advanced engineering mathematics

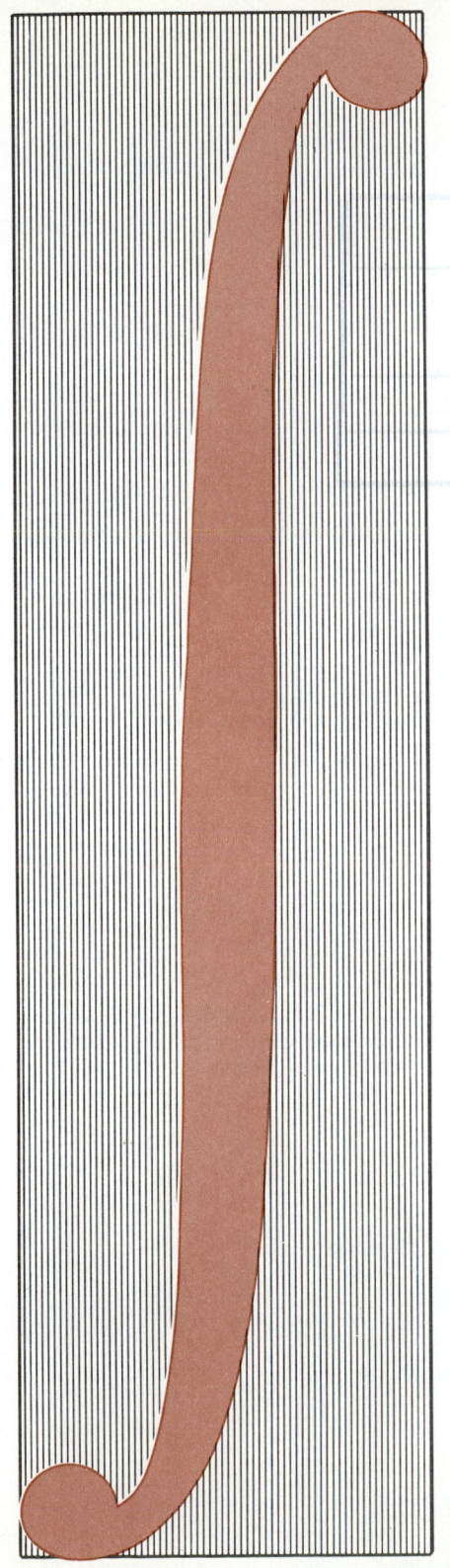

advanced engineering mathematics

STANLEY I. GROSSMAN
WILLIAM R. DERRICK
both of the University of Montana

HarperCollins*Publishers*

Sponsoring Editor: Peter Coveney
Project Editor: David Nickol
Text Design: Maria Carella
Cover Design: Joan Greenfield
Cover Illustration: Wassily Kandinsky. *Swinging 1925.* Oil on millboard,
  70.6 × 50.1 cm/27$\frac{3}{4}$ × 19$\frac{3}{4}$ in. Tate Gallery. Copyright ARS, NY/ADAGP 1987.
Text Art: Fine Line Illustrations, Inc.
Photo Research: Mira Schachne
Production Manager: Kewal K. Sharma
Compositor: TAPSCO, Inc.
Printer and Binder: R. R. Donnelley & Sons Company
Cover Printer: The Lehigh Press, Inc.

Advanced Engineering Mathematics

Library of Congress Cataloging-in-Publication Data

Grossman, Stanley I.
  Advanced Engineering Mathematics / Stanley I. Grossman, William R.
Derrick.
     p   cm.
  Includes index.
  ISBN 0-06-042534-2
    1. Engineering mathematics.   I. Derrick, William R.   II. Title.
TA330.G76 1988
510'. 2462—dc19

                                                                                    87-33423
                                                                                    CIP

93 9 8 7 6 5 4

Contents

Contents

2.3 Homogeneous Second-Order Linear Equations with Constant Coefficients *79*

2.4 Nonhomogeneous Equations: Method of Undetermined Coefficients *86*

2.5 Nonhomogeneous Equations: Variation of Parameters *93*

2.6 Euler Equations *97*

2.7 More on Electric Circuits *102*

2.8 Vibrational Motion: Simple Harmonic Motion *105*

2.9 Vibrational Motion: Damped Vibrations *110*

2.10 Vibrational Motion: Forced Vibrations *113*

2.11 Higher-Order Linear Differential Equations *118*

REVIEW EXERCISES FOR CHAPTER 2 *123*
APPENDIX TO CHAPTER 2 *124*

CHAPTER 3 **Systems of Linear Differential Equations** **127**

3.1 Method of Elimination for Linear Systems with Constant Coefficients *127*

3.2 Method of Determinants *136*

3.3 Electric Circuits with Several Loops *143*

3.4 Chemical Mixture and Population Biology Problems *149*

3.5 Mechanical Systems *154*

3.6 The Phase Plane *156*

3.7 Critical Points, Stability, and Phase Portraits for Linear Systems *163*

3.8 Stability of Nonlinear Systems *176*

REVIEW EXERCISES FOR CHAPTER 3 *185*

CHAPTER 4 **Laplace Transforms** **188**

4.1 Introduction: Definition and Basic Properties of the Laplace Transform *188*

4.2 Solving Initial Value Problems by Laplace Transform Methods *200*

4.3 Step Functions, Impulse Functions, and Periodic Functions *214*

CHAPTER 8   **Eigenvalues and Eigenvectors**   **453**

CHAPTER 9   **Vector Differential Calculus**   **507**

Preface

In writing a book about mathematics for engineers, the first problem one encounters is one of definition: What is engineering mathematics? The first part of the answer is easy. Each engineering student must acquire a basic understanding of one variable and multivariable calculus. He or she should be able to differentiate and integrate functions of one variable, take partial derivatives, and solve problems involving multiple integrals. The student should also appreciate the great applicability of calculus. These topics form the prerequisites for this text.

After calculus, engineering students study a variety of mathematical subjects. These include ordinary differential equations (taken by virtually all students), vector calculus (if not covered in the multivariable calculus course), linear algebra, partial differential equations, numerical analysis, and complex variables. Sometimes students take separate courses in each topic. Frequently, however, these subjects are covered in a year-long course under the general heading "engineering mathematics." It is at this course that this text is aimed.

The book is divided into six parts:

I. Ordinary Differential Equations Chapters 1–5 include a development of the theory of linear differential equations including discussions of linear systems, stability, Laplace transforms, and power series solutions. This development is accompanied by a wealth of applications in numerous scientific areas.

II. Linear Algebra Chapters 6–8 cover vector spaces, matrices, systems of equations, and eigenvalues and eigenvectors.

III. Vector Calculus Vector differential calculus is in Chapter 9; vector integral calculus appears in Chapter 10.

IV. Boundary Value Problems and Partial Differential Equations are discussed in Chapters 11 and 12. Included is a unique (for this kind of text) discussion of first-order partial differential equations.

V. Numerical Analysis Chapters 13 and 14 discuss a wide variety of topics, including numerical solutions of ordinary and partial differential equations and linear systems of equations. Here we also analyze (in Sections 13.6, 13.8, and 14.3) the error that is generated when a numerical technique is employed. This important topic is often slighted in comparable texts.

VI. Complex Variables Chapters 15–19 include analytic functions, complex integration, complex series, residues, and conformal mappings.

It would be very difficult to cover all 19 chapters in one course, and so a great deal of flexibility has been built in. The book has been written so that, wherever possible, each of the six parts is independent of the others. That is, the parts can be covered in any order. There are two exceptions. Some of the differential equations material in Chapters 1 and 2 is needed in the chapters (11 and 12) on boundary value problems and partial differential equations. In addition, it is useful to know something about differential equations, systems of equations, and eigenvalues before finding solutions numerically. So Sections 1.2, 1.3, and 1.4 are prerequisites for Section 13.5, Chapter 12 is a prerequisite for Section 13.9, Section 7.3 is needed for Sections 14.1 and 14.2, and Section 8.1 should be covered before Section 14.4.

The topics in this book constitute what is generally called undergraduate *applied mathematics*. Both words are important. This is a *mathematics* book. Whenever a theorem can be proved without invoking tools that are too sophisticated, it is proved. Mathematics with no mathematical proofs is no mathematics at all—just a grab bag of tricks.

However, students study this subject because of its great applicability in the physical sciences and engineering. It would be dishonest to separate the mathematics from the applications because, historically, much of the mathematics was developed in an attempt to solve certain types of physical problems. Thus, while not slighting the mathematics, we have written a book that stresses the great applicability of the topics being discussed. Our goal has been not only to teach mathematics, but to teach students how to *use* the mathematics in applied settings. The following list of features illustrates how these dual goals have been met:

- **Examples.** As students, we learned applied mathematics by reading examples and doing problems. *Advanced Engineering Mathematics* contains close to 700 examples—many more than are found in similar texts. Most examples contain all the algebraic and calculus steps needed to complete the solution. It is infuriating to be told that "it easily follows that. . ." when it is not at all obvious. Students have a right to see the "whole hand," so to speak, so that they always know how to get from A to B. In many instances, explanations are highlighted in color to make a step easier to follow.

- **Exercises.** The text contains over 5000 exercises—including both drill- and applied-type problems. More difficult problems are marked with an asterisk (*), and a few especially difficult ones are marked with a double asterisk (**). The exercises provide the most important learning tool in any undergraduate mathematics textbook. We stress to our students that no matter how well they think they understand our lectures or the textbook, they do not really know the material until they have worked problems. A vast difference exists between understanding someone else's solution and solving a new problem by yourself. Learning mathematics without doing problems is about as easy as learning to ski without going to the slopes.

• **Chapter Review Exercises.** At the end of each chapter, we have provided a collection of review exercises. Any student who can do these exercises can feel confident that he or she understands the material in the chapter.

• **Applications.** Most engineering mathematics textbooks contain applications to electric circuits and vibratory motions. We discuss these useful topics in a number of important places in the book. However, we provide many other applications in physics and engineering and apply the mathematics to topics in biology and economics as well. A partial list of applications follows:

Electric Circuits—in Sections 1.6, 2.7, 3.3, 4.4, and on pages 210 and 494
Vibratory Motion and Mechanic Systems—in Sections 2.8, 2.9, 2.10, 3.5, and on pages 133 and 454
Black Boxes and Transfer Functions—page 210
Carbon Dating—page 5
Chemical Flow—page 127
Chemical Mixture and Population Biology Problems—Section 3.4 (pages 149–153)
Compartmental Analysis—Section 1.7 (pages 46–54)
Competition and Predator-Prey Models (biology)—pages 498–499
Conservation of Energy—page 569
Coriolis Acceleration on the Earth—page 551
Coulomb's Law—page 651
Curves of Pursuit—Section 1.7 (pages 46–54)
An Epidemic Model (biology)—page 667
Escape Velocity—page 10
Fluid Flow—page 635, Section 19.5 (pages 1076–1081)
Free Fall—page 1
Full Wave Rectifier—page 695
Gravitational Potential—page 600
Heat Flow—page 649, Section 12.7 (pages 760–764), and Section 12.8 (pages 765–769)
Leontief Input-Output Model (economics)—page 373
Logistic Growth (biology)—page 11
A Model of Population Growth (biology)—Section 8.2 (pages 465–469)
Newton's Law of Cooling—page 4
A Projectile Under the Influence of the Earth's Rotation—page 239
Shocks in Gas Dynamics—page 738
The Vibrating String—Section 12.5 (pages 748–755)
Work—page 589

• **Accuracy.** Nothing is more infuriating to the student than to spend hours on a problem only to find that the answer in the back of the book is wrong. We have tried to ensure accuracy in the answer section in a number of ways. Each of us worked each problem and prepared an answer key. Then graduate students in New York and London were engaged to solve each problem. Solving applied mathematics problems in large numbers is a tedious task. We are grateful to the following individuals who put in uncounted hours helping us to provide the most accurate answers possible.

Chris Berry	Imperial College, University of London
Robert Bowles	University College London
Leon Gerber	St. John's University, New York
Elizabeth Jurisich	Courant Institute, New York University
Emma Tracey	Imperial College, University of London

Most numerical problems were solved with a Casio calculator. Minor differences in the last digit(s) may occur if you use a different calculator.

• **First-Order Partial Differential Equations.** Most engineering textbooks give a very cursory introduction to first-order partial differential equations. We delve deeper into this important topic. In Sections 12.1 and 12.2 we discuss *characteristics* and use them to solve both linear and certain non-linear equations.

• **Integral Tables.** In the real world, most nontrivial integrals are computed with the use of an integral table. The table in Appendix 1 of this book contains 220 entries. Every integral that comes up in this text can be found there.

• **Biographical Sketches.** Mathematics becomes more interesting if one knows something about the historical development of the subject. We chose seventeen mathematicians who contributed greatly to the understanding of differential equations. Biographies of these mathematicians appear on the pages indicated below:

Cauchy, 24	Poisson, 774
Bernoulli, 34	Lagrange, 809
Euler, 57	Cardano, 873
Laplace, 189	de Moivre, 888
Legendre, 283	Riemann, 907
Hamilton, 292	Weierstress, 989
Gauss, 366	Maclaurin, 999
Jacobi, 656	Dirichlet, 1083
Fourier, 678	

• **Answers and Other Aids.** The answers to most odd-numbered problems appear at the back of the book. In addition, we have prepared an Instructor's Solutions Manual that contains answers (and proofs) of the even-numbered exercises. The manual also contains detailed solutions to a great number of problems.

- **Numbering and Notation in the Text.** Numbering in the book is fairly standard. Within each section, examples, problems, theorems, and equations are numbered consecutively, starting with 1. Reference to an example, problem, theorem, or equation outside the section in which it appears is by chapter, section, and number. Thus, Example 4 in Section 2.3 is called, simply, Example 4 in that section, but outside the section it is referred to as Example 2.3.4. Frequently, a reference is accompanied by a page number to make it easier to find. As already mentioned, the more difficult problems are marked (*) or occasionally (**). Finally, the ends of examples or groups of examples are marked with a line and the ends of proofs are marked with a ■.

Acknowledgments

A number of reviewers provided useful suggestions for improving the quality of this text. In particular, we would like to thank the following:

Jim Buchanan	United States Naval Academy
John F. Cavalier	West Virginia Institute of Technology
James Dowdy	University of Virginia
William Firey	Oregon State University
Stuart Goldenberg	California State Polytechnic University, San Luis Obispo
Dar Ho	Georgia Institute of Technology
Allan M. Krall	Pennsylvania State University
Ray G. Langebartel	University of Illinois
Wayne Lewis	Texas Tech University
Deborah Franks Lochart	Michigan Technological University
David O. Lomen	University of Arizona
Richard MacCamy	Carnegie-Mellon University
Bernard Marshall	McGill University
Philip Miles	University of Wisconsin
William Miller	Worcester Polytechnic Institute
A. K. Mitra	Texas Tech University
John Douglas Moore	University of California, Santa Barbara
Robert Moreland	Texas Tech University
Norman Richert	University of Houston
John Scheick	Ohio State University
Klaus Schmitt	University of Utah
Michael Williams	Virginia Technological University

In addition, Professor Wayne Lewis and Professor Klaus Schmitt read page proofs for the entire book and caught a number of errors that might otherwise have found their way into print. We are grateful to Professors Lewis and Schmitt for helping to make this a cleaner textbook.

We wish to thank the following publishers for permission to use material that appears in this book:

Harcourt Brace Jovanovich for permission to use material from *Multivariate Calculus, Linear Algebra and Differential Equations,* Second Edition, 1986, by Stanley Grossman—in Chapters 9 and 10.

McGraw-Hill for permission to use (on page 43) an example in *Basic Circuit Theory,* 1969, by C. A. Desoer and E. S. Kuh.

Saunders Publishing Company for permission to use, in some of our biographical sketches, information that appears in their excellent history book *An Introduction to the History of Mathematics,* Fifth Edition, 1983, by Howard Eves.

Wadsworth Publishing Company for the use of material from *Elementary Linear Algebra,* Third Edition, 1987, by Stanley Grossman—in Chapters 6, 7, and 8.

Wadsworth International for permission to use complex variable material from *Complex Analysis and Applications,* Second Edition, 1984, by William Derrick.

West Publishing Co. for the use of material which appears in *Introduction to Differential Equations with Boundary Value Problems,* Third Edition, 1987, by William Derrick and Stanley Grossman.

A few of the biological exercises in this book first appeared in *Mathematics for the Biological Sciences,* by Stanley Grossman and James E. Turner, published by Macmillan in 1974. We are grateful to Professor Turner for permission to use this material.

Much of the manuscript was typed by Sandra Place in Hertfordshire, England. We are grateful to Mrs. Place for the great competence she brought to the job.

Finally, we are very grateful to the editorial and production staffs at Harper & Row for the skill they brought to this project. This was not an easy book to edit and typeset. The result is a tribute to their talent.

Stanley I. Grossman
William R. Derrick

advanced engineering mathematics

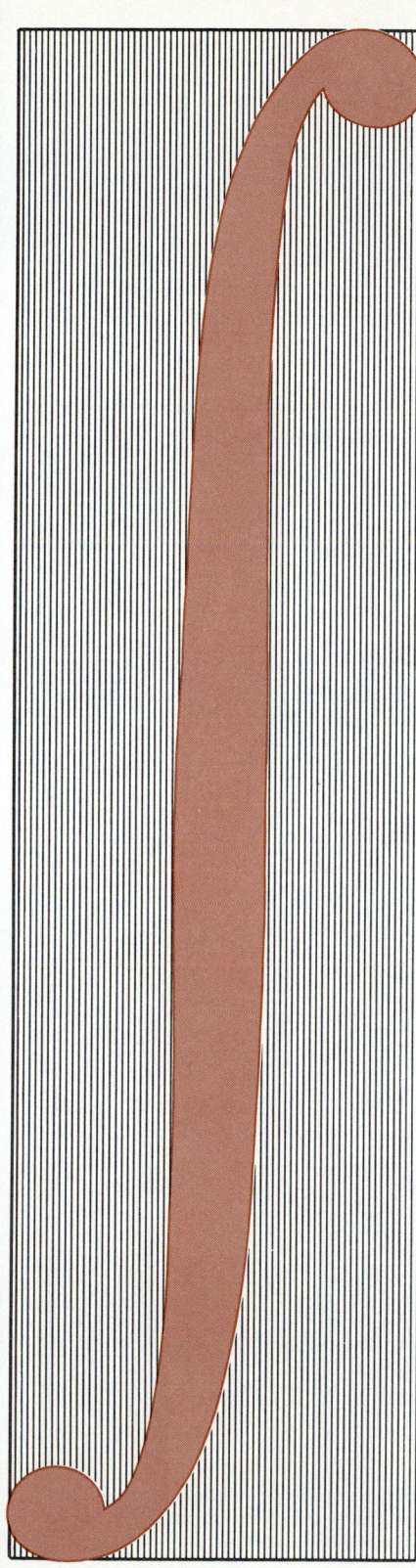

chapter 1

First-Order Differential Equations

1.1 Introduction

Many of the basic laws of the physical sciences and, more recently, of the biological and social sciences are formulated in terms of mathematical equations involving certain known and unknown quantities and their derivatives. Such equations are called **differential equations.** In this section, we shall show how some differential equations arise and how their solutions are obtained.

Before citing any examples, however, we should emphasize that in applying differential equations the most difficult problem often is to describe a real situation quantitatively. To do this, it is usually necessary to make simplifying assumptions that can be expressed in mathematical terms. Thus, for example, we initially describe the motion of a mass in space by assuming that (a) it is a point and (b) there is no friction or air resistance. These assumptions are not realistic, but the scientist can often glean valuable information from even highly idealized models that, once understood, can be modified to take other observable factors into account.

EXAMPLE 1: Free Fall

Newton's law of gravitation states that the magnitude of the gravitational force exerted by the earth on a body of mass m is proportional to its mass and inversely proportional to the square of its distance r from the center of the earth. Thus, if k denotes the constant of proportionality,

$$F = \frac{km}{r^2}.$$

By Newton's second law of motion

$$F = ma = m\frac{d^2r}{dt^2},$$

where $a = a(r)$ is the acceleration of the body when the distance to the earth is r. Hence, equating both forces and dividing by m, we get

$$\frac{d^2r}{dt^2} = \frac{k}{r^2}. \tag{1}$$

Note:† The velocity $v = dr/dt$ is negative because, as the object falls, its distance from the earth decreases. Moreover, the acceleration $a = dv/dt = d^2r/dt^2$ is also negative because as the object falls, its velocity gets more and more negative: that is, v decreases. Thus the constant k is negative.

We denote the acceleration of gravity at the surface of the earth (when $r = R$—the mean radius of the earth) by $a(R) = -g$. Then equation (1) leads to

$$-g = a(R) = \frac{k}{R^2},$$

so that $k = -gR^2$. Hence, we obtain the equation of motion

$$\frac{d^2r}{dt^2} = -g\frac{R^2}{r^2}, \tag{2}$$

where g is approximately 9.81 m/sec² (= 32.2 ft/sec²). If we substitute $r = R + h$, where h is the height of the body from the surface of the earth, then $dr/dt = dh/dt$ and equation (2) becomes

$$\frac{d^2h}{dt^2} = -g\frac{R^2}{(R+h)^2}.$$

When the height h is very small in comparison to the radius of the earth R, the ratio $R/(R + h)$ is very close to 1, so that the differential equation is well approximated by the usual equation found in calculus books:

$$\frac{d^2h}{dt^2} = -g. \tag{3}$$

Integrating both sides of equation (3) with respect to t, we obtain

$$h'(t) = -gt + C_1.$$

The constant C_1 can be determined by setting $t = 0$. We obtain $C_1 = h'(0)$, the initial velocity. Thus, the velocity of the body at any time t is given by the differential equation

$$h'(t) = -gt + h'(0). \tag{4}$$

Integrating once more with respect to t, we have

$$h(t) = -g\frac{t^2}{2} + h'(0)t + C_2.$$

The constant C_2 can also be found by setting $t = 0$. We obtain $C_2 = h(0)$, the initial height. Thus, the height of the body at any time t is

$$h(t) = -g\frac{t^2}{2} + h'(0)t + h(0). \tag{5}$$

---

† In this problem the force is directed toward the center of the earth. In what follows we shall only consider vectors directed in that or the opposite direction. These can be distinguished by signs: minus for vectors toward the center of the earth, plus in the opposite direction.

For example, if a ball is dropped from the top of a building 44.145 meters high, its initial velocity is $h'(0) = 0$ and its initial height is $h(0) = 4414.5$ cm. If we wish to find the length of time it takes for the ball to strike the ground, we substitute these values in equation (5) to obtain

$$0 = h(t) = -\frac{981t^2}{2} + 4414.5,$$

since the height at impact is zero. Solving for t, we have

$$490.5t^2 = 4414.5 \quad \text{or} \quad t^2 = 9.$$

Thus $t = \pm 3$. Since $t = -3$ has no physical significance, the answer is $t = 3$ sec.

The differential equation in Example 1 was solved directly by integration. If it were always possible to do this, then differential equations would be a direct application of integral calculus and there would be no need for separate chapters on differential equations. However, most differential equations that are solvable can only be solved by other techniques. One type of differential equation that we can solve is often discussed in a beginning calculus class.

EXAMPLE 2

Solve the differential equation

$$\frac{dy}{dx} = \alpha y, \qquad y(0) = 10 \tag{6}$$

where α is a constant.

SOLUTION
We rewrite (6)

$$\frac{dy}{y} = \alpha \, dx$$

and then integrate to obtain

$$\int \frac{dy}{y} = \int \alpha \, dx$$

or

$$\ln|y| = \alpha x + C. \qquad \text{(If ln } a = b, \text{ then } a = e^b.)$$

Hence

$$|y| = e^{\alpha x + C} = e^{\alpha x}e^C,$$

or

$$y = ke^{\alpha x}, \qquad \text{where } k = \pm e^C. \tag{7}$$

We check our answer:

If $y(x) = ke^{\alpha x}$, then $dy/dx = k(\alpha e^{\alpha x}) = \alpha(ke^{\alpha x}) = \alpha y$, so that $y = ke^{\alpha x}$ satisfies (6). That is, equation (6) has an infinite number of solutions, one for each real number k.

We can obtain a unique solution if we use the **initial condition,** $y(0) = 10$. Then

$$10 = ke^{\alpha \cdot 0} = ke^0 = k$$

so that $y = 10e^{\alpha x}$ is the unique solution to the **initial value problem** (6).

The method used to solve (6) in Example 2 is called **separation of variables,** since it involves placing the dependent and independent variables on different sides of the equation.

Exponential Growth and Decay

If $\alpha > 0$, we say that $e^{\alpha x}$ is **growing exponentially.** If $\alpha < 0$, it is **decaying exponentially** (see Fig. 1). Of course, if $\alpha = 0$, there is no growth and $y = e^0 = 1$ remains constant.

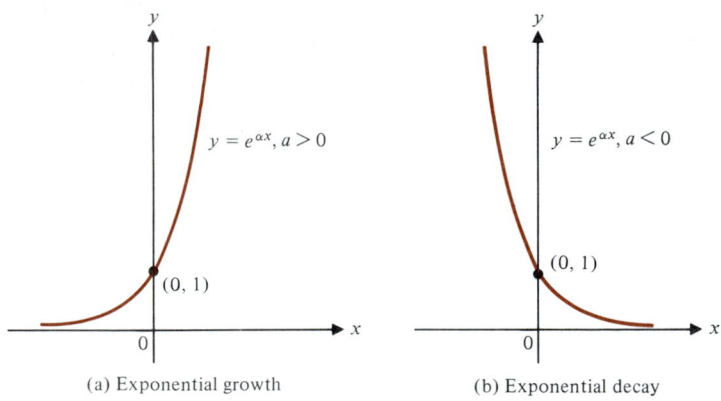

(a) Exponential growth

(b) Exponential decay

FIGURE 1

For a physical problem it would not make sense to have an infinite number of solutions. We can usually get around this difficulty, as in Example 2, by specifying the value of y for one particular value of x, say $y(x_0) = y_0$. This value is called an **initial condition** and it will give us a unique solution to the problem. We shall see this illustrated in the examples that follow.

Initial Condition

EXAMPLE 3: Newton's Law of Cooling

Newton's law of cooling states that the rate of change of the temperature difference between an object and its surrounding medium is proportional to the temperature difference. Let $\Delta(t)$ denote this temperature difference at any time t. Since a rate of change is expressed mathematically by a derivative, we may translate Newton's law of cooling into the equation

$$\frac{d\Delta}{dt} = \alpha \Delta. \tag{8}$$

α will be negative because the temperature difference is decreasing. From Example 2, we see that the unique solution to (8) is

$$\Delta(t) = \Delta(0)e^{\alpha t}. \tag{9}$$

General Solution

We call equation (9) the **general solution** to the differential equation (8), because every solution of (8) is of this form. Here $\Delta(0)$ is an arbitrary constant that denotes the temperature difference at time $t = 0$.

As a practical example, suppose that a pot of boiling water (100°C) is removed from the fire and allowed to cool at 20°C room temperature. Two minutes later, the temperature of the water in the pot is 80°C. What will be the temperature of the water 5 minutes after it has been removed from the fire?

Here the initial temperature difference is

$$\Delta(0) = 100° - 20° = 80°,$$

so that equation (9) becomes

$$\Delta(t) = 80e^{\alpha t}. \tag{10}$$

When $t = 2$ minutes, the temperature difference is

$$\Delta(2) = 80° - 20° = 60°C,$$

so if we substitute $t = 2$ into equation (10), we obtain

$$60 = 80e^{2\alpha},$$

or

$$e^{2\alpha} = \tfrac{60}{80} = \tfrac{3}{4}.$$

Taking the natural logarithm of both sides of this equation, we get

$$2\alpha = \ln(\tfrac{3}{4}),$$

or

$$\alpha = \tfrac{1}{2} \ln(\tfrac{3}{4}).$$

Note that α is negative (≈ -0.1438). This makes sense since the temperature difference must decrease as time increases.

Substituting this value of α into equation (10) and recalling that $e^{\ln x} = x$ and $a \ln b = \ln b^a$, we obtain

$$\Delta(t) = 80e^{[\ln(3/4)]t/2} = 80e^{\ln[(3/4)^{t/2}]}$$

$$= 80(\tfrac{3}{4})^{t/2}. \tag{11}$$

Particular Solution

Equation (11) is a **particular solution** of (7) since it is uniquely determined by the conditions specified for this particular situation.

Finally, to find the temperature of the water in the pot at $t = 5$ minutes, we first find the temperature difference by setting $t = 5$ in equation (11):

$$\Delta(5) = 80(\tfrac{3}{4})^{5/2} \approx 38.97.$$

Adding this difference to the room temperature of 20°C yields a temperature of approximately 58.97°C for the water in the pot 5 minutes after it has been removed from the fire.

EXAMPLE 4: Carbon Dating

Carbon dating is a technique used by archaeologists, geologists, and others who want to estimate the ages of certain artifacts and remains they uncover. The technique is based on certain properties of the carbon atom. In its natural state the nucleus of the carbon atom ^{12}C has 6 protons and 6 neutrons. Another *isotope* of carbon is ^{14}C, which has 2 additional neutrons in its nucleus. ^{14}C is *radioactive*. That is, it emits an electron and reaches the stable state ^{14}N. We make the assumption that the ratio of ^{14}C to ^{12}C in the atmosphere is constant. This assumption has been shown experimentally to be approximately valid, for although ^{14}C is constantly being lost through **radioactive decay** (as this process is often termed), new ^{14}C is constantly being produced by the cosmic bombardment of nitrogen in the upper atmosphere. Living plants and animals do not distinguish between ^{12}C and ^{14}C, so at the time of death the ratio of ^{12}C to ^{14}C in an organism is the same as the ratio in the atmosphere. However, this ratio changes after death since ^{14}C is converted to ^{14}N but no further ^{14}C is taken in.

It has been observed that ^{14}C decays at a rate proportional to its mass and that its **half-life** is approximately 5730 years.† That is, if a substance starts with 1 g of ^{14}C, then 5730 years later it would have $\frac{1}{2}$ g of ^{14}C, the other $\frac{1}{2}$ g having been converted to ^{14}N.

We may now pose a question typically asked by an archaeologist. The remains of an organism are unearthed and it is determined that the amount of ^{14}C present is 40% of what it would be for a similar living organism. What is the approximate age of the remains?

SOLUTION

Let $M(t)$ denote the mass of ^{14}C present in the remains. Then since ^{14}C decays at a rate proportional to its mass, we have

$$\frac{dM}{dt} = -\alpha M,$$

where α is the constant of proportionality. Then $M(t) = ce^{-\alpha t}$, where $c = M_0$, the initial amount of ^{14}C present. When $t = 0$, $M(0) = M_0$; when $t = 5730$ years, $M(5730) = \frac{1}{2}M_0$, since half the original amount of ^{14}C has been converted to ^{14}N. We can use this fact to solve for α since we have

$$\tfrac{1}{2}M_0 = M_0 e^{-\alpha \cdot 5730}, \quad \text{or} \quad e^{-5730\alpha} = \tfrac{1}{2}.$$

Thus

$$(e^{-\alpha})^{5730} = \tfrac{1}{2}, \quad \text{or} \quad e^{-\alpha} = (\tfrac{1}{2})^{1/5730}, \quad \text{and} \quad e^{-\alpha t} = (\tfrac{1}{2})^{t/5730},$$

so

$$M(t) = M_0(\tfrac{1}{2})^{t/5730}.$$

Now we are told that after t years (from the death of the organism to the present) $M(t) = 0.4M_0$, and we are asked to determine t. Then

$$0.4M_0 = M_0(\tfrac{1}{2})^{t/5730},$$

and taking natural logarithms (after dividing by M_0), we obtain

† This number was first determined in 1941 by the American chemist W. S. Libby, who based his calculations on the wood from sequoia trees, whose ages were determined by rings marking years of growth. Libby's method has come to be regarded as the archaeologist's absolute measuring scale. But in truth, this scale is flawed. Libby used the assumption that the atmosphere had at all times a constant amount of ^{14}C. Recently, however, the American chemist C. W. Ferguson of the University of Arizona deduced from his study of tree rings in 4000-year-old American giant sequoia trees that before 1500 B.C. the radiocarbon content of the atmosphere was higher than it was later. This result implied that objects from the pre-1500 B.C. era were much older than previously believed, because Libby's "clock" allowed for a smaller amount of ^{14}C than actually was present. For example, a find dated at 1800 B.C. was in fact from 2500 B.C. This fact has had a considerable impact on the study of prehistoric times. For a fascinating discussion of this subject, see Gerhard Herm, *The Celts,* St. Martin's Press, New York, 1975, pp. 90–92.

$$\ln 0.4 = \frac{t}{5730} \ln\left(\frac{1}{2}\right), \quad \text{or} \quad t = \frac{5730 \ln(0.4)}{\ln(\frac{1}{2})} \approx 7575 \text{ years.}$$

The carbon-dating method has been used successfully on numerous occasions. It was this technique that established that the Dead Sea scrolls were prepared and buried about two thousand years ago.

In Examples 2–4 we solved a very simple differential equation. Other, more elaborate differential equations can be solved by the method of separation of variables. We shall do this in Section 1.2. However, it should be apparent from these examples and the problems that follow that a large variety of applied problems can be solved by considering a simple differential equation. It is the purpose of this chapter to show you how to deal with other, more complicated differential equations arising in physical applications.

The physical examples given in this section are such that in each case we know that a solution exists. However, there is an inherent danger in confusing physical reality with the mathematical model given by the differential equation we are using to represent the real problem. It may well be that our reasoning was faulty, in which case the equations obtained might bear no connection with reality. Then solutions to the equations need not exist. We should also note that not all differential equations have solutions. For example, the equation

$$\left(\frac{dy}{dx}\right)^2 + 3 = 0$$

has no real-valued solutions, since $(dy/dx)^2 + 3 \geq 3$. On the other hand, the equation

$$\left(\frac{dy}{dx}\right)^2 + y^2 = 0$$

has $y = 0$ as its only real solution, whereas the equation

$$\frac{dy}{dx} + y = 0$$

has an infinite set of solutions $y = ce^{-x}$ for any constant c.

In Section 1.3 and elsewhere in this book we shall discuss conditions which ensure that differential equations do have solutions. Moreover, we shall show that if initial conditions are provided, then these solutions are unique.

Problems 1.1†

1. The growth rate of a bacteria population is proportional to its size. Initially the population is 10,000, while after 10 days its size is 25,000. What is the population size after 20 days? After 30 days?

2. In Problem 1 suppose instead that the population after 10 days is 6000. What is the population after 20 days? After 30 days?

† To complete most of these problems, you will need to use a hand calculator with $\boxed{\ln}$ and $\boxed{e^x}$ function keys. On some calculators e^x is obtained by pressing $\boxed{\text{INV}}$ $\boxed{\ln}$.

3. The population of a certain city grows 6% a year. If the population in 1970 was 250,000, what would be the population in 1980? In 2000?

4. When the air temperature is 70°F, an object cools from 170°F to 140°F in $\frac{1}{2}$ hr.

   (a) What will be the temperature after 1 hr?

   (b) When will the temperature be 90°F? [*Hint:* Use Newton's law of cooling.]

5. A hot coal (temperature 150°C) is immersed in ice water (temperature 0°C). After 30 sec the temperature of the coal is 60°C. Assume that the ice water is kept at 0°C.

   (a) What is the temperature of the coal after 2 min?

   (b) When will the temperature of the coal be 10°C?

** 6. The president and vice-president sit down for coffee. They are both served a cup of hot black coffee (at the same temperature). The president takes a container of cream and immediately adds it to his coffee, stirs it, and waits. The vice-president waits 10 minutes and then adds the same amount of cream (which has been kept cool) to her coffee and stirs it in. Then they both drink. Assuming that the temperature of the cream is lower than that of the air, who drinks the hotter coffee? [*Hint:* Use Newton's law of cooling. It is necessary to treat each case separately and to keep track of the volumes of coffee, cream, and the coffee-cream mixture.]†

7. An artifact contains 70% of a "normal" amount of ^{14}C. How old is the artifact?

8. Forty percent of a radioactive substance disappears in 100 years.

   (a) What is its half-life?

   (b) After how many years will 90% be gone?

9. Salt decomposes in water into sodium [Na^+] and chloride [Cl^-] ions at a rate proportional to its mass. Suppose there were 25 kg of salt initially and 15 kg after 10 hr.

   (a) How much salt would be left after one day?

   (b) After how many hours would there be less than $\frac{1}{2}$ kg of salt left?

10. X rays are absorbed into a uniform, partially opaque body as a function not of time but of penetration distance. The rate of change of the intensity $I(x)$ of the X ray is proportional to the intensity. Here x measures the distance of penetration. The more the X ray penetrates, the lower the intensity is. The constant of proportionality is the density D of the medium being penetrated.

   (a) Formulate a differential equation describing this phenomenon.

   (b) Solve for $I(x)$ in terms of x, D, and the initial (surface) intensity $I(0)$.

11. Radioactive beryllium is sometimes used to date fossils found in deep-sea sediment. The decay of beryllium satisfies the equation

   $$\frac{dA}{dt} = -\alpha A, \qquad \text{where} \quad \alpha = 1.5 \times 10^{-7}.$$

   What is the half-life of beryllium?

12. In a certain medical treatment a tracer dye is injected into the pancreas to measure its function rate. A normally active pancreas will secrete 4% of the dye each minute. A physician injects 0.3 g of the dye and 30 min later 0.1 g remains. How much dye would remain if the pancreas were functioning normally?

13. Atmospheric pressure is a function of altitude above sea level and is given by $dP/da = \beta P$, where β is a constant. The pressure is measured in millibars (mbar). At sea level ($a = 0$), $P(0)$ is 1013.25 mbar, which means that the atmosphere at sea level will support a column of mercury 1013.25 mm high at a standard temperature of 15°C. At an altitude of $a = 1500$ m, the pressure is 845.6 mbar.

   (a) What is the pressure at $a = 4000$ m?

   (b) What is the pressure at 10 km?

   (c) In California the highest and lowest points are Mount Whitney (4418 m) and Death Valley (86 m below sea level). What is the difference in their atmospheric pressures?

   (d) What is the atmospheric pressure at Mount Everest (elevation 8848 m)?

   (e) At what elevation is the atmospheric pressure equal to 1 mbar?

14. A bacteria population is known to grow exponentially. The following data were collected:

Number of days	Number of bacteria
5	936
10	2,190
20	11,986

† This is a famous old problem that keeps on popping up (with an ever-changing pair of characters) in books on games and puzzles in mathematics. The problem is hard and has stymied many a mathematician. Do not get frustrated if you cannot solve it. The trick is to write everything down and to keep track of all the variables. The fact that the air is warmer than the cream is critical. It should also be noted that guessing the correct answer is fairly easy. Proving that your guess is correct is what makes the problem difficult.

(a) What was the initial population?

(b) If the present growth rate were to continue, what would the population be after 60 days?

15. A bacteria population is declining exponentially. The following data were collected:

Number of hours	Number of bacteria
12	5969
24	3563
48	1269

(a) What was the initial population?

(b) How many bacteria are left after one week?

(c) When will there be no bacteria left? (i.e., when is $P(t) < 1$?)

16. A ball is thrown upward with an initial velocity v_0 m/sec from the top of a building h_0 m high. Find how high the ball travels and determine when it hits the ground for the following choices of v_0 and h_0; neglect air resistance and let $g = -9.8$ m/sec^2.

(a) $v_0 = 49$ m/sec, $h_0 = 539$ m

(b) $v_0 = 14$ m/sec, $h_0 = 21$ m

(c) $v_0 = 21$ m/sec, $h_0 = 175$ m

(d) $v_0 = 7$ m/sec, $h_0 = 56$ m

(e) $v_0 = 7.7$ m/sec, $h_0 = 42$ m

In Problems 17–20, use Newton's law of cooling to determine how long to bake a cake at the given oven temperature, assuming that it takes exactly 30 minutes to change 70°F dough into a 170°F cake in a 350°F oven.

17. 250°F

18. 400°F

19. 300°F

20. 200°F

1.2 Separation of Variables

In this section we use the technique of separation of variables to solve more complicated differential equations than the ones discussed in the last section.
    Consider the differential equation

$$\frac{dy}{dx} = f(x, y) \tag{1}$$

and suppose that the function $f(x, y)$ can be factored into a product

$$f(x, y) = g(x)h(y) \tag{2}$$

Separation of Variables

where $g(x)$ and $h(y)$ are each functions of only one variable. When this occurs, equation (1) can be solved by the method of **separation of variables.** To solve the equation, we substitute the product (2) into (1) to obtain

$$\frac{dy}{dx} = g(x)h(y),$$

or

$$\frac{1}{h(y)}\frac{dy}{dx} = g(x). \tag{3}$$

Integrating both sides of equation (3) with respect to x, we have

$$\int \frac{1}{h(y)}\frac{dy}{dx}dx = \int g(x)\,dx + C,$$

and by the change of variables procedure for integrals in calculus

$$\int \frac{1}{h(y)}\,dy = \int g(x)\,dx + C. \qquad (4)$$

If both integrals in equation (4) can be evaluated, a solution to the differential equation (1) is obtained. We illustrate this method with several examples.

EXAMPLE 1

Solve the differential equation $dx/dt = t\sqrt{1-x^2}$.

SOLUTION
We have

$$\frac{dx}{\sqrt{1-x^2}} = t\,dt \quad \text{or} \quad \int \frac{dx}{\sqrt{1-x^2}} = \int t\,dt + C.$$

Integration yields

$$\sin^{-1} x = \frac{t^2}{2} + C \quad \text{or} \quad x = \sin\!\left(\frac{t^2}{2} + C\right).$$

There are an infinite number of solutions, one for each value of C with $C \le \pi/2$. (What happens when $C > \pi/2$?) For certain initial conditions, there will be a unique solution. For example, suppose that $x(0) = 1/2$. Then

$$\frac{1}{2} = \sin\!\left(\frac{0^2}{2} + C\right) = \sin C \quad \text{and} \quad C = \sin^{-1}\frac{1}{2} = \frac{\pi}{6}.$$

Thus, the unique solution is

$$x(t) = \sin\!\left(\frac{t^2}{2} + \frac{\pi}{6}\right).$$

Note there is no solution if $x(0) = 2$, because the sine function takes values in the interval $[-1, 1]$.

EXAMPLE 2: Escape Velocity

In Example 1.1.1 on p. 1 we studied the motion of a body falling freely subject to the gravitational force of the earth. In that example we assumed that the height of the body was small in comparison to the radius R of the earth. However, if we wish to study the equation of motion for a communications satellite or an interplanetary vehicle, the distance r of the object from the center of the earth may be considerably larger than R. Thus, the approximation we made in obtaining equation (1.1.3) is no longer valid. Returning to equation (1.1.2),

$$\frac{d^2r}{dt^2} = -g\frac{R^2}{r^2}$$

and setting $v = dr/dt$, we see by the chain rule that

$$v = \frac{dr}{dt}$$

$$\downarrow$$

$$\frac{d^2r}{dt^2} = \frac{dv}{dt} = \frac{dv}{dr}\frac{dr}{dt} = v\frac{dv}{dr}.$$
(5)

Hence, equation (1.1.2) can be rewritten as

$$\frac{v\,dv}{dr} = -g\frac{R^2}{r^2},$$
(6)

where g and R are constant. Separating variables and integrating, we have

$$\int v\,dv = -gR^2 \int \frac{dr}{r^2} + C,$$

or

$$\frac{1}{2}v^2 = \frac{gR^2}{r} + C.$$

Assuming that the object is at the surface of the earth when $t = 0$, we get

$$\tfrac{1}{2}v(0)^2 = g\frac{R^2}{R} + C$$

or

$$C = \tfrac{1}{2}v(0)^2 - gR.$$

Thus,

$$v^2 = 2g\frac{R^2}{r} + v(0)^2 - 2gR.$$
(7)

For the object to escape the gravitational force of the earth, it is necessary that $v > 0$ for all time t. If we select $v(0) = \sqrt{2gR}$, the last two terms in equation (7) cancel, so that $v^2 > 0$ for all r. Observe that any smaller choice for $v(0)$ will allow the right side of equation (7) to be zero for some sufficiently large value of r. Thus, $v(0) = \sqrt{2gR} \approx 11.2$ km/sec is the initial velocity an object needs to escape the gravitational attraction of the earth. It is called the **escape velocity.**

The substitution we used in equation (5) can always be used to reduce an equation involving a second derivative into an equation involving only first derivatives, provided the independent variable does not appear explicitly in the equation.

EXAMPLE 3: Logistic Growth

Let $P(t)$ denote the population of a species at time t. The **growth rate per individual** of the population is defined as the growth of the population divided by the size of the population. Thus, for example, if the birth rate is 3.2 per hundred and the death rate is 1.8 per hundred, then the growth rate is $3.2 - 1.8 = 1.4$ per hundred $= 1.4/100 = 0.014$. We then write $dP/dt = 0.014P$.

Suppose that in a given population the average birth rate is a positive constant β. It is reasonable to assume that the average death rate is proportional to the number of individuals in the population. Greater populations mean greater crowding and more competition for food and territory. We call this constant of proportionality δ (which is greater than 0). Since dP/dt is the growth rate of the population, the growth rate per individual of the population is

$$\frac{1}{P}\frac{dP}{dt}.$$

Then the differential equation that governs the growth of this population is

$$\frac{1}{P}\frac{dP}{dt} = \beta - \delta P.$$

Multiplying both sides of this equation by P, we have

$$\frac{dP}{dt} = P(\beta - \delta P), \tag{8}$$

Logistic Equation and Logistic Growth

which is called the **logistic equation.**† The growth shown by this equation is called **logistic growth.**

Separating the variables, we have

$$\int \frac{dP}{P(\beta - \delta P)} = \int dt + C. \tag{9}$$

Using partial fractions, we find that

$$\frac{1}{P(\beta - \delta P)} = \frac{1}{\beta P} + \frac{\delta}{\beta(\beta - \delta P)}.$$

Substituting the right-hand side of this equation into equation (9) and integrating, we obtain

$$\frac{1}{\beta}\ln|P| - \frac{1}{\beta}\ln|\beta - \delta P| = t + C$$

or

$$\frac{1}{\beta}\ln\left|\frac{P}{\beta - \delta P}\right| = t + C. \tag{10}$$

Exponentiating both sides of this equation and denoting the arbitrary constant $\pm e^{\beta C}$ by C_1, we have

$$\left|\frac{P}{\beta - \delta P}\right| = e^{\beta t + \beta C} \qquad \text{and} \qquad \frac{P}{\beta - \delta P} = C_1 e^{\beta t}. \tag{11}$$

---

† The logistic equation is sometimes referred to as **Verhulst's equation** in honor of P. F. Verhulst (1804–1849), a Belgian mathematician who proposed this model for human population growth in 1838.

Setting $t = 0$, we find that

$$\frac{P(0)}{\beta - \delta P(0)} = C_1,$$

and substituting this value of C_1 into equation (11) yields

$$\frac{P(t)}{\beta - \delta P(t)} = \frac{P(0)}{\beta - \delta P(0)} e^{\beta t}.$$

Cross-multiplying and solving for $P(t)$, we obtain

$$P(t)[\beta - \delta P(0)] = P(0)[\beta - \delta P(t)]e^{\beta t}$$

or

$$\beta P(t) - \delta P(t)P(0) = \beta P(0)e^{\beta t} - \delta P(0)P(t)e^{\beta t}$$

so that

$$P(t)[\beta - \delta P(0) + \delta P(0)e^{\beta t}] = \beta P(0)e^{\beta t}$$

and, dividing top and bottom by $P(0)e^{\beta t}$,

$$P(t) = \frac{\beta P(0)e^{\beta t}}{\beta - \delta P(0) + \delta P(0)e^{\beta t}} = \frac{\beta}{\delta + \left[\dfrac{\beta}{P(0)} - \delta\right]e^{-\beta t}}. \tag{12}$$

Observe that since $\beta > 0$, as t increases the term $e^{-\beta t}$ approaches zero. Thus, the population approaches a limiting value of β/δ beyond which it cannot increase, since setting $P = \beta/\delta$ in equation (8) yields $dP/dt = 0$.

A Perspective: When Is a Differential Equation Separable?

In this section we saw how to solve a first-order differential equation when the equation was separable. However, it is not always clear that an equation is separable. For example, it is obvious that $f(x, y) = e^x \cos y$ is separable, but it is not so obvious that $f(x, y) = 2x^2 + y - x^2y + xy - 2x - 2$ is separable.† In this perspective, we give conditions that ensure that an equation is separable.‡

■ THEOREM 1

Suppose that $f(x, y) = g(x)h(y)$, where both g and h are differentiable. Then

$$f(x, y)f_{xy}(x, y) = f_x(x, y)f_y(x, y). \tag{13}$$

† $2x^2 + y - x^2y + xy - 2x - 2 = (1 + x - x^2)(y - 2)$.

‡ The results here are based on the paper "When is an ordinary differential equation separable?" by David Scott in *American Mathematical Monthly,* vol. 92 (1985): pp. 422–423.

Here the subscripts denote partial derivatives.

PROOF
Observe that

$$f_x(x, y) = g'(x)h(y)$$

$$f_y(x, y) = g(x)h'(y)$$

$$f_{xy}(x, y) = g'(x)h'(y).$$

Hence

$$f(x, y)f_{xy}(x, y) = g(x)h(y)g'(x)h'(y) = [g'(x)h(y)][g(x)h'(y)]$$

$$= f_x(x, y)f_y(x, y). \quad\blacksquare$$

It turns out that, under further conditions, if equation (13) holds, then $f(x, y)$ is separable. For what follows, D denotes an open disk in the xy-plane; that is, $D = \{(x, y): (x - a)^2 + (y - b)^2 < r^2\}$, where a, b, and r are real numbers and $r > 0$.

■ THEOREM 2
Suppose that in D, f, f_x, f_y, and f_{xy} exist and are continuous, $f(x, y) \neq 0$, and equation (13) holds. Then there are continuously differentiable functions $g(x)$ and $h(y)$ such that, for every $(x, y) \in D$,

$$f(x, y) = g(x)h(y). \tag{14}$$

PROOF
Since $f(x, y) \neq 0$ and f is continuous on D, f has the same sign on D. Assume $f(x, y) > 0$ for $(x, y) \in D$. A similar proof works if $f(x, y) < 0$ if f is replaced by $-f$. Now, from the quotient rule of differentiation,

equation (13)

$$\frac{\partial}{\partial y}\frac{f_x(x, y)}{f(x, y)} = \frac{f(x, y)f_{xy}(x, y) - f_x(x, y)f_y(x, y)}{f^2(x, y)} = 0.$$

When the partial derivative with respect to y of a function of x and y is identically zero, then that function must be a function of x only. Thus there is a function $\alpha(x)$ such that

$$\frac{f_x(x, y)}{f(x, y)} = \alpha(x).$$

Also, since $f(x, y) > 0$, $\ln f(x, y)$ is defined and

$$\frac{\partial}{\partial x}\ln f(x, y) = \frac{f_x(x, y)}{f(x, y)} = \alpha(x).$$

The function $\alpha(x)$ is continuous in D because it is the quotient of continuous functions and the function in the denominator is nonzero. Let $\beta(x) = \int \alpha(x)\, dx$. Then

$$\ln f(x, y) = \int \left[\frac{\partial}{\partial x} \ln f(x, y) \right] dx = \int \alpha(x)\, dx = \beta(x) + \gamma(y),$$

where γ is a function of y only. (The partial derivative with respect to x of a function of y only is zero. Thus $\gamma(y)$ represents the most general constant of integration.) Finally, let $g(x) = e^{\beta(x)}$ and $h(y) = e^{\gamma(y)}$. Then

$$f(x, y) = e^{\ln f(x,y)} = e^{\beta(x) + \gamma(y)} = e^{\beta(x)} e^{\gamma(y)} = g(x)h(y). \qquad \blacksquare$$

EXAMPLE 4

Let $f(x, y) = 2x^2 + y - x^2 y + xy - 2x - 2$. Then

$$f_x(x, y) = 4x - 2xy + y - 2$$

$$f_y(x, y) = 1 - x^2 + x$$

$$f_{xy}(x, y) = -2x + 1$$

$$f(x, y)f_{xy}(x, y) = (2x^2 + y - x^2 y + xy - 2x - 2)(-2x + 1)$$

$$= -4x^3 - xy + 2x^3 y - 3x^2 y + 6x^2 + 2x + y - 2$$

and

$$f_x(x, y)f_y(x, y) = (4x - 2xy + y - 2)(1 - x^2 + x)$$

$$= 2x - xy + y - 2 - 4x^3 + 2x^3 y - 3x^2 y + 6x^2.$$

Since the last two expressions are equal, we conclude by Theorem 2 that $f(x, y)$ is separable.

EXAMPLE 5

Let $f(x, y) = 1 + xy$. Then

$$f_x(x, y) = y$$

$$f_y(x, y) = x$$

$$f_{xy}(x, y) = 1$$

$$f(x, y)f_{xy}(x, y) = 1 + xy$$

and

$$f_x(x, y)f_y(x, y) = xy.$$

Since the last two expressions are unequal, we conclude that $f(x, y)$ is not separable.

Problems 1.2

In Problems 1–25 find the general solution by separating variables. If an initial condition is given, find the particular solution that satisfies that condition.

1. $\dfrac{dy}{dx} = \dfrac{e^x}{2y}$

2. $xy' = 3y, \quad y(2) = 5$

3. $\dfrac{dy}{dx} = \dfrac{e^y x}{e^y + x^2 e^y}$

4. $\dfrac{dx}{dy} = x\cos y, \quad x\!\left(\dfrac{\pi}{2}\right) = 1$

5. $\dfrac{dz}{dr} = r^2(1 + z^2)$

6. $\dfrac{dy}{dx} + y = y(xe^{x^2} + 1), \quad y(0) = 1$

7. $\dfrac{dP}{dQ} = P(\cos Q + \sin Q)$

8. $\dfrac{dy}{dx} = y^2(1 + x^2), \quad y(0) = 1$

9. $\dfrac{ds}{dt} + 2s = st^2, \quad s(0) = 1$

10. $\dfrac{dy}{dx} = \sqrt{1 - y^2}$

11. $(1 + x)\dfrac{dy}{dx} = -3y, \quad y(6) = 7$

12. $\dfrac{dx}{dt} + (\cos t)e^x = 0$

13. $\cot x\,\dfrac{dy}{dx} + y + 3 = 0$

14. $\dfrac{dx}{dt} = x(1 - \sin t), \quad x(0) = 1$

15. $\dfrac{dy}{dx} + \sqrt{\dfrac{1 - y^2}{1 - x^2}} = 0$

16. $(\tan y)\dfrac{dy}{dx} - \tan x = 0, \quad y(0) = 0$

17. $x^2\dfrac{dy}{dx} + y^2 = 0, \quad y(1) = 3$

18. $\dfrac{dy}{dx} = \dfrac{y^3 + 2y}{x^2 + 3x}, \quad y(1) = 1$

19. $e^x\!\left(\dfrac{dx}{dt} + 1\right) = 1, \quad x(0) = 1$

20. $\dfrac{ds}{dr} = \dfrac{s^2 + s - 2}{r^2 - 2r - 8}, \quad s(0) = 0$

21. $yy' = e^x$

22. $y' + y = y(xe^x + 1)$

23. $xy' = y(3 - x)$

24. $\dfrac{dy}{dx} = \dfrac{x^2 - xy - x + y}{xy - y^2}$

25. $\dfrac{dy}{dx} = \dfrac{x}{y} - \dfrac{x}{1 + y}, \quad y(0) = 1$

26. An object is falling in a vacuum with constant acceleration g. Express its velocity as a function of its height.

27. A rocket is launched from an initial position (x_0, y_0) with an initial speed v_0 and angle θ $(0 \le \theta \le \pi/2)$ measured from the horizontal. Find its horizontal and vertical coordinates $x(t)$ and $y(t)$ as functions of time. Assume that there is no air resistance and that gravity g is constant.

28. The economist Vilfredo Pareto (1848–1923) discovered that the rate of decrease of the number of people y having an income of at least x dollars in a stable economy is directly proportional to the number of such people and inversely proportional to their income. Obtain an expression (**Pareto's law**) for y in terms of x.

29. Assume that in addition to the gravitational attraction, a body falling close to the surface of the earth is subject to air resistance proportional to its velocity:

$$h'' = -g - ch', \qquad c > 0.$$

Find the velocity of the body at any time t, and determine its **terminal velocity** as $t \to \infty$.

* 30. A large open cistern filled with water has the shape of a hemisphere with radius 25 ft. The bowl has a circular hole of radius 1 ft in the bottom. By **Torricelli's law,**† water will flow out of the hole with the same speed it would attain in falling freely from the level of the water to the hole. How long will it take for all the water to flow from the cistern?

* 31. In Problem 30 find the shape of the cistern that would ensure that the water level drops at a constant rate.‡

** 32. On a certain day it began to snow early in the morning and the snow continued to fall at a constant rate. The velocity at which a snowplow is able to clear a road is inversely proportional to the height of the accumulated snow. The snowplow started at 11 a.m. and had cleared

---

† Evangelista Torricelli (1608–1647) was an Italian physicist.

‡ The ancient Egyptians (1380 B.C.) used water clocks based on this principle to tell time.

4 miles by 2 p.m. By 5 p.m. it had cleared another 2 miles. When did it start snowing?†

33. Table 1 shows data for the growth of yeast in a culture. Use equation (12) with $\beta = 0.55$ and $\delta = 8.3 \times 10^{-4}$ to calculate the predicted growth, and find the percentage error between observed and predicted values.

TABLE 1[1]

Time in hours	Observed yeast biomass
0	9.6
1	18.3
2	29.0
3	47.2
4	71.1
5	119.1
6	174.6
7	257.3
8	350.7
9	441.0
10	513.3
11	559.7
12	594.8
13	629.4
14	640.8
15	651.1
16	655.9
17	659.6
18	661.8

[1] Data from R. Pearl, "The growth of population," *Quarterly Review of Biology,* vol. 2 (1927), pp. 532–548.

* 34. **Obsidian dating‡** is a technique used by anthropologists that allows the dating of certain artifacts well beyond the reliable 40,000-year range of ^{14}C dating. Obsidian, a glassy volcanic rock, absorbs water from the atmosphere; the water seeps into the glass to form a hydration layer, which consists of a compound of water molecules and obsidian molecules. The depth of the hydration layer $x(t)$ beneath the surface is a function of the time t that has elapsed since the manufacture of the artifact. The velocity at which the hydration layer grows is inversely proportional to its depth. Find the depth of the layer for all time t.

35. A pond is shaped like a cone of radius r and depth d. Water flows into the pond at a constant rate i and is lost through evaporation at a rate proportional to the surface area.

   (a) Show that the volume $V(t)$ of water at time t satisfies the differential equation

   $$V' = i - k\pi\left(\frac{3rV}{\pi d}\right)^{2/3}.$$

   (b) Solve this equation.

   (c) What condition must be satisfied for the pond not to overflow?

** 36. (Snowplow chase)§ If a second snowplow starts at noon along the same path as the snowplow in Problem 32, when does it catch up to the first snowplow?

** 37. (Snowplow collision)¶ Suppose three identical snowplows start clearing the same road at 10 a.m., 11 a.m., and noon. If all three collide sometime after noon, when did it start snowing?

1.3 Classification of Differential Equations and Direction Fields

It should be apparent, if only from reading the examples in the previous section, that a great variety of types of differential equations can arise in the study of familiar phenomena. It is clearly necessary (and expedient) to study, independently, more restricted classes of these equations.

† Based on problem E275 of the Otto Dunkel Memorial Problem Book, *American Mathematical Monthly,* vol. 64 (1957), p. 54.

‡ We wish to thank H. W. Vayo for bringing to our attention the article by I. Friedman and R. L. Smith, "A new dating method using obsidian," *American Antiquity,* vol. 25 (1960), pp. 476–522.

§ This problem was first proposed by Fred Wan, *Applied Mathematics Notes* (January 1975), pp. 6–11.

¶ This problem was first proposed by M. S. Klamkin, *American Mathematical Monthly* 59 (January 1952), p. 42 (problem E963).

*Ordinary Differential Equation/
Partial Differential Equation*

The most obvious classification is based on the nature of the derivative(s) in the equation. A differential equation involving only ordinary derivatives (derivatives of functions of one variable) is called an **ordinary differential equation,** whereas one containing partial derivatives is called a **partial differential equation.** We shall postpone the further classification of partial differential equations until Chapter 12.

DEFINITION

Order

The **order** of a differential equation is defined as the order of the highest derivative appearing in the equation.

EXAMPLE 1

The following are examples of differential equations with indicated orders.

$$\textbf{(a)}\, dy/dx = ay \qquad\qquad\qquad \text{(first order)}$$
$$\textbf{(b)}\ \ x''(t) - 3x'(t) + x(t) = \cos t \quad \text{(second order)}$$
$$\textbf{(c)}\,(y^{(4)})^{3/5} - 2y'' = \cos x \qquad \text{(fourth order)}$$

Much of this book concerns solving differential equations. Thus, we need to define what we mean by a **solution.**

DEFINITION

Solution of an nth-*Order Differential
Equation*

A **solution of an nth-order differential equation** is a function that is n times differentiable and that satisfies the differential equation.

Symbolically, this means that a solution of the differential equation

$$F(x, y, y', \ldots, y^{(n)}) = 0 \qquad\qquad\qquad\qquad \textbf{(1)}$$

is a function $y(x)$, whose derivatives $y'(x)$, $y''(x)$, $\ldots$, $y^{(n)}(x)$ exist and satisfy the equation

$$F(x, y(x), y'(x), \ldots, y^{(n)}(x)) = 0$$

for all values of the independent variable x in any interval where (1) is defined.

EXAMPLE 2

The differential equation in Example 1.1.2 can be rewritten in the form $y' - \alpha y = 0$. It is easy to check that $y(x) = ke^{\alpha x}$ is a solution for all x and any real number k:

$$y'(x) - \alpha y(x) = (ke^{\alpha x})' - \alpha(ke^{\alpha x}) = \alpha ke^{\alpha x} - \alpha ke^{\alpha x} = 0.$$

As we saw in Sections 1.1 and 1.2, we are often interested in solving a first-order differential equation

$$\frac{dy}{dx} = f(x, y)$$

subject to a side condition

$$y(x_0) = y_0.$$

This is an example of an **initial value problem.** The side condition is called an **initial condition** and x_0 is called the **initial point.** More generally:

DEFINITION

Initial Value Problem

An **initial value problem** consists of a differential equation (of any order) together with a collection of initial conditions that must be satisfied by the solution of the differential equation and its derivatives at the initial point.

EXAMPLE 3

The following are examples of initial value problems.

(a) $dy/dx = 2y - 3x$, $y(0) = 2$. (Here $x = 0$ is the initial point.)
(b) $x''(t) + 5x'(t) + (\sin t)x(t) = 0$, $x(1) = 0$, $x'(1) = 7$. (Here $t = 1$ is the initial point.)

DEFINITION

Solution of an Initial Value Problem

We define a **solution of an** nth-order **initial value problem** as a function that is n times differentiable, satisfies the given differential equation, and satisfies the given initial conditions.

EXAMPLE 4

The function $y(x) = 2e^{3x}$ is a solution of the initial value problem

$$\frac{dy}{dx} = 3y, \qquad y(0) = 2$$

because $y(0) = 2e^{3\cdot 0} = 2e^0 = 2$ and

$$\frac{dy}{dx} = 2\frac{d}{dx}(e^{3x}) = 6e^{3x} = 3(2e^{3x}) = 3y.$$

Trivial Solution

If $y \equiv 0$ (that is, y identically equal to zero) is a solution to a differential equation on an interval I, then $y \equiv 0$ is called the **trivial solution** to that differential equation on I.

For example, $y \equiv 0$ is a solution to the differential equation in Example 4, although it is not a solution to the initial value problem because it does not satisfy the initial condition $[y(0) = 0$, not 2].

Implicit and Explicit Solutions

The solution to a differential equation may be given **implicitly** or **explicitly.** If we write the solution in the form $y = f(x)$, then our solution is given explicitly. If we write our solution in the form $f(x, y) = C$, a constant, then the solution is given implicitly.

EXAMPLE 5

(a) The solution $y = 2e^{3x}$ is an explicit solution of the differential equation in Example 4.

(b) Consider the differential equation

$$\frac{dy}{dx} = \frac{x}{y}$$

Then, separating variables, we obtain

$$y \, dy = x \, dx$$

$$\frac{y^2}{2} = \frac{x^2}{2} + c$$

$$y^2 - x^2 = 2c = C$$

Here the solution is given implicitly.

Note. We cannot here write y explicitly as a *function* of x because $y = \pm\sqrt{x^2 + C}$ is not a function.

DEFINITION

Boundary Value Problem

A **boundary value problem** consists of a differential equation and a collection of values that must be satisfied by the solution of the differential equation or its derivatives at no less than two different points.

EXAMPLE 6

The following are examples of boundary value problems.

(a) $\dfrac{d^2y}{dx^2} + 5xy = \cos x, \ y(0) = 0, \ y'(1) = 2$

(b) $\dfrac{dy}{dx} + 5xy = 0, \ y(0)y(1) = 2$

We shall postpone further consideration and discussion of boundary value problems until Chapter 11.

In each of Examples 1.1.1–1.1.4, we saw that once initial conditions were satisfied, the resulting initial value problem had a unique solution. [Note that two initial conditions were required in Example 1.1.1: the initial height $h(0)$ and the initial velocity $h'(0)$.] It is reasonable to ask if every initial value problem has a unique solution. Essentially, we are asking two questions:

1. Is there a solution to the problem?
2. If there is a solution, is it the only one?

As we shall see in the next two examples, the answer may be *no* to each question.

EXAMPLE 7

The initial value problem

$$\left(\frac{dy}{dx}\right)^2 + y^2 + 1 = 0, \qquad y(0) = 1$$

has no real-valued solutions, since the left-hand side is always positive for real-valued functions.

EXAMPLE 8

The initial value problem

$$\frac{dy}{dx} = xy^{1/3}, \qquad y(0) = 0 \tag{2}$$

has at least two solutions in the interval $-\infty < x < \infty$. Check that the functions

$$y \equiv 0 \quad \text{and} \quad y = \frac{1}{3\sqrt{3}} x^3$$

both satisfy the initial condition and the differential equation in (2).

Because of examples such as these, it is useful to obtain theorems that guarantee the existence of a unique solution. The following result, due originally to Cauchy† and developed in a more general form by Picard, is very popular because of the ease with which its hypotheses are checked. We shall not prove this result here.‡

■ **THEOREM 1**

Let f and $\partial f/\partial y$ be continuous in a rectangle R given by $a < x < b$, $c < y < d$ that contains the point (x_0, y_0) (see Fig. 1). Then, in an interval $x_0 - h < x < x_0 + h$ contained in $a < x < b$, there is a unique solution $y = y(x)$ of the initial value problem

$$\frac{dy}{dx} = f(x, y), \qquad y(x_0) = y_0.$$

■

† Cauchy gave a series of lectures at the École Polytechnique in Paris between 1820 and 1830. This theorem (and many other great results) was presented in the course of these lectures. The theorem was first published in 1835. See the biographical sketch on pages 24–25.

‡ For a proof of this and other results see W. R. Derrick and S. I. Grossman, *Introduction to Differential Equations with Boundary Value Problems*, 3rd ed., West Publishing Co., St. Paul, Minn., 1987, Appendix 3.

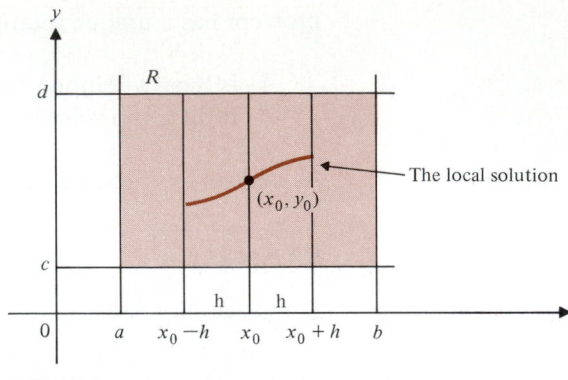

FIGURE 1

Local Existence-Uniqueness Theorem

Theorem 1 is called an **existence-uniqueness theorem,** because it provides criteria guaranteeing the existence of a unique solution. It requires that we check that f and $\partial f/\partial y$ are continuous functions in a rectangle R containing the initial point. In many of our examples, f and $\partial f/\partial y$ are continuous everywhere. When these criteria hold, the theorem guarantees that a unique solution will exist in some region containing the initial point (see Fig. 1). Since the guarantee is only for a small region around the initial point, we call such a theorem a **local** existence-uniqueness theorem. We illustrate how to utilize Theorem 1.

EXAMPLE 9

Does the initial value problem

$$\frac{dy}{dx} = x^2 + y^3, \qquad y(0) = 1$$

have a unique solution in a region around its initial point $(0, 1)$?

SOLUTION
Since $f(x, y) = x^2 + y^3$ and $\partial f/\partial y = 3y^2$ are continuous everywhere, they are certainly continuous in any rectangle R containing the initial point $(0, 1)$. Hence, a unique local solution exists.

EXAMPLE 10

If we look again at equation (2) in Example 8, we see that $\partial f/\partial y = x/3y^{2/3}$, which is not defined (and therefore not continuous) at the initial point $(0, 0)$. Thus, Theorem 1 does not apply to (2).

Theorem 1 is only one of many known existence-uniqueness theorems. Its conditions can be relaxed while still retaining its conclusions. We shall refer to a variety of existence-uniqueness theorems in subsequent sections. It can be shown that under certain easily stated conditions that apply to a wide variety of problems:

> There is a unique solution to an nth-order differential equation if the value of the unknown function and all its derivatives up to the $(n-1)$st are specified at a given point.

Although the major portion of this and the next chapter concerns methods for solving differential equations, a basic fact is that many differential equations arising in applications cannot be solved. That is, for many differential equations, it is impossible to express a solution in terms of elementary functions.

There are many ways to deal with this vexing problem. One is to look for numerical solutions. In other words, rather than look for a function $y(x)$ that solves the problem for every value of the independent variable x, we try to find an approximation at one or more values. This approach is discussed further in Section 1.8 and in Chapter 13.

Another approach is to try to describe, without solving the equation, how solutions "behave." Typical questions that can be asked in such a *qualitative* approach are:

1. Do solutions grow without bound as x increases?
2. Do solutions tend to zero?
3. Do solutions oscillate between certain values?

Much research on differential equations centers on finding answers to these questions. In the rest of this section, we describe one relatively simple way to obtain information about the solution to a differential equation. We shall say more about these questions in Chapter 3.

Direction Fields

Consider the first-order differential equation

$$y' = f(x, y). \tag{3}$$

Equation (3) contains a great deal of information. Under certain conditions the differential equation has a unique solution if we specify an initial condition. That is, there is a unique function $y(x)$ satisfying $y'(x) = f(x, y)$ and $y(x_0) = y_0$ for arbitrarily chosen numbers x_0 and y_0. The function $y(x)$ is a curve in the xy-plane. Even though we may not be able to find $y(x)$, *we know the slope of y at every point on the curve.* If the solution $y(x)$ passes through the point (x, y), then since $y' = f(x, y)$:

> The slope of the tangent line to the curve $y(x)$ at the point (x, y) is given by $f(x, y)$.

Direction Field

Thus, we know the direction of the solution curve $y(x)$ through any point in the xy-plane at which $f(x, y)$ is defined. The set of all these directions in the plane is called the **direction field** of the differential equation $y' = f(x, y)$. In many cases, we can use the direction field to sketch the solution to a differential equation without actually computing it.

EXAMPLE 11

Consider the initial value problem

$$y' = 2xy, \qquad y(0) = 1. \tag{4}$$

We have

$$y'(x) > 0, \qquad \text{if} \quad xy > 0$$

AUGUSTIN-LOUIS CAUCHY
1789–1857

Courtesy of the Granger Collection

Augustin-Louis Cauchy is considered the most outstanding mathematical analyst of the first half of the nineteenth century. He was born in Paris in 1789 and received his early education from his father. In secondary school he excelled at classical studies. Entering the École Polytechnique in 1805, Cauchy greatly impressed two of the greatest French mathematicians of the time: Joseph Lagrange (1736–1813) and Simon Laplace (1749–1827). Although Cauchy studied to be a civil engineer, he was persuaded by Lagrange and Laplace to accept a professorship of mathematics at the École Polytechnique.

Cauchy made many contributions to calculus. In his 1829 textbook *Leçons sur le calcul différentiel,* he gave the first reasonably clear definition of a limit and was the first to define the derivative as the limit of the difference quotient:

$$\frac{\Delta y}{\Delta x} = \frac{f(x + \Delta x) - f(x)}{\Delta x}.$$

He was also responsible for the modern definition of the definite integral as the limit of a sum.

Cauchy wrote extensively in both pure and applied mathematics. Only Euler wrote more. He contributed to many areas including real and complex function theory, determinants, probability theory, geometry, wave propagation theory, and infinite series. He provided the first rigorous proof of the existence of solutions to first-order differential equations.

Cauchy is credited with setting a new standard of rigor in mathematical publication. After Cauchy, it was much more difficult to publish a paper based on intuition; a strict adherence to formal proof was demanded.

The sheer volume of Cauchy's publication was overwhelming. When

and

$$y'(x) < 0, \qquad \text{if} \quad xy < 0.$$

Thus $y'(x) > 0$ in the first and third quadrants and $y'(x) < 0$ in the second and fourth quadrants. The direction field is sketched in Fig. 2. This figure consists of arrows with ends at a grid of points (x, y) and slopes $f(x, y) = 2xy$, where x and y are the coordinates of each grid point. Note that since x and y are positive in the first quadrant, the slopes of the tangent lines to any solution curve are positive, so that the solution curves increase and become steeper as x and y become larger. Along the axes, the solution is flat (has a horizontal tangent) because the derivative y' is zero. In the second quadrant, the slopes of the tangent lines are negative since $y' < 0$. Similar conditions apply in the third and fourth quadrants.

For the particular initial value problem we are considering in equation (4), we know that the solution curve must satisfy the initial condition $y(0) = 1$. Thus, the solution

the French Academy of Sciences began publishing its journal *Comptes Rendues* in 1835, Cauchy sent his work there to be published. Soon the printing bill for Cauchy's work alone became so large that the Academy placed a limit of four pages on each published paper. This rule is still in force today.

There are some unpleasant stories told of Cauchy. One of the most tragic had to do with the Norwegian mathematician Niels Henrik Abel (1802–1829). In 1826 Abel, who had already published some brilliant results, came to Paris in search of an academic position. He approached Cauchy and gave him an important paper he had just completed. Cauchy misplaced it. Abel wrote a friend, "Every beginner has a great deal of difficulty in getting noticed here. I have just finished an extensive treatise on a certain class of transcendental functions . . . but Mr. Cauchy scarcely deigned to glance at it." While waiting for a suitable position, Abel lived in an unheated apartment in Paris. In 1829, he died of tuberculosis. Ironically, a letter offering Abel a professorship of mathematics at the University of Berlin arrived two days after his death.

Cauchy was a political reactionary and a strong supporter of the Bourbons, the French kings who came to power in the years after the French revolution. When King Charles X, the last of the Bourbons, went into exile in 1830, Cauchy was forced to resign his position at the École Polytechnique. He was not allowed to return until 1848—and even then he refused to swear allegiance to the new government. He was also a religious bigot and spent much of his time attempting to convert others to his beliefs.

Cauchy was, however, a courageous defender of academic freedom. In 1843 he published a sharply worded letter in defence of freedom of conscience. This letter was partially responsible for the abolition of the oath of allegiance that Cauchy had so stubbornly refused to sign.

Cauchy died in 1857 at the age of 67 of a bronchial ailment. His last words were spoken to the Archbishop of Paris: "Men pass away, but their deeds abide."

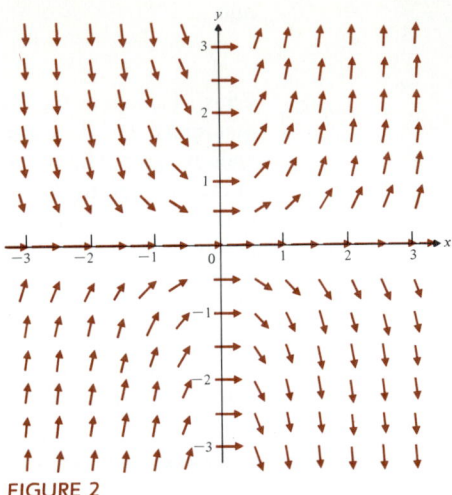

FIGURE 2

curve must pass through the point $(0, 1)$. Since that point is on the y-axis, the curve is initially flat and moves into the first quadrant with increasing values of x. As x increases, the solution curve begins to rise as $y'(x) > 0$ in the first quadrant. Hence $y > 1$ for all values of $x > 0$. On the other hand, if we allow x to be negative, the solution curve extends into the second quadrant. Since the slope is negative in this quadrant, the solution curve is decreasing as the curve moves to the right. Because xy becomes larger in absolute value as we move away from the y-axis, the curve becomes steeper. Putting this all together, we obtain the sketch in Fig. 3.

Of course, in this case the differential equation is easy to solve by separating variables:

$$\int \frac{dy}{y} = 2 \int x\,dx + C,$$

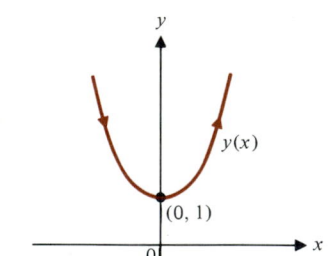

FIGURE 3

so that

$$\ln|y| = x^2 + C$$

or

$$y(x) = C_1 e^{x^2}.$$

Finally, since $y(0) = 1$, it follows that $C_1 = 1$ so that the solution we have found by using the direction field is

$$y = e^{x^2}.$$

EXAMPLE 12

Draw the direction field for the differential equation of logistic growth

$$\frac{dP}{dt} = P(\beta - \delta P)$$

where β and δ are positive constants.

SOLUTION

This is equation (1.2.8) (see p. 12). In Example 1.2.3 we found the solution

$$P(t) = \frac{\beta}{\delta + \left[\dfrac{\beta}{P(0)} - \delta\right]e^{-\beta t}}. \tag{5}$$

However, the direction field can be found without solving the equation. Since β and δ are positive, the following facts are evident:

$$P' < 0 \quad \text{if} \quad P < 0; \qquad\qquad P' = 0 \quad \text{if} \quad P = 0;$$

$$P' > 0 \quad \text{if} \quad 0 < P < \beta/\delta; \qquad P' = 0 \quad \text{if} \quad P = \beta/\delta;$$

$$P' < 0 \quad \text{if} \quad P > \beta/\delta.$$

Drawing arrows at the points (t, P) to indicate the slope of the tangent line of the solution through that point, we obtain the direction field shown in Fig. 4.

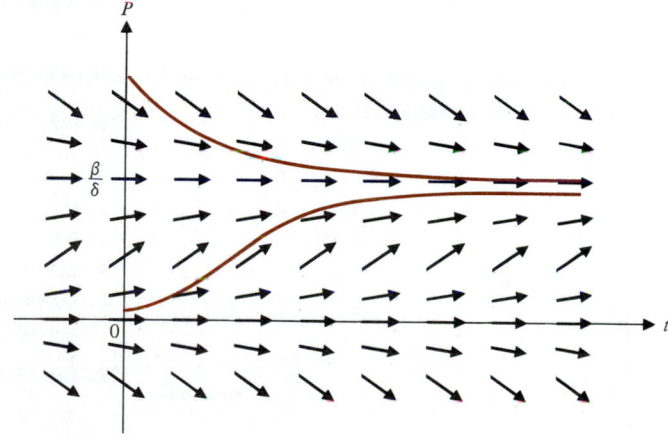

FIGURE 4 Direction field for $P' = P(\beta - \delta P)$.

In particular, any solution for which $P(0) > 0$, must approach the value β/δ, called the **carrying capacity** of the environment. This fact is confirmed by taking the limit at $t \to \infty$ of the solution given by (5).

Of course, if we can solve the differential equation directly, we need not plot the direction field. Nevertheless, direction fields provide a quick and useful, if crude, tool for getting an idea of the shape of the solution. And if a solution is not readily obtainable, direction fields provide an important first step in analyzing the behavior of solutions.

Problems 1.3

In Problems 1–7 state the order of each of the following differential equations.

1. $y' + ay = \sin^2 x$

2. $(d^2x/dt^2)^3 - 3x\, dx/dt = 4\cos t$

3. $s'''(t) - s''(t) = 0$

4. $d^5y/dx^5 = 0$

5. $y'' + y = 0$

6. $(dx/dt)^3 = x^5$

7. $x' - x^2 = 3x'''$

In Problems 8–13 state whether each of the following differential equations is an initial value or a boundary value problem.

8. $y'' + \omega^2 y = 0$, $y(0) = 0$, $y(1) = 1$ (here ω^2 is a constant)

9. $y'' + \omega^2 y = 0$, $y(0) = 0$, $y'(0) = 1$ (ω^2 constant)

10. $y'' + \omega^2 y = 0$, $y(0) = 0$, $y'(1) = 1$ (ω^2 constant)

11. $(dx/dt)^3 - 4x^2 = \sin t$, $x(0) = 3$

12. $y''' + 3y'' - (y')^2 + e^y = \sin x$, $y(0) = 0$, $y'(0) = 3$, $y''(0) = 5$

13. $y''' + 3y'' - (y')^2 + e^y = \sin x$, $y(0) = y(1) = 0$, $y'(0) = 2$

In Problems 14–18, verify that the given function or functions are solutions to the given differential equation.

14. $y'' + y = 0$; $y_1 = 2 \sin x$, $y_2 = -5 \cos x$

15. $y''' - y'' + y' - y = x$; $y_1 = e^x - x - 1$, $y_2 = 3 \cos x - x - 1$, $y_3 = \cos x + \sin x + e^x - x - 1$

16. $x^2 y'' - 2xy' + 2y = 0$; $y_1 = x$, $y_2 = x^2$, $y_3 = 2x - 3x^2$

17. $y'' - y = e^x$; $y_1 = \dfrac{x}{2} e^x$, $y_2 = \left(4 + \dfrac{x}{2}\right)e^x + 3e^{-x}$

18. $x^2 y'' + 5xy' + 4y = 0$; $y_1 = \dfrac{4 \ln x}{x^2}$, $y_2 = \dfrac{-6}{x^2}$ $(x > 0)$

19. By "guessing" that there is a solution to the equation
$$y'' - 4y' + 5y = 0$$
of the form
$$y = e^{ax} \cos bx,$$
find this solution. Can you "guess" a second solution?

20. By "guessing" that there is a solution to
$$y'' - 3y' - 4y = 0$$
of the form
$$y = e^{ax}$$
for some constant a, find two solutions of the equation.

21. Given that $y_1(x)$ and $y_2(x)$ are two solutions of the equation in Problem 20 check to see that $y_3(x) = c_1 y_1(x) + c_2 y_2(x)$ is also a solution, where c_1 and c_2 are arbitrary constants, by substituting y_3 into the differential equation.

22. Determine $\varphi(x)$ so that the functions $y = \sin \ln x$ and

$y = \cos \ln x (x > 0)$ are solutions of the differential equation
$$\lceil \varphi(x) y' \rceil' + \frac{y}{x} = 0.$$

23. Show that $\sin(1/x)$ and $\cos(1/x)$ are solutions of the differential equation
$$\frac{d}{dx}\left(x^2 \frac{dy}{dx}\right) + \frac{y}{x^2} = 0.$$

24. Verify that $y_1 = \sinh x$ and $y_2 = \cosh x$ are solutions of the differential equation $y'' - y = 0$. [*Hint:* $\cosh x = \frac{1}{2}(e^x + e^{-x})$; $\sinh x = \frac{1}{2}(e^x - e^{-x})$.]

25. Suppose that $\varphi(x)$ is a solution of the initial value problem $y'' + yy' = x^3$, with $y(-1) = 1$, $y'(-1) = 2$. Find $\varphi''(-1)$ and $\varphi'''(-1)$. [*Hint:* Differentiate.]

26. Let $\varphi(x)$ be a solution to $y' = x^2 + y^2$ with $y(1) = 2$. Find $\varphi'(1)$, $\varphi''(1)$, and $\varphi'''(1)$.

Direction Fields

27. (a) Plot the direction field for the differential equation
$$y' = y^{4/5}.$$
   (b) Plot the solution that satisfies $y(0) = 2$.
   (c) Plot the solution that satisfies $y(0) = -1$.

28. (a) Plot the direction field for the equation $y' = y^{3/5}$.
   (b) Plot the solution that satisfies $y(0) = 3$.
   (c) Plot the solution that satisfies $y(0) = -2$.

29. (a) Plot the direction field for the equation
$$y' = 2y(3 - y).$$
   (b) Plot the solution that satisfies $y(0) = -1$.
   (c) Plot the solution that satisfies $y(0) = 2$.
   (d) Plot the solution that satisfies $y(0) = 4$.

30. (a) Plot the direction field for the equation
$$y' = (y^2 - 4)(y - 4).$$
   (b) Plot the solution that satisfies $y(0) = -3$.
   (c) Plot the solution that satisfies $y(0) = -1$.
   (d) Plot the solution that satisfies $y(0) = 1$.
   (e) Plot the solution that satisfies $y(0) = 3$.
   (f) Plot the solution that satisfies $y(0) = 5$.

31. (a) Plot the direction field for the equation
$$y' = x^2 + y^2.$$
   (b) Plot the solution that satisfies $y(1) = 2$.

32. (a) Plot the direction field of the equation
$$y' = \frac{2xy}{1 + y^2}.$$

(b) Plot the solution that satisfies $y(1) = 1$.

(c) Plot the solution that satisfies $y(1) = -1$.

33. Plot the direction field of the equation

$$y' = \frac{4y - 5x}{y + x}.$$

34. (a) Plot the direction field of the equation

$$y' = 1 + x + y.$$

(b) Plot the solution that satisfies $y(0) = 1$.

35. (a) Plot the direction field of the equation

$$y' = e^{xy} - 1.$$

(b) Discuss the solution that passes through the origin.

(c) Compare the solution through $(0, 1)$ to that through $(0, -1)$.

1.4 Linear First-Order Differential Equations

Linear Differential Equation

DEFINITION

An nth-order differential equation is **linear** if it can be written in the form

$$\frac{d^n y}{dx^n} + a_{n-1}(x)\frac{d^{n-1} y}{dx^{n-1}} + \cdots + a_1(x)\frac{dy}{dx} + a_0(x)y = f(x).$$

Hence a first-order linear equation has the form

$$\frac{dy}{dx} + a(x)y = f(x),$$

while a second-order linear equation can be written as

$$\frac{d^2 y}{dx^2} + a(x)\frac{dy}{dx} + b(x)y = f(x).$$

The notation indicates that $a(x)$, $b(x)$, $f(x)$, etc. are functions of x alone.

DEFINITION

Nonlinear Differential Equation

Any differential equation that cannot be written in the form above is said to be **nonlinear.**

For example,

$$\frac{dy}{dx} = y^2$$

Homogeneous and Nonhomogeneous Differential Equations

is nonlinear. If the function $f(x)$ is the zero function, then the linear differential equation is said to be **homogeneous;** when $f(x) \neq 0$ we say the differential equation is **nonhomogeneous.**

Before dealing with the general nonhomogeneous first-order linear equation

$$\frac{dy}{dx} + a(x)y = f(x) \tag{1}$$

it is important to discuss the solution of the homogeneous equation

$$\frac{dy}{dx} + a(x)y = 0 \quad \text{or} \quad \frac{dy}{dx} = -a(x)y. \tag{2}$$

Separating variables, we have

$$\int \frac{dy}{y} = -\int a(x)\,dx + C$$

so that, integrating,

$$\ln|y| = -\int a(x)\,dx + C$$

and

$$y = C_1 e^{-\int a(x)\,dx} \qquad (C_1 = \pm e^C). \tag{3}$$

General Solution

Equation (3) is the **general solution** to equation (2). That is, it is the set of all solutions to the differential equation.

We have shown that the general solution to (2) is obtainable in terms of functions with which we are familiar whenever the antiderivative can be found. We illustrate this situation with two examples.

EXAMPLE 1

Solve the homogeneous differential equation

$$y' + 3y = 0$$

SOLUTION
From Example 1.1.2 we see that the general solution is

$$y = ke^{-3x}.$$

EXAMPLE 2

Solve the equation

$$\frac{dy}{dx} = -2xy, \qquad y(1) = 1. \tag{4}$$

SOLUTION
Separating variables, we have

$$\int \frac{dy}{y} = -2 \int x\,dx + C,$$

$$\ln|y| = -x^2 + C,$$

or

$$y = C_1 e^{-x^2}.$$

Finally, since $1 = y(1) = C_1 e^{-1}$, it follows that $C_1 = e$, so that the particular solution to the initial value problem (4) is

$$y = e^{1-x^2}.$$

We see that equation (3) can be written in the form

$$ye^{\int a(x)\,dx} = C \tag{5}$$

by multiplying both sides by $e^{\int a(x)\,dx}$. If we differentiate both sides of equation (5), we obtain

$$[y' + a(x)y]e^{\int a(x)\,dx} = 0 \tag{6}$$

since the derivative of an indefinite integral is the integrand. Notice that the expression in brackets is the left side of the original differential equation (2). We call the exponential

$$e^{\int a(x)\,dx} \tag{7}$$

Integrating Factor   an **integrating factor** for the linear equation (2).

We can capitalize on this fact to obtain the solution of the nonhomogeneous linear equation

$$\frac{dy}{dx} + a(x)y = f(x). \tag{8}$$

We multiply both sides of equation (8) by the integrating factor (7) to obtain

$$[y' + a(x)y]e^{\int a(x)\,dx} = f(x)e^{\int a(x)\,dx}. \tag{9}$$

But, as we saw in going from equation (5) to equation (6), the left side of equation (9) is the derivative of $ye^{\int a(x)\,dx}$. Thus

$$\frac{d}{dx}(ye^{\int a(x)\,dx}) = f(x)e^{\int a(x)\,dx}. \tag{10}$$

Integrating both sides of equation (10) with respect to x, we have

$$\int \frac{d}{dx}(ye^{\int a(x)\,dx})\,dx = \int f(x)e^{\int a(x)\,dx}\,dx + C$$

or

$$ye^{\int a(x)\,dx} = \int f(x)e^{\int a(x)\,dx}\,dx + C.$$

This can be written as

$$y = \left[\int f(x)e^{\int a(x)\,dx}\,dx + C\right]e^{-\int a(x)\,dx}. \tag{11}$$

Remark. Equation (11) provides an expression for the general solution of the first-order nonhomogeneous linear differential equation (8). It is usually better to go through the process of multiplying both sides of (8) by the integrating factor to obtain the solution than to try to memorize equation (11). We illustrate the use of integrating factors with several examples.

EXAMPLE 3

Solve the nonhomogeneous linear equation

$$y' = y + x^2, \qquad y(0) = 1. \tag{12}$$

SOLUTION

Rewriting equation (12) in the form

$$y' - y = x^2 \tag{13}$$

we see that $a(x) = -1$, so that the integrating factor is

$$e^{-\int dx} = e^{-x}.$$

Multiplying both sides of equation (13) by e^{-x}, we get

$$e^{-x}(y' - y) = x^2 e^{-x}$$

or

$$(ye^{-x})' = x^2 e^{-x}.$$

Integrating both sides, we have

$$ye^{-x} = \int x^2 e^{-x}\, dx + C$$

integrate by parts twice
$$\downarrow$$
$$= C - (x^2 + 2x + 2)e^{-x}$$

so that

$$y = Ce^x - (x^2 + 2x + 2).$$

Finally, setting $x = 0$, we get

$$1 = y(0) = C - 2,$$

implying that $C = 3$ and the solution of the initial value problem is

$$y = 3e^x - (x^2 + 2x + 2).$$

Useful Hint. When writing the solution, put in the constant of integration, C, as soon as you write $\int f(x)e^{\int a(x)\,dx}\,dx + C$. If you leave it out, you may forget it later and you will almost certainly get the wrong answer. For example, in Example 3 we wrote $ye^{-x} = \int x^2 e^{-x}\,dx + C$. If we had omitted the C, we would have ended up with $y = -x^2 + 2x + 2$, which does not satisfy $y(0) = 1$. In fact, without the Ce^x term it is impossible to find a solution that satisfies the initial condition.

EXAMPLE 4

Consider the equation $dy/dx = x^3 - 2xy$, where $y = 1$ when $x = 1$. Rewriting the equation as $dy/dx + 2xy = x^3$, we see that $a(x) = 2x$ and the integrating factor is $e^{\int a(x)\,dx} = e^{x^2}$. Thus multiplying both sides by e^{x^2} and integrating, we have

$$e^{x^2}y = \int x^3 e^{x^2}\,dx + C,$$

so that

$$y = e^{-x^2}\left[\int x^3 e^{x^2}\,dx + C\right].$$

We can integrate the integral by parts: Let $u = x^2$ and $dv = xe^{x^2}dx$. Then $du = 2x\,dx$, $v = \frac{1}{2}e^{x^2}$ and

$$\int x^3 e^{x^2}\,dx = \int x^2(xe^{x^2})\,dx = \frac{x^2 e^{x^2}}{2} - \int xe^{x^2}\,dx = e^{x^2}\left(\frac{x^2 - 1}{2}\right).$$

Thus replacing this term for the integral above, we have

$$y = e^{-x^2}\left[e^{x^2}\left(\frac{x^2 - 1}{2}\right) + C\right] = \frac{x^2 - 1}{2} + Ce^{-x^2}.$$

Setting $x = 1$ yields

$$1 = y(1) = Ce^{-1}.$$

Thus $c = e$ and the solution to the problem is

$$y = \tfrac{1}{2}(x^2 - 1) + e^{1-x^2}.$$

Bernoulli's Equation†

Certain nonlinear first-order equations can be reduced to linear equations by a suitable change of variables. The equation

$$\frac{dy}{dx} + a(x)y = f(x)y^n, \tag{14}$$

Bernoulli's Equation

which is known as **Bernoulli's Equation,** is of this type. Set $z = y^{1-n}$. Then $z' = (1 - n)y^{-n}y'$, so if we multiply both sides of equation (14) by $(1 - n)y^{-n}$, we obtain

$$(1 - n)y^{-n}y' + (1 - n)a(x)y^{1-n} = (1 - n)f(x)$$

or

$$\frac{dz}{dx} + (1 - n)a(x)z = (1 - n)f(x).$$

The equation is now linear and may be solved as before.

EXAMPLE 5

Solve

$$\frac{dy}{dx} - \frac{y}{x} = -\frac{5}{2}x^2 y^3. \tag{15}$$

† See the accompanying biographical sketch.

JAKOB BERNOULLI
(1654–1705)

Courtesy of Brown Brothers

One of the most distinguished families in the history of mathematics and science is the Bernoulli family of Switzerland, which, from the late seventeenth century on, produced an unusual number of capable mathematicians and scientists. The family record starts with the two brothers Jakob Bernoulli and Johann Bernoulli. These two men gave up earlier vocational interests and became mathematicians when Leibniz's papers began to appear in the *Acta eruditorum.* They were among the first mathematicians to realize the surprising power of the calculus and to apply the tool to a great diversity of problems. From 1687 until his death, Jakob occupied the mathematics chair at Basel University. The two brothers, often bitter rivals, maintained an almost constant exchange of ideas with Leibniz and with each other.

Among Jakob Bernoulli's contributions to mathematics are the early use of polar coordinates, the derivation in both rectangular and polar coordinates of the formula for the radius of curvature of a plane curve, the study of the catenary curve with extensions to strings of variable density and strings under the action of a central force, the study of a number of other higher plane curves, the discovery of the so-called **isochrone**—or curve along which a body will fall with uniform vertical velocity (it turned out to be a semicubical parabola with a vertical cusptangent), the determination of the form taken by an elastic rod fixed at one end and carrying a weight at the other, the form assumed by a flexible rectangular sheet having two opposite edges held horizontally fixed at the same height and loaded with a heavy liquid, and the shape of a rectangular sail filled with wind. He also proposed and discussed the problem of isoperimetric figures (planar closed paths of given species and fixed perimeter which include a maximum area) and was thus one of the first mathematicians to work in the calculus of variations. He was also one of the early students of mathematical probability; his book in this field, the *Ars conjectandi,* was published posthumously in 1713.

Several results in mathematics now bear Jakob Bernoulli's name. Among these are the *Bernoulli distribution* and *Bernoulli theorem* of statistics and probability theory, the *Bernoulli equation* met by every student in a first course in differential equations, the *Bernoulli numbers* and *Bernoulli polynomials* of number theory interest, and the *lemniscate of Bernoulli* encountered in many calculus courses. In Jakob Bernoulli's solution to the problem of the isochrone curve, which was published in the *Acta eruditorum* in 1690, we meet for the first time the word "integral" in a calculus sense. Leibniz had called the integral calculus *calculus summatorius*; in 1696 Leibniz and Johann Bernoulli agreed to call it *calculus integralis.*

Jakob Bernoulli was struck by the way the equiangular spiral reproduces itself under a variety of transformations and asked, in imitation of Archimedes, that such a spiral be engraved on his tombstone, along with the inscription *Eadem mutata resurgo* ("I shall arise the same, though changed").

SOLUTION

Here $n = 3$, so we let $z = y^{-2}$ and $z' = -2y^{-3}y'$ and multiply both sides of equation (15) by $-2y^{-3}$ to obtain

$$-2y^{-3}y' + \frac{2}{x}y^{-2} = 5x^2$$

or

$$z' + \frac{2z}{x} = 5x^2. \tag{16}$$

The integrating factor for this linear equation is

$$e^{2\int dx/x} = e^{2\ln|x|} = e^{\ln x^2} = x^2.$$

Multiplying both sides of equation (16) by x^2 we have

$$x^2 z' + 2xz = 5x^4$$

$$(x^2 z)' = 5x^4$$

so that

$$x^2 z = 5 \int x^4\, dx + C = x^5 + C.$$

Hence,

$$y^{-2} = z = x^3 + Cx^{-2}$$

or

$$y = \pm(x^3 + Cx^{-2})^{-1/2}.$$

A similar procedure can be used to solve

$$\frac{dy}{dx} + a(x)y = f(x)y \ln y. \tag{17}$$

We let $z = \ln y$. Then $z' = y'/y$, so that dividing equation (17) by y, we obtain the linear equation

$$\frac{dz}{dx} + a(x) = f(x)z.$$

Problems 1.4

In Problems 1–11, find the general solution for each equation. When an initial condition is given, find the particular solution that satisfies the condition.

1. $\dfrac{dx}{dt} = 3x$

2. $\dfrac{dy}{dx} + 22y = 0, \quad y(1) = 2$

3. $\dfrac{dx}{dt} = x + 1, \quad x(0) = 1$

4. $\dfrac{dy}{dx} + y = \sin x, \quad y(0) = 0$

5. $\dfrac{dx}{dy} - x \ln y = y^y$

6. $\dfrac{dy}{dx} + y = \dfrac{1}{1 + e^{2x}}$

7. $\dfrac{dy}{dx} - \dfrac{3}{x} y = x^3, \quad y(1) = 4$

8. $\dfrac{dx}{dt} + x \cot t = 2t \csc t$

9. $x' - 2x = t^2 e^{2t}$

10. $y' + \dfrac{2}{x} y = \dfrac{\cos x}{x^2}, \quad y(\pi) = 0$

11. $\dfrac{ds}{du} + s = ue^{-u} + 1$

12. Solve the equation

$$y - x\frac{dy}{dx} = \frac{dy}{dx} y^2 e^y$$

by reversing the roles of x and y (that is, treat x as the dependent variable).

13. Use the method shown in Problem 12 to solve

$$\frac{dy}{dx} = \frac{1}{e^{-y} - x}.$$

14. Find the solution of $dy/dx = 2(2x - y)$ that passes through the point $(0, -1)$.

15. In a study† on the rate at which education is being forgotten or made obsolete, the following linear first-order differential equation was used:

$$x' = 1 - kx,$$

where $x(t)$ denotes the education of an individual at time t and k is a constant given by the rate at which that education is being lost. Obtain an equation for x at time t.

16. Data collected in a botanical experiment‡ led to the differential equation

$$\frac{dI}{dw} = 0.088(2.4 - I).$$

Find the value of I as $w \to \infty$.

17. Assume that there exists an upper bound B for the size y of a crop in a given field. E. A. Mitscherlich proposed in 1939 the use of the linear differential equation

$$\frac{dy}{dt} = k(B - y)$$

as a model for agricultural growth. Find the general solution of this equation.

18. Suppose a population is growing at a rate proportional to its size and individuals are immigrating into the population at a constant rate.

(a) Find the linear differential equation governing this situation.

(b) Find its general solution.

19. Use the method we developed in this section for the solution of Bernoulli's equation (14) to solve the logistic equation (see Example 1.2.3)

$$\frac{dP}{dt} = P(\beta - \delta P).$$

§20. Let $N(t)$ be the biomass of a fish species in a given area of the ocean and suppose the rate of change of the biomass is governed by the logistic equation

$$\frac{dN}{dt} = rN\left(1 - \frac{N}{K}\right),$$

where K is the carrying capacity (see p. 27) for that species in that area. Assume that the rate at which fish are caught depends on the biomass. If E is the constant effort expended to harvest that species, then

(a) Find the resulting growth rate of the biomass.¶

(b) Solve the differential equation in part (a) by the Bernoulli method.

21. Suppose fish are harvested at a constant rate h independent of their biomass. Answer parts (a) and (b) of Problem 20 for this situation.

22. Find the effort E that will maximize the yield in Problem 20.

23. Show that if $h > rK/4$, in Problem 21 the species will become extinct regardless of the initial size of the biomass.

24. The differential equation governing the velocity v of an object of mass m subject to air resistance proportional to the instantaneous velocity is

$$m\frac{dv}{dt} = -mg - kv.$$

† L. Southwick and S. Zionts, "An Optimal-Control-Theory Approach to the Education Investment Decision," *Operations Research* 22 (1974): 1156–1174.

‡ R. L. Specht, "Dark Island Heath," *Australian Journal of Botany* 5 (1957): 137–172.

§ A number of other examples of models of renewable resources are given in C. W. Clark, *Mathematical Bioeconomics,* Wiley-Interscience, New York, 1976.

¶ This is called the **Schaefer model,** after the biologist M. B. Schaefer.

Solve the equation and determine the limiting velocity of the object as $t \to \infty$.

25. Repeat Problem 24 if air resistance is proportional to the square of the instantaneous velocity.

26. An infectious disease is introduced into a large population. The proportion of people who have been exposed to the disease increases with time. Suppose that $P(t)$ is the proportion of people who have been exposed to the disease within t years of its introduction. If $P'(t) = [1 - P(t)]/3$ and $P(0) = 0$, after how many years will the proportion have increased to 90 percent?

In Problems 27–32 find the general solution for each equation and a particular solution when an initial condition is given.

27. $\dfrac{dy}{dx} = -\dfrac{(6y^2 - x - 1)y}{2x}$

28. $y' = -y^3 x e^{-2x} + y$

29. $x\dfrac{dy}{dx} + y = x^4 y^3, \quad y(1) = 1$

30. $tx^2 \dfrac{dx}{dt} + x^3 = t \cos t$

31. $\dfrac{dy}{dx} + \dfrac{3}{x} y = x^2 y^2, \quad y(1) = 2$

32. $xyy' - y^2 + x^2 = 0$

1.5 Exact Equations

Total Differential

The **total differential** dg of a function of two variables $g(x, y)$ is defined by

$$dg = \frac{\partial g}{\partial x} dx + \frac{\partial g}{\partial y} dy.$$

EXAMPLE 1

Let $g(x, y) = x^2 y^3 + e^{4x} \sin y$. Compute the total differential dg.

SOLUTION

$$\frac{\partial g}{\partial x} = 2xy^3 + 4e^{4x} \sin y$$

and

$$\frac{\partial g}{\partial y} = 3x^2 y^2 + e^{4x} \cos y.$$

Hence

$$dg = (2xy^3 + 4e^{4x} \sin y)\, dx + (3x^2 y^2 + e^{4x} \cos y)\, dy.$$

We shall now use partial derivatives to solve ordinary differential equations. Suppose that we take the total differential of the equation $g(x, y) = c$:

$$dg = \frac{\partial g}{\partial x} dx + \frac{\partial g}{\partial y} dy = 0. \tag{1}$$

For example, the equation $xy = c$ has the total differential $y\, dx + x\, dy = 0$, which may be rewritten as the differential equation $y' = -y/x$. Reversing the situation, suppose that we start with the differential equation

$$M(x, y)\, dx + N(x, y)\, dy = 0. \tag{2}$$

If we can find a function $g(x, y)$ such that

$$\frac{\partial g}{\partial x} = M \quad \text{and} \quad \frac{\partial g}{\partial y} = N,$$

Exact Differential

then equation (2) becomes $dg = 0$, so that $g(x, y) = c$ is the general solution of equation (2). In this case $M\, dx + N\, dy$ is said to be an **exact differential,** and equation (2) is called an **exact differential equation.**

Cross-Derivative Test

It is very easy to determine whether a differential equation is exact by using the **cross-derivative test:** Let M, N, $\partial M/\partial y$, and $\partial N/\partial x$ be continuous over a rectangle. Then the equation $M(x, y)\, dx + N(x, y)\, dy = 0$ is exact if and only if†

$$\frac{\partial M}{\partial y} = \frac{\partial N}{\partial x} \tag{3}$$

If $M\, dx + N\, dy = 0$ is exact, then we can solve the differential equation (2) by finding the function g given above. The procedure for doing this is illustrated in Example 2.

EXAMPLE 2

Solve the equation

$$(1 - \sin x \tan y)\, dx + (\cos x \sec^2 y)\, dy = 0. \tag{4}$$

SOLUTION
Letting $M(x, y) = 1 - \sin x \tan y$ and $N(x, y) = \cos x \sec^2 y$, we have

$$\frac{\partial M}{\partial y} = -\sin x \sec^2 y = \frac{\partial N}{\partial x},$$

so the equation is exact. We now seek a function g of two variables such that $\partial g/\partial x = M$ and $\partial g/\partial y = N$. But if $\partial g/\partial x = M$, then

$$g(x, y) = \int M\, dx = \int (1 - \sin x \tan y)\, dx$$
$$= x + \cos x \tan y + h(y). \tag{5}$$

The "constant of integration" $h(y)$ occurring in (5) is an arbitrary function of y since we must introduce the most general term that vanishes under partial differentiation with respect to x. But

$$\cos x \sec^2 y = N(x, y) = \frac{\partial g}{\partial y} = \underbrace{\cos x \sec^2 y + h'(y)}$$

differentiating (5)
with respect to y

---

† Half of the proof is easy (see Problem 21). The other half is difficult. It can be found, for example, in R. C. Buck and E. F. Buck, *Advanced Calculus,* 3rd ed., McGraw-Hill, New York, 1978, p. 497.

This means that $h'(y) = 0$, so that $h(y) = k$, a constant. Thus the general solution to (5) is

$$g(x, y) = x + \cos x \tan y + k = C \text{ (another constant)}$$

or

$$x + \cos x \tan y = C_1$$

Multiplying by an Integrating Factor to Make an Equation Exact

It should be apparent that exact equations are comparatively rare, since the condition in equation (3) requires a precise balance of the functions M and N. For example,

$$(3x + 2y)\, dx + x\, dy = 0$$

is not exact. However, if we multiply the equation by x, then the new equation

$$(3x^2 + 2xy)\, dx + x^2\, dy = 0$$

is exact. The question we now must ask is: If

$$M(x, y)\, dx + N(x, y)\, dy = 0 \tag{6}$$

Integrating Factor

is not exact, under what conditions does an **integrating factor** $\mu(x, y)$ exist such that

$$\mu M\, dx + \mu N\, dy = 0$$

is exact? Surprisingly, the answer is, whenever equation (6) has a general solution $g(x, y) = c$. To see this, we solve equation (6) for dy/dx. By the chain rule

$$dg = \frac{\partial g}{\partial x} dx + \frac{\partial g}{\partial y} dy = \mu M\, dx + \mu N\, dy = 0$$

so that

$$\frac{dy}{dx} = -\frac{M}{N} = -\frac{\partial g/\partial x}{\partial g/\partial y},$$

from which it follows that

$$\frac{\partial g/\partial x}{M} = \frac{\partial g/\partial y}{N}.$$

Denote either side of the equation above by $\mu(x, y)$. Then

$$\frac{\partial g}{\partial x} = \mu M, \quad \frac{\partial g}{\partial y} = \mu N, \tag{7}$$

and (6) has at least one integrating factor μ. However, finding integrating factors is in general very difficult. Here is one procedure that is sometimes successful.

Since equation (7) indicates that $\mu M\,dx + \mu N\,dy = 0$ is exact, by equation we have

$$\mu\frac{\partial M}{\partial y} + M\frac{\partial \mu}{\partial y} = \frac{\partial}{\partial y}(\mu M) = \frac{\partial}{\partial x}(\mu N) = \mu\frac{\partial N}{\partial x} + N\frac{\partial \mu}{\partial x},$$

so that

$$\frac{1}{\mu}\left(N\frac{\partial \mu}{\partial x} - M\frac{\partial \mu}{\partial y}\right) = \frac{\partial M}{\partial y} - \frac{\partial N}{\partial x}. \qquad (8)$$

In case the integrating factor μ depends only on x, equation (8) becomes

$$\frac{1}{\mu}\frac{d\mu}{dx} = \frac{\partial M/\partial y - \partial N/\partial x}{N} = k(x, y) \qquad (9)$$

Since the left-hand side of this equation consists only of functions of x, $k(x, y)$ *must* also be a function of x. If this is indeed true, then μ can be found by separating the variables: $\mu(x) = e^{\int k(x)\,dx}$. A similar result holds if μ is a function of y alone, in which case

$$K = \frac{\partial M/\partial y - \partial N/\partial x}{-M}$$

is a function of y only. In this case, $\mu(y) = e^{\int K(y)\,dy}$ is the integrating factor.

EXAMPLE 3

Solve the equation

$$(3x^2 - y^2)\,dy - 2xy\,dx = 0.$$

SOLUTION
In this problem, $M = -2xy$ and $N = 3x^2 - y^2$, so that

$$\frac{\partial M}{\partial y} = -2x \quad \text{and} \quad \frac{\partial N}{\partial x} = 6x.$$

Then

$$K = \frac{\partial M/\partial y - \partial N/\partial x}{-M} = \frac{-4}{y},$$

so that

$$\mu = e^{-4\int y^{-1}\,dy} = e^{-4\,\ln|y|} = y^{-4}.$$

Multiplying the differential equation by y^{-4}, we obtain the exact equation

$$-\frac{2x}{y^3}\,dx + \left(\frac{3x^2 - y^2}{y^4}\right)dy = 0.$$

Integrating $M = -2x/y^3$ with respect to x and $N = (3x^2 - y^2)/y^4$ with respect to y, we get

$$-\int \frac{2x}{y^3}\,dx + h(y) = \int \frac{3x^2 - y^2}{y^4}\,dy + k(x),$$

or

$$-\frac{x^2}{y^3}+h(y)=-\frac{x^2}{y^3}+\frac{1}{y}+k(x).$$

Setting $k(x) = 0$ and $h(y) = 1/y$, we obtain the general solution

$$g(x, y) = \frac{1}{y} - \frac{x^2}{y^3} = c,$$

or

$$cy^3 - y^2 + x^2 = 0.$$

Problems 1.5

In Problems 1–11, verify that each given differential equation is exact and find the general solution. Find a particular solution when an initial condition is given.

1. $2xy\,dx + (x^2 + 1)\,dy = 0$

2. $[x\cos(x + y) + \sin(x + y)]\,dx + x\cos(x + y)\,dy = 0$, $y(1) = \pi/2 - 1$

3. $\left(4x^3y^3 + \dfrac{1}{x}\right)dx + \left(3x^4y^2 - \dfrac{1}{y}\right)dy = 0,\quad x(e) = 1$

4. $\left[\dfrac{\ln(\ln y)}{x} + \dfrac{2}{3}xy^3\right]dx + \left[\dfrac{\ln x}{y\ln y} + x^2y^2\right]dy = 0$

5. $(x - y\cos x)\,dx - \sin x\,dy = 0,\quad y(\pi/2) = 1$

6. $\cosh 2x\cosh 2y\,dx + \sinh 2x\sinh 2y\,dy = 0$

7. $(ye^{xy} + 4y^3)\,dx + (xe^{xy} + 12xy^2 - 2y)\,dy = 0$, $y(0) = 2$

8. $(3x^2\ln x + x^2 - y)\,dx - x\,dy = 0,\quad y(1) = 5$

9. $(2xy + e^y)\,dx + (x^2 + xe^y)\,dy = 0$

10. $(x^2 + y^2)\,dx + 2xy\,dy = 0,\quad y(1) = 1$

11. $\left(\dfrac{1}{x} - \dfrac{y}{x^2 + y^2}\right)dx + \left(\dfrac{x}{x^2 + y^2} - \dfrac{1}{y}\right)dy = 0$

In Problems 12–16, find an integrating factor for each differential equation and obtain the general solution.

12. $y\,dx + (y - x)\,dy = 0$

13. $(x^2 + y^2 + x)\,dx + y\,dy = 0$

14. $2y^2\,dx + (2x + 3xy)\,dy = 0$

15. $(x^2 + 2y)\,dx - x\,dy = 0$

16. $(x^2 + y^2)\,dx + (3xy)\,dy = 0$

17. Solve $xy\,dx + (x^2 + 2y^2 + 2)\,dy = 0$

18. Let $M = yF(xy)$ and $N = xG(xy)$. Show that $1/(xM - yN)$ is an integrating factor for

$$M\,dx + N\,dy = 0.$$

19. Use the result of Problem 18 to solve the equation

$$2x^2y^3\,dx + x^3y^2\,dy = 0.$$

20. Solve $(x^2 + y^2 + 1)\,dx - (xy + y)\,dy = 0$ [*Hint:* Try an integrating factor of the form $\mu(x, y) = (x + 1)^n$.]

21. Show that if $dg = M\,dx + N\,dy$, then $\partial M/\partial y = \partial N/\partial x$. Assume sufficient differentiability. [*Hint:* Use the equality of mixed second-order partial derivatives.]

1.6 Simple Electric Circuits

In this section we shall consider simple electric circuits containing a resistor and an inductor or capacitor in series with a source of electromotive force (emf). Such circuits are shown in Fig. 1(a) and (b), and their action can be understood easily without any special knowledge of electricity.

1. An electromotive force (emf) E (volts), usually a battery or generator, drives an electric charge Q (coulombs) and produces a current I (am-

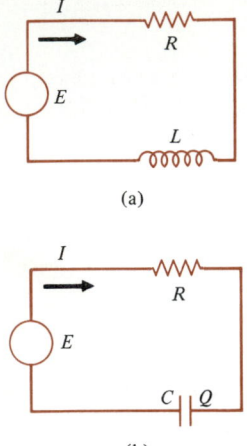

(a)

(b)

FIGURE 1

peres). The current is defined as the rate of flow of the charge, and we can write

$$I = \frac{dQ}{dt}. \tag{1}$$

2. A resistor of resistance R (ohms) is a component of the circuit that opposes the current, dissipating the energy in the form of heat. It produces a drop in voltage given by **Ohm's law:**

$$E_R = RI. \tag{2}$$

3. An inductor of inductance L (henrys) opposes any change in current by producing a voltage drop of

$$E_L = L \frac{dI}{dt}. \tag{3}$$

4. A capacitor of capacitance C (farads) stores charge. In so doing, it resists the flow of further charge, causing a drop in the voltage of

$$E_C = \frac{Q}{C}. \tag{4}$$

The quantities R, L, and C are usually constants associated with the particular component in the circuit; E may be a constant or a function of time. The fundamental principle guiding such circuits is given by **Kirchhoff's voltage law:**

The algebraic sum of all voltage drops around a closed circuit is zero.

In the circuit of Fig. 1(a) the resistor and the inductor cause voltage drops of E_R and E_L, respectively. The emf, however, *provides* a voltage of E (that is, a voltage drop of $-E$). Thus Kirchhoff's voltage law yields

$$E_R + E_L - E = 0.$$

Transposing E to the other side of the equation and using equations (2) and (3) to replace E_R and E_L, we have

$$L \frac{dI}{dt} + RI = E. \tag{5}$$

The following two examples illustrate the use of equation (5) in analyzing the circuit shown in Fig. 1(a).

EXAMPLE 1†

An inductance of 2 henrys (H) and a resistance of 10 ohms (Ω) are connected in series with an emf of 100 volts (V). If the current is zero when $t = 0$, what is the current at the end of 0.1 second?

SOLUTION

Since $L = 2$, $R = 10$, and $E = 100$, equation (5) and the initial current yield the initial value problem:

$$2\frac{dI}{dt} + 10I = 100, \quad I(0) = 0. \tag{6}$$

Dividing both sides of equation (6) by 2, we note that the resulting linear first-order equation has e^{5t} as an integrating factor, that is,

$$\frac{d}{dt}(e^{5t}I) = e^{5t}\left(\frac{dI}{dt} + 5I\right) = 50e^{5t}. \tag{7}$$

Integrating both sides of equation (7), we get

$$e^{5t}I(t) = 10e^{5t} + c,$$

or

$$I(t) = 10 + ce^{-5t}. \tag{8}$$

Setting $t = 0$ in equation (8) and using the initial condition $I(0) = 0$, we have

$$0 = I(0) = 10 + c,$$

which implies that $c = -10$. Substituting this value into (8), we obtain an equation for the current at all times t:

$$I(t) = 10(1 - e^{-5t}).$$

Thus, when $t = 0.1$, we have

$$I(0.1) = 10(1 - e^{-0.5}) \approx 3.93 \text{ amp.}$$

EXAMPLE 2

Suppose that the emf $E = 100 \sin 60t$ volts, but all other values remain the same as those given in Example 1. Then equation (5) yields

$$2\frac{dI}{dt} + 10I = 100 \sin 60t, \quad I(0) = 0. \tag{9}$$

Again dividing by 2 and multiplying both sides by the integrating factor e^{5t}, we have

$$\frac{d}{dt}(e^{5t}I) = e^{5t}\left(\frac{dI}{dt} + 5I\right) = 50e^{5t} \sin 60t. \tag{10}$$

Integrating both sides of (10) and using Formula 168 of the integral table we obtain

---

† This example is typical in electrical engineering. It is, for example, very similar to Exercise 12a in C. A. Desoer and E. S. Kuh, *Basic Circuit Theory*, McGraw-Hill, New York, 1969, p. 169.

$$I(t) = e^{-5t}\left[50\int (\sin 60t)e^{5t}\,dt + c\right]$$

$$= e^{-5t}\left[50e^{5t}\left(\frac{5\sin 60t - 60\cos 60t}{3625}\right) + c\right]$$

$$= \frac{2\sin 60t - 24\cos 60t}{29} + ce^{-5t}.$$

Thus setting $t = 0$, we find that $c = 24/29$ and

$$I(0.1) = \frac{2\sin 6 - 24\cos 6}{29} + \frac{24}{29}e^{-0.5} \approx -0.31 \text{ amp.}$$

For the circuit in Fig. 1(b) we have $E_R + E_C - E = 0$, or

$$RI + \frac{Q}{C} = E.$$

Using the fact that $I = dQ/dt$, we obtain the linear first-order equation

$$R\frac{dQ}{dt} + \frac{Q}{C} = E. \tag{11}$$

The next example illustrates how to use equation (11).

EXAMPLE 3

If a resistance of 2000 ohms and a capacitance of 5×10^{-6} farad (f) are connected in series with an emf of 100 volts, what is the current at $t = 0.1$ sec if $I(0) = 0.01$ ampere?

SOLUTION
Setting $R = 2000$, $C = 5 \times 10^{-6}$, and $E = 100$ in equation (11), we have

$$2000\left(\frac{dQ}{dt} + 100Q\right) = 100,$$

or

$$\frac{dQ}{dt} + 100Q = \frac{1}{20}, \tag{12}$$

from which we can determine $Q(0)$ since

$$\tfrac{1}{20} = Q'(0) + 100Q(0) = I(0) + 100Q(0).$$

Thus,

$$Q(0) = \tfrac{1}{100}[\tfrac{1}{20} - I(0)] = \tfrac{1}{100}[\tfrac{1}{20} - \tfrac{1}{100}]$$

$$= \tfrac{1}{100}(\tfrac{4}{100}) = 4 \times 10^{-4} \text{ coulombs.} \tag{13}$$

Multiplying both sides of equation (12) by the integrating factor e^{100t}, we get

$$\frac{d}{dt}(e^{100t}Q) = \frac{e^{100t}}{20},$$

and integrating this equation yields

$$e^{100t}Q = \frac{e^{100t}}{2000} + c.$$

Dividing both sides by e^{100t} gives us

$$Q(t) = \frac{1}{2000} + ce^{-100t},$$

and setting $t = 0$, we find that

$$c = Q(0) - \frac{1}{2000} = 4 \times 10^{-4} - 5 \times 10^{-4} = -10^{-4}.$$

Thus, the charge at all times t is

$$Q(t) = (5 - e^{-100t})/10^4,$$

and the current is

$$I(t) = Q'(t) = \tfrac{1}{100} e^{-100t}.$$

Thus $I(0.1) = 10^{-2}e^{-10} \approx 4.54 \times 10^{-7}$ amp.

Problems 1.6

In Problems 1–5, assume that the *RL* circuit shown in Fig. 1(a) has the given resistance, inductance, emf, and initial current. Find an expression for the current at all times t and calculate the current after 1 second.

1. $R = 10\ \Omega$, $L = 1$ H, $E = 12$ V, $I(0) = 0$ amp
2. $R = 8\ \Omega$, $L = 1$ H, $E = 6$ V, $I(0) = 1$ amp
3. $R = 50\ \Omega$, $L = 2$ H, $E = 100$ V, $I(0) = 0$ amp
4. $R = 10\ \Omega$, $L = 5$ H, $E = 10 \sin t$ V, $I(0) = 1$ amp
5. $R = 10\ \Omega$, $L = 10$ H, $E = e^t$ V, $I(0) = 0$ amp.

In Problems 6–10, use the given resistance, capacitance, emf, and initial charge in the *RC* circuit shown in Fig. 1(b). Find an expression for the charge at all times t.

6. $R = 1\ \Omega$, $C = 1$ f, $E = 12$ V, $Q(0) = 0$ coulomb
7. $R = 10\ \Omega$, $C = 0.001$ f, $E = 10 \cos 60t$ V, $Q(0) = 0$ coulomb
8. $R = 1\ \Omega$, $C = 0.01$ f, $E = \sin 60t$ V, $Q(0) = 0$ coulomb
9. $R = 100\ \Omega$, $C = 10^{-4}$ f, $E = 100$ V, $Q(0) = 1$ coulomb

10. $R = 200\ \Omega$, $C = 5 \times 10^{-5}$ f, $E = 1000$ V, $Q(0) = 1$ coulomb.

†11. The capacitor C in Fig. 2 is charged to 10 volts when the switch is closed. Obtain a differential equation for the capacitor voltage and find the voltage for all time t given that $R = 1000\ \Omega$ and $C = 10^{-6}$ f.

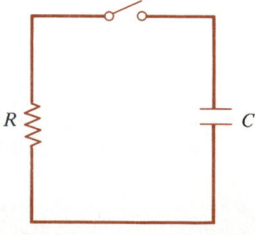

FIGURE 2

12. An inductance of 1 henry and a resistance of 2 ohms are connected in series with a battery of $6e^{-0.001t}$ volt. No current is flowing initially. When will the current measure 0.5 ampere?

† This example is Exercise 5.28 in Shearer *et al.*, *System Dynamics*, Addison-Wesley, Reading, Mass., 1971, p. 141. Reprinted with permission of Addison-Wesley Publishing Co.

13. A variable resistance $R = 1/(5 + t)$ ohms and a capacitance of 5×10^{-6} farad are connected in series with an emf of 100 volts. If $Q(0) = 0$, what is the charge on the capacitor after one minute?

14. In the RC circuit [Fig. 1(b)] with constant voltage E, how long will it take the current to decrease to one-half its original value?

15. Suppose that the voltage in an RC circuit is $E(t) = E_0 \cos \omega t$, where $2\pi/\omega$ is the period of the cycle. Assuming that the initial charge is zero, what are the charge and current as functions of R, C, ω, and t?

16. Show that the current in Problem 15 consists of two parts: a **steady-state** term that has a period of $2\pi/\omega$ and a **transient** term that tends to zero as t increases.

17. In Problem 16 show that if R is small, then the transient term can be quite large for small values of t. (This is why fuses can blow when a switch is flipped.)

18. Find the steady-state current, given that a resistance of 2000 ohms and a capacitance of 3×10^{-6} farad are connected in series with an alternating emf of $120 \cos 2t$ volts.

19. Find an expression for the current of a series RL circuit, where $R = 100\ \Omega$, $L = 2$ H, $I(0) = 0$, and the emf voltage satisfies

$$E = \begin{cases} 6, & \text{for} \quad 0 \leqslant t \leqslant 10, \\ 7 - e^{10-t}, & \text{for} \qquad t \geqslant 10. \end{cases}$$

20. Repeat Problem 19 with $R = 100/(1 + t)$, all other values remaining the same.

1.7 Further Applications: Compartmental Analysis and Curves of Pursuit

In this section we use the techniques developed earlier in the chapter to solve two interesting types of problems.

Compartmental Analysis

A complicated physical or biological process can often be divided into several distinct stages. The entire process can then be described by the interactions between the individual stages. Each such stage is called a **compartment** or pool, and the contents of each compartment are assumed to be well mixed. Material from one compartment is transferred to another and is immediately incorporated into the latter. Because of the name we have given to the stages, the entire process is called a **compartmental system.**† An **open** system is one in which there are inputs to or outputs from the system through one or more compartments. A system that is not open is said to be **closed.**

In this section we will investigate only the simplest such system: the one-compartment system. Additional work on more complicated systems will be found in later chapters.

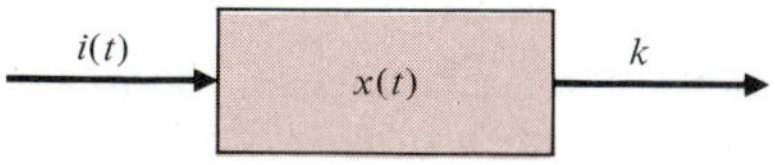

Figure 1 illustrates a one-compartment system consisting of a quantity $x(t)$ of material in the compartment, an input rate $i(t)$ at which material is being introduced into the system, and a **fractional transfer coefficient** k indicating the fraction of the material in the compartment that is being removed from the system per unit time. It is clear that the rate at which the quantity x is

FIGURE 1

---

† This name is frequently used in mathematical biology. Engineers refer to such systems as **block diagrams.**

changing depends on the difference between the input and output at any time t, leading to the differential equation

$$\frac{dx}{dt} = i(t) - kx(t). \tag{1}$$

As we saw in Section 1.4, this linear equation has the solution

$$x(t) = e^{-kt}\left[\int i(t)e^{kt}\,dt + c\right]. \tag{2}$$

This simple model applies to many different problems, as we shall illustrate below.

EXAMPLE 1

Strontium 90 (^{90}Sr) has a half-life of 25 years. If 10 grams of ^{90}Sr are initially placed in a sealed container, how many grams will remain after 10 years?

SOLUTION
Let $x(t)$ be the number of grams of ^{90}Sr at time t (years). Since the number of atoms present is very large, the number decaying per unit time is directly proportional to the number present at that time. The constant of proportionality k is the fractional transfer coefficient. Since there is no input, the equation involved is

$$\frac{dx}{dt} = -kx(t). \tag{3}$$

Equation (3) has the solution $x(t) = x_0 e^{-kt}$, where $x_0 = 10$ grams. To find k, we set $t = 25$ to obtain (since $5 = \frac{1}{2}(10)$ grams remain after 25 years)

$$5 = 10e^{-25k},$$

from which we find, after taking logarithms, that $k = (\ln 2)/25$. Thus

$$x(10) = 10e^{-(10\ln 2)/25}$$

$$= 10(2)^{-2/5} \approx 7.579\,\text{g}.$$

EXAMPLE 2

Consider a tank holding 100 gallons of water in which are dissolved 50 pounds of salt. Suppose that 2 gallons of brine, each containing 3 pounds of dissolved salt, run into the tank per minute, and the mixture, kept uniform by high-speed stirring, runs out of the tank at the rate of 2 gallons per minute. Find the amount of salt in the tank at any time t.

SOLUTION
Let $x(t)$ be the number of pounds of salt at the end of t minutes. Since each gallon of brine that enters the compartment (tank) contains 3 pounds of salt, we know that $i(t) = 6$. On the other hand, $k = 2/100$ since 2 of the 100 gallons in the tank are being removed each minute. Thus equation (1) becomes

$$\frac{dx}{dt} = 6 - \frac{2}{100}x \quad \text{or} \quad \frac{dx}{dt} + \frac{1}{50}x = 6.$$

Multiplying both sides by the integrating factor $e^{t/50}$, we get

$$(e^{t/50}x)' = 6e^{t/50},$$

which has the solution

$$x(t) = e^{-t/50}\left[6\int e^{t/50}\,dt + c\right],$$
$$= 300 + ce^{-t/50}.$$

At $t = 0$ we have

$$50 = x(0) = 300 + c,$$

so that

$$x(t) = 300 - 250e^{-t/50}.$$

Observe that x increases and the ratio of salt to water in the tank ($300/100 = 3$) approaches the ratio of salt to water in the input stream as time increases.

The fractional transfer coefficient k may be a function of time, as we shall see in the following example.

EXAMPLE 3

Suppose that, in Example 2, three gallons of brine, each containing 1 pound of salt, run into the tank each minute, and all other facts are the same. Now $i(t) = 3$, but since the quantity of brine in the tank increases with time, the fraction that is being transferred is $k = 2/(100 + t)$. The numerator of k is the number of gallons being removed, and $100 + t$ is the number of gallons in the tank at time t. The equation describing the system is

$$\frac{dx}{dt} = 3 - \frac{2x}{100 + t}$$

or

$$\frac{dx}{dt} + \frac{2x}{100 + t} = 3. \tag{4}$$

The integrating factor in this case is

$$e^{\int[2/(100+t)]\,dt} = e^{\ln(100+t)^2} = (100 + t)^2$$

so that we have, successively,

$$[(100 + t)^2 x]' = 3(100 + t)^2$$
$$(100 + t)^2 x = (100 + t)^3 + c$$

or

$$x(t) = (100 + t) + c(100 + t)^{-2}.$$

Setting $t = 0$, we find that $c = -50(100)^2$, so that

$$x(t) = 100 + t - 50(1 + t/100)^{-2}.$$

For example, after 100 minutes, we have

$$x(100) = 200 - 50/4 = 187.5 \text{ lb}$$

of salt in the tank.

The input function $i(t)$ may depend not only on time but also on the quantity present.

EXAMPLE 4

Systems with periodic inputs and fractional transfer coefficients often occur in biological processes due to the diurnal period of activity. For example, ACTH (adrenocorticotropic hormone) secretion by the anterior pituitary follows a 24-hour cycle, which drives the secretion of adrenal steroids in such a way that the levels of these steroids in the blood plasma peaks near 8:00 a.m. and is at a minimum near 8:00 p.m.

Let $k(t) = A + B \sin \omega t$, with $A > B$, in equation (1), which leads to the equation

$$\frac{dx}{dt} = i(t) - (A + B \sin \omega t)x. \tag{5}$$

Since

$$\int (A + B \sin \omega t)\, dt = At - \frac{B}{\omega} \cos \omega t + c,$$

we may use the integrating factor $e^{At+(B/\omega)(1-\cos \omega t)}$ on both sides of equation (5):

$$\frac{d}{dt}(e^{At+(B/\omega)(1-\cos \omega t)}x(t)) = e^{At+(B/\omega)(1-\cos \omega t)}[x' + (A + B \sin \omega t)x]$$

$$= e^{At+(B/\omega)(1-\cos \omega t)}i(t). \tag{6}$$

Integrating both sides of equation (6) from 0 to t, we have

$$e^{At+(B/\omega)(1-\cos \omega t)}x(t)|_0^t = \int_0^t i(t)e^{At+(B/\omega)(1-\cos \omega t)}\, dt,$$

or

$$x(t) = e^{-At-(B/\omega)(1-\cos \omega t)}\left[x(0) + \int_0^t i(t)e^{At+(B/\omega)(1-\cos \omega t)}\, dt\right]. \tag{7}$$

Since $1 - \cos \omega t = 2 \sin^2(\omega t/2)$, we can write equation (7) as

$$x(t) = e^{-At-2B\sin^2(\omega t/2)/\omega}\left[x(0) + \int_0^t i(t)e^{At+2B\sin^2(\omega t/2)/\omega}\, dt\right]. \tag{8}$$

If $i(t) = 0$, then $x(t)$ behaves as shown in Fig. 2, where $x(0)e^{-At}$ is an upper bound, and the factor $e^{-2B\sin^2(\omega t/2)/\omega}$ oscillates between $e^{-2B/\omega}$ and 1.

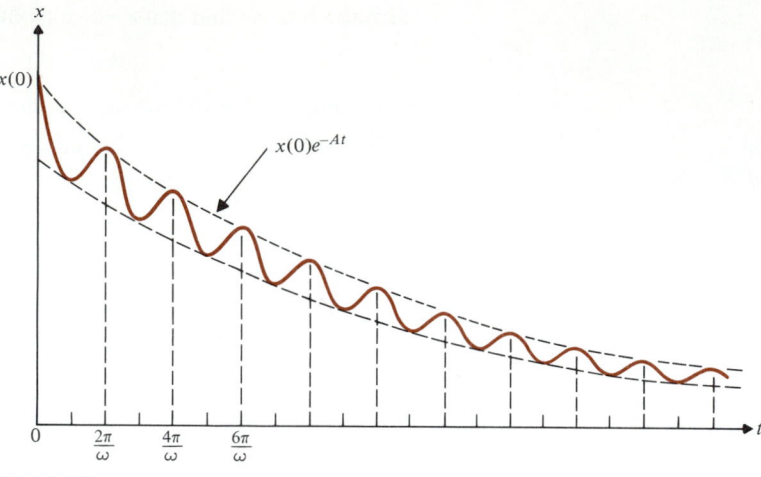

FIGURE 2

Curves of Pursuit†

Many interesting differential equations arise in studying the path of a pursuer in tracking prey.

EXAMPLE 5

Suppose that a hawk P at the point $(a, 0)$ spots a pigeon Q at the origin flying along the y-axis at a speed v. The hawk immediately flies toward the pigeon at a speed w. What will be the flight path of the hawk?

SOLUTION

Let time $t = 0$ at the instant the hawk starts flying toward the pigeon. After t seconds the pigeon will be at the point $Q = (0, vt)$ and the hawk at $P = (x, y)$. Since the line PQ is tangent to the path (see Fig. 3), we find that its slope is given by $y' = (y - vt)/x$, so that

$$xy' - y = -vt. \tag{9}$$

On the other hand, the length of the path traveled by the hawk can be computed by the formula for arc length of basic calculus

$$wt = \int ds = \int_x^a \sqrt{1 + (y')^2}\, dx. \tag{10}$$

Solving equations (9) and (10) for t and equating them, we have

$$\frac{y - xy'}{v} = \frac{1}{w} \int_x^a \sqrt{1 + (y')^2}\, dx. \tag{11}$$

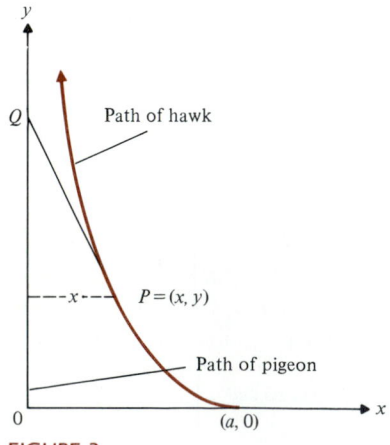

FIGURE 3

Path of hawk

$P = (x, y)$

Path of pigeon

$(a, 0)$

---

† An interesting discussion of this topic is contained in A. Bernhart, "Curves of pursuit II," *Scripta Mathematica*, vol. 23 (1957), pp. 49–66.

Differentiating both sides of equation (11) with respect to x yields

$$xy'' = \frac{v}{w}\sqrt{1+(y')^2}. \tag{12}$$

Setting $p = y'$, we find that equation (12) becomes

$$xp' = \frac{v}{w}\sqrt{1+p^2},$$

and we can separate the variables to obtain

$$\frac{dp}{\sqrt{1+p^2}} = \frac{v}{w}\frac{dx}{x}.$$

Integrating both sides of this equation (see Formula 22 in the integral table), we have

$$\ln(p + \sqrt{1+p^2}) = \frac{v}{w}\ln x + c.$$

Since $p = y' = 0$ when $x = a$ (the slope of the line PQ at $t = 0$ is zero), it follows that $c = -(v/w)\ln a$. Exponentiating both sides of this equation yields

$$p + \sqrt{1+p^2} = \left(\frac{x}{a}\right)^{v/w},$$

which, after some algebra, yields

$$\frac{dy}{dx} = p = \frac{1}{2}\left[\left(\frac{x}{a}\right)^{v/w} - \left(\frac{x}{a}\right)^{-v/w}\right]. \tag{13}$$

If we assume that the hawk flies faster than the pigeon ($w > v$), we may integrate equation (13) to obtain

$$y = \frac{a}{2}\left[\frac{(x/a)^{1+v/w}}{1+v/w} - \frac{(x/a)^{1-v/w}}{1-v/w}\right] + c.$$

Since $y = 0$ when $x = a$, we have

$$c = -\frac{a}{2}\left[\frac{1}{1+v/w} - \frac{1}{1-v/w}\right] = \frac{avw}{w^2-v^2}.$$

The hawk will catch the pigeon at $x = 0$ and $y = c = avw/(w^2 - v^2)$. The situation in which the hawk flies no faster than the pigeon ($w \leqslant v$) is discussed in Problems 12 and 13.

EXAMPLE 6

A destroyer is in a dense fog, which lifts for an instant, disclosing an enemy submarine on the surface four miles away. Suppose that the submarine dives immediately and proceeds at full speed in an unknown direction. What path should the destroyer select to be certain of passing directly over the submarine, if its velocity v is three times that of the submarine?

SOLUTION
Suppose that the destroyer has traveled three miles toward the place where the submarine was spotted. Then the submarine lies on the circle of radius one mile centered at where

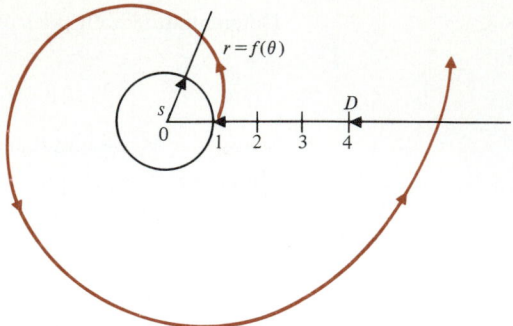

FIGURE 4

it was when spotted (see Fig. 4), since its velocity is one-third that of the destroyer. Since the location of the submarine can be described easily in polar coordinates, we make use of polar coordinates and assume that $r = f(\theta)$ is the path the destroyer should follow to be certain of passing over the submarine, regardless of the direction the latter chooses. Then the distance traveled by the submarine to the point where the paths will cross is $r - 1$, whereas that of the destroyer (which is three times longer) is given by the arc length formula in polar coordinates:

$$3(r - 1) = \int_0^\theta ds = \int_0^\theta \sqrt{(dr)^2 + (r\, d\theta)^2}$$

$$= \int_0^\theta \sqrt{(dr/d\theta)^2 + r^2}\, d\theta. \tag{14}$$

Differentiating both sides of equation (14) with respect to θ yields the differential equation

$$3r' = \sqrt{(r')^2 + r^2},$$

which simplifies to $8(r')^2 = r^2$. Taking the square roots of both sides and separating the variables, we have

$$\frac{dr}{r} = \frac{d\theta}{\sqrt{8}},$$

from which it follows that $\ln r = \theta/\sqrt{8} + c$, or

$$r = ce^{\theta/\sqrt{8}}. \tag{15}$$

Since $r = 1$ when $\theta = 0$, it follows that $c = 1$ and the path that the destroyer should follow is the spiral $r = e^{\theta/\sqrt{8}}$ after proceeding three miles toward where the submarine was spotted.

It should be noted that this path is not the only curve that the destroyer could follow. For example, suppose that the destroyer has gone six miles toward where the submarine was spotted (see Fig. 5). At this point, we can again follow a path $r = g(\theta)$. Since by now the submarine is two miles from the origin, the distance traveled by the submarine to where the paths will cross is $r - 2$, whereas the destroyer must go a distance

$$3(r - 2) = \int_{-\pi}^\theta \sqrt{(dr/d\theta)^2 + r^2}\, d\theta. \tag{16}$$

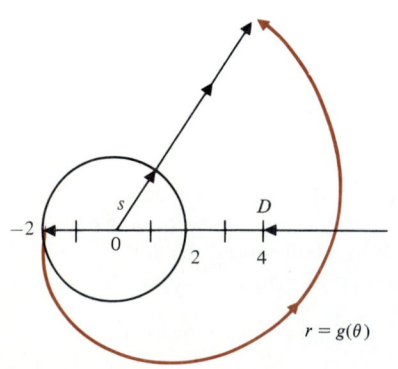

FIGURE 5

Equation (16) again leads to the general solution (15), but in this case $r = 2$ when $\theta = -\pi$, so that $c = 2e^{\pi/\sqrt{8}}$. Thus the spiral that the destroyer must follow is

$$r = 2e^{(\theta+\pi)/\sqrt{8}}.$$

Of course, the submarine captain can evade detection by not going at full speed or by following a curved trajectory.

Problems 1.7

1. Carbon 14 (^{14}C) has a half-life of 5730 years and is uniformly distributed in the atmosphere in the form of carbon dioxide. Living plants absorb carbon dioxide and maintain a fixed ratio of ^{14}C to the stable element ^{12}C. At death, the disintegration of ^{14}C changes this ratio. Compare the concentrations of ^{14}C in two identical pieces of wood, one of them freshly cut, the other 2000 years old.

2. Radioactive iodine ^{131}I is often used as a tracer in medicine. Suppose that a given dose Q_0 is injected into the bloodstream at time $t = 0$ and is evenly distributed in the entire bloodstream before any loss occurs. If the daily removal rate of the iodine by the kidney is k_1 percent, and k_2 percent by the thyroid gland, what percentage of the initial amount will still be in the blood after one day?

3. Suppose that an infected individual is introduced into a population of size N, all of whom are susceptible to the disease. If we assume that the rate of infection is proportional to the product of the numbers of infectives and susceptibles present, what will be the number of infections at any time t? Let k be the **specific infection rate**.

4. A tank initially contains 100 liters of fresh water. Brine containing 20 grams per liter of salt flows into the tank at the rate of 4 liters per minute, and the mixture, kept uniform by stirring, runs out at the same rate. How long will it take for the quantity of salt in the tank to become 1 kilogram?

5. Given the same data as in Problem 4, determine how long it will take for the quantity of salt in the tank to increase from 1 kilogram to $1\frac{1}{2}$ kilograms.

6. A tank contains 100 gallons of fresh water. Brine containing 2 pounds per gallon of salt runs into the tank at the rate of 4 gallons per minute, and the mixture, kept uniform by stirring, runs out at the rate of 2 gallons per minute. Find:

   (a) the amount of salt present when the tank has 120 gallons of brine;

   (b) the concentration of salt in the tank at the end of 20 minutes.

7. A tank contains 50 liters of water. Brine containing x grams per liter of salt enters the tank at the rate of 1.5 liters per minute. The mixture, thoroughly stirred, leaves the tank at the rate of 1 liter per minute. If the concentration is to be 20 grams per liter at the end of 20 minutes, what is the value of x?

8. A tank holds 500 gallons of brine. Brine containing 2 pounds per gallon of salt flows into the tank at the rate of 5 gallons per minute, and the mixture, kept uniform, flows out at the rate of 10 gallons per minute. If the maximum amount of salt is found in the tank at the end of 20 minutes, what was the initial salt content of the tank?

9. Phosphate excretion is at a minimum at 6:00 a.m. and rises to a peak at 6:00 p.m. If the rate of excretion is

$$\frac{1}{3} - \frac{1}{6}\cos\frac{\pi}{12}(t-6)$$

grams per hour at time t hours ($0 \le t \le 24$), the body contains 400 grams of phosphate, and the patient is only allowed to drink water, what is the amount of phosphate in the patient's body at all times t, for $0 \le t \le 24$?

10. Suppose in Problem 9 that the patient is allowed three meals during the day in such a way that the body takes in phosphate at a rate given by the formula

$$i(t) = \begin{cases} \frac{1}{3}\,\text{g/hr}, & 8 \le t \le 16, \\ 0\,\text{g/hr}, & \text{otherwise.} \end{cases}$$

Obtain a formula for the amount of phosphate in the patient's body at all times t. When is it at a maximum?

11. Given a one-compartment system with k constant and $i(t) = A + B\sin\omega t$, $A > B$, find a solution of the system. How does it differ from that of the system in which the input is constant and the fractional transfer coefficient is periodic [see equation (8)]?

12. Suppose that $v = w$ in Example 5. Prove that

$$y = \frac{a}{2}\left\{\frac{1}{2}\left[\left(\frac{x}{a}\right)^2 - 1\right] - \ln\frac{x}{a}\right\},$$

so that the hawk will never catch the pigeon. Using equations (9) and (13), show that the distance between the hawk and the pigeon is $(x^2 + a^2)/2a$ whenever the hawk is at the point (x, y) on the path. Thus the hawk will not come as close as $a/2$ to the pigeon.

13. Suppose that $v > w$ in Example 5. Show that

$$y = \frac{a}{2}\left[\frac{(x/a)^{1+(v/w)}-1}{1+(v/w)}+\frac{(a/x)^{(v/w)-1}-1}{(v/w)-1}\right],$$

so that the hawk will never catch the pigeon. Find the distance between the hawk and the pigeon in terms of the variable x.

14. Let the y-axis and the line $x = b$ be the banks of a river whose current has velocity v (in the negative y-direction). A man is at the origin, and his dog is at the point $(b, 0)$. When the man calls, the dog enters the river, swimming toward the man at a constant velocity $w(>v)$. What will be the path of the dog?

15. Where will the dog of Problem 14 land if $w = v$?

16. Show that the dog of Problem 14 will never land if $w < v$. Suppose that the man walks downriver at the velocity v while calling his dog. Will the dog now be able to land?

17. In Example 6 suppose the destroyer proceeds to where the

submarine was sighted, then turns 90° left and proceeds two miles before beginning the spiral search pattern. What is the equation of the path the destroyer should now follow?

18. Suppose the destroyer in Example 6 is only twice as fast as the submarine and the submarine is spotted when it is three miles away. Find a path that will guarantee the destroyer's passing over the submarine, assuming that both ships execute the same maneuvers as those given in the example.

19. Three snails at the corners of an equilateral triangle of side a begin to move with the same velocity, each toward the snail to its right. Centering the triangle at the origin with one vertex along the positive x-axis, find an equation for the slime path left by the snail that started on the x-axis.

20. Consider Problem 19 with four snails at the corners of the square $[0, a] \times [0, a]$. How far will the snails travel before they meet?

†* 21. A hawk is flying 100 feet below a sparrow that is 50 feet below an eagle. The sparrow flies straight forward horizontally, while both the eagle and the hawk fly directly toward the sparrow. The hawk flies twice as fast as the sparrow and reaches it at the same instant the eagle does. How far did each bird fly and how fast did the eagle fly?

1.8 Numerical Solutions of Differential Equations: Euler's Method

In Section 1.3 we discussed an elementary graphical technique for determining the solution of a differential equation. In this section we present a very elementary numerical method for "solving" differential equations.

Before presenting this numerical technique, it is useful to discuss when numerical methods could or should be employed. Such methods are used frequently when other methods are not applicable. Even when other methods do apply, there may be an advantage in having a numerical solution, as solutions in terms of more exotic special functions are sometimes difficult to interpret. There may also be computational advantages: the exact solution may be extremely tedious to obtain. Finally, in situations where the solution is given implicitly, a numerical solution will tell us more about the solution.

On the other hand, care must always be exercised in utilizing any numerical scheme, as the accuracy of the solution depends not only on the "correctness" of the numerical method being used but also on the precision of the device (hand calculator or computer) used for the computations.

From the general theory (see Section 1.3) we assume that the initial value problem

† Based on *American Mathematical Monthly*, vol. 40 (1933), pp. 436–437.

$$\frac{dy}{dx} = f(x, y), \qquad y(x_0) = y_0, \tag{1}$$

has a unique solution $y(x)$. The technique we will describe below approximates this solution $y(x)$ only at a finite number of points

$$x_0, \quad x_1 = x_0 + h, \quad x_2 = x_0 + 2h, \ldots, \quad x_n = x_0 + nh,$$

where h is some (nonzero) real number. The methods provide a value y_k that is an approximation to the exact value $y(x_k)$ for $k = 0, 1, \ldots, n$.

Euler's Method†

This procedure is crude but very simple. The idea is to approximate $y(x_1) = y_1$ assuming that $f(x, y)$ varies so little on the interval $x_0 \leqslant x \leqslant x_1$ that only a very small error is made by replacing it by the constant value $f(x_0, y_0)$. Integrating

$$\frac{dy}{dx} = f(x, y)$$

from x_0 to x_1, we obtain

$$y_1 - y_0 = y(x_1) - y(x_0)$$

$$= \int_{x_0}^{x_1} f(x, y) \, dx \approx f(x_0, y_0)(x_1 - x_0) \tag{2}$$

or, since $h = x_1 - x_0$,

$$y_1 = y_0 + hf(x_0, y_0).$$

Repeating the process with (x_1, y_1) to obtain y_2, etc., we obtain the **difference equation**

$$y_{n+1} = y_n + hf(x_n, y_n). \tag{3}$$

where $y_n \approx y(x_0 + nh)$, $n = 0, 1, 2, \ldots$. We shall solve equation (3) iteratively, that is, by first finding y_1, then using it to find y_2, and so on.

The geometric meaning of equation (3) is easily seen by considering the direction field of the differential equation (1): we are simply following the tangent to the solution curve passing through (x_n, y_n) for a small horizontal distance. Looking at Fig. 1, where the smooth curve is the unknown exact solution to the initial value problem (1), we see how equation (3) approximates the exact solution. Since $f(x_0, y_0)$ is the slope of the exact solution at (x_0, y_0), we follow this line to the point (x_1, y_1). Some solution to the differential equation passes through this point. We follow its tangent line at this point to reach (x_2, y_2), and so on. The differences Δ_k are errors at the kth stage in the process.

† See the biographical sketch on page 57.

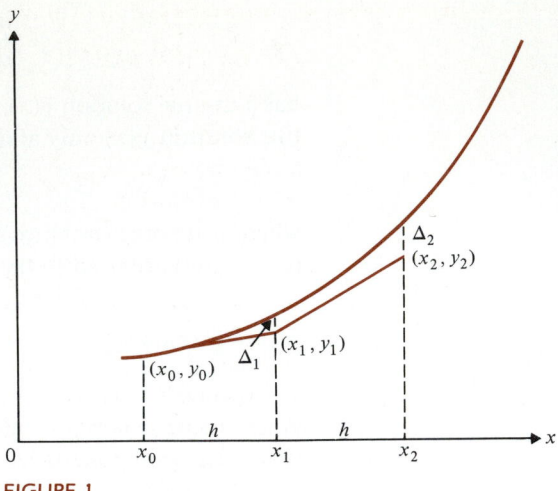

FIGURE 1

EXAMPLE 1

Use Euler's method to estimate $y(1)$, where $y(x)$ satisfies the initial value problem

$$\frac{dy}{dx} = y, \qquad y(0) = 1.$$

SOLUTION

From Section 1.1 we know that the solution to this problem is $y(x) = e^x$. Thus $y(1) = e^1 = e \approx 2.71828$. Let us apply Euler's method with $h = 1/5$ to this familiar problem to see if this answer is obtained.

We begin by dividing the interval $[0, 1]$ into five subintervals. Then $h = 1/5 = 0.2$ and $f(x_n, y_n) = y_n$. Thus we obtain successively

$$y_0 = y(0) = 1$$

$$y_1 = y_0 + 0.2y_0 = (1 + 0.2)y_0 = (1.2)y_0 = 1.2$$

$$y_2 = y_1 + 0.2y_1 = (1.2)y_1 = (1.2)^2$$

$$y_3 = y_2 + 0.2y_2 = (1.2)y_2 = (1.2)^3$$

$$y_4 = y_3 + 0.2y_3 = (1.2)y_3 = (1.2)^4$$

$$y_5 = y_4 + 0.2y_4 = (1.2)y_4 = (1.2)^5 \approx 2.48832.$$

Thus, with $h = 0.2$, Euler's method yields $y(1) \approx 2.48832$. Since we know that $y(1) = e \approx 2.71828$, our error is given by

$$\text{error} \approx 2.71828 - 2.48832 = 0.22996.$$

What happens if we double the number of subintervals? If $n = 10$, then $h = 0.1$ and, computing as before, we obtain $y(1) \approx y_{10} = (1.1)^{10} \approx 2.59374$. Thus the error is given by

$$\text{error} \approx 2.71828 - 2.59374 = 0.12454.$$

LEONHARD EULER
(1707–1783)

Courtesy of Brown Brothers

Leonhard Euler (pronounced "Oiler") was born in Basel, Switzerland. His father was a clergyman who had planned that his son should follow him into the ministry. The father was gifted at mathematics and, together with Johann Bernoulli, instructed young Leonhard in that subject as well as theology, astronomy, physics, medicine, and several Eastern languages.

In the early eighteenth century Catherine I of Russia, widow of the late Peter the Great, founded the St. Petersburg Academy. In 1727 Euler applied and was accepted for a chair in the faculty of medicine and physiology at the Academy. However, Catherine died the day Euler arrived in Russia and the Academy was plunged into turmoil. By 1730 Euler found himself in the chair of natural philosophy, from which he pursued his mathematical career. In 1741, accepting an invitation from Frederick the Great, Euler went to Berlin to head the Prussian Academy. Twenty-five years later he returned to St. Petersburg, where he died in 1783 at the age of 76.

Euler was the most prolific writer in the history of mathematics. He found results in virtually every branch of pure and applied mathematics. Although German was his native language, he wrote mostly in Latin and occasionally in French. His amazing productivity did not decline even when, in 1766, he became totally blind. During his lifetime Euler published 530 books and papers. When he died, he left so many unpublished manuscripts that the St. Petersburg Academy was still publishing his work in its *Proceedings* almost half a century later. His work enriched such diverse areas as hydraulics, celestial mechanics, lunar theory, and the theory of music, as well as mathematics.

Euler had a phenomenal memory. As a young man he memorized the entire Aeneid by Virgil (in Latin) and many years later could still recite the entire work. He was able to solve astonishingly complex mathematical problems in his head and is said to have solved in his head problems in astronomy that stymied Newton. The French Academician François Arago once commented that Euler could calculate without effort "just as men breathe, as eagles sustain themselves in the air."

Euler wrote in a mathematical language that is largely in use today. The following symbols were, among many others, first used by him: $f(x)$ for functional notation, e for the base of the natural logarithm, Σ for the summation sign, and i to denote the imaginary unit.

Euler's textbooks were models of clarity. They include the *Introductio in analysis infinitorum* (1748), the *Institutiones calculi differentialis* (1755), and the three-volume *Institutiones calculi integralis* (1768–1774). These and others of his works served as models for many of today's mathematics textbooks.

It is said that Euler did for mathematical analysis what Euclid did for geometry. It is no wonder that so many mathematicians who followed expressed the debt they owed to him.

In general, if $h = 1/n$, then

$$y_1 = y_0 + hy_0 = (1 + h)y_0 = 1 + h = 1 + \frac{1}{n}$$

$$y_2 = y_1\left(1 + \frac{1}{n}\right) = \left(1 + \frac{1}{n}\right)^2$$

$$y_3 = y_2\left(1 + \frac{1}{n}\right) = \left(1 + \frac{1}{n}\right)^3$$

$$\vdots$$

$$y_n = \left(1 + \frac{1}{n}\right)^n.$$

Thus

$$y(1) \approx \left(1 + \frac{1}{n}\right)^n.$$

Different values of $(1 + 1/n)^n$ are given in Table 1 (to 5 decimal places of accuracy). The numbers in the table should not be surprising. In fact, in elementary calculus the number e is *defined* by

$$e = \lim_{n \to \infty} \left(1 + \frac{1}{n}\right)^n.$$

Thus we have shown that $e = \lim_{n \to \infty} y_n$, where y_n is the nth iterate in Euler's method with $h = 1/n$. That is, y_n approximates $y(1)$ with better and better accuracy as n increases.

TABLE 1

n	$\left(1 + \dfrac{1}{n}\right)^n$
1	2
2	2.25
5	2.48832
10	2.59374
100	2.70481
1000	2.71692
10,000	2.71815
100,000	2.71827
1,000,000	2.71828

EXAMPLE 2

Find an approximate value for $y(1)$ if $y(x)$ satisfies the initial value problem

$$\frac{dy}{dx} = y + x^2, \qquad y(0) = 1. \tag{4}$$

Use five subintervals in your approximation.

SOLUTION

Here $h = 1/n = 1/5 = 0.2$ and we wish to find $y(1)$ by approximating the solution at $x = 0.0, 0.2, 0.4, 0.6, 0.8,$ and 1.0. We see that $f(x_n, y_n) = y_n + x_n^2$, and Euler's method [equation (3)] yields

$$y_{n+1} = y_n + hf(x_n, y_n) = y_n + h(y_n + x_n^2).$$

Since $y_0 = y(0) = 1$, we obtain

$$y_1 = y_0 + h(y_0 + x_0^2) = 1 + 0.2(1 + 0^2) = 1.2$$

$$y_2 = y_1 + h(y_1 + x_1^2) = 1.2 + 0.2[1.2 + (0.2)^2]$$

$$= 1.448 \approx 1.45$$

$$y_3 = y_2 + h(y_2 + x_2^2) = 1.45 + 0.2[1.45 + (0.4)^2] \approx 1.77$$

$$y_4 = y_3 + h(y_3 + x_3^2) = 1.77 + 0.2[1.77 + (0.6)^2] \approx 2.20$$

$$y_5 = y_4 + h(y_4 + x_4^2) = 2.20 + 0.2[2.20 + (0.8)^2] \approx 2.77.$$

We arrange our work as shown in Table 2. The value $y_5 = 2.77$, corresponding to $x_5 = 1.0$, is our approximate value for $y(1)$.

TABLE 2

x_n	y_n	$f(x_n, y_n) = y_n + x_n^2$	$y_{n+1} = y_n + h \cdot f(x_n, y_n)$
0.0	1.00	1.00	1.20
0.2	1.20	1.24	1.45
0.4	1.45	1.61	1.77
0.6	1.77	2.13	2.20
0.8	2.20	2.84	2.77
1.0	2.77		

Discretization Error

In Example 1.4.3 on p. 32 we showed that equation (4) has the exact solution $y = 3e^x - x^2 - 2x - 2$ (check that it does) so that $y(1) = 3e - 5 \approx 3.155$. Thus the Euler's method estimate was off by about 12 percent.† This is not surprising because we treated the derivative as a constant over intervals of length of 0.2. The error that arises in this way is called **discretization error,** because the "discrete" function $f(x_n, y_n)$ was substituted for the "continuously valued" function $f(x, y)$. It is usually true that if we reduce the step size h we can improve the accuracy of our answer, because then the "discretized" function $f(x_n, y_n)$ will be closer to the true value of $f(x, y)$ over the interval [0, 1]. This is illustrated in Fig. 2 with $h = 0.2$ and $h = 0.1$. Indeed, carrying out similar calculations with $h = 0.1$ yields an approximation of $y(1)$ of 2.94, which is a good deal more accurate (an error of about 7 percent).

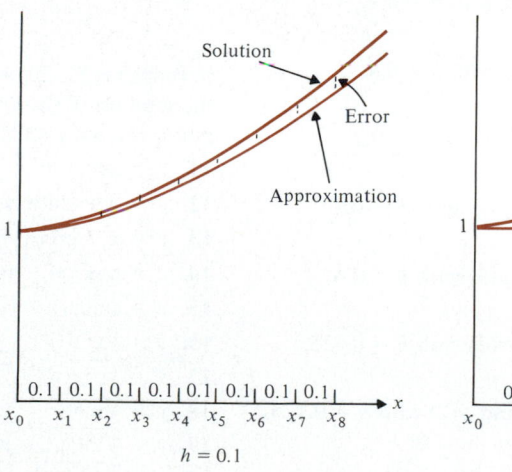

FIGURE 2

---

† Percent error $= \dfrac{3.155 - 2.77}{3.155} \approx 0.122 \approx 12\%.$

In general, reducing step size will improve accuracy. However, a warning must be attached to this. Reducing the step size will obviously increase the amount of work that must be done. Moreover, at every stage of the computation **round-off errors** are introduced. For example, in our calculations with $h = 0.2$, we rounded off the exact value 1.448 to the value 1.45 (correct to two decimal places). The rounded-off value was then used to calculate further values of y_n. It is not unusual for a computer solution of a more complicated differential equation to take several thousand individual computations, thus having several thousand round-off errors. In some problems the accumulated round-off error can be so large that the resulting computed solution will be sufficiently inaccurate to invalidate the result. Fortunately, this usually does not occur, since round-off errors can be positive or negative and tend to cancel one another out. This statement is made under the assumption (usually true) that the average of the round-off errors is zero. In any event, it should be clear that reducing the step size, thereby increasing the number of computations, is a procedure that should be carried out carefully. In general, each problem has an optimal step size, and a smaller than optimal step size will yield a greater error due to accumulated round-off errors.

Round-off Error

The material in this section was intended as a brief introduction to the numerical solution of differential equations. A far more extensive discussion of this topic is given in Chapter 13.

Problems 1.8

Solve Problems 1–10 by using Euler's method and the indicated value of h to two decimal places. Find the exact solution and compare.

1. $\dfrac{dy}{dx} = x + y$, $y(0) = 1$. Find $y(1)$ with $h = 0.2$.

2. $\dfrac{dy}{dx} = x - y$, $y(1) = 2$. Find $y(3)$ with $h = 0.4$.

3. $\dfrac{dy}{dx} = \dfrac{x-y}{x+y}$, $y(2) = 1$. Find $y(1)$ with $h = -0.2$.

4. $\dfrac{dy}{dx} = \dfrac{y}{x} + \left(\dfrac{y}{x}\right)^2$, $y(1) = 1$. Find $y(2)$ with $h = 0.2$.

5. $\dfrac{dy}{dx} = x\sqrt{1+y^2}$, $y(1) = 0$. Find $y(3)$ with $h = 0.4$.

6. $\dfrac{dy}{dx} = x\sqrt{1-y^2}$, $y(1) = 0$. Find $y(2)$ with $h = 0.125$.

7. $\dfrac{dy}{dx} = \dfrac{y}{x} - \dfrac{5}{2}x^2y^3$, $y(1) = \dfrac{1}{\sqrt{2}}$. Find $y(2)$ with $h = 0.125$.

8. $\dfrac{dy}{dx} = \dfrac{-y}{x} + x^2y^2$, $y(1) = \frac{2}{9}$. Find $y(3)$ with $h = \frac{1}{3}$.

9. $\dfrac{dy}{dx} = ye^x$, $y(0) = 2$. Find $y(2)$ with $h = 0.2$.

10. $\dfrac{dy}{dx} = xe^y$, $y(0) = 0$. Find $y(1)$ with $h = 0.1$.

In Problems 11–20, use Euler's method to graph approximately the solution of the given initial value problem by plotting the points (x_k, y_k) over the indicated range, where $x_k = x_0 + kh$.

11. $y' = xy^2 + y^3$, $y(0) = 1$, $h = 0.02$, $0 \le x \le 0.1$

12. $y' = x + \sin(\pi y)$, $y(1) = 0$, $h = 0.2$, $1 \le x \le 2$

13. $y' = x + \cos(\pi y)$, $y(0) = 0$, $h = 0.4$, $0 \le x \le 2$

14. $y' = \cos(xy)$, $y(0) = 0$, $h = \pi/4$, $0 \le x \le \pi$

15. $y' = \sin(xy)$, $y(0) = 1$, $h = \pi/4$, $0 \le x \le 2\pi$

16. $y' = \sqrt{x^2 + y^2}$, $y(0) = 1$, $h = 0.5$, $0 \le x \le 5$

17. $y' = \sqrt{y^2 - x^2}$, $y(0) = 1$, $h = 0.1$, $0 \le x \le 1$

18. $y' = \sqrt{x + y^2}$, $y(0) = 1$, $h = 0.2$, $0 \le x \le 1$

19. $y' = \sqrt{x + y^2}$, $y(1) = 2$, $h = -0.2$, $0 \le x \le 1$

20. $y' = \sqrt{x^2 + y^2}$, $y(1) = 5$, $h = -0.2$, $0 \le x \le 1$

(Optional)

1.9 Successive Approximations

In Sections 1.1–1.5 we discussed a number of techniques that can be used to solve first-order differential equations or initial value problems. In this section we shall describe an iterative technique, due to Picard, that provides an alternative approach to solving initial value problems. The main use of this method, however, is theoretical: it forms the basis for the proof of Picard's theorem (Theorem 1.3.1).

Consider the initial value problem

$$y' = f(x, y), \qquad y(x_0) = y_0. \tag{1}$$

If we integrate both sides of the differential equation in (1) from x_0 to x with respect to x, we have

$$\int_{x_0}^{x} y'(x)\, dx = \int_{x_0}^{x} f(x, y(x))\, dx$$

or

$$y(x) - y(x_0) = \int_{x_0}^{x} f(t, y(t))\, dt. \tag{2}$$

(We have changed the variable of integration on the right-hand side to t.)

Rewriting (2), we get [since $y(x_0) = y_0$]

$$y(x) = y_0 + \int_{x_0}^{x} f(t, y(t))\, dt. \tag{3}$$

Equation (3) is an alternative way of writing the initial value problem (1): Note that if we set $x = x_0$ in (3) then $y(x_0) = y_0$, since the integral is zero. Furthermore, if we differentiate both sides of (3) we obtain the differential equation

$$y'(x) = f(x, y(x)). \tag{4}$$

Thus (1) and (3) are equivalent ways of writing the same initial value problem.

Picard Iterations

We now define a sequence of functions $\{y_n(x)\}$, called **Picard† iterations**, by the successive formulas

$$y_0(x) = y_0$$

$$y_1(x) = y_0 + \int_{x_0}^{x} f(t, y_0(t))\, dt$$

$$y_2(x) = y_0 + \int_{x_0}^{x} f(t, y_1(t))\, dt$$

$$\vdots$$

$$y_n(x) = y_0 + \int_{x_0}^{x} f(t, y_{n-1}(t))\, dt. \tag{5}$$

† Emile Picard (1856–1941), one of the most eminent French mathematicians of the past century, made several outstanding contributions to mathematical analysis. Picard published the results we discuss in this section in 1890 and 1893 [*Journal de Mathématiques* (4), vol. 6 (1890), pp. 145–210; vol. 9 (1893), pp. 217–271].

It is possible to show that, under the conditions of Theorem 1.3.1 on p. 21, the Picard iterations defined by (5) converge uniformly to a solution of equation (3). We illustrate the process of this iteration by a simple example.

EXAMPLE 1

Consider the initial value problem

$$y'(x) = y(x), \qquad y(0) = 1. \tag{6}$$

As we know, equation (6) has the unique solution $y(x) = e^x$. In this case, the function $f(x, y)$ in equation (1) is given by $f(x, y(x)) = y(x)$, so the Picard iterations defined by (5) yield successively

$$y_0(x) = y_0 = 1,$$

$$y_1(x) = 1 + \int_0^x (1) \, dt = 1 + x,$$

$$y_2(x) = 1 + \int_0^x (1 + t) \, dt = 1 + x + \frac{x^2}{2},$$

$$y_3(x) = 1 + \int_0^x \left(1 + t + \frac{t^2}{2} \right) dt = 1 + x + \frac{x^2}{2!} + \frac{x^3}{3!},$$

and clearly,

$$y_n(x) = 1 + x + \frac{x^2}{2!} + \cdots + \frac{x^n}{n!} = \sum_{k=0}^{n} \frac{x^k}{k!}.$$

Hence

$$\lim_{n \to \infty} y_n(x) = \sum_{k=0}^{\infty} \frac{x^k}{k!} = e^x$$

since the series is the Maclaurin series of e^x.

You should not be fooled by the relative ease with which we obtained the solution in Example 1. In general, the Picard iterations quickly give rise to formidable integrals, which are often difficult or impossible to integrate.

EXAMPLE 2

If we apply Picard iterations to the initial value problem

$$y' = e^{x^2} y, \qquad y(0) = 1$$

we get

$$y_0(x) = y_0 = 1$$

$$y_1(x) = 1 + \int_0^x e^{t^2} \, dt$$

and this integral cannot be written in terms of elementary functions.

As stated earlier, these successive iterations are primarily a theoretical tool. In the next theorem we use a different technique to prove the uniqueness part of Theorem 1.3.1 for a linear first-order equation.

■ **THEOREM 1: Uniqueness Theorem**
Suppose that $a(x)$ and $f(x)$ are continuous functions. Then the linear initial value problem

$$\frac{dy}{dx} + a(x)y = f(x), \qquad y(x_0) = y_0 \tag{7}$$

has at most one solution.

Remark: This theorem does not claim that a solution exists. However, a global existence result for this equation has already been proved. One solution to equation (7) is given by formula (1.4.11) on p. 31 for an appropriate constant C. All integrals in equation (1.4.11) exist because $a(x)$ and $f(x)$ are assumed continuous.

PROOF
Suppose that both $y_1(x)$ and $y_2(x)$ satisfy (7). Let $y_3(x) = y_1(x) - y_2(x)$. We must show that $y_3(x) \equiv 0$. But, since y_1 and y_2 satisfy (7),

$$y_3' + a(x)y_3 = (y_1 - y_2)' + a(x)(y_1 - y_2)$$

$$= [y_1' + a(x)y_1] - [y_2' + a(x)y_2] = f(x) - f(x) = 0.$$

Also, $y_3(x_0) = y_1(x_0) - y_2(x_0) = y_0 - y_0 = 0$. Thus y_3 satisfies the initial value problem

$$\frac{dy}{dx} + a(x)y = 0, \qquad y(x_0) = 0. \tag{8}$$

Multiplying both sides of equation (8) by the integrating factor

$$e^{\int_{x_0}^x a(t)\, dt}$$

yields

$$[e^{\int_{x_0}^x a(t)\, dt} y_3(x)]' = e^{\int_{x_0}^x a(t)\, dt}[y_3' + a(x)y_3] = 0$$

so

$$e^{\int_{x_0}^x a(t)\, dt} y_3(x) = C \text{ (a constant)}.$$

Thus

$$y_3(x) = Ce^{-\int_{x_0}^x a(t)\, dt} \tag{9}$$

But the initial condition in (8) requires that

$$0 = y_3(x_0) = Ce^{-\int_{x_0}^{x_0} a(t)\, dt} = C$$

So $C = 0$. Substituting this value in (9) proves that $y_3 \equiv 0$. ■

Problems 1.9

1. Use Picard iterations to solve the initial value problem
$$y' = -y, \qquad y(0) = 1.$$

2. Do the same for the initial value problem
$$y' = 2y, \qquad y(0) = 5.$$

3. Repeat Problem 1 for the initial value problem
$$y' = x + y, \qquad y(0) = 0.$$

4. Repeat Problem 1 for the initial value problem
$$y' = 2xy, \qquad y(0) = 1.$$

Review Exercises for Chapter 1

In Exercises 1–4 find all solutions to the given differential equation. When an initial condition is given, find the particular solution that satisfies that condition.

1. $\dfrac{dy}{dx} = 3x$

2. $\dfrac{dx}{dt} = -2t, \quad x(0) = 4$

3. $\dfrac{dx}{dt} = \frac{1}{2}x, \quad x(0) = -3$

4. $\dfrac{dy}{dx} = 100y$

5. The relative annual rate of growth of a population is 15%. If the initial population is 10,000, what is the population after 5 years? After 10 years?

6. In Exercise 5, how long will it take for the population to double?

7. When a cake is taken out of the oven, its temperature is 125°C. Room temperature is 23°C. The temperature of the cake is 80°C after 10 min.

  (a) What will be its temperature after 20 min?

  (b) How long will the cake take to cool to 25°C?

8. An artifact contains 35% of the normal amount of ^{14}C. What is its approximate age?

9. What is the half-life of an exponentially decaying substance that loses 20% of its mass in one week?

10. How long will it take the substance in Exercise 9 to lose 75% of its mass? 95% of its mass?

11. (a) Plot the direction field for the differential equation $y' = y^{2/5}$.

  (b) Plot the solution that satisfies $y(0) = 5$.

  (c) Plot the solution that satisfies $y(0) = -2$.

12. (a) Plot the direction field for the differential equation $y' = -4xy/(1 + y^2)$.

  (b) Plot the solution that satisfies $y(2) = -1$.

  (c) Plot the solution that satisfies $y(2) = 3$.

13. J. H. Lambert (1728–1777) observed that very thin transparent layers of matter absorb light in direct proportion to the thickness of the layer and the amount of light incident on that layer. Express Lambert's law as a differential equation and solve it.

14. The **law of mass action** states that the rate of a chemical reaction (at constant temperature) is proportional to the product of the concentrations of the substances that are reacting. Consider the bimolecular reaction
$$Na + Cl \rightarrow NaCl,$$
where n moles per liter of sodium, Na, are combined with m moles per liter of chlorine, Cl, to produce salt. Let $x(t)$ be the number of moles per liter that have reacted after time t. Express the law of mass action as a differential equation, and solve it assuming $n \neq m$.

15. Repeat Exercise 14 assuming $n = m$.

16. A mountain climber starts from his base camp at 6:00 a.m. As he climbs, fatigue and oxygen deprivation take their toll so that the rate at which his elevation is increasing is inversely proportional to the elevation. At noon he is at an elevation of 19,000 ft, and at 2:00 p.m. he reaches the top of the mountain at 20,000 ft. How high was his base camp?

Find the general solution to each differential equation in Exercises 17–46. When an initial condition is given, find the particular solution that satisfies the condition.

17. $x\dfrac{dy}{dx} = y^2, \quad y(1) = 1$

18. $\dfrac{dy}{dx} = y\sqrt{1 - x}$

19. $\dfrac{dy}{dx} = \dfrac{\sqrt{1 - y^2}}{x}$

20. $x\dfrac{dy}{dx} = \tan y$, $\quad y(1) = \dfrac{\pi}{2}$

21. $\dfrac{dy}{dx} + x = x(y^2 + 1)$

22. $xy' = y(1 - 2y)$, $\quad y(1) = 2$

23. $yy' = \cos x$, $\quad y(\pi) = 0$　　**24.** $y' = xy(2 - 3y)$

25. $y' = x^2 y(1 - y)$　　　　　　　**26.** $y' = xy^2(1 - y)$

27. $y' - xy = 0$, $\quad y(1) = 1$　　**28.** $xy' - y = x$, $\quad y(1) = 1$

29. $y' - (\sin x)y = \sin x$　　　　**30.** $y' - \dfrac{1}{x}y = e^x$

31. $(1 + x^2)y' + xy = \sqrt{1+x^2}$

32. $y' - (\cos x)y = x^2$

33. $xy' - 2y = x^2$, $\quad y(1) = 1$

34. $xy' + (1 - x)y = xe^x$, $\quad y(1) = e$

35. $y' - xy = \begin{cases} 1, & x \leqslant 0 \\ 0, & x > 0 \end{cases}$

36. $y' + xy = \begin{cases} x, & x \leqslant 1 \\ 1, & x > 1 \end{cases}$

37. $y' = \dfrac{x - y}{x + 2y}$, $\quad y(0) = 1$ [*Hint:* Let $z = y/x$.]

38. $xy' = 2y + \sqrt{y^2 + x^2}$

39. $xy' = -y + \sqrt{xy + 1}$ [*Hint:* Let $z = xy$.]

40. $xy' = \sqrt{x^2 y^2 - 1} - y$, $\quad y(1) = 2$

41. $y' = y + xy^2$

42. $y' - xy = e^x y^3$

43. $(y - e^y \sec^2 x)\,dx + (x - e^y \tan x)\,dy = 0$

44. $(2x^2 y^3 - y^2)\,dx + (x^3 y^2 - x)\,dy = 0$

45. $\dfrac{dy}{dx} = \dfrac{3y^4 + 3x^2 y^3 - x^4 y}{xy^3 - 2x^5}$

46. $\dfrac{dy}{dx} + \dfrac{x + y\sqrt{x^2 + y^2}}{y + x\sqrt{x^2 + y^2}} = 0$

47. A paper mill is located next to a river having a constant flow of 1000 m³/sec that is situated at the only inlet to a lake having a volume of 10^9 m³. Assume that at time $t = 0$, the paper mill begins pumping pollutants into the river at the rate of 1 m³/sec, and the inflow and outflow of the lake are constant. How high a concentration of pollutant is there in the lake after 10 hr? after 100 hr? after 1 year?

48. Assume that the paper mill in Exercise 47 stops polluting the river at the end of 1 hour. Find an expression for the concentration of pollutant in the lake at all time t.

49. Assume the paper mill in Exercise 47 pollutes the river

for 1 hour each day. Find an expression for the concentration of pollutant in the lake at all time t. What is the maximum concentration that the pollution will reach in the lake?

50. A $20 \times 12 \times 8$ ft room contains five chain-smokers who are playing poker. An exhaust fan is removing 10 ft³/min of smoky air, which is replaced by pure air seeping in under the door. Each chain-smoker is contributing 0.1 ft³/min of smoke to the room. Find an expression for the concentration of smoke in the room at all time t, assuming the room contains no smoke at time $t = 0$.

51. A 6-ft chain weighing 10 lb per foot is placed on a frictionless table so that 1 ft of chain is hanging over the edge of the table. Find an equation describing the amount of chain still on the table for all time $t \geqslant 0$. (*Hint:* The mass of the chain that is falling changes with time.)

* 52. A power cable, hanging from fixed towers, has a weight of w lb per foot.

(a) Let $y(x)$ be the position of the cable x ft horizontally away from a tower and T_H be the horizontal component of tension in the cable. Show that

$$y'(x + \Delta x) - y'(x) = \dfrac{w}{T_H}\,\Delta s,$$

where Δs is the length of the cable over the horizontal interval of length Δx.

(b) Use part (a), the Pythagorean theorem, and limits to deduce the differential equation

$$y'' = \dfrac{w}{T_H}\sqrt{1 + (y')^2}. \qquad (1)$$

(c) Solve the differential equation in part (b) to obtain an expression for $y'(x)$, assuming $y'(0) = 0$.

53. Show that both $y = e^x$ and $y = e^{2x}$ are solutions of the second-order linear differential equation

$$y'' - 3y' + 2y = 0.$$

54. Consider the second-order linear equation

$$y'' + 5y' + 6y = 0. \qquad (2)$$

(a) Let $z = y' + 2y$. Show that (2) reduces to

$$z' + 3z = 0. \qquad (3)$$

(b) Solve (3), substitute z in the equation $y' + 2y = z$, and use the methods given in Section 1.4 to obtain the solution to equation (2).

55. Use the procedure outlined in Exercise 54 to find the general solution to

$$y'' + (a + b)y' + aby = 0,$$

where a and b are constants. [*Hint:* Let $z = y' + ay$.]

56. Use the method given in Exercise 54 to find the general solution to

$$y'' + 2ay' + a^2y = 0, \qquad a \text{ constant.}$$

57. Suppose a constant capacitor is connected in series to an emf whose voltage is a sine wave. Show that the current is 90° out of phase with the voltage.

58. Repeat Exercise 57 with the capacitor replaced by an inductor. What can you say in this case?

59. Newton's law of cooling can be used to estimate the time of death in a homicide.† Assume that at the time of death the body temperature was 98.6°F, when the corpse was discovered its temperature was 75°F, and 1 hour later it was 70°F. If the ambient temperature is 68°F, how soon before the corpse was discovered did the homicide occur?

60. **Stefan's law‡** of radiation states that the rate of change of temperature from a body at absolute temperature T is $T' = k(T^4 - T_0^4)$, where T_0 is the absolute temperature of the surrounding medium.

   (a) Solve this differential equation.

   * **(b)** Show that if $T - T_0$ is small compared to T_0, then Newton's law of cooling is a close approximation to Stefan's law.

61. Another equation that has been proposed to model population growth is the **Gompertz§** equation

$$\frac{dP}{dt} = P(a - b \ln P).$$

Find a solution to this equation and determine its behavior as $t \to \infty$.

62. A. G. W. Cameron,¶ in an article concerning the processes in the primitive solar nebula, obtained the first-order differential equation

$$\frac{dN}{dt} = \frac{aN^{5/6}}{(b - Bt)^{3/2}}$$

where a, b, and B are constants. Here $N(t)$ denotes, roughly, the number of interstellar particles that have accumulated

following the collapse of an interstellar cloud. Solve this equation.

63. A boat weighing 4000 lb is drifting away from a dock at 3 ft/sec. What is the minimum distance it will drift if a 180-lb crewman exerts a force equal to his weight on a rope tied to the bow of the boat?

64. L. L. Thurstone‖ used the separable differential equation

$$y' = \frac{2k}{\sqrt{m}}[y(1 - y)]^{3/2}$$

to describe the state of a learner $y(t)$ at time t, where k and m are positive constants depending on the learner and the complexity of the task, respectively. Find a solution to this equation.

In Exercises 65–70 use Euler's method and the given value of h to obtain an approximate solution at the indicated value of x.

65. $\dfrac{dy}{dx} = \dfrac{e^x}{y}$, $\quad y(0) = 2$. Find $y(3)$ with $h = \frac{1}{2}$.

66. $\dfrac{dy}{dx} = \dfrac{e^y}{x}$, $\quad y(1) = 0$. Find $y(\frac{1}{2})$ with $h = -0.1$.

67. $\dfrac{dy}{dx} = \dfrac{y}{\sqrt{1 + x^2}}$, $\quad y(0) = 1$. Find $y(3)$ with $h = \frac{1}{2}$.

68. $xy\dfrac{dy}{dx} = y^2 - x^2$, $\quad y(1) = 2$. Find $y(3)$ with $h = \frac{1}{2}$.

69. $\dfrac{dy}{dx} = y - xy^3$, $\quad y(0) = 1$. Find $y(3)$ with $h = \frac{1}{2}$.

70. $\dfrac{dy}{dx} = \dfrac{2xy}{3x^2 - y^2}$, $\quad y(-\frac{3}{8}) = -\frac{3}{4}$. Find $y(6)$ with $h = \frac{3}{8}$.

In Exercises 71–73 use Picard iterations to solve the given initial value problem.

71. $y' = y + 1$, $\quad y(0) = 0$

72. $y' = 3x^2y$, $\quad y(0) = 1$

73. $y' = y/x$, $\quad y(1) = 1$ [*Hint:* $x_0 = 1$ in equation (1.9.5).]

---

† D. A. Smith, "The homicide problem revisited," *Two Year College Mathematics Journal*, vol. 9 (1978), pp. 141–145.

‡ Named after the German physicist Josef Stefan (1835–1893).

§ Named after Benjamin Gompertz (1779–1865), an English mathematician.

¶ "Accumulation processes in the primitive solar nebula," *Icarus*, vol. 18 (1973), pp. 407–450.

‖ "The learning function," *Journal of General Psychology*, vol. 3 (1930), pp. 469–493.

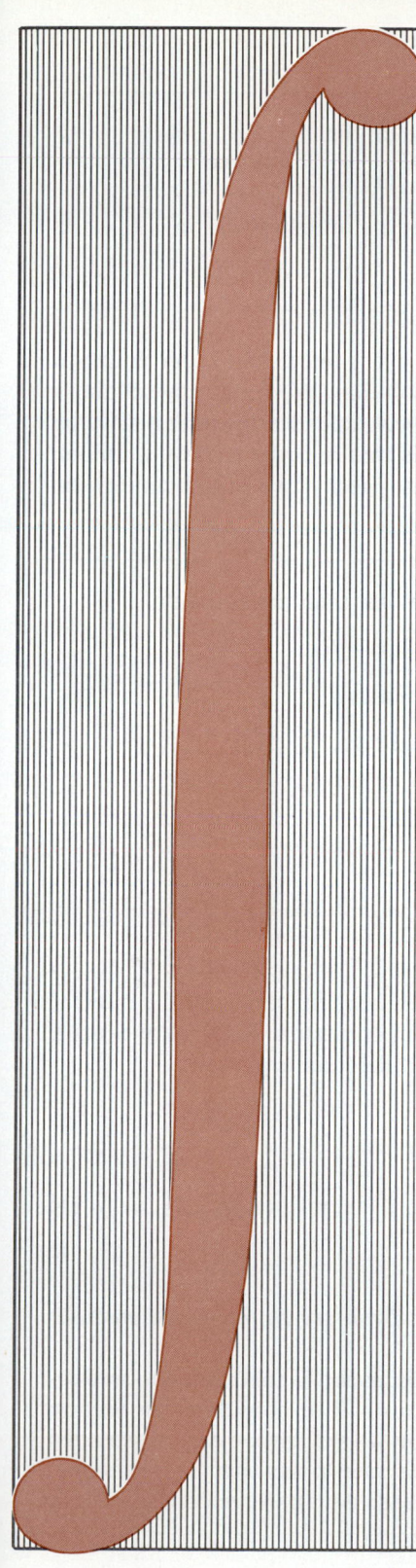

chapter 2

Second- and Higher-Order Linear Differential Equations

2.1 Theory of Linear Differential Equations

Although there is no procedure for explicitly solving arbitrary differential equations, systematic methods do exist for certain classes of differential equations. In this and the next section we shall study the theory of linear differential equations.

Recall that a differential equation is **linear** if it does not involve nonlinear functions (squares, exponentials, etc.) or products of the dependent variable and its derivatives. Thus $y'' + (x^3 \sin x)^5 y' + y = \cos x^3$ is linear, while $y'' + (y')^2 + y = 0$ is nonlinear. The most general first-order linear equation is

$$y'(x) + a(x)y(x) = f(x), \tag{1}$$

while the most general second-order linear equation can be written as

$$y''(x) + a(x)y'(x) + b(x)y(x) = f(x). \tag{2}$$

Here $a(x)$, $b(x)$, and $f(x)$ depend only on the independent variable x.

Equations (1) and (2) are special cases of the **general linear nth-order equation**

$$y^{(n)}(x) + a_{n-1}(x)y^{(n-1)}(x) + \cdots + a_1(x)y'(x) + a_0(x)y(x) = f(x).$$

In Section 1.4 we discussed the general first-order linear equation

$$y'(x) + a(x)y(x) = f(x).$$

If $a(x)$ and $f(x)$ are continuous, we saw that this equation has infinitely many solutions, since it involves an arbitrary constant [see equation (1.4.11) on p. 31]. This arbitrary constant can be determined if one condition $y(x_0) = y_0$ is given, as we saw in Examples 1.4.2 and 1.4.3. In this case, the equation has a unique solution. We can restate this basic fact as follows.

67

If $a(x)$ and $f(x)$ are continuous, then the equation

$$y'(x) + a(x)y(x) = f(x)$$

has one and only one solution that satisfies the initial condition

$$y(x_0) = y_0.$$

This is a very nice result, for it tells us that every linear first-order equation with a given initial condition has a unique solution. We need only set about finding it. It turns out that this special property holds for linear initial value problems of any order. The only difference is that in order to have a unique solution to a second-order equation, we must specify two initial conditions. Thus

---

If $a(x)$, $b(x)$, and $f(x)$ are continuous functions, then the equation

$$y''(x) + a(x)y'(x) + b(x)y(x) = f(x)$$

has a unique solution that satisfies the conditions

$$y(x_0) = y_0, \qquad y'(x_0) = y_1$$

for any real numbers x_0, y_0, and y_1.

In general, we have the following basic result.

■ **THEOREM 1: Existence-Uniqueness Theorem for Linear Initial Value Problems**

Let $a_0(x)$, $a_1(x)$, . . . , $a_{n-1}(x)$ and $f(x)$ be continuous functions on the interval $[x_1, x_2]$, and let $c_0, c_1, c_2, \ldots, c_{n-1}$ be n given constants. Then there exists a unique function $y(x)$ that satisfies the linear differential equation

$$y^{(n)} + a_{n-1}(x)y^{(n-1)} + a_{n-2}(x)y^{(n-2)} + \cdots + a_0(x)y = f(x)$$

on $[x_1, x_2]$ and the n initial conditions

$$y(x_0) = c_0, \quad y'(x_0) = c_1, \quad y''(x_0) = c_2, \ldots, y^{(n-1)}(x_0) = c_{n-1} \tag{3}$$

for some value x_0 in $[x_1, x_2]$.   ■

A proof of this theorem can be found in *Ordinary Differential Equations* by Jack K. Hale, Wiley, New York, 1969.

For simplicity we shall limit most of our discussion in this section to second-order linear equations. Higher-order equations will be considered in Section 2.11. We emphasize, however, that every result we shall prove can be extended to any linear differential equation. In the remainder of this chapter we assume, unless otherwise stated, that all functions in each equation are continuous for all real x.

DEFINITION

Homogeneous and Nonhomogeneous Differential Equations

If the function $f(x)$ is identically zero, we say that equation (2) is **homogeneous.** Otherwise, it is **nonhomogeneous.**

EXAMPLE 1

(a) The equation $y'' + 2xy' + 3y = 0$ is homogeneous.
(b) The equation $y'' + 2xy' + 3y = e^x$ is nonhomogeneous.

DEFINITION

Constant Coefficients

Variable Coefficients

If the coefficient functions $a(x)$ and $b(x)$ are constants, $a(x) = a$ and $b(x) = b,$ then the equation is said to have **constant coefficients.** As we shall see, linear differential equations with constant coefficients are the easiest to solve. If either $a(x)$ or $b(x)$ is not constant, the equation is said to have **variable coefficients.**

EXAMPLE 2

(a) The equation $y'' + 3y' - 10y = 0$ has constant coefficients.
(b) The equation $y'' + 3xy' - 10x^2y = 0$ has variable coefficients.

Before solving a second-order differential equation, it helps to know what we are seeking. A clue is provided by examining a first-order equation. Consider the equation

$$y' + 2y = 0.$$

In Section 1.1 we saw that one solution to this equation is $y = e^{-2x}$. In fact, $y = ce^{-2x}$ is a solution for any constant $c,$ and every solution to the equation has the form ce^{-2x}. We can summarize this result by noting that once we found one nonzero solution to the equation, we found all of the solutions, since every other solution is a constant multiple of this one solution.

It turns out that similar results hold for homogeneous second-order equations. The major difference is that now we have to find *two* solutions to (1) where neither solution is a multiple of the other. We now make these ideas more precise.

DEFINITION

Linear Combination

Let y_1 and y_2 be any two functions. By a **linear combination** of y_1 and y_2, we mean a function $y(x)$ that can be written in the form

$$y(x) = c_1 y_1(x) + c_2 y_2(x)$$

Linear Independence

for some constants c_1 and c_2. Two functions are **linearly independent** on an interval $[x_0, x_1]$ whenever the relation $c_1 y_1(x) + c_2 y_2(x) = 0$ for all x in $[x_0, x_1]$ implies that $c_1 = c_2 = 0$. Otherwise, they are **linearly dependent.** There is, however, an easier way to see that two functions y_1 and y_2 are linearly dependent. If $c_1 y_1(x) + c_2 y_2(x) = 0$ (where c_1 and c_2 are not both zero), we may suppose that $c_1 \neq 0$. Then, dividing the above expression by c_1, we obtain

$$y_1(x) + \frac{c_2}{c_1} y_2(x) = 0,$$

or

$$y_1(x) = -\frac{c_2}{c_1} y_2(x) = c y_2(x).$$

Therefore,

> Two functions are linearly dependent on the interval $[x_0, x_1]$ if and only if one of the functions is a constant multiple of the other.

Warning. The rule given above works only when dealing with two functions. When determining linear independence of the n functions $y_1, y_2, \ldots, y_n$ where $n > 2$, it is necessary to show that

$$c_1 y_1(x) + c_2 y_2(x) + \cdots + c_n y_n(x) = 0$$

holds only when $c_1 = c_2 = \cdots = c_n = 0$. We shall say more about this in Section 2.11.

Solutions of Homogeneous Equations

The notions of linear combination and linear independence are central to the theory of linear homogeneous equations, as illustrated by the results that follow. Consider the homogeneous second-order linear differential equation

$$y'' + a(x)y' + b(x)y = 0, \tag{4}$$

where $a(x)$ and $b(x)$ are continuous functions.

■ THEOREM 2
Two linearly independent solutions of equation (4) can always be found.   ■

(The proofs of this and the remaining theorems in this section are given in the appendix to this chapter.)

EXAMPLE 3

Verify that $y = e^{-5x}$ and $y = e^{2x}$ are linearly independent solutions to the equation

$$y'' + 3y' - 10y = 0.$$

SOLUTION

That e^{-5x} and e^{2x} are solutions is easy to show:

$$(e^{-5x})'' + 3(e^{-5x})' - 10e^{-5x} = 25e^{-5x} + 3(-5)e^{-5x} - 10e^{-5x} = 0;$$

$$(e^{2x})'' + 3(e^{2x})' - 10e^{2x} = 4e^{2x} + 3(2)e^{2x} - 10e^{2x} = 0.$$

Since $e^{-5x} = e^{-7x} \cdot e^{2x}$ and e^{-7x} is not a constant, we see that e^{-5x} and e^{2x} are linearly independent.

Every linear homogeneous second-order equation has two linearly independent solutions. The next two theorems show us that this is all the information we need. That is, once we have two linearly independent solutions, we can find all solutions.

Consider the homogeneous equation with continuous coefficients

$$y'' + a(x)y' + b(x)y = 0. \tag{5}$$

■ **THEOREM 3**

Let $y_1(x)$ and $y_2(x)$ be any two solutions of the homogeneous equation (5). Then any linear combination of them is also a solution of (5). ■

■ **THEOREM 4**

Let $y_1(x)$ and $y_2(x)$ be linearly independent solutions of (5) and let $y_3(x)$ be another solution of (5). Then there exist unique constants c_1 and c_2 such that

$$y_3(x) = c_1 y_1(x) + c_2 y_2(x). \qquad ■$$

In other words, any solution of (5) can be written as a linear combination of two given linearly independent solutions of (5).

We stress the importance of this theorem. It indicates that once we have found two linearly independent solutions y_1 and y_2 of equation (5), we have "essentially" found all the solutions of equation (5).

General Solution to a Linear Second-Order Equation

DEFINITION

The general solution of equation (5) is given by the linear combination

$$y(x) = c_1 y_1(x) + c_2 y_2(x),$$

where c_1 and c_2 are arbitrary constants, and y_1 and y_2 are two linearly independent solutions.

EXAMPLE 4

The general solution to $y'' + 3y' - 10y = 0$ is

$$y = c_1 e^{-5x} + c_2 e^{2x}. \qquad \text{see Example 3}$$

DEFINITION

Wronskian

Let $y_1(x)$ and $y_2(x)$ be any two solutions to equation (5). The **Wronskian†** of y_1 and y_2, $W(y_1, y_2)(x)$, is defined as

$$W(y_1, y_2)(x) = y_1(x)y_2'(x) - y_1'(x)y_2(x). \tag{6}$$

There is another way to write the Wronskian. This notation involves a 2×2 determinant. If you are unfamiliar with determinants, you can read about them in Section 7.7. We have

$$W(y_1, y_2)(x) = \begin{vmatrix} y_1(x) & y_2(x) \\ y_1'(x) & y_2'(x) \end{vmatrix} \tag{7}$$

Since

$$\begin{vmatrix} a & b \\ c & d \end{vmatrix} = ad - bc,$$

by the definition of a 2×2 determinant, we see that

$$\begin{vmatrix} y_1 & y_2 \\ y_1' & y_2' \end{vmatrix} = y_1 y_2' - y_2 y_1',$$

so that (6) and (7) denote the same thing.

Using the product rule of differentiation on equation (6) we see that

$$W'(y_1, y_2)(x) = y_1 y_2'' + y_1' y_2' - y_1' y_2' - y_1'' y_2$$

$$= y_1 y_2'' - y_1'' y_2.$$

Since y_1 and y_2 are solutions of equation (5),

$$y_1'' + ay_1' + by_1 = 0 \quad \text{and} \quad y_2'' + ay_2' + by_2 = 0.$$

Multiplying the first of these equations by y_2 and the second by y_1 and subtracting, we obtain

$$y_1 y_2'' - y_2 y_1'' + a(y_1 y_2' - y_2 y_1') = 0,$$

which is just

$$W' + aW = 0. \tag{8}$$

† Named after the Polish philosopher Josef M. H. Wronski (1778–1853).

From equation 1.4.3 (p. 30), we see that the solution of equation (8) can be written as

$$W(y_1, y_2)(x) = ce^{-\int a(x)dx} \tag{9}$$

Abel's Formula

for some arbitrary constant c. Equation (9) is known as **Abel's formula.**† Since an exponential is never zero, we see that $W(y_1, y_2)(x)$ is either always zero (when $c = 0$) or never zero (when $c \neq 0$). The importance of this fact is shown by the following theorem.

THEOREM 5

The solutions $y_1(x)$ and $y_2(x)$ of equation (5) are linearly independent on $[x_0, x_1]$ if and only if $W(y_1, y_2)(x) \neq 0$. ■

Theorem 5 is useful in at least three ways. First, it provides an easy way to determine whether or not two solutions are linearly independent. Second, it greatly simplifies the proof of Theorem 4—as we shall see in the appendix to this chapter. Third, the Wronskian can easily be extended to third- and higher-order equations with similar results. It is not easy to verify, directly, that three solutions are linearly independent, but the task is made easier by use of the Wronskian.

EXAMPLE 5

The functions $y_1 = e^{-5x}$ and $y_2 = e^{2x}$ are solutions to $y'' + 3y' - 10y = 0$. Then

$$W(y_1, y_2)(x) = \begin{vmatrix} e^{-5x} & e^{2x} \\ -5e^{-5x} & 2e^{2x} \end{vmatrix}$$

$$= e^{-5x}(2e^{2x}) - (-5e^{-5x})(e^{2x}) = 7e^{-3x}.$$

This is nonzero for every x, so y_1 and y_2 are linearly independent.

Solutions of Nonhomogeneous Equations

We now turn briefly to the nonhomogeneous equation

$$y'' + a(x)y' + b(x)y = f(x). \tag{10}$$

Let y_p be any particular solution to equation (10). If we know the general solution to the homogeneous equation

$$y'' + a(x)y' + b(x)y = 0, \tag{11}$$

we can find all solutions to equation (10).

† The tragic story of Niels Henrik Abel (1802–1829) is related in the biography of Cauchy on p. 25.

■ **THEOREM 6**

Let $y_p(x)$ be a particular solution to equation (10) and let $y^*(x)$ be any other solution. Then $y^*(x) - y_p(x)$ is a solution of equation (11); that is,

$$y^*(x) = c_1 y_1(x) + c_2 y_2(x) + y_p(x)$$

for some constants c_1 and c_2, where y_1, y_2 are two linearly independent solutions of equation (11). ■

Thus in order to find all solutions to the nonhomogeneous equation, we need only find one solution to the nonhomogeneous equation and the general solution of the homogeneous equation.

DEFINITION

General Solution to a Linear Nonhomogeneous Second-Order Equation

Let $y_p(x)$ be one solution to the nonhomogeneous equation (10) and let $y_1(x)$ and $y_2(x)$ be two linearly independent solutions to the homogeneous equation (11). Then the general solution to (10) is given by

$$y(x) = c_1 y_1(x) + c_2 y_2(x) + y_p(x)$$

where c_1 and c_2 are arbitrary constants.

EXAMPLE 6

It is easy to verify that $\frac{1}{2} t e^t$ is a particular solution of $x'' - x = e^t$. Two linearly independent solutions of $x'' - x = 0$ are given by $x_1 = e^t$ and $x_2 = e^{-t}$. The general solution is therefore $x(t) = \frac{1}{2} t e^t + c_1 e^t + c_2 e^{-t}$. Note that x_1 and x_2 are independent since $W(x_1, x_2) = e^t(-e^{-t}) - e^{-t}(e^t) = -2$.

In some important applications, the continuity requirements for the functions $a(x)$, $b(x)$, and $f(x)$ in equation (5) do not hold. For example, in the case of the equation

$$(x-1)y'' + x^2 y' + y = 0, \qquad y(0) = 1, \qquad y'(0) = 0,$$

the functions $a(x)$ and $b(x)$ are $x^2/(x-1)$ and $1/(x-1)$, respectively, which are discontinuous at $x = 1$. When this happens, some of our theoretical results do not apply (see problem 28). There are special techniques for handling some problems of this sort. We discuss some of these in Chapter 5.

Problems 2.1

In Problems 1–10, determine whether the given equation is linear or nonlinear. If it is linear, state whether it is homoge- neous or nonhomogeneous with constant or variable coeffi- cients.

1. $y'' + 2x^3 y' + y = 0$
2. $y'' + 2y' + y^2 = x$
3. $y'' + 3y' + yy' = 0$
4. $y'' + 3y' + 4y = 0$
5. $y'' + 3y' + 4y = \sin x$
6. $y'' + y(2 + 3y) = e^x$
7. $y'' + 4xy' + 2x^3 y = e^{2x}$
8. $y'' + \sin(xe^x)y' + 4xy = 0$
9. $3y'' + 16y' + 2y = 0$
10. $yy'y'' = 1$

In Problems 11–14, determine how many of the given functions are linearly independent on [0, 1].

11. $y_0 = 1$, $y_1 = 1 + x$, $y_2 = x^2$, $y_3 = x(1 - x)$, $y_4 = x$

12. $y_0 = \sin^2 x$, $y_1 = 1$, $y_2 = \sin x \cos x$, $y_3 = \cos^2 x$, $y_4 = \sin 2x$

13. $y_0 = 1 + x$, $y_1 = 1 - x$, $y_2 = 1$, $y_3 = x^2$, $y_4 = 1 + x^2$

14. $y_0 = 2\cos^2 x - 1$, $y_1 = \cos^2 x - \sin^2 x$, $y_2 = 1 - 2\sin^2 x$, $y_3 = \cos 2x$

In Problems 15–18, test each of the functions 1, x, x^2, and x^3 to see which functions satisfy the given differential equation. Then construct the *general solution* to the equation by writing a linear combination of the linearly independent solutions you have found.

15. $y'' = 0$
16. $y''' = 0$
17. $xy'' - y' = 0$
18. $x^2 y'' - 2xy' + 2y = 0$

19. Let $y_1(x)$ be a solution of the homogeneous equation

$$y'' + a(x)y' + b(x)y = 0$$

on the interval $\alpha \leqslant x \leqslant \beta$. Suppose that the curve y_1 is tangent to the x-axis at some point of this interval. Prove that y_1 must be identically zero.

20. Let $y_1(x)$ and $y_2(x)$ be two nontrivial solutions of the homogeneous equation

$$y'' + a(x)y' + b(x)y = 0$$

on the interval $\alpha \leqslant x \leqslant \beta$. Suppose $y_1(x_0) = y_2(x_0) = 0$ for some point $\alpha \leqslant x_0 \leqslant \beta$. Show that y_2 is a constant multiple of y_1.

21. (a) Show that $y_1(x) = \sin x^2$ and $y_2(x) = \cos x^2$ are linearly independent solutions of

$$xy'' - y' + 4x^3 y = 0.$$

(b) Calculate $W(y_1, y_2)$ and show that it is zero when $x = 0$. Does this result contradict Theorem 5? [*Hint:*

In Theorem 5, as elsewhere in this section, it is assumed that $a(x)$ and $b(x)$ are continuous.]

22. Show that x and x^3 are linearly independent solutions of $x^2 y'' - 3xy' + 3y = 0$, but their Wronskian is zero at $x = 0$. Does this contradict Theorem 5?

23. Show that

$$y_1(x) = \sin x \quad \text{and} \quad y_2(x) = 4\sin x - 2\cos x$$

are linearly independent solutions of $y'' + y = 0$. Write the solution $y_3(x) = \cos x$ as a linear combination of y_1 and y_2.

24. (a) Prove that $e^x \sin x$ and $e^x \cos x$ are linearly independent solutions of the equation

$$y'' - 2y' + 2y = 0.$$

(b) Find a solution that satisfies the conditions $y(0) = 1$, $y'(0) = 4$.

(c) Find another pair of linearly independent solutions.

25. Assume that some nonzero solution of

$$y'' + a(x)y' + b(x)y = 0, \quad y(0) = 0,$$

vanishes at some point x_1, where $x_1 > 0$. Prove that any other solution vanishes at $x = x_1$.

26. Define the function $s(x)$ to be the unique solution of the initial value problem

$$y'' + y = 0; \quad y(0) = 0, \quad y'(0) = 1,$$

and the function $c(x)$ as the solution of

$$y'' + y = 0; \quad y(0) = 1, \quad y'(0) = 0.$$

Without using trigonometry, prove that:

(a) $\dfrac{ds}{dx} = c(x)$;

(b) $\dfrac{dc}{dx} = -s(x)$;

(c) $s^2 + c^2 = 1$.

27. (a) Show that $y_1 = \sin \ln x^2$ and $y_2 = \cos \ln x^2$ are linearly independent solutions to

$$y'' + \frac{1}{x}y' + \frac{4}{x^2}y = 0 \qquad (x > 0).$$

(b) Calculate $W(y_1, y_2)(x)$.

28. Verify that the (infinite) family of functions

$$y = cx^3 + x$$

are all solutions of the initial value problem

$$x^2 y'' - 3xy' + 3y = 0, \qquad y(0) = 0,$$

$$y'(0) = 1.$$

Does this violate the existence-uniqueness theorem (Theorem 1)? Explain.

2.2 Using One Solution to Find Another (Reduction of Order Method)

As we saw in Theorem 2.1.4, it is easy to write down the general solution of the homogeneous equation

$$y'' + a(x)y' + b(x)y = 0, \tag{1}$$

provided we know two linearly independent solutions y_1 and y_2 of equation (1). The general solution is then given by

$$y = c_1 y_1 + c_2 y_2,$$

where c_1 and c_2 are arbitrary constants. Unfortunately, there is no general procedure for determining y_1 and y_2. However, a standard procedure does exist for finding y_2 when y_1 is known. This method is of considerable importance, since it is often possible to find one solution by inspecting the equation or by trial and error.

We assume that y_1 is a nonzero solution of equation (1) and seek another solution y_2 such that y_1 and y_2 are linearly independent. If it can be found, then since y_1 and y_2 are linearly independent, the ratio

$$\frac{y_2}{y_1} = v(x)$$

must be a nonconstant function of x, and $y_2 = vy_1$ must satisfy equation (1).

Now,

$$(vy_1)' = vy_1' + v'y_1$$

so

$$(vy_1)'' = (vy_1')' + (v'y_1)' = vy_1'' + v'y_1' + v'y_1' + v''y_1$$

$$= vy_1'' + 2v'y_1' + v''y_1$$

Since $y_2 = vy_1$ satisfies (1), we have

$$(vy_1)'' + a(x)(vy_1)' + bvy_1 = 0,$$

or

$$(v''y_1 + 2v'y_1' + vy_1'') + a(v'y_1 + vy_1') + bvy_1 = 0,$$

so that

$$y_1 v'' + (2y_1' + ay_1)v' + (y_1'' + ay_1' + by_1)v = 0. \tag{2}$$

That last term in parentheses in (2) must be zero, since y_1 is a solution to (1). Setting $z = v'$, equation (2) becomes

$$y_1 z' + (2y_1' + ay_1)z = 0. \tag{3}$$

Dividing by $y_1 z$, we can rewrite equation (3) in the form

$$\frac{z'}{z} = -\frac{2y_1'}{y_1} - a. \tag{4}$$

We can integrate this with respect to x to obtain

$$\ln z = -2 \ln y_1 - \int a(x)\, dx.$$

Thus,

$$z = e^{\ln z} = e^{-2\ln y_1 - \int a(x)\, dx} = e^{-2\ln y_1} e^{-\int a(x)\, dx}$$

$$e^{\ln u} = u$$

$$\downarrow$$

$$= e^{\ln y_1^{-2}} e^{-\int a(x)\, dx} = \frac{1}{y_1^2} e^{-\int a(x)\, dx}.$$

As $z = v'$, we have

$$z = v' = \frac{1}{y_1^2} e^{-\int a(x)\, dx}.$$

Since the exponential is never zero, v is nonconstant. To find v, we perform another integration and obtain

$$y_2 = v y_1 = y_1(x) \int \frac{e^{-\int a(x)\, dx}}{y_1^2(x)}\, dx. \tag{5}$$

EXAMPLE 1

$y_1 = x$ is a solution to

$$x^2 y'' - xy' + y = 0; \quad x > 0. \tag{6}$$

Find a second, linearly independent solution.

SOLUTION
We solve this problem in two ways.

METHOD 1
Let $y_2 = v y_1 = vx$. Then $y_2' = v'x + v$ and $y_2'' = v''x + 2v'$, so

$$x^2 y_2'' - xy_2' + y_2 = x^3 v'' + 2x^2 v' - x^2 v' - xv + xv = x^3 v'' + x^2 v' = 0.$$

Dividing both sides by x^2 we have

$$xv'' + v' = 0. \tag{7}$$

Set $u = v'$. Then (7) becomes

$$u' = -\frac{1}{x} u.$$

A solution to this linear first-order equation is

$$u(x) = e^{\int -(1/x)\,dx} = e^{-\ln x} = e^{\ln(1/x)} = \frac{1}{x}.$$

Since $u = v'$,

$$v(x) = \int u(x)\,dx = \int \frac{1}{x}dx = \ln x$$

and

$$y_2(x) = xv = x\ln x.$$

You should verify that y_2 is indeed a solution to (6).

METHOD 2

We use formula (5). To find y_2, we rewrite equation (6) as

$$y'' - \left(\frac{1}{x}\right)y' + \left(\frac{1}{x^2}\right)y = 0.$$

Then $a(x) = -(1/x)$

$$-\int a(x)\,dx = \int \frac{1}{x}dx = \ln x \quad \text{and} \quad e^{-\int a(x)\,dx} = e^{\ln x} = x$$

so that, from (5),

$$y_2 = x\int \frac{x}{x^2}dx = x\ln x.$$

Thus the general solution to (6) is

$$y = c_1 x + c_2 x\ln x, \qquad x > 0.$$

Remark: Formula (5) is there for your convenience. However, you should not lose sight of the process that led to this formula.

EXAMPLE 2

Consider the **Legendre equation of order one:**

$$(1 - x^2)y'' - 2xy' + 2y = 0, \qquad -1 < x < 1, \tag{8}$$

or

$$y'' - \frac{2x}{1 - x^2}y' + \frac{2y}{1 - x^2} = 0.$$

It is easy to verify that $y_1 = x$ is a solution. To find y_2, we note that $\int a(x)\,dx = \ln(1 - x^2)$, so that by equation (5) and partial fractions

$$y_2 = x \int \frac{e^{-\ln(1-x^2)}}{x^2} dx = x \int \frac{dx}{x^2(1-x^2)}$$

$$= x \int \left[\frac{1}{x^2} + \frac{1}{2}\left(\frac{1}{1+x} + \frac{1}{1-x} \right) \right] dx$$

$$= x \left[-\frac{1}{x} + \frac{1}{2}\ln\left(\frac{1+x}{1-x} \right) \right] = \frac{x}{2}\ln\left(\frac{1+x}{1-x} \right) - 1.$$

Note that in this example y_2 is defined in $-1 < x < 1$ even though $v(0)$ is undefined.

Problems 2.2

In each of Problems 1–10 a second-order differential equation and one solution $y_1(x)$ are given. Verify that $y_1(x)$ is indeed a solution and find a second, linearly independent solution.

1. $y'' - 2y' + y = 0$, $\quad y_1(x) = e^x$

2. $y'' + \left(\frac{3}{x} \right)y' = 0$, $\quad y_1(x) = 1$

3. $y'' - \left(\frac{2x}{1-x^2} \right)y' + \left(\frac{6}{1-x^2} \right)y = 0 \ (|x| < 1)$,

$y_1(x) = \frac{3x^2 - 1}{2}$

(This equation is called the **Legendre differential equation.** It is discussed in great detail in Section 5.5.)

4. $y'' - 2xy' + 2y = 0$, $\quad y_1(x) = x$. [*Hint:* Write your y_2 as an integral.]

5. $x^2 y'' + xy' - 4y = 0$, $\quad y_1(x) = x^2$. [*Hint:* Divide by x^2 first.]

6. $x^2 y'' - 2xy' + (x^2 + 2)y = 0 \ (x > 0)$, $\quad y_1 = x\sin x$

7. $xy'' + (2x - 1)y' - 2y = 0 \ (x > 0)$, $\quad y_1 = e^{-2x}$

8. $xy'' + (x - 1)y' + (3 - 12x)y = 0 \ (x > 0)$, $\quad y_1 = e^{3x}$

9. $xy'' - y' + 4x^3 y = 0 \ (x > 0)$, $\quad y_1 = \sin(x^2)$

10. $x^{1/3} y'' + y' + \left(\frac{1}{4}x^{-1/3} - \frac{1}{6x} - 6x^{-5/3} \right)y = 0$,

$y_1 = x^3 e^{-3x^{2/3}/4} \ (x > 0)$

11. The **Bessel differential equation** is given by

$$x^2 y'' + xy' + (x^2 - p^2)y = 0.$$

For $p = 1/2$, verify that $y_1(x) = (\sin x)/\sqrt{x}$ is a solution for $x > 0$. Find a second, linearly independent solution.

12. Letting $p = 0$ in the equation of Problem 11, we obtain the Bessel differential equation of index zero that we will study in Chapter 5. One solution is the Bessel function of order zero denoted by $J_0(x)$. In terms of $J_0(x)$, find a second, linearly independent solution.

13. The differential equation

$$y'' + a(xy' + y) = 0$$

arises in the study of turbulent flow in the wake of a cylinder.

(a) Verify that $y_1(x) = e^{-ax^2/2}$ is a solution of this equation.

(b) Find an expression for the general solution to this equation.

14. Verify that $y_1(x) = e^x$ is a solution to

$$xy'' - (x + n)y' + ny = 0,$$

with n a positive integer. Find a second, linearly independent solution.

2.3 Homogeneous Second-Order Linear Equations with Constant Coefficients

In this section we shall present a simple procedure for finding the general solution to the linear homogeneous equation with constant coefficients

$$y'' + ay' + by = 0. \tag{1}$$

We note that for the comparable first-order equation $y' + ay = 0$ (or $y' = -ay$) the general solution is $y(x) = ce^{-ax}$. It is then not implausible to "guess" that there may be a solution to equation (1) of the form $y(x) = e^{\lambda x}$ for some number λ (real or complex). Setting $y(x) = e^{\lambda x}$, we obtain $y' = \lambda e^{\lambda x}$ and $y'' = \lambda^2 e^{\lambda x}$ so that equation (1) yields

$$\lambda^2 e^{\lambda x} + a\lambda e^{\lambda x} + be^{\lambda x} = 0.$$

Since $e^{\lambda x} \neq 0$, we can divide this equation by $e^{\lambda x}$ to obtain

$$\lambda^2 + a\lambda + b = 0, \tag{2}$$

Characteristic Equation

where a and b are real numbers. Equation (2) is called the **characteristic equation** of the differential equation (1). It is clear that if λ satisfies equation (2), then $y(x) = e^{\lambda x}$ is a solution to equation (1).

Since the characteristic equation is quadratic, we have two roots:

$$\lambda_1 = \frac{-a + \sqrt{a^2 - 4b}}{2} \quad \text{and} \quad \lambda_2 = \frac{-a - \sqrt{a^2 - 4b}}{2}. \tag{3}$$

From equation (3) we see that there are three possible situations for the roots of λ_1 and λ_2 of the characteristic equation:

CASE 1
If $a^2 - 4b > 0$, then λ_1 and λ_2 are distinct real roots.

CASE 2
If $a^2 - 4b = 0$, then $\lambda_1 = \lambda_2$ is a real double root.

CASE 3
If $a^2 - 4b < 0$, then λ_1 and λ_2 are complex conjugate roots.

We shall examine each situation separately.

Roots Real and Unequal

CASE 1 $a^2 - 4b > 0$.
Then λ_1 and λ_2 are distinct real numbers [given by equation (3)] and $y_1(x) = e^{\lambda_1 x}$ and $y_2 = e^{\lambda_2 x}$ are distinct solutions.

These two solutions are linearly independent because

$$\frac{y_1}{y_2} = e^{(\lambda_1 - \lambda_2)x},$$

which is clearly not a constant when $\lambda_1 \neq \lambda_2$. Thus we have proved the following theorem.

■ **THEOREM 1**
If $a^2 - 4b > 0$, then the roots to the characteristic equation are real and unequal and the general solution to equation (1) is given by

$$y(x) = c_1 e^{\lambda_1 x} + c_2 e^{\lambda_2 x}, \qquad \text{(4)}$$

where c_1 and c_2 are arbitrary constants and λ_1 and λ_2 are the real roots of equation (2). ■

EXAMPLE 1

Consider the equation

$$y'' + 3y' - 10y = 0.$$

The characteristic equation is $\lambda^2 + 3\lambda - 10 = (\lambda - 2)(\lambda + 5) = 0$, and the roots are $\lambda_1 = 2$ and $\lambda_2 = -5$ (the order in which the roots are taken is irrelevant). The general solution is

$$y(x) = c_1 e^{2x} + c_2 e^{-5x}.$$

If we specify the initial conditions $y(0) = 1$ and $y'(0) = 3$, for example, then we can determine the constants c_1 and c_2. If we substitute $x = 0$ in the general solution, we obtain from the first initial condition, $y(0) = 1$,

$$c_1 e^0 + c_2 e^0 = c_1 + c_2 = 1.$$

Now, differentiate the general solution and substitute $x = 0$ to apply the second initial condition, $y'(0) = 3$, yielding

$$2c_1 e^0 - 5c_2 e^0 = 2c_1 - 5c_2 = 3.$$

Thus we obtain the system of simultaneous equations

$$c_1 + c_2 = 1,$$

$$2c_1 - 5c_2 = 3,$$

which have the unique solution $c_1 = \frac{8}{7}$ and $c_2 = -\frac{1}{7}$. The unique solution to the initial value problem is therefore

$$y(x) = \tfrac{1}{7}(8e^{2x} - e^{-5x}).$$

Roots Equal

CASE 2 $a^2 - 4b = 0$.
In this case equation (2) has the double root $\lambda_1 = \lambda_2 = -a/2$. Thus $y_1(x) = e^{-ax/2}$ is a solution of equation (1).

The procedure given in Section 2.2 can be used to find a second, linearly independent solution. Let $y_2 = vy_1 = ve^{-ax/2}$. Then

$$y_2' = v'e^{-ax/2} - \frac{a}{2}ve^{-ax/2},$$

$$y_2'' = v''e^{-ax/2} - av'e^{-ax/2} + \frac{a^2}{4}ve^{-ax/2},$$

so that if we substitute y_2 for y in equation (1) we get

$$\left(v'' - av' + \frac{a^2}{4}v\right)e^{-ax/2} + a\left(v' - \frac{a}{2}v\right)e^{-ax/2} + bve^{-ax/2} = 0. \qquad (5)$$

Dividing both sides by $e^{-ax/2}$, we obtain

$$v'' + \left(b - \frac{a^2}{4}\right)v = 0. \qquad (6)$$

But since $a^2 = 4b$, equation (6) becomes $v'' = 0$. Thus v' is a constant, which can be chosen to be 1 since we merely seek a solution, and $v = x$. Hence

$$y_2 = xe^{-ax/2}.$$

Check:

$$y_2' = e^{-ax/2}\left(1 - \frac{a}{2}x\right),$$

$$y_2'' = e^{-ax/2}\left(-a + \frac{a^2}{4}x\right),$$

$$y_2'' + ay_2' + by_2 = e^{-ax/2}\left(-a + \frac{a^2}{4}x + a - \frac{a^2}{2}x + bx\right)$$

$$= xe^{-ax/2}\left(-\frac{a^2}{4} + b\right) = 0 \qquad \text{because } a^2 - 4b = 0.$$

Since $y_2/y_1 = xe^{-ax/2}/e^{-ax/2} = x \neq$ constant, y_1 and y_2 are linearly independent and we have the following result:

■ THEOREM 2

If $a^2 - 4b = 0$, then the roots to the characteristic equation are equal and the general solution to equation (1) is given by

$$y(x) = c_1 e^{-(a/2)x} + c_2 x e^{-(a/2)x} = (c_1 + c_2 x)e^{-(a/2)x}, \qquad (7)$$

where c_1 and c_2 are arbitrary constants.   ■

EXAMPLE 2

Consider the equation

$$y'' - 6y' + 9 = 0.$$

The characteristic equation is $\lambda^2 - 6\lambda + 9 = (\lambda - 3)^2 = 0$, yielding the unique double root $\lambda_1 = -a/2 = 3$. The general solution is, therefore,

$$y(x) = c_1 e^{3x} + c_2 x e^{3x}.$$

Complex Conjugate Roots

CASE 3 $a^2 - 4b < 0$.

The roots of the characteristic equation to equation (1) are the **complex conjugates**

$$\lambda_1 = \alpha + i\beta, \qquad \lambda_2 = \alpha - i\beta,$$

where $\alpha = -a/2$, $\beta = \sqrt{4b - a^2}/2$, and $i = \sqrt{-1}$. Thus $y_1 = e^{\lambda_1 x}$ and $y_2 = e^{\lambda_2 x}$ are solutions to equation (1). However, in this case it is useful to recall that we proved that any linear combination of solutions is also a solution and instead consider the solutions

$$y_1^* = \frac{e^{\lambda_1 x} + e^{\lambda_2 x}}{2} \quad \text{and} \quad y_2^* = \frac{e^{\lambda_1 x} - e^{\lambda_2 x}}{2i}.$$

The number $\alpha + i\beta = \alpha + \beta\sqrt{-1}$ is called a **complex number.** We discuss complex numbers in great detail in Chapters 15–19. An introduction to complex numbers appears in Section 15.1. The Euler equations (derived in Section 15.7) state that

$$e^{i\theta} = \cos\theta + i\sin\theta$$

and

$$e^{-i\theta} = \cos\theta - i\sin\theta.$$

Using the Euler equations, we compute

$$e^{\lambda_1 x} = e^{\alpha x + i\beta x} = e^{\alpha x} e^{i\beta x} = e^{\alpha x}(\cos\beta x + i\sin\beta x)$$

and

$$e^{\lambda_2 x} = e^{\alpha x - i\beta x} = e^{\alpha x} e^{-i\beta x} = e^{\alpha x}(\cos\beta x - i\sin\beta x).$$

Finally, we observe that

$$y_1^* = \frac{e^{\lambda_1 x} + e^{\lambda_2 x}}{2} = e^{\alpha x}\cos\beta x \quad \text{and} \quad y_2^* = \frac{e^{\lambda_1 x} - e^{\lambda_2 x}}{2i} = e^{\alpha x}\sin\beta x.$$

Since $y_1^*/y_2^* = (e^{\alpha x}\cos\beta x)/(e^{\alpha x}\sin\beta x) = \cot\beta x$, $\beta \neq 0$, is not constant, y_1^* and y_2^* are linearly independent. We therefore have the following:

■ **THEOREM 3**

If $a^2 - 4b < 0$, then the characteristic equation has two complex roots and the general solution to equation (1) is given by

$$y(x) = e^{\alpha x}(c_1 \cos \beta x + c_2 \sin \beta x), \qquad (8)$$

where c_1 and c_2 are arbitrary constants and

$$\alpha = -\frac{a}{2}, \qquad \beta = \frac{\sqrt{4b - a^2}}{2}.$$

■

EXAMPLE 3

Let $y'' + y = 0$. Then the characteristic equation is $\lambda^2 + 1 = 0$ with roots $\pm i$. We have $\alpha = 0$ and $\beta = 1$ so that the general solution is

$$y(x) = c_1 \cos x + c_2 \sin x.$$

Equation of Harmonic Motion

This is the **equation of harmonic motion.**

EXAMPLE 4

Consider the equation $y'' + y' + y = 0$, $y(0) = 1$, $y'(0) = 3$. We have $\lambda^2 + \lambda + 1 = 0$ with roots $\lambda_1 = (-1 + i\sqrt{3})/2$ and $\lambda_2 = (-1 - i\sqrt{3})/2$. Then $\alpha = -\frac{1}{2}$ and $\beta = \sqrt{3}/2$, so that the general solution is

$$y(x) = e^{-x/2}\left(c_1 \cos \frac{\sqrt{3}}{2}x + c_2 \sin \frac{\sqrt{3}}{2}x\right).$$

To solve the initial value problem, we differentiate, set $x = 0$, and solve the simultaneous equations

$$c_1 = 1,$$

$$\frac{\sqrt{3}}{2}c_2 - \frac{1}{2}c_1 = 3.$$

Thus $c_1 = 1$, $c_2 = 7/\sqrt{3}$, and

$$y(x) = e^{-x/2}\left(\cos \frac{\sqrt{3}}{2}x + \frac{7}{\sqrt{3}} \sin \frac{\sqrt{3}}{2}x\right).$$

Problems 2.3

In Problems 1–32 find the general solution of each equation. When initial conditions are specified, give the particular solution that satisfies them.

1. $y'' - 4y = 0$

2. $x'' + x' - 6x = 0$, $x(0) = 0$, $x'(0) = 5$

3. $y'' - 3y' + 2y = 0$

4. $y'' + 5y' + 6y = 0$, $y(0) = 1$, $y'(0) = 2$

5. $4x'' + 20x' + 25x = 0$, $x(0) = 1$, $x'(0) = 2$

6. $y'' + 6y' + 9y = 0$

7. $x'' - x' - 6x = 0$, $x(0) = -1$, $x'(0) = 1$

8. $y'' - 8y' + 16y = 0$, $y(0) = 2$, $y'(0) = -1$

9. $y'' - 5y' = 0$

10. $y'' + 17y' = 0$, $y(0) = 1$, $y'(0) = 0$

11. $y'' + 2\pi y' + \pi^2 y = 0$

12. $y'' - 13y' + 42y = 0$

13. $z'' + 2z' - 15z = 0$

14. $w'' + 8w' + 12w = 0$

15. $y'' - 8y' + 16y = 0$, $y(0) = 1$, $y'(0) = 6$

16. $y'' + 2y' + y = 0$, $y(1) = 2/e$, $y'(1) = -3/e$

17. $y'' - 2y = 0$

18. $y'' + 6y' + 5y = 0$

19. $y'' - 5y = 0$, $y(0) = 3$, $y'(0) = -\sqrt{5}$

20. $y'' - 2y' - 2y = 0$, $y(0) = 1$, $y'(0) = 1 + 3\sqrt{3}$

21. $y'' + 2y' + 2y = 0$

22. $8y'' + 4y' + y = 0$, $y(0) = 0$, $y'(0) = 1$

23. $x'' + x' + 7x = 0$

24. $y'' + y' + 2y = 0$

25. $\dfrac{d^2 x}{d\theta^2} + 4x = 0$, $x\left(\dfrac{\pi}{4}\right) = 1$, $x'\left(\dfrac{\pi}{4}\right) = 3$

26. $y'' + y = 0$, $y(\pi) = 2$, $y'(\pi) = -1$

27. $y'' + \frac{1}{4}y = 0$, $y(\pi) = 1$, $y'(\pi) = -1$

28. $y'' + 6y' + 12y = 0$

29. $y'' + 2y' + 5y = 0$

30. $y'' + 2y' + 5y = 0$, $y(0) = 1$, $y'(0) = -3$

31. $y'' + 2y' + 2y = 0$, $y(\pi) = e^{-\pi}$, $y'(\pi) = -2e^{-\pi}$

32. $y'' + 2y' + 5y = 0$, $y(\pi) = e^{-\pi}$, $y'(\pi) = 3e^{-\pi}$

33. Suppose $z = c_1 z_1 + c_2 z_2$ is the general solution to equation (10). Explain why the quotient z'/z involves only one arbitrary constant. [*Hint:* Divide numerator and denominator by c_1.]

34. For arbitrary constants a, b, and c, find the substitution that changes the nonlinear equation

$$y' + ay^2 + by + c = 0$$

into a linear second-order equation with constant coefficients. What second-order equation is obtained?

Use the method in the box below to find the general solution to the Riccati equations in Problems 35–40. If an initial condition is specified, give the particular solution that satisfies that condition.

35. $y' + y^2 - 1 = 0$, $y(0) = -\frac{1}{3}$

36. $\dfrac{dx}{dt} + x^2 + 1 = 0$

37. $y' + y^2 - 2y + 1 = 0$

38. $y' + y^2 + 3y + 2 = 0$, $y(0) = 1$

39. $y' + y^2 - y - 2 = 0$

40. $y' + y^2 + 2y + 1 = 0$, $y(1) = 0$

The Riccati Equation†

Linear second-order differential equations may also be used in finding the solution to the **Riccati equation:**

$$y' + y^2 + a(x)y + b(x) = 0. \tag{9}$$

This nonlinear first-order equation frequently occurs in physical applications. To change equation (9) into a linear second-order equation, let $y = z'/z$. Then $y' = (z''/z) - (z'/z)^2$, so equation (9) becomes

$$\frac{z''}{z} - \left(\frac{z'}{z}\right)^2 + \left(\frac{z'}{z}\right)^2 + a(x)\left(\frac{z'}{z}\right) + b(x) = 0.$$

Multiplying by z, we obtain the linear second-order equation

$$z'' + a(x)z' + b(x)z = 0. \tag{10}$$

If the general solution to equation (10) can be found, the quotient $y = z'/z$ is the general solution to equation (9).

† Giacomo Riccati (1676–1754) was an Italian mathematician, physicist, and philosopher. Riccati was responsible for bringing much of Newton's work on calculus to the attention of Italian mathematicians.

2.4 Nonhomogeneous Equations: Method of Undetermined Coefficients

In this and the following section, we shall present methods for finding a particular solution to the nonhomogeneous linear equation

$$y'' + ay' + by = f(x). \tag{1}$$

First, however, we shall prove a very useful result concerning nonhomogeneous equations, called the **principle of superposition:**

Principle of Superposition

■ **THEOREM 1: Principle of Superposition**
Suppose the function $f(x)$ in (1) is a sum of two functions $f_1(x)$ and $f_2(x)$:

$$f(x) = f_1(x) + f_2(x).$$

If $y_1(x)$ is a solution of the equation

$$y'' + ay' + by = f_1(x) \tag{2}$$

and $y_2(x)$ is a solution of the equation

$$y'' + ay' + by = f_2(x), \tag{3}$$

then $y = y_1 + y_2$ is a solution of equation (1); that is, a solution of (1) is obtained by superimposing a solution of equation (3) on that of equation (2).

PROOF
Substituting $y = y_1 + y_2$ in the left-hand side of equation (1), we have

$$y'' + ay' + by = (y_1'' + y_2'') + a(y_1' + y_2') + b(y_1 + y_2)$$

$$= (y_1'' + ay_1' + by_1) + (y_2'' + ay_2' + by_2)$$

$$= f_1 + f_2 = f,$$

since y_1 and y_2 are solutions of equations (2) and (3), respectively. ■

Briefly, the principle of superposition tells us that if we can split the function $f(x)$ into a sum of two (or more) simpler expressions $f_k(x)$, then we can restrict our attention to solving the nonhomogeneous equations

$$y'' + ay' + by = f_k(x), \qquad k = 1, 2, \ldots, m, \tag{4}$$

because the solution to equation (1) is simply the sum of the solutions of these equations.

The method we shall present in this section *requires* that the function $f(x)$ in equation (1) be of *one* of the following three forms:

> **(i)** $P_n(x)$,
> **(ii)** $P_n(x)e^{\alpha x}$,   or
> **(iii)** $e^{\alpha x}[P_n(x) \cos \beta x + Q_n(x) \sin \beta x]$,

where $P_n(x)$ and $Q_n(x)$ are polynomials in x of degree n ($n \geq 0$). The method we present below can also be used if $f(x)$ is a sum of functions $f_k(x)$ of these three forms, since by the principle of superposition we can solve each of the equations in (4) and add the solutions together. **However, if any term of $f(x)$ is not of one of these three forms, we cannot use the method of this section.†**

Note that the three forms involve a multitude of situations. The following three functions are all of one of these forms:

(a) $2e^{3x}$ (the polynomial is the constant 2);
(b) $e^{4x} \cos x$ (here $P_n(x) = 1$ and $Q_n(x) = 0$ are both polynomials of degree zero);
(c) $x \cos x + \sin x$ (here $\alpha = 0$, $\beta = 1$, $P_n(x) = x + 0$, and $Q_n(x) = 0 \cdot x + 1$).

Method of Undetermined Coefficients

The **method of undetermined coefficients** assumes that the solution to equation (1) is exactly of the same "form" as $f(x)$. The technique requires that we replace each dependent variable y in (1) with an expression of the same form as $f(x)$ having polynomial terms with **undetermined coefficients.** If we compare both sides of the resulting equation, it is then possible to "determine" the unknown coefficients. The method will be illustrated by a number of examples.

EXAMPLE 1

Solve the following equation:

$$y'' - y = x^2. \tag{5}$$

SOLUTION
Since $f(x) = x^2$ is a polynomial of degree two, we "guess" that (5) has a solution $y_p(x)$ that is a polynomial of degree two. The most general polynomial of degree two is

$$y_p(x) = a + bx + cx^2.$$

Our problem is to *determine* the constants a, b, and c such that y_p is a solution to (5). After calculating that $y_p'' = 2c$, we substitute $y_p(x)$ into (5) to obtain

$$2c - (a + bx + cx^2) = x^2.$$

---

† Instead we must use the variation of parameters method, which will be described in Section 2.5. Note that forms (i) and (ii) are special cases of form (iii).

Equating coefficients, we have

Coefficient of constant term	Coefficient of x	Coefficient of x^2

$$2c - a = 0, \qquad -b = 0, \qquad -c = 1,$$

which immediately yields $a = -2$, $b = 0$, $c = -1$, and the particular solution

$$y_p(x) = -2 - x^2.$$

This particular solution is easily verified by substitution into (5). Finally, since the general solution of the homogeneous equation $y'' - y = 0$ is given by

$$y = c_1 e^x + c_2 e^{-x},$$

the general solution of (5) is

$$y = c_1 e^x + c_2 e^{-x} - 2 - x^2.$$

EXAMPLE 2

Solve

$$y'' - 3y' + 2y = e^x \sin x. \tag{6}$$

SOLUTION

Since $f(x)$ is of form (iii) with $P_n(x) = 0$ and $Q_n(x) = 1$, we "guess" that there is a solution to (6) of the form

$$y_p(x) = ae^x \sin x + be^x \cos x.$$

Then

$$y_p'(x) = (a - b)e^x \sin x + (a + b)e^x \cos x$$

and

$$y_p''(x) = 2ae^x \cos x - 2be^x \sin x.$$

Substituting these expressions into (6) we have

$$e^x(2a \cos x - 2b \sin x) - 3e^x[(a - b) \sin x + (a + b) \cos x]$$
$$+ 2e^x(a \sin x + b \cos x) = e^x \sin x.$$

Dividing both sides by e^x and equating the coefficients of $\sin x$ and $\cos x$, we have

$$2a - 3(a + b) + 2b = 0,$$
$$-2b - 3(a - b) + 2a = 1,$$

which yield $a = -\frac{1}{2}$ and $b = \frac{1}{2}$ so that

$$y_p = \frac{e^x}{2}(\cos x - \sin x).$$

Again, this result is easily verified by substitution. Finally, the general solution of (6) is

$$y = c_1 e^{2x} + c_2 e^x + \frac{e^x}{2}(\cos x - \sin x).$$

EXAMPLE 3

Solve $y'' + y = xe^{2x}$.

SOLUTION

Here $f(x)$ is of form (ii), where $P_n(x)$ is a polynomial of degree one, so we try a solution of the form

$$y_p(x) = e^{2x}(a + bx).$$

Then

$$y_p'(x) = e^{2x}(2a + b + 2bx), \qquad y_p''(x) = e^{2x}(4a + 4b + 4bx),$$

and substitution yields

$$e^{2x}(4a + 4b + 4bx) + e^{2x}(a + bx) = xe^{2x}.$$

Dividing both sides by e^{2x} and equating like powers of x, we obtain the equations

$$5a + 4b = 0, \qquad 5b = 1.$$

Thus $a = -\frac{4}{25}$, $b = \frac{1}{5}$, and a particular solution is

$$y_p(x) = \frac{e^{2x}}{25}(5x - 4).$$

Therefore, the general solution of this example is

$$y(x) = c_1 \sin x + c_2 \cos x + \frac{e^{2x}}{25}(5x - 4).$$

Difficulties arise in connection with problems of this type whenever any term of the guessed solution is a solution of the homogeneous equation

$$y'' + ay' + by = 0. \tag{7}$$

For example, in the equation

$$y'' + y = (1 + x + x^2) \sin x, \tag{8}$$

the function $f(x)$ is the sum of three functions, one of which ($\sin x$) is a solution to the homogeneous equation $y'' + y = 0$. As another example, in

$$y'' + y = (x + x^2) \sin x \tag{9}$$

the guessed solution is $y_p = (a_0 + a_1 x + a_2 x^2) \sin x + (b_0 + b_1 x + b_2 x^2) \cos x$ and $a_0 \sin x + b_0 \cos x$ is a solution to the homogeneous equation $y'' + y = 0$. When this situation occurs, the method of undetermined coefficients must be modified. To see why, consider the following example.

EXAMPLE 4

Find the solution to the equation

$$y'' - y = 2e^x. \tag{10}$$

SOLUTION

The general solution of $y'' - y = 0$ is

$$y(x) = c_1 e^x + c_2 e^{-x}.$$

Here $f(x) = 2e^x$ is a solution to the homogeneous equation. If we try to find a solution of the form Ae^x, we will get nowhere since Ae^x is a solution to the homogeneous equation for every constant A and, therefore, it cannot possibly be a solution to the nonhomogeneous equation.

What do we do? Recall that if λ was a double root of the characteristic equation for a homogeneous differential equation, then two solutions are $e^{\lambda x}$ and $xe^{\lambda x}$. This suggests that we try Axe^x, instead of Ae^x as a possible solution to (10). Thus, we consider a particular solution of the form

$$y_p = Axe^x.$$

Then

$$y_p' = Ae^x(x + 1), \qquad y_p''(x) = Ae^x(x + 2)$$

and

$$y_p'' - y_p = Ae^x(x + 2) - Axe^x = 2Ae^x = 2e^x.$$

Hence $A = 1$ and $y_p = xe^x$. Thus the general solution is

$$y(x) = c_1 e^x + c_2 e^{-x} + xe^x.$$

The preceding example suggests the following rule.

Modifications of the Method

If *any term* of the guessed solution $y_p(x)$ is a solution of the homogeneous equation (7), multiply $y_p(x)$ by x repeatedly until no term of the product $x^k y_p(x)$ is a solution of (7). Then use the product $x^k y_p(x)$ to solve equation (1).

EXAMPLE 5

Find the solution to

$$y'' + y = \cos x \tag{11}$$

that satisfies $y(0) = 2$ and $y'(0) = -3$.

SOLUTION

The general solution to $y'' + y = 0$ is $y = c_1 \cos x + c_2 \sin x$, and $f(x) = \cos x$ is a solution, so we must use the modification of the method to find a particular solution to (11). Ordinarily we would guess a solution of the form $y_p = A \cos x + B \sin x$. Instead we multiply by x and try a solution of the form

$$y_p = Ax \cos x + Bx \sin x.$$

Note that no term of y_p is a solution to $y'' + y = 0$. Then

$$y_p' = A \cos x - Ax \sin x + B \sin x + Bx \cos x = (A + Bx) \cos x + (B - Ax) \sin x$$
$$y_p'' = (2B - Ax) \cos x + (-2A - Bx) \sin x$$

and

From (11)
↓

$$\cos x = y_p'' + y_p = (-2A \sin x - Ax \cos x + 2B \cos x - Bx \sin x) + (Ax \cos x + Bx \sin x)$$

$$= -2A \sin x + 2B \cos x.$$

Therefore,

$$-2A = 0, \qquad 2B = 1, \qquad B = \tfrac{1}{2}$$

and

$$y_p = \tfrac{1}{2}x \sin x.$$

Thus the general solution to (11) is

$$y = c_1 \cos x + c_2 \sin x + \tfrac{1}{2}x \sin x.$$

We are not finished yet as initial conditions were given. We have

$$y' = -c_1 \sin x + c_2 \cos x + \tfrac{1}{2}x \cos x + \tfrac{1}{2} \sin x.$$

Then

$$y(0) = c_1 = 2 \quad \text{and} \quad y'(0) = c_2 = -3,$$

which yields the unique solution

$$y(x) = 2 \cos x - 3 \sin x + \tfrac{1}{2}x \sin x.$$

EXAMPLE 6

Find the general solution to

$$y'' + y = x \sin x.$$

SOLUTION

The guessed solution is $y_p = (Ax + B) \cos x + (Cx + D) \sin x$. Since $B \cos x + D \sin x$ solves $y'' + y = 0$, the modification is required. We therefore multiply by x and try a solution of the form

$$y_p = (Ax^2 + Bx) \cos x + (Cx^2 + Dx) \sin x$$

Then

$$y_p' = [Cx^2 + (2A + D)x + B] \cos x + [-Ax^2 + (2C - B)x + D] \sin x$$

$$y_p'' = [-Ax^2 + (4C - B)x + 2A + 2D] \cos x + [-Cx^2 - (4A + D)x + 2C - 2B] \sin x$$

and

given
↓

$$y_p'' + y_p = [4Cx + 2A + 2D] \cos x + [-4Ax + 2C - 2B] \sin x = x \sin x.$$

This yields $A = -\frac{1}{4}$, $B = 0$, $C = 0$, $D = \frac{1}{4}$ and the particular solution

$$y_p = -\tfrac{1}{4}x^2 \cos x + \tfrac{1}{4}x \sin x.$$

Thus the general solution is

$$y = (c_1 - \tfrac{1}{4}x^2) \cos x + (c_2 + \tfrac{1}{4}x) \sin x.$$

Let us now summarize the results of this section as follows: Consider the nonhomogeneous equation

$$y'' + ay' + by = f(x) \tag{12}$$

and the homogeneous equation

$$y'' + ay' + by = 0. \tag{13}$$

CASE 1
No term in the guessed solution $y_p(x)$ is a solution of equation (13). A particular solution of equation (12) will have the form $y_p(x)$ given by the table below:

$f(x)$	$y_p(x)$
$P_n(x)$	$a_0 + a_1 x + a_2 x^2 + \cdots + a_n x^n$
$P_n(x)e^{ax}$	$(a_0 + a_1 x + a_2 x^2 + \cdots + a_n x^n)e^{ax}$
$P_n(x)e^{ax} \sin bx$	
$+$	$(a_0 + a_1 x + a_2 x^2 + \cdots + a_n x^n)e^{ax} \sin bx +$
$Q_n(x)e^{ax} \cos bx$	$(c_0 + c_1 x + c_2 x + \cdots + c_n x^n)e^{ax} \cos bx$

CASE 2
If any term of $y_p(x)$ is a solution of equation (13), then multiply the appropriate function $y_p(x)$ of Case 1 by x^k, where k is the smallest integer such that no term in $x^k y_p(x)$ is a solution of equation (13).

Problems 2.4

In Problems 1–13, find the general solution of each given differential equation. If initial conditions are given, then find the particular solution that satisfies them.

1. $y'' + 4y = 3 \sin x$
2. $y'' - y' - 6y = 20e^{-2x}$, $y(0) = 0$, $y'(0) = 6$
3. $y'' - 3y' + 2y = 6e^{3x}$
4. $y'' + y' = 3x^2$, $y(0) = 4$, $y'(0) = 0$
5. $y'' - 2y' + y = -4e^x$
6. $y'' - 4y' + 4y = 6xe^{2x}$, $y(0) = 0$, $y'(0) = 3$
7. $y'' - 7y' + 10y = 100x$, $y(0) = 0$, $y'(0) = 5$
8. $y'' + y = 1 + x + x^2$
9. $y'' + y' = x^3 - x^2$

10. $y'' + 4y = 16x \sin 2x$
11. $y'' - 4y' + 5y = 2e^{2x} \cos x$
12. $y'' - y' - 2y = x^2 + \cos x$
13. $y'' + 6y' + 9y = 10e^{-3x}$

Use the principle of superposition to find the general solution of each of the equations in Problems 14–17.

14. $y'' + y = 1 + 2 \sin x$
15. $y'' - 2y' - 3y = x - x^2 + e^x$
16. $y'' + 4y = 3 \cos 2x - 7x^2$
17. $y'' + 4y' + 4y = xe^x + \sin x$

18. Show by the methods of this section that a particular solution of

$$y'' + 2ay' + b^2 y = A \sin \omega x \qquad (a, \omega > 0)$$

is given by

$$y = \frac{A \sin(\omega x - \alpha)}{\sqrt{(b^2 - \omega^2)^2 + 4\omega^2 a^2}},$$

where

$$\alpha = \tan^{-1} \frac{2a\omega}{(b^2 - \omega^2)}, \qquad (-\pi/2 < \alpha < \pi/2).$$

19. Let $f(x)$ be a polynomial of degree n. Show that, if $b \neq 0$, then there is always a solution that is a polynomial of degree n for the equation $y'' + ay' + by = f(x)$.

20. Use the method indicated in Problem 19 to find a particular solution of

$$y'' + 3y' + 2y = 9 + 2x - 2x^2.$$

In Problems 21–24 find particular solutions to the given differential equation.

21. $y'' + y = (x + x^2) \sin x$

22. $y'' - y' = x^2$

23. $y'' - 2y' + y = x^2 e^x$

24. $y'' - 4y' + 3y = x^3 e^{3x}$

25. In many physical problems (see Section 4.4) the nonhomogeneous term $f(x)$ is specified by different formulas in different intervals of x. Find a solution to the initial value problem

$$y'' + y = \begin{cases} x, & 0 \leq x \leq 1, \\ 1, & x \geq 1, \end{cases} \qquad y(0) = 0, \quad y'(0) = 1.$$

2.5 Nonhomogeneous Equations: Variation of Parameters†

In this section we shall consider a procedure, due to J. L. Lagrange (1736–1813), for finding a particular solution of any nonhomogeneous linear equation

$$y'' + a(x)y' + b(x)y = f(x), \tag{1}$$

where the functions $a(x)$, $b(x)$, and $f(x)$ are continuous. To use this method it is necessary to know the general solution $c_1 y_1(x) + c_2 y_2(x)$ of the homogeneous equation

$$y'' + a(x)y' + b(x)y = 0. \tag{2}$$

If $a(x)$ and $b(x)$ are constants, then the general solution to equation (2) can always be obtained by the methods of Section 2.3. If $a(x)$ and $b(x)$ are not both constants, it may be difficult to find this general solution; however, if one solution y_1 of equation (2) can be found, then the method of Section 2.2 will yield the general solution to equation (2).

Lagrange noticed that any particular solution y_p of equation (1) must have the property that y_p/y_1 and y_p/y_2 are not constants, suggesting that we look for a particular solution of equation (1) of the form

$$y(x) = c_1(x)y_1(x) + c_2(x)y_2(x). \tag{3}$$

This replacement of constants or parameters by variables gives the method its name. Differentiating equation (3), we obtain

$$y'(x) = c_1(x)y_1'(x) + c_2(x)y_2'(x) + c_1'(x)y_1(x) + c_2'(x)y_2(x).$$

† This procedure is also called the **variation of constants** method, or **Lagrange's method**.

To simplify this expression, it is convenient (but not necessary—see Problem 23) to set

$$c_1'(x)y_1(x) + c_2'(x)y_2(x) = 0. \tag{4}$$

Then

$$y'(x) = c_1(x)y_1'(x) + c_2(x)y_2'(x).$$

Differentiating once again, we obtain

$$y''(x) = c_1(x)y_1''(x) + c_2(x)y_2''(x) + c_1'(x)y_1'(x) + c_2'(x)y_2'(x).$$

Substitution of the expressions for $y(x)$, $y'(x)$ and $y''(x)$ into equation (1) yields

$$y'' + a(x)y' + b(x)y = c_1(x)(y_1'' + ay_1' + by_1) + c_2(x)(y_2'' + ay_2' + by_2)$$

$$+ c_1'y_1' + c_2'y_2'$$

$$= f(x).$$

But y_1 and y_2 are solutions to the homogeneous equation so that the equation above reduces to

$$c_1'y_1' + c_2'y_2' = f(x). \tag{5}$$

This gives a second equation relating $c_1'(x)$ and $c_2'(x)$, and we have the simultaneous equations

$$
\boxed{
\begin{aligned}
y_1 c_1' + y_2 c_2' &= 0, \\
y_1' c_1' + y_2' c_2' &= f(x).
\end{aligned}
}
\tag{6}
$$

The determinant of system (6) is the Wronskian, $W(y_1, y_2)(x)$:

$$\begin{vmatrix} y_1 & y_2 \\ y_1' & y_2' \end{vmatrix} = W(y_1, y_2)(x) \neq 0 \qquad \text{since } y_1 \text{ and } y_2 \text{ are linearly independent. See Theorem 2.1.5 on p. 73.}$$

Thus, for each value of x, $c_1'(x)$ and $c_2'(x)$ are uniquely determined and the problem has essentially been solved. We obtain, from (6),

$$y_1 y_2' c_1' + y_2 y_2' c_2' = 0 \qquad \text{first equation multiplied by } y_2'$$

$$\underline{y_1' y_2 c_1' + y_2 y_2' c_2' = y_2 f(x)} \qquad \text{second equation multiplied by } y_2$$

$$(y_1 y_2' - y_1' y_2)c_1' = -y_2 f(x)$$

or

$$c_1' = \frac{-y_2 f(x)}{W(y_1, y_2)(x)}.$$

A similar calculation yields an expression for c_2'. Thus, we obtain

$$c_1'(x) = \frac{-f(x)y_2(x)}{y_1(x)y_2'(x) - y_1'(x)y_2(x)} = \frac{-f(x)y_2(x)}{W(y_1, y_2)(x)}, \tag{7}$$

$$c_2'(x) = \frac{f(x)y_1(x)}{y_1(x)y_2'(x) - y_1'(x)y_2(x)} = \frac{f(x)y_1(x)}{W(y_1, y_2)(x)}. \tag{8}$$

Finally, if we can integrate c_1' and c_2', we can substitute c_1 and c_2 into equation (3) to obtain a particular solution to the nonhomogeneous equation. That is,

$$y_p(x) = c_1(x)y_1(x) + c_2(x)y_2(x),$$

where $c_1(x) = \int c_1'(x)\,dx$, $c_2(x) = \int c_2'(x)\,dx$, and $c_1'(x)$ and $c_2'(x)$ are given by (7) and (8).†

EXAMPLE 1

Solve $y'' - y = e^{2x}$ by the variation of parameters method.

SOLUTION

The solutions to the homogeneous equation are $y_1 = e^{-x}$ and $y_2 = e^x$. We obtain $W(y_1, y_2)(x) = 2$, so that equations (7) and (8) become

$$c_1'(x) = \frac{-e^{2x}e^x}{2} = \frac{-e^{3x}}{2}, \qquad c_2'(x) = \frac{e^{2x}e^{-x}}{2} = \frac{e^x}{2}.$$

Integrating these functions, we obtain $c_1(x) = -e^{3x}/6$ and $c_2(x) = e^x/2$. A particular solution is therefore

$$c_1(x)y_1(x) + c_2(x)y_2(x) = \frac{-e^{2x}}{6} + \frac{e^{2x}}{2} = \frac{e^{2x}}{3}$$

and the general solution is

$$y(x) = c_1 e^x + c_2 e^{-x} + \frac{e^{2x}}{3}.$$

EXAMPLE 2

Solve $y'' + y = \tan x$.

SOLUTION

The solutions to the homogeneous equation are $y_1 = \cos x$ and $y_2 = \sin x$. Also $W(y_1, y_2)(x) = 1$, so that equations (7) and (8) become

---

† It is not necessary to include arbitrary constants of integration in the computation of $c_1(x)$ and $c_2(x)$. Remember, by Theorem 2.1.6, we are only trying to find *a* particular solution. However, including arbitrary constants of integration will not lead to an incorrect answer: the constants will simply be absorbed by the solution to the homogeneous problem.

$$c_1'(x) = -\tan x \sin x = -\frac{\sin^2 x}{\cos x} = \frac{\cos^2 x - 1}{\cos x} = \cos x - \sec x,$$

$$c_2'(x) = \tan x \cos x = \sin x.$$

Hence

$$c_1(x) = \sin x - \ln|\sec x + \tan x|$$

and

$$c_2(x) = -\cos x.$$

Thus the particular solution is

$$y_p(x) = c_1(x)y_1(x) + c_2(x)y_2(x)$$

$$= \cos x \sin x - \cos x \ln|\sec x + \tan x| - \sin x \cos x$$

$$= -\cos x \ln|\sec x + \tan x|,$$

and the general solution is

$$y(x) = c_1 \cos x + c_2 \sin x - \cos x \ln|\sec x + \tan x|.$$

Example 2 illustrates that there are instances in which we cannot apply the method of undetermined coefficients. (Try to "guess" a solution in this case.) As a rule, the method of undetermined coefficients is easier to use if the function $f(x)$ is in the right form. However, the method of variation of parameters is far more general, since it will yield a solution whenever the functions c_1' and c_2' have known antiderivatives.

Problems 2.5

In Problems 1–20 find the general solution to each equation by the method of variation of parameters.

1. $y'' + 4y = 3 \sin x$

2. $y'' - y' - 6y = 20e^{-2x}$

3. $y'' - 3y' + 2y = 6e^{3x}$

4. $y'' + y' = 3x^2$

5. $y'' - 2y' + y = -4e^x$

6. $y'' - 4y' + 4y = 6xe^{2x}$

7. $y'' - 7y' + 10y = 100x$

8. $y'' + y = 1 + x + x^2$

9. $y'' + y' = x^3 - x^2$

10. $y'' + 4y = 16x \sin 2x$

11. $y'' - y' = \sec^2 x - \tan x$

12. $y'' + y = \cot x$

13. $y'' + 4y = \sec 2x$

14. $y'' + 4y = \sec x \tan x$

15. $y'' - 2y' + y = \dfrac{e^x}{(1-x)^2}$

16. $y'' - y = \sin^2 x$

17. $y'' - y = \dfrac{(2x-1)e^x}{x^2}$

18. $y'' - 3y' - 4y = \dfrac{e^{4x}(5x-2)}{x^3}$

19. $y'' - 4y' + 4y = \dfrac{e^{2x}}{(1+x)}$

20. $y'' + 2y' + y = e^{-x} \ln|x|$

21. Find a particular solution of

$$y'' + \frac{1}{x}y' - \frac{y}{x^2} = \frac{1}{x^2 + x^3} \qquad (x > 0),$$

given that two solutions of the associated homogeneous equation are $y_1 = x$ and $y_2 = 1/x$.

22. Find a particular solution of

$$y'' - \frac{2}{x}y' + \frac{2}{x^2}y = \frac{\ln|x|}{x} \qquad (x > 0),$$

given the two solutions $y_1 = x$ and $y_2 = x^2$ of the homogeneous equation.

* 23. This problem will show why there is no loss in generality in equation (4) by setting

$$c_1' y_1 + c_2' y_2 = 0.$$

Suppose instead that we let $c_1' y_1 + c_2' y_2 = z(x)$, with $z(x)$ an undetermined function of x.

(a) Show that we then obtain the system

$$c_1' y_1 + c_2' y_2 = z,$$

$$c_1' y_1' + c_2' y_2' = f - z' - az.$$

(b) Show that the system in part (a) has the solution

$$c_1' = \frac{-y_2 f}{W(y_1, y_2)} + \frac{(e^{\int a(x)\,dx} z y_2)'}{e^{\int a(x)\,dx} W(y_1, y_2)},$$

$$c_2' = \frac{y_1 f}{W(y_1, y_2)} - \frac{(e^{\int a(x)\,dx} z y_1)'}{e^{\int a(x)\,dx} W(y_1, y_2)}.$$

(c) Integrate by parts to show that

$$\int \frac{(e^{\int a(x)\,dx} z y_i)'}{e^{\int a(x)\,dx} W(y_1, y_2)}\,dx = \frac{z y_i}{W(y_1, y_2)}, \quad i = 1, 2.$$

(d) Conclude that the particular solution obtained by letting $c_1' y_1 + c_2' y_2 = z$ is identical to that obtained by assuming equation (4).

(e) Letting t be a dummy variable of integration, show the particular solution can always be represented by the integral

$$y_p(x) = \int^x \frac{y_2(x) y_1(t) - y_1(x) y_2(t)}{W(y_1, y_2)(t)} f(t)\,dt.$$

24. Verify that

$$y = \frac{1}{\omega} \int_0^x f(t) \sin \omega(x - t)\,dt$$

is a particular solution of $y'' + \omega^2 y = f(x)$.

25. Find a particular solution to the initial value problem

$$y'' - \omega^2 y = f(x), \qquad y(0) = y'(0) = 0.$$

[*Hint:* Use Problem 23.]

2.6 Euler Equations

For most linear second-order equations with variable coefficients it is impossible to write solutions in compact form in terms of elementary functions. In most cases it is necessary to use techniques such as the power series method (Chapter 5) to obtain information about solutions. However, there is one class of such equations that arise in applications for which closed-form solutions can be obtained. We discuss this class now.

An equation of the form

$$x^2 y'' + axy' + by = f(x), \qquad x \neq 0, \tag{1}$$

Euler Equation

is called an **Euler equation.**†

Note. Equation (1) can be written as

$$y'' + \frac{a}{x} y' + \frac{b}{x^2} y = \frac{f(x)}{x^2},$$

which is not defined for $x = 0$. This is why we make the restriction that $x \neq 0$.

We begin by solving the homogeneous Euler equation

$$x^2 y'' + axy' + by = 0, \qquad x \neq 0. \tag{2}$$

If we can find two linearly independent solutions to (2), then we can solve (1)

† See the biographical sketch on p. 57.

by the method of variation of parameters. There are two ways to solve equation (2). Each one involves a certain trick. We give one method here and leave the other method for the problem set (see Problem 18).

Our method involves guessing an appropriate solution to (2). We note that if $y = x^\lambda$ for some number λ, then $y' = \lambda x^{\lambda-1}$ and $y'' = \lambda(\lambda - 1)x^{\lambda-2}$. This is interesting because then $x^2 y''$, xy', and y all can be written as constant multiples of x^λ. Therefore, we guess that there is a solution having the form $y = x^\lambda$. Then, substituting this into equation (2), we obtain

$$\lambda(\lambda - 1)x^\lambda + a\lambda x^\lambda + bx^\lambda = x^\lambda[\lambda(\lambda - 1) + a\lambda + b] = 0.$$

Characteristic Equation

If $x \neq 0$, we can divide by x^λ to obtain the **characteristic equation**† for Euler's equation:

$$\lambda(\lambda - 1) + a\lambda + b = 0, \tag{3}$$

or

$$\lambda^2 + (a - 1)\lambda + b = 0. \tag{4}$$

As with constant-coefficient equations, there are three cases to consider.

CASE 1
The characteristic equation (4) has two distinct real roots.

EXAMPLE 1

Find the general solution to

$$x^2 y'' + 2xy' - 12y = 0, \qquad x \neq 0.$$

SOLUTION
The characteristic equation is

$$\lambda(\lambda - 1) + 2\lambda - 12 = \lambda^2 + \lambda - 12 = 0 = (\lambda + 4)(\lambda - 3)$$

with roots $\lambda_1 = -4$ and $\lambda_2 = 3$. Thus, two solutions (that are linearly independent) are

$$y_1 = x^{-4} = \frac{1}{x^4} \quad \text{and} \quad y_2 = x^3$$

so that the general solution is

$$y(x) = \frac{c_1}{x^4} + c_2 x^3.$$

In general, we have the following result:

† The method of characteristic equations is generally *not* applicable to equations with *variable* coefficients. The only reason it is applicable in this case is that the substitution in Problem 18 converts (1) into a second-order constant-coefficient equation with equation (3) as its characteristic equation.

■ **THEOREM 1**

If λ_1 and λ_2 are real and distinct, then the general solution to equation (2) is

$$y(x) = c_1 x^{\lambda_1} + c_2 x^{\lambda_2}, \qquad x \neq 0. \qquad (5)$$

■

CASE 2

The roots are real and equal ($\lambda_1 = \lambda_2$).

EXAMPLE 2

Find the general solution to

$$x^2 y'' - 3xy' + 4y = 0, \qquad x > 0. \qquad (6)$$

SOLUTION

The characteristic equation is

$$\lambda^2 - 4\lambda + 4 = (\lambda - 2)^2 = 0$$

with the single root $\lambda = 2$. Thus one solution is $y_1(x) = x^2$. To find a second solution, we use formula (2.2.5) on p. 77. First, we write (6) in the form

$$y'' - \frac{3}{x} y' + \frac{4}{x^2} y = 0.$$

Then, as in Section 2.2, $a(x) = -(3/x)$ so that

$$e^{-\int a(x)\, dx} = e^{\int (3/x)\, dx} = e^{3 \ln x} = x^3$$

and

$$y_2(x) = y_1(x) \int \frac{e^{-\int a(x)\, dx}}{y_1^2(x)}\, dx = x^2 \int \frac{x^3}{x^4}\, dx = x^2 \ln x.$$

Thus the general solution to equation (6) is

$$y(x) = c_1 x^2 + c_2 x^2 \ln x = x^2 (c_1 + c_2 \ln x).$$

■ **THEOREM 2**

If λ is the only root of the characteristic equation (4), then the general solution to (2) is

$$y(x) = x^\lambda (c_1 + c_2 \ln|x|). \qquad ■$$

CASE 3

The roots are complex conjugates ($\lambda_1 = \alpha + i\beta$, $\lambda_2 = \alpha - i\beta$).

EXAMPLE 3

Find the general solution of

$$x^2y'' + 5xy' + 13y = 0, \qquad x > 0. \tag{7}$$

SOLUTION

The characteristic equation is

$$\lambda^2 + 4\lambda + 13 = 0$$

and

$$\lambda = \frac{-4 \pm \sqrt{16 - 4(13)}}{2} = \frac{-4 \pm \sqrt{-36}}{2} = -2 \pm 3i.$$

Thus two linearly independent solutions are

$$y_1(x) = x^{-2+3i} \quad \text{and} \quad y_2(x) = x^{-2-3i}.$$

We can eliminate the imaginary exponents. First we note that

$$x^a = e^{\ln x^a} = e^{a \ln x}.$$

Then, by Euler's equations (see p. 83)

$$y_1(x) = (x^{-2})(x^{3i}) = x^{-2}e^{3i \ln x} = x^{-2}[\cos(3 \ln x) + i \sin(3 \ln x)]$$

and

$$y_2(x) = (x^{-2})(x^{-3i}) = x^{-2}e^{-3i \ln x} = x^{-2}[\cos(3 \ln x) - i \sin(3 \ln x)].$$

We now form two new solutions:

$$y_3(x) = \tfrac{1}{2}[y_1(x) + y_2(x)] = x^{-2} \cos(3 \ln x)$$

and

$$y_4(x) = \frac{1}{2i}[y_1(x) - y_2(x)] = x^{-2} \sin(3 \ln x).$$

These new solutions contain no complex numbers and are easier to work with. The general solution to (7) is

$$y(x) = x^{-2}[c_1 \cos(3 \ln x) + c_2 \sin(3 \ln x)].$$

■ THEOREM 3

If $\lambda_1 = \alpha + i\beta$ and $\lambda_2 = \alpha - i\beta$ are complex conjugate roots of the characteristic equation (4), then the general solution to (2) is

$$y(x) = x^\alpha[c_1 \cos(\beta \ln|x|) + c_2 \sin(\beta \ln|x|)]. \tag{8}$$

■

EXAMPLE 4

Find the general solution to

$$x^2 y'' + 2xy' - 12y = \sqrt{x}, \qquad x > 0. \tag{9}$$

SOLUTION

In Example 1 we found the homogeneous solutions

$$y_1 = x^{-4} \quad \text{and} \quad y_2 = x^3.$$

Then

$$W(y_1, y_2)(x) = \begin{vmatrix} x^{-4} & x^3 \\ -4x^{-5} & 3x^2 \end{vmatrix} = 3x^{-2} + 4x^{-2} = \frac{7}{x^2} \quad \text{and} \quad \frac{1}{W} = \frac{x^2}{7}.$$

We rewrite (9) in the standard form

$$y'' + \frac{2}{x} y' - \frac{12}{x^2} y = x^{-3/2}.$$

This is the form for which the variation of parameters formulas [formulas (2.5.7) and (2.5.8) on p. 95] apply. Then $f(x) = x^{-3/2}$ and we obtain

$$c_1'(x) = \frac{-f(x)y_2(x)}{W(y_1, y_2)(x)} = \frac{x^2}{7}(-x^{-3/2})(x^3) = \frac{-x^{7/2}}{7}$$

and

$$c_2'(x) = \frac{f(x)y_1(x)}{W(y_1, y_2)(x)} = \frac{x^2}{7}(x^{-3/2})(x^{-4}) = \tfrac{1}{7}x^{-7/2}.$$

Hence

$$c_1(x) = -\tfrac{1}{7} \cdot \tfrac{2}{9} x^{9/2}, \qquad c_2(x) = -\tfrac{1}{7} \cdot \tfrac{2}{5} x^{-5/2},$$

so that

$$y_p(x) = c_1(x)y_1(x) + c_2(x)y_2(x) = -\tfrac{1}{7}(\tfrac{2}{9}x^{9/2} \cdot x^{-4} + \tfrac{2}{5}x^{-5/2} \cdot x^3)$$

$$= \frac{-x^{1/2}}{7}(\tfrac{2}{9} + \tfrac{2}{5}) = -\tfrac{4}{45}x^{1/2}.$$

Thus the general solution is given by

$$y(x) = c_1 x^{-4} + c_2 x^3 - \tfrac{4}{45}x^{1/2}.$$

Problems 2.6

In Problems 1–17 find the general solution to the given Euler equation. Find the unique solution when initial conditions are given.

1. $x^2 y'' + xy' - y = 0$

2. $x^2 y'' - 5xy' + 9y = 0$

3. $x^2 y'' - xy' + 2y = 0$

4. $x^2 y'' - 2y = 0, \quad y(1) = 3, \quad y'(1) = 1$

5. $4x^2 y'' - 4xy' + 3y = 0, \quad y(1) = 0, \quad y'(1) = 1$

6. $x^2 y'' + 3xy' + 2y = 0$

7. $x^2 y'' - 3xy' + 3y = 0$

8. $x^2 y'' + 5xy' + 4y = 0, \quad y(1) = 1, \quad y'(1) = 3$

9. $x^2y'' + 5xy' + 5y = 0$

10. $4x^2y'' - 8xy' + 8y = 0$

11. $x^2y'' + 2xy' - 12y = 0$

12. $x^2y'' + xy' + y = 0$

13. $x^2y'' + 3xy' - 15y = 1/x$

14. $x^2y'' + 3xy' + y = 3x^6$

15. $x^2y'' - 5xy' + 9y = x^3$

16. $x^2y'' + 3xy' - 15y = x^2e^x$

17. $x^2y'' + xy' + y = 10$

* 18. Show that the homogeneous Euler equation (2) can be transformed into the constant-coefficient equation $y'' + (a - 1)y' + by = 0$ by making the substitution $x = e^t$ ($t = \ln x$). [*Hint:* By the chain rule

$$\frac{dy}{dt} = \frac{dy}{dx}\frac{dx}{dt} = x\frac{dy}{dx}$$

and

$$\frac{d^2y}{dt^2} = \frac{d}{dx}\left[x\frac{dy}{dx}\right]\frac{dx}{dt} = x^2\frac{d^2y}{dx^2} + x\frac{dy}{dx}.\right]$$

Use the method of Problem 18 to solve Problems 19–22.

19. $x^2y'' + 7xy' + 5y = x$

20. $x^2y'' + 3xy' - 3y = 5x^2$

21. $x^2y'' - 2y = \ln x,\quad (x > 0)$

22. $4x^2y'' - 4xy' + 3y = \sin \ln(-x)\quad (x < 0)$

23. The equation $xy'' + 4y' = 0$, $x > 0$, arises in astronomy.† Obtain its general solution.

2.7 More on Electric Circuits

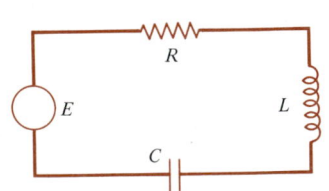

FIGURE 1

We shall make use of the electric circuit concepts developed in Section 1.6 and the methods of this chapter to study a simple electric circuit containing a resistor, an inductor, and a capacitor in series with an electromotive force (Fig. 1). Suppose that R, L, C, and E are constants. Applying Kirchhoff's law, we obtain

$$L\frac{dI}{dt} + RI + \frac{Q}{C} = E. \tag{1}$$

Since $dQ/dt = I$, we may differentiate equation (1) to get the second-order homogeneous differential equation

$$L\frac{d^2I}{dt^2} + R\frac{dI}{dt} + \frac{I}{C} = 0. \tag{2}$$

To solve this equation, we note that the characteristic equation

$$\lambda^2 + \frac{R}{L}\lambda + \frac{1}{CL} = 0$$

has the following roots:

$$\lambda_1 = \frac{-R + \sqrt{R^2 - 4L/C}}{2L}, \qquad \lambda_2 = \frac{-R - \sqrt{R^2 - 4L/C}}{2L},$$

or, rewriting the radical in dimensionless units,‡ we have

$$\lambda_1 = \frac{R}{2L}\left(-1 + \sqrt{1 - \frac{4L}{CR^2}}\right), \qquad \lambda_2 = \frac{R}{2L}\left(-1 - \sqrt{1 - \frac{4L}{CR^2}}\right). \tag{3}$$

Equation (2) may now be solved using the methods of Section 2.3.

† Z. Kopal, "Stress history of the moon and of terrestrial planets," *Icarus*, vol. 2 (1963), p. 381.

‡ 1 henry = 1 volt-s/amp; 1 farad = 1 coul/volt; 1 ohm = 1 volt/amp; 1 coul = 1 amp-s.

Remark: Equation (1) can be turned into a second-order differential equation even if E is a nondifferentiable function (like a square wave). Setting $I = dQ/dt$ and $dI/dt = d^2Q/dt^2$ in (1), we obtain

$$L\frac{d^2Q}{dt^2} + R\frac{dQ}{dt} + \frac{Q}{C} = E. \tag{1'}$$

EXAMPLE 1

Let $L = 1$ henry (H), $R = 100$ ohms (Ω), $C = 10^{-4}$ farads (f), and $E = 1000$ volts (V) in the circuit shown in Fig. 1. Suppose that no charge is present and no current is flowing at time $t = 0$ when E is applied. By equation (3) we see that the characteristic equation has the roots $\lambda_1 = -50 + 50\sqrt{3}i$ and $\lambda_2 = -50 - 50\sqrt{3}i$, since

$$R^2 - 4L/C = 10{,}000 - 4 \times 10^4 = -30{,}000 = -(100\sqrt{3})^2$$

and $R/2L = 50$. Thus

$$I(t) = e^{-50t}(c_1 \cos 50\sqrt{3}t + c_2 \sin 50\sqrt{3}t).$$

Applying the initial condition $I(0) = 0$, we have $c_1 = 0$. Hence

$$I(t) = c_2 e^{-50t} \sin 50\sqrt{3}t$$

and

$$I'(t) = 50c_2 e^{-50t}(\sqrt{3} \cos 50\sqrt{3}t - \sin 50\sqrt{3}t).$$

To establish the value of c_2, we must make use of equation (1) and the initial condition $Q(0) = 0$. Since

$$Q(t) = C\left(E - L\frac{dI}{dt} - RI\right)$$

$$= 10^{-4}[1000 - 50c_2 e^{-50t}(\sqrt{3} \cos 50\sqrt{3}t$$

$$- \sin 50\sqrt{3}t + 2 \sin 50\sqrt{3}t)]$$

$$= \frac{1}{10} - \frac{c_2}{200} e^{-50t}(\sin 50\sqrt{3}t + \sqrt{3} \cos 50\sqrt{3}t),$$

it follows that

$$Q(0) = \frac{1}{10} - \frac{c_2\sqrt{3}}{200} = 0 \quad \text{or} \quad c_2 = \frac{20}{\sqrt{3}}.$$

Hence

$$Q(t) = \frac{1}{10} - \frac{1}{10\sqrt{3}} e^{-50t}(\sin 50\sqrt{3}t + \sqrt{3} \cos 50\sqrt{3}t)$$

and

$$I(t) = \frac{20}{\sqrt{3}} e^{-50t} \sin 50\sqrt{3}t.$$

From these equations we observe that the current will rapidly damp out and the charge will rapidly approach its **steady-state value** of $\frac{1}{10}$ coulomb (coul). Here $I(t)$ is called the **transient current** (because its effect is brief).

EXAMPLE 2

Let the inductance, resistance, and capacitance in Example 1 remain the same, but suppose that $E = 962 \sin 60t$. By equation (1) we have

$$\frac{dI}{dt} + 100I + 10^4 Q = 962 \sin 60t, \tag{4}$$

and converting equation (4) so that all expressions are in terms of $Q(t)$, we obtain

$$\frac{d^2 Q}{dt^2} + 100\frac{dQ}{dt} + 10^4 Q = 962 \sin 60t. \tag{5}$$

It is evident that equation (5) has a particular solution of the form

$$Q_p(t) = A_1 \sin 60t + A_2 \cos 60t. \tag{6}$$

To determine the values A_1 and A_2, we substitute equation (6) into equation (5), obtaining the simultaneous equations

$$6400A_1 - 6000A_2 = 962,$$

$$6000A_1 + 6400A_2 = 0.$$

Thus $A_1 = \frac{2}{25}$, $A_2 = -\frac{3}{40}$, and since the general solution of the homogeneous equation is the same as that of equation (2), the general solution of equation (5) is

$$Q(t) = e^{-50t}(c_1 \cos 50\sqrt{3}t + c_2 \sin 50\sqrt{3}t) + \tfrac{2}{25} \sin 60t - \tfrac{3}{40} \cos 60t. \tag{7}$$

Differentiating (7), we obtain

$$I(t) = 50e^{-50t}[(\sqrt{3}c_2 - c_1) \cos 50\sqrt{3}t - (c_2 + \sqrt{3}c_1) \sin 50\sqrt{3}t] + \tfrac{24}{5} \cos 60t + \tfrac{9}{2} \sin 60t.$$

Setting $t = 0$ and using the initial conditions, we obtain

$$Q(0) = c_1 - \tfrac{3}{40} = 0, \qquad I(0) = 50(\sqrt{3}c_2 - c_1) + \tfrac{24}{5} = 0,$$

so that $c_1 = \frac{3}{40}$ and $c_2 = -21/1000\sqrt{3}$. Therefore

$$Q(t) = \frac{e^{-50t}}{1000}(75 \cos 50\sqrt{3}t - 7\sqrt{3} \sin 50\sqrt{3}t)$$

$$+ \frac{80 \sin 60t - 75 \cos 60t}{1000},$$

$$I(t) = -\frac{e^{-50t}}{5}(24 \cos 50\sqrt{3}t + 17\sqrt{3} \sin 50\sqrt{3}t)$$

$$+ \frac{48 \cos 60t + 45 \sin 60t}{10}.$$

Problems 2.7

1. In Example 1, let $L = 10$ H, $R = 250 \ \Omega$, $C = 10^{-3}$ f, and $E = 900$ V. With the same assumptions, calculate the current and charge for all values of $t \geq 0$.

2. In Problem 1, suppose instead that $E = 50 \cos 30t$. Find $Q(t)$ for $t \geq 0$.

In Problems 3–6, find the steady-state current in the *RLC* circuit of Fig. 1 where:

3. $L = 5$ henrys, $R = 10$ ohms, $C = 0.1$ farad,
$E = 25 \sin t$ volts.

4. $L = 10$ henrys, $R = 40$ ohms, $C = 0.025$ farad,
$E = 100 \cos 5t$ volts.

5. $L = 1$ henry, $R = 7$ ohms, $C = 0.1$ farad,
$E = 100 \sin 10t$ volts.

6. $L = 2.5$ henrys, $R = 10$ ohms, $C = 0.08$ farad,
$E = 100 \cos 5t$ volts.

Find the transient current in the *RLC* circuit of Fig. 1 for Problems 7–12.

7. Problem 3.   **8.** Problem 4.

9. Problem 5.   **10.** Problem 6.

11. $L = 20$ henrys, $R = 40$ ohms, $C = 10^{-3}$ farad,
$E = 500 \sin t$ volts.

12. $L = 24$ henrys, $R = 48$ ohms, $C = 0.375$ farad,
$E = 900 \cos 2t$ volts.

13. Given that $L = 1$ H, $R = 1200\ \Omega$, $C = 10^{-6}$ f, $I(0) = Q(0) = 0$, and $E = 100 \sin 600t$ volts, determine the transient current and the steady-state current.

14. Find the ratio of the current flowing in the circuit of Problem 13 to that which would be flowing if there were no resistance, at $t = 0.001$ second.

15. Consider the system governed by equation (1) for the case

where the resistance is zero and $E = E_0 \sin \omega t$. Show that the solution consists of two parts: a general solution with frequency $1/\sqrt{LC}$ and a particular solution with frequency ω. The frequency $1/\sqrt{LC}$ is called the **natural frequency** of the circuit. Note that if $\omega = 1/\sqrt{LC}$, then the particular solution disappears.

16. To allow for different variations of the voltage, let us assume in equation (1) that $E = E_0 e^{it}$ ($=E_0 \cos t + iE_0 \sin t$). Assume also, as in Problem 15, that $R = 0$. Finally, for simplicity assume that $E_0 = L = C = 1$. Then $1 = \omega = 1/\sqrt{LC}$.

(a) Show that equation (2) becomes

$$\frac{d^2 I}{dt^2} + I = e^{it}.$$

(b) Determine λ such that $I(t) = \lambda t e^{it}$ is a solution.

(c) Calculate the general solution and show that the magnitude of the current increases without bound as t increases. This phenomenon will produce **resonance.**

17. Let an inductance of L henrys, a resistance of R ohms, and a capacitance of C farads be connected in series with an emf of $E_0 \sin \omega t$ volts. Suppose $Q(0) = I(0) = 0$, and $4L > R^2 C$.

(a) Find the expressions for $Q(t)$ and $I(t)$.

(b) What value of ω will produce resonance?

18. Solve Problem 17 for $4L = R^2 C$.

19. Solve Problem 17 for $4L < R^2 C$.

2.8 Vibrational Motion: Simple Harmonic Motion

Differential equations were first studied in attempts to describe the motion of particles. As a simple example, consider a mass m attached to a coiled spring of stretched length l_0, the upper end of which is securely fastened (see Fig. 1).

We have denoted by the number zero the equilibrium position of the mass on the spring, that is, the point where the mass remains at rest. Suppose that the mass is given an initial displacement x_0 and an initial velocity v_0. Can we describe the future movement of the mass? To do so, we make the following assumptions about the force† exerted by the spring on the mass:

† The most common systems of units are given in the table below.

Systems of units	Force	Length	Mass	Time
International (SI)	newton (N)	meter (m)	kilogram (kg)	second (s)
English	pound (lb)	foot (ft)	slug	second (s) or (sec)

1 N = 1 kg-m/s² = 0.22481 lb; 1 m = 3.28084 ft;
1 kg = 0.06852 slug; 1 lb = 1 slug-ft/s² = 4.4482 N

1. All motion is along a vertical line through the center of gravity of the mass (which is then treated as if it were a point mass), and its direction is always from the mass toward the point of equilibrium.
2. At any time t the magnitude of the force exerted on the mass is proportional to the difference between the length l of the spring and its equilibrium length l_0. The positive constant of proportionality k is called the **spring constant,** and the principle above is known as **Hooke's law.**†

Spring Constant
Hooke's Law

Although it may seem that we are ignoring the effects of the gravitational force, that force was already compensated for by the elastic force in the spring when the mass attained its equilibrium position. Thus, in what follows in this section, it is not necessary to discuss the effects of the gravitational force.

Simple Harmonic Motion

Newton's second law of motion states that the force F acting on a particle moving with varying velocity v is equal to the time rate of change of the momentum mv and, since the mass is constant,

$$F = \frac{d(mv)}{dt} = ma = m\frac{d^2x}{dt^2}. \qquad \left(v = \frac{dx}{dt}\right)$$

Equating the two forces and applying Hooke's law, we have

$$m\frac{d^2x}{dt^2} = -kx, \tag{1}$$

where $x(t)$ denotes the displacement from equilibrium of the spring and is positive when the spring is stretched. The negative sign in (1) is due to the fact that the force always acts toward the equilibrium position and therefore is in the negative direction when x is positive.

Note that we have assumed that all other forces acting on the spring (friction, air resistance, etc.) can be ignored. Equation (1) yields the initial value problem

$$\frac{d^2x}{dt^2} + \frac{k}{m}x = 0, \qquad x(0) = x_0, \qquad x'(0) = v_0. \tag{2}$$

To find the solution of (2), we note that the characteristic equation has the complex root $\pm i\omega_0$, where $\omega_0 = \sqrt{k/m}$, leading to the general solution

$$x(t) = c_1 \cos \omega_0 t + c_2 \sin \omega_0 t.$$

Using the initial conditions, we find that $c_1 = x_0$ and $c_2 = v_0/\omega_0$, so that the solution of (2) is given by

$$x(t) = x_0 \cos \omega_0 t + (v_0/\omega_0) \sin \omega_0 t. \tag{3}$$

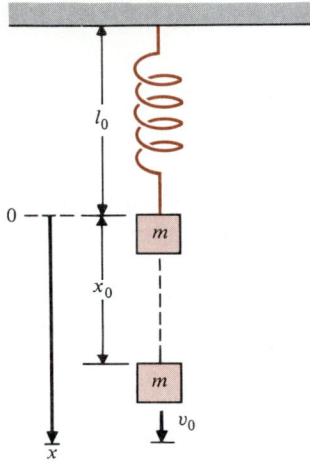

(positive direction is downward)

FIGURE 1

† Robert Hooke (1638–1703) was a British mathematician and physicist. He was one of the first scientists to state the **inverse square law:** the force of gravitational attraction between two bodies is inversely proportional to the square of the distance between them.

We would like to write $x(t)$ in the form

$$x(t) = A \sin(\omega_0 t + \varphi),$$

so that we can graph (and understand) the superposition of sinusoidal functions in (3). To do so we use the trigonometric formula for $\sin(x + y)$:

$$x(t) = A \sin(\omega_0 t + \varphi) = A \sin \omega_0 t \cos \varphi + A \cos \omega_0 t \sin \varphi$$

from (3)
↓

$$= x_0 \cos \omega_0 t + (v_0/\omega_0) \sin \omega_0 t.$$

Thus, equating coefficients of $\sin \omega_0 t$ and $\cos \omega_0 t$, we have

$$A \sin \varphi = x_0 \quad \text{and} \quad A \cos \varphi = v_0/\omega_0$$

and

$$x_0^2 + (v_0/\omega_0)^2 = A^2 \sin^2 \varphi + A^2 \cos^2 \varphi = A^2(\sin^2 \varphi + \cos^2 \varphi) = A^2$$

so that

$$A = \sqrt{x_0^2 + (v_0/\omega_0)^2}.$$

Also,

$$\cos \varphi = \frac{1}{A} \frac{v_0}{\omega_0} \quad \text{and} \quad \sin \varphi = \frac{x_0}{A}$$

so that

$$\tan \varphi = \frac{\sin \varphi}{\cos \varphi} = \frac{x_0/A}{v_0/\omega_0 A} = \frac{x_0 \omega_0}{v_0}.$$

Thus we may write equation (3) as

$$x(t) = A \sin(\omega_0 t + \varphi), \tag{4}$$

with

$$A = \sqrt{x_0^2 + (v_0/\omega_0)^2}$$

and

$$\varphi = \tan^{-1}(x_0 \omega_0 / v_0)$$

or

$$\varphi = \tan^{-1}(x_0 \omega_0 / v_0) + \pi.$$

Simple Harmonic Motion

Because of equation (4), the motion of the mass is called **simple harmonic motion,** since it is sinusoidal.

Amplitude

From this equation it is clear that the mass oscillates between the extreme positions $\pm A$; A is called the **amplitude** of the motion. Since the sine term has period $2\pi/\omega_0$, this is the time required for each complete oscillation. The **natural**

frequency f of the motion is the number of complete oscillations per unit time:†

$$f = \omega_0/2\pi. \tag{5}$$

Note that although the amplitude depends on the initial conditions, the frequency does not.

EXAMPLE 1

Suppose that $x_0 = 0.5$ m, $k = 0.4$ N/m, $m = 10$ kg, and $v_0 = 0.25$ m/s. Then $\omega_0 = \sqrt{k/m} = \sqrt{\frac{2}{50}} = \sqrt{0.04} = 0.2$ and equation (3) becomes

$$x(t) = 0.5 \cos 0.2t + \frac{0.25}{0.2} \sin 0.2t = 0.5 \cos 0.2t + 1.25 \sin 0.2t.$$

Now

$$A = \sqrt{0.5^2 + 1.25^2} = \sqrt{1.8125} \approx 1.3463$$

and

$$\varphi = \tan^{-1} \frac{(0.5)(0.2)}{0.25} = \tan^{-1} 0.4 \approx 0.3805 \text{ radians } (\approx 21.8°)$$

so that we may write

$$x(t) \approx 1.3463 \sin(0.2t + 0.3805) \text{ m}.$$

EXAMPLE 2

Consider a spring fixed at its upper end and supporting a weight of 10 pounds at its lower end. Suppose the 10-pound weight stretches the spring by 6 inches. Find the equation of motion of the weight if it is drawn to a position 4 inches below its equilibrium position and released.

SOLUTION

By Hooke's law, since a force of 10 lb stretches the spring by $\frac{1}{2}$ ft, $10 = k(\frac{1}{2})$ or $k = 20$ (lb/ft). We are given the initial values $x_0 = \frac{1}{3}$(ft) and $v_0 = 0$, so by equation (3) and the identity‡ $k/m = gk/w \approx 64 \text{ sec}^{-2}$, we obtain from (4)

$$x(t) = \tfrac{1}{3} \cos 8t \text{ ft}.$$

Thus the amplitude is $\frac{1}{3}$ ft (= 4 in.), and the frequency is $f = 4/\pi$ hertz.

Problems 2.8

In Problems 1–6, determine the equation of motion of a mass m attached to a coiled spring with spring constant k initially displaced a distance x_0 from equilibrium and released with velocity v_0 subject to no damping or external forces.

† Cycles/sec = hertz (Hz).

‡ The identity $w = mg$ may be used to convert weight to mass. Keep in mind that pounds and newtons are units of weight (force) whereas slugs and kilograms are units of mass. The gravitational constant $g = 9.81$ m/s² $= 32.2$ ft/sec² (approximately).

1. $m = 10$ kg, $k = 1000$ N/m, $x_0 = 1$ m, $v_0 = 0$,
2. $m = 10$ kg, $k = 10$ N/m, $x_0 = 0$, $v_0 = 1$ m/s,
3. $m = 10$ kg, $k = 10$ N/m, $x_0 = 3$ m, $v_0 = 4$ m/s,
4. $m = 1$ kg, $k = 16$ N/m, $x_0 = 4$ m, $v_0 = 0$,
5. $m = 1$ kg, $k = 25$ N/m, $x_0 = 0$ m, $v_0 = 3$ m/s,
6. $m = 9$ kg, $k = 1$ N/m, $x_0 = 4$ m, $v_0 = 1$ m/s,

7. One end of a rubber band is fixed at a point A. A 1-kg mass, attached to the other end, stretches the rubber band vertically to the point B in such a way that the length AB is 16 cm greater than the natural length of the band. If the mass is further drawn to a position 8 cm below B and released, what will be its velocity (if we neglect resistance) as it passes the position B?

8. If in Problem 7 the mass is released at a position 8 cm above B, what will be its velocity as it passes 1 cm above B?

9. A cylindrical block of wood of radius and height 1 ft and weighing 124.8 lb floats with its axis vertical in water (62.4 lb per ft³). If it is depressed so that the surface of the water is tangent to the block, and is then released, what will be its period of vibration and equation of motion? Neglect resistance. [*Hint:* The upward force on the block is equal to the weight of the water displaced by the block.]

* 10. A cubical block of wood, 1 ft on a side, is depressed so that its upper face lies along the surface of the water, and is then released. The period of vibration is found to be 1 sec. Neglecting resistance, what is the weight of the block of wood?

11. An ideal pendulum consists of a weightless rod of length l attached at one end to a frictionless hinge and supporting a body of mass m at the other end. Suppose the pendulum is displaced an angle θ_0 and released (see Fig. 2). The tangential acceleration of the ideal pendulum is $l\theta''$, and must be proportional, by Newton's second law of motion, to the tangential component of gravitational force.

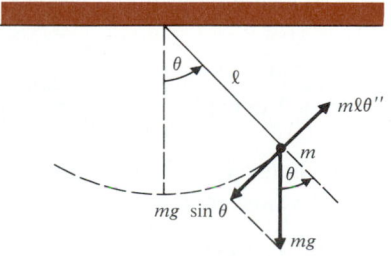

FIGURE 2

(a) Neglecting air resistance, show that the ideal pendulum satisfies the nonlinear initial value problem

$$l\frac{d^2\theta}{dt^2} = -g \sin\theta, \quad \theta(0) = \theta_0, \quad \theta'(0) = 0. \qquad (6)$$

(b) Assuming θ_0 is small, explain why (6) may be approximated by the linear initial value problem

$$\frac{d^2\theta}{dt^2} + \frac{g}{l}\theta = 0, \quad \theta(0) = \theta_0, \quad \theta'(0) = 0. \qquad (7)$$

(c) Solve (7) assuming that the rod is 6 inches long and that the initial displacement $\theta_0 = 0.5$ radian. What is the frequency of the pendulum?

12. A grandfather clock has a pendulum that is one meter long. The clock ticks each time the pendulum reaches the rightmost extent of its swing. Neglecting friction and air resistance, and assuming that the motion is small, determine how many times the clock ticks in one minute.

13. A turbine of unknown weight is placed on a spring-supported mounting platform of unknown spring constant. What is the natural frequency of the system if the turbine lowers the platform by $\frac{1}{8}$ in.?

* 14. A cubical block of wood 1 ft on a side and weighing 41.6 lb floats in water (62.4 lb/ft³). If it is depressed, so that its top just touches the surface of the water, and is then released, find its equation of motion, neglecting resistance.

15. Assume a straight tube has been bored through the center of the earth and a particle of weight w lb is dropped into the tube. If the radius of the earth is 3960 miles and the gravitational attraction is proportional to the distance from the center, find how long it will take (neglecting resistance) to drop halfway to the center. Will the particle pass through the tube to the other side?

16. A weight w is suspended by two springs, having spring constants k_1 and k_2, connected in series (see Fig. 3). What is the **effective** spring constant k for such a system? [*Hint:* No derivatives are required.]

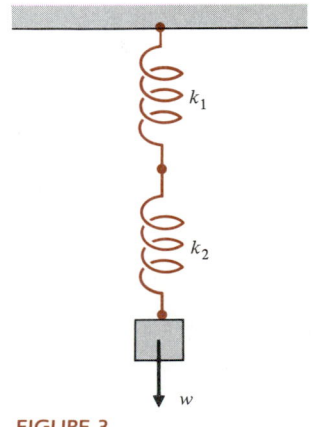

FIGURE 3

17. A small ball of weight w is placed in the middle of a tightly stretched perfectly flexible vertical string of length $2L$ and tension T_0. Show that for small lateral displacements the ball will undergo simple harmonic motion. What is its period? [*Note:* Neglect gravity.]

2.9 Vibrational Motion: Damped Vibrations

Throughout the discussion in Section 2.8 we made the assumption that there were no external forces acting on the spring. This assumption, however, is not very realistic. To take care of such things as friction in the spring and air resistance, we now assume that there is a damping force (that tends to slow things down), which can be thought of as the resultant of all external forces (except gravity) acting on the spring. It is reasonable to assume that the magnitude of the damping force is proportional to the velocity of the particle (for example, the slower the movement, the smaller the air resistance).† Therefore, we add the term $c(dx/dt)$, where c is the damping constant that depends on all external factors, to equation (2.8.1). The equation of motion then becomes

$$\frac{d^2x}{dt^2} = -\frac{k}{m}x - \frac{c}{m}\frac{dx}{dt}, \qquad x(0) = x_0, \qquad x'(0) = v_0 \tag{1}$$

or

$$\frac{d^2x}{dt^2} + \frac{c}{m}\frac{dx}{dt} + \frac{k}{m}x = 0, \qquad x(0) = x_0, \qquad x'(0) = v_0. \tag{2}$$

[Of course, since c depends on external factors, it may very well not be a constant at all but may vary with time and position. In that case c is really $c(t, x)$, and the equation becomes much harder to analyze than the constant-coefficient case.]

To study equation (2), we first find the roots of the characteristic equation:

$$\frac{-c \pm \sqrt{c^2 - 4mk}}{2m}. \tag{3}$$

The nature of the general solution will depend on the discriminant $\sqrt{c^2 - 4mk}$. If $c^2 > 4mk$, then both roots will be negative since $\sqrt{c^2 - 4mk} < c$. So in this case

$$x(t) = c_1 \exp\left(\frac{-c + \sqrt{c^2 - 4mk}}{2m}\right)t + c_2 \exp\left(\frac{-c - \sqrt{c^2 - 4mk}}{2m}\right)t \tag{4}$$

will become small as t becomes large whatever the initial conditions may be. Similarly, in the event the discriminant vanishes, then

$$x(t) = e^{(-c/2m)t}(c_1 + c_2 t), \tag{5}$$

and the solution has a similar behavior.

† In hydrodynamics the damping force is proportional to velocity squared.

EXAMPLE 1

A spring fixed at its upper end is stretched 6 inches by a 10-pound weight attached at its lower end. The spring-mass system is suspended in a viscous medium (such as oil or water) so that the system is subjected to a damping force of

$$5\frac{dx}{dt}\ (\text{lb}).$$

Describe the motion of the system if the weight is drawn down an additional 4 inches $(=\frac{1}{3}$ ft) and released.

SOLUTION

The differential equation is

$$\frac{d^2x}{dt^2} + 16\frac{dx}{dt} + 64x = 0 \qquad (6)$$

since $k/m = gk/w \approx 32(10/\frac{1}{2})/10 = 64\ \text{sec}^{-2}$ and $c/m = gc/w \approx 32(5)/10 = 16\ \text{sec}^{-1}$. The initial conditions are $x(0) = \frac{1}{3}$ and $x'(0) = 0$. Hence, the roots of the characteristic equation are

$$\frac{-16 \pm \sqrt{(16)^2 - 4(64)}}{2} = -8,$$

so that (6) has the general solution

$$x(t) = e^{-8t}(c_1 + c_2 t).$$

Applying the initial conditions yields

$$x(t) = \tfrac{1}{3}e^{-8t}(1 + 8t)\ \text{ft},$$

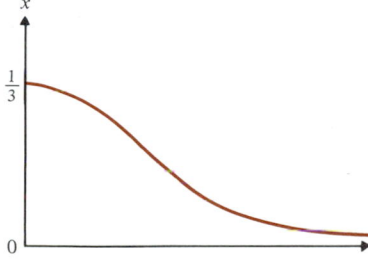

FIGURE 1

Critical Damping
Overdamped/Underdamped

which has the graph shown in Fig. 1. We observe that the solution does not oscillate. When this occurs we say the motion is **critically damped.** The exact value of c producing such damping, $c = 2\sqrt{mk}$, is called the **critical damping number.** A spring-weight system is called **overdamped** (or **underdamped**) if $c^2 > 4mk$ $(c^2 < 4mk)$.

If $c^2 < 4mk$, then the general solution is

$$x(t) = e^{(-c/2m)t}\left(c_1 \cos \frac{\sqrt{4mk - c^2}}{2m}t + c_2 \sin \frac{\sqrt{4mk - c^2}}{2m}t\right), \qquad (7)$$

which shows an oscillation with frequency

$$f = \frac{\sqrt{4mk - c^2}}{4\pi m}.$$

Damping Factor

The factor $e^{(-c/2m)t}$ is called the **damping factor.** Letting $c = 4$ (lb/(ft/sec)) in Example 1 leads to the general solution

$$x(t) = e^{-32t/5}(c_1 \cos \tfrac{24}{5}t + c_2 \sin \tfrac{24}{5}t)\ \text{ft}.$$

Note that $e^{-32t/5} \rightarrow 0$ as t tends to ∞, so that the damped motion decays to zero as time increases.

Using the initial values, we find that $c_1 = \frac{1}{3}$, $c_2 = \frac{4}{9}$, and the motion is illustrated in Fig. 2.

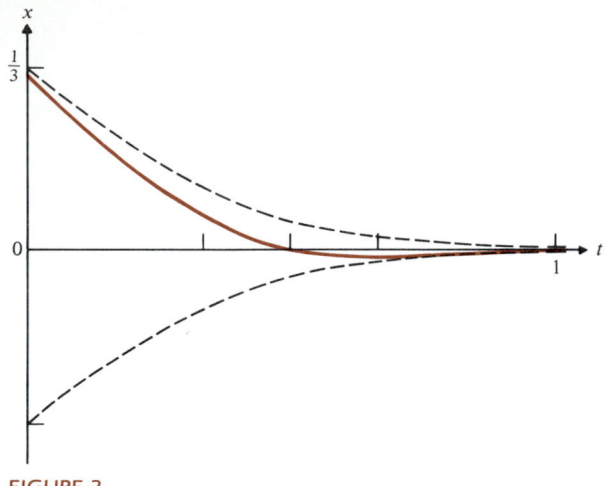

FIGURE 2

Damping forces also occur as the result of mechanical devices. A typical example is provided by shock absorbers in automobiles, where a piston is driven into a dashpot or damper (see Fig. 3). Mechanical systems involving torsional motion, such as brakes on automobiles, also experience damped vibrations (see Fig. 4). In this case the equation of motion is given by

$$I\frac{d^2\theta}{dt^2} + c\frac{d\theta}{dt} + k\theta = 0, \qquad \theta(0) = \theta_0, \qquad \theta'(0) = \theta'_0, \qquad \textbf{(8)}$$

where I is the moment of inertia, θ the angular displacement, $c\,d\theta/dt$ the damping torque, and $k\theta$ the elastic torque due to twisting of the shaft.

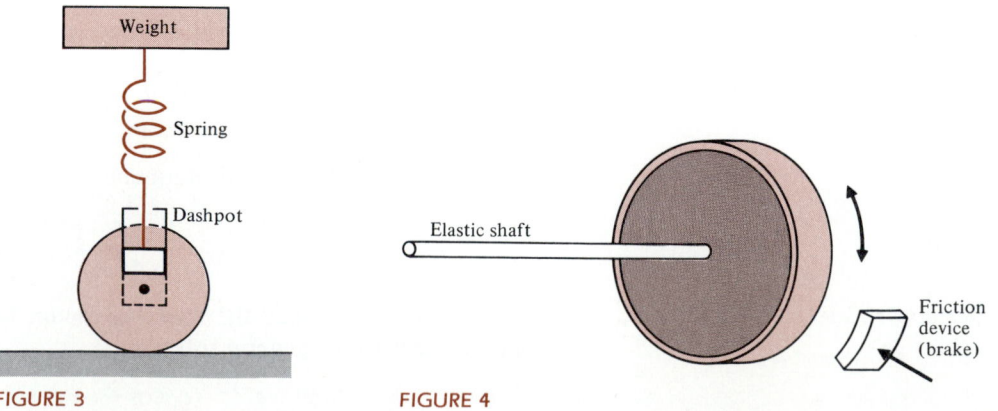

FIGURE 3   **FIGURE 4**

Problems 2.9

In Problems 1–6, determine the equation of motion of a mass m attached to a coiled spring with spring constant k initially displaced a distance x_0 from equilibrium and released with initial velocity v_0, subject to a damping constant c.

1. $m = 10$ kg, $k = 1000$ N/m, $x_0 = 1$ m, $v_0 = 0$, $c = 200$ kg/s $= 200$ N/(m/s).

2. $m = 10$ kg, $k = 10$ N/m, $x_0 = 0$, $v_0 = 1$ m/s, $c = 20$ kg/s.

3. $m = 10$ kg, $k = 10$ N/m, $x_0 = 3$ m, $v_0 = 4$ m/s, $c = 10\sqrt{5}$ kg/s.

4. $m = 1$ kg, $k = 16$ N/m, $x_0 = 4$ m, $v_0 = 0$, $c = 10$ kg/s.

5. $m = 1$ kg, $k = 25$ N/m, $x_0 = 0$ m, $v_0 = 3$ m/s, $c = 8$ kg/s.

6. $m = 9$ kg, $k = 1$ N/m, $x_0 = 4$ m, $v_0 = 1$ m/s, $c = 10$ kg/s.

7. A 10-g mass suspended from a spring vibrates freely, the resistance (in kg) being numerically equal to half the velocity (in m/s) at any instant. If the period of the motion is 8 s, what is the spring constant (in kg/s²)?

8. A weight w (lb) is suspended from a spring whose constant is 10 lb/ft. The motion of the weight is subject to a resistance (in lb) numerically equal to half the velocity (in ft/sec). If the motion is to have a 1-sec period, what are the possible values of w?

9. A weight of 48 lb hangs from a spring with spring constant

50 lb/ft. The damping in the system is 30% of critical. Determine the motion of the weight if it is pulled down 2 in. from its equilibrium position and released with an upward velocity of 1 ft/sec.

10. Answer Problem 9 if the damping is 90% of critical.

11. Answer Problem 9 if the damping is critical.

12. Answer Problem 9 if the damping is 150% of critical.

* 13. Answer Problem 9 if 48 lb of additional weight is suddenly applied to the system when it is at rest at equilibrium.

* 14. A weight, hung on an ideal spring in a vacuum (assume no air resistance) and set vibrating, has a period of 1 s. When the spring-mass system is placed in a viscous medium, where the resistance is proportional to the velocity, its (damped) period is found to be 2 s. Determine the differential equation corresponding to the damped vibrations.

* 15. A container weighing 2 lb is half filled with 4 lb of mercury. When hung on the end of a spring, it stretches it by 2 in. The period of oscillation is found to be 0.3 sec. If 4 lb more of mercury is added, the period becomes 0.4 sec. Can the resistance be proportional to the velocity in this situation?

** 16. The circular disk in Fig. 4 has a radius of 12 inches and weighs 100 lb. The observed frequency of torsional vibration is 2π rad/sec. When another body is attached to the same shaft, the observed frequency of torsional vibration is 2.4π rad/sec. Find the moment of inertia of the second body with respect to the axis of the shaft. [See (8).]

2.10 Vibrational Motion: Forced Vibrations

Free or Natural Vibrations

Forced Vibrations

The motion of the mass considered in Sections 2.8 and 2.9 is determined by the inherent forces of the spring-weight system and the natural forces acting on the system. Accordingly, the vibrations are called **free** or **natural vibrations.** We shall now assume that the mass is also subject to an external periodic force $F_0 \sin \omega t$, due to the motion of the object to which the upper end of the spring is attached (see Fig. 1). In this case the mass will undergo **forced vibrations.**

We saw in the previous sections that a spring-weight system, having a damping force proportional to the velocity, satisfies the initial value problem

$$m\frac{d^2x}{dt^2} = -kx - c\frac{dx}{dt}, \qquad x(0) = x_0, \qquad x'(0) = v_0.$$

If we subject such a system to an additional periodic external force

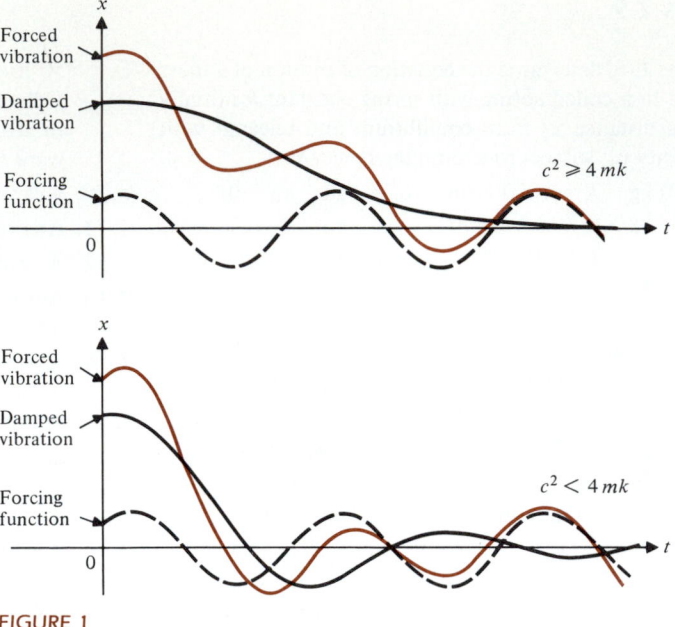

FIGURE 1

$F_0 \sin \omega t$, we obtain the nonhomogeneous second-order differential equation

$$m\frac{d^2x}{dt^2} = -kx - c\frac{dx}{dt} + F_0 \sin \omega t,$$

which we write in the form

$$\frac{d^2x}{dt^2} + \frac{c}{m}\frac{dx}{dt} + \frac{k}{m}x = \frac{F_0}{m}\sin \omega t. \tag{1}$$

By the method of undetermined coefficients, we know that $x(t)$ has a particular solution of the form

$$x_p(t) = b_1 \cos \omega t + b_2 \sin \omega t. \tag{2}$$

Substituting this function into equation (1) yields the simultaneous equations

$$(\omega_0^2 - \omega^2)b_1 + \frac{c\omega}{m}b_2 = 0,$$

$$-\frac{c\omega}{m}b_1 + (\omega_0^2 - \omega^2)b_2 = \frac{F_0}{m}, \tag{3}$$

where $\omega_0 = \sqrt{k/m}$, from which we obtain

$$b_1 = \frac{-F_0 c\omega}{m^2(\omega_0^2 - \omega^2)^2 + (c\omega)^2},$$

$$b_2 = \frac{F_0 m(\omega_0^2 - \omega^2)}{m^2(\omega_0^2 - \omega^2)^2 + (c\omega)^2}.$$

Using the same method we used to obtain equation (2.8.4), we have

$$x_p = A \sin(\omega t + \varphi),$$ **(4)**

where

$$A = \frac{F_0/k}{\sqrt{\left[1 - \left(\dfrac{\omega}{\omega_0}\right)^2\right]^2 + \left(2\dfrac{c}{c_0}\dfrac{\omega}{\omega_0}\right)^2}},$$

and

$$\tan \varphi = \frac{2\dfrac{c}{c_0}\dfrac{\omega}{\omega_0}}{\left(\dfrac{\omega}{\omega_0}\right)^2 - 1},$$

with $c_0 = 2m\omega_0$. Here A is the amplitude of the motion, φ is the **phase angle,** c/c_0 is the **damping ratio,** and ω/ω_0 is the **frequency ratio** of the motion.

The general solution is found by superimposing the periodic function (equation (4)) on the general solution (equations (2.9.4), (2.9.5), or (2.9.7)) of the homogeneous equation. Since the solution of the homogeneous equation damps out as t increases, the general solution will be very close to (4) for large values of t. Figure 1 illustrates two typical situations.

It is interesting to see what occurs if the damping constant c vanishes. There are two cases.

CASE 1
If $\omega^2 \neq \omega_0^2$, we superimpose the periodic function (4) on the general solution of the homogeneous equation $x'' + \omega_0^2 x = 0$, obtaining

$$x(t) = c_1 \cos \omega_0 t + c_2 \sin \omega_0 t + \frac{F_0/k}{1 - (\omega/\omega_0)^2} \sin \omega t.$$ **(5)**

Using the initial conditions, we find that

$$c_1 = x_0 \quad \text{and} \quad c_2 = \frac{v_0}{\omega_0} - \frac{(F_0/k)(\omega/\omega_0)}{1 - (\omega/\omega_0)^2}$$

so that

$$x(t) = A \sin(\omega_0 t + \varphi) + \frac{F_0/k}{1 - (\omega/\omega_0)^2} \sin \omega t,$$

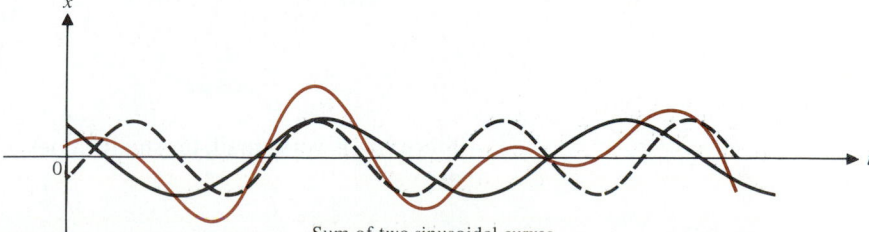

Sum of two sinusoidal curves

FIGURE 2

where

$$A = \sqrt{c_1^2 + c_2^2} \quad \text{and} \quad \tan \varphi = c_1/c_2.$$

Hence the motion in this case is simply the sum of two sinusoidal curves as illustrated in Fig. 2.

CASE 2
If $\omega^2 = \omega_0^2$, we must seek a particular solution of the form

$$x_p(t) = b_1 t \cos \omega t + b_2 t \sin \omega t, \tag{6}$$

since equation (2) is a solution of the homogeneous equation. (See form iii in our discussion of undetermined coefficients on p. 87.) Substituting equation (6) into

$$\frac{d^2x}{dt^2} + \frac{k}{m}x = \frac{F_0}{m} \sin \omega t,$$

we get

$$b_1 = -F_0/2m\omega \quad \text{and} \quad b_2 = 0,$$

so the general solution has the form

$$x(t) = c_1 \cos \omega t + c_2 \sin \omega t - \frac{F_0}{2m\omega} t \cos \omega t. \tag{7}$$

Resonance

Note that as t increases, the vibrations caused by the last term in equation (7) will increase without bound. The external force is said to be in **resonance** with the vibrating mass. It is evident that the displacement will become so large that the elastic limit of the spring will be exceeded, leading to fracture or to a permanent distortion in the spring.

Suppose that c is positive but very close to zero while $\omega^2 = \omega_0^2$. Note that equations (3) will yield $b_1 = -F_0/c\omega$ and $b_2 = 0$ when equation (2) is substituted in (1). Superimposing

$$x_p(t) = \frac{-F_0}{c\omega} \cos \omega t$$

on the solution of the homogeneous equation [see equation (2.9.7) on p. 111 since c^2 is very small] and letting $c_0 = 2m\omega_0$, we obtain

$$x(t) = e^{(-c/c_0)\omega_0 t}\left(c_1 \cos \omega_0 \sqrt{1 - \left(\frac{c}{c_0}\right)^2} t + c_2 \sin \omega_0 \sqrt{1 - \left(\frac{c}{c_0}\right)^2} t\right)$$

$$- \frac{F_0}{c\omega} \cos \omega t. \tag{8}$$

Since c/c_0 is very small, for small values of t we see that (8) can be approximated as

$$x(t) \approx c_1 \cos \omega t + c_2 \sin \omega t - \frac{F_0}{2m\omega}\left(\frac{2m}{c}\right) \cos \omega t,$$

which bears a marked resemblance to equation (7) *when (7) is evaluated at large values of t* (since $2m/c$ is large). Thus, the *damped* spring problem approaches resonance. This phenomenon is extremely important in engineering since resonance may produce undesirable effects such as metal fatigue and structural fracture, as well as desirable objectives such as sound and light amplification. An example of resonance occurred when a column of soldiers marched across Broughton bridge, near Manchester, England, in 1831. Their marching produced a periodic force closely approximating the natural frequency of the bridge, causing it to collapse.

Problems 2.10

In Problems 1–6, determine the equation of motion of a mass m attached to a coiled spring with spring constant k initially displaced a distance x_0 from equilibrium and released with velocity v_0 subject to a damping constant c and an external force $F_0 \sin \omega t$ N.

1. $m = 10$ kg, $k = 1000$ N/m, $x_0 = 1$ m, $v_0 = 0$,
   $c = 200$ kg/s, $F_0 = 1$ N, $\omega = 10$ rad/s.

2. $m = 10$ kg, $k = 10$ N/m, $x_0 = 0$, $v_0 = 1$ m/s,
   $c = 20$ kg/s, $F_0 = 1$ N, $\omega = 1$ rad/s.

3. $m = 10$ kg, $k = 10$ N/m, $x_0 = 3$ m, $v_0 = 4$ m/s,
   $c = 10\sqrt{5}$ kg/s, $F_0 = 1$ N, $\omega = 1$ rad/s.

4. $m = 1$ kg, $k = 16$ N/m, $x_0 = 4$ m, $v_0 = 0$,
   $c = 10$ kg/s, $F_0 = 4$ N, $\omega = 4$ rad/s.

5. $m = 1$ kg, $k = 25$ N/m, $x_0 = 0$ m, $v_0 = 3$ m/s,
   $c = 8$ kg/s, $F_0 = 1$ N, $\omega = 3$ rad/s.

6. $m = 9$ kg, $k = 1$ N/m, $x_0 = 4$ m, $v_0 = 1$ m/s,
   $c = 10$ kg/s, $F_0 = 2$ N, $\omega = \frac{1}{3}$ rad/s.

7. A 1-g mass is hanging at rest on a spring that is stretched 25 cm by the weight. The upper end of the spring is given the periodic force $0.01 \sin 2t$ N and air resistance has a magnitude (kg/s) 0.02162 times the velocity in meters per second. Find the equation of motion of the mass.

8. A 100-lb weight is suspended from a spring with spring constant 20 lb/in. When the system is vibrating freely, we observe that in consecutive cycles the amplitude decreases by 40%. If a force of $20 \cos \omega t$ N acts on the system, find the amplitude and phase shift of the resulting steady-state motion if $\omega = 9$ rad/sec.

9. Repeat Problem 8 for $\omega = 12$ rad/sec.

10. A particle of weight w moves along the x-axis under the influence of a force $F = -kx$. Friction on the particle is proportional to the force between the particle and the sur-

face on which it moves. Find the differential equation governing the motion of this particle.

* 11. Consider the forced vibrations of an undamped mechanical spring-weight system, where the external force is $F_0 \sin \omega t$ N.

   (a) Show that if $\omega \neq \omega_0 \ (= \sqrt{k/m})$, then the solution is given by

   $$x(t) = c_1 \cos \omega_0 t + c_2 \sin \omega_0 t + \frac{F_0}{k - m\omega^2} \sin \omega t.$$

   (b) Discuss what happens if ω is close, but not equal, to ω_0. [*Hint:* Use the procedure for finding equation (4) of Section 2.8. The phenomenon that occurs is called **beats** (see Fig. 3), and it occurs whenever an impressed frequency is close to a natural frequency of a mechanical system.]

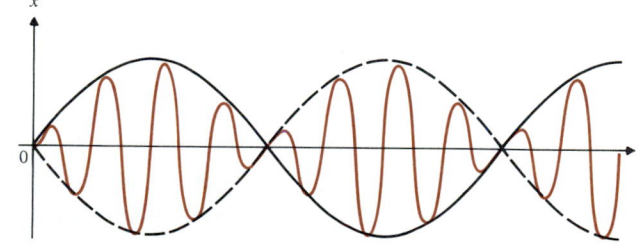

FIGURE 3

* 12. A 48-lb weight is suspended from a spring with spring constant 50 lb/in. In 10 cycles we see that the maximum displacement decreases by 50%. Suppose an external force of $F_0[\sin 15t + \sin 16t]$ N is applied to the spring-weight system. Will beats occur? Why?

2.11 Higher-Order Linear Differential Equations

In this section we extend the results of the chapter to linear differential equations of order higher than two. There is little theoretical difference between second-order and higher-order systems so we will be relatively brief. We shall state all results without proof.

The **general nonhomogeneous linear nth-order equation** is

$$y^{(n)}(x) + a_{n-1}(x)y^{(n-1)}(x) + \cdots + a_1(x)y'(x) + a_0(x)y(x) = f(x). \tag{1}$$

Associated Homogeneous Equation

The **associated homogeneous equation** is

$$y^{(n)}(x) + a_{n-1}(x)y^{(n-1)}(x) + \cdots + a_1(x)y'(x) + a_0(x)y(x) = 0. \tag{2}$$

In Theorem 2.1.1 on p. 68 we stated that equation (1) has a unique solution provided that all the functions in the equation are continuous and n initial conditions are specified. Now we shall concern ourselves with finding the general solutions to equations (1) and (2). To do so we shall follow the procedures we developed for solving second-order equations.

Linear Independence

We say that the functions $y_1, y_2, \ldots, y_n$ are **linearly independent** in $[x_0, x_1]$ if the following condition holds:

$$c_1 y_1(x) + c_2 y_2(x) + \cdots + c_n y_n(x) = 0 \text{ for all } x \text{ in } [x_0, x_1]$$

$$\text{implies that } c_1 = c_2 = \cdots = c_n = 0.$$

Otherwise the functions are **linearly dependent**. The expression $c_1 y_1 + c_2 y_2 + \cdots + c_n y_n$ is called a **linear combination** of the functions $y_1, y_2, \ldots, y_n$.

Linear Combination
Wronskian

The **Wronskian** of $y_1, y_2, \ldots, y_n$ is defined by

$$W(y_1, y_2, \ldots, y_n)(x) = \begin{vmatrix} y_1 & y_2 & \cdots & y_n \\ y_1' & y_2' & \cdots & y_n' \\ y_1'' & y_2'' & \cdots & y_n'' \\ \vdots & \vdots & & \vdots \\ y_1^{(n-1)} & y_2^{(n-1)} & \cdots & y_n^{(n-1)} \end{vmatrix} \tag{3}$$

■ **THEOREM 1**

Let $a_0, a_1, \ldots, a_{n-1}$ be continuous in $[x_0, x_1]$ and let $y_1, y_2, \ldots, y_n$ be n solutions of equation (2). Then

(i) $W(y_1, y_2, \ldots, y_n)(x)$ is either zero for all $x \in [x_0, x_1]$ or for no value of $x \in [x_0, x_1]$.

(ii) $y_1, y_2, \ldots, y_n$ are linearly independent if and only if

$$W(y_1, y_2, \ldots, y_n)(x) \neq 0. \qquad ■$$

EXAMPLE 1

The functions 1, x, and x^2 are solutions to the equation $y'''(x) = 0$. Determine whether they are linearly independent or dependent for all x.

SOLUTION

$$W(y_1, y_2, y_3) = \begin{vmatrix} 1 & x & x^2 \\ 0 & 1 & 2x \\ 0 & 0 & 2 \end{vmatrix} = 2 \neq 0$$

so the functions are linearly independent.

General Solution

Our procedure for solving equation (1) is as follows:

STEP 1
Find n linearly independent solutions, y_1, y_2, $\ldots$, y_n to the homogeneous equation (2).

STEP 2
Find one solution, y_p, to the nonhomogeneous equation (1).

General Solution

Then the **general solution** to (1) is given by

$$y(x) = c_1 y_1(x) + c_2 y_2(x) + \cdots + c_n y_n(x) + y_p \tag{4}$$

and

$$y(x) = c_1 y_1(x) + c_2 y_2(x) + \cdots + c_n y_n(x) \tag{5}$$

is the general solution to (2).

As in the case of second-order equations, we can generally find solutions to (2) only when the a_i's are constants. In this case, equations (1) and (2) are *Constant Coefficients* said to have **constant coefficients.**

The general nth-order linear, homogeneous constant-coefficient equation is

$$y^{(n)}(x) + a_{n-1} y^{(n-1)}(x) + \cdots + a_1 y'(x) + a_0 y(x) = 0. \tag{6}$$

We note that $(d^n/dx^n)e^{\lambda x} = \lambda^n e^{\lambda x}$. Thus, if we substitute $y = e^{\lambda x}$ into (6) and *Characteristic Equation* then divide by $e^{\lambda x}$, we obtain the **characteristic equation**

$$\lambda^n + a_{n-1}\lambda^{n-1} + \cdots + a_1\lambda + a_0 = 0 \tag{7}$$

Equation (7) has n roots λ_1, λ_2, $\ldots$, λ_n. Some of these roots may be real and distinct, real and equal, distinct complex conjugate pairs, or equal complex conjugate pairs. If a root λ_k (real or complex) occurs m times, we say that it has **multiplicity** m. The following rules tell us how to find the general solution to equation (6).

Procedure for Solving Linear Homogeneous Equations with Constant Coefficients

1. Obtain the characteristic equation (7).
2. Find the roots λ_1, λ_2, ..., λ_n of (7). (This is usually the most difficult step.)

Simple Root

3. For each real root λ_k of multiplicity 1 (**simple root**), one solution to (6) is $y_k = e^{\lambda_k x}$.
4. For each real root λ_k of multiplicity $m > 1$, m solutions to (6) are

$$y_1 = e^{\lambda_k x}, y_2 = xe^{\lambda_k x}, \ldots, y_m = x^{m-1}e^{\lambda_k x}.$$

5. If $\alpha + i\beta$ and $\alpha - i\beta$ are simple roots, then two solutions to (6) are

$$y_1 = e^{\alpha x} \cos \beta x \quad \text{and} \quad y_2 = e^{\alpha x} \sin \beta x.$$

6. If $\alpha + i\beta$ and $\alpha - i\beta$ are roots of multiplicity $m > 1$, then $2m$ solutions to (6) are

$$y_1 = e^{\alpha x} \cos \beta x, y_2 = xe^{\alpha x} \cos \beta x, \ldots, y_m = x^{m-1}e^{\alpha x} \cos \beta x,$$

$$y_{m+1} = e^{\alpha x} \sin \beta x, y_{m+2} = xe^{\alpha x} \sin \beta x, \ldots, y_{2m} = x^{m-1}e^{\alpha x} \sin \beta x.$$

7. If y_1, y_2, ..., y_n are the n solutions obtained in Steps 3, 4, 5, and 6, then y_1, y_2, ..., y_n are linearly independent and the general solution to (6) is given by

$$y(x) = c_1 y_1(x) + c_2 y_2(x) + \cdots + c_n y_n(x).$$

EXAMPLE 2

Find the general solution of

$$y''' - 3y'' - 10y' + 24y = 0.$$

SOLUTION

The characteristic equation is

$$\lambda^3 - 3\lambda^2 - 10\lambda + 24 = (\lambda - 2)(\lambda + 3)(\lambda - 4) = 0$$

with roots $\lambda_1 = 2$, $\lambda_2 = -3$, $\lambda_3 = 4$. Since these roots are real and distinct, three linearly independent solutions are

$$y_1 = e^{2x}, \qquad y_2 = e^{-3x}, \qquad y_3 = e^{4x}$$

and the general solution is

$$y(x) = c_1 e^{2x} + c_2 e^{-3x} + c_3 e^{4x}.$$

EXAMPLE 3

Find the general solution of

$$y^{(4)} - 4y''' + 6y'' - 4y' + y = 0.$$

SOLUTION

The characteristic equation is

$$\lambda^4 - 4\lambda^3 + 6\lambda^2 - 4\lambda + 1 = (\lambda - 1)^4 = 0$$

with the single root $\lambda = 1$ of multiplicity 4. Thus four linearly independent solutions are

$$y_1 = e^x, \qquad y_2 = xe^x, \qquad y_3 = x^2 e^x, \qquad y_4 = x^3 e^x$$

and the general solution is

$$y(x) = e^x (c_1 + c_2 x + c_3 x^2 + c_4 x^3).$$

EXAMPLE 4

Find the general solution of

$$y^{(5)} - 2y^{(4)} + 8y'' - 12y' + 8y = 0.$$

SOLUTION

The characteristic equation is

$$\lambda^5 - 2\lambda^4 + 8\lambda^2 - 12\lambda + 8 = 0.$$

This can be factored as

$$(\lambda + 2)(\lambda^2 - 2\lambda + 2)^2 = 0.$$

The solutions to $\lambda^2 - 2\lambda + 2 = 0$ are $\lambda = 1 \pm i$. Thus the roots are

$$\lambda_1 = -2 \text{ (simple)}, \qquad \lambda_2 = 1 + i, \quad \text{and} \quad \lambda_3 = 1 - i$$

with the complex roots λ_2 and λ_3 having multiplicity 2. Thus, five linearly independent solutions are

$$y_1 = e^{-2x}, \qquad y_2 = e^x \cos x, \qquad y_3 = xe^x \cos x, \qquad y_4 = e^x \sin x, \qquad y_5 = xe^x \sin x$$

and the general solution is

$$y(x) = c_1 e^{-2x} + (c_2 + c_3 x)e^x \cos x + (c_4 + c_5 x)e^x \sin x.$$

Remark. In solving the last three characteristic equations we made the factoring look easy. Finding roots of a polynomial of degree greater than two is, in general, very difficult.

How do we find a particular solution to the nonhomogeneous equation (1)? As with second-order equations, there are two methods: undetermined coefficients and variation of parameters. The method of undetermined coefficients is identical to the technique we used for second-order equations. The method of variation of parameters is discussed in Problems 29 and 30.

Finally, certain equations with variable coefficients can be solved. The higher-order Euler equation is discussed in Problems 31–34.

Problems 2.11

In Problems 1–16 find the general solution to the given equation. If initial conditions are given, find the particular solution that satisfies them.

1. $y^{(4)} + 2y'' + y = 0$

2. $y''' - y'' - y' + y = 0$

3. $y''' - 3y'' + 3y' - y = 0$, $y(0) = 1$, $y'(0) = 2$, $y''(0) = 3$

4. $x''' + 5x'' - x' - 5x = 0$

5. $y''' - 9y' = 0$, $y(0) = 3$, $y'(0) = 0$, $y''(0) = 18$

6. $y''' - 6y'' + 3y' + 10y = 0$

7. $y^{(4)} = 0$

8. $y^{(4)} - 9y'' = 0$

9. $y^{(4)} - 5y'' + 4y = 0$

10. $y^{(5)} - 2y''' + y' = 0$

11. $y^{(4)} - 4y'' = 0$, $y(0) = 1$, $y'(0) = 3$, $y''(0) = 0$, $y'''(0) = 16$

12. $y^{(4)} - 4y''' - 7y'' + 22y' + 24y = 0$

13. $y''' - y'' + y' - y = 0$

14. $y''' - 3y'' + 4y' - 2y = 0$, $y(0) = 1$, $y'(0) = 2$, $y''(0) = 3$

15. $y''' - 27y = 0$

16. $y^{(5)} + 2y''' + y' = 0$, $y(\pi/2) = 0$, $y'(\pi/2) = 1$, $y''(\pi/2) = 0$, $y'''(\pi/2) = -3$, $y^{(4)}(\pi/2) = 0$

17. Show that the solutions y_1, y_2, and y_3 of the linear third-order differential equation

$$y''' + a_1(x)y'' + a_2(x)y' + a_3(x)y = 0$$

that satisfy the conditions

$$y_1(x_0) = 1, \quad y_1'(x_0) = 0, \quad y_1''(x_0) = 0,$$
$$y_2(x_0) = 0, \quad y_2'(x_0) = 1, \quad y_2''(x_0) = 0,$$
$$y_3(x_0) = 0, \quad y_3'(x_0) = 0, \quad y_3''(x_0) = 1,$$

respectively, are linearly independent.

18. Show that *any* solution of

$$y''' + a_1(x)y'' + a_2(x)y' + a_3(x)y = 0$$

can be expressed as a linear combination of the solutions y_1, y_2, y_3 given in Problem 17. [*Hint:* If $y(x_0) = c_1$, $y'(x_0) = c_2$, and $y''(x_0) = c_3$, consider the linear combination $c_1 y_1 + c_2 y_2 + c_3 y_3$.]

* **19.** Consider the third-order equation

$$y''' + a(x)y'' + b(x)y' + c(x)y = 0$$

and let $y_1(x)$ and $y_2(x)$ be two linearly independent solutions. Define $y_3(x) = v(x)y_1(x)$ and assume that $y_3(x)$ is a solution to the equation.

(a) Find a second-order differential equation that is satisfied by v'.

(b) Show that $(y_2/y_1)'$ is a solution of this equation.

(c) Use the result of part (b) to find a second, linearly independent solution of the equation derived in part (a).

20. Consider the equation

$$y''' - \left(\frac{3}{x^2}\right)y' + \left(\frac{3}{x^3}\right)y = 0 \quad (x > 0).$$

(a) Show that $y_1(x) = x$ and $y_2(x) = x^3$ are two linearly independent solutions.

(b) Use the results of Problem 19 to get a third, linearly independent solution.

21. Consider the third-order equation

$$y''' + a(x)y'' + b(x)y' + c(x)y = 0,$$

where a, b, and c are continuous functions of x in some interval I. Prove that if $y_1(x)$, $y_2(x)$, and $y_3(x)$ are solutions to the equation, then so is any linear combination of them.

22. In Problem 21 let

$$W(y_1, y_2, y_3)(x) = \begin{vmatrix} y_1 & y_2 & y_3 \\ y_1' & y_2' & y_3' \\ y_1'' & y_2'' & y_3'' \end{vmatrix}.$$

(a) Show that W satisfies the differential equation $W'(x) = -a(x)W$.

(b) Prove that $W(y_1, y_2, y_3)(x)$ is either always zero or never zero.

23. (a) Prove that the solutions $y_1(x)$, $y_2(x)$, $y_3(x)$ of the equation in Problem 21 are linearly independent on $[x_0, x_1]$ if and only if $W(y_1, y_2, y_3) \neq 0$.

(b) Show that $\sin t$, $\cos t$, and e^t are linearly independent solutions of

$$y''' - y'' + y' - y = 0$$

on any interval (a, b) where $-\infty < a < b < \infty$.

24. Assume that $y_1(x)$ and $y_2(x)$ are two solutions to

$$y''' + a(x)y'' + b(x)y' + c(x)y = f(x).$$

Prove that $y_3(x) = y_1(x) - y_2(x)$ is a solution of the associated homogeneous equation.

In Problems 25–28 use the method of undetermined coefficients to find the general solution of the given equation.

25. $y''' - y'' - y' + y = e^x$

26. $y''' - y'' - y' + y = e^{-x}$

27. $y''' - 3y'' - 10y' + 24y = x + 3$

28. $y^{(4)} + 2y'' + y = 3\cos x$

* 29. Consider the third-order equation

$$y''' + ay'' + by' + cy = f(x). \qquad (8)$$

Let $y_1(x)$, $y_2(x)$, and $y_3(x)$ be three linearly independent solutions of the associated homogeneous equation. Assume that there is a solution of equation (8) of the form $y(x) = c_1(x)y_1(x) + c_2(x)y_2(x) + c_3(x)y_3(x)$.

(a) Following the steps used in deriving the variation of parameters procedure for second-order equations, derive a method for solving third-order equations.

(b) Find a particular solution of the equation

$$y''' - 2y' - 4y = e^{-x}\tan x.$$

30. Use the method derived in Problem 29 to find a particular solution of

$$y''' + 5y'' + 9y' + 5y = 2e^{-2x}\sec x.$$

In Problems 31–33 guess that there is a solution of the form $y = x^\lambda$ to solve the given Euler equation.

31. $x^3y''' + 2x^2y'' - xy' + y = 0$

32. $x^3y''' - 12xy' + 24y = 0$

33. $x^3y''' + 4x^2y'' + 3xy' + y = 0$

34. Show that the substitution $x = e^t$ can be used to solve the third-order Euler equation

$$x^3y''' + x^2y'' - 2xy' + 2y = 0.$$

Review Exercises for Chapter 2

In Exercises 1–5, a second-order differential equation and one solution $y_1(x)$ are given. Verify that $y_1(x)$ is indeed a solution and find a second linearly independent solution.

1. $y'' + 4y = 0$; $y_1(x) = \sin 2x$

2. $y'' - 6y' + 9y = 0$; $y_1(x) = e^{3x}$

3. $x^2y'' + xy' - 4y = 0$; $y_1(x) = x^2$

4. $y'' + \dfrac{1}{x}y' + \left(1 - \dfrac{1}{4x^2}\right)y = 0$; $y_1(x) = x^{-1/2}\sin x$

5. $(1 - x^2)y'' - 2xy' + 2y = 0$; $y_1(x) = x$

In Exercises 6–24, find the general solution to the given equation. If initial conditions are given, find the particular solution that satisfies them.

6. $y'' - 9y' + 20y = 0$

7. $y'' - 9y' + 20y = 0$; $y(0) = 3$, $y'(0) = 2$

8. $y'' - 3y' + 4y = 0$

9. $y'' - 3y' - 4y = 0$; $y(0) = 0$, $y'(0) = 1$

10. $y'' = 0$

11. $4y'' + 4y' + y = 0$

12. $y'' - 11y = 0$

13. $y'' - 2y' + 7y = 0$

14. $y'' - y' - 2y = \sin 2x$

15. $y''' - 6y'' + 11y' - 6y = 0$

16. $y'' - 2y' + y = xe^x$

17. $y'' - 2y' + y = x^2 - 1$; $y(0) = 2$, $y'(0) = 1$

18. $y'' + y = \sec x$, $0 < x < \dfrac{\pi}{2}$

19. $y'' - 2y' + y = \dfrac{2e^x}{x^3}$

20. $y'' + 4y' + 4y = e^{-2x}/x^2$; $x > 0$

21. $x^2y'' + 5xy' + 4y = 0$; $x > 0$

22. $x^2y'' - 2xy' + 3y = 0$; $x > 0$

23. $y''' + y'' - 8y' - 12y = 0$

24. $y^{(4)} + 8y'' + 4y = 0$

In Exercises 25–30 determine the equation of motion of a mass m attached to a coiled spring with spring constant k initially displaced a distance x_0 from equilibrium and released with velocity v_0 subject to:

(a) no damping or external forces,

(b) a damping constant c, but no external force,

(c) an external force $F_0\sin\omega t$, but no damping,

(d) both a damping constant c and external force $F_0\sin\omega t$.

25. $m = 20$ kg, $k = 1000$ N/m, $x_0 = 1$ m, $v_0 = 0$, $c = 200$ kg/s, $F_0 = 1$ N, $\omega = 10$ rad/s

26. $m = 25$ kg, $k = 40$ N/m, $x_0 = 0$, $v_0 = 1$ m/s,
    $c = 20$ kg/s, $F_0 = 1$ N, $\omega = 1$ rad/s

27. $m = 25$ kg, $k = 40$ N/m, $x_0 = 3$ m, $v_0 = 4$ m/s,
    $c = 10\sqrt{5}$ kg/s, $F_0 = 1$ N, $\omega = 1$ rad/s

28. $m = 1$ kg, $k = 36$ N/m, $x_0 = 4$ m, $v_0 = 0$,
    $c = 10$ kg/s, $F_0 = 4$ N, $\omega = 4$ rad/s

29. $m = 4$ kg, $k = 25$ N/m, $x_0 = 0$ m, $v_0 = 3$ m/s,
    $c = 8$ kg/s, $F_0 = 1$ N, $\omega = 3$ rad/s

30. $m = 9$ kg, $k = 81$ N/m, $x_0 = 4$ m, $v_0 = 1$ m/s,
    $c = 10$ kg/s, $F_0 = 2$ N, $\omega = \frac{1}{3}$ rad/s

31. Let an inductance of $L = 2$ henrys, a resistance of $R = 50$ ohms, and a capacitance $C = 10^{-4}$ farad be connected in series with an emf of $E = 1000$ volts. (see Fig. 2.7.1). Suppose no charge is present and no current is flowing at time $t = 0$, when E is applied. Find the transient and steady-state solutions for the charge Q at all times t.

In Exercises 32–35 find the steady-state current in the RLC circuit of Fig. 2.7.1, where

32. $L = 5$ henrys, $R = 20$ ohms, $C = 0.1$ farad,
    $E = 25 \sin t$ volts.

33. $L = 10$ henrys, $R = 240$ ohms, $C = 0.025$ farad,
    $E = 100 \cos 5t$ volts.

34. $L = 1$ henry, $R = 9$ ohms, $C = 0.1$ farad,
    $E = 100 \sin 10t$ volts.

35. $L = 2.5$ henrys, $R = 20$ ohms, $C = 0.08$ farad,
    $E = 100 \cos 5t$ volts.

36. A 27-lb weight hangs from a spring of spring constant 18 lb/in. During free motion the amplitude decreases to one-tenth in six complete cycles of the motion. Find the equation describing the motion of the spring-weight system.

37. A spring fixed at its upper end supports a 10-lb weight which stretches the spring by $\frac{1}{2}$ ft. An external periodic force of $2 \cos 8t$ lb is applied to the spring-weight system. Describe the motion, neglecting resistance.

38. Repeat Exercise 37 with a damping constant of $c = 0.01$ slugs/sec.

39. Repeat Exercise 38 with an external periodic force of $2 \sin \frac{65}{8} t$ lb.

Appendix to Chapter 2

Proofs of the Theorems of Section 2.1

We shall prove Theorems 2–6 in Section 2.1 in the order 3, 5, 2, 4, and 6.

■ **THEOREM 3 (p. 71)**
Let y_1 and y_2 be any two solutions of the homogeneous equation

$$y'' + a(x)y' + b(x)y = 0. \tag{1}$$

Then any linear combination of them is also a solution of (1).

PROOF
Let $y(x) = c_1 y_1(x) + c_2 y_2(x)$. Then

$$y'' + ay' + by = c_1 y_1'' + c_2 y_2'' + c_1 a y_1' + c_2 a y_2' + c_1 b y_1 + c_2 b y_2$$

$$= c_1(y_1'' + a y_1' + b y_1) + c_2(y_2'' + a y_2' + b y_2)$$

$$= c_1 \cdot 0 + c_2 \cdot 0 = 0$$

since y_1 and y_2 are solutions of the homogeneous equation (1).   ■

■ **THEOREM 5 (p. 73)**
The solutions y_1 and y_2 of equation (1) are linearly independent on $[x_0, x_1]$ if and only if $W(y_1, y_2)(x) \neq 0$.

PROOF

We first show that if $W(y_1, y_2)(x) = 0$, then y_1 and y_2 are linearly dependent. Select any point x_2 in $[x_0, x_1]$. Consider the system of equations

$$c_1 y_1(x_2) + c_2 y_2(x_2) = 0, \qquad c_1 y_1'(x_2) + c_2 y_2'(x_2) = 0, \tag{2}$$

where c_1 and c_2 are constants to be determined. Since the determinant of system (2) is

$$y_1(x_2) y_2'(x_2) - y_1'(x_2) y_2(x_2) = W(y_1, y_2)(x_2) = 0,$$

system (2) has a nontrivial solution (see Section 7.9); that is, there are constants c_1 and c_2 not both zero such that (2) holds. Define $y(x) = c_1 y_1(x) + c_2 y_2(x)$. By Theorem 3, $y(x)$ is a solution of equation (1). Then since c_1 and c_2 solve (2),

$$y(x_2) = c_1 y_1(x_2) + c_2 y_2(x_2) = 0$$

and

$$y'(x_2) = c_1 y_1'(x_2) + c_2 y_2'(x_2) = 0.$$

Thus $y(x)$ solves the initial value problem

$$y'' + a(x)y' + b(x)y = 0, \qquad y(x_2) = y'(x_2) = 0.$$

But this initial value problem also has the solution $y_3(x) = 0$ for all values of x in $x_0 \le x \le x_1$. By Theorem 1, the solution of this initial value problem is unique so that necessarily $y(x) = y_3(x) = 0$. Thus

$$y(x) = c_1 y_1(x) + c_2 y_2(x) = 0,$$

for all values of x in $x_0 \le x \le x_1$, which proves that y_1 and y_2 are linearly dependent.

We now assume that $W(y_1, y_2)(x) \ne 0$ in $[x_0, x_1]$ and shall prove that y_1 and y_2 are linearly independent. If y_1 and y_2 are not linearly independent, then there is a constant c such that $y_2 = cy_1$ or $y_1 = cy_2$. Assume that $y_2 = cy_1$. Then $y_2' = cy_1'$ and

$$W(y_1, y_2)(x) = y_1 y_2' - y_1' y_2 = y_1(cy_1') - y_1'(cy_1) = 0.$$

But this contradicts the assumption that $W \ne 0$. Hence the solutions y_1 and y_2 must be independent. ∎

THEOREM 2 (p. 70)

Two linearly independent solutions of equation (1) can always be found.

PROOF

The existence part of Theorem 1 guarantees that we can find a solution $y_1(x)$ to equation (1) satisfying

$$y_1(x_0) = 1 \quad \text{and} \quad y_1'(x_0) = 0.$$

Similarly, we can find a solution $y_2(x)$ to equation (1) satisfying

$$y_2(x_0) = 0 \quad \text{and} \quad y_2'(x_0) = 1.$$

Now

$$W(y_1, y_2)(x_0) = \begin{vmatrix} 1 & 0 \\ 0 & 1 \end{vmatrix} = 1 \neq 0$$

so y_1 and y_2 are linearly independent by Theorem 5. ∎

■ **THEOREM 4 (p. 71)**

Let $y_1(x)$ and $y_2(x)$ be linearly independent solutions to equation (1) and let $y_3(x)$ be another solution to (1). Then there exist unique constants c_1 and c_2 such that

$$y_3(x) = c_1 y_1(x) + c_2 y_2(x).$$

PROOF

Let $y_3(x_0) = a$ and $y_3'(x_0) = b$. Consider the linear system of equations in two unknowns c_1 and c_2:

$$y_1(x_0)c_1 + y_2(x_0)c_2 = a, \qquad y_1'(x_0)c_1 + y_2'(x_0)c_2 = b. \tag{3}$$

As we saw in the proof of Theorem 5, the determinant of this system is $W(y_1, y_2)(x_0)$, which is nonzero since the solutions are linearly independent. Thus there is a unique solution (c_1, c_2) to equation (2) and a solution $y^*(x) = c_1 y_1(x) + c_2 y_2(x)$ that satisfies the conditions $y^*(x_0) = a$ and $y^{*\prime}(x_0) = b$. Since every initial value problem has a unique solution (by Theorem 1), it must follow that $y_3(x) = y^*(x)$ on the interval $x_0 \leq x \leq x_1$, and so the proof is complete. ∎

■ **THEOREM 6 (p. 74)**

Let $y_p(x)$ be a solution of the nonhomogeneous equation

$$y'' + a(x)y' + b(x)y = f(x), \tag{4}$$

and let $y^*(x)$ be any other solution. Then $y^*(x) - y_p(x)$ is a solution to (1); that is,

$$y^*(x) = c_1 y_1(x) + c_2 y_2(x) + y_p(x)$$

for some constants c_1 and c_2, where y_1, y_2 are two linearly independent solutions of (1).

PROOF

We have

$$(y^* - y_p)'' + a(y^* - y_p)' + b(y^* - y_p)$$

$$= (y^{*\prime\prime} + ay^{*\prime} + by^*) - (y_p'' + ay_p' + by_p)$$

$$= f - f = 0.$$
∎

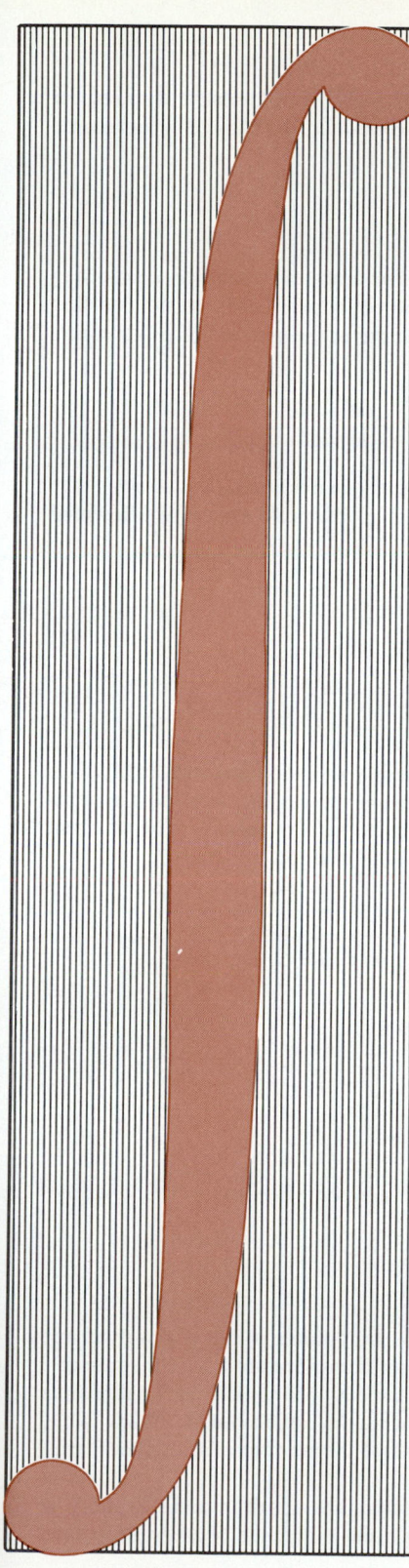

chapter 3

Systems of Linear Differential Equations

3.1 Method of Elimination for Linear Systems with Constant Coefficients

In the preceding chapters we discussed the problem of finding the solution to a single linear differential equation. In this chapter we shall discuss an elementary method for solving a system of simultaneous first-order linear differential equations by converting the system into a single, higher-order linear differential equation that may then be solved by the methods we have already seen. Systems of simultaneous differential equations arise in problems involving more than one unknown function, each of which is a function of a single independent variable (which often denotes time). For the sake of consistency throughout the remaining sections of this chapter, we shall denote the independent variable by t and the dependent variables by $x(t)$ and $y(t)$ or by the subscripted letters $x_1(t), x_2(t), \ldots, x_n(t)$.

Although some familiarity with the elementary properties of determinants (see Section 7.8) will be useful, no use of matrix methods in solving systems of simultaneous linear differential equations will be made in this chapter. A discussion of matrix methods for solving systems of linear differential equations can be found in Section 8.6. In this section we shall give some examples of how simple systems arise and shall describe two elementary procedures for finding their solution.

EXAMPLE 1

Suppose that a chemical solution flows from one container, at a rate proportional to its volume, into a second container. It flows out from the second container at a constant rate. Let $x(t)$ and $y(t)$ denote the volumes of solution in the first and second containers, respectively, at time t. (The containers may be, for example, cells, in which case we are describing a diffusion process across a cell wall.) To establish the necessary equations, we note that the change in volume equals the difference between input and output in each container. The rate of change in volume is the derivative of volume with respect to time. Since no chemical is flowing into the first container, the change in its volume equals the output:

$$\frac{dx}{dt} = -c_1 x,$$

where c_1 is a positive constant of proportionality. The amount of solution $c_1 x$ flowing out of the first container is the input of the second container. Let c_2 be the constant output of the second container. Then change in volume in the second container equals the difference between its input and output:

$$\frac{dy}{dt} = c_1 x - c_2.$$

System

Thus we can describe the flow of solution by means of two differential equations. Since more than one differential equation is involved, we have obtained a **system of differential equations:**

$$\frac{dx}{dt} = -c_1 x,$$

$$\frac{dy}{dt} = c_1 x - c_2,$$

(1)

where c_1 and c_2 are positive constants. By a solution of the system (1) we shall mean a pair of functions $(x(t), y(t))$ that simultaneously satisfy the two equations in (1). It is easy to solve this system by solving the two equations successively (for most systems this is not possible). If we denote the initial volumes in the two containers by $x(0)$ and $y(0)$, respectively, we see that the first equation has the solution

$$x(t) = x(0)e^{-c_1 t}.$$

(2)

Substituting equation (2) into the second equation of (1), we obtain

$$\frac{dy}{dt} = c_1 x(0)e^{-c_1 t} - c_2,$$

which, upon integration, yields the solution

$$y(t) = y(0) + x(0)(1 - e^{-c_1 t}) - c_2 t.$$

(3)

Equations (2) and (3) together constitute the unique solution of system (1) which satisfies the given initial conditions.

EXAMPLE 2

Let tank X contain 100 gallons of brine in which 100 pounds of salt is dissolved and tank Y contain 100 gallons of water. Suppose water flows into tank X at the rate of 2 gallons per minute, and the mixture flows from tank X into tank Y at 3 gallons per minute. From Y 1 gallon per minute is pumped back to X (establishing **feedback**) while 2 gallons per minute are flushed away. We wish to find the amount of salt in both tanks at all time t. (See Fig. 1.)

If we let $x(t)$ and $y(t)$ represent the number of pounds of salt in tanks X and Y at time t, and note that the change in weight equals the difference between input and

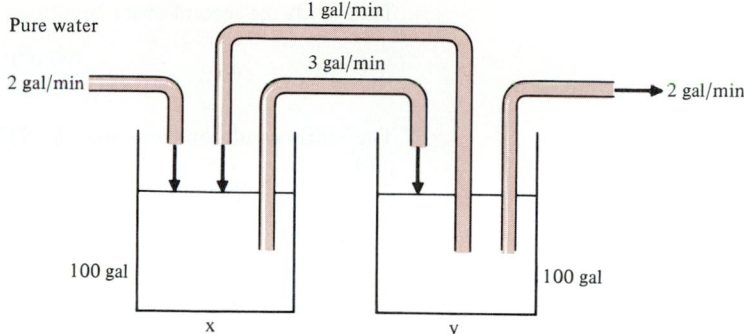

FIGURE 1

output, we can again derive a system of linear first-order equations. Tanks X and Y initially contain $x(0) = 100$ and $y(0) = 0$ pounds of salt, respectively, at time $t = 0$. The quantities $x/100$ and $y/100$ are, respectively, the amounts of salt contained in each gallon of water taken from tanks X and Y at time t. Three gallons are being removed from tank X and added to tank Y, while only one of the three gallons removed from tank Y is put in tank X. Thus we have the system

$$\frac{dx}{dt} = -3\frac{x}{100} + \frac{y}{100}, \qquad x(0) = 100$$

$$\frac{dy}{dt} = 3\frac{x}{100} - 3\frac{y}{100}, \qquad y(0) = 0. \tag{4}$$

Since both equations in system (4) involve both dependent variables, we cannot immediately solve for one of the variables, as we did in Example 1. Instead, we will use the operation of differentiation to eliminate one of the dependent variables. Suppose we begin by solving the second equation for $x(t)$ in terms of the dependent variable y and its derivative:

$$\frac{3}{100}x = \frac{3}{100}y + \frac{dy}{dt}$$

or

$$x = y + \frac{100}{3}\frac{dy}{dt}. \tag{5}$$

We now differentiate equation (5) and equate the result to the first equation in (4):

$$\frac{dx}{dt} = \frac{dy}{dt} + \frac{100}{3}\frac{d^2y}{dt^2} = \frac{-3}{100}x + \frac{y}{100}. \tag{6}$$

But, from the second equation of (4),

$$\frac{3}{100}x = \frac{dy}{dt} + \frac{3}{100}y$$

so, inserting this into the last expression in (6), we obtain

$$\frac{100}{3}\frac{d^2y}{dt^2} + \frac{dy}{dt} = \overbrace{-\frac{dy}{dt} - \frac{3}{100}y}^{\textstyle -\frac{3}{100}x} + \frac{1}{100}y.$$

This yields the second-order equation

$$\frac{100}{3}\frac{d^2y}{dt^2} + 2\frac{dy}{dt} + \frac{2y}{100} = 0.\qquad(7)$$

The initial conditions for equation (7) are obtained directly from the system (4), since $y(0) = 0$ and

$$y'(0) = 3\frac{x(0)}{100} - 3\frac{y(0)}{100} = 3.\qquad(8)$$

Multiplying both sides of equation (7) by $\frac{3}{100}$, we have the initial value problem

$$y'' + \frac{6}{100}y' + \frac{6}{(100)^2}y = 0,\qquad y(0) = 0,\quad y'(0) = 3.\qquad(9)$$

The characteristic equation for (9) is

$$\lambda^2 + \frac{6}{100}\lambda + \frac{6}{(100)^2} = 0$$

which has the roots

$$\lambda_{1,2} = \frac{-\dfrac{6}{100} \pm \sqrt{\dfrac{36}{(100)^2} - \dfrac{24}{(100)^2}}}{2} = \frac{-6 \pm \sqrt{12}}{200}$$

or

$$\lambda_1 = \frac{-3 + \sqrt{3}}{100}, \qquad \lambda_2 = \frac{-3 - \sqrt{3}}{100}.$$

Thus the general solution is

$$y(t) = c_1 e^{[(-3+\sqrt{3})t]/100} + c_2 e^{[(-3-\sqrt{3})t]/100}.$$

Using the initial conditions, we obtain the simultaneous equations

$$c_1 + c_2 = 0,$$

$$\frac{-3 + \sqrt{3}}{100}c_1 - \frac{3 + \sqrt{3}}{100}c_2 = 3.$$

These have the unique solution $c_1 = -c_2 = 50\sqrt{3}$. Hence

$$y(t) = 50\sqrt{3}[e^{[(-3+\sqrt{3})t]/100} - e^{[(-3-\sqrt{3})t/100]}]$$

and substituting this function into the right-hand side of equation (5) we obtain, after some algebra,

$$x(t) = 50[e^{[(-3+\sqrt{3})t]/100} + e^{[(-3-\sqrt{3})t]/100}].$$

Note that, as is evident from the problem, the amounts of salt in the two tanks approach zero as time tends to infinity.

Method of Elimination

The technique we used in solving Example 2 is called the **method of elimination,** since all but one of the dependent variables are eliminated by repeated differentiation. There are other methods for solving systems. One of these is discussed in the next section. A technique which makes use of matrix methods is discussed in Section 8.6.

We shall use the method of elimination to solve systems having the following form:

$$x'(t) = a_{11}x(t) + a_{12}y(t) + f_1(t),$$

$$y'(t) = a_{21}x(t) + a_{22}y(t) + f_2(t). \tag{10}$$

If $f_1(t) = f_2(t) \equiv 0$, then the system is said to be **homogeneous.** Otherwise it is **nonhomogeneous.** If a_{11}, a_{12}, a_{21}, and a_{22} are constants, then the system is said to have **constant coefficients.**

DEFINITION

Solution

A **solution** to system (10) is a pair of functions $(x(t), y(t))$ that are differentiable and satisfy the two equations in (10). The pair is also called a **vector solution.**

EXAMPLE 3

Find all solutions to the nonhomogeneous system

$$x' = 2x + y + t,$$

$$y' = x + 2y + t^2.$$

SOLUTION

Proceeding as in Example 2, we obtain

$$x'' = 2x' + y' + 1 = 2x' + (x + 2y + t^2) + 1$$

$$= 2x' + x + 2(x' - 2x - t) + t^2 + 1$$

or

$$x'' - 4x' + 3x = t^2 - 2t + 1 = (t - 1)^2. \tag{11}$$

The solution to the homogeneous part of equation (11) is $x(t) = c_1e^t + c_2e^{3t}$. Using the method of undetermined coefficients, we obtain the particular solution $x_p(t) = \frac{1}{3}t^2 + \frac{2}{9}t + \frac{11}{27}$, so the general solution of equation (11) is

$$x(t) = c_1e^t + c_2e^{3t} + \tfrac{1}{3}t^2 + \tfrac{2}{9}t + \tfrac{11}{27}.$$

As in Example 2, since $y = x' - 2x - t$, we obtain

$$y(t) = c_1e^t + 3c_2e^{3t} + \tfrac{2}{3}t + \tfrac{2}{9} - 2c_1e^t - 2c_2e^{3t} - \tfrac{2}{3}t^2 - \tfrac{4}{9}t - \tfrac{22}{27} - t$$

or

$$y(t) = -c_1e^t + c_2e^{3t} - \tfrac{2}{3}t^2 - \tfrac{7}{9}t - \tfrac{16}{27}.$$

EXAMPLE 4

Solve the equations

$$x' = \frac{3}{t}x + \frac{1}{t}y, \qquad t > 0$$

$$y' = -\frac{4}{t}x - \frac{1}{t}y, \qquad t > 0$$

SOLUTION

Differentiating the second equation and eliminating x we have

$$y'' = -\frac{4}{t}\left(\frac{3}{t}x + \frac{1}{t}y\right) + \frac{4}{t^2}x + \frac{1}{t^2}y - \frac{1}{t}y'$$

so that

$$y'' + \frac{1}{t}y' + \frac{3}{t^2}y = -\frac{8}{t^2}x = \frac{2}{t}\left(-\frac{4}{t}x\right) = \frac{2}{t}\left(y' + \frac{1}{t}y\right)$$

or

$$y'' - \frac{y'}{t} + \frac{y}{t^2} = 0.$$

But if we multiply this equation by t^2, we obtain the Euler equation

$$t^2 y'' - ty' + y = 0.$$

Using the method of Section 2.6, we find that

$$y(t) = c_1 t + c_2 t \ln t.$$

Then, from the second equation

$$\frac{4}{t}x = -y' - \frac{1}{t}y$$

and

$$x(t) = \frac{t}{4}(-c_1 - c_2 - c_2 \ln t - c_1 - c_2 \ln t)$$

$$= \frac{t}{4}(-2c_1 - c_2 - 2c_2 \ln t)$$

$$= \left(-\frac{c_1}{2} - \frac{c_2}{4}\right)t - \frac{c_2}{2}t \ln t.$$

A linear system of n first-order equations usually reduces to an nth-order linear differential equation, because it generally requires one differentiation to eliminate each variable $x_2, \ldots, x_n$ from the system.

EXAMPLE 5

Consider the mass-spring system of Fig. 2, which is a direct generalization of the system described in Section 2.8 (see p. 106). In this example we have two objects of mass m_1 and m_2 suspended by springs in series with spring constants k_1 and k_2. If the vertical displacements from equilibrium of the two point masses are denoted by $x_1(t)$ and $x_2(t)$, respectively, then using assumptions (1) and (2) (Hooke's law) on p. 106, we find that the net forces acting on the two masses are given by

$$F_1 = -k_1 x_1 + k_2(x_2 - x_1),$$

$$F_2 = -k_2(x_2 - x_1).$$

Here the positive direction is downward. Note that the first spring is compressed when $x_1 < 0$ and the second spring is compressed when $x_1 > x_2$. The equations of motion are

$$m_1 \frac{d^2 x_1}{dt^2} = -k_1 x_1 + k_2(x_2 - x_1) = -(k_1 + k_2)x_1 + k_2 x_2,$$

$$m_2 \frac{d^2 x_2}{dt^2} = -k_2(x_2 - x_1) = k_2 x_1 - k_2 x_2,$$

(12)

which comprise a system of two second-order linear differential equations with constant coefficients.

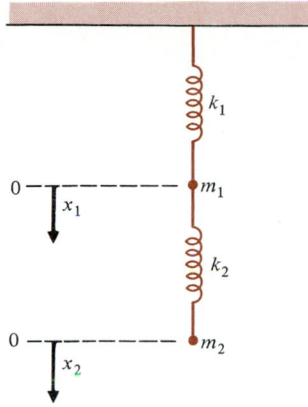

FIGURE 2

We shall now show that linear differential equations (and systems) of any order can be converted, by the introduction of new variables, into a system of first-order differential equations. This procedure is very important since it means that the study of first-order linear systems provides a unified theory for all linear differential equations and systems. From a practical point of view it means, for example, that once we know how to solve first-order linear systems with constant coefficients, we will be able to solve any constant-coefficient linear differential equation or system.

To rewrite system (12) as a first-order system, we define the new variables $x_3 = x_1'$ and $x_4 = x_2'$. Then $x_3' = x_1''$, $x_4' = x_2''$, and (12) can be expressed as the system of four first-order equations

$$x_1' = x_3,$$

$$x_2' = x_4,$$

$$x_1'' = x_3' = -\left(\frac{k_1 + k_2}{m_1}\right)x_1 + \left(\frac{k_2}{m_1}\right)x_2,$$

(13)

$$x_2'' = x_4' = \left(\frac{k_2}{m_2}\right)x_1 - \left(\frac{k_2}{m_2}\right)x_2.$$

If we wish, we can now use the method of elimination to reduce system (13) to a single fourth-order linear differential equation that can be solved by the techniques used earlier in this section.

To stress the importance of the concept of introducing variables to convert a higher-order equation or system into a system of first-order equations we prove:

■ THEOREM 1

The linear nth-order differential equation

$$x^{(n)} + a_1(t)x^{(n-1)} + a_2(t)x^{(n-2)} + \cdots + a_{n-1}(t)x' + a_n(t)x = f(t)$$

can be written as a system of n first-order linear equations.

PROOF

Define $x_1 = x$, $x_2 = x'$, $x_3 = x''$, ..., $x_n = x^{(n-1)}$. Then we have

$$x_1' = x_2,$$

$$x_2' = x_3,$$

$$\vdots$$

$$x_n' = -a_n x_1 - a_{n-1}x_2 - \cdots - a_1 x_n + f. \qquad ■$$

In some cases, Theorem 1 can be generalized to nonlinear differential equations. (See Problem 33.)

PROBLEMS 3.1

In Problems 1–15 find all solutions to the given system. If initial conditions are given, find the unique solution that satisfies them.

1. $x' = 4x - 3y$, $\quad y' = 5x - 4y$

2. $x' = 7x + 6y$, $\quad y' = 2x + 6y$

3. $x' = -x + y$, $\quad y' = -5x + 3y$

4. $x' = x + y$, $\quad y' = -x + 3y$; $\quad x(0) = 1$, $\quad y(0) = 3$

5. $x' = -4x - y$, $\quad y' = x - 2y$; $\quad x(0) = 2$, $\quad y(0) = -4$

6. $x' = 4x - 2y$, $\quad y' = 5x + 2y$; $\quad x(0) = 0$, $\quad y(0) = 1$

7. $x' = 4x - 3y$, $\quad y' = 8x - 6y$; $\quad x(0) = 1$, $\quad y(0) = 0$

8. $x' = x + 2y$, $\quad y' = 3x + 2y$

9. $x' = 8x - y$, $\quad y' = 4x + 12y$

10. $x' = 12x - 17y$, $\quad y' = 4x - 4y$

11. $x' = x + 2y + t - 1$, $\quad x(0) = 0$
$\quad y' = 3x + 2y - 5t - 2$, $\quad y(0) = 4$

12. $x' = x + y$, $\quad x(0) = 1$,
$\quad y' = y$, $\quad y(0) = 0$

13. $x' = 2x + y + 3e^{2t}$, $\quad y' = -4x + 2y + te^{2t}$

14. $x' = 3x + 3y + t$, $\quad y' = -x - y + 1$

15. $x' = 4x + y$, $x(\pi/4) = 0$,
$\quad y' = -8x + 8y$, $\quad y(\pi/4) = 1$

16. By elimination, find a solution to the following nonlinear system:
$\quad x' = x + \sin x \cos x + 2y$,
$\quad y' = (x + \sin x \cos x + 2y) \sin^2 x + x$.

In Problems 17–22, transform each given equation into a system of first-order equations. Do not solve the system.

17. $x'' - 6tx' + 3t^3 x = \cos t$

18. $x''' - x'' + (x')^2 - x^3 = t$

19. $x^{(4)} - \cos x(t) = t$

20. $x''' + xx'' - x'x^4 = \sin t$

21. $xx'x''x''' = t^5$

22. $x''' - 3x'' + 4x' - x = 0$

23. A mass m moves in xyz-space according to the following equations of motion:

$$mx'' = f(t, x, y, z),$$

$$my'' = g(t, x, y, z),$$

$$mz'' = h(t, x, y, z).$$

Transform these equations into a system of six first-order equations.

24. Consider the uncoupled system

$$x_1' = x_1, \qquad x_2' = x_2.$$

(a) What is the general solution of this system?

* (b) Show that there is no second-order equation equivalent to this system. [*Hint:* Show that any second-order equation has solutions that are not solutions of this system. This shows that first-order systems are more general than higher-order equations in the sense that any of the latter can be written as a first-order system, but not vice versa.]

Use the method of elimination to solve the systems in Problems 25 and 26.

25. $x_1' = x_1,$
$x_2' = 2x_1 + x_2 - 2x_3,$
$x_3' = 3x_1 + 2x_2 + x_3.$

26. $x_1' = x_1 + x_2 + x_3,$
$x_2' = 2x_1 + x_2 - x_3,$
$x_3' = -8x_1 - 5x_2 - 3x_3.$

27. In Example 2, when does tank Y contain a maximum amount of salt? How much salt is in tank Y at that time?

28. Suppose in Example 2 that the rate of flow from tank Y to tank X is two gallons per minute (instead of one) and all other facts are unchanged. Find the differential equations for the amount of salt in each tank at all times t.

29. Tank X contains 500 gallons of brine in which 500 lb of salt is dissolved. Tank Y contains 500 gal of water. Water flows into tank X at the rate of 30 gal/min, and the mixture flows into Y at the rate of 40 gal/min. From Y the solution is pumped back into X at the rate of 10 gal/min and into a third tank at the rate of 30 gal/min. Find the maximum amount of salt in Y. When does this concentration occur?

30. Suppose in Problem 29 that tank X contains 1000 gal of brine. Solve the problem, given that all other conditions are unchanged.

31. Consider the mass spring system illustrated in Fig. 3. Here three point masses are suspended in series by three springs

with spring constants k_1, k_2, and k_3, respectively. Formulate a system of second-order differential equations that describes this system.

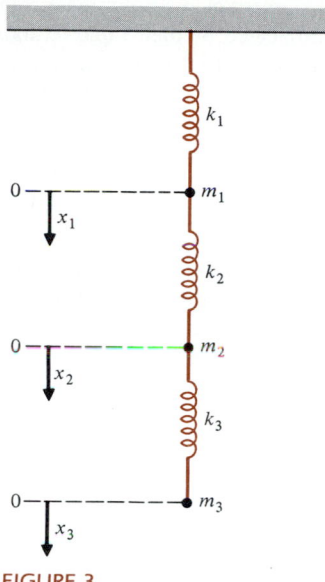

FIGURE 3

32. Find a single fourth-order linear differential equation in terms of the dependent variable x_1 for the system (13). Find a solution to the system if $m_1 = 1$ kg, $m_2 = 2$ kg, $k_1 = 5$ N/m, $k_2 = 4$ N/m.

33. Show that the differential equation

$$x^{(n)} = g(t, x, x', \dots, x^{(n-1)})$$

can be transformed into a system of n first-order equations.

34. In a study concerning the distribution of radioactive potassium, ^{42}K, between red blood cells and the plasma of the human blood, C. W. Sheppard and W. R. Martin† added ^{42}K to freshly drawn blood. They discovered that although the total amount of potassium (stable and radioactive) in the red cells and in the plasma remained practically constant during the experiment, the radioactivity was gradually transmitted from the plasma to the red cells. Thus the behavior of the radioactivity is that of a linear closed two-compartment system. If the fractional transfer coefficient (see p. 46) from the plasma to the cells is $k_{12} = 30.1$ percent per hour, while $k_{21} = 1.7$ percent per hour, and the initial radioactivity was 800 counts per minute in the plasma and 25 counts per minute in the red cells, what is the number of counts per minute in the red cells after 300 minutes?

---

† "Cation exchange between cells and plasma of mammalian blood," *Journal of General Physiology*, vol. 33 (1950), pp. 703–722.

35. The presence of temperature inversions and low wind speeds will often trap air pollutants in a mountain valley for an extended period of time. Gaseous sulfur compounds are often a significant air pollution problem, and their study is complicated by their rapid oxidation. Hydrogen sulfide, H_2S, oxidizes into sulfur dioxide, SO_2, which in turn oxidizes into a sulfate. The following model has been proposed† for determining the concentrations $x(t)$ and $y(t)$ of H_2S and SO_2, respectively, in a fixed airshed. Let

$$\frac{dx}{dt} = -\alpha x + \gamma, \qquad \frac{dy}{dt} = \alpha x - \beta y + \delta,$$

where the constants α and β are the conversion rates of H_2S into SO_2 and SO_2 into sulfate, respectively, and γ and δ are the production rates of H_2S and SO_2, respectively. Solve the equations sequentially and estimate the concentration levels that could be reached under a prolonged air pollution episode.

Electrical Networks

Electric networks are governed by the two laws of Kirchhoff:

> **1.** The algebraic sum of all voltage drops around any closed circuit is zero.
> **2.** The algebraic sum of the currents flowing into any junction in the network is zero.

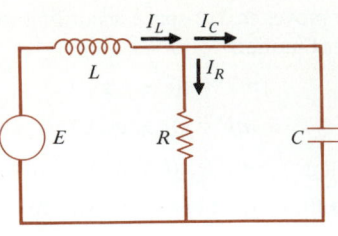

FIGURE 4

Consider the circuit in Fig. 4. There are two loops. By Kirchhoff's voltage law, we obtain

$$L\frac{dI_L}{dt} + RI_R = E,$$

$$\frac{Q_C}{C} - RI_R = 0.$$

Using the concepts of Section 1.6, solve Problems 36 and 37. Further examples of such circuits are given in Section 3.3.

36. Let $R = 50$ ohms, $L = 0.5$ henry, $C = 10^{-4}$ farads, and $E = 60$ volts in the network of Fig. 4. Suppose the currents I_R and I_L are both zero at time $t = 0$. Find the currents when $t = 1$ second.

37. Repeat Problem 36 for $E = 10 \sin t$ volts.

3.2 Method of Determinants

In this section we consider a more efficient method for solving the homogeneous system

$$x' = a_{11}x + a_{12}y,$$

$$y' = a_{21}x + a_{22}y,$$

$$(1)$$

where a_{11}, a_{12}, a_{21}, and a_{22} are constants.

Before doing so, we present a bit of theory. The results below are given without proof because they are similar to the results in Section 2.1.

† R. L. Bohac, "A mathematical model for the conversion of sulphur compounds in the Missoula Valley airshed," *Proceedings of the Montana Academy of Science* 34 (1974), 60–63.

Linear Combination

DEFINITION

The pair of functions $(x_3(t), y_3(t))$ is a **linear combination** of the pairs $(x_1(t), y_1(t))$ and $(x_2(t), y_2(t))$ if there exist constants c_1 and c_2 such that the following two equations hold simultaneously:

$$x_3(t) = c_1 x_1(t) + c_2 x_2(t),$$
$$y_3(t) = c_1 y_1(t) + c_2 y_2(t). \tag{2}$$

The next theorem is the systems analog of Theorem 2.1.3 on p. 71.

■ THEOREM 1

If the pairs $(x_1(t), y_1(t))$ and $(x_2(t), y_2(t))$ are solutions of the homogeneous system (1), then any linear combination of them is also a solution of the system (1). ■

EXAMPLE 1

Consider the system

$$x' = -x + 6y,$$
$$y' = x - 2y. \tag{3}$$

It is easy to verify that $(-2e^{-4t}, e^{-4t})$ and $(3e^t, e^t)$ are solutions of equations (3). Hence, by Theorem 1, the pair $(-2c_1 e^{-4t} + 3c_2 e^t, c_1 e^{-4t} + c_2 e^t)$ is a solution of equation (3) for any constants c_1 and c_2.

Linear Independence

DEFINITION

We define two pairs of functions $(x_1(t), y_1(t))$ and $(x_2(t), y_2(t))$ to be **linearly independent** on an interval I if, whenever the equations

$$c_1 x_1(t) + c_2 x_2(t) = 0,$$
$$c_1 y_1(t) + c_2 y_2(t) = 0 \tag{4}$$

hold for all values of t in I, then $c_1 = c_2 = 0$.

In Example 1, the two given pairs of solutions are linearly independent since $c_1 e^{-4t} + c_2 e^t$ vanishes for all t only when $c_1 = c_2 = 0$.

DEFINITION

Wronskian

Given two solutions $(x_1(t), y_1(t))$ and $(x_2(t), y_2(t))$ of the homogeneous system (1), we define the **Wronskian** of the two solutions by the following determinant:

$$W(t) = \begin{vmatrix} x_1(t) & y_1(t) \\ x_2(t) & y_2(t) \end{vmatrix} = x_1(t)y_2(t) - x_2(t)y_1(t). \tag{5}$$

We then have the next theorem.

■ **THEOREM 2**

If $W(t) \neq 0$ for every t in an interval I, then equations (2) are the **general solution** of the homogeneous system (1). That is, given any solution (x^*, y^*) of system (1), there exist constants c_1 and c_2 such that

$$x^* = c_1 x_1 + c_2 x_2,$$

$$y^* = c_1 y_1 + c_2 y_2. \tag{6}$$

■

Our main tool for solving second-order linear homogeneous equations with constant coefficients involved obtaining a characteristic equation by "guessing" that the solution had the form $y = e^{\lambda x}$.

Parallel to the method of Section 2.3, we guess that there is a solution to the system (1) of the form $(\alpha e^{\lambda t}, \beta e^{\lambda t})$, where α, β, and λ are constants yet to be determined. Substituting $x(t) = \alpha e^{\lambda t}$ and $y(t) = \beta e^{\lambda t}$ into equations (1), we obtain

$$x' = \alpha \lambda e^{\lambda t} = a_{11} \alpha e^{\lambda t} + a_{12} \beta e^{\lambda t},$$

$$y' = \beta \lambda e^{\lambda t} = a_{21} \alpha e^{\lambda t} + a_{22} \beta e^{\lambda t}.$$

After dividing by $e^{\lambda t}$, we obtain the linear system

$$(a_{11} - \lambda)\alpha + a_{12}\beta = 0,$$

$$a_{21}\alpha + (a_{22} - \lambda)\beta = 0. \tag{7}$$

We would like to find values for λ such that the system of equations (7) has a solution (α, β) where α and β are not both zero. According to the theory of determinants (see Section 7.9), such a solution will occur whenever the determinant of the system

$$D = \begin{vmatrix} a_{11} - \lambda & a_{12} \\ a_{21} & a_{22} - \lambda \end{vmatrix} = (a_{11} - \lambda)(a_{22} - \lambda) - a_{21}a_{12} \tag{8}$$

equals zero. Solving the equation $D = 0$, we obtain the quadratic equation

$$\lambda^2 - (a_{11} + a_{22})\lambda + (a_{11}a_{22} - a_{21}a_{12}) = 0. \tag{9}$$

DEFINITION

We define equation (9) as the **characteristic equation** of the system (1).

That we are using the same term again is no accident, as we shall now demonstrate. Suppose we differentiate the first equation in the system (1) and eliminate the function $y(t)$:

$$x'' = a_{11}x' + a_{12}(\overbrace{a_{21}x + a_{22}y}^{y'}).$$

Then

$$a_{12}y = x' - a_{11}x$$

$$x'' - a_{11}x' - a_{12}a_{21}x = a_{22}a_{12}y = a_{22}(\overbrace{x' - a_{11}x}),$$

and gathering like terms, we obtain the homogeneous equation

$$x'' - (a_{11} + a_{22})x' + (a_{11}a_{22} - a_{12}a_{21})x = 0. \tag{10}$$

The characteristic equation for equation (10) is exactly the same as equation (9). Hence the algebraic steps needed to obtain equation (10) can be avoided by setting the determinant $D = 0$.

As in Section 2.3, there are three cases to consider, depending on whether the two roots λ_1 and λ_2 of the characteristic equation are real and distinct, real and equal, or complex conjugates. We shall deal with the three cases separately. Rather than discuss the theory, we shall demonstrate each case with an example.

CASE 1: DISTINCT REAL ROOTS

EXAMPLE 2

Consider the system

$$x' = -x + 6y,$$
$$y' = x - 2y. \tag{11}$$

Here $a_{11} = -1$, $a_{12} = 6$, $a_{21} = 1$, $a_{22} = -2$, and equation (8) becomes

$$D = \begin{vmatrix} -1 - \lambda & 6 \\ 1 & -2 - \lambda \end{vmatrix} = (\lambda + 2)(\lambda + 1) - 6 = \lambda^2 + 3\lambda - 4 = 0,$$

which has the roots $\lambda_1 = -4$, $\lambda_2 = 1$.

At this point we can write down the general solution of *either* dependent variable:

$$x(t) = c_1 e^{-4t} + c_2 e^t, \tag{12}$$

or

$$y(t) = k_1 e^{-4t} + k_2 e^t. \tag{13}$$

It does not matter which expression, (12) or (13), we choose. However, once we pick one, we must determine the other dependent variable in terms of the first. For example, if we choose (12), then using the first equation in (11), we have

$$y = \tfrac{1}{6}(x' + x) = \tfrac{1}{6}(-4c_1 e^{-4t} + c_2 e^t + c_1 e^{-4t} + c_2 e^t)$$

$$= \tfrac{1}{6}(-3c_1 e^{-4t} + 2c_2 e^t) = -\tfrac{1}{2}c_1 e^{-4t} + \tfrac{1}{3}c_2 e^t.$$

Thus the general solution is

$$x(t) = c_1 e^{-4t} + c_2 e^t,$$

$$y(t) = -\tfrac{1}{2}c_1 e^{-4t} + \tfrac{1}{3}c_2 e^t,$$

which can be written as

$$(x(t), y(t)) = (c_1 e^{-4t} + c_2 e^t, -\tfrac{1}{2} c_1 e^{-4t} + \tfrac{1}{3} c_2 e^t).$$

CASE 2: ONE REAL ROOT

EXAMPLE 3

Consider the system

$$x' = -4x - y,$$

$$y' = x - 2y. \tag{14}$$

Equation (8) is

$$D = \begin{vmatrix} -4 - \lambda & -1 \\ 1 & -2 - \lambda \end{vmatrix} = (\lambda + 4)(\lambda + 2) + 1 = \lambda^2 + 6\lambda + 9 = 0,$$

which has the double root $\lambda_1 = \lambda_2 = -3$. Hence the general solution to (14) for x is given by

$$x(t) = c_1 e^{-3t} + c_2 t e^{-3t} = (c_1 + c_2 t) e^{-3t}.$$

From the first equation in (14),

$$y = -x' - 4x = -e^{-3t}(-3c_1 - 3c_2 t + c_2) - e^{-3t}(4c_1 + 4c_2 t)$$

$$= e^{-3t}(-c_1 - c_2 - c_2 t).$$

Thus the general solution to (14) is

$$(x(t), y(t)) = (e^{-3t}(c_1 + c_2 t), e^{-3t}(-c_1 - c_2 - c_2 t)).$$

CASE 3: COMPLEX CONJUGATE ROOTS

EXAMPLE 4

Consider the system

$$x' = 4x + y,$$

$$y' = -8x + 8y, \tag{15}$$

with

$$D = \begin{vmatrix} 4 - \lambda & 1 \\ -8 & 8 - \lambda \end{vmatrix} = \lambda^2 - 12\lambda + 40 = 0.$$

The roots of the characteristic equation are $\lambda_1 = 6 + 2i$ and $\lambda_2 = 6 - 2i$. Then the general solution to (15) for y is

$$y(t) = e^{6t}(c_1 \cos 2t + c_2 \sin 2t).$$

Using the second equation of (15), we have

$$8x = 8y - y' \quad \text{or} \quad x = y - \tfrac{1}{8}y',$$

so

$$x(t) = e^{6t}(c_1 \cos 2t + c_2 \sin 2t)$$

$$- \tfrac{1}{8}[6e^{6t}(c_1 \cos 2t + c_2 \sin 2t) + e^{6t}(-2c_1 \sin 2t + 2c_2 \cos 2t)]$$

$$\underbrace{\qquad\qquad\qquad\qquad\qquad\qquad\qquad\qquad\qquad\qquad\qquad}_{y'}$$

$$= e^{6t}[(c_1 - \tfrac{6}{8}c_1 - \tfrac{2}{8}c_2)\cos 2t + (c_2 - \tfrac{6}{8}c_2 + \tfrac{2}{8}c_1)\sin 2t]$$

$$= e^{6t}[\tfrac{1}{4}(c_1 - c_2)\cos 2t + \tfrac{1}{4}(c_1 + c_2)\sin 2t].$$

Thus the general solution to (15) is

$$(x(t), y(t)) = \left(\frac{e^{6t}}{4}[(c_1 - c_2)\cos 2t + (c_1 + c_2)\sin 2t], \; e^{6t}(c_1 \cos 2t + c_2 \sin 2t) \right).$$

We summarize below the results of Examples 2, 3, and 4.

To find the general solution of the system

$$x' = a_{11}x + a_{12}y,$$

$$y' = a_{21}x + a_{22}y,$$

in the case $a_{12} \neq 0$, first find two numbers λ_1 and λ_2 that satisfy the characteristic equation

$$\begin{vmatrix} a_{11} - \lambda & a_{12} \\ a_{21} & a_{22} - \lambda \end{vmatrix} = 0. \tag{16}$$

Case 1: If $\lambda_1 \neq \lambda_2$ are real, then

$$x(t) = c_1 e^{\lambda_1 t} + c_2 e^{\lambda_2 t}, \tag{17}$$

and $y(t)$ can be obtained from the equation

$$y = \frac{1}{a_{12}}(x' - a_{11}x). \tag{18}$$

Case 2: If $\lambda_1 = \lambda_2$, then

$$x(t) = e^{\lambda_1 t}(c_1 + c_2 t), \tag{19}$$

and $y(t)$ can be obtained from (18).

Case 3: If $\lambda_1 = a + ib$ and $\lambda_2 = a - ib$, then

$$x(t) = e^{at}(c_1 \cos bt + c_2 \sin bt), \tag{20}$$

and again $y(t)$ can be obtained from (18).

Remark 1 If $a_{12} = 0$ and $a_{21} \neq 0$, then $y(t)$ can be written in the form (17), (19), or (20) and $x(t)$ can then be obtained from the equation

$$x = \frac{1}{a_{21}}(y' - a_{22}y).\tag{21}$$

Remark 2 If $a_{12} = a_{21} = 0$, then the system is **uncoupled** and $x(t)$ and $y(t)$ can be obtained immediately:

$$x(t) = c_1 e^{a_{11}t} \quad \text{and} \quad y(t) = c_2 e^{a_{22}t}.$$

We close this section by noting that the method of determinants gives us a quick way to find the general solution to a homogeneous system of two equations in two unknown functions. The system can be modified to more than two equations but the algebra involved is often extremely tedious. It is an unfortunate fact that all methods for solving systems involving more than two unknown functions involve a great amount of algebra. For that reason, we shall, for the most part, confine ourselves to systems of two equations in two unknown functions. One method for obtaining solutions of larger systems is presented in Section 8.6.

Problems 3.2

In Problems 1–7, use the method of determinants to find two linearly independent solutions for each given system.

1. $x' = 4x - 3y,$
   $y' = 5x - 4y$

2. $x' = 7x + 6y,$
   $y' = 2x + 6y$

3. $x' = -x + y,$
   $y' = -5x + 3y$

4. $x' = x + y,$
   $y' = -x + 3y$

5. $x' = -4x - y,$
   $y' = x - 2y$

6. $x' = 4x - 2y,$
   $y' = 5x + 2y$

7. $x' = 4x - 3y,$
   $y' = 8x - 6y$

8. Consider the nonhomogeneous equations

$$x' = a_{11}x + a_{12}y + f_1,$$
$$y' = a_{21}x + a_{22}y + f_2.\tag{22}$$

Let (x_1, y_1) and (x_2, y_2) be two linearly independent solution pairs of the homogeneous system (1). Show that

$$x_p(t) = v_1(t)x_1(t) + v_2(t)x_2(t),$$

$$y_p(t) = v_1(t)y_1(t) + v_2(t)y_2(t),$$

is a particular solution of the system (22) if v_1 and v_2 satisfy the equations

$$v_1' x_1 + v_2' x_2 = f_1,$$
$$v_1' y_1 + v_2' y_2 = f_2.$$

This process for finding a particular solution of the nonhomogeneous system (22) is called the **variation of parameters method for systems.** Note the close parallel between this method and the method given in Section 2.5.

In Problems 9–13, use the variation of parameters method to find a particular solution for each given nonhomogeneous system.

9. $x' = 2x + y + 3e^{2t},$
   $y' = -4x + 2y + te^{2t}$

10. $x' = 3x + 3y + t,$
   $y' = -x - y + 1$

11. $x' = -2x + y,$
   $y' = -3x + 2y + 2\sin t$

12. $x' = -x + y + \cos t,$
   $y' = -5x + 3y$

13. $x' = 3x - 2y + t,$
   $y' = 2x - 2y + 3e^t$

* 14. Prove that two solutions of the homogeneous system (1) are linearly independent on $\mathbb{R}$ if and only if $W \neq 0$.

3.3 Electric Circuits with Several Loops

Kirchhoff's Laws

We shall make use of the concepts developed in Sections 1.6 and 2.7 to study electrical networks with two or more coupled closed circuits. The two fundamental principles governing such networks are the two laws of Kirchhoff:

> **i.** The algebraic sum of all voltage drops around any closed circuit is zero.
>
> **ii.** The algebraic sum of the currents flowing into any junction in the network is zero.

EXAMPLE 1

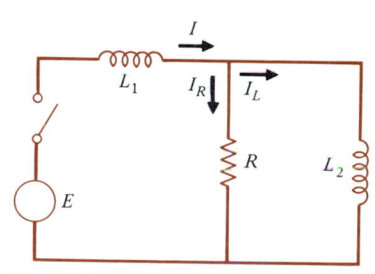

FIGURE 1

Consider the two-loop electric circuit in Fig. 1. Suppose we wish to find the current in each loop as a function of time, given that all currents are zero when the switch is closed at $t = 0$. Let I be the current flowing through the inductor L_1 and let I_L be the current flowing through the inductor L_2. By Kirchhoff's current law (ii), $I = I_R + I_L$, so that when we apply Kirchhoff's voltage law (i) to each loop, we obtain the system

$$L_1 \frac{dI}{dt} + RI_R = E, \tag{1}$$

$$L_2 \frac{dI_L}{dt} - RI_R = 0. \tag{2}$$

Replacing I by $I_R + I_L$ in (1), we have the system of simultaneous linear differential equations

$$L_1 \frac{dI_R}{dt} + L_1 \frac{dI_L}{dt} + RI_R = E, \tag{3}$$

$$L_2 \frac{dI_L}{dt} - RI_R = 0. \tag{4}$$

If we multiply (4) by L_1/L_2 we obtain

$$L_1 \frac{dI_L}{dt} = \frac{L_1 R}{L_2} I_R,$$

which we can substitute in (3) to eliminate the I_L variable, yielding

$$L_1 \frac{dI_R}{dt} + \left(\frac{L_1}{L_2} + 1\right) RI_R = E. \tag{5}$$

If L_1, L_2, and R are constant, we can solve the linear first-order differential equation by the method in Section 1.4. Assume that $L_1 = 1$ H, $L_2 = \frac{1}{2}$ H, $R = 20 \ \Omega$, and $E = 50$ V. Then

$$\frac{dI_R}{dt} + 60 I_R = 50.$$

Multiplying both sides by the integrating factor e^{60t}, we have

$$\frac{d}{dt}(e^{60t}I_R) = 50e^{60t},$$

and an integration yields

$$e^{60t}I_R = \tfrac{5}{6}e^{60t} + k_1.$$

Since $I_R(0) = 0$, it follows that $k_1 = -\tfrac{5}{6}$ and

$$I_R(t) = \tfrac{5}{6}(1 - e^{-60t}). \tag{6}$$

We can find I_L by substituting (6) into (4) to get

$$\frac{1}{2}\frac{dI_L}{dt} = \frac{100}{6}(1 - e^{-60t})$$

from which it follows that

$$I_L(t) = \frac{100}{3}\left(t + \frac{e^{-60t}}{60}\right) + k_2.$$

Setting $t = 0$, so that $I_L(0) = 0$, we obtain $k_2 = -\tfrac{5}{9}$ and

$$I_L(t) = \frac{100}{3}t + \frac{5}{9}(e^{-60t} - 1). \tag{7}$$

EXAMPLE 2

Consider the circuit in Fig. 2. There are two loops. By Kirchhoff's voltage law, we obtain

$$L\frac{dI_L}{dt} + RI_R = E, \tag{8}$$

$$\frac{Q_C}{C} - RI_R = 0. \tag{9}$$

Since $I = dQ/dt$, the second equation may be differentiated to obtain

$$\frac{I_C}{C} - R\frac{dI_R}{dt} = 0. \tag{10}$$

By Kirchhoff's current law, we have

$$I_L = I_C + I_R,$$

which, if substituted into equation (10), yields, together with equation (8), the nonhomogeneous system of linear first-order differential equations

$$\frac{dI_L}{dt} = -\frac{R}{L}I_R + \frac{E}{L},$$

$$\frac{dI_R}{dt} = \frac{I_L}{RC} - \frac{I_R}{RC}. \tag{11}$$

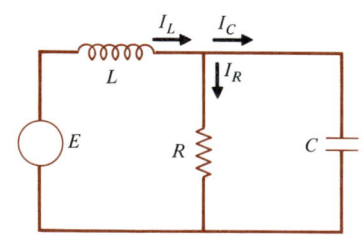

FIGURE 2

The characteristic equation of the associated homogeneous system is

$$D = \begin{vmatrix} -\lambda & -R/L \\ 1/RC & -\lambda - 1/RC \end{vmatrix} = \lambda(\lambda + 1/RC) + 1/LC$$

$$= \lambda^2 + \lambda/RC + 1/LC = 0. \tag{12}$$

The roots of equation (12) are $(-L \pm \sqrt{L^2 - 4R^2 LC})/2RLC$.

The value of the discriminant

$$L^2 - 4R^2 LC = L(L - 4R^2 C)$$

is now important. For simplicity assume that $R = 100$ ohms, $C = 1.5 \times 10^{-4}$ farads, $E = 100$ volts, and $L = 8$ henrys. Then

$$L - 4R^2 C = 8 - 4(100)^2 1.5(10^{-4}) = 2$$

so that the roots of (12) are $\lambda_1 = -50$ and $\lambda_2 = -\frac{50}{3}$.

Consider the homogeneous system

$$\frac{dI_L}{dt} = -\frac{R}{L} I_R, \tag{13}$$

$$\frac{dI_R}{dt} = \frac{I_L}{RC} - \frac{I_R}{RC}.$$

From equation (3.2.17) on p. 141, the general solution to (13) is given by

$$I_L(t) = c_1 e^{-50t} + c_2 e^{-(50/3)t}$$

$$I_R(t) = -\frac{L}{R} \frac{dI_L}{dt} = -\frac{8}{100} \left(-50 c_1 e^{-50t} - \frac{50}{3} c_2 e^{-(50/3)t} \right)$$

$$= 4 c_1 e^{-50t} + \frac{4}{3} c_2 e^{-(50/3)t}. \tag{14}$$

Since $E = 100$ volts is constant, when we use the method of undetermined coefficients, we assume a particular solution of the form

$$(I_L, I_R)_p = (A, B). \tag{15}$$

Substituting (15) into the nonhomogeneous system (11), we get

$$0 = -\frac{100}{8} B + \frac{100}{8},$$

$$0 = \frac{A - B}{(100)1.5 \times 10^{-4}},$$

so that $B = 1 = A$. Finally, we find the constants c_1 and c_2 by using the initial conditions $I_L(0) = 0 = I_R(0)$ in the equations

$$I_L = c_1 e^{-50t} + c_2 e^{-(50/3)t} + 1,$$

$$I_R = 4 c_1 e^{-50t} + \tfrac{4}{3} c_2 e^{-(50/3)t} + 1.$$

We obtain

$$I_L(0) = c_1 + c_2 + 1 = 0,$$

$$I_R(0) = 4 c_1 + \tfrac{4}{3} c_2 + 1 = 0,$$

with solution $c_1 = \frac{1}{8}$ and $c_2 = -\frac{9}{8}$. Thus the unique solution to the initial value problem is

$$(I_L, I_R) = (\tfrac{1}{8}e^{-50t} - \tfrac{9}{8}e^{-(50/3)t} + 1, \tfrac{1}{2}e^{-50t} - \tfrac{3}{2}e^{-(50/3)t} + 1).$$

EXAMPLE 3

If $L = 6$ henrys in Example 2 and all other facts remain the same, then $L - 4R^2C = 0$ and (12) has the double root $\lambda_1 = \lambda_2 = -1/2RC = -100/3$. Hence from equation (3.2.19) on p. 141, the general solution to the homogeneous system (13) is

$$I_L(t) = (c_1 + c_2 t)e^{-(100/3)t}$$

and

$$I_R(t) = -\frac{L}{R}\frac{dI_L}{dt} = -\frac{6}{100}e^{-(100/3)t}\left[\frac{-100}{3}(c_1 + c_2 t) + c_2\right]$$

$$= [(2c_1 - \tfrac{3}{50}c_2) + 2c_2 t]e^{-(100/3)t}.$$

As in Example 2, we find the particular solutions to (11) to be

$$(I_L, I_R) = (1, 1).$$

Hence the general solution to (11) is

$$I_L(t) = (c_1 + c_2 t)e^{-(100/3)t} + 1,$$

$$I_R(t) = [(2c_1 - \tfrac{3}{50}c_2) + 2c_2 t]e^{-(100/3)t} + 1.$$

Using the initial conditions, we have

$$I_L(0) = c_1 + 1 = 0,$$

$$I_R(0) = 2c_1 - \tfrac{3}{50}c_2 + 1 = 0,$$

or

$$c_1 = -1 \quad \text{and} \quad c_2 = -\tfrac{50}{3}.$$

Thus the unique solution to the initial value problem is

$$(I_L(t), I_R(t)) = ((-1 - \tfrac{50}{3}t)e^{-(100/3)t} + 1, (-1 - \tfrac{100}{3}t)e^{-(100/3)t} + 1).$$

EXAMPLE 4

Suppose, in Example 2, that $L = 3$ henrys and all other facts remain unchanged. Then $L - 4R^2C = -3$, so the characteristic equation (12) has the roots $\frac{100}{3}(-1 \pm i)$. Then, from equation (3.2.20) on p. 141, solutions to the homogeneous system (13) are

$$I_L(t) = e^{-(100/3)t}(c_1 \cos \tfrac{100}{3}t + c_2 \sin \tfrac{100}{3}t)$$

and

$$I_R(t) = \frac{L}{R}\left(-\frac{dI_L}{dt}\right) = \frac{3}{100}\left(-\frac{dI_L}{dt}\right)$$

$$= \tfrac{3}{100}e^{-(100/3)t}[\tfrac{100}{3}(c_1 \cos \tfrac{100}{3}t + c_2 \sin \tfrac{100}{3}t + c_1 \sin \tfrac{100}{3}t - c_2 \cos \tfrac{100}{3}t)]$$

$$= e^{-(100/3)t}[(c_1 - c_2) \cos \tfrac{100}{3}t + (c_1 + c_2) \sin \tfrac{100}{3}t].$$

The nonhomogeneous term in system (11) is E/L, a constant. Thus it is reasonable to find a particular solution to (11) of the form $I_L = A$ and $I_R = B$, where A and B are constants. Inserting these values into (11) we obtain, as in Example 2,

$$I_L = I_R = 1.$$

The general solution to (11) is, therefore,

$$I_L(t) = e^{-(100/3)t}(c_1 \cos \tfrac{100}{3}t + c_2 \sin \tfrac{100}{3}t) + 1$$

$$I_R(t) = e^{-(100/3)t}[(c_1 - c_2) \cos \tfrac{100}{3}t + (c_1 + c_2) \sin \tfrac{100}{3}t] + 1.$$

Finally, setting $I_R(0) = I_L(0) = 0$, we obtain

$$c_1 + 1 = 0 \quad \text{and} \quad c_1 - c_2 + 1 = 0$$

with solution $c_1 = -1$, $c_2 = 0$.

Thus the unique solution to our initial value problem is

$$I_L(t) = 1 - e^{-(100/3)t} \cos \tfrac{100}{3}t$$

$$I_R(t) = 1 - e^{-(100/3)t}(\cos \tfrac{100}{3}t + \sin \tfrac{100}{3}t).$$

Problems 3.3

1. Let $R = 100$ ohms, $L = 4$ henrys, $C = 10^{-4}$ farads and $E = 100$ volts in the network of Figure 1. Suppose the currents I_R and I_L are both zero at time $t = 0$. Find the currents when $t = 0.001$ second.

2. Let $L = 1$ henry in Problem 1 and suppose all the other facts are unchanged. Find the currents when $t = 0.001$ second and 0.1 second.

3. Let $L = 8$ henrys in Problem 1 and suppose all the other facts are unchanged. Find the currents when $t = 0.001$ second and 0.1 second.

4. Suppose $E = 100e^{-1000t}$ volts and all the other values are unchanged in Problem 1. Do:

   (a) Problem 1.

   (b) Problem 2.

   (c) Problem 3.

5. Repeat Problem 4 for $E = 100 \sin 60 \pi t$ volts.

6. Find the current at time t in each loop of the network in Fig. 3 given that $E = 100$ volts, $R = 10$ ohms, and $L = 10$ henrys.

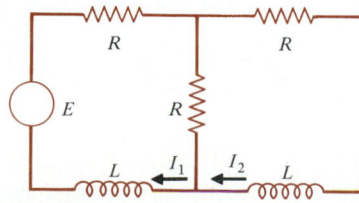

FIGURE 3

7. Repeat Problem 6 for $E = 10 \sin t$ volts.

8. Consider the air-core transformer network shown in Fig. 4 with $E = 10 \cos t$ volts, $R = 1$ ohm, $L = 2$ henrys and mutual inductance $L_* = -1$ henry (which depends on the relative modes of winding of the two coils involved). Treating the mutual inductance as an inductance for each circuit, find the two circuit currents at all times t assuming they are zero at $t = 0$.

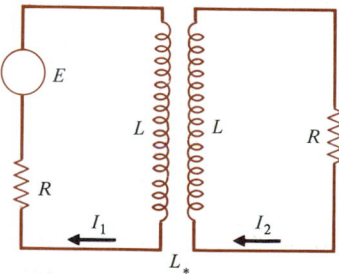

FIGURE 4

9. Consider the circuit shown in Fig. 5. Find the current in the top loop if all the resistances are $R = 100$ ohms and the inductances are $L = 1$ henry. Assume $E = 100$ volts and that there is no current flowing when the switch is closed at time $t = 0$.

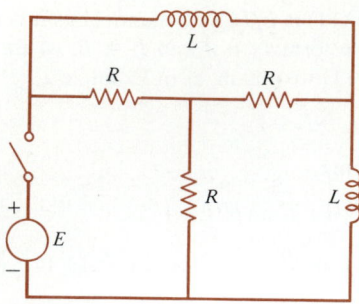

FIGURE 5

10. Repeat Problem 9 for the circuit in Fig. 6, where $C = 10^{-4}$ farads.

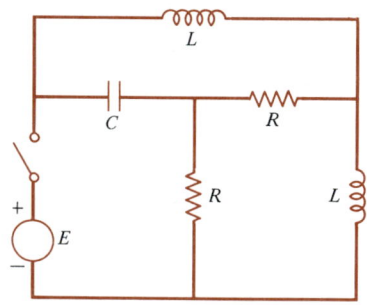

FIGURE 6

* **11.** Consider the **string insulator** shown in Fig. 7, and suppose that the voltage at $E_0 = A \cos \omega t$. Find the voltage at the kth junction E_k for each k. [*Hint:* Express E_k as a function of the voltages at previous junctions.]

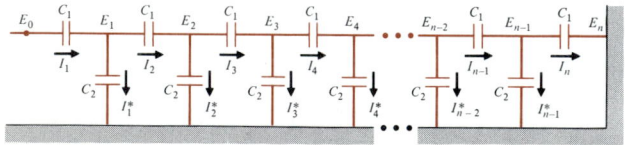

FIGURE 7

* **12.** Consider the **low-pass filter** in Fig. 8, where we *assume* that the voltage in the line is alternating and is given by $E_k = A_k \cos \omega t$, for fixed ω. Show that the only nontrivial solutions satisfy

$$\omega_N = \frac{2}{\sqrt{CL}} \sin \frac{N\pi}{2(n+1)}, \qquad N = 1, 2, \ldots, n.$$

(Thus waves of frequency higher than $\omega_n < 2/\sqrt{CL}$ are damped out as t increases.)

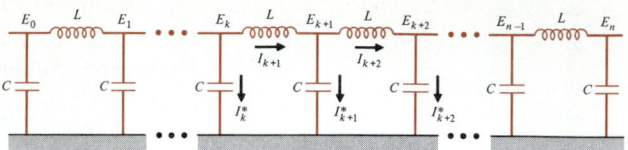

FIGURE 8

* **13.** Consider the network shown in Fig. 9, called a **high-pass filter.** Show that it damps out all waves below a certain **cutoff** frequency.

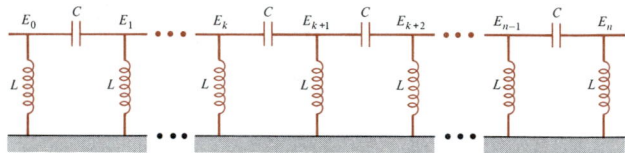

FIGURE 9

* **14.** The network shown in Fig. 10 is called a **band-pass filter,** since frequencies outside a certain band will be damped out. Find the two cutoff frequencies.

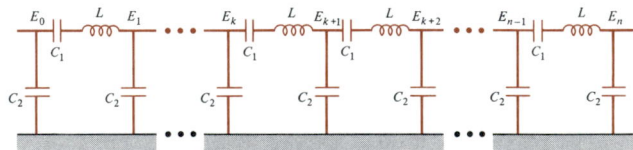

FIGURE 10

15. R. FitzHugh† proposed the electric circuit in Fig. 11 as a model for the transmission of current in a myelinated axon. Find the voltage at all junctions x_k in the network. (Assume $x_0 = A_0 \cos \omega t$.)

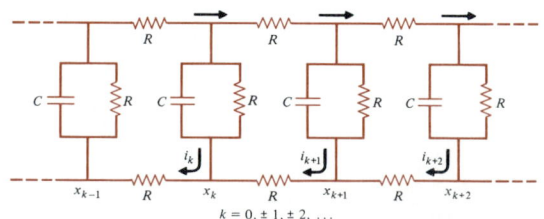

FIGURE 11

† "Computation of impulse initiation and saltatory conduction in myelinated fibres: theoretical basis of the velocity-diameter relation," *Biophysical Journal,* vol. 2 (1962), pp. 11–21.

3.4 Chemical Mixture and Population Biology Problems

In Section 3.1 we presented an example of a chemical mixture problem involving feedback. The exercise set of that section included similar situations, as well as applications of systems to medicine and air pollution. In this section we will present further examples related to these topics and to population biology. We solve the first example by the method of determinants.

EXAMPLE 1

Let tank X contain 10 gallons of brine in which 10 pounds of salt is dissolved and tank Y contain 20 gallons of water. Suppose water flows into tank X at the rate of 2 gallons per minute, and the mixture flows from tank X into tank Y at 4 gallons per minute. From Y 2 gallons are pumped back each minute to X (establishing feedback) while 2 gallons are flushed away. (a) Find the amount of salt in both tanks at all time t. (See Fig. 1.) (b) At what time is the amount of salt in tank Y a maximum?

SOLUTION

(a) If we let $x(t)$ and $y(t)$ represent the pounds of salt in tanks X and Y at time t and note that the change in weight equals the difference between input and output, we can again derive a system of linear first-order equations. Tanks X and Y initially contain $x(0) = 10$ and $y(0) = 0$ pounds of salt, respectively, at time $t = 0$. The quantities $x/10$ and $y/20$ are, respectively, the amounts of salt contained in each gallon of water taken from tanks X and Y at time t. Thus we have the system

$$\frac{dx}{dt} = -4\frac{x}{10} + 2\frac{y}{20}, \qquad x(0) = 10,$$

$$\frac{dy}{dt} = 4\frac{x}{10} - 4\frac{y}{20}, \qquad y(0) = 0. \tag{1}$$

From (1) we obtain the determinant

$$D = \begin{vmatrix} -\frac{2}{5} - \lambda & \frac{1}{10} \\ \frac{2}{5} & -\frac{1}{5} - \lambda \end{vmatrix} = (\lambda + \tfrac{1}{5})(\lambda + \tfrac{2}{5}) - \tfrac{1}{25} = 0,$$

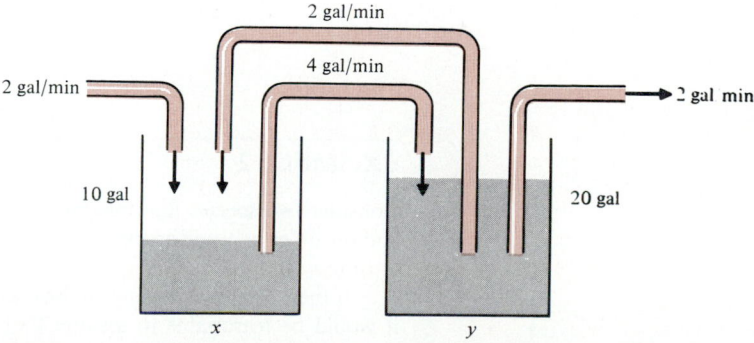

FIGURE 1

or

$$\lambda^2 + \tfrac{3}{5}\lambda + \tfrac{1}{25} = 0. \tag{2}$$

The characteristic equation (2) has the roots

$$\lambda_1 = \frac{-3 + \sqrt{5}}{10} \quad \text{and} \quad \lambda_2 = \frac{-3 - \sqrt{5}}{10}.$$

From equation (3.2.17) on p. 141,

$$x(t) = c_1 e^{[(-3+\sqrt{5})/10]t} + c_2 e^{[(-3-\sqrt{5})/10]t}.$$

From the first equation of (1),

$$y(t) = 10\frac{dx}{dt} + 4x$$

after some algebra
$$\downarrow$$
$$= (1 + \sqrt{5})c_1 e^{[(-3+\sqrt{5})/10]t} + (1 - \sqrt{5})c_2 e^{[(-3-\sqrt{5})/10]t}.$$

Using the initial conditions, we obtain

$$x(0) = c_1 + c_2 = 10,$$

$$y(0) = (1 + \sqrt{5})c_1 + (1 - \sqrt{5})c_2 = 0.$$

Then $c_1 = 5 - \sqrt{5}$, $c_2 = 5 + \sqrt{5}$, and the unique solution to the initial value problem is

$$(x(t), y(t)) = ((5 - \sqrt{5})e^{[(-3+\sqrt{5})/10]t} + (5 + \sqrt{5})e^{[(-3-\sqrt{5})/10]t},$$

$$4\sqrt{5}e^{[(-3+\sqrt{5})/10]t} - 4\sqrt{5}e^{[(-3-\sqrt{5})/10]t})$$

(b) $y(t)$ is a maximum when $y'(t) = 0$. But

$$y'(t) = 4\sqrt{5}\left[\frac{-3 + \sqrt{5}}{10}e^{[(-3+\sqrt{5})/10]t} + \frac{3 + \sqrt{5}}{10}e^{[(-3-\sqrt{5})/10]t}\right] = 0.$$

Multiplying through by $(10/4\sqrt{5})e^{(3/10)t}$ and simplifying, we obtain

$$(-3 + \sqrt{5})e^{(\sqrt{5}/10)t} = (-3 - \sqrt{5})e^{-(\sqrt{5}/10)t}$$

so that

$$e^{(\sqrt{5}/5)t} = \frac{-3 - \sqrt{5}}{-3 + \sqrt{5}} = \frac{7 + 3\sqrt{5}}{2}.$$

Hence

$$t = \sqrt{5} \ln\left(\frac{7 + 3\sqrt{5}}{2}\right) \approx 4.3 \text{ minutes.}$$

EXAMPLE 2

Consider two species that inhabit the same ecosystem. Denote one of the species by x and the other by y and assume that $x(t)$ and $y(t)$ are the sizes of the respective populations at time t.

  If the two populations were isolated from each other and resources were plentiful, it would be reasonable to assume that the growth rates of the populations would be proportional to their numbers:

$$\frac{dx}{dt} = ax \quad \text{and} \quad \frac{dy}{dt} = by,$$

where a and b are positive constants (the average growth rate per individual of each population).

However, if the two populations *compete* for the same resource (say they both consume the same forage), then the growth rate of each population is influenced by the size of the other population. In such a case we might arrive at the linear **competition system**

$$\frac{dx}{dt} = ax - Ay,$$

$$\frac{dy}{dt} = by - Bx. \tag{3}$$

The characteristic equation for the system (3) is

$$\lambda^2 - (a+b)\lambda + (ab - AB) = 0. \tag{4}$$

If (4) has the roots $\lambda_1 \neq \lambda_2$, then

$$x(t) = c_1 e^{\lambda_1 t} + c_2 e^{\lambda_2 t}$$

and

$$y(t) = \frac{1}{A}\left(ax - \frac{dx}{dt}\right) = \frac{1}{A}[(a - \lambda_1)c_1 e^{\lambda_1 t} + (a - \lambda_2)c_2 e^{\lambda_2 t}].$$

The constants c_1 and c_2 can be determined from the initial populations $x(0)$ and $y(0)$.

EXAMPLE 3

Many biological systems are controlled by the production of enzymes or hormones that stimulate or inhibit the secretion of some compound. For example, the pancreatic hormone glucagon stimulates the release of glucose from the liver to the plasma. A rise in blood glucose inhibits the secretion of glucagon but causes an increase in the production of the hormone insulin. Insulin, in turn, aids in the removal of glucose from the blood and in its conversion to glycogen in the muscle tissue. Let G and I be the deviations of plasma glucose and plasma insulin from the normal (fasting) level, respectively. We then have the system

$$\frac{dG}{dt} = -k_{11}G - k_{12}I,$$

$$\frac{dI}{dt} = k_{21}G - k_{22}I, \tag{5}$$

where the positive constants k_{ij} are model parameters, some of which may be determined experimentally. It is known that the system (5) exhibits a strongly damped oscillatory behavior, since direct injection of glucose into the blood will produce a fall of blood glucose to a level below fasting in about $1\frac{1}{2}$ hours followed by a rise slightly above the fasting level in about 3 hours. Hence, the characteristic equation of the system (5),

$$D = \begin{vmatrix} -k_{11} - \lambda & -k_{12} \\ k_{21} & -k_{22} - \lambda \end{vmatrix} = (k_{11} + \lambda)(k_{22} + \lambda) + k_{12}k_{21}$$

$$= \lambda^2 + (k_{11} + k_{22})\lambda + (k_{11}k_{22} + k_{12}k_{21}) = 0,$$

must have complex conjugate roots $-a \pm ib$, with $a = (k_{11} + k_{22})/2$ and $b = \sqrt{k_{12}k_{21} - (k_{11} - k_{22})^2/4}$, since only complex roots can lead to oscillatory behavior.

Thus, our solution has the form

$$(G(t), I(t)) = (e^{-at}(c_1 \cos bt + c_2 \sin bt), e^{-at}(d_1 \cos bt + d_2 \sin bt)),$$

where c_1, c_2, d_1, and d_2 are determined by the initial conditions and one of the equations in (5); and $b = 2\pi/3$ if we measure time in hours (since the period is 3 hours).

Assume now that the glucose injection was administered at a time when plasma insulin and glucose were at fasting levels and that the glucose was diffused completely in the blood before the insulin level began to increase ($t = 0$). Then $G(0) = G_0$ equals the ratio of the volume of glucose administered to blood volume, and $I(0) = 0$. Since $G(t)$ is at a maximum when $t = 0$, it follows that $c_1 = G_0$ and $c_2 = 0$. Hence

$$G(t) = G_0 e^{-at} \cos bt. \tag{6}$$

But

$$d_1 = I(0) = 0 \tag{7}$$

and from the first equation in (5)

$$\frac{dG}{dt} = G_0 e^{-at}(-a \cos bt - b \sin bt)$$

$$= -k_{11}G_0 e^{-at} \cos bt - k_{12}e^{-at}(d_2 \sin bt)$$

or, equating cosine and sine terms,

$$aG_0 = k_{11}G_0 \quad \text{and} \quad bG_0 = k_{12}d_2. \tag{8}$$

Thus

$$I(t) = \frac{bG_0}{k_{12}} e^{-at} \sin bt$$

and $k_{11} = a = (k_{11} + k_{22})/2$, so that $k_{11} = k_{22}$.

If the minimum level $G(\frac{3}{2})$ (< 0) is known, then by equation (6), $e^{-3a/2} = |G(\frac{3}{2})|/G_0$, so that

$$k_{11} = a = \frac{-2}{3} \ln \frac{|G(\frac{3}{2})|}{G_0}.$$

If we determine the plasma insulin at any given time $t_0 > 0$, we can then evaluate the parameters k_{12} and k_{21}.

Problems 3.4

1. Suppose in Example 1 that only $1\frac{1}{2}$ gallons per minute are pumped back from tank Y to tank X, while $2\frac{1}{2}$ gallons per minute are flushed away. If $2\frac{1}{2}$ gallons of water per minute flow into X, when does tank Y contain the maximum amount of salt? How much salt does it contain?

For simplicity, we assume that $k_1 = \frac{9}{2}m_1$, $k_2 = 2m_2$, and $7m_1 = 4m_2$. Then (3) becomes

see Section 7.9

$$0 = \begin{vmatrix} -\lambda & 0 & 1 & 0 \\ 0 & -\lambda & 0 & 1 \\ -8 & \frac{7}{2} & -\lambda & 0 \\ 2 & -2 & 0 & -\lambda \end{vmatrix} = \lambda^4 + 10\lambda^2 + 9 = (\lambda^2 + 9)(\lambda^2 + 1). \tag{4}$$

The four roots are $\lambda_1 = 3i$, $\lambda_2 = -3i$, $\lambda_3 = i$, $\lambda_4 = -i$. Hence, from the theory in Section 2.11 we know that there are constants a_1, a_2, a_3, a_4, b_1, b_2, b_3, and b_4 such that

$$x_1 = a_1 \cos 3t + a_2 \sin 3t + a_3 \cos t + a_4 \sin t,$$

$$x_2 = b_1 \cos 3t + b_2 \sin 3t + b_3 \cos t + b_4 \sin t. \tag{5}$$

Substituting (5) into (1), we get

$$a_1 = -\tfrac{7}{2}b_1, \qquad a_2 = -\tfrac{7}{2}b_2, \qquad a_3 = \tfrac{1}{2}b_3, \qquad a_4 = \tfrac{1}{2}b_4. \tag{6}$$

If initial conditions are given

$$x_1(0) = x_{10}, \qquad x_1'(0) = x_3(0) = x_{30},$$

$$x_2(0) = x_{20}, \qquad x_2'(0) = x_4(0) = x_{40},$$

we obtain the system of equations

$$a_1 \quad + a_3 \quad\quad = x_{10}$$

$$3a_2 \quad + a_4 = x_{30}$$

$$b_1 \quad + b_3 \quad\quad = x_{20}$$

$$3b_2 \quad + b_4 = x_{40}.$$

Together with (6), this system determines all the coefficients in (5):

$$b_1 = \frac{x_{20} - 2x_{10}}{8}, \qquad b_2 = \frac{x_{40} - 2x_{30}}{24},$$

$$b_3 = \frac{7x_{20} + 2x_{10}}{8}, \qquad b_4 = \frac{7x_{40} + 2x_{30}}{8}.$$

Problems 3.5

1. Show that the oscillations exhibited by the masses m_1 and m_2 in Example 1 can be represented by the superposition of *two* cosines.

2. Solve Example 1 with $k_1 = \frac{5}{2}m_1$, $k_2 = 2m_2$, and $3m_1 = 4m_2$.

3. Solve Example 1 with $k_1 = 3m_1/2$, $k_2 = 2m_2$, and $m_1 = 4m_2$.

4. When a water wave travels past any point, the water rises as the wave approaches the point and recedes as it moves past the point. If a ship is in the trough of a wave, its angular acceleration from the vertical is given by

$$\frac{d^2\psi}{dt^2} = -b^2\psi + A \sin \omega t. \tag{7}$$

(a) Rewrite (7) as a system of first-order differential equations.

(b) Solve the resulting system by the method of determinants.

* **5.** A straight slender shaft may be dynamically unstable when rotating at high speeds. When the period of rotation ω is nearly equal to one of the shaft's periods of lateral vibration, the shaft is said to **whirl.** A whirling shaft satisfies the differential equation

$$EI\frac{d^4y}{dx^4} = m\omega^2 y, \qquad (8)$$

where $y = y(x)$ is the distance of the shaft from its geometric axis, E is Young's modulus of elasticity, I is the moment of inertia of the shaft, and m is the mass. Assume that the shaft is hinged at both ends, has length L, and satisfies $y(0) = y(L) = y''(0) = y''(L) = 0$.

(a) Express (8) as a system of first-order differential equations.

(b) Show that nontrivial solutions exist if and only if $\omega = n^2\pi^2\sqrt{EI/m/L^2}$.

(c) Find the rotational speed necessary to produce whirling of a steel shaft $1\frac{1}{2}$ inches in diameter and 8 feet long. [*Hint:* $E = 4.32 \times 10^9$, $I = \pi r^4/4$, $m = 6/32.2$.]

6. Consider the coupled mechanical system shown in Fig. 2,

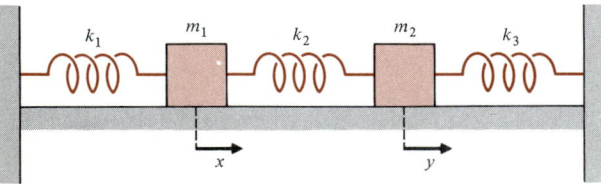

FIGURE 2

where two masses m_1 and m_2 rest on a frictionless plane and are attached to fixed walls by two springs with spring

constants k_1 and k_3. The masses are connected by a spring with spring constant k_2.

(a) Obtain the system of differential equations describing the motion of this mechanical system.

(b) Solve the system in part (a) and show that the motion of each mass is a superposition of two simple harmonic motions.

* **7.** Determine the equations of motion of a double pendulum consisting of two simple pendulums of masses m_1 and m_2 and lengths l_1 and l_2, respectively, shown in Fig. 3. (You may assume that the angular displacements are so small that $\sin \theta \approx \theta$ and $\sin \phi \approx \phi$.)

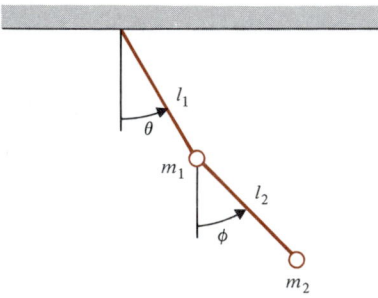

FIGURE 3

* **8.** Using the system in Problem 7, neglecting the terms containing $(\theta')^2$ and $(\phi')^2$, and replacing $\cos(\theta - \phi)$ by 1, show that one obtains the system

$$l_1(m_1 + m_2)\theta'' + m_2 l_2\phi'' + g(m_1 + m_2)\theta = 0$$
$$l_1\theta'' \qquad + l_2\phi'' \quad + g\phi \qquad = 0. \qquad (9)$$

9. Find the general solution of (9).

3.6 The Phase Plane

In the preceding sections we saw that there are large classes of differential equations and systems having solutions defined in some interval. However, if an equation is nonlinear, then there is usually no way to find its solution except by using numerical methods (as in Chapter 13). For this reason, it is necessary to seek methods for describing the nature of a solution without explicitly solving the equation.

First, it is necessary to ask, "what kind of information about a solution is it useful to have?" We indicate a partial answer by considering our old standby: second-order linear equations with constant coefficients.

EXAMPLE 1

Consider the three equations

  (a) $x'' + 3x' + 2x = 0$,
  (b) $x'' - 3x' + 2x = 0$,
  (c) $x'' + x = 0$.

The general solutions to these equations are

  (a) $x(t) = c_1 e^{-t} + c_2 e^{-2t}$,
  (b) $x(t) = c_1 e^{t} + c_2 e^{2t}$,
  (c) $x(t) = c_1 \cos t + c_2 \sin t$.

It is clear that all solutions of (a) approach zero as t tends to ∞, all solutions (except the zero solution) of (b) approach ∞ as t tends to ∞, and all solutions of (c) remain bounded and (except the zero solution) do not approach any constant as t tends to ∞. Furthermore, the solutions of (c) are periodic of period 2π. That is, $x(t + 2\pi) = x(t)$ for every real number t.

The solutions to nonlinear equations, too, may approach zero, become unbounded, or remain bounded as t becomes large. They also may be periodic. Much modern research in the theory of ordinary differential equations is concerned with finding conditions that will ensure that the solution of a nonlinear equation has one of these properties. There is no general method of analyzing all nonlinear equations. In the rest of this chapter we will discuss some of the oldest and most elementary ways of obtaining this information.

One of these methods is to consider the nonlinear equation as a perturbation of some linear equation; that is, attempt to approximate the nonlinear equation by a "related" linear equation. We shall illustrate this method with an example.

EXAMPLE 2

Consider the freely swinging (frictionless) pendulum of length l shown in Fig. 1. We indicated in Problem 2.8.11 (p. 109) that Newton's law of motion yields the nonlinear second-order equation

$$\frac{d^2\theta}{dt^2} + \omega^2 \sin \theta = 0, \tag{1}$$

where $\omega^2 = g/l$. However, since

$$\lim_{\theta \to 0} \frac{\sin \theta}{\theta} = 1,$$

we may approximate equation (1) for small values of θ by the linear equation

$$\frac{d^2\theta}{dt^2} + \omega^2\theta = 0. \tag{2}$$

The general solution of equation (2) is periodic:

$$\theta(t) = c_1 \cos \omega t + c_2 \sin \omega t.$$

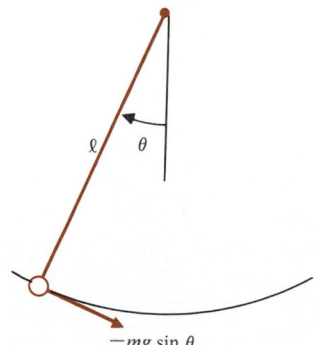

ℓ θ

$-mg \sin \theta$

FIGURE 1

How similar is this solution to the solution of equation (1)? Let $v = d\theta/dt$; then

$$\frac{d^2\theta}{dt^2} = \frac{dv}{dt} = \frac{dv}{d\theta}\frac{d\theta}{dt} = v\frac{dv}{d\theta} \tag{3}$$

so that

$$v\frac{dv}{d\theta} = -\omega^2 \sin \theta. \tag{4}$$

Separating variables and integrating we have

$$\int v\, dv = -\omega^2 \int \sin \theta\, d\theta + c$$

so that

$$v^2 = 2(c + \omega^2 \cos \theta).$$

Assuming that $v = 0$ when $\theta = \alpha$, we get

$$v = \frac{d\theta}{dt} = \omega\sqrt{2(\cos \theta - \cos \alpha)} \tag{5}$$

or

$$\int \frac{d\theta}{\sqrt{\cos \theta - \cos \alpha}} = \sqrt{2}\,\omega \int dt. \tag{6}$$

The indefinite integral of the left side of equation (6) cannot be obtained as an elementary function, so no direct way of comparing the solution of (6) to that of equation (2) is possible. We can, however, graph equation (5) for various values of α (see Fig. 2). We shall explain the implication of Fig. 2 in Section 3.8 (see Example 4 on p. 182).

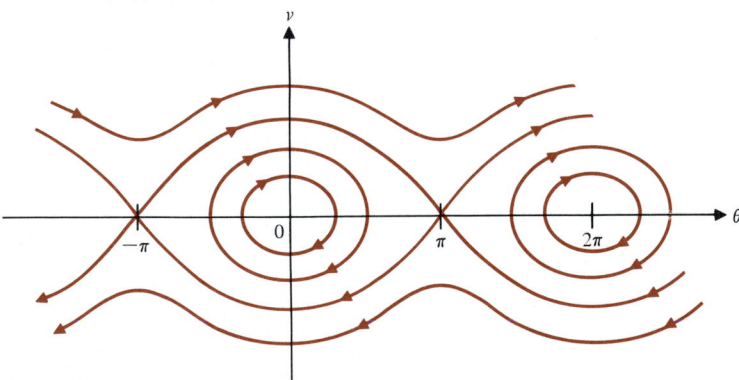

FIGURE 2

EXAMPLE 3: The Lotka-Volterra Equations

Assume that an ecosystem contains a predator species that feeds exclusively on a prey species and that the prey population has an ample food supply at all times. Let $y(t)$ and $x(t)$ denote the populations of the predator and prey species, respectively. Since food is readily available, the birth rate of the prey species is very likely to be a constant independent of time. The death rate, however, will certainly depend on the number of predators.

On the other hand, the birth rate of the predator species will be affected by the uncertain food supply, whereas its death rate may well be constant. We can write the growth rates per individual for the two species as a system

$$\frac{1}{x}\frac{dx}{dt} = a - by,$$

$$\frac{1}{y}\frac{dy}{dt} = \alpha x - \beta,$$

or

$$\frac{dx}{dt} = ax - bxy,$$

$$\frac{dy}{dt} = \alpha xy - \beta y, \tag{7}$$

Lotka-Volterra Equations

where a, b, α, and β are positive constants of proportionality. Equations (7) are called the **Lotka-Volterra equations.**

Although there is no explicit solution for the Lotka Volterra equations (7) (that is, no one has succeeded in finding formulas for the dependent variables x and y in terms of the independent variable t), it is possible to obtain a relation between the two populations x and y. Thus, even if we cannot determine what the populations will be at a specific time, we will know what influence one population has on the other. Hence, we seek an expression for y in terms of x. Using the chain rule of calculus, we have

$$\frac{dy}{dt} = \frac{dy}{dx}\frac{dx}{dt}.$$

Replacing dy/dt and dx/dt by their values in equation (7), we have

$$\alpha xy - \beta y = \frac{dy}{dx} \cdot (ax - bxy)$$

or

$$\frac{dy}{dx} = \frac{\alpha xy - \beta y}{ax - bxy} = \frac{y(\alpha x - \beta)}{x(a - by)}. \tag{8}$$

This equation is separable:

$$\int \frac{a - by}{y}dy = \int \frac{\alpha x - \beta}{x}dx$$

so that after we divide through and integrate,

$$a \ln y - by = \alpha x - \beta \ln x + c, \tag{9}$$

which can be graphed when the parameters a, b, c, α, and β are known. A graph for a given set of parameters is shown in Fig. 3.

There remains the problem of interpreting the graph in Fig. 3. First, the constant c is determined by substituting the initial populations $x(0)$ and $y(0)$ for x and y, respectively, into equation (9). Thus,

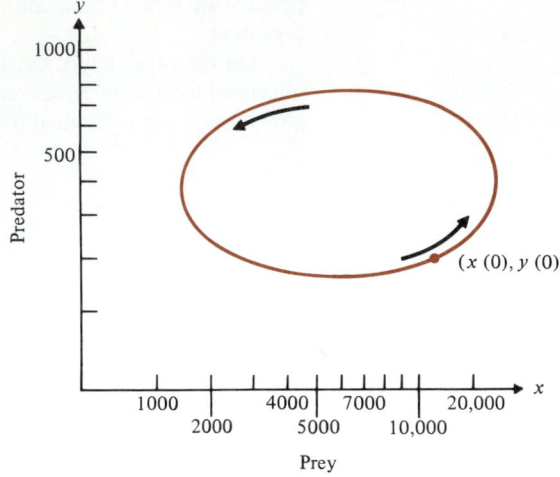

FIGURE 3

$$c = a \ln y(0) - by(0) - \alpha x(0) + \beta \ln x(0),$$

where a and α are the birth rates and b and β are the death rates of the prey and predator populations, respectively. The point $(x(0), y(0))$ is located somewhere on the closed curve shown in Fig. 3. As time increases, the point $(x(t), y(t))$ moves in a counterclockwise direction along the closed curve. This situation illustrates the following cycle:

1. As the number of prey increases, the prey are easier to catch; this leads to an increase in the predator population, until
2. the number of predators is so large that they severely deplete the quantity of prey, causing
3. a large number of deaths by starvation in the predator population, and allowing
4. reestablishment of the growth in the prey population.

The plane curves of Figs. 2 and 3 depict the relation between the dependent variables x and y. The independent variable t does not appear. The plane of the dependent variables is called the **phase plane** for the 2×2 system of differential equations. Each curve in the phase plane, determining a relation between the two dependent variables (θ, v and x, y), is called an **orbit**. A collection of orbits in a phase plane is sometimes called a **phase portrait** of the system. It is often possible to derive a great deal of information about the problem by examining the orbits in a phase portrait.

Phase Plane

Orbit
Phase Portrait

Although nonlinear equations and systems can take many different forms, they can be roughly classified into two different categories. To illustrate this, we consider the system of two first-order equations

$$x' = f(t, x, y),$$
$$y' = g(t, x, y),$$

(10)

where f and g are assumed to be continuously differentiable functions of t, x, and y over some region

$$D : a < t < b, \qquad c < x < d, \qquad e < y < h$$

in three-dimensional space. By a generalization of Theorem 1.3.1 on p. 21, this is enough to guarantee that there is a unique solution (defined over some interval in t) that passes through any initial point (t_0, x_0, y_0) in D.

DEFINITION

Autonomous and Nonautonomous Equations

The system (10) is said to be **autonomous** (time independent) if the functions f and g do not depend on t. Otherwise, equation (10) is said to be **nonautonomous**. Hence equation (10) is autonomous if $f(t, x, y) = f(x, y)$ and $g(t, x, y) = g(x, y)$.

EXAMPLE 4

The system

$$x' = -x^2 + y,$$
$$y' = -x + y^2,$$

is autonomous, whereas the system

$$x' = ty,$$
$$y' = -x$$

is nonautonomous.

EXAMPLE 5

Consider the equation of the harmonic oscillator

$$x'' + x = 0,$$

with initial conditions $x(0) = 1$, $x'(0) = 0$. We may write it as the autonomous system

$$x' = y, \qquad x(0) = 1,$$
$$y' = -x, \qquad y(0) = 0.$$

There are two ways to find the orbit. First, we observe that the unique solution of the initial value problem is the solution pair $(x, y) = (\cos t, -\sin t)$. Since $\cos^2 t + \sin^2 t = 1$, the orbit satisfies the equation

$$x^2 + y^2 = 1,$$

which is the unit circle in the xy-plane (see Fig. 4). The arrows in the figure indicate the direction in which the solutions move about the orbit as t increases. Note that as t increases, $\cos t$ (the x-coordinate) moves from 1 (when $t = 0$) to 0 (when $t = \pi/2$), to -1 (when $t = \pi$), to 0 (when $t = 3\pi/2$), and back to 1 (at $t = 2\pi$). Similarly, $-\sin t$ (the y-coordinate) moves from 0 to -1, to 0, to 1, and back to 0. This explains the direction indicated by the arrows in Fig. 4. Therefore, starting at the point (1, 0) (corresponding to $t = 0$), x decreases while y decreases.

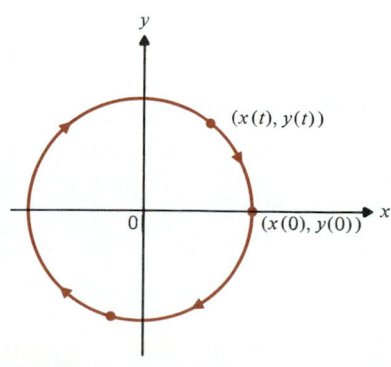

FIGURE 4

We can also find this orbit without solving the system. By the chain rule, we have

$$\frac{dy}{dx} = \frac{dy/dt}{dx/dt} = -\frac{x}{y}.$$

Separating the variables, we find that

$$y\,dy = -x\,dx$$

or, after an integration,

$$x^2 + y^2 = C.$$

To evaluate the constant C, we note that at $t = 0$, $x = 1$ and $y = 0$, so that $C = x^2(0) + y^2(0) = 1$. (Thus the radius of the orbit depends on the initial conditions.)

EXAMPLE 6

Again consider the harmonic oscillator with the new initial condition

$$x(t_0) = 1, \qquad y(t_0) = 0.$$

The unique solution pair is

$$(x(t), y(t)) = (\cos(t - t_0), -\sin(t - t_0)).$$

But then

$$x^2 + y^2 = \cos^2(t - t_0) + \sin^2(t - t_0) = 1,$$

which is the same orbit as in the previous example. In other words, the orbit is independent of the initial value of t [but not, of course, of the initial values $x(t_0)$ and $y(t_0)$]. This is a property shared by all autonomous systems. This fact will be assumed for the remainder of this chapter. Its proof can be found in most advanced differential equations texts.† The property does not hold for nonautonomous systems, as illustrated by the next example.

EXAMPLE 7

Consider the nonautonomous system

$$x' = \frac{1}{t}x, \qquad x(t_0) = 1,$$

$$y' = y, \qquad y(t_0) = 2.$$

Since these equations are uncoupled, it is easy to calculate the solution pair

$$(x(t), y(t)) = \left(\frac{t}{t_0}, 2e^{t-t_0}\right).$$

Since $t = t_0 x$, the orbit is

$$y = 2e^{t_0(x-1)}.$$

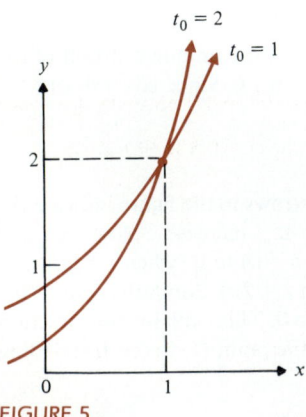

FIGURE 5

† See, for example, H. K. Wilson, *Ordinary Differential Equations*, Addison-Wesley, Reading, Mass., 1972.

Thus different initial values of t lead to different orbits. Figure 5 shows the orbits for $t_0 = 1$ and $t_0 = 2$.

Because of the properties of autonomous systems illustrated in Examples 5 and 6, we shall deal with them exclusively in the remainder of this chapter. As we saw earlier (see Example 2), autonomous systems can and do arise naturally in applications. A discussion of the properties of nonautonomous systems is more complicated (one reason being the necessity to worry about the initial value of t). Such a discussion can be found in many intermediate and advanced textbooks on differential equations.

Problems 3.6

In each of Problems 1–6, (a) find a related linear equation (as in Example 2), (b) find all constant solutions of the nonlinear equation, (c) solve the nonlinear equation, and (d) determine the behavior of the solutions to the nonlinear equation for values of $x(0)$ in the indicated range.

1. $x' = x - x^2, \quad -\infty < x(0) < \infty$
2. $x' = -2x + 3x^2, \quad x(0) \geq 0$
3. $x' = 2x + 3x^2, \quad x(0) \geq 0$
4. $x' = 2x - 3x^2, \quad -\infty < x(0) < \infty$
5. $x' = x(x - 1)(x - 2), \quad x(0) \geq 0$
6. $x' = -x(x - 1)(x - 2), \quad x(0) \geq 0$

7. **(a)** Draw the orbits for the initial value problem

$$x' = y, \qquad x(0) = a,$$
$$y' = -x, \qquad y(0) = b.$$

   (b) Show that these orbits are identical to those for the same system with the initial conditions $x(t_0) = a$, $y(t_0) = b$.

8. Show that the orbits for the equation $x'' + \omega^2 x = 0$ are ellipses centered at the origin.

9. Find the orbits for the system

$$x' = tx, \qquad x(t_0) = 1,$$
$$y' = -y, \qquad y(t_0) = 1,$$

   and graph these orbits for $t_0 = 0$, $t_0 = 1$, and $t_0 = 2$.

* 10. Suppose that the differential equations $x' = f(x, y)$ and $y' = g(x, y)$ have a unique solution whenever an initial condition is given for each variable. Show that no two orbits of the autonomous system

$$x' = f(x, y), \qquad y' = g(x, y) \tag{11}$$

   can ever intersect.

* 11. Use the result of Problem 10 to show that if (x_0, y_0) is a point having the property that $f(x_0, y_0) = g(x_0, y_0) = 0$ and if $(x(t), y(t))$ is a nonzero solution pair of the system (11) such that $f(x(t), y(t)) \neq 0$ for some value of t, then there is no value of t for which $(x(t), y(t)) = (x_0, y_0)$.

3.7 Critical Points, Stability, and Phase Portraits for Linear Systems

The general autonomous system of two first-order equations is given by

$$x' = f(x, y),$$
$$y' = g(x, y). \tag{1}$$

Critical Point or Stationary Point

DEFINITION

A point (x_0, y_0) is called a **critical point** (or **stationary point**) of the system (1) if

$$f(x_0, y_0) = g(x_0, y_0) = 0.$$

Any critical point (x_0, y_0) is a constant solution pair of (1), since the derivative of a constant is zero; thus

$$x_0' = 0 = f(x_0, y_0),\dagger$$

$$y_0' = 0 = g(x_0, y_0).$$

Point of Equilibrium

A critical point (x_0, y_0) of the system (1) is a **point of equilibrium,** since once we reach this point we can never leave it, the derivatives of both $x(t)$ and $y(t)$ being zero there. Physically, a critical point is often a point at which the potential energy is at a minimum. For instance, in Example 3.6.2, if we use the substitution $\mu = d\theta/dt$, we can write the system (1) of Section 3.6 (see p. 157) as

$$\theta' = \mu,$$

$$\mu' = -\omega^2 \sin \theta.$$

From Fig. 3.6.1, it is clear that the potential energy is a minimum when $\theta = 0$. The point $(0, 0)$ is a point of equilibrium of the system. Of course, there are other critical points (see the next example). This situation will be discussed in great detail in Example 3.8.4.

EXAMPLE 1

Consider the system

$$x' = y,$$

$$y' = -\omega^2 \sin x,$$

which was just obtained from equation (3.6.1). This system has infinitely many critical points, since $(k\pi, 0)$ is a critical point for all integers k.

EXAMPLE 2

Consider the system

$$x' = -x^2 + y,$$

$$y' = x - y^2.$$

The two critical points are $(0, 0)$ and $(1, 1)$. (Why?)

Stable Solution

Asymptotically Stable Solution
Unstable Solution

The notion of stability is central to any discussion of the behavior of differential equations. Roughly, a solution $\phi(t)$ to a system of equations is **stable** if whenever we start "close" to $\phi(t)$ we will stay close to $\phi(t)$ for all future values of t. It is **asymptotically stable** if it is stable and the solutions that start close to $\phi(t)$ approach $\phi(t)$ as t tends to ∞. Finally, $\phi(t)$ is **unstable** if it is not stable.

We already saw an example of a system that is stable. In Example 3.6.5 the orbits (see Fig. 3.6.4) were circles centered at the origin of radius $\sqrt{x^2(0) + y^2(0)}$. If the initial conditions are changed by a small amount, then

---

† The notation x_0' simply denotes the derivative of the constant function x_0.

the radius of the circular orbit is changed by a small amount, and thus the new orbit stays close to the original one.

To define "closeness" more precisely, we make use of the Pythagorean distance between two points in the plane. If (x_1, y_1) and (x_2, y_2) are the two points, then the distance between them is

$$d = [(x_1 - x_2)^2 + (y_1 - y_2)^2]^{1/2}.$$

We denote by $(x(t, x^*, y^*), y(t, x^*, y^*))$ the unique solution pair to the system (1) that satisfies the initial conditions

$$x(0) = x^*, \qquad y(0) = y^*.$$

Now we can give formal definitions of the above concepts.

DEFINITION

The **constant solution** (or critical point) (x_0, y_0) is said to be

1. **Stable** if for every number $\epsilon > 0$ there is a number $\delta > 0$ such that whenever

$$[(x_0 - x^*)^2 + (y_0 - y^*)^2]^{1/2} < \delta,$$

we have

$$\{[x_0 - x(t, x^*, y^*)]^2 + [y_0 - y(t, x^*, y^*)]^2\}^{1/2} < \epsilon,$$

for all $t \geq 0$ (That is, if you start close, you stay close);

2. **Asymptotically stable** if it is stable and there exists a number $A > 0$ such that whenever

$$[(x_0 - x^*)^2 + (y_0 - y^*)^2]^{1/2} < A,$$

we have

$$\lim_{t \to \infty} \{[x_0 - x(t, x^*, y^*)]^2 + [y_0 - y(t, x^*, y^*)]^2\} = 0$$

(That is, if you start close enough, the solution will approach the critical point as t tends to ∞);

3. **Unstable** if it is not stable   (That is, no matter how close to the constant solution you start, there are solutions that will move away from the constant solution).

The requirement that the solution be stable is part of the definition of asymptotic stability. There are examples, which we shall not cite here, of systems where all solutions eventually tend to zero but each gets very large, no matter how close to zero it starts. The zero solution for such a system is unstable.

The foregoing ideas are sketched in Fig. 1, where the critical point is taken to be the origin.

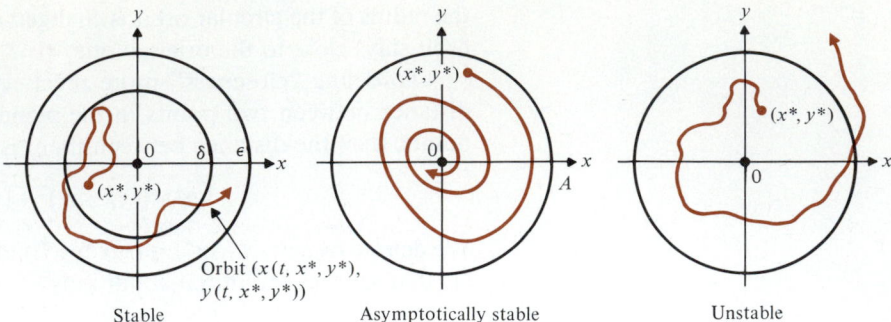

Stable   Asymptotically stable   Unstable

FIGURE 1

EXAMPLE 3

Consider again the harmonic oscillator

$$x' = y, \qquad y' = -x.$$

Clearly $(0, 0)$ is a constant solution. The solution pair with the initial conditions $x(0) = x^*$, $y(0) = y^*$ is

$$(x(t, x^*, y^*), y(t, x^*, y^*)) = (x^* \cos t + y^* \sin t, -x^* \sin t + y^* \cos t),$$

so that

$$(x(t, x^*, y^*) - 0)^2 + (y(t, x^*, y^*) - 0)^2$$

$$= (x^* \cos t + y^* \sin t)^2 + (-x^* \sin t + y^* \cos t)^2$$

$$= (x^{*2} + y^{*2}).$$

Thus $\delta = \epsilon$ satisfies the definition of stability and the zero solution is stable.

EXAMPLE 4

Consider the system

$$x' = y,$$

$$y' = -2x - 3y.$$

Again $(0, 0)$ is a critical point. The solution pair that satisfies $x(0) = x^*$, $y(0) = y^*$ is given by

$$(x(t, x^*, y^*), y(t, x^*, y^*))$$

$$= ((2x^* + y^*)e^{-t} - (x^* + y^*)e^{-2t}, -(2x^* + y^*)e^{-t} + 2(x^* + y^*)e^{-2t}),$$

which tends to zero as t tends to ∞ no matter how large x^* and y^* are. Hence the zero solution is asymptotically stable. As in this example, any situation in which A in Fig. 1(b) can be arbitrarily large yields the **global asymptotic stability** of the critical point.

Global Asymptotic Stability

EXAMPLE 5

The system

$$x' = y,$$

$$y' = -2x + 3y$$

has $(0, 0)$ as a critical point. However, the zero solution is unstable, since for any $x^* \neq 0$ and $y^* \neq 0$ (no matter how small), the solution pair is

$$(x(t, x^*, y^*), y(t, x^*, y^*))$$

$$= ((2x^* - y^*)e^t + (y^* - x^*)e^{2t}, (2x^* - y^*)e^t + 2(y^* - x^*)e^{2t}),$$

which becomes arbitrarily large as t tends to ∞.

The three examples considered above were all linear. The analysis of stability properties for nonlinear systems is much more difficult because, in general, solutions to such systems cannot be found. We shall consider a class of nonlinear systems in the next section, but in the remainder of this section we shall classify all possible linear systems† so that we may have some basis of comparison when we get to the nonlinear ones.

The linear system we shall consider is

$$\begin{aligned} x' &= a_{11}x + a_{12}y, \\ y' &= a_{21}x + a_{22}y, \end{aligned} \qquad a_{11}a_{22} - a_{12}a_{21} \neq 0, \qquad (2)$$

where the coefficients a_{ij} are real constants. As in Section 3.2, we derive the characteristic equation of the system

$$\lambda^2 - (a_{11} + a_{22})\lambda + (a_{11}a_{22} - a_{21}a_{12}) = 0, \qquad (3)$$

with roots λ_1 and λ_2.‡ The orbits of the system (2) will depend on the nature of these two roots. We shall therefore consider each case separately. We note that in all cases $(0, 0)$ will be the only constant solution (critical point) of the system (why?), so we shall restrict our attention to the nature of the orbits around the origin. This restriction also means that none of the roots of the characteristic equation (3) will be zero (see Problem 11).

In the following discussion, we shall examine the different possibilities for the roots of equation (3) and draw representative orbits of the system for each case. Thus, merely knowing the roots of the characteristic equation is enough to determine the nature of the orbits.

---

† Except those in which at least one of the roots of the characteristic equation (3) is zero.

‡ For those of you who cover the material in Chapter 8, λ_1 and λ_2 are the eigenvalues of the matrix

$$A = \begin{pmatrix} a_{11} & a_{12} \\ a_{21} & a_{22} \end{pmatrix}.$$

Although we avoid the eigenvalue terminology to make this chapter accessible to students who do not cover Chapter 8, the material should be regarded in terms of eigenvalues.

CASE 1 λ_1 *and* λ_2 *are real, distinct, and of the same sign.*
We may assume, for simplicity, that $\lambda_1 > \lambda_2$. Then, by Section 3.2, each solution pair has the form

$$(x(t), y(t)) = (c_1\alpha_1 e^{\lambda_1 t} + c_2\alpha_2 e^{\lambda_2 t}, c_1\beta_1 e^{\lambda_1 t} + c_2\beta_2 e^{\lambda_2 t}), \tag{4}$$

where c_1 and c_2 are arbitrary.

CASE 1(a) $\lambda_2 < \lambda_1 < 0$ *(both roots are negative).*
Clearly all solutions tend to $(0, 0)$ as t tends to ∞. First, we assume that $c_1 = 0$ and $c_2 \neq 0$. Then $y = (\beta_2/\alpha_2)x$, which means that the orbit is a straight line with slope β_2/α_2. If $c_1 \neq 0$ and $c_2 = 0$, then the situation is similar and we obtain the line $y = (\beta_1/\alpha_1)x$. To obtain the other orbits, we assume that c_1 and c_2 are both nonzero. Then

$$\frac{y(t)}{x(t)} = \frac{c_1\beta_1 e^{\lambda_1 t} + c_2\beta_2 e^{\lambda_2 t}}{c_1\alpha_1 e^{\lambda_1 t} + c_2\alpha_2 e^{\lambda_2 t}},$$

and dividing the numerator and denominator by $e^{\lambda_1 t}$, we have

$$\frac{y(t)}{x(t)} = \frac{c_1\beta_1 + c_2\beta_2 e^{(\lambda_2-\lambda_1)t}}{c_1\alpha_1 + c_2\alpha_2 e^{(\lambda_2-\lambda_1)t}} \to \frac{c_1\beta_1}{c_1\alpha_1} = \frac{\beta_1}{\alpha_1}$$

as t tends to ∞. Thus these orbits approach the origin with slope β_1/α_1. Similarly, as t tends to $-\infty$, all but two solutions become asymptotic to the lines with slope β_2/α_2. This situation is illustrated in Fig. 2 for three different values of the slopes β_1/α_1 and β_2/α_2. It is clear that in this case the zero solution is asymptotically stable. Here the origin is called a **stable node.**

Stable Node

CASE 1(b) $\lambda_1 > \lambda_2 > 0$ *(both roots are positive).*
Then all solutions (except the zero solution) approach ∞ as t tends to ∞. Hence the zero solution is unstable. The orbits are the same as in the previous case except that the direction of motion is reversed. The origin here is called an **unstable node.** As t tends to $-\infty$ all but two orbits approach zero with slope β_2/α_2, and as t tends to ∞ all but two orbits become asymptotic to lines with slope β_1/α_1 (see Fig. 2).

Unstable Node

CASE 2 λ_1 *and* λ_2 *are real with opposite signs.*
We assume that $\lambda_1 > 0 > \lambda_2$. The situation here is very different from Case 1. If $c_1 = 0$ and $c_2 \neq 0$, we obtain, as before,

$$\frac{y}{x} = \frac{\beta_2}{\alpha_2} \quad \text{or} \quad y = \frac{\beta_2}{\alpha_2}x.$$

As t tends to ∞, $x(t)$ and $y(t)$ approach zero. If $c_1 \neq 0$ and $c_2 = 0$, then $y = (\beta_1/\alpha_1)x$ and both x and y approach ∞ as t tends to ∞, and approach zero as t tends to $-\infty$. These orbits are sketched in Fig. 3. When both c_1 and c_2 are nonzero, the situation is more complicated. Again

$$\frac{y(t)}{x(t)} = \frac{c_1\beta_1 e^{\lambda_1 t} + c_2\beta_2 e^{\lambda_2 t}}{c_1\alpha_1 e^{\lambda_1 t} + c_2\alpha_2 e^{\lambda_2 t}} = \frac{c_1\beta_1 + c_2\beta_2 e^{(\lambda_2-\lambda_1)t}}{c_1\alpha_1 + c_2\alpha_2 e^{(\lambda_2-\lambda_1)t}}$$

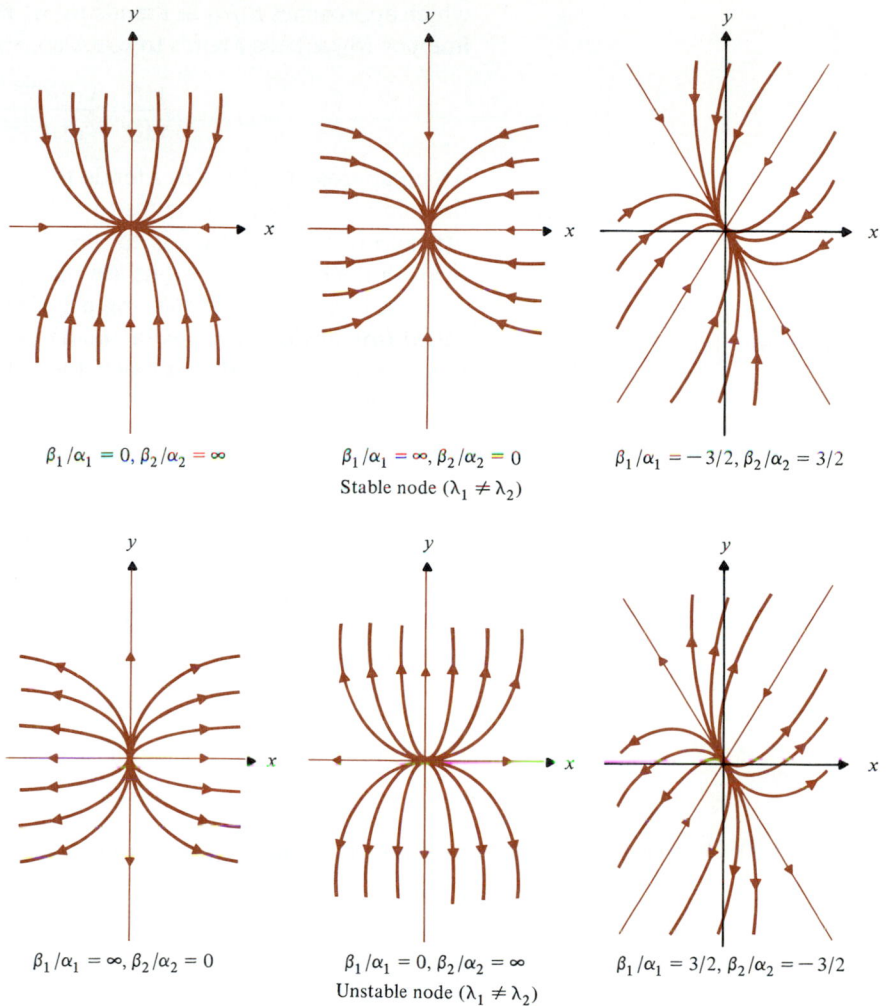

$\beta_1/\alpha_1 = 0,\ \beta_2/\alpha_2 = \infty$

$\beta_1/\alpha_1 = \infty,\ \beta_2/\alpha_2 = 0$
Stable node $(\lambda_1 \neq \lambda_2)$

$\beta_1/\alpha_1 = -3/2,\ \beta_2/\alpha_2 = 3/2$

$\beta_1/\alpha_1 = \infty,\ \beta_2/\alpha_2 = 0$

$\beta_1/\alpha_1 = 0,\ \beta_2/\alpha_2 = \infty$
Unstable node $(\lambda_1 \neq \lambda_2)$

$\beta_1/\alpha_1 = 3/2,\ \beta_2/\alpha_2 = -3/2$

FIGURE 2

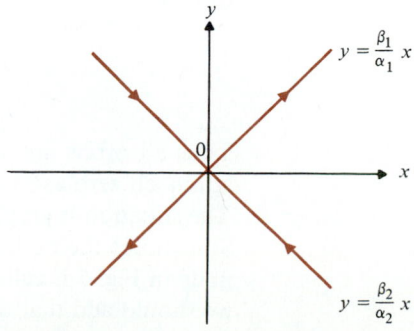

$y = \dfrac{\beta_1}{\alpha_1}\,x$

$y = \dfrac{\beta_2}{\alpha_2}\,x$

FIGURE 3

which approaches β_1/α_1 as t tends to ∞. Hence all orbits are asymptotic to the line $y = (\beta_1/\alpha_1)x$ as t tends to ∞. Also,

$$\frac{y(t)}{x(t)} = \frac{c_1\beta_1 e^{(\lambda_1-\lambda_2)t} + c_2\beta_2}{c_1\alpha_1 e^{(\lambda_1-\lambda_2)t} + c_2\alpha_2}$$

which approaches β_2/α_2 as t tends to $-\infty$. Hence all orbits are asymptotic to the line $y = (\beta_2/\alpha_2)x$ as t tends to $-\infty$. Finally, we observe that both $x(t)$ and $y(t)$ approach ∞ as t tends to $\pm\infty$, and, by uniqueness, no orbit can pass through the origin. The orbits are therefore as shown in Fig. 4.

It is clear here that the origin is unstable. In this situation, the origin is called (for obvious reasons) a **saddle point**. A saddle point has the property that exactly two orbits approach the origin and all others are "repelled" by it. The physical behavior corresponding to a saddle point is illustrated in Example 3.8.4.

CASE 3 $\lambda_1 = \lambda_2 = \lambda$.
Here $\lambda < 0$ or $\lambda > 0$.

CASE 3(a) $\lambda < 0$.
There are two ways in which the characteristic equation (3) can yield a double root. One possibility is

$$a_{11} = a_{22} \neq 0, \qquad a_{21} = a_{12} = 0. \tag{5}$$

Then the characteristic equation is

$$\lambda^2 - 2a_{11}\lambda + a_{11}^2 = 0,$$

and $\lambda = a_{11}$ is the double root. The system (2) is

$$x' = \lambda x,$$

$$y' = \lambda y.$$

The solutions are of the form

$$(x(t), y(t)) = (c_1 e^{\lambda t}, c_2 e^{\lambda t}),$$

so that

$$\frac{y}{x} = \frac{c_2}{c_1} \quad \text{or} \quad y = \frac{c_2}{c_1}x.$$

Thus all orbits are straight lines with slope c_2/c_1. Since $\lambda < 0$, all solutions approach zero as t tends to ∞, and the zero solution is asymptotically stable. The situation is graphed in Fig. 5. The origin in this case is also called a **node**. Sometimes the nodes shown in Fig. 2 are called **improper nodes,** whereas the node in Fig. 5 is called a **proper node**. We will not use this terminology. Also, we should add that the node of Fig. 5 is sometimes called a **star-shaped node**.

Saddle Point

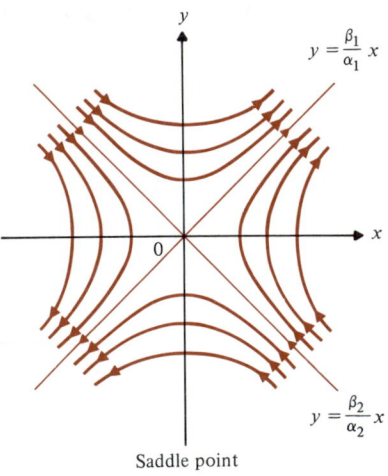

$y = \dfrac{\beta_1}{\alpha_1} x$

$y = \dfrac{\beta_2}{\alpha_2} x$

Saddle point

FIGURE 4

Improper Node
Proper Node
Star-Shaped Node

If $\lambda < 0$ is a double root but the equalities (5) do not hold, then the equations are coupled and the general solution is (from the technique summarized on p. 141),

$$(x(t), y(t)) = ([c_1\alpha_1 + c_2(\alpha_2 + \alpha_3 t)]e^{\lambda t}, [c_1\beta_1 + c_2(\beta_2 + \beta_3 t)]e^{\lambda t}). \tag{6}$$

Then

$$\frac{y}{x} = \frac{c_1\beta_1 + c_2\beta_2 + c_2\beta_3 t}{c_1\alpha_1 + c_2\alpha_2 + c_2\alpha_3 t} = \frac{c_1\beta_1/t + c_2\beta_2/t + c_2\beta_3}{c_1\alpha_1/t + c_2\alpha_2/t + c_2\alpha_3},$$

which approaches β_3/α_3 as t tends to $\pm\infty$. Since both $x(t)$ and $y(t)$ approach zero as t tends to ∞, the zero solution is asymptotically stable. Also all orbits are asymptotic to the line $y = (\beta_3/\alpha_3)x$ as t tends to $\pm\infty$, as illustrated in Fig. 6.

CASE 3(b) $\lambda > 0$.

The orbits are the same but with the arrows reversed, since now all solutions approach ∞ as t tends to ∞. (See Figs. 7 and 8.) In these two cases the origin is unstable. Cases 3(a) and 3(b) provide another situation in which nodes arise.

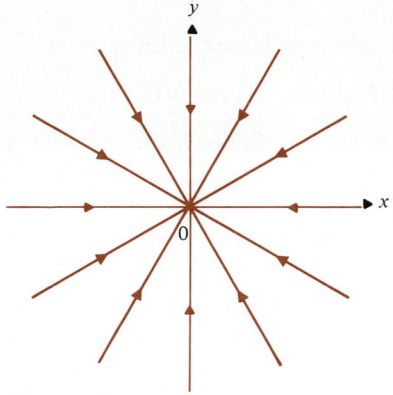

(Star–shaped) stable node ($\lambda_1 = \lambda_2$)

FIGURE 5

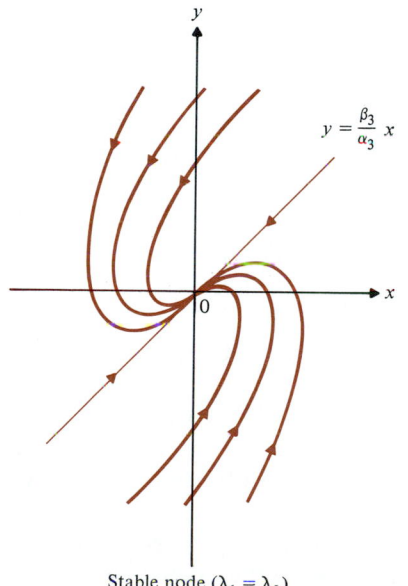

$y = \dfrac{\beta_3}{\alpha_3} x$

Stable node ($\lambda_1 = \lambda_2$)

FIGURE 6

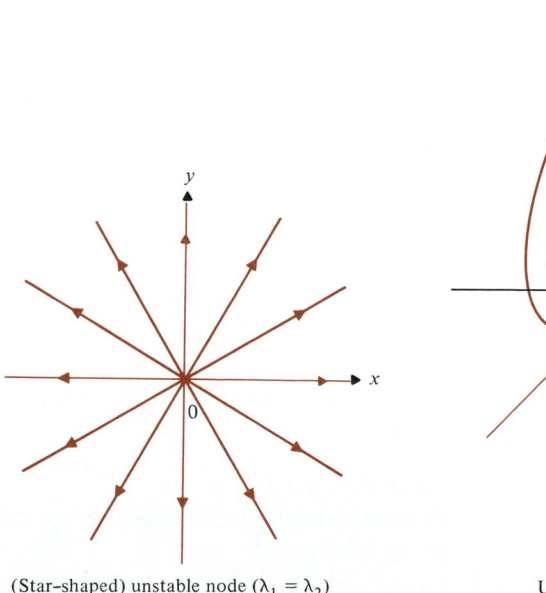

(Star–shaped) unstable node ($\lambda_1 = \lambda_2$)

FIGURE 7

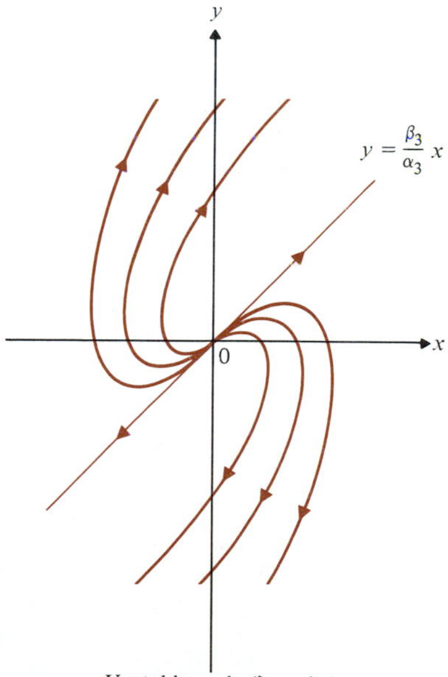

$y = \dfrac{\beta_3}{\alpha_3} x$

Unstable node ($\lambda_1 = \lambda_2$)

FIGURE 8

CASE 4 λ_1 *and* λ_2 *are complex conjugates but not pure imaginary.*
Then $\lambda_1 = a + ib$ and $\lambda_2 = a - ib$, where neither a nor b is zero.

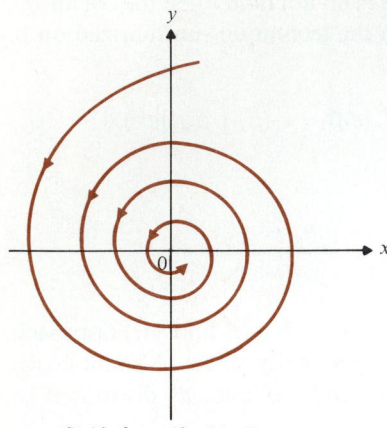

Stable focus (for $b > 0$)

FIGURE 9

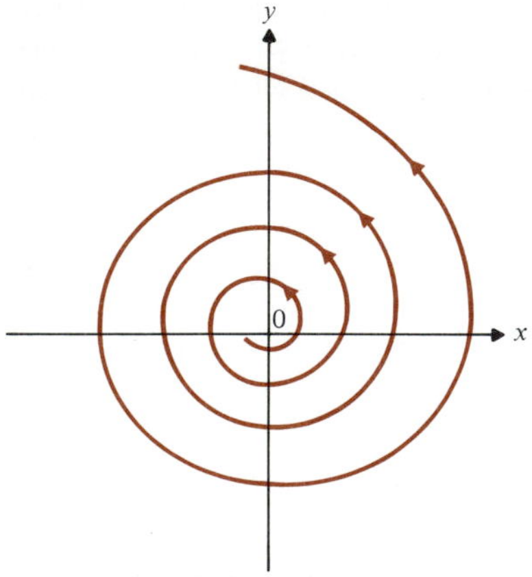

y

0

x

Unstable focus (for $b > 0$)

FIGURE 10

Stable Focus

CASE 4(a) $a < 0$.

By the technique summarized on p. 141, all solutions have the form

$$(x(t), y(t)) = (e^{at}[c_1(A_1 \cos bt - A_2 \sin bt) + c_2(A_1 \sin bt + A_2 \cos bt)],$$

$$e^{at}[c_1(B_1 \cos bt - B_2 \sin bt) + c_2(B_1 \sin bt + B_2 \cos bt)]). \quad (7)$$

To simplify the notation, we define

$$k_1 = c_1 A_1 + c_2 A_2, \qquad k_2 = -c_1 A_2 + c_2 A_1,$$

$$k_3 = c_1 B_1 + c_2 B_2, \qquad k_4 = -c_1 B_2 + c_2 B_1,$$

so that equation (7) becomes

$$(x(t), y(t)) = (e^{at}(k_1 \cos bt + k_2 \sin bt), e^{at}(k_3 \cos bt + k_4 \sin bt)). \quad (8)$$

We now define $A = \sqrt{k_1^2 + k_2^2}$ and $B = \sqrt{k_3^2 + k_4^2}$. Then we define α_1 and α_2 by

$$\cos \alpha_1 = \frac{k_1}{A}, \qquad \cos \alpha_2 = \frac{k_3}{B},$$

$$\sin \alpha_1 = -\frac{k_2}{A}, \qquad \sin \alpha_2 = -\frac{k_4}{B}, \quad (9)$$

so that $k_1 = A \cos \alpha_1$, $k_2 = -A \sin \alpha_1$, $k_3 = B \cos \alpha_2$, $k_4 = -B \sin \alpha_2$, and

$$(x(t), y(t)) = (Ae^{at}(\cos \alpha_1 \cos bt - \sin \alpha_1 \sin bt),$$

$$Be^{at}(\cos \alpha_2 \cos bt - \sin \alpha_2 \sin bt))$$

$$= (Ae^{at} \cos(bt + \alpha_1), Be^{at} \cos(bt + \alpha_2)). \quad (10)$$

Then

$$\frac{y}{x} = \frac{B \cos(bt + \alpha_1)}{A \cos(bt + \alpha_2)},$$

which is defined whenever $\cos(bt + \alpha_2) \neq 0$. Since this expression is periodic, it is clear that as t tends to ∞, the ratio y/x does not approach a limit, but the orbits must circle around the origin. Since $a < 0$, $x(t)$ and $y(t)$ approach zero as t tends to ∞. Hence the orbits must spiral in toward the origin. The zero is asymptotically stable, and the origin is called a **stable focus** (or **spiral point**). See Fig. 9.

CASE 4(b) $a > 0$.

Here the analysis is as before except that all solutions approach ∞ as t tends to ∞. See Fig. 10.

CASE 5 λ_1 *and* λ_2 *are pure imaginary.*

Let $\lambda_1 = ib$ and $\lambda_2 = -ib$, and use the same analysis as above with $a = 0$: We have

$$(x(t), y(t)) = (A \cos(bt + \alpha_1), B \cos(bt + \alpha_2)). \quad (11)$$

Clearly $x(t)$ and $y(t)$ are periodic with period $2\pi/b$ so that every orbit beginning at the point (x^*, y^*) when $t = t^*$ will return to the same point when $t = t^* + 2\pi/b$. Thus the orbits are closed curves. To get a feeling for the nature of these curves, we set $k_2 = k_3 = 0$ in equation (8) so that

$$x(t) = k_1 \cos bt \quad \text{and} \quad y(t) = k_4 \sin bt.$$

Then

$$\frac{x^2}{k_1^2} + \frac{y^2}{k_4^2} = \cos^2 bt + \sin^2 bt = 1,$$

which is the equation of an ellipse centered about $(0, 0)$ with the x- and y-axes as major and minor axes. If k_2 and k_3 are nonzero, we obtain ellipses with rotated major and minor axes. In this situation the zero solution is stable but not asymptotically stable, since solutions do not approach zero. Then the origin is called a **center**, which is illustrated in Fig. 11.

Center

The entire preceding analysis is summarized in the theorem below.

■ **THEOREM 1**

Consider the system

$$x' = a_{11}x + a_{12}y,$$

$$y' = a_{21}x + a_{22}y,$$

where the a_{ij} are real constants and $a_{11}a_{22} - a_{12}a_{21} \neq 0$ so that the origin $(0, 0)$ is the only critical point (see Problem 11). Let λ_1 and λ_2 be the two roots of the characteristic equation (3). Then

(a) the origin is stable if λ_1 and λ_2 are pure imaginary,
(b) the origin is asymptotically stable if Re $\lambda_1 < 0$ and Re $\lambda_2 < 0$,
(c) the origin is unstable in all other cases. Moreover, the behavior of the orbits near the origin is as indicated in Table 1.   ■

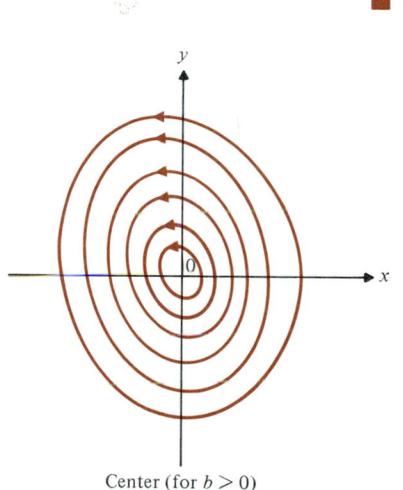

Center (for $b > 0$)

FIGURE 11

TABLE 1

λ_1, λ_2	Type of critical point
Real, distinct, negative	Stable node
Real, distinct, positive	Unstable node
Real, distinct, opposite signs	Saddle point (unstable)
Real, equal, negative	Stable node
Real, equal, positive	Unstable node
Complex conjugate, not pure imaginary, negative real parts	Stable focus
Complex conjugate, not pure imaginary, positive real parts	Unstable focus
Pure imaginary	Center (stable)

EXAMPLE 6

Consider the system

$$x' = x - 3y,$$
$$y' = x - y.$$

The characteristic equation is $\lambda^2 + 2 = 0$ with the roots $\pm\sqrt{2}i$. Therefore, the zero solution is stable and the origin is a center.

EXAMPLE 7

Consider the system

$$x' = 4x - y,$$
$$y' = 6x - 3y.$$

The characteristic equation is $\lambda^2 - \lambda - 6 = 0$ with the roots 3 and -2. Hence the origin is a saddle point and the zero solution is unstable.

EXAMPLE 8

Consider the system

$$x' = -4x - y,$$
$$y' = x - 2y.$$

The characteristic equation is $\lambda^2 + 6\lambda + 9 = 0$ with the double root $\lambda = -3$. Hence the origin is a stable node and the zero solution is asymptotically stable.

EXAMPLE 9

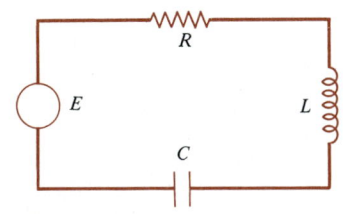

FIGURE 12

Consider the RLC circuit shown in Fig. 12. If $E = 0$, then from the discussion in Section 2.7 (see p. 102), we obtain the differential equation

$$\frac{d^2I}{dt^2} + \frac{R}{L}\frac{dI}{dt} + \frac{I}{CL} = 0 \tag{12}$$

for the description of this circuit. Writing the equation as a system, we obtain

$$I' = y,$$
$$y' = -\frac{1}{CL}I - \frac{R}{L}y.$$

The characteristic equation is

$$\lambda^2 + \frac{R}{L}\lambda + \frac{1}{CL} = 0$$

with the roots

$$\lambda_1 = \frac{-R + \sqrt{R^2 - 4L/C}}{2L} \quad \text{and} \quad \lambda_2 = \frac{-R - \sqrt{R^2 - 4L/C}}{2L} \tag{13}$$

There are three cases to consider, according to whether $R^2 - 4L/C$ is greater than zero, less than zero, or equal to zero.

1. If $R^2 - 4L/C > 0$, then λ_1 and λ_2 are real, distinct, and negative, so that the origin is a stable node.

2. If $R^2 - 4L/C = 0$, then $\lambda_1 = \lambda_2 = -R/2L$, so that the zero solution is again a stable node.

3. If $R^2 - 4L/C < 0$, then λ_1 and λ_2 are complex conjugates with negative real parts, so that the origin is a stable focus.

In all three cases the critical point $(0, 0)$ is asymptotically stable, and the transient current tends to zero as $t \to \infty$. Of course, the emf E is generally nonzero, but it affects only the steady-state current. By superposition, the current $I(t)$ is the sum of the steady-state and transient currents.

Problems 3.7

In Problems 1–8, describe the nature of the critical point $(0, 0)$ of each system and sketch the orbits.

1. $x' = 4x - 3y, \quad y' = 5x - 4y$
2. $x' = -2x + 3y, \quad y' = x - 3y$
3. $x' = -x + y, \quad y' = -5x + 3y$
4. $x' = x + y, \quad y' = -x - 3y$
5. $x' = -4x - 2y, \quad y' = 5x + 2y$
6. $x' = 4x - 3y, \quad y' = 8x + 5y$
7. $x' = 2x + y, \quad y' = -4x + 2y$
8. $x' = 2x, \quad y' = 2y$

9. Consider the system

$$x' = a_{11}x + a_{12}y,$$
$$y' = a_{21}x + a_{22}y.$$

Let $T = a_{11} + a_{22}$, $D = a_{11}a_{22} - a_{12}a_{21}$, and $C = T^2 - 4D$. Show that the origin is

(a) a node if $D > 0$ and $C \geq 0$;
(b) a saddle point if $D < 0$;
(c) a focus if $T \neq 0$ and $C < 0$;
(d) a center if $T = 0$ and $D > 0$.

10. Show that the zero solution in Problem 9 is

(a) asymptotically stable if $D > 0$ and $T < 0$;
(b) stable if $T = 0$ and $D > 0$;
(c) unstable if $T > 0$ or $D < 0$.

11. Show that $(0, 0)$ is the only critical point of the system (2) if and only if $\lambda = 0$ is not a root of the characteristic equation (3).

12. Let $L = 2$ henrys, $R = 50$ ohms, and $C = 0.0003$ farad. Locate graphically the position on the orbit of equation (12) at $t = 1$ if $I(0) = 0$ and $I'(0)$ is one of the values 10, 100, and 1000.

13. Let $L = 0.44$ henry, $R = 1200$ ohms, and $C = 10^{-6}$ farad. Locate graphically the position of the solution on the orbit at $t = 1$ if $I(0) = 0$ and $I'(0)$ is one of the values 10, 100, and 1000.

14. Consider the mass-spring system which is damped so that equation (2.9.2), p. 110 holds. That is,

$$x'' + \frac{c}{m}x' + \frac{k}{m}x = 0.$$

Following the analysis of Example 9, describe the behavior of the orbits near the origin for different positive values of m, c, and k.

15. Consider the system

$$x' = 0,$$
$$y' = -x + y.$$

(a) Solve the system.
(b) Show that $\lambda_1 = 0$ and $\lambda_2 = 1$ are the roots of the characteristic equation (3).
(c) Find all the critical points of the system.
(d) Show that all the orbits are straight lines.
(e) Show that the origin is an unstable equilibrium point.

16. Consider the system

$$x' = 0,$$
$$y' = -x - y.$$

(a) Solve the system.
(b) Show that the roots of equation (3) for this system are $\lambda_1 = 0$ and $\lambda_2 = -1$.
(c) Graph the orbits.
(d) Show that the zero solution is stable but not asymptotically stable.

3.8 Stability of Nonlinear Systems

In this section we shall discuss the stability properties of the autonomous non-linear system

$$x' = f(x, y),$$
$$y' = g(x, y) \tag{1}$$

where $f(0, 0) = g(0, 0) = 0$ so that the origin is a critical point. We assume that $f(x, y)$ and $g(x, y)$ possess continuous third partial derivatives so that they can each be expanded in a second-degree Taylor polynomial with a remainder term:†

$$f(x, y) = f(0,0) + \frac{\partial f}{\partial x}(0,0)x + \frac{\partial f}{\partial y}(0,0)y + \frac{\partial^2 f}{\partial x^2}(0,0)\frac{x^2}{2}$$
$$+ \frac{\partial^2 f}{\partial x \partial y}(0,0)xy + \frac{\partial^2 f}{\partial y^2}(0,0)\frac{y^2}{2} + \cdots,$$

$$g(x, y) = g(0,0) + \frac{\partial g}{\partial x}(0,0)x + \frac{\partial g}{\partial y}(0,0)y + \frac{\partial^2 g}{\partial x^2}(0,0)\frac{x^2}{2}$$
$$+ \frac{\partial^2 g}{\partial x \partial y}(0,0)xy + \frac{\partial^2 g}{\partial y^2}(0,0)\frac{y^2}{2} + \cdots, \tag{2}$$

where the omitted terms all involve polynomials in x and y of degree greater than 2. Now we define

$$a_{11} = \frac{\partial f}{\partial x}(0,0), \qquad a_{12} = \frac{\partial f}{\partial y}(0,0), \qquad a_{21} = \frac{\partial g}{\partial x}(0,0), \qquad a_{22} = \frac{\partial g}{\partial y}(0,0),$$

and assume that

$$a_{11}a_{22} - a_{12}a_{21} \neq 0 \tag{3}$$

so that $\lambda = 0$ is not a root of the characteristic equation (see Problem 3.7.11). Then, using the fact that $f(0, 0) = g(0, 0) = 0$, we can write $f(x, y)$ and $g(x, y)$ as

$$f(x, y) = a_{11}x + a_{12}y + f_1(x, y),$$
$$g(x, y) = a_{21}x + a_{22}y + g_1(x, y),$$

where

$$\lim_{x,y \to 0} \frac{f_1(x, y)}{\sqrt{x^2 + y^2}} = \lim_{x,y \to 0} \frac{g_1(x, y)}{\sqrt{x^2 + y^2}} = 0. \tag{4}$$

† Taylor's theorem for functions of two or more variables is discussed in Section 9.10.

The last property simply says that the point $(f_1(x, y), g_1(x, y))$ approaches the point $(0, 0)$ "faster" than the point (x, y) does. In one dimension this is easily visualized by noting, for example, that the function $f_1(x) = x^2$ goes to zero faster than x, or

$$\lim_{x \to 0} \frac{x^2}{x} = 0.$$

We now show that condition (4) holds for the function x^2. Once this has been done, it will be clear that the condition holds for xy, y^2, and higher powers of x and y (such as x^3, x^2y, xy^2, and y^3).

Now

$$\lim_{x,y \to 0} \frac{x^2}{\sqrt{x^2 + y^2}} = 0,$$

because

$$0 \le \frac{x^2}{\sqrt{x^2 + y^2}} \le \frac{x^2 + y^2}{\sqrt{x^2 + y^2}} = \sqrt{x^2 + y^2} \to 0,$$

as (x, y) tends to zero, and a similar proof applies for y^2. For xy use the inequality $2xy \le x^2 + y^2$, since $(x - y)^2 \ge 0$.

Note: Condition (3) will be satisfied if the determinant

$$\begin{vmatrix} \partial f / \partial x & \partial f / \partial y \\ \partial g / \partial x & \partial g / \partial y \end{vmatrix}$$

is nonzero at $(0, 0)$.

EXAMPLE 1

The system

$$x' = 2x + 3y + x^3,$$
$$y' = x - 2y - y^{3/2}$$

satisfies conditions (3) and (4) since (prove this)

$$\lim_{x,y \to 0} \frac{x^3}{\sqrt{x^2 + y^2}} = \lim_{x,y \to 0} \frac{y^{3/2}}{\sqrt{x^2 + y^2}} = 0.$$

In the rest of this section we shall consider the system

$$x' = a_{11}x + a_{12}y + f_1(x, y),$$
$$y' = a_{21}x + a_{22}y + g_1(x, y),$$

(5)

where equations (3) and (4) are satisfied, and the associated linear system

$$x' = a_{11}x + a_{12}y,$$

$$y' = a_{21}x + a_{22}y. \tag{6}$$

The following theorem enables us to determine the nature of the critical point (0, 0) of equation (5) by indicating the behavior of the solutions of the linear system (6) near the origin. The proof of this theorem is difficult and beyond the scope of this text.†

■ **THEOREM 1**
Let λ_1 and λ_2 be the roots of the characteristic equation of the linear system (6).
(i) The nonlinear system (5) has the same type of critical point at the origin as the linear system (6) whenever:

(a) $\lambda_1 \neq \lambda_2$ and (0, 0) is a node of the system (6),
(b) $\lambda_1 = \lambda_2$ and (0, 0) is not a star-shaped node of the system (6),
(c) (0, 0) is a saddle point of the system (6),
(d) (0, 0) is a focus of the system (6).

(ii) The origin is not necessarily the same type of critical point for the two systems:

(e) If $\lambda_1 = \lambda_2$ and (0, 0) is a star-shaped node of the system (6), then (0, 0) is either a node or a focus of the system (5).
(f) If (0, 0) is a center of the system (6), then (0, 0) is either a center or a focus of the system (5). ■

The next theorem relates the stability of the nonlinear system to that of the associated linear system. As before, the proof is omitted.

■ **THEOREM 2**
(i) If the zero solution of the system (6) is asymptotically stable, then the zero solution of the system (5) is asymptotically stable.
(ii) If the zero solution of the system (6) is unstable, then the zero solution of the system (5) is unstable.
(iii) If the zero solution of the system (6) is stable but not asymptotically stable, then the zero solution of the system (5) may be asymptotically stable, stable, or unstable. ■

† For a proof, see J. K. Hale, *Ordinary Differential Equations,* Wiley, New York, 1969, or the text by Wilson cited on p. 162.

Perturbation Theorems

Remark. Part (ii) of Theorem 2 holds even when equation (3) is not satisfied. We shall use this fact in Example 3.

Theorems 1 and 2 are often called **perturbation theorems,** because we may consider the nonlinear system (5) as a small perturbation of the linear system (6) near the origin.

EXAMPLE 2

In Example 3.6.3, we discussed the Lotka-Volterra equations as a model for the interaction of two competing species. A more complete model, including within-species competition is given by the system

$$x' = \beta_1 x - \delta_{11} x^2 - \delta_{12} xy,$$
$$y' = \beta_2 y - \delta_{21} xy - \delta_{22} y^2. \tag{7}$$

We can think of this system as a perturbation of the associated uncoupled linear system

$$x' = \beta_1 x,$$
$$y' = \beta_2 y. \tag{8}$$

The system (8) has the solution

$$(x(t), y(t)) = (c_1 e^{\beta_1 t}, c_2 e^{\beta_2 t}),$$

and

$$\lim_{x,y \to 0} \frac{x^2}{\sqrt{x^2 + y^2}} = \lim_{x,y \to 0} \frac{xy}{\sqrt{x^2 + y^2}} = \lim_{x,y \to 0} \frac{y^2}{\sqrt{x^2 + y^2}} = 0.$$

Hence Theorems 1 and 2 apply in the following cases:

1. If β_1 and β_2 are negative, then the critical point $(0, 0)$ is asymptotically stable for both systems (7) and (8). In both systems the origin is a stable node. This means that both populations will become extinct, in the absence of other factors, if the initial populations are small [that is, we start near $(0, 0)$]. Thus small initial populations cannot sustain themselves.
2. If β_1 and β_2 have opposite signs, then the origin in both cases is a saddle point, and the zero solution is unstable. This situation implies that one of the populations will become extinct while the other will grow without bound (since the asymptotes of the saddle point are the x- and y-axes).
3. If β_1 and β_2 are unequal and positive, then the origin in both cases is an unstable node. Hence both populations will increase if they start near $(0, 0)$. However, the orbits of (7) need not increase without bound, since as we move away from the origin the orbits may approach other critical points of (7) exhibiting a different type of behavior.

This is not a complete analysis of the system (7), since we can expect different and perhaps more interesting types of behavior near other critical points (see Problem 16). We shall illustrate a similar situation in the next example.

EXAMPLE 3

Consider the system

$$x' = -2xy = f(x, y),$$
$$y' = -x + y + xy - y^3 = g(x, y).$$ (9)

Setting $-2xy = -x + y + xy - y^3 = 0$, we find the three critical points $(0, 0)$, $(0, 1)$, and $(0, -1)$. We shall treat each of these separately.

CASE 1 $(0, 0)$.
The associated linear system is

$$x' = 0,$$
$$y' = -x + y.$$ (10)

The characteristic equation for (10) is $\lambda^2 - \lambda = 0$, with the roots $\lambda_1 = 0$, $\lambda_2 = 1$. Hence the origin of (10) is unstable, and therefore the zero solution of (9) is unstable by the remark following Theorem 2. [See Fig. 1].

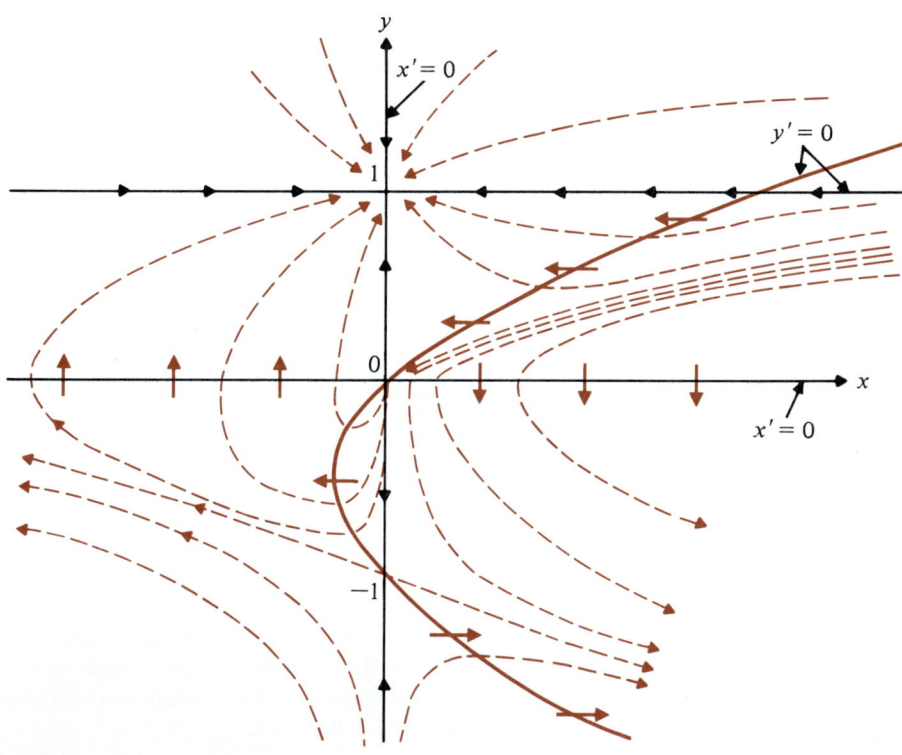

FIGURE 1

CASE 2 $(0, 1)$.
To obtain the associated linear system, we need to use Taylor's theorem to expand the right-hand sides of (9) around the point $(0, 1)$. Then

$$-2xy = f(0, 1) + \frac{\partial f}{\partial x}\bigg|_{(0, 1)} (x - 0) + \frac{\partial f}{\partial y}\bigg|_{(0, 1)} (y - 1) + \cdots$$

$$= -2x + \cdots [= -2x - 2x(y-1)]$$

and

$$-x + y + xy - y^3 = g(0, 1) + \frac{\partial g}{\partial x}\bigg|_{(0, 1)} (x-0) + \frac{\partial g}{\partial y}\bigg|_{(0, 1)} (y-1) + \cdots$$

$$= -2(y-1) + \cdots$$

$$[= -2(y-1) + x(y-1) - 3(y-1)^2 - (y-1)^3].$$

Selecting only the first-degree terms of these expansions, we obtain the associated linear system

$$x' = -2x,$$

$$y' = -2(y-1).$$

We define the new plane variables x and $z = y - 1$. Then the system is

$$x' = -2x,$$

$$z' = -2z$$

(11)

and the roots of the characteristic equation of (11) are $\lambda_1 = \lambda_2 = -2$. This means that the point $(0, 1)$ is a stable star-shaped node and the critical point $(0, 1)$ is an asymptotically stable solution of (9). By Theorem 1(e), the point $(0, 1)$ is either a stable node or a stable focus of (9). See Fig. 1. It turns out to be a node.

CASE 3 $(0, -1)$.
Expanding $f(x, y)$ and $g(x, y)$ around $(0, -1)$, we obtain

$$-2xy = f(0, -1) + \frac{\partial f}{\partial x}\bigg|_{(0, -1)} (x-0) + \frac{\partial f}{\partial y}\bigg|_{(0, -1)} (y+1) + \cdots$$

$$= 2x + \cdots [= 2x - 2x(y+1)]$$

and

$$-x + y + xy - y^3 = g(0, -1) + \frac{\partial g}{\partial x}\bigg|_{(0, -1)} (x-0) + \frac{\partial g}{\partial y}\bigg|_{(0, -1)} (y+1) + \cdots$$

$$= -2x - 2(y+1) + \cdots$$

$$[= -2x - 2(y+1) + x(y+1) + 3(y+1)^2 - (y+1)^3]$$

so that the associated linear system around the point $(0, -1)$ is

$$x' = 2x,$$

$$y' = -2x - 2(y+1).$$

(12)

Letting $z = y + 1$, we have $z' = y'$ and the roots of the characteristic equation for the system

$$x' = 2x,$$

$$z' = -2x - 2z$$

are $\lambda_1 = 2$, $\lambda_2 = -2$, so that the solution $(0, -1)$ is a saddle point of (9). This situation is illustrated in Figure 1, where we have drawn several orbits (the dotted lines) in the phase plane. The solid lines (including the axes) are the zero **isoclines** of the system,

Isoclines

that is, lines where $x' = 0$ or $y' = 0$. Observe that every critical point is located at the intersection of an $x' = 0$ isocline and a $y' = 0$ isocline [note that $y' = (1 - y)(y^2 + y - x)$].

EXAMPLE 4

In Example 3.6.2, we discussed the equation of motion of a frictionless pendulum given by

$$\theta'' + \omega^2 \sin \theta = 0.$$

Defining $\mu = \theta'$, we have the system

$$\theta' = \mu,$$
$$\mu' = -\omega^2 \sin \theta. \tag{13}$$

This system has an infinite number of critical points of the form $(k\pi, 0)$, $k = \pm 1, \pm 2, \pm 3, \ldots$. The associated linear system near the critical point $(0, 0)$ is

$$\theta' = \mu,$$
$$\mu' = -\omega^2 \theta, \tag{14}$$

since the Taylor series of $\sin \theta$ is

$$\sin \theta = \theta - \frac{\theta^3}{3!} + \frac{\theta^5}{5!} - \cdots.$$

This is the system of the familiar harmonic oscillator. The origin of (14) is a center, so we cannot conclude anything about the stability of the origin of (13). The results are more definitive for the critical point $(\pi, 0)$. Expanding $\sin \theta$ around $\theta = \pi$, we have

$$\sin \theta = \sin \pi + (\sin \theta)'|_{\theta = \pi}(\theta - \pi) + \cdots$$

$$= (\pi - \theta) - \frac{1}{3!}(\pi - \theta)^3 + \frac{1}{5!}(\pi - \theta)^5 - \cdots.$$

Thus the associated linear system is

$$\theta' = \mu,$$
$$\mu' = \omega^2(\theta - \pi). \tag{15}$$

Letting $z = \theta - \pi$, the roots of the characteristic equation $\lambda^2 - \omega^2 = 0$ are $\lambda_1 = \omega$, $\lambda_2 = -\omega$, so that $(\pi, 0)$ is a saddle point. Hence we may conclude that the critical point $(\pi, 0)$ of the system (13) is an unstable saddle point. Figure 3.6.1 illustrates intuitively what is happening. When $\theta = \pi$ the pendulum is pointing vertically upward, a position at which the potential energy is at a maximum and the kinetic energy is zero. Clearly such a position is unstable, and a small displacement will cause large deviations from the initial position. On the other hand, it is just as clear that $\theta = 0$ (the vertically downward position) is a stable center. An initial displacement of θ_0 of a frictionless pendulum will lead to periodic oscillations with a maximum displacement of θ_0. It is evident that all critical points of the form $(k\pi, 0)$ will be of the same type as $(0, 0)$ if k is even and of the type $(\pi, 0)$ if k is odd. Before drawing the orbits, we can show that the zero solution of (13) really is a center. By Theorem 1 it must be a center or a focus. But the system is **conservative** (there is no loss or gain of energy), whereas a stable focus would imply that the energy was decreasing to zero and an unstable focus would imply an increase in the total energy. Hence, a focus is ruled out, and the origin must be a center. This is all illustrated in Fig. 2.

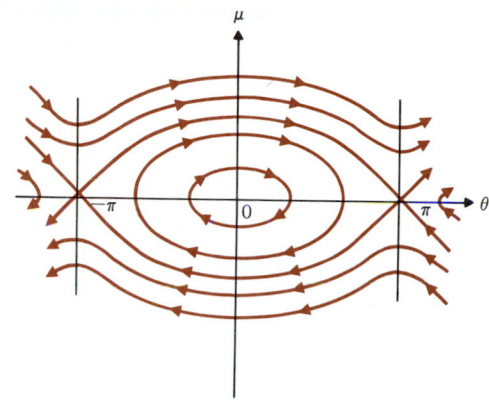

FIGURE 2

EXAMPLE 5

We shall consider the pendulum of the previous example, this time including the effect of friction. We assume that the frictional force is proportional to the angular velocity θ'. Denoting this constant of proportionality by $\epsilon > 0$, we obtain

$$\theta'' + \epsilon\theta' + \omega^2 \sin \theta = 0$$

or, equivalently,

$$\theta' = \mu,$$
$$\mu' = -\omega^2 \sin \theta - \epsilon\mu. \tag{16}$$

Again the origin is a critical point. However, in this case it is clear from physical considerations that the origin is asymptotically stable, since the friction will cause initial deviations from the vertical to damp out. We can prove this easily by considering the zero solution of the associated linear system

$$\theta' = \mu,$$
$$\mu' = -\omega^2\theta - \epsilon\mu. \tag{17}$$

The characteristic equation is $\lambda^2 + \epsilon\lambda + \omega^2 = 0$ with the roots

$$\lambda_1 = \frac{-\epsilon + \sqrt{\epsilon^2 - 4\omega^2}}{2} \quad \text{and} \quad \lambda_2 = \frac{-\epsilon - \sqrt{\epsilon^2 - 4\omega^2}}{2}$$

Hence the origin of (17) is a stable node (if $\epsilon > 2\omega$), a stable node (if $\epsilon = 2\omega$), or a stable focus (if $\epsilon < 2\omega$). In all cases, by Theorem 2, the origin of the system (16) is asymptotically stable. The last case is the most interesting. The "focal behavior" near the origin implies that the pendulum will continue to oscillate, but with constantly decreasing amplitudes.

EXAMPLE 6

Consider the *RLC* circuit of Example 3.7.9. Equation (3.7.12) is an idealized description of the circuit, since there may very well be nonlinear terms present (such as induced currents from other sources). To take this factor into account, we have the equation

$$I'' + \frac{R}{L}I' + \frac{1}{CL}I + f(I, I') = 0, \tag{18}$$

where $f(x, y)$ is a nonlinear function such that

$$f(0,0) = \frac{\partial f}{\partial x}\bigg|_{(0,0)} = \frac{\partial f}{\partial y}\bigg|_{(0,0)} = 0.$$

Equation (18) can be written in the form

$$I' = y,$$

$$y' = -\frac{1}{CL}I - \frac{R}{L}y - f(I, y).$$

The associated linear system is

$$I' = y,$$

$$y' = -\frac{1}{CL}I - \frac{R}{L}y. \tag{19}$$

But as we showed in Example 3.7.9, the zero solution of (19) is asymptotically stable, and so by Theorem 2 the zero solution of (18) is also asymptotically stable. Therefore, in analyzing the circuit, we may safely ignore the nonlinear terms for small initial values of I and I'.

Problems 3.8

In Problems 1 through 6, verify that in each case $(0, 0)$ is a critical point and determine the asymptotic behavior of solutions near that point.

1. $x' = 3 \sin x + e^y - 1, \quad y' = xy - y$

2. $x' = \ln(1 + y) + \cosh x - 1, \quad y' = \tan y + 2x$

3. $x' = 1 - y - e^{-x}, \quad y' = y - \sin x$

4. $x' = 2 \cos y - \frac{1}{2}y + e^x - 3 - \sin 2x, \quad y' = 2 \tan x - y$

5. $x' = (\sin x)(\cos x) + y^2, \quad y' = y^2 - x - x^3 + 3y$

6. $x' = -y + \epsilon x(1 - y^2), \quad \epsilon > 0, \quad y' = x$

In Problems 7–13, determine the critical points of each nonlinear equation; find the associated linear system for each of these critical points; determine, if possible, the nature of each critical point and its stability properties; and sketch the orbits near each such point.

7. $x' = x + x^3, \quad y' = y + y^3$

8. $x' = -\sin x + x^2, \quad y' = \sin y$

9. $x' = x - x^2 + xy, \quad y' = 2y - xy - 6y^2$

10. $x' = -e^y + 1, \quad y' = e^x - 1$

11. $x' = -xy^2 + y^2 - 7xy - x^2 - 6x, \quad y' = x^2 + y$

12. $x' = 1 - xy, \quad y' = x - y^3$

13. $x' = 2y, \quad y' = -2x - y + y^4$

14. (a) Convert the equation $x'' + ax' + bx + x^2 = 0$, $a, b > 0$, into a system.

(b) Show that the origin is a stable focus or node and the point $(-b, 0)$ is a saddle point.

(c) Conclude that the orbits have the form shown in Fig. 3.

(d) Sketch in the isoclines.

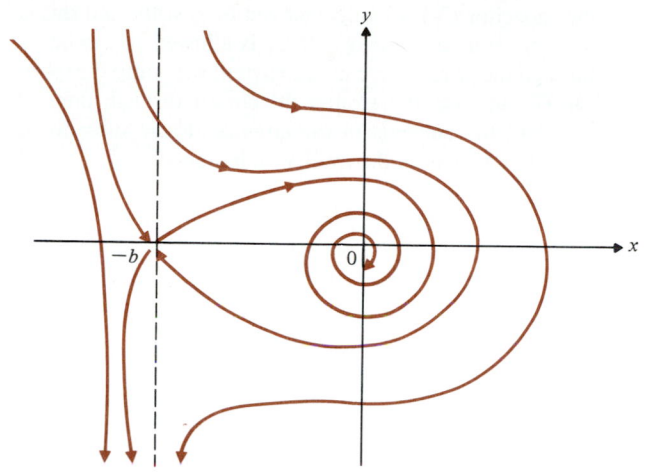

FIGURE 3

* **15.** The **Van der Pol equation**

$$x'' + \epsilon(x^2 - 1)x' + x = 0 \qquad (20)$$

arises in the study of a vacuum tube with three internal

elements (triode). Show that if $\epsilon < 0$, the origin is asymptotically stable. [It can also be shown that if $\epsilon > 0$, equation (20) has a periodic solution that is approached by all other solutions as t tends to ∞. This phenomenon is called a **limit cycle.†**]

16. Consider the following special case of the Lotka-Volterra equations (7):

$$\begin{aligned} x' &= -x - 2x^2 + xy, \\ y' &= -y + 7xy - 2y^2, \end{aligned} \qquad (21)$$

where x and y are measured in hundreds of organisms.

(a) Show that the system (21) has four critical points but that only two of them have any biological meaning (negative populations are not permissible).

(b) Show that if both populations are small initially, then both species will become extinct.

(c) Show that one biologically meaningful critical point is a saddle point, indicating that larger initial populations may lead to continued existence without the threat of extinction.

Review Exercises for Chapter 3

In Exercises 1–5 transform the equation into a first-order system.

1. $x'' + 4x = 0$

2. $x'' + 2x' + 5x = e^t$

3. $x''' - 6x'' + 2x' - 5x = 0$

4. $x'' - 3x' + 4t^2x = \sin t$

5. $xx'' + x'x''' = \ln t$

In Exercises 6–10, find the general solution for each system.

6. $x' = x + y, \quad y' = 9x + y$

7. $x' = x + 2y, \quad y' = 4x + 3y$

8. $x' = 4x - y, \quad y' = x + 2y$

9. $x' = 3x + 2y, \quad y' = -5x + y$

10. $x' = x - 4y, \quad y' = x + y$

11. Find the general solution to

$$\begin{aligned} x' &= -x - 3e^{-2t}, \\ y' &= -2x - y - 6e^{-2t}. \end{aligned}$$

12. Find the unique solution to

$$\begin{aligned} x' &= -4x - 6y + 9e^{-3t}, \ x(0) = -9, \\ y' &= x + y - 5e^{-3t}, \ y(0) = 4. \end{aligned}$$

13. A direct-current transmission line of length L, connecting a power source to a distant receiver, is subject to (i) drops in voltage due to the resistance of the line and (ii) leakage of current along the line due to imperfect insulation. If $E(x)$ and $I(x)$ are the voltage and current at a distance x from the power source, R is the resistance (ohms) per unit length of line, and G is the leakance (conductance) (mhos) per unit length, find

(a) A system of differential equations describing the voltage and current in the transmission line.

(b) The solution to part (a).

(c) The current and voltage at the end of the line.

14. J. P. Brady and C. Marmasse [*Psychological Record*, vol. 12 (1962), pp. 361–368] obtained the initial value problem

$$ay'' + y' + by = \beta, \qquad y(0) = \alpha, y'(0) = \beta$$

† See, for example, H. K. Wilson, *Ordinary Differential Equations,* Addison-Wesley, Reading, Mass., 1972.

to describe avoidance learning in rats. Here $y(t)$ is the value of the learning curve of the rat at time t, α and β are the initial values, and a and b are constants

(a) Obtain a system of first-order equations equivalent to the given initial value problem.

(b) Solve the system in part (a).

15. Tank X contains 150 gallons of pure water and tank Y 150 gallons of brine in which 60 pounds of salt is dissolved. Liquid circulates from each tank to the other at the rate of 9 gallons per minute, the mixture kept uniform in each tank by stirring.

(a) How much salt is in each tank at any time t?

(b) How much salt is in each tank as $t \to \infty$?

(c) When does tank X contain the maximum amount of salt?

16. Find the current I_2 in the indicated loop for the circuit in Fig. 1, where $R = 10$ ohms, and $L = 1$ henry, assuming $I_1 = I_2 = 0$ when the switch is closed at time $t = 0$.

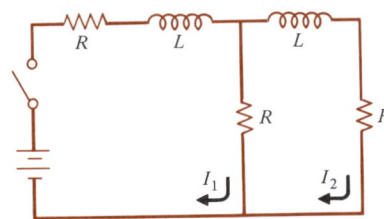

FIGURE 1

17. S. Grossberg [*Journal of Theoretical Biology* (1969), pp. 325–364] obtained the initial value problem

$$\frac{dx}{dt} = a(b - x) - cy, \qquad x(0) = b,$$

$$\frac{dy}{dt} = d(x - y) - (a + c)y, \qquad y(0) = \frac{bd}{a + d},$$

where a, b, c, d are positive constants. Show that y decays monotonically to a positive minimum.

18. A transformer having a spark gap in the primary circuit (see Fig. 2) is called an **oscillation transformer.** Suppose

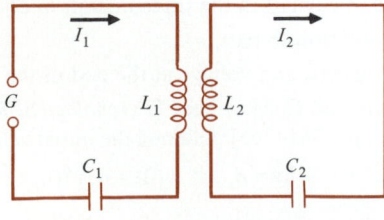

FIGURE 2

the capacitor C_1 is initially charged to e_0 volts, and this is the only source of energy. If C_1 is allowed to discharge through the primary circuit, the current will jump the spark gap G and continue around the circuit through the coil L_1. Find the currents in the circuits at any subsequent time. Let M denote the mutual inductance.

19. Using the information given in Fig. 3, derive a system of differential equations describing at any time the number of pounds of salt $x(t)$, $y(t)$, $z(t)$ in tanks X, Y, Z, respectively.

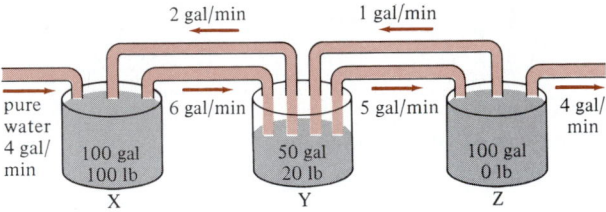

FIGURE 3

20. In problem 19:

(a) When does tank Y contain the maximum amount of salt?

(b) When does tank Z contain the maximum amount of salt?

In Exercises 21–24, describe the nature of the critical point $(0, 0)$ of each system and sketch the orbits.

21. $x' = -2x - 2y$, $\quad y' = -5x + y$

22. $x' = -12x + 7y$, $\quad y' = -7x + 2y$

23. $x' = 2x - y$, $\quad y' = 5x - 2y$

24. $x' = 3x + 2y$, $\quad y' = -5x + y$

In Exercises 25 and 26, verify that $(0, 0)$ is a critical point and determine the asymptotic behavior of solutions near that point.

25. $x' = 4(\sin x)e^{2y}$, $\quad y' = 2xy - 3y$

26. $x' = 3 \cos y - 2y - 4e^x + 1 - \sin x$, $y' = \tan x + y$

In Exercises 27–33, (a) determine the critical points, (b) find the associated linear system for each critical point, and (c) determine the nature of each critical point for each nonlinear system.

27. $x' = -x + 2x^3, \quad y' = 2y + y^3$

28. $x' = -x - x \sin x, \quad y' = \sin y$

29. $x' = -2x - xy + 2x^2, \quad y' = 5y + xy - 2y^2$

30. $x' = y, \quad y' = \epsilon(1 - x^2)y - x$

31. $x' = y - \epsilon(\frac{1}{3}x^3 - x), \quad y' = -x$

32. $x' = y, \quad y' = x^2(x^4 - 3)/(x^4 + 1)^2$

33. $x' = y, \quad y' = -x - x^3$

* 34. (Bendixon-DuLac Criterion)

The system

$$x' = F(x, y)$$

$$y' = G(x, y)$$

has no periodic solutions in a region Ω if there exists a continuously differentiable function $B(x, y)$ satisfying

$$(BF)_x + (BG)_y > 0$$

in D. Prove this.

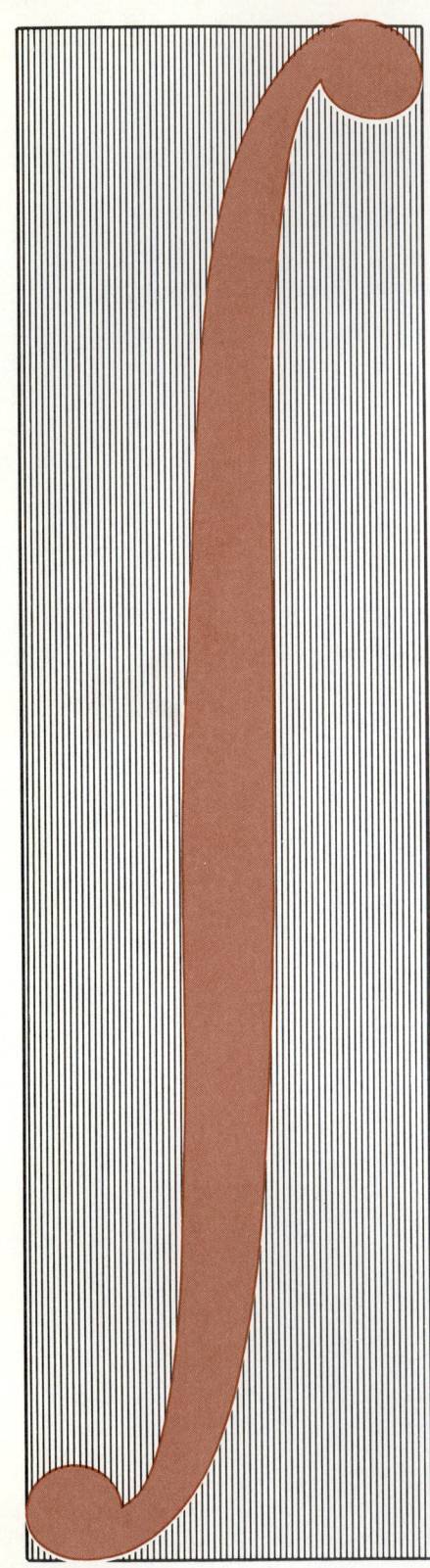

chapter 4

Laplace Transforms

4.1 Introduction: Definition and Basic Properties of the Laplace Transform

One of the most efficient methods of solving certain ordinary and partial differential equations is to use Laplace† transforms. The effectiveness of the Laplace transform is due to its ability to convert a differential equation into an algebraic equation, whose solution yields the solution of the differential equation when the transformation is reversed.

The Laplace transform is defined as a certain integral over the range from zero to infinity. We recall that such an integral is called an **improper integral.** Formally, if t_0 is a given real number, then for any function $f(t)$, we define

$$\int_{t_0}^{\infty} f(t)\, dt = \lim_{A \to \infty} \int_{t_0}^{A} f(t)\, dt.$$

If this limit exists and is finite, we say that the improper integral **converges.** Otherwise, it **diverges.**

EXAMPLE 1

Let $f(t) = e^{at}$, where $a \neq 0$. Then

$$\int_{0}^{\infty} e^{at}\, dt = \lim_{A \to \infty} \int_{0}^{A} e^{at}\, dt = \lim_{A \to \infty} \frac{1}{a} e^{at} \Big|_{0}^{A} = \lim_{A \to \infty} \frac{1}{a}(e^{aA} - 1).$$

Clearly this limit is finite (and is equal to $-1/a$) if $a < 0$ and diverges to $+\infty$ if $a > 0$.

---

† Named after the French mathematician P. S. Laplace (1749–1827). See the accompanying biographical sketch.

PIERRE-SIMON LAPLACE
(1749–1827)

Courtesy of Brown Brothers

Pierre-Simon Laplace was born of poor parents in 1749. His mathematical ability early won him good teaching posts, and as a political opportunist he ingratiated himself with whichever party happened to be in power during the uncertain days of the French Revolution.

His most outstanding work was done in the fields of celestial mechanics, probability, differential equations, and geodesy. He published two monumental works, *Traité de mécanique céleste* (five volumes, 1799–1825) and *Théorie analytique des probabilités* (1812), each of which was preceded by an extensive nontechnical exposition. The five-volume *Traité de mécanique céleste,* which earned him the title of "the Newton of France," embraced all previous discoveries in this field along with Laplace's own contributions and marked the author as the unrivaled master in the subject. It may be of interest to repeat a couple of anecdotes often told in connection with this work. When Napoleon teasingly remarked that God was not mentioned in his treatise, Laplace replied, "Sire, I did not need that hypothesis." And the American astronomer Nathaniel Bowditch, when he translated Laplace's treatise into English, remarked, "I never come across one of Laplace's 'Thus it plainly appears' without feeling sure that I have hours of hard work before me to fill up the chasm and find out and show how it plainly appears." Laplace's name is connected with the *nebular hypothesis* of cosmogony and with the so-called *Laplace equation* of potential theory (though neither of these contributions originated with Laplace), with the *Laplace transform* that later became the key to the operational calculus of Heaviside, and with the *Laplace expansion* of a determinant. Laplace died in 1827, exactly one hundred years after the death of Isaac Newton. According to one report, his last words were: "What we know is slight; what we don't know is immense."

The following story about Laplace is of interest and offers a valuable suggestion to one applying for a position. When Laplace arrived as a young man in Paris seeking a professorship of mathematics, he submitted his recommendations by prominent people to d'Alembert, but was not received. Returning to his lodgings, Laplace wrote d'Alembert a brilliant letter on the general principles of mechanics. This opened the door, and d'Alembert replied: "Sir, you notice that I paid little attention to your recommendations. You don't need any; you have introduced yourself better." A few days later Laplace was appointed professor of mathematics at the Military School of Paris.

Laplace was very generous to beginners in mathematical research. He called these beginners his stepchildren, and there are several instances in which he withheld publication of a discovery to allow a beginner the opportunity to publish first. Sadly, such generosity is rare in mathematics.

We close our brief account of Laplace with two quotations due to him. "All the effects of nature are only mathematical consequences of a small number of immutable laws." "In the final analysis, the theory of probability is only common sense expressed in numbers."

EXAMPLE 2

Let $f(t) = \cos t$. Then

$$\int_0^\infty \cos t \, dt = \lim_{A \to \infty} \int_0^A \cos t \, dt = \lim_{A \to \infty} \sin t \Big|_0^A = \lim_{A \to \infty} \sin A.$$

But $\sin A$ has no limit as $A \to \infty$. Therefore, $\int_0^\infty \cos t \, dt$ diverges, even though $-1 \leq \int_0^A \cos t \, dt \leq 1$ for every $A \geq 0$.

In the rest of this chapter we shall not rigorously calculate improper integrals. However, you should not forget that an improper integral is a special kind of limit.

DEFINITION

Let $f(t)$ be a real-valued function that is defined for $t \geq 0$. Suppose that $f(t)$ is multiplied by e^{-st} and the result is integrated with respect to t from zero to infinity. If the integral

$$F(s) = \mathcal{L}\{f(t)\} = \int_0^\infty e^{-st} f(t) \, dt \qquad (1)$$

Laplace Transform

converges, the resulting function of s† is called the **Laplace transform** of the function f. We stress that the original function f is a function of the variable t while its Laplace transform F is a function of the variable s.

It should now be apparent why the word **transform** is associated with this operation. The operation "transforms" the original function $f(t)$ into a new function $F(s) = \mathcal{L}\{f(t)\}$. Since we shall be interested in reversing the procedure, *Inverse Transform* we call the original function f the **inverse transform** of $F = \mathcal{L}\{f\}$ and denote it by $f = \mathcal{L}^{-1}\{F\}$. Note that the last statement translates into the mathematical symbols $\mathcal{L}^{-1}\{F\} = \mathcal{L}^{-1}\{\mathcal{L}\{f\}\} = f$. That is,

$$\text{if } \mathcal{L}\{f(t)\} = F(s), \text{ then } \mathcal{L}^{-1}\{F(s)\} = f(t). \qquad (2)$$

EXAMPLE 3

Let $f(t) = e^{at}$, where a is constant. Then

$$\mathcal{L}\{e^{at}\} = \int_0^\infty e^{-st} e^{at} \, dt = \frac{e^{-(s-a)t}}{a-s}\Big|_0^\infty,$$

† In general, the variable s is complex, but in this book we shall only consider real values of s.

which converges, if $s - a > 0$, to

$$\mathcal{L}\{e^{at}\} = \frac{1}{s-a}.$$

Thus the Laplace transform of the function e^{at} is the function $F(s) = 1/(s - a)$ for $s > a$. Note that we can also conclude, from (2), that $\mathcal{L}^{-1}\{F(s)\} = \mathcal{L}^{-1}\left\{\frac{1}{s-a}\right\} = e^{at}$.

The last example shows that $\mathcal{L}\{f(t)\}$ may not be defined for all values of s; but if it is defined, then it will exist for suitably large values of s. Indeed, $\mathcal{L}\{e^{at}\}$ is defined only for values $s > a$.

EXAMPLE 4

Let $f(t) = 1/t$. Then

$$\int_0^\infty \frac{e^{-st}}{t}\,dt = \int_0^1 \frac{e^{-st}}{t}\,dt + \int_1^\infty \frac{e^{-st}}{t}\,dt. \tag{3}$$

But for t in the interval $0 \le t \le 1$, $e^{-st} \ge e^{-s}$ if $s > 0$. Thus,

$$\int_0^\infty \frac{e^{-st}}{t}\,dt \ge e^{-s}\int_0^1 \frac{dt}{t} + \int_1^\infty \frac{e^{-st}}{t}\,dt.$$

However,

$$\int_0^1 t^{-1}\,dt = \lim_{A\to 0}\int_A^1 t^{-1}\,dt = \lim_{A\to 0} \ln t\,|_A^1$$

$$= \lim_{A\to 0}(\ln 1 - \ln A) = \lim_{A\to 0}(-\ln A) = +\infty.$$

Therefore, the integral (3) diverges, so that $f(t) = 1/t$ has no Laplace transform.

Examples 3 and 4 and the preceding discussion pinpoint the need for answers to the following questions:

1. Which functions f have Laplace transforms?
2. Can two functions f and g have the same Laplace transform?

In answering these questions, we shall be satisfied if a class of functions, large enough to contain virtually all functions that arise in practice, can be found for which the Laplace transforms exist and whose inverses are unique. We shall see in Section 4.2 that certain differential equations can be solved by first taking the Laplace transform of both sides of the equation, then finding the Laplace transform of the solution, and finally arriving at the answer by taking the unique inverse transform of the transform of the solution. The need for a unique inverse arises from the fact that we need to reverse the transformation to find the solution of a given problem and this step is not possible if the inverses within the given class of functions are not unique.

In order to give simple conditions that guarantee the existence of a Laplace transform, we shall require the following definitions.

Jump Discontinuity

A function f has a **jump discontinuity** at a point t^* if the function has different, finite limits as it approaches t^* from the left and from the right or if the two limits are equal but different from $f(t^*)$. Note that $f(t^*)$ may or may not be equal to either

$$\lim_{t \to t^{*+}} f(t) \quad \text{or} \quad \lim_{t \to t^{*-}} f(t).$$

Piecewise Continuous Function

In fact, $f(t^*)$ may not even be defined. A function f defined on $(0, \infty)$ is **piecewise continuous** if it is continuous on every finite interval $0 \le t \le b$, except possibly at finitely many points in each such finite interval where it has jump discontinuities. Such a function is illustrated in Fig. 1. The class of piecewise continuous functions includes every continuous function as well as many important discontinuous functions, such as the unit step function, square waves, and the staircase function, which we shall encounter later in this chapter.

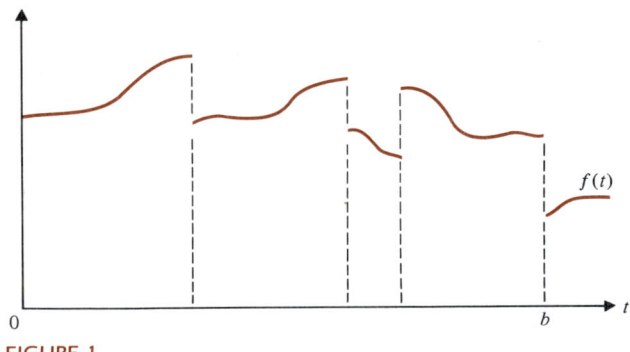

FIGURE 1

■ **THEOREM 1: Existence Theorem**

Let f be piecewise continuous on $t \ge 0$ and satisfy the condition

$$|f(t)| \le Me^{at} \tag{4}$$

for $t \ge T$ and for fixed nonnegative constants a, M, and T. Then $\mathcal{L}\{f(t)\}(s)$ exists for all $s > a$.

PROOF

Since f is piecewise continuous, $e^{-st}f(t)$ has a finite integral over any finite interval on $t \ge 0$, and

$$|\mathcal{L}\{f(t)\}| = \left| \int_0^\infty e^{-st}f(t)\,dt \right| \le \left| \int_0^T e^{-st}f(t)\,dt \right| + \int_T^\infty e^{-st}|f(t)|\,dt. \tag{5}$$

The first integral on the right-hand side of equation (5) exists, and by equation (4)

$$\int_T^\infty e^{-st}|f(t)|\,dt \le M \int_T^\infty e^{-(s-a)t}\,dt = \left. \frac{Me^{-(s-a)t}}{a-s} \right|_T^\infty,$$

which converges to $Me^{-(s-a)T}/(s-a)$ as $t \to \infty$ if $s > a$. Thus it follows by the comparison theorem of integrals that $\mathcal{L}\{f(t)\}$ exists. ∎

The conditions in Theorem 1 are easy to test. For example, $\mathcal{L}\{t^n\}$ exists because

$$t^n \le n!e^t, \qquad \text{for any } t > 0,$$

To see this observe that

$$e^t = 1 + t + \frac{t^2}{2!} + \cdots + \frac{t^n}{n!} + \cdots,$$

so that $t^n/n! \le e^t$, or $t^n \le n!e^t$, for $t > 0$. On the other hand

$$t^2 > \ln M + at$$

for sufficiently large t, regardless of the choice of M and a. Exponentiating both sides of this inequality, we get

$$e^{t^2} > e^{\ln M + at} = e^{\ln M}e^{at} = Me^{at}.$$

It is not hard to show that e^{t^2} has no Laplace transform.

It should also be noted that there are functions having Laplace transforms that do not satisfy the hypotheses of Theorem 1. For example, $f(t) = 1/\sqrt{t}$ is infinite at $t = 0$; but setting $x^2 = st$, we have

$$\mathcal{L}\{t^{-1/2}\} = \int_0^\infty e^{-st} t^{-1/2}\, dt = \frac{2}{\sqrt{s}} \int_0^\infty e^{-x^2}\, dx = \sqrt{\frac{\pi}{s}}$$

according to Formula 216 in the table of integrals.

The proof of uniqueness would take us too far afield at this point. However, it can be shown† that two functions having the same Laplace transform cannot differ at every point in an interval of positive length, although they may differ at several isolated points. Thus two different piecewise continuous functions having the same Laplace transform can differ only at isolated points. Such differences are generally of no importance in applications. Hence the Laplace transform has an essentially unique inverse. In particular, different continuous functions have different Laplace transforms.

One of the most important properties of the Laplace transform is stated in the following theorem.

■ **THEOREM 2: Linearity Property of the Laplace Transform**

If $\mathcal{L}\{f(t)\}$ and $\mathcal{L}\{g(t)\}$ exist, then $\mathcal{L}\{af(t) + bg(t)\}$ exists and

$$\mathcal{L}\{af(t) + bg(t)\} = a\mathcal{L}\{f(t)\} + b\mathcal{L}\{g(t)\}.$$

† See, for example, I. S. Sokolnikoff and R. M. Redheffer, *Mathematics of Physics and Modern Engineering,* McGraw-Hill, New York, 1966, p. 217.

PROOF
By definition

$$\mathcal{L}\{af(t)+bg(t)\} = \int_0^\infty e^{-st}[af(t)+bg(t)]\,dt$$

$$= a\int_0^\infty e^{-st}f(t)\,dt + b\int_0^\infty e^{-st}g(t)\,dt$$

$$= a\mathcal{L}\{f(t)\} + b\mathcal{L}\{g(t)\} = aF(s)+bG(s). \qquad \blacksquare$$

■ **COROLLARY: Linearity Property of the Inverse Laplace Transform**

If $F(s) = \mathcal{L}\{f(t)\}$ and $G(s) = \mathcal{L}\{g(t)\}$, then

$$\mathcal{L}^{-1}\{aF(s)+bG(s)\} = a\mathcal{L}^{-1}\{F(s)\} + b\mathcal{L}^{-1}\{G(s)\}$$

$$= af(t)+bg(t).$$

The proof of the corollary is left as an exercise (Problem 79).   ■

The linearity property allows us to deal with a linear equation term by term in order to obtain its Laplace transform.

It is essential that we begin to recognize the Laplace transforms of many different functions as quickly as possible. We list in Table 1 the transforms of seven basic functions. A much more extensive list can be found in Appendix 2. Note that, by definition, $0! = 1$.

TABLE 1

$f(t)$	$\mathcal{L}\{f(t)\}$	Domain of definition	
e^{at}	$\dfrac{1}{s-a}$	$s > a$ (real)	**(6)**
c (a constant)	$\dfrac{c}{s}$	$s > 0$	**(7)**
$t^n\ (n \geq 0)$	$\dfrac{n!}{s^{n+1}}$	$s > 0$	**(8)**
$\sin at$	$\dfrac{a}{s^2+a^2}$	$s > 0$	**(9)**
$\cos at$	$\dfrac{s}{s^2+a^2}$	$s > 0$	**(10)**
$\sinh at$	$\dfrac{a}{s^2-a^2}$	$s > |a|$	**(11)**
$\cosh at$	$\dfrac{s}{s^2-a^2}$	$s > |a|$	**(12)**

   Equation (6) has already been obtained in Example 3. The derivations of the other six transforms are given below. For equation (7), we have

$$\mathcal{L}\{c\} = \int_0^\infty e^{-st}c\,dt = \frac{ce^{-st}}{-s}\bigg|_0^\infty = \frac{c}{s}, \qquad \text{for } s > 0.$$

EXAMPLE 5

Let n be a positive integer. Then equation (8) is obtained by integrating

$$\mathcal{L}\{t^n\} = \int_0^\infty e^{-st}t^n\,dt$$

by parts with $u = t^n$ and $dv = e^{-st}\,dt$. We have $du = nt^{n-1}\,dt$ and $v = -(1/s)e^{-st}$:

$$\mathcal{L}\{t^n\} = \frac{-e^{-st}t^n}{s}\bigg|_0^\infty + \frac{n}{s}\int_0^\infty e^{-st}t^{n-1}\,dt = \frac{n}{s}\mathcal{L}\{t^{n-1}\}, \qquad \text{for } s > 0,$$

because $\lim_{t\to\infty}(e^{-st}t^n/s) = 0$, if $s > 0$, by n applications of L'Hôpital's rule. Thus

$$\mathcal{L}\{t^n\} = \frac{n}{s}\mathcal{L}\{t^{n-1}\} = \frac{n(n-1)}{s^2}\mathcal{L}\{t^{n-2}\} = \cdots = \frac{n!}{s^n}\mathcal{L}\{t^0\}.$$

But $t^0 = 1$, so by equation (7) we have $\mathcal{L}\{1\} = 1/s$ and $\mathcal{L}\{t^n\} = n!/s^{n+1}$.

EXAMPLE 6

The Laplace transform of $\sin at$ can also be obtained by integrating by parts. Let $u = \sin at$ and $dv = e^{-st}\,dt$; then $du = a\cos at\,dt$ and $v = -(1/s)e^{-st}$, so that

$$\mathcal{L}\{\sin at\} = \int_0^\infty e^{-st}\sin at\,dt = \frac{-e^{-st}\sin at}{s}\bigg|_0^\infty + \frac{a}{s}\int_0^\infty e^{-st}\cos at\,dt,$$

or

$$\mathcal{L}\{\sin at\} = \frac{a}{s}\mathcal{L}\{\cos at\}. \tag{13}$$

We must integrate this last integral again by parts with $u = \cos at$ and $dv = e^{-st}\,dt$, so that $du = -a\sin at\,dt$ and $v = -(1/s)e^{-st}$, yielding

$$\mathcal{L}\{\sin at\} = \frac{a}{s}\int_0^\infty e^{-st}\cos at\,dt = \frac{a}{s}\left[-\frac{e^{-st}\cos at}{s}\bigg|_0^\infty - \frac{a}{s}\int_0^\infty e^{-st}\sin at\,dt\right]$$

or

$$\mathcal{L}\{\sin at\} = \frac{a}{s^2} - \frac{a^2}{s^2}\mathcal{L}\{\sin at\}. \tag{14}$$

Moving the last term on the right-hand side of equation (14) to the left-hand side, we get

$$\left(1 + \frac{a^2}{s^2}\right)\mathcal{L}\{\sin at\} = \frac{a}{s^2} \quad \text{or} \quad \frac{s^2 + a^2}{s^2}\mathcal{L}\{\sin at\} = \frac{a}{s^2}.$$

Dividing both sides by $s^2/(s^2 + a^2)$, we obtain

$$\mathcal{L}\{\sin at\} = \frac{a}{s^2 + a^2}.$$

Equation (10) follows at once from equations (9) and (13) since

$$\mathcal{L}\{\cos at\} = \frac{s}{a}\mathcal{L}\{\sin at\} = \frac{s}{s^2 + a^2}.$$

Explain why these computations are valid only if $s > 0$.

Problem 41 illustrates another method for finding the Laplace transform of cos at and sin at by using the Euler equations (see Section 15.7): $e^{iat} = \cos(at) + i\sin(at)$.

EXAMPLE 7

Since sinh $at = (e^{at} - e^{-at})/2$, we can use Theorem 2 and equation (6) to obtain

$$\mathcal{L}\{\sinh at\} = \mathcal{L}\left\{\frac{e^{at} - e^{-at}}{2}\right\} = \frac{1}{2}(\mathcal{L}\{e^{at}\} - \mathcal{L}\{e^{-at}\})$$

$$= \frac{1}{2}\left(\frac{1}{s-a} - \frac{1}{s+a}\right) = \frac{a}{s^2 - a^2}.$$

Similarly,

$$\mathcal{L}\{\cosh at\} = \mathcal{L}\left\{\frac{e^{at} + e^{-at}}{2}\right\} = \frac{1}{2}(\mathcal{L}\{e^{at}\} + \mathcal{L}\{e^{-at}\})$$

$$= \frac{1}{2}\left(\frac{1}{s-a} + \frac{1}{s+a}\right) = \frac{s}{s^2 - a^2}.$$

EXAMPLE 8

Compute $\mathcal{L}\{3t^5 - t^8 + 4 - 5e^{2t} + 6\cos 3t\}$.

SOLUTION

Using the linearity property and the results in Table 1, we have

$$\mathcal{L}\{3t^5 - t^8 + 4 - 5e^{2t} + 6\cos 3t\}$$

$$= 3\mathcal{L}\{t^5\} - \mathcal{L}\{t^8\} + \mathcal{L}\{4\} - 5\mathcal{L}\{e^{2t}\} + 6\mathcal{L}\{\cos 3t\}$$

$$= 3\frac{5!}{s^6} - \frac{8!}{s^9} + \frac{4}{s} - \frac{5}{s-2} + \frac{6s}{s^2 + 9}$$

$$= \frac{360}{s^6} - \frac{40,320}{s^9} + \frac{4}{s} - \frac{5}{s-2} + \frac{6s}{s^2 + 9}.$$

Many other facts can be used to facilitate the computation of Laplace transforms. One of these is given below. Other facts will be given in the exercises and in the next three sections.

The next theorem presents a quick way to compute the Laplace transform $\mathcal{L}\{e^{at}f(t)\}$ when $\mathcal{L}\{f(t)\}$ is known.

■ **THEOREM 3: First Shifting Property**
of the Laplace Transform

Suppose that $F(s) = \mathcal{L}\{f(t)\}$ exists for $s > b$. If a is a real number, then

$$\mathcal{L}\{e^{at}f(t)\} = F(s-a), \qquad \text{for} \quad s > a+b. \tag{15}$$

Rewriting (15) in terms of the inverse Laplace transform, we have

$$\mathcal{L}^{-1}\{F(s-a)\} = e^{at}f(t). \tag{16}$$

PROOF
By definition,

$$\mathcal{L}\{e^{at}f(t)\} = \int_0^\infty e^{-st}e^{at}f(t)\,dt = \int_0^\infty e^{-(s-a)t}f(t)\,dt$$

$$= \mathcal{L}\{f(t)\}|_{s-a} = F(s-a),$$

if $s - a > b$ or $s > a + b$. As the formula suggests, to find $\mathcal{L}\{e^{at}f(t)\}$, we simply replace each s in $\mathcal{L}\{f(t)\}$ by $s - a$. ■

EXAMPLE 9

Compute $\mathcal{L}\{e^{2t}\cos 3t\}$.

SOLUTION
Since

$$F(s) = \mathcal{L}\{\cos 3t\} = \frac{s}{s^2+9}$$

and $a = 2$, we have

$$\mathcal{L}\{e^{2t}\cos 3t\} = F(s-2) = \frac{s-2}{(s-2)^2+9}.$$

If we apply the First Shifting Property to the Laplace transforms in Table 1, we obtain the results shown in Table 2 on page 198.

In Section 4.2 we will have frequent need to evaluate inverse Laplace transforms. The next example demonstrates the use of factoring in finding inverse Laplace transforms.

TABLE 2

$f(t)$	$\mathcal{L}\{f(t)\}$	Domain of definition
$e^{at}t^n$	$\dfrac{n!}{(s-a)^{n+1}}$	$s > a$
$e^{at}\sin bt$	$\dfrac{b}{(s-a)^2+b^2}$	$s > a$
$e^{at}\cos bt$	$\dfrac{s-a}{(s-a)^2+b^2}$	$s > a$
$e^{at}\sinh bt$	$\dfrac{b}{(s-a)^2-b^2}$	$s > a + |b|$
$e^{at}\cosh bt$	$\dfrac{s-a}{(s-a)^2-b^2}$	$s > a + |b|$

EXAMPLE 10

Compute

$$\mathcal{L}^{-1}\left\{\frac{s+9}{s^2+6s+13}\right\}.$$

SOLUTION

Example 9 provides a hint. We begin by completing the square in the denominator:

$$\frac{s+9}{s^2+6s+13} = \frac{s+9}{(s+3)^2+4}.$$

But $4 = 2^2$ and $s + 9 = (s + 3) + 6$, so that we can write

$$\frac{s+9}{s^2+6s+13} = \frac{(s+3)+6}{(s+3)^2+2^2}.$$

By the linearity property of the inverse Laplace transform (see the corollary to Theorem 2), the inverse transform of a sum equals the sum of the inverse transforms. Hence

$$\mathcal{L}^{-1}\left\{\frac{(s+3)+6}{(s+3)^2+2^2}\right\} = \mathcal{L}^{-1}\left\{\frac{(s+3)}{(s+3)^2+2^2}\right\} + 3\mathcal{L}^{-1}\left\{\frac{2}{(s+3)^2+2^2}\right\}$$

and using Table 2 (or the First Shifting Property and Table 1), we have

$$\mathcal{L}^{-1}\left\{\frac{(s+3)+6}{(s+3)^2+2^2}\right\} = e^{-3t}\cos 2t + 3e^{-3t}\sin 2t.$$

Problems 4.1

In Problems 1–40 find the Laplace transforms of the following functions, where a, b, and c are real constants. For what values of s are the transforms defined?

1. $5t + 2$

2. $7t - 8$

3. $9t^2 - 7$

4. $16t^2 - 4t$

5. $t^2 + 8t - 16$

6. $27t^3 - 9t + 4$

7. $\dfrac{t^3}{8} + \dfrac{t^2}{4} + \dfrac{t}{2} + 1$

8. $\dfrac{t^5}{120} + \dfrac{t^2}{6} + 1$

9. $at + b$

10. $at^2 + bt + c$

11. e^{5t+2}

12. e^{7t-8}

13. $e^{t/2}$

14. $e^{-t/3}$

15. $e^{-t-1/2}$

16. e^{at+b}

17. $\sin(3t)$

18. $\sin\dfrac{t}{2}$

19. $\cos(7t)$

20. $\cos(-t/3)$

21. $\sin(5t + 2)$

22. $\cos(7t - 8)$

23. $\cos(at + b)$

24. $\sin(at + b)$

25. $\cosh(t/2)$

26. $\sinh(-t/3)$

27. $\cosh(5t - 2)$

28. $\sinh(7t + 8)$

29. $\sinh(at + b)$

30. $\cosh(at + b)$

31. te^t

32. $t^2 e^{2t}$

33. $(t^3 - 1)e^{-t}$

34. $e^{3t}(t^2 + t)$

35. $e^t \sin t$

36. $e^{-t} \sin 2t$

37. $e^{4t}(\cos 2t)$

38. $e^{-t} \sinh 2t$

39. $e^{-t}(\sin t + \cos t)$

40. $e^{2t}(t + \cosh t)$

41. Using $e^{iat} = \cos at + i \sin at$ (see p. 83):
   (a) Show that $\mathcal{L}\{e^{iat}\} = 1/(s - ia)$, $s > 0$.
   (b) Show that $1/(s - ia) = (s + ia)/(s^2 + a^2)$.
   (c) Use parts (a) and (b) to derive (without integration by parts) the formulas for $\mathcal{L}\{\sin at\}$ and $\mathcal{L}\{\cos at\}$. [*Hint:* Equate real and imaginary parts.]

In Problems 42–55, find $f(t)$ where $F(s) = \mathcal{L}\{f(t)\}$ is given. If necessary, complete the square in the denominator.

42. $\dfrac{7}{s^2}$

43. $\dfrac{18}{s^3} + \dfrac{7}{s}$

44. $\dfrac{a_1}{s} + \dfrac{a_2}{s^2} + \dfrac{a_3}{s^3} + \cdots + \dfrac{a_{n+1}}{s^{n+1}}$

45. $\dfrac{s+1}{s^2+1}$

46. $\dfrac{7}{s-3}$

47. $\dfrac{s-2}{s^2-2}$

48. $\dfrac{s-2}{s^2+3}$

49. $\dfrac{1}{(s-1)^2}$

50. $\dfrac{3}{s^2+2s+2}$

51. $\dfrac{3}{s^2+4s+9}$

52. $\dfrac{s+12}{s^2+10s+35}$

53. $\dfrac{2s-1}{s^2+2s+8}$

54. $\dfrac{7s-8}{s^2+9s+25}$

55. $\dfrac{cs+d}{s^2+2as+b}$, $b > a^2 > 0$; a, b, c, d are real

In Problems 56–61, express each given hyperbolic function in terms of exponentials and apply the first shifting theorem to prove the given equality.

56. $\mathcal{L}\{\cosh^2 at\} = \dfrac{s^2 - 2a^2}{s(s^2 - 4a^2)}$

57. $\mathcal{L}\{\sinh^2 at\} = \dfrac{2a^2}{s(s^2 - 4a^2)}$

58. $\mathcal{L}\{\cosh at \sin at\} = \dfrac{a(s^2 + 2a^2)}{s^4 + 4a^4}$

59. $\mathcal{L}\{\cosh at \cos at\} = \dfrac{s^3}{s^4 + 4a^4}$

60. $\mathcal{L}\{\sinh at \sin at\} = \dfrac{2a^2 s}{s^4 + 4a^4}$

61. $\mathcal{L}\{\sinh at \cos at\} = \dfrac{a(s^2 - 2a^2)}{s^4 + 4a^4}$

Using the method above, find the Laplace transforms in Problems 62–67.

62. $\mathcal{L}\{\cosh at \cosh bt\}$

63. $\mathcal{L}\{\sinh at \sinh bt\}$

64. $\mathcal{L}\{\cosh at \sin bt\}$

65. $\mathcal{L}\{\cosh at \cos bt\}$

66. $\mathcal{L}\{\sinh at \sin bt\}$

67. $\mathcal{L}\{\sinh at \cos bt\}$

68. Suppose that $F(s) = \mathcal{L}\{f(t)\}$ exists for $s > a$. Show that
$$\mathcal{L}\{tf(t)\} = -F'(s), \qquad \text{for} \quad s > a. \tag{17}$$
[*Hint:* Assume that you can interchange the derivative and integral on the right-hand side of equation (17).]

69. Use equation (17) to show that
$$\mathcal{L}\{t^n f(t)\} = (-1)^n \frac{d^n}{ds^n} F(s), \qquad \text{for} \quad s > a. \tag{18}$$

Use equations (17) and (18) to compute the Laplace transform of the functions given in Problems 70–78. Assume a and b are real.

70. te^t

71. $t^3 e^{-t}$

72. $t \sin t$

73. $t^2 \cos 3t$

74. $te^t \sin t$

75. $te^{at} \cos bt$

76. $te^{at} \sin bt$

77. $3te^{-t} \cosh t$

78. $te^{-t} \sinh 2t$

79. Suppose that $f(t) = \mathcal{L}^{-1}\{F(s)\}$ and that $g(t) = \mathcal{L}^{-1}\{G(s)\}$. Prove the linearity property of the inverse Laplace transform:
$$af(t) + bg(t) = \mathcal{L}^{-1}\{aF(s) + bG(s)\},$$
where a and b are any real constants.

80. The **gamma function** is defined by
$$\Gamma(x) = \int_0^\infty e^{-u} u^{x-1}\, du, \quad x > 0.$$
   (a) Show that $\Gamma(x + 1) = \int_0^\infty e^{-u} u^x\, du$.
   (b) By integrating by parts, show that $\Gamma(x + 1) = x\Gamma(x)$.

(c) Show that $\Gamma(1) = 1$.

(d) Using the results of parts (b) and (c), show that if n is a positive integer, then $\Gamma(n + 1) = n!$

(e) By making the substitution $u = st$ in part (a), show that

$$\mathcal{L}\{t^x\} = \frac{\Gamma(x+1)}{s^{x+1}}, \qquad s > 0,\ x > -1.$$

81. It can be shown that $\Gamma(1/2) = \sqrt{\pi}$. Use this fact and the results of Problem 80 to compute

(a) $\mathcal{L}\{1/\sqrt{t}\}$

(b) $\mathcal{L}\{\sqrt{t}\}$

(c) $\mathcal{L}\{t^{5/2}\}$

4.2 Solving Initial Value Problems by Laplace Transform Methods

In this section we shall show how the theory developed in Section 4.1 can be applied to solve linear initial value problems. We shall see that the Laplace transform converts linear initial value problems with constant coefficients into algebraic equations whose solution is the Laplace transform of the solution to the initial value problem.

The most important property of Laplace transforms for solving differential equations concerns the transform of the derivative of a function f. We prove below that differentiation of f roughly corresponds to multiplication of the transform by s.

■ **THEOREM 1: Differentiation Property**

Let $f(t)$ satisfy the condition

$$|f(t)| \le Me^{at} \tag{1}$$

for $t \ge T$, for fixed nonnegative constants a, M, and T, and suppose that $f'(t)$ is piecewise continuous for $t \ge 0$. Then the Laplace transform of $f'(t)$ exists for all $s > a$, and

$$\mathcal{L}\{f'(t)\} = s\mathcal{L}\{f(t)\} - f(0). \tag{2}$$

PROOF

Since f is differentiable, it is also continuous. Hence it satisfies the conditions of the existence theorem, Theorem 4.1.1 on p. 192, and has a Laplace transform. Suppose, first, that $f'(t)$ is continuous on $t \ge 0$. Then integrating $\mathcal{L}\{f'(t)\}$ by parts, we set $u = e^{-st}$, $dv = f'(t)\,dt$ so that $du = -se^{-st}\,dt$, $v = f(t)$, and

$$\mathcal{L}\{f'(t)\} = \int_0^\infty e^{-st}f'(t)\,dt = e^{-st}f(t)\Big|_0^\infty + s\int_0^\infty e^{-st}f(t)\,dt. \tag{3}$$

Since $f(t)$ satisfies equation (1), the first term on the right-hand side in equation (3) vanishes at the upper limit when $s > a$, and by definition we obtain $\mathcal{L}\{f'(t)\} = s\mathcal{L}\{f(t)\} - f(0)$. When $f'(t)$ is piecewise continuous, the proof is similar. We simply break up the range of integration into parts on each of which $f'(t)$ is continuous and integrate by parts as in equation (3). All first terms will cancel out or vanish except $-f(0)$, and the second terms will combine to yield $s\mathcal{L}\{f(t)\}$. ∎

Theorem 1 may be extended to apply to piecewise continuous functions $f(t)$ (see Problem 50 at the end of this section).

Equation (2) may be applied repeatedly to obtain the Laplace transform of higher-order derivatives:

$$\mathcal{L}\{f''(t)\} = s\mathcal{L}\{f'(t)\} - f'(0) = s[s\mathcal{L}\{f(t)\} - f(0)] - f'(0)$$

or

$$\mathcal{L}\{f''(t)\} = s^2\mathcal{L}\{f(t)\} - sf(0) - f'(0). \tag{4}$$

Similarly,

$$\mathcal{L}\{f'''(t)\} = s^3\mathcal{L}\{f(t)\} - s^2f(0) - sf'(0) - f''(0),$$

leading by induction to the following extension of Theorem 1.

THEOREM 2

Let $f^{(k)}(t)$ satisfy equation (1) for $k = 0, 1, 2, \ldots, n - 1$ and suppose that $f^{(n)}(t)$ is piecewise continuous on $t \geq 0$. Then $\mathcal{L}\{f^{(n)}(t)\}$ exists and is given by

$$\mathcal{L}\{f^{(n)}(t)\} = s^n\mathcal{L}\{f(t)\} - s^{n-1}f(0) - s^{n-2}f'(0) - \cdots - f^{(n-1)}(0)$$

or

$$\mathcal{L}\{f^{(n)}(t)\} = s^n\mathcal{L}\{f(t)\} - \sum_{j=0}^{n-1} s^{n-j-1}f^{(j)}(0). \tag{5}$$

Theorems 1 and 2 are important because they are used to reduce the Laplace transform of a differential equation into an equation involving only the transform of the solution. Several such applications will be considered in this section. However, these theorems are also useful in determining the transforms of certain functions.

EXAMPLE 1

Compute $\mathcal{L}\{\sin^2 at\}$.

SOLUTION
Let $f(t) = \sin^2 at$. Then

$$f'(t) = 2a \sin at \cos at = a \sin 2at,$$

so

$$\frac{2a^2}{s^2 + 4a^2} = \mathcal{L}\{f'\} = s\mathcal{L}\{f\} - f(0).$$

Since $f(0) = 0$, it follows that

$$\mathcal{L}\{\sin^2 at\} = \frac{2a^2}{s(s^2 + 4a^2)}, \qquad s > 0.$$

EXAMPLE 2

Compute $\mathcal{L}\{t \sin at\}$.

SOLUTION
Suppose that $f(t) = t \sin at$. Then

$$f'(t) = \sin at + at \cos at, \qquad f''(t) = 2a \cos at - a^2 t \sin at.$$

Thus, since $f(0) = f'(0) = 0$,

$$2a\mathcal{L}\{\cos at\} - a^2\mathcal{L}\{f(t)\} = \mathcal{L}\{f''\} = s^2\mathcal{L}\{f\},$$

so that

$$(s^2 + a^2)\mathcal{L}\{f\} = 2a\mathcal{L}\{\cos at\} = \frac{2as}{s^2 + a^2}$$

or

$$\mathcal{L}\{f(t)\} = \frac{2as}{(s^2 + a^2)^2}, \qquad s > 0.$$

Alternatively, we can use the result of Problem 4.1.68 on p. 199 or Theorem 3 on p. 208. We have

$$\mathcal{L}\{t \sin at\} = -\frac{d}{ds}\mathcal{L}\{\sin at\} = -\frac{d}{ds}\left(\frac{a}{s^2 + a^2}\right) = \frac{2as}{(s^2 + a^2)^2}.$$

We now apply Theorems 1 and 2 to solve initial value problems. In what follows, we shall take Laplace transforms without worrying about their existence. Any solution so obtained must be checked by substitution into the original equation.

EXAMPLE 3

Find the solution of the initial value problem

$$y'' - 4y = 0, \qquad y(0) = 1, y'(0) = 2. \tag{6}$$

SOLUTION

Taking the Laplace transform of both sides of the differential equation in (6) and using the differentiation property, we transform equation (6) into the algebraic equation

$$[s^2 \mathcal{L}\{y\} - sy(0) - y'(0)] - 4\mathcal{L}\{y\} = [s^2 \mathcal{L}\{y\} - s - 2] - 4\mathcal{L}\{y\} = 0,$$

so that $(s^2 - 4)\mathcal{L}\{y\} = s + 2$ and

$$\mathcal{L}\{y\} = \frac{s+2}{s^2 - 4} = \frac{1}{s - 2}.$$

Thus, by Table 4.1.1 on p. 194, we have

$$y(t) = e^{2t},$$

which satisfies all the conditions in (6).

EXAMPLE 4

Solve the initial value problem

$$y'' + 4y = 0, \qquad y(0) = 1, \qquad y'(0) = 2. \tag{7}$$

SOLUTION

Using the differentiation property, we obtain

$$[s^2 \mathcal{L}\{y\} - sy(0) - y'(0)] + 4\mathcal{L}\{y\} = s^2 \mathcal{L}\{y\} - s - 2 + 4\mathcal{L}\{y\} = 0.$$

Solving for $\mathcal{L}\{y\}$, we have

$$\mathcal{L}\{y\} = \frac{s+2}{s^2 + 4} = \frac{s}{s^2 + 4} + \frac{2}{s^2 + 4}.$$

By reference to Table 4.1.1, we find that

$$y(t) = \cos 2t + \sin 2t,$$

which can readily be verified to be the solution of (7). In calculating the inverse transform, we used the linearity of $\mathcal{L}^{-1}$.

Example 4 indicates the necessity of writing $\mathcal{L}\{y\}$ as a linear combination of terms for which the inverse Laplace transforms are known.

EXAMPLE 5

Find the solution of the initial value problem

$$y'' - 3y' + 2y = 4t - 6, \qquad y(0) = 1, y'(0) = 3. \tag{8}$$

SOLUTION

Taking the Laplace transform of both sides and using the differentiation property, we have from Table 4.1.1,

$$[s^2 \mathcal{L}\{y\} - s - 3] - 3[s\mathcal{L}\{y\} - 1] + 2\mathcal{L}\{y\} = \frac{4}{s^2} - \frac{6}{s},$$

so that

$$(s^2 - 3s + 2)\mathcal{L}\{y\} = s + \frac{4}{s^2} - \frac{6}{s} = \frac{s^3 - 6s + 4}{s^2}.$$

Factoring the numerator and denominator, we obtain

$$\mathcal{L}\{y\} = \frac{s^3 - 6s + 4}{s^2(s^2 - 3s + 2)} = \frac{(s-2)(s^2 + 2s - 2)}{s^2(s-2)(s-1)} = \frac{s^2 + 2s - 2}{s^2(s-1)}.$$

But

$$\frac{s^2 + 2s - 2}{s^2(s-1)} = \frac{s^2}{s^2(s-1)} + \frac{2s - 2}{s^2(s-1)} = \frac{1}{s-1} + \frac{2}{s^2},$$

so that

$$\mathcal{L}\{y\} = \frac{1}{s-1} + \frac{2}{s^2}.$$

Using Table 4.1.1 we obtain the solution

$$y = e^t + 2t$$

to the initial value problem in equation (8).

We now discuss the most general second-order linear initial value problem with constant coefficients. Suppose that we wish to solve the nonhomogeneous differential equation with constant coefficients

$$y'' + ay' + by = f(t), \qquad y(0) = y_0, y'(0) = y_1. \tag{9}$$

The general existence-uniqueness theorem (Theorem 2.1.1 on p. 68) states that the initial value problem (9) will have a unique solution if $f(t)$ is continuous. Assuming this is the case, and taking the Laplace transforms of both sides, we obtain

$$\mathcal{L}\{y''\} + a\mathcal{L}\{y'\} + b\mathcal{L}\{y\} = \mathcal{L}\{f\}.$$

Now by Theorems 1 and 2 (differentiation properties) we have

$$[s^2 \mathcal{L}\{y\} - sy(0) - y'(0)] + a[s\mathcal{L}\{y\} - y(0)] + b\mathcal{L}\{y\} = \mathcal{L}\{f\}.$$

Then

$$[s^2 + as + b]\mathcal{L}\{y\} - [sy(0) + ay(0) + y'(0)] = \mathcal{L}\{f\},$$

so that

$$\mathcal{L}\{y\} = \frac{(s+a)y(0) + y'(0) + \mathcal{L}\{f\}}{s^2 + as + b}. \tag{10}$$

Three facts are evident from equation (10):

1. Initial conditions must be given.
2. The function f must have a Laplace transform.
3. We must be able to find $\mathcal{L}^{-1}$ of the right-hand side.

Thus Laplace transform methods are primarily intended for the solution of linear initial value problems with constant coefficients.

It should be clear that the major difficulty in solving problem (9) lies in finding the inverse transform of the right-hand side of equation (10). There is a general formula that provides the solution as an integral but we shall not discuss it in this text. Fortunately, many of the transforms you will encounter in solving initial value problems can be inverted using techniques from calculus. We illustrate with some examples.

EXAMPLE 6

Solve the initial value problem

$$y'' - 5y' + 4y = e^{2t}, \qquad y(0) = 1, y'(0) = 0.$$

SOLUTION

Making use of the differentiation property and Table 4.1.1, we have

$$[s^2\mathcal{L}\{y\} - sy(0) - y'(0)] - 5[s\mathcal{L}\{y\} - y(0)] + 4\mathcal{L}\{y\} = \mathcal{L}\{e^{2t}\}$$

or

$$[s^2\mathcal{L}\{y\} - s] - 5[s\mathcal{L}\{y\} - 1] + 4\mathcal{L}\{y\} = \frac{1}{s-2},$$

so that

$$(s^2 - 5s + 4)\mathcal{L}\{y\} = s - 5 + \frac{1}{s-2} = \frac{s^2 - 7s + 11}{s-2}.$$

Then

$$\mathcal{L}\{y\} = \frac{s^2 - 7s + 11}{(s-2)(s^2 - 5s + 4)} = \frac{s^2 - 7s + 11}{(s-2)(s-1)(s-4)}. \tag{11}$$

Review of Partial Fractions

At this point we pause. Remember that when you studied techniques of integration in calculus, you integrated functions like the right-hand side of equation (11) by using the method of *partial fractions*. This method is useful here. We seek constants A, B, and C such that

$$\frac{A}{s-2} + \frac{B}{s-1} + \frac{C}{s-4} = \frac{s^2 - 7s + 11}{(s-2)(s-1)(s-4)}. \tag{12}$$

Why? Because we know that $\mathcal{L}^{-1}\{1/(s-2)\} = e^{2t}$ so that $\mathcal{L}^{-1}\{A/(s-2)\} = Ae^{2t}$, and so on.

There is an easy method for finding these constants:

$$A = \frac{s^2 - 7s + 11}{(s-1)(s-4)}\bigg|_{s=2} = -\frac{1}{2};$$

$$B = \frac{s^2 - 7s + 11}{(s-2)(s-4)}\bigg|_{s=1} = \frac{5}{3};$$

$$C = \frac{s^2 - 7s + 11}{(s-2)(s-1)}\bigg|_{s=4} = -\frac{1}{6}.$$

Observe that we eliminate the denominator $(s - a)$ of each term on the left-hand side of equation (12) from the right-hand side of equation (12) and evaluate the resulting equation at $s = a$ to obtain the desired constant.

To understand why this procedure works, let us multiply both sides of equation (12) by $(s - 2)$. Then we have

$$A + (s-2)\left(\frac{B}{s-1} + \frac{C}{s-4}\right) = \frac{s^2 - 7s + 11}{(s-1)(s-4)}. \tag{13}$$

Setting $s = 2$ on both sides eliminates all but the constant A on the left-hand side of equation (13), and therefore

$$A = \frac{s^2 - 7s + 11}{(s-1)(s-4)}\bigg|_{s=2}.$$

Returning to our problem, we see that

$$\mathcal{L}\{y\} = \frac{(-\frac{1}{2})}{s-2} + \frac{(\frac{5}{3})}{s-1} + \frac{(-\frac{1}{6})}{s-4}$$

which implies, according to Table 4.1.1, that

$$y(t) = -\frac{e^{2t}}{2} + \frac{5e^t}{3} - \frac{e^{4t}}{6}.$$

EXAMPLE 7

Solve

$$y'' + 2y' + 2y = t, \qquad y(0) = y'(0) = 1.$$

SOLUTION

Using the differentiation property, we obtain

$$[s^2\mathcal{L}\{y\} - s - 1] + 2[s\mathcal{L}\{y\} - 1] + 2\mathcal{L}\{y\} = \frac{1}{s^2},$$

or

$$(s^2 + 2s + 2)\mathcal{L}\{y\} = \frac{1}{s^2} + s + 3 = \frac{s^3 + 3s^2 + 1}{s^2}$$

and

$$\mathcal{L}\{y\} = \frac{s^3 + 3s^2 + 1}{s^2(s^2 + 2s + 2)}. \tag{14}$$

The term $s^2 + 2s + 2$ can't be factored [since $2^2 - 4(1)(2) = -4 < 0$], so we write the right-hand side of equation (14) as

$$\frac{s^3 + 3s^2 + 1}{s^2(s^2 + 2s + 2)} = \frac{As + B}{s^2 + 2s + 2} + \frac{C}{s} + \frac{D}{s^2}. \tag{15}$$

Why do we do this? Because

$$\frac{s+1}{s^2 + 2s + 2} = \frac{s+1}{(s+1)^2 + 1},$$

so that

$$\mathcal{L}^{-1}\left\{\frac{s+1}{s^2 + 2s + 1}\right\} = \mathcal{L}^{-1}\left\{\frac{s+1}{(s+1)^2 + 1}\right\} = e^{-t}\cos t$$

by Table 4.1.2 on p. 198 as a consequence of the first shifting theorem. Similarly,

$$\mathcal{L}^{-1}\left\{\frac{1}{(s+1)^2 + 1}\right\} = e^{-t}\sin t.$$

Also, by Table 4.1.1 on p. 194,

$$\mathcal{L}^{-1}\left\{\frac{1}{s}\right\} = 1 \quad \text{and} \quad \mathcal{L}^{-1}\left\{\frac{1}{s^2}\right\} = t.$$

Further Review of Partial Fractions   There are tricks for finding the constants A, B, C, and D in equation (15), but these are more complicated than the method we used in Example 6. We can find these constants directly by combining terms:

$$\frac{(As + B)s^2 + C(s^2 + 2s + 2)s + D(s^2 + 2s + 2)}{s^2(s^2 + 2s + 1)} = \frac{s^3 + 3s^2 + 1}{s^2(s^2 + 2s + 2)},$$

so that, equating coefficients of like powers of s, we obtain:

$A + C$	$= 1$	these are the coefficients of s^3,
$B + 2C + D = 3$		these are the coefficients of s^2,
$2C + 2D = 0$		these are the coefficients of s,
$2D = 1$		these are the constant terms.

From the last equation we see that $D = \frac{1}{2}$, so working backward we obtain $C = -\frac{1}{2}$, $B = \frac{7}{2}$, and $A = \frac{3}{2}$. Then

$$y = \mathcal{L}^{-1}\left\{\frac{s^3 + 3s^2 + 1}{s^2(s^2 + 2s + 2)}\right\}$$

$$= \mathcal{L}^{-1}\left\{\frac{\frac{3}{2}s + \frac{7}{2}}{s^2 + 2s + 2} + \frac{(-\frac{1}{2})}{s} + \frac{(\frac{1}{2})}{s^2}\right\}$$

$$= \mathcal{L}^{-1}\left\{\frac{\frac{3}{2}(s+1)}{(s+1)^2 + 1} + \frac{2}{(s+1)^2 + 1} + \frac{(-\frac{1}{2})}{s} + \frac{(\frac{1}{2})}{s^2}\right\}$$

or

$$y = \frac{3}{2}e^{-t}\cos t + 2e^{-t}\sin t - \frac{1}{2} + \frac{1}{2}t,$$

which is the solution to our differential equation.

The methods used in the last two examples apply to the problem of inverting a Laplace transform obtained in trying to solve an initial value problem with constant coefficients. In some special cases, we can use these techniques to solve linear problems with variable coefficients. First, however, we need to prove the identities that were stated in Problems 68 and 69 in Section 4.1.

Consider the derivative

$$\frac{d}{ds}\mathcal{L}\{f(t)\} = \frac{d}{ds}\int_0^\infty e^{-st}f(t)\,dt. \tag{16}$$

If we reverse the order in which the operations of differentiation and integration are performed on the right-hand side of equation (16), we obtain

$$\frac{d}{ds}\mathcal{L}\{f(t)\} = \int_0^\infty \frac{d}{ds}e^{-st}f(t)\,dt = \int_0^\infty -te^{-st}f(t)\,dt = -\mathcal{L}\{tf(t)\}.$$

We have proved the following:

■ THEOREM 3

If $\mathcal{L}\{f(t)\}$ exists for $s > a$, then $\mathcal{L}\{tf(t)\}$ exists for $s > a$ and

$$\mathcal{L}\{tf(t)\} = -\frac{d}{ds}\mathcal{L}\{f(t)\}. \tag{17}$$

■

Using equation (17) repeatedly, we obtain

$$\mathcal{L}\{t^n f(t)\} = -\frac{d}{ds}\mathcal{L}\{t^{n-1}f(t)\} = (-1)^2\frac{d^2}{ds^2}\mathcal{L}\{t^{n-2}f(t)\}$$

$$= \cdots = (-1)^n\frac{d^n}{ds^n}\mathcal{L}\{f(t)\}. \tag{18}$$

Of course, it may not be legitimate to change the order of differentiation and integration in equation (16) (pathological examples do exist), but if the method succeeds in providing a correct solution to our problem, we need not be concerned. This lack of rigor reemphasizes the need of checking the final solution when solving problems by Laplace transform techniques.

The following example illustrates how equations (17) and (18) can be used in conjunction with the differentiation property to solve some initial value problems with variable coefficients.

EXAMPLE 8

Solve the equation with variable coefficients

$$ty'' - ty' - y = 0, \qquad y(0) = 0, y'(0) = 3.$$

SOLUTION

If we let $Y(s) = \mathcal{L}\{y(t)\}$, then by the differentiation property and equation (17)

$$\mathcal{L}\{ty''\} = -\frac{d}{ds}\mathcal{L}\{y''\} = -\frac{d}{ds}\{s^2 Y(s) - sy(0) - y'(0)\}$$

$$= -s^2 Y' - 2sY - y(0) = -s^2 Y' - 2sY$$

and

$$\mathcal{L}\{ty'\} = -\frac{d}{ds}\mathcal{L}\{y'\} = -\frac{d}{ds}\{sY - y(0)\} = -sY' - Y.$$

Substituting these expressions into the Laplace transform of the original equation yields

$$-s^2 Y' - 2sY + sY' + Y - Y = 0.$$

Rearranging and canceling terms, we have

$$(s^2 - s)Y' + 2sY = 0.$$

We now divide both sides by $s^2 - s = s(s - 1)$ to obtain

$$Y' + \frac{2}{s-1}Y = 0.$$

Separating variables,

$$\frac{dY}{Y} = -\frac{2}{s-1}ds,$$

and an integration yields

$$\ln|Y| = -2\ln|s-1| + c \quad \text{or} \quad Y(s) = \frac{c}{(s-1)^2}.$$

Thus, by Table 4.1.2,

$$y(t) = cte^t.$$

Note that $y(0) = 0$. To find c, we differentiate and use the second initial condition to obtain

$$3 = y'(0) = c(t+1)e^t|_{t=0} = c.$$

Thus, the unique solution to the initial value problem is given by

$$y(t) = 3te^t.$$

We caution the reader not to expect to be able to solve all variable coefficient equations by this method. It will work only when

(a) the coefficients $a_i(t)$ are polynomials in t,
(b) the differential equation involving $Y(s)$ can be solved, and
(c) the inverse transform of $Y(s)$ can be found.

It is rare that all these conditions can be met (see Problem 49).

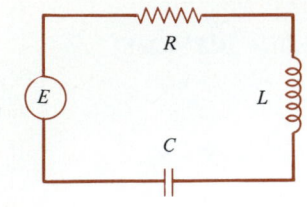

FIGURE 1

EXAMPLE 9

In Section 2.7 we applied Kirchhoff's law to obtain [see equation (2.7.1) on p. 102] the following differential equation relating the current (I), charge (Q), resistance (R), inductance (L), and capacitance (C) of the electric circuit shown in Fig. 1:

$$L\frac{dI}{dt} + RI + \frac{Q}{C} = E. \tag{19}$$

We then differentiated equation (19) using the fact that

$$\frac{dQ}{dt} = I \tag{20}$$

and, assuming that E was constant, obtained a homogeneous, second-order equation that we then solved. Now we make our model more realistic by assuming that the electromotive force (emf) $E = E(t)$ is a nonconstant function of time. Replacing each I in equation (19) by the identity in equation (20), we obtain

$$L\frac{dQ^2}{dt^2} + R\frac{dQ}{dt} + \frac{Q}{C} = E(t). \tag{21}$$

If $Q(0) = a$ and $Q'(0) = I(0) = b$, we can solve equation (21) by Laplace transform methods and use that solution to point out a feature common to many physical and biological systems.

We set $\hat{Q}(s) = \mathcal{L}\{Q(t)\}$, $\hat{E}(s) = \mathcal{L}\{E(t)\}$, and take the transform of both sides of equation (21), using the differentiation property, to obtain

$$L[s^2\hat{Q}(s) - as - b] + R[s\hat{Q}(s) - a] + \frac{1}{C}\hat{Q}(s) = \hat{E}(s),$$

or

$$\hat{Q}(s)\left(Ls^2 + Rs + \frac{1}{C}\right) = (Las + Lb + Ra) + \hat{E}(s). \tag{22}$$

Now let $U(s) = Ls^2 + Rs + 1/C$ and $V(s) = Las + Lb + Ra$ and rewrite equation (22) as

$$\hat{Q}(s) = \frac{\hat{E}(s) + V(s)}{U(s)} = \hat{E}(s)\left[\frac{1 + V(s)/\hat{E}(s)}{U(s)}\right]. \tag{23}$$

Defining $T(s) = (1 + V/\hat{E})/U$, we can rewrite equation (23) in the form

$$\hat{Q}(s) = \hat{E}(s)T(s). \tag{24}$$

Black Boxes and Transfer Functions

Black Box

Transfer Function

Engineers find the compartment or "black box" concept to be very useful in their work. In a **black box** something goes in (the input) and is transformed into something that comes out (the output). In the system under discussion we have the "black-box" setup shown in Fig. 2. In equation (24), we have a relationship between the transforms of the input and the output. The function $T(s)$ is called a **transfer function** and describes, precisely, the inner workings of the

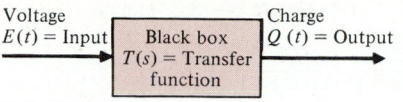

Voltage
$E(t)$ = Input | Black box $T(s)$ = Transfer function | Charge $Q(t)$ = Output

FIGURE 2

black box subject to the driving function $E(t)$. It tells us exactly what we get out in terms of what we put in and gives us a simple equation relating input to output. We shall exploit this concept in Sections 4.4 and 4.5.

EXAMPLE 10

Suppose the circuit in Fig. 1 is connected at $t = 0$ to the emf $E(t) = \cos t$ volts, and that $Q(0) = 0$ and $I(0) = 0$. Assume that $L = 1$ henry, $R = 6$ ohms, and $C = \frac{1}{9}$ farad. Then we have the initial value problem

$$\frac{dI}{dt} + 6I + 9Q = \cos t, \qquad Q(0) = 0, I(0) = 0.$$

Using the identity (20) we obtain

$$\frac{d^2Q}{dt^2} + 6\frac{dQ}{dt} + 9Q = \cos t, \qquad Q(0) = Q'(0) = 0, \tag{25}$$

and if we use the differentiation property of Laplace transforms, equation (25) becomes

$$s^2\hat{Q} + 6s\hat{Q} + 9\hat{Q} = \frac{s}{s^2 + 1}$$

or

$$\hat{Q}(s) \cdot [s^2 + 6s + 9] = \frac{s}{s^2 + 1}. \tag{26}$$

Comparing equations (22) and (26), we note that $U(s) = s^2 + 6s + 9 = (s + 3)^2$ and $V(s) = 0$, so that the transfer function $T(s) = 1/U(s)$. Solving for $\hat{Q}$ we have

$$\hat{Q}(s) = \frac{s}{s^2 + 1} \cdot \frac{1}{(s + 3)^2},$$

which we wish to write in partial fraction form:

$$\frac{s}{(s^2 + 1)(s + 3)^2} = \frac{As + B}{s^2 + 1} + \frac{C}{s + 3} + \frac{D}{(s + 3)^2}. \tag{27}$$

Multiplying both sides by $(s^2 + 1)(s + 3)^2$, we obtain

$$s = (As + B)(s + 3)^2 + C(s + 3)(s^2 + 1) + D(s^2 + 1)$$
$$= (A + C)s^3 + (6A + B + 3C + D)s^2 + (9A + 6B + C)s + (9B + 3C + D).$$

Equating like powers of s on both sides of the equation, we obtain the system

$$A \qquad + C \qquad = 0,$$
$$6A + B + 3C + D = 0,$$
$$9A + 6B + C \qquad = 1,$$
$$9B + 3C + D = 0. \tag{28}$$

Subtracting the first equation from the third and the fourth from the second, we get

$$8A + 6B = 1,$$
$$6A - 8B = 0,$$

from which we obtain $A = 0.08$, $B = 0.06$, $C = -0.08$, and $D = -0.3$. Hence

$$\hat{Q}(s) = \frac{0.08s}{s^2+1} + \frac{0.06}{s^2+1} - \frac{0.08}{s+3} - \frac{0.3}{(s+3)^2}$$

which, by Tables 4.1.1 and 4.1.2, yields

$$Q(t) = 0.08\cos t + 0.06\sin t - 0.08e^{-3t} - 0.3te^{-3t}. \tag{29}$$

Differentiating (29), we obtain the expression for the current:

$$I(t) = -0.08\sin t + 0.06\cos t - 0.06e^{-3t} + 0.9te^{-3t}.$$

Many other situations can be modeled using the idea of a black box and an appropriate transfer function relating output to input.

The following brief list of Laplace transforms will be useful in doing the exercises.

Short Table of Laplace Transforms

$f(t)$	$\mathscr{L}\{f(t)\}$	$f(t)$	$\mathscr{L}\{f(t)\}$
c	$\dfrac{c}{s}$	e^{at}	$\dfrac{1}{s-a}$
t^n	$\dfrac{n!}{s^{n+1}}$	$e^{at}t^n$	$\dfrac{n!}{(s-a)^{n+1}}$
$\sin bt$	$\dfrac{b}{s^2+b^2}$	$e^{at}\sin bt$	$\dfrac{b}{(s-a)^2+b^2}$
$\cos bt$	$\dfrac{s}{s^2+b^2}$	$e^{at}\cos bt$	$\dfrac{s-a}{(s-a)^2+b^2}$
$\sinh bt$	$\dfrac{b}{s^2-b^2}$	$e^{at}\sinh bt$	$\dfrac{b}{(s-a)^2-b^2}$
$\cosh bt$	$\dfrac{s}{s^2-b^2}$	$e^{at}\cosh bt$	$\dfrac{s-a}{(s-a)^2-b^2}$

$$\mathscr{L}\{f'(t)\} = s\mathscr{L}\{f(t)\} - f(0) \quad \mathscr{L}\{f''(t)\} = s^2\mathscr{L}\{f(t)\} - sf(0) - f'(0)$$

$$\mathscr{L}\{tf(t)\} = -\frac{d}{ds}\mathscr{L}\{f(t)\} \qquad \mathscr{L}\{t^nf(t)\} = (-1)^n\frac{d^n}{ds^n}\mathscr{L}\{f(t)\}$$

Problems 4.2

In Problems 1–20, solve the given initial value problems.

1. $y'' + y = 0$, $y(0) = 1$, $y'(0) = 0$
2. $y'' + y' = 0$, $y(0) = 0$, $y'(0) = 1$
3. $y'' - a^2y = 0$, $y(0) = A$, $y'(0) = B$
4. $y'' - ay' = 0$, $y(0) = 1$, $y'(0) = a$
5. $y'' + 2y' + 5y = 0$, $y(0) = y'(0) = 1$
6. $y'' - y' + y = 0$, $y(0) = y'(0) = 1$
7. $y'' - 4y' + 3y = 1$, $y(0) = 1$, $y'(0) = 4$
8. $y'' - 2y' - 3y = 5$, $y(0) = 0$, $y'(0) = 1$

9. $y'' - 9y = t$, $y(0) = 1$, $y'(0) = 2$
10. $y'' - 3y' - 4y = t^2$, $y(0) = 2$, $y'(0) = 1$
11. $y''' + y = 0$, $y(0) = y''(0) = 1$, $y'(0) = -1$
12. $y^{(4)} - y = 0$, $y(0) = y''(0) = 1$, $y'(0) = y'''(0) = 0$
13. $y^{(4)} - y = 0$, $y(0) = y''(0) = 0$, $y'(0) = y'''(0) = 1$
14. $y''' - 3y' - 2y = e^{2t}$, $y(0) = y'(0) = 0$, $y''(0) = 1$
15. $y'' + k^2y = \cos kt$, $y(0) = 0$, $y'(0) = k$ [*Hint:* Consider Example 2.]
16. $y'' + 4y = \cos t$, $y(0) = y'(0) = 0$

17. $y'' + a^2 y = \sin at$, $y(0) = a$, $y'(0) = a^2$

18. $y'' - y = te^t$, $y(0) = y'(0) = 1$

19. $y^{(4)} - y = \cos t$, $y(0) = y''(0) = 1$, $y'(0) = y'''(0) = 0$

20. $y^{(4)} - y = \sinh t$, $y(0) = y''(0) = 0$, $y'(0) = y'''(0) = 1$

In Problems 21–29, find the Laplace transform of each function by using the differentiation property or equation (18).

21. $\cos^2 at$

22. $t \cos at$

23. $t^2 \sin at$

24. $t^2 \cos at$

25. $t \sin^2 t$

26. $t \cos^2 t$

27. $t \sin^2 at$

28. $t^2 \cos^2 3t$

29. $t^2 \sin^2 2t$

30. By reversing the order of integration, show that if $F(s) = \mathcal{L}\{f(t)\}$, then

$$\int_s^\infty F(s)\, ds = \mathcal{L}\{f(t)/t\}. \qquad (30)$$

31. Let $g(t) = \int_0^t f(u)\, du$. Using calculus and the differentiation property (Theorem 1), show that

  *(a) $g'(t) = f(t)$ at all points of continuity of $f(t)$;

  (b) $\mathcal{L}\{f(t)\} = s\mathcal{L}\{g(t)\} - g(0)$;

  (c) $g(0) = 0$.

  Finally, using parts (b) and (c) conclude that

$$\mathcal{L}\left\{\int_0^t f(u)\, du\right\} = \frac{1}{s}\mathcal{L}\{f(t)\}. \qquad (31)$$

In Problems 32–44, use the results (30) and (31) of Problems 30 and 31 to compute the Laplace transform of the given function.

32. $\dfrac{\cos t - 1}{t}$

33. $\dfrac{\sin t}{t}$

34. $\dfrac{\sinh t}{t}$

35. $\dfrac{\sin 3t}{t}$

36. $\dfrac{\sinh kt}{t}$

37. $\dfrac{\sin kt}{t}$

38. $\dfrac{1 - \cos at}{t}$

39. $\dfrac{1 - \cosh at}{t}$

40. $\displaystyle\int_0^t \dfrac{\sin ku}{u}\, du$

41. $\displaystyle\int_0^t \dfrac{1 - \cosh au}{u}\, du$

42. $\displaystyle\int_0^t \dfrac{1 - \cos au}{u}\, du$

43. $\mathrm{erf}(t) = \dfrac{2}{\sqrt{\pi}} \displaystyle\int_0^t e^{-u^2}\, du$

44. $\dfrac{e^{-k^2/4t}}{\sqrt{\pi t}}$

Find the inverse Laplace transform of the functions in Problems 45–48. Use derivatives and integrals.

45. $\ln\left(1 + \dfrac{a^2}{s^2}\right)$

46. $\ln \dfrac{s - a}{s - b}$

47. $\arctan \dfrac{1}{s}$

* 48. $\dfrac{1}{s} \arctan \dfrac{1}{s}$

49. Consider the equation

$$y'' + ty = 0, \qquad y(0) = 0, \quad y'(0) = 1.$$

  (a) Obtain a differential equation for $Y(s) = \mathcal{L}\{y(t)\}$.

  (b) Solve the differential equation and find $Y(s)$. (Note that it is not possible to invert this transform by the methods we have discussed).

* 50. Let $f(t)$ be continuous, except for a jump discontinuity at $t = a\ (> 0)$, and let it satisfy all other conditions of Theorem 1. Prove that

$$\mathcal{L}\{f'(t)\} = s\mathcal{L}\{f(t)\} - f(0) - e^{-as}[f(a+0) - f(a-0)],$$

  where $f(a + 0) = \lim_{h\to 0^+} f(a + h)$ and $f(a - 0) = \lim_{h\to 0^-} f(a + h)$.

51. In Example 9, find $I(t)$ if $E(t) = \sin t$ volts, $Q(0) = I(0) = 0$, $L = 2$ henrys, $R = 20$ ohms, and $C = 0.02$ farad.

52. Let $L = 1$ henry, $R = 100$ ohms, $C = 10^{-4}$ farad, and $E = 1000 \sin t$ volts in the circuit in Fig. 1. Suppose that no charge and current is initially present. Find the current and charge at all times t.

53. Let L, R, and C be as in Problem 51 but let $E(t) = 10 \sin 10t$ volts. If $Q(0) = I(0) = 0$, find $Q(t)$.

54. A 50-kg mass is suspended from a spring with spring constant 20 N/m. When the system is vibrating freely, the maximum displacement of each consecutive cycle decreases by 20 percent. Assume a force equal to $10 \cos \omega t$ N acts on the system. Find the amplitude of the resultant steady-state motion if

  (a) $\omega = 8$ rad/s (radians per second)

  (b) $\omega = 10$ rad/s

  (c) $\omega = 12$ rad/s

  (d) $\omega = 14$ rad/s

  (e) $\omega = 18$ rad/s

4.3 Step Functions, Impulse Functions, and Periodic Functions

In Section 4.2 we saw how to solve linear differential equations in which the forcing function $f(t)$ was continuous. In a great number of applications, however, the forcing function is either a step function or an impulse function. In this section we shall define these terms and show how to compute the Laplace transforms of these two important types of functions. In Section 4.4 we shall give some examples of differential equations with discontinuous forcing functions.

Unit Step Function

Unit Step Function or HeavisideFunction

The following function, which is extremely important for practical applications, is known as the **unit step function** or **Heaviside function** (see Fig. 1a):

$$H(t) = \begin{cases} 0, & t < 0, \\ 1, & t > 0. \end{cases} \tag{1}$$

In particular, if a is any fixed constant, we can shift the Heaviside function by a units (see Fig. 1b) by defining

$$H(t-a) = \begin{cases} 0, & t < a, \\ 1, & t > a. \end{cases} \tag{2}$$

Then $H(t - a)$ has a jump discontinuity at $t = a$. Note that $H(t - a)$ is not defined at $t = a$.

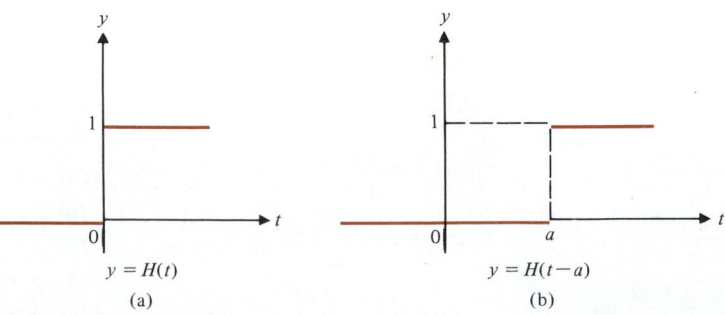

$y = H(t)$

(a)

$y = H(t-a)$

(b)

FIGURE 1

A practical use of such a function might be to model a light switch that is switched on a time $t = a$. For $a \geq 0$ and $s > 0$, we obtain

$$\mathcal{L}\{H(t-a)\} = \int_0^\infty e^{-st}H(t-a)\,dt = \int_a^\infty e^{-st}\,dt = \frac{e^{-as}}{s}. \tag{3}$$

The following theorem shows that multiplying a function by a unit step function has the effect of multiplying its transform by an exponential function.

■ **THEOREM 1: The Second Shifting Property of Laplace Transforms**
Let $a > 0$. Then

$$\mathcal{L}\{f(t-a)H(t-a)\} = e^{-as}\mathcal{L}\{f(t)\} \tag{4}$$

or, in terms of inverse transforms with $F(s) = \mathcal{L}\{f(t)\}$,

$$\mathcal{L}^{-1}\{e^{-as}F(s)\} = f(t-a)H(t-a). \tag{5}$$

PROOF
Using the definition and the substitution $x = t - a$, we find that

$$\mathcal{L}\{f(t-a)H(t-a)\} = \int_0^\infty e^{-st}f(t-a)H(t-a)\,dt = \int_a^\infty e^{-st}f(t-a)\,dt$$

$$\text{let } x = t - a$$

$$= \int_0^\infty e^{-s(x+a)}f(x)\,dx = e^{-as}\mathcal{L}\{f(t)\}.$$ ∎

The next three examples illustrate the use of the second shifting property.

EXAMPLE 1

$$\mathcal{L}\{\sin a(t-b)H(t-b)\} = e^{-bs}\mathcal{L}\{\sin at\} = \frac{ae^{-bs}}{(s^2+a^2)}.$$

EXAMPLE 2

Compute $\mathcal{L}\{f(t)\}$ when

$$f(t) = \begin{cases} e^t, & 0 \le t < 2\pi, \\ e^t + \cos t, & t > 2\pi. \end{cases}$$

SOLUTION
The function $f(t)$ has a jump discontinuity at $t = 2\pi$. We may write

$$f(t) = e^t + H(t - 2\pi) \cos(t - 2\pi),$$

since

$$H(t - 2\pi) \cos(t - 2\pi) = \begin{cases} 0, & \text{if } t < 2\pi, \\ \cos(t - 2\pi) = \cos t, & \text{if } t > 2\pi. \end{cases}$$

Thus, by the second shifting property (Theorem 1),

$$\mathcal{L}\{f(t)\} = \mathcal{L}\{e^t\} + e^{-2\pi s}\mathcal{L}\{\cos t\} = \frac{1}{s-1} + \frac{se^{-2\pi s}}{1 + s^2}.$$

EXAMPLE 3

Compute

$$\mathcal{L}^{-1}\left\{\frac{1 - e^{(-\pi s/2)}}{1 + s^2}\right\}.$$

SOLUTION
Observe that

$$\mathcal{L}^{-1}\left\{\frac{1 - e^{(-\pi s/2)}}{1 + s^2}\right\} = \mathcal{L}^{-1}\left\{\frac{1}{1 + s^2}\right\} - \mathcal{L}^{-1}\left\{\frac{e^{(-\pi s/2)}}{1 + s^2}\right\}$$

$$= \sin t - H(t - \pi/2) \sin(t - \pi/2)$$

$$= \sin t + H(t - \pi/2) \cos t.$$

The unit step function can be used as a building block in the construction of other functions. For example,

$$f_1(t) = H(t - a) - H(t - b), \qquad a < b,$$

is a square wave between a and b (see Fig. 2a), whereas

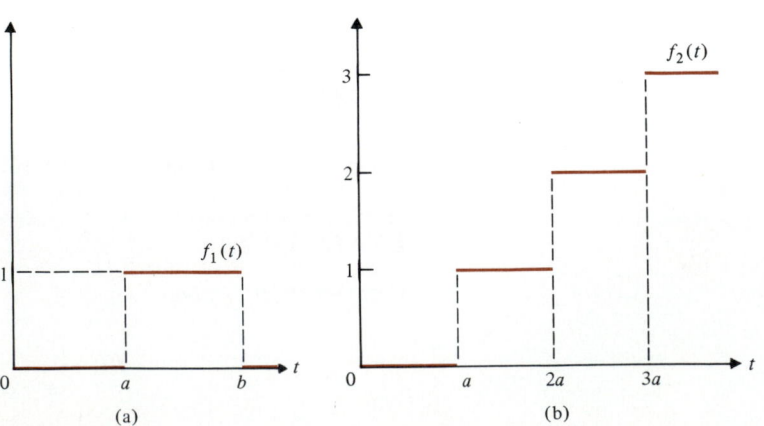

(a)

(b)

FIGURE 2

$$f_2(t) = H(t-a) + H(t-2a) + H(t-3a), \qquad a > 0,$$

yields a three-step staircase (see Fig. 2b). By the linearity property of Laplace transforms (Theorem 6.1.2), we obtain

$$\mathcal{L}\{f_1(t)\} = \frac{1}{s}(e^{-as} - e^{-bs})$$

and

$$\mathcal{L}\{f_2(t)\} = \frac{1}{s}(e^{-as} + e^{-2as} + e^{-3as}).$$

EXAMPLE 4

Compute the Laplace transform of the infinite staircase (obtained by continuing the staircase in Fig. 2b forever):

$$f(t) = H(t) + H(t-a) + H(t-2a) + H(t-3a) + \cdots, \qquad a > 0. \tag{6}$$

SOLUTION

Since $e^{-as} < 1$ if $as > 0$, we can use the formula for the sum of a geometric series

$$\sum_{n=0}^{\infty} x^n = 1 + x + x^2 + \cdots = \frac{1}{1-x}, \qquad |x| < 1. \tag{7}$$

Then, for $s > 0$, by equation (3),

$$\mathcal{L}\{f(t)\} = \frac{1}{s}(1 + e^{-as} + e^{-2as} + e^{-3as} + \cdots) = \frac{1}{s(1 - e^{-as})}. \tag{8}$$

EXAMPLE 5

Let $f(t)$ be the periodic square wave shown in Fig. 3. Then we can write $f(t)$ in the form

$$f(t) = H(t) - 2H(t-a) + 2H(t-2a) - 2H(t-3a) + \cdots,$$

from which it follows that

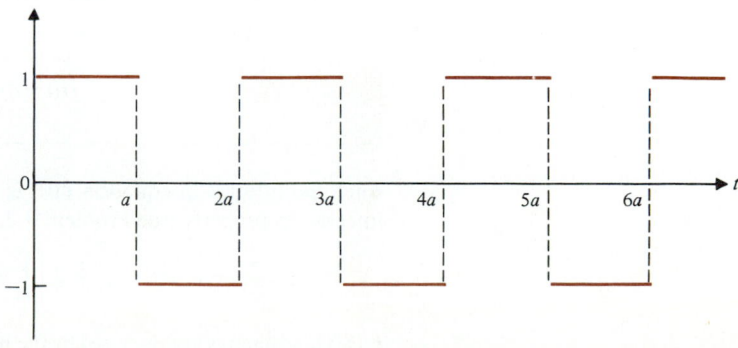

FIGURE 3

$$\mathcal{L}\{f(t)\} = \frac{1}{s}(1 - 2e^{-as} + 2e^{-2as} - 2e^{-3as} + \cdots) = \frac{1}{s}[2(1 - e^{-at} + e^{-2at} - \cdots) - 1]$$

set $x = -e^{-as}$ in (7)   multiply and divide by $e^{as/2}$

$$= \frac{1}{s}\left(\frac{2}{1 + e^{-as}} - 1\right) = \frac{1 - e^{-as}}{s(1 + e^{-as})} = \frac{1}{s}\left[\frac{e^{as/2} - e^{-as/2}}{e^{as/2} + e^{-as/2}}\right] = \frac{1}{s}\tanh\left(\frac{1}{2}as\right).$$

Unit Impulse Function

EXAMPLE 6

The **unit impulse function** (also called the **Dirac delta function**†) $\delta(t - a)$ is loosely described as a "function" that is zero everywhere except at $t = a$ and has the property that

$$\int_{-\infty}^{\infty} \delta(t - a)\,dt = 1. \tag{9}$$

As an illustration, we describe the Dirac delta function for $a = 0$. For any $\epsilon > 0$, consider the approximate delta functions (see Fig. 4)

$$\delta_\epsilon(t) = \begin{cases} \dfrac{1}{2\epsilon}, & -\epsilon < t < \epsilon, \\ 0, & |t| > \epsilon. \end{cases}$$

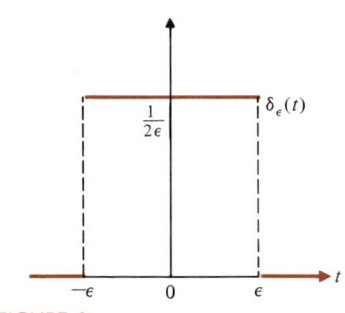

FIGURE 4

Clearly, $\delta_\epsilon(t)$ is piecewise continuous and

$$\int_{-\infty}^{\infty} \delta_\epsilon(t)\,dt = \int_{-\epsilon}^{\epsilon} \frac{1}{2\epsilon}\,dt = 1.$$

Then $\delta(t)$ may be defined by

$$\delta(t) = \lim_{\epsilon \to 0} \delta_\epsilon(t).$$

Of course, $\delta(t)$ is not a function. However, because it is the limit of piecewise continuous functions, we may treat $\delta(t)$ as if it were a legitimate function. We shall not prove this fact here, but shall nevertheless make use of it in all further discussions of the delta function.

Since $\delta(t - a)$ concentrates all its "mass" at $t = a$, we see that

$$H(t - a) = \int_{-\infty}^{t} \delta(u - a)\,du, \tag{10}$$

since the integral in equation (10) is zero if $t < a$ and equals 1 when $t > a$. By the integration property (see Problem 4.2.31), if $a \geq 0$,

$$\frac{1}{s}\mathcal{L}\{\delta(t - a)\} = \mathcal{L}\left\{\int_0^t \delta(u - a)\,du\right\} = \mathcal{L}\{H(t - a)\} = \frac{e^{-as}}{s}.$$

† This function was first discussed by the British physicist Paul A. M. Dirac (1902–1984) in 1932. In 1933 he received the Nobel prize in physics (jointly with E. Schrödinger) for his work in quantum theory.

Thus

$$\mathcal{L}\{\delta(t-a)\} = e^{-as}, \qquad \text{for} \quad a \geq 0 \quad \text{and} \quad s > 0. \tag{11}$$

The Heaviside function can be used in solving differential equations with discontinuous forcing functions:

EXAMPLE 7

Solve the initial value problem

$$y'' + y = f(t), \qquad y(0) = y'(0) = 0,$$

where

$$f(t) = \begin{cases} 1, & \text{for} \quad 0 \leq t \leq 1, \\ 0, & \text{for} \quad t > 1. \end{cases}$$

SOLUTION

We can rewrite $f(t)$ using Heaviside functions as $f(t) = H(t) - H(t - 1)$. Taking Laplace transforms, we get

$$s^2 \mathcal{L}\{y\} + \mathcal{L}\{y\} = \frac{1}{s} - \frac{e^{-s}}{s} \quad \text{or} \quad (s^2 + 1)\mathcal{L}\{y\} = \frac{1 - e^{-s}}{s}.$$

Hence

$$\mathcal{L}\{y\} = \frac{1 - e^{-s}}{s(s^2 + 1)},$$

and since

$$\frac{1}{s(s^2 + 1)} = \frac{1}{s} - \frac{s}{s^2 + 1},$$

we have

$$\mathcal{L}\{y\} = \frac{1}{s} - \frac{s}{s^2 + 1} - \frac{e^{-s}}{s} + \frac{se^{-s}}{s^2 + 1}.$$

Using the second shifting property (Theorem 1), we obtain

$$y(t) = 1 - \cos t - H(t - 1) + H(t - 1)\cos(t - 1)$$

$$= 1 - \cos t - H(t - 1)[1 - \cos(t - 1)].$$

The unit impulse function is sometimes the forcing function in an initial value problem.

EXAMPLE 8

Solve the differential equation

$$y'' + 2y' + y = \delta(t - 1), \qquad y(0) = 2, y'(0) = 3.$$

SOLUTION

Taking Laplace transforms and using equation (11) in Example 6, we obtain

$$(s^2 \mathcal{L}\{y\} - 2s - 3) + 2(s\mathcal{L}\{y\} - 2) + \mathcal{L}\{y\} = e^{-s}$$

or

$$(s^2 + 2s + 1)\mathcal{L}\{y\} = 2s + 7 + e^{-s}.$$

Hence, we get

$$\mathcal{L}\{y\} = \frac{2s + 7 + e^{-s}}{s^2 + 2s + 1} = \frac{2(s + 1)}{(s + 1)^2} + \frac{5}{(s + 1)^2} + \frac{e^{-s}}{(s + 1)^2}$$

$$= \frac{2}{(s + 1)} + \frac{5}{(s + 1)^2} + \frac{e^{-s}}{(s + 1)^2}.$$

Since $\mathcal{L}\{te^{-t}\} = (s + 1)^{-2}$, it follows from Theorem 1 that

$$\frac{e^{-s}}{(s + 1)^2} = e^{-s}\mathcal{L}\{te^{-t}\} = \mathcal{L}\{(t - 1)e^{-(t-1)}H(t - 1)\}.$$

Finally, the solution is

$$y(t) = 2e^{-t} + 5te^{-t} + (t - 1)e^{-(t-1)}H(t - 1)$$

$$= e^{-t}[2 + 5t + e(t - 1)H(t - 1)].$$

Laplace Transforms of Periodic Functions

Before going further, we prove a result that is very useful for finding the Laplace transform of a wide variety of functions.

■ **THEOREM 2: Periodicity Property of the Laplace Transform**

Let $f(t)$ be continuous in $[0, \omega]$ and periodic with period ω ($\omega > 0$), that is, $f(t + \omega) = f(t)$, for each $t \geq 0$. Then $f(t)$ has the Laplace transform

$$F(s) = \mathcal{L}\{f(t)\} = \frac{\int_0^\omega e^{-st}f(t)\,dt}{1 - e^{-\omega s}} \tag{12}$$

valid for every $s > 0$.

PROOF

By definition,

$$F(s) = \int_0^\infty e^{-st}f(t)\,dt = \int_0^\omega e^{-st}f(t)\,dt + \int_\omega^{2\omega} e^{-st}f(t)\,dt + \cdots$$

$$= \sum_{k=0}^\infty \int_{k\omega}^{(k+1)\omega} e^{-st}f(t)\,dt. \tag{13}$$

Now, making the substitution $u = t - k\omega$, we obtain

$$\int_{k\omega}^{(k+1)\omega} e^{-st}f(t)\,dt = \int_0^\omega e^{-s(u+k\omega)}f(u+k\omega)\,du$$

$$= e^{-sk\omega}\int_0^\omega e^{-su}f(u)\,du, \tag{14}$$

because of the periodicity of f. Thus, substituting equation (14) into equation (13) and using equation (7) with $x = e^{-s\omega}$, we have

$$F(s) = \sum_{k=0}^\infty e^{-sk\omega}\int_0^\omega e^{-su}f(u)\,du = \left[\int_0^\omega e^{-su}f(u)\,du\right]\sum_{k=0}^\infty (e^{-\omega s})^k$$

$$= \frac{\int_0^\omega e^{-su}f(u)\,du}{1 - e^{-\omega s}}.$$

Note that if $s > 0$, then $\omega s > 0$ and $e^{-\omega s} < 1$ so that the use of formula (7) is valid. ∎

EXAMPLE 9

Find the Laplace transform of the function

$$f(t) = |\sin at|, \qquad a > 0.$$

SOLUTION

Note that $f(t)$ has period $\omega = \pi/a$. By Theorem 2 we have

$$\mathcal{L}\{|\sin at|\} = \frac{\int_0^{\pi/a} e^{-st}\sin at\,dt}{1 - e^{-\pi s/a}}$$

since $|\sin at| = \sin at$ in $[0, \pi/a]$. Now using entry 168 in the table of integrals, we have

$$\int_0^{\pi/a} e^{-st}\sin at\,dt = \frac{e^{-st}}{s^2+a^2}(-s\sin at - a\cos at)\Big|_0^{\pi/a}$$

$$= \frac{a(e^{-\pi s/a} + 1)}{s^2 + a^2},$$

so that

$$\mathcal{L}\{|\sin at|\} = \frac{a}{s^2+a^2}\frac{1 + e^{-\pi s/a}}{1 - e^{-\pi s/a}},$$

which can be simplified by using hyperbolic functions to

$$\mathcal{L}\{|\sin at|\} = \frac{a}{s^2 + a^2}\coth\left(\frac{\pi s}{2a}\right).$$

EXAMPLE 10

Compute the Laplace transform of the periodic square wave

$$f(t) = H(t) - 2H(t-a) + 2H(t-2a) - 2H(t-3a) + \cdots, \qquad a > 0.$$

SOLUTION

We solved this problem in Example 5. We now solve it using Theorem 2 by noting that f is periodic of period $2a$ (see Fig. 3). We have, from (12),

$$F(s) = \frac{\int_0^{2a} e^{-st}f(t)\,dt}{1 - e^{-2as}}.$$

But $f(t) = 1$ for $0 < t < a$ and $f(t) = -1$ for $a < t < 2a$ so that

$$\int_0^{2a} e^{-st}f(t) = \int_0^a e^{-st}\,dt - \int_a^{2a} e^{-st}\,dt = -\frac{1}{s}e^{-st}\Big|_0^a + \frac{1}{s}e^{-st}\Big|_a^{2a}$$

$$= \frac{1}{s}(-e^{-as} + 1 + e^{-2as} - e^{-as}) = \frac{1}{s}(1 - e^{-as})^2.$$

Since $1 - e^{-2as} = (1 - e^{-as})(1 + e^{-as})$, we obtain

$$F(s) = \frac{1}{s}\frac{(1 - e^{-as})(1 - e^{-as})}{(1 - e^{-as})(1 + e^{-as})} = \frac{1 - e^{-as}}{s(1 + e^{-as})}.$$

We can write this in another way as

multiply top and bottom by $e^{as/2}$

$$\frac{1}{s}\frac{(1 - e^{-as})}{(1 + e^{-as})} = \frac{1}{s}\left(\frac{e^{as/2} - e^{-as/2}}{e^{as/2} + e^{-as/2}}\right) = \frac{1}{s}\tanh\frac{as}{2}.$$

Thus

$$\mathcal{L}\{f\} = \frac{1}{s}\tanh\frac{as}{2}.$$

EXAMPLE 11

Compute the Laplace transform of the sawtooth function given in Fig. 5.

SOLUTION

Clearly g is periodic of period $2a$. For $0 \le t \le a$, g is the line segment that passes through $(0, 0)$ and $(a, 1)$. The slope of this line is $1/a$ so $g(t) = (1/a)t$ for

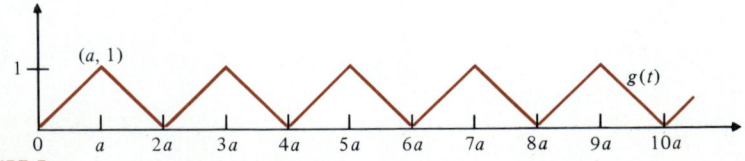

FIGURE 5

$0 \leq t \leq a$. For $a \leq t \leq 2a$, g is the line that passes through $(a, 1)$ and $(2a, 0)$. The slope of this line is $-1/a$ and the equation of the line is $g(t) = 2 - t/a$. Thus

$$g(t) = \begin{cases} \dfrac{1}{a}t, & 0 \leq t \leq a \\[2ex] 2 - \dfrac{t}{a}, & a \leq t \leq 2a \end{cases}$$

and, from (12),

$$\mathcal{L}\{g\} = \frac{\int_0^{2a} e^{-st}g(t)\,dt}{1 - e^{-2as}}. \tag{15}$$

But using entry 164 in the table of integrals

$$\int_0^{2a} e^{-st}g(t)\,dt = \frac{1}{a}\int_0^a te^{-st}\,dt + \frac{1}{a}\int_a^{2a}(2a-t)e^{-st}\,dt$$

$$= \frac{1}{a}\left[\frac{e^{-st}}{-s}\left(t+\frac{1}{s}\right)\right]\Big|_{t=0}^{t=a} + \frac{1}{a}\left[\frac{e^{-st}}{-s}\left(2a-t-\frac{1}{s}\right)\right]\Big|_a^{2a}$$

$$= \frac{1}{as^2}(1 - 2e^{-as} + e^{-2as}) = \frac{(1-e^{-as})^2}{as^2}.$$

We can complete the problem by substituting $2\sinh x = e^x - e^{-x}$ and $e^x + e^{-x} = 2\cosh x$ for $x = as/2$, but there is another way to obtain the solution. Observe that

$$g'(t) = \begin{cases} \dfrac{1}{a}, & 0 < t < a, \\[2ex] \dfrac{-1}{a}, & a < t < 2a. \end{cases}$$

That is, $g' = f/a$ where f is the periodic function of Example 10. Writing this another way, we see that

$$g(t) = \frac{1}{a}\int_0^t f(s)\,ds.$$

Then, from the result of Problem 4.2.31 (p. 213), we see that

$$\mathcal{L}\{g\} = \frac{1}{s}\mathcal{L}\left\{\frac{f}{a}\right\} = \frac{1}{s} \cdot \frac{1}{as}\tanh\frac{as}{2} = \frac{1}{as^2}\tanh\frac{as}{2}.$$

EXAMPLE 12

Solve the initial value problem

$$y'' + 2y' + 5y = f(t), \qquad y(0) = y'(0) = 0,$$

where $f(t)$ is the periodic square wave in Fig. 3 with $a = \pi$.

SOLUTION

Taking Laplace transforms and using the result in Example 5 (or 10), we have

$$(s^2 + 2s + 5)\mathcal{L}\{y\} = \frac{1 - e^{-\pi s}}{s(1 + e^{-\pi s})}$$

or

$$\mathcal{L}\{y\} = \frac{1}{s(s^2 + 2s + 5)} \frac{1 - e^{-\pi s}}{1 + e^{-\pi s}}.$$

But

$$\frac{1}{s(s^2 + 2s + 5)} = \frac{1}{5}\left[\frac{1}{s} - \frac{s + 2}{s^2 + 2s + 5}\right] = \frac{1}{5}\left[\frac{1}{s} - \frac{s + 2}{(s + 1)^2 + 2^2}\right]$$

and, by the geometric series in Example 4,

$$\frac{1 - e^{-\pi s}}{1 + e^{-\pi s}} = (1 - e^{-\pi s})(1 - e^{-\pi s} + e^{-2\pi s} - e^{-3\pi s} + \cdots)$$

$$= 1 - 2e^{-\pi s} + 2e^{-2\pi s} - 2e^{-3\pi s} + \cdots.$$

Thus,

$$\mathcal{L}\{y\} = \frac{1}{5}\left[\frac{1}{s} - \frac{s + 2}{(s + 1)^2 + 2^2}\right](1 - 2e^{-\pi s} + 2e^{-2\pi s} - 2e^{-3\pi s} + \cdots).$$

By the first and second shifting theorems we have

$$\mathcal{L}^{-1}\left\{\frac{1}{5}\left[\frac{1}{s} - \frac{(s + 1) + 1}{(s + 1)^2 + 2^2}\right]\right\} = \frac{1}{5}\left[1 - \overbrace{e^{-t}\left(\cos 2t + \frac{1}{2}\sin 2t\right)}^{\equiv g(t)}\right]$$

$$= \frac{1}{5}[1 - g(t)].$$

Then

$$\mathcal{L}^{-1}\left\{\frac{2}{5}\left[\frac{1}{s} - \frac{(s + 1) + 1}{(s + 1)^2 + 2^2}\right]e^{-k\pi s}\right\} = \frac{2}{5}[1 - g(t - k\pi)]H(t - k\pi).$$

But

$$g(t - k\pi) = e^{-(t - k\pi)}(\cos 2(t - k\pi) + \tfrac{1}{2}\sin 2(t - k\pi)) = e^{k\pi}g(t)$$

so that

$$y(t) = \tfrac{1}{5}[1 - g(t)] - \tfrac{2}{5}[1 - e^{\pi}g(t)]H(t - \pi) + \tfrac{2}{5}[1 - e^{2\pi}g(t)]H(t - 2\pi)$$

$$- \tfrac{2}{5}[1 - e^{3\pi}g(t)]H(t - 3\pi) + \cdots$$

$$= \tfrac{1}{5}[1 - 2H(t - \pi) + 2H(t - 2\pi) - 2H(t - 3\pi) + \cdots]$$

$$- \frac{g(t)}{5}[1 - 2e^{\pi}H(t - \pi) + 2e^{2\pi}H(t - 2\pi) - 2e^{3\pi}H(t - 3\pi) + \cdots].$$

Then

$$y(t) = \tfrac{1}{5}\{f(t) - g(t)[1 - 2e^{\pi}H(t - \pi) + 2e^{2\pi}H(t - 2\pi) - \cdots]\}.$$

Hence, if $n\pi < t < (n + 1)\pi$,

$$y(t) = \tfrac{1}{5}[(-1)^n - g(t)(1 - 2e^{\pi} + \cdots + (-1)^n 2e^{n\pi})]$$

$$= \frac{1}{5}\left\{(-1)^n - g(t)\left[2\left(\frac{1 + (-1)^n e^{(n+1)\pi}}{1 + e^{\pi}}\right) - 1\right]\right\}$$

$$= \frac{1}{5}\left[(-1)^n + g(t) - 2g(t)\left(\frac{1 + (-1)^n e^{(n+1)\pi}}{1 + e^{\pi}}\right)\right].$$

Problems 4.3

In Problems 1–8, use the second shifting theorem to prove the given equation.

1. $\mathcal{L}\{tH(t - 1)\} = e^{-s}\left(\dfrac{1}{s^2} + \dfrac{1}{s}\right)$

2. $\mathcal{L}\{t^2H(t - 1)\} = e^{-s}\left(\dfrac{2}{s^3} + \dfrac{2}{s^2} + \dfrac{1}{s}\right)$

3. $\mathcal{L}\{e^tH(t - 1)\} = \dfrac{e^{-(s-1)}}{s - 1}$

4. $\mathcal{L}\{e^{at}H(t - b)\} = \dfrac{e^{-b(s-a)}}{s - a}$

5. $\mathcal{L}\left\{\sin t \cdot H\left(t - \dfrac{\pi}{2}\right)\right\} = \dfrac{se^{-\pi s/2}}{s^2 + 1}$

6. $\mathcal{L}\{\cos a(t - b)H(t - b)\} = \dfrac{e^{-bs}s}{s^2 + a^2}$

7. $\mathcal{L}\{\sinh a(t - b)H(t - b)\} = \dfrac{e^{-bs}a}{s^2 - a^2}$

8. $\mathcal{L}\{\cosh a(t - b)H(t - b)\} = \dfrac{e^{-bs}s}{s^2 - a^2}$

In Problems 9–11, represent the graphed functions in terms of unit step functions and find their respective Laplace transforms.

9.

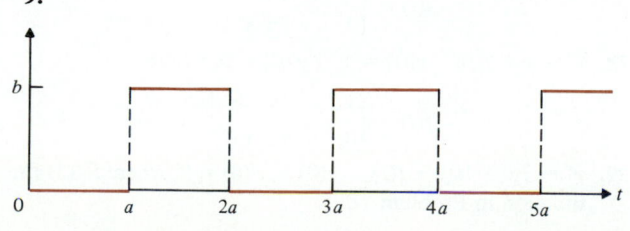

10.

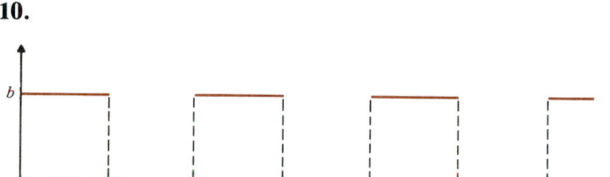

11.

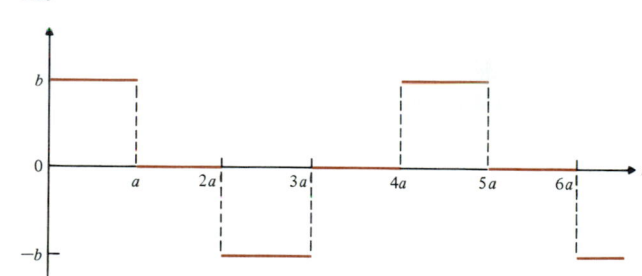

12. Let $g(t)$ be the function shown below.

    (a) Show that $g'(t)$ is piecewise continuous and that $g'(t) = f(t)$, where f is the step function of Problem 11 with $b = 1$.

    (b) Use the differentiation property to compute $\mathcal{L}\{g\}$.

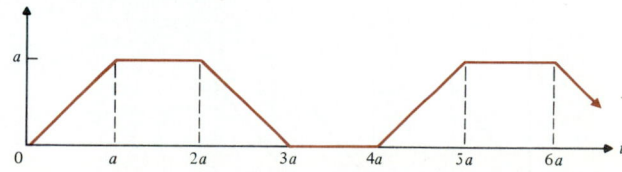

In Problems 13 and 14 use the method of Problem 12 to find the Laplace transform of the given functions.

13.

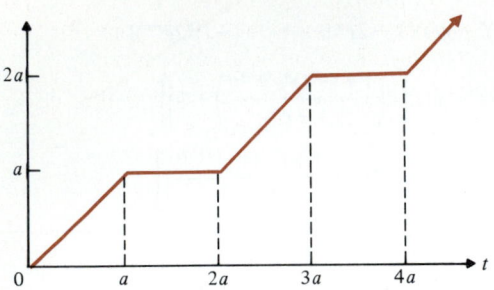

14.

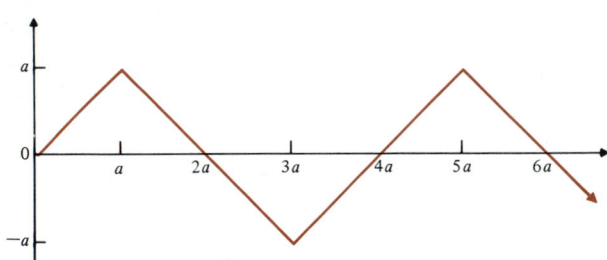

15. Compute the Laplace transform of the **half-wave rectifier** shown in the accompanying graph.

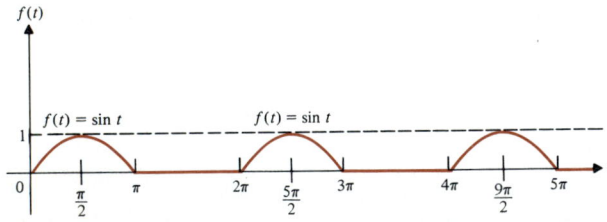

16. Compute the Laplace transform of the sawtooth wave given in the accompanying graph.

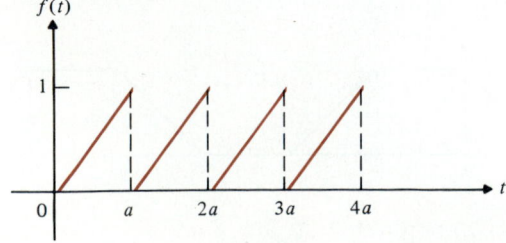

17. Let $f(t) = \begin{cases} \sin t, & t < 4\pi, \\ \sin t + \cos t, & t > 4\pi. \end{cases}$

Compute $\mathcal{L}\{f(t)\}$.

18. Let $f(t) = \begin{cases} \cos t, & t < 3\pi/2, \\ \cos t + \sin t, & t > 3\pi/2. \end{cases}$

Compute $\mathcal{L}\{f(t)\}$.

In Problems 19–24, compute the inverse of the given Laplace transform.

19. $\dfrac{e^{-\pi s}}{1 + s^2}$

20. $\dfrac{se^{-3\pi s}}{1 + s^2}$

21. $\dfrac{s - se^{-\pi s}}{1 + s^2}$

22. $\dfrac{1 - e^{-2s}}{s^2}$

23. $\dfrac{1 + e^{-4s}}{s^5}$

24. $\dfrac{e^{-2s}}{s^2 - 1}$ [*Hint:* Use partial fractions.]

25. Let $f(t) = \begin{cases} 1, & 0 < t < 1, \\ 3, & 1 < t < 7, \\ 5, & t > 7. \end{cases}$

Write $f(t)$ as a step function and compute $\mathcal{L}\{f(t)\}$.

26. Do the same for the function

$$f(t) = \begin{cases} -2, & 0 < t < 1, \\ 0, & 1 < t < 10, \\ 2, & t > 10. \end{cases}$$

In Problems 27–33 solve the given initial value problem with discontinuous forcing function $f(t)$. Draw a graph of the solution.

27. $y'' + y = f(t)$, $y(0) = y'(0) = 0$, where

$$f(t) = \begin{cases} t, & 0 < t < \pi \\ 0, & t > \pi \end{cases}$$

28. $y'' - y = f(t)$, $y(0) = 1$, $y'(0) = 0$, where

$$f(t) = \begin{cases} 1, & 0 < t < 1 \\ 0, & t > 1 \end{cases}$$

* 29. $y'' + 2y' + 10y = f(t)$, $y(0) = y'(0) = 0$, where $f(t)$ is the function in Problem 28.

* **30.** $y'' + 2y' + 5y = f(t)$, $y(0) = 0$, $y'(0) = 1$, where $f(t)$ is defined in Problem 27.

* **31.** $y'' + 2y' + 10y = f(t)$, $y(0) = 1$, $y'(0) = 0$, where $f(t)$ is defined in Problem 28.

32. $y'' - y' + 6y = \delta(t - 2)$, $y(0) = 1$, $y'(0) = -2$

33. $y'' - 4y' + 13y = \delta(t - 1)$, $y(0) = 0$, $y'(0) = 3$

* **34.** A model rocket of 1 kg mass blasts off from a playground field with a propulsion force of 10 N. Assume the motor runs out of propellant after 10 seconds, that the propellant has 500 grams mass, and that the propellant is consumed at a constant rate.

   (a) Write the equations of motion for this situation assuming a damping force (due to air resistance) proportional to the velocity of the rocket.

   (b) Solve the equation in part (a) using the Laplace transform methods of this section.

(Optional)

4.4 Some Differential Equations with Discontinuous Forcing Functions: Applications to Electrical Circuits

There are many physical models that give rise to differential equations with discontinuous forcing functions. In this section, we discuss several models involving electrical circuits. We begin with the *RLC* circuit discussed in Section 2.7.

Consider the differential equation

$$L\frac{dI}{dt} + RI + \frac{Q}{C} = E. \tag{1}$$

[See equation 2.7.1 on p. 102.] This is the equation that describes the circuit shown (again) in Fig. 1.

We assume that the voltage source, which may be a battery, is controlled by a switch that initially is turned off, and $Q(0) = I(0) = 0$. At some later time, t_1, the switch is turned on and the voltage is then equal to some constant value, E_0. In this situation we have

$$E(t) = E_0 H(t - t_1). \tag{2}$$

To find the current at all times $t > 0$, we use the technique of Example 4.2.9 on p. 210 and the fact that $s\hat{Q} = \hat{I}$ (since $dQ/dt = I$ and $Q(0) = 0$) to find that

$$\hat{I}(s) = T(s)\hat{E}(s) \tag{3}$$

where the transfer function is given by

$$T(s) = \frac{s/L}{s^2 + (R/L)s + (1/LC)}. \tag{4}$$

As before, the symbol ^ denotes the Laplace transform. From (2), we have

$$\hat{E}(s) = \mathcal{L}\{E_0 H(t - t_1)\} = \frac{E_0}{s}e^{-t_1 s}.$$

We can use this to solve for $\hat{I}(s)$ in equation (3).

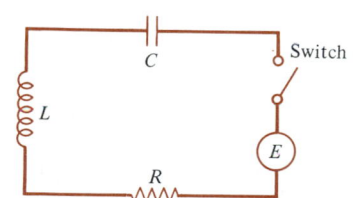

FIGURE 1

Remark. $E(t)$ is not defined at $t = t_1$ and so is not differentiable at that point. Therefore, the use of equation (3) is not strictly valid. However, we can get around this difficulty by observing that, according to equation (4.3.10), $H'(t - t_1) = \delta(t - t_1)$. Of course, $\delta(t - t_1)$ is not a function in the traditional sense but, since $\delta(t - t_1)$ has a Laplace transform [see equation (4.3.11)], we shall not worry about this difficulty.

EXAMPLE 1

An *RLC* circuit with $R = 10$ ohms, $L = 1$ henry, and $C = 0.01$ farad is hooked up to a battery that delivers a steady voltage of 20 volts when switched on. If the switch, initially off, is turned on after 10 seconds, find the current for all future values of t. Assume that Q and I are zero when the switch is turned on.

SOLUTION

Using equations (3) and (4) and the values given above, we have

$$\hat{I}(s) = \frac{s}{s^2 + 10s + 100}\left(\frac{E_0}{s}e^{-10s}\right)$$

$$= \frac{E_0 e^{-10s}}{s^2 + 10s + 100} = \frac{20e^{-10s}}{(s+5)^2 + 75}.$$

From the second shifting theorem, we obtain

$$\frac{e^{-10s}}{(s+5)^2 + 75} = \mathcal{L}\{f(t-10)H(t-10)\},$$

where $f(t)$ is the function whose Laplace transform is $[(s + 5)^2 + 75]^{-1}$. Then

$$f(t) = \mathcal{L}^{-1}\left\{\frac{1}{(s+5)^2 + 75}\right\}$$

$$= \frac{1}{\sqrt{75}}\mathcal{L}^{-1}\left\{\frac{\sqrt{75}}{(s+5)^2 + 75}\right\} = \frac{1}{\sqrt{75}}e^{-5t}\sin\sqrt{75}t.$$

Next,

$$f(t-10) = \frac{1}{\sqrt{75}}e^{-5(t-10)}\sin\sqrt{75}(t-10)$$

so, finally, we have

$$I(t) = \frac{20}{\sqrt{75}}e^{-5(t-10)}\sin\sqrt{75}(t-10)H(t-10).$$

Note that there is no current if $t < 10$.

EXAMPLE 2

Consider the *LC* circuit given in Fig. 2 with $I(0) = I'(0) = 0$. The voltage is given by

$$E(t) = \begin{cases} 25t, & 0 \le t \le 4, \\ 100, & t > 4. \end{cases}$$

Find the current for all values of $t \ge 0$.

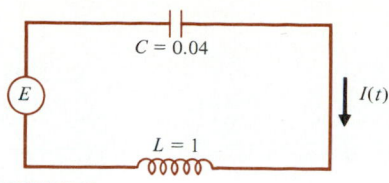

FIGURE 2

SOLUTION

Using equation (1) we have (since $1/0.04 = 25$)

$$\frac{dI}{dt} + 25Q = E$$

and, differentiating,

$$\frac{d^2I}{dt^2} + 25I = E'(t) = \begin{cases} 25, & 0 \le t \le 4, \\ 0, & t > 4, \end{cases}$$

or

$$\frac{d^2I}{dt^2} + 25I = 25 - 25H(t-4), \qquad I(0) = I'(0) = 0.$$

Then, taking Laplace transforms and using $I(0) = I'(0) = 0$, we have

$$s^2 \hat{I}(s) + 25\hat{I}(s) = \frac{25}{s} - \frac{25e^{-4s}}{s} = \frac{-25}{s}(e^{-4s} - 1)$$

so that

$$\hat{I}(s) = \frac{-25(e^{-4s} - 1)}{s(s^2 + 25)}.$$

Note that

$$\frac{25}{s(s^2 + 25)} = \frac{1}{s} - \frac{s}{s^2 + 25}$$

and

$$\hat{I}(s) = (e^{-4s} - 1)\left(\frac{-1}{s} + \frac{s}{s^2 + 25}\right)$$

$$= \frac{-e^{-4s}}{s} + e^{-4s}\frac{s}{s^2 + 25} + \frac{1}{s} - \frac{s}{s^2 + 25}.$$

Thus

$$I(t) = 1 - \cos 5t + H(t-4)[\cos 5(t-4) - 1].$$

EXAMPLE 3

Consider the parallel electric circuit shown in Fig. 3, where the arrows denote the direction of current flow over each component of the circuit. We assume that $I(t) = CE_0 \delta(t - t_1)$. Clearly there is no current except at $t = t_1$. Show that if $E(0) = 0$, then $I(t)$ yields enough current to charge the capacitor to the voltage E_0 immediately.

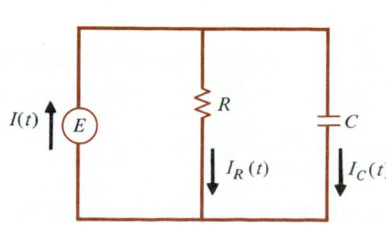

FIGURE 3

SOLUTION

We refer to Kirchhoff's laws given in Section 2.7. We have the system

$$E = RI_R, \qquad E = \frac{1}{C}Q_C, \tag{5}$$

or, differentiating,

$$E' = \frac{1}{C}\frac{dQ_C}{dt} = \frac{1}{C}I_C(t). \tag{6}$$

The total current is given by

$$I = I_R + I_C. \tag{7}$$

Thus

$$I(t) = I_R(t) + I_C(t) = \frac{E}{R} + CE'$$

or

$$E'(t) + \frac{1}{RC}E(t) = \frac{1}{C}I(t) = E_0\delta(t - t_1).$$

Then, taking transforms and using the fact that $E(0) = 0$, we have

$$s\hat{E}(s) + \frac{1}{RC}\hat{E}(s) = E_0 e^{-st_1}$$

or

$$\hat{E}(s) = \frac{E_0 e^{-st_1}}{s + 1/RC} = \mathcal{L}\{H(t - t_1)f(t - t_1)\},$$

where

$$\mathcal{L}\{f(t)\} = \frac{E_0}{s + 1/RC}.$$

Thus

$$f(t) = E_0 e^{-(1/RC)t},$$

and

$$E(t) = E_0 e^{-(1/RC)(t-t_1)}H(t - t_1).$$

Note that the voltage on the capacitor is zero before time $t = t_1$. At that time it jumps to the value E_0.

EXAMPLE 4

In Example 3, what happens to the current after time t_1?

SOLUTION
For $t > t_1$, $I(t) = 0$. Thus, for $t > t_1$ the only current is the current through the resistor and capacitor. From equation (5) we have, for $t > t_1$,

$$I_R(t) = \frac{1}{R}E(t) = \frac{1}{R}E_0 e^{-(1/RC)(t-t_1)}H(t - t_1) = \frac{E_0}{R}e^{-(1/RC)(t-t_1)}.$$

Thus the current through the resistor decreases exponentially after the time $t = t_1$; that is, the current "bleeds off" through the resistor after the instant at which the capacitor is charged.

Problems 4.4

1. Find the current for all t in the *RLC* circuit of Example 1 if $R = 20$ ohms, $L = \frac{1}{2}$ henry, and $C = 0.002$ farad, and a 50-volt battery, which is initially off, is turned on after 30 seconds.

2. Answer the question in Problem 1 if $R = 15$ ohms, $L = 2$ henrys, $C = 0.04$ farad, and E is a steady voltage battery of 25 volts that is turned on after 1 minute.

In Exercises 3–5, find the current for all values of t in the *LC* circuit of Example 2 using the given data. Graph the solution.

3. $L = 1$ H, $C = \frac{1}{16}$ f, $E(t) = \begin{cases} 16t, & 0 \le t \le 5, \\ 80, & t > 5 \end{cases}$ V.

4. $L = 1$ H, $C = 0.04$ f, $E(t) = \begin{cases} 0, & 0 \le t < 2, \\ 20t, & 2 \le t \le 4, \\ 80, & t > 4 \end{cases}$ V.

5. $L = 1$ H, $C = 0.1$ f, $E(t) = \begin{cases} 10t, & 0 \le t \le 2, \\ 20, & 2 \le t \le 4, \\ 20t, & t > 4 \end{cases}$ V.

6. In Example 3 find the voltage across the resistor after 20 seconds if $E_0 = 10$ volts, $t_1 = 10$ seconds, $R = 10$ ohms, and $C = 0.1$ farad.

7. An undamped spring (see p. 106) supports an object of 1-kg mass. The spring constant of the spring is 4 N/m. Suppose that a force $f(t)$ is applied to the object where

$$f(t) = \begin{cases} 2t, & 0 \le t < \pi/2, \\ 0, & t \ge \pi/2 \end{cases} \text{ N.}$$

  (a) Find the equation of motion of the mass.

  (b) Assuming that, initially, the mass is displaced downward 1 m before $f(t)$ is applied, find the position of the object for all $t > 0$.

8. Answer the questions in Problem 7 if

$$f(t) = \begin{cases} 0, & t < \pi/2, \\ 4, & t > \pi/2 \end{cases} \text{ N.}$$

4.5 The Transform of Convolution Integrals

It often occurs that in the process of solving a linear differential equation by transforms, we end up with a transform that is the product of two other transforms. Although we proved in Problem 4.1.79 that

$$\mathcal{L}^{-1}\{F + G\} = \mathcal{L}^{-1}\{F\} + \mathcal{L}^{-1}\{G\},$$

it is not true that $\mathcal{L}^{-1}\{FG\} = \mathcal{L}^{-1}\{F\}\mathcal{L}^{-1}\{G\}$. For example, if $F(s) = 1/s$ and $G(s) = 1/s^2$, then $F(s)G(s) = 1/s^3$, but

$$\mathcal{L}^{-1}\{FG\} = \mathcal{L}^{-1}\left\{\frac{1}{s^3}\right\} = \frac{t^2}{2}, \qquad \mathcal{L}^{-1}\{F\} = 1, \mathcal{L}^{-1}\left\{\frac{1}{s^2}\right\} = t$$

and, clearly, $\mathcal{L}^{-1}\{FG\} \ne \mathcal{L}^{-1}\{F\}\mathcal{L}^{-1}\{G\}$.

In this section we shall define the convolution of two functions f and g and show that $\mathcal{L}^{-1}\{FG\}$ is equal to the convolution of $\mathcal{L}^{-1}\{F\}$ and $\mathcal{L}^{-1}\{G\}$. We shall then apply this fact in a variety of ways.

■ **DEFINITION**

Convolution

If f and g are piecewise continuous functions, then the **convolution** of f and g, written $(f * g)$, is defined by

$$(f * g)(t) = \int_0^t f(t-u)g(u)\,du. \tag{1}$$

The notation $(f * g)(t)$ indicates that the convolution $f * g$ is a function of the independent variable t.

Using the change of variables $v = t - u$, we see that

$$(f * g)(t) = -\int_t^0 f(v)g(t-v)\,dv = \int_0^t g(t-v)f(v)\,dv = (g * f)(t). \tag{2}$$

Hence $(f * g)(t) = (g * f)(t)$, and we can take the convolution in either order without altering the result. We may now state the main result of this section.

■ **THEOREM 1: Convolution Theorem for Laplace Transforms**
Let $F(s) = \mathcal{L}\{f(t)\}$ and $G(s) = \mathcal{L}\{g(t)\}$. Then

$$\mathcal{L}\{(f * g)(t)\} = F(s)G(s).$$

PROOF
By definition

$$F(s)G(s) = \left(\int_0^\infty e^{-su}f(u)\,du\right)\left(\int_0^\infty e^{-sv}g(v)\,dv\right)$$

$$= \int_0^\infty \int_0^\infty e^{-s(u+v)}f(u)g(v)\,dv\,du. \tag{3}$$

If we make the change of variables $t = u + v$, with u fixed, then $dt = dv$ and the integral (3) is equal to

$$F(s)G(s) = \int_0^\infty \int_u^\infty e^{-st}f(u)g(t-u)\,dt\,du. \tag{4}$$

Changing the order of integration† and noting that

$$\int_0^\infty \int_u^\infty dt\,du = \int_0^\infty \int_0^t du\,dt$$

(see Fig. 1), we have the integral in (4) equal to

$$F(s)G(s) = \int_0^\infty \int_0^t e^{-st}f(u)g(t-u)\,du\,dt$$

$$= \int_0^\infty e^{-st}\left[\int_0^t g(t-u)f(u)\,du\right]dt$$

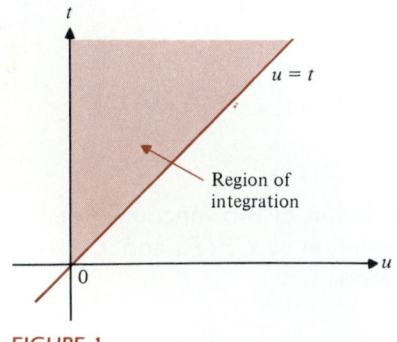

$u = t$

Region of integration

FIGURE 1

† This may not always be possible. Conditions under which reversing the order of integration is permissible are found in most advanced calculus texts.

$$= \int_0^\infty e^{-st}(g * f)(t)\, dt = \int_0^\infty e^{-st}(f * g)(t)\, dt$$

$$= \mathcal{L}\{f * g\}.$$

■ COROLLARY

If $F(s) = \mathcal{L}\{f(t)\}$ and $G(s) = \mathcal{L}\{g(t)\}$, then

$$\mathcal{L}^{-1}\{F(s)G(s)\} = (f * g)(t). \tag{5}$$

Our first applications of this theorem and corollary are in the computation of inverse transforms.

EXAMPLE 1

Compute $\mathcal{L}^{-1}\{s/(s^2 + 1)^2\}$.

SOLUTION
Since

$$\mathcal{L}\{\cos at\} = \frac{s}{s^2 + 1} \quad \text{and} \quad \mathcal{L}\{\sin t\} = \frac{1}{s^2 + 1},$$

we have

$$\mathcal{L}^{-1}\left\{\frac{s}{(s^2 + 1)^2}\right\} = \mathcal{L}^{-1}\left\{\frac{s}{s^2 + 1} \cdot \frac{1}{s^2 + 1}\right\}.$$

By Theorem 1, this is the convolution of $\cos t$ and $\sin t$. But

$$\sin t * \cos t = \int_0^t \sin(t - u) \cos u\, du$$

$$= \int_0^t (\sin t \cos u - \cos t \sin u) \cos u\, du$$

$$= \sin t \int_0^t \cos^2 u\, du - \cos t \int_0^t \sin u \cos u\, du,$$

$$= \left[\sin t\left(\frac{\sin u \cos u + u}{2}\right) - \cos t \frac{\sin^2 u}{2}\right]\Bigg|_{u=0}^{u=t}$$

$$= \frac{t \sin t}{2}.$$

Therefore,

$$\mathcal{L}^{-1}\left\{\frac{s}{(s^2 + 1)^2}\right\} = \frac{t \sin t}{2}. \tag{6}$$

EXAMPLE 2

Compute $\mathcal{L}^{-1}\{e^{-as}/s^{n+1}\}$, where $n \geq 1$ is an integer and $a > 0$ is a real number.

SOLUTION
Note that

$$\frac{1}{s^{n+1}} e^{-as} = \frac{1}{s^n} \frac{e^{-as}}{s} = \mathcal{L}\left\{\frac{t^{n-1}}{(n-1)!}\right\} \mathcal{L}\{H(t-a)\}.$$

Thus

$$\mathcal{L}^{-1}\left\{\frac{1}{s^{n+1}} e^{-as}\right\} = \int_0^t \frac{(t-u)^{n-1}}{(n-1)!} H(u-a)\, du.$$

If $t < a$, then $H(u - a) = 0$, but if $t > a$, then

$$\int_0^t \frac{(t-u)^{n-1}}{(n-1)!} H(u-a)\, du = \int_a^t \frac{(t-u)^{n-1}}{(n-1)!}\, du$$

$$= -\left.\frac{(t-u)^n}{n!}\right|_a^t = \frac{(t-a)^n}{n!}.$$

Thus

$$\mathcal{L}^{-1}\left\{\frac{1}{s^{n+1}} e^{-as}\right\} = \begin{cases} 0, & \text{if } t < a, \\ \dfrac{(t-a)^n}{n!}, & \text{if } t > a, \end{cases}$$

or

$$\mathcal{L}^{-1}\left\{\frac{1}{s^{n+1}} e^{-as}\right\} = \frac{(t-a)^n}{n!} H(t-a). \tag{7}$$

Note. This problem can also be solved using the second shifting theorem.

Convolutions are very useful in solving differential equations with discontinuous right-hand sides.

EXAMPLE 3

Solve the initial value problem

$$y'' + y = f(t), \qquad y(0) = 0, y'(0) = 1,$$

where

$$f(t) = \begin{cases} 1, & 0 < t < 1, \\ 0, & t > 1. \end{cases}$$

SOLUTION

Since $f(t) = H(t) - H(t-1)$, taking Laplace transforms of the differential equation we have

$$s^2 \mathcal{L}\{y\} - 1 + \mathcal{L}\{y\} = \frac{1 - e^{-s}}{s}$$

or

$$\mathcal{L}\{y\} = \frac{1 + s - e^{-s}}{s(s^2 + 1)} = \frac{1}{s} - \frac{s-1}{s^2+1} - \frac{e^{-s}}{s} \cdot \frac{1}{s^2+1}.$$

Using the convolution theorem for Laplace transforms (Theorem 1), we get

$$y(t) = 1 - \cos t + \sin t - \sin t * H(t-1).$$

But

$$\sin t * H(t-1) = \int_0^t \sin(t-u)H(u-1)\,du$$

and by the definition of the Heaviside function

$$\sin t * H(t-1) = H(t-1) \int_1^t \sin(t-u)\,du$$

$$= H(t-1) \cos(t-u) \Big|_1^t$$

$$= H(t-1)[1 - \cos(t-1)].$$

Thus

$$y(t) = 1 - \cos t + \sin t - H(t-1)[1 - \cos(t-1)].$$

Volterra Integral Equations

Although the convolution theorem is obviously very useful in calculating inverse transforms, it also has important applications in a very different area. In 1931 the Italian mathematician Vito Volterra[†] published a book that contained a fairly sophisticated model of population growth. It would be beyond the scope of this book to go into a derivation of Volterra's model. However, a central equation in this model is of the form

$$x(t) = f(t) + \int_0^t a(t-u)x(u)\,du. \tag{8}$$

Volterra Integral Equation

An equation of this type, where $f(t)$ and $a(t)$ can be assumed to be continuous, is called a **Volterra integral equation.** Since the publication of Volterra's papers,

† V. Volterra, *Leçons sur la Théorie Mathématique de la Lutte pour la Vie,* Gauthier-Villars, Paris, 1931.

many diverse phenomena in thermodynamics, electrical systems theory, nuclear reactor theory, and chemotherapy have been modeled with Volterra integral equations.

It is quite easy to see how Laplace transforms can be used to solve an equation in the form of equation (8). Taking transforms on both sides of equation (8), using the convolution theorem, and denoting transforms by the appropriate capital letters, we obtain

$$X(s) = F(s) + A(s)X(s)$$

or

$$X(s)[1 - A(s)] = X(s) - A(s)X(s) = F(s)$$

so that

$$X(s) = \frac{F(s)}{1 - A(s)}. \tag{9}$$

Looking at equation (9), we immediately see that if $F(s)$ and $A(s)$ are defined for $s \geq s_0$, then $X(s)$ is similarly defined so long as $A(s) \neq 1$. Once $X(s)$ is known, we may (if possible) calculate the solution $x(t) = \mathcal{L}^{-1}\{X(s)\}$.

EXAMPLE 4

Consider the integral equation

$$x(t) = t^2 + \int_0^t \sin(t - u)x(u)\, du. \tag{10}$$

Taking transforms, we have

$$X(s) = \frac{2}{s^3} + \frac{1}{s^2 + 1} \cdot X(s)$$

or

$$X(s) = \frac{2/s^3}{1 - 1/(s^2 + 1)} = \frac{2(s^2 + 1)}{s^5} = \frac{2}{s^3} + \frac{2}{s^5}.$$

Hence the solution to equation (10) is given by

$$x(t) = t^2 + \tfrac{1}{12}t^4.$$

Other applications of the very useful convolution theorem are given in the exercises. The student, however, should always keep in mind that the greatest difficulty in using any of these methods is that it is frequently difficult to calculate inverse transforms. Unfortunately, most problems that arise lead to inverting transforms that do not fit into familiar patterns. For this reason, methods have been devised for estimating such inverses. The interested reader should consult a more advanced book on Laplace transforms, such as the excellent book by Widder.[†]

† D. V. Widder, *The Laplace Transformation*, Princeton University Press, Princeton, N.J., 1941.

Problems 4.5

In Problems 1–5, find the Laplace transform of each given convolution integral.

1. $f(t) = \int_0^t (t-u)^3 \sin u \, du$

2. $f(t) = \int_0^t e^{-(t-u)} \cos 2u \, du$

3. $f(t) = \int_0^t (t-u)^3 u^5 \, du$

4. $f(t) = \int_0^t \sinh 4(t-u) \cosh 5u \, du$

5. $f(t) = \int_0^t e^{17(t-u)} u^{19} \, du$

In Problems 6–12, use the convolution theorem to calculate the inverse Laplace transforms of the given functions.

6. $F(s) = \dfrac{1}{s^2(s^2+a^2)}$

7. $F(s) = \dfrac{3}{s^4(s^2+1)}$

8. $F(s) = \dfrac{1}{(s^2+1)^2}$

9. $F(s) = \dfrac{1}{s(s^2+a^2)}$

10. $F(s) = \dfrac{1}{(s^2+1)^3}$

11. $F(s) = \dfrac{e^{-3s}}{s^3}$

12. $F(s) = \dfrac{e^{-10s}}{s^5}$

13. Solve the Volterra integral equation

$$x(t) = e^{-t} - 2\int_0^t \cos(t-u)x(u) \, du.$$

14. Solve the Volterra integral equation

$$x(t) = t + \frac{1}{6}\int_0^t (t-u)^3 x(u) \, du.$$

15. Find the solution of the initial value problem

$$y'' + 4y' + 13y = f(t), \qquad y(0) = y'(0) = 0,$$

where

$$f(t) = \begin{cases} 1, & t < \pi, \\ 0, & t > \pi. \end{cases}$$

16. Find the solution of the initial value problem

$$y'' - 2y' + 2y = f(t), \qquad y(0) = y'(0) = 0,$$

where

$$f(t) = \begin{cases} 0, & t < \pi/2, \\ 1, & \pi/2 < t < 3\pi/2, \\ 2, & t > 3\pi/2. \end{cases}$$

17. Use the convolution theorem to show that $\mathcal{L}\{\int_0^t f(u) \, du\} = F(s)/s$, where $F(s) = \mathcal{L}\{f(t)\}$. [*Hint:* $\int_0^t f(u) \, du = (1 * f)(t)$.]

18. Compute $\mathcal{L}\{\int_0^{t-a} f(u) \, du\}$. [*Hint:* $\int_0^{t-a} f(u) \, du = \int_0^t f(u)H(t - a - u) \, du$.]

4.6 Laplace Transform Methods for Systems

Laplace transform techniques are very useful for solving systems of differential equations with given initial conditions. Consider the system

$$\frac{dx}{dt} = a_{11}x + a_{12}y + f(t),$$

$$\frac{dy}{dt} = a_{21}x + a_{22}y + g(t),$$

\hfill (1)

with initial conditions $x(0) = x_0$, $y(0) = y_0$. Taking the Laplace transform of both equations in system (1) and letting the corresponding capital letters represent the Laplace transforms of the functions x, y, f, and g (that is, we let $X = L\{x\}$, $Y = L\{y\}$, and so on), we obtain

$$sX - x_0 = a_{11}X + a_{12}Y + F,$$

$$sY - y_0 = a_{21}X + a_{22}Y + G.$$

\hfill (2)

Gathering all the terms involving X and Y on the left-hand side, we obtain from (2) the system of simultaneous equations

$$(s - a_{11})X - a_{12}Y = F + x_0,$$

$$-a_{21}X + (s - a_{22})Y = G + y_0. \tag{3}$$

If $F + x_0$ and $G + y_0$ are not both identically zero, we may solve equations (3) simultaneously, obtaining

$$X = \frac{(s - a_{22})(F + x_0) + a_{12}(G + y_0)}{s^2 - (a_{11} + a_{22})s + (a_{11}a_{22} - a_{12}a_{21})},$$

$$Y = \frac{(s - a_{11})(G + y_0) + a_{21}(F + x_0)}{s^2 - (a_{11} + a_{22})s + (a_{11}a_{22} - a_{12}a_{21})}. \tag{4}$$

Note that the denominators of the fractions in equations (4) are identical to the characteristic polynomial for system (1) that we obtained in equation (3.2.9) on p. 138. Thus we may rewrite equations (4) in the form

$$X = \frac{(s - a_{22})(F + x_0) + a_{12}(G + y_0)}{(s - \lambda_1)(s - \lambda_2)},$$

$$Y = \frac{(s - a_{11})(G + y_0) + a_{21}(F + x_0)}{(s - \lambda_1)(s - \lambda_2)}, \tag{5}$$

where λ_1 and λ_2 are the solutions of the characteristic equation

$$\lambda^2 - (a_{11} + a_{22})\lambda + (a_{11}a_{22} - a_{12}a_{21}) = 0.$$

We can now find the solution (x, y) of system (1) with given initial conditions by inverting (if possible) equations (5).

It should be apparent from the discussion above that the Laplace transform allows us to convert a system of differential equations with given initial conditions into a system of simultaneous algebraic equations. This method clearly can be generalized to apply to systems of n linear first-order differential equations with constant coefficients, thus yielding a corresponding system of n simultaneous linear algebraic equations. As in the case of a single differential equation (see Section 4.2), Laplace transform methods are primarily intended for the solution of systems of linear differential equations with constant coefficients and given initial conditions. Any deviation from these conditions can so complicate the problem as to make no solution obtainable.

EXAMPLE 1

Consider the initial value problem

$$x' = x - y - e^{-t}, \qquad x(0) = 1,$$

$$y' = 2x + 3y + e^{-t}, \qquad y(0) = 0. \tag{6}$$

Using the differentiation property of Laplace transforms, we have

$$sX - 1 = X - Y - 1/(s+1),$$

$$sY = 2X + 3Y + 1/(s+1). \tag{7}$$

System (7) may be rewritten as

$$(s-1)X + Y = s/(s+1),$$

$$-2X + (s-3)Y = 1/(s+1), \tag{8}$$

from which we find that

$$X = \frac{s^2 - 3s - 1}{(s+1)[(s-2)^2 + 1]} = \frac{A_1}{s+1} + \frac{A_2(s-2) + A_3}{(s-2)^2 + 1},$$

$$Y = \frac{3s - 1}{(s+1)[(s-2)^2 + 1]} = \frac{B_1}{s+1} + \frac{B_2(s-2) + B_3}{(s-2)^2 + 1}. \tag{9}$$

Using the methods developed in Section 4.2, we obtain

$$A_1 = \tfrac{3}{10}, \qquad A_2 = \tfrac{7}{10}, \qquad A_3 = -\tfrac{11}{10},$$

$$B_1 = -\tfrac{2}{5}, \qquad B_2 = \tfrac{2}{5}, \qquad B_3 = \tfrac{9}{5}.$$

Therefore we have the solution

$$(x, y) = \left(\frac{3}{10} e^{-t} + \frac{e^{2t}}{10}(7 \cos t - 11 \sin t), -\frac{2}{5} e^{-t} + \frac{e^{2t}}{5}(2 \cos t + 9 \sin t) \right).$$

EXAMPLE 2: A Projectile under the Influence of the Earth's Rotation

Here we give a far more complicated model—a description of the motion of a particle projected from the earth where we do not ignore the effects of the earth's rotation on the particle.[†] To begin, we set up the needed three-dimensional coordinate system.

Assume that the positive x-axis points south, the positive y-axis points east, and the positive z-axis points in the direction opposite the direction of acceleration g due to gravity.

This model was studied by the German physicist B. M. Planck.[‡] Planck took the origin to have latitude β and the angular velocity of the earth to be ω. Assuming the mass of the particle to be negligible, Planck derived the following system of second-order equations:

$$\frac{d^2x}{dt^2} = 2\omega \sin \beta \frac{dy}{dt}, \tag{10}$$

$$\frac{d^2y}{dt^2} = -2\omega \left(\sin \beta \frac{dx}{dt} + \cos \beta \frac{dz}{dt} \right), \tag{11}$$

$$\frac{d^2z}{dt^2} = 2\omega \cos \beta \frac{dy}{dt} - g, \tag{12}$$

[†] This problem is very difficult computationally. Skip it if you balk at lengthy calculations. Unfortunately, real problems rarely have easy solutions. This problem is typical of those encountered beyond textbook examples.

[‡] *Einführung in die Allgemeine Mechanik*, 4th ed. (Leipzig: S. Hirzel, 1928), p. 81.

with

$$0 = x(0) = y(0) = z(0), \qquad x'(0) = u, \qquad y'(0) = v, \qquad z'(0) = w. \tag{13}$$

Taking transforms of both sides of equation (10), we obtain

$$s^2X - sx(0) - x'(0) = 2\omega \sin \beta[sY - y(0)],$$

and, using the conditions in equation (13),

$$s^2X - u = 2\omega(\sin \beta)sY,$$

or

$$s^2X - 2\omega(\sin \beta)sY = u. \tag{14}$$

Similarly (see Problem 13), we obtain

$$2\omega(\sin \beta)sX + s^2Y + 2\omega(\cos \beta)sZ = v \tag{15}$$

and

$$-2\omega(\cos \beta)sY + s^2Z = w - \frac{g}{s}. \tag{16}$$

Equations (14), (15), and (16) constitute a system of three equations in the three unknowns X, Y, and Z and have a unique solution if and only if the determinant of the system is nonzero. This determinant is

$$D = \begin{vmatrix} s^2 & -2\omega(\sin \beta)s & 0 \\ 2\omega(\sin \beta)s & s^2 & 2\omega(\cos \beta)s \\ 0 & -2\omega(\cos \beta)s & s^2 \end{vmatrix}$$

$$= s^6 + 4\omega^2(\sin^2 \beta)s^4 + 4\omega^2(\cos^2 \beta)s^4$$

$$= s^6 + 4\omega^2 s^4 = s^4(s^2 + 4\omega^2).$$

Thus for $s > 0$ the determinant is nonzero and the system has a unique solution. Using Cramer's rule (see Section 7.9), we obtain

$$X(s) = \frac{\begin{vmatrix} u & -2\omega(\sin \beta)s & 0 \\ v & s^2 & 2\omega(\cos \beta)s \\ w - \dfrac{g}{s} & -2\omega(\cos \beta)s & s^2 \end{vmatrix}}{D}$$

$$= \frac{us^4 - 4\omega^2 \sin \beta(\cos \beta)s^2\left(w - \dfrac{g}{s}\right) + 2\omega(\sin \beta)vs^3 + 4\omega^2(\cos^2 \beta)s^2u}{s^4(s^2 + 4\omega^2)}$$

$$= \frac{u}{s^2 + 4\omega^2} + v\left[\frac{2\omega \sin \beta}{s(s^2 + 4\omega^2)}\right] + u\left[\frac{4\omega^2 \cos^2 \beta}{s^2(s^2 + 4\omega^2)}\right]$$

$$- w\left[\frac{4\omega^2 \sin \beta \cos \beta}{s^2(s^2 + 4\omega^2)}\right] + g\left[\frac{4\omega^2 \sin \beta \cos \beta}{s^3(s^2 + 4\omega^2)}\right]. \tag{17}$$

Similarly (see Problem 14), we obtain

$$Y(s) = -u\left[\frac{2\omega\sin\beta}{s(s^2+4\omega^2)}\right] + v\left[\frac{1}{s^2+4\omega^2}\right]$$

$$- w\left[\frac{2\omega\cos\beta}{s(s^2+4\omega^2)}\right] + g\left[\frac{2\omega\cos\beta}{s^2(s^2+4\omega^2)}\right] \tag{18}$$

and

$$Z(s) = -u\left[\frac{4\omega^2\sin\beta\cos\beta}{s^2(s^2+4\omega^2)}\right] + v\left[\frac{2\omega\cos\beta}{s(s^2+4\omega^2)}\right]$$

$$+ w\left[\frac{1}{s^2+4\omega^2} + \frac{4\omega^2\sin^2\beta}{s^2(s^2+4\omega^2)}\right] - g\left[\frac{1}{s(s^2+4\omega^2)} + \frac{4\omega^2\sin^2\beta}{s^3(s^2+4\omega^2)}\right]. \tag{19}$$

Next (see Problem 15), equations (17), (18), and (19) can be inverted, and using the trigonometric identity $2\sin^2 A = 1 - \cos 2A$ we obtain

$$x(t) = \frac{u}{2\omega}(2\omega t\cos^2\beta + \sin^2\beta\sin 2\omega t) + \frac{v}{\omega}\sin\beta\sin^2\omega t$$

$$- \frac{w}{2\omega}\sin\beta\cos\beta(2\omega t - \sin 2\omega t) + \frac{g}{2\omega^2}\sin\beta\cos\beta(\omega^2 t^2 - \sin^2\omega t); \tag{20}$$

$$y(t) = -\frac{u}{\omega}\sin\beta\sin^2\omega t + \frac{v}{2\omega}\sin 2\omega t$$

$$- \frac{w}{\omega}\cos\beta\sin^2\omega t + \frac{g}{4\omega^2}\cos\beta(2\omega t - \sin 2\omega t); \tag{21}$$

$$z(t) = -\frac{u}{2\omega}\sin\beta\cos\beta(2\omega t - \sin 2\omega t) + \frac{v}{\omega}\cos\beta\sin^2\omega t$$

$$+ \frac{w}{2\omega}(2\omega t\sin^2\beta + \sin 2\omega t\cos^2\beta)$$

$$- \frac{g}{2\omega^2}(\omega^2 t^2\sin^2\beta + \cos^2\beta\sin^2\omega t). \tag{22}$$

These are the equations of motion of the projectile when the earth's rotation is taken into account. Finally, we note by L'Hôpital's rule that

$$\lim_{\omega\to 0}\frac{\sin 2\omega t}{2\omega} = \lim_{\omega\to 0}\frac{2t\cos\omega t}{2} = t$$

and

$$\lim_{\omega\to 0}\frac{\sin^2\omega t}{\omega} = \lim_{\omega\to 0}\frac{2t\sin\omega t\cos\omega t}{1} = 0.$$

Thus

$$\lim_{\omega\to 0}x(t) = ut, \qquad \lim_{\omega\to 0}y(t) = vt, \qquad \text{and} \qquad \lim_{\omega\to 0}z(t) = wt - \tfrac{1}{2}gt^2, \tag{23}$$

showing that equations (20), (21), and (22) reduce to the usual equations of motion when the earth's rotation is ignored.

Problems 4.6

In Problems 1–12 solve the initial value problems involving systems of differential equations by the Laplace transform method.

1. $x' = y$, $\quad x(0) = 1$
$y' = x$, $\quad y(0) = 0$

2. $x' = -3x + 4y$, $\quad x(0) = 3$
$y' = -2x + 3y$, $\quad y(0) = 2$

3. $x' = 4x - 2y$, $\quad x(0) = 2$
$y' = 5x + 2y$, $\quad y(0) = -2$

4. $x' = x - 2y$, $\quad x(0) = 1$
$y' = 4x + 5y$, $\quad y(0) = -2$

5. $x' = -3x + 4y + \cos t$, $\quad x(0) = 0$
$y' = -2x + 3y + t$, $\quad y(0) = 1$

6. $x' = 4x - 2y + e^t$, $\quad x(0) = 1$
$y' = 5x + 2y - t$, $\quad y(0) = 0$

7. $x' = x - 2y + t^2$, $\quad x(0) = 1$
$y' = 4x + 5y - e^t$, $\quad y(0) = -1$

8. $x'' + x + y = 0$, $\quad x(0) = x'(0) = 0$
$x' + y' = 0$, $\quad y(0) = 1$

9. $x'' = y + \sin t$, $\quad x(0) = 1$, $\quad x'(0) = 0$
$y'' = -x' + \cos t$, $\quad y(0) = -1$, $\quad y'(0) = -1$

10. $x'' = 2y + 2$, $\quad x(0) = 2$, $\quad x'(0) = 2$
$y' = -x + 5e^{2t} + 1$, $\quad y(0) = 1$

11. $x = z'$, $\quad x(0) = y(0) = z(0) = 1$
$x' + y + z = 1$
$-x + y' + z = 2 \sin t$

12. $x'' = y - z$, $\quad x(0) = x'(0) = 0$
$y'' = x' + z'$, $\quad y(0) = -1$, $\quad y'(0) = 1$
$z'' = -(1 + x + y)$, $\quad z(0) = 0$, $\quad z'(0) = 1$

13. Beginning with equations (11) and (12), use the differentiation property to obtain equations (15) and (16).

14. Use Cramer's rule in systems (14), (15), and (16) to obtain expressions (18) and (19) for Y and Z.

* 15. Invert the transforms in equations (17), (18), and (19) to solve for $x(t)$, $y(t)$, and $z(t)$.

Review Exercises for Chapter 4

In Exercises 1–17, find the Laplace transform of the given function.

1. $3t - 2$

2. $t^3 + 4t^2 - 2t + 1$

3. e^{2t-1}

4. $\cos(2t + 1)$

5. te^{-t}

6. $e^{-t} \cos 2t$

7. $\sinh(3t - 4)$

8. $t^3 e^{2t}$

9. $te^t \cos t$

10. $\cosh^2 2t$

11. $\cos^2 t$

12. $\int_0^t u \cos u \, du$

13. $t^2 \cos 2t$

14. $\cos t \cdot H(t - 2\pi)$

15. $\delta(t - 3)$

16. The function shown in the accompanying figure.

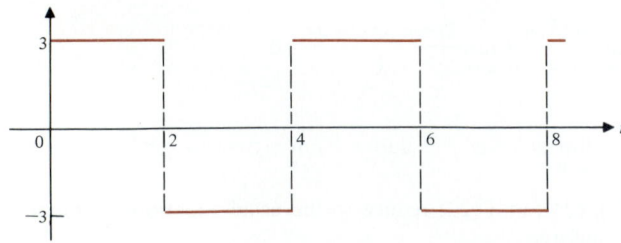

17. $f(t) = \begin{cases} \sin t, & t < 2\pi \\ \sin t + \cos t, & t > 2\pi \end{cases}$

In Exercises 18–32, find the inverse of the given Laplace transform.

18. $\dfrac{3}{s^2}$

19. $\dfrac{-14}{s}$

20. $\dfrac{s+2}{s^2+4}$

21. $\dfrac{1}{(s-2)^2}$

22. $\dfrac{s-3}{s^2-3}$

23. $\dfrac{1}{s^2+4s+5}$

24. $\dfrac{s+3}{s^2+8s+17}$

25. $\dfrac{s}{s^2-3s+2}$

26. $\dfrac{s+2}{s(s^2+4)}$

27. $\dfrac{s^2+8s-3}{(s^2+2s+1)(s^2+1)}$

28. $\ln \dfrac{s-3}{s-2}$

29. $\ln\left(1 + \dfrac{4}{s^2}\right)$

30. $\dfrac{se^{-(\pi/2)s}}{1+s^2}$

31. $\dfrac{s-se^{-(\pi/2)s}}{1+s^2}$

32. $\dfrac{e^{-3s}}{s^2-1}$

In Exercises 33–41, solve the given initial value problem by Laplace transform methods.

33. $y'' + y = 0,\quad y(0) = 1,\quad y'(0) = 3$

34. $y'' - 4y = 0,\quad y(0) = 2,\quad y'(0) = -5$

35. $y'' - 5y' + 6y = 0,\quad y(0) = 2,\quad y'(0) = 1$

36. $y'' + 4y' + 4y = 0,\quad y(0) = -1,\quad y'(0) = 2$

37. $y'' - y = te^{2t},\quad y(0) = 0,\quad y'(0) = 1$

38. $y'' + 9y = \sin t,\quad y(0) = y'(0) = 0$

39. $y'' + 2y' + 2y = H(t-3),\quad y(0) = y'(0) = 0$

40. $y' - 6y = H(t-2),\quad y(0) = 0$

41. $y'' - 4y' + 4y = \delta(t-1),\quad y(0) = 0,\quad y'(0) = 1$

In Exercises 42–44, find the Laplace transform of the given convolution integral.

42. $\int_0^t (t-u)^2 \sin u\, du$

43. $\int_0^t (t-u)^4 u^7\, du$

44. $\int_0^t e^{15(t-u)} u^{25}\, du$

45. Use the convolution theorem to compute the inverse Laplace transform of

(a) $F(s) = \dfrac{2}{s^3(s^2+1)}$

(b) $F(s) = \dfrac{1}{s(s^2+4)}$

46. Solve the initial value problem

$$y'' + 6y' + 18y = f(t),\qquad y(0) = y'(0) = 0,$$

where

$$f(t) = \begin{cases} 2, & t < \pi, \\ 0, & t > \pi. \end{cases}$$

47. If $f(0) = g(0) = 0$ show that

(a) $f' * g = f * g'$

(b) $(f * g)' = \frac{1}{2}(f' * g + f * g')$

48. Solve the system

$$x' = 3x + 2y, \qquad x(0) = 2$$

$$y' = -5x + y, \qquad y(0) = 1$$

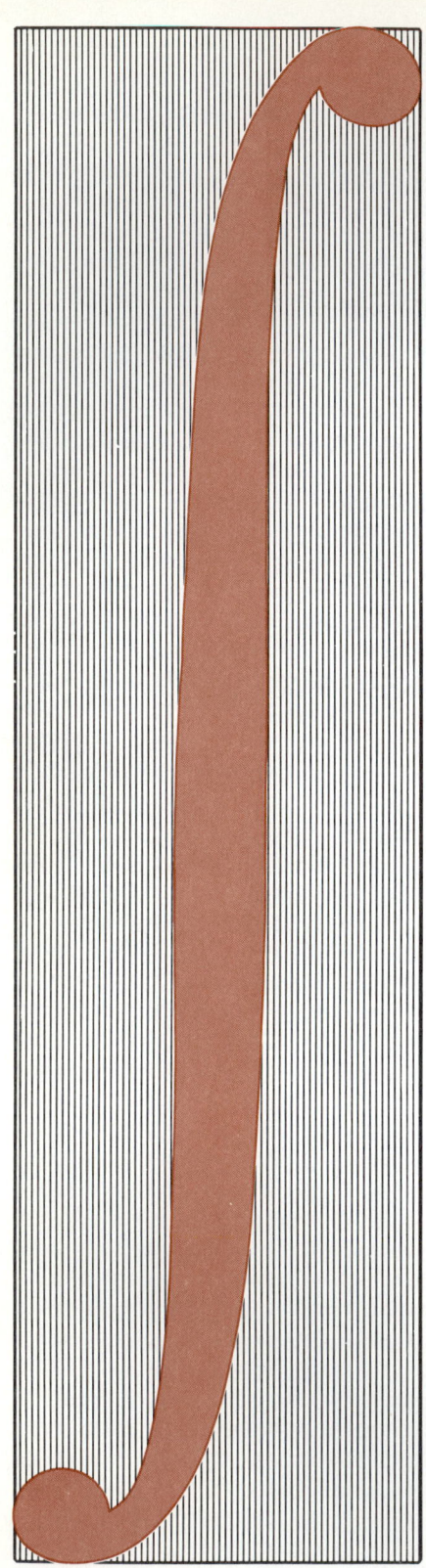

chapter 5

Power Series Solutions of Differential Equations

In Chapter 2 we studied several methods for solving second- and higher-order differential equations. With the exception of the Euler equation and a few equations in which one solution was easily guessed, the techniques applied only to linear differential equations with *constant coefficients*. The case of linear differential equations with *variable coefficients* is much more complicated. Unfortunately many of the most important differential equations in applied mathematics—for example, Bessel's equation and Legendre's equation—are of this type. In this chapter we shall consider a method for obtaining solutions to such equations. Since the solutions so obtained are in the form of power series, the procedure used is known as the *power series method*. The last two sections of the chapter are an introduction to Bessel functions and Legendre polynomials. In those sections we shall consider some of the properties of these special functions and a few of the standard procedures used in working with them.

5.1 The Power Series Method

In this section we shall review some of the basic properties of power series before discussing the power series method. We take it for granted that most readers will have received some background in power series in an earlier course in calculus. A **power series** in $(x - a)$ is an infinite series of the form

$$\sum_{n=0}^{\infty} c_n(x-a)^n = c_0 + c_1(x-a) + c_2(x-a)^2 + \cdots, \tag{1}$$

where $c_0, c_1, \ldots$ are constants, called the **coefficients** of the series, a is a constant called the **center** of the series, and x is an independent variable. In particular, a power series centered at zero ($a = 0$) has the form

$$\sum_{n=0}^{\infty} c_n x^n = c_0 + c_1 x + c_2 x^2 + c_3 x^3 + \cdots. \tag{2}$$

Note that polynomials are also power series, since they have this form.

A series of the form (1) can always be reduced to the form (2) by the substitution $X = x - a$. This substitution is merely a translation of the coordinate system. It is easy to see that the behavior of equation (2) near zero is exactly the same as the behavior of equation (1) near a. For this reason we need only study the properties of series of the form (2).

The most familiar power series are those that are obtained by the use of **Taylor's formula:**

Taylor's Formula

$$f(x) = \sum_{n=0}^{N} \frac{f^{(n)}(a)}{n!} (x - a)^n + R_N(x - a). \tag{3}$$

If the function $f(x)$ has derivatives of all orders at the point a, and for $|x - a|$ sufficiently small, the remainder term $R_N(x - a)$ tends to zero as N tends to infinity, we say that $f(x)$ is **analytic** at a and write

$$f(x) = \sum_{n=0}^{\infty} \frac{f^{(n)}(a)}{n!} (x - a)^n. \tag{4}$$

Taylor Series
Maclaurin Series

The series (4) is called the **Taylor series** of $f(x)$ at the point $x = a$. When $a = 0$, equation (4) is often called the **Maclaurin series** of $f(x)$. The following familiar expansions, valid for all x, may all be obtained by this method:

$$e^x = \sum_{n=0}^{\infty} \frac{x^n}{n!} = 1 + x + \frac{x^2}{2!} + \frac{x^3}{3!} + \cdots,$$

$$\sin x = \sum_{n=0}^{\infty} \frac{(-1)^n x^{2n+1}}{(2n+1)!} = x - \frac{x^3}{3!} + \frac{x^5}{5!} - \frac{x^7}{7!} + \cdots,$$

$$\cos x = \sum_{n=0}^{\infty} \frac{(-1)^n x^{2n}}{(2n)!} = 1 - \frac{x^2}{2!} + \frac{x^4}{4!} - \frac{x^6}{6!} + \cdots,$$

$$\sinh x = \sum_{n=0}^{\infty} \frac{x^{2n+1}}{(2n+1)!} = x + \frac{x^3}{3!} + \frac{x^5}{5!} + \frac{x^7}{7!} + \cdots,$$

$$\cosh x = \sum_{n=0}^{\infty} \frac{x^{2n}}{(2n)!} = 1 + \frac{x^2}{2!} + \frac{x^4}{4!} + \frac{x^6}{6!} + \cdots.$$

There are functions $f(x)$ that have derivatives of all orders at a given point a and yet are not analytic. In these cases, the remainder term $R_N(x - a)$ does not tend to zero as N tends to ∞. An example of such a function is given in Problem 17.

For the series (2) we compute

$$L = \lim_{n \to \infty} \frac{c_{n+1}}{c_n}. \tag{5}$$

If $L = 0$, the series converges for all x. This is the case for the Maclaurin series for e^x, since $c_{n+1}/c_n = 1/n + 1$. If $L = \infty$ then the series converges only for $x = 0$. For example, the series $\sum_{n=0}^{\infty} n!x^n$ converges only when $x = 0$ because $L = \lim_{n\to\infty} \dfrac{(n+1)!}{n!} = \lim_{n\to\infty}(n+1) = \infty$.

If $0 < L < \infty$, we let $R = \dfrac{1}{L}$. The series converges absolutely for $|x| < R$

Radius of Convergence

and diverges if $|x| > R$. The number R is called the **radius of convergence** of the power series (2). At $|x| = R$, the series may converge or diverge.

Remark. If the limit in (5) does not exist, we can obtain R by computing

$$\frac{1}{R} = \lim_{n\to\infty} \sup\{|c_n|^{1/n}, |c_{n+1}|^{1/(n+1)}, |c_{n+2}|^{1/(n+2)}, \ldots\}, \tag{6}$$

where sup{ } denotes the least upper bound.

EXAMPLE 1

Find the radius of convergence of the series $\sum_{n=0}^{\infty} \dfrac{x^n}{(n+1)2^n}$.

SOLUTION

$$L = \lim_{n\to\infty} \frac{1/(n+2)2^{n+1}}{1/(n+1)2^n} = \lim_{n\to\infty} \frac{(n+1)2^n}{(n+2)2^{n+1}} = \frac{1}{2}\lim_{n\to\infty}\frac{n+1}{n+2} = \frac{1}{2}.$$

Thus $R = 2$. If $x = 2$, then

$$\sum_{n=0}^{\infty} \frac{2^n}{(n+1)2^n} = \sum_{n=0}^{\infty}\frac{1}{n+1} = 1 + \frac{1}{2} + \frac{1}{3} + \cdots$$

diverges (this is the harmonic series). If $x = -2$, then

$$\sum_{n=0}^{\infty} \frac{(-2)^n}{(n+1)2^n} = \sum_{n=0}^{\infty}\frac{(-1)^n}{n+1} = 1 - \frac{1}{2} + \frac{1}{3} - \cdots$$

converges conditionally (to ln 2) by the alternating series test.

Suppose that $f(x) = \sum_{n=0}^{\infty} c_n x^n$ is a power series with radius of convergence R. Then, for $|x| < R$,

$$f'(x) = \sum_{n=0}^{\infty} nc_n x^{n-1} \quad \text{and} \quad \int f(x)\,dx = \sum_{n=0}^{\infty}\frac{c_n x^{n+1}}{n+1} + C$$

That is, *a power series can be differentiated and integrated term by term within its radius of convergence.* This is true because a power series converges *uniformly* in the interval $|x| \le r < R$. We shall discuss uniform convergence in Section 17.3.

The fundamental assumption made in solving a differential equation by the power series method is that the solution of the differential equation can be expressed in the form of a power series, say,

$$y = \sum_{n=0}^{\infty} c_n x^n = c_0 + c_1 x + c_2 x^2 + \cdots . \tag{7}$$

Power series expansions for y', y'', ... can then be obtained by differentiating equation (9) term by term:

$$y' = \sum_{n=1}^{\infty} n c_n x^{n-1} = c_1 + 2c_2 x + 3c_3 x^2 + \cdots , \tag{8}$$

$$y'' = \sum_{n=2}^{\infty} n(n-1) c_n x^{n-2} = 2c_2 + 3 \cdot 2 c_3 x + 4 \cdot 3 c_4 x^2 + \cdots , \tag{9}$$

and these series are substituted into the given differential equation. Adding the series substituted into the differential equation by collecting the terms involving like powers of x, we obtain an expression of the form

$$k_0 + k_1 x + k_2 x^2 + \cdots = \sum_{n=0}^{\infty} k_n x^n = 0, \tag{10}$$

where the coefficients k_0, k_1, k_2, ... are expressions involving the unknown coefficients c_0, c_1, c_2, ... Since equation (10) must hold for all values of x in some interval, all the coefficients $k_0, k_1, k_2, \ldots$ must vanish. From the equations

$$k_0 = 0, \qquad k_1 = 0, \qquad k_2 = 0, \ldots$$

it is then possible to determine successively the coefficients c_0, c_1, c_2, ... To illustrate that the power series method does provide the required solution, we shall solve three problems, two of which can be solved more easily by other methods.

EXAMPLE 2

Solve the initial value problem

$$y' = y + x^2, \qquad y(0) = 1. \tag{11}$$

SOLUTION
Inserting equations (7) and (8) into the differential equation, we have

$$c_1 + 2c_2 x + 3c_3 x^2 + 4c_4 x^3 + \cdots = (c_0 + c_1 x + c_2 x^2 + c_3 x^3 + \cdots) + x^2.$$

Collecting like powers of x yields

$$(c_1 - c_0) + (2c_2 - c_1)x + (3c_3 - c_2 - 1)x^2 + (4c_4 - c_3)x^3 + \cdots = 0.$$

Equating each of the coefficients to zero, we obtain the identities

$$c_1 - c_0 = 0, \qquad 2c_2 - c_1 = 0, \qquad 3c_3 - c_2 - 1 = 0, \qquad 4c_4 - c_3 = 0, \ldots ,$$

from which we find that

$$c_1 = c_0, \qquad c_2 = \frac{c_1}{2} = \frac{c_0}{2!}, \qquad c_3 = \frac{c_2 + 1}{3} = \frac{c_0 + 2}{3!}, \qquad c_4 = \frac{c_3}{4} = \frac{c_0 + 2}{4!}, \ldots .$$

With these values, equation (7) becomes

$$y = c_0 + c_0 x + \frac{c_0}{2!} x^2 + \frac{c_0 + 2}{3!} x^3 + \frac{c_0 + 2}{4!} x^4 + \frac{c_0 + 2}{5!} x^5 + \cdots$$

$$= (c_0 + 2) \left[1 + x + \frac{x^2}{2!} + \frac{x^3}{3!} + \frac{x^4}{4!} + \cdots \right] - 2 \left[1 + x + \frac{x^2}{2!} \right].$$

Looking carefully at the series in square brackets, we recognize the expansion for e^x, so we have the general solution

$$y = (c_0 + 2) e^x - x^2 - 2x - 2.$$

To solve the initial value problem, we set $x = 0$ to obtain,

$$1 = y(0) = c_0 + 2 - 2 = c_0.$$

Thus the solution of the initial value problem (11) is given by

$$y = 3e^x - x^2 - 2x - 2.$$

EXAMPLE 3

Solve the differential equation

$$y'' + y = 0. \tag{12}$$

SOLUTION
Using equations (7) and (9), we have

$$(2c_2 + 3 \cdot 2c_3 x + 4 \cdot 3c_4 x + \cdots) + (c_0 + c_1 x + c_2 x^2 + \cdots) = 0.$$

Gathering like powers of x yields

$$(2c_2 + c_0) + (3 \cdot 2c_3 + c_1) x + (4 \cdot 3c_4 + c_2) x^2 + \cdots = 0.$$

Setting each of the coefficients to zero, we obtain

$$2c_2 + c_0 = 0, \qquad 3 \cdot 2c_3 + c_1 = 0, \qquad 4 \cdot 3c_4 + c_2 = 0, \qquad 5 \cdot 4c_5 + c_3 = 0, \ldots$$

and

$$c_2 = -\frac{c_0}{2!}, \qquad c_3 = -\frac{c_1}{3!}, \qquad c_4 = -\frac{c_2}{4 \cdot 3} = \frac{c_0}{4!}, \qquad c_5 = -\frac{c_3}{5 \cdot 4} = \frac{c_1}{5!}, \ldots$$

Substituting these values into the power series (7) for y yields

$$y = c_0 + c_1 x - \frac{c_0}{2!} x^2 - \frac{c_1}{3!} x^3 + \frac{c_0}{4!} x^4 + \frac{c_1}{5!} x^5 + \cdots.$$

Splitting this series into two parts, we have

$$y = c_0 \left(1 - \frac{x^2}{2!} + \frac{x^4}{4!} - \cdots \right) + c_1 \left(x - \frac{x^3}{3!} + \frac{x^5}{5!} - \cdots \right).$$

Using the Maclaurin series for $\sin x$ and $\cos x$ we obtain the familiar general solution

$$y = c_0 \cos x + c_1 \sin x$$

to equation (12). We observe that in this case the power series method produces two arbitrary constants c_0, c_1.

So far we have considered only linear equation with constant coefficients. We turn now to linear equations with variable coefficients.

EXAMPLE 4

Solve the differential equation

$$y'' + xy' + y = 0. \tag{13}$$

SOLUTION

Using the power series method we obtain the equation

$$\sum_{n=2}^{\infty} n(n-1)c_n x^{n-2} + x \sum_{n=1}^{\infty} nc_n x^{n-1} + \sum_{n=0}^{\infty} c_n x^n = 0. \tag{14}$$

We use the summation notation in this example in order to develop the skill in manipulating power series that will be required later on. We would like to rewrite each of the sums in equation (14) so that the general term will contain the same power of x. Consider the first sum:

$$T_1 = \sum_{n=2}^{\infty} n(n-1)c_n x^{n-2}.$$

To obtain the exponent k in place of $n - 2$, we make the substitution $k = n - 2$. Then $n = k + 2$, so every place we see an n, we replace it with $k + 2$. Since n ranges from 2 to ∞ and $k = n - 2$, k ranges from $2 - 2 = 0$ to ∞ ($\infty - 2 = \infty$; explain why). Thus we have

$$T_1 = \sum_{k=0}^{\infty} (k+2)(k+1)c_{k+2} x^k.$$

Now, let $k = n$, so that

$$T_2 = x \sum_{n=1}^{\infty} nc_n x^{n-1} = \sum_{n=1}^{\infty} nc_n x \cdot x^{n-1} = \sum_{n=1}^{\infty} nc_n x^n = \sum_{k=1}^{\infty} kc_k x^k$$

and

$$T_3 = \sum_{n=0}^{\infty} c_n x^n = \sum_{k=0}^{\infty} c_k x^k.$$

Thus (14) becomes

$$\sum_{k=0}^{\infty} (k+2)(k+1)c_{k+2} x^k + \sum_{k=1}^{\infty} kc_k x^k + \sum_{k=0}^{\infty} c_k x^k = 0.$$

Note that the second sum can also be allowed to range from 0 to ∞, since $kc_k x^k = 0$ when $k = 0$.

Gathering like terms in x produces

$$\sum_{k=0}^{\infty} [(k+2)(k+1)c_{k+2} + (k+1)c_k]x^k = 0.$$

Recursion Formula

Setting the coefficients equal to zero, we obtain the general **recursion formula**

$$(k+2)(k+1)c_{k+2} + (k+1)c_k = 0.$$

Therefore $(k+2)c_{k+2} = -c_k$ or $c_{k+2} = -c_k/(k+2)$ and

$$c_2 = -\frac{c_0}{2}, \quad c_3 = -\frac{c_1}{3}, \quad c_4 = -\frac{c_2}{4} = \frac{c_0}{2 \cdot 4},$$

$$c_5 = -\frac{c_3}{5} = \frac{c_1}{3 \cdot 5}, \quad c_6 = -\frac{c_4}{6} = -\frac{c_0}{2 \cdot 4 \cdot 6}, \cdots.$$

Hence the power series for y can be written in the form

$$y = c_0 + c_1 x - \frac{c_0}{2}x^2 - \frac{c_1}{3}x^3 + \frac{c_0}{2 \cdot 4}x^4 + \frac{c_1}{3 \cdot 5}x^5 - \cdots$$

$$= c_0\left(1 - \frac{x^2}{2} + \frac{x^4}{2 \cdot 4} - \frac{x^6}{2 \cdot 4 \cdot 6} + \cdots\right) + c_1\left(x - \frac{x^3}{3} + \frac{x^5}{3 \cdot 5} - \frac{x^7}{3 \cdot 5 \cdot 7} + \cdots\right)$$

by separating the terms that involve c_0 and c_1. At this point we try to see whether we recognize the two series that have been obtained by the power series method. Frequently this is an unproductive task, but in this instance we are fortunate:

$$1 - \frac{x^2}{2} + \frac{x^4}{2 \cdot 4} - \frac{x^6}{2 \cdot 4 \cdot 6} + \cdots = 1 + \left(-\frac{x^2}{2}\right) + \frac{1}{2!}\left(-\frac{x^2}{2}\right)^2 + \frac{1}{3!}\left(-\frac{x^2}{2}\right)^3 + \cdots$$

or

$$y_1 = e^{-x^2/2}.$$

Although we cannot recognize the second series, knowing one solution allows us to determine the other. Substitute $y_2 = v(x)y_1$ into equation (13) (see Section 2.2). This yields

$$v''y_1 + v'(2y_1' + xy_1) + v(y_1'' + xy_1' + y_1) = 0.$$

The term in the last parentheses is zero since y_1 is a solution to equation (13). Setting $z = v'$, we obtain the *first*-order differential equation

$$y_1 z' + (2y_1' + xy_1)z = e^{-x^2/2}(z' - xz) = 0.$$

Separating variables we get $z = e^{x^2/2}$, so that the second solution to equation (13) is

$$y_2 = y_1 v = e^{-x^2/2} \int e^{x^2/2}\, dx.$$

Hence the general solution to equation (13) is given by

$$y = c_0 e^{-x^2/2} + c_1 e^{-x^2/2} \int e^{x^2/2}\, dx.$$

Note that this is as far as we can go since $\int e^{x^2/2}\, dx$ cannot be expressed in terms of elementary functions (these are the rational, trigonometric, exponential, and logarithmic functions).

Problems 5.1

In Problems 1–10 find the Taylor series centered at a and the radius of convergence for the given function.

1. $f(x) = e^x$, $\quad a = 1$

2. $f(x) = e^{-x}$, $\quad a = 0$

3. $f(x) = \cos x$, $\quad a = \pi/4$

4. $f(x) = \sinh x$, $\quad a = \ln 2$

5. $f(x) = e^{bx}$, $\quad a = -1$

6. $f(x) = xe^x$, $\quad a = 1$

7. $f(x) = x^2 e^{-x^2}$, $\quad a = 0$

8. $f(x) = \begin{cases} \dfrac{\sin x}{x}, & x \neq 0, \quad a = 0 \\ 1, & x = 0 \end{cases}$

9. $f(x) = (x - 1) \ln x$, $\quad a = 1$

10. $f(x) = \sin^2 x$, $\quad a = 0$

In Problems 11–14 derive the given Taylor series.

11. $\ln(1 + x) = x - \dfrac{x^2}{2} + \dfrac{x^3}{3} - \dfrac{x^4}{4} + \cdots, \quad |x| < 1$

12. $\sin^{-1} x = x + \dfrac{1}{2} \cdot \dfrac{x^3}{3} + \dfrac{1}{2} \cdot \dfrac{3}{4} \cdot \dfrac{x^5}{5} + \dfrac{1}{2} \cdot \dfrac{3}{4} \cdot \dfrac{5}{6} \cdot \dfrac{x^7}{7} + \cdots,$ $\quad |x| < 1$

13. $\ln x = (x - 1) - \dfrac{(x-1)^2}{2} + \dfrac{(x-1)^3}{3} - \dfrac{(x-1)^4}{4} + \cdots,$ $\quad 0 < x < 2$

14. $\dfrac{1}{2 - x} = 1 + (x - 1) + (x - 1)^2 + (x - 1)^3 + \cdots,$ $\quad 0 < x < 2$

15. Show that

$$\frac{1}{1 + x} = 1 - x + x^2 - x^3 + \cdots, \qquad |x| < 1$$

Then prove that

(a) $\ln(1 + x) = x - \dfrac{x^2}{2} + \dfrac{x^3}{3} - \dfrac{x^4}{4} + \cdots, \quad |x| < 1$

(b) $\tan^{-1} x = x - \dfrac{x^3}{3} + \dfrac{x^5}{5} - \dfrac{x^7}{7} + \cdots, \quad |x| < 1$

(c) $\dfrac{1}{(1 + x)^2} = 1 - 2x + 3x^2 - 4x^3 + \cdots, \quad |x| < 1$

16. Show that the series

$$\sum_{n=1}^{\infty} \frac{x^n}{n} = x + \frac{x^2}{2} + \frac{x^3}{3} + \frac{x^4}{4} + \cdots$$

diverges at $x = 1$ by proving that the partial sums satisfy the inequality

$$S_{2^k}(1) = \sum_{n=1}^{2^k} \frac{1}{n} \geq 1 + \frac{k}{2}.$$

(This exercise shows that even though the terms in a series may tend to zero, the series itself may diverge, in this case at $x = 1$.)

17. Consider the function

$$f(x) = \begin{cases} e^{-1/x^2}, & x \neq 0, \\ 0, & x = 0. \end{cases}$$

(a) Show that f has derivatives of all orders at $x = 0$ and that $f'(0) = f''(0) = \cdots = 0$.

(b) Conclude that $f(x)$ does not have a Taylor series expansion at $x = 0$, even though it is infinitely differentiable there. Thus f is not analytic at $x = 0$.

18. Using Taylor's formula, prove the **binomial formula**

$$(1 + x)^p = 1 + px + \frac{p(p-1)}{1 \cdot 2} x^2 + \frac{p(p-1)(p-2)}{1 \cdot 2 \cdot 3} x^3 + \cdots.$$

Where does it converge?

In Problems 19–34 find the general solution of each equation by the power series method. When initial conditions are specified, give the solution that satisfies them.

19. $y' = y - x$, $\quad y(0) = 2$

20. $y' = x^3 - 2xy$, $\quad y(0) = 1$

21. $y'' + y = x$

22. $y'' + 4y = 0$, $\quad y(0) = 1$, $\quad y'(0) = 0$

23. $(1 + x^2)y'' + 2xy' - 2y = 0$

24. $xy'' - xy' + y = e^x$, $\quad y(0) = 1$, $\quad y'(0) = 2$

25. $xy'' - x^2 y' + (x^2 - 2)y = 0$, $\quad y(0) = 0$, $\quad y'(0) = 1$

26. $(1 - x)y'' - y' + xy = 0$, $\quad y(0) = y'(0) = 1$

27. $y'' - 2xy' + 4y = 0$, $\quad y(0) = 1$, $\quad y'(0) = 0$

28. $(1 - x^2)y'' - xy' + y = 0$, $\quad y(0) = 0$, $\quad y'(0) = 1$

29. $y'' - xy' + y = -x \cos x$, $\quad y(0) = 0$, $\quad y'(0) = 2$

30. $y'' - xy' + xy = 0$, $\quad y(0) = 2$, $\quad y'(0) = 1$

31. $(1 - x)^2 y'' - (1 - x)y' - y = 0$, $\quad y(0) = y'(0) = 1$

32. $y'' - 2xy' + 2y = 0$

33. $y'' - 2xy' - 2y = x$, $\quad y(0) = 1$, $\quad y'(0) = -\tfrac{1}{4}$

34. $y'' - x^2 y = 0$

35. Airy's equation†

$$y'' - xy = 0$$

has applications in the theory of diffraction. Find the general solution of this equation.

36. Hermite's equation‡

$$y'' - 2xy' + 2py = 0,$$

where p is constant, arises in quantum mechanics in connection with the Schrödinger§ equation for a harmonic oscillator. Show that if p is a positive integer, one of the two linearly independent solutions of Hermite's equation is a polynomial, called the **Hermite polynomial $H_p(x)$**.

5.2 Ordinary and Singular Points

The power series method sometimes fails to yield a solution for one equation while working very well for an apparently similar equation.

EXAMPLE 1

Solve the equations

$$x^2 y'' + axy' + by = 0, \tag{1}$$

where

(i) $a = -2$, $b = 2$.
(ii) $a = -1$, $b = 1$.
(iii) $a = 1$, $b = 1$.

SOLUTION
Set

$$y = \sum_{n=0}^{\infty} c_n x^n, \qquad y' = \sum_{n=1}^{\infty} n c_n x^{n-1}, \qquad y'' = \sum_{n=2}^{\infty} n(n-1) c_n x^{n-2}.$$

Then, since $n(n-1) = 0$ at $n = 0$ and $n = 1$,

$$x^2 y'' + axy' + by = x^2 \sum_{n=0}^{\infty} n(n-1) c_n x^{n-2} + ax \sum_{n=0}^{\infty} n c_n x^{n-1} + b \sum_{n=0}^{\infty} c_n x^n$$

$$= \sum_{n=0}^{\infty} n(n-1) c_n x^n + \sum_{n=0}^{\infty} an c_n x^n + \sum_{n=0}^{\infty} b c_n x^n$$

$$= \sum_{n=0}^{\infty} [n(n-1) + an + b] c_n x^n = 0. \tag{2}$$

---

† Sir George Biddell Airy (1801–1892) was Lucasian Professor of Mathematics, director of the observatory, and Plumian Professor of Astronomy at Cambridge University in England until 1835. Then he was appointed director of the Greenwich Observatory (Astronomer Royal). He remained there until his retirement in 1881. He did much work in lunar and solar photography, planetary motion, optics, and other areas.

‡ Charles Hermite (1822–1901) was a French mathematician known for his contributions in algebra and number theory.

§ Erwin Schrödinger (1887–1961) was an Austrian physicist. He was awarded the Nobel prize in 1933 (jointly with P. A. M. Dirac) for his work in quantum mechanics.

(i) Substituting $a = -2$ and $b = 2$ into equation (2) yields

$$\sum_{n=0}^{\infty} (n^2 - 3n + 2)c_n x^n = 0.$$

Equating each of the coefficients to zero, we have $(n - 2)(n - 1)c_n = 0$, implying that $c_n = 0$ for all $n \neq 1$ or 2. Hence

$$y = c_1 x + c_2 x^2$$

is the general solution to equation (1) with $a = -2$ and $b = 2$.

(ii) Equation (2) yields, with $a = -1$ and $b = 1$,

$$\sum_{n=0}^{\infty} (n^2 - 2n + 1)c_n x^n = 0$$

so that $(n - 1)^2 c_n = 0$. Thus $c_n = 0$ for all $n \neq 1$, yielding the solution

$$y = c_1 x.$$

Since the general solution of a second-order linear differential equation involves *two* linearly independent solutions, the power series method has given us only half of the general solution. We can use the reduction of order method (Section 2.2) to find the other solution. Substitute $y = xv$ into equation (1) with $a = -1$ and $b = 1$:

$$x^2(xv)'' - x(xv)' + (xv) = 0 \quad \text{or} \quad x^3 v'' + x^2 v' = 0.$$

Setting $z = v'$, we obtain the first-order separable differential equation

$$x^3 \frac{dz}{dx} + x^2 z = 0.$$

Thus,

$$\frac{dz}{z} = -\frac{dx}{x}$$

so that $z = c/x$ and $v = c \ln|x|$. Hence the general solution to (ii) is

$$y = Ax + Bx \ln|x|.$$

(iii) Equation (2) gives us the series

$$\sum_{n=0}^{\infty} (n^2 + 1)c_n = 0.$$

Equating the coefficients to zero, we have $(n^2 + 1)c_n = 0$, so that $c_n = 0$ for every n. Hence, the power series method fails completely in helping us find the general solution

$$y = A \cos(\ln|x|) + B \sin(\ln|x|)$$

of equation (1) (check!).

It will be very useful if we can find the reason for this anomaly, because once we know the reason for the failure of the power series method, a method for avoiding this difficulty may become apparent. The main clue to the puzzle can be obtained by making the coefficient of the highest-order derivative equal to 1. Thus, if we write each of the second-order homogeneous equations in Example 1 in the form

$$y'' + a(x)y' + b(x)y = 0, \qquad (3)$$

the equations become

$$y'' + \frac{a}{x}y' + \frac{b}{x^2}y = 0.$$

Hence, neither of the terms $a(x)$ and $b(x)$ is defined at $x = 0$, so we certainly cannot expect to find a power series representation for $a(x)$ or $b(x)$ that converges in an open interval containing $x = 0$.

When $a(x)$ and $b(x)$ can both be represented by Maclaurin series that converge in an open interval containing $x = 0$, we have the following result.

■ THEOREM 1

There is a unique Maclaurin series $y(x)$ satisfying the initial value problem

$$y'' + a(x)y' + b(x)y = 0, \qquad y(0) = \alpha, \, y'(0) = \beta,$$

provided $a(x)$ and $b(x)$ can both be represented by Maclaurin series converging in an interval $|x| < R$. The power series $y(x)$ also converges in $|x| < R$. ■

The proof of this theorem is complicated and will not be given in this book.† Note, however, that it guarantees the success of the power series method whenever $a(x)$ and $b(x)$ are analytic at $x = 0$.

DEFINITION

Ordinary Point

We call $x = 0$ an **ordinary point** of the differential equation

$$y'' + a(x)y' + b(x)y = 0$$

Singular Point

when both $a(x)$ and $b(x)$ are analytic at $x = 0$. If $x = 0$ is not an ordinary point, it is called a **singular point** of the differential equation. Hence $x = 0$ is a singular point of equation (1) in Example 1, and it is an ordinary point of the differential equations in Examples 5.1.3–5.1.4.

EXAMPLE 2

Determine if the power series method yields a solution to each of the following equations.

$$(1 - x^2)y'' - 2xy' + 2y = 0. \qquad (4)$$

$$xy'' + 2y' + xy = 0. \qquad (5)$$

Is $x = 0$ an ordinary or a singular point of each equation?

SOLUTION

(i) Rewriting equation (4) in the form

$$y'' - \frac{2x}{1 - x^2}y' + \frac{2}{1 - x^2}y = 0,$$

† A proof may be found in E. A. Coddington, *An Introduction to Ordinary Differential Equations*, Prentice Hall, Englewood Cliffs, N.J., 1961, section 3.9.

we observe that

$$a(x) = \frac{-2x}{1-x^2} = -2x(1+x^2+x^4+\cdots),$$

$$b(x) = \frac{2}{1-x^2} = 2(1+x^2+x^4+\cdots)$$

are Maclaurin series that converge in the interval $|x| < 1$. Hence $x = 0$ is an ordinary point of equation (4) and the power series method will yield the general solution. Setting $y = \sum_{n=0}^{\infty} c_n x^n$ and multiplying the power series representations of y, y', and y'' by the coefficients of equation (4), we get

$$(1-x^2) \sum_{n=2}^{\infty} n(n-1)c_n x^{n-2} - 2x \sum_{n=1}^{\infty} nc_n x^{n-1} + 2 \sum_{n=0}^{\infty} c_n x^n = 0.$$

But

$$(1-x^2) \sum_{n=2}^{\infty} n(n-1)c_n x^{n-2} = \sum_{n=2}^{\infty} n(n-1)c_n x^{n-2} - \sum_{n=2}^{\infty} n(n-1)c_n x^n$$

$$\overset{\overset{\displaystyle k=n-2}{\downarrow}}{= \sum_{k=0}^{\infty} (k+2)(k+1)c_{k+2} x^k - \sum_{n=2}^{\infty} n(n-1)c_n x^n}$$

$$\overset{\overset{\displaystyle \substack{n(n-1)=0 \\ \text{at } n=0 \text{ and } n=1}}{\downarrow}}{= \sum_{k=0}^{\infty} (k+2)(k+1)c_{k+2} x^k - \sum_{n=0}^{\infty} n(n-1)c_n x^n}$$

$$\overset{\overset{\displaystyle k=n}{\downarrow}}{= \sum_{n=0}^{\infty} (n+2)(n+1)c_{n+2} x^n - \sum_{n=0}^{\infty} n(n-1)c_n x^n.}$$

Thus we obtain

$$\sum_{n=0}^{\infty} [(n+2)(n+1)c_{n+2} - n(n-1)c_n]x^n - \sum_{n=0}^{\infty} 2nc_n x^n + \sum_{n=0}^{\infty} 2c_n x^n = 0$$

or

$$\sum_{n=0}^{\infty} [(n+2)(n+1)c_{n+2} + (-n(n-1)-2n+2)c_n]x^n$$

$$= \sum_{n=0}^{\infty} [(n+2)(n+1)c_{n+2} + (-n^2-n+2)c_n]x^n$$

$$= \sum_{n=0}^{\infty} [(n+2)(n+1)c_{n+2} - (n+2)(n-1)c_n]x^n = 0$$

or

$$\sum_{n=0}^{\infty} (n+2)[(n+1)c_{n+2} - (n-1)c_n]x^n = 0. \qquad \textbf{(6)}$$

Most examples involving power series involve algebraic manipulations like the ones above. Here we have provided every detail. In subsequent examples we shall leave some of the details to you.

We note that equation (6) holds only if, for each n,

$$(n+1)c_{n+2} = (n-1)c_n \quad \text{or} \quad c_{n+2} = \frac{n-1}{n+1}c_n.$$

Setting $n = 1$ we immediately see that $c_3 = 0$. Hence $c_3 = c_5 = c_7 = \cdots = 0$. If even values of n are chosen, we have

$$c_2 = \frac{0-1}{0+1}c_0 = -c_0, \qquad c_4 = \frac{2-1}{2+1}c_2 = \frac{1}{3}c_2 = -\frac{1}{3}c_0,$$

$$c_6 = \frac{3}{5}c_4 = -\frac{3}{5}\cdot\frac{1}{3}c_0 = -\frac{1}{5}c_0, \qquad c_8 = \frac{5}{7}c_6 = -\frac{1}{7}c_0, \ldots,$$

and, choosing c_1 arbitrarily, we find the general solution

$$y = c_1 x + c_0\left(1 - x^2 - \frac{x^4}{3} - \frac{x^6}{5} - \cdots\right).$$

(ii) Here we have

$$y'' + \frac{2}{x}y' + y = 0$$

so that $a(x)$ is not defined at $x = 0$. Thus, $x = 0$ is a singular point of equation (5), and we should anticipate the possibility of trouble in using the power series method. We have $xy'' + 2y' + xy = 0$ and

$$x\sum_{n=2}^{\infty} n(n-1)c_n x^{n-2} + 2\sum_{n=1}^{\infty} nc_n x^{n-1} + x\sum_{n=0}^{\infty} c_n x^n = 0$$

or

$$\sum_{n=2}^{\infty} n(n-1)c_n x^{n-1} + \sum_{n=1}^{\infty} 2nc_n x^{n-1} + \sum_{n=0}^{\infty} c_n x^{n+1} = 0. \tag{7}$$

The first series is (setting $k = n - 1$)

$$\sum_{k=1}^{\infty} (k+1)kc_{k+1}x^k.$$

The second is

taking out the first term
$\downarrow$

$$\sum_{k=0}^{\infty} 2(k+1)c_{k+1}x^k = 2c_1 + \sum_{k=1}^{\infty} 2(k+1)c_{k+1}x^k.$$

Finally, setting $k = n + 1$ so that $n = k - 1$, the third series is

$$\sum_{k=1}^{\infty} c_{k-1}x^k,$$

and (7) becomes

$$2c_1 + \sum_{k=1}^{\infty} [(k+1)k + 2(k+1)]c_{k+1} + c_{k-1})x^k$$

$$= 2c_1 + \sum_{k=1}^{\infty} [(k+1)(k+2)c_{k+1} + c_{k-1}]x^k = 0$$

so that $c_1 = 0$ and

$$c_{k+1} = -\frac{c_{k-1}}{(k+1)(k+2)}, \text{ for } k \geq 1.$$

Thus $c_3 = c_5 = c_7 = \cdots = 0$. If odd values of k are chosen, we have

$$c_2 = -\frac{c_0}{3 \cdot 2} = -\frac{c_0}{3!}, \qquad c_4 = -\frac{c_2}{5 \cdot 4} = \frac{c_0}{5 \cdot 4 \cdot 3!} = \frac{c_0}{5!}, \cdots$$

and we obtain

$$y = c_0\left(1 - \frac{x^2}{3!} + \frac{x^4}{5!} - \frac{x^6}{7!} + \cdots\right) = \frac{c_0}{x}\left(x - \frac{x^3}{3!} + \frac{x^5}{5!} - \cdots\right) = c_0\frac{\sin x}{x}.$$

Hence, one solution to equation (5) is

$$y = c_0\frac{\sin x}{x}.$$

However, the power series method does not yield the general solution. We can find that solution by the method of reduction of order used in Example 5.1.4: setting $y = (v \sin x)/x$ we have, after some algebra,

$$(\sin x)v'' + 2(\cos x)v' = 0$$

or, letting $z = v'$,

$$\sin x\frac{dz}{dx} = -2z \cos x.$$

Hence

$$\frac{dz}{z} = -\frac{2 \cos x\,dx}{\sin x}$$

or

$$\ln|z| = -2 \ln|\sin x| + c$$

or

$$v' = \frac{k}{\sin^2 x} = k \csc^2 x.$$

Thus $v = -k \cot x$ so that a second solution of equation (5) is

$$y = \frac{\sin x}{x} \cdot \cot x = \frac{\sin x}{x} \cdot \frac{\cos x}{\sin x} = \frac{\cos x}{x}$$

and the general solution is

$$y = A\frac{\sin x}{x} + B\frac{\cos x}{x}.$$

Problems 5.2

In Problems 1–16, find two linearly independent power series about the ordinary point $x = 0$ that are solutions to the given differential equation.

1. $y'' - xy' + y = 0$ **2.** $y'' + xy' + y = 0$

3. $y'' - 3xy = 0$ **4.** $y'' - 2xy' + y = 0$

5. $y'' - xy' + xy = 0$ **6.** $y'' - x^2y = 0$

7. $y'' + x^2y' + 2xy = 0$ **8.** $y'' + x^2y' + xy = 0$

9. $(1 + x^2)y'' + 2xy' - 2y = 0$

10. $(1 - x)y'' - y' + xy = 0$

11. $(1 - x)^2y'' - (1 - x)y' - y = 0$

12. $(x^2 + 1)y'' - 6y = 0$

13. $(2x^2 + 1)y'' + 2xy' - 18y = 0$

14. $(x^2 + 2)y'' + 3xy' + y = 0$

15. $y'' - xy' + y = -x\cos x$

16. $y'' - 2xy' - 2y = x$

17. Find a solution to the initial value problem $y''' - xy = 0$ with $y(0) = 1$, $y'(0) = 0$, $y''(0) = 0$.

18. Solve **Airy's equation**

$$y'' - xy = 0, \qquad y(1) = 1, \qquad y'(1) = 0.$$

19. Solve the initial value problem $y'' - xy' - y = 0$ with $y(0) = 1$, $y'(0) = 0$.

In Problems 20–23 find the first four terms in the power series solution to the given initial value problem.

20. $y'' + (\sin x)y = 0$, $y(0) = 1$, $y'(0) = 0$

21. $y'' - e^x y = 0$, $y(0) = y'(0) = 1$

22. $y'' + (\cos x)y = 0$, $y(0) = 1$, $y'(0) = 0$

23. $y'' + (\cos x)y = 0$, $y(0) = 0$, $y'(0) = 1$

In Problems 24–32 use the power series method to obtain at least one solution about the singular point $x = 0$. Then use the reduction of order method to find the general solution.

24. $x^2y'' + 2xy' - 2y = 0$

25. $xy'' + (1 - 2x)y' - (1 - x)y = 0$

26. $xy'' + 2y' - xy = 0$

27. $x^2y'' + x(x - 1)y' - (x - 1)y = 0$

28. $xy'' + (1 - x)y' - y = 0$

29. $xy'' - (1 - x)y' - 2y = 0$

30. $x(x - 1)y'' - (1 - 3x)y' + y = 0$

31. $x(x - 1)y'' + 3y' - 2y = 0$

32. $x^2y'' - x(1 - x)y' + y = 0$

33. Does the power series method yield a solution to the equation
  (a) $x^2y' = y$?
  (b) $x^3y' = y$?

34. Show that the power series method fails for

$$x^2y'' + x^2y' + y = 0.$$

35. Show that the power series method fails for

$$x^3y'' + xy' + y = 0.$$

36. Show that the power series method fails for

$$x^4y'' + 2x^3y' - y = 0.$$

5.3 Frobenius' Method: The Indicial Equation

If we look at the solution of Example 5.2.2(ii) on p. 256, we see that the solution $(\cos x)/x$ was not obtained by the power series method. In fact, $(\cos x)/x$ cannot be written as a power series in x. However, it *can* be written as a power series in x times a power of x:

$$\frac{\cos x}{x} = x^{-1}\left(1 - \frac{x^2}{2!} + \frac{x^4}{4!} - \cdots\right).$$

This suggests that we should try to find solutions of the form

$$y = x^r(c_0 + c_1 x + c_2 x^2 + c_3 x^3 + \cdots), \tag{1}$$

where r is some real or complex number, whenever $x = 0$ is a singular point of the differential equation. For one class of singular points, this modification of the power series method does yield solutions.

DEFINITION

Regular Singular Point

We call $x = 0$ a **regular singular point** of the differential equation

$$y'' + a(x)y' + b(x)y = 0 \tag{2}$$

if both the functions $xa(x)$ and $x^2 b(x)$ have convergent Maclaurin series in an open interval containing $x = 0$. Observe that $x = 0$ is a regular singular point for the differential equation in Example 5.2.2(ii) because $xa(x) = 2$ and $x^2 b(x) = x^2$, both of which are convergent Maclaurin series with only one nonzero coefficient.

DEFINITION

Irregular Singular Point

A singular point that is not regular is called **irregular.** For example, the point $x = 0$ is an irregular singular point of the two equations

$$y'' + \frac{1}{x^2}y' + y = 0 \quad \text{and} \quad y'' + \frac{1}{x}y' + \frac{1}{x^3}y = 0$$

because in the first case $xa(x) = x^{-1}$, while in the second case $x^2 b(x) = x^{-1}$. The function of x^{-1} is not defined at $x = 0$ so it cannot have a convergent Maclaurin series.

Method of Frobenius

To simplify the explanation of the modified power series method, called the **method of Frobenius,**† we shall assume that $x = 0$ is a regular singular point of the equation

$$y'' + a(x)y' + b(x)y = 0 \tag{3}$$

and that equation (3) has a solution of the form

$$y = x^r(c_0 + c_1 x + c_2 x^2 + \cdots) = \sum_{n=0}^{\infty} c_n x^{r+n}, \qquad x > 0, \tag{4}$$

where r is some real or complex number. We can assume that $c_0 = 1$, since any constant multiple of a solution is again a solution of the differential equation. In addition, the choice $c_0 = 1$ simplifies much of the following discussion. The restriction $x > 0$ is necessary to prevent difficulties for certain values of r, such as $r = \frac{1}{2}$ and $-\frac{1}{4}$, since we are not interested in imaginary solutions. [If we need

† Georg Ferdinand Frobenius (1848–1917) was a German mathematician.

to find a solution valid for $x < 0$, we can change variables by substituting $X = -x$ into equation (3) and solve the resulting equation for $X > 0$.]

Since

$$y' = \sum_{n=0}^{\infty} c_n(r+n)x^{r+n-1}$$

and

$$y'' = \sum_{n=0}^{\infty} c_n(r+n)(r+n-1)x^{r+n-2},$$

equation (3) can be rewritten as

$$\sum_{n=0}^{\infty} c_n(r+n)(r+n-1)x^{r+n-2} + a(x)\sum_{n=0}^{\infty} c_n(r+n)x^{r+n-1} + b(x)\sum_{n=0}^{\infty} c_n x^{r+n} = 0$$

or, factoring an x and an x^2 in the second and third series, respectively, we obtain

$$\sum_{n=0}^{\infty} c_n[(r+n)(r+n-1) + (r+n)xa(x) + x^2b(x)]x^{r+n-2} = 0. \qquad \textbf{(5)}$$

As $x = 0$ is a regular singular point, both $xa(x)$ and $x^2b(x)$ can be expressed as convergent power series in x:

$$xa(x) = a_0 + a_1 x + a_2 x^2 + \cdots,$$

$$x^2b(x) = b_0 + b_1 x + b_2 x^2 + \cdots.$$

But $n \geq 0$, so that x^{r-2} is the smallest power of x in equation (5). Since the coefficients of a power series whose sum is zero must vanish, we have, for $n = 0$,

$$c_0[r(r-1) + a_0 r + b_0] = 0.$$

By hypothesis $c_0 = 1$, so that we obtain the **indicial equation**

Indicial Equation

$$\boxed{r(r-1) + a_0 r + b_0 = 0. \qquad \textbf{(6)}}$$

Exponents

whose roots r_1 and r_2, are called the **exponents** of the differential equation (3). In what follows we shall see that one of the solutions of equation (3) will always be of the form (4) and that there are three possible forms for the second linearly independent solution corresponding to the following cases:

CASE 1 r_1 and r_2 differ but not by an integer.

CASE 2 $r_1 = r_2$.

CASE 3 r_1 and r_2 differ by a nonzero integer.

We shall consider the three cases separately.

Case 1. r_1 and r_2 differ but not by an integer. This is the easiest case, since equation (3) will have two solutions, for $x > 0$, of the forms

$$y_1(x) = x^{r_1}(c_0 + c_1 x + c_2 x^2 + \cdots), \qquad c_0 = 1,$$

$$y_2(x) = x^{r_2}(c_0^* + c_1^* x + c_2^* x^2 + \cdots), \qquad c_0^* = 1.$$

That y_1 and y_2 are linearly independent follows easily from the fact that y_1/y_2 cannot be constant, since if it were, the roots r_1 and r_2 would coincide. The coefficients $c_1, c_2, \ldots$ are obtained by setting the coefficients of each power of x equal to zero in equation (5). We find $c_1^*, c_2^*, \ldots$ in a similar manner. The procedure is demonstrated in the following two examples.

EXAMPLE 1

Solve the Euler equation

$$y'' + \frac{1}{4x} y' + \frac{1}{8x^2} y = 0, \qquad x > 0. \tag{7}$$

SOLUTION
We substitute

$$y = x^r(c_0 + c_1 x + c_2 x^2 + \cdots) = x^r \sum_{n=0}^{\infty} c_n x^n = \sum_{n=0}^{\infty} c_n x^{r+n}$$

into equation (7) to obtain

$$\sum_{n=0}^{\infty}(r+n)(r+n-1)c_n x^{r+n-2} + \frac{1}{4x}\sum_{n=0}^{\infty}(r+n)c_n x^{r+n-1} + \frac{1}{8x^2}\sum_{n=0}^{\infty} c_n x^{r+n}$$

$$= \sum_{n=0}^{\infty}(r+n)(r+n-1)c_n x^{r+n-2} + \sum_{n=0}^{\infty}\frac{1}{4}(r+n)c_n x^{r+n-2} + \sum_{n=0}^{\infty}\frac{1}{8}c_n x^{r+n-2}$$

$$= \sum_{n=0}^{\infty} c_n\left[(r+n)(r+n-1) + \frac{1}{4}(r+n) + \frac{1}{8}\right]x^{r+n-2} = 0. \tag{8}$$

The indicial equation is obtained by setting the expression in brackets, for $n = 0$, equal to zero.

$$r(r-1) + \tfrac{1}{4}r + \tfrac{1}{8} = r^2 - \tfrac{3}{4}r + \tfrac{1}{8} = (r - \tfrac{1}{2})(r - \tfrac{1}{4}) = 0,$$

with roots $r = \tfrac{1}{4}$ and $r = \tfrac{1}{2}$ that do not differ by an integer. Assuming a solution of the form

$$y = x^{1/4}(c_0 + c_1 x + c_2 x^2 + \cdots),$$

equation (8) becomes (since $r = \frac{1}{4}$)

$$\sum_{n=0}^{\infty} c_n[(\tfrac{1}{4} + n)(\tfrac{1}{4} + n - 1) + (\tfrac{1}{4} + n)\tfrac{1}{4} + \tfrac{1}{8}]x^{(1/4)+n-2} = 0$$

or, after a bit of algebra,

$$\sum_{n=0}^{\infty} c_n\left(n^2 - \frac{n}{4}\right)x^{n-7/4} = 0.$$

Equating *all* terms of this series to zero, we get $n(n - \frac{1}{4})c_n = 0$, which holds only if $c_n = 0$, for $n > 0$. Thus $y_1(x) = c_0 x^{1/4} = x^{1/4}$.

To find the second solution we set $r = \frac{1}{2}$ in equation (8), obtaining

$$\sum_{n=0}^{\infty} c_n^*[(\tfrac{1}{2} + n)(\tfrac{1}{2} + n - 1) + (\tfrac{1}{2} + n)\tfrac{1}{4} + \tfrac{1}{8}]x^{(1/2)+n-2} = 0,$$

from which we get $n(n + \frac{1}{4})c_n^* = 0$. Thus $c_n^* = 0$ for $n > 0$, so that $y_2(x) = c_0^* \sqrt{x}$. Hence the general solution to equation (7) is

$$y = Ax^{1/4} + Bx^{1/2}.$$

The roots of the indicial equation may also be complex, as the following example illustrates.

EXAMPLE 2

Find the general solution of the equation

$$y'' + \frac{1}{x}y' + \frac{1}{x^2}y = 0, \qquad x > 0. \tag{9}$$

SOLUTION
Substituting $y = \sum_{n=0}^{\infty} c_n x^{r+n}$ into (9) yields

$$\sum_{n=0}^{\infty} c_n[(r + n)(r + n - 1) + (r + n) + 1]x^{r+n-2} = 0. \tag{10}$$

The indicial equation (obtained by setting $n = 0$ in the expression in brackets) is

$$r(r - 1) + r + 1 = r^2 + 1 = 0,$$

with the roots $r_1 = i$, $r_2 = -i$, which do not differ by an integer. Setting $r = i$ in equation (10), we have

$$\sum_{n=0}^{\infty} c_n[(i + n)(i + n - 1) + (i + n) + 1]x^{i+n-2} = 0.$$

Equating all coefficients of this series to zero, we have, after combining terms and using the fact that $(i + n)^2 = i^2 + 2in + n^2 = n^2 + 2in - 1$ (remember, $i^2 = -1$),

$$0 = c_n[(i + n)^2 + 1] = c_n(n^2 + 2in) = c_n n(n + 2i),$$

which holds only if $c_n = 0$ for $n > 0$. Now,

$$e^{\ln u} = u \quad \text{and} \quad e^{iu} = \cos u + i \sin u,$$

for any $u > 0$. This suggests that

$$x^i = e^{\ln x^i} = e^{i \ln x} = \cos \ln x + i \sin \ln x.$$

Thus, setting $c_0 = 1$ (any constant will give us a solution)

$$y_1(x) = x^i = e^{i(\ln x)} = [\cos(\ln x) + i \sin(\ln x)].$$

Similarly, substituting $r = -i$ into equation (10) yields the series

$$\sum_{n=0}^{\infty} c_n^*[(-i+n)(-i+n-1) + (-i+n) + 1]x^{-i+n-2} = 0,$$

whose coefficients satisfy the condition $c_n^*(n^2 - 2in) = 0$. Thus $c_n^* = 0$ for $n > 0$. Hence, setting $c_0^* = 1$

$$y_2(x) = x^{-i} = [\cos(\ln x) - i \sin(\ln x)].$$

Finally, since linear combinations of solutions are solutions, the real and imaginary parts of y_1 and y_2,

$$y_1^*(x) = \frac{1}{2}(y_1 + y_2) = \cos(\ln x), \qquad y_2^*(x) = \frac{1}{2i}(y_1 - y_2) = \sin(\ln x),$$

are solutions of equation (9). That y_1^* and y_2^* are linearly independent follows since $y_2^*/y_1^* = \tan(\ln x)$, which is nonconstant. Hence equation (9) has the general solution [see Example 5.2.1(iii) on p. 253]

$$y = A \cos(\ln x) + B \sin(\ln x), \qquad x > 0.$$

Case 2. $\mathbf{r_1 = r_2}$. Here we set $r = r_1$ and determine the coefficients $c_1, c_2,$ ... as in Case 1. We can then use the method of reduction of order (Section 2.2) to find the second linearly independent solution, since one solution is known. Consider the following example.

EXAMPLE 3

Use Frobenius' method to find the general solution of

$$x^2 y'' - xy' + y = 0, \qquad x > 0,$$

[see Example 5.2.1(ii)].

SOLUTION
Rewriting this equation in the form

$$y'' - \frac{1}{x}y' + \frac{1}{x^2}y = 0$$

and substituting $y = x^r(c_0 + c_1 x + c_2 x^2 + \cdots)$, we have

$$\sum_{n=0}^{\infty} c_n[(r+n)(r+n-1) - (r+n) + 1]x^{r+n-2} = 0. \tag{11}$$

Setting $n = 0$ we obtain the indicial equation

$$r(r-1) - r + 1 = r^2 - 2r + 1 = (r-1)^2 = 0,$$

with the double root $r = 1$. Then equation (11) yields (with $r = 1$)

$$\sum_{n=0}^{\infty} c_n[(n+1)n - (n+1) + 1]x^{n-1} = \sum_{n=0}^{\infty} n^2 c_n x^{n-1} = 0.$$

Hence $c_0 = 1$, $c_n = 0$ for $n > 0$, and one solution is $y_1 = x$. The second linearly independent solution, $y_2 = x \ln x$, is found as in Example 5.2.1(ii).

EXAMPLE 4

Solve the equation

$$y'' + y' + \frac{1}{4x^2} y = 0, \qquad x > 0. \tag{12}$$

SOLUTION

If we substitute $y = \sum_{n=0}^{\infty} c_n x^{r+n}$ into (12) we obtain

$$\sum_{n=0}^{\infty} c_n(r+n)(r+n-1)x^{r+n-2} + \sum_{n=0}^{\infty} c_n(r+n)x^{r+n-1} + \sum_{n=0}^{\infty} \tfrac{1}{4} c_n x^{r+n-2} = 0.$$

We need to express all sums in terms of the same power of x. If $k = n + 1$, then $n = k - 1$ and $x^{r+n-1} = x^{r+k-2}$. Thus, with $k = n + 1$, the second sum above can be written as

$$\sum_{k=1}^{\infty} c_{k-1}(r+k-1)x^{r+k-2} \overset{\text{setting } n = k}{=} \sum_{n=1}^{\infty} c_{n-1}(r+n-1)x^{r+n-2}$$

and we have (taking out the $n = 0$ terms from the first and third sums)

$$c_0[r(r-1) + \tfrac{1}{4}]x^{r-2} + \sum_{n=1}^{\infty} \{c_n[(r+n)(r+n-1) + \tfrac{1}{4}] + c_{n-1}(r+n-1)\}x^{r+n-2} = 0. \tag{13}$$

The indicial equation is

$$r(r-1) + \tfrac{1}{4} = r^2 - r + \tfrac{1}{4} = (r - \tfrac{1}{2})^2 = 0$$

which has the double root $r = \tfrac{1}{2}$. Substituting $r = \tfrac{1}{2}$ into (13) yields

$$\sum_{n=1}^{\infty} \{c_n[(n+\tfrac{1}{2})(n-\tfrac{1}{2}) + \tfrac{1}{4}] + c_{n-1}(n-\tfrac{1}{2})\}x^{n-3/2}$$

$$= \sum_{n=1}^{\infty} [n^2 c_n + (n-\tfrac{1}{2})c_{n-1}]x^{n-3/2} = 0.$$

This leads to the recurrence equation

$$c_n = -\frac{(n-\tfrac{1}{2})c_{n-1}}{n^2}, \qquad \text{for } n \geq 1.$$

Hence

$$c_1 = -\frac{c_0}{2}, \quad c_2 = -\frac{c_1(3/2)}{2^2} = \frac{3c_0}{2^2 \cdot 2^2}, \quad c_3 = -\frac{5c_2}{2 \cdot 3^2} = -\frac{3 \cdot 5 c_0}{2^3 \cdot 2^2 \cdot 3^2},$$

$$c_4 = -\frac{7c_3}{2 \cdot 4^2} = \frac{3 \cdot 5 \cdot 7 c_0}{2^4 \cdot 2^2 \cdot 3^2 \cdot 4^2}, \ldots,$$

so that

$$y_1(x) = x^{1/2}\left(c_0 - \frac{c_0}{2}x + \frac{3c_0}{2^2 \cdot 2^2}x^2 - \frac{3 \cdot 5 c_0}{2^3 \cdot 2^2 \cdot 3^2}x^3 + \frac{3 \cdot 5 \cdot 7 c_0}{2^4 \cdot 2^2 \cdot 3^2 \cdot 4^2}x^4 - \cdots\right)$$

$$= c_0 x^{1/2}\left[1 - \left(\frac{x}{2}\right) + \frac{3}{2^2}\left(\frac{x}{2}\right)^2 - \frac{3 \cdot 5}{2^2 \cdot 3^2}\left(\frac{x}{2}\right)^3 + \frac{3 \cdot 5 \cdot 7}{2^2 \cdot 3^2 \cdot 4^2}\left(\frac{x}{2}\right)^4 - \cdots\right]$$

$c_0 = 1$

$$\downarrow$$

$$= x^{1/2}\sum_{n=0}^{\infty} \frac{(2n)!}{(n!)^3}\left(\frac{-x}{4}\right)^n, \quad x > 0.$$

Without worrying whether this series converges we can use the method of reduction of order (Section 2.2) to produce the second linearly independent solution y_2. Recall that to find y_2 we set $y_2 = vy_1$, where y_1 is the solution above. Hence

$$y_2'' + y_2' + \frac{1}{4x^2}y_2 = v''y_1 + v'(2y_1' + y_1) + v\left(y_1'' + y_1' + \frac{1}{4x^2}y_1\right)$$

$$= v''y_1 + v'(2y_1' + y_1) = 0$$

because y_1 satisfies the differential equation. Thus

$$\frac{v''}{v'} = -2\frac{y_1'}{y_1} - 1 = \frac{-2\left(\frac{1}{2\sqrt{x}}\right)\left[1 - 3\left(\frac{x}{2}\right) + \frac{3 \cdot 5}{2^2}\left(\frac{x}{2}\right)^2 - \cdots\right]}{\sqrt{x}\left[1 - \left(\frac{x}{2}\right) + \frac{3}{2^2}\left(\frac{x}{2}\right)^2 - \cdots\right]} - 1.$$

After finding the first few terms by long division (carried out exactly as in the division of one polynomial by another), we have

$$\frac{v''}{v'} = \frac{-1}{x}\left(1 - x + \frac{x^2}{4} - \cdots\right) - 1 = \frac{-1}{x} - \frac{x}{4} + \cdots. \tag{14}$$

Integrating both sides of equation (14), we obtain

$$\ln v' = -\ln x - \frac{x^2}{8} + \cdots$$

or

$$v' = \frac{1}{x}\exp\left(-\frac{x^2}{8} + \cdots\right) = \frac{1}{x}\left[1 + \left(\frac{-x^2}{8} + \cdots\right) + \frac{1}{2!}\left(\frac{-x^2}{8} + \cdots\right)^2 + \cdots\right].$$

After expanding the exponential as a power series in x, we integrate once more and find that v has the form

$$v = \ln x - \frac{x^2}{16} + \cdots.$$

Then

$$y_2 = vy_1 = \left(\ln x - \frac{x^2}{16} + \cdots\right) \cdot \sqrt{x}\left[1 - \left(\frac{x}{2}\right) + \frac{3}{2^2}\left(\frac{x}{2}\right)^2 - \cdots\right]$$

$$= (\ln x)y_1 + \sqrt{x}\left(-\frac{x^2}{16} + \cdots\right),$$

and the general solution of equation (12) has the form

$$y(x) = \sqrt{x}\left\{(A + B \ln x)\left[1 - \left(\frac{x}{2}\right) + \cdots\right] + B\left(-\frac{x^2}{16} + \cdots\right)\right\}, \qquad x > 0.$$

Indeed, it is always true, when $r_1 = r_2$, that the general solution has the form

$$y(x) = x^r\left(\sum_{n=0}^{\infty} c_n x^n + \ln x \sum_{n=0}^{\infty} c_n^* x^n\right), \qquad x > 0.$$

Case 3. r_1 and r_2 differ by a nonzero integer. Suppose that $r_1 > r_2$. Then one solution of equation (3) will have the form

$$y_1 = x^{r_1}(c_0 + c_1 x + c_2 x^2 + \cdots), \quad c_0 = 1, x > 0,$$

as in Case 1. In some instances it is not possible to determine y_2 as was done in Case 1, because the procedure regenerates the same series expansion we obtained for y_1 (in this case, the first $r_1 - r_2$ coefficients c_n^* vanish). When this occurs, we proceed as in Case 2. These two possibilities are illustrated in the following two examples.

EXAMPLE 5

Solve **Bessel's equation of order $\frac{1}{2}$** (see Section 5.4):

$$x^2 y'' + xy' + [x^2 - (\tfrac{1}{2})^2]y = 0. \tag{15}$$

SOLUTION
We divide Bessel's equation through by x^2:

$$y'' + \frac{1}{x}y' + \left(1 - \frac{1}{4x^2}\right)y = 0.$$

Then, if $y = \sum_{n=0}^{\infty} c_n x^{r+n}$ we obtain

$$\sum_{n=0}^{\infty} c_n(r+n)(r+n-1)x^{r+n-2} + \sum_{n=0}^{\infty} c_n(r+n)x^{r+n-2} + \sum_{n=0}^{\infty} -\tfrac{1}{4}c_n x^{r+n-2} + \sum_{n=0}^{\infty} c_n x^{r+n} = 0.$$

The last sum can be written as $\sum_{n=2}^{\infty} c_{n-2} x^{r+n-2}$ (let $k = n + 2$) and we have, after taking out the $n = 0$ and $n = 1$ terms from the first three sums,

$$c_0[r(r-1) + r - \tfrac{1}{4}]x^{r-2} + c_1[r(r+1) + (r+1) - \tfrac{1}{4}]x^{r-1}$$

$$+ \sum_{n=2}^{\infty} \{c_n[(r+n)(r+n-1) + (r+n) - \tfrac{1}{4}] + c_{n-2}\}x^{r+n-2} = 0. \tag{16}$$

The indicial equation is

$$r(r-1)+r-\tfrac{1}{4}=r^2-\tfrac{1}{4}=(r-\tfrac{1}{2})(r+\tfrac{1}{2})=0.$$

Here the roots differ by the integer 1. Substituting $r=\tfrac{1}{2}$ in (16), we get

$$2c_1x^{-1/2}+\sum_{n=2}^{\infty}[c_n(n^2+n)+c_{n-2}]x^{n-3/2}=0.$$

Thus $c_1=0$ and

$$c_n=-\frac{c_{n-2}}{n(n+1)}, \qquad \text{for} \quad n\geq 2.$$

Thus all the coefficients with odd-numbered subscripts are zero, and

$$c_2=-\frac{c_0}{3!}, \quad c_4=-\frac{c_2}{4\cdot 5}=\frac{c_0}{5!}, \quad c_6=-\frac{c_4}{6\cdot 7}=-\frac{c_0}{7!}, \cdots,$$

and we have the solution

$$y_1(x)=\sqrt{x}\left(c_0-\frac{c_0}{3!}x^2+\frac{c_0}{5!}x^4-\frac{c_0}{7!}x^6+\cdots\right)$$

$$=c_0\sqrt{x}\left(1-\frac{x^2}{3!}+\frac{x^4}{5!}-\frac{x^6}{7!}+\cdots\right)$$

$$=\frac{1}{\sqrt{x}}\left(x-\frac{x^3}{3!}+\frac{x^5}{5!}-\frac{x^7}{7!}+\cdots\right)=\frac{\sin x}{\sqrt{x}}$$

since $c_0=1$. Now setting $r=-\tfrac{1}{2}$ in equation (16) we obtain (omitting some details)

$$\sum_{n=2}^{\infty}[c_n(n^2-n)+c_{n-2}]x^{n-5/2}=0$$

from which we get the recurrence relation

$$c_n=-\frac{c_{n-2}}{n(n-1)}.$$

Hence

$$c_2=-\frac{c_0}{2!}, \quad c_4=-\frac{c_2}{3\cdot 4}=\frac{c_0}{4!}, \quad c_6=-\frac{c_4}{5\cdot 6}=-\frac{c_0}{6!}, \cdots,$$

$$c_3=-\frac{c_1}{3!}, \quad c_5=-\frac{c_3}{4\cdot 5}=\frac{c_1}{5!}, \quad c_7=-\frac{c_5}{6\cdot 7}=-\frac{c_1}{7!}, \cdots,$$

and

$$y_2(x)=x^{-1/2}\left[c_0+c_1x-c_0\frac{x^2}{2!}-c_1\frac{x^3}{3!}+c_0\frac{x^4}{4!}+c_1\frac{x^5}{5!}-\cdots\right]$$

$$=\frac{1}{\sqrt{x}}(c_0\cos x+c_1\sin x), \qquad c_0=1.$$

Since the last term of y_2 is a multiple of y_1 and we are looking for linearly independent solutions, we may set $c_1 = 0$ to obtain

$$y_2 = \frac{\cos x}{\sqrt{x}}, \qquad x > 0.$$

That y_1 and y_2 are linearly independent follows from the fact that $y_1/y_2 = \tan x$, which is nonconstant. Thus the general solution of equation (15) is

$$y = \frac{A}{\sqrt{x}}\cos x + \frac{B}{\sqrt{x}}\sin x, \qquad x > 0.$$

EXAMPLE 6

Solve **Bessel's equation of order 1:**

$$x^2 y'' + xy' + (x^2 - 1)y = 0.$$

SOLUTION
Again, we first divide by x^2.

$$y'' + \frac{1}{x}y' + \left(1 - \frac{1}{x^2}\right)y = 0. \tag{17}$$

Substituting $y = \sum_{n=0}^{\infty} c_n x^{r+n}$ into (17) and changing the index of summation as in Example 5, we obtain

$$c_0[r(r-1) + r - 1]x^{r-2} + c_1[r(r+1) + (r+1) - 1]x^{r-1}$$

$$+ \sum_{n=2}^{\infty} \{c_n[(r+n)(r+n-1) + (r+n) - 1] + c_{n-2}\}x^{r+n-2} = 0. \tag{18}$$

The indicial equation is

$$r(r-1) + r - 1 = r^2 - 1 = 0$$

with roots $r_1 = 1$ and $r_2 = -1$ that differ by the integer 2. Setting $r = 1$ in (18) leads to

$$3c_1 + \sum_{n=2}^{\infty} [c_n(n^2 + 2n) + c_{n-2}]x^{n-1} = 0$$

so that $c_1 = 0$ and

$$c_n = -\frac{c_{n-2}}{n(n+2)}, \qquad \text{for } n \geq 2.$$

Thus all the coefficients with odd-numbered subscripts are zero, and

$$c_2 = -\frac{c_0}{2 \cdot 4}, \quad c_4 = -\frac{c_2}{4 \cdot 6} = \frac{c_0}{2 \cdot 4^2 \cdot 6}, \quad c_6 = \frac{-c_4}{6 \cdot 8} = \frac{-c_0}{2 \cdot 4^2 \cdot 6^2 \cdot 8}, \cdots.$$

Hence, since $c_0 = 1$,

$$y_1(x) = x\left(c_0 - \frac{c_0}{2 \cdot 4}x^2 + \frac{c_0}{2 \cdot 4^2 \cdot 6}x^4 - \frac{c_0}{2 \cdot 4^2 \cdot 6^2 \cdot 8}x^6 + \cdots\right)$$

$$= x\left(1 - \frac{1}{1!2!}\left(\frac{x}{2}\right)^2 + \frac{1}{2!3!}\left(\frac{x}{2}\right)^4 - \frac{1}{3!4!}\left(\frac{x}{2}\right)^6 + \cdots\right).$$

Now we shall see that the Frobenius method will not work for the second root $r_2 = -1$. Setting $r = -1$ in equation (18), we obtain

$$-c_1 x^{-2} + \sum_{n=2}^{\infty} [c_n(n^2 - 2n) + c_{n-2}]x^{n-3} = 0$$

so that $c_1 = 0$ and $n(n - 2)c_n = c_{n-2}$. Setting $n = 2$ we see that $c_0 = 0$, contradicting the assumption that $c_0 = 1$ (see p. 259). Note that all the coefficients will be zero, so the method fails for the second root. However, we do have *one* solution, indicating that the other solution can be obtained by the method of reduction of order:

$$\frac{v''}{v'} = \frac{-2y_1'}{y_1} - \frac{1}{x} = \frac{-3}{x} + \frac{x}{2} + \cdots .$$

Integrating both sides of this equation, we have

$$\ln v' = -3 \ln x + \frac{x^2}{4} + \cdots$$

or

$$v' = x^{-3} \exp\left(\frac{x^2}{4} + \cdots\right) = x^{-3} + \frac{1}{4}x^{-1} + \cdots .$$

Integrating once more, we get

$$v = -\tfrac{1}{2}x^{-2} + \tfrac{1}{4} \ln x + \cdots ,$$

so that

$$y_2 = vy_1 = \frac{1}{4}y_1 \ln x - \frac{1}{2}x^{-1} + \frac{x}{16} + \cdots , \qquad x > 0.$$

In general, if the method of Case 1 fails for r_2, the procedure above will yield the solution

$$y_2 = k_{-1}(\ln x)y_1 + x^{r_2}(k_0 + k_1 x + \cdots), \qquad x > 0,$$

where k_{-1} may equal zero.

We gather all the facts we have proved in this section in one theorem:

■ **THEOREM 1**

Let $x = 0$ be a regular singular point of the differential equation

$$y'' + a(x)y' + b(x)y = 0, \qquad x \neq 0, \tag{19}$$

and let r_1 and r_2 be the roots of the indicial equation

$$r(r - 1) + a_0 r + b_0 = 0,$$

where a_0 and b_0 are given by the power series expansions

$$xa(x) = a_0 + a_1 x + a_2 x^2 + \cdots ,$$

$$x^2 b(x) = b_0 + b_1 x + b_2 x^2 + \cdots .$$

Then equation (19) has two linearly independent solutions, y_1 and y_2, whose form depends on r_1 and r_2 as follows:

CASE 1

If r_1 and r_2 differ but not by an integer, then

$$y_1(x) = |x|^{r_1}\left(\sum_{n=0}^{\infty} c_n x^n\right), \qquad c_0 = 1,$$

$$y_2(x) = |x|^{r_2}\left(\sum_{n=0}^{\infty} c_n^* x^n\right), \qquad c_0^* = 1.$$

(The absolute-value signs are needed to avoid the assumption that $x > 0$.)

CASE 2

If $r_1 = r_2 = r$, then

$$y_1(x) = |x|^{r}\left(\sum_{n=0}^{\infty} c_n x^n\right), \qquad c_0 = 1,$$

$$y_2(x) = |x|^{r}\left(\sum_{n=1}^{\infty} c_n^* x^n\right) + y_1(x) \ln |x|.$$

CASE 3

If $r_1 - r_2$ is a positive integer, then

$$y_1(x) = |x|^{r_1}\left(\sum_{n=0}^{\infty} c_n x^n\right), \qquad c_0 = 1,$$

$$y_2(x) = |x|^{r_2}\left(\sum_{n=0}^{\infty} c_n^* x^n\right) + c_{-1}^* y_1(x) \ln |x|, \qquad c_0^* = 1,$$

and c_{-1}^* may equal zero.

Furthermore, if the power series expansions for $xa(x)$ and $x^2 b(x)$ are valid for $|x| < R$, then the solutions y_1 and y_2 are valid for $0 < |x| < R$. ■

Problems 5.3

In Problems 1–23 find the general solution to the given differential equation by the method of Frobenius.

1. $y'' + \dfrac{1}{2x} y' + \dfrac{1}{4x} y = 0$

2. $y'' + \dfrac{2(1-2x)}{x(1-x)} y' - \dfrac{2}{x(1-x)} y = 0$

3. $y'' + \dfrac{6}{x} y' + \left(\dfrac{6}{x^2} - 1\right) y = 0$

4. $y'' + \dfrac{4}{x} y' + \left(1 + \dfrac{2}{x^2}\right) y = 0$

5. $y'' + \dfrac{3}{x} y' + 4x^2 y = 0$

6. $(x-1)y'' - \left(\dfrac{4x^2 - 3x + 1}{2x}\right) y' + \left(\dfrac{2x^2 - x + 2}{2x}\right) y = 0$

7. $y'' + \dfrac{2}{x} y' - \dfrac{2}{x^2} y = 0$

8. $4xy'' + 2y' + y = 0$

9. $xy'' + 2y' + xy = 0$

10. $xy'' - y' + 4x^3 y = 0$

11. $xy'' + (1 - 2x)y' - (1 - x)y = 0$

12. $x(x + 1)^2 y'' + (1 - x^2)y' - (1 - x)y = 0$

13. $y'' - 2y' + \left(1 + \dfrac{1}{4x^2}\right)y = 0$

14. $x^2 y'' + Axy' + By = 0$, A and B constants.

15. $x(x - 1)y'' - (1 - 3x)y' + y = 0$

16. $x^2(x^2 - 1)y'' - x(x^2 + 1)y' + (x^2 + 1)y = 0$

17. $y'' + \dfrac{y'}{x} - y = 0$

18. $2xy'' - (x - 3)y' - y = 0$

19. $y'' + \dfrac{x+1}{2x}y' + \dfrac{3}{2x}y = 0$

20. $y'' + \dfrac{1}{2x}y' - \dfrac{x+1}{2x^2}y = 0$

21. $x^2 y'' + x(x - 1)y' - (x - 1)y = 0$

22. $xy'' - (3 + x)y' + 2y = 0$

23. $y'' + \dfrac{1}{4x^2}y = 0$

* 24. Prove that the second linearly independent solution of the equation $y'' + \dfrac{1}{x}y' + y = 0$ has the form

$$y_2 = y_1 \ln x + \sum_{n=1}^{\infty} \frac{(-1)^{n+1}}{2^{2n}(n!)^2}\left(1 + \frac{1}{2} + \cdots + \frac{1}{n}\right)x^{2n}, \qquad x > 0,$$

25. Consider the differential equation

$$y'' - \frac{1}{x^2}y' + \frac{1}{x^3}y = 0.$$

(a) Show that $x = 0$ is an irregular singular point of this equation.

(b) Use the fact that $y_1 = x$ is a solution to find a second independent solution.

(c) Show that the solution y_2 cannot be expressed as a series of the form (4). Thus this solution cannot be found by the method of Frobenius.

26. The differential equation $x^2 y'' + (4x - 1)y' + 2y = 0$ has $x = 0$ as an irregular singular point.

(a) Suppose that equation (4) is inserted into this equation. Show that $r = 0$ and the corresponding Frobenius method "solution" is $y = \sum_{n=0}^{\infty}(n + 1)!x^n$.

(b) Prove that the series above has radius of convergence $R = 0$. Hence even though a Frobenius series may formally satisfy a differential equation, it may not be a valid solution at an irregular singular point.

* 27. If $r_1 = r_2$, verify that the technique in Example 4 will always yield

$$v(x) = \ln x - (a_1 + 2c_1)x + \cdots$$

where

$$xa(x) = a_0 + a_1 x + a_2 x^2 + a_3 x^3 + \cdots$$

and $y_2 = vy_1$ with

$$y_1 = x^r(c_0 + c_1 x + c_2 x^2 + \cdots), \qquad c_0 = 1.$$

28. Show that the Frobenius method fails for the equation

$$x^4 y'' + 2x^3 y' - y = 0, \qquad x > 0,$$

which has $x = 0$ as an irregular singular point. Find a solution by assuming $y = \sum_{n=0}^{\infty} c_n x^{-n}$.

5.4 Bessel Functions

The differential equation

$$x^2 y'' + xy' + (x^2 - p^2)y = 0, \tag{1}$$

Bessel's Equation of Order p

which is known as **Bessel's equation of order p (≥ 0)**, is one of the most important differential equations in applied mathematics. The equation was investigated first in 1703 by Jakob Bernoulli (see p. 34), in connection with the oscillatory behavior of a hanging chain, and later by the German astronomer Friedrich Wilhelm Bessel (1784–1846) in his studies of planetary motion. Since then, Bessel functions have been used in the studies of elasticity, fluid motion, potential theory, diffusion, and the propagation of waves. We shall present a few applications of Bessel functions in Chapter 12 when we study the solutions of partial differential equations. (See, in particular, Section 12.9.)

Recall that in Section 5.3 we found the solution of Bessel's equation for $p = \frac{1}{2}$ and 1 (see Examples 5 and 6 there). In each of these examples, the method of Frobenius was an important tool, so we anticipate again the successful application of this procedure. We assume that a solution of the form

$$y(x) = \sum_{n=0}^{\infty} c_n x^{r+n}, \qquad x \neq 0, \qquad c_0 = 1 \tag{2}$$

exists for Bessel's equation of order p. We divide equation (1) by x^2:

$$y'' + \frac{1}{x} y' + \left(1 - \frac{p^2}{x^2}\right) y = 0$$

and then substitute (2) into this equation to obtain (after setting $k = n + 2$ in $\sum_{n=0}^{\infty} c_n x^{r+n}$),

$$\sum_{n=0}^{\infty} c_n (r+n)(r+n-1) x^{r+n-2} + \sum_{n=0}^{\infty} c_n (r+n) x^{r+n-2}$$

$$+ \sum_{n=0}^{\infty} -p^2 c_n x^{r+n-2} + \sum_{n=2}^{\infty} c_{n-2} x^{r+n-2} = 0$$

or

$$c_0 (r^2 - p^2) x^{r-2} + c_1 [(r+1)^2 - p^2] x^{r-1}$$

$$+ \sum_{n=2}^{\infty} \{c_n [(n+r)^2 - p^2] + c_{n-2}\} x^{r+n-2} = 0. \tag{3}$$

The indicial equation is $r^2 - p^2 = 0$ with roots $r_1 = p \, (\geq 0)$ and $r_2 = -p$. Setting $r = p$ in (3) yields

$$(1 + 2p) c_1 x^{p-1} + \sum_{n=2}^{\infty} [n(n+2p) c_n + c_{n-2}] x^{n+p-2} = 0$$

indicating that $c_1 = 0$ and

$$c_n = \frac{-c_{n-2}}{n(n+2p)}, \qquad \text{for } n \geq 2. \tag{4}$$

Hence all the coefficients with odd-numbered subscripts $c_{2j+1} = 0$, since by equation (4) they can all be expressed as a multiple of c_1. Letting $n = 2j + 2$, we see that the coefficients with even-numbered subscripts satisfy

$$c_{2(j+1)} = \frac{-c_{2j}}{2^2 (j+1)(p+j+1)}, \qquad \text{for } j \geq 0,$$

which yields

$$c_2 = \frac{-c_0}{2^2 (p+1)}, \qquad c_4 = \frac{-c_2}{2^2 \cdot 2(p+2)} = \frac{c_0}{2^4 2!(p+1)(p+2)},$$

$$c_6 = \frac{-c_4}{2^2 \cdot 3(p+3)} = \frac{-c_0}{2^6 3!(p+1)(p+2)(p+3)}, \ldots$$

Hence the series (2) becomes

$$y_1(x) = |x|^p \left[c_0 - \frac{c_0}{2^2(p+1)} x^2 + \frac{c_0}{2^4 2!(p+1)(p+2)} x^4 - \cdots \right]$$

$$= c_0 |x|^p \sum_{n=0}^{\infty} (-1)^n \frac{x^{2n}}{2^{2n} n!(p+1)(p+2) \cdots (p+n)}. \tag{5}$$

The Gamma Function

Gamma Function

To write equation (5) in a more compact form, we define the **gamma function** for all values $p > -1$:

$$\Gamma(p+1) = \int_0^\infty e^{-t} t^p \, dt.$$

Integrating $\Gamma(p + 1)$ by parts, we have

$$\Gamma(p+1) = \int_0^\infty e^{-t} t^p \, dt = \frac{e^{-t} t^{p+1}}{p+1} \Big|_0^\infty + \frac{1}{p+1} \int_0^\infty e^{-t} t^{p+1} \, dt.$$

The first expression on the right is zero, and the integral on the right side is $\Gamma(p + 2)$. We thus have the basic property of gamma functions:

$$\Gamma(p+2) = (p+1)\Gamma(p+1).$$

Since

$$\Gamma(1) = \int_0^\infty e^{-t} \, dt = -e^{-t} \Big|_0^\infty = 1,$$

it follows that $\Gamma(2) = \Gamma(1) = 1!$, $\Gamma(3) = 2\Gamma(2) = 2!$, ..., and in general $\Gamma(n + 1) = n!$ Thus, the gamma function is the extension to real numbers $p > -1$ of the factorial function.

It is customary in equation (5) to let $c_0 = [2^p \Gamma(p + 1)]^{-1}$. Then equation (5) becomes

$$J_p(x) = \left| \frac{x}{2} \right|^p \sum_{n=0}^{\infty} (-1)^n \frac{(x/2)^{2n}}{n! \Gamma(p+n+1)}, \qquad x \neq 0, \tag{6}$$

Bessel Function of the First Kind

which is known as the **Bessel function of the first kind of order p.** Thus $J_p(x)$ is the first solution of equation (1). It can be shown that the series $J_p(x)$ converges for all real x.

To find the second solution we must consider the difference $r_1 - r_2 = 2p$. By Case 1 of Section 5.3, if p is not a multiple of $\frac{1}{2}$, we will again be able to apply the method of Frobenius with $r = -p$ to find the second solution. The results of Examples 5.3.5 and 5.3.6 make it appear likely that we will obtain $\ln|x|$ terms only when p is an integer. Therefore, we set $r = -p$ in equation (3) and assume p is not an integer. Then we obtain

$$(1 - 2p)c_1 x^{-p-1} + \sum_{n=2}^{\infty} [n(n - 2p)c_n + c_{n-2}]x^{n-p-2} = 0, \qquad (7)$$

indicating that $c_1 = 0$ (if $p \neq \frac{1}{2}$) and

$$c_n = \frac{-c_{n-2}}{n(n - 2p)}. \qquad (8)$$

Note that when p is not a multiple of $\frac{1}{2}$, all the coefficients c_n with odd-numbered subscripts will be zero. If $p = (2m + 1)/2$, then c_{2m+1} is arbitrary and the recurrence relation (8) will yield the coefficients

$$c_{2m+3} = \frac{-c_{2m+1}}{2^2(p + 1)}, \quad c_{2m+5} = \frac{c_{2m+1}}{2^4 2!(p + 1)(p + 2)}, \dots$$

Thus, the odd-numbered coefficients generate the series

$$|x|^{-p}\left[c_{2m+1}x^{2m+1} - \frac{c_{2m+1}x^{2m+3}}{2^2(p + 1)} + \frac{c_{2m+1}x^{2m+5}}{2^4 2!(p + 1)(p + 2)} - \dots \right]$$

$$= c_{2m+1}|x|^p\left[1 - \frac{x^2}{2^2(p + 1)} + \frac{x^4}{2^4 2!(p + 1)(p + 2)} - \dots \right],$$

which is a multiple of $J_p(x)$. Thus, we can ignore the odd-numbered coefficients and concentrate on using equation (8) to calculate the coefficients with even-numbered subscripts. Then

$$c_2 = \frac{-c_0}{2^2(1 - p)}, \quad c_4 = \frac{-c_2}{2^2 \cdot 2(2 - p)} = \frac{c_0}{2^4 2!(1 - p)(2 - p)}, \dots,$$

and the second solution is the convergent series

$$J_{-p}(x) = \left|\frac{x}{2}\right|^{-p} \sum_{n=0}^{\infty} (-1)^n \frac{(x/2)^{2n}}{n!\Gamma(n - p + 1)}. \qquad (9)$$

To see that equations (6) and (9) are linearly independent, we obtain by long division

$$\frac{J_p(x)}{J_{-p}(x)} = \frac{|x/2|^p/\Gamma(p + 1) - |x/2|^{p+2}/2!\Gamma(p + 3) + \cdots}{|x/2|^{-p}/\Gamma(1 - p) - |x/2|^{2-p}/2!\Gamma(3 - p) + \cdots}$$

$$= \frac{|x/2|^{2p}}{\Gamma(1 + p)/\Gamma(1 - p)} + \frac{3p|x/2|^{2p+2}}{\Gamma(3 + p)/\Gamma(1 - p)} + \cdots,$$

which clearly is not a constant function. We have shown the following:

■ **THEOREM 1**

> If p is not an integer, then
>
> $$y(x) = AJ_p(x) + BJ_{-p}(x)$$
>
> is the general solution of Bessel's equation for all values $x \neq 0$.

■

If p is an integer, then the term $(n - 2p)$ in the recurrence relation (8) is zero for the integer $n = 2p$. Hence c_{2p-2} is zero and, iterating equation (8) repeatedly, we see that $c_{2p-2} = c_{2p-4} = \cdots = c_2 = c_0 = 0$. But this contradicts the assumed form (2) of the solution. Thus the method of Frobenius cannot be used when p is a positive integer, and the second linearly independent solution of equation (1) must be calculated by the method in Example 5.3.6 on p. 268. After an extremely long (but straightforward) calculation we obtain

$$y_2(x) = J_p(x) \ln|x| - \frac{1}{2}\left[\sum_{k=0}^{p-1}\frac{(p-k-1)!}{k!}\left(\frac{x}{2}\right)^{2k-p} + \frac{h_p}{p!}\left(\frac{x}{2}\right)^p\right.$$

$$\left. + \sum_{k=1}^{\infty}\frac{(-1)^k[h_k + h_{p+k}]}{k!(p+k)!}\left(\frac{x}{2}\right)^{2k+p}\right], \tag{10}$$

where

$$h_p = 1 + \frac{1}{2} + \frac{1}{3} + \cdots + \frac{1}{p} \tag{11}$$

and p is a positive integer.

It is customary to replace equation (10) by the linear combination of solutions

$$Y_p(x) = \frac{2}{\pi}[y_2(x) + (\gamma - \ln 2)J_p(x)], p = 0, 1, 2, \ldots,$$

where

$$\gamma = \lim_{p \to \infty}(h_p - \ln p) = 0.5772156649\ldots$$

Euler Constant
Neumann's Function

is the **Euler constant**. This particular solution is obviously independent of $J_p(x)$ and is called the **Bessel function of the second kind of order p** or **Neumann's**† **function of order p.** It is defined by the formula

† Carl G. Neumann (1832–1925) was a German mathematician.

$$Y_p(x) = \frac{2}{\pi} J_p(x) \left(\ln \frac{x}{2} + \gamma \right)$$

$$- \frac{1}{\pi} \left[\sum_{k=0}^{p-1} \frac{(p-k-1)!}{k!} \left(\frac{x}{2} \right)^{2k-p} + \frac{h_p}{p!} \left(\frac{x}{2} \right)^p + \sum_{k=1}^{\infty} (-1)^k \frac{[h_k + h_{p+k}]}{k!(p+k)!} \left(\frac{x}{2} \right)^{2k+p} \right]$$

for all integers $p = 0, 1, 2, \ldots$.

The function Y_p may be extended to all real numbers $p \geq 0$ (see Problem 24) by letting

$$Y_p(x) = \frac{1}{\sin p\pi} [J_p(x) \cos p\pi - J_{-p}(x)], \qquad p \neq 0, 1, 2, \ldots.$$

Using this definition of Y_p, we have the following result:

■ THEOREM 2

> The general solution of Bessel's equation of order p is
>
> $$y(x) = AJ_p(x) + BY_p(x), \qquad x \neq 0.$$

■

Graphs of the functions J_0, J_1, and J_2 and Y_0, Y_1, and Y_2 are given in Fig. 1.

Properties of Bessel's Functions

Now that we have the expansions for $J_p(x)$ and $Y_p(x)$, we can derive a number of important expressions involving Bessel functions and their derivatives. For simplicity we shall assume $x > 0$. The first two identities are immediate consequences of equation (6):

$$\frac{d}{dx} [x^p J_p(x)] = x^p J_{p-1}(x), \tag{12}$$

$$\frac{d}{dx} [x^{-p} J_p(x)] = -x^{-p} J_{p+1}(x). \tag{13}$$

To prove (12), we differentiate the product $x^p J_p$ term by term:

$$\frac{d}{dx} \sum_{n=0}^{\infty} (-1)^n \frac{2^p (x/2)^{2n+2p}}{n! \Gamma(p+n+1)} = \sum_{n=0}^{\infty} (-1)^n \frac{2^{p-1} (x/2)^{2n+2p-1} 2(n+p)}{n! \Gamma(p+n+1)}$$

$$= x^p \sum_{n=0}^{\infty} (-1)^n \frac{(x/2)^{2n+p-1}}{n! \Gamma(p+n)} = x^p J_{p-1}(x),$$

since $\Gamma(p + n + 1) = (p + n)\Gamma(p + n)$. The proof of equation (13) is similar

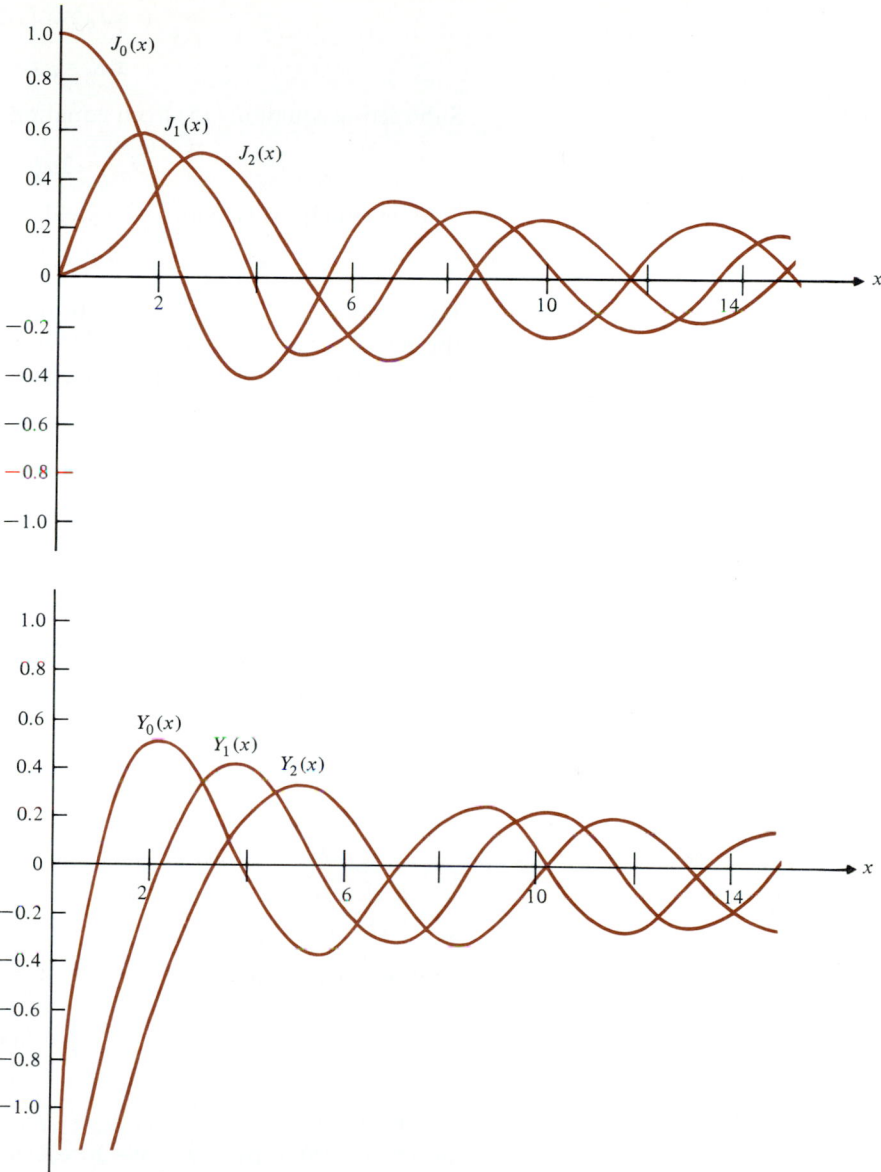

FIGURE 1

(see Problem 4). Expanding the left-hand sides of equations (12) and (13), we have

$$x^p J'_p + px^{p-1} J_p = x^p J_{p-1}$$

and

$$x^{-p} J'_p - px^{-p-1} J_p = -x^{-p} J_{p+1},$$

which may be simplified to yield the identities

$$xJ'_p = xJ_{p-1} - pJ_p, \tag{14}$$

$$xJ'_p = pJ_p - xJ_{p+1}. \tag{15}$$

Subtracting equation (15) from equation (14), we obtain the recursion relation

$$xJ_{p+1} - 2pJ_p + xJ_{p-1} = 0. \tag{16}$$

Adding equations (14) and (15) yields

$$2J'_p = J_{p-1} - J_{p+1}. \tag{17}$$

Equations (12)–(17) are extremely important in solving problems involving the Bessel functions, since they allow us to express Bessel functions of higher order in terms of lower-order functions.

EXAMPLE 1

Express $J_3(x)$ in terms of $J_0(x)$ and $J_1(x)$.

SOLUTION

We let $p = 2$ in equation (16):

$$xJ_3 = 4J_2 - xJ_1.$$

Applying formula (16) with $p = 1$ to J_2 yields

$$xJ_2 = 2J_1 - xJ_0.$$

Thus

$$J_3(x) = \frac{4}{x}J_2 - J_1 = \frac{4}{x^2}(2J_1 - xJ_0) - J_1 = \left(\frac{8}{x^2} - 1\right)J_1(x) - \frac{4}{x}J_0(x).$$

EXAMPLE 2

Evaluate the integral

$$\int x^4 J_1(x)\, dx. \tag{18}$$

SOLUTION

Integrating (18) by parts, we have, by equation (12),

$$\overset{\displaystyle (x^2 J_2)' = x^2 J_1}{\underset{\downarrow}{}}$$

$$\int x^2 [x^2 J_1(x)]\, dx = x^2(x^2 J_2) - \int x^2 J_2 \cdot 2x\, dx = x^4 J_2(x) - 2\int x^3 J_2\, dx.$$

Applying equation (12) to the last integral, we obtain

$$\int x^4 J_1(x)\, dx = x^4 J_2(x) - 2x^3 J_3(x) + C.$$

In general, an integral of the form

$$\int x^m J_n(x)\, dx,$$

where m and n are integers such that $m + n \geq 0$, can be completely integrated if $m + n$ is odd. But if $m + n$ is even, the result depends on the residual integral $\int J_0(x)\, dx$. It is not possible to reduce $\int J_0(x)\, dx$ [or $\int Y_0(x)\, dx$] any further, and for this reason values of the functions

$$\int_0^x J_0(x)\, dx \quad \text{and} \quad \int_0^x Y_0(x)\, dx$$

have been tabulated.†

EXAMPLE 3

Express $J_{3/2}(x)$ in terms of $\sin x$ and $\cos x$.

SOLUTION

Recall from Example 5.3.5 on p. 266 that the general solution of Bessel's equation of order $\frac{1}{2}$ can be written in terms of sines and cosines. Multiplying $y_1(x)$ and $y_2(x)$ by $c_0 = [\sqrt{2}\, \Gamma(\frac{3}{2})]^{-1}$, we have

$$J_{1/2}(x) = \frac{1}{\sqrt{2}\Gamma(\frac{3}{2})} \frac{\sin x}{\sqrt{x}} = \sqrt{\frac{2}{\pi x}} \sin x$$

and

$$J_{-1/2}(x) = \frac{1}{\sqrt{2}\Gamma(\frac{3}{2})} \frac{\cos x}{\sqrt{x}} = \sqrt{\frac{2}{\pi x}} \cos x,$$

since $\Gamma(\frac{3}{2}) = (\frac{1}{2})\Gamma(\frac{1}{2}) = \sqrt{\pi}/2$ (see entry 219 in the table of integrals). By equation (16),

$$J_{3/2}(x) = \frac{1}{x} J_{1/2} - J_{-1/2} = \sqrt{\frac{2}{\pi x}} \left(\frac{\sin x}{x} - \cos x \right).$$

Similar results hold for $Y_p(x)$ (see Problems 5, 6, and 7).

Many differential equations with variable coefficients can be reduced to Bessel equations.

EXAMPLE 4

Solve the equation

$$y'' + k^2 x y = 0. \tag{19}$$

---

† Milton Abramowitz and Irene A. Stegun, *Handbook of Mathematical Functions* (New York: Dover, 1965), Table 11.1, pp. 492–493.

SOLUTION

The following substitution will reduce equation (19) to a Bessel equation. Let $u = y/\sqrt{x}$ and $z = 2kx^{3/2}/3$. Then

$$\frac{du}{dz} = \frac{du/dx}{dz/dx} = \frac{y'}{kx} - \frac{y}{2kx^2}$$

and

$$\frac{d^2u}{dz^2} = \frac{\frac{d}{dx}\left(\frac{du}{dz}\right)}{dz/dx} = \frac{y''}{k^2x^{3/2}} - \frac{3y'}{2k^2x^{5/2}} + \frac{y}{k^2x^{7/2}}.$$

Hence

$$z^2\frac{d^2u}{dz^2} + z\frac{du}{dz} = \frac{4}{9}x^{3/2}y'' + \frac{1}{9}\frac{y}{\sqrt{x}},$$

and using equation (19) for y'', we obtain

$$z^2\frac{d^2u}{dz^2} + z\frac{du}{dz} = -\frac{4}{9}k^2x^3\left(\frac{y}{\sqrt{x}}\right) + \frac{1}{9}\frac{y}{\sqrt{x}} = -\left(z^2 - \frac{1}{9}\right)u$$

or

$$z^2\frac{d^2u}{dz^2} + z\frac{du}{dz} + \left(z^2 - \frac{1}{9}\right)u = 0, \qquad (20)$$

which is the Bessel equation of order $\frac{1}{3}$. Since equation (20) has the general solution

$$u(z) = AJ_{1/3}(z) + BJ_{-1/3}(z),$$

equation (19) has the solution

$$y(x) = \sqrt{x}[AJ_{1/3}(\tfrac{2}{3}kx^{3/2}) + BJ_{-1/3}(\tfrac{2}{3}kx^{3/2})].$$

Problems 5.4

1. Express $J_5(x)$ in terms of $J_0(x)$ and $J_1(x)$.

2. Express $J_{5/2}(x)$ in terms of $\sin x$ and $\cos x$.

3. Show that
   (a) $4J_p''(x) = J_{p+2}(x) - 2J_p(x) + J_{p-2}(x)$.
   (b) $-8J_p'''(x) = J_{p+3}(x) - 3J_{p+1}(x) + 3J_{p-1}(x) - J_{p-3}(x)$.

4. Prove that $[x^{-p}J_p(x)]' = -x^{-p}J_{p+1}(x)$.

5. Show that $[x^pY_p(x)]' = x^pY_{p-1}(x)$.

6. Prove that $[x^{-p}Y_p(x)]' = -x^{-p}Y_{p+1}(x)$.

7. Using the equations of Problems 5 and 6, show that
   (a) $x(Y_{p+1} + Y_{p-1}) = 2pY_p$.
   (b) $2Y_p' = Y_{p-1} - Y_{p+1}$.

8. Prove the following identities:
   (a) $\int J_1(x)\,dx = -J_0(x) + C$
   (b) $\int x^2J_1(x)\,dx = 2xJ_1(x) - x^2J_0(x) + C$
   (c) $\int xJ_0(x)\,dx = xJ_1(x) + C$
   (d) $\int x^3J_0(x)\,dx = (x^3 - 4x)J_1(x) + 2x^2J_0(x) + C$
   (e) $\int J_0(x)\cos x\,dx = xJ_0(x)\cos x + xJ_1(x)\sin x + C$
   (f) $\int J_0(x)\sin x\,dx = xJ_0(x)\sin x - xJ_1(x)\cos x + C$
   (g) $\int J_1(x)\cos x\,dx = xJ_1(x)\cos x - (x\sin x + \cos x)J_0(x) + C$
   (h) $\int J_1(x)\sin x\,dx = xJ_1(x)\sin x + (x\cos x - \sin x)J_0(x) + C$

9. Verify the following identities:

(a) $\int x^2 J_0(x)\,dx = x^2 J_1(x) + x J_0(x) - \int J_0(x)\,dx + C$

(b) $\int x^{-1} J_1(x)\,dx = -J_1(x) + \int J_0(x)\,dx + C$

(c) $\int x J_1(x)\,dx = -x J_0(x) + \int J_0(x)\,dx + C$

(d) $\int x^3 J_1(x)\,dx = 3x^2 J_1(x) - (x^3 - 3x)J_0(x) -$ $3\int J_0(x)\,dx + C$

10. Show that $\int J_0(\sqrt{x})\,dx = 2\sqrt{x}J_1(\sqrt{x}) + C$.

11. Show that $\int x J_0(x)\sin x\,dx = \frac{1}{3}\{x^2[J_0(x)\sin x - J_1(x)\cos x] + x J_1(x)\sin x\} + C$.

12. Show that $\int x J_1(x)\cos x\,dx = \frac{1}{3}\{x^2[J_1(x)\cos x - J_0(x)\sin x] + 2x J_1(x)\sin x\} + C$.

13. Show that $\int J_0(x)\,dx = 2[J_1(x) + J_3(x) + J_5(x) + \cdots] + C$.

In Problems 14–21 reduce each given equation to a Bessel equation and solve it.

14. $x^2 y'' + xy' + (a^2 x^2 - p^2)y = 0$

15. $4x^2 y'' + 4xy' + (x^2 - p^2)y = 0$

16. $x^2 y'' + xy' + 4(x^4 - p^2)y = 0$

17. $xy'' - y' + xy = 0$

18. $x^2 y'' + (x^2 + \frac{1}{4})y = 0$

19. $xy'' + (1 + 2k)y' + xy = 0$

20. $y'' + k^2 x^2 y = 0$

21. $y'' + k^2 x^4 y = 0$

22. Prove that the second linearly independent solution of the equation

$$y'' + \frac{1}{x}y' + y = 0$$

has the form

$$y_2(x) = J_0(x)\ln x$$

$$+ \sum_{n=1}^{\infty} \frac{(-1)^{n+1}}{2^{2n}(n!)^2}\left(1 + \frac{1}{2} + \cdots + \frac{1}{n}\right)x^{2n}, \qquad x > 0.$$

23. Obtain formula (10) for $p = 1$ by the methods of this section.

*24. Prove that if $p \geq 0$ is an integer,

$$\lim_{q \to p} \frac{J_q(x)\cos q\pi - J_{-q}(x)}{\sin q\pi} = Y_p(x).$$

This is the extension of Y_p to all real values $p \geq 0$.

25. (a) Expand $e^{(x/2)[t-(1/t)]}$ as a power series in t by multiplying the series for $e^{xt/2}$ and $e^{-x/2t}$.

(b) Show that the coefficient of t^n in the expansion obtained in part (a) is $J_n(x)$.

(c) Conclude that

$$e^{(x/2)[t-(1/t)]} = J_0(x) + \sum_{n=1}^{\infty} J_n(x)[t^n + (-t)^{-n}].$$

This function is called the **generating function** of the Bessel functions.

(d) Set $t = e^{i\theta}$ in the expression in part (c) and obtain the identities

$$\cos(x\sin\theta) = J_0(x) + 2\sum_{n=1}^{\infty} J_{2n}(x)\cos 2n\theta$$

and

$$\sin(x\sin\theta) = 2\sum_{n=1}^{\infty} J_{2n-1}(x)\sin(2n-1)\theta.$$

26. Show that $J_0(x) = (1/\pi)\int_0^\pi \cos(x\cos t)\,dt$.

27. Show that $J_n(x) = (1/\pi)\int_0^\pi \cos(nt - x\sin t)\,dt$.

28. **Modified Bessel functions.** The function $I_p(x) = i^{-p}J_p(ix)$, $i^2 = -1$ is called the **modified Bessel function of the first kind of order p.** Show that $I_p(x)$ is a solution of the differential equation $x^2 y'' + xy' - (x^2 + p^2)y = 0$, and obtain its series representation by the method of Frobenius.

29. Show that another solution of the differential equation in Problem 28 is the **modified Bessel function of the second kind of order p** (for p not an integer):

$$K_p(x) = \frac{\pi}{2\sin p\pi}[I_{-p}(x) - I_p(x)].$$

30. Prove (see Problem 25) that

$$e^{(x/2)[t+(1/t)]} = I_0(x) + \sum_{n=1}^{\infty} I_n(x)(t^n + t^{-n}).$$

31. Prove the following identities:

(a) $\dfrac{d}{dx}[x^p I_p(x)] = x^p I_{p-1}(x)$

(b) $\dfrac{d}{dx}[x^{-p} I_p(x)] = x^{-p} I_{p+1}(x)$

(c) $x(I_{p-1} - I_{p+1}) = 2p I_p$

32. (a) Prove for all integers $n > 0$ that

$$\int_0^\infty J_{n+1}(x)\,dx = \int_0^\infty J_{n-1}(x)\,dx$$

by means of equation (17), given the approximation

$$J_n(x) \approx \sqrt{\frac{2}{\pi x}}\cos\left(x - \frac{\pi}{4} - \frac{n\pi}{2}\right).$$

(b) Prove a similar fact for Y_n given the approximation

$$Y_n(x) \approx \sqrt{\frac{2}{\pi x}}\sin\left(x - \frac{\pi}{4} - \frac{n\pi}{2}\right).$$

(c) Show that $\int_0^\infty J_n(x)\,dx = 1$.

(d) Prove that $\int_0^\infty [J_n(x)/x]\,dx = 1/n$

33. In Section 2.8 we considered the vibrations of mass-spring systems whose springs had a constant elastic force per unit elongation (the spring constant k). In practice, the elastic force per unit elongation $k(t)$ of a spring decays with time. Suppose $k(t) = k_0 e^{-at} + k_1$, with $a > 0$, so that equation (2.8.2) becomes

$$\frac{d^2x}{dt^2} + \frac{k_0 e^{-at} + k_1}{m} x = 0, \qquad x(0) = x_0, \qquad x'(0) = v_0.$$

(a) Use the substitution $u = ce^{-at/2}$, $c = (2/a)\sqrt{k_0/m}$, to convert this differential equation into a Bessel equation.

(b) Solve the Bessel equation, and obtain a solution of the given initial value problem.

5.5 Legendre Polynomials

Legendre's Differential Equation

Another very important differential equation that arises in many applications (see, for example, Sections 11.9 and 12.9) is **Legendre's differential equation**

$$(1 - x^2)y'' - 2xy' + p(p+1)y = 0, \tag{1}$$

Legendre Function

where p is a given real number. Any solution of equation (1) is called a **Legendre function.**†

Dividing equation (1) by $(1 - x^2)$, we obtain

$$y'' - \frac{2x}{1-x^2}y' + \frac{p(p+1)}{1-x^2}y = 0$$

and we observe, using the geometric series

$$\frac{1}{1-x^2} = 1 + x^2 + x^4 + x^6 + \cdots,$$

that the coefficient functions

$$a(x) = \frac{-2x}{1-x^2} = -2x(1 + x^2 + x^4 + \cdots),$$

$$b(x) = \frac{p(p+1)}{1-x^2} = p(p+1)(1 + x^2 + x^4 + \cdots)$$

have convergent power series representations in the interval $|x| < 1$. By Theorem 5.2.1 on p. 254 it follows that (1) must have a power series representation valid in the interval $|x| < 1$. Substituting $y = \sum_{n=0}^\infty c_n x^n$, and its derivatives into equation (1), we have

$$(1 - x^2) \sum_{n=2}^\infty c_n n(n-1)x^{n-2} - 2x \sum_{n=1}^\infty c_n n x^{n-1} + p(p+1) \sum_{n=0}^\infty c_n x^n = 0$$

or

$$\sum_{n=0}^\infty \{(n+2)(n+1)c_{n+2} - c_n[n(n+1) - p(p+1)]\}x^n = 0. \tag{2}$$

† Named after the famous French mathematician Adrien Marie Legendre (1752–1833). See the accompanying biographical sketch.

ADRIEN-MARIE LEGENDRE
(1752–1833)

Courtesy of the Granger Collection

Adrien-Marie Legendre (1752–1833) is known in the history of elementary mathematics principally for his very popular *Eléments de géométrie,* in which he attempted a pedagogical improvement of Euclid's *Elements* by considerably rearranging and simplifying many of the propositions. This work was very favorably received in America and became the prototype of the geometry textbooks in this country. In fact, the first English translation of Legendre's geometry was made in 1819 by John Farrar of Harvard University. Three years later another English translation was made, by the famous Scottish writer Thomas Carlyle, who early in life was a teacher of mathematics. Carlyle's translation, as later revised by Charles Davies, and later still by J. H. Van Amringe, ran through thirty-three American editions. In later editions of his geometry, Legendre attempted to prove the parallel postulate (his Section 13.6). Legendre's chief work in higher mathematics centered on number theory, elliptic functions, the method of least squares, and integrals. He was also an assiduous computer of mathematical tables. Legendre's name is today connected with the second-order differential equation

$$(1 - x^2)y'' - 2xy' + n(n + 1)y = 0,$$

which is of considerable importance in applied mathematics. Functions satisfying this differential equation are called **Legendre functions** (of order n). When n is a nonnegative integer, the equation has polynomial solutions of special interest called **Legendre polynomials.** Legendre's name is also associated with the symbol $(c \mid p)$ of number theory. The **Legendre symbol** $(c \mid p)$ is equal to ± 1 according as the integer c, which is prime to p, is or is not a quadratic residue of the odd prime p. [For example, $(6 \mid 19) = 1$ since the congruence $x^2 \equiv 6 \pmod{19}$ has a solution, and $(39 \mid 47) = -1$ since the congruence $x^2 \equiv 39 \pmod{47}$ has no solution.]

In addition to his *Eléments de géométrie,* which appeared in 1794, Legendre published a two-volume 859-page work, *Essai sur la théorie des nombres* (1797–1798), which was the first treatise devoted exclusively to number theory. He later wrote a three-volume treatise, *Exercises du calcul intégral* (1811–1819), that, for comprehensiveness and authoritativeness, rivaled the similar work of Euler. Legendre later expanded parts of this work into another three-volume treatise, *Traité des fonctions elliptiques et des intégrals eulériennes* (1825–1832). In geodesy, Legendre achieved considerable fame for his triangulation of France.

Setting the coefficients of the sum (2) to zero, we obtain the recurrence relation

$$(n+2)(n+1)c_{n+2} = c_n(n^2 + n - p^2 - p) = c_n(n-p)(n+p+1).$$

Thus we have

$$c_{n+2} = -\frac{(p-n)(p+n+1)}{(n+2)(n+1)}c_n. \tag{3}$$

Therefore

$$c_2 = -\frac{p(p+1)}{2!}c_0, \quad c_3 = -\frac{(p-1)(p+2)}{3!}c_1,$$

$$c_4 = -\frac{(p-2)(p+3)}{4 \cdot 3}c_2 = \frac{(p-2)p(p+1)(p+3)}{4!}c_0, \cdots$$

Inserting these values for the coefficients into the power series expansion for $y(x)$ yields

$$y(x) = c_0 y_1(x) + c_1 y_2(x), \tag{4}$$

where

$$y_1(x) = 1 - p(p+1)\frac{x^2}{2!} + (p-2)p(p+1)(p+3)\frac{x^4}{4!} - \cdots, \tag{5}$$

$$y_2(x) = x - (p-1)(p+2)\frac{x^3}{3!} + (p-3)(p-1)(p+2)(p+4)\frac{x^5}{5!} - \cdots. \tag{6}$$

Dividing equation (6) by equation (5), we have

$$\frac{y_2(x)}{y_1(x)} = x + \frac{(p^2+p+1)}{3}x^3 + \cdots,$$

which obviously is nonconstant, implying that y_1 and y_2 are linearly independent. Thus equation (4) is the general solution of Legendre's equation (1) for $|x| < 1$.

In many applications the parameter p in Legendre's equation is a nonnegative integer. When this occurs, the right-hand side of equation (3) will vanish for $n = p$, implying that $c_{p+2} = c_{p+4} = c_{p+6} = \cdots = 0$. Thus one of the equations (5) or (6) reduces to a polynomial of degree p. (For even p, it is y_1; for odd p, it is y_2.) These polynomials, multiplied by an appropriate constant, are called the **Legendre polynomials.** It is customary to set

Legendre Polynomials

$$c_p = \frac{(2p)!}{2^p(p!)^2}, \qquad p = 0, 1, 2, \ldots, \tag{7}$$

so that by equation (3)

$$c_{p-2} = -\frac{p(p-1)}{2(2p-1)}c_p = -\frac{(2p-2)!}{2^p(p-1)!(p-2)!},$$

$$c_{p-4} = -\frac{(p-2)(p-3)}{4(2p-3)}c_{p-2} = \frac{(2p-4)!}{2^p 2!(p-2)!(p-4)!}, \cdots.$$

and in general

$$c_{p-2k} = \frac{(-1)^k(2p-2k)!}{2^p k!(p-k)!(p-2k)!}.$$

Then the **Legendre polynomials of degree p** are given by

$$P_p(x) = \sum_{k=0}^{M} \frac{(-1)^k(2p-2k)!}{2^p k!(p-k)!(p-2k)!} x^{p-2k}, \qquad p = 0, 1, 2, \ldots, \tag{8}$$

where M is the largest integer not greater than $p/2$. In particular, we have $P_0(x) = 1$, $P_1(x) = x$, $P_2(x) = \frac{1}{2}(3x^2 - 1)$, $P_3(x) = \frac{1}{2}(5x^3 - 3x)$, and $P_4(x) = \frac{1}{8}(35x^4 - 30x^2 + 3)$. As these results illustrate, as a consequence of the choice (7) of the value of c_p, we have $P_p(1) = 1$ and $P_p(-1) = (-1)^p$ for all integers $p \geq 0$.

To obtain an even more concise form than equation (8) for the Legendre polynomials, we observe that we can write

$$P_p(x) = \sum_{k=0}^{M} \frac{(-1)^k}{2^p k!(p-k)!} \frac{d^p}{dx^p}(x^{2p-2k}),$$

since

$$\frac{d^p}{dx^p}(x^{2p-2k}) = (2p-2k)\frac{d^{p-1}}{dx^{p-1}}(x^{2p-2k-1}) = \cdots$$

multiply and divide by $(p-2k)!$

$$= (2p-2k)\cdots(p-2k+1)x^{p-2k} = \frac{(2p-2k)!}{(p-2k)!}x^{p-2k}.$$

Hence

$$P_p(x) = \frac{1}{2^p p!}\frac{d^p}{dx^p}\sum_{k=0}^{M}(-1)^k\frac{p!}{k!(p-k)!}(x^2)^{p-k}.$$

We may now extend the range of this sum by letting k range from 0 to p. This extension will not affect the result, since the added terms are a polynomial of degree $<p$ so that the pth derivative will vanish. Thus

$$P_p(x) = \frac{1}{2^p p!}\frac{d^p}{dx^p}\sum_{k=0}^{p}\frac{p!}{k!(p-k)!}(x^2)^{p-k}(-1)^k,$$

and by the binomial formula we have

$$P_p(x) = \frac{1}{2^p p!}\frac{d^p}{dx^p}(x^2 - 1)^p, \qquad p = 0, 1, 2, \ldots. \tag{9}$$

Rodrigues' Formula

This formula, called **Rodrigues' formula**†, provides an easy way to compute successive Legendre polynomials.

† Named after the French mathematician and banker Olinde Rodrigues (1794–1851).

EXAMPLE 1

Show that $P_2(x) = \frac{1}{2}(3x^2 - 1)$.

SOLUTION
By Rodrigues' formula

$$P_2(x) = \frac{1}{2^2 2!} \frac{d^2}{dx^2}(x^4 - 2x^2 + 1) = \frac{1}{8}(12x^2 - 4) = \frac{1}{2}(3x^2 - 1).$$

We can use Rodrigues' formula to obtain several useful recurrence relations. Observe that

$$P'_{p+1} = \frac{d}{dx}\left[\frac{1}{2^{p+1}(p+1)!} \frac{d^{p+1}}{dx^{p+1}}(x^2 - 1)^{p+1} \right]$$

$$= \frac{d}{dx}\left\{ \frac{1}{2^p p!} \frac{d^p}{dx^p}[x(x^2 - 1)^p] \right\} = \frac{1}{2^p p!} \frac{d^{p+1}}{dx^{p+1}}[x(x^2 - 1)^p]. \qquad \textbf{(10)}$$

Taking the derivative of the term in brackets, we have

$$P'_{p+1} = \frac{1}{2^p p!} \frac{d^p}{dx^p}[(x^2 - 1)^p + 2px^2(x^2 - 1)^{p-1}]$$

$$= \frac{1}{2^p p!} \frac{d^p}{dx^p}[(2p + 1)(x^2 - 1)^p + 2p(x^2 - 1)^{p-1}]$$

$$= (2p + 1)P_p + P'_{p-1}, \qquad p = 1, 2, 3, \ldots .$$

We can get another recurrence relation from equation (10) if we consider the effect of repeated differentiations on a product of the form $x f(x)$. Note that

$$\frac{d}{dx}[x f(x)] = x \frac{d}{dx} f(x) + f(x), \qquad \frac{d^2}{dx^2}[x f(x)] = x \frac{d^2}{dx^2} f(x) + 2 \frac{d}{dx} f(x),$$

and in general

$$\frac{d^{p+1}}{dx^{p+1}}[x f(x)] = x \frac{d^{p+1}}{dx^{p+1}} f(x) + (p + 1) \frac{d^p}{dx^p} f(x). \qquad \textbf{(11)}$$

Applying equation (11) to the expression in brackets in equation (10), we obtain

$$P'_{p+1} = \frac{1}{2^p p!}\left[x \frac{d^{p+1}}{dx^{p+1}}(x^2 - 1)^p + (p + 1) \frac{d^p}{dx^p}(x^2 - 1)^p \right]$$

$$= xP'_p + (p + 1)P_p, \qquad p = 0, 1, 2, \ldots .$$

Thus we have proved the identities

$$(p + 1)P_p = P'_{p+1} - xP'_p, \qquad (2p + 1)P_p = P'_{p+1} - P'_{p-1}. \qquad \textbf{(12)}$$

Subtracting the first identity in (12) from the second one yields

$$pP_p = xP'_p - P'_{p-1}, \qquad p = 1, 2, \ldots. \tag{13}$$

Finally, we note that from (12) and (13) we can get

$$(p+1)P_{p+1} - (2p+1)xP_p + pP_{p-1}$$

$$= (xP'_{p+1} - P'_p) - x(P'_{p+1} - P'_{p-1}) + (P'_p - xP'_{p-1}) = 0,$$

so we can eliminate all derivatives and obtain the relation

$$(p+1)P_{p+1} + pP_{p-1} = (2p+1)xP_p, \qquad p = 1, 2, \ldots. \tag{14}$$

Equation (14) can be used to generate all the Legendre polynomials. We illustrate this iterative technique in the next example.

EXAMPLE 2

Starting with $P_0 = 1$ and $P_1 = x$, calculate the polynomials P_2, P_3, and P_4.

SOLUTION
By equation (14),

$$P_{p+1} = \frac{(2p+1)xP_p - pP_{p-1}}{p+1},$$

so that

$$P_2 = \frac{3xP_1 - P_0}{2} = \frac{3x^2 - 1}{2},$$

$$P_3 = \frac{5xP_2 - 2P_1}{3} = \frac{15x^3 - 5x - 4x}{6} = \frac{5x^3 - 3x}{2},$$

$$P_4 = \frac{7xP_3 - 3P_2}{4} = \frac{35x^4 - 21x^2 - 9x^2 + 3}{8} = \frac{35x^4 - 30x^2 + 3}{8}.$$

Problems 5.5

1. Calculate P_5, P_6, P_7, and P_8 by means of equation (14).
2. Prove that the series (5) has a radius of convergence $R = 1$.
3. Prove that the series (6) has a radius of convergence $R = 1$.
4. Calculate P_4 by means of Rodrigues' formula.
5. Prove that $P_{2p+1}(0) = 0$ for all integers $p \geq 0$.
6. Prove that $P_{2p}(0) = (-1)^p (2p)!/2^{2p}(p!)^2$ for all $p \geq 0$.
7. Prove that for all integers $p \geq 0$:
   (a) $P'_{2p}(0) = 0$

   (b) $P'_{2p+1}(0) = \dfrac{(-1)^p(2p+1)!}{2^{2p}(p!)^2}$

8. Show that for all integers $p > 0$:

   (a) $\displaystyle\int_0^1 P_p(x)\,dx = \frac{1}{p+1}P_{p-1}(0)$

   (b) $\int_0^1 P_{2p}(x)\,dx = 0$

   (c) $\displaystyle\int_0^1 P_{2p+1}(x)\,dx = (-1)^p \frac{(2p)!}{2^{2p+1}p!(p+1)!}$

   (d) Compute these integrals for $p = 0$

9. Consider the differential equation

$$y'' - 2xy' + 2py = 0, \tag{15}$$

which is known as **Hermite's equation.**†

(a) Use the method of Frobenius to show that all solutions of equation (15) are of the form

$$c_0\left[1 + \sum_{n=1}^{\infty} \frac{2^n(-p)(2-p)\cdots(2n-2-p)x^{2n}}{(2n)!}\right]$$

$$+ c_1\left[x + \sum_{n=1}^{\infty} \frac{2^n(1-p)(3-p)\cdots(2n-1-p)x^{2n+1}}{(2n+1)!}\right],$$

where c_0 and c_1 are arbitrary constants.

(b) Show that equation (15) has a polynomial solution of degree p for a nonnegative integer p. These polynomials, denoted by $H_p(x)$, are called the **Hermite polynomials of degree p.**

(c) Show that

$$H_p(x) = \sum_{n=0}^{M} \frac{(-1)^n p!(2x)^{p-2n}}{n!(p-2n)!},$$

where M is the greatest integer $\leq p/2$.

(d) Calculate H_0, H_1, H_2, H_3, and H_4.

10. Consider **Laguerre's equation**‡

$$xy'' + (1-x)y' + py = 0. \tag{16}$$

(a) Show that if p is a nonnegative integer, there is a polynomial solution to equation (16) of the form

$$L_p(x) = \sum_{n=0}^{p} \frac{(-1)^n p! x^n}{(p-n)!(n!)^2}.$$

The functions $L_p(x)$ are known as the **Laguerre polynomials.**

(b) Calculate $L_0(x)$, $L_1(x)$, $L_2(x)$, $L_3(x)$, and $L_4(x)$.

11. Use the binomial theorem (see Problem 5.1.18) to prove that

$$\frac{1}{\sqrt{1 - 2xz + z^2}} = P_0(x) + P_1(x)z + P_2(x)z^2$$

$$+ \cdots + P_n(x)z^n + \cdots,$$

where $P_n(x)$ is the nth Legendre polynomial. This identity is called the **generating function** for Legendre polynomials.

Review Exercises for Chapter 5

1. Show that for $|x - 1| < 1$,

$$\ln x = (x-1) - \tfrac{1}{2}(x-1)^2 + \tfrac{1}{3}(x-1)^3 \cdots.$$

2. Use Taylor's theorem to show that

$$\sin x = \sin a + \cos a(x - a)$$

$$- \frac{\sin a}{2!}(x-a)^2 - \frac{\cos a}{3!}(x-a)^3 + \cdots.$$

3. Verify that for $|x| < 1$,

$$\ln \frac{1+x}{1-x} = 2\left[x + \frac{x^3}{3} + \frac{x^5}{5} + \cdots + \frac{x^{2n-1}}{2n-1} + \cdots\right].$$

4. Show that

$$\ln \frac{x^2}{x^2 - 1} = \frac{1}{x^2} + \frac{1}{2x^4} + \frac{1}{3x^6} + \cdots,$$

for $|x| > 1$. [*Hint:* Consider the series $\ln(1 + x) = x - x^2/2 + x^3/3 - x^4/4 + \cdots$, $|x| < 1$.]

5. Use Taylor's theorem to show that

$$e^{\tan x} = 1 + x + \frac{x^2}{2!} + \frac{3x^3}{3!} + \frac{9x^4}{4!} + \frac{37x^5}{5!} + \cdots$$

for $|x| < \pi/2$.

In Problems 6–11 use power series to find the solution of the given initial value problem.

6. $xy'' + xy' - y = -e^{-x}$, $y(0) = 1$, $y'(0) = 0$

7. $xy'' - y' + (1-x)y = x^2(1-x)$, $y(0) = y'(0) = 1$

8. $(1+x)y'' + (2-x-x^2)y' + y = 0$, $y(0) = 0$, $y'(0) = 1$

9. $xy'' - 2y' + xy = x^2 - 2 - 2\sin x$, $y(0) = 0$, $y'(0) = 1$

10. $y'' - 3x^2y' - 6xy = 0$, $y(0) = 1$, $y'(0) = 0$

11. $y'' + 2xy' + 2y = 0$, $y(0) = 1$, $y'(0) = 0$

In Problems 12–17 use the method of Frobenius to solve the given differential equations.

---

† Charles Hermite (1822–1901) was a French mathematician known for his contributions to algebra and number theory.

‡ Edmond Laguerre (1834–1886) was a French mathematician whose research was in geometry and infinite series.

12. $4xy'' + 2y' - y = 0$

13. $2x(1 - 2x)y'' + (1 + 4x^2)y' - (1 + 2x)y = 0$

14. $2x(1 - 2x)y'' + (1 + x)y' - y = 0$

15. $x(1 - x)y'' + 2(1 - 2x)y' - 2y = 0$

16. $2x(1 - x)y'' + (1 - x)y' + 3y = 0$

17. $2x^2y'' + x(1 - x)y' - y = 0$

18. Gauss' hypergeometric equation is given by

$$x(1 - x)y'' + [c - (a + b + 1)x]y' - aby = 0,$$

where a, b, and c are constants. Show that it has a solution of the form

$$y_1(x) = 1 + \frac{ab}{c}x + \frac{a(a + 1)b(b + 1)}{c(c + 1)}\frac{x^2}{2!}$$

$$+ \frac{a(a + 1)(a + 2)b(b + 1)(b + 2)}{c(c + 1)(c + 2)}\frac{x^3}{3!} + \cdots .$$

This series is called the **hypergeometric series** and denoted by the symbol $F(a, b, c; x)$.

19. Prove, using the result of Problem 18, that

(a) $F(-a, b, b; -x) = (1 + x)^a$.

(b) $xF(1, 1, 2; -x) = \ln(1 + x)$.

20. Show that **Chebyshev's equation**†

$$(1 - x^2)y'' - xy' + p^2y = 0$$

has a polynomial solution when p is an integer.

21. Prove the following identities if $J_0(a) = J_0(b) = 0$:

(a) $\int_0^1 xJ_0(ax)J_0(bx)\, dx = 0, \quad a \neq b$

(b) $\int_0^1 xJ_0^2(ax)\, dx = \frac{1}{2}J_1^2(ax)$

(c) $x^2J_0'(x) = x[J_1(x) - J_0(x)] - (x^2 - 1)J_1(x)$

(d) $[xJ_0(x)J_1(x)]' = x[J_0^2(x) - J_1^2(x)]$

22. Find solutions in terms of Bessel functions for the following differential equations:

(a) $x^2y'' + (x^2 - 2)y = 0$

(b) $x^2y'' - xy' + (1 + x^2 - k^2)y = 0$

23. Using the method of Frobenius, find a solution to

$$x^2y'' - xy' + (x^2 + 1)y = 0.$$

24. Prove the following identities for integral values of n:

(a) $\int_{-1}^1 xP_n(x)P_{n-1}(x)\, dx = \frac{2n}{4n^2 - 1}$

(b) $\int_{-1}^1 (1 - x^2)[P_n'(x)]^2\, dx = \frac{2n(n + 1)}{2n + 1}$

(c) $(m + n + 1)\int_0^1 x^mP_n(x)\, dx = m\int_0^1 x^{m-1}P_{n-1}(x)\, dx$

25. Find a solution for the differential equation

$$y'' + 2(\cot x)y' + n(n + 1)y = 0.$$

[*Hint:* make the trigonometric substitution $t = \cos x$.]

26. Power series methods can also be used to obtain approximate solutions of nonlinear differential equations. In Problem 2.8.11 you showed that an ideal pendulum satisfies the nonlinear initial value problem

$$\frac{d^2\theta}{dt^2} + \frac{g}{l}\sin\theta = 0, \qquad \theta(0) = \theta_0, \qquad \theta'(0) = 0$$

(a) Show that the substitution $y = \theta$ and $x = \sqrt{g/l}\,t$ leads to the initial value problem

$$\frac{d^2y}{dx^2} + \sin y = 0, \qquad y(0) = \theta_0, \qquad y'(0) = 0.$$

* **(b)** Replace $\sin y$ by its Maclaurin series and substitute $y = \sum_{n=0}^\infty c_nx^n$ and show that

$$1 \cdot 2c_2 + \sin c_0 = 0$$

$$2 \cdot 3c_3 + c_1\cos c_0 = 0$$

$$3 \cdot 4c_4 + c_2\cos c_0 - \frac{c_1^2}{2}\sin c_0 = 0$$

$$4 \cdot 5c_5 + c_3\cos c_0 - c_1c_2\sin c_0 - \frac{c_1^3}{3}\cos c_0 = 0$$

(c) Conclude that the Taylor series of $\theta(t)$ has the form

$$\theta(t) = \theta_0 - \frac{g}{l}(\sin\theta_0)\frac{t^2}{2!} + \frac{g^2}{l^2}(\sin\theta_0\cos\theta_0)\frac{t^4}{4!} + (\text{terms in } t^6).$$

---

† Pafnuti L. Chebyshëv (1821–1894) was the leading Russian mathematician of his time.

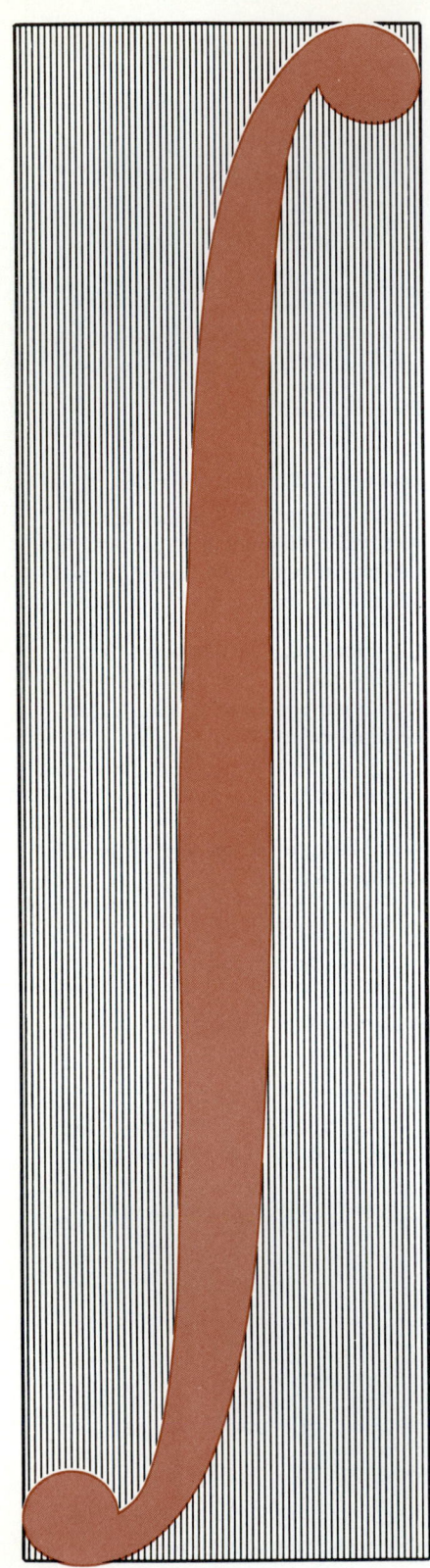

chapter 6

Vectors and Vector Spaces

6.0 Introduction

The study of vectors began essentially with the work of the great Irish mathematician Sir William Rowan Hamilton (1805–1865).† Hamilton knew that complex numbers could be represented as pairs of real numbers: $z = a + bi$ where a is the real part and b is the imaginary part (see Section 15.1). Rules for adding and multiplying these numbers were well known to mathematicians of the time (i.e., in the early nineteenth century). Complex numbers were used to study vectors and rotations in two-dimensional space. Hamilton devised an analogous system of numbers that could be used to study vectors and rotations in three-dimensional space. He considered ordered quadruples (a, b, c, d) of real numbers, which he called **quaternions,** and developed rules for adding and multiplying them.

After Hamilton's death, his work on quaternions was supplanted by the more adaptable work on vector analysis by the American mathematician and physicist Josiah Willard Gibbs (1839–1903) and the general treatment of ordered n-tuples by the German mathematician Hermann Grassmann (1809–1877).

Throughout Hamilton's life and for the remainder of the nineteenth century, there was considerable debate over the usefulness of quaternions and vectors. At the end of the century, the great British physicist Lord Kelvin wrote that quaternions "although beautifully ingenious, have been an unmixed evil to those who have touched them in any way . . . vectors . . . have never been of the slightest use to any creature."

But Kelvin was wrong. Today nearly all branches of classical and modern physics are represented using the language of vectors. Vectors are also used with increasing frequency in the social and biological sciences.

We begin this chapter with a review of vectors in space. We then discuss properties of real vector spaces and, especially, the vector space $\mathbb{R}^n$.

† See the accompanying biographical sketch on page 292.

6.1 Review of Vectors in the Plane and in Space

At the beginning of a course in multivariable calculus, every student studies vectors in the plane and in space. Thus we assume in this text that you are familiar with this material.

Besides being useful in the study of calculus, vectors in the plane and in space can be used to illustrate many of the properties of abstract vector spaces. For these reasons we provide in this and the next three sections a review of these vectors.

We stress that this is a review. Some proofs are omitted and there are fewer examples than usual. If this review is insufficient, we suggest that you reread the relevant material in your calculus book.

Vectors in the Plane: The Space $\mathbb{R}^2$

Vector
Components of a Vector

A **vector** $\mathbf{v}$ in the xy-plane is an ordered pair of real numbers (a, b). The numbers a and b are called the **components** of the vector $\mathbf{v}$. Here a is the x-coordinate and b is the y-coordinate. The **zero vector** is the vector $(0, 0)$. The set of all vectors in the plane is denoted $\mathbb{R}^2$.

Each vector in the plane can be depicted geometrically by drawing an arrow from $(0, 0)$ to (a, b). (See Fig. 1.)

Scalar

Since we shall often have to distinguish between real numbers and vectors (which, in this case, are pairs of real numbers), we shall use the term **scalar**† to denote a real number.

If $\mathbf{v} = (a, b)$, we find that the length or magnitude of the vector $\mathbf{v}$ is given by

$$|\mathbf{v}| = \textbf{magnitude of } \mathbf{v} = \sqrt{a^2 + b^2}. \tag{1}$$

This follows from the Pythagorean theorem (see Fig. 1). We have used the notation $|\mathbf{v}|$ to denote the magnitude of $\mathbf{v}$. Note that $|\mathbf{v}|$ is a *scalar*.

Direction

We now define the **direction** of the vector $\mathbf{v} = (a, b)$ to be the angle θ, measured counterclockwise in radians, that the vector makes with the positive x-axis. By convention, we choose θ such that $0 \leq \theta < 2\pi$. It follows from Fig. 1 that if $a \neq 0$, then

† The term *scalar* originated with Hamilton. His definition of the quaternion included what he called a *real part* and an *imaginary part*. In his paper "On Quaternions, or on a New System of Imaginaries in Algebra," in *Philosophical Magazine*, 3rd Ser., vol. 25 (1844), pp. 26–27, he wrote, "The algebraically *real* part may receive . . . all values contained on the one *scale* of progression of numbers from negative to positive infinity; we shall call it therefore the *scalar part*, or simply the *scalar* of the quaternion." In the same paper, Hamilton went on to define the imaginary part of his quaternion as the *vector* part. Although this was not the first usage of the word *vector*, it was the first time it was used in its present context. In fact, it is fair to say that the paper from which the above quotation was taken marks the beginning of modern vector analysis.

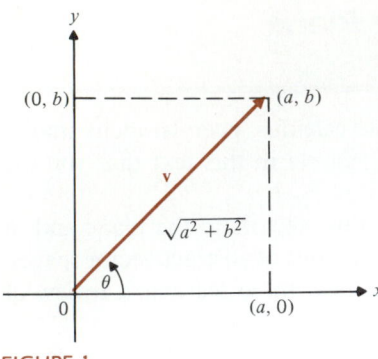

y

(0, b) --------- (a, b)

v

$\sqrt{a^2 + b^2}$

θ

0   (a, 0)   x

FIGURE 1

$$\tan \theta = \frac{b}{a}. \qquad (2)$$

EXAMPLE 1

Calculate the magnitudes and directions of (a) $(2, 2\sqrt{3})$ and (b) $(6, -6)$.

SOLUTION

(a) $|\mathbf{v}| = \sqrt{2^2 + (2\sqrt{3})^2} = \sqrt{16} = 4$; $\theta = \tan^{-1} \dfrac{2\sqrt{3}}{2} = \tan^{-1} \sqrt{3} = \pi/3.$

(b) $|\mathbf{v}| = \sqrt{6^2 + (-6)^2} = \sqrt{72} = 6\sqrt{2}.$

SIR WILLIAM ROWAN HAMILTON
(1805–1865)

Courtesy of the Granger Collection

Born in Dublin in 1805, where he spent most of his life, William Rowan Hamilton was without question Ireland's greatest mathematician. Hamilton's father (an attorney) and mother died when he was a small boy. His uncle, a linguist, took over the boy's education. By his fifth birthday Hamilton could read English, Hebrew, Latin, and Greek. By his thirteenth birthday he had mastered not only the languages of continental Europe, but also Sanskrit, Chinese, Persian, Arabic, Malay, Hindi, Bengali, and several others as well. Hamilton liked to write poetry, both as a child and as an adult, and his friends included the great English poets Samuel Taylor Coleridge and William Wordsworth. Hamilton's poetry was considered so bad, however, that it is fortunate that he developed other interests—especially in mathematics.

Although he enjoyed mathematics as a young boy, Hamilton's interest was greatly enhanced by a chance meeting at the age of 15 with Zerah Colburn, the American lightning calculator. Shortly afterward, Hamilton began to read important mathematical books of the time. In 1823, at the age of 17, he discovered an error in Simon Laplace's *Mécanique céleste* and wrote an impressive paper on the subject. A year later he entered Trinity College in Dublin.

Hamilton's university career was astonishing. At the age of 22, while still an undergraduate, he had so impressed the faculty that he was appointed Royal Astronomer of Ireland and Professor of Astronomy at the College. Shortly thereafter he wrote what is now considered a classic work on optics. Using only mathematical theory, he predicted conical refraction in certain types of crystals. Later this theory was confirmed by physicists. Largely because of this work, Hamilton was knighted in 1835.

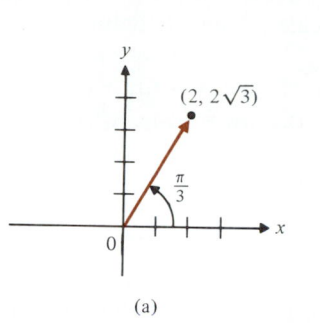

(a)

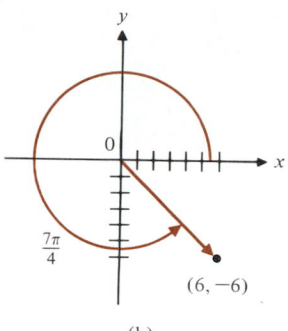

(b)

FIGURE 2

Since **v** is in the fourth quadrant and since $\tan^{-1}(-1) = -\pi/4$, $\theta = 2\pi - (\pi/4) = 7\pi/4$. The two vectors are sketched in Fig. 2.

Hamilton's first great purely mathematical paper appeared in 1833. In this work he described an algebraic way to manipulate pairs of real numbers. This work gives rules that are used today to add, subtract, multiply, and divide complex numbers. At first, however, Hamilton was unable to devise a multiplication for triples or n-tuples of numbers for $n > 2$. For 10 years he pondered this problem, and it is said that he solved it in an inspiration while walking on the Brougham Bridge in Dublin in 1843. The key was to discard the familiar commutative property of multiplication. The new objects he created were called *quaternions.* These were the precursors of what we now call *vectors.*

For the rest of his life, Hamilton spent most of his time developing the algebra of quaternions. He felt that they would have revolutionary significance in mathematical physics. His monumental work on this subject, *Treatise on Quaternions,* was published in 1853. Thereafter, he worked on an enlarged work, *Elements of Quaternions.* Although Hamilton died in 1865 before his *Elements* was completed, the work was published by his son in 1866.

Students of mathematics and physics know Hamilton in a variety of other contexts. In mathematical physics, for example, one encounters the Hamiltonian function, which often represents the total energy in a system, and the Hamilton-Jacobi differential equations of dynamics. In matrix theory, the Cayley-Hamilton theorem states that every matrix satisfies its own characteristic equation. [See Section 8.7.]

Despite the great work he was doing, Hamilton's final years were a torment to him. His wife was a semi-invalid and he was plagued by alcoholism. It is therefore gratifying to point out that, during these last years, the newly formed American National Academy of Sciences elected Sir William Rowan Hamilton to be its first foreign associate.

Sum of Two Vectors
Scalar Multiplication

Let $\mathbf{u} = (a_1, b_1)$ and $\mathbf{v} = (a_2, b_2)$ be two vectors in the plane and let α be a scalar. Then we define

(i) $\mathbf{u} + \mathbf{v} = (a_1 + a_2, b_1 + b_2)$, and
(ii) $\alpha\mathbf{u} = (\alpha a_1, \alpha b_1)$.

That is,

> to add two vectors, we add their corresponding components, and to multiply a vector by a scalar, we multiply each of its components by that scalar.

EXAMPLE 2

Let $\mathbf{u} = (1, 3)$ and $\mathbf{v} = (-2, 4)$. Calculate (a) $\mathbf{u} + \mathbf{v}$ and (b) $-3\mathbf{u} + 5\mathbf{v}$.

SOLUTION
(a) $\mathbf{u} + \mathbf{v} = (1 + (-2), 3 + 4) = (-1, 7)$.
(b) $-3\mathbf{u} + 5\mathbf{v} = -3(1, 3) + 5(-2, 4) = (-3, -9) + (-10, 20) = (-13, 11)$.

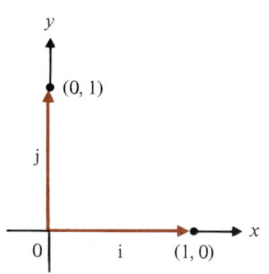

FIGURE 3

There are two special vectors in $\mathbb{R}^2$ which allow us to represent other vectors in $\mathbb{R}^2$ in a convenient way. We shall denote the vector $(1, 0)$ by the vector symbol $\mathbf{i}$ and the vector $(0, 1)$ by the vector symbol $\mathbf{j}$ (see Fig. 3). If (a, b) denotes any other vector in $\mathbb{R}^2$, then, since $(a, b) = a(1, 0) + b(0, 1)$, we may write

$$\mathbf{v} = (a, b) = a\mathbf{i} + b\mathbf{j}.$$

Moreover, any vector in $\mathbb{R}^2$ can be represented in a unique way in the form $a\mathbf{i} + b\mathbf{j}$ since the representation of (a, b) as a point in the plane is unique. (Put another way, a point in the xy-plane has one and only one x-coordinate and one and only one y-coordinate.)

Basis Vectors

Directed Line Segment

When the vector $\mathbf{v}$ is written in the form $\mathbf{v} = a\mathbf{i} + b\mathbf{j}$, we say that $\mathbf{v}$ *is resolved into its horizontal and vertical components* since, obviously, a is the horizontal component of $\mathbf{v}$ while b is its vertical component. The vectors $\mathbf{i}$ and $\mathbf{j}$ are called **basis vectors** for the vector space $\mathbb{R}^2$.

It is often useful to draw vectors whose initial points are not at $(0, 0)$. Let P and Q be two points in the plane. Then the **directed line segment** from P to Q, denoted by $\overrightarrow{PQ}$, is the straight line segment that extends from P to Q (see Fig. 4a). Note that the directed line segments $\overrightarrow{PQ}$ and $\overrightarrow{QP}$ are different since they point in opposite directions (Fig. 4b).

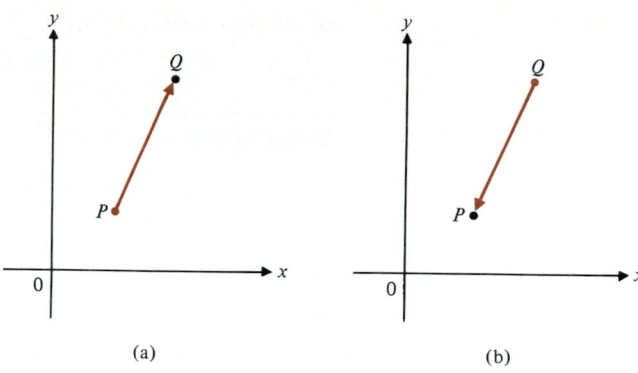

(a)   (b)

FIGURE 4

The point P in the directed line segment $\overrightarrow{PQ}$ (in Fig. 4a) is called the **initial point** of the segment and the point Q is called the **terminal point.** Two important properties of a directed line segment are its magnitude (length) and its direction.

Equivalent Vectors

If two directed line segments $\overrightarrow{PQ}$ and $\overrightarrow{RS}$ have the same magnitude and direction, we say that they are **equivalent** no matter where they are located with respect to the origin. Any directed line segment that is equivalent to the vector (a, b) is called a **representation** of the vector (a, b).

Now suppose that a vector v can be represented by the directed line segment $\overrightarrow{PQ}$ where $P = (a_1, b_1)$ and $Q = (a_2, b_2)$. (See Fig. 5.) If we label the point (a_2, b_1) as R, then we immediately see that

$$\mathbf{v} = \overrightarrow{PQ} = \overrightarrow{PR} + \overrightarrow{RQ}.$$

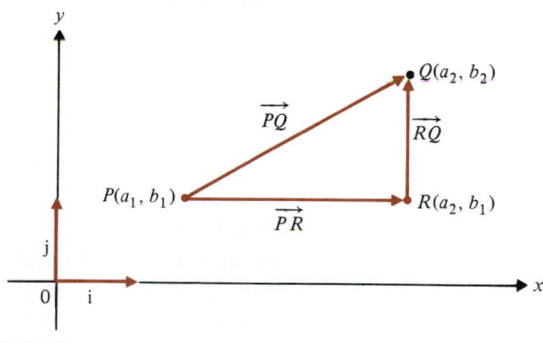

FIGURE 5

But the length of $\overrightarrow{PR}$ is $a_2 - a_1$ and since $\overrightarrow{PR}$ has the same direction as $\mathbf{i}$, we can write

$$\overrightarrow{PR} = (a_2 - a_1)\mathbf{i}.$$

Similarly,

$$\overrightarrow{RQ} = (b_2 - b_1)\mathbf{j}$$

and we may write

$$\mathbf{v} = (a_2 - a_1)\mathbf{i} + (b_2 - b_1)\mathbf{j}. \tag{3}$$

EXAMPLE 3

Resolve the vector represented by the directed line segment from $(-2, 3)$ to $(1, 5)$ into its vertical and horizontal components.

SOLUTION
Using (3), we have

$$\mathbf{v} = (a_2 - a_1)\mathbf{i} + (b_2 - b_1)\mathbf{j} = (1 - (-2))\mathbf{i} + (5 - 3)\mathbf{j} = 3\mathbf{i} + 2\mathbf{j}.$$

Unit Vector

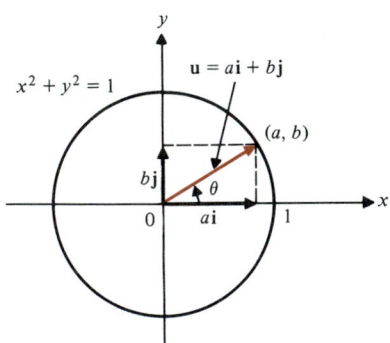

FIGURE 6

A **unit vector** $\mathbf{u}$ is a vector which has length 1. Let $\mathbf{u} = a\mathbf{i} + b\mathbf{j}$ be a unit vector. Then $|\mathbf{u}| = \sqrt{a^2 + b^2} = 1$, so $a^2 + b^2 = 1$ and $\mathbf{u}$ is a point on the unit circle (see Fig. 6). If θ is the direction of $\mathbf{u}$, we immediately see that $a = \cos \theta$ and $b = \sin \theta$. Thus any unit vector $\mathbf{u}$ can be written in the form

$$\mathbf{u} = (\cos \theta)\mathbf{i} + (\sin \theta)\mathbf{j} \tag{4}$$

where θ is the direction of $\mathbf{u}$. Finally,

Let $\mathbf{v}$ be any nonzero vector. Then $\mathbf{u} = \mathbf{v}/|\mathbf{v}|$ is the unit vector having the same direction as $\mathbf{v}$.

EXAMPLE 4

Find the unit vector having the same direction as $\mathbf{v} = 2\mathbf{i} - 3\mathbf{j}$.

SOLUTION
Here $|\mathbf{v}| = \sqrt{4 + 9} = \sqrt{13}$, so that $\mathbf{u} = \mathbf{v}/|\mathbf{v}| = (2/\sqrt{13})\mathbf{i} - (3/\sqrt{13})\mathbf{j}$ is the required unit vector.

Resultant

If more than one force is applied to an object, then we define the **resultant** of the forces applied to the object as the *vector sum* of these forces. We can think of the resultant as the *net* applied force.

EXAMPLE 5

A force of 3 newtons (N) is applied to the left side of an object, a 4 N force is applied from the bottom, and a force of 7 N is applied from an angle of $\pi/4$ to the horizontal. What is the resultant of forces applied to the object?

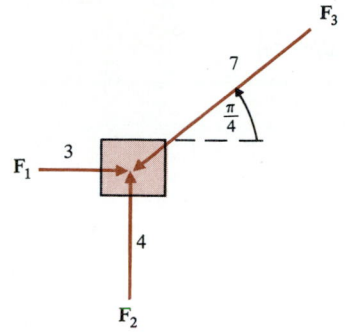

FIGURE 7

SOLUTION

The forces are indicated in Fig. 7. We write each force as a magnitude times a unit vector in the indicated direction. For convenience, we can think of the center of the object as being at the origin. Then $\mathbf{F}_1 = 3\mathbf{i}$, $\mathbf{F}_2 = 4\mathbf{j}$, and $\mathbf{F}_3 = -(7/\sqrt{2})(\mathbf{i} + \mathbf{j})$. This last follows from the fact that the vector $-(1/\sqrt{2})(\mathbf{i} + \mathbf{j})$ is a unit vector pointing toward the origin making an angle of $5\pi/4$ with the x-axis [see equation (4)]. Then the resultant (shown in Figs. 8a and 8b) is given by

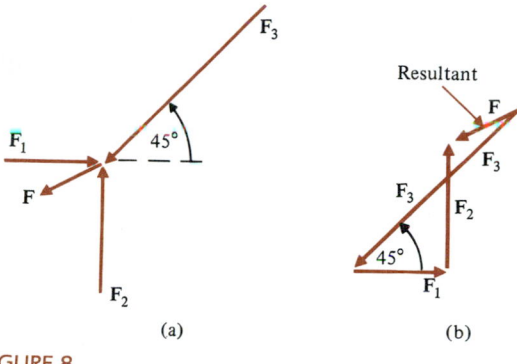

(a)   (b)

FIGURE 8

Vectors in Space: The Space $\mathbb{R}^3$

A **vector v** in space is an ordered triple of real numbers (a, b, c). The component a is the x-coordinate of **v**, b is the y-coordinate, and c is the z-coordinate. The **zero vector** is the vector $(0, 0, 0)$. The set of all vectors in space is denoted by $\mathbb{R}^3$.

The next result follows from the Pythagorean theorem. Let $\mathbf{v} = (a, b, c)$; then

$$|\mathbf{v}| = \mathbf{magnitude} \text{ of } \mathbf{v} = \sqrt{a^2 + b^2 + c^2}.$$

EXAMPLE 6

Let $\mathbf{v} = (1, 3, -2)$. Find $|\mathbf{v}|$.

SOLUTION
$|\mathbf{v}| = \sqrt{1^2 + 3^2 + (-2)^2} = \sqrt{14}$.

Let $\mathbf{u} = (x_1, y_1, z_1)$ and $\mathbf{v} = (x_2, y_2, z_2)$ be two vectors and let α be a real number (scalar). Then we define

$$\mathbf{u} + \mathbf{v} = (x_1 + x_2, y_1 + y_2, z_1 + z_2)$$

and

$$\alpha\mathbf{u} = (\alpha x_1, \alpha y_1, \alpha z_1).$$

Vector Space

The vectors in $\mathbb{R}^3$, with vector addition and scalar multiplication defined as above, form what we call a **vector space**. Indeed, we can show (see Problem 51) the following:

■ **THEOREM 1**

Let $\mathbf{u}$, $\mathbf{v}$, and $\mathbf{w}$ be any three vectors in space, let α and β be scalars, and let $\mathbf{0}$ denote the zero vector $(0, 0, 0)$. Then

(i) $\mathbf{u} + \mathbf{v} = \mathbf{v} + \mathbf{u}$	**(ii)** $\mathbf{u} + (\mathbf{v} + \mathbf{w}) = (\mathbf{u} + \mathbf{v}) + \mathbf{w}$
(iii) $\mathbf{v} + \mathbf{0} = \mathbf{v}$	**(iv)** $0\mathbf{v} = \mathbf{0}$
(v) $\alpha\mathbf{0} = \mathbf{0}$	**(vi)** $(\alpha\beta)\mathbf{v} = \alpha(\beta\mathbf{v})$
(vii) $\mathbf{v} + (-\mathbf{v}) = \mathbf{0}$	**(viii)** $(1)\mathbf{v} = \mathbf{v}$
(ix) $(\alpha + \beta)\mathbf{v} = \alpha\mathbf{v} + \beta\mathbf{v}$	**(x)** $\alpha(\mathbf{u} + \mathbf{v}) = \alpha\mathbf{u} + \alpha\mathbf{v}$
(xi) $|\alpha\mathbf{v}| = |\alpha||\mathbf{v}|$	**(xii)** $|\mathbf{u} + \mathbf{v}| \le |\mathbf{u}| + |\mathbf{v}|$

■

We can now formally define the direction of a vector in $\mathbb{R}^3$. We cannot define it as the angle θ the vector makes with the positive x-axis, since, for example, if $0 < \theta < \pi/2$, then there are an *infinite number* of vectors making the angle θ with the positive x-axis, and these together form a cone (see Fig. 9).

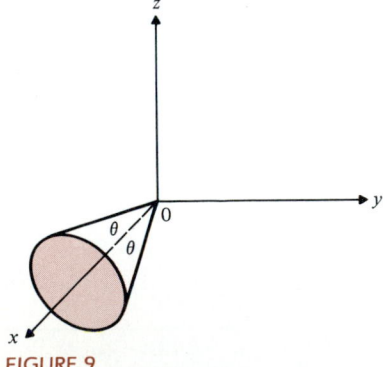

FIGURE 9

DEFINITION

Direction of a Vector

The **direction** of a nonzero vector $\mathbf{v}$ in $\mathbb{R}^3$ is defined as the unit vector $\mathbf{u} = \mathbf{v}/|\mathbf{v}|$.

Remark. We could have defined the direction of a vector $\mathbf{v}$ in $\mathbb{R}^2$ in this way. For if $\mathbf{u} = \mathbf{v}/|\mathbf{v}|$, then $\mathbf{u} = (\cos\theta, \sin\theta)$ where θ is the direction of $\mathbf{v}$ (according to the $\mathbb{R}^2$ definition).

Direction Cosines

Direction Angle

It would still be nice to define the direction of a vector in terms of some angles. Let **v** be the vector $\overrightarrow{OP}$ depicted in Fig. 10. We define α as the angle between **v** and the positive x-axis, β the angle between **v** and the positive y-axis, and γ the angle between **v** and the positive z-axis. The angles α, β, and γ are called the **direction angles** of the vector **v**. Then, from Fig. 10,

$$\cos \alpha = \frac{x_0}{|\mathbf{v}|}, \qquad \cos \beta = \frac{y_0}{|\mathbf{v}|}, \qquad \cos \gamma = \frac{z_0}{|\mathbf{v}|}. \tag{5}$$

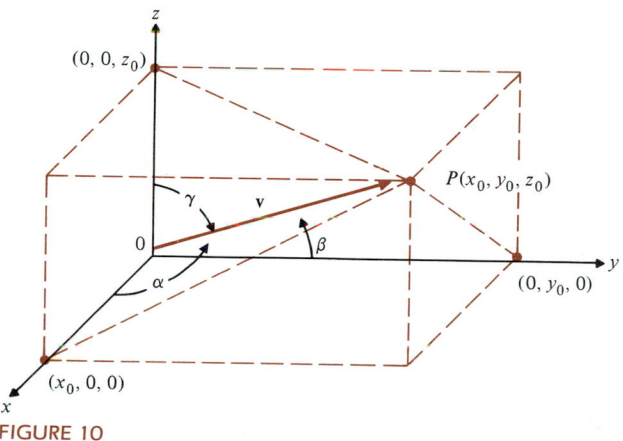

FIGURE 10

If **v** is a unit vector, then $|\mathbf{v}| = 1$ and

$$\cos \alpha = x_0, \qquad \cos \beta = y_0, \qquad \cos \gamma = z_0. \tag{6}$$

Direction Cosines

By definition, each of these three angles lies between 0 and π. The cosines of these angles are called the **direction cosines** of the vector **v**. Note, from (5), that

$$\cos^2 \alpha + \cos^2 \beta + \cos^2 \gamma = \frac{x_0^2 + y_0^2 + z_0^2}{|\mathbf{v}|^2} = \frac{x_0^2 + y_0^2 + z_0^2}{x_0^2 + y_0^2 + z_0^2} = 1. \tag{7}$$

If α, β, and γ are any three numbers, each between 0 and π, such that condition (7) is satisfied, then they determine a unique unit vector given by

$$\mathbf{u} = (\cos \alpha, \cos \beta, \cos \gamma).$$

Remark. If $\mathbf{v} = (a, b, c)$ and $|\mathbf{v}| \neq \mathbf{0}$, then the numbers a, b, and c are called **direction numbers** of the vector **v**.

EXAMPLE 7

Find the direction cosines of the vector $\mathbf{v} = (4, -1, 6)$.

SOLUTION

The direction of **v** is $\mathbf{v}/|\mathbf{v}| = \mathbf{v}/\sqrt{53} = (4/\sqrt{53}, -1/\sqrt{53}, 6/\sqrt{53})$. Then $\cos \alpha = 4/\sqrt{53} \approx 0.5494$, $\cos \beta = -1/\sqrt{53} \approx -0.1374$, and $\cos \gamma = 6/\sqrt{53} \approx 0.8242$. From these, we use a calculator to obtain $\alpha \approx 56.7° \approx 0.9891$ rad, $\beta \approx 97.9° \approx 1.709$ rad, and $\gamma = 34.5° \approx 0.6021$ rad.

Earlier we showed how any vector in the plane can be written in terms of the basis vectors **i** and **j**. Extending this idea to $\mathbb{R}^3$, we define

$$\mathbf{i} = (1, 0, 0), \qquad \mathbf{j} = (0, 1, 0), \qquad \mathbf{k} = (0, 0, 1). \qquad \textbf{(8)}$$

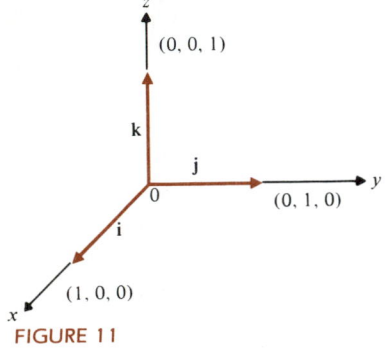

FIGURE 11

Here **i**, **j**, and **k** are unit vectors. The vector **i** lies along the x-axis, **j** along the y-axis, and **k** along the z-axis. These are sketched in Fig. 11. If $\mathbf{v} = (x, y, z)$ is any vector in $\mathbb{R}^3$, then

$$\mathbf{v} = (x, y, z) = (x, 0, 0) + (0, y, 0) + (0, 0, z) = x\mathbf{i} + y\mathbf{j} + z\mathbf{k}. \qquad \textbf{(9)}$$

That is, *any vector* **v** *in* $\mathbb{R}^3$ *can be written in a unique way in terms of the vectors* **i**, **j**, *and* **k**.

Let $P = (a_1, b_1, c_1)$ and $Q = (a_2, b_2, c_2)$. Then the representation $\overrightarrow{PQ}$ of **v** shows that

$$\mathbf{v} = (a_2 - a_1)\mathbf{i} + (b_2 - b_1)\mathbf{j} + (c_2 - c_1)\mathbf{k} \qquad \textbf{(10)}$$

EXAMPLE 8

Find a vector in space that can be represented by the directed line segment from $(2, -1, 4)$ to $(5, 1, -3)$.

SOLUTION

$\mathbf{v} = (5 - 2)\mathbf{i} + (1 - (-1))\mathbf{j} + (-3 - 4)\mathbf{k} = 3\mathbf{i} + 2\mathbf{j} - 7\mathbf{k}.$

EXAMPLE 9

Find a unit vector having the same direction as $\mathbf{v} = (2, 4, -3)$.

SOLUTION

Since $|\mathbf{v}| = \sqrt{2^2 + 4^2 + (-3)^2} = \sqrt{29}$, we have

$$\mathbf{u} = \mathbf{v}/|\mathbf{v}| = (2/\sqrt{29}, 4/\sqrt{29}, -3/\sqrt{29}).$$

Problems 6.1

In Problems 1–12 find the magnitude and direction of each vector in $\mathbb{R}^2$.

1. $\mathbf{v} = (4, 4)$
2. $\mathbf{v} = (-4, 4)$
3. $\mathbf{v} = (4, -4)$
4. $\mathbf{v} = (-4, -4)$
5. $\mathbf{v} = (\sqrt{3}, 1)$
6. $\mathbf{v} = (1, \sqrt{3})$
7. $\mathbf{v} = (-1, \sqrt{3})$
8. $\mathbf{v} = (1, -\sqrt{3})$
9. $\mathbf{v} = -\mathbf{i} - \sqrt{3}\mathbf{j}$
10. $\mathbf{v} = \mathbf{i} + 2\mathbf{j}$
11. $\mathbf{v} = -5\mathbf{i} + 8\mathbf{j}$
12. $\mathbf{v} = 11\mathbf{i} - 14\mathbf{j}$

In Problems 13–24 find the magnitude and direction cosines of each vector in $\mathbb{R}^3$.

13. $\mathbf{v} = (0, 3, 0)$
14. $\mathbf{v} = (-3, 0, 0)$
15. $\mathbf{v} = (0, 0, 14)$
16. $\mathbf{v} = (-2, 3, 0)$
17. $\mathbf{v} = (1, 0, 2)$
18. $\mathbf{v} = (1, 1, 1)$
19. $\mathbf{v} = \mathbf{i} - \mathbf{j} + \mathbf{k}$
20. $\mathbf{v} = -\mathbf{i} + \mathbf{j} - \mathbf{k}$
21. $\mathbf{v} = -\mathbf{i} - \mathbf{j} - \mathbf{k}$
22. $\mathbf{v} = 2\mathbf{i} + 5\mathbf{j} - 7\mathbf{k}$
23. $\mathbf{v} = -3\mathbf{i} + 3\mathbf{j} + 8\mathbf{k}$
24. $\mathbf{v} = -2\mathbf{i} - 3\mathbf{j} - 4\mathbf{k}$

In Problems 25–30 write the vector $\mathbf{v}$ that is represented by $\overrightarrow{PQ}$ in the form $a\mathbf{i} + b\mathbf{j}$ or $a\mathbf{i} + b\mathbf{j} + c\mathbf{k}$.

25. $P = (1, 2); \quad Q = (1, 3)$
26. $P = (2, 4); \quad Q = (-7, 4)$
27. $P = (5, 2); \quad Q = (-1, 3)$
28. $P = (8, -2); \quad Q = (-3, -3)$
29. $P = (2, -1, 4); \quad Q = (3, 0, 8)$
30. $P = (-1, 3, 6); \quad Q = (-3, -4, -7)$
31. Let $\mathbf{u} = (2, 3)$ and $\mathbf{v} = (-5, 4)$. Find
    (a) $3\mathbf{u}$
    (b) $\mathbf{u} + \mathbf{v}$
    (c) $\mathbf{v} - \mathbf{u}$
    (d) $2\mathbf{u} - 7\mathbf{v}$

In Problems 32–37 find a unit vector having the same direction as the given vector.

32. $\mathbf{v} = 2\mathbf{i} + 3\mathbf{j}$
33. $\mathbf{v} = \mathbf{i} - \mathbf{j}$
34. $\mathbf{v} = 3\mathbf{i} + 4\mathbf{j}$
35. $\mathbf{v} = \mathbf{i} + \mathbf{j} + \mathbf{k}$
36. $\mathbf{v} = 2\mathbf{i} - \mathbf{j} + 5\mathbf{k}$
37. $\mathbf{v} = 4\mathbf{i} - 9\mathbf{k}$

38. If $\mathbf{v} = a\mathbf{i} + b\mathbf{j} \neq \mathbf{0}$, show that $a/\sqrt{a^2 + b^2} = \cos\theta$ and $b/\sqrt{a^2 + b^2} = \sin\theta$ where θ is the direction of $\mathbf{v}$.

In Problems 39 and 40 find the resultant of the forces acting on an object. Then find the force that must be applied so that the object will remain at rest.

39. 5 N (from direction $\pi/3$), 5 N (from direction $2\pi/3$)
40. 7 N (from direction $\pi/60$), 7 N (from direction $\pi/3$), 14 N (from direction $5\pi/4$)

A vector $\mathbf{v}$ in $\mathbb{R}^2$ has a direction **opposite** to that of a vector $\mathbf{u}$ if $|\text{direction } \mathbf{v} - \text{direction } \mathbf{u}| = \pi$. In Problems 41–43 find a unit vector $\mathbf{v}$ which has a direction opposite the direction of the given vector $\mathbf{u}$.

41. $\mathbf{u} = \mathbf{i} + \mathbf{j}$
42. $\mathbf{u} = 2\mathbf{i} - 3\mathbf{j}$
43. $\mathbf{u} = -3\mathbf{i} + 4\mathbf{j}$

44. Show that there is no vector with direction angles $\pi/6$, $\pi/3$, and $\pi/4$.

45. Let $P = (2, 1, 4)$ and $Q = (3, -2, 8)$. Find a unit vector in the direction of $\overrightarrow{PQ}$.

In Problems 46–48 find a vector $\mathbf{v}$ in $\mathbb{R}^2$ having the given magnitude and direction.

46. $|\mathbf{v}| = 3; \quad \theta = \pi/6$
47. $|\mathbf{v}| = 8; \quad \theta = \pi/3$
48. $|\mathbf{v}| = 7; \quad \theta = \pi$

49. **Triangle inequality.** Show algebraically (i.e., strictly from the definitions of vector addition and magnitude) that for any two vectors $\mathbf{u}$ and $\mathbf{v}$ in $\mathbb{R}^2$, $|\mathbf{u} + \mathbf{v}| \leq |\mathbf{u}| + |\mathbf{v}|$. Explain geometrically why this result must be true.

50. Show that if neither $\mathbf{u}$ nor $\mathbf{v}$ is the zero vector, then $|\mathbf{u} + \mathbf{v}| = |\mathbf{u}| + |\mathbf{v}|$ if and only if $\mathbf{u}$ is a positive scalar multiple of $\mathbf{v}$.

51. Prove Theorem 1. [*Hint:* Write everything out in terms of the components of the vectors $\mathbf{u}$, $\mathbf{v}$, and $\mathbf{w}$.]

6.2 The Dot Product of Two Vectors and Projections

In Section 6.1 we showed how a vector can be multiplied by a scalar but not how two vectors can be multiplied. Actually, there are several ways to define the product of two vectors, and in this section we shall discuss one of them. We shall discuss other product operations in Section 6.3.

DEFINITION

Dot Product in $\mathbb{R}^2$

Let $\mathbf{u} = (a_1, b_1) = a_1\mathbf{i} + b_1\mathbf{j}$ and $\mathbf{v} = (a_2, b_2) = a_2\mathbf{i} + b_2\mathbf{j}$. Then the **dot product** of $\mathbf{u}$ and $\mathbf{v}$, denoted $\mathbf{u} \cdot \mathbf{v}$, is defined by

$$\mathbf{u} \cdot \mathbf{v} = a_1 a_2 + b_1 b_2. \tag{1}$$

Remark. The dot product of two vectors is a *scalar*. For this reason the dot product is often called the **scalar product**. It is also called the **inner product.**

EXAMPLE 1

If $\mathbf{u} = (1, 3)$, and $\mathbf{v} = (4, -7)$, then

$$\mathbf{u} \cdot \mathbf{v} = 1(4) + 3(-7) = 4 - 21 = -17.$$

DEFINITION

Dot Product in $\mathbb{R}^3$

If $\mathbf{u} = x_1\mathbf{i} + y_1\mathbf{j} + z_1\mathbf{k}$ and $\mathbf{v} = x_2\mathbf{j} + y_2\mathbf{j} + z_2\mathbf{k}$, then we define the **dot product** (or **scalar product** or **inner product**) by

$$\mathbf{u} \cdot \mathbf{v} = x_1 x_2 + y_1 y_2 + z_1 z_2. \tag{2}$$

As before, the dot product of two vectors is a *scalar*. Note that $\mathbf{i} \cdot \mathbf{i} = 1$, $\mathbf{j} \cdot \mathbf{j} = 1$, $\mathbf{k} \cdot \mathbf{k} = 1$, $\mathbf{i} \cdot \mathbf{j} = 0$, $\mathbf{j} \cdot \mathbf{k} = 0$, and $\mathbf{i} \cdot \mathbf{k} = 0$.

EXAMPLE 2

If $\mathbf{u} = 2\mathbf{i} - 3\mathbf{j} - 4\mathbf{k}$ and $\mathbf{v} = -3\mathbf{i} + \mathbf{j} - 2\mathbf{k}$, calculate $\mathbf{u} \cdot \mathbf{v}$.

SOLUTION
$\mathbf{u} \cdot \mathbf{v} = 2(-3) + (-3)(1) + (-4)(-2) = -1.$

■ THEOREM 1

For any vectors $\mathbf{u}, \mathbf{v}, \mathbf{w}$ and scalar α, we have

(i) $\mathbf{u} \cdot \mathbf{v} = \mathbf{v} \cdot \mathbf{u}$.

(ii) $(\mathbf{u} + \mathbf{v}) \cdot \mathbf{w} = \mathbf{u} \cdot \mathbf{w} + \mathbf{v} \cdot \mathbf{w}$.

(iii) $(\alpha\mathbf{u}) \cdot \mathbf{v} = \alpha(\mathbf{u} \cdot \mathbf{v})$.

(iv) $|\mathbf{u}| = \sqrt{\mathbf{u} \cdot \mathbf{u}}$.   ■

Angle Between Two Vectors

Let **u** and **v** be two nonzero vectors. Then the **angle** ϕ between **u** and **v** is defined as the smallest nonnegative angle between the representations of **u** and **v** that have the origin as their initial points. If $\mathbf{u} = \alpha\mathbf{v}$ for some scalar α, then we define $\phi = 0$ if $\alpha > 0$ and $\phi = \pi$ if $\alpha < 0$.

■ **THEOREM 2**

Let **u** and **v** be two nonzero vectors. If ϕ is the angle between them, then

$$\cos\phi = \frac{\mathbf{u} \cdot \mathbf{v}}{|\mathbf{u}||\mathbf{v}|}. \tag{3}$$

■

EXAMPLE 3

Find the cosine of the angle between the vectors $\mathbf{u} = 2\mathbf{i} + 3\mathbf{j}$ and $\mathbf{v} = -7\mathbf{i} + \mathbf{j}$.

SOLUTION

$\mathbf{u} \cdot \mathbf{v} = -14 + 3 = -11$, $|\mathbf{u}| = \sqrt{2^2 + 3^2} = \sqrt{13}$, and $|\mathbf{v}| = \sqrt{(-7)^2 + 1^2} = \sqrt{50}$, so that

$$\cos\phi = \frac{\mathbf{u} \cdot \mathbf{v}}{|\mathbf{u}||\mathbf{v}|} = \frac{-11}{\sqrt{13}\sqrt{50}} = \frac{-11}{\sqrt{650}} \approx -0.4315.$$

DEFINITION

Two nonzero vectors **u** and **v** are

Parallel and Orthogonal Vectors

(i) **parallel** if the angle between them is 0 or π, and
(ii) **orthogonal** (or **perpendicular**) if the angle between them is $\pi/2$.

■ **THEOREM 3**

Let **u** and **v** be two vectors in $\mathbb{R}^2$ or $\mathbb{R}^3$,

(i) **u** and **v** are parallel if and only if $\mathbf{v} = \alpha\mathbf{u}$ for some scalar α.
(ii) **u** and **v** are orthogonal if and only if $\mathbf{u} \cdot \mathbf{v} = 0$.　　■

EXAMPLE 4

Show that the vectors $\mathbf{u} = \mathbf{i} + 3\mathbf{j} - 4\mathbf{k}$ and $\mathbf{v} = -2\mathbf{i} - 6\mathbf{j} + 8\mathbf{k}$ are parallel.

SOLUTION

Here $(\mathbf{u} \cdot \mathbf{v})/(|\mathbf{u}||\mathbf{v}|) = -52/(\sqrt{26}\sqrt{104}) = -52/(\sqrt{26} \cdot 2\sqrt{26}) = -1$ so that $\cos\theta = -1$, $\theta = \pi$, and **u** and **v** are parallel (but have opposite directions). An easier way to see this is to note that $\mathbf{v} = -2\mathbf{u}$ so that, by Theorem 3, **u** and **v** are parallel.

Projection

A useful concept arising in the study of vectors is that of **projection** of one vector on another. Before we can discuss projections, we need to prove a theorem.

■ **THEOREM 4**

Let **v** be a nonzero vector. Then for any other vector **u**, the vector

$$w = u - [(u \cdot v)/|v|^2]v \text{ is orthogonal to } v.$$

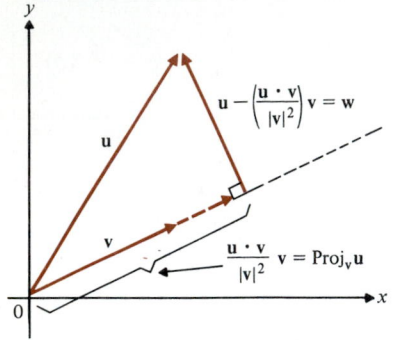

FIGURE 1

Projection of a Vector

PROOF

$$w \cdot v = \left(u - \frac{(u \cdot v)v}{|v|^2} \right) \cdot v = u \cdot v - \frac{(u \cdot v)(v \cdot v)}{|v|^2}$$

$$= u \cdot v - \frac{(u \cdot v)|v|^2}{|v|^2} = u \cdot v - u \cdot v = 0.$$

The vectors **u**, **v**, and **w** in $\mathbb{R}^2$ are shown in Fig. 1. ■

DEFINITION

Let **u** and **v** be nonzero vectors. Then the **projection of u onto v** is a vector, denoted Proj$_v$ **u**, which is defined by

$$\text{Proj}_v\, u = \frac{u \cdot v}{|v|^2} v = \left(\frac{u \cdot v}{|v|} \right) \frac{v}{|v|}. \tag{4}$$

$$\text{The \textbf{component} of } u \text{ in the direction } v \text{ is } (u \cdot v)/|v|. \tag{5}$$

Note that **v**/|**v**| is a unit vector in the direction of **v**.

Remark 1. From Fig. 1 and the fact that $\cos \phi = (u \cdot v)/(|u||v|)$, we find that

v and Proj$_v$ **u** have

    (i)   the same direction if $u \cdot v > 0$, and
  (ii) opposite directions if $u \cdot v < 0$.

Remark 2. If **u** and **v** are orthogonal, then $u \cdot v = 0$ so that Proj$_v$ **u** = **0**.

Remark 3. An alternative definition of projection is: if **u** and **v** are nonzero vectors, then Proj$_v$ **u** is the unique vector having the properties

---

(i) Proj$_v$ **u** is parallel to **v**, and
(ii) **u** − Proj$_v$ **u** is orthogonal to **v**.

EXAMPLE 5

Let **u** $= 2\mathbf{i} + 3\mathbf{j}$ and **v** $= \mathbf{i} + \mathbf{j}$. Calculate Proj$_v$ **u**.

SOLUTION
Proj$_v$ **u** $= (\mathbf{u}\cdot\mathbf{v})\mathbf{v}/|\mathbf{v}|^2 = [5/(\sqrt{2})^2]\mathbf{v} = (5/2)\mathbf{i} + (5/2)\mathbf{j}$.

EXAMPLE 6

Let **u** $= 2\mathbf{i} + 3\mathbf{j} + \mathbf{k}$ and **v** $= \mathbf{i} + 2\mathbf{j} - 6\mathbf{k}$. Find Proj$_v$ **u**.

SOLUTION
Here $(\mathbf{u}\cdot\mathbf{v})/|\mathbf{v}|^2 = \frac{2}{41}$ so that

$$\text{Proj}_v\ \mathbf{u} = \tfrac{2}{41}\mathbf{i} + \tfrac{4}{41}\mathbf{j} - \tfrac{12}{41}\mathbf{k}.$$

The component of **u** in the direction **v** is $(\mathbf{u}\cdot\mathbf{v})/|\mathbf{v}| = 2/\sqrt{41}$.

EXAMPLE 7: Work Done by a Force F

Assume a constant force **F** acts on an object, giving it a displacement **d**. Then the **work** W done by the force **F** in displacing the object is defined as the product of the distance moved times the length of the component of **F** in the direction of motion:

$$W = |\text{component of } \mathbf{F} \text{ along } \mathbf{d}| \times |\mathbf{d}|.$$

If we consider Fig. 2, we see that the component of **F** along **d** has length $|\mathbf{F}|\cos\phi$. Hence, by Theorem 2,

$$W = |\mathbf{F}||\mathbf{d}|\cos\phi = \mathbf{F}\cdot\mathbf{d}. \tag{6}$$

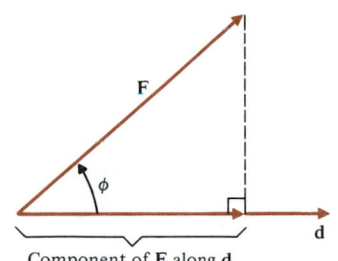

Component of **F** along **d**

FIGURE 2

EXAMPLE 8

A force of magnitude 4 N has the direction $\pi/3$. What is the work done in moving an object from the point $(1, 2)$ to the point $(5, 4)$ where distances are measured in meters?

SOLUTION
A unit vector with direction $\pi/3$ is given by **u** $= (\cos \pi/3)\mathbf{i} + (\sin \pi/3)\mathbf{j} = (1/2)\mathbf{i} + (\sqrt{3}/2)\mathbf{j}$. Thus **F** $= 4\mathbf{u} = 2\mathbf{i} + 2\sqrt{3}\mathbf{j}$. The displacement vector **d** is given by $(5 - 1)\mathbf{i} + (4 - 2)\mathbf{j} = 4\mathbf{i} + 2\mathbf{j}$. Thus

$$W = \mathbf{F}\cdot\mathbf{d} = (2\mathbf{i} + 2\sqrt{3}\mathbf{j})\cdot(4\mathbf{i} + 2\mathbf{j}) = (8 + 4\sqrt{3}) \approx 14.93 \text{ N-m}.$$

Problems 6.2

In Problems 1–10 calculate both the dot product of the two vectors and the cosine of the angle between them.

1. $\mathbf{u} = \mathbf{i} + \mathbf{j};\quad \mathbf{v} = \mathbf{i} - \mathbf{j}$
2. $\mathbf{u} = 3\mathbf{i};\quad \mathbf{v} = -7\mathbf{j}$
3. $\mathbf{u} = -5\mathbf{i};\quad \mathbf{v} = 18\mathbf{j}$
4. $\mathbf{u} = \alpha\mathbf{i};\quad \mathbf{v} = \beta\mathbf{j};\ \alpha,\ \beta$ real and nonzero
5. $\mathbf{u} = 2\mathbf{i} + 5\mathbf{j};\quad \mathbf{v} = 5\mathbf{i} + 2\mathbf{j}$
6. $\mathbf{u} = \mathbf{i} + \mathbf{j} + \mathbf{k};\quad \mathbf{v} = \mathbf{i} - \mathbf{k}$
7. $\mathbf{u} = \mathbf{i} - \mathbf{j} - \mathbf{k};\quad \mathbf{v} = \mathbf{i} + \mathbf{j} + 2\mathbf{k}$
8. $\mathbf{u} = 2\mathbf{i} + 3\mathbf{j};\quad \mathbf{v} = -3\mathbf{j} + 4\mathbf{k}$
9. $\mathbf{u} = 2\mathbf{i} - 5\mathbf{j} + 3\mathbf{k};\quad \mathbf{v} = -\mathbf{i} - 2\mathbf{j} + \mathbf{k}$
10. $\mathbf{u} = -3\mathbf{i} + 5\mathbf{j} - 7\mathbf{k};\quad \mathbf{v} = 4\mathbf{i} - 2\mathbf{j} + 6\mathbf{k}$

11. Show that for any real nonzero numbers α and β, the vectors $\mathbf{u} = \alpha\mathbf{i} + \beta\mathbf{j}$ and $\mathbf{v} = \beta\mathbf{i} - \alpha\mathbf{j}$ are orthogonal.
12. Let $\mathbf{u}, \mathbf{v}$, and $\mathbf{w}$ denote three arbitrary vectors. Explain why the product $\mathbf{u} \cdot \mathbf{v} \cdot \mathbf{w}$ *is not defined.*

In Problems 13–20 determine whether the given vectors are orthogonal, parallel, or neither. Then sketch each pair.

13. $\mathbf{u} = 3\mathbf{i} + 5\mathbf{j};\quad \mathbf{v} = -6\mathbf{i} - 10\mathbf{j}$
14. $\mathbf{u} = 2\mathbf{i} + 3\mathbf{j};\quad \mathbf{v} = 6\mathbf{i} - 4\mathbf{j}$
15. $\mathbf{u} = 2\mathbf{i} + 3\mathbf{j};\quad \mathbf{v} = 6\mathbf{i} + 4\mathbf{j}$
16. $\mathbf{u} = 2\mathbf{i} + 3\mathbf{j};\quad \mathbf{v} = -6\mathbf{i} + 4\mathbf{j}$
17. $\mathbf{u} = 7\mathbf{i};\quad \mathbf{v} = -23\mathbf{j}$
18. $\mathbf{u} = 2\mathbf{i} - 6\mathbf{j};\quad \mathbf{v} = -\mathbf{i} + 3\mathbf{j}$
19. $\mathbf{u} = \mathbf{i} + \mathbf{j};\quad \mathbf{v} = \alpha\mathbf{i} + \alpha\mathbf{j};\quad \alpha$ real
20. $\mathbf{u} = -2\mathbf{i} + 3\mathbf{j};\quad \mathbf{v} = -\mathbf{i} + 2\mathbf{j}$

21. Let $\mathbf{u} = 3\mathbf{i} + 4\mathbf{j}$ and $\mathbf{v} = \mathbf{i} + \alpha\mathbf{j}$. Determine α such that
    (a) $\mathbf{u}$ and $\mathbf{v}$ are orthogonal
    (b) $\mathbf{u}$ and $\mathbf{v}$ are parallel
    (c) the angle between $\mathbf{u}$ and $\mathbf{v}$ is $\pi/4$
    (d) the angle between $\mathbf{u}$ and $\mathbf{v}$ is $\pi/3$
22. Let $\mathbf{u} = -2\mathbf{i} + 5\mathbf{j}$ and $\mathbf{v} = \alpha\mathbf{i} - 2\mathbf{j}$. Determine α such that
    (a) $\mathbf{u}$ and $\mathbf{v}$ are orthogonal
    (b) $\mathbf{u}$ and $\mathbf{v}$ are parallel
    (c) the angle between $\mathbf{u}$ and $\mathbf{v}$ is $2\pi/3$
    (d) the angle between $\mathbf{u}$ and $\mathbf{v}$ is $\pi/3$
23. Let $P = (-3, 1, 7)$ and $Q = (8, 1, 7)$. Find a unit vector whose direction is opposite that of $\overrightarrow{PQ}$.

24. In Problem 23 find all points R for which $\overrightarrow{PR} \perp \overrightarrow{PQ}$.
* 25. Show that the set of points which satisfy the condition of Problem 24 and the condition $|\overrightarrow{PR}| = 1$ form a circle.

In Problems 26–38 let $\mathbf{u} = 2\mathbf{i} - 3\mathbf{j} + 4\mathbf{k}$, $\mathbf{v} = -2\mathbf{i} - 3\mathbf{j} + 5\mathbf{k}$, $\mathbf{w} = \mathbf{i} - 7\mathbf{j} + 3\mathbf{k}$, and $\mathbf{t} = 3\mathbf{i} + 4\mathbf{j} + 5\mathbf{k}$.

26. Calculate $\mathbf{u} + \mathbf{v}$.
27. Calculate $2\mathbf{u} - 3\mathbf{v}$.
28. Calculate $-18\mathbf{u}$.
29. Calculate $\mathbf{w} - \mathbf{u} - \mathbf{v}$.
30. Calculate $\mathbf{t} + 3\mathbf{w} - \mathbf{v}$.
31. Calculate $2\mathbf{u} - 7\mathbf{w} + 5\mathbf{v}$.
32. Calculate $2\mathbf{v} + 7\mathbf{t} - \mathbf{w}$.
33. Calculate $\mathbf{u} \cdot \mathbf{v}$.
34. Calculate $|\mathbf{w}|$.
35. Calculate $\mathbf{u} \cdot \mathbf{w} - \mathbf{w} \cdot \mathbf{t}$.
36. Calculate the angle between $\mathbf{u}$ and $\mathbf{w}$.
37. Calculate the angle between $\mathbf{t}$ and $\mathbf{w}$.
38. Calculate the angle between $\mathbf{v}$ and $\mathbf{t}$.

39. Show that the points $P = (3, 5, 6)$, $Q = (1, 2, 7)$, and $R = (6, 1, 0)$ are the vertices of a right triangle.
40. Show that the points $P = (3, 2, -1)$, $Q = (4, 1, 6)$, $R = (7, -2, 3)$, and $S = (8, -3, 10)$ are the vertices of a parallelogram.
41. Prove the **Cauchy-Schwarz inequality:** $|\mathbf{u} \cdot \mathbf{v}| \leq |\mathbf{u}||\mathbf{v}|$
42. Use the result of Problem 41 to prove the **triangle inequality:** $|\mathbf{u} + \mathbf{v}| \leq |\mathbf{u}| + |\mathbf{v}|$. When does equality hold?

In Problems 43–50 compute $\text{Proj}_\mathbf{v}\ \mathbf{u}$.

43. $\mathbf{u} = 3\mathbf{i};\quad \mathbf{v} = \mathbf{i} + \mathbf{j}$
44. $\mathbf{u} = -5\mathbf{j};\quad \mathbf{v} = \mathbf{i} + \mathbf{j}$
45. $\mathbf{u} = 2\mathbf{i} + \mathbf{j};\quad \mathbf{v} = \mathbf{i} - 2\mathbf{j}$
46. $\mathbf{u} = 2\mathbf{i} + 3\mathbf{j};\quad \mathbf{v} = 4\mathbf{i} + \mathbf{j}$
47. $\mathbf{u} = \mathbf{i} + \mathbf{j};\quad \mathbf{v} = 2\mathbf{i} - 3\mathbf{j}$
48. $\mathbf{u} = 2\mathbf{i} - \mathbf{j} - \mathbf{k};\quad \mathbf{v} = \mathbf{i} + \mathbf{j} + 3\mathbf{k}$
49. $\mathbf{u} = 4\mathbf{i} + 3\mathbf{j} + 2\mathbf{k};\quad \mathbf{v} = -7\mathbf{i} + 2\mathbf{j} + 5\mathbf{k}$
50. $\mathbf{u} = \mathbf{i} - \mathbf{k};\quad \mathbf{v} = \mathbf{j} + 4\mathbf{k}$

In Problems 51–55 find the work done when the force with given magnitude and direction moves an object from P to Q.

All distances are measured in meters. (Note that work can be negative.)

51. $|\mathbf{F}| = 3$ N; $\theta = 0$; $P = (2, 3)$; $Q = (1, 7)$

52. $|\mathbf{F}| = 2$ N; $\theta = \pi/2$; $P = (5, 7)$; $Q = (1, 1)$

53. $|\mathbf{F}| = 6$ N; $\theta = \pi/4$; $P = (2, 3)$; $Q = (-1, 4)$

54. $|\mathbf{F}| = 4$ N; $\theta = \pi/6$; $P = (-1, 2)$; $Q = (3, 4)$

55. $|\mathbf{F}| = 7$ N; $\theta = 2\pi/3$; $P = (4, -3)$; $Q = (1, 0)$

56. A force of 3 N acts in the direction of the vector with direction cosines $(1/\sqrt{6}, 1/\sqrt{3}, 1/\sqrt{2})$. Find the work done in moving the object from the point $(1, 2, 3)$ to the point $(2, 8, 11)$, where distance is measured in meters.

57. Find the work done when a force of 3 N acting in the direction of the vector $\mathbf{v} = \mathbf{i} + \mathbf{j} - \mathbf{k}$ moves an object from $(-1, 3, 4)$ to $(3, 7, -2)$, where distance is measured in meters.

In Problems 58–70 determine whether each pair of vectors is parallel, orthogonal, or neither.

58. $(2, -3)$; $(3, 2)$

59. $(2, -3)$; $(-3, 2)$

60. $(1, 2)$; $(2, 1)$

61. $(1, 2)$; $(2, 4)$

62. $(1, 2)$; $(-2, 4)$

63. $(1, 4, -7)$; $(2, 3, 2)$

64. $(1, -4, 7)$; $(2, 3, 2)$

65. $(1, 0, 1)$; $(3, 0, 3)$

66. $(1, 0, 1)$; $(-1, 0, -1)$

67. $(1, 0, 1)$; $(0, 0, 1)$

68. $(1, 0, 1)$; $(0, 1, 0)$

69. (a, b, c); $(-2a, -2b, -2c)$

70. $(a, b, 0)$; $(0, 0, c)$

71. Show that the vector $\mathbf{v} = a\mathbf{i} + b\mathbf{j}$ is orthogonal to the line $ax + by + c = 0$. [*Hint:* show that the line through $(0, 0)$ and (a, b) is orthogonal to the line $ax + by + c = 0$.]

72. Show that the vector $\mathbf{u} = b\mathbf{i} - a\mathbf{j}$ is parallel to the line $ax + by + c = 0$.

73. A triangle has vertices $(1, 3)$, $(4, -2)$, and $(-3, 6)$. Find the cosine of each of its angles.

74. A triangle has vertices (a_1, b_1), (a_2, b_2), and (a_3, b_3). Find a formula for the cosine of each of its angles.

75. Show that the diagonals of a parallelogram bisect each other.

76. Show that the diagonals of a rhombus are orthogonal.

77. Show that in a trapezoid, the line segment which joins the midpoints of the two sides that are not parallel is parallel to the parallel sides and has length equal to the average of the lengths of the parallel sides.

* 78. A solid polyhedron in space with exactly four vertices is called a **tetrahedron** (see Fig. 3). Let **P** represent the vector $\overrightarrow{OP}$, **Q** the vector $\overrightarrow{OQ}$, and so on. A line is drawn from each vertex to the centroid of the opposite side. Show that these four lines meet at the endpoint of the vector $\mathbf{v} = (\mathbf{P} + \mathbf{Q} + \mathbf{R} + \mathbf{S})/4$.

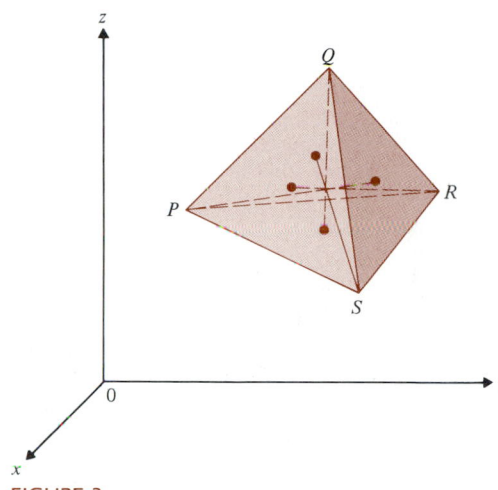

FIGURE 3

* 79. Let PQR be a triangle in $\mathbb{R}^3$. Show that if a force of N newtons moves an object around the triangle, the total work done by that force is zero.

* 80. Find the angle between the diagonal of a cube and the diagonal of one of its faces.

* 81. Let $ABCDEFG$ be a regular pyramid with the regular hexagon $ABCDEF$ as base and GH as altitude (see Fig. 4).

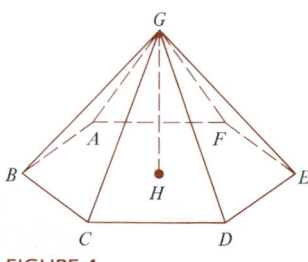

FIGURE 4

Define $\mathbf{u} = \overrightarrow{AB}$, $\mathbf{v} = \overrightarrow{AF}$, and $\mathbf{w} = \overrightarrow{HG}$. Then express in terms of $\mathbf{u}$, $\mathbf{v}$, and $\mathbf{w}$ the vectors $\overrightarrow{FE}$, $\overrightarrow{AE}$, and $\overrightarrow{AG}$.

6.3 The Cross Product of Two Vectors and Triple Products

To this point the only product of vectors which we have considered has been the dot or scalar product. We now define a new product, called the *cross product*† (or *vector product*), which is defined only in $\mathbb{R}^3$.

DEFINITION

Cross Product

Let $\mathbf{u} = a_1\mathbf{i} + b_1\mathbf{j} + c_1\mathbf{k}$ and $\mathbf{v} = a_2\mathbf{i} + b_2\mathbf{j} + c_2\mathbf{k}$. Then the **cross product (vector product)** of $\mathbf{u}$ and $\mathbf{v}$, denoted by $\mathbf{u} \times \mathbf{v}$, is a new vector defined by

$$\mathbf{u} \times \mathbf{v} = (b_1c_2 - c_1b_2)\mathbf{i} + (c_1a_2 - a_1c_2)\mathbf{j} + (a_1b_2 - b_1a_2)\mathbf{k}. \qquad (1)$$

Note that the result of the cross product is a vector, whereas the result of the dot product is a scalar.

Here the cross product seems to have been defined somewhat arbitrarily. There are many ways to define a vector product. Why was this definition chosen? We shall answer that question in this section by demonstrating some of the properties of the cross product and illustrating some of its uses.

EXAMPLE 1

Let $\mathbf{u} = \mathbf{i} - \mathbf{j} + 2\mathbf{k}$ and $\mathbf{v} = 2\mathbf{i} + 3\mathbf{j} - 4\mathbf{k}$. Calculate $\mathbf{w} = \mathbf{u} \times \mathbf{v}$.

SOLUTION

Using formula (1) we have

$$\mathbf{w} = [(-1)(-4) - (2)(3)]\mathbf{i} + [(2)(2) - (1)(-4)]\mathbf{j} + [(1)(3) - (-1)(2)]\mathbf{k}$$

$$= -2\mathbf{i} + 8\mathbf{j} + 5\mathbf{k}.$$

Note. In this example, $\mathbf{u} \cdot \mathbf{w} = \mathbf{v} \cdot \mathbf{w} = 0$. That is, $\mathbf{u} \times \mathbf{v}$ is orthogonal to both $\mathbf{u}$ and $\mathbf{v}$. As we shall see shortly, the cross product of $\mathbf{u}$ and $\mathbf{v}$ is always orthogonal to both $\mathbf{u}$ and $\mathbf{v}$.

Before continuing our discussion of the uses of the cross product, we remark that there is an easy way to calculate $\mathbf{u} \times \mathbf{v}$ if you are familiar with the elementary properties of 3×3 determinants. If you are not, we suggest that you either turn to Sections 7.7 and 7.8 or rely solely on the defining equation (1).

---

† The cross product was defined by Hamilton in one of a series of papers discussing his quaternions that were published in *Philosophical Magazine* between the years 1844 and 1850.

THEOREM 1

$$\mathbf{u} \times \mathbf{v} = \begin{vmatrix} \mathbf{i} & \mathbf{j} & \mathbf{k} \\ a_1 & b_1 & c_1 \\ a_2 & b_2 & c_2 \end{vmatrix}.$$

This expression is not a real determinant, but it does provide an easy way to remember the cross product.

EXAMPLE 2

Calculate $\mathbf{u} \times \mathbf{v}$ where $\mathbf{u} = 2\mathbf{i} + 4\mathbf{j} - 5\mathbf{k}$ and $\mathbf{v} = -3\mathbf{i} - 2\mathbf{j} + \mathbf{k}$.

SOLUTION

$$\mathbf{u} \times \mathbf{v} = \begin{vmatrix} \mathbf{i} & \mathbf{j} & \mathbf{k} \\ 2 & 4 & -5 \\ -3 & -2 & 1 \end{vmatrix} = (4-10)\mathbf{i} - (2-15)\mathbf{j} + (-4+12)\mathbf{k} = -6\mathbf{i} + 13\mathbf{j} + 8\mathbf{k}.$$

The following theorem summarizes some properties of the cross product.

THEOREM 2

Let $\mathbf{u}$, $\mathbf{v}$, and $\mathbf{w}$ be vectors in $\mathbb{R}^3$, and let α be a scalar. Then

Scalar Triple Product

(i) $\mathbf{u} \times \mathbf{0} = \mathbf{0} \times \mathbf{u} = \mathbf{0}$.

(ii) $\mathbf{u} \times \mathbf{v} = -(\mathbf{v} \times \mathbf{u})$. (The cross product is **anticommutative**.)

(iii) $(\alpha\mathbf{u} \times \mathbf{v}) = \alpha(\mathbf{u} \times \mathbf{v})$.

(iv) $\mathbf{u} \times (\mathbf{v} + \mathbf{w}) = (\mathbf{u} \times \mathbf{v}) + (\mathbf{u} \times \mathbf{w})$.

(v) $(\mathbf{u} \times \mathbf{v}) \cdot \mathbf{w} = \mathbf{u} \cdot (\mathbf{v} \times \mathbf{w}) = \begin{vmatrix} a_1 & b_1 & c_1 \\ a_2 & b_2 & c_2 \\ a_3 & b_3 & c_3 \end{vmatrix}$. (This is called the **scalar triple product** of $\mathbf{u}$, $\mathbf{v}$, and $\mathbf{w}$.)

(vi) $\mathbf{u} \cdot (\mathbf{u} \times \mathbf{v}) = \mathbf{v} \cdot (\mathbf{u} \times \mathbf{v}) = 0$. (That is, $\mathbf{u} \times \mathbf{v}$ is orthogonal to both $\mathbf{u}$ and $\mathbf{v}$.)

(vii) If $\mathbf{u}$ and $\mathbf{v}$ are parallel, then $\mathbf{u} \times \mathbf{v} = \mathbf{0}$.

PROOF

We prove (v), (vi), and (vii). The rest of the proof is left as an exercise (see Problems 56–59).

(v) Let $\mathbf{u} = a_1\mathbf{i} + b_1\mathbf{j} + c_1\mathbf{k}$, $\mathbf{v} = a_2\mathbf{i} + b_2\mathbf{j} + c_2\mathbf{k}$, and $\mathbf{w} = a_3\mathbf{i} + b_3\mathbf{j} + c_3\mathbf{k}$. Then

$$(\mathbf{u} \times \mathbf{v}) \cdot \mathbf{w} = [(b_1c_2 - c_1b_2)\mathbf{i} + (c_1a_2 - a_1c_2)\mathbf{j} + (a_1b_2 - b_1a_2)\mathbf{k}] \cdot (a_3\mathbf{i} + b_3\mathbf{j} + c_3\mathbf{k})$$

$$= b_1c_2a_3 - c_1b_2a_3 + c_1a_2b_3 - a_1c_2b_3 + a_1b_2c_3 - b_1a_2c_3.$$

$$\mathbf{u} \cdot (\mathbf{v} \times \mathbf{w}) = (a_1\mathbf{i} + b_1\mathbf{j} + c_1\mathbf{k}) \cdot [(b_2c_3 - c_2b_3)\mathbf{i} + (c_2a_3 - a_2c_3)\mathbf{j} + (a_2b_3 - b_2a_3)\mathbf{k}]$$

$$= a_1b_2c_3 - a_1c_2b_3 + b_1c_2a_3 - b_1a_2c_3 + c_1a_2b_3 - c_1b_2a_3$$

$$= (\mathbf{u} \times \mathbf{v}) \cdot \mathbf{w} = \begin{vmatrix} a_1 & b_1 & c_1 \\ a_2 & b_2 & c_2 \\ a_3 & b_3 & c_3 \end{vmatrix}.$$

(vi) We know that $\mathbf{u} \cdot (\mathbf{u} \times \mathbf{v}) = (\mathbf{u} \times \mathbf{v}) \cdot \mathbf{u}$ (since the dot product is commutative—see Theorem 6.2.1(i)). But, from parts (ii) and (v),

$$(\mathbf{u} \times \mathbf{v}) \cdot \mathbf{u} = \mathbf{u} \cdot (\mathbf{v} \times \mathbf{u}) = \mathbf{u} \cdot (-\mathbf{u} \times \mathbf{v}) = -\mathbf{u} \cdot (\mathbf{u} \times \mathbf{v}).$$

Thus $\mathbf{u} \cdot (\mathbf{u} \times \mathbf{v}) = -\mathbf{u} \cdot (\mathbf{u} \times \mathbf{v})$, which can occur only if $\mathbf{u} \cdot (\mathbf{u} \times \mathbf{v}) = 0$. A similar computation shows that $\mathbf{v} \cdot (\mathbf{u} \times \mathbf{v}) = 0$.

(vii) If $\mathbf{u}$ and $\mathbf{v}$ are parallel, then $\mathbf{v} = \alpha\mathbf{u}$ for some scalar α [Theorem 6.2.3(i)] so that

$$\mathbf{u} \times \mathbf{v} = \begin{vmatrix} \mathbf{i} & \mathbf{j} & \mathbf{k} \\ a_1 & b_1 & c_1 \\ \alpha a_1 & \alpha b_1 & \alpha c_1 \end{vmatrix} = 0$$

since a determinant with two proportional rows is zero (see Property 6 in Section 7.8, page 415.) ∎

EXAMPLE 3

Find two unit vectors orthogonal to $\mathbf{v}_1 = 3\mathbf{i} + 4\mathbf{j} - 2\mathbf{k}$ and $\mathbf{v}_2 = -3\mathbf{i} + 4\mathbf{j} + \mathbf{k}$.

SOLUTION
From Theorem 2(vi), two vectors orthogonal to $\mathbf{v}_1$ and $\mathbf{v}_2$ are $\mathbf{w}_1 = \mathbf{v}_1 \times \mathbf{v}_2$ and $\mathbf{w}_2 = \mathbf{v}_2 \times \mathbf{v}_1 = -\mathbf{v}_1 \times \mathbf{v}_2$. Then

$$\mathbf{w}_1 = \begin{vmatrix} \mathbf{i} & \mathbf{j} & \mathbf{k} \\ 3 & 4 & -2 \\ -3 & 4 & 1 \end{vmatrix} = 12\mathbf{i} + 3\mathbf{j} + 24\mathbf{k} \quad \text{and} \quad \mathbf{w}_2 = -12\mathbf{i} - 3\mathbf{j} - 24\mathbf{k}.$$

Since $|\mathbf{w}_1| = |\mathbf{w}_2| = \sqrt{12^2 + 3^2 + 24^2} = \sqrt{729} = 27$, the two required unit vectors are

$$\mathbf{u}_1 = \tfrac{12}{27}\mathbf{i} + \tfrac{3}{27}\mathbf{j} + \tfrac{24}{27}\mathbf{k} = \tfrac{4}{9}\mathbf{i} + \tfrac{1}{9}\mathbf{j} + \tfrac{8}{9}\mathbf{k} \quad \text{and} \quad \mathbf{u}_2 = -\tfrac{4}{9}\mathbf{i} - \tfrac{1}{9}\mathbf{j} - \tfrac{8}{9}\mathbf{k}.$$

■ **THEOREM 3**
If ϕ is the angle between $\mathbf{u}$ and $\mathbf{v}$, then

$$|\mathbf{u} \times \mathbf{v}| = |\mathbf{u}||\mathbf{v}| \sin \phi. \tag{2}$$

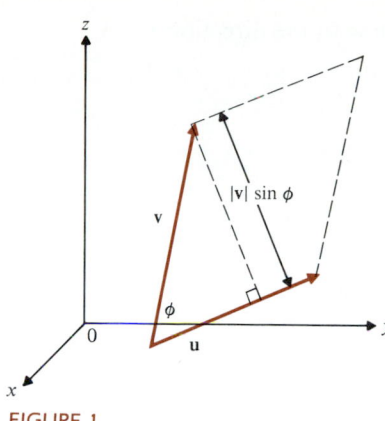

FIGURE 1

PROOF

It is straightforward to show (by comparing components) that

$$|\mathbf{u} \times \mathbf{v}|^2 = |\mathbf{u}|^2|\mathbf{v}|^2 - (\mathbf{u} \cdot \mathbf{v})^2 \tag{3}$$

(see Problem 62). Then since $(\mathbf{u} \cdot \mathbf{v})^2 = |\mathbf{u}|^2|\mathbf{v}|^2 \cos^2 \phi$ (from Theorem 6.2.2)

$$|\mathbf{u} \times \mathbf{v}|^2 = |\mathbf{u}|^2|\mathbf{v}|^2 - |\mathbf{u}|^2|\mathbf{v}|^2 \cos^2 \phi = |\mathbf{u}|^2|\mathbf{v}|^2(1 - \cos^2 \phi)$$

$$= |\mathbf{u}|^2|\mathbf{v}|^2 \sin^2 \phi$$

and the theorem follows after taking square roots of both sides. ∎

There is an interesting geometric interpretation of Theorem 3. The vectors **u** and **v** are sketched in Fig. 1 and can be thought of as two adjacent sides of a parallelogram. Then, from elementary geometry, we see that

$$\boxed{\text{area of the parallelogram} = |\mathbf{u}||\mathbf{v}| \sin \phi = |\mathbf{u} \times \mathbf{v}|. \tag{4}}$$

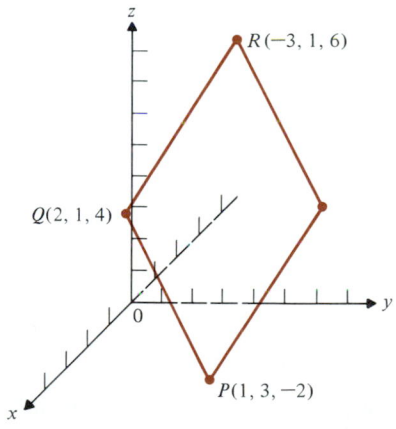

FIGURE 2

EXAMPLE 4

Find the area of the parallelogram with consecutive vertices at $P = (1, 3, -2)$, $Q = (2, 1, 4)$, and $R = (-3, 1, 6)$.

SOLUTION

The parallelogram is sketched in Fig. 2. We have

$$\text{area} = |\overrightarrow{PQ} \times \overrightarrow{QR}|$$

$$= |(\mathbf{i} - 2\mathbf{j} + 6\mathbf{k}) \times (-5\mathbf{i} + 2\mathbf{k})|$$

$$= \left| \begin{vmatrix} \mathbf{i} & \mathbf{j} & \mathbf{k} \\ 1 & -2 & 6 \\ -5 & 0 & 2 \end{vmatrix} \right| = |-4\mathbf{i} - 32\mathbf{j} - 10\mathbf{k}| = \sqrt{1140} \text{ square units.}$$

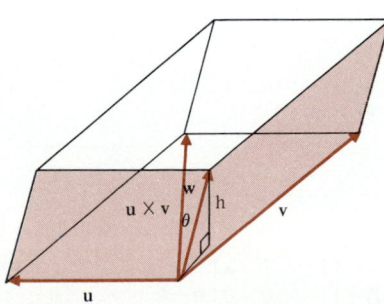

FIGURE 3

Geometric Interpretation of the Scalar Triple Product

Let **u**, **v**, and **w** be three vectors that are not in the same plane. Then they form the sides of a **parallelepiped** in space (see Fig. 3). Let us compute its volume. The base of the parallelepiped is a parallelogram. Its area, from (4), is equal to $|\mathbf{u} \times \mathbf{v}|$.

The vector $\mathbf{u} \times \mathbf{v}$ is orthogonal to both **u** and **v** and is therefore orthogonal to the parallelogram determined by **u** and **v**. The height of the parallelepiped, h, is measured along a vector orthogonal to the parallelogram. Therefore

h = absolute value of the component of **w** in the direction **u** × **v**

equation (6.2.5)
$$\downarrow$$
$$= \left| \frac{\mathbf{w} \cdot (\mathbf{u} \times \mathbf{v})}{|\mathbf{u} \times \mathbf{v}|} \right|.$$

Thus

$$\text{volume of parallelepiped} = \text{area of base} \times \text{height}$$

$$= |\mathbf{u} \times \mathbf{v}| \left[\frac{|\mathbf{w} \cdot (\mathbf{u} \times \mathbf{v})|}{|\mathbf{u} \times \mathbf{v}|} \right] = |\mathbf{w} \cdot (\mathbf{u} \times \mathbf{v})|$$

That is,

> The volume of the parallelepiped determined by the three vectors **u, v,** and **w** = $|(\mathbf{u} \times \mathbf{v}) \cdot \mathbf{w}|$ = absolute value of the scalar triple product of **u, v,** and **w**. **(5)**

We can derive an interesting and useful fact from (5). If **w** is in the plane of **u** and **v**, then **w** is perpendicular to **u** × **v**, which means that **w** · (**u** × **v**) = 0. Conversely, if (**u** × **v**) · **w** = 0, then **w** is perpendicular to (**u** × **v**), so **w** is in the plane determined by **u** and **v**. We conclude that

> three vectors **u, v,** and **w** are coplanar if and only if their scalar triple product is zero. **(6)**

We shall use this fact in the problem set in Section 6.4.

The Triple Cross Product and Other Identities

The vector **u** × (**v** × **w**) is called the **triple cross product** of **u, v,** and **w**. In the triple cross product, care must be taken because *the cross product is not associative.* That is,

> In general,
> $$\mathbf{u} \times (\mathbf{v} \times \mathbf{w}) \neq (\mathbf{u} \times \mathbf{v}) \times \mathbf{w}$$

EXAMPLE 5

$$\mathbf{i} \times (\mathbf{i} \times \mathbf{k}) = \mathbf{i} \times \begin{vmatrix} \mathbf{i} & \mathbf{j} & \mathbf{k} \\ 1 & 0 & 0 \\ 0 & 0 & 1 \end{vmatrix} = \mathbf{i} \times (-\mathbf{j}) = \begin{vmatrix} \mathbf{i} & \mathbf{j} & \mathbf{k} \\ 1 & 0 & 0 \\ 0 & -1 & 0 \end{vmatrix} = -\mathbf{k}$$

but

$$(\mathbf{i} \times \mathbf{i}) \times \mathbf{k} = \mathbf{0} \times \mathbf{k} = \mathbf{0}.$$

■ **THEOREM 4**

The following identities hold

$$\text{(i)} \quad \mathbf{u} \times (\mathbf{v} \times \mathbf{w}) = (\mathbf{u} \cdot \mathbf{w})\mathbf{v} - (\mathbf{u} \cdot \mathbf{v})\mathbf{w} \tag{7}$$

$$\text{(ii)} \quad (\mathbf{u} \times \mathbf{v}) \cdot (\mathbf{w} \times \mathbf{y}) = (\mathbf{u} \cdot \mathbf{w})(\mathbf{v} \cdot \mathbf{y}) - (\mathbf{u} \cdot \mathbf{y})(\mathbf{v} \cdot \mathbf{w}) \qquad \text{Lagrange's Identity} \tag{8}$$

$$\text{(iii)} \quad (\mathbf{u} \times \mathbf{v}) \times (\mathbf{w} \times \mathbf{y}) = [\mathbf{u} \cdot (\mathbf{v} \times \mathbf{y})]\mathbf{w} - [\mathbf{u} \cdot (\mathbf{v} \times \mathbf{w})]\mathbf{y} \tag{9}$$

PROOF

We prove (i) and leave the other two parts as an exercise (see Problems 60 and 61). Let $\mathbf{u} = u_1\mathbf{i} + u_2\mathbf{j} + u_3\mathbf{k}$, $\mathbf{v} = v_1\mathbf{i} + v_2\mathbf{j} + v_3\mathbf{k}$, and $\mathbf{w} = w_1\mathbf{i} + w_2\mathbf{j} + w_3\mathbf{k}$. Then

$$\mathbf{u} \times (\mathbf{v} \times \mathbf{w}) = \begin{vmatrix} \mathbf{i} & \mathbf{j} & \mathbf{k} \\ u_1 & u_2 & u_3 \\ v_2 w_3 - v_3 w_2 & v_3 w_1 - v_1 w_3 & v_1 w_2 - v_2 w_1 \end{vmatrix}$$

$$= (u_2 v_1 w_2 - u_2 v_2 w_1 - u_3 v_3 w_1 + u_3 v_1 w_3)\mathbf{i}$$

$$+ (u_3 v_2 w_3 - u_3 v_3 w_2 - u_1 v_1 w_2 + u_1 v_2 w_1)\mathbf{j}$$

$$+ (u_1 v_3 w_1 - u_1 v_1 w_3 - u_2 v_2 w_3 + u_2 v_3 w_2)\mathbf{k}$$

$$(\mathbf{u} \cdot \mathbf{w})\mathbf{v} = (u_1 w_1 + u_2 w_2 + u_3 w_3)(v_1\mathbf{i} + v_2\mathbf{j} + v_3\mathbf{k})$$

$$= (u_1 v_1 w_1 + u_2 v_1 w_2 + u_3 v_1 w_3)\mathbf{i} + (u_1 v_2 w_1 + u_2 v_2 w_2 + u_3 v_2 w_3)\mathbf{j}$$

$$+ (u_1 v_3 w_1 + u_2 v_3 w_2 + u_3 v_3 w_3)\mathbf{k}$$

$$(\mathbf{u} \cdot \mathbf{v})\mathbf{w} = (u_1 v_1 + u_2 v_2 + u_3 v_3)(w_1\mathbf{i} + w_2\mathbf{j} + w_3\mathbf{k})$$

$$= (u_1 v_1 w_1 + u_2 v_2 w_1 + u_3 v_3 w_1)\mathbf{i} + (u_1 v_1 w_2 + u_2 v_2 w_2 + u_3 v_3 w_2)\mathbf{j}$$

$$+ (u_1 v_1 w_3 + u_2 v_2 w_3 + u_3 v_3 w_3)\mathbf{k}$$

When we subtract and compare components we find that

$$\mathbf{u} \times (\mathbf{v} \times \mathbf{w}) = (\mathbf{u} \cdot \mathbf{w})\mathbf{v} - (\mathbf{u} \cdot \mathbf{v})\mathbf{w}. \qquad ■$$

Problems 6.3

In Problems 1–14 find the cross product $\mathbf{u} \times \mathbf{v}$.

1. $\mathbf{u} = \mathbf{i} - 2\mathbf{j}; \quad \mathbf{v} = 3\mathbf{k}$
2. $\mathbf{u} = 3\mathbf{i} - 7\mathbf{j}; \quad \mathbf{v} = \mathbf{i} + \mathbf{k}$
3. $\mathbf{u} = \mathbf{i} - \mathbf{j}; \quad \mathbf{v} = \mathbf{j} + \mathbf{k}$
4. $\mathbf{u} = -7\mathbf{k}; \quad \mathbf{v} = \mathbf{j} + 2\mathbf{k}$
5. $\mathbf{u} = -2\mathbf{i} + 3\mathbf{j}; \quad \mathbf{v} = 7\mathbf{i} + 4\mathbf{k}$

6. $\mathbf{u} = a\mathbf{i} + b\mathbf{j}; \quad \mathbf{v} = c\mathbf{i} + d\mathbf{j}$
7. $\mathbf{u} = a\mathbf{i} + b\mathbf{k}; \quad \mathbf{v} = c\mathbf{i} + d\mathbf{k}$
8. $\mathbf{u} = a\mathbf{j} + b\mathbf{k}; \quad \mathbf{v} = c\mathbf{i} + d\mathbf{k}$
9. $\mathbf{u} = 2\mathbf{i} - 3\mathbf{j} + \mathbf{k}; \quad \mathbf{v} = \mathbf{i} + 2\mathbf{j} + \mathbf{k}$
10. $\mathbf{u} = 3\mathbf{i} - 4\mathbf{j} + 2\mathbf{k}; \quad \mathbf{v} = 6\mathbf{i} - 3\mathbf{j} + 5\mathbf{k}$
11. $\mathbf{u} = -3\mathbf{i} - 2\mathbf{j} + \mathbf{k}; \quad \mathbf{v} = 6\mathbf{i} + 4\mathbf{j} - 2\mathbf{k}$

12. $\mathbf{u} = \mathbf{i} + 7\mathbf{j} - 3\mathbf{k};$ $\mathbf{v} = -\mathbf{i} - 7\mathbf{j} + 3\mathbf{k}$

13. $\mathbf{u} = \mathbf{i} - 7\mathbf{j} - 3\mathbf{k};$ $\mathbf{v} = -\mathbf{i} + 7\mathbf{j} - 3\mathbf{k}$

14. $\mathbf{u} = 2\mathbf{i} - 3\mathbf{j} + 5\mathbf{k};$ $\mathbf{v} = 3\mathbf{i} - \mathbf{j} - \mathbf{k}$

15. Find two unit vectors orthogonal to both $\mathbf{u} = 2\mathbf{i} - 3\mathbf{j}$ and $\mathbf{v} = 4\mathbf{j} + 3\mathbf{k}$.

16. Find two unit vectors orthogonal to both $\mathbf{u} = \mathbf{i} + \mathbf{j} + \mathbf{k}$ and $\mathbf{v} = \mathbf{i} - \mathbf{j} - \mathbf{k}$.

17. Use the cross product to find the sine of the angle ϕ between the vectors $\mathbf{u} = 2\mathbf{i} + \mathbf{j} - \mathbf{k}$ and $\mathbf{v} = -3\mathbf{i} - 2\mathbf{j} + 4\mathbf{k}$.

18. Use the dot product to calculate the cosine of the angle between the vectors of Problem 17. Then show that for the values you have calculated, $\sin^2 \phi + \cos^2 \phi = 1$.

In Problems 19–24 find the area of the parallelogram with the given adjacent vertices (see Fig. 2).

19. $(1, -2, 3); (2, 0, 1); (0, 4, 0)$

20. $(-2, 1, 1); (2, 2, 3); (-1, -2, 4)$

21. $(-2, 1, 0); (1, 4, 2); (-3, 1, 5)$

22. $(7, -2, -3); (-4, 1, 6); (5, -2, 3)$

23. $(a, 0, 0); (0, b, 0); (0, 0, c)$

24. $(a, b, 0); (a, 0, b); (0, a, b)$

25. Show that the area of the triangle PQR is given by $A = \frac{1}{2}|\overrightarrow{PQ} \times \overrightarrow{QR}|$.

26. Use the result of Problem 25 to calculate the area of the triangle with vertices at $(2, 1, -4)$, $(1, 7, 2)$, and $(3, -2, 3)$.

27. Calculate the area of the triangle with vertices at $(3, 1, 7)$, $(2, -3, 4)$, and $(7, -2, 4)$.

28. Calculate the area of the triangle with vertices at $(1, 0, 0)$, $(0, 1, 0)$, and $(0, 0, 1)$. Sketch this triangle.

In Problems 29–33 determine whether or not $\mathbf{u}$, $\mathbf{v}$, and $\mathbf{w}$ are coplanar.

29. $\mathbf{u} = \mathbf{i} + \mathbf{j},$ $\mathbf{v} = \mathbf{i} + \mathbf{k},$ $\mathbf{w} = 2\mathbf{i} + \mathbf{j} + \mathbf{k}$

30. $\mathbf{u} = \mathbf{i} - \mathbf{k},$ $\mathbf{v} = \mathbf{j} + \mathbf{k},$ $\mathbf{w} = \mathbf{i} + \mathbf{j}$

31. $\mathbf{u} = \mathbf{i} + \mathbf{j},$ $\mathbf{v} = \mathbf{j} + \mathbf{k},$ $\mathbf{w} = \mathbf{i} + \mathbf{k}$

32. $\mathbf{u} = 2\mathbf{i} - 3\mathbf{j} + 4\mathbf{k},$ $\mathbf{v} = -3\mathbf{i} + 5\mathbf{j} - 6\mathbf{k},$
    $\mathbf{w} = 7\mathbf{i} - 11\mathbf{j} + 14\mathbf{k}$

33. $\mathbf{u} = 3\mathbf{i} - \mathbf{j} + \mathbf{k},$ $\mathbf{v} = 2\mathbf{i} - 6\mathbf{j} - 2\mathbf{k},$ $\mathbf{w} = 5\mathbf{i} + 3\mathbf{j} - 4\mathbf{k}$

In Problems 34–41 compute the scalar triple product of the three vectors in the order given.

34. $\mathbf{i}, \mathbf{j}, \mathbf{k}$

35. $\mathbf{j}, \mathbf{i}, \mathbf{k}$

36. $\mathbf{i} + \mathbf{j}, \mathbf{i} - \mathbf{k}, \mathbf{j} + \mathbf{k}$

37. $\mathbf{i} - 2\mathbf{j}, 2\mathbf{j} + 5\mathbf{k}, -3\mathbf{i} + \mathbf{k}$

38. $\mathbf{i} + \mathbf{j}, 0, \mathbf{i} + 2\mathbf{j} + 3\mathbf{k}$

39. $2\mathbf{i} - \mathbf{j} + \mathbf{k}, \mathbf{k}, -\mathbf{i} + 3\mathbf{j} + 4\mathbf{k}$

40. $-\mathbf{i} - \mathbf{j} - \mathbf{k}, -\mathbf{j} - \mathbf{k}, -\mathbf{j}$

41. $2\mathbf{i} + 4\mathbf{j} - \mathbf{k}, 5\mathbf{i} - 7\mathbf{j} + 3\mathbf{k}, -4\mathbf{i} + 2\mathbf{j} - 3\mathbf{k}$

In Problems 42–48 compute the triple cross product in the order (first vector) × (second vector × third vector).

42. $\mathbf{i}, \mathbf{j}, \mathbf{j} + \mathbf{k}$

43. $\mathbf{i} + \mathbf{j}, \mathbf{j} + \mathbf{k}, \mathbf{i} + \mathbf{k}$

44. $\mathbf{i} - \mathbf{k}, \mathbf{k} - \mathbf{j}, \mathbf{i} + \mathbf{j}$

45. $2\mathbf{i} - \mathbf{j}, 3\mathbf{j} - 4\mathbf{k}, -\mathbf{i} + 2\mathbf{k}$

46. $\mathbf{i}, \mathbf{i} + \mathbf{j}, \mathbf{i} + \mathbf{j} + \mathbf{k}$

47. $\mathbf{i} - 2\mathbf{j} + 3\mathbf{k}, 4\mathbf{i} + \mathbf{j} - 3\mathbf{k}, -3\mathbf{i} - 2\mathbf{j} + \mathbf{k}$

48. $3\mathbf{i} - 2\mathbf{j} + 5\mathbf{k}, -4\mathbf{i} + 7\mathbf{k}, 2\mathbf{i} - 4\mathbf{j} - 3\mathbf{k}$

49. Calculate the volume of the parallelepiped determined by the vectors $\mathbf{u} = 2\mathbf{i} - \mathbf{j} + \mathbf{k}, \mathbf{v} = 3\mathbf{i} + 2\mathbf{j} - 2\mathbf{k}, \mathbf{w} = 3\mathbf{i} + 2\mathbf{j}$.

50. Calculate the volume of the parallelepiped determined by the vectors $\mathbf{u} = \mathbf{i} - \mathbf{j}, \mathbf{v} = 3\mathbf{i} + 2\mathbf{k},$ and $\mathbf{w} = -7\mathbf{j} + 3\mathbf{k}$.

51. Calculate the volume of the parallelepiped determined by the vectors $\overrightarrow{PQ}, \overrightarrow{PR},$ and $\overrightarrow{PS}$ where $P = (2, 1, -1), Q = (-3, 1, 4), R = (-1, 0, 2),$ and $S = (-3, -1, 5)$.

In Problems 52–55 let $\mathbf{u} = \mathbf{i} - 2\mathbf{j}, \mathbf{v} = 2\mathbf{i} + 3\mathbf{k}, \mathbf{w} = \mathbf{j} - \mathbf{k},$ and $\mathbf{y} = -\mathbf{i} + \mathbf{j} + \mathbf{k}$.

52. Show that $\mathbf{u} \times (\mathbf{v} \times \mathbf{w}) = (\mathbf{u} \cdot \mathbf{w})\mathbf{v} - (\mathbf{u} \cdot \mathbf{v})\mathbf{w}$.

53. Show that $(\mathbf{u} \times \mathbf{v}) \times \mathbf{w} = (\mathbf{u} \cdot \mathbf{w})\mathbf{v} - (\mathbf{v} \cdot \mathbf{w})\mathbf{u}$.

This result, together with the result of Problem 52, shows again that the cross product is not associative.

54. Show that $(\mathbf{u} \times \mathbf{v}) \cdot (\mathbf{w} \times \mathbf{y}) = (\mathbf{u} \cdot \mathbf{w})(\mathbf{v} \cdot \mathbf{y}) - (\mathbf{u} \cdot \mathbf{y})(\mathbf{v} \cdot \mathbf{w})$

55. Show that

$$(\mathbf{u} \times \mathbf{v}) \times (\mathbf{w} \times \mathbf{y}) = [\mathbf{u} \cdot (\mathbf{v} \times \mathbf{y})]\mathbf{w} - [\mathbf{u} \cdot (\mathbf{v} \times \mathbf{w})]\mathbf{y}$$

56. Show that $\mathbf{u} \times \mathbf{0} = \mathbf{0}$ for any vector $\mathbf{u}$.

57. Show that $\mathbf{u} \times \mathbf{v} = -(\mathbf{v} \times \mathbf{u})$. [*Hint:* Use Property 4 in Section 7.8, page 413.]

58. Show that $(\alpha\mathbf{u} \times \mathbf{v}) = \alpha(\mathbf{u} \times \mathbf{v})$. [*Hint:* Use Property 2 in Section 7.8.]

59. Show that $\mathbf{u} \times (\mathbf{v} + \mathbf{w}) = (\mathbf{u} \times \mathbf{v}) + (\mathbf{u} \times \mathbf{w})$. [*Hint:* Use Property 3 in Section 7.8.]

60. Prove identity (8). [*Hint:* Take the dot product of $\mathbf{y}$ and each side of (7). Then use Theorem 2(v).]

61. Prove identity (9). [*Hint:* Use (7).]

62. Prove that $|\mathbf{u} \times \mathbf{v}|^2 = |\mathbf{u}|^2|\mathbf{v}|^2 - (\mathbf{u} \cdot \mathbf{v})^2$.

6.4 Review of Lines and Planes in Space

In the plane $\mathbb{R}^2$ we can find the equation of a line if we know either two points on the line or one point and the slope of the line. In $\mathbb{R}^3$, our intuition tells us that the basic ideas are the same. Since two points determine a line, we should be able to calculate the equation of a line in space if we know two points on it. Alternatively, if we know one point and the direction of a line, we should also be able to find its equation.

Lines

We begin with two points $P = (x_1, y_1, z_1)$ and $Q = (x_2, y_2, z_2)$ on a line L. A vector parallel to L is a vector with representation $\overrightarrow{PQ}$. Thus [from equation (6.1.10) on p. 300]

$$\mathbf{v} = (x_2 - x_1)\mathbf{i} + (y_2 - y_1)\mathbf{j} + (z_2 - z_1)\mathbf{k} \tag{1}$$

is a vector parallel to L. Now let $R = (x, y, z)$ be another point on the line. Then $\overrightarrow{PR}$ is parallel to $\overrightarrow{PQ}$, which is parallel to $\mathbf{v}$, so that

$$\overrightarrow{PR} = t\mathbf{v} \tag{2}$$

for some real number t. Now look at Fig. 1. From this figure we have (in each of the three possible cases)

$$\overrightarrow{OR} = \overrightarrow{OP} + \overrightarrow{PR}. \tag{3}$$

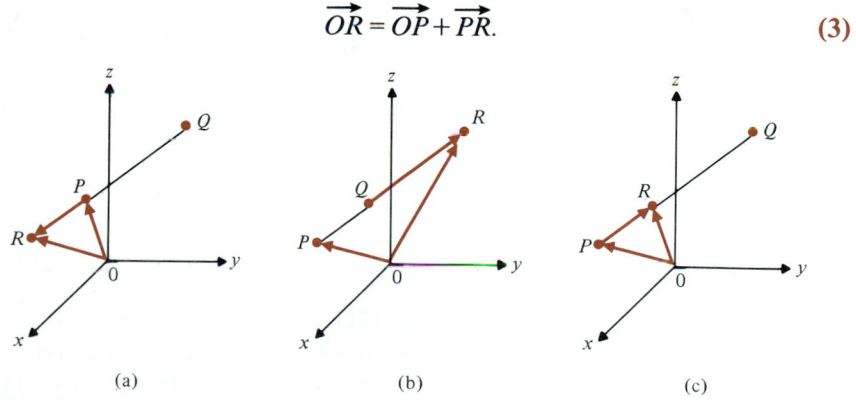

(a)   (b)   (c)

FIGURE 1

Combining (2) and (3), we get

$$\overrightarrow{PR} = \overrightarrow{OR} - \overrightarrow{OP} = t\mathbf{v}$$

or

$$\overrightarrow{OR} = \overrightarrow{OP} + t\mathbf{v} \tag{4}$$

Vector Equation

Equation (4) is called the **vector equation** of the line L. For if R is on L, then (4) is satisfied for some real number t. Conversely, if (4) is satisfied, then, reversing our steps, we see that $\overrightarrow{PR}$ is parallel to $\mathbf{v}$, which means that R is on L.

If we write out the components of Equation (4), we obtain

$$x\mathbf{i} + y\mathbf{j} + z\mathbf{k} = x_1\mathbf{i} + x_2\mathbf{j} + x_3\mathbf{k} + t(x_2 - x_1)\mathbf{i} + t(y_2 - y_1)\mathbf{j} + t(z_2 - z_1)\mathbf{k}$$

or

$$
\begin{aligned}
x &= x_1 + t(x_2 - x_1), \\
y &= y_1 + t(y_2 - y_1), \\
z &= z_1 + t(z_2 - z_1).
\end{aligned}
\tag{5}
$$

Parametric Equations

Equations (5) are called the **parametric equations** of a line.

Finally, solving for t in (5), and defining $x_2 - x_1 = a$, $y_2 - y_1 = b$, and $z_2 - z_1 = c$, we find that, if $abc \neq 0$,

$$\frac{x - x_1}{a} = \frac{y - y_1}{b} = \frac{z - z_1}{c}. \tag{6}$$

Symmetric Equations

Equations (6) are called the **symmetric equations** of the line. Here a, b, and c are direction numbers of the vector $\mathbf{v}$. Of course, equations (6) are valid only if a, b, and c are nonzero. The case in which one of these numbers is zero is discussed in Example 3.

EXAMPLE 1

Find a vector equation, parametric equations, and symmetric equations of the line L passing through the points $P = (2, -1, 6)$ and $Q = (3, 1, -2)$.

SOLUTION
First we calculate $\mathbf{v} = (3 - 2)\mathbf{i} + [1 - (-1)]\mathbf{j} + (-2 - 6)\mathbf{k} = \mathbf{i} + 2\mathbf{j} - 8\mathbf{k}$. Then, from (4), if $R = (x, y, z)$ is on the line, we obtain

$$\overrightarrow{OR} = x\mathbf{i} + y\mathbf{j} + z\mathbf{k} = \overrightarrow{OP} + t\mathbf{v} = 2\mathbf{i} - \mathbf{j} + 6\mathbf{k} + t(\mathbf{i} + 2\mathbf{j} - 8\mathbf{k}),$$

or

$$x = 2 + t, \qquad y = -1 + 2t, \qquad z = 6 - 8t.$$

Finally, since $a = 1$, $b = 2$, and $c = -8$, we find the symmetric equations

$$\frac{x - 2}{1} = \frac{y + 1}{2} = \frac{z - 6}{-8}. \tag{7}$$

To check this, we verify that $(2, -1, 6)$ and $(3, 1, -2)$ are indeed on the line. We have, after plugging these points into (7),

$$t = \frac{2-2}{1} = \frac{-1+1}{2} = \frac{6-6}{-8} = 0,$$

$$t = \frac{3-2}{1} = \frac{1+1}{2} = \frac{-2-6}{-8} = 1.$$

Other points on the line can be found. If $t = 3$, for example, we obtain

$$3 = \frac{x-2}{1} = \frac{y+1}{2} = \frac{z-6}{-8},$$

which yields the point $(5, 5, -18)$.

EXAMPLE 2

Find the symmetric equations of the line passing through the point $(1, -2, 4)$ and parallel to the vector $\mathbf{v} = \mathbf{i} + \mathbf{j} - \mathbf{k}$.

SOLUTION

We use formula (6) with $P = (x_1, y_1, z_1) = (1, -2, 4)$ and $\mathbf{v}$ as above so that $a = 1$, $b = 1$, and $c = -1$. This gives us

$$\frac{x-1}{1} = \frac{y+2}{1} = \frac{z-4}{-1}.$$

What happens if one of the direction numbers a, b, or c is zero?

EXAMPLE 3

Find the symmetric equation of the lines containing the points $P = (3, 4, -1)$ and $Q = (-2, 4, 6)$.

SOLUTION

Here $\mathbf{v} = -5\mathbf{i} + 7\mathbf{k}$ and $a = -5$, $b = 0$, $c = 7$. Then a parametric representation of the line is $x = 3 - 5t$, $y = 4$, and $z = -1 + 7t$. Solving for t, we find that

$$\frac{x-3}{-5} = \frac{z+1}{7} \quad \text{and} \quad y = 4.$$

The equation $y = 4$ is the equation of a plane parallel to the xz-plane, so we have obtained an equation of a line in that plane.

Planes

The equation of a line in space is obtained by specifying a point on the line and a vector *parallel* to this line. We can derive the equation of a plane in space by specifying a point in the plane and a vector orthogonal to every vector in the plane. This orthogonal vector is called a **normal vector** and is denoted by $\mathbf{n}$. (See Fig. 2.)

Normal Vector

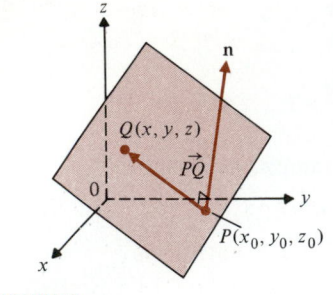

FIGURE 2

Let P be a point in space and let $\mathbf{n}$ be a given nonzero vector. Then the set of all points Q for which $\overrightarrow{PQ} \cdot \mathbf{n} = 0$ comprises a **plane** in $\mathbb{R}^3$.

Notation. We shall usually denote a plane by the symbol Π.

Let $P = (x_0, y_0, z_0)$ be a fixed point on a plane with normal vector $\mathbf{n} = a\mathbf{i} + b\mathbf{j} + c\mathbf{k}$. If $Q = (x, y, z)$ is any other point on the plane, then $\overrightarrow{PQ} = (x - x_0)\mathbf{i} + (y - y_0)\mathbf{j} + (z - z_0)\mathbf{k}$. Since $\overrightarrow{PQ}$ is perpendicular to $\mathbf{n}$, we have $\overrightarrow{PQ} \cdot \mathbf{n} = 0$. But this implies that

$$a(x - x_0) + b(y - y_0) + c(z - z_0) = 0. \tag{8}$$

A more common way to write the equation of a plane is easily derived from (8):

$$ax + by + cz = d, \tag{9}$$

where

$$d = ax_0 + by_0 + cz_0 = \overrightarrow{OP} \cdot \mathbf{n}.$$

EXAMPLE 4

Find the equation of the plane Π passing through the point $(2, 5, 1)$ having the normal vector $\mathbf{n} = \mathbf{i} - 2\mathbf{j} + 3\mathbf{k}$.

SOLUTION
From (8), we immediately obtain

$$(x - 2) - 2(y - 5) + 3(z - 1) = 0$$

or

$$x - 2y + 3z = -5.$$

EXAMPLE 5

Find the equation of the plane passing through the points $P = (1, 2, 1)$, $Q = (-2, 3, -1)$, and $R = (1, 0, 4)$.

SOLUTION
The vectors $\overrightarrow{PQ} = -3\mathbf{i} + \mathbf{j} - 2\mathbf{k}$ and $\overrightarrow{QR} = 3\mathbf{i} - 3\mathbf{j} + 5\mathbf{k}$ lie on the plane and are therefore orthogonal to the normal vector. Thus

$$\mathbf{n} = \overrightarrow{PQ} \times \overrightarrow{QR} = \begin{vmatrix} \mathbf{i} & \mathbf{j} & \mathbf{k} \\ -3 & 1 & -2 \\ 3 & -3 & 5 \end{vmatrix} = -\mathbf{i} + 9\mathbf{j} + 6\mathbf{k}$$

and we obtain

$$-(x - 1) + 9(y - 2) + 6(z - 1) = 0$$

or

$$-x + 9y + 6z = 23.$$

Note that if we choose another point, say Q, we get the equation $-(x + 2) + 9(y - 3) + 6(z + 1) = 0$, which also reduces to $-x + 9y + 6z = 23$.

Problems 6.4

In Problems 1–10 find a vector equation, parametric equations, and symmetric equations of the indicated line.

1. Containing $(2, 1, 3)$ and $(1, 2, -1)$
2. Containing $(1, -1, 1)$ and $(-1, 1, -1)$
3. Containing $(-4, 1, 3)$ and $(-4, 0, 1)$
4. Containing $(2, 3, -4)$ and $(2, 0, -4)$
5. Containing $(2, 2, 1)$ and parallel to $2\mathbf{i} - \mathbf{j} - \mathbf{k}$
6. Containing $(-1, -6, 2)$ and parallel to $4\mathbf{i} + \mathbf{j} - 3\mathbf{k}$
7. Containing (a, b, c) and parallel to $d\mathbf{i} + e\mathbf{j}$
8. Containing (a, b, c) and parallel to $d\mathbf{k}$
9. Containing $(4, 1, -6)$ and parallel to
$$(x - 2)/3 = (y + 1)/6 = (z - 5)/2$$
10. Containing $(3, 1, -2)$ and parallel to
$$(x + 1)/3 = (y + 3)/2 = (z - 2)/(-4)$$
11. Show that the lines
$$L_1: \frac{x - 3}{2} = \frac{y + 1}{4} = \frac{z - 2}{-1} \quad \text{and} \quad L_2: \frac{x - 3}{5} = \frac{y + 1}{-2} = \frac{z - 2}{2}$$
are orthogonal.
12. Let L_1 be given by
$$\frac{x - x_1}{a_1} = \frac{y - y_1}{b_1} = \frac{z - z_1}{c_1}$$
and L_2 be given by
$$\frac{x - x_1}{a_2} = \frac{y - y_1}{b_2} = \frac{z - z_1}{c_2}.$$
Show that L_1 is orthogonal to L_2 if and only if
$$a_1 a_2 + b_1 b_2 + c_1 c_2 = 0.$$
13. Show that the lines
$$L_1: \frac{x - 1}{1} = \frac{y + 3}{2} = \frac{z + 3}{3} \quad \text{and} \quad L_2: \frac{x - 3}{3} = \frac{y - 1}{6} = \frac{z - 8}{9}$$
are parallel.

Lines in $\mathbb{R}^3$ that do not have the same direction need not have a point in common.

14. Show that the lines $L_1: x = 1 + t$, $y = -3 + 2t$, $z = -2 - t$ and $L_2: x = 17 + 3s$, $y = 4 + s$, $z = -8 - s$ have the point $(2, -1, -3)$ in common.
15. Show that the lines $L_1: x = 2 - t$, $y = 1 + t$, $z = -2t$ and $L_2: x = 1 + s$, $y = -2s$, $z = 3 + 2s$ do not have a point in common.

16. Let L be given in its vector form $\overrightarrow{OR} = \overrightarrow{OP} + t\mathbf{v}$. Find a number t such that $\overrightarrow{OR}$ is perpendicular to $\mathbf{v}$.
17. Use the result of Problem 16 to find the distance between the line L (containing P and parallel to $\mathbf{v}$) and the origin when
   (a) $P = (2, 1, -4)$; $\quad \mathbf{v} = \mathbf{i} + \mathbf{j} + \mathbf{k}$
   (b) $P = (1, 2, -3)$; $\quad \mathbf{v} = 3\mathbf{i} - \mathbf{j} - \mathbf{k}$
   (c) $P = (-1, 4, 2)$; $\quad \mathbf{v} = -\mathbf{i} + \mathbf{j} + 2\mathbf{k}$

In Problems 18–21 find a line L orthogonal to the two given lines and passing through the given point.

18. $\dfrac{x + 2}{-3} = \dfrac{y - 1}{4} = \dfrac{z}{-5}$; $\quad \dfrac{x - 3}{7} = \dfrac{y + 2}{-2} = \dfrac{z - 8}{3}$; $\quad (1, -3, 2)$
19. $\dfrac{x - 2}{-4} = \dfrac{y + 3}{-7} = \dfrac{z + 1}{3}$; $\quad \dfrac{x + 2}{3} = \dfrac{y - 5}{-4} = \dfrac{z + 3}{-2}$;
   $(-4, 7, 3)$
20. $x = 3 - 2t$; $\quad y = 4 + 3t$; $\quad z = -7 + 5t$; $\quad x = -2 + 4s$,
   $y = 3 - 2s$, $z = 3 + s$; $\quad (-2, 3, 4)$
21. $x = 4 + 10t$, $\quad y = -4 - 8t$, $\quad z = 3 + 7t$, $\quad x = -2s$,
   $y = 1 + 4s$, $\quad z = -7 - 3s$; $\quad (4, 6, 0)$

* 22. Calculate the shortest distance between the lines
$$L_1: \frac{x - 2}{3} = \frac{y - 5}{2} = \frac{z - 1}{-1} \quad \text{and} \quad L_2: \frac{x - 4}{-4} = \frac{y - 5}{4} = \frac{z + 2}{1}.$$

[*Hint:* The shortest distance is measured along a vector $\mathbf{v}$ that is perpendicular to both L_1 and L_2. Let P be a point

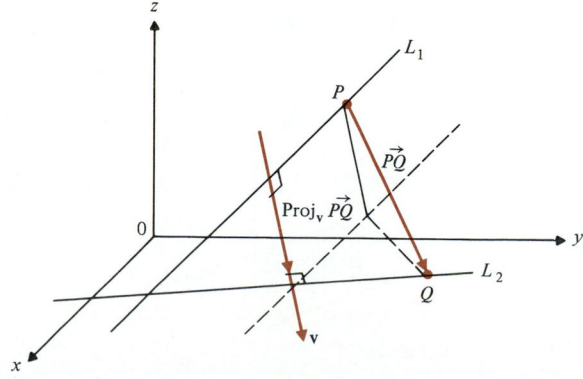

FIGURE 3

on L_1 and Q a point on L_2. Then the length of the projection of $\overrightarrow{PQ}$ on $\mathbf{v}$ is the distance between the lines, measured along a vector that is perpendicular to them both. See Fig. 3.]

* 23. Find the shortest distance between the lines

$$L_1: \frac{x+2}{3} = \frac{y-7}{-4} = \frac{z-2}{4} \quad \text{and} \quad L_2: \frac{x-1}{-3} = \frac{y+2}{4} = \frac{z+1}{1}$$

In Problems 24–37 find the equation of the plane.

24. $P = (0, 0, 0);\quad \mathbf{n} = \mathbf{i}$

25. $P = (0, 0, 0);\quad \mathbf{n} = \mathbf{j}$

26. $P = (0, 0, 0);\quad \mathbf{n} = \mathbf{k}$

27. $P = (1, 2, 3);\quad \mathbf{n} = \mathbf{i} + \mathbf{j}$

28. $P = (1, 2, 3);\quad \mathbf{n} = \mathbf{i} + \mathbf{k}$

29. $P = (1, 2, 3);\quad \mathbf{n} = \mathbf{j} + \mathbf{k}$

30. $P = (2, -1, 6);\quad \mathbf{n} = 3\mathbf{i} - \mathbf{j} + 2\mathbf{k}$

31. $P = (-4, -7, 5);\quad \mathbf{n} = -3\mathbf{i} - 4\mathbf{j} + \mathbf{k}$

32. $P = (-3, 11, 2);\quad \mathbf{n} = 4\mathbf{i} + \mathbf{j} - 7\mathbf{k}$

33. $P = (3, -2, 5);\quad \mathbf{n} = 2\mathbf{i} - 7\mathbf{j} - 8\mathbf{k}$

34. Containing $(1, 2, -4)$, $(2, 3, 7)$, and $(4, -1, 3)$

35. Containing $(-7, 1, 0)$, $(2, -1, 3)$, and $(4, 1, 6)$

36. Containing $(1, 0, 0)$, $(0, 1, 0)$, and $(0, 0, 1)$

37. Containing $(2, 3, -2)$, $(4, -1, -1)$, and $(3, 1, 2)$

* 38. Using Fig. 4, show that the distance from a point Q to a plane Π is given by

$$D = |\text{Proj}_{\mathbf{n}} \overrightarrow{PQ}| = \frac{|\overrightarrow{PQ} \cdot \mathbf{n}|}{|\mathbf{n}|}$$

where P is a point on the plane.

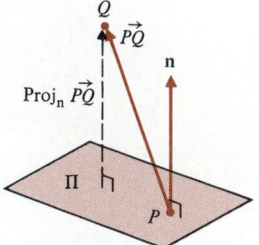

FIGURE 4

In Problems 39–42 find the distance from the given point to the given plane.

39. $(2, -1, 4);\quad 3x - y + 7z = 2$

40. $(4, 0, 1);\quad 2x - y + 8z = 3$

41. $(-7, -2, -2);\quad -2x + 8z = -5$

42. $(-3, 0, 2);\quad -3x + y + 5z = 0$

43. Prove that the distance between the plane

$$ax + by + cz = d$$

and the point (x_0, y_0, z_0) is given by

$$D = \frac{|ax_0 + by_0 + cz_0 - d|}{\sqrt{a^2 + b^2 + c^2}}$$

* 44. Find the distance between the lines $x = 2 - 3t$, $y = 1 + 2t$, $z = -2 - t$ and $x = 1 + 4s$, $y = -2 - s$, $z = 3 + s$.

* 45. Find the distance between the lines $x = -2 - 5t$, $y = -3 - 2t$, $z = 1 + 4t$ and $x = 2 + 3s$, $y = -1 + s$, $z = 3s$.

In Problems 46–48 find the equation of the set of all points of intersection of the two planes.

46. $\Pi_1: x - y + z = 2;\quad \Pi_2: 2x - 3y + 4z = 7$

47. $\Pi_1: 3x - y + 4z = 3;\quad \Pi_2: -4x - 2y + 7z = 8$

48. $\Pi_1: -2x - y + 17z = 4;\quad \Pi_2: 2x - y - z = -7$

The **angle between two planes** is defined as the acute angle between their normal vectors. In Problems 49–51 find the angle between the two planes.

49. The two planes of Problem 46

50. The two planes of Problem 47

51. The two planes of Problem 48

* 52. Let $\mathbf{u}$ and $\mathbf{v}$ be two nonparallel, nonzero vectors in a plane Π. Show that if $\mathbf{w}$ is any other vector in Π, then there exist scalars α and β such that $\mathbf{w} = \alpha\mathbf{u} + \beta\mathbf{v}$. This is called the **parametric representation** of the plane Π. [*Hint:* Draw a parallelogram in which $\alpha\mathbf{u}$ and $\beta\mathbf{v}$ form adjacent sides and the diagonal vector is $\mathbf{w}$.]

In Problems 53–57 determine whether the three given position vectors (that is, one endpoint at the origin) are coplanar [see statement (6.3.6)]. If they are coplanar, find the equation of the plane containing them.

53. $\mathbf{u} = 2\mathbf{i} - 3\mathbf{j} + 4\mathbf{k};\quad \mathbf{v} = 7\mathbf{i} - 2\mathbf{j} + 3\mathbf{k};\quad \mathbf{w} = 9\mathbf{i} - 5\mathbf{j} + 7\mathbf{k}$

54. $\mathbf{u} = -3\mathbf{i} + \mathbf{j} + 8\mathbf{k};\quad \mathbf{v} = -2\mathbf{i} - 3\mathbf{j} + 5\mathbf{k};\quad \mathbf{w} = 2\mathbf{i} + 14\mathbf{j} - 4\mathbf{k}$

55. $\mathbf{u} = 2\mathbf{i} + \mathbf{j} - 2\mathbf{k};\quad \mathbf{v} = 2\mathbf{i} - \mathbf{j} - 2\mathbf{k};\quad \mathbf{w} = 2\mathbf{i} - \mathbf{j} + 2\mathbf{k}$

56. $\mathbf{u} = 3\mathbf{i} - 2\mathbf{j} + \mathbf{k};\quad \mathbf{v} = \mathbf{i} + \mathbf{j} - 5\mathbf{k};\quad \mathbf{w} = -\mathbf{i} + 5\mathbf{j} - 16\mathbf{k}$

57. $\mathbf{u} = 2\mathbf{i} - \mathbf{j} - \mathbf{k};\quad \mathbf{v} = 4\mathbf{i} + 3\mathbf{j} + 2\mathbf{k};\quad \mathbf{w} = 6\mathbf{i} + 7\mathbf{j} + 5\mathbf{k}$

6.5 Vector Spaces

In Section 6.1 we reviewed the properties of vectors in the plane, $\mathbb{R}^2$, and in three-dimensional space, $\mathbb{R}^3$. We saw that vectors may be added and that they satisfy certain commutative and associative laws. Furthermore, if **x** is a vector in $\mathbb{R}^2$ or $\mathbb{R}^3$, then $\mathbf{x} + \mathbf{0} = \mathbf{x}$ and $\mathbf{x} + (-\mathbf{x}) = \mathbf{0}$. We can also multiply a vector by a scalar. Such scalar multiples satisfy distributive and associative laws, and $1\mathbf{x} = \mathbf{x}$. We call the sets $\mathbb{R}^2$ and $\mathbb{R}^3$ *vector spaces.*

In this section we shall generalize the concepts developed in Section 6.1 to the abstract world of real vector spaces. There is a great advantage in doing so. Once we have established a fact about vector spaces in general, we can apply that fact to *every* vector space. Otherwise, we would have to prove that fact again and again, once for each new vector space we encounter (and there is an endless supply of them).

Real Vector Space

DEFINITION

A **real vector space** V is a nonempty set of objects, called **vectors,** together with two operations called **addition** and **scalar multiplication** that satisfy the ten axioms listed below.

Notation If **x** and **y** are in V and if α is a real number, then we write $\mathbf{x} + \mathbf{y}$ for the sum of **x** and **y** and $\alpha\mathbf{x}$ for the scalar product of α and **x**.

Before we list the properties satisfied by vectors in a vector space, two things should be mentioned. First, while it might be helpful to think of $\mathbb{R}^2$ or $\mathbb{R}^3$ when dealing with a vector space, it often occurs that a vector space may appear to be very different from these comfortable spaces. (We shall see this shortly.) Second, the definition above is for a *real* vector space. The word "real" means that the scalars used are real numbers. It would be just as easy to define a *complex* vector space by using complex numbers instead of real ones. This chapter deals primarily with real vector spaces, but generalizations to other sets of scalars present little difficulty.

Axioms of a Vector Space

   i. If $\mathbf{x} \in V$ and $\mathbf{y} \in V$, then $\mathbf{x} + \mathbf{y} \in V$ (closure under addition).

   ii. For all **x**, **y**, and **z** in V, $(\mathbf{x} + \mathbf{y}) + \mathbf{z} = \mathbf{x} + (\mathbf{y} + \mathbf{z})$ (associative law of vector addition).

   iii. There is a vector $\mathbf{0} \in V$ such that for all $\mathbf{x} \in V$, $\mathbf{x} + \mathbf{0} = \mathbf{0} + \mathbf{x} = \mathbf{x}$ (**0** is called the additive identity).

   iv. If $\mathbf{x} \in V$, there is a vector $-\mathbf{x}$ in V such that $\mathbf{x} + (-\mathbf{x}) = \mathbf{0}$ ($-\mathbf{x}$ is called the additive inverse of **x**).

   v. If **x** and **y** are in V, then $\mathbf{x} + \mathbf{y} = \mathbf{y} + \mathbf{x}$ (commutative law of vector addition).

   vi. If $\mathbf{x} \in V$ and α is a scalar, then $\alpha\mathbf{x} \in V$ (closure under scalar multiplication).

vii. If $\mathbf{x}$ and $\mathbf{y}$ are in V and α is a scalar, then $\alpha(\mathbf{x} + \mathbf{y}) = \alpha\mathbf{x} + \alpha\mathbf{y}$ (first distributive law).

viii. If $\mathbf{x} \in V$ and α and β are scalars, then $(\alpha + \beta)\mathbf{x} = \alpha\mathbf{x} + \beta\mathbf{x}$ (second distributive law).

ix. If $\mathbf{x} \in V$ and α and β are scalars, then $\alpha(\beta\mathbf{x}) = \alpha\beta\mathbf{x}$ (associative law of scalar multiplication).

x. For every vector $\mathbf{x} \in V$, $1\mathbf{x} = \mathbf{x}$ (the scalar 1 is called a multiplicative identity).

There are many different kinds of vector spaces. For our purposes, the most important one is described below.

The Space $\mathbb{R}^n$

An **n-vector** $\mathbf{x}$ is an ordered set of n numbers which we write in the form

$$\mathbf{x} = (x_1, x_2, \ldots, x_n) \text{ or } \mathbf{x} = \begin{pmatrix} x_1 \\ x_2 \\ \vdots \\ x_n \end{pmatrix}. \text{ Then we define}$$

$$\mathbb{R}^n = \{(x_1, x_2, \ldots, x_n): \quad x_i \text{ is a real number for } i = 1, 2, \ldots, n\}.$$

In $\mathbb{R}^n$ we define vector addition and scalar multiplication just as we do for $\mathbb{R}^2$ and $\mathbb{R}^3$. In fact, $\mathbb{R}^n$ is the natural generalization of $\mathbb{R}^2$ or $\mathbb{R}^3$. That is, when we set $n = 2$ we get $\mathbb{R}^2$ and when we set $n = 3$ we get $\mathbb{R}^3$.

Vector Addition and Scalar Multiplication in $\mathbb{R}^n$

In $\mathbb{R}^n$, we define

$$(x_1, x_2, \ldots, x_n) + (y_1, y_2, \ldots, y_n) = (x_1 + y_1, x_2 + y_2, \ldots, x_n + y_n) \tag{1}$$

and

$$\alpha(x_1, x_2, \ldots, x_n) = (\alpha x_1, \alpha x_2, \ldots, \alpha x_n). \tag{2}$$

The following theorem generalizes Theorem 6.1.1 (p. 298). Its proof is left as an exercise.

■ THEOREM 1

$\mathbb{R}^n$ is a vector space. Every vector $\mathbf{x}$ in $\mathbb{R}^n$ can be written as a **row vector** $(x_1, x_2, \ldots, x_n)$ or a **column vector** $\begin{pmatrix} x_1 \\ x_2 \\ \vdots \\ x_n \end{pmatrix}$. The numbers $x_1, x_2, \ldots, x_n$ are called the **components** of $\mathbf{x}$. ■

Remark Whether we write $\mathbf{x}$ as a row vector or a column vector will depend on the use to which the vector is put. This will be clearer when we discuss matrix multiplication in Section 7.2. We note, however, that row vectors and column vectors are different things.

In much of Chapter 7 we shall deal with vectors in $\mathbb{R}^n$. However there are other interesting vector spaces. Here are some examples.

EXAMPLE 1

Let $V = \{0\}$. That is, V consists of the single number 0. Since

$$0 + 0 = 1 \cdot 0 = 0 + (0 + 0) = (0 + 0) + 0 = 0,$$

we see that V is a vector space. It is often referred to as a **trivial** vector space.

EXAMPLE 2

Let $V = \{1\}$. That is, V consists of the single number 1. This is *not* a vector space since it violates axiom (*i*)—the closure axiom. To see this we simply note that $1 + 1 = 2 \notin V$.

EXAMPLE 3

Let $V = \{(x, y): y = mx$, where m is a fixed real number and x is an arbitrary real number$\}$. That is, V consists of all points lying on the line $y = mx$ passing through the origin with slope m. Suppose that (x_1, y_1) and (x_2, y_2) are in V. Then $y_1 = mx_1$, $y_2 = mx_2$, and

$$(x_1, y_1) + (x_2, y_2) = (x_1, mx_1) + (x_2, mx_2) = (x_1 + x_2, mx_1 + mx_2)$$

$$= (x_1 + x_2, m(x_1 + x_2)) \in V.$$

Thus axiom (*i*) is satisfied. Axioms (*ii*), (*iii*), and (*v*) are obvious. Further,

$$-(x, mx) = (-x, -mx) = (-x, m(-x)) \in V$$

and

$$(x, mx) + (-x, m(-x)) = (0, 0) = \mathbf{0}$$

so that axiom (*iv*) is satisfied. The other axioms are easily verified and we see that the set of points in the plane lying on a straight line passing through the origin constitutes a vector space.

EXAMPLE 4

Let $V = \{(x, y): y = 2x + 1, x \in \mathbb{R}\}$. That is, V is the set of points lying on the line $y = 2x + 1$. V is *not* a vector space because closure is violated, as in Example 2. To see this, let us supose that (x_1, y_1) and (x_2, y_2) are in V. Then

$$(x_1, y_1) + (x_2, y_2) = (x_1 + x_2, y_1 + y_2).$$

If this last vector were in V, we would have

$$y_1 + y_2 = 2(x_1 + x_2) + 1 = 2x_1 + 2x_2 + 1.$$

But $y_1 = 2x_1 + 1$ and $y_2 = 2x_2 + 1$ so that

$$y_1 + y_2 = (2x_1 + 1) + (2x_2 + 1) = 2x_1 + 2x_2 + 2.$$

Hence we conclude that

$$(x_1 + x_2, y_1 + y_2) \notin V \quad \text{if} \quad (x_1, y_1) \in V \quad \text{and} \quad (x_2, y_2) \in V.$$

EXAMPLE 5

Let $V = \{(x, y, z): ax + by + cz = 0\}$. That is, V is the set of points in $\mathbb{R}^3$ lying on the plane passing through the origin with normal vector (a, b, c). Suppose (x_1, y_1, z_1) and (x_2, y_2, z_2) are in V. Then $(x_1, y_1, z_1) + (x_2, y_2, z_2) = (x_1 + x_2, y_1 + y_2, z_1 + z_2) \in V$ because $a(x_1 + x_2) + b(y_1 + y_2) + c(z_1 + z_2) = (ax_1 + by_1 + cz_1) + (ax_2 + by_2 + cz_2) = 0 + 0 = 0$; hence axiom (i) is satisfied. The other axioms are easily verified. Thus the set of points lying on a plane in $\mathbb{R}^3$ that passes through the origin comprises a vector space.

EXAMPLE 6

Let $V = P_n$, the set of polynomials with real coefficients of degree less than or equal to n. If $p \in P_n$, then

$$p(x) = a_n x^n + a_{n-1} x^{n-1} + \cdots + a_1 x + a_0$$

where each a_i is real. The sum $p(x) + q(x)$ is defined in the following way: If $q(x) = b_n x^n + b_{n-1} x^{n-1} + \cdots + b_1 x + b_0$, then

$$p(x) + q(x) = (a_n + b_n)x^n + (a_{n-1} + b_{n-1})x^{n-1} + \cdots + (a_1 + b_1)x + (a_0 + b_0).$$

The sum of two polynomials of degree less than or equal to n is another polynomial with degree less than or equal to n, so axiom (i) is satisfied. Properties (ii) and (v) to (x) are obvious. If we define the zero polynomial by $\mathbf{0} = 0x^n + 0x^{n-1} + \cdots + 0x + 0$, then $\mathbf{0} \in P_n$ and axiom (iii) is satisfied. Finally, letting $-p(x) = -a_n x^n - a_{n-1} x^{n-1} - \cdots - a_1 x - a_0$, we see that axiom (iv) holds, so P_n is a real vector space.

EXAMPLE 7

Let $V = C[0, 1] =$ the set of real-valued continuous functions defined on the interval $[0, 1]$. We define $(f + g)x = f(x) + g(x)$ and $(\alpha f)(x) = \alpha[f(x)]$. Since the sum of continuous functions is continuous, axiom (i) is satisfied and the other axioms can be verified with $\mathbf{0} =$ the zero function and $(-f)(x) = -f(x)$.

EXAMPLE 8: The Space $\mathbb{C}^n$

Let $V = \mathbb{C}^n = \{(c_1, c_2, \ldots, c_n): c_i$ is a complex number for $i = 1, 2, \ldots, n\}$ and the set of scalars is the set of complex numbers. It is not difficult to verify that $\mathbb{C}^n$, too, is a (complex) vector space if addition and scalar multiplication are given by (1) and (2).

As these examples suggest, there are many different kinds of vector spaces and many kinds of sets that are *not* vector spaces under standard addition and scalar multiplication operations. We now prove some elementary results about vector spaces.

THEOREM 2
Let V be a vector space. Then:

    i. $\alpha\mathbf{0} = \mathbf{0}$ for every real number α.
    ii. $0\mathbf{x} = \mathbf{0}$ for every $\mathbf{x} \in V$.
    iii. If $\alpha\mathbf{x} = \mathbf{0}$, then $\alpha = 0$ or $\mathbf{x} = \mathbf{0}$ (or both).
    iv. $(-1)\mathbf{x} = -\mathbf{x}$ for every $\mathbf{x} \in V$.

PROOF
    i. By axiom (*iii*), $\mathbf{0} + \mathbf{0} = \mathbf{0}$; and from axiom (*vii*),

$$\alpha(\mathbf{0} + \mathbf{0}) = \alpha\mathbf{0} + \alpha\mathbf{0} = \alpha\mathbf{0}. \tag{3}$$

Adding $-\alpha\mathbf{0}$ to both sides of the last equation in (3) and using the associative law (axiom *ii*), we obtain

$$[\alpha\mathbf{0} + \alpha\mathbf{0}] + (-\alpha\mathbf{0}) = \alpha\mathbf{0} + (-\alpha\mathbf{0}),$$

$$\alpha\mathbf{0} + [\alpha\mathbf{0} + (-\alpha\mathbf{0})] = \mathbf{0},$$

$$\alpha\mathbf{0} + \mathbf{0} = \mathbf{0},$$

$$\alpha\mathbf{0} = \mathbf{0}.$$

    ii. Essentially the same proof as used in part (*i*) works. We start with $0 + 0 = 0$ and use axiom (*viii*) to see that $0\mathbf{x} = (0 + 0)\mathbf{x} = 0\mathbf{x} + 0\mathbf{x}$ or $0\mathbf{x} + (-0\mathbf{x}) = 0\mathbf{x} + [0\mathbf{x} + (-0\mathbf{x})]$ or $\mathbf{0} = 0\mathbf{x} + \mathbf{0} = 0\mathbf{x}$.

    iii. Let $\alpha\mathbf{x} = \mathbf{0}$. If $\alpha \neq 0$, we multiply both sides of the equation by $1/\alpha$ to obtain $(1/\alpha)(\alpha\mathbf{x}) = (1/\alpha)\mathbf{0} = \mathbf{0}$ (by part *i*). But $(1/\alpha)(\alpha\mathbf{x}) = 1\mathbf{x} = \mathbf{x}$ (by axiom *ix*), so $\mathbf{x} = \mathbf{0}$.

    iv. We start with the fact that $1 + (-1) = 0$. Then, using part (*ii*), we obtain

$$\mathbf{0} = 0\mathbf{x} = [1 + (-1)]\mathbf{x} = 1\mathbf{x} + (-1)\mathbf{x} = \mathbf{x} + (-1)\mathbf{x}. \tag{4}$$

We add $-\mathbf{x}$ to both sides of (4) to obtain

$$\mathbf{0} + (-\mathbf{x}) = \mathbf{x} + (-1)\mathbf{x} + (-\mathbf{x}) = \mathbf{x} + (-\mathbf{x}) + (-1)\mathbf{x}$$

$$= \mathbf{0} + (-1)\mathbf{x} = (-1)\mathbf{x}.$$

Thus $-\mathbf{x} = (-1)\mathbf{x}$. Note that we were able to reverse the order of addition in the preceding equation by using the commutative law (axiom *v*).

Problems 6.5

In Problems 1–15 determine whether the given set and operations form a vector space. If not, list one axiom that does not hold.

1. $\{(x, y): y \leq 0; x, y \text{ real}\}$ with the usual addition and scalar multiplication of vectors.

2. The vectors in the plane lying in the first quadrant.

3. The set of vectors in $\mathbb{R}^3$ in the form (x, x, x).

4. The set of polynomials of degree 4 under the operations of Example 6.

5. The set consisting of the single vector $(0, 0)$ under the usual operations in $\mathbb{R}^2$.

6. The set of polynomials of degree $\leq n$ with zero constant term, under the operations of Example 6.

7. The set of polynomials of degree $\leq n$ with positive constant term a_0, under the operations of Example 6.

8. The set of continuous functions in $[0, 1]$ with $f(0) = 0$ and $f(1) = 0$ under the operations of Example 7.

9. The set of vectors in $\mathbb{R}^3$ lying on a line passing through the origin.

10. The set of vectors in $\mathbb{R}^3$ lying on the line $x = t + 1$, $y = 2t$, $z = t - 1$.

11. $\mathbb{R}^2$ with addition defined by $(x_1, y_1) + (x_2, y_2) = (x_1 + x_2 + 1, y_1 + y_2 + 1)$ and ordinary scalar multiplication.

12. The set of Problem 11 with scalar multiplication defined by $\alpha(x, y) = (\alpha + \alpha x - 1, \alpha + \alpha y - 1)$.

13. The set consisting of one object with addition defined by *object* + *object* = *object* and scalar multiplication defined by $\alpha(object) = object$.

14. The set of differentiable functions defined on $[0, 1]$ with the operations of Example 7.

15. The set of real numbers of the form $a + b\sqrt{2}$, where a and b are rational numbers, under the usual addition of real numbers and with scalar multiplication defined only for rational scalars.

16. Show that in a vector space the additive identity element is unique.

17. Show that in a vector space each vector has a unique additive inverse.

18. If **x** and **y** are vectors in a vector space V, show that there is a unique vector $\mathbf{z} \in V$ such that $\mathbf{x} + \mathbf{z} = \mathbf{y}$.

19. Show that the set of positive real numbers forms a vector space under the operations $x + y = xy$ and $\alpha x = x^\alpha$.

20. Show that the set of solutions of the homogeneous second-order differential equation

$$y''(x) + a(x)y'(x) + b(x)y(x) = 0,$$

where $a(x)$ and $b(x)$ are continuous functions, is a (real) vector space.

6.6 Subspaces

From the last section we know that $\mathbb{R}^2 = \{(x, y): x \in \mathbb{R} \text{ and } y \in \mathbb{R}\}$ is a vector space. In Example 6.5.3 we saw that $V = \{(x, y): y = mx\}$ is also a vector space. Moreover, it is clear that $V \subset \mathbb{R}^2$. That is, $\mathbb{R}^2$ has a subset that is also a vector space. In fact, all vector spaces have subsets that are also vector spaces.

DEFINITION

Subspace

Let H be a nonempty subset of a vector space V and suppose that H is itself a vector space under the operations of addition and scalar multiplication defined on V. Then H is said to be a **subspace** of V.

We shall encounter many examples of subspaces in this chapter. But first we prove a result that makes it relatively easy to determine whether a subset of V is indeed a subspace of V.

■ THEOREM 1

A nonempty subset H of the vector space V is a subspace of V if the two closure rules hold:

Rules for Checking Whether a Subset Is a Subspace

    i. If $\mathbf{x} \in H$ and $\mathbf{y} \in H$, then $\mathbf{x} + \mathbf{y} \in H$.
   ii. If $\mathbf{x} \in H$, then $\alpha\mathbf{x} \in H$ for every scalar α.

PROOF

To show that H is a vector space, we must show that axioms (i) to (x) on p. 321 hold under the operations of vector addition and scalar multiplication defined in V. The two closure operations (axioms i and vi) hold by hypothesis. Since vectors in H are also in V, the associative, commutative, distributive, and multiplicative identity laws (axioms ii, v, vii, $viii$, ix, and x) hold. Let $\mathbf{x} \in H$. Then $0\mathbf{x} \in H$ by hypothesis (ii). But by Theorem 6.5.2 (p. 325), (part ii), $0\mathbf{x} = \mathbf{0}$. Thus $\mathbf{0} \in H$ and axiom (iii) holds. Finally, by part (ii), $(-1)\mathbf{x} \in H$ for every $\mathbf{x} \in H$. By Theorem 6.5.2 (part iv), $-\mathbf{x} = (-1)\mathbf{x} \in H$ so that axiom (iv) also holds and the proof is complete. ∎

This theorem shows that to test whether a nonempty subset H of V is a subspace of V, it is only necessary to verify that:

$\mathbf{x} + \mathbf{y}$ and $\alpha\mathbf{x}$ are in H when $\mathbf{x}$ and $\mathbf{y}$ are in H and α is a scalar.

The preceding proof contains a fact that is important enough to mention explicitly:

             Every subspace of a vector space V contains $\mathbf{0}$.   **(1)**

This fact will often make it easy to see that a particular subset of V is *not* a vector space. That is, if a subset does not contain $\mathbf{0}$, then it is not a subspace.
    We now give some examples of subspaces.

EXAMPLE 1

For any vector space V, the subset $\{\mathbf{0}\}$ consisting of the zero vector alone is a subspace since $\mathbf{0} + \mathbf{0} = \mathbf{0}$ and $\alpha\mathbf{0} = \mathbf{0}$ for every real number α [part (i) of Theorem 6.5.2]. It is called the **trivial subspace.**

EXAMPLE 2

V is a subspace of itself for every vector space V.

Proper Subspace

The first two examples show that every vector space V contains two subspaces $\{\mathbf{0}\}$ and V (unless, of course, $V = \{\mathbf{0}\}$). It is more interesting to find other subspaces. Subspaces other than $\{\mathbf{0}\}$ and $\mathbf{V}$ are called **proper subspaces.**

EXAMPLE 3

Let $H = \{(x, y): y = mx\}$ (see Example 6.5.3, p. 323). Then, as we have already mentioned, H is a subspace of $\mathbb{R}^2$. It is true that the sets of vectors lying on straight lines through the origin are the only proper subspaces of $\mathbb{R}^2$.

EXAMPLE 4

Let $H = \{(x, y, z): x = at, y = bt, \text{ and } z = ct; a, b, c, t \text{ real}\}$. Then H consists of the vectors in $\mathbb{R}^3$ lying on a straight line passing through the origin. To see that H is a subspace of $\mathbb{R}^3$, we compute as follows: Let $\mathbf{x} = (at_1, bt_1, ct_1) \in H$ and $\mathbf{y} = (at_2, bt_2, ct_2) \in H$. Then $\mathbf{x} + \mathbf{y} = (a(t_1 + t_2), b(t_1 + t_2), c(t_1 + t_2)) \in H$ and $\alpha\mathbf{x} = (a(\alpha t_1), b(\alpha t_2), c(\alpha t_3)) \in H$. Thus H is a subspace of $\mathbb{R}^3$.

EXAMPLE 5

Let $\Pi = \{(x, y, z): ax + by + cz = 0; a, b, c \text{ real}\}$. Then, as we saw in Example 6.5.5, p. 324, Π is a vector space; thus Π is a subspace of $\mathbb{R}^3$.

We shall prove in Section 6.7 that sets of vectors lying on either lines or planes through the origin are the only proper subspaces of $\mathbb{R}^3$.

Before studying more examples, we note that *not every vector space has proper subspaces.*

EXAMPLE 6

Let H be a subspace of $\mathbb{R}$.† If $H \neq \{\mathbf{0}\}$, then H contains a nonzero real number α. Then, by axiom (vi), $1 = (1/\alpha)\alpha \in H$ and $\beta 1 = \beta \in H$ for every real number β. Thus if H is not the trivial subspace, then $H = \mathbb{R}$. That is, $\mathbb{R}$ has *no* proper subspace.

EXAMPLE 7

If P_n denotes the vector space of polynomials of degree $\leq n$ (Example 6.5.6, p. 324), and if $0 \leq m < n$, then P_m is a proper subspace of P_n, as can be verified.

EXAMPLE 8

$P_n[0, 1]\ddagger \subset C[0, 1]$ (see Example 6.5.7, p. 324) because every polynomial is continuous. P_n is a vector space for every integer n, so that each $P_n[0, 1]$ is a subspace of $C[0, 1]$.

---

† Note that $\mathbb{R}$ is a vector space over itself; that is, $\mathbb{R}$ is a vector space where the scalars are taken to be the reals.

‡ $P_n[0, 1]$ denotes the set of polynomials defined on the interval $[0, 1]$ of degree $\leq n$.

EXAMPLE 9

Let $C'[0, 1]$ denote the set of functions with continuous first derivatives defined on $[0, 1]$. Since every differentiable function is continuous, we have $C'[0, 1] \subset C[0, 1]$. The sum and scalar multiple of two differentiable functions are differentiable so $C'[0, 1]$ is a subspace of $C[0, 1]$. It is a proper subspace because not every continuous function is differentiable.

EXAMPLE 10

If $f \in C[0, 1]$, then $\int_0^1 f(x)\, dx$ exists. Let $H = \{f \in C[0, 1]: \int_0^1 f(x)\, dx = 0\}$. If $f \in H$ and $g \in H$, then $\int_0^1 [f(x) + g(x)]\, dx = \int_0^1 f(x)\, dx + \int_0^1 g(x)\, dx = 0 + 0 = 0$ and $\int_0^1 \alpha f(x)\, dx = \alpha \int_0^1 f(x)\, dx = 0$. Thus $f + g$ and αf are in H for every real number α. This shows that H is a proper subspace of $C[0, 1]$.

Problems 6.6

In Problems 1–13 determine whether the given subset H of the vector space V is a subspace of V.

1. $V = \mathbb{R}^2$; $H = \{(x, y): y \geq 0\}$
2. $V = \mathbb{R}^2$; $H = \{(x, y): x = y\}$
3. $V = \mathbb{R}^3$; H = the xy-plane
4. $V = \mathbb{R}^2$; $H = \{(x, y): x^2 + y^2 \leq 1\}$
5. $V = P_4$; $H = \{p \in P_4: \deg p = 4\}$
6. $V = P_4$; $H = \{p \in P_4: p(0) = 0\}$
7. $V = P_n$; $H = \{p \in P_n: p(0) = 0\}$
8. $V = P_n$; $H = \{p \in P_n: p(0) = 1\}$
9. $V = C[0, 1]$; $H = \{f \in C[0, 1]: f(0) = f(1) = 0\}$
10. $V = C[0, 1]$; $H = \{f \in C[0, 1]: f(0) = 2\}$
11. $V = C'[0, 1]$; $H = \{f \in C'[0, 1]: f'(0) = 0\}$
12. $V = C[a, b]$, where a and b are real numbers and $a < b$; $H = \{f \in C[a, b]: \int_a^b f(x)\, dx = 0\}$
13. $V = C[a, b]$; $H = \{f \in C[a, b]: \int_a^b f(x)\, dx = 1\}$

14. If $V = C[0, 1]$, let H_1 denote the subspace of Example 8 and H_2 denote the subspace of Example 9. Describe the set $H_1 \cap H_2$ and show that it is a subspace.

15. Let H_1 and H_2 be subspaces of a vector space V. Show that $H_1 \cap H_2$ is a subspace of V.

16. In Problem 15, show that $H_1 \cup H_2$ is not, in general, a subspace of V.

17. Let $H = \{(x, y, z, w): ax + by + cz + dw = 0\}$, where a, b, c, and d are real numbers not all zero. Show that H is a proper subspace of $\mathbb{R}^4$. H is called a *hyperplane* in $\mathbb{R}^4$.

18. Let $H = \{(x_1, x_2, \ldots, x_n): a_1 x_1 + a_2 x_2 + \cdots + a_n x_n = 0\}$, where $a_1, a_2, \ldots, a_n$ are real numbers not all zero. Show that H is a proper subspace of $\mathbb{R}^n$. H, as in Problem 17, is called a **hyperplane** in $\mathbb{R}^n$.

19. Let H_1 and H_2 be subspaces of a vector space V. Let $H_1 + H_2 = \{v: v = v_1 + v_2 \text{ with } v_1 \in H_1 \text{ and } v_2 \in H_2\}$. Show that $H_1 + H_2$ is a subspace of V.

20. Let v_1 and v_2 be two vectors in $\mathbb{R}^2$. Show that $H = \{v: v = av_1 + bv_2; a, b \text{ real}\}$ is a subspace of $\mathbb{R}^2$.

21. In Problem 20 show that if v_1 and v_2 are not collinear, then $H = \mathbb{R}^2$.

22. Let $v_1, v_2, \ldots, v_n$ be arbitrary vectors in a vector space V. Let $H = \{v \in V: v = a_1 v_1 + a_2 v_2 + \cdots + a_n v_n$, where $a_1, a_2, \ldots, a_n$ are scalars$\}$. Show that H is a subspace of V. H is called the subspace *spanned* by the vectors $v_1, v_2, \ldots, v_n$.

6.7 Linear Combinations, Linear Independence, Basis, and Dimension

In this section we present some ideas that are central to the study of vector spaces. We begin with a definition.

DEFINITION

Let $\mathbf{v}_1, \mathbf{v}_2, \ldots, \mathbf{v}_n$ be vectors in a vector space V. Then any expression of the form

$$a_1\mathbf{v}_1 + a_2\mathbf{v}_2 + \cdots + a_n\mathbf{v}_n \tag{1}$$

Linear Combination

where $a_1, a_2, \ldots, a_n$ are scalars is called a **linear combination** of the vectors $\mathbf{v}_1, \mathbf{v}_2, \ldots, \mathbf{v}_n$.

EXAMPLE 1

$\begin{pmatrix} -5 \\ 11 \end{pmatrix}$ is a linear combination of $\begin{pmatrix} 2 \\ 4 \end{pmatrix}$ and $\begin{pmatrix} 3 \\ -1 \end{pmatrix}$ because $\begin{pmatrix} -5 \\ 11 \end{pmatrix} = 2\begin{pmatrix} 2 \\ 4 \end{pmatrix} - 3\begin{pmatrix} 3 \\ -1 \end{pmatrix}$.

EXAMPLE 2

In $\mathbb{R}^3$, $\begin{pmatrix} -7 \\ 7 \\ 7 \end{pmatrix}$ is a linear combination of $\begin{pmatrix} -1 \\ 2 \\ 4 \end{pmatrix}$ and $\begin{pmatrix} 5 \\ -3 \\ 1 \end{pmatrix}$ since

$$\begin{pmatrix} -7 \\ 7 \\ 7 \end{pmatrix} = 2\begin{pmatrix} -1 \\ 2 \\ 4 \end{pmatrix} - \begin{pmatrix} 5 \\ -3 \\ 1 \end{pmatrix}.$$

EXAMPLE 3

In P_n every polynomial can be written as a linear combination of the monomials $1, x, x^2, \ldots, x^n$.

DEFINITION

Linear Dependence and Independence

Let $\mathbf{v}_1, \mathbf{v}_2, \ldots, \mathbf{v}_n$ be n vectors in a vector space V. Then the vectors are said to be **linearly dependent** if there exist n scalars $c_1, c_2, \ldots, c_n$ *not all zero* such that

$$c_1\mathbf{v}_1 + c_2\mathbf{v}_2 + \cdots + c_n\mathbf{v}_n = \mathbf{0} \tag{2}$$

If the vectors are not linearly dependent, they are said to be **linearly independent.**

In other words, the vectors $\mathbf{v}_1, \mathbf{v}_2, \ldots, \mathbf{v}_n$ are linearly dependent if some *nontrivial* (nonzero) linear combination of these vectors is equal to the zero vector. On the other hand, if the *only* linear combination of these vectors that equals the zero vector is the *trivial* linear combination (with $c_1 = c_2 = \cdots = c_n = 0$), then the vectors $\mathbf{v}_1, \mathbf{v}_2, \ldots, \mathbf{v}_n$ are linearly independent.

Note: By this definition the empty set is linearly independent. Also, if one of the vectors is the zero vector, then the set is linearly dependent (see Problem 6.8.42 on p. 349).

How do we determine whether a set of vectors is linearly dependent or independent? The case for two vectors is easy.

■ **THEOREM 1**
Two vectors are linearly dependent if and only if one is a scalar multiple of the other.

PROOF
First suppose that $v_2 = cv_1$ for some scalar $c \neq 0$. Then $cv_1 - v_2 = 0$ and v_1 and v_2 are linearly dependent. On the other hand, suppose that v_1 and v_2 are dependent. Then there are constants c_1 and c_2, not both zero, such that $c_1v_1 + c_2v_2 = 0$. If $c_1 \neq 0$, then, dividing by c_1, we obtain $v_1 + (c_2/c_1)v_2 = 0$ or

$$v_1 = \left(-\frac{c_2}{c_1}\right)v_2$$

That is, v_1 is a scalar multiple of v_2. If $c_1 = 0$ then $c_2 \neq 0$ and hence $v_2 = 0 = 0v_1$. ■

EXAMPLE 4

The vectors $v_1 = \begin{pmatrix} 2 \\ -1 \\ 0 \\ 3 \end{pmatrix}$ and $v_2 = \begin{pmatrix} -6 \\ 3 \\ 0 \\ -9 \end{pmatrix}$ are linearly dependent since $v_2 = -3v_1$.

EXAMPLE 5

The vectors $\begin{pmatrix} 1 \\ 2 \\ 4 \end{pmatrix}$ and $\begin{pmatrix} 2 \\ 5 \\ -3 \end{pmatrix}$ are linearly independent; if they were not, we would have

$\begin{pmatrix} 2 \\ 5 \\ -3 \end{pmatrix} = c\begin{pmatrix} 1 \\ 2 \\ 4 \end{pmatrix} = \begin{pmatrix} c \\ 2c \\ 4c \end{pmatrix}$. Then $2 = c$, $5 = 2c$, and $-3 = 4c$, which is clearly impossible for any number c.

In some situations the following theorem is useful.

■ **THEOREM 2**
In a vector space V the nonzero vectors $v_1, v_2, \ldots, v_n$ are linearly dependent if and only if at least one of them can be written as a linear combination of the vectors that precede it. That is, for some k, $1 < k \leq n$, there are scalars $\alpha_1, \alpha_2, \ldots, \alpha_{k-1}$ such that

$$v_k = \alpha_1v_1 + \alpha_2v_2 + \cdots + \alpha_{k-1}v_{k-1}. \tag{3}$$

PROOF
Suppose (3) holds. Then

$$\alpha_1\mathbf{v}_1 + \alpha_2\mathbf{v}_2 + \cdots + \alpha_{k-1}\mathbf{v}_{k-1} - \mathbf{v}_k + 0\mathbf{v}_{k+1} + \cdots + 0\mathbf{v}_n = \mathbf{0}$$

and we see, from (2), that the vectors are linearly dependent.

   Conversely, suppose that (2) holds. Let k be the largest integer such that $c_k \neq 0$ (k may equal n). Note that $k \geq 2$. For if $k = 1$, then (2) becomes $c_1\mathbf{v}_1 + 0\mathbf{v}_2 + \cdots + 0\mathbf{v}_n = \mathbf{0}$. Since $c_1 \neq 0$, we would conclude that $\mathbf{v}_1 = \mathbf{0}$, contradicting the assumption that the $\mathbf{v}_i$'s are nonzero. Thus (2) becomes

$$c_1\mathbf{v}_1 + c_2\mathbf{v}_2 + \cdots + c_{k-1}\mathbf{v}_{k-1} + c_k\mathbf{v}_k = \mathbf{0} \tag{4}$$

with $c_k \neq 0$. We divide both sides of (4) by c_k and rearrange to obtain

$$\mathbf{v}_k = -\frac{c_1}{c_k}\mathbf{v}_1 - \frac{c_2}{c_k}\mathbf{v}_2 - \cdots - \frac{c_{k-1}}{c_k}\mathbf{v}_{k-1}.$$

This completes the proof. ∎

EXAMPLE 6

Show that the vectors $(1, 2)$, $(-4, 3)$, and $(3, 1)$ in $\mathbb{R}^2$ are linearly dependent.

SOLUTION
We find constants α_1 and α_2 such that

$$(3, 1) = \alpha_1(1, 2) + \alpha_2(-4, 3). \tag{5}$$

Rewriting (5) we obtain

$$(\alpha_1 - 4\alpha_2, 2\alpha_1 + 3\alpha_2) = (3, 1)$$

or

$$\alpha_1 - 4\alpha_2 = 3$$

$$2\alpha_1 + 3\alpha_2 = 1$$

We multiply the first of these equations by 2 and then subtract:

$$2\alpha_1 - 8\alpha_2 = 6$$
$$\underline{2\alpha_1 + 3\alpha_2 = 1}$$
$$-11\alpha_2 = 5$$

So $\alpha_2 = -\frac{5}{11}$, $\alpha_1 = 3 + 4\alpha_2 = \frac{13}{11}$, and $(3, 1) = \frac{13}{11}(1, 2) - \frac{5}{11}(-4, 3)$. By Theorem 2, the three vectors are dependent.

   In order to determine whether a set of m vectors in $\mathbb{R}^n$ is linearly dependent or independent it is often necessary to solve a system of n linear equations in m unknowns. We saw a simple example of this in Example 6. In Section 7.4 we shall discuss a procedure for finding all solutions to such systems and shall give further examples of linear dependence and independence in $\mathbb{R}^n$. We shall prove the following theorem in Section 7.4 (Theorem 7.4.2, p. 381).

■ **THEOREM 3**

A set of m vectors in $\mathbb{R}^n$ is linearly dependent if $m > n$. ■

EXAMPLE 7

By Theorem 3, the four vectors $\begin{pmatrix} 1 \\ -1 \\ 2 \end{pmatrix}$, $\begin{pmatrix} 3 \\ 5 \\ 8 \end{pmatrix}$, $\begin{pmatrix} -2 \\ 4 \\ 7 \end{pmatrix}$, and $\begin{pmatrix} 2 \\ -3 \\ -4 \end{pmatrix}$ are linearly dependent in $\mathbb{R}^3$.

EXAMPLE 8

Show that the monomials $1, x, x^2, \ldots, x^n$ are linearly independent in P_n.

SOLUTION

If they are dependent, then by Theorem 2 there is an integer k and constants $\alpha_0, \alpha_1, \ldots, \alpha_{k-1}$ such that

$$x^k = \alpha_0 + \alpha_1 x + \cdots + \alpha_{k-1} x^{k-1}.$$

But this is impossible since the degree of the polynomial on the left is k while the degree of the polynomial on the right is $\leq k - 1$. Thus the monomials must be independent.

DEFINITION

Span

The vectors $\mathbf{v}_1, \mathbf{v}_2, \ldots, \mathbf{v}_n$ in a vector space V are said to **span** V if every vector in V can be written as a linear combination of them. That is, for every $\mathbf{v}$ in V, there are scalars $c_1, c_2, \ldots, c_n$ such that

$$\mathbf{v} = c_1 \mathbf{v}_1 + c_2 \mathbf{v}_2 + \cdots + c_n \mathbf{v}_n. \tag{6}$$

EXAMPLE 9

The vectors $\mathbf{i} = (1, 0, 0)$, $\mathbf{j} = (0, 1, 0)$, and $\mathbf{k} = (0, 0, 1)$ span $\mathbb{R}^3$.

EXAMPLE 10

The monomials $1, x, x^2, \ldots, x^n$ span P_n (by Example 3).

DEFINITION

Basis

A set of vectors $\{\mathbf{v}_1, \mathbf{v}_2, \ldots, \mathbf{v}_n\}$ forms a **basis** for V if:

   i. $\{\mathbf{v}_1, \mathbf{v}_2, \ldots, \mathbf{v}_n\}$ is linearly independent.
   ii. $\{\mathbf{v}_1, \mathbf{v}_2, \ldots, \mathbf{v}_n\}$ spans V.

EXAMPLE 11

In $\mathbb{R}^n$ we define

$$\mathbf{e}_1 = \begin{pmatrix} 1 \\ 0 \\ 0 \\ \cdot \\ \cdot \\ \cdot \\ 0 \end{pmatrix}, \mathbf{e}_2 = \begin{pmatrix} 0 \\ 1 \\ 0 \\ \cdot \\ \cdot \\ \cdot \\ 0 \end{pmatrix}, \mathbf{e}_3 = \begin{pmatrix} 0 \\ 0 \\ 1 \\ \cdot \\ \cdot \\ \cdot \\ 0 \end{pmatrix}, \ldots, \mathbf{e}_n = \begin{pmatrix} 0 \\ 0 \\ 0 \\ \cdot \\ \cdot \\ \cdot \\ 1 \end{pmatrix}.$$

If

$$\begin{pmatrix} 0 \\ 0 \\ \cdot \\ \cdot \\ \cdot \\ 0 \end{pmatrix} = \mathbf{0} = c_1\mathbf{e}_1 + c_2\mathbf{e}_2 + \cdots + c_n\mathbf{e}_n = \begin{pmatrix} c_1 \\ c_2 \\ \cdot \\ \cdot \\ \cdot \\ c_n \end{pmatrix},$$

then $c_1 = c_2 = \cdots = c_n = 0$, so the vectors $\mathbf{e}_1, \mathbf{e}_2, \ldots, \mathbf{e}_n$ are linearly independent.

Moreover, if $\mathbf{x} = \begin{pmatrix} x_1 \\ x_2 \\ \cdot \\ \cdot \\ \cdot \\ x_n \end{pmatrix} \in \mathbb{R}^n$, then $\mathbf{x} = x_1\mathbf{e}_1 + x_2\mathbf{e}_2 + \cdots + x_n\mathbf{e}_n$, so the vectors $\mathbf{e}_1$,

$\mathbf{e}_2, \ldots, \mathbf{e}_n$ span $\mathbb{R}^n$. We conclude that $\{\mathbf{e}_1, \mathbf{e}_2, \ldots, \mathbf{e}_n\}$ is a basis for $\mathbb{R}^n$. It is called the **standard basis** in $\mathbb{R}^n$.

EXAMPLE 12

The vectors $\begin{pmatrix} 1 \\ 2 \end{pmatrix}$ and $\begin{pmatrix} -1 \\ 1 \end{pmatrix}$ in $\mathbb{R}^2$ are linearly independent because neither one is a multiple of the other. Let $\begin{pmatrix} x \\ y \end{pmatrix}$ be a vector in $\mathbb{R}^2$. We can write $\begin{pmatrix} x \\ y \end{pmatrix}$ as a linear combination of $\begin{pmatrix} 1 \\ 2 \end{pmatrix}$ and $\begin{pmatrix} -1 \\ 1 \end{pmatrix}$. To see this, we find constants c_1 and c_2 such that

$$\begin{pmatrix} x \\ y \end{pmatrix} = c_1\begin{pmatrix} 1 \\ 2 \end{pmatrix} + c_2\begin{pmatrix} -1 \\ 1 \end{pmatrix}$$

$$\begin{pmatrix} x \\ y \end{pmatrix} = \begin{pmatrix} c_1 - c_2 \\ 2c_1 + c_2 \end{pmatrix}$$

or

$$c_1 - c_2 = x$$
$$\underline{2c_1 + c_2 = y}$$

(Adding) $$3c_1 = x + y$$

$$c_1 = \frac{x+y}{3}$$

$$c_2 = c_1 - x = \frac{x+y}{3} - x = \frac{-2x+y}{3}.$$

Thus

$$\binom{x}{y} = \frac{x+y}{3}\binom{1}{2} + \frac{-2x+y}{3}\binom{-1}{1}.$$

For example,

$$\binom{3}{-7} = \frac{3-7}{3}\binom{1}{2} + \frac{-6-7}{3}\binom{-1}{1} = \frac{-4}{3}\binom{1}{2} + \frac{-13}{3}\binom{-1}{1}.$$

Since the set $\left\{\binom{1}{2}, \binom{-1}{1}\right\}$ is linearly independent and spans $\mathbb{R}^2$, it is a basis for $\mathbb{R}^2$.

EXAMPLE 13

From Examples 8 and 10 we see that $\{1, x, x^2, \ldots, x^n\}$ is a basis for P_n.

In $\mathbb{R}^2$, the sets $\left\{\binom{1}{0}, \binom{0}{1}\right\}$ and $\left\{\binom{1}{2}, \binom{-1}{1}\right\}$ are bases. In fact, $\mathbb{R}^2$ has an infinite number of bases. We shall prove the following theorem on page 430.

■ **THEOREM 4**

Any set of n linearly independent vectors in $\mathbb{R}^n$ is a basis for $\mathbb{R}^n$. ■

Before we can define the dimension of a vector space, we need the following result. The proof is given in Section 7.4 (Theorem 7.4.5, p. 384).

■ **THEOREM 5**

If $\{\mathbf{u}_1, \mathbf{u}_2, \ldots, \mathbf{u}_m\}$ and $\{\mathbf{v}_1, \mathbf{v}_2, \ldots, \mathbf{v}_n\}$ are bases for the vector space V, then $m = n$; that is, any two bases in a vector space V have the same number of vectors. ■

DEFINITION

Dimension

Suppose that the nontrivial vector space V has a finite basis. Then the **dimension** of V is the number of vectors in a basis. In this case we write dim $V = n$. If V does not have a finite basis, then V is said to be **infinite-dimensional.** If $V = \{\mathbf{0}\}$, then V is said to be **zero-dimensional.**

EXAMPLE 14

From Theorem 4 or Example 11, we see that dim $\mathbb{R}^n = n$.

EXAMPLE 15

From Example 13 we see that dim $P_n = n + 1$.

EXAMPLE 16

$C[0, 1]$ is an infinite-dimensional vector space because the functions $1, x, x^2, x^3, \ldots$ are all in $C[0, 1]$ and form an infinite, linearly independent set. This means that $C[0, 1]$ cannot have a finite basis.

EXAMPLE 17

In Example 6.6.5 we saw that the plane $\Pi = \{(x, y, z); ax + by + cz = 0; a, b, c \text{ real}\}$ is a subspace of $\mathbb{R}^3$. What is its dimension?

SOLUTION

At least one of the numbers a, b, or c is nonzero. Suppose $c \neq 0$. Then, for $(x, y, z) \in \Pi$, $z = \dfrac{-a}{c} x - \dfrac{b}{c} y$ and we have

$$(x, y, z) = \left(x, y, \frac{-a}{c}x - \frac{b}{c}y\right) = x\left(1, 0, \frac{-a}{c}\right) + y\left(0, 1, \frac{-b}{c}\right).$$

Thus the vectors $\left(1, 0, \dfrac{-a}{c}\right)$ and $\left(0, 1, \dfrac{-b}{c}\right)$ span Π and, as they are clearly linearly independent, they form a basis for Π. Thus dim $\Pi = 2$.

EXAMPLE 18

Suppose that H is a two-dimensional subspace of $\mathbb{R}^3$. Show that all the vectors in H lie on the same plane.

SOLUTION

Let $\mathbf{v}_1 = (a_1, b_1, c_1)$ and $\mathbf{v}_2 = (a_2, b_2, c_2)$ be a basis for H. If $\mathbf{x} = (x, y, z) \in H$, then there exist real numbers s and t such that $\mathbf{x} = s\mathbf{v}_1 + t\mathbf{v}_2$ or $(x, y, z) = s(a_1, b_1, c_1) + t(a_2, b_2, c_2)$. Then

$$x = sa_1 + ta_2,$$
$$y = sb_1 + tb_2, \tag{7}$$
$$z = sc_1 + tc_2.$$

Let $\mathbf{v}_3 = (\alpha, \beta, \gamma) = \mathbf{v}_1 \times \mathbf{v}_2$. Then, from Theorem 6.3.2, part (vi) (p. 309), we have $\mathbf{v}_3 \cdot \mathbf{v}_1 = 0$ and $\mathbf{v}_3 \cdot \mathbf{v}_2 = 0$. Now, we calculate

$$\alpha x + \beta y + \gamma z = \alpha(sa_1 + ta_2) + \beta(sb_1 + tb_2) + \gamma(sc_1 + tc_2)$$
$$= (\alpha a_1 + \beta b_1 + \gamma c_1)s + (\alpha a_2 + \beta b_2 + \gamma c_2)t$$
$$= (\mathbf{v}_3 \cdot \mathbf{v}_1)s + (\mathbf{v}_3 \cdot \mathbf{v}_2)t = 0.$$

Thus if $(x, y, z) \in H$, then $\alpha x + \beta y + \gamma z = 0$, which shows that H is a plane passing through the origin with normal vector $\mathbf{v}_3 = \mathbf{v}_1 \times \mathbf{v}_2$.

We now turn to another way of finding subspaces of a vector space V.

Span of a Set of Vectors

DEFINITION

Let $\mathbf{v}_1$, $\mathbf{v}_2$, ..., $\mathbf{v}_n$ be a set of n vectors in a vector space V. The **span** of $\{\mathbf{v}_1, \mathbf{v}_2, \ldots, \mathbf{v}_n\}$ is the set of linear combinations of $\mathbf{v}_1$, $\mathbf{v}_2$, ..., $\mathbf{v}_n$. That is:

$$\text{span}\,\{\mathbf{v}_1, \mathbf{v}_2, \ldots, \mathbf{v}_n\} = \{\mathbf{v}: \mathbf{v} = a_1\mathbf{v}_1 + a_2\mathbf{v}_2 + \cdots + a_n\mathbf{v}_n\} \qquad (8)$$

where a_1, a_2, ..., a_n are scalars.

Remark: Note that the word *span* is used in two different ways in this section.

■ **THEOREM 6**
Span $\{\mathbf{v}_1, \mathbf{v}_2, \ldots, \mathbf{v}_n\}$ is a subspace of V.

PROOF
The proof is not difficult and is left as an exercise (see Problem 26). ■

The proof of the following useful theorem is also left as an exercise (see Problem 63).

■ **THEOREM 7**
Let H be a subspace of the finite dimensional vector space V. Then H is finite dimensional and

$$\dim H \leq \dim V$$

■

Theorem 7 has a number of interesting consequences. For example, we can use the theorem to find all proper subspaces of $\mathbb{R}^3$. Let H be such a subspace. Then dim $H = 0$, 1, 2, or 3. If dim $H = 0$, then $H = \{\mathbf{0}\}$. If dim $H = 3$, let $\{\mathbf{v}_1, \mathbf{v}_2, \mathbf{v}_3\}$ be a basis for H. Then $H = \text{span}\{\mathbf{v}_1, \mathbf{v}_2, \mathbf{v}_3\}$. But, by Theorem 4, $\{\mathbf{v}_1, \mathbf{v}_2, \mathbf{v}_3\}$ is a basis for $\mathbb{R}^3$, so $\mathbb{R}^3 = H$. That is, H is not a proper subspace of $\mathbb{R}^3$.

In Problems 52 and 53 you are asked to show that the only one-dimensional subspaces of $\mathbb{R}^3$ are lines passing through the origin. Combining this result with the results of Examples 17 and 18 we can conclude that

The only proper subspaces of $\mathbb{R}^3$ are sets of vectors lying on either lines or planes passing through the origin.

Problems 6.7

In Problems 1–8 determine whether the given pair of vectors
is linearly dependent or independent.

1. $\begin{pmatrix} 1 \\ 2 \end{pmatrix}, \begin{pmatrix} 2 \\ 1 \end{pmatrix}$
    2. $\begin{pmatrix} 1 \\ 2 \end{pmatrix}, \begin{pmatrix} -1 \\ 2 \end{pmatrix}$

3. $\begin{pmatrix} 1 \\ 2 \end{pmatrix}, \begin{pmatrix} 6 \\ 12 \end{pmatrix}$
    4. $\begin{pmatrix} 1 \\ 0 \\ 1 \end{pmatrix}, \begin{pmatrix} 1 \\ 1 \\ 1 \end{pmatrix}$

5. $\begin{pmatrix} 1 \\ 2 \\ 3 \end{pmatrix}, \begin{pmatrix} -2 \\ -4 \\ -6 \end{pmatrix},$
    6. $\begin{pmatrix} 1 \\ 0 \\ 1 \\ 0 \end{pmatrix}, \begin{pmatrix} 2 \\ 0 \\ 2 \\ 0 \end{pmatrix}$

7. $\begin{pmatrix} 1 \\ 1 \\ 1 \end{pmatrix}, \begin{pmatrix} 2.5 \\ 2.5 \\ 2.5 \end{pmatrix}$
    8. $\begin{pmatrix} 2 \\ 3 \\ 1 \end{pmatrix}, \begin{pmatrix} 0 \\ 0 \\ 0 \end{pmatrix}$

9. Show that the vectors $\begin{pmatrix} 1 \\ 0 \\ 0 \end{pmatrix}, \begin{pmatrix} 1 \\ 1 \\ 0 \end{pmatrix}, \begin{pmatrix} 1 \\ 1 \\ 1 \end{pmatrix}$ are linearly inde-

pendent.

10. Find conditions on the constants a, b, and c such that the

vectors $\begin{pmatrix} a \\ d \\ e \end{pmatrix}, \begin{pmatrix} 0 \\ b \\ f \end{pmatrix}, \begin{pmatrix} 0 \\ 0 \\ c \end{pmatrix}$ are linearly independent.

In Problems 11–17 determine whether the given set of vectors
are linearly dependent or independent.

11. In P_2: $1 - x, x$
12. In P_2: $-x, x^2 - 2x, 3x + 5x^2$
13. In P_2: $1 - x, 1 + x, x^2$
14. In P_3: $x, x^2 - x, x^3 - x$
15. In P_3: $2x, x^3 - 3, 1 + x - 4x^3, x^3 + 18x - 9$
16. In $C[0, 1]$: $\sin x, \cos x$
17. In $C[0, 1]$: $x, \sqrt{x}, \sqrt[3]{x}$

18. Show that any four polynomials in P_2 are linearly dependent.
19. Show that two polynomials cannot span P_2.
20. Show that any $n + 2$ polynomials in P_n are linearly dependent.
21. If $p_1, p_2, \ldots, p_m$ span P_n, show that $m \geq n + 1$.
22. Let S_1 and S_2 be two finite, linearly independent sets in a vector space V. Show that $S_1 \cap S_2$ is a linearly independent set.
23. Show that the infinite set $\{1, x, x^2, x^3, \ldots\}$ spans P, the vector space of polynomials.

24. Show that if $\mathbf{u}$ and $\mathbf{v}$ are in span $\{\mathbf{v}_1, \mathbf{v}_2, \ldots, \mathbf{v}_n\}$, then $\mathbf{u} + \mathbf{v}$ and $\alpha\mathbf{u}$ are in span $\{\mathbf{v}_1, \mathbf{v}_2, \ldots, \mathbf{v}_n\}$. [*Hint:* Using the definition of span write $\mathbf{u} + \mathbf{v}$ and $\alpha\mathbf{u}$ as linear combinations of $\mathbf{v}_1, \mathbf{v}_2, \ldots, \mathbf{v}_n$.]
25. Let $\mathbf{v}_1, \mathbf{v}_2, \ldots, \mathbf{v}_n, \mathbf{v}_{n+1}$ be $n + 1$ vectors that are in a vector space V. If $\mathbf{v}_1, \mathbf{v}_2, \ldots, \mathbf{v}_n$ span V, then prove that $\mathbf{v}_1, \mathbf{v}_2, \ldots, \mathbf{v}_n, \mathbf{v}_{n+1}$ also span V. That is, the addition of one (or more) vectors to a spanning set yields another spanning set.
26. Prove Theorem 6.
27. Let H be a subspace of V containing $\mathbf{v}_1, \mathbf{v}_2, \ldots, \mathbf{v}_n$. Show that span $\{\mathbf{v}_1, \mathbf{v}_2, \ldots, \mathbf{v}_n\} \subseteq H$. That is, span $\{\mathbf{v}_1, \mathbf{v}_2, \ldots, \mathbf{v}_n\}$ is the *smallest* subspace of V containing $\mathbf{v}_1, \mathbf{v}_2, \ldots, \mathbf{v}_n$.
28. Show that any subset of a set of linearly independent vectors is linearly independent.
29. Let $\{\mathbf{v}_1, \mathbf{v}_2, \ldots, \mathbf{v}_n\}$ be a linearly independent set. Show that the vectors $\mathbf{v}_1, \mathbf{v}_1 + \mathbf{v}_2, \mathbf{v}_1 + \mathbf{v}_2 + \mathbf{v}_3, \ldots, \mathbf{v}_1 + \mathbf{v}_2 + \cdots + \mathbf{v}_n$ are linearly independent.
30. Let $\mathbf{v}_1 = (x_1, y_1, z_1)$ and $\mathbf{v}_2 = (x_2, y_2, z_2)$ be in $\mathbb{R}^3$. Show that if $\mathbf{v}_2 = c\mathbf{v}_1$, then span $\{\mathbf{v}_1, \mathbf{v}_2\}$ is a line passing through the origin.
31. Let $\{\mathbf{v}_1, \mathbf{v}_2, \ldots, \mathbf{v}_n\}$ be a set of vectors having the property that the set $\{\mathbf{v}_i, \mathbf{v}_j\}$ is linearly dependent when $i \neq j$. Show that each vector in the set is a multiple of a single vector in the set.

In Problems 32–46 determine whether the given vectors constitute a basis.

32. In $\mathbb{R}^2$: $(2, 3), (3, 2)$
33. In $\mathbb{R}^2$: $(1, 4), (-2, -8)$
34. In $\mathbb{R}^2$: $(2, -3), (1, 2), (3, -5)$
35. In $\mathbb{R}^3$: $\begin{pmatrix} 1 \\ 0 \\ 2 \end{pmatrix}, \begin{pmatrix} 1 \\ 0 \\ 0 \end{pmatrix}, \begin{pmatrix} 3 \\ 0 \\ -7 \end{pmatrix}$
36. In $\mathbb{R}^3$: $\begin{pmatrix} 2 \\ -3 \\ 5 \end{pmatrix}, \begin{pmatrix} 4 \\ -2 \\ 0 \end{pmatrix}, \begin{pmatrix} 7 \\ 0 \\ 0 \end{pmatrix}$
37. In $\mathbb{R}^3$: $\begin{pmatrix} 1 \\ 1 \\ 1 \end{pmatrix}, \begin{pmatrix} 2 \\ 2 \\ 0 \end{pmatrix}, \begin{pmatrix} 3 \\ 0 \\ 0 \end{pmatrix}$
38. In $\mathbb{R}^3$: $\begin{pmatrix} 1 \\ 4 \\ 1 \end{pmatrix}, \begin{pmatrix} 3 \\ 5 \\ 2 \end{pmatrix}, \begin{pmatrix} 4 \\ 9 \\ 3 \end{pmatrix}$

39. In $\mathbb{R}^3$: $\begin{pmatrix} 2 \\ 1 \\ 6 \end{pmatrix}, \begin{pmatrix} 4 \\ -1 \\ 5 \end{pmatrix}, \begin{pmatrix} 12 \\ 3 \\ -18 \end{pmatrix}, \begin{pmatrix} 2/3 \\ 3/4 \\ 5/6 \end{pmatrix}$

40. In P_2: $1 - x^2, x$

41. In P_2: $-3x, 1 + x^2, x^2 - 5$

42. In P_2: $x^2 - 1, x^2 - 2, x^2 - 3$

43. In P_3: $1, 1 + x, 1 + x^2, 1 + x^3$

44. In P_3: $3, x^3 - 4x + 6, x^2$

45. $H = \{(x, y) \in \mathbb{R}^2: x + y = 0\}; (1, -1)$

46. $H = \{(x, y) \in \mathbb{R}^2: x + y = 0\}; (1, -1), (-3, 3)$

47. Find a basis in $\mathbb{R}^3$ for the set of vectors in the plane

$$2x - y - z = 0.$$

48. Find a basis in $\mathbb{R}^3$ for the set of vectors in the plane

$$3x - 2y + 6z = 0.$$

49. Find a basis in $\mathbb{R}^3$ for the set of vectors on the line

$$x/2 = y/3 = z/4.$$

50. Find a basis in $\mathbb{R}^3$ for the set of vectors on the line

$$x = 3t, y = -2t, z = t.$$

51. Show that the only proper subspaces of $\mathbb{R}^2$ are straight lines passing through the origin.

52. Show that $S = \{(x, y, z): x = at, y = bt, z = ct, \text{ where } a, b, \text{ and } c \text{ are real numbers}\}$ is a one-dimensional subspace of $\mathbb{R}^3$. S is the set of vectors lying on a straight line through the origin.

53. Suppose that H is a subspace of $\mathbb{R}^3$ with dim $H = 1$. Show that, for some numbers a, b, and c, $H = \{(x, y, z): x = at, y = bt, z = ct\}$.

54. Let H be a proper subspace of the finite-dimensional vector space V. Show that $0 < \dim H < \dim V$.

55. In $\mathbb{R}^4$ let $H = \{(x, y, z, w): ax + by + cz + dw = 0\}$, where $abcd \neq 0$.

  (a) Show that H is a subspace of $\mathbb{R}^4$.

  (b) Find a basis for H.

56. In $\mathbb{R}^n$ a **hyperplane** is a subspace of dimension $n - 1$. If H is a hyperplane in $\mathbb{R}^n$ show that

$$H = \{(x_1, x_2, \ldots, x_n): a_1x_1 + a_2x_2 + \cdots + a_nx_n = 0\}$$

where $a_1, a_2, \ldots, a_n$ are fixed real numbers, not all of which are zero.

57. Show that if $\{v_1, v_2, \ldots, v_n\}$ spans V, then $\dim V \leq n$.

58. Let H and K be subspaces of V such that $H \subseteq K$ and $\dim H = \dim K < \infty$. Show that $H = K$.

59. Let H and K be subspaces of V and define

$$H + K = \{\mathbf{h} + \mathbf{k}: \mathbf{h} \in H \text{ and } \mathbf{k} \in K\}.$$

  (a) Show that $H + K$ is a subspace of V.

  (b) If $H \cap K = \{\mathbf{0}\}$, show that $\dim (H + K) = \dim H + \dim K$.

* 60. If H is a subspace of the finite dimensional vector space V, show that there exists a unique subspace K of V such that **(a)** $H \cap K = \{\mathbf{0}\}$ and **(b)** $H + K = V$.

61. Show that two vectors $\mathbf{v}_1$ and $\mathbf{v}_2$ in $\mathbb{R}^2$ with endpoints at the origin are collinear if and only if

$$\dim \text{ span } \{\mathbf{v}_1, \mathbf{v}_2\} = 1.$$

62. Show that three vectors $\mathbf{v}_1$, $\mathbf{v}_2$, and $\mathbf{v}_3$ in $\mathbb{R}^3$ with endpoints at the origin are coplanar if and only if

$$\dim \text{ span } \{\mathbf{v}_1, \mathbf{v}_2, \mathbf{v}_3\} \leq 2.$$

* 63. Prove Theorem 7.

The following problems make use of material in Chapter 2.

64. Let V denote the subset of $C^{(2)}[0, 1]$ consisting of solutions to the linear homogeneous differential equation

$$y'' + a(x)y' + b(x)y = 0,$$

where $a(x)$ and $b(x)$ are continuous. Here $C^{(2)}[0, 1]$ denotes the set of twice continuously differentiable functions defined on $[0, 1]$.

  (a) Show that V is a subspace of $C[0, 1]$.

  (b) What is its dimension?

[See Problem 6.5.20 on p. 326.]

65. In Problem 64, let $H = \{y(x) \in V: y(0) = 0\}$.

  (a) Show that H is a subspace of V.

  (b) Find dim H.

In Problems 66–74 find a basis for the given vector space.

66. $V = \{x(t): x'' + x = 0\}$

67. $V = \{x(t): x'' - 4x' + 3x = 0\}$

68. $V = \{x(t): x'' + 3x' - 10x = 0\}$

69. $V = \{x(t): x'' - 4x' + 4x = 0\}$

70. $V = \{x(t): x'' + 6x' + 9x = 0\}$

71. $V = \{x(t): x'' + 4x = 0\}$

72. $V = \{x(t): x'' - 4x' + 5x = 0\}$

73. $V = \{x(t): x''' - 6x'' + 11x' - 6x = 0\}$

74. $V = \{x(t): x''' + 4x'' + x' + 4x = 0\}$

6.8 The Dot Product in $\mathbb{R}^n$, Orthonormal Bases, and Inner Product Spaces

In Section 6.2 we defined the dot product in $\mathbb{R}^2$ and $\mathbb{R}^3$. We can define the dot product in $\mathbb{R}^n$ in a similar way.

Dot Product in $\mathbb{R}^n$ (also known as the **inner product** or **scalar product**)

If $\mathbf{x} \in \mathbb{R}^n$ has the components $x_1, x_2, \ldots, x_n$ and $\mathbf{y} \in \mathbb{R}^n$ has the components $y_1, y_2, \ldots, y_n$ then the **dot product** of $\mathbf{x}$ and $\mathbf{y}$ is defined by

$$\mathbf{x} \cdot \mathbf{y} = x_1 y_1 + x_2 y_2 + \cdots + x_n y_n.$$

In this definition $\mathbf{x}$ and $\mathbf{y}$ can be row or column vectors. The dot product in $\mathbb{R}^n$ is usually written in one of three ways:

(I) $(x_1, x_2, \ldots, x_n) \cdot (y_1, y_2, \ldots, y_n) = x_1 y_1 + x_2 y_2 + \cdots + x_n y_n,$

(II) $\begin{pmatrix} x_1 \\ x_2 \\ \vdots \\ x_n \end{pmatrix} \cdot \begin{pmatrix} y_1 \\ y_2 \\ \vdots \\ y_n \end{pmatrix} = x_1 y_1 + x_2 y_2 + \cdots + x_n y_n,$

(III) $(x_1, x_2, \ldots, x_n) \cdot \begin{pmatrix} y_1 \\ y_2 \\ \vdots \\ y_n \end{pmatrix} = x_1 y_1 + x_2 y_2 + \cdots + x_n y_n.$

The third way will be used in Section 7.2 to define matrix multiplication.

EXAMPLE 1

Compute $(1, 3, -1, 2) \cdot \begin{pmatrix} 4 \\ 1 \\ 6 \\ 5 \end{pmatrix}$.

SOLUTION

$(1, 3, -1, 2) \cdot \begin{pmatrix} 4 \\ 1 \\ 6 \\ 5 \end{pmatrix} = 1 \cdot 4 + 3 \cdot 1 + (-1) \cdot 6 + 2 \cdot 5 = 11.$

In Theorem 6.7.4 we stated that n linearly independent vectors in $\mathbb{R}^n$ form a basis for $\mathbb{R}^n$. The most commonly used basis for $\mathbb{R}^n$ is the **standard basis** $E = \{\mathbf{e}_1, \mathbf{e}_2, \ldots, \mathbf{e}_n\}$, where

$$\mathbf{e}_j = (0, 0, \ldots, 0, 1, 0, \ldots, 0).$$

$$\uparrow$$

$$j\text{th position}$$

Among the main reasons this basis is preferred is that the vectors have two properties:

i. $\mathbf{e}_i \cdot \mathbf{e}_j = 0 \quad$ if $i \neq j$

ii. $\mathbf{e}_i \cdot \mathbf{e}_i = 1$

DEFINITION

Orthonormal Set in $\mathbb{R}^n$

The set of vectors $S = \{\mathbf{u}_1, \mathbf{u}_2, \ldots, \mathbf{u}_k\}$ in $\mathbb{R}^n$ is said to be an **orthonormal set** if

$$\mathbf{u}_i \cdot \mathbf{u}_j = 0 \qquad \text{if} \qquad i \neq j, \tag{1}$$

$$\mathbf{u}_i \cdot \mathbf{u}_i = 1. \tag{2}$$

If only equation (1) is satisfied, the set is called **orthogonal.**

Since we shall be working with the dot product extensively in this section, let us recall some basic facts. These facts generalize Theorem 6.2.1 on p. 302. The proofs follow easily from the definition of the dot product. We shall use these facts in the rest of this section without again mentioning them explicitly. If $\mathbf{u}$, $\mathbf{v}$, and $\mathbf{w}$ are in $\mathbb{R}^n$ and α is a real number, then

$$\mathbf{u} \cdot \mathbf{v} = \mathbf{v} \cdot \mathbf{u} \tag{3}$$

$$(\mathbf{u} + \mathbf{v}) \cdot \mathbf{w} = \mathbf{u} \cdot \mathbf{w} + \mathbf{v} \cdot \mathbf{w} \tag{4}$$

$$\mathbf{u} \cdot (\mathbf{v} + \mathbf{w}) = \mathbf{u} \cdot \mathbf{v} + \mathbf{u} \cdot \mathbf{w} \tag{5}$$

$$(\alpha\mathbf{u}) \cdot \mathbf{v} = \alpha(\mathbf{u} \cdot \mathbf{v}) \tag{6}$$

$$\mathbf{u} \cdot (\alpha\mathbf{v}) = \alpha(\mathbf{u} \cdot \mathbf{v}) \tag{7}$$

We now give another useful definition based on the fact that

$$\mathbf{v} \cdot \mathbf{v} = x_1^2 + x_2^2 + \cdots + x_n^2.$$

DEFINITION

Length or Norm of a Vector

If $\mathbf{v} \in \mathbb{R}^n$, then the **length** or **norm** of $\mathbf{v}$, written $|\mathbf{v}|$, is given by

$$|\mathbf{v}| = \sqrt{\mathbf{v} \cdot \mathbf{v}} \geq 0 \qquad\qquad (8)$$

with

$$|\mathbf{v}| = 0 \qquad \text{if and only if} \qquad \mathbf{v} = \mathbf{0}. \qquad (9)$$

EXAMPLE 2

If $\mathbf{v} = (x, y, z) \in \mathbb{R}^3$, then as we saw earlier,

$$|\mathbf{v}| = \sqrt{x^2 + y^2 + z^2}.$$

For example,

$$|(2, -1, 3)| = \sqrt{2^2 + (-1)^2 + 3^2} = \sqrt{14}.$$

We can now restate the definition of an orthonormal set.

> A set of vectors is orthonormal if each pair of them is orthogonal and each has length 1.

Orthonormal sets of vectors are reasonably easy to use. One reason is that any finite orthogonal set of nonzero vectors is linearly independent.

■ **THEOREM 1**

If $S = \{\mathbf{v}_1, \mathbf{v}_2, \ldots, \mathbf{v}_k\}$ is an orthogonal set of nonzero vectors, then S is linearly independent.

PROOF

Suppose that $c_1\mathbf{v}_1 + c_2\mathbf{v}_2 + \cdots + c_k\mathbf{v}_k = \mathbf{0}$. Then, for any $i = 1, 2, \ldots, k$,

$$0 = \mathbf{0} \cdot \mathbf{v}_i = (c_1\mathbf{v}_1 + c_2\mathbf{v}_2 + \cdots + c_i\mathbf{v}_i + \cdots + c_k\mathbf{v}_k) \cdot \mathbf{v}_i$$

$$= c_1(\mathbf{v}_1 \cdot \mathbf{v}_i) + c_2(\mathbf{v}_2 \cdot \mathbf{v}_i) + \cdots + c_i(\mathbf{v}_i \cdot \mathbf{v}_i) + \cdots + c_k(\mathbf{v}_k \cdot \mathbf{v}_i)$$

$$= c_1 0 + c_2 0 + \cdots + c_i|\mathbf{v}_i|^2 + \cdots + c_k 0 = c_i|\mathbf{v}_i|^2$$

Since $\mathbf{v}_i \neq 0$ by hypothesis, $|\mathbf{v}_i|^2 > 0$ and we have $c_i = 0$. This is true for $i = 1, 2, \ldots, k$ and the proof is complete. ■

We shall now see how *any* basis in $\mathbb{R}^n$ can be "turned into" an orthonormal basis. The method described below is called the **Gram–Schmidt orthonormalization process.†**

---

† Jörgen Pederson Gram (1850–1916) was a Danish actuary who was very interested in the science of measurement. Erhardt Schmidt (1876–1959) was a German mathematician.

■ **THEOREM 2: Gram–Schmidt Orthonormalization Process**

Let H be an m-dimensional subspace of $\mathbb{R}^n$. Then H has an orthonormal basis.†

PROOF

Let $S = \{\mathbf{v}_1, \mathbf{v}_2, \ldots, \mathbf{v}_m\}$ be a basis for H. We shall prove the theorem by constructing an orthonormal basis from the vectors in S. Before giving the steps in this construction we recall the fact that a linearly independent set of vectors does *not* contain the zero vector (see Problem 42).

STEP 1
Let

$$\mathbf{u}_1 = \frac{\mathbf{v}_1}{|\mathbf{v}_1|} \tag{10}$$

Then $\mathbf{u}_1 \cdot \mathbf{u}_1 = (\mathbf{v}_1/|\mathbf{v}_1|) \cdot (\mathbf{v}_1/|\mathbf{v}_1|) = (1/|\mathbf{v}_1|^2)(\mathbf{v}_1 \cdot \mathbf{v}_1) = 1$, so that $|\mathbf{u}_1| = 1$.

STEP 2
Let

$$\mathbf{v}_2' = \mathbf{v}_2 - (\mathbf{v}_2 \cdot \mathbf{u}_1)\mathbf{u}_1 \tag{11}$$

Then $\mathbf{v}_2' \cdot \mathbf{u}_1 = \mathbf{v}_2 \cdot \mathbf{u}_1 - (\mathbf{v}_2 \cdot \mathbf{u}_1)(\mathbf{u}_1 \cdot \mathbf{u}_1) = \mathbf{v}_2 \cdot \mathbf{u}_1 - \mathbf{v}_2 \cdot \mathbf{u}_1 = 0$, so that $\mathbf{v}_2'$ is orthogonal to $\mathbf{u}_1$. If $\mathbf{v}_2' = \mathbf{0}$, then $\mathbf{v}_2 - (\mathbf{v}_2 \cdot \mathbf{u}_1)\mathbf{u}_1 = 0$ so $\mathbf{v}_2$ is a multiple of $\mathbf{u}_1$ and is therefore a multiple of $\mathbf{v}_1$. This is impossible because the $\mathbf{v}$'s are linearly independent. Thus $\mathbf{v}_2' \neq \mathbf{0}$.

STEP 3
Let

$$\mathbf{u}_2 = \frac{\mathbf{v}_2'}{|\mathbf{v}_2'|} \tag{12}$$

Then clearly $\{\mathbf{u}_1, \mathbf{u}_2\}$ is an orthonormal set.

Suppose now that the vectors $\mathbf{u}_1, \mathbf{u}_2, \ldots, \mathbf{u}_k$ $(k < m)$ have been constructed and form an orthonormal set. We show how to construct $\mathbf{u}_{k+1}$.

STEP 4
Let

$$\mathbf{v}_{k+1}' = \mathbf{v}_{k+1} - (\mathbf{v}_{k+1} \cdot \mathbf{u}_1)\mathbf{u}_1 - (\mathbf{v}_{k+1} \cdot \mathbf{u}_2)\mathbf{u}_2 - \cdots - (\mathbf{v}_{k+1} \cdot \mathbf{u}_k)\mathbf{u}_k \tag{13}$$

Then, for $i = 1, 2, \ldots, k$,

$$\mathbf{v}_{k+1}' \cdot \mathbf{u}_i = \mathbf{v}_{k+1} \cdot \mathbf{u}_i - (\mathbf{v}_{k+1} \cdot \mathbf{u}_1)(\mathbf{u}_1 \cdot \mathbf{u}_i) - (\mathbf{v}_{k+1} \cdot \mathbf{u}_2)(\mathbf{u}_2 \cdot \mathbf{u}_i)$$

$$- \cdots - (\mathbf{v}_{k+1} \cdot \mathbf{u}_i)(\mathbf{u}_i \cdot \mathbf{u}_i) - \cdots - (\mathbf{v}_{k+1} \cdot \mathbf{u}_k)(\mathbf{u}_k \cdot \mathbf{u}_i)$$

But $\mathbf{u}_j \cdot \mathbf{u}_i = 0$ if $j \neq i$ and $\mathbf{u}_i \cdot \mathbf{u}_i = 1$. Thus

† Note that H may be $\mathbb{R}^n$ in this theorem. That is, $\mathbb{R}^n$ itself has an orthonormal basis.

$$\mathbf{v}'_{k+1} \cdot \mathbf{u}_i = \mathbf{v}_{k+1} \cdot \mathbf{u}_i - \mathbf{v}_{k+1} \cdot \mathbf{u}_i = 0$$

Hence $\{\mathbf{u}_1, \mathbf{u}_2, \ldots, \mathbf{u}_k, \mathbf{v}'_{k+1}\}$ is an orthogonal, linearly independent set and $\mathbf{v}'_{k+1} \neq \mathbf{0}$.

STEP 5

Let $\mathbf{u}_{k+1} = \mathbf{v}'_{k+1}/|\mathbf{v}'_{k+1}|$. Then clearly $\{\mathbf{u}_1, \mathbf{u}_2, \ldots, \mathbf{u}_k, \mathbf{u}_{k+1}\}$ is an orthonormal set and we continue in this manner until $k + 1 = m$ and the proof is complete. ∎

Remark. In $\mathbb{R}^2$ or $\mathbb{R}^3$ we can see, geometrically, what is happening. First we note that

$$\mathbf{v}'_2 = \mathbf{v}_2 - (\mathbf{v}_2 \cdot \mathbf{u}_1)\mathbf{u}_1 = \mathbf{v}_2 - \left(\mathbf{v}_2 \cdot \frac{\mathbf{v}_1}{|\mathbf{v}_1|}\right)\left(\frac{\mathbf{v}_1}{|\mathbf{v}_1|}\right) = \mathbf{v}_2 - \frac{(\mathbf{v}_2 \cdot \mathbf{v}_1)}{|\mathbf{v}_1|^2}\mathbf{v}_1$$

But, from equation (6.2.4) on p. 304, $[(\mathbf{v}_2 \cdot \mathbf{v}_1)/|\mathbf{v}_1|^2]\mathbf{v}_1$ is the projection of $\mathbf{v}_2$ on $\mathbf{v}_1$. Moreover, from Fig. 6.2.1, the vector $\mathbf{v}'_2 = \mathbf{v}_2 - \text{Proj}_{\mathbf{v}_1}\mathbf{v}_2$ is a vector orthogonal to $\mathbf{v}_1$.

Thus we see that the process we have described here uses a generalization of the notion of projection in $\mathbb{R}^2$ and $\mathbb{R}^3$.

EXAMPLE 3

Construct an orthonormal basis in $\mathbb{R}^3$ starting with the basis

$$\{\mathbf{v}_1, \mathbf{v}_2, \mathbf{v}_3\} = \left\{ \begin{pmatrix} 1 \\ 1 \\ 0 \end{pmatrix}, \begin{pmatrix} 0 \\ 1 \\ 1 \end{pmatrix}, \begin{pmatrix} 1 \\ 0 \\ 1 \end{pmatrix} \right\}.$$

SOLUTION

We first verify that the vectors are linearly independent (so we do, indeed, have a basis). If not, then from Theorem 6.7.2 there are constants a and b such that

$$\begin{pmatrix} 1 \\ 0 \\ 1 \end{pmatrix} = a\begin{pmatrix} 1 \\ 1 \\ 0 \end{pmatrix} + b\begin{pmatrix} 0 \\ 1 \\ 1 \end{pmatrix} = \begin{pmatrix} a \\ a+b \\ b \end{pmatrix}$$

or

$$a = 1$$
$$a + b = 0$$
$$b = 1$$

which is impossible since $a = 1$, $b = 1$, and $a + b = 2$, not 0. Now we compute $|\mathbf{v}_1| = \sqrt{2}$, so $\mathbf{u}_1 = \begin{pmatrix} 1/\sqrt{2} \\ 1/\sqrt{2} \\ 0 \end{pmatrix}$. Then

$$\mathbf{v}'_2 = \mathbf{v}_2 - (\mathbf{v}_2 \cdot \mathbf{u}_1)\mathbf{u}_1 = \begin{pmatrix} 0 \\ 1 \\ 1 \end{pmatrix} - \frac{1}{\sqrt{2}}\begin{pmatrix} 1/\sqrt{2} \\ 1/\sqrt{2} \\ 0 \end{pmatrix} = \begin{pmatrix} 0 \\ 1 \\ 1 \end{pmatrix} - \begin{pmatrix} 1/2 \\ 1/2 \\ 0 \end{pmatrix} = \begin{pmatrix} -1/2 \\ 1/2 \\ 1 \end{pmatrix}.$$

Since $|\mathbf{v}_2'| = \sqrt{3/2}$, $\mathbf{u}_2 = \sqrt{2/3}\begin{pmatrix} -1/2 \\ 1/2 \\ 1 \end{pmatrix} = \begin{pmatrix} -1/\sqrt{6} \\ 1/\sqrt{6} \\ 2/\sqrt{6} \end{pmatrix}$. Continuing, we have

$$\mathbf{v}_3' = \mathbf{v}_3 - (\mathbf{v}_3 \cdot \mathbf{u}_1)\mathbf{u}_1 - (\mathbf{v}_3 \cdot \mathbf{u}_2)\mathbf{u}_2$$

$$= \begin{pmatrix} 1 \\ 0 \\ 1 \end{pmatrix} - \frac{1}{\sqrt{2}}\begin{pmatrix} 1/\sqrt{2} \\ 1/\sqrt{2} \\ 0 \end{pmatrix} - \frac{1}{\sqrt{6}}\begin{pmatrix} -1/\sqrt{6} \\ 1/\sqrt{6} \\ 2/\sqrt{6} \end{pmatrix} = \begin{pmatrix} 1 \\ 0 \\ 1 \end{pmatrix} - \begin{pmatrix} 1/2 \\ 1/2 \\ 0 \end{pmatrix} - \begin{pmatrix} -1/6 \\ 1/6 \\ 2/6 \end{pmatrix} = \begin{pmatrix} 2/3 \\ -2/3 \\ 2/3 \end{pmatrix}$$

Finally $|\mathbf{v}_3'| = \sqrt{\frac{12}{9}} = 2/\sqrt{3}$, so that $\mathbf{u}_3 = \frac{\sqrt{3}}{2}\begin{pmatrix} 2/3 \\ -2/3 \\ 2/3 \end{pmatrix} = \begin{pmatrix} 1/\sqrt{3} \\ -1/\sqrt{3} \\ 1/\sqrt{3} \end{pmatrix}$. Thus the orthonormal

basis is $\left\{ \begin{pmatrix} 1/\sqrt{2} \\ 1/\sqrt{2} \\ 0 \end{pmatrix}, \begin{pmatrix} -1/\sqrt{6} \\ 1/\sqrt{6} \\ 2/\sqrt{6} \end{pmatrix}, \begin{pmatrix} 1/\sqrt{3} \\ -1/\sqrt{3} \\ 1/\sqrt{3} \end{pmatrix} \right\}$. This result should be checked.

Inner Product Spaces (Optional)

The space $\mathbb{R}^n$ is a vector space with an inner product (dot product) defined on it. It is one example of an inner product space. There are other important inner product spaces. Before giving a general definition we note that in $\mathbb{R}^n$, the inner product of two vectors is a real scalar. In other spaces (see Example 5 below) the inner product gives us a complex scalar. To include both cases, therefore, we assume in the following definition that the inner product of two vectors is a complex number. Since every real number is a complex number, this definition includes the real case as well.

Inner Product Space

The complex vector space V is called an **inner product space** if for every pair of vectors $\mathbf{u}$ and $\mathbf{v}$ in V, there is a unique complex number $(\mathbf{u}, \mathbf{v})$, called the **inner product** of $\mathbf{u}$ and $\mathbf{v}$, such that if $\mathbf{u}$, $\mathbf{v}$, and $\mathbf{w}$ are in V and $\alpha \in \mathbb{C}$, then

    i. $(\mathbf{v}, \mathbf{v}) \geq 0$
    ii. $(\mathbf{v}, \mathbf{v}) = 0$ if and only if $\mathbf{v} = \mathbf{0}$
    iii. $(\mathbf{u}, \mathbf{v} + \mathbf{w}) = (\mathbf{u}, \mathbf{v}) + (\mathbf{u}, \mathbf{w})$
    iv. $(\mathbf{u} + \mathbf{v}, \mathbf{w}) = (\mathbf{u}, \mathbf{w}) + (\mathbf{v}, \mathbf{w})$
    v. $(\mathbf{u}, \mathbf{v}) = \overline{(\mathbf{v}, \mathbf{u})}$
    vi. $(\alpha\mathbf{u}, \mathbf{v}) = \alpha(\mathbf{u}, \mathbf{v})$
    vii. $(\mathbf{u}, \alpha\mathbf{v}) = \bar{\alpha}(\mathbf{u}, \mathbf{v})$

The bar in conditions (v) and (vii) denotes the complex conjugate. (See Section 15.1 for a discussion of complex numbers.)

    Note. If $(\mathbf{u}, \mathbf{v})$ is real, then $\overline{(\mathbf{u}, \mathbf{v})} = (\mathbf{u}, \mathbf{v})$ and we can remove the bar in (v).

The **norm** of a vector **v** in an inner product space is denoted by $\|\mathbf{v}\|$ and given by

$$\|\mathbf{v}\| = (\mathbf{v}, \mathbf{v})^{1/2}$$

EXAMPLE 4

$\mathbb{R}^n$ is an inner product space with $(\mathbf{u}, \mathbf{v}) = \mathbf{u} \cdot \mathbf{v}$.

EXAMPLE 5

We defined the space $\mathbb{C}^n$ in Example 6.5.8. Let $\mathbf{x} = (x_1, x_2, \ldots, x_n)$ and $\mathbf{y} = (y_1, y_2, \ldots, y_n)$ be in $\mathbb{C}^n$. (Remember—this means that the x_i's and y_i's are complex numbers.) Then we define

$$(\mathbf{x}, \mathbf{y}) = x_1 \bar{y}_1 + x_2 \bar{y}_2 + \cdots + x_n \bar{y}_n. \tag{14}$$

To show that equation (14) defines an inner product we need some facts about complex numbers. If these are unfamiliar, refer to Section 15.1. For (i):

$$(\mathbf{x}, \mathbf{x}) = x_1 \bar{x}_1 + x_2 \bar{x}_2 + \cdots + x_n \bar{x}_n = |x_1|^2 + |x_2|^2 + \cdots + |x_n|^2.$$

Thus (i) and (ii) are satisfied since $|x_i|$ is a real number. Conditions (iii) and (iv) follow from the fact that $z_1(z_2 + z_3) = z_1 z_2 + z_1 z_3$ for any complex numbers z_1, z_2, and z_3. Condition (v) follows from the fact that $\overline{z_1 z_2} = \bar{z}_1 \bar{z}_2$ and $\bar{\bar{z}}_1 = z_1$ so that $x_1 \bar{y}_1 = \bar{x}_1 y_1$. Condition ($vi$) is obvious. For ($vii$): $(\mathbf{u}, \alpha \mathbf{v}) = \overline{(\alpha \mathbf{v}, \mathbf{u})} = \overline{(\overline{\alpha} \bar{\mathbf{v}}, \bar{\mathbf{u}})} = \bar{\alpha}(\bar{\mathbf{v}}, \bar{\mathbf{u}}) = \bar{\alpha}(\mathbf{u}, \mathbf{v})$. Here we used ($vi$) and ($v$).

EXAMPLE 6

In $\mathbb{C}^3$ let $\mathbf{x} = (1 + i, -3, 4 - 3i)$ and $\mathbf{y} = (2 - i, -i, 2 + i)$. Then

$$(\mathbf{x}, \mathbf{y}) = (1 + i)(\overline{2 - i}) + (-3)(\overline{-i}) + (4 - 3i)(\overline{2 + i})$$

$$= (1 + i)(2 + i) + (-3)(i) + (4 - 3i)(2 - i)$$

$$= (1 + 3i) - 3i + (5 - 10i) = 6 - 10i.$$

EXAMPLE 7

Suppose that $a < b$; let $V = C[a, b]$ and define

$$(f, g) = \int_a^b f(t) g(t) \, dt$$

We shall see that this is also an inner product.

(i) $(f, f) = \int_a^b f^2(t) \, dt \geq 0$. It is a basic theorem of calculus that if $f \in C[a, b]$, $f \geq 0$ on $[a, b]$, and $\int_a^b f(t) \, dt = 0$, then $f = 0$ on $[a, b]$. This proves (i) and (ii). (iii)–(vii) follow from basic facts about definite integrals.

EXAMPLE 8

Let $f(t) = t^2 \in C[0, 1]$ and $g(t) = (4 - t) \in C[0, 1]$. Then

$$(f, g) = \int_0^1 t^2(4 - t)\, dt = \int_0^1 (4t^2 - t^3)\, dt = \left(\frac{4t^3}{3} - \frac{t^4}{4}\right)\Big|_0^1 = \frac{13}{12}$$

Orthonormal Set

The set of vectors $\{\mathbf{v}_1, \mathbf{v}_2, \ldots, \mathbf{v}_n\}$ is an **orthonormal set** in an inner product space V if

$$(\mathbf{v}_i, \mathbf{v}_j) = 0 \quad \text{for} \quad i \neq j \tag{15}$$

and

$$|\mathbf{v}_i| = \sqrt{(\mathbf{v}_i, \mathbf{v}_i)} = 1 \tag{16}$$

If only (15) holds, the set is said to be **orthogonal.**

The Gram–Schmidt process can be used to find an orthonormal basis for any finite-dimensional inner product space. In Problem 31 you are asked to find an orthornormal basis for $P_2[0, 1]$.

In Chapter 11 we will discuss Fourier series. In solving problems using Fourier series, it is first necessary to find certain infinite, orthonormal sets.

EXAMPLE 9

Show that $\left\{\dfrac{1}{\sqrt{2\pi}}, \dfrac{\sin x}{\sqrt{\pi}}, \dfrac{\cos x}{\sqrt{\pi}}, \dfrac{\sin 2x}{\sqrt{\pi}}, \dfrac{\cos 2x}{\sqrt{\pi}}, \dfrac{\sin 3x}{\sqrt{\pi}}, \dfrac{\cos 3x}{\sqrt{\pi}}, \ldots\right\}$ is an orthonormal set

in $C[-\pi, \pi]$ with the inner product defined by $(f, g) = \int_{-\pi}^{\pi} f(x)g(x)\, dx$.

SOLUTION
Since

$$(f, g) = \int_{-\pi}^{\pi} f(x)g(x)\, dx,$$

we have

$$\|f\| = (f, f)^{1/2} = \left[\int_{-\pi}^{\pi} f^2(x)\, dx\right]^{1/2}.$$

Thus

$$\left\|\frac{1}{\sqrt{2\pi}}\right\| = \left(\int_{-\pi}^{\pi}\left(\frac{1}{\sqrt{2\pi}}\right)^2 dx\right)^{1/2} = \left(\int_{-\pi}^{\pi}\frac{1}{2\pi}\, dx\right)^{1/2} = 1,$$

entry 100 in the table
of integrals
↓

$$\left\|\frac{\sin kx}{\sqrt{\pi}}\right\|^2 = \frac{1}{\pi}\int_{-\pi}^{\pi}\sin^2 kx\, dx = \frac{1}{\pi}\left[\frac{x}{2} - \frac{\sin 2kx}{4k}\right]\Big|_{-\pi}^{\pi} = 1,$$

entry 115
↓

$$\left\|\frac{\cos kx}{\sqrt{\pi}}\right\|^2 = \frac{1}{\pi}\int_{-\pi}^{\pi}\cos^2 kx\, dx = \frac{1}{\pi}\left[\frac{x}{2} + \frac{\sin 2kx}{4k}\right]\Big|_{-\pi}^{\pi} = 1.$$

If $k \neq m$,

entry 124
↓

$$\left(\frac{\sin kx}{\sqrt{\pi}}, \frac{\cos mx}{\sqrt{\pi}}\right) = \frac{1}{\pi}\int_{-\pi}^{\pi}\sin kx \cos mx\, dx = \frac{1}{\pi}\left[-\frac{\cos(k-m)x}{2(k-m)} - \frac{\cos(k+m)x}{2(k+m)}\right]\Big|_{-\pi}^{\pi} = 0,$$

entry 103
↓

$$\left(\frac{\sin kx}{\sqrt{\pi}}, \frac{\sin mx}{\sqrt{\pi}}\right) = \frac{1}{\pi}\int_{-\pi}^{\pi}\sin kx \sin mx\, dx = \frac{1}{\pi}\left[\frac{\sin(k-m)x}{2(k-m)} - \frac{\sin(k+m)x}{2(k+m)}\right]\Big|_{-\pi}^{\pi} = 0,$$

entry 118
↓

$$\left(\frac{\cos kx}{\sqrt{\pi}}, \frac{\cos mx}{\sqrt{\pi}}\right) = \frac{1}{\pi}\int_{-\pi}^{\pi}\cos kx \cos mx\, dx = \frac{1}{\pi}\left[\frac{\sin(k-m)x}{2(k-m)} + \frac{\sin(k+m)x}{2(k+m)}\right]\Big|_{-\pi}^{\pi} = 0.$$

If $k = m$

entry 123
↓

$$\left(\frac{\sin kx}{\sqrt{\pi}}, \frac{\cos kx}{\sqrt{\pi}}\right) = \frac{1}{\pi}\int_{-\pi}^{\pi}\sin kx \cos kx\, dx = \frac{\sin^2 kx}{2k\pi}\Big|_{-\pi}^{\pi} = 0.$$

Finally observe that

$$\left(\frac{1}{\sqrt{2\pi}}, \frac{\sin kx}{\sqrt{\pi}}\right) = \frac{1}{\pi\sqrt{2}}\int_{-\pi}^{\pi}\sin kx\, dx = \frac{-1}{\pi\sqrt{2}}\left[\frac{\cos kx}{k}\right]\Big|_{-\pi}^{\pi} = 0,$$

$$\left(\frac{1}{\sqrt{2\pi}}, \frac{\cos kx}{\sqrt{\pi}}\right) = \frac{1}{\pi\sqrt{2}}\int_{-\pi}^{\pi}\cos kx\, dx = \frac{1}{\pi\sqrt{2}}\left[\frac{\sin kx}{k}\right]\Big|_{-\pi}^{\pi} = 0.$$

Problems 6.8

In Problems 1–5 compute the dot product of the given vectors.

1. $(1, 3, 2, -1), (4, 2, 5, 7)$

2. $\begin{pmatrix} 1 \\ 0 \\ 1 \\ 0 \\ 1 \end{pmatrix}, \begin{pmatrix} 0 \\ 1 \\ 0 \\ 1 \\ 0 \end{pmatrix}$
3. $\begin{pmatrix} 2 \\ 3 \\ 1 \\ 5 \\ 1 \end{pmatrix}, \begin{pmatrix} 3 \\ -1 \\ 2 \\ 4 \\ 6 \end{pmatrix}$

4. $(a, b, c, d, e), (b, c, d, e, a)$

5. $(1, 2, 3, \ldots, n), \left(1, \frac{1}{2}, \frac{1}{3}, \ldots, \frac{1}{n}\right)$

In Problems 6–10 compute the norm of the given vector.

6. $\begin{pmatrix} 1 \\ 2 \\ 3 \\ 4 \end{pmatrix}$
7. $\begin{pmatrix} 1 \\ 1/2 \\ 1/3 \\ 1/4 \end{pmatrix}$

8. $(0, 0, 1, 1, 0, 0)$ **9.** (a, b, c, d, e)

10. $(1, 1, 1, 1, 1, 1, 1, 1)$

In Problems 11–20 a vector space and a basis for that vector space are given. First verify that the vectors given really do form a basis. Then, starting with these vectors, construct an orthonormal basis.

11. $\mathbb{R}^2$; $\begin{pmatrix} 1 \\ 1 \end{pmatrix}$, $\begin{pmatrix} -1 \\ 1 \end{pmatrix}$

12. $\mathbb{R}^2$; $\begin{pmatrix} 2 \\ 3 \end{pmatrix}$, $\begin{pmatrix} 3 \\ 4 \end{pmatrix}$

13. $H = \{(x, y) \in \mathbb{R}^2 : x + y = 0\}$; $\begin{pmatrix} 1 \\ -1 \end{pmatrix}$

14. $H = \{(x, y) \in \mathbb{R}^2 : ax + by = 0, ab \neq 0\}$; $\begin{pmatrix} b \\ -a \end{pmatrix}$

15. $\mathbb{R}^2$; $\begin{pmatrix} a \\ b \end{pmatrix}$, $\begin{pmatrix} c \\ d \end{pmatrix}$, where $ad - bc \neq 0$

16. $\mathbb{R}^3$; $\begin{pmatrix} 1 \\ 2 \\ 3 \end{pmatrix}$, $\begin{pmatrix} 0 \\ 1 \\ 2 \end{pmatrix}$, $\begin{pmatrix} 0 \\ 0 \\ 5 \end{pmatrix}$

17. $\mathbb{R}^3$; $\begin{pmatrix} 1 \\ 2 \\ 0 \end{pmatrix}$, $\begin{pmatrix} 2 \\ 1 \\ 0 \end{pmatrix}$, $\begin{pmatrix} 1 \\ 0 \\ 2 \end{pmatrix}$

18. $\mathbb{R}^3$; $\begin{pmatrix} a \\ 0 \\ 0 \end{pmatrix}$, $\begin{pmatrix} b \\ c \\ 0 \end{pmatrix}$, $\begin{pmatrix} d \\ e \\ f \end{pmatrix}$, $acf \neq 0$

19. $\Pi = \{(x, y, z): 2x - y - z = 0\}$; $(1, 0, 2)$, $(0, 1, -1)$ [See Example 6.7.17.]

20. $L = \left\{(x, y, z): \dfrac{x}{2} = \dfrac{y}{3} = \dfrac{z}{4}\right\}$; $(2, 3, 4)$

In Problems 21–24 compute the inner product of the two vectors in $\mathbb{C}^n$.

21. $\begin{pmatrix} 1 \\ i \end{pmatrix}$, $\begin{pmatrix} -i \\ 1 \end{pmatrix}$

22. $\begin{pmatrix} 1 \\ 1+i \end{pmatrix}$, $\begin{pmatrix} 2-i \\ 1-i \end{pmatrix}$

23. $\begin{pmatrix} 1 \\ i \\ 1+i \end{pmatrix}$, $\begin{pmatrix} -i \\ 2-i \\ 5+2i \end{pmatrix}$

24. $\begin{pmatrix} 2i \\ i \\ -1 \\ 4+i \end{pmatrix}$, $\begin{pmatrix} -2-3i \\ 4+2i \\ i \\ 6 \end{pmatrix}$

25. Compute the norm of (a) $\begin{pmatrix} 2-i \\ 1+i \end{pmatrix}$ (b) $\begin{pmatrix} 1 \\ i \\ 1-i \\ 2+3i \end{pmatrix}$.

In Problems 26–30 compute (a) the inner product of the two functions in $C[0, 1]$ and (b) the norm of each function.

26. x, x^4

27. x, e^x

28. $1 + x$, $x^2 - 2$

29. e^x, $\sin x$

30. x^2, $\ln(x + 1)$

* 31. Construct an orthonormal basis for $P_2[0, 1]$, starting with the basis $\{1, x, x^2\}$ using the inner product of Example 7.

* 32. Find an orthonormal basis for $P_2[-1, 1]$. The polynomials you obtain are called **normalized Legendre polynomials**.

33. In $\mathbb{R}^2$, if $\mathbf{x} = \begin{pmatrix} x_1 \\ x_2 \end{pmatrix}$ and $\mathbf{y} = \begin{pmatrix} y_1 \\ y_2 \end{pmatrix}$, let $(\mathbf{x}, \mathbf{y})_* = x_1 y_1 + 3x_2 y_2$.

(a) Show that $(x, y)_*$ is an inner product on $\mathbb{R}^2$.

(b) Calculate $\left\| \begin{pmatrix} 2 \\ -3 \end{pmatrix} \right\|_*$.

34. If $\mathbf{u}_1, \mathbf{u}_2, \ldots, \mathbf{u}_n$ are orthonormal, show that

$$|\mathbf{u}_1 + \mathbf{u}_2 + \cdots + \mathbf{u}_n|^2 = |\mathbf{u}_1|^2 + |\mathbf{u}_2|^2 + \cdots + |\mathbf{u}_n|^2 = n$$

35. (a) Find a condition on the numbers a and b such that $\left\{\begin{pmatrix} a \\ b \end{pmatrix}, \begin{pmatrix} b \\ -a \end{pmatrix}\right\}$ and $\left\{\begin{pmatrix} a \\ b \end{pmatrix}, \begin{pmatrix} -b \\ a \end{pmatrix}\right\}$ form orthonormal bases in $\mathbb{R}^2$.

(b) Show that *any* orthonormal basis in $\mathbb{R}^2$ has one of these forms.

36. Prove the **Cauchy–Schwarz inequality** in $\mathbb{R}^n$: $|\mathbf{u} \cdot \mathbf{v}| \leq |\mathbf{u}||\mathbf{v}|$. [*Hint:* If $\mathbf{u} = \mathbf{0}$ or $\mathbf{v} = \mathbf{0}$, the result is obvious. If $\mathbf{u} \neq \mathbf{0}$ and $\mathbf{v} \neq \mathbf{0}$, use the fact that $|\mathbf{u}/|\mathbf{u}| - \mathbf{v}/|\mathbf{v}|| \geq 0$ and $|\mathbf{u}/|\mathbf{u}| + \mathbf{v}/|\mathbf{v}|| \geq 0$.

37. Show that, in Problem 36, $|\mathbf{u} \cdot \mathbf{v}| = |\mathbf{u}||\mathbf{v}|$ if and only if $\mathbf{u} = \lambda\mathbf{v}$ for some real number λ.

38. Using the result of Problem 37, prove that if $|\mathbf{u} + \mathbf{v}| = |\mathbf{u}| + |\mathbf{v}|$, then $\mathbf{u}$ and $\mathbf{v}$ are linearly dependent.

39. Using the result of Problem 36, prove the **triangle inequality**: $|\mathbf{u} + \mathbf{v}| \leq |\mathbf{u}| + |\mathbf{v}|$. [*Hint:* Expand $|\mathbf{u} + \mathbf{v}|^2$.]

40. Suppose that $\mathbf{x}_1, \mathbf{x}_2, \ldots, \mathbf{x}_k$ are vectors in $\mathbb{R}^n$ (not all zero) and

$$|\mathbf{x}_1 + \mathbf{x}_2 + \cdots + \mathbf{x}_k| = |\mathbf{x}_1| + |\mathbf{x}_2| + \cdots + |\mathbf{x}_k|$$

Show that dim span $\{\mathbf{x}_1, \mathbf{x}_2, \ldots, \mathbf{x}_k\} = 1$. [*Hint:* Use the results of Problems 38 and 39.]

41. Let $\{\mathbf{u}_1, \mathbf{u}_2, \ldots, \mathbf{u}_n\}$ be an orthonormal basis in $\mathbb{R}^n$ and let $\mathbf{v}$ be a vector in $\mathbb{R}^n$. Prove that $|\mathbf{v}|^2 = |\mathbf{v} \cdot \mathbf{u}_1|^2 + |\mathbf{v} \cdot \mathbf{u}_2|^2 + \cdots + |\mathbf{v} \cdot \mathbf{u}_n|^2$. This equality is called **Parseval's equality** in $\mathbb{R}^n$.

42. Let $\mathbf{v}_1, \mathbf{v}_2, \ldots, \mathbf{v}_k$ be a linearly independent set of vectors in $\mathbb{R}^n$. Show that $\mathbf{v}_i \neq \mathbf{0}$ for $i = 1, 2, \ldots, k$. [*Hint:* If $\mathbf{v}_i = \mathbf{0}$ for some i, show that $c_1\mathbf{v}_1 + c_2\mathbf{v}_2 + \cdots + c_i\mathbf{v}_i + \cdots + c_k\mathbf{v}_k = \mathbf{0}$ holds, where not all the c_i's are zero.]

Review Exercises for Chapter 6

In Exercises 1–6 find the magnitude and direction of the given vector.

1. $v = (3, 3)$

2. $v = -3i + 3j$

3. $v = (2, -2\sqrt{3})$

4. $v = (\sqrt{3}, 1)$

5. $v = -12i - 12j$

6. $v = i + 4j$

In Exercises 7–10 find the magnitude and the direction cosines of the given vector.

7. $v = 2i - k$

8. $v = 3j + 11k$

9. $v = i - 2j - 3k$

10. $v = -4i + j + 6k$

In Exercises 11–16 find a unit vector having the same direction as the given vector.

11. $v = i + j$

12. $v = -i + j$

13. $v = 2i + 5j$

14. $v = -7i + 3j$

15. $v = 3i + 4j$

16. $v = -2i - 2j$

17. Find a unit vector in the direction of $\overrightarrow{PQ}$, where $P = (3, -1, 2)$ and $Q = (-4, 1, 7)$.

18. Find a unit vector whose direction is opposite that of $\overrightarrow{PQ}$, where $P = (1, -3, 0)$ and $Q = (-7, 1, -4)$.

19. Let $u = 2i + 3j$ and $v = 4i + \alpha j$. Determine α such that

   (a) u and v are orthogonal

   (b) u and v are parallel

   (c) The angle between u and v is $\pi/4$

   (d) The angle between u and v is $\pi/6$

In Exercises 20–23 calculate $\text{Proj}_v u$.

20. $u = 14i; v = i + j$

21. $u = 14i, v = i - j$

22. $u = 3i - 2j; v = 3i + 2j$

23. $u = 3i + 2j; v = i - 3j$

In Exercises 24–27 find the work done when the force with given magnitude and direction moves an object from P to Q. All distances are measured in meters.

24. $|F| = 2$ N; $\theta = \pi/4$; $P = (1, 6)$; $Q = (2, 4)$

25. $|F| = 3$ N; $\theta = \pi/2$; $P = (3, -5)$; $Q = (2, 7)$

26. $|F| = 11$ N; $\theta = \pi/6$; $P = (-1, -2)$; $Q = (-7, -4)$

27. $|F| = 8$ N; $\theta = 2\pi/3$; $P = (-1, 4)$; $Q = (5, -6)$

In Exercises 28–35 let $u = i - 2j + 3k$, $v = -3i + 2j + 5k$, and $w = 2i - 4j + k$. Calculate:

28. $u - v$

29. $3v + 5w$

30. $\text{Proj}_v w$

31. $\text{Proj}_w u$

32. $2u - 4v + 7w$

33. $u \cdot w - w \cdot v$

34. The angle between u and v

35. The angle between v and w

36. Find the distance from the point $P = (3, -1, 2)$ to the line passing through the points $Q = (-2, -1, 6)$ and $R = (0, 1, -8)$.

37. Find the work done when a force of 4 N acting in the direction of the vector $v = -i + j + k$ moves an object from $(2, 1, -6)$ to $(3, 5, 8)$ (distance in meters).

In Exercises 38–41 find the cross product $u \times v$.

38. $u = 3i - j$; $v = 2i + 4k$

39. $u = 7j$; $v = i - k$

40. $u = 4i - j + 7k$; $v = -7i + j - 2k$

41. $u = -2i + 3j - 4k$; $v = -3i + j - 10k$

42. Find two unit vectors orthogonal to both $u = i - j + 3k$ and $v = -2i - 3j + 4k$.

43. Calculate the area of the parallelogram with the adjacent vertices $(1, 4, -2)$, $(-3, 1, 6)$, and $(1, -2, 3)$.

44. Calculate the area of the triangle with vertices at $(2, 1, 3)$, $(-4, 1, 7)$, and $(-1, -1, 3)$.

45. Calculate the volume of the parallelepiped determined by the vectors $i + j$, $2i - 3k$, and $2j + k$.

In Exercises 46–48 determine whether or not u, v, and w are coplanar.

46. $u = i + j + k$, $v = i - j + k$, $w = i + 3j + k$

47. $u = i + k$, $v = 2i - 3j$, $w = j + 2k$

48. $u = 2i - 3j + k$, $v = 5i + 3j - 4k$, $w = 3i + 6j - 5k$

In Exercises 49 and 50 compute the scalar triple product in the given order.

49. $i + 2j$, $2i - k$, $i + j + k$

50. $i - j - k$, $-3i - 2j + k$, $2i + 3j + 5k$

In Exercise 51 compute the triple cross product in the order (first vector) × (second vector × third vector).

51. $i - k$, $k - 2j$, $3i - 2j$

In Exercises 52–55 find a vector equation, parametric equations, and symmetric equations of the given line.

52. Containing $(3, -1, 4)$ and $(-1, 6, 2)$

53. Containing $(-4, 1, 0)$ and $(3, 0, 7)$

54. Containing $(3, 1, 2)$ and parallel to $3i - j - k$

55. Containing $(1, -2, -3)$ and parallel to

$$(x + 1)/5 = (y - 2)/(-3) = (z - 4)/2$$

56. Find the distance from the origin to the line passing through the point $(3, 1, 5)$ and having the direction $v = 2i - j + k$.

57. Find the equation of the line passing through $(-1, 2, 4)$ and orthogonal to $L_1: (x - 1)/4 = (y + 6)/3 = z/(-2)$ and $L_2: (x + 3)/5 = (y - 1)/1 = (z + 3)/4$.

In Exercises 58–60 find the equation of the plane containing the given point and orthogonal to the given normal vector.

58. $P = (1, 3, -2)$; $n = i + k$

59. $P = (1, -4, 6)$; $n = 2j - 3k$

60. $P = (-4, 1, 6)$; $n = 2i - 3j + 5k$

61. Find the equation of the plane containing the points $(-2, 4, 1)$, $(3, -7, 5)$, and $(-1, -2, -1)$.

62. Show that the position vectors $u = i - 2j + k$, $v = 3i + 2j - 3k$, and $w = 9i - 2j - 3k$ are coplanar and find the equation of the plane containing them.

In Exercises 63–72 determine whether the given set with the usual operations is a vector space. If so, determine its dimension. If it is finite-dimensional, find a basis for it.

63. The vectors (x, y, z) in $\mathbb{R}^3$ satisfying $x + 2y - z = 0$

64. The vectors (x, y, z) in $\mathbb{R}^3$ satisfying $x + 2y - z \leq 0$

65. The vectors (x, y, z, w) in $\mathbb{R}^4$ satisfying

$$x + y + z + w = 0$$

66. The set of polynomials of degree 4.

67. The set of polynomials of degree ≤ 4.

68. The set of polynomials of even degree ≤ 8.

69. The set of vectors in $\mathbb{R}^2$ with norm ≤ 1.

70. The set of complex numbers where the scalars are the real numbers.

71. $\{p \in P_3: p(0) = 0\}$

72. $\{p \in P_3: p(0) = 2\}$

In Problems 73–82 determine whether the given set of vectors is linearly dependent or independent.

73. $\begin{pmatrix} 1 \\ 2 \end{pmatrix}, \begin{pmatrix} 2 \\ 5 \end{pmatrix}$

74. $\begin{pmatrix} 3 \\ -1 \end{pmatrix}, \begin{pmatrix} -6 \\ 2 \end{pmatrix}$

75. $\begin{pmatrix} 2 \\ 3 \end{pmatrix}, \begin{pmatrix} 0 \\ 0 \end{pmatrix}$

76. $\begin{pmatrix} 1 \\ 4 \end{pmatrix}, \begin{pmatrix} 2 \\ -1 \end{pmatrix}, \begin{pmatrix} 3 \\ 5 \end{pmatrix}$

77. $\begin{pmatrix} 1 \\ 2 \\ 3 \end{pmatrix}, \begin{pmatrix} 5 \\ -1 \\ 1 \end{pmatrix}$

78. $\begin{pmatrix} 2 \\ -1 \\ 4 \end{pmatrix}, \begin{pmatrix} -4 \\ 2 \\ -8 \end{pmatrix}$

79. $\begin{pmatrix} 3 \\ 1 \\ 6 \end{pmatrix}, \begin{pmatrix} 0 \\ 2 \\ -5 \end{pmatrix}, \begin{pmatrix} 0 \\ 0 \\ 7 \end{pmatrix}$

80. $\begin{pmatrix} 1 \\ 0 \\ 6 \end{pmatrix}, \begin{pmatrix} -2 \\ 3 \\ 5 \end{pmatrix}, \begin{pmatrix} 4 \\ -1 \\ 8 \end{pmatrix}, \begin{pmatrix} -7 \\ 4 \\ 10 \end{pmatrix}$

81. In P_3: $1, 2 - x^2, 3 - x, 7x^2 - 8x$

82. In P_3: $1, 2 + x^3, 3 - x, 7x^2 - 8x$

In Problems 83–85 find the inner product of the given vectors.

83. $(1, 3, -5, 6), (2, 1, 5, 8)$

84. In $C[0, 1]$: $e^{-x}, \cos x$

85. In $C[0, 2\pi]$: $x, \sin 2x$

In Exercises 86–89 find an orthonormal basis for the given vector space.

86. $\mathbb{R}^2$ starting with the basis $\begin{pmatrix} 2 \\ 3 \end{pmatrix}, \begin{pmatrix} -1 \\ 4 \end{pmatrix}$

87. $\{(x, y, z) \in \mathbb{R}^3: x - y - z = 0\}$

88. $\{(x, y, z) \in \mathbb{R}^3: x = y = z\}$

89. $\{(x, y, z, w) \in \mathbb{R}^4: x = z \text{ and } y = w\}$

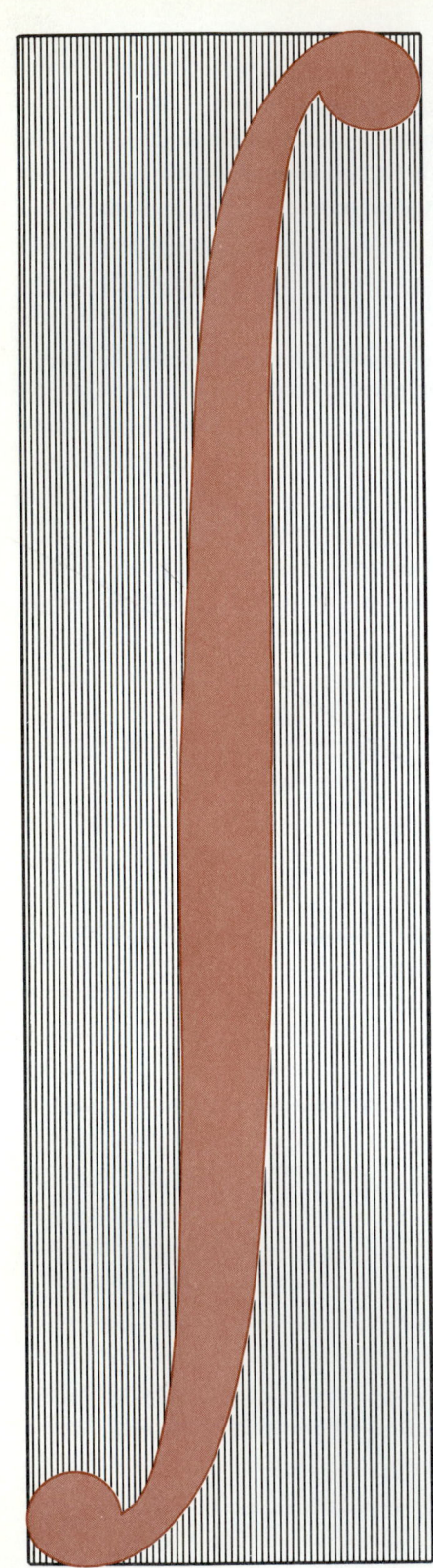

chapter 7

Matrices and Systems of Equations

7.1 Matrices

DEFINITION

An $m \times n$ **matrix**† A is a rectangular array of mn numbers arranged in a definite order in m rows and n columns:‡

$$A = \begin{pmatrix} a_{11} & a_{12} & \cdots & a_{1j} & \cdots & a_{1n} \\ a_{21} & a_{22} & \cdots & a_{2j} & \cdots & a_{2n} \\ \vdots & \vdots & & \vdots & & \vdots \\ a_{i1} & a_{i2} & \cdots & a_{ij} & \cdots & a_{in} \\ \vdots & \vdots & & \vdots & & \vdots \\ a_{m1} & a_{m2} & \cdots & a_{mj} & \cdots & a_{mn} \end{pmatrix}. \tag{1}$$

The number a_{ij} appearing in the ith row and jth column of A is called the ijth **component** of A. For convenience, the matrix A is written $A = (a_{ij})$. Usually, matrices will be denoted by capital letters.

If A is an $m \times n$ matrix with $m = n$, then A is called a **square matrix.** An $m \times n$ matrix with all components equal to zero is called the $m \times n$ **zero matrix** and is denoted O.

An $m \times n$ matrix is said to have the **size** $m \times n$. Two matrices $A = (a_{ij})$ and $B = (b_{ij})$ are **equal** if (i) they have the same size and (ii) corresponding components are equal.

---

† *Historical note:* The term "matrix" was first used in 1850 by the British mathematician James Joseph Sylvester (1814–1897) to distinguish matrices from determinants (which we shall discuss in Section 7.7). In fact, the term "matrix" was intended to mean "mother of determinants."

‡ As with vectors, we shall always assume, unless stated otherwise, that the numbers in a matrix are real.

EXAMPLE 1

The following are $m \times n$ matrices for various values of m and n:

i. $A = \begin{pmatrix} 1 & 3 \\ 4 & 2 \end{pmatrix}$, 2×2 (square) 　　　**ii.** $A = \begin{pmatrix} -1 & 3 \\ 4 & 0 \\ 1 & -2 \end{pmatrix}$, 3×2

iii. $\begin{pmatrix} -1 & 4 & 1 \\ 3 & 0 & 2 \end{pmatrix}$, 2×3 　　　**iv.** $\begin{pmatrix} 1 & 6 & -2 \\ 3 & 1 & 4 \\ 2 & -6 & 5 \end{pmatrix}$, 3×3 (square)

v. $\begin{pmatrix} 0 & 0 & 0 & 0 \\ 0 & 0 & 0 & 0 \end{pmatrix}$, 2×4 zero matrix

A vector is a special kind of matrix. The n-component row vector $(a_1, a_2, \ldots, a_n)$ is a $1 \times n$ matrix, whereas the n-component column vector

$$\begin{pmatrix} a_1 \\ a_2 \\ \vdots \\ a_n \end{pmatrix} \text{ is an } n \times 1 \text{ matrix.}$$

Matrices, like vectors, can be added and multiplied by scalars.

DEFINITION

Addition of Matrices

Let $A = (a_{ij})$ and $B = (b_{ij})$ be two $m \times n$ matrices. Then the sum of A and B is the $m \times n$ matrix $A + B$ given by

$$A + B = (a_{ij} + b_{ij}) = \begin{pmatrix} a_{11} + b_{11} & a_{12} + b_{12} & \cdots & a_{1n} + b_{1n} \\ a_{21} + b_{21} & a_{22} + b_{22} & \cdots & a_{2n} + b_{2n} \\ \vdots & \vdots & & \vdots \\ a_{m1} + b_{m1} & a_{m2} + b_{m2} & \cdots & a_{mn} + b_{mn} \end{pmatrix}. \quad (2)$$

That is, $A + B$ is the $m \times n$ matrix obtained by adding the corresponding components of A and B.

Warning. The sum of two matrices is defined only when both matrices have the same size. Thus, for example, it is not possible to add together the matrices $\begin{pmatrix} 1 & 2 & 3 \\ 4 & 5 & 6 \end{pmatrix}$ and $\begin{pmatrix} -1 & 0 \\ 2 & -5 \\ 4 & 7 \end{pmatrix}$.

EXAMPLE 2

$$\begin{pmatrix} 2 & 4 & -6 & 7 \\ 1 & 3 & 2 & 1 \\ -4 & 3 & -5 & 5 \end{pmatrix} + \begin{pmatrix} 0 & 1 & 6 & -2 \\ 2 & 3 & 4 & 3 \\ -2 & 1 & 4 & 4 \end{pmatrix} = \begin{pmatrix} 2 & 5 & 0 & 5 \\ 3 & 6 & 6 & 4 \\ -6 & 4 & -1 & 9 \end{pmatrix}$$

DEFINITION

Multiplication of a Matrix by a Scalar

If $A = (a_{ij})$ is an $m \times n$ matrix and if α is a scalar, then the $m \times n$ matrix αA is given by

$$\alpha A = (\alpha a_{ij}) = \begin{pmatrix} \alpha a_{11} & \alpha a_{12} & \cdots & \alpha a_{1n} \\ \alpha a_{21} & \alpha a_{22} & \cdots & \alpha a_{2n} \\ \vdots & \vdots & & \vdots \\ \alpha a_{m1} & \alpha a_{m2} & \cdots & \alpha a_{mn} \end{pmatrix} \qquad (3)$$

In other words, $\alpha A = (\alpha a_{ij})$ is the matrix obtained by multiplying each component of A by α.

EXAMPLE 3

Let $A = \begin{pmatrix} 1 & -3 & 4 & 2 \\ 3 & 1 & 4 & 6 \\ -2 & 3 & 5 & 7 \end{pmatrix}$.   Then $2A = \begin{pmatrix} 2 & -6 & 8 & 4 \\ 6 & 2 & 8 & 12 \\ -4 & 6 & 10 & 14 \end{pmatrix}$,

$$-3A = \begin{pmatrix} -3 & 9 & -12 & -6 \\ -9 & -3 & -12 & -18 \\ 6 & -9 & -15 & -21 \end{pmatrix}, \quad \text{and} \quad 0A = \begin{pmatrix} 0 & 0 & 0 & 0 \\ 0 & 0 & 0 & 0 \\ 0 & 0 & 0 & 0 \end{pmatrix}.$$

Matrices satisfy the algebraic properties listed in the following theorem. The proof is left as an exercise (see Problems 21–24).

■ THEOREM 1

Let A, B, and C be $m \times n$ matrices and let α be a scalar. Then:

i. $A + 0 = A$
ii. $0A = 0$
iii. $A + B = B + A$ (**commutative law for matrix addition**)
iv. $(A + B) + C = A + (B + C)$ (**associative law for matrix addition**)
v. $\alpha(A + B) = \alpha A + \alpha B$ (**distributive law for scalar multiplication**)
vi. $1A = A$

Note. The zero in part (i) of the theorem is the $m \times n$ zero matrix. In part (ii) the zero on the left is a scalar while the zero on the right is the $m \times n$ zero matrix. ■

Remark. In Section 6.5 (p. 321) we defined a vector space. Theorem 1 shows that the set of $m \times n$ matrices, denoted by M_{mn}, is a vector space. In Problem 31 you are asked to show that the dimension of M_{mn} is mn.

Problems 7.1

In Problems 1–12 perform the indicated computation with

$$A = \begin{pmatrix} 1 & 3 \\ 2 & 5 \\ -1 & 2 \end{pmatrix}, B = \begin{pmatrix} -2 & 0 \\ 1 & 4 \\ -7 & 5 \end{pmatrix}, \text{ and } C = \begin{pmatrix} -1 & 1 \\ 4 & 6 \\ -7 & 3 \end{pmatrix}.$$

1. $3A$

2. $A + B$

3. $A - C$

4. $2C - 5A$

5. $0B$ (0 is the scalar zero)

6. $-7A + 3B$

7. $A + B + C$

8. $C - A - B$

9. $2A - 3B + 4C$

10. $7C - B + 2A$

11. Find a matrix D such that $2A + B - D$ is the 3×2 zero matrix.

12. Find a matrix E such that $A + 2B - 3C + E$ is the 3×2 zero matrix.

In Problems 13–20 perform the indicated computation with

$$A = \begin{pmatrix} 1 & -1 & 2 \\ 3 & 4 & 5 \\ 0 & 1 & -1 \end{pmatrix}, B = \begin{pmatrix} 0 & 2 & 1 \\ 3 & 0 & 5 \\ 7 & -6 & 0 \end{pmatrix},$$

and $C = \begin{pmatrix} 0 & 0 & 2 \\ 3 & 1 & 0 \\ 0 & -2 & 4 \end{pmatrix}.$

13. $A - 2B$

14. $3A - C$

15. $A + B + C$

16. $2A - B + 2C$

17. $C - A - B$

18. $4C - 2B + 3A$

19. Find a matrix D such that $A + B + C + D$ is the 3×3 zero matrix.

20. Find a matrix E such that $3C - 2B + 8A - 4E$ is the 3×3 zero matrix.

21. Let $A = (a_{ij})$ be an $m \times n$ matrix and let $\bar{0}$ denote the $m \times n$ zero matrix. Show that $0A = \bar{0}$ and $\bar{0} + A = A$. Similarly, show that $1A = A$.

22. Let $A = (a_{ij})$ and $B = (b_{ij})$ be $m \times n$ matrices. Compute $A + B$ and $B + A$ and show that they are equal.

23. If α is a scalar and A and B are as in Problem 22, compute $\alpha(A + B)$ and $\alpha A + \alpha B$ and show that they are equal.

24. If $A = (a_{ij})$, $B = (b_{ij})$, and $C = (c_{ij})$ are $m \times n$ matrices, compute $(A + B) + C$ and $A + (B + C)$ and show that they are equal.

25. Consider the "graph" joining the four points in the figure. Construct a 4×4 matrix having the property that $a_{ij} = 0$

if point i is not connected (joined by a line segment) to point j and $a_{ij} = 1$ if point i is connected to point j.

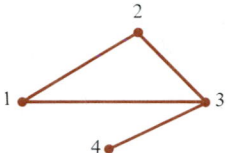

26. Do the same (this time constructing a 5×5 matrix) for the accompanying graph.

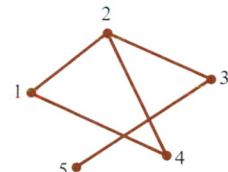

27. Nodal Matrix for an Electric Circuit. Consider the circuit sketched below.

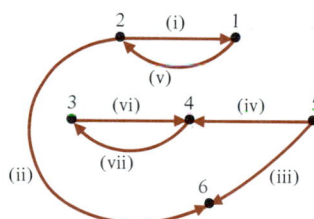

Each point or **node** in the circuit represents an electrical component. Each arrow indicates a connection or current flowing from one node to another. The circuit sketched has six nodes and seven connections. The **nodal matrix** N for this circuit is the 6×7 matrix having the following properties:

(a) $n_{ij} = 1$ if connection (j) leaves from node (i)

(b) $n_{ij} = -1$ if connection (j) enters node (i)

(c) $n_{ij} = 0$ otherwise

For example, $n_{47} = 1$ because connection (vii) leaves from node 4; $n_{37} = -1$ because connection (vii) enters node 3. Find the nodal matrix for the circuit.

28. Find the nodal matrix for the circuit drawn on the next page.

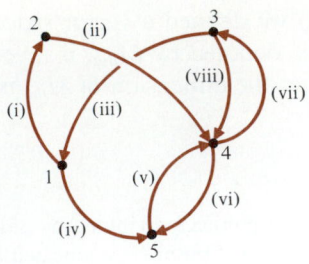

29. Show that the matrices $\begin{pmatrix} 1 & 0 & 0 \\ 0 & 0 & 0 \end{pmatrix}$, $\begin{pmatrix} 0 & 1 & 0 \\ 0 & 0 & 0 \end{pmatrix}$,

$\begin{pmatrix} 0 & 0 & 1 \\ 0 & 0 & 0 \end{pmatrix}$, $\begin{pmatrix} 0 & 0 & 0 \\ 1 & 0 & 0 \end{pmatrix}$, $\begin{pmatrix} 0 & 0 & 0 \\ 0 & 1 & 0 \end{pmatrix}$, and $\begin{pmatrix} 0 & 0 & 0 \\ 0 & 0 & 1 \end{pmatrix}$ are

linearly independent in M_{23}.

30. (a) Show that the matrices in Problem 29 form a basis for M_{23}.

    (b) What is the dimension of M_{23}?

31. Show that dim $M_{mn} = mn$. [*Hint:* Generalize the result of Problem 29.]

32. Let $H = \{A \in M_{22} : a_{11} = 0\}$.

    (a) Show that H is a subspace of M_{22}. **(b)** Find dim H.

7.2 Matrix Products and Systems of Equations

In this section we see how two matrices can be multiplied together. Quite obviously, we could define the product of two $m \times n$ matrices $A = (a_{ij})$ and $B = (b_{ij})$ to be the $m \times n$ matrix whose ijth component is $a_{ij}b_{ij}$. However, for just about all the important applications involving matrices, a different product is required. This matrix product is a generalization of the dot product for vectors.

DEFINITION

Matrix Product

Let $A = (a_{ij})$ be an $m \times n$ matrix whose ith row is denoted $\mathbf{a}_i$. Let $B = (b_{ij})$ be an $n \times p$ matrix whose jth column is denoted $\mathbf{b}_j$. Then the product of A and B is the $m \times p$ matrix $C = (c_{ij})$, where

$$c_{ij} = \mathbf{a}_i \cdot \mathbf{b}_j. \tag{1}$$

That is, the ijth element of the product AB is the dot product of the ith row $\mathbf{a}_i$ of A and the jth column $\mathbf{b}_j$ of B. Using the representation III for dot product given in Section 6.8 on p. 340, we write out the dot product:

$$\mathbf{a}_i \cdot \mathbf{b}_j = (a_{i1}, a_{i2}, \ldots, a_{in}) \cdot \begin{pmatrix} b_{1j} \\ b_{2j} \\ \vdots \\ b_{nj} \end{pmatrix} = a_{i1}b_{1j} + a_{i2}b_{2j} + \cdots + a_{in}b_{nj}$$

or

$$c_{ij} = a_{i1}b_{1j} + a_{i2}b_{2j} + \cdots + a_{in}b_{nj} = \sum_{k=1}^{n} a_{ik}b_{kj}. \qquad (2)$$

Warning. Two matrices can be multiplied together only if the number of columns of the first is equal to the number of rows of the second. Otherwise the vectors $\mathbf{a}_i$ and $\mathbf{b}_j$ will have different numbers of components and the dot product in equation (1) will not be defined.

EXAMPLE 1

If $A = \begin{pmatrix} 1 & 3 \\ -2 & 4 \end{pmatrix}$ and $B = \begin{pmatrix} 3 & -2 \\ 5 & 6 \end{pmatrix}$, calculate AB and BA.

SOLUTION
Since A and B are 2×2, the product matrices AB and BA will also be 2×2. Let $C = AB$; then

$$c_{11} = \mathbf{a}_1 \cdot \mathbf{b}_1 = (1 \quad 3) \cdot \begin{pmatrix} 3 \\ 5 \end{pmatrix} = 18,$$

$$c_{12} = \mathbf{a}_1 \cdot \mathbf{b}_2 = (1 \quad 3) \cdot \begin{pmatrix} -2 \\ 6 \end{pmatrix} = 16,$$

$$c_{21} = \mathbf{a}_2 \cdot \mathbf{b}_1 = (-2 \quad 4) \cdot \begin{pmatrix} 3 \\ 5 \end{pmatrix} = 14,$$

$$c_{22} = \mathbf{a}_2 \cdot \mathbf{b}_2 = (-2 \quad 4) \cdot \begin{pmatrix} -2 \\ 6 \end{pmatrix} = 28,$$

so that

$$C = AB = \begin{pmatrix} 18 & 16 \\ 14 & 28 \end{pmatrix}.$$

Similarly, leaving out the intermediate steps, we get

$$C^* = BA = \begin{pmatrix} 3 & -2 \\ 5 & 6 \end{pmatrix}\begin{pmatrix} 1 & 3 \\ -2 & 4 \end{pmatrix} = \begin{pmatrix} 3+4 & 9-8 \\ 5-12 & 15+24 \end{pmatrix} = \begin{pmatrix} 7 & 1 \\ -7 & 39 \end{pmatrix}.$$

Remark. Example 1 illustrates an important fact: Matrix products do not, in general, commute. That is, $AB \neq BA$ in general. It sometimes happens that $AB = BA$, but this will be the exception, not the rule. In fact, as the next example illustrates, it may occur that AB is defined while BA is not. Thus we must be careful of order when multiplying two matrices together.

EXAMPLE 2

Let $A = \begin{pmatrix} 2 & 0 & -3 \\ 4 & 1 & 5 \end{pmatrix}$ and $B = \begin{pmatrix} 7 & -1 & 4 & 7 \\ 2 & 5 & 0 & -4 \\ -3 & 1 & 2 & 3 \end{pmatrix}$. Calculate AB.

SOLUTION

We first note that A is a 2×3 matrix and B is a 3×4 matrix. Hence the number of columns of A equals the number of rows of B. The product AB is therefore defined and is a 2×4 matrix. Let $AB = C = (c_{ij})$. Then

$$c_{11} = (2 \quad 0 \quad -3) \cdot \begin{pmatrix} 7 \\ 2 \\ -3 \end{pmatrix} = 23 \qquad c_{12} = (2 \quad 0 \quad -3) \cdot \begin{pmatrix} -1 \\ 5 \\ 1 \end{pmatrix} = -5$$

$$c_{13} = (2 \quad 0 \quad -3) \cdot \begin{pmatrix} 4 \\ 0 \\ 2 \end{pmatrix} = 2 \qquad c_{14} = (2 \quad 0 \quad -3) \cdot \begin{pmatrix} 7 \\ -4 \\ 3 \end{pmatrix} = 5$$

$$c_{21} = (4 \quad 1 \quad 5) \cdot \begin{pmatrix} 7 \\ 2 \\ -3 \end{pmatrix} = 15 \qquad c_{22} = (4 \quad 1 \quad 5) \cdot \begin{pmatrix} -1 \\ 5 \\ 1 \end{pmatrix} = 6$$

$$c_{23} = (4 \quad 1 \quad 5) \cdot \begin{pmatrix} 4 \\ 0 \\ 2 \end{pmatrix} = 26 \qquad c_{24} = (4 \quad 1 \quad 5) \cdot \begin{pmatrix} 7 \\ -4 \\ 3 \end{pmatrix} = 39$$

Hence $AB = \begin{pmatrix} 23 & -5 & 2 & 5 \\ 15 & 6 & 26 & 39 \end{pmatrix}$. This completes the problem. Note that the product BA is *not* defined since the number of columns of B (four) is not equal to the number of rows of A (two).

EXAMPLE 3: Direct and Indirect Contact with a Contagious Disease

In this example we show how matrix multiplication can be used to model the spread of a contagious disease. Suppose that four individuals have contracted such a disease. This group has contacts with six people in a second group. We can represent these contacts, called *direct contacts,* by a 4×6 matrix. An example of such a matrix is given below:

DIRECT CONTACT MATRIX

First and second groups:

$$A = \begin{pmatrix} 0 & 1 & 0 & 0 & 1 & 0 \\ 1 & 0 & 0 & 1 & 0 & 1 \\ 0 & 0 & 0 & 1 & 1 & 0 \\ 1 & 0 & 0 & 0 & 0 & 1 \end{pmatrix}$$

Here we set $a_{ij} = 1$ if the ith person in the first group has made contact with the jth person in the second group. For example, the 1 in the 2, 4 position means that the second person in the first (infected) group has been in contact with the fourth person in the second group. Now suppose that a third group of five people has had a variety of direct contacts with individuals of the second group. We can also represent this by a matrix.

DIRECT CONTACT MATRIX
Second and third groups:

$$B = \begin{pmatrix} 0 & 0 & 1 & 0 & 1 \\ 0 & 0 & 0 & 1 & 0 \\ 0 & 1 & 0 & 0 & 0 \\ 1 & 0 & 0 & 0 & 1 \\ 0 & 0 & 0 & 1 & 0 \\ 0 & 0 & 1 & 0 & 0 \end{pmatrix}$$

Note that $b_{64} = 0$, which means that the sixth person in the second group has had no contact with the fourth person in the third group.

The *indirect* or *second-order* contacts between the individuals in the first and third groups is represented by the 4×5 matrix $C = AB$. To see this, observe that a person in group 3 can be infected from someone in group 2 who, in turn, has been infected by someone in group 1. For example, since $a_{24} = 1$ and $b_{45} = 1$, we see that, indirectly, the fifth person in group 3 has contact (through the fourth person in group 2) with the second person in group 1. The total number of indirect contacts between the second person in group 1 and the fifth person in group 3 is given by

$$c_{25} = a_{21}b_{15} + a_{22}b_{25} + a_{23}b_{35} + a_{24}b_{45} + a_{25}b_{55} + a_{26}b_{26}$$

$$= 1 \cdot 1 + 0 \cdot 0 + 0 \cdot 0 + 1 \cdot 1 + 0 \cdot 0 + 1 \cdot 0 = 2$$

We now compute.

INDIRECT CONTACT MATRIX
First and third groups:

$$C = AB = \begin{pmatrix} 0 & 0 & 0 & 2 & 0 \\ 1 & 0 & 2 & 0 & 2 \\ 1 & 0 & 0 & 1 & 1 \\ 0 & 0 & 2 & 0 & 1 \end{pmatrix}$$

We observe that only the second person in Group 3 has no indirect contacts with the disease. The fifth person in this group has $2 + 1 + 1 = 4$ indirect contacts.

We have seen that for matrix multiplication the commutative law does not hold. The next theorem shows that the associative law does hold.

■ **THEOREM 1: Associative Law for Matrix Multiplication**
Let $A = (a_{ij})$ be an $n \times m$ matrix, $B = (b_{ij})$ an $m \times p$ matrix, and $C = (c_{ij})$ a $p \times q$ matrix. Then the **associative law** holds:

$$A(BC) = (AB)C. \tag{3}$$

PROOF

Since A is $n \times m$ and B is $m \times p$, AB is $n \times p$. Thus $(AB)C = (n \times p) \times (p \times q)$ is an $n \times q$ matrix. Similarly BC is $m \times q$ and $A(BC)$ is $n \times q$ so that $(AB)C$ and $A(BC)$ are both of the same size. We must show that the ijth component of $(AB)C$ equals the ijth component of $A(BC)$. Define $D = (d_{ij}) = AB$. Then $d_{ij} = \sum_{k=1}^{m} a_{ik}b_{kj}$. The ijth component of $(AB)C = DC$ is $\sum_{l=1}^{p} d_{il}c_{lj} = \sum_{l=1}^{p} (\sum_{k=1}^{m} a_{ik}b_{kl})c_{lj} = \sum_{k=1}^{m} \sum_{l=1}^{p} a_{ik}b_{kl}c_{lj}$. Next we define $E = (e_{ij}) = BC$. Then $e_{ij} = \sum_{l=1}^{p} b_{il}c_{lj}$ and the ijth component of $A(BC) = AE$ is $\sum_{k=1}^{m} a_{ik}e_{kj} = \sum_{k=1}^{m} \sum_{l=1}^{p} a_{ik}b_{kl}c_{lj}$. Thus the ijth component of $(AB)C$ is equal to the ijth component of $A(BC)$. This proves the associative law. ■

From now on we shall write the product of three matrices simply as ABC. We can do this because $(AB)C = A(BC)$; thus we get the same answer no matter how the multiplication is carried out (provided that we do not commute any of the matrices).

The associative law can be extended to products with more factors. For example, if AB, BC, and CD are defined, then

$$ABCD = (AB)(CD) = A(BC)D = ((AB)C)D = A(B(CD)). \qquad (4)$$

■ **THEOREM 2: Distributive Laws for Matrix Multiplication**

If all the following sums and products are defined, then

$$A(B + C) = AB + AC \qquad (5)$$

and

$$(A + B)C = AC + BC. \qquad (6)$$

We prove the first distributive law [equation (5)]. The proof of the second one [equation (6)] is virtually identical and is therefore omitted. Let A be $n \times m$ and let B and C be $m \times p$. Then the kjth component of $B + C$ is $b_{kj} + c_{kj}$ and the ijth component of $A(B + C)$ is $\sum_{k=1}^{m} a_{ik}(b_{kj} + c_{kj}) = \sum_{k=1}^{m} a_{ik}b_{kj} + \sum_{k=1}^{m} a_{ik}c_{kj} = ij$th component of AB plus the ijth component of AC, and this proves equation (5). ■

Problems 7.2

In Problems 1–15 perform the indicated computation, if possible.

1. $\begin{pmatrix} 2 & 3 \\ -1 & 2 \end{pmatrix}\begin{pmatrix} 4 & 1 \\ 0 & 6 \end{pmatrix}$
    2. $\begin{pmatrix} 3 & -2 \\ 1 & 4 \end{pmatrix}\begin{pmatrix} -5 & 6 \\ 1 & 3 \end{pmatrix}$
    3. $\begin{pmatrix} 1 & -1 \\ 1 & 1 \end{pmatrix}\begin{pmatrix} -1 & 0 \\ 2 & 3 \end{pmatrix}$
    4. $\begin{pmatrix} -5 & 6 \\ 1 & 3 \end{pmatrix}\begin{pmatrix} 3 & -2 \\ 1 & 4 \end{pmatrix}$

5. $\begin{pmatrix} -4 & 5 & 1 \\ 0 & 4 & 2 \end{pmatrix}\begin{pmatrix} 3 & -1 & 1 \\ 5 & 6 & 4 \\ 0 & 1 & 2 \end{pmatrix}$

6. $\begin{pmatrix} 7 & 1 & 4 \\ 2 & -3 & 5 \end{pmatrix}\begin{pmatrix} 1 & 6 \\ 0 & 4 \\ -2 & 3 \end{pmatrix}$

7. $\begin{pmatrix} 1 & 6 \\ 0 & 4 \\ -2 & 3 \end{pmatrix}\begin{pmatrix} 7 & 1 & 4 \\ 2 & -3 & 5 \end{pmatrix}$

8. $\begin{pmatrix} 1 & 4 & -2 \\ 3 & 0 & 4 \end{pmatrix}\begin{pmatrix} 0 & 1 \\ 2 & 3 \end{pmatrix}$

9. $\begin{pmatrix} 1 & 4 & 6 \\ -2 & 3 & 5 \\ 1 & 0 & 4 \end{pmatrix}\begin{pmatrix} 2 & -3 & 5 \\ 1 & 0 & 6 \\ 2 & 3 & 1 \end{pmatrix}$

10. $\begin{pmatrix} 2 & -3 & 5 \\ 1 & 0 & 6 \\ 2 & 3 & 1 \end{pmatrix}\begin{pmatrix} 1 & 4 & 6 \\ -2 & 3 & 5 \\ 1 & 0 & 4 \end{pmatrix}$

11. $(1 \quad 4 \quad 0 \quad 2)\begin{pmatrix} 3 & -6 \\ 2 & 4 \\ 1 & 0 \\ -2 & 3 \end{pmatrix}$

12. $\begin{pmatrix} 3 & 2 & 1 & -2 \\ -6 & 4 & 0 & 3 \end{pmatrix}\begin{pmatrix} 1 \\ 4 \\ 0 \\ 2 \end{pmatrix}$

13. $\begin{pmatrix} 3 & -2 & 1 \\ 4 & 0 & 6 \\ 5 & 1 & 9 \end{pmatrix}\begin{pmatrix} 1 & 0 & 0 \\ 0 & 1 & 0 \\ 0 & 0 & 1 \end{pmatrix}$

14. $\begin{pmatrix} 1 & 0 & 0 \\ 0 & 1 & 0 \\ 0 & 0 & 1 \end{pmatrix}\begin{pmatrix} 3 & -2 & 1 \\ 4 & 0 & 6 \\ 5 & 1 & 9 \end{pmatrix}$

15. $\begin{pmatrix} a & b & c \\ d & e & f \\ g & h & j \end{pmatrix}\begin{pmatrix} 1 & 0 & 0 \\ 0 & 1 & 0 \\ 0 & 0 & 1 \end{pmatrix}$, where $a, b, c, d, e, f, g, h, j$ are

real numbers

16. Find a matrix $A = \begin{pmatrix} a & b \\ c & d \end{pmatrix}$ such that $A\begin{pmatrix} 2 & 3 \\ 1 & 2 \end{pmatrix} = \begin{pmatrix} 1 & 0 \\ 0 & 1 \end{pmatrix}$.

* 17. Let a_{11}, a_{12}, a_{21}, and a_{22} be given real numbers such that $a_{11}a_{22} - a_{12}a_{21} \neq 0$. Find numbers b_{11}, b_{12}, b_{21}, and b_{22} such that $\begin{pmatrix} a_{11} & a_{12} \\ a_{21} & a_{22} \end{pmatrix}\begin{pmatrix} b_{11} & b_{12} \\ b_{21} & b_{22} \end{pmatrix} = \begin{pmatrix} 1 & 0 \\ 0 & 1 \end{pmatrix}$.

18. Verify the associative law for multiplication for the matrices $A = \begin{pmatrix} 2 & -1 & 4 \\ 1 & 0 & 6 \end{pmatrix}$, $B = \begin{pmatrix} 1 & 0 & 1 \\ 2 & -1 & 2 \\ 3 & -2 & 0 \end{pmatrix}$,

and $C = \begin{pmatrix} 1 & 6 \\ -2 & 4 \\ 0 & 5 \end{pmatrix}$.

19. Let O be the $m \times n$ zero matrix and let A be an $n \times p$ matrix. Show that $OA = O_1$, where O_1 is the $m \times p$ zero matrix.

20. Verify the distributive law [equation (5)] for the matrices $A = \begin{pmatrix} 1 & 2 & 4 \\ 3 & -1 & 0 \end{pmatrix}$, $B = \begin{pmatrix} 2 & 7 \\ -1 & 4 \\ 6 & 0 \end{pmatrix}$, and $C = \begin{pmatrix} -1 & 2 \\ 3 & 7 \\ 4 & 1 \end{pmatrix}$.

21. Let A be a square matrix. Then A^2 is defined simply as AA. Calculate $\begin{pmatrix} 2 & -1 \\ 4 & 6 \end{pmatrix}^2$.

22. Calculate A^2, where $A = \begin{pmatrix} 1 & -2 & 4 \\ 2 & 0 & 3 \\ 1 & 1 & 5 \end{pmatrix}$.

23. Calculate A^3, where $A = \begin{pmatrix} -1 & 2 \\ 3 & 4 \end{pmatrix}$.

24. Calculate A^2, A^3, A^4, and A^5, where
$$A = \begin{pmatrix} 0 & 1 & 0 & 0 \\ 0 & 0 & 1 & 0 \\ 0 & 0 & 0 & 1 \\ 0 & 0 & 0 & 0 \end{pmatrix}.$$

25. An $n \times n$ matrix A has the property that its matrix product with any $n \times n$ matrix is the zero matrix. Prove that A is the zero matrix.

26. A **probability matrix** is a square matrix having two properties: (i) every component is nonnegative (≥ 0) and (ii) the sum of the elements in each row is 1. The following are probability matrices:
$$P = \begin{pmatrix} 1/4 & 1/4 & 1/2 \\ 0 & 1 & 0 \\ 1/3 & 1/3 & 1/3 \end{pmatrix} \text{ and } Q = \begin{pmatrix} 1 & 0 & 0 \\ 1/4 & 1/3 & 5/12 \\ 0 & 0 & 1 \end{pmatrix}.$$

Is QP a probability matrix?

27. Show that PQ is a probability matrix.

28. Let P be a probability matrix. Show that P^2 is a probability matrix.

* 29. Let P and Q be probability matrices of the same size. Prove that PQ is a probability matrix.

30. A round robin tennis tournament can be organized in the following way. Each of the n players plays all the others, and the results are recorded in an $n \times n$ matrix R as follows:
$$R_{ij} = \begin{cases} 1 & \text{if the } i\text{th player beats the } j\text{th player,} \\ 0 & \text{if the } i\text{th player loses to the } j\text{th player,} \\ 0 & \text{if } i = j. \end{cases}$$

The ith player is then assigned the score
$$S_i = \sum_{j=1}^{n} R_{ij} + \frac{1}{2}\sum_{j=1}^{n} (R^2)_{ij}.\dagger$$

$\dagger$ $(R^2)_{ij}$ is the ijth component of the matrix R^2.

(a) In a tournament between four players

$$R = \begin{pmatrix} 0 & 1 & 0 & 0 \\ 0 & 0 & 1 & 1 \\ 1 & 0 & 0 & 0 \\ 1 & 0 & 1 & 0 \end{pmatrix}.$$

Rank the players according to their scores.

(b) Interpret the meaning of the score.

31. Find two 2×2 matrices A and B such that $AB = 0$ but $A \neq 0$ and $B \neq 0$.

32. Find three 2×2 matrices A, B, and C such that $AB = AC$ but $B \neq C$.

* 33. The voltage-current relationship for a series impedance Z† is $\begin{pmatrix} U_1 \\ I_1 \end{pmatrix} = \begin{pmatrix} 1 & Z \\ 0 & 1 \end{pmatrix} \begin{pmatrix} U_2 \\ I_2 \end{pmatrix}$. For a shunt impedance it is $\begin{pmatrix} U_1 \\ I_1 \end{pmatrix} = \begin{pmatrix} 1 & 0 \\ 1/Z & 1 \end{pmatrix} \begin{pmatrix} U_2 \\ I_2 \end{pmatrix}$. Determine U_2 and I_2 as a function of U_1 and I_1 for the following networks:

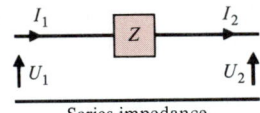

Series impedance

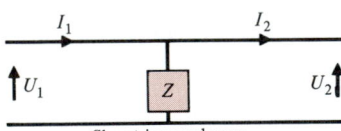

Shunt impendance

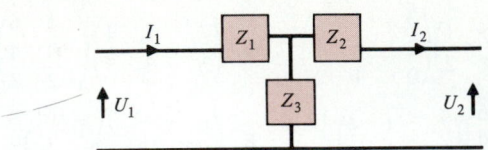

* 34. Determine the voltage-current relationship for the following network:

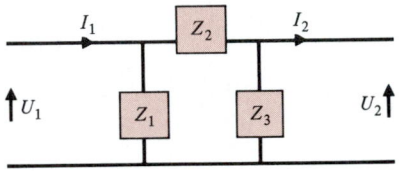

35. To describe electron spin, the following three matrices, called **Pauli matrices,** are used

$$X_1 = \frac{1}{2}\begin{pmatrix} 0 & 1 \\ 1 & 0 \end{pmatrix}, \quad X_2 = \frac{1}{2}\begin{pmatrix} 0 & -i \\ i & 0 \end{pmatrix}, \quad X_3 = \frac{1}{2}\begin{pmatrix} 1 & 0 \\ 0 & -1 \end{pmatrix},$$

(where Planck's constant is taken as unity). Find

(a) X_i^2 ($i = 1, 2, 3$)

(b) $X_i X_j + X_j X_i$ ($i, j = 1, 2, 3$).

7.3 Linear Systems of Equations

Consider the following system of two linear equations in the two unknowns x_1 and x_2:

$$a_{11}x_1 + a_{12}x_2 = b_1 \tag{1}$$
$$a_{21}x_1 + a_{22}x_2 = b_2$$

where a_{11}, a_{12}, a_{21}, a_{22}, b_1, and b_2 are given numbers. Each of these equations is the equation of a straight line (in the x_1x_2-plane instead of the xy-plane). The slope of the first line is $-a_{11}/a_{12}$; the slope of the second line is $-a_{21}/a_{22}$ (if $a_{12} \neq 0$ and $a_{22} \neq 0$). A **solution** to system (1) is a pair of numbers, denoted

Solution

† Problems 33, 34, and 35 are adapted from Alexander Graham, *Matrix Theory and Applications for Engineers and Mathematicians*, Ellis Herwood, Ltd., Chichester, England, 1979.

by (x_1, x_2), that satisfies both equations in (1). The questions that naturally arise are whether (1) has any solutions and, if so, how many solutions does it have?

These questions are easy to answer if we think about the problem geometrically. Each of the equations in (1) is the equation of a straight line. Thus there are three possibilities:

(i) The two lines intersect at one point and the system has exactly one solution.

(ii) The two lines are parallel and do not coincide. Then there is no point that lies on both lines and the system has no solution.

(iii) The two lines coincide. Then every point on the single line is a solution. That is, there are an infinite number of solutions.

EXAMPLE 1

Find all solutions to each system

$$\text{(a)} \quad x - y = 7 \qquad \text{(b)} \quad x - y = 7 \qquad \text{(c)} \quad x - y = 7$$
$$\quad\quad\; x + y = 5 \qquad\qquad 2x - 2y = 14 \qquad\qquad 2x - 2y = 10$$

SOLUTION

The lines represented in each system are sketched in Fig. 1.

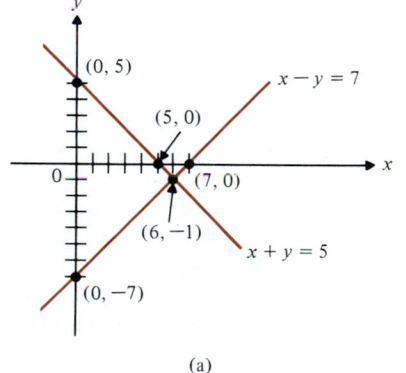

(a)

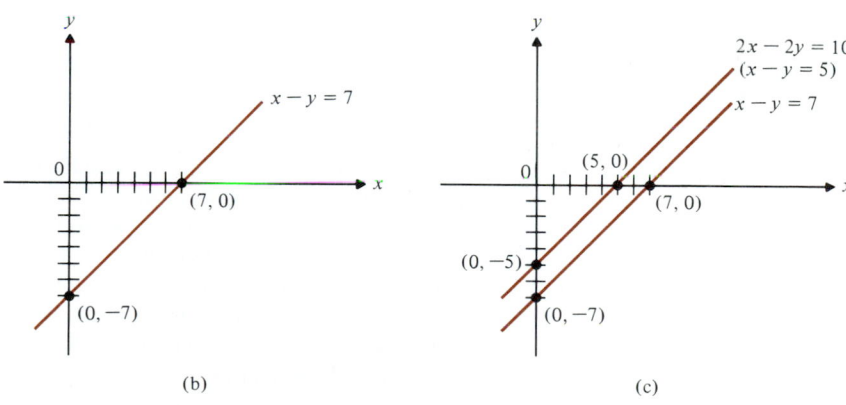

(b)   (c)

FIGURE 1

(a) The lines intersect at one point. By adding the equations together, we find the unique solution $x = 6$, $y = -1$.

(b) The two lines coincide (the second equation is twice the first). Every point on the line is a solution. From the first equation we see that $y = x - 7$. Thus all solutions have the form $(x, x - 7)$. For example, two solutions are $(0, -7)$ and $(3, -4)$.

(c) Now the lines are parallel and there is no solution. This follows directly by dividing the second equation by 2 to obtain $x - y = 5$. Evidently, it is impossible to find two numbers x and y such that both $x - y = 7$ and $x - y = 5$.

We now describe a method for finding all solutions (if any) to a system of m linear equations in n unknowns. In doing so we shall see that, like the 2×2 case, such a system has no solutions, one solution, or an infinite number of solutions. Before discussing the general method, we will look at some simple examples. Our procedure makes use of two simple algebraic facts.

Fact A: If $x = y$, then $cx = cy$ for any constant c.

Fact B: If $x = y$ and $z = w$, then $x + z = y + w$.

EXAMPLE 2

Solve the system

$$2x_1 + 4x_2 + 6x_3 = 18$$

$$4x_1 + 5x_2 + 6x_3 = 24 \qquad \text{(2)}$$

$$3x_1 + x_2 - 2x_3 = 4$$

SOLUTION

Here we seek three numbers x_1, x_2, and x_3 such that the three equations in (2) are satisfied. Our method of solution will be to simplify the equations until solutions can be readily identified. We begin by multiplying both sides of the first equation by $\frac{1}{2}$ to obtain a new first equation that has the same set of solutions as the original first equation. This is justified by Fact A. We obtain

$$x_1 + 2x_2 + 3x_3 = 9$$

$$4x_1 + 5x_2 + 6x_3 = 24 \qquad \text{(3)}$$

$$3x_1 + x_2 - 2x_3 = 4$$

Equivalent System

We stress that system (3) is **equivalent** to system (2). In other words, any solution to (2) is a solution to (3), and vice versa.

From Fact B, we can add two equations together to obtain a third, valid equation. This equation may replace either of the two equations used to obtain it in the system. When this is done, we obtain a system that is equivalent to the original system. That is, the set of solutions is unchanged. We begin simplifying system (3) by multiplying both sides of the first equation in (3) by -4 and adding this new equation to the second equation. This gives us

$$-4x_1 - 8x_2 - 12x_3 = -36$$

$$\underline{4x_1 + 5x_2 + 6x_3 = 24}$$

$$-3x_2 - 6x_3 = -12$$

The equation $-3x_2 - 6x_3 = -12$ is our new second equation and the system is now

$$x_1 + 2x_2 + 3x_3 = 9$$

$$-3x_2 - 6x_3 = -12$$

$$3x_1 + x_2 - 2x_3 = 4$$

We then multiply the first equation by -3 and add it to the third equation:

$$x_1 + 2x_2 + 3x_3 = 9$$
$$-3x_2 - 6x_3 = -12$$
$$-5x_2 - 11x_3 = -23$$

Note that in the last system the variable x_1 has been eliminated from the second and third equations. Next we divide the second equation by -3:

$$x_1 + 2x_2 + 3x_3 = 9$$
$$x_2 + 2x_3 = 4$$
$$-5x_2 - 11x_3 = -23$$

We multiply the second equation by -2 and add it to the first and then multiply the second equation by 5 and add it to the third:

$$x_1 \quad - x_3 = 1$$
$$x_2 + 2x_3 = 4$$
$$- x_3 = -3$$

We multiply the third equation by -1:

$$x_1 \quad - x_3 = 1$$
$$x_2 + 2x_3 = 4$$
$$x_3 = 3$$

Finally, we add the third equation to the first equation and add -2 times the third equation to the second equation to obtain the following system [which is equivalent to system (2)]:

$$x_1 \qquad = 4$$
$$x_2 \qquad = -2$$
$$x_3 = 3$$

Gauss-Jordan Elimination

Thus the point $(4, -2, 3)$ is the unique solution to the system (2). The method we have used here is called **Gauss-Jordan elimination.**†

Before going on to another example, let us summarize what we have done in this example:

i. We divided to make the coefficient of x_1 in the first equation equal to 1.
ii. We "eliminated" the x_1 terms in the second and third equations. That is, we made the coefficients of these terms equal to zero by multiplying

† Named after the great German mathematician Karl Friedrich Gauss (1777–1855) and the German engineer Wilhelm Jordan (1842–1899). See the biographical sketch of Gauss on the next page.

the first equation by appropriate numbers and then adding it to the second and third equations, respectively.

iii. We divided to make the coefficient of the x_2 term in the second equation equal to 1 and then proceeded to use the second equation to eliminate the x_2 terms in the first and third equations.

iv. We divided to make the coefficient of the x_3 term in the third equation equal to 1 and then proceeded to use the third equation to eliminate the x_3 terms in the first and second equations.

We emphasize that, at every step, we obtained systems that were equivalent. That is, each system had the same set of solutions as the one that preceded it. This follows from Facts A and B on p. 364.

Courtesy of the Granger Collection

CARL FRIEDRICH GAUSS
(1777–1855)

The greatest mathematician of the nineteenth century, Carl Friedrich Gauss is considered one of the three greatest mathematicians of all time—the others being Archimedes and Newton.

Gauss was born in Brunswick, Germany, in 1777. His father, a hard-working laborer who was exceptionally stubborn and did not believe in formal education, did what he could to keep Gauss from appropriate schooling. Fortunately for Carl (and for mathematics), his mother, while uneducated herself, encouraged her son in his studies and took considerable pride in his achievements until her death at the age of 97.

Gauss was a child prodigy. At the age of three, he found an error in his father's bookkeeping. A famous story tells of Carl, age 10, as a student in the local Brunswick school. The teacher there was known to assign tasks to keep his pupils busy. One day he asked his students to add the numbers from 1 to 100. Almost at once, Carl placed his slate face down with the words, "There it is." Afterwards, the teacher found that Gauss was the only one with the correct answer, 5050. Gauss had noticed that the numbers could be arranged in 50 pairs, each with the sum 101 (1 + 100, 2 + 99, and so on) and $50 \times 101 = 5050$. Later in life, Gauss joked that he could add before he could speak.

When Gauss was 15, the Duke of Brunswick noticed him and became his patron. The Duke helped him enter Brunswick College and, three years later, in 1795, to enter the University of Göttingen. Undecided between careers in mathematics and languages, Gauss chose mathematics after two remarkable discoveries. First, he invented the method of least squares a decade before the result was published by Legendre. Second, a month before his 19th birthday, he solved a problem whose solution had been sought for more than two thousand years. Gauss showed how to construct, using compass and ruler, a regular polygon with the number of sides not a multiple of 2, 3, or 5. On March 30, 1796, the day of this discovery, he began a diary, which contained as its first entry rules for construction of a 17-sided regular

Before solving other systems of equations, we introduce notation that makes it easier to write down each step in our procedure. The coefficients of the variables x_1, x_2, x_3 in system (1) can be written as the entries of a matrix A, called the **coefficient matrix** of the system:

Coefficient Matrix

$$A = \begin{pmatrix} 2 & 4 & 6 \\ 4 & 5 & 6 \\ 3 & 1 & -2 \end{pmatrix}. \tag{4}$$

Let the vectors **x** and **b** be defined by

$$\mathbf{x} = \begin{pmatrix} x_1 \\ x_2 \\ x_3 \end{pmatrix} \quad \text{and} \quad \mathbf{b} = \begin{pmatrix} 18 \\ 24 \\ 4 \end{pmatrix}.$$

polygon. The diary, which contains 146 statements of results in only 19 pages, is one of the most important documents in the history of mathematics.

After a short period at Göttingen, Gauss went to the University of Helmstädt and in 1798, at the age of 20, wrote his now famous doctoral dissertation. In it he gave the first mathematically rigorous proof of the fundamental theorem of algebra—that every polynomial of degree n has, counting multiplicities, exactly n roots. Many mathematicians, including Euler, Newton, and Lagrange, had attempted to prove this result.

Gauss made a great number of discoveries in physics as well as in mathematics. For example, in 1801 he used a new procedure to calculate, from very little data, the orbit of the planetoid Ceres. In 1833, he invented the electromagnetic telegraph with his colleague Wilhelm Weber (1804–1891). Although he did brilliant work in astronomy and electricity, it was Gauss's mathematical output that was astonishing. He made fundamental contributions to algebra and geometry. In 1811, he discovered a result that led to the development of complex variable theory by Cauchy. He is encountered in courses in matrix theory in the Gauss-Jordan method of elimination. Students of numerical analysis study Gaussian quadrature—a technique for numerical integration.

Gauss became Director of the Göttingen Observatory in 1807 and remained in that post until his death in 1855. Even after his death, his mathematical spirit remained to haunt nineteenth century mathematicians. Often it turned out that an important new result had been discovered earlier by Gauss and could be found in his unpublished notes.

In his mathematical writings, Gauss was a perfectionist and is probably the last mathematician who knew everything in his subject. Claiming that a cathedral was not a cathedral until the last piece of scaffolding was removed, he endeavored to make each of his published works complete, concise, and polished. He used a seal that pictured a tree carrying only a few fruit together with the motto *pauca sed matura* (few, but ripe). But Gauss also believed that mathematics must reflect the real world. At his death, Gauss was honored by a commemorative medal on which was inscribed "George V. King of Hanover to the Prince of Mathematicians."

Then system (1) can be written as

$$A\mathbf{x} = \mathbf{b}.$$

Augmented Matrix

In every step of Example 2, we wrote lots of x's. This is not necessary: Using matrix notation, system (2) can be written as the **augmented matrix**

$$\begin{pmatrix} 2 & 4 & 6 & | & 18 \\ 4 & 5 & 6 & | & 24 \\ 3 & 1 & -2 & | & 4 \end{pmatrix}. \tag{5}$$

For example, the first row in the augmented matrix (5) is read $2x_1 + 4x_2 + 6x_3 = 18$. Note that each row of the augmented matrix corresponds to one of the equations in the system.

We now introduce some terminology. We have seen that multiplying (or dividing) the sides of an equation by a nonzero number gives us an equivalent equation. Moreover, adding a multiple of one equation to another equation in a system gives us an equivalent system. Finally, if we interchange two equations in a system of equations, we obtain an equivalent system. These three operations, when applied to the rows of the augmented matrix representation of a system of equations, are called **elementary row operations.**

To sum up, the three elementary row operations applied to the augmented matrix representation of a system of equations are:

Elementary Row Operations

   i. Multiply (or divide) one row by a nonzero number.
  ii. Add a multiple of one row to another row.
 iii. Interchange two rows.

Row Reduction

The process of applying elementary row operations to simplify an augmented matrix is called **row reduction.**

Notation

   i. $M_i(c)$ stands for "multiply the ith row of a matrix by the number c."
  ii. $A_{i,j}(c)$ stands for "multiply the ith row by c and add it to the jth row."
 iii. $P_{i,j}$ stands for "interchange (permute) rows i and j."
 iv. $A \rightarrow B$ indicates that the augmented matrices A and B are equivalent; that is, the systems they represent have the same solution.

In Example 2 we saw that by using the elementary row operations (i) and (ii) several times we could obtain a system in which the solutions to the system were given explicitly. We now repeat the steps in Example 2, using the notation just introduced:

$$\begin{pmatrix} 2 & 4 & 6 & | & 18 \\ 4 & 5 & 6 & | & 24 \\ 3 & 1 & -2 & | & 4 \end{pmatrix} \xrightarrow{M_1(1/2)} \begin{pmatrix} 1 & 2 & 3 & | & 9 \\ 4 & 5 & 6 & | & 24 \\ 3 & 1 & -2 & | & 4 \end{pmatrix} \xrightarrow[A_{1,3}(-3)]{A_{1,2}(-4)} \begin{pmatrix} 1 & 2 & 3 & | & 9 \\ 0 & -3 & -6 & | & -12 \\ 0 & -5 & -11 & | & -23 \end{pmatrix}$$

$$\xrightarrow{M_2(-1/3)} \begin{pmatrix} 1 & 2 & 3 & | & 9 \\ 0 & 1 & 2 & | & 4 \\ 0 & -5 & -11 & | & -23 \end{pmatrix} \xrightarrow[A_{2,3}(5)]{A_{2,1}(-2)} \begin{pmatrix} 1 & 0 & -1 & | & 1 \\ 0 & 1 & 2 & | & 4 \\ 0 & 0 & -1 & | & -3 \end{pmatrix}$$

$$\xrightarrow{M_3(-1)} \begin{pmatrix} 1 & 0 & -1 & | & 1 \\ 0 & 1 & 2 & | & 4 \\ 0 & 0 & 1 & | & 3 \end{pmatrix} \xrightarrow[A_{3,2}(-2)]{A_{3,1}(1)} \begin{pmatrix} 1 & 0 & 0 & | & 4 \\ 0 & 1 & 0 & | & -2 \\ 0 & 0 & 1 & | & 3 \end{pmatrix}.$$

Again we can easily determine the solution $x_1 = 4$, $x_2 = -2$, $x_3 = 3$.

EXAMPLE 3

Solve the system

$$2x_1 + 4x_2 + 6x_3 = 18$$

$$4x_1 + 5x_2 + 6x_3 = 24$$

$$2x_1 + 7x_2 + 12x_3 = 30$$

SOLUTION

We observe that this system can be written as $A\mathbf{x} = \mathbf{b}$, where

$$A = \begin{pmatrix} 2 & 4 & 6 \\ 4 & 5 & 6 \\ 2 & 7 & 12 \end{pmatrix}, \quad \mathbf{x} = \begin{pmatrix} x_1 \\ x_2 \\ x_3 \end{pmatrix}, \quad \text{and} \quad \mathbf{b} = \begin{pmatrix} 18 \\ 24 \\ 30 \end{pmatrix}.$$

To solve, we proceed as in Example 2, first writing the system as an augmented matrix:

$$\begin{pmatrix} 2 & 4 & 6 & | & 18 \\ 4 & 5 & 6 & | & 24 \\ 2 & 7 & 12 & | & 30 \end{pmatrix}.$$

We then obtain, successively,

$$\xrightarrow{M_1(1/2)} \begin{pmatrix} 1 & 2 & 3 & | & 9 \\ 4 & 5 & 6 & | & 24 \\ 2 & 7 & 12 & | & 30 \end{pmatrix} \xrightarrow[A_{1,3}(-2)]{A_{1,2}(-4)} \begin{pmatrix} 1 & 2 & 3 & | & 9 \\ 0 & -3 & -6 & | & -12 \\ 0 & 3 & 6 & | & 12 \end{pmatrix}$$

$$\xrightarrow{M_2(-1/3)} \begin{pmatrix} 1 & 2 & 3 & | & 9 \\ 0 & 1 & 2 & | & 4 \\ 0 & 3 & 6 & | & 12 \end{pmatrix} \xrightarrow[A_{2,3}(-3)]{A_{2,1}(-2)} \begin{pmatrix} 1 & 0 & -1 & | & 1 \\ 0 & 1 & 2 & | & 4 \\ 0 & 0 & 0 & | & 0 \end{pmatrix}.$$

This is equivalent to the system of equations

$$x_1 \quad - \quad x_3 = 1$$

$$x_2 + 2x_3 = 4$$

This is as far as we can go. There are now only two equations in the three unknowns x_1, x_2, x_3 and there are an infinite number of solutions. To see this, let x_3 be chosen. Then $x_2 = 4 - 2x_3$ and $x_1 = 1 + x_3$. This will be a solution for any number x_3. We write these solutions in the vector form $(1 + x_3, 4 - 2x_3, x_3)$. For example, if $x_3 = 0$ we obtain the solution $(1, 4, 0)$. For $x_3 = 10$ we obtain the solution $(11, -16, 10)$.

EXAMPLE 4

Solve the system

$$2x_1 + 4x_2 + 6x_3 = 18$$

$$4x_1 + 5x_2 + 6x_3 = 24 \tag{6}$$

$$2x_1 + 7x_2 + 12x_3 = 40$$

SOLUTION

We use the augmented-matrix form and proceed exactly as in Example 3 to obtain, successively, the following systems. (Note how, in each step, we use either elementary row operation (i) or (ii).)

$$\begin{pmatrix} 2 & 4 & 6 & | & 18 \\ 4 & 5 & 6 & | & 24 \\ 2 & 7 & 12 & | & 40 \end{pmatrix} \xrightarrow{M_1(1/2)} \begin{pmatrix} 1 & 2 & 3 & | & 9 \\ 4 & 5 & 6 & | & 24 \\ 2 & 7 & 12 & | & 40 \end{pmatrix}$$

$$\xrightarrow[A_{1,3}(-2)]{A_{1,2}(-4)} \begin{pmatrix} 1 & 2 & 3 & | & 9 \\ 0 & -3 & -6 & | & -12 \\ 0 & 3 & 6 & | & 22 \end{pmatrix} \xrightarrow{M_2(-1/3)} \begin{pmatrix} 1 & 2 & 3 & | & 9 \\ 0 & 1 & 2 & | & 4 \\ 0 & 3 & 6 & | & 22 \end{pmatrix}$$

$$\xrightarrow[A_{2,3}(-3)]{A_{2,1}(-2)} \begin{pmatrix} 1 & 0 & -1 & | & 1 \\ 0 & 1 & 2 & | & 4 \\ 0 & 0 & 0 & | & 10 \end{pmatrix} \xrightarrow{M_3(1/10)} \begin{pmatrix} 1 & 0 & -1 & | & 1 \\ 0 & 1 & 2 & | & 4 \\ 0 & 0 & 0 & | & 1 \end{pmatrix}.$$

The last equation now reads $0x_1 + 0x_2 + 0x_3 = 1$, which is impossible since $0 \neq 1$. Thus system (6) has *no* solution.

We now consider the solution of a general system of m equations in n unknowns, given by

$$
\begin{aligned}
a_{11}x_1 + a_{12}x_2 + a_{13}x_3 + \cdots + a_{1n}x_n &= b_1 \\
a_{21}x_1 + a_{22}x_2 + a_{23}x_3 + \cdots + a_{2n}x_n &= b_2 \\
a_{31}x_1 + a_{32}x_2 + a_{33}x_3 + \cdots + a_{3n}x_n &= b_3 \\
\vdots \qquad \vdots \qquad \vdots \qquad \vdots \qquad \vdots \qquad \vdots \\
a_{m1}x_1 + a_{m2}x_2 + a_{m3}x_3 + \cdots + a_{mn}x_n &= b_m
\end{aligned}
\tag{7}
$$

This system can be written in the form $A\mathbf{x} = \mathbf{b}$ where

$$A = \begin{pmatrix} a_{11} & a_{12} & \cdots & a_{1n} \\ a_{21} & a_{22} & \cdots & a_{2n} \\ \vdots & \vdots & & \vdots \\ a_{m1} & a_{m2} & \cdots & a_{mn} \end{pmatrix}, \quad \mathbf{x} = \begin{pmatrix} x_1 \\ x_2 \\ \vdots \\ x_n \end{pmatrix}, \quad \text{and} \quad \mathbf{b} = \begin{pmatrix} b_1 \\ b_2 \\ \vdots \\ b_m \end{pmatrix}.$$

In system (7) all the a's and b's are given real numbers. The problem is to find all sets of n numbers, denoted by $(x_1, x_2, x_3, \ldots, x_n)$, that satisfy each of the m equations in (7). The number a_{ij} is the coefficient of the variable x_j in the ith equation.

We solve system (7) by writing the system as an augmented matrix $(A \mid \mathbf{b})$ and use the elementary row operations until the coefficient matrix A reaches *reduced row echelon form*.

DEFINITION

Reduced Row Echelon Form

A matrix is in **reduced row echelon form** if the following four conditions hold:

  i. All rows (if any) consisting entirely of zeros appear at the bottom of the matrix.
 ii. The first (starting from the left) number in any row not consisting entirely of zeros is 1. This first 1 is often called a **leading** 1.
iii. If two successive rows do not consist entirely of zeros, then the leading 1 in the lower row occurs farther to the right than the leading 1 in the higher row.
 iv. Any column containing the leading 1 in a row has zeros everywhere else.

EXAMPLE 5

The following matrices are in reduced row echelon form:

i. $\begin{pmatrix} 1 & 0 & 0 \\ 0 & 1 & 0 \\ 0 & 0 & 1 \end{pmatrix}$ ii. $\begin{pmatrix} 1 & 0 & 0 & 0 \\ 0 & 1 & 0 & 0 \\ 0 & 0 & 0 & 1 \end{pmatrix}$ iii. $\begin{pmatrix} 1 & 0 & 0 & 5 \\ 0 & 0 & 1 & 2 \end{pmatrix}$

iv. $\begin{pmatrix} 1 & 0 \\ 0 & 1 \end{pmatrix}$ v. $\begin{pmatrix} 1 & 0 & 2 & 5 \\ 0 & 1 & 3 & 6 \\ 0 & 0 & 0 & 0 \end{pmatrix}$

DEFINITION

Row Echelon Form

A matrix is in **row echelon form** if conditions (*i*), (*ii*), and (*iii*) hold in the preceding definition.

EXAMPLE 6

The following matrices are in row echelon form:

i. $\begin{pmatrix} 1 & 2 & 3 \\ 0 & 1 & 5 \\ 0 & 0 & 1 \end{pmatrix}$ ii. $\begin{pmatrix} 1 & -1 & 6 & 4 \\ 0 & 1 & 2 & -8 \\ 0 & 0 & 0 & 1 \end{pmatrix}$ iii. $\begin{pmatrix} 1 & 0 & 2 & 5 \\ 0 & 0 & 1 & 2 \end{pmatrix}$

iv. $\begin{pmatrix} 1 & 2 \\ 0 & 1 \end{pmatrix}$ v. $\begin{pmatrix} 1 & 3 & 2 & 5 \\ 0 & 1 & 3 & 6 \\ 0 & 0 & 0 & 0 \end{pmatrix}$

Remark 1. The difference between these two forms should be clear from the examples. In row echelon form, all the numbers below the first 1 in a row are zero. In reduced row echelon form, all the numbers above and below the first 1 in a row are zero. Thus reduced row echelon form is more exclusive. That is, every matrix in reduced row echelon form is in row echelon form, but not conversely.

Remark 2. We can always reduce a matrix to reduced row echelon form or row echelon form by performing elementary row operations. We saw this reduction to reduced row echelon form in Examples 2, 3, and 4.

We now describe a general method for reducing the coefficient matrix to reduced row echelon form. If $a_{11} = 0$, we permute the first row and the first row that has a nonzero coefficient for its x_1 term [elementary row operation (*iii*)]. We then divide the first row by a_{11} [elementary row operation (*i*)]. Next, we eliminate the x_1 term from all the other rows [using elementary row operation (*ii*)], and then continue the process above with the x_2 variable and rows 2 to m. The process is continued until one of three situations occurs:

> **i.** The last nonzero† equation reads $x_n = c$ for some constant c. Then there is either a unique solution or an infinite number of solutions to the system.
> **ii.** The last nonzero equation reads $a'_{ij}x_j + a'_{i,j+1}x_{j+1} + \cdots + a'_{i,j+k}x_k = c$ for some constant c where at least two of the a's are nonzero. That is, the last equation is a linear equation in two or more of the variables. Then there are an infinite number of solutions.
> **iii.** The last equation reads $0 = c$. Then there is no solution. In this case the system is called **inconsistent**. In cases (i) and (ii) the system is called **consistent.**

Inconsistent System

EXAMPLE 7

Solve the system

$$x_1 + 3x_2 - 5x_3 + x_4 = 4,$$

$$2x_1 + 5x_2 - 2x_3 + 4x_4 = 6.$$

SOLUTION

We write this system as an augmented matrix and row reduce:

$$\begin{pmatrix} 1 & 3 & -5 & 1 & | & 4 \\ 2 & 5 & -2 & 4 & | & 6 \end{pmatrix} \xrightarrow{A_{1,2}(-2)} \begin{pmatrix} 1 & 3 & -5 & 1 & | & 4 \\ 0 & -1 & 8 & 2 & | & -2 \end{pmatrix}$$

$$\xrightarrow{M_2(-1)} \begin{pmatrix} 1 & 3 & -5 & 1 & | & 4 \\ 0 & 1 & -8 & -2 & | & 2 \end{pmatrix} \xrightarrow{A_{2,1}(-3)} \begin{pmatrix} 1 & 0 & 19 & 7 & | & -2 \\ 0 & 1 & -8 & -2 & | & 2 \end{pmatrix}.$$

† The zero equation is the equation $0x_1 + 0x_2 + \cdots + 0x_n = 0$.

This is as far as we can go. The coefficient matrix is in reduced row echelon form—case (*ii*) above. There is evidently an infinite number of solutions. The variables x_3 and x_4 can be chosen arbitrarily. Then $x_2 = 2 + 8x_3 + 2x_4$ and $x_1 = -2 - 19x_3 - 7x_4$. All solutions are, therefore, represented by $(-2 - 19x_3 - 7x_4, 2 + 8x_3 + 2x_4, x_3, x_4)$. For example, if $x_3 = 1$ and $x_4 = 2$, we obtain the solution $(-35, 14, 1, 2)$.

As you will see if you do a lot of system solving, the computations can become very messy. It is a good rule of thumb to use a calculator whenever the fractions become unpleasant. It should be noted, however, that if computations are carried out on a computer or calculator, "round-off" errors can be introduced. This problem is discussed in Section 14.1.

We close this section with two examples illustrating how a system of linear equations can arise in a practical situation.

EXAMPLE 8

A model that is often used in economics is the **Leontief input-output model**.† Suppose an economic system has n industries. There are two kinds of demands on each industry. First there is the *external* demand from outside the system. If the system is a country, for example, then the external demand could be from another country. Second there is the demand placed on one industry by another industry in the same system. In the United States, for example, there is a demand on the output of the steel industry by the automobile industry.

Let e_i represent the external demand placed on the ith industry. Let a_{ij} represent the internal demand placed on the ith industry by the jth industry. More precisely, a_{ij} represents the number of units of the output of industry i needed to produce 1 unit of the output of industry j. Let x_i represent the output of industry i. Now we assume that the output of each industry is equal to its demand (that is, there is no overproduction). The total demand is equal to the sum of the internal and external demands. To calculate the internal demand on industry 2, for example, we note that industry 1 needs a_{21} units of the output of industry 2 to produce 1 unit of its output. If the output from industry 1 is x_1, then $a_{21}x_1$ is the total amount industry 1 needs from industry 2. Thus the total internal demand on industry 2 is $a_{21}x_1 + a_{22}x_2 + \cdots + a_{2n}x_n$.

We are led to the following system of equations obtained by equating the total demand with the output of each industry:

$$
\begin{aligned}
a_{11}x_1 + a_{12}x_2 + \cdots + a_{1n}x_n + e_1 &= x_1 \\
a_{21}x_1 + a_{22}x_2 + \cdots + a_{2n}x_n + e_2 &= x_2 \\
\vdots \qquad \vdots \qquad\quad \vdots \qquad \vdots \quad\; \vdots \\
a_{n1}x_1 + a_{n2}x_2 + \cdots + a_{nn}x_n + e_n &= x_n
\end{aligned}
\tag{8}
$$

† Named after American economist Wassily W. Leontief. This model was used in his pioneering paper "Qualitative Input and Output Relations in the Economic System of the United States" in *Review of Economic Statistics,* vol. 18 (1936), pp. 105–125. An updated version of this model appears in Leontief's book *Input-Output Analysis* (Oxford University Press, New York, 1966). Leontief won the Nobel Prize in economics in 1973 for his development of input-output analysis.

Or, rewriting (8) so that it looks like system (7), we get

$$
\begin{aligned}
(1-a_{11})x_1 - & \quad a_{12}x_2 - \cdots - & a_{1n}x_n = e_1 \\
-a_{21}x_1 + (1-a_{22})x_2 - & \cdots - & a_{2n}x_n = e_2 \\
\vdots \quad\quad & \vdots & \vdots \quad \vdots \\
-a_{n1}x_1 - & \quad a_{n2}x_2 - \cdots + (1-a_{nn})x_n = e_n
\end{aligned}
\tag{9}
$$

System (9) of n equations in n unknowns is very important in economic analysis.

EXAMPLE 9

In an economic system with three industries, suppose that the external demands are, respectively, 10, 25, and 20. Suppose that $a_{11} = 0.2$, $a_{12} = 0.5$, $a_{13} = 0.15$, $a_{21} = 0.4$, $a_{22} = 0.1$, $a_{23} = 0.3$, $a_{31} = 0.25$, $a_{32} = 0.5$, and $a_{33} = 0.15$. Find the output in each industry such that supply exactly equals demand.

SOLUTION

Here $n = 3$, $1 - a_{11} = 0.8$, $1 - a_{22} = 0.9$, and $1 - a_{33} = 0.85$. Then system (9) is

$$0.8x_1 - 0.5x_2 - 0.15x_3 = 10$$

$$-0.4x_1 + 0.9x_2 - 0.3x_3 = 25$$

$$-0.25x_1 - 0.5x_2 + 0.85x_3 = 20$$

Solving this system by using a calculator, we obtain successively (using five-decimal-place accuracy and Gauss-Jordan elimination)

$$
\left(\begin{array}{ccc|c}
0.8 & -0.5 & -0.15 & 10 \\
-0.4 & 0.9 & -0.3 & 25 \\
-0.25 & -0.5 & 0.85 & 20
\end{array}\right)
\xrightarrow{M_1(1/0.8)}
\left(\begin{array}{ccc|c}
1 & -0.625 & -0.1875 & 12.5 \\
-0.4 & 0.9 & -0.3 & 25 \\
-0.25 & -0.5 & 0.85 & 20
\end{array}\right)
$$

$$
\xrightarrow[A_{1,3}(0.25)]{A_{1,2}(0.4)}
\left(\begin{array}{ccc|c}
1 & -0.625 & -0.1875 & 12.5 \\
0 & 0.65 & -0.375 & 30 \\
0 & -0.65625 & 0.80313 & 23.125
\end{array}\right)
$$

$$
\xrightarrow{M_2(1/0.65)}
\left(\begin{array}{ccc|c}
1 & -0.625 & -0.1875 & 12.5 \\
0 & 1 & -0.57692 & 46.15385 \\
0 & 0.65625 & -0.80313 & 23.125
\end{array}\right)
$$

$$
\xrightarrow[A_{2,3}(0.65625)]{A_{2,1}(0.625)}
\left(\begin{array}{ccc|c}
1 & 0 & -0.54808 & 41.34615 \\
0 & 1 & -0.57692 & 46.15385 \\
0 & 0 & 0.42452 & 53.41346
\end{array}\right)
$$

$$
\xrightarrow{M_3(1/0.42452)}
\left(\begin{array}{ccc|c}
1 & 0 & -0.54808 & 41.34615 \\
0 & 1 & -0.57692 & 46.15385 \\
0 & 0 & 1 & 125.82106
\end{array}\right)
$$

$$
\xrightarrow[A_{3,2}(0.57692)]{A_{3,1}(0.54808)}
\left(\begin{array}{ccc|c}
1 & 0 & 0 & 110.30578 \\
0 & 1 & 0 & 118.74292 \\
0 & 0 & 1 & 125.82106
\end{array}\right).
$$

We conclude that the outputs needed for supply to equal demand are, approximately, $x_1 = 110$, $x_2 = 119$, and $x_3 = 126$.

Problems 7.3

In Problems 1–30 find all solutions (if any) to the given system by using Gauss-Jordan elimination.

1. $x_1 - 3x_2 = 4$
  $-4x_1 + 2x_2 = 6$

2. $2x_1 - x_2 = -3$
  $5x_1 + 7x_2 = 4$

3. $2x_1 - 8x_2 = 5$
  $-3x_1 + 12x_2 = 8$

4. $2x_1 - 8x_2 = 6$
  $-3x_1 + 12x_2 = -9$

5. $6x_1 + x_2 = 3$
  $-4x_1 - x_2 = 8$

6. $3x_1 + x_2 = 0$
  $2x_1 - 3x_2 = 0$

7. $4x_1 - 6x_2 = 0$
  $-2x_1 + 3x_2 = 0$

8. $5x_1 + 2x_2 = 3$
  $2x_1 + 5x_2 = 3$

9. $2x_1 + 3x_2 = 4$
  $3x_1 + 4x_2 = 5$

10. $ax_1 + bx_2 = c$
  $ax_1 - bx_2 = c$

11. $x_1 - 2x_2 + 3x_3 = 11$
  $4x_1 + x_2 - x_3 = 4$
  $2x_1 - x_2 + 3x_3 = 10$

12. $-2x_1 + x_2 + 6x_3 = 18$
  $5x_1 + 8x_3 = -16$
  $3x_1 + 2x_2 - 10x_3 = -3$

13. $3x_1 + 6x_2 - 6x_3 = 9$
  $2x_1 - 5x_2 + 4x_3 = 6$
  $-x_1 + 16x_2 - 14x_3 = -3$

14. $3x_1 + 6x_2 - 6x_3 = 9$
  $2x_1 - 5x_2 + 4x_3 = 6$
  $5x_1 + 28x_2 - 26x_3 = -8$

15. $x_1 + x_2 - x_3 = 7$
  $4x_1 - x_2 + 5x_3 = 4$
  $2x_1 + 2x_2 - 3x_3 = 0$

16. $x_1 + x_2 - x_3 = 7$
  $4x_1 - x_2 + 5x_3 = 4$
  $6x_1 + x_2 + 3x_3 = 18$

17. $x_1 + x_2 - x_3 = 7$
  $4x_1 - x_2 + 5x_3 = 4$
  $6x_1 + x_2 + 3x_3 = 20$

18. $x_1 - 2x_2 + 3x_3 = 0$
  $4x_1 + x_2 - x_3 = 0$
  $2x_1 - x_2 + 3x_3 = 0$

19. $x_1 + x_2 - x_3 = 0$
  $4x_1 - x_2 + 5x_3 = 0$
  $6x_1 + x_2 + 3x_3 = 0$

20. $2x_2 + 5x_3 = 6$
  $x_1 - 2x_3 = 4$
  $2x_1 + 4x_2 = -2$

21. $x_1 + 2x_2 - x_3 = 4$
  $3x_1 + 4x_2 - 2x_3 = 7$

22. $x_1 + 2x_2 - 4x_3 = 4$
  $-2x_1 - 4x_2 + 8x_3 = -8$

23. $x_1 + 2x_2 - 4x_3 = 4$
  $-2x_1 - 4x_2 + 8x_3 = -9$

24. $x_1 + 2x_2 - x_3 + x_4 = 7$
  $3x_1 + 6x_2 - 3x_3 + 3x_4 = 21$

25. $2x_1 + 6x_2 - 4x_3 + 2x_4 = 4$
  $x_1 - x_3 + x_4 = 5$
  $-3x_1 + 2x_2 - 2x_3 = -2$

26. $x_1 - 2x_2 + x_3 + x_4 = 2$
  $3x_1 + 2x_3 - 2x_4 = -8$
  $4x_2 - x_3 - x_4 = 1$
  $-x_1 + 6x_2 - 2x_3 = 7$

27. $x_1 - 2x_2 + x_3 + x_4 = 2$
  $3x_1 + 2x_3 - 2x_4 = -8$
  $4x_2 - x_3 - x_4 = 1$
  $5x_1 + 3x_3 - x_4 = -3$

28. $x_1 - 2x_2 + x_3 + x_4 = 2$
  $3x_1 + 2x_3 - 2x_4 = -8$
  $4x_2 - x_3 - x_4 = 1$
  $5x_1 + 3x_3 - x_4 = 0$

29. $x_1 + x_2 = 4$
  $2x_1 - 3x_2 = 7$
  $3x_1 + 2x_2 = 8$

30. $x_1 + x_2 = 4$
  $2x_1 - 3x_2 = 7$
  $3x_1 - 2x_2 = 11$

In Problems 31–39 determine whether the given matrix is in row echelon form (but not reduced row echelon form), reduced row echelon form, or neither.

31. $\begin{pmatrix} 1 & 1 & 0 \\ 0 & 1 & 1 \\ 0 & 0 & 1 \end{pmatrix}$

32. $\begin{pmatrix} 2 & 0 & 0 \\ 0 & 1 & 0 \\ 0 & 0 & -1 \end{pmatrix}$

33. $\begin{pmatrix} 1 & 0 & 1 & 0 \\ 0 & 1 & 1 & 0 \\ 0 & 0 & 0 & 0 \end{pmatrix}$

34. $\begin{pmatrix} 1 & 0 & 0 & 0 \\ 0 & 0 & 1 & 0 \\ 0 & 0 & 0 & 1 \end{pmatrix}$

35. $\begin{pmatrix} 0 & 1 & 0 & 0 \\ 1 & 0 & 0 & 0 \\ 0 & 0 & 0 & 0 \end{pmatrix}$

36. $\begin{pmatrix} 1 & 0 & 1 & 2 \\ 0 & 1 & 3 & 4 \end{pmatrix}$

37. $\begin{pmatrix} 1 & 0 \\ 0 & 1 \\ 0 & 0 \end{pmatrix}$

38. $\begin{pmatrix} 1 & 0 & 0 \\ 0 & 0 & 0 \\ 0 & 0 & 1 \end{pmatrix}$

39. $\begin{pmatrix} 1 & 0 & 0 & 4 \\ 0 & 1 & 0 & 5 \\ 0 & 1 & 1 & 6 \end{pmatrix}$

In Problems 40–45 use the elementary row operations to reduce the given matrices to row echelon form and reduced row echelon form.

40. $\begin{pmatrix} 1 & 1 \\ 2 & 3 \end{pmatrix}$

41. $\begin{pmatrix} -1 & 6 \\ 4 & 2 \end{pmatrix}$

42. $\begin{pmatrix} 1 & -1 & 1 \\ 2 & 4 & 3 \\ 5 & 6 & -2 \end{pmatrix}$ **43.** $\begin{pmatrix} 2 & -4 & 8 \\ 3 & 5 & 8 \\ -6 & 0 & 4 \end{pmatrix}$

44. $\begin{pmatrix} 2 & -4 & -2 \\ 3 & 1 & 6 \end{pmatrix}$ **45.** $\begin{pmatrix} 2 & -7 \\ 3 & 5 \\ 4 & -3 \end{pmatrix}$

46. In the Leontief input-output model of Example 8, suppose that there are three industries. Suppose further that $e_1 = 10$, $e_2 = 15$, $e_3 = 30$, $a_{11} = \frac{1}{3}$, $a_{12} = \frac{1}{2}$, $a_{13} = \frac{1}{6}$, $a_{21} = \frac{1}{4}$, $a_{22} = \frac{1}{4}$, $a_{23} = \frac{1}{8}$, $a_{31} = \frac{1}{12}$, $a_{32} = \frac{1}{3}$, and $a_{33} = \frac{1}{6}$. Find the output of each industry such that supply exactly equals demand.

* 47. Leontief used his model to analyze the 1958 American economy.† He divided the economy into 81 sectors and grouped them into six families of related sectors. For simplicity, we treat each family of sectors as a single sector so that we can treat the American economy as an economy with six industries. These industries are listed in Table 1. The input-output table, Table 2, gives internal demands in 1958 based on Leontief's figures. The units in the table are millions of dollars. Thus, for example, the number 0.173 in the 6, 5 position means that in order to produce $1 million worth of energy, it is necessary to provide $0.173 million = $173,000 worth of services. Finally, Leontief estimated the demands on the 1958 American economy (in millions of dollars), as listed in Table 3. In order to run the American economy in 1958 and meet all external demands, how many units in each of the six sectors had to be produced?

TABLE 1

Sector	Examples
Final nonmetal (FN)	Furniture, processed food
Final metal (FM)	Household appliances, motor vehicles
Basic metal (BM)	Machine-shop products, mining
Basic nonmetal (BN)	Agriculture, printing
Energy (E)	Petroleum, coal
Services (S)	Amusements, real estate

TABLE 2   INTERNAL DEMANDS IN 1958 U.S. ECONOMY

	FN	FM	BM	BN	E	S
FN	0.170	0.004	0	0.029	0	0.008
FM	0.003	0.295	0.018	0.002	0.004	0.016
BM	0.025	0.173	0.460	0.007	0.011	0.007
BN	0.348	0.037	0.021	0.403	0.011	0.048
E	0.007	0.001	0.039	0.025	0.358	0.025
S	0.120	0.074	0.104	0.123	0.173	0.234

TABLE 3   EXTERNAL DEMANDS ON 1958 U.S. ECONOMY (MILLIONS OF DOLLARS)

FN	$99,640
FM	$75,548
BM	$14,444
BN	$33,501
E	$23,527
S	$263,985

In Problems 48–53 write the given system in the form $A\mathbf{x} = \mathbf{b}$.

48. $2x_1 - x_2 = 3$
$4x_1 + 5x_2 = 7$

49. $x_1 - x_2 + 3x_3 = 11$
$4x_1 + x_2 - x_3 = -4$
$2x_1 - x_2 + 3x_3 = 10$

50. $3x_1 + 6x_2 - 7x_3 = 0$
$2x_1 - x_2 + 3x_3 = 1$

51. $4x_1 - x_2 + x_3 - x_4 = -7$
$3x_1 + x_2 - 5x_3 + 6x_4 = 8$
$2x_1 - x_2 + x_3 = 9$

52. $x_2 - x_3 = 7$
$x_1 + x_3 = 2$
$3x_1 + 2x_2 = -5$

53. $2x_1 + 3x_2 - x_3 = 0$
$-4x_1 + 2x_2 + x_3 = 0$
$7x_1 + 3x_2 - 9x_3 = 0$

54. Using Kirchhoff's laws (see p. 143): (a) Show that the equations for the unknown currents I_1, I_2, and I_3 in the figure on page 377 are

$$I_1 - I_2 - I_3 = 0$$
$$R_2 I_3 - R_1 I_2 = 0$$
$$R_1 I_2 + R_3 I_1 = E$$

(b) Determine the currents in terms of E, R_1, R_2, and R_3.

† The Structure of the U.S. Economy, *Scientific American*, Vol. 212, No. 4 (April, 1965), pp. 25–35.

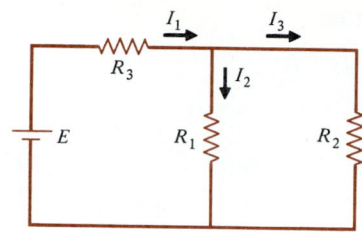

Using Kirchhoff's laws, find the currents in the following circuits in problems 55–56.

55.

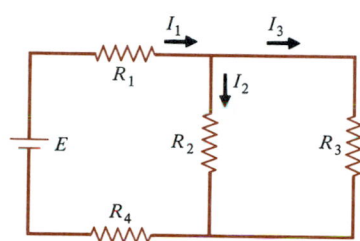

56.

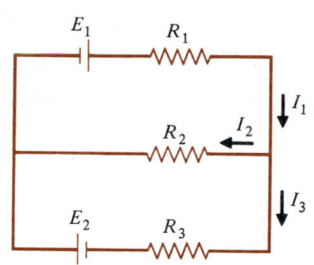

57. Show that $\begin{pmatrix} 1 & 2 \\ 0 & 1 \end{pmatrix}$ and $\begin{pmatrix} 1 & 4/3 \\ 0 & 1 \end{pmatrix}$ are row echelon forms for the matrix $\begin{pmatrix} 1 & 2 \\ 3 & 4 \end{pmatrix}$. That is, *the row echelon form of a matrix is not unique.*

58. Find two row echelon forms for the matrix $\begin{pmatrix} 4 & -2 \\ 5 & 7 \end{pmatrix}$.

7.4 Homogeneous Systems of Equations and Linear Independence

The general $m \times n$ system of linear equations [system (7.3.7), p. 370] is called **homogeneous** if all the constants $b_1, b_2, \ldots, b_m$ are zero. That is, the general homogeneous system is given by

$$\begin{aligned}
a_{11}x_1 + a_{12}x_2 + \cdots + a_{1n}x_n &= 0 \\
a_{21}x_1 + a_{22}x_2 + \cdots + a_{2n}x_n &= 0 \\
&\vdots \\
a_{m1}x_1 + a_{m2}x_2 + \cdots + a_{mn}x_n &= 0
\end{aligned} \qquad (1)$$

Homogeneous systems arise in a variety of ways. We shall see one of these in Examples 4 and 5. We begin by solving some homogeneous systems—again by the method of Gauss-Jordan elimination.

For the general linear system there are three possibilities: no solution, one solution, or an infinite number of solutions. For the general homogeneous system the situation is simpler. Since $x_1 = x_2 = \cdots = x_n = 0$ is always a solution (called the **trivial solution** or **zero solution**), there are only two possibilities: either the zero solution is the only solution or there are an infinite number of solutions in addition to the zero solution. Solutions other than the zero solution are called **nontrivial solutions.**

Trivial Solution

We observe that (1) can be written

$$Ax = 0,$$

where

$$A = \begin{pmatrix} a_{11} & a_{12} & \cdots & a_{1n} \\ a_{21} & a_{22} & \cdots & a_{2n} \\ \vdots & \vdots & & \vdots \\ a_{m1} & a_{m2} & \cdots & a_{mn} \end{pmatrix}, \qquad x = \begin{pmatrix} x_1 \\ x_2 \\ \vdots \\ x_n \end{pmatrix}, \qquad \text{and} \quad 0 = \begin{pmatrix} 0 \\ 0 \\ \vdots \\ 0 \end{pmatrix}. \quad m\text{ zeros}$$

EXAMPLE 1

Solve the homogeneous system

$$2x_1 + 4x_2 + 6x_3 = 0$$

$$4x_1 + 5x_2 + 6x_3 = 0$$

$$3x_1 + x_2 - 2x_3 = 0$$

SOLUTION

This is the homogeneous version of the system in Example 7.3.2, p. 364. It can be written in the form $Ax = 0$ where A and x are as in Example 7.3.2 and

$$0 = \begin{pmatrix} 0 \\ 0 \\ 0 \end{pmatrix}.$$

Reducing successively, we obtain (after dividing the first equation by 2)

$$\begin{pmatrix} 1 & 2 & 3 & | & 0 \\ 4 & 5 & 6 & | & 0 \\ 3 & 1 & -2 & | & 0 \end{pmatrix} \xrightarrow{\substack{A_{1,2}(-4) \\ A_{1,3}(-3)}} \begin{pmatrix} 1 & 2 & 3 & | & 0 \\ 0 & -3 & -6 & | & 0 \\ 0 & -5 & -11 & | & 0 \end{pmatrix} \xrightarrow{M_2(-1/3)} \begin{pmatrix} 1 & 2 & 3 & | & 0 \\ 0 & 1 & 2 & | & 0 \\ 0 & -5 & -11 & | & 0 \end{pmatrix}$$

$$\xrightarrow{\substack{A_{2,1}(-2) \\ A_{2,3}(5)}} \begin{pmatrix} 1 & 0 & -1 & | & 0 \\ 0 & 1 & 2 & | & 0 \\ 0 & 0 & -1 & | & 0 \end{pmatrix} \xrightarrow{M_3(-1)} \begin{pmatrix} 1 & 0 & -1 & | & 0 \\ 0 & 1 & 2 & | & 0 \\ 0 & 0 & 1 & | & 0 \end{pmatrix}$$

$$\xrightarrow{\substack{A_{3,1}(1) \\ A_{3,2}(-2)}} \begin{pmatrix} 1 & 0 & 0 & | & 0 \\ 0 & 1 & 0 & | & 0 \\ 0 & 0 & 1 & | & 0 \end{pmatrix}.$$

Thus the system has the unique solution $(0, 0, 0)$. That is, the system has only the trivial solution.

EXAMPLE 2

Solve the homogeneous system

$$x_1 + 2x_2 - x_3 = 0$$

$$3x_1 - 3x_2 + 2x_3 = 0$$

$$-x_1 - 11x_2 + 6x_3 = 0$$

SOLUTION

Using Gauss-Jordan elimination we obtain, successively,

$$\begin{pmatrix} 1 & 2 & -1 & | & 0 \\ 3 & -3 & 2 & | & 0 \\ -1 & -11 & 6 & | & 0 \end{pmatrix} \xrightarrow{\substack{A_{1,2}(-3) \\ A_{1,3}(1)}} \begin{pmatrix} 1 & 2 & -1 & | & 0 \\ 0 & -9 & 5 & | & 0 \\ 0 & -9 & 5 & | & 0 \end{pmatrix}$$

$$\xrightarrow{M_2(-1/9)} \begin{pmatrix} 1 & 2 & -1 & | & 0 \\ 0 & 1 & -5/9 & | & 0 \\ 0 & -9 & 5 & | & 0 \end{pmatrix} \xrightarrow{\substack{A_{2,1}(-2) \\ A_{2,3}(9)}} \begin{pmatrix} 1 & 0 & 1/9 & | & 0 \\ 0 & 1 & -5/9 & | & 0 \\ 0 & 0 & 0 & | & 0 \end{pmatrix}.$$

The augmented matrix is now in reduced row echelon form and, evidently, there are an infinite number of solutions given by $(-\frac{1}{9}x_3, \frac{5}{9}x_3, x_3)$. If $x_3 = 0$, for example, we obtain the trivial solution. If $x_3 = 1$ we obtain the solution $(-\frac{1}{9}, \frac{5}{9}, 1)$.

EXAMPLE 3

Solve the system

$$\begin{aligned} x_1 + x_2 - x_3 &= 0 \\ 4x_1 - 2x_2 + 7x_3 &= 0 \end{aligned} \qquad (2)$$

SOLUTION

Row-reducing, we obtain

$$\begin{pmatrix} 1 & 1 & -1 & | & 0 \\ 4 & -2 & 7 & | & 0 \end{pmatrix} \xrightarrow{A_{1,2}(-4)} \begin{pmatrix} 1 & 1 & -1 & | & 0 \\ 0 & -6 & 11 & | & 0 \end{pmatrix}$$

$$\xrightarrow{M_2(-1/6)} \begin{pmatrix} 1 & 1 & -1 & | & 0 \\ 0 & 1 & -11/6 & | & 0 \end{pmatrix} \xrightarrow{A_{2,1}(-1)} \begin{pmatrix} 1 & 0 & 5/6 & | & 0 \\ 0 & 1 & -11/6 & | & 0 \end{pmatrix}.$$

Thus an infinite number of solutions are given by $(-\frac{5}{6}x_3, \frac{11}{6}x_3, x_3)$. This is not surprising since system (2) contains three unknowns and only two equations.

In fact, if there are more unknowns than equations, the homogeneous system (1) will always have an infinite number of solutions. To see this, note that if there were only the trivial solution, then row reduction would lead us to the system

$$\begin{aligned} x_1 \qquad\qquad &= 0 \\ x_2 \qquad\quad &= 0 \\ \vdots \\ x_n &= 0 \end{aligned}$$

and, possibly, additional equations of the form $0 = 0$. But this system has at least as many equations as unknowns. Since row reduction does not change either the number of equations or the number of unknowns, we have a contradiction of our assumption that there were more unknowns than equations. Thus we have Theorem 1.

■ THEOREM 1

The homogeneous system (1) has an infinite number of solutions if $n > m$. That is, there are an infinite number of solutions if there are more unknowns than equations. ■

We can determine whether a set of vectors is linearly independent or dependent by solving an appropriate homogeneous system of equations.

EXAMPLE 4

Determine whether the vectors $\begin{pmatrix} 1 \\ -2 \\ 3 \end{pmatrix}$, $\begin{pmatrix} 2 \\ -2 \\ 0 \end{pmatrix}$, and $\begin{pmatrix} 0 \\ 1 \\ 7 \end{pmatrix}$ are linearly dependent or independent.

SOLUTION

Suppose that $c_1 \begin{pmatrix} 1 \\ -2 \\ 3 \end{pmatrix} + c_2 \begin{pmatrix} 2 \\ -2 \\ 0 \end{pmatrix} + c_3 \begin{pmatrix} 0 \\ 1 \\ 7 \end{pmatrix} = \mathbf{0} = \begin{pmatrix} 0 \\ 0 \\ 0 \end{pmatrix}$. Then, multiplying through and

adding, we have $\begin{pmatrix} c_1 + 2c_2 \\ -2c_1 - 2c_2 + c_3 \\ 3c_1 + 7c_3 \end{pmatrix} = \begin{pmatrix} 0 \\ 0 \\ 0 \end{pmatrix}$. This yields a homogeneous system of three

equations in the three unknowns c_1, c_2, and c_3:

$$c_1 + 2c_2 \qquad\quad = 0$$
$$-2c_1 - 2c_2 + c_3 = 0 \qquad\qquad \textbf{(3)}$$
$$3c_1 \qquad + 7c_3 = 0$$

Thus the vectors will be linearly dependent if and only if system (3) has nontrivial solutions. We write system (3) using an augmented matrix and then row-reduce:

$$\begin{pmatrix} 1 & 2 & 0 & | & 0 \\ -2 & -2 & 1 & | & 0 \\ 3 & 0 & 7 & | & 0 \end{pmatrix} \xrightarrow[A_{1,3}(-3)]{A_{1,2}(2)} \begin{pmatrix} 1 & 2 & 0 & | & 0 \\ 0 & 2 & 1 & | & 0 \\ 0 & -6 & 7 & | & 0 \end{pmatrix} \xrightarrow{M_2(1/2)} \begin{pmatrix} 1 & 2 & 0 & | & 0 \\ 0 & 1 & 1/2 & | & 0 \\ 0 & -6 & 7 & | & 0 \end{pmatrix}$$

$$\xrightarrow[A_{2,3}(6)]{A_{2,1}(-2)} \begin{pmatrix} 1 & 0 & -1 & | & 0 \\ 0 & 1 & 1/2 & | & 0 \\ 0 & 0 & 10 & | & 0 \end{pmatrix} \xrightarrow{M_3(1/10)} \begin{pmatrix} 1 & 0 & -1 & | & 0 \\ 0 & 1 & 1/2 & | & 0 \\ 0 & 0 & 1 & | & 0 \end{pmatrix}$$

$$\xrightarrow[A_{3,2}(-1/2)]{A_{3,1}(1)} \begin{pmatrix} 1 & 0 & 0 & | & 0 \\ 0 & 1 & 0 & | & 0 \\ 0 & 0 & 1 & | & 0 \end{pmatrix}.$$

The last system of equations reads $c_1 = 0$, $c_2 = 0$, $c_3 = 0$. Hence (3) has no nontrivial solutions and the given vectors are linearly independent.

EXAMPLE 5

Determine whether the vectors $\begin{pmatrix} 1 \\ -3 \\ 0 \end{pmatrix}$, $\begin{pmatrix} 3 \\ 0 \\ 4 \end{pmatrix}$, and $\begin{pmatrix} 11 \\ -6 \\ 12 \end{pmatrix}$ are linearly dependent or

independent.

SOLUTION

The equation $c_1\begin{pmatrix} 1 \\ -3 \\ 0 \end{pmatrix} + c_2\begin{pmatrix} 3 \\ 0 \\ 4 \end{pmatrix} + c_3\begin{pmatrix} 11 \\ -6 \\ 12 \end{pmatrix} = \begin{pmatrix} 0 \\ 0 \\ 0 \end{pmatrix}$ leads to the homogeneous system

$$c_1 + 3c_2 + 11c_3 = 0$$

$$-3c_1 \qquad - 6c_3 = 0 \tag{4}$$

$$4c_2 + 12c_3 = 0$$

Writing system (4) in augmented matrix form and row-reducing we obtain, successively,

$$\begin{pmatrix} 1 & 3 & 11 & | & 0 \\ -3 & 0 & -6 & | & 0 \\ 0 & 4 & 12 & | & 0 \end{pmatrix} \xrightarrow{A_{1,2}(3)} \begin{pmatrix} 1 & 3 & 11 & | & 0 \\ 0 & 9 & 27 & | & 0 \\ 0 & 4 & 12 & | & 0 \end{pmatrix}$$

$$\xrightarrow{M_2(1/9)} \begin{pmatrix} 1 & 3 & 11 & | & 0 \\ 0 & 1 & 3 & | & 0 \\ 0 & 4 & 12 & | & 0 \end{pmatrix} \xrightarrow[A_{2,3}(-4)]{A_{2,1}(-3)} \begin{pmatrix} 1 & 0 & 2 & | & 0 \\ 0 & 1 & 3 & | & 0 \\ 0 & 0 & 0 & | & 0 \end{pmatrix}.$$

We can stop here since we can see that system (4) has an infinite number of solutions. For example, the last augmented matrix reads

$$c_1 \qquad + 2c_3 = 0$$

$$c_2 + 3c_3 = 0$$

If we choose $c_3 = 1$, we have $c_2 = -3$ and $c_1 = -2$ so that, as is easily verified,

$-2\begin{pmatrix} 1 \\ -3 \\ 0 \end{pmatrix} - 3\begin{pmatrix} 3 \\ 0 \\ 4 \end{pmatrix} + \begin{pmatrix} 11 \\ -6 \\ 12 \end{pmatrix} = \begin{pmatrix} 0 \\ 0 \\ 0 \end{pmatrix}$ and the vectors are linearly dependent.

Theorem 1 can be used to prove an important theorem stated in Section 6.7.

■ **THEOREM 2 (Theorem 6.7.3).**
A set of m vectors in $\mathbb{R}^n$ is always linearly dependent if $m > n$.

PROOF
Let $\mathbf{v}_1, \mathbf{v}_2, \ldots, \mathbf{v}_m$ be m vectors in $\mathbb{R}^n$ and let us try to find constants $c_1, c_2, \ldots, c_m$ not all zero such that

$$c_1\mathbf{v}_1 + c_2\mathbf{v}_2 + \cdots + c_m\mathbf{v}_m = \mathbf{0}. \tag{5}$$

Let $\mathbf{v}_1 = \begin{pmatrix} a_{11} \\ a_{21} \\ \vdots \\ a_{n1} \end{pmatrix}, \mathbf{v}_2 = \begin{pmatrix} a_{12} \\ a_{22} \\ \vdots \\ a_{n2} \end{pmatrix}, \ldots, \mathbf{v}_m = \begin{pmatrix} a_{1m} \\ a_{2m} \\ \vdots \\ a_{nm} \end{pmatrix}$. Then equation (5) becomes

$$a_{11}c_1 + a_{12}c_2 + \cdots + a_{1m}c_m = 0$$
$$a_{21}c_1 + a_{22}c_2 + \cdots + a_{2m}c_m = 0$$
$$\vdots \qquad \vdots \qquad \qquad \vdots \qquad \vdots \tag{6}$$
$$a_{n1}c_1 + a_{n2}c_2 + \cdots + a_{nm}c_m = 0$$

System (6) is a homogeneous system of n equations in m unknowns with $m > n$. Then, by Theorem 1, since there are more unknowns than equations, the system has an infinite number of solutions. Thus there are scalars $c_1, c_2, \ldots, c_m$ not all zero that satisfy (6), which means that the vectors $v_1, v_2, \ldots, v_m$ are linearly dependent. ■

Let A be the $n \times m$ matrix with columns that are vectors $v_1, v_2, \ldots, v_m$ in $\mathbb{R}^n$. To determine whether the vectors are linearly dependent or independent, we write equation (5) and system (6). System (6) can be written as $Ac = 0$ with

$$c = \begin{pmatrix} c_1 \\ c_2 \\ \vdots \\ c_m \end{pmatrix}.$$ We therefore have the following:

■ **THEOREM 3**

The columns of the matrix A are linearly independent if and only if the only solution to the homogeneous system $Ax = 0$ is the trivial solution ($x = 0$). ■

There is a fundamental relationship between homogeneous and nonhomogeneous systems. Let A be an $m \times n$ matrix,

$$x = \begin{pmatrix} x_1 \\ x_2 \\ \vdots \\ x_n \end{pmatrix}, \quad b = \begin{pmatrix} b_1 \\ b_2 \\ \vdots \\ b_m \end{pmatrix}, \quad \text{and} \quad 0 = \begin{pmatrix} 0 \\ 0 \\ \vdots \\ 0 \end{pmatrix}.$$

m zeros

The general nonhomogeneous system can be written as

$$Ax = b. \tag{7}$$

Associated Homogeneous System   With A and x as in (7), we define the **associated homogeneous system** by

$$Ax = 0. \tag{8}$$

■ **THEOREM 4**

Let x_1 and x_2 be solutions of the nonhomogeneous system (7). Then their difference, $x_1 - x_2$, is a solution of the associated homogeneous system (8).

PROOF
$$A(x_1 - x_2) = Ax_1 - Ax_2 = b - b = 0. \qquad ■$$

■ **COROLLARY**

Let x be a particular solution to the nonhomogeneous system (7) and let y be another solution to (7). Then there exists a vector h that is a solution to the associated homogeneous system (8) such that

$$y = x + h. \tag{9}$$

PROOF

If $\mathbf{h}$ is defined by $\mathbf{h} = \mathbf{y} - \mathbf{x}$, then $\mathbf{h}$ solves (8) by Theorem 4 and $\mathbf{y} = \mathbf{x} + \mathbf{h}$. ∎

Theorem 4 and its corollary are very useful. They tell us that

> In order to find all solutions to the nonhomogeneous system (7), it is only necessary to find *one* solution to (7) and all solutions to the associated homogeneous system (8).

Remark. A very similar result holds for solutions of homogeneous and nonhomogeneous linear differential equations (see Problems 30 and 31). One of the many nice things about mathematics is that seemingly very different topics are closely interrelated.

EXAMPLE 6

Find all solutions to the nonhomogeneous system

$$x_1 + 2x_2 - x_3 = 2,$$
$$2x_1 + 3x_2 + 5x_3 = 5, \tag{10}$$
$$-x_1 - 3x_2 + 8x_3 = -1,$$

by using the Corollary given above.

SOLUTION

First we find one solution by row reduction:

$$\begin{pmatrix} 1 & 2 & -1 & | & 2 \\ 2 & 3 & 5 & | & 5 \\ -1 & -3 & 8 & | & -1 \end{pmatrix} \xrightarrow[A_{1,3}(1)]{A_{1,2}(-2)} \begin{pmatrix} 1 & 2 & -1 & | & 2 \\ 0 & -1 & 7 & | & 1 \\ 0 & -1 & 7 & | & 1 \end{pmatrix} \xrightarrow[A_{2,3}(-1)]{A_{2,1}(2)} \begin{pmatrix} 1 & 0 & 13 & | & 4 \\ 0 & -1 & 7 & | & 1 \\ 0 & 0 & 0 & | & 0 \end{pmatrix}.$$

We see that there are an infinite number of solutions. Setting $x_3 = 0$ (any other number would do), we obtain $x_1 = 4$ and $x_2 = -1$. So one particular solution is $\mathbf{x_p} = (4, -1, 0)$.

Row reduction of the associated homogeneous system leads to

$$\begin{pmatrix} 1 & 0 & 13 & | & 0 \\ 0 & -1 & 7 & | & 0 \\ 0 & 0 & 0 & | & 0 \end{pmatrix}.$$

Therefore all solutions to the homogeneous system satisfy

$$x_1 = -13x_3, \qquad x_2 = 7x_3,$$

or

$$\mathbf{x_h} = (x_1, x_2, x_3) = (-13x_3, 7x_3, x_3) = x_3(-13, 7, 1).$$

Thus each solution to (10) can be written

$$\mathbf{x} = \mathbf{x_p} + \mathbf{x_h} = (4, -1, 0) + x_3(-13, 7, 1)$$

for an appropriate value of x_3. For example, $x_3 = 0$ yields the solution $(4, -1, 0)$, while $x_3 = 2$ gives the solution $(-22, 13, 2)$.

We close this section by proving a theorem first stated in Section 6.7.

■ THEOREM 5 (THEOREM 6.7.5)

If $\{\mathbf{u}_1, \mathbf{u}_2, \ldots, \mathbf{u}_m\}$ and $\{\mathbf{v}_1, \mathbf{v}_2, \ldots, \mathbf{v}_n\}$ are bases for the vector space V, then $m = n$; that is, any two bases in a vector space V have the same number of vectors.

PROOF†

Let $S_1 = \{\mathbf{u}_1, \ldots, \mathbf{u}_m\}$ and $S_2 = \{\mathbf{v}_1, \ldots, \mathbf{v}_n\}$ be two bases for V. We must show that $m = n$. We prove this by showing that if $m > n$, then S_1 is a linearly dependent set, which contradicts the hypothesis that S_1 is a basis. This will show that $m \leq n$. The same proof will then show that $n \leq m$, and this will prove the theorem. Hence all we must show is that if $m > n$, then S_1 is dependent. Since S_2 constitutes a basis, we can write each $\mathbf{u}_i$ as a linear combination of $\mathbf{v}_i$'s. We have

$$
\begin{aligned}
\mathbf{u}_1 &= a_{11}\mathbf{v}_1 + a_{12}\mathbf{v}_2 + \cdots + a_{1n}\mathbf{v}_n \\
\mathbf{u}_2 &= a_{21}\mathbf{v}_1 + a_{22}\mathbf{v}_2 + \cdots + a_{2n}\mathbf{v}_n \\
&\;\;\vdots \qquad\;\; \vdots \qquad\;\; \vdots \qquad\qquad \vdots \\
\mathbf{u}_m &= a_{m1}\mathbf{v}_1 + a_{m2}\mathbf{v}_2 + \cdots + a_{mn}\mathbf{v}_n
\end{aligned}
\tag{11}
$$

To show that S_1 is dependent, we must find scalars $c_1, c_2, \ldots, c_m$, not all zero, such that

$$
c_1\mathbf{u}_1 + c_2\mathbf{u}_2 + \cdots + c_m\mathbf{u}_m = \mathbf{0}.
\tag{12}
$$

Inserting (11) into (12), we obtain

$$
c_1(a_{11}\mathbf{v}_1 + a_{12}\mathbf{v}_2 + \cdots + a_{1n}\mathbf{v}_n) + c_2(a_{21}\mathbf{v}_1 + a_{22}\mathbf{v}_2 + \cdots + a_{2n}\mathbf{v}_n)
$$
$$
+ \cdots + c_m(a_{m1}\mathbf{v}_1 + a_{m2}\mathbf{v}_2 + \cdots + a_{mn}\mathbf{v}_n) = \mathbf{0}.
\tag{13}
$$

Equation (13) can be rewritten as

$$
(a_{11}c_1 + a_{21}c_2 + \cdots + a_{m1}c_m)\mathbf{v}_1 + (a_{12}c_1 + a_{22}c_2 + \cdots + a_{m2}c_m)\mathbf{v}_2
$$
$$
+ \cdots + (a_{1n}c_1 + a_{2n}c_2 + \cdots + a_{mn}c_m)\mathbf{v}_n = \mathbf{0}.
\tag{14}
$$

But, since $\mathbf{v}_1, \mathbf{v}_2, \ldots, \mathbf{v}_n$ are linearly independent, we must have

$$
\begin{aligned}
a_{11}c_1 + a_{21}c_2 + \cdots + a_{m1}c_m &= 0 \\
a_{12}c_1 + a_{22}c_2 + \cdots + a_{m2}c_m &= 0 \\
\vdots \qquad\;\; \vdots \qquad\qquad \vdots \qquad\;\; \vdots \\
a_{1n}c_1 + a_{2n}c_2 + \cdots + a_{mn}c_m &= 0
\end{aligned}
\tag{15}
$$

---

† This proof is given for vector spaces with bases containing a finite number of vectors. We also treat the scalars as though they were real numbers. However, the proof works in the complex case as well.

System (15) is a homogeneous system of n equations in the m unknowns $c_1, c_2, \ldots, c_m$ and, since $m > n$, Theorem 1 tells us that the system has an infinite number of solutions. Thus there are scalars $c_1, c_2, \ldots, c_m$, not all zero, such that (12) is satisfied and therefore S_1 is a linearly dependent set. This contradiction proves that $m \leq n$ and, by exchanging the roles of S_1 and S_2, we can show that $n \leq m$ and the proof is complete. ∎

Problems 7.4

In Problems 1–13 find all solutions to the homogeneous systems.

1. $2x_1 - x_2 = 0$
$3x_1 + 4x_2 = 0$

2. $x_1 - 5x_2 = 0$
$-x_1 + 5x_2 = 0$

3. $x_1 + x_2 - x_3 = 0$
$2x_1 - 4x_2 + 3x_3 = 0$
$3x_1 + 7x_2 - x_3 = 0$

4. $x_1 + x_2 - x_3 = 0$
$2x_1 - 4x_2 + 3x_3 = 0$
$-x_1 - 7x_2 + 6x_3 = 0$

5. $x_1 + x_2 - x_3 = 0$
$2x_1 - 4x_2 + 3x_3 = 0$
$-5x_1 + 13x_2 - 10x_3 = 0$

6. $2x_1 + 3x_2 - x_3 = 0$
$6x_1 - 5x_2 + 7x_3 = 0$

7. $4x_1 - x_2 = 0$
$7x_1 + 3x_2 = 0$
$-8x_1 + 6x_2 = 0$

8. $x_1 - x_2 + 7x_3 - x_4 = 0$
$2x_1 + 3x_2 - 8x_3 + x_4 = 0$

9. $x_1 - 2x_2 + x_3 + x_4 = 0$
$3x_1 \qquad + 2x_3 - 2x_4 = 0$
$\qquad 4x_2 - x_3 - x_4 = 0$
$5x_1 \qquad + 3x_3 - x_4 = 0$

10. $-2x_1 \qquad\qquad + 7x_4 = 0$
$x_1 + 2x_2 - x_3 + 4x_4 = 0$
$3x_1 \qquad - x_3 + 5x_4 = 0$
$4x_1 + 2x_2 + 3x_3 \qquad = 0$

11. $2x_1 - x_2 = 0$
$3x_1 + 5x_2 = 0$
$7x_1 - 3x_2 = 0$
$-2x_1 + 3x_2 = 0$

12. $x_1 - 3x_2 = 0$
$-2x_1 + 6x_2 = 0$
$4x_1 - 12x_2 = 0$

13. $x_1 + x_2 - x_3 = 0$
$4x_1 - x_2 + 5x_3 = 0$
$-2x_1 + x_2 - 2x_3 = 0$
$3x_1 + 2x_2 - 6x_3 = 0$

14. Show that the homogeneous system
$$a_{11}x_1 + a_{12}x_2 = 0$$
$$a_{21}x_1 + a_{22}x_2 = 0$$
has an infinite number of solutions if and only if
$$a_{11}a_{22} - a_{21}a_{12} = 0.$$

15. Consider the system
$$2x_1 - 3x_2 + 5x_3 = 0$$
$$-x_1 + 7x_2 - x_3 = 0$$
$$4x_1 - 11x_2 + kx_3 = 0$$

For what value of k will the system have nontrivial solutions?

In Problems 16–21 determine whether the set of vectors is linearly dependent or independent.

16. $\begin{pmatrix} 1 \\ 0 \\ 1 \end{pmatrix}, \begin{pmatrix} 0 \\ 1 \\ 1 \end{pmatrix}, \begin{pmatrix} 1 \\ 1 \\ 0 \end{pmatrix}$

17. $\begin{pmatrix} -3 \\ 4 \\ 2 \end{pmatrix}, \begin{pmatrix} 7 \\ -1 \\ 3 \end{pmatrix}, \begin{pmatrix} 1 \\ 2 \\ 8 \end{pmatrix}$

18. $\begin{pmatrix} -3 \\ 4 \\ 2 \end{pmatrix}, \begin{pmatrix} 7 \\ -1 \\ 3 \end{pmatrix}, \begin{pmatrix} 1 \\ 1 \\ 8 \end{pmatrix}$

19. $\begin{pmatrix} 1 \\ -2 \\ 1 \\ 1 \end{pmatrix}, \begin{pmatrix} 3 \\ 0 \\ 2 \\ -2 \end{pmatrix}, \begin{pmatrix} 0 \\ 4 \\ -1 \\ -1 \end{pmatrix}, \begin{pmatrix} 5 \\ 0 \\ 3 \\ -1 \end{pmatrix}$

20. $\begin{pmatrix} 1 \\ -2 \\ 1 \\ 1 \end{pmatrix}, \begin{pmatrix} 3 \\ 0 \\ 2 \\ -2 \end{pmatrix}, \begin{pmatrix} 0 \\ 4 \\ -1 \\ 1 \end{pmatrix}, \begin{pmatrix} 5 \\ 0 \\ 3 \\ -1 \end{pmatrix}$

21. $\begin{pmatrix} 1 \\ -1 \\ 2 \end{pmatrix}, \begin{pmatrix} 4 \\ 0 \\ 0 \end{pmatrix}, \begin{pmatrix} -2 \\ 3 \\ 5 \end{pmatrix}, \begin{pmatrix} 7 \\ 1 \\ 2 \end{pmatrix}$

22. For what value(s) of α will the vectors $\begin{pmatrix} 1 \\ 2 \\ 3 \end{pmatrix}, \begin{pmatrix} 2 \\ -1 \\ 4 \end{pmatrix}, \begin{pmatrix} 3 \\ \alpha \\ 4 \end{pmatrix}$

be linearly dependent?

23. For what value(s) of α are the vectors $\begin{pmatrix} 2 \\ -3 \\ 1 \end{pmatrix}, \begin{pmatrix} -4 \\ 6 \\ -2 \end{pmatrix},$

$\begin{pmatrix} \alpha \\ 1 \\ 2 \end{pmatrix}$ linearly dependent? [*Hint:* Look carefully.]

In Problems 24–29 find all solutions to the given nonhomogeneous system by first finding one solution (if possible) and then finding all solutions to the associated homogeneous system.

24. $\quad x_1 - 3x_2 = 2$
$\quad -2x_1 + 6x_2 = -4$

25. $\quad x_1 - x_2 + x_3 = 6$
$\quad 3x_1 - 3x_2 + 3x_3 = 18$

26. $\quad x_1 - x_2 - x_3 = 2$
$\quad 2x_1 + x_2 + 2x_3 = 4$
$\quad x_1 - 4x_2 - 5x_3 = 2$

27. $\quad x_1 - x_2 - x_3 = 2$
$\quad 2x_1 + x_2 + 2x_3 = 4$
$\quad x_1 - 4x_2 - 5x_3 = 3$

28. $\quad x_1 + x_2 - x_3 + 2x_4 = 3$
$\quad 3x_1 + 2x_2 + x_3 - x_4 = 5$

29. $\quad x_1 - x_2 + x_3 - x_4 = -2$
$\quad -2x_1 + 3x_2 - x_3 + 2x_4 = 5$
$\quad 4x_1 - 2x_2 + 2x_3 - 3x_4 = 6$

30. Consider the linear homogeneous second-order differential equation

$$y''(x) + a(x)y'(x) + b(x)y(x) = 0, \qquad \text{(16)}$$

where $a(x)$ and $b(x)$ are continuous and the unknown function y is assumed to have a second derivative. Show that if y_1 and y_2 are solutions to (16), then $c_1 y_1 + c_2 y_2$ is a solution for any constants c_1 and c_2.

31. Suppose that y_p and y_q are solutions to the nonhomogeneous equation

$$y''(x) + a(x)y'(x) + b(x)y(x) = f(x).$$

Show that $y_p - y_q$ is a solution to the homogeneous equation (16).

7.5 The Inverse of a Square Matrix

In this section we define two kinds of matrices that are central to matrix theory. We begin with a simple example. Let $A = \begin{pmatrix} 2 & 5 \\ 1 & 3 \end{pmatrix}$ and $B = \begin{pmatrix} 3 & -5 \\ -1 & 2 \end{pmatrix}$. Then an easy computation shows that $AB = BA = I_2$, where $I_2 = \begin{pmatrix} 1 & 0 \\ 0 & 1 \end{pmatrix}$. The matrix I_2 is called the 2×2 *identity matrix.* The matrix B is called the *inverse* of A and is written A^{-1}.

DEFINITION

Identity Matrix

The $n \times n$ **identity matrix** is the $n \times n$ matrix with 1's down the **main diagonal**† and 0's everywhere else. That is,

$$I_n = (\delta_{ij}), \quad \text{where} \quad \delta_{ij} = \begin{cases} 1 & \text{if } i = j, \\ 0 & \text{if } i \neq j. \end{cases} \qquad \text{(1)}$$

---

† The main diagonal of $A = (a_{ij})$ consists of the components a_{11}, a_{22}, a_{33}, and so on. Unless otherwise stated, we shall refer to the main diagonal simply as the **diagonal**.

EXAMPLE 1

$$I_3 = \begin{pmatrix} 1 & 0 & 0 \\ 0 & 1 & 0 \\ 0 & 0 & 1 \end{pmatrix} \quad \text{and} \quad I_5 = \begin{pmatrix} 1 & 0 & 0 & 0 & 0 \\ 0 & 1 & 0 & 0 & 0 \\ 0 & 0 & 1 & 0 & 0 \\ 0 & 0 & 0 & 1 & 0 \\ 0 & 0 & 0 & 0 & 1 \end{pmatrix}.$$

■ THEOREM 1

Let A be a square $n \times n$ matrix. Then

$$AI_n = I_n A = A.$$

That is, I_n commutes with every $n \times n$ matrix and leaves it unchanged after multiplication on the left or right.

Note. I_n functions for $n \times n$ matrices the way the number 1 functions for real numbers (since $1 \cdot a = a \cdot 1 = a$ for every real number a).

PROOF
Let c_{ij} be the ijth element of AI_n. Then

$$c_{ij} = a_{i1}\delta_{1j} + a_{i2}\delta_{2j} + \cdots + a_{ij}\delta_{jj} + \cdots + a_{in}\delta_{nj}.$$

But, from (1), this sum is equal to a_{ij}. Thus $AI_n = A$. In a similar fashion we can show that $I_n A = A$, and this proves the theorem. ■

Notation. From now on we shall write any identity matrix simply as I, since if A is $n \times n$, the products IA and AI are defined only if I is also $n \times n$.

DEFINITION

Let A and B be $n \times n$ matrices. Suppose that

$$AB = BA = I.$$

Inverse of a Matrix

Then B is called the **inverse** of A and is written as A^{-1}. We then have

$$AA^{-1} = A^{-1}A = I.$$

Invertible Matrix

If A has an inverse, then A is said to be **invertible**.

Remark 1. From this definition it immediately follows that $(A^{-1})^{-1} = A$ if A is invertible.

Remark 2. This definition does *not* state that every square matrix has an inverse. In fact, many square matrices have no inverse. (See, for instance, Example 3 later in this section.)

In the last definition we defined *the* inverse of a matrix. This statement suggests that inverses are unique. This is indeed the case, as the following theorem shows.

■ **THEOREM 2**

If a square matrix A is invertible, then its inverse is unique.

PROOF

Suppose B and C are two inverses for A. We must show that $B = C$. By definition, we have $AB = BA = I$ and $AC = CA = I$. Then $B(AC) = BI = B$ and $(BA)C = IC = C$. But $B(AC) = (BA)C$ by the associative law of matrix multiplication. Hence $B = C$ and the theorem is proved. ■

Another important fact about inverses is given below.

■ **THEOREM 3**

Let A and B be invertible $n \times n$ matrices. Then AB is invertible and

$$(AB)^{-1} = B^{-1}A^{-1}.$$

PROOF

To prove this result, we refer to the definition of the inverse. That is, $B^{-1}A^{-1} = (AB)^{-1}$ if and only if $B^{-1}A^{-1}(AB) = (AB)(B^{-1}A^{-1}) = I$. But this follows since

$$\overset{\text{equation (4) on p. 360}}{(B^{-1}A^{-1})(AB) = B^{-1}(A^{-1}A)B = B^{-1}IB = B^{-1}B = I}$$

and

$$(AB)(B^{-1}A^{-1}) = A(BB^{-1})A^{-1} = AIA^{-1} = AA^{-1} = I.$$ ■

Consider the system of n equations in n unknowns

$$A\mathbf{x} = \mathbf{b},$$

and suppose that A is invertible. Then

$$A^{-1}A\mathbf{x} = A^{-1}\mathbf{b}, \qquad \text{we multiplied on the left by } A^{-1}$$

$$I\mathbf{x} = A^{-1}\mathbf{b}, \qquad A^{-1}A = I$$

$$\mathbf{x} = A^{-1}\mathbf{b}. \qquad I\mathbf{x} = \mathbf{x}$$

That is,

> If A is invertible, the system $A\mathbf{x} = \mathbf{b}$ has the unique solution $\mathbf{x} = A^{-1}\mathbf{b}$.
>
> (2)

This is one of the reasons we study matrix inverses.

Two basic questions come to mind once we have defined the inverse of a matrix:

Question 1. What matrices do have inverses?

Question 2. If a matrix has an inverse, how can we compute it?

We answer both questions in this section. Rather than starting by giving you what seems to be a set of arbitrary rules, we look first at what happens in the 2×2 case.

EXAMPLE 2

Let $A = \begin{pmatrix} 2 & -3 \\ -4 & 5 \end{pmatrix}$. Compute A^{-1} if it exists.

SOLUTION

Suppose that A^{-1} exists. We write $A^{-1} = \begin{pmatrix} x & y \\ z & w \end{pmatrix}$ and use the fact that $AA^{-1} = I$. Then

$$AA^{-1} = \begin{pmatrix} 2 & -3 \\ -4 & 5 \end{pmatrix}\begin{pmatrix} x & y \\ z & w \end{pmatrix} = \begin{pmatrix} 2x - 3z & 2y - 3w \\ -4x + 5z & -4y + 5w \end{pmatrix} = \begin{pmatrix} 1 & 0 \\ 0 & 1 \end{pmatrix}.$$

The last two matrices can be equal only if their corresponding components are equal. This means that

$$2x \quad\quad - 3z \quad\quad = 1 \tag{3}$$

$$2y \quad\quad - 3w = 0 \tag{4}$$

$$-4x \quad\quad + 5z \quad\quad = 0 \tag{5}$$

$$- 4y \quad\quad + 5w = 1 \tag{6}$$

This is a system of four equations in four unknowns. Note that there are two equations involving x and z only [equations (3) and (5)] and two equations involving y and w only [equations (4) and (6)]. We write these two systems in augmented matrix form:

$$\begin{pmatrix} 2 & -3 & | & 1 \\ -4 & 5 & | & 0 \end{pmatrix}, \tag{7}$$

$$\begin{pmatrix} 2 & -3 & | & 0 \\ -4 & 5 & | & 1 \end{pmatrix}. \tag{8}$$

Now, we know from Section 7.3 that if system (7) (in the variables x and z) has a unique solution, then Gauss-Jordan elimination of (7) will result in

$$\left(\begin{array}{cc|c} 1 & 0 & x \\ 0 & 1 & z \end{array}\right)$$

where (x, z) is the unique pair of numbers that satisfies $2x - 3z = 1$ and $-4x + 5z = 0$. Similarly, row reduction of (8) will result in

$$\left(\begin{array}{cc|c} 1 & 0 & y \\ 0 & 1 & w \end{array}\right)$$

where (y, w) is the unique pair of numbers that satisfies $2y - 3w = 0$ and $-4y + 5w = 1$.

Since the coefficient matrices in (7) and (8) are the same, we can perform the row reductions on the two augmented matrices simultaneously, by considering the new augmented matrix

$$\left(\begin{array}{cc|cc} 2 & -3 & 1 & 0 \\ -4 & 5 & 0 & 1 \end{array}\right). \tag{9}$$

If A is invertible, then the system defined by (3), (4), (5), and (6) has a unique solution and, by what we said above, Gauss-Jordan elimination will result in

$$\left(\begin{array}{cc|cc} 1 & 0 & x & y \\ 0 & 1 & z & w \end{array}\right).$$

We now carry out the computation, noting that the matrix on the left in (9) is A and the matrix on the right in (9) is I:

$$\left(\begin{array}{cc|cc} 2 & -3 & 1 & 0 \\ -4 & 5 & 0 & 1 \end{array}\right) \xrightarrow{M_1(1/2)} \left(\begin{array}{cc|cc} 1 & -3/2 & 1/2 & 0 \\ -4 & 5 & 0 & 1 \end{array}\right)$$

$$\xrightarrow{A_{1,2}(4)} \left(\begin{array}{cc|cc} 1 & -3/2 & 1/2 & 0 \\ 0 & -1 & 2 & 1 \end{array}\right)$$

$$\xrightarrow{M_2(-1)} \left(\begin{array}{cc|cc} 1 & -3/2 & 1/2 & 0 \\ 0 & 1 & -2 & -1 \end{array}\right)$$

$$\xrightarrow{A_{2,1}(3/2)} \left(\begin{array}{cc|cc} 1 & 0 & -5/2 & -3/2 \\ 0 & 1 & -2 & -1 \end{array}\right).$$

Thus $x = -\frac{5}{2}$, $y = -\frac{3}{2}$, $z = -2$, $w = -1$, and $A^{-1} = \left(\begin{array}{cc} -5/2 & -3/2 \\ -2 & -1 \end{array}\right)$. We still must check our answer. We have

$$AA^{-1} = \left(\begin{array}{cc} 2 & -3 \\ -4 & 5 \end{array}\right)\left(\begin{array}{cc} -5/2 & -3/2 \\ -2 & -1 \end{array}\right) = \left(\begin{array}{cc} 1 & 0 \\ 0 & 1 \end{array}\right)$$

and

$$A^{-1}A = \left(\begin{array}{cc} -5/2 & -3/2 \\ -2 & -1 \end{array}\right)\left(\begin{array}{cc} 2 & -3 \\ -4 & 5 \end{array}\right) = \left(\begin{array}{cc} 1 & 0 \\ 0 & 1 \end{array}\right).$$

Thus A is invertible and $A^{-1} = \begin{pmatrix} -5/2 & -3/2 \\ -2 & -1 \end{pmatrix}$.

EXAMPLE 3

Let $A = \begin{pmatrix} 1 & 2 \\ -2 & -4 \end{pmatrix}$. Calculate A^{-1} if it exists.

SOLUTION

If $A^{-1} = \begin{pmatrix} x & y \\ z & w \end{pmatrix}$ exists, then

$$AA^{-1} = \begin{pmatrix} 1 & 2 \\ -2 & -4 \end{pmatrix}\begin{pmatrix} x & y \\ z & w \end{pmatrix} = \begin{pmatrix} x+2z & y+2w \\ -2x-4z & -2y-4w \end{pmatrix} = \begin{pmatrix} 1 & 0 \\ 0 & 1 \end{pmatrix}.$$

This leads to the system

$$
\begin{array}{rcr}
x & + 2z & = 1 \\
y & + 2w & = 0 \\
-2x & - 4z & = 0 \\
-2y & - 4w & = 1
\end{array}
\tag{10}
$$

Using the same reasoning as in Example 2, we can write this system in the augmented matrix form $(A \mid I)$ and row-reduce:

$$\begin{pmatrix} 1 & 2 & \vline & 1 & 0 \\ -2 & -4 & \vline & 0 & 1 \end{pmatrix} \xrightarrow{A_{1,2}(2)} \begin{pmatrix} 1 & 2 & \vline & 1 & 0 \\ 0 & 0 & \vline & 2 & 1 \end{pmatrix}.$$

This is as far as we can go. The last line reads $0 = 2$ or $0 = 1$, depending on which of the two systems of equations (in x and z or in y and w) is being solved. Thus system (10) is inconsistent and A is not invertible.

The last two examples illustrate a procedure that always works when you are trying to find the inverse of a matrix.

Procedure for Computing the Inverse of a Square Matrix A

Step **1.** Write the augmented matrix $(A \mid I)$.

Step **2.** Use row reduction to reduce the matrix A to its reduced row echelon form.

Step **3.** Decide if A is invertible.

    (a) If A can be reduced to the identity matrix I, then A^{-1} will be the matrix to the right of the vertical bar.

    (b) If the row reduction of A leads to a row of zeros to the left of the vertical bar, then A is not invertible.

Remark. We can rephrase (a) and (b) as follows.

> *A square matrix A is invertible if and only if its reduced row echelon form is the identity matrix.*

Let $A = \begin{pmatrix} a_{11} & a_{12} \\ a_{21} & a_{22} \end{pmatrix}$. Then we define

$$\text{Determinant of } A = a_{11}a_{22} - a_{12}a_{21}. \qquad (11)$$

We abbreviate "determinant of A" by det A.

■ **THEOREM 4**

Let A be a 2×2 matrix. Then:

   i. A is invertible if and only if det $A \neq 0$.

   ii. If det $A \neq 0$, then

$$A^{-1} = \frac{1}{\det A}\begin{pmatrix} a_{22} & -a_{12} \\ -a_{21} & a_{11} \end{pmatrix}.\dagger \qquad (12)$$

PROOF

First suppose that det $A \neq 0$ and let $B = (1/\det A)\begin{pmatrix} a_{22} & -a_{12} \\ -a_{21} & a_{11} \end{pmatrix}$. Then

$$BA = \frac{1}{\det A}\begin{pmatrix} a_{22} & -a_{12} \\ -a_{21} & a_{11} \end{pmatrix}\begin{pmatrix} a_{11} & a_{12} \\ a_{21} & a_{22} \end{pmatrix}$$

$$= \frac{1}{a_{11}a_{22} - a_{12}a_{21}}\begin{pmatrix} a_{22}a_{11} - a_{12}a_{21} & 0 \\ 0 & -a_{21}a_{12} + a_{11}a_{22} \end{pmatrix} = \begin{pmatrix} 1 & 0 \\ 0 & 1 \end{pmatrix} = I.$$

Similarly, $AB = I$, which shows that A is invertible and that $B = A^{-1}$. We still must show that if A is invertible, then det $A \neq 0$. To do so, we consider the system

$$\begin{aligned} a_{11}x_1 + a_{12}x_2 &= b_1 \\ a_{21}x_1 + a_{22}x_2 &= b_2 \end{aligned} \qquad (13)$$

The system can be written in the form

$$A\mathbf{x} = \mathbf{b} \qquad (14)$$

---

† This formula can be obtained directly by applying our procedure for computing an inverse (see Problem 48).

with $\mathbf{x} = \begin{pmatrix} x_1 \\ x_2 \end{pmatrix}$ and $\mathbf{b} = \begin{pmatrix} b_1 \\ b_2 \end{pmatrix}$. Then, since A is invertible, we see from (2) that the system (14) has a unique solution given by

$$\mathbf{x} = A^{-1}\mathbf{b}.$$

But if the system has a unique solution, then the two lines whose equations are given in (13) are not parallel. That is, their slopes are unequal. In the $x_1 x_2$-plane (assuming that $a_{12} \neq 0$ and $a_{22} \neq 0$):

$$\text{slope of first line} = -\frac{a_{11}}{a_{12}},$$

and

$$\text{slope of second line} = -\frac{a_{21}}{a_{22}},$$

so

$$-\frac{a_{11}}{a_{12}} \neq -\frac{a_{21}}{a_{22}} \quad \text{or} \quad -a_{11}a_{22} \neq -a_{12}a_{21} \quad \text{or} \quad \det A = a_{11}a_{22} - a_{12}a_{21} \neq 0. \quad \blacksquare$$

Remark. In Problem 49 you are asked to complete the proof in the case $a_{12}a_{22} = 0$.

EXAMPLE 4

Let $A = \begin{pmatrix} 2 & -4 \\ 1 & 3 \end{pmatrix}$. Calculate A^{-1} if it exists.

SOLUTION
We find that $\det A = (2)(3) - (-4)(1) = 10$; hence A^{-1} exists. From Equation (12), we get

$$A^{-1} = \frac{1}{10}\begin{pmatrix} 3 & 4 \\ -1 & 2 \end{pmatrix} = \begin{pmatrix} 3/10 & 4/10 \\ -1/10 & 2/10 \end{pmatrix}.$$

Check.

$$A^{-1}A = \frac{1}{10}\begin{pmatrix} 3 & 4 \\ -1 & 2 \end{pmatrix}\begin{pmatrix} 2 & -4 \\ 1 & 3 \end{pmatrix} = \frac{1}{10}\begin{pmatrix} 10 & 0 \\ 0 & 10 \end{pmatrix} = \begin{pmatrix} 1 & 0 \\ 0 & 1 \end{pmatrix}$$

and

$$AA^{-1} = \begin{pmatrix} 2 & -4 \\ 1 & 3 \end{pmatrix}\begin{pmatrix} 3/10 & 4/10 \\ -1/10 & 2/10 \end{pmatrix} = \begin{pmatrix} 1 & 0 \\ 0 & 1 \end{pmatrix}.$$

EXAMPLE 5

Let $A = \begin{pmatrix} 1 & 2 \\ -2 & -4 \end{pmatrix}$. Calculate A^{-1} if it exists.

SOLUTION

We find that $\det A = (1)(-4) - (2)(-2) = -4 + 4 = 0$, so that A^{-1} does not exist, as we saw in Example 3.

The procedure described on p. 391 works for $n \times n$ matrices where $n > 2$. We illustrate this with a number of examples.

EXAMPLE 6

Let $A = \begin{pmatrix} 2 & 4 & 6 \\ 4 & 5 & 6 \\ 3 & 1 & -2 \end{pmatrix}$ (see Example 7.3.2 on p. 364). Calculate A^{-1} if it exists.

SOLUTION

We first put I next to A in an augmented matrix form

$$\begin{pmatrix} 2 & 4 & 6 & | & 1 & 0 & 0 \\ 4 & 5 & 6 & | & 0 & 1 & 0 \\ 3 & 1 & -2 & | & 0 & 0 & 1 \end{pmatrix}$$

and then carry out the row reduction.

$$\xrightarrow{M_1(1/2)} \begin{pmatrix} 1 & 2 & 3 & | & 1/2 & 0 & 0 \\ 4 & 5 & 6 & | & 0 & 1 & 0 \\ 3 & 1 & -2 & | & 0 & 0 & 1 \end{pmatrix} \xrightarrow[A_{1,3}(-3)]{A_{1,2}(-4)} \begin{pmatrix} 1 & 2 & 3 & | & 1/2 & 0 & 0 \\ 0 & -3 & -6 & | & -2 & 1 & 0 \\ 0 & -5 & -11 & | & -3/2 & 0 & 1 \end{pmatrix}$$

$$\xrightarrow{M_2(-1/3)} \begin{pmatrix} 1 & 2 & 3 & | & 1/2 & 0 & 0 \\ 0 & 1 & 2 & | & 2/3 & -1/3 & 0 \\ 0 & -5 & -11 & | & -3/2 & 0 & 1 \end{pmatrix} \xrightarrow[A_{2,3}(5)]{A_{2,1}(-2)} \begin{pmatrix} 1 & 0 & -1 & | & -5/6 & 2/3 & 0 \\ 0 & 1 & 2 & | & 2/3 & -1/3 & 0 \\ 0 & 0 & -1 & | & 11/6 & -5/3 & 1 \end{pmatrix}$$

$$\xrightarrow{M_3(-1)} \begin{pmatrix} 1 & 0 & -1 & | & -5/6 & 2/3 & 0 \\ 0 & 1 & 2 & | & 2/3 & -1/3 & 0 \\ 0 & 0 & 1 & | & -11/6 & 5/3 & -1 \end{pmatrix} \xrightarrow[A_{3,2}(-2)]{A_{3,1}(1)} \begin{pmatrix} 1 & 0 & 0 & | & -8/3 & 7/3 & -1 \\ 0 & 1 & 0 & | & 13/3 & -11/3 & 2 \\ 0 & 0 & 1 & | & -11/6 & 5/3 & -1 \end{pmatrix}.$$

Since A has now been reduced to I, we have

$$A^{-1} = \begin{pmatrix} -8/3 & 7/3 & -1 \\ 13/3 & -11/3 & 2 \\ -11/6 & 5/3 & -1 \end{pmatrix} = \frac{1}{6} \begin{pmatrix} -16 & 14 & -6 \\ 26 & -22 & 12 \\ -11 & 10 & -6 \end{pmatrix}.$$

We factor out 1/6 to make computations easier.

Check.

$$A^{-1}A = \frac{1}{6} \begin{pmatrix} -16 & 14 & -6 \\ 26 & -22 & 12 \\ -11 & 10 & -6 \end{pmatrix} \begin{pmatrix} 2 & 4 & 6 \\ 4 & 5 & 6 \\ 3 & 1 & -2 \end{pmatrix} = \frac{1}{6} \begin{pmatrix} 6 & 0 & 0 \\ 0 & 6 & 0 \\ 0 & 0 & 6 \end{pmatrix} = I.$$

We can also verify that $AA^{-1} = I$.

Warning. It is easy to make numerical errors in computing A^{-1}. Therefore it is essential to check the computations by verifying that $A^{-1}A = I$.

EXAMPLE 7

Let $A = \begin{pmatrix} 2 & 4 & 3 \\ 0 & 1 & -1 \\ 3 & 5 & 7 \end{pmatrix}$. Calculate A^{-1} if it exists.

SOLUTION

Proceeding as in Example 6 we obtain, successively, the following augmented matrices:

$$\left(\begin{array}{ccc|ccc} 2 & 4 & 3 & 1 & 0 & 0 \\ 0 & 1 & -1 & 0 & 1 & 0 \\ 3 & 5 & 7 & 0 & 0 & 1 \end{array}\right) \xrightarrow{M_1(1/2)} \left(\begin{array}{ccc|ccc} 1 & 2 & 3/2 & 1/2 & 0 & 0 \\ 0 & 1 & -1 & 0 & 1 & 0 \\ 3 & 5 & 7 & 0 & 0 & 1 \end{array}\right)$$

$$\xrightarrow{A_{1,3}(-3)} \left(\begin{array}{ccc|ccc} 1 & 2 & 3/2 & 1/2 & 0 & 0 \\ 0 & 1 & -1 & 0 & 1 & 0 \\ 0 & -1 & 5/2 & -3/2 & 0 & 1 \end{array}\right) \xrightarrow[A_{2,3}(1)]{A_{2,1}(-2)} \left(\begin{array}{ccc|ccc} 1 & 0 & 7/2 & 1/2 & -2 & 0 \\ 0 & 1 & -1 & 0 & 1 & 0 \\ 0 & 0 & 3/2 & -3/2 & 1 & 1 \end{array}\right)$$

$$\xrightarrow{M_3(2/3)} \left(\begin{array}{ccc|ccc} 1 & 0 & 7/2 & 1/2 & -2 & 0 \\ 0 & 1 & -1 & 0 & 1 & 0 \\ 0 & 0 & 1 & -1 & 2/3 & 2/3 \end{array}\right) \xrightarrow[A_{3,2}(1)]{A_{3,1}(-7/2)} \left(\begin{array}{ccc|ccc} 1 & 0 & 0 & 4 & -13/3 & -7/3 \\ 0 & 1 & 0 & -1 & 5/3 & 2/3 \\ 0 & 0 & 1 & -1 & 2/3 & 2/3 \end{array}\right).$$

Thus

$$A^{-1} = \begin{pmatrix} 4 & -13/3 & -7/3 \\ -1 & 5/3 & 2/3 \\ -1 & 2/3 & 2/3 \end{pmatrix}.$$

Check.

$$A^{-1}A = \begin{pmatrix} 4 & -13/3 & -7/3 \\ -1 & 5/3 & 2/3 \\ -1 & 2/3 & 2/3 \end{pmatrix} \begin{pmatrix} 2 & 4 & 3 \\ 0 & 1 & -1 \\ 3 & 5 & 7 \end{pmatrix} = \begin{pmatrix} 1 & 0 & 0 \\ 0 & 1 & 0 \\ 0 & 0 & 1 \end{pmatrix}.$$

EXAMPLE 8

Let $A = \begin{pmatrix} 1 & -3 & 4 \\ 2 & -5 & 7 \\ 0 & -1 & 1 \end{pmatrix}$. Calculate A^{-1} if it exists.

SOLUTION

Proceeding as before we obtain, successively,

$$\left(\begin{array}{ccc|ccc} 1 & -3 & 4 & 1 & 0 & 0 \\ 2 & -5 & 7 & 0 & 1 & 0 \\ 0 & -1 & 1 & 0 & 0 & 1 \end{array}\right) \xrightarrow{A_{1,2}(-2)} \left(\begin{array}{ccc|ccc} 1 & -3 & 4 & 1 & 0 & 0 \\ 0 & 1 & -1 & -2 & 1 & 0 \\ 0 & -1 & 1 & 0 & 0 & 1 \end{array}\right)$$

$$\xrightarrow[A_{2,3}(1)]{A_{2,1}(3)} \left(\begin{array}{ccc|ccc} 1 & 0 & 1 & -5 & 3 & 0 \\ 0 & 1 & -1 & -2 & 1 & 0 \\ 0 & 0 & 0 & -2 & 1 & 1 \end{array}\right).$$

This is as far as we can go. The matrix A *cannot* be reduced to the identity matrix and we can conclude that A is *not* invertible.

To understand why the last assertion in Example 8 is true, assume that we were trying to solve the system $A\mathbf{x} = \mathbf{b}$. Using Gauss-Jordan elimination, the steps above would yield an augmented matrix of the form

$$\left(\begin{array}{ccc|c} 1 & 0 & 1 & c_1 \\ 0 & 1 & -1 & c_2 \\ 0 & 0 & 0 & c_3 \end{array}\right).$$

If $c_3 = 0$, then by case (ii) of Section 7.3 (see p. 372) the system would have an infinite number of solutions. If $c_3 \neq 0$, by case (iii) there would be no solutions. In either case we rule out the possibility of having a unique solution. But if A is invertible, then $\mathbf{x} = A^{-1}\mathbf{b}$ is the *unique* solution to the system $A\mathbf{x} = \mathbf{b}$. Hence, we obtain the following important result:

> If in the row reduction of A to reduced row echelon form we end up with a row of zeros, then A is *not* invertible.

DEFINITION

Row Equivalent Matrices

Suppose that by elementary row operations we can transform the matrix A into the matrix B. Then A and B are said to be **row equivalent.**

The reasoning used above can be used to prove the following theorem (see Problem 40).

■ THEOREM 5

Let A be an $n \times n$ matrix.

    i. A is invertible if and only if A is row equivalent to the identity matrix I_n; that is, the reduced row echelon form of A is I_n.

    ii. A is invertible if and only if the system $A\mathbf{x} = \mathbf{b}$ has a unique solution for every n-vector $\mathbf{b}$.

    iii. If A is invertible, then this unique solution is given by $\mathbf{x} = A^{-1}\mathbf{b}$.

    iv. A is invertible if and only if the homogeneous system $A\mathbf{x} = \mathbf{0}$ has only the trivial solution $\mathbf{x} = \mathbf{0}$.   ■

EXAMPLE 9

Solve the system

$$2x_1 + 4x_2 + 3x_3 = 6,$$

$$x_2 - x_3 = -4,$$

$$3x_1 + 5x_2 + 7x_3 = 7.$$

SOLUTION

This system can be written as $A\mathbf{x} = \mathbf{b}$, where $A = \begin{pmatrix} 2 & 4 & 3 \\ 0 & 1 & -1 \\ 3 & 5 & 7 \end{pmatrix}$ and $\mathbf{b} = \begin{pmatrix} 6 \\ -4 \\ 7 \end{pmatrix}$. In Example 7 we saw that A^{-1} exists and

$$A^{-1} = \begin{pmatrix} 4 & -13/3 & -7/3 \\ -1 & 5/3 & 2/3 \\ -1 & 2/3 & 2/3 \end{pmatrix}.$$

Thus the unique solution is given by

$$\mathbf{x} = \begin{pmatrix} x_1 \\ x_2 \\ x_3 \end{pmatrix} = A^{-1}\mathbf{b} = \begin{pmatrix} 4 & -13/3 & -7/3 \\ -1 & 5/3 & 2/3 \\ -1 & 2/3 & 2/3 \end{pmatrix} \begin{pmatrix} 6 \\ -4 \\ 7 \end{pmatrix} = \begin{pmatrix} 25 \\ -8 \\ -4 \end{pmatrix}.$$

In this chapter we have seen that the concepts of solving a system of equations, finding the inverse of a matrix, and determining linear independence or dependence are somehow related. The theorem below describes this relationship. A partial proof follows.

■ **THEOREM 6: Summing Up Theorem—View 1**

Let A be an $n \times n$ matrix. Then each of the following six statements implies the other five (that is, if one is true, all are true):

   i. A is invertible.
  ii. The only solution to the homogeneous system $A\mathbf{x} = \mathbf{0}$ is the trivial solution ($\mathbf{x} = \mathbf{0}$).
 iii. The system $A\mathbf{x} = \mathbf{b}$ has a unique solution for every n-vector $\mathbf{b}$.
 iv. A is row equivalent to the $n \times n$ identity matrix I_n.
  v. The rows (and columns) of A are linearly independent.
 vi. $\det A \neq 0$. (So far, $\det A$ is only defined if A is a 2×2 matrix.)

PROOF

We have already seen that statements (i) and (iii) are equivalent [Theorem 5 (part ii)], that (i) and (iv) are equivalent [Theorem 5 (part i)], and that (ii) and (v) [for columns] are equivalent (Theorem 7.4.3). We shall see that (ii) and (iv) are equivalent. Suppose that (ii) holds. That is, suppose that $A\mathbf{x} = \mathbf{0}$ has only the trivial solution $\mathbf{x} = \mathbf{0}$. If we write out this system we obtain

$$\begin{matrix} a_{11}x_1 + a_{12}x_2 + \cdots + a_{1n}x_n = 0 \\ a_{21}x_1 + a_{22}x_2 + \cdots + a_{2n}x_n = 0 \\ \vdots \qquad \vdots \qquad\qquad \vdots \\ a_{n1}x_1 + a_{n2}x_2 + \cdots + a_{nn}x_n = 0 \end{matrix} \tag{15}$$

If A were not equivalent to I_n, then row reduction of the augmented matrix associated with (15) would leave us with a row of zeros. But if, say, the last row is zero, then the last equation reads $0 = 0$. Then, the homogeneous system reduces to one with $n - 1$ equations in n unknowns which, by Theorem 7.4.1 on p. 380 has an infinite number of solutions. But we assumed that $\mathbf{x} = \mathbf{0}$ was the only solution to system (15). This contradiction shows that A is row equivalent to I_n. Conversely, suppose that (*iv*) holds; that is, suppose that A is row equivalent to I_n. Then by Theorem 5 (part *i*), A is invertible and by Theorem 5 (part *iii*) the unique solution to $A\mathbf{x} = \mathbf{0}$ is $\mathbf{x} = A^{-1}\mathbf{0} = \mathbf{0}$. Thus (*ii*) and (*iv*) are equivalent. In Theorem 4 we showed that (*i*) and (*vi*) are equivalent in the 2×2 case. In Section 7.9 we shall show the equivalence of (*ii*) and (*vi*) and that if (*v*) is true for columns, it is also true for rows. ∎

Remark. We could add another statement to the theorem. Suppose the system $A\mathbf{x} = \mathbf{b}$ has a unique solution. Let R be a matrix in row echelon form that is row equivalent to A. Then R cannot have a row of zeros because if it did, it could not be reduced to the identity matrix.† Thus the row echelon form of A must look like this:

$$\begin{pmatrix} 1 & r_{12} & r_{13} & \cdots & r_{1n} \\ 0 & 1 & r_{23} & \cdots & r_{2n} \\ 0 & 0 & 1 & \cdots & r_{3n} \\ \vdots & \vdots & \vdots & & \vdots \\ 0 & 0 & 0 & \cdots & 1 \end{pmatrix}. \tag{16}$$

That is, R is a matrix with 1's down the diagonal and 0's below it. We thus have Theorem 7.

■ THEOREM 7

If any one of the statements in Theorem 6 holds, then the row echelon form of A has the form of matrix (16). ∎

We have seen that in order to verify that $B = A^{-1}$, we have to check that $AB = BA = I$. It turns out that only half this work has to be done.

■ THEOREM 8

Let A and B be $n \times n$ matrices. Then A is invertible and $B = A^{-1}$ if

(i) $BA = I$ **or** **(ii)** $AB = I$.

Remark. This theorem simplifies the work in checking that one matrix is the inverse of another.

PROOF

i. We assume that $BA = I$. Consider the homogeneous system $A\mathbf{x} = \mathbf{0}$. Multiplying both sides of this equation on the left by B, we obtain

$$BA\mathbf{x} = B\mathbf{0}. \tag{17}$$

---

† Note that if the ith row of R contains only zeros, then the homogeneous system $R\mathbf{x} = \mathbf{0}$ contains more unknowns than equations (since the ith equation is the zero equation) and the system has an infinite number of solutions. But then $A\mathbf{x} = \mathbf{0}$ has an infinite number of solutions, which is a contradiction of our assumption.

But $BA = I$ and $B0 = 0$, so (17) becomes $I\mathbf{x} = \mathbf{0}$ or $\mathbf{x} = \mathbf{0}$. This shows that $\mathbf{x} = \mathbf{0}$ is the only solution to $A\mathbf{x} = \mathbf{0}$ and, by Theorem 6 (parts i and ii), this means that A is invertible. We still have to show that $B = A^{-1}$. But

$$B = BI = B(AA^{-1}) = (BA)A^{-1} = IA^{-1} = A^{-1}.$$

ii. Let $AB = I$. Then, from part (i), $A = B^{-1}$. From the definition of the inverse of a matrix this means that $AB = BA = I$, which proves that A is invertible and that $B = A^{-1}$. This completes the proof. ■

PROBLEMS 7.5

In Problems 1–15 determine whether the given matrix is invertible. If it is, calculate the inverse.

1. $\begin{pmatrix} 2 & 1 \\ 3 & 2 \end{pmatrix}$

2. $\begin{pmatrix} -1 & 6 \\ 2 & -12 \end{pmatrix}$

3. $\begin{pmatrix} 0 & 1 \\ 1 & 0 \end{pmatrix}$

4. $\begin{pmatrix} 1 & 1 \\ 3 & 3 \end{pmatrix}$

5. $\begin{pmatrix} a & a \\ b & b \end{pmatrix}$

6. $\begin{pmatrix} 1 & 1 & 1 \\ 0 & 2 & 3 \\ 5 & 5 & 1 \end{pmatrix}$

7. $\begin{pmatrix} 3 & 2 & 1 \\ 0 & 2 & 2 \\ 0 & 0 & -1 \end{pmatrix}$

8. $\begin{pmatrix} 1 & 1 & 1 \\ 0 & 1 & 1 \\ 0 & 0 & 1 \end{pmatrix}$

9. $\begin{pmatrix} 1 & 6 & 2 \\ -2 & 3 & 5 \\ 7 & 12 & -4 \end{pmatrix}$

10. $\begin{pmatrix} 3 & 1 & 0 \\ 1 & -1 & 2 \\ 1 & 1 & 1 \end{pmatrix}$

11. $\begin{pmatrix} 2 & -1 & 4 \\ -1 & 0 & 5 \\ 19 & -7 & 3 \end{pmatrix}$

12. $\begin{pmatrix} 1 & 2 & 3 \\ 1 & 1 & 2 \\ 0 & 1 & 2 \end{pmatrix}$

13. $\begin{pmatrix} 1 & 1 & 1 & 1 \\ 1 & 2 & -1 & 2 \\ 1 & -1 & 2 & 1 \\ 1 & 3 & 3 & 2 \end{pmatrix}$

14. $\begin{pmatrix} 1 & 0 & 2 & 3 \\ -1 & 1 & 0 & 4 \\ 2 & 1 & -1 & 3 \\ -1 & 0 & 5 & 7 \end{pmatrix}$

15. $\begin{pmatrix} 1 & -3 & 0 & -2 \\ 3 & -12 & -2 & -6 \\ -2 & 10 & 2 & 5 \\ -1 & 6 & 1 & 3 \end{pmatrix}$

16. Show that if A, B, and C are invertible matrices, then ABC is invertible and $(ABC)^{-1} = C^{-1}B^{-1}A^{-1}$.

17. If $A_1, A_2, \ldots, A_m$ are invertible $n \times n$ matrices, show that $A_1 A_2 \cdots A_m$ is invertible and calculate its inverse.

18. Show that the matrix $\begin{pmatrix} 3 & 4 \\ -2 & -3 \end{pmatrix}$ is equal to its own inverse.

19. Show that the matrix $\begin{pmatrix} a_{11} & a_{12} \\ a_{21} & a_{22} \end{pmatrix}$ is equal to its own inverse if $A = \pm I$ or if $a_{11} = -a_{22}$ and $a_{21} a_{12} = 1 - a_{11}^2$.

20. Find the output vector $\mathbf{x}$ in the Leontief input-output model (see p. 373) if $n = 3$, $\mathbf{e} = \begin{pmatrix} 30 \\ 20 \\ 40 \end{pmatrix}$, and
$$A = \begin{pmatrix} 1/5 & 1/5 & 0 \\ 2/5 & 2/5 & 3/5 \\ 1/5 & 1/10 & 2/5 \end{pmatrix}.$$

* 21. Suppose that A is $n \times m$ and B is $m \times n$ so that AB is $n \times n$. Show that AB is not invertible if $n > m$. [*Hint:* Show that there is a nonzero vector $\mathbf{x}$ such that $AB\mathbf{x} = \mathbf{0}$ and then apply Theorem 6.]

* 22. Use the methods of this section to find the inverses of the following matrices with complex entries:

a. $\begin{pmatrix} i & 2 \\ 1 & -i \end{pmatrix}$

b. $\begin{pmatrix} 1-i & 0 \\ 0 & 1+i \end{pmatrix}$

c. $\begin{pmatrix} 1 & i & 0 \\ -i & 0 & 1 \\ 0 & 1+i & 1-i \end{pmatrix}$

23. Show that for every real number θ the matrix $\begin{pmatrix} \sin\theta & \cos\theta & 0 \\ \cos\theta & -\sin\theta & 0 \\ 0 & 0 & 1 \end{pmatrix}$ is invertible and find its inverse.

24. Calculate the inverse of $A = \begin{pmatrix} 2 & 0 & 0 \\ 0 & 3 & 0 \\ 0 & 0 & 4 \end{pmatrix}$.

25. A square matrix $A = (a_{ij})$ is called **diagonal** if all its elements off the main diagonal are zero. That is, $a_{ij} = 0$ if $i \neq j$. (The matrix of Problem 24 is diagonal.) Show that a diagonal matrix is invertible if and only if each of its diagonal components is nonzero.

26. Let

$$A = \begin{pmatrix} a_{11} & 0 & \cdots & 0 \\ 0 & a_{22} & \cdots & 0 \\ & & \ddots & \\ 0 & 0 & \cdots & a_{nn} \end{pmatrix}$$

be a diagonal matrix such that each of its diagonal components is nonzero. Calculate A^{-1}.

27. Calculate the inverse of $A = \begin{pmatrix} 2 & 1 & -1 \\ 0 & 3 & 4 \\ 0 & 0 & 5 \end{pmatrix}$.

28. Show that the matrix $A = \begin{pmatrix} 1 & 0 & 0 \\ -2 & 0 & 0 \\ 4 & 6 & 1 \end{pmatrix}$ is not invertible.

* 29. A square matrix is called **upper (lower) triangular** if all its elements below (above) the main diagonal are zero. (The matrix of Problem 27 is upper triangular and the matrix of Problem 28 is lower triangular.) Show that an upper or lower triangular matrix is invertible if and only if each of its diagonal elements is nonzero.

* 30. Show that the inverse of an invertible upper triangular matrix is upper triangular. [*Hint:* First prove the result for a 3×3 matrix.]

In Problems 31 and 32 a matrix is given. In each case show that the matrix is not invertible by finding a nonzero vector **x** such that $A\mathbf{x} = \mathbf{0}$.

31. $\begin{pmatrix} 2 & -1 \\ -4 & 2 \end{pmatrix}$

32. $\begin{pmatrix} 1 & -1 & 3 \\ 0 & 4 & -2 \\ 2 & -6 & 8 \end{pmatrix}$

33. A factory for the construction of quality furniture has two divisions: a machine shop, where the parts of the furniture are fabricated, and an assembly and finishing division, where the parts are put together into the finished product. Suppose there are 12 employees in the machine shop and 20 in the assembly and finishing division and that each employee works an 8-hour day. Suppose further that the factory produces only two products: chairs and tables. A chair requires $\frac{384}{17}$ hours of machine shop time and $\frac{480}{17}$ hours of assembly and finishing time. A table requires $\frac{240}{17}$ hours of machine shop time and $\frac{640}{17}$ hours of assembly and finishing time. Assuming that there is an unlimited demand for these products and that the manufacturer wishes to

keep all employees busy, how many chairs and how many tables can this factory produce each day?

34. A witch's magic cupboard contains 10 oz of ground four-leaf clovers and 14 oz of powdered mandrake root. The cupboard will replenish itself automatically provided she uses up exactly all her supplies. A batch of love potion requires $3\frac{1}{13}$ oz of ground four-leaf clovers and $2\frac{2}{13}$ oz of powdered mandrake root. One recipe of a well-known (to witches) cure for the common cold requires $5\frac{5}{13}$ oz of four-leaf clovers and $10\frac{10}{13}$ oz of mandrake root. How much of the love potion and the cold remedy should the witch make in order to use up the supply in the cupboard exactly?

35. A farmer feeds his cattle a mixture of two types of feed. One standard unit of type A feed supplies a steer with 10% of its minimum daily requirement of protein and 15% of its requirement of carbohydrates. Type B feed contains 12% of the requirement of protein and 8% of the requirement of carbohydrates in a standard unit. If the farmer wishes to feed his cattle exactly 100% of their minimum daily requirement of protein and carbohydrates, how many units of each type of feed should he give a steer each day?

Problems 36–39 deal with the Leontief input-output model.

36. In Example 7.3.8 (p. 373) we described the Leontief input-output model.

(a) Write system (8) on p. 373 in the form $A\mathbf{x} + \mathbf{e} = \mathbf{x}$.

(b) Write this system as $(I - A)\mathbf{x} = \mathbf{e}$.

The matrix A of internal demands is called the **technology matrix** and the matrix $I - A$ is called the **Leontief matrix**.

* (c) Find the technology and Leontief matrices for the model of the U.S. economy given in Problem 7.3.47 on p. 376.

* 37. In Problem 36, compute $(I - A)^{-1}$ and use it to obtain the solution to the problem posed in Problem 7.3.47.

38. A much simplified version of an input-output table for the 1958 Israeli economy divides that economy into three sectors—agriculture, manufacturing, and energy—with the following result.†

	Agriculture	Manufacturing	Energy
Agriculture	0.293	0	0
Manufacturing	0.014	0.207	0.017
Energy	0.044	0.010	0.216

† Wassily Leontief, *Input-Output Economics*, Oxford University Press, New York, 1966, pp. 54–57.

(a) How many units of agricultural production are required to produce one unit of agricultural output?

(b) How many units of agricultural production are required to produce 200,000 units of agricultural output?

(c) How many units of agricultural product go into the production of 50,000 units of energy?

(d) How many units of energy go into the production of 50,000 units of agricultural products?

39. Continuing Problem 38 exports (in thousands of Israeli pounds) in 1958 were

Agriculture	13,213
Manufacturing	17,597
Energy	1,786

(a) Compute the technology and Leontief matrices.

(b) Determine the number of Israeli pounds worth of agricultural products, manufactured goods, and energy required to run this model of the Israeli economy and export the stated value of products.

40. Prove parts (*i*) and (*ii*) of Theorem 5.

In Problems 41–47 compute the row echelon form of the given matrix and use it to determine directly whether the given matrix is invertible.

41. The matrix of Problem 1.

42. The matrix of Problem 4.

43. The matrix of Problem 7.

44. The matrix of Problem 9.

45. The matrix of Problem 11.

46. The matrix of Problem 13.

47. The matrix of Problem 14.

48. Let $A = \begin{pmatrix} a_{11} & a_{12} \\ a_{21} & a_{22} \end{pmatrix}$ and assume that $a_{11}a_{22} - a_{12}a_{21} \neq 0$.

Derive formula (12) by row reducing the augmented matrix

$$\begin{pmatrix} a_{11} & a_{12} & | & 1 & 0 \\ a_{21} & a_{22} & | & 0 & 1 \end{pmatrix}.$$

49. Assume that system (13) has a unique solution.

(a) Show that $|a_{12}| + |a_{22}| \neq 0$.

(b) If $a_{12} = 0$, show that $a_{11}a_{22} \neq 0$.

(c) If $a_{22} = 0$, show that $a_{12}a_{21} \neq 0$.

7.6 The Transpose of a Matrix; Symmetric Matrices

Corresponding to every matrix is another matrix which, as we shall see in Section 7.8, has properties very similar to those of the original matrix.

DEFINITION

Transpose

Let $A = (a_{ij})$ be an $m \times n$ matrix. Then the **transpose** *of A*, written A^t, is the $n \times m$ matrix obtained by interchanging the rows and columns of A. Succinctly, we may write $A^t = (a_{ji})$. In other words,

$$\text{if} \quad A = \begin{pmatrix} a_{11} & a_{12} & \cdots & a_{1n} \\ a_{21} & a_{22} & \cdots & a_{2n} \\ \vdots & \vdots & & \vdots \\ a_{m1} & a_{m2} & \cdots & a_{mn} \end{pmatrix}, \quad \text{then}$$

$$A^t = \begin{pmatrix} a_{11} & a_{21} & \cdots & a_{m1} \\ a_{12} & a_{22} & \cdots & a_{m2} \\ \vdots & \vdots & & \vdots \\ a_{1n} & a_{2n} & \cdots & a_{mn} \end{pmatrix}. \tag{1}$$

Simply put, the ith row of A is the ith column of A^t and the jth column of A is the jth row of A^t.

EXAMPLE 1

Find the transposes of the matrices

$$A = \begin{pmatrix} 2 & 3 \\ 1 & 4 \end{pmatrix}, \qquad B = \begin{pmatrix} 2 & 3 & 1 \\ -1 & 4 & 6 \end{pmatrix}, \qquad C = \begin{pmatrix} 1 & 2 & -6 \\ 2 & -3 & 4 \\ 0 & 1 & 2 \\ 2 & -1 & 5 \end{pmatrix}.$$

SOLUTION

Interchanging the rows and columns of each matrix, we obtain

$$A^t = \begin{pmatrix} 2 & 1 \\ 3 & 4 \end{pmatrix}, \qquad B^t = \begin{pmatrix} 2 & -1 \\ 3 & 4 \\ 1 & 6 \end{pmatrix}, \qquad C^t = \begin{pmatrix} 1 & 2 & 0 & 2 \\ 2 & -3 & 1 & -1 \\ -6 & 4 & 2 & 5 \end{pmatrix}.$$

Note, for example, that 4 is the component in row 2 and column 3 of C while 4 is the component in row 3 and column 2 of C^t. That is, the 23 element of C is the 32 element of C^t.

■ **THEOREM 1**

Suppose $A = (a_{ij})$ is an $n \times m$ matrix and $B = (b_{ij})$ is an $m \times p$ matrix. Then:

  i. $(A^t)^t = A$. (2)
  ii. $(AB)^t = B^t A^t$ (3)
  iii. If A and B are $n \times m$, then $(A + B)^t = A^t + B^t$. (4)

PROOF

  i. This follows directly from the definition of the transpose.
  ii. First we note that AB is an $n \times p$ matrix, so $(AB)^t$ is $p \times n$. Also, B^t is $p \times m$ and A^t is $m \times n$, so $B^t A^t$ is $p \times n$. Thus both matrices in equation (3) have the same size. Now the ijth element of AB is $\sum_{k=1}^m a_{ik} b_{kj}$ and this is the jith element of $(AB)^t$. Let $C = B^t$ and $D = A^t$. Then the ijth element c_{ij} of C is b_{ji} and the ijth element d_{ij} of D is a_{ji}. Thus the jith element of CD = the jith element of $B^t A^t$ = $\sum_{k=1}^m c_{jk} d_{ki}$ = $\sum_{k=1}^m b_{kj} a_{ik}$ = $\sum_{k=1}^m a_{ik} b_{kj}$ = the jith element of $(AB)^t$. This completes the proof of part (ii).
  iii. This part is left as an exercise (see Problem 11). ■

The transpose plays an important role in matrix theory. We shall see in succeeding chapters that A and A^t have many properties in common. Since columns of A^t are rows of A, we shall be able to use facts about the transpose to conclude that many things that are true about the rows of a matrix are true about its columns.

We conclude this section with an important definition.

Symmetric Matrix

DEFINITION

The $n \times n$ (square) matrix A is called **symmetric** if $A^t = A$. That is, if a_{ij} denotes the ijth component of A, then A is symmetric if $a_{ij} = a_{ji}$ for $i, j = 1, 2, \ldots, n$.

EXAMPLE 2

The following four matrices are symmetric:

$$I, \quad A = \begin{pmatrix} 1 & 2 \\ 2 & 3 \end{pmatrix}, \quad B = \begin{pmatrix} 1 & -4 & 2 \\ -4 & 7 & 5 \\ 2 & 5 & 0 \end{pmatrix}, \quad C = \begin{pmatrix} -1 & 2 & 4 & 6 \\ 2 & 7 & 3 & 5 \\ 4 & 3 & 8 & 0 \\ 6 & 5 & 0 & -4 \end{pmatrix}.$$

Problems 7.6

In Problems 1–10 find the transpose of the given matrix.

1. $\begin{pmatrix} -1 & 4 \\ 6 & 5 \end{pmatrix}$

2. $\begin{pmatrix} 3 & 0 \\ 1 & 2 \end{pmatrix}$

3. $\begin{pmatrix} 2 & 3 \\ -1 & 2 \\ 1 & 4 \end{pmatrix}$

4. $\begin{pmatrix} 2 & -1 & 0 \\ 1 & 5 & 6 \end{pmatrix}$

5. $\begin{pmatrix} 1 & 2 & 3 \\ -1 & 0 & 4 \\ 1 & 5 & 5 \end{pmatrix}$

6. $\begin{pmatrix} 1 & 2 & 3 \\ 2 & 4 & -5 \\ 3 & -5 & 7 \end{pmatrix}$

7. $\begin{pmatrix} 1 & 0 & 1 & 0 \\ 0 & 1 & 0 & 1 \end{pmatrix}$

8. $\begin{pmatrix} 2 & -1 \\ 2 & 4 \\ 1 & 6 \\ 1 & 5 \end{pmatrix}$

9. $\begin{pmatrix} a & b & c \\ d & e & f \\ g & h & j \end{pmatrix}$

10. $\begin{pmatrix} 0 & 0 & 0 \\ 0 & 0 & 0 \end{pmatrix}$

11. Let A and B be $n \times m$ matrices. Show, using the definition of the transpose, that $(A + B)^t = A^t + B^t$.

12. Find numbers α and β such that $\begin{pmatrix} 2 & \alpha & 3 \\ 5 & -6 & 2 \\ \beta & 2 & 4 \end{pmatrix}$ is symmetric.

13. If A and B are symmetric $n \times n$ matrices, prove that $A + B$ is symmetric.

14. If A and B are symmetric $n \times n$ matrices, show that $(AB)^t = BA$.

15. For any matrix A, show that the product matrix AA^t is defined and is a symmetric matrix.

16. Show that every diagonal matrix (see Problem 7.5.25, p. 399) is symmetric.

17. Show that the transpose of every upper triangular matrix (see Problem 7.5.29) is lower triangular.

18. A square matrix is called **skew-symmetric** if $A^t = -A$ (that is, $a_{ij} = -a_{ji}$). Which of the following matrices are skew-symmetric?

a. $\begin{pmatrix} 1 & -6 \\ 6 & 0 \end{pmatrix}$

b. $\begin{pmatrix} 0 & -6 \\ 6 & 0 \end{pmatrix}$

c. $\begin{pmatrix} 2 & -2 & -2 \\ 2 & 2 & -2 \\ 2 & 2 & 2 \end{pmatrix}$

d. $\begin{pmatrix} 0 & 1 & -1 \\ -1 & 0 & 2 \\ 1 & -2 & 0 \end{pmatrix}$

19. Let A and B be $n \times n$ skew-symmetric matrices. Show that $A + B$ is skew-symmetric.

20. If A is skew-symmetric, show that every component on the main diagonal of A is zero.

21. If A and B are skew-symmetric $n \times n$ matrices, show that $(AB)^t = BA$, so that AB is symmetric if and only if A and B commute.

* 22. Let $A = \begin{pmatrix} a_{11} & a_{12} \\ a_{21} & a_{22} \end{pmatrix}$ be a matrix with nonnegative entries having the properties that (i) $a_{11}^2 + a_{21}^2 = 1$ and $a_{12}^2 + a_{22}^2 = 1$ and (ii) $\begin{pmatrix} a_{11} \\ a_{21} \end{pmatrix} \cdot \begin{pmatrix} a_{12} \\ a_{22} \end{pmatrix} = 0$. Show that A is invertible and that $A^{-1} = A^t$.

7.7 The Determinant of a Square Matrix

Let $A = \begin{pmatrix} a_{11} & a_{12} \\ a_{21} & a_{22} \end{pmatrix}$ be a 2×2 matrix. In Section 7.5 we defined the determinant of A by

$$\det A = a_{11}a_{22} - a_{12}a_{21}. \tag{1}$$

We shall often denote $\det A$ by

$$|A| = \begin{vmatrix} a_{11} & a_{12} \\ a_{21} & a_{22} \end{vmatrix}.$$

We showed that A is invertible if and only if $\det A \neq 0$. As we shall see, this important theorem is valid for $n \times n$ matrices.

In this section and in Sections 7.8 and 7.9, we develop some of the basic properties of determinants and see how they can be used to calculate inverses and solve systems of n linear equations in n unknowns.

We shall define the determinant of an $n \times n$ matrix *inductively*. In other words, we use our knowledge of a 2×2 determinant to define a 3×3 determinant, use this to define a 4×4 determinant, and so on. We start by defining a 3×3 determinant.†

DEFINITION

3 × 3 Determinant

Let $A = \begin{pmatrix} a_{11} & a_{12} & a_{13} \\ a_{21} & a_{22} & a_{23} \\ a_{31} & a_{32} & a_{33} \end{pmatrix}$. Then the **3 × 3 determinant** of the matrix A is given by

$$\det A = |A| = a_{11}\begin{vmatrix} a_{22} & a_{23} \\ a_{32} & a_{33} \end{vmatrix} - a_{12}\begin{vmatrix} a_{21} & a_{23} \\ a_{31} & a_{33} \end{vmatrix} + a_{13}\begin{vmatrix} a_{21} & a_{22} \\ a_{31} & a_{32} \end{vmatrix}. \tag{2}$$

Note the minus sign before the second term on the right side of (2).

EXAMPLE 1

Let $A = \begin{pmatrix} 3 & 5 & 2 \\ 4 & 2 & 3 \\ -1 & 2 & 4 \end{pmatrix}$. Calculate $|A|$.

---

† There are several ways to define a determinant and this is one of them. It is important to realize that "det" is a function which assigns a *number* to a *square* matrix.

SOLUTION

$$|A| = \begin{vmatrix} 3 & 5 & 2 \\ 4 & 2 & 3 \\ -1 & 2 & 4 \end{vmatrix} = 3\begin{vmatrix} 2 & 3 \\ 2 & 4 \end{vmatrix} - 5\begin{vmatrix} 4 & 3 \\ -1 & 4 \end{vmatrix} + 2\begin{vmatrix} 4 & 2 \\ -1 & 2 \end{vmatrix}$$

$$= 3 \cdot 2 - 5 \cdot 19 + 2 \cdot 10 = -69.$$

EXAMPLE 2

Calculate $\begin{vmatrix} 2 & -3 & 5 \\ 1 & 0 & 4 \\ 3 & -3 & 9 \end{vmatrix}$.

SOLUTION

$$\begin{vmatrix} 2 & -3 & 5 \\ 1 & 0 & 4 \\ 3 & -3 & 9 \end{vmatrix} = 2\begin{vmatrix} 0 & 4 \\ -3 & 9 \end{vmatrix} - (-3)\begin{vmatrix} 1 & 4 \\ 3 & 9 \end{vmatrix} + 5\begin{vmatrix} 1 & 0 \\ 3 & -3 \end{vmatrix}$$

$$= 2 \cdot 12 + 3(-3) + 5(-3) = 0.$$

Before defining $n \times n$ determinants, we first note that in equation (2), $\begin{pmatrix} a_{22} & a_{23} \\ a_{32} & a_{33} \end{pmatrix}$ is the matrix obtained by deleting the first row and first column of A; $\begin{pmatrix} a_{21} & a_{23} \\ a_{31} & a_{33} \end{pmatrix}$ is the matrix obtained by deleting the first row and second column of A; and $\begin{pmatrix} a_{21} & a_{22} \\ a_{31} & a_{32} \end{pmatrix}$ is the matrix obtained by deleting the first row and third column of A. If we denote these three matrices by M_{11}, M_{12}, and M_{13}, respectively, and if $A_{11} = \det M_{11}$, $A_{12} = -\det M_{12}$, and $A_{13} = \det M_{13}$, then equation (2) can be written

$$\det A = |A| = a_{11}A_{11} + a_{12}A_{12} + a_{13}A_{13}. \qquad \textbf{(3)}$$

DEFINITION

Minor

Let A be an $n \times n$ matrix and let M_{ij} be the $(n-1) \times (n-1)$ matrix obtained from A by deleting the ith row and jth column of A. M_{ij} is called the **ijth minor** of A.

EXAMPLE 3

Let $A = \begin{pmatrix} 2 & -1 & 4 \\ 0 & 1 & 5 \\ 6 & 3 & -4 \end{pmatrix}$. Find M_{13} and M_{32}.

SOLUTION

Deleting the first row and third column of A, we obtain $M_{13} = \begin{pmatrix} 0 & 1 \\ 6 & 3 \end{pmatrix}$. Similarly, by eliminating the third row and second column we obtain $M_{32} = \begin{pmatrix} 2 & 4 \\ 0 & 5 \end{pmatrix}$.

Cofactor

DEFINITION

Let A be an $n \times n$ matrix. The **ijth cofactor** of A, denoted by A_{ij}, is given by

$$A_{ij} = (-1)^{i+j} \det M_{ij}. \tag{4}$$

That is, the ijth cofactor of A is obtained by taking the determinant of the ijth minor and multiplying it by $(-1)^{i+j}$. Note that

$$(-1)^{i+j} = \begin{cases} 1 & \text{if } i+j \text{ is even,} \\ -1 & \text{if } i+j \text{ is odd.} \end{cases}$$

Remark. The definition of cofactor makes sense because we are going to define an $n \times n$ determinant with the assumption that we already know what an $(n-1) \times (n-1)$ determinant is.

EXAMPLE 4

In Example 3 we have

$$A_{13} = (-1)^{1+3}|M_{13}| = \begin{vmatrix} 0 & 1 \\ 6 & 3 \end{vmatrix} = -6,$$

$$A_{32} = (-1)^{3+2}|M_{32}| = -\begin{vmatrix} 2 & 4 \\ 0 & 5 \end{vmatrix} = -10.$$

We now consider the general $n \times n$ matrix. Here

$$A = \begin{pmatrix} a_{11} & a_{12} & \cdots & a_{1n} \\ a_{21} & a_{22} & \cdots & a_{2n} \\ \vdots & \vdots & & \vdots \\ a_{n1} & a_{n2} & \cdots & a_{nn} \end{pmatrix}. \tag{5}$$

DEFINITION

$n \times n$ *Determinant*

Let A be an $n \times n$ matrix. Then the **determinant** of A, written $\det A$ or $|A|$, is given by

$$\det A = |A| = a_{11}A_{11} + a_{12}A_{12} + a_{13}A_{13} + \cdots + a_{1n}A_{1n}$$

$$= \sum_{k=1}^{n} a_{1k}A_{1k}. \tag{6}$$

Expansion by Cofactors

The expression on the right side of (6) is called an **expansion by cofactors.**

Remark. In equation (6) we defined the determinant by expanding by cofactors using components of A in the first row. We shall see in the next theorem that we get the same answer if we expand by cofactors in any row or column. First, we give an example of the use of equation (6).

EXAMPLE 5

Calculate $\det A$, where

$$A = \begin{pmatrix} 1 & 3 & 5 & 2 \\ 0 & -1 & 3 & 4 \\ 2 & 1 & 9 & 6 \\ 3 & 2 & 4 & 8 \end{pmatrix}.$$

SOLUTION

$$\begin{vmatrix} 1 & 3 & 5 & 2 \\ 0 & -1 & 3 & 4 \\ 2 & 1 & 9 & 6 \\ 3 & 2 & 4 & 8 \end{vmatrix} = a_{11}A_{11} + a_{12}A_{12} + a_{13}A_{13} + a_{14}A_{14} = 1\begin{vmatrix} -1 & 3 & 4 \\ 1 & 9 & 6 \\ 2 & 4 & 8 \end{vmatrix} - 3\begin{vmatrix} 0 & 3 & 4 \\ 2 & 9 & 6 \\ 3 & 4 & 8 \end{vmatrix}$$

$$+ 5\begin{vmatrix} 0 & -1 & 4 \\ 2 & 1 & 6 \\ 3 & 2 & 8 \end{vmatrix} - 2\begin{vmatrix} 0 & -1 & 3 \\ 2 & 1 & 9 \\ 3 & 2 & 4 \end{vmatrix} = 1(-92) - 3(-70) + 5(2) - 2(-16) = 160.$$

The proof of the following important theorem is given in Section 7.8.

■ **THEOREM 1: Basic Theorem**

Let the $n \times n$ matrix A be given by (5). Then

$$\det A = a_{i1}A_{i1} + a_{i2}A_{i2} + \cdots + a_{in}A_{in} = \sum_{k=1}^{n} a_{ik}A_{ik} \tag{7}$$

for $i = 1, 2, \ldots, n$. That is, we can calculate $\det A$ by expanding by cofactors in *any* row of A. Furthermore:

$$\det A = a_{1j}A_{1j} + a_{2j}A_{2j} + \cdots + a_{nj}A_{nj} = \sum_{k=1}^{n} a_{kj}A_{kj}. \tag{8}$$

Since the jth column of A is $\begin{pmatrix} a_{1j} \\ a_{2j} \\ \vdots \\ a_{nj} \end{pmatrix}$, equation (8) indicates that we can calculate det A by expanding by cofactors in any column of A.

EXAMPLE 6

Returning to Example 5, we note that there is a zero in the a_{21} entry of A. Thus the task of calculating det A is simplified by expanding along row 2 or column 1. To expand along column 1, we get

$$\begin{vmatrix} 1 & 3 & 5 & 2 \\ 0 & -1 & 3 & 4 \\ 2 & 1 & 9 & 6 \\ 3 & 2 & 4 & 8 \end{vmatrix} = 1 \cdot A_{11} + 0 \cdot A_{21} + 2 \cdot A_{31} + 3 \cdot A_{41}$$

$$= 1 \begin{vmatrix} -1 & 3 & 4 \\ 1 & 9 & 6 \\ 2 & 4 & 8 \end{vmatrix} + 2 \begin{vmatrix} 3 & 5 & 2 \\ -1 & 3 & 4 \\ 2 & 4 & 8 \end{vmatrix} - 3 \begin{vmatrix} 3 & 5 & 2 \\ -1 & 3 & 4 \\ 1 & 9 & 6 \end{vmatrix}$$

$$= 1(-92) + 2(84) - 3(-28) = 160.$$

It is clear that calculating the determinant of an $n \times n$ matrix can be tedious. To calculate a 4×4 determinant, we must usually calculate four 3×3 determinants. To calculate a 5×5 determinant, we must usually calculate five 4×4 determinants—which is the same as calculating twenty 3×3 determinants. Fortunately, techniques exist for greatly simplifying these computations. Some of these methods are discussed in the next section. There are, however, some matrices whose determinants can easily be calculated.

DEFINITION

Triangular Matrix

A square matrix is called **upper triangular** if all its components below the diagonal are zero. It is **lower triangular** if all its components above the diagonal are zero. A matrix is called **diagonal** if all its elements not on the diagonal are zero; that is, $A = (a_{ij})$ is upper triangular if $a_{ij} = 0$ for $i > j$, lower triangular if $a_{ij} = 0$ for $i < j$, and diagonal if $a_{ij} = 0$ for $i \neq j$.

EXAMPLE 7

The matrices $A = \begin{pmatrix} 2 & 1 & 7 \\ 0 & 2 & -5 \\ 0 & 0 & 1 \end{pmatrix}$ and $B = \begin{pmatrix} -2 & 3 & 0 & 1 \\ 0 & 0 & 2 & 4 \\ 0 & 0 & 1 & 3 \\ 0 & 0 & 0 & -2 \end{pmatrix}$ are upper triangular;

$C = \begin{pmatrix} 5 & 0 & 0 \\ 2 & 3 & 0 \\ -1 & 2 & 4 \end{pmatrix}$ and $D = \begin{pmatrix} 0 & 0 \\ 1 & 0 \end{pmatrix}$ are lower triangular; I and $E = \begin{pmatrix} 2 & 0 & 0 \\ 0 & -7 & 0 \\ 0 & 0 & -4 \end{pmatrix}$ are diagonal. Note that a diagonal matrix is both upper and lower triangular.

EXAMPLE 8

Let

$$A = \begin{pmatrix} a_{11} & 0 & 0 & 0 \\ a_{21} & a_{22} & 0 & 0 \\ a_{31} & a_{32} & a_{33} & 0 \\ a_{41} & a_{42} & a_{43} & a_{44} \end{pmatrix}$$

be lower triangular. Compute det A.

SOLUTION

$$\det A = a_{11}A_{11} + 0A_{12} + 0A_{13} + 0A_{14} = a_{11}A_{11}$$

$$= a_{11} \begin{vmatrix} a_{22} & 0 & 0 \\ a_{32} & a_{33} & 0 \\ a_{41} & a_{42} & a_{43} \end{vmatrix}$$

$$= a_{11} a_{22} \begin{vmatrix} a_{33} & 0 \\ a_{43} & a_{44} \end{vmatrix}$$

$$= a_{11} a_{22} a_{33} a_{44}.$$

Example 8 can easily be generalized to prove the following.

■ **THEOREM 2**

Let $A = (a_{ij})$ be an upper or lower triangular $n \times n$ matrix. Then

$$\det A = a_{11} a_{22} a_{33} \cdots a_{nn}.$$

That is: *The determinant of a triangular matrix equals the product of its diagonal components.* ■

EXAMPLE 9

The determinants of the six matrices in Example 7 are $|A| = 2 \cdot 2 \cdot 1 = 4$; $|B| = (-2)(0)(1)(-2) = 0$; $|C| = 5 \cdot 3 \cdot 4 = 60$; $|D| = 0$; $|I| = 1$; $|E| = (2)(-7)(-4) = 56$.

Problems 7.7

In Problems 1–20 calculate the determinant.

1. $\begin{vmatrix} 1 & 0 & 3 \\ 0 & 1 & 4 \\ 2 & 1 & 0 \end{vmatrix}$

2. $\begin{vmatrix} -1 & 1 & 0 \\ 2 & 1 & 4 \\ 1 & 5 & 6 \end{vmatrix}$

3. $\begin{vmatrix} 3 & -1 & 4 \\ 6 & 3 & 5 \\ 2 & -1 & 6 \end{vmatrix}$

4. $\begin{vmatrix} -1 & 0 & 6 \\ 0 & 2 & 4 \\ 1 & 2 & -3 \end{vmatrix}$

5. $\begin{vmatrix} -2 & 3 & 1 \\ 4 & 6 & 5 \\ 0 & 2 & 1 \end{vmatrix}$

6. $\begin{vmatrix} 5 & -2 & 1 \\ 6 & 0 & 3 \\ -2 & 1 & 4 \end{vmatrix}$

7. $\begin{vmatrix} -2 & 3 & 6 \\ 4 & 1 & 8 \\ -2 & 0 & 0 \end{vmatrix}$

8. $\begin{vmatrix} 2 & -1 & 3 \\ 4 & 0 & 6 \\ 5 & -2 & 3 \end{vmatrix}$

9. $\begin{vmatrix} 1 & -1 & 2 & 4 \\ 0 & -3 & 5 & 6 \\ 1 & 4 & 0 & 3 \\ 0 & 5 & -6 & 7 \end{vmatrix}$

10. $\begin{vmatrix} 2 & -3 & 1 & 4 \\ 0 & -2 & 0 & 0 \\ 3 & 7 & -1 & 2 \\ 4 & 1 & -3 & 8 \end{vmatrix}$

11. $\begin{vmatrix} 1 & 1 & -1 & 0 \\ -3 & 4 & 6 & 0 \\ 2 & 5 & -1 & 3 \\ 4 & 0 & 3 & 0 \end{vmatrix}$

12. $\begin{vmatrix} 3 & -1 & 2 & 1 \\ 4 & 3 & 1 & -2 \\ -1 & 0 & 2 & 3 \\ 6 & 2 & 5 & 2 \end{vmatrix}$

13. $\begin{vmatrix} 2 & 0 & 0 & 0 \\ 0 & 0 & 3 & 0 \\ 0 & -1 & 0 & 0 \\ 0 & 0 & 0 & 4 \end{vmatrix}$

14. $\begin{vmatrix} 0 & a & 0 & 0 \\ b & 0 & 0 & 0 \\ 0 & 0 & 0 & c \\ 0 & 0 & d & 0 \end{vmatrix}$

15. $\begin{vmatrix} 1 & 2 & 0 & 0 \\ 3 & -2 & 0 & 0 \\ 0 & 0 & 1 & -5 \\ 0 & 0 & 7 & 2 \end{vmatrix}$

16. $\begin{vmatrix} a & b & 0 & 0 \\ c & d & 0 & 0 \\ 0 & 0 & a & -b \\ 0 & 0 & c & d \end{vmatrix}$

17. $\begin{vmatrix} 2 & -1 & 0 & 4 & 1 \\ 3 & 1 & -1 & 2 & 0 \\ 3 & 2 & -2 & 5 & 1 \\ 0 & 0 & 4 & -1 & 6 \\ 3 & 2 & 1 & -1 & 1 \end{vmatrix}$

18. $\begin{vmatrix} 1 & -1 & 2 & 0 & 0 \\ 3 & 1 & 4 & 0 & 0 \\ 2 & -1 & 5 & 0 & 0 \\ 0 & 0 & 0 & 2 & 3 \\ 0 & 0 & 0 & -1 & 4 \end{vmatrix}$

19. $\begin{vmatrix} a & 0 & 0 & 0 & 0 \\ 0 & 0 & b & 0 & 0 \\ 0 & 0 & 0 & 0 & c \\ 0 & 0 & 0 & d & 0 \\ 0 & e & 0 & 0 & 0 \end{vmatrix}$

20. $\begin{vmatrix} 2 & 5 & -6 & 8 & 0 \\ 0 & 1 & -7 & 6 & 0 \\ 0 & 0 & 0 & 4 & 0 \\ 0 & 2 & 1 & 5 & 1 \\ 4 & -1 & 5 & 3 & 0 \end{vmatrix}$

21. Show that if A and B are diagonal $n \times n$ matrices, then det $AB = $ det A det B.

* 22. Show that if A and B are lower triangular matrices, then det $AB = $ det A det B.

23. Show that, in general, it is not true that $\det(A + B) = $ det $A + $ det B.

24. Show that if A is triangular, then det $A \neq 0$ if and only if all the diagonal components of A are nonzero.

25. Prove Theorem 2 for a lower triangular matrix.

* 26. We say that the vectors $\begin{pmatrix} 1 \\ 0 \end{pmatrix}$ and $\begin{pmatrix} 0 \\ 1 \end{pmatrix}$ *generate the area* 1 in the plane since if we construct a square with three of its vertices at $(0, 0)$, $(1, 0)$, and $(0, 1)$, we see that the area is 1. (See Fig. 1a.) More generally, if $\begin{pmatrix} x_1 \\ y_1 \end{pmatrix}$ and $\begin{pmatrix} x_2 \\ y_2 \end{pmatrix}$ are two linearly independent 2-vectors, then they generate an area defined to be the area of the parallelogram with three of its four vertices at $(0, 0)$, (x_1, y_1), and (x_2, y_2). (See Fig. 1b.) Let A be a 2×2 matrix. If k denotes the area generated by $\begin{pmatrix} x_1 \\ y_1 \end{pmatrix}$ and $\begin{pmatrix} x_2 \\ y_2 \end{pmatrix}$, where $\begin{pmatrix} x_1 \\ y_1 \end{pmatrix} = A\begin{pmatrix} 1 \\ 0 \end{pmatrix}$ and $\begin{pmatrix} x_2 \\ y_2 \end{pmatrix} = A\begin{pmatrix} 0 \\ 1 \end{pmatrix}$, show that $k = |\det A|$.

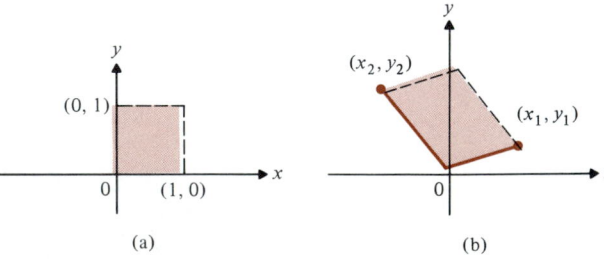

(a)   (b)

FIGURE 1

** 27. Let $\mathbf{u}_1$ and $\mathbf{u}_2$ be two 2-vectors and let $\mathbf{v}_1 = A\mathbf{u}_1$ and $\mathbf{v}_2 = A\mathbf{u}_2$. Show that

(area generated by $\mathbf{v}_1$ and $\mathbf{v}_2$)

$$= \text{(area generated by } \mathbf{u}_1 \text{ and } \mathbf{u}_2)|\det A|.$$

This provides a geometric interpretation of the determinant.

7.8 Properties of Determinants

In this section we shall list and prove some properties of the determinant. We shall also give a proof of the *basic theorem* (Theorem 7.7.1), that

$$\det A = a_{i1}A_{i1} + a_{i2}A_{i2} + \cdots + a_{in}A_{in} \tag{1}$$

or

$$\det A = a_{1j}A_{1j} + a_{2j}A_{2j} + \cdots + a_{nj}A_{nj} \tag{2}$$

for the $n \times n$ matrix $A = (a_{ij})$. We shall see that these properties can be used to reduce greatly the work involved in evaluating a determinant.

■ **PROPERTY 1**

If any row or column of A is the zero vector, then $\det A = 0$.

PROOF

Suppose the ith row of A contains all zeros. That is, $a_{ij} = 0$ for $j = 1, 2, \ldots, n$. Then $\det A = a_{i1}A_{i1} + a_{i2}A_{i2} + \cdots + a_{in}A_{in} = 0 + 0 + \cdots + 0 = 0$. The same proof works if the jth column is the zero vector. ■

EXAMPLE 1

It is easy to verify that

$$\begin{vmatrix} 2 & 3 & 5 \\ 0 & 0 & 0 \\ 1 & -2 & 4 \end{vmatrix} = 0 \quad \text{and} \quad \begin{vmatrix} -1 & 3 & 0 & 1 \\ 4 & 2 & 0 & 5 \\ -1 & 6 & 0 & 4 \\ 2 & 1 & 0 & 1 \end{vmatrix} = 0.$$

■ **PROPERTY 2**

If the ith row or the jth column of A is multiplied by the constant c, then $\det A$ is multiplied by c. That is, if we call this new matrix B, then

$$|B| = \begin{vmatrix} a_{11} & a_{12} & \cdots & a_{1n} \\ a_{21} & a_{22} & \cdots & a_{2n} \\ \vdots & \vdots & & \vdots \\ ca_{i1} & ca_{i2} & \cdots & ca_{in} \\ \vdots & \vdots & & \vdots \\ a_{n1} & a_{n2} & \cdots & a_{nn} \end{vmatrix} = c \begin{vmatrix} a_{11} & a_{12} & \cdots & a_{1n} \\ a_{21} & a_{22} & \cdots & a_{2n} \\ \vdots & \vdots & & \vdots \\ a_{i1} & a_{i2} & \cdots & a_{in} \\ \vdots & \vdots & & \vdots \\ a_{n1} & a_{n2} & \cdots & a_{nn} \end{vmatrix} = c|A|. \tag{3}$$

PROOF

To prove (3) we expand in the ith row of A to obtain

$$\det B = ca_{i1}A_{i1} + ca_{i2}A_{i2} + \cdots ca_{in}A_{in}$$

$$= c(a_{i1}A_{i1} + a_{i2}A_{i2} + \cdots + a_{in}A_{in}) = c \det A.$$

A similar proof works for columns. ■

EXAMPLE 2

Let $A = \begin{pmatrix} 1 & -1 & 2 \\ 3 & 1 & 4 \\ 0 & -2 & 5 \end{pmatrix}$. Then $\det A = 16$. If we multiply the second row by 4, we have

$B = \begin{pmatrix} 1 & -1 & 2 \\ 12 & 4 & 16 \\ 0 & -2 & 5 \end{pmatrix}$ and $\det B = 64 = 4 \det A$. If the third column is multiplied by

-3 we obtain $C = \begin{pmatrix} 1 & -1 & -6 \\ 3 & 1 & -12 \\ 0 & -2 & -15 \end{pmatrix}$ and $\det C = -48 = -3 \det A$.

Remark. Using Property 2 we can prove (see Problem 28) the following interesting fact: For any scalar α and $n \times n$ matrix A, $\det \alpha A = \alpha^n \det A$.

■ **PROPERTY 3**
Let

$$A = \begin{pmatrix} a_{11} & a_{12} & \cdots & a_{1j} & \cdots & a_{1n} \\ a_{21} & a_{22} & \cdots & a_{2j} & \cdots & a_{2n} \\ \vdots & \vdots & & \vdots & & \vdots \\ a_{n1} & a_{n2} & \cdots & a_{nj} & \cdots & a_{nn} \end{pmatrix},$$

$$B = \begin{pmatrix} a_{11} & a_{12} & \cdots & \alpha_{1j} & \cdots & a_{1n} \\ a_{21} & a_{22} & \cdots & \alpha_{2j} & \cdots & a_{2n} \\ \vdots & \vdots & & \vdots & & \vdots \\ a_{n1} & a_{n2} & \cdots & \alpha_{nj} & \cdots & a_{nn} \end{pmatrix},$$

and $\quad C = \begin{pmatrix} a_{11} & a_{12} & \cdots & a_{1j}+\alpha_{1j} & \cdots & a_{1n} \\ a_{21} & a_{22} & \cdots & a_{2j}+\alpha_{2j} & \cdots & a_{2n} \\ \vdots & \vdots & & \vdots & & \vdots \\ a_{n1} & a_{n2} & \cdots & a_{nj}+\alpha_{nj} & \cdots & a_{nn} \end{pmatrix}.$

Then

$$\det C = \det A + \det B. \tag{4}$$

In other words, suppose that A, B, and C are identical except for the jth column and that the jth column of C is the sum of the jth columns of A and B. Then $\det C = \det A + \det B$. The same statement is true for rows.

PROOF

We expand det C in the jth column to obtain

$$\det C = (a_{1j} + \alpha_{1j})A_{1j} + (a_{2j} + \alpha_{2j})A_{2j} + \cdots + (a_{nj} + \alpha_{nj})A_{nj}$$

$$= (a_{1j}A_{1j} + a_{2j}A_{2j} + \cdots + a_{nj}A_{nj})$$

$$+ (\alpha_{1j}A_{1j} + \alpha_{2j}A_{2j} + \cdots + \alpha_{nj}A_{nj}) = \det A + \det B. \quad \blacksquare$$

EXAMPLE 3

Let $A = \begin{pmatrix} 1 & -1 & 2 \\ 3 & 1 & 4 \\ 0 & -2 & 5 \end{pmatrix}$, $B = \begin{pmatrix} 1 & -6 & 2 \\ 3 & 2 & 4 \\ 0 & 4 & 5 \end{pmatrix}$, and $C = \begin{pmatrix} 1 & -1-6 & 2 \\ 3 & 1+2 & 4 \\ 0 & -2+4 & 5 \end{pmatrix} =$

$\begin{pmatrix} 1 & -7 & 2 \\ 3 & 3 & 4 \\ 0 & 2 & 5 \end{pmatrix}$. Then det $A = 16$, det $B = 108$, and det $C = 124 = $ det $A + $ det B.

■ PROPERTY 4

Interchanging any two rows (or columns) of A has the effect of multiplying det A by -1.

PROOF

We prove the statement for rows and assume first that two adjacent rows are interchanged. That is, we assume that the ith and $(i + 1)$st rows are interchanged. Let

$$A = \begin{pmatrix} a_{11} & a_{12} & \cdots & a_{1n} \\ a_{21} & a_{22} & \cdots & a_{2n} \\ \vdots & \vdots & & \vdots \\ a_{i1} & a_{i2} & \cdots & a_{in} \\ a_{i+1,1} & a_{i+1,2} & \cdots & a_{i+1,n} \\ \vdots & \vdots & & \vdots \\ a_{n1} & a_{n2} & & a_{nn} \end{pmatrix} \quad \text{and}$$

$$B = \begin{pmatrix} a_{11} & a_{12} & \cdots & a_{1n} \\ a_{21} & a_{22} & \cdots & a_{2n} \\ \vdots & \vdots & & \vdots \\ a_{i+1,1} & a_{i+1,2} & \cdots & a_{i+1,n} \\ a_{i1} & a_{i2} & \cdots & a_{in} \\ \vdots & \vdots & & \vdots \\ a_{n1} & a_{n2} & \cdots & a_{nn} \end{pmatrix}.$$

Then, expanding det A in its ith row and det B in its $(i + 1)$st row, we obtain

$$\det A = a_{i1}A_{i1} + a_{i2}A_{i2} + \cdots + a_{in}A_{in},$$

$$\det B = a_{i1}B_{i+1,1} + a_{i2}B_{i+1,2} + \cdots + a_{in}B_{i+1,n}.$$

(5)

Here $A_{ij} = (-1)^{i+j}|M_{ij}|$, where M_{ij} is obtained by crossing off the ith row and jth column of A. Notice now that if we cross off the $(i + 1)$st row and jth column of B, we obtain the same M_{ij}. Thus

$$B_{i+1,j} = (-1)^{i+1+j}|M_{ij}| = -(-1)^{i+j}|M_{ij}| = -A_{ij}$$

so that, from equations (5), $\det B = -\det A$.

Now suppose that $i < j$ and that the ith and jth rows are to be interchanged. We can do this by interchanging adjacent rows several times. It will take $j - i$ interchanges to move row j into the ith row. Then row i will be in the $(i + 1)$st row and it will take an additional $j - i - 1$ interchanges to move row i into the jth row. Finally, the total number of interchanges of adjacent rows is $(j - i) + (j - i - 1) = 2j - 2i - 1$, which is odd. Thus $\det A$ is multiplied by -1 an odd number of times, which is what we needed to show. ■

EXAMPLE 4

Let $A = \begin{pmatrix} 1 & -1 & 2 \\ 3 & 1 & 4 \\ 0 & -2 & 5 \end{pmatrix}$. By interchanging the first and third rows we obtain $B = \begin{pmatrix} 0 & -2 & 5 \\ 3 & 1 & 4 \\ 1 & -1 & 2 \end{pmatrix}$. By interchanging the first and second columns of A we obtain $C = \begin{pmatrix} -1 & 1 & 2 \\ 1 & 3 & 4 \\ -2 & 0 & 5 \end{pmatrix}$. Then, by direct calculation, we find that $\det A = 16$ and $\det B = \det C = -16$.

■ PROPERTY 5

If A has two equal rows or columns, then $\det A = 0$.

PROOF

Suppose the ith and jth rows of A are equal. By interchanging these rows we get a matrix B having the property that $\det B = -\det A$ (from Property 4). But since row i = row j, interchanging them gives us the same matrix. Thus $A = B$ and $\det A = \det B = -\det A$. Thus $2 \det A = 0$, which can happen only if $\det A = 0$. ■

EXAMPLE 5

$$\begin{vmatrix} 2 & 3 \\ 2 & 3 \end{vmatrix} = 2(3) - 2(3) = 0.$$

■ **PROPERTY 6**

If one row (column) of A is a constant multiple of another row (column), then $\det A = 0$.

PROOF

Assume $(a_{j1}, a_{j2}, \ldots, a_{jn}) = c(a_{i1}, a_{i2}, \ldots, a_{in})$; that is, the jth row of A is c times the ith row. By Property 2, $\det A$ is c times the determinant of a matrix whose ith row and jth row are equal. But by Property 5, the latter matrix has determinant equal to zero. Hence $\det A = 0$. The proof is similar for columns. ■

EXAMPLE 6

$$\begin{vmatrix} 2 & 4 & 1 & 12 \\ -1 & 1 & 0 & 3 \\ 0 & -1 & 9 & -3 \\ 7 & 3 & 6 & 9 \end{vmatrix} = 0 \text{ since the fourth column is three times the second column.}$$

■ **PROPERTY 7**

If a multiple of one row (column) of A is added to another row (column) of A, then the determinant is unchanged.

PROOF

Let B be the matrix obtained by adding k times the ith row of A to the jth row of A; that is,

$$(b_{j1}, b_{j2}, \ldots, b_{jn}) = (a_{j1} + ka_{i1}, a_{j2} + ka_{i2}, \ldots, a_{jn} + ka_{in}).$$

Then, by Property 3,

$$\det B = \det A + \det C,$$

where the jth row of C is equal to k times its ith row. By Property 6, $\det C = 0$ and the proof is complete. ■

EXAMPLE 7

Let $A = \begin{pmatrix} 1 & -1 & 2 \\ 3 & 1 & 4 \\ 0 & -2 & 5 \end{pmatrix}$. Then $\det A = 16$. If we multiply the third row by 4 and add it to the second row, we obtain a new matrix B given by

$$B = \begin{pmatrix} 1 & -1 & 2 \\ 3 + 4(0) & 1 + 4(-2) & 4 + 5(4) \\ 0 & -2 & 5 \end{pmatrix} = \begin{pmatrix} 1 & -1 & 2 \\ 3 & -7 & 24 \\ 0 & -2 & 5 \end{pmatrix}$$

and $\det B = 16 = \det A$.

The properties we have discussed make it much easier to evaluate high-order determinants. We simply row-reduce the matrix whose determinant we wish to calculate until the determinant is in an easily evaluated form. The most common goal will be to use Property 7 repeatedly until either (*i*) the new determinant has a row (column) of zeros or one row (column) is a multiple of another row (column)—in which case the determinant is zero—or (*ii*) the new matrix is triangular so that its determinant is the product of its diagonal elements. Alternatively, make all the components except one in one row or column equal to zero and then expand in that row or column.

EXAMPLE 8

Calculate

$$|A| = \begin{vmatrix} 1 & 3 & 5 & 2 \\ 0 & -1 & 3 & 4 \\ 2 & 1 & 9 & 6 \\ 3 & 2 & 4 & 8 \end{vmatrix}.$$

SOLUTION (See Example 7.7.5.)
There is already a zero in the first column, so it is simplest to reduce other elements in the first column to zero. We then continue to reduce, aiming for a triangular matrix:

Multiply the first row by -2 and add it to the third row and multiply the first row by -3 and add it to the fourth row.

$$|A| = \begin{vmatrix} 1 & 3 & 5 & 2 \\ 0 & -1 & 3 & 4 \\ 0 & -5 & -1 & 2 \\ 0 & -7 & -11 & 2 \end{vmatrix}$$

Multiply the second row by -5 and -7 and add it to the third and fourth rows, respectively.

$$= \begin{vmatrix} 1 & 3 & 5 & 2 \\ 0 & -1 & 3 & 4 \\ 0 & 0 & -16 & -18 \\ 0 & 0 & -32 & -26 \end{vmatrix}$$

Factor out -16 from the third row (using Property 2).

$$= -16 \begin{vmatrix} 1 & 3 & 5 & 2 \\ 0 & -1 & 3 & 4 \\ 0 & 0 & 1 & \frac{9}{8} \\ 0 & 0 & -32 & -26 \end{vmatrix}$$

Multiply the third row by 32 and add it to the fourth row.

$$= -16 \begin{vmatrix} 1 & 3 & 5 & 2 \\ 0 & -1 & 3 & 4 \\ 0 & 0 & 1 & \frac{9}{8} \\ 0 & 0 & 0 & 10 \end{vmatrix}$$

Now we have an upper triangular matrix and $|A| = -16(1)(-1)(1)(10) = (-16)(-10) = 160$.

EXAMPLE 9

Calculate

$$|A| = \begin{vmatrix} -2 & 1 & 0 & 4 \\ 3 & -1 & 5 & 2 \\ -2 & 7 & 3 & 1 \\ 3 & -7 & 2 & 5 \end{vmatrix}.$$

SOLUTION

There are a number of ways to proceed here and it is not apparent which way will get us the answer most quickly. However, since there is already one zero in the first row, we begin our reduction in that row.

Multiply the second column by 2 and −4 and add it to the first and fourth columns, respectively.

$$|A| = \begin{vmatrix} 0 & 1 & 0 & 0 \\ 1 & -1 & 5 & 6 \\ 12 & 7 & 3 & -27 \\ -11 & -7 & 2 & 33 \end{vmatrix}$$

Interchange the first two columns.

$$= - \begin{vmatrix} 1 & 0 & 0 & 0 \\ -1 & 1 & 5 & 6 \\ 7 & 12 & 3 & -27 \\ -7 & -11 & 2 & 33 \end{vmatrix}$$

Multiply the second column by −5 and −6 and add it to the third and fourth columns, respectively.

$$= - \begin{vmatrix} 1 & 0 & 0 & 0 \\ -1 & 1 & 0 & 0 \\ 7 & 12 & -57 & -99 \\ -7 & -11 & 57 & 99 \end{vmatrix}$$

Since the fourth column is now a multiple of the third column (column 4 = $\frac{99}{57}$ × column 3), we see that $|A| = 0$.

There are three additional facts about determinants that will be very useful to us.

■ THEOREM 1

Let A be an $n \times n$ matrix. Then

$$a_{i1}A_{j1} + a_{i2}A_{j2} + \cdots + a_{in}A_{jn} = 0 \qquad \text{if } i \neq j. \tag{6}$$

Note. From Theorem 7.7.1, the sum in equation (6) equals det A if $i = j$.

PROOF

Let B be the matrix obtained by replacing the jth row of A by the ith row of A. Then, since two rows of B are equal, det $B = 0$. But $B = A$ except in the jth row. Thus if we calculate det B by expanding in the jth row of B, we obtain the sum in (6) and the theorem is proved. Note that when we expand in the jth row, the jth row is deleted in computing the cofactors of B. Thus $B_{jk} = A_{jk}$ for $k = 1, 2, \ldots, n$. ■

■ THEOREM 2

Let A be an $n \times n$ matrix. Then

$$\det A = \det A^t. \tag{7}$$

PROOF

This proof uses mathematical induction. If you are unfamiliar with this important method of proof, refer to Appendix 3. We first prove the theorem in the case $n = 2$. If

$$|A| = \begin{vmatrix} a_{11} & a_{12} \\ a_{21} & a_{22} \end{vmatrix} = a_{11}a_{22} - a_{12}a_{21}$$

then

$$|A^t| = \begin{vmatrix} a_{11} & a_{21} \\ a_{12} & a_{22} \end{vmatrix} = a_{11}a_{22} - a_{21}a_{12} = |A|$$

so the theorem is true for $n = 2$. Next we assume the theorem to be true for $(n - 1) \times (n - 1)$ matrices and prove it for $n \times n$ matrices. This will prove the theorem. Let $B = A^t$. Then

$$|A| = \begin{vmatrix} a_{11} & a_{12} & \cdots & a_{1n} \\ a_{21} & a_{22} & \cdots & a_{2n} \\ \vdots & \vdots & & \vdots \\ a_{n1} & a_{n2} & \cdots & a_{nn} \end{vmatrix} \quad \text{and} \quad |A^t| = |B| = \begin{vmatrix} a_{11} & a_{21} & \cdots & a_{n1} \\ a_{12} & a_{22} & \cdots & a_{n2} \\ \vdots & \vdots & & \vdots \\ a_{1n} & a_{2n} & \cdots & a_{nn} \end{vmatrix}.$$

We expand $|A|$ in the first row and expand $|B|$ in the first column. This gives us

$$|A| = a_{11}A_{11} + a_{12}A_{12} + \cdots + a_{1n}A_{1n},$$

$$|B| = a_{11}B_{11} + a_{12}B_{21} + \cdots + a_{1n}B_{n1}.$$

We need to show that $A_{1k} = B_{k1}$ for $k = 1, 2, \ldots, n$. But $A_{1k} = (-1)^{1+k}|M_{1k}|$ and $B_{k1} = (-1)^{k+1}|N_{k1}|$, where M_{1k} is the $1k$th minor of A and N_{k1} is the $k1$st minor of B. Then

$$|M_{1k}| = \begin{vmatrix} a_{21} & a_{22} & \cdots & a_{2,k-1} & a_{2,k+1} & \cdots & a_{2n} \\ a_{31} & a_{32} & \cdots & a_{3,k-1} & a_{3,k+1} & \cdots & a_{3n} \\ \vdots & \vdots & & \vdots & \vdots & & \vdots \\ a_{n1} & a_{n2} & \cdots & a_{n,k-1} & a_{n,k+1} & \cdots & a_{nn} \end{vmatrix}$$

and

$$|N_{k1}| = \begin{vmatrix} a_{21} & a_{31} & \cdots & a_{n1} \\ a_{22} & a_{32} & \cdots & a_{n2} \\ \vdots & \vdots & & \vdots \\ a_{2,k-1} & a_{3,k-1} & \cdots & a_{n,k-1} \\ a_{2,k+1} & a_{3,k+1} & \cdots & a_{n,k+1} \\ \vdots & \vdots & & \vdots \\ a_{2n} & a_{3n} & \cdots & a_{nn} \end{vmatrix}.$$

Clearly $M_{1k} = N_{k1}^t$, and since both are $(n-1) \times (n-1)$ matrices, the induction hypothesis tells us that $|M_{1k}| = |N_{k1}|$. Thus $A_{1k} = B_{k1}$ and the proof is complete. ■

EXAMPLE 10

$$|A| = \begin{vmatrix} 1 & 2 \\ 3 & 4 \end{vmatrix} = 1(4) - 2(3) = -2$$

and

$$|A^t| = \begin{vmatrix} 1 & 3 \\ 2 & 4 \end{vmatrix} = 1(4) - 2(3) = -2.$$

■ **THEOREM 3**

Let A and B be $n \times n$ matrices. Then

$$\det AB = \det A \det B. \qquad (8)$$

That is, *the determinant of the product is the product of the determinants.*

PROOF

First assume that A is the diagonal matrix

$$A = D = \begin{pmatrix} d_1 & 0 & \cdots & 0 \\ 0 & d_2 & \cdots & 0 \\ \vdots & \vdots & & \vdots \\ 0 & 0 & \cdots & d_n \end{pmatrix}.$$

Then using Property 2 repeatedly, we have

$$\det DB = \begin{vmatrix} d_1 b_{11} & d_1 b_{12} & \cdots & d_1 b_{1n} \\ d_2 b_{21} & d_2 b_{22} & \cdots & d_2 b_{2n} \\ \vdots & \vdots & & \vdots \\ d_n b_{n1} & d_n b_{n2} & \cdots & d_n b_{nn} \end{vmatrix}$$

$$= d_1 \begin{vmatrix} b_{11} & b_{12} & \cdots & b_{1n} \\ d_2 b_{21} & d_2 b_{22} & \cdots & d_2 b_{2n} \\ \vdots & & & \\ d_n b_{n1} & d_n b_{n2} & \cdots & d_n b_{nn} \end{vmatrix}$$

$$= d_1 d_2 \cdot \ldots \cdot d_n \begin{vmatrix} b_{11} & b_{12} & \cdots & b_{1n} \\ b_{21} & b_{22} & \cdots & b_{2n} \\ \vdots & \vdots & & \vdots \\ b_{n1} & b_{n2} & \cdots & b_{nn} \end{vmatrix}$$

$$= d_1 \cdot d_2 \cdot \ldots \cdot d_n \det B$$

$$= \det D \det B.$$

Now observe that we can use the elementary row operations (ii) and (iii) (see p. 368) to reduce a general matrix A to a diagonal matrix D. [Elementary row operation (i) is only used in the Gauss-Jordan elimination to obtain the reduced row echelon form.] By Properties 4 and 7 we see that if we only use elementary row operations (ii) and (iii) the determinant does not change, except for a sign reversal when two rows are interchanged. Note that the same elimination steps reduce AB to DB, with *exactly* the same effect on its determinant. Thus,

$$\det AB = \pm \det DB = \pm \det D \det B = \det A \det B,$$

since the result has been shown to hold when D is diagonal. ■

EXAMPLE 11

Verify equation (8) for $A = \begin{pmatrix} 1 & -1 & 2 \\ 3 & 1 & 4 \\ 0 & -2 & 5 \end{pmatrix}$ and $B = \begin{pmatrix} 1 & -2 & 3 \\ 0 & -1 & 4 \\ 2 & 0 & -2 \end{pmatrix}$.

SOLUTION

$\det A = 16$ and $\det B = -8$. We calculate

$$AB = \begin{pmatrix} 1 & -1 & 2 \\ 3 & 1 & 4 \\ 0 & -2 & 5 \end{pmatrix}\begin{pmatrix} 1 & -2 & 3 \\ 0 & -1 & 4 \\ 2 & 0 & -2 \end{pmatrix} = \begin{pmatrix} 5 & -1 & -5 \\ 11 & -7 & 5 \\ 10 & 2 & -18 \end{pmatrix}$$

and $\det AB = -128 = (16)(-8) = \det A \det B$.

Proof of the Basic Theorem (Optional)

PROOF

We show that

$$\det A = \sum_{k=1}^{n} a_{ik} A_{ik} \tag{9}$$

by mathematical induction. For the 2×2 matrix $A = \begin{pmatrix} a_{11} & a_{12} \\ a_{21} & a_{22} \end{pmatrix}$, we first expand the first row by cofactors: $\det A = a_{11}A_{11} + a_{12}A_{12} = a_{11}(a_{22}) + a_{12}(-a_{21}) = a_{11}a_{22} - a_{12}a_{21}$. Similarly, expanding in the second row, we obtain $a_{21}A_{21} + a_{22}A_{22} = a_{21}(-a_{12}) + a_{22}(a_{11}) = a_{11}a_{22} - a_{12}a_{21}$. Thus we get the same result by expanding in any row of a 2×2 matrix and this proves equality (9) in the 2×2 case.

We now assume that equality (9) holds for all $(n-1) \times (n-1)$ matrices. We must show that it holds for $n \times n$ matrices. Our procedure will be to expand by cofactors in the first and ith rows and show that the expansions are identical. If we expand in the first row, then a typical term in the cofactor expansion is

$$a_{1k}A_{1k} = (-1)^{1+k}a_{1k}|M_{1k}|, \qquad k = 1, 2, \ldots, n. \tag{10}$$

Note that this is the only place in the expansion of $|A|$ that the term a_{1k} occurs since another typical term is $a_{1m}A_{1m} = (-1)^{1+m}|M_{1m}|$, $k \neq m$, and M_{1m} is obtained by deleting the first row and mth column of A (and a_{1k} is in the first row of A). Since M_{1k} is an $(n-1) \times (n-1)$ matrix, we can, by the induction hypothesis, calculate $|M_{1k}|$ by expanding in the ith row of A (which is the $(i-1)$st row of M_{1k}). A typical term in this expansion is

$$a_{il}\text{ (cofactor of } a_{il} \text{ in } M_{1k}) \qquad (k \neq l). \tag{11}$$

For the reasons outlined above, this is the only term in the expansion of $|M_{1k}|$ in the ith row of A that contains the term a_{il}. Substituting (11) into (10), we find that

$$(-1)^{1+k}a_{1k}a_{il}\text{ (cofactor of } a_{il} \text{ in } M_{1k}) \qquad (k \neq l) \tag{12}$$

is the only occurrence of the term $a_{1k}a_{il}$ in the cofactor expansion of det A in the first row.

Now if we expand by cofactors in the ith row of A (where $i \neq 1$), a typical term is

$$(-1)^{i+l}a_{il}|M_{il}|, \qquad l = 1, 2, \ldots, n, \tag{13}$$

and a typical term in the expansion of $|M_{il}|$ in the first row of M_{il} is

$$a_{1k}\text{ (cofactor of } a_{1k} \text{ in } M_{il}) \qquad (k \neq l) \tag{14}$$

and, inserting (14) in (13), we find that the only occurrence of the term $a_{il}a_{1k}$ in the expansion of det A along its ith row is

$$(-1)^{i+l}a_{1k}a_{il}\text{ (cofactor of } a_{1k} \text{ in } M_{il}) \qquad (k \neq l). \tag{15}$$

If we can show that the expressions in (12) and (15) are the same, then (1) will be proved, for the term in (12) is the only occurrence of $a_{1k}a_{il}$ in the first row expansion, the term in (15) is the only occurrence of $a_{1k}a_{il}$ in the ith row expansion, and k, i, and l are arbitrary. This will show that the sums of the terms in the first and ith row expansions are the same.

Now let $M_{1i,kl}$ denote the $(n-2) \times (n-2)$ matrix obtained by deleting the first and ith rows and kth and lth columns of A. (This is called a **second-order minor** of A.) We first suppose that $k < l$. Then

$$M_{1k} = \begin{pmatrix} a_{21} & \cdots & a_{2,k-1} & a_{2,k+1} & \cdots & a_{2l} & \cdots & a_{2n} \\ \vdots & & \vdots & \vdots & & \vdots & & \vdots \\ a_{i1} & \cdots & a_{i,k-1} & a_{i,k+1} & \cdots & a_{il} & \cdots & a_{in} \\ \vdots & & \vdots & \vdots & & \vdots & & \vdots \\ a_{n1} & \cdots & a_{n,k-1} & a_{n,k+1} & \cdots & a_{nl} & \cdots & a_{nn} \end{pmatrix}, \tag{16}$$

$$M_{il} = \begin{pmatrix} a_{11} & \cdots & a_{1k} & \cdots & a_{1,l-1} & a_{1,l+1} & \cdots & a_{1n} \\ \vdots & & \vdots & & \vdots & \vdots & & \vdots \\ a_{i-1,1} & \cdots & a_{i-1,k} & \cdots & a_{i-1,l-1} & a_{i-1,l+1} & \cdots & a_{i-1,n} \\ a_{i+1,1} & \cdots & a_{i+1,k} & \cdots & a_{i+1,l-1} & a_{i+1,l+1} & \cdots & a_{i+1,n} \\ \vdots & & \vdots & & \vdots & \vdots & & \vdots \\ a_{n1} & \cdots & a_{nk} & \cdots & a_{n,l-1} & a_{n,l+1} & \cdots & a_{nn} \end{pmatrix}. \tag{17}$$

From (16) and (17), we see that

$$\text{Cofactor of } a_{il} \text{ in } M_{1k} = (-1)^{(i-1)+(l-1)}|M_{1i,kl}|, \tag{18}$$

$$\text{Cofactor of } a_{1k} \text{ in } M_{il} = (-1)^{1+k}|M_{1i,kl}|. \tag{19}$$

Thus (12) becomes

$$(-1)^{1+k}a_{1k}a_{il}(-1)^{(i-1)+(l-1)}|M_{1i,kl}| = (-1)^{i+k+l-1}a_{1k}a_{il}|M_{1i,kl}| \tag{20}$$

and (15) becomes

$$(-1)^{i+l}a_{1k}a_{il}(-1)^{1+k}|M_{1i,kl}| = (-1)^{i+k+l+1}a_{1k}a_{il}|M_{1i,kl}|. \tag{21}$$

But $(-1)^{i+k+l-1} = (-1)^{i+k+l+1}$, so the right sides of equations (20) and (21) are equal. Hence expressions (12) and (15) are equal and (1) is proved in the case $k < l$. If $k > l$, then, by similar reasoning, we find that

$$\text{Cofactor of } a_{il} \text{ in } M_{1k} = (-1)^{(i-1)+l}|M_{1i,kl}|,$$

$$\text{Cofactor of } a_{1k} \text{ in } M_{il} = (-1)^{1+(k-1)}|M_{1i,kl}|,$$

so that (12) becomes

$$(-1)^{1+k}a_{1k}a_{il}(-1)^{(i-1)+l}|M_{1i,kl}| = (-1)^{i+k+l}a_{1k}a_{il}|M_{1i,kl}|$$

and (15) becomes

$$(-1)^{i+l}a_{1k}a_{il}(-1)^{1+(k-1)}|M_{1i,kl}| = (-1)^{i+k+l}a_{1k}a_{il}|M_{1i,kl}|$$

This completes the proof of equation (9).

To prove equation (7.7.8) on p. 408 (expansion in columns), we go through a similar process. If we expand in the kth and lth columns, we find that the only occurrences of the term $a_{1k}a_{il}$ will be given by (12) and (15). (See Problems 38 and 39.) This shows that the expansion by cofactors in any two columns is the same and that each is equal to the expansion along any row. This completes the proof. ∎

Problems 7.8

In Problems 1–20 evaluate the determinant by using the methods of this section.

1. $\begin{vmatrix} 3 & -5 \\ 2 & 6 \end{vmatrix}$

2. $\begin{vmatrix} 4 & 1 \\ 0 & -3 \end{vmatrix}$

3. $\begin{vmatrix} -1 & 0 & 2 \\ 3 & 1 & 4 \\ 2 & 0 & -6 \end{vmatrix}$

4. $\begin{vmatrix} 2 & 1 & -1 \\ 3 & -2 & 0 \\ 5 & 1 & 6 \end{vmatrix}$

5. $\begin{vmatrix} -3 & 2 & 4 \\ 1 & -1 & 2 \\ -1 & 4 & 0 \end{vmatrix}$

6. $\begin{vmatrix} 0 & -2 & 3 \\ 1 & 2 & -3 \\ 4 & 0 & 5 \end{vmatrix}$

7. $\begin{vmatrix} -2 & 3 & 6 \\ 4 & 1 & 8 \\ -2 & 0 & 0 \end{vmatrix}$

8. $\begin{vmatrix} -2 & 3 & 6 \\ 4 & 1 & 8 \\ -2 & 1 & 0 \end{vmatrix}$

9. $\begin{vmatrix} -2 & 3 & 6 \\ 4 & 1 & 8 \\ -2 & 0 & 1 \end{vmatrix}$

10. $\begin{vmatrix} 2 & -1 & 3 \\ 4 & 0 & 6 \\ 5 & -2 & 3 \end{vmatrix}$

11. $\begin{vmatrix} 2 & -1 & 3 \\ 4 & 0 & 6 \\ 5 & 0 & 3 \end{vmatrix}$

12. $\begin{vmatrix} 2 & -1 & 3 \\ 4 & 0 & 0 \\ 5 & -2 & 3 \end{vmatrix}$

13. $\begin{vmatrix} 1 & -1 & 2 & 4 \\ 0 & -3 & 5 & 6 \\ 1 & 4 & 0 & 3 \\ 0 & 5 & -6 & 7 \end{vmatrix}$

14. $\begin{vmatrix} 2 & -3 & 1 & 4 \\ 0 & -2 & 0 & 0 \\ 3 & 7 & -1 & 2 \\ 4 & 1 & -3 & 8 \end{vmatrix}$

15. $\begin{vmatrix} 1 & 1 & -1 & 0 \\ -3 & 4 & 6 & 0 \\ 2 & 5 & -1 & 3 \\ 4 & 0 & 3 & 0 \end{vmatrix}$

16. $\begin{vmatrix} 3 & -1 & 2 & 1 \\ 4 & 3 & 1 & -2 \\ -1 & 0 & 2 & 3 \\ 6 & 2 & 5 & 2 \end{vmatrix}$

17. $\begin{vmatrix} 0 & 2 & 0 & 0 \\ 0 & 0 & 3 & 0 \\ -1 & 0 & 0 & 0 \\ 0 & 0 & 0 & 4 \end{vmatrix}$

18. $\begin{vmatrix} 1 & 2 & 0 & 0 \\ 3 & -2 & 0 & 0 \\ 0 & 0 & 1 & -5 \\ 0 & 0 & 7 & 2 \end{vmatrix}$

19. $\begin{vmatrix} 2 & -1 & 0 & 4 & 1 \\ 3 & 1 & -1 & 2 & 0 \\ 3 & 2 & -2 & 5 & 1 \\ 0 & 0 & 4 & -1 & 6 \\ 3 & 2 & 1 & -1 & 1 \end{vmatrix}$

20. $\begin{vmatrix} 1 & -2 & 3 & -5 & 7 \\ 2 & 0 & -1 & -5 & 6 \\ 4 & 7 & 3 & -9 & 4 \\ 3 & 1 & -2 & -2 & 3 \\ -5 & -1 & 3 & 7 & -9 \end{vmatrix}$

In Problems 21–27 compute the determinant assuming that

$$\begin{vmatrix} a_{11} & a_{12} & a_{13} \\ a_{21} & a_{22} & a_{23} \\ a_{31} & a_{32} & a_{33} \end{vmatrix} = 8.$$

21. $\begin{vmatrix} a_{31} & a_{32} & a_{33} \\ a_{21} & a_{22} & a_{23} \\ a_{11} & a_{12} & a_{13} \end{vmatrix}$

22. $\begin{vmatrix} a_{31} & a_{32} & a_{33} \\ a_{11} & a_{12} & a_{13} \\ a_{21} & a_{22} & a_{23} \end{vmatrix}$

23. $\begin{vmatrix} a_{11} & a_{12} & a_{13} \\ 2a_{21} & 2a_{22} & 2a_{23} \\ a_{31} & a_{32} & a_{33} \end{vmatrix}$

24. $\begin{vmatrix} -3a_{11} & -3a_{12} & -3a_{13} \\ 2a_{21} & 2a_{22} & 2a_{23} \\ 5a_{31} & 5a_{32} & 5a_{33} \end{vmatrix}$

25. $\begin{vmatrix} a_{11} & 2a_{13} & a_{12} \\ a_{21} & 2a_{23} & a_{22} \\ a_{31} & 2a_{33} & a_{32} \end{vmatrix}$

26. $\begin{vmatrix} a_{11} - a_{12} & a_{12} & a_{13} \\ a_{21} - a_{22} & a_{22} & a_{23} \\ a_{31} - a_{32} & a_{32} & a_{33} \end{vmatrix}$

27. $\begin{vmatrix} 2a_{11} - 3a_{21} & 2a_{12} - 3a_{22} & 2a_{13} - 3a_{23} \\ a_{31} & a_{32} & a_{33} \\ a_{21} & a_{22} & a_{23} \end{vmatrix}$

28. Using Property 2, show that if α is a number and A is an $n \times n$ matrix, then $\det \alpha A = \alpha^n \det A$.

* 29. Show that

$$\begin{vmatrix} 1+x_1 & x_2 & x_3 & \cdots & x_n \\ x_1 & 1+x_2 & x_3 & \cdots & x_n \\ x_1 & x_2 & 1+x_3 & \cdots & x_n \\ \vdots & \vdots & \vdots & & \vdots \\ x_1 & x_2 & x_3 & \cdots & 1+x_n \end{vmatrix}$$

$$= 1 + x_1 + x_2 + \cdots + x_n.$$

* 30. If A is an $n \times n$ skew-symmetric matrix, show that $\det A = (-1)^n \det A$.

31. Using the result of Problem 30, show that if A is a skew-symmetric $n \times n$ matrix and n is odd, then $\det A = 0$.

32. A matrix A is called **orthogonal** if A is invertible and $A^{-1} = A^t$. Show that if A is orthogonal, then $\det A = \pm 1$.

** 33. Let Δ denote the triangle in the plane with vertices at (x_1, y_1), (x_2, y_2), and (x_3, y_3). Show that the area of the triangle is given by

$$\text{Area of } \Delta = \pm \frac{1}{2} \begin{vmatrix} 1 & x_1 & y_1 \\ 1 & x_2 & y_2 \\ 1 & x_3 & y_3 \end{vmatrix}.$$

Under what circumstances will this determinant equal zero?

** 34. Three lines, no two of which are parallel, determine a triangle in the plane. Suppose that the lines are given by

$$a_{11}x + a_{12}y + a_{13} = 0,$$

$$a_{21}x + a_{22}y + a_{23} = 0,$$

$$a_{31}x + a_{32}y + a_{33} = 0.$$

Show that the area determined by the lines is

$$\frac{\pm 1}{2A_{13}A_{23}A_{33}} \begin{vmatrix} A_{11} & A_{12} & A_{13} \\ A_{21} & A_{22} & A_{23} \\ A_{31} & A_{32} & A_{33} \end{vmatrix}.$$

Here A_{ij} is the ijth cofactor of the matrix of coefficients for the system of equations.

35. The 3×3 Vandermonde† determinant is given by

$$D_3 = \begin{vmatrix} 1 & 1 & 1 \\ a_1 & a_2 & a_3 \\ a_1^2 & a_2^2 & a_3^2 \end{vmatrix}.$$

Show that $D_3 = (a_2 - a_1)(a_3 - a_1)(a_3 - a_2)$.

36. $D_4 = \begin{vmatrix} 1 & 1 & 1 & 1 \\ a_1 & a_2 & a_3 & a_4 \\ a_1^2 & a_2^2 & a_3^2 & a_4^2 \\ a_1^3 & a_2^3 & a_3^3 & a_4^3 \end{vmatrix}$ is the 4×4 Vandermonde de-

terminant. Show that

$$D_4 = (a_2 - a_1)(a_3 - a_1)(a_4 - a_1)(a_3 - a_2)(a_4 - a_2)(a_4 - a_3).$$

** 37. (a) Define the $n \times n$ Vandermonde determinant D_n.

(b) Show that $D_n = \prod_{\substack{i=1 \\ j>i}}^{n} (a_j - a_i)$, where $\prod$ stands for the word "product." Note that the product in Problem 36 can be written $\prod_{\substack{i=1 \\ j>i}}^{4} (a_j - a_i)$.

38. Show that if A is expanded along its kth column, then the only occurrence of the term $a_{1k}a_{il}$ is given by equation (12).

39. Show that if A is expanded along its lth column, then the only occurrence of the term $a_{1k}a_{il}$ is given by equation (15).

7.9 Determinants and Inverses; Cramer's Rule

In this section we shall discuss a method for finding the inverse using determinants. This method is not very useful for computational purposes if the matrix is large, but it is useful theoretically. We shall also complete our task of summing up the properties of a matrix that we have developed in this chapter. We begin with a simple result.

■ **THEOREM 1**

If A is invertible, then $\det A \neq 0$ and

$$\det A^{-1} = \frac{1}{\det A}. \tag{1}$$

PROOF
From Theorems 7.7.2 and 7.8.3, we have

$$1 = \det I = \det AA^{-1} = \det A \det A^{-1}. \tag{2}$$

If $\det A$ were equal to zero, then equation (2) would read $1 = 0$. Thus $\det A \neq 0$ and $\det A^{-1} = 1/\det A$. ■

† A. T. Vandermonde (1735–1796) was a French mathematician.

Before using determinants to calculate inverses, we need to define the *adjoint* of a matrix $A = (a_{ij})$. Let $B = (A_{ij})$ be the matrix of cofactors of A. (Remember that a cofactor is a number.) Then

$$B = \begin{pmatrix} A_{11} & A_{12} & \cdots & A_{1n} \\ A_{21} & A_{22} & \cdots & A_{2n} \\ \vdots & \vdots & & \vdots \\ A_{n1} & A_{n2} & \cdots & A_{nn} \end{pmatrix}. \tag{3}$$

DEFINITION

The Adjoint

Let A be an $n \times n$ matrix and let B, given by (3), denote the matrix of its cofactors. Then the **adjoint** of A, written adj A, is the transpose of the $n \times n$ matrix B; that is,

$$\text{adj } A = B^t = \begin{pmatrix} A_{11} & A_{21} & \cdots & A_{n1} \\ A_{12} & A_{22} & \cdots & A_{n2} \\ \vdots & \vdots & & \vdots \\ A_{1n} & A_{2n} & \cdots & A_{nn} \end{pmatrix}. \tag{4}$$

Note. The adjoint is sometimes called the **comatrix** of A.

EXAMPLE 1

Let $A = \begin{pmatrix} 2 & 4 & 3 \\ 0 & 1 & -1 \\ 3 & 5 & 7 \end{pmatrix}$. Compute adj A.

SOLUTION

We have $A_{11} = \begin{vmatrix} 1 & -1 \\ 5 & 7 \end{vmatrix} = 12$, $A_{12} = -\begin{vmatrix} 0 & -1 \\ 3 & 7 \end{vmatrix} = -3$, $A_{13} = -3$, $A_{21} = -13$, $A_{22} = 5$,

$A_{23} = 2$, $A_{31} = -7$, $A_{32} = 2$, and $A_{33} = 2$. Thus

$$B = \begin{pmatrix} 12 & -3 & -3 \\ -13 & 5 & 2 \\ -7 & 2 & 2 \end{pmatrix} \text{ and adj } A = B^t = \begin{pmatrix} 12 & -13 & -7 \\ -3 & 5 & 2 \\ -3 & 2 & 2 \end{pmatrix}.$$

Warning. In taking the adjoint of a matrix, do not forget to transpose the matrix of cofactors.

THEOREM 2

Let A be an $n \times n$ matrix. Then

$$(A)(\text{adj } A) = \begin{pmatrix} \det A & 0 & 0 & \cdots & 0 \\ 0 & \det A & 0 & \cdots & 0 \\ 0 & 0 & \det A & \cdots & 0 \\ \vdots & \vdots & \vdots & & \vdots \\ 0 & 0 & 0 & \cdots & \det A \end{pmatrix} = (\det A)I. \qquad (5)$$

PROOF

Let $C = (c_{ij}) = (A)(\text{adj } A)$. Then

$$C = \begin{pmatrix} a_{11} & a_{12} & \cdots & a_{1n} \\ a_{21} & a_{22} & \cdots & a_{2n} \\ \vdots & \vdots & & \vdots \\ a_{n1} & a_{n2} & \cdots & a_{nn} \end{pmatrix} \begin{pmatrix} A_{11} & A_{21} & \cdots & A_{n1} \\ A_{12} & A_{22} & \cdots & A_{n2} \\ \vdots & \vdots & & \vdots \\ A_{1n} & A_{2n} & \cdots & A_{nn} \end{pmatrix}. \qquad (6)$$

We have

$$c_{ij} = (i\text{th row of } A) \cdot (j\text{th column of adj } A)$$

$$= (a_{i1} a_{i2} \cdots a_{in}) \cdot \begin{pmatrix} A_{j1} \\ A_{j2} \\ \vdots \\ A_{jn} \end{pmatrix}.$$

Thus

$$c_{ij} = a_{i1} A_{j1} + a_{i2} A_{j2} + \cdots + a_{in} A_{jn}. \qquad (7)$$

Now if $i = j$, the sum in (7) equals $a_{i1} A_{i1} + a_{i2} A_{i2} + \cdots + a_{in} A_{in}$, which is the expansion of $\det A$ in the ith row of A. On the other hand, if $i \neq j$ then from Theorem 7.8.1 the sum in (7) equals zero. Thus

$$c_{ij} = \begin{cases} \det A & \text{if } i = j, \\ 0 & \text{if } i \neq j. \end{cases}$$

This proves the theorem ∎

We can now state the main result.

THEOREM 3

Let A be an $n \times n$ matrix. Then A is invertible if and only if $\det A \neq 0$. If $\det A \neq 0$, then

$$A^{-1} = \frac{1}{\det A} \text{ adj } A. \qquad (8)$$

Note that Theorem 7.5.4 for 2×2 matrices is a special case of this theorem.

PROOF
If A is invertible, then det $A \neq 0$ by Theorem 1. If det $A \neq 0$, then, from Theorem 2,

$$(A)\left(\frac{1}{\det A}\operatorname{adj} A\right) = \frac{1}{\det A}[A(\operatorname{adj} A)] = \frac{1}{\det A}(\det A)I = I.$$

But, by Theorem 7.5.8, if $AB = I$, then $B = A^{-1}$. Thus $(1/\det A)\operatorname{adj} A = A^{-1}$. ∎

EXAMPLE 2

Let $A = \begin{pmatrix} 2 & 4 & 3 \\ 0 & 1 & -1 \\ 3 & 5 & 7 \end{pmatrix}$. Determine whether A is invertible and calculate A^{-1} if it is.

SOLUTION
Since det $A = 3 \neq 0$, we see that A is invertible. From Example 1,

$$\operatorname{adj} A = \begin{pmatrix} 12 & -13 & -7 \\ -3 & 5 & 2 \\ -3 & 2 & 2 \end{pmatrix}.$$

Thus

$$A^{-1} = \frac{1}{3}\begin{pmatrix} 12 & -13 & -7 \\ -3 & 5 & 2 \\ -3 & 2 & 2 \end{pmatrix} = \begin{pmatrix} 4 & -13/3 & -7/3 \\ -1 & 5/3 & 2/3 \\ -1 & 2/3 & 2/3 \end{pmatrix}.$$

Check.

$$A^{-1}A = \frac{1}{3}\begin{pmatrix} 12 & -13 & -7 \\ -3 & 5 & 2 \\ -3 & 2 & 2 \end{pmatrix}\begin{pmatrix} 2 & 4 & 3 \\ 0 & 1 & -1 \\ 3 & 5 & 7 \end{pmatrix} = \frac{1}{3}\begin{pmatrix} 3 & 0 & 0 \\ 0 & 3 & 0 \\ 0 & 0 & 3 \end{pmatrix} = I.$$

Consider the system of n equations in n unknowns $A\mathbf{x} = \mathbf{b}$, and suppose that det $A \neq 0$. Then the system has the unique solution $\mathbf{x} = A^{-1}\mathbf{b}$. We can develop a method for finding that solution without row reduction and without computing A^{-1}. (See, however, the remark following Example 3).

Let $D = \det A$. We define n new matrices:

$$A_1 = \begin{pmatrix} b_1 & a_{12} & \cdots & a_{1n} \\ b_2 & a_{22} & \cdots & a_{2n} \\ \vdots & \vdots & & \vdots \\ b_n & a_{n2} & \cdots & a_{nn} \end{pmatrix}, \quad A_2 = \begin{pmatrix} a_{11} & b_1 & \cdots & a_{1n} \\ a_{21} & b_2 & \cdots & a_{2n} \\ \vdots & \vdots & & \vdots \\ a_{n1} & b_n & \cdots & a_{nn} \end{pmatrix}, \quad \ldots,$$

$$A_n = \begin{pmatrix} a_{11} & a_{12} & \cdots & b_1 \\ a_{21} & a_{22} & \cdots & b_2 \\ \vdots & \vdots & & \vdots \\ a_{n1} & a_{n2} & \cdots & b_n \end{pmatrix}.$$

That is, A_i is the matrix obtained by replacing the ith column of A with $\mathbf{b}$. Finally, let $D_1 = \det A_1$, $D_2 = \det A_2$, $\ldots$, $D_n = \det A_n$.

■ **THEOREM 4: Cramer's Rule†**

Let A be an $n \times n$ matrix and suppose that $\det A \neq 0$. Then the unique solution to the system $A\mathbf{x} = \mathbf{b}$ is given by

$$x_1 = \frac{D_1}{D},\, x_2 = \frac{D_2}{D},\, \ldots,\, x_i = \frac{D_i}{D},\, \ldots,\, x_n = \frac{D_n}{D}. \tag{9}$$

PROOF

The solution to $A\mathbf{x} = \mathbf{b}$ is $\mathbf{x} = A^{-1}\mathbf{b}$. But

$$A^{-1}\mathbf{b} = \frac{1}{D}(\operatorname{adj} A)\mathbf{b} = \frac{1}{D}\begin{pmatrix} A_{11} & A_{21} & \cdots & A_{n1} \\ A_{12} & A_{22} & \cdots & A_{n2} \\ \vdots & \vdots & & \vdots \\ A_{1n} & A_{2n} & \cdots & A_{nn} \end{pmatrix}\begin{pmatrix} b_1 \\ b_2 \\ \vdots \\ b_n \end{pmatrix}.$$

Now $(\operatorname{adj} A)\mathbf{b}$ is an n-vector, the jth component of which is

$$(A_{1j}\, A_{2j} \cdots A_{nj}) \cdot \begin{pmatrix} b_1 \\ b_2 \\ \vdots \\ b_n \end{pmatrix} = b_1 A_{1j} + b_2 A_{2j} + \cdots + b_n A_{nj}. \tag{10}$$

Consider the matrix A_j:

$$A_j = \begin{pmatrix} a_{11} & a_{12} & \cdots & b_1 & \cdots & a_{1n} \\ a_{21} & a_{22} & \cdots & b_2 & \cdots & a_{2n} \\ \vdots & \vdots & & \vdots & & \vdots \\ a_{n1} & a_{n2} & \cdots & b_n & \cdots & a_{nn} \end{pmatrix}. \tag{11}$$

$$\uparrow$$
$$j\text{th column}$$

Using the basic theorem (Theorem 7.7.1) and expanding A_j by cofactors along the jth column, we get

$$D_j = |A_j| = b_1 A_{1j} + b_2 A_{2j} + \cdots + b_n A_{nj}.$$

But this is the same as the right side of (10). Thus the jth component of $(\operatorname{adj} A)\mathbf{b}$ is D_j and we have

† Named for the Swiss mathematician Gabriel Cramer (1704–1752). Cramer published the rule in 1750 in his *Introduction to the Analysis of Lines of Algebraic Curves.* Actually, there is much evidence to suggest that the rule was known as early as 1729 to Colin Maclaurin (1698–1746), who was probably the most outstanding British mathematician in the years following the death of Newton.

$$\mathbf{x} = \begin{pmatrix} x_1 \\ x_2 \\ \vdots \\ x_n \end{pmatrix} = A^{-1}\mathbf{b} = \frac{1}{D}(\text{adj } A)\mathbf{b} = \frac{1}{D}\begin{pmatrix} D_1 \\ D_2 \\ \vdots \\ D_n \end{pmatrix} = \begin{pmatrix} D_1/D \\ D_2/D \\ \vdots \\ D_n/D \end{pmatrix}$$

and the proof is complete.   ◼

EXAMPLE 3

Using Cramer's rule solve the system

$$2x_1 + 4x_2 + 6x_3 = 18,$$

$$4x_1 + 5x_2 + 6x_3 = 24,$$

$$3x_1 + x_2 - 2x_3 = 4. \qquad (12)$$

SOLUTION
We have solved this before—using row reduction in Example 7.3.2 on p. 364. We could also solve it by calculating A^{-1} (Example 7.5.6, p. 394) and then finding $A^{-1}\mathbf{b}$. We now solve it by using Cramer's rule. First we have

$$D = \begin{vmatrix} 2 & 4 & 6 \\ 4 & 5 & 6 \\ 3 & 1 & -2 \end{vmatrix} = 6 \neq 0,$$

so that system (12) has a unique solution. Then $D_1 = \begin{vmatrix} 18 & 4 & 6 \\ 24 & 5 & 6 \\ 4 & 1 & -2 \end{vmatrix} = 24$, $D_2 = \begin{vmatrix} 2 & 18 & 6 \\ 4 & 24 & 6 \\ 3 & 4 & -2 \end{vmatrix} = -12$, and $D_3 = \begin{vmatrix} 2 & 4 & 18 \\ 4 & 5 & 24 \\ 3 & 1 & 4 \end{vmatrix} = 18$. Hence $x_1 = \dfrac{D_1}{D} = \dfrac{24}{6} = 4$, $x_2 = \dfrac{D_2}{D} = -\dfrac{12}{6} = -2$, and $x_3 = \dfrac{D_3}{D} = \dfrac{18}{6} = 3$.

Remark. If $n > 3$, Gauss-Jordan elimination is a much more efficient method than Cramer's rule. It is also easier to program.

The results of Sections 7.7–7.9 enable us to extend our summing up theorem, last seen on p. 397.

◼ **THEOREM 5: Summing-Up Theorem—View 2**
Let A be an $n \times n$ matrix. Then each of the following six statements implies the other five. (That is, if one is true, all are true.)

  i. A is invertible.
  ii. The only solution to the homogeneous system $A\mathbf{x} = \mathbf{0}$ is the trivial solution ($\mathbf{x} = \mathbf{0}$).
  iii. The system $A\mathbf{x} = \mathbf{b}$ has a unique solution for every n-vector $\mathbf{b}$.
  iv. A is row equivalent to the $n \times n$ identity matrix I_n.

    v. The rows (and columns) of A are linearly independent.

   vi. det $A \neq 0$.

PROOF

In Theorem 7.5.6 we proved the equivalence of parts (i), (ii), (iii), and (iv). The equivalence of parts (i) and (vi) is proved in Theorem 3 of this section. We saw, in Theorem 7.4.3, that (ii) and (v) are equivalent for columns. However the rows of an $n \times n$ matrix are linearly independent if and only if its columns are linearly independent. This is true because the rows of A are the columns of A^t, and det $A =$ det A^t. ∎

EXAMPLE 4

Determine whether the vectors $\begin{pmatrix} 1 \\ -2 \\ 3 \end{pmatrix}$, $\begin{pmatrix} 4 \\ 1 \\ 5 \end{pmatrix}$, and $\begin{pmatrix} 1 \\ 0 \\ 2 \end{pmatrix}$ are linearly independent or dependent.

SOLUTION

Let $A = \begin{pmatrix} 1 & 4 & 1 \\ -2 & 1 & 0 \\ 3 & 5 & 2 \end{pmatrix}$. Then $|A| = \begin{vmatrix} 1 & 4 & 1 \\ -2 & 1 & 0 \\ 3 & 5 & 2 \end{vmatrix} = 5 \neq 0$, and hence the vectors (which are the columns of A) are linearly independent.

EXAMPLE 5

Determine whether the vectors $\begin{pmatrix} -2 \\ 4 \\ 5 \end{pmatrix}$, $\begin{pmatrix} 3 \\ 1 \\ 0 \end{pmatrix}$, and $\begin{pmatrix} 4 \\ 6 \\ 5 \end{pmatrix}$ are linearly independent or dependent.

SOLUTION

Proceeding as in Example 4, we have $\begin{vmatrix} -2 & 3 & 4 \\ 4 & 1 & 6 \\ 5 & 0 & 5 \end{vmatrix} = 0$; thus the vectors are linearly dependent.

We close this section by proving a theorem cited in Section 6.7.

■ **THEOREM 6 (THEOREM 6.7.4)**

Any set of n linearly independent vectors in $\mathbb{R}^n$ is a basis for $\mathbb{R}^n$.

PROOF

Let $\mathbf{v}_1 = \begin{pmatrix} a_{11} \\ a_{21} \\ \vdots \\ a_{n1} \end{pmatrix}$, $\mathbf{v}_2 = \begin{pmatrix} a_{12} \\ a_{22} \\ \vdots \\ a_{n2} \end{pmatrix}$, $\ldots$, $\mathbf{v}_n = \begin{pmatrix} a_{1n} \\ a_{2n} \\ \vdots \\ a_{nn} \end{pmatrix}$ be linearly independent.

We show that $\{v_1, v_2, \ldots, v_n\}$ spans $\mathbb{R}^n$. This will prove that they are a basis.

Let $\mathbf{v} = \begin{pmatrix} x_1 \\ x_2 \\ \vdots \\ x_n \end{pmatrix}$ be a vector in $\mathbb{R}^n$. We must show that there exist scalars c_1, $c_2, \ldots, c_n$ such that

$$\mathbf{v} = c_1\mathbf{v}_1 + c_2\mathbf{v}_2 + \cdots + c_n\mathbf{v}_n.$$

That is,

$$\begin{pmatrix} x_1 \\ x_2 \\ \vdots \\ x_n \end{pmatrix} = c_1 \begin{pmatrix} a_{11} \\ a_{21} \\ \vdots \\ a_{n1} \end{pmatrix} + c_2 \begin{pmatrix} a_{12} \\ a_{22} \\ \vdots \\ a_{n2} \end{pmatrix} + \cdots + c_n \begin{pmatrix} a_{1n} \\ a_{2n} \\ \vdots \\ a_{nn} \end{pmatrix}. \tag{13}$$

In (13), we multiply through, add, and equate components to obtain a system of n equations in the n unknowns $c_1, c_2, \ldots, c_n$:

$$\begin{aligned} a_{11}c_1 + a_{12}c_2 + \cdots + a_{1n}c_n &= x_1 \\ a_{21}c_1 + a_{22}c_2 + \cdots + a_{2n}c_n &= x_2 \\ &\vdots \\ a_{n1}c_1 + a_{n2}c_2 + \cdots + a_{nn}c_n &= x_n \end{aligned} \tag{14}$$

We write (14) as $A\mathbf{c} = \mathbf{v}$, where

$$A = \begin{pmatrix} a_{11} & a_{12} & \cdots & a_{1n} \\ a_{21} & a_{22} & \cdots & a_{2n} \\ \vdots & \vdots & & \vdots \\ a_{n1} & a_{n2} & \cdots & a_{nn} \end{pmatrix} \quad \text{and} \quad \mathbf{c} = \begin{pmatrix} c_1 \\ c_2 \\ \vdots \\ c_n \end{pmatrix}.$$

But system (14) has a unique solution if and only if $\det A \neq 0$. And $\det A \neq 0$ because the columns of A are linearly independent. (This all follows from Theorem 5.) Thus there is a unique vector $\mathbf{c}$ satisfying system (14) and the theorem is proved. ∎

Remark 1. This theorem not only shows that $\mathbf{v}$ can be written as a linear combination of the independent vectors $\mathbf{v}_1, \mathbf{v}_2, \ldots, \mathbf{v}_n$ but also that this can be done in *only one way* (since the solution vector $\mathbf{c}$ is unique).

Remark 2. It is also the case that any n vectors in $\mathbb{R}^n$ that span $\mathbb{R}^n$ form a basis for $\mathbb{R}^n$. See Problem 33.

Problems 7.9

In Problems 1–12 use the methods of this section to determine whether the given matrix is invertible. If so, compute the inverse.

1. $\begin{pmatrix} 3 & 2 \\ 1 & 2 \end{pmatrix}$

2. $\begin{pmatrix} 3 & 6 \\ -4 & -8 \end{pmatrix}$

3. $\begin{pmatrix} 0 & 1 \\ 1 & 0 \end{pmatrix}$

4. $\begin{pmatrix} 1 & 1 & 1 \\ 0 & 2 & 3 \\ 5 & 5 & 1 \end{pmatrix}$

5. $\begin{pmatrix} 3 & 2 & 1 \\ 0 & 2 & 2 \\ 0 & 1 & -1 \end{pmatrix}$

6. $\begin{pmatrix} 1 & 1 & 1 \\ 0 & 1 & 1 \\ 0 & 0 & 1 \end{pmatrix}$

7. $\begin{pmatrix} 1 & 2 & 3 \\ 1 & 1 & 2 \\ 0 & 1 & 2 \end{pmatrix}$

8. $\begin{pmatrix} 3 & 1 & 0 \\ 1 & -1 & 2 \\ 1 & 1 & 1 \end{pmatrix}$

9. $\begin{pmatrix} 2 & -1 & 4 \\ -1 & 0 & 5 \\ 19 & -7 & 3 \end{pmatrix}$

10. $\begin{pmatrix} 1 & 6 & 2 \\ -2 & 3 & 5 \\ 7 & 12 & -4 \end{pmatrix}$

11. $\begin{pmatrix} 1 & 1 & 1 & 1 \\ 1 & 2 & -1 & 2 \\ 1 & -1 & 2 & 1 \\ 1 & 3 & 3 & 2 \end{pmatrix}$

12. $\begin{pmatrix} 1 & -3 & 0 & -2 \\ 3 & -12 & -2 & -6 \\ -2 & 10 & 2 & 5 \\ -1 & 6 & 1 & 3 \end{pmatrix}$

In Problems 13–21 solve the given system by using Cramer's rule.

13. $x_1 + 4x_2 = 13$
$4x_1 + 2x_2 = 10$

14. $3x_1 - x_2 = 0$
$4x_1 + 2x_2 = 5$

15. $x_1 + x_2 + x_3 = 1$
$2x_1 + 3x_2 + 4x_3 = 3$
$4x_1 + 9x_2 + 16x_3 = 11$

16. $x_1 + x_2 + x_3 = 6$
$2x_1 - x_2 = 0$
$2x_1 + x_3 = 3$

17. $2x_1 + 2x_2 + x_3 = 7$
$x_1 + 2x_2 - x_3 = 0$
$-x_1 + x_2 + 3x_3 = 1$

18. $2x_1 + 5x_2 - x_3 = -1$
$4x_1 + x_2 + 3x_3 = 3$
$-2x_1 + 2x_2 = 0$

19. $2x_1 + x_2 - x_3 = 4$
$x_1 + x_3 = 2$
$-x_2 + 5x_3 = 1$

20. $x_1 + x_2 + x_3 + x_4 = 6$
$2x_1 - x_3 - x_4 = 4$
$3x_3 + 6x_4 = 3$
$x_1 - x_4 = 5$

21. $x_1 - x_4 = 7$
$2x_2 + x_3 = 2$
$4x_1 - x_2 = -3$
$3x_3 - 5x_4 = 2$

22. Show that an $n \times n$ matrix A is invertible if and only if A^t is invertible.

23. For $A = \begin{pmatrix} 1 & 1 \\ 2 & 5 \end{pmatrix}$, verify that $\det A^{-1} = 1/\det A$.

24. For $A = \begin{pmatrix} 1 & -1 & 3 \\ 4 & 1 & 6 \\ 2 & 0 & -2 \end{pmatrix}$, verify that $\det A^{-1} = 1/\det A$.

25. For what values of α is the matrix $\begin{pmatrix} \alpha & -3 \\ 4 & 1-\alpha \end{pmatrix}$ not invertible?

26. For what values of α does the matrix
$\begin{pmatrix} -\alpha & \alpha-1 & \alpha+1 \\ 1 & 2 & 3 \\ 2-\alpha & \alpha+3 & \alpha+7 \end{pmatrix}$ not have an inverse?

27. Suppose that the $n \times n$ matrix A is not invertible. Show that $(A)(\mathrm{adj}\,A)$ is the zero matrix.

* 28. Consider the triangle below.

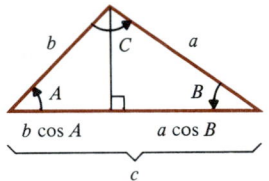

(a) Show, using elementary trigonometry, that
$$b\cos A + a\cos B = c$$
$$c\cos A + a\cos C = b$$
$$c\cos B + b\cos C = a$$

(b) If the system of part (a) is thought of as a system of three equations in the three unknowns $\cos A$, $\cos B$, and $\cos C$, show that the determinant of the system is nonzero.

(c) Use Cramer's rule to solve for $\cos C$.

(d) Use part (c) to prove the **law of cosines:**
$$c^2 = a^2 + b^2 - 2ab\cos C.$$

In Problems 29–31, use determinants to determine whether the given vectors are linearly dependent or independent.

29. $\begin{pmatrix} 2 \\ -1 \\ 4 \end{pmatrix}, \begin{pmatrix} 1 \\ 6 \\ 2 \end{pmatrix}, \begin{pmatrix} 0 \\ 0 \\ 1 \end{pmatrix}$

30. $\begin{pmatrix} 1 \\ -3 \\ 2 \\ 1 \end{pmatrix}, \begin{pmatrix} 4 \\ 0 \\ 1 \\ 2 \end{pmatrix}, \begin{pmatrix} 0 \\ 3 \\ -1 \\ 4 \end{pmatrix}, \begin{pmatrix} 5 \\ -6 \\ 4 \\ -1 \end{pmatrix}$

31. $\begin{pmatrix} 3 \\ 0 \\ 1 \\ 0 \end{pmatrix}, \begin{pmatrix} 5 \\ 1 \\ -1 \\ 2 \end{pmatrix}, \begin{pmatrix} 6 \\ 0 \\ 1 \\ 0 \end{pmatrix}, \begin{pmatrix} 0 \\ 0 \\ 1 \\ 1 \end{pmatrix}$

32. Show that the n n-vectors are linearly independent.

$$\begin{pmatrix} 1 \\ 0 \\ 0 \\ \vdots \\ 0 \end{pmatrix}, \begin{pmatrix} 0 \\ 1 \\ 0 \\ \vdots \\ 0 \end{pmatrix}, \begin{pmatrix} 0 \\ 0 \\ 1 \\ \vdots \\ 0 \end{pmatrix}, \dots, \begin{pmatrix} 0 \\ 0 \\ 0 \\ \vdots \\ 1 \\ 0 \end{pmatrix}, \begin{pmatrix} 0 \\ 0 \\ 0 \\ \vdots \\ 1 \end{pmatrix}.$$

33. Suppose that the n vectors $\mathbf{v}_1, \mathbf{v}_2, \dots, \mathbf{v}_n$ in $\mathbb{R}^n$ span $\mathbb{R}^n$. Prove that $\{\mathbf{v}_1, \mathbf{v}_2, \dots, \mathbf{v}_n\}$ is a basis for $\mathbb{R}^n$. [*Hint:* Let A be the matrix whose columns are $\mathbf{v}_1, \mathbf{v}_2, \dots, \mathbf{v}_n$. Show that the system $A\mathbf{x} = \mathbf{b}$ has a solution for every $\mathbf{b} \in \mathbb{R}^n$. Use this fact to show that there is a matrix B such that $AB = I$. Conclude that A is invertible so its columns are linearly independent].

7.10 The Rank and Nullity of a Matrix

In Section 6.7 we introduced the notion of linear independence. We showed in Theorem 7.9.5 that if A is an $n \times n$ matrix with det $A \neq 0$, then the columns and the rows of A form sets of linearly independent vectors. However, if A is a square matrix with det $A = 0$, or if A is not a square matrix, then these results tell us little about the number of linearly independent rows or columns of A. In this section we shall fill in this gap.

DEFINITION

Rank

Let A be an $m \times n$ matrix. Then the **rank** of A, denoted by $\rho(A)$, is the maximum number of linearly independent rows of A.

EXAMPLE 1

Find $\rho(A)$ if $A = \begin{pmatrix} 1 & 2 & 3 \\ 2 & 4 & 6 \\ 3 & 6 & 9 \end{pmatrix}$.

SOLUTION

The second row is twice the first and the third row is three times the first. Thus A has one linearly independent row and $\rho(A) = 1$.

EXAMPLE 2

Find $\rho(A)$ if $A = \begin{pmatrix} 3 & 5 & 2 \\ 4 & 2 & 3 \\ -1 & 2 & 4 \end{pmatrix}$.

SOLUTION

In Example 7.7.1 on p. 404 we computed det $A = -69$. By the summing-up theorem (p. 429), the rows of A are linearly independent. Thus $\rho(A) = 3$.

EXAMPLE 3

Find $\rho(A)$ if $A = \begin{pmatrix} 1 & 3 & -1 & 2 \\ 4 & 1 & 5 & 6 \end{pmatrix}$.

SOLUTION

Since the second row is not a multiple of the first, the two rows of A are linearly independent so $\rho(A) = 2$.

We can generalize Example 2 to obtain the following result.

■ **THEOREM 1**

Let A be an $n \times n$ matrix. Then $\rho(A) = n$ if and only if $\det A \neq 0$. ■

DEFINITION

Let $\mathbf{r}_1, \mathbf{r}_2, \ldots, \mathbf{r}_m$ denote the rows of A. Then the **row space** of A is given by

$$R_A = \mathrm{span}\{\mathbf{r}_1, \mathbf{r}_2, \ldots, \mathbf{r}_m\}.$$

Row Space

Hence

$$\rho(A) = \dim R_A.$$

■ **THEOREM 2**

If A is row equivalent to B, then $\rho(A) = \rho(B)$.

PROOF

Let C be the matrix obtained by performing an elementary row operation on A. We show that $\rho(A) = \rho(C)$. This will complete the proof since B is obtained from A by performing a number of elementary row operations on A.

CASE 1

Interchange two rows of A. Then $R_A = R_C$ because the rows of A and C are the same (just written in a different order).

CASE 2

Multiply the ith row of A by $c \neq 0$. If the rows of A are $\{\mathbf{r}_1, \mathbf{r}_2, \ldots, \mathbf{r}_i, \ldots, \mathbf{r}_m\}$, then the rows of C are $\{\mathbf{r}_1, \mathbf{r}_2, \ldots, c\mathbf{r}_i, \ldots, \mathbf{r}_m\}$. Obviously, $c\mathbf{r}_i = c(\mathbf{r}_i)$ and $\mathbf{r}_i = (1/c)c\mathbf{r}_i$. Thus each row of C is a multiple of one row of A and vice versa. This means that each row of C is in the span of the rows of A and vice versa. We have

$$R_A \subseteq R_C \quad \text{and} \quad R_C \subseteq R_A, \quad \text{so } R_C = R_A.$$

CASE 3

Multiply the ith row of A by $c \neq 0$ and add it to the jth row. Now the rows of C are $\{\mathbf{r}_1, \mathbf{r}_2, \ldots, \mathbf{r}_i, \ldots, c\mathbf{r}_i + \mathbf{r}_j, \ldots, \mathbf{r}_m\}$. Here

$$\mathbf{r}_j = \underbrace{(c\mathbf{r}_i + \mathbf{r}_j)}_{j\text{th row of } C} - c\underbrace{(\mathbf{r}_i)}_{i\text{th row of } C}$$

so each row of A can be written as a linear combination of the rows of C and vice versa. Then, as before,

$$R_A \subseteq R_C \quad \text{and} \quad R_C \subseteq R_A, \quad \text{so } R_C = R_A.$$

In each case we have $R_A = R_C$ so $\rho(A) = \dim R_A = \dim R_C = \rho(C)$. ■

EXAMPLE 4

Determine the rank and row space of $A = \begin{pmatrix} 1 & -1 & 3 \\ 2 & 0 & 4 \\ -1 & -3 & 1 \end{pmatrix}$.

SOLUTION

We row-reduce to obtain a simpler matrix:

$$\begin{pmatrix} 1 & -1 & 3 \\ 2 & 0 & 4 \\ -1 & -3 & 1 \end{pmatrix} \xrightarrow{\begin{subarray}{c} A_{1,2}(-2) \\ A_{1,3}(1) \end{subarray}} \begin{pmatrix} 1 & -1 & 3 \\ 0 & 2 & -2 \\ 0 & -4 & 4 \end{pmatrix} \xrightarrow{M_2(1/2)} \begin{pmatrix} 1 & -1 & 3 \\ 0 & 1 & -1 \\ 0 & -4 & 4 \end{pmatrix} \xrightarrow{A_{2,3}(4)} \begin{pmatrix} 1 & -1 & 3 \\ 0 & 1 & -1 \\ 0 & 0 & 0 \end{pmatrix} = B.$$

Since B has two independent rows, we have $\rho(B) = \rho(A) = 2$ and

$$R_A = \text{span}\{(1, -1, 3), (0, 1, -1)\}.$$

Column Space

DEFINITION

Let $\mathbf{c}_1, \mathbf{c}_2, \ldots, \mathbf{c}_n$ be the columns of A. Then the **column space** of A is

$$C_A = \text{span}\{\mathbf{c}_1, \mathbf{c}_2, \ldots, \mathbf{c}_n\}.$$

■ **THEOREM 3**

Let A be an $m \times n$ matrix. Then

$$\dim C_A = \dim R_A = \rho(A).$$

That is, the number of linearly independent columns of A is equal to the number of linearly independent rows of A.

PROOF

As before, we denote the rows of A by $\mathbf{r}_1, \mathbf{r}_2, \ldots, \mathbf{r}_m$, and let $k = \dim R_A$. Let $S = \{\mathbf{s}_1, \mathbf{s}_2, \ldots, \mathbf{s}_k\}$ be a basis for R_A. Then every row vector of A can be written as a linear combination of the vectors in S, and we have, for some constants α_{ij}:

$$\begin{aligned}
\mathbf{r}_1 &= \alpha_{11}\mathbf{s}_1 + \alpha_{12}\mathbf{s}_2 + \cdots + \alpha_{1k}\mathbf{s}_k \\
\mathbf{r}_2 &= \alpha_{21}\mathbf{s}_1 + \alpha_{22}\mathbf{s}_2 + \cdots + \alpha_{2k}\mathbf{s}_k \\
&\vdots \qquad \vdots \qquad \vdots \qquad\qquad \vdots \\
\mathbf{r}_m &= \alpha_{m1}\mathbf{s}_1 + \alpha_{m2}\mathbf{s}_2 + \cdots + \alpha_{mk}\mathbf{s}_k
\end{aligned} \tag{1}$$

Now, the jth component of $\mathbf{r}_i$ is a_{ij}. Let s_{ij} denote the jth component of $\mathbf{s}_i$. Then, if we equate the jth components of both sides of (1), we obtain

$$a_{1j} = \alpha_{11} s_{1j} + \alpha_{12} s_{2j} + \cdots + \alpha_{1k} s_{kj}$$
$$a_{2j} = \alpha_{21} s_{1j} + \alpha_{22} s_{2j} + \cdots + \alpha_{2k} s_{kj}$$
$$\vdots$$
$$a_{mj} = \alpha_{m1} s_{1j} + \alpha_{m2} s_{2j} + \cdots + \alpha_{mk} s_{kj}$$

or

$$\begin{pmatrix} a_{1j} \\ a_{2j} \\ \vdots \\ a_{mj} \end{pmatrix} = s_{1j} \begin{pmatrix} \alpha_{11} \\ \alpha_{21} \\ \vdots \\ \alpha_{m1} \end{pmatrix} + s_{2j} \begin{pmatrix} \alpha_{12} \\ \alpha_{22} \\ \vdots \\ \alpha_{m2} \end{pmatrix} + \cdots + s_{kj} \begin{pmatrix} \alpha_{1k} \\ \alpha_{2k} \\ \vdots \\ \alpha_{mk} \end{pmatrix}. \tag{2}$$

Let α_i denote the vector $\begin{pmatrix} \alpha_{1i} \\ \alpha_{2i} \\ \vdots \\ \alpha_{mi} \end{pmatrix}$. Then since the left-hand side of (2) is the

jth column of A, we see that we can write every column of A as a linear combination of $\alpha_1, \alpha_2, \ldots, \alpha_k$, which means that $\alpha_1, \alpha_2, \ldots, \alpha_k$ span C_A and

$$\dim C_A \le k = \dim R_A. \tag{3}$$

But equation (3) holds for any matrix A. In particular it holds for A^t. But $C_{A^t} = R_A$ and $R_{A^t} = C_A$. Thus, since, from (3), $\dim C_{A^t} \le \dim R_{A^t}$, we have

$$\dim R_A \le \dim C_A. \tag{4}$$

Combining (3) and (4) completes the proof. ∎

EXAMPLE 5

From Example 4 and Theorem 2 we know that the matrix $A = \begin{pmatrix} 1 & -1 & 3 \\ 2 & 0 & 4 \\ -1 & -3 & 1 \end{pmatrix}$ has two

linearly independent columns. You should verify that

$$\begin{pmatrix} 3 \\ 4 \\ 1 \end{pmatrix} = 2 \begin{pmatrix} 1 \\ 2 \\ -1 \end{pmatrix} - \begin{pmatrix} -1 \\ 0 \\ -3 \end{pmatrix}.$$

We next show how the notion of rank can be used to solve linear systems of equations. Again we consider the system of m equations in n unknowns

$$a_{11}x_1 + a_{12}x_2 + \cdots + a_{1n}x_n = b_1$$
$$a_{21}x_1 + a_{22}x_2 + \cdots + a_{2n}x_n = b_2$$
$$\vdots \qquad \vdots \qquad \qquad \vdots \qquad \vdots \tag{5}$$
$$a_{m1}x_1 + a_{m2}x_2 + \cdots + a_{mn}x_n = b_m$$

which we write as $A\mathbf{x} = \mathbf{b}$. We use the symbol $(A, \mathbf{b})$ to denote the $m \times (n + 1)$ augmented matrix obtained (as in Section 7.3) by adjoining the vector $\mathbf{b}$ to A.

■ THEOREM 4

The system $A\mathbf{x} = \mathbf{b}$ has at least one solution if and only if $\mathbf{b} \in C_A$. This will occur if and only if A and the augmented matrix $(A, \mathbf{b})$ have the same rank.

PROOF

If $\mathbf{c}_1, \mathbf{c}_2, \ldots, \mathbf{c}_n$ are the columns of A, then we can write system (5) as

$$x_1\mathbf{c}_1 + x_2\mathbf{c}_2 + \cdots + x_n\mathbf{c}_n = \mathbf{b}. \tag{6}$$

System (6) will have a solution if and only if $\mathbf{b}$ can be written as a linear combination of the columns of A. That is, to have a solution we must have $\mathbf{b} \in C_A$. If $\mathbf{b} \in C_A$, then $(A, \mathbf{b})$ has the same number of linearly independent columns as A so that A and $(A, \mathbf{b})$ have the same rank. If $\mathbf{b} \notin C_A$, then $\rho(A, \mathbf{b}) = \rho(A) + 1$ and the system has no solutions. This completes the proof. ■

EXAMPLE 6

Determine whether the system

$$2x_1 + 4x_2 + 6x_3 = 18$$

$$4x_1 + 5x_2 + 6x_3 = 24$$

$$2x_1 + 7x_2 + 12x_3 = 40$$

has solutions.

SOLUTION

Let $A = \begin{pmatrix} 2 & 4 & 6 \\ 4 & 5 & 6 \\ 2 & 7 & 12 \end{pmatrix}$. Then we row-reduce to obtain, successively,

$$\xrightarrow{M_1(1/2)} \begin{pmatrix} 1 & 2 & 3 \\ 4 & 5 & 6 \\ 2 & 7 & 12 \end{pmatrix} \xrightarrow[A_{1,3}(-2)]{A_{1,2}(-4)} \begin{pmatrix} 1 & 2 & 3 \\ 0 & -3 & -6 \\ 0 & 3 & 6 \end{pmatrix}$$

$$\xrightarrow{M_2(-1/3)} \begin{pmatrix} 1 & 2 & 3 \\ 0 & 1 & 2 \\ 0 & 3 & 6 \end{pmatrix} \xrightarrow[A_{2,3}(-3)]{A_{2,1}(-2)} \begin{pmatrix} 1 & 0 & -1 \\ 0 & 1 & 2 \\ 0 & 0 & 0 \end{pmatrix}.$$

Thus $\rho(A) = 2$. Similarly, we row-reduce $(A, \mathbf{b})$ to obtain

$$\begin{pmatrix} 2 & 4 & 6 & | & 18 \\ 4 & 5 & 6 & | & 24 \\ 2 & 7 & 12 & | & 40 \end{pmatrix} \xrightarrow{M_1(1/2)} \begin{pmatrix} 1 & 2 & 3 & | & 9 \\ 4 & 5 & 6 & | & 24 \\ 2 & 7 & 12 & | & 40 \end{pmatrix}$$

$$\xrightarrow[A_{1,3}(-2)]{A_{1,2}(-4)} \begin{pmatrix} 1 & 2 & 3 & | & 9 \\ 0 & -3 & -6 & | & -12 \\ 0 & 3 & 6 & | & 22 \end{pmatrix}$$

$$\xrightarrow{M_2(-1/3)} \begin{pmatrix} 1 & 2 & 3 & | & 9 \\ 0 & 1 & 2 & | & 4 \\ 0 & 3 & 6 & | & 22 \end{pmatrix} \xrightarrow[A_{2,3}(-3)]{A_{2,1}(-2)} \begin{pmatrix} 1 & 0 & -1 & | & 1 \\ 0 & 1 & 2 & | & 4 \\ 0 & 0 & 0 & | & 10 \end{pmatrix}.$$

It is easy to see that the last three columns of the last matrix are linearly independent. Thus $\rho(A, \mathbf{b}) = 3$ and there are no solutions to the system.

EXAMPLE 7

Determine whether the system

$$x_1 - x_2 + 2x_3 = 4$$

$$2x_1 + x_2 - 3x_3 = -2$$

$$4x_1 - x_2 + x_3 = 6$$

has solutions.

SOLUTION

Let $A = \begin{pmatrix} 1 & -1 & 2 \\ 2 & 1 & -3 \\ 4 & -1 & 1 \end{pmatrix}$. Then $\det A = 0$, so $\rho(A) < 3$. Since the first column is not a multiple of the second, we see that the first two columns are linearly independent; hence $\rho(A) = 2$. To compute $\rho(A, \mathbf{b})$, we row-reduce:

$$\begin{pmatrix} 1 & -1 & 2 & \big| & 4 \\ 2 & 1 & -3 & \big| & -2 \\ 4 & -1 & 1 & \big| & 6 \end{pmatrix} \xrightarrow{\substack{A_{1,2}(-2) \\ A_{1,3}(-4)}} \begin{pmatrix} 1 & -1 & 2 & \big| & 4 \\ 0 & 3 & -7 & \big| & -10 \\ 0 & 3 & -7 & \big| & -10 \end{pmatrix} \xrightarrow{A_{2,3}(-1)} \begin{pmatrix} 1 & -1 & 2 & \big| & 4 \\ 0 & 3 & -7 & \big| & -10 \\ 0 & 0 & 0 & \big| & 0 \end{pmatrix}.$$

We see that $\rho(A, \mathbf{b}) = 2$ and there are an infinite number of solutions to the system. (Also note that if there were a unique solution, we would have $\det A \neq 0$.)

In Section 7.4 we discussed homogeneous systems of equations. We saw that a homogeneous system has either one solution, the trivial solution, or an infinite number of solutions. On p. 382 we remarked that every solution to system (5) can be written as $\mathbf{x} + \mathbf{h}$ where $\mathbf{x}$ is a particular solution to (5) and $\mathbf{h}$ is some solution of the associated homogeneous system. We now analyze the homogeneous system more closely.

DEFINITION
Let A be an $m \times n$ matrix. We define

$$N_A = \{\mathbf{x} \in \mathbb{R}^n : A\mathbf{x} = \mathbf{0}\}. \tag{7}$$

Kernel or Nullspace

N_A is called the **kernel** or **nullspace** of the matrix A.

■ **THEOREM 5**
N_A is a subspace of $\mathbb{R}^n$.

PROOF
Let $\mathbf{x}_1$ and $\mathbf{x}_2$ be in N_A. Then

$$A(\alpha\mathbf{x}_1 + \beta\mathbf{x}_2) = \alpha A\mathbf{x}_1 + \beta A\mathbf{x}_2 = \alpha \cdot \mathbf{0} + \beta \cdot \mathbf{0} = \mathbf{0}.$$

Thus $\alpha\mathbf{x}_1 + \beta\mathbf{x}_2 \in N_A$ and, by Theorem 6.6.1, N_A is a subspace of $\mathbb{R}^n$. ■

Nullity

DEFINITION

The **nullity** of a matrix, denoted by $\nu(A)$, is the dimension of the kernel of A. That is,

$$\nu(A) = \dim N_A.$$

Consider the homogeneous system

$$
\begin{aligned}
a_{11}x_1 + a_{12}x_2 + \cdots + a_{1n}x_n &= 0 \\
a_{21}x_1 + a_{22}x_2 + \cdots + a_{2n}x_n &= 0 \\
&\;\;\vdots \\
a_{m1}x_1 + a_{m2}x_2 + \cdots + a_{mn}x_n &= 0
\end{aligned}
\tag{8}
$$

We write this system as

$$A\mathbf{x} = \mathbf{0}. \tag{9}$$

Evidently, if $\mathbf{x}$ is a solution to (9), then $\mathbf{x} \in N_A$. Also, N_A has a basis consisting of k vectors in $\mathbb{R}^n$ where $k = \nu(A)$. Let $\mathbf{v}_1, \mathbf{v}_2, \ldots, \mathbf{v}_k$ be a basis for N_A. Then, if $\mathbf{x}$ solves (9), there are unique scalars $c_1, c_2, \ldots, c_k$ such that

$$\mathbf{x} = c_1\mathbf{v}_1 + c_2\mathbf{v}_2 + \cdots + c_k\mathbf{v}_k. \tag{10}$$

Moreover, if $c_1, c_2, \ldots, c_k$ are k numbers, then (10) is a solution to (9). We conclude that

the number of arbitrary constants in the solution of $A\mathbf{x} = \mathbf{0}$ is equal to $\nu(A)$. (11)

EXAMPLE 8

Consider the system $A\mathbf{x} = \mathbf{0}$ where $A = \begin{pmatrix} 2 & -1 & 3 \\ 4 & -2 & 6 \\ -6 & 3 & -9 \end{pmatrix}$. We compute N_A and $\nu(A)$.

To compute N_A, we row-reduce:

$$
\begin{pmatrix} 2 & -1 & 3 & | & 0 \\ 4 & -2 & 6 & | & 0 \\ -6 & 3 & -9 & | & 0 \end{pmatrix}
\xrightarrow{\substack{A_{1,2(-2)} \\ A_{1,3(3)}}}
\begin{pmatrix} 2 & -1 & 3 & | & 0 \\ 0 & 0 & 0 & | & 0 \\ 0 & 0 & 0 & | & 0 \end{pmatrix}.
$$

Thus $\begin{pmatrix} x \\ y \\ z \end{pmatrix} \in N_A$ if $2x - y + 3z = 0$, or $y = 2x + 3z$. Here x and z are arbitrary so, from

(11), $\nu(A) = 2$. Choose $x = 1$, $z = 0$ and $x = 0$, $z = 1$ to obtain $\begin{pmatrix} 1 \\ 2 \\ 0 \end{pmatrix}$ and $\begin{pmatrix} 0 \\ 3 \\ 1 \end{pmatrix}$. Since these are linearly independent, they form a basis for N_A. To see this more clearly, let $\begin{pmatrix} x \\ y \\ z \end{pmatrix}$ be in N_A. Then $y = 2x + 3z$ so

$$\begin{pmatrix} x \\ y \\ z \end{pmatrix} = \begin{pmatrix} x \\ 2x + 3z \\ z \end{pmatrix} = x\begin{pmatrix} 1 \\ 2 \\ 0 \end{pmatrix} + z\begin{pmatrix} 0 \\ 3 \\ 1 \end{pmatrix}.$$

The next theorem ties together the notions of rank and nullity. We omit the proof.†

■ **THEOREM 6**

Let A be an $m \times n$ matrix. Then

$$\rho(A) + \nu(A) = n.$$

■

EXAMPLE 9

In Example 4 we saw that $\rho(A) = \rho\begin{pmatrix} 1 & -1 & 3 \\ 2 & 0 & 4 \\ -1 & -3 & 1 \end{pmatrix} = 2$, so $\nu(A) = 3 - \rho(A) = 3 - 2 = 1$.

You should verify that

$$N_A = \text{span}\left\{ \begin{pmatrix} -2 \\ 1 \\ 1 \end{pmatrix} \right\}.$$

The results of this section allow us to restate our Summing-Up Theorem—last seen on p. 429.

■ **THEOREM 7: Summing-Up Theorem—View 3**

Let A be an $n \times n$ matrix. Then the following eight statements are equivalent. That is, if one is true, all are true.

    i. A is invertible.

    ii. The only solution to the homogeneous system $A\mathbf{x} = \mathbf{0}$ is the trivial solution ($\mathbf{x} = \mathbf{0}$).

    iii. The system $A\mathbf{x} = \mathbf{b}$ has a unique solution for every n-vector $\mathbf{b}$.

    iv. A is row equivalent to the $n \times n$ identity matrix I_n.

† For a proof see S. I. Grossman, *Elementary Linear Algebra,* 3rd ed., Wadsworth, Belmont, Calif., 1987, p. 211.

v. The rows (and columns) of A are linearly independent.

vi. $\det A \neq 0$.

vii. $\nu(A) = 0$.

viii. $\rho(A) = n$.

Moreover, if one of the above fails to hold, then for every vector $\mathbf{b} \in \mathbb{R}^n$, the system $A\mathbf{x} = \mathbf{b}$ has either no solution or an infinite number of solutions. It has an infinite number of solutions if and only if $\rho(A) = \rho((A, \mathbf{b}))$.

PROOF

We proved the equivalence of (i)–(vi) in Theorem 7.9.5. Statements (vi) and (viii) are equivalent because of Theorem 1. Finally, according to Theorem 6, statements (vii) and (viii) are equivalent. ∎

Problems 7.10

In Problems 1–15 find the rank and nullity of the given matrix.

1. $\begin{pmatrix} 1 & 2 \\ 3 & 4 \end{pmatrix}$

2. $\begin{pmatrix} 1 & -1 & 2 \\ 3 & 1 & 0 \end{pmatrix}$

3. $\begin{pmatrix} -1 & 3 & 2 \\ 2 & -6 & -4 \end{pmatrix}$

4. $\begin{pmatrix} 1 & -1 & 2 \\ 3 & 1 & 4 \\ -1 & 0 & 4 \end{pmatrix}$

5. $\begin{pmatrix} 1 & -1 & 2 \\ 3 & 1 & 4 \\ 5 & -1 & 8 \end{pmatrix}$

6. $\begin{pmatrix} -1 & 2 & 1 \\ 2 & -4 & -2 \\ -3 & 6 & 3 \end{pmatrix}$

7. $\begin{pmatrix} 1 & -1 & 2 & 3 \\ 0 & 1 & 4 & 3 \\ 1 & 0 & 6 & 6 \end{pmatrix}$

8. $\begin{pmatrix} 1 & -1 & 2 & 3 \\ 0 & 1 & 4 & 3 \\ 1 & 0 & 6 & 5 \end{pmatrix}$

9. $\begin{pmatrix} 2 & 3 \\ -1 & 1 \\ 4 & 7 \end{pmatrix}$

10. $\begin{pmatrix} 1 & -1 & 2 & 3 \\ 0 & 1 & 0 & 1 \\ 1 & 0 & 1 & 0 \\ 0 & 0 & 0 & 1 \end{pmatrix}$

11. $\begin{pmatrix} 1 & -1 & 2 & 1 \\ -1 & 0 & 1 & 2 \\ 1 & -2 & 5 & 4 \\ 2 & -1 & 1 & -1 \end{pmatrix}$

12. $\begin{pmatrix} 1 & -1 & 2 & 3 \\ -2 & 2 & -4 & -6 \\ 2 & -2 & 4 & 6 \\ 3 & -3 & 6 & 9 \end{pmatrix}$

13. $\begin{pmatrix} -1 & -1 & 0 & 0 \\ 0 & 0 & 2 & 3 \\ 4 & 0 & -2 & 1 \\ 3 & -1 & 0 & 4 \end{pmatrix}$

14. $\begin{pmatrix} 3 & 0 & 0 \\ 0 & 0 & 0 \\ 0 & 0 & 6 \end{pmatrix}$

15. $\begin{pmatrix} 1 & 2 & 3 \\ 0 & 0 & 4 \\ 0 & 0 & 6 \end{pmatrix}$

In Problems 16–22 find a basis for the row space and kernel of the given matrix.

16. The matrix of Problem 2

17. The matrix of Problem 5

18. The matrix of Problem 6

19. The matrix of Problem 8

20. The matrix of Problem 11

21. The matrix of Problem 12

22. The matrix of Problem 13

In Problems 23–26 use Theorem 4 to determine whether the given system has any solutions.

23.
$$x_1 + x_2 - x_3 = 7$$
$$4x_1 - x_2 + 5x_3 = 4$$
$$6x_1 + x_2 + 3x_3 = 20$$

24.
$$x_1 + x_2 - x_3 = 7$$
$$4x_1 - x_2 + 5x_3 = 4$$
$$6x_1 + x_2 + 3x_3 = 18$$

25.
$$x_1 - 2x_2 + x_3 + x_4 = 2$$
$$3x_1 \quad + 2x_3 - 2x_4 = -8$$
$$4x_2 - x_3 - x_4 = 1$$
$$5x_1 \quad + 3x_3 - x_4 = -3$$

26.
$$x_1 - 2x_2 + x_3 + x_4 = 2$$
$$3x_1 \quad + 2x_3 - 2x_4 = -8$$
$$4x_2 - x_3 - x_4 = 1$$
$$5x_1 \quad + 3x_3 - x_4 = 0$$

27. Show that the rank of a diagonal matrix is equal to the number of nonzero components on the diagonal.

28. Let A be an upper triangular $n \times n$ matrix with zeros on the diagonal. Show that $\rho(A) < n$.

29. Show that for any matrix A, $\rho(A) = \rho(A^t)$.

30. Show that if A is an $m \times n$ matrix and $m < n$, then (**a**) $\rho(A) \leq m$ and (**b**) $\nu(A) \geq n - m$.

31. Let A be an $m \times n$ matrix and let B and C be invertible $m \times m$ and $n \times n$ matrices, respectively. Prove that $\rho(A) = \rho(BA) = \rho(AC)$. That is, multiplying a matrix by an invertible matrix does not change its rank.

32. Let A and B be $m \times n$ and $n \times p$ matrices, respectively. Show that $\rho(AB) \leq \min(\rho(A), \rho(B))$.

33. Let A be a 5×7 matrix with rank 5. Show that the linear system $A\mathbf{x} = \mathbf{b}$ has at least one solution for every 5-vector $\mathbf{b}$.

* 34. Let A and B be $m \times n$ matrices. Show that if $\rho(A) = \rho(B)$, then there exist invertible matrices C and D such that $B = CAD$.

35. If $B = CAD$, where C and D are invertible, prove that $\rho(A) = \rho(B)$.

36. Suppose that any k rows of A are linearly independent while any $k + 1$ rows of A are linearly dependent. Show that $\rho(A) = k$.

37. If A is an $n \times n$ matrix, show that $\rho(A) < n$ if and only if there is a vector $\mathbf{x} \in \mathbb{R}^n$ such that $\mathbf{x} \neq \mathbf{0}$ and $A\mathbf{x} = \mathbf{0}$.

38. Let A be an $m \times n$ matrix. Suppose that for every $\mathbf{y} \in \mathbb{R}^m$ there is an $\mathbf{x} \in \mathbb{R}^n$ such that $A\mathbf{x} = \mathbf{y}$. Show that $\rho(A) = m$.

7.11 Linear Transformations

In this section we discuss a special class of functions, called *linear transformations,* which occur with great frequency in linear algebra and other branches of mathematics. They are also important in a wide variety of applications. Before defining a linear transformation, let us study two simple examples to see what can happen.

EXAMPLE 1

In $\mathbb{R}^2$, define a function T by the formula $T\begin{pmatrix} x \\ y \end{pmatrix} = \begin{pmatrix} x \\ -y \end{pmatrix}$. Geometrically, T takes a vector in $\mathbb{R}^2$ and reflects it about the x-axis. This is illustrated in Fig. 1. Once we have given our basic definition, we shall see that T is a linear transformation from $\mathbb{R}^2$ into $\mathbb{R}^2$.

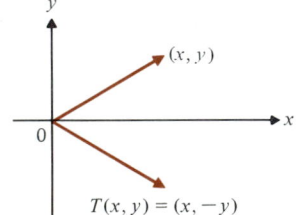

FIGURE 1

$T(x, y) = (x, -y)$

EXAMPLE 2

Suppose the vector $\mathbf{v} = \begin{pmatrix} x \\ y \end{pmatrix}$ in the xy-plane is rotated through an angle of θ (measured in degrees or radians) in the counterclockwise direction. Call the new rotated vector $\mathbf{v}' = \begin{pmatrix} x' \\ y' \end{pmatrix}$. Then, as in Fig. 2, if r denotes the length of $\mathbf{v}$ (which is unchanged by rotation),

$$x = r\cos\alpha, \qquad y = r\sin\alpha,$$
$$x' = r\cos(\theta + \alpha), \qquad y' = r\sin(\theta + \alpha).\dagger$$

But $r\cos(\theta + \alpha) = r\cos\theta\cos\alpha - r\sin\theta\sin\alpha$, so that

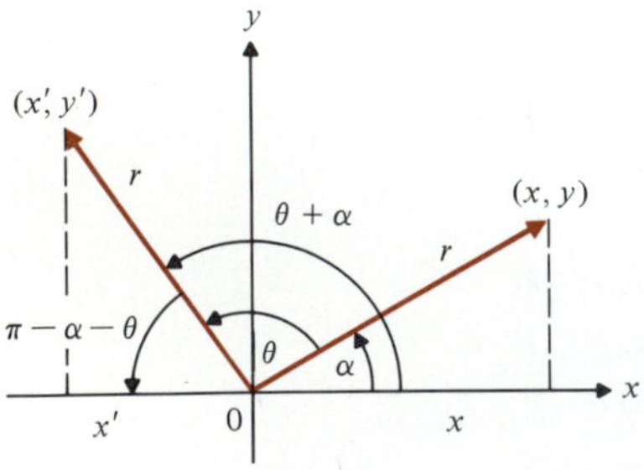

$$x' = x\cos\theta - y\sin\theta. \qquad (1)$$

FIGURE 2

† These follow from the standard definitions of $\cos\theta$ and $\sin\theta$ as the x and y coordinates of a point on the unit circle. If (x, y) is a point on the circle centered at the origin of radius r, then $x = r\cos\phi$ and $y = r\sin\phi$, where ϕ is the angle the vector (x, y) makes with the positive x-axis.

Similarly, $r \sin(\theta + \alpha) = r \sin \theta \cos \alpha + r \cos \theta \sin \alpha$ or

$$y' = x \sin \theta + y \cos \theta. \tag{2}$$

Let

$$A_\theta = \begin{pmatrix} \cos \theta & -\sin \theta \\ \sin \theta & \cos \theta \end{pmatrix}. \tag{3}$$

Then, from (1) and (2), we see that $A_\theta \begin{pmatrix} x \\ y \end{pmatrix} = \begin{pmatrix} x' \\ y' \end{pmatrix}$. The linear transformation $T: \mathbb{R}^2 \to \mathbb{R}^2$ defined by $Tv = A_\theta v$, where A_θ is given by (3), is called a **rotation transformation.** As we shall see, this is also a linear transformation.

We now give the basic definition.

DEFINITION

Linear Transformation

Let V and W be vector spaces. A **linear transformation** T from V into W is a function that assigns to each vector $\mathbf{v} \in V$ a unique vector $T\mathbf{v} \in W$ and that satisfies, for each $\mathbf{u}$ and $\mathbf{v}$ in V and each scalar α,

$$T(\mathbf{u} + \mathbf{v}) = T\mathbf{u} + T\mathbf{v} \tag{4}$$

$$T(\alpha\mathbf{v}) = \alpha T\mathbf{v} \tag{5}$$

Notation. We write $T: V \to W$ to indicate that T takes V into W.

Terminology. Linear transformations are often called **linear operators.** Functions that satisfy (4) and (5) are called **linear functions.**

To see how to verify that the transformations in Examples 1 and 2 are linear, select any two vectors

$$\mathbf{u} = \begin{pmatrix} u_1 \\ u_2 \end{pmatrix} \quad \text{and} \quad \mathbf{v} = \begin{pmatrix} v_1 \\ v_2 \end{pmatrix}$$

in $\mathbb{R}^2$. Then, for Example 1, $T\begin{pmatrix} x \\ y \end{pmatrix} = \begin{pmatrix} x \\ -y \end{pmatrix}$ so

$$T(\mathbf{u} + \mathbf{v}) = T\begin{pmatrix} u_1 + v_1 \\ u_2 + v_2 \end{pmatrix} = \begin{pmatrix} u_1 + v_1 \\ -u_2 - v_2 \end{pmatrix} = \begin{pmatrix} u_1 \\ -u_2 \end{pmatrix} + \begin{pmatrix} v_1 \\ -v_2 \end{pmatrix} = T\mathbf{u} + T\mathbf{v}$$

and

$$T(\alpha\mathbf{u}) = T\begin{pmatrix} \alpha u_1 \\ \alpha u_2 \end{pmatrix} = \begin{pmatrix} \alpha u_1 \\ -\alpha u_2 \end{pmatrix} = \alpha\begin{pmatrix} u_1 \\ -u_2 \end{pmatrix} = \alpha T\mathbf{u}.$$

Similarly,

$$A_\theta(\mathbf{u} + \mathbf{v}) = \begin{pmatrix} \cos\theta & -\sin\theta \\ \sin\theta & \cos\theta \end{pmatrix}\begin{pmatrix} u_1 + v_1 \\ u_2 + v_2 \end{pmatrix}$$

$$= \begin{pmatrix} \cos\theta & -\sin\theta \\ \sin\theta & \cos\theta \end{pmatrix}\begin{pmatrix} u_1 \\ u_2 \end{pmatrix} + \begin{pmatrix} \cos\theta & -\sin\theta \\ \sin\theta & \cos\theta \end{pmatrix}\begin{pmatrix} v_1 \\ v_2 \end{pmatrix}$$

$$= A_\theta\mathbf{u} + A_\theta\mathbf{v}$$

and

$$A_\theta(\alpha\mathbf{u}) = \begin{pmatrix} \cos\theta & -\sin\theta \\ \sin\theta & \cos\theta \end{pmatrix}\begin{pmatrix} \alpha u_1 \\ \alpha u_2 \end{pmatrix} = \alpha\begin{pmatrix} \cos\theta & -\sin\theta \\ \sin\theta & \cos\theta \end{pmatrix}\begin{pmatrix} u_1 \\ u_2 \end{pmatrix}$$

$$= \alpha A_\theta\mathbf{u}.$$

EXAMPLE 3: Matrix Multiplication as a Linear Transformation

Let A be an $m \times n$ matrix and define $T: \mathbb{R}^n \to \mathbb{R}^m$ by $T\mathbf{x} = A\mathbf{x}$. Since $A(\mathbf{x} + \mathbf{y}) = A\mathbf{x} + A\mathbf{y}$ and $A(\alpha\mathbf{x}) = \alpha A\mathbf{x}$ if $\mathbf{x}$ and $\mathbf{y}$ are in $\mathbb{R}^n$, we see that T is a linear transformation. Thus: *Every $m \times n$ matrix A gives rise to a linear transformation from $\mathbb{R}^n$ into $\mathbb{R}^m$.* We shall see that a certain converse is true: *Every linear transformation between finite-dimensional vector spaces can be represented by a matrix.*

EXAMPLE 4: Zero Transformation

Let V and W be vector spaces and define $T: V \to W$ by $T\mathbf{v} = \mathbf{0}$. Then $T(\mathbf{v}_1 + \mathbf{v}_2) = \mathbf{0} = \mathbf{0} + \mathbf{0} = T\mathbf{v}_1 + T\mathbf{v}_2$ and $T(\alpha\mathbf{v}) = \mathbf{0} = \alpha\mathbf{0} = \alpha T\mathbf{v}$. Here T is called the **zero transformation**.

EXAMPLE 5: Identity Transformation

Let V be a vector space and define $I: V \to V$ by $I\mathbf{v} = \mathbf{v}$. Here I is obviously a linear transformation. It is called the **identity transformation** or **identity operator**.

EXAMPLE 6: Orthogonal Projection Transformation

Let H be a proper subspace of $\mathbb{R}^n$. We define the **orthogonal projection transformation** $P: V \to H$ by

$$P\mathbf{v} = \text{Proj}_H\, \mathbf{v}.$$

Let $\{\mathbf{u}_1, \mathbf{u}_2, \ldots, \mathbf{u}_k\}$ be an orthonormal basis for H. Then from the proof of the Gram-Schmidt process on p. 343, we see that

$$P\mathbf{v} = (\mathbf{v} \cdot \mathbf{u}_1)\mathbf{u}_1 + (\mathbf{v} \cdot \mathbf{u}_2)\mathbf{u}_2 + \cdots + (\mathbf{v} \cdot \mathbf{u}_k)\mathbf{u}_k.$$

Since $(\mathbf{v}_1 + \mathbf{v}_2) \cdot \mathbf{u} = \mathbf{v}_1 \cdot \mathbf{u} + \mathbf{v}_2 \cdot \mathbf{u}$ and $(\alpha\mathbf{v}) \cdot \mathbf{u} = \alpha(\mathbf{v} \cdot \mathbf{u})$, we see that P is a linear transformation.

EXAMPLE 7

Let $T: \mathbb{R}^3 \to \mathbb{R}^3$ be defined by $T\begin{pmatrix} x \\ y \\ z \end{pmatrix} = \begin{pmatrix} x \\ y \\ 0 \end{pmatrix}$. Then T is the projection operator taking

a vector in space and projecting it into the xy-plane. Similarly, $T\begin{pmatrix} x \\ y \\ z \end{pmatrix} = \begin{pmatrix} x \\ 0 \\ z \end{pmatrix}$ projects a

vector in space into the xz-plane.

EXAMPLE 8

Let $J: C[0, 1] \to \mathbb{R}$ be defined by $Jf = \int_0^1 f(x)\, dx$. Since $\int_0^1 [f(x) + g(x)]\, dx = \int_0^1 f(x)\, dx + \int_0^1 g(x)\, dx$ and $\int_0^1 \alpha f(x)\, dx = \alpha \int_0^1 f(x)\, dx$ if f and g are continuous, we see that J is linear. For example, $J(x^3) = \frac{1}{4}$. J is called an **integral operator.**

EXAMPLE 9

Let $D: C^{(1)}[0, 1] \to C[0, 1]$ be defined by $Df = f'$. Since $(f + g)' = f' + g'$ and $(\alpha f)' = \alpha f'$ if f and g are differentiable, we see that D is linear. D is called a **differential operator.**

Warning. Not every function that looks linear actually is linear. For example, define $T: \mathbb{R} \to \mathbb{R}$ by $Tx = 2x + 3$. Then $\{(x, Tx): x \in \mathbb{R}\}$ is a straight line in the xy-plane. But T is not linear since $T(x + y) = 2(x + y) + 3 = 2x + 2y + 3$ and $Tx + Ty = (2x + 3) + (2y + 3) = 2x + 2y + 6$. The only linear functions from $\mathbb{R}$ to $\mathbb{R}$ are functions of the form $f(x) = mx$ for some real number m. Thus, among all straight line functions, the only ones that are linear are the ones that pass through the origin.

■ **THEOREM 1**

Let $T: V \to W$ be a linear transformation. Then for all vectors $\mathbf{u}, \mathbf{v}, \mathbf{v}_1, \mathbf{v}_2, \ldots,$ $\mathbf{v}_n$ in V and all scalars $\alpha_1, \alpha_2, \ldots, \alpha_n$:

    i. $T(\mathbf{0}) = \mathbf{0}$
    ii. $T(\mathbf{u} - \mathbf{v}) = T\mathbf{u} - T\mathbf{v}$
    iii. $T(\alpha_1\mathbf{v}_1 + \alpha_2\mathbf{v}_2 + \cdots + \alpha_n\mathbf{v}_n) = \alpha_1 T\mathbf{v}_1 + \alpha_2 T\mathbf{v}_2 + \cdots + \alpha_n T\mathbf{v}_n$

Note. In part (*i*) the $\mathbf{0}$ on the left is the zero vector in V while the $\mathbf{0}$ on the right is the zero vector in W.

PROOF
    i. $T(\mathbf{0}) = T(\mathbf{0} + \mathbf{0}) = T(\mathbf{0}) + T(\mathbf{0})$. Thus $\mathbf{0} = T(\mathbf{0}) - T(\mathbf{0}) = T(\mathbf{0}) + T(\mathbf{0}) - T(\mathbf{0}) = T(\mathbf{0})$.
    ii. $T(\mathbf{u} - \mathbf{v}) = T[\mathbf{u} + (-1)\mathbf{v}] = T\mathbf{u} + T[(-1)\mathbf{v}] = T\mathbf{u} + (-1)T\mathbf{v} = T\mathbf{u} - T\mathbf{v}$.

iii. We prove this part by induction (see Appendix 3). For $n = 2$, we get $T(\alpha_1\mathbf{v}_1 + \alpha_2\mathbf{v}_2) = T(\alpha_1\mathbf{v}_1) + T(\alpha_2\mathbf{v}_2) = \alpha_1 T\mathbf{v}_1 + \alpha_2 T\mathbf{v}_2$. Thus the equation holds for $n = 2$. We assume that it holds for $n = k$ and prove it for $n = k + 1$: $T(\alpha_1\mathbf{v}_1 + \alpha_2\mathbf{v}_2 + \cdots + \alpha_k\mathbf{v}_k + \alpha_{k+1}\mathbf{v}_{k+1}) = T(\alpha_1\mathbf{v}_1 + \alpha_2\mathbf{v}_2 + \cdots + \alpha_k\mathbf{v}_k) + T(\alpha_{k+1}\mathbf{v}_{k+1})$, and using the induction assumption for $n = k$, this is equal to $(\alpha_1 T\mathbf{v}_1 + \alpha_2 T\mathbf{v}_2 + \alpha_k T\mathbf{v}_k) + \alpha_{k+1} T\mathbf{v}_{k+1}$, which is what we wanted to show. This completes the proof. ■

Remark. Note that part (*ii*) of Theorem 1 is a special case of part (*iii*).

An important fact about linear transformations is that they are completely determined by what they do to basis vectors.

■ **THEOREM 2**

Let V be a finite-dimensional vector space with basis $B = \{\mathbf{v}_1, \mathbf{v}_2, \ldots, \mathbf{v}_n\}$. Let $\mathbf{w}_1, \mathbf{w}_2, \ldots, \mathbf{w}_n$ be n vectors in W. Suppose that T_1 and T_2 are two linear transformations from V to W such that $T_1\mathbf{v}_i = T_2\mathbf{v}_i = \mathbf{w}_i$ for $i = 1, 2, \ldots, n$. Then for any vector $\mathbf{v} \in V$, $T_1\mathbf{v} = T_2\mathbf{v}$. That is, $T_1 = T_2$.

PROOF

Since B is a basis for V, there exists a unique set of scalars $\alpha_1, \alpha_2, \ldots, \alpha_n$ such that $\mathbf{v} = \alpha_1\mathbf{v}_1 + \alpha_2\mathbf{v}_2 + \cdots + \alpha_n\mathbf{v}_n$. Then, from part (*iii*) of Theorem 1,

$$T_1\mathbf{v} = T_1(\alpha_1\mathbf{v}_1 + \alpha_2\mathbf{v}_2 + \cdots + \alpha_n\mathbf{v}_n) = \alpha_1 T_1\mathbf{v}_1 + \alpha_2 T_1\mathbf{v}_2 + \cdots + \alpha_n T_1\mathbf{v}_n$$

$$= \alpha_1\mathbf{w}_1 + \alpha_2\mathbf{w}_2 + \cdots + \alpha_n\mathbf{w}_n.$$

Similarly,

$$T_2\mathbf{v} = T_2(\alpha_1\mathbf{v}_1 + \alpha_2\mathbf{v}_2 + \cdots + \alpha_n\mathbf{v}_n) = \alpha_1 T_2\mathbf{v}_1 + \alpha_2 T_2\mathbf{v}_2 + \cdots + \alpha_n T_2\mathbf{v}_n$$

$$= \alpha_1\mathbf{w}_1 + \alpha_2\mathbf{w}_2 + \cdots + \alpha_n\mathbf{w}_n.$$

Thus $T_1\mathbf{v} = T_2\mathbf{v}$. ■

Theorem 2 tells us that if $T: V \to W$ is finite-dimensional, then we need to know only what T does to basis vectors in V. This determines T completely. To see this let $\mathbf{v}_1, \mathbf{v}_2, \ldots, \mathbf{v}_n$ be a basis in V and let $\mathbf{v}$ be another vector in V. Then $\mathbf{v} = \alpha_1\mathbf{v}_1 + \alpha_2\mathbf{v}_2 + \cdots + \alpha_n\mathbf{v}_n$, so

$$T\mathbf{v} = \alpha_1 T\mathbf{v}_1 + \alpha_2 T\mathbf{v}_2 + \cdots + \alpha_n T\mathbf{v}_n.$$

Thus we can compute $T\mathbf{v}$ for any vector $\mathbf{v} \in V$ if we know $T\mathbf{v}_1, T\mathbf{v}_2, \ldots, T\mathbf{v}_n$.

Another question arises: If $\mathbf{w}_1, \mathbf{w}_2, \ldots, \mathbf{w}_n$ are n vectors in W, does there exist a linear transformation T such that $T\mathbf{v}_i = \mathbf{w}_i$ for $i = 1, 2, \ldots, n$? The answer is yes, as the next theorem shows.

■ **THEOREM 3**

Let V be a finite-dimensional vector space with basis $B = \{\mathbf{v}_1, \mathbf{v}_2, \ldots, \mathbf{v}_n\}$. Let W be a vector space containing the n vectors $\mathbf{w}_1, \mathbf{w}_2, \ldots, \mathbf{w}_n$. Then there exists a unique linear transformation $T: V \to W$ such that $T\mathbf{v}_i = \mathbf{w}_i$ for $i = 1, 2, \ldots, n$.

PROOF

Define a function T as follows:

   i. $T\mathbf{v}_i = \mathbf{w}_i$.

   ii. If $\mathbf{v} = \alpha_1\mathbf{v}_1 + \alpha_2\mathbf{v}_2 + \cdots + \alpha_n\mathbf{v}_n$, then

$$T\mathbf{v} = \alpha_1\mathbf{w}_1 + \alpha_2\mathbf{w}_2 + \cdots + \alpha_n\mathbf{w}_n. \tag{6}$$

Because B is a basis for V, T is defined for every $\mathbf{v} \in V$; and since W is a vector space, $T\mathbf{v} \in W$. Thus it only remains to show that T is linear. But this follows directly from equation (6). For if $\mathbf{u} = \alpha_1\mathbf{v}_1 + \alpha_2\mathbf{v}_2 + \cdots + \alpha_n\mathbf{v}_n$ and $\mathbf{v} = \beta_1\mathbf{v}_1 + \beta_2\mathbf{v}_2 + \cdots + \beta_n\mathbf{v}_n$, then

$$T(\mathbf{u} + \mathbf{v}) = T[(\alpha_1 + \beta_1)\mathbf{v}_1 + (\alpha_2 + \beta_2)\mathbf{v}_2 + \cdots + (\alpha_n + \beta_n)\mathbf{v}_n]$$

$$= (\alpha_1 + \beta_1)\mathbf{w}_1 + (\alpha_2 + \beta_2)\mathbf{w}_2 + \cdots + (\alpha_n + \beta_n)\mathbf{w}_n = (\alpha_1\mathbf{w}_1 + \alpha_2\mathbf{w}_2$$

$$+ \cdots + \alpha_n\mathbf{w}_n) + (\beta_1\mathbf{w}_1 + \beta_2\mathbf{w}_2 + \cdots + \beta_n\mathbf{w}_n) = T\mathbf{u} + T\mathbf{v}.$$

Similarly $T(\alpha\mathbf{v}) = \alpha T\mathbf{v}$, so T is linear. The uniqueness of T follows from Theorem 2 and the theorem is proved. ■

Remark. In Theorems 2 and 3 the vectors $\mathbf{w}_1, \mathbf{w}_2, \ldots, \mathbf{w}_n$ need not be distinct or even independent. Moreover, we emphasize that the theorems are true if V is any finite-dimensional vector space, not just $\mathbb{R}^n$. Note also that W does not have to be finite-dimensional.

EXAMPLE 10

Find a linear transformation from $\mathbb{R}^2$ into the plane

$$W = \left\{ \begin{pmatrix} x \\ y \\ z \end{pmatrix} : 2x - y + 3z = 0 \right\}.$$

SOLUTION

From Example 6.7.17 we know that W is a two-dimensional subspace of $\mathbb{R}^3$. W has basis vectors $\mathbf{w}_1 = \begin{pmatrix} 1 \\ 2 \\ 0 \end{pmatrix}$ and $\mathbf{w}_2 = \begin{pmatrix} 0 \\ 3 \\ 1 \end{pmatrix}$. Using the standard basis in $\mathbb{R}^2$, $\mathbf{v}_1 = \begin{pmatrix} 1 \\ 0 \end{pmatrix}$ and $\mathbf{v}_2 = \begin{pmatrix} 0 \\ 1 \end{pmatrix}$, we define the linear transformation T by $T\begin{pmatrix} 1 \\ 0 \end{pmatrix} = \begin{pmatrix} 1 \\ 2 \\ 0 \end{pmatrix}$ and $T\begin{pmatrix} 0 \\ 1 \end{pmatrix} = \begin{pmatrix} 0 \\ 3 \\ 1 \end{pmatrix}$. Then, as

the discussion following Theorem 2 shows, T is completely determined. For example,

$$T\begin{pmatrix} 5 \\ -7 \end{pmatrix} = T\left[5\begin{pmatrix} 1 \\ 0 \end{pmatrix} - 7\begin{pmatrix} 0 \\ 1 \end{pmatrix} \right] = 5T\begin{pmatrix} 1 \\ 0 \end{pmatrix} - 7T\begin{pmatrix} 0 \\ 1 \end{pmatrix} = 5\begin{pmatrix} 1 \\ 2 \\ 0 \end{pmatrix} - 7\begin{pmatrix} 0 \\ 3 \\ 1 \end{pmatrix} = \begin{pmatrix} 5 \\ -11 \\ -7 \end{pmatrix}.$$

We have seen that if A is an $m \times n$ matrix, then the transformation $T\mathbf{x} = A\mathbf{x}$, from $\mathbb{R}^n$ into $\mathbb{R}^m$, is linear. We shall now prove the converse that was promised in Example 3. Recall, from Theorem 3, that we need to know only what T does to a basis of the domain of the transformation.

■ **THEOREM 4**

Let $T: \mathbb{R}^n \rightarrow \mathbb{R}^m$ be a linear transformation. Then there exists a unique $m \times n$ matrix A_T such that

$$T\mathbf{x} = A_T\mathbf{x} \qquad \text{for every } \mathbf{x} \in \mathbb{R}^n. \tag{7}$$

PROOF

Let $\mathbf{w}_1 = T\mathbf{e}_1$, $\mathbf{w}_2 = T\mathbf{e}_2, \ldots, \mathbf{w}_n = T\mathbf{e}_n$. Let A_T be the matrix whose columns are $\mathbf{w}_1, \mathbf{w}_2, \ldots, \mathbf{w}_n$. Then $A_T\mathbf{e}_i = \mathbf{w}_i$. Now take any vector $\mathbf{x}$ in $\mathbb{R}^n$,

$$\mathbf{x} = x_1\mathbf{e}_1 + x_2\mathbf{e}_2 + \cdots + x_n\mathbf{e}_n,$$

and observe that by Theorem 1

$$T\mathbf{x} = x_1 T\mathbf{e}_1 + x_2 T\mathbf{e}_2 + \cdots + x_n T\mathbf{e}_n = x_1\mathbf{w}_1 + x_2\mathbf{w}_2 + \cdots + x_n\mathbf{w}_n.$$

But

$$A_T\mathbf{x} = x_1 A_T\mathbf{e}_1 + x_2 A_T\mathbf{e}_2 + \cdots + x_n A_T\mathbf{e}_n = x_1\mathbf{w}_1 + x_2\mathbf{w}_2 + \cdots + x_n\mathbf{w}_n$$

so that $T\mathbf{x} = A_T\mathbf{x}$. Uniqueness is guaranteed by Theorem 3. ■

Transformation Matrix

DEFINITION

The matrix A_T in Theorem 4 is called the **transformation matrix** *corresponding to T.*

EXAMPLE 11

Find the transformation matrix A_T corresponding to the projection of a vector in $\mathbb{R}^3$ onto the xy-plane.

SOLUTION

Here $T\begin{pmatrix} x \\ y \\ z \end{pmatrix} = \begin{pmatrix} x \\ y \\ 0 \end{pmatrix}$. In particular, $T\begin{pmatrix} 1 \\ 0 \\ 0 \end{pmatrix} = \begin{pmatrix} 1 \\ 0 \\ 0 \end{pmatrix}$, $T\begin{pmatrix} 0 \\ 1 \\ 0 \end{pmatrix} = \begin{pmatrix} 0 \\ 1 \\ 0 \end{pmatrix}$, and $T\begin{pmatrix} 0 \\ 0 \\ 1 \end{pmatrix} = \begin{pmatrix} 0 \\ 0 \\ 0 \end{pmatrix}$. Thus $A_T = \begin{pmatrix} 1 & 0 & 0 \\ 0 & 1 & 0 \\ 0 & 0 & 0 \end{pmatrix}$. Note that $A_T\begin{pmatrix} x \\ y \\ z \end{pmatrix} = \begin{pmatrix} 1 & 0 & 0 \\ 0 & 1 & 0 \\ 0 & 0 & 0 \end{pmatrix}\begin{pmatrix} x \\ y \\ z \end{pmatrix} = \begin{pmatrix} x \\ y \\ 0 \end{pmatrix}$.

Problems 7.11

In Problems 1–29 determine whether the given transformation from V to W is linear. If T is linear, find the matrix A_T such that $T\mathbf{x} = A_T\mathbf{x}$.

1. $T: \mathbb{R}^2 \to \mathbb{R}^2; \ T\begin{pmatrix} x \\ y \end{pmatrix} = \begin{pmatrix} x \\ 0 \end{pmatrix}$

2. $T: \mathbb{R}^2 \to \mathbb{R}^2; \ T\begin{pmatrix} x \\ y \end{pmatrix} = \begin{pmatrix} 1 \\ y \end{pmatrix}$

3. $T: \mathbb{R}^3 \to \mathbb{R}^2; \ T\begin{pmatrix} x \\ y \\ z \end{pmatrix} = \begin{pmatrix} x \\ y \end{pmatrix}$

4. $T: \mathbb{R}^3 \to \mathbb{R}^2; \ T\begin{pmatrix} x \\ y \\ z \end{pmatrix} = \begin{pmatrix} 0 \\ y \end{pmatrix}$

5. $T: \mathbb{R}^3 \to \mathbb{R}^2; \ T\begin{pmatrix} x \\ y \\ z \end{pmatrix} = \begin{pmatrix} 1 \\ z \end{pmatrix}$

6. $T: \mathbb{R}^2 \to \mathbb{R}^2; \ T\begin{pmatrix} x \\ y \end{pmatrix} = \begin{pmatrix} x^2 \\ y^2 \end{pmatrix}$

7. $T: \mathbb{R}^2 \to \mathbb{R}^2; \ T\begin{pmatrix} x \\ y \end{pmatrix} = \begin{pmatrix} y \\ x \end{pmatrix}$

8. $T: \mathbb{R}^2 \to \mathbb{R}^2; \ T\begin{pmatrix} x \\ y \end{pmatrix} = \begin{pmatrix} x+y \\ x-y \end{pmatrix}$

9. $T: \mathbb{R}^2 \to \mathbb{R}; \ T\begin{pmatrix} x \\ y \end{pmatrix} = xy$

10. $T: \mathbb{R}^n \to \mathbb{R}; \ T\begin{pmatrix} x_1 \\ x_2 \\ \vdots \\ x_n \end{pmatrix} = x_1 + x_2 + \cdots + x_n$

11. $T: \mathbb{R} \to \mathbb{R}^n; \ T(x) = \begin{pmatrix} x \\ x \\ \vdots \\ x \end{pmatrix}$

12. $T: \mathbb{R}^4 \to \mathbb{R}^2; \ T\begin{pmatrix} x \\ y \\ z \\ w \end{pmatrix} = \begin{pmatrix} x+z \\ y+w \end{pmatrix}$

13. $T: \mathbb{R}^4 \to \mathbb{R}^2; \ T\begin{pmatrix} x \\ y \\ z \\ w \end{pmatrix} = \begin{pmatrix} xz \\ yw \end{pmatrix}$

14. $T: M_{nn} \to M_{nn}; \ T(A) = AB$, where B is a fixed $n \times n$ matrix

15. $T: M_{nn} \to M_{nn}; \ T(A) = A^t A$

16. $T: M_{mn} \to M_{mp}; \ T(A) = AB$, where B is a fixed $n \times p$ matrix

17. $T: D_n \to D_n; \ T(D) = D^2$ (D_n is the set of $n \times n$ diagonal matrices)

18. $T: D_n \to D_n; \ T(D) = I + D$

19. $T: P_2 \to P_1; \ T(a_0 + a_1 x + a_2 x^2) = a_0 + a_1 x$

20. $T: P_2 \to P_1; \ T(a_0 + a_1 x + a_2 x^2) = a_1 + a_2 x$

21. $T: \mathbb{R} \to P_n; \ T(a) = a + ax + ax^2 + \cdots + ax^n$

22. $T: P_2 \to P_4; \ T(p(x)) = [p(x)]^2$

23. $T: C[0, 1] \to C[0, 1]; \ (Tf)(x) = f^2(x)$

24. $T: C[0, 1] \to C[0, 1]; \ (Tf)(x) = f(x) + 1$

25. $T: C[0, 1] \to \mathbb{R}; \ Tf = \int_0^1 f(x)g(x)\,dx$, where g is a fixed function in $C[0, 1]$

26. $T: C^{(1)}[0, 1] \to C[0, 1]; \ Tf = (fg)'$, where g is a fixed function in $C^{(1)}[0, 1]$

27. $T: C[0, 1] \to C[1, 2]; \ (Tf)(x) = f(x - 1)$

28. $T: C[0, 1] \to \mathbb{R}; \ Tf = f(\tfrac{1}{2})$

29. $T: M_{nn} \to \mathbb{R}; \ T(A) = \det A$

30. Let $T: \mathbb{R}^2 \to \mathbb{R}^2$ be given by $T(x, y) = (-x, -y)$. Describe T geometrically.

31. Let T be a linear transformation from $\mathbb{R}^2 \to \mathbb{R}^3$ such that

$$T\begin{pmatrix} 1 \\ 0 \end{pmatrix} = \begin{pmatrix} 1 \\ 2 \\ 3 \end{pmatrix} \text{ and } T\begin{pmatrix} 0 \\ 1 \end{pmatrix} = \begin{pmatrix} -4 \\ 0 \\ 5 \end{pmatrix}.$$ Find **(a)** $T\begin{pmatrix} 2 \\ 4 \end{pmatrix}$ and **(b)** $T\begin{pmatrix} -3 \\ 7 \end{pmatrix}$.

32. In Example 2: **(a)** Find the rotation matrix A_θ when $\theta = \pi/6$. **(b)** What happens to the vector $\begin{pmatrix} -3 \\ 4 \end{pmatrix}$ if it is rotated through an angle of $\pi/6$ in the counterclockwise direction?

33. Let $A_\theta = \begin{pmatrix} \cos\theta & -\sin\theta & 0 \\ \sin\theta & \cos\theta & 0 \\ 0 & 0 & 1 \end{pmatrix}$. Describe geometrically the linear transformation $T: \mathbb{R}^3 \to \mathbb{R}^3$ given by $T\mathbf{x} = A_\theta \mathbf{x}$.

34. Answer the question in Problem 33 for

$$A_\theta = \begin{pmatrix} \cos\theta & 0 & -\sin\theta \\ 0 & 1 & 0 \\ \sin\theta & 0 & \cos\theta \end{pmatrix}.$$

35. Suppose that, in a real vector space V, T satisfies $T(\mathbf{x} + \mathbf{y}) = T\mathbf{x} + T\mathbf{y}$ and $T(\alpha\mathbf{x}) = \alpha T\mathbf{x}$ for $\alpha \geq 0$. Show that T is linear.

36. Find a linear transformation $T: M_{33} \to M_{22}$.

37. If T is a linear transformation from V to W, show that

$$T(-\mathbf{x}) = -T\mathbf{x}$$

38. If T is a linear transformation from V to W, show that $T0 = 0$. Are the two zero vectors here the same?

39. Let V be an inner product space and let $\mathbf{u}_0 \in V$ be fixed. Let $T: V \to \mathbb{R}$ (or $\mathbb{C}$) be defined by $T\mathbf{v} = (\mathbf{v}, \mathbf{u}_0)$. Show that T is linear.

* 40. Show that if V is a complex inner product space and $T: V \to \mathbb{C}$ is defined by $T\mathbf{v} = (\mathbf{u}_0, \mathbf{v})$ for a fixed vector $\mathbf{u}_0 \in V$, then T is not linear.

41. Let V be an inner product space with the finite-dimensional subspace H. Let $\{\mathbf{u}_1, \mathbf{u}_2, \ldots, \mathbf{u}_k\}$ be a basis for H. Show that $T: V \to H$ defined by $T\mathbf{v} = (\mathbf{v}, \mathbf{u}_1)\mathbf{u}_1 + (\mathbf{v}, \mathbf{u}_2)\mathbf{u}_2 + \cdots + (\mathbf{v}, \mathbf{u}_n)\mathbf{u}_n$ is a linear transformation.

42. Let $T: V \to W$ be a linear transformation, let $\{\mathbf{v}_1, \mathbf{v}_2, \ldots, \mathbf{v}_n\}$ be a basis for V, and suppose that $T\mathbf{v}_i = \mathbf{0}$ for $i = 1, 2, \ldots, n$. Show that T is the zero transformation.

43. In Problem 42, suppose that $W = V$ and $T\mathbf{v}_i = \mathbf{v}_i$ for $i = 1, 2, \ldots, n$. Show that T is the identity operator.

44. Find all linear transformations from $\mathbb{R}^2$ into $\mathbb{R}^2$ such that the line $y = 0$ is carried into the line $x = 0$.

* 45. Find all linear transformations from $\mathbb{R}^2$ into $\mathbb{R}^2$ that carry the line $y = ax$ into the line $y = bx$.

Review Exercises for Chapter 7

In Exercises 1–14, find all solutions (if any) to the given systems.

1.
$$3x_1 + 6x_2 = 9$$
$$-2x_1 + 3x_2 = 4$$

2.
$$3x_1 + 6x_2 = 9$$
$$2x_1 + 4x_2 = 6$$

3.
$$3x_1 - 6x_2 = 9$$
$$-2x_1 + 4x_2 = 6$$

4.
$$x_1 + x_2 + x_3 = 2$$
$$2x_1 - x_2 + 2x_3 = 4$$
$$-3x_1 + 2x_2 + 3x_3 = 8$$

5.
$$x_1 + x_2 + x_3 = 0$$
$$2x_1 - x_2 + 2x_3 = 0$$
$$-3x_1 + 2x_2 + 3x_3 = 0$$

6.
$$x_1 + x_2 + x_3 = 2$$
$$2x_1 - x_2 + 2x_3 = 4$$
$$-x_1 + 4x_2 + x_3 = 2$$

7.
$$x_1 + x_2 + x_3 = 2$$
$$2x_1 - x_2 + 2x_3 = 4$$
$$-x_1 + 4x_2 + x_3 = 3$$

8.
$$x_1 + x_2 + x_3 = 0$$
$$2x_1 - x_2 + 2x_3 = 0$$
$$-x_1 + 4x_2 + x_3 = 0$$

9.
$$2x_1 + x_2 - 3x_3 = 0$$
$$4x_1 - x_2 + x_3 = 0$$

10.
$$x_1 + x_2 = 0$$
$$2x_1 + x_2 = 0$$
$$3x_1 + x_2 = 0$$

11.
$$x_1 + x_2 = 1$$
$$2x_1 + x_2 = 3$$
$$3x_1 + x_2 = 4$$

12.
$$x_1 + x_2 + x_3 + x_4 = 4$$
$$2x_1 - 3x_2 - x_3 + 4x_4 = 7$$
$$-2x_1 + 4x_2 + x_3 - 2x_4 = 1$$
$$5x_1 - x_2 + 2x_3 + x_4 = -1$$

13.
$$x_1 + x_2 + x_3 + x_4 = 0$$
$$2x_1 - 3x_2 - x_3 + 4x_4 = 0$$
$$-2x_1 + 4x_2 + x_3 - 2x_4 = 0$$
$$5x_1 - x_2 + 2x_3 + x_4 = 0$$

14.
$$x_1 + x_2 + x_3 + x_4 = 0$$
$$2x_1 - 3x_2 - x_3 + 4x_4 = 0$$
$$-2x_1 + 4x_2 + x_3 - 2x_4 = 0$$

15. Find the distance from the point $(3, -2)$ to the line $x - 2y = 6$.

16. Solve the system represented by $\begin{pmatrix} 1 & 4 & | & 2 \\ 3 & 7 & | & 5 \end{pmatrix}$.

17. Solve the system represented by $\begin{pmatrix} 1 & 2 & -4 & | & 4 \\ -2 & -4 & 8 & | & -8 \end{pmatrix}$.

18. Solve the system represented by $\begin{pmatrix} 1 & 2 & -4 & | & 4 \\ -2 & -4 & 8 & | & -9 \end{pmatrix}$.

19. Solve the homogeneous system represented by
$$\begin{pmatrix} 1 & -2 & 3 & | & 0 \\ 4 & 1 & -1 & | & 0 \\ 2 & -1 & 3 & | & 0 \end{pmatrix}.$$

20. Solve the homogeneous system represented by
$$\begin{pmatrix} 1 & 1 & -1 & | & 0 \\ 4 & -1 & 5 & | & 0 \\ 6 & 1 & 3 & | & 0 \end{pmatrix}.$$

21. Solve the system represented by the augmented matrix
$$\begin{pmatrix} 1 & 3 & -2 & 1 & | & 3 \\ 2 & -6 & 4 & -1 & | & 2 \\ 4 & 12 & -8 & 2 & | & 4 \\ -3 & 0 & 6 & -2 & | & -8 \end{pmatrix}.$$

22. Solve the homogeneous system represented by the augmented matrix $\begin{pmatrix} 1 & 2 & -3 & 5 & 4 & | & 0 \\ -2 & 4 & 7 & -3 & 5 & | & 0 \\ -4 & 0 & 13 & -13 & -3 & | & 0 \end{pmatrix}.$

23. Three chemicals are combined to form three grades of fertilizer. A unit of grade I fertilizer requires 10 kg of chemical A, 30 of B, and 60 of C. A unit of grade II requires 20 kg of A, 30 of B, and 50 of C. A unit of grade III requires 50 kg of A and 50 of C. If 1600 kg of A, 1200 of B, and 3200 of C are available, how many units of the three grades should be produced to use all available supplies? [*Hint:* To solve this problem, first write the resulting system in the form $A\mathbf{x} = \mathbf{b}$.]

In Exercises 24–31 perform the indicated computations.

24. $3\begin{pmatrix} -2 & 1 \\ 0 & 4 \\ 2 & 3 \end{pmatrix}$

25. $\begin{pmatrix} 1 & 0 & 3 \\ 2 & -1 & 6 \end{pmatrix} + \begin{pmatrix} 2 & 0 & 4 \\ -2 & 5 & 8 \end{pmatrix}$

26. $5\begin{pmatrix} 2 & 1 & 3 \\ -1 & 2 & 4 \\ -6 & 1 & 5 \end{pmatrix} - 3\begin{pmatrix} -2 & 1 & 4 \\ 5 & 0 & 7 \\ 2 & -1 & 3 \end{pmatrix}$

27. $\begin{pmatrix} 2 & 3 \\ -1 & 4 \end{pmatrix}\begin{pmatrix} 5 & -1 \\ 2 & 7 \end{pmatrix}$

28. $\begin{pmatrix} 2 & 3 & 1 & 5 \\ 0 & 6 & 2 & 4 \end{pmatrix}\begin{pmatrix} 5 & 7 & 1 \\ 2 & 0 & 3 \\ 1 & 0 & 0 \\ 0 & 5 & 6 \end{pmatrix}$

29. $\begin{pmatrix} 2 & 3 & 5 \\ -1 & 6 & 4 \\ 1 & 0 & 6 \end{pmatrix}\begin{pmatrix} 0 & -1 & 2 \\ 3 & 1 & 2 \\ -7 & 3 & 5 \end{pmatrix}$

30. $\begin{pmatrix} 1 & 0 & 3 & -1 & 5 \\ 2 & 1 & 6 & 2 & 5 \end{pmatrix}\begin{pmatrix} 7 & 1 \\ 2 & 3 \\ -1 & 0 \\ 5 & 6 \\ 2 & 3 \end{pmatrix}$

31. $\begin{pmatrix} 1 & -1 & 2 \\ 3 & 5 & 6 \\ 2 & 4 & -1 \end{pmatrix}\begin{pmatrix} 2 \\ 1 \\ 3 \end{pmatrix}$

32. Verify the associative law of matrix multiplication for the matrices $A = \begin{pmatrix} 2 & 3 & 1 \\ 0 & 4 & 6 \end{pmatrix}$, $B = \begin{pmatrix} 1 & 0 & 2 \\ 0 & 3 & 3 \\ 5 & 1 & -1 \end{pmatrix}$, and $C = \begin{pmatrix} 5 & 6 \\ -1 & 2 \\ 0 & 1 \end{pmatrix}$.

In Exercises 33–37 calculate the row echelon form and the inverse of the given matrix (if the inverse exists).

33. $\begin{pmatrix} 2 & 3 \\ -1 & 4 \end{pmatrix}$

34. $\begin{pmatrix} -1 & 2 \\ 2 & -4 \end{pmatrix}$

35. $\begin{pmatrix} 1 & 2 & 0 \\ 2 & 1 & -1 \\ 3 & 1 & 1 \end{pmatrix}$

36. $\begin{pmatrix} -1 & 2 & 0 \\ 4 & 1 & -3 \\ 2 & 5 & -3 \end{pmatrix}$

37. $\begin{pmatrix} 2 & 0 & 4 \\ -1 & 3 & 1 \\ 0 & 1 & 2 \end{pmatrix}$

In Exercises 38–40 first write the system in the form $A\mathbf{x} = \mathbf{b}$, then calculate A^{-1}, and, finally, use matrix multiplication to obtain the solution vector.

38. $x_1 - 3x_2 = 4$
$2x_1 + 5x_2 = 7$

39. $x_1 + 2x_2 \quad\quad = 3$
$2x_1 + x_2 - x_3 = -1$
$3x_1 + x_2 + x_3 = 7$

40. $2x_1 \quad\quad + 4x_3 = 7$
$-x_1 + 3x_2 + x_3 = -4$
$x_2 + 2x_3 = 5$

In Exercises 41–46 calculate the transpose of the given matrix and determine whether the matrix is symmetric, skew-symmetric,† or neither.

41. $\begin{pmatrix} 2 & 3 & 1 \\ -1 & 0 & 2 \end{pmatrix}$

42. $\begin{pmatrix} 4 & 6 \\ 6 & 4 \end{pmatrix}$

43. $\begin{pmatrix} 2 & 3 & 1 \\ 3 & -6 & -5 \\ 1 & -5 & 9 \end{pmatrix}$

44. $\begin{pmatrix} 0 & 5 & 6 \\ -5 & 0 & 4 \\ -6 & -4 & 0 \end{pmatrix}$

45. $\begin{pmatrix} 1 & -1 & 4 & 6 \\ -1 & 2 & 5 & 7 \\ 4 & 5 & 3 & -8 \\ 6 & 7 & -8 & 9 \end{pmatrix}$

46. $\begin{pmatrix} 0 & 1 & -1 & 1 \\ -1 & 0 & 1 & -2 \\ 1 & 1 & 0 & 1 \\ 1 & -2 & -1 & 0 \end{pmatrix}$

† We remind you that A is skew-symmetric if $A^t = -A$.

In Exercises 47–50 solve the system by using Cramer's rule.

47. $2x_1 - x_2 = 3$

$3x_1 + 2x_2 = 5$

48. $x_1 - x_2 + x_3 = 7$

$2x_1 \qquad - 5x_3 = 4$

$3x_2 - x_3 = 2$

49. $2x_1 + 3x_2 - x_3 = 5$

$-x_1 + 2x_2 + 3x_3 = 0$

$4x_1 - x_2 + x_3 = -1$

50. $x_1 \qquad - x_3 + x_4 = 7$

$2x_2 + 2x_3 - 3x_4 = -1$

$4x_1 - x_2 - x_3 \qquad = 0$

$-2x_1 + x_2 + 4x_3 \qquad = 2$

In Exercises 51–58 calculate the determinant.

51. $\begin{vmatrix} -1 & 2 \\ 0 & 4 \end{vmatrix}$

52. $\begin{vmatrix} -3 & 5 \\ -7 & 4 \end{vmatrix}$

53. $\begin{vmatrix} 1 & -2 & 3 \\ 0 & 4 & 5 \\ 0 & 0 & 6 \end{vmatrix}$

54. $\begin{vmatrix} 5 & 0 & 0 \\ 6 & 2 & 0 \\ 10 & 100 & 6 \end{vmatrix}$

55. $\begin{vmatrix} 1 & -1 & 2 \\ 3 & 4 & 2 \\ -2 & 3 & 4 \end{vmatrix}$

56. $\begin{vmatrix} 3 & 1 & -2 \\ 4 & 0 & 5 \\ -6 & 1 & 3 \end{vmatrix}$

57. $\begin{vmatrix} 1 & -1 & 2 & 3 \\ 4 & 0 & 2 & 5 \\ -1 & 2 & 3 & 7 \\ 5 & 1 & 0 & 4 \end{vmatrix}$

58. $\begin{vmatrix} 3 & 15 & 17 & 19 \\ 0 & 2 & 21 & 60 \\ 0 & 0 & 1 & 50 \\ 0 & 0 & 0 & -1 \end{vmatrix}$

In Exercises 59–64 use determinants to calculate the inverse (if one exists).

59. $\begin{pmatrix} -3 & 4 \\ 2 & 1 \end{pmatrix}$

60. $\begin{pmatrix} 3 & -5 & 7 \\ 0 & 2 & 4 \\ 0 & 0 & -3 \end{pmatrix}$

61. $\begin{pmatrix} 1 & -1 & 2 \\ 3 & 1 & 4 \\ 5 & -1 & 8 \end{pmatrix}$

62. $\begin{pmatrix} 1 & 1 & 1 \\ 1 & 0 & 1 \\ 0 & 1 & 1 \end{pmatrix}$

63. $\begin{pmatrix} 2 & 1 & 0 & 0 \\ 0 & -1 & 3 & 0 \\ 1 & 0 & 0 & -2 \\ 3 & 0 & -1 & 0 \end{pmatrix}$

64. $\begin{pmatrix} 3 & -1 & 2 & 4 \\ 1 & 1 & 0 & 3 \\ -2 & 4 & 1 & 5 \\ 6 & -4 & 1 & 2 \end{pmatrix}$

65. Using determinants, determine whether each set of vectors is linearly dependent or independent.

(a) $\begin{pmatrix} 1 \\ 5 \\ 2 \end{pmatrix}; \begin{pmatrix} 3 \\ 0 \\ 4 \end{pmatrix}; \begin{pmatrix} -5 \\ 5 \\ 6 \end{pmatrix}$

(b) $(2, 1, 4); (3, -2, 6); (-1, -4, -2)$

In Exercises 66–71 find the kernel, nullity, and rank of the given matrix.

66. $A = \begin{pmatrix} 1 & -2 \\ -2 & 4 \end{pmatrix}$

67. $A = \begin{pmatrix} 1 & -1 & 3 \\ 2 & 0 & 4 \\ 0 & -2 & 2 \end{pmatrix}$

68. $A = \begin{pmatrix} 1 & -1 & 2 \\ 0 & 1 & 4 \\ 1 & -1 & 0 \end{pmatrix}$

69. $A = \begin{pmatrix} 2 & 4 & -2 \\ -1 & -2 & 1 \end{pmatrix}$

70. $A = \begin{pmatrix} 2 & 3 \\ -1 & 2 \\ 4 & 6 \end{pmatrix}$

71. $A = \begin{pmatrix} 1 & -1 & 2 & 3 \\ 0 & 1 & -1 & 0 \\ 1 & -2 & 3 & 3 \\ 2 & -3 & 5 & 6 \end{pmatrix}$

In Exercises 72–77 determine whether the given transformation from V to W is linear.

72. $T: \mathbb{R}^2 \to \mathbb{R}^2; T(x, y) = (0, -y)$

73. $T: \mathbb{R}^3 \to \mathbb{R}^3; T(x, y, z) = (1, y, z)$

74. $T: \mathbb{R}^2 \to \mathbb{R}^2; T(x, y) = x/y$

75. $T: P_1 \to P_2; (Tp)(x) = xp(x)$

76. $T: P_2 \to P_2; (Tp)(x) = 1 + p(x)$

77. $T: C[0, 1] \to \mathbb{R}; (Tf)(x) = f(1)$

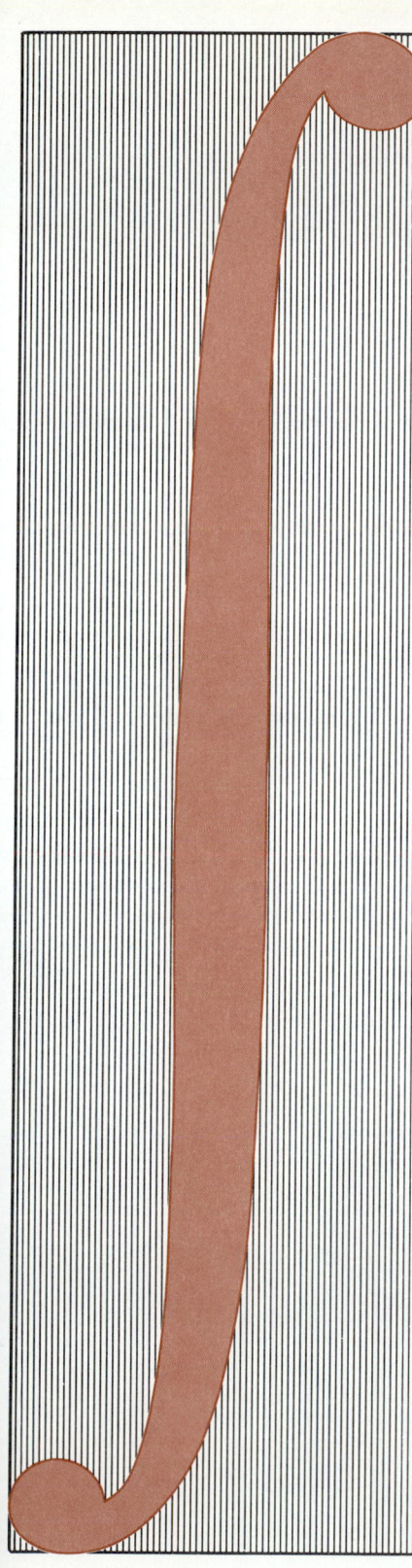

chapter 8

Eigenvalues and Eigenvectors

8.1 Eigenvalues and Eigenvectors

Let A be an $n \times n$ matrix. In a great variety of applications (several of which are given in this chapter), it is useful to find a vector $\mathbf{v}$ in $\mathbb{R}^n$ such that $A\mathbf{v}$ and $\mathbf{v}$ are multiples of each other. That is, we seek a vector $\mathbf{v}$ and a scalar λ such that

$$A\mathbf{v} = \lambda\mathbf{v}. \tag{1}$$

If $\mathbf{v} \neq \mathbf{0}$ and λ satisfy (1), then λ is called a *eigenvalue* of A and $\mathbf{v}$ is called an *eigenvector* of A corresponding to the eigenvalue λ. The purpose of this chapter is to investigate properties of eigenvalues and eigenvectors.

Let A be an $n \times n$ matrix with real† components. The number λ (real or complex) is called an **eigenvalue** of A if there is a *nonzero* vector $\mathbf{v}$ with real or complex entries such that

$$A\mathbf{v} = \lambda\mathbf{v}.$$

The vector $\mathbf{v} \neq 0$ is called an **eigenvector of A corresponding to the eigenvalue** λ.

Note. *Eigen* is the German word for "own" or "proper." Eigenvalues are also called **proper values, characteristic values,** or **latent roots** and eigenvectors are called **proper vectors, characteristic vectors** or **latent vectors.**

Remark. As we shall see (for example, in Example 7), a matrix with real components can have complex eigenvalues and eigenvectors. That is why, in the definition, we have asserted that λ and the components of $\mathbf{v}$ may be complex.

† This definition is also valid if A has complex components, but since the matrices we deal with have real components, the definition is sufficient for our purposes.

EXAMPLE 1

Let

$$A = \begin{pmatrix} 10 & -18 \\ 6 & -11 \end{pmatrix}.$$

Then

$$A\begin{pmatrix} 2 \\ 1 \end{pmatrix} = \begin{pmatrix} 10 & -18 \\ 6 & -11 \end{pmatrix}\begin{pmatrix} 2 \\ 1 \end{pmatrix} = \begin{pmatrix} 2 \\ 1 \end{pmatrix}.$$

Thus $\lambda_1 = 1$ is an eigenvalue of A with corresponding eigenvector $\mathbf{v}_1 = \begin{pmatrix} 2 \\ 1 \end{pmatrix}$. Similarly,

$$A\begin{pmatrix} 3 \\ 2 \end{pmatrix} = \begin{pmatrix} 10 & -18 \\ 6 & -11 \end{pmatrix}\begin{pmatrix} 3 \\ 2 \end{pmatrix} = \begin{pmatrix} -6 \\ -4 \end{pmatrix} = -2\begin{pmatrix} 3 \\ 2 \end{pmatrix},$$

so $\lambda_2 = -2$ is an eigenvalue of A with corresponding eigenvector $\mathbf{v}_2 = \begin{pmatrix} 3 \\ 2 \end{pmatrix}$. As we soon see, $\lambda_1 = 1$ and $\lambda_2 = -2$ are the only eigenvalues of A.

EXAMPLE 2

Let $A = I$. Then for any $\mathbf{v}$, $A\mathbf{v} = I\mathbf{v} = \mathbf{v}$. Thus $\lambda = 1$ is the only eigenvalue of A and every nonzero vector $\mathbf{v}$ is an eigenvector of I.

EXAMPLE 3: A Mass-Spring System

Consider the mass-spring depicted in Fig. 1. In Section 3.5 we obtained the following system of differential equations governing the behavior of this system (see p. 154):

$$x_1' = x_3,$$

$$x_2' = x_4,$$

$$x_1'' = x_3' = -\left(\frac{k_1 + k_2}{m_1}\right)x_1 + \left(\frac{k_2}{m_1}\right)x_2,$$

$$x_2'' = x_4' = \left(\frac{k_2}{m_2}\right)x_1 - \left(\frac{k_2}{m_2}\right)x_2.$$

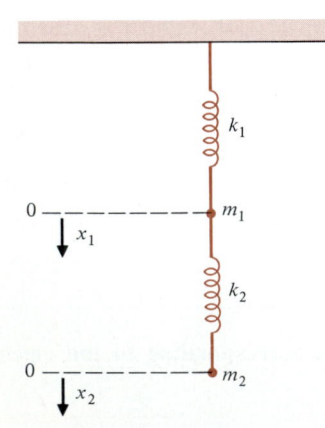

FIGURE 1

Let

$$\mathbf{x} = \begin{pmatrix} x_1 \\ x_2 \\ x_3 \\ x_4 \end{pmatrix} \quad \text{and} \quad A = \begin{pmatrix} 0 & 0 & 1 & 0 \\ 0 & 0 & 0 & 1 \\ -\left(\dfrac{k_1 + k_2}{m_1}\right) & \dfrac{k_2}{m_1} & 0 & 0 \\ \dfrac{k_2}{m_2} & -\dfrac{k_2}{m_2} & 0 & 0 \end{pmatrix}$$

Then the system can be written as

$$\mathbf{x}' = A\mathbf{x}. \tag{2}$$

A solution to the scalar equation $x' = ax$ is $x(t) = e^{at}$. Thus it is reasonable to seek a solution to (4) of the form $\mathbf{x}(t) = e^{\lambda t}\mathbf{c}$ where $\mathbf{c}$ is a constant vector. Inserting $\mathbf{x}(t) = e^{\lambda t}\mathbf{c}$ into both sides of (2), we obtain

$$\mathbf{x}'(t) = \lambda e^{\lambda t}\mathbf{c} = A\mathbf{x} = e^{\lambda t}A\mathbf{c}.$$

Dividing by $e^{\lambda t}$ (which is nonzero) we obtain

$$\lambda\mathbf{c} = A\mathbf{c}.$$

That is, $\mathbf{x}(t) = e^{\lambda t}\mathbf{c}$ is a solution to (2) whenever $\mathbf{c}$ is a eigenvector of A with corresponding eigenvalue λ.

We shall return to this problem in Section 8.6.

We compute the eigenvalues and eigenvectors of many matrices in this section. But first we need to prove some facts that can simplify our computations.

Suppose that λ is an eigenvalue of A. Then there exists a nonzero vector

$$\mathbf{v} = \begin{pmatrix} x_1 \\ x_2 \\ \vdots \\ x_n \end{pmatrix} \neq \mathbf{0}$$

such that $A\mathbf{v} = \lambda\mathbf{v} = \lambda I\mathbf{v}$.

Rewriting this, we have

$$(\lambda I - A)\mathbf{v} = \mathbf{0}. \tag{3}$$

If A is an $n \times n$ matrix, equation (3) is a homogeneous system of n equations in the unknowns $x_1, x_2, \ldots, x_n$. Since, by assumption, the system has nontrivial solutions, we conclude that $\det(\lambda I - A) = 0$ (see Theorem 7.9.5 on p. 429).

Conversely, if $\det(\lambda I - A) = 0$, then (3) has nontrivial solutions and λ is an eigenvalue of A. On the other hand, if $\det(\lambda I - A) \neq 0$, then (3) has only the solution $\mathbf{v} = \mathbf{0}$, so λ is *not* an eigenvalue of A. Summing up these facts, we have the following.

■ THEOREM 1

Let A be an $n \times n$ matrix. Then λ is an eigenvalue of A if and only if

$$p(\lambda) = \det(\lambda I - A) = 0. \tag{4}$$

■

Characteristic Equation
Characteristic Polynomial

Equation (4) is called the **characteristic equation** of A and $p(\lambda)$ is called the **characteristic polynomial** of A. Compare this with equation (3.2.9) on p. 138.

As becomes apparent in the examples, $p(\lambda)$ is a polynomial of degree n in λ; for example, if

$$A = \begin{pmatrix} a & b \\ c & d \end{pmatrix},$$

then

$$\lambda I - A = \begin{pmatrix} \lambda & 0 \\ 0 & \lambda \end{pmatrix} - \begin{pmatrix} a & b \\ c & d \end{pmatrix} = \begin{pmatrix} \lambda - a & -b \\ -c & \lambda - d \end{pmatrix}$$

and

$$p(\lambda) = \det(\lambda I - A) = (\lambda - a)(\lambda - d) - bc$$
$$= \lambda^2 - (a + d)\lambda + (ad - bc).$$

Similarly, if

$$A = \begin{pmatrix} a_{11} & a_{12} & \cdots & a_{1n} \\ a_{21} & a_{22} & \cdots & a_{2n} \\ \vdots & \vdots & & \vdots \\ a_{n1} & a_{n2} & \cdots & a_{nn} \end{pmatrix},$$

then

$$p(\lambda) = \det(\lambda I - A) = \begin{vmatrix} \lambda - a_{11} & -a_{12} & \cdots & -a_{1n} \\ -a_{21} & \lambda - a_{22} & \cdots & -a_{2n} \\ \vdots & \vdots & & \vdots \\ -a_{n1} & -a_{n2} & \cdots & \lambda - a_{nn} \end{vmatrix},$$

and $p(\lambda)$ can be written in the form

$$p(\lambda) = \lambda^n + b_{n-1}\lambda^{n-1} + \cdots + b_1\lambda + b_0. \tag{5}$$

By the fundamental theorem of algebra (which we prove in Section 16.6), any polynomial of degree n with real or complex coefficients has exactly n roots (counting multiplicities). By this we mean, for example, that the polynomial $(\lambda - 1)^5$ has five roots, all equal to the number 1. Thus $p(\lambda)$ can be factored as follows:

$$p(\lambda) = (\lambda - \lambda_1)^{r_1}(\lambda - \lambda_2)^{r_2} \cdots (\lambda - \lambda_m)^{r_m}, \tag{6}$$

where $m \geq 1$, $\lambda_1, \lambda_2, \ldots, \lambda_m$ are real or complex numbers [the roots of $p(\lambda)$], and $r_1, r_2, \ldots, r_m$ are positive integers with $r_1 + r_2 + \cdots + r_m = n$. The factorization (6) is unique except for the order of factors. Since $p(\lambda_i) = 0$, $i = 1, 2, \ldots, m$, each λ_i is an eigenvalue of A. The number r_i is called the **algebraic multiplicity** of λ_i. Also, $p(\lambda) \neq 0$ if λ is not in the set $\{\lambda_1, \lambda_2, \ldots, \lambda_m\}$, so this set contains all the distinct eigenvalues of A. If the algebraic multiplicity of λ_i is 1, then λ_i is called a **simple eigenvalue**.

Algebraic Multiplicity

Simple Eigenvalue

■ THEOREM 2

Let v_1 and v_2 be eigenvectors of A corresponding to the same eigenvalue λ. Then $v = c_1v_1 + c_2v_2 \neq 0$ is an eigenvector of A corresponding to λ; that is,

> any nonzero linear combination of eigenvectors corresponding to the same eigenvalue is an eigenvector corresponding to that eigenvalue.

PROOF

$$Av = A(c_1v_1 + c_2v_2) = c_1Av_1 + c_2Av_2 = c_1\lambda v_1 + c_2\lambda v_2$$

$$= \lambda(c_1v_1 + c_2v_2) = \lambda v.$$ ■

According to Theorem 2, corresponding to each eigenvalue of a square matrix A there are an infinite number of eigenvectors. Our aim is to find as many *linearly independent* eigenvectors as possible corresponding to each eigenvalue. The following result is very useful and is given without proof.

■ THEOREM 3

If λ is a simple eigenvalue of A (that is, the algebraic multiplicity of λ is 1), then there is only one linearly independent eigenvector corresponding to λ. ■

We now prove another useful result.

■ THEOREM 4

Let A be an $n \times n$ matrix and let $\lambda_1, \lambda_2, \ldots, \lambda_m$ be distinct eigenvalues of A with corresponding eigenvectors $v_1, v_2, \ldots, v_m$. Then $v_1, v_2, \ldots, v_m$ are linearly independent; that is, eigenvectors corresponding to distinct eigenvalues are linearly independent.

PROOF

We prove this by mathematical induction. We start with $m = 2$. Suppose that

$$c_1v_1 + c_2v_2 = 0. \tag{7}$$

Then, multiplying both sides of equation (7) by A, we have

$$0 = A(c_1v_1 + c_2v_2) = c_1Av_1 + c_2Av_2$$

or

$$c_1\lambda_1v_1 + c_2\lambda_2v_2 = 0. \tag{8}$$

We then multiply (7) by λ_1 and subtract it from (8) to obtain

$$(c_1\lambda_1v_1 + c_2\lambda_2v_2) - (c_1\lambda_1v_1 + c_2\lambda_1v_2) = 0$$

or

$$c_2(\lambda_2 - \lambda_1)v_2 = 0.$$

Since $\mathbf{v}_2 \neq \mathbf{0}$ (by the definition of an eigenvector) and since $\lambda_2 \neq \lambda_1$, we conclude that $c_2 = 0$. Then inserting $c_2 = 0$ in (7), we obtain $c_1 = 0$, which proves the theorem in the case $m = 2$. Now suppose that the theorem is true for $m = k$; that is, assume that any k eigenvectors corresponding to distinct eigenvalues are linearly independent. We prove the theorem for $m = k + 1$. So we assume that

$$c_1\mathbf{v}_1 + c_2\mathbf{v}_2 + \cdots + c_k\mathbf{v}_k + c_{k+1}\mathbf{v}_{k+1} = \mathbf{0}. \tag{9}$$

Then, multiplying both sides of equation (9) by A and using the fact that $A\mathbf{v}_i = \lambda_i\mathbf{v}_i$, we obtain

$$c_1\lambda_1\mathbf{v}_1 + c_2\lambda_2\mathbf{v}_2 + \cdots + c_k\lambda_k\mathbf{v}_k + c_{k+1}\lambda_{k+1}\mathbf{v}_{k+1} = \mathbf{0}. \tag{10}$$

We multiply both sides of (9) by λ_{k+1} and subtract it from (10):

$$c_1(\lambda_1 - \lambda_{k+1})\mathbf{v}_1 + c_2(\lambda_2 - \lambda_{k+1})\mathbf{v}_2 + \cdots + c_k(\lambda_k - \lambda_{k+1})\mathbf{v}_k = \mathbf{0}.$$

But, by the induction assumption, $\mathbf{v}_1, \mathbf{v}_2, \ldots, \mathbf{v}_k$ are linearly independent. Thus

$$c_1(\lambda_1 - \lambda_{k+1}) = c_2(\lambda_2 - \lambda_{k+1}) = \cdots = c_k(\lambda_k - \lambda_{k+1}) = 0,$$

and, since $\lambda_i \neq \lambda_{k+1}$ for $i = 1, 2, \ldots, k$, we conclude that $c_1 = c_2 = \cdots = c_k = 0$. But, from (9), this means that $c_{k+1} = 0$. Thus the theorem is true for $m = k + 1$ and the proof is complete. ∎

We now proceed to calculate eigenvalues and corresponding eigenspaces. We do this using a three-step procedure:

To Compute Eigenvalues and Eigenvectors

  a. Find $p(\lambda) = \det(\lambda I - A)$. (11)
  b. Find the roots $\lambda_1, \lambda_2, \ldots, \lambda_m$ of $p(\lambda) = 0$. (12)
  c. Corresponding to each eigenvalue λ_i, find the set of solutions to the homogeneous system $(\lambda_i I - A)\mathbf{v} = \mathbf{0}$ or $(A - \lambda_i I)\mathbf{v} = \mathbf{0}$.

  (13)

Remark. Step b is often the hardest one to carry out.

EXAMPLE 4

Let

$$A = \begin{pmatrix} 4 & 2 \\ 3 & 3 \end{pmatrix}.$$

Then

$$\det(\lambda I - A) = \begin{vmatrix} \lambda - 4 & -2 \\ -3 & \lambda - 3 \end{vmatrix} = (\lambda - 4)(\lambda - 3) - 6$$

$$= \lambda^2 - 7\lambda + 6 = (\lambda - 1)(\lambda - 6) = 0.$$

Thus the eigenvalues of A are $\lambda_1 = 1$ and $\lambda_2 = 6$. For $\lambda_1 = 1$, we solve $(A - I)\mathbf{v} = \mathbf{0}$ or

$$\begin{pmatrix} 3 & 2 \\ 3 & 2 \end{pmatrix}\begin{pmatrix} x_1 \\ x_2 \end{pmatrix} = \begin{pmatrix} 0 \\ 0 \end{pmatrix}.$$

Clearly, any eigenvector corresponding to $\lambda_1 = 1$ satisfies $3x_1 + 2x_2 = 0$, or $x_1 = -2x_2/3$. One such eigenvector is

$$\mathbf{v}_1 = \begin{pmatrix} 2 \\ -3 \end{pmatrix}.$$

According to Theorem 3, we cannot find any other linearly independent eigenvector corresponding to $\lambda = 1$.

Similarly, the equation $(A - 6I)\mathbf{v} = \mathbf{0}$ means that

$$\begin{pmatrix} -2 & 2 \\ 3 & -3 \end{pmatrix}\begin{pmatrix} x_1 \\ x_2 \end{pmatrix} = \begin{pmatrix} 0 \\ 0 \end{pmatrix},$$

or $x_1 = x_2$. Thus $\mathbf{v}_2 = \begin{pmatrix} 1 \\ 1 \end{pmatrix}$ is an eigenvector corresponding to $\lambda_2 = 6$. Note that $\mathbf{v}_1$ and $\mathbf{v}_2$ are linearly independent by Theorem 4.

EXAMPLE 5

Let

$$A = \begin{pmatrix} 1 & -1 & 4 \\ 3 & 2 & -1 \\ 2 & 1 & -1 \end{pmatrix}.$$

Then

$$\det(\lambda I - A) = \begin{vmatrix} \lambda - 1 & 1 & -4 \\ -3 & \lambda - 2 & 1 \\ -2 & -1 & \lambda + 1 \end{vmatrix} = \lambda^3 - 2\lambda^2 - 5\lambda + 6$$

$$= (\lambda - 1)(\lambda + 2)(\lambda - 3) = 0.$$

Thus the eigenvalues of A are $\lambda_1 = 1$, $\lambda_2 = -2$, and $\lambda_3 = 3$. Since each eigenvalue is simple, we need only find one eigenvector corresponding to each eigenvalue. Corresponding to $\lambda = 1$, we have

$$(A - I)\mathbf{v} = \begin{pmatrix} 0 & -1 & 4 \\ 3 & 1 & -1 \\ 2 & 1 & -2 \end{pmatrix}\begin{pmatrix} x_1 \\ x_2 \\ x_3 \end{pmatrix} = \begin{pmatrix} 0 \\ 0 \\ 0 \end{pmatrix}. \tag{14}$$

We write out the equations in (14):

$$0x_1 - x_2 + 4x_3 = 0$$

$$3x_1 + x_2 - x_3 = 0 \tag{15}$$

$$2x_1 + x_2 - 2x_3 = 0$$

We solve this system by Gauss–Jordan elimination (see Section 7.3):

$$\begin{pmatrix} 0 & -1 & 4 & | & 0 \\ 3 & 1 & -1 & | & 0 \\ 2 & 1 & -2 & | & 0 \end{pmatrix} \xrightarrow[A_{1,3}(1)]{A_{1,2}(1)} \begin{pmatrix} 0 & -1 & 4 & | & 0 \\ 3 & 0 & 3 & | & 0 \\ 2 & 0 & 2 & | & 0 \end{pmatrix}$$

$$\xrightarrow{M_2(1/3)} \begin{pmatrix} 0 & -1 & 4 & | & 0 \\ 1 & 0 & 1 & | & 0 \\ 2 & 0 & 2 & | & 0 \end{pmatrix} \xrightarrow{A_{2,3}(-2)} \begin{pmatrix} 0 & -1 & 4 & | & 0 \\ 1 & 0 & 1 & | & 0 \\ 0 & 0 & 0 & | & 0 \end{pmatrix}.$$

This denotes the system of equations

$$-x_2 + 4x_3 = 0,$$

$$x_1 \quad + \quad x_3 = 0.$$

Thus $x_1 = -x_3$, $x_2 = 4x_3$, and an eigenvector corresponding to $\lambda = 1$ is

$$\mathbf{v}_1 = \begin{pmatrix} -1 \\ 4 \\ 1 \end{pmatrix}.$$

Because of Theorem 3, we know that any other eigenvector of A corresponding to $\lambda = 1$ is a multiple of $\mathbf{v}_1$.

For $\lambda_2 = -2$, we have

$$[A - (-2I)]\mathbf{v} = (A + 2I)\mathbf{v} = \mathbf{0}$$

or

$$\begin{pmatrix} 3 & -1 & 4 \\ 3 & 4 & -1 \\ 2 & 1 & 1 \end{pmatrix}\begin{pmatrix} x_1 \\ x_2 \\ x_3 \end{pmatrix} = \begin{pmatrix} 0 \\ 0 \\ 0 \end{pmatrix}.$$

This leads to

$$\begin{pmatrix} 3 & -1 & 4 & | & 0 \\ 3 & 4 & -1 & | & 0 \\ 2 & 1 & 1 & | & 0 \end{pmatrix} \xrightarrow[A_{1,3}(1)]{A_{1,2}(4)} \begin{pmatrix} 3 & -1 & 4 & | & 0 \\ 15 & 0 & 15 & | & 0 \\ 5 & 0 & 5 & | & 0 \end{pmatrix}$$

$$\xrightarrow{M_2(1/15)} \begin{pmatrix} 3 & -1 & 4 & | & 0 \\ 1 & 0 & 1 & | & 0 \\ 5 & 0 & 5 & | & 0 \end{pmatrix} \xrightarrow[A_{2,3}(-5)]{A_{2,1}(-4)} \begin{pmatrix} -1 & -1 & 0 & | & 0 \\ 1 & 0 & 1 & | & 0 \\ 0 & 0 & 0 & | & 0 \end{pmatrix}.$$

Thus $x_2 = -x_1$, $x_3 = -x_1$, and an eigenvector corresponding to $\lambda = -2$ is

$$\mathbf{v}_2 = \begin{pmatrix} 1 \\ -1 \\ -1 \end{pmatrix}.$$

Finally, for $\lambda_3 = 3$, we have

$$(A - 3I)\mathbf{v} = \begin{pmatrix} -2 & -1 & 4 \\ 3 & -1 & -1 \\ 2 & 1 & -4 \end{pmatrix}\begin{pmatrix} x_1 \\ x_2 \\ x_3 \end{pmatrix} = \begin{pmatrix} 0 \\ 0 \\ 0 \end{pmatrix}$$

and

$$\begin{pmatrix} -2 & -1 & 4 & | & 0 \\ 3 & -1 & -1 & | & 0 \\ 2 & 1 & -4 & | & 0 \end{pmatrix} \xrightarrow[A_{3,2}(1)]{A_{3,1}(1)} \begin{pmatrix} 0 & 0 & 0 & | & 0 \\ 5 & 0 & -5 & | & 0 \\ 2 & 1 & -4 & | & 0 \end{pmatrix}$$

$$\xrightarrow{M_2(1/5)} \begin{pmatrix} 0 & 0 & 0 & | & 0 \\ 1 & 0 & -1 & | & 0 \\ 2 & 1 & -4 & | & 0 \end{pmatrix} \xrightarrow{A_{2,3}(-4)} \begin{pmatrix} 0 & 0 & 0 & | & 0 \\ 1 & 0 & -1 & | & 0 \\ -2 & 1 & 0 & | & 0 \end{pmatrix}.$$

Thus $x_3 = x_1$, $x_2 = 2x_1$, and an eigenvector corresponding to $\lambda = 3$ is

$$\mathbf{v}_3 = \begin{pmatrix} 1 \\ 2 \\ 1 \end{pmatrix}.$$

Remark. In this and every other example, there are always an infinite number of choices for each eigenvector. We arbitrarily choose a simple one by setting one or more of the x_i's equal to 1. Other normalizations are sometimes useful.

In the last example we found three eigenvectors that, according to Theorem 4, must be linearly independent. There is an easy way to verify this. The following result was proved in Section 7.9.

THEOREM 5 (Theorem 7.9.5)
The $n \times 1$ vectors $\mathbf{v}_1, \mathbf{v}_2, \ldots, \mathbf{v}_n$ in $\mathbb{R}^n$ are linearly independent if and only if the determinant of the matrix whose columns are $\mathbf{v}_1, \mathbf{v}_2, \ldots, \mathbf{v}_n$ is nonzero.

EXAMPLE 6

In Example 5 we can verify that the vectors

$$\mathbf{v}_1 = \begin{pmatrix} -1 \\ 4 \\ 1 \end{pmatrix}, \quad \mathbf{v}_2 = \begin{pmatrix} 1 \\ -1 \\ -1 \end{pmatrix}, \quad \text{and} \quad \mathbf{v}_3 = \begin{pmatrix} 1 \\ 2 \\ 1 \end{pmatrix}$$

are linearly independent because

$$\begin{vmatrix} -1 & 1 & 1 \\ 4 & -1 & 2 \\ 1 & -1 & 1 \end{vmatrix} = -6 \neq 0.$$

EXAMPLE 7

Let

$$A = \begin{pmatrix} 3 & -5 \\ 1 & -1 \end{pmatrix}.$$

Then

$$\det(\lambda I - A) = \begin{vmatrix} \lambda - 3 & 5 \\ -1 & \lambda + 1 \end{vmatrix} = \lambda^2 - 2\lambda + 2 = 0.$$

Then

$$\left.\begin{array}{l}\lambda_1\\\lambda_2\end{array}\right\} = \frac{-(-2) \pm \sqrt{4 - 4(1)(2)}}{2} = \frac{2 \pm \sqrt{-4}}{2} = \frac{2 \pm 2i}{2} = 1 \pm i.$$

Thus $\lambda_1 = 1 + i$ and $\lambda_2 = 1 - i$. Then

$$[A - (1+i)I]\mathbf{v} = \begin{pmatrix} 2-i & -5 \\ 1 & -2-i \end{pmatrix}\begin{pmatrix} x_1 \\ x_2 \end{pmatrix} = \begin{pmatrix} 0 \\ 0 \end{pmatrix}$$

and we obtain $(2 - i)x_1 - 5x_2 = 0$ and $x_1 + (-2 - i)x_2 = 0$.† Thus $x_1 = (2 + i)x_2$, which yields the eigenvector (corresponding to $\lambda_1 = 1 + i$)

$$\mathbf{v}_1 = \begin{pmatrix} 2+i \\ 1 \end{pmatrix}.$$

Similarly,

$$[A - (1-i)I]\mathbf{v} = \begin{pmatrix} 2+i & -5 \\ 1 & -2+i \end{pmatrix}\begin{pmatrix} x_1 \\ x_2 \end{pmatrix} = \begin{pmatrix} 0 \\ 0 \end{pmatrix},$$

or $x_1 + (-2 + i)x_2 = 0$, which yields $x_1 = (2 - i)x_2$, and we obtain the eigenvector (corresponding to $\lambda_2 = 1 - i$)

$$\mathbf{v}_2 = \begin{pmatrix} 2-i \\ 1 \end{pmatrix}.$$

Remark. This example illustrates that a real matrix may have complex eigenvalues and eigenvectors. It should be pointed out that some texts define eigenvalues of real matrices as the *real* roots of the characteristic equation. With this definition, the matrix of the last example has *no* eigenvalues. This might make the computations simpler, but it also significantly reduces the usefulness of the theory of eigenvalues and eigenvectors. We shall see an important illustration of the use of complex eigenvalues in Section 8.6 in our computation of matrix solutions.

EXAMPLE 8

Let

$$A = \begin{pmatrix} 4 & 1 \\ 0 & 4 \end{pmatrix}.$$

Then

$$\det(\lambda I - A) = \begin{vmatrix} \lambda - 4 & -1 \\ 0 & \lambda - 4 \end{vmatrix} = (\lambda - 4)^2 = 0,$$

---

† Note that $(-2-i)\begin{pmatrix} 2-i \\ 1 \end{pmatrix} = \begin{pmatrix} -5 \\ -2-i \end{pmatrix}.$

so $\lambda = 4$ is an eigenvalue of algebraic multiplicity 2 and we have

$$(A - 4I)\mathbf{v} = \begin{pmatrix} 0 & 1 \\ 0 & 0 \end{pmatrix}\begin{pmatrix} x_1 \\ x_2 \end{pmatrix} = \begin{pmatrix} x_2 \\ 0 \end{pmatrix}.$$

Thus $x_2 = 0$ and x_1 is arbitrary. Therefore the only linearly independent eigenvector is

$$\mathbf{v} = \begin{pmatrix} 1 \\ 0 \end{pmatrix}.$$

Here is an instance where a 2×2 matrix has only one linearly independent eigenvector.

Observe that zero is an eigenvalue of A if and only if

$$p(0) = \det(0I - A) = \det(-A) = (-1)^n \det A = 0.$$

Thus, only singular matrices can have zero eigenvalues. This observation permits one final extension of the Summing-Up Theorem (see Theorem 7.10.7, p. 440).

■ **THEOREM 6: Summing-Up Theorem—View 4**
Let A be an $n \times n$ matrix. Then the following statements are equivalent. That is, if one is true, all are true.

    i. A is invertible.
    ii. The only solution to the homogeneous system $A\mathbf{x} = \mathbf{0}$ is the trivial solution ($\mathbf{x} = \mathbf{0}$).
    iii. The system $A\mathbf{x} = \mathbf{b}$ has a unique solution for every n-vector $\mathbf{b}$.
    iv. A is row equivalent to the $n \times n$ identity matrix I_n.
    v. The rows (and columns) of A are linearly independent.
    vi. $\det A \neq 0$.
    vii. $\nu(A) = 0$.
    viii. $\rho(A) = n$.
    ix. Zero is *not* an eigenvalue of A.

Problems 8.1

In Problems 1–20 calculate the eigenvalues and eigenvectors of the given matrix. If the algebraic multiplicity of an eigenvalue is greater than 1, determine the number of linearly independent eigenvectors that correspond to it.

1. $\begin{pmatrix} -2 & -2 \\ -5 & 1 \end{pmatrix}$

2. $\begin{pmatrix} -12 & 7 \\ -7 & 2 \end{pmatrix}$

3. $\begin{pmatrix} 2 & -1 \\ 5 & -2 \end{pmatrix}$

4. $\begin{pmatrix} -3 & 0 \\ 0 & -3 \end{pmatrix}$

5. $\begin{pmatrix} -3 & 2 \\ 0 & -3 \end{pmatrix}$

6. $\begin{pmatrix} 3 & 2 \\ -5 & 1 \end{pmatrix}$

7. $\begin{pmatrix} 1 & -1 & 0 \\ -1 & 2 & -1 \\ 0 & -1 & 1 \end{pmatrix}$

8. $\begin{pmatrix} 1 & 1 & -2 \\ -1 & 2 & 1 \\ 0 & 1 & -1 \end{pmatrix}$

9. $\begin{pmatrix} 5 & 4 & 2 \\ 4 & 5 & 2 \\ 2 & 2 & 2 \end{pmatrix}$

10. $\begin{pmatrix} 1 & 2 & 2 \\ 0 & 2 & 1 \\ -1 & 2 & 2 \end{pmatrix}$

11. $\begin{pmatrix} 0 & 1 & 0 \\ 0 & 0 & 1 \\ 1 & -3 & 3 \end{pmatrix}$

12. $\begin{pmatrix} -3 & -7 & -5 \\ 2 & 4 & 3 \\ 1 & 2 & 2 \end{pmatrix}$

13. $\begin{pmatrix} 1 & -1 & -1 \\ 1 & -1 & 0 \\ 1 & 0 & -1 \end{pmatrix}$

14. $\begin{pmatrix} 7 & -2 & -4 \\ 3 & 0 & -2 \\ 6 & -2 & -3 \end{pmatrix}$

15. $\begin{pmatrix} 4 & 6 & 6 \\ 1 & 3 & 2 \\ -1 & -5 & -2 \end{pmatrix}$

16. $\begin{pmatrix} 4 & 1 & 0 & 1 \\ 2 & 3 & 0 & 1 \\ -2 & 1 & 2 & -3 \\ 2 & -1 & 0 & 5 \end{pmatrix}$

17. $\begin{pmatrix} a & 0 & 0 & 0 \\ 0 & a & 0 & 0 \\ 0 & 0 & a & 0 \\ 0 & 0 & 0 & a \end{pmatrix}$

18. $\begin{pmatrix} a & b & 0 & 0 \\ 0 & a & 0 & 0 \\ 0 & 0 & a & 0 \\ 0 & 0 & 0 & a \end{pmatrix}$; $b \neq 0$

19. $\begin{pmatrix} a & b & 0 & 0 \\ 0 & a & c & 0 \\ 0 & 0 & a & 0 \\ 0 & 0 & 0 & a \end{pmatrix}$; $bc \neq 0$

20. $\begin{pmatrix} a & b & 0 & 0 \\ 0 & a & c & 0 \\ 0 & 0 & a & d \\ 0 & 0 & 0 & a \end{pmatrix}$; $bcd \neq 0$

21. Show that for any real numbers a and b, the matrix $A = \begin{pmatrix} a & b \\ -b & a \end{pmatrix}$ has the eigenvectors $\begin{pmatrix} 1 \\ i \end{pmatrix}$ and $\begin{pmatrix} 1 \\ -i \end{pmatrix}$.

In Problems 22–28 assume that the matrix A has the eigenvalues $\lambda_1, \lambda_2, \ldots, \lambda_k$.

22. Show that the eigenvalues of A^t are $\lambda_1, \lambda_2, \ldots, \lambda_k$.

23. Show that the eigenvalues of αA are $\alpha\lambda_1, \alpha\lambda_2, \ldots, \alpha\lambda_k$.

24. Show that A^{-1} exists if and only if $\lambda_1\lambda_2 \cdots \lambda_k \neq 0$.

* 25. If A^{-1} exists, show that the eigenvalues of A^{-1} are $1/\lambda_1, 1/\lambda_2, \ldots, 1/\lambda_k$.

26. Show that the matrix $A - \alpha I$ has the eigenvalues $\lambda_1 - \alpha, \lambda_2 - \alpha, \ldots, \lambda_k - \alpha$.

* 27. Show that the eigenvalues of A^2 are $\lambda_1^2, \lambda_2^2, \ldots, \lambda_k^2$.

* 28. Show that the eigenvalues of A^m are $\lambda_1^m, \lambda_2^m, \ldots, \lambda_k^m$ for $m = 1, 2, 3, \ldots$.

29. Let λ be an eigenvalue of A with corresponding eigenvector $\mathbf{v}$. Let $p(\lambda) = a_0 + a_1\lambda + a_2\lambda^2 + \cdots + a_n\lambda^n$. Define the matrix $p(A)$ by $p(A) = a_0 I + a_1 A + a_2 A^2 + \cdots + a_n A^n$. Show that $p(A)\mathbf{v} = p(\lambda)\mathbf{v}$.

30. Using the result of Problem 29, show that if $\lambda_1, \lambda_2, \ldots, \lambda_k$ are eigenvalues of A, then $p(\lambda_1), p(\lambda_2), \ldots, p(\lambda_k)$ are eigenvalues of $p(A)$.

31. Show that if A is an upper triangular matrix, then the eigenvalues of A are the diagonal components of A.

32. **Eigenspace.** Let λ be an eigenvalue of the $n \times n$ matrix A. Prove that $\mathbf{0}$ and the set of eigenvectors of A corresponding to λ is a subspace of $\mathbb{R}^n$. This subspace is called the **eigenspace** of λ and is denoted E_λ.

33. **Geometric Multiplicity.** The **geometric multiplicity** of an eigenvalue λ is defined as the dimension of the eigenspace E_λ. Show that if an $n \times n$ matrix A has n distinct eigenvalues, then the geometric multiplicity of each eigenvalue is 1.

34. Let $A = \begin{pmatrix} 3 & 2 & 4 \\ 2 & 0 & 2 \\ 4 & 2 & 3 \end{pmatrix}$. (a) Show that $\lambda = -1$ is an eigenvalue of algebraic multiplicity 2. (b) Find a basis for E_{-1}. (c) Show that $\dim E_{-1} = 2$. That is, in this case, the geometric multiplicity equals the algebraic multiplicity.

35. Let $A = \begin{pmatrix} -5 & -5 & -9 \\ 8 & 9 & 18 \\ -2 & -3 & -7 \end{pmatrix}$. (a) Show that $\lambda = -1$ is an eigenvalue of algebraic multiplicity 3. (b) Find a basis for E_{-1}. (c) Show that the geometric multiplicity of -1 is 1.

36. Let $A = \begin{pmatrix} -1 & -3 & -9 \\ 0 & 5 & 18 \\ 0 & -2 & -7 \end{pmatrix}$. (a) Show that $\lambda = -1$ is an eigenvalue of algebraic multiplicity 3. (b) Find a basis for E_{-1}. (c) Show that the geometric multiplicity of -1 is 2.

37. Let $A = -I$. (a) Show that $\lambda = -1$ is an eigenvalue of algebraic and geometric multiplicity 3.

Note: Problems 35, 36, 37 illustrate the fact that if λ is an eigenvalue of algebraic multiplicity 3, then there can be one, two or three linearly independent eigenvectors corresponding to λ. A similar result is true if the algebraic multiplicity is n, where $n > 1$.

38. Let $A_1 = \begin{pmatrix} 2 & 0 & 0 & 0 \\ 0 & 2 & 0 & 0 \\ 0 & 0 & 2 & 0 \\ 0 & 0 & 0 & 2 \end{pmatrix}$, $A_2 = \begin{pmatrix} 2 & 1 & 0 & 0 \\ 0 & 2 & 0 & 0 \\ 0 & 0 & 2 & 0 \\ 0 & 0 & 0 & 2 \end{pmatrix}$,

$A_3 = \begin{pmatrix} 2 & 1 & 0 & 0 \\ 0 & 2 & 1 & 0 \\ 0 & 0 & 2 & 0 \\ 0 & 0 & 0 & 2 \end{pmatrix}$, and $A_4 = \begin{pmatrix} 2 & 1 & 0 & 0 \\ 0 & 2 & 1 & 0 \\ 0 & 0 & 2 & 1 \\ 0 & 0 & 0 & 2 \end{pmatrix}$.

Show that, for each matrix, $\lambda = 2$ is an eigenvalue of algebraic multiplicity 4. In each case, compute the geometric multiplicity of $\lambda = 2$.

* 39. Let A be a real $n \times n$ matrix. Show that if λ_1 is a complex eigenvalue of A with eigenvector $\mathbf{v}_1$, then $\bar{\lambda}_1$ is an eigenvalue of A with eigenvector $\bar{\mathbf{v}}_1$.

* **40.** A **probability matrix** is an $n \times n$ matrix having two properties:

  i. $a_{ij} \geq 0$ for every i and j.
  ii. The sum of the components in every column is 1.

Prove that 1 is an eigenvalue of every probability matrix.

41. Find the eigenvalues of the Pauli matrices (see Problem 7.2.35)

$$X_1 = \frac{1}{2}\begin{pmatrix} 0 & 1 \\ 1 & 0 \end{pmatrix}, \quad X_2 = \frac{1}{2}\begin{pmatrix} 0 & -i \\ i & 0 \end{pmatrix} \quad \text{and} \quad X_3 = \frac{1}{2}\begin{pmatrix} 1 & 0 \\ 0 & -1 \end{pmatrix}.$$

(Optional)

8.2 A Model of Population Growth

In this section we show how the theory of eigenvalues and eigenvectors can be used to analyze a model of the growth of a bird population.† We begin by discussing a simple model of population growth. We assume that a certain species grows at a constant rate; that is, the population of the species after one time period (which could be an hour, a week, a month, a year, etc.) is a constant multiple of the population in the previous time period. One way this could happen, for example, is that each generation is distinct and each organism produces r offspring and then dies. If p_n denotes the population after the nth time period, we would have

$$p_n = r p_{n-1}.$$

For example, this model might describe a bacteria population where, at a given time, an organism splits into two separate organisms. Then $r = 2$. Let p_0 denote the initial population. Then $p_1 = r p_0$, $p_2 = r p_1 = r(r p_0) = r^2 p_0$, $p_3 = r p_2 = r(r^2 p_0) = r^3 p_0$, and so on, so that

$$p_n = r^n p_0. \tag{1}$$

From this model we see that the population increases without bound if $r > 1$ and decreases to zero if $r < 1$. If $r = 1$ the population remains at the constant value p_0.

This model is, evidently, very simplistic. One obvious objection is that the number of offspring produced depends, in many cases, on the ages of the adults. For example, in a human population the average female adult over 50 would certainly produce fewer children than the average 21-year-old female. To deal with this difficulty, we introduce a model which allows for age groupings with different fertility rates.

We now look at a model of population growth for a species of birds. In this bird population we assume that the number of female birds equals the number of males. Let $p_{j,n-1}$ denote the population of juvenile (immature) females in the $(n-1)$st year and let $p_{a,n-1}$ denote the number of adult females in the $(n-1)$st year. Some of the juvenile birds will die during the year. We assume that a certain proportion α of the juvenile birds survive to become adults in the spring of the nth year. Each surviving female bird produces eggs later in

† The material in this section is based on a paper by D. Cooke: "A 2×2 matrix model of population growth," *Mathematical Gazette*, vol. 61, no. 416, June, 1977, pp. 120–123.

the spring, which hatch to produce, on the average, k juvenile female birds in the following spring. Adults also die, and the proportion of adults that survive from one spring to the next is β.

This constant survival rate of birds is not just a simplistic assumption. It appears to be the case with most of the natural bird populations that have been studied. This means that the adult survival rate of many bird species is independent of age. Perhaps few birds in the wild survive long enough to exhibit the effects of old age. Moreover, in many species the number of offspring seems to be uninfluenced by the age of the mother.

In the notation introduced above, $p_{j,n}$ and $p_{a,n}$ represent, respectively, the populations of juvenile and adult females in the nth year. Putting together all the information given, we arrive at the following 2×2 system:

$$p_{j,n} = \qquad\qquad kp_{a,n-1}$$
$$p_{a,n} = \alpha p_{j,n-1} + \beta p_{a,n-1}$$

(2)

or

$$\mathbf{p}_n = A\mathbf{p}_{n-1},$$

(3)

where $\mathbf{p}_n = \begin{pmatrix} p_{j,n} \\ p_{a,n} \end{pmatrix}$ and $A = \begin{pmatrix} 0 & k \\ \alpha & \beta \end{pmatrix}$. It is clear, from (3), that $\mathbf{p}_1 = A\mathbf{p}_0$, $\mathbf{p}_2 = A\mathbf{p}_1 = A(A\mathbf{p}_0) = A^2\mathbf{p}_0, \ldots,$ and so on. Hence

$$\mathbf{p}_n = A^n\mathbf{p}_0,$$

(4)

where $\mathbf{p}_0$ is the vector of initial populations of juvenile and adult females.

Equation (4) is like equation (1), but now we are able to distinguish between the survival rates of juvenile and adult birds.

EXAMPLE 1

Let $A = \begin{pmatrix} 0 & 2 \\ 0.3 & 0.5 \end{pmatrix}$. This means that each adult female produces two female offspring and, since the number of males is assumed equal to the number of females, at least four eggs—and probably many more, since losses among fledglings are likely to be high. From the model, it is apparent that α and β lie in the interval $[0, 1]$. Since juvenile birds are not as likely as adults to survive, we must have $\alpha < \beta$.

In Table 1 we assume that, initially, there are 10 female (and 10 male) adults and no juveniles. The computations were done on a computer, but the work would not be too onerous if done on a hand calculator. For example, $\mathbf{p}_1 = \begin{pmatrix} 0 & 2 \\ 0.3 & 0.5 \end{pmatrix}\begin{pmatrix} 0 \\ 10 \end{pmatrix} = \begin{pmatrix} 20 \\ 5 \end{pmatrix}$, so that $p_{j,1} = 20$, $p_{a,1} = 5$, the total female population after 1 year is 25, and the ratio of juvenile to adult females is 4 to 1. In the second year, $\mathbf{p}_2 = \begin{pmatrix} 0 & 2 \\ 0.3 & 0.5 \end{pmatrix}\begin{pmatrix} 20 \\ 5 \end{pmatrix} = \begin{pmatrix} 10 \\ 8.5 \end{pmatrix}$, which we round down to $\begin{pmatrix} 10 \\ 8 \end{pmatrix}$ since we cannot have $8\frac{1}{2}$ adult birds. Table 1

TABLE 1

Year n	No. of juveniles $p_{j,n}$	No. of adults $p_{a,n}$	Total female population T_n in nth year	$p_{j,n}/p_{a,n}$ [1]	T_n/T_{n-1} [1]
0	0	10	10	0	—
1	20	5	25	4.00	2.50
2	10	8	18	1.18	0.74
3	17	7	24	2.34	1.31
4	14	8	22	1.66	0.96
5	17	8	25	2.00	1.13
10	22	12	34	1.87	1.06
11	24	12	36	1.88	1.07
12	25	13	38	1.88	1.06
20	42	22	64	1.88	1.06

[1] The figures in these columns were obtained before the numbers in the previous columns were rounded. Thus, for example, in year 2, $p_{j,2}/p_{a,2} = 10/8.5 \approx 1.176470588 \approx 1.18$.

shows the ratios $p_{j,n}/p_{a,n}$ and the ratios T_n/T_{n-1} of the total number of females in successive years.

In Table 1 it appears that the ratio $p_{j,n}/p_{a,n}$ is approaching the constant 1.88 while the total population seems to be increasing at a constant rate of 6 percent a year. Let us see if we can determine why this is the case. If we calculate the eigenvalues of $A = \begin{pmatrix} 0 & 2 \\ 0.3 & 0.5 \end{pmatrix}$, we obtain the characteristic equation

$$\lambda^2 - 0.5\lambda - 0.6 = 0$$

with roots

$$\lambda = \frac{0.5 \pm \sqrt{0.25 + 2.4}}{2} = \frac{0.5 \pm \sqrt{2.65}}{2} \approx \begin{cases} 1.06, \\ -0.56. \end{cases}$$

Obtaining an eigenvalue $\lambda_1 \approx 1.06$ is certainly interesting, since this is approximately the growth rate of the population noted in Table 1. An eigenvector $\mathbf{v}_1$ corresponding to $\lambda_1 \approx 1.06$ is obtained by solving

$$(A - 1.06I)\mathbf{v}_1 = \begin{pmatrix} -1.06 & 2 \\ 0.3 & -0.56 \end{pmatrix}\begin{pmatrix} x_1 \\ x_2 \end{pmatrix} = \begin{pmatrix} 0 \\ 0 \end{pmatrix}$$

so that $1.06x_1 = 2x_2$. Setting $x_2 = 1$ we get $x_1 = 2/1.06 \approx 1.88$ so that $\mathbf{v}_1 = \begin{pmatrix} 1.88 \\ 1 \end{pmatrix}$.

Obtaining 1.88 as the top entry in $\mathbf{v}_1$ is also interesting, as this is the ratio of juvenile to female adults in Table 1. We now show that these observations are not an accident, but a fundamental property of problems where the population at time n, $\mathbf{p}_n$, is obtained by multiplying the initial population, $\mathbf{p}_0$, n times by a matrix A.

For simplicity we shall restrict our considerations to 2×2 matrices, although the results we obtain also apply to $n \times n$ matrices. Suppose that A has the real distinct eigenvalues λ_1 and λ_2 with corresponding eigenvectors $\mathbf{v}_1$ and $\mathbf{v}_2$. Since $\mathbf{v}_1$ and $\mathbf{v}_2$ are linearly independent, we can express the initial population as a linear combination of $\mathbf{v}_1$ and $\mathbf{v}_2$:

$$\mathbf{p}_0 = a_1\mathbf{v}_1 + a_2\mathbf{v}_2 \tag{5}$$

for some real numbers a_1 and a_2. Then (4) becomes

$$\mathbf{p}_n = A^n(a_1\mathbf{v}_1 + a_2\mathbf{v}_2). \tag{6}$$

But $A\mathbf{v}_1 = \lambda_1\mathbf{v}_1$ and $A^2\mathbf{v}_1 = A(A\mathbf{v}_1) = A(\lambda_1\mathbf{v}_1) = \lambda_1 A\mathbf{v}_1 = \lambda_1(\lambda_1\mathbf{v}_1) = \lambda_1^2\mathbf{v}_1$. Thus we can see that $A^n\mathbf{v}_1 = \lambda_1^n\mathbf{v}_1$, $A^n\mathbf{v}_2 = \lambda_2^n\mathbf{v}_2$, and, from (6),

$$\mathbf{p}_n = a_1\lambda_1^n\mathbf{v}_1 + a_2\lambda_2^n\mathbf{v}_2. \tag{7}$$

Now assume that $|\lambda_1| > |\lambda_2|$ (this is certainly the case in Example 1). We can then rewrite (7) as

$$\mathbf{p}_n = \lambda_1^n\left[a_1\mathbf{v}_1 + \left(\frac{\lambda_2}{\lambda_1}\right)^n a_2\mathbf{v}_2\right]. \tag{8}$$

Since $|\lambda_2/\lambda_1| < 1$, it is apparent that $(\lambda_2/\lambda_1)^n$ gets very small as n gets large. Thus, for n large,

$$\mathbf{p}_n \approx a_1\lambda_1^n\mathbf{v}_1 \tag{9}$$

This means that, in the long run, the age distribution stabilizes and is proportional to $\mathbf{v}_1$. Each age group will change by a factor of λ_1 each year. Thus—in the long run—equation (4) acts just like equation (1). In the short term—that is, before "stability" is reached—the numbers oscillate. The magnitude of this oscillation depends on the magnitude of λ_2/λ_1 (which is negative, thus explaining the oscillation).

Remark. In the preceding computations precision was lost because we rounded to only two decimal places of accuracy. Much greater accuracy is obtained by using a hand calculator or computer. For example, using a hand calculator, we easily calculate $\lambda_1 \approx 1.06394103$, $\lambda_2 \approx -0.5639410298$, $\mathbf{v}_1 \approx \begin{pmatrix} 1 \\ 0.531970515 \end{pmatrix}$, $\mathbf{v}_2 \approx \begin{pmatrix} 1 \\ -0.2819705149 \end{pmatrix}$, and the ratio of $p_{j,n}$ to $p_{a,n}$ is seen to be $1/0.531970515 \approx 1.879803432$.

It is remarkable just how much information is available from a simple computation of eigenvalues. It is of great interest to know whether a population will ultimately increase or decrease. It will increase if $\lambda_1 > 1$, and decrease if $|\lambda_1| < 1$. More will be said about this in Chapter 14.

Before we close this section we indicate two limitations of this model:

i. Birth and death rates often change from year to year and are particularly dependent on the weather. This model assumes a constant environment.

ii. Ecologists have found that for many species birth and death rates vary with the size of the population. In particular, a population cannot grow when it reaches a certain size due to the effects of limited food resources and overcrowding. It is obvious that a population cannot grow indefinitely at a constant rate. Otherwise that population would overrun the earth.

Problems 8.2

In Problems 1–3 find the numbers of juvenile and adult female birds after 1, 2, 5, 10, 19, and 20 years. Then find the long-term ratios of $p_{j,n}$ to $p_{a,n}$ and T_n to T_{n-1}. [*Hint:* Use equations (7) and (9) and a hand calculator and round to three decimals.]

1. $\mathbf{p}_0 = \begin{pmatrix} 0 \\ 12 \end{pmatrix}$; $k = 3$, $\alpha = 0.4$, $\beta = 0.6$

2. $\mathbf{p}_0 = \begin{pmatrix} 0 \\ 15 \end{pmatrix}$; $k = 1$, $\alpha = 0.3$, $\beta = 0.4$

3. $\mathbf{p}_0 = \begin{pmatrix} 0 \\ 20 \end{pmatrix}$; $k = 4$, $\alpha = 0.7$, $\beta = 0.8$

4. Show that if $\alpha = \beta$ and $\alpha > \frac{1}{2}$, then the bird population will always increase in the long run if at least one female offspring on the average is produced by each female adult.

5. Show that, in the long run, the ratio $p_{j,n}/p_{a,n}$ approaches the limiting value k/λ_1.

6. Suppose we divide the adult birds into two age groups: those 1–5 years old and those more than 5 years old. Assume that the survival rate for birds in the first group is β while in the second group it is γ (and $\beta > \gamma$). Assume that the birds in the first group are equally divided as to age. (That is, if there are 100 birds in the group, then 20 are 1 year old, 20 are 2 years old, and so on.) Formulate a 3×3 matrix model for this situation.

8.3 Similar Matrices and Diagonalization

In this section we describe an interesting and useful relationship that can hold between two matrices.

Similar Matrices

DEFINITION
Two $n \times n$ matrices A and B are said to be **similar** if there exists an invertible $n \times n$ matrix C such that

$$B = C^{-1}AC. \tag{1}$$

Similarity Transformation

The function defined by (1) which takes the matrix A into the matrix B is called a **similarity transformation.**

Note. $C^{-1}(A_1 + A_2)C = C^{-1}A_1C + C^{-1}A_2C$ and $C^{-1}(\alpha A)C = \alpha C^{-1}AC$, so that the function defined by (1) is, in fact, a linear transformation. This explains the use of the word "transformation" in the definition above.

The purpose of this section is to show that (*i*) similar matrices have several important properties in common and (*ii*) most matrices are similar to diagonal matrices.

EXAMPLE 1

Let $A = \begin{pmatrix} 2 & 1 \\ 0 & -1 \end{pmatrix}$, $B = \begin{pmatrix} 4 & -2 \\ 5 & -3 \end{pmatrix}$, and $C = \begin{pmatrix} 2 & -1 \\ -1 & 1 \end{pmatrix}$. Then $CB = \begin{pmatrix} 2 & -1 \\ -1 & 1 \end{pmatrix}\begin{pmatrix} 4 & -2 \\ 5 & -3 \end{pmatrix} = \begin{pmatrix} 3 & -1 \\ 1 & -1 \end{pmatrix}$ and $AC = \begin{pmatrix} 2 & 1 \\ 0 & -1 \end{pmatrix}\begin{pmatrix} 2 & -1 \\ -1 & 1 \end{pmatrix} = \begin{pmatrix} 3 & -1 \\ 1 & -1 \end{pmatrix}$. Thus $CB =$

AC. Since det $C = 1 \neq 0$, C is invertible; and since $CB = AC$, we have $C^{-1}CB = C^{-1}AC$ or $B = C^{-1}AC$. This shows that A and B are similar.

Note. If $AC = CB$, it is not necessary to compute C^{-1} to show that A and B are similar. It is only necessary to know that C is nonsingular.

■ **THEOREM 1**

If A and B are similar $n \times n$ matrices, then A and B have the same characteristic equation and, therefore, have the same eigenvalues.

PROOF

A and B are similar, $B = C^{-1}AC$ and, since det $AB =$ det A det B,

$$\det(B - \lambda I) = \det(C^{-1}AC - \lambda I) = \det[C^{-1}AC - C^{-1}(\lambda I)C]$$

$$= \det[C^{-1}(A - \lambda I)C] = \det(C^{-1})\det(A - \lambda I)\det(C)$$

$$= \det(C^{-1})\det(C)\det(A - \lambda I) = \det(C^{-1}C)\det(A - \lambda I)$$

$$= \det I \det(A - \lambda I) = \det(A - \lambda I).$$

This means that A and B have the same characteristic equation and, since eigenvalues are roots of the characteristic equation, they have the same eigenvalues with the same algebraic multiplicities. ■

EXAMPLE 2

We check that the matrices A and B, which were shown to be similar in Example 1, have the same eigenvalues:

$$\det(A - \lambda I) = \begin{vmatrix} 2 - \lambda & 1 \\ 0 & -1 - \lambda \end{vmatrix} = \lambda^2 - \lambda - 2 = (\lambda - 2)(\lambda + 1) = 0,$$

$$\det(B - \lambda I) = \begin{vmatrix} 4 - \lambda & -2 \\ 5 & -3 - \lambda \end{vmatrix} = \lambda^2 - \lambda - 2 = (\lambda - 2)(\lambda + 1) = 0.$$

In both cases, the eigenvalues are 2 and -1.

In a variety of applications it is quite useful to "diagonalize" a matrix A—that is, to find, if possible, a diagonal matrix similar to A.

DEFINITION

Diagonalizable Matrix

An $n \times n$ matrix A is **diagonalizable** if there is at least one diagonal matrix D such that A is similar to D.

Remark. If D is a diagonal matrix, then its eigenvalues are its diagonal components. If A is similar to D, then A and D have the same eigenvalues (by Theorem 1). Putting these two facts together, we observe that if A is diagonalizable, then A is similar to a diagonal matrix whose diagonal components are the eigenvalues of A.

The next theorem tells us when a matrix is diagonalizable.

■ **THEOREM 2**

An $n \times n$ matrix A is diagonalizable if and only if it has n linearly independent eigenvectors. In that case, the diagonal matrix D similar to A is given by

$$D = \begin{pmatrix} \lambda_1 & 0 & 0 & \cdots & 0 \\ 0 & \lambda_2 & 0 & \cdots & 0 \\ 0 & 0 & \lambda_3 & \cdots & 0 \\ \vdots & \vdots & \vdots & & \vdots \\ 0 & 0 & 0 & \cdots & \lambda_n \end{pmatrix} \tag{2}$$

where $\lambda_1, \lambda_2, \ldots, \lambda_n$ are the eigenvalues of A. If C is a matrix whose columns are linearly independent eigenvectors of A, then

$$D = C^{-1}AC. \tag{3}$$

PROOF

We first assume that A has n linearly independent eigenvectors $\mathbf{v}_1, \mathbf{v}_2, \ldots, \mathbf{v}_n$ corresponding to the (not necessarily distinct) eigenvalues $\lambda_1, \lambda_2, \ldots, \lambda_n$. Let

$$\mathbf{v}_1 = \begin{pmatrix} c_{11} \\ c_{21} \\ \vdots \\ c_{n1} \end{pmatrix} \quad \mathbf{v}_2 = \begin{pmatrix} c_{12} \\ c_{22} \\ \vdots \\ c_{n2} \end{pmatrix}, \ldots, \mathbf{v}_n = \begin{pmatrix} c_{1n} \\ c_{2n} \\ \vdots \\ c_{nn} \end{pmatrix}$$

and let

$$C = \begin{pmatrix} c_{11} & c_{12} & \cdots & c_{1n} \\ c_{21} & c_{22} & \cdots & c_{2n} \\ \vdots & \vdots & & \vdots \\ c_{n1} & c_{n2} & \cdots & c_{nn} \end{pmatrix}.$$

Then C is invertible since its columns are linearly independent. Now

$$AC = A(\mathbf{v}_1 \quad \mathbf{v}_2 \quad \cdots \quad \mathbf{v}_n) = (A\mathbf{v}_1 \quad A\mathbf{v}_2 \quad \cdots \quad A\mathbf{v}_n)$$

$$= (\lambda_1\mathbf{v}_1 \quad \lambda_2\mathbf{v}_2 \quad \cdots \quad \lambda_n\mathbf{v}_n) = CD,$$

where

$$CD = \begin{pmatrix} c_{11} & c_{12} & \cdots & c_{1n} \\ c_{21} & c_{22} & \cdots & c_{2n} \\ \vdots & \vdots & & \vdots \\ c_{n1} & c_{n2} & \cdots & c_{nn} \end{pmatrix} \begin{pmatrix} \lambda_1 & 0 & \cdots & 0 \\ 0 & \lambda_2 & \cdots & 0 \\ \vdots & \vdots & & \vdots \\ 0 & 0 & \cdots & \lambda_n \end{pmatrix}.$$

Thus

$$AC = CD \qquad (4)$$

and, since C is invertible, we can multiply both sides of (4) on the left by C^{-1} to obtain

$$D = C^{-1}AC \qquad (5)$$

This proves that if A has n linearly independent eigenvectors, then A is diagonalizable. Conversely, suppose that A is diagonalizable. That is, suppose that (5) holds for some invertible matrix C. Let $\mathbf{v}_1, \mathbf{v}_2, \ldots, \mathbf{v}_n$ be the columns of C. Then $AC = CD$ and, reversing the arguments above, we immediately see that $A\mathbf{v}_i = \lambda_i \mathbf{v}_i$ for $i = 1, 2, \ldots, n$. Thus $\mathbf{v}_1, \mathbf{v}_2, \ldots, \mathbf{v}_n$ are eigenvectors of A and are linearly independent because C is invertible. ∎

Notation. To indicate that D is a diagonal matrix with diagonal components $\lambda_1, \lambda_2, \ldots, \lambda_n$, we write $D = \text{diag}(\lambda_1, \lambda_2, \ldots, \lambda_n)$.

Theorem 2 has a useful corollary that follows immediately from Theorem 8.1.4 on p. 457.

■ **COROLLARY**

If the $n \times n$ matrix A has n distinct eigenvalues, then A is diagonalizable. ∎

Remark. If the real coefficients of a polynomial of degree n are picked at random, then, with probability 1, the polynomial will have n distinct roots. It is not difficult to see, intuitively, why this is so. If $n = 2$, for example, then the equation $\lambda^2 + a\lambda + b = 0$ has equal roots if and only if $a^2 = 4b$—not a likely event if a and b are chosen at random. We can, of course, write down polynomials having roots of algebraic multiplicity greater than 1, but these polynomials are exceptional. Thus, without attempting to be mathematically precise, it is fair to say that *most* polynomials have distinct roots. Hence *most* matrices have distinct eigenvalues and, as we stated at the beginning of the section, *most* matrices are diagonalizable.

EXAMPLE 3

Let $A = \begin{pmatrix} 4 & 2 \\ 3 & 3 \end{pmatrix}$. In Example 8.1.4 on p. 458 we found the two linearly independent eigenvectors $\mathbf{v}_1 = \begin{pmatrix} 2 \\ -3 \end{pmatrix}$ and $\mathbf{v}_2 = \begin{pmatrix} 1 \\ 1 \end{pmatrix}$. Then, setting $C = \begin{pmatrix} 2 & 1 \\ -3 & 1 \end{pmatrix}$, we find that

$$C^{-1}AC = \frac{1}{5}\begin{pmatrix} 1 & -1 \\ 3 & 2 \end{pmatrix}\begin{pmatrix} 4 & 2 \\ 3 & 3 \end{pmatrix}\begin{pmatrix} 2 & 1 \\ -3 & 1 \end{pmatrix}$$

$$= \frac{1}{5}\begin{pmatrix} 1 & -1 \\ 3 & 2 \end{pmatrix}\begin{pmatrix} 2 & 6 \\ -3 & 6 \end{pmatrix} = \frac{1}{5}\begin{pmatrix} 5 & 0 \\ 0 & 30 \end{pmatrix} = \begin{pmatrix} 1 & 0 \\ 0 & 6 \end{pmatrix},$$

which is the matrix whose diagonal components are the eigenvalues of A.

EXAMPLE 4

Let $A = \begin{pmatrix} 1 & -1 & 4 \\ 3 & 2 & -1 \\ 2 & 1 & -1 \end{pmatrix}$. In Example 8.1.5 on p. 459 we computed the three linearly

independent eigenvectors $\mathbf{v}_1 = \begin{pmatrix} -1 \\ 4 \\ 1 \end{pmatrix}$, $\mathbf{v}_2 = \begin{pmatrix} 1 \\ -1 \\ -1 \end{pmatrix}$, and $\mathbf{v}_3 = \begin{pmatrix} 1 \\ 2 \\ 1 \end{pmatrix}$. Then $C =$

$\begin{pmatrix} -1 & 1 & 1 \\ 4 & -1 & 2 \\ 1 & -1 & 1 \end{pmatrix}$ is invertible and

$$AC = \begin{pmatrix} 1 & -1 & 4 \\ 3 & 2 & -1 \\ 2 & 1 & -1 \end{pmatrix} \begin{pmatrix} -1 & 1 & 1 \\ 4 & -1 & 2 \\ 1 & -1 & 1 \end{pmatrix} = \begin{pmatrix} -1 & -2 & 3 \\ 4 & 2 & 6 \\ 1 & 2 & 3 \end{pmatrix}$$

$$= \begin{pmatrix} -1 & 1 & 1 \\ 4 & -1 & 2 \\ 1 & -1 & 1 \end{pmatrix} \begin{pmatrix} 1 & 0 & 0 \\ 0 & -2 & 0 \\ 0 & 0 & 3 \end{pmatrix} = CD,$$

where D is the matrix whose diagonal components are the eigenvalues of A.

Remark. Since there are an infinite number of ways to choose an eigenvector, there are an infinite number of ways to choose the diagonalizing matrix C. The only advice is to choose the eigenvectors and matrix C that are, arithmetically, the easiest to work with. This usually means that you should insert as many 0's and 1's as possible.

EXAMPLE 5

Let $A = \begin{pmatrix} 4 & 1 \\ 0 & 4 \end{pmatrix}$. In Example 8.1.8 on p. 462 we saw that A did *not* have two linearly independent eigenvectors. Suppose that A were diagonalizable (in contradiction to Theorem 2). Then $D = \begin{pmatrix} 4 & 0 \\ 0 & 4 \end{pmatrix}$ and there would be an invertible matrix C such that $C^{-1}AC = D$. Multiplying this equation on the left by C and on the right by C^{-1}, we find that $A = CDC^{-1} = C\begin{pmatrix} 4 & 0 \\ 0 & 4 \end{pmatrix}C^{-1} = C(4I)C^{-1} = 4CIC^{-1} = 4CC^{-1} = 4I = \begin{pmatrix} 4 & 0 \\ 0 & 4 \end{pmatrix} = D$. But $A \neq D$, so no such C exists. In particular, A is not diagonizable.

We have seen that many matrices are similar to diagonal matrices. However, two questions remain:

i. Is it possible to determine whether a given matrix is diagonalizable without computing eigenvalues and eigenvectors?

ii. What do we do if A is not diagonalizable?

We shall give a partial answer to the first question in the next section. The second question can be answered by computing the **Jordan canonical form** of a matrix. This topic is discussed in a number of linear algebra books. We do not discuss Jordan canonical form.†

Problems 8.3

In Problems 1–15 determine whether the given matrix A is diagonalizable. If it is, find a matrix C such that $C^{-1}AC = D$.

1. $\begin{pmatrix} -2 & -2 \\ -5 & 1 \end{pmatrix}$

2. $\begin{pmatrix} 3 & -1 \\ -2 & 4 \end{pmatrix}$

3. $\begin{pmatrix} 2 & -1 \\ 5 & -2 \end{pmatrix}$

4. $\begin{pmatrix} 3 & -5 \\ 1 & -1 \end{pmatrix}$

5. $\begin{pmatrix} 3 & 2 \\ -5 & 1 \end{pmatrix}$

6. $\begin{pmatrix} 1 & -1 & 0 \\ -1 & 2 & -1 \\ 0 & -1 & 1 \end{pmatrix}$

7. $\begin{pmatrix} 1 & 1 & -2 \\ -1 & 2 & 1 \\ 0 & 1 & -1 \end{pmatrix}$

8. $\begin{pmatrix} 2 & 1 & 0 \\ 0 & 0 & 1 \\ 0 & 0 & 0 \end{pmatrix}$

9. $\begin{pmatrix} 3 & 0 & 0 \\ 0 & 0 & 1 \\ 0 & 0 & 2 \end{pmatrix}$

10. $\begin{pmatrix} 3 & -1 & -1 \\ 1 & 1 & -1 \\ 1 & -1 & 1 \end{pmatrix}$

11. $\begin{pmatrix} 7 & -2 & -4 \\ 3 & 0 & -2 \\ 6 & -2 & -3 \end{pmatrix}$

12. $\begin{pmatrix} 4 & 6 & 6 \\ 1 & 3 & 2 \\ -1 & -5 & -2 \end{pmatrix}$

13. $\begin{pmatrix} -3 & -7 & -5 \\ 2 & 4 & 3 \\ 1 & 2 & 2 \end{pmatrix}$

14. $\begin{pmatrix} -2 & -2 & 0 & 0 \\ -5 & 1 & 0 & 0 \\ 0 & 0 & 2 & -1 \\ 0 & 0 & 5 & -2 \end{pmatrix}$

15. $\begin{pmatrix} 4 & 1 & 0 & 1 \\ 2 & 3 & 0 & 1 \\ -2 & 1 & 2 & -3 \\ 2 & -1 & 0 & 5 \end{pmatrix}$

16. Show that if A is similar to B and B is similar to C, then A is similar to C.

17. Let $A = \begin{pmatrix} 1 & 2 & 0 \\ 0 & 2 & 0 \\ 0 & 0 & 1 \end{pmatrix}$.

   (a) Show that the only eigenvalues of A are 1 and 2.

   (b) Show that A is diagonalizable.

This example shows that a 3×3 matrix (which is not the

identity matrix) may be diagonalizable even though it does not have three distinct eigenvalues.

** 18. If A is similar to B, show that $\rho(A) = \rho(B)$ and $\nu(A) = \nu(B)$. [*Hint:* See Section 7.10. First prove that if C is invertible, then $\nu(CA) = \nu(A)$ by showing that $\mathbf{x} \in N_A$ if and only if $\mathbf{x} \in N_{CA}$. Next prove that $\rho(AC) = \rho(A)$ by showing that $R_A = R_{AC}$. Conclude that $\rho(AC) = \rho(CA) = \rho(A)$. Finally, use the fact that C^{-1} is invertible to show that $\rho(C^{-1}AC) = \rho(A)$.]

19. If A is similar to B, show that A^n is similar to B^n for any positive integer n.

20. If A is similar to B, show that $\det A = \det B$.

21. Let $D = \begin{pmatrix} 1 & 0 \\ 0 & -1 \end{pmatrix}$. Compute D^{20}.

22. Let $A = \begin{pmatrix} 3 & -4 \\ 2 & -3 \end{pmatrix}$. Compute A^{20}. [*Hint:* Find a C such that $A = CDC^{-1}$, where D is diagonal and show that $A^{20} = CD^{20}C^{-1}$.]

23. Suppose that $C^{-1}AC = D$. Show that for any integer n, $A^n = CD^nC^{-1}$. This gives an easy way to compute powers of a diagonalizable matrix.

24. Use the result of Problem 23 to compute A^{10}, where $A = \begin{pmatrix} 3 & 2 & 4 \\ 2 & 0 & 2 \\ 4 & 2 & 3 \end{pmatrix}$.

25. Let A be an $n \times n$ matrix whose characteristic equation is $(\lambda - c)^n = 0$. Show that A is diagonalizable if and only if $A = cI$.

26. If A is diagonalizable, show that $\det A = \lambda_1 \lambda_2 \cdots \lambda_n$, where $\lambda_1, \lambda_2, \ldots, \lambda_n$ are the eigenvalues of A.

* 27. Let A and B be real $n \times n$ matrices with distinct eigenvalues. Prove that $AB = BA$ if and only if A and B have the same eigenvectors.

---

† An introduction to Jordan canonical form can be found in S. I. Grossman, *Elementary Linear Algebra*, 3rd ed., Wadsworth, Belmont, Calif., 1987, Section 6.6.

8.4 Symmetric Matrices and Orthogonal Diagonalization

In this section we shall see that symmetric matrices have a number of important properties. In particular, we shall show that any symmetric matrix has n linearly independent real eigenvectors and therefore, by Theorem 8.3.2, is diagonalizable. We begin by proving that the eigenvalues of a real symmetric matrix are real.

■ **THEOREM 1**

Let A be a real $n \times n$ symmetric matrix. Then the eigenvalues of A are real.

PROOF (Optional)

Let λ be an eigenvalue of A with eigenvector $\mathbf{v}$; that is, $A\mathbf{v} = \lambda\mathbf{v}$. Here $\mathbf{v}$ is a vector in $\mathbb{C}^n$, so its entries may be complex numbers. In $\mathbb{C}^n$ we define the dot product as follows:

$$\mathbf{x} \bullet \mathbf{y} = x_1 \bar{y}_1 + x_2 \bar{y}_2 + \cdots + x_n \bar{y}_n = \sum_{i=1}^{n} x_i \bar{y}_i, \tag{1}$$

where $\bar{y}_i$ denotes the complex conjugate of y_i. In particular, for $\mathbf{x} = \mathbf{y}$ we obtain

$$\mathbf{x} \bullet \mathbf{x} = \sum_{i=1}^{n} x_i \bar{x}_i = \sum_{i=1}^{n} |x_i|^2. \tag{2}$$

From (1) it follows that

$$(\alpha\mathbf{x}) \bullet \mathbf{y} = \alpha(\mathbf{x} \bullet \mathbf{y}) \tag{3}$$

and

$$\mathbf{x} \bullet (\alpha\mathbf{y}) = \bar{\alpha}(\mathbf{x} \bullet \mathbf{y}). \tag{4}$$

Since λ is an eigenvalue of A with corresponding eigenvector $\mathbf{v}$, we have, from (3),

$$A\mathbf{v} \bullet \mathbf{v} = (\lambda\mathbf{v}) \bullet \mathbf{v} = \lambda(\mathbf{v} \bullet \mathbf{v}). \tag{5}$$

The ith component of the vector $A\mathbf{v}$ is $\sum_{j=1}^{n} a_{ij} v_j$. Thus

$$A\mathbf{v} \bullet \mathbf{v} = \sum_{i=1}^{n} \left(\sum_{j=1}^{n} a_{ij} v_j \right) \bar{v}_i = \sum_{i=1}^{n} \sum_{j=1}^{n} a_{ij} v_j \bar{v}_i. \tag{6}$$

Now the jth component of $A\mathbf{v}$ can be written $\sum_{i=1}^{n} a_{ji} v_i$. Thus

$$\mathbf{v} \bullet A\mathbf{v} = \sum_{j=1}^{n} v_j \overline{\sum_{i=1}^{n} a_{ji} v_i}, \tag{7}$$

and

$$\overline{\sum_{i=1}^{n} a_{ji} v_i} = \overline{a_{j1} v_1 + a_{j2} v_2 + \cdots + a_{jn} v_n} = \overline{a_{j1}} \bar{v}_1 + \overline{a_{j2}} \bar{v}_2 + \cdots + \overline{a_{jn}} \bar{v}_n = \sum_{i=1}^{n} \overline{a_{ji}} \bar{v}_i.$$

But A is real and symmetric so $\overline{a_{ji}} = a_{ji} = a_{ij}$ and we have

$$\overline{\sum_{i=1}^{n} a_{ji} v_i} = \sum_{i=1}^{n} a_{ij} \bar{v}_i. \tag{8}$$

Inserting (8) into (7) we find that

$$\mathbf{v} \cdot A\mathbf{v} = \sum_{j=1}^{n} v_j \sum_{i=1}^{n} a_{ij} \bar{v}_i = \sum_{j=1}^{n} \sum_{i=1}^{n} a_{ij} v_j \bar{v}_i. \tag{9}$$

The double sums in (6) and (9) are the same because, as each sum is finite, we get the same result whether we sum first with respect to i or j. Thus

$$A\mathbf{v} \cdot \mathbf{v} = \mathbf{v} \cdot A\mathbf{v}. \tag{10}$$

But, from (4) and (5),

$$\mathbf{v} \cdot A\mathbf{v} = \mathbf{v} \cdot \lambda\mathbf{v} = \bar{\lambda}(\mathbf{v} \cdot \mathbf{v}) \quad \text{and} \quad A\mathbf{v} \cdot \mathbf{v} = \lambda(\mathbf{v} \cdot \mathbf{v}).$$

So, from (10),

$$\lambda(\mathbf{v} \cdot \mathbf{v}) = \bar{\lambda}(\mathbf{v} \cdot \mathbf{v}). \tag{11}$$

But $(\mathbf{v} \cdot \mathbf{v}) = |\mathbf{v}|^2 \neq 0$, since $\mathbf{v}$ is an eigenvector. Thus we can divide both sides of (11) by $(\mathbf{v} \cdot \mathbf{v})$ to obtain

$$\lambda = \bar{\lambda}. \tag{12}$$

If $\lambda = a + ib$, then $\bar{\lambda} = a - ib$ and, from (12), we have

$$a + ib = a - ib,$$

which can hold only if $b = 0$. This shows that $\lambda = a$; hence λ is real and the proof is complete. ■

We saw in Theorem 8.1.4 on p. 457 that eigenvectors corresponding to different eigenvalues are linearly independent. For symmetric matrices the result is stronger: *Eigenvectors of a symmetric matrix corresponding to different eigenvalues are orthogonal.*

■ **THEOREM 2**
Let A be a real symmetric $n \times n$ matrix. If λ_1 and λ_2 are distinct eigenvalues with corresponding real eigenvectors $\mathbf{v}_1$ and $\mathbf{v}_2$, then $\mathbf{v}_1$ and $\mathbf{v}_2$ are orthogonal.

PROOF
We compute

$$A\mathbf{v}_1 \cdot \mathbf{v}_2 = \lambda_1 \mathbf{v}_1 \cdot \mathbf{v}_2 = \lambda_1(\mathbf{v}_1 \cdot \mathbf{v}_2). \tag{13}$$

In Problem 27 you are asked to show that

$$A\mathbf{v}_1 \cdot \mathbf{v}_2 = \mathbf{v}_1 \cdot A^t \mathbf{v}_2. \tag{14}$$

Then

$$Av_1 \cdot v_2 = v_1 \cdot A^t v_2 = v_1 \cdot Av_2 = v_1 \cdot (\lambda_2 v_2) = \lambda_2 (v_1 \cdot v_2). \qquad (15)$$

Combining (13) and (15) we have $\lambda_1(v_1 \cdot v_2) = \lambda_2(v_1 \cdot v_2)$ and since $\lambda_1 \neq \lambda_2$, we conclude that $v_1 \cdot v_2 = 0$. This is what we wanted to show. ∎

We now state the main result of this section. Its proof, which is difficult (and optional), is given at the end of this section.

THEOREM 3
Let A be a real symmetric $n \times n$ matrix. Then A has n real orthonormal eigenvectors. ∎

Before continuing, we need a new definition.

Orthogonal Matrix

DEFINITION
The $n \times n$ matrix Q is called **orthogonal** if Q is invertible and

$$Q^{-1} = Q^t.$$

Orthogonal matrices are not difficult to find, according to the next theorem.

THEOREM 4
The $n \times n$ matrix Q is orthogonal if and only if the columns of Q form an orthonormal basis for $\mathbb{R}^n$.

PROOF
Let

$$Q = \begin{pmatrix} a_{11} & a_{12} & \cdots & a_{1n} \\ a_{21} & a_{22} & \cdots & a_{2n} \\ \vdots & \vdots & & \vdots \\ a_{n1} & a_{n2} & \cdots & a_{nn} \end{pmatrix}.$$

Then

$$Q^t = \begin{pmatrix} a_{11} & a_{21} & \cdots & a_{n1} \\ a_{12} & a_{22} & \cdots & a_{n2} \\ \vdots & \vdots & & \vdots \\ a_{1n} & a_{2n} & \cdots & a_{nn} \end{pmatrix}.$$

Let $B = (b_{ij}) = Q^t Q$. Then

$$b_{ij} = a_{1i}a_{1j} + a_{2i}a_{2j} + \cdots + a_{ni}a_{nj} = c_i \cdot c_j,$$

where c_i denotes the ith column of Q. If the columns of Q are orthonormal, then

$$b_{ij} = \begin{cases} 0 & \text{if} \quad i \neq j, \\ 1 & \text{if} \quad i = j. \end{cases} \tag{16}$$

That is, $B = I$. Conversely if $Q^t = Q^{-1}$, then $B = I$, so that (16) holds and the columns of Q are orthonormal. This completes the proof. ■

EXAMPLE 1

You should verify that the vectors $\begin{pmatrix} 1/\sqrt{2} \\ 1/\sqrt{2} \\ 0 \end{pmatrix}$, $\begin{pmatrix} -1/\sqrt{6} \\ 1/\sqrt{6} \\ 2/\sqrt{6} \end{pmatrix}$, $\begin{pmatrix} 1/\sqrt{3} \\ -1/\sqrt{3} \\ 1/\sqrt{3} \end{pmatrix}$ form an orthonormal

basis in $\mathbb{R}^3$. Thus the matrix $Q = \begin{pmatrix} 1/\sqrt{2} & -1/\sqrt{6} & 1/\sqrt{3} \\ 1/\sqrt{2} & 1/\sqrt{6} & -1/\sqrt{3} \\ 0 & 2/\sqrt{6} & 1/\sqrt{3} \end{pmatrix}$ is an orthogonal matrix. To

check this we note that

$$Q^t Q = \begin{pmatrix} 1/\sqrt{2} & 1/\sqrt{2} & 0 \\ -1/\sqrt{6} & 1/\sqrt{6} & 2/\sqrt{6} \\ 1/\sqrt{3} & -1/\sqrt{3} & 1/\sqrt{3} \end{pmatrix} \begin{pmatrix} 1/\sqrt{2} & -1/\sqrt{6} & 1/\sqrt{3} \\ 1/\sqrt{2} & 1/\sqrt{6} & -1/\sqrt{3} \\ 0 & 2/\sqrt{6} & 1/\sqrt{3} \end{pmatrix} = \begin{pmatrix} 1 & 0 & 0 \\ 0 & 1 & 0 \\ 0 & 0 & 1 \end{pmatrix}.$$

Theorem 3 tells us that if A is symmetric, then $\mathbb{R}^n$ has a basis $B = \{\mathbf{u}_1, \mathbf{u}_2, \ldots, \mathbf{u}_n\}$ consisting of orthonormal eigenvectors of A. Let Q be the matrix whose columns are $\mathbf{u}_1, \mathbf{u}_2, \ldots, \mathbf{u}_n$. Then Q is an orthogonal matrix. This leads to the following definition.

DEFINITION

Orthogonally Diagonalizable Matrix   An $n \times n$ matrix A is said to be **orthogonally diagonalizable** if there exists an orthogonal matrix Q such that

$$Q^t A Q = D, \tag{17}$$

where $D = \text{diag}(\lambda_1, \lambda_2, \ldots, \lambda_n)$ and $\lambda_1, \lambda_2, \ldots, \lambda_n$ are the eigenvalues of A.

Note. Remember that Q is orthogonal if $Q^t = Q^{-1}$; hence (17) could be written as $Q^{-1} A Q = D$.

■ **THEOREM 5**

Let A be a real $n \times n$ matrix. Then A is orthogonally diagonalizable if and only if A is symmetric.

PROOF

Let A be symmetric. Then, by Theorem 3, A is orthogonally diagonalizable with Q the matrix whose columns are the orthonormal eigenvectors given in Theorem 3. Conversely, suppose that A is orthogonally diagonalizable. Then there exists an orthogonal matrix Q such that $Q^tAQ = D$. Multiplying this equation on the left by Q and on the right by Q^t and using the fact that $Q^tQ = QQ^t = I$, we obtain

$$A = QDQ^t.$$

Then $A^t = (QDQ^t)^t = (Q^t)^tD^tQ^t = QDQ^t = A$. Thus A is symmetric and the theorem is proved. In the last series of equations we used the facts that $(AB)^t = B^tA^t$, $(A^t)^t = A$ (see Theorem 7.6.1, p. 402), and $D^t = D$ for any diagonal matrix D. ∎

Before giving examples, we provide the following three-step procedure for finding the orthogonal matrix Q that diagonalizes the symmetric matrix A. Recall the definition of eigenspace given on p. 464.

Procedure for Finding a Diagonalizing Matrix Q

   i. Find a basis for each eigenspace of A.
  ii. Find an orthonormal basis for each eigenspace of A by using the Gram–Schmidt process.
 iii. Write Q as the matrix whose columns are the orthonormal eigenvectors obtained in step (ii).

EXAMPLE 2

Let $A = \begin{pmatrix} 1 & -2 \\ -2 & 3 \end{pmatrix}$. Then the characteristic equation of A is

$$\det(A - \lambda I) = \begin{vmatrix} 1-\lambda & -2 \\ -2 & 3-\lambda \end{vmatrix} = \lambda^2 - 4\lambda - 1 = 0,$$

which has the roots $\lambda = (4 \pm \sqrt{20})/2 = (4 \pm 2\sqrt{5})/2 = 2 \pm \sqrt{5}$. For $\lambda_1 = 2 - \sqrt{5}$, we obtain

$$(A - \lambda I)\mathbf{v} = \begin{pmatrix} -1+\sqrt{5} & -2 \\ -2 & 1+\sqrt{5} \end{pmatrix}\begin{pmatrix} x_1 \\ x_2 \end{pmatrix} = \begin{pmatrix} 0 \\ 0 \end{pmatrix}.$$

An eigenvector is $\mathbf{v}_1 = \begin{pmatrix} 2 \\ -1+\sqrt{5} \end{pmatrix}$; $|\mathbf{v}_1| = \sqrt{2^2 + (-1+\sqrt{5})^2} = \sqrt{10 - 2\sqrt{5}}$. Thus $\mathbf{u}_1 = \dfrac{1}{\sqrt{10-2\sqrt{5}}}\begin{pmatrix} 2 \\ -1+\sqrt{5} \end{pmatrix}$. Next, for $\lambda_2 = 2 + \sqrt{5}$, we compute

$$(A - \lambda I)\mathbf{v} = \begin{pmatrix} -1-\sqrt{5} & -2 \\ -2 & 1-\sqrt{5} \end{pmatrix}\begin{pmatrix} x_1 \\ x_2 \end{pmatrix} = \begin{pmatrix} 0 \\ 0 \end{pmatrix} \text{ and } \mathbf{v}_2 = \begin{pmatrix} 1-\sqrt{5} \\ 2 \end{pmatrix}.$$

Note that $\mathbf{v}_1 \cdot \mathbf{v}_2 = 0$ (which must be true according to Theorem 2). Then $|\mathbf{v}_2| = \sqrt{10 - 2\sqrt{5}}$, so that $\mathbf{u}_2 = \dfrac{1}{\sqrt{10 - 2\sqrt{5}}} \begin{pmatrix} 1 - \sqrt{5} \\ 2 \end{pmatrix}$. Finally

$$Q = \frac{1}{\sqrt{10 - 2\sqrt{5}}} \begin{pmatrix} 2 & 1 - \sqrt{5} \\ -1 + \sqrt{5} & 2 \end{pmatrix}, \quad Q^t = \frac{1}{\sqrt{10 - 2\sqrt{5}}} \begin{pmatrix} 2 & -1 + \sqrt{5} \\ 1 - \sqrt{5} & 2 \end{pmatrix}$$

and

$$Q^t A Q = \frac{1}{10 - 2\sqrt{5}} \begin{pmatrix} 2 & -1 + \sqrt{5} \\ 1 - \sqrt{5} & 2 \end{pmatrix} \begin{pmatrix} 1 & -2 \\ -2 & 3 \end{pmatrix} \begin{pmatrix} 2 & 1 - \sqrt{5} \\ -1 + \sqrt{5} & 2 \end{pmatrix}$$

$$= \frac{1}{10 - 2\sqrt{5}} \begin{pmatrix} 2 & -1 + \sqrt{5} \\ 1 - \sqrt{5} & 2 \end{pmatrix} \begin{pmatrix} 4 - 2\sqrt{5} & -3 - \sqrt{5} \\ -7 + 3\sqrt{5} & 4 + 2\sqrt{5} \end{pmatrix}$$

$$= \frac{1}{10 - 2\sqrt{5}} \begin{pmatrix} 30 - 14\sqrt{5} & 0 \\ 0 & 10 + 6\sqrt{5} \end{pmatrix} = \begin{pmatrix} 2 - \sqrt{5} & 0 \\ 0 & 2 + \sqrt{5} \end{pmatrix}.$$

EXAMPLE 3

Let $A = \begin{pmatrix} 5 & 4 & 2 \\ 4 & 5 & 2 \\ 2 & 2 & 2 \end{pmatrix}$. Then A is symmetric and $\det(A - \lambda I) =$

$\begin{pmatrix} 5 - \lambda & 4 & 2 \\ 4 & 5 - \lambda & 2 \\ 2 & 2 & 2 - \lambda \end{pmatrix} = -(\lambda - 1)^2(\lambda - 10)$. Corresponding to $\lambda = 1$ we compute the

linearly independent eigenvectors $\mathbf{v}_1 = \begin{pmatrix} -1 \\ 1 \\ 0 \end{pmatrix}$ and $\mathbf{v}_2 = \begin{pmatrix} -1 \\ 0 \\ 2 \end{pmatrix}$. Corresponding to $\lambda =$

10 we find that $\mathbf{v}_3 = \begin{pmatrix} 2 \\ 2 \\ 1 \end{pmatrix}$. To find Q, we apply the Gram–Schmidt process to $\{\mathbf{v}_1, \mathbf{v}_2\}$,

a basis for E_1. Since $|\mathbf{v}_1| = \sqrt{2}$, we set $\mathbf{u}_1 = \begin{pmatrix} -1/\sqrt{2} \\ 1/\sqrt{2} \\ 0 \end{pmatrix}$. Next,

$$\mathbf{v}_2' = \mathbf{v}_2 - (\mathbf{v}_2 \cdot \mathbf{u}_1)\mathbf{u}_1 = \begin{pmatrix} -1 \\ 0 \\ 2 \end{pmatrix} - \frac{1}{\sqrt{2}} \begin{pmatrix} -1/\sqrt{2} \\ 1/\sqrt{2} \\ 0 \end{pmatrix} = \begin{pmatrix} -1 \\ 0 \\ 2 \end{pmatrix} - \begin{pmatrix} -1/2 \\ 1/2 \\ 0 \end{pmatrix} = \begin{pmatrix} -1/2 \\ -1/2 \\ 2 \end{pmatrix}.$$

Then $|\mathbf{v}_2| = \sqrt{18/4} = 3\sqrt{2}/2$ and $\mathbf{u}_2 = \dfrac{2}{3\sqrt{2}} \begin{pmatrix} -1/2 \\ -1/2 \\ 2 \end{pmatrix} = \begin{pmatrix} -1/3\sqrt{2} \\ -1/3\sqrt{2} \\ 4/3\sqrt{2} \end{pmatrix}$. We check this by noting

that $\mathbf{u}_1 \cdot \mathbf{u}_2 = 0$. Finally, we have $\mathbf{u}_3 = \mathbf{v}_3/|\mathbf{v}_3| = \tfrac{1}{3}\mathbf{v}_3 = \begin{pmatrix} 2/3 \\ 2/3 \\ 1/3 \end{pmatrix}$. We can check this too by

noting that $\mathbf{u}_1 \cdot \mathbf{u}_3 = 0$ and $\mathbf{u}_2 \cdot \mathbf{u}_3 = 0$. Thus

$$Q = \begin{pmatrix} -1/\sqrt{2} & -1/3\sqrt{2} & 2/3 \\ 1/\sqrt{2} & -1/3\sqrt{2} & 2/3 \\ 0 & 4/3\sqrt{2} & 1/3 \end{pmatrix}$$

and

$$Q^t A Q = \begin{pmatrix} -1/\sqrt{2} & 1/\sqrt{2} & 0 \\ -1/3\sqrt{2} & -1/3\sqrt{2} & 4/3\sqrt{2} \\ 2/3 & 2/3 & 1/3 \end{pmatrix} \begin{pmatrix} 5 & 4 & 2 \\ 4 & 5 & 2 \\ 2 & 2 & 2 \end{pmatrix} \begin{pmatrix} -1/\sqrt{2} & -1/3\sqrt{2} & 2/3 \\ 1/\sqrt{2} & -1/3\sqrt{2} & 2/3 \\ 0 & 4/3\sqrt{2} & 1/3 \end{pmatrix}$$

$$= \begin{pmatrix} -1/\sqrt{2} & 1/\sqrt{2} & 0 \\ -1/3\sqrt{2} & -1/3\sqrt{2} & 4/3\sqrt{2} \\ 2/3 & 2/3 & 1/3 \end{pmatrix} \begin{pmatrix} -1/\sqrt{2} & -1/3\sqrt{2} & 20/3 \\ 1/\sqrt{2} & -1/3\sqrt{2} & 20/3 \\ 0 & 4/3\sqrt{2} & 10/3 \end{pmatrix}$$

$$= \begin{pmatrix} 1 & 0 & 0 \\ 0 & 1 & 0 \\ 0 & 0 & 10 \end{pmatrix}.$$

Conjugate Transpose

Hermitian Matrix

Unitary Martix

In this section we have proved results for real symmetric matrices. If $A = (a_{ij})$ is a complex matrix, then the **conjugate transpose** of A, denoted A^*, is defined by the requirement that the ijth element of $A^* = \bar{a}_{ji}$. The matrix A is called **Hermitian** if $A^* = A$. It turns out that Theorems 1, 2, and 3 are also true for Hermitian matrices except that the eigenvectors need not be real. Moreover, if we define a **unitary** matrix to be a complex matrix U with $U^* = U^{-1}$, then, using the proof of Theorem 5, we can show that a Hermitian matrix is "unitarily" diagonalizable. We leave all these facts as exercises (see Problems 21–23).

EXAMPLE 4

Let

$$A = \begin{pmatrix} 2 & 3-i & 4+2i \\ 2+i & 5-2i & 7+3i \\ 4-2i & -3+7i & 5-2i \end{pmatrix}.$$

Find A^*.

SOLUTION

$$A^* = \bar{A}^t = \begin{pmatrix} 2 & 3+i & 4-2i \\ 2-i & 5+2i & 7-3i \\ 4+2i & -3-7i & 5+2i \end{pmatrix}^t = \begin{pmatrix} 2 & 2-i & 4+2i \\ 3+i & 5+2i & -3-7i \\ 4-2i & 7-3i & 5+2i \end{pmatrix}.$$

EXAMPLE 5

The following matrices are Hermitian:

$$A = \begin{pmatrix} 2 & 3+2i \\ 3-2i & 5 \end{pmatrix} \quad \text{and} \quad B = \begin{pmatrix} 2 & -1+5i & 7+2i \\ -1-5i & 4 & 3-5i \\ 7-2i & 3+5i & 7 \end{pmatrix}.$$

EXAMPLE 6

Show that

$$A = \begin{pmatrix} \dfrac{1}{2} & \dfrac{\sqrt{3}}{2}i \\ -\dfrac{\sqrt{3}}{2}i & -\dfrac{1}{2} \end{pmatrix}$$

is unitary.

SOLUTION

First we note that $A^* = A$. Then

$$A^*A = \begin{pmatrix} \dfrac{1}{2} & \dfrac{\sqrt{3}}{2}i \\ -\dfrac{\sqrt{3}}{2}i & -\dfrac{1}{2} \end{pmatrix} \begin{pmatrix} \dfrac{1}{2} & \dfrac{\sqrt{3}}{2}i \\ -\dfrac{\sqrt{3}}{2}i & -\dfrac{1}{2} \end{pmatrix} = \begin{pmatrix} 1 & 0 \\ 0 & 1 \end{pmatrix}$$

so $A^* = A^{-1}$ and A is unitary.

The main results of this section can be restated as follows for complex matrices:

■ THEOREM 1′

The eigenvalues of a Hermitian matrix are real. ■

DEFINITION

Unitarily Diagonalizable Matrix

An $n \times n$ complex matrix A is **unitarily diagonalizable** if there exists a unitary matrix U such that $U^*AU = D$ where $D = \text{diag}(\lambda_1, \lambda_2, \ldots, \lambda_n)$ and λ_1, λ_2, $\ldots$, λ_n are the eigenvalues of A.

■ THEOREM 5′

Let A be a complex $n \times n$ matrix. Then A is unitarily diagonalizable if and only if A is Hermitian. ■

We conclude this section with a proof of Theorem 3.

PROOF OF THEOREM 3 (Optional)

We prove that to every eigenvalue λ of algebraic multiplicity k, there correspond k orthonormal eigenvectors. This step, combined with Theorem 2, will be sufficient. Let $\mathbf{u}_1$ be an eigenvector of A corresponding to λ_1. We can assume that $|\mathbf{u}_1| = 1$. We can also assume that $\mathbf{u}_1$ is real because λ_1 is real and $\mathbf{u}_1 \in N_{A-\lambda_1 I}$, the kernel of the real matrix $A - \lambda_1 I$. This kernel is a subspace of $\mathbb{R}^n$

by Theorem 7.10.5 on p. 438. Next we note that $\{\mathbf{u}_1\}$ can be expanded into a basis $\{\mathbf{u}_1, \mathbf{v}_2, \mathbf{v}_3, \ldots, \mathbf{v}_n\}$ for $\mathbb{R}^n$ and, by the Gram–Schmidt process, we can turn this basis into the orthonormal basis $\{\mathbf{u}_1, \mathbf{u}_2, \ldots, \mathbf{u}_n\}$. Let Q be the orthogonal matrix whose columns are $\mathbf{u}_1, \mathbf{u}_2, \ldots, \mathbf{u}_n$. For convenience of notation we write $Q = (\mathbf{u}_1, \mathbf{u}_2, \ldots, \mathbf{u}_n)$. Now Q is invertible and $Q^t = Q^{-1}$, so A is similar to $Q^t A Q$ and, by Theorem 8.3.1 on p. 470, $Q^t A Q$ and A have the same characteristic polynomial: $|Q^t A Q - \lambda I| = |A - \lambda I|$. But,

$$
Q^t = \begin{pmatrix} \mathbf{u}_1^t \\ \mathbf{u}_2^t \\ \vdots \\ \mathbf{u}_n^t \end{pmatrix}
$$

so that

$$
Q^t A Q = \begin{pmatrix} \mathbf{u}_1^t \\ \mathbf{u}_2^t \\ \vdots \\ \mathbf{u}_n^t \end{pmatrix} A (\mathbf{u}_1 \quad \mathbf{u}_2 \quad \cdots \quad \mathbf{u}_n) = \begin{pmatrix} \mathbf{u}_1^t \\ \mathbf{u}_2^t \\ \vdots \\ \mathbf{u}_n^t \end{pmatrix} (A\mathbf{u}_1 \quad A\mathbf{u}_2 \quad \cdots \quad A\mathbf{u}_n)
$$

$$
= \begin{pmatrix} \mathbf{u}_1^t \\ \mathbf{u}_2^t \\ \vdots \\ \mathbf{u}_n^t \end{pmatrix} (\lambda_1 \mathbf{u}_1 \quad A\mathbf{u}_2 \quad \cdots \quad A\mathbf{u}_n) = \begin{pmatrix} \lambda_1 & \mathbf{u}_1^t A\mathbf{u}_2 & \cdots & \mathbf{u}_1^t A\mathbf{u}_n \\ 0 & \mathbf{u}_2^t A\mathbf{u}_2 & \cdots & \mathbf{u}_2^t A\mathbf{u}_n \\ \vdots & \vdots & & \vdots \\ 0 & \mathbf{u}_n^t A\mathbf{u}_2 & \cdots & \mathbf{u}_n^t A\mathbf{u}_n \end{pmatrix}.
$$

The zeros appear because $\mathbf{u}_1^t \mathbf{u}_j = \mathbf{u}_1 \cdot \mathbf{u}_j = 0$ if $j \neq 1$. Now $[Q^t A Q]^t = Q^t A^t (Q^t)^t = Q^t A Q$. Thus $Q^t A Q$ is symmetric, which means that there must be zeros in the first row of $Q^t A Q$ to match the zeros in the first column. Thus

$$
Q^t A Q = \begin{pmatrix} \lambda_1 & 0 & 0 & \cdots & 0 \\ 0 & q_{22} & q_{23} & \cdots & q_{2n} \\ 0 & q_{32} & q_{33} & \cdots & q_{3n} \\ \vdots & \vdots & \vdots & & \vdots \\ 0 & q_{n2} & q_{n3} & \cdots & q_{nn} \end{pmatrix}
$$

and

$$
|Q^t A Q - \lambda I| = \begin{vmatrix} \lambda_1 - \lambda & 0 & 0 & \cdots & 0 \\ 0 & q_{22} - \lambda & q_{23} & \cdots & q_{2n} \\ 0 & q_{32} & q_{33} - \lambda & \cdots & q_{3n} \\ \vdots & \vdots & \vdots & & \vdots \\ 0 & q_{n2} & q_{n3} & \cdots & q_{nn} - \lambda \end{vmatrix}
$$

$$
= (\lambda_1 - \lambda) \begin{vmatrix} q_{22} - \lambda & q_{23} & \cdots & q_{2n} \\ q_{32} & q_{33} - \lambda & \cdots & q_{3n} \\ \vdots & \vdots & & \vdots \\ q_{n2} & q_{n3} & \cdots & q_{nn} - \lambda \end{vmatrix} = (\lambda - \lambda_1)|M_{11}(\lambda)|,
$$

where $M_{11}(\lambda)$ is the 1, 1 minor of $Q^tAQ - \lambda I$. If $k = 1$, there is nothing to prove. If $k > 1$, then $|A - \lambda I|$ contains the factor $(\lambda - \lambda_1)^2$ and, therefore, $|Q^tAQ - \lambda I|$ also contains the factor $(\lambda - \lambda_1)^2$. Thus $|M_{11}(\lambda)|$ contains the factor $\lambda - \lambda_1$, which means that $|M_{11}(\lambda_1)| = 0$. This means that the last $n - 1$ columns of $Q^tAQ - \lambda_1 I$ are linearly dependent. Since the first column of $Q^tAQ - \lambda_1 I$ is the zero vector, $Q^tAQ - \lambda_1 I$ contains at most $n - 2$ linearly independent columns. In other words, $\rho(Q^tAQ - \lambda_1 I) \le n - 2$. But $Q^tAQ - \lambda_1 I$ and $A - \lambda_1 I$ are similar; hence, by Problem 8.3.18, $\rho(A - \lambda_1 I) \le n - 2$. Therefore $\nu(A - \lambda_1 I) \ge 2$, which means that $N_{A-\lambda_1 I}$ contains at least two linearly independent eigenvectors. If $k = 2$, we are done. If $k > 2$, then we take two orthonormal vectors $\mathbf{u}_1, \mathbf{u}_2$ in $N_{A-\lambda_1 I}$ and expand them into a new orthonormal basis $\{\mathbf{u}_1, \mathbf{u}_2, \ldots, \mathbf{u}_n\}$ for $\mathbb{R}^n$ and define $P = (\mathbf{u}_1, \mathbf{u}_2, \ldots, \mathbf{u}_n)$. Then, exactly as before, we show that

$$
P^tAP - \lambda I = \begin{pmatrix} \lambda_1 - \lambda & 0 & 0 & 0 & \cdots & 0 \\ 0 & \lambda_1 - \lambda & 0 & 0 & \cdots & 0 \\ 0 & 0 & \beta_{33} - \lambda & \beta_{34} & \cdots & \beta_{3n} \\ 0 & 0 & \beta_{43} & \beta_{44} - \lambda & \cdots & \beta_{4n} \\ \vdots & \vdots & \vdots & \vdots & & \vdots \\ 0 & 0 & \beta_{n3} & \beta_{n4} & \cdots & \beta_{nn} - \lambda \end{pmatrix}
$$

Since $k > 2$, we show, as before, that the determinant of the matrix in brackets is zero when $\lambda = \lambda_1$, which shows that $\rho(P^tAP - \lambda_1 I) \le n - 3$, so that $\nu(P^tAP - \lambda_1 I) = \nu(A - \lambda_1 I) \ge 3$. Then dim $N_{A-\lambda_1 I} \ge 3$ and so on. We can continue this process for dim $N_{A-\lambda_1 I} = k$ steps. Finally, for each $N_{A-\lambda_i I}$ we can find an orthonormal basis. This completes the proof. ∎

Problems 8.4

In Problems 1–8 find an orthogonal matrix that diagonalizes the given symmetric matrix.

1. $\begin{pmatrix} 3 & 4 \\ 4 & -3 \end{pmatrix}$

2. $\begin{pmatrix} 2 & 1 \\ 1 & 2 \end{pmatrix}$

3. $\begin{pmatrix} 1 & -1 \\ -1 & 1 \end{pmatrix}$

4. $\begin{pmatrix} 1 & -1 & -1 \\ -1 & 1 & -1 \\ -1 & -1 & 1 \end{pmatrix}$

5. $\begin{pmatrix} -1 & 2 & 2 \\ 2 & -1 & 2 \\ 2 & 2 & 1 \end{pmatrix}$

6. $\begin{pmatrix} 1 & -1 & 0 \\ -1 & 2 & -1 \\ 0 & -1 & 1 \end{pmatrix}$

7. $\begin{pmatrix} 3 & 2 & 2 \\ 2 & 2 & 0 \\ 2 & 0 & 4 \end{pmatrix}$

8. $\begin{pmatrix} 1 & -1 & 0 & 0 \\ -1 & 0 & 0 & 0 \\ 0 & 0 & 0 & 0 \\ 0 & 0 & 0 & 2 \end{pmatrix}$

9. Let Q be a symmetric orthogonal matrix. Show that if λ is an eigenvalue of Q, then $\lambda = \pm 1$.

10. A is **orthogonally similar** to B if there exists an orthogonal matrix Q such that $B = Q^tAQ$. Suppose that A is ortho-gonally similar to B and that B is orthogonally similar to C. Show that A is orthogonally similar to C.

11. Show that if $Q = \begin{pmatrix} a & b \\ c & d \end{pmatrix}$ is orthogonal, then $b = \pm c$. [*Hint:* Write out the equations that result from the equation $Q^tQ = I$.]

12. Suppose that A is a real symmetric matrix every one of whose eigenvalues is zero. Show that A is the zero matrix.

13. Show that if a real 2×2 matrix A has eigenvectors that are orthogonal, then A is symmetric.

14. Let A be a real skew-symmetric matrix ($A^t = -A$). Prove that every eigenvalue of A is of the form $i\alpha$, where α is a real number. That is, prove that every eigenvalue of A is a pure imaginary number.

In Problems 15–18 find the conjugate transpose, A^*, of the given matrix.

15. $A = \begin{pmatrix} 1+i & 2 \\ 5+2i & 3 \end{pmatrix}$

16. $A = \begin{pmatrix} -1+3i & 2+5i \\ -7+2i & 5+6i \end{pmatrix}$

17. $A = \begin{pmatrix} 2 & 3 & 5 \\ -i & 2i & 5i \\ 1+i & -1+i & 2-3i \end{pmatrix}$

18. $A = \begin{pmatrix} 3 & 2-i & 4+2i \\ 2+i & 5 & -2-i \\ 4-2i & -2+i & -3 \end{pmatrix}$

19. Show that the diagonal components of a Hermitian matrix are real.

20. A matrix A is **skew-Hermitian** if $A^* = -A$. Show that the diagonal components of a skew-Hermitian matrix are pure imaginary.

* **21.** Show that the eigenvalues of a complex $n \times n$ Hermitian matrix are real. [*Hint:* Use the fact that in $\mathbb{C}^n$, $A\mathbf{x} \cdot \mathbf{y} = \mathbf{x} \cdot A^*\mathbf{y}$.]

* **22.** If A is an $n \times n$ Hermitian matrix, show that eigenvectors corresponding to different eigenvalues are orthogonal.

** **23.** By repeating the proof of Theorem 3, except that $\bar{\mathbf{v}}_i^t$ replaces $\mathbf{v}_i^t$ where appropriate, show that any $n \times n$ Hermitian matrix has n orthonormal eigenvectors.

24. Find a unitary matrix U such that U^*AU is diagonal, where
$$A = \begin{pmatrix} 1 & 1-i \\ 1+i & 0 \end{pmatrix}.$$

25. Do the same for $A = \begin{pmatrix} 2 & 3-3i \\ 3+3i & 5 \end{pmatrix}$.

26. Prove that the determinant of a Hermitian matrix is real.

27. Show that for any matrix A,
$$A\mathbf{v}_1 \cdot \mathbf{v}_2 = \mathbf{v}_1 \cdot A^t\mathbf{v}_2.$$

8.5 Quadratic Forms and Conic Sections

In this section we use the material of Section 8.4 to discover information about the graphs of quadratic equations. Quadratic equations and quadratic forms, which are defined below, arise in a variety of ways. For example, we can use quadratic forms to obtain information about the conic sections in $\mathbb{R}^2$ (circles, parabolas, ellipses, hyperbolas) and extend this theory to describe certain surfaces, called *quadric surfaces,* in $\mathbb{R}^3$. These topics are discussed later in the section. Although we shall not discuss it in this text, quadratic forms also are used in a number of applications ranging from a description of cost functions in economics to an analysis of the control of a rocket traveling in space.

Quadratic Equation

DEFINITION

i. A **quadratic equation in two variables with no linear terms** is an equation of the form

$$ax^2 + bxy + cy^2 = d \qquad (1)$$

where $|a| + |b| + |c| \neq 0$.

Quadratic Form

ii. A **quadratic form in two variables** is an expression of the form

$$F(x, y) = ax^2 + bxy + cy^2 \qquad (2)$$

where $|a| + |b| + |c| \neq 0$.

Obviously quadratic equations and quadratic forms are closely related. We begin our analysis of quadratic forms with a simple example.

Consider the quadratic form $F(x, y) = x^2 - 4xy + 3y^2$. Let $\mathbf{v} = \begin{pmatrix} x \\ y \end{pmatrix}$ and $A = \begin{pmatrix} 1 & -2 \\ -2 & 3 \end{pmatrix}$. Then

$$A\mathbf{v} \cdot \mathbf{v} = \begin{pmatrix} 1 & -2 \\ -2 & 3 \end{pmatrix}\begin{pmatrix} x \\ y \end{pmatrix} \cdot \begin{pmatrix} x \\ y \end{pmatrix} = \begin{pmatrix} x - 2y \\ -2x + 3y \end{pmatrix} \cdot \begin{pmatrix} x \\ y \end{pmatrix}$$

$$= (x^2 - 2xy) + (-2xy + 3y^2) = x^2 - 4xy + 3y^2 = F(x, y).$$

Thus we have "represented" the quadratic form $F(x, y)$ by the symmetric matrix A in the sense that

$$F(x, y) = A\mathbf{v} \cdot \mathbf{v}. \tag{3}$$

Conversely, if A is a symmetric matrix, then equation (3) defines a quadratic form $F(x, y) = A\mathbf{v} \cdot \mathbf{v}$.

We can represent $F(x, y)$ by many matrices but only one symmetric matrix. To see this, let $A = \begin{pmatrix} 1 & a \\ b & 3 \end{pmatrix}$, where $a + b = -4$. Then $A\mathbf{v} \cdot \mathbf{v} = F(x, y)$. If $A = \begin{pmatrix} 1 & 3 \\ -7 & 3 \end{pmatrix}$, for example, then $A\mathbf{v} = \begin{pmatrix} x + 3y \\ -7x + 3y \end{pmatrix}$ and $A\mathbf{v} \cdot \mathbf{v} = x^2 - 4xy + 3y^2$. If, however, we insist that A be symmetric, then we must have $a + b = -4$ and $a = b$. This pair of equations has the unique solution $a = b = -2$.

If $F(x, y) = ax^2 + bxy + cy^2$ is a quadratic form, let

$$A = \begin{pmatrix} a & b/2 \\ b/2 & c \end{pmatrix}. \tag{4}$$

Then

$$A\mathbf{v} \cdot \mathbf{v} = \left[\begin{pmatrix} a & b/2 \\ b/2 & c \end{pmatrix}\begin{pmatrix} x \\ y \end{pmatrix}\right] \cdot \begin{pmatrix} x \\ y \end{pmatrix} = \begin{pmatrix} ax + (b/2)y \\ (b/2)x + cy \end{pmatrix} \cdot \begin{pmatrix} x \\ y \end{pmatrix}$$

$$= ax^2 + bxy + cy^2 = F(x, y).$$

Now let us return to the quadratic equation (1). Using (3), we can write (1) as

$$A\mathbf{v} \cdot \mathbf{v} = d \tag{5}$$

where A is symmetric. By Theorem 8.4.5 on p. 478, there is an orthogonal matrix Q such that $Q^t AQ = D$, where $D = \text{diag}(\lambda_1, \lambda_2)$ and λ_1 and λ_2 are the (real, because A is symmetric) eigenvalues of A. Then $A = QDQ^t$ (remember that $Q^t = Q^{-1}$) and (5) can be written

$$(QDQ^t \mathbf{v}) \cdot \mathbf{v} = d. \tag{6}$$

But, by equation (8.4.14), $A\mathbf{v} \cdot \mathbf{y} = \mathbf{v} \cdot A^t \mathbf{y}$. Thus

$$Q(DQ^t \mathbf{v}) \cdot \mathbf{v} = DQ^t \mathbf{v} \cdot Q^t \mathbf{v} \tag{7}$$

so that (6) reads

$$[DQ^t \mathbf{v}] \cdot Q^t \mathbf{v} = d. \tag{8}$$

Let $\mathbf{v}' = Q^t \mathbf{v}$. Then $\mathbf{v}'$ is a 2-vector and (8) becomes

$$D\mathbf{v}' \cdot \mathbf{v}' = d. \tag{9}$$

Let us look at (9) more closely. We can write $\mathbf{v}' = \begin{pmatrix} x' \\ y' \end{pmatrix}$. Since a diagonal matrix is symmetric, (9) defines a quadratic form $F'(x', y')$ in the variables x' and y'. If $D = \begin{pmatrix} a' & 0 \\ 0 & c' \end{pmatrix}$, then $D\mathbf{v}' = \begin{pmatrix} a' & 0 \\ 0 & c' \end{pmatrix}\begin{pmatrix} x' \\ y' \end{pmatrix} = \begin{pmatrix} a'x' \\ c'y' \end{pmatrix}$ and

$$F'(x', y') = D\mathbf{v}' \cdot \mathbf{v}' = \begin{pmatrix} a'x' \\ c'y' \end{pmatrix} \cdot \begin{pmatrix} x' \\ y' \end{pmatrix} = a'x'^2 + c'y'^2.$$

That is: $F'(x', y')$ *is a quadratic form with the $x'y'$ term missing.* Hence equation (9) is a quadratic equation in the new variables x', y' with the $x'y'$ term missing.

Let us take another look at the matrix Q. Since Q is real and orthogonal, $1 = \det QQ^{-1} = \det QQ^t = \det Q \det Q^t = \det Q \det Q = (\det Q)^2$. Thus $\det Q = \pm 1$. If $\det Q = -1$, we can interchange the rows of Q to make the determinant of this new Q equal to 1. Then it can be shown (see Problem 36) that $Q = \begin{pmatrix} \cos \theta & -\sin \theta \\ \sin \theta & \cos \theta \end{pmatrix}$ for some number θ with $0 \le \theta < 2\pi$. But, from Example 7.11.2 on p. 443, this means that Q is a rotation matrix. We have therefore proved the following theorem.

■ **THEOREM 1: Principal Axes Theorem in $\mathbb{R}^2$**

Let

$$ax^2 + bxy + cy^2 = d \tag{10}$$

be a quadratic equation in the variables x and y. Then there exists a number θ in $[0, 2\pi)$ such that equation (10) can be written in the form

$$a'x'^2 + c'y'^2 = d \tag{11}$$

where x', y' are the axes obtained by rotating the x- and y-axes through an angle of θ in the counterclockwise direction. Moreover, the numbers a' and c'

are the eigenvalues of the matrix $A = \begin{pmatrix} a & b/2 \\ b/2 & c \end{pmatrix}$. The x'- and y'-axes are called the **principal axes** of the graph of the quadratic equation (10). ∎

We can use Theorem 1 to identify three important conic sections. Recall that the **standard equations** of a circle, ellipse, and hyperbola are:

$$\text{Circle:} \qquad x^2 + y^2 = r^2 \tag{12}$$

$$\text{Ellipse:} \qquad \frac{x^2}{a^2} + \frac{y^2}{b^2} = 1 \tag{13}$$

$$\text{Hyperbola:} \quad \left\{ \begin{array}{l} \dfrac{x^2}{a^2} - \dfrac{y^2}{b^2} = 1 \qquad (14) \\[1em] \text{or} \\[1em] \dfrac{y^2}{a^2} - \dfrac{x^2}{b^2} = 1 \qquad (15) \end{array} \right.$$

EXAMPLE 1

Identify the conic section whose equation is

$$x^2 - 4xy + 3y^2 = 6. \tag{16}$$

SOLUTION

We can write (16) in the form

$$A\mathbf{x} \cdot \mathbf{x} = \begin{pmatrix} 1 & -2 \\ -2 & 3 \end{pmatrix}\begin{pmatrix} x \\ y \end{pmatrix} \cdot \begin{pmatrix} x \\ y \end{pmatrix} = 6.$$

To identify the conic, we need only obtain the eigenvalues of A. $\det(A - \lambda I) = (1 - \lambda)(3 - \lambda) - 4 = \lambda^2 - 4\lambda - 1 = 0$, so that $\lambda = 2 \pm \sqrt{5}$. Thus (16) can be written in the new variables as

$$(2 - \sqrt{5})x'^2 + (2 + \sqrt{5})y'^2 = 6,$$

or, as the standard hyperbola,

$$\frac{y'^2}{6/(2 + \sqrt{5})} - \frac{x'^2}{6/(\sqrt{5} - 2)} = 1. \tag{17}$$

More work is required if we want to determine the angle θ through which the axes must be rotated to obtain (17). In Example 8.4.2 on p. 479 we saw that A can be diagonalized to $D = \begin{pmatrix} 2 - \sqrt{5} & 0 \\ 0 & 2 + \sqrt{5} \end{pmatrix}$ by using the orthogonal matrix

$$Q = \frac{1}{\sqrt{10 - 2\sqrt{5}}}\begin{pmatrix} 2 & 1 - \sqrt{5} \\ -1 + \sqrt{5} & 2 \end{pmatrix}, \qquad \det Q = 1.$$

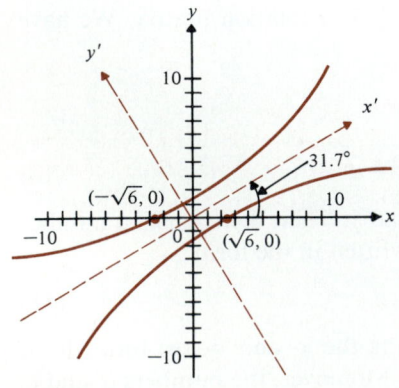

FIGURE 1

Using Problem 36 and the upper left entry of Q, we get

$$\cos \theta = \frac{2}{\sqrt{10 - 2\sqrt{5}}} \approx 0.85065.$$

Thus θ is in the first quadrant and, using a table (or a calculator), we find that $\theta \approx 0.5536$ rad $\approx 31.7°$. Thus (16) is the equation of a standard hyperbola rotated through an angle of $31.7°$ (see Fig. 1).

EXAMPLE 2

Identify the conic section whose equation is

$$5x^2 - 2xy + 5y^2 = 4. \tag{18}$$

SOLUTION

Here $A = \begin{pmatrix} 5 & -1 \\ -1 & 5 \end{pmatrix}$ so that

$$\det(A - \lambda I) = (5 - \lambda)^2 - 1 = \lambda^2 - 10\lambda + 24 = (\lambda - 6)(\lambda - 4) = 0,$$

or $\lambda_1 = 6$ and $\lambda_2 = 4$. Thus, (18) can be rewritten in the rotated plane as

$$6x'^2 + 4y'^2 = 4$$

or

$$\frac{x'^2}{(2/3)} + y'^2 = 1.$$

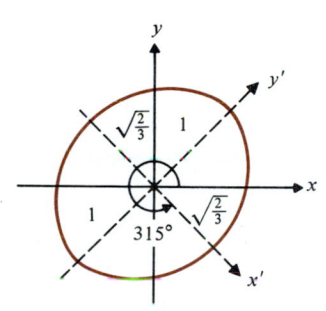

$$5x^2 - 2xy + 5y^2 = 4$$

FIGURE 2

Hence (18) is an ellipse. We leave to the reader the task of showing that the angle of rotation is $315°$. (See Fig. 2.)

EXAMPLE 3

Identify the conic section whose equation is

$$-5x^2 + 2xy - 5y^2 = 4 \tag{19}$$

SOLUTION

Rewriting (19) as $5x^2 - 2xy + 5y^2 = -4$, and using Example 2, we get

$$\frac{x'^2}{(2/3)} + y'^2 = -1. \tag{20}$$

But the left-hand side of (20) is nonnegative for all real numbers x' and y', implying that no real numbers x and y satisfy (19). Conic sections that are not ellipses, hyperbolas, or parabolas are called **degenerate conics**. Quadratic equations with no graph are sometimes called **imaginary conics**.

Writing the conic section

$$ax^2 + bxy + cy^2 = d \tag{21}$$

in the form $A\mathbf{x} \cdot \mathbf{x} = d$, with $A = \begin{pmatrix} a & b/2 \\ b/2 & c \end{pmatrix}$, we see that the characteristic polynomial of A is given by

$$(a-\lambda)(c-\lambda) - \frac{b^2}{4} = \lambda^2 - (a+c)\lambda + \left(ac - \frac{b^2}{4}\right).$$

Letting

$$\lambda^2 - (a+c)\lambda + \left(ac - \frac{b^2}{4}\right) = (\lambda - \lambda_1)(\lambda - \lambda_2),$$

it follows that $\lambda_1\lambda_2 = ac - b^2/4 = \det A$. In the rotated coordinates, (21) becomes

$$\lambda_1 x'^2 + \lambda_2 y'^2 = d. \tag{22}$$

If λ_1 and λ_2 have the same sign, then (21) defines an ellipse (or a circle) or a degenerate conic as in Examples 2 and 3. If λ_1 and λ_2 have opposite signs, then (21) is the equation of a hyperbola (as in Example 1). We can therefore prove the following.

■ **THEOREM 2**

If $A = \begin{pmatrix} a & b/2 \\ b/2 & c \end{pmatrix}$, then the quadratic equation (21) with $d \neq 0$ is the equation of:

   i. A hyperbola if $\det A < 0$.
   ii. An ellipse, circle, or degenerate conic section if $\det A > 0$.
   iii. A pair of straight lines or a degenerate conic section if $\det A = 0$.
   iv. If $d = 0$, then (21) is the equation of two straight lines if $\det A \neq 0$ and the equation of a single line if $\det A = 0$.

PROOF

We have already shown why (*i*) and (*ii*) are true. To prove part (*iii*), suppose that $\det A = 0$. Then, by our Summing-Up Theorem (Theorem 8.1.6), $\lambda = 0$ is an eigenvalue of A and equation (22) reads $\lambda_1 x'^2 = d$ or $\lambda_2 y'^2 = d$. If $\lambda_1 x'^2 = d$ and $d/\lambda_1 > 0$, then $x_1' = \pm\sqrt{d/\lambda_1}$ is the equation of two straight lines in the xy-plane. If $d/\lambda_1 < 0$, then we have $x'^2 < 0$ (which is impossible) and we obtain a degenerate conic. The same facts hold if $\lambda_2 y'^2 = d$. Part (*iv*) is left as an exercise (see Problem 37). ■

The methods described above can be used to analyze quadratic equations in more than two variables. We give one example below.

EXAMPLE 4

Consider the quadratic equation

$$5x^2 + 8xy + 5y^2 + 4xz + 4yz + 2z^2 = 100. \tag{23}$$

If $A = \begin{pmatrix} 5 & 4 & 2 \\ 4 & 5 & 2 \\ 2 & 2 & 2 \end{pmatrix}$ and $\mathbf{v} = \begin{pmatrix} x \\ y \\ z \end{pmatrix}$, then (23) can be written in the form

$$A\mathbf{v} \cdot \mathbf{v} = 100. \qquad (24)$$

From Example 8.4.3 on p. 480,

$$Q^t A Q = D = \begin{pmatrix} 1 & 0 & 0 \\ 0 & 1 & 0 \\ 0 & 0 & 10 \end{pmatrix}, \text{ where } Q = \begin{pmatrix} -1/\sqrt{2} & -1/3\sqrt{2} & 2/3 \\ 1/\sqrt{2} & -1/3\sqrt{2} & 2/3 \\ 0 & 4/3\sqrt{2} & 1/3 \end{pmatrix}.$$

Let

$$\mathbf{v}' = \begin{pmatrix} x' \\ y' \\ z' \end{pmatrix} = Q^t \mathbf{v} = \begin{pmatrix} -1/\sqrt{2} & 1/\sqrt{2} & 0 \\ -1/3\sqrt{2} & -1/3\sqrt{2} & 4/3\sqrt{2} \\ 2/3 & 2/3 & 1/3 \end{pmatrix} \begin{pmatrix} x \\ y \\ z \end{pmatrix}$$

$$= \begin{pmatrix} (-1/\sqrt{2})x + (1/\sqrt{2})y \\ -(1/3\sqrt{2})x - (1/3\sqrt{2})y + (4/3\sqrt{2})z \\ (2/3)x + (2/3)y + (1/3)z \end{pmatrix}.$$

Then, as before, $A = QDQ^t$ and $A\mathbf{v} \cdot \mathbf{v} = QDQ^t\mathbf{v} \cdot \mathbf{v} = DQ^t\mathbf{v} \cdot Q^t\mathbf{v} = D\mathbf{v}' \cdot \mathbf{v}'$. Thus (24) can be written in the new variables x', y', z' as $D\mathbf{v}' \cdot \mathbf{v}' = 100$ or

$$x'^2 + y'^2 + 10z'^2 = 100. \qquad (25)$$

In $\mathbb{R}^3$, the surface defined by (25) is called an **ellipsoid** (see Fig. 3).

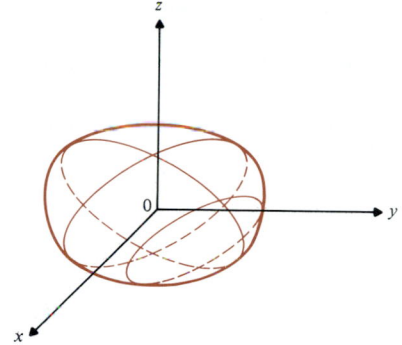

z

y

0

x

FIGURE 3

There is a great variety of three-dimensional surfaces of the form $A\mathbf{v} \cdot \mathbf{v} = d$, where $\mathbf{v} \in \mathbb{R}^3$. Such surfaces are called **quadric surfaces.**

We close this section by noting that quadratic forms can be defined in any number of variables.

DEFINITION

Quadratic Form

Let $\mathbf{v} = \begin{pmatrix} x_1 \\ x_2 \\ \vdots \\ x_n \end{pmatrix}$ and let A be a symmetric $n \times n$ matrix. Then a **quadratic form**

in $x_1, x_2, \ldots, x_n$ is an expression of the form

$$F(x_1, x_2, \ldots, x_n) = A\mathbf{v} \cdot \mathbf{v}. \qquad (26)$$

■

Problems 8.5

In Problems 1–13 write the quadratic equation in the form $A\mathbf{v} \cdot \mathbf{v} = d$ (where A is a symmetric matrix) and eliminate the xy-term by rotating the axes through an angle of θ. Write the equation in terms of the new variables and identify the conic section obtained.

1. $3x^2 - 2xy - 5 = 0$

2. $4x^2 + 4xy + y^2 = 9$

3. $4x^2 + 4xy - y^2 = 9$

4. $xy = 1$

5. $xy = a;\ a > 0$

6. $4x^2 + 2xy + 3y^2 + 2 = 0$

7. $xy = a;\ a < 0$

8. $x^2 + 4xy + 4y^2 - 6 = 0$

9. $-x^2 + 2xy - y^2 = 0$

10. $2x^2 + xy + y^2 = 4$

11. $3x^2 - 6xy + 5y^2 = 36$

12. $x^2 - 3xy + 4y^2 = 1$

13. $6x^2 + 5xy - 6y^2 + 7 = 0$

14. What are the possible forms of the graph of
$$ax^2 + bxy + cy^2 = 0?$$

In Problems 15–18 write the quadratic form in new variables x', y', and z' so that no cross-product terms (xy, xz, yz) are present.

15. $x^2 - 2xy + y^2 - 2xz - 2yz + z^2$

16. $-x^2 + 4xy - y^2 + 4xz + 4yz + z^2$

17. $3x^2 + 4xy + 2y^2 + 4xz + 4z^2$

18. $x^2 - 2xy + 2y^2 - 2yz + z^2$

In Problems 19–21 find a symmetric matrix A such that the quadratic form can be written in the form $A\mathbf{x} \cdot \mathbf{x}$.

19. $x_1^2 + 2x_1x_2 + x_2^2 + 4x_1x_3 + 6x_2x_3 + 3x_3^2 + 7x_1x_4 - 2x_2x_4 + x_4^2$

20. $x_1^2 - x_2^2 + x_1x_3 - x_2x_4 + x_4^2$

21. $3x_1^2 - 7x_1x_2 - 2x_2^2 + x_1x_3 - x_2x_3 + 3x_3^2 - 2x_1x_4 + x_2x_4 - 4x_3x_4 - 6x_4^2 + 3x_1x_5 - 5x_3x_5 + x_4x_5 - x_5^2$

22. Suppose that for some nonzero value of d, the graph of $ax^2 + bxy + cy^2 = d$ is a hyperbola. Show that the graph is a hyperbola for any other nonzero value of d.

23. Show that if $a \neq c$, the xy-term in quadratic equation (1) will be eliminated by rotation through an angle θ if θ is given by $\cot 2\theta = (a - c)/b$.

24. Show that if $a = c$ in Problem 23, then the xy-term will be eliminated by a rotation through an angle of either $\pi/4$ or $-\pi/4$.

* 25. Suppose that a rotation converts $ax^2 + bxy + cy^2$ into $a'(x')^2 + b'(x'y') + c'(y')^2$. Show that:

a. $a + c = a' + c'$

b. $b^2 - 4ac = b'^2 - 4a'c'$

* 26. A quadratic form $F(\mathbf{x}) = F(x_1, x_2, \ldots, x_n)$ is said to be **positive definite** if $F(\mathbf{x}) \geq 0$ for every $\mathbf{x} \in \mathbb{R}^n$ and $F(\mathbf{x}) = 0$ if and only if $\mathbf{x} = \mathbf{0}$. Show that F is positive definite if and only if the symmetric matrix A associated with F has positive eigenvalues.

27. A quadratic form $F(\mathbf{x})$ is said to be **positive semidefinite** if $F(\mathbf{x}) \geq 0$ for every $\mathbf{x} \in \mathbb{R}^n$. Show that F is positive semidefinite if and only if the eigenvalues of the symmetric matrix associated with F are all nonnegative.

The definitions of **negative definite** and **negative semidefinite** are the definitions in Problems 26 and 27 with ≤ 0 replacing ≥ 0. A quadratic form is **indefinite** if it is none of the above. In Problems 28–35 determine whether the given quadratic form is positive definite, positive semidefinite, negative definite, negative semidefinite, or indefinite.

28. $3x^2 + 2y^2$

29. $-3x^2 - 3y^2$

30. $3x^2 - 2y^2$

31. $x^2 + 2xy + 2y^2$

32. $x^2 - 2xy + 2y^2$

33. $x^2 - 4xy + 3y^2$

34. $-x^2 + 4xy - 3y^2$

35. $-x^2 + 2xy - 2y^2$

* 36. Let $Q = \begin{pmatrix} a & b \\ c & d \end{pmatrix}$ be a real orthogonal matrix with $\det Q = 1$. Define the number $\theta \in [0, 2\pi)$:

a. If $a \geq 0$ and $c > 0$, then $\theta = \cos^{-1} a$ $(0 < \theta \leq \pi/2)$.

b. If $a \geq 0$ and $c < 0$, then $\theta = 2\pi - \cos^{-1} a$ $(3\pi/2 \leq \theta < 2\pi)$.

c. If $a \leq 0$ and $c > 0$, then $\theta = \cos^{-1} a$ $(\pi/2 \leq \theta < \pi)$.

d. If $a \leq 0$ and $c < 0$, then $\theta = 2\pi - \cos^{-1} a$ $(\pi < \theta \leq 3\pi/2)$.

e. If $a = 1$ and $c = 0$, then $\theta = 0$.

f. If $a = -1$ and $c = 0$, then $\theta = \pi$.

(Here $\cos^{-1} x \in [0, \pi]$ for $x \in [-1, 1]$.) With θ chosen as above, show that

$$Q = \begin{pmatrix} \cos \theta & -\sin \theta \\ \sin \theta & \cos \theta \end{pmatrix}.$$

37. Prove, using formula (22), that equation (21) is the equation of two straight lines in the xy-plane when $d = 0$ and $\det A \neq 0$. If $\det A = d = 0$, show that equation (21) is the equation of a single line.

8.6 Matrix Differential Equations

In Chapter 1, we studied the linear first-order differential equation

$$x'(t) = ax(t). \tag{1}$$

We saw that this equation arises in a number of contexts: growth rates of populations, radioactive decay, and monetary investments, for example. We found that the only solutions to (1) are of the form

$$x(t) = x_0 e^{at}, \tag{2}$$

where $x_0 = x(0)$ is the **initial value** of x.

In Chapter 3, we considered the system of n differential equations in n unknown functions:

$$
\begin{aligned}
x_1'(t) &= a_{11}x_1(t) + a_{12}x_2(t) + \cdots + a_{1n}x_n(t) \\
x_2'(t) &= a_{21}x_1(t) + a_{22}x_2(t) + \cdots + a_{2n}x_n(t) \\
&\ \ \vdots \qquad\quad \vdots \qquad\quad \vdots \qquad\qquad \vdots \\
x_n'(t) &= a_{n1}x_1(t) + a_{n2}x_2(t) + \cdots + a_{nn}x_n(t)
\end{aligned}
\tag{3}
$$

where the a_{ij}'s are real numbers. System (3) is called an $n \times n$ **first-order system of linear differential equations.**

Now, let

$$
\mathbf{x}(t) = \begin{pmatrix} x_1(t) \\ x_2(t) \\ \vdots \\ x_n(t) \end{pmatrix}.
$$

Here $\mathbf{x}(t)$ is called a **vector function.** We define

$$
\mathbf{x}'(t) = \begin{pmatrix} x_1'(t) \\ x_2'(t) \\ \vdots \\ x_n'(t) \end{pmatrix}.
$$

Then, if we define the $n \times n$ matrix

$$
A = \begin{pmatrix}
a_{11} & a_{12} & \cdots & a_{1n} \\
a_{21} & a_{22} & \cdots & a_{2n} \\
\vdots & \vdots & & \vdots \\
a_{n1} & a_{n2} & \cdots & a_{nn}
\end{pmatrix},
$$

system (3) can be written as

$$\mathbf{x}'(t) = A\mathbf{x}(t). \tag{4}$$

■ **THEOREM 1**

Let λ be an eigenvalue of A with corresponding eigenvector $\mathbf{v}$. Then $\mathbf{x}(t) = e^{\lambda t}\mathbf{v}$ is a solution to (4).

PROOF

If $\mathbf{x}(t) = e^{\lambda t}\mathbf{v}$, then

$$\overset{\lambda \text{ is an eigenvalue}}{\downarrow}$$

$$\mathbf{x}'(t) = \lambda e^{\lambda t}\mathbf{v} = e^{\lambda t}(\lambda \mathbf{v}) = e^{\lambda t}(A\mathbf{v}) = A(e^{\lambda t}\mathbf{v}) = A\mathbf{x}(t). \qquad ■$$

There is a theorem (which we shall not prove here) that states that the system (3) or (4) has n linearly independent solutions.† If A has n linearly independent eigenvectors $\mathbf{v}_1, \mathbf{v}_2, \ldots, \mathbf{v}_n$ corresponding to the eigenvalues $\lambda_1, \lambda_2, \ldots, \lambda_n$ (not necessarily distinct), then n linearly independent solutions to (4) are

$$e^{\lambda_1 t}\mathbf{v}_1, e^{\lambda_2 t}\mathbf{v}_2, \ldots, e^{\lambda_n t}\mathbf{v}_n.$$

General Solution

Then the **general solution** to (4) is, in this case, defined by

$$\mathbf{x}(t) = c_1 e^{\lambda_1 t}\mathbf{v}_1 + c_2 e^{\lambda_2 t}\mathbf{v}_2 + \cdots + c_n e^{\lambda_n t}\mathbf{v}_n.$$

If A does not have n linearly independent eigenvectors, then some other technique must be used to find the general solution to (4). For 2×2 systems our technique can fail only when A has one eigenvalue of algebraic multiplicity 2 but geometric multiplicity 1 (i.e., only one linearly independent eigenvector). A method for dealing with this situation is suggested in Problem 25.

EXAMPLE 1: An Electric Circuit

Consider the two-loop electric circuit given in Fig. 1. In Example 3.3.2 on p. 144 we found that the system of equations describing this circuit is

$$\frac{dI_L}{dt} = -\frac{R}{L}I_R + \frac{E}{L},$$

$$\frac{dI_R}{dt} = \frac{I_L}{RC} - \frac{I_R}{RC}. \qquad (5)$$

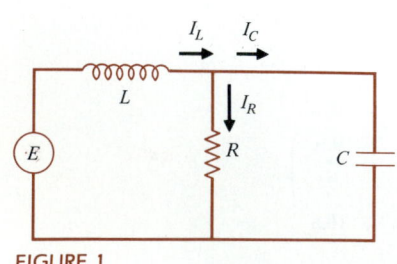

FIGURE 1

Find the current through the resistor and the inductor at all times t if $R = 100$ ohms, $C = 1.5 \times 10^{-4}$ farad, $L = 8$ henrys, $E = 0$, $I_L(0) = 0.1$ amp, and $I_R(0) = 0.2$ amp.

---

† See W. Derrick and S. Grossman, *Introduction to Differential Equations with Boundary Value Problems,* 3rd ed., West Publishing Co., St. Paul, Minn., 1987, Appendix 3.

SOLUTION

Using the values given, the system (5) can be written as

$$\frac{d\mathbf{I}(t)}{dt} = \frac{d}{dt}\begin{pmatrix} I_L(t) \\ I_R(t) \end{pmatrix} = \begin{pmatrix} 0 & -\dfrac{R}{L} \\ \dfrac{1}{RC} & -\dfrac{1}{RC} \end{pmatrix}\mathbf{I}(t) + \begin{pmatrix} \dfrac{E}{L} \\ 0 \end{pmatrix}, \qquad \mathbf{I}(0) = \begin{pmatrix} I_L(0) \\ I_R(0) \end{pmatrix},$$

or

$$\frac{d\mathbf{I}(t)}{dt} = \begin{pmatrix} 0 & -25/2 \\ 200/3 & -200/3 \end{pmatrix}\mathbf{I}(t), \qquad \mathbf{I}(0) = \begin{pmatrix} 0.1 \\ 0.2 \end{pmatrix}.$$

You should verify that the eigenvalues of $A = \begin{pmatrix} 0 & -25/2 \\ 200/3 & -200/3 \end{pmatrix}$ are $\lambda_1 = -50$, $\lambda_2 =$

$-50/3$ with corresponding eigenvectors $\mathbf{v}_1 = \begin{pmatrix} 1 \\ 4 \end{pmatrix}$ and $\mathbf{v}_2 = \begin{pmatrix} 3 \\ 4 \end{pmatrix}$. Thus the general solution

is

$$\mathbf{I}(t) = c_1 e^{-50t}\begin{pmatrix} 1 \\ 4 \end{pmatrix} + c_2 e^{-(50/3)t}\begin{pmatrix} 3 \\ 4 \end{pmatrix} = \begin{pmatrix} c_1 e^{-50t} + 3c_2 e^{-(50/3)t} \\ 4c_1 e^{-50t} + 4c_2 e^{-(50/3)t} \end{pmatrix}.$$

But

$$I(0) = \begin{pmatrix} c_1 + 3c_2 \\ 4c_1 + 4c_2 \end{pmatrix} = \begin{pmatrix} 0.1 \\ 0.2 \end{pmatrix}.$$

The solution to this system of two linear equations in two unknowns is $c_1 = c_2 = 0.025$. Thus the unique solution to the initial value problem is

$$\mathbf{I}(t) = \begin{pmatrix} 0.025e^{-50t} + 0.075e^{-(50/3)t} \\ 0.1e^{-50t} + 0.1e^{-(50/3)t} \end{pmatrix}.$$

EXAMPLE 2: A Mass-Spring System

In Section 3.5 and in Example 8.1.3 we discussed the mass-spring system depicted in Fig. 2. We found that the equations governing this system are

$$x_1' = x_3,$$

$$x_2' = x_4,$$

$$x_3' = -\left(\frac{k_1 + k_2}{m_1}\right)x_1 + \left(\frac{k_2}{m_1}\right)x_2,$$

$$x_4' = \left(\frac{k_2}{m_2}\right)x_1 - \left(\frac{k_2}{m_2}\right)x_2.$$

(6)

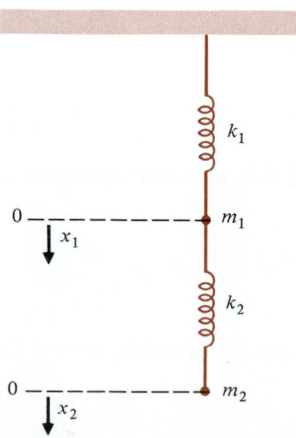

FIGURE 2

Suppose that $k_1 = 9m_1/2$, $k_2 = 2m_2$, and $7m_1 = 4m_2$. Find the unique solution to (6) that satisfies $x_1(0) = 1$, $x_1'(0) = 0$, $x_2(0) = 3$, $x_2'(0) = 0$.

SOLUTION

We write (6) as $\mathbf{x}'(t) = A\mathbf{x}(t)$ where $\mathbf{x}(t) = \begin{pmatrix} x_1(t) \\ x_2(t) \\ x_3(t) \\ x_4(t) \end{pmatrix}$,

$$A = \begin{pmatrix} 0 & 0 & 1 & 0 \\ 0 & 0 & 0 & 1 \\ -\left(\dfrac{k_1+k_2}{m_1}\right) & \dfrac{k_2}{m_1} & 0 & 0 \\ \dfrac{k_2}{m_2} & -\dfrac{k_2}{m_2} & 0 & 0 \end{pmatrix} = \begin{pmatrix} 0 & 0 & 1 & 0 \\ 0 & 0 & 0 & 1 \\ -8 & 7/2 & 0 & 0 \\ 2 & -2 & 0 & 0 \end{pmatrix}, \text{ and } \mathbf{x}(0) = \begin{pmatrix} 1 \\ 3 \\ 0 \\ 0 \end{pmatrix}.$$

From the computation on p. 155 we see that the eigenvalues of A are $\pm i$, $\pm 3i$. We compute corresponding eigenvectors. If $\lambda = i$, we obtain

$$\begin{pmatrix} -i & 0 & 1 & 0 \\ 0 & -i & 0 & 1 \\ -8 & 7/2 & -i & 0 \\ 2 & -2 & 0 & -i \end{pmatrix} \begin{pmatrix} u_1 \\ u_2 \\ u_3 \\ u_4 \end{pmatrix} = \begin{pmatrix} 0 \\ 0 \\ 0 \\ 0 \end{pmatrix}.$$

We row reduce:

$$\begin{pmatrix} -i & 0 & 1 & 0 & | & 0 \\ 0 & -i & 0 & 1 & | & 0 \\ -8 & 7/2 & -i & 0 & | & 0 \\ 2 & -2 & 0 & -i & | & 0 \end{pmatrix} \xrightarrow{M_1(i)} \begin{pmatrix} 1 & 0 & i & 0 & | & 0 \\ 0 & -i & 0 & 1 & | & 0 \\ -8 & 7/2 & -i & 0 & | & 0 \\ 2 & -2 & 0 & -i & | & 0 \end{pmatrix}$$

$$\xrightarrow[A_{1,4}(-2)]{A_{1,3}(8)} \begin{pmatrix} 1 & 0 & i & 0 & | & 0 \\ 0 & -i & 0 & 1 & | & 0 \\ 0 & 7/2 & 7i & 0 & | & 0 \\ 0 & -2 & -2i & -i & | & 0 \end{pmatrix} \xrightarrow{M_2(i)} \begin{pmatrix} 1 & 0 & i & 0 & | & 0 \\ 0 & 1 & 0 & i & | & 0 \\ 0 & 7/2 & 7i & 0 & | & 0 \\ 0 & -2 & -2i & -i & | & 0 \end{pmatrix}$$

$$\xrightarrow[A_{2,4}(2)]{A_{2,3}(-7/2)} \begin{pmatrix} 1 & 0 & i & 0 & | & 0 \\ 0 & 1 & 0 & i & | & 0 \\ 0 & 0 & 7i & -7i/2 & | & 0 \\ 0 & 0 & -2i & i & | & 0 \end{pmatrix} \xrightarrow{M_3(-i/7)} \begin{pmatrix} 1 & 0 & i & 0 & | & 0 \\ 0 & 1 & 0 & i & | & 0 \\ 0 & 0 & 1 & -1/2 & | & 0 \\ 0 & 0 & -2i & i & | & 0 \end{pmatrix}$$

$$\xrightarrow[A_{3,4}(2i)]{A_{3,1}(-i)} \begin{pmatrix} 1 & 0 & 0 & i/2 & | & 0 \\ 0 & 1 & 0 & i & | & 0 \\ 0 & 0 & 1 & -1/2 & | & 0 \\ 0 & 0 & 0 & 0 & | & 0 \end{pmatrix}.$$

Thus $u_3 = \frac{1}{2}u_4$, $u_2 = -iu_4$, $u_1 = -(i/2)u_4$ and, setting $u_4 = 2$, we obtain the eigenvector

$$\mathbf{v}_1 = \begin{pmatrix} -i \\ -2i \\ 1 \\ 2 \end{pmatrix}.$$

From the result of Problem 8.1.39 on p. 464, an eigenvector corresponding to $-i = \bar{i}$ is

$$\mathbf{v}_2 = \bar{\mathbf{v}}_1 = \begin{pmatrix} i \\ 2i \\ 1 \\ 2 \end{pmatrix}.$$

Similar computations yield

$$\lambda = 3i \qquad \qquad \lambda = -3i$$

$$\mathbf{v}_3 = \begin{pmatrix} 7i \\ -2i \\ -21 \\ 6 \end{pmatrix} \quad \text{and} \quad \mathbf{v}_4 = \begin{pmatrix} -7i \\ 2i \\ -21 \\ 6 \end{pmatrix}.$$

The general solution is

$$\mathbf{x}(t) = c_1 e^{it} \begin{pmatrix} -i \\ -2i \\ 1 \\ 2 \end{pmatrix} + c_2 e^{-it} \begin{pmatrix} i \\ 2i \\ 1 \\ 2 \end{pmatrix} + c_3 e^{3it} \begin{pmatrix} 7i \\ -2i \\ -21 \\ 6 \end{pmatrix} + c_4 e^{-3it} \begin{pmatrix} -7i \\ 2i \\ -21 \\ 6 \end{pmatrix}. \tag{7}$$

Setting $t = 0$ in (7), we obtain the system

$$-ic_1 + ic_2 + 7ic_3 - 7ic_4 = 1$$

$$-2ic_1 + 2ic_2 - 2ic_3 + 2ic_4 = 3$$

$$c_1 + c_2 - 21c_3 - 21c_4 = 0$$

$$2c_1 + 2c_2 + 6c_3 + 6c_4 = 0$$

The unique solution to this system is

$$c_1 = \frac{23}{32} i, \quad c_2 = -\frac{23}{32} i, \quad c_3 = \frac{i}{32}, \quad c_4 = -\frac{i}{32}.$$

Inserting these values into (7) and adding the vectors, we obtain

$$\mathbf{x}(t) = \frac{i}{32} \begin{pmatrix} -23ie^{it} - 23ie^{-it} + 7ie^{3it} + 7ie^{-3it} \\ -46ie^{it} - 46ie^{-it} - 2ie^{3it} - 2ie^{-3it} \\ 23e^{it} - 23e^{-it} - 21e^{3it} + 21e^{-3it} \\ 46e^{it} - 46e^{-it} + 6e^{3it} - 6e^{-3it} \end{pmatrix}$$

$$= \frac{1}{32} \begin{pmatrix} 23(e^{it} + e^{-it}) - 7(e^{3it} + e^{-3it}) \\ 46(e^{it} + e^{-it}) + 2(e^{3it} + e^{-3it}) \\ 23i(e^{it} - e^{-it}) - 21i(e^{3it} - e^{-3it}) \\ 46i(e^{it} - e^{-it}) + 6i(e^{3it} - e^{-3it}) \end{pmatrix}. \tag{8}$$

From the Euler equations (see p. 83),

$$e^{it} + e^{-it} = 2 \cos t, \qquad e^{3it} + e^{-3it} = 2 \cos 3t,$$

$$e^{it} - e^{-it} = 2i \sin t, \qquad e^{3it} - e^{-3it} = 2i \sin 3t.$$

Inserting these values into (8), we obtain the solution

$$\mathbf{x}(t) = \begin{pmatrix} (23/16) \cos t - (7/16) \cos 3t \\ (23/8) \cos t + (1/8) \cos 3t \\ (-23/16) \sin t + (21/16) \sin 3t \\ (-23/8) \sin t - (3/8) \sin 3t \end{pmatrix} = \begin{pmatrix} x_1(t) \\ x_2(t) \\ x_3(t) \\ x_4(t) \end{pmatrix},$$

where $x_1(t)$ = displacement of m_1, $x_2(t)$ = displacement of m_2, $x_3(t)$ = velocity of m_1, and $x_4(t)$ = velocity of m_2.

We now apply our computations to a simple biological model of population growth. Suppose that in an ecosystem there are two interacting species S_1 and S_2. We denote the populations of the species at time t by $x_1(t)$ and $x_2(t)$. One system governing the relative growth of the two species is

$$x_1'(t) = ax_1(t) + bx_2(t),$$

$$x_2'(t) = cx_1(t) + dx_2(t).$$

We can interpret the constants a, b, c, and d as follows. If the species are *competing*, then it is reasonable to have $b < 0$ and $c < 0$. This is true because increases in the population of one species will slow the growth of the other. A second model is a *predator-prey* relationship. If S_1 is the prey and S_2 is the predator (S_2 eats S_1), then it is reasonable to have $b < 0$ and $c > 0$ since an increase in the predator species will cause a decrease in the prey species, while an increase in the prey species will cause an increase in the predator species (since it will have more food). Finally, in a *symbiotic* relationship (a partnership of different species), we would likely have $b > 0$ and $c > 0$. Of course, the constants a, b, c, and d depend on a wide variety of factors including available food, time of year, climate, limits due to overcrowding, other competing species, and so on. We shall analyze two different models by using the material in this section. We assume that t is measured in years.

EXAMPLE 3: A Competition Model

Consider the system

$$x_1'(t) = 3x_1(t) - x_2(t),$$

$$x_2'(t) = -2x_1(t) + 2x_2(t).$$

Here, an increase in the population of one species causes a decline in the growth rate of another. Suppose that the initial populations are $x_1(0) = 90$ and $x_2(0) = 150$. Find the populations of both species for $t > 0$.

SOLUTION

We have $A = \begin{pmatrix} 3 & -1 \\ -2 & 2 \end{pmatrix}$. The eigenvalues of A are $\lambda_1 = 1$ and $\lambda_2 = 4$ with corresponding eigenvectors $\mathbf{v}_1 = \begin{pmatrix} 1 \\ 2 \end{pmatrix}$ and $\mathbf{v}_2 = \begin{pmatrix} 1 \\ -1 \end{pmatrix}$. Then the general solution is

$$\mathbf{x}(t) = \begin{pmatrix} x_1(t) \\ x_2(t) \end{pmatrix} = c_1 e^t \begin{pmatrix} 1 \\ 2 \end{pmatrix} + c_2 e^{4t} \begin{pmatrix} 1 \\ -1 \end{pmatrix} = \begin{pmatrix} c_1 e^t + c_2 e^{4t} \\ 2c_1 e^t - c_2 e^{4t} \end{pmatrix}.$$

Setting $t = 0$ we have

$$\mathbf{x}(0) = \begin{pmatrix} c_1 + c_2 \\ 2c_1 - c_2 \end{pmatrix} = \begin{pmatrix} 90 \\ 150 \end{pmatrix},$$

from which we find that $c_1 = 80$ and $c_2 = 10$ so that

$$\mathbf{x}(t) = \begin{pmatrix} 80e^t + 10e^{4t} \\ 160e^t - 10e^{4t} \end{pmatrix}.$$

For example, after 6 months ($t = \frac{1}{2}$ year), $x_1(t) = 80e^{1/2} + 10e^2 \approx 206$ individuals, while $x_2(t) = 160e^{1/2} - 10e^2 \approx 190$ individuals. More significantly, $160e^t - 10e^{4t} = 0$ when $16e^t = e^{4t}$ or $16 = e^{3t}$ or $3t = \ln 16$ and $t = (\ln 16)/3 \approx 2.77/3 \approx 0.92$ years $\approx$ 11 months. Thus the second species will be eliminated after only 11 months even though it started with a larger population. In Problems 18 and 19 you are asked to show that neither population will be eliminated if $x_2(0) = 2x_1(0)$ and that the first population will be eliminated if $x_2(0) > 2x_1(0)$. Thus, as was well known to Darwin, survival in this very simple model depends on the relative sizes of the competing species when competition begins.

EXAMPLE 4: A Predator-Prey Model

Consider the predator-prey model governed by the system

$$x'_1(t) = x_1(t) + x_2(t),$$

$$x'_2(t) = -x_1(t) + x_2(t).$$

If the initial populations are $x_1(0) = x_2(0) = 1000$, determine the populations of the two species for $t > 0$.

SOLUTION

Here $A = \begin{pmatrix} 1 & 1 \\ -1 & 1 \end{pmatrix}$ with characteristic equation $\lambda^2 - 2\lambda + 2 = 0$, complex roots $\lambda_1 = 1 + i$ and $\lambda_2 = 1 - i$, and eigenvectors $\mathbf{v}_1 = \begin{pmatrix} 1 \\ i \end{pmatrix}$ and $\mathbf{v}_2 = \begin{pmatrix} 1 \\ -i \end{pmatrix}$. Then

$$\mathbf{x}(t) = \begin{pmatrix} c_1 e^{(1+i)t} + c_2 e^{(1-i)t} \\ ic_1 e^{(1+i)t} - ic_2 e^{(1-i)t} \end{pmatrix},$$

$$\mathbf{x}(0) = \begin{pmatrix} c_1 + c_2 \\ ic_1 - ic_2 \end{pmatrix} = \begin{pmatrix} 1000 \\ 1000 \end{pmatrix},$$

so $c_1 = 500(1 - i)$, $c_2 = 500(1 + i)$, and

$$\mathbf{x}(t) = 500 \begin{pmatrix} (1-i)e^{(1+i)t} + (1+i)e^{(1-i)t} \\ (1+i)e^{(1+i)t} + (1-i)e^{(1-i)t} \end{pmatrix}.$$

Now,

$$(1-i)e^{(1+i)t} + (1+i)e^{(1-i)t} = (1-i)e^t e^{it} + (1+i)e^t e^{-it}$$

$$= (1-i)e^t(\cos t + i \sin t) + (1+i)e^t(\cos t - i \sin t)$$

$$= e^t(2 \cos t + 2 \sin t)$$

and

$$(1+i)e^{(1+i)t} + (1-i)e^{(1-i)t} = e^t(2 \cos t - 2 \sin t).$$

Thus

$$\mathbf{x}(t) = \begin{pmatrix} 1000e^t(\cos t + \sin t) \\ 1000e^t(\cos t - \sin t) \end{pmatrix}.$$

The prey species is eliminated when $1000e^t(\cos t - \sin t) = 0$ or when $\sin t = \cos t$. The first positive solution of this last equation is $t = \pi/4 \approx 0.7854$ year ≈ 9.4 months.

Problems 8.6

In Problems 1–9 find the general solution of the system $\mathbf{x}'(t) = A\mathbf{x}(t)$.

1. $A = \begin{pmatrix} -2 & -2 \\ -5 & 1 \end{pmatrix}$

2. $A = \begin{pmatrix} 3 & -1 \\ -2 & 4 \end{pmatrix}$

3. $A = \begin{pmatrix} 2 & -1 \\ 5 & -2 \end{pmatrix}$

4. $A = \begin{pmatrix} 3 & -5 \\ 1 & -1 \end{pmatrix}$

5. $A = \begin{pmatrix} 1 & -1 \\ 1/3 & -1/6 \end{pmatrix}$

6. $A = \begin{pmatrix} -2 & 1 \\ 5 & 2 \end{pmatrix}$

7. $A = \begin{pmatrix} -12 & 8 \\ -6 & 2 \end{pmatrix}$

8. $A = \begin{pmatrix} 1 & 1 & -2 \\ -1 & 2 & 1 \\ 0 & 1 & -1 \end{pmatrix}$

9. $A = \begin{pmatrix} 4 & 6 & 6 \\ 1 & 3 & 2 \\ -1 & -5 & -2 \end{pmatrix}$

10. Find I_L and I_R for the circuit in Fig. 1 if $E = 0$, $R = 200$ ohms, $C = 4 \times 10^{-4}$ farad, $L = 10$ henrys, $I_L(0) = 0.5$ amp, and $I_R(0) = 0$.

11. Answer Problem 10 if $I_L(0) = 0$, $I_R(0) = 0.5$ amp, and all other values remain the same.

12. Find the current at time t in each loop of the circuit shown below given that $E = 0$, $R = 10$ ohms, $L = 10$ henrys, $I_1(0) = 0.05$ amp, and $I_2(0) = 0.2$ amp.

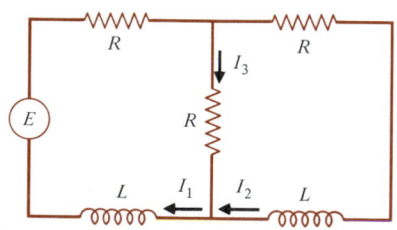

13. Solve Example 2 with $k_1 = 13m_1/2$, $k_2 = 2m_2$, and $3m_1 = 4m_2$.

14. Solve Example 2 with $k_1 = 6m_1$, $k_2 = 2m_2$, and $m_1 = m_2$.

* 15. A rotating, straight slender shaft may be dynamically unstable at high speeds. When the period ω of rotation is nearly equal to one of its periods of lateral vibration, the shaft is then said to **whirl**. A whirling shaft satisfies the differential equation

$$EI\frac{d^4y}{dx^4} = m\omega^2 y, \qquad (9)$$

where $y = y(x)$ is the distance of the shaft from its geometric axis, E denotes Young's modulus of elasticity, I is the moment of inertia of the shaft, and m is the mass. Assume the shaft is hinged at both ends and satisfies

$$y(0) = y(L) = y''(0) = y''(L) = 0.$$

(a) Express (9) as a system of first-order differential equations.

(b) Show that nontrivial solutions exist if and only if $\omega = n^2\pi^2\sqrt{EI/m}/L^2$.

(c) Find the rotational speed necessary to produce whirling of a steel shaft $1\frac{1}{2}$ inches in diameter and 8 feet long. [*Hint:* $E = 4.32 \times 10^9$, $I = \pi r^4/4$, $m = 6/32.2$.]

* 16. Consider the coupled mechanical system shown below, where two masses m_1 and m_2 rest on a frictionless plane and are attached to fixed walls by two springs with spring constants k_1 and k_3. The masses m_1 and m_2 are connected by a spring with spring constant k_2.

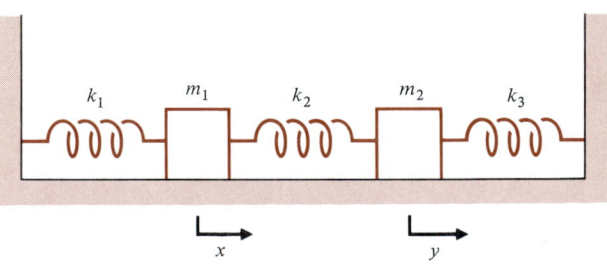

(a) Obtain the system of differential equations describing the motion of this mechanical system.

(b) Solve the system in part (a), and show that the motion of each mass is a superposition of two simple harmonic motions (called the **normal modes** of vibration).

17. **A Model of Symbiosis.** Consider the symbiotic model governed by the system

$$x_1'(t) = -\tfrac{1}{2}x_1(t) + x_2(t),$$
$$x_2'(t) = \tfrac{1}{4}x_1(t) - \tfrac{1}{2}x_2(t).$$

Note that in this model the population of each species increases proportionally to the population of the other and decreases proportionally to its own population. Suppose that $x_1(0) = 200$ and $x_2(0) = 500$. Determine the population of each species for $t > 0$.

18. In Example 3, show that if the initial vector $\mathbf{x}(0) = \begin{pmatrix} a \\ 2a \end{pmatrix}$, where a is a constant, then both populations grow at a rate proportional to e^t.

19. In Example 3, show that if $x_2(0) > 2x_1(0)$, then the first population will be eliminated.

20. In a water desalinization plant there are two tanks of water. Suppose that tank 1 contains 1000 liters of brine in which 1000 kg of salt is dissolved and tank 2 contains 1000 liters of pure water. Suppose that water flows into tank 1 at the rate of 20 liters per minute and the mixture flows from tank 1 into tank 2 at a rate of 30 liters per minute. From tank 2, 10 liters are pumped back to tank 1 (establishing **feedback**) while 20 liters are flushed away. Find the amount of salt in both tanks at all times t. [*Hint:* Write the information as a 2×2 system and let $x_1(t)$ and $x_2(t)$ denote the amount of salt in each tank.]

21. Consider the *second-order differential equation*

$$x''(t) + ax'(t) + bx(t) = 0.$$

(a) Letting $x_1(t) = x(t)$ and $x_2(t) = x'(t)$, write the preceding equation as a first-order system in the form of equation (4), where A is a 2×2 matrix.

(b) Show that the characteristic equation of A is

$$\lambda^2 + a\lambda + b = 0.$$

In Problems 22–24 use the result of Problem 21 to solve the given equation.

22. $x'' + 5x' + 6x = 0; \quad x(0) = 1, \quad x'(0) = 0$

23. $x'' + 6x' + 5x = 0; \quad x(0) = 1, \quad x'(0) = 2$

24. $x'' - 3x' - 10x = 0; \quad x(0) = 3, \quad x'(0) = 2$

25. Consider the system $\mathbf{x}' = A\mathbf{x}$ where $A = \begin{pmatrix} 7 & -1 \\ 9 & 1 \end{pmatrix}$.

(a) Show that $\lambda = 4$ is an eigenvalue of A of algebraic multiplicity 2, with only one linearly independent eigenvector.

(b) Find a vector $\mathbf{v}_1$ such that $\mathbf{x}_1(t) = e^{4t}\mathbf{v}_1$ is a solution to the system.

(c) Find two vectors $\mathbf{u}$ and $\mathbf{w}$ such that $\mathbf{x}_2(t) = e^{4t}\mathbf{u} + te^{4t}\mathbf{w}$ is a second, linearly independent solution to the system. [*Hint:* Using the fact that $\mathbf{x}_2'(t) = A\mathbf{x}_2(t)$, show that you obtain the system

$$4\mathbf{u} + \mathbf{w} = A\mathbf{u},$$
$$4\mathbf{w} = A\mathbf{w}$$

so that $\mathbf{w}$ is an eigenvector of A (and is therefore a multiple of $\mathbf{v}_1$.]

(d) Write the general solution to the system.

In Problems 26 and 27 use the technique suggested in Problem 25 to obtain the general solution to $\mathbf{x}'(t) = A\mathbf{x}(t)$.

26. $A = \begin{pmatrix} -10 & -7 \\ 7 & 4 \end{pmatrix}$

27. $A = \begin{pmatrix} -12 & 7 \\ -7 & 2 \end{pmatrix}$.

28. **A Predator-Prey Model.** Consider the following system in which species 1 is the prey and species 2 is the predator:

$$x_1'(t) = 2x_1(t) - x_2(t),$$
$$x_2'(t) = x_1(t) + 4x_2(t).$$

Find the populations of the two species for $t > 0$ if the initial populations are $x_1(0) = 500$ and $x_2(0) = 100$.

29. In Problem 28 show that the first population will become extinct in α years, where $\alpha = x_1(0)/[x_1(0) + x_2(0)]$.

8.7 The Cayley-Hamilton Theorem

There are many useful results concerning the eigenvalues of a matrix besides the ones given in the last section. In this section we shall discuss an important one that says that any matrix satisfies its own characteristic equation. First, we need to define what we mean by a **matrix polynomial**.

Matrix Polynomial

Let $p(x) = x^n + a_{n-1}x^{n-1} + \cdots + a_1x + a_0$ be a polynomial and let A be an $n \times n$ matrix. Then powers of A are defined and we define

$$p(A) = A^n + a_{n-1}A^{n-1} + \cdots + a_1A + a_0I \tag{1}$$

EXAMPLE 1

Let $A = \begin{pmatrix} -1 & 4 \\ 3 & 7 \end{pmatrix}$ and $p(x) = x^2 - 5x + 3$. Then

$$P(A) = A^2 - 5A + 3I = \begin{pmatrix} 13 & 24 \\ 18 & 61 \end{pmatrix} + \begin{pmatrix} 5 & -20 \\ -15 & -35 \end{pmatrix} + \begin{pmatrix} 3 & 0 \\ 0 & 3 \end{pmatrix} = \begin{pmatrix} 21 & 4 \\ 3 & 29 \end{pmatrix}.$$

Expression (1) is a polynomial with scalar coefficients defined for a matrix variable. We can also define a polynomial with *square matrix* coefficients by

$$Q(\lambda) = B_0 + B_1\lambda + B_2\lambda^2 + \cdots + B_n\lambda^n. \tag{2}$$

If A is a matrix of the same size, then we define

$$Q(A) = B_0 + B_1A + B_2A^2 + \cdots + B_mA^m. \tag{3}$$

When we use equation (3) we must be careful about the order in which we write the various matrices since matrices do not commute under multiplication.

THEOREM 1

If $P(\lambda)$ and $Q(\lambda)$ are polynomials in the scalar variable λ with square matrix coefficients and if $P(\lambda) = Q(\lambda)(A - \lambda I)$, then $P(A) = O$, the zero matrix.

PROOF

If $Q(\lambda)$ is given by equation (2), then

$$P(\lambda) = (B_0 + B_1\lambda + B_2\lambda^2 + \cdots + B_n\lambda^n)(A - \lambda I)$$

$$= B_0A + B_1A\lambda + B_2A\lambda^2 + \cdots + B_nA\lambda^n$$

$$- B_0\lambda - B_1\lambda^2 - B_2\lambda^3 - \cdots - B_n\lambda^{n+1}. \tag{4}$$

Then, substituting A for λ in (4), we obtain

$$P(A) = B_0A + B_1A^2 + B_2A^3 + \cdots + B_nA^{n+1}$$

$$- B_0A - B_1A^2 - B_2A^3 - \cdots - B_nA^{n+1} = O. \quad \blacksquare$$

Note. We cannot prove this theorem by substituting $\lambda = A$ to obtain $P(A) = Q(A)(A - A) = 0$. This is because it is possible to find polynomials $P(\lambda)$

and $Q(\lambda)$ with matrix coefficients such that $F(\lambda) = P(\lambda)Q(\lambda)$ but $F(A) \neq P(A)Q(A)$. (See Problem 10.)

We can now state the main theorem.

■ **THEOREM 2: The Cayley-Hamilton Theorem†**

Every square matrix satisfies its own characteristic equation. That is, if $p(\lambda) = 0$ is the characteristic equation of A, then $p(A) = O$.

PROOF

We have

$$p(\lambda) = \det(A - \lambda I) = \begin{vmatrix} a_{11} - \lambda & a_{12} & \cdots & a_{1n} \\ a_{21} & a_{22} - \lambda & \cdots & a_{2n} \\ \vdots & \vdots & & \vdots \\ a_{n1} & a_{n2} & \cdots & a_{nn} - \lambda \end{vmatrix}.$$

Clearly, any cofactor of $(A - \lambda I)$ is a polynomial in λ. Thus the adjoint of $A - \lambda I$ (see the definition on p. 425) is an $n \times n$ matrix each of whose components is a polynomial in λ. That is:

$$\text{adj}(A - \lambda I) = \begin{pmatrix} p_{11}(\lambda) & p_{12}(\lambda) & \cdots & p_{1n}(\lambda) \\ p_{21}(\lambda) & p_{22}(\lambda) & \cdots & p_{2n}(\lambda) \\ \vdots & \vdots & & \vdots \\ p_{n1}(\lambda) & p_{22}(\lambda) & \cdots & p_{nn}(\lambda) \end{pmatrix}.$$

This means that we can think of $\text{adj}(A - \lambda I)$ as a polynomial, $Q(\lambda)$, in λ with $n \times n$ matrix coefficients. For example,

$$\begin{pmatrix} -\lambda^2 - 2\lambda + 1 & 2\lambda^2 - 7\lambda - 4 \\ 4\lambda^2 + 5\lambda - 2 & -3\lambda^2 - \lambda + 3 \end{pmatrix} = \begin{pmatrix} -1 & 2 \\ 4 & -3 \end{pmatrix}\lambda^2 + \begin{pmatrix} -2 & -7 \\ 5 & -1 \end{pmatrix}\lambda + \begin{pmatrix} 1 & -4 \\ -2 & 3 \end{pmatrix}.$$

Now, from Theorem 7.9.3 on p. 426,

$$\det(A - \lambda I)I = [\text{adj}(A - \lambda I)][A - \lambda I] = Q(\lambda)(A - \lambda I). \tag{5}$$

But $\det(A - \lambda I)I = p(\lambda)I$. If

$$p(\lambda) = \lambda^n + a_{n-1}\lambda^{n-1} + \cdots + a_1\lambda + a_0,$$

then we define

$$P(\lambda) = p(\lambda)I = \lambda^n I + a_{n-1}\lambda^{n-1}I + \cdots + a_1\lambda I + a_0 I.$$

Thus, from (5), we have $P(\lambda) = Q(\lambda)(A - \lambda I)$. Finally, from Theorem 1, $P(A) = O$. This completes the proof. ■

---

† Named after Sir William Rowan Hamilton (discussed in Chapter 6—see his biography on p. 292) and Arthur Cayley (1821–1895). Cayley published the first discussion of this famous theorem in 1858. Independently, Hamilton discovered the result in his work on quaternions.

EXAMPLE 2

Let $A = \begin{pmatrix} 1 & -1 & 4 \\ 3 & 2 & -1 \\ 2 & 1 & -1 \end{pmatrix}$. In Example 8.1.5 on p. 459 we computed the characteristic equation $\lambda^3 - 2\lambda^2 - 5\lambda + 6 = 0$. Now we compute

$$A^2 = \begin{pmatrix} 6 & 1 & 1 \\ 7 & 0 & 11 \\ 3 & -1 & 8 \end{pmatrix}, \quad A^3 = \begin{pmatrix} 11 & -3 & 22 \\ 29 & 4 & 17 \\ 16 & 3 & 5 \end{pmatrix}$$

and

$$A^3 - 2A^2 - 5A + 6I = \begin{pmatrix} 11 & -3 & 22 \\ 29 & 4 & 17 \\ 16 & 3 & 5 \end{pmatrix} + \begin{pmatrix} -12 & -2 & -2 \\ -14 & 0 & -22 \\ -6 & 2 & -16 \end{pmatrix}$$

$$+ \begin{pmatrix} -5 & 5 & -20 \\ -15 & -10 & 5 \\ -10 & -5 & 5 \end{pmatrix} + \begin{pmatrix} 6 & 0 & 0 \\ 0 & 6 & 0 \\ 0 & 0 & 6 \end{pmatrix}$$

$$= \begin{pmatrix} 0 & 0 & 0 \\ 0 & 0 & 0 \\ 0 & 0 & 0 \end{pmatrix}.$$

When A is an invertible matrix, we can use the Cayley-Hamilton theorem to find A^{-1}. Let

$$p(\lambda) = \lambda^n + a_{n-1}\lambda^{n-1} + \cdots + a_1\lambda + a_0 = 0$$

be the characteristic equation of A. Then, by the Cayley-Hamilton theorem,

$$p(A) = A^n + a_{n-1}A^{n-1} + \cdots + a_1A + a_0I = O$$

and

$$A^{-1}p(A) = A^{n-1} + a_{n-1}A^{n-2} + \cdots + a_2A + a_1I + a_0A^{-1} = O.$$

Thus

$$A^{-1} = \frac{1}{a_0}(-A^{n-1} - a_{n-1}A^{n-2} - \cdots - a_2A - a_1I). \tag{6}$$

Note that $a_0 \neq 0$ because $a_0 = \det A$ (why?) and we assumed that A was invertible.

EXAMPLE 3

Let $A = \begin{pmatrix} 1 & -1 & 4 \\ 3 & 2 & -1 \\ 2 & 1 & -1 \end{pmatrix}$. Then $p(\lambda) = \lambda^3 - 2\lambda^2 - 5\lambda + 6$. Here $n = 3$, $a_2 = -2$, $a_1 = -5$, $a_0 = 6$, and

$$A^{-1} = \frac{1}{6}(-A^2 + 2A + 5I)$$

$$= \frac{1}{6}\left[\begin{pmatrix} -6 & -1 & -1 \\ -7 & 0 & -11 \\ -3 & 1 & -8 \end{pmatrix} + \begin{pmatrix} 2 & -2 & 8 \\ 6 & 4 & -2 \\ 4 & 2 & -2 \end{pmatrix} + \begin{pmatrix} 5 & 0 & 0 \\ 0 & 5 & 0 \\ 0 & 0 & 5 \end{pmatrix}\right]$$

$$= \frac{1}{6}\begin{pmatrix} 1 & -3 & 7 \\ -1 & 9 & -13 \\ 1 & 3 & -5 \end{pmatrix}.$$

Note that we computed A^{-1} with a single division, only one calculation of a determinant (in order to find $p(\lambda) = \det(A - \lambda I)$), and one matrix multiplication (to compute A^2).

Problems 8.7

In Problems 1–9: **(a)** Find the characteristic equation $p(\lambda) = 0$ of the given matrix; **(b)** verify that $p(A) = 0$; **(c)** use part (b) to compute A^{-1}.

1. $\begin{pmatrix} -2 & -2 \\ -5 & 1 \end{pmatrix}$

2. $\begin{pmatrix} 2 & -1 \\ 5 & -2 \end{pmatrix}$

3. $\begin{pmatrix} 1 & -1 & 0 \\ -1 & 2 & -1 \\ 0 & -1 & 1 \end{pmatrix}$

4. $\begin{pmatrix} 1 & 2 & 2 \\ 0 & 2 & 1 \\ -1 & 2 & 2 \end{pmatrix}$

5. $\begin{pmatrix} 0 & 1 & 0 \\ 0 & 0 & 1 \\ 1 & -3 & 3 \end{pmatrix}$

6. $\begin{pmatrix} -3 & -7 & -5 \\ 2 & 4 & 3 \\ 1 & 2 & 2 \end{pmatrix}$

7. $\begin{pmatrix} 2 & -1 & 3 \\ 4 & 1 & 6 \\ 1 & 5 & 3 \end{pmatrix}$

8. $\begin{pmatrix} 1 & 0 & 1 & 0 \\ 2 & -1 & 0 & 2 \\ -1 & 0 & 0 & 1 \\ 4 & 1 & -1 & 0 \end{pmatrix}$

9. $\begin{pmatrix} a & b & 0 & 0 \\ 0 & a & c & 0 \\ 0 & 0 & a & d \\ 0 & 0 & 0 & a \end{pmatrix}$; $bcd \neq 0$

10. Let $P(\lambda) = B_0 + B_1\lambda$ and $Q(\lambda) = C_0 + C_1\lambda$, where B_0, B_1, C_0, and C_1 are $n \times n$ matrices.
   a. Compute $F(\lambda) = P(\lambda)Q(\lambda)$.
   b. Let A be an $n \times n$ matrix. Show that $F(A) = P(A)Q(A)$ if A commutes with both C_0 and C_1.

Review Exercises for Chapter 8

In Exercises 1–6 calculate the eigenvalues and eigenvectors of the given matrix.

1. $\begin{pmatrix} -8 & 12 \\ -6 & 10 \end{pmatrix}$

2. $\begin{pmatrix} 2 & 5 \\ 0 & 2 \end{pmatrix}$

3. $\begin{pmatrix} 1 & 0 & 0 \\ 3 & 7 & 0 \\ -2 & 4 & -5 \end{pmatrix}$

4. $\begin{pmatrix} 1 & -1 & 0 \\ 1 & 2 & 1 \\ -2 & 1 & -1 \end{pmatrix}$

5. $\begin{pmatrix} 5 & -2 & 0 & 0 \\ 4 & -1 & 0 & 0 \\ 0 & 0 & 3 & -1 \\ 0 & 0 & 2 & 3 \end{pmatrix}$

6. $\begin{pmatrix} -2 & 1 & 0 \\ 0 & -2 & 1 \\ 0 & 0 & -2 \end{pmatrix}$

In Exercises 7–15 determine whether the given matrix A is diagonalizable. If it is, find a matrix C such that $C^{-1}AC = D$. If A is symmetric, find an orthogonal matrix Q such that $Q^t A Q = D$.

7. $\begin{pmatrix} -18 & -15 \\ 20 & 17 \end{pmatrix}$

8. $\begin{pmatrix} 17/2 & 9/2 \\ -15 & -8 \end{pmatrix}$

9. $\begin{pmatrix} 1 & 1 & 1 \\ -1 & -1 & 0 \\ -1 & 0 & -1 \end{pmatrix}$

10. $\begin{pmatrix} 4 & 2 & 0 \\ 2 & 4 & 0 \\ 0 & 0 & -3 \end{pmatrix}$

11. $\begin{pmatrix} -3 & 2 & 1 \\ -7 & 4 & 2 \\ -5 & 3 & 2 \end{pmatrix}$

12. $\begin{pmatrix} 8 & 0 & 12 \\ 0 & -2 & 0 \\ 12 & 0 & -2 \end{pmatrix}$

13. $\begin{pmatrix} 2 & 2 & 0 \\ 2 & 2 & 0 \\ 0 & 0 & -3 \end{pmatrix}$

14. $\begin{pmatrix} 4 & 2 & -2 & 2 \\ 1 & 3 & 1 & -1 \\ 0 & 0 & 2 & 0 \\ 1 & 1 & -3 & 5 \end{pmatrix}$

15. $\begin{pmatrix} 3 & 4 & -4 & 0 \\ 0 & -1 & 0 & 0 \\ 0 & 0 & -1 & 0 \\ 0 & -4 & 4 & 3 \end{pmatrix}$

In Exercises 16–20 identify the conic section and write it in new variables with the xy term absent.

16. $xy = -4$

17. $4x^2 + 2xy + 2y^2 = 8$

18. $4x^2 - 3xy + y^2 = 1$

19. $3y^2 - 2xy - 5 = 0$

20. $x^2 - 4xy + 4y^2 + 1 = 0$

21. Write the quadratic form $2x^2 + 4xy + 2y^2 - 3z^2$ in new variables x', y', and z' so that no cross-product terms are present.

In Problems 22–25 find the general solution of each system.

22. $\begin{aligned} x_1' &= 4x_1 + 2x_2 \\ x_2' &= 3x_1 + 3x_2 \end{aligned}$

23. $\begin{aligned} x_1' &= -8x_1 + 12x_2 \\ x_2' &= -6x_1 + 10x_2 \end{aligned}$

24. $\begin{aligned} x_1' &= 3x_1 - 3x_2 \\ x_2' &= x_1 + x_2 \end{aligned}$

25. $\begin{aligned} x_1' &= x_1 - x_2 + 4x_3 \\ x_2' &= 3x_1 + 2x_2 - x_3 \\ x_3' &= 2x_1 + x_2 - x_3 \end{aligned}$

26. Use the Cayley-Hamilton theorem to compute the inverse of

$$A = \begin{pmatrix} 2 & 3 & 1 \\ -1 & 1 & 0 \\ -2 & -1 & 4 \end{pmatrix}.$$

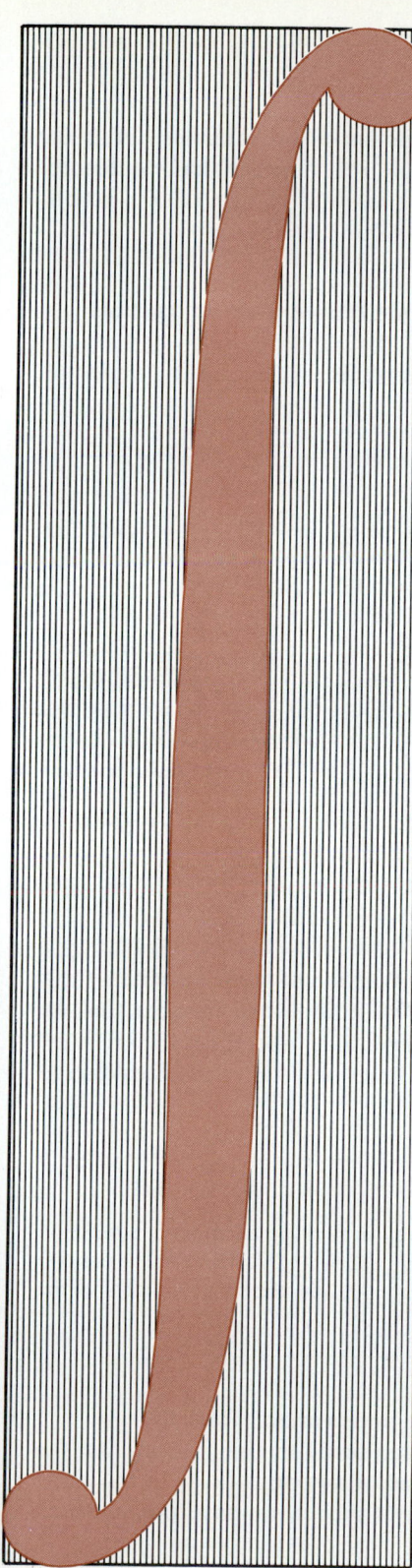

chapter 9

Vector Differential Calculus

9.1 Scalar Functions of Several Variables

For most of the functions we have encountered so far in this book, we have been able to write $y = f(x)$. This means that we could write the variable y explicitly in terms of the single variable x. However, in a great variety of applications it is necessary to write the quantity of interest in terms of two or more variables. We have already encountered this situation. For example, the volume of a right circular cylinder is given by

$$V = \pi r^2 h, \tag{1}$$

where r is the radius of the cylinder and h is its height. That is, V is a function of the *two* variables r and h.

As a second example, consider the **ideal gas law,** which relates pressure, volume, and temperature for an ideal gas: We have

$$PV = nRT,$$

where P is the pressure of the gas, V is the volume, T is the absolute temperature (i.e., in degrees Kelvin), n is the number of moles of the gas, and R is a constant. Solving for P, we find that

$$P = nRT/V. \tag{2}$$

That is, we can write P as a function of the *three* variables n, T, and V.

There are many examples like those cited above. It is probably fair to say that very few physical, biological, or economic quantities depend on one variable alone. Often we write these quantities in terms of one variable simply because functions of only one variable are the easiest functions to handle.

Function of Two Variables

DEFINITION

Let D be a subset of $\mathbb{R}^2$. Then a **real-valued function of two variables** f is a rule that assigns to each point (x, y) in D a unique real number that we denote $f(x, y)$. The set D is called the **domain** of f. The set $\{f(x, y): (x, y) \in D\}$, which is the set of values the function f takes on, is called the **range** of f.

Terminology. A real-valued function is often called a **scalar function.**

Remark. We emphasize that the domain of f is a subset of $\mathbb{R}^2$, while the range is a subset of $\mathbb{R}$, the real numbers.

EXAMPLE 1

Find the domain and range of the function f given by $f(x, y) = \sqrt{4 - x^2 - y^2}$, and find $f(0, 1)$ and $f(-1, 1)$.

SOLUTION

The function f is defined when the expression under the square root sign is nonnegative. Thus $D = \{(x, y): x^2 + y^2 \le 4\}$. This is the disk centered at the origin with radius 2. Because x^2 and y^2 are nonnegative, $4 - x^2 - y^2$ is largest when $x = y = 0$. Thus the largest value of $\sqrt{4 - x^2 - y^2} = \sqrt{4} = 2$. Since $x^2 + y^2 \le 4$, the least value of $4 - x^2 - y^2$ is 0, taken when $x^2 + y^2 = 4$ (at all points on the circle $x^2 + y^2 = 4$). Thus the range of f is the closed interval $[0, 2]$. Finally,

$$f(0, 1) = \sqrt{4 - 0^2 - 1^2} = \sqrt{3} \qquad \text{and} \qquad f(-1, 1) = \sqrt{4 - (-1)^2 - 1^2} = \sqrt{2}.$$

Function of Three Variables

DEFINITION

Let D be a subset of $\mathbb{R}^3$. Then a **real-valued function of three variables** f is a rule that assigns to each point (x, y, z) in D a unique real number that we denote $f(x, y, z)$. The set D is called the **domain** of f, and the set $\{f(x, y, z): (x, y, z) \in D\}$, which is the set of values the function f takes on, is called the **range** of f.

EXAMPLE 2

The formula for the pressure P in an ideal gas given in (2) is an example of a function of three variables: $P(n, T, V) = RnT/V$. The domain consists of all points in the positive octant of $\mathbb{R}^3$: $D = \{(n, T, V): n \ge 0, T \ge 0, V > 0\}$, and the range consists of the nonnegative real numbers.

We shall define a function of more than three variables shortly.

We now turn to a discussion of the graph of a function. Recall that the graph of a function of one variable f is the set of all points (x, y) in the plane such that $y = f(x)$. Using this definition as our model, we have the following definition.

Graph of a Function of
Two Variables

DEFINITION

The **graph** of a function f of two variables x and y is the set of all points (x, y, z) in $\mathbb{R}^3$ such that $z = f(x, y)$. The graph of a function of two variables is an example of a **surface** in $\mathbb{R}^3$.

EXAMPLE 3

What is the graph of the function

$$z = f(x, y) = \sqrt{1 - x^2 - y^2}? \tag{3}$$

SOLUTION

We first note that $z \geq 0$. Then squaring both sides of (3), we have $z^2 = 1 - x^2 - y^2$, or $x^2 + y^2 + z^2 = 1$. This equation is the equation of the unit sphere centered at $(0, 0, 0)$. However, since $z \geq 0$, the graph of f is the upper hemisphere.

It is often very difficult to graph a function $z = f(x, y)$ by plotting points in space. Traditional curve plotting techniques in three dimensions are tedious and, except in the easiest cases, often yield poor results. Frequently, the best one can do is describe **cross sections** of the surface by holding one coordinate constant and letting the other vary.

The advent of computers with graphing capabilities provides an exciting new technology for obtaining graphs of such surfaces. Two computer-drawn surfaces are given in Fig. 1.

The situation becomes much more complicated when we try to sketch the graph of a function of three variables $w = f(x, y, z)$. We would need *four dimensions* to sketch such a surface. But we are limited to three dimensions, and we see that we have reached the point where three-dimensional geometry fails us. We are *not* saying that curves and surfaces in four-dimensional space do not exist; they do—only that we are not able to sketch them.

We have said that it is usually very difficult to sketch the graph of a function of two variables. Fortunately, there is a way to describe the graph of such a function in two dimensions. The idea for what we are about to do comes from cartographers. Cartographers have the problem of indicating three-dimensional features (such as mountains and valleys) on a two-dimensional surface. They

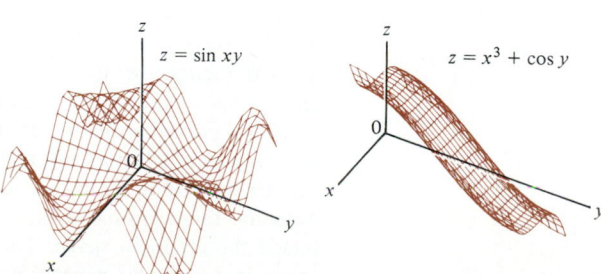

FIGURE 1

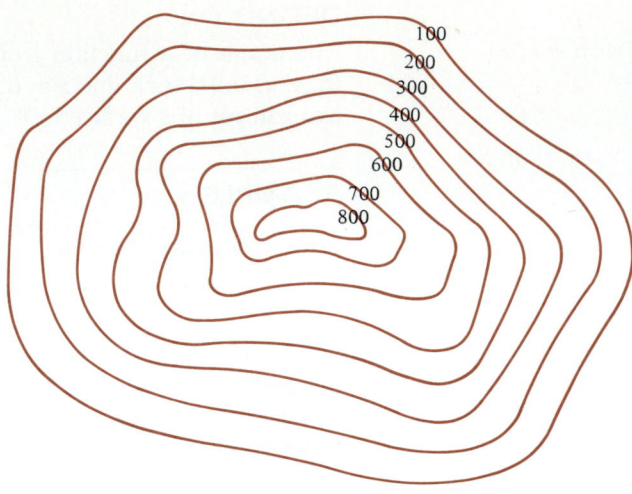

FIGURE 2

solve the problem by drawing a **contour map (topographic map),** which is a map in which points of constant elevation are joined to form curves, called **contour curves.** The closer together these contour curves are drawn, the steeper is the terrain. A portion of a typical contour map is sketched in Fig. 2.

Level Curves

We can use the same idea to depict the function $z = f(x, y)$ graphically. If z is fixed, then the equation $f(x, y) = z$ frequently defines a curve in the xy-plane, called a **level curve.**

    We can think of a level curve as the projection of a cross section of the surface lying in a plane specified by the value of z parallel to the xy-plane. In other words, a level curve is the projection of the intersection of the surface $z = f(x, y)$ with the plane $z = c$. This idea is best illustrated with some examples.

EXAMPLE 4

Sketch the level curves of $z = x^2 + y^2$.

SOLUTION

If $z > 0$, then $z = a^2$ for some positive number $a > 0$. Hence all level curves are circles of the form $x^2 + y^2 = a^2$. The number a^2 can be thought of as the "elevation" of points on a level curve. Some of these curves are sketched in Fig. 3. Each level curve encloses a projection of a "slice" of the actual graph of the function in three dimensions. In this case each circle is the projection onto the xy-plane of a part of the surface in space. Actually, this example is especially simple because we can, without much difficulty, sketch the graph in space. The equation $z = x^2 + y^2$ is the equation of an elliptic paraboloid (actually, a circular paraboloid). It is sketched in Fig. 4. In this easy case we can see that if we slice this surface parallel to the xy-plane, we obtain the circles whose

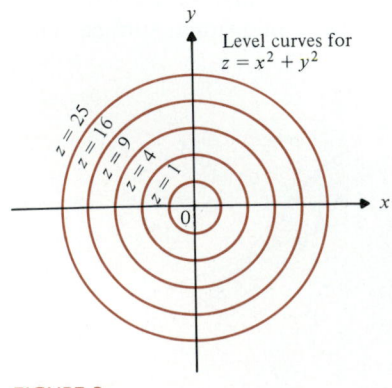

Level curves for $z = x^2 + y^2$

FIGURE 3

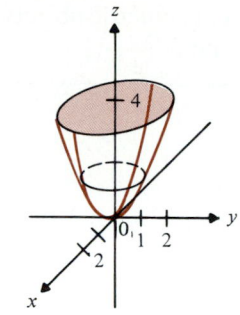

FIGURE 4

Function of n Variables

projections onto the xy-plane are the level curves. In most cases, of course, we will not be able to sketch the graph in space easily, so we will have to rely on our "contour map" sketch in $\mathbb{R}^2$. If $z = 0$, the level curve consists of a single point (the origin). There are no level curves for $z < 0$ since $x^2 + y^2 \geq 0$.

Most of the real-valued (scalar) functions encountered in this book are functions of one, two, or three variables. However, for completeness, we define below a function of n variables.

DEFINITION

Let Ω be a subset of $\mathbb{R}^n$. A (scalar-valued) **function of n variables** f is a rule that assigns to each vector $\mathbf{x} = (x_1, x_2, \ldots, x_n)$ in Ω a unique real number which we denote $f(\mathbf{x})$ or $f(x_1, x_2, \ldots, x_n)$. The set Ω is called the **domain** of f and is denoted dom f. The set $\{f(\mathbf{x}): \mathbf{x} \in \Omega\}$, which is the set of values the function f takes on, is called the **range** of f.

Notation. We will write

$$f: \mathbb{R}^n \to \mathbb{R}$$

to indicate that f is a real-valued function whose domain is a subset of $\mathbb{R}^n$.

EXAMPLE 5

Let $\Omega = \mathbb{R}^5$ and let $f(x_1, x_2, x_3, x_4, x_5) = x_1 x_2 x_3 x_4 x_5$. Then the range of f is $\mathbb{R}$.

Usually, as in $\mathbb{R}^2$, when the domain Ω is not given, we take the domain of f to be the largest subset of $\mathbb{R}^n$ for which the expression $f(x)$ makes sense.

EXAMPLE 6

In $\mathbb{R}^4$, let

$$f(\mathbf{x}) = \frac{1}{x_1^2 + x_2^2 + x_3^2 + x_4^2 - 1}.$$

Then $f(\mathbf{x})$ is defined except when $x_1^2 + x_2^2 + x_3^2 + x_4^2 = 1$. This is the equation of the **unit sphere** in $\mathbb{R}^4$. Thus the domain of $f = \{(x_1, x_2, x_3, x_4): x_1^2 + x_2^2 + x_3^2 + x_4^2 \neq 1\}$. It is a bit more difficult to find the range of f. We note that $x_1^2 + x_2^2 + x_3^2 + x_4^2 - 1$ can be made as close to zero—from either side—as desired. Thus $f(\mathbf{x})$ can take on arbitrarily large values. Since $x_1^2 + x_2^2 + x_3^2 + x_4^2 - 1 \geq -1$, $f(\mathbf{x})$ cannot take on negative values larger than -1 (to the right of -1). In addition, $f(\mathbf{x}) \neq 0$ for any vector $\mathbf{x}$. Finally, if $|\mathbf{x}|$ is large (see p. 342 for the definition of the norm, $|\mathbf{x}|$), then $f(\mathbf{x})$ is small and positive. Thus the range of $f = (-\infty, -1] \cup (0, \infty) = \mathbb{R} - (-1, 0]$.

As in Example 6, we can define spheres and other geometric objects in $\mathbb{R}^n$, where $n > 3$, by analogy with known objects in $\mathbb{R}^2$ and $\mathbb{R}^3$.

Problems 9.1

In Problems 1–35, find the domain and range of the indicated function.

1. $f(x, y) = \sqrt{x^2 + y^2}$

2. $f(x, y) = \sqrt{1 + x + y}$

3. $f(x, y) = \dfrac{x}{y}$

4. $f(x, y) = \sqrt{1 - x^2 - 4y^2}$

5. $f(x, y) = \sqrt{1 - x^2 + 4y^2}$

6. $f(x, y) = \sin(x + y)$

7. $f(x, y) = e^x + e^y$

8. $f(x, y) = \dfrac{1}{(x^2 - y^2)^{3/2}}$

9. $f(x, y) = \tan(x - y)$

10. $f(x, y) = \sqrt{\dfrac{x + y}{x - y}}$

11. $f(x, y) = \sqrt{\dfrac{x - y}{x + y}}$

12. $f(x, y) = \sin^{-1}(x + y)$

13. $f(x, y) = \cos^{-1}(x - y)$

14. $f(x, y) = \dfrac{y}{|x|}$

15. $f(x, y) = \dfrac{x^2 - y^2}{x + y}$

16. $f(x, y) = \ln(1 + x^2 - y^2)$

* 17. $f(x, y) = \left(\dfrac{x}{2y}\right) + \left(\dfrac{2y}{x}\right)$

18. $f(x, y, z) = x + y + z$

19. $f(x, y, z) = \sqrt{x + y + z}$

20. $f(x, y, z) = \dfrac{1}{\sqrt{x^2 + y^2 + z^2}}$

21. $f(x, y, z) = \dfrac{1}{\sqrt{x^2 - y^2 + z^2}}$

22. $f(x, y, z) = \dfrac{1}{\sqrt{x^2 - y^2 - z^2}}$

23. $f(x, y, z) = \sqrt{-x^2 - y^2 - z^2}$

24. $f(x, y, z) = \ln(x - 2y - 3z + 4)$

25. $f(x, y, z) = \dfrac{xy}{z}$

26. $f(x, y, z) = \sin(x + y - z)$

27. $f(x, y, z) = \sin^{-1}(x + y - z)$

28. $f(x, y, z) = \ln(x + y - z)$

29. $f(x, y, z) = \tan^{-1}\left(\dfrac{x + z}{y}\right)$

30. $f(x, y, z) = e^{xy + z}$

31. $f(x, y, z) = \dfrac{e^x + e^y}{e^z}$

32. $f(x, y, z) = xyz$

33. $f(x, y, z) = \dfrac{1}{xyz}$

34. $f(x, y, z) = \dfrac{x}{y + z}$

35. $f(x, y, z) = \sin x + \cos y + \sin z$

In Problems 36–42, sketch the graph of the given function.

36. $z = 4x^2 + 4y^2$

37. $y = x^2 + 4z^2$

38. $x = 4z^2 - 4y^2$

39. $z = x^2 - 4y^2$

40. $z = \sqrt{x^2 + 4y^2 + 4}$

41. $y = \sqrt{x^2 - 4z^2 + 4}$

42. $x = \sqrt{4 - z^2 - 4y^2}$

In Problems 43–50, describe the level curves of the given function and sketch these curves for the given values of z.

43. $z = \sqrt{1 + x + y}$; $z = 0, 1, 5, 10$

44. $z = \dfrac{x}{y}$; $z = -3, -1, 1, 3, 5$

45. $z = \sqrt{1 - x^2 - 4y^2}$; $z = 0, \frac{1}{4}, \frac{1}{2}, 1$

46. $z = \sqrt{1 + x^2 - y}$; $z = 0, 1, 2, 5$

47. $z = \cos^{-1}(x - y)$; $z = 0, \dfrac{\pi}{6}, \dfrac{\pi}{3}, \dfrac{\pi}{2}$

48. $z = \sqrt{\dfrac{x + y}{x - y}}$; $z = 0, 1, 2, 5$

49. $z = \tan(x + y)$; $z = -1, 0, 1, \sqrt{3}$

50. $z = \tan^{-1}(x - y^2)$; $z = 0, \pi/6, \pi/4$

51. The temperature T at any point on an object in the plane is given by $T(x, y) = 20 + x^2 + 4y^2$. Sketch the isothermal (constant temperature) curves for $T = 50$, $T = 60$, and $T = 70$ degrees.

52. The voltage at a point (x, y) on a metal plate placed in the xy-plane is given by $V(x, y) = \sqrt{1 - 4x^2 - 9y^2}$. Sketch the equipotential (constant voltage) curves for $V = 1.0$ V, $V = 0.5$ V, and $V = 0.25$ V.

53. A manufacturer earns $P(x, y) = 100 + 2x^2 + 3y^2$ dollars each year for producing x and y units, respectively, of two products. Sketch the constant profit curves for $P = \$100$, $P = \$200$, and $P = \$1000$.

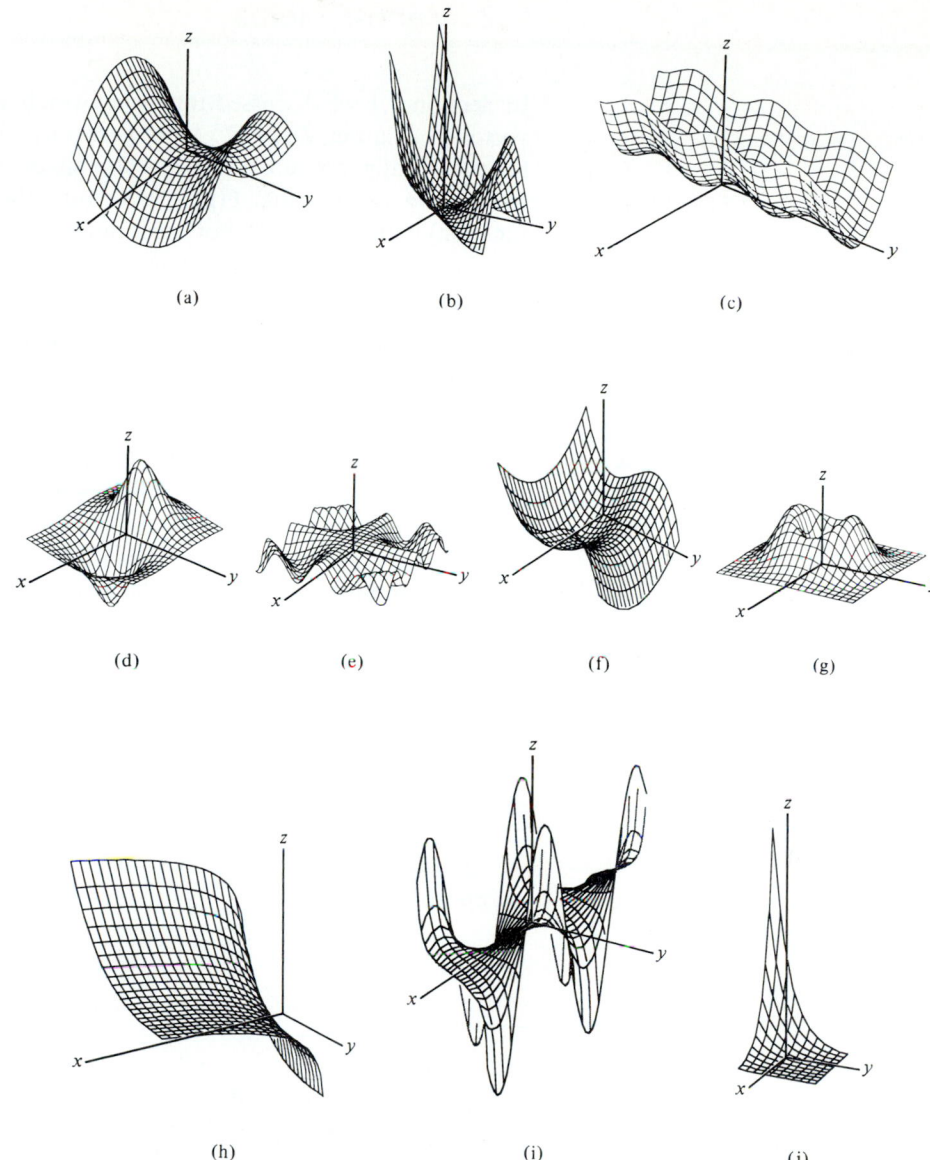

(a)

(b)

(c)

(d)

(e)

(f)

(g)

(h)

(i)

(j)

In Problems 54–63 a function of two variables is given. Match the function to one of ten computer-drawn graphs given below.

54. $z = x^2 + \sin y$

55. $z = \ln x + e^{-y}$

56. $z = \cos xy$

57. $z = y^2 - x^2$

58. $z = x^2 - y^3 - x + 2y - 2$

59. $z = (y - \frac{3}{2}x^2)(y - \frac{1}{2}x^2)$

60. $z = e^{-(x+y)}$

61. $z = \sin x \tan y, \quad -\pi/2 < y < \pi/2$

62. $z = \dfrac{1}{(x+0.15)^2 + y^2 + 0.2} - \dfrac{1}{(x-0.15)^2 + y^2 + 0.2}$

63. $z = (2x^2 + 3y^2)e^{1-x^2-y^2}$

9.2 Vector Fields

In Section 9.1 we discussed functions which assigned a real number to each vector in a subset, D, of $\mathbb{R}^n$. More generally, suppose we assign a vector $\mathbf{F(p)}$ to each point $\mathbf{p}$ in a subset D of a vector space. We call this assignment a **vector field** on D and say that $\mathbf{F(p)}$ is a **vector-valued function** (or, simply, **vector function**) with domain D. For most purposes D will be a subset of $\mathbb{R}^2$ or $\mathbb{R}^3$.

Vector Field in $\mathbb{R}^2$

DEFINITION

Let D be a subset of $\mathbb{R}^2$. Then $\mathbf{F}$ is a **vector field** in D if $\mathbf{F}$ assigns to every $\mathbf{x}$ in D a unique vector $\mathbf{F(x)}$ in $\mathbb{R}^2$.

Remark. Simply put, a vector field in $\mathbb{R}^2$ is a function whose domain is a subset of $\mathbb{R}^2$ and whose range is a subset of $\mathbb{R}^2$. Two vector fields in $\mathbb{R}^2$ are sketched in Fig. 1. The meaning of this sketch is that to every point $\mathbf{x}$ in D a unique vector $\mathbf{F(x)}$ is assigned. That is, the function value $\mathbf{F(x)}$ is represented by an arrow with $\mathbf{x}$ at the tail of the arrow.

This definition can be extended, in an obvious way, to $\mathbb{R}^3$ or $\mathbb{R}^n$.

Vector Field in $\mathbb{R}^3$

DEFINITION

Let D be a subset of $\mathbb{R}^3$. Then $\mathbf{F}$ is a **vector field** in D if $\mathbf{F}$ assigns to each vector $\mathbf{x}$ in D a unique vector $\mathbf{F(x)}$ in $\mathbb{R}^3$.

Remark. A vector field in $\mathbb{R}^3$ is a function whose domain and range are subsets of $\mathbb{R}^3$.

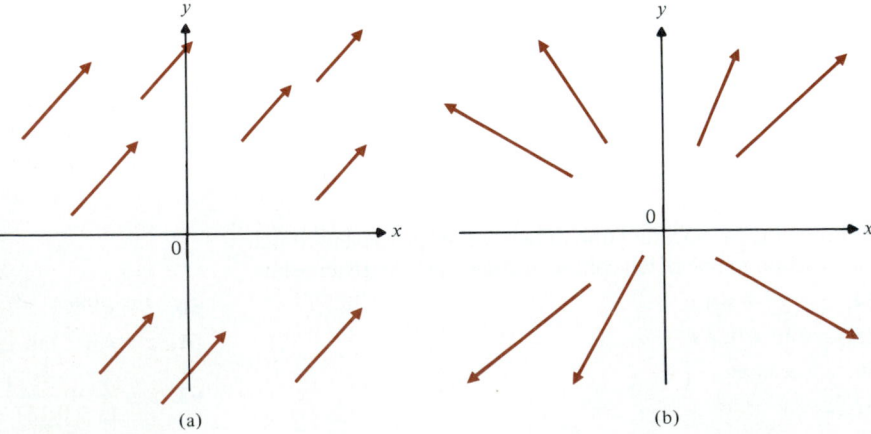

(a)   (b)

FIGURE 1

EXAMPLE 1: Gravitational Field

Let m_1 represent the mass of a (relatively) fixed object in space and let m_2 denote the mass of an object moving near the fixed object. Then the magnitude of the gravitational force between the objects is given by

$$|\mathbf{F}| = G\frac{m_1 m_2}{r^2}, \tag{1}$$

where r is the distance between the objects and G is a universal constant. If we assume that the first object is at the origin, then we may denote the position of the second object by $\mathbf{x}$, and then, since $r = |\mathbf{x}|$, (1) can be written

$$|\mathbf{F}| = \frac{Gm_1 m_2}{|\mathbf{x}|^2}. \tag{2}$$

Also, the force acts toward the origin, that is, in the direction opposite to that of the position vector $\mathbf{x}$. Therefore

$$\text{direction of } \mathbf{F} = -\frac{\mathbf{x}}{|\mathbf{x}|}, \tag{3}$$

so from (2) and (3),

$$\mathbf{F}(\mathbf{x}) = \frac{\alpha\mathbf{x}}{|\mathbf{x}|^3}, \tag{4}$$

where $\alpha = -Gm_1 m_2$.

The vector field (4) is called a **gravitational field.** We can sketch this vector field without much difficulty because for every $\mathbf{x} \neq \mathbf{0} \in \mathbb{R}^3$, $\mathbf{F}(\mathbf{x})$ points toward the origin. The sketch appears in Fig. 2. Note, too, that the vectors have increased magnitudes as we get closer to the origin.

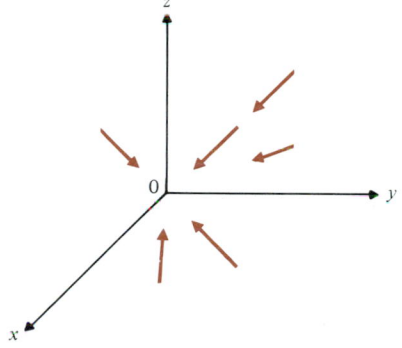

FIGURE 2

EXAMPLE 2

Let $\mathbf{F}$ be the vector field in $\mathbb{R}^2$ given by

$$\mathbf{F} = \frac{2(x^2 - y^2 - 1)}{[(x+1)^2 + y^2][(x-1)^2 + y^2]}\mathbf{i} + \frac{4xy}{[(x+1)^2 + y^2][(x-1)^2 + y^2]}\mathbf{j}. \tag{5}$$

It can be shown that $\mathbf{F}$ is the electric field in the plane caused by two infinite straight wires that are perpendicular to the xy-plane and that pass through the points $(1, 0)$ and $(-1, 0)$. It is assumed that the wires are uniformly charged with electricity and that the charge of one is opposite to the charge of the other and of the same magnitude. This field is sketched in Fig. 3. (Also, see Problem 25).

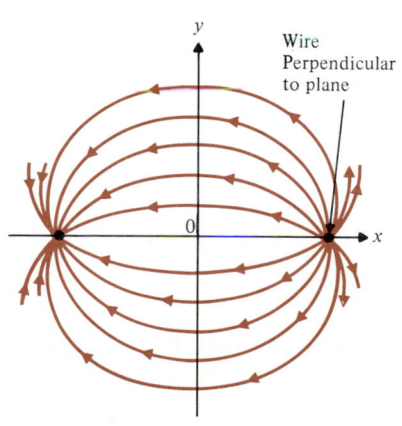

Wire
Perpendicular
to plane

FIGURE 3

EXAMPLE 3: Velocity Field of a Rotating Body

Suppose a body is rotating in space around an axis of rotation (see Fig. 4). We define a vector $\mathbf{w}$ as follows: $\mathbf{w}$ has the direction of the axis of rotation such that if we look from the initial point of $\mathbf{w}$ to its terminal point, the motion appears clockwise. (This is the **right hand rule** where the fingers point in the direction of motion and the thumb, held at a right angle, points in the direction of $\mathbf{w}$.) Now

$$\text{angular speed } \omega = \frac{\text{tangential speed of a point } \mathbf{x}(t)}{\text{distance from } \mathbf{x} \text{ to the axis of rotation}}. \tag{6}$$

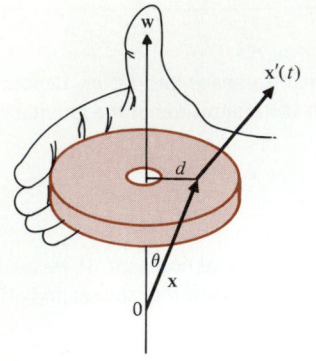

FIGURE 4

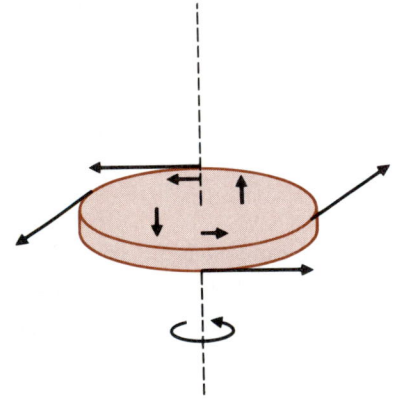

FIGURE 5

Note that if $\mathbf{x}(t)$ represents the position vector, then the velocity vector is $\mathbf{x}'(t)$ and its speed is $|\mathbf{x}'(t)|$. Thus if d denotes the distance of a point $\mathbf{x}$ from the axis of rotation, we have, from (6),

$$\omega = \frac{|\mathbf{x}'(t)|}{d}. \tag{7}$$

Then we define the magnitude of $\mathbf{w}$ to be ω. That is,

$$|\mathbf{w}| = \omega. \tag{8}$$

Let θ denote the angle between $\mathbf{x}$ and $\mathbf{w}$. Then from Fig. 4

$$\sin\theta = \frac{d}{|\mathbf{x}|} \quad \text{or} \quad d = |\mathbf{x}|\sin\theta,$$

and

$$\omega d = |\mathbf{x}|\,\omega\sin\theta = |\mathbf{x}|\,|\mathbf{w}|\sin\theta = |\mathbf{x}\times\mathbf{w}|.$$

Thus, from (7),

$$|\mathbf{x}'(t)| = |\mathbf{x}\times\mathbf{w}|. \tag{9}$$

Now, again from Fig. 4, we see that the velocity vector $\mathbf{x}'$ is perpendicular to both $\mathbf{x}$ and $\mathbf{w}$, so that the direction of $\mathbf{v}=\mathbf{x}'$ is the direction of $\mathbf{w}\times\mathbf{x}$. Combining this result with (9), we obtain

$$\mathbf{v} = \mathbf{w}\times\mathbf{x}.$$

We can write this **velocity field** as

$$\mathbf{v}(x, y, z) = \mathbf{w}\times(x\mathbf{i} + y\mathbf{j} + z\mathbf{k}). \tag{10}$$

This field is sketched in Fig. 5.

EXAMPLE 4

Suppose that the axis of rotation in Example 3 is the y-axis and that $\mathbf{w}$ points in the positive y-direction. Then

$$\mathbf{w} = \omega\mathbf{j} \quad \text{and} \quad \mathbf{w}\times\mathbf{x} = \begin{vmatrix} \mathbf{i} & \mathbf{j} & \mathbf{k} \\ 0 & \omega & 0 \\ x & y & z \end{vmatrix} = \omega z\mathbf{i} - \omega x\mathbf{k} = \omega(z\mathbf{i} - x\mathbf{k}).$$

Note. In Examples 3 and 4, ω is not necessarily constant. If ω is constant, the body is said to have **constant angular speed.**

Problems 9.2

In Problems 1–21, compute the vector field $(\partial f/\partial x, \partial f/\partial y)$ or $(\partial f/\partial x, \partial f/\partial y, \partial f/\partial z)$ of the given function.

1. $f(x, y) = (x + y)^2$

2. $f(x, y) = e^{\sqrt{xy}}$

3. $f(x, y) = \cos(x - y)$

4. $f(x, y) = \ln(2x - y + 1)$

5. $f(x, y) = \sqrt{x^2 + y^3}$

6. $f(x, y) = \tan^{-1} \dfrac{y}{x}$

7. $f(x, y) = y \tan(y - x)$

8. $f(x, y) = x^2 \sinh y$

9. $f(x, y) = \sec(x + 3y)$

10. $f(x, y) = \dfrac{x - y}{x + y}$

11. $f(x, y) = \dfrac{x^2 - y^2}{x^2 + y^2}$

12. $f(x, y) = \dfrac{e^{x^2} - e^{-y^2}}{3y}$

13. $f(x, y, z) = xyz$

14. $f(x, y, z) = \sin x \cos y \tan z$

15. $f(x, y, z) = \dfrac{x^2 - y^2 + z^2}{3xy}$

16. $f(x, y, z) = x \ln y - z \ln x$

17. $f(x, y, z) = xy^2 + y^2 z^3$

18. $f(x, y, z) = (y - z)e^{x+2y+3z}$

19. $f(x, y, z) = x \sin y \ln z$

20. $f(x, y, z) = \dfrac{x - z}{\sqrt{1 - y^2 + x^2}}$

21. $f(x, y, z) = x \cosh z - y \sin x$

22. Find an equation describing a gravitational field in $\mathbb{R}^2$.

23. Find the velocity field of a rotating body whose axis of rotation is the z-axis with $\mathbf{w}$ pointing in the direction of the positive z-axis. Assume that ω is constant.

24. Answer the question of Problem 23 after replacing the z-axis with the x-axis.

25. Show that the force field given in Example 2 is the gradient (see p. 561) of the function

$$F(x, y) = \ln \dfrac{\sqrt{(x - 1)^2 + y^2}}{\sqrt{(x + 1)^2 + y^2}}.$$

Draw figures (similar to Fig. 2) of the vector fields in $\mathbb{R}^2$ given by the vector functions in Problems 26–29.

26. $\mathbf{F}(\mathbf{x}) = \dfrac{\mathbf{x}}{|\mathbf{x}|}$

27. $\mathbf{F}(\mathbf{x}) = \dfrac{\mathbf{x} - \mathbf{x}_0}{|\mathbf{x} - \mathbf{x}_0|^2}$

28. $\mathbf{F}(\mathbf{x}) = \dfrac{\mathbf{x}_0 - \mathbf{x}}{|\mathbf{x} - \mathbf{x}_0|}$

29. $\mathbf{F}(\mathbf{x}) = \dfrac{-\mathbf{x}}{|\mathbf{x}|^{1/2}}.$

9.3 Derivative of a Vector Function

In Section 9.2 we defined a vector field $\mathbf{F}$ by stating that it is a rule that assigns a vector $\mathbf{F}(\mathbf{x})$ to every point $\mathbf{x}$ in a subset D of some vector space. In this section we shall review the differential calculus of such vector functions when D is a subset of $\mathbb{R}$. Then $\mathbf{x} = t$, a real number, and the vectors $\mathbf{f}(t)$ belong to $\mathbb{R}^n$. Everything that happens in $\mathbb{R}^n$ is a natural extension of behavior of vectors in $\mathbb{R}^2$. Thus we begin our discussion with the case $n = 2$.

Since $\mathbf{f}(t)$ is a vector in $\mathbb{R}^2$, we can express it in terms of its projections on the two coordinate axes in the form

$$\mathbf{f}(t) = f_1(t)\mathbf{i} + f_2(t)\mathbf{j}. \tag{1}$$

The projections $f_1(t)$ and $f_2(t)$ are real-valued functions of the real variable t. The notions of continuity and differentiability of $\mathbf{f}$ will be developed in terms of the continuity and differentiability of f_1 and f_2. Restating our definition of a vector function in terms of the coordinate system in $\mathbb{R}^2$, we have the following.

DEFINITION

Vector Function in $\mathbb{R}^2$

Let f_1 and f_2 be functions of the real variable t. Then for all values of t for which $f_1(t)$ and $f_2(t)$ are defined, we define the **vector-valued function $\mathbf{f}$** by

$$\mathbf{f}(t) = (f_1(t), f_2(t)) = f_1(t)\mathbf{i} + f_2(t)\mathbf{j}. \tag{2}$$

The **domain** of **f** is the intersection of the domains of f_1 and f_2.

Remark. We stress that a vector function is a special case of a vector field: the case in which the domain is a subset of $\mathbb{R}$.

EXAMPLE 1

Let $\mathbf{f}(t) = f_1(t)\mathbf{i} + f_2(t)\mathbf{j} = (1/t)\mathbf{i} + \sqrt{t+1}\,\mathbf{j}$. Find the domain of **f**.

SOLUTION

The domain of **f** is the set of all t for which f_1 and f_2 are defined. Since $f_1(t)$ is defined for $t \neq 0$ and $f_2(t)$ is defined for $t \geq -1$, we see that the domain of **f** is the set

$$\{t : t \geq -1 \text{ and } t \neq 0\}.$$

Let **f** be a vector function. Then for each t in the domain of **f**, the endpoint of the vector $f_1(t)\mathbf{i} + f_2(t)\mathbf{j}$ is a point (x, y) in the xy-plane, where

$$x = f_1(t) \quad \text{and} \quad y = f_2(t). \tag{3}$$

Plane Curves and Parametric Equations

DEFINITION

Suppose that the interval $[a, b]$ is in the domain of the function **f** and that both f_1 and f_2 are continuous in $[a, b]$. Then the set of points $(f_1(t), f_2(t))$ for $a \leq t \leq b$ is called a **plane curve** C. Equation (2) is called the **vector equation** of C, while equations (3) are called the **parametric equations** or **parametric representation** of C. In this context the variable t is called a **parameter.**

Remark. We shall usually refer to a plane curve simply as a **curve.**

EXAMPLE 2

Describe the curve given by the vector equation

$$\mathbf{f}(t) = (\cos t)\mathbf{i} + (\sin t)\mathbf{j}, \qquad 0 \leq t \leq 2\pi. \tag{4}$$

SOLUTION

We see that for every t, $|\mathbf{f}(t)| = 1$ since $|\mathbf{f}(t)| = \sqrt{\cos^2 t + \sin^2 t} = 1$. Moreover, if we write the curve in its parametric representation, we find that

$$x = \cos t, \qquad y = \sin t, \tag{5}$$

and since $\cos^2 t + \sin^2 t = 1$, we have

$$x^2 + y^2 = 1,$$

which is the equation of the unit circle. This curve is sketched in Fig. 1. Note that in the sketch the parameter t represents both the length of the arc from $(1, 0)$ to the endpoint of the vector and the angle (measured in radians) the vector makes with the positive x-axis. The representation $x^2 + y^2 = 1$ is called the *Cartesian equation* of the curve given by (5).

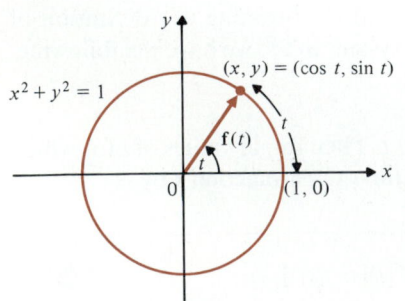

FIGURE 1

Remark: As t increases from 0 to 2π, we move around the unit circle in the counterclockwise direction. If t were instead restricted to the range $0 \le t \le \pi$, then we would not get the entire circle. Rather, we would stop at the point $(\cos \pi, \sin \pi) = (-1, 0)$, which would give us the upper semicircle only.

DEFINITION

Cartesian Equation of a Plane Curve

A **Cartesian equation** of the curve $\mathbf{f}(t) = x(t)\mathbf{i} + y(t)\mathbf{j}$ is an equation relating the variables x and y only.

EXAMPLE 3

Describe and sketch the curve given parametrically by $x = t + 3$, $y = t^2 - t + 2$.

SOLUTION

With problems of this type the easiest thing to do is to write t as a function of x or y, if possible. Since $x = t + 3$, we immediately see that $t = x - 3$ and $y = t^2 - t + 2 = (x - 3)^2 - (x - 3) + 2 = x^2 - 7x + 14$. This is the Cartesian equation of the curve and is the equation of a parabola. It is sketched in Fig. 2. Note that in the Cartesian equation of the parabola, the parameter t does not appear.

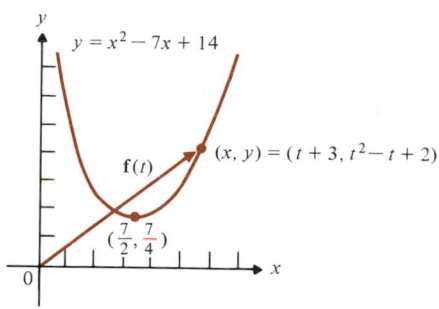

FIGURE 2

These last two examples show that vector functions provide a useful means of describing parametric representations of curves. Since most of the curves we will want to consider consist of a single piece, we will want the coordinate functions $x(t)$ and $y(t)$ [or $f_1(t)$ and $f_2(t)$] to be continuous (why?). Since continuity of real-valued functions of a real variable is defined by a limit, we must extend this notion to vector functions.

DEFINITION

Let $\mathbf{f}(t) = f_1(t)\mathbf{i} + f_2(t)\mathbf{j}$. Let t_0 be any real number or $+\infty$, or $-\infty$. If $\lim_{t \to t_0} f_1(t)$ and $\lim_{t \to t_0} f_2(t)$ both exist, then we define

$$\lim_{t \to t_0} \mathbf{f}(t) = [\lim_{t \to t_0} f_1(t)]\mathbf{i} + [\lim_{t \to t_0} f_2(t)]\mathbf{j}.$$

That is, *the **limit of a vector function** is determined by the limits of its component functions.* Thus in order to calculate the limit of a vector function, it is only necessary to calculate two ordinary limits.

EXAMPLE 4

Let $\mathbf{f}(t) = [(\sin t)/t]\mathbf{i} + [\ln(3 + t)]\mathbf{j}$. Calculate $\lim_{t\to 0} \mathbf{f}(t)$.

SOLUTION

$$\lim_{t\to 0} \mathbf{f}(t) = \left[\lim_{t\to 0} \frac{\sin t}{t}\right]\mathbf{i} + [\lim_{t\to 0} \ln(3+t)]\mathbf{j} = \mathbf{i} + (\ln 3)\mathbf{j}$$

DEFINITION

Continuity of a Vector Function

$\mathbf{f}$ is **continuous** at t_0 if the component functions f_1 and f_2 are continuous at t_0. Thus, $\mathbf{f}$ is continuous at t_0 if

    (i) $\mathbf{f}$ is defined at t_0.

    (ii) $\lim_{t\to t_0} \mathbf{f}(t)$ exists.

    (iii) $\lim_{t\to t_0} \mathbf{f}(t) = \mathbf{f}(t_0)$.

DEFINITION

Differentiable Vector Function

Let f be defined at t. Then $\mathbf{f}$ is **differentiable** at t if

$$\lim_{\Delta t\to 0} \frac{\mathbf{f}(t + \Delta t) - \mathbf{f}(t)}{\Delta t} \tag{6}$$

exists and is finite. The vector function $\mathbf{f}'$ defined by

$$\mathbf{f}'(t) = \frac{d\mathbf{f}}{dt} = \lim_{\Delta t\to 0} \frac{\mathbf{f}(t + \Delta t) - \mathbf{f}(t)}{\Delta t} \tag{7}$$

is called the **derivative** of $\mathbf{f}$, and the domain of $\mathbf{f}'$ is the set of all t such that the limit in (6) exists.

DEFINITION

Differentiability in an Open Interval

The vector function $\mathbf{f}$ is **differentiable** on the open interval I if $\mathbf{f}'(t)$ exists for every t in I.

Before giving examples of the calculation of derivatives, we state a theorem that makes this calculation no more difficult than the calculation of "ordinary" derivatives. The proof is left as an exercise. (See Problem 57.)

■ **THEOREM 1**

If $\mathbf{f}(t) = f_1(t)\mathbf{i} + f_2(t)\mathbf{j}$, then at any value t for which $f'_1(t)$ and $f'_2(t)$ exist,

$$\mathbf{f}'(t) = f'_1(t)\mathbf{i} + f'_2(t)\mathbf{j}.$$

That is, the derivative of a vector function is determined by the derivatives of its component functions. ■

EXAMPLE 5

Let $\mathbf{f}(t) = (\cos t)\mathbf{i} + e^{2t}\mathbf{j}$. Calculate $\mathbf{f}'(t)$.

SOLUTION

$$\mathbf{f}'(t) = \frac{d}{dt}(\cos t)\mathbf{i} + \frac{d}{dt}e^{2t}\mathbf{j} = -(\sin t)\mathbf{i} + 2e^{2t}\mathbf{j}.$$

Once we know how to calculate the first derivative of $\mathbf{f}$, we can calculate higher derivatives as well.

DEFINITION

Second Derivative

If the function $\mathbf{f}'$ is differentiable at t, we define the **second derivative** of $\mathbf{f}$ to be the derivative of $\mathbf{f}'$. That is,

$$\mathbf{f}'' = (\mathbf{f}')'. \tag{8}$$

Geometric Interpretation of $\mathbf{f}'$

We now seek a **geometric interpretation for** $\mathbf{f}'$. As we have seen, the set of vectors $\mathbf{f}(t) = f_1(t)\mathbf{i} + f_2(t)\mathbf{j}$ for t in the domain of $\mathbf{f}$ describe a curve C in $\mathbb{R}^2$. Assume $\mathbf{f}'(t) \neq \mathbf{0}$ and consider

$$\mathbf{f}'(t) = \lim_{\Delta t \to 0} \frac{1}{\Delta t}[\mathbf{f}(t + \Delta t) - \mathbf{f}(t)]. \tag{9}$$

The denominator in the difference quotient in (9) is a number, while the numerator is a secant vector (see Fig. 3a). Observe that the direction of this secant vector approaches that of the tangent line to C at $\mathbf{f}(t)$ as $\Delta t \to 0$ (see Fig. 3b). Hence $\mathbf{f}'(t)$ has the same direction as the tangent line to C at $\mathbf{f}(t)$. We call $\mathbf{f}'(t)$

Tangent Vector

the **tangent vector** to C at $\mathbf{f}(t)$.

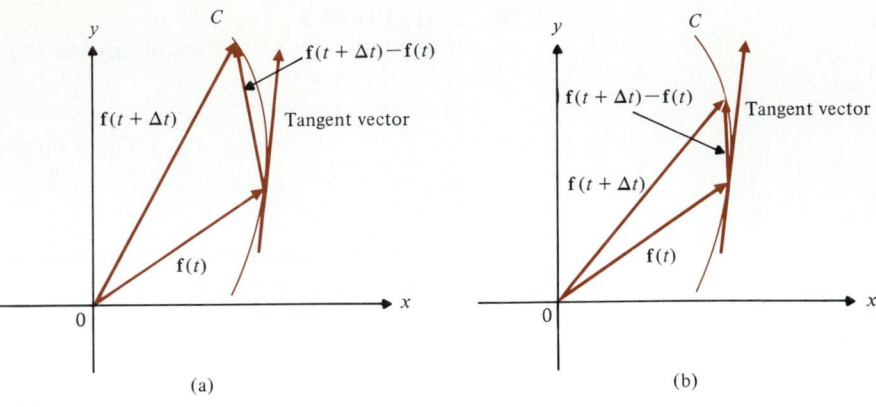

FIGURE 3

DEFINITION

Unit Tangent Vector

It is sometimes useful to calculate a **unit tangent vector** to a curve C. This vector is a tangent vector with a magnitude of 1. The unit tangent vector is usually denoted $\mathbf{T}$ and can be calculated by the formula

$$\mathbf{T}(t) = \frac{\mathbf{f}'(t)}{|\mathbf{f}'(t)|} \tag{10}$$

for any number t so long as $\mathbf{f}'(t) \neq 0$. This follows since $\mathbf{f}'(t)$ is a tangent vector and $\mathbf{f}'/|\mathbf{f}'|$ is a unit vector.

Vector Function in $\mathbb{R}^3$

From our discussion above it is natural to define a **vector function f in $\mathbb{R}^3$** of a real variable t by

$$\mathbf{f}(t) = (f_1(t), f_2(t), f_3(t)) = f_1(t)\mathbf{i} + f_2(t)\mathbf{j} + f_3(t)\mathbf{k}. \tag{11}$$

Curve in $\mathbb{R}^3$

If f_1, f_2, and f_3 are continuous over an interval I, than as t varies over I, the set of points traced out by the end of the vector $\mathbf{f}$ is called a **curve** in 3-space. In particular, $\mathbf{f}$ is differentiable if and only if its component functions f_1, f_2, and f_3 are differentiable, and in that case

$$\mathbf{f}'(t) = f'_1(t)\mathbf{i} + f'_2(t)\mathbf{j} + f'_3(t)\mathbf{k}. \tag{12}$$

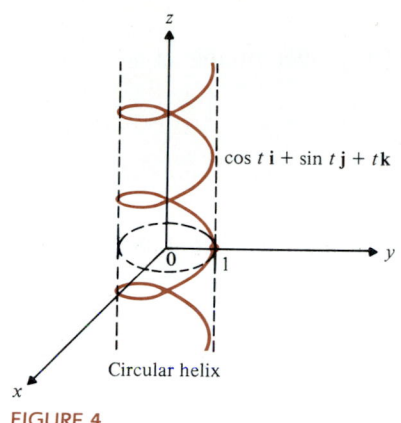

$\cos t\,\mathbf{i} + \sin t\,\mathbf{j} + t\mathbf{k}$

Circular helix

FIGURE 4

EXAMPLE 6

The curve $\mathbf{f}(t) = \cos t\,\mathbf{i} + \sin t\,\mathbf{j} + t\mathbf{k}$ is called a **circular helix.** (a) Sketch the curve. (b) Compute its derivative.

SOLUTION

(a) The curve $\cos t\,\mathbf{i} + \sin t\,\mathbf{j}$ in the xy-plane is a parameterization of the unit circle. Thus, as t increases, the x- and y-coordinates "wind" around the unit circle while z increases. We therefore obtain the curve in Fig. 4, which lies on the vertical circular cylinder with base on the unit circle.
(b) $\mathbf{f}'(t) = -\sin t\,\mathbf{i} + \cos t\,\mathbf{j} + \mathbf{k}$.

We now extend the definitions in this section to $\mathbb{R}^n$.

DEFINITION

Vector-Valued Function

Let $f_1, f_2, \ldots, f_n$ be functions of the real variable t. Then for all values of t for which $f_1(t), f_2(t), \ldots, f_n(t)$ are defined, the **vector-valued function f** is given by

$$\mathbf{f}(t) = (f_1(t), f_2(t), \ldots, f_n(t)). \tag{13}$$

The functions $f_1, f_2, \ldots, f_n$ are called the **component functions** of **f.**

Notation. We shall denote a vector-valued function by

$$\mathbf{f}: \mathbb{R} \to \mathbb{R}^n.$$

DEFINITION

Continuity and Limits

The function **f** given by (13) is **continuous** at t_0 if each component function is continuous at t_0. In addition,

$$\lim_{t \to t_0} \mathbf{f}(t) = \left(\lim_{t \to t_0} f_1(t), \lim_{t \to t_0} f_2(t), \ldots, \lim_{t \to t_0} f_n(t) \right)$$

if $\lim_{t \to t_0} f_i(t)$ exists for $i = 1, 2, \ldots, n$.

DEFINITION

Curve in $\mathbb{R}^n$
Closed Curve

Suppose that the interval $[a, b]$ is in the domain of each of the functions $f_1, f_2, \ldots, f_n$. If $\mathbf{f}: \mathbb{R} \to \mathbb{R}^n$ is given by (13) and is continuous at each t, then the set of vectors

$$C = \{\mathbf{f}(t): a \leq t \leq b\} \tag{14}$$

is called a **curve in $\mathbb{R}^n$.** If $\mathbf{f}(a) = \mathbf{f}(b)$, we say that the curve is **closed.**

DEFINITION

Let $\mathbf{f}(t) = (f_1(t), f_2(t), \ldots, f_n(t))$ where each $f_i(t)$ is differentiable at t_0.

Differentiability

(i) Then **f** is **differentiable** at t_0 and

$$\mathbf{f}'(t_0) = (f'_1(t_0), f'_2(t_0), \ldots, f'_n(t_0)).$$

(ii) If f'_i is continuous at t_0 for $i = 1, 2, \ldots, n$, then f is said to be **continuously differentiable** at t_0.

EXAMPLE 7

If $\mathbf{f}(t) = (t, t^2, \ldots, t^n)$, compute $\mathbf{f}'(t)$.

SOLUTION

$$\mathbf{f}'(t) = (1, 2t, 3t^2, \ldots, nt^{n-1}).$$

EXAMPLE 8

If $\mathbf{f}(t) = (c_1, c_2, \ldots, c_n)$, a constant vector, then $\mathbf{f}'(t) = (0, 0, \ldots, 0) = \mathbf{0}$.

EXAMPLE 9

If $\mathbf{f}(t) = f(t)\mathbf{x}$ where $\mathbf{x} = (x_1, x_2, \ldots, x_n)$ is a constant vector and f is a scalar function, then

$$\mathbf{f}'(t) = \frac{d}{dt}(f(t)x_1, f(t)x_2, \ldots, f(t)x_n) = (f'(t)x_1, f'(t)x_2, \ldots, f'(t)x_n) = f'(t)\mathbf{x}.$$

In particular, if $f(t) = t$, then

$$\frac{d}{dt}(t\mathbf{x}) = \mathbf{x}.$$

DEFINITION

Unit Tangent Vector

If $\mathbf{f}'(t) \neq \mathbf{0}$, we define the **unit tangent vector** in $\mathbb{R}^n$ by

$$\mathbf{T}(t) = \frac{\mathbf{f}'(t)}{|\mathbf{f}'(t)|}. \tag{15}$$

EXAMPLE 10

Compute the unit tangent vector to the curve of Example 7 at $t = 1$.

SOLUTION

In Example 7, $\mathbf{f}'(t) = (1, 2t, 3t^2, \ldots, nt^{n-1})$ so $\mathbf{f}'(1) = (1, 2, 3, \ldots, n)$ and

$$|\mathbf{f}'(1)| = \sqrt{1^2 + 2^2 + \cdots + n^2} = \sqrt{\frac{n(n+1)(2n+1)}{6}}.\dagger$$

Thus

$$T(1) = \frac{\sqrt{6}}{\sqrt{n(n+1)(2n+1)}}(1, 2, 3, \ldots, n).$$

For example, if $n = 5$, we obtain

$$\mathbf{T}(1) = \sqrt{\frac{6}{330}}(1, 2, 3, 4, 5) = \frac{1}{\sqrt{55}}(1, 2, 3, 4, 5).$$

The familiar rules of differentiation yield similar rules for vector functions. The proofs are left to the reader.

■ **THEOREM 2**

Let $\mathbf{f} : \mathbb{R} \to \mathbb{R}^n$, and $\mathbf{g} : \mathbb{R} \to \mathbb{R}^n$ be differentiable on an open interval (a, b). Let the scalar function h be differentiable on (a, b). Let α be a scalar and let $\mathbf{v}$ be a constant vector in $\mathbb{R}^n$. Then

(i) $\mathbf{f} + \mathbf{g}$ is differentiable and

$$\frac{d}{dt}(\mathbf{f} + \mathbf{g}) = \frac{d\mathbf{f}}{dt} + \frac{d\mathbf{g}}{dt} = \mathbf{f}' + \mathbf{g}'.$$

(ii) $\alpha\mathbf{f}$ is differentiable and

$$\frac{d}{dt}\alpha\mathbf{f} = \alpha\frac{d\mathbf{f}}{dt} = \alpha\mathbf{f}'.$$

(iii) $\mathbf{v} \cdot \mathbf{f}$ is differentiable and

$$\frac{d}{dt}\mathbf{v} \cdot \mathbf{f} = \mathbf{v} \cdot \frac{d\mathbf{f}}{dt} = \mathbf{v} \cdot \mathbf{f}'$$

---

† This formula can be proved by mathematical induction.

(iv) $h\mathbf{f}$ is differentiable and

$$\frac{d}{dt}h\mathbf{f} = h\frac{d\mathbf{f}}{dt} + \frac{dh}{dt}\mathbf{f} = h\mathbf{f}' + h'\mathbf{f}.$$

(v) $\mathbf{f} \cdot \mathbf{g}$ is differentiable and

$$\frac{d}{dt}\mathbf{f}\cdot\mathbf{g} = \mathbf{f}\cdot\frac{d\mathbf{g}}{dt} + \frac{d\mathbf{f}}{dt}\cdot\mathbf{g} = \mathbf{f}\cdot\mathbf{g}' + \mathbf{f}'\cdot\mathbf{g}.$$

■

Problems 9.3

In Problems 1–8, find the domain of each vector-valued function.

1. $\mathbf{f}(t) = \dfrac{1}{t}\mathbf{i} + \dfrac{1}{t-1}\mathbf{j}$

2. $\mathbf{f}(t) = \sqrt{t}\,\mathbf{i} + \dfrac{1}{t}\mathbf{j}$

3. $\mathbf{f}(s) = \dfrac{1}{s^2-1}\mathbf{i} + (s^2-1)\mathbf{j}$

4. $\mathbf{f}(u) = e^{1/u}\mathbf{i} + e^{-1/(u+1)}\mathbf{j}$

5. $\mathbf{f}(s) = (\ln s)\mathbf{i} + \ln(1-s)\mathbf{j}$

6. $\mathbf{f}(r) = (\sin r)\mathbf{i} + r\mathbf{j}$

7. $\mathbf{f}(t) = (\sec t)\mathbf{i} + (\csc t)\mathbf{j}$

8. $\mathbf{f}(w) = (\tan w)\mathbf{i} + (\cot w)\mathbf{j}$

In Problems 9–20, find the Cartesian equation of each curve and then sketch the curve in the xy-plane.

9. $\mathbf{f}(t) = t^2\mathbf{i} + 2t\mathbf{j}$

10. $\mathbf{f}(t) = (2t-3)\mathbf{i} + t^2\mathbf{j}$

11. $\mathbf{f}(t) = t^2\mathbf{i} + t^3\mathbf{j}$

12. $\mathbf{f}(t) = 3(\sin t)\mathbf{i} + 3(\cos t)\mathbf{j}$

13. $\mathbf{f}(t) = (2t-1)\mathbf{i} + (4t+3)\mathbf{j}$

14. $\mathbf{f}(t) = 2(\cosh t)\mathbf{i} + 2(\sinh t)\mathbf{j}$

15. $\mathbf{f}(t) = (t^4+t^2+1)\mathbf{i} + t^2\mathbf{j}$

16. $\mathbf{f}(t) = t^2\mathbf{i} + t^8\mathbf{j}$

17. $\mathbf{f}(t) = t^3\mathbf{i} + (t^9-1)\mathbf{j}$

18. $\mathbf{f}(t) = t\mathbf{i} + e^t\mathbf{j}$

19. $\mathbf{f}(t) = e^t\mathbf{i} + t^2\mathbf{j}$

20. $\mathbf{f}(t) = (t^2+t-3)\mathbf{i} + \sqrt{t}\,\mathbf{j}$

In Problems 21–30, calculate the first and second derivatives of the given vector function.

21. $\mathbf{f}(t) = t\mathbf{i} - t^5\mathbf{j}$

22. $\mathbf{f}(t) = (1+t^2)\mathbf{i} + \dfrac{2}{t}\mathbf{j}$

23. $\mathbf{f}(t) = (\sin 2t)\mathbf{i} + (\cos 3t)\mathbf{j}$

24. $\mathbf{f}(t) = \dfrac{t}{1+t}\mathbf{i} - \dfrac{1}{\sqrt{t}}\mathbf{j}$

25. $\mathbf{f}(t) = (\ln t)\mathbf{i} + e^{3t}\mathbf{j}$

26. $\mathbf{f}(t) = e^t(\sin t)\mathbf{i} + e^t(\cos t)\mathbf{j}$

27. $\mathbf{f}(t) = (\tan t)\mathbf{i} + (\sec t)\mathbf{j}$

28. $\mathbf{f}(t) = (\tan^{-1} t)\mathbf{i} + (\sin^{-1} t)\mathbf{j}$

29. $\mathbf{f}(t) = (\ln \cos t)\mathbf{i} + (\ln \sin t)\mathbf{j}$

30. $\mathbf{f}(t) = (\cosh t)\mathbf{i} + (\sinh t)\mathbf{j}$

In Problems 31–40, find the unit tangent vector to the given curve for the given value of t.

31. $\mathbf{f}(t) = t^2\mathbf{i} + t^3\mathbf{j}$; $t = 1$

32. $\mathbf{f}(t) = t\mathbf{i} + \dfrac{1}{t}\mathbf{j}$; $t = 1$

33. $\mathbf{f}(t) = (\cos t)\mathbf{i} + (\sin t)\mathbf{j}$; $t = 0$

34. $\mathbf{f}(t) = (\cos t)\mathbf{i} + (\sin t)\mathbf{j}$; $t = \pi/2$

35. $\mathbf{f}(t) = (\cos t)\mathbf{i} + (\sin t)\mathbf{j}$; $t = \pi/4$

36. $\mathbf{f}(t) = (\cos t)\mathbf{i} + (\sin t)\mathbf{j};\quad t = 3\pi/4$

37. $\mathbf{f}(t) = (\tan t)\mathbf{i} + (\sec t)\mathbf{j};\quad t = 0$

38. $\mathbf{f}(t) = (\ln t)\mathbf{i} + e^{2t}\mathbf{j};\quad t = 1$

39. $\mathbf{f}(t) = \dfrac{t}{t+1}\mathbf{i} + \dfrac{t+1}{t}\mathbf{j};\quad t = 2$

40. $\mathbf{f}(t) = \dfrac{t+1}{t}\mathbf{i} + \dfrac{t}{t+1}\mathbf{j};\quad t = 2$

41. The equation $(x^2/a^2) + (y^2/b^2) = 1$ is the equation of an ellipse. Show that the curve given by the vector equation $\mathbf{f}(t) = a(\cos t)\mathbf{i} + b(\sin t)\mathbf{j}$ is an ellipse.

42. A cannonball is shot upward from ground level at an angle of 45° with an initial speed of 1300 ft/sec. Find a parametric representation of the path of the cannonball. Then find the Cartesian equation of this path.

43. How many feet (horizontally) does the cannonball in Problem 42 travel before it hits the ground?

44. How many feet (vertically) does the cannonball in Example 42 travel before it hits the ground?

45. An object is thrown down from the top of a 150-m building at an angle of 30° (below the horizontal) with an initial velocity of 100 m/sec. Determine a parametric representation of the path of the object. [*Hint:* Draw a picture.]

46. When the object in Problem 45 hits the ground, how far is it from the base of the building?

47. A point is located 25 cm from the center of a wheel 1 m in diameter. Find the parametric representation of the curve traced by that point as the wheel rolls.

48. Answer the question in Problem 47 if the point is located on the circumference of the wheel.

49. The ellipse $(x^2/a^2) + (y^2/b^2) = 1$ can be written parametrically as $x = a \cos \theta$, $y = b \sin \theta$. Find a unit tangent vector to the ellipse at $\theta = \pi/4$.

50. Find a unit tangent vector to the **cycloid**
$$\mathbf{f}(\alpha) = r(\alpha - \sin \alpha)\mathbf{i} + r(1 - \cos \alpha)\mathbf{j} \text{ for } \alpha = 0.$$

51. Find a unit tangent vector to the cycloid of Problem 50 for $\alpha = \pi/2$.

52. Find a unit tangent vector to the cycloid of Problem 50 for $\alpha = \pi/3$.

In Problems 53–56 compute the unit tangent vector to the given curve.

53. $\mathbf{f}(t) = (1, t, t^2, t^3)$ at $t = 1$

54. $\mathbf{f}(t) = (t^2, t^4, \ldots, t^{2n})$ at $t = 1$

55. $\mathbf{f}(t) = (\sin t, \cos t, \sin t, \cos t, \sin t, \cos t)$ at $t = \pi/6$

56. $\mathbf{f}(t) = (e^t, e^{2t}, \ldots, e^{10t})$ at $t = 0$

57. Prove Theorem 1. [*Hint:* $\mathbf{f}'(t) = \lim\limits_{\Delta t \to 0} \dfrac{1}{\Delta t}[\mathbf{f}(t + \Delta t) - \mathbf{f}(t)].$]

9.4 Curves and Arc Length

In the last section we saw that a curve C in $\mathbb{R}^2$ can be defined parametrically by a vector function $\mathbf{f}(t) = x(t)\mathbf{i} + y(t)\mathbf{j} = (x(t), y(t))$ of a real variable t. We also stated that $\mathbf{f}'(t)$ is a tangent vector to C at $\mathbf{f}(t)$. Our first goal in this section will be to justify this designation and to indicate some of the problems that may occur.

Consider the curves shown in Fig. 1. The curves shown have:

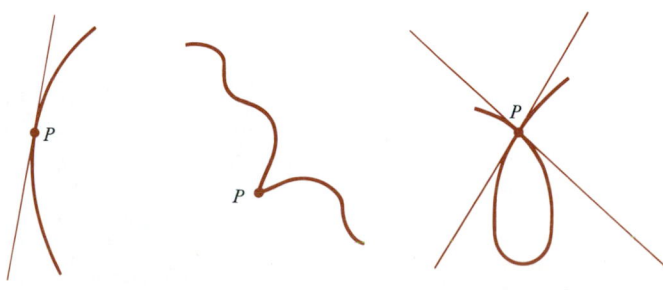

(a) Unique tangent at P (b) No tangent at P (c) Two tangents at P

FIGURE 1

(a) a unique tangent line at P,

(b) no tangent line at P, and

(c) two (or more) tangent lines at P.

A condition which ensures that there is at least one tangent line at each point is that

$$f'_1 \text{ and } f'_2 \text{ exist} \quad \text{and} \quad [f'_1(t)]^2 + [f'_2(t)]^2 \neq 0. \tag{1}$$

Remark. Condition (1) simply states that the derivatives of f_1 and f_2 are not zero at the same value of t. Hence $\mathbf{f}'(t) \neq \mathbf{0}$.

■ **THEOREM 1**

Let (x_0, y_0) be on the curve C given by $x = f_1(t)$ and $y = f_2(t)$ where f_1 and f_2 have continuous derivatives. If the curve passes through (x_0, y_0) when $t = t_0$,† then the slope m of a line tangent to C at (x_0, y_0) is given by

$$m = \lim_{t \to t_0} \frac{f'_2(t)}{f'_1(t)}, \tag{2}$$

provided that this limit exists.

PROOF

We refer to Fig. 2. From the figure we see that

$$\frac{dy}{dx} = \lim_{\Delta t \to 0} \frac{f_2(t_0 + \Delta t) - f_2(t_0)}{f_1(t_0 + \Delta t) - f_1(t_0)} \overset{\text{L'Hôpital's rule}}{=} \lim_{\Delta t \to 0} \frac{f'_2(t_0 + \Delta t)}{f'_1(t_0 + \Delta t)} \overset{\text{let } t = t_0 + \Delta t}{=} \lim_{t \to t_0} \frac{f'_2(t)}{f'_1(t)}. \quad ■$$

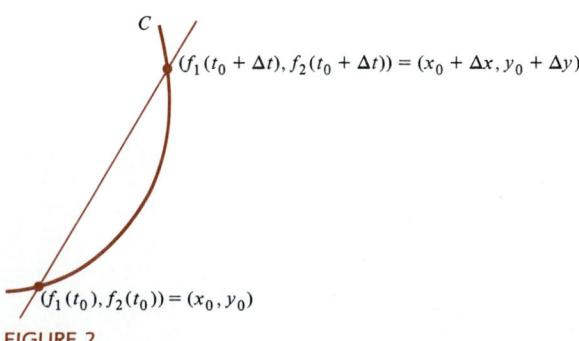

FIGURE 2

† The curve may pass through the point (x_0, y_0) for other values of t as well, and therefore it may have other tangent lines at that point, as in Fig. 1c.

Remark. In L'Hôpital's rule in the last proof we are differentiating with respect to the variable Δt.

The following consequence of Theorem 1 is immediate:

■ **COROLLARY**

If $f_2'(t_0)$ exists and $f_1'(t_0) \neq 0$, then under the hypotheses of Theorem 1

$$m = \frac{f_2'(t_0)}{f_1'(t_0)}. \tag{3}$$

■

Using the corollary, we find that the tangent line to C at t_0 is given by

$$\frac{y - y_0}{x - x_0} = \frac{f_2'(t_0)}{f_1'(t_0)},$$

or

$$f_2'(t_0)(x - x_0) - f_1'(t_0)(y - y_0) = 0. \tag{4}$$

Note that if $f_1'(t_0) = 0$, then equation (4) is still valid, and we obtain

$$f_2'(t_0)(x - x_0) = 0. \tag{5}$$

Since by assumption $[f_1'(t_0)]^2 + [f_2'(t_0)]^2 \neq 0$, the assumption $f_1'(t_0) = 0$ implies that $f_2'(t_0) \neq 0$. Thus we can divide by $f_2'(t_0)$ in (5) to obtain the equation

$$x - x_0 = 0, \quad \text{or} \quad x = x_0.$$

In this case the tangent line is vertical. If $f_2'(t_0) = 0$, then C has a horizontal tangent at t_0.

In sum, if condition (1) holds, then

> **(i)** C has a vertical tangent line at t_0 if
>
> $$\left.\frac{dx}{dt}\right|_{t=t_0} = f_1'(t_0) = 0 \quad \text{and} \quad f_2'(t_0) \neq 0.$$
>
> **(ii)** C has a horizontal tangent line at t_0 if
>
> $$\left.\frac{dy}{dt}\right|_{t=t_0} = f_2'(t_0) = 0 \quad \text{and} \quad f_1'(t_0) \neq 0.$$

(6)

EXAMPLE 1

Find the equation of a line (or lines) tangent to the curve $x = e^t$, $y = e^{-t}$ at the point $(1, 1)$.

SOLUTION

The point $(1, 1)$ is reached only when $t = 0$. Then

$$\frac{dy}{dx} = \frac{d(e^{-t})/dt}{d(e^t)/dt} = -\frac{e^{-t}}{e^t}\bigg|_{t=0} = -1,$$

and the equation of the tangent line is

$$\frac{y-1}{x-1} = -1, \quad \text{or} \quad y = -x + 2.$$

In Example 9.3.2 we showed that

$$\mathbf{f}(t) = (\cos t)\mathbf{i} + (\sin t)\mathbf{j}, \qquad 0 \le t \le 2\pi$$

is a parametric representation for the unit circle. Note that in this case the length of the vector $\mathbf{f}(t)$ is constant. Vector functions of constant length have the following important property.

◼ THEOREM 2

Let $\mathbf{f}(t) = f_1\mathbf{i} + f_2\mathbf{j}$ be a differentiable vector function such that $|\mathbf{f}(t)| = \sqrt{f_1^2(t) + f_2^2(t)}$ is constant. Then

$$\mathbf{f} \cdot \mathbf{f}' = 0. \tag{7}$$

PROOF

Suppose $|\mathbf{f}(t)| = C$, a constant. Then

$$\mathbf{f} \cdot \mathbf{f} = f_1^2 + f_2^2 = |\mathbf{f}|^2 = C^2,$$

so that

$$\frac{d}{dt}(\mathbf{f} \cdot \mathbf{f}) = \frac{d}{dt}C^2 = 0.$$

But

$$\frac{d}{dt}(\mathbf{f} \cdot \mathbf{f}) = \mathbf{f} \cdot \mathbf{f}' + \mathbf{f}' \cdot \mathbf{f} = 2\mathbf{f} \cdot \mathbf{f}' = 0, \quad \text{so} \quad \mathbf{f} \cdot \mathbf{f}' = 0. \quad ◼$$

There is an interesting geometric application of Theorem 2. Let $\mathbf{f}(t)$ be a differentiable vector function. For all t for which $\mathbf{f}'(t) \ne 0$, we let $\mathbf{T}(t)$ denote the unit tangent vector to the curve $\mathbf{f}(t)$. Then since $|\mathbf{T}(t)| = 1$ by the definition of a *unit* tangent vector on page 522, we have, from Theorem 2 [assuming that $\mathbf{T}'(t)$ exists],

$$\mathbf{T}(t) \cdot \mathbf{T}'(t) = 0. \tag{8}$$

That is, $\mathbf{T}'(t)$ is *orthogonal* to $\mathbf{T}(t)$. Recalling that a line perpendicular to a tangent line is called a **normal line,** we call the vector $\mathbf{T}'$ a **normal vector** to the curve **f.** Finally, whenever $\mathbf{T}'(t) \neq 0$, we can define the **unit normal vector** to the curve **f** at t in $\mathbb{R}^2$ as

Unit Normal Vector

$$\mathbf{n}(t) = \frac{\mathbf{T}'(t)}{|\mathbf{T}'(t)|}. \tag{9}$$

EXAMPLE 2

Calculate a unit normal vector to the curve $\mathbf{f}(t) = [(t^3/3) - t]\mathbf{i} + t^2\mathbf{j}$ at $t = 3$.

SOLUTION
Here $\mathbf{f}'(t) = (t^2 - 1)\mathbf{i} + 2t\mathbf{j}$ and

$$|\mathbf{f}'(t)| = \sqrt{(t^2-1)^2 + 4t^2} = \sqrt{t^4 - 2t^2 + 1 + 4t^2} = \sqrt{t^4 + 2t^2 + 1} = t^2 + 1.$$

Thus

$$\mathbf{T}(t) = \frac{\mathbf{f}'(t)}{|\mathbf{f}'(t)|} = \frac{t^2-1}{t^2+1}\mathbf{i} + \frac{2t}{t^2+1}\mathbf{j}.$$

Then

$$\mathbf{T}'(t) = \frac{d}{dt}\left(\frac{t^2-1}{t^2+1}\right)\mathbf{i} + \frac{d}{dt}\left(\frac{2t}{t^2+1}\right)\mathbf{j} = \frac{4t}{(t^2+1)^2}\mathbf{i} + \frac{2-2t^2}{(t^2+1)^2}\mathbf{j}.$$

Finally,

$$|\mathbf{T}'(t)| = \left\{\left[\frac{4t}{(t^2+1)^2}\right]^2 + \left[\frac{2-2t^2}{(t^2+1)^2}\right]^2\right\}^{1/2} = \frac{1}{(t^2+1)^2}(16t^2 + 4 - 8t^2 + 4t^4)^{1/2}$$

$$= \frac{1}{(t^2+1)^2}\sqrt{4t^4 + 8t^2 + 4} = \frac{2}{(t^2+1)},$$

so that

$$\mathbf{n}(t) = \frac{\mathbf{T}'(t)}{|\mathbf{T}'(t)|} = \frac{t^2+1}{2}\left[\frac{4t}{(t^2+1)^2}\mathbf{i} + \frac{2-2t^2}{(t^2+1)^2}\mathbf{j}\right] = \frac{2t}{t^2+1}\mathbf{i} + \frac{1-t^2}{t^2+1}\mathbf{j}.$$

At $t = 3$, $\mathbf{T}(t) = \frac{4}{5}\mathbf{i} + \frac{3}{5}\mathbf{j}$ and $\mathbf{n} = \frac{3}{5}\mathbf{i} - \frac{4}{5}\mathbf{j}$. Note that $\mathbf{T}(3) \cdot \mathbf{n}(3) = 0$. This is sketched in Fig. 3.

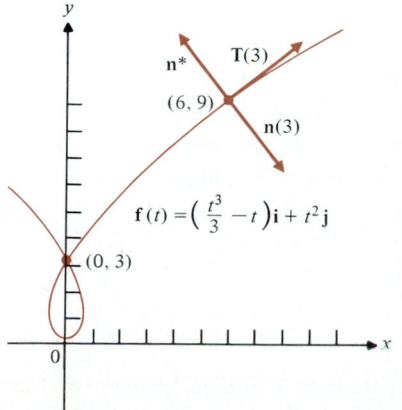

FIGURE 3

In Fig. 3 we see that there are two vectors perpendicular to $\mathbf{T}$ at the point P: one pointing "inward" ($\mathbf{n}$) and one pointing "outward" ($\mathbf{n}^*$). In general, the situation is as shown in Fig. 4. How do we know which of the two vectors is $\mathbf{n}$?

Since $\mathbf{n} = \mathbf{T}'/|\mathbf{T}'|$, consider the derivative

$$\mathbf{T}'(t) = \lim_{\Delta t \to 0} \frac{\mathbf{T}(t + \Delta t) - \mathbf{T}(t)}{\Delta t}.$$

Assuming that $\Delta t > 0$, the difference quotient suggests that we pay special attention to the vector $\mathbf{T}(t + \Delta t) - \mathbf{T}(t)$. If we translate $\mathbf{T}(t + \Delta t)$ so that it begins at the same point $\mathbf{T}(t)$ does, we observe that the vector $\mathbf{T}(t + \Delta t) - \mathbf{T}(t)$ points in the direction of the convex side of the curve (see Fig. 4). Following these tangent curves back as Δt decreases to zero, it is apparent that the vector $\mathbf{T}(t + \Delta t) - \mathbf{T}(t)$ always points in that direction. Hence, if $\mathbf{T}' \neq 0$,

> *The unit normal vector* $\mathbf{n}$ *always points in the direction of the convex side of the curve.*

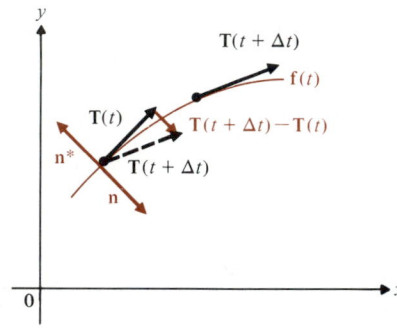

FIGURE 4

Rectifiable Curve
Arc Length

DEFINITION

Let the curve C be given parametrically by $\mathbf{f}(t) = x(t)\mathbf{i} + y(t)\mathbf{j}$, where $\mathbf{P}_0 = \mathbf{f}(t_0)$. Suppose $\mathbf{f}$ has a continuous derivative in the interval $[t_0, b]$ and suppose that for every t_1 in $[t_0, b]$, $\int_{t_0}^{t_1} |\mathbf{f}'(t)|\, dt$ exists. Then $\mathbf{f}$ is said to be **rectifiable** in the interval $[t_0, b]$, and the **arc length** of the curve $\mathbf{f}(t)$ in the interval $[t_0, t_1]$ is given by

$$s(t_1) = \int_{t_0}^{t_1} |\mathbf{f}'(t)|\, dt, \tag{10}$$

where $t_0 \leq t_1 \leq b$. That is, $s(t_1)$ is the length along the curve C from $\mathbf{P}_0$ to $\mathbf{P}_1 = \mathbf{f}(t_1)$. (See Fig. 5.)

A proof that formula (10) yields the length of the curve independent of the given parameterization can be obtained from geometric principles.† However, it is not difficult to give an intuitive idea of what is going on. Consider an arc of the curve between t_0 and $t_0 + \Delta t$ (see Fig. 6). First, we note that as $\Delta t \to 0$, the ratio of the length of the secant line L to the length of the arc Δs between the points $(x(t_0), y(t_0))$ and $(x(t_0 + \Delta t), y(t_0 + \Delta t))$ approaches 1. That is,

$$\lim_{\Delta t \to 0} \frac{\Delta s}{L} = 1.$$

---

† See, for example, R. C. Buck and E. F. Buck, *Advanced Calculus* 3rd Edition, McGraw-Hill, New York, 1978, pp. 404–405.

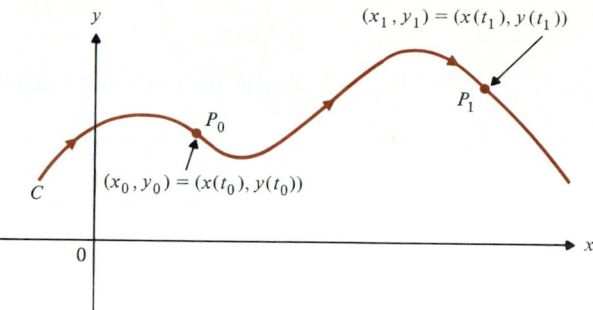

FIGURE 5

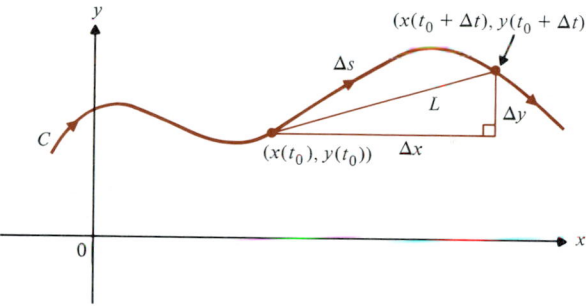

FIGURE 6

Then since $L = \sqrt{\Delta x^2 + \Delta y^2}$, we have

$$\frac{\Delta s}{\Delta t} = \left(\frac{\Delta s}{L}\right)\left(\frac{L}{\Delta t}\right) = \left(\frac{\Delta s}{L}\right)\left(\frac{\sqrt{\Delta x^2 + \Delta y^2}}{\Delta t}\right) = \left(\frac{\Delta s}{L}\right)\sqrt{\left(\frac{\Delta x}{\Delta t}\right)^2 + \left(\frac{\Delta y}{\Delta t}\right)^2}$$

and

$$\frac{ds}{dt} = \lim_{\Delta t \to 0}\frac{\Delta s}{\Delta t} = \lim_{\Delta t \to 0}\frac{\Delta s}{L} \cdot \lim_{\Delta t \to 0}\sqrt{\left(\frac{\Delta x}{\Delta t}\right)^2 + \left(\frac{\Delta y}{\Delta t}\right)^2} = \sqrt{\left(\frac{dx}{dt}\right)^2 + \left(\frac{dy}{dt}\right)^2}.$$

The last step is justified by the continuity of dx/dt and dy/dt.† Note, in particular, that

$$\frac{ds}{dt} = \sqrt{\left(\frac{dx}{dt}\right)^2 + \left(\frac{dy}{dt}\right)^2} = |\mathbf{f}'(t)|. \tag{11}$$

† See, for example, R. C. Buck and E. F. Buck, *Advanced Calculus,* 3*rd* Edition, McGraw-Hill, New York, 1978, p. 404.

EXAMPLE 3

Calculate the length of the arc of the curve $\mathbf{f}(t) = (2t - t^2)\mathbf{i} + \frac{8}{3}t^{3/2}\mathbf{j}$ between $t = 1$ and $t = 3$.

SOLUTION
$\mathbf{f}'(t) = (2 - 2t)\mathbf{i} + 4\sqrt{t}\,\mathbf{j}$ and

$$|\mathbf{f}'(t)| = \sqrt{4 - 8t + 4t^2 + 16t} = 2\sqrt{t^2 + 2t + 1} = 2(t+1),$$

so

$$s = \int_1^3 2(t+1)\,dt = (t^2 + 2t)\Big|_1^3 = 12.$$

The facts that we have developed concerning arc length and tangent vectors in $\mathbb{R}^2$ extend by identical proofs to $\mathbb{R}^3$ (or $\mathbb{R}^n$):

> **(i)** $\mathbf{T}(t) = \mathbf{f}'(t)/|\mathbf{f}'(t)|$, for $\mathbf{f}'(t) \neq 0$.
> **(ii)** If $|\mathbf{f}(t)| = C$ over an interval I, then $\mathbf{f} \cdot \mathbf{f}' = 0$ on I.
> **(iii)** $ds/dt = |\mathbf{f}'(t)|$.
> **(iv)** $s(t_1) = \int_{t_0}^{t_1} |\mathbf{f}'(t)|\,dt$.

EXAMPLE 4

Let $\mathbf{f}(t) = (\cos t)\mathbf{i} + (\sin t)\mathbf{j} + t\mathbf{k}$. (This curve is called a **circular helix**. It is sketched in Fig. 9.3.4.)

> **(a)** Calculate $\mathbf{T}(t)$ at $t = \pi/3$.
> **(b)** Find the length of the arc from $t = 0$ to $t = 4$.

SOLUTION
(a) $\mathbf{f}'(t) = -(\sin t)\mathbf{i} + (\cos t)\mathbf{j} + \mathbf{k}$ and

$$|\mathbf{f}'(t)| = \sqrt{\sin^2 t + \cos^2 t + 1} = \sqrt{2},$$

so that

$$\mathbf{T}(t) = -\frac{\sin t}{\sqrt{2}}\mathbf{i} + \frac{\cos t}{\sqrt{2}}\mathbf{j} + \frac{1}{\sqrt{2}}\mathbf{k}.$$

At $t = \pi/3$,

$$\mathbf{T}\left(\frac{\pi}{3}\right) = -\frac{\sqrt{3}}{2\sqrt{2}}\mathbf{i} + \frac{1}{2\sqrt{2}}\mathbf{j} + \frac{1}{\sqrt{2}}\mathbf{k}.$$

(b) Since $ds/dt = |\mathbf{f}'(t)| = \sqrt{2}$,

$$s(4) = \int_0^4 \sqrt{2}\,dt = 4\sqrt{2}.$$

Problems 9.4

In Problems 1–20, find the length of the arc over the given interval or the length of the closed curve.

1. $x = t^3$; $y = t^2$; $1 \le t \le 4$

2. $x = \cos 2\theta$; $y = \sin 2\theta$; $0 \le \theta \le \pi/2$

3. $x = t^3 + 1$; $y = 3t^2 + 2$; $0 \le t \le 2$

4. $x = 1 + t$; $y = (1 + t)^{3/2}$; $0 \le t \le 1$

5. $x = \dfrac{1}{\sqrt{t + 1}}$; $y = \dfrac{t}{2(t + 1)}$; $0 \le t \le 4$

6. $x = e^t \cos t$; $y = e^t \sin t$; $0 \le t \le \pi/2$

7. $x = \sin^2 t$; $y = \cos^2 t$; $0 \le t \le \pi/2$

8. The hypocycloid of four cusps $x = a \cos^3 \theta$, $y = a \sin^3 \theta$, $a > 0$. [*Hint:* Calculate the length in the first quadrant and multiply by 4.]

* 9. The cardioid $r = a(1 + \sin \theta)$. [*Hint:* $x = r \cos \theta$, $y = r \sin \theta$

$$\int \sqrt{1 + \sin \theta}\, d\theta = \int \sqrt{1 + \sin \theta} \cdot \frac{\sqrt{1 - \sin \theta}}{\sqrt{1 - \sin \theta}}\, d\theta.$$

Pay attention to signs.]

10. One arc of the cycloid $x = a(\theta - \sin \theta)$, $y = a(1 - \cos \theta)$, $a > 0$.

11. $x = t^3$; $y = t^2$; $-1 \le t \le 1$ [*Hint:* $\sqrt{t^2} = -t$ for $t < 0$.]

12. $r = a \sin \theta$; $0 \le \theta \le \pi/2$, $a > 0$

13. $r = a \cos \theta$; $0 \le \theta \le \pi$, $a > 0$

* 14. $r = a\theta$; $0 \le \theta \le 2\pi$, $a > 0$

15. $r = e^\theta$; $0 \le \theta \le 3$

16. $r = \theta^2$; $0 \le \theta \le \pi$

17. $r = 6 \cos^2(\theta/2)$; $0 \le \theta \le \pi/2$

18. $r = \sin^3(\theta/3)$; $0 \le \theta \le \pi/2$ [*Hint:* $\sin^2(\theta/3) = \frac{1}{2}(1 - \cos(2\theta/3))$.]

19. $\mathbf{f}(t) = e^t(\sin t)\mathbf{i} + e^t(\cos t)\mathbf{j}$; $0 \le t \le \pi/2$

20. $\mathbf{f}(t) = 3(\cos \theta)\mathbf{i} + 3(\sin \theta)\mathbf{j}$; $0 \le \theta \le 2\pi$

In Problems 21–34, find $\mathbf{T}(t)$, $\mathbf{n}(t)$, and the particular vectors $\mathbf{T}$ and $\mathbf{n}$ when $t = t_0$. Then sketch the curve near $t = t_0$ and include the vectors $\mathbf{T}$ and $\mathbf{n}$ in your sketch.

21. $\mathbf{f} = (\cos 3t)\mathbf{i} + (\sin 3t)\mathbf{j}$; $t_0 = 0$

22. $\mathbf{f} = (\cos 5t)\mathbf{i} + (\sin 5t)\mathbf{j}$; $t_0 = \pi/2$

23. $\mathbf{f} = 2(\cos 4t)\mathbf{i} + 2(\sin 4t)\mathbf{j}$; $t_0 = \pi/4$

24. $\mathbf{f} = -3(\cos 10t)\mathbf{i} - 3(\sin 10t)\mathbf{j}$; $t_0 = \pi$

25. $\mathbf{f} = 8(\cos t)\mathbf{i} + 8(\sin t)\mathbf{j}$; $t_0 = \pi/4$

26. $\mathbf{f} = 4t\mathbf{i} + 2t^2\mathbf{j}$; $t_0 = 1$

27. $\mathbf{f} = (2 + 3t)\mathbf{i} + (8 - 5t)\mathbf{j}$; $t_0 = 3$

28. $\mathbf{f} = (4 - 7t)\mathbf{i} + (-3 + 5t)\mathbf{j}$; $t_0 = -5$

29. $\mathbf{f} = (a + bt)\mathbf{i} + (c + dt)\mathbf{j}$, a, b, c, d real; any t_0

30. $\mathbf{f} = 2t\mathbf{i} + (e^{-t} + e^t)\mathbf{j}$; $t_0 = 0$

31. $\mathbf{f} = (t - \cos t)\mathbf{i} + (1 - \sin t)\mathbf{j}$; $t_0 = \pi/2$

32. $\mathbf{f} = (t - \cos t)\mathbf{i} + (1 - \sin t)\mathbf{j}$; $t_0 = \pi$

33. $\mathbf{f} = (t - \cos t)\mathbf{i} + (1 - \sin t)\mathbf{j}$; $t_0 = \pi/4$

34. $\mathbf{f} = (\ln \sin t)\mathbf{i} + (\ln \cos t)\mathbf{j}$; $t_0 = \pi/6$

In Problems 35–40, find the unit tangent vector $\mathbf{T}$ for the given value of t.

35. $\mathbf{f}(t) = t\mathbf{i} + t^2\mathbf{j} + t^3\mathbf{k}$; $t = 1$

36. $\mathbf{f}(t) = t^3\mathbf{i} + t^5\mathbf{j} + t^7\mathbf{k}$; $t = 1$

37. $\mathbf{f}(t) = t\mathbf{i} + e^t\mathbf{j} + e^{-t}\mathbf{k}$; $t = 0$

38. $\mathbf{f}(t) = t^2\mathbf{i} + t^2\mathbf{j} + t^{5/2}\mathbf{k}$; $t = 4$

39. $\mathbf{f}(t) = 4(\cos 2t)\mathbf{i} + 9(\sin 2t)\mathbf{j} + t\mathbf{k}$; $t = \pi/4$

40. $\mathbf{f}(t) = (\cosh t)\mathbf{i} + (\sinh t)\mathbf{j} + t^2\mathbf{k}$; $t = 0$

41. Find the arc length of the curve $\mathbf{f} = 2(\cos 3t)\mathbf{i} + 2(\sin 3t)\mathbf{j} + t^2\mathbf{k}$ between $t = 0$ and $t = 10$.

42. Find the arc length of the curve $\mathbf{f} = e^t(\cos 2t)\mathbf{i} + e^t(\sin 2t)\mathbf{j} + e^t\mathbf{k}$ between $t = 1$ and $t = 4$.

43. Find the arc length of the curve $\mathbf{f} = \frac{2}{3}t^3\mathbf{i} + (1 + t^{9/2})\mathbf{j} + (1 - t^{9/2})\mathbf{k}$ between $t = 0$ and $t = 2$.

44. The parametric representation of the ellipse $(x^2/a^2) + (y^2/b^2) = 1$ is given by $x = a \cos \theta$, $y = b \sin \theta$. Find an integral that represents the length of the circumference of an ellipse but do not try to evaluate it. The integral you obtain is called an *elliptic integral* and it arises in a variety of physical applications. It cannot be integrated (except numerically) unless $a = b$.

45. Find the arc length of the curve

$$\mathbf{f}(t) = (1, t, 2t, 3t, \ldots, nt) \qquad \text{for } t \in [1, 5].$$

9.5 Tangent Vector, Curvature, and Torsion

In Section 9.3 we saw that the derivative $\mathbf{f}'(t)$ of a continuously differentiable vector function $\mathbf{f}(t)$ has a geometric interpretation: it is a tangent vector at any point on the curve C defined by $\mathbf{f}(t)$. This interpretation was justified in Section 9.4, where additional facts were provided about unit tangent vectors and the arc length of curves in $\mathbb{R}^2$ and $\mathbb{R}^n$. In all of this work we assumed that the curve C is defined by a continuous vector function $\mathbf{f}(t)$, where t is an arbitrary parameter. In many situations it is convenient to use the arc length s as a parameter: Let

Arc Length as a Parameter

$$\mathbf{f}(s) = x(s)\mathbf{i} + y(s)\mathbf{j} \tag{1}$$

be a point on the curve C described by $\mathbf{f}$ that is s units along the curve C from a fixed reference point $P_0 = x(0)\mathbf{i} + y(0)\mathbf{j}$. The point P_0 may be chosen arbitrarily. By convention, the orientation corresponding to increasing values of s along C is called the **positive orientation** of C. Clearly, there are two ways of orienting C; the transformation from one to the other involves changing the parameter s to $-s$.

EXAMPLE 1

What is the arc length parametrization of the circle centered at the origin with radius r? Take $P_0 = (r, 0)$.

SOLUTION
Suppose we want to give this circle its usual counterclockwise orientation; then every point P_θ is given by

$$P_\theta = (r \cos \theta)\mathbf{i} + (r \sin \theta)\mathbf{j},$$

where $0 \le \theta \le 2\pi$; θ is measured in radians. By the definition of radian measure, P_θ is $r\theta$ units away from P_0 along the circle. Thus $s = r\theta$ so that $\theta = s/r$. Then

$$\mathbf{f}(s) = r[\cos(s/r)\mathbf{i} + \sin(s/r)\mathbf{j}], \qquad 0 \le s \le 2\pi r,$$

is the required arc length parametrization.

EXAMPLE 2

Let $\mathbf{f}(t) = (2t - t^2)\mathbf{i} + \frac{8}{3}t^{3/2}\mathbf{j}$, $t \ge 0$. Write this curve with arc length as a parameter.

SOLUTION
Suppose that the fixed point is $P_0 = (0, 0)$ when $t = 0$. Then from Example 9.4.3,

$$ds/dt = |f'(t)| = 2(t + 1),$$

so that

$$s = \int_0^t \frac{ds}{du}\,du = \int_0^t 2(u + 1)\,du = t^2 + 2t.$$

This leads to the equations

$$t^2 + 2t - s = 0 \quad \text{and} \quad t = \frac{-2 + \sqrt{4 + 4s}}{2} = \sqrt{1 + s} - 1.$$

We took the positive square root here since it is assumed that t starts at 0 and increases.

Then

$$x = 2t - t^2 = 4\sqrt{1+s} - 4 - s,$$

$$y = \tfrac{8}{3}t^{3/2} = \tfrac{8}{3}(\sqrt{1+s} - 1)^{3/2},$$

and we obtain

$$\mathbf{f}(s) = (4\sqrt{1+s} - 4 - s)\mathbf{i} + \tfrac{8}{3}(\sqrt{1+s} - 1)^{3/2}\mathbf{j}.$$

As Example 2 illustrates, writing **f** explicitly with arc length as a parameter can be tedious (or, more often, impossible) to do in terms of elementary functions.

There is an interesting and important relationship between position vectors, tangent vectors, and normal vectors that becomes apparent when we use s as a parameter.

THEOREM 1

If the curve C is parametrized by $\mathbf{f}(s) = x(s)\mathbf{i} + y(s)\mathbf{j}$, where s is arc length and x and y have continuous derivatives with $\mathbf{f}'(s) \neq \mathbf{0}$, then the unit tangent vector **T** is given by

$$\mathbf{T}(s) = \frac{d\mathbf{f}}{ds}. \tag{2}$$

PROOF

With *any* parametrization of C, the unit tangent vector is given by [see equation (9.3.10)]

$$\mathbf{T}(t) = \frac{\mathbf{f}'(t)}{|\mathbf{f}'(t)|}.$$

Choosing $t = s$ yields

$$\mathbf{T}(s) = \frac{d\mathbf{f}/ds}{|d\mathbf{f}/ds|}.$$

But from equation (9.4.11), $|\mathbf{f}'(t)| = ds/dt$, so that

$$\left|\frac{d\mathbf{f}}{ds}\right| = \left|\frac{d\mathbf{f}/dt}{ds/dt}\right| = \left|\frac{\mathbf{f}'(t)}{ds/dt}\right| = 1.$$

Theorem 1 is quite useful in that it provides a check of our calculation of the parametrization in terms of arc length. For if

$$\mathbf{f}(s) = x(s)\mathbf{i} + y(s)\mathbf{j},$$

then

$$T = \frac{df}{ds} = \frac{dx}{ds}i + \frac{dy}{ds}j.$$

But $|T| = 1$, so that $|T|^2 = 1$, which implies that

$$\left(\frac{dx}{ds}\right)^2 + \left(\frac{dy}{ds}\right)^2 = 1. \tag{3}$$

We can apply this result in Example 2. We have

$$x(s) = 4\sqrt{1+s} - 4 - s \quad \text{and} \quad y(s) = \tfrac{8}{3}(\sqrt{1+s} - 1)^{3/2},$$

so

$$\frac{dx}{ds} = \frac{2}{\sqrt{1+s}} - 1, \qquad \frac{dy}{ds} = \frac{2}{\sqrt{1+s}}(\sqrt{1+s} - 1)^{1/2},$$

and

$$\left(\frac{dx}{ds}\right)^2 + \left(\frac{dy}{ds}\right)^2 = \frac{4}{1+s} - \frac{4}{\sqrt{1+s}} + 1 + \frac{4}{1+s}(\sqrt{1+s} - 1) = 1,$$

as expected.

If the curve C is represented by a twice continuously differentiable vector function $f(s)$ parametrized by arc length, then the rate of change of the unit tangent vector with respect to the arc length has a geometric meaning.

DEFINITION

Curvature

(i) The **curvature** of C, denoted by $\kappa(s)$, is given by

$$\kappa(s) = |T'(s)| = |f''(s)|. \tag{4}$$

Note. Observe that, in (4), κ is written with arc length as the parameter.

Radius of Curvature

(ii) The **radius of curvature** $\rho(s)$ is defined by

$$\rho(s) = 1/\kappa(s), \qquad \text{if } \kappa(s) > 0. \tag{5}$$

Remark 1. The curvature is a measure of *how fast* the curve turns as we move along it with unit speed. The radius of curvature is the radius of the circle having that *constant* curvature. (We verify this in Example 3, below.)

Remark 2. If $\kappa(s) = 0$, we say that the radius of curvature is *infinite*. To understand this idea, note that if $\kappa(s) = 0$, then the "curve" does not bend and so is a straight line (see Example 4). A straight line can be thought of as an arc of a circle with *infinite* radius.

EXAMPLE 3

What is the radius of curvature of a circle of radius r?

SOLUTION
Using Example 1 and the parametrization

$$\mathbf{f}(s) = r\left[\cos\left(\frac{s}{r}\right)\mathbf{i} + \sin\left(\frac{s}{r}\right)\mathbf{j}\right],$$

we see that

$$\mathbf{f}''(s) = -\frac{1}{r}\left[\cos\left(\frac{s}{r}\right)\mathbf{i} + \sin\left(\frac{s}{r}\right)\mathbf{j}\right],$$

so that

$$\rho(s) = \frac{1}{\kappa(s)} = \frac{1}{|\mathbf{f}''(s)|} = \frac{r}{\sqrt{\cos^2(s/r) + \sin^2(s/r)}} = r.$$

It will be convenient to express the curvature of a curve C defined by an arbitrarily parametrized vector function $\mathbf{f}(t)$.

■ **THEOREM 2**

If C is defined by $\mathbf{f}(t) = x(t)\mathbf{i} + y(t)\mathbf{j}$, with $\mathbf{f}'(t) \neq \mathbf{0}$, then the curvature at the point $\mathbf{f}(t)$ is

$$\kappa(t) = \frac{|x'y'' - y'x''|}{[(x')^2 + (y')^2]^{3/2}} = \frac{\sqrt{|\mathbf{f}'|^2|\mathbf{f}''|^2 - (\mathbf{f}' \cdot \mathbf{f}'')^2}}{|\mathbf{f}'(t)|^3}. \tag{6}$$

PROOF
Since $ds/dt = |\mathbf{f}'(t)|$ and $\mathbf{T}(t) = \mathbf{f}'(t)/|\mathbf{f}'(t)|$ we have

$$\kappa(s) = \left|\frac{d\mathbf{T}}{ds}\right| = \frac{|d\mathbf{T}/dt|}{|ds/dt|} = \frac{1}{|\mathbf{f}'(t)|}\left|\frac{d}{dt}\left(\frac{\mathbf{f}'(t)}{|\mathbf{f}'(t)|}\right)\right| = \frac{\|\mathbf{f}'(t)|\mathbf{f}''(t) - g(t)\mathbf{f}'(t)|}{|\mathbf{f}'(t)|^3},$$

where

$$g(t) = \frac{d}{dt}|\mathbf{f}'(t)| = \frac{d}{dt}[(x')^2 + (y')^2]^{1/2} = \frac{x'x'' + y'y''}{\sqrt{(x')^2 + (y')^2}} = \frac{\mathbf{f}' \cdot \mathbf{f}''}{|\mathbf{f}'|}.$$

Thus

$$\kappa(s) = \sqrt{(|\mathbf{f}'|\mathbf{f}'' - g\mathbf{f}') \cdot (|\mathbf{f}'|\mathbf{f}'' - g\mathbf{f}')}/|\mathbf{f}'|^3$$

$$= \sqrt{|\mathbf{f}'|^2|\mathbf{f}''|^2 - 2g|\mathbf{f}'|(\mathbf{f}' \cdot \mathbf{f}'') + g^2|\mathbf{f}'|^2}/|\mathbf{f}'|^3$$

$$= \sqrt{|\mathbf{f}'|^2|\mathbf{f}''|^2 - (\mathbf{f}' \cdot \mathbf{f}'')^2}/|\mathbf{f}'|^3. \qquad ∎$$

Another useful expression for the curvature is (see Problem 36)

$$\kappa(t) = \frac{|\mathbf{f}' \times \mathbf{f}''|}{|\mathbf{f}'|^3}.$$

EXAMPLE 4

Show that for a straight line $\kappa(t) = 0$.

SOLUTION

A line through the points $\mathbf{x}_1$ and $\mathbf{x}_2$ is represented parametrically by $\mathbf{f}(t) = \mathbf{x}_1 + t(\mathbf{x}_2 - \mathbf{x}_1)$. Then $\mathbf{f}'(t) = \mathbf{x}_2 - \mathbf{x}_1$ and $\mathbf{f}''(t) = \mathbf{0}$, so that by (6)

$$\kappa(t) = 0.$$

If $y = f(x)$ is the Cartesian equation of the curve C then the curvature, now denoted by $\kappa(x)$, is much simpler. Here the parametric representation of C is

$$x = t, \qquad y = f(t),$$

so that $x' = 1$ and $x'' = 0$. Thus

$$\kappa(x) = \frac{|y''|}{[1 + (y')^2]^{3/2}}. \qquad (7)$$

Recall from equation (9.4.9) that the unit normal vector to the curve defined by $\mathbf{f}$ at t in $\mathbb{R}^2$ is given by

$$\mathbf{n} = \mathbf{T}'/|\mathbf{T}'|.$$

If the curve, $\mathbf{f}(s)$, is parametrized by arc length then

$$\mathbf{n}(s) = \mathbf{T}'(s)/|\mathbf{T}'(s)|$$

so that by (4)

$$\mathbf{T}'(s) = \kappa(s)\mathbf{n}(s). \qquad (8)$$

The definition we have given for curvature also is valid in $\mathbb{R}^3$ (and $\mathbb{R}^n$), but we are faced with a problem in trying to define the unit normal vector, because there is a circle [or $(n - 2)$-dimensional surface] of unit vectors orthogonal to $\mathbf{T}$. Instead, we have

DEFINITION

The **principal unit normal vector n** in $\mathbb{R}^3$ (or $\mathbb{R}^n$) is defined by

$$\mathbf{n}(s) = \frac{1}{\kappa(s)}\frac{d\mathbf{T}}{ds}. \tag{9}$$

Remark. Having defined a unit normal vector, we must show that it is orthogonal to the unit tangent vector in order to justify its name.

■ **THEOREM 3**

$\mathbf{n}(s) \perp \mathbf{T}(s)$.

PROOF

Since $1 = \mathbf{T} \cdot \mathbf{T}$, we differentiate both sides with respect to s to obtain

$$0 = \mathbf{T} \cdot \frac{d\mathbf{T}}{ds} + \frac{d\mathbf{T}}{ds} \cdot \mathbf{T} = 2\mathbf{T} \cdot \frac{d\mathbf{T}}{ds} = (2\kappa)\mathbf{T} \cdot \mathbf{n}. \quad ■$$

EXAMPLE 5

Find the curvature and principal unit normal vector, $\kappa(s)$ and $\mathbf{n}(s)$, for the circular helix of Example 9.4.4, p. 534, at $t = \pi/3$.

SOLUTION

Since $ds/dt = \sqrt{2}$, we have $s = \sqrt{2}t$ and $t = s/\sqrt{2}$, so that

$$\mathbf{T}(s) = -\frac{\sin(s/\sqrt{2})}{\sqrt{2}}\mathbf{i} + \frac{\cos(s/\sqrt{2})}{\sqrt{2}}\mathbf{j} + \frac{1}{\sqrt{2}}\mathbf{k}$$

and

$$\frac{d\mathbf{T}}{ds} = -\frac{\cos(s/\sqrt{2})}{2}\mathbf{i} - \frac{\sin(s/\sqrt{2})}{2}\mathbf{j}.$$

Hence

$$\kappa(s) = \sqrt{\frac{\cos^2(s/\sqrt{2})}{4} + \frac{\sin^2(s/\sqrt{2})}{4}} = \frac{1}{2}$$

and

$$\mathbf{n}(s) = \frac{1}{\kappa(s)}\frac{d\mathbf{T}}{ds} = -\left(\cos\frac{s}{\sqrt{2}}\right)\mathbf{i} - \left(\sin\frac{s}{\sqrt{2}}\right)\mathbf{j}.$$

At $s = \sqrt{2}t = \sqrt{2}\pi/3,$

$$\mathbf{n}\!\left(\frac{\sqrt{2}\pi}{3}\right) = -\frac{1}{2}\mathbf{i} - \frac{\sqrt{3}}{2}\mathbf{j}.$$

Having defined two orthogonal vectors in $\mathbb{R}^3$ we define a third so as to form a right-handed triple of unit vectors, called the **trihedron** of C, at the given point.

DEFINITION

Binormal Vector

The **binormal vector B** to the curve **f** is defined by

$$\mathbf{B} = \mathbf{T} \times \mathbf{n}. \tag{10}$$

From this definition we see that **B** is orthogonal to both **T** and **n**. Moreover, since $\mathbf{T} \perp \mathbf{n}$, the angle θ between **T** and **n** is $\pi/2$, and

$$|\mathbf{B}| = |\mathbf{T} \times \mathbf{n}| = |\mathbf{T}||\mathbf{n}| \sin\theta = 1.$$

For Example 5 we have (since $t = s/\sqrt{2}$)

$$\mathbf{B} = \mathbf{T} \times \mathbf{n} = \begin{vmatrix} \mathbf{i} & \mathbf{j} & \mathbf{k} \\ -\dfrac{\sin t}{\sqrt{2}} & \dfrac{\cos t}{\sqrt{2}} & \dfrac{1}{\sqrt{2}} \\ -\cos t & -\sin t & 0 \end{vmatrix} = \frac{\sin t}{\sqrt{2}}\mathbf{i} - \frac{\cos t}{\sqrt{2}}\mathbf{j} + \frac{1}{\sqrt{2}}\mathbf{k}. \tag{11}$$

Since $\mathbf{B} \cdot \mathbf{T} = 0$ and $\mathbf{B} \cdot \mathbf{n} = 0$, differentiating the first identity with respect to arc length yields, by (9),

$$0 = \mathbf{B}' \cdot \mathbf{T} + \mathbf{B} \cdot \mathbf{T}' = \mathbf{B}' \cdot \mathbf{T} + \kappa\mathbf{B} \cdot \mathbf{n},$$

so that $\mathbf{B}' \cdot \mathbf{T} = 0$. If $\mathbf{B}' \neq \mathbf{0}$ then it is also perpendicular to **B** (differentiate $\mathbf{B} \cdot \mathbf{B} = 1$), so that $\mathbf{B}'$ is perpendicular to the plane spanned by **T** and **B**. Hence

$$\mathbf{B}' = -\tau\mathbf{n}, \tag{12}$$

Torsion

with τ a scalar function called the **torsion** of C at the given point. The torsion measures the rate of twisting of the curve C since $|\mathbf{B}'| = |-\tau\mathbf{n}| = |\tau|$. Observe that if we take the dot product of both sides of (12) with **n**, we get

$$\tau = \tau\mathbf{n} \cdot \mathbf{n} = -\mathbf{B}' \cdot \mathbf{n}. \tag{13}$$

EXAMPLE 6

Compute the torsion of the circular helix of Example 5 at $t = \dfrac{\pi}{3}$.

SOLUTION

Using (11) and differentiating, we have

$$\mathbf{B}' = \frac{\cos t}{\sqrt{2}}\mathbf{i} + \frac{\sin t}{\sqrt{2}}\mathbf{j} = \frac{1}{\sqrt{2}}\left(\frac{1}{2}\mathbf{i} + \frac{\sqrt{3}}{2}\mathbf{j}\right) \qquad \text{at } t = \frac{\pi}{3}.$$

But, from Example 5, $\mathbf{n}(\sqrt{2}\pi/3) = -\tfrac{1}{2}\mathbf{i} - (\sqrt{3}/2)\mathbf{j}$. Thus, from (13),

$$\tau = -\frac{1}{2\sqrt{2}}(\mathbf{i} + \sqrt{3}\mathbf{j})\cdot\left(-\frac{1}{2}\mathbf{i} - \frac{\sqrt{3}}{2}\mathbf{j}\right) = \frac{1}{4\sqrt{2}}(1 + 3) = \frac{1}{\sqrt{2}}.$$

Summary	**Frenet† Formulas**
Unit Tangent Vector $$\mathbf{T} = \frac{\mathbf{f}'(t)}{|\mathbf{f}'(t)|} = \frac{d\mathbf{f}}{ds}$$ *Principal Unit Normal Vector* $$\mathbf{n} = \frac{\mathbf{T}'(s)}{|\mathbf{T}'(s)|}$$ *Unit Binormal Vector* $$\mathbf{B} = \mathbf{T} \times \mathbf{n}$$	$$\mathbf{T}' = \kappa\mathbf{n}$$ $$\mathbf{B}' = -\tau\mathbf{n}$$ $$\mathbf{n}' = \tau\mathbf{B} - \kappa\mathbf{T}$$ (see Problem 37)

Problems 9.5

In Problems 1–20, find the curvature and radius of curvature for each curve. Sketch the unit tangent vector and the circle of curvature at the given point.

1. $\mathbf{f} = 2 \cos t\mathbf{i} + 2 \sin t\mathbf{j};\quad t = \pi/4$
2. $\mathbf{f} = 2 \cos t\mathbf{i} + 2 \sin t\mathbf{j};\quad t = \pi/27$
3. $\mathbf{f} = t\mathbf{i} + t^2\mathbf{j};\quad t = 1$
4. $\mathbf{f} = 3 \sin t\mathbf{i} + 4 \cos t\mathbf{j};\quad t = 0$
5. $\mathbf{f} = 3 \sin t\mathbf{i} + 4 \cos t\mathbf{j};\quad t = \pi/2$
6. $\mathbf{f} = 3 \sin t\mathbf{i} + 4 \cos t\mathbf{j};\quad t = \pi/4$

7. $\mathbf{f} = (\cos t + t \sin t)\mathbf{i} + (\sin t - t \cos t)\mathbf{j};\quad t = \pi/6$
8. $y = x^2;\ (0, 0)$
9. $y = x^2;\ (1, 1)$
10. $xy = 1;\ (1, 1)$
11. $y = e^x;\ (0, 1)$
12. $y = e^x;\ (1, e)$
13. $y = \ln x;\ (1, 0)$
14. $y = \cos x;\ (\pi/3, \tfrac{1}{2})$

† Named after the French mathematician J.-F. Frenet (1816–1900).

15. $y = ax^2 + bx + c$; $(0, c)$, $a \neq 0$

16. $y = \ln \cos x$; $(\pi/4, \ln(1/\sqrt{2}))$

17. $y = \sqrt{1 - x^2}$; $(0, 1)$

* 18. $y = \sin^{-1} x$; $(1, \pi/2)$

19. $x = \cos y$; $(0, \pi/2)$

20. $x = y^3$; $(1, 1)$

21. At what point on the parabola $y = ax^2$ is the curvature a maximum?

22. At what point on the curve $y = \ln x$ is the curvature a maximum?

23. For what value of t in the interval $[0, \pi/2]$ is the curvature of the curve $\mathbf{f}(t) = a(\cos^3 t)\mathbf{i} + a(\sin^3 t)\mathbf{j}$ a minimum? For what value is it a maximum?

* 24. Let $r = f(\theta)$ be the equation of a curve in polar coordinates. Show that

$$\kappa(\theta) = \frac{|r^2 + 2(dr/d\theta)^2 - r\, d^2r/d\theta^2|}{[r^2 + (dr/d\theta)^2]^{3/2}}.$$

In Problems 25–32, find the unit tangent vector $\mathbf{T}$, the unit normal vector $\mathbf{n}$, the binormal vector $\mathbf{B}$, the curvature κ, and the torsion τ at the given value of t.

25. $\mathbf{f} = a(\sin t)\mathbf{i} + a(\cos t)\mathbf{j} + t\mathbf{k}$; $t = \pi/4$; $a > 0$

26. $\mathbf{f} = a(\sin t)\mathbf{i} + a(\cos t)\mathbf{j} + t\mathbf{k}$; $t = \pi/6$; $a > 0$

27. $\mathbf{f} = a(\cos t)\mathbf{i} + b(\sin t)\mathbf{j} + t\mathbf{k}$; $t = 0$; $a > 0, b > 0$, $a \neq b$

28. $\mathbf{f} = a(\cos t)\mathbf{i} + b(\sin t)\mathbf{j} + t\mathbf{k}$; $t = \pi/2$; $a > 0, b > 0$, $a \neq b$

29. $\mathbf{f} = t\mathbf{i} + t^2\mathbf{j} + t^3\mathbf{k}$; $t = 0$

30. $\mathbf{f} = e^t(\cos 2t)\mathbf{i} + e^t(\sin 2t)\mathbf{j} + e^t\mathbf{k}$; $t = 0$

31. $\mathbf{f} = e^t(\cos 2t)\mathbf{i} + e^t(\sin 2t)\mathbf{j} + e^t\mathbf{k}$; $t = \pi/4$

32. $\mathbf{f} = \mathbf{i} + t\mathbf{j} + t^2\mathbf{k}$; $t = 1$

33. Let $\mathbf{f} = a(\cos t)\mathbf{i} + a(\sin t)\mathbf{j} + t\mathbf{k}$ (a circular helix). Show that the angle between the unit tangent vector $\mathbf{T}$ and the z-axis is constant.

34. Show, by writing out the component functions, that

$$(\mathbf{f} \times \mathbf{g})' = (\mathbf{f}' \times \mathbf{g}) + (\mathbf{f} \times \mathbf{g}').$$

35. Show that the curvature of a straight line in space is zero.

36. Prove that $\kappa = |\mathbf{f}' \times \mathbf{f}''|/|\mathbf{f}'|^3$.

37. Show that $d\mathbf{n}/ds = -\kappa\mathbf{T} + \tau\mathbf{B}$. [*Hint:* Use the fact that $\mathbf{n} = \mathbf{B} \times \mathbf{T}$.]

38. Show that the torsion of a plane curve is zero.

39. Show that $\tau = \mathbf{T} \cdot (\mathbf{n} \times \mathbf{n}')$.

40. Prove that $\tau = [\mathbf{f}' \cdot (\mathbf{f}'' \times \mathbf{f}''')]/\kappa^2$.

9.6 Tangential and Normal Components of Acceleration

Suppose that an object is moving in the plane. Then we can describe its motion parametrically by the vector function

$$\mathbf{f}(t) = f_1(t)\mathbf{i} + f_2(t)\mathbf{j}. \tag{1}$$

In this context $\mathbf{f}$ is called the **position vector** of the object, and the curve described by $\mathbf{f}$ is called the **trajectory** of the object. We then have the following definition.

DEFINITION
If $\mathbf{f}'$ and $\mathbf{f}''$ exist, then

Velocity Vector and Acceleration Vector

(i) $\mathbf{v}(t) = \mathbf{f}'(t) = f_1'(t)\mathbf{i} + f_2'(t)\mathbf{j}$ \hfill (2)

is called the **velocity vector** of the moving object at time t, and

(ii) $\mathbf{a}(t) = \dfrac{d\mathbf{v}}{dt} = \mathbf{f}''(t) = f_1''(t)\mathbf{i} + f_2''(t)\mathbf{j}$ \hfill (3)

is called the **acceleration vector** of the object.

This definition is, of course, not surprising. It simply extends to the vector case our notion of velocity as the derivative of position and acceleration as the derivative of velocity.

DEFINITION

Speed and Acceleration Scalar

(i) The **speed** $v(t)$ of a moving object is the magnitude of the velocity vector.

(ii) The **acceleration scalar** $a(t)$ is the magnitude of the acceleration vector.

Remark 1. Since we have already shown that $|\mathbf{f}'(t)| = ds/dt$ [equation (9.4.11)], we have, since $v(t) = |\mathbf{v}(t)| = |\mathbf{f}'(t)|$,

$$v(t) = \frac{ds}{dt}. \tag{4}$$

Remark 2. Although $\mathbf{a}(t)$ is the derivative of $\mathbf{v}(t)$, it is *not true* in general that $a(t)$ is the derivative of the speed $v(t)$. For example, consider the motion along the unit circle given by

$$\mathbf{f}(t) = (\cos t)\mathbf{i} + (\sin t)\mathbf{j}.$$

Then

$$\mathbf{v}(t) = -(\sin t)\mathbf{i} + (\cos t)\mathbf{j} \quad \text{and} \quad \mathbf{a}(t) = -(\cos t)\mathbf{i} - (\sin t)\mathbf{j}.$$

But $v(t) = |\mathbf{v}(t)| = 1$, so that $dv/dt = 0$ and $a(t) = |\mathbf{a}(t)| = 1$, which is, evidently, not equal to dv/dt.

Remark 3. It follows from Theorem 9.4.2 that *if speed is constant, then the velocity and acceleration vectors are orthogonal.*

DEFINITION

Momentum

The **momentum P** of a particle at any time t is a vector defined as the product of the mass m of the particle and its velocity $\mathbf{v}$. That is,

$$\mathbf{P} = m\mathbf{v}. \tag{5}$$

Newton's Second Law of Motion

Newton's second law of motion states that the *rate of change of momentum of a moving object is equal to the resultant force and is in the direction of that force.* That is,

$$\mathbf{F} = \frac{d\mathbf{P}}{dt}. \tag{6}$$

If the mass of the object is constant, then using (5) and (6), we have the familiar law

$$\mathbf{F} = \frac{d}{dt}(m\mathbf{v}) = m\frac{d\mathbf{v}}{dt} = m\mathbf{a}. \tag{7}$$

If mass is not constant, then (7) becomes

$$\mathbf{F} = \frac{d}{dt}(m\mathbf{v}) = m\frac{d\mathbf{v}}{dt} + \mathbf{v}\frac{dm}{dt} = m\mathbf{a} + \mathbf{v}\frac{dm}{dt}. \tag{8}$$

EXAMPLE 1: Central Force and Centripetal Acceleration

An object of constant mass m moves in the elliptical orbit given by $\mathbf{f}(t) = a(\cos \alpha t)\mathbf{i} + b(\sin \alpha t)\mathbf{j}$. Find the force acting on the object at any time t.

SOLUTION

We easily find that $\mathbf{a}(t) = \mathbf{f}''(t) = -\alpha^2 a(\cos \alpha t)\mathbf{i} - \alpha^2 b(\sin \alpha t)\mathbf{j} = -\alpha^2\mathbf{f}(t)$. Since m is constant, $\mathbf{F} = m\mathbf{a} = -m\alpha^2\mathbf{f}(t)$. Thus the force always acts in the direction opposite to the direction of the position vector (thereby pointing *toward* the origin) and has a magnitude proportional to the distance of the object from the origin. Such a force is called a **central force** (see Fig. 1). By definition a central force is a force that is directed toward or away from a fixed point. The acceleration $\mathbf{a}(t) = -\alpha^2\mathbf{f}(t)$ is called **centripetal acceleration**. This is acceleration, due to rotation, that is directed toward the center. It is also referred to as **radial acceleration**.

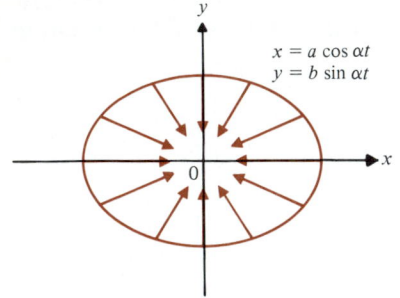

$x = a \cos \alpha t$
$y = b \sin \alpha t$

FIGURE 1 Central force

Conservation of Momentum

If the vector sum of forces acting on a system is zero, then, since $\mathbf{0} = \mathbf{F} = \mathbf{P}'(t)$, we find that the *total momentum of the system is constant*. This fact is called the **principle of the conservation of linear momentum.** We now see how that principle can be applied to the motion of a rocket.

Let the mass of a rocket be denoted by $m(t)$, and suppose that the rocket is traveling with velocity $\mathbf{v}(t)$. The rocket is propelled by gas emissions. We assume that the constant exhaust velocity of the gas ejected from the rocket, relative to the rocket, is $\mathbf{u}$. Then the total velocity of the gas (relative to the earth, say) is $\mathbf{u} + \mathbf{v}$. The rate of change of momentum on the rocket is

$$\mathbf{P}'(t) = \frac{d}{dt}(m\mathbf{v}) = m\frac{d\mathbf{v}}{dt} + \mathbf{v}\frac{dm}{dt}.$$

As the rocket loses mass, the exhaust gains it, so the rate of change of momentum of the gas acting on the rocket† is given by

$$-(\mathbf{u} + \mathbf{v})\frac{dm}{dt}.$$

Here dm/dt is the rate at which exhaust gas accumulates. The total force acting on the rocket-gas system is the sum of the rates of change of momentum for each of the two components of the system; that is,

---

† The rest of the column of gas no longer acts on the rocket.

$$\mathbf{F} = m\frac{d\mathbf{v}}{dt} + \mathbf{v}\frac{dm}{dt} - \mathbf{u}\frac{dm}{dt} - \mathbf{v}\frac{dm}{dt} = m\frac{d\mathbf{v}}{dt} - \mathbf{u}\frac{dm}{dt},$$

or

$$\mathbf{F} + \mathbf{u}\frac{dm}{dt} = m\frac{d\mathbf{v}}{dt}. \tag{9}$$

The term $\mathbf{u}\,dm/dt$ is called the **thrust** of the rocket. Note that $dm/dt < 0$ (since the rocket is losing mass), so that the thrust has the direction *opposite* to that of the velocity $\mathbf{u}$ of the exhaust gases.

Now if the rocket is free of any gravitational field, then there are no external forces acting on the system, so $\mathbf{F} = 0$ and (9) becomes

$$\mathbf{u}\frac{dm}{dt} = m\frac{d\mathbf{v}}{dt}. \tag{10}$$

or

$$\frac{d\mathbf{v}}{dt} = \frac{\mathbf{u}}{m}\frac{dm}{dt},$$

and integration yields

$$\mathbf{v}(t) - \mathbf{v}_0 = \mathbf{u}\int_{m_0}^{m}\frac{dm}{m} = \mathbf{u}\ln m\Big|_{m_0}^{m} = -\mathbf{u}\ln\left(\frac{m_0}{m}\right). \tag{11}$$

Here $\mathbf{v}_0$ is the initial velocity vector and m_0 is the initial mass of the rocket and the fuel combined. Thus, if we ignore all gravitational fields, we can calculate its velocity at any time if we know the velocity of the escaping gases and the proportion of gases that have been expelled.

EXAMPLE 2

A 1000-kg rocket carrying 2000 kg of fuel is motionless in space. Its engine starts and its exhaust velocity is 0.7 km/sec. What is its speed when all its fuel is consumed?

SOLUTION
Here $\mathbf{v}_0 = \mathbf{0}$. Then

$$|\mathbf{v}| = |\mathbf{u}|\ln\frac{m_0}{m} = 0.7\ln\frac{3000}{1000} = 0.7\ln 3 \approx 0.769\,\text{km/sec}$$

$$\approx 2769\,\text{km/hr}\,(\approx 1720\,\text{mi/hr}).$$

There is an interesting relationship between curvature and acceleration vectors. If a particle is moving along the curve C with position vector

$$\mathbf{f}(t) = x(t)\mathbf{i} + y(t)\mathbf{j},$$

then the acceleration vector is given by

$$\mathbf{a}(t) = \frac{d^2x}{dt^2}\mathbf{i} + \frac{d^2y}{dt^2}\mathbf{j}. \tag{12}$$

The representation (12) resolves **a** into its horizontal and vertical components. However, there is another representation that is often more useful. Imagine yourself driving on the highway. If the car in which you are riding accelerates forward, you are pressed to the back of your seat. If it turns sharply to one side, you are thrown to the other. Both motions are due to acceleration. The second force is related to the rate at which the car turns, which is, of course, related to the curvature of the road. Thus we would like to express the acceleration vector as a component in the direction of motion and a component somehow related to the curvature of the path. How do we do so?

■ **THEOREM 1**

Let the curve C be given by the vector function $\mathbf{f}(t)$. Then the acceleration vector $\mathbf{a}(t) = \mathbf{f}''(t)$ can be written as

$$\mathbf{a} = \frac{d^2s}{dt^2}\mathbf{T} + \left(\frac{ds}{dt}\right)^2 \kappa\,\mathbf{n} \tag{13}$$

where **T** and **n** are the unit tangent and unit normal vectors to the curve, s is the arc length measured from the point reached when $t = 0$, and κ is the curvature.

Note. Since $v = ds/dt$ is the speed, we can write (13) as

$$\mathbf{a} = \frac{dv}{dt}\mathbf{T} + v^2\kappa\,\mathbf{n} = \frac{dv}{dt}\mathbf{T} + \frac{v^2}{\rho}\mathbf{n}. \tag{14}$$

Thus the tangential component of acceleration is $a_{\mathrm{T}} = dv/dt$ and the normal component of acceleration is $a_{\mathrm{n}} = v^2/\rho$.

PROOF OF THEOREM 1
We have

$$\mathbf{v} = \frac{d\mathbf{f}}{dt} = \frac{d\mathbf{f}}{ds}\frac{ds}{dt} = \mathbf{T}\frac{ds}{dt}$$

(from Theorem 9.5.1, $\mathbf{T} = d\mathbf{f}/ds$). Then

$$\mathbf{a} = \frac{d\mathbf{v}}{dt} = \frac{d}{dt}\left(\mathbf{T}\frac{ds}{dt}\right) = \frac{d^2s}{dt^2}\mathbf{T} + \frac{ds}{dt}\frac{d\mathbf{T}}{dt}.$$

But

$$\frac{ds}{dt}\frac{d\mathbf{T}}{dt} = \frac{ds}{dt}\left(\frac{d\mathbf{T}}{ds}\frac{ds}{dt}\right) = \left(\frac{ds}{dt}\right)^2\frac{d\mathbf{T}}{ds} = \left(\frac{ds}{dt}\right)^2 \kappa\,\mathbf{n}.$$

The last step follows from the definition of the principal unit normal (on p. 541), and the theorem is proved. ◼

EXAMPLE 3

A 1500-kg race car is driven at a speed of 150 km/hr on an unbanked circular race track of radius 250 m. What frictional force must be exerted by the tires on the road surface to keep the car from skidding?

SOLUTION

The frictional force exerted by the tires must be equal to the component of the force (due to acceleration) normal to the circular race track. That is,

$$F = mv^2\kappa = \frac{mv^2}{\rho} = (1500\,\text{kg})\frac{(150{,}000\,\text{m})^2}{(3600\,\text{sec})^2}\cdot\frac{1}{250\,\text{m}}$$

$$= 10{,}416.7\,(\text{kg})(\text{m})/\text{sec}^2 = 10{,}416.7\,\text{N}.$$

EXAMPLE 4

Coefficient of Friction

Let the car of Example 3 have the **coefficient of friction** μ. That is, the maximum frictional force that can be exerted by the car on the road surface is μmg, where mg is the **normal force** of the car on the road (the force of the car on the road due to gravity). What is the minimum value μ can take in order that the car not slide off the road?

SOLUTION

We must have $\mu mg \geq 10{,}416.7$ N. But $\mu mg = \mu(9.81)(1500)$, so we obtain

$$\mu \geq \frac{10{,}416.7}{(9.81)(1500)} \approx 0.71.$$

All of the properties we have developed above in $\mathbb{R}^2$ hold in $\mathbb{R}^3$ (and also in $\mathbb{R}^n$). Let

$$\mathbf{f}(t) = f_1(t)\mathbf{i} + f_2(t)\mathbf{j} + f_3(t)\mathbf{k};$$

then for the trajectory $\mathbf{f}$ we have

DEFINITION

If $\mathbf{f}'$ and $\mathbf{f}''$ exist, then

(i) $\mathbf{v}(t) = \mathbf{f}'(t) = f'_1(t)\mathbf{i} + f'_2(t)\mathbf{j} + f'_3(t)\mathbf{k}$

is called the **velocity vector**.

Velocity and Acceleration Vectors in $\mathbb{R}^3$

(ii) $\mathbf{a}(t) = \dfrac{d\mathbf{v}}{dt} = \mathbf{f}''(t) = f_1''(t)\mathbf{i} + f_2''(t)\mathbf{j} + f_3''(t)\mathbf{k}$

is called the **acceleration vector.**

Speed and Acceleration Scalar in $\mathbb{R}^3$

(iii) The **speed** of the object $v(t)$ is the magnitude of the velocity vector:

$$v(t) = |\mathbf{v}(t)|.$$

(iv) The **acceleration scalar** $a(t)$ is the magnitude of the acceleration vector:

$$a(t) = |\mathbf{a}(t)|.$$

(v) $\mathbf{a}(t) = \dfrac{dv}{dt}\mathbf{T} + \dfrac{v^2}{\rho}\mathbf{n}.$

The proof in $\mathbb{R}^3$ of **(v)** is identical to the proof of Theorem 1.

EXAMPLE 5

Consider a disk in space centered at the origin that rotates around the z-axis with constant angular speed ω. See Fig. 2. Let M be a particle that moves along a radial line toward the edge of the disk with constant speed v_0. If $\mathbf{x}(t) = (x(t), y(t))$ denotes the position of the particle at time t, we have

$$\mathbf{x}(t) = v_0 t\mathbf{r}, \tag{15}$$

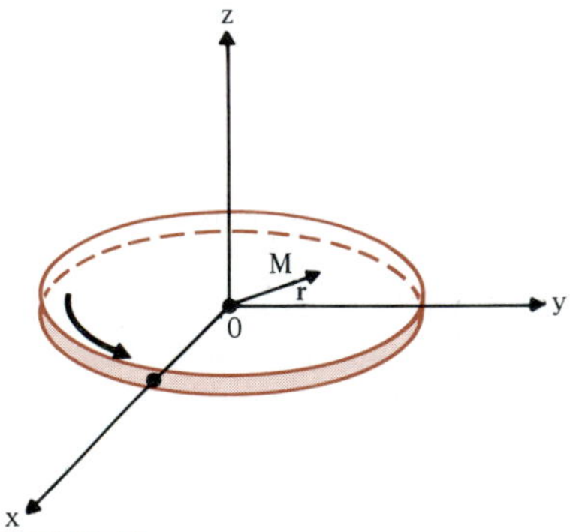

where $\mathbf{r}$ is a unit vector directed away from the origin. Since $\mathbf{r}$ is moving (with the disk) with angular speed ω, we have

$$\mathbf{r}(t) = \cos \omega t\,\mathbf{i} + \sin \omega t\,\mathbf{j}. \tag{16}$$

Thus

$$\mathbf{v}(t) = \mathbf{x}'(t) = v_0 t\mathbf{r}' + v_0\mathbf{r}$$

and

$$\mathbf{a}(t) = v_0 t\mathbf{r}'' + 2v_0\mathbf{r}'. \tag{17}$$

From (16),

$$\mathbf{r}''(t) = -\omega^2(\cos \omega t\,\mathbf{i} + \sin \omega t\,\mathbf{j}) = -\omega^2\mathbf{r}(t),$$

so

$$v_0 t\mathbf{r}''(t) = -\omega^2 v_0 t\mathbf{r} = -\omega^2\mathbf{x}(t).$$

That is, the term $v_0 t\mathbf{r}''$ in (17) is the centripetal acceleration of the particle (see Example 1). This is the acceleration directed toward the center of the disk.

The other expression in (17) is more surprising. The term

$$\mathbf{a}_c = 2v_0\mathbf{r}'$$

Coriolis Acceleration

is called the **Coriolis acceleration.**† It results from both the rotation of the disk and the motion of the particle on the disk. Its direction is in the direction of a tangent vector to the disk because $\mathbf{r}'$ is orthogonal to the radial vector $\mathbf{r}$ (explain why).

FIGURE 2

---

† Named after the French physicist and engineer Gaspard Coriolis (1792–1843). Coriolis first described this type of acceleration in a paper published in 1835.

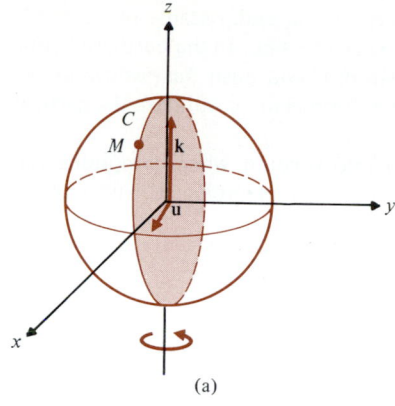

(a)

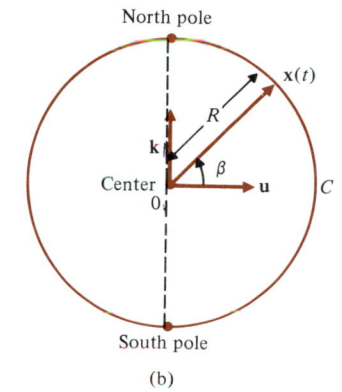

(b)

FIGURE 3

EXAMPLE 6: Coriolis Acceleration on the Earth

Suppose that a particle M of mass m moves along a meridian† C of the earth with constant angular speed β (relative to the earth). Find the acceleration of the particle, taking into account the rotation of the earth.

SOLUTION

The motion is depicted in Fig. 3a. In Fig. 3b we depict the meridian C, inside which a coordinate system is drawn. In the figure, $\mathbf{k}$ is the unit vector drawn from the center of the earth in the direction of the North pole and $\mathbf{u}$ is a unit vector lying in the plane of C and orthogonal to $\mathbf{k}$ (and, therefore, in the xy-plane). If C were in the xy-plane, the motion of M on C would be given by

$$\mathbf{x}(t) = R \cos \beta t\, \mathbf{i} + R \sin \beta t\, \mathbf{j} \tag{18}$$

since the constant angular speed is β. In our case we are in a plane with basis vectors $\mathbf{u}$ and $\mathbf{k}$ (instead of $\mathbf{i}$ and $\mathbf{j}$) so the position vector of M on C is given by

$$\mathbf{x}(t) = R \cos \beta t\, \mathbf{u} + R \sin \beta t\, \mathbf{k}, \tag{19}$$

where R is the radius of the earth.‡ The vector $\mathbf{k}$ is on the axis of rotation of the earth so it does not change as the earth rotates. However, the vector $\mathbf{u}$ rotates and, since $\mathbf{u}$ lies in the xy-plane,

$$\mathbf{u}(t) = \cos \omega t\, \mathbf{i} + \sin \omega t\, \mathbf{j}, \tag{20}$$

where ω is the angular speed of the earth. If time is measured in hours, then $\omega \approx 2\pi/24 = \pi/12$ since the earth rotates 2π radians every 24 hours (approximately).

Now we compute, from (19),

$$\mathbf{v}(t) = \mathbf{x}'(t) = R \cos \beta t\, \mathbf{u}'(t) - \beta R \sin \beta t\, \mathbf{u}(t) + \beta R \cos \beta t\, \mathbf{k},$$

$$\mathbf{a}(t) = \mathbf{x}''(t) = R \cos \beta t\, \mathbf{u}''(t) - 2\beta R \sin \beta t\, \mathbf{u}' - \beta^2 R \cos \beta t\, \mathbf{u} - \beta^2 R \sin \beta t\, \mathbf{k}. \tag{21}$$

But, from (20),

$$\mathbf{u}''(t) = -\omega^2 \mathbf{u}(t)$$

so, from (21) and (19)

$$\mathbf{a}(t) = -\omega^2 R \cos \beta t\, \mathbf{u}(t) - 2\beta R \sin \beta t\, \mathbf{u}'(t) - \beta^2 \mathbf{x}(t). \tag{22}$$

The vector $-\omega^2 R \cos \beta t\, \mathbf{u}$ points toward the earth's axis of rotation and is the centripetal acceleration due to the rotation of the earth. The vector $-\beta^2 \mathbf{x}$ points toward the earth's center and is the centripetal acceleration due to the rotation of M along the great circle containing C. The middle vector in (22), $-2\beta R \sin \beta t\, \mathbf{u}'$, is denoted by $\mathbf{a}_c$. It is tangent to the surface of the earth and is again called Coriolis acceleration. We make two observations: First,

$$|\mathbf{a}_c| = 2|\beta|R|\sin \beta t|.$$

From Fig. 3b we see that $\beta = 0$ at the equator, $\beta = \pi/2$ at the north pole, and $\beta = -\pi/2$ at the South pole. Thus the magnitude of $\mathbf{a}_c$ is largest at the poles and is zero at the equator.

---

† A meridian is half of a great circle on the earth (i.e. a semicircle centered at the center of the earth) that ends at the poles.

‡ R varies between 3950 miles $\simeq 6357$ km at each pole to 3963 miles ≈ 6378 km at the equator. However, in this example we shall assume R is constant.

Coriolis Force

Since the particle has mass m, the **Coriolis force** is ma_c and, because of Newton's third law of motion, the particle will experience a force of $-ma_c$. In the northern hemisphere $\sin \beta t > 0$ and the force $-ma_c = 2\beta mR \sin \beta t\ \mathbf{u}'$ will push the particle to the right. In the southern hemisphere $\sin \beta t < 0$ and the Coriolis force will push the particle to the left.

Therefore, the Coriolis force must be taken into account when computing the trajectories of rockets or missiles. It is also important to meteorologists who wish to track air flow under changes in pressure.

Problems 9.6

In Problems 1–15, the position vector of a moving particle is given. For the indicated value of t, calculate the velocity vector, the acceleration vector, the speed, and the acceleration scalar. Then sketch the portion of the trajectory showing the velocity and acceleration vectors.

1. $\mathbf{f} = (\cos 3t)\mathbf{i} + (\sin 3t)\mathbf{j}$; $t = 0$

2. $\mathbf{f} = (\cos 5t)\mathbf{i} + (\sin 5t)\mathbf{j}$; $t = \pi/2$

3. $\mathbf{f} = 2(\cos 4t)\mathbf{i} + 2(\sin 4t)\mathbf{j}$; $t = \pi/6$

4. $\mathbf{f} = -3(\cos 10t)\mathbf{i} - 3(\sin 10t)\mathbf{j}$; $t = \pi$

5. $\mathbf{f} = 4t\mathbf{i} + 2t^2\mathbf{j}$; $t = 1$

6. $\mathbf{f} = (2 + 3t)\mathbf{i} + (8 - 5t)\mathbf{j}$; $t = 3$

7. $\mathbf{f} = (4 - 7t)\mathbf{i} + (-3 + 5t)\mathbf{j}$; $t = -5$

8. $\mathbf{f} = (a + bt)\mathbf{i} + (c + dt)\mathbf{j}$; $t = t_0$, a, b, c, d, real

9. $\mathbf{f} = 2t\mathbf{i} + (e^{-t} + e^t)\mathbf{j}$; $t = 0$

10. $\mathbf{f} = t^2\mathbf{i} + t^3\mathbf{j} + t^4\mathbf{k}$; $t = 2$

11. $\mathbf{f} = (\cos t)\mathbf{i} + (\sin t)\mathbf{j} + t^4\mathbf{k}$; $t = 1$

12. $\mathbf{f} = (\ln t)\mathbf{i} + \dfrac{1}{t}\mathbf{j} + \dfrac{1}{t^2}\mathbf{k}$; $t = 1$

13. $\mathbf{f} = (\cosh t)\mathbf{i} + (\sinh t)\mathbf{j} + t\mathbf{k}$; $t = 0$

14. $\mathbf{f} = e^t\mathbf{i} + e^{-t}\mathbf{j} + \sqrt{t}\mathbf{k}$; $t = 4$

15. $\mathbf{f} = \sqrt{t}\mathbf{i} + \sqrt[3]{t}\mathbf{j} + \sqrt[4]{t}\mathbf{k}$; $t = 1$

* 16. A bullet is shot from a gun with an initial velocity of 1200 m/sec. The gun is inclined at an angle of 45°. Find (a) the total distance traveled by the bullet, (b) the horizontal distance traveled by the bullet, (c) the maximum height it reaches, and (d) the speed of the bullet at impact. Assume that the gun is held at ground level.

* 17. Answer the questions of Problem 16 if the angle of inclination of the gun is 60° and the initial velocity is 3000 ft/sec.

18. A man is standing at the top of a 200-m building. He throws a ball horizontally with an initial speed of 20 m/sec.

(a) How far does the ball travel in its path?

(b) How far from the base of the building does the ball hit the ground?

(c) At what angle with the horizontal does the ball hit the ground?

(d) With what speed does the ball hit the ground?

19. An airplane, flying horizontally at a height of 1500 m and with a speed of 450 km/hr, releases a bomb.

(a) How long does it take the bomb to hit the ground?

(b) How far does it travel in the horizontal direction?

(c) What is its speed of impact?

20. A 2000-kg rocket carrying 3000 kg of fuel is initially motionless in space. Its exhaust velocity is 1 km/sec. What is its velocity when all its fuel is consumed?

In Problems 21–28, find the tangential and normal components of acceleration for each of the given position vectors.

21. $\mathbf{f} = (\cos 2t)\mathbf{i} + (\sin 2t)\mathbf{j}$

22. $\mathbf{f} = 2(\cos t)\mathbf{i} + 3(\sin t)\mathbf{j}$

23. $\mathbf{f} = t\mathbf{i} + t^2\mathbf{j}$

24. $\mathbf{f} = t\mathbf{i} + (\cos t)\mathbf{j}$

25. $\mathbf{f} = (t^3 - 3t)\mathbf{i} + (t^2 - 1)\mathbf{j}$

26. $\mathbf{f} = e^{-t}\mathbf{i} + e^t\mathbf{j}$

27. $\mathbf{f} = t^2\mathbf{i} + t^3\mathbf{j}$

28. $\mathbf{f} = (\sin t^2)\mathbf{i} + (\cos t^2)\mathbf{j}$

29. Show that if a particle is moving at a constant speed, then the tangential component of acceleration is zero.

30. Suppose that the driver of the car of Examples 3 and 4 reduces his speed by a factor of M. Show that the frictional force needed to keep the car from skidding is reduced by a factor of M^2.

31. A truck traveling at 80 km/hr and weighing 10,000 kg is moving on a curved unbanked stretch of track. The equation of the curved section is the parabola $y = x^2 - x$ meters.

What is the frictional force exerted by the wheels of the truck on the track at the "point" (0, 0)?

32. If the coefficient of friction for the truck in Problem 31 is 2.5, what is the maximum speed it can achieve at the point (0, 0) without going off the road?

33. If the race car of Example 3 is placed on a track with half the radius of the original one, how much slower would it have to be driven so as not to increase the normal component of acceleration?

* 34. A woman swings a rope attached to a bucket containing 3 kg of water. The pail rotates in the vertical plane in a circular path with a radius of 1 m. What is the smallest number of revolutions that must be made every minute in order that the water stay in the pail? [*Hint:* Determine the centripetal acceleration.]

35. Show in Example 1 that the magnitude a_r of the centripetal acceleration is given by $a_r = v^2/r$, where v is the speed and r is the distance of the particle from the origin.

In Problems 36–39 compute the Coriolis acceleration when the motion $v_0 t \mathbf{r}$ in Example 5 is replaced by the given displacement vector.

36. $5 v_0 t \mathbf{r}$

37. $v_0 t^2 \mathbf{r}$

38. $v_0 t^5 \mathbf{r}$

39. $\dfrac{v_0}{1+t} \mathbf{r}$

* 40. Compute the Coriolis acceleration if a particle moves at constant angular speed on a great circle C of the earth that does not pass through the poles. [*Hint:* Write the motion on C in terms of two orthogonal vectors $\mathbf{u}$ and $\mathbf{v}$ lying in the plane of C. Neither vector is constant because neither one lies on the axis of rotation.]

9.7 The Chain Rule and the Mean Value Theorem

In this section we discuss some properties of functions of several variables that we will need in the following sections. For simplicity we will discuss these concepts for functions of two variables; the generalization to functions of three or more variables will be readily apparent. Some familiarity by the student with the concepts of a first course in calculus will be assumed.

Open and Closed Disks; Open Set

(i) The **open disk** D_r, centered at (x_0, y_0) with radius r is the subset of $\mathbb{R}^2$ given by

$$\{(x, y) : (x - x_0)^2 + (y - y_0)^2 < r^2\}.$$

(ii) The **closed disk** centered at (x_0, y_0) with radius r is the subset of $\mathbb{R}^2$ given by

$$\{(x, y) : (x - x_0)^2 + (y - y_0)^2 \leq r^2\}.$$

(iii) The **boundary** of the open or closed disk defined in (i) or (ii) is the circle

$$\{(x, y) : (x - x_0)^2 + (y - y_0)^2 = r^2\}.$$

(iv) A **neighborhood** of a point (x_0, y_0) in $\mathbb{R}^2$ is any set containing an open disk centered at (x_0, y_0).

(v) A set Ω in $\mathbb{R}^2$ is said to be **open** if for every point (x_0, y_0) in Ω, there is an open disk D_r centered at (x_0, y_0) such that D_r is wholly contained in Ω.

Remark. In this definition the words "open" and "closed" have meanings very similar to their meanings in the terms "open interval" and "closed interval." An open interval does not contain its endpoints. An open disk does not contain any point on its boundary. Similarly, a closed interval contains all its boundary points, as does a closed disk. (See Fig. 1.)

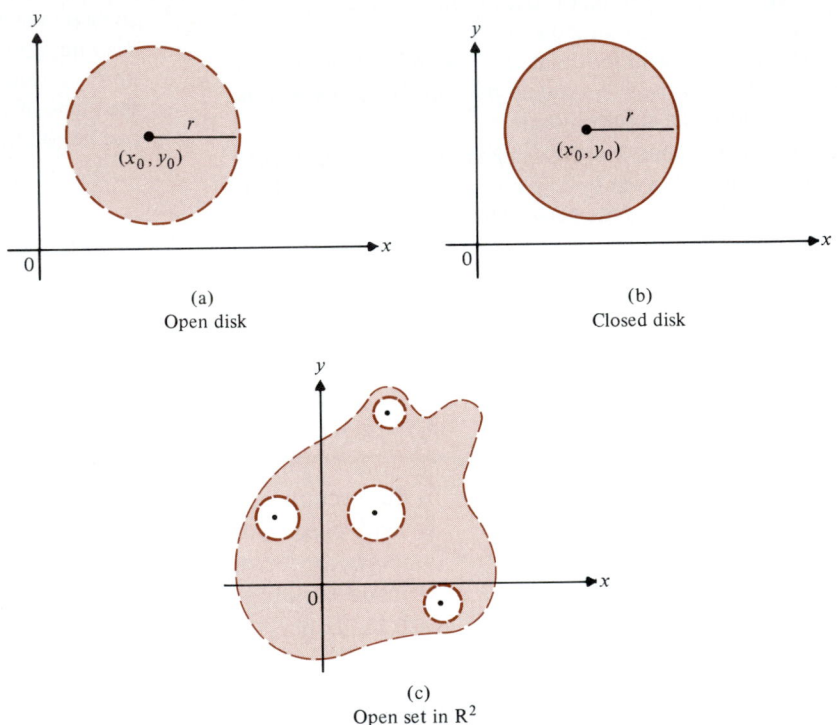

(a)
Open disk

(b)
Closed disk

(c)
Open set in $\mathbb{R}^2$

FIGURE 1

We briefly review the notions of limit, continuity, and partial differentiation. Intuitively, we say that $f(x, y)$ approaches the limit L as (x, y) approaches (x_0, y_0) if $f(x, y)$ gets arbitrarily "close" to L as (x, y) approaches (x_0, y_0). We define this notion precisely below.

DEFINITION

Let $f(x, y)$ be defined in a neighborhood of (x_0, y_0) but not necessarily at (x_0, y_0) itself. Then the **limit** of $f(x, y)$ as (x, y) approaches (x_0, y_0) is L, written

Limit

$$\lim_{(x,y) \to (x_0,y_0)} f(x, y) = L,$$

if for every number $\epsilon > 0$, there is a number $\delta > 0$ such that $|f(x, y) - L| < \epsilon$ for every $(x, y) \neq (x_0, y_0)$ in the open disk centered at (x_0, y_0) with radius δ. This definition is illustrated in Fig. 2.

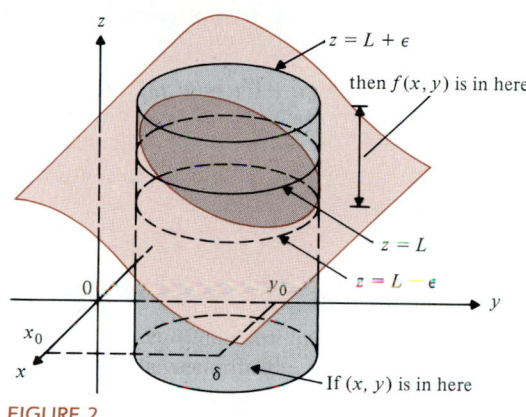

FIGURE 2

DEFINITION

Continuity

(i) Let $f(x, y)$ be defined at every point (x, y) in a neighborhood of (x_0, y_0). Then f is **continuous** at (x_0, y_0) if

$$\lim_{(x,y) \to (x_0, y_0)} f(x, y) = f(x_0, y_0).$$

Otherwise, f is said to be **discontinuous** at (x_0, y_0).

(ii) f is **continuous in a subset** S of $\mathbb{R}^2$ if it is continuous at every point (x, y) in S.

DEFINITION

Let $z = f(x, y)$.

Partial Derivatives

(i) The **partial derivative of f with respect to** x is the function

$$f_x = \frac{\partial z}{\partial x} = \frac{\partial f}{\partial x} = \lim_{\Delta x \to 0} \frac{f(x + \Delta x, y) - f(x, y)}{\Delta x}. \tag{1}$$

$\partial f / \partial x$ is defined at every point (x, y) in the domain of f where the limit (1) exists.

(ii) The **partial derivative of f with respect to** y is the function

$$f_y = \frac{\partial z}{\partial y} = \frac{\partial f}{\partial y} = \lim_{\Delta y \to 0} \frac{f(x, y + \Delta y) - f(x, y)}{\Delta y}. \tag{2}$$

$\partial f / \partial y$ is defined at every point (x, y) in the domain of f where the limit (2) exists.

Remark. This definition allows us to calculate partial derivatives in the same way we calculate ordinary derivatives by allowing only one of the variables to vary. It also allows us to use all the formulas from one-variable calculus.

DEFINITION

Second Partial Derivatives

A **second partial derivative** is defined as a partial derivative of a first partial derivative.

$$f_{xx} = (f_x)_x, \qquad f_{xy} = (f_x)_y,$$
$$f_{yx} = (f_y)_x, \qquad f_{yy} = (f_y)_y.$$

The derivatives f_{xy} and f_{yx} are called the **mixed second partials.**

The proof of the following theorem can be found in most intermediate calculus texts.†

■ **THEOREM 1: Equality of Mixed Partials**
Suppose that f, f_x, f_y, f_{xy}, and f_{yx} are all continuous at (x_0, y_0). Then

$$f_{xy}(x_0, y_0) = f_{yx}(x_0, y_0). \tag{3}$$

■

Several chain rules hold for partial derivatives.

■ **THEOREM 2: Chain Rule**
Let $z = f(x, y)$ be continuous and have continuous first partial derivatives at every point of an open disk D. Suppose $x = x(t)$ and $y = y(t)$ are differentiable functions for all t in some open interval I. Assume further that for each t in I, the point $(x(t), y(t))$ lies in D. Then $z = f(x(t), y(t))$ is a differentiable function for all t in I and

$$\frac{dz}{dt} = \frac{\partial z}{\partial x}\frac{dx}{dt} + \frac{\partial z}{\partial y}\frac{dy}{dt} = f_x\frac{dx}{dt} + f_y\frac{dy}{dt}. \tag{4}$$

† See, for example, S. Grossman, *Multivariable Calculus, Linear Algebra, and Differential Equations,* 2nd ed., Harcourt Brace Jovanovich, San Diego, CA, 1986, p. 195.

PROOF

Let t belong to I and choose Δt so small that $t + \Delta t$ also belongs to I. Define

$$\Delta x = x(t + \Delta t) - x(t), \qquad \Delta y = y(t + \Delta t) - y(t)$$

and

$$\Delta z = f(x(t) + \Delta x, y(t) + \Delta y) - f(x(t), y(t)). \tag{5}$$

Adding and subtracting terms we may rewrite (5) as

$$\Delta z = [f(x + \Delta x, y + \Delta y) - f(x, y + \Delta y)] + [f(x, y + \Delta y) - f(x, y)].$$

Applying the mean value theorem† for a function of a single variable to each term in brackets, we obtain

$$\Delta z = \Delta x \frac{\partial f}{\partial x}\bigg|_{(x_0, y + \Delta y)} + \Delta y \frac{\partial f}{\partial y}\bigg|_{(x, y_0)},$$

where x_0 lies between $x(t)$ and $x(t) + \Delta x$ and y_0 lies between $y(t)$ and $y(t) + \Delta y$. Dividing both sides by Δt and taking the limit as $\Delta t \to 0$, we have

$$\lim_{\Delta t \to 0} \frac{\Delta z}{\Delta t} = \lim_{\Delta t \to 0} \left[\frac{\Delta x}{\Delta t} \frac{\partial f}{\partial x}\bigg|_{(x_0, y + \Delta y)} + \frac{\Delta y}{\Delta t} \frac{\partial f}{\partial y}\bigg|_{(x, y_0)} \right].$$

Since x and y are differentiable and f_x and f_y are continuous in D, we obtain

$$\frac{dz}{dt} = \frac{dx}{dt} \frac{\partial f}{\partial x} + \frac{dy}{dt} \frac{\partial f}{\partial y}. \qquad \blacksquare$$

Theorem 2 (and its proof) can be immediately extended:

■ **THEOREM 3**

Let $z = f(x, y)$ be continuous and have continuous first partial derivatives at every point of an open disk D. Suppose $x = x(r, s)$ and $y = y(r, s)$ have first partial derivatives in an open disk D^*, such that for any point (r, s) in D^*, the point $(x(r, s), y(r, s))$ lies in D. Then the function $z = f(x(r, s), y(r, s))$ has first partials with respect to r and s in D^* and

$$\frac{\partial z}{\partial r} = \frac{\partial z}{\partial x} \frac{\partial x}{\partial r} + \frac{\partial z}{\partial y} \frac{\partial y}{\partial r} \tag{6}$$

$$\frac{\partial z}{\partial s} = \frac{\partial z}{\partial x} \frac{\partial x}{\partial s} + \frac{\partial z}{\partial y} \frac{\partial y}{\partial s}. \tag{7}$$

$\blacksquare$

---

† The mean value theorem of one-variable calculus says that if f is continuous on $[a, b]$ and differentiable on (a, b), then there is a number c in (a, b) such that $f(b) - f(a) = f'(c)(b - a)$.

EXAMPLE 1

The radius of a right circular cone is increasing at a rate of 3 cm/sec, while its height is increasing at a rate of 5 cm/sec. How fast is the volume increasing when $r = 15$ cm and $h = 25$ cm?

SOLUTION

We are asked to find dV/dt, where $V = \frac{1}{3}\pi r^2 h$. But

$$\frac{dV}{dt} = \frac{\partial V}{\partial r}\frac{dr}{dt} + \frac{\partial V}{\partial h}\frac{dh}{dt} = \frac{2}{3}\pi rh(3) + \frac{1}{3}\pi r^2(5)$$

$$= 2\pi rh + \frac{5}{3}\pi r^2 = \pi[2 \cdot 15 \cdot 25 + \frac{5}{3}(15)^2]$$

$$= 1125\pi \text{ cm}^3/\text{sec}.$$

EXAMPLE 2

According to the ideal gas law, the pressure, volume, and absolute temperature of n moles of an ideal gas are related by

$$PV = nRT,$$

where R is a constant. Suppose that the volume of an ideal gas is increasing at a rate of 10 cm^3/min and the pressure is decreasing at a rate of 0.3 N/cm^2/min. How is the temperature of the gas changing when the volume of 5 mol of a gas is 100 cm^3 and the pressure is 2 N/cm^2?

SOLUTION

We have $T = PV/nR$, where $n = 5$. Then

$$\frac{dT}{dt} = \frac{\partial T}{\partial P}\frac{dP}{dt} + \frac{\partial T}{\partial V}\frac{dV}{dt} = \frac{V}{nR}(-0.3) + \frac{P}{nR}(10) = \frac{100}{5R}(-0.3) + \frac{2}{5R}(10)$$

$$= \frac{-2}{R} \,{}^\circ\text{K/min}.$$

As we saw in the proof of Theorem 2, the mean value theorem for a differentiable function $f(x)$ of a single variable states that

$$f(x + h) - f(x) = hf'(c), \tag{8}$$

where c lies between x and $x + h$. We can extend this result to functions of two variables. We first need a definition.

DEFINITION

Line Segment

Let $\mathbf{x}$ and $\mathbf{y}$ be two vectors in $\mathbb{R}^n$. Then the **line segment** L joining $\mathbf{x}$ and $\mathbf{y}$ is the set defined by

$$L = \{\mathbf{v} : \mathbf{v} = t\mathbf{x} + (1 - t)\mathbf{y} \qquad \text{for } 0 \le t \le 1\}.$$

Note: Both **x** and **y** are on *L*, as is seen by setting $t = 0$ and $t = 1$ in the equation for *L*.

■ **THEOREM 4: Mean Value Theorem**

Let $f(x, y)$ be continuous and have continuous first partial derivatives in an open disk *D*. Let (x_0, y_0) and $(x_0 + h, y_0 + k)$ be points in *D*. Then there is a point (x^*, y^*) on the line segment joining (x_0, y_0) to $(x_0 + h, y_0 + k)$ such that

$$f(x_0 + h, y_0 + k) - f(x_0, y_0) = h\frac{\partial f}{\partial x}(x^*, y^*) + k\frac{\partial f}{\partial y}(x^*, y^*). \tag{9}$$

PROOF

Define $F(t) = f(x_0 + th, y_0 + tk)$, $0 \le t \le 1$, then

$$F(0) = f(x_0, y_0) \quad \text{and} \quad F(1) = f(x_0 + h, y_0 + k).$$

Now *F* depends only on the single variable *t*, and we have

$$f(x_0 + h, y_0 + k) - f(x_0, y_0) = F(1) - F(0). \tag{10}$$

Then observe that from the definition of $F(t)$, $F(t)$ satisfies all the hypotheses of the mean value theorem since $x(t) = x_0 + th$ and $y(t) = y_0 + tk$ are differentiable functions in *t* and $f(x, y)$ has continuous first partials. Thus there is a number t_0 in $(0, 1)$ such that

$$F(1) - F(0) = F'(t_0) = \frac{dF}{dt} = f_x\frac{dx}{dt} + f_y\frac{dy}{dt}. \tag{11}$$

$$\uparrow$$
$$\text{by (4)}$$

But

$$\frac{dx}{dt} = \frac{d}{dt}(x_0 + th) = h, \qquad \frac{dy}{dt} = \frac{d}{dt}(y_0 + tk) = k \tag{12}$$

so that if we substitute (12) in (11) this together with (10) yields the desired equation (9), and the proof is complete. ■

Problems 9.7

In Problems 1–4 use the definition to verify the given limit.

1. $\lim\limits_{(x,y)\to(1,2)} (3x + y) = 5$

2. $\lim\limits_{(x,y)\to(3,-1)} (x - 7y) = 10$

3. $\lim\limits_{(x,y)\to(5,-2)} (ax + by) = 5a - 2b$

4. $\lim\limits_{(x,y)\to(1,1)} \dfrac{x}{y} = 1$

In Problems 5–8 show that the indicated limit does not exist. Explain why. [*Hint:* Consider the quotient along various curves that approach the origin.]

5. $\lim\limits_{(x,y)\to(0,0)} \dfrac{x + y}{x - y}$

6. $\lim\limits_{(x,y)\to(0,0)} \dfrac{xy}{x^2 - y^2}$

7. $\displaystyle\lim_{(x,y)\to(0,0)}\frac{x^2-y^2}{x^2+y^2}$

8. $\displaystyle\lim_{(x,y,z)\to(0,0,0)}\frac{yx+z^2}{x^2+y^2+z^2}$

9. Find a function $g(x)$ such that the function

$$f(x, y) = \begin{cases} \dfrac{x^2-y^2}{x-y}, & x \neq y \\ g(x), & x = y \end{cases}$$

is continuous at every point in $\mathbb{R}^2$.

10. Find a number c such that the function

$$f(x, y) = \begin{cases} \dfrac{3xy}{\sqrt{x^2+y^2}}, & (x, y) \neq (0,0) \\ c, & (x, y) = (0,0) \end{cases}$$

is continuous at the origin.

In Problems 11–14 calculate $\partial z/\partial x$ and $\partial z/\partial y$.

11. $z = x^2 y$

12. $z = \sin(x^2 + y^3)$

13. $z = \sqrt{xy + 2y^3}$

14. $z = \ln(x^3 y^5 - 2)$

15. Let $f(x, y) = \begin{cases} \dfrac{x+y}{x-y}, & x \neq y \\ 0, & x = y \end{cases}$

    (a) Show that f is not continuous at $(0, 0)$. [*Hint:* Show that $\lim_{(x,y)\to(0,0)} f(x, y)$ does not exist.]

    (b) Do $f_x(0, 0)$ and $f_y(0, 0)$ exist?

16. A **partial differential equation** is an equation involving partial derivatives. Show that the function $z = f(x, y) = e^{(x+\sqrt{3}y)/4} - 4x - 2y - 4 - 2\sqrt{3}$ satisfies the partial differential equation $\partial z/\partial x + \sqrt{3}\partial z/\partial y - z = 4x + 2y$.

In Problems 17–20, use the chain rule to calculate dz/dt. Check your answer by first writing z or w as a function of t and then differentiating.

17. $z = xy$, $x = e^t$, $y = e^{2t}$

18. $z = x^2 + y^2$, $x = \cos t$, $y = \sin t$

19. $z = \dfrac{y}{x}$, $x = t^2$, $y = t^3$

20. $z = e^x \sin y$, $x = \sqrt{t}$, $y = \sqrt[3]{t}$

In Problems 21–24 use the chain rule to calculate the indicated partial derivatives.

21. $z = xy$; $x = r + s$; $y = r - s$; $\partial z/\partial r$ and $\partial z/\partial s$

22. $z = x^2 + y^2$; $x = \cos(r + s)$; $y = \sin(r - s)$; $\partial z/\partial r$ and $\partial z/\partial s$

23. $z = \dfrac{y}{x}$; $x = e^r$; $y = e^s$; $\dfrac{\partial z}{\partial r}$ and $\dfrac{\partial z}{\partial s}$

24. $z = \sin\dfrac{y}{x}$; $x = \dfrac{r}{s}$; $y = \dfrac{s}{r}$; $\dfrac{\partial z}{\partial r}$ and $\dfrac{\partial z}{\partial s}$

25. Let $w = f(x, y, z)$ be differentiable and let $x = r \cos \theta$, $y = r \sin \theta$, and $z = t$. Calculate $\partial w/\partial r$, $\partial w/\partial\theta$, and $\partial w/\partial t$. (These are cylindrical coordinates.)

26. If a particle is falling in a fluid, then according to *Stokes' law* the velocity of the particle is given by

$$V = \frac{2g}{9}(\rho_P - \rho_f)\frac{r^2}{\eta},$$

where g is the acceleration due to gravity, ρ_P is the density of the particle, ρ_f is the density of the fluid, r is the radius of the particle (in centimeters), and η is the absolute viscosity of the liquid. Calculate $V\rho_P$, $V\rho_f$, V_r, and V_η.

27. The pressure of 8 mol of an ideal gas is decreasing at a rate of 0.4 N/cm²/min, while the temperature is decreasing at a rate of 0.5 °K/min. How fast is the volume of the gas changing when $V = 1000$ cm³ and $P = 3$ N/cm²? Is the volume increasing or decreasing? Use the value $R = 8.314$ J/mol °K.

28. The angle A of a triangle ABC is increasing at a rate of $3°$/sec, the side AB is increasing at a rate of 1 cm/sec, and the side AC is decreasing at a rate of 2 cm/sec. How fast is the side BC changing when $A = 30°$, $AB = 10$ cm, and $AC = 24$ cm? Is the length of BC increasing or decreasing? [*Hint:* Use the law of cosines and convert degrees to radians.]

29. The partial differential equation

$$\frac{\partial^2 y}{\partial t^2} = c^2 \frac{\partial^2 y}{\partial x^2}$$

is called the *wave equation*. Show that if f is any twice differentiable function, then $y(x, t) = \frac{1}{2}[f(x - ct) + f(x + ct)]$ is a solution to this equation.

* 30. Let $z = f(x, y)$ be differentiable. Write the expression $(\partial z/\partial x)^2 + (\partial z/\partial y)^2$ in terms of polar coordinates r and θ.

* 31. Let $z = f(x, y)$, let $x = x(r, s)$, and let $y = y(r, s)$. Show that

$$\frac{\partial^2 z}{\partial r^2} = \frac{\partial}{\partial x}\left(\frac{\partial f}{\partial x}\frac{\partial x}{\partial r} + \frac{\partial f}{\partial y}\frac{\partial y}{\partial r}\right)\left(\frac{\partial x}{\partial r}\right) + \frac{\partial}{\partial y}\left(\frac{\partial f}{\partial x}\frac{\partial x}{\partial r} + \frac{\partial f}{\partial y}\frac{\partial y}{\partial r}\right)\left(\frac{\partial y}{\partial r}\right).$$

** **32.** The partial differential equation

$$\frac{\partial^2 f}{\partial x^2} + \frac{\partial^2 f}{\partial y^2} = 0.$$

is called *Laplace's equation*. If we write (x, y) in polar coordinates ($x = r \cos \theta$, $y = r \sin \theta$), show that Laplace's equation becomes

$$\frac{\partial^2 f}{\partial r^2} + \frac{1}{r^2} \frac{\partial^2 f}{\partial \theta^2} + \frac{1}{r} \frac{\partial f}{\partial r} = 0.$$

[*Hint:* Write $\partial^2 f/\partial x^2$ and $\partial^2 f/\partial y^2$ in terms of r and θ, using the result of Problem 31.]

9.8 Differentiability, Gradients, and Directional Derivatives

In Section 9.7 we discussed some of the properties of the partial derivatives of a differentiable scalar field f. In this section we study a vector field that can be associated to any such scalar field. For simplicity, we again limit consideration to functions of two or three variables.

The Gradient in $\mathbb{R}^2$

DEFINITION

Let f be a function of two variables such that f_x and f_y exist at a point $\mathbf{x} = (x, y)$. Then the **gradient** of f at $\mathbf{x}$, denoted $\nabla f(\mathbf{x})$, or grad $\mathbf{f}$, is given by

$$\nabla f(\mathbf{x}) = f_x(x, y)\mathbf{i} + f_y(x, y)\mathbf{j}. \tag{1}$$

That is, the gradient is the vector of first partial derivatives of f at $\mathbf{x}$.

Note that the gradient of f is a vector field, as defined on p. 514. That is, for every point $\mathbf{x}$ in $\mathbb{R}^2$ for which $\nabla f(\mathbf{x})$ is defined, $\nabla f(\mathbf{x})$ is a vector in $\mathbb{R}^2$.

The Gradient in $\mathbb{R}^3$

DEFINITION

Let f be a function of three variables whose first partial derivatives exist at the point $\mathbf{x} = (x, y, z)$. Then the gradient of f is given by the vector field

$$\nabla f(\mathbf{x}) = f_x(\mathbf{x})\mathbf{i} + f_y(\mathbf{x})\mathbf{j} + f_z(\mathbf{x})\mathbf{k}. \tag{2}$$

Remark. The gradient of f is denoted ∇f, which is read "del" f. This symbol, an inverted Greek delta, was first used in the 1850s, although the name "del" first appeared in print only in 1901. The symbol ∇ is also called *nabla*. This name is used because someone once suggested to the Scottish mathematician Peter Guthrie Tait (1831–1901) that ∇ looks like an Assyrian harp, the Assyrian name of which is nabla.† Tait, incidentally, was one of the mathematicians who helped carry on Hamilton's development of the theory of quaternions and vectors in the nineteenth century.

EXAMPLE 1

Let $f(x, y) = xy$. Then

$$\nabla f(x, y) = f_x \mathbf{i} + f_y \mathbf{j} = y\mathbf{i} + x\mathbf{j}.$$

Let f and g be differentiable scalar fields whose first partial derivatives exist in a neighborhood of $\mathbf{x}$. Then for every scalar α it is easy to see, using the definition of gradient, that αf and $f + g$ are differentiable at $\mathbf{x}$, and

> **(i)** $\nabla(\alpha f) = \alpha \nabla f$, and
> **(ii)** $\nabla(f + g) = \nabla f + \nabla g$.   (3)

Using the terminology of Section 7.11, we see that the gradient is a linear operator. Linear operators play an extremely important role in functional analysis as well as in the theory of vector spaces.

If we use the notation

$$\Delta \mathbf{x} = (\Delta x, \Delta y) = \Delta x \mathbf{i} + \Delta y \mathbf{j},$$

then we have

$$\nabla f(\mathbf{x}) \cdot \Delta \mathbf{x} = (f_x \mathbf{i} + f_y \mathbf{j}) \cdot (\Delta x \mathbf{i} + \Delta y \mathbf{j}) = f_x(x, y)\Delta x + f_y(x, y)\Delta y.$$

Also,

$$f(x + \Delta x, y + \Delta y) = f(\mathbf{x} + \Delta \mathbf{x}).$$

In Section 9.7 (p. 559) we proved a mean value theorem for functions of two variables. We showed that if f is continuous with continuous first partial derivatives in an open disk D and if (x_0, y_0) and $(x_0 + h, y_0 + k)$ are in D, then there is a point (x^*, y^*) on the line segment joining (x_0, y_0) and $(x_0 + h, y_0 + k)$ such that

$$f(x_0 + h, y_0 + k) - f(x_0, y_0) = h \frac{\partial f}{\partial x}(x^*, y^*) + k \frac{\partial f}{\partial y}(x^*, y^*).$$

Setting $\mathbf{x} = (x_0, y_0)$, $\Delta x = (h, k)$, and $x^* = (x^*, y^*)$, we can rewrite this result as

---

† Fortunately, most (but certainly not all) of the mathematical terms currently in use have more to do with the objects they describe. We might further point out that ∇ is also called *atled,* which is delta spelled backward.

$$f(\mathbf{x} + \Delta\mathbf{x}) - f(\mathbf{x}) = \nabla f(\mathbf{x}^*) \cdot \Delta\mathbf{x}, \tag{4}$$

where $\mathbf{x}^*$ is a point on the line segment joining $\mathbf{x}$ to $\mathbf{x} + \Delta\mathbf{x}$. If the first partial derivatives of f are continuous functions, then the vector field ∇f is continuous and we can rewrite (4) as

$$f(\mathbf{x} + \Delta\mathbf{x}) - f(\mathbf{x}) = \nabla f(\mathbf{x}) \cdot \Delta\mathbf{x} + g(\Delta\mathbf{x}), \tag{5}$$

where

$$\frac{g(\Delta\mathbf{x})}{|\Delta\mathbf{x}|} = [\nabla f(\mathbf{x}^*) - \nabla f(\mathbf{x})] \cdot \frac{\Delta\mathbf{x}}{|\Delta\mathbf{x}|} \to 0 \tag{6}$$

as $\Delta\mathbf{x} \to \mathbf{0}$, since $\mathbf{x}^* \to \mathbf{x}$. This suggests the following definition of differentiability of a scalar field:

DEFINITION

Differentiability

Let f be a function of two variables that is defined in a neighborhood of a point $\mathbf{x} = (x, y)$. Let $\Delta\mathbf{x} = (\Delta x, \Delta y)$. If $f_x(x, y)$ and $f_y(x, y)$ are continuous, then f is **differentiable** at $\mathbf{x}$ if there is a scalar function g such that

$$f(\mathbf{x} + \Delta\mathbf{x}) - f(\mathbf{x}) = \nabla f(\mathbf{x}) \cdot \Delta\mathbf{x} + g(\Delta\mathbf{x}), \tag{7}$$

where

$$\lim_{\Delta\mathbf{x} \to \mathbf{0}} \frac{g(\Delta\mathbf{x})}{|\Delta\mathbf{x}|} = 0. \tag{8}$$

In your one-variable calculus course you saw that every differentiable function is continuous but not every continuous function is differentiable. For example, $f(x) = |x|$ is continuous at 0, but $f'(0)$ is undefined since

$$f'(x) = \begin{cases} 1, & x > 0, \\ -1, & x < 0. \end{cases}$$

We now show that the same statements can be made about a function of two or more variables.

The following theorem follows immediately from (7) by letting $\Delta\mathbf{x} \to \mathbf{0}$.

■ **THEOREM 1**

If f is differentiable at $\mathbf{x}$, then f is continuous at $\mathbf{x}$. ■

It is not difficult to show that continuity does not imply differentiability.

For example, $f(x, y) = |x| + y$ is continuous but not differentiable at $(0, 0)$. We can make an even stronger statement. The function

$$f(x, y) = \begin{cases} \dfrac{xy}{x^2 + y^2}, & (x, y) \neq (0, 0), \\[2mm] 0, & (x, y) = (0, 0) \end{cases}$$

is not even continuous at $(0, 0)$ although both f_x and f_y exist at $(0, 0)$ (see Problem 28). Thus *existence of the partial derivatives does not ensure differentiability.* However, the following result can be proved.†

■ **THEOREM 2**

If f and its first partial derivatives are defined and continuous at (x_0, y_0), then f is differentiable at (x_0, y_0). ■

Let $\Delta\mathbf{x}$ be parallel to the x-axis: $\Delta\mathbf{x} = (\Delta x, 0)$. Then (7) becomes

$$\frac{f(\mathbf{x} + \Delta\mathbf{x}) - f(\mathbf{x})}{|\Delta\mathbf{x}|} = \nabla f(\mathbf{x}) \bullet \frac{\Delta\mathbf{x}}{|\Delta\mathbf{x}|} + \frac{g(\Delta\mathbf{x})}{|\Delta\mathbf{x}|}. \tag{9}$$

Taking the limit of both sides of (9) as $|\Delta\mathbf{x}| = \Delta x \to 0$ we get

$$\lim_{\Delta x \to 0} \frac{f(\mathbf{x} + \Delta\mathbf{x}) - f(\mathbf{x})}{|\Delta\mathbf{x}|} = \nabla f(\mathbf{x}) \bullet (1, 0) = f_x(\mathbf{x}).$$

Similarly, if $\Delta\mathbf{x} = (0, \Delta y)$, the limit of both sides of (9) is $f_y(\mathbf{x})$. Thus (7) provides a way of obtaining the rate of change of f along directions parallel to the coordinate axes. However, it *also* provides a way of finding the rate of change of f in *any* direction. Let $\Delta\mathbf{x}$ be any vector, and set $t = |\Delta\mathbf{x}|$. Then

$$\Delta\mathbf{x} = |\Delta\mathbf{x}| \frac{\Delta\mathbf{x}}{|\Delta\mathbf{x}|} = t\mathbf{u}, \tag{10}$$

where $\mathbf{u}$ is a unit vector. Replacing $\Delta\mathbf{x}$ in (9) by (10) we get

$$\frac{f(\mathbf{x} + t\mathbf{u}) - f(\mathbf{x})}{t} = \nabla f(\mathbf{x}) \bullet \mathbf{u} + \frac{g(t\mathbf{u})}{t}. \tag{11}$$

Note that $f(\mathbf{x} + t\mathbf{u})$ is the value of f at a point t units away from $\mathbf{x}$ in the direction $\mathbf{u}$ (see Fig. 1). If we let $z(t) = f(\mathbf{x} + t\mathbf{u})$, then (11) becomes

$$\frac{z(t) - z(0)}{t} = \nabla f(\mathbf{x}) \bullet \mathbf{u} + \frac{g(t\mathbf{u})}{t}. \tag{12}$$

Taking the limit as $t \to 0$ of both sides of (12) we get

$$z'(0) = \frac{dz}{dt}\bigg|_{t=0} = \nabla f(\mathbf{x}) \bullet \mathbf{u},$$

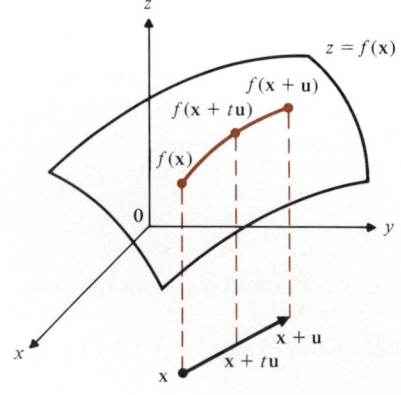

FIGURE 1

† See R. C. Buck and E. F. Buck, *Advanced Calculus,* 3rd ed., McGraw-Hill, New York, 1978, p. 131.

which is the instantaneous rate of change of f in the direction **u**. This leads to the following definition.

DEFINITION

Directional Derivative

Let f be differentiable at a point $\mathbf{x} = (x, y)$ in $\mathbb{R}^2$ and let **u** be a unit vector. Then the **directional derivative of f in the direction u,** denoted $f'_{\mathbf{u}}(\mathbf{x})$, is given by

$$f'_{\mathbf{u}}(\mathbf{x}) = \nabla f(\mathbf{x}) \cdot \mathbf{u}. \qquad (13)$$

Remark 1. Note that if $\mathbf{u} = \mathbf{i}$, then $\nabla f \cdot \mathbf{u} = \partial f / \partial x$ and (13) reduces to the partial derivative $\partial f / \partial x$. Similarly, if $\mathbf{u} = \mathbf{j}$, then (13) reduces to $\partial f / \partial y$.

Remark 2. The definition of directional derivative makes sense if f is a function of three variables. Then, of course, **u** is a unit vector in $\mathbb{R}^3$.

EXAMPLE 2

Let $z = f(x, y) = xy^2$. Calculate the directional derivative of f in the direction of the vector $\mathbf{v} = 2\mathbf{i} + 3\mathbf{j}$ at the point $(4, -1)$.

SOLUTION

A unit vector in the direction **v** is $\mathbf{u} = (2/\sqrt{13})\mathbf{i} + (3/\sqrt{13})\mathbf{j}$. Also, $\nabla f = y^2\mathbf{i} + 2xy\mathbf{j}$. Thus

$$f'_{\mathbf{u}}(x, y) = \nabla f(\mathbf{x}) \cdot \mathbf{u} = \frac{2y^2}{\sqrt{13}} + \frac{6xy}{\sqrt{13}} = \frac{2y^2 + 6xy}{\sqrt{13}}.$$

At $(4, -1)$, $f'_{\mathbf{u}}(4, -1) = -22/\sqrt{13}$.

EXAMPLE 3

Let $f(x, y, z) = x \ln y - e^{xz^3}$. Calculate the directional derivative of f in the direction of the vector $\mathbf{v} = \mathbf{i} - \mathbf{j} + 3\mathbf{k}$. Evaluate this derivative at the point $(-5, 1, -2)$.

SOLUTION

A unit vector in the direction of **v** is $\mathbf{u} = (1/\sqrt{11})\mathbf{i} - (1/\sqrt{11})\mathbf{j} + (3/\sqrt{11})\mathbf{k}$, and

$$\nabla f = (\ln y - z^3 e^{xz^3})\mathbf{i} + \frac{x}{y}\mathbf{j} - 3xz^2 e^{xz^3}\mathbf{k}.$$

Thus

$$f'_{\mathbf{u}}(\mathbf{x}) = \nabla f(\mathbf{x}) \cdot \mathbf{u} = \frac{\ln y - z^3 e^{xz^3} - (x/y) - 9xz^2 e^{xz^3}}{\sqrt{11}},$$

and at $(-5, 1, -2)$

$$f'_{\mathbf{u}}(-5, 1, -2) = \frac{5 + 188e^{40}}{\sqrt{11}}.$$

There is an interesting geometric interpretation of the directional derivative. By the definition on p. 304, the projection of ∇f on $\mathbf{u}$ is given by

$$\text{Proj}_{\mathbf{u}}\ \nabla f = \frac{\nabla f \cdot \mathbf{u}}{|\mathbf{u}|^2}\,\mathbf{u},$$

and since $\mathbf{u}$ is a unit vector, the component of ∇f in the direction $\mathbf{u}$ is given by

$$\frac{\nabla f \cdot \mathbf{u}}{|\mathbf{u}|^2} = \nabla f \cdot \mathbf{u}.$$

Thus the *directional derivative of f in the direction $\mathbf{u}$ is the component of the gradient of f in the direction $\mathbf{u}$.* This is illustrated in Fig. 2.

We now derive another remarkable property of the gradient. Recall that $\mathbf{u} \cdot \mathbf{v} = |\mathbf{u}|\,|\mathbf{v}| \cos \theta$, where θ is the smallest angle between the vectors $\mathbf{u}$ and $\mathbf{v}$. Thus the directional derivative of f in the direction $\mathbf{u}$ can be written as

$$f'_{\mathbf{u}}(\mathbf{x}) = \nabla f(\mathbf{x}) \cdot \mathbf{u} = |\nabla f(\mathbf{x})|\,|\mathbf{u}| \cos \theta, \tag{14}$$

or since $\mathbf{u}$ is a unit vector,

$$f'_{\mathbf{u}}(\mathbf{x}) = |\nabla f(\mathbf{x})| \cos \theta. \tag{15}$$

Now $\cos \theta = 1$ when $\theta = 0$, which occurs when $\mathbf{u}$ has the direction of ∇f. Similarly, $\cos \theta = -1$ when $\theta = \pi$, which occurs when $\mathbf{u}$ has the direction of $-\nabla f$. Also, $\cos \theta = 0$ when $\theta = \pi/2$. Thus since $-1 \le \cos \theta \le 1$, equation (14) implies the following important result.

■ **THEOREM 3**

Let f be differentiable. Then f increases most rapidly in the direction of its gradient and decreases most rapidly in the direction opposite to that of its gradient. Furthermore, f is stationary in a direction perpendicular to its gradient.   ■

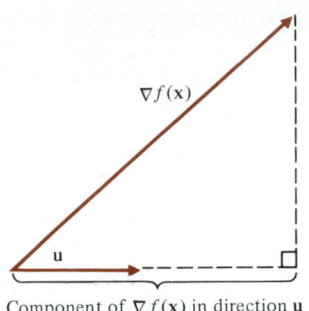

$\nabla f(\mathbf{x})$

$\mathbf{u}$

Component of $\nabla f(\mathbf{x})$ in direction $\mathbf{u}$

FIGURE 2

EXAMPLE 4

The distribution of voltage on a metal plate is given by

$$V = 50 - x^2 - 4y^2.$$

(a) At the point $(1, -2)$, in what direction does the voltage increase most rapidly?
(b) In what direction does it decrease most rapidly?
(c) What is the magnitude of this increase or decrease?
(d) In what direction does it not change?

SOLUTION

$$\nabla V = V_x \mathbf{i} + V_y \mathbf{j} = -2x\mathbf{i} - 8y\mathbf{j}. \text{ At } (1, -2),\ \nabla V = -2\mathbf{i} + 16\mathbf{j}.$$

(a) The voltage increases most rapidly as we move in the direction of $-2\mathbf{i} + 16\mathbf{j}$.
(b) It decreases most rapidly in the direction of $2\mathbf{i} - 16\mathbf{j}$.

(c) The magnitude of the increase or decrease is $\sqrt{2^2 + 16^2} = \sqrt{260}$.

(d) A unit vector perpendicular to ∇V is $(16\mathbf{i} + 2\mathbf{j})/\sqrt{260}$. The voltage is constant in this or the opposite direction.

EXAMPLE 5

In Example 4, describe the path of a particle that starts at the point $(1, -2)$ and moves in the direction of greatest voltage increase.

SOLUTION

The path of the particle will be that of the gradient. That is, at each point of the path, the gradient at that point is tangent to the path. If the particle follows the path $\mathbf{f}(t) = x(t)\mathbf{i} + y(t)\mathbf{j}$, then since the direction of the path is $\mathbf{f}'(t) = x'(t)\mathbf{i} + y'(t)\mathbf{j}$ and since this direction is also given by $\nabla V = -2x\mathbf{i} - 8y\mathbf{j}$, we must have

$$x'(t) = -2x(t) \quad \text{and} \quad y'(t) = -8y(t).$$

From Section 1.1 (see p. 3), the solutions to these differential equations are

$$x(t) = c_1 e^{-2t} \quad \text{and} \quad y(t) = c_2 e^{-8t}.$$

But $x(0) = 1$ and $y(0) = -2$, so that

$$x(t) = e^{-2t} \quad \text{and} \quad y(t) = -2e^{-8t}.$$

Then, since $e^{-8t} = (e^{-2t})^4$, we see that the particle moves along the path

$$y = -2x^4.$$

Remark. Technically, a direction is a unit vector, so we should choose the direction $(-2x\mathbf{i} - 8y\mathbf{j})/\sqrt{4x^2 + 64y^2}$ in our computations. But this choice would not change the final answer.

Chain Rule Revisited

In Theorem 9.7.2 we proved the chain rule for a function of two variables: Let $z = f(x, y)$ be continuous with continuous partial derivatives at every point of an open disk D. Suppose that $x = x(t)$ and $y = y(t)$ are differentiable functions for all t in some open interval I and that $(x(t), y(t)) \in D$ if $t \in I$. Then $z = f(x(t), y(t))$ is a differentiable function of t for every t in I and

$$\frac{dz}{dt} = \frac{\partial z}{\partial x}\frac{dx}{dt} + \frac{\partial z}{\partial y}\frac{dy}{dt} = f_x\frac{dx}{dt} + f_y\frac{dy}{dt}. \tag{16}$$

Now if $\mathbf{x}(t) = (x(t), y(t))$ then $\mathbf{x}'(t) = \left(\dfrac{dx}{dt}, \dfrac{dy}{dt}\right)$, so, since $\nabla f = (f_x, f_y)$, (16) can be written as

$$\frac{dz}{dt} = \nabla f(\mathbf{x}) \cdot \mathbf{x}'(t). \tag{17}$$

Theorem 9.7.2 can be proved in the same way for a function f of n variables $x_1(t), x_2(t), \ldots, x_n(t)$ assuming that $x_i(t)$ is differentiable for t in I and that $\mathbf{x} = (x_1(t), x_2(t), \ldots, x_n(t)) \in D$ if $t \in I$. We assume that f and each of its partial derivatives are continuous for every $\mathbf{x}$ in D. We then have the following theorem.

■ **THEOREM 4: Vector Form of the Chain Rule in $\mathbb{R}^n$**

Under the assumptions above, $y = f(\mathbf{x}(t))$ is a differentiable function of t for every t in I and

$$\frac{dy}{dt} = \nabla f(\mathbf{x}) \cdot \mathbf{x}'(t).$$

■

We now show how gradients arise in a fundamental way in physics. Let $\mathbf{F}(\mathbf{x})$ be any vector field in $\mathbb{R}^2$ (or $\mathbb{R}^3$). Is $\mathbf{F}$ the gradient of some scalar field f? That is, do we have

$$\mathbf{F} = \nabla f? \tag{18}$$

Conservative Vector Field
Potential Function

If $\mathbf{F} = -\nabla f$ for some function f, then $\mathbf{F}$ is said to be a **conservative vector field** and f is called a **potential function** for $\mathbf{F}$. The reason for this terminology will be made clear shortly. [If $\mathbf{F} = \nabla f$, then $\mathbf{F} = -\nabla(-f)$, so that the introduction of the minus sign does not cause any problem.]

Now let $\mathbf{x}(t) = x(t)\mathbf{i} + y(t)\mathbf{j}$ be a differentiable curve and suppose that a particle of mass m moves along it. Suppose further that the force acting on the particle at any time t is given by $\mathbf{F}(\mathbf{x}(t))$, where $\mathbf{F}$ is assumed to be a conservative vector field. By Newton's second law,

$$\mathbf{F}(\mathbf{x}(t)) = m\mathbf{a}(t) = m\mathbf{x}''(t).$$

But since $\mathbf{F}$ is conservative, (18) implies that $\mathbf{F}(\mathbf{x}) = -\nabla f(\mathbf{x})$ for some differentiable function f. Then we have

$$-\nabla f(\mathbf{x}(t)) = m\mathbf{x}''(t),$$

or

$$m\mathbf{x}'' + \nabla f(\mathbf{x}) = \mathbf{0}. \tag{19}$$

We now take the dot product of both sides of (19) with $\mathbf{x}'$ to obtain

$$m\mathbf{x}' \cdot \mathbf{x}'' + \nabla f(\mathbf{x}) \cdot \mathbf{x}' = 0. \tag{20}$$

But by the product rule,

$$\frac{d}{dx}|\mathbf{x}'(t)|^2 = \frac{d}{dt}(\mathbf{x}'(t) \cdot \mathbf{x}'(t)) = \mathbf{x}'(t) \cdot \mathbf{x}''(t) + \mathbf{x}''(t) \cdot \mathbf{x}'(t) = 2\mathbf{x}'(t) \cdot \mathbf{x}''(t), \tag{21}$$

and by equation (17)

$$\frac{d}{dt}f(\mathbf{x}(t)) = \nabla f(\mathbf{x}(t)) \cdot \mathbf{x}'(t) \tag{22}$$

Using (21) and (22) in (20), we obtain

$$\frac{d}{dt}\left[\frac{1}{2}m|\mathbf{x}'|^2 + f(\mathbf{x}(t))\right] = 0,$$

which implies that

$$\tfrac{1}{2}m|\mathbf{x}'|^2 + f(\mathbf{x}(t)) = C, \tag{23}$$

Kinetic Energy

Potential Energy

where C is a constant. This is one of the versions of the **law of conservation of energy.** The term $\frac{1}{2}m|\mathbf{x}'|^2 = \frac{1}{2}m|\mathbf{v}|^2$ is called the **kinetic energy** of the particle, and the term $f(\mathbf{x}(t))$ is called the **potential energy** of the particle. Equation (23) tells us simply that if the force function $\mathbf{F}$ is conservative, then the total energy of the system is constant and, moreover, the potential function f of $\mathbf{F}$ represents the potential energy of the system.

The **principle of the conservation of energy** states that energy may be transformed from one form to another but cannot be created or destroyed; that is, the total energy is constant. Thus it seems reasonable that force fields in classical physics are conservative (although since the work of Einstein, it has been found that energy can be transformed into mass and vice versa, so that there are forces that are not conservative). One example of a conservative force is given by the force of gravitational attraction. Let m_1 represent the mass of a (relatively) fixed object in space and let m_2 denote the mass of an object moving near the fixed object. Then the magnitude of the gravitational force between the objects is given by

$$|\mathbf{F}| = G\frac{m_1 m_2}{r^2}, \tag{24}$$

where r is the distance between the objects and G is a universal constant. If we assume that the first object is at the origin, then we may denote the position of the second object by $\mathbf{x}(t)$, and then since $r = |\mathbf{x}|$, (24) can be written

$$|\mathbf{F}| = \frac{Gm_1 m_2}{|\mathbf{x}|^2}. \tag{25}$$

Also, the force acts toward the origin, that is, in the direction opposite to that of the positive vector $\mathbf{x}$. Therefore

$$\mathbf{F} = -\frac{Gm_1 m_2}{|\mathbf{x}|^3}\mathbf{x}, \tag{26}$$

where $|\mathbf{x}| = r = \sqrt{x^2 + y^2 + z^2}$. Now consider the scalar field $f(\mathbf{x}) = c/r$:

$$f_x = \frac{\partial}{\partial x}(cr^{-1}) = -cr^{-2}\frac{\partial r}{\partial x} = \frac{-c}{x^2 + y^2 + z^2} \cdot \frac{x}{\sqrt{x^2 + y^2 + z^2}} = \frac{-cx}{r^3}.$$

Similarly, $f_y = -cy/r^3$ and $f_z = -cz/r^3$. Thus

$$\nabla f(\mathbf{x}) = -\frac{c\mathbf{x}}{r^3}$$

so that

$$\mathbf{F} = \nabla\left(\frac{Gm_1m_2}{|\mathbf{x}|}\right). \tag{27}$$

Problems 9.8

In Problems 1–15, calculate the directional derivative of the given function at the given point in the direction of the given vector **v**.

1. $f(x, y) = xy$ at $(2, 3)$; $\mathbf{v} = \mathbf{i} + 3\mathbf{j}$

2. $f(x, y) = 2x^2 - 3y^2$ at $(1, -1)$; $\mathbf{v} = -\mathbf{i} + 2\mathbf{j}$

3. $f(x, y) = \ln(x + 3y)$ at $(2, 4)$; $\mathbf{v} = \mathbf{i} + \mathbf{j}$

4. $f(x, y) = ax^2 + by^2$ at (c, d); $\mathbf{v} = \alpha\mathbf{i} + \beta\mathbf{j}$

5. $f(x, y) = \tan^{-1}\dfrac{y}{x}$ at $(2, 2)$; $\mathbf{v} = 3\mathbf{i} - 2\mathbf{j}$

6. $f(x, y) = \dfrac{x-y}{x+y}$ at $(4, 3)$; $\mathbf{v} = -\mathbf{i} - 2\mathbf{j}$

7. $f(x, y) = xe^y + ye^x$ at $(1, 2)$; $\mathbf{v} = \mathbf{i} + \mathbf{j}$

8. $f(x, y) = \sin(2x + 3y)$ at $(\pi/12, \pi/9)$; $\mathbf{v} = -2\mathbf{j} + 3\mathbf{j}$

9. $f(x, y, z) = xy + yz + xz$ at $(1, 1, 1)$; $\mathbf{v} = \mathbf{i} + \mathbf{j} + \mathbf{k}$

10. $f(x, y, z) = xy^3z^5$ at $(-3, -1, 2)$; $\mathbf{v} = -\mathbf{i} - 2\mathbf{j} + \mathbf{k}$

11. $f(x, y, z) = \ln(x + 2y + 3z)$ at $(1, 2, 0)$; $\mathbf{v} = 2\mathbf{i} + \mathbf{j} - \mathbf{k}$

12. $f(x, y, z) = xe^{yz}$ at $(2, 0, -4)$; $\mathbf{v} = -\mathbf{i} + 2\mathbf{j} + 5\mathbf{k}$

13. $f(x, y, z) = x^2y^3 + z\sqrt{x}$ at $(1, -2, 3)$; $\mathbf{v} = 5\mathbf{j} + \mathbf{k}$

14. $f(x, y, z) = e^{-(x^2+y^2+z^2)}$ at $(1, 1, 1)$; $\mathbf{v} = \mathbf{i} + 3\mathbf{j} - 5\mathbf{k}$

15. $f(x, y, z) = \dfrac{1}{\sqrt{x^2+y^2+z^2}}$ at $(-1, 2, 3)$; $\mathbf{v} = \mathbf{i} - \mathbf{j} + \mathbf{k}$

16. The voltage (potential) at any point on a metal structure is given by

$$v(x, y, z) = \frac{1}{0.02 + \sqrt{x^2+y^2+z^2}}.$$

At the point $(1, -1, 2)$, in what direction does the voltage increase most rapidly?

17. The temperature at any point in a solid metal ball centered at the origin is given by

$$T(x, y, z) = 100e^{-(x^2+y^2+z^2)}.$$

(a) Where is the ball hottest?

(b) Show that at any point (x, y, z) on the ball, the direction of greatest increase in temperature is a vector pointing toward the origin.

18. The temperature distribution of a solid ball centered at the origin is given by

$$T(x, y, z) = \frac{100}{x^2+y^2+z^2+1}.$$

(a) Where is the ball hottest?

(b) Find the direction of greatest decrease of temperature at the point $(3, -1, 2)$.

(c) Find the direction of greatest increase in temperature. Does this vector point toward the origin?

19. The temperature distribution on a plate is given by

$$T(x, y) = 1 - \frac{x^2}{a^2} - \frac{y^2}{b^2}.$$

Find the path of a heat-seeking particle (i.e., a particle that always moves in the direction of greatest increase in temperature) if it starts at the point (a, b).

20. Find the path of the particle in Problem 19 if it starts at the point $(-a, b)$.

21. The height of a mountain is given by $h(x, y) = 3000 - 2x^2 - y^2$, where the x-axis points south, the y-axis points east, and all distances are measured in meters. Suppose that a mountain climber is at the point $(30, -20, 800)$.

(a) If the climber moves in the southwest direction, will she ascend or descend?

(b) In what direction should the climber move so as to ascend most rapidly?

22. Show that if f and g are differentiable functions of three variables, then fg is differentiable and

$$\nabla(fg) = f\nabla g + g\nabla f.$$

* 23. Show that $\nabla f \equiv 0$ if and only if f is constant.

24. Show that if $\nabla f = \nabla g$, then there is a constant c for which $f(x, y) = g(x, y) + c$. [*Hint:* Use the results of Problem 23.]

* 25. What is the most general function f such that $\nabla f(\mathbf{x}) = \mathbf{x}$ for every $\mathbf{x}$ in $\mathbb{R}^2$?

* 26. Let $f(x, y) = \begin{cases} (x^2 + y^2) \sin \dfrac{1}{\sqrt{x^2 + y^2}}, & (x, y) \neq (0,0) \\ 0, & (x, y) = (0,0). \end{cases}$

    (a) Calculate $f_x(0, 0)$ and $f_y(0, 0)$.

    (b) Explain why f_x and f_y are *not* continuous at $(0, 0)$.

    (c) Show that f is differentiable at $(0, 0)$.

27. Suppose that f is a differentiable function of one variable and g is a differentiable function of three variables. Show that $f \circ g$ is differentiable and $\nabla(f \circ g) = f'(g)\nabla g$.

* 28. Let

$$f(x, y) = \begin{cases} \dfrac{xy}{x^2 + y^2}, & (x, y) \neq (0,0), \\ 0, & (x, y) = (0,0). \end{cases}$$

Show that f_x and f_y exist at $(0, 0)$ but that f is not continuous at $(0, 0)$. [*Hint:* Show that if $(x, y) \to (0, 0)$ along the line $y = x$, then $f(x, y) \to \frac{1}{2}$ but that $f(x, y) \to -\frac{1}{2}$ along the line $y = -x$. Conclude that $\lim_{(x,y)\to(0,0)} f(x, y)$ does not exist.]

29. Show that the force given by $\mathbf{F}(\mathbf{x}) = -\alpha\mathbf{x}/|\mathbf{x}|^2$ is conservative by finding a potential function for it.

30. Do the same as in Problem 29 for the force given by $\mathbf{F}(\mathbf{x}) = -\alpha\mathbf{x}/|\mathbf{x}|^4$.

31. If α is a constant, show that the force $\mathbf{F}(\mathbf{x}) = -\alpha\mathbf{x}/|\mathbf{x}|^k$ is conservative.

32. Show that the force $\mathbf{F}(x, y) = y\mathbf{i} + x\mathbf{j}$ is conservative.

33. Show that if $\mathbf{F}$ and $\mathbf{G}$ are conservative, then $\alpha\mathbf{F} + \beta\mathbf{G}$ is also conservative for any constants α and β.

* 34. Show that the force $\mathbf{F}(x, y) = y\mathbf{i} - x\mathbf{j}$ is *not* conservative [*Hint:* Assume that $\mathbf{F} = -\nabla f$, so that $-y = \partial f/\partial x$ and $x = \partial f/\partial y$. Then observe the mixed partials are unequal.]

9.9. Tangent Planes, Normal Lines, and Gradients

Let $z = f(x, y)$ be a function of two variables. As we have seen, the graph of f is a surface in $\mathbb{R}^3$. More generally, the graph of the equation $F(x, y, z) = 0$ is a surface in $\mathbb{R}^3$. The surface $F(x, y, z) = 0$ is called **differentiable** at a point (x_0, y_0, z_0) if $\partial F/\partial x$, $\partial F/\partial y$, and $\partial F/\partial z$ all exist and are continuous at (x_0, y_0, z_0). In $\mathbb{R}^2$, a differentiable curve has a unique tangent line at each point. In $\mathbb{R}^3$ a differentiable surface in $\mathbb{R}^3$ has a unique tangent plane at each point at which $\partial F/\partial x$, $\partial F/\partial y$, and $\partial F/\partial z$ are not all zero. We will formally define what we mean by a tangent plane to a surface after a bit, although it should be easy enough to visualize (see Fig. 1). We note here that not every surface has a

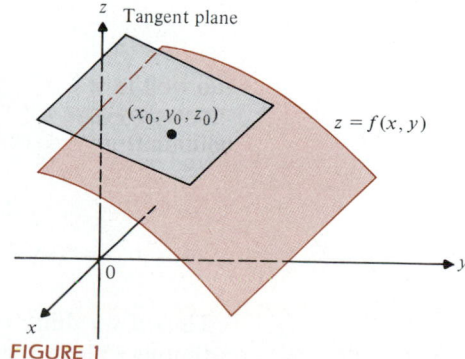

FIGURE 1

unique tangent plane at every point. For example, the cone $z = \sqrt{x^2 + y^2}$ does not have a unique tangent plane at the origin (see Fig. 2).

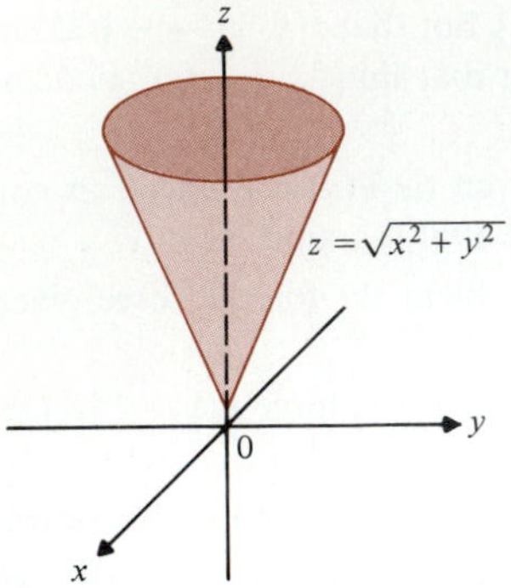

Assume that the surface S given by $F(x, y, z) = 0$ is differentiable. Let C be any curve lying on S. That is, C can be given parametrically by $\mathbf{g}(t) = x(t)\mathbf{i} + y(t)\mathbf{j} + z(t)\mathbf{k}$. Recall the chain rule (given on p. 556) and assume that all derivatives exist. Then

$$\frac{d}{dt}F(x, y, z) = \frac{\partial F}{\partial x}\frac{dx}{dt} + \frac{\partial F}{\partial y}\frac{dy}{dt} + \frac{\partial F}{\partial z}\frac{dz}{dt}.$$

Since $\nabla F = (\partial F/\partial x, \partial F/\partial y, \partial F/\partial z)$ and $\mathbf{g}'(t) = (x'(t), y'(t), z'(t))$ we can rewrite the chain rule as

$$\frac{dF}{dt} = \nabla F \cdot \mathbf{g}'(t)$$

or

$$F'(t) = \nabla F \cdot \mathbf{g}'(t). \tag{1}$$

But since $F(x(t), y(t), z(t)) = 0$ for all t [since $(x(t), y(t), z(t))$ is on S], we see that $F'(t) = 0$ for all t. But $\mathbf{g}'(t)$ is tangent to the curve C for every number t. Thus (1) implies the following:

> *The gradient of F at a point $\mathbf{x}_0 = (x_0, y_0, z_0)$ on S is orthogonal to the tangent vector at $\mathbf{x}_0$ to any curve C on S passing through $\mathbf{x}_0$.*

This statement is illustrated in Fig. 3.

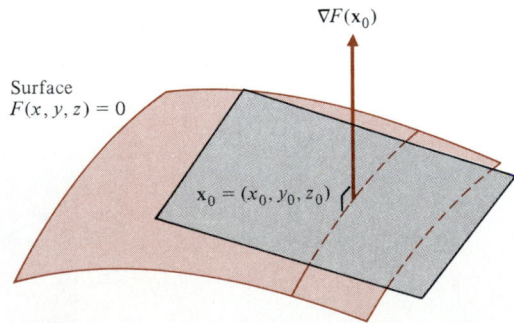

$\nabla F(\mathbf{x}_0)$

Surface
$F(x, y, z) = 0$

$\mathbf{x}_0 = (x_0, y_0, z_0)$

FIGURE 3

Thus if we think of all the vectors tangent to a surface at a point $\mathbf{x}_0$ as constituting a plane, then $\nabla F(\mathbf{x}_0)$ is a *normal* vector to that plane. This motivates the following definition.

FIGURE 2

$z = \sqrt{x^2 + y^2}$

DEFINITION

Let F be differentiable at $\mathbf{x}_0 = (x_0, y_0, z_0)$ and let the surface S be defined by $F(x, y, z) = 0$. Suppose that $|\nabla F(\mathbf{x}_0)| \neq 0$.

Tangent Plane and Normal Line

(i) The **tangent plane** to S at $\mathbf{x}_0$ is the plane passing through the point $\mathbf{x}_0$ with normal vector $\nabla F(\mathbf{x}_0)$.

(ii) The **normal line** to S at $\mathbf{x}_0$ is the line passing through $\mathbf{x}_0$ having the same direction as $\nabla F(\mathbf{x}_0)$.

EXAMPLE 1

Find the equation of the tangent plane and symmetric equations of the normal line to the ellipsoid $x^2 + (y^2/4) + (z^2/9) = 3$ at the point $(1, 2, 3)$.

SOLUTION

Since $F(x, y, z) = x^2 + (y^2/4) + (z^2/9) - 3 = 0$, we have

$$\nabla F = \frac{\partial F}{\partial x}\mathbf{i} + \frac{\partial F}{\partial y}\mathbf{j} + \frac{\partial F}{\partial z}\mathbf{k} = 2x\mathbf{i} + \frac{y}{2}\mathbf{j} + \frac{2z}{9}\mathbf{k}.$$

Then $\nabla F(1, 2, 3) = 2\mathbf{i} + \mathbf{j} + \frac{2}{3}\mathbf{k}$, and the equation of the tangent plane is (see p. 318)

$$2(x-1) + (y-2) + \tfrac{2}{3}(z-3) = 0,$$

or

$$2x + y + \tfrac{2}{3}z = 6.$$

The normal line is given by (see p. 316)

$$\frac{x-1}{2} = y - 2 = \tfrac{3}{2}(z-3).$$

The situation is even simpler if we can write the surface in the form $z = f(x, y)$. That is, the surface is the graph of a function of two variables. Then $F(x, y, z) = f(x, y) - z = 0$, so that

$$F_x = f_x, \qquad F_y = f_y, \qquad F_z = -1,$$

and the normal vector $\mathbf{N}$ to the tangent plane is

$$\mathbf{N} = f_x(x_0, y_0)\mathbf{i} + f_y(x_0, y_0)\mathbf{j} - \mathbf{k}. \tag{2}$$

Remark. One interesting consequence of this fact is that if $z = f(x, y)$ and if $\nabla f(x_0, y_0) = 0$, then *the tangent plane to the surface at $(x_0, y_0, f(x_0, y_0))$ is parallel to the xy-plane (i.e., it is horizontal)*. This occurs because at $(x_0, y_0, f(x_0, y_0))$, $\mathbf{N} = (\partial f/\partial x)\mathbf{i} + (\partial f/\partial y)\mathbf{j} - \mathbf{k} = \nabla f - \mathbf{k} = -\mathbf{k}$. Thus the z-axis is normal to the tangent plane.

EXAMPLE 2

Find the tangent plane and normal line to the surface $z = x^3 y^5$ at the point $(2, 1, 8)$.

SOLUTION

$\mathbf{N} = (\partial z/\partial x)\mathbf{i} + (\partial z/\partial y)\mathbf{j} - \mathbf{k} = 3x^2 y^5 \mathbf{i} + 5x^3 y^4 \mathbf{j} - \mathbf{k} = 12\mathbf{i} + 40\mathbf{j} - \mathbf{k}$ at $(2, 1, 8)$. Then the tangent plane is given by

$$12(x - 2) + 40(y - 1) - (z - 8) = 0,$$

or

$$12x + 40y - z = 56.$$

Symmetric equations of the normal line are

$$\frac{x - 2}{12} = \frac{y - 1}{40} = \frac{z - 8}{-1}.$$

We can write the equation of the tangent plane to a surface $z = f(x, y)$ so that it looks like the equation of the tangent line to a curve in $\mathbb{R}^2$. This will further illustrate the connection between the derivative of a function of one variable and the gradient. Recall that if P is a point on a plane and $\mathbf{N}$ is a normal vector, then if Q denotes any other point on the plane, the equation of the plane can be written

$$\overrightarrow{PQ} \cdot \mathbf{N} = 0. \tag{3}$$

In this case, since $z = f(x, y)$, a point on the surface takes the form $(x, y, z) = (x, y, f(x, y))$. Then $\mathbf{N} = f_x \mathbf{i} + f_y \mathbf{j} - \mathbf{k}$ and the equation of the tangent plane at $(x_0, y_0, f(x_0, y_0))$ becomes, using (3),

$$0 = [(x, y, z) - (x_0, y_0, z_0)] \cdot (f_x, f_y, -1)$$

$$= (x - x_0, y - y_0, z - z_0) \cdot (f_x, f_y, -1)$$

$$= (x - x_0)f_x + (y - y_0)f_y - (z - z_0). \tag{4}$$

We can rewrite (4) as

$$z = f(x_0, y_0) + (x - x_0)f_x + (y - y_0)f_y. \tag{5}$$

Denote (x_0, y_0) by $\mathbf{x}_0$ and (x, y) by $\mathbf{x}$. Then (5) can be written as

$$z = f(\mathbf{x}_0) + (\mathbf{x} - \mathbf{x}_0) \cdot \nabla f(\mathbf{x}_0). \tag{6}$$

Recall that if $y = f(x)$ is differentiable at x_0, then the equation of the tangent line to the curve at the point $(x_0, f(x_0))$ is given by

$$\frac{y - f(x_0)}{x - x_0} = f'(x_0),$$

or

$$y = f(x_0) + (x - x_0)f'(x_0). \tag{7}$$

This similarity between (6) and (7) illustrates quite vividly the importance of the gradient vector of a function of several variables as the generalization of the derivative of a function of one variable.

We can use the facts derived in this section to define a special directional derivative. Since the gradient vector is normal to the level curve $f(x, y) = c$ for any constant c, we say that the directional derivative of f in the direction of the gradient is the **normal derivative** of f and is denoted by $\partial f/\partial n$. We then have, from equation (9.8.15) on p. 566,

Normal Derivative

$$\text{normal derivative} = \frac{\partial f}{\partial n} = |\nabla f|. \tag{8}$$

EXAMPLE 3

Let $f(x, y) = xy^2$. Calculate the normal derivative and evaluate it at the point $(3, -2)$.

SOLUTION

$\nabla f = y^2 \mathbf{i} + 2xy\mathbf{j}$. Then $\partial f/\partial n = |\nabla f| = \sqrt{y^4 + 4x^2y^2}$. At $(3, -2)$, $\partial f/\partial n = \sqrt{16 + 144} = \sqrt{160}$.

Problems 9.9

In Problems 1–16, find the equation of the tangent plane and symmetric equations of the normal line to the given surface at the given point.

1. $x^2 + y^2 + z^2 = 1$; $(1, 0, 0)$
2. $x^2 + y^2 + z^2 = 1$; $(0, 1, 0)$
3. $x^2 + y^2 + z^2 = 1$; $(0, 0, 1)$
4. $x^2 - y^2 + z^2 = 1$; $(1, 1, 1)$
5. $\dfrac{x^2}{a^2} + \dfrac{y^2}{b^2} + \dfrac{z^2}{c^2} = 3$; (a, b, c)
6. $\dfrac{x^2}{a^2} + \dfrac{y^2}{b^2} + \dfrac{z^2}{c^2} = 3$; $(-a, b, -c)$
7. $x^{1/2} + y^{1/2} + z^{1/2} = 6$; $(4, 1, 9)$
8. $ax + by + cz = d$; $\left(\dfrac{1}{a}, \dfrac{1}{b}, \dfrac{d-2}{c}\right)$
9. $xyz = 4$; $(1, 2, 2)$
10. $xy^2 - yz^2 + zx^2 = 1$; $(1, 1, 1)$
11. $4x^2 - y^2 - 5z^2 = 15$; $(3, 1, -2)$
12. $xe^y - ye^z = 1$; $(1, 0, 0)$
13. $\sin xy - 2 \cos yz = 0$; $(\pi/2, 1, \pi/3)$
14. $x^2 + y^2 + 4x + 2y + 8z = 7$; $(2, -3, -1)$
15. $e^{xyz} = 5$; $(1, 1, \ln 5)$
16. $\sqrt{\dfrac{x+y}{z-1}} = 1$; $(1, 1, 3)$

In Problems 17–24, write the equation of the tangent plane and find the symmetric equations of the normal line to the given surface at the given point.

17. $z = xy^2$; $(1, 1, 1)$
18. $z = \ln(x - 2y)$; $(3, 1, 0)$
19. $z = \sin(2x + 5y)$; $(\pi/8, \pi/20, 1)$
20. $z = \sqrt{\dfrac{x+y}{x-y}}$; $(5, 4, 3)$
21. $z = \tan^{-1}\dfrac{y}{x}$; $\left(-2, 2, -\dfrac{\pi}{4}\right)$

22. $z = \sinh xy^2$; $(0, 3, 0)$

23. $z = \sec(x - y)$; $(\pi/2, \pi/6, 2)$

24. $z = e^x \cos y + e^y \cos x$; $(\pi/2, 0, e^{\pi/2})$

* 25. Find the two points of intersection of the surface $z = x^2 + y^2$ and the line
$$\frac{x-3}{1} = \frac{y+1}{-1} = \frac{z+2}{-2}.$$

26. At each of the points of intersection found in Problem 25, find the cosine of the angle between the given line and the normal line to the surface.

* 27. Show that every line normal to the surface of a sphere passes through the center of the sphere.

28. Show that every line normal to the cone $z^2 = ax^2 + ay^2$ intersects the z-axis.

29. Let f be a differentiable function of one variable and let $z = yf(y/x)$. Show that all tangent planes to the surface defined by this equation have a point in common.

30. The *angle between two surfaces* at a point of intersection is defined to be the angle between their normal lines. Show that if two surfaces $F(x, y, z) = 0$ and $G(x, y, z) = 0$ intersect at right angles at a point $\mathbf{x}_0$, then
$$\nabla F(\mathbf{x}_0) \bullet \nabla G(\mathbf{x}_0) = 0.$$

31. Show that the sum of the squares of the x-, y-, and z-intercepts of any plane tangent to the surface $x^{2/3} + y^{2/3} + z^{2/3} = a^{2/3}$ is constant.

32. The equation $F(x, y) = 0$ defines a curve in $\mathbb{R}^2$. Show that if F is differentiable, then $\nabla F(x, y)$ is normal to the curve at every point.

33. Use the result of Problem 32 to find the equation of the tangent line to a curve $F(x, y) = 0$ at a point (x_0, y_0).

In Problems 34–39, use the results of Problems 32 and 33 to find a normal vector and the equation of the tangent line to the curve at the given point.

34. $xy = 5$; $(1, 5)$

35. $x^2 + xy + y^2 + 3x - 5y = 16$; $(1, -2)$

36. $\dfrac{x+y}{x-y} = 7$; $(4, 3)$

37. $xe^{xy} = 1$; $(1, 0)$

38. $\dfrac{x^2}{4} + \dfrac{y^2}{16} = 1$; $(\sqrt{2}, 2\sqrt{2})$

39. $\tan(x + y) = 1$; $(\pi/4, 0)$

40. Show that at any point (x, y), $y \neq 0$, the curve $x/y = a$ is orthogonal to the curve $x^2 + y^2 = r^2$ for any constants a and r.

In Problems 41–44 calculate the normal derivative at the given point.

41. $f(x, y) = x + 2y$ at $(1, 4)$

42. $f(x, y) = e^{x+3y}$ at $(1, 0)$

43. $f(x, y) = \tan^{-1} \dfrac{y}{x}$ at $(-1, 1)$

44. $f(x, y) = \sqrt{\dfrac{x-y}{x+y}}$ at $(3, 1)$

(Optional)   **9.10   Taylor's Theorem in *n* Variables**

In this section we shall extend Taylor's theorem to a function of n variables. More precisely, we shall indicate how a function $f(\mathbf{x})$ can be approximated by an mth-degree polynomial $p_m(\mathbf{x})$ with a remainder term. First we need a definition.

Let $f: \Omega \to \mathbb{R}$ be a mapping defined on the open set $\Omega \subset \mathbb{R}^n$. The partial derivative

$$\frac{\partial^m f}{\partial x_1^{k_1} \partial x_2^{k_2} \cdots \partial x_n^{k_n}}, \qquad \text{where } k_1 + k_2 + \cdots + k_n = m, \tag{1}$$

The Class $C^{(m)}$ (Ω)

is called an **mth order partial derivative**. The function f is said to be of **class $C^{(m)}(\Omega)$** [written $f \in C^{(m)}(\Omega)$] if f and all its partial derivatives of order $\leq m$ are continuous in Ω.

The result we want is obtained by repeated application of the chain rule (Theorem 9.8.4 on p. 568). We assume that the function $f: \mathbb{R}^n \to \mathbb{R}$ is defined on an open set Ω and that the line segment joining $\mathbf{x}$ and $\mathbf{x}_0$ in $\mathbb{R}^n$ is in Ω (see the definition on p. 558). We define the vector $\mathbf{h}$ by

$$\mathbf{h} = \mathbf{x} - \mathbf{x}_0. \tag{2}$$

Keep in mind that $\mathbf{h} = (h_1, h_2, \ldots, h_n)$ is an n-vector. Let

$$g(t) = f(\mathbf{x}_0 + t\mathbf{h}). \tag{3}$$

Note that g is defined for $t \in [0, 1]$ because $\{\mathbf{x}: \mathbf{x} = \mathbf{x}_0 + t\mathbf{h}, 0 \le t \le 1\}$ is the line segment joining $\mathbf{x}_0$ and $\mathbf{x}$. We assume that f is of class $C^{(m)}(\Omega)$. Then g has m continuous derivatives, which we now compute. We first note that

$$\frac{d}{dt}(\mathbf{x}_0 + t\mathbf{h}) = \mathbf{h} \tag{4}$$

since $\mathbf{h}$ is assumed to be constant (see Example 9.3.9 on p. 524). Thus, by the chain rule,

$$g'(t) = \frac{d}{dt} f(\mathbf{x}_0 + t\mathbf{h}) = \nabla f(\mathbf{x}_0 + t\mathbf{h}) \cdot \mathbf{h}. \tag{5}$$

Writing out the terms in the scalar product in (5) with $f_i = \partial f / \partial x_i$, we have

$$g'(t) = \sum_{i=1}^{n} f_i(\mathbf{x}_0 + t\mathbf{h}) h_i. \tag{6}$$

The idea now is to compute higher-order derivatives of g and then to apply Taylor's theorem of one variable to g. Consider the function $f_i(\mathbf{x}_0 + t\mathbf{h})$. We can apply the chain rule again to obtain

$$\frac{d}{dt} f_i(\mathbf{x}_0 + t\mathbf{h}) = \nabla f_i(\mathbf{x}_0 + t\mathbf{h}) \cdot \mathbf{h}. \tag{7}$$

But $\nabla f_i = (f_{i1}, f_{i2}, \ldots, f_{in})$, so that, from (7),

$$\frac{d}{dt} f_i(\mathbf{x}_0 + t\mathbf{h}) = \sum_{j=1}^{n} f_{ij}(\mathbf{x}_0 + t\mathbf{h}) h_j. \tag{8}$$

Note that the subscripts on f represent partial derivatives, whereas those on the vector $\mathbf{h}$ represent the components of $\mathbf{h}$. Now, from (6),

$$g''(t) = \sum_{i=1}^{n} \frac{d}{dt} f_i(\mathbf{x}_0 + t\mathbf{h}) h_i. \tag{9}$$

So inserting (8) into (9) yields

$$g''(t) = \sum_{i=1}^{n} \left[\sum_{j=1}^{n} f_{ij}(\mathbf{x}_0 + t\mathbf{h}) h_j \right] h_i. \tag{10}$$

This is getting complicated, but if you have come this far, you can probably see the pattern. We have

$$g'''(t) = \sum_{i=1}^{n} \left\{ \sum_{j=1}^{n} \left[\sum_{k=1}^{n} f_{ijk}(\mathbf{x}_0 + t\mathbf{h})h_k \right] h_j \right\} h_i \tag{11}$$

which we write in the abbreviated form (with each index running over the set $1, 2, \ldots, n$)

$$g'''(t) = \sum_{i,j,k=1}^{n} f_{ijk}(\mathbf{x}_0 + t\mathbf{h})h_i h_j h_k. \tag{12}$$

Finally, we have

$$g^{(m)}(t) = \sum_{i_1, i_2, \ldots, i_m = 1}^{n} f_{i_1 i_2 i_3 \ldots i_m}(\mathbf{x}_0 + t\mathbf{h})h_{i_1} h_{i_2} \cdots h_{i_m}. \tag{13}$$

Formula (13) is formidable, but it is useful in deriving Taylor's theorem for n variables.

■ **THEOREM 1**
Let $f: \mathbb{R}^n \to \mathbb{R}$ be of class $C^{(m+1)}(\Omega)$ and let the line segment joining $\mathbf{x}_0$ and $\mathbf{x}$ be in Ω. Then if $\mathbf{x} = (x_1, x_2, \ldots, x_n)$ and $\mathbf{x}_0 = (x_1^{(0)}, x_2^{(0)}, \ldots, x_n^{(0)})$,

$$f(\mathbf{x}) = f(\mathbf{x}_0) + \sum_{i=1}^{n} f_i(\mathbf{x}_0)(x_i - x_i^{(0)}) + \frac{1}{2!} \sum_{i=1}^{n} \sum_{j=1}^{n} f_{ij}(\mathbf{x}_0)(x_i - x_i^{(0)})(x_j - x_j^{(0)})$$

$$+ \frac{1}{3!} \sum_{i,j,k=1}^{n} f_{ijk}(\mathbf{x}_0)(x_i - x_i^{(0)})(x_j - x_j^{(0)})(x_k - x_k^{(0)}) + \cdots$$

$$+ \frac{1}{m!} \sum_{i_1, i_2, \ldots, i_m = 1}^{n} f_{i_1 i_2 \ldots i_m}(\mathbf{x}_0)(x_{i_1} - x_{i_1}^{(0)})(x_{i_2} - x_{i_2}^{(0)}) \cdots (x_{i_m} - x_{i_m}^{(0)}) + R_m(\mathbf{x})$$

$$\tag{14}$$

where

$$R_m(\mathbf{x}) = \tag{15}$$

$$\frac{1}{(m+1)!} \sum_{i_1, i_2, \ldots, i_{m+1} = 1}^{n} f_{i_1 i_2 \ldots i_{m+1}}(\mathbf{x}_0 + c\mathbf{h})(x_{i_1} - x_{i_1}^{(0)})(x_{i_2} - x_{i_2}^{(0)}) \cdots (x_{i_{m+1}} - x_{i_{m+1}}^{(0)})$$

for some number c in $(0, 1)$.

PROOF
Since $g(t)$ has $m + 1$ continuous derivatives on $[0, 1]$ we have, from Taylor's theorem,

$$g(1) = g(0) + g'(0) + \frac{1}{2!}g''(0) + \cdots + \frac{1}{m!}g^{(m)}(0) + \frac{1}{(m+1)!}g^{(m+1)}(c) \tag{16}$$

for some c in $(0, 1)$. But, from (2) and (3),

$$g(1) = f(\mathbf{x}_0 + \mathbf{h}) = f(\mathbf{x}_0 + \mathbf{x} - \mathbf{x}_0) = f(\mathbf{x}) \quad \text{and} \quad g(0) = f(\mathbf{x}_0).$$

Also, $h_i = (x_i - x_i^{(0)})$, so inserting (6), (10), (12), and (13) into (16) gives the desired result. ∎

Notation. We shall denote the polynomial given in (14) by $p_m(\mathbf{x})$.

We can write Taylor's theorem for n variables in a more compact notation that will illustrate again how the gradient vector generalizes the derivative of a function of one variable.

We start with equation (5):

$$g'(t) = \nabla f(\mathbf{x}_0 + t\mathbf{h}) \cdot \mathbf{h}.$$

We write this as

$$g'(t) = (\mathbf{h} \cdot \nabla) f.$$

We introduce the notation $(\mathbf{h} \cdot \nabla)^n f$ recursively:

$$(\mathbf{h} \cdot \nabla)^2 f = \mathbf{h} \cdot \nabla (\mathbf{h} \cdot \nabla) f,$$

$$(\mathbf{h} \cdot \nabla)^3 f = \mathbf{h} \cdot \nabla [(\mathbf{h} \cdot \nabla)^2 f], \quad \text{and so on.}$$

But

$$
\begin{aligned}
(\mathbf{h} \cdot \nabla)^2 f &= \mathbf{h} \cdot \nabla \left[\overset{\text{from (6)}}{\underset{i=1}{\overset{n}{\sum}} f_i(\mathbf{x}_0 + t\mathbf{h}) h_i} \right] \\
&= \sum_{i=1}^{n} \left[\sum_{j=1}^{n} f_{ij}(\mathbf{x}_0 + t\mathbf{h}) h_j \right] h_i = g''(t).
\end{aligned}
$$

This formula holds for $0 \le t \le 1$. Similarly,

$$(\mathbf{h} \cdot \nabla)^3 f = g'''(t),$$

and so on. In each case f is evaluated at $\mathbf{x}_0 + t\mathbf{h}$.

Now, $\mathbf{h} = \mathbf{x} - \mathbf{x}_0$, $f(\mathbf{x}) = f(\mathbf{x}_0 + \mathbf{h}) = f(\mathbf{x}_0 + t\mathbf{h})$ for $t = 1$ and $f(\mathbf{x}_0) = f(\mathbf{x}_0 + 0\mathbf{h})$. Thus, using our new notation in equations (14) and (15), we have the following alternative form of Taylor's theorem:

■ **THEOREM 1′ (Taylor's Theorem)**

Let f be defined in the disk $C = \{\mathbf{x} : |\mathbf{x} - \mathbf{x}_0| < r\}$ and suppose all the partial derivatives of order $\le n$ are continuous and partial derivatives of order $n + 1$ exist in D. Then

$$f(\mathbf{x}_0 + \mathbf{h}) = \sum_{k=0}^{n} \frac{(\mathbf{h} \cdot \nabla)^k f(\mathbf{x}_0)}{k!} + \frac{(\mathbf{h} \cdot \nabla)^{n+1} f(\mathbf{x}_0 + c\mathbf{h})}{(n+1)!} \qquad \textbf{(17)}$$

where $0 < c < 1$. ■

Remark. Just as in the one-variable case, it is essential to compute all the gradients (derivatives) at an *arbitrary point* $\mathbf{x}$ in D. Once all the required terms $(\mathbf{h} \cdot \nabla)^k f(\mathbf{x})$, $k = 1, \ldots, n + 1$, have been calculated, we evaluate them at $\mathbf{x}_0$.

Taylor's theorem is not terribly difficult to apply, but it does become tedious when computing higher-order terms of a function of n variables. We shall therefore keep our examples simple.

EXAMPLE 1

Compute the first three terms of Taylor's theorem for the function $f(x, y) = x^2 y$ at $\mathbf{x}_0 = (x_1^0, x_2^0) = (1, 2)$.

SOLUTION

The first term is $f(\mathbf{x}_0) = 2$. The second term is

$$\sum_{i=1}^{2} f_i(\mathbf{x}_0)(x_i - x_i^0) = f_1(\mathbf{x}_0)(x - 1) + f_2(\mathbf{x}_0)(y - 2).$$

Since $f_1 = f_x = 2xy$ and $f_2 = f_y = x^2$ at $\mathbf{x}_0$, we get

$$4(x - 1) + (y - 2).$$

The third term is (since $f_{12} = f_{xy} = f_{yx} = f_{21}$)

$$\frac{1}{2!} \sum_{i=1}^{2} \sum_{j=1}^{2} f_{ij}(\mathbf{x}_0)(x_i - x_i^0)(x_j - x_j^0)$$

$$= \frac{1}{2!}[f_{11}(\mathbf{x}_0)(x - 1)^2 + 2f_{12}(\mathbf{x}_0)(x - 1)(y - 2) + f_{22}(\mathbf{x}_0)(y - 2)^2]$$

or, as $f_{11} = 2y$, $f_{12} = 2x$, and $f_{22} = 0$,

$$\frac{1}{2!}[4(x - 1)^2 + 4(x - 1)(y - 2)] = 2(x - 1)^2 + 2(x - 1)(y - 2).$$

Remark. It is true that if $f(\mathbf{x})$ is a polynomial of degree m, then $f(\mathbf{x}) = p_m(\mathbf{x})$. We will not attempt to prove this fact. The proof is not conceptually difficult, but it involves messy computations (see Problems 17, 18, and 19).

EXAMPLE 2

Compute $p_2(\mathbf{x})$ for $\mathbf{x}_0 = 0$ and $f(\mathbf{x}) = e^{x_1 + x_2 + \cdots + x_n}$.

SOLUTION

$f(\mathbf{x}_0) = e^0 = 1$. The second term in (14) is

$$\sum_{i=1}^{n} f_i(0)x_i. \tag{18}$$

Now $f_i(\mathbf{x}) = e^{x_1 + x_2 + \cdots + x_n}$, so that $f_i(0) = 1$ and (18) becomes

$$\sum_{i=1}^{n} x_i = x_1 + x_2 + \cdots + x_n.$$

The third term in (14) is

$$\frac{1}{2!}\sum_{i=1}^{n}\sum_{j=1}^{n}f_{ij}(0)x_ix_j. \tag{19}$$

Again $f_{ij}(0) = 1$ and (19) becomes

$$\frac{1}{2!}\sum_{i=1}^{n}\sum_{j=1}^{n}x_ix_j = \frac{1}{2!}\sum_{i=1}^{n}x_i\sum_{j=1}^{n}x_j = \frac{1}{2!}\left(\sum_{i=1}^{n}x_i\right)^2 = \frac{1}{2!}(x_1 + x_2 + \cdots + x_n)^2.$$

Thus

$$p_2(\mathbf{x}) = 1 + (x_1 + x_2 + \cdots + x_n) + \frac{1}{2!}(x_1 + x_2 + \cdots + x_n)^2.$$

We can obtain this answer immediately from the one-dimensional Maclaurin formula for the exponential

$$e^x = 1 + x + \frac{x^2}{2!} + \frac{x^3}{3!} + \cdots.$$

Letting $x = x_1 + x_2 + \cdots + x_n$, we have

$$e^{x_1 + x_2 + \cdots + x_n} \approx 1 + (x_1 + x_2 + \cdots + x_n)$$

$$+ \frac{(x_1 + x_2 + \cdots + x_n)^2}{2!} + \cdots + \frac{(x_1 + x_2 + \cdots + x_n)^k}{k!} + \cdots$$

Taylor Polynomials and the Binomial Theorem

Before doing any more examples, we illustrate how the computation of Taylor polynomials can be made a bit easier. We first suppose that $n = 2$ so that $\mathbf{x} = (x, y)$ and $\mathbf{x}_0 = (x_0, y_0)$. Let us compute $p_2(x, y)$ with $x_1 = x$, $x_2 = y$, $x_1^{(0)} = x_0$ and $x_2^{(0)} = y_0$. Also $f_1(\mathbf{x}_0) = f_x(x_0, y_0)$ and $f_2(\mathbf{x}_0) = f_y(x_0, y_0)$. Equation (14) then becomes

$$f(x, y) = f(x_0, y_0) + f_x(x_0, y_0)(x - x_0) + f_y(x_0, y_0)(y - y_0)$$

$$+ \frac{1}{2!}[f_{xx}(x_0, y_0)(x - x_0)^2 + 2f_{xy}(x_0, y_0)(x - x_0)(y - y_0) + f_{yy}(x, y)(y - y_0)^2]$$

$$+ R_3(x, y). \tag{20}$$

Here we have used the fact that $f_{xy} = f_{yx}$ (Theorem 9.7.1).

The linear term is easy to remember. What about the quadratic term? Recall that

$$(a + b)^2 = a^2 + 2ab + b^2. \tag{21}$$

Compare this expression with the quadratic terms in brackets in equation (20)! What about the cubic term? The binomial theorem states that

$$(a + b)^3 = a^3 + 3a^2b + 3ab^2 + b^3. \tag{22}$$

If we compute the third degree term in (14), again with $n = 2$, we obtain

$$\frac{1}{3!}[f_{xxx}(x_0, y_0)(x - x_0)^3 + 3f_{xxy}(x_0, y_0)(x - x_0)^2(y - y_0)$$

$$+ 3f_{xyy}(x_0, y_0)(x - x_0)(y - y_0)^2 + f_{yyy}(y - y_0)^3]. \qquad (23)$$

Compare expressions (22) and (23).

These results for the quadratic and cubic terms suggest the following rule, which we offer without proof:

If $n = 2$, the kth degree term in the Taylor polynomial of f at (x_0, y_0) is obtained as follows:

STEP (i)
Expand $(a + b)^k$ by the binomial theorem.

STEP (ii)
A typical term in the expansion is

$$\binom{k}{j}a^{k-j}b^j = \frac{k!}{j!(k-j)!}a^{k-j}b^j.$$

Write the term

$$\binom{k}{j}f_{\underbrace{xx \cdots x}_{k-j \text{ times}}\underbrace{yy \cdots y}_{j \text{ times}}}(x - x_0)^{k-j}(y - y_0)^j. \qquad (24)$$

There will be $k + 1$ such terms (for $j = 0, 1, 2, \ldots, k$)

STEP (iii)
The kth degree term in the Taylor polynomial is

$$\frac{1}{k!}[\text{sum of the } k + 1 \text{ terms obtained in step (ii)}]$$

Note. This procedure can be extended to $n > 2$. For example,

$$(a + b + c)^2 = a^2 + 2ab + 2ac + b^2 + 2bc + c^2.$$

Then, as can be verified, the quadratic term in (14) at $\mathbf{x}_0 = (x_0, y_0, z_0)$ for $n = 3$ is

$$\frac{1}{2!}[f_{xx}(x - x_0)^2 + 2f_{xy}(x - x_0)(y - y_0) + 2f_{xz}(x - x_0)(z - z_0)$$

$$+ f_{yy}(y - y_0)^2 + 2f_{yz}(y - y_0)(z - z_0) + f_{zz}(z - z_0)^2]. \qquad (25)$$

In (25), f_{xx} denotes $f_{xx}(x_0, y_0, z_0)$, f_{xz} denotes $f_{xz}(x_0, y_0, z_0)$, and so on.

After a few computations, it becomes apparent that computing a Taylor polynomial is not much more difficult than computing expressions of the form $(x_1 + x_2 + \cdots + x_n)^k$.

EXAMPLE 3

Compute $p_2(\mathbf{x})$ for $f(\mathbf{x}) = \sin(2x + y)$ at $\mathbf{x}_0 = (\pi/6, \pi/3)$.

SOLUTION
We see that, at $\mathbf{x}_0$, $2x + y = 2\pi/3$ so

$$f(x_0, y_0) = \sin\left(\frac{2\pi}{3}\right) = \frac{\sqrt{3}}{2}.$$

But $f_x(x, y) = 2\cos(2x + y)$ and $f_y(x, y) = \cos(2x + y)$, so that

$$f_x\left(\frac{\pi}{6}, \frac{\pi}{3}\right) = -1, \quad f_y\left(\frac{\pi}{6}, \frac{\pi}{3}\right) = -\frac{1}{2}$$

Next

$$f_{xx}(x, y) = -4\sin(2x + y) \quad \text{and} \quad f_{xx}\left(\frac{\pi}{6}, \frac{\pi}{3}\right) = -2\sqrt{3}$$

$$f_{xy}(x, y) = -2\sin(2x + y) \quad \text{and} \quad 2f_{xy}\left(\frac{\pi}{6}, \frac{\pi}{3}\right) = -2\sqrt{3}$$

$$f_{yy}(y, y) = -\sin(2x + y) \quad \text{and} \quad f_{yy}\left(\frac{\pi}{6}, \frac{\pi}{3}\right) = -\frac{\sqrt{3}}{2}$$

Then, from (20), we have

$$p_2(x, y) = \frac{\sqrt{3}}{2} - \left(x - \frac{\pi}{6}\right) - \frac{1}{2}\left(y - \frac{\pi}{3}\right)$$

$$+ \frac{1}{2}\left[-2\sqrt{3}\left(x - \frac{\pi}{6}\right)^2 - 2\sqrt{3}\left(x - \frac{\pi}{6}\right)\left(y - \frac{\pi}{3}\right) - \frac{\sqrt{3}}{2}\left(y - \frac{\pi}{3}\right)^2\right].$$

$$= \frac{\sqrt{3}}{2} - \left(x - \frac{\pi}{6}\right) - \frac{1}{2}\left(y - \frac{\pi}{3}\right) - \sqrt{3}\left(x - \frac{\pi}{6}\right)^2$$

$$- \sqrt{3}\left(x - \frac{\pi}{6}\right)\left(y - \frac{\pi}{3}\right) - \frac{\sqrt{3}}{4}\left(y - \frac{\pi}{3}\right)^2.$$

Problems 9.10

The **linearization** of a function $f: \mathbb{R}^n \to \mathbb{R}$ at $\mathbf{x}_0$ is the first-degree Taylor polynomial that approximates that function at $\mathbf{x}_0$. In Problems 1–7 find the linearization of the given function.

1. $\sin(x + y)$; $\mathbf{x}_0 = \mathbf{0}$
2. $\cos(3x - 2y)$; $\mathbf{x}_0 = \mathbf{0}$
3. $e^{x_1 - 4x_2 + x_3}$; $\mathbf{x}_0 = \mathbf{0}$
4. $\ln(1 + x_1 + x_2 + \cdots + x_n)$; $\mathbf{x}_0 = \mathbf{0}$

5. $\sqrt{x + y}$; $\mathbf{x}_0 = (2, 2)$
6. $\sqrt{x_1 + x_2 + x_3 + x_4}$; $\mathbf{x}_0 = (1, 1, 1, 1)$
7. $\frac{x}{y} + \frac{y}{x}$; $\mathbf{x}_0 = (2, 1)$

In Problems 8–16 find the Taylor polynomial of degree m that approximates the given function at the given point.

8. $f(x, y) = \ln(x + 2y)$; $\mathbf{x}_0 = (1, 0)$; $m = 2$

9. $f(x, y) = \sin(x^2 + y^2)$; $\mathbf{x}_0 = (0, 0)$; $m = 2$

10. $f(x_1, x_2, x_3, x_4) = \sin(x_1 + x_2 + x_3 + x_4)$;

$$\mathbf{x}_0 = \left(\frac{\pi}{8}, \frac{\pi}{8}, \frac{\pi}{8}, \frac{\pi}{8}\right); \quad m = 2$$

11. $f(x, y, z) = \sin xyz$; $\mathbf{x}_0 = \mathbf{0}$; $m = 9$

12. $f(x_1, x_2, \ldots, x_n) = \cos(x_1 x_2 \cdots x_n)$; $\mathbf{x}_0 = \mathbf{0}$; $m = 4n$

13. $f(x, y) = e^x \sin y$; $\mathbf{x}_0 = \mathbf{0}$; $m = 2$

14. $f(x, y) = \sin x \cos y$; $\mathbf{x}_0 = \mathbf{0}$; $m = 2$

15. $f(x, y) = e^x \sin y$; $\mathbf{x}_0 = \left(2, \frac{\pi}{4}\right)$; $m = 2$

16. $f(x, y) = \sin x \cos y$; $\mathbf{x}_0 = \left(\frac{\pi}{2}, \pi\right)$; $m = 3$

17. Verify that the second-degree Taylor polynomial of xy around $\mathbf{0}$ is xy.

18. Verify that the fifth-degree Taylor polynomial of x^2yz^2 around $\mathbf{0}$ is x^2yz^2.

* 19. (a) Show that if $f(x, y)$ is a polynomial of degree m, then its mth-degree Taylor polynomial $p_m(\mathbf{x})$ around $\mathbf{0}$ is equal to f.

   (b) What can you say about the mth-degree Taylor polynomial around a value $\mathbf{x}_0 \neq \mathbf{0}$?

20. Write out the third-degree Taylor polynomial for a function of three variables. [*Hint:* First compute $(a + b + c)^3$.]

21. Write out the fifth-degree term in the Taylor polynomial for a function of two variables.

Review Exercises for Chapter 9

In Exercises 1–8, find the Cartesian equation of each curve and then sketch the curve in the xy-plane.

1. $\mathbf{f}(t) = t\mathbf{i} + 2t\mathbf{j}$

2. $\mathbf{f}(t) = (2t - 6)\mathbf{i} + t^2\mathbf{j}$

3. $\mathbf{f}(t) = t^2\mathbf{i} + (2t - 6)\mathbf{j}$

4. $\mathbf{f}(t) = t^2\mathbf{i} + t^4\mathbf{j}$

5. $\mathbf{f}(t) = (\cos 4t)\mathbf{i} + (\sin 4t)\mathbf{j}$

6. $\mathbf{f}(t) = 4(\sin t)\mathbf{i} + 9(\sin t)\mathbf{j}$

7. $\mathbf{f}(t) = t^6\mathbf{i} + t^2\mathbf{j}$

8. $\mathbf{f}(t) = e^t(\cos t)\mathbf{i} + e^t(\sin t)\mathbf{j}$

In Exercises 9–16, find the slope of the line tangent to the given curve for the given value of the parameter, and then find all points at which the curve has vertical and horizontal tangents.

9. $x = t^3$; $y = 6t$; $t = 1$

10. $x = t^7$; $y = t^8 - 5$; $t = 2$

11. $x = \sin 5\theta$; $y = \cos 5\theta$; $\theta = \pi/3$

12. $x = \cos^2 \theta$; $y = -3\theta$; $\theta = \pi/4$

13. $x = \cosh t$; $y = \sinh t$; $t = 0$

14. $x = \dfrac{2}{\theta}$; $y = -3\theta$; $\theta = 10\pi$

15. $x = 3 \cos \theta$; $y = 4 \sin \theta$; $\theta = \pi/3$

16. $x = 3 \cos \theta$; $y = -4 \sin \theta$; $\theta = 2\pi/3$

17. Find the slope of the line tangent to the polar curve $r = -3 \cos \theta$ for $\theta = \pi/3$.

18. Find the slope of the line tangent to the polar curve $r = \sin 2\theta$ for $\theta = \pi/8$.

19. Find the slope of the line tangent to the *spiral of Archimedes* $r = \theta$ for $\theta = \pi$.

20. Calculate the equation of the line that is tangent to the curve $x = t^2 - t - 2$, $y = t^2 - t + 1$ at the point $(4, 7)$.

In Exercises 21–24, find the first and second derivatives of the given vector functions.

21. $\mathbf{f}(t) = 2t\mathbf{i} - t^3\mathbf{j}$

22. $\mathbf{f}(t) = \left(\dfrac{1}{t^2}\right)\mathbf{i} + 3\sqrt{t}\,\mathbf{j}$

23. $\mathbf{f}(t) = (\cos 5t)\mathbf{i} + 2(\sin 4t)\mathbf{j}$

24. $\mathbf{f}(t) = (\tan t)\mathbf{i} + (\cot t)\mathbf{j}$

In Exercises 25–30, find the unit tangent and unit normal vectors to the given curve for the given value of t.

25. $\mathbf{f}(t) = t^4\mathbf{i} + t^5\mathbf{j}$; $t = 1$

26. $\mathbf{f}(t) = (\cos 2t)\mathbf{i} + (\sin 2t)\mathbf{j}$; $t = \pi/6$

27. $\mathbf{f}(t) = (\sin 5t)\mathbf{i} + (\cos 5t)\mathbf{j}$; $t = \pi/30$

28. $\mathbf{f}(t) = (\cosh 2t)\mathbf{i} + (\sinh 2t)\mathbf{j}$; $t = 0$

29. $\mathbf{f}(t) = (\ln t)\mathbf{i} + \sqrt{t}\,\mathbf{j}$; $t = 1$

30. $\mathbf{f}(t) = (\tan t)\mathbf{i} + (\cot t)\mathbf{j}$; $t = \pi/3$

In Exercises 31–36, find the length of the arc over the given interval or the length of the closed curve.

31. $x = \cos 4\theta$; $y = \sin 4\theta$; $0 \leq \theta \leq \pi/12$

32. $x = e^t \sin t$; $y = e^t \cos t$; $0 \leq t \leq \pi/2$

33. $r = 2(1 + \cos \theta)$

34. $r = 5 \sin \theta$

35. $r = \theta^2$; $1 \le \theta \le 5$

36. $r = 2\theta$; $0 \le \theta \le 20\pi$

In Exercises 37–40, find parametric equations for the given curve in terms of the arc length s measured from the point reached when $t = 0$. Verify your answer by using equation (9.5.3).

37. $\mathbf{f} = 3t\mathbf{i} + 4t^{3/2}\mathbf{j}$

38. $\mathbf{f} = \frac{2}{9}t^{9/2}\mathbf{i} + \frac{1}{3}t^3\mathbf{j}$

39. $\mathbf{f} = 2(\cos 3t)\mathbf{i} + 2(\sin 3t)\mathbf{j}$

40. $\mathbf{f} = e^t(\sin t)\mathbf{i} + e^t(\cos t)\mathbf{j}$

In Exercises 41–43, calculate all second partial derivatives and show that all pairs of mixed partials are equal.

41. $f(x, y) = xy^3$

42. $f(x, y) = \sqrt{x^2 - y^2}$

43. $f(x, y) = \dfrac{x + y}{x - y}$

In Exercises 44–51, calculate the gradient of the given function at the given point.

44. $f(x, y) = x^2 - y^3$; $(1, 2)$

45. $f(x, y) = \tan^{-1}\dfrac{y}{x}$; $(-1, -1)$

46. $f(x, y) = \dfrac{x - y}{x + y}$; $(3, 2)$

47. $f(x, y) = \cos(x - 2y)$; $\left(\dfrac{\pi}{2}, \dfrac{\pi}{6}\right)$

48. $f(x, y, z) = xy + yz^3$; $(1, 2, -1)$

49. $f(x, y, z) = \dfrac{x - y}{3z}$; $(2, 1, 4)$

50. $f(x, y, z) = \dfrac{1}{\sqrt{x^2 + y^2 + z^2}}$; (a, b, c)

51. $f(x, y, z) = e^{-(x^2 + y^3 + z^4)}$; $(0, -1, 1)$

In Exercises 52–60, use the chain rule to calculate the indicated derivative.

52. $z = 2xy$; $x = \cos t$, $y = \sin t$; dz/dt

53. $z = \sin^{-1}\dfrac{y}{x}$; $x = 1 + t$, $y = t^2$; $\dfrac{dz}{dt}$

54. $w = \ln(1 - x - 2y + 3z)$; $x = e^t \sin t$, $y = e^t \cos t$, $z = t^2$; dw/dt

55. $z = \dfrac{y}{x}$; $x = r - s$, $y = r + s$; $\dfrac{\partial z}{\partial s}$

56. $z = xy^3$; $x = \dfrac{r}{s}$, $y = \dfrac{s^2}{r}$; $\dfrac{\partial z}{\partial r}$

57. $z = \sin(x - y)$; $x = e^{r+s}$, $y = e^{r-s}$; $\partial z/\partial s$

58. $w = xyz$; $x = rs$, $y = \dfrac{r}{s}$, $z = s^2 r^3$; $\dfrac{\partial w}{\partial r}$ and $\dfrac{\partial w}{\partial s}$

59. $w = x^3 y + y^3 z$; $x = rst$, $y = \dfrac{rs}{t}$, $z = \dfrac{rt}{s}$; $\dfrac{\partial w}{\partial s}$ and $\dfrac{\partial w}{\partial t}$

60. $w = \ln(x + 2y + 5z)$; $x = e^{r+s+t}$, $y = \sqrt{rst^2}$, $z = \dfrac{1}{\sqrt{r + s + t}}$; $\dfrac{\partial w}{\partial r}$ and $\dfrac{\partial w}{\partial t}$

61. Show that $\displaystyle\lim_{(x,y) \to (0,0)} \dfrac{xy}{y^2 - x^2}$ does not exist.

62. Show that $\displaystyle\lim_{(x,y) \to (0,0)} \dfrac{y^2 - 2x}{y^2 + 2x}$ does not exist.

63. Show that $\displaystyle\lim_{(x,y) \to (0,0)} \dfrac{4xy^3}{x^2 + y^4} = 0$.

64. Show that $\displaystyle\lim_{(x,y,z) \to (0,0,0)} \dfrac{zy^2 + x^3}{x^2 + y^2 + z^2} = 0$.

In Exercises 65–74, find the curvature and radius of curvature for each curve. Sketch the unit tangent vector and the circle of curvature at the given point.

65. $\mathbf{f} = (\cos 2t)\mathbf{i} + (\sin 2t)\mathbf{j}$; $t = \pi/3$

66. $\mathbf{f} = t^2\mathbf{i} + 2t\mathbf{j}$; $t = 2$

67. $\mathbf{f} = 4(\cos t)\mathbf{i} + 9(\sin t)\mathbf{j}$; $t = \pi/4$

68. $y = 2x^2$; $(0, 0)$

69. $xy = 1$; $(2, \frac{1}{2})$

70. $y = e^{-x}$; $(1, 1/e)$

71. $y = \sqrt{x}$; $(4, 2)$

72. $r = 3 \cos 2\theta$; $\theta = \pi/6$

73. $r = 1 + \sin \theta$; $\theta = \pi/2$

74. $r = 3\theta$; $\theta = \pi$

In Exercises 75–78, find the tangential and normal components of acceleration for each of the given position vectors.

75. $\mathbf{f} = 2(\sin t)\mathbf{i} + 2(\cos t)\mathbf{j}$

76. $\mathbf{f} = 4(\cos t)\mathbf{i} + 9(\sin t)\mathbf{j}$

77. $\mathbf{f} = 3t^2\mathbf{i} + 2t^3\mathbf{j}$

78. $\mathbf{f} = (\cos t^2)\mathbf{i} + (\sin t^2)\mathbf{j}$

79. A 1300-kg race car is driven at a speed of 175 km/hr on an unbanked circular race track of radius 65 m. What frictional force must be exerted by the tires on the road surface to keep the car from skidding?

80. A 1500-kg rocket carrying 1000 kg of fuel starts from a position motionless in space. Its exhaust velocity is 1.5 km/sec. What is its velocity when all its fuel is consumed?

81. The rocket in Exercise 80 is launched from the earth, and gas is emitted at a constant rate of 40 kg/sec.

   (a) Find the net initial force acting on the rocket.

   (b) Find the net force just before all the fuel is expended.

82. In Exercise 81, what is the minimum thrust needed to get the rocket off the ground?

In Exercises 83–88, calculate the directional derivative of the given function at the given point, in the direction of the given vector **v**.

83. $f(x, y) = \dfrac{y}{x}$ at $(1, 2)$; $\mathbf{v} = \mathbf{i} - \mathbf{j}$

84. $f(x, y) = 3x^2 - 4xy$ at $(3, -1)$; $\mathbf{v} = 2\mathbf{i} + 5\mathbf{j}$

85. $f(x, y) = \tan^{-1} \dfrac{y}{x}$ at $(1, -1)$; $\mathbf{v} = -3\mathbf{i} + 2\mathbf{j}$

86. $f(x, y, z) = xy^2 - zy^3$ at $(1, 2, 3)$; $\mathbf{v} = \mathbf{i} - \mathbf{j} + 2\mathbf{k}$

87. $f(x, y, z) = \dfrac{1}{\sqrt{x^2 + y^2 + z^2}}$ at $(1, -1, 2)$; $\mathbf{v} = -2\mathbf{i} + \mathbf{j} - 3\mathbf{k}$

88. $f(x, y, z) = e^{-(x+y^2-xz)}$ at $(1, 0, -1)$; $\mathbf{v} = 2\mathbf{i} + 5\mathbf{j} + \mathbf{k}$

89. Show that the force $\mathbf{f}(\mathbf{x}) = -3\mathbf{x}/|\mathbf{x}|^5$ is conservative and find a potential function for **f**.

90. Show that the force $\mathbf{f}(x, y) = y^2\mathbf{i} - x^2\mathbf{j}$ is not conservative.

In Exercises 91–96 find the equation of the tangent plane and symmetric equations of the normal line to the given surface at the point given.

91. $x^2 + y^2 + z^2 = 3$; $(1, 1, 1)$

92. $x^{1/2} + y^{3/2} + z^{1/2} = 3$; $(1, 0, 4)$

93. $3x - y + 5z = 15$; $(-1, 2, 4)$

94. $xy^2 = yz^3$; $(1, 1, 1)$

95. $xyz = 6$; $(-2, 1, -3)$

96. $\sqrt{\dfrac{x-y}{y+z}} = \dfrac{1}{2}$; $(2, 1, 3)$

97. Find the third-degree Taylor polynomial of $\cos(x + 2y)$ at $\mathbf{x}_0 = \mathbf{0}$.

98. Find the second-degree Taylor polynomial of
$$\ln(1 + 5x - 4y) \text{ at } \mathbf{x}_0 = \mathbf{0}.$$

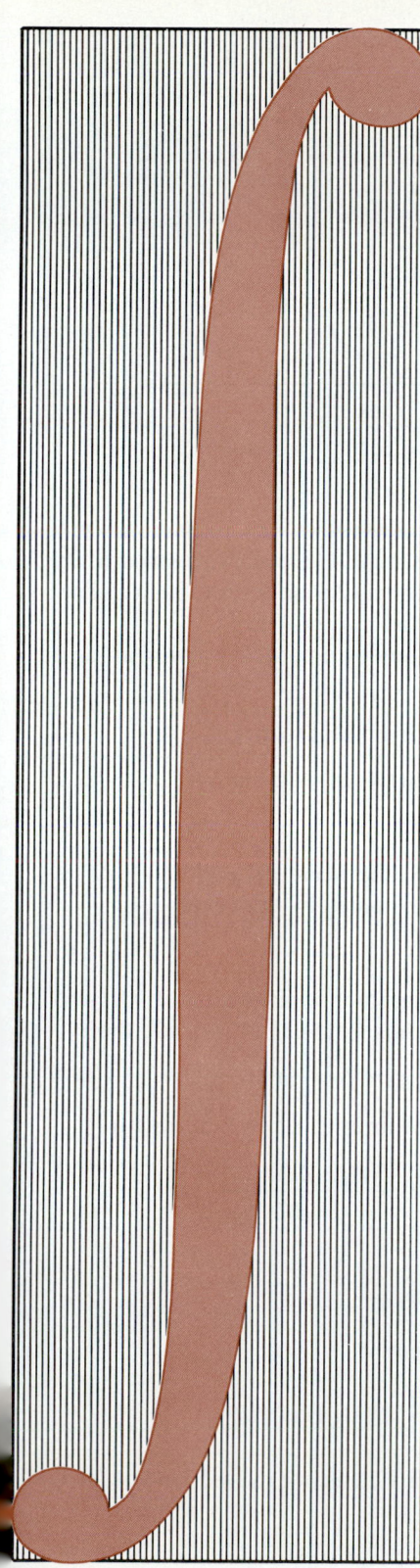

chapter 10

Vector Integral Calculus

In this chapter we define the concepts of line and surface integrals and consider several important applications of these integrals that occur frequently in physics and engineering. We shall show that a line integral is a natural extension of the definite integral, and the surface integral is a generalization of the double integral of calculus.

10.1 Line Integrals

In Section 6.2 we showed that if a constant force $\mathbf{F}$ is applied to a particle that moves along a vector $\mathbf{d}$, then the work done by the force on the particle is given by

$$W = \mathbf{F} \cdot \mathbf{d}. \tag{1}$$

We now calculate the work done when a particle moves along a curve C. In doing so, we will define an important concept in applied mathematics—the *line integral*.

Suppose that a curve in the plane is given parametrically by

$$C: \quad \mathbf{x}(t) = x(t)\mathbf{i} + y(t)\mathbf{j}. \tag{2}$$

If a force is applied to a particle moving along C, then such a force will have magnitude and direction, so the force will be a vector function of the vector $\mathbf{x}(t)$. That is, the force will be a vector field. We write

$$\mathbf{F}(\mathbf{x}) = \mathbf{F}(x, y) = P(x, y)\mathbf{i} + Q(x, y)\mathbf{j} \tag{3}$$

where P and Q are scalar-valued functions. We assume that $\mathbf{F}$ is defined in a region R so that our computations are valid for any curve C lying in the region R (see Fig. 1). The problem is to determine the work done when a particle moves on C from a point $\mathbf{x}(a)$ to a point $\mathbf{x}(b)$ subject to the force $\mathbf{F}$ given by

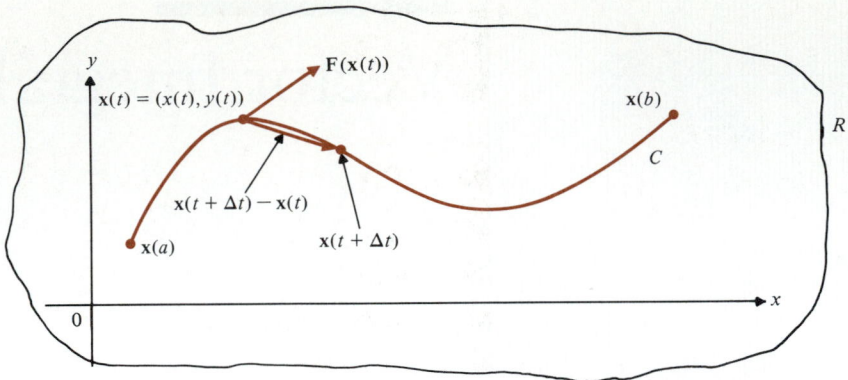

FIGURE 1

Piecewise Smooth Curve

(3). We will assume in our discussion that the curve C is **piecewise smooth.** By that we mean that the functions $x(t)$ and $y(t)$ in (2) are continuously differentiable or that x' and y' exist and are piecewise continuous.

To motivate the definition of the work done by a particle moving along a curve C, let

$$W(t) = \text{work done in moving from } \mathbf{x}(a) \text{ to } \mathbf{x}(t). \qquad \textbf{(4)}$$

Then the work done in moving from $\mathbf{x}(t)$ to $\mathbf{x}(t + \Delta t)$ is given by

$$W(t + \Delta t) - W(t). \qquad \textbf{(5)}$$

Now if Δt is small, then the part of the curve between $\mathbf{x}(t)$ and $\mathbf{x}(t + \Delta t)$ is "close" to a straight line and so can be approximated by the vector

$$\mathbf{x}(t + \Delta t) - \mathbf{x}(t). \qquad \textbf{(6)}$$

If Δt is small and if $\mathbf{F}(\mathbf{x})$ is continuous, then the force applied between $\mathbf{x}(t)$ and $\mathbf{x}(t + \Delta t)$ is approximately equal to $\mathbf{F}(\mathbf{x}(t))$. Thus by (1), if Δt is small, then

$$W(t + \Delta t) - W(t) \approx \mathbf{F}(\mathbf{x}(t)) \bullet [\mathbf{x}(t + \Delta t) - \mathbf{x}(t)]. \qquad \textbf{(7)}$$

We divide both sides of (7) by Δt and take the limit as $\Delta t \to 0$ to obtain

$$\lim_{\Delta t \to 0} \frac{W(t + \Delta t) - W(t)}{\Delta t} = \lim_{\Delta t \to 0} \left\{ \mathbf{F}(\mathbf{x}(t)) \bullet \frac{[\mathbf{x}(t + \Delta t) - \mathbf{x}(t)]}{\Delta t} \right\},$$

or

$$W'(t) = \mathbf{F}(\mathbf{x}(t)) \bullet \mathbf{x}'(t). \qquad \textbf{(8)}$$

Integrating both sides of (8) from a to b, we get

$$W(b) - W(a) = \int_a^b W'(t)\, dt = \int_a^b \mathbf{F}(\mathbf{x}(t)) \bullet \mathbf{x}'(t)\, dt.$$

But $W(a) = 0$, since no work is done if the particle is not moved, and $W(b)$ is the total work done (denoted by W). Hence

$$W = \int_a^b \mathbf{F}(\mathbf{x}(t)) \cdot \mathbf{x}'(t)\, dt. \tag{9}$$

Equation (9) is our definition of the work done by a particle moving along a curve C from $\mathbf{x}(a)$ to $\mathbf{x}(b)$. Note that this definition involves the chosen parametrization (2) of the curve C. Thus, it is reasonable to ask if a different parametrization of C yields the same value for W. It turns out (see Problem 21) that the line integral (9) is *independent* of the parametrization; that is, the same value of W is obtained for any parametrization of C. Thus, we can write equation (9) as

$$W = \int_C \mathbf{F}(\mathbf{x}) \cdot \mathbf{dx}. \tag{10}$$

The symbol $\int_C$ is read "the integral along the curve C." The integral in (10) is called a *line integral of* $\mathbf{F}$ *over* C.

Remark. If $\mathbf{F}$ acts along C and C lies along the x-axis, then C is given by $\mathbf{x}(t) = x(t)\mathbf{i} + 0\mathbf{j}$, $\mathbf{F}(\mathbf{x}) = F(x)\mathbf{i}$ and $\mathbf{x}'(t) = x'(t)\mathbf{i}$, so that (10) becomes

$$\int_C \mathbf{F}(\mathbf{x}) \cdot d\mathbf{x} = \int_a^b (F(x(t))\mathbf{i}) \cdot (x'(t)\mathbf{i})\, dt = \int_a^b F(x(t))x'(t)\, dt = \int_{x(a)}^{x(b)} F(x)\, dx,$$

which is our usual definite integral.

Remark. Since $\mathbf{F}$ is given by (3), we can write equation (9) as

$$W = \int_a^b [P(x(t), y(t))\, x'(t) + Q(x(t), y(t))\, y'(t)]\, dt. \tag{11}$$

EXAMPLE 1

A particle is moving along the parabola $y = x^2$ subject to a force given by the vector field $2xy\mathbf{i} + (x^2 + y^2)\mathbf{j}$. How much work is done in moving from the point $(1, 1)$ to the point $(3, 9)$ if forces are measured in newtons and distances are measured in meters?

SOLUTION

The curve C is given parametrically by

$$\mathbf{x}(t) = t\mathbf{i} + t^2\mathbf{j} \qquad \text{between} \qquad t = 1 \quad \text{and} \quad t = 3.$$

We therefore have $x(t) = t$ and $y(t) = t^2$. Then $P = 2xy = 2t^3$ and $Q = x^2 + y^2 = t^2 + t^4$. Also, $x'(t) = 1$ and $y'(t) = 2t$, so by (11)

$$W = \int_1^3 [(2t^3)1 + (t^2 + t^4)(2t)]\, dt = \int_1^3 (4t^3 + 2t^5)\, dt = 322\tfrac{2}{3} \text{ joules.}$$

EXAMPLE 2

Two electrical charges of like polarity (i.e., both positive or both negative) will repel each other. If a charge of α coulombs is placed at the origin and a charge of 1 coulomb of the same polarity is at the point (x, y), then the force of repulsion is given by

$$F(x, y) = \frac{\alpha x}{(x^2 + y^2)^{3/2}}\mathbf{i} + \frac{\alpha y}{(x^2 + y^2)^{3/2}}\mathbf{j}.$$

How much work is done by the force on the 1-coulomb charge as the charge moves on the straight line from $(1, 0)$ to $(3, -2)$?

SOLUTION
The straight line is the line $y = 1 - x$, or, parametrically,

$$\mathbf{x}(t) = x(t)\mathbf{i} + y(t)\mathbf{j} = t\mathbf{i} + (1 - t)\mathbf{j} \qquad \text{for} \qquad 1 \le t \le 3.$$

Then

$$P = \frac{\alpha t}{[(t^2 + (1 - t)^2]^{3/2}} = \frac{\alpha t}{(2t^2 - 2t + 1)^{3/2}}$$

and

$$Q = \frac{\alpha(1 - t)}{[t^2 + (1 - t)^2]^{3/2}} = \frac{\alpha(1 - t)}{(2t^2 - 2t + 1)^{3/2}}.$$

Then, since $x'(t) = 1$ and $y'(t) = -1$,

$$W = \alpha \int_1^3 \left\{ \frac{t}{(2t^2 - 2t + 1)^{3/2}}(1) + \frac{(1 - t)}{(2t^2 - 2t + 1)^{3/2}}(-1) \right\} dt$$

$$= \alpha \int_1^3 \frac{2t - 1}{(2t^2 - 2t + 1)^{3/2}}\, dt = \frac{\alpha}{2} \int_1^3 \frac{4t - 2}{(2t^2 - 2t + 1)^{3/2}}\, dt$$

$$= -\alpha(2t^2 - 2t + 1)^{-1/2} \Big|_1^3 = \alpha\left(1 - \frac{1}{\sqrt{13}}\right).$$

EXAMPLE 3

How much work is done by the force on the charge in Example 2 if it moves in the counterclockwise direction along the semicircle $x^2 + y^2 = a^2$, $y \ge 0$?

SOLUTION
Here

$$\mathbf{x}(t) = a(\cos t)\mathbf{i} + a(\sin t)\mathbf{j}, \qquad 0 \le t \le \pi.$$

Then

$$P = \frac{\alpha x}{(x^2 + y^2)^{3/2}} = \frac{a\alpha \cos t}{(a^2 \cos^2 t + a^2 \sin^2 t)^{3/2}} = \frac{a\alpha \cos t}{a^3} = \frac{\alpha \cos t}{a^2}$$

and

$$Q = \frac{\alpha y}{(x^2 + y^2)^{3/2}} = \frac{\alpha \sin t}{a^2}.$$

Since $x'(t) = -a \sin t$ and $y'(t) = a \cos t$,

$$W = \frac{\alpha}{a} \int_0^\pi [\cos t(-\sin t) + \sin t(\cos t)] \, dt = 0.$$

This result is no surprise since in this example the force **F** and the direction of the curve **x**′ are orthogonal. Thus the component of the force in the direction of motion is zero, which means that the force does no work.

We used the notion of work to motivate the discussion of the line integral. We now give a general definition of the line integral in the plane.

DEFINITION

Let P and Q be continuous on a set S containing the smooth (or piecewise smooth) curve C given by

$$C: \quad \mathbf{x}(t) = x(t)\mathbf{i} + y(t)\mathbf{j}, \qquad t \in [a, b].$$

Let the vector field **F** be given by

$$\mathbf{F}(x, y) = P(x, y)\mathbf{i} + Q(x, y)\mathbf{j}.$$

Line Integral in the Plane

Then the **line integral** of **F** over C is given by

$$\int_C \mathbf{F}(\mathbf{x}) \cdot d\mathbf{x} = \int_a^b [P(x(t), y(t))x'(t) + Q(x(t), y(t))y'(t)] \, dt. \qquad (12)$$

Remark. If C is piecewise smooth but not smooth, then C is made up of a number of "sections," each of which is smooth. Since C is continuous, these sections are joined. Some typical piecewise smooth curves are sketched in Fig. 2. If C is made up of the n smooth curves $C_1, C_2, \ldots, C_n$, then

$$\int_C \mathbf{F}(\mathbf{x}) \cdot d\mathbf{x} = \int_{C_1} \mathbf{F}(\mathbf{x}) \cdot d\mathbf{x} + \int_{C_2} \mathbf{F}(\mathbf{x}) \cdot d\mathbf{x} + \cdots + \int_{C_n} \mathbf{F}(\mathbf{x}) \cdot d\mathbf{x}. \qquad (13)$$

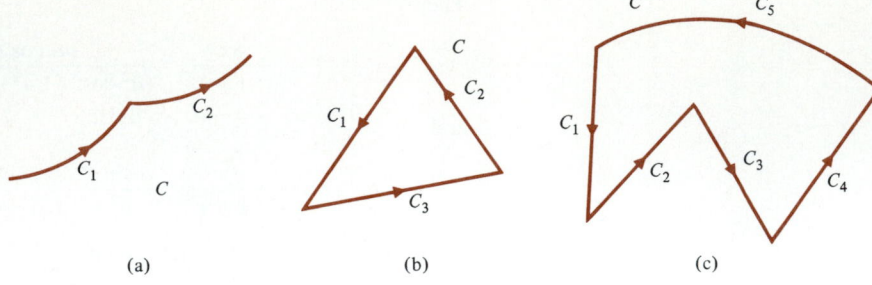

(a)   (b)   (c)

FIGURE 2

EXAMPLE 4

Calculate $\int_C \mathbf{F(x)} \cdot \mathbf{dx}$, where $\mathbf{F}(x, y) = xy\mathbf{i} + ye^x\mathbf{j}$ and C is the rectangle joining the points $(0, 0)$, $(2, 0)$, $(2, 1)$, and $(0, 1)$ if C is traversed in the counterclockwise direction.

SOLUTION

The rectangle is sketched in Fig. 3, and it is made up of four smooth curves (straight lines). We have

$$C_1: \quad \mathbf{x}(t) = t\mathbf{i}, \qquad\qquad 0 \le t \le 2$$

$$C_2: \quad \mathbf{x}(t) = 2\mathbf{i} + t\mathbf{j}, \qquad 0 \le t \le 1$$

$$C_3: \quad \mathbf{x}(t) = (2-t)\mathbf{i} + \mathbf{j}, \qquad 0 \le t \le 2$$

$$C_4: \quad \mathbf{x}(t) = (1-t)\mathbf{j}, \qquad 0 \le t \le 1.$$

Note, for example, that on C_3, $t = 0$ corresponds to the point $2\mathbf{i} + \mathbf{j} = (2, 1)$ and $t = 2$ corresponds to $(2 - 2)\mathbf{i} + \mathbf{j} = (0, 1)$. Thus as t increases, we do move along C_3 in the direction indicated by the arrow in Fig. 3. This illustrates why our parametrization of the rectangle is correct.

Now

$$\int_C \mathbf{F(x)} \cdot \mathbf{dx} = \int_{C_1} \mathbf{F(x)} \cdot \mathbf{dx} + \int_{C_2} \mathbf{F(x)} \cdot \mathbf{dx} + \int_{C_3} \mathbf{F(x)} \cdot \mathbf{dx} + \int_{C_4} \mathbf{F(x)} \cdot \mathbf{dx}.$$

On C_1, $x = t$ and $y = 0$, so that $xy = 0$, $ye^x = 0$, and

$$\int_{C_1} \mathbf{F(x)} \cdot \mathbf{dx} = 0.$$

On C_2, $x = 2$, $y = t$, $x' = 0$, $y' = 1$, and $ye^x = te^2$, so that

$$\int_{C_2} \mathbf{F(x)} \cdot \mathbf{dx} = \int_0^1 te^2 \, dt = \frac{e^2}{2}.$$

On C_3, $x = (2 - t)$, $y = 1$, $x' = -1$, $y' = 0$, and $xy = 2 - t$, so that

$$\int_{C_3} \mathbf{F(x)} \cdot \mathbf{dx} = \int_0^2 (2-t)(-1) \, dt = -2.$$

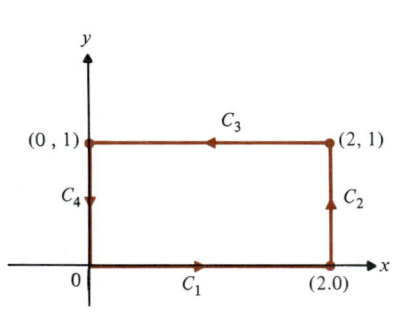

FIGURE 3

On C_4, $x = 0$, $y = 1 - t$, $x' = 0$, $y' = -1$, $xy = 0$, and $ye^x = 1 - t$, so that

$$\int_{C_4} \mathbf{F(x)} \cdot \mathbf{dx} = \int_0^1 (1 - t)(-1)\, dt = -\frac{1}{2}.$$

Hence

$$\int_C \mathbf{F(x)} \cdot \mathbf{dx} = 0 + \frac{e^2}{2} - 2 - \frac{1}{2} = \frac{e^2}{2} - \frac{5}{2} \approx 1.2.$$

Problems 10.1

In Problems 1–14, calculate $\int_C \mathbf{F(x)} \cdot \mathbf{dx}$.

1. $\mathbf{F}(x, y) = x^2\mathbf{i} + y^2\mathbf{j}$; C is the straight line from $(0, 0)$ to $(2, 4)$.

2. $\mathbf{F}(x, y) = x^2\mathbf{i} + y^2\mathbf{j}$; C is the parabola $y = x^2$ from $(0, 0)$ to $(2, 4)$.

3. $\mathbf{F}(x, y) = xy\mathbf{i} + (y - x)\mathbf{j}$; C is the line $y = 2x - 4$ from $(1, -2)$ to $(2, 0)$.

4. $\mathbf{F}(x, y) = xy\mathbf{i} + (y - x)\mathbf{j}$; C is the curve $y = \sqrt{x}$ from $(0, 0)$ to $(1, 1)$.

5. $\mathbf{F}(x, y) = xy\mathbf{i} + (y - x)\mathbf{j}$; C is the unit circle in the counterclockwise direction.

6. $\mathbf{F}(x, y) = xy\mathbf{i} + (y - x)\mathbf{j}$; C is the triangle joining the points $(0, 0)$, $(0, 1)$, and $(1, 0)$ in the counterclockwise direction.

7. $\mathbf{F}(x, y) = xy\mathbf{i} + (y - x)\mathbf{j}$; C is the triangle joining the points $(0, 0)$, $(1, 0)$, and $(1, 1)$ in the counterclockwise direction.

8. $\mathbf{F}(x, y) = e^x\mathbf{i} + e^y\mathbf{j}$; C is the curve of Problem 4.

9. $\mathbf{F}(x, y) = (x^2 + 2y)\mathbf{i} - y^2\mathbf{j}$; C is the part of the ellipse $x^2 + 9y^2 = 9$ joining the points $(0, -1)$ and $(0, 1)$ in the clockwise direction.

10. $\mathbf{F}(x, y) = (\cos x)\mathbf{i} - (\sin y)\mathbf{j}$; C is the curve of Problem 9.

11. $\mathbf{F}(x, y) = e^{x+y}\mathbf{i} + e^{x-y}\mathbf{j}$; C is the curve of Problem 6.

12. $\mathbf{F}(x, y) = e^{x+y}\mathbf{i} + e^{x-y}\mathbf{j}$; C is the curve of Problem 7.

13. $\mathbf{F}(x, y) = (y/x^2)\mathbf{i} + (x/y^2)\mathbf{j}$; C is the straight line segment from $(2, 1)$ to $(4, 6)$.

14. $\mathbf{F}(x, y) = (\ln x)\mathbf{i} + (\ln y)\mathbf{j}$; C is the curve $\mathbf{x}(t) = 2t\mathbf{i} + t^3\mathbf{j}$ for $1 \le t \le 4$.

In Problems 15–19 forces are given in newtons and distances are given in meters.

15. Calculate the work done when a force field $\mathbf{F}(x, y) = x^3\mathbf{i} + xy\mathbf{j}$ moves a particle from the point $(0, 1)$ to the point $(1, e^{\pi/2})$ along the curve $\mathbf{x}(t) = (\sin t)\mathbf{i} + e^t\mathbf{j}$.

16. Calculate the work done when the force field $\mathbf{F}(x, y) = xy\mathbf{i} + (2x^3 - y)\mathbf{j}$ moves a particle around the unit circle in the counterclockwise direction.

17. What is the work done if the particle in Problem 16 is moved in the clockwise direction?

18. Calculate the work done by the force field $\mathbf{F}(x, y) = -y^2x\mathbf{i} + 2x\mathbf{j}$ when a particle is moved around the ellipse $(x^2/a^2) + (y^2/b^2) = 1$ in the counterclockwise direction.

19. Calculate the work done by the force field $\mathbf{F}(x, y) = 2xy\mathbf{i} + y^2\mathbf{j}$ when a particle is moved around the triangle of Problem 7.

20. What is the work done on the 1-coulomb particle of Example 2 if it moves along the line $y = 2x - 3$ from $(1, -1)$ to $(2, 1)$?

* 21. Let $g: [\alpha, \beta] \to [a, b]$ be an increasing differentiable function such that $g(\alpha) = a$ and $g(\beta) = b$. Suppose a curve C in the plane is given parametrically by

$$C: \quad \mathbf{x}(t) = x(t)\mathbf{i} + y(t)\mathbf{j}, \qquad a \le t \le b.$$

(a) Show that C is also represented parametrically by

$$\bar{\mathbf{x}}(u) = x(g(u))\mathbf{i} + y(g(u))\mathbf{j}, \qquad \alpha \le u \le \beta.$$

(b) Verify that

$$\int_a^b \mathbf{F(x}(t)) \cdot \mathbf{x}'(t)\, dt = \int_\alpha^\beta \mathbf{F}(\bar{\mathbf{x}}(u)) \cdot \bar{\mathbf{x}}'(u)\, du.$$

(c) Conclude that the line integral (10) is independent of parametrization.

10.2 Gradient Fields and Independence of Path

As we saw in Section 10.1, a line integral is evaluated by reducing it to a definite integral. *The value of a line integral of a function generally depends not only on the endpoints but also on the given path of integration.*

EXAMPLE 1

Evaluate the integral $\int_C \mathbf{F}(\mathbf{x}) \cdot d\mathbf{x}$ for $\mathbf{F}(x, y) = xy\mathbf{i} + x\mathbf{j}$, where

    (a) C is the straight line from $(0, 0)$ to $(1, 1)$,

    (b) C is the parabola $y = x^2$ from $(0, 0)$ to $(1, 1)$.

SOLUTION

(a) C is given by $x = t$, $y = t$, or $\mathbf{x}(t) = t\mathbf{i} + t\mathbf{j}$ so that $\mathbf{x}'(t) = \mathbf{i} + \mathbf{j}$, $xy = t^2$, and

$$\int_C \mathbf{F}(\mathbf{x}) \cdot d\mathbf{x} = \int_0^1 [(t^2\mathbf{i} + t\mathbf{j}) \cdot (\mathbf{i} + \mathbf{j})]\, dt$$

$$= \int_0^1 (t^2 + t)\, dt = \frac{t^3}{3} + \frac{t^2}{2}\Big|_0^1 = \frac{5}{6}.$$

(b) Here C is parametrized by $x = t$, $y = t^2$, or $\mathbf{x}(t) = t\mathbf{i} + t^2\mathbf{j}$ so that $\mathbf{x}'(t) = \mathbf{i} + 2t\mathbf{j}$ and

$$\int_C \mathbf{F}(\mathbf{x}) \cdot d\mathbf{x} = \int_0^1 [(t^3\mathbf{i} + t\mathbf{j}) \cdot (\mathbf{i} + 2t\mathbf{j})]\, dt$$

$$= \int_0^1 (t^3 + 2t^2)\, dt = \left(\frac{t^4}{4} + \frac{2t^3}{3}\right)\Big|_0^1 = \frac{11}{12}.$$

Although dependence of the value of the line integral on the path of integration is the usual situation, we shall now show that for certain types of vector functions the value of the line integral will depend only on the endpoints of the path of integration.

We first illustrate what we have in mind with an example.

EXAMPLE 2

Let $\mathbf{F}(x, y) = y\mathbf{i} + x\mathbf{j}$. Calculate $\int_C \mathbf{F}(\mathbf{x}) \cdot d\mathbf{x}$, where C is as follows:

    (a) The straight line from $(0, 0)$ to $(1, 1)$.

    (b) The parabola $y = x^2$ from $(0, 0)$ to $(1, 1)$.

    (c) The curve $\mathbf{x}(t) = t^{3/2}\mathbf{i} + t^5\mathbf{j}$ from $(0, 0)$ to $(1, 1)$.

SOLUTION

(a) C is given by $\mathbf{x}(t) = t\mathbf{i} + t\mathbf{j}$. Then $\mathbf{x}'(t) = \mathbf{i} + \mathbf{j}$ and $\mathbf{F} \cdot \mathbf{x}' = (t\mathbf{i} + t\mathbf{j}) \cdot (\mathbf{i} + \mathbf{j}) = 2t$, so

$$\int_C \mathbf{F}(\mathbf{x}) \cdot d\mathbf{x} = \int_0^1 2t\, dt = 1.$$

(b) C is given by $\mathbf{x}(t) = t\mathbf{i} + t^2\mathbf{j}$. Then $\mathbf{x}'(t) = \mathbf{i} + 2t\mathbf{j}$ and $\mathbf{F} \cdot \mathbf{x}' = (t^2\mathbf{i} + t\mathbf{j}) \cdot (\mathbf{i} + 2t\mathbf{j}) = 3t^2$, so

$$\int_C \mathbf{F}(\mathbf{x}) \cdot d\mathbf{x} = \int_0^1 3t^2\, dt = 1.$$

(c) Here $\mathbf{x}'(t) = \frac{3}{2}\sqrt{t}\,\mathbf{i} + 5t^4\mathbf{j}$ and $\mathbf{F}(x, y) = t^5\mathbf{i} + t^{3/2}\mathbf{j}$, so

$$\mathbf{F} \cdot \mathbf{x}' = (t^5\mathbf{i} + t^{3/2}\mathbf{j}) \cdot (\tfrac{3}{2}\sqrt{t}\,\mathbf{i} + 5t^4\mathbf{j}) = \tfrac{3}{2}t^{11/2} + 5t^{11/2} = \tfrac{13}{2}t^{11/2}.$$

Then

$$\int_C \mathbf{F} \cdot \mathbf{x}'\,dt = \int_0^1 \tfrac{13}{2}t^{11/2}\,dt = 1.$$

Before stating our next result, we give three definitions.

DEFINITION

Connected Set

A set Ω in $\mathbb{R}^2$ is **(arcwise) connected** if any two points in Ω can be joined by a piecewise smooth curve lying entirely in Ω.

DEFINITION

Open Set

A set Ω in $\mathbb{R}^2$ is **open** if for every point $\mathbf{x} \in \Omega$, there is an open disk D with center at $\mathbf{x}$ that is contained in Ω.

DEFINITION

Region

A **region** Ω in $\mathbb{R}^2$ is an open, connected set.

These definitions are illustrated in Fig. 1.

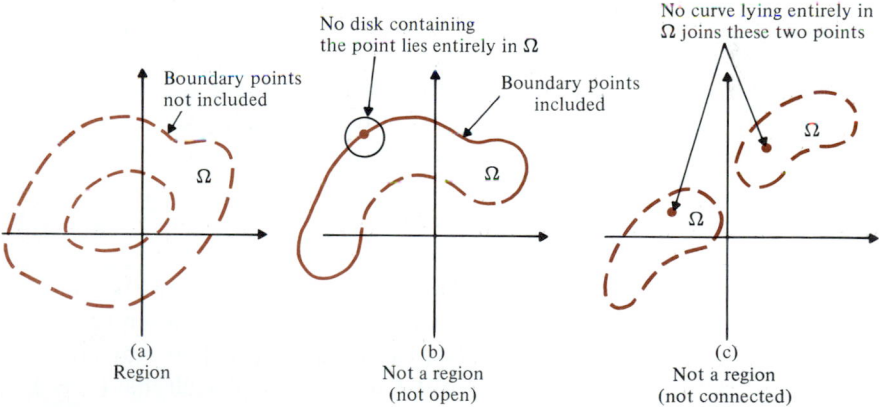

(a)
Region

(b)
Not a region
(not open)

(c)
Not a region
(not connected)

FIGURE 1

Independence of Path

In Example 2 we saw that on three very different curves, we obtained the same answer as we moved between the points $(0, 0)$ and $(1, 1)$. In fact, as we shall show in a moment, we get the same answer if we integrate certain functions along any piecewise smooth curve C joining these two points. When this happens, we say that the line integral is **independent of the path**. A condition that ensures that a line integral is independent of the path over which it is integrated is given below.

■ **THEOREM 1**

Let **F** be continuous in a region Ω in $\mathbb{R}^2$. Then **F** is the gradient of a differentiable function f if and only if for any piecewise smooth curve C lying in Ω, the line integral $\int_C \mathbf{F}(\mathbf{x}) \cdot d\mathbf{x}$ is independent of the path.

PROOF

We first assume that $\mathbf{F}(\mathbf{x}) = \nabla f$ for some differentiable function f. Recall the vector form of the chain rule:

$$\frac{d}{dt} f(\mathbf{x}(t)) = \nabla f(\mathbf{x}(t)) \cdot \mathbf{x}'(t) \tag{1}$$

[see equation (9.8.17)]. Let C given by $\mathbf{x}(t)$: $a \le t \le b$ be any curve such that $\mathbf{x}(a) = \mathbf{x}_0$, and $\mathbf{x}(b) = \mathbf{x}_1$. We will assume that C is smooth. Otherwise, we could write the line integral in the form (10.1.13) and treat the integral over each smooth curve C_i separately. Using (1), we have

$$\int_C \mathbf{F}(\mathbf{x}) \cdot d\mathbf{x} = \int_a^b \mathbf{F}(\mathbf{x}(t)) \cdot \mathbf{x}'(t)\, dt = \overset{\overset{\displaystyle \mathbf{F}=\nabla f}{\swarrow}}{\int_a^b} \nabla f(\mathbf{x}(t)) \cdot \mathbf{x}'(t)\, dt$$

$$= \int_a^b \frac{d}{dt} f(\mathbf{x}(t))\, dt = f(\mathbf{x}(b)) - f(\mathbf{x}(a)) = f(\mathbf{x}_1) - f(\mathbf{x}_0).$$

This proves that the line integral is independent of the path since $f(\mathbf{x}_1) - f(\mathbf{x}_0)$ does not depend on the particular curve chosen.

We now assume that $\int_C \mathbf{F}(\mathbf{x}) \cdot d\mathbf{x}$ is independent of the path and prove that $\mathbf{F} = \nabla f$ for some differentiable function f. Let $\mathbf{x}_0$ be a fixed point in Ω and let $\mathbf{x}$ be any other point in Ω. Since Ω is connected, there is at least one piecewise smooth path C joining $\mathbf{x}_0$ and $\mathbf{x}$, with C wholly contained in Ω. We define a function f by

$$f(\mathbf{x}) = \int_C \mathbf{F}(\mathbf{x}) \cdot d\mathbf{x}.$$

This function is well defined because, by hypothesis, $\int_C \mathbf{F}(\mathbf{x}) \cdot d\mathbf{x}$ is the same no matter what path is chosen between $\mathbf{x}_0$ and $\mathbf{x}$. Write $\mathbf{x}_0 = (x_0, y_0)$ and $\mathbf{x} = (x, y)$. Since Ω is open, there is an open disk D centered at (x, y) that is contained in Ω. Choose $\Delta x > 0$ such that $(x + \Delta x, y) \in D$, and let C_1 be the horizontal line segment joining (x, y) to $(x + \Delta x, y)$. The situation is depicted in Fig. 2. We see that $C \cup C_1$ is a path joining (x_0, y_0) to $(x + \Delta x, y)$, so that

$$f(x + \Delta x, y) = \int_C \mathbf{F}(\mathbf{x}) \cdot d\mathbf{x} + \int_{C_1} \mathbf{F}(\mathbf{x}) \cdot d\mathbf{x} = f(x, y) + \int_{C_1} \mathbf{F}(\mathbf{x}) \cdot d\mathbf{x}. \tag{2}$$

A parametrization for C_1 (which is a horizontal line) is

$$\mathbf{x}(t) = (x + t\,\Delta x, y), \qquad 0 \le t \le 1$$

and

$$\mathbf{x}'(t) = (\Delta x, 0).$$

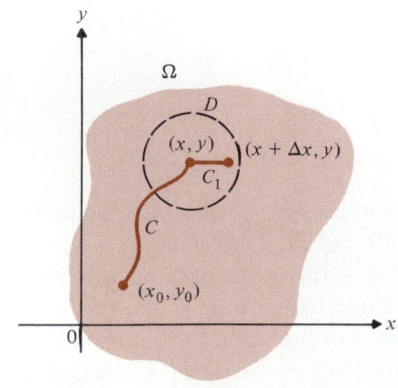

FIGURE 2

Suppose that $\mathbf{F}(\mathbf{x}) = P(x, y)\mathbf{i} + Q(x, y)\mathbf{j}$. Then on C_1

$$\mathbf{F}(\mathbf{x}) \cdot d\mathbf{x} = P(x + t\Delta x, y)\, \Delta x. \tag{3}$$

From (2) and (3) we compute

$$f(x + \Delta x, y) - f(x, y) = \int_{C_1} \mathbf{F}(\mathbf{x}) \cdot d\mathbf{x} = \int_0^1 P(x + t\, \Delta x, y)\, \Delta x\, dt,$$

so that

$$\frac{f(x + \Delta x, y) - f(x, y)}{\Delta x} = \int_0^1 P(x + t\, \Delta x, y)\, dt. \tag{4}$$

In (4), $P(x + t\, \Delta x, y)$ is a function of one variable (t), so we may apply the mean value theorem for integrals to see that there is a number $\bar{x}$ with $x < \bar{x} < x + \Delta x$ such that

$$\int_0^1 P(x + t\, \Delta x, y)\, dt = P(\bar{x}, y) \int_0^1 dt = P(\bar{x}, y).$$

Thus taking the limits as $\Delta x \to 0$ on both sides of (4), we have

P is continuous and $\bar{x} \to x$ as $\Delta x \to 0$

$$\frac{\partial f(\mathbf{x})}{\partial x} = \lim_{\Delta x \to 0} \frac{f(x + \Delta x, y) - f(x, y)}{\Delta x} = \lim_{\Delta x \to 0} P(\bar{x}, y) = P(x, y).$$

In a similar manner, we can show that $(\partial f/\partial y)(\mathbf{x}) = Q(x, y)$. This shows that $\mathbf{F} = \nabla f$, and f is differentiable, by Theorem 9.8.2, since the first partials are continuous functions. ■

In proving this theorem, we also proved the following corollary.

■ **COROLLARY 1**

Suppose that $\mathbf{F}$ is a continuous in a region Ω in $\mathbb{R}^2$ and $\mathbf{F} = \nabla f$ for some differentiable function f. Then for any piecewise smooth curve C in Ω starting at the point $\mathbf{x}_0$ and ending at the point $\mathbf{x}_1$,

$$\int_C \mathbf{F}(\mathbf{x}) \cdot d\mathbf{x} = f(\mathbf{x}_1) - f(\mathbf{x}_0); \tag{5}$$

that is, the value of the integral depends only on the endpoints of the path. ■

Remark. This corollary is really the line integral analog of the fundamental theorem of calculus. It says that we can evaluate the line integral of a gradient field by evaluating at two points the function for which $\mathbf{F}$ is the gradient.

Remark. In Theorem 1 it is important that **F** be continuous on Ω, not only on C.

EXAMPLE 2 (Continued)

Since $\mathbf{F}(\mathbf{x}) = y\mathbf{i} + x\mathbf{j} = \nabla(xy)$, we immediately find that for any curve C starting at $(0, 0)$ and ending at $(1, 1)$

$$\int_C \mathbf{F}(\mathbf{x}) \bullet \mathbf{dx} = xy|_{(1,1)} - xy|_{(0,0)} = 1 - 0 = 1.$$

There is another important consequence of Theorem 1. Let C be a closed curve (i.e., $\mathbf{x}_0 = \mathbf{x}_1$).

> If **F** is continuous in a region Ω and if **F** is the gradient of a differentiable function f, then for any closed curve C lying in Ω,
>
> $$\int_C \mathbf{F}(\mathbf{x}) \bullet \mathbf{dx} = f(\mathbf{x}_0) - f(\mathbf{x}_0) = 0. \tag{5}'$$

We now state a general result that tells us whether **F** is a gradient of some function f (i.e., whether **F** is conservative). Its proof is difficult and is omitted.†

■ **THEOREM 2**

Let $\mathbf{F}(x, y) = P(x, y)\mathbf{i} + Q(x, y)\mathbf{j}$ and suppose that P, Q, $\partial P/\partial y$, and $\partial Q/\partial x$ are continuous in an open disk D centered at (x, y). Then in D, **F** is the gradient of a function f if and only if‡

$$\frac{\partial P}{\partial y} = \frac{\partial Q}{\partial x}. \tag{6}$$

■

EXAMPLE 3

Show that the vector field

$$\mathbf{F}(x, y) = \left(4x^3y^3 + \frac{1}{x}\right)\mathbf{i} + \left(3x^4y^2 - \frac{1}{y}\right)\mathbf{j}$$

is conservative in any open disk in the first quadrant, and find all functions f for which $\mathbf{F} = \nabla f$.

---

† For a proof see R. C. Buck and E. F. Buck, *Advanced Calculus, 3rd* Ed, McGraw-Hill, New York, 1978, p. 497.

‡ We suggested a proof of part of this theorem in Problem 1.5.21 on p. 41. More will be said about this theorem on p. 612.

SOLUTION

$P(x, y) = 4x^3y^3 + (1/x)$ and $Q(x, y) = 3x^4y^2 - (1/y)$, so

$$\frac{\partial P}{\partial y} = 12x^3y^2 = \frac{\partial Q}{\partial x}.$$

If $\nabla f = \mathbf{F}$, then $\partial f/\partial x = P$, so

$$f(x, y) = \int \left(4x^3y^3 + \frac{1}{x}\right) dx = x^4y^3 + \ln|x| + g(y).$$

Differentiating with respect to y, we have

$$Q = \frac{\partial f}{\partial y} = 3x^4y^2 + g'(y) = 3x^4y^2 - \frac{1}{y}.$$

Thus $g'(y) = -1/y$, $g(y) = -\ln|y| + C$, and finally,

$$f(x, y) = x^4y^3 + \ln|x| - \ln|y| + C = x^4y^3 + \ln\left|\frac{x}{y}\right| + C.$$

Finally, observe that since $\mathbf{F}$ is conservative in any open disk in the first quadrant,

$$\int_C \mathbf{F}(\mathbf{x}) \cdot \mathbf{dx}$$

is independent of path for any curve C in the first quadrant.

■ **COROLLARY 2**

Let $\mathbf{F}(x, y) = P(x, y)\mathbf{i} + Q(x, y)\mathbf{j}$. Then if P, Q, $\partial P/\partial y$, and $\partial Q/\partial x$ are all continuous in an open disk D containing C and if $\partial P/\partial y = \partial Q/\partial x$, then $\int_C \mathbf{F}(\mathbf{x}) \cdot \mathbf{dx}$ is independent of the path. ■

Why are we being so careful to restrict the domain of the functions in Theorem 2 and Corollary 2? The reason is that $\partial P/\partial y = \partial Q/\partial x$ may not force the field $\mathbf{F} = P\mathbf{i} + Q\mathbf{j}$ to be conservative if the domain is more general.

EXAMPLE 4

Let $\mathbf{F}(\mathbf{x}) = \dfrac{y}{x^2+y^2}\mathbf{i} - \dfrac{x}{x^2+y^2}\mathbf{j}$. Show that $\mathbf{F}$ satisfies condition (6), but is not conservative in the region $0 < x^2 + y^2 < 4$.

SOLUTION

To see that (6) holds, observe that

$$\frac{\partial P}{\partial y} = \frac{\partial}{\partial y}\left(\frac{y}{x^2+y^2}\right) = \frac{x^2-y^2}{(x^2+y^2)^2} \quad \text{and} \quad \frac{\partial Q}{\partial x} = \frac{\partial}{\partial x}\left(\frac{-x}{x^2+y^2}\right) = \frac{x^2-y^2}{(x^2+y^2)^2}.$$

However, if we let C be the unit circle parametrized by $\mathbf{x}(t) = (\cos t)\mathbf{i} + (\sin t)\mathbf{j}$, $0 \le t \le 2\pi$, we have $\mathbf{x}'(t) = (-\sin t)\mathbf{i} + (\cos t)\mathbf{j}$ and

$$\int_C \mathbf{F}(\mathbf{x}) \cdot d\mathbf{x} = \int_0^{2\pi} [(\sin t)\mathbf{i} - (\cos t)\mathbf{j}] \cdot [(-\sin t)\mathbf{i} + (\cos t)\mathbf{j}] \, dt = -\int_0^{2\pi} dt = -2\pi.$$

$$\uparrow$$
$$x = \cos t$$
$$y = \sin t$$

Thus $\mathbf{F}$ is not conservative, because if it were equation (5)′ would guarantee that $\int_C \mathbf{F}(\mathbf{x}) \cdot d\mathbf{x} = 0$. The problem in this example is that $\mathbf{F}$ is not defined at the origin. This "hole" in the domain of definition of the vector field causes the difficulty. More will be said about this in Section 10.4.

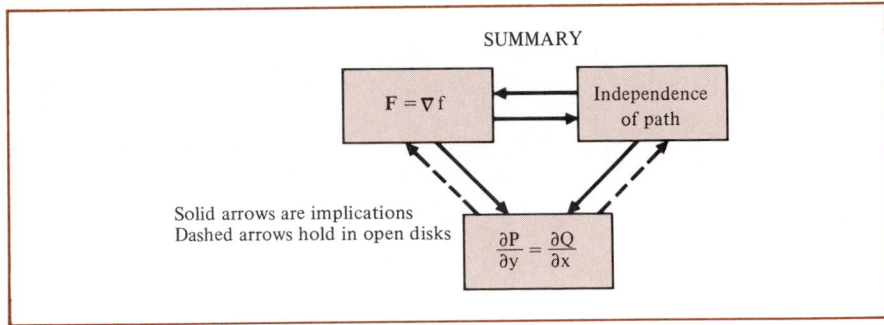

There is a physical interpretation of these results:

> *The work done by a conservative force field as it moves a particle completely around a closed path is zero.*

This also follows from the fact that if $\mathbf{F}$ is conservative, then the work done in moving an object between two points depends only on the points, and not on the path taken between the points.

Note that if the domain is an open disk, we need only verify that $\partial P/\partial y = \partial Q/\partial x$ to determine that the field is conservative (also see p. 612).

Scalar Potentials

Let $\mathbf{F}$ be a conservative vector field. Then $\mathbf{F} = -\nabla f$† for some differentiable function f. We call f a **scalar potential.**

EXAMPLE 5: Gravitational Potential

In Section 9.8 (p. 569) we saw that the magnitude of the gravitational force between two objects, M_1 and M_2, of masses m_1 and m_2 is given by

$$|\mathbf{F}| = \frac{G m_1 m_2}{|\mathbf{x}|^2},$$

† $-\nabla f = \nabla(-f)$.

where it is assumed that one mass is at the origin and the other is $r = |\mathbf{x}|$ units away. The gravitational force acts along a line joining the two bodies, so if we treat the first mass as a point mass at the origin, then $\mathbf{F}$ acts in the direction of a unit vector $-\mathbf{u}$ pointing from M_2 toward the origin (see Fig. 3). That is,

$$\mathbf{F}(\mathbf{x}) = -\frac{Gm_1 m_2}{|\mathbf{x}|^2}\mathbf{u} = -\frac{Gm_1 m_2}{|\mathbf{x}|^3}\mathbf{x}. \tag{7}$$

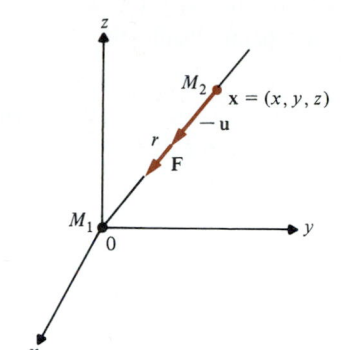

FIGURE 3

Note that $\mathbf{F}$ is a *central force* (see p. 546). The **gravitational potential** at $\mathbf{x}$ is defined as the work required to bring a unit mass from infinity to $\mathbf{x}$. On p. 570 we showed that $\mathbf{F} = \nabla(Gm_1 m_2/|\mathbf{x}|)$ so $\mathbf{F}$ is conservative and its scalar potential is $Gm_1 m_2/|\mathbf{x}|$. We show that this is equal to its gravitational potential. This is given by

$$\text{gravitational potential} = W = \int_C \mathbf{F}(\mathbf{x}) \cdot d\mathbf{x},$$

where C is the ray extending from ∞ to $\mathbf{x}$. Since $\mathbf{F}$ is conservative, we have

$$W = f(\mathbf{x}) - f(\infty) = \frac{Gm_1 m_2}{|\mathbf{x}|}.$$

Here we have arbitrarily assigned zero to the potential at infinity. This makes sense since $\lim_{|\mathbf{x}| \to \infty} f(\mathbf{x}) = 0$.

We note that we can never measure potential, only potential *difference*. Potential and potential difference are the same when, as here, the potential at one of the points takes the value zero.

Problems 10.2

In Problems 1–14, determine if the given vector field is conservative, and if so find all functions f for which $\nabla f = \mathbf{F}$.

1. $\mathbf{F}(x, y) = 2xy\mathbf{i} + (x^2 + 1)\mathbf{j}$

2. $\mathbf{F}(x, y) = (4x^3 - ye^{xy})\mathbf{i} + (\tan y - xe^{xy})\mathbf{j}$

3. $\mathbf{F}(x, y) = (4x^2 - 4y^2)\mathbf{i} + (8xy - \ln y)\mathbf{j}$

4. $\mathbf{F}(x, y) = [x\cos(x + y) + \sin(x + y)]\mathbf{i} + x\cos(x + y)\mathbf{j}$

5. $\mathbf{F}(x, y) = 2x(\cos y)\mathbf{i} + x^2(\sin y)\mathbf{j}$

6. $\mathbf{F}(x, y) = \left[\dfrac{\ln(\ln y)}{x} + \dfrac{2}{3}xy^3\right]\mathbf{i} + \left(\dfrac{\ln x}{y\ln y} + x^2 y^2\right)\mathbf{j}$

7. $\mathbf{F}(x, y) = (x - y\cos x)\mathbf{i} - (\sin x)\mathbf{j}$

8. $\mathbf{F}(x, y) = e^{x^2 y}\mathbf{i} + e^{x^2 y}\mathbf{j}$

9. $\mathbf{F}(x, y) = (3x\ln x + x^5 - y)\mathbf{i} - x\mathbf{j}$

10. $\mathbf{F}(x, y) = \left[\left(\dfrac{1}{x^2}\right) + y^2\right]\mathbf{i} + (2xy)\mathbf{j}$

11. $\mathbf{F}(x, y) = \left(\tan^2 x - \dfrac{y}{x^2 + y^2}\right)\mathbf{i} + \left(\dfrac{x}{x^2 + y^2} - e^y\right)\mathbf{j}$

12. $\mathbf{F}(x, y) = [\cos(x + 3y) - x^4]\mathbf{i} + \left[3\cos(x + 3y) - \left(\dfrac{2}{y}\right)\right]\mathbf{j}$

13. $\mathbf{F}(x, y) = (x^2 + y^2 + 1)\mathbf{i} - (xy + y)\mathbf{j}$

14. $\mathbf{F}(x, y) = \left[-\left(\dfrac{1}{x^3}\right) + 4x^3 y\right]\mathbf{i} + (\sin y + \sqrt{y} + x^4)\mathbf{j}$

Consider the vector field

$$\mathbf{F}(x, y, z) = P(x, y, z)\mathbf{i} + Q(x, y, z)\mathbf{j} + R(x, y, z)\mathbf{k}.$$

$\mathbf{F}$ is conservative if there is a differentiable function f such that

$$\nabla f(x, y, z) = \mathbf{F}(x, y, z).$$

If P, Q, R, $\partial P/\partial y$, $\partial P/\partial z$, $\partial Q/\partial x$, $\partial Q/\partial z$, $\partial R/\partial x$, and $\partial R/\partial y$ are continuous in an open ball, then, analogously to Theorem 2, $\mathbf{F}$ is conservative if and only if

$$\frac{\partial P}{\partial y} = \frac{\partial Q}{\partial x}, \qquad \frac{\partial R}{\partial x} = \frac{\partial P}{\partial z}, \qquad \frac{\partial Q}{\partial z} = \frac{\partial R}{\partial y}.$$

In Problems 15–18, use the result above to test if $\mathbf{F}$ is conservative. If so, find all functions f for which $\nabla f = \mathbf{F}$.

15. $\mathbf{F} = \mathbf{i} + \mathbf{j} + \mathbf{k}$

16. $\mathbf{F} = yz\mathbf{i} + xz\mathbf{j} + xy\mathbf{k}$

17. $\mathbf{F} = \left[\left(\dfrac{y}{z}\right)+x^2\right]\mathbf{i} + \left[\left(\dfrac{x}{z}\right)-\sin y\right]\mathbf{j} + \left[\cos z - \left(\dfrac{xy}{z^2}\right)\right]\mathbf{k}$

18. $\mathbf{F} = \left(\dfrac{1}{x+2y+3z}-3x\right)\mathbf{i} + \left(\dfrac{2}{x+2y+3z}+y^2\right)\mathbf{j}$

$\quad + \left(\dfrac{3}{x+2y+3z}-\tan^{-1} z\right)\mathbf{k}$

In Problems 19–26, show that $\mathbf{F}$ is conservative and use Corollary 1 to calculate $\int_C \mathbf{F}(\mathbf{x}) \bullet d\mathbf{x}$, where C is any smooth curve starting at $\mathbf{x}_0$ and ending at $\mathbf{x}_1$.

19. $\mathbf{F}(x, y) = 2xy\mathbf{i} + (x^2 + 1)\mathbf{j}$; $\mathbf{x}_0 = (0, 1)$, $\mathbf{x}_1 = (2, 3)$

20. $\mathbf{F}(x, y) = (4x^2 - 4y^2)\mathbf{i} + (y - 8xy)\mathbf{j}$; $\mathbf{x}_0 = (-1, 1)$,
$\quad \mathbf{x}_1 = (4, e)$

21. $\mathbf{F}(x, y) = [x \cos(x + y) + \sin(x + y)]\mathbf{i} + x \cos(x + y)\mathbf{j}$;
$\quad \mathbf{x}_0 = (0, 0)$, $\mathbf{x}_1 = (\pi/6, \pi/3)$

22. $\mathbf{F}(x, y) = [x + y^2]\mathbf{i} + 2xy\mathbf{j}$; $\mathbf{x}_0 = (1, 4)$, $\mathbf{x}_1 = (3, 2)$

23. $\mathbf{F}(x, y) = 2x(\cos y)\mathbf{i} - x^2(\sin y)\mathbf{j}$; $\mathbf{x}_0 = (0, \pi/2)$,
$\quad \mathbf{x}_1 = (\pi/2, 0)$

24. $\mathbf{F}(x, y) = (2xy^3 - 2)\mathbf{i} + (3x^2y^2 + \cos y)\mathbf{j}$; $\mathbf{x}_0 = (1, 0)$,
$\quad \mathbf{x}_1 = (0, -\pi)$

25. $\mathbf{F}(x, y) = e^y\mathbf{i} + xe^y\mathbf{j}$; $\mathbf{x}_0 = (0, 0)$, $\mathbf{x}_1 = (5, 7)$

26. $\mathbf{F}(x, y) = (\cosh x)(\cosh y)\mathbf{i} + (\sinh x)(\sinh y)\mathbf{j}$;
$\quad \mathbf{x}_0 = (0, 0)$, $\mathbf{x}_1 = (1, 2)$

27. Show that $\mathbf{F} = P\mathbf{i} + Q\mathbf{j} + R\mathbf{k}$ is conservative in the open unit ball $x^2 + y^2 + z^2 < 1$, assuming P, Q, and R have continuous first partials, if and only if $\nabla \times \mathbf{F} = 0$.

28. If a particle moves around a circle with constant angular speed ω, then the radial force accelerating the particle away from the center is called **centrifugal force** and is given by

$$\mathbf{F}_C = \omega^2 |\mathbf{x}|\mathbf{u},$$

where $\mathbf{x}$ is the position of the particle and $\mathbf{u}$ is a unit vector pointing away from the center. The **potential,** f_C, of the centrifugal force is defined as the work done in moving the particle from $\mathbf{0}$ to $\mathbf{x}$. Assuming that $f_C(\mathbf{0}) = 0$, show that

$$f_C = -\tfrac{1}{2}\omega^2 |\mathbf{x}|^2.$$

10.3 Review of Double Integrals

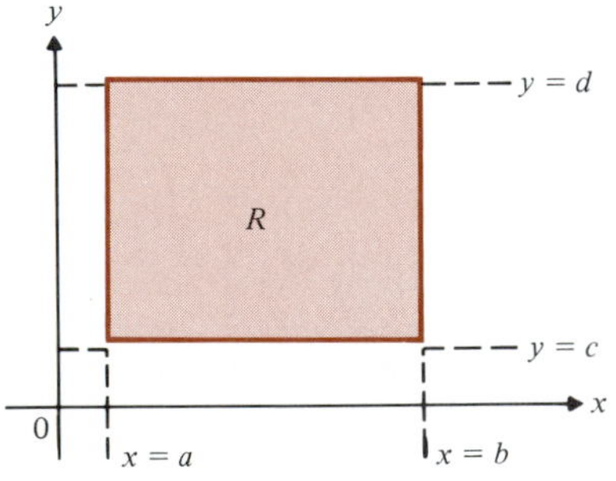

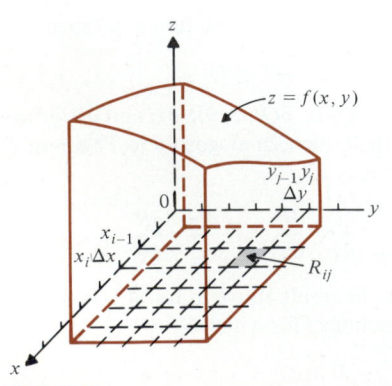

FIGURE 1

In the remaining sections of this chapter we shall need the concept of double integrals. The reader will be familiar with this concept from calculus; this section will constitute a brief review of this topic.

Let R denote the rectangle in $\mathbb{R}^2$ given by

$$R = \{(x, y) : a \le x \le b \text{ and } c \le y \le d\}. \tag{1}$$

This rectangle is sketched in Fig. 1.

Recall that, by definition,

$$\int_a^b f(x)\, dx = \lim_{\Delta x \to 0} \sum_{i=1}^n f(x_i^*)\, \Delta x, \tag{2}$$

where the interval $[a, b]$ is partitioned into n subintervals of length Δx [$= (b - a)/n$] and the limit in (2) is independent of the points x_i^* in the n subintervals.

To define a double integral of the function $z = f(x, y)$ over the rectangle R, we partition R as in Fig. 2 and choose a point (x_i^*, y_j^*) in each subrectangle. A typical subrectangle is sketched in Fig. 3. We define $\Delta s = \sqrt{(\Delta x)^2 + (\Delta y)^2}$ (see Fig. 3). Finally, let $\Delta A = \Delta x\, \Delta y$.

DEFINITION

Let $z = f(x, y)$ and let the rectangle R be given by (1). Let $\Delta A = \Delta x\, \Delta y$. Suppose that

$$\lim_{\Delta s \to 0} \sum_{i=1}^n \sum_{j=1}^m f(x_i^*, y_j^*)\, \Delta A$$

FIGURE 2

Double Integral over a Rectangle

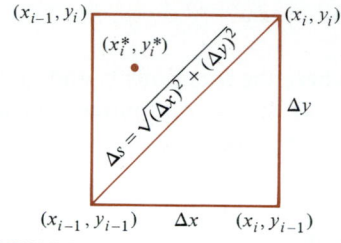

FIGURE 3

exists and is independent of the way in which the points (x_i^*, y_j^*) are chosen. Then the **double integral of** f **over** R, written $\iint_R f(x, y)\, dA$, is defined by

$$\iint\limits_R f(x, y)\, dA = \lim_{\Delta s \to 0} \sum_{i=1}^{n} \sum_{j=1}^{m} f(x_i^*, y_j^*)\, \Delta A. \tag{3}$$

If the limit in (3) exists, then the function f is said to be **integrable** over R.

Note. $\iint_R f(x, y)\, dA$ is a number, not a function. This is similar to the one-dimensional case where the definite integral $\int_a^b f(x)\, dx$ is a number.

■ **THEOREM 1: Existence of the Double Integral over a Rectangle†**
If f is continuous on R, then f is integrable over R. ■

Consider the bounded regions in Fig. 4, where $g_1(x)$, $g_2(x)$, $h_1(y)$, and $h_2(y)$ are continuous. Since each Ω is bounded, we can draw a rectangle R around it. Let f be defined over Ω. We then define a new function F by

$$F(x, y) = \begin{cases} f(x, y), & \text{for } (x, y) \text{ in } \Omega \\ 0, & \text{for } (x, y) \text{ in } R \text{ but not in } \Omega. \end{cases} \tag{4}$$

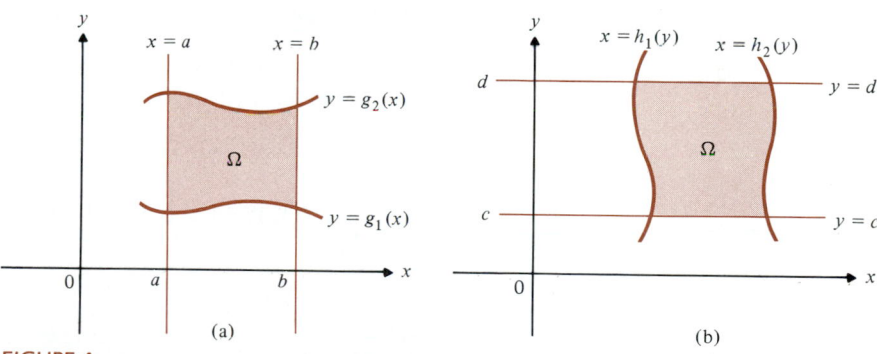

FIGURE 4

Integrability over a Region

DEFINITION

Let f be defined for (x, y) in Ω and let F be defined by (4). Then we write

$$\iint\limits_{\Omega} f(x, y)\, dA = \iint\limits_{R} F(x, y)\, dA \tag{5}$$

† See, for example, R. C. Buck and E. F. Buck, *Advanced Calculus,* 3rd ed., McGraw-Hill, New York, 1978, p. 169.

if the integral on the right exists. In this case we say that f is **integrable over** Ω.

■ **THEOREM 2: Existence of the Double Integral over a More General Region**

Let Ω be one of the regions depicted in Fig. 4 where the functions g_1 and g_2 or h_1 and h_2 are continuous. Let F be defined by (4). If f is continuous over Ω, then f is integrable over Ω and its integral is given by (5).   ■

Mean Value Theorem for Double Integrals

■ **THEOREM 3**

Let Ω be one of the regions in Figure 4 and suppose that f is continuous over Ω. Let A_Ω denote the area of Ω. Then there exists a point (x_0, y_0) in Ω such that

$$\iint_\Omega f(x, y)\, dA = f(x_0, y_0) A_\Omega.$$

■

Iterated Integrals: Repeated Integrals

We can compute some double integrals by integrating over one variable at a time. If R is given by (1), then

$$\iint_R f(x, y)\, dA = \int_a^b \left\{ \int_c^d f(x, y)\, dy \right\} dx = \int_c^d \left\{ \int_a^b f(x, y)\, dx \right\} dy. \tag{6}$$

EXAMPLE 1

Compute $I = \iint_R (x + 2y)\, dA$ where $R = \{(x, y) : 1 \le x \le 2 \text{ and } 3 \le y \le 5\}$.

SOLUTION

$$I = \iint_R (x + 2y)\, dA = \int_1^2 \left[\int_3^5 (x + 2y)\, dy \right] dx$$

In computing the bracketed integral, we treat x as a constant

$$= \int_1^2 \left[(xy + y^2) \Big|_{y=3}^{y=5} \right] dx = \int_1^2 [(5x + 25) - (3x + 9)]\, dx$$

$$= \int_1^2 (2x + 16)\, dx = (x^2 + 16x) \Big|_1^2 = 19.$$

The same answer is obtained by integrating first with respect to x.

If Ω is one of the regions in Fig. 4, we can compute the double integral as follows:

■ **THEOREM 4**

Let f be continuous over a region Ω given in Fig. 4a or 4b.

(i) If Ω is of the form of Fig. 4a, where g_1 and g_2 are continuous, then

$$\iint_{\Omega} f(x, y)\, dA = \int_a^b \int_{g_1(x)}^{g_2(x)} f(x, y)\, dy\, dx. \tag{7}$$

(ii) If Ω is of the form of Fig. 4b, where h_1 and h_2 are continuous, then

$$\iint_{\Omega} f(x, y)\, dA = \int_c^d \int_{h_1(y)}^{h_2(y)} f(x, y)\, dx\, dy. \tag{8}$$

■

Volume

If f is nonnegative over Ω, then the double integral in (7) or (8) represents the **volume** of the solid bounded above by f and lying over the region Ω.

EXAMPLE 2

Find the volume of the solid under the surface $z = x^2 + y^2$ and lying above the region

$$\Omega = \{(x, y): 0 \le x \le 1 \quad \text{and} \quad x^2 \le y \le \sqrt{x}\}.$$

SOLUTION
Ω is sketched in Fig. 5. We see that $0 \le x \le 1$ and $x^2 \le y \le \sqrt{x}$. Then using (7), we have

$$V = \int_0^1 \int_{x^2}^{\sqrt{x}} (x^2 + y^2)\, dy\, dx = \int_0^1 \left\{ \left(x^2 y + \frac{y^3}{3} \right) \Big|_{x^2}^{\sqrt{x}} \right\} dx$$

$$= \int_0^1 \left\{ \left(x^2 \sqrt{x} + \frac{(\sqrt{x})^3}{3} \right) - \left(x^2 (x^2) + \frac{(x^2)^3}{3} \right) \right\} dx$$

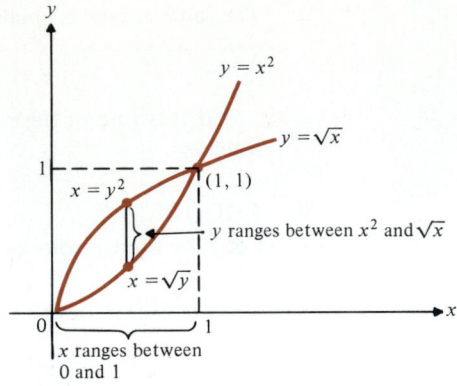

FIGURE 5

$$= \int_0^1 \left(x^{5/2} + \frac{x^{3/2}}{3} - x^4 - \frac{x^6}{3} \right) dx$$

$$= \left(\frac{2x^{7/2}}{7} + \frac{2x^{5/2}}{15} - \frac{x^5}{5} - \frac{x^7}{21} \right)\Big|_0^1 = \frac{2}{7} + \frac{2}{15} - \frac{1}{5} - \frac{1}{21} = \frac{18}{105}.$$

We can calculate this integral in another way. We note that x varies between the curves $x = y^2$ and $x = \sqrt{y}$. Then using (8), since $0 \le y \le 1$ and $y^2 \le x \le \sqrt{y}$, we have

$$V = \int_0^1 \int_{y^2}^{\sqrt{y}} (x^2 + y^2) \, dx \, dy,$$

which is also equal to 18/105.

Reversing the Order of Integration

EXAMPLE 3

Evaluate $\int_1^2 \int_1^{x^2} (x/y) \, dy \, dx$.

SOLUTION

$$\int_1^2 \int_1^{x^2} \frac{x}{y} \, dy \, dx = \int_1^2 \left\{ x \ln y \Big|_1^{x^2} \right\} dx = \int_1^2 x \ln x^2 \, dx = \int_1^2 2x \ln x \, dx$$

It is necessary to use integration by parts to complete the problem. Setting $u = \ln x$ and $dv = 2x \, dx$, we have $du = (1/x) \, dx$, $v = x^2$, and

$$\int_1^2 2x \ln x \, dx = x^2 \ln x \Big|_1^2 - \int_1^2 x \, dx = 4 \ln 2 - \frac{x^2}{2}\Big|_1^2 = 4 \ln 2 - \frac{3}{2}.$$

There is an easier way to calculate the double integral. We simply **reverse the order of integration**. The region of integration is sketched in Figure 6. If we want to integrate first with respect to x, we note that we can describe the region by

$$\Omega = \{(x, y): \sqrt{y} \le x \le 2 \text{ and } 1 \le y \le 4 \}.$$

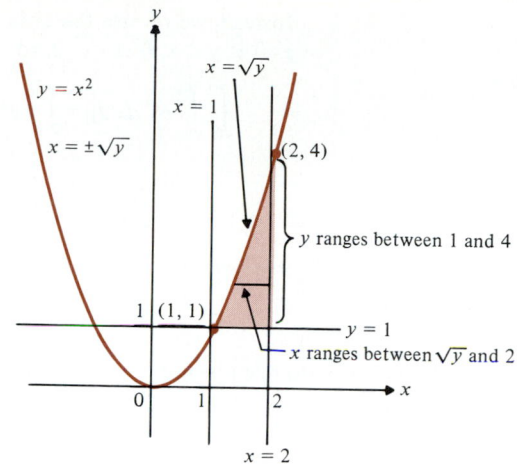

FIGURE 6

Then

$$\int_1^2 \int_1^{x^2} \frac{x}{y}\,dy\,dx = \iint_\Omega \frac{x}{y}\,dA = \int_1^4 \int_{\sqrt{y}}^2 \frac{x}{y}\,dx\,dy = \int_1^4 \left\{ \frac{x^2}{2y}\bigg|_{\sqrt{y}}^2 \right\} dy$$

$$= \int_1^4 \left(\frac{2}{y} - \frac{1}{2}\right) dy = \left(2\ln y - \frac{y}{2}\right)\bigg|_1^4 = 2\ln 4 - \frac{3}{2}$$

$$= 4\ln 2 - \frac{3}{2}.$$

Note that in this case it is easier to integrate first with respect to x.

EXAMPLE 4

Compute $\int_0^2 \int_y^2 e^{x^2}\,dx\,dy$.

SOLUTION

The region of integration is sketched in Fig. 7. We first observe that the double integral cannot be evaluated directly since it is impossible to find an antiderivative for e^{x^2}.

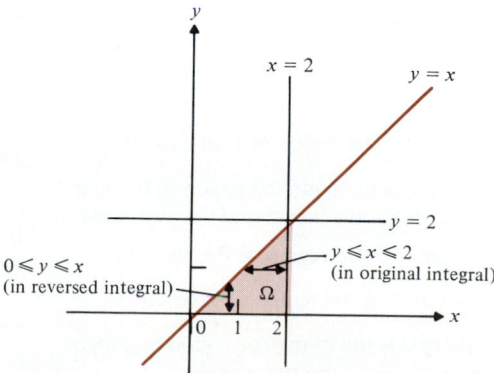

FIGURE 7

Instead, we reverse the order of integration. From Fig. 6 we see that Ω can be written as $0 \leq y \leq x$, $0 \leq x \leq 2$, so

$$\int_0^2 \int_y^2 e^{x^2}\, dx\, dy = \iint_\Omega e^{x^2}\, dA = \int_0^2 \int_0^x e^{x^2}\, dy\, dx = \int_0^2 \left(ye^{x^2}\Big|_{y=0}^{y=x}\right) dx = \int_0^2 xe^{x^2}\, dx$$

$$= \frac{1}{2}e^{x^2}\Big|_0^2 = \frac{1}{2}(e^4 - 1).$$

Problems 10.3

In Problems 1–23, evaluate the given double integral.

1. $\displaystyle\int_0^1 \int_0^2 xy^2\, dx\, dy$

2. $\displaystyle\int_{-1}^3 \int_2^4 (x^2 - y^3)\, dy\, dx$

3. $\displaystyle\int_2^5 \int_0^4 e^{(x-y)}\, dx\, dy$

4. $\displaystyle\int_0^1 \int_{x^2}^x x^3 y\, dy\, dx$

5. $\displaystyle\int_2^4 \int_{1+y}^{2+3y} (x - y^2)\, dx\, dy$

6. $\displaystyle\int_{\pi/4}^{\pi/3} \int_{\sin x}^{\cos x} (x + 2y)\, dy\, dx$

7. $\displaystyle\int_0^3 \int_{-\sqrt{9-y^2}}^{\sqrt{9-y^2}} x^2 y\, dx\, dy$

8. $\displaystyle\int_1^2 \int_{y^5}^{3y^5} \frac{1}{x}\, dx\, dy$

9. $\iint_\Omega (x^2 + y^2)\, dA$, where $\Omega = \{(x, y): 1 \leq x \leq 2$ and $-1 \leq y \leq 1\}$.

10. $\iint_\Omega 2xy\, dA$, where $\Omega = \{(x, y): 0 \leq x \leq 4$ and $1 \leq y \leq 3\}$.

11. $\iint_\Omega (x - y)^2\, dA$, where $\Omega = \{(x, y): -2 \leq x \leq 2$ and $0 \leq y \leq 1\}$.

12. $\iint_\Omega \sin(2x + 3y)\, dA$, where $\Omega = \{(x, y): 0 \leq x \leq \pi/6$ and $0 \leq y \leq \pi/18\}$.

13. $\iint_\Omega xe^{(x^2+y)}\, dA$, where Ω is the region of Problem 10.

14. $\iint_\Omega (x - y^2)\, dA$, where Ω is the bounded region in the first quadrant between the x-axis, the y-axis, and the unit circle.

15. $\iint_\Omega (x^2 + y)\, dA$, where Ω is the region of Problem 14.

16. $\iint_\Omega (x^3 - y^3)\, dA$, where Ω is the region of Problem 14.

17. $\iint_\Omega (x + 2y)\, dA$, where Ω is the triangular region bounded by the lines $y = x$, $y = 1 - x$, and the y-axis.

18. $\iint_\Omega e^{x+2y}\, dA$, where Ω is the region of Problem 17.

19. $\iint_\Omega (x^2 + y)\, dA$, where Ω is the bounded region in the first quadrant between the parabolas $y = x^2$ and $y = 1 - x^2$.

20. $\iint_\Omega (1/\sqrt{y})\, dA$, where Ω is the region of Problem 19.

21. $\iint_\Omega (y/\sqrt{x^2+y^2})\, dA$, where $\Omega = \{(x, y): 1 \leq x \leq y$ and $1 \leq y \leq 2\}$.

22. $\iint_\Omega [e^{-y}/(1 + x^2)]\, dA$, where Ω is the first quadrant.

23. $\iint_\Omega (x + y)\, e^{-(x+y)}\, dA$, where Ω is the first quadrant.

In Problems 24–33, (a) sketch the region over which the integral is taken. Then (b) change the order of integration, and (c) evaluate the given integral.

24. $\displaystyle\int_0^2 \int_{-1}^3 dx\, dy$

25. $\displaystyle\int_0^4 \int_{-5}^8 (x + y)\, dy\, dx$

26. $\displaystyle\int_2^4 \int_1^y \frac{y^3}{x^3}\, dx\, dy$

27. $\displaystyle\int_0^1 \int_0^x dy\, dx$

28. $\displaystyle\int_0^1 \int_x^1 dy\, dx$

29. $\displaystyle\int_0^{\pi/2} \int_0^{\cos y} y\, dx\, dy$

30. $\displaystyle\int_0^2 \int_0^{\sqrt{4-y^2}} (4 - x^2)^{3/2}\, dx\, dy$

31. $\displaystyle\int_0^1 \int_{\sqrt{x}}^{\sqrt[3]{x}} (1 + y^6)\, dy\, dx$

32. $\displaystyle\int_0^1 \int_{\sqrt{y}}^1 \sqrt{3 - x^3}\, dx\, dy$

33. $\displaystyle\int_0^\infty \int_x^\infty \frac{1}{(1 + y^2)^{7/5}}\, dy\, dx$

In Problems 34–40 find the volume of the given solid.

34. The solid bounded by the plane $x + y + z = 3$ and the three coordinate planes.

35. The solid bounded by the planes $x = 0$, $x + 2y + z = 6$, $z = 0$, and $x - 2y + z = 6$.

36. The solid bounded by the cylinders $x^2 + y^2 = 4$ and $y^2 + z^2 = 4$.

37. The solid bounded by the cylinder $x^2 + z^2 = 1$ and the planes $y = 0$ and $y = 2$.

* 38. The ellipsoid $x^2 + 4y^2 + 9z^2 = 36$.

39. The solid bounded above by the sphere $x^2 + y^2 + z^2 = 9$ and below by the plane $z = \sqrt{5}$.

40. The solid bounded by the planes $y = 0$, $y = x$, and the cylinder $x + z^2 = 2$.

10.4 Green's Theorem in the Plane

In this section we state a result that gives an important relationship between line integrals and double integrals.

Let Ω be a region in the plane (a typical region is sketched in Fig. 1). The curve (indicated by the arrows) that goes around the edge of Ω in the direction that keeps Ω on the left (sometimes called the *counterclockwise* direction) is called the **boundary of** Ω and is denoted $\partial\Omega$. Let

$$\mathbf{F}(x, y) = P(x, y)\mathbf{i} + Q(x, y)\mathbf{j}.$$

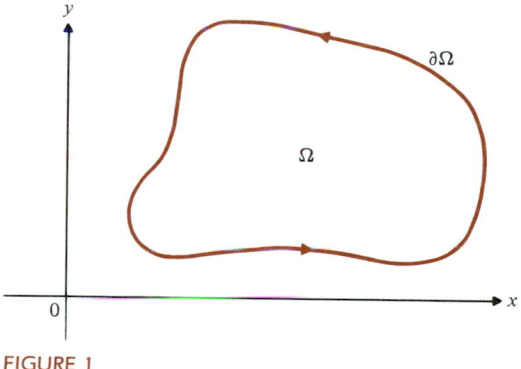

FIGURE 1

If the curve $\partial\Omega$ is given by

$$\partial\Omega: \quad \mathbf{x}(t) = x(t)\mathbf{i} + y(t)\mathbf{j},$$

we can write

$$\mathbf{F}(\mathbf{x}) \cdot \mathbf{dx} = P\,dx + Q\,dy. \tag{1}$$

We then denote the line integral of $\mathbf{F}$ around $\partial\Omega$ by

$$\oint_{\partial\Omega} P\,dx + Q\,dy. \tag{2}$$

The symbol $\oint_{\partial\Omega}$ indicates that $\partial\Omega$ is a closed curve around which we integrate in the counterclockwise direction (the direction of the arrow).

DEFINITION

A curve C is called **simple** if it does not cross itself. That is, suppose C is given by

$$C: \quad \mathbf{x}(t) = x(t)\mathbf{i} + y(t)\mathbf{j}, \quad t \in [a, b].$$

Then C is simple if and only if $\mathbf{x}(t_1) \neq \mathbf{x}(t_2)$ whenever $t_1 \neq t_2$. A closed curve is simple if it is simple except at the endpoints (see Fig. 2).

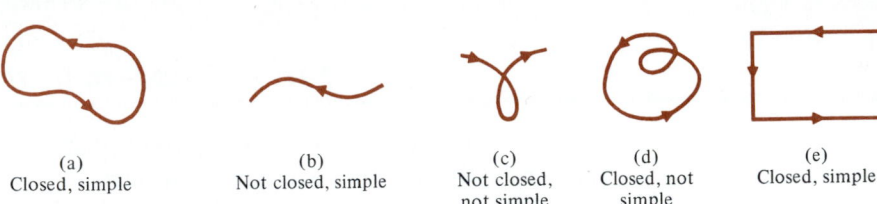

(a)   (b)   (c)   (d)   (e)
Closed, simple   Not closed, simple   Not closed,   Closed, not   Closed, simple
                                          not simple   simple

FIGURE 2

DEFINITION

A region Ω in the xy-plane is called **simply connected** if it has the following property: If C is a simple closed curve contained in Ω, then every point in the region enclosed by C is also in Ω. Intuitively, a region is simply connected if it has no holes (see Fig. 3).

Every point on this
line segment is removed

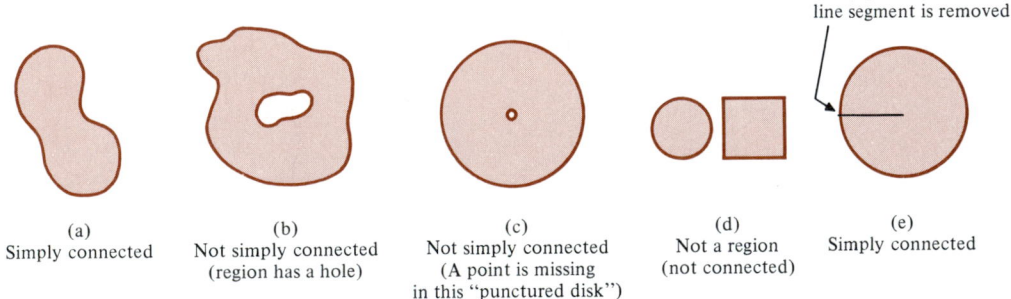

(a)   (b)   (c)   (d)   (e)
Simply connected   Not simply connected   Not simply connected   Not a region   Simply connected
                  (region has a hole)   (A point is missing   (not connected)
                                      in this "punctured disk")

FIGURE 3

We are now ready to state the principal result of this section.

■ **THEOREM 1: Green's Theorem† in the Plane**

Let Ω be a simply connected region in the xy-plane bounded by a piecewise smooth curve $\partial\Omega$. Let P and Q be continuous with continuous first partials in an open disk containing Ω. Then

---

† Named after George Green (1793–1841), a British mathematician and physicist, who wrote an essay in 1828 on electricity and magnetism that contained this important theorem. Green was the self-educated son of a baker. His 1828 essay was published for private circulation. It was largely overlooked until it was rediscovered by Lord Kelvin in 1846. The theorem was independently discovered by the Russian mathematician Michel Ostrogradski (1801–1861), and to this day the theorem is known in the Soviet Union as *Ostrogradski's theorem*. It is also known as *Gauss's theorem*.

$$\oint_{\partial\Omega} P\,dx + Q\,dy = \iint_{\Omega} \left(\frac{\partial Q}{\partial x} - \frac{\partial P}{\partial y}\right) dx\,dy. \tag{3}$$

Remark. This theorem shows how the line integral of a function around the boundary of a region is related to a double integral over that region.

PARTIAL PROOF OF GREEN'S THEOREM
We prove Green's theorem in the case in which Ω takes the simple form given in Fig. 4. The region Ω can be written as

$$\{(x, y): a \le x \le b, g_1(x) \le y \le g_2(x)\}, \tag{4}$$

or

$$\{(x, y): c \le y \le d, h_1(y) \le x \le h_2(y)\}. \tag{5}$$

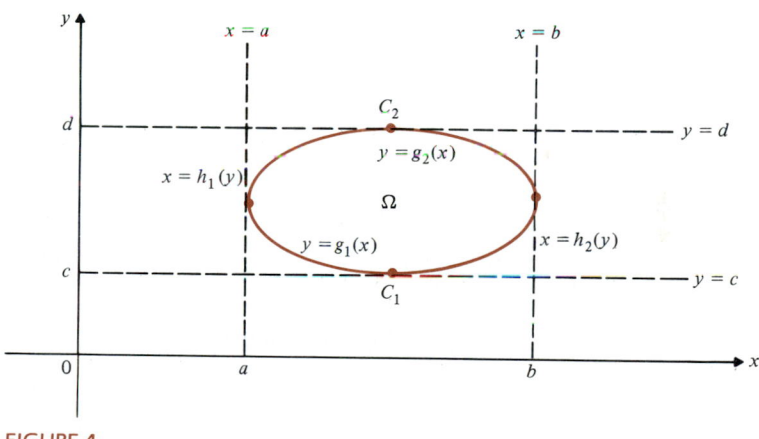

FIGURE 4

We first calculate

$$\iint_{\Omega} \frac{\partial P}{\partial y}\,dx\,dy = \int_a^b \left\{\int_{g_1(x)}^{g_2(x)} \frac{\partial P}{\partial y}\,dy\right\} dx = \int_a^b \left\{P(x, y)\Big|_{y=g_1(x)}^{y=g_2(x)}\right\} dx$$

$$= \int_a^b [P(x, g_2(x)) - P(x, g_1(x))]\,dx. \tag{6}$$

Now

$$\oint_{\partial\Omega} P\,dx = \int_{C_1} P\,dx + \int_{C_2} P\,dx, \tag{7}$$

where C_1 is the graph of $g_1(x)$ from $x = a$ to $x = b$ and C_2 is the graph of $g_2(x)$ from $x = b$ to $x = a$ (note the order). Since on C_1, $y = g_1(x)$, we have

$$\int_{C_1} P(x, y)\, dx = \int_a^b P(x, g_1(x))\, dx. \tag{8}$$

Similarly,

$$\int_{C_2} P(x, y)\, dx = \int_b^a P(x, g_2(x))\, dx = -\int_a^b P(x, g_2(x))\, dx. \tag{9}$$

Thus

$$\oint_{\partial\Omega} P\, dx = \int_a^b P(x, g_1(x))\, dx - \int_a^b P(x, g_2(x))\, dx$$

$$= -\int_a^b [P(x, g_2(x)) - P(x, g_1(x))]\, dx. \tag{10}$$

Comparing (6) and (10), we find that

$$\iint_\Omega -\frac{\partial P}{\partial y}\, dx\, dy = \oint_{\partial\Omega} P\, dx. \tag{11}$$

Similarly, using the representation (5), we have

$$\iint_\Omega \frac{\partial Q}{\partial x}\, dx\, dy = \int_c^d \left\{ \int_{h_1(y)}^{h_2(y)} \frac{\partial Q}{\partial x}\, dx \right\} dy$$

$$= \int_c^d [Q(h_2(y), y) - Q(h_1(y), y)]\, dy,$$

which by analogous reasoning yields the equation

$$\iint_\Omega \frac{\partial Q}{\partial x}\, dx\, dy = \oint_{\partial\Omega} Q\, dy. \tag{12}$$

Adding (11) and (12) completes the proof of the theorem in the special case that Ω can be written in the form (4) and (5).†   ∎

Remark. In Example 10.2.4 we considered a vector field $\mathbf{F} = P\mathbf{i} + Q\mathbf{j}$, where $\partial P/\partial y = \partial Q/\partial x$ but $\mathbf{F}$ was not conservative. The difficulty was due to the fact that the domain of definition of $\mathbf{F}$ was not simply connected. Theorem 10.2.2 can be extended to any simply connected region Ω. Then, by Green's theorem

---

† For a proof of Green's theorem for more general regions, see R. C. Buck and E. F. Buck, *Advanced Calculus*, 3rd Ed. McGraw-Hill, New York, 1978, p. 479–483.

$$\oint_{\partial\Omega} \mathbf{F}(\mathbf{x}) \cdot \mathbf{dx} = \oint_{\partial\Omega} P\,dx + Q\,dy = \iint_{\Omega} \left(\frac{\partial Q}{\partial x} - \frac{\partial P}{\partial y}\right) dx\,dy = 0,$$

for a conservative field $\mathbf{F}$. [See equation (10.2.5′), p. 598.]

EXAMPLE 1

Evaluate $\oint_{\partial\Omega} xy\,dx + (x - y)\,dy$, where Ω is the rectangle $\{(x, y): 0 \le x \le 1, 1 \le y \le 3\}$.

SOLUTION

$P(x, y) = xy$, $Q(x, y) = x - y$, $\partial Q/\partial x = 1$, and $\partial P/\partial y = x$, so

$$\oint_{\partial\Omega} xy\,dx + (x - y)\,dy = \int_0^1 \int_1^3 (1 - x)\,dy\,dx = \int_0^1 \left\{ (1 - x)y \Big|_1^3 \right\} dx$$

$$= \int_0^1 2(1 - x)\,dx = 1.$$

EXAMPLE 2

Evaluate $\oint_C (x^3 + y^3)\,dx + (2y^3 - x^3)\,dy$, where C is the unit circle.

SOLUTION

We first note that $C = \partial\Omega$, where Ω is the unit disk. Next, we have $(\partial Q/\partial x) - (\partial P/\partial y) = -3x^2 - 3y^2 = -3(x^2 + y^2)$. Thus

$$\oint_C (x^3 + y^3)\,dx + (2y^3 - x^3)\,dy = -3 \iint_{\Omega} (x^2 + y^2)\,dx\,dy$$

Converting to polar coordinates
$$\downarrow$$
$$= -3 \int_0^{2\pi} \int_0^1 (r^2) r\,dr\,d\theta$$

$$= -3 \int_0^{2\pi} \left\{ \frac{r^4}{4} \Big|_0^1 \right\} d\theta$$

$$= -\frac{3}{4} \int_0^{2\pi} d\theta = -\frac{3\pi}{2}.$$

Green's theorem can be useful for calculating area. Recall that

$$\text{area enclosed by } \Omega = \iint_{\Omega} dA. \tag{13}$$

But by Green's theorem,

$$\iint_\Omega dA = \oint_{\partial\Omega} x\,dy = \oint_{\partial\Omega} (-y)\,dx = \frac{1}{2}\oint_{\partial\Omega} [(-y)\,dx + x\,dy] \qquad \textbf{(14)}$$

(explain why). Any of the line integrals in (14) can be used to calculate area.

EXAMPLE 3

Use Green's theorem to calculate the area enclosed by the ellipse

$$(x^2/a^2) + (y^2/b^2) = 1.$$

SOLUTION
The ellipse can be written parametrically as

$$\mathbf{x}(t) = a(\cos t)\mathbf{i} + b(\sin t)\mathbf{j}, \qquad 0 \le t \le 2\pi.$$

Then using the first line integral in (14), we obtain

$$A = \oint_{\partial\Omega} x\,dy = \int_0^{2\pi} (a\cos t)\,d(b\sin t)$$

$$= \int_0^{2\pi} (a\cos t)b\cos t\,dt = ab\int_0^{2\pi}\cos^2 t\,dt$$

$$= \frac{ab}{2}\int_0^{2\pi}(1 + \cos 2t)\,dt = \pi ab.$$

Note how much easier this calculation is than the direct evaluation of $\iint_A dx\,dy$, where A denotes the area enclosed by the ellipse.

Curl and Divergence in $\mathbb{R}^2$

There are two very interesting and important vector interpretations of Green's theorem.

DEFINITION
Let $\mathbf{F}(x, y) = P(x, y)\mathbf{i} + Q(x, y)\mathbf{j}$ be a vector field in the plane.

(i) The **curl** of $\mathbf{F}$ is given by

$$\text{curl }\mathbf{F} = \frac{\partial Q}{\partial x} - \frac{\partial P}{\partial y}. \qquad \textbf{(15)}$$

(ii) The **divergence** of $\mathbf{F}$, denoted div $\mathbf{F}$, is given by

$$\text{div } \mathbf{F} = \frac{\partial P}{\partial x} + \frac{\partial Q}{\partial y}. \tag{16}$$

Before explaining these terms more fully, we will write Green's theorem in two equivalent vector forms.

First, recall from equation (9.5.2) on p. 537 that if $\mathbf{T}$ denotes the unit tangent vector to a curve $\mathbf{x}(t) = x(t)\mathbf{i} + y(t)\mathbf{j}$ and if s denotes the parameter of arc length, then

$$\mathbf{T} = \frac{d\mathbf{x}}{ds},$$

or

$$d\mathbf{x} = \mathbf{T}\, ds. \tag{17}$$

Now let $\mathbf{T}$ denote the unit tangent vector to $\partial\Omega$. Then if $\mathbf{F} = P\mathbf{i} + Q\mathbf{j}$, we can write, using (17),

$$\oint_{\partial\Omega} P\, dx + Q\, dy = \oint_{\partial\Omega} \mathbf{F} \cdot d\mathbf{x} = \oint_{\partial\Omega} \mathbf{F} \cdot \mathbf{T}\, ds. \tag{18}$$

Then applying Green's theorem and using (15) and (18), we obtain the following theorem.

■ **THEOREM 2: First Vector Form of Green's Theorem†**
Under the hypotheses of Theorem 1,

$$\oint_{\partial\Omega} \mathbf{F} \cdot \mathbf{T}\, ds = \iint_{\Omega} \text{curl } \mathbf{F}\, dx\, dy. \tag{19}$$

■

Now from the expression

$$\oint_{\partial\Omega} P\, dx + Q\, dy = \iint_{\Omega} \left(\frac{\partial Q}{\partial x} - \frac{\partial P}{\partial y} \right) dx\, dy,$$

† This form of Green's theorem is sometimes called **Stokes's theorem in the plane.**

we may replace P by $-Q$ and Q by P to obtain

$$\oint_{\partial\Omega} -Q\,dx + P\,dy = \iint_{\Omega} \left(\frac{\partial P}{\partial x} + \frac{\partial Q}{\partial y}\right) dx\,dy. \qquad (20)$$

If $dx\,\mathbf{i} + dy\,\mathbf{j}$ represents the vector $\mathbf{T}\,ds$, then $dy\,\mathbf{i} - dx\,\mathbf{j}$ represents the vector $\mathbf{n}\,ds$, where $\mathbf{n}$ is the unit normal vector to the curve. This is easy to see since $(dx\,\mathbf{i} + dy\,\mathbf{j}) \cdot (dy\,\mathbf{i} - dx\,\mathbf{j}) = 0$ and both $\mathbf{T}\,ds$ and $\mathbf{n}\,ds$ have the same magnitude ds. Thus the left-hand side of (20) can be written

$$\oint_{\partial\Omega} -Q\,dx + P\,dy = \oint_{\partial\Omega} \mathbf{F} \cdot \mathbf{n}\,ds. \qquad (21)$$

Using (16) and (21) in (20) yields the next theorem.

■ **THEOREM 3: Second Vector Form of Green's Theorem†**
Under the hypothesis of Theorem 1,

$$\oint_{\partial\Omega} \mathbf{F} \cdot \mathbf{n}\,ds = \iint_{\Omega} \operatorname{div} \mathbf{F}\,dx\,dy. \qquad (22)$$

■

Circulation and Flux

There are interesting physical interpretations of the two vector forms of Green's theorem. Let $\mathbf{F}(x, y)$ denote the direction and rate of flow of a fluid at a point (x, y) in the plane. The integral

$$\oint_{\partial\Omega} \mathbf{F} \cdot \mathbf{T}\,ds$$

is the integral of the component of the flow in the direction tangent to the boundary of Ω and is called the **circulation** of $\mathbf{F}$ around the boundary of Ω (see Fig. 5). By (19) and the mean value theorem for double integrals (p. 604), we have

$$\oint_{\partial\Omega} \mathbf{F} \cdot \mathbf{T}\,ds = \iint_{\Omega} \operatorname{curl} \mathbf{F}\,dx\,dy = [\operatorname{curl} \mathbf{F}(x_0, y_0)] \text{ (area of } \Omega), \qquad (23)$$

where $(x_0, y_0) \in \Omega$. If Ω is small, then curl $\mathbf{F}$ is nearly constant and curl $\mathbf{F}(x, y) \approx$ curl $\mathbf{F}(x_0, y_0)$ for $(x, y) \in \Omega$. From (23) it appears that the curl represents

---

† This form of Green's theorem is sometimes called the **divergence theorem** or **Gauss's Theorem in the plane.**

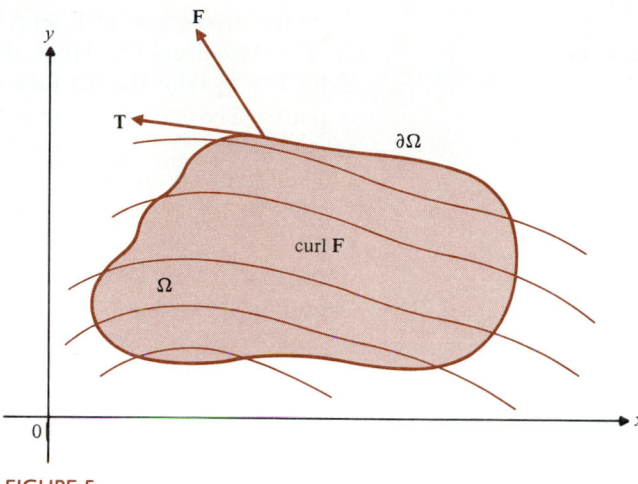

FIGURE 5

the circulation per unit area at the point (x, y). If curl $\mathbf{F} = 0$ for every (x, y) in Ω, then the fluid flow $\mathbf{F}$ is called **irrotational**.

Now $\oint_{\partial\Omega} \mathbf{F} \cdot \mathbf{n} \, ds$ is the component of flow in the direction of the outward normal to $\partial\Omega$, and is called the **flux** across $\partial\Omega$ (see Fig. 6). The flux is the rate at which fluid is flowing across the boundary of Ω from inside Ω. If Ω is small, then using (22), we have

$$\oint_{\partial\Omega} \mathbf{F} \cdot \mathbf{n} \, ds = \iint_{\Omega} \text{div } \mathbf{F} \, dx \, dy \overset{\overset{\text{mean value theorem}}{\downarrow}}{=} [\text{div } \mathbf{F}(x_0, y_0)](\text{area of } \Omega)$$

$$\approx [\text{div } \mathbf{F}(x, y)](\text{area of } \Omega). \tag{24}$$

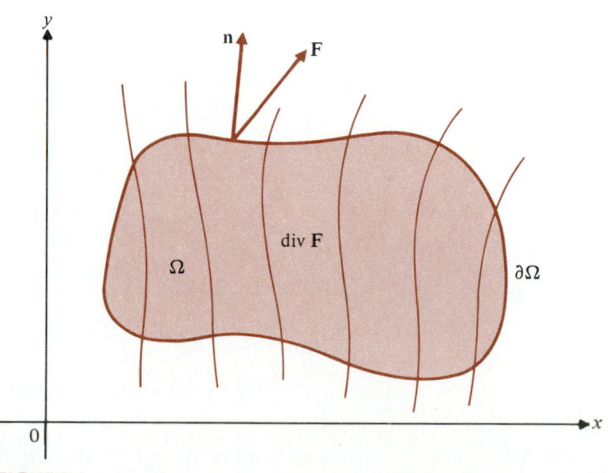

FIGURE 6

Thus the divergence of **F** represents the net rate of flow away from (x, y). If div **F** $= 0$ for every (x, y) in Ω, then the flow **F** is called **incompressible,** because if div **F** $= 0$, then the net flow into the region is 0. That is, fluid is not accumulating. The accumulation of fluid in a fixed region would cause it to become compressed.

We shall say more about divergence and curl in Section 10.6.

Before leaving this section we note that there are other ways to write the curl and divergence, and therefore there are other ways to write Green's theorem in the plane. Very loosely speaking, we can think of the gradient symbol ∇ as the vector function $(\partial/\partial x, \partial/\partial y)$. Then if **F** $= (P, Q)$,

$$\nabla \times \mathbf{F} = \begin{vmatrix} \mathbf{i} & \mathbf{j} & \mathbf{k} \\ \dfrac{\partial}{\partial x} & \dfrac{\partial}{\partial y} & 0 \\ P & Q & 0 \end{vmatrix} = \left(\frac{\partial Q}{\partial x} - \frac{\partial P}{\partial y} \right)\mathbf{k} = (\text{curl } \mathbf{F})\mathbf{k}.$$

Thus curl **F** $= (\text{curl } \mathbf{F})\mathbf{k} \cdot \mathbf{k} = (\nabla \times \mathbf{F}) \cdot \mathbf{k}$ (since $\mathbf{k} \cdot \mathbf{k} = 1$). For this reason curl **F** is often written in the plane (with the **k** omitted) as

$$\text{curl } \mathbf{F} = \nabla \times \mathbf{F}. \tag{25}$$

Similarly,

$$\nabla \cdot \mathbf{F} = \left(\frac{\partial}{\partial x}, \frac{\partial}{\partial y} \right) \cdot (P, Q) = \frac{\partial P}{\partial x} + \frac{\partial Q}{\partial y} = \text{div } \mathbf{F},$$

and we write

$$\text{div } \mathbf{F} = \nabla \cdot \mathbf{F}. \tag{26}$$

Then using (25), (26), and Theorems 2 and 3, we have the following forms of Green's theorem:

$$\oint_{\partial\Omega} \mathbf{F} \cdot \mathbf{T}\, ds = \iint_{\Omega} (\nabla \times \mathbf{F}) \cdot \mathbf{k}\, dx\, dy \tag{27}$$

and

$$\oint_{\partial\Omega} \mathbf{F} \cdot \mathbf{n}\, ds = \iint_{\Omega} \nabla \cdot \mathbf{F}\, dx\, dy. \tag{28}$$

Problems 10.4

In Problems 1–15, find the line integral by using Green's theorem.

1. $\oint_{\partial\Omega} 3y\,dx + 5x\,dy$; $\Omega = \{(x, y): 0 \le x \le 1, 0 \le y \le 1\}$.

2. $\oint_{\partial\Omega} ay\,dx + bx\,dy$; Ω is the region of Problem 1.

3. $\oint_{\partial\Omega} e^x \cos y\,dx + e^x \sin y\,dy$; Ω is the region enclosed by the triangle with vertices at $(0, 0)$, $(1, 0)$, and $(0, 1)$.

4. The integral of Problem 3, where Ω is the region enclosed by the triangle with vertices at $(0, 0)$, $(1, 0)$, and $(1, 1)$.

5. The integral of Problem 3, where Ω is the region enclosed by the rectangle with vertices at $(0, 0)$, $(2, 0)$, $(2, 1)$, and $(0, 1)$.

6. $\oint_{\partial\Omega} 2xy\,dx + x^2\,dy$; Ω is the unit disk.

7. $\oint_{\partial\Omega}(x^2 + y^2)\,dx - 2xy\,dy$; Ω is the unit disk.

8. $\oint_{\partial\Omega}(1/y)\,dx + (1/x)\,dy$; Ω is the region bounded by the lines $y = 1$ and $x = 16$ and the curve $y = \sqrt{x}$.

9. $\oint_{\partial\Omega} \cos y\,dx + \cos x\,dy$; Ω is the region enclosed by the rectangle $\{(x, y): 0 \le x \le \pi/4, 0 \le y \le \pi/3\}$.

10. $\oint_{\partial\Omega} x^2 y\,dx - xy^2\,dy$; Ω is the disk $x^2 + y^2 \le 9$.

11. $\oint_{\partial\Omega} y \ln x\,dy$; $\Omega = \{(x, y): 1 \le y \le 3, e^y \le x \le e^{y^3}\}$.

12. $\oint_{\partial\Omega} \sqrt{1 + y^2}\,dx$; $\Omega = \{(x, y): -1 \le y \le 1, y^2 \le x \le 1\}$.

13. $\oint_{\partial\Omega} ay\,dx + bx\,dy$; Ω is a region of the type (4), (5).

14. $\oint_{\partial\Omega} e^x \sin y\,dx + e^x \cos y\,dy$; Ω is the region enclosed by the ellipse $(x^2/a^2) + (y^2/b^2) = 1$.

15. $\oint_{\partial\Omega}(-4x/\sqrt{1+y^2})\,dx + (2x^2y/(1 + y^2)^{3/2})\,dy$; Ω is a region of the type (4), (5).

16. Use one of the line integrals in (14) to calculate the area enclosed by the circle $\mathbf{x}(t) = a(\cos t)\mathbf{i} + a(\sin t)\mathbf{j}$.

17. Use Green's theorem to calculate the area enclosed by the triangle with vertices at (a_1, b_1), (a_2, b_2), and (a_3, b_3), assuming that the three points are not collinear.

18. Use Green's theorem to calculate the area of the quadrilateral with vertices at $(0, 0)$, $(2, 1)$, $(-1, 3)$, and $(4, 4)$.

19. Use Green's theorem to calculate the area of the quadrilateral with vertices at (a_1, b_1), (a_2, b_2), (a_3, b_3), and (a_4, b_4), assuming that no three of the points are collinear

and that no point is within the triangle whose vertices are the other three points.

In Problems 20–26, calculate (a) curl $\mathbf{F}$, (b) $\oint_{\partial\Omega} \mathbf{F} \cdot \mathbf{T}\,ds$, (c) div $\mathbf{F}$, and (d) $\oint_{\partial\Omega} \mathbf{F} \cdot \mathbf{n}\,ds$.

20. $\mathbf{F}(x, y) = x^2\mathbf{i} + y^2\mathbf{j}$; Ω is the region of Problem 1.

21. $\mathbf{F}(x, y) = y^2\mathbf{i} + x^2\mathbf{j}$; Ω is the region of Problem 1.

22. $\mathbf{F}(x, y) = ay\mathbf{i} + bx\mathbf{j}$; Ω is the region of Problem 1.

23. $\mathbf{F}(x, y) = y^3\mathbf{i} + x^3\mathbf{j}$; Ω is the unit disk.

24. $\mathbf{F}(x, y) = x\mathbf{i} + y\mathbf{j}$; Ω is the unit disk.

25. $\mathbf{F}(x, y) = y\mathbf{i} - x\mathbf{j}$; Ω is the unit disk.

26. $\mathbf{F}(x, y) = xy\mathbf{i} + (y^2 - x^2)\mathbf{j}$; Ω is the region of Problem 3.

27. Let $\partial\Omega$ be the ellipse $(x^2/a^2) + (y^2/b^2) = 1$. Let $\mathbf{F}(x, y)$ be the vector field $-x\mathbf{i} - y\mathbf{j}$ that, at any point (x, y), points toward the origin. Show that $\oint_{\partial\Omega} \mathbf{F} \cdot \mathbf{T}\,ds = 0$.

* 28. Let Ω be the disk $x^2 + y^2 \le a^2$. Show that
$$\oint_{\partial\Omega} \alpha\sqrt{x^2 + y^2}\,dx + \beta\sqrt{x^2 + y^2}\,dy = 0.$$

29. Let Ω be as in Problem 28 and suppose that g is continuously differentiable. Show that
$$\oint_{\partial\Omega} \alpha g(x^2 + y^2)\,dx + \beta g(x^2 + y^2)\,dy = 0.$$

30. Show that the vector flow $\mathbf{F}(x, y) = x\mathbf{i} + y\mathbf{j}$ is irrotational.

31. Show that the vector flow $\mathbf{F}(x, y) = \sin x\,e^x\mathbf{i} + y^{5/2}\mathbf{j}$ is irrotational.

32. Show that for any continuously differentiable functions f and g, the vector flow $\mathbf{F}(x, y) = f(x)\mathbf{i} + g(y)\mathbf{j}$ is irrotational.

33. Show that the following vector flow is incompressible.
$$\mathbf{F}(x, y) = y\sqrt{x^2 + y^2}\mathbf{i} - x\sqrt{x^2 + y^2}\mathbf{j}$$

34. Let g be as in Problem 29. Show that the vector flow $\mathbf{F}(x, y) = -yg(x^2 + y^2)\mathbf{i} + xg(x^2 + y^2)\mathbf{j}$ is incompressible.

35. Let $\mathbf{F}(x, y) = y/(x^2 + y^2)\mathbf{i} - x/(x^2 + y^2)\mathbf{j}$.
    (a) Show that curl $\mathbf{F} = 0$.
    (b) Show that $\oint_C \mathbf{F} \cdot \mathbf{T}\,ds \ne 0$ if C is the unit circle oriented counterclockwise.
    (c) Explain why the results of (a) and (b) do not contradict Theorem 2.

10.5 Surface Integrals and Flux

In Section 9.4 we found a formula for the length of a plane curve [given by $y = f(x)$] by showing that the length of a small "piece" of the curve was approximately equal to $\sqrt{1 + [f'(x)]^2}\,\Delta x$. We now define the area of a surface

$z = f(x, y)$ that lies over a region Ω in the xy-plane. We define it by analogy with the arc length formula

DEFINITION

Lateral Surface Area

Let f be continuous with continuous partial derivatives in the region Ω in the xy-plane. Then the **lateral surface area** σ of the graph of f over Ω is defined by

$$\sigma = \iint\limits_{\Omega} \sqrt{1 + f_x^2(x, y) + f_y^2(x, y)}\, dA. \tag{1}$$

Remark. The assumption that f is continuously differentiable over Ω ensures that the integral in (1) exists.

To see why (1) is chosen as the definition of lateral surface area, consider the following justification:

Derivation of the Formula for Surface Area. We begin by calculating the surface area $\Delta\sigma$ over a rectangle ΔA with sides Δx and Δy. The situation is depicted in Fig. 1, in which it is assumed that $f(x, y) > 0$ for (x, y) in Ω. We assume that f has continuous partial derivatives over Ω. If Δx and Δy are small, then the region $PQSR$ in space has, approximately, the shape of a parallelogram. Thus by equation (6.3.4), p. 311:

$$\Delta\sigma \approx \text{area of parallelogram} = |\vec{PQ} \times \vec{PR}|. \tag{2}$$

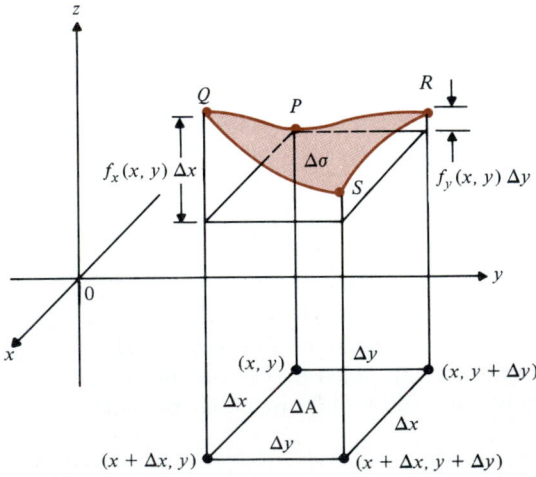

FIGURE 1

Now

$$\vec{PQ} = (x + \Delta x,\, y,\, f(x + \Delta x,\, y)) - (x,\, y,\, f(x,\, y))$$

$$= (\Delta x,\, 0,\, f(x + \Delta x,\, y) - f(x,\, y)).$$

But if Δx is small, then

$$\frac{f(x + \Delta x, y) - f(x, y)}{\Delta x} \approx f_x(x, y),$$

so that

$$f(x + \Delta x, y) - f(x, y) \approx f_x(x, y)\,\Delta x$$

and

$$\vec{PQ} \approx (\Delta x, 0, f_x(x, y)\,\Delta x). \tag{3}$$

Similarly,

$$\vec{PR} = (x, y + \Delta y, f(x, y + \Delta y)) - (x, y, f(x, y))$$

$$= (0, \Delta y, f(x, y + \Delta y) - f(x, y)),$$

and if Δy is small,

$$\vec{PR} \approx (0, \Delta y, f_y(x, y)\,\Delta y). \tag{4}$$

Thus from (3) and (4),

$$\vec{PQ} \times \vec{PQ} \approx \begin{vmatrix} \mathbf{i} & \mathbf{j} & \mathbf{k} \\ \Delta x & 0 & f_x(x, y)\,\Delta x \\ 0 & \Delta y & f_y(x, y)\,\Delta y \end{vmatrix}$$

$$= -f_x(x, y)\,\Delta x\,\Delta y\mathbf{i} - f_y(x, y)\,\Delta x\,\Delta y\mathbf{j} + \Delta x\,\Delta y\mathbf{k}$$

$$= (-f_x(x, y)\mathbf{i} - f_y(x, y)\mathbf{j} + \mathbf{k})\,\Delta x\,\Delta y,$$

so that from (2)

$$\Delta\sigma \approx \sqrt{f_x^2(x, y) + f_y^2(x, y) + 1}\,\overbrace{\Delta x\,\Delta y}^{=\,\Delta A}. \tag{5}$$

Finally, adding up the surface area over rectangles that partition Ω and taking a limit yields (1).

Remark. We emphasize that formula (1) is a definition. A definition is not something that we have to prove, of course, but it is essential that it should give the desired answers in all cases where the concept of surface area can be established independently.

DEFINITION

Smooth Surface

The surface $z = f(x, y)$ is called **smooth** at a point (x_0, y_0, z_0) if $\partial f/\partial x$ and $\partial f/\partial y$ are continuous at (x_0, y_0). If the surface is smooth at all points in the domain of f, we speak of it as a **smooth surface.** That is, if f is continuously differentiable, then the surface is smooth.

A surface integral is very much like a double integral. Suppose that the surface $z = f(x, y)$ is smooth for (x, y) in a bounded region Ω in the xy-plane.

Since $f(x, y)$ is continuous on Ω, we find from the definition of the double integral and Theorem 10.3.1 that

$$\iint\limits_{\Omega} f(x, y)\, dA = \lim_{\Delta s \to 0} \sum_{i=1}^{n} \sum_{j=1}^{m} f(x_i^*, y_j^*)\, \Delta x\, \Delta y \tag{6}$$

exists, where $\Delta s = \sqrt{\Delta x^2 + \Delta y^2}$ and the limit is independent of the way in which the points (x_i^*, y_j^*) are chosen in the rectangle R_{ij}. We stress that in (6) the quantity $\Delta x\, \Delta y$ represents the *area* of the rectangle R_{ij}.

The double integral (6) is an integral over a region in the plane. A surface integral, which we will soon define, is an integral over a surface in space. Suppose we wish to integrate the function $F(x, y, z)$ over the surface S given by $z = f(x, y)$ where $(x, y) \in \Omega$ and, as before, Ω is a bounded region in the xy-plane. Such a surface is sketched in Fig. 2.

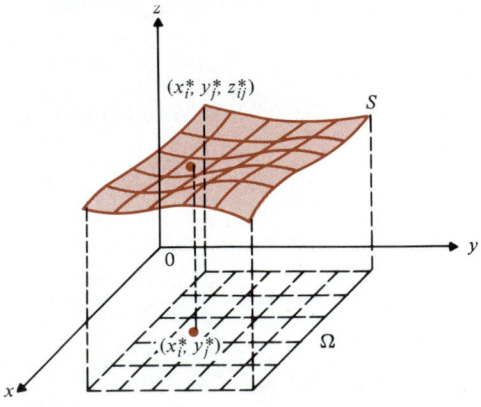

FIGURE 2

We partition Ω into rectangles (and parts of rectangles) as before. This procedure provides a partition of S into mn "subsurfaces" S_{ij}, where

$$S_{ij} = \{(x, y, z): z = f(x, y) \text{ and } (x, y) \in R_{ij}\}.$$

We choose a point in each S_{ij}. Such a point will have the form (x_i^*, y_j^*, z_{ij}^*), where $z_{ij}^* = f(x_i^*, y_j^*)$ and $(x_i^*, y_j^*) \in R_{ij}$. We let $\Delta \sigma_{ij}$ denote the surface area of S_{ij}. This is analogous to the notation $\Delta x\, \Delta y$ as the area of the rectangle R_{ij}. Then we write the double sum

$$\sum_{i=1}^{n} \sum_{j=1}^{m} F(x_i^*, y_j^*, z_{ij}^*)\, \Delta \sigma_{ij}$$

and consider

$$\lim_{\Delta s \to 0} \sum_{i=1}^{n} \sum_{j=1}^{m} F(x_i^*, y_j^*, z_{ij}^*)\, \Delta \sigma_{ij}. \tag{7}$$

DEFINITION

Surface Integral

Suppose the limit in (7) exists and is independent of the way the surface S is partitioned and the way in which the points (x_i^*, y_j^*, z_{ij}^*), are chosen in S_{ij}. Then F is said to be **integrable** over S and the **surface integral** of F over S, denoted by $\iint_S F(x, y, z)\, d\sigma$, is given by

$$\iint_S F(x, y, z)\, d\sigma = \lim_{\Delta s \to 0} \sum_{i=1}^{n} \sum_{j=1}^{m} F(x_i^*, y_j^*, z_{ij}^*)\, \Delta \sigma_{ij}. \qquad (8)$$

Three central questions are raised by this definition.

    (i) Under what conditions does the surface integral of F over S exist?

    (ii) How can we compute the surface integral if it does exist?

   (iii) What good is a surface integral?

We shall answer the first two questions now and answer the third one by giving some applications of the surface integral later in the section.

    The key to evaluating the limit in (8) is to note that, from formula (5),

$$\Delta \sigma_{ij} \approx \sqrt{f_x^2(x_i^*, y_j^*) + f_y^2(x_i^*, y_j^*) + 1}\; \Delta x\, \Delta y. \qquad (9)$$

Then inserting (9) into the limit in (8) and noting that $z_{ij}^* = f(x_i^*, y_j^*)$, we have

$$\lim_{\Delta s \to 0} \sum_{i=1}^{n} \sum_{j=1}^{m} F(x_i^*, y_j^*, z_{ij}^*)\, \Delta \sigma_{ij} \qquad (10)$$

$$= \lim_{\Delta s \to 0} \sum_{i=1}^{n} \sum_{j=1}^{m} F(x_i^*, y_j^*, f(x_i^*, y_j^*)) \sqrt{f_x^2(x_i^*, y_j^*) + f_y^2(x_i^*, y_j^*) + 1}\; \Delta x\, \Delta y.$$

Now if $z = f(x, y)$ is a smooth surface over Ω, then f_x and f_y are continuous over Ω, so that $\sqrt{f_x^2 + f_y^2 + 1}$ is also continuous over Ω. Furthermore, if $F(x, y, z)$ is continuous for (x, y, z) on a region containing S, then by Theorem 10.3.1

$$\iint_\Omega F(x, y, f(x, y)) \sqrt{f_x^2(x, y) + f_y^2(x, y) + 1}\; dA$$

exists and is equal to the right-hand limit in (10). We therefore have the following important result.

■ THEOREM 1

Let S: $z = f(x, y)$ be a smooth surface for (x, y) in the bounded region Ω in the xy-plane. Then if F is continuous on a region containing S, F is integrable over S, and

$$\iint\limits_{S} F(x, y, z)\, d\sigma = \iint\limits_{\Omega} F(x, y, f(x, y))\, \sqrt{f_x^2(x, y) + f_y^2(x, y) + 1}\; dA. \qquad \textbf{(11)}$$

Remark. Using (11), we can compute a surface integral over S by transforming it into an ordinary double integral over the region Ω that is the projection of S into the xy-plane.

Remark. If $F(x, y, z) = 1$ in (11), then (11) reduces to

$$\sigma = \iint\limits_{S} d\sigma = \iint\limits_{\Omega} \sqrt{f_x^2(x, y) + f_y^2(x, y) + 1}\; dA.$$

This is the formula for surface area.

EXAMPLE 1

Compute $\iint_S (x^2 + y^2 + 3z^2)\, d\sigma$, where S is the part of the circular paraboloid $z = x^2 + y^2$ with $x^2 + y^2 \le 9$.

SOLUTION

Here $f(x, y) = x^2 + y^2$, so that $f_x = 2x$, $f_y = 2y$, and $d\sigma = \sqrt{1 + 4x^2 + 4y^2}\; dx\, dy$. Thus

$$I = \iint\limits_{S} (x^2 + y^2 + 3z^2)\, d\sigma = \iint\limits_{\Omega} [x^2 + y^2 + 3(x^2 + y^2)^2]\, \sqrt{1 + 4x^2 + 4y^2}\; dA,$$

where Ω is the disk (in the xy-plane) $x^2 + y^2 \le 9$. The problem is greatly simplified by the use of polar coordinates. We have, using $x^2 + y^2 = r^2$,

$$I = \int_0^{2\pi} \int_0^3 [r^2 + 3(r^2)^2]\, \sqrt{1 + 4r^2}\; r\, dr\, d\theta$$

$$= \int_0^{2\pi} \int_0^3 (r^3 + 3r^5)\, \sqrt{1 + 4r^2}\; dr\, d\theta$$

$$= 2\pi \int_0^3 (r^3 + 3r^5)\, \sqrt{1 + 4r^2}\; dr.$$

There are several ways to complete the evaluation of this integral. One way is to integrate by parts twice (start with $u = r^2 + 3r^4$). Another way is to make the substitution $r = \frac{1}{2} \tan \varphi$. The result is

$$I = 2\pi \left\{ 21(37)^{3/2} + \frac{1}{120}[1 - 55(37)^{5/2}] + \frac{1}{280}(37^{7/2} - 1) \right\} \approx 12{,}629.4.$$

There are many applications for which it is necessary to compute a surface integral. For example, suppose that for (x, y) in a region Ω in the xy-plane, $z = f(x, y)$ is the equation of a thin metallic surface in space. Suppose further that the density of the surface varies and is given by $\rho(x, y, z)$ for (x, y, z) on the surface. Our problem is to compute the total mass of the sheet. We do so by first partitioning Ω in the usual way. This procedure leads to a partition of the metallic surface into subsurfaces, as before. We again denote this subsurface S_{ij}:

$$S_{ij} = \{(x, y, z): z = f(x, y) \text{ and } (x, y) \in R_{ij}\}.$$

If $\Delta\sigma_{ij}$ denotes the surface area of S_{ij}, if $\Delta\sigma_{ij}$ is small, and if $\rho(x, y, z)$ is continuous, then $\rho(x, y, z)$ is approximately constant on S_{ij}. If (x_i^*, y_j^*) is a point in R_{ij}, then $(x_i^*, y_j^*, z_{ij}^*) = (x_i^*, y_j^*, f(x_i^*, y_j^*))$ is a point on S_{ij}, and

$$\rho(x, y, z) \approx \rho(x_i^*, y_j^*, f(x_i^*, y_j^*)) \tag{12}$$

for (x, y, z) on S_{ij}. Now if area density is constant, then the mass μ of an object with area σ and density ρ is given by

$$\mu = \rho\sigma. \tag{13}$$

Thus if μ_{ij} denotes the mass of S_{ij}, then combining (12) and (13), we have

$$\mu_{ij} \approx \rho(x_i^*, y_j^*, f(x_i^*, y_j^*)) \, \Delta\sigma_{ij},$$

and adding up the masses of the mn "subsurfaces" S_{ij}, we obtain

$$\mu \approx \sum_{i=1}^{n} \sum_{j=1}^{m} \rho(x_i^*, y_j^*, f(x_i^*, y_j^*)) \, \Delta\sigma_{ij}. \tag{14}$$

Finally, we let $\Delta s \to 0$, as before. This gives us

$$\mu = \lim_{\Delta s \to 0} \sum_{i=1}^{n} \sum_{j=1}^{m} \rho(x_i^*, y_j^*, f(x_i^*, y_j^*)) \, \Delta\sigma_{ij} = \iint_S \rho(x, y, f(x, y)) \, d\sigma. \tag{15}$$

EXAMPLE 2

A metallic dome has the shape of a hemisphere centered at the origin with radius 4 m. Its area density at a point (x, y, z) in space is given by $\rho(x, y, z) = 25 - x^2 - y^2$ kilograms per square meter. Find the total mass of the dome.

SOLUTION

The equation of the hemisphere is $x^2 + y^2 + z^2 = 16$ with $z \geq 0$. Thus $z = f(x, y) = \sqrt{16 - x^2 - y^2}$ (for $x^2 + y^2 \leq 16$),

$$f_x = \frac{-x}{\sqrt{16 - x^2 - y^2}}, \qquad f_y = \frac{-y}{\sqrt{16 - x^2 - y^2}},$$

and

$$\sqrt{f_x^2 + f_y^2 + 1} = \sqrt{\frac{x^2}{16 - x^2 - y^2} + \frac{y^2}{16 - x^2 - y^2} + 1}$$

$$= \sqrt{\frac{x^2 + y^2 + (16 - x^2 - y^2)}{16 - x^2 - y^2}} = \frac{4}{\sqrt{16 - x^2 - y^2}}.$$

Then using polar coordinates, we have

$$\overset{\Omega \text{ is the circle centered}}{\underset{\downarrow}{\text{at } (0,0) \text{ with radius } 4}}$$

$$\mu = \iint_S \rho(x, y, z) \, d\sigma = \int_0^{2\pi} \int_0^4 (25 - r^2) \left(\frac{4}{\sqrt{16 - r^2}} \right) r \, dr \, d\theta$$

$$= 2\pi \int_0^4 (25 - r^2) \left(\frac{4}{\sqrt{16 - r^2}} \right) r \, dr.$$

Now let $16 - r^2 = u^2$, so that $25 - r^2 = 9 + u^2$ and $r \, dr = -u \, du$. Then the integral becomes

$$2\pi \int_4^0 (9 + u^2) \left(\frac{4}{u} \right) (-u \, du) = 8\pi \int_0^4 (9 + u^2) \, du$$

$$= (8\pi) \left(\frac{172}{3} \right) = \frac{1376\pi}{3} \approx 1441 \text{ kg.}$$

We can derive another way to represent a surface integral. First we need to define the orientation of a surface.

Orientation of a Surface

Consider the smooth surface $z = f(x, y)$. We can write this surface in two ways:

$$F(x, y, z) = f(x, y) - z = 0 \quad \text{and} \quad G(x, y, z) = z - f(x, y) = 0. \tag{16}$$

From Section 9.9 (p. 572) we know that the gradient vectors ∇F and ∇G are normal to the surface at every point on the surface. Then, if ∇F and $\nabla G \neq 0$,

$$\mathbf{n}_1 = \frac{\nabla F}{|\nabla F|} \quad \text{and} \quad \mathbf{n}_2 = \frac{\nabla G}{|\nabla G|} \tag{17}$$

are unit normal vectors. In fact, it is evident from the way F and G are defined that $\mathbf{n}_2 = -\mathbf{n}_1$.

We choose one of these normal vectors, denote it by **n,** and call it the **outward unit normal vector.** The direction of **n** is called the **positive normal direction** to the surface at a point.

Outward Unit Normal Vector

Remark. If S is a closed surface such as the surface of a ball, then, by convention, we choose **n** so that it points away from the region bounded by the surface. This explains the use of the term *outward* unit normal vector.

Orientable Surface

Let P_0 be a point on the surface S. Let C be any closed curve on S that passes through P_0. Then S is said to be **orientable** if the outward unit normal vector at P_0 is preserved as the outward unit normal is displaced continuously around C until it returns to P_0.

If a surface is orientable, then the choice of the outward unit normal vector **n** determines a positive direction on S by displacing **n** along curves lying in S.

Most of the surfaces we consider in this text are orientable. The most famous example of a nonorientable surface is given by the Möbius strip.† To construct a Möbius strip, take a long rectangular strip of paper, twist one of the shorter ends once, and paste the shorter ends together (as in Fig. 3). In our terminology, the Möbius strip is not orientable because a normal vector moving once around the closed curve indicated by the dotted line will change direction.

In this text we will write the surface $z = f(x, y)$ as $G(x, y, z) = z - f(x, y) = 0$. Then

$$\nabla G = G_x\mathbf{i} + G_y\mathbf{j} + G_z\mathbf{k} = -f_x\mathbf{i} - f_y\mathbf{j} + \mathbf{k}$$

and

$$\mathbf{n} = \frac{\nabla G}{|\nabla G|} = \frac{-f_x\mathbf{i} - f_y\mathbf{j} + \mathbf{k}}{\sqrt{f_x^2 + f_y^2 + 1}}. \tag{18}$$

This vector determines our orientation (positive direction).

We now define the angle γ to be the acute angle between **n** and the positive z-axis. This angle is depicted in Fig. 4. Since **k** is a unit vector having the direction of the positive z-axis, we have, from Theorem 6.2.2 on p. 303,

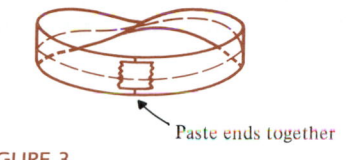

FIGURE 3

Paste ends together

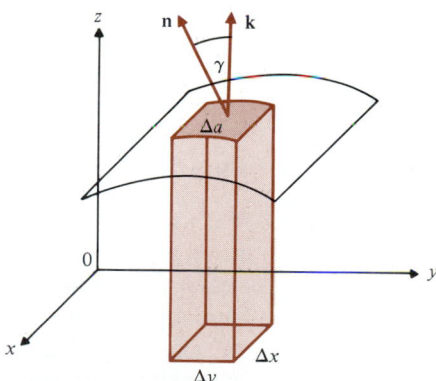

FIGURE 4

$$\cos \gamma = \frac{\mathbf{n} \cdot \mathbf{k}}{|\mathbf{n}|\,|\mathbf{k}|} = \frac{1}{\sqrt{f_x^2 + f_y^2 + 1}} \tag{19}$$

and

$$\sec \gamma = \sqrt{f_x^2 + f_y^2 + 1}. \tag{20}$$

† Named after the German mathematician August Ferdinand Möbius (1790–1868). Möbius, who was a student of Gauss, did important work in geometry, mechanics, and number theory.

Hence we have, from (11),

$$\iint_S F(x, y, z)\, d\sigma = \iint_\Omega F(x, y, f(x, y))\, \sec \gamma(x, y)\, dx\, dy. \qquad \textbf{(21)}$$

EXAMPLE 3

Compute $\iint_S (x + 2y + 3z)\, d\sigma$, where S is the part of the plane $2x - y + z = 3$, that lies above the triangular region Ω in the xy-plane bounded by the x- and y-axes and the line $y = 1 - 2x$.

SOLUTION

Here S is the plane $2x - y + z = 3$. From Section 9.9 we know that $\mathbf{N} = 2\mathbf{i} - \mathbf{j} + \mathbf{k}$ is normal to this surface, so $\mathbf{n} = \mathbf{N}/|\mathbf{N}| = (1/\sqrt{6})(2\mathbf{i} - \mathbf{j} + \mathbf{k})$ is the outward unit normal vector. Then $\cos \gamma = \mathbf{n} \cdot \mathbf{k} = 1/\sqrt{6}$ and $\sec \gamma = \sqrt{6}$, so, by Fig. 5,

$$\iint_S (x + 2y + 3z)\, d\sigma = \iint_\Omega [(x + 2y) + 3(3 - 2x + y)]\sqrt{6}\, dx\, dy,$$

$$= \sqrt{6} \int_0^{1/2} \int_0^{1-2x} (9 - 5x + 5y)\, dy\, dx = \sqrt{6} \int_0^{1/2} \left(9y - 5xy + \frac{5y^2}{2}\right)\Big|_0^{1-2x} dx$$

$$= \sqrt{6} \int_0^{1/2} \left[9(1 - 2x) - 5x(1 - 2x) + \frac{5}{2}(1 - 2x)^2\right] dx$$

$$= \sqrt{6} \int_0^{1/2} \left(20x^2 - 33x + \frac{23}{2}\right) dx = \frac{59}{24}\sqrt{6}.$$

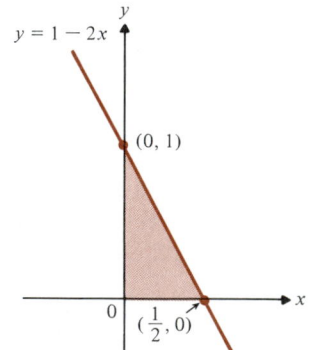

$y = 1 - 2x$

$(0, 1)$

0 $\left(\frac{1}{2}, 0\right)$

FIGURE 5

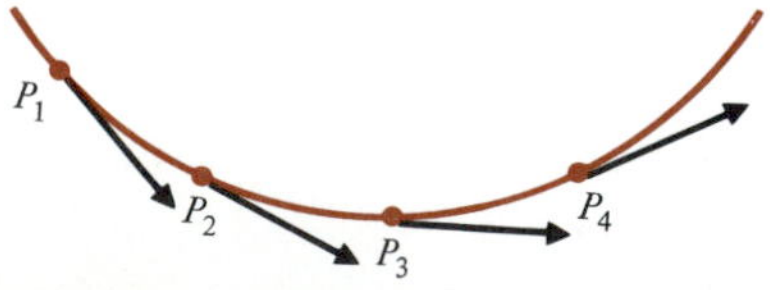

FIGURE 6 A streamline

Streamline

Flux

Flux

We now give another application of the surface integral and interpret the integral in yet another way. Suppose that a smooth surface is immersed in a continuous vector field. For example, this could be a permeable surface immersed in a fluid or a conductor surrounded by an electric field. We denote the vector field by $\mathbf{v}(x, y, z)$ and assume that it is a flow field; that is, we assume that something is moving. For example, the fluid could be in motion. As a particle moves through the fluid, it traces out a **streamline.** One streamline is illustrated in Fig. 6. As the particle moves from P_1 to P_2 to P_3, it traces out a path that is determined by the vectors $\mathbf{v}(x, y, z)$ at each point through which it passes. That is, $\mathbf{v}$ is tangent to the streamline at every point. Thus, if the particle is at the point (x_0, y_0, z_0), then $\mathbf{v}(x_0, y_0, z_0)$ determines where the particle will be in the next instant of time. Loosely speaking, the **flux** across a piece of surface is proportional to a weighted count of the streamlines that cut across the surface. The weight factor is the cosine of the angle between the normal direction and the streamline direction at each crossing.

A typical surface immersed in a vector field is sketched in Fig. 7a. In Fig. 7b we have sketched the flux across a small subsurface. The flux across a point on the surface can be thought of as the component of the vector field in the direction normal to the surface at that point. From the definition (see p. 304), this **normal component** of the vector field is given by

Normal Component

$$\frac{\mathbf{v} \cdot \mathbf{n}}{|\mathbf{n}|} = \mathbf{v} \cdot \mathbf{n}$$

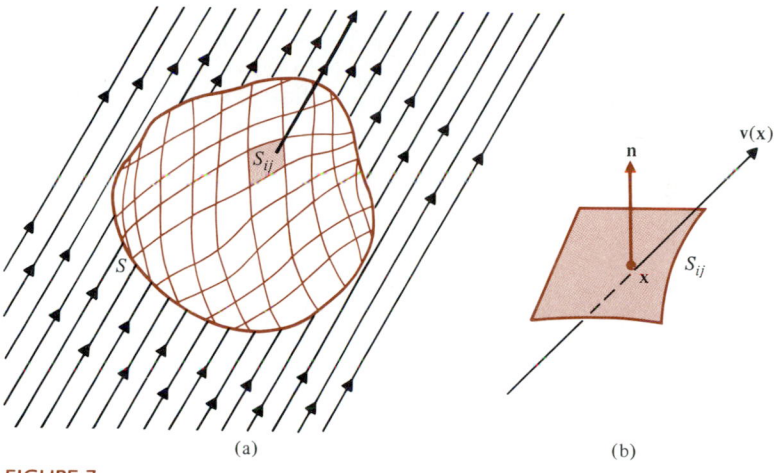

(a)   (b)

FIGURE 7

since $|\mathbf{n}| = 1$. If the subsurface S_{ij} is very small, then because $\mathbf{v}$ is continuous, $\mathbf{v} \cdot \mathbf{n}$ will vary little over S_{ij} and the flux across S_{ij} will be approximately equal to $\mathbf{v}(x_i^*, y_j^*, f(x_i^*, y_j^*)) \cdot \mathbf{n}$ times the surface area of S_{ij}. Thus

$$\text{flux across } S_{ij} \approx \mathbf{v}(x_i^*, y_j^*, f(x_i^*, y_j^*)) \cdot \mathbf{n} \Delta \sigma_{ij}.$$

Then

$$\text{total flux} \approx \sum_{i=1}^{n} \sum_{j=1}^{n} \mathbf{v}(x_i^*, y_j^*, f(x_i^*, y_j^*)) \cdot \mathbf{n} \Delta \sigma_{ij}.$$

This result leads to the following formal definition.

DEFINITION

Flux of a Vector Field over a Surface

Let S be a smooth surface with the vector field $\mathbf{v}(x, y, z)$ continuous on S. Then the **flux** of $\mathbf{v}$ over S is defined by

$$\text{flux of } \mathbf{v} \text{ over } S = \iint_S \mathbf{v} \cdot \mathbf{n} \, d\sigma. \tag{22}$$

The following theorem tells us how flux can be computed.

■ **THEOREM 2**

Let $S: z = f(x, y)$ for $(x, y) \in \Omega$ be a smooth surface and let $\mathbf{n}(x, y, z)$ be its outward unit normal vector. Let $\mathbf{v} = \mathbf{v}(x, y, z)$ be a vector field continuous on S. If

$$\mathbf{v}(x, y, z) = P(x, y, z)\mathbf{i} + Q(x, y, z)\mathbf{j} + R(x, y, z)\mathbf{k}, \qquad (23)$$

then

$$\text{flux of } \mathbf{v} \text{ over } S = \iint_S \mathbf{v} \bullet \mathbf{n} \, d\sigma = \iint_\Omega (-Pf_x - Qf_y + R) \, dx \, dy. \qquad (24)$$

PROOF

Using (18) and (20), we find that

$$\mathbf{v} \bullet \mathbf{n} = (P\mathbf{i} + Q\mathbf{j} + R\mathbf{k}) \bullet \left(\frac{-f_x \mathbf{i} - f_y \mathbf{j} + \mathbf{k}}{\sqrt{f_x^2 + f_y^2 + 1}} \right)$$

$$= \frac{1}{\sqrt{f_x^2 + f_y^2 + 1}} (-Pf_x - Qf_y + R)$$

$$= \frac{-Pf_x - Qf_y + R}{\sec \gamma}. \qquad (25)$$

Now from (21)

$$\iint_S \mathbf{v} \bullet \mathbf{n} \, d\sigma = \iint_\Omega \mathbf{v} \bullet \mathbf{n} \sec \gamma \, dx \, dy. \qquad (26)$$

But from (25)

$$\mathbf{v} \bullet \mathbf{n} \sec \gamma = -Pf_x - Qf_y + R. \qquad (27)$$

Inserting (27) into the right-hand integral in (26) completes the proof of the theorem.   ■

EXAMPLE 4

Find the flux across the conical surface $z = \sqrt{x^2 + y^2}$, $x^2 + y^2 \neq 0$, where $x^2 + y^2 \leq 1$, $x \geq 0$, $y \geq 0$, if the velocity field is given by $\mathbf{v} = x^2\mathbf{i} + y^2\mathbf{j} + z\mathbf{k}$.

SOLUTION

We have $f_x = x/\sqrt{x^2 + y^2}$ and $f_y = y/\sqrt{x^2 + y^2}$, so

$$\text{flux} = \iint_\Omega (-Pf_x - Qf_y + R) \, dx \, dy$$

$$= \iint_\Omega \left(-\frac{x^3}{\sqrt{x^2+y^2}} - \frac{y^3}{\sqrt{x^2+y^2}} + \sqrt{x^2+y^2} \right) dx\, dy$$

Using polar coordinates

$$= \int_0^{\pi/2} \int_0^1 \left(-\frac{r^3 \cos^3 \theta}{r} - \frac{r^3 \sin^3 \theta}{r} + r \right) r\, dr\, d\theta$$

$$= \int_0^{\pi/2} \int_0^1 [-r^3(\cos^3 \theta + \sin^3 \theta) + r^2]\, dr\, d\theta$$

$$= \int_0^{\pi/2} \left\{ \left[-\frac{r^4}{4}(\cos^3 \theta + \sin^3 \theta) + \frac{r^3}{3} \right] \Big|_0^1 \right\} d\theta$$

$$= \int_0^{\pi/2} \frac{1}{3}\, d\theta - \frac{1}{4}\int_0^{\pi/2} (\cos^3 \theta + \sin^3 \theta)\, d\theta = \frac{\pi}{6} - \frac{1}{3}.$$

Problems 10.5

In Problems 1–9, find the area of the part of the surface that lies over the given bounded region Ω.

1. $z = x + 2y$; $\Omega = \{(x, y): 0 \le x \le y,\ 0 \le y \le 2\}$

2. $z = 4x + 7y$; Ω = region between $y = x^2$ and $y = x^5$

3. $z = ax + by$; Ω = upper half of unit disk

4. $z = y^2$; $\Omega = \{(x, y): 0 \le x \le 2,\ 0 \le y \le 4\}$

* 5. $z = 3 + x^{2/3}$; $\Omega = \{(x, y): -1 \le x \le 1,\ 1 \le y \le 2\}$

6. $z = (x^4/4) + (1/8x^2)$; $\Omega = \{(x, y): 1 \le x \le 2,$ $0 \le y \le 5\}$

7. $z = \frac{1}{3}(y^2 + 2)^{3/2}$; $\Omega = \{(x, y):$ $-4 \le x \le 7, 0 \le y \le 3\}$

8. $z = 2 \ln(1 + y)$; $\Omega = \{(x, y): 0 \le x \le 2,\ 0 \le y \le 1\}$

* 9. $(z + 1)^2 = 4x^3$; $\Omega = \{(x, y): 0 \le x \le 1,\ 0 \le y \le 2\}$

* 10. Let $z = f(x, y)$ be the equation of a plane (i.e., $z = ax + by + c$). Show that over the region Ω, the area of the plane is given by

$$\sigma = \iint_\Omega \sec \gamma\, dA,$$

where γ is the angle between the normal vector **N** to the plane and the positive z-axis. [*Hint:* Show, using the dot product, that

$$\cos \gamma = \frac{\mathbf{N} \cdot \mathbf{k}}{|\mathbf{N}|} = \frac{1}{\sqrt{1 + a^2 + b^2}}.]$$

In Problems 11–27, evaluate the surface integral over the given surface.

11. $\iint_S x\, d\sigma$, where S: $z = x^2, 0 \le x \le 1, 0 \le y \le 2$.

12. $\iint_S y\, d\sigma$, where S is as in Problem 11.

13. $\iint_S x^2\, d\sigma$, where S is as in Problem 11.

14. $\iint_S (x^2 - 2y^2)\, d\sigma$, where S is as in Problem 11.

15. $\iint_S \sqrt{1 + 4z}\, d\sigma$, where S is as in Problem 11.

* 16. $\iint_S x\, d\sigma$, where S is the hemisphere $x^2 + y^2 + z^2 = 4$, $z \ge 0, x^2 + y^2 \le 4$.

* 17. $\iint_S xy\, d\sigma$, where S is as in Problem 16.

18. $\iint_S (x + y)\, d\sigma$, where S is the plane $x + 2y - 3z = 4$; $0 \le x \le 1, 1 \le y \le 2$.

19. $\iint_S yz\, d\sigma$, where S is as in Problem 18.

20. $\iint_S z^2\, d\sigma$, where S is as in Problem 18.

* 21. $\iint_S (x^2 + y^2 + z^2)\, d\sigma$, where S is the part of the plane $x - y = 4$ that lies inside the cylinder $y^2 + z^2 = 4$.

22. $\iint_S \cos z\, d\sigma$, where S is the plane $2x + 3y + z = 1$ for $0 \le x \le 1$ and $-1 \le y \le 2$.

* 23. $\iint_S z\, d\sigma$, where S is the tetrahedron bounded by the coordinate planes and the plane $4x + 8y + 2z = 16$.

24. $\iint_S |x|\, d\sigma$, where S is the hemisphere $x^2 + y^2 + z^2 = 4$; $x \ge 0, y^2 + z^2 \le 4$.

25. $\iint_S z\, d\sigma$, where S is the surface of Problem 24.

26. $\iint_S z^2\, d\sigma$, where S is the hemisphere $x^2 + y^2 + z^2 = 9$, $y \geq 0$, $x^2 + z^2 \leq 9$.

27. $\iint_S x^2\, d\sigma$, where S is the surface of Problem 26.

28. Find the mass of a triangular metallic sheet with corners at $(1, 0, 0)$, $(0, 1, 0)$, and $(0, 0, 1)$ if its density is constant.

29. Find the mass of the sheet of Problem 28 if its density is proportional to x^2.

30. Find the mass of a metallic sheet in the shape of the hemisphere $x^2 + y^2 + z^2 = 9$, $z \geq 0$, $x^2 + y^2 \leq 9$, if its density is proportional to its distance from the origin.

31. **(a)** Show that $\iint_S (x^2 + y^2)\, d\sigma$ over the hemisphere $x^2 + y^2 + z^2 = r^2$, $z \geq 0$, $x^2 + y^2 \leq r^2$, is equal to $\frac{4}{3}\pi r^4$.

   (b) Show that $\iint_S x^2\, d\sigma = \iint_S y^2\, d\sigma = \iint_S z^2\, d\sigma = \frac{1}{3}\iint_S (x^2 + y^2 + z^2)\, d\sigma$.

   (c) Explain why the last integral is equal to $\frac{2}{3}\pi r^4$, without performing any integration.

   (d) Use (c) to explain why $\iint_S (x^2 + y^2)\, d\sigma = \frac{4}{3}\pi r^4$, without performing any integration.

In Problems 32–40, compute the flux $\iint_S \mathbf{v} \cdot \mathbf{n}\, d\sigma$ for the given surface lying in the given vector field.

32. $S: z = xy; \quad 0 \leq x \leq 1, 0 \leq y \leq 2; \quad \mathbf{v} = x^2 y\mathbf{i} - z\mathbf{j}$

33. $S: z = 4 - x - y; \quad x \geq 0, y \geq 0, z \geq 0;$
$\mathbf{v} = -3x\mathbf{i} - y\mathbf{j} + 3z\mathbf{k}$

34. $S: x^2 + y^2 + z^2 = 1; \quad z \geq 0; \quad \mathbf{v} = x\mathbf{i} + y\mathbf{j} + z\mathbf{k}$

35. $S: x^2 + y^2 + z^2 = 1; \quad y \geq 0; \quad \mathbf{v} = x\mathbf{i} + y\mathbf{j} + z\mathbf{k}$

36. $S: x^2 + y^2 + z^2 = 1; \quad x \leq 0; \quad \mathbf{v} = x\mathbf{i} + y\mathbf{j} + z\mathbf{k}$

37. $S: z = \sqrt{x^2 + y^2}; \quad x^2 + y^2 \leq 1; \quad \mathbf{v} = x\mathbf{i} - y\mathbf{j} + xy\mathbf{k}$

38. $S: x = \sqrt{y^2 + z^2}; \quad y^2 + z^2 \leq 1; \quad \mathbf{v} = y\mathbf{i} - z\mathbf{j} + yz\mathbf{k}$

* 39. S: region bounded by $y = 1$ and $y = \sqrt{x^2 + z^2}$; $x^2 + z^2 \leq 1; \quad \mathbf{v} = x\mathbf{i} - z\mathbf{j} + xz\mathbf{k}$

40. S: unit sphere; $\quad \mathbf{v} = x\mathbf{i} + y\mathbf{j} + z\mathbf{k}$

41. Show that if $\mathbf{v} = a\mathbf{i} + b\mathbf{j} + c\mathbf{k}$, where a, b, and c are constants, then $\iint_S \mathbf{v} \cdot \mathbf{n}\, d\sigma = 0$, where S is the sphere $x^2 + y^2 + z^2 = r^2$.

42. Find the flux of the vector field $\mathbf{v} = xz\mathbf{i} + y^2\mathbf{j} - xy^3 z^2\mathbf{k}$ across the surface of the sphere S in Problem 41.

43. Find the flux of the vector field of Problem 42 over the tetrahedron formed by the coordinate planes and the plane $x + y + z = 1$.

10.6 Divergence and Curl of a Vector Field in $\mathbb{R}^3$

In Section 10.4 we defined the divergence of $F(x, y) = P(x, y)\mathbf{i} + Q(x, y)\mathbf{j}$ by

$$\operatorname{div} \mathbf{F} = \frac{\partial P}{\partial x} + \frac{\partial Q}{\partial y} \tag{1}$$

and the curl of $\mathbf{F}$ by

$$\operatorname{curl} \mathbf{F} = \frac{\partial Q}{\partial x} - \frac{\partial P}{\partial y}. \tag{2}$$

We also wrote

$$\operatorname{div} \mathbf{F} = \nabla \cdot \mathbf{F} \tag{3}$$

and

$$\operatorname{curl} \mathbf{F} = \nabla \times \mathbf{F}, \tag{4}$$

where ∇, the gradient symbol, is regarded as the vector $(\partial/\partial x, \partial/\partial y)$. Moreover, we used both div $\mathbf{F}$ and curl $\mathbf{F}$ to give alternative versions of Green's theorem in the plane.

In this section we define the divergence and curl of a vector field in $\mathbb{R}^3$, and in Sections 10.7 and 10.8 we shall show how these are used in the statement of two very important theorems about surface integrals.

Let the function $F(x, y, z)$ be given. Then the gradient of F is given by

$$\nabla F = \frac{\partial F}{\partial x}\mathbf{i} + \frac{\partial F}{\partial y}\mathbf{j} + \frac{\partial F}{\partial z}\mathbf{k}. \tag{5}$$

We can think of the gradient as a function that takes a differentiable function of (x, y, z) into a vector field in $\mathbb{R}^3$. We write this function, symbolically, as

$$\nabla = \frac{\partial}{\partial x}\mathbf{i} + \frac{\partial}{\partial y}\mathbf{j} + \frac{\partial}{\partial z}\mathbf{k}. \tag{6}$$

The operator in (6) is a useful device for writing things down. For example, (5) can be written as

$$\nabla F = \left(\frac{\partial}{\partial x}\mathbf{i} + \frac{\partial}{\partial y}\mathbf{j} + \frac{\partial}{\partial z}\mathbf{k}\right)F = \frac{\partial F}{\partial x}\mathbf{i} + \frac{\partial F}{\partial y}\mathbf{j} + \frac{\partial F}{\partial z}\mathbf{k}.$$

We now define the divergence and curl of a vector field $\mathbf{F}$ in $\mathbb{R}^3$ given by

$$\mathbf{F}(x, y, z) = P(x, y, z)\mathbf{i} + Q(x, y, z)\mathbf{j} + R(x, y, z)\mathbf{k} \tag{7}$$

DEFINITION

Divergence and Curl

Let the vector field $\mathbf{F}$ be given by (7), where P, Q, and R are differentiable. Then the **divergence** of $\mathbf{F}$ (div $\mathbf{F}$) and **curl** of $\mathbf{F}$ (curl $\mathbf{F}$) are given by

$$\text{div } \mathbf{F} = \frac{\partial P}{\partial x} + \frac{\partial Q}{\partial y} + \frac{\partial R}{\partial z} \tag{8}$$

and

$$\text{curl } \mathbf{F} = \left(\frac{\partial R}{\partial y} - \frac{\partial Q}{\partial z}\right)\mathbf{i} + \left(\frac{\partial P}{\partial z} - \frac{\partial R}{\partial x}\right)\mathbf{j} + \left(\frac{\partial Q}{\partial x} - \frac{\partial P}{\partial y}\right)\mathbf{k}. \tag{9}$$

Note. In $\mathbb{R}^3$, div $\mathbf{F}$ is a scalar function and curl $\mathbf{F}$ is a vector field.

Before giving examples of divergence and curl, we derive an easy way to remember how to compute them.

◾ THEOREM 1

Let ∇ be given by (6) and let the differentiable vector field $\mathbf{F}$ be given by (7). Then

> (i)   div $\mathbf{F} = \nabla \cdot \mathbf{F}$ (ii)   curl $\mathbf{F} = \nabla \times \mathbf{F}$.   (10)

PROOF

(i) $\nabla \cdot \mathbf{F} = \left(\dfrac{\partial}{\partial x} \mathbf{i} + \dfrac{\partial}{\partial y} \mathbf{j} + \dfrac{\partial}{\partial z} \mathbf{k} \right) \cdot (P\mathbf{i} + Q\mathbf{j} + R\mathbf{k})$

$$= \dfrac{\partial P}{\partial x} + \dfrac{\partial Q}{\partial y} + \dfrac{\partial R}{\partial z} = \text{div } \mathbf{F}.$$

(ii) $\nabla \times \mathbf{F} = \begin{vmatrix} \mathbf{i} & \mathbf{j} & \mathbf{k} \\ \dfrac{\partial}{\partial x} & \dfrac{\partial}{\partial y} & \dfrac{\partial}{\partial z} \\ P & Q & R \end{vmatrix}$

$$= \left(\dfrac{\partial R}{\partial y} - \dfrac{\partial Q}{\partial z} \right) \mathbf{i} + \left(\dfrac{\partial P}{\partial z} - \dfrac{\partial R}{\partial x} \right) \mathbf{j} + \left(\dfrac{\partial Q}{\partial x} - \dfrac{\partial P}{\partial y} \right) \mathbf{k} = \text{curl } \mathbf{F}. \quad \blacksquare$$

EXAMPLE 1

Compute the divergence and curl of $\mathbf{F}(x, y, z) = xy\mathbf{i} + (z^2 - 2y)\mathbf{j} + \cos yz\mathbf{k}$.

SOLUTION

$$\text{div } \mathbf{F} = \dfrac{\partial}{\partial x}(xy) + \dfrac{\partial}{\partial y}(z^2 - 2y) + \dfrac{\partial}{\partial z}(\cos yz)$$

$$= y - 2 - y \sin yz$$

and

$$\text{curl } \mathbf{F} = \begin{vmatrix} \mathbf{i} & \mathbf{j} & \mathbf{k} \\ \dfrac{\partial}{\partial x} & \dfrac{\partial}{\partial y} & \dfrac{\partial}{\partial z} \\ xy & z^2 - 2y & \cos yz \end{vmatrix}$$

$$= \left[\dfrac{\partial}{\partial y} \cos yz - \dfrac{\partial}{\partial z}(z^2 - 2y) \right] \mathbf{i} + \left[\dfrac{\partial}{\partial z} xy - \dfrac{\partial}{\partial x} \cos yz \right] \mathbf{j}$$

$$+ \left[\dfrac{\partial}{\partial x}(z^2 - 2y) - \dfrac{\partial}{\partial y} xy \right] \mathbf{k}$$

$$= (-z \sin yz - 2z)\mathbf{i} - x\mathbf{k}.$$

EXAMPLE 2

Compute the divergence and curl of

$$\mathbf{F}(x, y, z) = yz\mathbf{i} + xz\mathbf{j} + xy\mathbf{k}.$$

SOLUTION

$$\text{div } \mathbf{F} = \frac{\partial}{\partial x}yz + \frac{\partial}{\partial y}xz + \frac{\partial}{\partial z}xy = 0.$$

$$\text{curl } \mathbf{F} = \begin{vmatrix} \mathbf{i} & \mathbf{j} & \mathbf{k} \\ \dfrac{\partial}{\partial x} & \dfrac{\partial}{\partial y} & \dfrac{\partial}{\partial z} \\ yz & xz & xy \end{vmatrix}$$

$$= \left(\frac{\partial}{\partial y}xy - \frac{\partial}{\partial z}xz\right)\mathbf{i} + \left(\frac{\partial}{\partial z}yz - \frac{\partial}{\partial x}xy\right)\mathbf{j} + \left(\frac{\partial}{\partial x}xz - \frac{\partial}{\partial y}yz\right)\mathbf{k}$$

$$= (x - x)\mathbf{i} + (y - y)\mathbf{j} + (z - z)\mathbf{k} = \mathbf{0}.$$

As in $\mathbb{R}^2$, the curl of a vector field $\mathbf{F}$ represents the circulation per unit area at the point (x, y, z). If curl $\mathbf{F} = \mathbf{0}$ for every (x, y, z) in some region W in $\mathbb{R}^3$, then the fluid flow $\mathbf{F}$ is called **irrotational.** The divergence of $\mathbf{F}$ at a point (x, y, z) represents the net rate of flow away from (x, y, z). If div $\mathbf{F} = 0$ for every (x, y, z) in W, then the flow $\mathbf{F}$ is called **incompressible** or **solenoidal.** The vector field in Example 2 is both irrotational and incompressible. Let us examine these ideas more closely.

Physical Interpretation of Divergence and Curl

Divergence. Suppose a fluid is flowing through the small volume $dx\,dy\,dz$ depicted in Fig. 1. Let $\mathbf{v} = v_x\mathbf{i} + v_y\mathbf{j} + v_z\mathbf{k}$ and ρ denote, respectively, the velocity and the density of the fluid at the origin. Fluid is flowing in all directions. We consider the positive x-direction first. The fluid flowing into this volume per unit time through the face $EFGH$ is

$$\begin{array}{l}\text{rate of flow in} \\ \text{(face } EFGH)\end{array} = \rho v_x|_{x=0}\,dy\,dz. \tag{11}$$

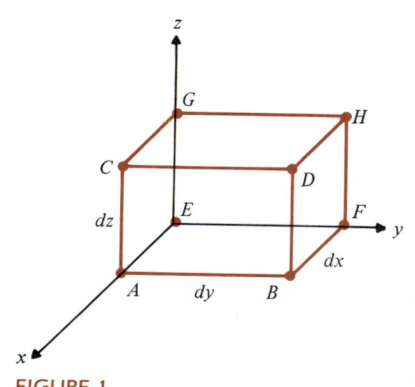

FIGURE 1

The components ρv_y and ρv_z of the flow are tangential to this face and contribute nothing to the flow through this face. The rate of flow out (in the positive x-direction) through the face $ABCD$ is

$$\begin{array}{l}\text{rate of flow out} \\ \text{(face } ABCD)\end{array} = \rho v_x|_{x=dx}\,dy\,dz. \tag{12}$$

By the mean-value theorem, we have

$$\rho v_x|_{x=dx}\,dy\,dz - \rho v_x|_{x=0}\,dy\,dz = \left[\frac{\partial}{\partial x}(\rho v_x)\,dx\right]dy\,dz, \tag{13}$$

where the partial derivative is evaluated at some point in $(0, dx)$. Hence

$$\begin{array}{l}\text{net flow in} \\ \text{positive } x\text{-direction} \\ \text{at the point } (x, y, z)\end{array} = \text{flow out} - \text{flow in} = \frac{\partial}{\partial x}(\rho v_x)\,dx\,dy\,dz.$$

Similar results hold in the positive y- and z-directions and we have

$$\frac{\text{net flow}}{\text{per unit time}} = \left[\frac{\partial}{\partial x}(\rho v_x) + \frac{\partial}{\partial y}(\rho v_y) + \frac{\partial}{\partial z}(\rho v_z)\right] dx\, dy\, dz = \text{div}(\rho \mathbf{v})\, dx\, dy\, dz.$$

Therefore the net flow of the compressible fluid from the volume element $dx\, dy\, dz$ per unit volume per unit time is $\text{div}(\rho \mathbf{v})$. This is why we call it **divergence.**

Continuity Equation

A direct result of our calculations is the **continuity equation**

$$\frac{\partial \rho}{\partial t} + \text{div}(\rho \mathbf{v}) = 0, \tag{14}$$

which states that a net flow from the volume results in a decreased density inside the volume.

The divergence appears in a wide variety of physical problems, ranging from a probability current density in quantum mechanics to neutron leakage in a nuclear reactor.

Curl. Consider the circulation of fluid around the rectangular loop in the xy-plane drawn in Fig. 2. Let the velocity vector $\mathbf{v} = v_x \mathbf{i} + v_y \mathbf{j}$. Now

Circulation = flow along 1 + flow along 2 + flow along 3 + flow along 4.

Recall that distance is the integral of velocity. Thus, treating v_x and v_y as constants (since dx and dy are small), we have

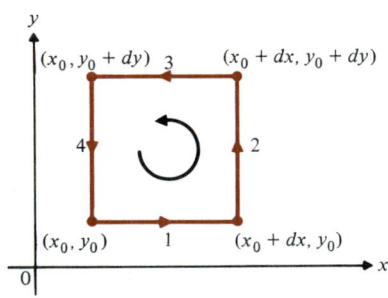

FIGURE 2

$$\text{flow along } 1 = \int_{x_0}^{x_0+dx} v_x(x_0, y_0)\, dx = v_x(x_0, y_0)\, dx,$$

$$\text{flow along } 2 = \int_{y_0}^{y_0+dy} v_y(x_0 + dx, y_0)\, dy = v_y(x_0 + dx, y_0)\, dy,$$

$$\text{flow along } 3 = \int_{x_0+dx}^{x_0} v_x(x_0, y_0 + dy)\, dx$$

$$= -\int_{x_0}^{x_0+dx} v_x(x_0, y_0 + dy)\, dx = -v_x(x_0, y_0 + dy)\, dx,$$

$$\text{flow along } 4 = \int_{y_0+dy}^{y_0} v_y(x_0, y_0)\, dy = -v_y(x_0, y_0)\, dy.$$

Using Maclaurin expansions around (x_0, y_0), we have

$$v_y(x_0 + t, y_0) = v_y(x_0, y_0) + \frac{\partial v_y}{\partial x}(x_0, y_0)t + \text{higher order terms}$$

and

$$v_y(x_0 + dx, y_0) = v_y(x_0, y_0) + \frac{\partial v_y}{\partial x}(x_0, y_0)\, dx + \text{higher order terms}$$

so

$$\text{flow along } 2 \approx v_y(x_0, y_0)\, dy + \frac{\partial v_y}{\partial x}(x_0, y_0)\, dx\, dy.$$

Similarly,

$$v_x(x_0, y_0 + dy) = v_x(x_0, y_0) + \frac{\partial v_x}{\partial y}\, dy + \text{higher order terms}$$

and

$$\text{flow along } 3 \approx -v_x(x_0, y_0)\, dx - \frac{\partial v_x}{\partial y}(x_0, y_0)\, dx\, dy.$$

Adding the circulations along 1, 2, 3, and 4 we have

$$\text{circulation}_{1234} = \left(\frac{\partial v_y}{\partial x} - \frac{\partial v_x}{\partial y}\right) dx\, dy$$

or, dividing by $dx\, dy$,

$$\begin{array}{l}\text{circulation} \\ \text{per unit area} = \text{curl } \mathbf{v}(x_0, y_0). \\ \text{at } (x_0, y_0)\end{array}$$

In principle, curl $\mathbf{v}$ at (x_0, y_0) could be determined by inserting a paddle wheel into the moving fluid at the point (x_0, y_0). The rotation of the paddle wheel would be a measure of the curl.

In Section 10.2 we stated that $\mathbf{F}(x, y) = P(x, y)\mathbf{i} + Q(x, y)\mathbf{j}$ is the gradient of a function f in a simply connected region if $\partial P/\partial y = \partial Q/\partial x$. If $\mathbf{F}(x, y, z) = P(x, y, z)\mathbf{i} + Q(x, y, z)\mathbf{j} + R(x, y, z)\mathbf{k,}$ then there is a condition that can be used to check whether there is a differentiable function f such that $\mathbf{F} = \nabla f$. We give this result without proof.†

■ **THEOREM 2**
Let $\mathbf{F}(x, y, z) = P(x, y, z)\mathbf{i} + Q(x, y, z)\mathbf{j} + R(x, y, z)\mathbf{k}$ and suppose that P, Q, R, $\partial P/\partial y$, $\partial P/\partial z$, $\partial Q/\partial x$, $\partial Q/\partial z$, $\partial R/\partial x$, and $\partial R/\partial y$ are continuous in a simply connected region. Then $\mathbf{F}$ is the gradient of a differentiable function f if and only if

$$\frac{\partial P}{\partial y} = \frac{\partial Q}{\partial x}, \qquad \frac{\partial R}{\partial x} = \frac{\partial P}{\partial z}, \qquad \frac{\partial Q}{\partial z} = \frac{\partial R}{\partial y}. \tag{15}$$

■

Using (15) and (9), we obtain the following interesting result.

† For a proof, see R. C. Buck and E. F. Buck, *Advanced Calculus,* 3rd Ed., McGraw-Hill, New York, 1978, p. 497–498.

■ **THEOREM 3**

The differentiable vector field $\mathbf{F}$ is the gradient of a function f in a simply connected region if and only if curl $\mathbf{F} = \mathbf{0}$.   ■

Problems 10.6

In Problems 1–10, compute the divergence and curl of the given vector field.

1. $\mathbf{F}(x, y, z) = x^2\mathbf{i} + y^2\mathbf{j} + z^2\mathbf{k}$

2. $\mathbf{F}(x, y, z) = (\sin y)\mathbf{i} + (\sin z)\mathbf{j} + (\sin x)\mathbf{k}$

3. $\mathbf{F}(x, y, z) = a\mathbf{i} + b\mathbf{j} + c\mathbf{k}$; a, b, c constants

4. $\mathbf{F}(x, y, z) = \sqrt{1 + x^2 + y^2}\mathbf{i} + \sqrt{1 + x^2 + y^2}\mathbf{j} + z^4\mathbf{k}$

5. $\mathbf{F}(x, y, z) = xy\mathbf{i} + yz\mathbf{j} + xz\mathbf{k}$

6. $\mathbf{F}(x, y, z) = (y^2 + z^2)\mathbf{i} + (x^2 + z^2)\mathbf{j} + (x^2 + y^2)\mathbf{k}$

7. $\mathbf{F}(x, y, z) = e^{yz}\mathbf{i} + e^{xz}\mathbf{j} + e^{xy}\mathbf{k}$

8. $\mathbf{F}(x, y, z) = e^{xy}\mathbf{i} + e^{yz}\mathbf{j} + e^{xz}\mathbf{k}$

9. $\mathbf{F}(x, y, z) = \dfrac{x}{y}\mathbf{i} + \dfrac{y}{z}\mathbf{j} + \dfrac{z}{x}\mathbf{k}$

10. $\mathbf{F}(x, y, z) = \sqrt{y+z}\,\mathbf{i} + \sqrt{x+z}\,\mathbf{j} + \sqrt{x+y}\,\mathbf{k}$

11. Let f, g, and h be differentiable functions of two variables. Show that the vector field $\mathbf{F}(x, y, z) = f(y, z)\mathbf{i} + g(x, z)\mathbf{j} + h(x, y)\mathbf{k}$ is incompressible.

In Problems 12–18, assume that all given functions are differentiable.

12. Show that $\operatorname{div}(\mathbf{F} + \mathbf{G}) = \operatorname{div}\mathbf{F} + \operatorname{div}\mathbf{G}$.

13. If $f = f(x, y, z)$ is a scalar function, show that
$$\operatorname{div}(f\mathbf{F}) = f\operatorname{div}\mathbf{F} + \nabla f \cdot \mathbf{F}.$$

14. Show that $\operatorname{curl}(\mathbf{F} + \mathbf{G}) = \operatorname{curl}\mathbf{F} + \operatorname{curl}\mathbf{G}$.

15. If f is as in Problem 13, show that
$$\operatorname{curl}(f\mathbf{F}) = f\operatorname{curl}\mathbf{F} + \nabla f \times \mathbf{F}.$$

16. If f is as in Problem 13, show that curl grad $f = \operatorname{curl}(\nabla f) = \mathbf{0}$.

17. Show that div curl $\mathbf{F} = 0$.

18. Show that $\operatorname{div}(\mathbf{F} \times \mathbf{G}) = \mathbf{G} \cdot \operatorname{curl}\mathbf{F} - \mathbf{F} \cdot \operatorname{curl}\mathbf{G}$.

The **Laplacian** of a twice-differentiable scalar function $f = f(x, y, z)$, denoted by $\nabla^2 f$, is defined by

$$\text{Laplacian of } f = \nabla^2 f = \frac{\partial^2 f}{\partial x^2} + \frac{\partial^2 f}{\partial y^2} + \frac{\partial^2 f}{\partial z^2}. \qquad \textbf{(16)}$$

In Problems 19–22, compute $\nabla^2 f$.

19. $f(x, y, z) = xyz$

20. $f(x, y, z) = x^2 + y^2 + z^2$

21. $f(x, y, z) = \dfrac{1}{\sqrt{x^2 + y^2 + z^2}}$

22. $f(x, y, z) = 2x^2 + 5y^2 + 3z^2$

23. A function that satisfies the equation $\nabla^2 f = 0$, called **Laplace's equation,** is called **harmonic.** Which of the functions in Problems 19–22 are harmonic?

24. Show that $\nabla^2 f = \operatorname{div}(\operatorname{grad} f)$.

25. The **Laplacian** of a vector field $\mathbf{F}$ is given by $\nabla^2\mathbf{F} = \nabla^2 P\mathbf{i} + \nabla^2 Q\mathbf{j} + \nabla^2 R\mathbf{k}$, where $\mathbf{F} = P\mathbf{i} + Q\mathbf{j} + R\mathbf{k}$. Show that curl curl $\mathbf{F} = \nabla \operatorname{div}\mathbf{F} - \nabla^2\mathbf{F}$.

26. Let $\mathbf{F}$ be the vector field of Problem 1. Verify that
$$\operatorname{div}\operatorname{curl}\mathbf{F} = 0.$$

27. Let f be the function given in Problem 21. Verify that curl grad $f = \mathbf{0}$.

28. It is true, although we will not attempt to prove it, that if $\operatorname{div}\mathbf{F} = 0$ in a simply connected region, then $\mathbf{F} = \operatorname{curl}\mathbf{G}$ for some vector field $\mathbf{G}$. Let
$$\mathbf{F} = x\mathbf{i} + \frac{y}{2}\mathbf{j} - \frac{3}{2}z\mathbf{k}.$$

   (a) Verify that $\operatorname{div}\mathbf{F} = 0$.

   (b) Find a vector field $\mathbf{G}$ such that $\mathbf{F} = \operatorname{curl}\mathbf{G}$.

29. If curl $\mathbf{F} = \mathbf{0}$ in $\mathbb{R}^3$, then $\mathbf{F} = \nabla f$ for some function f. Let $\mathbf{F} = 4xyz\mathbf{i} + 2x^2 z\mathbf{j} + 2x^2 y\mathbf{k}$.

   (a) Verify that curl $\mathbf{F} = \mathbf{0}$.

   (b) Find a function f such that $\mathbf{F} = \nabla f$.

30. The electrostatic field of a point charge q is
$$\mathbf{E} = \frac{q}{4\pi\epsilon_0} \cdot \frac{\mathbf{r}}{r^2}.$$

   Calculate div $\mathbf{E}$. What happens at the origin?

31. Show that the gravitational force $\mathbf{F} = Gm_1 m_2\mathbf{x}/|\mathbf{x}|^3$ (see p. 515) is irrotational.

32. Show that if $\mathbf{v}$ is irrotational, then $\mathbf{v} \times \mathbf{x}$ is incompressible (solenoidal).

* 33. The vector potential $\mathbf{P}$ of a magnetic dipole, dipole moment $\mathbf{m}$, is given by
$$\mathbf{P}(\mathbf{x}) = \frac{\mu_0}{4\pi}\left(\mathbf{m} \times \frac{\mathbf{x}}{|\mathbf{x}|^3}\right).$$

Show that the magnetic induction $\mathbf{B} = \text{curl } \mathbf{P}$ is given by

$$\mathbf{B} = \frac{\mu_0}{4\pi} \frac{3\mathbf{x}(\mathbf{x} \cdot \mathbf{m}) - \mathbf{m}}{|\mathbf{x}|^3}.$$

34. The velocity of a two-dimensional flow of liquid is given by $\mathbf{v} = u(x, y)\mathbf{i} - v(x, y)\mathbf{j}$. If the liquid is incompressible and the flow is irrotational show that

$$\frac{\partial u}{\partial x} = \frac{\partial v}{\partial y} \quad \text{and} \quad \frac{\partial u}{\partial y} = -\frac{\partial v}{\partial x}.$$

These are the **Cauchy-Riemann** equations. We shall discuss them further in Section 15.6.

10.7 Stokes's Theorem

In Section 10.4 we discussed Green's theorem in the plane, one vector form of which was given by (see Theorem 10.4.2)

$$\int_{\partial\Omega} \mathbf{F} \cdot \mathbf{T} \, ds = \iint_{\Omega} \text{curl } \mathbf{F} \, dx \, dy, \tag{1}$$

where $\partial\Omega$ is the piecewise smooth boundary of a region Ω in the xy-plane. We now generalize this result.

If C is a closed curve enclosing the region Ω in the plane, then by *traversing C in the positive sense* we mean moving around C so that Ω is always on the left. This motion corresponds to moving in a counterclockwise direction (see Fig. 1).

Remark. Like the definition of the outward unit normal vector, the words "counterclockwise" and "left" are undefined when applied to closed curves in $\mathbb{R}^3$. To see why, picture a circle in a plane in $\mathbb{R}^3$. You can look at the circle from either side of the plane. The counterclockwise direction from one side would appear as the clockwise direction from the other side. (Take a large piece of translucent paper, place it between yourself and a friend, and ask him or her to draw a circle in the counterclockwise direction. It will appear clockwise to you.) Thus the "positive sense" in which a curve is traversed is arbitrary. However, once we have chosen a positive or counterclockwise direction, the outward unit normal vector can be unambiguously defined: At any point on a smooth surface, we choose a tangent vector $\mathbf{u}$ lying in the unique tangent plane at that point. We then obtain a second tangent vector $\mathbf{v}$ by rotating $\mathbf{u}$ 90° in the counterclockwise direction while remaining in the tangent plane. Finally, we define the direction of $\mathbf{n}$ to be the direction of $\mathbf{u} \times \mathbf{v}$ (so that the vectors $\mathbf{u}$, $\mathbf{v}$, and $\mathbf{n}$ form a right-handed system).

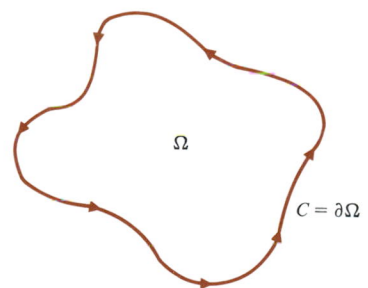

FIGURE 1

Let $S: z = f(x, y)$ be a smooth surface for $(x, y) \in \Omega$, where Ω is bounded. We assume that the boundary of S, denoted ∂S, is a piecewise smooth, simple closed curve in $\mathbb{R}^3$. The positive direction on ∂S corresponds to the positive direction of $\partial\Omega$, where $\partial\Omega$ is the projection of ∂S into the xy-plane. This orientation is illustrated in Fig. 2.

We now state the second major result in vector calculus (the first was Green's theorem).

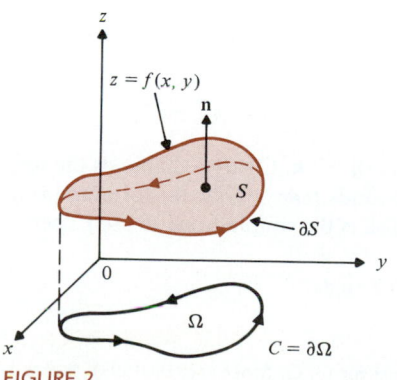

FIGURE 2

■ **THEOREM 1: Stokes's Theorem†**

Let $\mathbf{F}(x, y, z) = P(x, y, z)\mathbf{i} + Q(x, y, z)\mathbf{j} + R(x, y, z)\mathbf{k}$ be continuously differentiable on a bounded region W in space that contains the smooth surface S, and let ∂S be a piecewise smooth, simple closed curve traversed in the positive sense. Then

$$\oint_{\partial S} \mathbf{F} \cdot \mathbf{T} \, ds = \iint_S \text{curl } \mathbf{F} \cdot \mathbf{n} \, d\sigma, \qquad (2)$$

where $\mathbf{T}$ is the unit tangent vector to the curve ∂S at a point (x, y, z) and $\mathbf{n}$ is the outward unit normal vector to the surface S at the point (x, y, z). ■

The proof of this theorem is difficult and is left to the end of this section.

Remark 1. Note the similarity between Stokes's theorem and the first vector form of Green's theorem.

Remark 2. If ∂S is given parametrically by $\mathbf{x}(t) = x(t)\mathbf{i} + y(t)\mathbf{j} + z(t)\mathbf{k}$, then by Theorem 9.5.1 on p. 537

$$\mathbf{T} = \frac{d\mathbf{x}}{ds}, \qquad \text{or} \qquad d\mathbf{x} = \mathbf{T} \, ds. \qquad (3)$$

Thus Stokes's theorem can be written as

$$\oint_{\partial S} \mathbf{F} \cdot d\mathbf{x} = \iint_S \text{curl } \mathbf{F} \cdot \mathbf{n} \, d\sigma. \qquad (4)$$

EXAMPLE 1

Use Stokes's theorem to evaluate $\oint_C \mathbf{F} \cdot d\mathbf{x}$, where $\mathbf{F}(x, y, z) = (z - 2y)\mathbf{i} + (3x - 4y)\mathbf{j} + (z + 3y)\mathbf{k}$ and C is the unit circle in the plane $z = 2$.

SOLUTION

C is given parametrically by $\mathbf{x}(t) = (\cos t)\mathbf{i} + (\sin t)\mathbf{j} + 2\mathbf{k}$. Clearly, C bounds the unit disk $x^2 + y^2 \leq 1$, $z = 2$ in the plane $z = 2$. (It bounds many surfaces—for example, a hemisphere with this circle as its base. But the disk is the simplest one to use.) Then

$$\oint_C \mathbf{F} \cdot d\mathbf{x} = \iint_S \text{curl } \mathbf{F} \cdot \mathbf{n} \, d\sigma.$$

---

† Named after the British mathematician and physicist Sir G. G. Stokes (1819–1903). Stokes is also known as one of the first to discuss the notion of uniform convergence (in 1848).

We compute

$$\text{curl } \mathbf{F} = \begin{vmatrix} \mathbf{i} & \mathbf{j} & \mathbf{k} \\ \dfrac{\partial}{\partial x} & \dfrac{\partial}{\partial y} & \dfrac{\partial}{\partial z} \\ z - 2y & 3x - 4y & z + 3y \end{vmatrix} = 3\mathbf{i} + \mathbf{j} + 5\mathbf{k},$$

and $\mathbf{n} = \mathbf{k}$ (since $\mathbf{k}$ is normal to any vector lying in a plane parallel to the xy-plane). Thus

$$\text{curl } \mathbf{F} \cdot \mathbf{n} = 5,$$

so that

$$\oint_C \mathbf{F} \cdot \mathbf{dx} = 5 \iint_S d\sigma.$$

But $\iint_S d\sigma =$ the surface area of $S =$ the area of the unit disk $= \pi$. Thus

$$\oint_C \mathbf{F} \cdot \mathbf{dx} = 5\pi.$$

From Theorem 10.6.3, in a simply connected region $\mathbf{F}$ is the gradient of a function f if and only if curl $\mathbf{F} = \mathbf{0}$. This result provides a proof of the fact that

$$\oint_C \mathbf{F} \cdot \mathbf{dx} = 0$$

if $\mathbf{F}$ is the gradient of a function f, because

$$\oint_C \mathbf{F} \cdot \mathbf{dx} = \iint_S \text{curl } \mathbf{F} \cdot \mathbf{n} \, d\sigma = 0,$$

where S is any smooth surface whose boundary is C.

We can combine several results to obtain the following theorem.

■ THEOREM 2

Let $\mathbf{F} = P(x, y, z)\mathbf{i} + Q(x, y, z)\mathbf{j} + R(x, y, z)\mathbf{k}$ be continuously differentiable on a simply connected region S in space. Then the following conditions are equivalent. That is, if one is true, all are true.

(i) $\mathbf{F}$ is the gradient of a differentiable function f.

(ii) $\dfrac{\partial P}{\partial y} = \dfrac{\partial Q}{\partial x}, \dfrac{\partial R}{\partial x} = \dfrac{\partial P}{\partial z},$ and $\dfrac{\partial Q}{\partial z} = \dfrac{\partial R}{\partial y}.$

(iii) $\oint_C \mathbf{F} \cdot \mathbf{dx} = 0$ for every piecewise smooth, simple closed curve C lying in S.

(iv) $\oint_C \mathbf{F} \cdot \mathbf{dx}$ is independent of path.

(v) Curl $\mathbf{F} = \mathbf{0}$.

Many interesting physical results follow from Stokes's theorem.

EXAMPLE 2: Ampère's Law

Suppose a steady current is flowing in a wire with an electric current density given by the vector field **i**. It is known in physics that such a current will set up a magnetic field, which is usually denoted **B**. In Fig. 3 we see a collection of compass needles near a wire carrying (a) no current and (b) a very strong current. A special case of one of the famous Maxwell equations† states that

$$\text{curl } \mathbf{B} = \mathbf{i}. \tag{5}$$

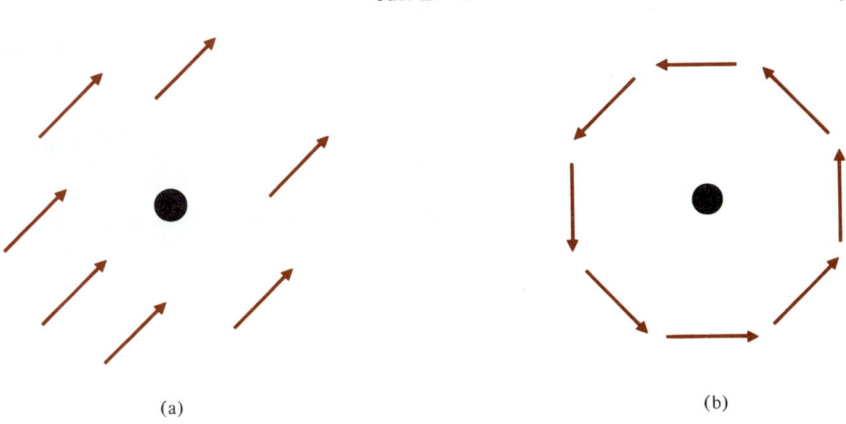

(a)   (b)

FIGURE 3

From this equation we can deduce an important physical law. Let S be a surface with smooth boundary ∂S. Then $\oint_{\partial S} \mathbf{B} \cdot \mathbf{T} \, ds$ is defined as the **circulation** of the magnetic field around ∂S. By Stokes's theorem

$$\oint_{\partial S} \mathbf{B} \cdot \mathbf{T} \, ds = \iint_S \text{curl } \mathbf{B} \cdot \mathbf{n} \, d\sigma.$$

But by (5), $\iint_S \text{curl } \mathbf{B} \cdot \mathbf{n} \, d\sigma = \iint_S \mathbf{i} \cdot \mathbf{n} \, d\sigma$, so

$$\oint_{\partial S} \mathbf{B} \cdot \mathbf{T} \, ds = \iint_S \mathbf{i} \cdot \mathbf{n} \, d\sigma. \tag{6}$$

In other words, (6) states that *the total current flowing through an electric field in a surface S is equal to the circulation of the magnetic field induced by the current around the boundary of S*. This important result is known as **Ampère's law.**

We close this section with a proof of Stokes's theorem.

PROOF OF STOKES'S THEOREM (OPTIONAL)
We have $\mathbf{F} = P\mathbf{i} + Q\mathbf{j} + R\mathbf{k}$ and

$$\text{curl } \mathbf{F} = \left(\frac{\partial R}{\partial y} - \frac{\partial Q}{\partial z}\right)\mathbf{i} + \left(\frac{\partial P}{\partial z} - \frac{\partial R}{\partial x}\right)\mathbf{j} + \left(\frac{\partial Q}{\partial x} - \frac{\partial P}{\partial y}\right)\mathbf{k}. \tag{7}$$

† James Maxwell (1831–1879) was a British physicist. He formulated four equations, known as **Maxwell's equations,** which describe electromagnetic phenomena.

The surface S is given by S: $z = f(x, y)$ for (x, y) in Ω. Then by equation (10.5.24), p. 630,

$$\iint_S \text{curl } \mathbf{F} \cdot \mathbf{n} \, d\sigma =$$

$$\iint_\Omega \left[\left(\frac{\partial R}{\partial y} - \frac{\partial Q}{\partial z} \right)\left(-\frac{\partial f}{\partial x} \right) + \left(\frac{\partial P}{\partial z} - \frac{\partial R}{\partial x} \right)\left(-\frac{\partial f}{\partial y} \right) + \left(\frac{\partial Q}{\partial x} - \frac{\partial P}{\partial y} \right) \right] dx\, dy. \tag{8}$$

Suppose that $\mathbf{x}(t) = x(t)\mathbf{i} + y(t)\mathbf{j}$ for $t_0 \le t \le t_1$ parametrizes $\partial\Omega$. Then ∂S is parametrized by

$$\mathbf{x}(t) = x(t)\mathbf{i} + y(t)\mathbf{j} + f(x(t), y(t))\mathbf{k}, \qquad t_0 \le t \le t_1.$$

Thus

$$\oint_{\partial S} \mathbf{F} \cdot \mathbf{T} \, ds = \oint_{\partial S} \mathbf{F} \cdot \mathbf{dx} = \int_{t_0}^{t_1} \left(P\frac{dx}{dt} + Q\frac{dy}{dt} + R\frac{dz}{dt} \right) dt. \tag{9}$$

By the chain rule,

$$\frac{dz}{dt} = \frac{\partial z}{\partial x}\frac{dx}{dt} + \frac{\partial z}{\partial y}\frac{dy}{dt}. \tag{10}$$

Substituting (10) into (9) yields

$$\oint_{\partial S} \mathbf{F} \cdot \mathbf{T} \, ds = \int_{t_0}^{t_1} \left[\left(P + R\frac{\partial z}{\partial x} \right)\frac{dx}{dt} + \left(Q + R\frac{\partial z}{\partial y} \right)\frac{dy}{dt} \right] dt$$

$$= \oint_{\partial\Omega} \left(P + R\frac{\partial z}{\partial x} \right) dx + \left(Q + R\frac{\partial z}{\partial y} \right) dy. \tag{11}$$

We apply Green's theorem to (11) to obtain

$$\oint_{\partial S} \mathbf{F} \cdot \mathbf{T} \, ds = \iint_\Omega \left\{ \frac{\partial[Q + R(\partial z/\partial y)]}{\partial x} - \frac{\partial[P + R(\partial z/\partial x)]}{\partial y} \right\} dx\, dy. \tag{12}$$

We now apply the chain rule again. This is a bit complicated. For example,

$$\frac{\partial}{\partial x} Q(x, y, z) = \frac{\partial Q}{\partial x} + \frac{\partial Q}{\partial y}\frac{\partial y}{\partial x} + \frac{\partial Q}{\partial z}\frac{\partial z}{\partial x} = \frac{\partial Q}{\partial x} + \frac{\partial Q}{\partial z}\frac{\partial z}{\partial x},$$

since y is not a function of x (so that $\partial y/\partial x = 0$). Also

$$\frac{\partial}{\partial x}\left(R\frac{\partial z}{\partial y} \right) = \left[\frac{\partial}{\partial x} R(x, y, z) \right]\frac{\partial z}{\partial y} + R\frac{\partial}{\partial x}\left\{ \frac{\partial[z(x, y)]}{\partial y} \right\}$$

$$= \frac{\partial R}{\partial x}\frac{\partial z}{\partial y} + \frac{\partial R}{\partial z}\frac{\partial z}{\partial x}\frac{\partial z}{\partial y} + R\frac{\partial^2 z}{\partial x\, \partial y} + R\frac{\partial^2 z}{\partial y^2}\frac{\partial y}{\partial x},$$

and the last term is zero because y is not a function of x. Differentiating inside the right-hand integral in (12), we obtain

$$\oint_{\partial S} \mathbf{F} \cdot \mathbf{T} \, ds = \iint_{\Omega} \left[\left(\frac{\partial Q}{\partial x} + \frac{\partial Q}{\partial z} \frac{\partial z}{\partial x} + \frac{\partial R}{\partial x} \frac{\partial z}{\partial y} + \frac{\partial R}{\partial z} \frac{\partial z}{\partial x} \frac{\partial z}{\partial y} + R \frac{\partial^2 z}{\partial x \, \partial y} \right) \right.$$

$$\left. - \left(\frac{\partial P}{\partial y} + \frac{\partial P}{\partial z} \frac{\partial z}{\partial y} + \frac{\partial R}{\partial y} \frac{\partial z}{\partial x} + \frac{\partial R}{\partial z} \frac{\partial z}{\partial y} \frac{\partial z}{\partial x} + R \frac{\partial^2 z}{\partial y \, \partial x} \right) \right] dz \, dy$$

After terms cancel

$$= \iint_{\Omega} \left(\frac{\partial Q}{\partial x} + \frac{\partial Q}{\partial z} \frac{\partial z}{\partial x} + \frac{\partial R}{\partial x} \frac{\partial z}{\partial y} - \frac{\partial P}{\partial y} - \frac{\partial P}{\partial z} \frac{\partial z}{\partial y} - \frac{\partial R}{\partial y} \frac{\partial z}{\partial x} \right) dx \, dy. \qquad (13)$$

After rearranging and noting that $\partial f/\partial x = \partial z/\partial x$ and $\partial f/\partial y = \partial z/\partial y$, we see that the integrals in (8) and (13) are identical. Thus

$$\iint_S \operatorname{curl} \mathbf{F} \cdot \mathbf{n} \, d\sigma = \oint_{\partial S} \mathbf{F} \cdot \mathbf{T} \, ds.$$

■

Note. We have proved this theorem only under the simplifying assumption that the surface can be written in the form $z = f(x, y)$. The proof for more general surfaces is best left to an advanced calculus book.

Problems 10.7

In Problems 1–8, evaluate the line integral by using Stokes's theorem.

1. $\oint_C \mathbf{F} \cdot d\mathbf{x}$, where $\mathbf{F}(x, y, z) = (x + y)\mathbf{i} + (z - 2x + y)\mathbf{j} + (y - z)\mathbf{k}$ and C is the unit circle in the plane $z = 5$.

2. $\oint_C \mathbf{F} \cdot d\mathbf{x}$, where $\mathbf{F}(x, y, z) = ax\mathbf{i} + by\mathbf{j} + cz\mathbf{k}$ and C is the curve of Problem 1.

3. $\oint_C \mathbf{F} \cdot d\mathbf{x}$, where $\mathbf{F}$ is as in Example 1 and C is the boundary of the triangle joining the points $(2, 0, 0)$, $(0, 2, 0)$, and $(0, 0, 2)$.

4. $\oint_C \mathbf{F} \cdot d\mathbf{x}$, where $\mathbf{F}$ is as in Example 1 and C is the boundary of the triangle joining the points $(d, 0, 0)$, $(0, d, 0)$, and $(0, 0, d)$.

5. $\oint_C y^2 x^3 \, dx + 2xyz^3 \, dy + 3xy^2 z^2 \, dz$, where C is given parametrically by $\mathbf{x}(t) = 2 \cos t \mathbf{i} + 3\mathbf{j} + 2 \sin t \mathbf{k}, 0 \le t \le 2\pi$.

6. $\oint_C \mathbf{F} \cdot d\mathbf{x}$, where $\mathbf{F} = 2y(x - z)\mathbf{i} + (x^2 + z^2)\mathbf{j} + y^3 \mathbf{k}$, C is the square $0 \le x \le 3, 0 \le y \le 3, z = 4$.

7. $\oint_C e^x \, dx + x \sin y \, dy + (y^2 - x^2) \, dz$, where C is the equilateral triangle formed by the intersection of the plane $x + y + z = 3$ with the three coordinate planes.

8. $\oint_C \mathbf{F} \cdot \mathbf{T} \, ds$, where $\mathbf{F} = -3y\mathbf{i} + 3x\mathbf{j} + \mathbf{k}$ and C is the circle $x^2 + y^2 = 1, z = 3$.

9. Show that Green's theorem is really a special case of Stokes's theorem.

In Problems 10–13, verify Stokes's theorem for the given S and $\mathbf{F}$.

10. $\mathbf{F} = z^2 \mathbf{i} + x^2 \mathbf{j} + y^2 \mathbf{k}$; S is the part of the plane $x + y + z = 1$ lying in the first octant (i.e., $x \ge 0, y \ge 0, z \ge 0$).

11. $\mathbf{F} = y^2 \mathbf{i} + z^2 \mathbf{j} + x^2 \mathbf{k}$ and S is the part of the plane $x + 2y + 3z = 6$ lying in the first octant.

12. $\mathbf{F} = 2y\mathbf{i} + x^2 \mathbf{j} + 3x\mathbf{k}$; S is the hemisphere $x^2 + y^2 + z^2 = 16, z \ge 0$.

13. $\mathbf{F} = y\mathbf{i} - x\mathbf{j} - z\mathbf{k}$; S is the circle $x^2 + y^2 \le 9, z = 2$.

14. Let S be a sphere. Use Stokes's theorem to show that $\iint_S \operatorname{curl} \mathbf{F} \cdot \mathbf{n} \, d\sigma = 0$.

* 15. Let S be a smooth closed surface bounding a simply connected region. Show that $\iint_S \operatorname{curl} \mathbf{F} \cdot \mathbf{n} \, d\sigma = 0$ for any continuously differentiable vector field $\mathbf{F}$.

* 16. Let $\mathbf{E}$ be an electric field and let $\mathbf{B}(t)$ be a magnetic field in space induced by $\mathbf{E}$. One of Maxwell's equations states that $\operatorname{curl} \mathbf{E} = -\partial \mathbf{B}/\partial t$. Let S be a surface in space with smooth boundary C. We define

voltage drop around $C = \oint_C \mathbf{E} \cdot d\mathbf{x}$.

Prove **Faraday's law,** which states that the voltage drop around C is equal to the time rate of decrease of the magnetic flux through S. [*Warning:* At one point it is necessary to interchange the order of differentiation and integration. A proof that this manipulation is "legal" requires techniques from advanced calculus; so just assume that it can be done.]

* **17.** A magnetic induction **B** is generated by electric current in a ring of radius R. Show that the magnitude of the vector potential **P** (**B** = curl **P**) is given by

$$|\mathbf{P}| = \varphi/2\pi R,$$

where φ is the total magnetic flux passing through the ring. [*Hint:* **P** is tangential to the ring.]

10.8 The Divergence Theorem

In Section 10.4 we gave the second vector form of Green's theorem (Theorem 10.4.3):

$$\oint_{\partial\Omega} \mathbf{F} \cdot \mathbf{n} \, ds = \iint_{\Omega} \operatorname{div} \mathbf{F} \, dx \, dy, \tag{1}$$

where $\partial\Omega$ is the piecewise smooth boundary of a region Ω in the xy-plane. In this section we extend this theorem to obtain the third major result in vector integral calculus: the **divergence theorem.**

Green's theorem gives us a relationship between a line integral in $\mathbb{R}^2$ around a closed curve and a double integral over the region in $\mathbb{R}^2$ enclosed by that curve. Stokes's theorem shows a relationship between a line integral in $\mathbb{R}^3$ around a closed curve and a surface integral over a surface that has the closed curve as the boundary. As we shall see, the divergence theorem gives us a relationship between a surface integral over a closed surface and a triple integral over the solid bounded by that surface.

Let S be a surface that forms the complete boundary of a solid W in space. A typical region and its boundary are sketched in Fig. 1. We assume that S is smooth or piecewise smooth. This assumption ensures that $\mathbf{n}(x, y, z)$, the outward unit normal to S, is continuous or piecewise continuous as a function of (x, y, z).

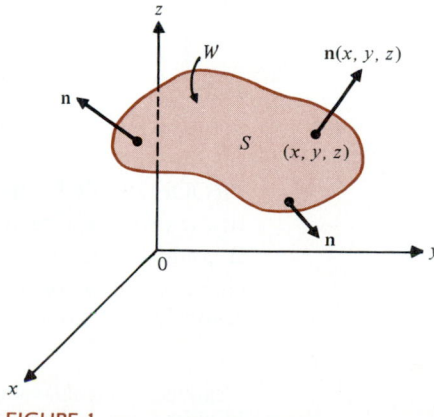

FIGURE 1

■ **THEOREM 1: The Divergence Theorem†**

Let W be a solid in $\mathbb{R}^3$ completely bounded by the smooth or piecewise smooth surface S. Let $\mathbf{F}$ be a smooth vector field on W, and let $\mathbf{n}$ denote the outward unit normal to S. Then

$$\iint_S \mathbf{F} \cdot \mathbf{n} \, d\sigma = \iiint_W \operatorname{div} \mathbf{F} \, dx \, dy \, dz. \tag{2}$$

■

Remark 1. Just as Green's theorem equates a line integral to an ordinary double integral, the divergence theorem equates a surface integral to an ordinary triple integral.

Remark 2. If $\mathbf{F} = P\mathbf{i} + Q\mathbf{j} + R\mathbf{k}$, then we can write the divergence theorem in the form

$$\iint_S (P \cos \alpha + Q \cos \beta + R \cos \gamma) \, d\sigma = \iiint_W \operatorname{div} \mathbf{F} \, dx \, dy \, dz, \tag{3}$$

where α, β, and γ are the angles that $\mathbf{n}$ makes with the positive coordinate axes. To see why this is so, observe that we have from equation (10.5.19), p. 627,

$$\cos \gamma = \mathbf{n} \cdot \mathbf{k} \qquad (\mathbf{n} \text{ and } \mathbf{k} \text{ are unit vectors}).$$

Similarly, we have

$$\cos \beta = \mathbf{n} \cdot \mathbf{j} \quad \text{and} \quad \cos \alpha = \mathbf{n} \cdot \mathbf{i}.$$

Thus

$$\mathbf{n} = (\cos \alpha)\mathbf{i} + (\cos \beta)\mathbf{j} + (\cos \gamma)\mathbf{k}$$

and

$$\mathbf{F} \cdot \mathbf{n} = P \cos \alpha + Q \cos \beta + R \cos \gamma.$$

PROOF OF THE DIVERGENCE THEOREM

In the proof of Green's theorem we assumed (see Fig. 10.4.4) that the region Ω could be enclosed by two curves—an upper curve $y = g_2(x)$ and a lower curve $y = g_1(x)$ or a left curve $x = h_1(y)$ and a right curve $x = h_2(y)$. This is equivalent to saying that any line parallel to the x- or y-axis crosses $\partial\Omega$ in at

† This theorem is also known as **Gauss's theorem,** named after the German mathematician Carl Friedrich Gauss (1777–1855). See the biographical sketch of Gauss on p. 366.

most two points. In order to prove the divergence theorem, we assume much the same thing. That is, we assume that W can be enclosed by an upper surface and a lower surface so that any line parallel to one of the coordinate axes crosses S in at most two points. This assumption is not necessary; smoothness of S is sufficient. But it allows us to give a reasonably simple proof.

From (3) it is sufficient to prove that

$$\iint_S P \cos \alpha \, d\sigma = \iiint_W \frac{\partial P}{\partial x} \, dx \, dy \, dz, \tag{4}$$

$$\iint_S Q \cos \beta \, d\sigma = \iiint_W \frac{\partial Q}{\partial y} \, dx \, dy \, dz, \tag{5}$$

and

$$\iint_S R \cos \gamma \, d\sigma = \iiint_W \frac{\partial R}{\partial z} \, dx \, dy \, dz. \tag{6}$$

The proofs of (4), (5), and (6) are virtually identical, so we will prove (6) only and leave (4) and (5) to you. By the assumption stated above, we can think of S as "walnut shaped." That is, S consists of two surfaces: an upper surface S_2: $z = f_2(x, y)$ and a lower surface S_1: $z = f_1(x, y)$. This is depicted in Fig. 2. We then have

$$\iint_S R \cos \gamma \, d\sigma = \iint_{S_1} R \cos \gamma \, d\sigma + \iint_{S_2} R \cos \gamma \, d\sigma.$$

Now from equation (10.5.21) on p. 628,

$$\iint_{S_2} R(x, y, z) \cos \gamma \, d\sigma = \iint_\Omega R(x, y, f_2(x, y)) \cos \gamma \sec \gamma \, dx \, dy$$

$$= \iint_\Omega R(x, y, f_2(x, y)) \, dx \, dy, \tag{7}$$

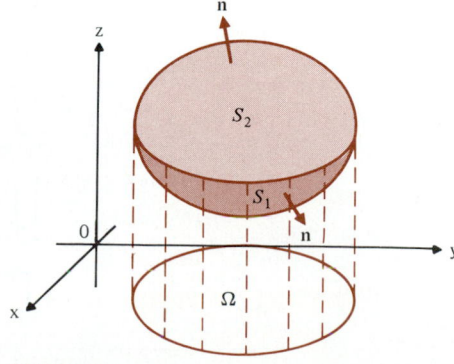

FIGURE 2

where Ω is the projection of S_2 into the xy-plane. On S_1 the situation is a bit different because the outer normal $\mathbf{n}$ on S_1 points downward and we assumed in deriving equation (10.5.21) that $\mathbf{n}$ points upward (so that the angle γ between $\mathbf{n}$ and the positive z-axis is an acute angle). We solve this problem by using $-\mathbf{n}$ in (10.5.21) so that on S_1 we have

$$\iint_{S_1} R(x, y, z) \cos \gamma \, d\sigma = -\iint_{\Omega} R(x, y, f_1(x, y)) \cos \gamma \sec \gamma \, dx \, dy$$

$$= -\iint_{\Omega} R(x, y, f_1(x, y)) \, dx \, dy.$$

Thus

$$\iint_{S} R \cos \gamma \, ds = \iint_{\Omega} [R(x, y, f_2(x, y)) - R(x, y, f_1(x, y))] \, dx \, dy$$

$$= \iint_{\Omega} \left(\int_{f_1(x,y)}^{f_2(x,y)} \frac{\partial R}{\partial z} \, dz \right) dx \, dy = \iiint_{W} \frac{\partial R}{\partial z} \, dz \, dx \, dy,$$

and under the restrictions already mentioned, this completes the proof. ■

EXAMPLE 1

Compute $\iint_S \mathbf{F} \cdot \mathbf{n} \, d\sigma$, where $\mathbf{F}(x, y, z) = x^2\mathbf{i} + 2y\mathbf{j} + 4z^2\mathbf{k}$ and S is the surface of the cylinder $x^2 + y^2 \le 4, 0 \le z \le 2$.

SOLUTION
div $\mathbf{F} = 2x + 2 + 8z$, so

$$\iint_{S} \mathbf{F} \cdot \mathbf{n} \, d\sigma = \iiint_{W} (2x + 2 + 8z) \, dz \, dx \, dy$$

$$= \iint_{x^2+y^2\le4} \int_{0}^{2} (2x + 2 + 8z) \, dz \, dx \, dy$$

$$= \iint_{x^2+y^2\le4} \left[(2xz + 2z + 4z^2) \Big|_{z=0}^{z=2} \right] dx \, dy$$

$$= \iint_{x^2+y^2\le4} (4x + 20) \, dx \, dy = \int_{0}^{2\pi} \int_{0}^{2} (4r \cos \theta + 20) r \, dr \, d\theta$$

$$= 4 \int_{0}^{2\pi} \int_{0}^{2} (r^2 \cos \theta + 5r) \, dr \, d\theta = 4 \int_{0}^{2\pi} \left(\frac{8}{3} \cos \theta + 10 \right) d\theta = 80\pi.$$

At the end of Section 10.4 and in Section 10.6 we stated that the divergence of **F** at a point (x, y) represents the net rate of flow away from (x,y). We can see this more clearly in light of the divergence theorem. Let us divide this region W into a set of small subregions W_i with boundaries S_i. Since div **F** is continuous, if W_i is sufficiently small, div **F** is approximately constant, and by the mean-value theorem

$$\iiint_{W_i} \text{div } \mathbf{F} \, dx \, dy \, dz = \text{div } \mathbf{F}(\mathbf{x}) \iiint_{W_i} dx \, dy \, dz$$

$$= (\text{div } \mathbf{F}(\mathbf{x})) \text{ volume } W_i, \tag{8}$$

for some point **x** in W_i. But

$$\iiint_{W_i} \text{div } \mathbf{F} \, dx \, dy \, dz = \iint_{S_i} \mathbf{F} \cdot \mathbf{n} \, d\sigma, \tag{9}$$

which is the flux of **F** through S_i. Thus by (8) and (9), at a point $\mathbf{x} = (x, y, z)$

$$\text{div } \mathbf{F}(x, y, z) = \frac{\iiint_{W_i} \text{div } \mathbf{F} \, dx \, dy \, dz}{\text{volume of } W_i} = \frac{\iint_{S_i} \mathbf{F} \cdot \mathbf{n} \, d\sigma}{\text{volume of } W_i}. \tag{10}$$

This last term can be thought of as *flux per unit volume* or net average rate of flow away from (x, y, z).

We can use the right side of (10) to represent the average rate of flow out of any region W. Assuming that the flow is steady and the fluid is incompressible, if the value of

$$\frac{1}{\text{volume } W} \iint_{S} \mathbf{F} \cdot \mathbf{n} \, d\sigma \tag{11}$$

is not zero, then there must be **sources** or **sinks** in W, that is, points where fluid is produced or removed.

Let $\mathbf{x}_0$ be a source (or sink) in the region W. If we use a limiting process to shrink W down to the point $\mathbf{x}_0$, it follows from (10) that the limit is div $\mathbf{F}(\mathbf{x}_0)$. Thus, the divergence at the point $\mathbf{x}_0$ of the vector field of a steady incompressible flow is the **source** (or **sink**) **intensity.**

EXAMPLE 2: Heat Flow

Heat in a solid flows in the direction of decreasing temperature. Experiments have shown that the flow rate **F** is proportional to the gradient of the temperature T, that is,

$$\mathbf{F} = -k \nabla T, \tag{12}$$

where the constant k is the thermal conductivity of the solid. Let W be a region in the solid and let S be its boundary surface. Suppose the heat flow is steady (this occurs, for example, when identical conditions have been maintained for a long period of time), and no heat sources or sinks are in W. Then the average flow rate through S must be zero. Hence, by (9)

$$0 = \iint\limits_{S} \mathbf{F} \cdot \mathbf{n}\, d\sigma = \iiint\limits_{W} \text{div } \mathbf{F}\, dx\, dy\, dz$$

$$= -k \iiint\limits_{W} \nabla \cdot \nabla T\, dx\, dy\, dz.$$

By (10.6.16) we have

$$-k \iiint\limits_{W} \nabla^2 T\, dx\, dy\, dz = 0.\dagger \qquad (13)$$

If we assume no region in the solid contains heat sources or sinks, (13) will hold for every region W in the solid. Assuming that the integrand is continuous, the only way this could happen would be if the integrand is zero everywhere in the solid, that is,

$$\nabla^2 T = 0. \qquad (14)$$

Equation (14) is called **Laplace's equation.** It is of fundamental importance in steady flows. Methods for finding solutions to Laplace's equation will be discussed in Chapter 12.

EXAMPLE 3: Green's Formulas

Let $\mathbf{F} = u\nabla v$, where u and v are scalar functions. Then

$$\text{div } \mathbf{F} = \nabla \cdot (u\nabla v) = u\nabla^2 v + \nabla u \cdot \nabla v$$

and

$$\mathbf{F} \cdot \mathbf{n} = u(\nabla v \cdot \mathbf{n}).$$

Using the divergence theorem (2), we have **Green's first formula:**

$$\iiint\limits_{W} (u\nabla^2 v + \nabla u \cdot \nabla v)\, dx\, dy\, dz = \iint\limits_{S} u(\nabla v \cdot \mathbf{n})\, d\sigma. \qquad (15)$$

Interchanging the roles of u and v in (15) and subtracting the result from (15), we obtain **Green's second formula**

$$\iiint\limits_{W} (u\nabla^2 v - v\nabla^2 u)\, dx\, dy\, dz = \iint\limits_{S} (u\nabla v - v\nabla u) \cdot \mathbf{n}\, d\sigma. \qquad (16)$$

† Recall [equation (10.6.16) on p. 638] that $\nabla^2 f = \dfrac{\partial^2 f}{\partial x^2} + \dfrac{\partial^2 f}{\partial y^2} + \dfrac{\partial^2 f}{\partial z^2}$ is called the **Laplacian** of f.

EXAMPLE 4

Suppose f is a solution of Laplace's equation $\nabla^2 f = 0$ in a domain R. Let W be a region in R whose piecewise smooth boundary S lies entirely in R. Suppose $f \equiv 0$ on S. Then by (15), with $u = v = f$,

$$\iiint_W |\nabla f|^2 \, dx \, dy \, dz = \iiint_W [f \nabla^2 f + \nabla f \cdot \nabla f] \, dx \, dy \, dz = \iint_S f(\nabla f \cdot \mathbf{n}) \, d\sigma = 0.$$

Since ∇f is continuous, this means $\nabla f = \mathbf{0}$ in W. Thus f is constant on W, and since it is 0 on S, it must be zero on $W \cup S$.

One consequence of this example is that two solutions of Laplace's equation that agree on S must be identical on $W \cup S$, since their difference satisfies the hypotheses above.

EXAMPLE 5: Coulomb's Law and Gauss's Law

According to **Coulomb's law**† the force of attraction (or repulsion) of the point charge q_1 on q_2 is given by

$$\mathbf{E} = \frac{1}{4\pi\epsilon_0} \frac{q_1 q_2}{r^2} \mathbf{u}, \tag{17}$$

where r is the distance between the charges and $\mathbf{u}$ is a unit vector directed from q_1 to q_2. The constant ϵ_0 is the **permittivity** of free space ($1/4\pi\epsilon_0 = 9 \times 10^9$ N $-$ m^2/C^2, where C denotes coulombs).

Let S denote a closed surface enclosing a point charge q. If Φ_E denotes the flux of the electric field surrounding q and through S, then **Gauss's law** states that

$$\epsilon_0 \Phi_E = q$$

or, using equation (10.5.24), on p. 630,

$$\epsilon_0 \iint_S \mathbf{E} \cdot \mathbf{n} \, d\sigma = q. \tag{18}$$

Gauss's law is one of the four Maxwell equations. We show here that Gauss's law can be derived from Coulomb's law.

Since q is enclosed by S, we can enclose q in a sphere S_r centered at q with arbitrarily small radius r. (See Fig. 3.) We can join S and S_r by a hollow tube of arbitrarily small radius to form a single, simply connected surface S_c. Then, by the divergence theorem,

$$\iint_{S_c} \mathbf{E} \cdot \mathbf{n} \, d\sigma = \iiint_W \operatorname{div} \mathbf{E} \, dx \, dy \, dz,$$

where W is the region between S and S_r. This region encloses no charge so it contains no sources or sinks. Thus div $E = 0$ and we have

$$\iint_{S_c} \mathbf{E} \cdot \mathbf{n} \, d\sigma = 0.$$

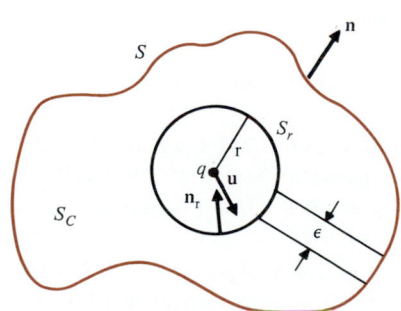

FIGURE 3

† Named after the French physicist Charles de Coulomb (1736–1806). Coulomb verified his law experimentally using a torsion balance. He wrote up his results in a famous 1785 memoir to the French Academy of Sciences.

But $S_c = S - S_r$ so we have

$$\iint_S \mathbf{E} \cdot \mathbf{n} \, d\sigma - \iint_{S_r} \mathbf{E} \cdot \mathbf{n}_r \, d\sigma = 0.$$

We ignore the contribution along the tube because the radius of the tube is arbitrarily small. The minus sign is there because $\mathbf{n}_r$ points inward. Thus

$$\iint_S \mathbf{E} \cdot \mathbf{n} \, d\sigma = \iint_{S_r} \mathbf{E} \cdot \mathbf{n}_r \, d\sigma. \qquad (19)$$

Now, from Coulomb's law, a unit point charge at the origin will produce the electric field given by

$$\mathbf{E} = \frac{q}{4\pi\epsilon_0 r^2} \mathbf{u}$$

Thus

$$\iint_{S_r} \mathbf{E} \cdot \mathbf{n}_r \, d\sigma = \frac{q}{4\pi\epsilon_0 r^2} \iint_{S_r} \mathbf{u} \cdot \mathbf{n}_r \, d\sigma \qquad (20)$$

Moreover, since $\mathbf{u}$ points toward the surface of S_r, $\mathbf{u}$ and $\mathbf{n}_r$ are parallel so $\mathbf{u} \cdot \mathbf{n}_r = 1$ and (20) becomes

$$\iint_S \mathbf{E} \cdot \mathbf{n} \, d\sigma = \frac{q}{4\pi\epsilon_0 r^2} \iint_{S_r} d\sigma = \frac{q}{4\pi\epsilon_0 r^2} \cdot 4\pi r^2 = \frac{q}{\epsilon_0},$$

$$\overset{\text{surface area}}{\underset{\text{of } S_r}{\downarrow}}$$

from which Gauss's law follows.

Problems 10.8

In Problems 1–15, evaluate the surface integral by using the divergence theorem.

1. $\iint_S \mathbf{F} \cdot \mathbf{n} \, d\sigma$, where $\mathbf{F} = x\mathbf{i} + y\mathbf{j} + z\mathbf{k}$ and S is the unit sphere.

2. $\iint_S \mathbf{F} \cdot \mathbf{n} \, d\sigma$, where $F = x^2\mathbf{i} + y^2\mathbf{j} + z^2\mathbf{k}$ and S is the unit sphere.

3. $\iint_S \mathbf{F} \cdot \mathbf{n} \, d\sigma$, where $\mathbf{F}$ is as in Problem 1 and S is the cylinder $x^2 + y^2 \le 4$, $0 \le z \le 3$.

4. $\iint_S \mathbf{F} \cdot \mathbf{n} \, d\sigma$, where $\mathbf{F} = y\mathbf{i} + z\mathbf{j} + x\mathbf{k}$ and S is as in Problem 3.

5. $\iint_S \mathbf{F} \cdot \mathbf{n} \, d\sigma$, where $\mathbf{F} = (y^2 + z^2)^{3/2}\mathbf{i} + \sin(x^2 - z^5)^{4/3}\mathbf{j} + e^{x^2 - y^3}\mathbf{k}$ and S is the ellipsoid

$$\frac{x^2}{a^2} + \frac{y^2}{b^2} + \frac{z^2}{c^2} = 1, \qquad abc \neq 0.$$

6. $\iint_S \mathbf{F} \cdot \mathbf{n} \, d\sigma$, where $\mathbf{F} = x\mathbf{i} + y\mathbf{j} + z\mathbf{k}$ and S is the surface of the unit cube $0 \le x \le 1$, $0 \le y \le 1$, $0 \le z \le 1$.

7. $\iint_S \mathbf{F} \cdot \mathbf{n} \, d\sigma$, where $\mathbf{F} = x^2\mathbf{i} + y^2\mathbf{j} - xy\mathbf{k}$ and S is as in Problem 6.

8. $\iint_S \mathbf{F} \cdot \mathbf{n} \, d\sigma$, where $\mathbf{F} = xyz\mathbf{i} + yz\mathbf{j} + z\mathbf{k}$ and S is as in Problem 6.

9. $\iint_S \mathbf{F} \cdot \mathbf{n} \, d\sigma$, where $\mathbf{F} = 2x\mathbf{i} + 3y\mathbf{j} + z\mathbf{k}$ and S is the boundary of the hemisphere $x^2 + y^2 + z^2 = 9$, $z \ge 0$.

10. $\iint_S \mathbf{F} \cdot \mathbf{n} \, d\sigma$, where $\mathbf{F} = x^2\mathbf{i} + y^2\mathbf{j} + z^2\mathbf{k}$ and S is as in Problem 9.

11. $\iint_S \mathbf{F} \cdot \mathbf{n} \, d\sigma$, where $\mathbf{F} = xy\mathbf{i} + y^2\mathbf{j} + yz\mathbf{k}$ and S is the boundary of the tetrahedron with vertices at $(0, 0, 0)$, $(1, 0, 0)$, $(0, 1, 0)$, and $(0, 0, 1)$.

12. $\iint_S \mathbf{F} \cdot \mathbf{n} \, d\sigma$, where $\mathbf{F} = y^2\mathbf{i} + x^2\mathbf{j} + z^2\mathbf{k}$ and S is as in Problem 11.

13. $\iint_S \mathbf{F} \cdot \mathbf{n} \, d\sigma$, where $\mathbf{F} = x(1 - \sin y)\mathbf{i} + (y - \cos y)\mathbf{j} + z\mathbf{k}$ and S is as in Problem 11.

14. $\iint_S \mathbf{F} \cdot \mathbf{n} \, d\sigma$, where $\mathbf{F} = x\mathbf{i} + y\mathbf{j} + z\mathbf{k}$ and S is the surface of the region bounded by the parabolic cylinder $z = 1 - y^2$, the plane $x + z = 2$, and the xy- and yz-planes.

15. $\iint_S \mathbf{F} \cdot \mathbf{n} \, d\sigma$, where $\mathbf{F} = (x^2 + e^{y\cos z})\mathbf{i} + (xy - \tan z^{1/3})\mathbf{j} + (x - y^{3/5})^{2/9}\mathbf{k}$ and S is as in Problem 14.

16. If $\mathbf{F}$ is twice continuously differentiable, prove that $\iint_S \text{curl } \mathbf{F} \cdot \mathbf{n} \, d\sigma = 0$ for any smooth closed surface S.

17. Show that $\iint_S \mathbf{F} \cdot \mathbf{n} \, d\sigma = 0$ if $\mathbf{F}$ is a constant vector field and S is a closed smooth surface.

18. If $\mathbf{F} = x\mathbf{i} + y\mathbf{j} + z\mathbf{k}$ and W is a solid with closed smooth boundary S, show that

$$\text{volume of } W = \frac{1}{3} \iint_S \mathbf{F} \cdot \mathbf{n} \, d\sigma.$$

19. Suppose that a vector field $\mathbf{v}$ is tangent to a closed smooth surface S, where S is the boundary of a solid W. Show that $\iiint_W \text{div } \mathbf{v} \, dx \, dy \, dz = 0$.

* 20. Prove Coulomb's law by assuming Gauss's law.

** 21. Let C be a path in a changing magnetic field. According to **Faraday's law,**

$$\oint_C \mathbf{E} \cdot \mathbf{dx} = -\frac{\partial \Phi}{\partial t},$$

where $\Phi = \iint_S \mathbf{B} \cdot \mathbf{n} \, d\sigma$ is the magnetic flux and $\mathbf{E}$ is an electric field. Show that

$$\text{curl } E = -\frac{\partial B}{\partial t}.$$

This equation is Maxwell's third equation.

10.9 Changing Variables in Multiple Integrals and the Jacobian

In many places in this and the preceding chapter we found it useful to evaluate a double integral by first converting to polar coordinates:

$$\iint_\Omega f(x, y) \, dx \, dy = \iint_\Omega f(r \cos \theta, r \sin \theta) r \, dr \, d\theta. \tag{1}$$

In this section we show how, under certain conditions, it is possible to convert from one set of coordinates to another in double and triple integrals. Much of what we do is a generalization of the formula for changing variables in ordinary definite integrals. Let us recall that formula now. If $x = g(u)$ is a differentiable one-to-one function, then

$$\int_a^b f(x) \, dx = \int_c^d f(g(u))g'(u) \, du, \tag{2}$$

where $c = g^{-1}(a)$ and $d = g^{-1}(b)$.

Now suppose we wish to change the variables of integration in the double integral

$$\iint_\Omega f(x, y) \, dx \, dy. \tag{3}$$

We assume that the new variables are called u and v and that they are related to the old variables x and y by the relations

$$x = g(u, v) \quad \text{and} \quad y = h(u, v). \tag{4}$$

The functional relationship described by (4) is called a **mapping** from the uv-plane into the xy-plane. We shall assume that there is a region Σ in the uv-plane that gets mapped **onto** Ω by the mapping described by (4) and the mapping is **one-to-one.** That is:

(i) For every $(x, y) \in \Omega,$ there is a $(u, v) \in \Sigma$ such that $x = g(u, v)$ and $y = h(u, v)$.

(ii) If $g(u_1, v_1) = g(u_2, v_2)$ and $h(u_1, v_1) = h(u_2, v_2)$, then $u_1 = u_2$ and $v_1 = v_2$.

Condition (ii) is the natural extension of the definition of one-to-one for functions of two variables. We illustrate what is going on in Fig. 1.

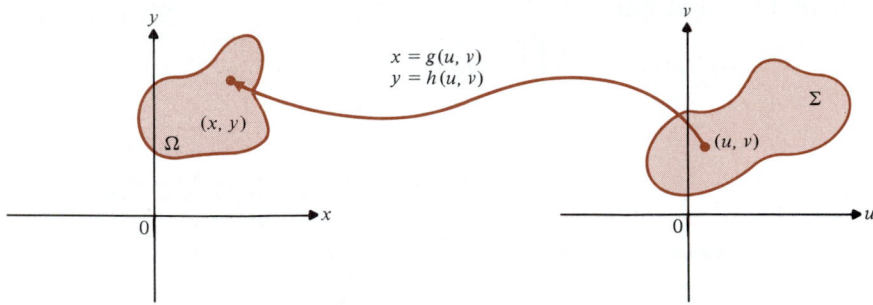

FIGURE 1

EXAMPLE 1

Let $x = u^2 - v^2$ and $y = 3uv$. We shall compute the image of the vertical line $u = k$ in the uv-plane under the mapping. Substituting k for u in the equations given above, we have

$$x = k^2 - v^2 \quad \text{and} \quad y = 3kv.$$

Thus $v = y/3k$ and

$$x = k^2 - \frac{y^2}{9k^2}. \tag{5}$$

For every $k \neq 0$, equation (5) is the equation of a parabola in the xy-plane. Moreover, if $v = c$, a constant, then $x = u^2 - c^2$, $y = 3uc$, $u = y/3c$, and

$$x = \frac{y^2}{9c^2} - c^2, \tag{6}$$

which is also a parabola for $c \neq 0$. Thus the functions given above map straight lines into parabolas. Some of these curves are sketched in Fig. 2.

We now give an important definition.

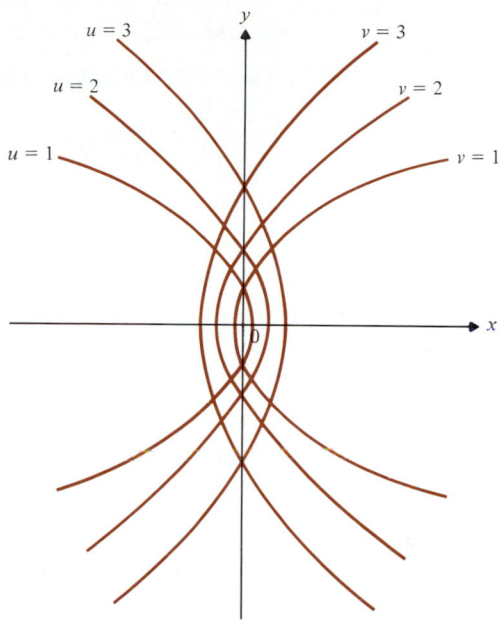

FIGURE 2

DEFINITION

Jacobian

Let $x = g(u, v)$ and $y = h(u, v)$ be differentiable. Then the **Jacobian**† of x and y with respect to u and v, denoted $\partial(x, y)/\partial(u, v)$, is

$$\frac{\partial(x, y)}{\partial(u, v)} = \frac{\partial x}{\partial u}\frac{\partial y}{\partial v} - \frac{\partial x}{\partial v}\frac{\partial y}{\partial u} = \begin{vmatrix} \dfrac{\partial x}{\partial u} & \dfrac{\partial x}{\partial v} \\[2mm] \dfrac{\partial y}{\partial u} & \dfrac{\partial y}{\partial v} \end{vmatrix}. \tag{7}$$

■

EXAMPLE 2

If $x = u^2 - v^2$ and $y = 3uv$ as in Example 1, then $\partial x/\partial u = 2u$, $\partial x/\partial v = -2v$, $\partial y/\partial u = 3v$, and $\partial y/\partial v = 3u$, so

$$\begin{vmatrix} \dfrac{\partial x}{\partial u} & \dfrac{\partial x}{\partial v} \\[2mm] \dfrac{\partial y}{\partial u} & \dfrac{\partial y}{\partial v} \end{vmatrix} = \begin{vmatrix} 2u & -2v \\ 3v & 3u \end{vmatrix} = 6(u^2 + v^2).$$

† Named after the German mathematician Carl Gustav Jacob Jacobi (1804–1851). See the accompanying biographical sketch.

EXAMPLE 3

Let $x = r \cos \theta$ and $y = r \sin \theta$. Then

$$\frac{\partial x}{\partial r} = \cos \theta, \qquad \frac{\partial x}{\partial \theta} = -r \sin \theta, \qquad \frac{\partial y}{\partial r} = \sin \theta, \qquad \frac{\partial y}{\partial \theta} = r \cos \theta,$$

so

$$\frac{\partial(x, y)}{\partial(r, \theta)} = \begin{vmatrix} \cos \theta & -r \sin \theta \\ \sin \theta & r \cos \theta \end{vmatrix}$$

$$= r \cos^2 \theta + r \sin^2 \theta = r(\sin^2 \theta + \cos^2 \theta) = r.$$

CARL GUSTAV JACOB JACOBI
(1804–1851)

Courtesy of Historical Pictures Service

The son of a prosperous banker, Carl Gustav Jacob Jacobi was born in Potsdam, Prussia, in 1804. He was educated at the University of Berlin, where he received his doctorate in 1824. In 1827, he was appointed Extraordinary Professor of Mathematics at the University of Königsberg. Jacobi taught at Königsberg until 1842, when he returned to Berlin under a pension from the Prussian government. He remained in Berlin until his death in 1851.

A prolific writer of mathematical treatises, Jacobi was best known in his time for his results in the theory of elliptic functions. Today, however, he is most remembered for his work on determinants. He was one of the two most creative developers of determinant theory, the other being Cauchy. In 1829, Jacobi published a paper on algebra that contained the notation for the Jacobian that we use today. In 1841 he published an extensive treatise titled *De determinantibus functionalibus,* which was devoted to results about the Jacobian. Jacobi showed the relationship between the Jacobian of functions of several variables and the derivative of a function of one variable. He also showed that *n* functions of *n* variables are linearly independent if and only if their Jacobian is not identically zero.

In addition to being a fine mathematician, Jacobi was considered the greatest teacher of mathematics of his generation. He inspired and influenced an astonishing number of students. To dissuade his students from mastering great amounts of mathematics before setting off to do their own research, Jacobi often remarked, "Your father would never have married, and you would not be born, if he had insisted on knowing all the girls in the world before marrying one."

Jacobi believed strongly in research in pure mathematics and frequently defended it against the claim that research should always be applicable to something. He once said, "The real end of science is the honor of the human mind."

To obtain our main result we need assumptions (i) and (ii) cited earlier in this section. We need also to assume the following:

(iii) $C_1 = \partial\Omega$ and $C_2 = \partial\Sigma$ are simple closed curves, and as (u, v) moves once about C_2 in the positive direction, $(x, y) = (g(u, v), h(u, v))$ moves once around C_1 in the positive or negative direction.

(iv) All second order partial derivatives are continuous.

■ **THEOREM 1: Change of Variables in a Double Integral**

If assumptions (i), (ii), (iii), and (iv) hold, then

$$\iint\limits_{\Omega} f(x, y)\, dx\, dy = \pm \iint\limits_{\Sigma} f[g(u, v), h(u, v)] \frac{\partial(x, y)}{\partial(u, v)}\, du\, dv. \qquad (8)$$

The plus (minus) sign is taken if as (u, v) moves around C_2 in the positive direction $(x, y) = (g(u, v), h(u, v))$ moves around C_1 in the positive (negative) direction.

PROOF

Define $F(x, y) = \int_{x_0}^{x} f(t, y)\, dt$. By the fundamental theorem of calculus, F is continuous and

$$\frac{\partial F}{\partial x} = f(x, y). \qquad (9)$$

Then by Green's theorem

$$\iint\limits_{\Omega} f(x, y)\, dx\, dy = \iint\limits_{\Omega} \frac{\partial F}{\partial x}\, dx\, dy = \oint_{C_1} F\, dy. \qquad (10)$$

We wish to write $\oint_{C_1} F\, dy$ as a line integral in the uv-plane. Let $\mathbf{u} = u(t)\mathbf{i} + v(t)\mathbf{j}$ be a parametric representation of C_2 in the uv-plane for $a \le t \le b$. Then by assumption (iii), $\mathbf{x} = x(t)\mathbf{i} + y(t)\mathbf{j} = g(u(t), v(t))\mathbf{i} + h(u(t), v(t))\mathbf{j}$, $a \le t \le b$, is a parametric representation for C_1. The only difference is that this representation may traverse C_1 in either the positive or the negative direction. Thus

$$\oint_{C_1} F(x, y)\, dy = \int_{a}^{b} F(g(u(t), v(t)), h(u(t), v(t))) \frac{dy}{dt}\, dt. \qquad (11)$$

Now by the chain rule and the fact that $y = h(u, v)$,

$$\frac{dy}{dt} = \frac{\partial h}{\partial u} u'(t) + \frac{\partial h}{\partial v} v'(t). \qquad (12)$$

To simplify notation, let $\bar{F}(u, v) = F(g(u, v), h(u, v))$. Then if we substitute (12) into (11), we obtain

$$\oint_{C_1} F(x, y)\, dy = \int_a^b \bar{F}(u(t), v(t)) \left[\frac{\partial h}{\partial u} u'(t) + \frac{\partial h}{\partial v} v'(t) \right] dt$$

$$= \int_a^b \left[\bar{F} \frac{\partial h}{\partial u} u'(t) + \bar{F} \frac{\partial h}{\partial v} v'(t) \right] dt. \tag{13}$$

Using the definition of the line integral, we can write (13) as

$$\oint_{C_1} F(x, y)\, dy = \pm \oint_{C_2} \bar{F} \frac{\partial h}{\partial u} du + \bar{F} \frac{\partial h}{\partial v} dv, \tag{14}$$

where the $\pm$ depends on whether (x, y) traverses C_1 in the same direction as, or in the direction opposite to, that in which (u, v) traverses C_2 as t goes from a to b. Let

$$P = \bar{F} \frac{\partial h}{\partial u} \quad \text{and} \quad Q = \bar{F} \frac{\partial h}{\partial v}. \tag{15}$$

Then by Green's theorem

$$\oint_{C_2} P\, du + Q\, dv = \iint_\Sigma \left(\frac{\partial Q}{\partial u} - \frac{\partial P}{\partial v} \right) du\, dv. \tag{16}$$

But from the product rule

$$\frac{\partial Q}{\partial u} = \frac{\partial \bar{F}}{\partial u} \frac{\partial h}{\partial v} + \bar{F} \frac{\partial^2 h}{\partial u \partial v} \tag{17}$$

and

$$\frac{\partial P}{\partial v} = \frac{\partial \bar{F}}{\partial v} \frac{\partial h}{\partial u} + \bar{F} \frac{\partial^2 h}{\partial v \partial u}, \tag{18}$$

so that from (14)–(18),

Equal because mixed partials are continuous

$$\oint_{C_1} F\, dy = \pm \iint_\Sigma \left(\frac{\partial \bar{F}}{\partial u} \frac{\partial h}{\partial v} + \bar{F} \frac{\partial^2 h}{\partial u \partial v} - \frac{\partial \bar{F}}{\partial v} \frac{\partial h}{\partial u} - \bar{F} \frac{\partial^2 h}{\partial v \partial u} \right) du\, dv$$

$$= \pm \iint_\Sigma \left(\frac{\partial \bar{F}}{\partial u} \frac{\partial h}{\partial v} - \frac{\partial \bar{F}}{\partial v} \frac{\partial h}{\partial u} \right) du\, dv. \tag{19}$$

But

$$\frac{\partial \bar{F}}{\partial u} = \frac{\partial \bar{F}}{\partial x} \frac{\partial x}{\partial u} + \frac{\partial \bar{F}}{\partial y} \frac{\partial y}{\partial u} = \frac{\partial \bar{F}}{\partial x} \frac{\partial g}{\partial u} + \frac{\partial \bar{F}}{\partial y} \frac{\partial h}{\partial u},$$

and similarly for $\partial \bar{F}/\partial v$. Thus

$$\int_{C_1} F\, dy = \pm \iint_{\Sigma} \left\{ \left(\frac{\partial \bar{F}}{\partial x}\frac{\partial g}{\partial u} + \frac{\partial \bar{F}}{\partial y}\frac{\partial h}{\partial u} \right)\frac{\partial h}{\partial v} - \left(\frac{\partial \bar{F}}{\partial x}\frac{\partial g}{\partial v} + \frac{\partial \bar{F}}{\partial y}\frac{\partial h}{\partial v} \right)\frac{\partial h}{\partial u} \right\} du\, dv$$

$$= \pm \iint_{\Sigma} \frac{\partial \bar{F}}{\partial x}\left(\frac{\partial g}{\partial u}\frac{\partial h}{\partial v} - \frac{\partial g}{\partial v}\frac{\partial h}{\partial u} \right) du\, dv.$$

But $\partial \bar{F}/\partial x = f(g(u, v), h(u, v))$ and $(\partial g/\partial u)(\partial h/\partial v) - (\partial g/\partial v)(\partial h/\partial u) = \partial(x, y)/\partial(u, v)$. Thus

$$\int_{C_1} F\, dy = \pm \iint_{\Sigma} f(g(u, v), h(u, v)) \frac{\partial(x, y)}{\partial(u, v)}\, du\, dv. \tag{20}$$

Combining (10) with (20) completes the proof. ■

Remark. Conditions (i) and (ii) given on p. 654 are often difficult to verify. It can be shown that these conditions hold locally if all partial derivatives are continuous in Ω and the Jacobian $\partial(x, y)/\partial(u, v)$ is not zero on Ω.

EXAMPLE 4

We can use Theorem 1 to obtain the polar coordinate formula (1) very easily. For if $x = r \cos \theta$ and $y = r \sin \theta$, then $\partial(x, y)/\partial(r, \theta) = r$ by Example 3, and (1) follows immediately from (8).

EXAMPLE 5

Let Ω be the region in the upper half of the xy-plane bounded by the parabolas $y^2 = 9 - 9x$ and $y^2 = 9 + 9x$ and by the x-axis. Compute $\iint_{\Omega} (x + y)\, dx\, dy$ by making the change of variables $x = u^2 - v^2$, $y = 3uv$.

SOLUTION

We saw in Example 1 that this mapping takes straight lines in the uv-plane into parabolas in the xy-plane. For example, if $y^2 = 9 - 9x$, then $y^2 = 9u^2v^2 = 9 - 9x$, or $u^2v^2 = 1 - x = 1 - u^2 + v^2$, or $u^2(1 + v^2) = u^2 + u^2v^2 = 1 + v^2$, and $u^2 = 1$, so $u = \pm 1$. Similarly, if $y^2 = 9 + 9x$, then $v = \pm 1$. Since, in Ω, $y \geq 0$, we must have $uv \geq 0$, so that u and v have the same sign. We will choose $u = v = 1$ for reasons to be made clear shortly. Note also that if $u = 0$, then $y = 0$ and $x = -v^2 \leq 0$, so that the positive v-axis in the uv-plane is mapped into the negative x-axis in the xy-plane. Similarly, the positive u-axis in the uv-plane is mapped into the positive x-axis in the xy-plane. The situation is sketched in Fig. 3. The reason we chose $u = v = 1$ is that moving around Σ in the positive direction corresponds to moving around Ω in the positive direction. [Try it: Take the path $v = 0$ (the u-axis) to $u = 1$ to $v = 1$ to $u = 0$.] Thus the integral around the "parabolic" region given in the problem can be reduced to an integral around a square. Also, as we computed in Example 2,

$$\frac{\partial(x, y)}{\partial(u, v)} = 6(u^2 + v^2).$$

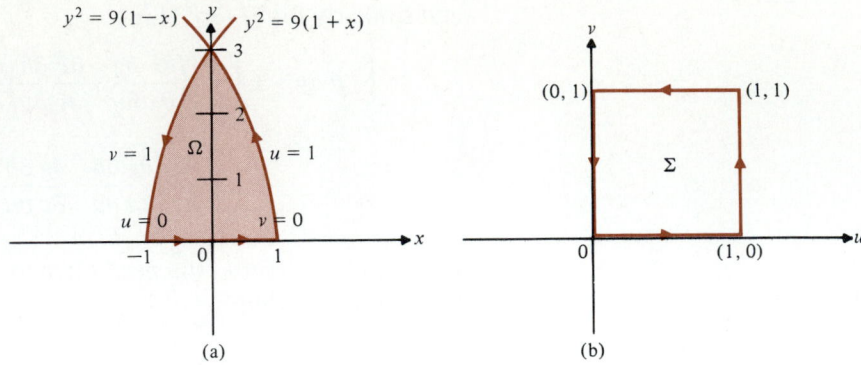

FIGURE 3

Finally, $x + y = u^2 + 3uv - v^2$, so

$$\iint_\Omega (x+y)\,dx\,dy = 6 \int_0^1 \int_0^1 (u^2 + 3uv - v^2)(u^2 + v^2)\,du\,dv$$

$$= 6 \int_0^1 \int_0^1 (u^4 + 3u^3v + 3uv^3 - v^4)\,du\,dv = \frac{9}{2}.$$

For triple integrals there is a result analogous to the one we have proven for double integrals. We will state this result without proof.

Let

$$x = g(u, v, w) \qquad y = h(u, v, w), \qquad z = j(u, v, w). \tag{21}$$

Jacobian in $\mathbb{R}^3$

We define the **Jacobian** of the transformation from a region U in a uvw-space to a region W in xyz-space as

$$\text{Jacobian} = \frac{\partial(x, y, z)}{\partial(u, v, w)} = \begin{vmatrix} \dfrac{\partial x}{\partial u} & \dfrac{\partial x}{\partial v} & \dfrac{\partial x}{\partial w} \\[2mm] \dfrac{\partial y}{\partial u} & \dfrac{\partial y}{\partial v} & \dfrac{\partial y}{\partial w} \\[2mm] \dfrac{\partial z}{\partial u} & \dfrac{\partial z}{\partial v} & \dfrac{\partial z}{\partial w} \end{vmatrix}. \tag{22}$$

Then under hypotheses similar to the ones made in Theorem 1, we have the following theorem.

■ **THEOREM 2: Change of Variables in a Triple Integral**

$$\iiint\limits_{W} F(x, y, z)\,dx\,dy\,dz$$

$$= \pm \iiint\limits_{U} F(g(u, v, w), h(u, v, w), j(u, v, w)) \frac{\partial(x, y, z)}{\partial(u, v, w)}\,du\,dv\,dw$$

■

EXAMPLE 6

Let $x = \rho \sin \varphi \cos \theta$, $y = \rho \sin \varphi \sin \theta$, and $z = \rho \cos \varphi$. These are spherical coordinates. Then

$$\frac{\partial(x, y, z)}{\partial(\rho, \varphi, \theta)} = \begin{vmatrix} \dfrac{\partial x}{\partial \rho} & \dfrac{\partial x}{\partial \varphi} & \dfrac{\partial x}{\partial \theta} \\[2mm] \dfrac{\partial y}{\partial \rho} & \dfrac{\partial y}{\partial \varphi} & \dfrac{\partial y}{\partial \theta} \\[2mm] \dfrac{\partial z}{\partial \rho} & \dfrac{\partial z}{\partial \varphi} & \dfrac{\partial z}{\partial \theta} \end{vmatrix} = \begin{vmatrix} \sin \varphi \cos \theta & \rho \cos \varphi \cos \theta & -\rho \sin \varphi \sin \theta \\ \sin \varphi \sin \theta & \rho \cos \varphi \sin \theta & \rho \sin \varphi \cos \theta \\ \cos \varphi & -\rho \sin \varphi & 0 \end{vmatrix}$$

Expanding in the last row
↓

$$= \cos \varphi \begin{vmatrix} \rho \cos \varphi \cos \theta & -\rho \sin \varphi \sin \theta \\ \rho \cos \varphi \sin \theta & \rho \sin \varphi \cos \theta \end{vmatrix}$$

$$+ \rho \sin \theta \begin{vmatrix} \sin \varphi \cos \theta & -\rho \sin \varphi \sin \theta \\ \sin \varphi \sin \theta & \rho \sin \varphi \cos \theta \end{vmatrix}$$

$$= \cos \varphi(\rho^2 \cos \varphi \sin \varphi \cos^2 \theta + \rho^2 \sin \varphi \cos \varphi \sin^2 \theta)$$

$$+ \rho \sin \varphi(\rho \sin^2 \varphi \cos^2 \theta + \rho \sin^2 \varphi \sin^2 \theta)$$

$$= \cos \varphi(\rho^2 \sin \varphi \cos \varphi) + \rho \sin \varphi(\rho \sin^2 \varphi)$$

$$= \rho^2 \sin \varphi(\cos^2 \varphi + \sin^2 \varphi) = \rho^2 \sin \varphi.$$

Thus

$$\iiint\limits_{W} f(x, y, z)\,dx\,dy\,dz = \iiint\limits_{U} f(\rho \sin \varphi \cos \theta, \rho \sin \varphi \sin \theta, \rho \cos \theta)\rho^2 \sin \varphi\,d\rho\,d\varphi\,d\theta.$$

Problems 10.9

In Problems 1–20, compute the Jacobian of the given transformation.

1. $x = u + v, \quad y = u - v$

2. $x = u^2 - v^2, \quad y = u^2 + v^2$

3. $x = u^2 - v^2, \quad y = 2uv$

4. $x = \sin u, \quad y = \cos v$

5. $x = u + 3v - 1, \quad y = 2u + 4v + 6$

6. $x = v - 2u, \quad y = u + 2v$

7. $x = au + bv, \quad y = bu - av$

8. $x = e^v, \quad y = e^u$

9. $x = ue^v, \quad y = ve^u$

* 10. $x = u^v, \quad y = v^u$

11. $x = \ln(u + v), \quad y = \ln uv$

12. $x = \tan u, \quad y = \sec v$

13. $x = u \sec v, \quad y = v \csc u$

14. $x = u \ln v, \quad y = v \ln u$

15. $x = u + v + w, \quad y = u - v - w, \quad z = -u + v + w$

16. $x = au + bv + cw, \quad y = au - bv - cw, \quad z = -au + bv + cw$

17. $x = u^2 + v^2 + w^2, \quad y = u + v + w, \quad z = uvw$

18. $x = u \sin v, \quad y = v \cos w, \quad z = w \sin u$

19. $x = e^u, \quad y = e^v, \quad z = e^w$

20. $x = u \ln(v + w), \quad y = v \ln(u + w), \quad z = w \ln(u + v)$

In Problems 21–25, transform the integral in (x, y) to an integral in (u, v) by using the given transformation. You need not evaluate the integral.

21. $\int_0^1 \int_y^1 xy \, dx \, dy; \quad x = u - v, \quad y = u + v.$

22. $\int \int_\Omega e^{(x+y)/(x-y)} \, dx \, dy$, where Ω is the region in the first quadrant between the lines $x + y = 1$ and $x + y = 2$; $x = u + v, \ y = u - v.$

23. $\int \int_\Omega y \, dx \, dy$, where Ω is the region $7x^2 + 6\sqrt{3}x(y - 1) + 13(y - 1)^2 \le 16$; use the transformation $x = \sqrt{3}u + (\frac{1}{2})v$, $y = 1 - u + (\sqrt{3}/2)v.$

24. $\int_0^1 \int_0^x (x^2 + y^2) \, dx \, dy; \quad x = v, \ y = u.$

* 25. $\int \int_\Omega (y - x) \, dy \, dx$, where Ω is the region bounded by $y = 2$, $y = x$, and $x = -y^2$; $x = v - u^2, \ y = u + v.$

26. Let Ω be the region in the first quadrant of the xy-plane bounded by the hyperbolas $xy = 1$, $xy = 2$, and the lines $x = y$ and $x = 4y$. Compute $\int \int_\Omega x^2 y^2 \, dx \, dy$ by setting $x = u$ and $y = u/v.$

27. Let W be the solid enclosed by the ellipsoid $(x^2/a^2) + (y^2/b^2) + (z^2/c^2) = 1$. Then

$$\text{volume of } W = \int \int \int_W dx \, dy \, dz.$$

Compute this volume by making the transformation $x = au, \ y = bv, \ z = cw$. [*Hint:* In uvw-space you'll obtain a sphere.]

28. Compute $\int \int \int_W (xy + xz + yz) \, dx \, dy \, dz$, where W is the region of Problem 27.

Review Exercises for Chapter 10

In Exercises 1–6, compute the gradient vector of the given function.

1. $f(x, y) = (x + y)^3$

2. $f(x, y) = \sin(x + 2y)$

3. $f(x, y) = \sqrt{xy}$

4. $f(x, y) = \dfrac{x + y}{x - y}$

5. $f(x, y, z) = x^2 + y^2 + z^2$

6. $f(x, y, z) = xyz$

In Exercises 7 and 8, sketch the given vector field.

7. $\mathbf{F}(x, y) = (x^2 + y^2)\mathbf{i} - 2xy\mathbf{j}$

8. $\mathbf{F}(x, y, z) = x\mathbf{i} - y\mathbf{j} - z\mathbf{k}$

In Exercises 9–11, calculate $\int_C \mathbf{F}(\mathbf{x}) \cdot d\mathbf{x}.$

9. $\mathbf{F}(x, y) = x^2\mathbf{i} + y^2\mathbf{j};$ C is the curve $y = x^{3/2}$ from $(0, 0)$ to $(1, 1).$

10. $\mathbf{F}(x, y) = x^2 y\mathbf{i} - xy^2\mathbf{j};$ C is the unit circle in the counterclockwise direction.

11. $\mathbf{F}(x, y) = 3xy\mathbf{i} - y\mathbf{j};$ C is the triangle joining the points $(0, 0), (1, 1),$ and $(0, 1)$ in the counterclockwise direction.

12. Calculate the work done when the force field $F(x, y) = x^2 y i + (y^3 + x^3)j$ moves a particle around the unit circle in the counterclockwise direction.

13. Calculate the work done when the force field $F(x, y) = 3(x - y)i + x^5j$ moves a particle around the triangle of Exercise 11.

14. Show that $F(x, y) = 3x^2y^2i + 2x^3yj$ is conservative, and calculate $\int_C F(x) \cdot dx$, where C starts at $(1, 2)$ and ends at $(3, -1)$.

15. Show that $F(x, y) = e^{xy}(1 + xy)i + x^2e^{xy}j$ is conservative, and calculate $\int_C F(x) \cdot dx$, where C starts at $(-1, 0)$ and ends at $(0, 1)$.

16. Evaluate $\oint_{\partial\Omega} 2y\, dx + 4x\, dy$, where $\Omega = \{(x, y): 0 \le x \le 2, 0 \le y \le 2\}$.

17. Evaluate $\oint_{\partial\Omega} x^2y\, dx + xy^2\, dy$, where Ω is the region enclosed by the triangle of Exercise 11.

18. Evaluate $\oint_{\partial\Omega} (x^3 + y^3)\, dx + (x^2y^2)\, dy$, where Ω is the unit disk.

19. Evaluate $\oint_{\partial\Omega} \sqrt{1 + x^2}\, dy$; $\Omega = \{(x, y): -1 \le x \le 1, x^2 \le y \le 1\}$.

20. Evaluate

$$\oint_{\partial\Omega} \frac{2x^2 + y^2 - xy}{(x^2 + y^2)^{1/2}} dx + \frac{xy - x^2 - 2y^2}{(x^2 + y^2)^{1/2}} dy,$$

where Ω is an open disk centered at the origin with the origin deleted.

21. Let $F(x, y) = xy^2i + x^2yj$ and let Ω denote the disk of radius 2 centered at $(0, 0)$. Calculate (a) curl F, (b) $\oint_{\partial\Omega} F \cdot T\, ds$, (c) div F, and (d) $\oint_{\partial\Omega} F \cdot n\, ds$.

22. Do Problem 21 for $F(x, y) = y^2i - x^2j$, where Ω is the region enclosed by the triangle of Exercise 11.

23. Show that the vector flow $F(x, y) = (\cos x^2)i + e^yj$ is irrotational.

24. Show that the vector flow $F(x, y) = -(x^2y + y^3)i + (x^3 + xy^2)j$ is incompressible.

In Exercises 25 and 26, evaluate $\int_C F(x) \cdot dx$.

25. $F(x, y, z) = xi + yj + zk$; C is the curve $x(t) = t^3i + t^2j + tk$ from $(0, 0, 0)$ to $(1, 1, 1)$.

26. $F(x, y, z) = x^2i + y^2j + z^2k$; C is the helix $x(t) = (\sin t)i + (\cos t)j + 2tk$ from $(0, 1, 0)$ to $(1, 0, \pi)$.

27. Show that $F(x, y, z) = -(y/z)i - (x/z)j + (xy/z^2)k$ is conservative in the set $z > 0$, and use that fact to evaluate $\int_C F(x) \cdot dx$, where C is a piecewise smooth curve joining the points $(1, 1, 1)$ and $(2, -1, 3)$ and not crossing the xy-plane.

In Exercises 28–33, evaluate the surface integral over the given surface.

28. $\iint_S y\, d\sigma$, where $S: z = y^2$, $0 \le x \le 2, 0 \le y \le 1$.

29. $\iint_S (x^2 + y^2)\, d\sigma$, where S is as in Exercise 28.

30. $\iint_S y^2\, d\sigma$, where S is the hemisphere $x^2 + y^2 + z^2 = 9$, $z \ge 0, x^2 + y^2 \le 9$.

31. $\iint_S xz\, d\sigma$, where S is as in Exercise 30.

32. $\iint_S y\, d\sigma$, where S is the boundary of the tetrahedron bounded by the coordinate planes and the plane $x + y + z = 1$.

33. $\iint_S (x^2 + y^2 + z^2)\, d\sigma$, where S is the part of the plane $y - x = 3$ that lies inside the cylinder $y^2 + z^2 = 9$.

34. Find the mass of the triangular metallic sheet with corners at $(1, 0, 0), (0, 1, 0)$, and $(0, 0, 1)$ if its density is proportional to y^2.

35. Find the mass of a metallic sheet in the shape of the hemisphere $x^2 + y^2 + z^2 = 1$, $y \ge 0, x^2 + z^2 \le 1$, if its density is proportional to the distance from the y-axis.

In Exercises 36–39, compute the flux $\iint_S v \cdot n\, d\sigma$ for the given surface lying in the given vector field.

36. $S: z = 2xy$; $0 \le x \le 1$, $0 \le y \le 4$; $v = xy^2i - 2zj$

37. $S: x^2 + y^2 + z^2 = 1$; $z \ge 0$; $v = xi + 2yj + 3zk$

38. $S: y = \sqrt{x^2 + z^2}$; $x^2 + z^2 \le 4$; $v = xi - xzj + zk$

39. S: unit sphere; $v = x^2i + y^2j + z^2k$

In Exercises 39–45, compute the divergence and curl of the given vector field.

40. $F(x, y, z) = xi + yj + zk$

41. $F(x, y, z) = (x - y)i + (y - z)j + (z - x)k$

42. $F(x, y, z) = yzi + xzj + xyk$

43. $F(x, y, z) = \ln xi + \ln yj + \ln zk$

44. $F(x, y, z) = e^{yz}i + e^{xz}j + e^{xy}k$

45. $F(x, y, z) = \cos yi + \cos xj + \cos zk$

In Exercises 46–48, evaluate the line integral by using Stokes's theorem.

46. $\oint_C F \cdot dx$, where $F(x, y, z) = (x + 2y)i + (y - 3z)j + (z - x)k$ and C is the unit circle in the plane $z = 2$.

47. $\oint_C F \cdot dx$, where $F = xi + yj + zk$ and C is the boundary of the triangle joining the points $(1, 0, 0), (0, 1, 0)$, and $(0, 0, 1)$.

48. $\oint_C F \cdot dx$, where $F = y^2i + x^2j + z^2k$ and C is the boundary of the part of the plane $x + y + z = 1$ lying in the first octant.

49. Compute $\iint_S \text{curl } \mathbf{F} \cdot \mathbf{n} \, d\sigma$, where S is the unit sphere and $\mathbf{F} = e^{xy}\mathbf{i} + \tan^{-1} z\mathbf{j} + (x + y + z)^{7/3} z^2\mathbf{k}$.

In Exercises 50–55, evaluate the surface integral by using the divergence theorem.

50. $\iint_S \mathbf{F} \cdot \mathbf{n} \, d\sigma$, where $\mathbf{F} = x\mathbf{i} + 2y\mathbf{j} + 3z\mathbf{k}$ and S is the unit sphere.

51. $\iint_S \mathbf{F} \cdot \mathbf{n} \, d\sigma$, where $\mathbf{F} = ax\mathbf{i} + by\mathbf{j} + cz\mathbf{k}$ and S is the unit sphere.

52. $\iint_S \mathbf{F} \cdot \mathbf{n} \, d\sigma$, where $\mathbf{F}$ is as in Exercise 50 and S is the cylinder $x^2 + y^2 = 9, 0 \le z \le 6$.

53. $\iint_S \mathbf{F} \cdot \mathbf{n} \, d\sigma$, where $\mathbf{F} = (y^3 - z)\mathbf{i} + x^2 e^z\mathbf{j} + \sin xy\mathbf{k}$ and S is the ellipsoid $(x^2/4) + (y^2/16) + (z^2/25) = 1$.

54. $\iint_S \mathbf{F} \cdot \mathbf{n} \, d\sigma$, where $\mathbf{F}$ is as in Exercise 50 and S is the boundary of the unit cube.

55. $\iint_S \mathbf{F} \cdot \mathbf{n} \, d\sigma$, where $\mathbf{F} = xz\mathbf{i} + xy\mathbf{j} + xyz\mathbf{k}$ and S is the unit cube in the first octant.

In Exercises 56–63, compute the Jacobian of the given transformation.

56. $x = u + 2v, \quad y = 2u - v$

57. $x = u^3 - v^3, \quad y = u^3 + v^3$

58. $x = v \ln u, \quad y = u \ln v$

59. $x = ve^u, \quad y = ue^v$

60. $x = \dfrac{u}{v}, \quad y = \dfrac{v}{u}$

61. $x = v \tan u, \quad y = \tan uv$

62. $x = u + v + w, \quad y = u - 2v + 3w, \quad z = -2u + v - 5w$

63. $x = vw, \quad y = uw, \quad z = uv$

64. Transform the integral $\int_0^1 \int_x^1 xy \, dx \, dy$ by making the transformation $x = u + v, y = u - v$.

65. Transform $\iint_\Omega e^{(x-y)/(x+y)} \, dx \, dy$, where Ω is the region in the first quadrant between the lines $x + y = 2$ and $x + y = 3$, by making the transformation $u = x - y$ and $v = x + y$. Then evaluate the integral.

chapter 11

Fourier Series and Boundary Value Problems

11.1 Introduction

Until now we have considered differential equations whose solutions are determined by initial conditions. In this chapter we shall discuss another equally important way to specify a particular solution of a differential equation, namely, specifying certain values of the function and its derivatives at two or more points. The differential equation together with the given conditions at two or more points is called a *boundary value problem*. (This was defined earlier in Section 1.3.)

In this section we shall present some examples of boundary value problems, together with a discussion of some of the powerful methods that are used to solve them. We shall see how the concepts of eigenvalues and eigenfunctions arise naturally in these situations and discuss the importance of Fourier series techniques in solving nonhomogeneous boundary value problems.

Since a boundary value problem requires that conditions be given at two or more points, we can assume the differential equation involved to be at least of second order, because a first-order equation is usually completely specified by a single condition. (This follows from the existence-uniqueness results cited in Chapter 2.)

EXAMPLE 1

Consider the simple harmonic motion of a mass m attached to a coiled spring with spring constant k. Applying Newton's law of motion and Hooke's law to this situation, we obtained in Section 2.8 [see p. 106], the differential equation

$$x'' + p^2 x = 0, \qquad p^2 = \frac{k}{m} > 0, \tag{1}$$

where $x = x(t)$ denotes the displacement at time t of the mass from its equilibrium position. Suppose the mass is initially ($t = 0$) at its equilibrium position when it is given

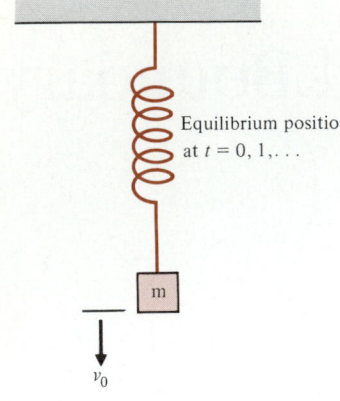

Equilibrium position
at $t = 0, 1, \ldots$

m

v_0

FIGURE 1

Eigenfunctions
Eigenvalues

an unknown initial velocity $v_0 > 0$. How can we guarantee that the mass will again be at its equilibrium position after precisely one second? Obviously, such a mechanism provides a simple timepiece. (See Fig. 1.)

Our purpose now is to discover some of the properties of the unknown solution $x(t)$. If the time is measured in seconds, then we have the boundary conditions

$$x(0) = x(1) = 0. \tag{2}$$

The boundary conditions together with the differential equation (1) specify the boundary value problem we must solve.

We recall from Chapter 2 that to find the general solution of the differential equation (1) we need only solve the characteristic equation

$$\lambda^2 + p^2 = 0.$$

Since the roots of the characteristic equation are $\lambda = \pm ip$, the general solution of equation (1) is given by

$$x(t) = A \cos pt + B \sin pt. \tag{3}$$

Setting $t = 0$ in equation (3) and using the boundary condition $x(0) = 0$, we observe that $A = 0$, so that the solution must consist only of multiples of $\sin pt$. For the boundary condition $x(1) = 0$ to hold we must also have

$$B \sin p = x(1) = 0.$$

There are two cases to consider: either $B = 0$ or $\sin p = 0$. If $B = 0$, then the solution (3) is constantly zero, which means that the mass remains at its equilibrium position at all times t. This is impossible, since the mass was given an initial velocity $v_0 > 0$. Therefore, $\sin p = 0$, which is possible only if p is a nonzero multiple of π:

$$p = \pm \pi, \pm 2\pi, \pm 3\pi, \ldots.$$

[If $p = 0$, we see that (3) is again constantly zero.] Thus we have arrived at a surprising conclusion: there are infinitely many solutions of the form

$$x(t) = B \sin (n\pi)t, \tag{4}$$

where n is a nonzero integer, for the boundary value problem (1), (2). Furthermore, the constant B cannot be specified without additional information. Indeed, B depends on v_0 and n, since

$$v_0 = x'(0) = n\pi B$$

or $B = v_0/n\pi$.

Actually, in doing the calculations above, we used one fact that was not part of the boundary value problem (1), (2). This was the assumption, given in the statement of the problem, that $v_0 > 0$. This assumption was used to disallow the *trivial solutions* $x(t) \equiv 0$ for all t of the boundary value problem (1), (2). However, the results we have obtained illustrate one of the basic facts about boundary value problems: *Nontrivial solutions will exist only for certain values of the parameter p.* All such nontrivial solutions (4) are called **eigenfunctions** of the problem, and the corresponding values of p^2 that yield these eigenfunctions are called **eigenvalues** of the problem. Rephrasing our work in this context, we see that

Eigenfunctions: $\quad x(t) = \left(\dfrac{v_0}{n\pi}\right) \sin n\pi t \tag{5}$

is an eigenfunction of this problem corresponding to the eigenvalue

$$\text{Eigenvalues:} \qquad \frac{k}{m} = p^2 = (n\pi)^2. \tag{6}$$

It is interesting to discover the physical significance of this solution. Suppose we are given a spring with (fixed) spring constant k. Solving equation (6) for m, we obtain the masses

$$m_n = \frac{k}{n^2\pi^2}, \qquad n = 1, 2, 3, \ldots, \tag{7}$$

each corresponding to a different value of n. Since k and π are fixed quantities, it is *only* for these masses that a solution of the boundary value problem will exist. For any other choice, no matter what initial velocity is given, the mass will not be back at its equilibrium position after exactly one second. Thus the construction of our clock depends on the selection of one of the masses (7). Furthermore, equation (5) indicates that if the mass m_n is chosen, it will be at its equilibrium position n times during the time period $0 < t \leq 1$. Since increasing n decreases both the mass m_n and the amplitude ($v_0/n\pi$) of the oscillations, we arrive at the conclusion that *the lighter the mass, the more rapid the oscillation.* This fact is easy to observe physically.

We stress an important point: In the mechanical problems we discussed in Chapters 1–3, we were asked to solve an initial value problem. We knew that for any mass m and spring constant k, the problem had a unique solution if the initial mass, $x(0)$, and velocity, $x'(0)$, were specified.

Here the situation is very different. The boundary value problem may have nontrivial solutions only for particular values of m and k. Thus, in a sense, the constants m and k appear as variables. Of course, any particular mass may be placed on any particular spring, however, the boundary conditions $x(0) = x(1) = 0$ will be satisfied only for some masses on some springs.

EXAMPLE 2

A population is subject to an influenza epidemic. Let $s(t)$ and $i(t)$ be, respectively, the numbers of susceptible and infected individuals at time t. Suppose that the rate of change of the number of susceptible individuals is proportional to the number of infected individuals, whereas the rate of change of the number of infected individuals is proportional to the number of susceptible individuals. (The difference between the changes in the numbers of infected and susceptible individuals consists of those who have gained immunity or died.) The epidemic lasts a certain period of time, say two months. [With this assumption, we can set $i(0) = i(2) = 0$.] We wish to discover the number of infected individuals at all time t (in months).

From the hypotheses above, we obtain the system of equations

$$\frac{ds}{dt} = -pi,$$

$$\frac{di}{dt} = qs, \tag{8}$$

where p and q are unknown positive constants of proportionality. Differentiating the second of the equations in the system (8) with respect to t, and substituting the first equation for ds/dt, we obtain the second-order equation

$$\frac{d^2i}{dt^2} = q\frac{ds}{dt} = -qpi$$

or

$$i'' + k^2 i = 0, \qquad k^2 = pq, \tag{9}$$

with the boundary conditions

$$i(0) = i(2) = 0. \tag{10}$$

This boundary value problem is very similar to that of Example 1. We again find the general solution

$$i(t) = A \cos kt + B \sin kt.$$

Setting $t = 0$ and using the first boundary condition, we find that $0 = i(0) = A$. The second boundary condition $t = 2$ implies that

$$B \sin 2k = i(2) = 0.$$

Since we are not interested in the trivial solution $i(t) \equiv 0$, it follows that $B \neq 0$, so that $2k$ must be a nonzero multiple of π. Hence the eigenvalues of this problem are

$$k^2 = \left(\frac{\pi}{2}\right)^2, (\pi)^2, \left(\frac{3\pi}{2}\right)^2, \ldots, \left(\frac{n\pi}{2}\right)^2, \ldots,$$

and the corresponding eigenfunctions $i_k(t)$ are given by

$$i_k(t) = B \sin \frac{n\pi t}{2}, \qquad k^2 = \left(\frac{n\pi}{2}\right)^2. \tag{11}$$

If the number of infected individuals is assumed to be positive during the entire epidemic $0 < t < 2$, then $n = 1$. This is a reasonable limitation since negative numbers of infected individuals do not have a physical meaning.

EXAMPLE 3

We return to Example 1, but assume that the spring-mass system is also subject to a periodic external force $K \sin \omega t$, where K, $\omega > 0$. Using the theory of forced vibrations developed in Section 2.10, we obtain the nonhomogeneous second-order equation

$$m\frac{d^2x}{dt^2} = -kx + K \sin \omega t.$$

Using the same boundary conditions as before, we are led to the nonhomogeneous boundary value problem

$$x'' + \frac{k}{m}x = \frac{K}{m} \sin \omega t, \qquad x(0) = x(1) = 0. \tag{12}$$

There are two cases to consider: either $\omega = n\pi$ for some integer n or $\omega \neq n\pi$ for every integer n. We assume the latter condition and leave it to the reader to find out what happens if $\omega = n\pi$ for some integer n (see Problem 12).

First, we use the method of undetermined coefficients (Section 2.4) to find a particular solution. We suppose that

$$x(t) = A \sin \omega t + B \cos \omega t.$$

Since $x(t)$ must satisfy the boundary condition $x(0) = 0$, we see that $0 = x(0) = B$. Hence, we have

$$x(t) = A \sin \omega t. \tag{13}$$

Differentiating (13) twice and substituting into (12) we obtain

$$-A\omega^2 \sin \omega t + \frac{k}{m} A \sin \omega t = \frac{K}{m} \sin \omega t.$$

Canceling the $\sin \omega t$ on both sides of this equation and solving for A, we obtain

$$A = \frac{K/m}{(k/m) - \omega^2} = \frac{K}{k - m\omega^2}.$$

The particular function we have obtained by the method of undetermined coefficients,

$$x_p(t) = \frac{K}{k - m\omega^2} \sin \omega t, \tag{14}$$

does not satisfy the boundary condition $x_p(1) = 0$. To see this observe that

$$x_p(1) = \frac{K}{k - m\omega^2} \sin \omega \neq 0,$$

because $\sin \omega = 0$ only when $\omega = n\pi$ for some integer n.

So what can we do? What we do next is one reason we study Fourier series.

STEP 1
Observe that $x_n(t) = \sin n\pi t$ satisfies $x(0) = x(1) = 0$ [$\sin n\pi t$ is an eigenfunction of the homogeneous problem (1), (2)].

STEP 2
Try to form an infinite linear combination of the functions $\sin n\pi t$ to obtain a function that solves (12).

We shall see using the methods in Section 11.4 that the trigonometric series

$$2\pi \sin \omega \sum_{n=1}^{\infty} \frac{(-1)^n n}{\omega^2 - (n\pi)^2} \sin n\pi t, \tag{15}$$

called a *Fourier series*, converges to the function $\sin \omega t$ at all points in $0 \leq t < 1$, and to zero at $t = 0$ and $t = 1$. Making use of this fact, we seek a solution of the form

$$x(t) = 2\pi \sin \omega \sum_{n=1}^{\infty} \frac{(-1)^n n c_n}{\omega^2 - (n\pi)^2} \sin n\pi t. \tag{16}$$

If (16) is substituted into (12), then it is possible to show that $c_n = K/(k - mn^2\pi^2)$. With these values of c_n we can obtain the general solution to (12).

Summary

(i) Linear boundary value problems like (1), (2) or (9), (10) usually have solutions for specified values of the parameter in the equation. These values are called *eigenvalues* and the corresponding solutions are called *eigenfunctions.*

(ii) We can solve certain nonhomogeneous boundary value problems by writing solutions as infinite series or **Fourier series** of eigenfunctions of the corresponding homogeneous problems.

At this point you may wonder whether very many functions can be represented as a Fourier series. The question is significant, since forcing functions can be quite arbitrary. Actually, we shall find that the class of functions that have Fourier series representations includes almost every practical forcing function. Although the Fourier series (16) may appear complicated, we shall discover that it is really quite easy to work with once we have become more familiar with it. We shall see that the *n*th term in the series behaves essentially like the *n*th coordinate of a vector. This phenomenon, called *orthogonality,* will be one of the central themes of this chapter.

In the next three sections we shall discuss how to express many different kinds of functions as Fourier series. In Section 11.8 we shall solve a number of nonhomogeneous boundary value problems.

Two things should be added here. First, many different kinds of nonlinear boundary value problems arise in applications. These are more difficult to solve and we shall not do so in this introductory text.

Finally, it should be noted that boundary value problems can involve differential equations of order greater than two. For example, in studying the deflection of a beam, one will arrive at the fourth-order boundary value problem

$$\frac{d^4 y}{dx^4} = k^2 y,$$

with specified boundary conditions $y(0)$, $y'(0)$, $y(L)$, and $y'(L)$, where L is the length of the beam.

Problems 11.1

1. What eigenfunctions and eigenvalues will result in Example 1 if the boundary condition $x(1) = 0$ is replaced by $x(T) = 0$, where T is some fixed nonzero constant?

2. Consider the boundary value problem

$$\frac{d^2 x}{dt^2} - kx(t) = 0, \quad x(0) = x(1) = 0,$$

where k is any real number. What are the eigenvalues and eigenfunctions of this problem?

3. Answer Problem 2 given that the boundary condition $x(1) = 0$ is replaced by $x(T) = 0$ for some fixed nonzero constant T.

4. Answer Problem 2 given that the boundary condition $x(0) = 0$ is replaced by $x(-1) = 0$, all other conditions remaining the same.

5. Find the eigenfunctions and eigenvalues of the boundary value problem

$$x''(t) + k^2 x(t) = 0, \quad x(-T) = x(T) = 0,$$

where T is a fixed nonzero constant.

6. Find the eigenvalues and eigenfunctions of

$$x''(t) + k^2 x(t) = 0, \qquad x(0) = x'(1) = 0.$$

7. Answer Problem 6 if the condition $x'(1) = 0$ is replaced by $x'(T) = 0$ for some fixed nonzero constant T.

8. Answer Problem 6 if the conditions $x(0) = x'(1) = 0$ are replaced by $x'(0) = x(1) = 0$.

9. Answer Problem 6 if the conditions $x(0) = x'(1) = 0$ are replaced by $x'(0) = x'(1) = 0$.

* 10. Find the eigenvalues and eigenvectors of the boundary value problem

$$(x^2 y')' + \lambda y = 0, \qquad y(1) = y'(e) = 0.$$

* 11. Answer Problem 10 if the boundary conditions are replaced by $y'(1) = y(e) = 0$.

* 12. In Example 3, show that if $\omega = n_0 \pi$ and $m \neq k/n_0^2 \pi^2$, then the boundary value problem (12) has the general solution

$$x(t) = A \sin n\pi t + \frac{Kn^2}{k(n^2 - n_0^2)} \sin n_0 \pi t.$$

11.2 Orthogonal Sets of Functions

Let $f_1(x)$ and $f_2(x)$ be two functions that are continuous on the interval $[a, b]$.

DEFINITION

Orthogonal Functions

> $f_1(x)$ and $f_2(x)$ are said to be **orthogonal** on $[a, b]$ if
>
> $$\int_a^b f_1(x) f_2(x) \, dx = 0. \tag{1}$$

DEFINITION

Norm

The **norm** of f_1, denoted $\|f_1\|$, is defined by

$$\|f_1\| = \left(\int_a^b f_1^2(x) \, dx \right)^{1/2}. \tag{2}$$

DEFINITION

Orthogonal and Orthonormal Set of Functions

The infinite set of continuous functions $f_1(x), f_2(x), f_3(x), \ldots$ is said to be **orthogonal** on $[a, b]$ if $f_n(x)$ and $f_m(x)$ are orthogonal whenever $n \neq m$. The set is **orthonormal** if it is orthogonal and $\|f_n\| = 1$ for $n = 1, 2, 3, \ldots$.

EXAMPLE 1

The functions $f_n(x) = \sin nx$, $n = 1, 2, 3, \ldots$, are an orthogonal set of functions on the interval $-\pi \leq x \leq \pi$. To see this, we compute, for $n \neq m$,

Entry 103 in Appendix 1

$$\int_{-\pi}^{\pi} f_n(x)f_m(x)\,dx = \int_{-\pi}^{\pi} \sin nx \, \sin mx \, dx = \left[\frac{\sin(n-m)x}{2(n-m)} - \frac{\sin(m+n)x}{2(m+n)}\right]\Big|_{-\pi}^{\pi} = 0,$$

since $\sin k\pi = 0$ for any integer k. Also

Entry 100

$$\|f_n\| = \left[\int_{-\pi}^{\pi} \sin^2 nx\,dx\right]^{1/2} = \left[\left(\frac{x}{2} - \frac{\sin 2nx}{4n}\right)\Big|_{-\pi}^{\pi}\right]^{1/2} = \sqrt{\pi}.$$

Thus the functions

$$\frac{\sin x}{\sqrt{\pi}}, \frac{\sin 2x}{\sqrt{\pi}}, \frac{\sin 3x}{\sqrt{\pi}}, \ldots$$

form an orthonormal set over the interval $[-\pi, \pi]$.

EXAMPLE 2

The functions

$$\frac{1}{\sqrt{2\pi}}, \frac{\cos x}{\sqrt{\pi}}, \frac{\sin x}{\sqrt{\pi}}, \frac{\cos 2x}{\sqrt{\pi}}, \frac{\sin 2x}{\sqrt{\pi}}, \ldots$$

form an orthonormal set of functions on the interval $-\pi \leq x \leq \pi$. We have already demonstrated the orthonormality of the $\sin nx/\sqrt{\pi}$ terms. For $n \neq m$

Entry 118

$$\int_{-\pi}^{\pi} \frac{\cos nx}{\sqrt{\pi}} \frac{\cos mx}{\sqrt{\pi}}\,dx = \frac{1}{\pi}\left[\frac{\sin(n-m)x}{2(n-m)} + \frac{\sin(n+m)x}{2(n+m)}\right]\Big|_{-\pi}^{\pi} = 0$$

and

Entry 124

$$\int_{-\pi}^{\pi} \frac{\sin mx}{\sqrt{\pi}} \frac{\cos nx}{\sqrt{\pi}}\,dx = \frac{1}{\pi}\left[\frac{-\cos(m-n)x}{2(m-n)} - \frac{\cos(m+n)x}{2(m+n)}\right]\Big|_{-\pi}^{\pi} = 0$$

because $\cos[(m \pm n)\pi] = \cos[(m \pm n)(-\pi)]$. For $m = n$

Entry 115

$$\int_{-\pi}^{\pi} \frac{\cos^2 nx}{\pi}\,dx = \frac{1}{\pi}\left[\frac{x}{2} + \frac{\sin 2nx}{4n}\right]\Big|_{-\pi}^{\pi} = 1,$$

but

Entry 123

$$\int_{-\pi}^{\pi} \frac{\cos nx}{\sqrt{\pi}} \frac{\sin nx}{\sqrt{\pi}}\,dx = \frac{1}{\pi}\left[\frac{\sin^2 nx}{2n}\right]\Big|_{-\pi}^{\pi} = 0.$$

Also,

$$\int_{-\pi}^{\pi} \frac{1}{\sqrt{2\pi}} \frac{\cos nx}{\sqrt{\pi}}\,dx = \frac{1}{\pi\sqrt{2}} \frac{\sin nx}{n}\Big|_{-\pi}^{\pi} = 0$$

and

$$\int_{-\pi}^{\pi} \frac{1}{\sqrt{2\pi}} \frac{\sin nx}{\sqrt{\pi}}\,dx = \frac{1}{\pi\sqrt{2}}\left(\frac{-\cos nx}{n}\right)\Big|_{-\pi}^{\pi} = \frac{1}{n\pi\sqrt{2}}[\cos(-n\pi) - \cos n\pi] = 0.$$

Finally,

$$\left\|\frac{1}{\sqrt{2\pi}}\right\| = \left[\int_{-\pi}^{\pi}\left(\frac{1}{\sqrt{2\pi}}\right)^2 dx\right]^{1/2} = \left[\frac{1}{2\pi}\int_{-\pi}^{\pi} dx\right]^{1/2} = \left(\frac{1}{2\pi}\cdot 2\pi\right)^{1/2} = 1$$

and

$$\left\|\frac{\cos nx}{\sqrt{\pi}}\right\| = \left[\frac{1}{\pi}\int_{-\pi}^{\pi}\cos^2 nx\,dx\right]^{1/2} \overset{\text{Entry 115}}{\underset{\downarrow}{=}} \left[\frac{1}{\pi}\left(\frac{x}{2}+\frac{\sin 2nx}{4n}\right)\Big|_{-\pi}^{\pi}\right]^{1/2} = 1.$$

Some important sets of functions $\{f_n(x)\}$ that occur in applications are not orthogonal, but have the property that for some nontrivial continuous nonnegative function $w(x)$ the integral

$$\int_a^b w(x)f_n(x)f_m(x)\,dx = 0 \qquad\qquad (3)$$

Orthogonality and Orthonormality with Respect to a Weight Function

whenever $n \neq m$. If this happens, the set $\{f_n(x)\}$ is said to be **orthogonal with respect to the weight function** $w(x)$ **on the interval** $a \leq x \leq b$. The **weighted norm** of f_n is defined as

$$\|f_n\|_w = \left[\int_a^b w(x)f_n(x)^2\,dx\right]^{1/2}, \qquad\qquad (4)$$

and $\{f_n(x)\}$ is said to be **orthonormal with respect to the weight function** $w(x)$, if in addition to being orthogonal with respect to $w(x)$, it satisfies the condition $\|f_n\|_w = 1$ for all n.

EXAMPLE 3

Consider **Laguerre's equation**†

$$(xe^{-x}y')' + ne^{-x}y = 0. \qquad\qquad (5)$$

This equation was discussed in Problem 5.5.10 on p. 288. [Equation (5) is obtained from equation (5.5.16) by multiplying both sides of (5.5.16) by e^{-x}.] In Problem 5.5.10 you were asked to show that for each integer n, (5) has a polynomial solution of degree n, called a **Laguerre polynomial**. For each n, denote the corresponding Laguerre polynomial by $y_n(x)$. Consider the set of Laguerre polynomials $\{y_n(x)\}$, $n = 1, 2, 3, \ldots$,

---

† Edmond Laguerre (1834–1886), a French mathematician, made many contributions to geometry and the theory of infinite series.

and suppose $m \neq n$. Then we have the equations

$$(xe^{-x}y'_n)' + ne^{-x}y_n = 0,$$

$$(xe^{-x}y'_m)' + me^{-x}y_m = 0.$$

We multiply the first equation by y_m and the second by $-y_n$ and add them to obtain

$$y_m(xe^{-x}y'_n)' - y_n(xe^{-x}y'_m)' = e^{-x}(m-n)y_ny_m. \tag{6}$$

Integrating both sides of equation (6) on the interval $0 \leqslant x \leqslant \infty$, we have

$$\int_0^\infty [y_m(xe^{-x}y'_n)' - y_n(xe^{-x}y'_m)'] \, dx = (m-n)\int_0^\infty e^{-x}y_ny_m \, dx. \tag{7}$$

But by the product formula for differentiation,

$$[xe^{-x}(y_my'_n - y_ny'_m)]' = (xe^{-x}y'_n)'y_m + xe^{-x}y'_ny'_m - (xe^{-x}y'_m)'y_n - xe^{-x}y'_my'_n$$

$$= y_m(xe^{-x}y'_n)' - y_n(xe^{-x}y'_m)'.$$

Thus equation (7) becomes

$$(m-n)\int_0^\infty e^{-x}y_ny_m \, dx = xe^{-x}(y_my'_n - y_ny'_m)\Big|_0^\infty = 0, \quad m \neq n,$$

because $f(x) = x(y_my'_n - y_ny'_m)$ is a polynomial (since each term is a polynomial) so that the quotient $f(x)/e^x$ tends to zero as $x \to \infty$. This may be verified by applying L'Hôpital's rule as many times as the degree of f to the quotient $f(x)/e^x$:

$$\lim_{x \to \infty} \frac{f(x)}{e^x} = \lim_{x \to \infty} \frac{f'(x)}{e^x} = \cdots = \lim_{x \to \infty} \frac{c}{e^x} = 0.$$

Therefore, the set of functions $\{y_n\}$, $n = 1, 2, \ldots$, is orthogonal with respect to the weight function e^{-x} on $0 \leqslant x < \infty$.

Problems 11.2

In Problems 1–8, show that each given set of functions is orthogonal on the given interval and determine the corresponding orthonormal set.

1. $\{\cos nx\}, \quad n = 0, 1, 2, \ldots; \quad 0 \leqslant x \leqslant 2\pi$

2. $\left\{\sin \dfrac{n\pi x}{T}\right\}, \quad n = 1, 2, 3, \ldots; \quad -T \leqslant x \leqslant T$

3. $\left\{\cos \dfrac{2n\pi x}{T}\right\}, \quad n = 0, 1, 2, \ldots; \quad 0 \leqslant x \leqslant T$

4. $\{\sin 2nx\}, \quad n = 1, 2, 3, \ldots; \quad 0 \leqslant x \leqslant \pi$

5. $\{\cos 2nx\}, \quad n = 0, 1, 2, \ldots; \quad 0 \leqslant x \leqslant \pi$

6. $\{\sin 3nx\}, \quad n = 1, 2, 3, \ldots; \quad -\pi \leqslant x \leqslant \pi$

7. $\{\cos 3nx\}, \quad n = 0, 1, 2, \ldots; \quad |x| \leqslant \pi$

8. $\{\sin 2nx, \cos 2nx\}, \quad n = 1, 2, 3, \ldots; \quad |x| \leqslant \pi$

9. **Hermite polynomials.†** The functions

$$H_0 = 1, \quad H_n(x) = (-1)^n e^{x^2} \frac{d^n}{dx^n} e^{-x^2}, n = 1, 2, 3, \ldots,$$

are called **Hermite polynomials** (see Problem 5.5.9 on p. 288). Prove that:

(a) $H_1(x) = 2x, \quad H_2(x) = 4x^2 - 2,$
    $H_3(x) = 8x^3 - 12x.$

(b) The Hermite polynomials satisfy the relation

$$H_{n+1}(x) = 2xH_n(x) - H'_n(x).$$

† Charles Hermite (1822–1901), a French mathematician, was known for his work in number theory.

(c) $H_n(x)$ is a solution of **Hermite's equation**

$$y'' - 2xy' + 2ny = 0.$$

(d) The set of functions $\{H_n(x)\}$ is orthogonal with respect to the weight function $\exp(-x^2)$ on the interval $-\infty < x < \infty$.

* 10. The norm of a function $f(x)$ on an interval $a \le x \le b$,

$$\|f\| = \left(\int_a^b f^2(x)\, dx \right)^{1/2},$$

is a generalization of the notion of the distance of a point (x, y) in the plane from the origin:

$$d(x, y) = \sqrt{x^2 + y^2}.$$

Assume that f and g are continuous functions on $a \le x \le b$. Prove that the norm satisfies the following three properties:

(a) $\|f\| \ge 0$, where the equality holds if and only if $f(x) \equiv 0$;

(b) $\|af\| = |a| \cdot \|f\|$ for any constant a;

(c) $\|f + g\| \le \|f\| + \|g\|$ (triangle inequality).

[*Hint:* (a) Use the fact that if $f(x_0) \ne 0$, then by continuity $f^2(x) \ge \epsilon > 0$ for all x in some interval $x_0 - \delta \le x \le x_0 + \delta$, where $\delta > 0$. (c) Show that for any two real numbers α and β

$$0 \le \|\alpha f - \beta g\|^2 = \alpha^2 \|f\|^2 - 2\alpha\beta (f, g) + \beta^2 \|g\|^2.$$

Then let $\alpha = \|g\|$ and $\beta = \|f\|$, showing that

$$(f, g) = \int_a^b f(x) g(x)\, dx \le \|f\| \cdot \|g\|.$$

This is called **Cauchy's inequality.** Finally, use the last inequality to show that

$$\|f + g\|^2 \le (\|f\| + \|g\|)^2.]$$

11.3 The Fourier Representation of a Periodic Function

Sound travels in waves. Some sound waves can be described mathematically by the equation of simple harmonic motion (see p. 107)

$$x(t) = A_1 \sin(\omega_0 t + \varphi_1),$$

where $x(t)$ is the displacement from equilibrium, A_1 is the amplitude, $\omega_0/2\pi$ is the frequency (in cycles per second), and φ_1 is the initial phase. We can write this equation as

$$x(t) = A_1 \sin \omega_0 t \cos \varphi_1 + A_1 \sin \varphi_1 \cos \omega_0 t$$

or

$$x(t) = a_1 \cos \omega_0 t + b_1 \sin \omega_0 t$$

where $a_1 = A \sin \varphi_1$ and $b_1 = A \cos \varphi_1$.

When the middle C is struck on a piano, the basic or fundamental frequency of the tone emitted is 264 cycles/sec. However, other tones are emitted as well, because of resonance. These tones, called **overtones** or **harmonics,** have frequencies which are integral multiples of the basic frequency $\dfrac{\omega_0}{2\pi}, \dfrac{2\omega_0}{2\pi}, \dfrac{3\omega_0}{2\pi}, \dfrac{4\omega_0}{2\pi}, \ldots$. The richness of musical sound is related to the number of harmonics that can be detected by the human ear. The larger the amplitude of each harmonic, the more likely it is to be detected.

A tone of frequency $n\omega_0/2\pi$ can be described by

$$x(t) = A_n \sin(n\omega_0 t + \varphi_n) = a_n \cos n\omega_0 + b_n \sin n\omega_0.$$

Thus a sound wave, together with all its harmonics, can be represented as the infinite sum

$$x(t) = a_1 \cos \omega_0 t + b_1 \sin \omega_0 t + a_2 \cos 2\omega_0 t + b_2 \sin 2\omega_0 t + \cdots$$

$$= \sum_{n=1}^{\infty} (a_n \cos n\omega_0 t + b_n \sin n\omega_0 t).$$

The series written above is called a *trigonometric series*.

We shall see in this section that most periodic functions can be written as trigonometric series. When this is done, we will be able to see the harmonics of the function. This has significance in many different kinds of physical problems.

DEFINITION

Periodic Function

The function $f(x)$ is said to be **periodic of period** T if there is a number $T > 0$ such that

$$f(x + T) = f(x) \qquad \text{for every real number } x. \tag{1}$$

Period

The smallest positive number T for which (1) holds is called the **period** of f.

EXAMPLE 1

$\sin x$ is periodic of period 2π because $\sin(x + 2\pi) = \sin x$.

EXAMPLE 2

$\cos ax$ is periodic of period $2\pi/a$ because $\cos a(x + 2\pi/a) = \cos(ax + 2\pi) = \cos ax$.

EXAMPLE 3

The "square-wave" function sketched in Fig. 1 is periodic of period 2.

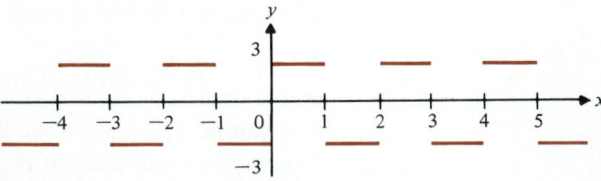

FIGURE 1

An infinite series of the form

$$\frac{a_0}{2} + \sum_{n=1}^{\infty} (a_n \cos nx + b_n \sin nx), \tag{2}$$

Trigonometric Series

where a_n and b_n are constants, is called a **trigonometric series.** The function defined by (2) is periodic of period 2π (at most). For example, for each n

$$\cos n(x + 2\pi) = \cos(nx + 2n\pi) = \cos nx \cos 2n\pi - \sin nx \sin 2n\pi = \cos nx.$$

Thus, if the series (2) converges to a periodic function $F(x)$ on the interval $[-\pi, \pi]$, then it converges for all real x, since

$$F(x) = \frac{a_0}{2} + \sum_{n=1}^{\infty} (a_n \cos nx + b_n \sin nx) \tag{3}$$

is a periodic function of period 2π.

In what follows we shall assume that $F(x)$ is **continuous** or **piecewise continuous** in an interval $[a, b]$. This means that F is continuous at all but a finite number of points in $[a, b]$ and that, at each point of discontinuity, the one-sided limits of F

$$F(x + 0) = \lim_{h \to 0^+} F(x + h) \quad \text{and} \quad F(x - 0) = \lim_{h \to 0^+} F(x - h)$$

both exist and are finite. If $F(x + 0) \neq F(x - 0)$ we say that F has a **jump discontinuity** at x (see Fig. 2a).

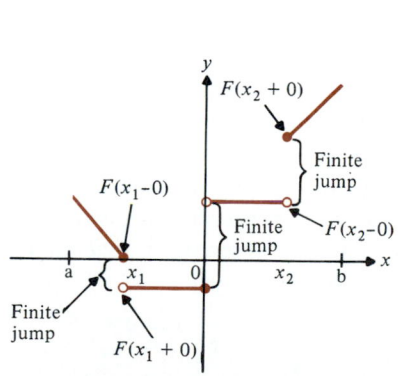

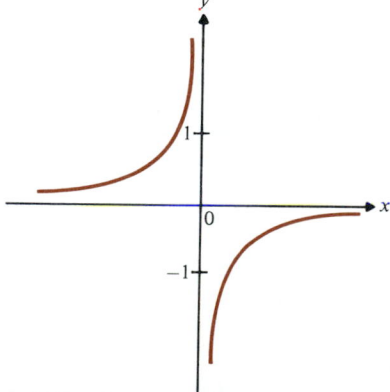

(a) Function piecewise continuous in $[a, b]$
Jump discontinuities at x_1, 0, and x_2

(b) Function not piecewise continuous in $[-1, 1]$; values become infinite as $x \to 0$

FIGURE 2

DEFINITION

Fourier Series
Fourier Coefficients
Euler Formulas

Let $F(x)$ be periodic of period 2π and be piecewise continuous in $[-\pi, \pi]$. Suppose $F(x)$ can be written as a trigonometric series (3). Then (3) is called a **Fourier series** for $F(x)$. The constants a_n and b_n are called the **Fourier coefficients** of $F(x)$ and are given by the **Euler formulas:**

$$a_n = \frac{1}{\pi} \int_{-\pi}^{\pi} F(x) \cos nx \, dx, \qquad b_n = \frac{1}{\pi} \int_{-\pi}^{\pi} F(x) \sin nx \, dx. \qquad \textbf{(4)}$$

Observe that since $\cos 0 = 1$

$$a_0 = \frac{1}{\pi} \int_{-\pi}^{\pi} F(x) \, dx$$

Courtesy of Historical Pictures Service

JEAN BAPTISTE JOSEPH FOURIER
(1768–1830)

One of the greatest mathematicians of the nineteenth century was the Frenchman Jean Baptiste Joseph Fourier.

Fourier was born in Auxerre in 1768 and died in Paris in 1830. The son of a tailor, he was orphaned at the age of eight and educated in a military school conducted by the Benedictines, where he was given a lectureship in mathematics before his 21st birthday. He assisted in the promotion of the French Revolution and was rewarded by a chair at the École Polytechnique. He resigned from this position so that he could accompany Napoleon on the Egyptian expedition. In 1798 he was appointed governor of Lower Egypt. Following the British victories and the capitulation of the French in 1801, Fourier returned to France and was made prefect of Grenoble. It was while at Grenoble that he started his experiments on heat.

In 1807, Fourier presented a paper before the French Academy of Sciences that initiated a new and highly fruitful chapter in the history of mathematics. The paper dealt with the practical problem of the flow of heat in metallic rods, plates, and solid bodies. In the course of the presentation of the paper, Fourier made the startling claim that any function, defined in a finite closed interval by an arbitrarily drawn graph, can be resolved into a sum of sine and cosine functions. To be more explicit, he claimed that *any* function whatever, no matter how capriciously it is defined in the interval $(-\pi, \pi)$, can be represented in that interval by

$$\frac{a_0}{2} + \sum_{n=1}^{\infty} (a_n \cos nx + b_n \sin nx),$$

where the a's and the b's are suitable real numbers. Such a series is known as a **trigonometric series,** and it was not new to the mathematicians of the time. Indeed, a number of more or less well behaved functions had been shown to be representable by such a series. But Fourier claimed that *any* function defined in $(-\pi, \pi)$ can be so represented. The scientists at the Academy were very skeptical of Fourier's claim, and the paper, which was

but $b_0 = 0$ since $\sin 0 = 0$. To see why the coefficients a_n and b_n are given by (4), observe that

$$\frac{1}{\pi} \int_{-\pi}^{\pi} F(x) \cos kx \, dx = \frac{1}{\pi} \int_{-\pi}^{\pi} \left(\frac{a_0}{2} + \sum_{n=1}^{\infty} a_n \cos nx + b_n \sin nx \right) \cos kx \, dx.$$

If term-by-term integration is permissible,† we have

† Most advanced calculus textbooks contain proofs that term-by-term integration is permissible if series (3) converges *uniformly* on $a \leq x \leq b$. Uniform convergence is discussed in Section 17.3.

judged by Lagrange, Laplace, and Legendre, was rejected. However, to encourage Fourier to develop his ideas more carefully, the French Academy made the problem of heat propagation the subject of a grand prize to be awarded in 1812. Fourier submitted a revised paper in 1811, which was judged by a group containing, among others, the former three judges, and the paper won the prize, though it was criticized for lack of rigor and so was not recommended for publication in the Academy's *Mémoires*.

Resentful, Fourier continued his researches on heat, and, in 1822, after a move to Paris in 1816, he published one of the great classics of mathematics, his *Théorie analytique de la chaleur* (Analytical Theory of Heat). Two years after the publication of his great work, Fourier became secretary of the French Academy, and, in that capacity, was able to have his 1811 paper published in its original form in the Academy's *Mémoires*.

Though it has been shown that Fourier's claim that *any* function can be represented by a trigonometric series (or **Fourier series,** as they are commonly called today) is too extravagant, the class of functions so representable is very broad indeed. The Fourier series have proved to be highly valuable in such fields of study as acoustics, optics, electrodynamics, thermodynamics, and many others.

Lord Kelvin (William Thomson, 1824–1907) claimed that his whole career in mathematical physics was influenced by Fourier's work on heat, and Clerk Maxwell (1831–1879) pronounced Fourier's treatise "a great mathematical poem."

A sad story is told about Fourier and his interest in heat. It seems that from his experience in Egypt, and maybe his work on heat, he became convinced that desert heat is the ideal condition for good health. He accordingly clothed himself in many layers of garments and lived in rooms of unbearably high temperature. It has been said by some that this obsession with heat hastened his death, by heart disease, so that he died, thoroughly cooked, in his sixty-third year.

Perhaps Fourier's most quoted sentence (it appeared in his early work on the mathematical theory of heat) is: "The deep study of nature is the most fruitful source of mathematical discovery."

$$\frac{1}{\pi} \int_{-\pi}^{\pi} F(x) \cos kx \, dx = \frac{a_0}{2\pi} \int_{-\pi}^{\pi} \cos kx \, dx$$

$$+ \sum_{n=1}^{\infty} \left(\frac{a_n}{\pi} \int_{-\pi}^{\pi} \cos nx \cos kx \, dx + \frac{b_n}{\pi} \int_{-\pi}^{\pi} \sin nx \cos kx \, dx \right). \qquad (5)$$

However, as we saw in Example 11.2.2, the functions

$$\frac{1}{\sqrt{2\pi}}, \quad \frac{\cos x}{\sqrt{\pi}}, \quad \frac{\sin x}{\sqrt{\pi}}, \quad \frac{\cos 2x}{\sqrt{\pi}}, \quad \frac{\sin 2x}{\sqrt{\pi}}, \ldots$$

form an orthonormal set of functions on the interval $-\pi \le x \le \pi$. Hence all of the integrals in (5) are zero except

$$\frac{a_k}{\pi} \int_{-\pi}^{\pi} \cos^2 kx \, dx = a_k, \text{ if } k \neq 0, \quad \text{or} \quad \frac{a_0}{2\pi} \int_{-\pi}^{\pi} 1 \, dx = a_0.$$

Thus the first of the Euler formulas has been verified. The second Euler formula is proved similarly.

Remark. There is nothing special about the period of $F(x)$. Suppose that instead of having period 2π, the function F has period T:

$$F(x + T) = F(x).$$

Then define the function

$$G(x) = F\left(\frac{xT}{2\pi}\right) = F(y),$$

where $y = xT/2\pi$, and note that

$$G(x + 2\pi) = F\left(\frac{(x + 2\pi)T}{2\pi}\right) = F\left(\frac{xT}{2\pi} + T\right) = F\left(\frac{xT}{2\pi}\right) = G(x).$$

If there is a Fourier series (2) that converges to $G(x)$, then

$$G(x) = \frac{a_0}{2} + \sum_{n=1}^{\infty} (a_n \cos nx + b_n \sin nx) \qquad (6)$$

on the interval $-\pi \le x \le \pi$; the substitution $x = 2\pi y/T$ can be used in (6) to obtain a Fourier series for $F(y)$:

$$F(y) = G(2\pi y/T) = \frac{a_0}{2} + \sum_{n=1}^{\infty} \left(a_n \cos \frac{2n\pi y}{T} + b_n \sin \frac{2n\pi y}{T} \right).$$

The change of variables above can also be used to show the orthogonality of the functions

$$1, \quad \cos \frac{2\pi y}{T}, \quad \sin \frac{2\pi y}{T}, \quad \cos \frac{4\pi y}{T}, \quad \sin \frac{4\pi y}{T}, \ldots$$

over the interval $|y| \le T/2$.

So far we have concentrated only on properties of the Fourier series (3). We have not discussed how to go about determining the Fourier series for a given periodic function $F(x)$. The task is not difficult: we merely compute the coefficients a_n and b_n using the Euler formulas (4) and substitute these values in (3). The result is called the **Fourier series corresponding to** $F(x)$. The remaining question is whether the Fourier series so obtained actually converges to $F(x)$. If the Fourier series does converge to $F(x)$, we call it a **representation** of $F(x)$.

The class of periodic functions $F(x)$ that can be represented by Fourier series is very large, so large, in fact, that it originally aroused a big controversy. The following theorem gives sufficient conditions for almost all conceivable practical applications.

■ **THEOREM 1: Fourier Convergence Theorem**

Let $F(x)$ be a periodic function with period T and such that $F(x)$ and $F'(x)$ are piecewise continuous on the interval $-T/2 \le x \le T/2$. Then $F(x)$ has a Fourier series

$$F(x) = \frac{a_0}{2} + \sum_{n=1}^{\infty} \left(a_n \cos \frac{2n\pi x}{T} + b_n \sin \frac{2n\pi x}{T} \right), \tag{7}$$

whose coefficients are given by the **Euler formulas**

$$a_n = \frac{2}{T} \int_{-T/2}^{T/2} F(x) \cos \frac{2n\pi x}{T} \, dx, \qquad n = 0, 1, 2, \ldots, \tag{8}$$

$$b_n = \frac{2}{T} \int_{-T/2}^{T/2} F(x) \sin \frac{2n\pi x}{T} \, dx, \qquad n = 1, 2, \ldots. \tag{9}$$

The Fourier series (7) converges to $F(x)$ at all points where F is continuous, and to $[F(x + 0) + F(x - 0)]/2$ at all points of jump discontinuity of F. ■

Remark. Note that $[F(x + 0) + F(x - 0)]/2$ is the average of the right- and left-hand limits at the point x. At any point of continuity, both of these values coincide, so (7) converges to $[F(x + 0) + F(x - 0)]/2$ for all x in the interval $-T/2 \le x \le T/2$. Note also that

$$a_0 = \frac{2}{T} \int_{-T/2}^{T/2} F(x) \, dx$$

since $\cos 0 = 1$.

Although the conditions given in this theorem guarantee the existence and convergence of a Fourier series for the function $F(x)$, they are *not* the most general of such conditions. Moreover, the convergence of a Fourier series to a function $F(x)$ does not imply that F satisfies the conditions given in this theorem. In summary, the conditions given in this theorem are neither necessary nor sufficient conditions. Furthermore, even with these limitations, the proof of the theorem as stated is too complicated to be presented here.†

Although the theory of Fourier series is complicated, the application of these series is not difficult. It should be clear from Theorem 1 that Fourier series apply to a much wider class of functions than Taylor series do, since discontinuous functions cannot have a Taylor series representation. It is also useful to consider some functions for which Theorem 1 *does not* apply.

EXAMPLE 4

The functions $F_1(x) = 1/x$ and
$$F_2(x) = (-1)^n, \qquad \frac{1}{n+1} < |x| \leq \frac{1}{n}, \qquad n = 1, 2, 3, \ldots,$$
do not satisfy the hypotheses of Theorem 1 in the interval $-1 \leq x \leq 1$, since neither function is piecewise continuous. The function F_1 has an infinite jump discontinuity at $x = 0$, whereas F_2 has an infinite number of discontinuities in $-1 \leq x \leq 1$.

Let us now illustrate the procedure involved in obtaining the Fourier series of a function.

EXAMPLE 5

Find the Fourier series of the periodic function (see Fig. 3)
$$f(x) = (-1)^n k, \qquad n < x < n + 1. \tag{10}$$

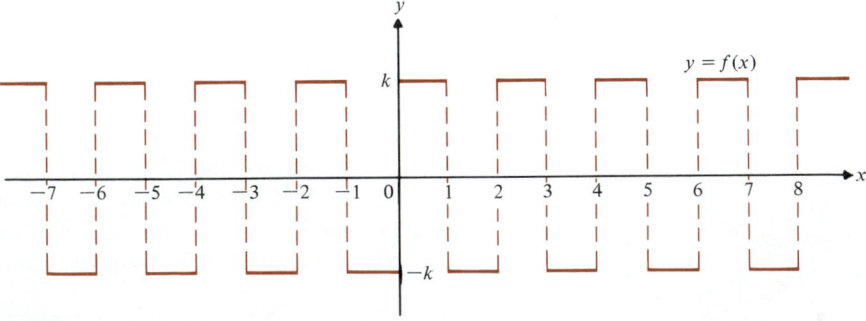

FIGURE 3

† Proofs of this theorem can be found in most books on advanced calculus or complex variables. See, for example, W. Kaplan, *Advanced Calculus,* Addison-Wesley, Reading, Mass., 1973, p. 484; or W. R. Derrick, *Complex Analysis and Applications,* 2nd ed., Wadsworth, Belmont, Calif., 1984, p. 267.

Functions of this type occur as off-on controls in mechanical systems. Observe that $f(x)$ has period $T = 2$; by equation (8), with $T = 2$,

$$a_n = \int_{-1}^{1} f(x) \cos n\pi x \, dx = -k \int_{-1}^{0} \cos n\pi x \, dx + k \int_{0}^{1} \cos n\pi x \, dx.$$

If $n = 0$, we obtain $a_0 = -k + k = 0$; and if $n \neq 0$,

$$a_n = \frac{-k}{n\pi} \sin n\pi x \Big|_{-1}^{0} + \frac{k}{n\pi} \sin n\pi x \Big|_{0}^{1} = 0.$$

Similarly, since $\cos(-t) = \cos t$,

$$b_n = -k \int_{-1}^{0} \sin n\pi x \, dx + k \int_{0}^{1} \sin n\pi x \, dx$$

$$= \frac{k}{n\pi} \cos n\pi x \Big|_{-1}^{0} - \frac{k}{n\pi} \cos n\pi x \Big|_{0}^{1} = \frac{2k}{n\pi}(1 - \cos n\pi),$$

which is zero for even n and equals $4k/n\pi$ for odd n. Thus

$$f(x) = \frac{4k}{\pi}\left(\sin \pi x + \frac{1}{3} \sin 3\pi x + \frac{1}{5} \sin 5\pi x + \cdots \right). \tag{11}$$

The graphs of $f(x)$ and the first three partial sums of equation (11) are shown in Fig. 4.

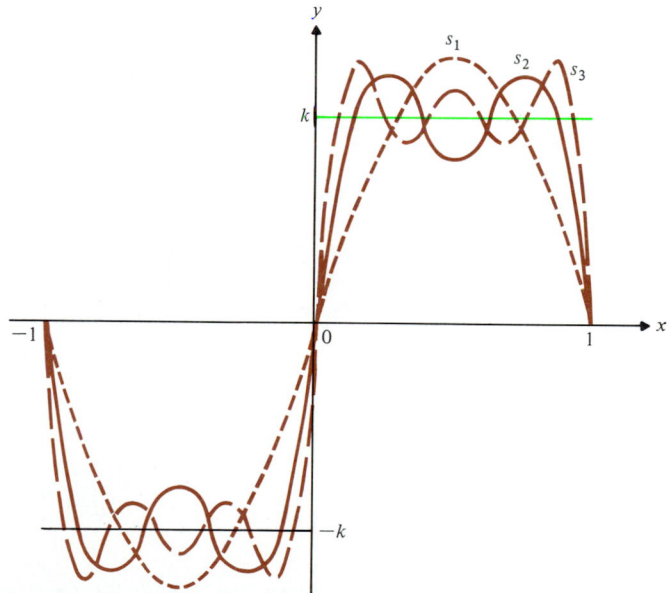

FIGURE 4

As a bonus for our previous work, note that if we set $x = \frac{1}{2}$ in (11), we obtain the series

$$k = \frac{4k}{\pi}\left(1 - \frac{1}{3} + \frac{1}{5} - \frac{1}{7} + \cdots \right),$$

or

$$\frac{\pi}{4} = 1 - \frac{1}{3} + \frac{1}{5} - \frac{1}{7} + \cdots,$$

since $\sin \pi/2 = 1$, $\sin 3\pi/2 = -1$, and so on. This is a famous result that Leibniz obtained by means of a complicated geometrical construction.

Odd Function

Also, observe that the function $f(x)$ in this example satisfies the condition $f(-x) = -f(x)$. Functions having this property are said to be **odd** functions. In particular, observe that all functions of the form $\sin n\pi x$ are odd functions.

EXAMPLE 6

Consider the sawtooth function

$$f(x) = \begin{cases} x+1, & -1 \leq x \leq 0, \\ -x+1, & 0 \leq x \leq 1, \end{cases} \qquad f(x+2) = f(x),$$

shown in Fig. 5a. Again $T = 2$ and

$$a_n = \int_{-1}^0 (x+1) \cos n\pi x \, dx + \int_0^1 (-x+1) \cos n\pi x \, dx.$$

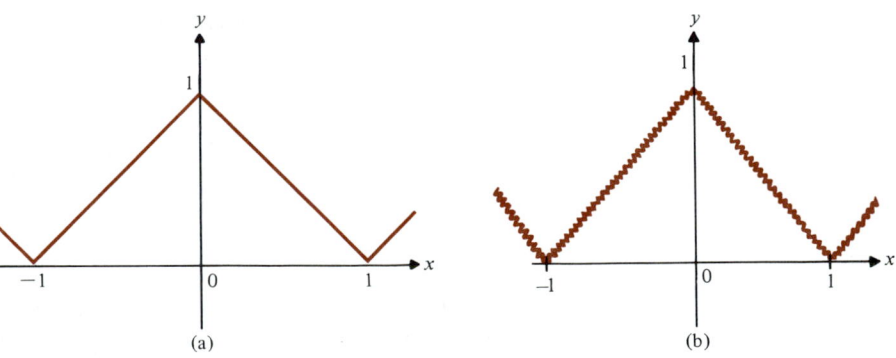

(a)   (b)

FIGURE 5

If $n = 0$, we have

$$a_0 = \int_{-1}^0 (x+1) \, dx + \int_0^1 (-x+1) \, dx = \left(\frac{x^2}{2} + x\right)\Big|_{-1}^0 + \left(x - \frac{x^2}{2}\right)\Big|_0^1 = 1,$$

and integrating by parts for $n \neq 0$, we obtain

$$a_n = \frac{(x+1)}{n\pi} \sin n\pi x \Big|_{-1}^0 - \frac{1}{n\pi} \int_{-1}^0 \sin n\pi x \, dx$$

$$+ \frac{(1-x)}{n\pi} \sin n\pi x \Big|_0^1 + \frac{1}{n\pi} \int_0^1 \sin n\pi x \, dx$$

$$= \frac{1}{(n\pi)^2} \cos n\pi x \Big|_{-1}^0 - \frac{1}{(n\pi)^2} \cos n\pi x \Big|_0^1 = \frac{2}{(n\pi)^2} (1 - \cos n\pi).$$

But $\cos n\pi = (-1)^n$. Thus a_n is zero for even $n \neq 0$ and equals $4/(n\pi)^2$ for odd n. Similarly,

$$b_n = \frac{-(x+1)}{n\pi} \cos n\pi x \bigg|_{-1}^{0} + \frac{1}{n\pi} \int_{-1}^{0} \cos n\pi x \, dx$$

$$+ \frac{(x-1)}{n\pi} \cos n\pi x \bigg|_{0}^{1} - \frac{1}{n\pi} \int_{0}^{1} \cos n\pi x \, dx$$

$$= \frac{-1}{n\pi} + \frac{1}{(n\pi)^2} \sin n\pi x \bigg|_{-1}^{0} + \frac{1}{n\pi} - \frac{1}{(n\pi)^2} \sin n\pi x \bigg|_{0}^{1} = 0.$$

Thus

$$f(x) = \frac{1}{2} + \frac{4}{\pi^2} \left(\cos \pi x + \frac{1}{3^2} \cos 3\pi x + \frac{1}{5^2} \cos 5\pi x + \frac{1}{7^2} \cos 7\pi x + \cdots \right). \tag{12}$$

A computer-drawn sketch of the first four terms in (12) is given in Fig. 5b. Incidentally, note that if we set $x = 1$, then

$$0 = \frac{1}{2} - \frac{4}{\pi^2} \left(1 + \frac{1}{3^2} + \frac{1}{5^2} + \frac{1}{7^2} + \cdots \right),$$

or

$$\frac{\pi^2}{8} = 1 + \frac{1}{3^2} + \frac{1}{5^2} + \frac{1}{7^2} + \cdots .$$

We observe that if term-by-term differentiation of (12) is valid, we obtain

$$f'(x) = \frac{-4}{\pi} \left(\sin \pi x + \frac{1}{3} \sin 3\pi x + \frac{1}{5} \sin 5\pi x + \cdots \right),$$

which is the same series as (11) for $k = -1$. An explanation for this fact is found by noting that $f'(x) = (-1)^{n+1}$ for $n < x < n + 1$.

Finally, note that this function satisfies the condition $f(-x) = f(x)$. All such functions are called **even** functions. In particular, $\cos n\pi x$ is an even function for every integer n.

Even Function

EXAMPLE 7

Consider the function

$$f(x) = \begin{cases} k \sin Tx, & 0 < x < \pi/T, \\ 0, & -\pi/T < x < 0, \end{cases} \quad f\left(x + \frac{2\pi}{T}\right) = f(x),$$

which is obtained by passing a sinusoidal voltage $k \sin Tx$ through a half-wave rectifier (see Fig. 6). Here the period is $2\pi/T$, so

$$a_n = \frac{T}{\pi} \int_{0}^{\pi/T} k \sin Tx \cos nTx \, dx, \qquad a_0 = \frac{T}{\pi} \int_{0}^{\pi/T} k \sin Tx \, dx = \frac{2k}{\pi}.$$

For $n \neq 1$

Formula 124 in Appendix 1

$$a_n = \overset{\downarrow}{\frac{Tk}{\pi}} \left[\frac{-\cos(T - Tn)x}{2(T - Tn)} - \frac{\cos(T + nT)x}{2(T + Tn)} \right]\Bigg|_0^{\pi/T}$$

$$= \frac{k}{2\pi} \left[\frac{\cos(n-1)\pi - 1}{n - 1} - \frac{\cos(n+1)\pi - 1}{n + 1} \right], \qquad \text{for } n \neq 1,$$

and

$$a_1 = \frac{kT}{2\pi} \int_0^{\pi/T} \sin 2Tx\,dx = \frac{kT}{2\pi} \left(\frac{1 - \cos 2\pi}{2T} \right) = 0.$$

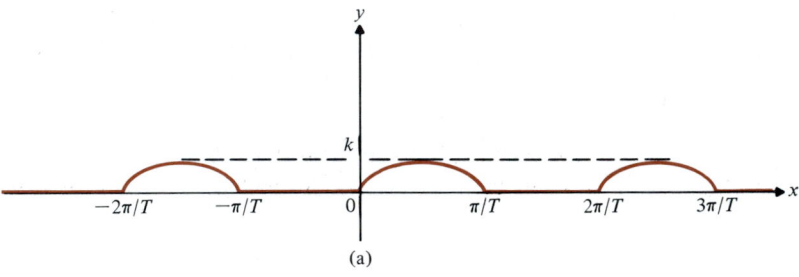

(a)

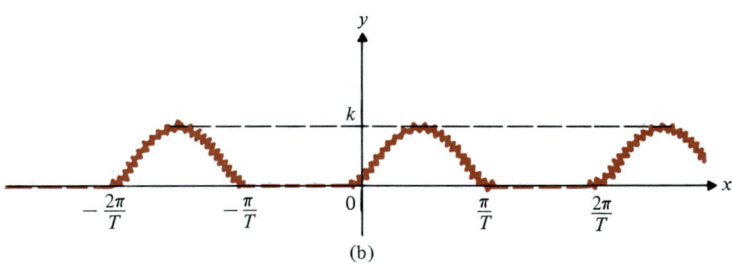

(b)

FIGURE 6

When n is odd, $a_n = 0$, and when n is even,

$$a_n = \frac{k}{\pi} \left(\frac{1}{n+1} - \frac{1}{n-1} \right) = \frac{-2k}{\pi(n^2 - 1)}.$$

Similarly, for $n \neq 1$,

$$b_n = \frac{T}{\pi} \int_0^{\pi/T} k \sin Tx \sin nTx\,dx$$

Formula 103

$$\overset{\downarrow}{=} \frac{k}{2\pi} \left[\frac{\sin(n-1)Tx}{n - 1} - \frac{\sin(n+1)Tx}{n + 1} \right]\Bigg|_0^{\pi/T} = 0;$$

and if $n = 1$, by Formula 100 of Appendix 1,

$$b_1 = \frac{k}{2\pi} (Tx - \sin Tx \cos Tx)\Bigg|_0^{\pi/T} = \frac{k}{2}.$$

Thus, since only b_1 and the evenly subscripted terms $a_{2n} = -2k/\pi(4n^2 - 1)$ are nonzero,

$$f(x) = \frac{k}{\pi} + \frac{k}{2}\sin Tx - \frac{2k}{\pi}\sum_{n=1}^{\infty}\frac{\cos 2nTx}{4n^2 - 1}. \tag{13}$$

Incidentally, if we set $x = \pi/2T$, another intriguing identity results:

$$k = \frac{k}{\pi} + \frac{k}{2} - \frac{2k}{\pi}\sum_{n=1}^{\infty}\frac{\cos n\pi}{4n^2 - 1},$$

or

$$\frac{\pi}{2} - 1 = 2\left(\frac{1}{1\cdot 3} - \frac{1}{3\cdot 5} + \frac{1}{5\cdot 7} - \frac{1}{7\cdot 9} + \cdots\right).$$

Note that in this example the Fourier series (13) contains both sine and cosine terms. A computer-drawn sketch of the first four terms in (13) is given in Fig. 6b.

Although we have concentrated in this section on trigonometric series, it is important to mention that other orthogonal sets of functions often provide useful series expansions. Let $\{f_n(x)\}$ be an orthogonal set of functions on an interval $a \le x \le b$ and suppose that the function $F(x)$ can be represented in terms of the functions $f_n(x)$ by a convergent series

$$F(x) = \sum_{n=1}^{\infty}c_nf_n(x) = c_1f_1(x) + c_2f_2(x) + \cdots. \tag{14}$$

Generalized Fourier Series

Series (14) is called a **generalized Fourier series of** $F(x)$ and the coefficients c_1, $c_2, \ldots$ are called the **generalized Fourier coefficients of** $F(x)$ **with respect to the orthogonal set of functions** $\{f_n(x)\}$. It is easy to determine the constants $c_1, c_2,$ $\ldots$ by the following procedure. Fixing m, we multiply both sides of (14) by $f_m(x)$ and integrate the result over the interval $a \le x \le b$:

$$\int_a^b F(x)f_m(x)\,dx = \sum_{n=1}^{\infty}c_n\int_a^b f_n(x)f_m(x)\,dx = \sum_{n=1}^{\infty}c_n(f_n, f_m).$$

Note that we have again assumed term-by-term integration to be permissible—an assumption that sometimes is not valid. Since $\int_a^b f_n(x)f_m(x)\,dx = 0$ whenever $n \ne m$, we have

$$\int_a^b F(x)f_m(x)\,dx = c_m\|f_m\|^2,$$

so that

$$\boxed{c_m = \frac{1}{\|f_m\|^2}\int_a^b F(x)f_m(x)\,dx, \qquad m = 1, 2, \ldots. \tag{15}}$$

Moreover, we can also use orthogonal sets of functions $\{f_n(x)\}$ **with respect to a weight function** $w(x)$ on an interval $a \le x \le b$. If $F(x)$ can be represented by a generalized Fourier series in these weighted orthogonal functions,

$$F(x) = \sum_{n=1}^{\infty} c_n f_n(x), \tag{16}$$

then the general Fourier coefficients c_k can be determined by multiplying both sides of (16) by $w(x)f_k(x)$ and integrating over the interval $a \le x \le b$:

$$\int_a^b w(x)F(x)f_k(x)\,dx = \sum_{n=1}^{\infty} c_n \int_a^b w(x)f_n(x)f_k(x)\,dx = c_k\|f_k\|_w^2,$$

or

$$c_k = \frac{\int_a^b F(x)f_k(x)w(x)\,dx}{\|f_k\|_w^2}. \tag{17}$$

Problems 11.3

1. Find the smallest positive period for each of the functions $\cos 2x$, $\sin \pi x$, $\cos(2\pi nx/T)$, $\sin 2k\pi x$.

2. Show that a constant function is periodic with any period $T > 0$.

3. Suppose that $f(x)$ has period T. What is the period of $f(ax/b)$?

4. Prove that a convergent infinite series of functions of period T is periodic of period T.

In Problems 5–13 find the Fourier series of each function $f(x)$ of period 2π, where one period is defined, and accurately plot the first three partial sums

$$\frac{a_0}{2} + \sum_{n=1}^{k} (a_n \cos nx + b_n \sin nx), \qquad k = 1, 2, 3.$$

5. $f(x) = x, \quad |x| < \pi$

6. $f(x) = \begin{cases} 0, & -\pi < x < 0 \\ 1, & 0 < x < \pi \end{cases}$

7. $f(x) = x^2, \quad |x| < \pi$

8. $f(x) = \begin{cases} 0, & -\pi < x < 0 \\ x, & 0 < x < \pi \end{cases}$

9. $f(x) = |x|, \quad |x| < \pi$

10. $f(x) = \begin{cases} x, & -\pi < x < 0 \\ x - \pi, & 0 < x < \pi \end{cases}$

11. $f(x) = \begin{cases} -1, & -\pi < x < -1 \\ x, & -1 < x < 1 \\ 1, & 1 < x < \pi \end{cases}$

12. $f(x) = \begin{cases} \pi + x, & -\pi < x < 0 \\ \pi - x, & 0 < x < \pi \end{cases}$

13. $f(x) = e^x, \quad |x| < \pi$

In Problems 14–21 find the Fourier series of each function $f(x)$ of period T, where one of the periods is defined.

14. $f(x) = x, \quad |x| < 1, \quad T = 2$

15. $f(x) = x, \quad 0 < x < 2, \quad T = 2$

16. $f(x) = x, \quad 0 < x < 3, \quad T = 3$

17. $f(x) = x^2, \quad |x| < 1, \quad T = 2$

18. $f(x) = x^2, \quad 0 < x < 2, \quad T = 2$

19. $f(x) = \begin{cases} 0, & 0 < x < 1, \\ 1, & 1 < x < 2, \quad T = 2 \end{cases}$

20. $f(x) = \begin{cases} 0, & 0 < x < 1, \\ x - 1, & 1 < x < 2, \quad T = 2 \end{cases}$

21. $f(x) = \begin{cases} x, & 0 < x < 1, \\ 1, & 1 < x < 2, \quad T = 2 \end{cases}$

22. Find the Fourier series of the periodic function of period 2π

$$f(x) = \frac{x^2}{4}, |x| < \pi,$$

and use this series to verify the identities:

$$\frac{\pi^2}{6} = 1 + \frac{1}{2^2} + \frac{1}{3^2} + \frac{1}{4^2} + \frac{1}{5^2} + \cdots,$$

$$\frac{\pi^2}{12} = 1 - \frac{1}{2^2} + \frac{1}{3^2} - \frac{1}{4^2} + \frac{1}{5^2} - \cdots,$$

$$\frac{\pi^2}{8} = 1 + \frac{1}{3^2} + \frac{1}{5^2} + \frac{1}{7^2} + \frac{1}{9^2} + \cdots.$$

11.4 Half-Range Expansions

In Section 11.3 we defined the concepts of odd and even functions. We repeat: A function $f(x)$ is **odd** if it satisfies the condition

Odd Function

$$f(-x) = -f(x) \tag{1}$$

and **even** if

Even Function

$$f(-x) = f(x). \tag{2}$$

As we saw, the function $\sin n\pi x$ is odd, whereas $\cos n\pi x$ is even for $n = 1, 2, 3, \dots$.

Knowing that a function $f(x)$ is even or odd can help us avoid unnecessary work in computing the Fourier coefficients of $f(x)$. This claim is based on the following facts:

■ **THEOREM 1**

   a. The product of two even or two odd functions is even.
   b. The product of an even and an odd function is odd.
   c. If $g(x)$ is an odd function, then

$$\int_{-T/2}^{0} g(x)\,dx = -\int_{0}^{T/2} g(x)\,dx \tag{3}$$

and, for every $T > 0$,

$$\int_{-T/2}^{T/2} g(x)\,dx = 0. \tag{4}$$

   d. If $g(x)$ is an even function, then

$$\int_{-T/2}^{0} g(x)\,dx = \int_{0}^{T/2} g(x)\,dx$$

and

$$\int_{-T/2}^{T/2} g(x)\,dx = 2\int_{0}^{T/2} g(x)\,dx. \tag{5}$$

PROOF

   a. Let $f(x)$ and $g(x)$ be even functions. By (2),

$$f(-x)g(-x) = f(x)g(x),$$

indicating that their product is even. Now suppose that they are both odd functions. By (1),

$$f(-x)g(-x) = [-f(x)][-g(x)] = f(x)g(x),$$

so again their product is even.
   b. If $f(x)$ is even and $g(x)$ is odd, then

$$f(-x)g(-x) = f(x)[-g(x)] = -f(x)g(x),$$

so the product is odd.

c. Let $g(x)$ be an odd function and let $x = -t$, so that $dx = -dt$. Then

$$\int_{-T/2}^{0} g(x)\,dx = -\int_{T/2}^{0} g(-t)\,dt = \int_{0}^{T/2} g(-t)\,dt = -\int_{0}^{T/2} g(t)\,dt,$$

so we have

$$\int_{-T/2}^{T/2} g(x)\,dx = \int_{-T/2}^{0} g(x)\,dx + \int_{0}^{T/2} g(x)\,dx = 0.$$

d.

$$\int_{-T/2}^{0} g(x)\,dx = -\int_{T/2}^{0} g(-t)\,dt = \int_{0}^{T/2} g(t)\,dt. \qquad ■$$

What precisely is the effect of (4) and (5) in computing the Fourier coefficients of an even or odd periodic function $F(x)$? If $F(x)$ is even and of period T, then $F(x)\sin(2n\pi x/T)$ is odd and of period T by Theorem 1(b). Hence, by (4),

$$\int_{-T/2}^{T/2} F(x)\sin\frac{2n\pi x}{T}\,dx = 0,$$

and all the Fourier coefficients b_n are zero when $F(x)$ is even. Similarly, if $F(x)$ is odd and of period T, then $F(x)\cos(2n\pi x/T)$ is odd and, by (4),

$$\int_{-T/2}^{T/2} F(x)\cos\frac{2n\pi x}{T}\,dx = 0,$$

indicating that all the Fourier coefficients a_n are zero. Equation (5) often simplifies the computation of the nonzero coefficients. Thus we have the following result.

■ **THEOREM 2**

Let $F(x)$ be a periodic function of period T that has a Fourier series. If $F(x)$ is an even function, then all the Fourier coefficients b_n are zero, whereas if $F(x)$ is an odd function, then the Fourier coefficients a_n are zero. ■

Remark. The Fourier series of an even function,

$$F(x) = \frac{a_0}{2} + \sum_{n=1}^{\infty} a_n \cos\frac{2n\pi x}{T}, \qquad (6)$$

Fourier Cosine Series

is called a **Fourier cosine series,** whereas that of an odd function,

$$F(x) = \sum_{n=1}^{\infty} b_n \sin \frac{2n\pi x}{T}, \tag{7}$$

Fourier Sine Series

is said to be a **Fourier sine series.** Observe that Theorem 2 implies that for even and odd functions, we need to calculate only half of the coefficients that are generally required.

EXAMPLE 1

Find the Fourier series of the periodic function of period 2π given by

$$F(x) = |x|, \qquad |x| \le \pi.$$

SOLUTION

Since $F(-x) = |-x| = |x| = F(x)$ is even, we need only calculate a Fourier cosine series. By (5),

$$a_n = \frac{1}{\pi} \int_{-\pi}^{\pi} |x| \cos nx \, dx = \frac{2}{\pi} \int_0^{\pi} x \cos nx \, dx.$$

Integrating by parts, we have

$$a_n = \frac{2}{\pi} \left(\frac{x \sin nx}{n} \Big|_0^{\pi} - \frac{1}{n} \int_0^{\pi} \sin nx \, dx \right)$$

$$= \frac{2 \cos nx}{\pi n^2} \Big|_0^{\pi}$$

$$= \frac{2}{\pi n^2} [(-1)^n - 1], \qquad \text{for } n \ge 1.$$

Also, by (5),

$$a_0 = \frac{1}{\pi} \int_{-\pi}^{\pi} |x| \, dx = \frac{2}{\pi} \int_0^{\pi} x \, dx = \frac{x^2}{\pi} \Big|_0^{\pi} = \pi,$$

so

$$F(x) = \frac{\pi}{2} + \sum_{n=1}^{\infty} \frac{2}{\pi n^2} [(-1)^n - 1] \cos nx,$$

or, since $a_{2n} = 0$,

$$|x| = \frac{\pi}{2} - \frac{4}{\pi} \sum_{k=0}^{\infty} \frac{\cos(2k+1)x}{(2k+1)^2}, \qquad |x| \le \pi. \tag{8}$$

In particular, setting $x = 0$ in (8) we have

$$0 = \frac{\pi}{2} - \frac{4}{\pi} \sum_{k=0}^{\infty} \frac{1}{(2k+1)^2},$$

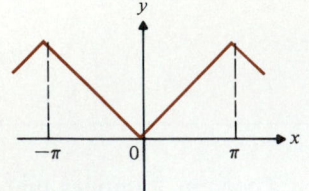

$$\frac{\pi^2}{8} = \sum_{k=0}^{\infty} \frac{1}{(2k+1)^2} = 1 + \frac{1}{3^2} + \frac{1}{5^2} + \cdots. \tag{9}$$

FIGURE 1

EXAMPLE 2

What is the Fourier series of the periodic function

$$F(x) = x, \qquad |x| < \tau?$$

SOLUTION

Since $F(-x) = -x = -F(x)$ is odd, all the coefficients $a_n = 0$. By Theorem 1(a) and (5), since F has period $T = 2\tau$

$$b_n = \frac{1}{\tau} \int_{-\tau}^{\tau} x \sin \frac{n\pi x}{\tau} dx = \frac{2}{\tau} \int_0^{\tau} x \sin \frac{n\pi x}{\tau} dx.$$

Integrating by parts, we have

$$b_n = \frac{2}{\tau} \left[\frac{-\tau x \cos(n\pi x/\tau)}{n\pi} \bigg|_0^{\tau} + \frac{\tau}{n\pi} \int_0^{\tau} \cos \frac{n\pi x}{\tau} dx \right]$$

$$= \frac{2}{\tau} \left[\frac{-(-1)^n \tau^2}{n\pi} + \left(\frac{\tau}{n\pi}\right)^2 \sin \frac{n\pi x}{\tau} \bigg|_0^{\tau} \right]$$

$$= \frac{-2(-1)^n \tau}{n\pi}.$$

Hence

$$x = \frac{-2\tau}{\pi} \sum_{n=1}^{\infty} \frac{(-1)^n}{n} \sin \frac{n\pi x}{\tau}, \qquad |x| < \tau. \tag{10}$$

In particular, setting $x = \tau/2$ in (10) we have

$$\frac{\tau}{2} = \frac{-2\tau}{\pi} \sum_{n=1}^{\infty} \frac{(-1)^n}{n} \sin \frac{n\pi}{2},$$

or

$$\frac{\pi}{4} = 1 - \frac{1}{3} + \frac{1}{5} - \frac{1}{7} + \cdots. \tag{11}$$

In many problems of physics and engineering there is a practical need to apply a Fourier series to a nonperiodic function $F(x)$ that is defined only on the interval $0 < x < \tau$. Because of physical or mathematical considerations, it may be permissible to *extend* $F(x)$ over the interval $-\tau < x < \tau$, making it periodic of period $T = 2\tau$. Figure 2 illustrates two such extensions. The **odd extension** in Fig. 2b has a Fourier *sine* series; the **even extension** in Fig. 2c has a Fourier *cosine* series.

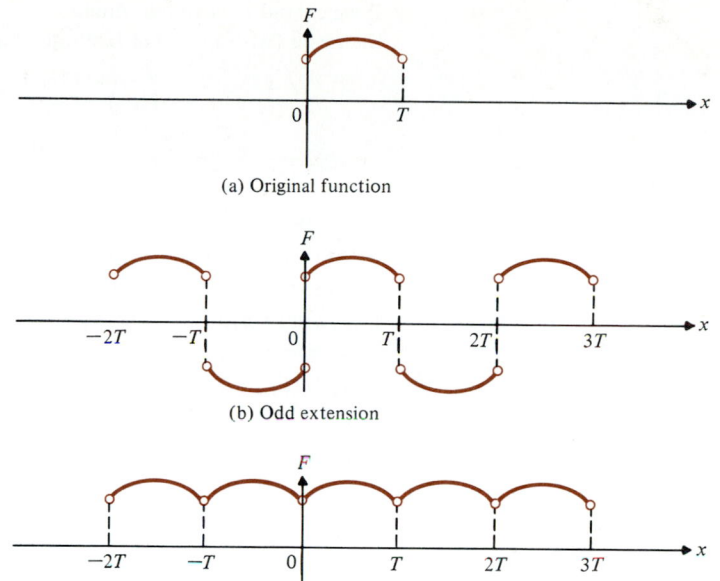

(a) Original function

(b) Odd extension

(c) Even extension

FIGURE 2

EXAMPLE 3

Let $F(x) = x$ on $0 \leq x \leq \pi$. If we extend F as an even function of period 2π, we obtain the graph in Fig. 3c. Hence, for the even extension, $F_e(x) = |x|$ on $|x| \leq \pi$, so the result of Example 1 holds and

$$F_e(x) = |x| = \frac{\pi}{2} - \frac{4}{\pi} \sum_{k=0}^{\infty} \frac{\cos(2k+1)x}{(2k+1)^2}, \qquad |x| \leq \pi.$$

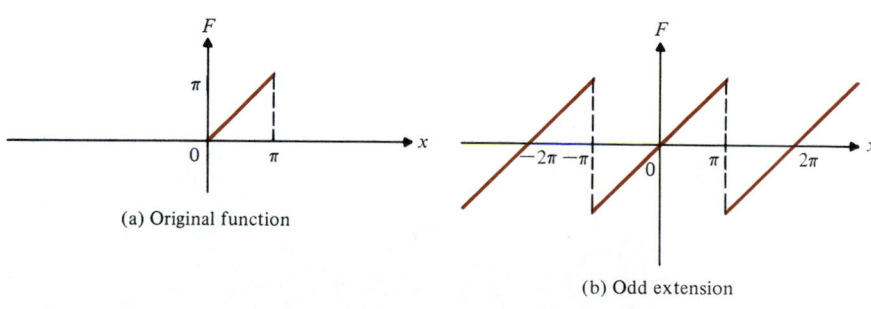

(a) Original function

(b) Odd extension

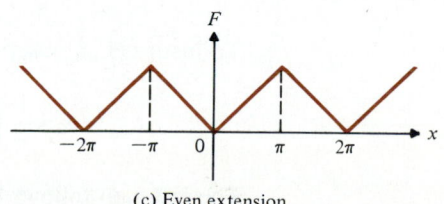

(c) Even extension

FIGURE 3

If we extend F as an odd function of period 2π (Fig. 3b), then $F_0(x) = x$ on $|x| < \pi$ and the result (with $\tau = \pi$) of Example 2 applies:

$$F_0(x) = x = -2 \sum_{n=1}^{\infty} \frac{(-1)^n}{n} \sin nx, \qquad |x| < \pi.$$

Since the Fourier series is based on a function $F(x)$ defined on only half of the interval of periodicity, the Fourier series for either an even or an odd extension is called a **half-range expansion.** Using (4) and (5) we obtain the following result.

■ **THEOREM 3**

Let $F(x)$ be a function defined on the interval $0 \le x \le \tau$. Then an **even half-range expansion** of $F(x)$ is given by

Even Half-Range Expansion

$$F(x) = \frac{a_0}{2} + \sum_{n=1}^{\infty} a_n \cos \frac{n\pi x}{\tau},$$

where

$$a_n = \frac{2}{\tau} \int_0^{\tau} F(x) \cos \frac{n\pi x}{\tau} dx, \qquad n = 0, 1, 2, \ldots, \qquad \textbf{(12)}$$

and an **odd half-range expansion** is given by

Odd Half-Range Expansion

$$F(x) = \sum_{n=1}^{\infty} b_n \sin \frac{n\pi x}{\tau},$$

where

$$b_n = \frac{2}{\tau} \int_0^{\tau} F(x) \sin \frac{n\pi x}{\tau} dx, \qquad n = 1, 2, \ldots. \qquad \textbf{(13)}$$

■

To justify (12), observe that, by the Euler formulas,

$$a_n = \frac{1}{\tau} \int_{-\tau}^{\tau} F(x) \cos \frac{n\pi x}{\tau} dx = \frac{2}{\tau} \int_0^{\tau} F(x) \cos \frac{n\pi x}{\tau} dx.$$

The last step follows from Theorem 1d and the fact that $F(x)$ and $\cos(n\pi x/\tau)$ are even. Formula (13) can be proved in a similar way.

EXAMPLE 4: Full Wave Rectifier

A full wave rectifier passes the positive peaks of an incoming sine wave and inverts the negative peaks. Thus the current can be written (see Fig. 4)

$$i(t) = \begin{cases} \sin \omega t, & 0 < \omega t < \pi, \\ -\sin \omega t, & -\pi < \omega t < 0. \end{cases} \tag{14}$$

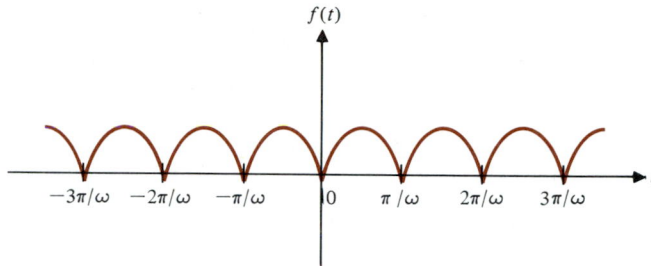

FIGURE 4

We can use a Fourier series to show how well the rectifier approaches direct current (i.e., is constant). Since $i(t)$ is even, we can use (12) to write the Fourier cosine series for the current. Here $\tau = \pi/\omega$; so substituting this value and $n = 0$ in (12), we get

$$a_0 = \frac{2\omega}{\pi} \int_0^{\pi/\omega} \sin \omega t \, dt = \frac{-2}{\pi} \cos \omega t \Big|_0^{\pi/\omega} = \frac{4}{\pi}.$$

If $n \neq 1$, (12) yields

$$a_n = \frac{2\omega}{\pi} \int_0^{\pi/\omega} \sin \omega t \cos n\omega t \, dt$$

entry 124 in Appendix 1

$$= \frac{2\omega}{\pi} \left[-\frac{\cos \omega(1-n)t}{2\omega(1-n)} - \frac{\cos \omega(1+n)t}{2\omega(1+n)} \right]_0^{\pi/\omega}$$

$$= \frac{2\omega}{\pi} \left[\frac{1 - \cos(n-1)\pi}{2\omega(1-n)} + \frac{1 - \cos(n+1)\pi}{2\omega(1+n)} \right].$$

If n is odd, $n - 1$ and $n + 1$ are even, so $\cos(n - 1)\pi = \cos(n + 1)\pi = 1$ and $a_n = 0$. If n is even, $n - 1$ and $n + 1$ are odd, so $\cos(n - 1)\pi = \cos(n + 1)\pi = -1$ and

$$a_n = \frac{2\omega}{\pi} \left[\frac{2}{2\omega(1-n)} + \frac{2}{2\omega(1+n)} \right] = \frac{2}{\pi} \left[\frac{1}{1-n} + \frac{1}{1+n} \right] = \frac{-4}{\pi(n^2 - 1)}.$$

Substituting $n = 2m$, we obtain

$$i(t) = \frac{2}{\pi} - \frac{4}{\pi} \sum_{m=1}^{\infty} \frac{\cos 2m\omega t}{4m^2 - 1}.$$

The original frequency ω has been eliminated. The lowest surviving harmonic has frequency 2ω and amplitude $4/3\pi$. Higher harmonics (of frequencies 4ω, 6ω, . . . , $2m\omega$, . . .) have amplitudes that fall off proportionally to $1/m^2$. Thus the full wave rectifier does a fairly good job of approximating direct current. Whether it is good enough depends on the particular use to which it is put.

In Problems 1–10 determine whether each function is even, odd, or neither.

1. x^2

2. x^3

3. $x \sin x$

4. e^x

5. $x + x^3$

6. $x + x^2$

7. $\ln x$

8. $\sin x + \cos x$

9. $F(x) = \begin{cases} x, & -T < x \le 0, \\ 0, & 0 < x < T, \ F(x + 2T) = F(x) \end{cases}$

10. $F(x) = \begin{cases} x^2, & -T < x < 0, \\ -x^2, & 0 < x < T, \ F(x + 2T) = F(x) \end{cases}$

In Problems 11–20 find the Fourier series of the given functions, assuming that they are periodic. Pay particular attention to whether each function is even or odd.

11. $F(x) = x^2, \ |x| \le \pi$

12. $F(x) = x^3, \ |x| \le \pi$

13. $F(x) = |x^3|, \ |x| \le \pi$

14. $F(x) = |\sin x|, \ |x| \le \pi$

15. $F(x) = \begin{cases} x + T, & -T < x < 0 \\ -x + T, & 0 < x < T \end{cases}$

16. $F(x) = \begin{cases} -x - T, & -T < x < 0 \\ x - T, & 0 < x < T \end{cases}$

17. $F(x) = \begin{cases} -x - T, & -T < x < 0 \\ -x + T, & 0 < x < T \end{cases}$

18. $F(x) = \begin{cases} x + T, & -T < x < 0 \\ x - T, & 0 < x < T \end{cases}$

19. $F(x) = \begin{cases} (x + 1)^2, & -1 < x < 0 \\ (x - 1)^2, & 0 < x < 1 \end{cases}$

20. $F(x) = \begin{cases} (x + 1)^2, & -1 < x < 0 \\ -(x - 1)^2, & 0 < x < 1 \end{cases}$

Extend each of the functions in Problems 21–30 as both an even and an odd periodic function. Sketch the resulting graphs and compute their corresponding Fourier cosine or sine series.

21. $F(x) = k, \ 0 < x < 1$

22. $F(x) = x^2, \ 0 < x < 1$

23. $F(x) = x^3, \ 0 < x < 1$

24. $F(x) = e^x, \ 0 < x < 1$

25. $F(x) = \begin{cases} 1, & 0 < x < 1 \\ 0, & 1 < x < 2 \end{cases}$

26. $F(x) = \begin{cases} 1, & 0 < x < 1 \\ 1/2, & 1 < x < 2 \end{cases}$

27. $F(x) = \begin{cases} 1, & 0 < x < 1 \\ -x + 2, & 1 < x < 2 \end{cases}$

28. $F(x) = \begin{cases} x, & 0 < x < 1 \\ 1, & 1 < x < 2 \end{cases}$

29. $F(x) = \begin{cases} x^2, & 0 < x < 1 \\ 1, & 1 < x < 2 \end{cases}$

30. $F(x) = \begin{cases} x^2, & 0 < x < 1/2 \\ 1, & 1/2 < x < 1 \end{cases}$

31. Using the formula $e^{ix} = \cos x + i \sin x$, show that

(a)
$$\cos nx = \frac{1}{2}(e^{inx} + e^{-inx}),$$

$$\sin nx = \frac{1}{2i}(e^{inx} - e^{-inx});$$

(b) the Fourier series

$$F(x) = \frac{a_0}{2} + \sum_{n=1}^{\infty} (a_n \cos nx + b_n \sin nx)$$

can be written in the **complex form**

$$F(x) = \sum_{n=-\infty}^{\infty} c_n e^{inx};$$

(c) the **complex Fourier coefficients** c_n in (b) are given by

$$c_n = \frac{1}{2\pi} \int_{-\pi}^{\pi} F(x) e^{-inx} \, dx, \quad n = 0, \pm 1, \pm 2, \ldots.$$

32. Show that the complex Fourier coefficients of an even function are real.

33. Show that the complex Fourier coefficients of an odd function are pure imaginary.

11.5 Fourier Integrals

The Fourier series provides a powerful method for representing a periodic function in terms of a series of orthogonal functions. Illustrations of their use were provided in Section 11.1 and further examples will be given in Chapter 12. However, nonperiodic functions arise naturally in many practical problems, so it would be very advantageous if we could generalize the Fourier series technique to extend to such functions.

An example of a nonperiodic function is provided by a single unrepeated pulse (see Fig. 1c). Observe that this single pulse can be obtained by beginning with a periodic function consisting of identical pulses each a distance $\lambda = 1$ unit apart and letting the distance λ between pulses tend to ∞.

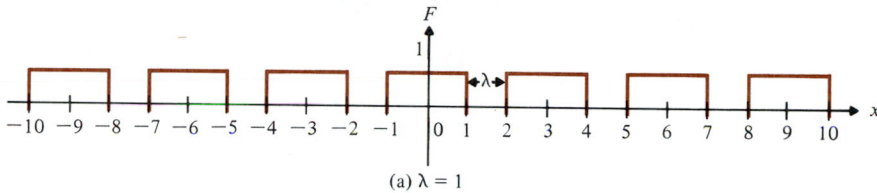

(a) $\lambda = 1$

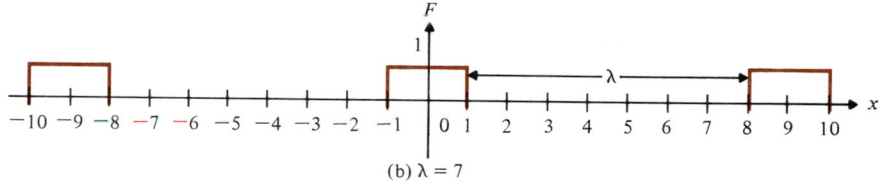

(b) $\lambda = 7$

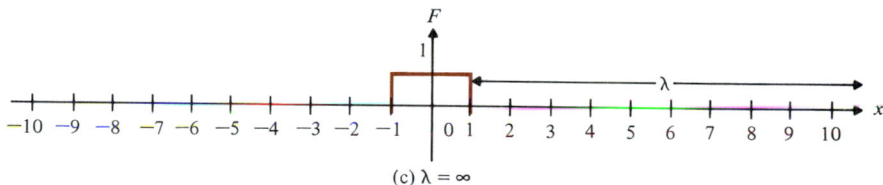

(c) $\lambda = \infty$

Pulse Train of Decreasing Frequency

FIGURE 1. Pulse train of decreasing frequency.

We might now be able to discover the "Fourier series" for the function in Fig. 1c by obtaining the Fourier series for the function in Fig. 1a and letting $\lambda \to \infty$. We do this in the following example.

EXAMPLE 1

We can define the periodic functions in Figs. 1a and 1b by

$$F_\lambda(x) = \begin{cases} 1, & |x| < 1, \\ 0, & 1 < |x| < 1 + \lambda, \end{cases} \tag{1}$$

where λ is the distance between pulses. Then the period of $F_\lambda(x)$ is $\lambda + 2$, so half the period is $1 + \lambda/2$. Since $F_\lambda(x)$ is even, it has a Fourier cosine series with

Formula (12) on p. 694
$$a_n = \frac{4}{\lambda + 2} \int_0^{1+\lambda/2} F_\lambda(x) \cos\left(\frac{n\pi x}{1 + \lambda/2}\right) dx$$

$$= \frac{4}{\lambda + 2} \int_0^1 \cos\left(\frac{2n\pi x}{\lambda + 2}\right) dx$$

since $F_\lambda(x) = 0$ on $1 < |x| < 1 + \lambda$. Hence

$$a_n = \frac{4}{\lambda+2}\left[\frac{(\lambda+2)\,\sin[2n\pi x/(\lambda+2)]}{2n\pi}\right]\bigg|_0^1 = \frac{2}{n\pi}\sin\left(\frac{2n\pi}{\lambda+2}\right), \qquad n \neq 0,$$

and

$$a_0 = \frac{4}{\lambda+2}.$$

Thus

$$F_\lambda(x) = \frac{2}{\lambda+2} + \frac{2}{\pi}\sum_{n=1}^{\infty}\frac{1}{n}\sin\left(\frac{2n\pi}{\lambda+2}\right)\cos\left(\frac{2n\pi x}{\lambda+2}\right). \tag{2}$$

Now set $t_n = 2n\pi/(\lambda+2)$ and observe that $\Delta t = t_{n+1} - t_n = 2\pi/(\lambda+2)$ for $n = 0, 1, 2,$ .... Hence (2) can be rewritten as

$$F_\lambda(x) = \frac{\Delta t}{\pi} + \frac{2}{\pi}\sum_{n=1}^{\infty}\frac{\sin(t_n)}{t_n}\cos(t_n x)\,\Delta t. \tag{3}$$

If we take the limit of the right side of (3) as $\lambda \to \infty$, and note that $\Delta t \to 0$, we obtain

$$\lim_{\lambda\to\infty}F_\lambda(x) = \frac{2}{\pi}\lim_{\Delta t\to 0}\sum_{n=1}^{\infty}\frac{\sin(t_n)}{t_n}\cos(t_n x)\,\Delta t \tag{4}$$

since the first term on the right side of (3) vanishes. The right side of (4) defines an (improper) definite integral where the integrand is evaluated at the left endpoint of the interval $[t_n, t_{n+1}]$ of length Δt. Hence

$$F(x) = \lim_{\lambda\to\infty}F_\lambda(x) = \frac{2}{\pi}\int_0^{\infty}\frac{\sin t\,\cos(tx)}{t}\,dt. \tag{5}$$

Thus, if this procedure holds, we have represented $F(x)$ (the function in Fig. 1c) as an improper integral (5).

We now adapt the procedure in Example 1 to an arbitrary periodic function F_λ of period 2λ. Assume $F_\lambda(x)$ can be represented by a Fourier series

$$F_\lambda(x) = \frac{a_0}{2} + \sum_{n=1}^{\infty}\left[a_n\cos\frac{n\pi x}{\lambda} + b_n\sin\frac{n\pi x}{\lambda}\right], \tag{6}$$

where

$$a_n = \frac{1}{\lambda}\int_{-\lambda}^{\lambda}F_\lambda(x)\cos\frac{n\pi x}{\lambda}\,dx \tag{7}$$

and

$$b_n = \frac{1}{\lambda}\int_{-\lambda}^{\lambda}F_\lambda(x)\sin\frac{n\pi x}{\lambda}\,dx.$$

Substitute $t_n = n\pi/\lambda$ and $\Delta t = t_{n+1} - t_n = \pi/\lambda$ in (6) and (7) to obtain

$$F_\lambda(x) = \frac{1}{2\lambda} \int_{-\lambda}^{\lambda} F_\lambda(x)\,dx + \frac{1}{\pi} \sum_{n=1}^{\infty} \left[\cos(t_n x) \int_{-\lambda}^{\lambda} F_\lambda(x) \cos(t_n u)\,du \right.$$

$$\left. + \sin(t_n x) \int_{-\lambda}^{\lambda} F_\lambda(u) \sin(t_n u)\,du \right] \Delta t. \tag{8}$$

If we let $\lambda \to \infty$, then $\Delta t \to 0$ and we have

$$F(x) = \lim_{\lambda \to \infty} F_\lambda(x) = \lim_{\lambda \to \infty} \frac{1}{2\lambda} \int_{-\lambda}^{\lambda} F_\lambda(x)\,dx$$

$$+ \frac{1}{\pi} \lim_{\substack{\lambda \to \infty \\ (\Delta t \to 0)}} \left\{ \sum_{n=1}^{\infty} \left[\cos(t_n x) \int_{-\lambda}^{\lambda} F_\lambda(u) \cos(t_n u)\,du \right. \right.$$

$$\left. \left. + \sin(t_n x) \int_{-\lambda}^{\lambda} F_\lambda(u) \sin(t_n u)\,du \right] \Delta t \right\}. \tag{9}$$

If

$$\int_{-\infty}^{\infty} |F(x)|\,dx < \infty, \tag{10}$$

that is, if $F(x)$ is absolutely integrable on $\mathbb{R}$, then the first integral in (9) is zero and

$$F(x) = \frac{1}{\pi} \int_0^{\infty} \left[\cos tx \int_{-\infty}^{\infty} F(u) \cos(tu)\,du + \sin tx \int_{-\infty}^{\infty} F(u) \sin(tu)\,du \right] dt.$$

This may be rewritten in the form

$$F(x) = \frac{1}{\pi} \int_0^{\infty} [A(t) \cos tx + B(t) \sin tx]\,dt, \tag{11}$$

where

$$A(t) = \int_{-\infty}^{\infty} F(u) \cos tu\,du,$$

$$B(t) = \int_{-\infty}^{\infty} F(u) \sin tu\,du. \tag{12}$$

Fourier Integral

Equations (12) resemble Euler's formulas; the representation (11), (12) of $F(x)$ is called the **Fourier integral** of $F(x)$.

The heuristic approach we used to develop the Fourier integral provides an indication of the following result, but is not a proof of the validity of Fourier integral representations. The proof of the following theorem can be found in various places.† Results using more general hypotheses can also be found.

■ **THEOREM 1 (Fourier Integral)**

If $F(x)$ and $F'(x)$ are piecewise continuous and the integral (10) exists and is finite, then $F(x)$ can be represented by the Fourier integral (11), (12). At points of discontinuity of $F(x)$, the value of the Fourier integral equals the average of the left- and right-hand limits of $F(x)$:

$$\frac{1}{\pi} \int_0^\infty [A(t)\cos tx + B(t)\sin tx]\,dt = \frac{F(x+0)+F(x-0)}{2}. \qquad (13)$$

EXAMPLE 1 (Continued)

The function $F(x)$ shown in Fig. 1c is discontinuous at $x = \pm 1$ and satisfies the hypotheses in Theorem 1. Thus we have **Dirichlet's discontinuous function**

$$\int_0^\infty \frac{\sin t \cos tx}{t}\,dt = \begin{cases} \pi/2, & \text{for } 0 \le x < 1, \\ \pi/4, & \text{for } x = 1, \\ 0, & \text{for } x > 1. \end{cases} \qquad (14)$$

Dirichlet Integral

In particular, when $x = 0$ we have **Dirichlet's integral**

$$\int_0^\infty \frac{\sin t}{t}\,dt = \frac{\pi}{2}.$$

EXAMPLE 2

We have seen that partial sums of the Fourier series of a function $F(x)$ provide curves that approximate the function $F(x)$. Similarly, in the case of Fourier integrals, definite integrals of the form

$$\frac{1}{\pi} \int_0^N [A(t)\cos tx + B(t)\sin tx]\,dt \qquad (15)$$

provide approximations of the nonperiodic function $F(x)$. For example, we would expect that for N large,

$$\int_0^N \frac{\sin t \cos tx}{t}\,dt \approx \begin{cases} \pi/2, & \text{for } 0 \le x < 1, \\ \pi/4, & \text{for } x = 1, \\ 0, & \text{for } x > 1. \end{cases}$$

† For example, see W. R. Derrick, *Complex Analysis and Applications*, 2nd ed., Wadsworth, Belmont, Calif., 1984, p. 273.

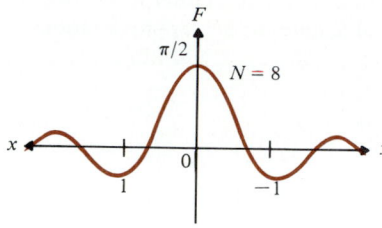

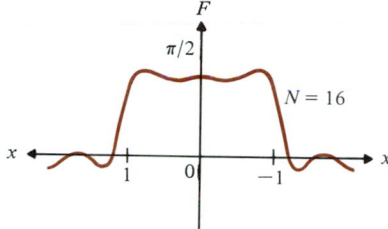

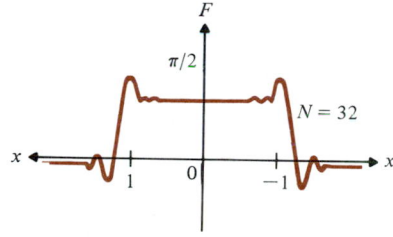

FIGURE 2

Figure 2 shows various approximations of $F(x)$ for differing values of N. Observe the small oscillations that occur as the curve approaches the points of discontinuity of $F(x)$. One might expect that these small oscillations would damp out as N increases, since the integral converges to $F(x)$ as $N \to \infty$. Instead, as N increases the oscillations *remain* but are shifted closer to the points of discontinuity $x = \pm 1$. This is called the **Gibbs phenomenon**.† To see why this occurs, use the trigonometric identities

$$\sin(t + tx) = \sin t \cos tx + \cos t \sin tx,$$

$$\sin(t - tx) = \sin t \cos tx - \cos t \sin tx,$$

to rewrite the integral in the form

$$\int_0^N \frac{\sin t \cos tx}{t}\,dt = \frac{1}{2}\int_0^N \frac{\sin(t+tx)}{t}\,dt + \frac{1}{2}\int_0^N \frac{\sin(t-tx)}{t}\,dt. \tag{16}$$

Substitute $u = t(1 + x)$ in the first integral and $u = t(x - 1)$ in the second to obtain

$$\int_0^N \frac{\sin t \cos tx}{t}\,dt = \frac{1}{2}\int_0^{N(1+x)} \frac{\sin u\,du}{u} - \frac{1}{2}\int_0^{N(x-1)} \frac{\sin u\,du}{u}$$

$$= \frac{1}{2}\int_{N(x-1)}^{N(x+1)} \frac{\sin u}{u}\,du. \tag{17}$$

The integrand is shown as the solid line and the **sine integral**

$$\text{Si}(z) = \int_0^z \frac{\sin u}{u}\,du \tag{18}$$

as the dotted line in Fig. 3. Since

$$\int_0^N \frac{\sin t \cos tx}{t}\,dt = \tfrac{1}{2}[\text{Si}(N(1 + x)) - \text{Si}(N(x - 1))],$$

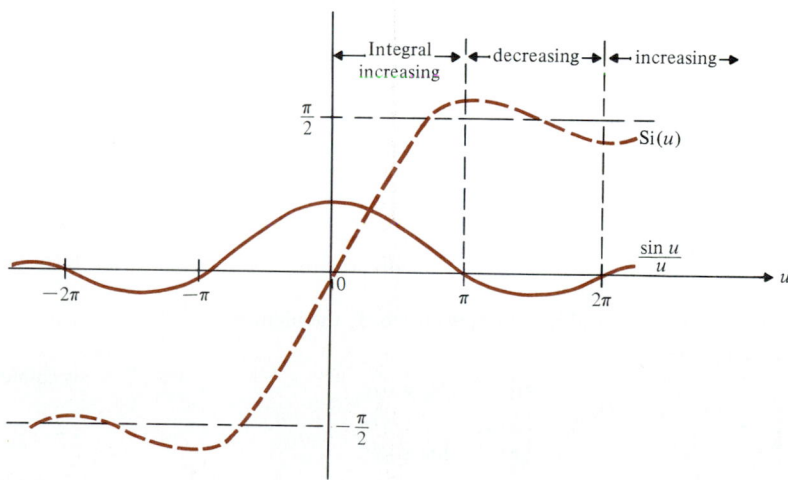

FIGURE 3

† Josiah Willard Gibbs (1839–1903) was an American mathematical physicist. He first wrote about the phenomenon named after him in an 1899 article in *Nature*. The phenomenon was explained in 1906 by the French mathematician Maxime Bôcher.

increasing N merely shifts the oscillations by selecting values of x closer to the discontinuities at ± 1. The Gibbs phenomenon is a general feature of such approximations.

Problems 11.5

In Problems 1–12, find the Fourier integral of the given function.

1. $F(x) = \begin{cases} x, & |x| < 1, \\ 0, & |x| > 1 \end{cases}$

2. $F(x) = \begin{cases} x^3, & |x| < 1, \\ 0, & |x| > 1 \end{cases}$

3. $F(x) = \begin{cases} \operatorname{sgn} x, & |x| < 1,\dagger \\ 0, & |x| > 1 \end{cases}$

4. $F(x) = \begin{cases} 1, & |x| < 1, \\ 1/2, & 1 < |x| < 2, \\ 0, & |x| > 2 \end{cases}$

5. $F(x) = \dfrac{1}{1 + x^2}$

6. $F(x) = \begin{cases} \cos x, & |x| < \pi/2, \\ 0, & |x| > \pi/2 \end{cases}$

7. $F(x) = \begin{cases} \sin x, & |x| < \pi, \\ 0, & |x| > \pi \end{cases}$

8. $F(x) = \begin{cases} \sin x, & |x| < \pi/2, \\ 0, & |x| > \pi/2 \end{cases}$

9. $F(x) = \begin{cases} \sinh x, & |x| < T, \\ 0, & |x| > T \end{cases}$

10. $F(x) = \begin{cases} \cosh x, & |x| < T, \\ 0, & |x| > T \end{cases}$

11. $F(x) = \begin{cases} |x|, & |x| < 1, \\ 2 - |x|, & 1 < |x| < 2, \\ 0, & |x| > 2 \end{cases}$

12. $F(x) = \begin{cases} 1, & |x| < 1, \\ 2 - |x|, & 1 < |x| < 2, \\ 0, & |x| > 2 \end{cases}$

Using the Fourier integral representation, verify the identities in Problems 13–16.

13. $\displaystyle\int_0^\infty \dfrac{\cos tx + t \sin tx}{1 + t^2}\, dt = \begin{cases} 0 & \text{for } x < 0, \\ \pi/2 & \text{for } x = 0, \\ \pi e^{-x} & \text{for } x > 0 \end{cases}$

[*Hint:* Use (11).]

14. $\displaystyle\int_0^\infty \dfrac{1 - \cos kt}{t} \sin tx\, dt = \begin{cases} \pi/2, & 0 < x < k, \\ \pi/4, & x = k, \\ 0, & x > k \end{cases}$

[*Hint:* See Problems 17 and 18.]

15. $\displaystyle\int_0^\infty \dfrac{\cos tx}{t^2 + k^2}\, dt = \dfrac{\pi e^{-kx}}{2k}$ for $k > 0$, $x \geq 0$

16. $\displaystyle\int_0^\infty \dfrac{t \sin tx}{t^2 + k^2}\, dt = \dfrac{\pi e^{-kx}}{2}$ for $k > 0$, $x > 0$

17. Prove that if $F(x)$ is an even function then $B(t) = 0$ in (12).

18. Prove that if $F(x)$ is an odd function then $A(t) = 0$ in (12).

19. Show that $F(x) = 1$, $x > 0$, cannot be represented by a Fourier integral.

20. (a) Show that (11) can be rewritten in the form

$$F(x) = \frac{1}{\pi} \int_0^\infty \left[\int_{-\infty}^\infty F(u) \cos(tx - tu)\, du \right] dt. \qquad \textbf{(19)}$$

(b) Show the integral in brackets in (19) is an even function of t so that $F(x)$ can be written as

$$F(x) = \frac{1}{2\pi} \int_{-\infty}^\infty \left[\int_{-\infty}^\infty F(u) \cos t(x - u)\, du \right] dt. \qquad \textbf{(20)}$$

(c) Show that

$$\int_{-\infty}^\infty \left[\int_{-\infty}^\infty F(u) \sin t(x - u)\, du \right] dt = 0.$$

(d) Use parts **(b)** and **(c)** to rewrite

$$F(x) = \frac{1}{2\pi} \int_{-\infty}^\infty \left[\int_{-\infty}^\infty F(u) e^{it(x-u)}\, du \right] dt.$$

This is the **complex form** of the Fourier integral. The term (equivalent to Euler's formulas)

$$c(t) = \frac{1}{\sqrt{2\pi}} \int_{-\infty}^\infty F(u) e^{-itu}\, du$$

is called the **inverse Fourier transform** of $F(x)$ and

$$F(x) = \frac{1}{\sqrt{2\pi}} \int_{-\infty}^\infty c(t) e^{itx}\, dt$$

is called the **Fourier transform** of $c(t)$.

$\dagger \operatorname{sgn} x = \begin{cases} 1, & \text{if } x > 0, \\ -1, & \text{if } x < 0. \end{cases}$

11.6 Sturm-Liouville Problems

Examples 11.1.1 and 11.1.2, as well as several of the homogeneous boundary value problems we shall encounter in Chapter 12, belong to a wide class of problems whose eigenfunctions and eigenvalues have particularly nice properties. We develop a small part of the fascinating theory of this class of problems in this and the next two sections.

Consider the differential equation

$$[r(x)y']' + [p(x) + \lambda q(x)]y = 0, \tag{1}$$

where $r(x)$, $r'(x)$, $p(x)$, and $q(x)$ are continuous functions on some interval $a \leq x \leq b$ and λ is a real parameter. Many of the differential equations of applied mathematics, such as Legendre's equation and Bessel's equation, can be written in the form (1). Suppose that we impose the boundary conditions

$$a_1 y(a) - a_2 y'(a) = 0,$$
$$b_1 y(b) - b_2 y'(b) = 0, \tag{2}$$

at the endpoints of the interval and require that at least one coefficient in each equation in (2) be nonzero. Equation (1), together with the boundary conditions (2), is known as a **Sturm-Liouville problem.**†

Sturm-Liouville Problem

Observe, as in Example 11.1.1, that any Sturm-Liouville problem has the trivial solution $y \equiv 0$. For certain values of the parameter λ, nontrivial solutions of the Sturm-Liouville problem may exist. All such nontrivial solutions are called **eigenfunctions**‡ of the problem, and the corresponding values of λ that yield these solutions are called **eigenvalues** of the problem.

Eigenfunctions, Eigenvalues

EXAMPLE 1

Find the eigenvalues and eigenfunctions of the Sturm-Liouville problem

$$y'' + \lambda y = 0, \qquad y(0) = y'(1) = 0. \tag{3}$$

SOLUTION

Setting $r(x) \equiv q(x) \equiv 1$ and $p(x) \equiv 0$, we see that equation (3) has the form of equation (1), and letting $a = 0$, $b = 1$, $a_1 = -b_2 = 1$, and $a_2 = b_1 = 0$, we have the right type of boundary conditions. Thus (3) is a Sturm-Liouville problem.

---

† The Swiss mathematician J. C. F. Sturm (1803–1855) made many significant contributions to the theory of differential equations. Joseph Liouville (1809–1882), a French mathematician, was noted for his work in complex analysis.

‡ Any constant multiple of an eigenfunction is also an eigenfunction.

The roots of the characteristic equation are $\pm\sqrt{-\lambda}$, so we obtain the general solutions

$$y(x) = \begin{cases} c_1 e^{\sqrt{-\lambda}x} + c_2 e^{-\sqrt{-\lambda}x}, & \text{if } \lambda < 0, \\ c_1 + c_2 x, & \text{if } \lambda = 0, \\ c_1 \cos\sqrt{\lambda}x + c_2 \sin\sqrt{\lambda}x, & \text{if } \lambda > 0. \end{cases}$$

Thus we need to examine three cases:

CASE 1
If $\lambda < 0$, the boundary conditions $y(0) = 0$ and $y'(1) = 0$ yield, after a short computation, the homogeneous system of equations

$$c_1 + c_2 = 0,$$

$$\sqrt{-\lambda}e^{\sqrt{-\lambda}}c_1 - \sqrt{-\lambda}e^{-\sqrt{-\lambda}}c_2 = 0.$$

It is easy to verify that the only solution to this system is $c_1 = c_2 = 0$, which indicates that problem (3) has only trivial solutions if $\lambda < 0$.

CASE 2
If $\lambda = 0$, the boundary condition $y(0) = 0$ implies that $c_1 = 0$, and since $y' \equiv c_2$, the second condition forces c_2 to vanish. Again (3) has only a trivial solution.

CASE 3
If $\lambda > 0$, the condition $y(0) = 0$ implies that $c_1 = 0$, so the solution to the problem must have the form

$$y(x) = c_2 \sin\sqrt{\lambda}x. \tag{4}$$

Differentiating (4), we have $y'(x) = c_2\sqrt{\lambda} \cos\sqrt{\lambda}x$, and setting $x = 1$, we obtain the equation $c_2\sqrt{\lambda} \cos\sqrt{\lambda} = 0$. Hence in this case we have nontrivial solutions whenever $\cos\sqrt{\lambda}$ is zero. This occurs whenever the constant $\sqrt{\lambda}$ is one of the numbers

$$\frac{\pi}{2}, \frac{3\pi}{2}, \frac{5\pi}{2}, \frac{7\pi}{2}, \ldots;$$

that is, whenever λ takes on one of the values

$$\frac{\pi^2}{4}, \frac{9\pi^2}{4}, \frac{25\pi^2}{4}, \frac{49\pi^2}{4}, \ldots. \tag{5}$$

The values (5) are the eigenvalues of (3), and the eigenfunctions corresponding to the eigenvalues $\lambda_k = (2k-1)^2\pi^2/4$, $k = 1, 2, 3, \ldots$ have the form

$$y_k(x) = A \sin\frac{(2k-1)\pi x}{2}, \qquad A \neq 0.$$

EXAMPLE 2

Consider the Sturm-Liouville problem

$$(xy')' + \frac{\lambda}{x}y = 0, \qquad y(1) = y(e) = 0. \tag{6}$$

If we write $(xy')'$ as $xy'' + y'$ and multiply the differential equation on both sides by x, we obtain the Euler equation considered in Section 2.6. We therefore assume that $y = x^r$ is a solution of (6) and find, after a short calculation, that

$$(rx^r)' + \lambda x^{r-1} = (r^2 + \lambda)x^{r-1} = 0.$$

Since x does not vanish on $1 \le x \le e$, it follows that $r = \pm\sqrt{-\lambda}$. Again we must deal with three cases:

CASE 1

If $\lambda < 0$, we have a general solution of the form

$$y(x) = c_1 x^{\sqrt{-\lambda}} + c_2 x^{-\sqrt{-\lambda}}.$$

The boundary conditions yield the homogeneous equations

$$c_1 + c_2 = 0,$$

$$e^{\sqrt{-\lambda}}c_1 + e^{-\sqrt{-\lambda}}c_2 = 0.$$

If we set $c_2 = -c_1$ in the second equation, we obtain

$$2c_1 \sinh \sqrt{-\lambda} = c_1(e^{\sqrt{-\lambda}} - e^{-\sqrt{-\lambda}}) = 0,$$

and since $\sinh \sqrt{-\lambda} \ne 0$ if $\lambda \ne 0$, it follows that $c_1 = c_2 = 0$. Thus (6) has only trivial solutions when $\lambda < 0$.

CASE 2

If $\lambda = 0$, equation (6) reduces to the equation $(xy')' = 0$. When we integrate both sides, we obtain $xy' = c_1$, so $y' = c_1/x$ and $y(x) = c_1 \ln x + c_2$. The condition $y(1) = 0$ implies that $c_2 = 0$. Then $y(e) = c_1 \ln e = c_1 = 0$. Again we have only trivial solutions.

CASE 3

If $\lambda > 0$, we use the identity $e^{i\theta} = \cos \theta + i \sin \theta$ (see equation (15.7.5), p. 912) to write

$$y_1(x) = x^{\sqrt{-\lambda}} = (e^{\ln x})^{i\sqrt{\lambda}} = \cos(\sqrt{\lambda} \ln x) + i \sin(\sqrt{\lambda} \ln x),$$

$$y_2(x) = x^{-\sqrt{-\lambda}} = (e^{\ln x})^{-i\sqrt{\lambda}} = \cos(\sqrt{\lambda} \ln x) - i \sin(\sqrt{\lambda} \ln x).$$

If we let $y_1^* = (y_1 + y_2)/2$ and $y_2^* = (y_1 - y_2)/2i$, we may write the general solution of (6) in the form

$$y(x) = c_1 \cos(\sqrt{\lambda} \ln x) + c_2 \sin(\sqrt{\lambda} \ln x), \qquad \lambda > 0.$$

Setting $x = 1$, we find that the condition $y(1) = 0$ implies that $c_1 = 0$, so

$$y(x) = c_2 \sin(\sqrt{\lambda} \ln x).$$

Finally, if we set $x = e$, we have $0 = y(e) = c_2 \sin \sqrt{\lambda}$. Therefore (6) has a nontrivial solution whenever $\sqrt{\lambda} = k\pi$, $k = 1, 2, 3, \ldots$. Thus the numbers $\lambda_k = k^2\pi^2$, $k = 1, 2, 3$, $\ldots$ are all eigenvalues of (6), and the eigenfunctions corresponding to λ_k have the form

$$y_k(x) = A \sin(k\pi \ln x).$$

Note that it is not necessary to write this solution in the form $A \sin(k\pi \ln |x|)$, because we are interested only in values of x in the interval $1 \le x \le e$. Thus $x > 0$ and $\ln|x| = \ln x$.

EXAMPLE 3

Consider the Sturm-Liouville problem

$$y'' + \lambda y = 0, \qquad y(0) + y'(0) = 0, \qquad y(1) = 0.$$

Then, as in Example 1,

$$y(x) = \begin{cases} c_1 e^{\sqrt{-\lambda}\,x} + c_2 e^{-\sqrt{-\lambda}\,x}, & \text{if } \lambda < 0, \\ c_1 + c_2 x, & \text{if } \lambda = 0, \\ c_1 \cos \sqrt{\lambda}\,x + c_2 \sin \sqrt{\lambda}\,x, & \text{if } \lambda > 0. \end{cases}$$

It is not difficult to verify that the boundary conditions imply that $c_1 = c_2 = 0$ if $\lambda < 0$. If $\lambda = 0$, we have

$$y(0) = c_1, \qquad y'(0) = c_2, \qquad y(1) = c_1 + c_2,$$

and so both boundary conditions imply that $c_1 + c_2 = 0$. We therefore obtain the eigenfunctions

$$y_0(x) = c_1(1 - x).$$

If $\lambda > 0$, then

$$y(0) = c_1, \qquad y'(0) = \sqrt{\lambda}\,c_2, \qquad y(1) = c_1 \cos \sqrt{\lambda} + c_2 \sin \sqrt{\lambda}.$$

The boundary conditions imply that

$$c_1 + \sqrt{\lambda}\,c_2 = 0,$$

$$c_1 \cos \sqrt{\lambda} + c_2 \sin \sqrt{\lambda} = 0.$$

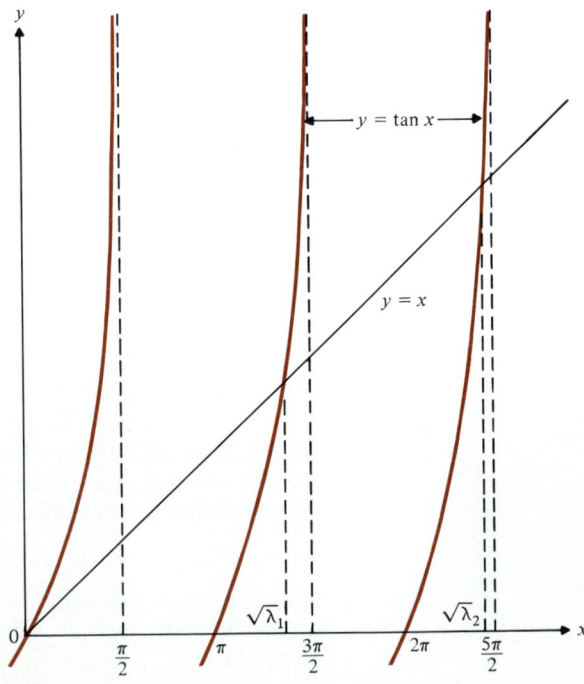

FIGURE 1

Setting $c_1 = -\sqrt{\lambda}c_2$, we find that the second equation becomes

$$-c_2\sqrt{\lambda} \cos \sqrt{\lambda} + c_2 \sin \sqrt{\lambda} = 0,$$

or, dividing by $c_2 \cos \sqrt{\lambda}$ and simplifying,

$$\tan \sqrt{\lambda} = \sqrt{\lambda}.$$

We can easily see that there are an infinite number of real eigenvalues by plotting the curves $y = \tan x$ and $y = x$. The eigenvalues are the squares of the x-values of the points of intersection. From Fig. 1 it seems clear that the square roots of the eigenvalues get closer and closer to the vertical asymptotes of $\tan x$; that is,

$$\sqrt{\lambda_n} \approx \left(\frac{2n+1}{2}\right)\pi \qquad \text{or} \qquad \lambda_n \approx \frac{(2n+1)^2}{4}\pi^2.$$

The eigenfunctions are

$$y_n(x) = A(\sin \sqrt{\lambda_n}x - \sqrt{\lambda_n} \cos \sqrt{\lambda_n}x), \qquad n \geq 1, \qquad A \neq 0.$$

Note. The first positive solution of $\tan x = x$ is $x_1 \approx 4.4934$. Thus the first eigenvalue is $\lambda_1 = x_1^2 \approx 20.1906$.

Problems 11.6

Find the eigenvalues and corresponding eigenfunctions of the given Sturm-Liouville problems.

1. $y'' + \lambda y = 0, \quad y(0) = y(\pi) = 0$
2. $y'' + \lambda y = 0, \quad y(0) = y'(\pi) = 0$
3. $y'' + \lambda y = 0, \quad y(-\pi/2) = y(\pi/2) = 0$
4. $y'' + \lambda y = 0, \quad y'(0) = y'(\pi) = 0$
5. $(xy')' + \dfrac{\lambda}{x}y = 0, \quad y(1) = y(e^2) = 0$

6. $(x^2y')' + \lambda y = 0, \quad y(1) = y(e) = 0$
7. $(x^2y')' + \dfrac{\lambda^2}{x^2}y = 0, \quad y(\frac{1}{2}) = y(1) = 0 \quad$ [*Hint:* Try $y_1 = \sin(\lambda/x)$.]
8. $\left(\dfrac{1}{2x}y'\right)' - 2x\lambda y = 0, \quad y(1) = y(\sqrt{2}) = 0.$

11.7 The Orthogonality Theorem

In this section we investigate four basic properties of the eigenvalues and eigenfunctions of the Sturm-Liouville problem mentioned in the last section:

$$[r(x)y']' + [p(x) + \lambda q(x)]y = 0, \tag{1}$$

with the boundary conditions

$$a_1 y(a) - a_2 y'(a) = 0, \qquad b_1 y(b) - b_2 y'(b) = 0, \tag{2}$$

where at least one coefficient in each equation in (2) is nonzero.

We assume throughout this section that the real-valued functions $r(x)$,

$r'(x)$, $p(x)$, and $q(x)$ in (1) are continuous on the interval $a \leq x \leq b$ and that $r(x)$ and $q(x)$ are positive on $a < x < b$.

Sturm-Liouville problems with these restrictions have the following four basic properties:

---

i. They have infinitely many eigenvalues.

ii. The eigenvalues are real (so we need not worry about complex eigenvalues).

iii. Each eigenvalue has a single linearly independent eigenfunction (simplifying our task of finding eigenfunctions).

iv. Eigenfunctions corresponding to different eigenvalues are orthogonal with respect to the weight function $q(x)$ (allowing us to construct generalized Fourier series in terms of these eigenfunctions).

Not all of the restrictions are required for each of these properties, but all four hold when the restrictions apply.

We prove only part (iv). Parts (ii) and (iii) are left as exercises (see Problems 8 and 11); part (i) is beyond the scope of this book.

We begin by showing that the eigenfunctions of this problem are orthogonal with respect to the weight function $q(x)$.

■ **THEOREM 1: Orthogonality Theorem**

Let the real-valued functions $r(x)$, $r'(x)$, $p(x)$, and $q(x)$ of equation (1) be continuous on the interval $a \leq x \leq b$, and let $y_m(x)$ and $y_n(x)$ be eigenfunctions corresponding to distinct eigenvalues λ_m and λ_n of the Sturm-Liouville problem (1), (2). Then y_m and y_n are orthogonal with respect to the weight function $q(x)$.

PROOF

The functions y_m and y_n satisfy the equations

$$(ry'_m)' + (p + \lambda_m q)y_m = 0,$$

$$(ry'_n)' + (p + \lambda_n q)y_n = 0.$$

Multiplying the first equation by y_n and the second by $-y_m$ and adding the resulting equations together yields

$$y_n(ry'_m)' - y_m(ry'_n)' = (\lambda_n - \lambda_m)qy_m y_n.$$

Integrating all terms from a to b and the first two terms by parts, we obtain

$$\left(y_n r y'_m \big|_a^b - \int_a^b r y'_m y'_n \, dx \right) - \left(y_m r y'_n \big|_a^b - \int_a^b r y'_n y'_m \, dx \right) = (\lambda_n - \lambda_m) \int_a^b q y_m y_n \, dx.$$

(3)

Canceling the two identical integrals on the left-hand side, we obtain

$$r(b)[y_n(b)y'_m(b) - y_m(b)y'_n(b)] - r(a)[y_n(a)y'_m(a) - y_m(a)y'_n(a)]$$

$$= (\lambda_n - \lambda_m) \int_a^b q y_m y_n \, dx. \qquad (4)$$

By hypothesis, at least one of the constants a_1, a_2 in (2) is nonzero, so we have either

$$y(a) = \frac{a_2}{a_1} y'(a) \qquad \text{or} \qquad y'(a) = \frac{a_1}{a_2} y(a). \qquad (5)$$

If the first of these equations is true then it is satisfied by both $y_m(a)$ and $y_n(a)$, so the terms in the second set of brackets in (4) become

$$y_n(a)y'_m(a) - y_m(a)y'_n(a) = \frac{a_2}{a_1} y'_n(a)y'_m(a) - \frac{a_2}{a_1} y'_m(a)y'_n(a) = 0.$$

Similarly if the second equation in (5) is true, then the terms in the second set of brackets are zero. In the same way, the second boundary condition in (2) causes the terms in the first set of brackets in (4) to be zero. Thus

$$(\lambda_n - \lambda_m) \int_a^b q y_m y_n \, dx = 0,$$

and since $\lambda_n \neq \lambda_m$, the proof is complete. ∎

Remark. Observe that if $r(a) = 0$, we do not need the first boundary condition in (2) to prove the orthogonality theorem. Similarly, the second boundary condition is not required if $r(b) = 0$. Finally, if $r(a) = r(b)$, we can also obtain the conclusion of the orthogonality theorem by assuming the two-point conditions

$$y(a) = y(b), \qquad y'(a) = y'(b) \qquad (6)$$

instead of (2).† The proof of this fact is obvious, since the quantities in brackets in (4) are now identical, so that the left-hand side of (4) is zero. Furthermore, *if the functions $r(x)$, $p(x)$, and $q(x)$ are periodic with period $(b - a)$, then any eigenfunction of this problem is also periodic with period $(b - a)$* (see Problem 6).

EXAMPLE 1

Consider the Sturm-Liouville problem

$$y'' + \lambda y = 0, \qquad y(0) = y'(1) = 0.$$

---

† Note that property (iii) fails if equations (2) are replaced by the mixed two-point conditions (6) (see Problem 10).

As we saw in Example 11.6.1, the eigenfunctions of this problem have the form

$$\sin\frac{\pi}{2}x,\ \sin\frac{3\pi}{2}x,\ \sin\frac{5\pi}{2}x,\dots,\ \sin\frac{(2k-1)}{2}\pi x,\dots.$$

Theorem 1 implies that these functions are orthogonal, a fact that we established directly in Section 11.3.

EXAMPLE 2

Using the orthogonality theorem, it is easy to verify that the Legendre polynomials form an orthogonal set of functions. Since $[(1-x^2)y']' = (1-x^2)y'' - 2xy'$, we can write Legendre's equation (see Section 5.5) in the Sturm-Liouville form

$$[(1-x^2)y']' + \lambda y = 0, \qquad \lambda = n(n+1).$$

Since $r(x) = 1 - x^2$ vanishes at $x = \pm 1$, no boundary conditions are needed for the theorem to apply; and since $q(x) \equiv 1$, we immediately have

$$(P_m, P_n) = \int_{-1}^{1} P_m(x)P_n(x)\, dx = 0, \qquad \text{if } m \neq n.$$

The endpoints a and b in the orthogonality theorem need not be finite provided the improper integrals in (3) all exist. We illustrate this situation with the following example.

EXAMPLE 3: The Laguerre Equation

$$(xe^{-x}y')' + ne^{-x}y = 0$$

is of the Sturm-Liouville type, and $r(x) = xe^{-x}$ vanishes at $x = 0$. By L'Hôpital's rule,

$$\lim_{x\to\infty} r(x) = \lim_{x\to\infty} \frac{x}{e^x} = \lim_{x\to\infty} \frac{1}{e^x} = 0.$$

Thus for the Laguerre polynomials $\{y_n(x)\}$ no boundary conditions are required to apply the orthogonality theorem, and the Laguerre polynomials $\{y_n(x)\}$ form an orthogonal set of functions with respect to the weight function $q(x) = e^{-x}$ (compare this proof to that in Example 11.2.3):

$$\int_0^{\infty} e^{-x}y_m(x)y_n(x)\, dx = 0 \qquad \text{if } m \neq n.$$

In every example we have seen so far, the eigenvalues of the given Sturm-Liouville problem are real, not because the examples were carefully chosen, but because the situation holds for every Sturm-Liouville problem. The proof of this fact, stated below, is very similar to the proof of Theorem 1 and is therefore left as an exercise (see Problem 8).

■ THEOREM 2

The eigenvalues of the Sturm-Liouville problem (1), (2) are all real. ■

Finally, we have observed that in every example there are an infinite number of eigenvalues, and that corresponding to each eigenvalue there is only one linearly independent eigenfunction; that is, if y_1 and y_2 are two eigenfunctions corresponding to the eigenvalue λ, then there is a constant c such that $y_1 = cy_2$. When the latter condition holds, the eigenvalue is said to be **simple**. These results are summarized in the theorem below, whose proof, however, is beyond the scope of this text.† (The simplicity of the eigenvalues can be proved; see Problem 11.)

Simple Eigenvalue

■ THEOREM 3

The eigenvalues of the Sturm-Liouville problem (1), (2) are simple. Moreover, there are an infinite number of them, which can be arranged in an increasing order

$$\lambda_1 < \lambda_2 < \lambda_3 < \cdots < \lambda_k < \cdots,$$

where λ_k tends to ∞ as k tends to ∞. ■

All three theorems of this section are needed for the solution of non-homogeneous boundary value problems in Section 11.8.

Problems 11.7

In Problems 1–5 verify the implications of Theorems 1, 2, and 3 for each given Sturm-Liouville problem.

1. $y'' + \lambda y = 0$, $\quad y(0) = y(\pi) = 0$
2. $y'' + \lambda y = 0$, $\quad y(0) = y'(\pi) = 0$
3. $y'' + \lambda y = 0$, $\quad y(-\pi/2) = y(\pi/2) = 0$
4. $(xy')' + (\lambda/x)y = 0$, $\quad y(1) = y(e^2) = 0$
5. $(x^2y')' + (\lambda/x^2)y = 0$, $\quad y(\frac{1}{2}) = y(1) = 0$

* 6. Prove that if $r(x)$, $p(x)$, and $q(x)$ are continuous periodic functions of period $(b - a)$, then all eigenfunctions of the boundary value problem (1), (6) are periodic with period $(b - a)$.

7. **a.** Show that the Hermite equation [see Problem 11.2.9(c)] can be written in the form

$$(e^{-x^2}y')' + 2ne^{-x^2}y = 0.$$

b. Using the orthogonality theorem, prove that the Hermite polynomials are orthogonal with respect to the weight function $\exp(-x^2)$ on $-\infty < x < \infty$.

8. Prove that the eigenvalues of the Sturm-Liouville problem (1), (2) are all real by carrying out the following steps:

a. Show that if $\lambda = \alpha + i\beta$ is an eigenvalue with the corresponding eigenfunction $y_\lambda(x) = u(x) + iv(x)$, then $\bar\lambda = \alpha - i\beta$ is an eigenvalue with the corresponding eigenfunction $\bar y_\lambda(x) = u(x) - iv(x)$.

b. Using the same proof as that of Theorem 1, show that

$$(\lambda - \bar\lambda) \int_a^b y_\lambda(x)\bar y_\lambda(x)q(x)\,dx = 0.$$

c. Use the result of part (b) of this problem and the fact that $y_\lambda \bar y_\lambda = u^2 + v^2$ to show that $\beta = 0$.

9. The boundary value problem

$$y'' - \frac{\lambda}{x}y' + \frac{\lambda}{x^2}y = 0, \qquad y(1) = y(2) = 0,$$

is not a Sturm-Liouville problem. Show that none of the eigenvalues is real.

† See, for example, the excellent book by R. Courant and D. Hilbert, *Methods of Mathematical Physics*, Wiley, New York, 1953, vol. I, chap. 6.

10. Consider the boundary value problem

$$y'' + \lambda y = 0, \qquad y(-1) = y(1), \qquad y'(-1) = y'(1).$$

a. Explain why it is not a Sturm-Liouville problem.

b. Calculate the eigenvalues of this problem.

c. Show that corresponding to each nonzero eigenvalue there are two linearly independent eigenfunctions; that is, show that the eigenvalues are not simple.

11. Prove that the eigenvalues of a Sturm-Liouville problem are simple. [*Hint:* Assume that y_1 and y_2 are two eigenfunctions corresponding to the eigenvalue λ. Use the boundary conditions (2) to show that the Wronskian $W(y_1, y_2)(x) = 0$ for $a \le x \le b$.]

12. Show that any second-order linear differential equation can be written in the form (1).

11.8 Nonhomogeneous Boundary Value Problems

In Example 11.1.3 (p. 668) we considered a nonhomogeneous boundary value problem given by equation (11.1.12). At that time we stated that the solution of (11.1.12), for $\omega \ne n\pi$, could be expressed by the Fourier sine series (11.1.16). In this section we justify that statement and unite the seemingly unrelated concepts of Fourier series and Sturm-Liouville problems. We discover that the solutions of certain nonhomogeneous boundary value problems can be obtained as generalized Fourier series in the eigenfunctions of the associated Sturm-Liouville problem.

Eigenfunction Expansions

This method of **eigenfunction expansions** is one of the main techniques used in solving nonhomogeneous boundary value problems. Furthermore, it provides the motivation for the very powerful method of **Green's functions** that we discuss at the end of this section, which is extremely useful for solving such problems.

Suppose that we wish to solve the *nonhomogeneous* differential equation

$$[r(x)w']' + p(x)w = f(x) \tag{1}$$

on $a \le x \le b$ with the boundary conditions

$$a_1 w(a) - a_2 w'(a) = 0, \qquad b_1 w(b) - b_2 w'(b) = 0. \tag{2}$$

We first consider a different boundary value problem,

$$[r(x)y']' + [p(x) + \lambda q(x)]y = 0, \tag{3}$$

with identical boundary conditions,

$$a_1 y(a) - a_2 y'(a) = 0, \qquad b_1 y(b) - b_2 y'(b) = 0, \tag{4}$$

and assume that its eigenvalues and eigenfunctions are known. We are free to choose $q(x)$ in any way we wish provided we are able to determine the eigenfunctions of (3), (4). Different choices of $q(x)$ yield different sets of eigenfunctions, and some sets may be easier to work with than others. We assume that $r(x)$, $r'(x)$, $p(x)$, and $q(x)$ are continuous and that r and q are positive on $a \le x \le b$. By Theorem 11.7.3 and the orthogonality theorem, we know that the problem (3), (4) has an infinite orthogonal set $\{y_k\}$ of eigenfunctions with respect to the weight function $q(x)$.

The object now is to express the solution $w(x)$ of (1), (2) (if one exists) as a generalized Fourier series [see equation (11.3.16)] of the form

$$w(x) = \sum_{n=1}^{\infty} c_n y_n, \tag{5}$$

where the functions y_n are eigenfunctions of (3), (4). Of course, since we are proceeding formally, when the process is complete we will have to check that our result is indeed a solution. As we saw in Section 11.3, the problem of determining which functions can be represented in this way is not easy. Although the proof is beyond the scope of this book, *any piecewise continuous function $w(x)$ with piecewise continuous derivative $w'(x)$ has a representation (5) at all points of continuity, provided that a and b are finite.*† (At a point of discontinuity, the generalized Fourier series converges to $[w(x + 0) + w(x - 0)]/2$.)

To obtain the formal representation (5), we proceed in a manner similar to the proof of the orthogonality theorem. Multiplying (1) by y and (3) by $-w$, we obtain the equation

$$y(rw')' - w(ry')' = fy + \lambda qyw.$$

Integrating both sides from a to b and the left-hand side by parts, we have

$$\left(yrw' \Big|_a^b - \int_a^b rw'y' \, dx \right) - \left(wry' \Big|_a^b - \int_a^b ry'w' \, dx \right) = \int_a^b (fy + \lambda qyw) \, dx.$$

The integrals on the left-hand side cancel out, yielding the equation

$$r(b)[y(b)w'(b) - w(b)y'(b)] - r(a)[y(a)w'(a) - w(a)y'(a)] = \int_a^b (fy + \lambda qyw) \, dx. \tag{6}$$

Letting $\lambda = \lambda_k$ and $y = y_k$, we see, as in the proof of the orthogonality theorem, that the left-hand side of (6) is zero. Thus

$$\lambda_k \int_a^b qwy_k \, dx = -\int_a^b fy_k \, dx,$$

and since, by equation (11.3.17) on p. 688,

$$c_k = \frac{\int_a^b qwy_k \, dx}{\int_a^b qy_k^2 \, dx},$$

we finally obtain

$$c_k = -\frac{\int_a^b fy_k \, dx}{\lambda_k \int_a^b qy_k^2 \, dx}. \tag{7}$$

† Much more general theorems are known. See, for example, E. A. Coddington and N. Levinson, *Theory of Ordinary Differential Equations,* McGraw-Hill, New York, 1955, p. 199.

In particular, if the eigenfunctions y_k have been chosen so that they are ortho-normal with respect to q, then the denominator in the right-hand side of (7) reduces to just the eigenvalue λ_k, and we have the representation

$$w(x) = -\sum_{n=1}^{\infty}\left(\lambda_n^{-1}\int_a^b fy_n\,dx\right)y_n(x) \qquad (8)$$

for the solution of the boundary value problem (1), (2). We now apply this result to justify the solution of Example 11.1.3 [see equation (11.1.16)].

EXAMPLE 1

Suppose that we again consider the forced vibrations of a spring-mass system, with spring constant k, attached to an external periodic force $K \sin \omega t$. For simplicity, we ignore the damping term and merely consider the nonhomogeneous equation

$$x'' + \frac{k}{m}x = \frac{K}{m}\sin \omega t. \qquad (9)$$

We again suppose that the object is at its equilibrium position initially and when $t = 1$ second, and we seek a solution to the nonhomogeneous boundary value problem.

To apply the eigenfunction expansion (8), we consider the Sturm-Liouville problem

$$y'' + (\lambda + 1)\frac{k}{m}y = 0, \qquad y(0) = y(1) = 0. \qquad (10)$$

The eigenvalues of (10) are obtained by setting $\sqrt{(\lambda+1)k/m}$ equal to a multiple of π; hence

$$\lambda_n = \frac{mn^2\pi^2}{k} - 1, \qquad n = 1, 2, 3, \ldots.$$

The corresponding eigenfunctions have the form

$$y_n = A \sin n\pi t, \qquad n = 1, 2, 3, \ldots,$$

and are orthonormal with respect to $q(x) = k/m$ if $A = \sqrt{2m/k}$. By equation (7), the Fourier coefficients c_n are given by

$$c_n = -\sqrt{\frac{2m}{k}} \cdot \frac{K}{m\lambda_n}\int_0^1 \sin \omega t \sin n\pi t \, dt.$$

Using Formula 103 of Appendix 1 and the identity

$$\sin(\omega \pm n\pi)t = \sin \omega t \cos n\pi t \pm \cos \omega t \sin n\pi t,$$

we have

$$c_n = -\sqrt{\frac{2m}{k}} \cdot \frac{K}{2m\lambda_n}\left[\frac{\sin(\omega - n\pi)t}{\omega - n\pi} - \frac{\sin(\omega + n\pi)t}{\omega + n\pi}\right]\Bigg|_0^1$$

$$= -\sqrt{\frac{2m}{k}} \cdot \frac{K}{m\lambda_n}\sin \omega \cos n\pi \frac{n\pi}{\omega^2 - n^2\pi^2}, \qquad \omega \neq n\pi.$$

If $\omega = n\pi$, then

$$c_n = -\sqrt{\frac{2m}{k}} \cdot \frac{K}{2m\lambda_n},$$

and all the other Fourier coefficients vanish by the orthogonality of the eigenfunctions. Thus the solution $x(t)$ must be stated for two cases:

(a) If $\omega \neq n\pi$, $n = 1, 2, 3, \ldots$, then by (8) and the identity $\cos n\pi = (-1)^n$,

$$x(t) = -\frac{2K\pi \sin \omega}{k} \sum_{n=1}^{\infty} \frac{(-1)^n n}{\lambda_n(\omega^2 - n^2\pi^2)} \sin n\pi t.$$

This solution is identical to equation (11.1.16).

(b) Suppose that $\omega = n\pi$ for some positive integer n. Then using (8) and the value of λ_n, we obtain the particular solution

$$x_p(t) = -\frac{K}{k\lambda_n} \sin n\pi t = \frac{K \sin n\pi t}{k - mn^2\pi^2}.$$

This solution is valid provided that the spring constant $k \neq mn^2\pi^2$.

The general solution is now obtained as in Example 11.1.3. When $k = mn^2\pi^2$, there is *no* solution. Observe that in this case the eigenvalue,

$$\lambda_n = \frac{mn^2\pi^2}{k} - 1,$$

is zero.

The following result is beyond the scope of this book, but it provides the necessary information to decide whether or not there is a solution in the case above.

■ **THEOREM 1†**

Let one of the eigenvalues λ_k of the problem in (3), (4) be zero. Then the nonhomogeneous boundary value problem in (1), (2) has a solution if and only if

$$\int_a^b f(t) y_k(t)\, dt = 0.$$

■

EXAMPLE 2

Consider the boundary value problem

$$(xw')' + \frac{w}{x} = \frac{1}{x}, \qquad w(1) = w(e) = 0. \tag{11}$$

To apply the method of eigenfunction expansion (8), we let $p(x) = q(x) = f(x) = 1/x$ and consider the Sturm-Liouville problem

$$(xy')' + \frac{(1+\lambda)}{x} y = 0, \qquad y(1) = y(e) = 0. \tag{12}$$

† See, for example, E. A. Coddington and N. Levinson, *Theory of Ordinary Differential Equations*, McGraw-Hill, New York, 1955, p. 294.

By Example 11.6.2, we know that the eigenvalues $\lambda^* = 1 + \lambda$ of (12) are given by

$$1 + \lambda_k = \lambda_k^* = k^2\pi^2, \qquad k = 1, 2, 3, \ldots,$$

so $\lambda_k = k^2\pi^2 - 1$ and the eigenfunctions are of the form

$$y_k = A \sin(k\pi \ln x).$$

Using the substitution $u = k\pi \ln x$, $du = (k\pi/x)\, dx$, we have

$$\int_1^e \frac{1}{x} \sin^2(k\pi \ln x)\, dx = \frac{1}{k\pi} \int_0^{k\pi} \sin^2 u\, du = \frac{u - \cos u \sin u}{2k\pi}\bigg|_0^{k\pi} = \frac{1}{2}.$$

Letting $A = \sqrt{2}$, we obtain an orthonormal set of eigenfunctions. By (7), the coefficients c_k of the generalized Fourier series are given by

$$c_k = -\lambda_k^{-1} \int_1^e \frac{\sqrt{2}}{x} \sin(k\pi \ln x)\, dx = \frac{\sqrt{2}\lambda_k^{-1}}{k\pi} \cos(k\pi \ln x)\bigg|_1^e$$

$$= \frac{\sqrt{2}(\cos k\pi - 1)}{k\pi\lambda_k} = \begin{cases} 0, & k \text{ even,} \\[2mm] \dfrac{-2\sqrt{2}}{k\pi(k^2\pi^2 - 1)}, & k \text{ odd.} \end{cases}$$

Hence the solution of problem (11) is given by the series

$$w(x) = -4 \sum_{n=0}^{\infty} \frac{\sin[(2n+1)\pi \ln x]}{[(2n+1)\pi]^3 - [(2n+1)\pi]}.$$

A number of further examples of the use of eigenfunction expansions will be found in Chapter 12, where they are used in the solution of certain partial differential equations.

The methods we have presented in this chapter are not the only techniques that can be used to solve boundary value problems. A very powerful procedure, attributed to George Green (1793–1841), involves the representation of the solution of boundary value problem (1), (2) in the form

$$w(x) = \int_a^b K(x, t)f(t)\, dt. \tag{13}$$

Actually, in light of the eigenfunction expansion (8), a representation of the form (13) is not surprising, for if we use t as the variable of integration and interchange the sum and integral in (8), we obtain the expression

$$\int_a^b \left[-\sum_{n=1}^{\infty} \frac{y_n(t)y_n(x)}{\lambda_n} \right] f(t)\, dt.$$

Green's Function

Hence the **Green's function** $K(x, t)$ may be presumed to equal

$$K(x, t) = -\sum_{n=1}^{\infty} \frac{y_n(t)y_n(x)}{\lambda_n}. \tag{14}$$

All the assumptions we made above can be justified.† Problems 8–12 indicate another method of obtaining the Green's function $K(x, t)$.

Problems 11.8

In Problems 1–5 use the method of eigenfunction expansions to solve the given nonhomogeneous boundary value problems.

1. $y'' + y = x, \quad y(0) = y(\pi) = 0$

2. $y'' + 2y = \cos x, \quad y(0) = y(1) = 0$

3. $y'' + 4y = x^2, \quad y(0) = y'(1) = 0$

4. $(xy')' + \dfrac{1}{x} y = \ln x, \quad y'(1) = y'(e^{2\pi}) = 0$

5. $(xy')' + \dfrac{3}{x} y = \dfrac{1}{x} \sin(\ln x), \quad y(1) = y(2) = 0$

* 6. Consider the nonhomogeneous problem

$$y'' + \lambda y = f(x), \qquad y(0) = y(1) = 0.$$

Show that if $f(x)$ is continuous, then there is a unique solution to this problem if and only if λ is *not* an eigenvalue of the associated homogeneous equation $y'' + \lambda y = 0$, $y(0) = y(1) = 0$.

* 7. Prove the claim in Problem 6 for

$$y'' + \lambda y = f(x), \qquad y(0) = y(1), \qquad y'(0) = y'(1).$$

Consider the differential equation

$$(rw')' + qw = 0.$$

It is easy to find a nontrivial solution w_1 of this equation with $w_1(a) = 0$, and a nontrivial solution w_2 with $w_2(b) = 0$. Define the function

$$k(x, t) = \begin{cases} w_1(x)w_2(t), & a < x < t, \\ w_1(t)w_2(x), & t < x < b. \end{cases}$$

8. Ignoring any difficulties at $t = x$, show that

$$\frac{d}{dx} \int_a^b k(x, t)f(t)\, dt = \int_a^b \frac{\partial}{\partial x} k(x, t)f(t)\, dt.$$

9. Show that

$$\frac{d}{dx}\left[r(x)\frac{d}{dx}\int_a^b k(x, t)f(t)\, dt \right]$$

$$= r(x)f(x)W(x) - q(x) \int_a^b k(x, t)f(t)\, dt,$$

where W is the Wronskian of the functions w_1 and w_2.

10. Using the result of Problem 9, conclude that

$$w_3(x) = \int_a^b k(x, t)f(t)\, dt$$

satisfies the equation

$$[r(x)w_3']' + q(x)w_3 = f(x)r(x)W(x).$$

11. Show that

$$\frac{d}{dx}(r(x)W(x)) = 0,$$

and conclude that $r(x)W(x) = C$, a constant.

12. Prove that

$$w(x) = C^{-1}w_3(x)$$

is a solution of the boundary value problem

$$(r(x)w')' + q(x)w = f(x), \qquad w(a) = w(b) = 0.$$

Hence conclude that the Green's function is given by $K(x, t) = C^{-1}k(x, t)$.

13. Using the ideas of Problems 8–12, find the Green's function for the boundary value problem

$$w''(x) - w(x) = f(x), \qquad w(0) = w(1) = 0.$$

14. Find the Green's function for the boundary value problem

$$w''(x) + k^2 w(x) = f(x), \qquad w(0) = w(a) = 0.$$

* 15. Generalize the procedure used in Problems 8–12 to obtain the Green's function for the boundary value problem

$$(rw')' + qw = f, \qquad w(0) = w'(1) = 0.$$

Apply the generalized procedure of Problem 15 to Problems 16–18.

16. $w''(x) = f(x), \quad w(0) = w'(1) = 0$

17. $w''(x) + 4w(x) = f(x), \quad w(0) = w'(1) = 0$

18. $(xw')' = f(x), \quad w(1) = w'(e) = 0$

† See, for example, E. A. Coddington and N. Levinson, *Theory of Ordinary Differential Equations*, McGraw-Hill, New York, 1955.

(Optional)   ## 11.9 Mean-Square Approximation

We have seen that if $F(x)$ is periodic of period $2T$ and if $F(x)$ can be represented by a trigonometric series, then

$$F(x) = \frac{a_0}{2} + \sum_{n=1}^{\infty} \left(a_n \cos \frac{n\pi x}{T} + b_n \sin \frac{n\pi x}{T} \right), \tag{1}$$

where the coefficients $a_0, a_1, b_1, a_2, b_2, \ldots$ are given by the Euler formulas (11.3.8) and (11.3.9).

In a practical situation no one adds up the infinite number of terms in (1). Rather, it is necessary to stop at some number N. We then have an **approximation** of $F(x)$:

$$F(x) \approx \frac{a_0}{2} + \sum_{n=1}^{N} \left(a_n \cos \frac{n\pi x}{T} + b_n \sin \frac{n\pi x}{T} \right) = P_N(x). \tag{2}$$

Trigonometric Polynomial

The function $P_N(x)$ is called a **trigonometric polynomial**. A question that naturally arises is

Is P_N the "best" approximation of the form (2) for $F(x)$?

In order to answer the question, we have to decide how to measure the "goodness" of an approximation. Here are four commonly used measures: Let lub denote the *least upper bound* of a set of numbers. (According to the completeness axiom, every bounded set of numbers has a least upper bound.)

$$\epsilon_a = \text{absolute error} = \underset{|x| \le T}{\text{lub}} \, |F(x) - P_N(x)|, \tag{3}$$

$$\epsilon_{as} = \text{absolute square error} = \underset{|x| \le T}{\text{lub}} \, [F(x) - P_N(x)]^2, \tag{4}$$

$$\epsilon_m = \text{mean error} = \int_{-T}^{T} |F(x) - P_N(x)| \, dx, \tag{5}$$

$$\epsilon_{ms} = \text{mean-square error} = \int_{-T}^{T} [F(x) - P_N(x)]^2 \, dx. \tag{6}$$

The first two measure error at each point x in $[-T, T]$ and take the least upper bound of all these values. The sequence $\{P_N(x)\}$ converges to $F(x)$ at each point x if the term (3) or (4) approaches 0 as $N \to \infty$.

If we require ϵ_a (or ϵ_{as}) to be small, we shall require also that $\epsilon_m/2T$ (or $\epsilon_{ms}/2T$) be small, because

$$\epsilon_m = \int_{-T}^{T} |F(x) - P_N(x)| \, dx \le \epsilon_a \int_{-T}^{T} dx = 2\epsilon_a T.$$

In many situations it may not matter that $|F(x) - P_N(x)|$ [or $(F(x) - P_N(x))^2$] be large on small sets of x-values. Then mean error or mean-square error is used. If either of the terms (5) or (6) approaches zero as $N \to \infty$, then we say that $P_N(x)$ **converges in the mean** to $F(x)$.

Convergence in Mean

When dealing with Fourier series it is much easier to use mean-square error as our measure. With this measure, we shall answer the question posed earlier as follows:

> $F(x)$ is "best" approximated in the mean-square sense by the polynomial (2) in which a_n and b_n are the Fourier coefficients
>
> $$a_n = \frac{1}{T} \int_{-T}^{T} F(x) \cos \frac{n\pi x}{T} dx \quad \text{and} \quad b_n = \frac{1}{T} \int_{-T}^{T} F(x) \sin \frac{n\pi x}{T} dx. \qquad (7)$$

We prove this result in a more general setting. Before going further, reread the material on generalized Fourier series that begins on p. 687. Let $f_1(x), f_2(x), \ldots, f_N(x), \ldots$ be an orthonormal set of functions on the interval $[a, b]$ (see p. 671). That is,

$$\int_a^b f_n(x)f_m(x)\, dx = \begin{cases} 0, & \text{if } n \neq m, \\ 1, & \text{if } n = m. \end{cases} \qquad (8)$$

Let $F(x)$ be a continuous function on $[a, b]$. We seek the polynomial

$$P_N(x) = a_1 f_1(x) + a_2 f_2(x) + \cdots + a_N f_N(x)$$

that best approximates $F(x)$ in the mean-square sense. That is, we seek coefficients $a_1, a_2, \ldots, a_N$ such that

$$\epsilon_{ms} = \int_a^b [F(x) - (a_1 f_1(x) + a_2 f_2(x) + \cdots + a_N f_N(x))]^2\, dx \qquad (9)$$

is a minimum. We expand the integrand in (9):

$$\epsilon_{ms} = \underbrace{\int_a^b [F(x)]^2\, dx}_{①} - \underbrace{2 \int_a^b [a_1 f_1(x) + a_2 f_2(x) + \cdots + a_N f_N(x)] F(x)\, dx}_{②}$$

$$+ \underbrace{\int_a^b [a_1 f_1(x) + a_2 f_2(x) + \cdots + a_N f_N(x)]^2\, dx.}_{③} \qquad (10)$$

We consider each of the three integrals in (10):

① $\quad\displaystyle \int_a^b [F(x)]^2\, dx = \|F\|^2.$

② $\quad\displaystyle \int_a^b a_m f_m(x) F(x)\, dx = a_m \int_a^b f_m(x) F(x)\, dx = a_m c_m$

Formula (15) on p. 687
with $\|f_m\|^2 = 1$ $\downarrow$
$= a_m$ times the mth generalized Fourier coefficient of $F(x)$ with respect to the orthonormal set $\{f_n(x)\}$.

Thus

$$2 \int_a^b [a_1 f_1(x) + \cdots + a_N f_N(x)] F(x) \, dx = 2[a_1 c_1 + a_2 c_2 + \cdots + a_N c_N].$$

③ There are two kinds of terms when we expand the integrand:

(a)
$$\int_a^b a_m^2 f_m^2(x) \, dx = a_m^2 \int_a^b f_m^2(x) \overset{\text{from (8)}}{=} a_m^2,$$

(b)
$$\int_a^b 2 a_k a_m f_k(x) f_m(x) \, dx = 2 a_k a_m \int_a^b f_k(x) f_m(x) \, dx \overset{\text{from (8)}}{=} 0, \qquad k \neq m.$$

Thus

$$\epsilon_{ms} = \|F\|^2 + \sum_{m=1}^N (a_m^2 - 2 a_m c_m).$$

But, completing the square, we find that

$$a_m^2 - 2 a_m c_m = (a_m - c_m)^2 - c_m^2,$$

so

$$\epsilon_{ms} = \|F\|^2 - \sum_{m=1}^N c_m^2 + \sum_{m=1}^N (a_m - c_m)^2. \tag{11}$$

Equation (11) enables us to prove a number of interesting results.

■ **THEOREM 1: Fourier Coefficients Are Best in the Mean-Square Sense**

Let $\{f_n(x)\}$ be an orthonormal set of functions in the interval $[a, b]$ and suppose that $F(x)$ is continuous on $[a, b]$. Define

$$c_m = \int_a^b F(x) f_m(x) \, dx. \tag{12}$$

Let $P_N(x) = c_1 f_1(x) + c_2 f_2(x) + \cdots + c_N f_N(x)$. If $a_1, a_2, \ldots, a_N$ is another set of constants, let

$$q_N(x) = a_1 f_1(x) + a_2 f_2(x) + \cdots + a_n f_N(x). \tag{13}$$

Then

$$\int_a^b [F(x) - P_N(x)]^2 \, dx \leq \int_a^b [F(x) - q_N(x)]^2 \, dx.$$

Fourier Polynomial

That is, the **Fourier polynomial** $P_N(x)$ gives the least mean-square error among all functions that can be written in the form (13).

PROOF

In (11), $\|F\|^2$ and $\sum_{m=1}^{N} c_m^2$ are constant. The last term, $\sum_{m=1}^{N} (a_m - c_m)^2$, is nonnegative and is minimized when $a_m = c_m$, for $m = 1, 2, \ldots, N$. In this case $\epsilon_{ms} = \|F\|^2 - \sum_{m=1}^{N} c_m^2$. ∎

■ **COROLLARY**

Statement (7) is valid. ■

■ **THEOREM 2: Bessel's Inequality**

Under the assumptions of Theorem 1, the series $\sum_{m=1}^{\infty} c_m^2$ converges and

$$\sum_{m=1}^{\infty} c_m^2 \leq \int_a^b [F(x)]^2 \, dx. \tag{14}$$

PROOF

Let $a_m = c_m$ in (11). Then

$$0 \leq \int_a^b [F(x) - P_N(x)]^2 \, dx = \|F\|^2 - \sum_{m=1}^{N} c_m^2$$

or

$$\sum_{m=1}^{N} c_m^2 \leq \|F\|^2 = \int_a^b [F(x)]^2 \, dx.$$

Let $N \to \infty$ to obtain (14). ■

Least Squares Polynomial Approximation

In Section 5.5 we discussed **Legendre's differential equation**

$$(1 - x^2)y'' - 2xy' + p(p+1)y = 0. \tag{15}$$

We showed (see p. 284) that if p is a nonnegative integer, then (15) has at least one solution that is a polynomial of degree p. This solution is called a **Legendre polynomial,** denoted by $P_p(x)$. We saw on p. 285 that

$$P_p(x) = \sum_{k=0}^{M} \frac{(-1)^k (2p - 2k)!}{2^p k! (p-k)! (p-2k)!} x^{p-2k}, \qquad p = 0, 1, 2, \ldots, \tag{16}$$

where M is the largest integer not greater than $p/2$. In particular, $P_0(x) = 1$, $P_1(x) = x$, $P_2(x) = \frac{1}{2}(3x^2 - 1)$, $P_3(x) = \frac{1}{2}(5x^3 - 3x)$, and $P_4(x) = \frac{1}{8}(35x^4 - 30x^2 + 3)$. In Problems 16–18 you are asked to show that $\{P_n(x)\}$ is an orthogonal set on the interval $[-1, 1]$. (Also see Example 11.7.2.) Let

Equation (15) on p. 687

$$\downarrow$$
$$c_m = \frac{1}{\|P_m\|^2} \int_{-1}^{1} F(x)P_m(x)\, dx. \tag{17}$$

Then, by Theorem 1, the Nth degree polynomial that best approximates $F(x)$ over $[-1, 1]$ in the mean-square sense is given by

$$q_N(x) = c_1 P_1(x) + c_2 P_2(x) + \cdots + c_N P_N(x). \tag{18}$$

Least Squares Polynomial Approximation

The polynomial $q_N(x)$ is called the **least squares polynomial approximation** to $F(x)$ on the interval $[-1, 1]$.

EXAMPLE 1

Find the least squares approximation to e^x on $[-1, 1]$ by a straight line, that is, by a polynomial of degree 1.

SOLUTION
$P_0 = 1$ and $\|P_0\|^2 = \int_{-1}^{1} 1\, dx = 2$; $P_1(x) = x$ and $\|P_1\|^2 = \int_{-1}^{1} x^2\, dx = \frac{2}{3}$. Then

$$c_0 = \frac{1}{2}\int_{-1}^{1} e^x\, dx = \frac{e^1 - e^{-1}}{2} = \sinh 1,$$

$$c_1 = \frac{3}{2}\int_{-1}^{1} xe^x\, dx = \frac{3}{2}(x-1)e^x\Big|_{-1}^{1} = \frac{3}{e}.$$

Thus

$$q_1(x) = (\sinh 1)P_0(x) + \frac{3}{e}P_1(x) = \sinh 1 + \frac{3x}{e}.$$

Note that $q_1(x)$ is *not* the first two terms of the Maclaurin expansion of e^x $(=1 + x + x^2/2! + x^3/3! + \cdots)$.

Problems 11.9

1. Find constants a_0, a_1, and a_2 such that $\int_{-\pi}^{\pi} [1 - P_2(x)]^2\, dx$ is a minimum, where $P_2(x) = \frac{a_0}{2} + a_1 \cos x + a_2 \sin 2x$.

2. Find constants a_1, a_2, and a_3 such that $\int_{0}^{\pi} [1 - P_3(x)]^2\, dx$ is a minimum, where $P_3(x) = a_1 \sin x + a_2 \sin 2x + a_3 \sin 3x$.

3. Find constants a_1, a_2, and a_3 such that $\int_{-1}^{1} [x - P_3(x)]^2\, dx$ is a minimum, where $P_3(x) = a_1 \sin \pi x + a_2 \sin 2\pi x + a_3 \sin 3\pi x$.

4. Find constants a_0, a_1, a_2, and a_3 such that

$$\int_{0}^{1} [x - P_3(x)]^2\, dx$$

is a minimum, where $P_3(x) = \frac{a_0}{2} + a_1 \cos \pi x + a_2 \cos 2\pi x + a_3 \cos 3\pi x$.

5. Find a polynomial $P_3(x) = \frac{a_0}{2} + a_1 \cos \pi x + a_2 \cos 2\pi x + a_3 \cos 3\pi x$ that best approximates x^2 in the mean-square sense in the interval $[-1, 1]$.

6. Let $\{f_n(x)\}$ denote the functions $\{1, \sin x, \cos x, \sin 2x, \cos 2x, \ldots\}$. Let $F(x)$ be continuous in $[-\pi, \pi]$ and suppose that a_n and b_n are given by formulas (11.3.8) and (11.3.9) on p. 680. Prove that

$$\frac{a_0^2}{2} + \sum_{n=1}^{\infty} a_n^2 + \sum_{n=1}^{\infty} b_n^2 \le \frac{1}{\pi}\int_{-\pi}^{\pi} [F(x)]^2\, dx.$$

In Problems 7–9 represent each polynomial as a linear combination of Legendre polynomials.

7. x^4

8. $x^5 - x + 3$

9. $9x^3 - 8x^2 + 7x - 6$

In Problems 10–14 find a quadratic polynomial that best approximates the given function in the least squares sense on the interval $[-1, 1]$.

10. e^x

11. $\sin x$

12. $\cos x$

13. $\sinh x$

14. $\dfrac{1}{1+x^2}$

15. Given that $p(x)$ is a polynomial of degree $n \, (\geq 1)$ and

$$\int_{-1}^{1} x^k p(x)\,dx = 0 \qquad \text{for } k = 0, 1, 2, \ldots, n-1,$$

show that $p(x) = cP_n(x)$ for some constant c.

16. Use Rodrigues' formula [(5.5.9) on p. 285] to show that

$$\int_{-1}^{1} P_n(x)P_m(x)\,dx = \frac{1}{2^n n!} \int_{-1}^{1} P_m(x)\frac{d^n}{dx^n}[(x^2-1)^n]\,dx.$$

17. (a) Use the result of Problem 16 to show that

$$\int_{-1}^{1} P_n(x)P_m(x)\,dx = \frac{P_m(x)}{2^n n!} \frac{d^{n-1}}{dx^{n-1}}[(x^2-1)^n]\Bigg|_{-1}^{1}$$

$$-\frac{1}{2^n n!}\int_{-1}^{1} P'_m(x)\frac{d^{n-1}}{dx^{n-1}}[(x^2-1)^n]\,dx.$$

(b) Show that the first term above is zero.

18. (a) Use the result of Problem 17 and repeated integration by parts to show that

$$\int_{-1}^{1} P_n(x)P_m(x)\,dx = \frac{(-1)^n}{2^n n!}\int_{-1}^{1} P_m^{(n)}(x)(x^2-1)^n\,dx.$$

(b) Explain why $\int_{-1}^{1} P_n(x)P_m(x)\,dx = 0$ if $n \neq m$.

* 19. Use the result of Problem 18 with $m = n$ to show that

$$\|P_n\|^2 = \frac{2}{2n+1}.$$

[*Hint:* Make the substitution $x = \sin\theta$ and use entry 218 in Appendix 1.]

Review Exercises for Chapter 11

In Exercises 1–4 show that the given set of functions is orthogonal on the given interval and determine the corresponding orthonormal set.

1. $\{\sin nx\}, \quad n = 1, 2, 3, \ldots, \quad 0 \leq x \leq \pi$

2. $\{\cos n\pi x\}, \quad n = 0, 1, 2, \ldots, \quad 0 \leq x \leq 2$

3. $\{\sin 2n\pi x, \cos 2n\pi x\}, \quad n = 1, 2, \ldots, \quad 0 \leq x \leq 1$

4. $\{\sin n\pi x\}, \quad n = 1, 2, 3, \ldots, \quad 0 \leq x \leq 1$

Expand each of the functions in Exercises 5–8 in a Fourier series. Examine each series at all the points of discontinuity.

5. $f(x) = x \sin x, \quad 0 < x < 2\pi$

6. $f(x) = \sqrt{1 - \cos x}, \quad |x| < \pi$

7. $f(x) = |\sin x|, \quad |x| < \pi$

8. $f(x) = \begin{cases} x, & 0 < x < \pi \\ 0, & \pi < x < 2\pi \end{cases}$

Find Fourier sine and cosine series for the functions in Exercises 9–12 by extending the given function on the interval $[0, a]$ to an odd or even function on $[-a, a]$.

9. $f(x) = x^2, \quad 0 < x < \pi$

10. $f(x) = e^x, \quad 0 < x < \pi$

11. $f(x) = x, \quad 0 < x < 1$

12. $f(x) = 2 - x, \quad 0 < x < 2$

13. Find a Fourier sine series for $f(x) = x^2 - 2$ in $0 < x < 2$ and use this series to obtain a series for π^3.

In Exercises 14–19 find the values or approximate values of the eigenvalues and corresponding eigenfunctions of the given Sturm-Liouville problems.

14. $y'' - \lambda y = 0, \quad y'(0) = y(1) = 0$

15. $y'' + \lambda y = 0, \quad y'(0) = y(\pi) = 0$

16. $y'' - \lambda y = 0, \quad y(0) + y'(0) = 0, \quad y(1) = 0$

17. $y'' + \lambda y = 0, \quad y(0) = 0, \quad y(1) + y'(1) = 0$

18. $y'' + (-9 + \lambda)y = 0, \quad y'(0) = y'(1) = 0$

19. $y'' + (-9 + \lambda)y = 0, \quad y(0) = y(1) + y'(1) = 0$

20. Show that the generalized Laguerre equation

$$xy'' + (a + 1 - x)y' + ny = 0$$

is of Sturm-Liouville type, and determine the weight function for which the resulting polynomials are orthogonal.

21. Show that Chebyshev's equation

$$(1 - x^2)y'' - xy' + n^2 y = 0$$

is of Sturm-Liouville type, and determine the weight function for which the resulting polynomials are orthogonal.

22. Solve $y'' + y = e^x$, in $0 < x < 1$, with $y(0) = 0$, $y'(1) = 0$

23. Solve $y'' - y = 2x$, in $0 < x < 1$, with $y'(0) = 0$, $y(1) = 0$

24. Solve $y'' - y = e^x$, in $0 < x < 1$, with $y(0) = 1$, $y'(1) = 0$

25. Solve $[(1 + x)^2 y']' - y = x$, in $0 < x < 1$, with $y(0) = y(1) = 0$.

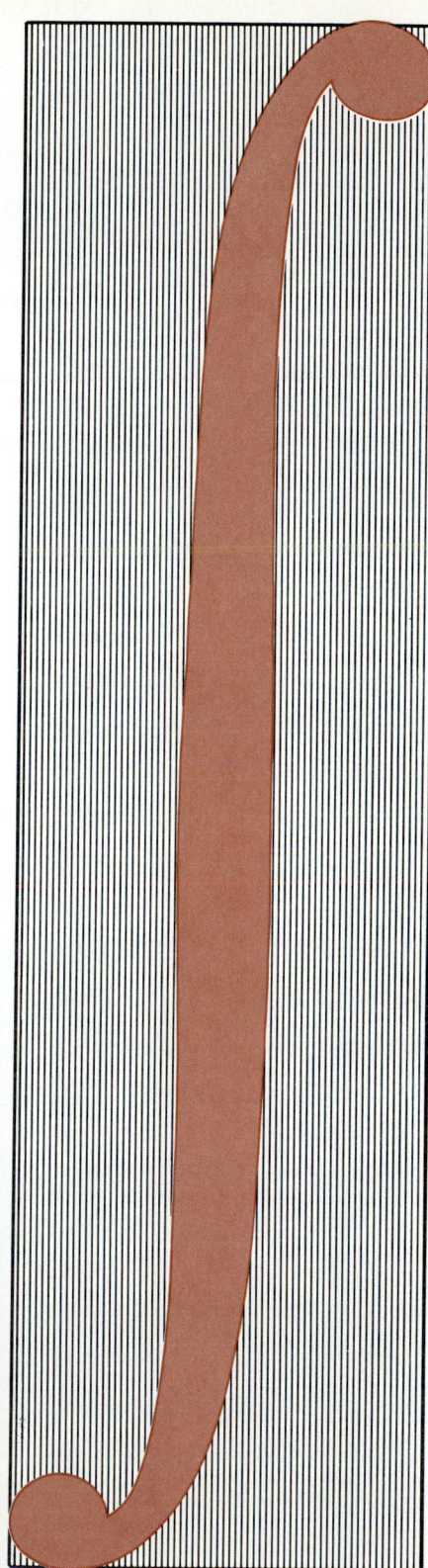

chapter 12

Partial Differential Equations

A partial differential equation is an equation involving a function of two or more variables and some of its partial derivatives. Thus the crucial difference between partial and ordinary differential equations is the number of independent variables involved in the equation.

In this chapter we present a brief introduction to partial differential equations. Since the general theory is much too difficult to be presented here, we merely examine a few simple examples. Because these examples have a number of important practical applications, even this cursory treatment is of significant value.

12.1 First-Order Linear Partial Differential Equations

In this section we are concerned with first-order linear partial differential equations in two independent variables, that is, equations of the form

$$a(x, y)\frac{\partial z}{\partial x} + b(x, y)\frac{\partial z}{\partial y} + c(x, y)z = f(x, y), \tag{1}$$

where a, b, and c are functions of x and y. Equation (1) is **linear** since it does not involve any nonlinear functions of the dependent variable z and its partial derivatives; that is, equation (1) is of **first degree** in z and its partial derivatives.

When a, b, and c are constants, that is, when the coefficients of (1) are constant, the partial differential equation can be solved by rotating the axes. Let

$$\tan \alpha = b/a \tag{2}$$

and set

$$u = x \cos \alpha + y \sin \alpha,$$
$$v = -x \sin \alpha + y \cos \alpha. \tag{3}$$

Using (3) to change the independent variables x and y in (1), we obtain a new differential equation in u and v that is an ordinary differential equation, because the partial derivative with respect to v is absent. We illustrate this with an example.

EXAMPLE 1

Consider the partial differential equation

$$\frac{\partial z}{\partial x} + \frac{\partial z}{\partial y} - z = 0. \tag{4}$$

Since $a = b = 1$, it follows from (2) that $\tan \alpha = 1$ and $\alpha = \pi/4$ radians, so $\cos \alpha = \sin \alpha = 1/\sqrt{2}$. Thus, by equation (3),

$$u = \frac{x+y}{\sqrt{2}}, \qquad v = \frac{-x+y}{\sqrt{2}}, \tag{5}$$

and we have (using the chain rule)

$$\frac{\partial z}{\partial x} = \frac{\partial z}{\partial u} \frac{\partial u}{\partial x} + \frac{\partial z}{\partial v} \frac{\partial v}{\partial x} = \frac{1}{\sqrt{2}} \left(\frac{\partial z}{\partial u} - \frac{\partial z}{\partial v} \right),$$

$$\frac{\partial z}{\partial y} = \frac{\partial z}{\partial u} \frac{\partial u}{\partial y} + \frac{\partial z}{\partial v} \frac{\partial v}{\partial y} = \frac{1}{\sqrt{2}} \left(\frac{\partial z}{\partial u} + \frac{\partial z}{\partial v} \right). \tag{6}$$

Substituting the equations in (6) into equation (4), we get

$$\sqrt{2} \frac{\partial z}{\partial u} - z = 0. \tag{7}$$

Although equation (7) is still a partial differential equation in the independent variables u and v, we treat the variable v as if it were a parameter and solve the first-order linear differential equation

$$\sqrt{2} z' - z = 0, \qquad \text{where} \quad z' = \frac{dz}{du}.$$

Since $(1/z)\, dz = (1/\sqrt{2})\, du$, we get

$$\ln|z| = \frac{u}{\sqrt{2}} + g(v),$$

where the last term is an arbitrary differentiable function of the parameter v (which vanishes whenever we take its partial derivative with respect to u). Simplifying, we have

$$z = e^{u/\sqrt{2}} G(v), \qquad G(v) = e^{g(v)}.$$

Replacing u and v by the substitutions in (5), we get

$$z = e^{(x+y)/2} G\left(\frac{y-x}{\sqrt{2}}\right). \tag{8}$$

That this is, indeed, a solution of equation (4) for *any* differentiable function G is easily checked by performing the operations indicated in that equation. We call such a solution a **general solution** for the partial differential equation.

General Solution

Up to now we have stressed the similarity between first-order ordinary and partial differential equations. We encounter a substantial difference when we consider initial value problems. The solution of an ordinary differential equation is completely determined by prescribing a value for it at a single point. This is generally not true for partial differential equations. In order to obtain a *unique* solution, it is usually necessary to prescribe values for the solution on an entire *line*. A suitable initial condition is

$$z(x, 0) = z_0(x)$$

for all x, where z_0 is a differentiable function. A solution satisfying this initial condition is then easily obtained by setting $y = 0$ in equation (8). For example, suppose we are given the initial condition

$$z(x, 0) = \frac{x}{2} + e^x. \tag{9}$$

Setting $y = 0$ in (8) and substituting the initial condition for the left-hand side of (8), we have

$$\frac{x}{2} + e^x = e^{x/2} G\left(\frac{-x}{\sqrt{2}}\right),$$

or

$$G\left(\frac{-x}{\sqrt{2}}\right) = \frac{x}{2} e^{-x/2} + e^{x/2}.$$

Setting $w = -x/\sqrt{2}$, we have $x = -\sqrt{2}\, w$ and we obtain the exact form of the function G:

$$G(w) = \frac{-w}{\sqrt{2}} e^{w/\sqrt{2}} + e^{-w/\sqrt{2}}.$$

Replacing this function for G in (8) yields

$$z = e^{(x+y)/2}\left[\left(\frac{x-y}{2}\right) e^{(y-x)/2} + e^{(x-y)/2}\right] = \left(\frac{x-y}{2}\right) e^y + e^x,$$

which is the solution to the initial value problem given by equations (4) and (9).

Cauchy Problem

A problem like that in Example 1, in which we require that the solution satisfy a line of initial conditions, is called a **Cauchy problem** or an **initial value problem**. The solution of a Cauchy problem can be visualized as a surface $z = z(x, y)$ in three-dimensional Euclidean space. Thus we can use three-dimensional analytic geometry to increase our understanding of the solution.

We call a surface $z = z(x, y)$ that is the solution of a Cauchy problem an **integral surface.** This name is used because "integrations" are required in solving (partial) differential equations. If we write the equation of an integral surface in the form

$$F(x, y, z) = z(x, y) - z = 0, \tag{10}$$

we can find the tangent plane to the integral surface by taking the total differential

$$z_x \, dx + z_y \, dy - dz = 0.$$

The vector $\mathbf{N} = z_x \mathbf{i} + z_y \mathbf{j} - \mathbf{k}$ is normal to the integral surface (10) at any point (see p. 573). Rewriting (1) in the form

$$a(x, y)z_x + b(x, y)z_y = g(x, y, z), \tag{11}$$

where $g(x, y, z) = f(x, y) - c(x, y)z$, we see that the vector

$$\mathbf{V} = a\mathbf{i} + b\mathbf{j} + g\mathbf{k} \tag{12}$$

is orthogonal to $\mathbf{N}$ since

$$\mathbf{V} \cdot \mathbf{N} = az_x + bz_y - g = 0.$$

Hence $\mathbf{V}$ is tangent to the integral surface and lies in the tangent plane at every point (see Fig. 1). Thus the first-order partial differential equation provides the geometric *requirement* that any integral surface through a given point be tangent to the vector $\mathbf{V}$. This means that if we begin at some point specified by the initial condition (which must belong to the integral surface) and move in the direction of the *known* tangent vector $\mathbf{V}$, we move along a curve lying entirely on the integral surface $F(x, y, z) = 0$. This curve is called a **characteristic.** The integral surface can often be described in terms of these characteristics.

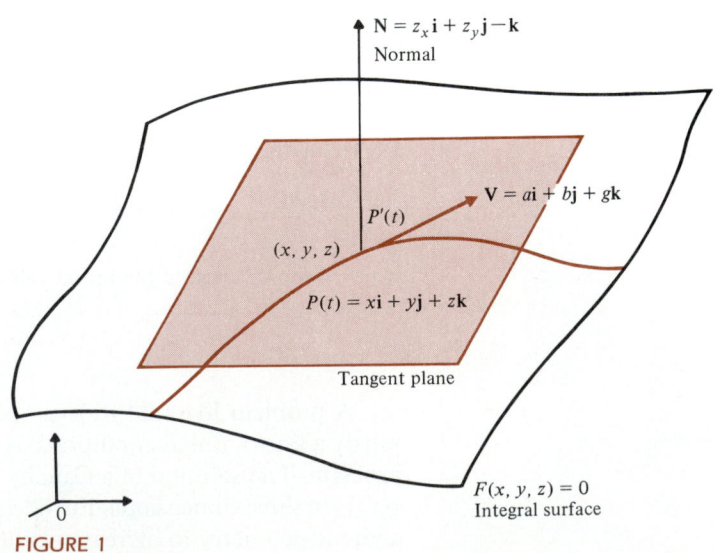

FIGURE 1

Points on a characteristic curve can be described parametrically by an expression of the form

$$\mathbf{P}(t) = x(t)\mathbf{i} + y(t)\mathbf{j} + z(t)\mathbf{k}. \tag{13}$$

If we differentiate (13) with respect to t, we obtain a tangent vector. This vector must belong to the tangent plane of the integral surface at the given point. Furthermore, it must be proportional to $\mathbf{V}$, since the characteristic is obtained by moving in the $\mathbf{V}$ direction. Thus the coordinates of $\mathbf{P}'(t)$ and $\mathbf{V}$ are proportional; that is,

$$\frac{dx/dt}{a(x, y)} = \frac{dy/dt}{b(x, y)} = \frac{dz/dt}{g(x, y, z)},$$

or

$$\frac{dx}{a} = \frac{dy}{b} = \frac{dz}{g}. \tag{14}$$

Equations (14) provide a pair of first-order ordinary differential equations:

$$\frac{dy}{dx} = \frac{b(x, y)}{a(x, y)}, \qquad \frac{dz}{dx} = \frac{g(x, y, z)}{a(x, y)},$$

or $\tag{15}$

$$\frac{dx}{dy} = \frac{a(x, y)}{b(x, y)}, \qquad \frac{dz}{dy} = \frac{g(x, y, z)}{b(x, y)}.$$

The solutions of these equations determine the characteristics of the partial differential equation. The following example illustrates the use of characteristics.

EXAMPLE 2

Consider the Cauchy problem of Example 1:

$$\frac{\partial z}{\partial x} + \frac{\partial z}{\partial y} = z, \qquad z(x, 0) = \frac{x}{2} + e^x. \tag{16}$$

Since $a = b = 1$, and $g = z$, by (14) we have

$$\frac{dx}{1} = \frac{dy}{1} = \frac{dz}{z},$$

and the system of differential equations (15) is

$$\frac{dx}{dy} = 1, \qquad \frac{dz}{dy} = z, \tag{17}$$

with solutions

$$x = y + c_1, \qquad z = c_2 e^y. \tag{18}$$

Setting $y = 0$ in (18), we have $x = c_1$ and $z(x, 0) = c_2$. In the xz-plane we have

$$c_2 = z(x, 0) = \overset{\overset{\text{initial condition}}{\downarrow}}{\frac{x}{2}} + e^x = \frac{c_1}{2} + e^{c_1}. \tag{19}$$

Equation (19) determines the relation between the parameters c_1 and c_2. Rewriting (18) as

$$c_1 = x - y, \qquad c_2 = z e^{-y},$$

we have, by (19),

$$z e^{-y} = c_2 = \frac{c_1}{2} + e^{c_1} = \frac{x - y}{2} + e^{x-y},$$

or

$$z = \left(\frac{x - y}{2}\right) e^y + e^x.$$

Notice that this is the solution to the Cauchy problem that was obtained in Example 1.

In the next example we show that the method of characteristics applies to Cauchy problems with variable coefficients.

EXAMPLE 3

Solve the initial-value problem

$$z_x + y z_y = 2xz, \qquad z(0, y) = y - y^2. \tag{20}$$

SOLUTION

In equation (14) we have $a = 1$, $b(x, y) = y$, and $g(x, y, z) = 2xz$. Thus the characteristics are determined by the ordinary differential equations

$$\frac{dx}{1} = \frac{dy}{y} = \frac{dz}{2xz},$$

or

$$dx = \frac{dy}{y}, \qquad 2x \, dx = \frac{dz}{z}.$$

These differential equations have the solutions

$$x + c_1 = \ln|y|, \qquad x^2 + c_2 = \ln|z|,$$

or

$$y = c_3 e^x, \qquad z = c_4 e^{x^2}. \tag{21}$$

Setting $x = 0$, we have, from (20) and (21), $y = c_3$ and

$$c_4 = z(0, y) = y - y^2 = c_3 - c_3^2. \tag{22}$$

Using (21), we have

$$c_3 = ye^{-x}, \qquad c_4 = ze^{-x^2}. \tag{23}$$

Combining (22) and (23), we obtain

$$ze^{-x^2} = c_4 = c_3 - c_3^2 = ye^{-x} - (ye^{-x})^2,$$

or

$$z = ye^{x^2-x} - y^2e^{x^2-2x}. \tag{24}$$

To check that (24) is the solution to (20), observe that $z(0, y) = y - y^2$,

$$z_x = (2x - 1)ye^{x^2-x} - (2x - 2)y^2e^{x^2-2x},$$

$$yz_y = ye^{x^2-x} - 2y^2e^{x^2-2x},$$

and

$$2xz = 2xye^{x^2-x} - 2xy^2e^{x^2-2x}.$$

Solution of a Cauchy Problem

To solve the Cauchy problem

partial differential equation:

$$a(x, y)\frac{\partial z}{\partial x} + b(x, y)\frac{\partial z}{\partial y} = g(x, y, z),$$

initial condition: $z(x, 0) = z_0(x),$

by the method of characteristics, perform the following four steps:

STEP 1
Write the differential equations

$$\frac{dx}{a(x, y)} = \frac{dy}{b(x, y)} = \frac{dz}{g(x, y, z)}.$$

STEP 2
Solve (if possible) the first differential equation

$$\frac{dx}{dy} = \frac{a(x, y)}{b(x, y)}$$

for x in terms of y, obtaining $x = x(y, c_1)$. Substitute this expression (if necessary) for x in the second differential equation

$$\frac{dz}{dy} = \frac{g(x, y, z)}{b(x, y)}$$

and solve for z in terms of y, obtaining $z = z(y, c_2)$.

STEP 3

Set $y = 0$ in the two solutions in step 2 and substitute the resulting expressions into the initial condition $z = z_0(x)$, obtaining an equation involving only c_1 and c_2:

$$z(0, c_2) = z_0(x(0, c_1)).$$

STEP 4

Use the solutions in step 2 to eliminate the constants c_1 and c_2 in the final equation in step 3. The resulting equation in x, y, and z is the solution of the Cauchy problem.

Problems 12.1

In Problems 1–8 use the method of rotation of axes (as in Example 1) to solve the given Cauchy problem.

1. $\dfrac{\partial z}{\partial x} - \dfrac{\partial z}{\partial y} + \sqrt{2}z = 0, \quad z(x, 0) = x$

2. $\dfrac{\partial z}{\partial x} - \dfrac{\partial z}{\partial y} - \sqrt{2}z = 0, \quad z(x, 0) = x^2$

3. $\dfrac{\partial z}{\partial x} + \dfrac{\partial z}{\partial y} + \sqrt{2}z = 0, \quad z(x, 0) = x + e^x$

4. $\dfrac{\partial z}{\partial x} + \dfrac{\partial z}{\partial y} - \sqrt{2}z = 0, \quad z(0, y) = e^y - y$

5. $3\dfrac{\partial z}{\partial x} - 4\dfrac{\partial z}{\partial y} + 2z = 7, \quad z(x, 0) = e^x$

6. $\dfrac{\partial z}{\partial x} + \dfrac{\partial z}{\partial y} + z = 2, \quad z(x, 0) = \sin x$

7. $\dfrac{\partial z}{\partial x} + \dfrac{\partial z}{\partial y} - z = e^x, \quad z(x, 0) = 0$

8. $\dfrac{\partial z}{\partial x} - \dfrac{\partial z}{\partial y} + z = y, \quad z(x, 0) = x^2$

In Problems 9–22 use the method of characteristics to solve the given Cauchy problem.

9. Problem 1

10. Problem 2

11. Problem 3

12. Problem 4

13. Problem 5

14. Problem 6

15. Problem 7

16. Problem 8

17. $xz_x + z_y = 1, \quad z(1, y) = e^{-y}$

18. $xz_x + yz_y = z, \quad z(1, y) = y^2$

19. $z_x + yz_y = z, \quad z(x, 1) = xe^{-x}$

20. $xz_x + y^{-1}z_y = 1, \quad z(x, 0) = 5 - x$

21. $z_x + \dfrac{z_y}{2y} - z = 2, \quad z(x, 0) = \sin x - 2$

22. $xz_x + z_y - xz = x, \quad z(e, y) = y - 2$

* 23. Adapt the method of characteristics to the Cauchy problem

$$z_x + z_y = z, \quad z(y, y) = e^y.$$

Discuss your result.

The following two problems show that Cauchy problems may have no solutions or infinitely many solutions.

* 24. Consider the Cauchy problem

$$z_x + z_y = 1, \quad z(y, y) = e^y.$$

Show that it has no solution.

25. Show that the Cauchy problem

$$z_x + z_y = 1, \quad z(y, y) = y + 7$$

has infinitely many solutions.

26. Let f be a real-valued function on $\mathbb{R}^2$ that is **homogeneous of degree** n; that is,

$$f(tx, ty) = t^n f(x, y), \tag{i}$$

for all t real and (x, y) in $\mathbb{R}^2$. Show that $z = f(x, y)$ satisfies the partial differential equation

$$xz_x + yz_y = nz.$$

[*Hint:* Differentiate (i) with respect to t and then set $t = 1$.]

12.2 Initial Value Problems for Quasi-Linear First-Order Equations

In Section 12.1 we developed the method of characteristics and applied it to linear first-order partial differential equations. In this section we show that this method applies to an even wider class of Cauchy problems. We also discuss some difficulties that arise when using this technique.

A first-order partial differential equation of the form

$$a(x, y, z)\frac{\partial z}{\partial x} + b(x, y, z)\frac{\partial z}{\partial y} = g(x, y, z) \tag{1}$$

Quasi-linear Equation

is called **quasi-linear.** Note that since a and b may involve the dependent variable z, quasi-linear equations need not be linear in z and its first partial derivatives. The set of quasi-linear partial differential equations of the first order includes the linear first-order equations as a subset. We now apply the method of characteristics to quasi-linear equations.

EXAMPLE 1

Solve the quasi-linear Cauchy problem

$$zz_x + yz_y = x, \qquad z(0, y) = \frac{1}{y}. \tag{2}$$

SOLUTION
Proceeding as in Section 12.1, we have

$$\frac{dx}{z} = \frac{dy}{y} = \frac{dz}{x}.$$

Then $x\,dx = z\,dz$, so $z^2 = x^2 + c_1$ and $z = \sqrt{x^2 + c_1}$. Also

$$\frac{dy}{y} = \frac{dx}{z} = \frac{dx}{\sqrt{x^2 + c_1}}.$$

Using Formula 67 in Appendix 1 we obtain

$$\ln|y| = c_2 + \ln|x + \sqrt{x^2 + c_1}| = c_2 + \ln|x + z|,$$

or

$$y = c_3(x + z).$$

Setting $x = 0$, we have

$$z(0, y) = \frac{1}{y} \text{ is given}$$

$$c_3 = \frac{y}{z(0, y)} \xrightarrow{\quad} \frac{1}{z^2(0, y)} = \frac{1}{c_1}.$$

Therefore

$$\frac{y}{x+z} = c_3 = \frac{1}{c_1} = \frac{1}{z^2 - x^2},$$

or

$$x + z = y(z^2 - x^2). \tag{3}$$

It is easy to check that (3) satisfies the initial condition. To see that it satisfies the quasi-linear equation, we differentiate implicitly first with respect to x and then with respect to y:

$$1 + z_x = 2yzz_x - 2xy,$$

$$z_y = (z^2 - x^2) + 2yzz_y.$$

Thus

$$z_x = \frac{1 + 2xy}{2yz - 1} \quad \text{and} \quad z_y = \frac{z^2 - x^2}{1 - 2yz}.$$

Then,

$$\overset{\text{equation (3)}}{\underset{\downarrow}{}}$$

$$zz_x + yz_y = \frac{z + 2xyz - y(z^2 - x^2)}{2yz - 1} = \frac{z + 2xyz - (x + z)}{2yz - 1} = x.$$

So far we have concentrated on situations where no difficulties are encountered in using the method of characteristics. The next examples provide a sample of various types of technical difficulties that may arise.

EXAMPLE 2

Find the general solution of the quasi-linear partial differential equation

$$zz_x + z_y = 0. \tag{4}$$

SOLUTION

If we follow the technique of Example 1 formally, we obtain the system of ordinary differential equations

$$\frac{dx}{z} = \frac{dy}{1} = \frac{dz}{0}.$$

Division by zero is not allowed. However, if we look again at the development leading to equation (12.1.14) on p. 729, we see that all that is required is that the coordinates of the tangent $\mathbf{P}'(t)$ be proportional to those of $\mathbf{V}$. This is equivalent to the system

$$\frac{dx}{z} = \frac{dy}{1} \quad \text{and} \quad dz = 0. \tag{5}$$

Hence $z = c_2$ and $x/z = y + c_1$ (since z is constant).

If initial conditions for z are prescribed in the form $z(x, 0) = f(x)$, with f an arbitrary differentiable function, then setting $y = 0$, we obtain

$$\begin{array}{c} \text{set } y = 0 \text{ in} \\ x = z(y + c_1) \\ \downarrow \end{array}$$

$$c_2 = z(x, 0) = f(x) = f(c_1 z) = f(c_1 c_2).$$

Hence

$$z = f\left(\left[\frac{x}{z} - y\right]z\right) = f(x - yz)$$

is the general solution of any such Cauchy problem. Observe that

$$z_x = f'(x - yz)(1 - yz_x), \qquad z_y = f'(x - yz)(-z - yz_y);$$

therefore

$$z_x = \frac{f'}{1 + yf'}, \qquad z_y = \frac{-zf'}{1 + yf'},$$

from which (4) follows immediately.

EXAMPLE 3

Consider the partial differential equation

$$z_x + 2z_y = 0.$$

Here

$$\frac{dx}{1} = \frac{dy}{2}, \qquad dz = 0,$$

so that

$$2\,dx = dy, \qquad dz = 0,$$

and

$$2x + c_1 = y, \qquad z = c_2.$$

The constants c_1 and c_2 are arbitrary, so the characteristics are given by all horizontal lines (i.e., lines in a plane parallel to the xy-plane) parallel to $y = 2x$. Assume that $c_2 = f(c_1)$; then

$$z = f(y - 2x) \tag{6}$$

is the general solution of the partial differential equation. We now consider what effect various initial conditions have on (6).

a. If $z(x, x) = e^x$ is the initial condition, then setting $y = x$ in (6) we have

$$f(-x) = e^x \qquad \text{or} \qquad f(x) = e^{-x}.$$

Thus the solution of the corresponding Cauchy problem is $z = e^{2x-y}$.

b. If $z(x, 2x) = e^x$ is the initial condition, then when we set $y = 2x$ in (6) we obtain $e^x = f(0)$. This is impossible, since x is variable. The problem here is that the initial condition is defined on a characteristic ($y = 2x$) on which z *must be* *constant,* because of the partial differential equation. Thus we have *no* solution in this case.

c. If $z(x, 2x) = k$ (a constant), setting $y = 2x$ we have $f(0) = k$. In this case, *any* differentiable function satisfying $f(0) = k$ is a solution. For example, if $f(t) = at^n + k$, then $z = a(y - 2x)^n + k$ is a solution to the Cauchy problem for every number a and integer $n > 0$.

It is clear from (b) and (c) above that if the initial condition is given along a characteristic, we cannot guarantee the existence or uniqueness of a solution. It can be shown† that the Cauchy problem (1) with initial condition $z(x_0(t), y_0(t)) = z_0(t)$, $\alpha \leq t \leq \beta$, has a unique solution in a disk centered at $(x_0(t_0), y_0(t_0))$, where t_0 is the initial value of t, provided

$$a(x_0(t), y_0(t), z_0(t)) \frac{dy_0}{dt} \neq b(x_0(t), y_0(t), z_0(t)) \frac{dx_0}{dt}.$$

EXAMPLE 4

A partial differential equation that arises in the one-dimensional motion of an ideal gas or fluid, because of the conservation of mass, is

$$\rho_t + (\rho v)_x = 0, \tag{7}$$

where ρ, v, x, and t represent density, velocity, displacement, and time in the flow of the gas or fluid. For certain (**isentropic‡**) flows, we are interested in solutions where the density is a function of the velocity; that is, where $\rho = f(v)$. In this case, (7) becomes

$$f'(v)v_t + [f(v) + vf'(v)]v_x = 0,$$

or

$$\left[v + \frac{f(v)}{f'(v)}\right]v_x + v_t = 0. \tag{8}$$

For simplicity, assume that $f(v) = a + bv$. Then we get

$$\left(\frac{a}{b} + 2v\right)v_x + v_t = 0. \tag{9}$$

The characteristics for (9) are determined by the system

$$\frac{dx}{(a/b) + 2v} = \frac{dt}{1}, \qquad dv = 0.$$

Hence v is constant on each characteristic, and since

$$\frac{dx}{dt} = \frac{a}{b} + 2v,$$

dx/dt is also constant on characteristics:

$$x = \left(2c_2 + \frac{a}{b}\right)t + c_1, \qquad v = c_2.$$

---

† See G. C. Zachmanoglou and D. W. Thoe, *Introduction to Partial Differential Equations with Applications,* Williams & Wilkins, Baltimore, 1976.

‡ **Isentropic** means having constant entropy. **Entropy** (for a gas) is a scalar field that is proportional to the potential temperature.

Thus the characteristics are straight lines that are parallel in each plane $v = c_2$. Since c_1 and c_2 are arbitrary, we can assume that $c_2 = g(c_1)$, so

$$v = g\left(x - \left(2v + \frac{a}{b}\right)t\right) \tag{10}$$

is the general solution to (9).

Now suppose the initial condition is

$$v(x, 0) = \begin{cases} \epsilon(1 + \cos x), & |x| < \pi, \\ 0, & |x| > \pi. \end{cases}$$

Then $g(x) = v(x, 0)$, and

$$v(x, t) = \begin{cases} \epsilon\left[1 + \cos\left(x - \left(2v + \frac{a}{b}\right)t\right)\right], & \text{if } \left|x - \left(2v + \frac{a}{b}\right)t\right| < \pi, \\ 0, & \text{if } \left|x - \left(2v + \frac{a}{b}\right)t\right| > \pi. \end{cases}$$

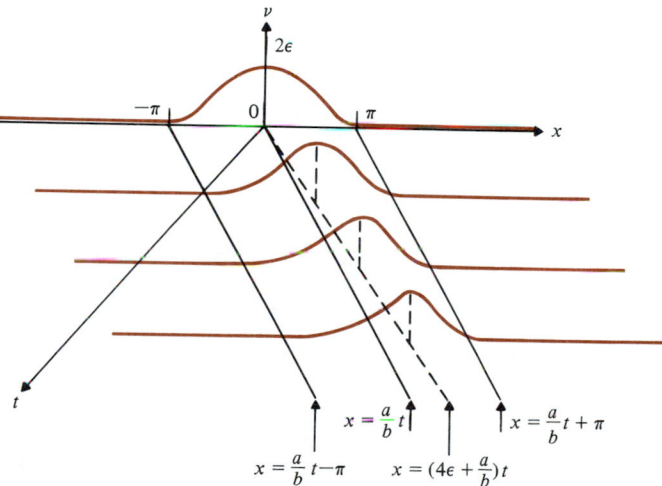

FIGURE 1

If we look at Fig. 1 we see that the initial velocity wave propagates along the characteristics $x = (at/b) \pm \pi$ for $v = 0$, but that the *shape* of the wave changes, since the maximum of 2ϵ is located above the line $x = (4\epsilon + a/b)t$ in the base plane. More will be said about this example in Section 12.3.

Problems 12.2

In Problems 1–20 find either the general solution or, if initial conditions are given, the solution to the Cauchy problem.

1. $z_x + z_y = 0$

2. $z_x - z_y = 0$

3. $xz_x + z_y = 0$

4. $zz_x - z_y = 0$

5. $xz_x + zz_y = 1$

6. $zz_x + z_y = z^2$

7. $z_x + z^2 z_y = 0$, $z(x, 0) = \cos x$

8. $z_x + zz_y = 0$, $z(0, y) = \cosh y$

9. $x^2 z_x + y^2 z_y = z^2$, $z(x, 2x) = 1$

10. $xz_x = yz_y$, $z(x, x) = x^2$

11. $zz_x + yz_y = x$, $z(x, 1) = 2x$

12. $(y + z)z_x + yz_y = 0$

13. $zz_x + z_y = 0,\quad z(x, 0) = x$

14. $z^2 z_x + z_y = 0$

15. $xz_x + yz_y = z^2$

16. $(1 + x)z_x - yz_y = 1 - z^2$

17. $xz_x + yz_y = 1 - z^2,\quad z(x, 1) = e^x$

18. $zz_x + zz_y = z^2$

19. $yz_x + zz_y = xy$

20. $zz_x + zyz_y = y$

21. The quasi-linear equation

$$(1 + \tfrac{3}{2}z)z_x + z_y = 0$$

arises in the study of gravity water waves in shallow water.† Find a general solution.

* 22. Traffic flow on a highway can be modeled as a one-dimensional flow along the x-axis (as the highway). Let $\rho(x, t)$ be the density (cars per unit length) at the point x of the highway at time t, and let $q(x, t)$ be the rate at which cars pass the point x per unit time at time t.

(a) Assuming that there are no sources or sinks of cars at any point of the highway and that the functions ρ and q are sufficiently smooth, show that ρ and q satisfy the conservation equation

$$\rho_t + q_x = 0.$$

(b) Assume that the flow rate depends only on the density, $q = f(\rho)$, for some smooth function f. What quasi-linear equation is obtained?

(c) Find the general solution of the equation in part (b) if $f(\rho) = c\rho(1 - \rho/k)$.

(Optional)

12.3 Applications to the Theory of Shocks in Gas Dynamics

In Example 12.2.4 we saw how a certain type of physical application gives rise to quasi-linear Cauchy problems. In this section we study the one-dimensional, time-dependent flow of a compressible fluid under constant pressure p. Let u, ρ, and E represent the fluid velocity, density, and internal energy per unit volume. The basic equations of gas dynamics‡ are

$$u_t + uu_x = 0,$$

$$\rho_t + (\rho u)_x = 0, \tag{1}$$

$$E_t + (Eu)_x + pu_x = 0.$$

The substitution $s = E + p$ allows us to replace the third equation in (1) by

$$s_t + (su)_x = 0, \tag{2}$$

since p is constant. Note that (2) has the same form as the second equation in (1). Finally, to convert this into a Cauchy problem, we assume the initial conditions

$$u(x, 0) = f(x), \qquad \rho(x, 0) = g(x), \qquad s(x, 0) = h(x), \tag{3}$$

where f, g, and h have sufficient smoothness for what is yet to come.

If we solve the first equation in (1), which is independent of ρ and s, we can use the solution u to solve the remaining *linear* partial differential equations.

---

† Robert L. Street, *Partial Differential Equations*, Brooks/Cole, Monterey, Calif., 1973.

‡ See, for example, S. I. Pai, *Magnetogas-Dynamics and Plasma Dynamics*, Prentice-Hall, Englewood Cliffs, N.J., 1962.

As in Section 12.2, the characteristics of the first equation in (1) are given by the system of differential equations

$$\frac{dt}{1} = \frac{dx}{u} \quad \text{and} \quad du = 0.$$

Here the characteristics are $u = c_2$ and $ut = x + c_1$ (since u is constant along the characteristic), in three-dimensional txu-space. Applying the first set of initial conditions in (3), we have (with $t = 0$)

$$c_2 = u(x, 0) = f(x) = f(-c_1).$$

Hence we obtain the general solution

$$u = f(x - tu). \tag{4}$$

To check that this is indeed a solution, observe that, by the chain rule,

$$u_t = (-u - tu_t)f'(x - tu) \quad \text{or} \quad (1 + tf')u_t = -uf', \tag{5}$$

$$u_x = (1 - tu_x)f'(x - tu) \quad \text{or} \quad (1 + tf')u_x = f'. \tag{6}$$

Thus the first equation in (1) is satisfied if $1 + tf' \neq 0$. Observe that

$$1 + tf'(x - tu) > 0 \tag{7}$$

for very small values of t.

We may rewrite (5) and (6) as

$$u_t = \frac{-uf'(x - tu)}{1 + tf'(x - tu)}, \qquad u_x = \frac{f'(x - tu)}{1 + tf'(x - tu)}. \tag{8}$$

Notice that u_t and u_x tend to ∞ if $1 + tf'$ tends to zero. When $1 + tf'$ becomes zero, we shall show below that the solution u develops a discontinuity known as a **shock.** Shocks are a common phenomenon in fluid and gas dynamics.

Shocks

EXAMPLE 1

Consider the general solution (12.2.10) of Example 12.2.4:

$$v = g\left(x - \left(2v + \frac{a}{b}\right)t\right).$$

In that example we let the initial condition be

$$v(x, 0) = \begin{cases} \epsilon(1 + \cos x), & |x| < \pi, \\ 0, & |x| > \pi, \end{cases}$$

so that

$$v(x, t) = \begin{cases} \epsilon(1 + \cos \eta), & |\eta| < \pi, \\ 0, & |\eta| > \pi, \end{cases}$$

where $\eta = x - (2v + a/b)t$. Hence

$$v_x = -\epsilon(1 - 2tv_x) \sin \eta,$$

or

$$v_x = \frac{-\epsilon \sin \eta}{1 - 2t\epsilon \sin \eta}. \tag{9}$$

Since ϵ is constant, the shock first appears for $\eta = \pi/2$ ($\sin \eta = 1$) when $t = 1/2\epsilon$. At that value of t, the wave in the x direction exhibits a vertical tangent at height $v = \epsilon$. After that, v is not a function: for example, if $t_0 = \pi/6\epsilon$, then when

$$x_0 = \left(3\epsilon + \frac{a}{b}\right)t_0 + \frac{\pi}{3} = \left(2\epsilon + \frac{a}{b}\right)t_0 + \frac{\pi}{2}$$

it follows that v takes on both the values $3\epsilon/2$ and ϵ, since

$$v = \epsilon\left\{1 + \cos\left[2\left(\frac{3\epsilon}{2} - v\right)t_0 + \frac{\pi}{3}\right]\right\} \text{ and } v = \epsilon\left\{1 + \cos\left[2(\epsilon - v)t_0 + \frac{\pi}{2}\right]\right\}.$$

Hence the waves shown in Fig. 12.2.1 occur only as long as $t < 1/2\epsilon$.

To understand why shocks occur, consider equation (4). Select any fixed number x_0 and calculate $u_0 = f(x_0)$. The points lying on the intersection of the two surfaces

$$x - tu_0 = x_0 \quad \text{and} \quad u = u_0 \tag{10}$$

satisfy (4), since $u = u_0 = f(x_0) = f(x - tu_0)$. Thus the straight line defined by (10) lies on the solution surface. In particular, this means that along the line

$$x - tu_0 = x_0 \tag{11}$$

in the xt-plane, which passes through the point $(x_0, 0)$, the solution u to the Cauchy problem is constant and equal to $u_0 = f(x_0)$.

Now suppose we pick a different number x_1, calculate $u_1 = f(x_1) \neq u_0$, and consider the line

$$x - tu_1 = x_1. \tag{12}$$

As long as lines (11) and (12), for arbitrary choices of x_0 and x_1, do not intersect in the half plane $t > 0$, the solution exists for all $t > 0$. But if two of these lines intersect in $t > 0$ (see Fig. 1), then at the point of intersection we have a difficulty: by (11) the solution u must equal u_0, whereas (12) requires that it equal $u_1 \neq u_0$. Thus the solution u cannot exist at the point of intersection given by $x = x_1 + tu_1 = x_0 + tu_0$, or

$$t = \frac{x_1 - x_0}{u_0 - u_1}, \qquad x = \frac{u_0 x_1 - x_0 u_1}{u_0 - u_1}. \tag{13}$$

In particular, the solution does not exist for

$$t \geq \frac{x_1 - x_0}{u_0 - u_1}.$$

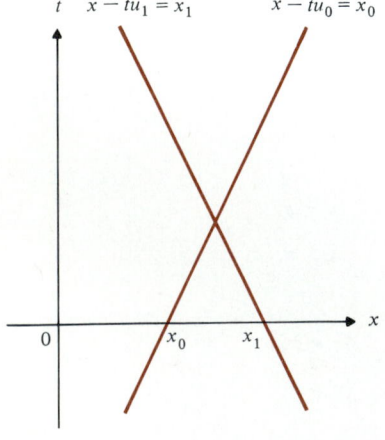

FIGURE 1

Although the existence of shocks presents a technical difficulty, there are ways of handling this obstacle by generalizing the meaning of a solution of a partial differential equation.† These concepts are beyond the scope of this book.

Returning to the original Cauchy problem of equations (1)–(3), we write the second equation in (1) as

$$\rho_t + u\rho_x + \rho u_x = 0. \tag{14}$$

Since

$$u_x = \frac{f'(x-tu)}{1+tf'(x-tu)}, \qquad u = f(x-tu),$$

we try a solution for ρ of the form

$$\rho = \frac{G(x-tu)}{1+tf'(x-tu)} \tag{15}$$

so that all terms in (14) have a denominator of $(1+tf')^2$. Then (14) becomes

$$\rho_t = \frac{(1+tf')G'(-u-tu_t) - G(f' + tf''(-u-tu_t))}{(1+tf')^2},$$

$$u\rho_x = u\left[\frac{(1+tf')G'(1-tu_x) - Gtf''(1-tu_x)}{(1+tf')^2}\right],$$

$$\rho u_x = \frac{Gf'}{(1+tf')^2}.$$

We find that

$$\rho_t + u\rho_x + \rho u_x = \frac{-tG'(1+tf')(u_t + uu_x) + t^2 Gf''(u_t + uu_x)}{(1+tf')^2} = 0,$$

since $u_t + uu_x = 0$ by (1). Setting $t = 0$ in (15), we get

$$\rho(x, 0) = G(x),$$

implying, by the second initial condition, that $G = g$ and that the general solution of the density equation is

$$\rho = \frac{g(x-tu)}{1+tf'(x-tu)}.$$

Similarly, since (2) is of the same form as the second equation in (1), we get

$$s = \frac{h(x-tu)}{1+tf'(x-tu)},$$

from which it follows that

$$E = \frac{h(x-tu)}{1+tf'(x-tu)} - p.$$

---

† See the survey article by P. D. Lax, "The formation and decay of shock waves," *American Mathematical Monthly*, vol. 79 (1972), pp. 227–241.

Problems 12.3

1. Solve the Cauchy problem

$$uu_x + u_t = 0, \quad u(x, 0) = x.$$

Do shocks occur in this problem for $t \geq 0$?

2. Repeat Problem 1 for the Cauchy problem

$$u^2 u_x + u_t = 0, \qquad u(x, 0) = x.$$

[*Hint:* Use the Maclaurin series of $\sqrt{1 + a}$.]

3. Repeat Problem 1 for

$$uu_x + u_t = 0, \quad u(x, 0) = x^2.$$

4. For the Cauchy problem of equations (1)–(3), show that

if $f(x)$ is nondecreasing, then shocks never occur for $t \geq 0$.

5. For the Cauchy problem of equations (1)–(3), show that if $f(x)$ is strictly decreasing over any interval, then a shock eventually develops at some positive t. Derive a formula for the time and location of the shock.

* 6. Discuss whether shocks occur for the Cauchy problem

$$u_t + F(u)u_x = 0, \quad u(x, 0) = f(x).$$

* 7. Discuss whether shocks occur for the Cauchy problem

$$u_t + (F(u))_x = 0, \quad u(x, 0) = f(x).$$

12.4 Classification of Linear Second-Order Equations

In this section we are concerned only with second-order linear partial differential equations of two independent variables with constant coefficients, that is, equations of the form

$$a\frac{\partial^2 w}{\partial x^2}(x, y) + 2b\frac{\partial^2 w}{\partial x \partial y}(x, y) + c\frac{\partial^2 w}{\partial y^2}(x, y) + k\frac{\partial w}{\partial x}(x, y)$$

$$+ m\frac{\partial w}{\partial y}(x, y) + nw(x, y) = f(x, y), \qquad \textbf{(1)}$$

where a, b, c, k, m, and n are constants (the number 2 in front of the coefficient b simplifies later computations). Since we want to consider second-order equations, we assume that at least one of the coefficients a, b, or c is nonzero. If so, equation (1) is called a **second-order** equation. The procedure for solving second-order equations is much more complicated than that for first-order equations. Fortunately, as we show in this section, the general second-order equation (1) can be transformed into one of four standard simpler forms:

$$\frac{\partial^2 z}{\partial x^2} + \frac{\partial^2 z}{\partial y^2} + \lambda z = G(x, y) \qquad \text{(elliptic)}$$

$$\frac{\partial^2 z}{\partial x^2} - \frac{\partial^2 z}{\partial y^2} + \lambda z = G(x, y) \qquad \text{(hyperbolic)}$$

$$\frac{\partial^2 z}{\partial x^2} + \lambda\frac{\partial z}{\partial y} = G(x, y) \qquad \text{(parabolic)}$$

$$\frac{\partial^2 z}{\partial x^2} + \lambda z = G(x, y) \qquad \text{(degenerate)}$$

λ constant.

In addition to being easier to solve, these standard forms arise naturally in a number of important applications. In Sections 12.5 and 12.6 we shall study a particular hyperbolic equation, called the **wave equation,** that arises in modeling small transverse vibrations in a tightly stretched string. In Sections 12.7 and 12.8 we shall consider a parabolic equation, called the **heat equation,** that is used to analyze heat conduction in thin round bars. In Sections 12.8 and 12.9 we shall study a particular case of an elliptic equation, called **Laplace's equation,** that is used in determining steady-state flows (of fluids, heat, etc.).

Initially, it is convenient to simplify equation (1) by another rotation of axes. Using the substitution

$$u = x \cos \alpha + y \sin \alpha,$$
$$v = -x \sin \alpha + y \cos \alpha, \tag{2}$$

with $\tan 2\alpha = 2b/(a - c)$, we can eliminate the mixed second partial $\partial^2 w/\partial u \, \partial v$ and obtain an equation of the form

$$A \frac{\partial^2 w}{\partial u^2} + C \frac{\partial^2 w}{\partial v^2} + K \frac{\partial w}{\partial u} + M \frac{\partial w}{\partial v} + nw$$

$$= f(u \cos \alpha - v \sin \alpha, u \sin \alpha + v \cos \alpha) = F(u, v), \tag{3}$$

where

$$A = a \cos^2 \alpha + 2b \sin \alpha \cos \alpha + c \sin^2 \alpha,$$

$$C = a \sin^2 \alpha - 2b \sin \alpha \cos \alpha + c \cos^2 \alpha,$$

$$K = k \cos \alpha + m \sin \alpha,$$

$$M = m \cos \alpha - k \sin \alpha.$$

To see why this happens, observe that

$$\frac{\partial w}{\partial x} = \frac{\partial w}{\partial u} \frac{\partial u}{\partial x} + \frac{\partial w}{\partial v} \frac{\partial v}{\partial x} = \cos \alpha \frac{\partial w}{\partial u} - \sin \alpha \frac{\partial w}{\partial v},$$

$$\frac{\partial w}{\partial y} = \frac{\partial w}{\partial u} \frac{\partial u}{\partial y} + \frac{\partial w}{\partial v} \frac{\partial v}{\partial y} = \sin \alpha \frac{\partial w}{\partial u} + \cos \alpha \frac{\partial w}{\partial v},$$

and

$$\frac{\partial^2 w}{\partial x^2} = \frac{\partial}{\partial u} \left(\frac{\partial w}{\partial x} \right) \frac{\partial u}{\partial x} + \frac{\partial}{\partial v} \left(\frac{\partial w}{\partial x} \right) \frac{\partial v}{\partial x}$$

$$= \cos \alpha \left(\cos \alpha \frac{\partial^2 w}{\partial u^2} - \sin \alpha \frac{\partial^2 w}{\partial u \, \partial v} \right) - \sin \alpha \left(\cos \alpha \frac{\partial^2 w}{\partial u \, \partial v} - \sin \alpha \frac{\partial^2 w}{\partial v^2} \right)$$

$$= \cos^2 \alpha \frac{\partial^2 w}{\partial u^2} - 2 \sin \alpha \cos \alpha \frac{\partial^2 w}{\partial u \, \partial v} + \sin^2 \alpha \frac{\partial^2 w}{\partial v^2}.$$

Similarly,

$$\frac{\partial^2 w}{\partial x \partial y} = \sin \alpha \cos \alpha \frac{\partial^2 w}{\partial u^2} + (\cos^2 \alpha - \sin^2 \alpha) \frac{\partial^2 w}{\partial u \partial v} - \sin \alpha \cos \alpha \frac{\partial^2 w}{\partial v^2},$$

and

$$\frac{\partial^2 w}{\partial y^2} = \sin^2 \alpha \frac{\partial^2 w}{\partial u^2} + 2 \sin \alpha \cos \alpha \frac{\partial^2 w}{\partial u \partial v} + \cos^2 \alpha \frac{\partial^2 w}{\partial v^2}.$$

Then

$$a \frac{\partial^2 w}{\partial x^2} + 2b \frac{\partial^2 w}{\partial x \partial y} + c \frac{\partial^2 w}{\partial y^2} = A \frac{\partial^2 w}{\partial u^2} + 2B \frac{\partial^2 w}{\partial u \partial v} + C \frac{\partial^2 w}{\partial v^2},$$

where

$$B = (c - a) \sin \alpha \cos \alpha + b(\cos^2 \alpha - \sin^2 \alpha).$$

By the double angle formulas of trigonometry,

$$B = \frac{(c-a)}{2} \sin 2\alpha + b \cos 2\alpha = b \cos 2\alpha \left[\frac{(c-a)}{2b} \tan 2\alpha + 1 \right].$$

Since

$$\tan 2\alpha = \frac{2b}{a-c},$$

it follows that $B = 0$. An easy but messy calculation (see Problem 21) shows that

$$b^2 - ac = B^2 - AC = -AC. \tag{4}$$

Elliptic, Hyperbolic, and Parabolic Equations

By analogy with analytic geometry, we call the partial differential equation (1) an **elliptic** equation if $b^2 - ac < 0$, that is, if A and C have the same sign. If A and C have opposite signs, or equivalently, if $b^2 - ac > 0$, we say that (1) is a **hyperbolic** equation. Finally, if $b^2 - ac = 0$, (1) is called a **parabolic** equation. In summary, (1) is called

$$\left. \begin{array}{l} \text{hyperbolic} \\ \text{parabolic} \\ \text{elliptic} \end{array} \right\} \quad \text{if } b^2 - ac \left\{ \begin{array}{l} > 0, \\ = 0, \\ < 0. \end{array} \right. \tag{5}$$

EXAMPLE 1

Classify the second-order linear partial differential equation

$$\frac{\partial^2 w}{\partial x^2} + 2 \frac{\partial^2 w}{\partial x \partial y} + 4 \frac{\partial^2 w}{\partial y^2} + 7 \frac{\partial w}{\partial x} + w = 0. \tag{6}$$

SOLUTION

Here $a = b = 1$ and $c = 4$, so $b^2 - ac = -3$. Thus the partial differential equation (6) is elliptic.

Once we have eliminated the mixed second partial term, we can, for suitable values of α and β, use the substitution

$$w = e^{\alpha u + \beta v} z$$

to eliminate *at least one* of the first partials in (3). Note that by the product rule of differentiation,

$$\frac{\partial w}{\partial u} = \alpha w + e^{\alpha u + \beta v} \frac{\partial z}{\partial u}$$

and

$$\frac{\partial w}{\partial v} = \beta w + e^{\alpha u + \beta v} \frac{\partial z}{\partial v},$$

so that

$$\frac{\partial^2 w}{\partial u^2} = \alpha^2 w + 2\alpha e^{\alpha u + \beta v} \frac{\partial z}{\partial u} + e^{\alpha u + \beta v} \frac{\partial^2 z}{\partial u^2}$$

and

$$\frac{\partial^2 w}{\partial v^2} = \beta^2 w + 2\beta e^{\alpha u + \beta v} \frac{\partial z}{\partial v} + e^{\alpha u + \beta v} \frac{\partial^2 z}{\partial v^2}.$$

Equation (3) becomes

$$A \frac{\partial^2 z}{\partial u^2} + C \frac{\partial^2 z}{\partial v^2} + (2\alpha A + K) \frac{\partial z}{\partial u} + (2\beta C + M) \frac{\partial z}{\partial v}$$

$$+ (\alpha^2 A + \beta^2 C + \alpha K + \beta M + n) z = e^{-(\alpha u + \beta v)} F(u, v). \qquad (7)$$

If A and $C \neq 0$, then choosing $\alpha = -K/2A$ and $\beta = -M/2C$ allows us to eliminate the first partial terms, yielding the partial differential equation

$$A \frac{\partial^2 z}{\partial u^2} + C \frac{\partial^2 z}{\partial v^2} + \left(n - \frac{K^2}{4A} - \frac{M^2}{4C} \right) z = e^{-(\alpha u + \beta v)} F(u, v). \qquad (8)$$

If either A or C is zero, we cannot eliminate one of the first partial terms.

A change of scale in the independent variables provides a final simplification: If A and $C \neq 0$, set $u = \sqrt{|A|}\, x$ and $v = \sqrt{|C|}\, y$, so that (8) becomes

$$\frac{A}{|A|} \frac{\partial^2 z}{\partial x^2} + \frac{C}{|C|} \frac{\partial^2 z}{\partial y^2} + \left(n - \frac{K^2}{4A} - \frac{M^2}{4C} \right) z = G(x, y). \qquad (9)$$

From the calculations above, it follows that (1) can be transformed into one of the following four standard forms:

**Standard Forms for a Second-Order,
Linear Partial Differential Equation
with Constant Coefficients**

$$\frac{\partial^2 z}{\partial x^2} + \frac{\partial^2 z}{\partial y^2} + \lambda z = G(x, y) \qquad \text{(elliptic)} \qquad \textbf{(10)}$$

$$\frac{\partial^2 z}{\partial x^2} - \frac{\partial^2 z}{\partial y^2} + \lambda z = G(x, y) \qquad \text{(hyperbolic)} \qquad \textbf{(11)}$$

$$\frac{\partial^2 z}{\partial x^2} + \lambda \frac{\partial z}{\partial y} = G(x, y) \qquad \text{(parabolic)} \qquad \textbf{(12)}$$

$$\frac{\partial^2 z}{\partial x^2} + \lambda z = G(x, y) \qquad \text{(degenerate)} \qquad \textbf{(13)}$$

λ constant.

Remark. Equation (10) occurs when A and C in (9) are nonzero and have the same sign. Equation (11) occurs when A and C are nonzero and have opposite signs. Equation (12) arises when $C = 0$ and $AM \neq 0$, by choosing $\alpha = -K/2A$ and $\beta = (K^2/4A - n)/M$; here $\lambda = M/A$. Equation (13) occurs when $C = M = 0$; here $\lambda = n - K^2/4A$.

The degenerate case may be treated as an ordinary differential equation with parameter v, in a manner very similar to Example 12.1.1. For this reason, we do not consider this case any further. In the next three sections we consider examples of elliptic, hyperbolic, and parabolic equations, showing how each arises in practice and indicating what steps can be taken to obtain a solution.

EXAMPLE 2

To transform the equation

$$\frac{\partial^2 w}{\partial x^2} + 4\frac{\partial^2 w}{\partial x \partial y} + \frac{\partial^2 w}{\partial y^2} + \frac{\partial w}{\partial x} = 0$$

into standard form, observe that $b^2 - ac = 3$, since $a = c = 1$ and $b = 2$. Therefore the equation is hyperbolic. Since $a - c = 0$, $\tan 2\alpha$ is infinite, so $\alpha = 45° = \pi/4$. Substituting this value into (3), we obtain

$$3\frac{\partial^2 w}{\partial u^2} - \frac{\partial^2 w}{\partial v^2} + \frac{1}{\sqrt{2}}\frac{\partial w}{\partial u} - \frac{1}{\sqrt{2}}\frac{\partial w}{\partial v} = 0.$$

Setting

$$w = z \exp\left[-\frac{\sqrt{2}}{12}(u + 3v)\right]$$

(or $\alpha = -K/2A = -1/6\sqrt{2} = -\sqrt{2}/12$ and $\beta = -M/2C = -1/2\sqrt{2} = -\sqrt{2}/4$), we obtain from (8)

$$3\frac{\partial^2 z}{\partial u^2} - \frac{\partial^2 z}{\partial v^2} + \frac{z}{12} = 0.$$

Finally, letting $u = \sqrt{3}\,x$ and $v = y$, we get the hyperbolic equation

$$\frac{\partial^2 z}{\partial x^2} - \frac{\partial^2 z}{\partial y^2} + \frac{z}{12} = 0.$$

Problems 12.4

In Problems 1–10 classify the given equation as elliptic, parabolic, hyperbolic, or degenerate second-order partial differential equations.

1. $\dfrac{\partial^2 w}{\partial x^2} - \dfrac{\partial^2 w}{\partial y^2} = 0$

2. $\dfrac{\partial^2 w}{\partial x^2} + \dfrac{\partial^2 w}{\partial y^2} + w = 0$

3. $w_{xx} + 2w_{xy} + w_{yy} + w_x + w_y = 0$

4. $w_{xx} + w_{xy} + 4w_{yy} + 5w_x = 2$

5. $w_{xx} + 2w_{xy} + 2w_{yy} = 0$

6. $w_{xx} + 2w_{xy} + w_{yy} + w = 0$

7. $w_{xx} + 2w_{xy} + 4w_{yy} + 5w = 0$

8. $w_{xx} + 2w_{xy} - w_{yy} + w_y = 0$

9. $3w_{xx} - 5w_{xy} + 2w_{yy} = \sin(xy)$

10. $7w_{xy} + 5w_{yy} + 8w = x/y$

In Problems 11–20 transform the given equation into standard form.

11. $4\dfrac{\partial^2 w}{\partial x^2} + 3\dfrac{\partial^2 w}{\partial y^2} - w = 0$

12. $7\dfrac{\partial^2 w}{\partial x^2} + 4\dfrac{\partial^2 w}{\partial y^2} + 3w = 2$

13. $\dfrac{\partial^2 w}{\partial x^2} - \dfrac{\partial^2 w}{\partial y^2} + 2\dfrac{\partial w}{\partial x} - 5\dfrac{\partial w}{\partial y} + 7w = 0$

14. $\dfrac{\partial^2 w}{\partial x^2} + 2\dfrac{\partial^2 w}{\partial x\partial y} + 2\dfrac{\partial^2 w}{\partial y^2} = 0$

15. $\dfrac{\partial^2 w}{\partial x^2} + 2\dfrac{\partial^2 w}{\partial x\partial y} + \dfrac{\partial^2 w}{\partial y^2} + w = 0$

16. $\dfrac{\partial^2 w}{\partial x^2} + 2\dfrac{\partial^2 w}{\partial x\partial y} - \dfrac{\partial^2 w}{\partial y^2} = 0$

17. $\dfrac{\partial^2 w}{\partial x^2} + 2\dfrac{\partial^2 w}{\partial x\partial y} + 4\dfrac{\partial^2 w}{\partial y^2} + 5w = 0$

18. $\dfrac{\partial^2 w}{\partial x^2} + 2\dfrac{\partial^2 w}{\partial x\partial y} + \dfrac{\partial^2 w}{\partial y^2} + \dfrac{\partial w}{\partial y} = 0$

19. $\dfrac{\partial^2 w}{\partial x^2} - 4\dfrac{\partial^2 w}{\partial x\partial y} + \dfrac{\partial^2 w}{\partial y^2} + \dfrac{\partial w}{\partial y} = 0$

20. $w_{xx} - 4w_{xy} + 2w_{yy} + w_x - 3w_y + w = 0$

21. Obtain equation (4) by a direct calculation.

The classification of a partial differential equation can be done even if the coefficients in (1) are not constants. In such a case we divide the plane into the regions where the **discriminant**

$$\Delta = b^2 - ac$$

is everywhere positive, everywhere negative, or everywhere zero. For example

$$w_{xx} + xw_{yy} = 0$$

has discriminant $\Delta = -x$, so the partial differential equation is hyperbolic in the region $\{(x, y): x < 0\}$, elliptic in $\{(x, y): x > 0\}$, and parabolic in $\{(x, y): x = 0\}$. In Problems 22–27 describe the regions in the plane where the given equation is elliptic, parabolic, or hyperbolic.

22. $w_{xx} - xw_{yy} = 0$

23. $w_{xx} - xyw_{yy} = 0$

24. $x^2 w_{xx} + w_{yy} = xy$

25. $w_{xx} + (x^2 - y^2)w_{yy} = 0$

26. $w_{xx} + 2xw_{xy} + w_{yy} + 5w = 0$

27. $y\dfrac{\partial^2 w}{\partial x^2} - 2\dfrac{\partial^2 w}{\partial x\partial y} + e^x\dfrac{\partial^2 w}{\partial y^2} + y^2\dfrac{\partial w}{\partial y} - w = 0$

12.5 The Vibrating String: d'Alembert's Method

Consider a string that is tightly stretched between two fixed points 0 and L on the x-axis (see Fig. 1). Suppose that the string is pulled back vertically a distance that is very small compared to the length L and released at time $t = 0$, causing it to vibrate. Our problem is to determine the displacement $y(x, t)$ of the point on the string that is x units away from the end 0, at any time t.

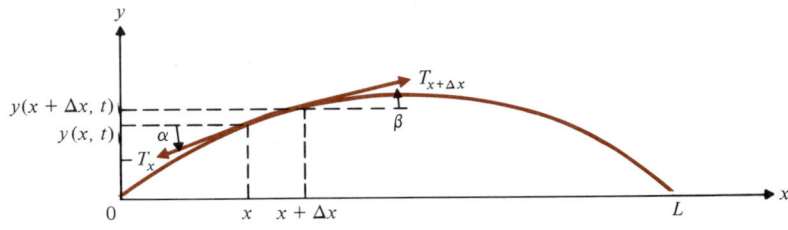

FIGURE 1

To avoid making our equation too complicated, we make two simplifying assumptions:

a. Our "ideal" string has uniform mass m per unit horizontal length and offers no resistance to bending.

b. The tension T in the string is so large that the gravitational force on the string may be neglected.

Consider a small segment of length Δx. By Newton's second law of motion, the total force acting on this piece of string is equal to the mass of the string multiplied by its acceleration:

$$F = ma = (m\Delta x)\frac{\partial^2 y}{\partial t^2}. \tag{1}$$

We assume in this equation that the string is moving only in the xy-plane and that each particle in the string moves only vertically.

Let T_x and $T_{x+\Delta x}$ be the tension vectors at the endpoints of the given segment. These forces are applied tangentially (see Fig. 1) since the string offers no resistance to bending. Since there is no motion in the x-direction, the x-components of the tension vectors must coincide:

$$T_{x+\Delta x}\cos \beta = T_x \cos \alpha \equiv T. \tag{2}$$

Thus T is constant, since x and Δx are arbitrary. Similarly, the difference in the y-components of the tension vectors must equal the total force acting on the string. Then by equation (1),

$$T_{x+\Delta x}\sin \beta - T_x \sin \alpha = m\Delta x\frac{\partial^2 y}{\partial t^2}. \tag{3}$$

Dividing each term in (3) by the corresponding term in (2), we have

$$\frac{T_{x+\Delta x}\sin\beta}{T_{x+\Delta x}\cos\beta} - \frac{T_x\sin\alpha}{T_x\cos\alpha} = \frac{m}{T}\Delta x\frac{\partial^2 y}{\partial t^2},$$

or

$$\tan\beta - \tan\alpha = \frac{m}{T}\Delta x\frac{\partial^2 y}{\partial t^2}. \tag{4}$$

Since

$$\tan\alpha = \frac{\partial y}{\partial x}\Big|_x \quad \text{and} \quad \tan\beta = \frac{\partial y}{\partial x}\Big|_{x+\Delta x},$$

we may rewrite (4) in the form

$$\frac{1}{\Delta x}\left[\frac{\partial y}{\partial x}\Big|_{x+\Delta x} - \frac{\partial y}{\partial x}\Big|_x\right] = \frac{m}{T}\frac{\partial^2 y}{\partial t^2}.$$

Letting Δx tend to zero, we obtain in the limit the equation

$$\frac{\partial^2 y}{\partial x^2} = \frac{m}{T}\frac{\partial^2 y}{\partial t^2}.$$

Since m/T is positive, it is clear that this is a hyperbolic partial differential equation. The equation, written in the form

$$\frac{\partial^2 y}{\partial t^2} = c^2\frac{\partial^2 y}{\partial x^2}, \qquad c^2 = \frac{T}{m} = \frac{\text{tension}}{\text{mass per unit length}}, \tag{5}$$

Wave Equation

is often called the one-dimensional **wave equation,** where the constant c^2 indicates that the coefficient is positive.

As yet we have not made use of the fact that the string is fixed at its endpoints. We may write these boundary conditions as

$$y(0, t) = y(L, t) = 0, \qquad t \geq 0. \tag{6}$$

In addition, we have not taken into account the initial distortion of the string and the fact that it was at rest when released. These initial conditions can be written as

$$y(x, 0) = f(x), \qquad 0 \leq x \leq L, \tag{7}$$

$$\frac{\partial y}{\partial t}\Big|_{t=0} = 0, \tag{8}$$

where $f(x)$ is a continuous function depicting the original distortion (see Examples 1 and 2, following), and $(\partial y/\partial t)(x, t)$ is the velocity of the point x units away from the origin at time t.

The direct method we use to solve this problem is due to d'Alembert.† Since it is only rarely possible to apply this technique, we develop two more applicable methods in this chapter.

We begin by defining the variables

$$u = x + ct, \qquad v = x - ct, \tag{9}$$

and transforming the wave equation (5) into one involving variables u and v. By the chain rule, we have

$$\frac{\partial y}{\partial t} = \frac{\partial y}{\partial u}\frac{\partial u}{\partial t} + \frac{\partial y}{\partial v}\frac{\partial v}{\partial t} = c\left(\frac{\partial y}{\partial u} - \frac{\partial y}{\partial v}\right), \tag{10}$$

where we used (9) to obtain the partial derivatives $\partial u/\partial t = c = -\partial v/\partial t$. Taking the partial derivative with respect to t of (10), we obtain

$$\frac{\partial^2 y}{\partial t^2} = c\frac{\partial}{\partial t}\left(\frac{\partial y}{\partial u} - \frac{\partial y}{\partial v}\right) = c\left[\frac{\partial}{\partial u}\left(\frac{\partial y}{\partial u} - \frac{\partial y}{\partial v}\right)\frac{\partial u}{\partial t} + \frac{\partial}{\partial v}\left(\frac{\partial y}{\partial u} - \frac{\partial y}{\partial v}\right)\frac{\partial v}{\partial t}\right]$$

$$= c^2\left[\left(\frac{\partial^2 y}{\partial u^2} - \frac{\partial^2 y}{\partial u\,\partial v}\right) - \left(\frac{\partial^2 y}{\partial v\,\partial u} - \frac{\partial^2 y}{\partial v^2}\right)\right].$$

Since the mixed second partials are equal, we finally have

$$\frac{\partial^2 y}{\partial t^2} = c^2\left(\frac{\partial^2 y}{\partial u^2} - 2\frac{\partial^2 y}{\partial u\,\partial v} + \frac{\partial^2 y}{\partial v^2}\right).$$

Similarly, we find that

$$\frac{\partial^2 y}{\partial x^2} = \left(\frac{\partial^2 y}{\partial u^2} + 2\frac{\partial^2 y}{\partial u\,\partial v} + \frac{\partial^2 y}{\partial v^2}\right).$$

Substituting these results into (5) and canceling like terms yield the equation

$$\frac{\partial^2 y}{\partial u\,\partial v} = 0, \tag{11}$$

since $4c^2 \neq 0$. It is now easy to solve (11) by performing two successive integrations. Integrating with respect to u, we obtain

$$\frac{\partial y}{\partial v} = g'(v), \tag{12}$$

where g' is an unknown function in v. From (12) it is evident that g' is the partial derivative, with respect to v, of the solution y—which is the reason for using the prime in our notation. Integrating (12) with respect to v yields

$$y = g(v) + h(u), \tag{13}$$

d'Alembert's Solution

where h is an unknown function in u. Substituting (9) into (13) yields **d'Alembert's solution,**

---

† The French mathematician Jean Le Rond d'Alembert (1717–1783) is known for his contributions in mechanics.

$$y(x, t) = g(x - ct) + h(x + ct), \tag{14}$$

of the wave equation (5).

Note that each function $g(x - ct)$ and $h(x + ct)$ is separately a solution of the wave equation (5). Consider the solution $g(x - ct)$. If we set $t = 0$, the initial distortion of the string is given by $y(x, 0) = g(x)$. At a fixed time later, its distortion is $y(x, t) = g(x - ct)$. Thus the shape of the string can be obtained from the graph of the function $g(x)$ by moving an interval of length L to the right with velocity c. For this reason, $g(x - ct)$ is called a **traveling wave.** Similarly, $h(x + ct)$ is a traveling wave that moves to the left with velocity c. Hence d'Alembert's solution is a superposition of two traveling waves, one moving to the right, the other to the left, with the same velocity c.

Traveling Wave

The functions g and h can be determined from the initial conditions. Setting $t = 0$, we find that

$$f(x) = y(x, 0) = g(x) + h(x). \tag{15}$$

Noting that g and h are functions of one variable, we may use the chain rule to differentiate (14) with respect to t. Setting $t = 0$, we obtain

$$0 = \frac{\partial y}{\partial t}\bigg|_{t=0} = -cg'(x - ct) + ch'(x + ct)|_{t=0} = c[h'(x) - g'(x)].$$

From the last equation we known that $h' = g'$, so

$$h(x) = g(x) + k,$$

where k is a constant. Replacing this function for h in (15) yields $f(x) = 2g(x) + k$ or $g = (f - k)/2$. Hence $h = (f + k)/2$, so the solution of the problem finally becomes

$$y(x, t) = \tfrac{1}{2}[f(x - ct) + f(x + ct)]. \tag{16}$$

Note that since the boundary conditions (6) must be satisfied by (16), setting $x = 0$ in (16) yields

$$0 = \tfrac{1}{2}[f(-ct) + f(ct)],$$

or

$$f(-ct) = -f(ct).$$

Thus $f(x)$ is odd. This result may seem curious, since $f(x)$ has been defined only on the interval $0 \le x \le L$. However, it causes no difficulty, since we may extend f to the interval $-L \le x \le L$ by defining

$$f(-x) = -f(x) \qquad \text{for} \qquad -L \le x \le 0.$$

This extension is always possible as $f(0) = 0$. Similarly, setting $x = L$ and using the fact that f is odd, we find that $0 = \tfrac{1}{2}[f(L - ct) + f(L + ct)]$ or

$$f(L + ct) = -f(L - ct) = f(ct - L),$$

so that f has period $2L$. Again, we can extend $f(x)$ to the entire real line by letting

$$f(x + 2L) = f(x),$$

for all x. This extension of $f(x)$ to the entire real line seems very reasonable if we return to the notion of traveling waves. It is clear, from equation (16), that we must immediately leave the interval $0 \leq x \leq L$ when $t > 0$. Now suppose we move $f(x)$ gradually to the right. Since $y(0, t) = y(L, t) = 0$ for all t, any positive contribution at 0 or L from the traveling wave $f(x - ct)$ must be offset by an equal negative contribution from $f(x + ct)$ [which is obtained by moving $f(x)$ the same distance to the left], and vice versa. In effect, this procedure amounts to a reflection of wave forms at the boundary. (See Fig. 2.)

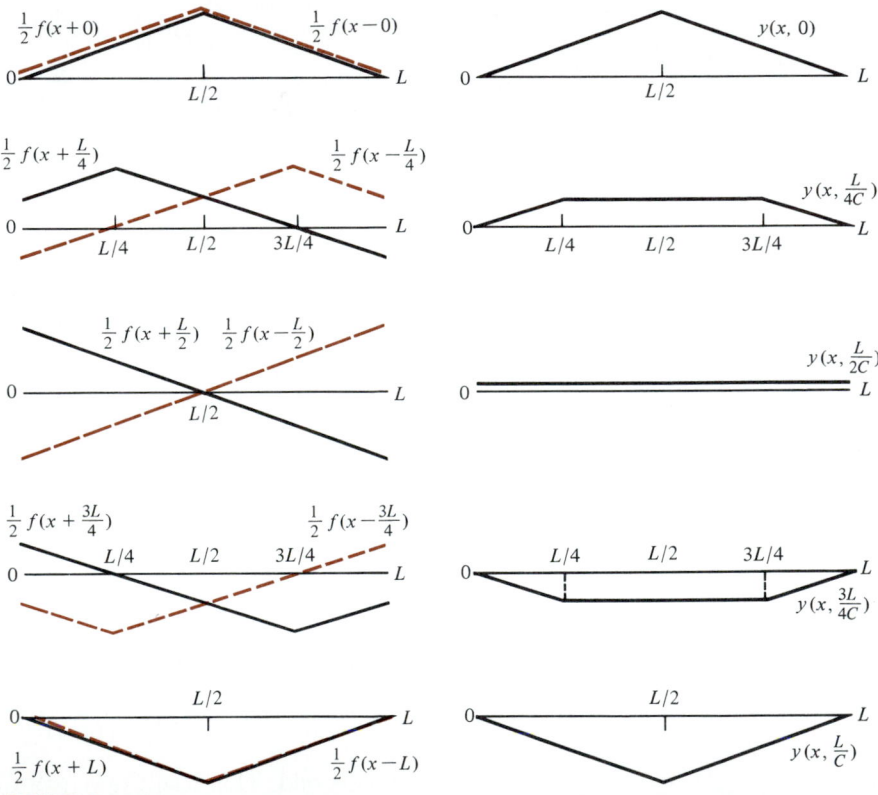

FIGURE 2. Solution $y(x, t)$ obtained as a superposition of a wave traveling to the right (dashed) and a wave traveling to the left (solid).

By Theorem 11.4.2, the function $f(x)$ has a Fourier sine series—a fact we shall refer to in the next section. (The continuity of f is guaranteed by the fact that the string is unbroken. We assume that f' is piecewise continuous, since that is what it is in most practical situations, such as in the next example.)

EXAMPLE 1

Suppose that we pluck the string at its center (see Fig. 3), a distance $y(L/2, 0) = y_0$ meters (y_0 is assumed to be small). We can assume that the distortion consists of two straight lines from $(0, 0)$ to $(L/2, y_0)$ and from $(L/2, y_0)$ to $(L, 0)$. Hence

$$f(x) = \begin{cases} \dfrac{2y_0}{L}x, & 0 \le x \le \dfrac{L}{2}, \\[2mm] \dfrac{2y_0}{L}(L-x), & \dfrac{L}{2} \le x \le L, \end{cases}$$

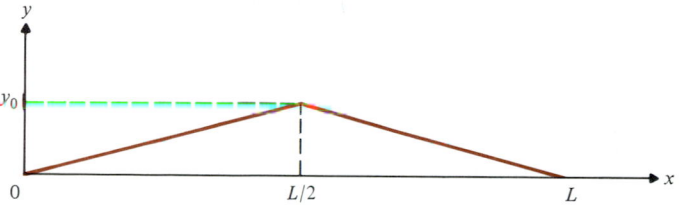

FIGURE 3

which has the piecewise continuous derivative

$$f'(x) = \begin{cases} \dfrac{2y_0}{L}, & 0 \le x < \dfrac{L}{2}, \\[2mm] \dfrac{-2y_0}{L}, & \dfrac{L}{2} < x \le L. \end{cases}$$

If the string is 4 meters long, has a mass of 0.08 kilogram, and is subject to a constant tension of 2 newtons ($=$kg-m/s^2), we obtain the solution

$$y(x, t) = \tfrac{1}{2}[f(x - 10t) + f(x + 10t)],$$

since (remember that m = mass per unit length)

$$c^2 = \frac{T}{m} = \frac{2}{0.08/4} = 100 \,(\text{meters/second})^2.$$

To discover the exact location of a point 1 meter from the end 0 after exactly 1 second, we set $x = t = 1$. Then

$$y(1, 1) = \tfrac{1}{2}[f(-9) + f(11)],$$

and since f has period $2L = 8$ and is odd, we get

$$y(1, 1) = \frac{1}{2}[f(-1) + f(3)] = \frac{1}{2}[f(3) - f(1)]$$

$$= \frac{1}{2}\left(\frac{2y_0}{L} - \frac{2y_0}{L}\right) = \frac{1}{2}\left(\frac{2y_0}{4} - \frac{2y_0}{4}\right) = 0.$$

EXAMPLE 2

Suppose that $f(x) = y_0 \sin (\pi x/L)$. By equation (16),

$$y(x, t) = \frac{y_0}{2}\left[\sin \frac{\pi}{L}(x - ct) + \sin \frac{\pi}{L}(x + ct)\right],$$

and by the addition formulas of trigonometry, this reduces to

$$y(x, t) = y_0 \sin \frac{\pi x}{L} \cos \frac{\pi ct}{L}.$$

Thus, if we fix our attention on the point x_0 units from the left, we see that it oscillates with a period $2L/c$ and an amplitude $y_0 \sin(\pi x_0/L)$.

Problems 12.5

In Problems 1–8 use d'Alembert's solution (16) to find the displacement $y(x, t)$ for each given function $f(x)$, point x, and time t. Assume that $c = 5$ and $L = 6$ meters.

1. $f(x) = 0.01 \sin(\pi x/L)$, $x = 2$, $t = 1$

2. $f(x) = 0.1 \sin(\pi x/L)$, $x = 3$, $t = 2$

3. $f(x) = 0.01 \sin(2\pi x/L)$, $x = 3$, $t = 1$

4. $f(x) = 0.1 \sin(2\pi x/L)$, $x = 2$, $t = 2$

5. $f(x) = \begin{cases} 0.1x, & 0 \le x \le 3, \\ 0.1(6 - x), & 3 \le x \le 6 \end{cases}$ $x = 2$, $t = 2$

6. $f(x) = \begin{cases} 0.1x, & 0 \le x \le 2, \\ 0.2(3 - x), & 2 \le x \le 4, \\ 0.1(x - 6), & 4 \le x \le 6 \end{cases}$ $x = 3$, $t = 5$

7. $f(x) = \begin{cases} 0.1x, & 0 \le x \le 2, \\ 0.2, & 2 \le x \le 4, \\ 0.1(6 - x), & 4 \le x \le 6 \end{cases}$ $x = 3$, $t = 5$

8. $f(x) = \begin{cases} 0.01x^2, & 0 \le x \le 3, \\ 0.01(6 - x)^2, & 3 \le x \le 6 \end{cases}$ $x = 2$, $t = 2$

9. Assume, instead of the initial conditions (7) and (8), that a string is initially in its equilibrium position and is set vibrating by each of its points x being given an initial velocity $f'(x)$, where f is a differentiable function. Prove that the solution is given by

$$y(x, t) = \frac{1}{2c}[f(x + ct) - f(x - ct)].$$

Using the result of Problem 9 and setting $c = 5$ and $L = 6$ meters, do Problems 10–15 for the given functions $f'(x)$ and values x and t.

10. $f'(x) = 0.01 \sin(\pi x/L)$, $x = 2$ meters, $t = 3$ seconds

11. $f'(x) = 0.1 \sin(2\pi x/L)$, $x = 1$ meter, $t = 1$ second

12. $f'(x) = 3(6x - x^2)$, $x = 3$ meters, $t = 0.01$ second

13. $f'(x) = 0.1 \sin^2(\pi x/L)$, $x = 1$ meter, $t = 0.1$ second

14. $f'(x) = \begin{cases} 0.01x, & 0 \le x \le 3, \\ 0.01(6 - x), & 3 \le x \le 6 \end{cases}$ $x = 2$, $t = 1$

15. $f'(x) = \begin{cases} 0.01(3 - x), & 0 \le x \le 3, \\ 0.01(x - 3), & 3 \le x \le 6 \end{cases}$ $x = 2$, $t = 1$

16. Suppose that we subject a vibrating string to both an initial displacement $y(x, 0) = f(x)$, $0 \le x \le L$, and an initial velocity $y_t(x, 0) = g'(x)$, $0 \le x \le L$. Show that the solution to this initial value problem is

$$y(x, t) = \frac{1}{2}[f(x - ct) + f(x + ct)]$$

$$+ \frac{1}{2c}[g(x + ct) - g(x - ct)].$$

17. Use the result of Problem 16 to solve the initial value problem

$$y_{tt} = y_{xx}, \quad |x| < \infty, \quad t > 0$$

$$y(x, 0) = \sin \frac{\pi x}{2} = y_t(x, 0).$$

18. Suppose that we subject a vibrating string to a damping force that is proportional at each instant to the velocity at each point. Show that the resulting hyperbolic differential equation is of the form

$$\frac{\partial^2 y}{\partial t^2} - k\frac{\partial y}{\partial t} = c^2 \frac{\partial^2 y}{\partial x^2}.$$

* 19. The distortionless solution of the transmission-line problem of a telegraph cable satisfies the partial differential equation

$$v_{xx} = \alpha v_{tt} + \beta v_t + \gamma v, \quad 0 < x < \infty, \quad t > 0,$$

where $\alpha = LC$, $\beta = RC + GL$, $\gamma = RG$, and v, L, R, C, and G are the voltage, inductance, resistance, capacitance, and conductance between the line and ground. What is the form of the solution?

12.6 Separation of Variables: The Wave Equation

One of the simplest techniques for solving ordinary differential equations consists of separating the variables. Although we now have *two* independent variables, we can nevertheless adapt the technique to all partial differential equations of the forms (12.4.10)–(12.4.13) whenever the function $F(u, v) \equiv 0$. The method is best explained by means of an example.

Suppose that we again consider the problem of a vibrating string given by the wave equation

$$\frac{\partial^2 y}{\partial t^2} = c^2 \frac{\partial^2 y}{\partial x^2}, \qquad c^2 = \frac{T}{m}, \tag{1}$$

with boundary conditions

$$y(0, t) = y(L, t) = 0, \qquad t \geq 0, \tag{2}$$

and initial conditions

$$y(x, 0) = f(x), \qquad \frac{\partial y}{\partial t}\bigg|_{t=0} = 0, \qquad 0 \leq x \leq L. \tag{3}$$

We begin by seeking a solution of the form

$$y(x, t) = X(x)T(t), \tag{4}$$

where X is a function of x alone and T is a function involving only the independent variable t. Of course there is no guarantee that such a solution exists. In fact, as we will see later, separation is often not possible. Nevertheless, we can try it out and see if it works. Using primes to denote differentiation, we observe that

$$\frac{\partial^2 y}{\partial t^2} = \frac{\partial^2}{\partial t^2}[XT] = XT'' \qquad \text{and} \qquad \frac{\partial^2 y}{\partial x^2} = X''T,$$

so that equation (1) becomes

$$XT'' = c^2 X''T. \tag{5}$$

It is now possible to separate the variables and write equation (5) as

$$\frac{T''}{T} = c^2 \frac{X''}{X}. \tag{6}$$

Initially we had no assurance that it would be possible to separate the variables as we have done. Should such a separation prove impossible, this method would not apply and we would have to turn to other techniques.

Now suppose we fix t and allow x to vary. Since the left-hand side of equation (6) is constant, the right-hand side must be constant for all values of x. Consequently, the left-hand side is constant *regardless* of the value of t, and we have

$$\frac{T''}{T} = c^2 \frac{X''}{X} = k,$$

where k is constant. From this set of equalities we obtain two ordinary differential equations

$$T'' - kT = 0, \tag{7}$$

$$X'' - \frac{k}{c^2} X = 0. \tag{8}$$

Since our solution must satisfy the boundary conditions (2), if a solution of the form (4) exists, we must have the equations

$$X(0)T(t) = X(L)T(t) = 0 \qquad \text{for all } t.$$

If $T(t) \equiv 0$, then $y \equiv 0$ and the string is constantly at rest, which is impossible if $f(x) \not\equiv 0$. For $f \not\equiv 0$, T must be nonzero for at least one value of t, which implies that $X(0) = X(L) = 0$. Hence equation (8) becomes a Sturm-Liouville problem with these boundary conditions. Thus the problem of solving the partial differential equation (1) has been reduced to solving two ordinary differential equations, (7) and (8). The constant k is arbitrary, but its value must be the same for both equations. Using the characteristic equation (see p. 80) $\lambda^2 - k = 0$, we can obtain the general solution

$$T(t) = \begin{cases} a_1 e^{\sqrt{k}t} + a_2 e^{-\sqrt{k}t}, & \text{if } k > 0, \\[2mm] a_1 t + a_2, & \text{if } k = 0, \\[2mm] a_1 \cos \sqrt{-k}t + a_2 \sin \sqrt{-k}t, & \text{if } k < 0. \end{cases}$$

Similarly, we also have the general solution

$$X(x) = \begin{cases} b_1 e^{\sqrt{k}x/c} + b_2 e^{-\sqrt{k}x/c}, & \text{if } k > 0, \\[2mm] b_1 x + b_2, & \text{if } k = 0, \\[2mm] b_1 \cos \dfrac{\sqrt{-k}x}{c} + b_2 \sin \dfrac{\sqrt{-k}x}{c}, & \text{if } k < 0. \end{cases}$$

Setting $x = 0$ and $x = L$ in the general solution yields, for $k > 0$, the homogeneous system of equations

$$X(0) = b_1 \qquad + b_2 \qquad = 0,$$

$$X(L) = b_1 e^{\sqrt{k}L/c} + b_2 e^{-\sqrt{k}L/c} = 0.$$

This system has a nonzero solution if and only if the determinant of its coefficients is zero. But

$$\begin{vmatrix} 1 & 1 \\ e^{\sqrt{k}L/c} & e^{-\sqrt{k}L/c} \end{vmatrix} = e^{-\sqrt{k}L/c} - e^{\sqrt{k}L/c} = -2 \sinh \frac{\sqrt{k}L}{c} \neq 0,$$

so that $X(x) \equiv 0$ for $k > 0$. Then $y \equiv 0$, which is impossible if $f(x) \not\equiv 0$. For $k = 0$, the condition $X(0) = 0$ implies that $b_2 = 0$, while $X(L) = 0$ requires that $b_1 = 0$, again leading to the zero solution. So we are left with the remaining possibility that k is negative, and we set $k = -r^2$. This yields the general solution

$$X(x) = b_1 \cos \frac{rx}{c} + b_2 \sin \frac{rx}{c}.$$

Letting $x = 0$ and $x = L$, we have

$$X(0) = b_1 = 0 \quad \text{and} \quad X(L) = b_2 \sin \frac{rL}{c} = 0.$$

To prevent our again having the zero solution, we must choose an r such that rL/c is a positive multiple of π:

$$\frac{rL}{c} = n\pi \qquad \text{or} \qquad r = \frac{n\pi c}{L}.$$

Thus we obtain an infinite set of solutions,

$$X_n(x) = b_2 \sin \frac{n\pi x}{L}, \tag{9}$$

each associated with the choice

$$k = -\frac{n^2\pi^2 c^2}{L^2}. \tag{10}$$

Hence the corresponding solutions for T are

$$T_n(t) = a_1 \cos \frac{n\pi ct}{L} + a_2 \sin \frac{n\pi ct}{L}.$$

Finally, we obtain an infinite set of solutions,

$$y_n(x, t) = X_n(x)T_n(t) = \left(B_n \cos \frac{n\pi ct}{L} + B_n^* \sin \frac{n\pi ct}{L} \right) \sin \frac{n\pi x}{L}, \tag{11}$$

where B_n and B_n^* are constants, that satisfy the partial differential equation (1) and the boundary conditions (2). In the language of Chapter 11 we see that we have found the *eigenvalues* (10) and the corresponding *eigenfunctions* (11) of the boundary value problem (1), (2).

We now try to select a solution that satisfies the initial conditions (3). It is extremely likely that no one solution (11) will satisfy (3). However, any finite sum of solutions (11) is again a solution of the boundary value problem (by the *principle of superposition*), so we can try to find a finite linear combination of these solutions that satisfies the initial conditions (3). Generally, even this will fail, so as a last resort we try an infinite series of the solutions (11):

$$y(x, t) = \sum_{n=1}^{\infty} y_n(x, t) = \sum_{n=1}^{\infty} \left(B_n \cos \frac{n\pi ct}{L} + B_n^* \sin \frac{n\pi ct}{L} \right) \sin \frac{n\pi x}{L}. \qquad (12)$$

There is no guarantee at this point that such a series converges, but since its form is very similar to that of the Fourier series (11.3.7) (p. 680), there is every reason to be optimistic. Furthermore, if (12) converges and the series obtained by the formal term-by-term partial differentiations $\partial/\partial t$, $\partial^2/\partial t^2$, $\partial/\partial x$, and $\partial^2/\partial x^2$ all converge uniformly, then (12) is again a solution of the boundary value problem (1), (2). If, in addition, we can satisfy the initial conditions (3), then we have solved the vibrating string problem.

Setting $t = 0$ in equation (12) and using the first of equations (3), we have

$$y(x, 0) = \sum_{n=1}^{\infty} B_n \sin \frac{n\pi x}{L} = f(x). \qquad (13)$$

We again assume that $f(x)$ and $f'(x)$ are piecewise continuous (see Section 12.5). The infinite series (13) is a Fourier sine series, which requires that the function $f(x)$ be odd [$f(-x) = -f(x)$], and periodic with period $2L$. As in Section 12.5, we can extend the function $f(x)$, defined on the interval $0 \leq x \leq L$, in such a way that it is odd and periodic with period $2L$. Since $f(0) = 0$, we let $f(-x) = -f(x)$ on $-L \leq x \leq 0$ and require that $f(x + 2L) = f(x)$ for all real numbers x. Then the coefficients B_n are given by the Euler formula

$$B_n = \frac{1}{L} \int_{-L}^{L} f(x) \sin \frac{n\pi x}{L} dx = \frac{2}{L} \int_{0}^{L} f(x) \sin \frac{n\pi x}{L} dx, \qquad n = 1, 2, 3, \ldots. \qquad (14)$$

The last equality holds because $f(x)$ and $\sin(n\pi x/L)$ are odd and their product is even [see equation (11.4.5), p. 689]. Taking the partial derivative of (12) with respect to t and using the second of the initial conditions (3), we have

$$\frac{\partial y}{\partial t} = \frac{\pi c}{L} \sum_{n=1}^{\infty} \left(nB_n^* \cos \frac{n\pi ct}{L} - nB_n \sin \frac{n\pi ct}{L} \right) \sin \frac{n\pi x}{L} \Bigg|_{t=0} = 0,$$

or

$$\sum_{n=1}^{\infty} nB_n^* \sin \frac{n\pi x}{L} = 0.$$

Again we obtain an expression of the form (14) for B_n^*, but this time the function involved is zero. Hence all the coefficients B_n^* vanish. Thus solution (12) reduces to

$$y(x, t) = \sum_{n=1}^{\infty} B_n \cos \frac{n\pi ct}{L} \sin \frac{n\pi x}{L}, \qquad (15)$$

where the coefficients B_n are given by (14). It is possible to write series (15) in a more compact form by using the trigonometric identity

$$2 \sin A \cos B = \sin(A + B) + \sin(A - B).$$

Then

$$y(x, t) = \frac{1}{2} \sum_{n=1}^{\infty} B_n \left[\sin \frac{n\pi}{L}(x+ct) + \sin \frac{n\pi}{L}(x-ct) \right]$$

$$= \frac{1}{2} \left[\sum_{n=1}^{\infty} B_n \sin \frac{n\pi}{L}(x+ct) + \sum_{n=1}^{\infty} B_n \sin \frac{n\pi}{L}(x-ct) \right],$$

and the series can be evaluated by substituting $x + ct$ and $x - ct$, respectively, for the variable x in (13). Hence

$$y(x, t) = \tfrac{1}{2}[f(x+ct) + f(x-ct)],$$

which is precisely the result we obtained directly in Section 12.5.

EXAMPLE 1

Let a vibrating string have length π and let $c^2 = 1$. Suppose that the initial velocity is zero and the initial distortion $f(x) = \pi x - x^2$. Instead of using d'Alembert's solution, we here obtain an expansion of the form (15) by calculating the coefficients B_n.

From (14) we have

$$B_n = \frac{2}{\pi} \int_0^\pi (\pi x - x^2) \sin nx \, dx.$$

We integrate by parts twice:

$$B_n = \frac{2}{\pi} \left[-(\pi x - x^2) \frac{\cos nx}{n} \Big|_0^\pi + \frac{1}{n} \int_0^\pi (\pi - 2x) \cos nx \, dx \right]$$

$$= \frac{2}{n\pi} \left[(\pi - 2x) \frac{\sin nx}{n} \Big|_0^\pi + \frac{2}{n} \int_0^\pi \sin nx \, dx \right]$$

$$= \frac{4}{n^2\pi} \left(-\frac{\cos nx}{n} \Big|_0^\pi \right) = \frac{4}{n^3\pi}[1 - (-1)^n].$$

Hence the solution is

$$y(x, t) = \frac{4}{\pi} \sum_{n=1}^{\infty} \frac{[1 - (-1)^n]}{n^3} \cos nt \sin nx$$

$$= \frac{8}{\pi} \left(\cos t \sin x + \frac{1}{3^3} \cos 3t \sin 3x + \frac{1}{5^3} \cos 5t \sin 5x + \cdots \right)$$

$$= \frac{8}{\pi} \sum_{n=0}^{\infty} \frac{\cos(2n+1)t \sin(2n+1)x}{(2n+1)^3}.$$

Problems 12.6

In Problems 1–6 use the method of separation of variables to find the displacement $y(x, t)$ in the wave equation for each initial condition $y(x, 0) = f(x)$.

1. $f(x) = 0.1 \sin(\pi x/L), \quad 0 \le x \le L$

2. $f(x) = 0.01 \sin(2\pi x/L), \quad 0 \le x \le L$

3. $f(x) = 0.1x(L - x), \quad 0 \le x \le L$

4. $f(x) = 0.1x^2(L - x), \quad 0 \le x \le L$

5. $f(x) = \begin{cases} 0.1x, & 0 \le x \le 3 \\ 0.1(6 - x), & 3 \le x \le 6 \end{cases}$

6. $f(x) = \begin{cases} 0.1x, & 0 \le x \le 2 \\ 0.2, & 2 \le x \le 4 \\ 0.1(6 - x), & 4 \le x \le 6 \end{cases}$

In Problems 7–14 reduce the given equations to pairs of ordinary differential equations by the method of separation of variables.

7. $\dfrac{\partial^2 y}{\partial x^2} + \dfrac{\partial y}{\partial t} = 0$

8. $\dfrac{\partial^2 y}{\partial x^2} + x\dfrac{\partial y}{\partial t} = 0$

9. $\dfrac{\partial^2 y}{\partial x \partial t} + \dfrac{\partial y}{\partial t} = 0$

10. $z_{xx} = y^2 z_{yy}$

11. $z_{yy} + xyz_x = 0$

12. $z_{xx} - z_{yy} = z$

13. $u_{rr} + \dfrac{1}{r} u_r + \dfrac{1}{r^2} u_{\theta\theta} = 0, \quad$ where $u = u(r, \theta)$

14. $u_{rr} + \dfrac{1}{r} u_r + \dfrac{1}{r^2} u_{\theta\theta} + u_{zz} = 0, \quad$ where $u = u(r, \theta, z)$

For Problems 15–20 do the following:

(a) Find every solution of the given equation by the technique of separation of variables.

(b) Find the nontrivial solutions if we require the solutions to satisfy the boundary conditions

$$y(0, t) = y(\pi, t) = 0.$$

(c) Using the nontrivial solutions in part (b), indicate whether it would be possible to find a solution that also satisfies the initial condition

$$y(x, 0) = f(x), \quad 0 < x < \pi,$$

for $f(x) = x(\pi - x)$.

15. $\dfrac{\partial^2 y}{\partial x^2} + \dfrac{1}{x^2}\dfrac{\partial^2 y}{\partial t^2} = 0$

16. $\dfrac{\partial^2 y}{\partial x \partial t} = y$

17. $\dfrac{\partial^2 y}{\partial x^2} + \dfrac{\partial^2 y}{\partial t^2} = 0$

18. $\dfrac{\partial^2 y}{\partial x^2} - 2\dfrac{\partial y}{\partial x} = -\dfrac{\partial y}{\partial t}$

19. $\dfrac{\partial^2 y}{\partial t^2} = \dfrac{\partial^2 y}{\partial x^2} + y$

20. $\dfrac{\partial^2 y}{\partial t^2} = x^2\dfrac{\partial^2 y}{\partial x^2}$

12.7 Heat Flow and the Heat Equation

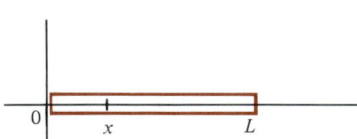

FIGURE 1

Consider a cylindrical rod of length L and radius R (so that the cross-sectional area is $A = \pi R^2$) composed of a uniform heat-conducting material. We assume that heat can enter and leave the rod only through its ends; that is, the lateral surface of the rod is completely insulated. Let x measure the distance along the rod (see Fig. 1) and let $T(x, t)$ denote the temperature at time t at a point x units along the rod (we assume that the temperature is uniform across any cross-sectional area of the rod). Let ρ be the **density** of the rod (mass per unit volume), which is assumed to be constant. The **specific heat** c of the rod is defined as the amount of heat (joules, BTUs) that must be supplied to raise the temperature of one unit mass (kilograms, pounds) of the rod by one degree (Celsius, Fahrenheit) from a standard temperature.

Consider a section of the rod between x and $x + dx$. Since mass = volume $\times$ density, the mass between these two points is $\rho A \, dx$. In order to change the temperature of the rod between these two points from 0 to $T(x, t)$ degrees, we must supply $T(x, t) \cdot c\rho A \cdot dx$ units of heat. Thus, between any two points x_0 and x_1, the heat energy contained in the rod at time t is

$$Q(t) = \int_{x_0}^{x_1} T(x, t) c\rho A \, dx. \tag{1}$$

By the law of conservation of energy, if there are no heat sources within the rod, the heat energy in any part of the rod can increase or decrease only because of a lateral flow of heat through the boundaries of that part of the rod. This **heat flux,** written as $Q_F(x, t)$, is the quantity of heat energy per unit time passing through a unit area in the cross section x units from the left-hand end in a positive (rightward) direction. Thus, to find the rate of change of the heat energy between the x_0 and x_1 cross sections, we need only take the difference

$$\frac{dQ}{dt} = AQ_F(x_0, t) - AQ_F(x_1, t), \tag{2}$$

where the first term on the right-hand side is the heat energy flowing in and the second is that flowing out (see Fig. 2).

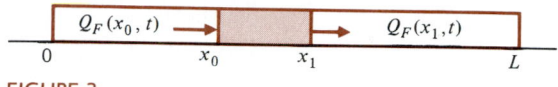

FIGURE 2

It is an empirical law of physics that the heat flux at any point is proportional to the **temperature gradient** $\partial T(x, t)/\partial x$ at that point. The constant of proportionality is called the **thermal conductivity** of the rod and is denoted by κ. Since heat flows in the direction of decreasing temperature, we have

$$Q_F(x, t) = -\kappa \frac{\partial T}{\partial x}(x, t),$$

where $\kappa > 0$. Since no heat sources are in this part of the rod, we have

$$AQ_F(x_0, t) - AQ_F(x_1, t) = A \left[\kappa \frac{\partial T}{\partial x}(x_1, t) - \kappa \frac{\partial T}{\partial x}(x_0, t) \right]$$

$$= \int_{x_0}^{x_1} \frac{\partial}{\partial x} \left[\kappa \frac{\partial T}{\partial x}(x, t) \right] A \, dx. \tag{3}$$

Substituting the right-hand side of (3) for the right-hand side of (2) and differentiating (1), we have

$$\int_{x_0}^{x_1} \frac{\partial}{\partial x} \left(\kappa \frac{\partial T}{\partial x} \right) A \, dx = \frac{dQ}{dt}$$

$$= \int_{x_0}^{x_1} c\rho \frac{\partial T}{\partial t} A \, dx,$$

or

$$\int_{x_0}^{x_1} \left[c\rho \frac{\partial T}{\partial t} - \frac{\partial}{\partial x} \left(\kappa \frac{\partial T}{\partial x} \right) \right] A \, dx = 0. \tag{4}$$

Equation (4) must hold in any interval $x_0 \leq x \leq x_1$, so if we assume that the functions involved are all continuous, the expression in brackets must vanish. To see this, note that if it were positive (negative) at some point, then by continuity it would be positive (negative) on some interval, in which case (4) would not hold. Thus, for all x in the interval $[0, L]$,

$$cp \frac{\partial T}{\partial t} = \frac{\partial}{\partial x}\left(\kappa \frac{\partial T}{\partial x}\right), \tag{5}$$

and if we assume that κ is constant,

$$\frac{\partial T}{\partial t} = \delta \frac{\partial^2 T}{\partial x^2}, \qquad \delta = \frac{\kappa}{c\rho}. \tag{6}$$

Heat Equation

Equation (6) is called the **heat (conduction) equation** and is clearly parabolic. The constant δ is positive and measures the **diffusivity** of the material of the rod. To completely specify our problem, we select the case in which the boundary conditions are

$$T(0, t) = T(L, t) = 0, \qquad t \geq 0, \tag{7}$$

and the initial condition is

$$T(x, 0) = f(x), \qquad 0 \leq x \leq L, \tag{8}$$

where f is a given function. In physical terms, we are keeping the ends of the rod at zero temperature and letting $f(x)$ denote the initial temperature at any point x of the rod. We can modify the problem by selecting other, possibly time-dependent, boundary conditions (see Problems 7–9).

We begin to solve the heat equation by assuming the existence of a solution to the problem of the form

$$T(x, t) = X(x)\mathcal{T}(t).$$

Substituting this equation into (6) yields

$$X\mathcal{T}' = \delta X''\mathcal{T},$$

or

$$\frac{\mathcal{T}'}{\delta\mathcal{T}} = \frac{X''}{X}. \tag{9}$$

The function on the left-hand side depends only on t, so it is constant for fixed t and arbitrary x. On the other hand, the function on the right side of (9) depends only on x and is constant for fixed x and arbitrary t. The only way these situations can hold simultaneously is for each function to be constant, say k. Then we obtain the pair of ordinary differential equations

$$\mathcal{T}' - k\delta\mathcal{T} = 0, \tag{10}$$

$$X'' - kX = 0. \tag{11}$$

The boundary conditions (7) may be written as

$$X(0)\mathcal{T}(t) = X(L)\mathcal{T}(t) = 0,$$

implying that $X(0) = X(L) = 0$ unless the rod has zero initial temperature at every point. If we ignore this uninteresting case, we note that the boundary value problem (11), $X(0) = X(L) = 0$, is almost identical to the boundary value problem for X in the case of the vibrating string [see equation (12.6.8)]. Paralleling the development of Section 12.6, we see that the only nonzero solutions for the boundary value problem (11) with $X(0) = X(L) = 0$ arise for negative values of k. Setting $k = -r^2$, we find that k is an eigenvalue of the problem only if r is a multiple of π/L, and the eigenfunctions corresponding to $-(n\pi/L)^2$ have the form

$$X_n(x) = A \sin \frac{n\pi x}{L}.$$

Setting $k = -(n\pi/L)^2$ in (10), we see that this first-order equation has the general solution

$$\mathcal{T}_n(t) = Be^{-(n\pi/L)^2\delta t}.$$

Hence we consider an infinite series of the form

$$T(x, t) = \sum_{n=1}^{\infty} X_n(x)\mathcal{T}_n(t) = \sum_{n=1}^{\infty} B_n \sin \frac{n\pi x}{L} e^{-(n\pi/L)^2\delta t}. \tag{12}$$

Setting $t = 0$ and making use of the initial condition (8), we have

$$T(x, 0) = \sum_{n=1}^{\infty} B_n \sin \frac{n\pi x}{L} = f(x).$$

Thus, in order that separation of variables work, $f(x)$ must be representable as a Fourier sine series, implying that f must be extended as an odd, piecewise continuous function of period $2L$ with piecewise continuous derivatives. When we make this extension, the coefficients B_n are given by

$$B_n = \frac{1}{L} \int_{-L}^{L} f(x) \sin \frac{n\pi x}{L} dx = \frac{2}{L} \int_{0}^{L} f(x) \sin \frac{n\pi x}{L} dx, \qquad n = 1, 2, 3, \ldots. \tag{13}$$

The solution (12) is completely determined and the series must converge, since $T(x, 0)$ converges and the exponential factors are less than 1 for all $t > 0$. Observe that the exponential factors in (12) cause $T(x, t)$ to approach zero as t tends to infinity.

EXAMPLE 1

Suppose that the rod has $L = \pi$, $\delta = 1$, and initial temperature $f(x) = \sin 2x$. By the orthogonality of the set of functions $\{\sin nx\}$ (see Example 11.2.1), the coefficients B_n are all zero for $n \neq 2$. For $n = 2$, we have, from equation (13) and from Example 11.2.1,

$$B_2 = \frac{1}{\pi} \int_{-\pi}^{\pi} \sin^2 2x \, dx = 1.$$

Thus the solution of this problem is given by

$$T(x, t) = (\sin 2x)e^{-4t}.$$

EXAMPLE 2

Let the rod have length 1, $\delta = 1$, and initial temperature $f(x) = x(1 - x^2)$. Observe that $f(x)$ is an odd function for all $|x| \leq 1$, since

$$f(-x) = -x[1 - (-x)^2] = -x(1 - x^2) = -f(x).$$

Extend f to the reals $\mathbb{R}$ periodically by letting $f(x + 2) = f(x)$. Then

$$B_n = 2 \int_0^1 f(x) \sin n\pi x \, dx = 2 \int_0^1 (x - x^3) \sin n\pi x \, dx.$$

We integrate by parts three times to obtain (the details should now be familiar)

$$B_n = \frac{12(-1)^{n+1}}{n^3 \pi^3}.$$

Hence, by (12), the solution to this initial-value problem is

$$T(x, t) = \frac{12}{\pi^3} \sum_{n=1}^{\infty} \frac{(-1)^{n+1} \sin n\pi x e^{-n^2 \pi^2 t}}{n^3}.$$

Problems 12.7

In Problems 1–6 find the temperature $T(x, t)$ in an insulated rod π units long whose ends are kept at $0°C$, where the initial temperature is $f(x)$.

1. $f(x) = x(\pi - x)$

2. $f(x) = x(\pi^2 - x^2)$

3. $f(x) = x^2(\pi - x)$

4. $f(x) = x \cos \dfrac{x}{2}$

5. $f(x) = \begin{cases} x, & 0 \leq x \leq \pi/2 \\ 0, & \pi/2 < x \leq \pi \end{cases}$

6. $f(x) = \begin{cases} 0, & 0 \leq x < \pi/3 \\ 1, & \pi/3 \leq x < 2\pi/3 \\ 0, & 2\pi/3 \leq x \leq \pi \end{cases}$

7. Suppose that the ends of the rod are kept at different constant temperatures:

$$T(0, t) = T_1, \quad T(L, t) = T_2.$$

Find the temperature at x as t tends to ∞; that is, find the **steady-state** temperature. [*Hint:* The steady-state temperature is *not* time dependent.]

8. Denote the steady-state temperature by $T_s(x)$ and define the **transient** temperature in the rod by

$$T_t(x, t) = T(x, t) - T_s(x).$$

Verify that $T_t(x, t)$ is given by equation (12). This shows that we need only superimpose the steady-state temperature on the solution (12) to obtain the solution for nonzero boundary conditions.

9. Suppose that the ends of the rod are kept at the temperatures

$$T(0, t) = T_1 e^{-c^2 \delta t}, \quad T(L, t) = T_2 e^{-c^2 \delta t}.$$

For what initial conditions $f(x)$ can a solution be found by the separation of variables? What happens if cL is a multiple of π?

* 10. Consider the nonhomogeneous heat equation

$$\frac{\partial u}{\partial t} - k \frac{\partial^2 u}{\partial x^2} = A(x, t),$$

where $A(x, t)$ is given by

$$A(x, t) = \sum_{n=1}^{\infty} a_n(t) \sin \frac{n\pi x}{L}$$

for known functions $a_n(t)$. Suppose that initial and boundary conditions are given by

$$u(0, t) = u(L, t) = 0, \quad t \geq 0,$$

$$u(x, 0) = f(x), \quad 0 \leq x \leq L,$$

with $f(x)$ known. Find a Fourier series for the solution $u(x, t)$. [*Hint:* Assume that B_n in (12) is a function of t.]

12.8 Two-Dimensional Heat Flow and Laplace's Equation

Consider a rectangular plane sheet of heat-conducting material of uniform thickness θ, density ρ, specific heat c, and thermal conductivity κ. We may suppose that the set of points (x, y) with $0 \leq x \leq L$ and $0 \leq y \leq M$ is a face of the sheet. We assume that the faces of the sheet are insulated and that heat can enter and leave only through the edges of the sheet.

Select any interior point (x, y) on the face and consider a rectangular region *ABCD* whose corner coordinates are given in Fig. 1. It is reasonable to assume the heat flow in each plane parallel to the face of the sheet to be identical, so that the flow is two-dimensional. In the discussion below we assume that the distances Δx and Δy are very small.

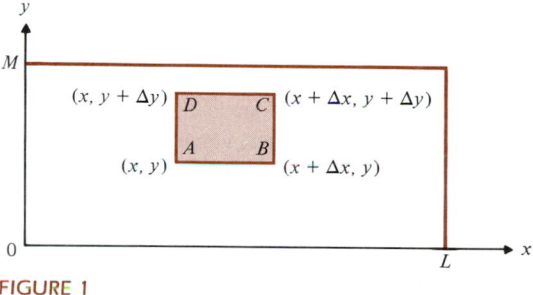

FIGURE 1

The rate of change of heat energy in the sheet *ABCD* at any time t is approximately (see Section 12.7)

$$c\rho\theta\,\Delta x\,\Delta y\frac{\partial T}{\partial t}, \tag{1}$$

where $T = T(x, y, t)$ is the temperature at any point (x, y) at time t. Since Δx and Δy are small, we assume that the heat flux along each edge is constant. Then the heat energy passing through *ABCD* in the vertical direction is

$$\theta\,\Delta x[Q_F(x, y, t) - Q_F(x, y + \Delta y, t)], \tag{2}$$

where Q_F is the heat flux. In the horizontal direction the heat energy is

$$\theta\,\Delta y[Q_F(x, y, t) - Q_F(x + \Delta x, y, t)]. \tag{3}$$

In Section 12.7 we mentioned that the heat flux at any point is proportional to the temperature gradient at that point: $Q_F(x, y, t) = -\kappa|\nabla T(x, y, t)|$. Adding (2) and (3), we get another approximation of the rate of change of heat energy in the sheet *ABCD* by writing

$$\kappa\theta\,\Delta y\left[\frac{\partial T}{\partial x}(x + \Delta x, y, t) - \frac{\partial T}{\partial x}(x, y, t)\right] + \kappa\theta\,\Delta x\left[\frac{\partial T}{\partial y}(x, y + \Delta y, t) - \frac{\partial T}{\partial y}(x, y, t)\right]. \tag{4}$$

Equating (4) and (1) and dividing by $\Delta x\, \Delta y$ yield

$$c\rho\theta\frac{\partial T}{\partial t} = \kappa\theta\frac{\left[\dfrac{\partial T}{\partial x}(x+\Delta x, y, t) - \dfrac{\partial T}{\partial x}(x, y, t)\right]}{\Delta x} + \kappa\theta\frac{\left[\dfrac{\partial T}{\partial y}(x, y+\Delta y, t) - \dfrac{\partial T}{\partial y}(x, y, t)\right]}{\Delta y}.$$

Passing to the limit as Δx and Δy both approach zero, we obtain

$$c\rho\theta\frac{\partial T}{\partial t} = \kappa\theta\left[\frac{\partial^2 T}{\partial x^2} + \frac{\partial^2 T}{\partial y^2}\right],$$

which may be written in the form

$$\frac{\partial^2 T}{\partial x^2} + \frac{\partial^2 T}{\partial y^2} = \frac{1}{\delta}\frac{\partial T}{\partial t}, \qquad \delta = \frac{\kappa}{c\rho} \tag{5}$$

Two-Dimensional Heat Equation

where $\delta = \kappa/c\rho$ is the **diffusivity.** Equation (5) is called the **two-dimensional heat equation.**

   If the edges of the sheet are kept at constant temperatures, and time is allowed to increase to infinity, the temperature at any given point stabilizes, that is, steady-state conditions are attained. The steady-state temperature T at a point on the sheet then depends only on its position and not on time. Thus $\partial T/\partial t$ is zero throughout the sheet and equation (5) becomes

Laplace's Equation

$$\frac{\partial^2 T}{\partial x^2} + \frac{\partial^2 T}{\partial y^2} = 0. \tag{6}$$

This elliptic partial differential equation, commonly referred to as **Laplace's equation,** arises in many other problems of applied mathematics. We now give an example of the solution of Laplace's equation by the method of separation of variables.

EXAMPLE 1

We begin by studying the steady-state heat flow problem in the rectangular sheet of size $L \times M$ to which Laplace's equation applies. Suppose that the upper horizontal edge is kept at 100°C, while the other three edges are kept at 0°C. We may write these boundary conditions for $T(x, y)$ as follows:

$$T(0, y) = T(L, y) = 0, \qquad 0 < y < M, \tag{7}$$

$$T(x, 0) = 0, \qquad T(x, M) = 100, \qquad 0 < x < L. \tag{8}$$

Since we are seeking the steady-state solution, time is not a factor in this problem, and therefore no initial condition makes sense.

Setting $T(x, y) = X(x)Y(y)$, we find that (6) becomes

$$X''Y + XY'' = 0,$$

which may be written in separated form as

$$\frac{X''}{X} = -\frac{Y''}{Y}. \tag{9}$$

The left-hand side of (9) is constant for fixed x and arbitrary y. Hence Y''/Y must be constant. Letting k be this constant, we obtain the pair of ordinary differential equations

$$X'' - kX = 0,$$

$$Y'' + kY = 0. \tag{10}$$

From (7) we have

$$X(0)Y(y) = X(L)Y(y) = 0,$$

and $Y(y) \not\equiv 0$ [since otherwise $T(x, y) \equiv 0$, contradicting the second equation of (8)], so that $X(0) = X(L) = 0$. Thus again we have the situation encountered in Section 12.7, and the eigenvalues are all negative. Setting $k = -r^2$, we find that rL must be a multiple of π in order that we have a nonzero solution of the boundary value problem for X. Hence $r = n\pi/L$, and the functions

$$X_n(x) = A \sin \frac{n\pi x}{L}, \qquad n = 1, 2, 3, \dots, \tag{11}$$

are the eigenfunctions of the problem. Setting $k = -(n\pi/L)^2$ in the second equation of (10) yields the general solution

$$Y_n(y) = B_1 e^{n\pi y/L} + B_2 e^{-n\pi y/L}.$$

The first condition of (8) implies that $Y(0) = 0$, thus $B_1 = -B_2$, and we can rewrite Y_n as

$$Y_n(y) = B \sinh \frac{n\pi y}{L}. \tag{12}$$

As in Section 12.7, to enlarge the class of possible solutions, we consider an infinite sum of products of the terms (11) and (12):

$$T(x, y) = \sum_{n=1}^{\infty} c_n \sin \frac{n\pi x}{L} \sinh \frac{n\pi y}{L}. \tag{13}$$

Evaluating equation (13) at any point (x, M), we obtain, by (8),

$$100 = \sum_{n=1}^{\infty} \left(c_n \sinh \frac{n\pi M}{L} \right) \sin \frac{n\pi x}{L}, \qquad 0 < x < L,$$

which again is a Fourier sine series, requiring that we extend the boundary condition $T(x, M)$ to the interval $-L \leq x \leq L$ as an odd piecewise continuous periodic function of period $2L$ with a piecewise continuous derivative. The function

$$f(x) = \begin{cases} 100, & 0 < x < L, \\ -100, & -L < x < 0, \end{cases}$$

with $f(x + 2L) = f(x)$ satisfies these conditions. Hence $c_n \sinh(n\pi M/L)$ must equal the nth Fourier (sine) coefficient of $f(x)$:

$$c_n \sinh \frac{n\pi M}{L} = \frac{1}{L} \int_{-L}^{L} f(x) \sin \frac{n\pi x}{L} dx = \frac{200}{L} \int_{0}^{L} \sin \frac{n\pi x}{L} dx$$

$$= \frac{200}{n\pi}(1 - \cos n\pi) = \frac{200}{n\pi}[1 - (-1)^n].$$

Hence $c_{2k} = 0$, while

$$c_{2k+1} = \frac{400}{\pi(2k+1)\sinh[(2k+1)\pi M/L]},$$

and the solution is given by

$$T(x, y) = \frac{400}{\pi}\left[\frac{\sin(\pi x/L)\sinh(\pi y/L)}{\sinh(\pi M/L)} + \frac{\sin(3\pi x/L)\sinh(3\pi y/L)}{3\sinh(3\pi M/L)} + \cdots\right]$$

$$= \frac{400}{\pi}\sum_{n=1}^{\infty}\frac{\sin[(2n-1)\pi x/L]\sinh[(2n-1)\pi y/L]}{(2n-1)\sinh[(2n-1)\pi M/L]}.$$

We may use this formula to compute the steady-state temperature at any point in the rectangle. For example,

$$T\left(\frac{L}{2}, \frac{M}{2}\right) = \frac{400}{\pi}\sum_{n=1}^{\infty}\frac{\sin[(2n-1)\pi/2]\sinh[(2n-1)\pi M/2L]}{(2n-1)\sinh[(2n-1)\pi M/L]}.$$

Since

$$\frac{\sinh(A/2)}{\sinh A} = \frac{e^{A/2} - e^{-A/2}}{e^A - e^{-A}} = \frac{1}{e^{A/2} + e^{-A/2}} = \frac{1}{2\cosh(A/2)},$$

we obtain

$$T\left(\frac{L}{2}, \frac{M}{2}\right) = \frac{200}{\pi}\left[\frac{1}{\cosh(\pi M/2L)} - \frac{1}{3\cosh(3\pi M/2L)} + \frac{1}{5\cosh(5\pi M/2L)} - \cdots\right],$$

which, for given L and M, can be approximated to any desired degree of accuracy.

Problems 12.8

1. A square plate with sides of length L has both faces insulated. The upper horizontal edge is kept at 50°C, while all the other edges are at 0°C. Find the steady-state temperatures at the points $(L/2, L/2)$ and $(L/4, L/4)$.

2. If in Problem 1 the temperatures along the edge $y = L$ are given by

$$T(x, L) = x(L - x),$$

and all the other conditions remain the same, what is $T(x, y)$?

3. If in Problem 1 the temperatures along the upper horizontal edge are

$$T(x, L) = 100 \sin(\pi x/L),$$

with all the other conditions remaining unchanged, what is $T(x, y)$?

4. Suppose that the temperature along the edge $y = L$ in Problem 2 is changed to

$$T(x, L) = x(L^2 - x^2),$$

with all the other conditions remaining the same. Find the temperature at the points $(L/4, L/4)$ and $(L/4, 3L/4)$.

Assume in Problems 5–8 that all other previous conditions are unchanged.

5. Repeat Problem 4 with $T(x, L) = x^2(L - x)$.

6. Repeat Problem 2 with $T(x, L) = x \cos(\pi x/2L)$.

7. Repeat Problem 2 with $T(x, L) = 10x \sin \pi x/L$.

8. Repeat Problem 4 with $T(x, L) = 100 \sin^2(\pi x/L)$.

9. A rectangular plate of length $2L$ and height L has both faces insulated. The upper edge is kept at 50°C and the lower edge at 0°C, while the vertical edges have temperature

$$T(-L, y) = T(L, y) = 50 \sin \frac{\pi y}{2L}, \quad 0 \le y \le L.$$

Find the temperature at any point in the plate.

10. Repeat Problem 9 with

$$T(-L, y) = T(L, y) = 50y/L, \quad 0 \le y \le L,$$

all other conditions remaining unchanged.

11. Repeat Problem 10 with the upper edge now changed to

$$T(x, L) = 50(x/L)^2,$$

all other conditions remaining unchanged.

* 12. Extend the development in this section to a solid cube of side L.

  (a) What is the three-dimensional heat equation for such a situation?

  (b) What is the steady-state equation for part (a)?

  (c) Assume that all faces, except $z = L$, of the cube are kept at 0°C, while the $z = L$ face is kept at 100°C. What is the steady-state solution in the form of a Fourier series?

12.9 Laplace's Equation in Polar and Spherical Coordinates

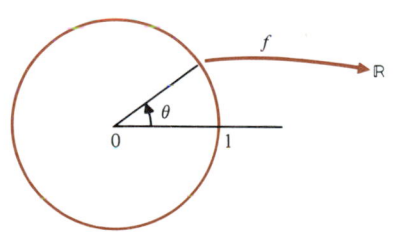

FIGURE 1

We now consider the steady-state heat flow problem in the disk D of radius 1 centered at the origin. It should be clear from the nature of the situation that the problem will be simplified if we can use polar coordinates (r, θ) instead of rectangular coordinates (x, y). Assume that the boundary of the disk is kept at the temperature

$$T(1, \theta) = f(\theta), \qquad 0 \le \theta \le 2\pi, \tag{1}$$

where the function $f(\theta)$ is periodic with period 2π and has a continuous derivative (see Fig. 1). Now we need only write Laplace's equation, $\partial^2 T/\partial x^2 + \partial^2 T/\partial y^2 = 0$, in polar coordinates. Recall the transformation formulas

$$r = \sqrt{x^2 + y^2}, \qquad x = r \cos \theta,$$

$$\theta = \tan^{-1} \frac{y}{x}, \dagger \qquad y = r \sin \theta.$$

By the chain rule, we have

$$\frac{\partial T}{\partial x} = \frac{\partial T}{\partial r} \frac{\partial r}{\partial x} + \frac{\partial T}{\partial \theta} \frac{\partial \theta}{\partial x} = \frac{\partial T}{\partial r} \cos \theta - \frac{\partial T}{\partial \theta} \frac{\sin \theta}{r},$$

since

$$\frac{\partial r}{\partial x} = \frac{\partial}{\partial x} \sqrt{x^2 + y^2} = \frac{x}{\sqrt{x^2 + y^2}} = \frac{x}{r} = \cos \theta,$$

† This formula is valid in the right half-plane. See p. 881 for the left half-plane.

and

$$\frac{\partial \theta}{\partial x} = \frac{-y/x^2}{1 + (y/x)^2} = \frac{-y}{r^2} = \frac{-\sin \theta}{r}.$$

Then

$$\frac{\partial^2 T}{\partial x^2} = \frac{\partial}{\partial r}\left(\frac{\partial T}{\partial r}\frac{\partial r}{\partial x} + \frac{\partial T}{\partial \theta}\frac{\partial \theta}{\partial x}\right)\frac{\partial r}{\partial x} + \frac{\partial}{\partial \theta}\left(\frac{\partial T}{\partial r}\frac{\partial r}{\partial x} + \frac{\partial T}{\partial \theta}\frac{\partial \theta}{\partial x}\right)\frac{\partial \theta}{\partial x}$$

$$= \frac{\partial^2 T}{\partial r^2}\cos^2 \theta - \frac{\partial^2 T}{\partial r \partial \theta}\frac{\sin \theta \cos \theta}{r} + \frac{\partial T}{\partial \theta}\frac{\sin \theta \cos \theta}{r^2}$$

$$- \frac{\partial^2 T}{\partial \theta \partial r}\frac{\sin \theta \cos \theta}{r} + \frac{\partial T}{\partial r}\frac{\sin^2 \theta}{r} + \frac{\partial^2 T}{\partial \theta^2}\frac{\sin^2 \theta}{r^2} + \frac{\partial T}{\partial \theta}\frac{\cos \theta \sin \theta}{r^2}.$$

Similarly,

$$\frac{\partial^2 T}{\partial y^2} = \frac{\partial}{\partial r}\left(\frac{\partial T}{\partial r}\sin \theta + \frac{\partial T}{\partial \theta}\frac{\cos \theta}{r}\right)\frac{\partial r}{\partial y} + \frac{\partial}{\partial \theta}\left(\frac{\partial T}{\partial r}\sin \theta + \frac{\partial T}{\partial \theta}\frac{\cos \theta}{r}\right)\frac{\partial \theta}{\partial y}$$

$$= \frac{\partial^2 T}{\partial r^2}\sin^2 \theta + \frac{\partial^2 T}{\partial r \partial \theta}\frac{\sin \theta \cos \theta}{r} - \frac{\partial T}{\partial \theta}\frac{\cos \theta \sin \theta}{r^2}$$

$$+ \frac{\partial^2 T}{\partial \theta \partial r}\frac{\sin \theta \cos \theta}{r} + \frac{\partial T}{\partial r}\frac{\cos^2 \theta}{r} + \frac{\partial^2 T}{\partial \theta^2}\frac{\cos^2 \theta}{r^2} - \frac{\partial T}{\partial \theta}\frac{\sin \theta \cos \theta}{r^2}.$$

Adding the two equations, we obtain

Laplace's Equation in Polar Coordinates

$$\frac{\partial^2 T}{\partial x^2} + \frac{\partial^2 T}{\partial y^2} = \frac{\partial^2 T}{\partial r^2} + \frac{1}{r^2}\frac{\partial^2 T}{\partial \theta^2} + \frac{1}{r}\frac{\partial T}{\partial r} = 0, \tag{2}$$

which is **Laplace's equation in polar coordinates.**

Now we let $T(r, \theta) = R(r)\Theta(\theta)$ and transform (2) into

$$R''\Theta + \frac{1}{r^2}R\Theta'' + \frac{1}{r}R'\Theta = 0,$$

or

$$\left(R'' + \frac{1}{r}R'\right)\Theta = -\frac{R}{r^2}\Theta''.$$

Separating the variables, we have

$$\frac{r^2 R'' + r R'}{R} = -\frac{\Theta''}{\Theta} = k,$$

since each side of the first equality is a function of only one variable. Hence we obtain two ordinary differential equations:

$$\Theta'' + k\Theta = 0, \tag{3}$$

and

$$r^2 R'' + rR' - kR = 0. \tag{4}$$

Since $f(\theta) = T(1, \theta) = R(1)\Theta(\theta)$, it is clear that $\Theta(\theta)$ must be periodic with period 2π. We assume that f is not identically zero [see Problem 12 for the case where $f(\theta) \equiv 0$], so $R(1) \neq 0$. Observe that $\Theta(0) = \Theta(2\pi)$ and $\Theta'(0) = \Theta'(2\pi)$. By the remark following the orthogonality theorem (p. 709) and Problem 11.7.6, the boundary value problem in Θ has periodic eigenfunctions of period 2π. Thus the eigenvalues must be $k = n^2$, with the corresponding eigenfunctions

$$\Theta_n(\theta) = a_1 \cos n\theta + a_2 \sin n\theta, \qquad n = 0, 1, 2, 3, \ldots.$$

Equation (4) is an Euler equation (see Section 2.6) and can be solved by letting $R = r^\lambda$. Letting $k = n^2$, we see that (4) becomes

$$r^\lambda[\lambda(\lambda - 1) + \lambda - n^2] = r^\lambda(\lambda^2 - n^2) = 0,$$

and the roots of the equation $\lambda^2 - n^2 = 0$ are $\lambda = \pm n$. Hence the general solution of (4) with $k = n^2$ is

$$R_0(r) = b_1 + b_2 \ln r, \qquad \text{for } n = 0,$$

and

$$R_n(r) = b_1 r^n + b_2 r^{-n}, \qquad n = 1, 2, 3, \ldots.$$

Since we want R_n to exist for all values in the range $0 \leq r \leq 1$, b_2 must vanish and $R_n(r) = b_1 r^n$, $n = 0, 1, 2, 3, \ldots$. Thus we seek a solution in the form of the infinite series

$$T(r, \theta) = \frac{A_0}{2} + \sum_{n=1}^{\infty} r^n (A_n \cos n\theta + B_n \sin n\theta). \tag{5}$$

To evaluate the constants A_n, B_n, we note that

$$f(\theta) = T(1, \theta) = \frac{A_0}{2} + \sum_{n=1}^{\infty} A_n \cos n\theta + B_n \sin n\theta. \tag{6}$$

Hence (6) is the Fourier series of f, and the coefficients A_n and B_n are the Fourier coefficients of f given by the Euler formulas

$$A_n = \frac{1}{\pi} \int_{-\pi}^{\pi} f(\phi) \cos n\phi \, d\phi, \qquad B_n = \frac{1}{\pi} \int_{-\pi}^{\pi} f(\phi) \sin n\phi \, d\phi.$$

Since (6) converges and $r \leq 1$, series (5) also converges. For example, if $f(\theta) =$

$\cos \theta$, all the coefficients except A_1 vanish, and $A_1 = 1$. Hence the temperature is given by

$$T(r, \theta) = r \cos \theta$$

at all points in the unit disk.

It is interesting to note what occurs if we replace the coefficients A_n and B_n in (5) by the Euler formulas and interchange the sums and integrals (see Problem 9):

$$T(r, \theta) = \frac{1}{2\pi} \int_{-\pi}^{\pi} f(\phi)\, d\phi$$

$$+ \sum_{n=1}^{\infty} r^n \left[\frac{\cos n\theta}{\pi} \int_{-\pi}^{\pi} f(\phi) \cos n\phi\, d\phi + \frac{\sin n\theta}{\pi} \int_{-\pi}^{\pi} f(\phi) \sin n\phi\, d\phi\right]$$

$$= \frac{1}{\pi} \int_{-\pi}^{\pi} f(\phi)\left[\frac{1}{2} + \sum_{n=1}^{\infty} r^n(\cos n\theta \cos n\phi + \sin n\theta \sin n\phi)\right] d\phi.$$

Since

$$\cos n(\theta - \phi) = \cos n\theta \cos n\phi + \sin n\theta \sin n\phi,$$

we obtain

$$T(r, \theta) = \frac{1}{\pi} \int_{-\pi}^{\pi} f(\phi)\left[\frac{1}{2} + \sum_{n=1}^{\infty} r^n \cos n(\theta - \phi)\right] d\phi.$$

But

$$\cos n(\theta - \phi) = \frac{e^{in(\theta-\phi)} + e^{-in(\theta-\phi)}}{2},$$

so

$$\sum_{n=1}^{\infty} r^n \cos n(\theta - \phi) = \frac{1}{2} \sum_{n=1}^{\infty} r^n[e^{in(\theta-\phi)} + e^{-in(\theta-\phi)}]$$

$$= \frac{1}{2}\left\{\sum_{n=1}^{\infty} [re^{i(\theta-\phi)}]^n + \sum_{n=1}^{\infty} [re^{-i(\theta-\phi)}]^n\right\}.$$

The complex number $e^{it} = \cos t + i \sin t$ can be represented as a vector of length $\sqrt{(\cos t)^2 + (\sin t)^2} = 1$ (see p. 882), so $re^{i(\theta-\phi)}$ and $re^{i(\phi-\theta)}$ both have length equal to $r(<1)$. Using the geometric series, we have

$$\sum_{n=1}^{\infty} r^n \cos n(\theta - \phi) = \frac{1}{2}\left[\frac{re^{i(\theta-\phi)}}{1 - re^{i(\theta-\phi)}} + \frac{re^{-i(\theta-\phi)}}{1 - re^{-i(\theta-\phi)}}\right]$$

$$= \frac{1}{2}\left\{\frac{r[e^{i(\theta-\phi)} + e^{-i(\theta-\phi)}] - 2r^2}{r^2 - r[e^{i(\theta-\phi)} + e^{-i(\theta-\phi)}] + 1}\right\}$$

$$= \frac{r \cos(\theta - \phi) - r^2}{r^2 - 2r \cos(\theta - \phi) + 1}.$$

Thus

$$\frac{1}{2} + \sum_{n=1}^{\infty} r^n \cos n(\theta - \phi) = \frac{1}{2} + \frac{r \cos(\theta - \phi) - r^2}{r^2 + 1 - 2r \cos(\theta - \phi)}$$

$$= \frac{1 - r^2}{2[r^2 + 1 - 2r \cos(\theta - \phi)]}$$

and

Poisson Integral Formula

$$T(r, \theta) = \frac{1 - r^2}{2\pi} \int_{-\pi}^{\pi} \frac{f(\phi)}{r^2 + 1 - 2r \cos(\theta - \phi)} \, d\phi. \qquad (7)$$

Equation (7), the **Poisson integral formula,**† is valid for all values $r < 1$ (see Problem 9). It indicates that the temperature at any interior point (r, θ) of the unit disk may be obtained by integrating the boundary temperatures according to formula (7). In particular, if $r = 0$, then the temperature at the center of the disk is

$$T(0, \theta) = \frac{1}{2\pi} \int_{-\pi}^{\pi} f(\phi) \, d\phi;$$

that is, the temperature at the center is the integral average of the boundary temperatures. This fact, often called the **mean value theorem,** holds for all functions that satisfy Laplace's equation on the unit disk $r \leq 1$.

EXAMPLE 1

In this example we consider the three-dimensional Laplace equation

$$\frac{\partial^2 T}{\partial x^2} + \frac{\partial^2 T}{\partial y^2} + \frac{\partial^2 T}{\partial z^2} = 0 \qquad (8)$$

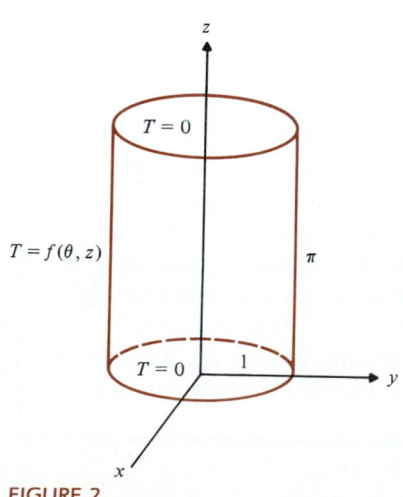

$T = 0$

$T = f(\theta, z)$

π

$T = 0$ 1

FIGURE 2

associated with the steady-state heat flow problem in a solid.

Suppose that the solid is a rectangular cylinder of height π and radius 1 (Fig. 2). Since such a solid is easily described in cylindrical coordinates, it is convenient to obtain Laplace's equation (8) in terms of the coordinates (r, θ, z), where

$$x = r \cos \theta, \qquad y = r \sin \theta, \qquad z = z.$$

† Named in honor of Siméon Denis Poisson (1781–1840), a French mathematician and physicist. See the accompanying biographical sketch.

Courtesy of the David Smith Collection

SIMÉON POISSON
(1781–1840)

Siméon Poisson was born in Pithiviers, France in 1781 and died in Paris in 1840. He was educated by his father, a private soldier who on retirement received a small administrative post in his village and, when the French Revolution broke out, assumed the governing of the place. Relatives wished to press the young Poisson, much against his own wishes, into medicine. The education was undertaken by an uncle, who started the boy off with pricking veins in cabbage leaves with a lancet. When he had perfected himself in this, he was graduated to cutting on blisters. But in almost the first case in which he did this by himself, his patient died within a few hours. Although the doctors assured him that "The event was a very common one," he vowed to have nothing more to do with the profession.

Strong mathematical interests led Poisson in 1798 to enter the École Polytechnique to study the subject, where his abilities impressed Lagrange and Laplace. Upon graduation he was made a lecturer at the École Polytechnique. The rest of his life was spent in various government posts and professorships. Somewhat of a socialist, he remained a staunch republican until 1815, when he joined the Legitimists. Because of this allegiance, he became a baron in 1825, and was made a peer of France, in 1837, by Louis Philippe.

Poisson's mathematical publications were numerous, numbering between 300 and 400. Two of his best known treatises are *Traité de mécanique* (2 vols., 1811, 1833) and *Recherches sur la probabilité des jugements* (1837). In his papers he considered such matters as the mathematical theory of electricity and magnetism, physical astronomy, the attraction of ellipsoids, definite integrals, series, and the theory of elasticity. The student encounters *Poisson's constant* (in electricity), the *Poisson ratio* (in elasticity), *Poisson's integral* and *Poisson's equation* (in potential theory), and *Poisson's law* (in probability theory).

A droll story links Poisson to one of his professional interests. When a boy, he was put in the care of a nurse. One day, when his father came to see him, the nurse had gone out and left the youngster suspended by his straps to a nail in the wall—to protect the boy, the nurse said, from the disease and dirt of the floor. Poisson said that his gymnastic efforts when thus suspended caused him to swing back and forth, and it was in this way that he early became familiar with the pendulum, the study of which occupied much of his later life.

Poisson once remarked: "Life is good for only two things, discovering mathematics and teaching mathematics." He excelled in both pursuits.

Obviously we are simply writing the xy-coordinates in polar form. Thus the calculations we did earlier still apply, and Laplace's equation becomes [from equation (2)]

$$\frac{\partial^2 T}{\partial r^2} + \frac{1}{r}\frac{\partial T}{\partial r} + \frac{1}{r^2}\frac{\partial^2 T}{\partial \theta^2} + \frac{\partial^2 T}{\partial z^2} = 0. \tag{9}$$

We now apply the method of separation of variables to the function T and, accordingly, set $T(r, \theta, z) = R(r)\Theta(\theta)Z(z)$. Equation (9) becomes

$$R''\Theta Z + \frac{1}{r}R'\Theta Z + \frac{1}{r^2}R\Theta''Z + R\Theta Z'' = 0,$$

or

$$[(R'' + r^{-1}R')\Theta + r^{-2}R\Theta'']Z = -R\Theta Z''.$$

Separating out the z-variables, we have

$$\frac{R'' + r^{-1}R'}{R} + r^{-2}\frac{\Theta''}{\Theta} = -\frac{Z''}{Z} = k_1,$$

since the left-hand side of the equation is a function of r and θ, whereas the right-hand side is a function of z alone. Thus

$$Z'' + k_1 Z = 0,$$

and

$$\frac{R'' + r^{-1}R'}{R} = k_1 - r^{-2}\frac{\Theta''}{\Theta}.$$

Multiplying both sides of the last equation by r^2, we have

$$\frac{r^2 R'' + r R'}{R} = k_1 r^2 - \frac{\Theta''}{\Theta},$$

or

$$\frac{r^2 R'' + r R'}{R} - k_1 r^2 = -\frac{\Theta''}{\Theta} = k_2,$$

since each side is a function of only one variable. Hence we also obtain the ordinary differential equations

$$\Theta'' + k_2 \Theta = 0$$

and

$$r^2 R'' + r R' - (k_2 + k_1 r^2)R = 0. \tag{10}$$

If we now assume the boundary conditions

$$T(r, \theta, 0) = T(r, \theta, \pi) = 0, \qquad T(1, \theta, z) = f(\theta, z),$$

where f is a continuously differentiable function in θ and z, we can translate the conditions into terms of R, Θ, and Z:

$$R(r)\Theta(\theta)Z(0) = R(r)\Theta(\theta)Z(\pi) = 0,$$

$$R(1)\Theta(\theta)Z(z) = f(\theta, z).$$

Since R and Θ are not constantly zero [assuming $f(\theta, z) \neq 0$], it follows that $Z(0) = Z(\pi) = 0$. Also, since f is continuous, it must be periodic with period 2π in θ (since f is the temperature on the surface of the cylinder). Thus $\Theta(\theta)$ is periodic with period 2π, so $\Theta(0) = \Theta(2\pi)$ and $\Theta'(0) = \Theta'(2\pi)$. Finally, we know that $R(1) \neq 0$ and that $R(0)$ is finite. As in (3), the eigenvalues of the boundary value problem

$$\Theta'' + k_2\Theta = 0, \qquad \Theta(0) = \Theta(2\pi), \qquad \Theta'(0) = \Theta'(2\pi),$$

are $k_2 = m^2$, $m = 0, 1, 2, 3, \ldots$, with the corresponding eigenfunctions

$$\Theta_m(\theta) = A_1 \cos m\theta + A_2 \sin m\theta.$$

Similarly, the boundary value problem

$$Z'' + k_1 Z = 0, \qquad Z(0) = Z(\pi) = 0,$$

has the eigenvalues $k_1 = n^2$, $n = 1, 2, 3, \ldots$, and the corresponding eigenfunctions

$$Z_n(z) = B \sin nz.$$

Setting $k_1 = n^2$ and $k_2 = m^2$ in (10), we obtain

$$r^2 R'' + rR' - (n^2 r^2 + m^2)R = 0, \tag{11}$$

which may be transformed into Bessel's equation by the substitution $\rho = inr$. To verify this fact, we note that

$$\frac{dR}{d\rho} = \frac{dR}{dr}\frac{dr}{d\rho} = \frac{1}{in} R'(r),$$

and

$$\frac{d^2R}{d\rho^2} = \frac{d}{d\rho}\left(\frac{1}{in}R'\right) = \frac{1}{(in)^2} R''(r) = -\frac{1}{n^2} R''(r).$$

Substituting these values into equation (11), we obtain

$$-n^2 r^2 \frac{d^2R}{d\rho^2} + inr\frac{dR}{d\rho} - (n^2 r^2 + m^2)R = 0,$$

or

$$\rho^2 \frac{d^2R}{d\rho^2} + \rho\frac{dR}{d\rho} + (\rho^2 - m^2)R = 0. \tag{12}$$

This is Bessel's equation of order m (see p. 271). By Theorem 5.4.2 (p. 276), equation (12) has the general solution

$$R(\rho) = C_1 J_m(\rho) + C_2 Y_m(\rho), \qquad \rho \neq 0.$$

And since $R(0)$ is finite, the constant $C_2 = 0$, because the Bessel function of the second kind, Y_m, involves a $\ln \rho$ term that approaches $-\infty$ as ρ approaches zero. We are, therefore, led to the double series

$$T(r, \theta, z) = \sum_{m=0}^{\infty} \sum_{n=1}^{\infty} (a_{mn} \cos m\theta + b_{mn} \sin m\theta) \sin nz \, J_m(inr).$$

The coefficients a_{mn} and b_{mn} can be determined by setting $r = 1$ and requiring that

$$f(\theta, z) = \sum_{m=0}^{\infty} \left[\left(\sum_{n=1}^{\infty} a_{mn} J_m(in) \sin nz \right) \cos m\theta + \left(\sum_{n=1}^{\infty} b_{mn} J_m(in) \sin nz \right) \sin m\theta \right].$$

$$\text{(13)}$$

If we define

$$a_m(z) = \sum_{n=1}^{\infty} a_{mn} J_m(in) \sin nz$$

and $\qquad\qquad\qquad\qquad\qquad\qquad\qquad\qquad\qquad\qquad\qquad$ (14)

$$b_m(z) = \sum_{n=1}^{\infty} b_{mn} J_m(in) \sin nz,$$

we see that (13) becomes

$$f(\theta, z) = \sum_{m=0}^{\infty} a_m(z) \cos m\theta + b_m(z) \sin m\theta,$$

which is a Fourier series in θ, where z is an arbitrary parameter. By the Euler formulas, it follows that

$$a_m(z) = \frac{1}{\pi} \int_{-\pi}^{\pi} f(\theta, z) \cos m\theta \, d\theta, \qquad b_m(z) = \frac{1}{\pi} \int_{-\pi}^{\pi} f(\theta, z) \sin m\theta \, d\theta.$$

If the functions $a_m(z)$ and $b_m(z)$ can be calculated, we can then use the Fourier sine series (14) in z to calculate the coefficients a_{mn} and b_{mn}:

$$a_{mn} = \frac{2}{\pi J_m(in)} \int_0^{\pi} a_m(z) \sin nz \, dz,$$

$$b_{mn} = \frac{2}{\pi J_m(in)} \int_0^{\pi} b_m(z) \sin nz \, dz.$$

Although the formal procedure delineated above is straightforward, the calculations involved are, to say the least, very tedious. In Problem 8 the reader is asked to develop a similar procedure that leads to the use of series of Legendre polynomials.

Problems 12.9

1. Consider the steady-state heat flow problem on the unit disk D [see equation (2)] and suppose that the boundary condition (1) is given by $f(\theta) = |\theta - \pi|$. Find the Fourier series solution (5) of this problem.

2. Repeat Problem 1 with $f(\theta) = \cos 2\theta$.

3. Find the Poisson integral solution (7) of Problem 1. What is the temperature at the origin? What is the temperature at the point $(r, \theta) = (\frac{1}{2}, 0)$?

4. Find the Poisson integral solution for Problem 2. What is the temperature at the origin?

5. Consider the steady-state heat flow problem on the unit disk and suppose that $f(\theta) = \sin^2 \theta$. Find the temperature at the point $r = \frac{1}{2}, \theta = 0°$ by

   (a) the eigenfunction expansion method;

   (b) the Poisson integral formula.

6. Repeat Problem 5 with $f(\theta) = \sin \theta - \cos \theta$.

7. Repeat Problem 5 with $f(\theta) = \theta(2\pi - \theta)$.

* 8. (a) Show that the three-dimensional Laplace equation (8) may be written in spherical coordinates r, θ, ϕ ($x = r \cos \theta \sin \phi, y = r \sin \theta \sin \phi, z = r \cos \phi$) in the form

$$\frac{\partial^2 T}{\partial r^2} + \frac{2}{r}\frac{\partial T}{\partial r} + \frac{1}{r^2}\frac{\partial^2 T}{\partial \phi^2} + \frac{\cot \phi}{r^2}\frac{\partial T}{\partial \phi} + \frac{1}{r^2 \sin^2 \phi}\frac{\partial^2 T}{\partial \theta^2} = 0. \tag{15}$$

   (b) Separate the variables in equation (15) to obtain the ordinary differential equations

$$\Theta'' + k_1 \Theta = 0,$$

$$R'' + \left(\frac{2}{r}\right)R' + \left(\frac{k_2}{r^2}\right)R = 0,$$

$$[(\sin \phi)\Phi']' - \left(\frac{k_2 \sin^2 \phi + k_1}{\sin \phi}\right)\Phi = 0.$$

   (c) Assume the boundary condition $T(1, \theta, \phi) = f(\theta, \phi)$ for a sphere of radius 1, where f is continuously differentiable in θ and ϕ and periodic with period 2π in θ. Find the eigenfunctions of the three boundary value problems thus defined. [*Hint:* Use Legendre polynomials.]

   (d) Obtain the eigenfunction expansion for $T(r, \theta, \phi)$.

** 9. Justify the steps involved in obtaining the Poisson integral formula (7) from the solution (5) of the steady-state heat equation in the disk $r \leq 1$. [*Hint:* Show that series (5) converges uniformly in the disk $r \leq r_0 < 1$.]

* 10. Generalize the problem in Fig. 1 to obtain the steady-state temperature in a disk of radius R centered at the origin, and prove that the corresponding Poisson integral formula is

$$T(r, \theta) = \frac{R^2 - r^2}{2\pi} \int_{-\pi}^{\pi} \frac{f(\phi)\, d\phi}{R^2 + r^2 - 2rR\cos(\theta - \phi)}, \quad r < R,$$

where the continuously differentiable function $f(\phi)$ describes the boundary temperature in the direction $0 \leq \phi < 2\pi$.

* 11. Let D be any bounded polygon in the xy-plane, and let (x_0, y_0) be a point interior to D. Let u be a real-valued function defined on D that has continuous derivatives of at least order two.

   (a) Show that if u attains its maximal value M at (x_0, y_0), then

$$\left.\frac{\partial^2 u}{\partial x^2} + \frac{\partial^2 u}{\partial y^2}\right|_{(x_0,y_0)} \leq 0.$$

   (b) Show that if

$$\frac{\partial^2 u}{\partial x^2} + \frac{\partial^2 u}{\partial y^2} > 0$$

   at all interior points of D, then u does not attain its maximal value at any interior point of D.

   (c) Suppose that

$$\frac{\partial^2 u}{\partial x^2} + \frac{\partial^2 u}{\partial y^2} \geq 0 \tag{16}$$

   at all interior points of D. Show that the function

$$w(x, y) = u(x, y) + \epsilon[(x - x_0)^2 + (y - y_0)^2]$$

   does not attain its maximal value in D at (x_0, y_0), for any value $\epsilon > 0$.

   (d) Conclude that any function satisfying equation (16) at all interior points of D must attain its maximal value on the boundary of D. (This is called the **maximum principle**.)

* 12. Use the maximum principle obtained in Problem 11 to show that any function u that satisfies Laplace's equation attains both its maximum and its minimum on the boundary of D. (Physically, this means that, in a steady-state heat flow problem, the maximum and minimum temperatures occur on the boundary of the region.)

Vibrations of a Drum. The following problems show how to find the equation of the deflections of a two-dimensional vibrating membrane. This generalizes the problem of the vibrating string to two spatial dimensions.

13. (a) If a drum of radius R, fixed at its boundary, is struck (see Fig. 3), the amplitude $u(x, y, t)$ of its vibration satisfies the two-dimensional wave equation

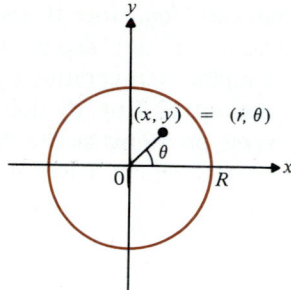

FIGURE 3

$$\frac{\partial^2 u}{\partial t^2} = c^2\left(\frac{\partial^2 u}{\partial x^2} + \frac{\partial^2 u}{\partial y^2}\right),$$

where $c^2 = T/\rho$, T is the tension, and ρ is the area density.

(b) Show that if $x = r \cos\theta$ and $y = r \sin\theta$, the two-dimensional wave equation can be written

$$\frac{\partial^2 u}{\partial t^2} = c^2\left(\frac{\partial^2 u}{\partial r^2} + \frac{1}{r}\frac{\partial u}{\partial r} + \frac{1}{r^2}\frac{\partial^2 u}{\partial \theta^2}\right). \tag{17}$$

14. Show that if the vibrations do not depend on θ (that is, if there is radial symmetry) then (17) reduces to

$$\frac{\partial^2 u}{\partial t^2} = c^2\left(\frac{\partial^2 u}{\partial r^2} + \frac{1}{r}\frac{\partial u}{\partial r}\right).$$

15. Write $u(r, t) = U(r)V(t)$. Show that $U(r)$ satisfies the differential equation

$$U'' + \frac{1}{r}U' + \lambda^2 U = 0 \tag{18}$$

for some constant λ. [*Hint:* Because the drum is fixed at its boundary, we must have $u(R, t) = 0$ for all $t \geq 0$.]

16. (a) Make the substitution $r = z/\lambda$ in (18) and show that U satisfies Bessel's differential equation of order zero

$$\frac{\partial^2 U}{\partial z^2} + \frac{1}{z}\frac{\partial U}{\partial z} + U = 0. \tag{19}$$

(b) Using the material in Section 5.4, show that the general solution to (19) is

$$U(z) = c_1 J_0(z) + c_2 Y_0(z) \tag{20}$$

where J_0 and Y_0 are Bessel functions of the first and second kind, respectively.

* (c) Explain why c_2 must be zero in (20).

17. Set $c_1 = 1$ and $c_2 = 0$ in (20).

(a) Explain why $U(R)$ must equal 0.

(b) It can be shown that $J_0(z)$ has an infinite number of zeros $s_1, s_2, s_3, \ldots$ and $s_n \to \infty$ as $n \to \infty$.† Show that

$$U_n(r) = J_0\left(\frac{s_n}{R}r\right)$$

satisfies (18) and the condition $U(R) = 0$.

18. (a) Show that $V(t)$ satisfies the equation

$$V'' + k^2 V = 0 \qquad \text{where } k = \lambda c. \tag{21}$$

(b) Show that if $\lambda_n = s_n/R$ and $k_n = \lambda_n c$, then

$$V_n(t) = a_n \cos k_n t + b_n \sin k_n t \tag{22}$$

satisfies (21).

** 19. Suppose that the drum has initial deflection and velocity given by

$$u(r, 0) = f_1(r), \quad \text{and} \quad u_t(r, 0) = f_2(r).$$

(a) Show that the **Bessel series**

$$u(r, t) = \sum_{n=1}^{\infty} U_n(r)V_n(t) = \sum_{n=1}^{\infty} (a_n \cos k_n t + b_n \sin k_n t)J_0\left(\frac{s_n r}{R}\right) \tag{23}$$

satisfies equation (17) and the condition $u(R, t) = 0$.

(b) Find formulas for a_n and b_n in (23). [*Hint:* Assume that f_1 and f_2 are continuous and that the Bessel functions $\left\{J_0\left(\frac{s_n}{R}r\right)\right\}$, $n = 1, 2, \ldots$ are orthogonal on $[0, R]$.]

12.10 Laplace Transform Methods for Partial Differential Equations

The one-dimensional heat equation

$$\frac{\partial T}{\partial t} = \kappa \frac{\partial^2 T}{\partial x^2}$$

† The first four zeros are $s_1 \approx 2.405$, $s_2 \approx 5.520$, $s_3 \approx 8.654$ and $s_4 \approx 14.931$.

is sometimes called the **thermal diffusion equation,** since it describes the flow of heat along a uniform rod. The same equation may also be used to describe one-dimensional fluid flow, since we can replace temperature by mass (or concentration) in the considerations of equation (12.7.6). By doing this, we can use the heat equation to study such diverse problems as the dispersal of pollutants in the atmosphere or the flow of a chemical across a membrane. In some cases, the fluid flow is aided (or retarded) by other forces; for example, the motion of a pollutant emitted from a smokestack involves both mixing (**diffusion**) of the pollutant in the air and transport (**convection**) of the pollutant by air currents. The change in the concentration T in the interval from x_0 to x_1, due to convection, is

$$v[T(x_0, t) - T(x_1, t)],$$

where v is the velocity of the current. Adding this term to equation (12.7.3) (p. 761), we obtain

$$\frac{dQ}{dt} = \int_{x_0}^{x_1} \frac{\partial}{\partial x}\left(\kappa \frac{\partial T}{\partial x} - vT\right)A\,dx.$$

Differentiating equation (12.7.1), we finally have, by considerations similar to those in Section 12.7,

$$c\rho \frac{\partial T}{\partial t} = \frac{\partial}{\partial x}\left(\kappa \frac{\partial T}{\partial x}\right) - v\frac{\partial T}{\partial x}. \tag{1}$$

Diffusion Equation

Equation (1) is generally called the **diffusion equation.** If κ is zero, we call (1) a **pure convection;** if $v = 0$, a **pure diffusion.**

The method of separation of variables is again very useful in solving diffusion equations. However, in some situations the boundary and initial conditions are such that it is not possible to obtain a solution by this technique. Consider the following example of a pure diffusion.

EXAMPLE 1

Suppose that κ is constant and $v = 0$ in (1). Then (1) reduces to the thermal diffusion equation

$$\frac{\partial T}{\partial t} = \kappa \frac{\partial^2 T}{\partial x^2}. \tag{2}$$

Assume that we are given the initial condition

$$T(x, 0) = 0, \qquad 0 \leq x < \infty, \tag{3}$$

and the boundary conditions

$$\lim_{x \to +\infty} T(x, t) = \lim_{x \to +\infty} \frac{\partial T}{\partial x}(x, t) = 0, \tag{4}$$

$$T(0, t) = T_0(\neq 0), \qquad 0 < t. \tag{5}$$

These conditions could be used to describe the heat flow in an infinitely long rod when a constant source of heat is applied at $x = 0$, or to describe the concentration of a pollutant at varying distances from a smokestack, in the absence of winds, with the stack constantly emitting pollutants, for all time $t > 0$. If we let $T(x, t) = X(x)\mathcal{T}(t)$, we have

$$X\mathcal{T}' = \kappa X''\mathcal{T} \qquad \text{or} \qquad \frac{\mathcal{T}'}{\kappa \mathcal{T}} = \frac{X''}{X} = c,$$

from which we obtain the two differential equations

$$\mathcal{T}' = c\kappa \mathcal{T} \qquad \text{and} \qquad X'' - cX = 0. \tag{6}$$

The first differential equation in (6) has the general solution $\mathcal{T}(t) = Ce^{c\kappa t}$. Using the initial condition (3), we observe that

$$0 = T(x, 0) = X(x)\mathcal{T}(0) = CX(x).$$

Hence either $C = 0$, implying that $\mathcal{T} \equiv 0$, or $X \equiv 0$. In either case $T \equiv 0$, contradicting the second boundary condition (5). Thus, there is no solution of the form $T(x, t) = X(x)\mathcal{T}(t)$ for this problem. It is therefore apparent that other techniques must be developed to treat problems of this type.

The effectiveness of Laplace transform methods for ordinary differential equations suggests that they might be useful in this context. The rest of this section is devoted to a discussion of how transformation techniques may be applied to partial differential equations. It should be noted, however, that the effectiveness of Laplace transforms is subject to several limitations. The equation *must* be linear and should have constant coefficients, and there must be appropriate initial conditions. Even when these conditions are met, there is no guarantee of success; and even if a solution can be obtained, it may be easier to obtain by other methods.

To apply Laplace transforms to a partial differential equation, we must begin by considering the ranges of the independent variables. We here illustrate the entire procedure as it relates to Example 1. In this problem, the independent variables x and t may both assume all values in the range 0 to ∞. This property is very desirable, since the definition of the Laplace transform of a function requires its integration over this range [see equation (4.1.1)]. At this point both variables look promising. We recall, however, that the differentiation property (Theorem 4.2.1, p. 200) of Laplace transforms requires an additional initial condition for each order of the derivative involved. In this case, (2) contains first derivatives in t and second derivatives in x. Since $(\partial T/\partial x)(0, t)$ is not known, there are not enough data for the x variable. Thus we treat x as a parameter and consider (2) as an ordinary differential equation with t as the only independent variable. Let

$$\mathcal{L}\{T(x, t)\} = \int_0^\infty e^{-st} T(x, t)\, dt.$$

Then $\mathcal{L}\{T\}$ is a function of s and the parameter x. Using the differentiation property of Laplace transforms and the initial condition (3), we may transform (2) into the ordinary differential equation

$$s\mathcal{L}\{T\} - 0 = \mathcal{L}\left\{\frac{\partial T}{\partial t}\right\} = \kappa\mathcal{L}\left\{\frac{\partial^2 T}{\partial x^2}\right\} = \kappa\frac{\partial}{\partial x^2}\mathcal{L}\{T\}. \tag{7}$$

Note that we have interchanged the operations of taking the Laplace transform and differentiating with respect to x. This exchange may not be valid, but the objection can be ignored if the method succeeds in producing a solution to the problem. Hence it is essential that we verify any result obtained by this method.

It is important now to interpret the boundary conditions (4) and (5) for $\mathcal{L}(T)$ whenever possible. Note that (5) yields

$$\mathcal{L}\{T\}(0, s) = \int_0^\infty e^{-st}T(0, t)\,dt = \frac{T_0}{s}. \tag{8}$$

Also, after warning that the indicated operations may not always be valid, we obtain

$$\lim_{x \to +\infty} \mathcal{L}\{T\}(x, s) = \lim_{x \to +\infty} \int_0^\infty e^{-st}T(x, t)\,dt$$

$$= \int_0^\infty e^{-st}\left[\lim_{x \to +\infty} T(x, t)\right]dt = 0. \tag{9}$$

Now we treat s as a parameter and x as the independent variable. Setting $z = \mathcal{L}\{T\}$, we may rewrite (7) as

$$\frac{d^2 z}{dx^2} = \frac{s}{\kappa}z, \tag{10}$$

where s is fixed. The characteristic equation for the differential equation (10) has the roots $\pm\sqrt{s/\kappa}$, so z has the general solution

$$z(x) = c_1 e^{\sqrt{s/\kappa}x} + c_2 e^{-\sqrt{s/\kappa}x}.$$

Since $s > 0$, the coefficient c_1 must be zero, because the term $e^{\sqrt{s/\kappa}x}$ tends to infinity as x approaches $+\infty$, violating the identity (9). Setting $x = 0$, we find that $c_2 = T_0/s$ by (8), so we finally have

$$\mathcal{L}\{T\} = z(x) = \frac{T_0}{s}e^{-\sqrt{s/\kappa}x}. \tag{11}$$

Again we treat x as a parameter and seek the inverse Laplace transform of (11). By Formula 43 of Appendix 2, with $r = x/\sqrt{\kappa}$, we see that

$$T(x, t) = T_0\left(1 - \frac{2}{\sqrt{\pi}}\int_0^{x/2\sqrt{\kappa t}} e^{-u^2}\,du\right). \tag{12}$$

To check that (12) is indeed the solution of Example 1, we note from Formula 216 of Appendix 1 that

$$\int_0^\infty e^{-u^2}\, du = \frac{\sqrt{\pi}}{2}.$$ (13)

The integral in (12) tends to $\sqrt{\pi}/2$ if t approaches zero or if x approaches $+\infty$. Thus (3) and the first limit of (4) hold. If $x = 0$, we obtain (5). Differentiating (12) with respect to x, we have

$$\frac{\partial T}{\partial x} = -\frac{T_0}{\sqrt{\pi \kappa t}} e^{-(x^2/4\kappa t)},$$

which vanishes as x approaches $+\infty$, and differentiating again, we obtain

$$\kappa \frac{\partial^2 T}{\partial x^2} = \frac{T_0 x}{2\sqrt{\pi \kappa}} \frac{e^{-(x^2/4\kappa t)}}{t^{3/2}} = \frac{\partial T}{\partial t}.$$

Thus in this case the Laplace transform method does yield a solution.

EXAMPLE 2

We here consider the vibrating string problem of Section 12.5 given by the wave equation

$$\frac{\partial^2 y}{\partial t^2} = c^2 \frac{\partial^2 y}{\partial x^2},$$ (14)

with the boundary conditions

$$y(0, t) = y(L, t) = 0$$ (15)

and initial conditions

$$y(x, 0) = f(x), \qquad \left.\frac{\partial y}{\partial t}\right|_{(x,0)} = 0.$$ (16)

Since $0 \le x \le L$, we select t as our independent variable, for the Laplace transform. By the differentiation property [see equation (4.2.4), p. 201], we transform (14) into

$$s^2 \mathcal{L}\{y\} - sf(x) - 0 = c^2 \frac{\partial^2}{\partial x^2} \mathcal{L}\{y\}.$$

Thus we have the nonhomogeneous second-order ordinary differential equation

$$c^2 Y'' - s^2 Y = -sf(x),$$ (17)

where $Y(x) = \mathcal{L}\{y\}$ and s is a parameter. By (15), we have $Y(0) = Y(L) = 0$. Our problem thus reduces to a nonhomogeneous boundary value problem. It may then be solved by the methods of Section 11.8. To simplify matters, we assume that $f(x) = y_0 \sin(\pi x/L)$ and so avoid having to use eigenfunction expansions to find the solution Y. A particular solution of (17) is of the form

$$Y_p = A \cos \frac{\pi x}{L} + B \sin \frac{\pi x}{L}.$$

Hence

$$-A\left(\frac{c^2\pi^2}{L^2}+s^2\right)\cos\frac{\pi x}{L}-B\left(\frac{c^2\pi^2}{L^2}+s^2\right)\sin\frac{\pi x}{L}=-sy_0\sin\frac{\pi x}{L},$$

implying that $A = 0$ and $B = sy_0/[s^2 + (c^2\pi^2/L^2)]$. Thus the general solution of (17) is

$$Y = ae^{sx/c}+be^{-sx/c}+\frac{sy_0}{s^2+(c^2\pi^2/L^2)}\sin\frac{\pi x}{L}. \tag{18}$$

Setting $x = 0$ and $x = L$ in (18), we obtain the homogeneous simultaneous equations

$$a \quad + b \quad = 0,$$
$$ae^{sL/c}+be^{-sL/c}=0.$$

Since the determinant of the coefficients on the left-hand side of (19) is nonzero, $a = b = 0$ and (18) reduces to

$$\mathcal{L}\{y\}=Y=\frac{sy_0}{s^2+(c^2\pi^2/L^2)}\sin\frac{\pi x}{L}.$$

Solving for y, we obtain

$$y=y_0\sin\frac{\pi x}{L}\cos\frac{\pi ct}{L},$$

which agrees with the solution of Example 12.5.2, obtained by d'Alembert's method.

Problems 12.10

In Problems 1–6 obtain the solution of the given boundary and initial condition problem by the method of Laplace transforms. Be sure to verify your solution.

1. $\dfrac{\partial^2 y}{\partial x^2}=4\dfrac{\partial^2 y}{\partial t^2},\quad t>0,\quad x>0$

$y(x, 0) = 0,\quad \left.\dfrac{\partial y}{\partial t}\right|_{(x,0)}=1,\quad x\ge 0$

$y(0, t) = t,\quad \lim\limits_{x\to\infty} y(x, t)$ is finite

2. $\dfrac{\partial^2 y}{\partial x^2}=16\dfrac{\partial^2 y}{\partial t^2},\quad t>0,\quad x>0$

$y(x, 0) = 0,\quad \left.\dfrac{\partial y}{\partial t}\right|_{(x,0)}=1,\quad x\ge 0$

$y(0, t) = \sin t,\quad \lim\limits_{x\to\infty} y(x, t)$ is finite

3. $\dfrac{\partial T}{\partial t}=\delta\dfrac{\partial^2 T}{\partial x^2},\quad t>0,\quad x>0$

$T(x, 0) = 0,\quad 0\le x<\infty,\quad T(0, t)=1,\quad t>0$

$\lim\limits_{x\to\infty} T(x, t)=\lim\limits_{x\to\infty}\dfrac{\partial T}{\partial x}(x, t)=0,\quad t>0$

4. $\dfrac{\partial T}{\partial t}=\delta\dfrac{\partial^2 T}{\partial x^2},\quad t>0,\quad x>0$

$T(x, 0) = e^{-x},\quad x\ge 0,\quad T(0, t)=T_0(\ne 0)$

$\lim\limits_{x\to\infty} T(x, t)=\lim\limits_{x\to\infty}\dfrac{\partial T}{\partial x}(x, t)=0,\quad t>0$

5. $\dfrac{\partial T}{\partial t}=\delta\dfrac{\partial^2 T}{\partial x^2},\quad t>0,\quad x>0$

$T(x, 0) = 0,\quad x\ge 0,$

$T(0, t) = \sin\omega t$

$\lim\limits_{x\to\infty} T(x, t)=\lim\limits_{x\to\infty}\dfrac{\partial T}{\partial x}(x, t)=0,\quad t>0$

6. $\dfrac{\partial T}{\partial t}=\delta\dfrac{\partial^2 T}{\partial x^2}+\mu\dfrac{\partial T}{\partial x},\quad t>0,\quad x>0$

$T(x, 0) = 0,\quad x\ge 0$

$T(0, t) = T_0(\ne 0),$

$\lim\limits_{x\to\infty} T(x, t)=\lim\limits_{x\to\infty}\dfrac{\partial T}{\partial x}(x, t)=0,\quad t>0$

Review Exercises for Chapter 12

In Exercises 1–4 find the solution $z(x, y)$ of each partial differential equation that satisfies the given condition.

1. $z_x + z_y - z = e^y, \quad z(x, 0) = x$

2. $z_x + z_y - 2z = 0, \quad z(x, 0) = e^x$

3. $z_x + z_y - z = 1 - x, \quad z(x, 0) = x + 1$

4. $z_x - z_y + z = 0, \quad z(x, 0) = \sin x$

In Exercises 5–8 find the temperature in an insulated rod a units long whose ends are kept at 0°C, where the initial temperature is given by

5. $f(x) = x(a - x)$

6. $f(x) = x^2(a^2 - x^2)$

7. $f(x) = \begin{cases} x, & 0 \le x \le a/2, \\ a - x, & a/2 \le x \le a \end{cases}$

8. $f(x) = \begin{cases} x^2, & 0 \le x \le a/2, \\ 0, & a/2 < x \le a \end{cases}$

Find general solutions of the equations in Exercises 9–12 by separation of variables.

9. $\dfrac{\partial^2 y}{\partial x^2} = \dfrac{t^2}{x^2} \dfrac{\partial^2 y}{\partial t^2}$

10. $\dfrac{\partial^2 y}{\partial x^2} + \dfrac{\partial^2 y}{\partial x \partial t} = \dfrac{\partial y}{\partial t}$

11. $\dfrac{\partial^2 y}{\partial t^2} = \dfrac{\partial^2 y}{\partial x^2} - y$

12. $\dfrac{\partial^2 y}{\partial x^2} + \dfrac{\partial y}{\partial x} = \dfrac{\partial y}{\partial t}$

13. A tightly stretched string fixed at $x = 0$ and $x = a$ is initially displaced to the position $y(x, 0) = 2 \sin^3(\pi x/a)$. At time $t = 0$, it is released from rest from this position. Find the displacement of any point on the string at any time t.

14. Repeat Exercise 13 if the initial displacement is $y(x, 0) = a^2 x - x^3$.

15. The vibrating string in Exercise 13 is subjected to a damping force that is proportional to the velocity at each point and instant in time.

　(a) Find the differential equation that the damped string satisfies.

　(b) Solve the equation by separation of variables.

　* (c) If the initial displacement is $y(x, 0) = f(x)$, $0 \le x \le a$, express the solution as an infinite series.

16. A rectangular plate is bounded by the lines $x = 0$, $y = 0$, $x = 100$, $y = 50$. Its surfaces are insulated and the temperature along the upper edge is given by $T(x, 50) = x(100 - x)$, while the other edges are kept at 0°C. Find the steady-state temperature.

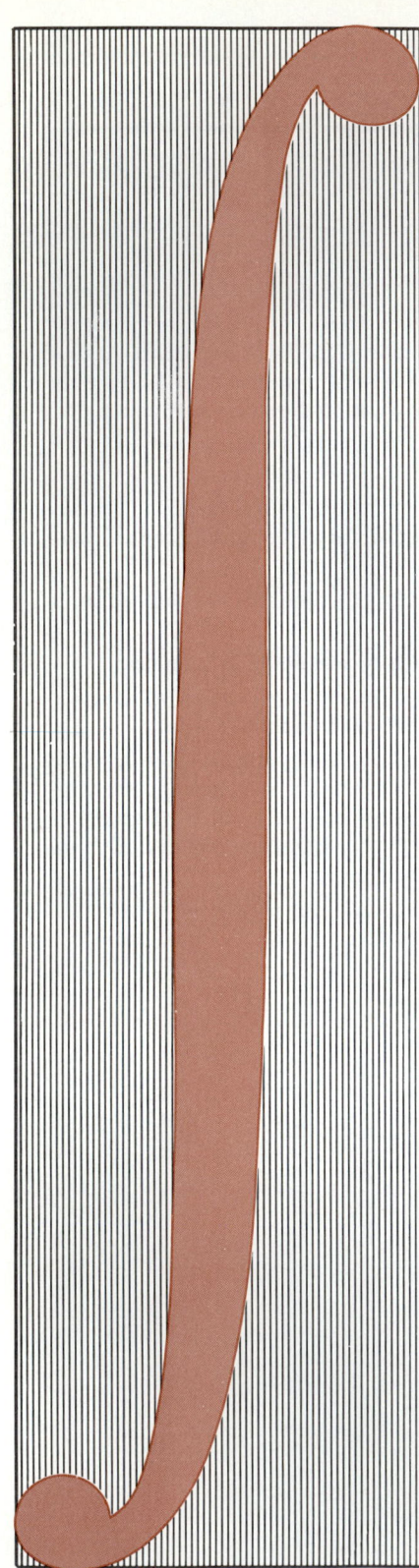

chapter 13

Numerical Analysis

13.1 Introduction

In every chapter of this book we have performed numerical computations: solving differential equations, using Laplace transforms, applying the power series method, solving systems of equations, inverting matrices. With few exceptions, we have limited our examples to problems involving a small number of variables—not because most applications have only two or three variables but because the computations would have been too tedious otherwise.

With the recent and widespread use of calculators and computers, the situation has been altered. The remarkable strides made in the last few years in the theory of numerical methods for solving certain computational problems have made it possible to perform, quickly and accurately, the calculations mentioned in the first paragraph.

The use of the computer presents new difficulties, however. Computers do not store numbers such as $\frac{2}{3}$, $7\frac{3}{8}$, $\sqrt{2}$, and π. Rather, most computers use what is called **floating-point arithmetic.** In this system every number is represented in the form

$$x = \pm 0.d_1 d_2 \cdots d_k \times 10^n \qquad (1)$$

where d_1, d_2, . . . , d_k are single-digit integers and n is an integer. Any number written in this form is called a **floating-point number.** In equation (1), the number $\pm 0.d_1 d_2 \cdots d_k$ is called the **mantissa** and the number n is called the **exponent.** The number k is called the **number of significant digits in the expression.**

If the number of significant digits were unlimited, then we would have no problem. Almost every time numbers are introduced into the computer, however, errors begin to accumulate. This can happen in one of two ways:

i. **Truncation:** All significant digits after k digits are simply "cut off." For example, if truncation is used, $\frac{2}{3} = 0.666666\ldots$ is stored (with $k = 8$) as $\frac{2}{3} = 0.66666666 \times 10^0$.

ii. **Rounding:** If $d_{k+1} \geq 5$, then 1 is added to d_k and the resulting number is truncated. Otherwise, the number is simply truncated. For example, with rounding (and $k = 8$), $\frac{2}{3}$ is stored as $\frac{2}{3} = 0.66666667 \times 10^0$.

Individual round-off or truncation errors do not seem very significant. When thousands of computational steps are involved, however, the **accumulated** round-off error can be devastating. Thus, in discussing any numerical scheme, it is necessary to know not only whether you will get the right answer, theoretically, but also how badly the round-off errors will accumulate. To keep track of things, we define two types of error. If x is the actual value of a number and if x^* is the number that appears in the computer, then the **absolute error** ε_a is defined by

Absolute Error

$$\varepsilon_a = |x^* - x|. \tag{2}$$

More interesting in most situations is the **relative error** ε_r, defined by

Relative Error

$$\varepsilon_r = \left| \frac{x^* - x}{x} \right|. \tag{3}$$

Much of numerical analysis is concerned with questions of **convergence** and **stability.** If x is the answer to a problem and our computational method gives us approximating values x_n, then the method converges if, theoretically, x_n approaches x as n gets large. If it can be shown that the round-off errors will not accumulate in such a way as to make the answer unreliable, then the method is stable.

It is easy to give an example of a procedure in which round-off error can be quite large. Suppose we wish to compute $y = 1/(x - 0.66666665)$. For $x = \frac{2}{3}$, if the computer truncates, then $x = 0.66666666$ and $y = 1/0.00000001 = 10^8 = 10 \times 10^7$. If the computer rounds, then $x = 0.66666667$ and $y = 1/0.00000002 = 5 \times 10^7$. The difference here is enormous. The correct answer is $1/(\frac{2}{3} - \frac{66666665}{100000000}) = 60,000,000 = 6 \times 10^7$.

In this and the next chapter we shall examine several numerical procedures for solving the problems we have studied in other chapters of this book. Whenever possible, we shall also discuss convergence and stability. This is but a very superficial view of numerical methods, however; entire books and courses are devoted to the subject. For a more exhaustive view, you are encouraged to consult the following references:

1. W. Cheney and D. Kincaid, *Numerical Mathematics and Computing,* Brooks/Cole Publishing Co., Monterey, Calif., 1980.
2. S. D. Conte and C. deBoor, *Elementary Numerical Analysis: An Algorithmic Approach,* 3rd ed., McGraw-Hill, New York, 1980.

3. P. J. Davis and P. Rabinowitz, *Methods of Numerical Integration*, Academic Press, New York, 1975.

4. C. W. Gear, *Numerical Initial Value Problems in Ordinary Differential Equations*, Prentice-Hall, Englewood Cliffs, N.J., 1971.

5. P. Henrici, *Discrete Variable Methods in Ordinary Differential Equations*, Wiley, New York, 1962.

6. F. B. Hildebrand, *Introduction to Numerical Analysis*, McGraw-Hill, New York, 1974.

7. R. D. Richtmyer, *Difference Methods for Initial Value Problems*, Wiley (Interscience), New York, 1957.

8. J. H. Wilkinson, *Rounding Errors in Algebraic Processes*, Prentice-Hall, Englewood Cliffs, N.J., 1963.

13.2 Newton's Method for Solving Equations

Consider the equation

$$f(x) = 0, \tag{1}$$

where f is assumed to be differentiable in some interval $[a, b]$. It is often important to calculate the *roots* of equation (1), that is, the values of x that satisfy the equation. For example, if $f(x)$ is a polynomial of degree 5, say, then the roots of $f(x)$ could be as in Fig. 1.

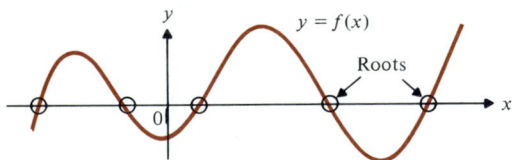

FIGURE 1

The intermediate value theorem states that if f is continuous on $[a, b]$ and k is some number between $f(a)$ and $f(b)$, then there is at least one number x in (a, b) such that $f(x) = k$. That is, f takes on all values between $f(a)$ and $f(b)$. We can use the intermediate value theorem as a crude tool to estimate the roots of $f(x) = 0$. Suppose that f is continuous, $f(a) < 0$ and $f(b) > 0$; then there is at least one number c in (a, b) such that $f(c) = 0$. If $f'(x) > 0$ on (a, b) [or $f'(x) < 0$ on (a, b)] then f is increasing (or decreasing) and there can be only one solution to $f(x) = 0$ in (a, b). We have proved:

■ **THEOREM 1**
Suppose that f' is continuous in $[a, b]$, and

(i) $f(a)$ and $f(b)$ have different signs,
(ii) $|f'(x)| \neq 0$ on $[a, b]$.

Then the equation $f(x) = 0$ has a unique solution c on $[a, b]$. ∎

We can make use of Theorem 1 to find roots by trial and error, but there is often a much better way.

In the seventeenth century Newton discovered a method for estimating a solution, or root, by defining a sequence of numbers that become successively closer and closer to the root sought. His method is best illustrated graphically. Let $y = f(x)$ as in Fig. 2. A number x_0 is chosen arbitrarily. We then locate the point $(x_0, f(x_0))$ on the graph and draw the tangent line to the curve at that point. Next, we follow the tangent line down until it hits the x-axis. The point of intersection of the tangent line and the x-axis is called x_1. We then repeat the process to arrive at the next point, x_2. On the graph we have labeled the solution to $f(x) = 0$ as s. That is, $f(s) = 0$. For our graph at least, it seems as if the points $x_0, x_1, x_2, \cdots$ are approaching the point $x = s$. In fact, this happens for quite a few functions, and the rate of approach to the solution is quite rapid.

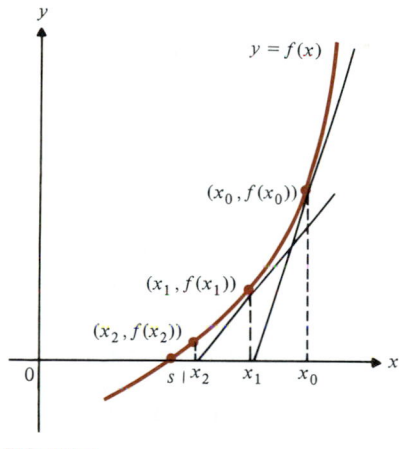

FIGURE 2

Having briefly looked at a graphical representation of Newton's method, let us next develop a formula for giving us x_1 from x_0, x_2 from x_1, and so on. The slope of the tangent line at the point $(x_0, f(x_0))$ is $f'(x_0)$. Two points on this line are $(x_1, 0)$ and $(x_0, f(x_0))$. Therefore

$$\frac{0 - f(x_0)}{x_1 - x_0} = f'(x_0). \tag{2}$$

Solving (2) for x_1 gives us

$$x_1 = x_0 - \frac{f(x_0)}{f'(x_0)}. \tag{3}$$

Similarly,

$$x_2 = x_1 - \frac{f(x_1)}{f'(x_1)}. \tag{4}$$

In general, we obtain

$$x_{n+1} = x_n - \frac{f(x_n)}{f'(x_n)}. \tag{5}$$

This last step tells us how to obtain the $(n + 1)$st point if the nth point is given, as long as $f'(x_n) \neq 0$ so that (5) is defined. Thus if we start with a given value x_0, we can obtain $x_1, x_2, x_3, x_4, \ldots$. Formula (5) is called **Newton's formula.** The set of numbers $x_0, x_1, x_2, x_3, \ldots$ form a sequence. Under conditions that we describe shortly, the sequence will converge to a number s such that $f(s) = 0$. That is,

$$\lim_{n \to \infty} x_n = s. \tag{6}$$

Before stating the theorem that guarantees when the sequence defined by (5) converges to a solution of $f(x) = 0$, we give two simple examples.

EXAMPLE 1

Let $r > 1$. Formulate a rule for calculating the square root of r, and find $\sqrt{2}$.

SOLUTION
We must find an x such that $x = \sqrt{r}$, or $x^2 = r$, or $x^2 - r = 0$. Let $f(x) = x^2 - r$. Then if $f(s) = 0$, s will be a square root of r because

$$s^2 - r = 0 \quad \text{or} \quad s^2 = r.$$

By Newton's formula, since $f'(x) = 2x$, we have

$$x_{n+1} = x_n - \frac{f(x_n)}{f'(x_n)} = x_n - \frac{(x_n^2 - r)}{2x_n} = \frac{2x_n^2 - x_n^2 + r}{2x_n} = \frac{1}{2}\left(x_n + \frac{r}{x_n}\right).\dagger \tag{7}$$

Selecting any value for x_0, we use (7) to generate the sequence $x_0, x_1, x_2, \ldots$, which we hope will converge to $\sqrt{r}$. Setting $r = 2$ in (7), we get

$$x_{n+1} = \frac{1}{2}\left(x_n + \frac{2}{x_n}\right). \tag{8}$$

Using a calculator, we obtain the sequence in Table 1, starting with $x_0 = 1$. An answer correct to nine decimal places is obtained after only four steps! We are limited in accuracy only by the fact that our calculator can display only 10 digits.

---

† Note that this formula requires only the basic arithmetic operations of addition, subtraction, multiplication, and division. These are the operations most easily performed in a calculator, which is why many calculators use this formula, or a simple modification of it, to find square roots.

TABLE 1

n	x_n	$\dfrac{2}{x_n}$	$x_n + \dfrac{2}{x_n}$	$x_{n+1} = \dfrac{1}{2}\left(x_n + \dfrac{2}{x_n}\right)$
0	1.0	2.0	3.0	1.5
1	1.5	1.333333333	2.833333333	1.416666667
2	1.416666667	1.411764706	2.828431373	1.414215686
3	1.414215686	1.414211438	2.828427125	1.414213562
4	1.414213562	1.414213562	2.828427125	1.414213562

EXAMPLE 2

Formulate a rule for calculating the kth root of a given number r, and use it to find $\sqrt[3]{17}$.

SOLUTION

We must find an x such that $x = r^{1/k}$, or $x^k = r$, or $f(x) = x^k - r = 0$. Then $f'(x) = kx^{k-1}$, and

$$x_{n+1} = x_n - \frac{x_n^k - r}{kx_n^{k-1}} = x_n - \frac{1}{k}\frac{x_n^k}{x_n^{k-1}} + \frac{r}{kx_n^{k-1}} = \left(1 - \frac{1}{k}\right)x_n + \frac{r}{kx_n^{k-1}}. \qquad (9)$$

To find $\sqrt[3]{17}$, we set $k = 3$ and $r = 17$, obtaining

$$x_{n+1} = \frac{2}{3}x_n + \frac{17}{3x_n^2}.$$

Since $\sqrt[3]{17}$ is between 2 and 3, we choose $x_0 = 2$ (3 would do just as well) and obtain the values tabulated in Table 2. The last number x_5 is correct to nine decimal places. Again, the rapid convergence of Newton's method is illustrated.

TABLE 2

n	x_n	$\dfrac{2}{3}x_n$	x_n^2	$\dfrac{17}{3x_n^2}$	$x_{n+1} = \dfrac{2}{3}x_n + \dfrac{17}{3x_n^2}$
0	2.0	1.333333333	4.0	1.416666667	2.75
1	2.75	1.833333333	7.5625	0.7493112948	2.582644628
2	2.582644628	1.721763085	6.670053275	0.8495684267	2.571331512
3	2.571331512	1.714221008	6.611745745	0.8570605836	2.571281592
4	2.571281592	1.714187728	6.611489025	0.8570938627	2.571281591

When using Newton's method, three questions naturally arise.

1. Is there a unique solution to $f(x) = 0$?
2. Does the sequence of Newton iterates converge to it?
3. If so, how fast does it converge?

The first question has already been answered in Theorem 1. To see how

Theorem 1 is used, observe that if we set $[a, b] = [2, 3]$ in Example 2, then since

$$f(x) = x^3 - 17, \qquad f(2) = -9, \qquad f(3) = 10,$$

and (i) is satisfied. Also, $f'(x) = 3x^2$ and $|f'(x)| \neq 0$ on $[2, 3]$. Thus there is a unique solution.

We now state a remarkable theorem that answers questions (2) and (3); a partial proof of the theorem is given at the end of this section.

Assume that f, f', and f'' are continuous in $[a, b]$. Then there are numbers m and M such that $|f'(x)| \geq m$ on $[a, b]$ and $|f''(x)| \leq M$ on $[a, b]$. Moreover, if $|f'(x)| \neq 0$ in $[a, b]$ then $m > 0$.

■ **THEOREM 2: Quadratic Convergence Theorem**

Let f, f', and f'' be continuous on $[a, b]$. Suppose that

  (i) $f(a)$ and $f(b)$ have different signs,
  (ii) $|f'(x)| \geq m > 0$,
  (iii) $|f''(x)| \leq M$,
  (iv) $M/m \leq 4$,
  (v) $x_0 \in [a, b]$.

Then

  (a) The sequence of Newton iterates $x_0, x_1, x_2, \cdots$ converges to the unique solution s of $f(x) = 0$ in $[a, b]$.
  (b) The convergence is quadratic. That is, each iterate is accurate to twice as many decimal places as the one that precedes it.   ■

EXAMPLE 3

In Example 2, $f(x) = x^3 - 17$, $|f'(x)| = 3x^2 \geq 3 \cdot 2^2 = 12$ on $[2, 3]$, and $|f''(x)| = 6x \leq 18$ on $[2, 3]$. Thus $m = 12$, $M = 18$, and $M/m = \frac{3}{2} < 4$. Thus Newton's method converges quadratically to $\sqrt[3]{17} \approx 2.571281591$. Now, from Table 2,

$|x_1 - \sqrt[3]{17}| = |2.75 - 2.571281591| = 0.178718409$ correct to 0 decimal places,

$|x_2 - \sqrt[3]{17}| = |2.582644628 - 2.571281591|$

$\qquad = 0.011363037$ correct to 1 decimal place (almost 2),

$|x_3 - \sqrt[3]{17}| = |2.571331512 - 2.571281591|$

$\qquad = 0.000049921$ correct to 4 decimal places,

$|x_3 - \sqrt[3]{17}| = |2.571281592 - 2.571281591|$

$\qquad = 0.000000001$ correct to 8 decimal places.

This illustrates nicely the quadratic convergence of Newton's method.

EXAMPLE 4

Find the real roots of $p(x) = x^3 + x^2 + 7x - 3$.

SOLUTION

We differentiate to find that $p'(x) = 3x^2 + 2x + 7$. This polynomial has no real roots (explain why), and, since $p'(0) = 7$, we conclude that $p(x)$ is an increasing function and has exactly one real root. It is graphed in Fig. 3. Now

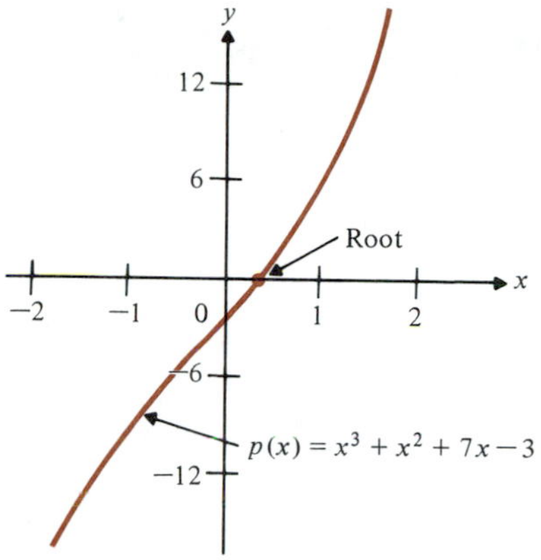

$$p(0) = -3 \quad \text{and} \quad p(1) = 6,$$

so the root is in the interval $[0, 1]$. We compute

$$|p'(x)| = 3x^2 + 2x + 7 \geq 7 = m \quad \text{on} \quad [0, 1],$$

and

$$|p''(x)| = 6x + 2 \leq 8 = M \quad \text{on} \quad [0, 1].$$

Since $M/m = \frac{8}{7} < 4$, we conclude that Newton's method converges quadratically to the unique root of $p(x) = 0$. We have

$$x_{n+1} = x_n - \frac{p(x_n)}{p'(x_n)} = x_n - \frac{x_n^3 + x_n^2 + 7x_n - 3}{3x_n^2 + 2x_n + 7}$$

$$= \frac{x_n(3x_n^2 + 2x_n + 7) - (x_n^3 + x_n^2 + 7x_n - 3)}{3x_n^2 + 2x_n + 7} = \frac{2x_n^3 + x_n^2 + 3}{3x_n^2 + 2x_n + 7}.$$

If we choose $x_0 = 0$, we obtain the results in Table 3. The root is $s = 0.3970992165$, correct to 10 decimal places.

FIGURE 3

$p(x) = x^3 + x^2 + 7x - 3$

TABLE 3

n	x_n	$2x_n^3 + x_n^2 + 3$	$3x_n^2 + 2x_n + 7$	$x_{n+1} = \dfrac{2x_n^3 + x_n^2 + 3}{3x_n^2 + 2x_n + 7}$
0	0	3.0	7.0	0.4285714286
1	0.4285714286	3.341107872	8.408163265	0.3973647712
2	0.3973647712	3.283385572	8.268425827	0.3970992352
3	0.3970992352	3.282923214	8.267261878	0.3970992165
4	0.3970992165	3.282923182	8.267261796	0.3970992165

EXAMPLE 5

Find all roots (if any) of the equation $\sin x = x/2$ in the interval $(0, 2\pi]$.

SOLUTION

The graphs of $y = \sin x$ and $y = x/2$ are given in Fig. 4. From these it is evident that there is exactly one real root of the equation $\sin x = x/2$ in $(0, 2\pi]$, and it lies in the interval $[\pi/2, \pi]$.

Setting $f(x) = \sin x - x/2$, we have $f'(x) = \cos x - 1/2$, and Newton's method provides the rule

$$x_{n+1} = x_n - \frac{f(x_n)}{f'(x_n)} = x_n - \frac{\sin x_n - x_n/2}{\cos x_n - 1/2} = x_n - \frac{2\sin x_n - x_n}{2\cos x_n - 1}.$$

Since we know that the root is in $[\pi/2, \pi]$, we can choose this interval to be our interval. We now verify the four conditions in Theorem 2.

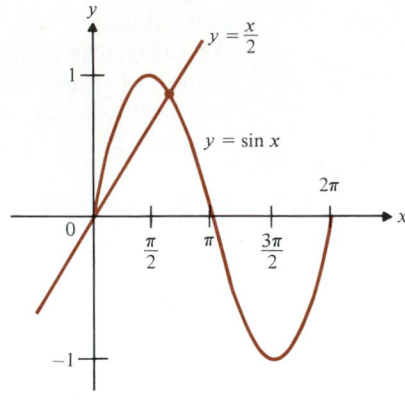

FIGURE 4

(i) $f(\pi/2) = 1 - \pi/4 > 0$, and $f(\pi) = 0 - \pi/2 < 0$.

(ii) $f'(x) = \cos x - \frac{1}{2}$, and $\cos x$ decreases from 0 to -1 in $[\pi/2, \pi]$, so the smallest value for $|f'(x)| = |\cos(\pi/2) - \frac{1}{2}| = |0 - \frac{1}{2}| = \frac{1}{2} = m$.

(iii) $f''(x) = -\sin x$ so $|f''(x)| \le 1 = M$.

(iv) $M/m = 1/\frac{1}{2} = 2 < 4$.

Thus the Newton iterates will converge for any $x_0 \in [\pi/2, \pi]$.

We start with $x_0 = 2$ and carry out the iteration given in Table 4. We find that the unique solution (correct to 10 significant figures) is $x_n = 1.895494267$ radians ($\approx 108.6°$).

TABLE 4

n	x_n	$\sin x_n$	$2 \sin x_n - x_n$	$\cos x_n$
0	2	0.9092974268	−0.1814051463	−0.4161468365
1	1.900995594	0.9459777536	−0.0090400868	−0.3242315372
2	1.895511645	0.9477415894	−0.0000284662	−0.3190389940
3	1.895494267	0.9477471335	0.0000000001	−0.3190225241

n	$2 \cos x_n - 1$	$\dfrac{2 \sin x_n - x_n}{2 \cos x_n - 1}$	$x_{n+1} = x_n - \dfrac{2 \sin x_n - x_n}{2 \cos x_n - 1}$
0	−1.832293673	0.0990044058	1.900995594
1	−1.648463074	0.0054839486	1.895511645
2	−1.638077988	0.0000173778	1.895494267
3	−1.638045048	−0.0000000003	1.895494267

Let us make some final remarks about Newton's method. First, if x_n is a solution to $f(x) = 0$, then

$$x_{n+1} = x_n - \frac{f(x_n)}{f'(x_n)} = x_n - 0 = x_n.$$

This result explains why, in Tables 1, 3, and 4, the value of x_n did not change after the desired accuracy was reached. Second, it is not necessary to check the conditions of Theorems 1 and 2 every time Newton's method is used. However, failure to do so could result in one of the following problems:

(i) The sequence generated by (5) could grow without bound, never even getting close to a solution.

(ii) The sequence could converge to a solution, but there could be *other* solutions in the interval [a, b] that would be missed. This problem can often be avoided if a graph is drawn that gives an idea about where (approximately) the roots are located.

(iii) The sequence generated by (5) could get into a closed nonconverging loop.

PARTIAL PROOF OF THE CONVERGENCE THEOREM (OPTIONAL)

According to Theorem 1, conditions (i) and (ii) imply that there is a unique number s in (a, b) such that $f(s) = 0$. Let ϵ_n denote the error of the nth iterate; that is, let

$$\epsilon_n = |x_n - s|. \tag{10}$$

Suppose that x_n is accurate to k decimal places. Then

$$\epsilon_n \leq \tfrac{1}{2} \times 10^{-k}. \tag{11}$$

We shall show that x_{n+1} is accurate to $2k$ decimal places, that is

$$\epsilon_{n+1} = |x_{n+1} - s| \leq \tfrac{1}{2} \times 10^{-2k}.$$

By Newton's formula

$$\epsilon_{n+1} = |s - x_{n+1}| = \left| s - \left(x_n - \frac{f(x_n)}{f'(x_n)} \right) \right| = \left| s - x_n + \frac{f(x_n)}{f'(x_n)} \right|. \tag{12}$$

Now apply Taylor's theorem to $f(s)$ at x_n:

$$0 = f(s) = f(x_n) + \frac{f'(x_n)}{1!}(s - x_n) + \frac{f''(c_n)}{2!}(s - x_n)^2, \tag{13}$$

where c_n is a number between x_n and s. Condition (ii) implies that $f'(x_n) \neq 0$, so we can divide both ends of (13) by $f'(x_n)$ and rearrange terms to obtain

$$0 = \frac{f(x_n)}{f'(x_n)} + (s - x_n) + \frac{1}{2}\frac{f''(c_n)}{f'(x_n)}(s - x_n)^2$$

or

$$s - x_n + \frac{f(x_n)}{f'(x_n)} = \frac{-1}{2}\frac{f''(c_n)}{f'(x_n)}(s - x_n)^2. \tag{14}$$

Taking the absolute value of both sides and using (10) and (12), we have

$$\left| s - x_n + \frac{f(x_n)}{f'(x_n)} \right| = \frac{1}{2}\frac{|f''(c_n)|}{|f'(x_n)|}|s - x_n|^2$$

or

$$\epsilon_{n+1} = \frac{1}{2}\frac{|f''(c_n)|}{|f'(x_n)|}\,\epsilon_n^2. \tag{15}$$

By conditions (ii), (iii), and (iv), $|f'(x_n)| \geq m$, $|f''(c_n)| \leq M$, and $M/m \leq 4$ so (15) becomes

$$\epsilon_{n+1} \leq \frac{1}{2}\frac{M}{m}\,\epsilon_n^2 \leq \frac{4}{2}\,\epsilon_n^2 = 2\epsilon_n^2. \tag{16}$$

Using (11) it now follows that

$$\epsilon_{n+1} \leq 2(\tfrac{1}{2}\times 10^{-k})^2 = \tfrac{1}{2}\times 10^{-2k},$$

which is what we wanted to prove. ∎

Remark. This is not a complete proof of convergence because we assumed that some term in the sequence $x_0, x_1, x_2, \cdots$ was accurate to at least one decimal place. Indeed, it suffices to show that some iterate x_n is within $\frac{1}{3}$ of s, implying $\epsilon_n < \frac{1}{3}$, because from (16)

$$\epsilon_{n+3} \leq 2\epsilon_{n+2}^2 \leq 2(2\epsilon_{n+1}^2)^2 \leq 8(2\epsilon_n^2)^4 = 128\epsilon_n^8 < \frac{128}{3^8} < \frac{1}{2}\times 10^{-1}.$$

Of course, we still haven't proved that $\epsilon_n < \frac{1}{3}$.

Problems 13.2

1. Find an interval over which Theorem 2 applies to the calculation of $\sqrt{90}$ by solving $x^2 - 90 = 0$.

2. Using Newton's method, find a formula for finding reciprocals without dividing (the reciprocal of x is $1/x$).

3. Show that there is no interval $[a, b]$ for which Theorem 2 applies when $f(x) = x^2 + 5x + 7$. Explain why this must be the case. Try Newton's method with $x_0 = 0$; what happens?

4. Try to use Newton's method to find a solution to $\sin x = x$ on $[\pi/4, \pi]$. What happens? Which of the hypotheses of Theorem 2 are violated? Why is the method doomed to fail?

In Problems 5–16, calculate all answers to as many decimal places of accuracy as are displayed by your calculator or computer.

5. Use the result of Problem 1 to choose a good initial approximation for $\sqrt{90}$. Continue using Newton's method.

6. Use the formula you obtained for Problem 2 to approximate $1/7$ and $1/81$.

7. Approximate $\sqrt[4]{25}$ by using Newton's method.

8. Approximate $\sqrt[5]{10}$ by using Newton's method.

9. Use Newton's method to approximate the roots of $x^2 - 7x + 5 = 0$. Compare your results with the values obtained by using the quadratic formula. [*Hint:* Two calculations are necessary.]

10. Approximate all solutions of the equation $x^3 - 6x^2 - 15x + 4 = 0$. [*Hint:* There are three roots. Draw a sketch first.]

11. Approximate all solutions of the equation $x^3 - 8x^2 + 2x - 15 = 0$.

12. Approximate all solutions of the equation $x^3 + 3x^2 - 24x - 40 = 0$.

13. (a) Show graphically that there is a unique solution to $\cos x = x$ in the interval $[0, \pi/2]$.

    (b) Use Newton's method to approximate the solution.

14. Use Newton's method to estimate the unique solution in the interval $(0, 3\pi/2)$ to the equation $x = \tan x$.

15. Find the smallest positive root of $4\cos x = e^x$.

16. Show by means of a graph that there is only one solution to the equation $x = e^{-x}$ and determine this solution.

* **17.** The derivative $F'(x_n)$ can be approximated by the difference quotient

$$\frac{F(x_n) - F(x_{n-1})}{x_n - x_{n-1}}.$$

Clearly, this quotient converges to the derivative as the difference between successive iterations approaches zero. Using the difference quotient instead of the derivative in Newton's method, derive a second-order difference equa-

tion that defines successive iterates. (*Note:* This defines a method, known as **regula falsi,** for the numerical solution of equations. The method is useful when calculation of derivatives is undesirable.)

* **18.** Using the method indicated in Problem 17, formulate algorithms for calculating square roots and cube roots and calculate $\sqrt{15}$ and $\sqrt[3]{6}$. Compare these computations to those of Newton's method. (Note that two initial choices must be made.)

13.3 Numerical Integration

Consider the problem of evaluating

$$\int_0^1 \sqrt{1 + x^3}\, dx \quad \text{or} \quad \int_0^1 e^{x^2}\, dx. \tag{1}$$

Since both $\sqrt{1 + x^3}$ and e^{x^2} are continuous in [0, 1], we know that both the definite integrals given here exist. They represent the areas under the curves $y = \sqrt{1 + x^3}$ and $y = e^{x^2}$ for x between 0 and 1. The problem is that none of the methods of calculus will enable us to find the antiderivative of $\sqrt{1 + x^3}$ or e^{x^2} because neither antiderivative can be expressed in terms of finitely many of the functions we know.

In fact, there are many continuous functions whose antiderivative cannot be expressed in terms of finitely many functions we know. In those cases we cannot use the fundamental theorem of calculus to evaluate a definite integral. Nevertheless, it may be very important to approximate the value of such an integral. For that reason many methods have been devised to approximate the value of a definite integral to as many decimal places as are deemed necessary. All these techniques come under the heading of **numerical integration.** We shall not discuss this vast subject in great generality here. Rather, we shall introduce two reasonably effective methods for estimating a definite integral: the **trapezoidal rule** and **Simpson's rule.**†

We know that

$$\int_a^b f(x)\, dx = \lim_{\Delta x \to 0} [f(x_1^*)\, \Delta x + f(x_2^*)\, \Delta x + \cdots + f(x_n^*)\, \Delta x]$$

$$= \lim_{\Delta x \to 0} \sum_{i=1}^n f(x_i^*)\, \Delta x. \tag{2}$$

In other words, when the lengths of the subintervals in a partition of [a, b] are small, the sum in the right-hand side of (2) gives us an approximation to the

† One reasonably elementary book that gives a more complete discussion of numerical integration is by S. D. Conte and C. deBoor, *Elementary Numerical Analysis: An Algorithmic Approach,* 3rd ed., McGraw-Hill, New York, 1980.

integral. Here the area is approximated by a sum of areas of rectangles. This is the way the definite integral was approximated in your calculus course. We now develop a more efficient way to approximate the integral.

Trapezoidal Rule

Let f be as in Fig. 1 and let us partition the interval $[a, b]$ by the equally spaced points

$$a = x_0 < x_1 < x_2 < \cdots < x_{i-1} < x_i < \cdots < x_n = b, \tag{3}$$

where $x_i - x_{i-1} = \Delta x = (b - a)/n$. In Fig. 1 we have indicated that the area under the curve can be approximated by the sum of the areas of n trapezoids.

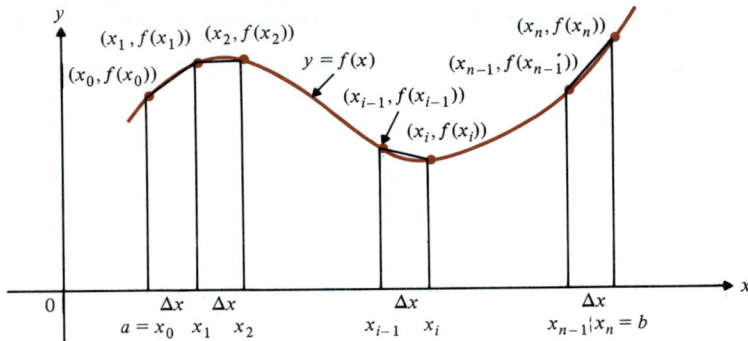

FIGURE 1

One typical trapezoid is sketched in Fig. 2. The area of the trapezoid is the area of the rectangle plus the area of the triangle. But the area of the rectangle is $f(x_i)\,\Delta x$, and the area of the triangle is $\frac{1}{2}[f(x_{i-1}) - f(x_i)]\,\Delta x$, so that

$$\text{area of trapezoid} = \overbrace{f(x_i)\Delta x}^{\text{Area of rectangle}} + \overbrace{\tfrac{1}{2}[f(x_{i-1}) - f(x_i)]\Delta x}^{\text{Area of triangle}}$$

$$= \tfrac{1}{2}[f(x_{i-1}) + f(x_i)]\,\Delta x.$$

Then

$$\int_a^b f(x)\,dx \approx \text{sum of the areas of the trapezoids}$$

$$= \tfrac{1}{2}[f(x_0) + f(x_1)]\,\Delta x + \tfrac{1}{2}[f(x_1) + f(x_2)]\,\Delta x + \cdots$$

$$+ \tfrac{1}{2}[f(x_{n-2}) + f(x_{n-1})]\,\Delta x + \tfrac{1}{2}[f(x_{n-1}) + f(x_n)]\,\Delta x,$$

or

$$\int_a^b f(x)\,dx \approx \tfrac{1}{2}\,\Delta x[f(x_0) + 2f(x_1) + 2f(x_2) + \cdots + 2f(x_{n-1}) + f(x_n)]. \tag{4}$$

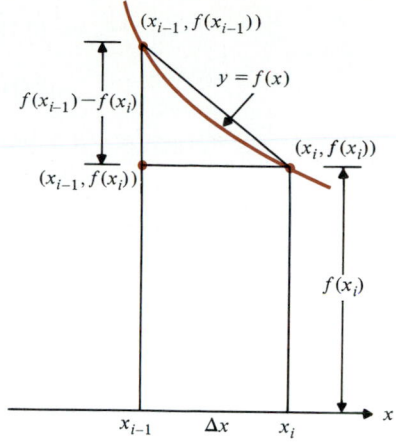

FIGURE 2

The approximation formula (4) is called the **trapezoidal rule** for numerical integration. Note that since $\Delta x = (b - a)/n$, we can write (4) as

Trapezoidal Rule

$$\int_a^b f(x)\,dx \approx \frac{b-a}{2n}[f(x_0) + 2f(x_1) + 2f(x_2) + \cdots + 2f(x_{n-1}) + f(x_n)]. \quad (5)$$

EXAMPLE 1

Estimate $\int_1^2 (1/x)\,dx$ by using the trapezoidal rule first with $n = 5$ and then with $n = 10$.

SOLUTION
(i) Here $n = 5$ and

$$\Delta x = \frac{b-a}{n} = \frac{2-1}{5} = \frac{1}{5} = 0.2.$$

Then $x_0 = 1$, $x_1 = 1.2$, $x_2 = 1.4$, $x_3 = 1.6$, $x_4 = 1.8$, and $x_5 = 2$. From (4)

$$\int_1^2 \frac{1}{x}\,dx \approx \frac{1}{2}\,\Delta x[f(x_0) + 2f(x_1) + 2f(x_2) + 2f(x_3) + 2f(x_4) + f(x_5)]$$

$$= \frac{0.2}{2}\left(\frac{1}{1} + \frac{2}{1.2} + \frac{2}{1.4} + \frac{2}{1.6} + \frac{2}{1.8} + \frac{1}{2}\right)$$

$$\approx 0.1(1 + 1.6667 + 1.4286 + 1.25 + 1.1111 + 0.5)$$

$$= 0.1(6.9564) \approx 0.6956.$$

(ii) Now $n = 10$ and $\Delta x = 1/10 = 0.1$, so that $x_0 = 1$, $x_1 = 1.1, \ldots, x_9 = 1.9$, and $x_{10} = 2$. Thus

$$\int_1^2 \frac{1}{x}\,dx \approx \frac{1}{2}(0.1)\left[1 + \frac{2}{1.1} + \frac{2}{1.2} + \frac{2}{1.3} + \frac{2}{1.4} + \frac{2}{1.5} + \frac{2}{1.6} + \frac{2}{1.7} + \frac{2}{1.8} + \frac{2}{1.9} + \frac{1}{2}\right]$$

$$\approx 0.05[1 + 1.8182 + 1.6667 + 1.5385 + 1.4286 + 1.3333 + 1.25$$

$$+ 1.1765 + 1.1111 + 1.0526 + 0.5]$$

$$= 0.05[13.8755] = 0.6938.$$

We can check our calculations by integrating:

$$\int_1^2 \frac{1}{x}\,dx = \ln x \Big|_1^2 = \ln 2 - \ln 1 = \ln 2 \approx 0.6931.$$

We can see that by increasing the number of intervals, we increase the accuracy of our answer. This, of course, is not surprising. However, we are naturally led to ask what kind of accuracy we can expect by using the trapezoidal rule. In general, two kinds of errors are encountered when we use a numerical method to integrate. The first kind is the kind we have already seen—the error obtained by approximating the curve between the points $(x_{i-1}, f(x_{i-1}))$ and $(x_i, f(x_i))$ by the straight line joining those points. Since we now consider the function at a finite or *discrete* number of points, the error incurred by this approximation is called **discretization error**. However, there is another kind of error we will usually encounter. As we saw in Example 1, we rounded our calculations to four decimal places. Each such "rounding" led to an error in our calculation. The accumulated effect of this rounding is called **round-off error**. Note that as we increase the number of intervals in our calculation, we improve the accuracy of our approximation to the area under the curve. This, evidently, has the effect of reducing the discretization error. On the other hand, an increase in the number of subintervals leads to an increase in the number of computations, which, in turn, leads to an increase in the accumulated round-off error. In fact, there is a delicate balance between these two types of errors and often there is an "optimal" number of intervals to be chosen so as to minimize the total error. Round-off error depends on the type of device used for the computations (pencil and paper, hand calculator, computer, etc.) and will not be discussed further here. However, we can give a formula for estimating the discretization error incurred in using the trapezoidal rule.

Let the sum in (5) be denoted by T and let ϵ_n^T denote the discretization error when n subintervals are used. For example,

Discretization Error

Round-off Error

$$\epsilon_1^T = \int_a^b f(x)\,dx - \frac{b-a}{2}[f(x_0) + f(x_1)], \qquad x_0 = a, x_1 = b,$$

and

$$\epsilon_n^T = \int_a^b f(x)\,dx - \frac{b-a}{2n}[f(x_0) + 2f(x_1) + \cdots + 2f(x_{n-1}) + f(x_n)]$$

$$= \int_a^b f(x)\,dx - \frac{b-a}{2n}\sum_{k=0}^{n-1}[f(x_{k+1}) + f(x_k)], \qquad x_0 = a, x_n = b. \qquad \textbf{(6)}$$

Let F be an antiderivative of f:

$$F(t) = \int_a^t f(x)\,dx. \tag{7}$$

The Taylor series for F is

$$F(a+h) = F(a) + F'(a)h + F''(a)\frac{h^2}{2!} + F'''(a)\frac{h^3}{3!} + \cdots$$

and $F(a) = 0$. Thus, since $F' = f$ by the fundamental theorem of calculus, (7) yields

$$\int_a^{a+h} f(x)\,dx = f(a)h + f'(a)\frac{h^2}{2!} + f''(a)\frac{h^3}{3!} + \cdots. \tag{8}$$

Now consider ϵ_1^T in (5). If $h = b - a$, we have

$$\epsilon_1^T = \int_a^b f(x)\,dx - \frac{b-a}{2}[f(a) + f(a+h)]. \tag{9}$$

Expanding the last term in (9) by its Taylor series and applying (8) to the integral in (9), we obtain

$$\epsilon_1^T = \left[f(a)h + f'(a)\frac{h^2}{2!} + f''(a)\frac{h^3}{3!} + \cdots \right]$$

$$- \frac{h}{2}\left[f(a) + \left(f(a) + f'(a)h + f''(a)\frac{h^2}{2!} + \cdots \right) \right]$$

$$= -\frac{1}{12} f''(a)h^3 + \cdots.$$

Applying a similar analysis to each interval $[x_k, x_{k+1}]$ in (6) yields

$$\int_{x_k}^{x_{k+1}} f(x)\,dx - \frac{h}{2}[f(x_{k+1}) + f(x_k)] = -\frac{1}{12} f''(x_k)h^3 + \cdots,$$

where $h = (b-a)/n$. Hence (6) becomes

$$\epsilon_n^T = -\frac{h^3}{12}\left[\sum_{k=0}^{n-1} f''(x_k) \right] + \cdots, \tag{10}$$

and an application of the intermediate value theorem (see Problem 46) allows us to express the average

$$\frac{1}{n} \sum_{k=0}^{n-1} f''(x_k)$$

by the value of f'' at some point c in the interval $[a, b]$. Thus

$$\epsilon_n^T \approx -\frac{h^3}{12}\left[\sum_{k=0}^{n-1} f''(x_k) \right] = -\frac{h^2}{12}\frac{(b-a)}{n} \sum_{k=0}^{n-1} f''(x_k)$$

$$= -\frac{h^2}{12}(b-a) f''(c). \tag{11}$$

■ **THEOREM 1**

Let f be defined on $[a, b]$ and let f' and f'' be continuous on $[a, b]$. Then there is a number c in (a, b) such that

$$\epsilon_n^T = -\frac{(b-a)}{12}(\Delta x)^2 f''(c).$$ ■

■ **COROLLARY 1: Error Bound for Trapezoidal Rule**

If $|f''(x)| \le M$ for all x in $[a, b]$, then

$$|\epsilon_n^T| \le M\frac{(b-a)^3}{12n^2}.$$ (12)

PROOF
From (11)

$$|\epsilon_n^T| = \frac{b-a}{12}\Delta x^2 |f''(c)| \le \frac{b-a}{12}\left(\frac{b-a}{n}\right)^2 M = \frac{M(b-a)^3}{12n^2}.$$ ■

Remark. The expression in (12) gives the maximum possible value of the error. Often, as in the next example, the actual error will be quite a bit less. You may think of the error bound as a *guarantee* that the error will not be any greater.

EXAMPLE 2

Use the trapezoidal rule to estimate $\int_0^2 e^{x^2}\, dx$ with a maximum error of 1.

SOLUTION
We must choose n large enough so that $|\epsilon_n^T| \le 1$. For $f(x) = e^{x^2}$, we have $f'(x) = 2xe^{x^2}$ and $f''(x) = (2 + 4x^2)e^{x^2}$. Since this function is an increasing function, its maximum over the interval $[0, 2]$ occurs at 2. Then $M = f''(2) = 18e^4 \approx 983$. Hence from (12)

$$|\epsilon_n^T| \le \frac{M(b-a)^3}{12n^2} \le \frac{(983)2^3}{12n^2} \approx \frac{655}{n^2}.$$

We need $655/n^2 \le 1$, or $n^2 \ge 655$, or $n \ge \sqrt{655}$. The smallest n that meets this requirement is $n = 26$. Hence we use the trapezoidal rule with $n = 26$ and $\Delta x = (b - a)/n = 2/26 = 1/13$. We have $x_0 = 0$, $x_1 = 1/13$, $x_2 = 2/13$, $\ldots$, $x_{25} = 25/13$, and $x_{26} = 26/13 = 2$. Then

$$\int_0^2 e^{x^2}\, dx \approx \tfrac{1}{2} \cdot \tfrac{1}{13}[e^0 + 2e^{(1/13)^2} + 2e^{(2/13)^2} + \cdots + 2e^{(25/13)^2} + e^{(26/13)^2}]$$

$$= \tfrac{1}{26}(430.564) \approx 16.560.$$

Values of the function $\int_0^x e^{t^2}\, dt$ have been tabulated. To six decimal places, the correct value of $\int_0^2 e^{t^2}\,dt$ is 16.452627. Thus our answer is actually correct to within 0.11. The next method we discuss will enable us to calculate this integral with greater accuracy and about the same amount of work.

Simpson's Rule

We now derive a second method for estimating a definite integral. Look at the three sketches in Fig. 3. In Fig. 3a the area under the curve $y = f(x)$ over the interval $[x_i, x_{i+2}]$ is approximated by rectangles, where the height of each rectangle is the value of the function at an endpoint of an interval. In Fig. 3b we have depicted the trapezoidal approximation to this area. The "top" of the first trapezoid is the straight line joining the *two* points $(x_i, f(x_i))$ and $(x_{i+1}, f(x_{i+1}))$ and the "top" of the second trapezoid is given in an analogous manner. In Fig. 3c we are approximating the required area by drawing a figure whose "top" is the parabola passing through the *three* points $(x_i, f(x_i))$, $(x_{i+1}, f(x_{i+1}))$, and $(x_{i+2}, f(x_{i+2}))$. As we shall see, this method will give us a better approximation to the area under the curve. First, we need to calculate the area depicted in Fig. 3c.

The proof of the following theorem is left as an exercise (see Problem 45).

■ **THEOREM 2**
The area A_{i+2} bounded by the parabola† passing through the points $(x_i, f(x_i))$, $(x_{i+1}, f(x_{i+1}))$, and $(x_{i+2}, f(x_{i+2}))$, the lines $x = x_i$ and $x = x_{i+2}$, and the x-axis (where $x_{i+1} - x_i = x_{i+2} - x_{i+1} = \Delta x$) is given by

$$A_{i+2} = \tfrac{1}{3}\,\Delta x[f(x_i) + 4f(x_{i+1}) + f(x_{i+2})]. \tag{13}$$

■

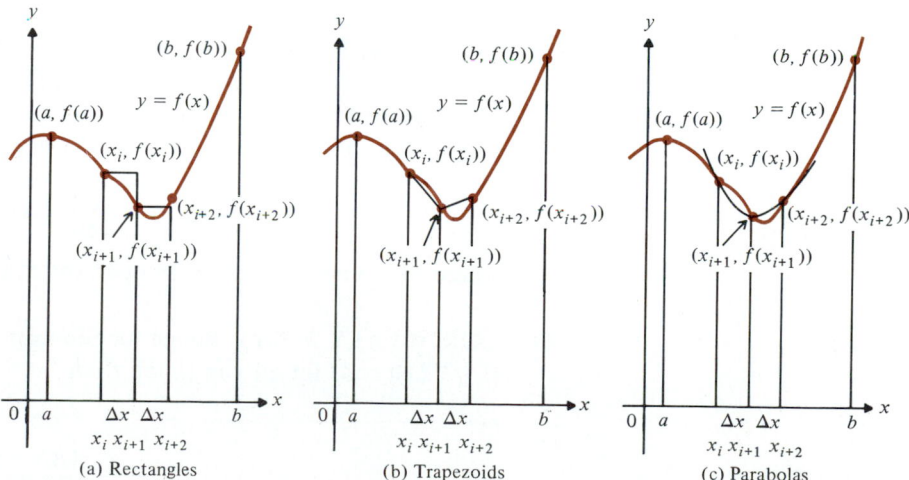

FIGURE 3

(a) Rectangles (b) Trapezoids (c) Parabolas

† In Fig. 3c we assume that f is positive over the interval $[x_i, x_{i+2}]$. The method we are about to develop does *not* require that f be positive. However, the method is easier to motivate if we make this assumption.

Now suppose that the interval $[a, b]$ is divided into $2n$ subintervals of equal lengths $\Delta x = (b - a)/2n$. Then from (13) we have

$$\int_a^b f(x)\,dx \approx A_2 + A_4 + A_6 + \cdots + A_{2n}$$

$$= \tfrac{1}{3}\Delta x[f(x_0) + 4f(x_1) + f(x_2)] + \tfrac{1}{3}\Delta x[f(x_2) + 4f(x_3) + f(x_4)]$$

$$+ \cdots + \tfrac{1}{3}\Delta x[f(x_{2n-2}) + 4f(x_{2n-1}) + f(x_{2n})],$$

or

Simpson's Rule

$$\int_a^b f(x)\,dx \approx \tfrac{1}{3}\Delta x[f(x_0) + 4f(x_1) + 2f(x_2) + 4f(x_3)$$

$$+ 2f(x_4) + \cdots + 2f(x_{2n-2}) + 4f(x_{2n-1}) + f(x_{2n})]. \qquad \textbf{(14)}$$

The approximation in (14) is called **Simpson's rule**† (or the **parabolic rule**) for approximating a definite integral. From (14) we see that there is about the same amount of work needed to estimate an integral by using Simpson's rule as there is by using the trapezoidal rule with the same number of subintervals. However, the discretization error in Simpson's rule is usually a good deal less, as is suggested in the following theorem, whose proof (similar to that of Theorem 1) can be found in the text cited on p. 797.

Remark. It is important to note that the formula for the trapezoidal rule has n subintervals and the formula for Simpson's rule has $2n$ subintervals. With $2n$ subintervals we have n parabolas, the areas under which we must compute. That is why we use the number $2n$ instead of n in using Simpson's rule.

■ **THEOREM 3**
Let f be defined on $[a, b]$ and let f', f'', f''', and $f^{(4)}$ be continuous on $[a, b]$. Then the discretization error ϵ_{2n}^S of Simpson's rule (14), using $2n$ equally spaced subintervals of length $\Delta x = (b - a)/2n$, is given by

$$\epsilon_{2n}^S = -\frac{(b-a)}{180}(\Delta x)^4 f^{(4)}(c),$$

where c is some number in the open interval (a, b).   ■

■ **COROLLARY 2: Error Bound for Simpson's Rule**
If $|f^{(4)}(x)| \leq M$ for all x in $[a, b]$, then

$$|\epsilon_{2n}^S| \leq \frac{M(b-a)^5}{2880n^4}. \qquad \textbf{(15)}$$

† Named after the British mathematician Thomas Simpson (1710–1761), who published the result in his *Mathematical Dissertations on Physical and Analytical Subjects* in 1743.

PROOF

$$|\epsilon_{2n}^S| = \frac{b-a}{180}(\Delta x)^4|f^{(4)}(c)| \le \left(\frac{b-a}{180}\right)\left(\frac{b-a}{2n}\right)^4 M$$

$$\le \frac{M(b-a)^5}{(180)(16n^4)} = \frac{M(b-a)^5}{2880n^4}.$$

∎

EXAMPLE 3

Use Simpson's rule to estimate $\int_1^2 (1/x)\,dx$ by using 10 subintervals. What is the maximum error in your estimate? Compare this error with the exact answer of ln 2.

SOLUTION

Here we have $\Delta x = \frac{1}{10}$ and $n = 5$ ($2n = 10$), so that from (14)

$$\int_1^2 \frac{1}{x}\,dx \approx \frac{1}{30}\left(\frac{1}{1}+\frac{4}{1.1}+\frac{2}{1.2}+\frac{4}{1.3}+\frac{2}{1.4}+\frac{4}{1.5}+\frac{2}{1.6}+\frac{4}{1.7}+\frac{2}{1.8}+\frac{4}{1.9}+\frac{1}{2}\right)$$

$$\approx \frac{1}{30}(1 + 3.636364 + 1.666667 + 3.076923 + 1.428571 + 2.666667$$

$$+ 1.25 + 2.352941 + 1.111111 + 2.105263 + 0.5)$$

$$= \frac{1}{30}(20.794507) \approx 0.693150.$$

To six decimal places, ln 2 = 0.693147, so our answer is very accurate indeed. To calculate the maximum possible error, we first need to calculate $f^{(4)}$. But $f(x) = 1/x$, $f'(x) = -1/x^2$, $f''(x) = 2/x^3$, $f'''(x) = -6/x^4$, and $f^{(4)}(x) = 24/x^5$. Over the interval [1, 2], $|24/x^5| \le 24$, so that $M = 24$. Then we use formula (15) with $M = 24$ and $n = 5$ (so that $2n = 10$) to obtain

$$|\epsilon_{2n}^S| \le \frac{24}{(2880)5^4} = \frac{24}{(2880)(625)} = \frac{24}{1,800,000} \approx 0.0000133.$$

Our actual error is approximately $0.693150 - 0.693147 = 0.000003$, about one-fourth the maximum possible discretization error. Notice that in this example Simpson's rule gives a far more accurate answer than the trapezoidal rule using the same number of subintervals (10) and similar amounts of calculation.

EXAMPLE 4

Use Simpson's rule to estimate $\int_0^2 e^{x^2}\,dx$ with a maximum error of 0.1.

SOLUTION

If $f(x) = e^{x^2}$, we have already calculated (in Example 3) that $f''(x) = (2 + 4x^2)e^{x^2}$. Then $f'''(x) = (12x + 8x^3)e^{x^2}$ and $f^{(4)}(x) = (12 + 48x^2 + 16x^4)e^{x^2}$. This is an increasing function for x in [0, 2], so that $M = f^{(4)}(2) = 460e^4 \approx 25,115$. Since $(b - a)^5 = 2^5 = 32$, we must choose n such that

$$|\epsilon_{2n}^S| \le \frac{M(b-a)^5}{2880n^4} \approx \frac{(25,115)(32)}{2880n^4} \approx \frac{279}{n^4} < 0.1 = \frac{1}{10}.$$

We then need $n^4 > 2790$, or $n > 2790^{1/4}$. The smallest integer n that satisfies this inequality is $n = 8$. Thus to obtain the required accuracy, we use Simpson's rule with $2n = 16$ subintervals. Then $\Delta x = (b - a)/2n = \frac{2}{16} = \frac{1}{8}$, and we have

$$\int_0^2 e^{x^2}\, dx \approx \frac{1}{3} \cdot \frac{1}{8}(e^0 + 4e^{(1/8)^2} + 2e^{(2/8)^2} + 4e^{(3/8)^2} + \cdots + 2e^{(14/8)^2}$$

$$+ 4e^{(15/8)^2} + e^{(16/8)^2}) \approx \frac{1}{24}(395.0118) \approx 16.4588.$$

Since the correct value is 16.452627 (correct to six decimal places), our answer is really correct to within 0.007.

Notice how in the calculation of $\int_0^2 e^{x^2}\, dx$, Simpson's rule gives us more accuracy with fewer calculations than the trapezoidal rule does.

There are many other methods that can be used to approximate definite integrals. For example, there are methods in which the points $x_0, x_1, \ldots, x_n$ are *not* equally spaced. One such method is called **Gaussian quadrature.** We shall not discuss this very useful method here except to note that it can be found in any introductory book on numerical analysis (see the references cited in Section 13.1).

The problems at the end of this section can all be done reasonably quickly by using a scientific hand calculator. For more accuracy it may be necessary to evaluate a function at hundreds, or even thousands, of points. This problem is a manageable one only if you have access to a high-speed computer or a programmable calculator. If, in fact, you do have such access, you should write a computer program to estimate an integral using Simpson's rule and then use it to calculate each of the integrals in the problem set to at least six decimal places of accuracy.

Problems 13.3

In Problems 1–15, estimate the given definite integral by using (i) the trapezoidal rule and (ii) Simpson's rule over the given number of intervals. Then (iii) use error formula (12) to obtain a bound for the error of the trapezoidal approximation and (iv) use error formula (15) to obtain a bound for the error of the approximation using Simpson's rule. Then (v) calculate the integral exactly. Finally, compare the actual errors in your computations with the maximum possible errors found in (iii) and (iv).

1. $\int_0^1 x\, dx$; 4 intervals

2. $\int_{-2}^2 x\, dx$; 6 intervals

3. $\int_0^1 x^2\, dx$; 4 intervals

4. $\int_0^1 e^x\, dx$; 4 intervals

5. $\int_0^2 e^x\, dx$; 6 intervals

6. $\int_1^2 \frac{1}{x^2}\, dx$; 6 intervals

7. $\int_1^2 \sqrt{x}\, dx$; 8 intervals

8. $\int_0^3 \frac{1}{\sqrt{1+x}}\, dx$; 6 intervals

9. $\int_0^{\pi/2} \sin x\, dx$; 4 intervals

10. $\int_0^{\pi/2} \cos x\, dx$; 4 intervals

11. $\int_0^{\pi/3} \tan x\, dx$; 8 intervals

12. $\int_2^5 \frac{x}{\sqrt{x^2+1}}\, dx$; 6 intervals

13. $\int_1^e \ln x\, dx$; 6 intervals

14. $\int_0^1 \frac{1}{1+x^2}\, dx$; 10 intervals

15. $\int_0^1 e^x \sin x\, dx$; 8 intervals

In Problems 16–30, approximate the given integral by using (i) the trapezoidal rule and (ii) Simpson's rule, with the indicated number of subintervals.

16. $\int_0^1 \sqrt{x + x^2}\, dx$; 4 intervals

17. $\int_1^2 \frac{\sin x}{x}\, dx$; 6 intervals

18. $\int_1^3 \frac{x}{\sin x}\, dx$; 6 intervals

19. $\int_0^1 e^{\sqrt{x}}\, dx$; 6 intervals

20. $\int_0^1 e^{x^3}\, dx$; 8 intervals

21. $\int_0^\pi \sin x^2\, dx$; 8 intervals

22. $\int_1^2 \sqrt{\ln x}\, dx$; 10 intervals

23. $\int_{-1}^1 e^{-x^2}\, dx$; 10 intervals

24. $\int_0^1 \sqrt{1 + x^3}\, dx$; 10 intervals

25. $\int_0^1 \frac{dx}{\sqrt{1 + x^3}}$; 10 intervals

26. $\int_0^1 xe^{x^3}\, dx$; 10 intervals

27. $\int_0^1 \ln(1 + e^x)\, dx$; 8 intervals

28. $\int_1^3 \frac{x^2}{\sqrt[3]{1 + x}}\, dx$; 10 intervals

29. $\int_0^1 \sinh x^2\, dx$; 10 intervals

30. $\int_0^{\pi/4} \sin(\tan x)\, dx$; 10 intervals

In Problems 31–36, find a bound for the discretization error by using the trapezoidal rule and Simpson's rule.

31. The integral of Problem 20

32. The integral of Problem 21

33. The integral of Problem 23

34. The integral of Problem 24

35. The integral of Problem 26

36. The integral of Problem 27

37. The integral $(1/\sqrt{2\pi}) \int_{-a}^a e^{-x^2/2}\, dx$ is very important in probability theory. Using Simpson's rule, estimate $(1/\sqrt{2\pi}) \int_{-1}^1 e^{-x^2/2}\, dx$ with an error of less than 0.01 [*Hint:* Show that $\int_{-a}^a e^{-x^2/2}\, dx = 2 \int_0^a e^{-x^2/2}\, dx$.]

38. Estimate $(1/\sqrt{2\pi}) \int_{-5}^5 e^{-x^2/2}\, dx$ with an error of less than 0.01. [*Hint:* See Problem 37.]

* 39. (a) Estimate $(1/\sqrt{2\pi}) \int_{-50}^{50} e^{-x^2/2}\, dx$ with an error of less than 0.1.

   (b) Can you guess what happens to $(1/\sqrt{2\pi}) \int_{-N}^N e^{-x^2/2}\, dx$ as N grows without bound?

40. We know that

$$\int_0^1 \frac{dx}{1 + x^2} = \tan^{-1} x \Big|_0^1 = \frac{\pi}{4}.$$

Thus

$$\pi = 4 \int_0^1 \frac{dx}{1 + x^2}.$$

   (a) With Simpson's rule how many subintervals does it take to estimate π with an error of less than 0.0001?

   (b) Using this number of subintervals, give your estimate of π.

41. How many subintervals would it take to estimate ln 2 by using the trapezoidal rule applied to the integral $\int_1^2 (1/x)\, dx$ with an error of less than 10^{-10}?

42. How many subintervals would it take to perform the estimate in Problem 41 by using Simpson's rule?

43. Estimate $\int_{1/2}^1 J_{1/2}(x)\, dx$ where $J_{1/2}(x) = \sqrt{\dfrac{2}{\pi x}} \sin x$ is the Bessel function of order $\frac{1}{2}$.

44. Show that Simpson's rule provides the exact answer for $\int_a^b p_3(x)\, dx$ if p_3 is a polynomial of degree three or less.

45. Prove Theorem 2. [*Hint:* Show that both the shaded area in Fig. 3c and $\frac{1}{3} \Delta x[f(x_i) + 4f(x_{i+1}) + f(x_{i+2})]$ are equal to $a(6x_i^2 + 12x_i\, \Delta x + 8\, \Delta x)^2 + b(6x_i + 6\, \Delta x) + 6c$ if the parabola is written as $y = ax^2 + bx + c$.]

46. Show that if f'' is continuous on $[a, b]$ and $a = x_0 < x_1 < \cdots < x_n = b$, then $\dfrac{1}{n} \sum_{k=0}^{n-1} f''(x_k) = f''(c)$ for some c in $[a, b]$.

(Optional)　　13.4 Polynomial Interpolation

Taylor's theorem shows us how to approximate a differentiable function by a polynomial. In some applications it is more important to find a polynomial which is *equal to* the function f at a certain number of specified points in some interval over which f is defined. Such a polynomial is said to **interpolate** the function f between these points. For example, in Fig. 1 the fourth-degree polynomial P_4 interpolates the function f at the points x_1, x_2, x_3, and x_4. The existence of an interpolating polynomial is given in the theorem below.

■ THEOREM 1

Let the function f be defined on some interval $[a, b]$ and let $x_0, x_1, x_2, \ldots, x_n$ be $n + 1$ *distinct points* of $[a, b]$. Then there exists a *unique* polynomial $q(x)$ of degree less than or equal to n such that

$$q(x_k) = f(x_k) \quad \text{for} \quad k = 0, 1, 2, \ldots, n. \tag{1}$$

This polynomial is called the **Lagrange†** **interpolating polynomial.**

Lagrange Interpolating Polynomial

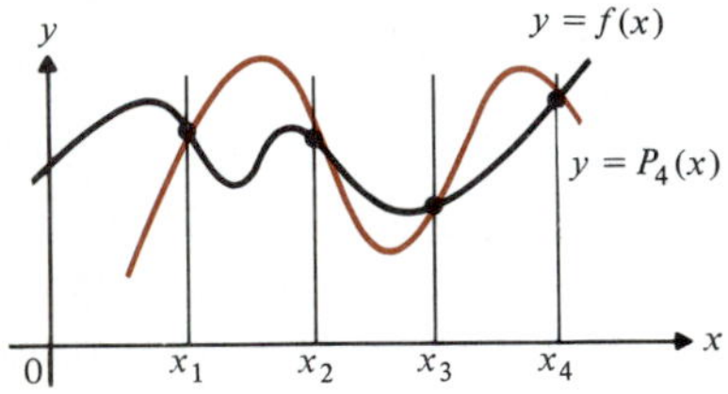

FIGURE 1

PROOF
Define the polynomials $r_k(x)$, $k = 0, 1, \ldots, n$, by

$$r_k(x) = \frac{(x - x_0)(x - x_1) \cdots (x - x_{k-1})(x - x_{k+1}) \cdots (x - x_n)}{(x_k - x_0)(x_k - x_1) \cdots (x_k - x_{k-1})(x_k - x_{k+1}) \cdots (x_k - x_n)}. \tag{2}$$

We note the following facts about each r_k:

(i) r_k is a polynomial of degree n [the denominator of the quotient in equation (2) is a constant],
(ii) $r_k(x_k) = 1$, and
(iii) $r_k(x_j) = 0$ if $j \neq k$.

We now define

$$q(x) = f(x_0)r_0(x) + f(x_1)r_1(x) + \cdots + f(x_n)r_n(x). \tag{3}$$

Since $q(x)$ is a sum of polynomials of degree n, $q(x)$ is of degree $\leq n$. Moreover, using facts (ii) and (iii) above, we see that $q(x_k) = f(x_k)$ for each k, $k = 0, 1, 2, \ldots, n$, and this proves that $q(x)$ is *an* interpolating polynomial. We still must prove that this polynomial is unique. Suppose that $q(x)$ and $s(x)$ are two interpolating polynomials. Let $u(x) = q(x) - s(x)$. Then $u(x)$ is a polynomial of degree $\leq n$ and $u(x_k) = q(x_k) - s(x_k) = f(x_k) - f(x_k) = 0$ for $k = 0, 1, 2, \ldots, n$. But, from the fundamental theorem of algebra (proved in Section 16.6) any nonzero polynomial of degree n has at most n zeros. Since $u(x)$ has the $n + 1$ distinct zeros $x_0, x_1, \ldots, x_n$, we must conclude that $u(x) = 0$ for every x so that $q(x) = s(x)$, thereby proving the uniqueness of the interpolating polynomial. ■

† See the biography of Lagrange on p. 809.

JOSEPH LOUIS LAGRANGE
(1736–1813)

The Granger Collection

Joseph Louis Lagrange was one of the two greatest mathematicians of the eighteenth century—the other being Leonhard Euler (see page 57). Born in 1736 in Turin, Italy, Lagrange was the youngest of eleven children of French and Italian parents and the only one to survive to adulthood. Educated in Turin, Joseph Louis became a professor of mathematics in the military academy there when he was still quite young.

When Euler left his post at the court of Frederick the Great in Berlin in 1766, he recommended that Lagrange be appointed his successor. Accepting Euler's advice, Frederick wrote to Lagrange that "the greatest king in Europe" wished to invite to his court "the greatest mathematician in Europe." Lagrange accepted and remained in Berlin for twenty years. Afterwards, he accepted a post at the Ecole Polytechnique in France.

Lagrange had a deep influence on nineteenth and twentieth century mathematics. He is perhaps best known as the first great mathematician to attempt to make calculus mathematically rigorous. His major work in this area was his 1797 paper "Théorie des fonctions analytiques contenant les principes du calcul différentiel." In this work, Lagrange tried to make calculus more logical—rather than more useful. His key idea was to represent a function $f(x)$ by a Taylor series. For example, we can write $1/(1 - x) = 1 + x + x^2 + \cdots + x^n + \cdots$ (a result that can be obtained by long division). Lagrange multiplied the coefficient of x^n by $n!$ and called the result the nth *derived function* of $1/(1 - x)$ at $x = 0$. This is the origin of the word *derivative.* The notation $f'(x), f''(x), \ldots$ was first used by Lagrange as was the form of the remainder term. Beginning in the 1750s, he invented the calculus of variations. He made significant contributions to ordinary differential equations, partial differential equations, numerical analysis, number theory, and algebra. In 1788 he published his *Mécanique Analytique,* which contained the equations of motion of a dynamical system. Today these equations are known as *Lagrange's equations.*

Lagrange lived in France during the French revolution. In 1790 he was placed on a committee to reform weights and measures and later became the head of a related committee that, in 1799, recommended the adoption of the system that we know today as the *metric system.* Despite his work for the revolution, however, Lagrange was disgusted by its cruelties. After the great French chemist Lavoisier was guillotined, Lagrange exclaimed, "It took the mob only a moment to remove his head; a century will not suffice to reproduce it."

In his later years, Lagrange was often lonely and depressed. When he was 56, the 17-year-old daughter of his friend the astronomer P. C. Lemonier was so moved by his unhappiness that she proposed to him. The resulting marriage apparently turned out to be ideal for both.

TABLE 1

k	0	1	2	3
x_k	1	2	3	4
$f(x_k)$	1	$\frac{1}{2}$	$\frac{1}{3}$	$\frac{1}{4}$

EXAMPLE 1

Find the polynomial of degree ≤ 3 which interpolates the function $y = 1/x$ at the values 1, 2, 3, and 4.

SOLUTION
We refer to Table 1. Using (2) we have

$$r_0(x) = \frac{(x-x_1)(x-x_2)(x-x_3)}{(x_0-x_1)(x_0-x_2)(x_0-x_3)} = \frac{(x-2)(x-3)(x-4)}{(1-2)(1-3)(1-4)} = -\frac{1}{6}(x-2)(x-3)(x-4)$$

$$r_1(x) = \frac{(x-x_0)(x-x_2)(x-x_3)}{(x_1-x_0)(x_1-x_2)(x_1-x_3)} = \frac{(x-1)(x-3)(x-4)}{(2-1)(2-3)(2-4)} = \frac{1}{2}(x-1)(x-3)(x-4)$$

$$r_2(x) = \frac{(x-x_0)(x-x_1)(x-x_3)}{(x_2-x_0)(x_2-x_1)(x_2-x_3)} = \frac{(x-1)(x-2)(x-4)}{(3-1)(3-2)(3-4)} = -\frac{1}{2}(x-1)(x-2)(x-4)$$

$$r_3(x) = \frac{(x-x_0)(x-x_1)(x-x_2)}{(x_3-x_0)(x_3-x_1)(x_3-x_2)} = \frac{(x-1)(x-2)(x-3)}{(4-1)(4-2)(4-3)} = \frac{1}{6}(x-1)(x-2)(x-3).$$

Then the interpolating polynomial is of third degree and is given by

$$q(x) = r_0(x)f(x_0) + r_1(x)f(x_1) + r_2(x)f(x_2) + r_3(x)f(x_3)$$

$$= -\frac{1}{6}(x-2)(x-3)(x-4) + \frac{1}{4}(x-1)(x-3)(x-4)$$

$$- \frac{1}{6}(x-1)(x-2)(x-4) + \frac{1}{24}(x-1)(x-2)(x-3)$$

$$= -\frac{1}{24}(x^3 - 10x^2 + 35x - 50).$$

(We've spared you the missing algebraic steps.) This is illustrated in Fig. 2.

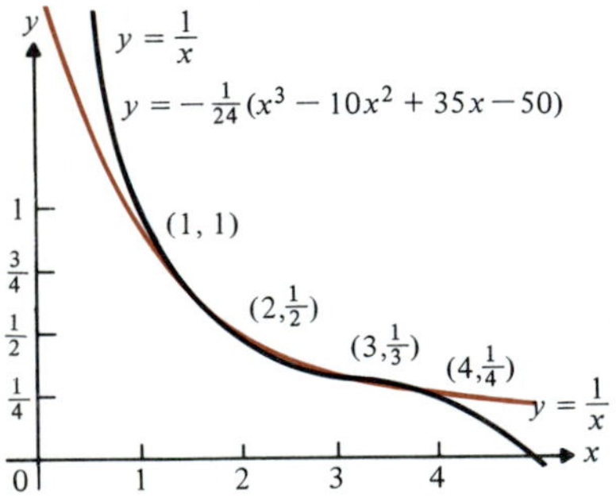

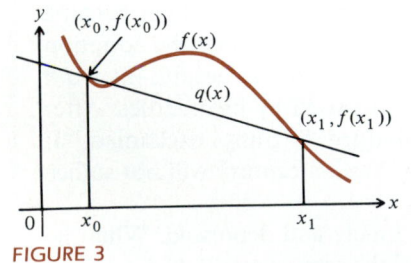

$y = \frac{1}{x}$

$y = -\frac{1}{24}(x^3 - 10x^2 + 35x - 50)$

$(1, 1)$

$(2, \frac{1}{2})$

$(3, \frac{1}{3})$ $(4, \frac{1}{4})$

$y = \frac{1}{x}$

FIGURE 2

Linear Interpolation

When $n = 1$ in equations (2) and (3) we obtain the formula for **linear interpolation**. In this case we interpolate at the two points x_0 and x_1. Then

$$r_0(x) = \frac{x - x_1}{x_0 - x_1}, \qquad r_1(x) = \frac{x - x_0}{x_1 - x_0},$$

and

$$q(x) = f(x_0)\left(\frac{x - x_1}{x_0 - x_1}\right) + f(x_1)\left(\frac{x - x_0}{x_1 - x_0}\right) = \frac{(x_1 - x)f(x_0) + (x - x_0)f(x_1)}{x_1 - x_0}. \qquad (4)$$

This is seen to be the equation of the straight line passing through the points $(x_0, f(x_0))$ and $(x_1, f(x_1))$. The situation is illustrated in Fig. 3.

FIGURE 3

$(x_0, f(x_0))$

$f(x)$

$q(x)$

$(x_1, f(x_1))$

x_0 x_1

EXAMPLE 2

Using the values $\sin 35° \approx 0.5736$ and $\sin 40° \approx 0.6428$, use linear interpolation to find an approximate value for $\sin 37°$.

SOLUTION

The equation of the straight line joining the points $(x_0, f(x_0)) = (35°, 0.5736)$ and $(x_1, f(x_1)) = (40°; 0.6428)$ is given by

$$q(x) = \frac{(40 - x)(0.5736) + (x - 35)(0.6428)}{40 - 35}$$

$$= (0.01384)x + 0.0892.$$

Substituting $x = 37°$ yields

$$q(37°) = (0.01384)(37) + (0.0892) \approx 0.6013.$$

The correct value of $\sin 37°$ is 0.6018 to four decimal places.

The preceding examples lead to the question—how good an approximation is given by the interpolating polynomial? Intuitively, it seems that the farther we are from an interpolating point [a point of the form $(x_k, f(x_k))$], the greater the distance between the interpolating polynomial and the original function. A precise expression for the error is given by the following theorem.

■ **THEOREM 2**

Suppose that $f, f', \ldots, f^{(n+1)}$ are continuous over the interval $[a, b]$. Let x be any number in $[a, b]$ and let I_x denote the smallest closed interval containing the numbers $x, x_0, x_1, \ldots, x_n$. Then there is a number x^* in I_x such that

$$f(x) - q(x) = \frac{Q(x)}{(n+1)!} f^{(n+1)}(x^*), \tag{5}$$

where $Q(x)$ is defined by $Q(x) = (x - x_0)(x - x_1)(x - x_2) \cdots (x - x_n)$ and $q(x)$ is the Lagrange interpolating polynomial for $f(x)$ which interpolates at the points $(x_0, f(x_0)), (x_1, f(x_1)), \ldots, (x_n, f(x_n))$.

Remark. We cannot use formula (5) to calculate the actual "error" $f(x) - q(x)$ since the number x^* is not known. However, as with Taylor's formula, it is often possible to use the formula to obtain a bound on the maximum difference between a function and its interpolating polynomial.

PROOF OF THEOREM 2

If $x = x_k$ for some k, then there is nothing to prove as both sides of (5) are equal to zero. Therefore suppose that $x \neq x_k$ and define $\alpha = [f(x) - q(x)]/Q(x)$ (since x is fixed, α is a constant). Define the function $S(t)$ by

$$S(t) = f(t) - q(t) - \alpha Q(t). \tag{6}$$

If $t = x_k$ for some k, then

$$S(x_k) = f(x_k) - q(x_k) - \alpha Q(x_k) = 0.$$

Similarly,

$$S(x) = f(x) - q(x) - \left[\frac{f(x) - q(x)}{Q(x)}\right] Q(x) = 0.$$

Recall that Rolle's theorem states that if f and f' are continuous on $[a, b]$ and if $f(a) = f(b) = 0$, then there is a number c in (a, b) such that $f'(c) = 0$. Now, the function S is zero for at least $n + 2$ numbers in the interval I_x. Rolle's theorem implies that S' has at least $n + 1$ zeros in that interval (one zero of S' between every two consecutive zeros of S). Then S'' has at least n zeros, S''' has at least $n - 1$ zeros and, finally, $S^{(n+1)}$ has at least one zero at a number, which we shall call x^*, in the interval (a, b). If we differentiate both sides of (6) $n + 1$ times, we obtain

$$S^{(n+1)}(t) = f^{(n+1)}(t) - q^{(n+1)}(t) - \alpha Q^{(n+1)}(t). \tag{7}$$

But since $Q(t) = t^{n+1} + \cdots$, we have $Q^{(n+1)}(t) = (n + 1)!$. Thus substituting $t = x^*$ in (7), we have

$$S^{(n+1)}(x^*) = 0 = f^{(n+1)}(x^*) - q^{(n+1)}(x^*) - \alpha(n + 1)!.$$

But $q^{(n+1)}(x) = 0$ since $q(x)$ is a polynomial of degree at most n and we obtain

$$f^{(n+1)}(x^*) = \alpha(n + 1)! = \frac{f(x) - q(x)}{Q(x)}(n + 1)!.$$

Finally, after rearranging terms, we find that

$$f(x) - q(x) = \frac{Q(x)}{(n + 1)!} f^{(n+1)}(x^*),$$

which is what we want to prove.   ■

EXAMPLE 3

If there is only one interpolating point $(x_0, f(x_0))$, then $n = 0$, $q(x) = f(x_0)$, $Q(x) = (x - x_0)$ and, from (5), we obtain

$$f(x) - f(x_0) = (x - x_0)f'(x^*), \tag{8}$$

where x^* is in the smallest interval containing x and x_0. This is one of the intervals $[x_0, x]$ or $[x, x_0]$. Does this look familiar? It should, for it is simply the mean value theorem of differential calculus.

EXAMPLE 4

If there are two interpolating points we obtain the linear interpolating polynomial given by (4). Then the error is given by

$$f(x) - q(x) = \frac{(x - x_0)(x - x_1)}{2} f''(x^*).$$

Suppose that $|f''(x)| \leq M$ for some constant M. Define the function h by

$$h(x) = \frac{Q(x)}{2} = \frac{(x - x_0)(x - x_1)}{2} = \frac{x^2 - (x_0 + x_1)x + x_0 x_1}{2}$$

and we see that $h'(x) = x - [(x_0 + x_1)/2]$ which is zero when $x = (x_0 + x_1)/2$. In the interval $[x_0, x_1]$, $h(x) \leq 0$. Since $h''(x) = 1 > 0$, $h(x)$ has a minimum at $x = (x_0 + x_1)/2$. Since $h(x)$ is negative, this means that $|h(x)|$ has a maximum at

$(x_0 + x_1)/2$ in the interval $[x_0, x_1]$. If $x = (x_0 + x_1)/2$, then $h(x) = -(x_1 - x_0)^2/8$ and we obtain the bound

$$|f(x) - q(x)| \le \frac{(x_1 - x_0)^2}{8} M. \tag{9}$$

EXAMPLE 5

In Example 2, we have $x_0 = 35° \approx 0.61087$ rad and $x_1 = 40° \approx 0.69813$ rad. Since $f(x) = \sin x, f''(x) = -\sin x$ and $|f''(x)| \le 1$. Then from (9) we obtain

$$|f(x) - q(x)| \le \frac{(0.69813 - 0.61087)^2}{8} \approx 0.00095.$$

In our calculation the actual error was approximately $0.6018 - 0.6013 = 0.0005$.

EXAMPLE 6: Error in Cubic Interpolation

In Example 1 we approximated the function $f(x) = 1/x$ by a cubic polynomial $q(x) = -(1/24)(x^3 - 10x^2 + 35x - 50)$. How well does $q(x)$ approximate $f(x)$ over the interval $[1, 4]$?

SOLUTION

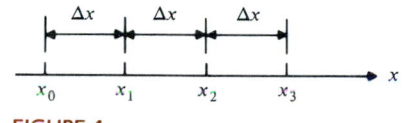

FIGURE 4

We will derive a formula for an upper bound on the maximum error of cubic interpolation. Let the four numbers $x_0, x_1, x_2,$ and x_3 be equally spaced as in Fig. 4. Here $Q(x) = (x - x_0)(x - x_1)(x - x_2)(x - x_3)$. It is not difficult to show that the maximum of $|Q(x)|$ over the interval $[x_0, x_3]$ occurs at $x^* = x_0 + 0.381966\Delta x$ and that $|Q(x^*)| = (\Delta x)^4$ (see Problem 14). Then if M denotes the maximum value of $f^{(4)}(x)$ on $[x_0, x_3]$, we obtain

$$|f(x) - q(x)| \le \frac{1}{4!} (\Delta x)^4 M. \tag{10}$$

Returning to our example, since $f(x) = 1/x$, we have $f'(x) = -(1/x^2)$, $f''(x) = 2/x^3$, $f'''(x) = -(6/x^4)$ and $f^{(4)}(x) = 24/x^5$ which is bounded by $M = 24$ on the interval $[1, 4]$. Since $\Delta x = 1$ here, we see that

$$|f(x) - q(x)| \le 1.$$

For example, if $x = \frac{3}{2}$, then $f(\frac{3}{2}) = \frac{2}{3} \approx 0.6667$ and $q(\frac{3}{2}) \approx 0.6927$ with an error of 0.026. If $x = \frac{5}{2}$, then $f(x) = \frac{2}{5} = 0.4$ and $q(\frac{5}{2}) \approx 0.3906$ with an error of 0.0094. Here the actual error is much less than the maximum error as calculated by the result of Theorem 2. Actually, the error bound obtained here is unusually large. If we had instead chosen $\Delta x = 0.1$, we would have obtained the error bound

$$|f(x) - q(x)| \le (0.1)^4 = 0.0001.$$

Thus the cubic polynomial which interpolates the function $f(x) = 1/x$ at the numbers $1, 1.1, 1.2,$ and 1.3 gives values of $1/x$ correct to at least 3 decimal places for any x in the interval $[1, 1.3]$.

Problems 13.4

In Problems 1–10 find the Lagrange polynomial which interpolates the given function at the given points. Graph the function and its interpolating polynomial over the interval of interpolation.

1. $f(x) = \sin x$; $x = 0$; $x = \pi/2$
2. $f(x) = \ln x$; $x = 1$; $x = 3$
3. $f(x) = \sqrt{1+x}$; $x = 0$; $x = 3$; $x = 8$
4. $f(x) = 1/x^2$; $x = 1$; $x = 2$; $x = 3$
5. $f(x) = e^x$; $x = -1$; $x = 0$; $x = 1$
6. $f(x) = \sin x$; $x = 0$; $x = \pi/6$; $x = \pi/3$; $x = \pi/2$
7. $f(x) = \tan x$; $x = 0$; $x = \pi/6$; $x = \pi/4$
8. $f(x) = x^4$; $x = 0$; $x = 1$
9. $f(x) = x^4$; $x = 0$; $x = \frac{1}{2}$; $x = 1$
10. $f(x) = x^4$; $x = 0$; $x = \frac{1}{3}$; $x = \frac{2}{3}$; $x = 1$.

11. Given that $\ln 2 \approx 0.693147$ and that $\ln 3 \approx 1.09861$, use linear interpolation to estimate $\ln 2.3$. What is the maximum error of your estimate?

12. In addition to the information in Problem 11, assume that $\ln 2.5 \approx 0.916291$. Estimate $\ln 2.3$ using quadratic interpolation. What is the maximum error of your estimate?

13. Given that $\ln \frac{7}{3} \approx 0.847298$ and $\ln \frac{8}{3} \approx 0.980829$, and using the information in Problem 11, find the cubic polynomial which interpolates $\ln x$ at the numbers 2, $2\frac{1}{3}$, $2\frac{2}{3}$, and 3. Use this to estimate $\ln 2.3$. What is the maximum error now?

14. Show that the maximum value of
$$|(x - x_0)(x - x_1)(x - x_2)(x - x_3)|$$
occurs at $x^* = x_0 + 0.381996\Delta x$ and is equal to $(\Delta x)^4$ assuming that $x_1 = x_0 + \Delta x$, $x_2 = x_0 + 2\,\Delta x$ and $x_3 = x_0 + 3\,\Delta x$.

* 15. **Quartic interpolation.** Consider the problem of interpolating the function f at the five equally spaced values $x_k = x_0 + k\,\Delta x$, $k = 0, 1, 2, 3, 4$.

(a) Find the maximum value over the interval $[x_0, x_4]$ of $|Q(x)| = |(x - x_0)(x - x_1)(x - x_2)(x - x_3)(x - x_4)|$.
(b) If $|f^{(5)}(x)| \leq M$ on $[x_0, x_4]$, find a bound on the interpolation error $|f(x) - q(x)|$ on $[x_0, x_4]$ where $q(x)$ is the fourth-degree Lagrange interpolating polynomial.

16. Given that $\sqrt{\frac{5}{4}} \approx 1.11803$, $\sqrt{\frac{3}{2}} \approx 1.22474$, $\sqrt{\frac{7}{4}} \approx 1.32288$, and $\sqrt{2} \approx 1.41421$, find the quartic (fourth-degree) polynomial that interpolates $f(x) = \sqrt{x}$ over the interval $[1, 2]$ at the five equally spaced points 1, $\frac{5}{4}$, $\frac{3}{2}$, $\frac{7}{4}$, 2.

17. Use the polynomial obtained in Problem 16 to estimate $\sqrt{1.2}$ and $\sqrt{1.7}$. What is an upper bound on the error in each of these estimates?

* 18. Let $q(x)$ be the polynomial that interpolates $f(x)$ at the $n + 1$ points $x_0, x_1, \ldots, x_n$. Show that if $x \neq x_k$, $q(x)$ can be written
$$q(x) = Q(x)\left[\frac{f(x_0)}{(x - x_0)Q'(x_0)} + \frac{f(x_1)}{(x - x_1)Q'(x_1)} \right.$$
$$\left. + \cdots + \frac{f(x_n)}{(x - x_n)Q'(x_n)} \right].$$

19. Use the result of Problem 18 and L'Hôpital's rule to show that for each k, $\lim_{x \to x_k} q(x) = f(x_k)$.

20. Show that if f is a polynomial of degree n, then the Lagrange polynomial which interpolates f at $n + 1$ points is f itself.

21. Find the interpolating polynomials which interpolate $f(x) = \sin x$ at 2, 3, and 4 equally spaced points in the interval $[0, \pi/2]$. Calculate an error bound for each of these polynomials. Then find actual errors by using the polynomials to estimate $\sin \frac{1}{2}$ and $\sin 1$.

13.5   Euler's Methods for First-Order Differential Equations

In Section 1.8 we described Euler's method of approximating the solution of the first-order initial value problem

$$\frac{dy}{dx} = f(x, y), \qquad y(x_0) = y_0, \tag{1}$$

at the points

$$x_0, x_1 = x_0 + h, x_2 = x_0 + 2h, \ldots, x_n = x_0 + nh,$$

for h some nonzero real number. **Euler's method,** defined by the iterative formula (see Fig. 1)

Euler's Method

$$y_{n+1} = y_n + hf(x_n, y_n) \tag{2}$$

with $y_0 = y(x_0)$, approximates the solution by following the tangent to the solution curve passing through (x_n, y_n) for a small horizontal distance. The following example illustrates how the work can be organized in tabular form.

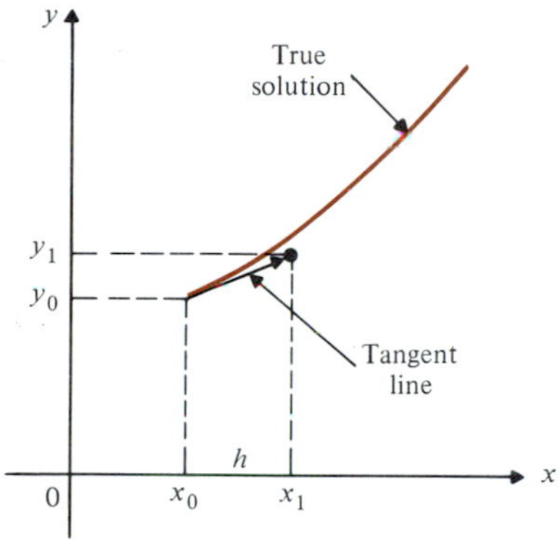

FIGURE 1

EXAMPLE 1

Find an approximate value for the solution of the initial value problem

$$y' = x + y^2, \qquad y(1) = 0,$$

at $x = 1, 1.1, 1.2, 1.3, 1.4, 1.5.$

SOLUTION

We arrange our work in columns, in Table 1: the first column contains the values of x at $1, 1.1, \ldots, 1.5$; the second, the values of y beginning with the initial condition; the

TABLE 1   EULER METHOD WITH $h = 0.1$, FOR
$y' = x + y^2$, $y(0) = 1$

x_n	y_n	$f(x_n, y_n)$	$y_n + hf(x_n, y_n)$
1.00	0.00	1.00	0.10
1.10	0.10	1.11	0.21
1.20	0.21	1.24	0.34
1.30	0.34	1.41	0.48
1.40	0.48	1.63	0.64
1.50	0.64		

TABLE 2   EULER METHOD WITH $h = 0.05$

x_n	y_n	$f(x_n, y_n)$	$y_n + hf(x_n, y_n)$
1	0	1	0.05
1.05	0.05	1.0525	0.10263
1.1	0.10263	1.1105	0.15815
1.15	0.15815	1.1750	0.21690
1.2	0.21690	1.2470	0.27925
1.25	0.27925	1.3280	0.34565
1.3	0.34565	1.4195	0.41663
1.35	0.41663	1.5236	0.49281
1.4	0.49281	1.6429	0.57495
1.45	0.57495	1.7806	0.66398
1.5	0.66398		

third, the computed value of $f(x, y) = x + y^2$ for the given x and y in that row; and the last, the computed value of y_{n+1} according to equation (2). The value in the last column is then transferred to the y entry in the next row to compute the entries in the third and fourth columns.

As we saw in Section 1.8, we can usually reduce the discretization error by reducing the step size. For example, if instead of a step size of $h = 0.1$ we used $h = 0.05$, our approximate solution would be that shown in Table 2, while $h = 0.025$ would yield Table 3. Observe that in each case, the approximate value of $y(1.5)$ increases. This is the behavior that one would anticipate if the solution curves in the direction field are

TABLE 3   EULER METHOD WITH $h = 0.025$

x_n	y_n	$f(x_n, y_n)$	$y_n + h \cdot f(x_n, y_n)$
1.000	0.00000	1.0000	0.02500
1.025	0.02500	1.0256	0.05064
1.050	0.05064	1.0526	0.07695
1.075	0.07695	1.0809	0.10398
1.100	0.10398	1.1108	0.13175
1.125	0.13175	1.1424	0.16031
1.150	0.16031	1.1757	0.18970
1.175	0.18970	1.2110	0.21997
1.200	0.21997	1.2484	0.25118
1.225	0.25118	1.2881	0.28339
1.250	0.28339	1.3303	0.31664
1.275	0.31664	1.3753	0.35103
1.300	0.35103	1.4232	0.38661
1.325	0.38661	1.4745	0.42347
1.350	0.42347	1.5293	0.46170
1.375	0.46170	1.5882	0.50140
1.400	0.50140	1.6514	0.54269
1.425	0.54269	1.7195	0.58568
1.450	0.58568	1.7930	0.63050
1.475	0.63050	1.8725	0.67732
1.500	0.67732		

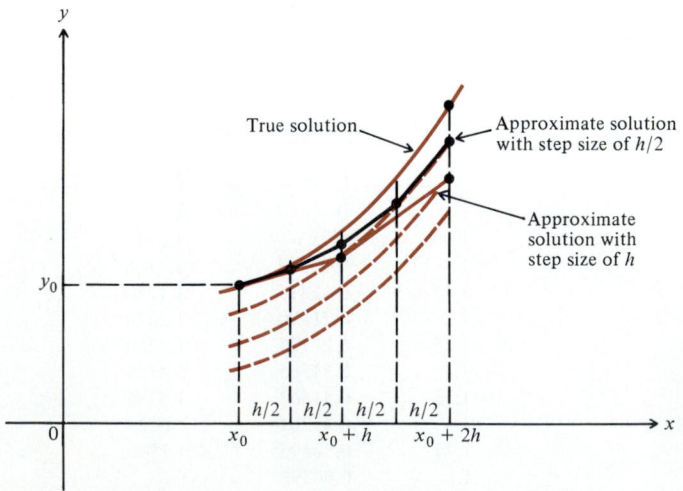

FIGURE 2

all increasing in the interval $1 \leq x \leq 1.5$. Figure 2 illustrates this assertion: the approximation with step size $h/2$ lies above the one of step size h, because the slope of the solution curves near the true solution at $x_0 + h/2$ exceeds the slope at x_0. Thus, two steps of size $h/2$ incur less error from the true solution than one step of size h.

Improved Euler Method

The **improved Euler method** was developed to reduce the discretization error that occurs in the Euler method. The procedure *averages* the slopes at the left and right endpoints of each step and uses this "average" slope to compute the next y-value. To see why this is a reasonable procedure, integrate both sides of equation (1) between x_0 and x_1:

$$y(x_1) - y(x_0) = \int_{x_0}^{x_1} \frac{dy}{dx} dx = \int_{x_0}^{x_1} f(x, y(x)) \, dx$$

or

$$y(x_1) = y(x_0) + \int_{x_0}^{x_1} f(x, y(x)) \, dx. \tag{3}$$

If we approximate the integral in (3) by the *trapezoidal rule* with two points, we obtain for $h = x_1 - x_0$

$$y(x_1) \approx y(x_0) + \frac{h}{2} [f(x_0, y(x_0)) + f(x_1, y(x_1))]. \tag{4}$$

Since $y(x_1)$ is not known, we shall replace it by the value found by Euler's method, which we call z_1; then (4) can be replaced by the system of equations

$$z_1 = y_0 + hf(x_0, y_0),$$

$$y_1 = y_0 + \frac{h}{2} [f(x_0, y_0) + f(x_1, z_1)].$$

Figure 3 illustrates the effect of averaging the slopes at the endpoints of the interval and provides a geometric interpretation of each step of the general procedure:

Improved Euler Method

$$z_{n+1} = y_n + hf(x_n, y_n),$$

$$y_{n+1} = y_n + \frac{h}{2} [f(x_n, y_n) + f(x_{n+1}, z_{n+1})]. \tag{5}$$

EXAMPLE 2

Consider the initial value problem in Example 1.8.2:

$$\frac{dy}{dx} = y + x^2, \qquad y(0) = 1. \tag{6}$$

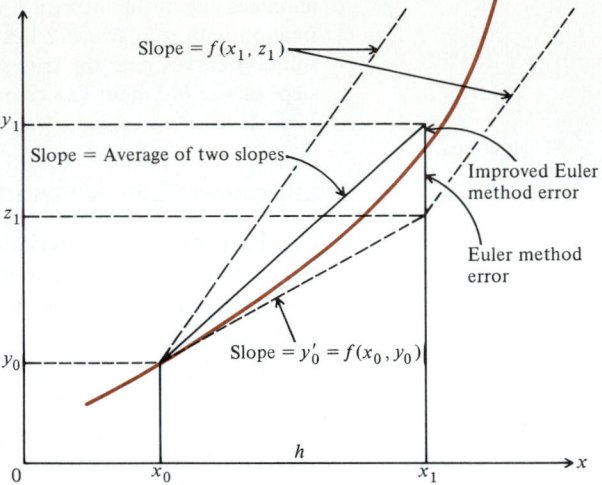

FIGURE 3. Improved Euler method

We wish to determine $y(1)$. We saw in Section 1.8 that with $h = 0.2$ the Euler method yields the approximation $y(1) \approx 2.77$, which is off by about 12% from the exact solution $y(1) = 3e - 5 \approx 3.1548$. Using $x_0 = 0$, $y_0 = 1$, and $h = 0.2$, equation (5) yields

$$z_1 = y_0 + hf(x_0, y_0) = 1 + 0.2(1 + 0^2) = 1.2,$$

$$y_1 = y_0 + \frac{h}{2}[f(x_0, y_0) + f(x_1, z_1)] = 1 + 0.1[(1 + 0^2) + (1.2 + 0.2^2)]$$

$$= 1 + 0.1[2.24] = 1.224 \approx 1.22,$$

$$z_2 = y_1 + hf(x_1, y_1) = 1.22 + 0.2[1.22 + (0.2)^2] = 1.472 \approx 1.47,$$

$$y_2 = y_1 + \frac{h}{2}[f(x_1, y_1) + f(x_2, z_2)]$$

$$= 1.22 + 0.1[1.22 + (0.2)^2 + 1.47 + (0.4)^2] = 1.509 \approx 1.51,$$

and so on. Table 4 shows the approximate values of the solution of equation (6); the error from the exact solution this time is less than $\frac{1}{2}\%$.

If we reduce the step size to $h = 0.1$, we obtain the results in Table 5. The actual value obtained using the improved Euler method to five significant figures (see Table 6) is $y(1) \approx 3.1504$, an error of 0.14%.

TABLE 4

x_n	y_n	$f(x_n, y_n)$ $= y_n + x_n^2$	z_{n+1}	$f(x_{n+1}, z_{n+1})$ $= z_{n+1} + x_{n+1}^2$	y_{n+1}
0.0	1.00	1.00	1.20	1.24	1.22
0.2	1.22	1.26	1.47	1.63	1.51
0.4	1.51	1.67	1.84	2.20	1.90
0.6	1.90	2.26	2.35	2.99	2.43
0.8	2.43	3.07	3.04	4.04	3.14
1.0	3.14				

TABLE 5   IMPROVED EULER METHOD WITH $h = 0.1$,
FOR $y' = y + x^2$, $y(0) = 1$

x_n	y_n	$f(x_n, y_n)$	z_{n+1}	$f(x_{n+1}, z_{n+1})$	y_{n+1}
0.00	1.00	1.00	1.10	1.11	1.11
0.10	1.11	1.12	1.22	1.26	1.22
0.20	1.22	1.26	1.35	1.44	1.36
0.30	1.36	1.45	1.50	1.66	1.52
0.40	1.52	1.68	1.68	1.93	1.70
0.50	1.70	1.95	1.89	2.25	1.91
0.60	1.91	2.27	2.13	2.62	2.15
0.70	2.15	2.64	2.41	3.05	2.43
0.80	2.43	3.07	2.74	3.55	2.77
0.90	2.77	3.58	3.12	4.12	3.15
1.00	3.15				

TABLE 6   A COMPARISON OF RESULTS USING THE EULER AND
IMPROVED EULER METHODS WITH $h = 0.2$ AND
$h = 0.1$, FOR THE INITIAL VALUE PROBLEM
$y' = y + x^2$, $y(0) = 1$

x	Euler	Improved Euler	Exact
$h = 0.2$			
0.0	1.000000	1.00000000	1.00000000
0.2	1.200000	1.22400000	1.22420828
0.4	1.448000	1.51408000	1.51547409
0.6	1.769600	1.90237760	1.90635640
0.8	2.195520	2.42810067	2.43662279
1.0	2.762624	3.13908282	3.15484548
$h = 0.1$			
0.0	1.00000000	1.00000000	1.00000000
0.1	1.10000000	1.10550000	1.10551275
0.2	1.21100000	1.22412750	1.22420828
0.3	1.33610000	1.35936089	1.35957642
0.4	1.47871000	1.51504378	1.51547409
0.5	1.64258100	1.69542338	1.69616381
0.6	1.83183910	1.90519283	1.90635640
0.7	2.05102301	2.14953808	2.15125812
0.8	2.30512531	2.43418958	2.43662279
0.9	2.59963784	2.76547948	2.76880934
1.0	2.94060163	3.15040483	3.15484548

EXAMPLE 3

We apply the improved Euler method to the initial value problem in Example 1:

$$y' = x + y^2, \qquad y(1) = 0,$$

with $h = 0.1$. The values in Table 7 are obtained. Here $y(1) \approx 0.6918$ to four decimal places (see Table 8). If we change the step size to $h = 0.05$ we obtain $y(1) \approx 0.6916$, while $h = 0.025$ yields $y(1) \approx 0.6915$, suggesting that the calculation in Table 8 probably has an error of not more than 0.0003 ($< 0.05\%$).

TABLE 7   IMPROVED EULER METHOD WITH $h = 0.1$,
          FOR $y' = x + y^2$, $y(1) = 0$

x_n	y_n	$f(x_n, y_n)$	z_{n+1}	$f(x_{n+1}, z_{n+1})$	y_{n+1}
1.00	0.00	1.00	0.10	1.11	0.11
1.10	0.11	1.11	0.22	1.25	0.22
1.20	0.22	1.25	0.35	1.42	0.36
1.30	0.36	1.43	0.50	1.65	0.51
1.40	0.51	1.66	0.68	1.96	0.69
1.50	0.69				

TABLE 8   COMPARISON OF RESULTS USING THE
          EULER AND IMPROVED EULER
          METHODS WITH $h = 0.1$ AND $h = 0.05$,
          FOR THE INITIAL VALUE PROBLEM
          $y' = x + y^2$, $y(1) = 0$

x	Euler	Improved Euler
$h = 0.1$		
1.0	0.000000000	0.000000000
1.1	0.100000000	0.105500000
1.2	0.211000000	0.223402573
1.3	0.335452100	0.356966908
1.4	0.476704911	0.510823653
1.5	0.639429668	0.691781574
$h = 0.05$		
1.00	0.000000000	0.000000000
1.05	0.050000000	0.051312500
1.10	0.102625000	0.105398434
1.15	0.158151595	0.162573808
1.20	0.216902191	0.223209962
1.25	0.279254519	0.287746157
1.30	0.345653673	0.356705840
1.35	0.416627496	0.430717867
1.40	0.492806420	0.510544562
1.45	0.574949328	0.597119307
1.50	0.663977664	0.691597674

In the next section we shall discuss the discretization error of Euler's method. A similar study could be done for the improved Euler method, but we shall not do so in this text.

Problems 13.5

In Problems 1–10, solve each problem exactly using the methods of Chapter 1. Then (a) use the Euler method and the indicated value of h, and (b) use the improved Euler method and the given value of h. Compare the accuracy of the two methods with the exact answer.

1. $\dfrac{dy}{dx} = x + y$, $y(0) = 1$. Find $y(1)$ with $h = 0.2$.

2. $\dfrac{dy}{dx} = x - y$, $y(1) = 2$. Find $y(3)$ with $h = 0.4$.

3. $\dfrac{dy}{dx} = \dfrac{x - y}{x + y}$, $y(2) = 1$. Find $y(1)$ with $h = -0.2$.

4. $\dfrac{dy}{dx} = \dfrac{y}{x} + \left(\dfrac{y}{x}\right)^2$, $y(1) = 1$. Find $y(2)$ with $h = 0.2$.

5. $\dfrac{dy}{dx} = x\sqrt{1+y^2}$, $y(1) = 0$. Find $y(3)$ with $h = 0.4$.

6. $\dfrac{dy}{dx} = x\sqrt{1-y^2}$, $y(1) = 0$. Find $y(2)$ with $h = 0.125$.

7. $\dfrac{dy}{dx} = \dfrac{y}{x} - \dfrac{5}{2}x^2y^3$, $y(1) = \dfrac{1}{\sqrt{2}}$. Find $y(2)$ with $h = 0.125$.

8. $\dfrac{dy}{dx} = \dfrac{-y}{x} + x^2y^2$, $y(1) = \dfrac{2}{9}$. Find $y(3)$ with $h = \frac{1}{3}$.

[*Hint:* This is a Bernoulli equation.]

9. $\dfrac{dy}{dx} = ye^x$, $y(0) = 2$. Find $y(2)$ with $h = 0.2$.

10. $\dfrac{dy}{dx} = xe^y$, $y(0) = 0$. Find $y(1)$ with $h = 0.1$.

In Problems 11–20, use (a) the Euler method, or (b) the improved Euler method to graph approximately the solution of the given initial value problem by plotting the points (x_k, y_k) over the indicated range, where $x_k = x_0 + kh$.

11. $y' = xy^2 + y^3$, $y(0) = 1$, $h = 0.02$, $0 \leqslant x \leqslant 0.1$

12. $y' = x + \sin(\pi y)$, $y(1) = 0$, $h = 0.2$, $1 \leqslant x \leqslant 2$

13. $y' = x + \cos(\pi y)$, $y(0) = 0$, $h = 0.4$, $0 \leqslant x \leqslant 2$

14. $y' = \cos(xy)$, $y(0) = 0$, $h = \pi/4$, $0 \leqslant x \leqslant \pi$

15. $y' = \sin(xy)$, $y(0) = 1$, $h = \pi/4$, $0 \leqslant x \leqslant 2\pi$

16. $y' = \sqrt{x^2 + y^2}$, $y(0) = 1$, $h = 0.5$, $0 \leqslant x \leqslant 5$

17. $y' = \sqrt{y^2 - x^2}$, $y(0) = 1$, $h = 0.1$, $0 \leqslant x \leqslant 1$

18. $y' = \sqrt{x + y^2}$, $y(0) = 1$, $h = 0.2$, $0 \leqslant x \leqslant 1$

19. $y' = \sqrt{x + y^2}$, $y(1) = 2$, $h = -0.2$, $0 \leqslant x \leqslant 1$

20. $y' = \sqrt{x^2 + y^2}$, $y(1) = 5$, $h = -0.2$, $0 \leqslant x \leqslant 1$

21. Let

$$\frac{dy}{dx} = \frac{x-y}{x+y}, \quad y(2) = 0.$$

Use the improved Euler method to approximate $y(1)$ with $h = -0.2$. Compare your answer with the exact value and explain why the method failed.

22. Use the improved Euler method to approximate $y(3)$ with $h = 0.4$ for the initial value problem

$$\frac{dy}{dx} + \frac{3}{x}y = x^2y^2, \quad y(1) = 2.$$

Compare your answer to the exact answer and explain why the numerical technique failed.

23. Repeat Problem 22 for the initial value problem

$$\frac{dy}{dx} + \frac{y}{x} = x^3y^3, \quad y(1) = 1.$$

* 24. Let $y' = e^{xy}$, $y(0) = 1$, and obtain a value for $y(4)$ with $h = 0.5$ using the improved Euler method. What difficulties are encountered? How much confidence do you have in your answer?

13.6 An Error Analysis for Euler's Method

(Optional)

In this section we discuss only the discretization errors encountered in the use of Euler's method. Round-off errors depend not only on the method and the number of steps in the calculation, but also on the type of instrument (hand-calculator, computer, pencil and paper, etc.) used for computing the answer. Round-off error will not be discussed in this section (it will be discussed in Section 13.8), although it never should be ignored.

Let us again consider the first-order initial value problem

$$y' = f(x, y), \qquad y(x_0) = y_0, \tag{1}$$

and use the iteration scheme

$$y_{n+1} = y_n + hf(x_n, y_n), \tag{2}$$

where h is a fixed step size.

We shall assume for the remainder of this section that $f(x, y)$ possesses continuous first partial derivatives. Then, on any finite interval, $\partial f(x, y)/\partial y$ is bounded by some constant that we denote by L (a continuous function is always bounded on a closed, bounded interval). Since $y'(x) = f(x, y)$, we obtain, by the chain rule,

$$y''(x) = \frac{\partial f}{\partial x}(x, y) + \frac{\partial f}{\partial y}(x, y)y'(x),$$

which must be continuous since it is the sum of continuous functions. Hence $y''(x)$ must be bounded on the interval $x_0 \leq x \leq a$. So we assume that $|y''(x)| < M$ for some positive constant M.

We now wish to estimate the error ϵ_n at the nth step of the iteration defined by equation (2). Since $y(x_n)$ is the exact value of the solution $y(x)$ at the point $x_n = x_0 + nh$, and y_n is the approximate value at that point, the error at the nth step is given by

$$\epsilon_n = y_n - y(x_n). \tag{3}$$

Note that $y_0 = y(x_0)$, so that $\epsilon_0 = 0$.

Now $y(x_{n+1}) = y(x_n + h)$ and $y''(x)$ is continuous. So we may use Taylor's theorem with remainder to obtain

$$y(x_{n+1}) = y(x_n + h) = y(x_n) + hy'(x_n) + \frac{h^2}{2}y''(\xi_n), \tag{4}$$

where $x_n \leq \xi_n \leq x_{n+1}$. We may now state the main result of this section.

■ **THEOREM 1**

Let $f(x, y)$ have continuous first partial derivatives and let y_n be the approximate solution of equation (1) generated by Euler's method [equation (2)]. Suppose that $y(x)$ is defined and the inequalities

$$\left|\frac{\partial f}{\partial y}(x, y)\right| < L \quad \text{and} \quad |y''(x)| < M$$

hold on the bounded interval $x_0 \leq x \leq a$. Then the error $\epsilon_n = y_n - y(x_n)$ satisfies the inequality

$$|\epsilon_n| \leq \frac{hM}{2L}(e^{(x_n - x_0)L} - 1) = \frac{hM}{2L}(e^{nhL} - 1). \tag{5}$$

In particular, since $x_n - x_0 \leq a - x_0$ (which is finite), every $|\epsilon_n|$ tends to zero as h tends to zero.

PROOF
Subtracting (4) from (2) yields

$$y_{n+1} - y(x_{n+1}) = y_n - y(x_n) + h[f(x_n, y_n) - y'(x_n)] - \frac{h^2}{2}y''(\xi_n)$$

or

$$\epsilon_{n+1} = \epsilon_n + h[f(x_n, y_n) - f(x_n, y(x_n))] - \frac{h^2}{2} y''(\xi_n). \tag{6}$$

By the mean value theorem of differential calculus,

$$f(x_n, y_n) - f(x_n, y(x_n)) = \frac{\partial f}{\partial y}(x_n, \hat{y}_n)[y_n - y(x_n)] = \frac{\partial f}{\partial y}(x_n, \hat{y}_n)\epsilon_n, \tag{7}$$

where $\hat{y}_n$ is between y_n and $y(x_n)$. We substitute (7) into (6) to obtain

$$\epsilon_{n+1} = \epsilon_n + h\frac{\partial f}{\partial y}(x_n, \hat{y}_n)\epsilon_n - \frac{h^2}{2} y''(\xi_n). \tag{8}$$

But $|\partial f/\partial y| \leq L$ and $|y''| \leq M$, so that taking the absolute value of both sides of (8) and using the triangle inequality, we obtain

$$|\epsilon_{n+1}| \leq |\epsilon_n| + hL|\epsilon_n| + \frac{h^2}{2} M = (1 + hL)|\epsilon_n| + \frac{h^2}{2} M. \tag{9}$$

We now consider the first-order difference equation

$$r_{n+1} = (1 + hL)r_n + \frac{h^2}{2} M, \qquad r_0 = 0, \tag{10}$$

and claim that if r_n is the solution to (10), then $|\epsilon_n| \leq r_n$. We show this by induction. It is true for $n = 0$, since $\epsilon_0 = r_0 = 0$. We assume it is true for $n = k$ and prove it for $n = k + 1$. That is, we assume that $|\epsilon_m| \leq r_m$, for $m = 0, 1, \ldots, k$. Then

$$r_{k+1} = (1 + hL)r_k + \frac{h^2}{2} M \geq (1 + hL)|\epsilon_k| + \frac{h^2}{2} M \geq |\epsilon_{k+1}|,$$

and the claim is proved. [The last step follows from equation (9).] We can solve the constant-coefficient first-order nonhomogeneous difference equation (10) by proceeding inductively: $r_1 = h^2M/2 = [(1 + hL) - 1]hM/2L$, $r_2 = [(1 + hL) + 1]h^2M/2 = [(1 + hL)^2 - 1]hM/2L$, and, using the identity $(1 + x)^n - 1 = (1 + x)[(1 + x)^{n-1} - 1] + x$,

$$r_n = \frac{hM}{2L}(1 + hL)^n - \frac{hM}{2L}.$$

Now, $e^{hL} = 1 + hL + h^2L^2/2! + \cdots$, so that

$$1 + hL \leq e^{hL} \quad \text{and} \quad (1 + hL)^n \leq (e^{hL})^n = e^{nhL}$$

Thus

$$|\epsilon_n| \leq r_n \leq \frac{hM}{2L} e^{nhL} - \frac{hM}{2L} = \frac{hM}{2L}(e^{nhL} - 1). \tag{11}$$

But $x_n = x_0 + nh$, so that $x_n - x_0 = nh$ and (11) becomes

$$|\epsilon_n| \le \frac{hM}{2L}(e^{(x_n-x_0)L} - 1),$$

and the theorem is proved.   ■

Theorem 1 not only shows that the errors get small as h tends to zero but also tells us *how fast* the errors decrease. If we define the constant k by

$$k = \frac{M}{2L}|e^{(a-x_0)L} - 1|, \tag{12}$$

then we have

$$|\epsilon_n| \le kh. \tag{13}$$

Thus the error is bounded by a *linear* function of h. (Note that $|\epsilon_n|$ is bounded by a term that depends only on h, not on n.) Roughly speaking, this implies that the error decreases at a rate proportional to the decrease in the step size. If, for example, we halve the step size, then we can expect at least to halve the error. Actually, since the estimates used in arriving at equation (13) are very crude, we can often do better, as in Example 1.8.2, where we halved the step size and decreased the error by a factor of four. Nevertheless, it is useful to have an upper bound for the error. It should be noted, however, that this bound may be difficult to evaluate, since it is frequently difficult to find a bound for $y''(x)$.

EXAMPLE 1

Consider the equation $y' = y$, $y(0) = 1$. We have

$$f(x, y) = y \quad \text{and} \quad \left|\frac{\partial f}{\partial y}(x, y)\right| = 1 = L.$$

Since the solution of the problem is $y(x) = e^x$, we have $|y''| \le e^1 = M$ on the interval $0 \le x \le 1$. Then equation (12) becomes

$$k = \frac{e}{2}|e - 1| = \frac{e^2 - e}{2} \approx 2.34,$$

so that

$$|\epsilon_n| \le 2.34h.$$

Therefore, using a step size of $h = 0.1$, say, we can expect to have an error at each step of less than 0.234 (see Table 1). We note that the greatest actual error is about half of the maximum possible error according to equation (13).

It turns out that it is possible to derive error estimates like equation (11) or (13) for every method we shall discuss in this chapter for solving differential

TABLE 1

x_n	$y_n = y'_n$	$y_{n+1} = y_n + hy'_n$	$y(x_n) = e^{x_n}$	$\epsilon_n = y_n - y(x_n)$
0.0	1.00	1.10	1.00	0.00
0.1	1.10	1.21	1.11	−0.01
0.2	1.21	1.33	1.22	−0.01
0.3	1.33	1.46	1.35	−0.02
0.4	1.46	1.61	1.49	−0.03
0.5	1.61	1.77	1.65	−0.04
0.6	1.77	1.95	1.82	−0.05
0.7	1.95	2.15	2.01	−0.06
0.8	2.15	2.37	2.23	−0.08
0.9	2.37	2.61	2.46	−0.09
1.0	2.61		2.72	−0.11

equations numerically. Actually, to derive these estimates would take us beyond the scope of this book,† but we should mention that for the Runge-Kutta method, which will be discussed in Section 13.7, the discretization error ϵ_n is of the form

$$|\epsilon_n| \leqslant kh^4,$$

for some appropriate constant k. Thus halving the step size, for example, has the effect of decreasing the bound on the error by a factor of $2^4 = 16$. However, the price for this greater accuracy is to have to calculate $f(x, y)$ at four points [see p. 826] for each step in the iteration.

Problems 13.6

1. Consider the differential equation $y' = -y$, $y(0) = 1$. We wish to find $y(1)$.

   (a) Calculate an upper bound on the error of Euler's method as a function of h.

   (b) Calculate this bound for $h = 0.1$ and $h = 0.2$.

   (c) Perform the iterations for $h = 0.2$ and $h = 0.1$ and compare the actual error with the maximum error.

2. Consider the equation of Problem 1. If we ignore round-off error, how many iterations would have to be performed in order to guarantee that the calculation of $y(1)$ obtained by Euler's method be correct to:

   (a) five decimal places?

   (b) six decimal places?

3. Answer the questions in Problem 2 for the equation $y' = 3y - x^2$, $y(1) = 2$, to find $y(1.5)$.

13.7 The Runge-Kutta Method

This powerful method gives accurate results without a large number of steps (that is, without the need to make the step size too small). The efficiency is obtained by using a version of Simpson's rule in evaluating the integral in

$$y(x_1) = y(x_0) + \int_{x_0}^{x_1} f(x, y(x))\, dx. \tag{1}$$

† For a more detailed analysis, see, for example, C. W. Gear, *Numerical Initial Value Problems in Ordinary Differential Equations,* Prentice-Hall, Englewood Cliffs, N.J., 1971.

As before, the method estimates the solution at x_1 of the initial value problem

$$y' = f(x, y), \qquad y(x_0) = y_0. \tag{2}$$

If $f(x, y(x))$ were a function of x alone, then *Simpson's rule* (p. 804) would give

$$\int_{x_0}^{x_0+h} f(x)\, dx \approx \frac{h/2}{3}\left[f(x_0) + 4f\left(x_0 + \frac{h}{2}\right) + f(x_0 + h)\right]$$

$$\approx \frac{1}{6}\left[hf(x_0) + 4hf\left(x_0 + \frac{h}{2}\right) + hf(x_0 + h)\right]. \tag{3}$$

We modify this rule in the following way (a theoretical justification will be provided later in this section) to adjust for the fact that f also depends on y:

$$\int_{x_0}^{x_0+h} f(x, y(x))\, dx \approx \tfrac{1}{6}[m_1 + 2m_2 + 2m_3 + m_4], \tag{4}$$

where

$$m_1 = hf(x_0, y_0),$$

$$m_2 = hf\left(x_0 + \frac{h}{2}, y_0 + \frac{m_1}{2}\right),$$

$$m_3 = hf\left(x_0 + \frac{h}{2}, y_0 + \frac{m_2}{2}\right),$$

$$m_4 = hf(x_0 + h, y_0 + m_3). \tag{5}$$

Note that if f depends only on x, then $m_2 = m_3$ and (4) becomes Simpson's rule. Setting $x_1 = x_0 + h$, $x_2 = x_0 + 2h$, ..., $x_n = x_0 + nh$, we obtain the following recursive algorithm, called the **Runge-Kutta method:**†

Runge-Kutta Method

$$y_{n+1} = y_n + \tfrac{1}{6}(m_1 + 2m_2 + 2m_3 + m_4), \tag{6}$$

where

$$m_1 = hf(x_n, y_n),$$

$$m_2 = hf\left(x_n + \frac{h}{2}, y_n + \frac{m_1}{2}\right),$$

$$m_3 = hf\left(x_n + \frac{h}{2}, y_n + \frac{m_2}{2}\right),$$

$$m_4 = hf(x_n + h, y_n + m_3). \tag{7}$$

† Named after the German mathematicians Carl Runge (1856–1927) and Wilhelm Kutta (1867–1944).

Note that the values m_1/h, m_2/h, m_3/h, and m_4/h are four slopes between $x_n \leqslant x \leqslant x_{n+1}$, so that (6) is a weighted average of these slopes—a procedure similar to the one we used in the improved Euler method. The following two examples will illustrate the use of (6) and (7).

EXAMPLE 1

Consider the initial value problem in Example 13.5.2 (and Example 1.8.2) where we wish to evaluate $y(1)$ given that

$$\frac{dy}{dx} = y + x^2, \qquad y(0) = 1.$$

If we apply the Runge-Kutta method with $h = 1$, then $y(1) = y_1$, which is obtained by first calculating the expressions in (7):

$$m_1 = f(0, 1) = 1,$$

$$m_2 = f(\tfrac{1}{2}, \tfrac{3}{2}) = \tfrac{7}{4},$$

$$m_3 = f(\tfrac{1}{2}, \tfrac{15}{8}) = \tfrac{17}{8},$$

$$m_4 = f(1, \tfrac{25}{8}) = \tfrac{33}{8}.$$

Thus, using (6),

$$y_1 = 1 + \tfrac{1}{6}(1 + \tfrac{7}{2} + \tfrac{17}{4} + \tfrac{33}{8}) = \tfrac{151}{48} \approx 3.146.$$

In one step this method got us even closer to the correct value than the improved Euler method did in five steps.

If we use five steps ($h = 0.2$) to calculate $y(1)$, we can arrange our work in tabular form as shown in Table 1. The $y(1)$ entry, when displayed to four decimal places is $y(1) \approx 3.1548$, which is the exact answer to that many significant figures (see Table 2).

TABLE 1   RUNGE-KUTTA METHOD WITH $h = 0.2$, FOR $y' = y + x^2$, $y(0) = 1$

x_n	y_n	m_1	m_2	m_3	m_4
0.00	1.00	0.20	0.22	0.22	0.25
0.20	1.22	0.25	0.29	0.29	0.34
0.40	1.52	0.34	0.39	0.39	0.45
0.60	1.91	0.45	0.52	0.53	0.62
0.80	2.44	0.62	0.71	0.72	0.83
1.00	3.15				

TABLE 2   COMPARISON OF RESULTS USING THE EULER, IMPROVED EULER, AND RUNGA-KUTTA METHODS WITH $h = 0.2$ AND $h = 0.1$, FOR THE INITIAL VALUE PROBLEM $y' = y + x^2$, $y(0) = 1$

x	Euler	Improved Euler	Runge-Kutta	Exact
$h = 0.2$				
0.0	1.000000	1.00000000	1.00000000	1.00000000
0.2	1.200000	1.22400000	1.22420667	1.22420828
0.4	1.448000	1.51408000	1.51546869	1.51547409
0.6	1.769600	1.90237760	1.90634413	1.90635640
0.8	2.195520	2.42810067	2.43659938	2.43662279
1.0	2.762624	3.13908282	3.15480515	3.15484548

TABLE 2   *Continued*

x	Euler	Improved Euler	Runge-Kutta	Exact
$h = 0.1$				
0.0	1.00000000	1.00000000	1.00000000	1.00000000
0.1	1.10000000	1.10550000	1.10551271	1.10551275
0.2	1.21100000	1.22412750	1.22420815	1.22420828
0.3	1.33610000	1.35936089	1.35957618	1.35957642
0.4	1.47871000	1.51504378	1.51547370	1.51547409
0.5	1.64258100	1.69542338	1.69616320	1.69616381
0.6	1.83183910	1.90519283	1.90635551	1.90635640
0.7	2.05102301	2.14953808	2.15125689	2.15125812
0.8	2.30512531	2.43418958	2.43662112	2.43662279
0.9	2.59963784	2.76547948	2.76880714	2.76880934
1.0	2.94060163	3.15040483	3.15484264	3.15484548

EXAMPLE 2

Consider the initial value problem

$$y' = x + y^2, \qquad y(1) = 0$$

(see Examples 13.5.1 and 13.5.3). We wish to find approximate values of the solution at $x = 1$, 1.1, 1.2, 1.3, 1.4, and 1.5. Setting $h = 0.1$ and applying the Runge-Kutta method, (6) and (7), we obtain the values in Table 3. If we display the $y(1.5)$ entry to four significant figures, we obtain $y(1.5) \approx 0.6915$ (see Table 4).

TABLE 3   RUNGE-KUTTA METHOD WITH $h = 0.1$, FOR $y' = x + y^2$, $y(1) = 0$

x_n	y_n	m_1	m_2	m_3	m_4
1.00	0.00	0.10	0.11	0.11	0.11
1.10	0.11	0.11	0.12	0.12	0.12
1.20	0.22	0.12	0.13	0.13	0.14
1.30	0.36	0.14	0.15	0.15	0.17
1.40	0.51	0.17	0.18	0.18	0.20
1.50	0.69				

TABLE 4   COMPARISON OF RESULTS USING THE EULER, IMPROVED EULER, AND RUNGE-KUTTA METHODS WITH $h = 0.1$ AND $h = 0.05$, FOR THE INITIAL-VALUE PROBLEM $y' = x + y^2$, $y(1) = 0$

x	Euler	Improved Euler	Runge-Kutta
$h = 0.1$			
1.0	0.000000000	0.000000000	0.000000000
1.1	0.100000000	0.105500000	0.105360367
1.2	0.211000000	0.223402573	0.223135961
1.3	0.335452100	0.356966908	0.356601567
1.4	0.476704911	0.510823653	0.510424419
1.5	0.639429668	0.691781574	0.691496736

TABLE 4 *Continued*

x	Euler	Improved Euler	Runge-Kutta
$h = 0.05$			
1.00	0.000000000	0.000000000	0.000000000
1.05	0.050000000	0.051312500	0.051293290
1.10	0.102625000	0.105398434	0.105360321
1.15	0.158151595	0.162573808	0.162517274
1.20	0.216902191	0.223209962	0.223135802
1.25	0.279254519	0.287746157	0.287655681
1.30	0.345653673	0.356705840	0.356601169
1.35	0.416627496	0.430717867	0.430602377
1.40	0.492806420	0.510544562	0.510423556
1.45	0.574949328	0.597119307	0.597001050
1.50	0.663977664	0.691597674	0.691494999

In Euler's method we used one value of the derivative $y' = f(x, y)$ for each iteration. In the improved Euler's method we used two values. In the Runge-Kutta formulas we made use of four values of the derivative for each iteration. We shall show how to find the four "best" values in a sense to be made precise later.

To begin, we need to recall the Taylor series expansions of functions of one or two variables:

$$y(x_0 + h) = y(x_0) + hy'(x_0) + \frac{h^2}{2!} y''(x_0) + \frac{h^3}{3!} y'''(x_0) + \cdots, \tag{8}$$

$$f(x_0 + mh, y_0 + nh) = f(x_0, y_0) + h(mf_x + nf_y)$$

$$+ \frac{h^2}{2!} (m^2 f_{xx} + 2mn f_{xy} + n^2 f_{yy})$$

$$+ \frac{h^3}{3!} (m^3 f_{xxx} + 3m^2 n f_{xxy} + 3mn^2 f_{xyy} + n^3 f_{yyy})$$

$$+ \cdots, \tag{9}$$

where the partials are all evaluated at the point (x_0, y_0). Since

$$y' = f(x, y), \tag{10}$$

we find that

$$y'' = f_x + (f_y)y' = f_x + ff_y,$$

$$y''' = f_{xx} + 2ff_{xy} + f^2 f_{yy} + f_y(f_x + ff_y),$$

and so on. (Note that $f_{xy} = f_{yx}$ when these derivatives are continuous.) Thus (8) can be written in the form

$$y_1 - y_0 = hf + \frac{h^2}{2} (f_x + ff_y) + \frac{h^3}{6} [f_{xx} + 2ff_{xy} + f^2 f_{yy} + f_y(f_x + ff_y)] + \cdots. \tag{11}$$

The main idea is somehow to select several points (x, y) so that the Taylor series expansions [equation (9)] of the corresponding $f(x, y)$ terms coincide with the terms on the right-hand side of equation (11). Suppose we let

$$m_1 = hf(x_0, y_0),$$

$$m_2 = hf(x_0 + nh, y_0 + nm_1),$$

$$m_3 = hf(x_0 + ph, y_0 + pm_2),$$

$$m_4 = hf(x_0 + qh, y_0 + qm_3).$$

(The reason for doing this will be made clear shortly.) Using (9), we may write these values as

$$m_1 = hf,$$

$$m_2 = h\left[f + nh(f_x + ff_y) + \frac{(nh)^2}{2}(f_{xx} + 2ff_{xy} + f^2f_{yy}) + \cdots\right],$$

$$m_3 = h\left\{f + ph(f_x + ff_y) + \frac{h^2}{2}[p^2(f_{xx} + 2ff_{xy} + f^2f_{yy}) + 2npf_y(f_x + ff_y)] + \cdots\right\},$$

$$m_4 = h\left\{f + qh(f_x + ff_y) + \frac{h^2}{2}[q^2(f_{xx} + 2ff_{xy} + f^2f_{yy}) + 2pqf_y(f_x + ff_y)] + \cdots\right\},$$

where all functions are evaluated at the point (x_0, y_0).

We now consider an expression of the form

$$am_1 + bm_2 + cm_3 + dm_4$$

and try to equate it to the right-hand side of (11). This will have the effect of giving us a numerical scheme that agrees with the solution to (10) up to and including third-order terms. Then the error will be no greater than terms like kh^4, and so on. Matching like expressions, we find that

$$
\begin{array}{ll}
\text{coefficient of } hf & a + b + c + d = 1, \\
\text{coefficient of } h^2(f_x + ff_y) & bn + cp + dq = \tfrac{1}{2}, \\
\text{coefficient of } \tfrac{1}{2}h^3(f_{xx} + 2ff_{xy} + f^2f_{yy}) & bn^2 + cp^2 + dq^2 = \tfrac{1}{3}, \\
\text{coefficient of } h^3[f_y(f_x + ff_y)] & cnp + dpq = \tfrac{1}{6}.
\end{array}
\qquad (12)
$$

Now any solution of these equations will produce a method in which there is no error up to the third-order terms. Suppose we taken $n = p = \tfrac{1}{2}$ and $q = 1$. Then (12) reduces to the system of equations

$$a + b + c + d = 1,$$

$$b + c + 2d = 1,$$

$$3b + 3c + 12d = 4,$$

$$3c + 6d = 2,$$

which has the solution $a = d = \frac{1}{6}$, $b = c = \frac{1}{3}$. Thus the Runge-Kutta formula (6) and (7) agrees with $y_1 - y_0$ for all terms up to and including the terms in h^3. Actually, with quite a bit more work, one can show that they agree in the h^4 terms, too. Thus the error (if any) involves only terms in h^5 and higher. Hence, for small h, we should expect to get very good results.

Other formulas are also readily derivable. Suppose we choose $n = \frac{1}{3}$, $p = \frac{2}{3}$, $q = 1$. Then (12) yields

$$a + b + c + d = 1,$$
$$2b + 4c + 6d = 3,$$
$$b + 4c + 9d = 3,$$
$$4c + 12d = 3,$$

which has a solution $a = d = \frac{1}{8}$, $b = c = \frac{3}{8}$. The formula, known as the **Kutta-Simpson $\frac{3}{8}$-rule,** may be written as

$$y_1 = y_0 + \frac{1}{8}(m_1 + 3m_2 + 3m_3 + m_4),$$

where

$$m_1 = hf(x_0, y_0),$$
$$m_2 = hf(x_0 + h/3, y_0 + m_1/3),$$
$$m_3 = hf(x_0 + 2h/3, y_0 + 2m_2/3),$$
$$m_4 = hf(x_0 + h, y_0 + m_3).$$

$$(13)$$

Similarly, the choice $n = \frac{1}{3}$, $p = \frac{2}{3}$, $q = 1$ also yields a solution $a = c = 0$, $b = \frac{3}{4}$, $d = \frac{1}{4}$; hence

$$y_1 = y_0 + \frac{1}{4}(3m_2 + m_4),$$

where m_2 and m_4 are defined as in (13). Since the number of possible choices of n, p, and q is infinite, the reader may be amused by solving (12) for a, b, c, and d with whatever values of n, p, and q that he or she selects.

Problems 13.7

In Problems 1–10, solve each problem exactly using the methods of Chapter 1. Then use the Runge-Kutta method with the given h. Compare the accuracy of the method with the exact answer.

1. $\dfrac{dy}{dx} = x + y$, $\quad y(0) = 1$. Find $y(1)$ with $h = 0.2$.

2. $\dfrac{dy}{dx} = x - y$, $\quad y(1) = 2$. Find $y(3)$ with $h = 0.4$.

3. $\dfrac{dy}{dx} = \dfrac{x - y}{x + y}$, $\quad y(2) = 1$. Find $y(1)$ with $h = -0.2$.

4. $\dfrac{dy}{dx} = \dfrac{y}{x} + \left(\dfrac{y}{x}\right)^2$, $\quad y(1) = 1$. Find $y(2)$ with $h = 0.2$.

5. $\dfrac{dy}{dx} = x\sqrt{1 + y^2}$, $\quad y(1) = 0$. Find $y(3)$ with $h = 0.4$.

6. $\dfrac{dy}{dx} = x\sqrt{1 - y^2}$, $\quad y(1) = 0$. Find $y(2)$ with $h = 0.125$.

7. $\dfrac{dy}{dx} = \dfrac{y}{x} - \dfrac{5}{2}x^2 y^3$, $\quad y(1) = \dfrac{1}{\sqrt{2}}$. Find $y(2)$ with $h = 0.125$.

8. $\dfrac{dy}{dx} = \dfrac{-y}{x} + x^2y^2$, $y(1) = \dfrac{2}{9}$. Find $y(3)$ with $h = \frac{1}{3}$.

9. $\dfrac{dy}{dx} = ye^x$, $y(0) = 2$. Find $y(2)$ with $h = 0.2$.

10. $\dfrac{dy}{dx} = xe^y$, $y(0) = 0$. Find $y(1)$ with $h = 0.1$.

In Problems 11–20, use the Runge-Kutta method to graph approximately the solution of the given initial value problem by plotting the points (x_k, y_k) over the indicated range, where $x_k = x_0 + kh$.

11. $y' = xy^2 + y^3$, $y(0) = 1$, $h = 0.02$, $0 \leqslant x \leqslant 0.1$

12. $y' = x + \sin(\pi y)$, $y(1) = 0$, $h = 0.2$, $1 \leqslant x \leqslant 2$

13. $y' = x + \cos(\pi y)$, $y(0) = 0$, $h = 0.4$, $0 \leqslant x \leqslant 2$

14. $y' = \cos(xy)$, $y(0) = 0$, $h = \pi/4$, $0 \leqslant x \leqslant \pi$

15. $y' = \sin(xy)$, $y(0) = 1$, $h = \pi/4$, $0 \leqslant x \leqslant 2\pi$

16. $y' = \sqrt{x^2 + y^2}$, $y(0) = 1$, $h = 0.5$, $0 \leqslant x \leqslant 5$

17. $y' = \sqrt{y^2 - x^2}$, $y(0) = 1$, $h = 0.1$, $0 \leqslant x \leqslant 1$

18. $y' = \sqrt{x + y^2}$, $y(0) = 1$, $h = 0.2$, $0 \leqslant x \leqslant 1$

19. $y' = \sqrt{x + y^2}$, $y(1) = 2$, $h = -0.2$, $0 \leqslant x \leqslant 1$

20. $y' = \sqrt{x^2 + y^2}$, $y(1) = 5$, $h = -0.2$, $0 \leqslant x \leqslant 1$

21. Let

$$\frac{dy}{dx} = \frac{x - y}{x + y}, \quad y(2) = 0$$

Use the Runge-Kutta method to approximate $y(1)$ with

$h = -0.2$. Compare your answer with the exact value and explain why the method failed.

22. Use the Runge-Kutta method to approximate $y(3)$ with $h = 0.4$ for the initial value problem

$$\frac{dy}{dx} + \frac{3}{x}y = x^2y^2, \quad y(1) = 2.$$

Compare your answer to the exact answer and explain why the numerical technique failed.

23. Repeat Problem 22 for the initial value problem

$$\frac{dy}{dx} + \frac{y}{x} = x^3y^3, \quad y(1) = 1.$$

24. Let $y' = e^{xy}$, $y(0) = 1$, and obtain a value for $y(4)$ with $h = 0.5$ using the Runge-Kutta method. What difficulties are encountered? How much confidence do you have in your answer?

25. Set $d = 0$ in equation (12) and let $n = \frac{1}{3}$, $p = \frac{2}{3}$. Then find the coefficients a, b, and c for these choices. The resulting formula is called **Heun's formula.**

26. Set $a = d = 0$, $b = \frac{3}{4}$, $c = \frac{1}{4}$ in (12) and find n and p.

27. Set $a = \frac{2}{8}$, $b = c = \frac{3}{8}$, $d = 0$ in (12) and find n and p.

** 28. Prove that the Runge-Kutta formula agrees with equation (11) up to and including the h^4 terms.

29. Explain why the choice $n = \frac{1}{3}$, $p = \frac{2}{3}$, $q = 1$ in (12) yielded two sets of solutions a, b, c, and d. Are other solutions possible? Does the choice $n = p = \frac{1}{2}$, $q = 1$ allow for multiple solutions?

13.8 Propagation of Round-Off Error: An Example of Numerical Instability

(Optional)

In this section we shall show how a theoretically very accurate method can produce results that are useless. A **multistep method** is a method (such as the improved Euler method) that involves information about the solution at more than one point. Consider the multistep method given by the equation

$$y_{n+1} = y_{n-1} + 2hf(x_n, y_n). \tag{1}$$

Here it is necessary to use both the nth and the $(n - 1)$st iterate to obtain the $(n + 1)$st iterate. It can be shown† that this method has the following error estimate:

$$|\epsilon_n| = |y_n - y(x_n)| \leqslant kh^2.$$

† See S. D. Conte, *Elements of Numerical Analysis*, Section 6.6, McGraw-Hill, New York, 1965.

Since the error for Euler's method is $|\epsilon_n| \le kh$, we would theoretically expect more accuracy in solving an initial value problem by using equation (1) than by using Euler's method. However, this does not always turn out to be the case.

EXAMPLE 1

Consider the initial value problem

$$y' = -y + 2, \qquad y(0) = 1.$$

The solution to this equation is easily obtained: $y(x) = 2 - e^{-x}$. Let us obtain $y(5)$ by Euler's method and the method of equation (1). To use the latter, we need two initial values y_0 and y_1. Since we know the solution, we use the exact value $y_1 = y(x_1) = 2 - e^{-x_1}$. Table 1 illustrates the computation with a step size $h = 0.25$. The second column is the correct value of $y(x_n)$ to four decimal places. Column three gives the Euler iterates, and column four gives the iterates obtained by the two-step method (1). Column five is the Euler error, $\epsilon_n^{(E)} = y_n^{(E)} - y(x_n)$, and column six is the error of the two-step method, $\epsilon_n^{(2s)} = y_n^{(2s)} - y(x_n)$.

TABLE 1

x_n	$y(x_n) = 2 - e^{-x_n}$	$y_n^{(E)} = y_{n-1}^{(E)}$ $+ h(2 - y_{n-1}^{(E)})$	$y_n^{(2s)} = y_{n-2}^{(2s)}$ $+ 2h(2 - y_{n-1}^{(2s)})$	$\epsilon_n^{(E)}$	$\epsilon_n^{(2s)}$
0.00	1.0000	1.0000	1.0000	0.0000	0.0000
0.25	1.2212	1.2500	1.2212	0.0288	0.0000
0.50	1.3935	1.4375	1.3894	0.0440	−0.0041
0.75	1.5276	1.5781	1.5265	0.0505	−0.0011
1.00	1.6321	1.6836	1.6262	0.0515	−0.0059
1.25	1.7135	1.7627	1.7134	0.0492	−0.0001
1.50	1.7769	1.8220	1.7695	0.0453	−0.0074
1.75	1.8262	1.8665	1.8287	0.0403	+0.0025
2.00	1.8647	1.8999	1.8552	0.0352	−0.0095
2.25	1.8946	1.9249	1.9011	0.0303	+0.0065
2.50	1.9179	1.9437	1.9047	0.0258	−0.0132
2.75	1.9361	1.9578	1.9488	0.0217	+0.0127
3.00	1.9502	1.9684	1.9303	0.0182	−0.0199
3.25	1.9612	1.9763	1.9837	0.0151	+0.0225
3.50	1.9698	1.9822	1.9385	0.0124	−0.0313
3.75	1.9765	1.9867	2.0145	0.0102	+0.0380
4.00	1.9817	1.9900	1.9313	0.0083	−0.0504
4.25	1.9857	1.9925	2.0489	0.0068	+0.0632
4.50	1.9889	1.9944	1.9069	0.0055	−0.0820
4.75	1.9913	1.9958	2.0955	0.0045	+0.1042
5.00	1.9933	1.9969	1.8952	0.0036	−0.0981

Numerical Instability

It is evident that the two-step method (1) produces a smaller error for small values of x_n than Euler's method. However, as x_n increases, the error in Euler's method decreases, whereas the error in the two-step method not only increases but does so with oscillating sign. This phenomenon is called **numerical instability**. As we shall see, it is due to a propagation of round-off errors.

Let us now explain what will lead to this instability. In the example, $f(x_n, y_n) = -y_n + 2$, so that (1) is

$$y_{n+1} = y_{n-1} + 2h(2 - y_n)$$

or

$$y_{n+1} + 2hy_n - y_{n-1} = 4h, \qquad y_0 = 1. \tag{2}$$

Difference Equation

This is a linear nonhomogeneous **difference equation.** It can be solved in a manner very similar to that used in solving linear nonhomogeneous differential equations. First, we find the general solution of the homogeneous difference equation

$$y_{n+1} + 2hy_n - y_{n-1} = 0. \tag{3}$$

To do this we guess that there is a solution of the form $y_n = \lambda^n$. If we insert $y_n = \lambda^n$ into (3), we obtain

$$\lambda^{n-1}(\lambda^2 + 2h\lambda - 1) = 0.$$

Assuming $\lambda \neq 0$ (otherwise we have a trivial solution) we see that the two roots of the **characteristic equation**

$$\lambda^2 + 2h\lambda - 1 = 0 \tag{4}$$

are

$$\lambda_1 = \frac{-2h + \sqrt{4h^2 + 4}}{2} = -h + \sqrt{1 + h^2}$$

and

$$\lambda_2 = -h - \sqrt{1 + h^2}.$$

Hence (3) has the **general solution**

$$y_n = c_1\lambda_1^n + c_2\lambda_2^n.$$

Finally, we need to find *any* solution of the nonhomogeneous equation (2). Since $4h$ is constant, we try the equivalent of the method of undetermined coefficients: Set $y_m = A$, then (2) becomes

$$A + 2hA - A = 4h$$

so that $A = 2$. Hence, the general solution of (2) is given by

$$y_n = c_1\lambda_1^n + c_2\lambda_2^n + 2. \tag{5}$$

By the binomial theorem (see Problem 5.1.18 on p. 251),

$$(1 + h^2)^{1/2} = 1 + \tfrac{1}{2}h^2 - \tfrac{1}{8}h^4 + \tfrac{1}{16}h^6 - \cdots,$$

where the omitted terms are higher powers of h. Hence the roots (4) of the characteristic equation can be written as

$$\lambda_1 = 1 - h + \alpha(h) \quad \text{and} \quad \lambda_2 = -1 - h - \alpha(h), \tag{6}$$

where

$$\alpha(h) = \frac{h^2}{2} - \frac{h^4}{8} + \frac{h^6}{16} - \cdots.$$

Substituting (6) into (5) yields

$$y_n = c_1[1 - h + \alpha(h)]^n + c_2(-1)^n[1 + h + \alpha(h)]^n + 2. \tag{7}$$

From calculus we know that

$$\lim_{k \to \infty} \left(1 + \frac{1}{k}\right)^k = \lim_{h \to 0}(1 + h)^{1/h} = e.$$

Therefore, since $x_n = 0 + nh = nh$, we have

$$\lim_{h \to 0}(1 - h)^n = \lim_{h \to 0}(1 - h)^{x_n/h} = e^{-x_n} \quad \text{and} \quad \lim_{h \to 0}(1 + h)^n = e^{x_n}.$$

Hence as $h \to 0$, we may ignore the higher-order terms $\alpha(h)$ in (7) to obtain

$$y_n \approx c_1 e^{-x_n} + 2 + c_2(-1)^n e^{x_n}. \tag{8}$$

Here lies the problem. The exact solution of the problem requires that $c_1 = -1$ and $c_2 = 0$. However, even a small round-off error may cause c_2 to be nonzero and this error will grow exponentially while the real solution is approaching the constant two. This is the phenomenon we observed in Table 1. Note that the $(-1)^n$ in (8) causes the errors to oscillate (as we also observed).

The problem arose because we approximated a *first*-order differential equation by a *second*-order difference equation. Such approximations do not always lead to this kind of instability, but it is a possibility that cannot be ignored. In general, to analyze the effectiveness of a given method, we must not only estimate the discretization error, but also show that the method is not numerically unstable (that is, it is **numerically stable**) for the given problem.

13.9 Solution of Partial Differential Equations by Finite Differences

There are a number of different numerical methods for solving partial differential equations. In this section we shall present one of these methods: solution by the use of **finite differences**. This method has the advantage of always being applicable, whether the partial differential equation is linear or not.

Before continuing, we stress that the numerical solution of partial differential equations is a vast and highly technical topic. Entire books are devoted to it. Here we are providing only a very rudimentary introduction to this interesting subject.

Since a partial differential equation is an equation involving partial derivatives, this method involves replacing the partial derivatives by a *finite* number of *differences* of function values.

Suppose that $z = z(x, y)$ is a differentiable function of two variables defined in some region $G \subset \mathbb{R}^2$. For simplicity, we shall assume that G is a rectangle containing the square grid of points $P_{i,j}$ spaced a distance h apart (see Fig. 1) in the horizontal and vertical direction. (This is *not* essential.†) We can then

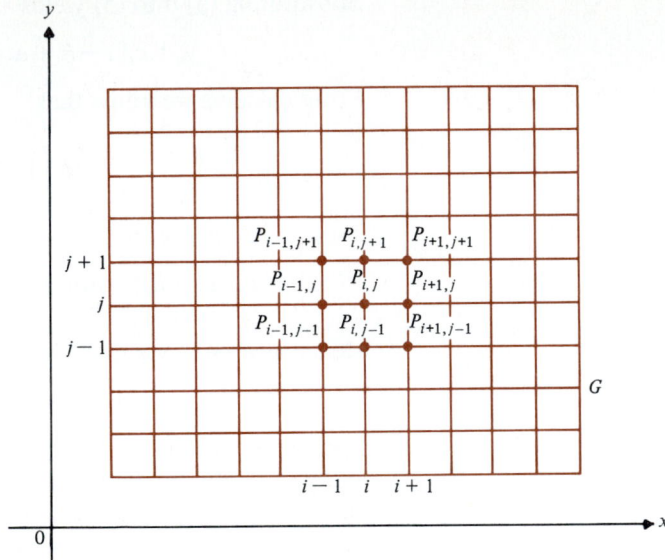

FIGURE 1. A square grid

approximate the partial derivative $\partial z/\partial x$ at the point $P_{i,j}$ in three ways: If $z_{i,j}$ denotes the value of z at $P_{i,j}$, then

$$\frac{\partial z}{\partial x}\bigg|_{i,j} \approx \frac{1}{h}(z_{i+1,j} - z_{i,j}), \tag{1}$$

$$\frac{\partial z}{\partial x}\bigg|_{i,j} \approx \frac{1}{h}(z_{i,j} - z_{i-1,j}), \tag{2}$$

or

$$\frac{\partial z}{\partial x}\bigg|_{i,j} \approx \frac{1}{2h}(z_{i+1,j} - z_{i-1,j}). \tag{3}$$

Equations (1)–(3) are all elementary consequences of the Taylor series expansions of z:

$$z_{i+1,j} = z_{i,j} + h\left(\frac{\partial z}{\partial x}\bigg|_{i,j}\right) + \frac{h^2}{2!}\left(\frac{\partial^2 z}{\partial x^2}\bigg|_{i,j}\right) + h^3(\text{other terms}) \tag{4}$$

† The reader is referred to R. D. Richtmyer, *Difference Methods for Initial Value Problems*, Wiley (Interscience), New York, 1957, for methods for handling nonrectangular regions and grids.

and

$$z_{i-1,j} = z_{i,j} + (-h)\left(\frac{\partial z}{\partial x}\bigg|_{i,j}\right) + \frac{h^2}{2!}\left(\frac{\partial^2 z}{\partial x^2}\bigg|_{i,j}\right) - h^3(\text{other terms}). \qquad (5)$$

Writing (4) in the form

$$z_{i+1,j} - z_{i,j} = h\left(\frac{\partial z}{\partial x}\bigg|_{i,j}\right) + h^2(\text{other terms}) \qquad (6)$$

and dividing both sides by h yields (1). The truncation error in (1) involves terms in h, since we are omitting the expression "h(other terms)" after dividing both sides of (6) by h. Similarly, (2) is obtained from (5) with a truncation error in terms of h. Equation (3) is obtained by subtracting (5) from (4). Note that we obtain

$$z_{i+1,j} - z_{i-1,j} = 2h\left(\frac{\partial z}{\partial x}\bigg|_{i,j}\right) + h^3(\text{other terms}), \qquad (7)$$

so that when we divide both sides of (7) by $2h$ to obtain (3), the truncation error is in terms of h^2.

Although the truncation error in (3) is in terms of h^2, this does not mean that its use will always provide a more accurate answer than (1) or (2). Indeed, examples exist for which (1) and (2) yield good answers while (3) leads to nonsense (see Problem 9). Similarly,

$$\frac{\partial z}{\partial y}\bigg|_{i,j} \approx \frac{1}{h}(z_{i,j+1} - z_{i,j}), \qquad (8)$$

$$\frac{\partial z}{\partial y}\bigg|_{i,j} \approx \frac{1}{h}(z_{i,j} - z_{i,j-1}), \qquad (9)$$

and

$$\frac{\partial z}{\partial y}\bigg|_{i,j} \approx \frac{1}{2h}(z_{i,j+1} - z_{i,j-1}). \qquad (10)$$

Second partial derivatives can be obtained by approximating the second partial by first partials and the latter by finite differences of the $z_{i,j}$:

$$\frac{\partial^2 z}{\partial x^2}\bigg|_{i,j} \approx \frac{1}{h}\left[\left(\frac{\partial z}{\partial x}\bigg|_{i+1,j}\right) - \left(\frac{\partial z}{\partial x}\bigg|_{i,j}\right)\right]$$

$$\approx \frac{1}{h}\left(\frac{z_{i+1,j} - z_{i,j}}{h} - \frac{z_{i,j} - z_{i-1,j}}{h}\right)$$

$$= \frac{1}{h^2}(z_{i+1,j} - 2z_{i,j} + z_{i-1,j}). \qquad (11)$$

Similarly,

$$\frac{\partial^2 z}{\partial y^2}\bigg|_{i,j} \approx \frac{1}{h^2}(z_{i,j+1} - 2z_{i,j} + z_{i,j-1}), \tag{12}$$

while using (3)

$$\frac{\partial^2 z}{\partial x \partial y}\bigg|_{i,j} \approx \frac{1}{2h}\left[\left(\frac{\partial z}{\partial y}\bigg|_{i+1,j}\right) - \left(\frac{\partial z}{\partial y}\bigg|_{i-1,j}\right)\right]$$

$$\approx \frac{1}{2h}\left(\frac{z_{i+1,j+1} - z_{i+1,j-1}}{2h} - \frac{z_{i-1,j+1} - z_{i-1,j-1}}{2h}\right)$$

$$= \frac{1}{4h^2}(z_{i+1,j+1} - z_{i+1,j-1} - z_{i-1,j+1} + z_{i-1,j-1}). \tag{13}$$

The following example illustrates the use of these approximations.

EXAMPLE 1

Find an approximate solution for the one-dimensional heat equation

$$\frac{\partial u}{\partial t} = \frac{\partial^2 u}{\partial x^2}, \qquad 0 \leq x \leq 8, \qquad 0 \leq t \leq 16, \tag{14}$$

with boundary and initial conditions

$$u(x, 0) = x(8 - x), \qquad 0 \leq x \leq 8,$$

$$u(0, t) = 0 = u(16, t), \qquad 0 \leq t \leq 16. \tag{15}$$

SOLUTION
Suppose we decide to use a square grid with points spaced $h = 2$ units apart. We then obtain the grid shown in Fig. 2. The values of u on the x-axis and on the vertical sides

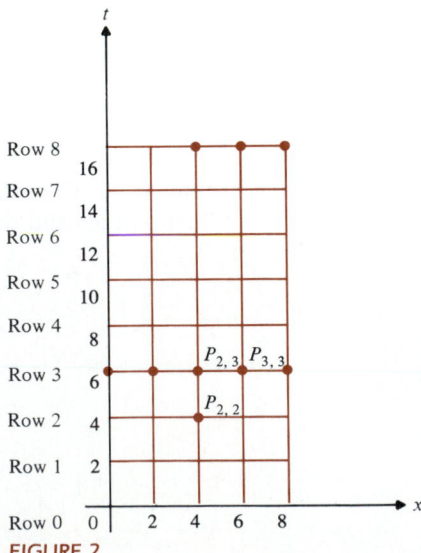

FIGURE 2

are determined by the boundary and initial conditions. If we denote the u value of the point (ih, jh) by $u_{i,j}$, we have

$$u_{1,0} = 2(8-2) = 12, \qquad u_{2,0} = 16, \qquad u_{3,0} = 12$$

and

$$u_{0,j} = u_{4,j} = 0 \qquad \text{for} \qquad j = 0, 1, \ldots, 8.$$

The problem now is to determine $u_{i,j}$ at all the remaining grid points.

Since we are using the first subscript of $u_{i,j}$ to indicate the horizontal position and the second subscript to indicate vertical position, we can use (8) to rewrite the left side of (14) as

$$\left.\frac{\partial u}{\partial t}\right|_{i,j} \approx \frac{1}{h}(u_{i,j+1} - u_{i,j}) = \tfrac{1}{2}(u_{i,j+1} - u_{i,j}). \tag{16}$$

Similarly, the right side of (14) may be expressed by using (11) as

$$\left.\frac{\partial^2 u}{\partial x^2}\right|_{i,j} \approx \frac{1}{h^2}(u_{i+1,j} - 2u_{i,j} + u_{i-1,j}) = \tfrac{1}{4}(u_{i+1,j} - 2u_{i,j} + u_{i-1,j}). \tag{17}$$

Equating (16) and (17), we have

$$\tfrac{1}{2}(u_{i,j+1} - u_{i,j}) = \tfrac{1}{4}(u_{i+1,j} - 2u_{i,j} + u_{i-1,j})$$

or

$$u_{i,j+1} = \tfrac{1}{2}(u_{i+1,j} + u_{i-1,j}). \tag{18}$$

Observe that all the terms on the right side of (18) belong to row j, while the left side is on row $j + 1$. Since the values on row 0 are given by the initial condition, we can use (18) to obtain the values on row 1, then apply it again to obtain the values on row 2, and so forth. The calculation is shown in Table 1, where the value of $u_{i,j}$ appears in the ith column and jth row. For example $u(2, 10)$ is, by Fig. 2, equal to the value $u_{1,5}$ in column 1 and row 5; $u(2, 10) \approx u_{1,5} = 2$.

TABLE 1

	column 0	column 1	column 2	column 3	column 4
row 8	0	0.75	1	0.75	0
row 7	0	1	1.5	1	0
row 6	0	1.5	2	1.5	0
row 5	0	2	3	2	0
row 4	0	3	4	3	0
row 3	0	4	6	4	0
row 2	0	6	8	6	0
row 1	0	8	12	8	0
row 0	0	12	16	12	0

EXAMPLE 2

Consider the boundary value problem

$$\frac{\partial u}{\partial t} = \frac{1}{2}\frac{\partial^2 u}{\partial x^2}, \qquad 0 \le x \le 12, \qquad 0 \le t \le 12, \tag{19}$$

with boundary and initial conditions

$$u(x, 0) = \tfrac{1}{4}x(15 - x), \qquad 0 \le x \le 12,$$

$$u(0, t) = 0, \qquad u(12, t) = 9, \qquad 0 \le t \le 12.$$

If we use a square grid with points spaced $h = 3$ units apart, we have

$$u_{1,0} = \tfrac{1}{4}(3)(15 - 3) = 9, \qquad u_{2,0} = 13.5, \qquad u_{3,0} = 13.5$$

$$u_{0,j} = 0 \quad \text{and} \quad u_{4,j} = 9 \qquad \text{for} \quad j = 0, 1, 2, 3, 4.$$

TABLE 2

	column 0	column 1	column 2	column 3	column 4
row 4	0	6.604	10.639	11.104	9.0
row 3	0	7.083	11.292	11.583	9.0
row 2	0	7.625	12.000	12.125	9.0
row 1	0	8.250	12.750	12.750	9.0
row 0	0	9.000	13.500	13.500	9.0

Table 2 holds the results of our calculations. Here $u(9, 9) \approx u_{3,3} = 11.583$. The results are based on approximations similar to those in Example 1. Equation (19) becomes

$$\tfrac{1}{3}(u_{i,j+1} - u_{i,j}) = \tfrac{1}{2} \cdot \tfrac{1}{9}(u_{i+1,j} - 2u_{i,j} + u_{i-1,j})$$

or

$$u_{i,j+1} = \tfrac{1}{6}u_{i+1,j} + \tfrac{2}{3}u_{i,j} + \tfrac{1}{6}u_{i-1,j}. \tag{20}$$

From what we have done in Examples 1 and 2, it is apparent that the partial differential equation

$$\frac{\partial u}{\partial t} = \frac{\partial^2 u}{\partial x^2}$$

can be approximated by the finite difference equation

$$\frac{1}{h}(u_{i,j+1} - u_{i,j}) = \frac{1}{h^2}(u_{i+1,j} - 2u_{i,j} + u_{i-1,j})$$

or

$$u_{i,j+1} = \frac{1}{h}u_{i+1,j} + \left(1 - \frac{2}{h}\right)u_{i,j} + \frac{1}{h}u_{i-1,j}. \tag{21}$$

It can be shown† that (21) provides a useful approximation as long as $h \ge 2$. For h smaller than 2, (21) yields unbounded oscillations as time increases. Thus square meshes are useful only for very coarse (large h) approximations to the heat equation. For finer approximations we must employ a rectangular grid

---

† See W. F. Ames, *Numerical Methods for Partial Differential Equations*, Academic Press, New York, 1977, p. 45.

with t intervals of size Δt and x intervals of size Δx, with convergence guaranteed when

$$0 < \frac{\Delta t}{(\Delta x)^2} \le \frac{1}{2}.$$

The best results are obtained when $\Delta t/(\Delta x)^2 = \frac{1}{6}$.

Examples 1 and 2 provided an illustration of a *forward* difference equation, because we obtained each row explicitly based on the values of the previous row. The next example illustrates a situation where an explicit determination is more difficult, and iterative techniques may be useful.

EXAMPLE 3

Consider Laplace's equation

$$\frac{\partial^2 z}{\partial x^2} + \frac{\partial^2 z}{\partial y^2} = 0, \qquad 0 \le x \le 1, \qquad 0 \le y \le 1, \tag{22}$$

with boundary data

$$z(0, y) = z(1, y) = z(x, 1) = 0, \qquad 0 \le x \le 1, \qquad 0 \le y \le 1,$$

$$z(x, 0) = f(x), \qquad 0 < x < 1. \tag{23}$$

Using (11) and (12), we may approximate (22) on a square grid with points spaced h units apart by

$$\frac{1}{h^2}(z_{i+1, j} - 2z_{i, j} + z_{i-1, j}) + \frac{1}{h^2}(z_{i, j+1} - 2z_{i, j} + z_{i, j-1}) = 0,$$

or

$$z_{i, j} = \tfrac{1}{4}(z_{i+1, j} + z_{i-1, j} + z_{i, j+1} + z_{i, j-1}). \tag{24}$$

Equation (24) tells us that $z_{i, j}$ is the average of the four surrounding points. Note that this implies that no maxima or minima may exist at interior points of the grid; that is, all extrema must lie on the boundary of the grid.

Assume that the grid involves rows 0 to n and columns 0 to n [$(n + 1)^2$ grid points]. We have been given in (23) all of the values of the grid points on the boundary. The problem is to find the values of all the interior grid points.

For simplicity assume that $n = 4$, so that the nine points $z_{i, j}$, $1 \le i \le 3$, $1 \le j \le 3$, are the only values we need to determine (see Fig. 3). The points on the boundary of the grid are determined by the boundary conditions (24):

$$z_{01} = f(\tfrac{1}{4}), \qquad z_{02} = f(\tfrac{1}{2}), \qquad z_{03} = f(\tfrac{3}{4}),$$

and all the rest are zero.

Observe that we cannot immediately obtain z_{11} because z_{12} and z_{21} are not known. Similarly, z_{12} depends on the unknown values z_{11}, z_{13}, and z_{22}. Writing (24) as

$$z_{i+1, j} + z_{i-1, j} + z_{i, j+1} + z_{i, j-1} - 4z_{ij} = 0 \tag{25}$$

and taking all the *known* terms to the right side, we obtain nine equations in nine unknowns:

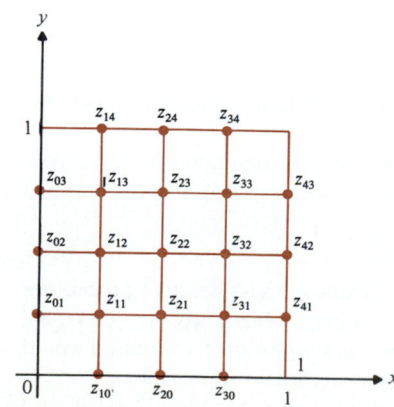

FIGURE 3

$$z_{21} + z_{12} - 4z_{11} = -z_{10} - z_{01}$$

$$z_{11} + z_{22} + z_{13} - 4z_{12} = -z_{02}$$

$$z_{12} + z_{23} - 4z_{13} = -z_{03} - z_{14}$$

$$z_{11} + z_{22} + z_{31} - 4z_{21} = -z_{20}$$

$$z_{12} + z_{21} + z_{23} + z_{32} - 4z_{22} = 0$$

$$z_{13} + z_{22} + z_{33} - 4z_{23} = -z_{24}$$

$$z_{21} + z_{32} - 4z_{31} = -z_{30} - z_{41}$$

$$z_{22} + z_{31} + z_{33} - 4z_{32} = -z_{42}$$

$$z_{23} + z_{32} - 4z_{33} = -z_{34} - z_{43}. \tag{26}$$

We can rewrite the linear system (26) in matrix notation as

$$
\overset{A}{
\begin{pmatrix}
-4 & 1 & 0 & 1 & 0 & 0 & 0 & 0 & 0 \\
1 & -4 & 1 & 0 & 1 & 0 & 0 & 0 & 0 \\
0 & 1 & -4 & 0 & 0 & 1 & 0 & 0 & 0 \\
1 & 0 & 0 & -4 & 1 & 0 & 1 & 0 & 0 \\
0 & 1 & 0 & 1 & -4 & 1 & 0 & 1 & 0 \\
0 & 0 & 1 & 0 & 1 & -4 & 0 & 0 & 1 \\
0 & 0 & 0 & 1 & 0 & 0 & -4 & 1 & 0 \\
0 & 0 & 0 & 0 & 1 & 0 & 1 & -4 & 1 \\
0 & 0 & 0 & 0 & 0 & 1 & 0 & 1 & -4
\end{pmatrix}}
\overset{\mathbf{z}}{
\begin{pmatrix}
z_{11} \\ z_{12} \\ z_{13} \\ z_{21} \\ z_{22} \\ z_{23} \\ z_{31} \\ z_{32} \\ z_{33}
\end{pmatrix}}
=
\overset{\mathbf{b}}{
\begin{pmatrix}
-z_{01} \\ -z_{02} \\ -z_{03} \\ 0 \\ 0 \\ 0 \\ 0 \\ 0 \\ 0
\end{pmatrix}}
\tag{27}
$$

or $A\mathbf{z} = \mathbf{b}$. (Most of the entries in $\mathbf{b}$ are zero since most of the boundary values are zero.) Note that the matrix A has a very definite pattern and contains a large number of zeros. Matrices with large numbers of zeros are called **sparse matrices**. Observe, also, that A is **diagonally dominant**. That is, the absolute value of the diagonal number in each row is greater than the sum of the absolute values of the other numbers in that row.

At this point two methods may be used to solve (27):

(i) We can use Gaussian elimination to obtain the solution explicitly (see Sections 7.3 and 14.1).

(ii) We can use the Jacobi or Gauss-Seidel method (or some other iterative technique) to approximate the solution by iteration (these methods will be explained in Section 14.2).

For our problem, either procedure would be adequate. (The iterative procedures converge because A is diagonally dominant.) For a larger problem, say $n = 1000$, the matrix would contain a million entries. Gaussian elimination in such a situation would present enormous difficulties. Iterative techniques are often used in this case.

Problems 13.9

In Problems 1–8 approximate the solution to the given boundary value problem by the methods of this section for a square grid of points spaced h units apart.

1. $\dfrac{\partial T}{\partial t} = \dfrac{\partial^2 T}{\partial x^2}$, $0 < x < 20$, $0 < t < 24$

$T(x, 0) = |10 - x|$, $0 \le x \le 20$

$T(0, t) = T(20, t) = 0$, $0 \le t \le 24$, $h = 4$

2. Repeat Problem 1 with $T(x, 0) = (10 - x)^2, 0 \le x \le 20$.

3. $\dfrac{\partial T}{\partial t} = \dfrac{\partial^2 T}{\partial x^2} + 2\,\dfrac{\partial T}{\partial x}$, $0 < x < 20$, $0 < t < 20$

$T(x, 0) = |10 - x|$, $0 \le x \le 20$

$T(0, t) = T(20, t) = 0$, $0 \le t \le 20$, $h = 4$

4. Repeat Problem 3 with $h = 5$.

5. Repeat Problem 3 with the boundary and initial conditions

$T(x, 0) = \tfrac{1}{2}x$, $0 \le x \le 20$

$T(0, t) = 0$, $T(20, t) = 10$, $0 \le t \le 20$, $h = 4$

6. $\dfrac{\partial^2 z}{\partial x^2} + \dfrac{\partial^2 z}{\partial y^2} = 0$, $0 < x < 1$, $0 < y < 1$

$z(x, 0) = x$, $z(1, y) = 1 - y$

$z(x, 1) = 0 = z(0, y)$, $h = \tfrac{1}{3}$

7. Repeat Problem 6 with $h = \tfrac{1}{4}$.

8. Repeat Problem 6 with $h = \tfrac{1}{4}$ and $z(x, 0) = x^2$.

9. Consider the heat equation

$$\frac{\partial u}{\partial t} = \frac{\partial^2 u}{\partial x^2}$$

(a) Use (10) and (11) to obtain the finite difference approximation of the heat equation with a square grid of size h.

(b) Use the substitution $u_{ij} = e^{(aj + \sqrt{-1}\,bi)h}$ in the difference equation in part (a) and show that

$$\sinh(ah) = -\frac{4}{h}\sin^2\!\left(\frac{bh}{2}\right).$$

(c) Use part (b) to show that

$$|e^{ah}| > 1,$$

implying that $|u_{ij}|$ grows without bound as j increases, regardless of the initial conditions. [This shows the approximation in part (a) is unstable, because any round-off error leads to arbitrarily large values for j sufficiently large. The computed solution would then bear very little resemblance to the exact solution.]

Review Exercises for Chapter 13

In Exercises 1–6, solve the given initial value problem using the methods of Chapter 1. Then use the

(a) improved Euler method or

(b) Runge-Kutta method

and the given value of h to obtain an approximate solution at the indicated value of x. Compare the numerical answer with the exact answer.

1. $\dfrac{dy}{dx} = \dfrac{e^x}{y}$, $y(0) = 2$. Find $y(3)$ with $h = \tfrac{1}{2}$.

2. $\dfrac{dy}{dx} = \dfrac{e^y}{x}$, $y(1) = 0$. Find $y(\tfrac{1}{2})$ with $h = -0.1$.

3. $\dfrac{dy}{dx} = \dfrac{y}{\sqrt{1 + x^2}}$, $y(0) = 1$. Find $y(3)$ with $h = \tfrac{1}{2}$.

4. $xy\dfrac{dy}{dx} = y^2 - x^2$, $y(1) = 2$. Find $y(3)$ with $h = \tfrac{1}{2}$.

5. $\dfrac{dy}{dx} = y - xy^3$, $y(0) = 1$. Find $y(3)$ with $h = \tfrac{1}{2}$.

6. $\dfrac{dy}{dx} = \dfrac{2xy}{3x^2 - y^2}$, $y(-\tfrac{3}{8}) = -\tfrac{3}{4}$. Find $y(6)$ with $h = \tfrac{3}{8}$.

7. Consider the differential equation in Exercise 6 with the initial condition $y(0) = -1$. Use the Runge-Kutta method or the improved Euler method to calculate $y(6)$ with $h = 1$. Why does the numerical solution differ from the exact answer?

8. Suppose the initial condition in Exercise 4 is $y(1) = 1$. Can any of the methods of Sections 13.5 and 13.7 provide the correct answer for $y(3)$?

9. Consider the initial value problem

$$y' = 1 + y^2, \qquad y(0) = 0.$$

Can any of the methods in Sections 13.5 or 13.7 be used to obtain $y(2)$?

In Exercises 10–17, use the trapezoidal rule (T) or Simpson's rule (S) to estimate the given integral with the given number of subintervals.

10. $\int_0^1 e^{-x} \, dx$;   T, $n = 10$

11. $\int_0^1 e^{-x} \, dx$;   S, $n = 10$

12. $\int_0^1 e^{x^3} \, dx$;   T, $n = 4$

13. $\int_0^1 e^{x^3} \, dx$;   S, $n = 4$

14. $\int_0^1 \dfrac{dx}{\sqrt{1+x^4}}$;   T, $n = 6$

15. $\int_0^1 \dfrac{dx}{\sqrt{1+x^4}}$;   S, $n = 6$

16. $\int_0^{\pi/2} \cos \sqrt{x} \, dx$; S, $n = 6$

17. $\int_{\pi/6}^{\pi/2} \ln(\sin x) \, dx$; S, $n = 8$

18. How many subintervals are needed in Exercise 12 to obtain a discretization error less than 0.01?

19. How many subintervals are needed in Exercise 13 to obtain a discretization error less than 0.00001?

20. Use Simpson's rule to estimate $\int_1^2 (1/x^2) \, dx$ with an error of less than 0.0001. Compare your answer with the actual answer, which is easily obtained by integration.

21. Answer the questions in Exercise 18 for the integral $\int_1^2 \ln x \, dx$.

In Exercises 22–27 find the Lagrange polynomial which interpolates the given function at the given points. Graph the function and its interpolating polynomial over the interval of interpolation.

22. $f(x) = \sin x$; $x = 0$; $x = \pi/4$; $x = \pi/2$

23. $f(x) = e^x$; $x = 0$; $x = 1$; $x = 2$

24. $f(x) = 1/(1 + x^2)$; $x = 0$; $x = 1$; $x = 2$

25. $f(x) = x^5$; $x = 0$; $x = 1$; $x = -1$

26. $f(x) = x^5$; $x = 0$; $x = 1$; $x = 2$; $x = 3$

27. $f(x) = x^5$; $x = \frac{1}{2}$; $x = 2.7$; $x = 5.3$; $x = 7.8$; $x = 10.4$; $x = 17.6$.

28. Given that $\sqrt[3]{2} \approx 1.259921$, use linear interpolation to estimate $\sqrt[3]{1.7}$. What is the maximum error of your estimate?

29. In Exercise 28, use the fact that $\sqrt[3]{1.5} \approx 1.144714$ together with quadratic interpolation to estimate $\sqrt[3]{1.7}$. What is the maximum error of your estimate?

30. Solve the boundary value problem

$$\frac{\partial u}{\partial t} = \frac{\partial^2 u}{\partial x^2}, \qquad 0 < x < 100, \qquad 0 < t < 50$$

$$u(x, 0) = x, \qquad u(0, t) = 0, \qquad u(100, t) = 100,$$

by approximating the partial differential equation by a finite difference equation. Let $h = 10$.

31. Approximate the solution of the boundary value problem

$$\frac{\partial^2 z}{\partial x^2} + \frac{\partial^2 z}{\partial y^2} = z, \qquad 0 < x < 1, \qquad 0 < y < 1$$

with

$$z(x, 0) = x, \qquad z(1, y) = 1 - y^2$$

$$z(x, 1) = 0, \qquad z(0, y) = 0, \qquad h = \tfrac{1}{4}$$

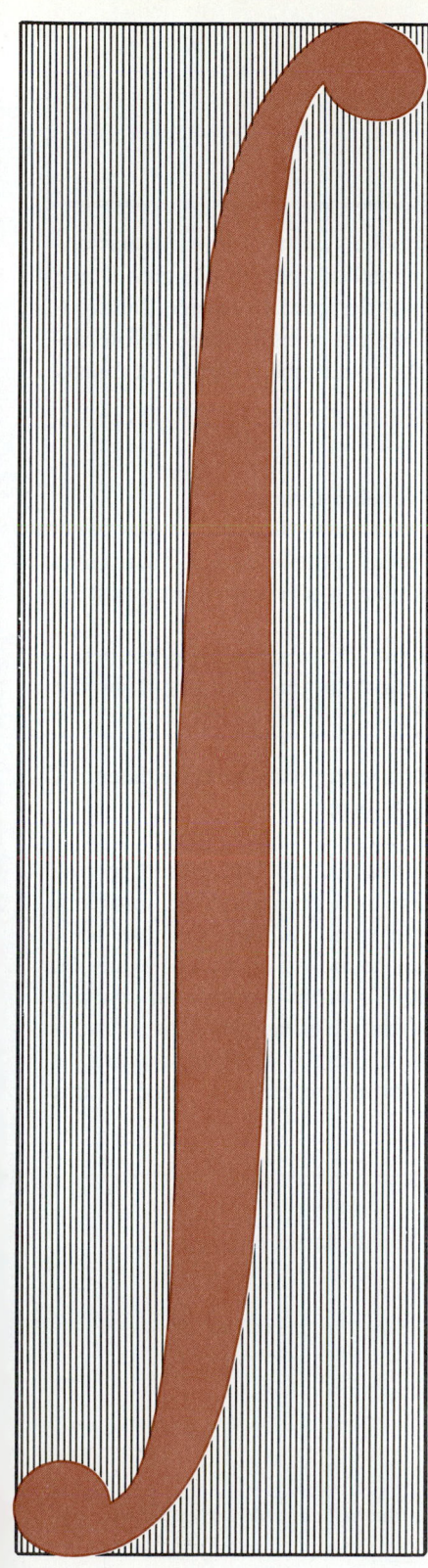

chapter 14

Numerical Methods for Matrices

In this chapter we discuss some of the numerical techniques that are available for solving systems of linear equations, finding eigenvalues, and inverting matrices. Although we shall limit our examples to problems involving a small number of variables, the methods we shall describe apply to systems involving large numbers of variables. Of course, as the number of variables increases, so does the number of calculations, leading to situations where the accumulated round-off error can be devastating. Some of these problems will be described in this chapter.

We shall use the notions of absolute error, ϵ_a, and relative error, ϵ_r, defined in Section 13.1. We urge the reader to become familiar with these ideas before proceeding with this chapter.

The techniques described in this chapter provide a superficial view of the methods available for dealing with matrices. Entire books, such as those listed below, are devoted to the subject.

Numerical Linear Algebra References

1. E. K. Blum, *Numerical Analysis and Computation: Theory and Practice,* Addison-Wesley, Reading, Mass., 1972.
2. R. L. Burden and J. D. Faires, *Numerical Analysis,* 3rd ed., Prindle, Weber and Schmidt, Boston, 1983.
3. S. D. Conte, *Elementary Numerical Analysis,* 2nd ed., McGraw-Hill, New York, 1972.
4. D. K. Faddeev and V. N. Faddeeva, *Computational Methods of Linear Algebra,* Freeman, San Francisco, 1963.
5. L. Fox, *An Introduction to Numerical Linear Algebra,* Oxford University Press, New York, 1965.

14.1 Systems of Linear Equations: Gaussian Elimination and Ill Conditioning

It is not difficult to program a computer to solve a system of linear equations by the Gaussian or Gauss-Jordan elimination method used in this text. There is, however, a variation of the method that was designed to reduce the accumulated round-off error in solving an $n \times n$ system of equations. Before describing this method, recall the definition of the row echelon form of a matrix (see p. 371).

DEFINITION

Row Echelon Form

Let A be an $m \times n$ matrix. Then A is in **row echelon form** if:

   i. All rows consisting entirely of zeros appear at the bottom of the matrix.
   ii. The first (starting from the left) number in any row not consisting entirely of zeros is a 1.
   iii. If two successive rows do not consist entirely of zeros, then the first 1 in the lower row occurs farther to the right than the first 1 in the higher row.

Note. If A is an $n \times n$ upper triangular matrix with 1's on the main diagonal, then it is easy to verify that A is in row echelon form.

From Chapter 7, it is apparent that any matrix can be reduced to row echelon form by Gaussian elimination. There is a computational problem with this method, however. If we divide by a small number that has been rounded, the result could contain a significant round-off error. For example, $1/0.00075 \approx 1333.3$ while $1/0.0008 = 1250.0$. To avoid this problem, we use a method called **Gaussian elimination with partial pivoting.** The idea is always to divide by the largest component in a column, thereby avoiding, so far as is possible, the type of error illustrated above. We describe the method with a simple example.

Gaussian Elimination With Partial Pivoting

EXAMPLE 1

Solve the following system by Gaussian elimination with partial pivoting:

$$\begin{aligned} x_1 - x_2 + x_3 &= 1, \\ -3x_1 + 2x_2 - 3x_3 &= -6, \\ 2x_1 - 5x_2 + 4x_3 &= 5. \end{aligned}$$

SOLUTION

STEP 1.
Write the system in augmented matrix form. From the first column with nonzero com-

Pivot

ponents (called the *pivot column*), select the component with the *largest absolute value*. This component is called the **pivot**:

$$\text{pivot} \longrightarrow \begin{pmatrix} 1 & -1 & 1 & | & 1 \\ \boxed{-3} & 2 & -3 & | & -6 \\ 2 & -5 & 4 & | & 5 \end{pmatrix}.$$

STEP 2.

Rearrange the rows to move the pivot to the top:

$$\begin{pmatrix} \boxed{-3} & 2 & -3 & | & -6 \\ 1 & -1 & 1 & | & 1 \\ 2 & -5 & 4 & | & 5 \end{pmatrix}$$

(first and second rows were interchanged).

STEP 3.

Divide the first row by the pivot:

$$\begin{pmatrix} 1 & -2/3 & 1 & | & 2 \\ 1 & -1 & 1 & | & 1 \\ 2 & -5 & 4 & | & 5 \end{pmatrix}$$

(first row divided by -3).

STEP 4.

Add multiples of the first row to the other rows to make all the other components in the pivot column equal to zero:

$$\begin{pmatrix} 1 & -2/3 & 1 & | & 2 \\ 0 & -1/3 & 0 & | & -1 \\ 0 & -11/3 & 2 & | & 1 \end{pmatrix}$$

(first row multiplied by -1 and -2 and added to the second and third rows).

STEP 5.

Delete the first row and column and perform steps 1–4 on the resulting *submatrix:*

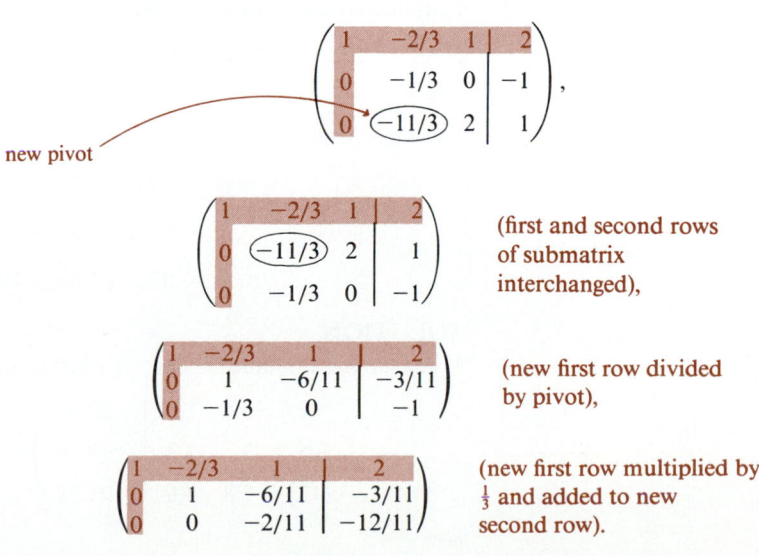

new pivot

(first and second rows of submatrix interchanged),

(new first row divided by pivot),

(new first row multiplied by $\frac{1}{3}$ and added to new second row).

STEP 6.

Continue in this manner until the matrix is in row echelon form:

$$\left(\begin{array}{ccc|c} 1 & -2/3 & 1 & 2 \\ 0 & 1 & -6/11 & -3/11 \\ 0 & 0 & \boxed{-2/11} & -12/11 \end{array}\right),$$

new pivot

$$\left(\begin{array}{ccc|c} 1 & -2/3 & 1 & 2 \\ 0 & 1 & -6/11 & -3/11 \\ 0 & 0 & 1 & 6 \end{array}\right)$$ (divided new first row by pivot).

STEP 7.

Use **back substitution** to find the solution (if any) to the system. Evidently, we have $x_3 = 6$. Then $x_2 - \frac{6}{11}x_3 = -\frac{3}{11}$ or

$$x_2 = -\frac{3}{11} + \frac{6}{11}x_3 = -\frac{3}{11} + \frac{6}{11}(6) = 3.$$

Finally, $x_1 - \frac{2}{3}x_2 + x_3 = 2$ or

$$x_1 = 2 + \frac{2}{3}x_2 - x_3 = 2 + \frac{2}{3}(3) - 6 = -2.$$

The unique solution is given by the vector $(-2, 3, 6)$.

Remark. **Complete pivoting** involves finding the component in A with largest absolute value, not just the component in the first nonzero column. The problem with this method is that it usually involves relabeling variables when the columns are interchanged to bring the pivot to the first column. For this reason the partial pivoting method described above is more popular.

We now examine the partial pivoting method applied to a computationally more difficult system. Calculations were done on a hand-calculator and were rounded to six significant digits.

EXAMPLE 2

Solve the system

$$2x_1 - 3.5x_2 + x_3 = 22.35,$$

$$-5x_1 + 3x_2 + 3.3x_3 = -9.08,$$

$$12x_1 + 7.8x_2 + 4.6x_3 = 21.38.$$

SOLUTION

Using the steps outlined above, we obtain, successively:

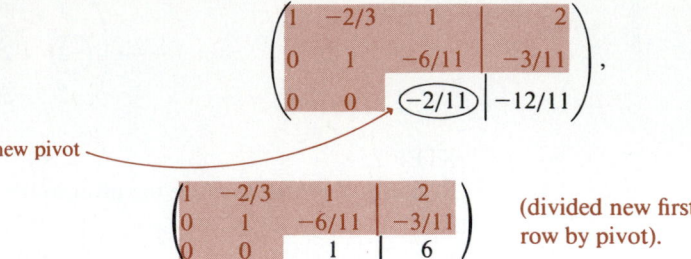

pivot

$$\xrightarrow{M_1(1/12)} \begin{pmatrix} 1 & 0.65 & 0.383333 & \bigm| & 1.78167 \\ -5 & 3 & 3.3 & \bigm| & -9.08 \\ 2 & -3.5 & 1 & \bigm| & 22.35 \end{pmatrix}$$

$$\xrightarrow[A_{1,3}(-2)]{A_{1,2}(5)} \begin{pmatrix} 1 & 0.65 & 0.383333 & \bigm| & 1.78167 \\ 0 & \boxed{6.25} & 5.21667 & \bigm| & -0.17165 \\ 0 & -4.8 & 0.233334 & \bigm| & 18.7867 \end{pmatrix}$$

new pivot

$$\xrightarrow{M_2(1/6.25)} \begin{pmatrix} 1 & 0.65 & 0.383333 & \bigm| & 1.78167 \\ 0 & 1 & 0.834667 & \bigm| & -0.027464 \\ 0 & -4.8 & 0.233334 & \bigm| & 18.7867 \end{pmatrix}$$

$$\xrightarrow{A_{2,3}(4.8)} \begin{pmatrix} 1 & 0.65 & 0.383333 & \bigm| & 1.78167 \\ 0 & 1 & 0.834667 & \bigm| & -0.027464 \\ 0 & 0 & \boxed{4.23974} & \bigm| & 18.6549 \end{pmatrix}$$

new pivot

$$\xrightarrow{M_3(1/4.23974)} \begin{pmatrix} 1 & 0.65 & 0.383333 & \bigm| & 1.78167 \\ 0 & 1 & 0.834667 & \bigm| & -0.027464 \\ 0 & 0 & 1 & \bigm| & 4.40001 \end{pmatrix}.$$

The matrix is now in row echelon form. Using back substitution, we obtain

$$x_3 = 4.40001,$$

$$x_2 = -0.027464 - 0.834667x_3 = -0.027464 - (0.834667)(4.40001)$$

$$= -3.70001,$$

$$x_1 = 1.78167 - (0.65)(x_2) - (0.383333)x_3 = 1.78167 - (0.65)(-3.70001)$$

$$-(0.383333)(4.40001) = 2.50001.$$

The correct solution is $x_1 = 2.5$, $x_2 = -3.7$, and $x_3 = 4.4$. Our answers are very accurate indeed.

Remark. This example illustrates the fact that it is tedious and inefficient to use this method without a calculator—especially if several significant digits of accuracy are required.

The next example shows how pivoting can significantly improve the answers. Here we round to only three significant digits, thereby introducing greater round-off errors.

EXAMPLE 3

Consider the system

$$\begin{aligned} 0.0002x_1 - 0.00031x_2 + 0.0017x_3 &= 0.00609, \\ 5x_1 \qquad\quad - 7x_2 \qquad + 6x_3 &= 7, \\ 8x_1 \qquad\quad + 6x_2 \qquad + 3x_3 &= 2. \end{aligned}$$

The exact solution is $x_1 = -2$, $x_2 = 1$, $x_3 = 4$. Let us first solve the system by Gaussian elimination without pivoting, rounding to three significant figures.

$$\begin{pmatrix} 0.0002 & -0.00031 & 0.0017 & | & 0.00609 \\ 5 & -7 & 6 & | & 7 \\ 8 & 6 & 3 & | & 2 \end{pmatrix} \xrightarrow{M_1(1/0.0002)} \begin{pmatrix} 1 & -1.55 & 8.5 & | & 30.5 \\ 5 & -7 & 6 & | & 7 \\ 8 & 6 & 3 & | & 2 \end{pmatrix}$$

$$\xrightarrow[A_{1,3}(-8)]{A_{1,2}(-5)} \begin{pmatrix} 1 & -1.55 & 8.5 & | & 30.5 \\ 0 & 0.75 & -36.5 & | & -146 \\ 0 & 18.4 & -65 & | & -242 \end{pmatrix} \xrightarrow{M_2(1/0.75)} \begin{pmatrix} 1 & -1.55 & 8.5 & | & 30.5 \\ 0 & 1 & -48.7 & | & -195 \\ 0 & 18.4 & -65 & | & -242 \end{pmatrix}$$

$$\xrightarrow{A_{2,3}(-18.4)} \begin{pmatrix} 1 & -1.55 & 8.5 & | & 30.5 \\ 0 & 1 & -48.7 & | & -195 \\ 0 & 0 & 831 & | & 3346 \end{pmatrix} \xrightarrow{M_3(1/831)} \begin{pmatrix} 1 & -1.55 & 8.5 & | & 30.5 \\ 0 & 1 & -48.7 & | & -195 \\ 0 & 0 & 1 & | & 4.03 \end{pmatrix}.$$

This yields

$$x_3 = 4.03,$$

$$x_2 = -195 + (48.7)(4.03) = 1.26,$$

$$x_1 = 30.5 + (1.55)(1.26) - 8.5(4.03) = -1.8.$$

Here the errors are significant. The relative errors, given as percentages, are

$$x_1: \quad \varepsilon_r = \left| \frac{-0.2}{2} \right| = 10\%,$$

$$x_2: \quad \varepsilon_r = \left| \frac{0.26}{1} \right| = 26\%,$$

$$x_3: \quad \varepsilon_r = \left| \frac{0.03}{4} \right| = 0.75\%.$$

Let us now repeat the procedure *with* pivoting. We obtain (with the pivots circled):

$$\begin{pmatrix} 0.0002 & -0.00031 & 0.0017 & | & 0.00609 \\ 5 & -7 & 6 & | & 7 \\ 8 & 6 & 3 & | & 2 \end{pmatrix}$$

$$\xrightarrow{P_{1,3}} \begin{pmatrix} \boxed{8} & 6 & 3 & | & 2 \\ 5 & -7 & 6 & | & 7 \\ 0.0002 & -0.00031 & 0.0017 & | & 0.00609 \end{pmatrix}$$

$$\xrightarrow{M_1(1/8)} \begin{pmatrix} 1 & 0.75 & 0.375 & | & 0.25 \\ 5 & -7 & 6 & | & 7 \\ 0.0002 & -0.00031 & 0.0017 & | & 0.00609 \end{pmatrix}$$

$$\xrightarrow[A_{1,3}(-0.0002)]{A_{1,2}(-5)} \begin{pmatrix} 1 & 0.75 & 0.375 & | & 0.25 \\ 0 & \boxed{-10.8} & 4.13 & | & 5.75 \\ 0 & -0.00046 & 0.00163 & | & 0.00604 \end{pmatrix}$$

$$\xrightarrow{M_2(-1/10.8)} \begin{pmatrix} 1 & 0.75 & 0.375 & | & 0.25 \\ 0 & 1 & -0.382 & | & -0.532 \\ 0 & -0.00046 & 0.00163 & | & 0.00604 \end{pmatrix}$$

$$A_{2,3}(0.00046) \longrightarrow \left(\begin{array}{ccc|c} 1 & 0.75 & 0.375 & 0.25 \\ 0 & 1 & -0.382 & -0.532 \\ 0 & 0 & \boxed{0.00145} & 0.0058 \end{array} \right)$$

$$M_3(1/0.00145) \longrightarrow \left(\begin{array}{ccc|c} 1 & 0.75 & 0.375 & 0.25 \\ 0 & 1 & -0.382 & -0.532 \\ 0 & 0 & 1 & 4.00 \end{array} \right).$$

Hence

$$x_3 = 4.00,$$

$$x_2 = -0.532 + (0.382)(4.00) = 0.996,$$

$$x_1 = 0.25 - 0.75(0.996) - (0.375)(4.00) = -2.00.$$

Thus, with pivoting and three-significant-digit rounding, x_1 and x_3 are obtained exactly and x_2 is obtained with the relative error of $0.004/1 = 0.4\%$.

Before leaving this section, we note that there are some matrices for which a small round-off error can have disastrous results. Such matrices are called **ill-conditioned.**

EXAMPLE 4

Consider the system

$$x_1 + x_2 = 1,$$

$$x_1 + 1.005x_2 = 0.$$

The solution is easily seen to be $x_1 = 201$, $x_2 = -200$. If, with or without pivoting, the coefficients are rounded to three significant digits, we obtain the system

$$x_1 + x_2 = 1,$$

$$x_1 + 1.01x_2 = 0,$$

with the solution $x_1 = 101$, $x_2 = -100$. Note that by rounding we introduced a relative error of $0.005/1.005 \approx 0.5\%$ in one coefficient but this induced an error of about 50% in the final answer!

There are techniques for recognizing and dealing with ill-conditioned matrices. Some of these are discussed in the references listed at the beginning of this chapter.

Problems 14.1

In Problems 1–4 solve the given system by Gaussian elimination with partial pivoting. Use a hand-calculator and round to six significant digits at every step.

1.
$$2x_1 - x_2 + x_3 = 0.3$$
$$-4x_1 + 3x_2 - 2x_3 = -1.4$$
$$3x_1 - 8x_2 + 3x_3 = 0.1$$

2. $4.7x_1 + 1.81x_2 + 2.6x_3 = -5.047$
$-3.4x_1 - 0.25x_2 + 1.1x_3 = 11.495$
$12.3x_1 + 0.06x_2 + 0.77x_3 = 7.9684$

3. $-7.4x_1 + 3.61x_2 + 8.04x_3 = 25.1499$
$12.16x_1 - 2.7x_2 - 0.891x_3 = 3.2157$
$-4.12x_1 + 6.63x_2 - 4.38x_3 = -36.1383$

4. $4.1x_1 - 0.7x_2 + 8.3x_3 + 3.9x_4 = -4.22$
$2.6x_1 + 8.1x_2 + 0.64x_3 - 0.8x_4 = 37.452$
$-5.3x_1 - 0.2x_2 + 7.4x_3 - 0.55x_4 = -25.73$
$0.8x_1 - 1.3x_2 + 3.6x_3 + 1.6x_4 = -7.7$

In Problems 5 and 6 solve the system by Gaussian elimination with and without pivoting by rounding to three significant figures. Then solve the system exactly and compute the relative errors of all six computed values.

5. $0.1x_1 + 0.05x_2 + 0.2x_3 = 1.3$
$12x_1 + 25x_2 - 3x_3 = 10$
$-7x_1 + 8x_2 + 15x_3 = 2$

6. $0.02x_1 + 0.03x_2 - 0.04x_3 = -0.04$
$16x_1 + 2x_2 + 4x_3 = 0$
$50x_1 + 10x_2 + 8x_3 = 6$

7. Show that the system

$$x_1 + x_2 = 50,$$
$$x_1 + 1.026x_2 = 20,$$

is ill-conditioned if rounding is done to three significant digits. What is the approximate relative error in each answer induced by rounding?

8. Use Gaussian elimination rounding to three significant digits to solve

$$-0.0001x_1 + x_2 = 2,$$
$$-x_1 + x_2 = 3.$$

(This shows that even a well-conditioned problem may yield poor answers if pivoting is avoided.)

14.2 Systems of Linear Equations: Iterative Methods

In the last section we developed a method for solving linear systems *directly*. That is, we carried out a fixed number of steps that led to a single answer. In numerical analysis, this procedure is the exception rather than the rule. A much more common procedure is called *iteration*. With iteration, the idea is to come up with a *sequence* of approximations to the answer. If things work well, this sequence will converge to the correct answer in the sense that each term or *iterate* in the sequence (after a certain stage) is a better approximation to the answer than the ones that precede it. This was the type of procedure that was used in Newton's formula of Section 13.2.

There are two commonly used iterative techniques for solving a system of equations $A\mathbf{x} = \mathbf{b}$: the **Jacobi**† **method** and the **Gauss-Seidel**‡ **method.** These methods are used under certain special circumstances. If A is ill-conditioned, for example, then, as we have seen, certain direct techniques fail. If the matrix A has a large number of zeros (A is then called a *sparse matrix*), iterative techniques will often provide better results with less work. The two methods do not always converge, however. After describing them, we shall examine some conditions under which the methods always converge. In the following discussion we assume that $\det A \neq 0$ so that the system has a unique solution.

---

† Carl Gustav Jacobi (1804–1851) was a brilliant and versatile German mathematician. See the biography of Jacobi on p. 656.

‡ We encountered the great Karl Fredrich Gauss in Chapter 1. P. L. V. Seidel (1821–1896) was a German mathematician. See the biography of Gauss on p. 366.

Jacobi Iteration

Let us illustrate the method by solving a particular system. First we note that since $\det A \neq 0$, A has no zero columns. Thus, by possibly rearranging the rows of A we can get a new coefficient matrix A' with nonzero diagonal components (see Problem 14). Hence we assume that for the $n \times n$ matrix $A = (a_{ij})$, $a_{ii} \neq 0$ for $i = 1, 2, \ldots, n$.

EXAMPLE 1

Solve the system

$$
\begin{aligned}
4.4x_1 - 2.3x_2 + 0.7x_3 &= -7.43, \\
0.8x_1 + 2.5x_2 + 1.1x_3 &= 12.17, \\
-1.6x_1 + 0.4x_2 - 5.2x_3 &= 26.12.
\end{aligned}
\tag{1}
$$

SOLUTION
The following computations are carried out to five significant digits.

STEP 1
Rewrite system (1) so that, in the ith equation, x_i is written in terms of the other variables:

$$
\begin{aligned}
x_1 &= -\frac{7.43}{4.4} + \frac{2.3}{4.4}x_2 - \frac{0.7}{4.4}x_3 = -1.6886 + 0.52273x_2 - 0.15909x_3, \\
x_2 &= \frac{12.17}{2.5} - \frac{0.8}{2.5}x_1 - \frac{1.1}{2.5}x_3 = 4.868 - 0.32x_1 - 0.44x_3, \\
x_3 &= -\frac{26.12}{5.2} - \frac{1.6}{5.2}x_1 + \frac{0.4}{5.2}x_2 = -5.0231 - 0.30769x_1 + 0.076923x_2.
\end{aligned}
\tag{2}
$$

STEP 2
Arbitrarily choose an initial approximation to the solution: $x_1^{(0)}$, $x_2^{(0)}$, $x_3^{(0)}$. If no other information is available, choose $x_1^{(0)} = x_2^{(0)} = x_3^{(0)} = 0$.

STEP 3
Substitute these initial values into the right-hand side of (2) to obtain the new approximation $x_1^{(1)}$, $x_2^{(1)}$, $x_3^{(1)}$:

$$
\begin{aligned}
x_1^{(1)} &= -1.6886 + 0 - 0 = -1.6886, \\
x_2^{(1)} &= 4.868 - 0 - 0 = 4.868, \\
x_3^{(1)} &= -5.0231 - 0 + 0 = -5.0231.
\end{aligned}
\tag{3}
$$

STEP 4
Use the values computed in step 3 to compute $x_1^{(2)}$, $x_2^{(2)}$, $x_3^{(2)}$ and continue in this fashion to generate the sequences $\{x_1^{(n)}\}$, $\{x_2^{(n)}\}$, $\{x_3^{(n)}\}$.

$$
\begin{aligned}
x_1^{(2)} &= -1.6886 + 0.52273x_2^{(1)} - 0.15909x_3^{(1)} \\
&= -1.6886 + 0.52273(4.868) - 0.15909(-5.0231) = 1.6552,
\end{aligned}
$$

$$x_2^{(2)} = 4.868 - 0.32x_1^{(1)} - 0.44x_3^{(1)}$$

$$= 4.868 - 0.32(-1.6886) - 0.44(-5.0231) = 7.6185,$$

$$x_3^{(2)} = -5.0231 - 0.30769x_1^{(1)} + 0.076923x_2^{(1)}$$

$$= -5.0231 - 0.30769(-1.6886) + 0.076923(4.868) = -4.1291.$$

Continuing in this fashion, we obtain Table 1 (rounded to five figures).

TABLE 1

Iterate	$x_1^{(n)}$	$x_2^{(n)}$	$x_3^{(n)}$
0	0	0	0
1	−1.6886	4.868	−5.0231
2	1.6552	7.6185	−4.1291
3	2.9507	6.1551	−4.9464
4	2.3158	6.1002	−5.4575
5	2.3684	6.5282	−5.2664
6	2.5617	6.4273	−5.2497
7	2.5063	6.3581	−5.3169
8	2.4808	6.4054	−5.3052
9	2.5037	6.4084	−5.2937
10	2.5034	6.3960	−5.3005
11	2.4980	6.3991	−5.3014
12	2.4998	6.4013	−5.2995
13	2.5006	6.3998	−5.2999
14	2.4999	6.3998	−5.3002
15	2.5000	6.4001	−5.3000

It appears that the sequences are converging to the values $x_1 = 2.5$, $x_2 = 6.4$, $x_3 = -5.3$. That these values are the solution can be verified by direct substitution. Note that, at least for this problem, the Jacobi iterates converge, but they converge rather slowly. The Gauss-Seidel method, which we now describe, can improve the speed of convergence.

Gauss-Seidel Iteration

If you look closely at the steps in Jacobi iteration, you will note some inefficiency in step 3. In computing $x_2^{(n)}$, we have already computed a new value for $x_1^{(n)}$ but have instead used the old value $x_1^{(n-1)}$. If the iterates are converging, it makes sense to use the latest available information. Thus, the steps in this method are identical to that of the Jacobi method, except that in step 3 we use the most recent value of each variable.

EXAMPLE 2

Solve the system of Example 1 by using the Gauss-Seidel method.

SOLUTION
We start, as before, with $x_1^{(0)} = x_2^{(0)} = x_3^{(0)} = 0$. Then, as before,

$$x_1^{(1)} = -1.6886 + 0 + 0 = -1.6886.$$

But the next step is different. Using this new approximation to x_1, we obtain

$$x_2^{(1)} = 4.868 - 0.32x_1^{(1)} - 0.44x_3^{(0)} = 4.868 - 0.32(-1.6886) = 5.4084.$$

Now we have new approximations for both x_1 and x_2. Using these values in system (2), we have

$$x_3^{(1)} = -5.0231 - 0.30769x_1^{(1)} + 0.076923x_2^{(1)}$$

$$= -5.0231 - 0.30769(-1.6886) + 0.076923(5.4084) = -4.0875.$$

Continuing in this way (always using the latest approximations), we obtain Table 2.

TABLE 2

Iterate	$x_1^{(n)}$	$x_2^{(n)}$	$x_3^{(n)}$
0	0	0	0
1	-1.6886	5.4084	-4.0875
2	1.7888	6.0941	-5.1047
3	2.3091	6.3752	-5.2432
4	2.4780	6.3820	-5.2946
5	2.4898	6.4009	-5.2968
6	2.5000	6.3986	-5.3001
7	2.4993	6.4003	-5.2998
8	2.5002	6.3998	-5.3001
9	2.5000	6.4000	-5.3000
10	2.5000	6.4000	-5.3000

Again we conclude that $x_1 = 2.5$, $x_2 = 6.4$, $x_3 = -5.3$. Note that the Gauss-Seidel iterations converge more rapidly than the Jacobi iterations.

Warning. It is usually (but not always) true that the Gauss-Seidel method is more efficient than the Jacobi method. In Example 7 we encounter a system for which the Jacobi iterates converge (slowly) but the Gauss-Seidel iterates diverge. The converse is also possible (see Problem 13).

Convergence

As mentioned above, these two methods do not always give a converging sequence of iterates. We cite below several conditions which ensure that the iterates converge. The proofs are beyond the scope of this text; for a good discussion of this problem, consult the book by Fox cited earlier.

DEFINITION
Let

$$A = \begin{pmatrix} a_{11} & a_{12} & \cdots & a_{1n} \\ a_{21} & a_{22} & \cdots & a_{2n} \\ \vdots & \vdots & & \vdots \\ a_{n1} & a_{n2} & \cdots & a_{nn} \end{pmatrix}.$$

Then A is **strictly diagonally dominant** if, in every row, the absolute value of the diagonal component is greater than the sum of the absolute values of the off-diagonal components. That is:

$$|a_{ii}| > |a_{i1}| + |a_{i2}| + \cdots + |a_{i,i-1}| + |a_{i,i+1}| + \cdots + |a_{in}| = \sum_{\substack{j=1 \\ j \neq i}}^{n} |a_{ij}| \qquad (4)$$

for $i = 1, 2, \ldots, n$.

EXAMPLE 3

The matrix $A = \begin{pmatrix} 4.4 & -2.3 & 0.7 \\ 0.8 & 2.5 & 1.1 \\ -1.6 & 0.4 & -5.2 \end{pmatrix}$ is strictly diagonally dominant because

$$|4.4| > |-2.3| + |0.7| = 3,$$

$$|2.5| > \ |0.8| + |1.1| = 1.9,$$

and

$$|-5.2| > |-1.6| + |0.4| = 2.$$

EXAMPLE 4

Consider the system

$$x_1 + 3x_2 - x_3 = \ \ 6,$$

$$4x_1 - x_2 + x_3 = \ \ 5,$$

$$x_1 + x_2 - 7x_3 = -9. \qquad (5)$$

The matrix $A = \begin{pmatrix} 1 & 3 & -1 \\ 4 & -1 & 1 \\ 1 & 1 & -7 \end{pmatrix}$ is *not* strictly diagonally dominant. However, if we interchange the first two equations in system (5) (which, of course, does not change the solutions), then the matrix of the rearranged system is $\begin{pmatrix} 4 & -1 & 1 \\ 1 & 3 & -1 \\ 1 & 1 & -7 \end{pmatrix}$, which *is* strictly diagonally dominant.

The importance of systems with strictly diagonally dominant coefficient matrices is given in the following theorem. Its proof is suggested in Problem 18.

■ **THEOREM 1**

If A is strictly diagonally dominant, then both the Jacobi and the Gauss-Seidel iterations converge to the unique solution of $A\mathbf{x} = \mathbf{b}$ for any vector $\mathbf{b}$. ■

Note. Since the matrix of Examples 1 and 2 is strictly diagonally dominant, by this theorem we know before doing any computations that both sequences of iterates will converge.

Remark. As we shall see in Example 6, there are matrices that are *not* strictly diagonally dominant but for which both sequences of iterates will converge.

Let A be an $n \times n$ matrix. Let L denote the $n \times n$ matrix consisting of the components of A below the main diagonal and zero everywhere else; D is the $n \times n$ matrix with the same diagonal components as A and zero everywhere else; U is the matrix consisting of the components of A above the main diagonal and zeros everywhere else. Here L, D, and U are called the **lower triangular,** the **diagonal,** and the **upper triangular** parts of A, respectively. We have, clearly,

$$A = L + D + U. \tag{6}$$

EXAMPLE 5

Let $A = \begin{pmatrix} 1 & 2 & 3 \\ 4 & 5 & 6 \\ 7 & 8 & 9 \end{pmatrix}$. Then $L = \begin{pmatrix} 0 & 0 & 0 \\ 4 & 0 & 0 \\ 7 & 8 & 0 \end{pmatrix}$, $D = \begin{pmatrix} 1 & 0 & 0 \\ 0 & 5 & 0 \\ 0 & 0 & 9 \end{pmatrix}$, and $U = \begin{pmatrix} 0 & 2 & 3 \\ 0 & 0 & 6 \\ 0 & 0 & 0 \end{pmatrix}$.

With this notation we have the following theorem.

■ **THEOREM 2**

Let $r(A)$ denote the absolute value of the eigenvalue of A with largest absolute value. Then, referring to the system $A\mathbf{x} = \mathbf{b}$ with $\det A \neq 0$:

i. The Jacobi iterates will converge if and only if

$$r[D^{-1}(L + U)] < 1. \tag{7}$$

ii. The Gauss-Seidel iterates will converge if and only if

$$r[(D + L)^{-1}U] < 1. \tag{8}$$

■

EXAMPLE 6

Let $A = \begin{pmatrix} 2 & 3 \\ 1 & 4 \end{pmatrix}$; then $L = \begin{pmatrix} 0 & 0 \\ 1 & 0 \end{pmatrix}$, $D = \begin{pmatrix} 2 & 0 \\ 0 & 4 \end{pmatrix}$, $U = \begin{pmatrix} 0 & 3 \\ 0 & 0 \end{pmatrix}$, $D^{-1} = \begin{pmatrix} 1/2 & 0 \\ 0 & 1/4 \end{pmatrix}$,

$D^{-1}(L + U) = \begin{pmatrix} 0 & 3/2 \\ 1/4 & 0 \end{pmatrix}$, and the eigenvalues of $D^{-1}(L + U)$ are $\pm\sqrt{\frac{3}{8}}$. Thus

$r[D^{-1}(L + U)] = \sqrt{\frac{3}{8}}$. Similarly, we find that $(D + L)^{-1}U = \begin{pmatrix} 0 & 3/2 \\ 0 & -3/8 \end{pmatrix}$ with eigenvalues

0 and $-\frac{3}{8}$. Thus $r[(D + L)^{-1}U] = \frac{3}{8}$. This provides an example of a matrix that is not strictly diagonally dominant but for which both the Jacobi and the Gauss-Seidel iterates will converge.

EXAMPLE 7

Let $A = \begin{pmatrix} 1 & 0 & 1 \\ -1 & 1 & 0 \\ 1 & 2 & -3 \end{pmatrix}$. Then $L = \begin{pmatrix} 0 & 0 & 0 \\ -1 & 0 & 0 \\ 1 & 2 & 0 \end{pmatrix}$, $D = \begin{pmatrix} 1 & 0 & 0 \\ 0 & 1 & 0 \\ 0 & 0 & -3 \end{pmatrix}$, and $U = \begin{pmatrix} 0 & 0 & 1 \\ 0 & 0 & 0 \\ 0 & 0 & 0 \end{pmatrix}$. We find that $D^{-1} = \begin{pmatrix} 1 & 0 & 0 \\ 0 & 1 & 0 \\ 0 & 0 & -1/3 \end{pmatrix}$ and $D^{-1}(L + U) =$

$\begin{pmatrix} 1 & 0 & 0 \\ 0 & 1 & 0 \\ 0 & 0 & -1/3 \end{pmatrix}\begin{pmatrix} 0 & 0 & 1 \\ -1 & 0 & 0 \\ 1 & 2 & 0 \end{pmatrix} = \begin{pmatrix} 0 & 0 & 1 \\ -1 & 0 & 0 \\ -1/3 & -2/3 & 0 \end{pmatrix}$.

The characteristic equation of $D^{-1}(L + U)$ is $\lambda^3 + \lambda/3 - \frac{2}{3} = 0$ with approximate roots $\lambda_1 \approx 0.748$, $\lambda_2 \approx -0.374 + 0.868i$, and $\lambda_3 \approx -0.374 - 0.868i$. We have $|\lambda_1| \approx 0.748$, $|\lambda_2| = |\lambda_3| \approx \sqrt{0.374^2 + 0.868^2} \approx 0.945$. Thus $r[D^{-1}(L + U)] \approx 0.945$ and *the Jacobi iterates will converge.* On the other hand, we find that

$$D + L = \begin{pmatrix} 1 & 0 & 0 \\ -1 & 1 & 0 \\ 1 & 2 & -3 \end{pmatrix} \quad \text{and}$$

$$(D + L)^{-1}U = \begin{pmatrix} 1 & 0 & 0 \\ 1 & 1 & 0 \\ 1 & 2/3 & -1/3 \end{pmatrix}\begin{pmatrix} 0 & 0 & 1 \\ 0 & 0 & 0 \\ 0 & 0 & 0 \end{pmatrix} = \begin{pmatrix} 0 & 0 & 1 \\ 0 & 0 & 1 \\ 0 & 0 & 1 \end{pmatrix}.$$

The characteristic equation of $(D + L)^{-1}U$ is $\lambda^2(\lambda - 1) = 0$ so that $\lambda_1 = \lambda_2 = 0$, $\lambda_3 = 1$, and $r[(D + L)^{-1}U] = 1$. Thus *the Gauss-Seidel iterates diverge.*

Remark. It can be further shown that the *rate* of convergence in the two methods depends on the values of $r[D^{-1}(L + U)]$ and $r[(D + L)^{-1}U]$. The smaller the value of r, the faster will be the rate of convergence.

Remark. Let $|A|$ denote the **max-row sum norm** of A:

$$|A| = \max_{1 \le i \le n} \sum_{j=1}^{n} |a_{ij}|. \tag{9}$$

It is possible to prove that, for any $n \times n$ matrix A,

$$r(A) \le |A|. \tag{10}$$

The following result then follows directly from Theorem 2.

■ **THEOREM 3**

Let A, D, L, and U be as in Theorem 2. Then:

i. The sequence of Jacobi iterates converges if

$$|D^{-1}(L+U)| < 1. \tag{11}$$

ii. The sequence of Gauss-Seidel iterates converges if

$$|(D+L)^{-1}U| < 1. \tag{12}$$

■

EXAMPLE 8

Let $A = \begin{pmatrix} 1 & 0 & 1/2 \\ 1/2 & 1 & 1 \\ 1/2 & 0 & 1 \end{pmatrix}$. Then A is not strictly diagonally dominant, but we can still

show that the Gauss-Seidel iterates will converge. For $D + L = \begin{pmatrix} 1 & 0 & 0 \\ 1/2 & 1 & 0 \\ 1/2 & 0 & 1 \end{pmatrix}$,

$(D+L)^{-1} = \begin{pmatrix} 1 & 0 & 0 \\ -1/2 & 1 & 0 \\ -1/2 & 0 & 1 \end{pmatrix}$ and $(D+L)^{-1}U = \begin{pmatrix} 1 & 0 & 0 \\ -1/2 & 1 & 0 \\ -1/2 & 0 & 1 \end{pmatrix}\begin{pmatrix} 0 & 0 & 1/2 \\ 0 & 0 & 1 \\ 0 & 0 & 0 \end{pmatrix} =$

$\begin{pmatrix} 0 & 0 & 1/2 \\ 0 & 0 & 3/4 \\ 0 & 0 & -1/4 \end{pmatrix}$. Thus $|(D+L)^{-1}U| = \frac{3}{4} < 1$.

Problems 14.2

In Problems 1–6 determine whether the given matrix is strictly diagonally dominant.

1. $\begin{pmatrix} 2 & 1 \\ 1 & 2 \end{pmatrix}$

2. $\begin{pmatrix} 3 & 3 \\ 4 & 5 \end{pmatrix}$

3. $\begin{pmatrix} 1 & 1/2 & 1/2 \\ 1/2 & 1 & 1/2 \\ 1/2 & 1/2 & 1 \end{pmatrix}$

4. $\begin{pmatrix} 1 & 1/2 & 1/3 \\ 1/2 & 1 & -1/3 \\ -1/2 & -1/3 & 1 \end{pmatrix}$

5. $\begin{pmatrix} 3 & -2 & 0 \\ 1 & -4 & 2 \\ -3 & 1 & -5 \end{pmatrix}$

6. $\begin{pmatrix} 6 & -2 & 3 \\ -3 & 5 & 2 \\ -2 & -4 & -7 \end{pmatrix}$

In Problems 7–12 solve the given system by using the Jacobi method and the Gauss-Seidel method. Carry out your computations until $|x_1^{(n)} - x_1^{(n-1)}|/|x_1^{(n)}|$ is smaller than the number given in parentheses. Start with all initial approximations equal to zero and use five significant figures.

7. $2x_1 - x_2 = 7$
    $3x_1 + 5x_2 = 4$ (0.01)

8. $3.3x_1 - 2.7x_2 = -0.6$
    $-4.2x_1 + 8.3x_2 = 11.95$ (0.001)

9. $3x_1 - x_2 + x_3 = 4$
    $2x_1 + 5x_2 + 2x_3 = -5$ (0.01)
    $x_1 + 2x_2 + 4x_3 = 20$

10. $3.8x_1 + 1.6x_2 + 0.9x_3 = 3.72$
    $-0.7x_1 + 5.4x_2 + 1.6x_3 = 3.16$ (0.001)
    $1.5x_1 + 1.1x_2 - 3.2x_3 = 43.78$

11. $5.2x_1 + 3.1x_2 - 1.6x_3 = 1.64$
    $1.7x_1 + 2.4x_2 + 0.3x_3 = 20.42$ (0.001)
    $-6.3x_1 - 3.7x_2 - 12.6x_3 = 0.27$

12. $-3.1x_1 + 1.9x_2 - 0.77x_3 = -12.806$
    $0.9x_1 - 2.4x_2 + 1.06x_3 = 12.165$ (0.0001)
    $7.6x_1 - 3.9x_2 + 16.5\ x_3 = 27.931$

13. Consider the system

$$x_1 + \tfrac{1}{2}x_2 + \tfrac{1}{2}x_3 = 2$$

$$\tfrac{1}{2}x_1 + x_2 + \tfrac{1}{2}x_3 = 2$$

$$\tfrac{1}{2}x_1 + \tfrac{1}{2}x_2 + x_3 = 2$$

(a) Show that the matrix of the system is not strictly diagonally dominant.

(b) Starting with $x_1^{(0)} = x_2^{(0)} = x_3^{(0)} = 0.8$, show that the Jacobi iterates oscillate back and forth between the values 0.8 and 1.2. That is, show that the sequence of Jacobi iterates diverges.

(c) Show that the Gauss-Seidel iterates converge to the solution $x_1 = x_2 = x_3 = 1$ by computing eight iterates and rounding to five significant figures.

* (d) Explain the results of parts (b) and (c) in light of Theorem 2.

* 14. Let A be an $n \times n$ matrix with det $A \neq 0$. Show that it is always possible to rearrange the rows of A so that the diagonal components of A are all nonzero.

15. Let A be a diagonal matrix with det $A \neq 0$. Show that $r[D^{-1}(L + U)] = r[(D + L)^{-1}U] = 0$.

16. Let A be an invertible upper or lower triangular matrix. Show that the sequences of Jacobi and Gauss-Seidel iterates always converge.

17. Let $A = \begin{pmatrix} a & b \\ c & d \end{pmatrix}$. Show that both the Jacobi and the Gauss-Seidel iterates converge if and only if $|bc/ad| < 1$. This shows that it is impossible to find an example of a 2×2 system where one sequence of iterates converges while the other does not.

* 18. Use part (i) of Theorem 3 to show that if A is strictly diagonally dominant, then the Jacobi iterates converge.

14.3 Error Analysis—or When Do We Stop?

In solving problems by iteration, there is always the question of determining when to stop. There are two ways to make this decision. First, we can agree to stop after a fixed number of iterations, say 10 or 20. But since we do not know how many iterations it will take to get a reasonably accurate answer, this method is not very useful.

A better device is to stop when the relative error ε_r is sufficiently small. Remember:

$$\varepsilon_r = \left| \frac{x^* - x}{x} \right|, \tag{1}$$

where x is the exact solution and x^* is the approximation. Of course, we cannot compute ε_r exactly since we do not know the exact answer x. (If we did, we wouldn't have any problem in the first place.) We can, however, for many numerical schemes, estimate the relative error in an iteration scheme by the formula

$$\varepsilon_r^{(n)} = \left| \frac{x^{(n)} - x^{(n-1)}}{x^{(n)}} \right|. \tag{2}$$

Formula (2) can be explained in the following way: If we know that the scheme converges, then the iterate $x^{(n)}$ is getting closer and closer to the "correct" answer x. Thus the absolute error $\varepsilon_a = |x^{(n)} - x|$ is approaching zero. But then, since $x^{(n)} \approx x$, we have $|x^{(n)} - x^{(n-1)}| \approx |x - x^{(n-1)}|$. This means that formula (2) approximates the true relative error

$$\varepsilon_r = \left| \frac{x^{(n-1)} - x}{x} \right|.$$

Thus we can agree to stop when $\varepsilon_r^{(n)}$ is smaller than some agreed upon value ε. Typically we make $\varepsilon = 0.1, 0.01, 0.001$, or some similar value. As an alternative, we can stop when $x_1^{(n)}$ and $x_1^{(n-1)}$ agree to a certain number of decimal places.

In Example 14.2.1, suppose we agree to iterate until the estimated relative error $\varepsilon_r^{(n)}$ in the computation of x_1 is less than 0.01. From Table 14.2.1 we get Table 1.

TABLE 1

Iterate	$x_1^{(n)}$	$\|x_1^{(n)} - x_1^{(n-1)}\|$	$\varepsilon_r^{(n)} = \left\| \frac{x_1^{(n)} - x_1^{(n-1)}}{x_1^{(n)}} \right\|$
0	0		
1	−1.6886	1.6886	1
2	1.6552	3.3438	2.02020
3	2.9507	1.2955	0.43905
4	2.3158	0.6349	0.27416
5	2.3684	0.0526	0.02221
6	2.5617	0.1933	0.07546
7	2.5063	0.0554	0.02210
8	2.4808	0.0255	0.01028
9	2.5037	0.0229	0.00915
10	2.5034	0.0003	0.00012

Here we would stop after the ninth iterate. This would give us the estimate $x_1 \approx 2.5037$. Since $x_1 = 2.5$, the true relative error is $0.0037/2.5 = 0.00148$. It is apparent that this method gives us only a crude measure of the relative error. It is easy, however, to compute the approximations $\varepsilon_r^{(n)}$; and, under conditions of convergence, they do provide a reasonable measure of how close we are getting to the right answer.

This method applies to *any* iterative procedure. For example, we can also use the estimated relative error to decide how small a step size we need when using the Euler methods or Runge-Kutta procedures. If we are working with Example 13.7.2

$$y' = x + y^2, \qquad y(1) = 0, \tag{3}$$

we can evaluate $y(1.5)$ for different step sizes h. Using Euler's method we obtain the values in Table 2. As we can see, the value of $y(1.5) \approx 0.6915$ obtained

TABLE 2 RELATIVE ERROR USING EULER'S
         METHOD

n	Step size h	$y_n(1.5)$	$\epsilon_r^{(n)} = \left\| \dfrac{y_n - y_{n-1}}{y_n} \right\|$
0	0.500	0.5000	—
1	0.250	0.5781	0.1351
2	0.100	0.6394	0.0959
3	0.050	0.6640	0.0370
4	0.025	0.6773	0.0196
5	0.010	0.6857	0.0123
6	0.005	0.6886	0.0042

using the improved Euler method or Runge-Kutta procedure is certainly within the relative error $(0.6915 - 0.6886 = 0.0029)$.

14.4 Computing Eigenvalues and Eigenvectors

As we have seen, the computation of eigenvalues and eigenvectors for a given matrix A is important for a variety of applications. It is tempting to estimate eigenvalues by first finding the characteristic polynomial $p(\lambda) = \det(\lambda I - A)$ and then estimating, directly, the roots of $p(\lambda)$. There are two problems with this approach. First, polynomials are often ill-conditioned; that is, a small round-off error in the coefficients of the polynomial can lead to large errors in the roots. Second, even if the coefficients of $p(\lambda)$ are exact, it is still difficult to find all the roots of a polynomial. For these reasons a number of techniques have been devised for computing eigenvalues and eigenvectors directly. The first of these is used to compute the eigenvalue of largest absolute value.

DEFINITION

Dominant Eigenvalue and Eigenvector

Let $\lambda_1, \lambda_2, \ldots, \lambda_n$ be the eigenvalues of A. Then the **eigenvalue** λ_1 is **dominant** if

$$|\lambda_1| > |\lambda_i| \qquad \text{for } i = 2, \ldots, n. \tag{1}$$

If $\mathbf{v}_1$ is an eigenvector of A corresponding to λ_1, then $\mathbf{v}_1$ is called a **dominant eigenvector.**

EXAMPLE 1

If the eigenvalues of A are $-4, -2, 1, 3$, then -4 is dominant.

EXAMPLE 2

If the eigenvalues of A are $-5, 3, 5$, then A has no dominant eigenvalue since $|-5| = |5|$.

Observe that if A is a real matrix, its complex eigenvalues occur in conjugate pairs. Thus, if A has a dominant eigenvalue, that eigenvalue cannot be complex, because its absolute value (if complex) would *equal* that of its conjugate.

We now describe a method, called the **power method**, for computing the dominant eigenvalue and eigenvector of a matrix.

The Power Method

Let $\lambda_1, \lambda_2, \ldots, \lambda_n$ be the eigenvalues of A and suppose that

$$|\lambda_1| > |\lambda_2| \geq |\lambda_3| \geq \cdots \geq |\lambda_n|. \tag{2}$$

That is, λ_1 is the dominant eigenvalue. Suppose further that A is diagonalizable; that is, A has n linearly independent eigenvectors $\mathbf{u}_1, \mathbf{u}_2, \ldots, \mathbf{u}_n$. Let $\mathbf{x}_0$ be a nonzero vector in $\mathbb{R}^n$. There are constants $c_1, c_2, \ldots, c_n$ such that

$$\mathbf{x}_0 = c_1\mathbf{u}_1 + c_2\mathbf{u}_2 + \cdots + c_n\mathbf{u}_n. \tag{3}$$

We assume that $c_1 \neq 0$. Define a sequence of iterates by the formula

$$\mathbf{x}_{n+1} = A\mathbf{x}_n. \tag{4}$$

Then

$$\mathbf{x}_1 = A\mathbf{x}_0 = c_1 A\mathbf{u}_1 + c_2 A\mathbf{u}_2 + \cdots + c_n A\mathbf{u}_n$$

$$= c_1\lambda_1\mathbf{u}_1 + c_2\lambda_2\mathbf{u}_2 + \cdots + c_n\lambda_n\mathbf{u}_n.$$

Continuing to multiply by powers of A, we find that

$$\mathbf{x}_2 = A\mathbf{x}_1 = A^2\mathbf{x}_0 = A(A\mathbf{x}_0) = c_1\lambda_1 A\mathbf{u}_1 + c_2\lambda_2 A\mathbf{u}_2 + \cdots + c_n\lambda_n A\mathbf{u}_n$$

$$= c_1\lambda_1^2\mathbf{u}_1 + c_2\lambda_2^2\mathbf{u}_2 + \cdots + c_n\lambda_n^2\mathbf{u}_n,$$

$$\vdots$$

$$\mathbf{x}_k = A^k\mathbf{x}_0 = c_1\lambda_1^k\mathbf{u}_1 + c_2\lambda_2^k\mathbf{u}_2 + \cdots + c_n\lambda_n^k\mathbf{u}_n, \tag{5}$$

or

$$\mathbf{x}_k = A^k\mathbf{x}_0 = \lambda_1^k\left[c_1\mathbf{u}_1 + c_2\left(\frac{\lambda_2}{\lambda_1}\right)^k\mathbf{u}_2 + \cdots + c_n\left(\frac{\lambda_n}{\lambda_1}\right)^k\mathbf{u}_n\right]. \tag{6}$$

Since $|\lambda_i| < |\lambda_1|$ for $i = 2, 3, \ldots, n$, we see that $|\lambda_i/\lambda_1|^k$ approaches zero as k increases. Therefore we may write

$$\mathbf{x}_k = A^k\mathbf{x}_0 \approx \lambda_1^k c_1\mathbf{u}_1. \tag{7}$$

Suppose $\mathbf{u}_1 = \begin{pmatrix} a_1 \\ a_2 \\ \vdots \\ a_n \end{pmatrix}$. Then $\lambda_1^k c_1 \mathbf{u}_1 = \begin{pmatrix} \lambda_1^k c_1 a_1 \\ \lambda_1^k c_1 a_2 \\ \vdots \\ \lambda_1^k c_1 a_n \end{pmatrix}$.

Now let a_j be nonzero. Then we form the quotient

$$\alpha_j^{(k+1)} = \frac{j\text{th component of } A^{k+1}\mathbf{x}_0}{j\text{th component of } A^k\mathbf{x}_0} \approx \frac{\lambda_1^{k+1} c_1 a_j}{\lambda_1^k c_1 a_j} = \lambda_1. \tag{8}$$

This gives us a method for computing λ_1. We simply look at the ratio of the jth components of $\mathbf{x}_{k+1}$ and $\mathbf{x}_k$ and let k get large. Moreover, once we have found λ_1, we also know an eigenvector corresponding to λ_1, because by equation (7),

$$\mathbf{x}_k \approx \lambda_1^k c_1 \mathbf{u}_1 \tag{9}$$

is an eigenvector corresponding to λ_1 since $\lambda_1^k c_1$ is a scalar and $\mathbf{u}_1$ is a corresponding eigenvector.

Warning. The power method described above will work only if A has a dominant eigenvalue.†

EXAMPLE 3

Use the power method to find the dominant eigenvalue and eigenvector of $A = \begin{pmatrix} -4 & -5 \\ 1 & 2 \end{pmatrix}$.

SOLUTION

Here $\mathbf{x}_0$ is arbitrary so we choose a simple value for it: $\mathbf{x}_0 = \begin{pmatrix} 1 \\ 1 \end{pmatrix}$. Then

$$\mathbf{x}_1 = A\mathbf{x}_0 = \begin{pmatrix} -4 & -5 \\ 1 & 2 \end{pmatrix}\begin{pmatrix} 1 \\ 1 \end{pmatrix} = \begin{pmatrix} -9 \\ 3 \end{pmatrix}, \qquad \alpha_1^{(1)} = \frac{-9}{1} = -9, \qquad \alpha_2^{(1)} = \frac{3}{1} = 3,$$

$$\mathbf{x}_2 = A\mathbf{x}_1 = \begin{pmatrix} -4 & -5 \\ 1 & 2 \end{pmatrix}\begin{pmatrix} -9 \\ 3 \end{pmatrix} = \begin{pmatrix} 21 \\ -3 \end{pmatrix}, \qquad \alpha_1^{(2)} = \frac{21}{-9} \approx -2.3333, \qquad \alpha_2^{(2)} = \frac{-3}{3} = -1.$$

Continuing in this fashion we obtain Table 1. All results are rounded to five significant figures.

---

† It can be shown that the power method will work even when A is not diagonalizable. In that case, however, convergence is at a slower rate.

TABLE 1

Iterate	$\mathbf{x}_k$ (as a row vector)	$\alpha_1^{(k)}$	$\alpha_2^{(k)}$
0	$(1, 1)$	—	—
1	$(-9, 3)$	-9	3
2	$(21, -3)$	-2.3333	-1
3	$(-69, 15)$	-3.2857	-5
4	$(201, -39)$	-2.9130	-2.6
5	$(-609, 123)$	-3.0299	-3.1538
6	$(1821, -363)$	-2.9901	-2.9512
7	$(-5469, 1095)$	-3.0033	-3.0165
8	$(16401, -3279)$	-2.9989	-2.9945
9	$(-49209, 9843)$	-3.0004	-3.0018

It appears that $\alpha_1^{(k)}$ and $\alpha_2^{(k)}$ are converging to -3—which, as is easily verified, is the dominant eigenvalue of A (the other one is $\lambda_2 = 1$). Moreover, we see that $\mathbf{v}_1 = \begin{pmatrix} -49209 \\ 9843 \end{pmatrix}$ is approximately equal to an eigenvector of A. To simplify this vector, we "normalize" it by dividing through by its largest component (in absolute value) $-49{,}209$ to obtain $\mathbf{v}_1' \approx \begin{pmatrix} 1 \\ -0.20002 \end{pmatrix} \approx \begin{pmatrix} 1 \\ -1/5 \end{pmatrix}$. You should verify that $\begin{pmatrix} 1 \\ -1/5 \end{pmatrix}$ is an eigenvector of A corresponding to the eigenvalue $\lambda_1 = -3$.

The Power Method with Scaling

In the last example we saw that the iterates grew very rapidly in size. To prevent this, we do what we did to complete the problem: We *normalize* or *scale* the vector $\mathbf{x}_k$ by dividing it by its largest component (in absolute value). If we call the new scaled iterate $\mathbf{x}_k'$, then:

$$\mathbf{x}_{k+1} = A\mathbf{x}_k'. \tag{10}$$

This new method is called the **power method with scaling.** It will give us an eigenvector $\mathbf{u}$ with largest component 1 and we can then find the dominant eigenvalue by solving the equation $A\mathbf{u} = \lambda_1 \mathbf{u}$ for λ_1.

EXAMPLE 4

Redo Example 3 by using the power method with scaling.

SOLUTION

If $\mathbf{x}_0 = \begin{pmatrix} 1 \\ 1 \end{pmatrix}$, then $\mathbf{x}_1 = \begin{pmatrix} -9 \\ 3 \end{pmatrix}$ as before, and

$$\mathbf{x}_1' = -\frac{1}{9}\begin{pmatrix} -9 \\ 3 \end{pmatrix} = \begin{pmatrix} 1 \\ -1/3 \end{pmatrix};$$

then

$$\mathbf{x}_2 = A\mathbf{x}_1' = \begin{pmatrix} -4 & -5 \\ 1 & 2 \end{pmatrix}\begin{pmatrix} 1 \\ -1/3 \end{pmatrix} = \begin{pmatrix} -7/3 \\ 1/3 \end{pmatrix},$$

so that

$$\mathbf{x}_2' = -\frac{3}{7}\begin{pmatrix} -7/3 \\ 1/3 \end{pmatrix} = \begin{pmatrix} 1 \\ -1/7 \end{pmatrix} = \begin{pmatrix} 1 \\ -0.14286 \end{pmatrix}.$$

We carry out further iterations in Table 2.

TABLE 2

Iterate	$\mathbf{x}_k$	$\mathbf{x}_k'$ (normalized)
0	(1, 1)	(1, 1)
1	(−9, 3)	(1, −0.33333)
2	(−2.3333, 0.33333)	(1, −0.14286)
3	(−3.2857, 0.71428)	(1, −0.21739)
4	(−2.9131, 0.56522)	(1, −0.19403)
5	(−3.0299, 0.61194)	(1, −0.20197)
6	(−2.9902, 0.59606)	(1, −0.19934)
7	(−3.0033, 0.60132)	(1, −0.20022)
8	(−2.9989, 0.59956)	(1, −0.19993)
9	(−3.0004, 0.60014)	(1, −0.20002)

As before, we can conclude that $\mathbf{v} = \begin{pmatrix} 1 \\ -0.2 \end{pmatrix}$ is an eigenvector of A corresponding to λ_1. Then $A\mathbf{v} = \lambda_1\mathbf{v}$ or $\begin{pmatrix} -4 & -5 \\ 1 & 2 \end{pmatrix}\begin{pmatrix} 1 \\ -0.2 \end{pmatrix} = \begin{pmatrix} \lambda_1 \\ -0.2\lambda_1 \end{pmatrix}$, which yields $\begin{pmatrix} -3 \\ 0.6 \end{pmatrix} = \begin{pmatrix} \lambda_1 \\ -0.2\lambda_1 \end{pmatrix}$. Hence $\lambda_1 = -3$.

Note. In Table 2 the first component of $\mathbf{x}_k$ tends to -3. This must be the case since, for k large,

$$\mathbf{x}_k = A\mathbf{x}_{k-1}' \approx \lambda_1\mathbf{x}_{k-1}'.$$

But the first component of $\mathbf{x}_{k-1}'$ is 1. Hence the first component of $\mathbf{x}_k \approx \lambda_1$. This means that we can find λ_1 directly from the table.

Deflation

The power method has the obvious drawback of giving us only the dominant eigenvalue. There are many ways to compute other eigenvalues. We examine one method here. First we need the following result.

THEOREM 1

Let $\lambda_1, \lambda_2, \ldots, \lambda_n$ be the eigenvalues of A. Let λ_1 be the dominant eigenvalue with eigenvector $\mathbf{u}_1$. Let $\mathbf{v}$ be a column vector such that $\mathbf{u}_1 \cdot \mathbf{v} = 1$. If the matrix B is given by

$$B = A - \lambda_1 \mathbf{u}_1 \mathbf{v}^t, \tag{11}$$

then the eigenvalues of B are $\{0, \lambda_2, \lambda_3, \ldots, \lambda_n\}$.†

The proof of this theorem can be found on p. 239 in the book by Blum referenced at the beginning of this chapter.

EXAMPLE 5

In Example 3 we had $A = \begin{pmatrix} -4 & -5 \\ 1 & 2 \end{pmatrix}$, $\lambda_1 = -3$, and $\mathbf{u}_1 = \begin{pmatrix} 1 \\ -1/5 \end{pmatrix}$. If $\mathbf{v} = \begin{pmatrix} 1/2 \\ -5/2 \end{pmatrix}$, then $\mathbf{u}_1 \cdot \mathbf{v} = 1$ and

$$B = A - \lambda_1 \mathbf{u}_1 \mathbf{v}^t = \begin{pmatrix} -4 & -5 \\ 1 & 2 \end{pmatrix} + 3 \begin{pmatrix} 1 \\ -1/5 \end{pmatrix}(1/2, -5/2)$$

$$= \begin{pmatrix} -4 & -5 \\ 1 & 2 \end{pmatrix} + 3 \begin{pmatrix} 1/2 & -5/2 \\ -1/10 & 1/2 \end{pmatrix}$$

$$= \begin{pmatrix} -4 & -5 \\ 1 & 2 \end{pmatrix} + \begin{pmatrix} 3/2 & -15/2 \\ -3/10 & 3/2 \end{pmatrix} = \begin{pmatrix} -5/2 & -25/2 \\ 7/10 & 7/2 \end{pmatrix}.$$

Then $\det B = 0$; hence zero is an eigenvalue of B, as expected. The other eigenvalue of B is the second eigenvalue of A. We compute this by the power method with scaling. Starting with $x_0 = \begin{pmatrix} 1 \\ 1 \end{pmatrix}$, we obtain

$$\mathbf{x}_1 = B\mathbf{x}_0 = \begin{pmatrix} -2.5 & -12.5 \\ 0.7 & 3.5 \end{pmatrix}\begin{pmatrix} 1 \\ 1 \end{pmatrix} = \begin{pmatrix} -15 \\ 4.2 \end{pmatrix}.$$

Then

$$\mathbf{x}_1' = \begin{pmatrix} 1 \\ -0.28 \end{pmatrix}$$

and

$$\mathbf{x}_2 = B\mathbf{x}_1' = \begin{pmatrix} -2.5 & -12.5 \\ 0.7 & 3.5 \end{pmatrix}\begin{pmatrix} 1 \\ -0.28 \end{pmatrix} = \begin{pmatrix} 1 \\ -0.28 \end{pmatrix}.$$

Thus, without further ado, we see that $\begin{pmatrix} 1 \\ -0.28 \end{pmatrix}$ is an eigenvector of B (but not of A) corresponding to the eigenvalue $\lambda_2 = 1$. Therefore the eigenvalues of $\begin{pmatrix} -4 & -5 \\ 1 & 2 \end{pmatrix}$ are -3 and 1.

---

† Note that since $\mathbf{u}_1$ is an $n \times 1$ matrix (a column vector) and $\mathbf{v}$ is also an $n \times 1$ matrix, then $\mathbf{v}^t$ is an $1 \times n$ matrix and $\mathbf{u}_1 \mathbf{v}^t$ is an $n \times n$ matrix.

Note. It is not a coincidence that $B\mathbf{x}_0$ is an eigenvector of B if B is a 2×2 matrix—this is *always* the case. (See Problem 14 for a suggestion as to why this is so.)

EXAMPLE 6

Compute the eigenvalues of $A = \begin{pmatrix} 4 & -1 & 1 \\ -1 & 3 & -2 \\ 1 & -2 & 3 \end{pmatrix}$ by the power method with scaling and deflation.

SOLUTION

Let $\mathbf{x}_0 = \begin{pmatrix} 1 \\ 1 \\ 1 \end{pmatrix}$. Then $\mathbf{x}_1 = A\mathbf{x}_0 = \begin{pmatrix} 4 \\ 0 \\ 2 \end{pmatrix}$ and $\mathbf{x}_1' = \begin{pmatrix} 1 \\ 0 \\ 0.5 \end{pmatrix}$. Similarly, $\mathbf{x}_2 = A\mathbf{x}_1' = \begin{pmatrix} 4.5 \\ -2 \\ 2.5 \end{pmatrix}$ and

$\mathbf{x}_2' = \begin{pmatrix} 1 \\ -0.44444 \\ 0.55556 \end{pmatrix}$. Continuing in this manner we obtain the values in Table 3.

TABLE 3

Iterate	$\mathbf{x}_k$	$\mathbf{x}_k'$	$\alpha_k = $ 1st component of $\mathbf{x}_k$
0	(1, 1, 1)	(1, 1, 1)	1
1	(4, 0, 2)	(1, 0, 0.5)	4
2	(4.5, −2, 2.5)	(1, −0.44444, 0.55556)	4.5
3	(5, −3.4444, 3.5556)	(1, −0.68888, 0.71112)	5
4	(5.4, −4.4889, 4.5111)	(1, −0.83128, 0.83539)	5.4
5	(5.6667, −5.1646, 5.1687)	(1, −0.91139, 0.91212)	5.6667
6	(5.8235, −5.5584, 5.5591)	(1, −0.95448, 0.95460)	5.8235
7	(5.9091, −5.7726, 5.7728)	(1, −0.97690, 0.97693)	5.9091
8	(5.9538, −5.8846, 5.8846)	(1, −0.98838, 0.98838)	5.9538
9	(5.9768, −5.9419, 5.9419)	(1, −0.99416, 0.99416)	5.9768
10	(5.9883, −5.9708, 5.9708)	(1, −0.99708, 0.99708)	5.9883

It appears that the α_k's are converging to $\lambda_1 = 6$ with corresponding eigenvector

$\mathbf{u}_1 = \begin{pmatrix} 1 \\ -1 \\ 1 \end{pmatrix}$. You should verify this. Next we find a vector $\mathbf{v}$ such that $\mathbf{u}_1 \cdot \mathbf{v} = 1$. One ob-

vious choice is $\mathbf{v} = \begin{pmatrix} 1/3 \\ -1/3 \\ 1/3 \end{pmatrix}$. Then

$$\mathbf{u}_1 \mathbf{v}^t = \begin{pmatrix} 1 \\ -1 \\ 1 \end{pmatrix}(1/3, -1/3, 1/3) = \begin{pmatrix} 1/3 & -1/3 & 1/3 \\ -1/3 & 1/3 & -1/3 \\ 1/3 & -1/3 & 1/3 \end{pmatrix},$$

so that

$$B = A - \lambda_1 \mathbf{u}_1 \mathbf{v}^t = \begin{pmatrix} 4 & -1 & 1 \\ -1 & 3 & -2 \\ 1 & -2 & 3 \end{pmatrix} - 6\begin{pmatrix} 1/3 & -1/3 & 1/3 \\ -1/3 & 1/3 & -1/3 \\ 1/3 & -1/3 & 1/3 \end{pmatrix}$$

$$= \begin{pmatrix} 4 & -1 & 1 \\ -1 & 3 & -2 \\ 1 & -2 & 3 \end{pmatrix} - \begin{pmatrix} 2 & -2 & 2 \\ -2 & 2 & -2 \\ 2 & -2 & 2 \end{pmatrix} = \begin{pmatrix} 2 & 1 & -1 \\ 1 & 1 & 0 \\ -1 & 0 & 1 \end{pmatrix}.$$

We see that $\det B = 0$; hence zero is an eigenvalue of B. To find the dominant eigenvalue of B, we again use the power method with scaling. The results are tabulated in Table 4.

TABLE 4

Iterate	$\mathbf{x}_k$	$\mathbf{x}'_k$	$\alpha_k = $ 1st component of $\mathbf{x}_k$
0	$(1, 1, 1)$	$(1, 1, 1)$	1
1	$(2, 2, 0)$	$(1, 1, 0)$	2
2	$(3, 2, -1)$	$(1, 0.66667, -0.33333)$	3
3	$(3, 1.6667, -1.3333)$	$(1, 0.55557, -0.44443)$	3
4	$(3, 1.5556, -1.4444)$	$(1, 0.51853, -0.48147)$	3
5	$(3, 1.5185, -1.4815)$	$(1, 0.50617, -0.49383)$	3
6	$(3, 1.5062, -1.4938)$	$(1, 0.50207, -0.49793)$	3
7	$(3, 1.5021, -1.4979)$	$(1, 0.50070, -0.49930)$	3

Now it seems that the iterates are converging to $\lambda_2 = 3$ and $\mathbf{u}'_2 = (1, 1/2, -1/2)$. Again this can be verified. Although $\mathbf{u}_2$ is an eigenvector of both A and B, this is not always the case. (In Example 5, for instance, $\begin{pmatrix} 1 \\ -0.28 \end{pmatrix}$ was an eigenvector of B but not of A.)

Finally, we use deflation again to find the last eigenvalue of B (and therefore of A). We set $\mathbf{v}_1 = \begin{pmatrix} 1 \\ 0 \\ 0 \end{pmatrix}$. Then $\mathbf{u}_2 \cdot \mathbf{v}_1 = 1$, $\mathbf{u}_2 \mathbf{v}'_1 = \begin{pmatrix} 1 \\ 1/2 \\ -1/2 \end{pmatrix}(1, 0, 0) = \begin{pmatrix} 1 & 0 & 0 \\ 1/2 & 0 & 0 \\ -1/2 & 0 & 0 \end{pmatrix}$ and

$$C = B - \lambda_2 \mathbf{u}_2 \mathbf{v}'_1 = \begin{pmatrix} 2 & 1 & -1 \\ 1 & 1 & 0 \\ -1 & 0 & 1 \end{pmatrix} - \begin{pmatrix} 3 & 0 & 0 \\ 3/2 & 0 & 0 \\ -3/2 & 0 & 0 \end{pmatrix} = \begin{pmatrix} -1 & 1 & -1 \\ -1/2 & 1 & 0 \\ 1/2 & 0 & 1 \end{pmatrix}.$$ We shall omit the iteration, which shows that the dominant eigenvalue of C is $\lambda_3 = 1$. Thus the eigenvalues of A are 6, 3, and 1.

The power method together with deflation provides a reasonable way to find the eigenvalues of A if no two eigenvalues of A have the same absolute value and if each approximation to an eigenvalue is a good one. If, for example, λ_1 is inaccurate, then the computation of λ_2 by deflation could be a good deal more inaccurate.

There are many other ways to compute, numerically, the eigenvalues of a square matrix. One method that works fairly well on a symmetric matrix is called **Jacobi's method.** The idea is to compute a sequence of orthogonal matrices whose diagonal components approach the eigenvalues of A. Many of the references listed at the beginning of the chapter discuss that method. Finally, we note that the decision "when to stop" can be made, as in the last section, by computing approximate values of the relative error $\varepsilon_r^{(n)}$.

Problems 14.4

In Problems 1–6 estimate the dominant eigenvalue and eigenvector of A by using the power method with scaling.

1. $\begin{pmatrix} -2 & -2 \\ -5 & 1 \end{pmatrix}$

2. $\begin{pmatrix} 8 & 3 \\ -3 & -2 \end{pmatrix}$

3. $\begin{pmatrix} -22.3 & -32 \\ 12 & 17.7 \end{pmatrix}$

4. $\begin{pmatrix} 1 & -1 & 4 \\ 3 & 2 & -1 \\ 2 & 1 & -1 \end{pmatrix}$

5. $\begin{pmatrix} 3 & 2 & 4 \\ 2 & 0 & 2 \\ 4 & 2 & 3 \end{pmatrix}$ **6.** $\begin{pmatrix} 5 & 4 & 2 \\ 4 & 5 & 2 \\ 2 & 2 & 2 \end{pmatrix}$

7. Use the power method to estimate the dominant eigenvalue of $A = \begin{pmatrix} 1 & 7 \\ 6 & 3 \end{pmatrix}$:

(a) Rounding to five significant figures, continue the iterations until the estimated relative error $\varepsilon_r^{(n)} < 0.001$.

(b) Compute the dominant eigenvalue exactly. What is the exact value of ε_r?

8. For the matrix $A = \begin{pmatrix} -16.32 & 13 \\ 8 & 4.79 \end{pmatrix}$, follow the steps of Problem 7. Use six significant figures.

9. Show that the iterates of the power method fail to converge for the matrix $A = \begin{pmatrix} -3 & 5 \\ -2 & 3 \end{pmatrix}$. Explain why.

10. Repeat Problem 9 for the matrix $A = \begin{pmatrix} 2 & -1 \\ 5 & -2 \end{pmatrix}$.

In Problems 11–13 use deflation to find the other eigenvalues.

11. For the matrix of Problem 1

12. For the matrix of Problem 3

13. For the matrix of Problem 4

14. Let $A = \begin{pmatrix} a & 0 \\ 0 & b \end{pmatrix}$. Show that, for any 2-vector $\mathbf{x}_0$, $B\mathbf{x}_0$ is an eigenvector of B, where B is defined by equation (11).

Review Exercises for Chapter 14

In Exercises 1–4 reduce the given matrix to row echelon form.

1. $\begin{pmatrix} 2 & -4 & 6 \\ 1 & -3 & 5 \\ -4 & 9 & -13 \end{pmatrix}$

2. $\begin{pmatrix} 1 & 3 & 5 \\ 2 & 4 & 6 \\ -1 & -2 & 7 \end{pmatrix}$

3. $\begin{pmatrix} 1 & 3 & 5 \\ 2 & 4 & 6 \\ -1 & -2 & -3 \end{pmatrix}$

4. $\begin{pmatrix} 1 & 2 & 3 \\ 4 & 5 & 6 \\ 7 & 8 & 9 \end{pmatrix}$

In Exercises 5 and 6 solve the given system by Gaussian elimination with partial pivoting. Round to six significant digits at every step.

5.
$3.6x_1 + 8.2x_2 - 6.4x_3 = 1.26$
$-4.5x_1 - 5.9x_2 + 0.3x_3 = 2.57$
$0.7x_1 + 3.6x_2 - 4.8x_3 = 2.15$

6.
$1.3x_1 - 9.6x_2 + 5.35x_3 = 0.515$
$-12x_1 - 15x_2 + 3.8x_3 = -71.966$
$1.06x_1 - 22.2x_2 + 9.93x_3 = 1.809$

In Exercises 7–10 determine whether the given matrix is strictly diagonally dominant.

7. $\begin{pmatrix} -1 & 1/2 & 1/3 \\ -5/6 & 1 & 0 \\ 2 & 3/2 & -4 \end{pmatrix}$

8. $\begin{pmatrix} -1 & 1/3 & -1/3 \\ -5/6 & 1 & 1/6 \\ 2 & 3/2 & -4 \end{pmatrix}$

9. $\begin{pmatrix} 4 & -1 & 0 & 0 \\ -1 & 4 & -1 & 0 \\ 0 & -1 & 4 & -1 \\ 0 & 0 & -1 & 4 \end{pmatrix}$

10. $\begin{pmatrix} 2 & 1 & 0 & 0 \\ 1 & 2 & 1 & 0 \\ 0 & 1 & 2 & 1 \\ 0 & 0 & 1 & 2 \end{pmatrix}$

In Exercises 11 and 12 solve the given system by using the Jacobi and Gauss-Seidel methods. Carry out the iterations until the estimated relative error $\varepsilon_r^{(n)}$ is smaller than the number given in parentheses. Use six significant digits in all computations.

11.
$2.7x_1 - 0.9x_2 + 1.3x_3 = 6.98$
$-0.3x_1 + x_2 + 0.4x_3 = -2.77$ (0.001)
$4x_1 - 3.3x_2 + 9.6x_3 = 21.79$

12.
$42.31x_1 + 8.62x_2 + 19.4x_3 = -2.2502$
$-4.73x_1 + 80.4x_2 - 37.2x_3 = 3.5402$ (0.0001)
$8.37x_1 + 30.9x_2 - 57.4x_3 = -24.0858$

In Exercises 13–15 estimate the dominant eigenvalue and eigenvector by using the power method with scaling.

13. $\begin{pmatrix} 8 & -2 \\ 4 & 2 \end{pmatrix}$

14. $\begin{pmatrix} -6 & 3 \\ 6 & 1 \end{pmatrix}$

15. $\begin{pmatrix} 1 & -1 & 0 \\ -1 & 2 & -1 \\ 0 & -1 & 1 \end{pmatrix}$

16. Use deflation to find the second eigenvalue of the matrix of Exercise 14.

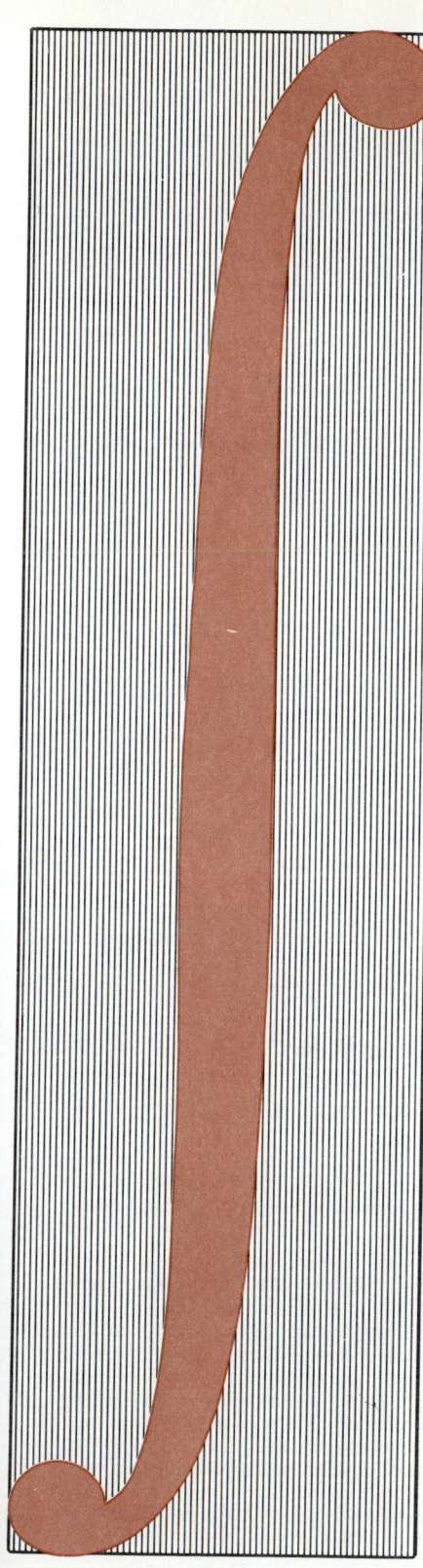

chapter 15

Analytic Functions of a Complex Variable

Complex numbers were first proposed by Girolamo Cardano,† in a monumental treatise on the solution of cubic and quartic equations entitled *Ars Magna,* in 1545. To appreciate the audacity of this proposal, one must realize that the concept of negative numbers had just gained acceptance and there was still controversy about their properties. Cardano's "fictitious" quantities were ignored by most mathematicians until the mathematical genius Carl Friedrich Gauss gave them their present name and used them in proving the Fundamental Theorem of Algebra, which states that every nonconstant polynomial has at least one zero. In this chapter and the four that follow, we shall study properties of complex numbers and complex-valued functions of a complex variable. We shall see that the theory of functions of a complex variable extends the concepts of calculus to the complex plane. In so doing, differentiation and integration acquire new depth and elegance, and the two-dimensional nature of the complex plane yields many results useful in applied mathematics.

15.1 Complex Numbers

In algebra we encounter the problem of finding the roots of the polynomial

$$\lambda^2 + a\lambda + b = 0. \tag{1}$$

To find the roots, we use the quadratic formula to obtain

$$\lambda = \frac{-a \pm \sqrt{a^2 - 4b}}{2}.$$

If $a^2 - 4b > 0$, there are two real roots. If $a^2 - 4b = 0$, we obtain the single

† See the biographical sketch of Cardano on p. 873.

root (of multiplicity 2) $\lambda = -a/2$. To deal with the case $a^2 - 4b < 0$, we introduce the **imaginary number**†

$$i = \sqrt{-1}, \qquad \text{so that} \quad i^2 = -1.$$

Then for $a^2 - 4b < 0$

$$\sqrt{a^2 - 4b} = \sqrt{(4b - a^2)(-1)} = \sqrt{4b - a^2}\,\sqrt{-1} = \sqrt{4b - a^2}\,i,$$

and the two roots of (1) are given by

$$\lambda_1 = -\frac{a}{2} + \frac{\sqrt{4b - a^2}}{2}i \quad \text{and} \quad \lambda_2 = -\frac{a}{2} - \frac{\sqrt{4b - a^2}}{2}i.$$

EXAMPLE 1

Find the roots of the quadratic equation $\lambda^2 + 2\lambda + 5 = 0$.

SOLUTION

We have $a = 2$, $b = 5$, and $a^2 - 4b = -16$. Thus $\sqrt{a^2 - 4b} = \sqrt{-16} = \sqrt{16}\,\sqrt{-1} = 4i$, and the roots are

$$\lambda_1 = \frac{-2 + 4i}{2} = -1 + 2i \quad \text{and} \quad \lambda_2 = -1 - 2i.$$

DEFINITION

Complex Number

A **complex number** is a number of the form

$$z = x + yi, \tag{2}$$

Real Part
Imaginary Part
Cartesian Form

where x and y are real numbers. x is called the **real part** of z and is denoted Re z. y is called the **imaginary part** of z and is denoted Im z. Representation (2) is sometimes called the **Cartesian form** of the complex number z.

---

† You should not be troubled by the term "imaginary." It's just a name. The British mathematician Alfred North Whitehead, in the chapter on imaginary numbers in his *Introduction to Mathematics,* wrote:

> At this point it may be useful to observe that a certain type of intellect is always worrying itself and others by discussion as to the applicability of technical terms. Are the incommensurable numbers properly called numbers? Are the positive and negative numbers really numbers? Are the imaginary numbers imaginary, and are they numbers?—are types of such futile questions. Now, it cannot be too clearly understood that, in science, technical terms are names arbitrarily assigned, like Christian names to children. There can be no question of the names being right or wrong. They may be judicious or injudicious; for they can sometimes be so arranged as to be easy to remember, or so as to suggest relevant and important ideas. But the essential principle involved was quite clearly enunciated in Wonderland to Alice by Humpty Dumpty, when he told her, apropos of his use of words, 'I pay them extra and make them mean what I like'. So we will not bother as to whether imaginary numbers are imaginary, or as to whether they are numbers, but will take the phrase as the arbitrary name of a certain mathematical idea, which we will now endeavour to make plain.

Remark. If $y = 0$ in equation (2), then $z = x$ is a real number. In this context we can regard the set of real numbers as a subset of the set of complex numbers.

EXAMPLE 2

The following are complex numbers.

(a) $z = 1 + 2i$ (Re $z = 1$, Im $z = 2$)
(b) $z = -3 + 5i$ (Re $z = -3$, Im $z = 5$)
(c) $z = 4$ (Re $z = 4$, Im $z = 0$)
(d) $z = -8i$ (Re $z = 0$, Im $z = -8$)

GIROLAMO CARDANO
(1501–1576)

Courtesy of the Granger Collection

Girolamo Cardano is one of the most extraordinary characters in the history of mathematics. He was born in Pavia, Italy, in 1501 as the illegitimate son of a jurist and developed into a man of passionate contrasts. He commenced his turbulent professional life as a doctor, studying, teaching, and writing mathematics while practicing his profession. He once traveled as far as Scotland, and, upon his return to Italy, he successively held important chairs at the Universities of Pavia and Bologna. He was imprisoned for a time for heresy because he published a horoscope of Christ's life. Resigning his chair in Bologna, he moved to Rome and became a distinguished astrologer, receiving a pension as astrologer to the papal court. He died in Rome in 1576, by his own hand, one story says, so as to fulfill his earlier astrological prediction of the date of his death. Many stories are told of his wickedness, as when in a fit of rage he cut off the ears of his younger son. Some of the stories could be exaggerations of his enemies, and it may be that he has been maligned. His autobiography, of course, supports this viewpoint.

One of the most gifted and versatile men of his time, Cardano wrote a number of works on arithmetic, astronomy, physics, medicine, and other subjects. His greatest work is his *Ars Magna,* the first great Latin treatise devoted solely to algebra. Here notice is taken of negative roots of an equation and some attention is paid to computations with imaginary numbers. There also occurs a crude method for obtaining an approximate value of a root of an equation of any degree. There is evidence that he was familiar with "Descartes' rule of signs." An inveterate gambler, Cardano wrote a gambler's manual in which some interesting questions on probability are considered.

Notation. We shall denote the set of complex numbers by the symbol $\mathbb{C}$. Thus

$$\mathbb{C} = \{z : z = x + yi \quad \text{where} \quad x, y \in \mathbb{R}\}.$$

Recall that the real numbers satisfy the following five rules of algebra, called the **field axioms.**

Field Axioms

1. **Commutative laws**

   $$a + b = b + a \quad \text{and} \quad ab = ba.$$

2. **Associative laws**

   $$(a + b) + c = a + (b + c) \quad \text{and} \quad (ab)c = a(bc).$$

3. **Distributive laws**

   $$a(b + c) = ab + ac \quad \text{and} \quad (a + b)c = ac + bc.$$

4. **Identities.** The **additive identity** 0 and **multiplicative identity** 1 satisfy

   $$a + 0 = a = 0 + a \quad \text{and} \quad a \cdot 1 = a = 1 \cdot a.$$

5. **Inverses.** Each real number a has an **additive inverse** $(-a)$, and, if $a \neq 0$, a **multiplicative inverse** a^{-1} satisfying

   $$a + (-a) = 0 = (-a) + a \quad \text{and} \quad aa^{-1} = 1 = a^{-1}a.$$

The complex numbers can be shown to satisfy the same axioms once we have defined the addition, scalar multiplication, and multiplication of complex numbers. Let $z_1 = x_1 + y_1 i$ and $z_2 = x_2 + y_2 i$.

Addition of Complex Numbers

$$z_1 + z_2 = (x_1 + x_2) + (y_1 + y_2)i \tag{3}$$

That is, the sum of two complex numbers is obtained by adding their corresponding real and imaginary parts.

EXAMPLE 3

Let $z_1 = 3 + 2i$ and $z_2 = 7 - 5i$. Then

$$z_1 + z_2 = (3 + 7) + (2 - 5)i = 10 - 3i.$$

Scalar Multiplication of a Complex Number

If $a \in \mathbb{R}$ and $z = x + yi \in \mathbb{C}$, then

$$az = ax + ayi. \tag{4}$$

EXAMPLE 4

If $z = 7 - 5i$, then

$$4z = 28 - 20i \quad \text{and} \quad -5z = -35 + 25i.$$

Multiplication of Complex Numbers

$$z_1 z_2 = (x_1 x_2 - y_1 y_2) + (x_1 y_2 + x_2 y_1)i. \tag{5}$$

Remark. Rule (5) makes sense because $i^2 = -1$. To see this we multiply in the "ordinary" way, treating i as an algebraic symbol:

$$(x_1 + y_1 i)(x_2 + y_2 i) = x_1 x_2 + x_1 y_2 i + x_2 y_1 i + y_1 y_2 \overset{\overset{\displaystyle = -1}{\downarrow}}{i^2}$$

$$= x_1 x_2 - y_1 y_2 + (x_1 y_2 + x_2 y_1)i.$$

EXAMPLE 5

Let $z = 2 + 3i$ and $w = 5 - 4i$. Compute zw.

SOLUTION

$$zw = (2 + 3i)(5 - 4i) = (2)(5) + 2(-4i) + (3i)(5) + (3i)(-4i)$$

$$= 10 - 8i + 15i - 12i^2 = 10 + 7i + 12$$

$$= 22 + 7i.$$

We now return to the axioms listed on p. 874. You are asked to show (in Problems 46–48) that complex numbers satisfy the commutative, associative, and distributive laws that real numbers satisfy. Also, it is easy to see that $0 + z = z + 0 = z$ and $z \cdot 1 = z \cdot 1 = z$ for any complex number z. What about the additive and multiplicative inverses? The additive inverse is easy:

$$z + (-z) = (x + iy) + (-x - iy) = 0 + i0 = 0.$$

To define the multiplicative inverse, we need to be able to divide complex numbers. Before we do this we need a definition.

DEFINITION

Conjugate of a Complex Number

Let $z = x + iy$. Then the **conjugate** of z, denoted $\bar{z}$, is given by

$$\bar{z} = x - iy. \tag{6}$$

EXAMPLE 6

If $z = 3 + 5i$, then $\bar{z} = 3 - 5i$.

In algebraic manipulations, it is useful to use the conjugate because of the following algebraic fact:

$$z\bar{z} = (x + iy)(x - iy) = x^2 - xyi + xyi + y^2 = x^2 + y^2. \tag{7}$$

We can use this fact to find the multiplicative inverse of a complex number. We illustrate the procedure with an example.

EXAMPLE 7

Find the inverse of $3 + 5i$.

SOLUTION

To obtain $(3 + 5i)^{-1}$, we multiply and divide by the conjugate $3 - 5i$ to obtain

$$(3 + 5i)^{-1} = \frac{1}{3 + 5i} = \frac{3 - 5i}{(3 + 5i)(3 - 5i)}$$

$$\overset{\text{from (7)}}{\underset{\downarrow}{=}} \frac{3 - 5i}{3^2 + 5^2} = \frac{3 - 5i}{34} = \frac{3}{34} - \frac{5}{34}i.$$

EXAMPLE 8

Express the quotient $\dfrac{1 - 2i}{3 - 4i}$ as a complex number.

SOLUTION

Multiplying numerator and denominator by the complex conjugate $3 - (-4i) = 3 + 4i$ of the denominator, we have

$$\frac{1-2i}{3-4i} = \frac{(1-2i)(3+4i)}{(3-4i)(3+4i)} = \frac{3-6i+4i-8i^2}{9+16} = \frac{(3+8)-2i}{25}$$

$$= \frac{11}{25} - \frac{2}{25}i.$$

We mention two other identities involving the conjugate:

$$z + \bar{z} = (x+iy)+(x-iy) = 2x = 2 \operatorname{Re} z, \tag{8}$$

and

$$z - \bar{z} = (x+iy)-(x-iy) = 2iy = 2i \operatorname{Im} z. \tag{9}$$

It is not difficult to show (see Problem 35) that

$$\bar{z} = z \text{ if and only if } z \text{ is real.} \tag{10}$$

Pure Imaginary Number

If $z = \beta i$ with β real, then z is said to be **pure imaginary.** We can then show (see Problem 36) that

$$\bar{z} = -z \text{ if and only if } z \text{ is pure imaginary.} \tag{11}$$

If $z_1 = x_1 + iy_1$ and $z_2 = x_2 + iy_2$, then

$$\overline{z_1 + z_2} = \overline{(x_1+x_2)+i(y_1+y_2)} = (x_1+x_2) - i(y_1+y_2)$$

$$= (x_1 - iy_1) + (x_2 - iy_2) = \bar{z}_1 + \bar{z}_2.$$

Thus, the complex conjugate of a sum of complex numbers is the sum of their conjugates:

$$\overline{z_1 + z_2} = \bar{z}_1 + \bar{z}_2.$$

Similarly, we can show (see Problems 42–44) that

$$\overline{z_1 - z_2} = \bar{z}_1 - \bar{z}_2,$$

$$\overline{z_1 z_2} = \bar{z}_1 \bar{z}_2,$$

and, if $z_2 \neq 0$,

$$\overline{(z_1/z_2)} = \bar{z}_1 / \bar{z}_2.$$

We can plot complex numbers in the xy-plane by plotting Re z along the x-axis and Im z along the y-axis. Thus, each complex number can be thought of as a point or vector in the xy-plane. With this representation, the xy-plane is called the **complex plane.** Some representative points are plotted in Fig. 1.

Figure 2 depicts a representative value of z and $\bar{z}$.

Complex Plane

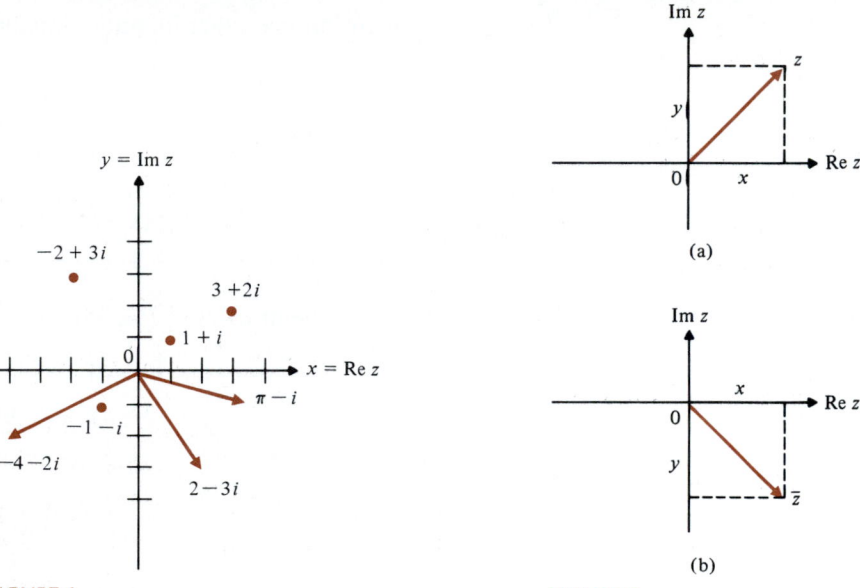

FIGURE 1   FIGURE 2

Absolute Value of a Complex Number

DEFINITION

For $z = x + iy$, we define the **magnitude** or **length** or **absolute value** of z, denoted by $|z|$, by

$$|z| = \sqrt{x^2 + y^2}. \tag{12}$$

From (7), we see that $|z| = \sqrt{z\bar{z}}$ or

$$z\bar{z} = |z|^2. \tag{13}$$

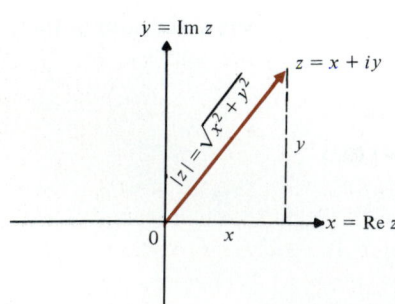

FIGURE 3

The magnitude of a complex number is easily depicted in the complex plane (see Fig. 3). We see, from the Pythagorean theorem, that $|z| = \sqrt{x^2 + y^2}$ is equal to the length of the hypotenuse of the right triangle with sides x and y.

EXAMPLE 9

Let $z = 7 - 2i$. Compute $|z|$.

SOLUTION

$$|z| = \sqrt{7^2 + (-2)^2} = \sqrt{49 + 4} = \sqrt{53}.$$

Problems 15.1

In Problems 1–28 express the given quantity as a single complex number.

1. $(2 - 3i) + (7 - 4i)$

2. $3(4 + i) - 5(-3 + 6i)$

3. $(1 + i)(1 - i)$

4. $(2 - 3i)(4 + 7i)$

5. $(-3 + 2i)(7 + 3i)$

6. $(1 + i)(1 + i)$

7. $\dfrac{1 + i}{1 - i}$

8. $\dfrac{1 - i}{1 + i}$

9. $\dfrac{2 + i}{3 - 4i}$

10. $\dfrac{3 - 4i}{2 + i}$

11. $\dfrac{3 - 2i}{4 + i}$

12. $\dfrac{4 + 5i}{1 - i}$

13. $(2 + i)(2 - i)$

14. $\dfrac{2 + i}{2 - i}$

15. $(2 + i)(5i)$

16. $\dfrac{2 + i}{5i}$

17. $\dfrac{5i}{2 + i}$

18. $\dfrac{5}{2 + i}$

19. $(1 + i)(1 + i)(1 + i)$

20. $\dfrac{7 + 9i}{-4 + 6i}$

21. $(1 - i)^2$

22. $(1 - i)^3$

23. $(1 - 2i)^2$

24. $i^2(1 + i)^3$

25. $\dfrac{2 + i}{3 - i} - \dfrac{4 + i}{1 + 2i}$

26. $\dfrac{3 + 2i}{1 + i} + \dfrac{5 - 2i}{-1 + i}$

27. $(1 + i)(1 + 2i)(1 + 3i)$

28. $(1 - i)(1 - 2i)(1 - 3i)$

In Problems 29–34 find the conjugate of the given complex number.

29. i

30. -7

31. $3 - 8i$

32. $-4 + 12i$

33. $5 + 6i$

34. $-2 - i$

35. Show that $z = x + iy$ is real if and only if $z = \bar{z}$. [*Hint:* If $z = \bar{z}$, show that $y = 0$.]

36. Show that $z = x + iy$ is pure imaginary if and only if $z = -\bar{z}$. [*Hint:* If $z = -\bar{z}$, show that $x = 0$.]

37. Prove that $\mathrm{Re}(iz) = -\mathrm{Im}\, z$.

38. Prove that $\mathrm{Re}(z) = \mathrm{Im}(iz)$.

39. Prove that if $z_1 z_2 = 0$, then $z_1 = 0$ or $z_2 = 0$.

40. Show that if $\mathrm{Im}\, z > 0$, then $\mathrm{Im}(1/z) < 0$.

* 41. Suppose $z_1 + z_2$ and $z_1 z_2$ are both negative real numbers. Prove that z_1 and z_2 must be real.

In Problems 42–44, let $z_1 = x_1 + iy_1$ and $z_2 = x_2 + iy_2$.

42. Show that $\overline{z_1 - z_2} = \bar{z}_1 - \bar{z}_2$.

43. Show that $\overline{z_1 z_2} = \bar{z}_1 \bar{z}_2$.

44. Show that $\overline{(z_1/z_2)} = \bar{z}_1/\bar{z}_2$, where $z_2 \neq 0$.

* 45. Let $p(z) = z^n + a_{n-1}z^{n-1} + a_{n-2}z^{n-2} + \cdots + a_1 z + a_0$, with $a_0, a_1, \ldots, a_{n-1}$ real numbers. Show that if $p(z) = 0$, then $p(\bar{z}) = 0$. That is: **The non-real roots of polynomials with real coefficients occur in complex conjugate pairs.**

46. Show that $z_1 + z_2 = z_2 + z_1$ and $z_1 z_2 = z_2 z_1$ for any complex numbers z_1 and z_2.

47. Show that $(z_1 + z_2) + z_3 = z_1 + (z_2 + z_3)$ and $(z_1 z_2)z_3 = z_1(z_2 z_3)$ for any complex numbers z_1, z_2, and z_3.

48. Show that $z_1(z_2 + z_3) = z_1 z_2 + z_1 z_3$ and $(z_1 + z_2)z_3 = z_1 z_3 + z_2 z_3$ for any complex numbers z_1, z_2, and z_3.

In his book *Ars Magna,* Girolamo Cardano included a method for finding the roots of the general cubic equation

$$z^3 + pz^2 + qz + r = 0$$

that had been discovered by Nicolo Tartaglia.

* 49. Show that the substitution $w = z + p/3$ reduces the general cubic to an equation of the form $w^3 + aw + b = 0$.

* 50. Show that the roots of the equation in Problem 49 are

$$w = A + B, \; -\frac{A + B}{2} + \frac{A - B}{2}\sqrt{3}i, \; -\frac{A + B}{2} - \frac{A - B}{2}\sqrt{3}i,$$

where $A = \sqrt[3]{-\dfrac{b}{2}+D}$, $B = \sqrt[3]{-\dfrac{b}{2}-D}$, and

$$D = \sqrt{\dfrac{b^2}{4}+\dfrac{a^3}{27}}.$$

* 51. Show that complex numbers are needed even for finding the *real* roots of $w^3 - 19w + 30 = 0$ by Tartaglia's method.

15.2 Polar Representation of a Complex Number and the Triangle Inequality

In Section 15.1 we saw that a complex number could be viewed as a vector. The length of the vector is given by the absolute value of z:

$$|z| = \sqrt{x^2 + y^2}.$$

Its direction is given by its polar angle θ between the line $0z$ and the positive x-axis. From Fig. 1 we see that $r = |z|$ is the distance from z to the origin and that when the angle θ is in the first or fourth quadrant, then

$$\theta = \tan^{-1}\frac{y}{x}. \tag{1}$$

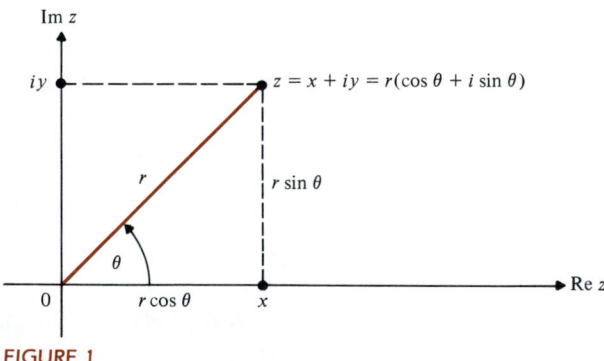

FIGURE 1

DEFINITION

Argument

We call the polar angle θ the **argument** of z and denote it by arg z. Note that the argument arg z is **multivalued** because the angles $\theta \pm 2\pi$, $\theta \pm 4\pi$, ... can also be used to locate the point z in polar coordinates. By convention, we call the value of the argument that lies in the interval

$$-\pi < \theta \le \pi \tag{2}$$

Principal Value

the **principal value** of the argument, and denote it by Arg z. Thus, the notation arg z is not specific, as it includes multiples of 2π, whereas the notation

$$\text{Arg } z + 2\pi k, \qquad k \text{ an integer,}$$

is used to identify a specific angle.

We know that tan θ is positive in the first and third quadrants and is negative in the second and fourth quadrants. Also,

$$\tan(\theta \pm \pi) = \frac{\sin(\theta \pm \pi)}{\cos(\theta \pm \pi)} = \frac{-\sin\theta}{-\cos\theta} = \tan\theta.$$

By definition, $\tan^{-1}\left(\dfrac{y}{x}\right)$ is a number in the interval $\left(\dfrac{-\pi}{2}, \dfrac{\pi}{2}\right)$. From these facts we conclude the following:

Values of θ = Arg z = Arg($x + iy$)

quadrant of θ	values of x and y	value of θ	range of θ
I	$x > 0, y > 0$	$\tan^{-1}\dfrac{y}{x}$	$0 < \theta < \dfrac{\pi}{2}$
II	$x < 0, y > 0$	$\pi + \tan^{-1}\dfrac{y}{x}$	$\dfrac{\pi}{2} < \theta < \pi$
III	$x < 0, y < 0$	$\tan^{-1}\dfrac{y}{x} - \pi$	$-\pi < \theta < -\dfrac{\pi}{2}$
IV	$x > 0, y < 0$	$\tan^{-1}\dfrac{y}{x}$	$-\dfrac{\pi}{2} < \theta < 0$

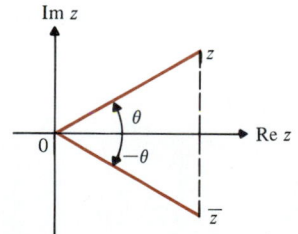

Im z

Re z

θ

$-\theta$

z

$\bar{z}$

0

FIGURE 2

From Fig. 2 we see that

$$|\bar{z}| = \sqrt{x^2 + (-y)^2} = \sqrt{x^2 + y^2} = |z|, \qquad (3)$$

and

$$\arg \bar{z} = -\arg z.\dagger \qquad (4)$$

† Equation (4) is a *multivalued* identity: each value of the left side corresponds to some value of the right side.

We can use $|z|$ and arg z to obtain a more convenient way of representing complex numbers. From Fig. 1 it is evident that if $z = x + iy$, $r = |z|$, and $\theta = $ arg z, then

$$x = r \cos \theta \quad \text{and} \quad y = r \sin \theta.$$

DEFINITION

Polar Form of a Complex Number

If $z = x + iy$, $r = |z|$, and $\theta = $ arg z, then the **polar form** of z is given by

$$z = r(\cos \theta + i \sin \theta).\dagger \qquad (5)$$

Note the following important facts:

$$|\cos \theta + i \sin \theta| = \sqrt{\cos^2 \theta + \sin^2 \theta} = \sqrt{1} = 1 \qquad (6)$$

and

$$\bar{z} = r[\cos(-\theta) + i \sin(-\theta)] = r(\cos \theta - i \sin \theta). \qquad (7)$$

EXAMPLE 1

Determine the polar forms of the following complex numbers, using only the principal value of their arguments (see Fig. 3):

  (a) 1
  (b) −1
  (c) i
  (d) $1 + i$
  (e) $-1 - \sqrt{3}\, i$
‡ (f) $-2 + 7i$.

SOLUTION

  (a) From Fig. 3a we see that arg $1 = 2\pi k$, k any integer. The principal value Arg $1 = 0$. Since $|1| = 1$, we have the polar form

$$1 = 1(\cos 0 + i \sin 0).$$

† Engineering books often use the notations $r \underline{/\theta}$ and r cis θ for the polar representation of z.
‡ ▤ This symbol means a calculator is needed to do the problem.

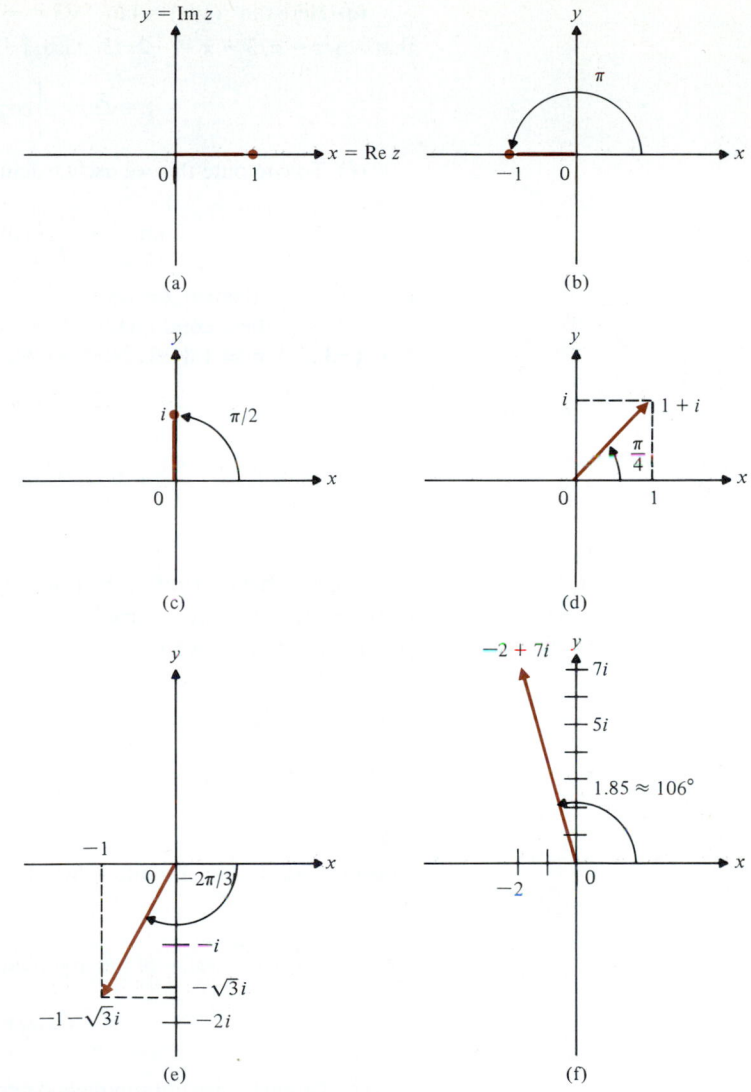

FIGURE 3

(b) Since $\text{Arg}(-1) = \pi$ (Fig. 3b) and $|-1| = 1$, we have

$$-1 = 1(\cos \pi + i \sin \pi).$$

(c) From Fig. 3c we see that the principal value of arg i is $\text{Arg}\, i = \pi/2$. Since $|i| = \sqrt{0^2 + 1^2} = 1$, it follows that

$$i = 1\left(\cos \frac{\pi}{2} + i \sin \frac{\pi}{2}\right).$$

(d) $\text{Arg}(1 + i) = \pi/4$ and $|1 + i| = \sqrt{1^2 + 1^2} = \sqrt{2}$, so that

$$1 + i = \sqrt{2}\left(\cos \frac{\pi}{4} + i \sin \frac{\pi}{4}\right).$$

(e) Here $\tan^{-1}(y/x) = \tan^{-1}\sqrt{3} = \pi/3$. However, Arg z is in the third quadrant, so that Arg $z = \pi/3 - \pi = -2\pi/3$. Also, $|-1 - i\sqrt{3}| = \sqrt{1^2 + (\sqrt{3})^2} = \sqrt{1+3} = 2$, so that

$$-1 - \sqrt{3}\,i = 2\left[\cos\left(-\frac{2\pi}{3}\right) + i\sin\left(\frac{-2\pi}{3}\right)\right].$$

(f) To compute this we use a calculator:

$$\tan^{-1}\left(-\frac{7}{2}\right) = \tan^{-1}(-3.5) \approx -1.2925.$$

But $\tan^{-1} x$ is defined for angles in the interval $(-\pi/2, \pi/2)$. Since (from Fig. 3f) $-2 + 7i$ is in the second quadrant, we see that the principle value of arg z is Arg $z = \tan^{-1}(-3.5) + \pi \approx 1.8491$. Next, we see that

$$|-2 + 7i| = \sqrt{(-2)^2 + 7^2} = \sqrt{53}.$$

Hence

$$-2 + 7i \approx \sqrt{53}(\cos 1.8491 + i\sin 1.8491).$$

We emphasize that each complex number (except 0) has an **infinite number** of polar forms—one for each value of the argument. We can speak about **the** polar form only if we limit ourselves to the principal value of the argument.

EXAMPLE 2

Find all polar representations of $1 - i$.

SOLUTION

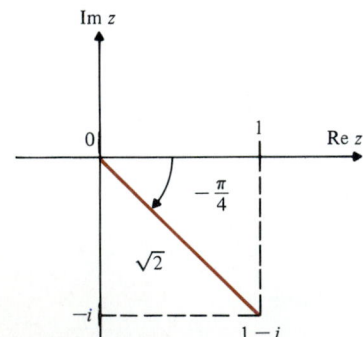

FIGURE 4 Polar representation of $1 - i$

Consider Fig. 4. The absolute value of $1 - i$ is

$$|1 - i| = \sqrt{1^2 + (-1)^2} = \sqrt{2},$$

while the principal value of the argument $1 - i$ is

$$\mathrm{Arg}(1 - i) = -\frac{\pi}{4}.$$

Since polar angles are not uniquely determined, the argument is

$$\arg(1 - i) = \frac{-\pi}{4} + 2\pi k,$$

where k is any integer. Thus, the polar representation of $1 - i$ is

$$1 - i = \sqrt{2}\left[\cos\left(\frac{-\pi}{4} + 2\pi k\right) + i\sin\left(\frac{-\pi}{4} + 2\pi k\right)\right].$$

We can use the geometric representation of a complex number in polar form to obtain a number of interesting results. The first makes use of the absolute value and the fact that a straight line segment is the curve of shortest length between two points.

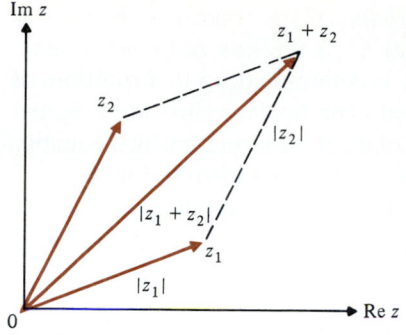

FIGURE 5 Triangle inequality:

$$|z_1 + z_2| \leqslant |z_1| + |z_2|$$

The Triangle Inequality

$$|z_1 + z_2| \leqslant |z_1| + |z_2|. \tag{8}$$

The triangle inequality follows by observing that the lengths $|z_1|$ and $|z_2|$ of the sides of the shaded triangle in Fig. 5 exceed $|z_1 + z_2|$. The triangle inequality can also be proved algebraically (see Problem 23).

EXAMPLE 3

We note that $6 + 2i = (2 + 3i) + (4 - i)$. To verify the triangle inequality for these three vectors, we observe that

$$|6 + 2i| = \sqrt{40} < \sqrt{13} + \sqrt{17} = |2 + 3i| + |4 - i|.$$

The multiplication of two complex numbers z and w has interesting geometric interpretations when we write both numbers in their polar representations. Let $\theta = \arg z$ and $\phi = \arg w$. Writing z and w in their polar representations, we have

$$z = |z|(\cos \theta + i \sin \theta) \quad \text{and} \quad w = |w|(\cos \phi + i \sin \phi).$$

Then

$$zw = |z||w|(\cos \theta + i \sin \theta)(\cos \phi + i \sin \phi)$$

$$= |z||w|[(\cos \theta \cos \phi - \sin \theta \sin \phi) + i(\sin \theta \cos \phi + \cos \theta \sin \phi)]$$

and by the addition formulas of trigonometry,

$$zw = |z||w|[\cos(\theta + \phi) + i \sin(\theta + \phi)]. \tag{9}$$

Since

$$|\cos(\theta + \phi) + i \sin(\theta + \phi)| = 1,$$

equation (9) yields

$$|zw| = |z||w| \tag{10}$$

and

$$\arg zw = \arg z + \arg w. \tag{11}$$

Hence, the length of the vector zw is the *product* of the lengths of the vectors z and w, whereas the argument of the vector zw is the *sum* of the arguments of the vectors z and w. Since the argument is determined up to a multiple of 2π, equation (11) is interpreted to mean that if particular values are assigned to any two of the terms, then there is a value of the third term for which equality holds. The geometric construction of the product zw is shown in Fig. 6.

Division of complex numbers leads to the following equation:

$$\frac{z}{w} = \frac{z\bar{w}}{w\bar{w}} = \frac{|z|(\cos\theta + i\sin\theta)|\bar{w}|(\cos\phi - i\sin\phi)}{|w|^2}, \qquad w \neq 0.$$

Now $|\bar{w}| = |w|$, and we obtain by the addition formulas of trigonometry

$$\frac{z}{w} = \frac{|z|}{|w|}[\cos(\theta - \phi) + i\sin(\theta - \phi)].$$

Hence,

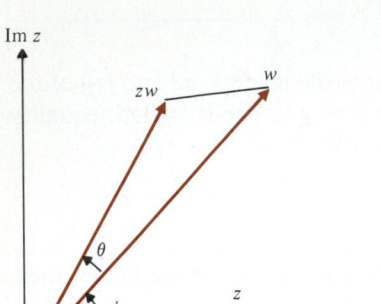

FIGURE 6 Complex multiplication. Triangle 01z is similar to triangle Ow(zw)

$$\left|\frac{z}{w}\right| = \frac{|z|}{|w|} \tag{12}$$

and

$$\arg\left(\frac{z}{w}\right) = \arg z - \arg w, \tag{13}$$

with equation (13) subject to an interpretation similar to that of equation (11).

Problems 15.2

In Problems 1–20 find the absolute value, argument, and polar representation of the given complex number. Use a calculator where indicated.

1. i

2. $-i$

3. $1 - i$

4. $-1 + i$

5. $-1 - i$

6. $1 + \sqrt{3}i$

7. $-1 + \sqrt{3}i$

8. $\sqrt{3} - i$

9. $-\sqrt{3} - i$

10. $-3 + 4i$

11. $4 + 3i$

12. $5 - 12i$

13. $2 + 7i$

14. $2 - i$

15. $5 + 2i$

16. $12 + 5i$

17. $7 - 2i$

18. $50 + 50i$

19. $-100 + 100i$

20. $30 - 30\sqrt{3}i$

21. For any complex number z_0 and real number $a > 0$, describe $\{z : |z - z_0| = a\}$.

22. Describe $\{z : |z - z_0| \le a\}$, where z_0 and a are as in Problem 21.

23. Prove the triangle inequality algebraically. [*Hint:* Show that $|z_1 + z_2|^2 \le (|z_1| + |z_2|)^2$ by using the fact that Re $z_1\bar{z}_2 \le |z_1\bar{z}_2| = |z_1||z_2|$.]

24. Show that the circle of radius 1 centered at the origin (the unit circle) is the set of points in the complex plane that satisfy $|z| = 1$.

25. Prove that

$$\left|\frac{z - a}{1 - \bar{a}z}\right| < 1$$

if $|z| < 1$ and $|a| < 1$.

26. Show that the triangle inequality is an equality for nonzero numbers z_1 and z_2 if and only if $\arg z_1 = \arg z_2$.

27. Show that if z_0 is a root of a polynomial $P(z)$ with real coefficients, then $\bar{z}_0$ is also a root of $P(z)$.

28. Prove that $|z - w| \geq \|z| - |w\|$. [*Hint*: Use the triangle inequality.]

15.3 De Moivre's Theorem and Roots

In this section we derive a result that is very useful for finding rational powers of complex numbers. We begin with the following identity [equation (9) on p. 885]:

If $\theta = \arg z$ and $\phi = \arg w$, then

$$zw = |z\|w|[\cos(\theta + \phi) + i \sin(\theta + \phi)]. \tag{1}$$

Let $z = w$ in (1). Then we obtain, since $\theta = \phi$,

$$z^2 = |z|^2[\cos(2\theta) + i \sin(2\theta)].$$

Setting $w = z^2$ in (1), we obtain

$$z(z^2) = |z\|z|^2[\cos(\theta + 2\theta) + i \sin(\theta + 2\theta)]$$

or

$$z^3 = |z|^3[\cos(3\theta) + i \sin(3\theta)].$$

Since $z = |z|(\cos \theta + i \sin \theta)$, we have shown that

$$(\cos \theta + i \sin \theta)^2 = \cos(2\theta) + i \sin(2\theta)$$

and

$$(\cos \theta + i \sin \theta)^3 = \cos(3\theta) + i \sin(3\theta).$$

Continuing this process, we obtain **De Moivre's theorem,†**

De Moivre's Theorem

$$(\cos \theta + i \sin \theta)^n = \cos n\theta + i \sin n\theta. \tag{2}$$

where n is a positive integer. De Moivre's theorem has many useful applications.

EXAMPLE 1

Calculate $(1 - i)^{23}$.

SOLUTION

We could multiply $1 - i$ by itself 23 times to obtain the answer, but, using De Moivre's theorem, we can find the answer quite easily. We saw in Example 15.2.2 on p. 884 that

$$1 - i = \sqrt{2}\left[\cos\left(\frac{-\pi}{4} + 2\pi k\right) + i \sin\left(\frac{-\pi}{4} + 2\pi k\right)\right].$$

† See the biographical sketch of De Moivre on p. 888.

Using the principal value of the argument, we have

$$(1 - i) = \sqrt{2}\left[\cos\left(\frac{-\pi}{4}\right) + i \sin\left(\frac{-\pi}{4}\right)\right].$$

Then, by De Moivre's theorem,

$$(1 - i)^{23} = (\sqrt{2})^{23}\left[\cos\left(\frac{-\pi}{4}\right) + i \sin\left(\frac{-\pi}{4}\right)\right]^{23}$$

$$= 2^{23/2}\left[\cos\left(\frac{-23\pi}{4}\right) + i \sin\left(\frac{-23\pi}{4}\right)\right].$$

ABRAHAM DE MOIVRE
(1667–1754)

Courtesy of the Granger Collection

Abraham De Moivre was a French Huguenot who moved to the more congenial political climate of London after the revocation of the Edict of Nantes in 1685. He earned his living in England by private tutoring, and he became an intimate friend of Isaac Newton.

De Moivre is particularly noted for his work *Annuities upon Lives,* which played an important role in the history of actuarial mathematics, his *Doctrine of Chances,* which contained much new material on the theory of probability, and his *Miscellanea analytica,* which contributed to recurrent series, probability, and analytic trigonometry. De Moivre is credited with the first treatment of the probability integral,

$$\int_0^\infty e^{-x^2}\,dx = \frac{\sqrt{\pi}}{2},$$

and of (essentially) the normal frequency curve

$$y = ce^{-hx^2}, \qquad c \text{ and } h \text{ constants,}$$

so important in the study of statistics. The misnamed Stirling's formula, which says that for very large n

$$n! \approx (2\pi n)^{1/2} e^{-n} n^n,$$

is due to De Moivre and is highly useful for approximating factorials of large numbers. The formula

$$(\cos x + i \sin x)^n = \cos nx + i \sin nx, \qquad i = \sqrt{-1},$$

known by De Moivre's name and found in many theory of equations textbooks, was familiar to De Moivre for the case where n is a positive integer. This formula has become the keystone of analytic trigonometry.

An interesting fable is often told of De Moivre's death. According to the story, De Moivre noticed that each day he required a quarter of an hour more sleep than on the preceding day. When this arithmetic progression reached 24 hours, De Moivre passed away.

But $-23\pi/4 = \pi/4 - 6\pi$, and $\sin(\pi/4) = \cos(\pi/4) = 1/\sqrt{2}$, so we get

$$(1-i)^{23} = 2^{23/2}\left(\frac{1}{\sqrt{2}} + \frac{i}{\sqrt{2}}\right) = 2^{11}(1+i) = 2048(1+i).$$

EXAMPLE 2

Compute an approximate value for $(2 + 5i)^{10}$.

SOLUTION

$|2 + 5i| = \sqrt{2^2 + 5^2} = \sqrt{29}$ and the principal value of $\arg(2 + 5i) = \tan^{-1}\frac{5}{2} \approx 1.19028995$. Then, by De Moivre's theorem,

$$(2+5i)^{10} \approx (\sqrt{29})^{10}(\cos 1.19028995 + i \sin 1.19028995)^{10}$$

$$= (29)^5(\cos 11.9028995 + i \sin 11.9028995)$$

$$\approx (29)^5[0.7878593 + i(-0.6158553)]$$

$$\approx (1.62 - 1.26i) \times 10^7.$$

De Moivre's theorem can also be used to find the roots of a complex number. If z is an nth root of the complex number w, then

$$z^n = w.$$

To find z set

$$z = |z|(\cos\theta + i\sin\theta) \quad \text{and} \quad w = |w|(\cos\phi + i\sin\phi),$$

where $\theta = \arg z$ and $\phi = \arg w$. Then, from De Moivre's theorem, we have

$$|z|^n(\cos n\theta + i\sin n\theta) = |w|(\cos\phi + i\sin\phi).$$

Thus, we may take

$$|z| = |w|^{1/n}$$

and

$$\theta = \frac{1}{n}\arg w = \frac{1}{n}(\text{Arg } w + 2\pi k), \qquad k = 0, \pm 1, \pm 2, \dots. \qquad (3)$$

Although equation (3) provides infinitely many values for θ, only n different polar angles are obtained because

$$\frac{2\pi(k+n)}{n} = \frac{2\pi k}{n} + 2\pi,$$

so the polar angles repeat every n integers. Therefore, we restrict our attention to the n polar angles

$$\theta = \frac{1}{n}(\text{Arg } w + 2\pi k), \qquad k = 0, 1, \dots, n-1.$$

EXAMPLE 3

Find the six 6th roots of 64.

SOLUTION

If $w = 64$, then $|w| = 64$, $|w|^{1/6} = 2$, and arg $w = 0 \pm 2k\pi$. Then

$$\frac{1}{n} \arg w = \frac{1}{6}(2k\pi), \qquad k = 0, 1, 2, 3, 4, 5,$$

$$= 0, \frac{\pi}{3}, \frac{2\pi}{3}, \pi, \frac{4\pi}{3}, \frac{5\pi}{3}.$$

Hence the six 6th roots of 64 are

$$z_0 = 2(\cos 0 + i \sin 0) = 2,$$

$$z_1 = 2\left(\cos \frac{\pi}{3} + i \sin \frac{\pi}{3}\right) = 1 + \sqrt{3}\ i,$$

$$z_2 = 2\left(\cos \frac{2\pi}{3} + i \sin \frac{2\pi}{3}\right) = -1 + \sqrt{3}\ i,$$

$$z_3 = 2(\cos \pi + i \sin \pi) = -2,$$

$$z_4 = 2\left(\cos \frac{4\pi}{3} + i \sin \frac{4\pi}{3}\right) = -1 - \sqrt{3}\ i,$$

$$z_5 = 2\left(\cos \frac{5\pi}{3} + i \sin \frac{5\pi}{3}\right) = 1 - \sqrt{3}\ i.$$

Problems 15.3

In Problems 1–10, use De Moivre's theorem to express each number in the form $x + iy$, where x and y are real.

1. $(1 + i)^{29}$

2. $(-1 + i)^{17}$

3. $(-1 - i)^{36}$

4. $(2 + 2i)^{12}$

5. $(\sqrt{3} + i)^{15}$

6. $(-\sqrt{3} + i)^{13}$

7. $(-1 - \sqrt{3}\ i)^{10}$

8. $(1 + 2i)^{12}$

9. $(-3 + 4i)^8$

10. $(\frac{1}{2} + \frac{1}{3}i)^6$

Find all the solutions of the following equations in Problems 11–20.

11. $z^2 = i$

12. $z^2 = 1 + i$

13. $z^2 = 2 - i$

14. $-z^2 = \sqrt{3} + i$

15. $z^3 = 2 + i$

16. $z^3 = 1 + \sqrt{3}i$

17. $z^4 = i$

18. $z^4 = -1$

19. $z^8 = 1$

20. $z^4 = 3 + 2i$

21. Derive expressions for $\cos 4\theta$ and $\sin 4\theta$ by comparing the De Moivre formula and the expansion of $(\cos \theta + i \sin \theta)^4$.

22. Prove De Moivre's formula by mathematical induction. [*Hint:* Recall the trigonometric identities $\cos(x + y) = \cos x \cos y - \sin x \sin y$ and $\sin(x + y) = \sin x \cos y + \cos x \sin y$.]

23. The n roots of the equation $z^n = 1$ are called **nth roots of unity.** Show that the nth roots of unity are given by

$$z_k = \cos\left(\frac{2\pi k}{n}\right) + i \sin\left(\frac{2\pi k}{n}\right), \qquad k = 0, 1, \dots, n - 1.$$

24. Let z_k be any nth root of unity. Prove that

$$1 + z_k + z_k^2 + \cdots + z_k^{n-1} = 0, \qquad z_k \neq 1.$$

25. If $1, z_1, z_2, \dots, z_{n-1}$ are the nth roots of unity, show that

$$(z - z_1)(z - z_2) \cdots \cdots (z - z_{n-1}) = 1 + z + z^2 + \cdots + z^{n-1}.$$

* 26. Let $z = x + iy$. For $y \neq 0$ we define sgn $y = y/|y|$; that is,

$$\text{sgn } y = \begin{cases} 1, & y > 0, \\ -1, & y < 0. \end{cases}$$

Show that the two values for $\sqrt{z}$ are given by

$$\sqrt{z} = \pm[\sqrt{(|z|+x)/2} + \text{sgn } y\sqrt{(|z|-x)/2}\ i].$$

In Problems 27–31 use the result of Problem 26 to compute the two square roots of the given number.

27. $\sqrt{3i}$

28. $\sqrt{2+3i}$

29. $\sqrt{-1+i}$

30. $\sqrt{-4-5i}$

31. $\sqrt{2-i}$

* 32. By minimizing the expression $\sum_{k=1}^{n}(|a_k| - \lambda|z_k|)^2$, where $a_1, \ldots, a_n, z_1, \ldots, z_n$ are complex numbers, for arbitrary real λ, show that

$$\left(\sum_{k=1}^{n}|a_k z_k|\right)^2 \leq \left(\sum_{k=1}^{n}|a_k|^2\right)\left(\sum_{k=1}^{n}|z_k|^2\right).$$

* 33. Prove **Lagrange's identity:**

$$\left|\sum_{k=1}^{n}a_k z_k\right|^2 = \left(\sum_{k=1}^{n}|a_k|^2\right)\left(\sum_{k=1}^{n}|z_k|^2\right) - \sum_{1 \leq j < k \leq n}|a_j \bar{z}_k - a_k \bar{z}_j|^2.$$

* 34. **Enestrom-Kakeya theorem.** Let $P(z)$ be a polynomial with real coefficients,

$$P(z) = a_n z^n + a_{n-1}z^{n-1} + \cdots + a_1 z + a_0,$$

satisfying $a_0 > a_1 > \cdots > a_n > 0$. Prove that all roots of $P(z)$ satisfy $|z| > 1$. [*Hint:* Apply the triangle inequality to $(1-z)P(z) = a_0 - [(a_0 - a_1)z + (a_1 - a_2)z^2 + \cdots + (a_{n-1} - a_n)z^n + a_n z^{n+1}]$.]

15.4 Sets in the Complex Plane

ε-Neighborhood

Let z_0 be a complex number. Then an **ε-neighborhood** of z_0 is the set of all points z whose distance from z_0 is less than ϵ, that is, all z satisfying $|z - z_0| < \epsilon$ (see Fig. 1). Pictorially, this is the interior of a disk centered at z_0 of radius ϵ.

Interior Point
Boundary Point
Interior, Exterior,
Complement, Boundary

Let S be a set of points in the complex plane $\mathbb{C}$. The point z_0 is said to be an **interior point** of S if some ε-neighborhood of z_0 is contained entirely in S; the set of all interior points of S is called the **interior** of S and is denoted by Int S. The **complement** of S is the set $\mathbb{C} - S$ of all points not in S. The set Int($\mathbb{C} - S$) is referred to as the **exterior** of S. A point z_0 is a **boundary point** of S if every ε-neighborhood of z_0 contains points in S and points not in S. Note that every boundary point of S is not in the interior or exterior of S. The set of all boundary points of S is called the **boundary** of S (see Fig. 2).

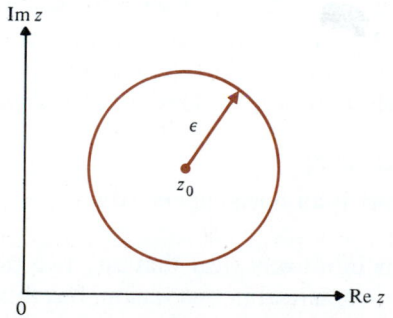

FIGURE 1 An ε-neighborhood of z_0

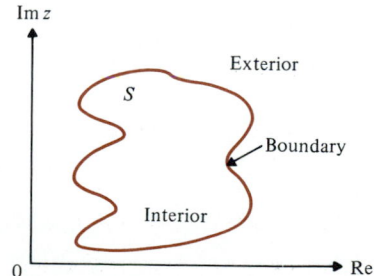

FIGURE 2 Interior, exterior, and boundary of a set

EXAMPLE 1: Open Unit Disk

Let S_0 be the set of all points z such that $|z| < 1$. Find the interior, boundary, and exterior of the set S_0. This set is called the **open unit disk.**

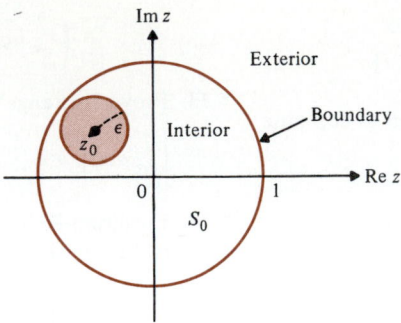

FIGURE 3

Interior, boundary, and exterior of the open, unit disk $|z| < 1$

SOLUTION

Let z_0 be any point in S_0. Note that the disk $|z - z_0| < \epsilon$ lies entirely in S_0 whenever $\epsilon < 1 - |z_0|$. Thus, every point of S_0 is an interior point. Similarly, every point z_0 satisfying $|z_0| > 1$ is exterior to S_0. If $|z_0| = 1$, then every ϵ-neighborhood of z_0 will contain points in S_0 and points not in S_0. Hence, the boundary of S_0 consists of all points on the circle $|z| = 1$, the interior of S_0 is the set $|z| < 1$, and the exterior of S_0 is the set of all points satisfying $|z| > 1$ (see Fig. 3).

Open Set

Closed Set

Bounded Set, Unbounded Set

A set is **open** if all its points are interior points; that is, $S = \text{Int } S$ when S is open. Thus, the set S_0 in our example above is an open set. The complement of an open set is said to be **closed.** For example, the set T of all points z such that $|z| \geq 1$ is closed. Many sets are neither open nor closed.

A set S is said to be **bounded** if there is a positive real number R such that all z in S satisfy $|z| < R$. If this condition does not hold, we say that S is **unbounded.** For example, the set S_0 in our earlier example is bounded, but $T = \{z : |z| \geq 1\}$ is unbounded.

Connected Set

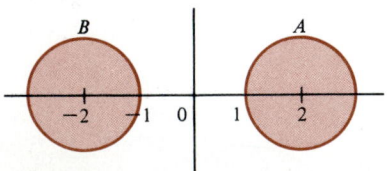

FIGURE 4 $A \cup B$ not connected

A set S is **connected** if it cannot be represented as the union of two nonempty disjoint sets A and B, neither containing a boundary point of the other. Intuitively, what this says is that S consists of a single piece. For example S_0 is connected, but the set of all z for which $|z - 2| < 1$ or $|z + 2| < 1$ is not connected, as we can let A be the set of all z such that $|z - 2| < 1$, and B be the set of all z for which $|z + 2| < 1$ (see Fig. 4). Then A and B are disjoint open sets, and each set does not contain a boundary point of the other (why?).

DEFINITION

Region

A **region**† is an open connected set.

It is intuitively clear that any two points in a region can be joined by a polygon contained in that region, but this fact requires verification. The proof can be found in most standard complex variables texts.‡ The following theorem is illustrated in Fig. 5.

† Many books call an open connected set a **domain.** We will avoid this usage to prevent possible confusion when describing the domain of definition of a complex function.

‡ See, for example. W. R. Derrick, *Complex Analysis and Applications,* 2nd ed., Wadsworth, Belmont, Calif., 1984.

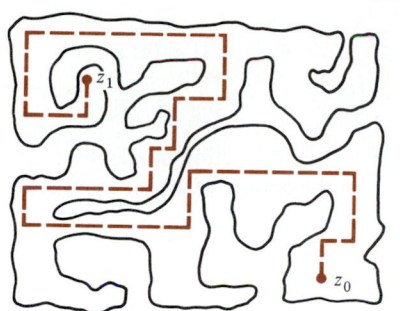

FIGURE 5 Polygon joining two points z_0, z_1

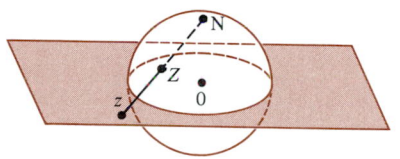

FIGURE 6 The Riemann sphere

Simply Connected Region

THEOREM 1

Any two points of a region can be joined by a polygon that lies in the region.

For many purposes it is useful to extend the system $\mathbb{C}$ of complex numbers by including a point at infinity, denoted by the symbol ∞. This new set is called the **extended complex plane** $\mathcal{M}$, and the point ∞ satisfies the following algebraic rules:

$$a + \infty = \infty + a = \infty, \qquad \frac{a}{\infty} = 0, \qquad a \neq \infty,$$

$$b \cdot \infty = \infty \cdot b = \infty, \qquad \frac{b}{0} = \infty, \qquad b \neq 0.$$

As a geometric model for $\mathcal{M}$ we use the $x_1^2 + x_2^2 + x_3^2 = 1$ unit sphere in three-dimensional space. We associate to each point z in the plane that point Z in the sphere where the ray originating from the north pole N and passing through z intersects the sphere. Thus, N corresponds to ∞ (see Fig. 6) and ϵ-neighborhoods of N on the unit sphere correspond to neighborhoods of the point at infinity. This model is called the **Riemann sphere** and the point correspondence is referred to as stereographic projection. It can be shown that

> all straight lines in $\mathbb{C}$ correspond to circles passing through ∞ in $\mathcal{M}$.†

A region is **simply connected** if its complement in $\mathcal{M}$ is connected. This implies that a simply connected region has no "holes" in it. For example, the set S_0 in our earlier example is simply connected, but the set of all z satisfying $0 < |z| < 1$ is not, since the origin forms a "hole" for this set (see Fig. 7). This region is called a **punctured disk.**

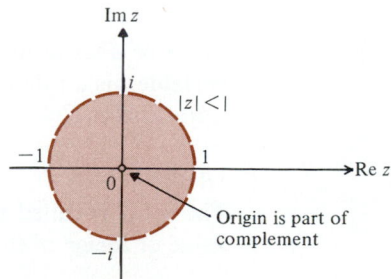

FIGURE 7 $0 < |z| < 1$ (punctured disk)

† Other common notations for $\mathcal{M}$ are S, Σ, $\mathbb{C} \cup \{\infty\}$.

Problems 15.4

In Problems 1–10, classify the sets according to the terms *open, closed, bounded, connected,* and *simply connected.* Graph each set and determine its boundary.

1. $|z + 3| < 2$
2. $|\text{Re } z| < 1$
3. $|\text{Im } z| > 1$
4. $0 < |z - 1| \leqslant 1$
5. $|z| \leqslant \text{Re } z + 2$
6. $|z - 1| - |z + 1| > 2$
7. $|z + 1| + |z + i| \geqslant 2$
8. $|z - 1| < \text{Im } z$
9. $2\sqrt{2} < |z - 1| + |z + 1| < 3$
10. $\|z - i| - |z + i\| < 1$

* 11. The closure of a set S is the intersection of all closed sets containing S. Prove that the closure of a connected set is connected.

15.5 Continuous Functions of a Complex Variable

The study of one-variable calculus involves defining a number of concepts and then using these definitions to obtain important properties and useful applications. If you look back at your calculus book you will find the following central ideas discussed in roughly the order given below:

   (i) function,
   (ii) limit of a function,
  (iii) continuous function,
  (iv) derivative of a function,
   (v) integral of a function.

In this section we study continuous complex-valued functions of a complex variable. In so doing we shall first, as in one-variable calculus, have to define what we mean by a function and what we mean by a limit. We shall see that these definitions are very similar to those in calculus.

In Section 15.6 we shall compute derivatives of complex-valued functions. Here we observe that differentiation of a complex function has some different properties from differentiation of a real-valued function.

We begin with a basic definition.

DEFINITION

Complex-Valued Function of a Complex Variable

Let S be a set in the complex plane. A **complex-valued function of a complex variable** f is a rule that assigns a unique complex number w to each complex number z in S.

DEFINITION

Domain
Image

The set S is called the **domain** of f. We write $w = f(z)$ and say that w is the **value** or **image** of the function at the point z in the domain of the function.

Both z and w are complex numbers, so we may write

$$z = x + iy \quad \text{and} \quad w = u + iv,$$

where x, y, u, and v are real numbers. Then we obtain

$$w = u(z) + iv(z) = u(x, y) + iv(x, y). \qquad (1)$$

That is, a complex-valued function of a complex variable consists of a **pair** of real-valued functions of two real variables.

Remark. In Section 9.2 (see p. 514) we discussed vector fields in $\mathbb{R}^2$. We defined a vector field as a function or mapping from a subset Ω of $\mathbb{R}^2$ into $\mathbb{R}^2$. That is, if $\mathbf{f}$ is a vector field in $\mathbb{R}^2$, then for every $(x, y) \in \Omega \subset \mathbb{R}^2$, $\mathbf{f}(x, y) = (u(x, y), v(x, y))$ is in $\mathbb{R}^2$. Thus we can see that a complex-valued function can be thought of as a vector field mapping $\mathbb{R}^2$ (with the notation of the complex plane) into $\mathbb{R}^2$.

EXAMPLE 1

Let $w = f(z) = z^2$.
  (a) Compute $f(3 + i)$ and $f(-2 + 5i)$.
  (b) Express $w = z^2$ as a pair of real-valued functions of two real variables.

SOLUTION
  (a)
  $$f(3 + i) = (3 + i)^2 = (3 + i)(3 + i) = 9 + 6i + i^2 = 8 + 6i,$$
  $$f(-2 + 5i) = (-2 + 5i)^2 = 4 - 20i + (5i)^2 = -21 - 20i.$$

  (b) Setting $z = x + iy$, we obtain
  $$w = z^2 = (x + iy)^2 = (x^2 - y^2) + i(2xy).$$

Thus, $u(x, y) = x^2 - y^2$ and $v(x, y) = 2xy$.

Real-valued functions of a real variable $y = f(x)$ can be described geometrically by a graph in the xy-plane. No such convenient representation is possible for $w = f(z)$, as it would require four dimensions, two for each complex variable. Instead, information about the function is displayed by drawing separate complex planes for the variables z and w and indicating correspondences between points or sets of points in the two planes (see Fig. 1). The function f is said to be a **mapping** of the set S in the z-plane into the w-plane.

Mapping

Recall that a real-valued function f is **one-to-one** if $f(x) = f(y)$ implies that $x = y$. For example, the function $f(x) = x^3$ is one-to-one, whereas $f(x) = x^2$ is not $[(-2)^2 = 2^2 = 4$ but $-2 \neq 2]$. A real-valued function f is **onto** a set Y if for every y in Y, there is an x in the domain of f such that $y = f(x)$. That is, f is onto Y if every point in Y can be obtained as the image of a point in the domain of f.

Similar definitions can be given for complex-valued functions.

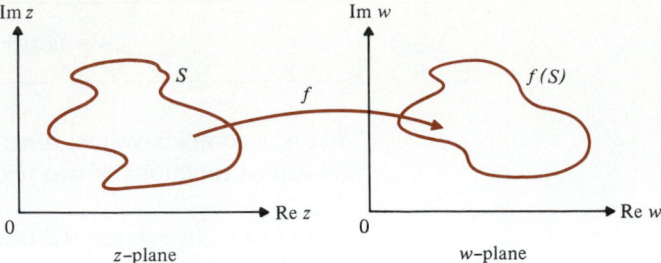

FIGURE 1 A mapping $w = f(z)$

DEFINITION

One-to-One and Onto Functions

A function f mapping a set S into a set S', $f:S \to S'$, is said to be **one-to-one** if $f(z_1) = f(z_2)$ only for $z_1 = z_2$; it is said to be **onto** if $S' = f(S)$ where $f(S)$ is the set of all values assumed by f on the set S.† That is, for every $w \in S'$, there is at least one number $z \in S$ such that $f(z) = w$.

Image Set

We call $f(S)$ the **image set** of S under the mapping f.

EXAMPLE 2

Analyze the function $w = 3z$.

SOLUTION
Setting $z = x + iy$, we obtain

$$w = u + iv = 3x + i(3y).$$

Thus, $u = 3x$, $v = 3y$, and each nonzero vector in the z-plane is stretched into a vector, with the same argument but three times its length, in the w-plane. Since any point $a + ib$ in the w-plane is the image of the point $(a/3) + i(b/3)$ in the z-plane, the function $w = 3z$ is onto. The function is also one-to-one, since $3z_1 = 3z_2$ only when $z_1 = z_2$.

EXAMPLE 3

Describe the image set of the function $w = z^3$ defined on the disk $|z| < 2$, and state if this mapping is one-to-one.

SOLUTION
Writing each point of the disk in its polar representation

$$z = r(\cos \theta + i \sin \theta),$$

where $0 \leqslant r = |z| < 2$ and $-\pi < \theta \leqslant \pi$, we obtain

$$w = z^3 = r^3(\cos 3\theta + i \sin 3\theta).$$

Hence, each argument is tripled, indicating that the disk $|z| < 2$ is mapped onto $|w| < 8$ with each point of $0 < |w| < 8$ the image of *three* points of $0 < |z| < 2$. For example, $z = 1, (-1 \pm \sqrt{3}i)/2$ all map to $w = 1$. Hence, the mapping is not one-to-one. (See Fig. 2.) Compare this with the real case $y = x^3$.

---

† One-to-one functions are often called **injections** and onto functions are referred to as **surjections**. A function that is both an injection and a surjection is called a **bijection**.

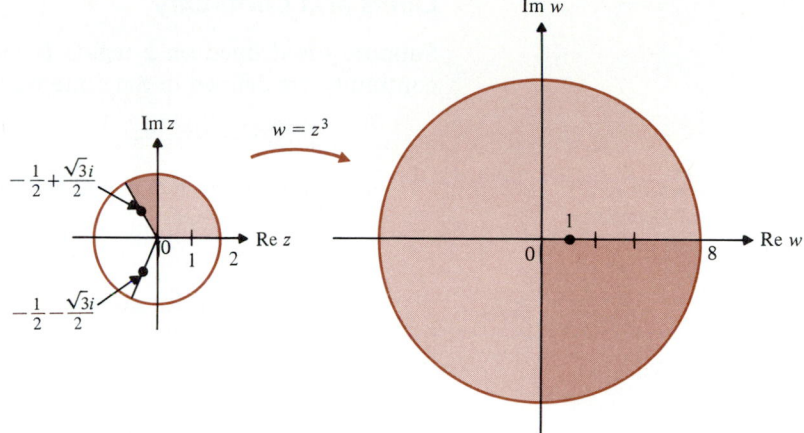

FIGURE 2 The mapping $w = z^3$

EXAMPLE 4

Determine if the function

$$w = \frac{z-1}{z-2}$$

is one-to-one, and state where the function is defined.

SOLUTION

Suppose z_1 and z_2 yield the same value of w:

$$\frac{z_1 - 1}{z_1 - 2} = \frac{z_2 - 1}{z_2 - 2}.$$

Cross-multiplying, we have

$$(z_1 - 1)(z_2 - 2) = (z_2 - 1)(z_1 - 2)$$

or

$$z_1 z_2 - 2z_1 - z_2 + 2 = z_1 z_2 - z_1 - 2z_2 + 2.$$

Canceling like terms and gathering all the terms involving z_1 on one side and z_2 on the other, we obtain $z_1 = z_2$, implying that the function is one-to-one.

The answer to the second question depends on what values of w we wish to allow. If we restrict w to the complex plane $\mathbb{C}$, then the function is not defined at $z = 2$, since the denominator vanishes. However, if we allow w to assume all values in the extended plane $\mathcal{M}$, then the function can be defined on $\mathcal{M}$ (see p. 893), with $z = 2$ mapped to $w = \infty$. The image of the point at ∞ is obtained by evaluating

$$w = \frac{z-1}{z-2} = \frac{1 - 1/z}{1 - 2/z}$$

as z tends to ∞. Thus, $z = \infty$ is mapped to $w = 1$ (see Problem 36).

Limits and Continuity

Suppose f is defined on a region G and a is a point in G. Then limits and continuity are defined in the same way as in the real variable case.

DEFINITION

Limit and Continuity of a Complex-Valued Function

(a) The function $f(z)$ is said to have **limit** A as z approaches a,

$$\lim_{z \to a} f(z) = A,$$

provided that for every $\epsilon > 0$ there exists a number $\delta > 0$ such that

$$|f(z) - A| < \epsilon \tag{2}$$

whenever $0 < |z - a| < \delta$.

(b) The function $f(z)$ is said to be **continuous** at a if and only if

$$\lim_{z \to a} f(z) = f(a) \tag{3}$$

(see Fig. 3). A continuous function is one that is continuous at all points where it is defined.

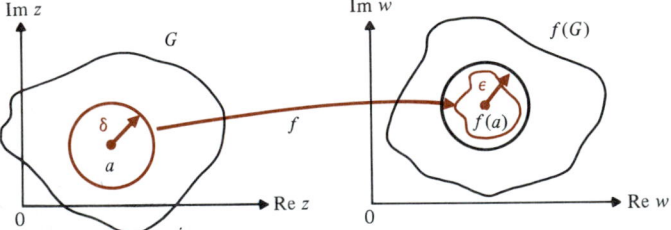

FIGURE 3 Continuity of f at a.

Geometrically, the definition of a limit states that given $\epsilon > 0$, there is $\delta > 0$ such that $f(z)$ will take values in an ϵ-neighborhood of A if z takes values in a δ-neighborhood of a, except possibly the value a.

EXAMPLE 5

Prove that $\lim\limits_{z \to 3} \dfrac{z-1}{z-2} = 2$.

SOLUTION

We must show that for every $\epsilon > 0$, there is a $\delta > 0$ such that

$$\left| \frac{z-1}{z-2} - 2 \right| < \epsilon \qquad \text{whenever} \qquad 0 < |z - 3| < \delta.$$

To find δ we work backward. That is, we assume that

$$\left| \frac{z-1}{z-2} - 2 \right| < \epsilon, \tag{4}$$

and determine the value of δ that will be required for this inequality. Observe that

$$\left|\frac{z-1}{z-2}-2\right| = \left|\frac{(z-1)-2(z-2)}{z-2}\right| = \left|\frac{3-z}{z-2}\right| < \frac{\delta}{|z-2|}, \tag{5}$$

since we are assuming that $0 < |z-3| < \delta$, with δ yet to be determined in terms of ϵ. If $\delta < \frac{1}{2}$, using the triangle inequality

$$1 = |1-(3-z)+(3-z)| \leqslant |1-(3-z)| + |3-z| < |z-2| + \delta$$

so that

$$|z-2| > 1 - \delta > 1 - \tfrac{1}{2} = \tfrac{1}{2}.$$

Hence, using (5) we have

$$\left|\frac{z-1}{z-2}-2\right| < 2\delta.$$

Thus, given any small number $\epsilon > 0$, if we select $\delta < \min(\tfrac{1}{2}, \tfrac{1}{2}\epsilon)$, we obtain

$$\left|\frac{z-1}{z-2}-2\right| < \epsilon.$$

EXAMPLE 6

Show that $\lim\limits_{z \to 0} \dfrac{z^2}{|z|^2}$ does not exist.

SOLUTION
Let $z = x + iy$. Then

$$\frac{z^2}{|z|^2} = \frac{(x+iy)^2}{x^2+y^2} = \frac{x^2-y^2+2xyi}{x^2+y^2}. \tag{6}$$

Now, there are many ways to approach 0. If we approach along the real axis, then $y = \text{Im } z = 0$ and we obtain [setting $y = 0$ in (6)]

$$\lim_{\substack{z \to 0 \\ y=0}} \frac{z^2}{|z|^2} = \lim_{x \to 0} \frac{x^2}{x^2} = \lim_{x \to 0} 1 = 1.$$

On the other hand, if we approach along the imaginary axis, then $x = \text{Re } z = 0$ and we have

$$\lim_{\substack{z \to 0 \\ x=0}} \frac{z^2}{|z|^2} = \lim_{y \to 0} \frac{-y^2}{y^2} = \lim_{y \to 0} -1 = -1.$$

Thus we get different answers depending on how we approach the origin. To prove that the limit cannot exist, we note that we have shown that in any open disk centered at the origin, there are points at which $z^2/|z|^2$ takes on the values $+1$ and -1. Hence $z^2/|z|^2$ cannot have a limit as $z \to 0$.

Example 6 leads to the following general rule:

Rule for Nonexistence of a Limit

$\lim\limits_{z \to z_0} f(z)$ does not exist if there are two different paths for z to approach z_0 which give different values for this limit.

Since the definition of a limit of a complex-valued function of a complex variable is identical to that of a real-valued function of a real variable and absolute values behave as in the real case, precisely the same rule of limits apply. The verifications of the following properties are exact analogs of the usual proofs of elementary calculus. We omit the proofs.

Rules of Limits

Let $\lim\limits_{z \to a} f(z) = A$ and $\lim\limits_{z \to a} g(z) = B$.

Then

(i) $\lim\limits_{z \to a} [f(z) \pm g(z)] = A \pm B$, the limit of a sum or difference of two functions is the sum or difference of their limits.

(ii) $\lim\limits_{z \to a} f(z)g(z) = AB$, the limit of a product is the product of the limits.

(iii) $\lim\limits_{z \to a} \dfrac{f(z)}{g(z)} = \dfrac{A}{B}$, for $B \neq 0$. The limit of a quotient is the quotient of the limits if $B \neq 0$.

The rules of limits can be used to prove that every polynomial function in z

$$f(z) = a_n z^n + a_{n-1} z^{n-1} + \cdots + a_1 z + a_0$$

is continuous in $\mathbb{C}$. Note that the identity function

$$f(z) = z$$

is clearly continuous at any point by letting $\delta = \epsilon$. Applying the second rule of limits repeatedly, we see that $f(z) = z^n$ is continuous for every positive integer n. Every constant function $f(z) = c$ is trivially continuous, since all δ-neighborhoods of any point are mapped into any ϵ-neighborhood of c. Again, applying the second rule of limits, we see that $f(z) = a_n z^n$ is continuous. Finally, using the first rule of limits repeatedly, we see that all polynomials are contin-

Rational Function

uous. Indeed, using the third rule of limits, we find that all quotients of polynomials (called **rational functions**)

$$F(z) = \frac{a_n z^n + \cdots + a_1 z + a_0}{b_m z^m + \cdots + b_1 z + b_0}$$

are continuous at those points where the denominator does not vanish. It also follows from the rules of limits that the sum $f(z) + g(z)$ and the product $f(z)g(z)$ of two continuous functions are continuous, and the quotient $f(z)/g(z)$ is defined and continuous at all points where $g(z)$ does not vanish.

EXAMPLE 7

Determine if the function

$$f(z) = \begin{cases} \dfrac{z^2 - 1}{z - 1}, & z \neq 1, \\ \\ 3, & z = 1, \end{cases}$$

is continuous.

SOLUTION

Clearly, f is continuous in the set $z \neq 1$, since the denominator is nonzero. Thus, the only point where we still need to check the continuity is $z = 1$. However,

$$\lim_{z \to 1} \frac{z^2 - 1}{z - 1} = 2$$

because

$$\frac{z^2 - 1}{z - 1} = \frac{(z - 1)(z + 1)}{z - 1} = z + 1$$

if $z \neq 1$. But by definition, $f(1) = 3 \neq 2$. Hence, f is not continuous.

Problems 15.5

Use the ϵ-δ definition of a limit to verify the limits in Problems 1–10.

1. $\lim\limits_{z \to 1} 2z = 2$

2. $\lim\limits_{z \to i} iz = -1$

3. $\lim\limits_{z \to -i} z + i = 0$

4. $\lim\limits_{z \to i} z^2 + 1 = 0$

5. $\lim\limits_{z \to 1+i} 2z - 3 = -1 + 2i$

6. $\lim\limits_{z \to 1+i} z^2 = 2i$

7. $\lim\limits_{z \to 2} \dfrac{z^2 - 4}{z - 2} = 4$

8. $\lim\limits_{z \to -i} \dfrac{z^2 + 1}{z + i} = -2i$

9. $\lim\limits_{z \to 1} \dfrac{z^3 - 1}{z - 1} = 3$

10. $\lim\limits_{z \to 2} \dfrac{z^2 - 3z + 2}{z - 2} = 1$

11. Show that $\lim\limits_{z \to 0} \dfrac{z}{|z|}$ does not exist.

12. Show that $\lim\limits_{z \to 0} \dfrac{z - \bar{z}}{z + \bar{z}}$ does not exist.

13. Show that $\lim\limits_{z \to 0} \dfrac{|z|^2}{z} = 0$.

14. Determine whether $\lim\limits_{z \to 0} \dfrac{z^3}{|z|^3}$ exists. If so, calculate it.

In Problems 15–20 use the limit theorems to compute the given limits.

15. $\lim\limits_{z \to i} (z^3 - 2z + 3)$

16. $\lim\limits_{z \to 2-i} \dfrac{z+i}{z-i}$

17. $\lim\limits_{z \to -1+i} \dfrac{z^2}{1-z}$

18. $\lim\limits_{z \to 1+i} (z^2 + 2z + 3 + 4i)$

19. $\lim\limits_{z \to 3+i} \dfrac{i}{z^2 - 1}$

20. $\lim\limits_{z \to 1-2i} \dfrac{z^2 - 1}{z^2 + 1}$

31. $f(z) = \dfrac{|z|^2}{z}$

32. $f(z) = \dfrac{(\operatorname{Re} z)(\operatorname{Im} z)}{|z|^2}$

33. $f(z) = \dfrac{(\operatorname{Re} z)^2 - (\operatorname{Im} z)^2}{|z|^2}$

Prove that the functions in Problems 21–24 are continuous in $\mathbb{C}$:

21. $w = \operatorname{Re} z$

22. $w = \operatorname{lm} z$

23. $w = \bar{z}$

24. $w = |z|$

Suppose $f(z)$ is a continuous function on a region G. Prove that the functions in Problems 25–28 are continuous on G.

25. $\operatorname{Re} f(z)$

26. $\operatorname{lm} f(z)$

27. $|f(z)|$

28. $f(\bar{z})$

29. At what points is the function

$$f(z) = \begin{cases} \dfrac{z^3 - 1}{z^2 - 1}, & z \neq \pm 1, \\[2mm] \dfrac{3}{2}, & z = \pm 1, \end{cases}$$

continuous?

Prove that each function in Problems 30–33 is continuous for $z \neq 0$. Can the function be defined to make it continuous at $z = 0$?

30. $f(z) = \dfrac{z \operatorname{Re} z}{|z|^2}$

34. Prove that every function of the form

$$w = \frac{z-a}{z-b}, \qquad a \neq b,$$

is a one-to-one mapping of the extended plane $\mathcal{M}$ onto itself.

35. Prove that every function of the form

$$w = \frac{az+b}{cz+d}, \qquad ad \neq bc,$$

is a one-to-one mapping of $\mathcal{M}$ onto itself.

36. The function $f(z)$ has the limit A as z approaches ∞,

$$\lim_{z \to \infty} f(z) = A,$$

if for every $\epsilon > 0$ there exists a number $\delta > 0$ such that

$$|f(z) - A| < \epsilon \qquad \text{whenever } |z| > \delta.$$

Use this definition to prove that

$$\lim_{z \to \infty} \frac{z-1}{z-2} = 1.$$

* 37. Suppose the coefficients of the polynomial

$$P(z) = a_n z^n + \cdots + a_1 z + a_0$$

satisfy $|a_0| \geqslant |a_1| + |a_2| + \cdots + |a_n|$. Prove that $P(z)$ has no roots in the unit disk $|z| < 1$. [*Hint:* Note that

$$|P(z)| \geqslant |a_0| - [|a_1|\,|z| + \cdots + |a_n|\,|z|^n].$$

Compare this result with Problem 15.3.34.]

15.6 Analytic Functions and the Cauchy-Riemann Equations

The derivative of a complex-valued function of a complex variable is defined in precisely the same way as the real-valued case of elementary calculus.

DEFINITION

Derivative and Analyticity

The **derivative** f' of f at a is given by

$$f'(a) = \lim_{h \to 0} \frac{f(a+h) - f(a)}{h}, \tag{1}$$

Analytic Function

Entire Function

provided the limit exists. The function f is said to be **analytic** (or **holomorphic**) on the region G if it has a derivative at each point of G, and f is said to be **entire** if it is analytic on all of $\mathbb{C}$.

Observe that in the definition above, h is a complex number, as is the quotient $[f(a+h) - f(a)]/h$. Thus, for the derivative to exist, it is necessary that the quotient above tend to a unique complex number $f'(a)$ independent of the manner in which h approaches zero.

EXAMPLE 1

Compute $f'(z)$ where $f(z) = z^2$.

SOLUTION

$$f'(z) = \lim_{h \to 0} \frac{(z+h)^2 - z^2}{h} = \lim_{z \to 0} \frac{z^2 + 2zh + h^2 - z^2}{h}$$

$$= \lim_{z \to 0} \frac{2zh + h^2}{h} = \lim_{h \to 0} \frac{h(2z+h)}{h} = \lim_{h \to 0}(2z+h) = 2z.$$

This result is not surprising. It shows that the derivative of z^2 is $2z$, for every complex number z. In particular, we see that z^2 is analytic on $\mathbb{C}$, that is, $f(z) = z^2$ is an entire function.

Before continuing, we make a useful observation:

If f has a derivative at a, then f is continuous at a.

The proof of this fact is contained in the following computation:

$$\lim_{h \to 0} f(a+h) = \lim_{h \to 0} \left\{ \left[\frac{f(a+h) - f(a)}{h} \right] \cdot h + f(a) \right\} = f'(a) \cdot 0 + f(a) = f(a).$$

Many of the differentiation rules for one-variable calculus hold for a complex-valued function. We give four of these rules here. We assume that f and g are differentiable at a and c is a complex number. Then $f + g$, cf, and $f \cdot g$ are differentiable at a and

(i) $(f + g)'(a) = f'(a) + g'(a)$.   The derivative of the sum of two functions is the sum of their derivatives

(ii) $(cf)'(a) = cf'(a)$.   scalar multiple rule

(iii) $(fg)'(a) = f(a)g'(a) + g(a)f'(a)$.   product rule

Moreover, if $g(a) \neq 0$, then f/g is differentiable at a and

(iv) $\left(\dfrac{f}{g}\right)'(a) = \dfrac{g(a)f'(a) - f(a)g'(a)}{g^2(a)}$.   quotient rule

Finally, consider the composite function $(f \circ g)(z) = f(g(z))$. If g is differentiable at a and f is differentiable at $g(a)$, then

(v) $(f \circ g)'(a) = f'(g(a))g'(a)$.   chain rule

The proofs are identical to those in any elementary calculus text and so are omitted.

In Example 1 we showed that $(z^2)' = 2z$. In fact, if n is a positive integer, then

$$(z^n)' = nz^{n-1}. \qquad (2)$$

This follows exactly as the comparable result for the real-valued function $f(x) = x^n$. Using (2) and the sum and scalar multiple rules [(i) and (ii)], we can see that

$$p(z) = a_n z^n + a_{n-1} z^{n-1} + \cdots + a_1 z + a_0$$

has the derivative

$$p'(z) = na_n z^{n-1} + (n-1)a_{n-1} z^{n-2} + \cdots + 2a_2 z + a_1 \qquad (3)$$

for every complex number z. That is,

every polynomial is an entire function.

Moreover, using the quotient rule, we see that the rational function $p(z)/q(z)$ is analytic in any region that does not contain points at which $q(z) = 0$.

EXAMPLE 2

(a) Compute the derivative of $f(z) = \dfrac{z+3}{z^2+1}$.

(b) What is the largest region in which f is analytic?

SOLUTION

(a) By the quotient rule,

$$f'(z) = \frac{(z^2+1)(z+3)' - (z+3)(z^2+1)'}{(z^2+1)^2} = \frac{z^2+1-(z+3)(2z)}{(z^2+1)^2} = \frac{-z^2-6z+1}{(z^2+1)^2}$$

provided $z^2 + 1 \neq 0$.

(b) The derivative is defined as long as $z^2 + 1 \neq 0$, or $z \neq \pm i$. Thus f is analytic over $S = \{z \in \mathbb{C} : z \neq \pm i\}$.

We have seen that there are many similarities between differentiation for functions of real variables and differentiation for functions of a complex variable. However, there is a fundamental difference between them, which we now explore. Let $z = (x, y)$ and suppose that h is real. Then

$$f'(z) = \lim_{h \to 0} \frac{f(x+h, y) - f(x, y)}{h} = \frac{\partial f}{\partial x}(z) = f_x(z). \tag{4}$$

But if $h = ik$ is purely imaginary, then

$$f'(z) = \lim_{k \to 0} \frac{f(x, y+k) - f(x, y)}{ik} = \frac{1}{i} \frac{\partial f}{\partial y}(z) = -if_y(z). \tag{5}$$

Thus, the existence of a complex derivative forces the function to satisfy the partial differential equation

$$f_x = -if_y.$$

Writing $f(z) = u(z) + iv(z)$, where u and v are real-valued functions of a complex variable, and equating the real parts and imaginary parts of

$$u_x + iv_x = f_x = -if_y = v_y - iu_y,$$

Cauchy-Riemann Equations   we obtain the **Cauchy-Riemann† differential equations**

$$u_x = v_y, \qquad v_x = -u_y. \tag{6}$$

We have proved:

---

† See the biography of Cauchy on p. 24 and the biography of Riemann on page 907.

■ **THEOREM 1: Necessary Conditions for Differentiability**

If the function $f(z) = u(z) + iv(z)$ has a derivative at the point z, then the first partial derivatives of u and v, with respect to x and y, exist and satisfy the Cauchy-Riemann equations (6). ■

EXAMPLE 3

Let $f(z) = z^2 = (x^2 - y^2) + 2xyi$. Since f is entire, $u = x^2 - y^2$ and $v = 2xy$ must satisfy the Cauchy-Riemann equations. Observe that

$$u_x = 2x = v_y \quad \text{and} \quad -u_y = 2y = v_x.$$

The Cauchy-Riemann equations have many uses. For example, a function is *not* differentiable at a point at which they do not hold.

EXAMPLE 4

Let $f(z) = |z|^2 = x^2 + y^2$. Then $u = x^2 + y^2$, $v = 0$ and $u_x = 2x$, $u_y = 2y$, $v_x = 0 = v_y$, so f satisfies the Cauchy-Riemann equations only at 0. Thus f is not differentiable for $|z| > 0$. However, f has a derivative when $z = 0$, because

$$f'(0) = \lim_{h \to 0} \frac{|h|^2}{h} = \lim_{h \to 0} \frac{h\bar{h}}{h} = \lim_{h \to 0} \bar{h} = 0.$$

At this point one might ask whether the Cauchy-Riemann equations are enough to guarantee the existence of a derivative at a given point. The following example by D. Menchoff shows that this is not the case. Let

$$f(z) = \begin{cases} \dfrac{z^5}{|z|^4}, & z \neq 0, \\ 0 & z = 0. \end{cases}$$

Then

$$\frac{f(z)}{z} = \left(\frac{z}{|z|} \right)^4, \qquad z \neq 0,$$

so that on the real axis $f(z)/z = (x/|x|)^4 = 1$, while on the line $y = x$, we have $z = x + iy = x(1 + i)$, so that

$$\frac{f(z)}{z} = \left(\frac{x(1+i)}{|x| \, |1+i|} \right)^4 = \frac{(1+i)^4}{(\sqrt{2})^4} = \frac{-4}{4} = -1.$$

Thus

$$\lim_{h \to 0} \frac{f(0 + h) - f(0)}{h} = \lim_{h \to 0} \frac{f(h)}{h}$$

GEORG FRIEDRICH RIEMANN
(1826–1866)

Courtesy of the Granger Collection

Georg Friedrich Riemann was born in 1826 in the village of Breselenz, Hanover. His father was a Lutheran pastor. Throughout his life Riemann was exceedingly shy and in frail health. Although his family was by no means wealthy, Riemann was able to get a good education, both at the University of Berlin and at the University of Göttingen. His work at Göttingen culminated with a brilliant thesis in the area of functions of a complex variable. At Göttingen he worked under the greatest mathematician of the nineteenth century, Karl Friedrich Gauss (1777–1855).

In 1854 Riemann was appointed *Privatdozent* (official but unpaid lecturer) at the University of Göttingen. According to the custom of the day, he was asked to give a probationary lecture. The result was one of the greatest papers of comparable size ever presented in the history of mathematics. The title of the lecture was "Über die Hypothesen welche der Geometrie zu Grunde liegen" ("On the Hypotheses Which Lie at the Foundation of Geometry"). Rather than discussing a specific example, it urged a global view of geometry that revolutionized the study of that subject. After this lecture, and perhaps for the only time in a career that spanned approximately sixty years, Gauss paid compliments to the work of someone else.

Riemann's great paper was presented in 1854 but was not published until 1868. One of the central ideas in the paper was that geometry could be discussed in general curved spaces rather than only in the sphere (which is one specific curved space). Albert Einstein made use of this idea in his general theory of relativity.

Riemann made important contributions to many other areas of mathematics and theoretical physics. He is famous for having clarified the concept of the definite integral. It is this definition, now known as the *Riemann integral,* that is the basis for the integration chapter in all basic calculus texts.

In 1859 Riemann became a full professor at the University of Göttingen, having become an assistant professor only two years earlier. The chair that he occupied was previously held by another great German mathematician, Peter Gustav Lejeune Dirichlet (1805–1859). Dirichlet, in turn, had succeeded Gauss. Riemann's career was cut short in 1866, in Italy, where he had gone to seek a cure for tuberculosis. He died at the age of 39.

does not exist and f does not have a derivative at $z = 0$; but expanding the expression for f yields

$$u(x, 0) = x, \qquad u(0, y) = 0 = v(x, 0), \qquad v(0, y) = y.$$

Hence,

$$u_x(0,0) = 1 = v_y(0,0), \qquad -u_y(0,0) = 0 = v_x(0,0),$$

and the Cauchy-Riemann equations hold. That is, a function may fail to have a derivative at a point even though the Cauchy-Riemann equations may hold at that point.

What is missing? Recall that a real-valued function f of two variables may fail to be differentiable at (x_0, y_0) even though $f_x(x_0, y_0)$ and $f_y(x_0, y_0)$ both exist. However, if both partial derivatives are continuous at (x_0, y_0), then f is differentiable at (x_0, y_0) (see Theorem 9.8.2). This suggests the missing step. The proof of the following theorem is difficult and is omitted.†

■ THEOREM 2: Sufficient Conditions for Differentiability

Let $f(z) = u(x, y) + iv(x, y)$, defined in some region G containing the point z_0, have *continuous* first partial derivatives, with respect to x and y, satisfying the Cauchy-Riemann equations at z_0. Then $f'(z_0)$ exists. ■

EXAMPLE 5

Show that the function

$$f(z) = e^{x^2 - y^2}(\cos 2xy + i \sin 2xy)$$

is entire.

SOLUTION

We must check that the first partials of

$$u = e^{x^2 - y^2} \cos 2xy \quad \text{and} \quad v = e^{x^2 - y^2} \sin 2xy$$

are continuous and satisfy the Cauchy-Riemann equations at all points of $\mathbb{C}$. We have

$$u_x = 2e^{x^2 - y^2}(x \cos 2xy - y \sin 2xy) = v_y$$

and

$$-u_y = 2e^{x^2 - y^2}(y \cos 2xy + x \sin 2xy) = v_x.$$

Furthermore, they are continuous functions in $\mathbb{C}$, so $f(z)$ is entire.

EXAMPLE 6

Describe the region of analyticity of the function

$$f(z) = \frac{(x - 1) - iy}{(x - 1)^2 + y^2} = u + iv.$$

---

† A proof may be found in W. R. Derrick, *Complex Analysis and Applications,* 2nd ed., Wadsworth, Belmont, Calif., 1984, pp. 41–42.

SOLUTION

The first partials of $u = \operatorname{Re} f$ and $v = \operatorname{Im} f$ satisfy

$$u_x = \frac{y^2 - (x-1)^2}{[(x-1)^2 + y^2]^2} = v_y$$

and

$$u_y = \frac{-2y(x-1)}{[(x-1)^2 + y^2]^2} = -v_x.$$

These functions are continuous for all $z \neq 1$. Note that $f(z)$ is not defined at $z = 1$. Hence, $f(z)$ is analytic for all $z \neq 1$, and by (4),

$$f'(z) = u_x + iv_x = \frac{y^2 - (x-1)^2}{[(x-1)^2 + y^2]^2} + \frac{2y(x-1)}{[(x-1)^2 + y^2]^2} i.$$

EXAMPLE 7

Compute $f'(z)$, where $f(z) = e^x(\cos y + i \sin y)$.

SOLUTION

Here

$$u = e^x \cos y, \qquad v = e^x \sin y,$$

$$u_x = e^x \cos y = v_y, \qquad v_x = e^x \sin y = -u_y.$$

Thus $f_x = u_x + iv_x = f'(z) = f(z)$. The function $e^x(\cos y + i \sin y)$ is called the **complex exponential function.** We shall discuss this function in great detail in the next section.

In the real-variable case of elementary calculus, we know that when the derivative of a function is zero on some interval, then the function is constant in that interval. A similar result holds for complex variables.

■ **THEOREM 3: Zero Derivative Theorem**

Let f be analytic on a region G and $f'(z) = 0$ at each z in G. Then f is constant on G. The same conclusion holds if either $\operatorname{Re} f$, $\operatorname{Im} f$, $|f|$, or $\arg f$ is constant on G.

PROOF

Since $f'(z) = u_x(z) + iv_x(z)$, the vanishing of the derivative implies $u_x = v_y$, $v_x = -u_y$ are all zero. Thus, u and v are constant on lines parallel to the coordinate axes, and since G is polygonally connected (see the theorem on p. 893) $f = u + iv$ is constant on G.

If u (or v) is constant, $v_x = -u_y = 0 = u_x = v_y$, implying $f'(z) = u_x(z) + iv_x(z) = 0$ and f is constant.

If $|f|$ is constant, so is $|f|^2 = u^2 + v^2$, implying that

$$uu_x + vv_x = 0, \qquad uu_y + vv_y = vu_x - uv_x = 0.$$

Solving these two equations for u_x, v_x, we have $u_x = v_x = 0$ unless the determinant $u^2 + v^2$ vanishes. Since $|f|^2 = u^2 + v^2$ is constant, if $u^2 + v^2 = 0$ at a single point, then it is constantly zero and f vanishes identically. Otherwise the derivative vanishes and f is constant.

If $\arg f = c$, then $f(G)$ lies on the line

$$v = (\tan c) \cdot u,$$

unless $u \equiv 0$, in which case we are done. But $(1 - i \tan c)f$ is analytic and

$$\mathrm{Im}(1 - i \tan c)f = v - (\tan c)u = 0,$$

implying that $(1 - i \tan c)f$ is constant. Thus, so is f.   ■

Problems 15.6

In Problems 1–10 compute the derivative of each function and determine whether the function is entire. If not, find the points at which the derivative does not exist.

1. $f(z) = z^3 - 2z$

2. $f(z) = z^5 - 2z + 4$

3. $f(z) = \dfrac{1}{z}$

4. $f(z) = \dfrac{z+1}{z-1}$

5. $f(z) = \dfrac{z-1}{z^3+1}$

6. $f(z) = \dfrac{z+1}{z^2}$

7. $f(z) = \dfrac{z^2+1}{z^2-1}$

8. $f(z) = z - \dfrac{2}{z^2}$

9. $f(z) = \dfrac{4}{(1-z)^2}$

10. $f(z) = \dfrac{z^3}{(z^2+1)^4}$

In Problems 11–18 show that each function is entire and compute its derivative.

11. $f(z) = (x + y) + i(y - x)$

12. $f(z) = (x^2 - 2xy - y^2) + i(x^2 + 2xy - y^2)$

13. $f(z) = (x^2 + 2x - y^2) + 2y(x + 1)i$

14. $f(z) = \cos x \cosh y - i \sin x \sinh y$

15. $f(z) = \sin x \cosh y + i \cos x \sinh y$

16. $f(z) = e^{x^2-y^2}(\cos 2xy + i \sin 2xy)$

17. $f(z) = (x^3 - 3xy^2) + i(3x^2y - y^3)$

18. $f(z) = \sin(x^2 - y^2) \cosh(2xy) + i \cos(x^2 - y^2) \sinh(2xy)$

Using the Cauchy-Riemann equations, prove that the functions in Problems 19–22 do not have a derivative at any point in $\mathbb{C}$.

19. $f(z) = \bar{z}$

20. $f(z) = \mathrm{Re}\, z$

21. $f(z) = \mathrm{Im}\, z$

22. $f(z) = |z|$

Using the Cauchy-Riemann equations and the definition of a derivative, determine where the functions in Problems 23–26 have a derivative.

23. $f(z) = \bar{z}^2$

24. $f(z) = (\mathrm{Re}\, z)^2$

25. $f(z) = \bar{z}\, \mathrm{Re}\, z$

26. $f(z) = z\, \mathrm{Im}\, z$

In Problems 27–30 determine regions over which the given function is analytic and compute the derivatives in those regions.

27. $f(z) = \dfrac{x}{x^2+y^2} - i\dfrac{y}{x^2+y^2}$

28. $f(z) = \dfrac{-x}{(y-1)^2+x^2} - i\left[\dfrac{y^2-3y+2+x^2}{(y-1)^2+x^2}\right]$

29. $f(z) = \sin\left(\dfrac{x}{x^2+y^2}\right)\cosh\left(\dfrac{y}{x^2+y^2}\right)$
$\qquad - i\cos\left(\dfrac{x}{x^2+y^2}\right)\sinh\left(\dfrac{y}{x^2+y^2}\right)$

* 30. $f(z) = \dfrac{1}{2}\ln(x^2 + y^2) + i\tan^{-1}\dfrac{y}{x}$

31. Show that at $z = 0$ the function

$$f(z) = \begin{cases} \dfrac{\bar{z}^3}{|z|^2}, & z \neq 0, \\ 0, & z = 0, \end{cases}$$

satisfies the Cauchy-Riemann equations but does not have a derivative.

32. Show that the function

$$f(z) = \begin{cases} e^{-1/z^4}, & z \neq 0, \\ 0, & z = 0, \end{cases}$$

satisfies the Cauchy-Riemann equations at $z = 0$ but does not have a derivative at that point.

* **33.** If u and v are expressed in terms of polar coordinates (r, θ), show that the Cauchy-Riemann equations can be written in the form

$$\frac{\partial u}{\partial r} = \frac{1}{r}\frac{\partial v}{\partial \theta}, \qquad \frac{1}{r}\frac{\partial u}{\partial \theta} = -\frac{\partial v}{\partial r}, \qquad r \neq 0.$$

[*Hint:* Recall that $x = r \cos \theta$ and $y = -r \sin \theta$. Then use the chain rule for real-valued functions. For example,
$$\frac{\partial u}{\partial \theta} = \frac{\partial u}{\partial x}\frac{\partial x}{\partial \theta} + \frac{\partial u}{\partial y}\frac{\partial y}{\partial \theta}.]$$

34. Prove that the function

$$f(z) = r^5(\cos 5\theta + i \sin 5\theta)$$

satisfies the Cauchy-Riemann equations in polar form for all $z \neq 0$.

* **35.** If all the zeros of a polynomial $P(z)$ have negative real parts, prove that the same is true for all the zeros of $P'(z)$. [*Hint:* Factor $P(z)$ and consider $P'(z)/P(z)$.]

36. If $f(z) = u + iv$ and $\bar{f} = u - iv$ are both analytic, prove that f is constant.

37. Let $f(z) = u + iv$ be entire and suppose $u \cdot v$ is constant. Prove that f is constant.

38. If $f(z) = u + iv$ is entire and $v = u^2$, then show that f is constant.

39. If $f = u + iv$ is entire and $u^2 = v^2$, then prove that f is constant.

40. Suppose the analytic function f is real-valued on the region G. Prove that f is constant on G.

* **41.** Let $f(z) = z^3$, $z_1 = 1$, and $z_2 = i$. Prove there is no point z_0 on the line segment from z_1 to z_2 such that

$$f(z_2) - f(z_1) = f'(z_0)(z_2 - z_1).$$

This shows that the mean value theorem for real functions does not extend to complex functions.

42. If $z = x + iy$, show that no entire function has the function $f(z) = x$ as its derivative.

15.7 The Complex Exponential

In Section 15.6 we saw that polynomials and rational functions in a real variable yield analytic functions when the real variable is replaced by z. This is by no means an isolated example. In fact, all elementary functions in calculus, such as exponentials, logarithms, and trigonometric functions, give rise to analytic functions when suitably extended to the complex plane. In the next three sections we shall define extensions of these elementary functions and indicate some of their properties.

We begin with the exponential e^x. We wish to define a function $f(z) = e^z$ that is analytic and coincides with the real exponential function when z is real.

From calculus, we recall that the main property of e^x is that it is its own derivative: $(e^x)' = e^x$. Thus, viewed as a differential equation, it satisfies the initial value problem

$$f'(z) = f(z), \qquad f(0) = 1. \tag{1}$$

We want $f(z) = e^z$ to satisfy (1), coincide with e^x when $y = 0$, and behave in accordance with the laws of exponents:

$$e^z = e^{x+iy} = e^x e^{iy}. \tag{2}$$

Of course, at this point, there is no guarantee that we will be able to define such a function. However, these considerations suggest that the complex exponential, if it exists, may have the form

$$f(z) = e^x p(y), \tag{3}$$

where $p(0) = 1$. The problem now is to determine an appropriate function $p(y)$.

As it turns out, we have already encountered an entire function that satisfies all of these properties. In Example 15.6.7 we showed that

$$f(z) = e^x(\cos y + i \sin y) \tag{4}$$

is entire and satisfies $f'(z) = f(z)$. Clearly $f(z)$ has the form indicated in (3) and $f(0) = e^0(\cos 0 + i \sin 0) = 1$. Thus, if we define

Euler's Formula

$$e^{iy} = \cos y + i \sin y, \tag{5}$$

(called **Euler's formula**) the law of exponents (2) will also hold, and (4) will coincide with e^x when $y = 0$.

The Complex Exponential

The **complex exponential,** given by

$$e^z = e^x(\cos y + i \sin y), \tag{6}$$

is a nonzero entire function satisfying $(e^z)' = e^z$ and $e^0 = 1$. Moreover $e^z \neq 0$, since neither e^x nor $\cos y + i \sin y$ is zero.

Euler's formula is very useful. Note that

$$e^{2\pi i} = \cos(2\pi) + i \sin(2\pi) = 1,$$

$$e^{\pi i} = \cos \pi + i \sin \pi = -1,$$

$$e^{\pi i/2} = \cos\frac{\pi}{2} + i \sin\frac{\pi}{2} = i,$$

$$e^{-\pi i/2} = \cos\left(\frac{-\pi}{2}\right) + i \sin\left(\frac{-\pi}{2}\right) = -i.$$

The polar representation of a complex number, given in Section 15.2, can be simplified by using the complex exponential:

$$z = r(\cos \theta + i \sin \theta), \tag{7}$$

where $r = |z|$ and $\theta = \arg z$. But, from (5), $\cos \theta + i \sin \theta = e^{i\theta}$ so that

$$z = re^{i\theta} = |z|e^{i \arg z}. \tag{8}$$

If $z_1 = x_1 + iy_1$ and $z_2 = x_2 + iy_2$, the addition formulas of trigonometry imply that

$$e^{z_1}e^{z_2} = e^{x_1}e^{x_2}(\cos y_1 + i \sin y_1)(\cos y_2 + i \sin y_2)$$

$$= e^{x_1+x_2}[(\cos y_1 \cos y_2 - \sin y_1 \sin y_2) + i(\sin y_1 \cos y_2 + \cos y_1 \sin y_2)]$$

$$= e^{x_1+x_2}[\cos(y_1 + y_2) + i \sin(y_1 + y_2)]$$

$$= e^{x_1+x_2}e^{i(y_1+y_2)} = e^{z_1+z_2}.$$

Hence

$$e^{z_1}e^{z_2} = e^{z_1+z_2}. \tag{9}$$

Furthermore, since

$$e^{z_1-z_2}e^{z_2} = e^{z_1-z_2+z_2} = e^{z_1},$$

it follows that

$$e^{z_1-z_2} = e^{z_1}/e^{z_2}. \tag{10}$$

Using the sum-of-exponents formula repeatedly, we obtain $e^{nz} = (e^z)^n$. This identity provides a quick proof of De Moivre's theorem by letting $z = e^{i\theta}$:

$$(\cos \theta + i \sin \theta)^n = (e^{i\theta})^n = e^{in\theta} = \cos n\theta + i \sin n\theta, \tag{11}$$

for $n = 0, \pm 1, \pm 2, \dots$.

EXAMPLE 1

Compute $(1 - i)^{23}$.

SOLUTION
$\text{Arg}(1 - i) = \tan^{-1}(-1) = -\pi/4$ and $|1 - i| = \sqrt{2}$. Thus, using (11), we obtain

$$\text{since } e^{6\pi i} = (e^{2\pi i})^3 = 1^3 = 1$$
$$\downarrow$$
$$(1-i)^{23} = (\sqrt{2}e^{-\pi i/4})^{23} = 2^{23/2}e^{-23\pi i/4} = 2^{23/2}e^{i(-23\pi/4 + 24\pi/4)}$$

$$= 2^{23/2}e^{\pi i/4} = 2^{11}(\sqrt{2}e^{\pi i/4})$$

$$= 2^{11}(1 + i).$$

EXAMPLE 2

Find all the complex numbers z that satisfy $e^z = -1$.

SOLUTION

Let $z = x + iy$ and write both sides of $e^z = -1$ in polar form:

$$e^x e^{iy} = |-1| e^{i \arg(-1)}.$$

Since $|-1| = 1$, we have $e^x = 1$, or $x = 0$. Also $\arg(-1) = \pi + 2n\pi$, $n = 0, \pm 1, \pm 2, \ldots,$ so that

$$y = (2n + 1)\pi, \qquad n = 0, \pm 1, \pm 2, \ldots.$$

Finally, we have

$$z = x + iy = (2n + 1)\pi i, \qquad n = 0, \pm 1, \pm 2, \ldots.$$

EXAMPLE 3

Find all complex numbers z that satisfy

$$e^{4z} = 1 - \sqrt{3} i.$$

SOLUTION

$|1 - \sqrt{3} i| = 2$ and $\arg(1 - \sqrt{3} i) = -\pi/3 + 2n\pi$, $n = 0, \pm 1, \ldots.$ Thus

$$1 - \sqrt{3} i = 2 e^{(-\pi/3 + 2n\pi)i}, \qquad n = 0, \pm 1, \pm 2, \ldots.$$

Then, if $z = x + iy$, $e^{4z} = e^{4x} e^{4yi}$, so that

$$|e^{4z}| = e^{4x} = |1 - \sqrt{3} i| = 2.$$

Thus

$$4x = \ln 2 \quad \text{and} \quad x = \tfrac{1}{4} \ln 2 = \ln 2^{1/4}.$$

Also

$$e^{4yi} = e^{(-\pi/3 + 2n\pi)i}$$

so that

$$4yi = \left(-\frac{\pi}{3} + 2n\pi \right) i$$

and

$$y = -\frac{\pi}{12} + \frac{n\pi}{2}, \qquad n = 0, \pm 1, \pm 2, \ldots.$$

Therefore, the answer is

$$z = \ln 2^{1/4} + \left(-\frac{\pi}{12} + \frac{n\pi}{2} \right) i, \qquad n = 0, \pm 1, \pm 2, \ldots.$$

The real-valued function e^x has the important property of being one-to-one. That is, if $e^{x_1} = e^{x_2}$ then $x_1 = x_2$. This means that the function e^x has an inverse, which we denote by $\ln x$. The important point here is that we are able to define the logarithm $\ln x$ *because* e^x is one-to-one.

We shall define the complex logarithm in Section 15.9. However, this definition will not be immediate as in the real case because e^z is *not* one-to-one. This follows from the periodicity of e^z, since $e^z = e^{z+2n\pi i}$ for $n = 0, \pm 1, \pm 2, \ldots$. Thus, if we wish to obtain a one-to-one function using e^z, we must restrict its domain.

In equation (15.2.2) on p. 880 we restricted the argument of a complex number to lie in the interval $-\pi < \theta \le \pi$. In order to make e^z one-to-one, it is natural, therefore, to restrict y to the interval $-\pi < \theta \le \pi$. Hence

e^z is one-to-one in the region

$$S = \{z = x + iy : -\infty < x < \infty, -\pi < y \le \pi\}.$$

Principal Region

The infinite strip sketched in Fig. 1 is called the **principal region** of the function e^z.

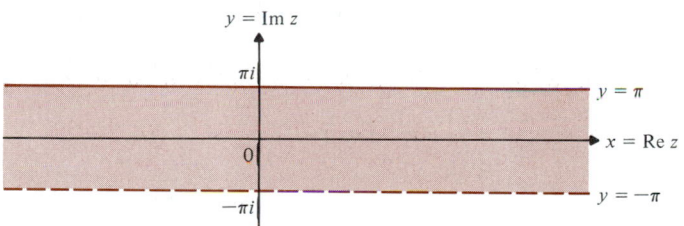

FIGURE 1 Principal region for e^z

Problems 15.7

In Problems 1–17 express each number in the form $x + iy$.

1. $e^{3\pi i}$

2. $e^{i\pi/4}$

3. $e^{i\pi/3}$

4. $e^{-i\pi/6}$

5. $e^{-2\pi i/3}$

6. $e^{5\pi i/6}$

7. $e^{3+(\pi/2)i}$

8. $e^{2-4\pi i/3}$

9. $e^{-1+(\pi/6)i}$

10. $e^{(1+\pi i)/2}$

11. $e^{-1+\pi i/4}$

12. $e^{(-1+\pi i)/4}$

13. $e^{(6+3\pi i)/2}$

14. $e^{7\pi i/2}$

15. $e^{50\pi i}$

16. $e^{2001\pi i}$

17. $e^{-(2+\pi i/6)}$

In Problems 18–26 find all the complex numbers z that satisfy the given equation.

18. $e^z = 1$

19. $e^z = i$

20. $e^z = 2i$

21. $e^{iz} = 2$

22. $e^{iz} = -1$

23. $e^z = -\sqrt{3} + i$

24. $e^{2z} = 4 + 4i$

25. $e^{3z} = 6 - 6i$

26. $e^{10z} = -5 - 5i$

27. Obtain all values of $e^{\pi i k/2}$, k an integer

28. Show that $(\overline{e^z}) = e^{\bar{z}}$.

In Problems 29–36, calculate each number using equation (8) and De Moivre's theorem.

29. $(1 + i)^{29}$

30. $(-1 + i)^{17}$

31. $(-1 - i)^{36}$

32. $(2 + 2i)^{12}$

33. $(\sqrt{3} + i)^{15}$ **34.** $(-\sqrt{3} + i)^{13}$

35. $(1 - \sqrt{3}i)^{14}$ **36.** $\left(\dfrac{1}{\sqrt{2}} - \dfrac{1}{\sqrt{2}} i\right)^{19}$

In Problems 37–40, find the sums using De Moivre's theorem.

37. $1 + \cos x + \cos 2x + \cdots + \cos nx$

38. $\cos x + \cos 3x + \cos 5x + \cdots + \cos(2n - 1)x$

39. $\sin x + \sin 2x + \sin 3x + \cdots + \sin nx$

40. $\sin x + \sin 3x + \sin 5x + \cdots + \sin(2n - 1)x$

41. If $f(z)$ is entire, show that $e^{f(z)}$ is entire and find its derivative.

42. Express all the nth roots of 1 in exponential form.

43. Compute all the 5th roots of 1.

44. Recall the series

$$\cos x = 1 - \frac{x^2}{2!} + \frac{x^4}{4!} - \frac{x^6}{6!} + \cdots,$$

and

$$\sin x = x - \frac{x^3}{3!} + \frac{x^5}{5!} - \frac{x^7}{7!} + \cdots.$$

Show that if $x = i\theta$ is inserted into the series

$$e^U = 1 + U + \frac{U^2}{2!} + \frac{U^3}{3!} + \cdots,$$

we obtain

$$e^{i\theta} = \cos\theta + i\sin\theta.$$

* 45. Prove that e^z is the only analytic solution to the complex differential equation $f'(z) = f(z), f(0) = 1$.

15.8 Trigonometric and Hyperbolic Functions

The complex exponential can be used to define the complex trigonometric functions. Since $e^{iy} = \cos y + i \sin y$ and $e^{-iy} = \cos y - i \sin y$, it follows that

$$\frac{e^{iy} + e^{-iy}}{2} = \frac{(\cos y + i \sin y) + (\cos y - i \sin y)}{2} = \cos y$$

and

$$\frac{e^{iy} - e^{-iy}}{2i} = \frac{(\cos y + i \sin y) - (\cos y - i \sin y)}{2i} = \sin y.$$

We extend these definitions to the complex plane as follows.

The Complex Trigonometric Functions

$$\cos z = \frac{e^{iz} + e^{-iz}}{2}, \qquad \sin z = \frac{e^{iz} - e^{-iz}}{2i}. \tag{1}$$

These functions are entire as they are sums of entire functions and satisfy

$$(\cos z)' = \frac{ie^{iz} - ie^{-iz}}{2} = -\frac{e^{iz} - e^{-iz}}{2i} = -\sin z, \tag{2}$$

$$(\sin z)' = \frac{ie^{iz} + ie^{-iz}}{2i} = \frac{e^{iz} + e^{-iz}}{2} = \cos z. \tag{3}$$

In addition, these functions are periodic of period 2π since, for example,

$$\cos(z + 2\pi) = \frac{e^{i(z+2\pi)} + e^{-i(z+2\pi)}}{2} = \frac{e^{iz+2\pi i} + e^{-iz-2\pi i}}{2} \overset{e^{2\pi i}=1}{\downarrow} \frac{e^{iz} + e^{-iz}}{2} = \cos z.$$

EXAMPLE 1

Compute $\cos(2 + 3i)$.

SOLUTION

$$\cos(2 + 3i) = \frac{e^{i(2+3i)} + e^{-i(2+3i)}}{2} = \frac{e^{-3+2i} + e^{3-2i}}{2}$$

$$= \frac{e^{-3}(\cos 2 + i \sin 2) + e^{3}(\cos 2 - i \sin 2)}{2}$$

$$= \frac{(e^3 + e^{-3}) \cos 2}{2} - \frac{(e^3 - e^{-3}) \sin 2}{2} i \approx -4.19 - 9.11i.$$

The other four trigonometric functions, defined in terms of the sine and cosine functions by the usual relations

$$\tan z = \frac{\sin z}{\cos z}, \qquad \cot z = \frac{\cos z}{\sin z},$$

$$\sec z = \frac{1}{\cos z}, \qquad \csc z = \frac{1}{\sin z}, \tag{4}$$

are analytic except where their denominators vanish, and satisfy the standard rules of differentiation

$$(\tan z)' = \sec^2 z, \qquad (\sec z)' = \sec z \tan z,$$

$$(\cot z)' = -\csc^2 z, \qquad (\csc z)' = -\csc z \cot z. \tag{5}$$

We verify two of these:

$$(\tan z)' = \left(\frac{\sin z}{\cos z}\right)' \overset{\text{quotient rule}}{\downarrow} \frac{\cos z(\sin z)' - \sin z(\cos z)'}{\cos^2 z}$$

$$= \frac{\cos z(\cos z) - \sin z(-\sin z)}{\cos^2 z}$$

see equation (6) below

$$= \frac{\cos^2 z + \sin^2 z}{\cos^2 z} \overset{\downarrow}{=} \frac{1}{\cos^2 z} = \sec^2 z.$$

chain rule

$$(\sec z)' = \left(\frac{1}{\cos z}\right)' = [(\cos z)^{-1}]' \overset{\downarrow}{=} -(\cos z)^{-2}(-\sin z)$$

$$= \frac{\sin z}{\cos^2 z} = \frac{1}{\cos z} \cdot \frac{\sin z}{\cos z} = \sec z \tan z.$$

All the usual trigonometric identities are still valid in complex variables, the proofs depending on properties of the exponential. For example,

$$\cos^2 z + \sin^2 z = \tfrac{1}{4}[(e^{iz} + e^{-iz})^2 - (e^{iz} - e^{-iz})^2] = 1, \tag{6}$$

and

$$\cos z_1 \cos z_2 - \sin z_1 \sin z_2$$

$$= \frac{e^{iz_1} + e^{-iz_1}}{2} \frac{e^{iz_2} + e^{-iz_2}}{2} - \frac{e^{iz_1} - e^{-iz_1}}{2i} \frac{e^{iz_2} - e^{-iz_2}}{2i}$$

$$= \frac{2e^{iz_1}e^{iz_2} + 2e^{-iz_1}e^{-iz_2}}{4} = \cos(z_1 + z_2). \tag{7}$$

Recall the definitions of the real hyperbolic sine and cosine functions:

$$\sinh x = \frac{e^x - e^{-x}}{2} \quad \text{and} \quad \cosh x = \frac{e^x + e^{-x}}{2}.$$

We can use these functions to obtain some interesting identities.
From the definition of $\cos z$ we have

$$\cos z = \cos(x + iy) = \frac{e^{-y+ix} + e^{y-ix}}{2}$$

$$= \tfrac{1}{2}e^{-y}(\cos x + i \sin x) + \tfrac{1}{2}e^{y}(\cos x - i \sin x)$$

$$= \left(\frac{e^y + e^{-y}}{2}\right)\cos x - i\left(\frac{e^y - e^{-y}}{2}\right)\sin x.$$

Thus,

$$\cos z = \cos x \cosh y - i \sin x \sinh y. \tag{8}$$

Similarly, we find

$$\sin z = \sin x \cosh y + i \cos x \sinh y. \tag{9}$$

When dealing with a function, it is useful to know when the function vanishes (that is, the points at which it takes the value 0). The zeros of $\cos x$ are the numbers $x = \pi/2 + n\pi$, $n = 0, \pm1, \pm2, \ldots$. The zeros of $\sin x$ are multiples of π. It turns out that these are the only zeros of the complex functions $\cos z$ and $\sin z$, respectively.

THEOREM 1

The real zeros of $\sin z$ and $\cos z$ are their only zeros.

PROOF

If $\sin z = 0$, equation (9) shows we must have

$$\sin x \cosh y = 0 \quad \text{and} \quad \cos x \sinh y = 0.$$

But $\cosh y \geq 1$, implying that the first term vanishes only when $\sin x = 0$, that is, when $x = 0, \pm\pi, \pm2\pi, \ldots$. However, for these values $\cos x$ does not vanish. Hence, we must have $\sinh y = 0$, or $y = 0$. Thus

$$\sin z = 0 \quad \text{implies} \quad z = n\pi, \qquad n \text{ an integer.}$$

This statement also applies to $\tan z$, and in like manner we find

$$\cos z = 0 \quad \text{implies} \quad z = (n + \tfrac{1}{2})\pi, \qquad n \text{ an integer.} \qquad \blacksquare$$

The complex hyperbolic functions are defined by extending the real definitions to the complex plane.

The Complex Hyperbolic Functions

$$\sinh z = \frac{e^z - e^{-z}}{2}, \qquad \cosh z = \frac{e^z + e^{-z}}{2}. \tag{10}$$

The other four complex hyperbolic functions are defined as in the real case. Again, all the usual identities and rules of differentiation apply to the complex hyperbolic functions (see Problems 47–49). Note, moreover, that

$$\sinh iz = \frac{e^{iz} - e^{-iz}}{2} = i\left(\frac{e^{iz} - e^{-iz}}{2i}\right) = i \sin z \tag{11}$$

and

$$\cosh iz = \frac{e^{iz} + e^{-iz}}{2} = \cos z. \qquad (12)$$

Thus, the complex hyperbolic functions are intimately related to the complex trigonometric functions, as multiplying by i simply rotates every vector in $\mathbb{C}$ counterclockwise by 90° (since Arg $i = \pi/2$). Hence, the zeros of sinh z and cosh z are pure imaginary.

Finally, since $f(y) = e^{iy}$ is periodic of period 2π, it follows that

sinh z and cosh z are periodic of period $2\pi i$.

Problems 15.8

In Problems 1–16, express each of the numbers in the form $x + iy$. First express the answer in terms of trigonometric and hyperbolic functions of a real variable. Then use a calculator, if necessary, to obtain an answer accurate to three decimal places.

1. sin i
2. cos($-i$)
3. cos(1 + i)
4. sin(1 − i)
5. tan $2i$
6. sec $2i$
7. sin(2 + 5i)
8. tan(1 + i)
9. cosh(1 + i)
10. sinh πi
11. cosh($2\pi i$)
12. cosh($5\pi i$)
13. cosh($\pi i/4$)
14. sinh($\pi i/4$)
15. tan $\pi i/4$
16. sec $\pi i/4$

17. Compute $|\sin z|$.
18. Compute $|\cos z|$.
19. Compute $|\tan z|$.
20. Compute $|\sec z|$.
21. Write tan z in the form $u(x, y) + iv(x, y)$.
22. Do the same as in Problem 21 for the function sec z.

In Problems 23–30 find all complex numbers that solve the given equation.

23. $\cos z = \sin z$
24. $\cos z = -i \sin z$

25. $\sin z = 2$
26. $\cos z = \frac{1}{2}$
27. $\sinh z = 0$
28. $\cosh z = 0$
29. $\cosh z = 2$
30. $\cosh z = i$

31. Are there any points z where sinh $z = \cosh z$?
32. Show that $\overline{\sin z} = \sin \bar{z}$.
33. Show that $\overline{\cos z} = \cos \bar{z}$.

In Problems 34–41 prove the given identities.

34. $\sin(z_1 \pm z_2) = \sin z_1 \cos z_2 \pm \cos z_1 \sin z_2$
35. $\cos(z_1 - z_2) = \cos z_1 \cos z_2 + \sin z_1 \sin z_2$
36. $\sin(-z) = -\sin z$ and $\cos(-z) = \cos z$
37. $\sin 2z = 2 \sin z \cos z$
38. $\cos 2z = \cos^2 z - \sin^2 z$
39. $\tan 2z = \dfrac{2 \tan z}{1 - \tan^2 z}$
40. $|\sin z|^2 = \sin^2 x + \sinh^2 y$
41. $|\cos z|^2 = \cos^2 x + \sinh^2 y$

Prove the identities in Problems 42–46.

42. $\cosh^2 z - \sinh^2 z = 1$, $\cosh(-z) = \cosh z$,
  $\sinh(-z) = -\sinh z$
43. $\sinh(z_1 + z_2) = \sinh z_1 \cosh z_2 + \cosh z_1 \sinh z_2$

44. $\cosh(z_1 + z_2) = \cosh z_1 \cosh z_2 + \sinh z_1 \sinh z_2$

45. $i \sinh z = \sin iz, \quad \cosh z = \cos iz, \quad i \tanh z = \tan iz$

46. $|\sinh z|^2 = \sinh^2 x + \sin^2 y, \quad |\cosh z|^2 = \sinh^2 x + \cos^2 y$

Prove the rules of differentiation given in Problems 47–49.

47. $(\sinh z)' = \cosh z, \quad (\cosh z)' = \sinh z$

48. $(\tanh z)' = \operatorname{sech}^2 z, \quad (\coth z)' = -\operatorname{csch}^2 z$

49. $(\operatorname{sech} z)' = -\operatorname{sech} z \tanh z, \quad (\operatorname{csch} z)' = -\operatorname{csch} z \coth z$

50. Find all the zeros of $\sinh z$ and $\cosh z$.

51. Verify that $e^z = \cosh z + \sinh z$.

52. Verify that $e^{iz} = \cos z + i \sin z$.

15.9 The Complex Logarithm and Complex Powers

The real-valued functions $f(x) = e^x$ and $g(x) = \ln x \; (= \log_e x)$ are inverse functions. That is,

$$(f \circ g)(x) = e^{\ln x} = x, \qquad \text{if } x > 0, \tag{1}$$

and

$$(g \circ f)(x) = \ln e^x = x. \tag{2}$$

Both functions are one-to-one so that each has an inverse. That is why, in the study of one-variable calculus, it is possible to start with e^x and define $\ln x$ as the inverse of e^x. Alternatively, one can start with $\ln x$ (defined as $\ln x = \int_1^x (1/t) \, dt$) and then obtain e^x as its inverse.

The situation is more complicated when defining the complex logarithm. The problem is that the complex exponential e^z is *not* one-to-one. In fact, since e^z is periodic of period $2\pi i$, there are infinitely many solutions to the equation $e^z = w$, for any nonzero complex number w. For example, as we saw in Example 15.7.2 on p. 914, all numbers of the form $z = (2n + 1)\pi i$ are solutions to the equation $e^z = -1$. Thus, if we wanted to define an inverse function at -1, we would have to choose from the values $\pi i, 3\pi i, 5\pi i$, and so on.

We can avoid this problem by restricting the domain of the exponential function e^z. We shall do so in a moment. Let us first determine the form the complex logarithm must take.

In the real case, if $y = \ln x$, then

$$x = e^y. \tag{3}$$

Now, let z be a nonzero complex number and suppose that there is a complex number w that satisfies the equation

$$z = e^w. \tag{4}$$

If $r = |z|$, $\theta = \arg z$, and $w = u + iv$, we obtain

$$z = re^{i\theta} = e^{u+iv} = e^u e^{iv}. \tag{5}$$

Equating real and imaginary parts of (5), we obtain

$$r = e^u \quad \text{and} \quad v = \theta.$$

Since u is a real number, we know that $r = e^u$ if and only if $u = \ln r$. Thus

$$w = \ln r + i\theta = \ln|z| + i \arg z. \qquad (6)$$

Note that $\ln|z|$ is defined as long as $z \neq 0$. Moreover, $\arg z$ is infinitely valued, so that $\ln|z| + i \arg z$ is multiply defined.

DEFINITION

If $z \neq 0$, then the **logarithm** of z, denoted by $\log z$, is defined by

Complex Logarithm

$$\boxed{\log z = \ln|z| + i \arg z. \qquad (7)}$$

Multifunction

The complex logarithm is not a function because it is infinitely valued. We will call any such multivalued expression a **multifunction.**

EXAMPLE 1

Compute all values of $\log(-1)$.

SOLUTION

Since $|-1| = 1$ and $\arg(-1) = (2n + 1)\pi$, $n = 0, \pm 1, \pm 2, \ldots$, we have

$$\log(-1) = \ln 1 + i \arg(-1) = (2n + 1)\pi i, \qquad n = 0, \pm 1, \pm 2, \ldots.$$

EXAMPLE 2

Compute all values of $\log(1 + i)$.

SOLUTION

Since $|1 + i| = \sqrt{2}$ and $\arg(1 + i) = \pi/4 + 2n\pi$, $n = 0, \pm 1, \pm 2, \ldots$, we have

$$\log(1 + i) = \ln\sqrt{2} + (2n + \tfrac{1}{4})\pi i,$$
$$= \tfrac{1}{2} \ln 2 + (2n + \tfrac{1}{4})\pi i, \qquad n = 0, \pm 1, \pm 2, \ldots.$$

It is evident that $\log z$ would be a function (that is, single-valued) if the domain of $\arg z$ were restricted.

DEFINITION

The **principal value** of $\log z$, denoted by Log z, is defined as

Principal Value of the Complex Logarithm

$$\boxed{\operatorname{Log} z = \ln|z| + i \operatorname{Arg} z, \qquad -\pi < \operatorname{Arg} z \leq \pi. \qquad (8)}$$

Log z is a function whose domain is $\mathbb{C} - \{0\}$. Note, however, that Log z is not continuous on Arg $z = \pi$. (See Problem 40.)

EXAMPLE 3

From Examples 1 and 2 we see that

$$\text{Log}(-1) = \pi i \quad \text{and} \quad \text{Log}(1+i) = \tfrac{1}{2}\ln 2 + \frac{\pi}{4}i.$$

We can think of the complex logarithm log z as an infinite number of functions, each of which has a restricted domain. For example, we can write

$$f_0(z) = \text{Log } z = \ln|z| + i \text{ Arg } z, \qquad -\pi < \arg z \le \pi,$$

$$f_1(z) = \ln|z| + i \arg z, \qquad -3\pi < \arg z \le -\pi,$$

$$f_2(z) = \ln|z| + i \arg z, \qquad \pi < \arg z \le 3\pi,$$

$$f_3(z) = \ln|z| + i \arg z, \qquad -5\pi < \arg z \le -3\pi,$$

$$f_4(z) = \ln|z| + i \arg z, \qquad 3\pi < \arg z \le 5\pi,$$

and so on. These functions are called **branches** of the logarithm. The values $\pm\pi, \pm3\pi, \pm5\pi, \ldots$ of the argument of z all correspond to the negative real axis. They are the values of the argument at which we switch from one branch to another. The rays

$$\arg z = \pm\pi, \pm3\pi, \ldots$$

are called **branch cuts** of the logarithm. In any region that does not include a branch cut, the corresponding branch of the logarithm is analytic. We shall prove this shortly.

The domain of analyticity for the function Log z is sketched in Fig. 1.

Remark. It is possible to define log z so that it is analytic anywhere we wish except at $z = 0$. To do so it is necessary to move the branch cuts to other locations. For example we can move the branch cut to the negative imaginary axis, so that log z becomes the infinite collection of functions

$$g_0(z) = \ln|z| + i \arg z, \qquad -\frac{\pi}{2} < \arg z \le \frac{3\pi}{2},$$

$$g_1(z) = \ln|z| + i \arg z, \qquad \frac{3\pi}{2} < \arg z \le \frac{7\pi}{2},$$

$$g_2(z) = \ln|z| + i \arg z, \qquad \frac{-5\pi}{2} < \arg z \le \frac{-\pi}{2}.$$

Now the logarithm is analytic along the rays $\arg z = \pm\pi, \pm3\pi, \ldots$.

Branch Cut

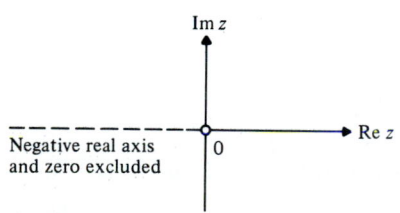

Negative real axis
and zero excluded

FIGURE 1

■ **THEOREM 1**

Log z is analytic in the region $G = \{z : |z| > 0, -\pi < \arg z < \pi\}$.

PROOF

For simplicity assume $|\text{Arg } z| < \pi/2$ (so that $\tan^{-1}(y/x)$ is defined). Then

$$u + iv = \text{Log } z = \ln|z| + i \text{ Arg } z = \tfrac{1}{2}\ln(x^2 + y^2) + i \tan^{-1}(y/x).$$

We next compute

$$u_x = \frac{x}{x^2 + y^2}, \qquad v_x = \frac{1}{1 + (y/x)^2}\left(\frac{-y}{x^2}\right) = \frac{-y}{x^2 + y^2},$$

$$u_y = \frac{y}{x^2 + y^2}, \qquad v_y = \frac{1}{1 + (y/x)^2}\left(\frac{1}{x}\right) = \frac{x}{x^2 + y^2}.$$

We see that $u_x = v_y$ and $u_y = -v_x$ so the Cauchy-Riemann equations hold. Moreover, the partial derivatives are continuous, so Log z is analytic in G. ■

Remark 1. The same proof works for each of the branches of log z.

Remark 2. The proof is free of restrictions if we use the polar form of the Cauchy-Riemann equations. In Problem 15.6.33 on p. 911 you were asked to derive these equations:

$$\frac{\partial u}{\partial r} = \frac{1}{r}\frac{\partial v}{\partial \theta} \quad \text{and} \quad \frac{1}{r}\frac{\partial u}{\partial \theta} = -\frac{\partial v}{\partial r}.$$

If

$$w = \text{Log } z = \ln|z| + i \text{ Arg } z = \ln r + i\theta,$$

then $u = \ln r$, $v = \theta$,

$$\frac{\partial u}{\partial r} = \frac{1}{r}, \qquad \frac{\partial v}{\partial r} = 0,$$

$$\frac{\partial u}{\partial \theta} = 0, \qquad \frac{\partial v}{\partial \theta} = 1,$$

and the Cauchy-Riemann equations are easily seen to be satisfied in G.

Remark 3. In Theorem 1, we must exclude $\arg z = \pm\pi$. In Problem 40 you are asked to show that Log z is *not* continuous on the negative real axis.

■ **THEOREM 2**

The derivative of log z is $1/z$ for any branch of log z.

PROOF

Let $w = \log z$. Then

$$z = e^w \tag{9}$$

We know from Remark 1 that w' exists. Thus we can use the chain rule to differentiate both sides of (9) with respect to z to obtain

$$1 = w'e^w \quad \text{or} \quad w' = \frac{1}{e^w} = \frac{1}{z}. \qquad \blacksquare$$

Complex Powers

Let a and x be positive real numbers and let

$$y = x^a.$$

Then

$$\ln y = a \ln x$$

and, exponentiating,

$$y = e^{\ln y} = e^{a \ln x}.$$

Thus

$$x^a = e^{a \ln x} \qquad \text{if } a, x > 0. \tag{10}$$

We use equation (10) to define complex exponentials.

Complex Powers

DEFINITION

The **complex power** z^a is defined by

$$\boxed{z^a = e^{a \log z}, \qquad a \text{ complex}, z \neq 0. \tag{11}}$$

The complex power z^a has an infinite number of branches because $\log z$ has an infinite number of branches. Thus z^a is a multifunction. The branch of z^a corresponding to Log z is called the **principal branch** of z^a.

EXAMPLE 4

Compute all values of i^i.

SOLUTION

$i^i = e^{i \log i}$. Now

$$\log i = \ln|i| + i \arg i = \ln 1 + i\left(\frac{\pi}{2} + 2n\pi\right) = \left(\frac{\pi}{2} + 2n\pi\right)i.$$

Thus

$$i^i = e^{i(\pi/2 + 2n\pi)i} = e^{-\pi/2 - 2n\pi}, \qquad n = 0, \pm 1, \pm 2, \ldots.$$

The principal value of i^i (obtained by setting $n = 0$) is the real number $e^{-\pi/2} \approx 0.208$.

EXAMPLE 5

Describe the branches of $z^{1/2}$.

SOLUTION

Let $\theta = \text{Arg } z$ and $r = |z|$. Then

$$z^{1/2} = e^{(\log z)/2} = e^{(\ln r + i[\theta + 2\pi n])/2}$$

$e^{\ln r^{1/2}} = r^{1/2}$

$$\overset{\downarrow}{=} \sqrt{r}\left[\cos\left(\frac{\theta + 2\pi n}{2}\right) + i \sin\left(\frac{\theta + 2\pi n}{2}\right)\right].$$

For the principal branch of $z^{1/2}$ we set $n = 0$, obtaining

$$f(z) = \text{principal branch of } z^{1/2} = \sqrt{r}\left[\cos\frac{\theta}{2} + i \sin\frac{\theta}{2}\right].$$

The branch where $\pi < \text{arg } z \leq 3\pi$ (set $n = 1$) gives

$$g(z) = z^{1/2}\big|_{\pi < \text{arg } z \leq 3\pi} = \sqrt{r}\left[\cos\left(\frac{\theta}{2} + \pi\right) + i \sin\left(\frac{\theta}{2} + \pi\right)\right]$$

$$= -\sqrt{r}\left[\cos\left(\frac{\theta}{2}\right) + i \sin\left(\frac{\theta}{2}\right)\right] = -f(z).$$

Thus $f(z)$ and $g(z)$ give us the two square roots of z. There are, of course, other branches, but each of these is equal to either $f(z)$ or $g(z)$. For example, in the interval $-7\pi < \text{arg } z \leq -5\pi$, $\text{arg } z = \theta - 6\pi$, so that

$$h(z) = z^{1/2}\big|_{-7\pi < \text{arg } z \leq -5\pi} = \sqrt{r}[\cos \tfrac{1}{2}(\theta - 6\pi) + i \sin \tfrac{1}{2}(\theta - 6\pi)]$$

$$= \sqrt{r}[\cos(\tfrac{1}{2}\theta - 3\pi) + i \sin(\tfrac{1}{2}\theta - 3\pi)] = g(z).$$

Similarly, it is easy to show that

$$j(z) = z^{1/2}\big|_{5\pi < \text{arg } z \leq 7\pi} = f(z).$$

We can use the facts in this section to prove a number of logarithmic and exponential identities. We prove five of them here.

■ **THEOREM 3**

Let z_1, z_2, a, and b be nonzero complex numbers. Then, allowing for different multiples of 2π in the arguments,† we have

(a) $\log(z_1 z_2) = \log z_1 + \log z_2$.

(b) $\log(z_1/z_2) = \log z_1 - \log z_2$.

(c) $z^a z^b = z^{a+b}$.

(d) $z^a/z^b = z^{a-b}$.

(e) z^a is analytic in each branch and $(z^a)' = az^{a-1}$.

† By this we mean the following, for each of these multivalued identities: if values are chosen for all but one term in each equation, then a value exists for the remaining term for which equality holds.

PROOF

(a) $\log z_1 = \ln|z_1| + i \arg z_1$ and $\log z_2 = \ln|z_2| + i \arg z_2$, so that

$$\log(z_1 z_2) = \ln|z_1 z_2| + i \arg z_1 z_2$$

see equation (11) on p. 885
$$\downarrow$$
$$= \ln|z_1| + \ln|z_2| + i(\arg z_1 + \arg z_2)$$

$$= (\ln|z_1| + i \arg z_1) + (\ln|z_2| + i \arg z_2)$$

$$= \log z_1 + \log z_2.$$

(b) This follows since $\ln|z_1/z_2| = \ln|z_1| - \ln|z_2|$ and $\arg(z_1/z_2) = \arg z_1 - \arg z_2$.

(c) Let $w_1 = z^a$ and $w_2 = z^b$. Then

$$w_1 = e^{a \log z}, \qquad w_2 = e^{b \log z},$$

and

$$z^a z^b = w_1 w_2 = e^{a \log z} e^{b \log z} = e^{(a+b)\log z} = z^{a+b}.$$

(d) $\dfrac{z^a}{z^b} = \dfrac{w_1}{w_2} = \dfrac{e^{a \log z}}{e^{b \log z}} = e^{(a-b) \log z} = z^{a-b}.$

(e) z^a is analytic because it is the composition of analytic functions (the composition of analytic functions is analytic by the chain rule). Also,

chain rule
$$\downarrow$$
$$(z^a)' = (e^{a \log z})' = e^{a \log z}(a \log z)' = z^a \cdot \frac{a}{z}$$

from (c)
$$= a\frac{z^a \downarrow}{z^1} = az^{a-1}. \qquad \blacksquare$$

Warning Care must be taken when working with the principal branch of the logarithm, Log z, as properties of logarithms may not apply. For example,

$$\text{Log } i = \ln|i| + i \text{ Arg } i = i\pi/2,$$

$$\text{Log}(-1 + i) = \ln|-1 + i| + i \text{ Arg}(-1 + i)$$

$$= \ln \sqrt{2} + i\frac{3\pi}{4},$$

but

$$\text{Log}[i(-1 + i)] = \text{Log}(-1 - i)$$

$$= \ln|-1 - i| + i \text{ Arg}(-1 - i)$$

$$= \ln \sqrt{2} - i\frac{3\pi}{4},$$

so that

$$\text{Log}[i(-1+i)] \neq \text{Log } i + \text{Log}(-1 \pm i).$$

The problem here is that the sum of the angles $\pi/2$ and $3\pi/4$ is $5\pi/4$, which is not the principal value $-3\pi/4$ of the argument of $i(-1+i) = -1 - i$.

Inverse Trigonometric Multifunctions

We can use the complex logarithm to define the **inverse trigonometric multifunctions.** For example, let

$$z = \sin w = \frac{e^{iw} - e^{-iw}}{2i}.$$

Multiplying both sides of this equation by $2ie^{iw}$ yields

$$2ize^{iw} = e^{2iw} - 1 \quad \text{or} \quad e^{2iw} - 2ize^{iw} - 1 = 0.$$

This is a quadratic equation in the variable e^{iw}. The solutions are

$$e^{iw} = \frac{2iz \pm \sqrt{(2iz)^2 + 4}}{2} = iz \pm (1 - z^2)^{1/2}.$$

Hence

$$iw = \log[iz \pm (1 - z^2)^{1/2}] \quad \text{or} \quad w = \frac{1}{i} \log[iz \pm (1 - z^2)^{1/2}].$$

To decide which value of the square roots should be chosen, recall from calculus that

$$\frac{d}{dx} \sin^{-1} x = \frac{1}{(1 - x^2)^{1/2}} \quad \text{and} \quad \frac{d}{dx} \cos^{-1} x = \frac{-1}{(1 - x^2)^{1/2}}.$$

It is not hard to check that

$$\frac{d}{dz}\left(\frac{1}{i} \log[iz + (1 - z^2)^{1/2}]\right) = \frac{1}{(1 - z^2)^{1/2}}$$

and

$$\frac{d}{dz}\left(\frac{1}{i} \log[iz - (1 - z^2)^{1/2}]\right) = \frac{-1}{(1 - z^2)^{1/2}}.$$

This leads to the following definitions.

Complex Inverse Trigonometric Multifunctions

DEFINITION

The **inverse trigonometric multifunctions** $\sin^{-1} z$ and $\cos^{-1} z$ are defined by

$$\sin^{-1} z = \frac{1}{i} \log[iz + (1 - z^2)^{1/2}] \tag{12}$$

$$\cos^{-1} z = \frac{1}{i} \log[iz - (1 - z^2)^{1/2}]. \tag{13}$$

Since $\log z$ is a multifunction, so are $\sin^{-1} z$ and $\cos^{-1} z$. Although we shall not do so here, it is possible to prove that these functions are analytic in regions that do not include the **branch points** $z = \pm 1$.

Problems 15.9

In Problems 1–14 find (a) all values of the given logarithm and (b) the principal value.

1. $\log i$

2. $\log(-i)$

3. $\log 1$

4. $\log(1 - i)$

5. $\log(-1 + i)$

6. $\log(1 + \sqrt{3}i)$

7. $\log(\sqrt{3} + i)$

8. $\log(-1 - \sqrt{3}i)$

9. $\log(1 - \sqrt{3}i)$

10. $\log(-\sqrt{3} + i)$

11. $\log(2 + 3i)$

12. $\log(-4 + 6i)$

13. $\log(i/4)$

14. $\log(-8 - 2i)$

In Problems 15–24 find (a) all values and (b) the principal value of the given expression.

15. 1^i

16. 1^{2i}

17. 2^i

18. i^{-i}

19. $(1 + i)^i$

20. $(-1 + i)^i$

21. $(1 + i)^{1+i}$

22. $(-1 + \sqrt{3}i)^i$

23. $(1 - \sqrt{3}i)^{1+i}$

24. $(\sqrt{3} + i)^{3+4i}$

25. Show that $\mathrm{Log}(-1 - i) - \mathrm{Log}\, i \neq \mathrm{Log}\left(\dfrac{-1 - i}{i}\right)$.

26. Show that $\mathrm{Log}(i^3) \neq 3\,\mathrm{Log}\, i$.

Problems 27–39 involve multifunctions. Interpret equality as was done in Theorem 3.

27. Prove that $\log z^a = a \log z$, a complex $\neq 0$, $z \neq 0$. (Here we identify arguments that differ by multiples of 2π.)

28. Is 1 raised to any power always equal to 1?

29. Explain why $\cos^{-1} z = \dfrac{1}{i} \log[z + (z^2 - 1)^{1/2}]$.

30. Show that $\tan^{-1} z = \dfrac{i}{2} \log\!\left(\dfrac{i + z}{i - z}\right)$, $z \neq \pm i$.

31. Prove that $\sinh^{-1} z = \log[z + (z^2 + 1)^{1/2}]$.

32. Prove that $\cosh^{-1} z = \log[z + (z^2 - 1)^{1/2}]$.

33. Prove that $\tanh^{-1} z = \frac{1}{2} \log\!\left(\dfrac{1 + z}{1 - z}\right)$, $z \neq \pm 1$.

34. Prove that $(\sin^{-1} z)' = (1 - z^2)^{-1/2}$, $z \neq \pm 1$.

35. Prove that $(\cos^{-1} z)' = -(1 - z^2)^{-1/2}$, $z \neq \pm 1$.

36. Prove that $(\tan^{-1} z)' = \dfrac{1}{1 + z^2}$, $z \neq \pm i$.

37. Prove that $(\sinh^{-1} z)' = (1 + z^2)^{-1/2}$, $z \neq \pm i$.

38. Prove that $(\cosh^{-1} z)' = (z^2 - 1)^{-1/2}$, $z \neq \pm 1$.

39. Find the flaw in the following argument:

$$i = (-1)^{1/2} = [(-1)^3]^{1/2} = (-1)^{3/2} = i^3 = -i.$$

40. Explain why $\mathrm{Log}\, z$ is not continuous on the negative real axis. [*Hint:* Consider $\lim_{\theta \to \pi^+} \mathrm{Log}(re^{i\theta})$ and $\lim_{\theta \to \pi^-} \mathrm{Log}(re^{i\theta})$.]

Review Exercises for Chapter 15

In Exercises 1–7 perform the indicated operation.

1. $(5 - i) + (6 - 2i)$

2. $(4 + i)(3 - 2i)$

3. $(-1 + 5i)(\frac{1}{2} - \frac{1}{3}i)$

4. $\dfrac{2 + i}{4 - i}$

5. $\dfrac{2 + 3i}{4 - 2i}$

6. $(1 + i)^2$

7. $(1 - i)(1 - 2i)(1 - 3i)$

In Exercises 8–12 find the absolute value, the principal value of the argument, and the polar representation of the given number.

8. $2i$

9. -4

10. $-4 - 4i$

11. $2\sqrt{3} + 2i$

12. $-2 - 2\sqrt{3}i$

In Exercises 13–16 compute the given quantity.

13. $(1 - i)^{50}$

14. $(\frac{1}{2} - \frac{1}{2}i)^{12}$

15. $\left(\dfrac{\sqrt{3} - i}{2}\right)^{10}$

16. $(-2 - 2i)^8$

In Exercises 17–20 find all solutions to the given equation.

17. $z^4 = 1$

18. $z^2 = 4 + i$

19. $z^2 = 2 - 3i$

20. $z^3 = \sqrt{3} + i$

21. Graph the set $|z - 4| < 3$ and give its boundary points.

22. Graph the set $|\text{Re } z| > 2$ and give its boundary points.

23. Graph the set $|\text{Im } z| < 4$ and give its boundary points.

24. Graph the set $\pi/6 < \arg z < \pi/2$ and give its boundary points.

25. Compute $\lim\limits_{z \to 1+2i} \dfrac{z}{1 + z^2}$.

26. Compute the derivative of $f(z) = z/(z^2 + 1)$.

In Exercises 27–30 show that the given function is entire by verifying the Cauchy-Riemann equations and compute its derivative.

27. $f(z) = (z + 2)^2$

28. $f(z) = (x - y)^2 - 2y^2 + [(x + y)^2 - 2y^2]i$

29. $f(z) = e^{2x}(\cos 2y + i \sin 2y)$

30. $f(z) = e^{-3x}(\cos 3y - i \sin 3y)$

31. Show that the function $f(z) = (2x^2 - 2y^2) - 4xyi$ is differentiable only at the origin.

32. Determine where, if anywhere, $f(z) = z \text{ Re } z$ is differentiable.

In Exercises 33–42, write each number in the form $x + iy$.

33. $e^{4\pi i}$

34. $e^{-i\pi/3}$

35. $e^{22\pi i}$

36. $e^{4+(\pi/2)i}$

37. $e^{-3+(5\pi/6)i}$

38. $\cos i$

39. $\sin(1 - i)$

40. $\tan(1 - i)$

41. $\cosh(\pi i/6)$

42. $\sinh(\sqrt{3} + i)$

In Exercises 43–52 find all the complex numbers z that satisfy the given equation.

43. $e^{2z} = 1$

44. $e^z = -i$

45. $e^z = -1 + i$

46. $e^{4z} = 8 + 8\sqrt{3}i$

47. $e^{iz} = i$

48. $\cos 2z = \sin 2z$

49. $\cos z = 3$

50. $\cos z = i$

51. $\sinh z = 1$

52. $\cosh z = 1$

In Exercises 53–60 find (a) all values of the given expression and (b) the principal value.

53. $\log 2i$

54. $\log 2$

55. $\log(-2\sqrt{3} - 2i)$

56. $\log(5 - 5i)$

57. 1^{3i}

58. $(-i)^i$

59. $(2 + 2i)^{2i}$

60. $(1 - i)^{1-i}$

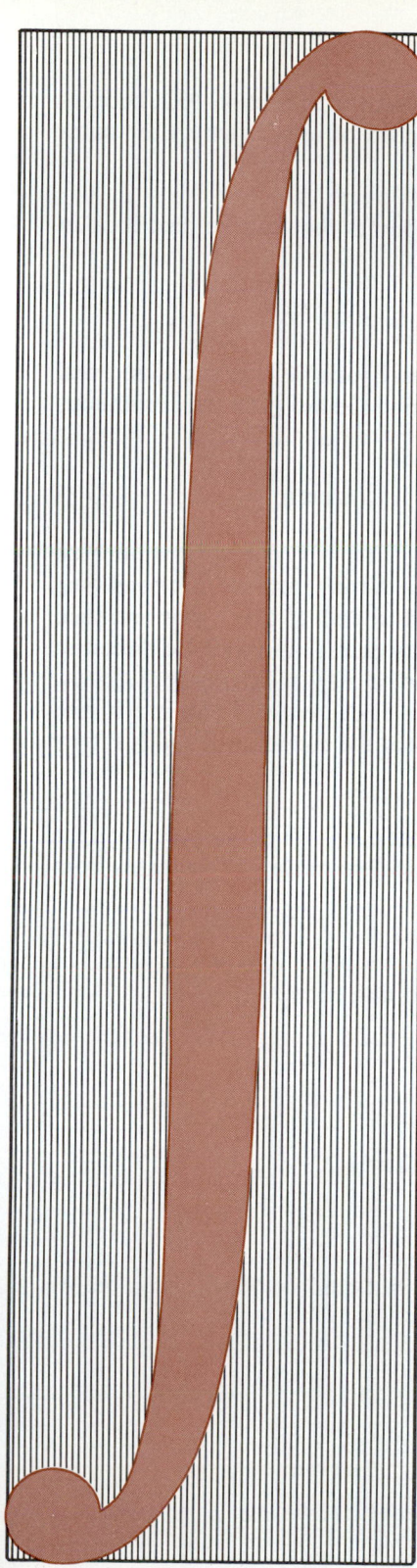

chapter 16

Complex Integration

Integration is an important and useful concept in elementary calculus. The two-dimensional nature of the complex plane suggests the consideration of integrals along arbitrary curves in $\mathbb{C}$ instead of only on segments of the real axis. These "line integrals" have interesting and unusual properties when the function being integrated is analytic. Complex integration is one of the most beautiful and elegant theories in mathematics. We begin by discussing line integrals in the complex plane.

16.1 Line Integrals in the Complex Plane

In Section 10.1 we defined the line integral in $\mathbb{R}^2$. We repeat our definition here:

DEFINITION

Let P and Q be continuous functions on a set S containing the smooth (or piecewise smooth) curve C given by

$$C : \mathbf{x}(t) = x(t)\mathbf{i} + y(t)\mathbf{j}, \qquad t \in [a, b].$$

Let the vector field $\mathbf{F}$ be given by

$$\mathbf{F}(x, y) = P(x, y)\mathbf{i} + Q(x, y)\mathbf{j}.$$

Then the **line integral** of $\mathbf{F}$ over C is given by

$$\int_C \mathbf{F}(\mathbf{x}) \cdot \mathbf{dx} = \int_a^b [P(x(t), y(t))x'(t) + Q(x(t), y(t))y'(t)] \, dt. \tag{1}$$

In Section 10.1 we showed that this definition arose from a calculation of work and we gave a variety of examples of the calculation of line integrals. At this point we suggest that you read (or reread) the material in Section 10.1, because the idea of the line integral is central to the idea of integration in the complex plane.

We now extend these ideas to the complex plane.

A **curve** or **arc** or **path** C is any set of points that can be described in parametric form by

$$C : x = x(t), \qquad y = y(t), \qquad a \leqslant t \leqslant b,$$

with $x(t)$, $y(t)$ continuous functions of the real variable t in the closed real interval $[a, b]$. In the complex plane we describe the arc C by the continuous complex-valued function of a real variable

$$C : z = z(t) = x(t) + iy(t), \qquad a \leqslant t \leqslant b.$$

The arc C is said to be **smooth** if the function $z'(t) = x'(t) + iy'(t)$ is nonzero and continuous on $a \leqslant t \leqslant b$. A **piecewise smooth (pws)** arc is an arc consisting of a finite number of smooth arcs joined end to end. If C is a pws arc, then $x(t)$ and $y(t)$ are continuous, but their derivatives $x'(t)$ and $y'(t)$ are piecewise continuous. An arc is a **simple**, or **Jordan**, arc if $z(t_1) = z(t_2)$ only if $t_1 = t_2$, that is, if it is non-self-intersecting. An arc is a **closed curve** if $z(a) = z(b)$ and a **Jordan curve** if it is closed and simple except at the endpoints a and b. Figure 1 illustrates some of these concepts.

Arc or Path in the Complex Plane
Smooth Arc
Piecewise Smooth Arc

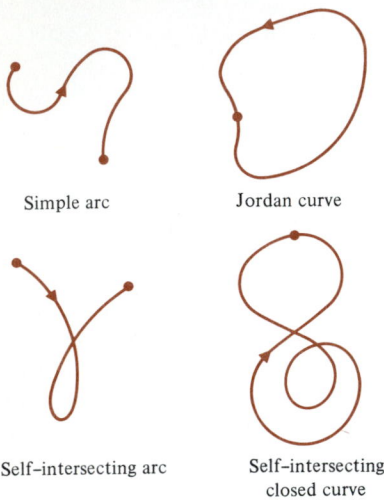

Simple arc   Jordan curve

Self–intersecting arc   Self–intersecting closed curve

FIGURE 1 Arcs and curves

Remark. As in Chapter 10, we assume that $a < b$. Then the arrows on the curves in Fig. 1 indicate the direction in which we move as t increases from a to b.

Remark. Jordan curves separate the complex plane into two regions. The bounded region is called the **interior** of the Jordan curve.

Jordan Arc
Interior

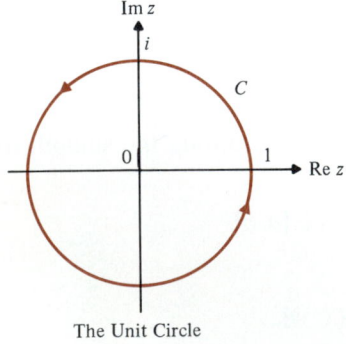

The Unit Circle

FIGURE 2

EXAMPLE 1

Sketch the arc given by the parametrization

$$C : z(t) = e^{it}, \qquad 0 \leq t \leq 2\pi.$$

SOLUTION

Since $(e^{it})' = ie^{it} \neq 0$, the arc C is smooth. Note that $|e^{it}| = 1$ and $e^0 = e^{2\pi i} = 1$. Hence, C is a parametrization of the unit circle traversed in a counterclockwise direction. We see that C is a Jordan curve (see Fig. 2), with $|z| < 1$ as its interior.

EXAMPLE 2

Sketch the arc given parametrically by

$$C : z(t) = \begin{cases} 1 - i(1 - t), & 0 \leq t \leq 1, \\ 1 + t - i, & -1 \leq t \leq 0. \end{cases}$$

SOLUTION

We first check for continuity. The only potential problem is at $t = 0$. We find that

$$\lim_{t \to 0^+} z(t) = \lim_{t \to 0^+} [(1 - i(1 - t)] = 1 - i$$

and

$$\lim_{t \to 0^-} z(t) = \lim_{t \to 0^-} (1 + t - i) = 1 - i.$$

Thus $z(t)$ is continuous at 0 and, therefore, it is continuous in the entire interval $[-1, 1]$.

Next we check for smoothness. We compute the derivative $z'(t)$ by differentiating the definition of $z(t)$ with respect to t on each interval:

$$z'(t) = \begin{cases} i, & 0 < t < 1, \\ 1, & -1 < t < 0. \end{cases}$$

We see that $z'(0)$ does not exist but that $z'(t)$ exists and is continuous (in fact, is piecewise constant) for $0 < |t| < 1$. Thus, although C is not smooth, it is piecewise smooth. In each of the intervals $[-1, 0]$ and $[0, 1]$ the graph of C is a straight line [since $z'(t)$ is constant]. The easiest way to sketch a straight line is to plot two points on it. We find that $z(-1) = -i$, $z(0) = 1 - i$, and $z(1) = 1$. This leads to the graph in Fig. 3. We see from this graph that C is a simple arc.

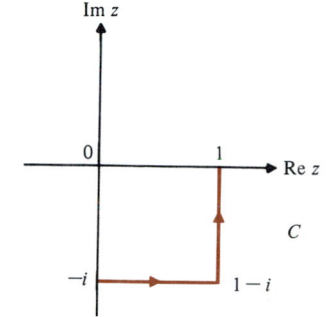

FIGURE 3

Orientation

It will be useful when discussing line integrals to consider the **orientation** of a Jordan curve. A Jordan curve is parametrized in its **positive sense** if its interior is kept to the left as the curve is traversed with increasing t. For example, the parametrization $z(t) = e^{it} = \cos t + i \sin t$, $0 \le t \le 2\pi$, parametrizes $|z| = 1$ in its positive sense, whereas $z(t) = e^{-it}$, $0 \le t \le 2\pi$, does not. This is illustrated in Fig. 4.

We are now ready to define the line integral in the complex plane.

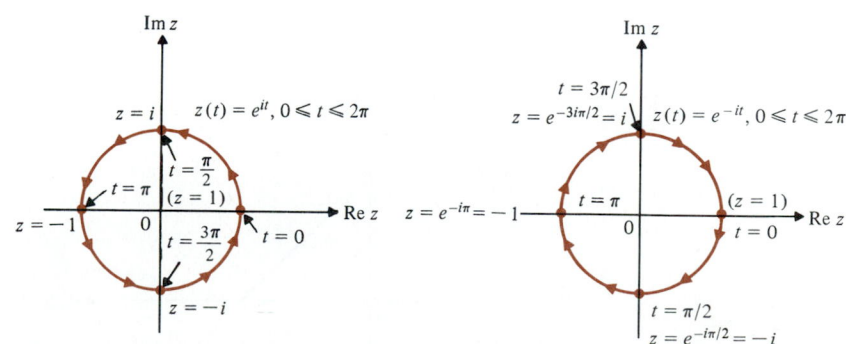

(a) Positively oriented Jordan curve
(interior on the left as t increases).

(b) Jordan curve negatively oriented
(interior on the right as t increases).

FIGURE 4

Let C be a smooth arc in the complex plane and let the complex function $f(z)$ be continuous on C. We use the parametrization of C to define the line integral of f on C in terms of two real integrals. If the two real integrals can be evaluated, a value can be assigned to the line integral of f.

Line Integral in the Complex Plane

DEFINITION

Let $C : z = z(t)$, $a \leqslant t \leqslant b$ be a smooth arc and $f(z) = u(z) + iv(z)$ be continuous on C. Then, the line integral of f on C is given by

$$\int_C f(z)\,dz = \int_a^b f(z(t))z'(t)\,dt$$

$$= \int_a^b [u(z(t)) + iv(z(t))][x'(t) + iy'(t)]\,dt$$

$$= \int_a^b [u(z(t))x'(t) - v(z(t))y'(t)]\,dt$$

$$+ i\int_a^b [u(z(t))y'(t) + v(z(t))x'(t)]\,dt.$$

The line integral over a pws arc C is obtained by applying the definition above to each of the finitely many closed intervals on which $z(t)$ is smooth and summing the results.

EXAMPLE 3

Evaluate $\int_C x\,dz$ along
    (a) the smooth arc (line segment) C shown in Fig. 5a.
    (b) the pws arc C^* shown in Fig. 5b.

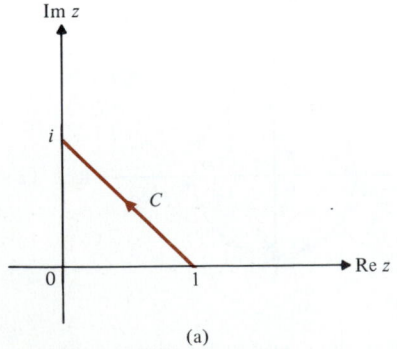

(a)

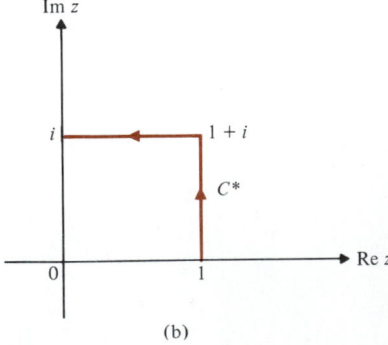

(b)

FIGURE 5

SOLUTION

(a) We first must find a parametrization for the line segment joining the points 1 and i. To do so we find the equation of the straight line joining the points $(1, 0)$ and $(0, 1)$. We have

$$\frac{y}{x-1} = \text{slope} = -1 \quad \text{or} \quad y = 1 - x.$$

Since the y values increase from 0 to 1 as the arc C is traversed from 1 to i, we set

$$y(t) = t, \qquad 0 \leq t \leq 1,$$

$$x(t) = 1 - y(t) = 1 - t,$$

so that the arc C is given by

$$C : z(t) = x(t) + iy(t) = (1 - t) + it, \qquad 0 \leq t \leq 1.$$

Then

$$z'(t) = -1 + i$$

and

$$\int_C x\,dz = \int_0^1 \overbrace{x(t)}^{f(z(t))} z'(t)\,dt = \int_0^1 (1-t)(-1+i)\,dt$$

$$= (-1+i)\int_0^1 (1-t)\,dt = (-1+i)\left[t - \frac{t^2}{2}\bigg|_0^1\right] = \frac{-1+i}{2}.$$

(b) We must now parametrize the two lines from 1 to $1 + i$ and from $1 + i$ to i. On the first line, x is constant and y increases from 0 to 1. On the second, y is constant and x increases from 0 to 1. Thus, on the first line, we have

$$z(t) = 1 + it, \qquad 0 \leq t \leq 1.$$

On the second line, it is convenient to have t increase from 1 to 2. This leads to

$$z(t) = (2 - t) + i, \qquad 1 \leq t \leq 2.$$

Putting these together, we obtain

$$C^* : z(t) = \begin{cases} 1 + it, & 0 \leq t \leq 1, \\ (2 - t) + i, & 1 \leq t \leq 2. \end{cases}$$

Then

$$z'(t) = \begin{cases} i, & 0 < t < 1, \\ -1, & 1 < t < 2. \end{cases}$$

Hence C^* is piecewise smooth but not smooth. To obtain the integral, we integrate over each interval on which C^* is smooth and add the results. Keep in mind that for $0 \leq t \leq 1$, $x(t) = 1$ and for $1 \leq t \leq 2$, $x(t) = 2 - t$. We obtain

$$\int_{C^*} x\,dz = \int_0^1 i\,dt + \int_1^2 (2-t)(-1)\,dt = -\frac{1}{2} + i.$$

Independence of Parametrization

A question arises: Since arcs and curves have many different parametrizations, will we get different values for the line integral of f when we use different parametrizations of C? Fortunately, the answer is no. To justify this assertion, consider Example 3b.

If we choose a different parametrization for C^*, say

$$C^* : z(t) = \begin{cases} 1 + i \ln t, & 1 \leqslant t \leqslant e, \\ 2 - \dfrac{t}{e} + i, & e \leqslant t \leqslant 2e, \end{cases}$$

we have

$$z'(t) = \begin{cases} i/t, & 1 \leqslant t \leqslant e, \\ -1/e, & e \leqslant t \leqslant 2e, \end{cases}$$

and

$$\int_{C^*} x\, dz = \int_1^e \frac{i}{t}\, dt + \int_e^{2e} \left(2 - \frac{t}{e}\right)\left(\frac{-1}{e}\right) dt = -\frac{1}{2} + i.$$

This is exactly the value obtained in Example 3b using a different parametrization for C^*.

Hence, the line integral is independent of the two parametrizations of C^*. This is always the case when the change of parameters is piecewise differentiable, as can be seen by using the change-of-variable formula of integral calculus. (See Problem 30.)

Remark. In Example 3 we evaluated the line integral of the function $f(z) = \operatorname{Re} z = x$ over two different arcs C and C^* that join the same two points 1 and i. Observe that we obtained different results for these two integrations. This shows that the complex line integral is generally *not* independent of path. More will be said about this in Section 16.3.

EXAMPLE 4

Calculate

$$\int_C z\, dz \quad \text{and} \quad \int_{C^*} z\, dz,$$

where C and C^* are the two arcs shown in Fig. 5.

SOLUTION
Using the parametrization of C from Example 3, we get

$$\int_C z\, dz = \int_0^1 [(1-t) + it](-1+i)\, dt = -\int_0^1 dt + i\int_0^1 (1-2t)\, dt = -1.$$

Parametrizing C^* as in Example 3, we have

$$\int_{C^*} z \, dz = \int_0^1 (1 + it)i \, dt + \int_1^2 [(2 - t) + i](-1) \, dt$$

$$= i \int_0^1 dt - \int_0^1 t \, dt + \int_1^2 (t - 2) \, dt - i \int_1^2 dt = -1.$$

Observe that both paths yield the same result.

EXAMPLE 5

Let C be the unit circle with positive orientation and let n be an integer. Compute $\int_C z^n \, dz$.

SOLUTION
The curve C may be given parametrically by

$$C : z(t) = e^{it}, \qquad 0 \le t \le 2\pi.$$

Then $z'(t) = ie^{it}$, $z^n = e^{int}$, and we have

$$\int_C z^n \, dz = \int_0^{2\pi} ie^{int} e^{it} \, dt = i \int_0^{2\pi} e^{i(n+1)t} \, dt.$$

There are two cases to consider: $n \ne -1$ and $n = -1$. If $n \ne -1$, then by Euler's formula

$$i \int_0^{2\pi} e^{i(n+1)t} \, dt = i \int_0^{2\pi} [\cos(n+1)t + i\sin(n+1)t] \, dt$$

$$= i \left[\frac{\sin(n+1)t}{n+1} - \frac{i\cos(n+1)t}{n+1} \right] \Big|_0^{2\pi}$$

$$= \left[\frac{\cos(n+1)t}{n+1} + \frac{i\sin(n+1)t}{n+1} \right] \Big|_0^{2\pi} = \frac{1}{n+1} e^{i(n+1)t} \Big|_0^{2\pi} = 0$$

because of the periodicity of e^{it}.

If $n = -1$, then $e^{i(n+1)t} = e^0 = 1$ and

$$i \int_0^{2\pi} e^{i(n+1)t} \, dt = i \int_0^{2\pi} dt = 2\pi i.$$

Thus

$$\int_{|z|=1} z^n \, dz = \begin{cases} 0, & \text{if } n \ne -1, \\ 2\pi i, & \text{if } n = -1. \end{cases}$$

Problems 16.1

In Problems 1–10 two points in the complex plane are given. Find a parametrization of the straight line segment extending from the first point to the second point.

1. $0, 1$

2. $1, 0$

3. $-i, i$

4. $1, i$

5. $i, 1$

6. $1 + i, 1 - i$

7. $1 - i, 1 + i$

8. $3 - 2i, -1 + 4i$

9. $-2 + 5i, 3 - 7i$

10. $4 + 2i, -1 - 3i$

11. Show that the parametrization

$$C: z(t) = a \cos t + ib \sin t, 0 \leqslant t \leqslant 2\pi,$$

describes the ellipse

$$\frac{x^2}{a^2} + \frac{y^2}{b^2} = 1.$$

In Problems 12–20, determine pws parametrizations for the indicated arcs or curves.

12. Semicircle from 1 to −1

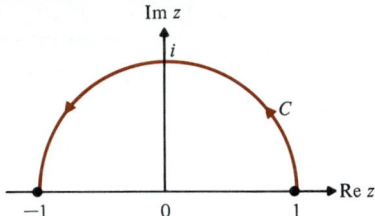

13. Triangle

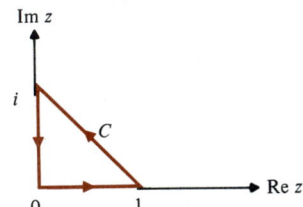

14. Square

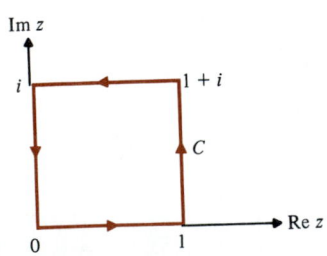

15. Rectangle

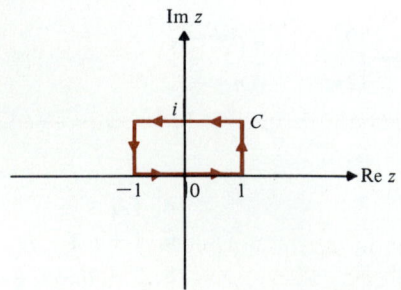

16. Rectangle

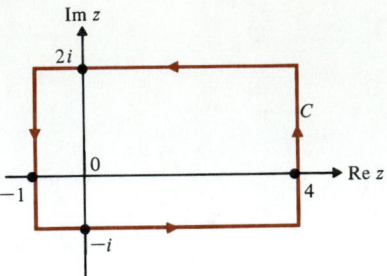

17. Diamond

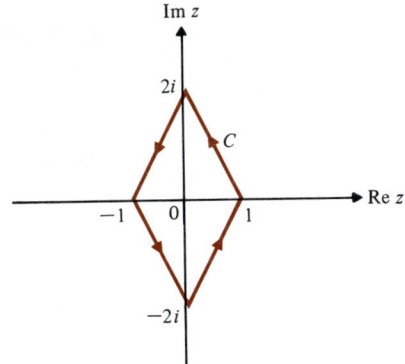

18. Circle

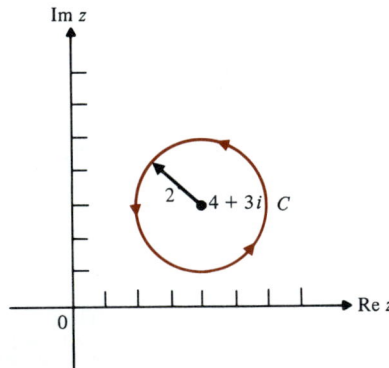

19. Semicircle

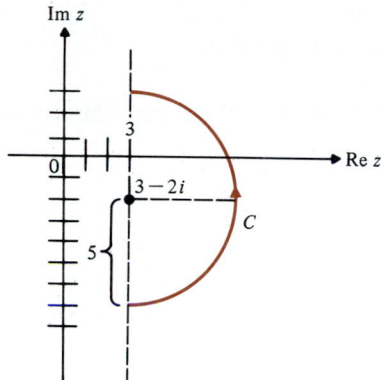

20. Barbell beginning at 1

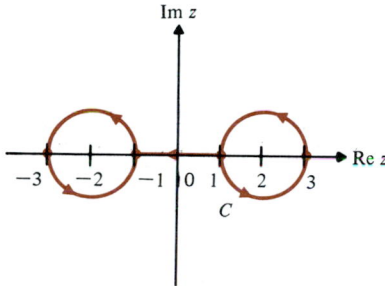

21. Let C be the circle centered at z_0 with radius r. Show that

$$\int_C \frac{dz}{z - z_0} = 2\pi i.$$

22. If $n \neq -1$ is an integer and C is the curve of Problem 21, show that

$$\int_C (z - z_0)^n\, dz = 0.$$

23. Show that $z'(t)$ can be interpreted as a vector tangent to the arc $C : z = z(t)$ at all points where $z'(t)$ is nonzero.

Evaluate the integrals $\int x\, dz$, $\int y\, dz$, $\int \bar{z}\, dz$ along the paths given in Problems 24–26.

24. The directed line segment from 0 to $1 - i$

25. Around the circle $|z| = 1$

26. Around the circle $|z - a| = R$

27. Evaluate $\int_C y\, dz$, where C is the straight line joining 1 to i.

28. Evaluate $\int_C y\, dz$, where C is the arc in the first quadrant along $|z| = 1$ joining 1 to i.

29. Evaluate $\int_C y\, dz$, where C is the arc along the coordinate axes joining 1 to i.

* 30. Let the curve C be given by the parametrization

$$z = z(t), \qquad a \le t \le b.$$

Suppose $t = t(T)$ is a differentiable function mapping $A \le T \le B$ onto $a \le t \le b$ with $t' > 0$. Show that $\int_C f(z)\, dz$ yields the same value using the parametrization $z = z(t(T))$, $A \le T \le B$, as it does with the first parametrization of C.

16.2 The Fundamental Theorem of Calculus

Although the techniques of line integrals in Section 16.1 yield a useful way to calculate complex line integrals, the procedure is more complicated than necessary for certain complex-valued functions. In this section we shall describe a more efficient technique.

We recall the **fundamental theorem of calculus** for a real-valued function of one variable:

Fundamental Theorem of Calculus

If a real-valued function $f(x)$ is continuous in an interval $a \le x \le b$, then $f(x)$ possesses antiderivatives in that interval. If $F(x)$ is any antiderivative of $f(x)$ in

$a \leqslant x \leqslant b$, then

$$\int_a^b f(x)\,dx = F(b) - F(a). \qquad (1)$$

Example 16.1.3 shows that the fundamental theorem cannot be extended to complex line integrals, because

$$\int_C x\,dz = \frac{-1+i}{2} \quad \text{and} \quad \int_{C*} x\,dz = -\frac{1}{2} + i,$$

and both arcs begin at 1 and end at i (see Figure 16.1.5). Since $f(z) = \text{Re } z = x$ is a continuous function, it is clear that we cannot obtain a theorem similar to the fundamental theorem of calculus for *all* continuous complex-valued functions $f(z)$. Suppose, instead, that we consider only those continuous functions $f(z)$ that are derivatives of an analytic function $F = U + iV$ in some simply connected region Ω containing the smooth arc C. Then, by definition,

$$\int_C f(z)\,dz = \int_C F'(z)\,dz = \int_a^b F'(z(t))z'(t)\,dt.$$

By the chain rule of calculus, we have

$$\int_a^b F'(z(t))z'(t)\,dt = \int_a^b \frac{d}{dt}[F(z(t))]\,dt$$

$$= \int_a^b \frac{d}{dt}[U(z(t))]\,dt + i\int_a^b \frac{d}{dt}[V(z(t))]\,dt.$$

Applying the fundamental theorem of calculus to each of these real integrals, we obtain

$$\int_C f(z)\,dz = [U(z(b)) - U(z(a))] + i[V(z(b)) - V(z(a))]$$

$$= F(z(b)) - F(z(a)).$$

We can easily extend this result to pws arcs by adding the results obtained from the smooth subarcs. Since the result depends only on the endpoints of each smooth subarc, we have proved the following theorem.

■ **THEOREM 1: Fundamental Theorem (of Calculus) for a Complex Line Integral**

If $F(z)$ is an analytic function with a continuous derivative $f(z) = F'(z)$ in a simply connected region Ω containing the pws arc $C : z = z(t)$, $a \leqslant t \leqslant b$, then

$$\int_C f(z)\,dz = F(z(b)) - F(z(a)). \qquad (2)$$

Since the integral depends only on the endpoints of the arc C, *the integral is independent of path.* Thus, the same result is obtained for *any* pws arc in Ω with these endpoints. *For pws closed curves C, the fundamental theorem yields*

$$\int_C f(z)\,dz = 0, \tag{3}$$

since $F(z(\beta)) = F(z(\alpha))$.   ■

EXAMPLE 1

Calculate

$$\int_C z\,dz \quad \text{and} \quad \int_{C*} z\,dz,$$

where C and $C*$ are the two arcs shown in Fig. 16.1.5.

SOLUTION
The continuous function $f(z) = z$ is the derivative of the entire function $F(z) = z^2/2$. Thus if we use the fundamental theorem, and integrate along any pws arc C beginning at 1 and ending at i, we get

$$\int_C z\,dz = \frac{z^2}{2}\Big|_1^i = \frac{i^2 - 1}{2} = -1.$$

(Compare this work with Example 16.1.4.)

EXAMPLE 2

Let $P(z)$ be any polynomial and C be a pws arc. Show that:
  (a) $\int_C P(z)\,dz = 0$ if C is a closed curve,
  (b) $\int_C P(z)\,dz$ depends only on the endpoints of C.

SOLUTION
Every polynomial $P(z)$ is continuous in the complex plane. Furthermore, if

$$P(z) = a_n z^n + a_{n-1} z^{n-1} + \cdots + a_1 z + a_0,$$

then $P(z)$ is the derivative of the analytic polynomial

$$Q(z) = \frac{a_n z^{n+1}}{n+1} + \frac{a_{n-1} z^n}{n} + \cdots + \frac{a_1 z^2}{2} + a_0 z.$$

Thus, the fundamental theorem is satisfied and parts (a) and (b) hold.

EXAMPLE 3

Since $\cos z$ is entire and has the antiderivative $\sin z$, we have

$$\int_{-i}^{i} \cos z \, dz = \sin z \Big|_{-i}^{i} = 2 \sin i = 2i \sinh(1),$$

and along any pws closed curve C,

$$\int_{C} \cos z \, dz = 0.$$

EXAMPLE 4

Compute

$$\int_{|z|=1} z^n \, dz, \qquad n = -1, 0, 1, 2, \dots. \tag{4}$$

SOLUTION
If $n \neq -1$, then the continuous function

$$z^n = \frac{d}{dz}\left(\frac{z^{n+1}}{n+1}\right), \qquad n = 0, 1, 2, \dots,$$

and since $C : |z| = 1$ is a smooth closed curve, (3) yields

$$\int_{|z|=1} z^n \, dz = 0, \qquad n = 0, 1, 2, \dots.$$

However, for $n = -1$, we have the integral

$$\int_{|z|=1} \frac{dz}{z},$$

and the antiderivatives of $1/z$ are logarithms. In Section 15.9 we saw that each logarithm function has a restricted domain. For example, the principal value of the logarithm (see p. 922) is given by

$$\text{Log } z = \ln|z| + i \text{ Arg } z, \qquad -\pi < \text{Arg } z \leqslant \pi.$$

In this case the arc $|z| = 1$ is *not closed* because although the numbers $e^{\pi i}$ and $e^{-\pi i}$ are equal to -1, the values of log $e^{\pi i}$ and log $e^{-\pi i}$ are unequal. If we integrate along C from $e^{-\pi i}$ to $e^{\pi i}$ and apply the fundamental theorem (since the region is simply connected), we obtain

$$\int_{|z|=1} \frac{dz}{z} = (\ln|z| + i \arg z) \Big|_{e^{-\pi i}}^{e^{\pi i}} = \pi i - (-\pi i) = 2\pi i.$$

The computation here may be made completely legitimate by defining log z so that it is analytic on the branch cuts (see p. 923).

Problems 16.2

In Problems 1–10 compute the line integral by using the fundamental theorem.

1. $\int_C z^2 \, dz$ where C is the directed line segment from 1 to i.

2. $\int_C \frac{1}{2} \, dz$ where C is the directed line segment from i to 2.

3. $\int_C e^{2z} \, dz$ where C is the quarter of the unit circle extending from 1 to i.

4. $\int_C (z^2 + 3z + 2)\, dz$ where C is the part of the rectangle in Problem 16.1.15 that excludes the line from -1 to 1.

5. $\int_C \sin z\, dz$ where C is the semicircle of Problem 16.1.12.

6. $\int_C (z^2 + 1/z^2)\, dz$ where C is the directed line segment from $1 + i$ to $1 - i$.

7. $\int_C (z^3 - 2z + 3)\, dz$ where C is the arc sketched below.

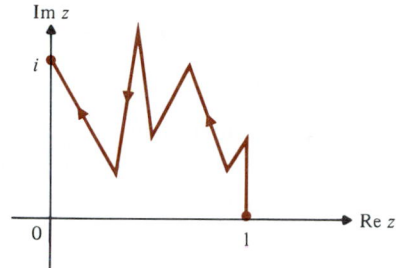

8. $\int_C \cos z\, dz$ where C is the directed line segment from i to 0.

9. $\int_C \sinh z\, dz$ where C is the semicircle in Problem 16.1.19.

10. $\int_C \cosh z\, dz$ where C is as in Problem 9.

Use parametrizations of the arcs to evaluate the integrals in Problems 11–17. Confirm your answer by using the fundamental theorem.

11. Evaluate the integral $\int_C (z - a)^n\, dz$, n an integer, around the circle $|z - a| = R$. (The answer for $n = -1$ differs from the rest.)

12. Evaluate $\int_C e^z\, dz$, where C is the straight-line path joining 1 to i.

13. Evaluate $\int_C e^z\, dz$, whre C is the path in the first quadrant along the circle $|z| = 1$ joining 1 to i.

14. Evaluate $\int_C e^z\, dz$, where C is the path along the coordinate axes joining 1 to i.

15. Evaluate $\int_{-i}^{i} e^{\pi z}\, dz$, along the directed line segment from $-i$ to i.

16. Evaluate $\int_{-1}^{i} \sinh(az)\, dz$, along the directed line segment from -1 to i.

17. Evaluate $\int_{1}^{i} (z - 1)^3\, dz$, along the directed line segment from 1 to i.

18. If C is the ellipse $z(t) = a \cos t + ib \sin t$, $0 \leqslant t \leqslant 2\pi$, $a^2 - b^2 = 1$, show that

$$\int_C \frac{dz}{\sqrt{1 - z^2}} = \pm 2\pi,$$

depending on which value of the radical is taken. [*Hint:* $1 - z^2(t) = [z'(t)]^2$.]

19. Let $C_1 : z(t) = e^{it}$ and $C_2 : z(t) = e^{-it}$, $0 \leqslant t \leqslant \pi$. Evaluate $\int dz/z^2$ along each curve.

20. Evaluate $\int \text{Log}\, z\, dz$ along each curve given in Problem 19.

21. Evaluate $\int \sqrt{z}\, dz$ along each curve given in Problem 19. [Use the principal branch of $\sqrt{z}$.]

16.3 Cauchy's Theorem

In Examples 16.1.5 and 16.2.4 we found that the line integral of a polynomial along a pws closed curve vanishes but that

$$\int_{|z|=1} \frac{dz}{z} = 2\pi i.$$

Note that the function $1/z$ is not analytic at the origin. Could it be that the line integral of a function along a pws Jordan curve vanishes when the function is analytic on and *inside* the curve? Surprisingly, that is correct.

We shall prove that the line integral along a pws Jordan curve is zero if we assume that the derivative of the analytic function in the integrand is continuous inside the pws Jordan curve. This is not an unreasonable requirement, since the derivative of every analytic function we have encountered is analytic.

Let $f = u + iv$ be analytic in a bounded, simply connected region Ω and let C be a pws Jordan curve in Ω. We rewrite the integral of f along C in the form

$$\int_C f(z)\,dz = \int_C (u + iv)(dx + i\,dy) = \int_C u\,dx - v\,dy + i\int_C v\,dx + u\,dy.$$

If f' is continuous on Ω, then the first partials u_x, u_y, v_x, and v_y are continuous. Applying Green's theorem (p. 610) to the two line integrals on the right, with R the region interior to C, we get

$$\int_C f(z)\,dz = -\int\int_R (v_x + u_y)\,dx\,dy + i\int\int_R (u_x - v_y)\,dx\,dy. \qquad (1)$$

The first partials satisfy the Cauchy-Riemann equations $u_x = v_y$ and $u_y = -v_x$, since f is analytic. Hence, both integrands on the right side are zero. Under the assumption that $f'(z)$ is continuous on Ω, we have proved the following theorem.

■ **THEOREM 1: Cauchy's Theorem**

Let the function $f(z)$ be analytic in a simply connected region Ω and let C be a pws Jordan curve in Ω. Then

$$\int_C f(z)\,dz = 0. \qquad (2)$$

■

The drawback to our proof is the assumption that $f'(z)$ is continuous on Ω. However, although we shall not prove this here, it is the case that this condition is unnecessary. That is, every analytic function has an analytic derivative.†

EXAMPLE 1

Let $p(z)$ be a polynomial. Then, since every polynomial is entire, we have

$$\int_C p(z)\,dz = 0,$$

where C is a pws Jordan curve.

EXAMPLE 2

$\int_C e^{z^2}\,dz = 0$ for every pws Jordan curve C because e^{z^2} is entire.

---

† For a proof of this and related facts, see W. Derrick, *Complex Analysis and Applications*, 2nd ed., Wadsworth, Belmont, Calif., 1984, Section 2.5. See Theorem 16.5.2.

EXAMPLE 3

Let $p(z)/q(z)$ be a rational function. Let C be any pws Jordan curve whose interior does not contain a root of $q(z)$. Then

$$\int_C \frac{p(z)}{q(z)} \, dz = 0.$$

Remark. In applying Cauchy's theorem it is necessary to check that $f(z)$ is analytic on a simply connected region Ω containing C. Practically, this means that $f(z)$ is analytic at every point on and *inside* C. If this condition does not hold, then the conclusion of the theorem might not hold. For example, $f(z) = 1/z$ is analytic everywhere except at 0. In Example 16.1.5 we showed that

$$\int_{|z|=1} \frac{1}{z} \, dz = 2\pi i.$$

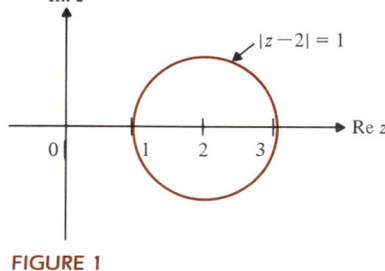

FIGURE 1

This does not contradict Cauchy's theorem because $1/z$ is not analytic inside the unit circle. On the other hand,

$$\int_{|z-2|=1} \frac{1}{z} \, dz = 0$$

since $|z - 2| = 1$ is the equation of a circle of radius 1 centered at 2 (see Fig. 1) and $1/z$ is analytic on and inside this circle.

Next we state a theorem that gives four basic properties of complex line integrals. The proofs of the first three parts follow from the definition of the line integral in Section 16.1.

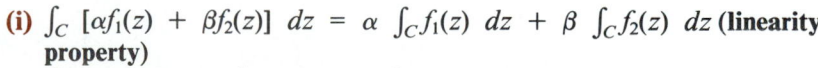

FIGURE 2

■ **THEOREM 2**

Let $f(z)$, $f_1(z)$, and $f_2(z)$ be analytic in a region Ω, let C, C_1, and C_2 be paths in Ω, and let α and β be complex numbers. Then

(i) $\int_C [\alpha f_1(z) + \beta f_2(z)] \, dz = \alpha \int_C f_1(z) \, dz + \beta \int_C f_2(z) \, dz$ **(linearity property)**

(ii) $\int_{C_1+C_2} f(z) \, dz = \int_{C_1} f(z) \, dz + \int_{C_2} f(z) \, dz$, where $C_1 + C_2$ is the path obtaining by traversing first C_1 followed by C_2 (see Fig. 2).

(iii) $\int_{-C} f(z) \, dz = -\int_C f(z) \, dz$, where $-C$ is the path obtained by traversing the arc C in the reverse direction (see Fig. 3).

(iv)
$$\left| \int_C f(z) \, dz \right| \leq \int_C |f(z(t))| \, |z'(t)| \, dt. \tag{3}$$

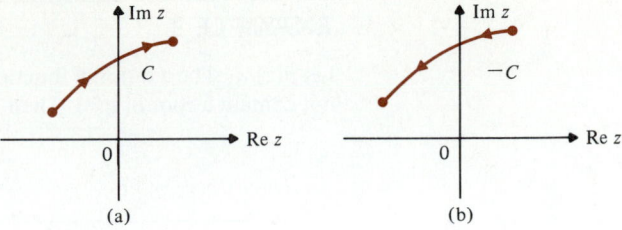

FIGURE 3

Remark. We define

$$|dz| = |dx + i\,dy| = \sqrt{(dx)^2 + (dy)^2} = ds, \tag{4}$$

so that $|dz|$ is the **arc length differential.** Then, since $dz = z'(t)\,dt$, we have $|dz| = |z'(t)|\,dt$ and (3) becomes

$$\left| \int_C f(z)\,dz \right| \le \int_C |f(z)|\,|dz|. \tag{5}$$

EXAMPLE 4

Show that

$$\left| \int_{|z|=1} e^z\,dz \right| \le 2\pi e.$$

SOLUTION
From part (iv) of Theorem 2 we have

$$\left| \int_{|z|=1} e^z\,dz \right| \le \int_{|z|=1} |e^z|\,|dz|.$$

Since $|e^z| = e^x \le e$ for all points $z = x + iy$ on the unit circle.

$$\int_{|z|=1} |e^z|\,|dz| \le e \int_{|z|=1} |dz|.$$

Since $|dz|$ is the arc length differential, the integral on the right-hand side is merely the length of the unit circle. Hence

$$\left| \int_{|z|=1} e^z\,dz \right| \le 2\pi e.$$

In fact, it is clear that

$$\left| \int_{|z|=1} e^z\,dz \right| < 2\pi e,$$

since $|e^z|$ attains the value e only at $z = 1$.

Independence of Path

In Section 10.2 (p. 596) we saw that a great variety of line integrals in the plane are independent of path. That is, if (x_1, y_1) and (x_2, y_2) are two points in the plane, then $\int_C \mathbf{F(x)} \cdot \mathbf{dx}$ is the same for any piecewise smooth curve joining (x_1, y_1) and (x_2, y_2).

For complex line integrals, independence of path follows quite easily from Cauchy's theorem. Let C be a path joining two points z_1 and z_2 in the complex plane and suppose that $f(z)$ is analytic in a simply connected region Ω containing C (see Fig. 4a).

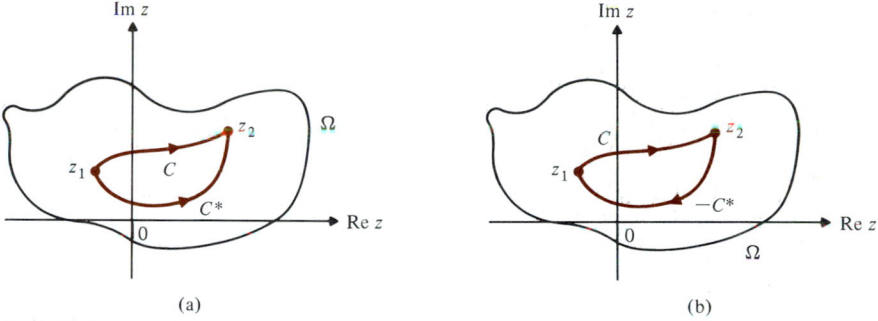

(a)   (b)

FIGURE 4

We choose another path C^* joining z_1 to z_2 with C^* in Ω. Then, as in Fig. 4b, the curve $C - C^*$ is a simple closed curve on and inside which $f(z)$ is analytic. By Cauchy's theorem,

$$\int_{C-C^*} f(z)\,dz = 0.$$

But, by Theorem 2, parts (ii) and (iii),

$$0 = \int_{C-C^*} f(z)\,dz = \int_C f(z)\,dz + \int_{-C^*} f(z)\,dz = \int_C f(z)\,dz - \int_{C^*} f(z)\,dz.$$

That is,

$$\int_C f(z)\,dz = \int_{C^*} f(z)\,dz. \tag{6}$$

We have shown that $\int_C f(z)\,dz$ is the same for any path in Ω from z_1 to z_2. The result is true even if C and C^* intersect at other points. Look at Fig. 5. Here C and C^* also intersect at z_3. We can write $C = C_1 + C_2$ and $C^* = C_1^* + C_2^*$ as in Fig. 5. Then from (6), since C_1 and C_1^* are arcs in Ω joining z_1 to z_3 and C_2 and C_2^* join z_3 to z_2,

$$\int_{C_1} f(z)\,dz = \int_{C_1^*} f(z)\,dz \quad \text{and} \quad \int_{C_2} f(z)\,dz = \int_{C_2^*} f(z)\,dz.$$

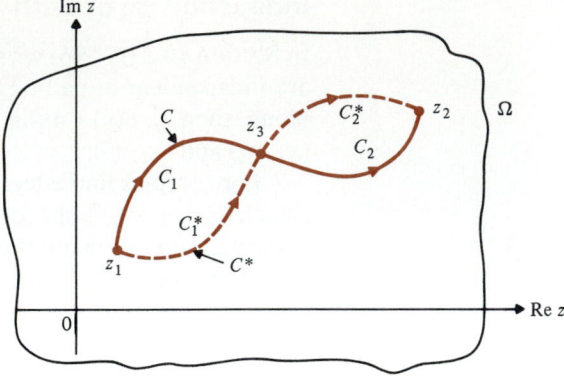

FIGURE 5

Using Theorem 2(ii) again, we have

$$\int_C f(z)\,dz = \int_{C_1} f(z)\,dz + \int_{C_2} f(z)\,dz$$

$$= \int_{C_1^*} f(z)\,dz + \int_{C_2^*} f(z)\,dz = \int_{C^*} f(z)\,dz.$$

A similar proof can be used to show that the integral is independent of path no matter how many times the paths intersect (if they intersect at a finite number of points) and a different proof (which we omit) holds for an infinite number of points of intersection. We have proved:

■ **THEOREM 3: Independence of Path Theorem**
Let $f(z)$ be analytic in a simply connected region Ω and let z_1 and z_2 be two points in Ω. If C and C^* are two pws paths lying in Ω, joining z_1 to z_2, then the line integral depends only on the endpoints and we can write

$$\int_C f(z)\,dz = \int_{z_1}^{z_2} f(z)\,dz = \int_{C^*} f(z)\,dz.$$

■

We use the independence of path to prove that analytic functions have analytic antiderivatives.

■ **THEOREM 4: Antiderivative Theorem**
Let $f(z)$ be analytic in a simply connected region Ω. Then there exists an analytic function $F(z)$ such that $F'(z) = f(z)$ for every z in Ω.

PROOF

Choose z_0 in Ω and define the function

$$F(z) = \int_{z_0}^{z} f(w)\, dw$$

for every z in Ω. By independence of path, $F(z)$ is well defined. We now show that $F(z)$ is analytic and that $F'(z) = f(z)$. For fixed z in Ω, choose h small enough so that $z + h \in \Omega$. Then

$$F(z+h) - F(z) = \int_{z_0}^{z+h} f(w)\, dw - \int_{z_0}^{z} f(w)\, dw$$

Theorem 2(ii)
$$\downarrow$$
$$= \int_{z}^{z+h} f(w)\, dw + \int_{z_0}^{z} f(w)\, dw - \int_{z_0}^{z} f(w)\, dw$$

$$= \int_{z}^{z+h} f(w)\, dw.$$

Observe that

$$f(z)h = f(z) \int_{z}^{z+h} dw = \int_{z}^{z+h} f(z)\, dw \quad \text{or} \quad f(z) = \frac{1}{h} \int_{z}^{z+h} f(z)\, dw$$

so that

$$\frac{F(z+h) - F(z)}{h} = \frac{1}{h} \int_{z}^{z+h} f(w)\, dw$$

add and subtract $f(z)$
$$\downarrow$$
$$= f(z) + \frac{1}{h} \int_{z}^{z+h} [f(w) - f(z)]\, dw.$$

Thus

$$\frac{F(z+h) - F(z)}{h} - f(z) = \frac{1}{h} \int_{z}^{z+h} [f(w) - f(z)]\, dw.$$

We will be done if we can show that

$$\lim_{h \to 0} \frac{1}{h} \int_{z}^{z+h} [f(w) - f(z)]\, dw = 0.$$

Since f is analytic, it is continuous. For each $\epsilon > 0$, choose a $\delta > 0$ such that $|f(w) - f(z)| < \epsilon$ if $|w - z| < \delta$. Then, for $|h| < \delta$,

Theorem 2(iv)
$$\left| \frac{1}{h} \int_{z}^{z+h} [f(w) - f(z)]\, dw \right| \overset{\downarrow}{\leq} \frac{1}{|h|} \int_{z}^{z+h} |f(w) - f(z)|\, |dw| < \frac{\epsilon}{|h|} \int_{z}^{z+h} |dw| = \epsilon.$$

Hence $\lim_{h \to 0} \left[\dfrac{F(z+h) - F(z)}{h} - f'(z) \right] = 0$ and the theorem is proved. ∎

EXAMPLE 5

Compute $\int_1^i z^2\,dz$.

SOLUTION

$z^3/3$ is an antiderivative for z^2 so

$$\int_1^i z^2\,dz = \frac{z^3}{3}\bigg|_1^i = \frac{1}{3}(i^3 - 1) = \frac{1}{3}(-i - 1).$$

EXAMPLE 6

Compute $\int_{\pi/3}^i \cos z\,dz$.

SOLUTION

$\sin z$ is an antiderivative for $\cos z$ so

$$\int_{\pi/3}^i \cos z\,dz = \sin z\bigg|_{\pi/3}^i = \sin i - \sin \pi/3 \overset{\text{formula (9) on p. 919}}{=} -\sqrt{3}/2 + i\sinh 1.$$

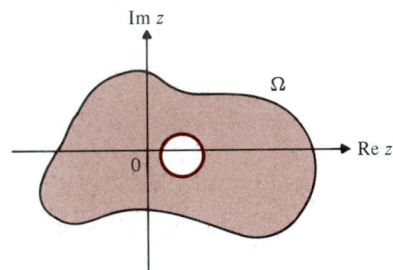

Im z

Ω

0 Re z

FIGURE 6 Doubly connected region

Cauchy's Theorem in a Region with Holes

Consider the region Ω in Fig. 6. It is not simply connected because it has a hole. Such a region is called **doubly connected.** If it has two holes it is **triply connected,** and so on. Cauchy's theorem can be extended to such regions as follows:

◼ **THEOREM 5: Cauchy's Theorem on a Multiply Connected Region**

Let the inside of the pws Jordan curve C_0 contain the disjoint pws Jordan curves $C_1, \ldots, C_n$, none of which is contained inside another. Suppose $f(z)$ is analytic in a region Ω containing the set S consisting of all points on and inside C_0 but not inside C_k, $k = 1, \ldots, n$. Then

$$\int_{C_0} f(z)\,dz = \sum_{k=1}^n \int_{C_k} f(z)\,dz. \tag{7}$$

PROOF

We can always find disjoint pws arcs L_k, $k = 0, \ldots, n$, joining C_k to C_{k+1} (with L_n joining C_n to C_0) such that two pws Jordan curves are formed, each lying in some simply connected subregion of Ω. (See Fig. 7.) By Cauchy's theorem, the integral of $f(z)$ on these curves, each traversed in the positive sense, vanishes. But the total contribution of these two curves is equivalent to traversing C_0 in

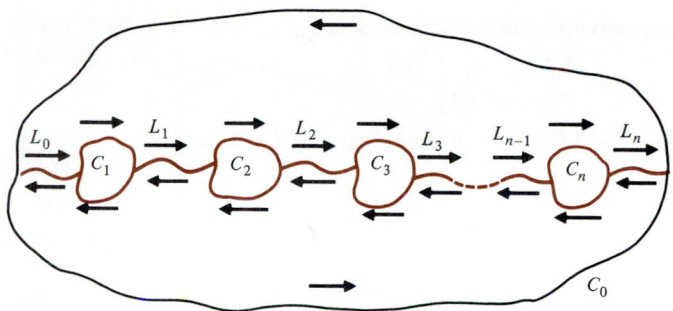

FIGURE 7 A multiply connected domain

the positive sense, $C_1, \ldots, C_n$ in the negative (opposite) sense, and $L_0, \ldots, L_n$ in opposite directions. Thus, the integrals on the arcs L_k cancel out and

$$0 = \int_{C_0 - \sum\limits_{k=1}^{n} C_k} f(z)\,dz = \int_{C_0} f(z)\,dz - \sum_{k=1}^{n} \int_{C_k} f(z)\,dz.$$ ∎

EXAMPLE 7

Letting $w = z - z_0$, we have $dw = dz$ so that for $r < 1$

$$\int_{|z-z_0|=r} \frac{dz}{z - z_0} = \int_{|w|=r} \frac{dw}{w} = \int_{|w|=1} \frac{dw}{w} = 2\pi i.$$

Theorem 5   Example 16.1.5

Problems 16.3

In problems 1–10 compute the integral by using Theorem 3.

1. $\int_0^i (z^3 + 2z + 1)\,dz$

2. $\int_{1+i}^{1-i} (2z + 3)\,dz$

3. $\int_0^i \cos z\,dz$

4. $\int_{-i\pi}^{i\pi} \sin\!\left(\frac{z}{2}\right) dz$

5. $\int_0^{\pi i/8} \sinh 2z\,dz$

6. $\int_0^{\pi i/4} e^z\,dz$

7. $\int_0^{2\pi i/3} z e^{z^2}\,dz$

8. $\int_{2i}^{3i} (z^3 - 1)^2\,dz$

9. $\int_{-\pi i/2}^{\pi i/2} z \cos z^2\,dz$

10. $\int_1^i e^z \sin z\,dz$

11. Show that $\displaystyle\int_{|z-2|=3} \frac{dz}{z - 2} = 2\pi i.$

12. Compute $\displaystyle\int_{|z|=2} \frac{dz}{z - 3}.$

13. Compute $\displaystyle\int_{|z|=3} \frac{dz}{z(z - 1)}.$ [*Hint:* Use partial fractions.]

In Problems 14–20 integrate each function around the positively oriented unit circle.

14. $f(z) = z$

15. $f(z) = z^2$

16. $f(z) = \dfrac{1}{z^2 - 4}$

17. $f(z) = \dfrac{1}{z^2 + 4}$

18. $f(z) = \dfrac{z}{z^2 - 4}$

19. $f(z) = e^{4z}$

20. $f(z) = z \sin z.$

* 21. Show that

$$\int_{|z|=1} \frac{\text{Log } z}{z}\,dz = 0,$$

even though $(\text{Log } z)/z$ is not analytic on $|z| \le 1$. What result is obtained if we integrate

$$\int_C \frac{\log z}{z}\,dz$$

on $C: z(t) = e^{it}, 0 \le t \le 2\pi$? Explain.

22. Without computing the integral, show that

$$\left| \int_{|z|=2} \frac{dz}{z^2 + 1} \right| \leq \frac{4\pi}{3}.$$

23. If C is the semicircle $|z| = R$, $|\arg z| \leq \pi/2$, $R > 1$, show that

$$\left| \int_C \frac{\operatorname{Log} z}{z^2} \, dz \right| \leq \frac{\pi}{R} \left(\ln R + \frac{\pi}{2} \right)$$

and hence that the value of the integral tends to zero as $R \to \infty$.

* 24. Let C_1 denote the line segment from 0 to $1 + i$ and C_2 denote the parabola $(y = x^2)$ from 0 to $1 + i$. Compute

$$\int_{C_1} \bar{z} \, dz \quad \text{and} \quad \int_{C_2} \bar{z} \, dz.$$

Does this result contradict the independence of path theorem? Explain.

** 25. Morera's Theorem. Prove the following converse of Cauchy's theorem: Suppose that $f(z)$ is continuous in a simply connected region Ω and that $\int_C f(z) \, dz = 0$ for every pws closed curve in Ω. Then $f(z)$ is analytic in Ω.

16.4 Using Cauchy's Theorem to Compute Real Integrals

(Optional)

In this section we show how Cauchy's theorem can be used to compute real integrals. The techniques employed here have evolved over a long period of time and will likely seem difficult. When you read these examples, keep in mind that the solutions are not at all obvious to anyone who encounters them for the first time. However, the techniques are useful and understanding them is certainly worth the effort.

EXAMPLE 1: The Poisson Kernel†

Show that

$$\frac{1}{2\pi} \int_0^{2\pi} \frac{R^2 - r^2}{R^2 - 2Rr \cos \theta + r^2} \, d\theta = 1, \qquad 0 < r < R. \tag{1}$$

The integrand appearing in this integral is called the **Poisson kernel**. The Poisson kernel has many useful properties; we see some of these in Section 19.6.

SOLUTION

Consider the quotient

$$f(z) = \frac{R + z}{R - z}. \tag{2}$$

† See the discussion of the Poisson integral formula in Section 12.9, p. 773.

If z is on the circle centered at the origin with radius r, we have $z = re^{i\theta}$ for $0 \le \theta < 2\pi$ and

$$f(z) = \frac{R + re^{i\theta}}{R - re^{i\theta}} = \frac{(R + re^{i\theta})(R - re^{-i\theta})}{(R - re^{i\theta})(R - re^{-i\theta})} = \frac{R^2 - r^2 + rR(e^{i\theta} - e^{-i\theta})}{R^2 - r^2 - rR(e^{i\theta} + e^{-i\theta})}$$

equation (15.8.1) on p. 916

$$\downarrow$$
$$= \frac{R^2 - r^2 + 2irR \sin \theta}{R^2 - 2rR \cos \theta + r^2}. \tag{3}$$

Then

$$\operatorname{Re} f(z) = \frac{R^2 - r^2}{R^2 - 2rR \cos \theta + r^2} = \text{the Poisson kernel.} \tag{4}$$

Also, since $z = re^{i\theta}$ and r is fixed,

$$\frac{dz}{d\theta} = rie^{i\theta} = iz,$$

or

$$\frac{dz}{iz} = d\theta \qquad \text{on } |z| = r. \tag{5}$$

Hence, from (4) and (5)

$$\frac{1}{2\pi} \int_0^{2\pi} \frac{R^2 - r^2}{R^2 - 2rR \cos \theta + r^2} d\theta = \frac{1}{2\pi} \int_{|z|=r} \operatorname{Re}\left(\frac{R + z}{R - z}\right) \frac{dz}{iz}$$

and, since $1/2\pi$ and $d\theta = dz/iz$ are real,

$$\frac{1}{2\pi} \int_0^{2\pi} \frac{R^2 - r^2}{R^2 - 2rR \cos \theta + r^2} d\theta = \operatorname{Re}\left(\frac{1}{2\pi i} \int_{|z|=r} \frac{R + z}{R - z} \frac{dz}{z}\right).$$

But by partial fractions,

$$\frac{1}{2\pi i} \int_{|z|=r} \frac{R + z}{z(R - z)} dz = \frac{1}{2\pi i} \int_{|z|=r} \left(\frac{1}{z} + \frac{2}{R - z}\right) dz = \frac{1}{2\pi i} \int_{|z|=r} \frac{dz}{z} + \frac{1}{2\pi i} \int_{|z|=r} \frac{2dz}{R - z}.$$

The function $2/(R - z)$ is analytic on and inside $|z| = r$ since $r < R$. Thus, by Cauchy's theorem,

$$\int_{|z|=r} \frac{2dz}{R - z} = 0.$$

Finally, using Theorem 16.3.5, we have

$$\frac{1}{2\pi i} \int_{|z|=r} \frac{dz}{z} = \frac{1}{2\pi i} \int_{|z|=1} \frac{dz}{z} = 1.$$

Thus

$$\frac{1}{2\pi} \int_0^{2\pi} \frac{R^2 - r^2}{R^2 - 2rR \cos \theta + r^2} d\theta = \operatorname{Re}\left(\frac{1}{2\pi i} \int_{|z|=r} \frac{R + z}{R - z} \frac{dz}{z}\right) = \operatorname{Re} 1 = 1.$$

EXAMPLE 2

Show that

$$\int_{-\infty}^{\infty} e^{-x^2} \cos 2bx\, dx = \sqrt{\pi} e^{-b^2}.$$

Remark. This formula cannot be obtained from any technique of elementary calculus.

SOLUTION

Consider the piecewise smooth curve sketched in Fig. 1. The entire function $f(z) = e^{-z^2}$ is analytic on the region $S = \{z : |x| \le a, 0 \le y \le b\}$ enclosed by this curve C. Thus, by Cauchy's theorem

$$\oint_C e^{-z^2}\, dz = 0. \tag{6}$$

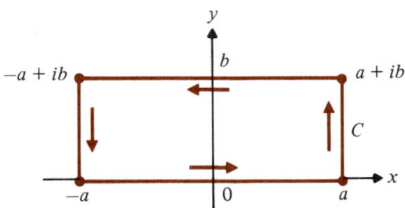

FIGURE 1

C consists of four line segments:

$$\text{from } -a \text{ to } a: \qquad z = x,$$
$$\text{from } a \text{ to } a + ib: \qquad z = a + iy, 0 \le y \le b,$$
$$\text{from } a + ib \text{ to } -a + ib: \qquad z = x + ib, a \ge x \ge -a,$$
$$\text{from } -a + ib \text{ to } -a: \qquad z = -a + iy, b \ge y \ge 0.$$

Using this parametrization, (6) becomes

$$0 = \int_{-a}^{a} e^{-x^2}\, dx + \int_{0}^{b} e^{-(a+iy)^2} i\, dy + \int_{a}^{-a} e^{-(x+ib)^2}\, dx + \int_{b}^{0} e^{-(-a+iy)^2} i\, dy.$$

Now,

$$\int_{0}^{b} e^{-(a+iy)^2} i\, dy + \int_{b}^{0} e^{-(-a+iy)^2} i\, dy = i \int_{0}^{b} [e^{-(a+iy)^2} - e^{-(-a+iy)^2}]\, dy$$

$$= i \int_{0}^{b} e^{-a^2+y^2}(e^{-2iay} - e^{2iay})\, dy = e^{-a^2} \int_{0}^{b} e^{y^2} \frac{(e^{2iay} - e^{-2iay})}{i}\, dy = 2e^{-a^2} \int_{0}^{b} e^{y^2} \sin 2ay\, dy.$$

Also,

$$\int_{a}^{-a} e^{-(x+ib)^2}\, dx = -\int_{-a}^{a} e^{-x^2+b^2} e^{-2ibx}\, dx$$

$$= -e^{b^2} \int_{-a}^{a} e^{-x^2} \cos 2bx\, dx + ie^{b^2} \int_{-a}^{a} e^{-x^2} \sin 2bx\, dx. \tag{7}$$

Observe that $g(x) = e^{-x^2} \sin 2bx$ is an odd function; that is, $g(-x) = -g(x)$. By Theorem 11.4.1(c) $\int_{-a}^{a} g(x)\,dx = 0$ if g is odd. Thus the second integral in (7) is zero and we have

$$0 = \int_{-a}^{a} e^{-x^2}\,dx - e^{b^2} \int_{-a}^{a} e^{-x^2} \cos 2bx\,dx + 2e^{-a^2} \int_{0}^{b} e^{y^2} \sin 2ay\,dy. \tag{8}$$

Using polar coordinates, we can calculate

$$\left(\int_{-\infty}^{\infty} e^{-x^2}\,dx\right)^2 = \int_{-\infty}^{\infty} e^{-x^2}\,dx \cdot \int_{-\infty}^{\infty} e^{-y^2}\,dy$$

$$= \int_{-\infty}^{\infty} \int_{-\infty}^{\infty} e^{-(x^2+y^2)}\,dx\,dy$$

$$= \int_{0}^{2\pi} \int_{0}^{\infty} e^{-r^2} r\,dr\,d\theta = \pi,$$

so that

$$\int_{-\infty}^{\infty} e^{-x^2}\,dx = \sqrt{\pi}. \tag{9}$$

The identity in (9) indicates that the first two integrals in (8) are convergent as $a \to \infty$ because

$$\left| \int_{-a}^{a} e^{-x^2} \cos 2bx\,dx \right| \le \int_{-a}^{a} e^{-x^2} |\cos 2bx|\,dx \le \int_{-a}^{a} e^{-x^2}\,dx.$$

The last term in (8) tends to zero as $a \to \infty$ because the integral is bounded,

$$\left| \int_{0}^{b} e^{y^2} \sin ay\,dy \right| \le \int_{0}^{b} e^{y^2} |\sin ay|\,dy \le b e^{b^2},$$

and $e^{-a^2} \to 0$. Thus, we have from (8) and (9)

$$\int_{-\infty}^{\infty} e^{-x^2} \cos 2bx\,dx = e^{-b^2} \int_{-\infty}^{\infty} e^{-x^2}\,dx = \sqrt{\pi}\,e^{-b^2}.$$

Remark. As stated at the beginning of this section, computations in these two examples were neither easy nor obvious. They involved techniques and observations that were discovered over a long period of time. However, the examples are important because they embody the essence of applied mathematics: A new result is found (in this case Cauchy's theorem). Then old techniques are combined with this new result to obtain the solutions to a number of new problems.

Problems 16.4

* **1.** Prove that

$$\int_{0}^{\pi/2} e^{a \cos t} \cos(t + a \sin t)\,dt = \frac{\sin a}{a}, \qquad a > 0,$$

by integrating e^z along the Jordan curve in the first quadrant composed of the quarter circle of $|z| = a$ and the line segments from ia to 0 and 0 to a.

* 2. Show that

$$\int_0^T e^{at} \cos bt\, dt = \frac{e^{aT}(a \cos bT + b \sin bT) - a}{a^2 + b^2},$$

and

$$\int_0^T e^{at} \sin bt\, dt = \frac{e^{aT}(a \sin bT - b \cos bT) + b}{a^2 + b^2},$$

by integrating $f(z) = e^z$ along the line segment joining 0 to $(a + ib)T$.

* 3. Show that

$$\int_0^T \sin at \cosh bt\, dt$$

$$= \frac{b \sin aT \sinh bT - a \cos aT \cosh bT + a}{a^2 + b^2},$$

and

$$\int_0^T \cos at \sinh bt\, dt$$

$$= \frac{b \cos aT \cosh bT + a \sin aT \sinh bT - b}{a^2 + b^2},$$

by integrating $f(z) = \sin z$ along the line segment from 0 to $(a + ib)T$.

* 4. Prove that

$$\int_0^\infty \frac{\sin(x^2)}{x}\, dx = \frac{\pi}{4}$$

by integrating $\dfrac{1}{z} e^{iz^2}$ over the curve shown in Fig. 2.

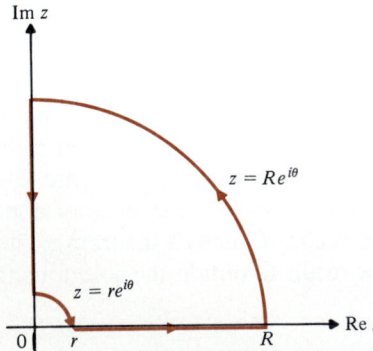

FIGURE 2

5. Assuming $0 < b < 1$ and applying Cauchy's theorem to the function $f(z) = (1 + z^2)^{-1}$ along the boundary of the rectangle in Fig. 1, show that

$$\int_{-\infty}^\infty \frac{(1 - b^2 + x^2)\, dx}{(1 - b^2 + x^2)^2 + 4x^2 b^2} = \pi.$$

6. Prove that

$$\int_{-\infty}^\infty e^{-kx^2} \cos ax\, dx = \sqrt{\frac{\pi}{k}}\, e^{-a^2/4k}, \qquad k > 0,\ a \text{ real},$$

by using the same procedure as in Example 2 with the function $f(z) = e^{-kz^2}$. Check your answer by changing variables.

7. Prove that

$$\int_{-\infty}^\infty \frac{(1 - b^2 + x^2) \cos kx + 2xb \sin kx}{(1 - b^2 + x^2)^2 + 4x^2 b^2}\, dx = e^{kb} \int_{-\infty}^\infty \frac{\cos kx}{1 + x^2}\, dx,$$

$$\int_{-\infty}^\infty \frac{(1 - b^2 + x^2) \sin kx - 2xb \cos kx}{(1 - b^2 + x^2)^2 + 4x^2 b^2}\, dx = 0$$

with $0 < b < 1$ and k real.

8. Let $0 < b < 2^{-1/2}$ and show that

$$\int_{-\infty}^\infty \frac{\mathrm{Re}(1 + (x - ib)^4)}{|1 + (x + ib)^4|^2}\, dx = \int_{-\infty}^\infty \frac{dx}{1 + x^4}.$$

9. Prove that

$$\int_0^\infty e^{-x^2} \sin 2xb\, dx = e^{-b^2} \int_0^b e^{x^2}\, dx, \qquad b > 0,$$

by integrating around a suitable rectangle.

* 10. Prove the equalities, known as the **Fresnel integrals,**

$$\int_0^\infty \cos x^2\, dx = \int_0^\infty \sin x^2\, dx = \frac{\sqrt{\pi}}{2\sqrt{2}}$$

by applying Cauchy's theorem to the function $f(z) = e^{-z^2}$ along the boundary of the sector $0 \le |z| \le R$, $0 \le \arg z \le \pi/4$.

* 11. Show that

$$\int_0^\infty e^{-x^2} \cos(x^2)\, dx = \frac{\sqrt{\pi}}{4} \sqrt{\sqrt{2} + 1},$$

$$\int_0^\infty e^{-x^2} \sin(x^2)\, dx = \frac{\sqrt{\pi}}{4} \sqrt{\sqrt{2} - 1},$$

by integrating e^{-z^2} along the boundary of the sector $0 \le |z| \le R$, $0 \le \arg z \le \pi/8$.

* 12. Prove Dirichlet's integral

$$\int_0^\infty \frac{\sin x}{x}\, dx = \frac{\pi}{2}$$

by integrating $f(z) = e^{iz}/z$ along the boundary of the set $r \le |z| \le R$, $0 \le \arg z \le \pi$. (See example 11.5.1.)

16.5 The Cauchy Integral Formula

Cauchy made the remarkable discovery that the values of an analytic function inside a pws Jordan curve are completely determined by its values on that curve.

■ **THEOREM 1: The Cauchy Integral Formula**

Let $f(z)$ be analytic on a simply connected region Ω containing the pws Jordan curve C. Then if z_0 is inside C

$$f(z_0) = \frac{1}{2\pi i} \int_C \frac{f(z)}{z - z_0}\, dz. \tag{1}$$

PROOF

Fix z_0. Then, since $f(z)$ is continuous at z_0, given $\epsilon > 0$, there exists a closed disk $|z - z_0| \le r$ lying inside C for which $|f(z) - f(z_0)| < \epsilon$. (See Fig. 1.) Since $f(z)/(z - z_0)$ is analytic on a region containing the points on and inside C satisfying $|z - z_0| \ge r$, the Cauchy theorem on multiply connected regions (Theorem 16.3.5) implies that

$$\frac{1}{2\pi i} \int_C \frac{f(z)}{z - z_0}\, dz = \frac{1}{2\pi i} \int_{|z-z_0|=r} \frac{f(z)}{z - z_0}\, dz.$$

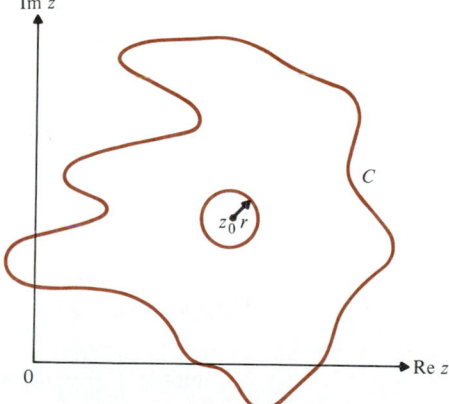

FIGURE 1 The Cauchy integral formula

But $f(z) = f(z_0) + [f(z) - f(z_0)]$, so that

$$\int_{|z-z_0|=r} \frac{f(z)}{z - z_0}\, dz = f(z_0) \int_{|z-z_0|=r} \frac{dz}{z - z_0} + \int_{|z-z_0|=r} \frac{f(z) - f(z_0)}{z - z_0}\, dz.$$

By Example 16.3.7, the first integral on the right-hand side equals $2\pi i$, and we have

$$\int_{|z-z_0|=r} \frac{f(z)}{z-z_0}\,dz = 2\pi i f(z_0) + \int_{|z-z_0|=r} \frac{f(z)-f(z_0)}{z-z_0}\,dz.$$

Using part (iv) of Theorem 16.3.2 (see p. 945),

$$\left|\int_{|z-z_0|=r} \frac{f(z)}{z-z_0}\,dz - 2\pi i f(z_0)\right| = \left|\int_{|z-z_0|=r} \frac{f(z)-f(z_0)}{z-z_0}\,dz\right|$$

$$\leq \int_{|z-z_0|=r} \frac{|f(z)-f(z_0)|}{|z-z_0|}\,|dz| \leq \epsilon \int_{|z-z_0|=r} \frac{|dz|}{|z-z_0|} \overset{|z-z_0|=r}{=} \frac{\epsilon}{r}\int_{|z-z_0|=r} |dz| = 2\pi\epsilon,$$

since $|dz|$ is the arc length differential and the last integral is merely the perimeter of the circle $|z - z_0| = r$. We have shown that

$$\left|\int_{|z-z_0|=r} \frac{f(z)}{z-z_0}\,dz - 2\pi i f(z_0)\right| < 2\pi\epsilon.$$

Since ϵ can be chosen arbitrarily close to 0, the proof is complete. ∎

EXAMPLE 1

Compute $\displaystyle\int_{|z|=2} \frac{z^3+2z-5}{z-1}\,dz.$

SOLUTION
Since $f(z) = z^3 + 2z - 5$ is entire, by Cauchy's integral formula

$$\frac{1}{2\pi i}\int_{|z|=2} \frac{z^3+2z-5}{z-1}\,dz = f(1) = 1^3 + 2\cdot 1 - 5 = -2.$$

Hence

$$\int_{|z|=2} \frac{z^3+2z-5}{z-1}\,dz = -4\pi i.$$

EXAMPLE 2

Compute $\displaystyle\int_C \frac{e^{z^6}}{z+i}\,dz,$ where $-i$ is inside the Jordan curve C.

SOLUTION
Since $f(z) = e^{z^6}$ is entire, by Cauchy's integral formula

$$\int_C \frac{e^{z^6}}{z+i}\,dz = 2\pi i f(-i) = 2\pi i e^{(-i)^6} = 2\pi i e^{-1} = \frac{2\pi i}{e}.$$

EXAMPLE 3

Integrate

$$\int_C \frac{\cos z}{z^3 + z}\,dz$$

over the given curves: (a) $C: |z| = 2$, (b) $C: |z| = \frac{1}{2}$, and (c) $C: |z - i/2| = 1$.

SOLUTION

(a) $C: |z| = 2$. We have

by partial fractions

$$\frac{1}{z^3 + z} = \frac{1}{z(z^2 + 1)} = \frac{1}{z(z+i)(z-i)} = \frac{1}{z} - \frac{1}{2(z+i)} - \frac{1}{2(z-i)}.$$

Now the points 0, $-i$, and i are all inside the circle $|z| = 2$, so we apply Theorem 16.3.5 and the Cauchy integral formula three times to obtain

$$\int_C \frac{\cos z}{z^3 + z}\,dz = \int_C \frac{\cos z}{z}\,dz - \frac{1}{2}\int_C \frac{\cos z}{z+i}\,dz - \frac{1}{2}\int_C \frac{\cos z}{z-i}\,dz$$

$$\cosh(1) = \cosh(-1) \text{ and} \\ \text{formula (15.8.12) on p. 920}$$

$$= 2\pi i\left[\cos(0) - \frac{1}{2}\cos(-i) - \frac{1}{2}\cos i\right] = 2\pi i[1 - \cosh(1)].$$

(b) $C: |z| = \frac{1}{2}$. As $\cos z/(z^2 + 1)$ is analytic on and inside C, the integral equals $2\pi i$ times its value at $z = 0$, that is,

$$\int_C \frac{\cos z}{z^3 + z}\,dz = 2\pi i\left(\frac{\cos z}{z^2 + 1}\right)\Bigg|_{z=0} = 2\pi i.$$

(c) $C: |z - i/2| = 1$. Since $\cos z/(z + i)$ is analytic on and inside C, by partial fractions we have

$$\frac{1}{z(z - i)} = i\left(\frac{1}{z} - \frac{1}{z - i}\right),$$

so that

$$\int_C \frac{\cos z}{z^3 + z}\,dz = 2\pi i\left[i\left(\frac{\cos(0)}{i}\right) - i\left(\frac{\cos i}{2i}\right)\right] = 2\pi i\left[1 - \frac{1}{2}\cosh(1)\right].$$

Of course, we can do all three examples utilizing the partial fraction decomposition in part (a), since the corresponding integrals vanish when the points 0 or $\pm i$ lie outside C.

Derivatives of Analytic Functions

The Cauchy integral formula can be used to prove the following remarkable fact: all derivatives of an analytic function are analytic. That is, an analytic function is infinitely differentiable. The following theorem not only indicates why this is true but also tells us the form these derivatives take.

■ **THEOREM 2: Cauchy's Formula for Derivatives†**

Let $f(z)$ be analytic on a simply connected region Ω containing the pws Jordan curve C. Then, if z_0 is inside C, all derivatives of f exist and are analytic at z_0. Moreover,

$$f'(z_0) = \frac{1!}{2\pi i} \int_C \frac{f(z)}{(z - z_0)^2} \, dz, \tag{2}$$

$$f''(z_0) = \frac{2!}{2\pi i} \int_C \frac{f(z)}{(z - z_0)^3} \, dz, \tag{3}$$

$$\vdots$$

$$f^{(n)}(z_0) = \frac{n!}{2\pi i} \int_C \frac{f(z)}{(z - z_0)^{n+1}} \, dz. \tag{4}$$

■

Remark. Equations (2)–(4) are obtained by repeated differentiation with respect to z_0 of both sides of the Cauchy integral formula (1). The proof of Theorem 2 requires verification of the validity of differentiating under the integral sign.

EXAMPLE 4

Integrate

$$\int_C \frac{\cos z}{z^2(z - 1)} \, dz$$

over (a) $C : |z| = \frac{1}{3}$, (b) $C : |z - 1| = \frac{1}{3}$, and (c) $C : |z| = 2$.

SOLUTION

(a) $C : |z| = \frac{1}{3}$. In this case $f(z) = \cos z/(z - 1)$ is analytic on and inside C, so by Cauchy's theorem for derivatives we obtain

$$\int_C \frac{(\cos z)/(z - 1)}{z^2} \, dz = \int_C \frac{f(z)}{z^2} \, dz = 2\pi i f'(0).$$

But

$$f'(z) = \frac{-(z - 1)\sin z - \cos z}{(z - 1)^2} \quad \text{and} \quad f'(0) = -1,$$

† For a complete proof see W. R. Derrick, *Complex Analysis and Applications,* 2nd ed., Wadsworth, Belmont, Calif., 1984, p. 111.

implying that

$$\int_C \frac{\cos z}{z^2(z-1)}\,dz = -2\pi i.$$

(b) $C : |z - 1| = \frac{1}{3}$. Since C does not enclose 0, $z^{-2} \cos z$ is analytic on and inside C and the integral equals $2\pi i$ times the value of $z^{-2} \cos z$ at $z = 1$, that is,

$$\int_C \frac{z^{-2} \cos z}{z-1}\,dz = 2\pi i \cos(1).$$

(c) $C : |z| = 2$. There are two ways to do this problem. First, as we see in Fig. 2, $f(z) = (\cos z)/z^2(z - 1)$ is analytic in the triply connected region bounded by the three circles. Thus, by Theorem 16.3.5 (with $n = 2$) on p. 950 and parts (a) and (b),

$$\int_{|z|=2} \frac{\cos z}{z^2(z-1)}\,dz = \int_{|z|=1/3} \frac{\cos z}{z^2(z-1)}\,dz + \int_{|z-1|=1/3} \frac{\cos z}{z^2(z-1)}\,dz = 2\pi i[-1 + \cos 1].$$

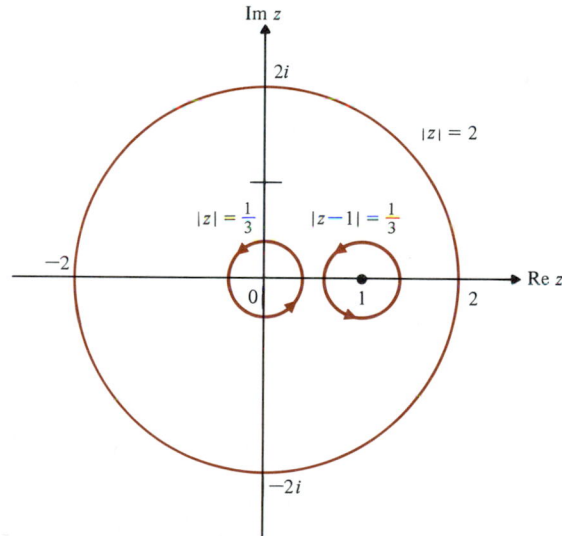

FIGURE 2

Another approach is to use partial fractions:

$$\int_{|z|=2} \frac{\cos z}{z^2(z-1)}\,dz = \int_{|z|=2} \cos z \left(\frac{1}{z-1} - \frac{1}{z} - \frac{1}{z^2} \right)\,dz.$$

Applying Cauchy's formula to $(\cos z)/(z - 1)$ and $(\cos z)/z$ and Cauchy's formula for the first derivative to $(\cos z)/z^2$, we have

$$\int_{|z|=2} \frac{\cos z}{z^2(z-1)}\,dz = 2\pi i \cos 1 - 2\pi i \cos 0 - 2\pi i[(\cos z)'|_{z=0}] = 2\pi i[\cos 1 - 1].$$

Problems 16.5

In Problems 1–14 use Cauchy's integral formula to evaluate the given integral over the given curve. All integrals are taken in the positive (counterclockwise) sense.

1. $\int_{|z-1|=4} \dfrac{z}{z+1}\, dz$

2. $\int_{|z|=1} \dfrac{z^3+3z+1}{z}\, dz$

3. $\int_{|z-2|=3} \dfrac{e^z}{z-1}\, dz$

4. $\int_{|z|=10} \dfrac{\sin 2z}{z}\, dz$

5. $\int_{|z|=1} \dfrac{\cos z}{z-\pi/4}\, dz$

6. $\int_{|z|=1} \dfrac{\cos z}{z+\pi/4}\, dz$

7. $\int_{|z|=2} \dfrac{z}{z^2-1}\, dz$

8. $\int_{|z|=5} \dfrac{z}{z^2+1}\, dz$

9. $\int_{|z|=1/2} \dfrac{z^2+2z+3}{(z-1)(z+2)}\, dz$

10. $\int_{|z|=3/2} \dfrac{z^2+2z+3}{(z-1)(z+2)}\, dz$

11. $\int_{|z|=5/2} \dfrac{z^2+2z+3}{(z-1)(z+2)}\, dz$

12. $\int_{|z|=1} \dfrac{e^z}{z^2+5z}\, dz$

13. $\int_{|z|=10} \dfrac{e^z}{z^2+5z}\, dz$

14. $\int_{|z|=1} \dfrac{4\cosh z}{4z^2+\pi^2}\, dz$

In Problems 15–28 use Cauchy's integral formula for derivatives to compute the given integral. As before, all curves are traversed in the positive sense.

15. $\int_{|z|=3} \dfrac{z^3}{(z+1)^2}\, dz$

16. $\int_{|z|=5} \dfrac{z^2+1}{(z-i)^2}\, dz$

17. $\int_{|z|=4} \dfrac{e^z}{(z+\pi i)^3}\, dz$

18. $\int_{|z|=1} \dfrac{e^z \cos z}{z^2}\, dz$

19. $\int_{|z-1|=4} \dfrac{e^{z^2}}{(z+i)^2}\, dz$

20. $\int_C \dfrac{\sin(z+i)}{(z-2i)^3}\, dz$; C: a square enclosing $2i$

21. $\int_C \dfrac{\sin z + \cos z}{(z-\pi/6)^5}\, dz$; C: an ellipse enclosing $\pi/6$

22. $\int_{|z|=2} \dfrac{e^z}{z^2(z+1)}\, dz$ [*Hint:* Use partial fractions.]

23. $\int_{|z-1|=\frac{3}{2}} \dfrac{z^3+1}{(z+1)^2(z^2+1)}\, dz$

24. $\int_{|z-1|=3} \dfrac{z^3+1}{(z+1)^2(z^2+1)}\, dz$

25. $\int_{|z|=5} \dfrac{\sinh z}{(z+1)^2(z-1)^2}\, dz$

26. $\int_{|z-1|=2} \dfrac{\cos z}{(z-1)^3}\, dz$

27. $\int_{|z-1|=2} \dfrac{\sin z}{(z^2+1)^2}\, dz$

28. $\int_{|z+1|=4} \dfrac{\cosh z}{(z+1)^3}\, dz$

In Problems 29–31, evaluate

$$\int_C \frac{dz}{(z-a)(z-b)^2}$$

by decomposing the integrand into partial fractions.

29. If a and b lie inside C.

30. If a lies inside and b lies outside C.

31. If b lies inside and a lies outside C.

* 32. Let $f(z)$ be analytic and bounded by M in $|z| \le R$. Prove that

$$|f^{(n)}(z)| \le \frac{MRn!}{(R-|z|)^{n+1}}, \qquad |z| < R.$$

* 33. If $f(z)$ is analytic in $|z| < 1$ and $|f(z)| \le (1-|z|)^{-1}$, show that

$$|f^{(n)}(0)| \le (n+1)!\left(1+\frac{1}{n}\right)^n.$$

* 34. Can an analytic function $f(z)$ satisfy $|f^{(n)}(z)| > n!n^n$, for all positive integers n, at some point z? Explain.

* 35. Compute

$$\int_{|z|=1} \frac{e^{kz^n}}{z}\, dz,$$

for n a positive integer. Then show that

$$\int_0^{2\pi} e^{k\cos n\theta} \cos(k \sin n\theta)\, d\theta = 2\pi.$$

* 36. The Legendre polynomial $P_n(z)$ is defined by

$$P_n(z) = \frac{1}{2^n n!} \frac{d^n}{dz^n}[(z^2-1)^n].$$

Using Cauchy's formula for derivatives, show that

$$P_n(z_0) = \frac{1}{2\pi i} \int_C \frac{(z^2-1)^n\, dz}{2^n(z-z_0)^{n+1}},$$

where z_0 is inside the pws Jordan curve C.

* 37. Let $P(z)$ be a polynomial none of whose roots lies on the pws Jordan curve C. Show that

$$\frac{1}{2\pi i}\int_C \frac{P'(z)}{P(z)}\,dz$$

equals the number of roots of $P(z)$ inside C including multiplicities. [*Hint:* Write $P(z)$ as a product of terms.]

16.6 Liouville's Theorem, the Maximum Principle, and the Fundamental Theorem of Algebra

In this section we discuss three useful consequences of Cauchy's integral formula and its extension to higher derivatives. We first need a simple consequence of Cauchy's integral formula.

■ **THEOREM 1: Gauss's Mean Value Theorem**
Let $f(z)$ be analytic in the disk $|z - z_0| < R$. Then

$$f(z_0) = \frac{1}{2\pi}\int_0^{2\pi} f(z_0 + re^{i\theta})\,d\theta, \qquad 0 < r < R. \tag{1}$$

Remark. Gauss's theorem states that $f(z_0)$ is the average of the values of $f(z)$ on any circle of radius r centered at z_0 that is contained in the region of analyticity of $f(z)$ (see Fig. 1).

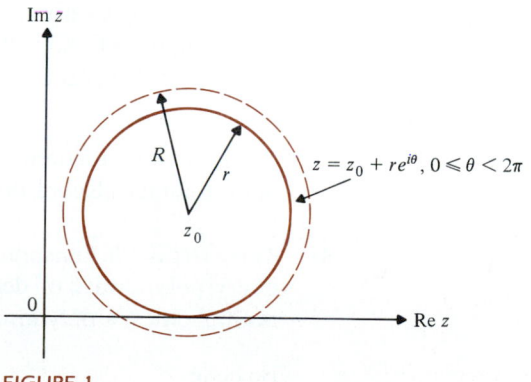

FIGURE 1

PROOF
The Cauchy integral formula states that

$$f(z_0) = \frac{1}{2\pi i}\int_{|z-z_0|=r} \frac{f(z)}{z - z_0}\,dz. \tag{2}$$

If $z = z_0 + re^{i\theta}$, $0 \le \theta < 2\pi$, then $dz/d\theta = ire^{i\theta}$, $z - z_0 = re^{i\theta}$, and (2) becomes

$$f(z_0) = \frac{1}{2\pi i} \int_0^{2\pi} \frac{f(z_0 + re^{i\theta})}{re^{i\theta}} ire^{i\theta}\, d\theta = \frac{1}{2\pi} \int_0^{2\pi} f(z_0 + re^{i\theta})\, d\theta,$$

for all $0 < r < R$.　　　　　　　　　　　　　　　　　　　　　　■

THEOREM 2: Cauchy's Estimate

Let $f(z)$ be analytic and satisfy $|f(z)| \le M$ in $|z - z_0| \le r$. Then

$$|f^{(n)}(z_0)| \le \frac{Mn!}{r^n}.$$

PROOF

By Cauchy's theorem for derivatives, we have

$$|f^{(n)}(z_0)| = \left| \frac{n!}{2\pi i} \int_{|z-z_0|=r} \frac{f(z)}{(z-z_0)^{n+1}}\, dz \right| \le \frac{n!}{2\pi} \int_{|z-z_0|=r} \frac{|f(z)|}{|z-z_0|^{n+1}}\, |dz|$$

$$\le \frac{Mn!}{2\pi r^{n+1}} \int_{|z-z_0|=r} |dz| = \frac{Mn!}{r^n}.$$

　　　　　　　　　　　　　　　　　　　　　　　　　　　■

We now state the first of our three main results.

THEOREM 3: Liouville's Theorem

An entire function cannot be bounded on all of the complex plane unless it is a constant.

PROOF

Suppose $f(z)$ is entire and bounded by M. Then at any point z_0 in the complex plane, Cauchy's estimate implies that $|f'(z_0)| \le M/r$. But r can be made arbitrarily large, so $f'(z_0) = 0$ at all z_0 in the complex plane. Hence, $f(z)$ is constant in the complex plane.　　　　　　　　　　　　　　　　　　■

　　　　Liouville's theorem yields an easy verification of a very important theorem of elementary algebra that is usually stated without proof.

THEOREM 4: Fundamental Theorem of Algebra

Every polynomial of degree greater than zero has a root. In fact, counting multiplicities, a polynomial of degree n has exactly n roots.

PROOF

Suppose $P(z) = a_n z^n + a_{n-1} z^{n-1} + \cdots + a_1 z + a_0$ is not zero for any value z. Then the function $f(z) = 1/P(z)$ is entire. Furthermore, $|f(z)|$ approaches zero as $|z|$ tends to infinity because

$$|f(z)| = \frac{1}{|z|^n |a_n + a_{n-1}/z + \cdots + a_0/z^n|}.$$

Therefore, $|f(z)|$ is bounded for all z. By Liouville's theorem, $f(z)$ and consequently $P(z)$ are constant, contradicting the hypothesis that $n > 0$. Thus $P(z)$ has at least one root.

To show that $P(z)$ actually has n roots (including multiple roots), we observe that by the fundamental theorem of algebra, it has at least one root, say, z_0. Thus,

$$P(z) = P(z) - P(z_0)$$

$$= a_n(z^n - z_0^n) + a_{n-1}(z^{n-1} - z_0^{n-1}) + \cdots + a_1(z - z_0)$$

$$= (z - z_0)Q(z),$$

where $Q(z)$ is a polynomial in z of degree $n - 1$. If $n - 1 > 0$, then $Q(z)$ has a root. Continuing in this fashion, we can extract n factors from $P(z)$; thus $P(z)$ has exactly n roots. ∎

We next prove one of the most useful theorems in the theory of analytic functions.

THEOREM 5: Maximum Principle

If $f(z)$ is analytic and nonconstant in a region Ω, then $|f(z)|$ has no maximum in Ω.

PROOF

Suppose there is a point z_0 in Ω satisfying $|f(z)| \le |f(z_0)|$ for all z in Ω. A region is open by definition so, since $z_0 \in \Omega$, there is a number $r > 0$ such that $|z - z_0| \le r$ lies in Ω. Then by Gauss's mean value theorem,

$$f(z_0) = \frac{1}{2\pi} \int_0^{2\pi} f(z_0 + re^{it})\, dt;$$

that is, the value at the center of the circle equals the integral average of its values on the circle. By assumption, $|f(z_0 + re^{it})| \le |f(z_0)|$, and if strict inequality holds for some value t, it must hold, by the continuity of $|f(z)|$, on an arc of the circle. But then

$$|f(z_0)| \le \frac{1}{2\pi} \int_0^{2\pi} |f(z_0 + re^{it})|\, dt < \frac{1}{2\pi} \int_0^{2\pi} |f(z_0)|\, dt = |f(z_0)|,$$

a contradiction. So $|f(z_0 + re^{it})| = |f(z_0)|$ for $0 \le t \le 2\pi$, and since the procedure holds on all circles $|z - z_0| = s$, $0 < s \le r$, $|f(z)|$ is constant on the disk $|z - z_0| \le r$.

Theorem 15.4.1 states that any point z^* in Ω can be joined to z_0 by a polygonal path C interior to Ω. If ϵ is the distance between C and the boundary of Ω, then $|z - z_n| \le r$ lies in Ω whenever $z_n \in C$ and $r < \epsilon$. The idea now is to draw a sequence of disks $|z - z_n| \le r$, $n = 0, 1, 2, \ldots, N$, such that $0 < |z_n - z_{n-1}| \le r/2$, for $n = 1, 2, \ldots, N$, with $z_N = z^*$. The argument above guarantees that $|f(z)|$ is constant on each of these disks. Hence $|f(z^*)| = |f(z_0)|$, so $|f(z)|$ is constant on Ω. But this contradicts the hypothesis. Thus $|f(z)|$ has no maximum in Ω. ∎

We denote by $\bar{\Omega}$ the set $\Omega \cup \partial\Omega$. Since $\bar{\Omega}$ contains its boundary points, $\bar{\Omega}$ is a closed set. We now state a useful corollary to the maximum principle.

■ **COROLLARY to Maximum Principle**
Let $f(z)$ be analytic in a bounded region Ω and continuous on $\bar{\Omega}$. Then $|f(z)|$ attains its maximum on the boundary of Ω.

PROOF
Since $\bar{\Omega}$ is closed and bounded and $|f(z)|$ is continuous on $\bar{\Omega}$, a theorem of ordinary calculus states that $|f(z)|$ attains a maximum somewhere on $\bar{\Omega}$. By the maximum principle, it cannot be in Ω, so it must be on the boundary of Ω. ∎

Problems 16.6

1. Show that if $f(z)$ is entire and $f(z)$ is not a constant function, then for any positive numbers K and N, there exists a z with $|z| > K$ such that $|f(z)| > N$.

2. Suppose that $f(z)$ is entire and, for sufficiently large z, $|f(z)| < A|z|$ where A is a real number. Show that $f(z) = az + b$ for some numbers a and b. [Hint: Use the result of Problem 16.5.32 on p. 962.]

3. Suppose that $f(z)$ is entire and, for sufficiently large z, $|f(z)| < A|z|^2$. Show that $f(z)$ is a quadratic polynomial.

4. Prove that an entire function satisfying $|f(z)| < A|z|^n$ for some n and all sufficiently large $|z|$ must be a polynomial.

5. Let $f(z)$ be analytic in $|z| < 1$ and satisfy $f(0) = 0$. Define $F(z) = f(z)/z$ for all z in $0 < |z| < 1$. What value can be given to $F(0)$ to make $F(z)$ analytic in $|z| < 1$? [Hint: Apply Cauchy's theorem for derivatives to $f(z)$ on $|z| = r < 1$. Then show that the resulting function, analytic in $|z| < r$, coincides with F on $0 < |z| < r$. Use partial fractions.]

6. Using the results in the problem above and the maximum principle, prove **Schwarz's lemma:** Let $f(z)$ be analytic for $|z| < 1$ and satisfy the conditions $f(0) = 0$ and $|f(z)| \leq 1$. Then $|f(z)| \leq |z|$ and $|f'(0)| \leq 1$, with equality only if $f(z) = e^{i\theta}z$ for some fixed real θ.

7. Show that in Schwarz's lemma, $|f(z)| \leq 1$ for $|z| < 1$ implies that $|f'(0)| \leq 1$ regardless of the value of $f(0)$.

8. Prove the **minimum principle.** Let $f(z)$ be analytic in a bounded region Ω and continuous and nonzero on $\bar{\Omega}$. Then $|f(z)|$ attains its minimum on the boundary of Ω. [Hint: Apply the maximum principle to the function $g(z) = 1/f(z)$.]

9. Give an example to show why the nonzero condition is necessary for the validity of the minimum principle.

* 10. Let $f(z)$ be analytic and nonconstant in $|z| < R$ and denote by $M(r)$ the maximum of $|f(z)|$ on $|z| = r$. Prove that $M(r)$ is strictly increasing for $0 \leq r < R$.

* 11. Prove that if $f(z)$ is analytic and nonconstant in the bounded region Ω, is continuous in $\bar{\Omega}$, and has constant absolute value on the boundary of Ω, then it must have at least one zero in Ω.

** 12. Prove the **three-circles theorem:** If $f(z)$ is analytic in a region containing the annulus $0 < r_1 \leq |z| \leq r_2$ and satisfies the inequalities $|f(z)| \leq M_1$ on $|z| = r_1$ and $|f(z)| \leq M_2$ on $|z| = r_2$, then the maximum of $|f(z)|$ on $|z| = r$, $r_1 \leq r \leq r_2$, is at most equal to
$$M_1^{(\ln(r_2/r))/(\ln(r_2/r_1))} \cdot M_2^{(\ln(r/r_1))/(\ln(r_2/r_1))}.$$

16.7 Harmonic Functions

In Section 12.9 we discussed Laplace's equation
$$u_{xx} + u_{yy} = 0.$$

This equation is of fundamental importance in physics, arising in connection with heat and fluid flows as well as gravitational and electrostatic fields. For example, the steady state temperature u in a heat flow is the real part of an analytic function $w = u + iv$. By the Cauchy-Riemann equations, we have formally

$$u_{xx} = (u_x)_x = (v_y)_x = (v_x)_y = (-u_y)_y = -u_{yy}$$

and Laplace's equation holds for u. Similarly, it holds for v.

Harmonic Function

Any real-valued function $u(x, y)$ with continuous second partial derivatives that satisfies Laplace's equation on a region Ω is said to be **harmonic** on Ω. Thus we have

■ **THEOREM 1**

Let $f(z) = u(z) + iv(z)$ be analytic on the region Ω. Then both of the real-valued functions $u(z)$ and $v(z)$ are harmonic on Ω. ■

EXAMPLE 1

Show that the function $u(x, y) = x^2 - y^2$ is harmonic on the complex plane.

SOLUTION

There are two ways to solve this problem,. First, by direct calculation, the function u has continuous first and second partials:

$$u_x = 2x, \qquad u_y = -2y$$

$$u_{xx} = 2, \qquad u_{xy} = u_{yx} = 0, \qquad u_{yy} = -2.$$

Since

$$u_{xx} + u_{yy} = 2 - 2 = 0,$$

u is harmonic. Second, observe that $u = \mathrm{Re}(z^2)$. Since $f(z) = z^2$ is entire, it follows that its real part u is harmonic in $\mathbb{C}$.

EXAMPLE 2

Show that the function $v(x, y) = e^x \sin y$ is harmonic in the complex plane.

SOLUTION

Observe that v has continuous partials of all orders and

$$v_x = e^x \sin y, \qquad v_y = e^x \cos y,$$

$$v_{xx} = e^x \sin y, \qquad v_{yy} = -e^x \sin y,$$

so $v_{xx} + v_{yy} = 0$ and v is harmonic. Alternatively, note that $v = \mathrm{Im}(e^z)$.

■ **THEOREM 2**

Let $u(z)$ be a real-valued function that is harmonic on a simply connected region Ω. Then the line integral

$$v(z) = \int_C u_x \, dy - u_y \, dx,$$

where C is any pws arc in Ω joining z_0 to z, is harmonic on Ω, and the function $f(z) = u(z) + iv(z)$ is analytic on Ω. We call $v(z)$ a **harmonic conjugate** of $u(z)$.

PROOF

Let $F(z) = u_x - iu_y$. Then, if we write $U = u_x$ and $V = -u_y$, we see that the function $F = U + iV$ satisfies the Cauchy-Riemann equations:

$$\overset{\overset{\textstyle u \text{ is harmonic}}{\downarrow}}{U_x = (u_x)_x = u_{xx} = -u_{yy} = (-u_y)_y = V_y,}$$

$$U_y = (u_x)_y = u_{xy} = u_{yx} = -(-u_{yx}) = -(-u_y)_x = -V_x.$$

Thus $F(z)$ is analytic on Ω. Since Ω is simply connected, we can use the fundamental theorem to define an analytic antiderivative of $F(z)$:

$$f(z) = u(z_0) + \int_C F(z)\,dz = u(z_0) + \int_C (u_x - iu_y)(dx + idy)$$

$$= u(z_0) + \int_C (u_x\,dx + u_y\,dy) + i\int_C (u_x\,dy - u_y\,dx).$$

The first integrand is the total differential of the function $u = u(x, y)$, so that

$$f(z) = u(z_0) + \int_C du + i\int_C u_x\,dy - u_y\,dx$$

$$= u(z) + i\int_C u_x\,dy - u_y\,dx.$$

Thus, we have constructed an analytic function $f(z)$ having $u(z)$ as its real part. This means that the integral

$$v(z) = \int u_x\,dy - u_y\,dx, \tag{1}$$

defined up to an arbitrary constant, is harmonic on Ω. ∎

We will illustrate the use of the integral above in determining harmonic conjugates in the following examples.

EXAMPLE 3

Find the harmonic conjugate on the complex plane of the function

$$u = x^2 - y^2.$$

SOLUTION

The function u was shown to be harmonic on the complex plane in Example 1. Thus the harmonic conjugate $v(x, y)$ must satisfy

$$v(z) = \int u_x \, dy - u_y \, dx = \int 2x \, dy + 2y \, dx$$

$$= 2 \int d(xy) = 2xy + k, \qquad k \text{ constant.}$$

Note that $v(z) = \text{Im}(z^2 + ik)$.

EXAMPLE 4

Is there an analytic function $f = u + iv$ for which

$$u = \frac{x}{x^2 + y^2}?$$

SOLUTION

Observe that u is not defined at the origin. Thus, we shall seek a harmonic conjugate of u on some simply connected subset of $\mathbb{C} - \{0\}$. First we show that u is harmonic:

$$u_x = \frac{y^2 - x^2}{(x^2 + y^2)^2}, \qquad u_y = \frac{-2xy}{(x^2 + y^2)^2}, \qquad \text{and} \qquad u_{xx} = \frac{2x^3 - 6xy^2}{(x^2 + y^2)^3} = -u_{yy}.$$

Thus $u_{xx} + u_{yy} = 0$, and u is harmonic. A harmonic conjugate is given by

$$v(z) = \int u_x \, dy - u_y \, dx = \int \frac{(y^2 - x^2)\, dy + 2xy \, dx}{(x^2 + y^2)^2} = \int v_x \, dx + v_y \, dy.$$

Hence

$$v_x = \frac{2xy}{(x^2 + y^2)^2}, \qquad v = \int v_x \, dx = \frac{-y}{x^2 + y^2} + g(y), \qquad \text{and} \qquad v_y = \frac{y^2 - x^2}{(x^2 + y^2)^2} + g'(y).$$

Thus $g'(y) = 0$, so $g(y)$ is constant and

$$v(z) = \int d\left(\frac{-y}{x^2 + y^2}\right) = \frac{-y}{x^2 + y^2}$$

up to an arbitrary constant. Note that the function

$$f(z) = u + iv = \frac{x - iy}{x^2 + y^2} = \frac{\bar{z}}{|z|^2} = \frac{1}{z}$$

is analytic for $z \neq 0$.

The correspondence between analytic and harmonic functions yields many important properties for the latter.

■ THEOREM 3: Maximum Principle for Harmonic Functions

If $u(z)$ is harmonic and nonconstant in a simply connected region Ω, then $u(z)$ has no maximum or minimum in Ω.

PROOF

Constructing a conjugate harmonic function $v(z)$, we have that $f = u + iv$ is analytic in Ω. Likewise

$$F(z) = e^{f(z)} = e^{u+iv}$$

is analytic in Ω, and $|F(z)| = e^{u(z)}$. Since $F(z)$ is nonzero in Ω, applying the maximum and minimum principles† for analytic functions to F, it follows that e^u has no maximum or minimum in Ω. Since the real function e^u is an increasing function of u, the proof is complete. ■

■ **THEOREM 4: Mean Value Theorem for Harmonic Functions**

If $u(z)$ is harmonic in $|z - z_0| < R$, then

$$u(z_0) = \frac{1}{2\pi} \int_0^{2\pi} u(z_0 + re^{i\theta})\, d\theta, \qquad 0 < r < R.$$

PROOF

Construct a harmonic conjugate $v(z)$ so that $f = u + iv$ is analytic on $|z - z_0| < R$. Gauss's mean value theorem (Theorem 16.6.1) states that

$$f(z_0) = \frac{1}{2\pi} \int_0^{2\pi} f(z_0 + re^{i\theta})\, d\theta, \qquad 0 < r < R.$$

Taking the real part of both sides yields the desired equation. ■

Harmonic functions are important because Laplace's equation arises so often in applications. In Chapter 19 we shall show how the facts about harmonic functions developed in this section can help us to solve a variety of interesting physical problems.

Problems 16.7

In Problems 1–8 show that the given function is harmonic.

1. $u(x, y) = 5$

2. $u(x, y) = 3x + 12y$

3. $u(x, y) = -2x + 4y$

4. $u(x, y) = e^x \cos y$

5. $u(x, y) = 2xy$

6. $u(x, y) = \sin x \sinh y$

7. $u(x, y) = x^3 - 3xy^2$

8. $u(x, y) = e^{x^2 - y^2} \sin(2xy)$

In Problems 9–15 determine whether the given function is the real or imaginary part of an analytic function.

9. $u(x, y) = \sin x \cosh y$

10. $u(x, y) = \sqrt{x^2 + y^2}$

11. $u(x, y) = \dfrac{x}{y}$

12. $u(x, y) = \ln(x^2 + y^2)$

13. $u(x, y) = e^{y/x}$

14. $u(x, y) = e^{2x} \cos 2y$

15. $u(x, y) = e^{3x} \sin 2y$

Find the conjugates of the harmonic functions given in Problems 16–26.

16. $u(x, y) = xy$

17. $v(x, y) = xy$

18. $u(x, y) = -2x + 5y$

19. $v(x, y) = -2x + 5y$

20. $u(x, y) = x^2 - (y - 1)^2$

21. $u(x, y) = \frac{1}{2} \ln(x^2 + y^2)$

22. $v(x, y) = (x + 2)^2 - y^2$

23. $v(x, y) = x^2 - y^2 + 2y$

24. $u(x, y) = \cosh x \cos y$

25. $u(x, y) = \tan^{-1} \dfrac{2xy}{x^2 - y^2}$

---

† The maximum principle is Theorem 16.6.5 and the minimum principle is given in Problem 16.6.8.

26. $u(x, y) = \dfrac{x(x-1)+y^2}{(x-1)^2+y^2}$

27. If u and v are harmonic functions, show that $au + bv$ is also harmonic, where a and b are real constants. Show that uv is harmonic if u and v are conjugate harmonic functions.

28. Show that $\ln|f(z)|$ is harmonic whenever $f(z)$ is analytic and nonzero.

29. Show that the maximum principle holds for multiply connected regions.

* 30. Prove that $\int_0^\pi \ln \sin \theta \, d\theta = -\pi \ln 2$. [*Hint:* Apply the mean value theorem to $\ln|1 + z|$ in $|z| \le r < 1$, and let $r \to 1^-$.]

Review Exercises for Chapter 16

In Exercises 1–5 find a parametrization of the given arc.

1. The line segment joining 2 to $-2i$.

2. The line segment joining $2 - i$ to $3 + 4i$.

3. The semicircle from $-i$ to i.

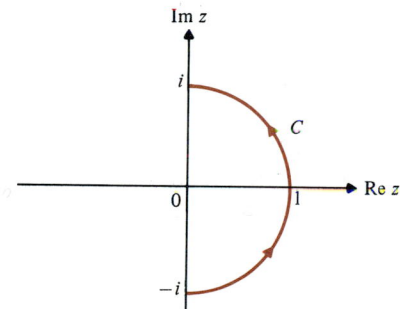

4. The triangle

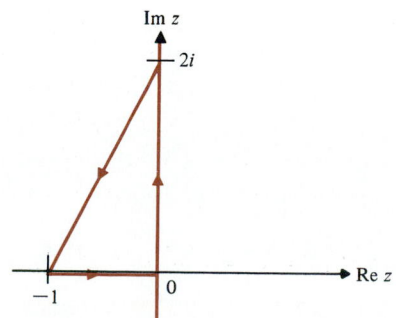

5. The rectangle

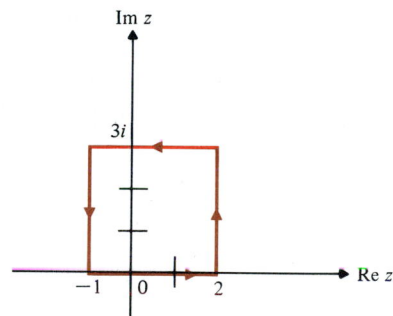

In Exercises 6–40 compute the integral.

6. $\displaystyle\int_2^{2i} z^2 \, dz$

7. $\displaystyle\int_i^1 \frac{3}{z} \, dz$, for any arc in the first quadrant.

8. $\displaystyle\int_{-i}^i \cos z \, dz$

9. $\displaystyle\int_C e^{z^2} \, dz$, where C is the triangle of Exercise 4.

10. $\displaystyle\int_{|z+1|=2} \frac{dz}{z+2}$

11. $\displaystyle\int_C x \, dz$, where C is the line segment from $-i$ to 1

12. $\displaystyle\int_C \bar{z} \, dz$, where C is the semicircle of Exercise 3

13. $\displaystyle\int_{|z|=1} z^2 \, dz$

14. $\displaystyle\int_{|z|=1} \operatorname{Im} \bar{z}\, dz$

15. $\displaystyle\int_{|z|=3} \frac{dz}{z^2+4}$

16. $\displaystyle\int_C \frac{2z+1}{z^2+z-6}\, dz$, where C is the square with edges at 0, 1, $1+i$, and i

17. $\displaystyle\int_{|z|=5} \frac{\sin z}{z}\, dz$

18. $\displaystyle\int_{|z-2|=1} \frac{e^{2z}}{z}\, dz$

19. $\displaystyle\int_{|z-1|=1} \frac{dz}{z^2-1}$

20. $\displaystyle\int_{|z-3|=1} \frac{dz}{z^2-1}$

21. $\displaystyle\int_{|z-1|=3} \frac{dz}{z^2-1}$

22. $\displaystyle\int_0^i (z^4+z)\, dz$

23. $\displaystyle\int_{1-i}^{1+i} (z-5)\, dz$

24. $\displaystyle\int_{-i\pi/2}^{i\pi/2} \cos z\, dz$

25. $\displaystyle\int_0^i z e^{z^2}\, dz$

26. $\displaystyle\int_C \frac{3}{z}\, dz$, where C is the semicircle in Exercise 3

27. $\displaystyle\int_{|z-1|=1} \frac{e^{3z}}{(z+1)^3}\, dz$

28. $\displaystyle\int_{|z+1|=1} \frac{e^{3z}}{(z+1)^3}\, dz$

29. $\displaystyle\int_{|z|=2} \frac{dz}{(z-i)^2}$

30. $\displaystyle\int_{|z|=2} \frac{dz}{(z+i)^2}$

31. $\displaystyle\int_{|z|=5} \frac{\cos 2z}{z}\, dz$

32. $\displaystyle\int_{|z|=1} \frac{2z^2-5z+4}{(z+2)(z+4)}\, dz$

33. $\displaystyle\int_{|z|=3} \frac{2z^2-5z+4}{(z+2)(z+4)}\, dz$

34. $\displaystyle\int_{|z|=5} \frac{2z^2-5z+4}{(z+2)(z+4)}\, dz$

35. $\displaystyle\int_{|z|=5} \frac{2z^2-5z+4}{(z+2)^2}\, dz$

36. $\displaystyle\int_{|z|=5} \frac{2z^2-5z+4}{(z+3)^2}\, dz$

37. $\displaystyle\int_{|z|=2} \frac{e^{5z}}{(z-i)^4}\, dz$

38. $\displaystyle\int_C \frac{\sin 2z}{(z+\pi/2)^6}\, dz$; C is a square centered at $-\pi/2$

39. $\displaystyle\int_C \frac{\cosh z^2}{(z+i)^3}\, dz$; C is a triangle containing $-i$

40. $\displaystyle\int_{|z|=1} \frac{\sinh(z^3+3)\operatorname{Log}(z+5)}{(z-2)^{10}}\, dz$

41. Show that

$$\int_0^T \cos at \cosh bt\, dt = \frac{a \sin aT \cosh bT + b \cos aT \sinh bT}{a^2+b^2},$$

by integrating $f(z) = \cos z$ along the line segment from 0 to $(a+ib)T$.

42. Compute $\displaystyle\int_0^T \sin at \sinh bt\, dt$.

In Exercises 43–48 show that the given function u is harmonic and compute (a) a harmonic conjugate and (b) an analytic function f for which u is its real part.

43. $u(x, y) = -10xy$

44. $u(x, y) = 2(x^2-y^2)$

45. $u(x, y) = xy(x^2-y^2)$

46. $u(x, y) = \cos x \cosh y$

47. $u(x, y) = 3\ln(x^2+y^2)$, $x^2+y^2 > 0$

48. $u(x, y) = x^4 - 6x^2y^2 + y^4$

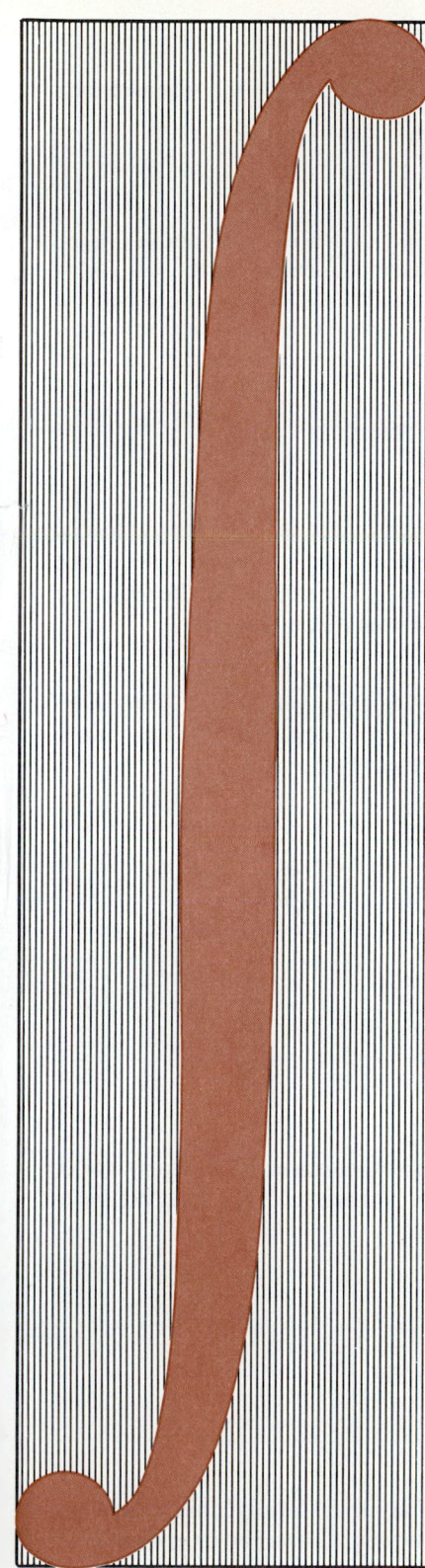

chapter 17

Infinite Series

The study of infinite series is an important part of a calculus course. Series can be used in a variety of applications—for example, to solve certain differential equations (as in Chapter 5) and to estimate functions and integrals numerically. As we shall see, series representations of complex-valued functions are even more important. They play a central role in the theory of functions of a complex variable and have a surprising number of applications.

In this chapter we shall study complex series and, in Chapter 18, we shall provide a number of interesting applications. In Appendix A at the end of this chapter we discuss real infinite sequences and series, a topic usually covered in calculus. If you are unfamiliar with these notions, you should refer to Appendix A or any calculus text.

17.1 Complex Sequences

According to a popular dictionary,† a *sequence* is "the following of one thing after another." In mathematics we could define a sequence intuitively as a succession of numbers that never terminates. The numbers in the sequence are called the *terms* of the sequence. In a sequence there is one term for each positive integer.

EXAMPLE 1

Consider the sequence

$$\underset{\substack{\text{1st} \\ \text{term}}}{\dfrac{1}{2}}, \quad \underset{\substack{\text{2nd} \\ \text{term}}}{\dfrac{1}{4}}, \quad \underset{\substack{\text{3rd} \\ \text{term}}}{\dfrac{1}{8}}, \quad \underset{\substack{\text{4th} \\ \text{term}}}{\dfrac{1}{16}}, \quad \underset{\substack{\text{5th} \\ \text{term}}}{\dfrac{1}{32}}, \dots, \underset{\substack{n\text{th} \\ \text{term}}}{\dfrac{1}{2^n}}, \dots$$

† *The Random House Dictionary,* Ballantine Books, New York, 1978.

We see that there is one term for each positive integer. The terms in this sequence form an infinite set of numbers, which we write as

$$A = \left\{ \frac{1}{2}, \frac{1}{4}, \frac{1}{8}, \ldots, \frac{1}{2^n}, \ldots \right\}. \tag{1}$$

That is, the set A consists of all numbers of the form $1/2^n$, where n is a positive integer. There is another way to describe this set. We define the function f by the rule $f(n) = 1/2^n$, where the domain of f is the set of positive integers. Then the set A is precisely the set of values taken by the function f.

In general, we have the following formal definition.

DEFINITION

Sequence

A **sequence** of numbers is a function whose domain is the set of positive integers. The values taken by the function are called **terms** of the sequence. A sequence of real numbers is called a **real sequence;** however, if any number in the sequence is complex, we say that we have a **sequence of complex numbers.**

Notation. We shall often denote the terms of a sequence by z_n. Thus if the function given in the definition above is f, then $z_n = f(n)$. With this notation, *we can denote the set of values taken by the sequence by $\{z_n\}$.*

EXAMPLE 2

The following are sequences:

$$\textbf{(a)} \ \{z_n\} = \left\{ \frac{1}{n} \right\}, \qquad \textbf{(b)} \ \{z_n\} = \{i\sqrt{n}\}, \qquad \textbf{(c)} \ \{z_n\} = \left\{ \frac{e^{in}}{n!} \right\}.$$

(a) is a real sequence, while (b) and (c) are complex sequences.

DEFINITION

Finite Limit of a Sequence

The sequence $\{z_n\}$ has a **finite limit** L, a complex number, if, for every $\epsilon > 0$, there exists an integer $N > 0$ such that if $n \geq N$, then $|z_n - L| < \epsilon$. We write $z_n \to L$ or

$$\lim_{n \to \infty} z_n = L. \tag{2}$$

The idea behind this definition is sketched in Fig. 1.

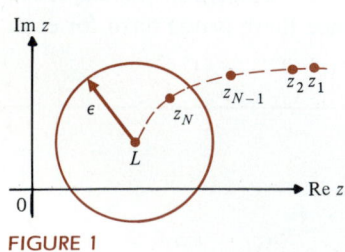

FIGURE 1

DEFINITION

Convergence of a Sequence

If the limit in (2) exists and if L is finite, we say that the sequence **converges** or is **convergent.** Otherwise, we say that the sequence **diverges** or is **divergent.**

EXAMPLE 3

The sequence $\{1/2^n\}$ is convergent since, $\lim_{n\to\infty} 1/2^n = \lim_{n\to\infty}(1/2)^n = 0$.

EXAMPLE 4

The sequence $\{r^n\}$ is divergent for $r > 1$ since $\lim_{n\to\infty} r^n = \infty$ if $r > 1$.

EXAMPLE 5

The sequence $\{(-i)^n\}$ is divergent since the values z_n alternate between ± 1 and $\pm i$ but do not stay close to any fixed number as n becomes large.

EXAMPLE 6

Does the sequence $\{e^{in}/n\}$ converge or diverge?

SOLUTION
Note that $|e^{in}| = 1$, so that $e^{in}/n \to 0$. Hence the sequence converges to zero.

THEOREM 1
Let $\{z_n\} = \{a_n + ib_n\}$ be a complex sequence. Then $\{z_n\}$ converges if and only if each real sequence $\{a_n\}$ and $\{b_n\}$ converges. Moreover, if $a_n \to a$ and $b_n \to b$, then

$$L = \lim_{n\to\infty} z_n = \lim_{n\to\infty}(a_n + ib_n) = a + ib. \tag{3}$$

PROOF
Suppose that $\{a_n\}$ and $\{b_n\}$ both converge and let $\epsilon > 0$ be given. Choose N_1 so that $|a_n - a| < \epsilon/\sqrt{2}$ if $n \geq N_1$ and choose N_2 so that $|b_n - b| < \epsilon/\sqrt{2}$ if $n \geq N_2$. Then, if $n \geq \max\{N_1, N_2\}$, we have

$$|z_n - L| = |(a_n + ib_n) - (a + ib)| = |(a_n - a) + i(b_n - b)|$$

$$= \sqrt{(a_n - a)^2 + (b_n - b)^2} \leq \sqrt{\left(\frac{\epsilon}{\sqrt{2}}\right)^2 + \left(\frac{\epsilon}{\sqrt{2}}\right)^2} = \epsilon.$$

Thus

$$\lim_{n\to\infty}(a_n + ib_n) = a + ib.$$

Conversely, suppose that $z_n \to a + ib$. Choose N so that $|z_n - (a + ib)| < \epsilon$ if $n \geq N$. Then

$$|a_n - a| \leq \sqrt{(a_n - a)^2 + (b_n - b)^2} = |z_n - (a + ib)| < \epsilon$$

so that $a_n \to a$. A similar computation shows that $b_n \to b$. ■

EXAMPLE 7

The sequence $z_n = 1/n + n^{1/n}i \to i$ because $1/n \to 0$ and $n^{1/n} \to 1$.

EXAMPLE 8

The sequence $z_n = e^{in\theta}$ diverges if $0 < \theta < 2\pi$ because $e^{in\theta} = \cos n\theta + i \sin n\theta$ and both of the sequences $\{\cos n\theta\}$ and $\{\sin n\theta\}$ diverge.

Bounded Sequence

The sequence $\{z_n\}$ is **bounded** if there is a real number $M > 0$ such that $|z_n| \le M$ for every $n \ge 1$. If the sequence is not bounded, it is said to be **unbounded.**

Graphically, a sequence $\{z_n\}$ is bounded if each z_n lies in a disk centered at the origin of radius M. This is illustrated in Fig. 2.

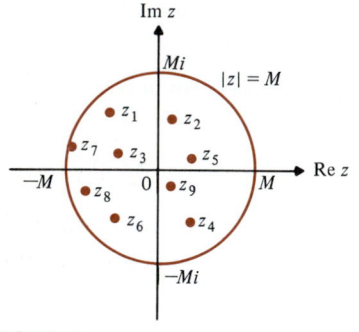

FIGURE 2

EXAMPLE 9

The sequence $\{e^{in\theta}\}$ is bounded because $|e^{in\theta}| = 1$ for every n.

EXAMPLE 10

The sequence $\{(1 - i)^n\}$ is unbounded because $|1 - i| = \sqrt{2}$ and $|(1 - i)^n| = 2^{n/2} \to \infty$ as $n \to \infty$.

■ THEOREM 2

If the sequence $\{z_n\}$ is convergent, then it is bounded. Hence, if a sequence is unbounded, it must be divergent.

PROOF
Suppose $z_n \to L$. For any $\epsilon > 0$ we can find a number N such that $|z_n - L| < \epsilon$ for $n \ge N$. Hence all but finitely many terms of the sequence lie in the disk $|z_n - L| < \epsilon$. If we choose a disk of radius

$$R = \max[|z_1|, |z_2|, \ldots, |z_{N-1}|, |L| + \epsilon],$$

every term of the sequence lies in the disk $|z| \le R$. Thus $\{z_n\}$ is bounded. ■

Cauchy Sequence

A sequence $\{z_n\}$ is a **Cauchy sequence** if, for every $\epsilon > 0$, there is an N such that whenever $n, m > N$, we have

$$|z_n - z_m| < \epsilon. \tag{4}$$

■ THEOREM 3[†]

A sequence is convergent if and only if it is a Cauchy sequence.   ■

† For a proof see R. C. Buck and E. F. Buck, *Advanced Calculus,* 3rd ed., McGraw-Hill, New York, 1978, p. 63.

EXAMPLE 11

Prove that the sequence $\{(i)^n/n\}$ is a Cauchy sequence.

SOLUTION

Observe that if $m > n$, by the triangle inequality

$$|z_n - z_m| = \left| \frac{i^n}{n} - \frac{i^m}{m} \right| \le \frac{1}{n} + \frac{1}{m} < \frac{2}{n}.$$

Thus, if we select $N > 2/\epsilon$ it follows that for $n \ge N$

$$|z_n - z_m| < \frac{2}{n} \le \frac{2}{N} < \epsilon.$$

Problems 17.1

In Problems 1–16 write the first four terms of the given sequence. Determine if the sequence is bounded or unbounded and convergent or divergent. If convergent, find the limit.

1. $\{z_n\} = \left\{ \frac{3}{n} + 5n \right\}$

2. $\{z_n\} = \{e^{n\pi i}\}$

3. $\{z_n\} = \{e^{-n\pi i/6}\}$

4. $\{z_n\} = \{\sin n\pi i\}$

5. $\{z_n\} = \{\cosh n\pi i/2\}$

6. $\{z_n\} = \{\text{Log } in\}$

7. $\{z_n\} = \{(-i)^n/\sqrt{n}\}$

8. $\{z_n\} = \{1 - i^{-n}\}$

9. $\{z_n\} = \{1/(in)^n\}$

10. $\{z_n\} = \{e^{\pi in}/n^2\}$

11. $\{z_n\} = \left\{ \frac{n^2}{n+i} \right\}$

12. $\{z_n\} = \left\{ \frac{n+i}{n^2} \right\}$

13. $\{z_n\} = \left\{ \frac{in}{n+1} \right\}$

14. $\{z_n\} = \{i^n/n\}$

15. $\{z_n\} = \{\sin in/n\}$

16. $\{z_n\} = \left\{ \frac{n+i}{n+1} \right\}$.

17. Show that if a sequence converges, its limit is unique.

18. Suppose $z_n \to 0$ and $|w_n| \le |z_n|$ for each n. Prove that $w_n \to 0$.

Determine if the sequences in Problems 19–22 are Cauchy sequences. If Cauchy, what do they converge to?

19. $\{z_n\} = \{e^{in}/n\}$

20. $\{z_n\} = \{e^{in}\}$

21. $\{z_n\} = \left\{ \frac{1-i^n}{n} \right\}$

22. $\{z_n\} = \{e^{i/n}\}$

17.2 Complex Series

DEFINITION

Let $\{z_n\}$ be a sequence. Then the infinite sum

$$\sum_{k=1}^{\infty} z_k = z_1 + z_2 + z_3 + \cdots \tag{1}$$

Infinite Series
Term
Partial Sum

is called an **infinite series** (or, simply, **series**). Each z_k in (1) is called a **term** of the series. The **partial sums** of the series are given by

$$S_n = \sum_{k=1}^{n} z_k, \qquad n = 1, 2, 3, \ldots \tag{2}$$

Convergent Series

The term S_n is called the **nth partial sum** of the series. If the sequence of partial sums $\{S_n\}$ converges to L, then we say that the series $\sum_{k=1}^{\infty} z_k$ **converges** to L and we write

$$\sum_{k=1}^{\infty} z_k = L. \tag{3}$$

Divergent Series

Otherwise, we say that the series $\sum_{k=1}^{\infty} z_k$ **diverges**.

■ **THEOREM 1: Necessary Condition for Convergence**
If $\sum_{k=1}^{\infty} z_k$ converges, then $z_n \to 0$ as $n \to \infty$.

PROOF
By definition of the nth partial sum, $z_n = S_n - S_{n-1}$. But

$$\lim_{n \to \infty} z_n = \lim_{n \to \infty} S_n - \lim_{n \to \infty} S_{n-1} = \sum_{k=1}^{\infty} z_k - \sum_{k=1}^{\infty} z_k = 0. \qquad ■$$

■ **COROLLARY**
If $z_n \nrightarrow 0$ as $n \to \infty$, then $\sum_{k=1}^{\infty} z_k$ diverges.   ■

EXAMPLE 1

$\sum_{k=1}^{\infty} e^{ik\theta}$ diverges for every real θ because $|e^{ik\theta}| = 1$ so that $e^{ik\theta} \nrightarrow 0$.

■ **THEOREM 2: Geometric Series**
The geometric series $\sum_{k=0}^{\infty} z^k$ converges if $|z| < 1$ and diverges if $|z| \geq 1$. Moreover, if $|z| < 1$, then

$$\sum_{k=0}^{\infty} z^k = 1 + z + z^2 + z^3 + \cdots = \frac{1}{1-z}. \tag{4}$$

PROOF

The partial sum S_n is given by

$$S_n = 1 + z + z^2 + \cdots + z^{n-1} + z^n.$$

Also,

$$zS_n = z + z^2 + \cdots + z^n + z^{n+1}.$$

Subtracting these two expressions, we have

$$S_n - zS_n = (1 - z)S_n = 1 - z^{n+1}.$$

If $z \neq 1$, we obtain, by division,

$$S_n = \frac{1 - z^{n+1}}{1 - z} = \frac{1}{1 - z} - \frac{z^{n+1}}{1 - z}, \tag{5}$$

where the number z is fixed. If $|z| < 1$, then

$$|z^{n+1}| = |z|^{n+1} \to 0 \qquad \text{as} \qquad n \to \infty$$

so that $S_n \to 1/(1 - z)$. If $|z| > 1$, then $|z^{n+1}| = |z|^{n+1} \to \infty$ as $n \to \infty$. If $|z| = 1$, then $z = e^{i\theta}$ and $z^n = e^{in\theta} \not\to 0$ so, by the corollary to Theorem 1, the series diverges for $|z| \geq 1$. ■

EXAMPLE 2

$\sum_{k=0}^{\infty} (e^{i\theta}/2)^k$ converges because $|e^{i\theta}/2| = \frac{1}{2}$.

Furthermore, by (4),

$$\sum_{k=0}^{\infty} \left(\frac{e^{i\theta}}{2}\right)^k = \frac{1}{1 - \frac{1}{2}e^{i\theta}} = \frac{2}{2 - e^{i\theta}} \cdot \frac{2 - e^{-i\theta}}{2 - e^{-i\theta}}$$

$$= \frac{2(2 - e^{-i\theta})}{4 - 2(e^{i\theta} + e^{-i\theta}) + 1} = \frac{2[(2 - \cos\theta) + i\sin\theta]}{5 - 4\cos\theta}.$$

Note that if $\theta = 0$, $e^{i\theta} = 1$ and we obtain $\sum_{k=0}^{\infty}(\frac{1}{2})^k = 2$. Similarly, for $\theta = \pi/2$, $e^{i\pi/2} = i$ and the series becomes

$$\sum_{k=0}^{\infty} \left(\frac{i}{2}\right)^k = \frac{4 + 2i}{5}.$$

Similar computations apply for any value of θ.

DEFINITION

Absolute Convergence

The series $\sum_{k=1}^{\infty} z_k$ **converges absolutely** if $\sum_{k=1}^{\infty} |z_k|$ is a convergent real series.

■ **THEOREM 3**

If $\sum_{k=1}^{\infty} z_k$ converges absolutely, then it converges.

Remark. The converse of this theorem is false. For example $\sum_{k=1}^{\infty} (-1)^k/k$ converges but does not converge absolutely.

PROOF

The series $\sum_{k=1}^{\infty} |z_k|$ converges so the partial sums

$$\{T_n\} = \left\{ \sum_{k=1}^{n} |z_k| \right\}$$

converge. By Theorem 17.1.3, $\{T_n\}$ is a Cauchy sequence. Thus, for $\epsilon > 0$, there is an N such that $n, m > N$ implies that

$$|T_n - T_m| < \epsilon.$$

Now, let $S_n = \sum_{k=1}^{n} z_k$ be the nth partial sum of $\sum_{k=1}^{\infty} z_n$. Then, if N is as above, for $n > m > N$ we have

$$|S_n - S_m| = \left| \sum_{k=1}^{n} z_k - \sum_{k=1}^{m} z_k \right| = |z_{m+1} + z_{m+2} + \cdots + z_n|$$

triangle inequality
$$\downarrow$$
$$\leq |z_{m+1}| + |z_{m+2}| + \cdots + |z_n|$$

$$= \sum_{k=1}^{n} |z_k| - \sum_{k=1}^{m} |z_k| = |T_n - T_m| < \epsilon.$$

Thus $\{S_n\}$ is a Cauchy sequence and so, by Theorem 17.1.3, $\{S_n\}$ converges. By definition this means that $\sum_{k=1}^{\infty} z_k$ converges. ■

Theorem 3 provides us with a powerful tool for proving convergence of a complex series for the following reason: $\sum |z_k|$ is a real series with nonnegative terms and so all the tests given in Appendix A of this chapter for such series can be used. We cite some of these here:

■ **THEOREM 4: Comparison Test**

Let $\sum z_k$ be a complex series and let $\sum a_k$ be a convergent real series with nonnegative terms. If $|z_k| \leq a_k$ for all k, then $\sum z_k$ converges absolutely. ■

EXAMPLE 3

Show that $\sum_{k=1}^{\infty} 1/(k + i)^2$ converges absolutely.

SOLUTION

$|k + i| = \sqrt{k^2 + 1}$ so that

$$\frac{1}{|k + i|^2} = \frac{1}{k^2 + 1} < \frac{1}{k^2}$$

and $\sum_{k=1}^{\infty} 1/k^2$ converges, by the integral test (see Theorem 3 in Appendix A) or Problem 19.

■ THEOREM 5: Ratio Test

Suppose that $z_n \neq 0$ for all n sufficiently large and that $\lim_{n \to \infty} |z_{n+1}/z_n| = L$.

(i) If $L < 1$, then $\sum_{k=1}^{\infty} z_k$ converges absolutely.

(ii) If $L > 1$, then $\sum_{k=1}^{\infty} z_k$ diverges.

(iii) If $L = 1$, then $\sum_{k=1}^{\infty} z_k$ may converge or diverge and the ratio test is inconclusive; some other test must be used.

PROOF

(i) Let $a_n = \left| \dfrac{z_{n+1}}{z_n} \right|$ so that $\lim_{n \to \infty} a_n = L < 1$. Pick ϵ so that $L + \epsilon < 1$. By definition of the limit, there exists a number $N > 0$ such that $a_n < L + \epsilon$ if $n \geq N$. This means that for every $k \geq N$

$$\left| \frac{z_{k+1}}{z_k} \right| < L + \epsilon \quad \text{or} \quad |z_{k+1}| < |z_k|(L + \epsilon).$$

Thus

$$|z_{N+2}| < |z_{N+1}|(L + \epsilon) < |z_N|(L + \epsilon)^2$$

and, for $k \geq N$,

$$|z_k| = |z_{k-N+N}| < |z_N|(L + \epsilon)^{k-N}.$$

Hence

$$\sum_{k=1}^{\infty} |z_k| = \sum_{k=1}^{N} |z_k| + \sum_{k=N+1}^{\infty} |z_k| < \sum_{k=1}^{N} |z_k| + |z_N| \sum_{k=N+1}^{\infty} (L + \epsilon)^{k-N}.$$

The first sum is a finite sum of terms and the second sum is a constant times a converging geometric series (since $L + \epsilon < 1$). Thus $\sum_{k=1}^{\infty} z_k$ converges absolutely.

(ii) For $L > 1$, pick ϵ such that $L - \epsilon > 1$. Then choose M so that $a_k > L - \epsilon$ if $k \geq M$. Then, proceeding as in part (i), we find that for $k \geq M$

$$|z_k| > |z_M|(L - \epsilon)^{k-M}.$$

But $|L - \epsilon| > 1$ so $(L - \epsilon)^{k-M} \to \infty$ as $k \to \infty$. Thus $z_k \not\to 0$ as $k \to \infty$ and, by the corollary to Theorem 1, $\sum_{k=1}^{\infty} z_k$ diverges.

(iii) Observe that both of the series

$$\sum_{k=1}^{\infty} \frac{1}{k} \quad \text{and} \quad \sum_{k=1}^{\infty} \frac{(-1)^k}{k}$$

have

$$\lim_{n \to \infty} \left| \frac{z_{n+1}}{z_n} \right| = \lim_{n \to \infty} \frac{n}{n+1} = 1$$

but the first diverges while the second converges. (See Table 1 in Appendix A.)

■

EXAMPLE 4

Determine whether $\sum\limits_{k=0}^{\infty} \dfrac{(7+2i)^k}{k!}$ converges or diverges.

SOLUTION

$$\left|\frac{z_{n+1}}{z_n}\right| = \frac{|7+2i|^{n+1}}{(n+1)!}\frac{n!}{|7+2i|^n} = \frac{|7+2i|}{n+1} \to 0$$

as $n \to \infty$, so the series converges. As we shall see in Example 17.3.9, this series converges to e^{7+2i}.

■ **THEOREM 6: Root Test**

Suppose that $\lim_{n\to\infty} |z_n|^{1/n} = L$.

(i) If $L < 1$, $\sum_{k=1}^{\infty} z_k$ converges absolutely.

(ii) If $L > 1$, $\sum_{k=1}^{\infty} z_k$ diverges.

(iii) If $L = 1$, the series may either converge or diverge.

PROOF

(i) Choose ϵ such that $L + \epsilon < 1$. Then, there exists an N such that $|z_k|^{1/k} < L + \epsilon$ for $k \geq N$ and

$$\sum_{k=1}^{\infty} |z_k| < \sum_{k=1}^{N-1} |z_k| + \sum_{k=N}^{\infty} (L+\epsilon)^k = \sum_{k=1}^{N-1} |z_k| + (L+\epsilon)^N \sum_{k=0}^{\infty} (L+\epsilon)^k$$

$$= \sum_{k=1}^{N-1} |z_k| + \frac{(L+\epsilon)^N}{1-(L+\epsilon)}$$

since the series in $L + \epsilon$ is geometric. Both terms on the right are finite so the series $\sum z_k$ converges absolutely.

(ii) Here $|z_n| > (L - \epsilon)^k \not\to 0$ as in the proof of the ratio test.

(iii) Observe that

$$\lim_{n\to\infty} \left(\frac{1}{n^a}\right)^{1/n} = \lim_{n\to\infty} e^{-(a \ln n)/n} \overset{\text{L'Hôpital's rule}}{=} \lim_{n\to\infty} e^{-(a/n)/1} = e^0 = 1$$

so that $L = 1$ for both the divergent series $\sum 1/n$ and the convergent series $\sum 1/n^2$ (see Problem 19).

Hence, we see that if $L = 1$, the root test is inconclusive. ■

Problems 17.2

In Problems 1–6 find the sum of the given geometric series.

1. $\sum\limits_{k=0}^{\infty} (\tfrac{1}{2} + \tfrac{1}{2}i)^k$
2. $\sum\limits_{k=0}^{\infty} (-\tfrac{1}{2} + \tfrac{1}{2}i)^k$
3. $\sum\limits_{k=0}^{\infty} (\tfrac{1}{2} - \tfrac{1}{2}i)^k$
4. $\sum\limits_{k=0}^{\infty} (-\tfrac{1}{2} - \tfrac{1}{2}i)^k$

5. $\displaystyle\sum_{k=0}^{\infty}\left(\frac{e^{\pi i/6}}{3}\right)^k$

6. $\displaystyle\sum_{k=0}^{\infty}\left(\frac{2e^{i\theta}}{3}\right)^k$

In Problems 7–17 determine convergence or divergence of the given series.

7. $\displaystyle\sum_{k=1}^{\infty}\frac{(2+8i)^k}{k!}$

8. $\displaystyle\sum_{k=1}^{\infty}\frac{(80+120i)^k}{k!}$

9. $\displaystyle\sum_{k=1}^{\infty}\frac{(-1)^k(50-2i)^{2k+1}}{(2k+1)!}$

10. $\displaystyle\sum_{k=1}^{\infty}\frac{(-1)^k(-2+5i)^{2k}}{(2k)!}$

11. $\displaystyle\sum_{k=1}^{\infty}e^{k\pi i/2}$

12. $\displaystyle\sum_{k=0}^{\infty}e^{-k\pi i/20}$

13. $\displaystyle\sum_{k=0}^{\infty}\frac{3^k e^{50k\pi^2 i}}{4^k}$

14. $\displaystyle\sum_{k=1}^{\infty}\frac{e^{\pi ki}}{k^{3/2}}$

15. $\displaystyle\sum_{k=1}^{\infty}\frac{1}{(k+2i)^2+3}$

16. $\displaystyle\sum_{k=1}^{\infty}\frac{1}{(k+i)^2\,\text{Log}(3+k\pi i)}$

17. $\displaystyle\sum_{k=1}^{\infty}\left(\frac{k}{k+20i}\right)^k$

18. Prove that the series $\sum_{k=1}^{\infty}1/k(k+1)$ converges. [*Hint:* Use the fact that $1/k(k+1)=1/k-1/(k+1)$.]

19. Prove that the series $\sum_{k=1}^{\infty}1/k^2$ converges. [*Hint:* Use Problem 18 and the comparison test.]

Divergent Series (optional). It is sometimes interesting to perform calculations with diverging complex series.† Consider the geometric sum

$$1+z+z^2+\cdots=\frac{1}{1-z}.$$

20. By setting $z=e^{i\theta}$, $0<\theta<2\pi$, show formally that

$$1+e^{i\theta}+e^{2i\theta}+\cdots=\tfrac{1}{2}+\tfrac{1}{2}i\cot\tfrac{1}{2}\theta.$$

21. Using the result of Problem 20, show formally that for $0<\theta<2\pi$

$$\tfrac{1}{2}+\cos\theta+\cos 2\theta+\cdots=0$$

and

$$\sin\theta+\sin 2\theta+\cdots=\tfrac{1}{2}\cot\tfrac{1}{2}\theta.$$

22. Proceeding as in Problem 20, deduce that

$$\sin\theta-\sin 2\theta+\cdots=\tfrac{1}{2}\tan\tfrac{1}{2}\theta,$$

and

$$\tfrac{1}{2}-\cos\theta+\cos 2\theta-\cdots=0$$

for $-\pi<\theta<\pi$.

17.3 Complex Power Series

In the last section we discussed series of complex numbers. Now we discuss series of functions.

DEFINITION

Power Series

(i) A **power series** in z is a series of the form

$$\sum_{k=0}^{\infty}c_k z^k=c_0+c_1 z+c_2 z^2+\cdots+c_n z^n+\cdots. \tag{1}$$

(ii) A **power series** in $(z-z_0)$ is a series of the form

$$\sum_{k=0}^{\infty}c_k(z-z_0)^k=c_0+c_1(z-z_0)+c_2(z-z_0)^2+\cdots+c_n(z-z_0)^n+\cdots, \tag{2}$$

where z_0 is a complex number.

A power series in $(z-z_0)$ can be converted to a power series in w by the change of variables $w=z-z_0$. Then $\sum_{k=0}^{\infty}c_k(z-z_0)^k=\sum_{k=0}^{\infty}c_k w^k$. For example, consider

† The following set of problems was adapted from the fascinating book *Divergent Series* by G. H. Hardy, Clarendon Press, Oxford, 1949, Section 1.2.

$$\sum_{k=0}^{\infty} \frac{(z-3)^k}{k!} \tag{3}$$

If $w = z - 3$, then the power series in $(z - 3)$ given by (3) can be written as

$$\sum_{k=0}^{\infty} \frac{w^k}{k!},$$

which is a power series in w. Thus, it suffices to discuss power series in z.

DEFINITION

Convergence and Divergence of a Power Series

    (i) A power series is said to **converge** at z if the series of complex numbers $\sum_{k=0}^{\infty} c_k z^k$ converges. Otherwise, it is said to **diverge** at z.

    (ii) A power series is said to converge in a set D in the complex plane if it converges for every complex number z in D.

EXAMPLE 1

For what complex numbers does the power series

$$\sum_{k=0}^{\infty} \frac{z^k}{3^k} = 1 + \frac{z}{3} + \frac{z^2}{3^2} + \frac{z^3}{3^3} + \cdots \tag{4}$$

converge?

SOLUTION
The nth term in this series is $z^n/3^n$. Using the root test, we find that

$$L = \lim_{n \to \infty} \left| \frac{z^n}{3^n} \right|^{1/n} = \lim_{n \to \infty} \left| \frac{z}{3} \right| = \left| \frac{z}{3} \right|.$$

Thus, the series converges absolutely if $|z| < 3$ and diverges if $|z| > 3$. If $|z| = 3$, then $z = 3e^{i\theta}$ and $z^k/3^k = e^{ik\theta}$. But $\sum_{k=0}^{\infty} e^{ik\theta}$ is a diverging geometric series from Theorem 17.2.1. Thus $\sum_{k=0}^{\infty} z^k/3^k$ converges if $|z| < 3$ and diverges if $|z| \geq 3$.

DEFINITION

Circle and Radius of Convergence

If the series $\sum_{k=0}^{\infty} c_k(z - z_0)^k$ converges for $|z - z_0| < R$ and diverges for $|z - z_0| > R$, where $0 < R < \infty$, then the circle $|z - z_0| = R$ is called the **circle of convergence**. R is called the **radius of convergence**.

Based on the results of Section 17.2 for the ratio and root tests, we obtain the following theorem:

■ THEOREM 1
Consider the power series $\sum_{k=0}^{\infty} c_k z^k$ and suppose that $\lim_{n \to \infty} |c_{n+1}/c_n|$ exists and is equal to L or that $\lim_{n \to \infty} |c_n|^{1/n}$ exists and is equal to L.

(i) If $L = \infty$, then $R = 0$ and the series converges only at $z = 0$.

(ii) If $L = 0$, then $R = \infty$ and the series converges for all values of z.

(iii) If $0 < L < \infty$, then $R = 1/L$ and the series converges if $|z| < R$ and diverges if $|z| > R$. At $|z| = R$, the series may converge or diverge.

PROOF

By the ratio test, the series $\sum_{n=0}^{\infty} c_n z^n$ converges if

$$1 > \lim_{n \to \infty} \left| \frac{c_{n+1} z^{n+1}}{c_n z^n} \right| = |z| \lim_{n \to \infty} \left| \frac{c_{n+1}}{c_n} \right| = |z| L$$

and diverges if $1 < |z|L$. Hence, convergence occurs (for $L \neq 0, \infty$) if $|z| < 1/L = R$, and divergence occurs when $|z| > 1/L = R$. The proof for the root test is similar and is left to the reader. ∎

EXAMPLE 2

Find the circle of convergence for the series $\sum_{k=0}^{\infty} z^k/(k + 1)$.

SOLUTION

By the ratio test we have

$$\frac{1}{R} = L = \lim_{n \to \infty} \left| \frac{1/(n+1)}{1/n} \right| = \lim_{n \to \infty} \frac{n}{n+1} = 1.$$

Hence $|z| = 1$ is the circle of convergence, and the series converges (absolutely) in $|z| < 1$ and diverges in $|z| > 1$.

If $|z| = 1$ then $z = e^{i\theta}$ and

$$\sum_{k=0}^{\infty} \frac{z^k}{k+1} = \sum_{k=0}^{\infty} \frac{e^{ik\theta}}{k+1}.$$

But $|e^{ik\theta}/k + 1| = 1/(k + 1)$ and $\sum_{k=0}^{\infty} 1/(k + 1)$ is the diverging harmonic series so that, for any value of θ, $\sum_{k=0}^{\infty} e^{ik\theta}/(k + 1)$ does not converge absolutely. Note that if $\theta = 0$ we obtain the diverging harmonic series, while for $\theta = \pi$ we obtain the convergent alternating series

$$\sum_{k=0}^{\infty} \frac{e^{ik\pi}}{k+1} = 1 - \tfrac{1}{2} + \tfrac{1}{3} - \tfrac{1}{4} + \cdots.$$

EXAMPLE 3

Find the circle of convergence of the power series $\sum_{k=0}^{\infty} 2^k z^k/\ln(k + 2)$.

SOLUTION

Here $c_n = 2^n/\ln(n + 2)$ and

$$L = \lim_{n \to \infty} \left| \frac{c_{n+1}}{c_n} \right| = \lim_{n \to \infty} \left| \frac{2^{n+1}/\ln(n+3)}{2^n/\ln(n+2)} \right| = 2 \lim_{n \to \infty} \frac{\ln(n+2)}{\ln(n+3)} = 2.$$

Thus $R = 1/L = \tfrac{1}{2}$ and the circle of convergence is $|z| = \tfrac{1}{2}$.

EXAMPLE 4

Find the radius of convergence of the power series $\sum_{k=0}^{\infty} z^{2k} = 1 + z^2 + z^4 + \cdots$.

SOLUTION

$$\sum_{k=0}^{\infty} z^{2k} = 1 + 0 \cdot z + 1 \cdot z^2 + 0 \cdot z^3 + 1 \cdot z^4 + 0 \cdot z^5 + 1 \cdot z^6 + \cdots.$$

This example illustrates the pitfalls of blindly applying formulas. We have $c_0 = 1$, $c_1 = 0$, $c_2 = 1$, $c_3 = 0, \ldots$. Thus the ratio c_{n+1}/c_n is 0 if n is even and is undefined if n is odd. If we try to apply the root test we have $|c_1| = 0$, $\sqrt{|c_2|} = 1$, $\sqrt[3]{|c_3|} = 0$, $\sqrt[4]{|c_4|} = 1$, $\ldots$, an oscillating sequence that doesn't have a limit. In such cases we may apply Hadamard's theorem which follows. (The proof is omitted.)

■ THEOREM 2: Hadamard's Theorem†

Suppose that the sequence $\{|c_n|^{1/n}\}$ does not converge.

(i) If the sequence is bounded, then the radius of convergence of the power series $\sum_{k=0}^{\infty} c_k(z - z_0)^k$ is given by $R = 1/M$, where M is the largest limit of the convergent subsequences of $\{|c_n|^{1/n}\}$.

(ii) If the sequence is unbounded, then $R = 0$ and the sequence converges only for $z = z_0$. ■

For Example 4, note there are two convergent subsequences: one consisting only of zeros, the other only of ones. Thus $M = 1$ and the radius of convergence for the series in Example 4 is $R = 1/M = 1$.

EXAMPLE 5

Find the radius of convergence of the series

$$1 + 3z + 4z^2 + 27z^3 + \cdots + 2^{2k}z^{2k} + 3^{2k+1}z^{2k+1} + \cdots.$$

SOLUTION

Here $c_{2k} = 2^{2k}$ and $c_{2k+1} = 3^{2k+1}$. Then

$$\left| \frac{c_{2k+1}}{c_{2k}} \right| = \frac{3^{2k+1}}{2^{2k}} = 3 \cdot \left(\frac{3}{2} \right)^{2k} \to \infty, \qquad \text{as} \qquad k \to \infty,$$

but

$$\left| \frac{c_{2k}}{c_{2k-1}} \right| = \frac{2^{2k}}{3^{2k-1}} = 3 \left(\frac{2}{3} \right)^{2k} \to 0 \qquad \text{as} \qquad k \to \infty$$

so the ratio test does not work. We compute $|c_{2k}|^{1/2k} = (2^{2k})^{1/2k} = 2$, $|c_{2k+1}|^{1/(2k+1)} = 3$, so that the largest limit is $M = 3$ and $R = \frac{1}{3}$.

† Jacques Hadamard (1865–1963) was a French mathematician who is best known for his proof of the prime number theorem in 1896.

So far we have only been interested in the convergence or divergence of power series. Assuming that the power series $\sum_{k=0}^{\infty} c_k z^k$ converges in $|z| < R$, where R is the radius of convergence of the series, we know that the power series represents some complex-valued function in the disk $|z| < R$. Is this function analytic in $|z| < R$? If it is, how do we go about differentiating and integrating it? The answer to these questions requires the following concepts.

We say that the series $L = \sum_{k=0}^{\infty} c_k z^k$ converges *at* z_0 if, for every $\epsilon > 0$, there is a number N such that $|\sum_{k=0}^{n} c_k z_0^k - L| < \epsilon$ if $n \geq N$. However, this N may depend on both ϵ and z_0. When it depends on z_0, the type of convergence we have is called **pointwise convergence**. However, if N depends only on ϵ and not on z_0, then we have **uniform convergence**.

Pointwise Convergence

DEFINITION

Uniform Convergence

The series $\sum_{k=0}^{\infty} f_k(z)$ **converges uniformly** to $f(z)$ on a region Ω if, for every $\epsilon > 0$, there exists a number $N > 0$ such that

$$\left| f(z) - \sum_{k=0}^{n} f_k(z) \right| < \epsilon, \qquad \text{if } n > N, \tag{5}$$

for all z in Ω.

Remark. We stress that the N in (5) works for *every* z in Ω.

EXAMPLE 6

Show that $\sum_{k=0}^{\infty} z^k$ converges uniformly to $1/(1 - z)$ in $\Omega = \{z : |z| \leq \frac{1}{2}\}$.

SOLUTION
We know (see Theorem 17.2.2) that

$$\left| \frac{1}{1-z} - \sum_{k=0}^{N} z^k \right| = \left| \sum_{k=0}^{\infty} z^k - \sum_{k=0}^{N} z^k \right|$$

$$= \left| \sum_{k=N+1}^{\infty} z^k \right| \leq \sum_{k=N+1}^{\infty} |z|^k$$

$$\leq \left(\frac{1}{2}\right)^{N+1} \sum_{k=0}^{\infty} \left(\frac{1}{2}\right)^k = \frac{(1/2)^{N+1}}{1 - \frac{1}{2}}$$

$$= \left(\frac{1}{2}\right)^N,$$

since $|z| \leq \frac{1}{2}$. For any $\epsilon > 0$, we can always find an integer N such that $(\frac{1}{2})^N < \epsilon$. This N depends only on ϵ, so the convergence is uniform in Ω.

Remark. The definition of uniform convergence of a function is always used together with a region Ω. In Example 6 we used the fact that $|z| \le \frac{1}{2}$ in Ω to prove the uniformity. Without this requirement we would not have been able to say that

$$\sum_{k=N+1}^{\infty} |z|^k \le \sum_{k=N+1}^{\infty} \left(\frac{1}{2}\right)^k = \left(\frac{1}{2}\right)^{N+1} \sum_{k=0}^{\infty} \left(\frac{1}{2}\right)^k.$$

For example, if we use the region $\Omega^* = \{z : |z| < 1\}$ we will not be able to prove uniform convergence because

$$\sum_{k=N+1}^{\infty} |z|^k = \frac{|z|^{N+1}}{1 - |z|} \to \infty \qquad \text{as} \qquad |z| \to 1^-.$$

The importance of uniform convergence is provided by the following result, whose proof is given in Appendix B at the end of this chapter.

■ **THEOREM 3: Weierstrass's Theorem†**

The sum of a uniformly convergent series of *analytic* functions in Ω is analytic and may be differentiated or integrated term by term. That is, if $\sum_{k=0}^{\infty} f_k(z)$ converges uniformly to $f(z)$ on a region Ω, and each $f_k(z)$ is analytic, then

$$f'(z) = \sum_{k=0}^{\infty} f'_k(z) \tag{6}$$

for any z in Ω. Moreover, for any pws closed curve C in Ω,

$$\int_C f(z)\,dz = \sum_{k=0}^{\infty} \int_C f_k(z)\,dz. \tag{7}$$

■

Although this theorem is stated in a very general way, we are not interested in series of arbitrary analytic functions $f_k(z)$. The functions we want to consider are $f_k(z) = c_k(z - z_0)^k$, so that the series is the power series

$$\sum_{k=0}^{\infty} c_k(z - z_0)^k.$$

In this case, the following result has important consequences.

† See the accompanying biographical sketch.

KARL THEODOR WILHELM WEIERSTRASS
(1815–1897)

Courtesy of the Granger Collection

It is generally thought that a potential mathematician of the first rank, in order to succeed, must start serious mathematical studies at an early age and must not be dulled by an inordinate amount of elementary teaching. Karl Theodor Wilhelm Weierstrass, who was born in Ostenfelde, Germany, in 1815, is an outstanding exception to these two general rules. A misdirected youth spent in studying the law and finance gave Weierstrass a late start in mathematics, and it was not until he was forty that he finally emancipated himself from secondary teaching by obtaining an instructorship at the University of Berlin, and another eight years passed before, in 1864, he was awarded a full professorship at the university and could finally devote all his time to advanced mathematics. Weierstrass never regretted the years he spent in elementary teaching, and he later carried over his remarkable pedagogical abilities into his university work, becoming probably the greatest teacher of advanced mathematics that the world has yet known.

Weierstrass wrote a number of early papers on hyperelliptic integrals, Abelian functions, and algebraic differential equations, but his widest known contribution to mathematics is his construction of the theory of complex functions by means of power series. This, in a sense, was an extension to the complex plane of the idea earlier attempted by Lagrange, but Weierstrass carried it through with absolute rigor. Weierstrass showed particular interest in entire functions and in functions defined by infinite products. He discovered uniform convergence and started the so-called arithmetization of analysis, or the reduction of the principles of analysis to real number concepts. A large number of his mathematical findings were introduced to the mathematical world, not through publication by him, but through notes taken of his lectures. He was very generous in allowing students and others to carry out, and receive credit for, investigations of many of his mathematical gems.

In algebra, Weierstrass was perhaps the first to give a so-called postulational definition of a determinant. He defined the determinant of a square matrix A as a polynomial in the elements of A which is homogeneous and linear in the elements of each row of A, which merely changes sign when two rows of A are permuted, and which reduces to 1 when A is the corresponding identity matrix. He also contributed to the theory of bilinear and quadratic forms.

Weierstrass was a very influential teacher, and his meticulously prepared lectures established an ideal for many future mathematicians; "Weierstrassian rigor" became synonymous with "extremely careful reasoning." Weierstrass was "the mathematical conscience par excellence," and he became known as "the father of modern analysis." He died in Berlin in 1897, just one hundred years after the first publication, in 1797 by Lagrange, of an attempt to rigorize the calculus.

■ **THEOREM 4: Abel's Theorem†**

Let $R > 0$ be the radius of convergence of the power series $\sum_{k=0}^{\infty} c_k(z - z_0)^k$. Then the series converges uniformly in any disk $|z - z_0| \le r$ with $r < R$.

PROOF

Choose s with $r < s < R$. By Theorem 2, $R = 1/M$ where M is the largest limit of the subsequences of the sequence $\{|c_n|^{1/n}\}$. Since $s < R$, $1/s > 1/R = M$. Choose δ such that $1/s > M + \delta$. Then, by the definition of the largest limit of the subsequences, there exists an N such that $|c_n|^{1/n} < M + \delta < 1/s$ for every $n \ge N$ and, therefore,

$$|c_n| < \frac{1}{s^n} \qquad \text{for} \qquad n \ge N$$

Hence, for $|z - z_0| \le r$,

$$\sum_{k=N}^{\infty} |c_k|\,|z - z_0|^k < \sum_{k=N}^{\infty} \left(\frac{r}{s}\right)^k = \left(\frac{r}{s}\right)^N \sum_{k=0}^{\infty} \left(\frac{r}{s}\right)^k = \frac{(r/s)^N}{1 - (r/s)}$$

since $r/s < 1$. This last term approaches 0 as $N \to \infty$. Thus the sum $\sum_{k=N}^{\infty} |c_k|\,|z - z_0|^k$ can be made less than ε by choosing N large enough. This N depends only on ε and not on the particular value of z in the disk $|z - z_0| \le r$. Thus, the convergence is uniform.   ■

If we apply Theorem 4 in conjunction with Weierstrass's Theorem 3 we obtain the following:

■ **THEOREM 5**

Let the series

$$f(z) = \sum_{k=0}^{\infty} c_k(z - z_0)^k \tag{8}$$

have a positive radius of convergence (i.e., $R > 0$). Then

   (i) $f(z)$ is analytic for $|z - z_0| < R$.

   (ii)
$$f'(z) = \sum_{k=1}^{\infty} kc_k(z - z_0)^{k-1} \tag{9}$$

and the power series (9) has radius of convergence R. Indeed,

† Niels Henrik Abel (1802–1829) was a Norwegian mathematician who published some brilliant results in mathematical analysis in a career curtailed by tragedy. For an account of his last year, read the biography of Cauchy on p. 24.

$$f''(z) = \sum_{k=2}^{\infty} k(k-1)c_k(z-z_0)^{k-2}, \tag{10}$$

$$f'''(z) = \sum_{k=3}^{\infty} k(k-1)(k-2)c_k(z-z_0)^{k-3}, \tag{11}$$

$$\vdots$$

$$f^{(n)}(z) = \sum_{k=n}^{\infty} k(k-1)(k-2)\cdots(k-n+1)c_k(z-z_0)^{k-n}. \tag{12}$$

(iii)

$$g(z) = \sum_{k=0}^{\infty} \frac{c_k(z-z_0)^{k+1}}{k+1} + K \tag{13}$$

is an antiderivative for $f(z)$. That is, $g'(z) = f(z)$ for every z with $|z| < R$. Moreover, the radius of convergence for the series (13) is R.

PROOF
(ii) Being the derivative of an analytic function, $f'(z)$ is analytic and, by Theorem 3, can be differentiated term by term. ∎

EXAMPLE 7

We know, from Theorem 17.2.2, that

$$\frac{1}{1-z} = \sum_{k=0}^{\infty} z^k = 1 + z + z^2 + \cdots, \qquad |z| < 1.$$

Thus

$$\left(\frac{1}{1-z}\right)' = \frac{1}{(1-z)^2} = \sum_{k=1}^{\infty} kz^{k-1} = 1 + 2z + 3z^2 + \cdots, \qquad |z| < 1.$$

Both series have radius of convergence 1. Differentiating once more, we have

$$\left(\frac{1}{1-z}\right)'' = \frac{2}{(1-z)^3} = \sum_{k=2}^{\infty} k(k-1)z^{k-2} = 2 + 6z + 12z^2 + \cdots, \qquad |z| < 1,$$

also with radius of convergence 1.

A power series can also be integrated term by term, so we can also obtain results with this tool.

EXAMPLE 8

Since $\dfrac{d}{dz}\log(1-z) = \dfrac{-1}{1-z}$, we can integrate the series in Example 7 term by term to obtain

$$\log(1-z) = K - z - \frac{z^2}{2} - \frac{z^3}{3} - \cdots, \qquad |z| < 1.$$

The choice of K will depend on the branch of the logarithm that we select. For example,

$$\text{Log}(1-z) = -z - \frac{z^2}{2} - \frac{z^3}{3} - \cdots, \qquad |z| < 1,$$

since Log $1 = 0$ (set $z = 0$).

EXAMPLE 9

The series

$$f(z) = 1 + z + \frac{z^2}{2!} + \frac{z^3}{3!} + \cdots = \sum_{k=0}^{\infty} \frac{z^k}{k!}$$

converges for every complex number z since $L = \lim_{n \to \infty} \dfrac{|z|^{n+1}/(n+1)!}{|z|^n/n!} = \lim_{n \to \infty} \dfrac{|z|}{n+1} = 0$. But

$$f'(z) = \frac{d}{dz} 1 + \frac{d}{dz} z + \frac{d}{dz} \frac{z^2}{2!} + \cdots = 1 + z + \frac{z^2}{2!} + \cdots = f(z).$$

Also, $f(0) = 1$. Thus $f(z)$ satisfies the initial value problem

$$f'(z) = f(z), \qquad f(0) = 1.$$

From the discussion in Section 15.7, we know that the unique solution to this initial value problem is $f(z) = e^z$. Thus

$$e^z = 1 + z + \frac{z^2}{2!} + \frac{z^3}{3!} + \cdots = \sum_{k=0}^{\infty} \frac{z^k}{k!}. \tag{14}$$

Remark. In the last example we found a power series representation for e^z. Could there be another one? The answer is no, as shown by the following theorem.

■ **THEOREM 6: Uniqueness Theorem for Power Series**

Suppose that $\sum_{k=0}^{\infty} c_k z^k$ and $\sum_{k=0}^{\infty} d_k z^k$ have the same radius of convergence $R > 0$ and converge to the same value for each z with $|z| < R$. Then the series are identical. That is, $c_k = d_k$ for $k = 0, 1, 2, \ldots$.

PROOF
Since

$$c_0 + c_1 z + c_2 z^2 + \cdots = d_0 + d_1 z + d_2 z^2 + \cdots \tag{15}$$

we may set $z = 0$ to see that $c_0 = d_0$. Differentiating both sides of (15), we obtain

$$c_1 + 2c_2 z + 3c_3 z^2 + \cdots = d_1 + 2d_2 z + 3d_3 z^2 + \cdots. \tag{16}$$

Setting $z = 0$ in (16) yields $c_1 = d_1$. Repeating the same process, we see that $c_n = d_n$ for every n. ∎

Remark. It is important to note that although term-by-term differentiation is valid inside the circle of convergence, it need not be at points on the circle of convergence where the series converges. The following example illustrates this anomaly.

EXAMPLE 10

Consider the power series

$$f(z) = \sum_{k=1}^{\infty} \frac{z^k}{k^2}. \tag{17}$$

Using the ratio test, we have

$$\frac{1}{R} = \lim_{n \to \infty} \left| \frac{c_{n+1}}{c_n} \right| = \lim_{n \to \infty} \left| \frac{n}{n+1} \right|^2 = 1,$$

so that $f(z)$ is convergent and analytic for $|z| < 1$. Notice that

$$f(1) = \sum_{k=1}^{\infty} \frac{1}{k^2}$$

is convergent (see Problem 17.2.19). However, term-by-term differentiation

$$f'(z) = \sum_{k=1}^{\infty} \frac{z^{k-1}}{k}$$

is not valid at $z = 1$, because if it were

$$f'(1) = \sum_{k=1}^{\infty} \frac{1}{k} \tag{18}$$

would exist. However, the series in (18) is the divergent harmonic series.

Problems 17.3

1. Find a power series expansion for $1/(1 + z^2)$ valid for $|z| < 1$.

2. Recall from Problem 15.9.36 on p. 929 that the derivative of $\tan^{-1} z$ is $1/(1 + z^2)$. Use this fact and the result of Problem 1 to obtain a power series expansion for $\tan^{-1} z$.

3. Use the result of Problem 2 to obtain an estimate of π that is accurate to two decimal places. [*Hint:* Remember that $\tan^{-1} 1 = \pi/4$.]

4. Use the result of Example 8 to estimate the following quantities to two decimal places of accuracy.

(a) Log 0.5

(b) Log 1.6

(c) $\text{Log}(1 + i/2)$

(d) $\text{Log}(\frac{1}{3} - \frac{1}{4}i)$.

In Problems 5–24 find the radius of convergence for each power series.

5. $\sum\limits_{k=0}^{\infty} \dfrac{z^k}{6^k}$

6. $\sum\limits_{k=0}^{\infty} \dfrac{z^k}{(1+i)^k}$

7. $\sum\limits_{k=0}^{\infty} (4z)^k$

8. $\sum\limits_{k=0}^{\infty} \dfrac{(z+1)^k}{k^2+1}$

9. $\sum\limits_{k=0}^{\infty} \dfrac{(z-i)^k}{(ik)^3+3}$

10. $\sum\limits_{k=0}^{\infty} \dfrac{(z+20i)^k}{k!}$

11. $\sum\limits_{k=2}^{\infty} \dfrac{z^k}{k \ln k}$

12. $\sum\limits_{k=0}^{\infty} z^{5k}$

13. $\sum\limits_{k=1}^{\infty} \dfrac{z^k}{k^k}$

14. $\sum\limits_{k=1}^{\infty} \dfrac{kz^k}{\ln(k+1)}$

15. $\sum\limits_{k=2}^{\infty} \dfrac{(z+1-i)^k}{(\ln k)^k}$

16. $\sum\limits_{k=1}^{\infty} \dfrac{(2i)^k z^k}{k^4}$

17. $\sum\limits_{k=0}^{\infty} \dfrac{(2z+5-2i)^k}{(4i)^{2k}}$

18. $\sum\limits_{k=1}^{\infty} \left(\dfrac{k}{3}\right)^k z^k$

19. $\sum\limits_{k=1}^{\infty} k^k(z+1)^k$

20. $\sum\limits_{k=1}^{\infty} \dfrac{\ln k(z+i)^k}{k-i}$

* 21. $\sum\limits_{k=1}^{\infty} \dfrac{k^k}{k!} z^k$

* 22. $\sum\limits_{k=1}^{\infty} \dfrac{k!}{k^k} z^k$

23. $\sum\limits_{k=0}^{\infty} c_k z^k$ where $c_k = \begin{cases} 2, & \text{if } k \text{ is even} \\ 3, & \text{if } k \text{ is odd} \end{cases}$

24. $1 + 2z + 3z^2 + z^3 + 2z^4 + 3z^5 + z^6 + 2z^7 + 3z^8 + \cdots$

In Problems 25–30 use the series obtained in Example 7 and substitution to obtain a series for the given function. Determine the radius of convergence for this new series.

25. $f(z) = \dfrac{1}{1+z} + \dfrac{1}{1-z}$

26. $f(z) = \dfrac{1}{1+z} + \dfrac{1}{2+z}$

27. $f(z) = \dfrac{1}{(1+z)^2}$

28. $f(z) = \dfrac{z}{1-z^2}$

29. $f(z) = \dfrac{1}{(z-1)(z+2)}$

30. $f(z) = \dfrac{1}{(1+z)^2} + \dfrac{1}{1-z}$

In Problems 31–42 prove or disprove that the given series converges uniformly in the given region.

31. $\sum\limits_{k=0}^{\infty} z^{3k}, \quad |z| \le \frac{3}{4}$

32. $\sum\limits_{k=1}^{\infty} \left(\dfrac{z}{k}\right)^k, \quad |z| \le 1$

33. $\sum\limits_{k=1}^{\infty} \dfrac{z^k}{(6i)^k}, \quad |z| \le 5$

34. $\sum\limits_{k=0}^{\infty} \dfrac{(z+1)^k}{3^k}, \quad |z+1| \le \frac{1}{4}$

35. $\sum\limits_{k=1}^{\infty} \dfrac{z^k}{k^2}, \quad |z| \le \frac{7}{8}$

36. $\sum\limits_{k=1}^{\infty} \left(\dfrac{z}{k^2}\right)^k, \quad |z| \le 1000$

37. $\sum\limits_{k=2}^{\infty} \dfrac{z^k}{(\ln k)^k}, \quad |z| \le 1$

38. $\sum\limits_{k=1}^{\infty} \dfrac{z^k}{k^2+|z|}, \quad |z| \le 1$

39. $\sum\limits_{k=0}^{\infty} e^{-kz}, \quad \text{Re } z \ge \delta > 0$

40. $\sum\limits_{k=1}^{\infty} \dfrac{[1+(-1)^k]}{k} z^k, \quad |z| \le 0.95$

41. $\sum\limits_{k=1}^{\infty} z^k e^{ik|z|}, \quad |z| \le 0.999$

42. $\sum\limits_{k=1}^{\infty} \dfrac{\sin kz}{k}; \quad |z| \le 0.5$

43. Define the complex Bessel function $J_0(z)$ by

$$J_0(z) = \sum\limits_{k=0}^{\infty} [(-1)^k/(k!)^2](z/2)^{2k}.$$

(a) What is the radius of convergence of this series?

(b) Show that $J_0(z)$ satisfies the differential equation

$$z^2 J_0''(z) + z J_0'(z) + z^2 J_0(z) = 0.$$

In Problems 44–46, let $g(t)$ be a continuous complex-valued function on $0 \le t \le 1$ and define

$$f(z) = \int_0^1 g(t) e^{zt} \, dt.$$

44. Show that, for fixed z, the series

$$g(t) e^{zt} = \sum\limits_{n=0}^{\infty} \dfrac{z^n t^n g(t)}{n!}$$

converges uniformly on $0 \le t \le 1$.

45. Prove that $f(z)$ is entire. [*Hint:* Use Problem 44 to interchange sum and integral.]

46. Prove that $f'(z) = \int_0^1 tg(t) e^{zt} \, dt$.

47. Let $g(t)$ be continuous on $0 \le t \le 1$ and define

$$f(z) = \int_0^1 g(t) \sin(zt)\, dt.$$

Prove that $f(z)$ is entire and find its derivative.

48. Let $g(t)$ be continuous on $0 \le t \le 1$ and define

$$f(z) = \int_0^1 \frac{g(t)}{1 - zt}\, dt, \qquad |z| < 1.$$

Prove that $f(z)$ is analytic in $|z| < 1$ and find its derivative.

17.4 Taylor Series

Taylor series† are first encountered at the end of a one-variable calculus course. The basic result is given by the following theorem:

■ **THEOREM 1**

If the function $f(x)$ can be represented by an infinite series $f(x) = \sum_{k=0}^{\infty} a_k x^k$, then f has derivatives of all orders at 0 and $a_k = f^{(k)}(0)/k!$. More generally, if $f(x) = \sum_{k=0}^{\infty} a_k (x - x_0)^k$, then f has derivatives of all orders at x_0 and $a_k = f^{(k)}(x_0)/k!$. In this case

$$f(x) = f(x_0) + f'(x_0)(x - x_0) + \frac{f''(x_0)(x - x_0)^2}{2!} + \cdots$$

$$= \sum_{k=0}^{\infty} \frac{f^{(k)}(x_0)(x - x_0)^k}{k!}. \tag{1}$$

■

This theorem is quite useful but it presents one severe difficulty. The theorem says that *if* a function can be represented as a power series, then the function is infinitely differentiable and the series takes the form (1). For example, $f(x) = e^x$ is infinitely differentiable, with $f^{(k)}(x) = e^x$ for every k, so that for $x_0 = 0$, $f^{(k)}(0) = 1$, and the series (1) takes the form

$$e^x = 1 + x + \frac{x^2}{2!} + \frac{x^3}{3!} + \cdots.$$

However, Theorem 1 does *not* prove that this expansion is correct. It only states that *if* e^x can be written as a power series around $x_0 = 0$, then the series must take the form given above. To prove that e^x can be written as a power series

---

† Named after the English mathematician Brook Taylor (1685–1731), a pupil of Isaac Newton, who published what we now call *Taylor's formula* in *Methodus Incrementorum* in 1715. There was a considerable controversy over whether Taylor's discovery was, in fact, a plagiarism of an earlier result of the Swiss mathematician Johann Bernoulli (1667–1748).

some other argument must be used. One such argument was given in Example 17.3.9.

To see the difficulty more clearly, consider the function

$$f(x) = \begin{cases} e^{-1/x^2}, & \text{if } x \neq 0, \\ 0, & \text{if } x = 0. \end{cases}$$

By using L'Hôpital's rule repeatedly, it is not difficult to show that $\lim_{x \to 0} f^{(k)}(x) = 0$ for $k = 1, 2, 3, \ldots$. Thus, f is infinitely differentiable and $f^{(k)}(0) = 0$ for $k = 1, 2, 3, \ldots$. Evidently, then, f *cannot* be written as a convergent power series around $x_0 = 0$. For if it could, we would have, by Theorem 1,

$$f(x) = f(0) + f'(0)x + \frac{f''(0)x^2}{2!} + \frac{f'''(0)x^3}{3!} + \cdots$$

$$= 0 + 0 + 0 + \cdots = 0,$$

which is false.

The beauty of complex Taylor series is that this difficulty does not arise. That is, if f is analytic at z_0, then $f(z)$ can be written as a convergent power series about $z = z_0$ and the series has the form (1) (with z's instead of x's).

■ **THEOREM 2: Taylor's Theorem**

Let $f(z)$ be analytic in the region Ω containing the point z_0. Then the representation

$$f(z) = f(z_0) + \frac{f'(z_0)}{1!}(z - z_0) + \cdots + \frac{f^{(n)}(z_0)}{n!}(z - z_0)^n + \cdots$$

$$= \sum_{k=0}^{\infty} \frac{f^{(k)}(z_0)(z - z_0)^k}{k!} \tag{2}$$

holds in all disks $|z - z_0| < r$ contained in Ω. Series (2) is called the **Taylor series representation** for $f(z)$.

PROOF

Let z be any point of the open disk $|\zeta - z_0| < r$ contained, with its boundary, in Ω and use the Cauchy integral formula (p. 957) to express $f(z)$ as an integral.

$$f(z) = \frac{1}{2\pi i} \int_{|\zeta - z_0| = r} \frac{f(\zeta)}{\zeta - z} d\zeta. \tag{3}$$

Now

$$\zeta - z = (\zeta - z_0) \left[1 - \frac{z - z_0}{\zeta - z_0} \right] \quad \text{and} \quad \left| \frac{z - z_0}{\zeta - z_0} \right| < 1. \tag{4}$$

Recall that

$$\frac{1}{1-r} = 1 + r + r^2 + \cdots + r^{n-1} + \frac{r^n}{1-r}. \tag{5}$$

Hence, using (4) and (5) in (3), we have

$$f(z) = \frac{1}{2\pi i} \int_{|\zeta - z_0| = r} \frac{f(\zeta)}{\zeta - z_0} \frac{1}{[1 - (z - z_0)/(\zeta - z_0)]} d\zeta$$

$$= \frac{1}{2\pi i} \int_{|\zeta - z_0| = r} \frac{f(\zeta)}{\zeta - z_0} \left[1 + \frac{z - z_0}{\zeta - z_0} + \cdots + \left(\frac{z - z_0}{\zeta - z_0} \right)^{n-1} + Q_n \right] d\zeta, \tag{6}$$

where

$$Q_n(z) = \frac{(z - z_0)^n}{(\zeta - z)(\zeta - z_0)^{n-1}}.$$

But by Cauchy's formula for derivatives (Theorem 16.5.2 on p. 960) we have

$$\frac{1}{2\pi i} \int_{|\zeta - z_0| = r} \frac{f(\zeta)(z - z_0)^k}{(\zeta - z_0)^{k+1}} d\zeta = \frac{(z - z_0)^k}{k!} \frac{k!}{2\pi i} \int_{|\zeta - z_0| = r} \frac{f(\zeta)}{(\zeta - z_0)^{k+1}} d\zeta$$

$$= \frac{f^{(k)}(z_0)(z - z_0)^k}{k!} \tag{7}$$

for $k = 0, 1, \ldots, n - 1$, so that (6) becomes

$$f(z) = f(z_0) + \frac{f'(z_0)}{1!}(z - z_0) + \cdots + \frac{f^{(n-1)}(z_0)}{(n-1)!}(z - z_0)^{n-1} + R_n, \tag{8}$$

where

$$R_n = \frac{(z - z_0)^n}{2\pi i} \int_{|\zeta - z_0| = r} \frac{f(\zeta)}{(\zeta - z)(\zeta - z_0)^n} d\zeta. \tag{9}$$

Let $\rho = |z - z_0| < r$. Then

Problem (15.2.28),
p. 887
↓

$$|z - \zeta| = |z - z_0 + z_0 - \zeta| \geq |z_0 - \zeta| - |z - z_0| = r - \rho$$

for all ζ on $|\zeta - z_0| = r$. Thus, $1/|\zeta - z| \leq 1/(r - \rho)$.

Let M = maximum of $|f(\zeta)|$ for $|\zeta - z_0| = r$ (see Fig. 1). Then, from (9)

$$|R_n| \leq \frac{|z - z_0|^n}{2\pi} \int_{|\zeta - z_0| = r} \frac{|f(\zeta)|}{|\zeta - z||\zeta - z_0|^n} |d\zeta|$$

$$\leq \frac{\rho^n}{2\pi} \int_{|\zeta - z_0| = r} \frac{M}{(r - \rho)r^n} |d\zeta| = \frac{\rho^n}{2\pi} \frac{2\pi r M}{(r - \rho)r^n} = \frac{rM}{r - \rho} \left(\frac{\rho}{r} \right)^n.$$

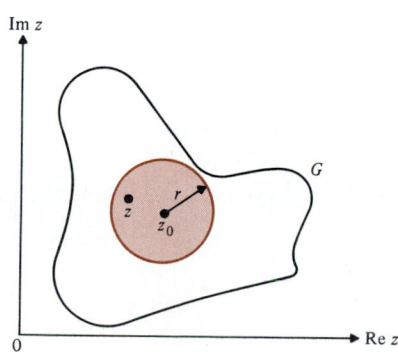

FIGURE 1 The disk $|\zeta - z_0| < r$ contained in Ω

Since $\rho/r < 1$, $|R_n| \to 0$ as $n \to \infty$. Therefore, from (8),

$$f(z) = \lim_{n \to \infty} \left[\sum_{k=0}^{n-1} \frac{f^{(k)}(z_0)(z - z_0)^k}{k!} + R_n \right]$$

$$= \sum_{k=0}^{\infty} \frac{f^{(k)}(z_0)(z - z_0)^k}{k!}.$$

■

Maclaurin Series

DEFINITION

When $z_0 = 0$, series (2) is called the **Maclaurin† series representation** for $f(z)$.

EXAMPLE 1

Since e^z is entire, we know that it has a Maclaurin series, which we have already computed (in Example 17.3.9):

$$e^z = 1 + z + \frac{z^2}{2!} + \frac{z^3}{3!} + \cdots. \tag{10}$$

Series (10) is valid for all z.

EXAMPLE 2

Find the Maclaurin series for $\cos z$.

SOLUTION
We have

$$f(z) = \cos z, \qquad f(0) = 1,$$
$$f'(z) = -\sin z, \qquad f'(0) = 0,$$
$$f''(z) = -\cos z, \qquad f''(0) = -1,$$
$$f'''(z) = \sin z, \qquad f'''(0) = 0,$$
$$f^{(4)}(z) = \cos z, \qquad f^{(4)}(0) = 1, \text{ and so on.}$$

Thus

$$\cos z = 1 - \frac{z^2}{2!} + \frac{z^4}{4!} - \frac{z^6}{6!} + \cdots = \sum_{k=0}^{\infty} (-1)^k \frac{z^{2k}}{(2k)!}. \tag{11}$$

EXAMPLE 3

Find the Maclaurin series for $\sin z$.

† See the accompanying biographical sketch.

COLIN MACLAURIN
(1698–1746)

Courtesy of the Granger Collection

Considered the finest British mathematician of the generation after Newton, Colin Maclaurin was certainly one of the best mathematicians of the eighteenth century.

Born in Scotland, Maclaurin was a mathematical prodigy and entered Glasgow University at the age of eleven. By the age of nineteen he was a professor of mathematics in Aberdeen and later obtained a post at the University of Edinburgh.

Maclaurin is best known for the *Maclaurin series,* which is the Taylor series in the case $x_0 = 0$. He used this series in his 1742 work *Treatise of Fluxions.* (Maclaurin acknowledged that the series had first been used by Taylor in 1715.) The *Treatise of Fluxions* was most significant in that it presented the first logical description of Newton's method of fluxions. This work was written to defend Newton from the attacks of the powerful Bishop George Berkeley (1685–1753). Berkeley was troubled (as are many of today's calculus students) by the idea of a quotient that in the limit takes the form 0/0. This, of course, is what we obtain when we take a derivative. Berkeley wrote:

And what are these fluxions? The velocities of evanescent increments. And what are these same evanescent increments? They are neither finite quantities nor quantities infinitely small nor yet nothing. May we not call them ghosts of departed quantities?

Maclaurin answered Berkeley by using geometrical arguments. Later, Newton's calculus was put on an even firmer footing by the work of Lagrange in 1797 (see p. 808).

Maclaurin made many other contributions to mathematics—especially in the areas of geometry and algebra. He published his *Geometria organica* when only 21 years old. His posthumous work *Treatise of Algebra,* published in 1748, contained many important results, including the well-known *Cramer's rule* for solving a system of equations (Cramer published the result in 1750).

In 1745, when "Bonnie Prince Charlie" marched against Edinburgh, Maclaurin helped defend the city. When the city fell, Maclaurin escaped, fleeing to York, where he died in 1746 at the age of 48.

SOLUTION

Differentiating series (11) and multiplying both sides by -1, we obtain

$$\sin z = z - \frac{z^3}{3!} + \frac{z^5}{5!} - \frac{z^7}{7!} + \cdots = \sum_{k=0}^{\infty} \frac{(-1)^k z^{2k+1}}{(2k+1)!}. \tag{12}$$

Series (12) is *a* series expansion of $\sin z$. By the uniqueness theorem for power series (Theorem 17.3.6 on p. 992) we know that it must be *the* Maclaurin series for $\sin z$.

EXAMPLE 4

Find the Taylor series for $\text{Log } z$ around $z_0 = 1$.

SOLUTION

First we note that as $\text{Log } 0$ is not defined, any series we come up with will have a radius of convergence of at most 1. We compute

$$f(z) = \text{Log } z, \qquad f(1) = 0,$$
$$f'(z) = \frac{1}{z}, \qquad f'(1) = 1,$$
$$f''(z) = \frac{-1}{z^2}, \qquad f''(1) = -1,$$
$$f'''(z) = \frac{2}{z^3}, \qquad f'''(1) = 2,$$
$$f^{(4)}(z) = \frac{-6}{z^4}, \qquad f^{(4)}(1) = -6,$$

and, in general,

$$f^{(k)}(z) = \frac{(-1)^{k-1}(k-1)!}{z^k}, \qquad f^{(k)}(1) = (-1)^{k-1}(k-1)!$$

Thus

$$\text{Log } z = 0 + (z-1) - \frac{(z-1)^2}{2!} + \frac{2(z-1)^3}{3!} - \frac{3!(z-1)^4}{4!} + \frac{4!(z-1)^5}{5!} - \cdots$$

$$= (z-1) - \frac{(z-1)^2}{2} + \frac{(z-1)^3}{3} - \frac{(z-1)^4}{4} + \cdots$$

$$= \sum_{k=1}^{\infty} \frac{(-1)^{k-1}(z-1)^k}{k}$$

or

$$\text{Log } z = -\sum_{k=1}^{\infty} \frac{(1-z)^k}{k}. \tag{13}$$

The reader should verify that the radius of convergence of this series is 1.

In dealing with multivalued functions like $\log z$, it is necessary to specify the branch on which we are working. The series expansion for $\text{Log } z$ around 1 obtained in the last example is correct because $\text{Arg } 1 = 0$ and the principal branch of $\log z$, $\text{Log } z$, is defined for $-\pi \leq \text{Arg } z < \pi$. Other branches will lead to other Taylor series.

EXAMPLE 5

Find the Taylor series for $\log z$ about $1 = z_0 = e^{2\pi i}$.

SOLUTION

Here $\arg z_0 = 2\pi$ and we must use the branch of $\log z$ given by (see p. 923)

$$\log z = \ln|z| + i \arg z, \qquad \pi < \arg z \leq 3\pi.$$

The derivative of this branch of $\log z$ is $1/z$ and so the values of the derivatives of $\log z$ evaluated at $1 = e^{2\pi i}$ will be the same as before. However, here

$$f(e^{2\pi i}) = \log 2\pi i = i \arg e^{2\pi i} = 2\pi i \neq \text{Log } 1 = 0.$$

Thus the Taylor series now will have an aditional $2\pi i$ term and we have, from (13) and the fact that $e^{2\pi i} = 1$,

$$\log z = 2\pi i + \sum_{k=1}^{\infty} \frac{(-1)^{i+1}(z-1)^k}{k}, \qquad \pi < \arg z \leq 3\pi.$$

EXAMPLE 6

Find the analytic solution of the differential equation

$$f''(z) - 2zf'(z) - 2f(z) = 0$$

with initial conditions $f(0) = 1$, $f'(0) = 0$.

SOLUTION

The solution $f(z)$ has the Maclaurin series

$$f(z) = c_0 + c_1 z + c_2 z^2 + \cdots + c_n z^n + \cdots, \qquad c_0 = 1.$$

We differentiate twice to obtain

$$f'(z) = c_1 + 2c_2 z + 3c_3 z^2 + \cdots + nc_n z^{n-1} + \cdots, \qquad c_1 = 0,$$

$$f''(z) = 2c_2 + 6c_3 z + \cdots + (n+2)(n+1)c_{n+2} z^n + \cdots.$$

Then, gathering terms with like powers,

$$f''(z) - 2zf'(z) - 2f(z) = \sum_{n=0}^{\infty} [(n+2)(n+1)c_{n+2} - 2(n+1)c_n]z^n = 0,$$

so $(n+1)[(n+2)c_{n+2} - 2c_n] = 0$, for $n = 0, 1, 2, \ldots$. The recursion equation $c_{n+2} = 2c_n/(n+2)$ implies that the general analytic solution of the differential equation is

$$f(z) = c_0\left(1 + z^2 + \frac{z^4}{2!} + \frac{z^6}{3!} + \cdots\right)$$

$$+ c_1 z\left(1 + \frac{1}{3!}(2z)^2 + \frac{2}{5!}(2z)^4 + \frac{3}{7!}(2z)^6 + \cdots\right).$$

Since $c_0 = 1$ and $c_1 = 0$, we obtain the entire function

$$f(z) = 1 + z^2 + \frac{z^4}{2!} + \frac{z^6}{3!} + \cdots = e^{z^2}$$

as the only analytic solution to the initial value problem.

Problems 17.4

In Problems 1–10 find the Maclaurin series for the given function and determine its radius of convergence.

1. $\sin 2z$ **2.** $\cos z^2$

3. $\sinh z$ **4.** $\cosh z$

5. $\dfrac{1}{1 - z^4}$ **6.** e^{iz}

7. e^{-z^3} **8.** $\text{Log}(1 + z^2)$

9. $\text{Log}(1 + iz)$ **10.** $\tan^{-1} z$

11. $\dfrac{\sin z}{z}$

12. ze^{-z}

13. $z^2 e^{z^2}$

14. $\sin^2 z$ [*Hint:* $\sin^2 z = (1 - \cos 2z)/2$.]

15. $\cos^2 z$

In Exercises 16–25, expand the given functions in a Taylor series about z_0. Indicate the largest disk where the representation is valid.

16. $f(z) = \dfrac{1}{1 - z}$, $z_0 = -1$ **17.** $f(z) = \dfrac{1}{1 - z}$, $z_0 = i$

18. $f(z) = \cos z$, $z_0 = \dfrac{\pi}{2}$ **19.** $f(z) = \sin z$, $z_0 = \dfrac{\pi}{2}$

20. $f(z) = \dfrac{1}{z}$, $z_0 = 1$ **21.** $f(z) = \text{Log } z$, $z_0 = i$

22. $f(z) = \log z$, $z_0 = -1$ **23.** $f(z) = \log z$, $z_0 = 2e^{3\pi i}$

24. $f(z) = \sinh z$, $z_0 = 1$

25. $f(z) = (z - 1)\text{Log } z$, $z_0 = 1$

26. Find the first four nonzero terms of the Taylor series for $\csc z$ at $\pi/2$.

* 27. Find the first three nonzero terms of the Maclaurin series for $\text{Log}(\cos z)$.

28. Differentiate the Maclaurin series for $\sinh z$ and show that it is equal to the Maclaurin series for $\cosh z$.

29. Prove the **binomial theorem** for complex α:

$$(1 + z)^\alpha = 1 + \frac{\alpha}{1}z + \frac{\alpha(\alpha - 1)}{2!}z^2 + \frac{\alpha(\alpha - 1)(\alpha - 2)}{3!}z^3 + \cdots$$

$$= 1 + \sum_{k=1}^{\infty} \frac{\alpha(\alpha - 1) \cdots (\alpha - k + 1)}{k!} z^k, \qquad |z| < 1.$$

30. Show that for any complex number z

$$1 + \frac{z}{2} + \frac{z(z - 1)}{2^2 2!} + \cdots$$

$$+ \frac{z(z - 1) \cdots (z - n + 1)}{2^n n!} + \cdots = \left(\frac{3}{2}\right)^z.$$

[*Hint:* Use the binomial series.]

31. Use the binomial theorem to find the Maclaurin series for $\sqrt[4]{1 + z}$.

32. Use the result of Problem 31 to obtain a Maclaurin series for $\sqrt[4]{1 + z^3}$.

33. Use the result of Problem 32 to estimate $\int_0^{0.5} \sqrt[4]{1 + x^3}\, dx$ to four significant figures.

34. Using the technique suggested in Problems 31–33, estimate $\int_0^{1/4} (1 + \sqrt{x})^{3/5}\, dx$ to four significant figures.

35. The **error function** (which arises in mathematical statistics) is defined by

$$\text{erf}(z) = \frac{2}{\sqrt{\pi}} \int_0^z e^{-t^2}\, dt.$$

(a) Find a Maclaurin series for $\text{erf}(z)$ by integrating the Maclaurin series for e^{-z^2}.

(b) Use the series obtained in (a) to estimate, with an error < 0.0001, erf(1) and erf($\frac{1}{2}$).

📖 **36.** The **complementary error function** is defined by

$$\text{erfc}(z) = 1 - \text{erf}(z) = 1 - \frac{2}{\sqrt{\pi}} \int_0^z e^{-t^2}\,dt = \frac{2}{\sqrt{\pi}} \int_z^\infty e^{-t^2}\,dt.$$

Find a Maclaurin series for erfc(z) and use it to estimate erfc(1) and erfc($\frac{1}{2}$) with a maximum error of 0.0001. Note that for large values of z, erfc(z) can be estimated by integrating the last integral by parts.

* 37. The **sine integral** is defined by

$$\text{Si}(z) = \int_0^z \frac{\sin t}{t}\,dt.$$

(a) Show that Si(z) is defined and continuous for all complex z.

(b) Find a Maclaurin series expansion for Si(z).

(c) Estimate Si(1) and Si($\frac{1}{2}$) with a maximum error of 0.0001.

38. Let $f(z) = \tan z$.

(a) Compute $f(0)$ and $f'(0)$.

(b) Show that $f'(z) = 1 + f^2(z)$.

(c) Use the result of part (b) to compute $f''(z)$ in terms of $f(z)$ and $f'(z)$ and find $f''(0)$.

(d) Compute $f'''(z)$ and $f'''(0)$.

(e) Compute $f^{(4)}(0), f^{(5)}(0), f^{(6)}(0)$, and $f^{(7)}(0)$ and use these to find the first four nonzero terms in the Taylor series for $\tan z$.

39. Bernoulli numbers. Let

$$f(z) = \begin{cases} \dfrac{z}{e^z - 1}, & z \neq 0 \\ 1, & z = 0 \end{cases}.$$

(a) Show that $f(z)$ is analytic at 0.

(b) Write

$$f(z) = B_0 + B_1\frac{z}{1!} + B_2\frac{z^2}{2!} + B_3\frac{z^3}{3!} + \cdots = \sum_{k=0}^\infty \frac{B_k z^k}{k!}.$$

The coefficients $B_0, B_1, B_2, \ldots$ are called **Bernoulli numbers.**

(c) Compute $B_0, B_1, B_2, B_3, B_4, B_5, B_6$, and B_7 by writing $f(z)(e^z - 1) = z$ and taking the product of the series for $f(z)$ and $e^z - 1$.

** 40. (a) Show that

$$\tan z = \frac{2i}{e^{2iz} - 1} - \frac{4i}{e^{4iz} - 1} - i.$$

[*Hint:* Use formula (15.8.1) on p. 916.]

(b) Using the series expansion for $1/(1 + z^2)$, the binomial theorem, and the series in Problem 39b, show that

$$\tan z = \sum_{k=1}^\infty (-1)^{k+1} B_{2k} \frac{2^{2k}(2^{2k} - 1)}{(2k)!} z^{2k-1}.$$

41. Euler numbers

(a) Show that if $\sec z = \sum_{k=0}^\infty c_k z^k$, then $c_1 = c_3 = c_5 = \cdots = 0$.

(b) The **Euler numbers** E_{2k} are defined by $E_{2k} = (-1)^k (2k)! c_{2k}$ where c_{2k} are as in part (a). Compute E_0, E_2, E_4, E_6, and E_8.

42. Find the most general power series (involving two arbitrary constants) satisfying the differential equation

$$f''(z) + f(z) = 0.$$

43. Find a Maclaurin series satisfying the differential equation $f'(z) = 1 + zf(z)$ with the initial condition $f(0) = 0$. What is its radius of convergence?

44. Determine the general Maclaurin series solution of the differential equation $zf''(z) + f'(z) + zf(z) = 0$ and show that it is entire. [*Hint:* Prove $\sqrt[n]{n!} \to \infty$ as $n \to \infty$, since e^z is entire.]

45. Find the general Maclaurin series solution of the differential equation $(1 - z^2)f''(z) - 2zf'(z) + n(n + 1)f(z) = 0$.

* 46. Use a Maclaurin series to solve the functional equation $f(z^2) = z + f(z)$. Where does the series converge?

47. Suppose $f(z)$ and $g(z)$ are analytic in a neighborhood of z_0 and $f(z_0) = g(z_0) = 0$ while $g'(z_0) \neq 0$. Prove **L'Hôpital's theorem**

$$\lim_{z \to z_0} \frac{f(z)}{g(z)} = \frac{f'(z_0)}{g'(z_0)}.$$

17.5 Laurent Series

A series of the form

$$c_0 + \frac{c_1}{z} + \frac{c_2}{z^2} + \cdots + \frac{c_n}{z^n} + \cdots = \sum_{k=-\infty}^0 c_k z^k \qquad \textbf{(1)}$$

can be considered a power series in the variable $1/z$. If R is its radius of convergence, the series will converge absolutely whenever $|1/z| < R$ or $|z| > 1/R$. The convergence is uniform in every region $|z| \geq \rho$, $\rho > 1/R$, and the series diverges for $|z| < 1/R$. Thus the series represents an analytic function in $|z| > 1/R$.

Annulus

An **annulus** is defined as the ring-shaped region between two circles. Specifically, if $|z - z_0| = r$ and $|z - z_0| = R$ are two circles with $r < R$, then the open region between them is the **open annulus** A given by

$$A = \{z : r < |z - z_0| < R\}. \tag{2}$$

This annulus is sketched in Fig. 1.

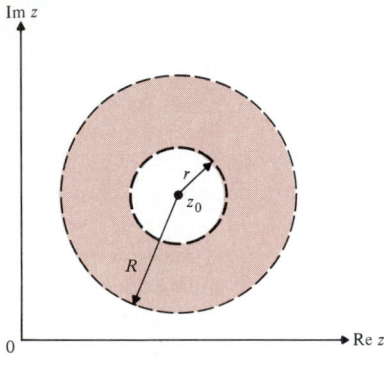

Im z

r

z_0

R

0 ——→ Re z

FIGURE 1 Open annulus

Laurent Series

Suppose that $\sum_{k=-\infty}^{-1} c_k z^k$ converges in the region $|z| > r$ and the power series $\sum_{k=0}^{\infty} c_k z^k$ converges in the disk $|z| < R$. If $r < R$, then both series converge in the open annulus $r < |z| < R$. In this case the sum is written

$$\sum_{k=-\infty}^{\infty} c_k z^k. \tag{3}$$

Series (3) represents an analytic function in this annulus. Similarly,

$$\sum_{k=-\infty}^{\infty} c_k(z - z_0)^k$$

represents an analytic function in the annulus $r < |z - z_0| < R$. We call all expansions of this type **Laurent**† **series**. Conversely, we now prove that a function analytic in an annulus $r < |z - z_0| < R$ can be expanded into a Laurent series.

Laurent series

■ **THEOREM 1: Laurent's Theorem**
If $f(z)$ is analytic in the annulus $0 \leq r < |z - z_0| < R \leq \infty$, then it can be expanded in a Laurent series

$$f(z) = \sum_{k=-\infty}^{\infty} c_k(z - z_0)^k, \tag{4}$$

where

$$c_k = \frac{1}{2\pi i} \int_{|\zeta - z_0| = \rho} \frac{f(\zeta) \, d\zeta}{(\zeta - z_0)^{k+1}}, \qquad k = 0, \pm 1, \pm 2, \ldots, r < \rho < R. \tag{5}$$

† Pierre Alphonse Laurent (1813–1854) was a French mathematician and engineer. He published the result now known as *Laurent's theorem* in 1843.

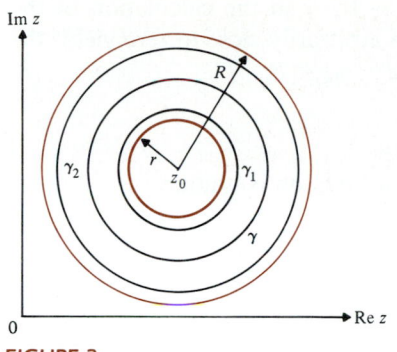

Im z

γ_2

γ_1

R

r

z_0

γ

Re z

0

FIGURE 2

PROOF

Let γ_1 and γ_2 denote the circles $|\zeta - z_0| = r + \epsilon$ and $|\zeta - z_0| = R - \epsilon$, respectively, with $0 < \epsilon < (R - r)/2$ (see Fig. 2).

By the Cauchy integral formula,

$$f(z) = \frac{1}{2\pi i} \int_{\gamma_2} \frac{f(\zeta)\, d\zeta}{\zeta - z} - \frac{1}{2\pi i} \int_{\gamma_1} \frac{f(\zeta)\, d\zeta}{\zeta - z}$$

for all z satisfying $r + \epsilon < |z - z_0| < R - \epsilon$.

Proceeding exactly as we did in proving Taylor's theorem, the first integral becomes

$$\frac{1}{2\pi i} \int_{\gamma_2} \frac{f(\zeta)\, d\zeta}{\zeta - z} = \sum_{k=0}^{\infty} c_k (z - z_0)^k$$

with

$$c_k = \frac{1}{2\pi i} \int_{\gamma_2} \frac{f(\zeta)\, d\zeta}{(\zeta - z_0)^{k+1}}, \qquad k = 0, 1, 2, \ldots.$$

For the second integral observe that

$$-\frac{1}{\zeta - z} = \frac{1}{(z - z_0) - (\zeta - z_0)} = \frac{1}{(z - z_0)[1 - (\zeta - z_0)/(z - z_0)]}. \qquad (6)$$

But $|(\zeta - z_0)/(z - z_0)| < 1$ since $|\zeta - z_0| < |z - z_0|$ on γ_1. Thus, by the geometric series and (6),

$$-\frac{1}{\zeta - z} = \frac{1}{z - z_0} \sum_{k=0}^{\infty} \left(\frac{\zeta - z_0}{z - z_0} \right)^k.$$

Furthermore, this geometric series converges uniformly by Abel's theorem (Theorem 17.3.4). By Weierstrass's theorem (Theorem 17.3.3) we may integrate the series term by term:

$$-\frac{1}{2\pi i} \int_{\gamma_1} \frac{f(\zeta)}{\zeta - z}\, d\zeta = \frac{1}{2\pi i} \int_{\gamma_1} \frac{f(\zeta)}{z - z_0} \left[\sum_{k=0}^{\infty} \left(\frac{\zeta - z_0}{z - z_0} \right)^k \right] d\zeta$$

$$= \sum_{k=0}^{\infty} \frac{1}{(z - z_0)^{k+1}} \cdot \frac{1}{2\pi i} \int_{\gamma_1} f(\zeta)(\zeta - z_0)^k\, d\zeta. \qquad (7)$$

If we define

$$c_{-k-1} = \frac{1}{2\pi i} \int_{\gamma_1} \frac{f(\zeta)}{(\zeta - z_0)^{-k}}\, d\zeta, \qquad (8)$$

then (7) becomes

$$-\frac{1}{2\pi i} \int_{\gamma_1} \frac{f(\zeta)}{\zeta - z}\, d\zeta = \sum_{k=-\infty}^{-1} c_k (z - z_0)^k.$$

Finally, since $f(\zeta)/(\zeta - z_0)^{n+1}$ is analytic inside and on $\gamma - \gamma_1$ or $\gamma_2 - \gamma$, where γ is the circle $|\zeta - z_0| = \rho$, $r < \rho < R$, Cauchy's theorem on a multiply connected region implies that we may replace γ_1 or γ_2 by γ in the calculation of the coefficients c_k. Noticing that ϵ can be chosen arbitrarily close to zero yields the desired representation on the annulus $r < |z - z_0| < R$. ∎

■ **THEOREM 2: Uniqueness of Laurent Series**

The Laurent series representation for a given function is unique.

PROOF

If $f(z)$ had representations

$$f(z) = \sum_{k=-\infty}^{\infty} a_k(z-z_0)^k, \quad \text{and} \quad f(z) = \sum_{k=-\infty}^{\infty} b_k(z-z_0)^k,$$

then multiplying by $(z - z_0)^n$, for any integer n, and integrating along $|z - z_0| = \rho$ would yield by uniform convergence

$$\sum_{k=-\infty}^{\infty} a_k \cdot \int_C (z-z_0)^{n+k}\, dz = \sum_{k=-\infty}^{\infty} b_k \cdot \int_C (z-z_0)^{n+k}\, dz.$$

Since all powers of $z - z_0$ except $(z - z_0)^{-1}$ have analytic antiderivatives in $r < |z - z_0| < R$, their integrals vanish by the fundamental theorem. Thus $2\pi i a_{-n-1} = 2\pi i b_{-n-1}$, implying $a_n = b_n$ for all integers n. ∎

The Laurent coefficients c_n are not often obtained by use of their integral formulas. We give examples of the techniques employed in avoiding this computation.

EXAMPLE 1

Obtain the Laurent series for $z^{-2} \cos z$.

SOLUTION

Since $\cos z = \sum_{k=0}^{\infty} \dfrac{(-1)^k z^{2k}}{(2k)!} = 1 - \dfrac{z^2}{2!} + \dfrac{z^4}{4!} - \dfrac{z^6}{6!} + \cdots$, we can divide each term by z^2 to obtain

$$\frac{\cos z}{z^2} = \frac{1}{z^2} - \frac{1}{2!} + \frac{z^2}{4!} - \frac{z^4}{6!} + \cdots + \sum_{k=-1}^{\infty} (-1)^{k+1} \frac{z^{2k}}{(2k+2)!}, \qquad 0 < |z| < \infty. \qquad (9)$$

Remark. Example 1 illustrates the value of the uniqueness theorem. Without it, we would know that the series (9) is *a* Laurent series for $z^{-2} \cos z$, but not necessarily the one obtained by repeated use of formula (5). However, because of Theorem 2 we can conclude that the series (9) is *the* Laurent series for $z^{-2} \cos z$.

EXAMPLE 2

Compute the Laurent series for e^{1/z^2}.

SOLUTION
Replacing z by $1/z^2$ in the series

$$e^z = \sum_{k=0}^{\infty} \frac{z^k}{k!}$$

yields

$$e^{1/z^2} = \sum_{k=0}^{\infty} \frac{z^{-2k}}{k!} = \sum_{k=-\infty}^{0} \frac{z^{2k}}{(-k)!}, \qquad 0 < |z|.$$

EXAMPLE 3

Consider the function $(z^2 - 3z + 2)^{-1}$. It is analytic everywhere except at $z = 1, 2$. Find its Laurent series on the regions (a) $1 < |z| < 2$, (b) $|z| < 1$, (c) $|z| > 2$, and (d) $0 < |z - 1| < 1$.

SOLUTION
(a) In the annulus $1 < |z| < 2$, writing

$$\frac{1}{z^2 - 3z + 2} = \frac{1}{(z-1)(z-2)} = \frac{1}{z-2} - \frac{1}{z-1},$$

we may expand the fractions in the form

$$\frac{1}{z-2} - \frac{1}{z-1} = \frac{-1/2}{1 - z/2} - \frac{1/z}{1 - 1/z}$$

$$= -\frac{1}{2} \sum_{k=0}^{\infty} \left(\frac{z}{2}\right)^k - \frac{1}{z} \sum_{k=0}^{\infty} \left(\frac{1}{z}\right)^k.$$

The first series converges for $|z| < 2$ and the second converges for $|1/z| < 1$ or $|z| > 1$. Thus both series converge in the annulus $1 < |z| < 2$ and we have

$$\frac{1}{(z-1)(z-2)} = -\sum_{k=-\infty}^{-1} z^k - \sum_{k=0}^{\infty} \frac{z^k}{2^{k+1}}.$$

(b) For $|z| < 1$, we expand the expression as

$$\frac{1}{z-2} - \frac{1}{z-1} = \frac{-1/2}{1 - z/2} + \frac{1}{1 - z} = -\frac{1}{2} \sum_{k=0}^{\infty} \left(\frac{z}{2}\right)^k + \sum_{k=0}^{\infty} z^k.$$

Hence, since both series converge when $|z| < 1$,

$$\frac{1}{(z-1)(z-2)} = \sum_{k=0}^{\infty} \left(1 - \frac{1}{2^{k+1}}\right) z^k, \qquad |z| < 1.$$

(c) Here $|z| > 2$ and

$$\frac{1}{z-2} - \frac{1}{z-1} = \frac{1/z}{1 - 2/z} - \frac{1/z}{1 - 1/z}$$

$$= \frac{1}{z} \sum_{k=0}^{\infty} \left(\frac{2}{z}\right)^k - \frac{1}{z} \sum_{k=0}^{\infty} \left(\frac{1}{z}\right)^k = \sum_{k=0}^{\infty} (2^k - 1) z^{-k-1}$$

$$= \sum_{k=1}^{\infty} (2^{k-1} - 1) z^{-k} = \sum_{k=-\infty}^{-1} \left(\frac{1}{2^{k+1}} - 1\right) z^k.$$

(d) In $0 < |z - 1| < 1$, we obtain

$$\frac{1}{z-2} - \frac{1}{z-1} = -\frac{1}{z-1} - \frac{1}{1-(z-1)} = -\frac{1}{z-1} - \sum_{k=0}^{\infty} (z-1)^k.$$

Hence, in $0 < |z - 1| < 1$,

$$\frac{1}{(z-1)(z-2)} = -\frac{1}{z-1} - \frac{1}{1-(z-1)} = -\sum_{k=-1}^{\infty} (z-1)^k.$$

Problems 17.5

In Problems 1–8 find the Laurent series for the given function that converges in the annulus $0 < |z| < \infty$.

1. $\dfrac{e^z}{z}$　　　　　　　　　**2.** $\sin \dfrac{1}{z}$

3. $z \cos \dfrac{1}{z^2}$　　　　　　　**4.** $\dfrac{\sin 2z}{z^3}$

5. $\dfrac{e^{z^2}}{z^{10}}$　　　　　　　　　**6.** $\dfrac{e^{1/z^2}}{z^{10}}$

7. $ze^{1/z}$　　　　　*** 8.** $\sin z \sin \dfrac{1}{z}$

(*Hint:* Multiply their series.)

Find the Laurent series of the function $(z^2 + z)^{-1}$ in the regions given in Problems 9–11.

9. $0 < |z| < 1$

10. $0 < |z - 1| < 1$

11. $1 < |z - 1| < 2$

Represent the function $(z^3 - z)^{-1}$ as a Laurent series in the regions given in Problems 12–15.

12. $0 < |z| < 1$

13. $1 < |z|$

14. $0 < |z - 1| < 1$

15. $1 < |z - 1| < 2$

Find the Laurent series of the function $z/(z^2 + z - 2)$ in the regions given in Problems 16–21.

16. $|z| < 1$　　　　　　**17.** $0 < |z - 1| < 3$

18. $0 < |z + 2| < 3$　　　**19.** $1 < |z| < 2$

20. $|z| > 2$　　　　　　　　　**21.** $|z + 2| > 3$

Find the Laurent series of the functions given in Problems 22–24 in the region $0 < |z - 1| < 1$.

22. $\dfrac{1}{z-1} \sin \dfrac{1}{z}$

23. $\dfrac{1}{z} \sin \dfrac{1}{z-1}$

24. $z \sin \dfrac{1}{z-1}$

* 25. Find a Laurent series for $e^{z+1/z}$. [*Hint:* $e^{z+1/z} = e^z e^{1/z}$.]

* 26. Find a Laurent series for $\sin(z + 1/z)$. [*Hint:* $\sin(a + b) = \sin a \cos b + \cos a \sin b$.]

27. Suppose $f(z)$ is analytic and bounded by M in $r < |z - z_0| < R$. Show that the coefficients of its Laurent series satisfy

$$|a_n| \leqslant MR^{-n}, \quad |a_{-n}| \leqslant Mr^n, \quad n = 0, 1, 2, \ldots.$$

Suppose $r = 0$. Can $f(z_0)$ be defined in such a way that $f(z)$ is analytic in $|z - z_0| < R$?

28. Bessel's function $J_n(z)$ is defined as the nth coefficient $(n \geqslant 0)$ of the Laurent series of the function

$$e^{(z/2)(\zeta - 1/\zeta)} = \sum_{n=-\infty}^{\infty} J_n(z) \zeta^n.$$

Show that

$$J_n(z) = \frac{1}{\pi} \int_0^{\pi} \cos(n\theta - z \sin \theta) \, d\theta.$$

* 29. Using the coefficient of the Laurent series of $e^{1/z}$ in $|z| > 0$, show that

$$\frac{1}{2\pi} \int_{-\pi}^{\pi} e^{\cos \theta} \cos(\sin \theta - n\theta) \, d\theta = \frac{1}{n!}, \qquad n = 0, 1, 2, \ldots.$$

[*Hint:* Integrate on $|z| = 1$.]

* 30. Evaluate the integral

$$\int_{-\pi}^{\pi} (\cos \theta)^m \cos n\theta \, d\theta, \qquad m, n \text{ integers},$$

by comparing the coefficients of the Laurent series of $(z + 1/z)^m$ with its binomial expansion.

* 31. Find the Laurent series of $\csc z$ in $\pi < |z| < 2\pi$.

17.6 Zeros and Singularities

In this section we classify the zeros and singularities of analytic functions.

Zeros of Analytic Functions

Finding the zeros of a function is an important problem in applied mathematics. Taylor's theorem can be used to show that the zeros of a nonconstant analytic function are isolated. That is, around each such zero we can find a disk such that there is no other zero in the disk. This is illustrated in Fig. 1. To prove this result we first state a theorem that is a consequence of Taylor's theorem.

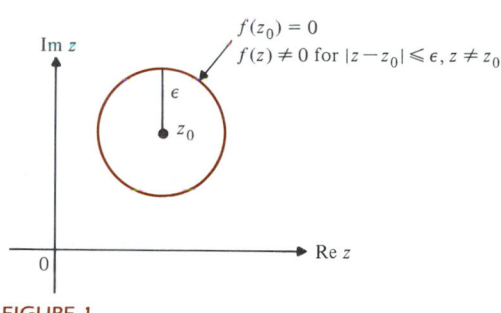

FIGURE 1

■ THEOREM 1

If $f(z)$ is analytic in a region Ω containing the point z_0, and $f^{(n)}(z_0) = 0$ for $n = 1, 2, \ldots$, then $f(z)$ is constant in Ω.

PROOF

By Taylor's theorem, $f(z) = f(z_0)$ for all z in any disk $|\zeta - z_0| < r$ contained in Ω. Let $g(z) = f(z) - f(z_0)$. Then g is analytic in Ω and $g^{(n)}(z) = 0$, $n = 0, 1, 2, \ldots$, for all z in this disk. By a proof identical to that in the maximum principle (see p. 965) it follows that $g^{(n)}(z) = 0$, $n = 0, 1, 2, \ldots$ for all z in Ω. But, by Taylor's theorem or the zero derivative theorem (p. 909) it follows that $g(z) \equiv 0$ on Ω. Thus $f(z) = f(z_0)$ for all z in Ω. ■

Order of a Zero

DEFINITION

Let $f(z)$ be a nonconstant analytic function with $f(z_0) = 0$. Then, from Theorem 1, there is a positive integer n such that $f^{(n)}(z_0) \neq 0$. The **order of the zero of f at z_0** is defined as the smallest integer n such that $f^{(n)}(z_0) \neq 0$.

EXAMPLE 1

Find the order of the zero of $f(z) = 2z(e^z - 1)$ at $z = 0$.

SOLUTION
We compute

$$f(z) = 2z(e^z - 1), \qquad\qquad f(0) = 0,$$

$$f'(z) = 2ze^z + 2(e^z - 1), \qquad f'(0) = 0,$$

$$f''(z) = 2ze^z + 4e^z, \qquad\qquad f''(0) = 4 \neq 0.$$

Thus the order of the zero is 2 since $f''(0)$ is the first nonzero derivative. This is also clear from the Maclaurin series of $f(z)$:

$$2z(e^z - 1) = 2z \sum_{n=1}^{\infty} \frac{z^n}{n!} = 2z^2 \left(1 + \frac{z}{2!} + \frac{z^2}{3!} + \cdots \right).$$

■ **THEOREM 2**

The zeros of a nonconstant function which is analytic in a region Ω are isolated.

PROOF
Suppose that $f(z_0) = 0$ and the order of the zero is n. Then $f^{(n)}(z_0) \neq 0$ and

$$f(z_0) = f'(z_0) = \cdots = f^{(n-1)}(z_0) = 0.$$

We therefore use equations (17.4.8) and (17.4.9) on p. 997 to write

$$f(z) = (z - z_0)^n f_n(z), \qquad f_n(z) = \frac{1}{2\pi i} \int_{|\zeta - z_0| = r} \frac{f(\zeta)}{(\zeta - z)(\zeta - z_0)^n} \, d\zeta,$$

with $f_n(z)$ analytic inside the disk $|\zeta - z_0| \leq r$ contained in Ω by Cauchy's theorem for derivatives. Furthermore,

$$f_n(z_0) = \frac{1}{2\pi i} \int_{|\zeta - z_0| = r} \frac{f(\zeta)}{(\zeta - z_0)^{n+1}} \, d\zeta = \frac{f^{(n)}(z_0)}{n!} \neq 0.$$

Since $f_n(z)$ is continuous and $f_n(z_0) \neq 0$, there is an $\epsilon > 0$ such that $f_n(z) \neq 0$ if $|z - z_0| < \epsilon$. Then $f(z) = (z - z_0)^n f_n(z) \neq 0$ for $0 < |z - z_0| < \epsilon$, and the theorem is proved. ■

EXAMPLE 2

Show that there is no function analytic in $|z| < 2$ satisfying the condition

$$f\left(\frac{1}{n} \right) = \frac{(-1)^n}{n}, \qquad n = 1, 2, 3, \ldots.$$

SOLUTION

First observe that if such an analytic function exists

$$f(0) = f\left(\lim_{n\to\infty} \frac{1}{n}\right) = \lim_{n\to\infty} f\left(\frac{1}{n}\right) = \lim_{n\to\infty} \frac{(-1)^n}{n} = 0,$$

since f is continuous at 0. Define the function $F(z) = z - f(z)$. Evidently $F(z)$ is a nonconstant analytic function satisfying $F(0) = 0$ and $F(1/2m) = 0$, for $m = 1, 2, 3, \ldots$. Hence, $z = 0$ is a zero of $F(z)$ that is not isolated, contradicting Theorem 2.

Isolated Singularities

Isolated Singularity

A function $f(z)$, analytic in a region $0 < |z - z_0| < R$, but not analytic or even necessarily defined at z_0, is said to have an **isolated singularity** at z_0. These singularities are classified into three categories: removable singularities, poles, and essential singularities.

DEFINITION

Removable Singularities

Removable singularities are those where it is possible to assign a complex number to $f(z_0)$ in such a way that $f(z)$ becomes analytic in $|z - z_0| < R$.

EXAMPLE 3

The function $f(z) = (\sin z)/z$ is not defined at 0. However, we have

$$\frac{\sin z}{z} = \frac{1}{z}\left(z - \frac{z^3}{3!} + \frac{z^5}{5!} - \frac{z^7}{7!} + \cdots\right) = 1 - \frac{z^2}{3!} + \frac{z^4}{5!} - \frac{z^7}{7!} + \cdots.$$

The Maclaurin series for $(\sin z)/z$ converges for all z and takes the value 1 when $z = 0$. Hence, the function

$$g(z) = \begin{cases} \dfrac{\sin z}{z}, & z \neq 0, \\ 1, & z = 0, \end{cases}$$

is equal to $f(z)$ for $z \neq 0$ and is entire. Thus we see that $f(z)$ has a removable singularity at 0.

In Example 3 we saw that $f(z) \to 1$ as $z \to 0$. The following theorem generalizes this observation.

■ THEOREM 3

Suppose that $f(z)$ is analytic in $0 < |z - z_0| < R$. Then $f(z)$ has a removable singularity at z_0 if and only if $\lim_{z\to z_0} f(z) = L$ exists. In that case

$$g(z) = \begin{cases} f(z), & z \neq z_0, \\ L, & z = z_0, \end{cases} \tag{1}$$

is analytic in $|z - z_0| < R$.

PROOF

If $f(z)$ has a removable singularity at z_0, then a value L can be assigned to $f(z_0)$ making f analytic in $|z - z_0| < R$. But then f is continuous on $|z - z_0| < R$ so that $\lim_{z \to z_0} f(z) = f(z_0) = L$.

Conversely, suppose that $L = \lim_{z \to z_0} f(z)$ exists. We show that $g(z)$, as defined by (1), is analytic in $|z - z_0| < R$. We know that $g(z)$ is analytic in $0 < |z - z_0| < R$ and is continuous on $|z - z_0| \le r < R$ [continuity at 0 follows from the definition of $g(z_0) = L$]. Thus, by the corollary to the maximum principle on p. 966, $g(z)$ is bounded on $|z - z_0| \le r$. Using the definition of the Laurent coefficients in Laurent's theorem, Theorem 17.5.1, we have for $k > 0$

$$c_{-k} = \frac{1}{2\pi i} \int_{|z-z_0|=r} g(z)(z - z_0)^{k-1}\, dz. \tag{2}$$

Then

$$|c_{-k}| \le \frac{1}{2\pi} \int_{|z-z_0|=r} |g(z)||z - z_0|^{k-1}\, |dz| \le \frac{Mr^{k-1}}{2\pi} \cdot 2\pi r = Mr^k,$$

where M is a bound for g on $|z - z_0| \le r$. Since r can be chosen arbitrarily small, $c_{-k} = 0$ for $k = 1, 2, \ldots$. Thus the Laurent series for $g(z)$ reduces to a Taylor series, which implies g is analytic in $|z - z_0| < R$. ■

DEFINITION

Poles

If $f(z)$ has an isolated singularity at $z = z_0$ and $\lim_{z \to z_0} f(z) = \infty$,† then z_0 is called a **pole** of $f(z)$.

To analyze the behavior of $f(z)$ at a pole, let $g(z) = 1/f(z)$. Since $\lim_{z \to z_0} g(z) = 0$, $g(z)$ has a removable singularity at z_0. Moreover, since $f(z)$ is analytic and nonconstant in a neighborhood of z_0, z_0 is an isolated zero of the nonconstant analytic function $g(z)$, by Theorem 2. Let n be the order of the zero of $g(z)$ at z_0. Then $f(z)$ is said to have a **pole of order n at z_0**.

Pole of Order n

Now, since z_0 is a zero of order n of $g(z)$ and since this zero is isolated, it follows that there is a number $r > 0$ such that $g(z) \ne 0$ for $0 < |z - z_0| < r$. By definition of the order of a zero, we have

$$g(z) = (z - z_0)^n g_n(z) \tag{3}$$

where $g_n(z)$ is analytic and nonzero in $|z - z_0| < r \le R$. If $f_n(z) = 1/g_n(z)$, then (3) becomes

$$f(z) = \frac{f_n(z)}{(z - z_0)^n} \tag{4}$$

where $f_n(z)$ is analytic in $|z - z_0| < r$.

---

† By this we simply mean that the limit is unbounded; that is, given any $M > 0$ there exists an $\epsilon > 0$ such that $|f(z)| > M$ whenever $0 < |z - z_0| < \epsilon$.

Moreover, since $f_n(z)$ is analytic in $|z - z_0| < r$, $f_n(z)$ has a Taylor expansion

$$f_n(z) = \sum_{k=0}^{\infty} b_k(z - z_0)^k,$$

so that

$$f(z) = (z - z_0)^{-n} f_n(z) = \sum_{k=0}^{\infty} b_k(z - z_0)^{k-n}$$

$$= \sum_{k=-n}^{\infty} c_k(z - z_0)^k, \qquad \text{where } c_k = b_{k+n}.$$

We summarize our results:

Suppose that $f(z)$ has an isolated singularity at z_0 and that $f(z)$ is analytic in $0 < |z - z_0| < R$. If

$$\lim_{z \to z_0} f(z) = \infty,$$

then $f(z)$ has a pole at z_0. The order of this pole is the smallest integer n such that $f_n(z) = (z - z_0)^n f(z)$ is analytic at z_0. In particular, $f(z)$ has a convergent Laurent series given by

$$f(z) = \sum_{k=-n}^{\infty} c_k(z - z_0)^k, \tag{5}$$

which is valid in $0 < |z - z_0| < r$.

EXAMPLE 4

The function $f(z) = (\cos z)/z^2$ has a pole of order 2 at 0

$$\frac{\cos z}{z^2} = \frac{1}{z^2}\left(1 - \frac{z^2}{2!} + \frac{z^4}{4!} - \frac{z^6}{6!} + \cdots \right) = \frac{1}{z^2} - \frac{1}{2!} + \frac{z^2}{4!} - \frac{z^4}{6!} + \cdots$$

because z^{-2} is the first nonzero term in the Laurent series for $f(z)$.

EXAMPLE 5

Determine the orders of the poles of

$$f(z) = \frac{1}{z^2(z + i)^7} + \frac{1}{z^4(z + i)^3}.$$

SOLUTION

$f(z)$ has a pole of order 4 at 0 and a pole of order 7 at $-i$. To see this more clearly, we observe that

$$z^4 f(z) = \frac{z^2}{(z + i)^7} + \frac{1}{(z + i)^3},$$

which is analytic at 0, and

$$(z+i)^7 f(z) = \frac{1}{z^2} + \frac{(z+i)^4}{z^4},$$

which is analytic at $-i$.

Note that

$$z^3 f(z) = \frac{z}{(z+i)^7} + \frac{1}{z(z+i)^3}$$

has a singularity at $z = 0$, and a similar result holds for $(z+i)^6 f(z)$.

DEFINITION

Meromorphic Function

A function analytic in a region Ω, except for poles, is said to be **meromorphic** in Ω.

The functions in Examples 4 and 5 are meromorphic functions.

If $f(z)$ and $g(z)$ are analytic in G, and $g(z)$ is not identically zero, then the singularities of the quotient $f(z)/g(z)$ agree with the zeros of $g(z)$. They are poles whenever $f(z)$ is nonzero or has a zero of order less than that of $g(z)$; otherwise they are removable singularities. Extending $f(z)/g(z)$ by continuity over the removable singularities, we obtain a meromorphic function in G. For example, $f(z) = \tan z = \sin z / \cos z$ is meromorphic in $\mathbb{C}$ with poles at $z = (k + \frac{1}{2})\pi$, $k = 0, \pm 1, \pm 2, \ldots$, and $z = \infty$ is an accumulation point of poles.

DEFINITION

Essential Singularities

Essential singularities are all isolated singularities that are not removable or poles. In this case $f(z)$ does not have a limit as $z \to z_0$, and infinitely many of the coefficients c_n, $n = 1, 2, 3, \ldots$, of its Laurent series about z_0 do not vanish, since otherwise z_0 is a pole or a removable singularity.

EXAMPLE 6

The function $e^{1/z} = 1 + \frac{1}{z} + \frac{1}{2!z^2} + \frac{1}{3!z^3} + \cdots$ has an essential singularity at $z = 0$.

Remark. We have classified all isolated singularities into three groups. However singularities are not necessarily isolated. For example, the function

$$f(z) = \frac{1}{\sin(1/z)}$$

has singularities at 0 and wherever the denominator is zero, that is, whenever $1/z = n\pi$ or $z = 1/n\pi$, $n = \pm 1, \pm 2, \pm 3, \ldots$. Thus $z = 0$ is not an isolated singularity since for every $\epsilon > 0$, there are an infinite number of singularities in the region $|z| < \epsilon$.

The behavior of a function in an ϵ-neighborhood of an essential singularity is very complicated, as the following result demonstrates.

■ **THEOREM 4: Weierstrass-Casorati Theorem**

An analytic function approaches any given value arbitrarily closely in any ϵ-neighborhood of an essential singularity.

PROOF

If the theorem is false, we can find a complex number A and a $\delta > 0$ such that $|f(z) - A| > \delta$ in every neighborhood $0 < |z - z_0| < \epsilon$ of the essential singularity z_0. Then

$$\left| \frac{f(z) - A}{z - z_0} \right| > \frac{\delta}{|z - z_0|} \to \infty \qquad \text{as } z \to z_0,$$

implying that $g(z) = [f(z) - A]/(z - z_0)$ has a pole at z_0. Thus, $g(z)$ is meromorphic in $|z - z_0| < \epsilon$. But then so is $f(z) = A + (z - z_0)g(z)$, contradicting the hypothesis that z_0 is an essential singularity. ■

In fact, more can be shown, although the proof is difficult and will not be given here:

■ **THEOREM 5: Picard's Theorem**

An analytic function assumes every complex number, with possibly one exception, infinitely often in any ϵ-neighborhood of an essential singularity. ■

Problems 17.6

In Problems 1–10 find the zeros of each function. What is the order of each zero?

1. z^3

2. $z^2 + 2z + 1$

3. $(z^2 + 1)^4$

4. $z^3 e^z$

5. $z^3 \sin z$

6. $\text{Log } z^3$

7. $z^5 - 1$

8. $z^2(\cos z - 1)$ (at 0 only)

9. $6 \sin z^2 + z^2(z^4 - 6)$ (at 0 only)

10. $z - \tan z$ (at 0 only).

11. Prove that if two functions analytic on a region Ω coincide on a sequence of points in Ω that converges to a point in Ω, then they coincide everywhere in Ω.

Determine if there exists a function analytic in $|z| < 2$ assuming, at the points $z = 1/n$, $n = 1, 2, 3, \ldots$, the values given in Problems 12–15.

12. $0, 1, 0, -1, 0, 1, 0, -1, \ldots$

13. $1, 0, \frac{1}{3}, 0, \frac{1}{5}, 0, \frac{1}{7}, 0, \frac{1}{9}, 0, \ldots$

14. $1, \frac{2}{3}, \frac{3}{5}, \frac{4}{7}, \frac{5}{9}, \frac{6}{11}, \frac{7}{13}, \frac{8}{15}, \ldots$

15. $\frac{1}{2}, -\frac{1}{2}, \frac{1}{3}, -\frac{1}{3}, \frac{1}{4}, -\frac{1}{4}, \frac{1}{5}, -\frac{1}{5}, \ldots$

16. Give an example of two functions that agree at infinitely many points in a region Ω, yet are different.

17. Show that if f is a nonconstant analytic function in Ω, the set of points z satisfying $f(z) = \alpha$, α in $\mathbb{C}$, does not have an accumulation point in Ω.

In Problems 18–43 find and classify the singularities of each function.

18. $\dfrac{z^2}{z}$

19. $\dfrac{z}{z^2}$

20. $\dfrac{z + i}{z^2 + 1}$

21. $\dfrac{z + 1}{z^2 + 1}$

22. $\dfrac{z^2 - 3z + 2}{z^3 - 6z^2 + 11z - 6}$

23. $\dfrac{1}{\sin z}$

24. $\tan z$

25. $\sin \dfrac{1}{z}$

26. $\dfrac{e^z}{1+z^2}$

27. $e^{z-1/z}$

28. $ze^{1/z}$

29. $z^{-4}\cos z$

30. $e^{z/(1-z)}$

31. $\cot\dfrac{1}{z}$

32. $\cot\dfrac{1}{z}-\dfrac{1}{z}$

33. $\cot z-\dfrac{2}{z}$

34. $\dfrac{\cot z}{z^2}$

35. $\dfrac{\sinh z}{z^{10}}$

36. $\dfrac{e^{iz}}{z^4}$

37. $\tanh z$

38. $\dfrac{\cos z}{z+\pi/2}$

39. $\dfrac{1}{\sin^2 z}$

40. $\dfrac{1}{\cos^2 z}$

41. $\cot z-\dfrac{1}{z}$

42. $\dfrac{1-e^z}{1+e^z}$

43. $e^{\tan(1/z)}$

44. Show that every singularity of a rational function is either removable or a pole.

45. Prove that $f(z)$ has a removable singularity at z_0 if and only if $\lim_{z\to z_0}(z-z_0)f(z)=0$.

46. Construct a function having a removable singularity at $z=-1$, a pole of order 3 at $z=0$, and an essential sin-

gularity at $z=1$. Then find its Laurent series in $0<|z|<1$.

Show that each of the integrands in Problems 47–50 has a removable singularity at $z=0$. Remove the singularity and obtain the Maclaurin series of each integral.

47. $\text{Si}(z)=\displaystyle\int_0^z\frac{\sin\zeta}{\zeta}\,d\zeta$

48. $C(z)=\displaystyle\int_0^z\frac{\cos\zeta-1}{\zeta}\,d\zeta$

49. $E(z)=\displaystyle\int_0^z\frac{e^\zeta-1}{\zeta}\,d\zeta$

50. $L(z)=\displaystyle\int_0^z\frac{\text{Log}(1+\zeta)}{\zeta}\,d\zeta$

51. Show that the function $f(z)=e^{1/z}$ assumes every value except 0 infinitely often in any ϵ-neighborhood of $z=0$.

52. Prove that an entire function having a nonessential singularity at ∞ must be a polynomial. What kind of singularity do e^z, $\sin z$, and $\cos z$ have at ∞?

53. Show that a function meromorphic in $\mathcal{M}$ must be the quotient of two polynomials.

54. Prove that an entire function that omits the values 0 and 1 is constant. [*Hint:* Use Picard's theorem.]

Review Exercises for Chapter 17

In Exercises 1–6 determine convergence or divergence of the given series.

1. $\displaystyle\sum_{k=1}^\infty\frac{(4-13i)^k}{k!}$

2. $\displaystyle\sum_{k=1}^\infty\frac{e^{\pi ki/2}}{k}$

3. $\displaystyle\sum_{k=1}^\infty\frac{1}{(k-i)^2+i}$

4. $\displaystyle\sum_{k=1}^\infty\frac{k+i-3}{k^3-i+2}$

5. $\displaystyle\sum_{k=1}^\infty\frac{2^k e^{30\pi^3 k^2 i}}{k!}$

6. $\displaystyle\sum_{k=1}^\infty\frac{(-ik)^k}{k!}$

In Exercises 7–12 find the circle of convergence for each power series.

7. $\displaystyle\sum_{k=0}^\infty\frac{z^k}{4^k}$

8. $\displaystyle\sum_{k=0}^\infty\frac{z^k}{(2-3i)^k}$

9. $\displaystyle\sum_{k=0}^\infty\frac{(z-i)^k}{k^2+4}$

10. $\displaystyle\sum_{k=0}^\infty\frac{(z-i)^k}{k!}$

11. $\displaystyle\sum_{k=1}^\infty\frac{(z+1)^k}{k^k}$

12. $\displaystyle\sum_{k=1}^\infty\left(\frac{k}{4i}\right)^k(z-2i)^k$

In Exercises 13–16 find the sum of the geometric series.

13. $\displaystyle\sum_{k=0}^\infty\left(\frac{2}{3}i\right)^k$

14. $\displaystyle\sum_{k=0}^\infty\left(\frac{e^{\pi i/4}}{4}\right)^k$

15. $\displaystyle\sum_{k=0}^\infty\left(\frac{1}{3}+\frac{1}{3}i\right)^k$

16. $\displaystyle\sum_{k=0}^\infty\left(\frac{1-i}{\sqrt{5}}\right)^k$

In Problems 17–24 find Taylor series for the given function at the given value z_0.

17. $f(z) = e^{z^4}, \quad z_0 = 0$

18. $f(z) = \cos 2z, \quad z_0 = 0$

19. $f(z) = \sin iz, \quad z_0 = 0$

20. $f(z) = z^3 e^{z^2}, \quad z_0 = 0$

21. $f(z) = \text{Log}(1 - iz), \quad z_0 = 0$

22. $f(z) = 1/z, \quad z_0 = i$

23. $f(z) = \sin z, \quad z_0 = \pi$

24. $f(z) = \text{Log } z, \quad z_0 = i/2$

25. Find the first four nonzero terms of the Taylor series for $\cot z$ at $\pi/2$. What is its radius of convergence?

26. Find a Maclaurin series for $\sqrt[3]{1 + z^2}$.

In Exercises 27–30 prove that the given series converges uniformly in the given region.

27. $\displaystyle\sum_{k=0}^{\infty} \frac{(z - 2 + i)^k}{k!}, \quad |z - (2 - i)| \le 50$

28. $\displaystyle\sum_{k=1}^{\infty} \frac{\cos k^5 |z|}{k^4}, \quad z \in \mathbb{C}$

29. $\displaystyle\sum_{k=1}^{\infty} \frac{e^{-2i|z|}}{k^3}, \quad z \in \mathbb{C}$

30. $\displaystyle\sum_{k=1}^{\infty} \frac{z^k e^{-2i|z|}}{2^k}, \quad |z| \le 1.8$

In Exercises 31–34 find a Laurent series for the given function that converges for $0 < |z| < \infty$.

31. $\dfrac{e^{z^2}}{z^5}$

32. $z^3 \sin \dfrac{1}{z}$

33. $z^2 e^{-1/z}$

34. $\dfrac{\cos 4z}{z^6}.$

In Exercises 35–40 find a Laurent series for the given function in the given region.

35. $\dfrac{1}{z^2 + 4z}, \quad 0 < |z| < 4$

36. $\dfrac{1}{z^2 + 4z}, \quad 0 < |z + 4| < 4$

37. $\dfrac{1}{z^2 + 4z}, \quad 4 < |z - 4| < 8$

38. $\dfrac{z}{z^2 - 5z + 6}, \quad |z| < 2$

39. $\dfrac{z}{z^2 - 5z + 6}, \quad 0 < |z - 2| < 1$

40. $\dfrac{z}{z^2 - 5z + 6}, \quad 1 < |z - 3|$

41. Find a Laurent series for $z \cos \dfrac{1}{z - 2}$ valid in $0 < |z - 2| < 2$.

In Exercises 42–45 find the orders of the zeros of each function.

42. $z^3 e^z$

43. $z^3 - 6z^2 + 12z - 8$

44. $(z^2 - 1)^5$

45. $z^5(\cos z - 1)$

In Exercises 46–54 find and classify the singularities of each function.

46. $\dfrac{\sin^2 z}{z^2}$

47. $\dfrac{\sin^2 z}{z^3}$

48. $\dfrac{z^2 - z - 6}{z^3 + 4z^2 - 11z - 30}$

49. $\sin \dfrac{1}{z^2}$

50. $z \csc z$

51. $\dfrac{1}{z - 1} + \dfrac{1}{(z - 2)^2} + \dfrac{1}{(z^2 + 1)^{10}}$

52. $\dfrac{\cosh z}{z^5}$

53. $\dfrac{e^{iz^2}}{z^5}$

54. $\dfrac{1}{\sin^3 z}$

Appendix A: Review of Real Sequences and Series

The notions of convergence, divergence, and limit of a sequence defined in Section 17.1 apply to real sequences with no change, except that everything now takes place on the real line $\mathbb{R}$ instead of the complex plane $\mathbb{C}$. The first real difference arises with the notion of **boundedness,** where the ordering of the real numbers permits a sharper definition.

DEFINITION

Boundedness

(i) The sequence $\{a_n\}$ is **bounded above** if there is a number M_1 such that $a_n \leq M_1$ for every positive integer n. M_1 is called an **upper bound** for $\{a_n\}$.

(ii) It is **bounded below** if there is a number M_2 such that $M_2 \leq a_n$ for every positive integer n. M_2 is called a **lower bound** for $\{a_n\}$.

(iii) It is **bounded** if there is a number $M > 0$ such that $|a_n| \leq M$ for every positive integer n. M is called a **bound** for $\{a_n\}$.

(iv) If the sequence is not bounded, it is called **unbounded.**

Just as in the complex case,

■ THEOREM 1

Every convergent sequence is bounded.   ■

The converse of Theorem 1 is *not* true. That is, it is not true that every bounded sequence is convergent. For example, the sequences $\{(-1)^n\}$ and $\{\sin n\}$ are both bounded *and* divergent. Since boundedness alone does not ensure convergence, we need some other property.

DEFINITION

Monotonicity

(i) The sequence $\{a_n\}$ is **monotone increasing** if $a_n \leq a_{n+1}$ for every $n \geq 1$.

(ii) The sequence $\{a_n\}$ is **monotone decreasing** if $a_n \geq a_{n+1}$ for every $n \geq 1$.

(iii) The sequence $\{a_n\}$ is **monotonic** if it is either monotone increasing or monotone decreasing.

DEFINITION

Strict Monotonicity

(i) The sequence $\{a_n\}$ is **strictly increasing** if $a_n < a_{n+1}$ for every $n \geq 1$.

(ii) The sequence $\{a_n\}$ is **strictly decreasing** if $a_n > a_{n+1}$ for every $n \geq 1$.

(iii) The sequence $\{a_n\}$ is **strictly monotonic** if it is either strictly increasing or strictly decreasing.

Using these notions, we obtain

■ THEOREM 2

A bounded monotonic real sequence is convergent. ■

Completeness Axiom

The **completeness axiom** of the real number system states that every set of real numbers that is bounded above has a **least upper bound** and that every set of real numbers that is bounded below has a **greatest lower bound.** In particular, *if the sequence $\{a_n\}$ is bounded above and increasing, then it converges to its least upper bound. Similarly, if $\{a_n\}$ is bounded below and decreasing, then it converges to its greatest lower bound.*

Series of real numbers are defined in precisely the same fashion as series of complex numbers: Their convergence depends on the existence of a limit for the sequence of partial sums. In addition to the properties (such as the comparison, ratio, and root tests) that we developed in Section 17.2, real series also satisfy the following tests:

■ THEOREM 3: The Integral Test

Let f be a function that is continuous, positive, and decreasing for all $x \geq 1$. Then the series

$$\sum_{k=1}^{\infty} f(k) = f(1) + f(2) + f(3) + \cdots + f(n) + \cdots$$

converges if $\int_1^{\infty} f(x)\, dx$ converges, and diverges if $\int_1^n f(x)\, dx \to \infty$ as $n \to \infty$. ■

■ THEOREM 4: Limit Comparison Test

Let $\sum_{k=1}^{\infty} a_k$ and $\sum_{k=1}^{\infty} b_k$ be series with positive terms. If there is a number $c > 0$ such that

$$\lim_{k \to \infty} \frac{a_k}{b_k} = c,$$

then either both series converge or both series diverge. ■

DEFINITION

Alternating Series

A series in which successive terms have opposite signs is called an **alternating series.**

■ THEOREM 5: Alternating Series Test

Let $\{a_k\}$ be a decreasing sequence of positive numbers such that $\lim_{k \to \infty} a_k = 0$. Then the alternating series $\sum_{k=1}^{\infty} (-1)^{k+1} a_k = a_1 - a_2 + a_3 - a_4 + \cdots$ converges. ■

Conditional Convergence

DEFINITION

An alternating series is said to be **conditionally convergent** if it is convergent but not absolutely convergent.

■ THEOREM 6

If $S = \sum_{k=1}^{\infty} (-1)^{k+1} a_k$ is a convergent alternating series with $\{a_k\}$ monotone decreasing, then for any n,

$$|S - S_n| \leq |a_{n+1}|. \qquad ■$$

There is one fascinating fact about an alternating series that is conditionally but not absolutely convergent:

> By reordering the terms of a conditionally convergent alternating series, the new series of rearranged terms can be made to converge to *any* real number.

In the following table we summarize the convergence tests and other useful facts that apply to real series.

TABLE 1   TESTS OF CONVERGENCE

Test	Description	Examples and comments
Convergence test for a geometric series	$\sum_{k=0}^{\infty} r^k$ converges to $1/(1-r)$ if $|r| < 1$ and diverges if $|r| \geq 1$	$\sum_{k=0}^{\infty} (\frac{1}{2})^k$ converges to 2; $\sum_{k=0}^{\infty} 2^k$ diverges
Look at the terms of the series—the limit test	If $|a_k|$ does not converge to 0, then $\sum a_n$ diverges	If $a_k \to 0$, then $\sum_0^{\infty} a_k$ may converge ($\sum_{k=0}^{\infty} 1/k^2$) or it may not (the harmonic series $\sum_{k=0}^{\infty} 1/k$)
Comparison test	If $0 \leq a_k \leq b_k$ and $\sum b_k$ converges, then $\sum a_k$ converges. If $a_k \geq b_k \geq 0$ and $\sum b_k$ diverges, then $\sum a_k$ diverges	It is not necessary that $a_k \leq b_k$ or $a_k \geq b_k$ for *all* k, only for $k \geq N$ for some integer N; convergence or divergence of a series is not affected by the values of the first few terms
Integral test	If $a_k = f(k) \geq 0$, then $\sum_{k=1}^{\infty} a_k$ converges if $\int_1^{\infty} f(x)\,dx$ converges and $\sum_{k=1}^{\infty} a_k$ diverges if $\int_1^{\infty} f(x)\,dx$ diverges	Use this test whenever $f(x)$ can easily be integrated
$\sum_{k=1}^{\infty} 1/k^a$	$\sum_{k=1}^{\infty} 1/k^a$ diverges if $0 \leq a \leq 1$ and converges if $a > 1$	

TABLE 1   *(Continued)*

Test	Description	Examples and comments
Limit comparison test	If $a_k > 0$, $b_k > 0$ and there is a number $c > 0$ such that $\lim_{k \to \infty} a_k/b_k = c$, then either both series converge or both series diverge	Use the limit comparison test when a series $\sum b_k$ can be found such that (a) it is known whether $\sum b_k$ converges or diverges and (b) it appears that a_k/b_k has an easily computed limit; (b) will be true, for instance, when $a_k = 1/p(k)$ and $b_k = 1/q(k)$ where $p(k)$ and $q(k)$ are polynomials
Ratio test	If $a_k > 0$ and $\lim_{n \to \infty} a_{n+1}/a_n = L$, then $\sum_{k=1}^{\infty} a_k$ converges if $L < 1$ and diverges when $L > 1$	This is often the easiest test to apply; note that if $L = 1$, then the series may either converge ($\sum 1/k^2$) or diverge ($\sum 1/k$)
Root test	If $a_k > 0$ and $\lim_{n \to \infty} (a_n)^{1/n} = L$, then $\sum_{k=1}^{\infty} a_k$ converges if $L < 1$ and diverges if $L > 1$	If $L = 1$, the series may either converge ($\sum 1/k^2$) or diverge ($\sum 1/k$); the root test is the hardest test to apply; it is most useful when a_k is something raised to the kth power [$\sum 1/(\ln k)^k$, for example]
Alternating series test	$\sum (-1)^k a_k$ with $a_k \geq 0$ converges if (a) $a_k \to 0$ as $k \to \infty$ and (b) $\{a_k\}$ is a decreasing sequence; also, $\sum (-1)^k a_k$ diverges if $\lim_{k \to \infty} a_k \neq 0$	This test can only be applied when the terms are alternately positive and negative; if there are two or more positive (or negative) terms in a row, then try another test.
Absolute convergence test for a series with both positive and negative terms	$\sum a_k$ converges absolutely if $\sum |a_k|$ converges	To determine whether $\sum |a_k|$ converges, try any of the tests that apply to series with nonnegative terms

Appendix B: Proof of Weierstrass's Theorem

Before proving Weierstrass's Theorem (see p. 1022) we need to prove the following result which is a converse to Cauchy's theorem.

■ **THEOREM 1: Morera's Theorem**
If $f(z)$ is continuous in a simply connected region Ω and satisfies

$$\int_C f(z)\, dz = 0 \qquad \qquad \textbf{(1)}$$

for all pws closed curves C in Ω, then $f(z)$ is analytic in Ω.

PROOF

Choose a point z_0 in Ω and define

$$F(z) = \int_{z_0}^{z} f(w)\, dw \qquad (2)$$

for all z in Ω. We show first that $F(z)$ is well defined. Note that we cannot use the independence of path theorem because we are not assuming that $f(z)$ is analytic. Suppose that C_1 and C_2 are two paths joining z_0 to z as in Fig. 1.

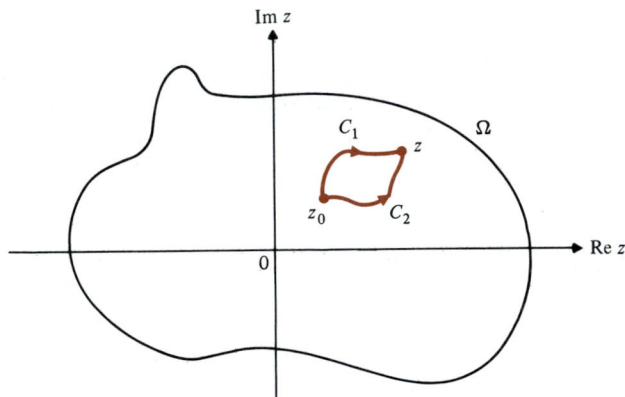

FIGURE 1

Then $C_1 - C_2$ is a pws curve in Ω so, by hypothesis,

$$0 = \int_{C_1-C_2} f(z)\, dz = \int_{C_1} f(z)\, dz + \int_{-C_2} f(z)\, dz = \int_{C_1} f(z)\, dz - \int_{C_2} f(z)\, dz.$$

Thus

$$\int_{C_1} f(z)\, dz = \int_{C_2} f(z)\, dz \qquad (3)$$

and the function defined by (2) is well defined. The rest of the proof is patterned after that of the antiderivative theorem, 16.3.4, which relies only on the continuity of f. In that proof we showed that $F(z)$ defined by (2) is analytic in Ω and that $F'(z) = f(z)$. Finally, from Theorem 16.5.2, we conclude that $f(z)$ is analytic in Ω, since it is the derivative of an analytic function. ■

■ **THEOREM 2: Weierstrass's Theorem** (see p. 988)

The sum of a uniformly convergent series of analytic functions in Ω is analytic and may be differentiated and integrated term by term in Ω.

PROOF

Let $f(z) = \sum_{k=1}^{\infty} f_k(z)$ with each $f_k(z)$ analytic in the region Ω. By uniform convergence, given $\epsilon > 0$, there is a positive integer K such that

$$\left| f(z) - \sum_{k=1}^{k^*} f_k(z) \right| < \frac{\epsilon}{3}, \tag{4}$$

for every $k^* > K$ and all z in Ω. By continuity, for any z_0 in Ω and fixed $k^* > K$, there is a $\delta > 0$ such that

$$\left| \sum_{k=1}^{k^*} f_k(z) - \sum_{k=1}^{k^*} f_k(z_0) \right| \le \sum_{k=1}^{k^*} |f_k(z) - f_k(z_0)| < \frac{\epsilon}{3}, \tag{5}$$

whenever $|z - z_0| < \delta$ for z in Ω. By the triangle inequality and inequalities (4) and (5)

$$|f(z) - f(z_0)| \le \left| f(z) - \sum_{k=1}^{k^*} f_k(z) \right| + \left| \sum_{k=1}^{k^*} f_k(z) - \sum_{k=1}^{k^*} f_k(z_0) \right|$$

$$+ \left| \sum_{k=1}^{k^*} f_k(z_0) - f(z_0) \right| < \frac{\epsilon}{3} + \frac{\epsilon}{3} + \frac{\epsilon}{3} = \epsilon,$$

for any z in Ω such that $|z - z_0| < \delta$. Thus f is continuous in Ω.

By Cauchy's theorem, for any pws closed curve C lying in a disk contained in Ω,

$$\int_C f_k(z)\,dz = 0, \qquad k = 1, 2, \ldots, n, \ldots$$

so that

$$\left| \int_C f(z)\,dz \right| = \left| \int_C f(z)\,dz - \sum_{k=1}^{n} \int_C f_k(z)\,dz \right|$$

$$\le \int_C \left| f(z) - \sum_{k=1}^{n} f_k(z) \right| |dz|. \tag{6}$$

Choose $\epsilon > 0$. Then by uniform convergence there is an N such that if $n \ge N$, $|f(z) - \sum_{k=1}^{n} f_k(z)| < \epsilon$ if $n \ge N$. Thus, from (6),

$$\left| \int_C f(z)\,dz \right| < \epsilon L, \tag{7}$$

where L is the length of C. Since ϵ can be made arbitrarily small, we have $|\int_C f(z)\,dz| = 0$, so that by Morera's theorem we see that $f(z)$ is analytic in Ω. In particular, the estimate in (6) and (7) shows that on any pws arc C in Ω

$$\int_C f(z)\,dz = \sum_{k=1}^{\infty} \int_C f_k(z)\,dz. \tag{8}$$

Now, by Cauchy's formula for derivatives

$$f'(z) = \frac{1}{2\pi i} \int_C \frac{f(\zeta)}{(\zeta - z)^2}\,d\zeta$$

and

$$f'_n(z) = \frac{1}{2\pi i} \int_C \frac{f_n(\zeta)}{(\zeta - z)^2} d\zeta.$$

Pick z_0 in Ω and choose r so that $|z - z_0| < r/2$. Consider the circle $|\zeta - z_0| = r$ (see Fig. 2). Then

$$\left| f'(z) - \sum_{k=1}^{n} f'_k(z) \right| = \left| \frac{1}{2\pi i} \int_{|\zeta - z_0| = r} \frac{f(\zeta) - \sum_{k=1}^{n} f_k(\zeta)}{(\zeta - z)^2} d\zeta \right|. \tag{9}$$

From the triangle inequality we have

$$|\zeta - z| = |(\zeta - z_0) - (z - z_0)| \geq \|\zeta - z_0| - |z - z_0\|$$

$$> \left| r - \frac{r}{2} \right| = \frac{r}{2}.$$

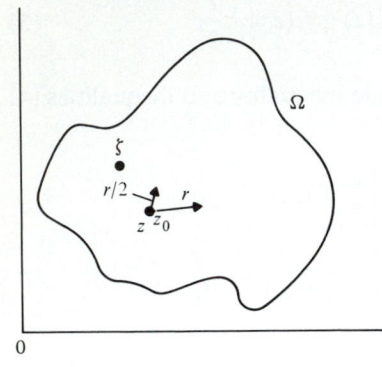

FIGURE 2

Thus $1/|\zeta - z| \leq 2/r$ and

$$\frac{1}{|\zeta - z|^2} \leq \frac{4}{r^2}.$$

Thus, choosing $n \geq N$, we have from (9)

$$\left| f'(z) - \sum_{k=1}^{n} f'_k(z) \right| \leq \frac{1}{2\pi} \int_{|\zeta - z_0| = r} \frac{|f(\zeta) - \sum_{k=1}^{n} f_k(\zeta)|}{|\zeta - z|^2} |d\zeta|$$

$$\leq \frac{\epsilon}{2\pi} \cdot \frac{4}{r^2} \cdot 2\pi r = \frac{4\epsilon}{r}.$$

Since ϵ is arbitrary and N does not depend on z, we see that $\sum f'_n(z)$ converges uniformly to $f'(z)$ on $|z - z_0| < r/2$ and, by extension, to all of Ω (explain why). Thus the proof is complete. ∎

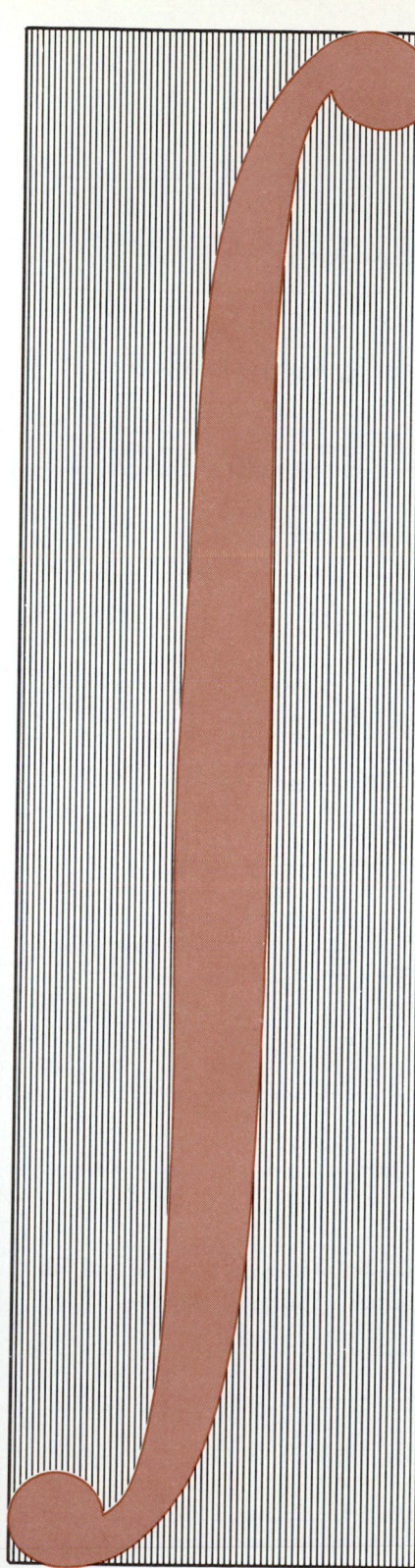

chapter 18

Residues and Contour Integration

18.1 Residues

In Section 17.5 (p. 1004) we proved that if $f(z)$ is analytic in the annulus $0 < |z - z_0| < R$, then it can be uniquely expanded into a Laurent series

$$f(z) = \sum_{k=-\infty}^{\infty} c_k(z - z_0)^k. \tag{1}$$

DEFINITION: Residue

The coefficient c_{-1} in the series (1) is called the **residue** of $f(z)$ at z_0 and is denoted by $\operatorname{Res}_{z_0} f(z)$. It is given by (see equation (17.5.5) on p. 1004)

$$\operatorname{Res}_{z_0} f(z) = c_{-1} = \frac{1}{2\pi i} \int_{|z-z_0|=\rho} f(z)\, dz, \qquad 0 < \rho < R. \tag{2}$$

Let C be any pws Jordan curve contained in the annulus $0 < |z - z_0| < R$ that encloses z_0. Then it follows from (2) and the Cauchy theorem on a multiply connected region (see p. 950) that

$$\int_C f(z)\, dz = 2\pi i c_{-1} = 2\pi i \operatorname{Res}_{z_0} f(z). \tag{3}$$

Residues can also be obtained directly from a known Laurent series.

EXAMPLE 1

Compute $\operatorname{Res}_0 e^{1/z}$.

SOLUTION
It is obvious that 0 is a singularity of the function

$$e^{1/z} = 1 + \frac{1}{z} + \frac{1}{2!z^2} + \frac{1}{3!z^3} + \cdots.$$

The residue at 0 is the coefficient of the $1/z$ term. Hence $\text{Res}_0\, e^{1/z} = 1$.

EXAMPLE 2

Compute $\int_C e^{1/z}\, dz$ where C is a pws Jordan curve having the origin in its interior.

SOLUTION
The only singularity of $e^{1/z}$ occurs at 0. Thus, from (3) and the result of Example 1,

$$\int_C e^{1/z}\, dz = 2\pi i\, \text{Res}_0\, e^{1/z} = 2\pi i(1) = 2\pi i.$$

EXAMPLE 3

Find $\text{Res}_1\, f(z)$ where $f(z) = (z-1)^2 \sin \dfrac{1}{z-1}$.

SOLUTION
Observe that

$$(z-1)^2 \sin \frac{1}{z-1} = (z-1)^2 \left[\frac{1}{z-1} - \frac{1}{3!(z-1)^3} + \frac{1}{5!(z-1)^5} - \cdots \right]$$

$$= (z-1) - \frac{1}{3!(z-1)} + \frac{1}{5!(z-1)^3} - \cdots.$$

Thus, $\text{Res}_1\, f(z) = -\dfrac{1}{3!} = -\dfrac{1}{6}$.

The integral formula in (3) can be generalized:

■ **THEOREM 1: The Residue Theorem**
Let $f(z)$ be analytic in a region Ω containing the set of all points inside and on a pws Jordan curve C except for a finite number of singularities $z_1, \ldots, z_k$ inside C. Then

$$\int_C f(z)\, dz = 2\pi i \sum_{n=1}^{k} \text{Res}_{z_n}\, f(z). \tag{4}$$

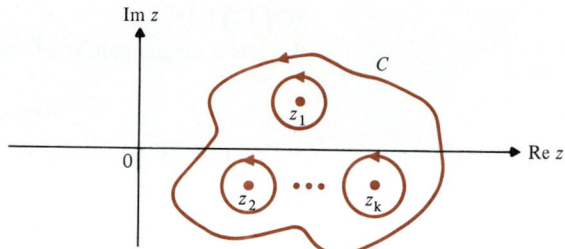

FIGURE 1

PROOF
We can draw circles $|z - z_n| = r_n$ (>0), $n = 1, \ldots, k$, inside C such that the disks $|z - z_n| \le r_n$ are disjoint from each other. (See Fig. 1.) By Cauchy's theorem on a multiply connected domain

$$\int_C f(z)\,dz = \sum_{n=1}^{k} \int_{|z-z_n|=r_n/2} f(z)\,dz. \tag{5}$$

Since $f(z)$ has the unique singularity z_n in $|z - z_n| \le r_n/2$, formula (2) applies and we have

$$\text{Res}_{z_n} f(z) = c_{-1} = \frac{1}{2\pi i} \int_{|z-z_n|=r_n/2} f(z)\,dz, \qquad n = 1, \ldots, k. \tag{6}$$

Putting (5) and (6) together proves the theorem. ∎

To be able to use the residue theorem, we need to develop an efficient way to find residues.

Suppose that $f(z)$ has a simple pole at z_0. Then $(z - z_0)f(z)$ is an analytic function representable by a Taylor series, because

$$(z - z_0)f(z) = (z - z_0)\left[\frac{c_{-1}}{z - z_0} + c_0 + c_1(z - z_0) + c_2(z - z_0)^2 + \cdots\right]$$

$$= c_{-1} + c_0(z - z_0) + c_1(z - z_0)^2 + \cdots. \tag{7}$$

Taking the limit as $z \to z_0$ in (7), we find that $\text{Res}_{z_0} f(z) = c_{-1}$. Hence

if $f(z)$ has a simple pole at z_0, then
$$\text{Res}_{z_0} f(z) = \lim_{z \to z_0}(z - z_0)f(z). \tag{8}$$

EXAMPLE 4

Find $\text{Res}_{\pi/3} f(z)$ where $f(z) = (\cos z)/(z - \pi/3)$.

SOLUTION

$f(z)$ has a simple pole at $\pi/3$ so, from (8),

$$\text{Res}_{\pi/3}\, f(z) = \lim_{z \to \pi/3} \frac{(z - \pi/3)\cos z}{z - \pi/3}$$

$$= \lim_{z \to \pi/3}\, \cos z = \cos \pi/3 = \tfrac{1}{2}.$$

EXAMPLE 5

Evaluate the integral

$$\int_{|z-1/2|=1} \frac{e^z}{z^3 - z}\, dz.$$

SOLUTION

The singularities of the integrand occur at $z = 0, \pm 1$. Hence, only the residues at the simple poles at 0 and 1 need to be calculated since only 0 and 1 are inside the circle $|z - \tfrac{1}{2}| = 1$, while -1 is not. Then

$$\text{Res}_0\, \frac{e^z}{z(z^2 - 1)} = \lim_{z \to 0} \frac{e^z}{z^2 - 1} = -1$$

and

$$\text{Res}_1\, \frac{e^z}{z(z^2 - 1)} = \lim_{z \to 1} \frac{(z - 1)e^z}{z(z^2 - 1)}$$

$$= \lim_{z \to 1} \frac{e^z}{z(z + 1)} = \frac{e}{2}.$$

Thus,

$$\int_{|z-1/2|=1} \frac{e^z}{z^3 - z}\, dz = 2\pi i\left(\frac{e}{2} - 1\right) = \pi i(e - 2).$$

If $f(z)$ has a pole of order n at z_0, then

$$f(z) = \frac{c_{-n}}{(z - z_0)^n} + \cdots + \frac{c_{-1}}{z - z_0} + c_0 + c_1(z - z_0) + \cdots.$$

Thus, $(z - z_0)^n f(z)$ is analytic in a neighborhood of z_0 and

$$(z - z_0)^n f(z) = c_{-n} + c_{-n+1}(z - z_0) + \cdots$$

$$+ c_{-1}(z - z_0)^{n-1} + c_0(z - z_0)^n + \cdots. \tag{9}$$

Differentiating (9) $n - 1$ times, we get

$$\frac{d^{n-1}}{dz^{n-1}}\left[(z - z_0)^n f(z)\right] = (n - 1)!\, c_{-1} + n!\, c_0(z - z_0) + \cdots. \tag{10}$$

Thus

$$(n-1)!c_{-1} = (n-1)! \operatorname{Res}_{z_0} f(z) = \lim_{z \to z_0} \frac{d^{n-1}}{dz^{n-1}} [(z-z_0)^n f(z)], \qquad (11)$$

so that we have

---

if $f(z)$ has a pole of order n at z_0, then

$$\operatorname{Res}_{z_0} f(z) = \frac{1}{(n-1)!} \lim_{z \to z_0} \frac{d^{n-1}}{dz^{n-1}} [(z-z_0)^n f(z)]. \qquad (12)$$

EXAMPLE 6

Find $\operatorname{Res}_{-2} f(z)$ for $f(z) = z^2 e^z/(z+2)^3$.

SOLUTION

Note that $f(z)$ has a pole of order 3 at $z_0 = -2$. Then, differentiating successively,

$$\frac{d^2}{dz^2} [(z+2)^3 f(z)] = \frac{d^2}{dz^2} [z^2 e^z] = \frac{d}{dz} [(z^2 + 2z)e^z] = e^z(z^2 + 4z + 2).$$

Thus, by (12),

$$\operatorname{Res}_{-2} f(z) = \frac{1}{2!} \lim_{z \to -2} e^z(z^2 + 4z + 2) = \tfrac{1}{2} e^{-2}(-2) = -e^{-2}.$$

EXAMPLE 7

Compute the residue at all singularities of $g(z) = (\sin z)/(z^3 - z)$.

SOLUTION

Since

$$\frac{\sin z}{z^3 - z} = \frac{\sin z}{z(z-1)(z+1)}$$

$g(z)$ has a removable singularity at 0 (because $\lim_{z \to 0} \sin z/z = 1$) and simple poles at 1 and -1. Thus the Laurent series of g at 0 is a Taylor series, implying that

$$\operatorname{Res}_0 g(z) = \lim_{z \to 0} \frac{\sin z}{z^2 - 1} = 0,$$

Also

$$\operatorname{Res}_1 g(z) = \lim_{z \to 1} \frac{\sin z}{z(z+1)} = \frac{\sin 1}{2},$$

$$\operatorname{Res}_{-1} g(z) = \lim_{z \to -1} \frac{\sin z}{z(z-1)} = \frac{\sin(-1)}{2} = \frac{-\sin 1}{2}.$$

The situation in the last example is not an isolated incident, but a general principle: An analytic function $f(z)$ having a removable singularity at z_0 can always be represented by a Taylor series in a neighborhood of z_0. Thus

---

if $f(z)$ has a removable singularity at z_0, then

$$\text{Res}_{z_0}\, f(z) = 0. \qquad\qquad (13)$$

EXAMPLE 8

Find the residue at all singularities of $h(z) = z/(\sin z)$.

SOLUTION

Note that $h(z)$ has a removable singularity at $z = 0$ so that by (13) $\text{Res}_0\, h(z) = 0$. Moreover, $h(z)$ has simple poles at $z_0 = \pi k$, $k = \pm 1, \pm 2, \ldots$. Since $\sin(z - \pi k) = (-1)^k \sin z$, the residues at πk are given by

$$\text{Res}_{\pi k}\, h(z) = \lim_{z \to \pi k} \frac{(z - \pi k)z}{\sin z} = \lim_{z \to \pi k} \frac{(z - \pi k)(-1)^k z}{\sin(z - \pi k)} = (-1)^k \pi k, \qquad k = 0, \pm 1, \pm 2, \ldots.$$

We make one final observation about calculating residues. Suppose that the quotient $h(z) = f(z)/g(z)$ has a simple pole at z_0. Then $g(z_0) = 0$, since z_0 is a simple zero of g, and

$$\text{Res}_{z_0}\, h(z) = \lim_{z \to z_0}(z - z_0)\frac{f(z)}{g(z)} = \lim_{z \to z_0} \frac{f(z)}{[g(z) - g(z_0)]/(z - z_0)} = \frac{f(z_0)}{g'(z_0)}.$$

That is,

---

if $\dfrac{f(z)}{g(z)}$ has a simple pole at z_0, then

$$\text{Res}_{z_0} \frac{f(z)}{g(z)} = \frac{f(z_0)}{g'(z_0)}. \qquad\qquad (14)$$

EXAMPLE 9

Compute $\text{Res}_0\, h(z)$ where $h(z) = (z^3 + 2z + 5)/(e^z - 1)$.

SOLUTION

$e^z - 1 = z + z^2/2! + \cdots$ so that $h(z)$ has a simple pole at 0. Since $g'(z) = e^z$, by (14) we see that

$$\text{Res}_0\, h(z) = \frac{(z^3 + 2z + 5)|_0}{e^z|_0} = 5.$$

EXAMPLE 10

Compute $\displaystyle\int_{|z|=1} \frac{z^3 + 2z + 5}{e^z - 1}\, dz$.

SOLUTION

The only singularity of the function $\dfrac{z^3 + 2z + 5}{e^z - 1}$ in $\mathbb{C}$ is a simple pole at 0. From Example 9, $\operatorname{Res}_0 h(z) = 5$. Thus, from (4),

$$\int_{|z|=1} \frac{z^3 + 2z + 5}{e^z - 1}\, dz = 2\pi i(5) = 10\pi i.$$

EXAMPLE 11

Show that $\displaystyle\int_{|z|=2} \frac{e^z}{(z^2 - 1)^2}\, dz = \frac{\pi i}{e}$.

SOLUTION

Since $(z^2 - 1)^2 = [(z + 1)(z - 1)]^2 = (z + 1)^2(z - 1)^2$, the function $e^z/(z^2 - 1)^2$ has poles of order 2 at 1 and -1 and both are inside $|z| = 2$. Then

$$\operatorname{Res}_1 f(z) = \lim_{z \to 1} \frac{d}{dz}\left[\frac{(z-1)^2 e^z}{(z^2-1)^2}\right] = \lim_{z \to 1} \frac{d}{dz}\left[\frac{e^z}{(z+1)^2}\right]$$

$$= \lim_{z \to 1} \frac{(z+1)^2 e^z - 2e^z(z+1)}{(z+1)^4} = 0,$$

$$\operatorname{Res}_{-1} f(z) = \lim_{z \to -1} \frac{d}{dz}\left[\frac{(z+1)^2 e^z}{(z^2-1)^2}\right] = \lim_{z \to -1} \frac{d}{dz}\left[\frac{e^z}{(z-1)^2}\right]$$

$$= \lim_{z \to -1} \frac{(z-1)^2 e^z - 2e^z(z-1)}{(z-1)^4} = \frac{1}{2e}.$$

Therefore

$$\int_{|z|=2} \frac{e^z}{(z^2 - 1)^2}\, dz = 2\pi i\left(0 + \frac{1}{2e}\right) = \frac{\pi i}{e}.$$

Problems 18.1

In Problems 1–20 compute the residue at each singularity of the given function.

1. $f(z) = \dfrac{1}{1+z}$

2. $f(z) = \dfrac{z^2}{1-z}$

3. $f(z) = \dfrac{z^2}{1-z^2}$

4. $f(z) = \dfrac{z}{1+z^2}$

5. $f(z) = \dfrac{e^{z^2}}{z}$

6. $f(z) = \dfrac{e^z}{z^2}$

7. $f(z) = \dfrac{\cos z}{z^2}$

8. $f(z) = \dfrac{e^z}{z^{100}}$

9. $f(z) = \dfrac{e^z}{z^3 - z}$

10. $f(z) = \dfrac{\sin z^2}{z^2}$

11. $f(z) = \dfrac{\cos^2 z}{z}$

12. $f(z) = \dfrac{3}{e^z - 1}$

13. $f(z) = \dfrac{4}{z^2 - 5z + 4}$

14. $f(z) = \dfrac{\sin z}{(z-1)^2}$

15. $f(z) = \dfrac{z^3 + 2z + 5}{z^3 - 3z^2 + 2z}$

16. $f(z) = \dfrac{z^2 + 1}{z^3}$

17. $f(z) = \dfrac{z}{(z^2 + 1)^2}$

18. $f(z) = ze^{1/z}$

19. $f(z) = z \cos \dfrac{1}{z}$

20. $f(z) = (z - 1)e^{1/z}$

In Problems 21–50 evaluate the given integral.

21. $\displaystyle\int_{|z|=1} \dfrac{e^z}{z}\, dz$

22. $\displaystyle\int_{|z|=2} \dfrac{dz}{1+z}$

23. $\displaystyle\int_{|z|=1} \dfrac{e^{z^2}}{z}\, dz$

24. $\displaystyle\int_{|z|=1} \cot z\, dz$

25. $\displaystyle\int_{|z|=1/2} \csc z\, dz$

26. $\displaystyle\int_{|z|=1} \dfrac{e^z}{z^{100}}\, dz$

27. $\displaystyle\int_{|z-1|=1} \dfrac{\sin z}{(z-1)^2}\, dz$

28. $\displaystyle\int_{|z|=3} \dfrac{dz}{z^2 - 7z + 10}$

29. $\displaystyle\int_{|z|=2} \dfrac{dz}{(z+1)^2(z-3)^2}$

30. $\displaystyle\int_{|z-2|=3} \dfrac{\cos z}{z-2}\, dz$

31. $\displaystyle\int_{|z|=2} \dfrac{z^3}{z^2 - 1}\, dz$

32. $\displaystyle\int_{|z|=2} \dfrac{z^3}{z^2 + 1}\, dz$

33. $\displaystyle\int_{|z|=2} \dfrac{e^{2z}}{z^3 + z}\, dz$

34. $\displaystyle\int_{|z-1|=2} \dfrac{e^z}{z(z-1)(z+5)^3}\, dz$

35. $\displaystyle\int_{|z|=10} \dfrac{dz}{z^3 - 5z}$

36. $\displaystyle\int_{|z|=1} \dfrac{\sin z}{z^2 - \pi^2/36}\, dz$

37. $\displaystyle\int_{|z|=1} \dfrac{(3z+1)}{9z^3 + z}\, dz$

38. $\displaystyle\int_{|z|=5} \dfrac{(3z+1)^2}{9z^3 + z}\, dz$

39. $\displaystyle\int_{|z|=1/2} \dfrac{\sin z}{(z-1)^2(z^2+4)}\, dz$

40. $\displaystyle\int_{|z|=3/2} \dfrac{\sin z}{(z-1)^2(z^2+4)}\, dz$

41. $\displaystyle\int_{|z|=5/2} \dfrac{\sin z}{(z-1)^2(z^2+4)}\, dz$

42. $\displaystyle\int_{|z|=3} \tan^2 z\, dz$

43. $\displaystyle\int_{|z+1|=2} \dfrac{dz}{z^4 + 1}$

44. $\displaystyle\int_{|z|=2} \dfrac{z^5}{(z^3 + 1)^2}\, dz$

45. $\displaystyle\int_{|z|=4} \dfrac{z^2 - 2z - 8}{(z-1)(z+2)(z-3)^2(z+5)}\, dz$

46. $\displaystyle\int_{|z|=3/2} \dfrac{z^3 - 2z^2 + 8z + 5}{z^6 + z^5 - 2z^4}\, dz$

47. $\displaystyle\int_{|z|=1} \tan 4\pi z\, dz$

48. $\displaystyle\int_{|z|=1} e^z \csc \dfrac{\pi}{2} z\, dz$

49. $\displaystyle\int_{|z|=1} \dfrac{\cosh z}{9z^2 - 1}\, dz$

50. $\displaystyle\int_{|z|=1} \dfrac{\cosh z}{9z^2 + 1}\, dz$

** 51. Prove the **inside-outside theorem.** Let

$$F(z) = \dfrac{a_n z^n + a_{n-1} z^{n-1} + \cdots + a_1 z + a_0}{b_m z^m + b_{m-1} z^{m-1} + \cdots + b_1 z + b_0},$$

where $m \geq n + 2$. Then if S is the set of singularities of $F(z)$ inside the pws Jordan curve C, and S^* is the set of singularities of $F(z)$ outside C,

$$\int_C F(z)\, dz = \begin{cases} 2\pi i \displaystyle\sum_S \operatorname{Res} F(z), \\[2ex] -2\pi i \displaystyle\sum_{S^*} \operatorname{Res} F(z). \end{cases}$$

52. Evaluate the integral

$$\int_{|z|=1} \dfrac{z+a}{z^n(z+b)}\, dz, \qquad |b| > 1.$$

Evaluate each integral in Problems 53–56 where n is a large positive integer.

53. $\displaystyle\int_{|z-1|=2} \dfrac{dz}{z^n(z^2 + 1)}$

54. $\displaystyle\int_{|z-1|=\sqrt{5}/2} \dfrac{dz}{z^n(z^2 + 1)}$

55. $\displaystyle\int_{|z-i|=3/2} \dfrac{dz}{z^n(z^2 + 1)}$

56. $\displaystyle\int_{|z-i|=1/2} \dfrac{dz}{z^n(z^2 + 1)}$

57. Suppose that $p(z)$ and $q(z)$ are polynomials. Show that all the residues of the function $[p(z)/q(z)]'$ are zero. [*Hint:* Consider the partial fractions decomposition of $p(z)/q(z)$.]

18.2 Evaluation of Real Definite Integrals in Sines and Cosines

In this and the next section we show how the residue theorem can be used to evaluate real definite integrals.

Consider an integral of the form

$$\int_0^{2\pi} F(\cos\theta, \sin\theta)\, d\theta, \tag{1}$$

where $F(s, t) = p(s, t)/q(s, t)$ is a rational function of s and t. To integrate (1), we make the substitution

$$z = e^{i\theta}.$$

Then, as θ goes from 0 to 2π, z traverses the unit circle in the positive direction. Moreover,

$$\cos\theta = \frac{1}{2}(e^{i\theta} + e^{-i\theta}) = \frac{1}{2}\left(z + \frac{1}{z}\right), \tag{2}$$

$$\sin\theta = \frac{1}{2i}(e^{i\theta} - e^{-i\theta}) = \frac{1}{2i}\left(z - \frac{1}{z}\right), \tag{3}$$

and

$$\frac{dz}{d\theta} = ie^{i\theta} = iz \quad \text{or} \quad d\theta = \frac{dz}{iz}. \tag{4}$$

Thus, we have proved the following theorem:

■ **THEOREM 1**

$$\int_0^{2\pi} F(\cos\theta, \sin\theta)\, d\theta = \int_{|z|=1} F\left[\frac{1}{2}\left(z + \frac{1}{z}\right), \frac{1}{2i}\left(z - \frac{1}{z}\right)\right]\frac{dz}{iz}. \tag{5}$$

■

EXAMPLE 1

Show that

$$\int_0^{\pi} \frac{d\theta}{a + b\cos\theta} = \frac{\pi}{\sqrt{a^2 - b^2}}, \qquad a > b > 0.$$

SOLUTION

Since $\cos\theta$ takes on the same values on $[\pi, 2\pi]$ as it does on $[0, \pi]$, the integral above equals

$$\frac{1}{2}\int_0^{2\pi} \frac{d\theta}{a + b\cos\theta}.$$

From (2),

$$a + b \cos \theta = a + \frac{b}{2}\left(z + \frac{1}{z}\right) = \frac{bz^2 + 2az + b}{2z},$$

so that

$$\frac{1}{2} \int_0^{2\pi} \frac{d\theta}{a + b \cos \theta} = \frac{1}{i} \int_{|z|=1} \frac{dz}{bz^2 + 2az + b}.$$

Now, by the quadratic formula, the roots of $bz^2 + 2az + b = 0$ are

$$z = \frac{-2a \pm \sqrt{4a^2 - 4b^2}}{2b} = \frac{-a \pm \sqrt{a^2 - b^2}}{b}.$$

Both roots are real since $a > b$ and

$$\left| \frac{-a - \sqrt{a^2 - b^2}}{b} \right| = \left| \frac{a + \sqrt{a^2 - b^2}}{b} \right| > \left| \frac{a}{b} \right| > 1.$$

Let $p = (-a + \sqrt{a^2 - b^2})/b$ and $q = (-a - \sqrt{a^2 - b^2})/b$. Then $pq = 1$ and since $|q| > 1$, it follows that $|p| < 1$. Thus p is the only zero of the denominator lying inside the unit circle; it is a simple pole. Thus, by the residue theorem,

$$\frac{1}{i} \int_{|z|=1} \frac{dz}{bz^2 + 2az + b} = \frac{1}{i} 2\pi i \, \text{Res}_p \frac{1}{bz^2 + 2az + b}.$$

But $b(z - p)(z - q) = bz^2 + 2az + b$ so that the residue is equal to

$$\lim_{z \to p} \frac{z - p}{b(z - p)(z - q)} = \frac{1}{b(p - q)} = \frac{1}{2\sqrt{a^2 - b^2}},$$

and

$$\int_0^{\pi} \frac{d\theta}{a + b \cos \theta} = 2\pi \cdot \frac{1}{2\sqrt{a^2 - b^2}} = \frac{\pi}{\sqrt{a^2 - b^2}}.$$

EXAMPLE 2

Evaluate $\displaystyle\int_0^{2\pi} \frac{\cos^2 \theta}{2 + \sin \theta} \, d\theta$.

SOLUTION
Using (2), (3), and (4) we have

$$\int_0^{2\pi} \frac{\cos^2 \theta}{2 + \sin \theta} \, d\theta = \frac{1}{i} \int_{|z|=1} \frac{[\frac{1}{2}(z + 1/z)]^2}{2 + (1/2i)(z - 1/z)} \frac{dz}{z}$$

$$= \frac{1}{i} \int_{|z|=1} \frac{(2i/4)(z^2 + 2 + 1/z^2)}{4i + (z - 1/z)} \frac{dz}{z}$$

$$= \frac{1}{2} \int_{|z|=1} \frac{z^4 + 2z^2 + 1}{z^2(z^2 + 4iz - 1)} \, dz.$$

The zeros of the denominator are 0 (of order 2) and $(-2 \pm \sqrt{3})i$. But $|(-2 - \sqrt{3})i| > 1$ so, inside the unit circle, the integrand has a pole of order 2 at 0 and a simple pole at $(-2 + \sqrt{3})i$. The residues are

$$\text{Res}_0\, f(z) = \lim_{z \to 0} \frac{d}{dz}\left(\frac{z^4 + 2z^2 + 1}{z^2 + 4zi - 1}\right)$$

$$= \lim_{z \to 0} \frac{(z^2 + 4zi - 1)(4z^3 + 4z) - (z^4 + 2z^2 + 1)(2z + 4i)}{(z^2 + 4zi - 1)^2}$$

$$= -4i = \frac{4}{i}$$

and

$$\text{Res}_{(-2+\sqrt{3})i}\, f(z) = \lim_{z \to (-2+\sqrt{3})i} \frac{z^4 + 2z^2 + 1}{z^2(z + 2i + \sqrt{3}i)}$$

$$= \frac{84 - 48\sqrt{3}}{(-7 + 4\sqrt{3})(2\sqrt{3}i)} = \frac{1}{i}\left(\frac{84 - 48\sqrt{3}}{24 - 14\sqrt{3}}\right).$$

Thus

$$\int_0^{2\pi} \frac{\cos^2 \theta}{2 + \sin \theta}\, d\theta = 2\pi i \cdot \frac{1}{2i}\left(\frac{84 - 48\sqrt{3}}{24 - 14\sqrt{3}} + 4\right) = \pi\left(\frac{180 - 104\sqrt{3}}{24 - 14\sqrt{3}}\right) \approx 0.536\pi.$$

Problems 18.2

In Problems 1–8 evaluate the integral.

1. $\displaystyle\int_0^{2\pi} \frac{d\theta}{2 + \sin \theta}$

2. $\displaystyle\int_0^{2\pi} \frac{d\theta}{4 + \cos \theta}$

3. $\displaystyle\int_0^{2\pi} \frac{\cos^2 \theta}{5 + 4 \cos \theta}\, d\theta$

4. $\displaystyle\int_0^{2\pi} \frac{\sin^2 \theta}{2 + \cos \theta}\, d\theta$

5. $\displaystyle\int_0^{2\pi} \frac{\cos^2 \theta}{6 - 2 \sin \theta}\, d\theta$

6. $\displaystyle\int_0^{2\pi} \frac{\sin \theta}{4 - \cos \theta}\, d\theta$

7. $\displaystyle\int_0^{2\pi} \frac{\cos 3\theta}{5 - 4 \cos \theta}\, d\theta$

8. $\displaystyle\int_0^{2\pi} \frac{\sin^2 \theta}{13 - 12 \sin 2\theta}\, d\theta.$

Prove the identities in Problems 9–17. In Problems 14–16 n is a nonnegative integer.

9. $\displaystyle\int_0^{\pi/2} \frac{d\theta}{a + \sin^2 \theta} = \frac{\pi}{2\sqrt{a^2 + a}}, \quad a > 0$

10. $\displaystyle\int_0^{\pi/2} \frac{d\theta}{(a + \sin^2 \theta)^2} = \frac{\pi(2a + 1)}{4\sqrt{(a^2 + a)^3}}, \quad a > 0$

11. $\displaystyle\int_0^{2\pi} \frac{d\theta}{a^2 \cos^2 \theta + b^2 \sin^2 \theta} = \frac{2\pi}{ab}, \quad a, b > 0$

12. $\displaystyle\int_0^{2\pi} \frac{d\theta}{(a^2 \cos^2 \theta + b^2 \sin^2 \theta)^2} = \frac{\pi(a^2 + b^2)}{a^3 b^3}, \quad a, b > 0$

13. $\displaystyle\int_0^{2\pi} \frac{d\theta}{1 - 2a \cos \theta + a^2} = \begin{cases} \dfrac{2\pi}{1 - a^2}, & \text{if } |a| < 1 \\[2mm] \dfrac{2\pi}{a^2 - 1}, & \text{if } |a| > 1 \end{cases}$

14. $\displaystyle\int_0^{2\pi} \cos^n \theta\, d\theta = \begin{cases} \dfrac{n!\pi}{2^{n-1}[(n/2)!]^2}, & \text{if } n \text{ is even} \\[2mm] 0, & \text{if } n \text{ is odd} \end{cases}$

15. $\displaystyle\int_0^{2\pi} (a \cos \theta + b \sin \theta)^n\, d\theta =$

$$\begin{cases} \dfrac{n!\pi}{2^{n-1}[(n/2)!]^2} \cdot \sqrt{(a^2 + b^2)^n}, & n \text{ even} \\[2mm] 0, & n \text{ odd} \end{cases} \quad a, b \text{ real}$$

16. $\displaystyle\int_0^{2\pi} e^{\cos \theta} \cos(n\theta - \sin \theta)\, d\theta = \frac{2\pi}{n!}$

* 17. $\displaystyle\int_0^{2\pi} \cot(\theta + ib)\, d\theta = -2\pi i \operatorname{sign} b, \quad b \text{ real and nonzero}$

18.3 Evaluation of Improper Real Integrals

In Theorem 18.2.1 the interval of integration was automatically transformed into a closed curve, allowing us to apply the residue theorem. In the next application this is not possible, so instead we replace the given curve by a closed curve such that, in the limit, the values of the integrals agree.

■ **THEOREM 1**

Suppose $F(z)$ is the quotient of two polynomials in z such that

(i) $F(z)$ has no poles on the real axis, and
(ii) the degree of the denominator exceeds the degree of the numerator by at least 2.

Then

$$\int_{-\infty}^{\infty} F(x) \cos ax\, dx = \mathrm{Re}\left[2\pi i \sum_{\mathrm{Im}\, z > 0} \mathrm{Res}\, F(z)e^{iaz} \right], \qquad a \ge 0 \qquad (1)$$

$$\int_{-\infty}^{\infty} F(x) \sin ax\, dx = \mathrm{Im}\left[2\pi i \sum_{\mathrm{Im}\, z > 0} \mathrm{Res}\, F(z)e^{iaz} \right], \qquad a \ge 0 \qquad (2)$$

The sum in (1) and (2) is taken on all the poles of $F(z)$ in the upper half plane $\mathrm{Im}\, z > 0$.†

PROOF

Let C be the closed curve obtained by taking the line segment $(-R, R)$ on the real axis followed by the semicircle $z = Re^{i\theta}$, $0 \le \theta \le \pi$. Since $F(z)$ is the quotient of polynomials, its poles, and hence those of $F(z)e^{iaz}$, occur only at zeros of the denominator and thus are finite in number. If R is chosen sufficiently large, all poles of $F(z)$ in the upper half plane will lie inside C (see Fig. 1). Then the residue theorem implies

$$2\pi i \sum_{\mathrm{Im}\, z > 0} \mathrm{Res}\, F(z)e^{iaz} = \int_C F(z)e^{iaz}\, dz$$

$$= \int_{-R}^{R} F(x)e^{iax}\, dx + \int_0^{\pi} F(Re^{i\theta})e^{ia\,Re^{i\theta}} i\, Re^{i\theta}\, d\theta. \qquad (3)$$

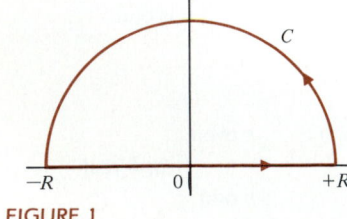

FIGURE 1

By hypothesis (ii) and the argument on p. 964, $|z^2 F(z)|$ is bounded by a

---

† Note that the improper integrals in (1) and (2) are Fourier integrals (see p. 699).

constant M at all points of the upper half plane not inside C. This means that on C

$$|R^2 e^{2i\theta} F(Re^{i\theta})| \le M \quad \text{or} \quad |F(Re^{i\theta})| \le \frac{M}{R^2}.$$

Furthermore,

$$|e^{ia\,Re^{i\theta}}| = |e^{iaR(\cos\theta + i\sin\theta)}|$$

$$= |e^{iaR\cos\theta}|\,|e^{-aR\sin\theta}| = e^{-aR\sin\theta}$$

and

$$|iRe^{i\theta}| = R.$$

Thus

$$\left| \int_0^\pi F(Re^{i\theta}) e^{ia\,Re^{i\theta}} i\,Re^{i\theta}\,d\theta \right| \le \frac{M}{R} \int_0^\pi e^{-aR\sin\theta}\,d\theta \le \frac{M\pi}{R},$$

since $e^{-aR\sin\theta} \le 1$. This shows that

$$\lim_{R\to\infty} \left| \int_0^\pi F(Re^{i\theta}) e^{iaRe^{i\theta}} i\,Re^{i\theta}\,d\theta \right| = 0.$$

By (ii) and the comparison theorem of improper integrals in calculus it follows that

$$\int_{-\infty}^\infty F(x) \cos ax\,dx \quad \text{and} \quad \int_{-\infty}^\infty F(x) \sin ax\,dx, \qquad a \ge 0,$$

both converge. Letting $R \to \infty$ in (3), we have

$$\int_{-\infty}^\infty F(x) e^{iax}\,dx = 2\pi i \sum_{\text{Im } z>0} \text{Res } F(z) e^{iaz}, \qquad a \ge 0,$$

from which the result follows by taking the real and imaginary parts of both sides. ■

EXAMPLE 1

Show that $\displaystyle\int_0^\infty \frac{\cos ax}{x^2+b^2}\,dx = \frac{\pi e^{-ab}}{2b}, \qquad a \ge 0, b > 0.$

SOLUTION
Here $F(z) = 1/(z^2 + b^2)$ with poles at $\pm ib$. Only ib is in the upper half plane. Moreover, the degree of the denominator is two more than the degree of the numerator so Theorem 1 applies. Since $(\cos ax)/(x^2 + b^2)$ is an even function, we obtain from (1)

$$\int_0^\infty \frac{\cos ax}{x^2+b^2}\,dx = \frac{1}{2}\int_{-\infty}^\infty \frac{\cos ax}{x^2+b^2}\,dx = \frac{1}{2}\,\text{Re}\left[2\pi i\,\text{Res}_{ib}\,\frac{e^{iaz}}{z^2+b^2}\right]$$

$$= \text{Re}\left[\pi i \lim_{z\to ib}\frac{e^{iaz}}{z+ib}\right] = \text{Re}\left[\pi i\frac{e^{-ab}}{2ib}\right] = \frac{\pi e^{-ab}}{b}.$$

Since

$$\text{Im}\,\frac{\pi}{b}e^{-ab}=0,$$

we see that

$$\int_{-\infty}^{\infty}\frac{\sin ax}{x^2+b^2}\,dx=0.$$

This also follows from the fact that $(\sin ax)/(x^2+b^2)$ is an odd function.

EXAMPLE 2

Compute $\displaystyle\int_0^{\infty}\frac{dx}{x^6+1}$.

SOLUTION
The function $1/(x^6+1)$ is even so that

$$\int_0^{\infty}\frac{dx}{x^6+1}=\frac{1}{2}\int_{-\infty}^{\infty}\frac{dx}{x^6+1}.$$

The poles of $1/(z^6+1)$ are the sixth roots of $-1=e^{\pi i}$. These are the simple poles $e^{\pi i/6}$, $e^{-\pi i/6}$, $e^{\pi i/2}$, $e^{-\pi i/2}$, $e^{5\pi i/6}$, $e^{-5\pi i/6}$. Only the poles $e^{\pi i/6}$, $e^{\pi i/2}=i$, and $e^{5\pi i/6}$ are in the upper half plane. From formula (18.1.14) on p. 1030, we have

$$\text{Res}_{e^{\pi i/6}}\frac{1}{z^6+1}=\frac{1}{(z^6+1)'}\bigg|_{e^{\pi i/6}}=\frac{1}{6z^5}\bigg|_{e^{\pi i/6}}=\frac{1}{6}e^{-5\pi i/6},$$

$$\text{Res}_i\frac{1}{z^6+1}=\frac{1}{6z^5}\bigg|_i=\frac{1}{6i^5}=\frac{1}{6i}=-\frac{i}{6},$$

$$\text{Res}_{e^{5\pi i/6}}\frac{1}{z^6+1}=\frac{1}{6}[e^{5\pi i/6}]^{-5}=\frac{1}{6}e^{-25\pi i/6}=\frac{1}{6}e^{-\pi i/6}.$$

Applying (1) with $a=0$, we obtain

$$\frac{1}{2}\int_{-\infty}^{\infty}\frac{dx}{x^6+1}=\text{Re}\,\frac{1}{2}\cdot\frac{1}{6}\cdot 2\pi i[e^{-5\pi i/6}-i+e^{-\pi i/6}]$$

$$=\text{Re}\,\frac{\pi}{6}i\left(-\frac{\sqrt{3}}{2}-\frac{1}{2}i-i+\frac{\sqrt{3}}{2}-\frac{1}{2}i\right)$$

$$=\text{Re}\,\frac{\pi}{6}i(-2i)=\frac{\pi}{3}.$$

The proof of Theorem 1 does not hold if the degree of the denominator of $F(x)$ is only one more than the degree of the numerator. However, we have the following result:

■ THEOREM 2
Suppose that $F(z)$ is the quotient of two polynomials in z such that

(i) $F(z)$ has no poles on the real axis, and

(ii)′ the degree of the denominator is one more than the degree of the numerator.

Then equations (1) and (2) hold for $a > 0$.

PROOF

We first observe that we cannot use the comparison theorem to obtain convergence of the integral

$$\int_{-\infty}^{\infty} F(x)e^{iax}\,dx, \qquad a > 0.$$

In fact, we must prove that

$$\int_{-X_1}^{X_2} F(x)e^{iax}\,dx, \qquad a > 0,$$

has a limit as X_1 and X_2 tend independently to ∞. Let C be the boundary of the rectangle with vertices at the points $-X_1$, X_2, $X_2 + iY$, $-X_1 + iY$, the constants X_1, X_2, Y chosen large enough that the poles of $F(z)$ in the upper half plane lie inside C (see Fig. 2).

Condition (ii)′ now shows $|zF(z)|$ bounded by M at all points in $y > 0$ not inside C. This means that if Im $z > 0$ and z is outside C, $|F(z)| \leq M/|z|$.

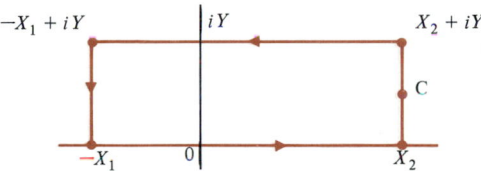

FIGURE 2

On the path from X_2 to $X_2 + iY$, $z = X_2 + iy$, $0 \leq y \leq Y$, and

$$\left| \int_{X_2}^{X_2 + iY} F(z)e^{iaz}\,dz \right| \leq \int_0^Y |F(X_2 + iy)||e^{iaX_2}|e^{-ay}\,dy$$

$$\leq \int_0^Y \frac{Me^{-ay}}{|X_2 + iy|}\,dy.$$

But $|X_2 + iy| \geq |X_2|$ so $1/|X_2 + iy| \leq 1/X_2$ and the last integral is bounded by

$$\frac{M}{X_2} \int_0^Y e^{-ay}\,dy = \frac{-M}{aX_2}e^{-ay}\Big|_0^Y = \frac{M}{aX_2}(1 - e^{-aY}) < \frac{M}{aX_2}.$$

Similarly, the integral on the line segment joining $-X_1 + iY$ to $-X_1$ is bounded by M/aX_1. Next, the line segment from $X_2 + iY$ to $-X_1 + iY$ is parametrized by $z = x + iY$, as x goes from X_2 to $-X_1$. Thus

$$\left| \int_{X_2 + iY}^{-X_1 + iY} F(z)e^{iaz}\,dz \right| \leq \frac{Me^{-aY}}{Y} \int_{-X_1}^{X_2} dx = \frac{Me^{-aY}}{Y}(X_1 + X_2).$$

Using the residue theorem and the triangle inequality, we have

$$\left| \int_{-X_1}^{X_2} F(x)e^{iax}\,dx - 2\pi i \sum_{\text{Im } z>0} \text{Res } F(z)e^{iaz} \right| < M\left[\frac{1}{aX_1} + \frac{1}{aX_2} + \frac{e^{-aY}}{Y}(X_1 + X_2) \right].$$

First letting $Y \to \infty$ and then letting X_1 and X_2 tend independently to ∞ yields the result. ■

EXAMPLE 3

Show that

$$\int_0^\infty \frac{x \sin ax}{x^2 + b^2}\,dx = \frac{\pi}{2} e^{-ab}, \qquad a > 0, \quad b > 0.$$

SOLUTION
Conditions (i) and (ii)′ apply to $F(z) = z/(z^2 + b^2)$, so that

$$\int_{-\infty}^\infty \frac{x \sin ax}{x^2 + b^2}\,dx = \text{Im}\left(2\pi i \ \text{Res}_{ib} \frac{ze^{iaz}}{z^2 + b^2} \right) = \text{Im}\left(2\pi i \lim_{z\to ib} \frac{ze^{iaz}}{z + ib} \right) = \text{Im}\left(2\pi i \frac{ibe^{-ab}}{2ib} \right) = \pi e^{-ab}.$$

The result follows because the integrand is an even function.

Problems 18.3

In Problems 1–16 compute the integral.

1. $\displaystyle \int_0^\infty \frac{dx}{x^2 + 1}$

2. $\displaystyle \int_{-\infty}^\infty \frac{1}{x^4 + 1}\,dx$

3. $\displaystyle \int_{-\infty}^\infty \frac{x}{x^4 + 1}\,dx$

4. $\displaystyle \int_0^\infty \frac{x^2}{x^4 + 1}\,dx$

5. $\displaystyle \int_{-\infty}^\infty \frac{x \cos 2x}{x^2 + 1}\,dx$

6. $\displaystyle \int_0^\infty \frac{x \sin x}{(x^2 + 1)^2}\,dx$

7. $\displaystyle \int_0^\infty \frac{x^2 \cos 2x}{(x^2 + 1)^2}\,dx$

8. $\displaystyle \int_0^\infty \frac{x^3 \sin 2x}{(x^2 + 1)^2}\,dx$

9. $\displaystyle \int_0^\infty \frac{x^2 \cos x}{x^4 + 1}\,dx$

10. $\displaystyle \int_0^\infty \frac{x^3 \sin x}{(x^2 + 1)(x^2 + 4)}\,dx$

11. $\displaystyle \int_{-\infty}^\infty \frac{x^2}{(1 + x^2)^2}\,dx$

12. $\displaystyle \int_0^\infty \frac{x^3 \sin x}{x^4 + 5x^2 + 6}\,dx$

13. $\displaystyle \int_{-\infty}^\infty \frac{x^2 \cos x}{x^4 + 9}\,dx$

14. $\displaystyle \int_{-\infty}^\infty \frac{x}{x^6 + 1}\,dx$

15. $\displaystyle \int_{-\infty}^\infty \frac{x^2}{x^6 + 1}\,dx$

16. $\displaystyle \int_{-\infty}^\infty \frac{x^3}{x^6 + 1}\,dx$

In Problems 17–26 verify the given identity.

17. $\displaystyle \int_{-\infty}^\infty \frac{x\,dx}{(x^2 + 2x + 2)^2} = -\frac{\pi}{2}$

18. $\displaystyle \int_{-\infty}^\infty \frac{x^2\,dx}{(x^2 + 2x + 2)^2} = \pi$

19. $\displaystyle \int_0^\infty \frac{x^2\,dx}{(x^2 + a^2)^2} = \frac{\pi}{4a}, \quad a > 0$

20. $\displaystyle \int_{-\infty}^\infty \frac{dx}{(x^2 + a^2)(x^2 + b^2)} = \frac{\pi}{ab(a + b)}, \quad a, b > 0$

21. $\displaystyle \int_{-\infty}^\infty \frac{dx}{(x^2 + 1)^{n+1}} = \frac{(2n)!\pi}{2^{2n}(n!)^2}, \quad n \text{ a nonnegative integer}$

22. $\displaystyle \int_{-\infty}^\infty \frac{\cos ax\,dx}{(x^2 + b^2)^2} = \frac{\pi(1 + ab)e^{-ab}}{2b^3}, \quad a \geqslant 0, b > 0$

23. $\displaystyle \int_{-\infty}^\infty \frac{x^3 \sin ax\,dx}{(x^2 + b^2)^2} = \frac{\pi}{2}(2 - ab)e^{-ab}, \quad a, b > 0$

24. $\displaystyle \int_0^\infty \frac{\cos ax}{x^4 + b^4}\,dx = \frac{\pi}{2b^3} e^{-(ab)/\sqrt{2}} \sin\left(\frac{ab}{\sqrt{2}} + \frac{\pi}{4} \right), \quad a \geqslant 0, b > 0$

25. $\displaystyle \int_0^\infty \frac{x \sin ax}{x^4 + b^4}\,dx = \frac{\pi}{2b^2} e^{-(ab)/\sqrt{2}} \sin \frac{ab}{\sqrt{2}}, \quad a \geqslant 0, b > 0$

26. $\displaystyle \int_0^\infty \frac{x^3 \sin ax}{x^4 + b^4}\,dx = \frac{\pi}{2} e^{-(ab)/\sqrt{2}} \cos \frac{ab}{\sqrt{2}}, \quad a, b > 0$

(Optional)

18.4 Using the Residue Theorem to Sum Convergent Series

The residue theorem can, in certain circumstances, be used to find the sum of a convergent series. We have the following theorem.

■ **THEOREM 1**
Let $P(z)$ be a polynomial of degree ≥ 2 with zeros $z_1, z_2, \ldots, z_n$, none of which is an integer. Then

$$\sum_{k=-\infty}^{\infty} \frac{1}{P(k)} = -\pi \sum_{j=1}^{n} \operatorname{Res}_{z_j} \frac{\cot \pi z}{P(z)}, \tag{1}$$

$$\sum_{k=-\infty}^{\infty} \frac{(-1)^k}{P(k)} = -\pi \sum_{j=1}^{n} \operatorname{Res}_{z_j} \frac{\csc \pi z}{P(z)}. \tag{2}$$

PROOF
We shall prove (1) here and leave the proof of (2) as an exercise (see Problem 13). Let N be a positive integer sufficiently large so that all the zeros of $P(z)$ are in the interior of the circle C_N: $|z| = N + \frac{1}{2}$. Then

$$\int_{C_N} \frac{\pi \cot \pi z}{P(z)} \, dz = 2\pi i \sum_{\text{inside } C_N} \operatorname{Res} \frac{\pi \cot \pi z}{P(z)}.$$

The residues of $(\cot \pi z)/P(z)$ come in two varieties. First are the zeros of $P(z)$: $z_1, z_2, \ldots, z_n$. In addition, $\cot \pi z = (\cos \pi z)/(\sin \pi z)$ has simple poles at the zeros of $\sin \pi z$ in C_N. These occur at

$$-N, -N-1, \ldots, -1, 0, 1, 2, \ldots, N-1, N.$$

Let us compute the residue at the integer $z = k$. We have

$$\operatorname{Res}_k \frac{\pi \cos \pi z}{P(z) \sin \pi z} = \lim_{z \to k} \frac{(z-k)\pi \cos \pi z}{P(z) \sin \pi z}$$

$$= \lim_{z \to k} \frac{\pi \cos \pi z}{P(z)} \lim_{z \to k} \frac{z-k}{\sin \pi z} \overset{\text{L'Hôpitals' rule}}{=} \frac{\pi \cos k\pi}{P(k)} \lim_{z \to k} \frac{1}{\pi \cos \pi z}$$

$$= \frac{1}{P(k)}.$$

Thus

$$\int_{C_N} \frac{\pi \cot \pi z}{P(z)} \, dz = 2\pi i \left[\pi \sum_{j=1}^{n} \operatorname{Res}_{z_j} \frac{\cot \pi z}{P(z)} + \sum_{k=-N}^{N} \frac{1}{P(k)} \right]. \tag{3}$$

Note that it was important here that none of the zeros of $P(z)$ was an integer. This means that none of the numbers $z_1, z_2, \ldots, z_n$ is a pole of $\cot \pi z$.

We now prove that

$$\lim_{N \to \infty} \int_{C_N} \frac{\pi \cot \pi z}{P(z)} dz = 0. \tag{4}$$

The function $\cot \pi z$ is analytic in the region $\Omega = \mathbb{C} - \bigcup_{k=-\infty}^{\infty} \{|z - k| \leq \varepsilon\}$. Moreover, if we choose $\varepsilon < \frac{1}{2}$, then $C_N \subset \Omega$. By the maximum principle it follows that $\cot \pi z$ is bounded by M on Ω and therefore on C_N. By the argument on p. 964

$$\left| \frac{z^2}{P(z)} \right| \leq M_1$$

on every C_N for N sufficiently large. Selecting N large, we have

$$\left| \int_{C_N} \frac{\pi \cot \pi z}{P(z)} dz \right| \leq \pi \int_{C_N} \left| \frac{z^2}{P(z)} \right| \frac{|\cot \pi z|}{|z|^2} |dz|$$

$$\leq \pi M_1 M \int_{C_N} \frac{1}{|z|^2} |dz| \overset{\substack{|z| = N + \frac{1}{2} \text{ on } C_N \quad \text{length of } C_N \\ \downarrow \qquad \overbrace{\qquad\qquad}}}{\leq} \frac{\pi M_1 M}{(N + \frac{1}{2})^2} 2\pi(N + \tfrac{1}{2})$$

$$= \frac{2\pi^2 M_1 M}{N + \frac{1}{2}} \to 0$$

as $N \to \infty$. Using (4), we may take the limit as $N \to \infty$ in (3) to obtain (1). ∎

EXAMPLE 1

Compute $L = \sum_{k=0}^{\infty} \frac{1}{k^2 + 1}$.

SOLUTION
Since $k^2 + 1$ is even,

$$L = \sum_{k=-\infty}^{0} \frac{1}{k^2 + 1}$$

and

$$\sum_{k=-\infty}^{\infty} \frac{1}{k^2 + 1} = \sum_{k=-\infty}^{0} \frac{1}{k^2 + 1} + \sum_{k=0}^{\infty} \frac{1}{k^2 + 1} - 1 = 2L - 1$$

(we subtracted 1 because the term obtained by setting $k = 0$ is added in twice).

Now $1/(z^2 + 1)$ has the poles $\pm i$. Thus, from (1),

$$\sum_{k=-\infty}^{\infty} \frac{1}{k^2+1} = -\pi \left[\operatorname{Res}_i \frac{\cot \pi z}{z^2+1} + \operatorname{Res}_{-i} \frac{\cot \pi z}{z^2+1} \right]$$

$$= -\pi \left[\frac{\cot \pi i}{2i} + \frac{\cot(-\pi i)}{-2i} \right]$$

$$= i\pi \cot \pi i = \pi \coth \pi.$$

Thus

$$2L - 1 = \pi \coth \pi$$

and

$$L = \sum_{k=0}^{\infty} \frac{1}{k^2+1} = \frac{1}{2} + \frac{\pi}{2} \coth \pi \approx 2.0767.$$

Problems 18.4

In Problems 1–12 find the sum of the given series.

1. $\displaystyle\sum_{k=0}^{\infty} \frac{1}{k^2+4}$

2. $\displaystyle\sum_{k=0}^{\infty} \frac{1}{k^2+a^2}, \; a > 0$

3. $\displaystyle\sum_{k=0}^{\infty} \frac{1}{k^4+1}$

4. $\displaystyle\sum_{k=-\infty}^{\infty} \frac{1}{(3k+1)^4}$

5. $\displaystyle\sum_{k=0}^{\infty} \frac{(-1)^k}{k^2+1}$

6. $\displaystyle\sum_{k=0}^{\infty} \frac{(-1)^k}{(4k+2)^3}$

7. $\displaystyle\sum_{k=-\infty}^{\infty} \frac{1}{(3k+2)^2}$

8. $\displaystyle\sum_{k=-\infty}^{\infty} \frac{(-1)^k}{(4k+1)^3}$

9. $\displaystyle\sum_{k=-\infty}^{\infty} \frac{1}{k^4-2}$

10. $\displaystyle\sum_{k=0}^{\infty} \frac{(-1)^k}{k^4+a^4}, \; a > 0$

11. $\displaystyle\sum_{k=-\infty}^{\infty} \frac{1}{(k-a)^2}, \; a > 0$ not an integer

12. $\displaystyle\sum_{k=0}^{\infty} \frac{1}{(2k^2+1)^2}$

13. Prove part (b) of Theorem 1.

18.5 The Argument Principle

Another application of the residue theorem, useful in determining the number of zeros and poles of a meromorphic function, is the following result.

■ **THEOREM 1: The Argument Principle**
Let $w = f(z)$ be meromorphic in the simply connected region Ω and let C be a pws Jordan curve in Ω that avoids the zeros and poles of $f(z)$. Then

$$\frac{1}{2\pi i} \int_C \frac{f'(z)}{f(z)} \, dz = Z - P = \text{number of zeros of } f(z) \text{ minus}$$
$$\text{number of poles of } f(z), \text{ including}$$
$$\text{multiplicities, lying inside } C. \tag{1}$$

PROOF

If a is a zero of order k of $f(z)$, then we write $f(z) = (z - a)^k f_0(z)$, with $f_0(z)$ analytic and nonzero in an ϵ-neighborhood of a. Thus,

$$f'(z) = k(z-a)^{k-1} f_0(z) + (z-a)^k f'_0(z)$$

and

$$\frac{f'(z)}{f(z)} = \frac{k}{z-a} + \frac{f'_0(z)}{f_0(z)}.$$

Since f'_0/f_0 is analytic in an ϵ-neighborhood of a, we see that f'/f has a pole of order 1 with residue k at $z = a$.

On the other hand, if a is a pole of order h of $f(z)$, then $f(z) = f_0(z)/(z-a)^h$ with $f_0(z)$ again nonzero and analytic in an ϵ-neighborhood of a. In this case

$$f'(z) = \frac{(z-a)^h f'_0(z) - h(z-a)^{h-1} f_0(z)}{(z-a)^{2h}} = \frac{f'_0(z)}{(z-a)^h} - \frac{h f_0(z)}{(z-a)^{h+1}}$$

so that

$$\frac{f'(z)}{f(z)} = \frac{-h}{z-a} + \frac{f'_0(z)}{f_0(z)}$$

has a pole of order 1 with residue $(-h)$ at $z = a$. By the residue theorem, it follows that

$$\frac{1}{2\pi i} \int_C \frac{f'(z)}{f(z)} \, dz = Z - P,$$

where Z is the sum of all the orders k of the zeros of $f(z)$, and P is the sum of all the orders h of the poles of $f(z)$, lying inside C. ∎

Remark. If we let $w = f(z)$, then the argument principle can be rewritten as

$$\frac{1}{2\pi i} \int_{f(C)} \frac{dw}{w} = \frac{1}{2\pi i} \int_C \frac{f'(z)\, dz}{f(z)} = Z - P. \tag{2}$$

But

$$\int_{f(C)} \frac{dw}{w} = \log w = \ln|w| + i \arg w$$

and $f(C)$ is a closed curve, so that by the fundamental theorem of calculus (see Theorem 16.2.1 on p. 940) the real parts cancel out, leaving only the difference in argument of the starting and ending points of the curve $f(C)$. Thus, the first integral in (2) measures the number of times $f(C)$ winds (counterclockwise) around 0; in other words, it measures the *variation of the argument* of $f(z)$ as z traverses the curve C. This is what gives the theorem its name.

EXAMPLE 1

Compute $\int_{|z|=10} \tan z \, dz$.

SOLUTION

$$\tan z = \frac{\sin z}{\cos z} = -\frac{(-\sin z)}{\cos z} = -\frac{f'(z)}{f(z)}$$

where $f(z) = \cos z$. Thus, by the argument principle,

$$\int_{|z|=10} \tan z \, dz = -2\pi i \text{ (number of zeros of } \cos z \text{ minus number of poles of } \cos z \text{ inside } |z| = 10).$$

Now $\cos z$ has no poles since it is entire. Its zeros are at $(n + \frac{1}{2})\pi$, $n = 0, \pm 1, \pm 2, \ldots$. Inside the circle $|z| = 10$, it has zeros at $\pi/2$, $-\pi/2$, $3\pi/2$, $-3\pi/2$, $5\pi/2$, and $-5\pi/2$ ($7\pi/2 \approx 11$ is outside the circle). We count 6 zeros so

$$\int_{|z|=10} \tan z \, dz = -12\pi i.$$

A most useful application of the argument principle is the following result.

THEOREM 2: Rouché's Theorem

Let $f(z)$ and $g(z)$ be analytic in a simply connected region Ω. If $|f(z)| > |g(z) - f(z)|$ at all points of the pws Jordan curve C lying in Ω then $f(z)$ and $g(z)$ have the same number of zeros inside C.

PROOF

The hypothesis $|f(z)| > |g(z) - f(z)|$ forces both functions to be nonzero on C; thus C avoids the poles and zeros of $F(z) = g(z)/f(z)$. However, for all z on C,

$$\left| \frac{g(z)}{f(z)} - 1 \right| < 1.$$

Thus $F(C)$ does not wind around 0, so the argument principle implies that $F(z)$ has the same number of zeros as it has poles inside C. But these correspond to the zeros of $g(z)$ and $f(z)$, respectively, so the proof is complete. ■

EXAMPLE 2

Find the number of roots of the equation $z^4 + 5z + 1 = 0$ lying inside the circle $|z| = 1$.

SOLUTION

Let $f(z) = 5z$ and $g(z) = z^4 + 5z + 1$. Then, by the triangle inequality,

$$|g(z) - f(z)| \leq |z|^4 + 1 < |5z| = |f(z)|$$

on $|z| = 1$. Since $f(z)$ has one zero inside $|z| = 1$, so does $g(z)$. On the other hand, letting $f(z) = z^4$, we have

$$|g(z) - f(z)| = |5z + 1| \leqslant 11 < 16 = |z|^4 = |f(z)|$$

on $|z| = 2$. Thus, $g(z)$ has four zeros inside $|z| = 2$, three of which lie in the annulus $1 < |z| < 2$, since no zeros lie on $|z| = 1$.

EXAMPLE 3

Show that $z - e^z + a = 0$, $a > 1$, has one root in the left half plane.

SOLUTION

Let $f(z) = z + a$ and $g(z) = z - e^z + a$. For $z = iy$ or $|z| = R > 2a$, $x < 0$, we have

$$|g(z) - f(z)| = e^{\text{Re }z} \leqslant 1 < a \leqslant |f(z)|,$$

and $f(z)$ has only one root (at $z = -a$), so the proof is complete.

EXAMPLE 4

Find the number of roots of the equation

$$z^4 + iz^3 + 3z^2 + 2iz + 2 = 0$$

lying in the upper half plane.

SOLUTION

Let $f(z) = z^4 + 3z^2 + 2 = (z^2 + 2)(z^2 + 1)$ and $g(z) = z^4 + iz^3 + 3z^2 + 2iz + 2$. For $z = x$ or $|z| = R \geqslant 2$, we have

$$|g(z) - f(z)| = |z||z^2 + 2| < |z^2 + 1||z^2 + 2| = |f(z)|,$$

so $g(z)$ has two roots in the upper half plane.

EXAMPLE 5

Find the number of roots of the equation

$$7z^3 - 5z^2 + 4z - 2 = 0$$

in the disk $|z| \leqslant 1$.

SOLUTION

If we multiply the equation by $z + 1$, we obtain

$$7z^4 + 2z^3 - z^2 + 2z - 2 = 0.$$

Letting $f(z) = 7z^4$ and $g(z) = 7z^4 + 2z^3 - z^2 + 2z - 2$, we find by the triangle inequality that

$$|g(z) - f(z)| \leqslant 2|z|^3 + |z|^2 + 2|z| + 2 < 7|z|^4 = |f(z)|,$$

whenever $|z| = 1 + \epsilon$, $\epsilon > 0$. Hence, $g(z)$ has four roots in $|z| \leqslant 1$, implying that the original equation had three roots in the closed unit disk.

Problems 18.5

In Problems 1–6, use the argument principle to compute the integral.

1. $\displaystyle\int_{|z|=8} \cot z \, dz$

2. $\displaystyle\int_{|z|=5/2} \frac{3z^2 - 12z + 11}{z^3 - 6z^2 + 11z - 6} \, dz$

3. $\displaystyle\int_{|z|=1} \frac{e^z}{e^z - 1} \, dz$

4. $\displaystyle\int_{|z|=4} \frac{e^z}{e^z + 1} \, dz$

5. $\displaystyle\int_{|z|=5} \frac{2z^3 - z}{z^4 - z^2 - 2} \, dz$

6. $\displaystyle\int_{|z|=10} \tan \frac{z}{4} \, dz$

In Problems 7–10 find the number of roots of the given equation inside the circle $|z| = 1$.

7. $z^5 + 8z + 10 = 0$

8. $z^8 - 2z^5 + z^3 - 8z^2 + 3 = 0$

9. $z^6 + 3z^5 - 2z^2 + 2z - 9 = 0$

10. $z^7 - 7z^6 + 4z^3 - 1 = 0$

11. How many of the roots of the equations given in Problems 7–10 lie inside $|z| = 2$?

12. How many roots of the equation

$$3z^4 - 6iz^3 + 7z^2 - 2iz + 2 = 0$$

lie in the upper half plane?

13. How many roots of the equation

$$z^6 + z^5 - 6z^4 - 5z^3 + 10z^2 + 5z - 5 = 0$$

lie in the right half plane?

14. Find the number of roots of the equation

$$9z^4 + 7z^3 + 5z^2 + z + 1 = 0$$

lying in the disk $|z| \leq 1$.

15. How many roots of the equation

$$z^4 + 2z^3 - 3z^2 - 3z + 6 = 0$$

lie in the disk $|z| \leq 1$?

16. Use Rouché's theorem to prove the fundamental principle of algebra.

* 17. Suppose that $f(z)$ is analytic on and inside a pws Jordan curve C and that $f(z)$ is real for z on C. Prove that $f(z)$ is constant.

Review Exercises for Chapter 18

In Exercises 1–12 compute the residue at each singularity of the given function.

1. $f(z) = \dfrac{1}{z - 2}$

2. $f(z) = \dfrac{z^2}{3 - z}$

3. $f(z) = \dfrac{\cos 2z}{z^2}$

4. $f(z) = \dfrac{\sin 2z}{z^2}$

5. $f(z) = \dfrac{\sin z}{(z + 2)^2}$

6. $f(z) = z^2 e^{2/z}$

7. $f(z) = \dfrac{2}{z^2 + 9}$

8. $f(z) = \dfrac{2}{z^2 - 9}$

9. $f(z) = \dfrac{3z + 2}{(z^4 + 3z^2 + 2)}$

10. $f(z) = \dfrac{\sin^3 z}{z^3}$

11. $f(z) = \dfrac{z^4 - 3z^2 + 3}{z^2 - 5z + 6}$

12. $f(z) = \dfrac{e^{z^2}}{e^z - 1}$

In Exercises 13–24 evaluate the given integral.

13. $\displaystyle\int_{|z|=1} \frac{e^{2z}}{z} \, dz$

14. $\displaystyle\int_{|z|=4} \frac{dz}{1 + z}$

15. $\displaystyle\int_{|z|=1} \frac{dz}{z^2 - 5z + 6}$

16. $\displaystyle\int_{|z|=5/2} \frac{dz}{z^2 - 5z + 6}$

17. $\displaystyle\int_{|z|=7/2} \frac{dz}{z^2 - 5z + 6}$

18. $\displaystyle\int_{|z-2|=2} \frac{\sin z}{(z - 2)^2} \, dz$

19. $\displaystyle\int_{|z|=2} \frac{z^4}{z^2 + 1} \, dz$

20. $\displaystyle\int_{|z|=4} \frac{e^z}{z^3 + 4z} \, dz$

21. $\displaystyle\int_{|z|=1} \frac{\sin z}{(z - 2)^2(z^2 + 9)} \, dz$

22. $\displaystyle\int_{|z|=5/2} \frac{\sin z}{(z - 2)^2(z^2 + 9)} \, dz$

23. $\displaystyle\int_{|z|=7/2} \frac{\sin z}{(z - 2)^2(z^2 + 9)} \, dz$

24. $\displaystyle\int_{|z|=5} \cot^2 z \, dz$

In Exercises 25–33 evaluate the integral.

25. $\displaystyle\int_0^{2\pi} \frac{d\theta}{2+\cos\theta}$

26. $\displaystyle\int_0^{2\pi} \frac{d\theta}{3-2\sin\theta}$

27. $\displaystyle\int_0^{2\pi} \frac{\sin^2\theta}{2-\cos\theta}\,d\theta$

28. $\displaystyle\int_0^{\infty} \frac{\cos 2x}{x^2+1}\,dx$

29. $\displaystyle\int_0^{\infty} \frac{dx}{x^4+16}$

30. $\displaystyle\int_0^{\infty} \frac{x^2}{(1+x^2)^2}\,dx$

31. $\displaystyle\int_0^{\infty} \frac{x\sin x}{x^2+4}\,dx$

32. $\displaystyle\int_0^{\infty} \frac{x^2\cos x}{x^4+3x^2+2}\,dx$

33. $\displaystyle\int_{-\infty}^{\infty} \frac{x^2\sin^2 x}{x^4+16}\,dx$

In Exercises 34–37 find the sum of the given series.

34. $\displaystyle\sum_{k=0}^{\infty} \frac{1}{k^2+9}$

35. $\displaystyle\sum_{k=0}^{\infty} \frac{(-1)^k}{k^2+4}$

36. $\displaystyle\sum_{k=-\infty}^{\infty} \frac{(-1)^k}{(6k+1)^2}$

37. $\displaystyle\sum_{k=-\infty}^{\infty} \frac{1}{(6k+1)^2}$

38. Compute $\displaystyle\int_{|z|=5} \sec z\,dz$

39. Compute $\displaystyle\int_{|z|=6} \frac{z^3-2z}{z^4-4z^2}\,dz$

40. Find the number of roots of $z^7 - 4z^3 + z - 1 = 0$ inside the unit circle.

* 41. Show that if z is not a pole of $\csc z$, then

$$\csc z = \frac{1}{z} + 2\sum_{k=1}^{\infty} \frac{(-1)^k z}{z^2 - k^2\pi^2}.$$

* 42. (a) Show that the series in Exercise 41 converges uniformly.

(b) Integrate the series term by term to obtain

$$\text{Log}\tan\tfrac{1}{2}z = K + \text{Log}\,\tfrac{1}{2}z + \sum_{k=1}^{\infty}(-1)^k \text{Log}\left(1 - \frac{z^2}{k^2\pi^2}\right).$$

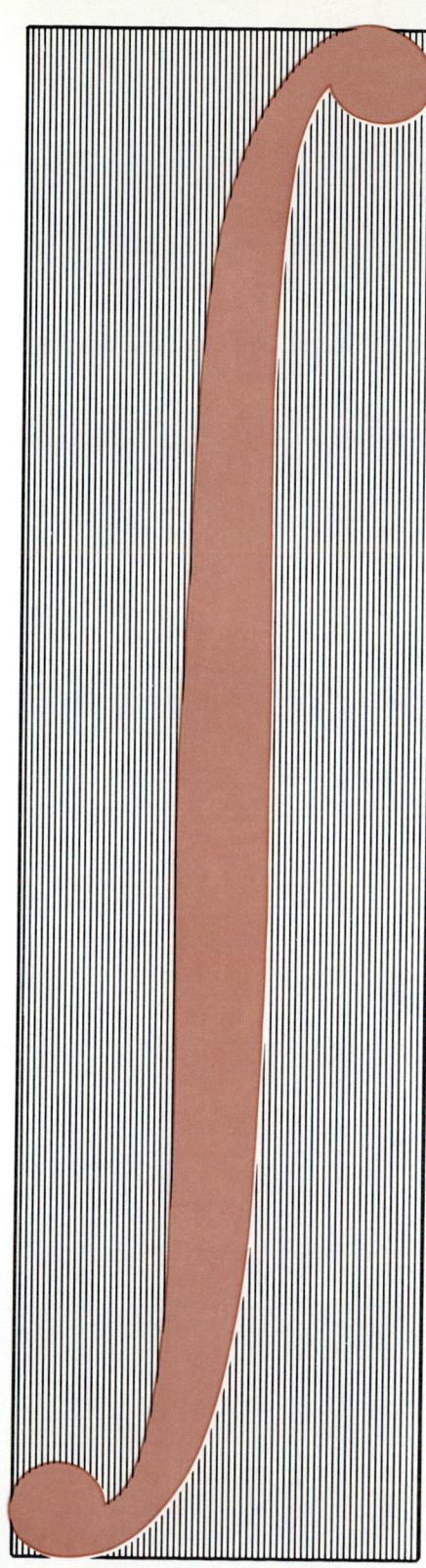

chapter 19

Conformal Mapping

19.1 The Complex Function Viewed as a Mapping

In Chapters 15–18 we looked at a complex function one point at a time. That is, we started with a point z_0 in the domain of $f(z)$ and computed $f(z_0)$. In this chapter, we shall look at a complex function $w = f(z)$, as a mapping from the z-plane to the w-plane (see Section 15.5, p. 896). We shall describe how certain sets in the z-plane are transformed in the w-plane by the mapping.

EXAMPLE 1

Describe the images of lines and circles in the z-plane under the mapping $w = z^2$.

SOLUTION
We have

$$w = u + iv = (x + iy)^2 = x^2 - y^2 + 2xyi.$$

Thus

$$u = x^2 - y^2 \quad \text{and} \quad v = 2xy. \tag{1}$$

We begin by considering the images of straight lines in the z-plane. Straight lines parallel to the imaginary axis have the equation Re $z = x = c$ and straight lines parallel to the y-axis have the equation Im $z = y = c$. If $x = c$, then, from (1)

$$u = c^2 - y^2, \qquad v = 2cy.$$

If $c \neq 0$, we obtain

$$y = \frac{v}{2c} \qquad \text{so that} \quad u = c^2 - \frac{v^2}{4c^2}.$$

The equation $u = c^2 - v^2/4c^2$ is the equation of a parabola in the w-plane with focus at the origin, directrix the line $u = 2c^2$, u-axis as the axis, and opening to the left. The

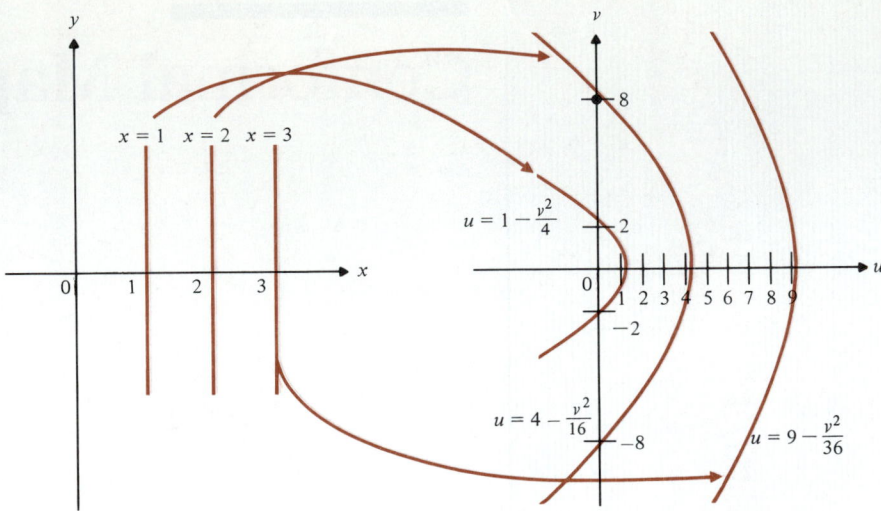

FIGURE 1

images of the lines $x = 1$, $x = 2$, and $x = 3$ are sketched in Fig. 1. Note that the image of $x = -c$ is the same as the image of $x = c$. The image of $x = 0$ is the negative real axis including the origin.

Next, if $y = c$, then

$$u = x^2 - c^2, \qquad v = 2cx, \qquad x = \frac{v}{2c},$$

and

$$u = \frac{v^2}{4c^2} - c^2.$$

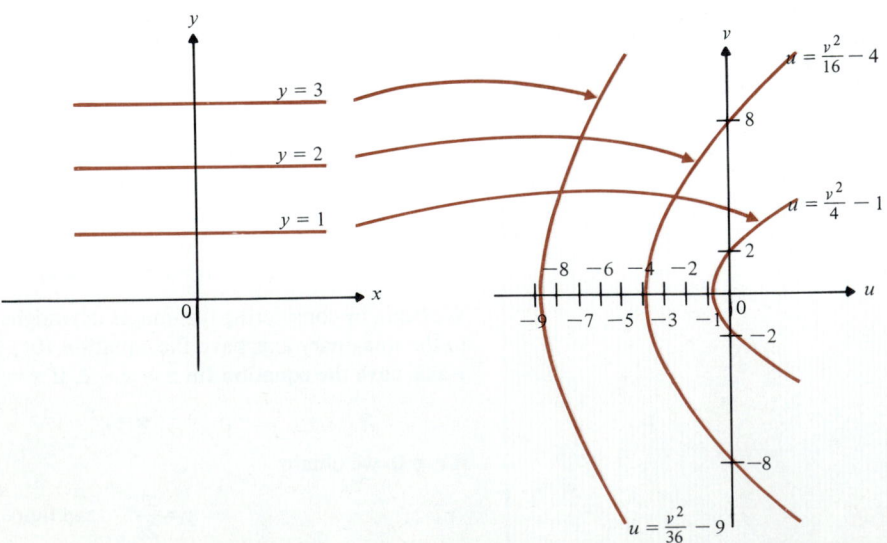

FIGURE 2

This is a parabola with focus at the origin, directrix the lines $u = -2c^2$, axis the u-axis, and opening to the right. The images of the lines $y = 1$, $y = 2$, and $y = 3$ are sketched in Fig. 2. The image of $y = 0$ is the positive real axis including the origin.

The images of other straight lines can be found. For example, if $y = x$, then $u = 0$ and $v = 2x^2 \geq 0$. Thus the image of $y = x$ is the half line $u = 0$, $v \geq 0$ (see Fig. 3).

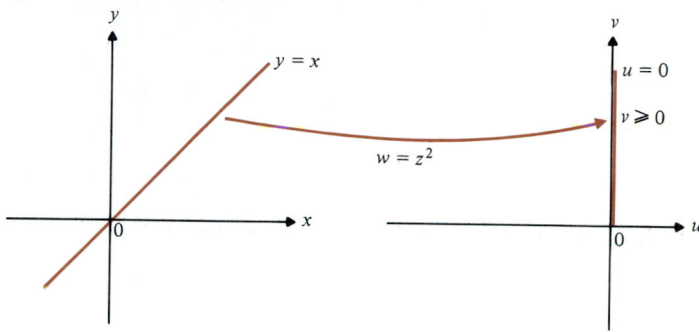

FIGURE 3

We can look at this mapping in another way.

If $z = re^{i\theta}$, then $w = r^2 e^{2i\theta}$. That is, under the mapping $w = z^2$, $|z|$ is squared and arg z is doubled. This means, for example, that the image of a circle centered at the origin is another circle centered at the origin in the w-plane with squared radius (see Fig. 4). Moreover, the image circle is traversed twice since $0 \leq$ arg $z \leq 2\pi$ implies that $0 \leq$ arg $z^2 \leq 4\pi$. Note that this mapping is not one-to-one since every nonzero complex number is the square of two distinct numbers.

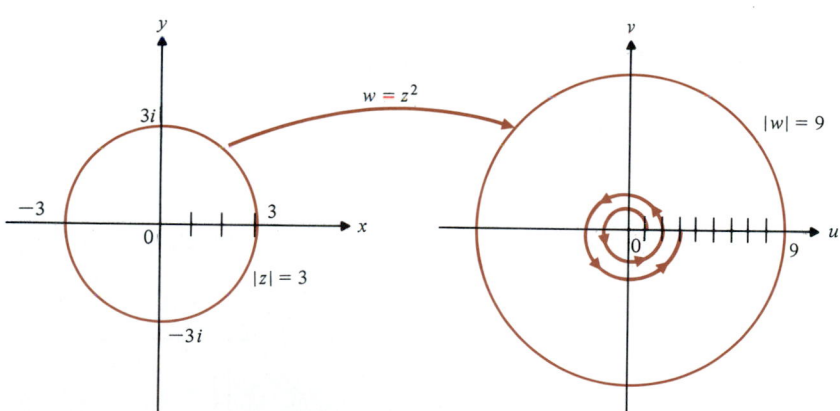

FIGURE 4

We can also describe the image of a region under this mapping. We give some examples graphically in Figs. 5, 6, and 7.

Note: We have to choose which part of the w-plane to shade: the "inside" of the parabola or the "outside." We shade the inside because 0 satisfies Re $z \leq 1$ and $f(0) = 0^2 = 0$ is the image of 0 and is "inside" the parabola. A test point will help you determine which of two regions to shade.

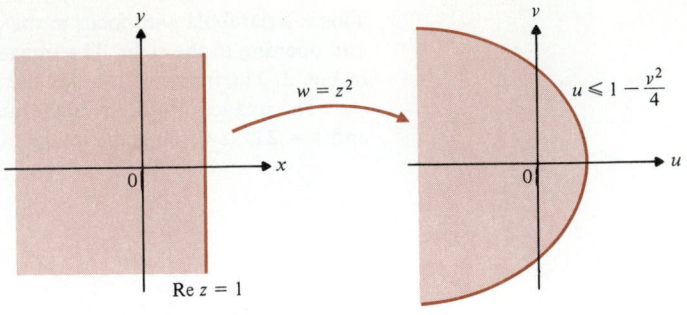

(a) Re $z \leq 1$

(b) Image of Re $z \leq 1$

FIGURE 5

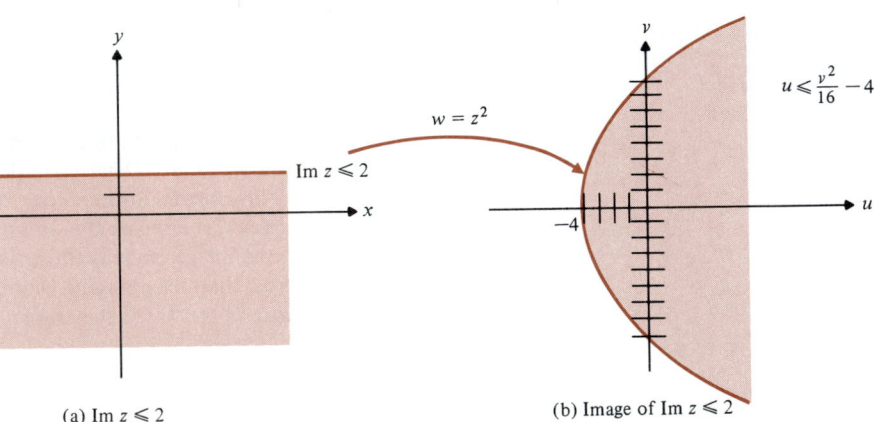

(a) Im $z \leq 2$

(b) Image of Im $z \leq 2$

FIGURE 6

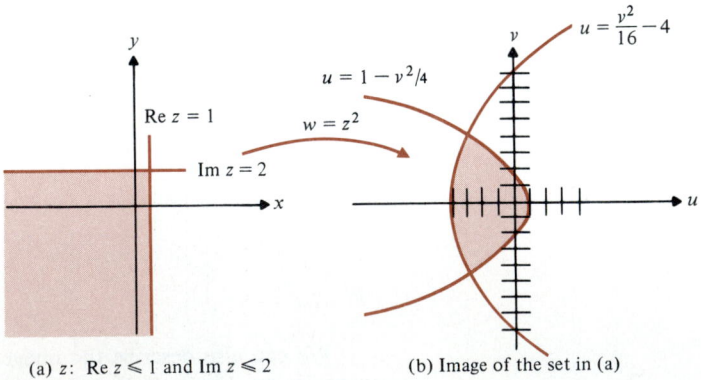

(a) z: Re $z \leq 1$ and Im $z \leq 2$

(b) Image of the set in (a)

FIGURE 7

EXAMPLE 2

What are the images of straight lines parallel to the coordinate axes under the mapping $w = e^z$?

SOLUTION

$$w = e^z = e^x e^{iy} = e^x(\cos y + i \sin y)$$

so

$$u = e^x \cos y \quad \text{and} \quad v = e^x \sin y.$$

Moreover, $|w| = e^x$ and arg $w = y$.

First consider the line Re $z = x = c$. Then

$$|w| = e^c = k, \text{ a constant,} \quad \text{and} \quad \text{arg } w = y,$$

so that the image of the line $x = c$ is a circle of radius e^c centered at the origin. Moreover, the circle is traversed once in the positive sense each time y increases 2π units (see Fig. 8).

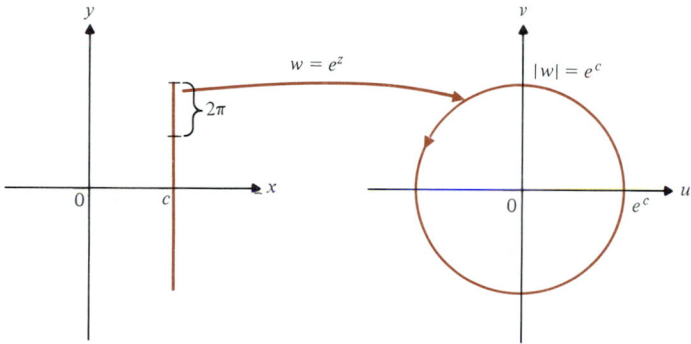

FIGURE 8

If $y = c$, then $w = e^x e^{ic}$ so arg $w = c$, a constant, and $|w| = e^x$ takes values in the open interval $(0, \infty)$. This is sketched in Fig. 9. Note that the image of the point $(0, c)$ is e^{ic}. The mapping is not one-to-one because $e^{z+2\pi ni} = e^z$, for every integer n.

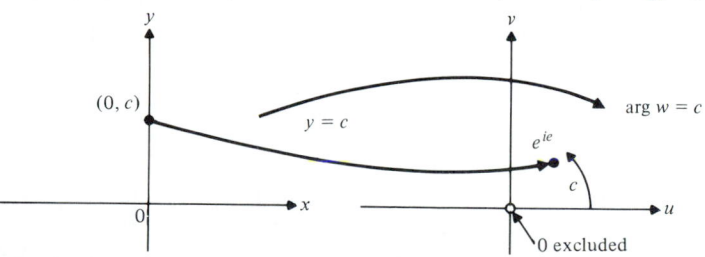

FIGURE 9

EXAMPLE 3

Find the images of circles and straight lines under the mapping $w = \dfrac{1}{z}$.

SOLUTION
Suppose that

$$x = \text{Re } z = \frac{z + \bar{z}}{2} = c \neq 0. \tag{2}$$

Since $w = \dfrac{1}{z}$, $z = \dfrac{1}{w}$ and $\bar{z} = \dfrac{1}{\bar{w}}$ so (2) can be rewritten as

$$\frac{1}{w} + \frac{1}{\bar{w}} = 2c \quad \text{or} \quad \frac{\bar{w} + w}{2} = cw\bar{w} = c|w|^2$$

so that

$$\text{Re } w = c|w|^2.$$

Thus

$$u = c(u^2 + v^2) \quad \text{or} \quad u^2 - \frac{u}{c} + v^2 = 0.$$

Completing the squares we obtain

$$\left(u - \frac{1}{2c}\right)^2 + v^2 = \left(\frac{1}{2c}\right)^2$$

which is the equation of a circle centered at $1/2c$ with radius $1/2c$. If $x = 0$, then $w = 1/iy = -i/y$, and we obtain the line $u = 0$. Analogously, the image of $y = c$ is the circle $u^2 + (v - 1/2c)^2 = 1/4c^2$. These are illustrated in Figs. 10 and 11.

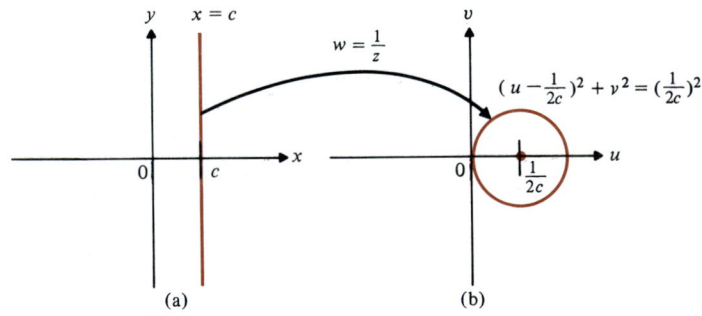

FIGURE 10

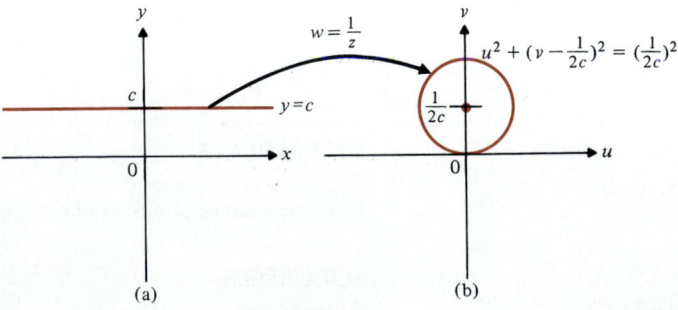

FIGURE 11.

Problems 19.1

1. Show that the image of the line $y = -2x + 1$ under the transformation $w = z^2$ is a parabola, and graph it in the w-plane.

2. Show that the image of the line $y = mx + b$ under the transformation $w = z^2$ is a parabola. [*Hint:* The equation $Ax^2 + Bxy + Cy^2 + Dx + Ey + F = 0$ is the equation of a parabola if and only if $B^2 - 4AC = 0$.]

3. Consider the mapping $w = \cos z$.

   (a) Show that it is not one-to-one.

   (b) Show that it is onto.

Problems 4–8 refer to the transformation $w = \cos z$.

4. Show that the image of the line $x = (2k + 1)\pi/2$ is the line $u = 0$.

5. Show that the image of the line $x = 2k\pi$ is part of the line $v = 0$. Which part?

6. Show that the image of the line $x = c$ where $c \neq 2k\pi$ or $(2k + 1)\pi/2$ is one branch of a hyperbola. Find the equation of this image in the w-plane.

7. Show that the image of the line $y = 0$ is a line segment. Determine that segment.

8. Show that the image of the line $y = c \neq 0$ is an ellipse. Find its equation.

9. What is the image of the line $x = 2$ under the transformation $w = z^3$?

10. Find the image of the line $y = -1$ under the transformation $w = z^3$.

In Problems 11–20 find and sketch the image of the given region under the mapping $w = z^2$.

11. $\{z : \operatorname{Im} z > -1\}$

12. $\{z : \operatorname{Re} z > 1\}$

13. $\{z : \operatorname{Re} z \geq 0 \text{ and } \operatorname{Im} z \geq 0\}$

14. $\{z : \operatorname{Re} z \geq 0 \text{ and } \operatorname{Im} z \leq 0\}$

15. $\{z : \operatorname{Re} z \leq 0 \text{ and } \operatorname{Im} z \geq 0\}$

16. $\{z : \operatorname{Re} z \leq 0 \text{ and } \operatorname{Im} z \leq 0\}$

17. $\{z : \operatorname{Re} z > 3 \text{ and } \operatorname{Im} z < -2\}$

18. $\{z : \operatorname{Re} z < 1 \text{ and } \operatorname{Im} z \geq 1\}$

19. $\{z : |z| \leq 2\}$

20. $\{z : |z| > 1\}$

In Problems 21–32 find and sketch the image of the following sets under the mapping $w = e^z$.

21. $\{z : \operatorname{Re} z < 1\}$

22. $\{z : \operatorname{Im} z \geq 2\}$

23. $\{z : \operatorname{Im} z \leq -1\}$

24. $\{z : \operatorname{Re} z > 0\}$

25. $\{z : \operatorname{Re} z > 0 \text{ and } \operatorname{Im} z > 0\}$

26. $\{z : \operatorname{Re} z > 0 \text{ and } \operatorname{Im} z < 0\}$

27. $\{z : \operatorname{Re} z < 0 \text{ and } \operatorname{Im} z > 0\}$

28. $\{z : \operatorname{Re} z < 0 \text{ and } \operatorname{Im} z < 0\}$

29. $\{z : \operatorname{Re} z \geq 2 \text{ and } \operatorname{Im} z \leq 2\}$

30. $\{z : \operatorname{Re} z < -1 \text{ and } \operatorname{Im} z > 3\}$

31. $\{z : |z| < 1\}$

32. $\{z : |z| \geq 1\}$

In Problems 33–40 find and sketch the image of the following sets under the mapping $w = \sin z$.

33. $\left\{z : \operatorname{Re} z = (2k + 1)\dfrac{\pi}{2}\right\}$

34. $\{z : \operatorname{Re} z = 2k\pi\}$

35. $\left\{z : \operatorname{Im} z = (2k + 1)\dfrac{\pi}{2}\right\}$

36. $\{z : \operatorname{Im} z = 2k\pi\}$

37. $\{z : \operatorname{Re} z = c, c \neq k\pi \text{ or } (2k + 1)\pi/2\}$

38. $\{z : \operatorname{Im} z = c, c \neq k\pi \text{ or } (2k + 1)\pi/2\}$

39. $\{z : \operatorname{Re} z < 1 \text{ and } \operatorname{Im} z \geq 2\}$

40. $\{z : \operatorname{Re} z > -2 \text{ and } \operatorname{Im} z > 1\}$

41. Show that the image of the line segment $\{z : \operatorname{Im} z = c, 2k\pi \leq x < (2k + 2)\pi\}$ under the mapping $w = \sin z$ is the ellipse given by $|w + 1| + |w - 1| = 2 \cosh c$.

42. Show that the image of the interior of the rectangle with corners at $\pm\pi$ and $\pm\pi + ic$ under the mapping $w = \sin z$ is the interior of the ellipse obtained in Problem 41 except for a slit from -1 to 1 and a slit along the negative v-axis.

43. Show that the image of a straight line or circle under the transformation $w = 1/z$ is a straight line or circle by carrying out the following steps:

   (a) Show that every straight line or circle in the xy-plane has the equation $Ax^2 + Ay^2 + Cx + Dy + E = 0$.

   (b) Write the equation in (a) in terms of z and $\bar{z}$.

   (c) Substitute $z = 1/w$ and $\bar{z} = 1/\bar{w}$ in the equation obtained in part (b).

   (d) Write the equation obtained in part (c) in terms of u and v and show that this is the equation of a straight line or circle.

In Problems 44–52 find the image of the given set under the transformation $w = 1/z$.

44. $\{z : \text{Re } z > 0\}$

45. $\{z : \text{Im } z < 1\}$

46. $\{z : y = x\}$

47. $\{z : y = 2x + 1\}$

48. $\{z : |z| = 1\}$

49. $\{z : |z| \geq 2\}$

50. $\{z : |z - 1| = 2\}$

51. $\{z : |z + 1| = 1\}$

52. $\{z : |z| \leq 3\}$

In Problems 53–62 determine whether the given function is one-to-one.

53. $w = z^3$

54. $w = z + 3$

55. $w = \dfrac{1}{z}$

56. $w = 2z - 1$

57. $w = z^2 - 2z + 1$

58. $w = \dfrac{z + i}{z - i}$

59. $w = 2i$

60. $w = z^4$

61. $w = z^5$

62. $w = z^{1/2}$

63. Find an analytic function such that the image of the set $\{|z| : \text{Im } z \geq 0\}$ is $\{|w| : \text{Im } w \geq 1\}$ and $f(0) = 1 + i$.

64. Find an analytic function such that the image of the set $\{|z| : \text{Re } z \leq 2\}$ is $\{|w| : \text{Re } w \geq 0\}$ and $f(2) = i$.

65. Find the image of $\{z : y = ax + b\}$ under the transformation $w = 3iz + 2$.

19.2 Properties of Conformal Mappings

Let the curve C be given by $C : z = z(t)$, $a \leq t \leq b$. We have defined (see p. 933) the **orientation** of C as the direction in which we move along C as t increases. This is illustrated in Fig. 1.

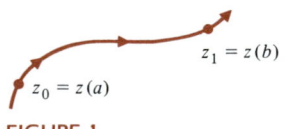
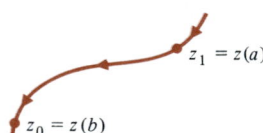

FIGURE 1

Let C_1 and C_2 be two oriented smooth Jordan arcs that intersect at a point z_0. Because they are smooth, each has a unique tangent line at z_0. The **angle** between C_1 and C_2 at z_0 is defined as the angle θ between the tangent lines. (See Fig. 2.) The way we measure the angle determines the **orientation** or **sense** of the angle. We can measure it from T_1 to T_2 (as in Fig. 2) or from T_2 to T_1.

Angle Between Two Curves
Sense of an Angle

DEFINITION

Conformal Mapping at a Point

A mapping $w = f(z)$ is **conformal** at z_0 if the following condition holds: Let C_1 and C_2 be two smooth curves that intersect at z_0. Then

> angle between C_1 and C_2 at z_0 = angle between $f(C_1)$ and $f(C_2)$ at $f(z_0)$ with both angles measured in the same sense.

That is, *a mapping is conformal if it is both angle and sense preserving.*

A conformal map at z_0 is illustrated in Fig. 3.

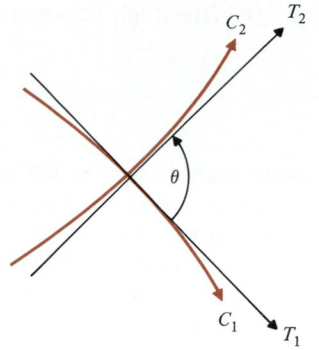

FIGURE 2

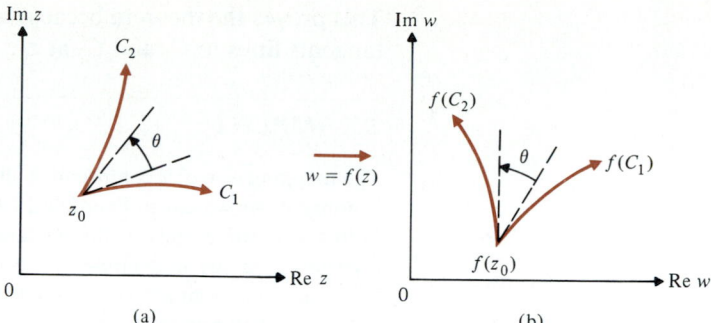

FIGURE 3 A conformal map at z_0

Conformal Function

A function $w = f(z)$ is said to be **conformal in a region** Ω if it is conformal at every point z_0 in Ω. The following theorem tells us that an analytic function is conformal at every point at which its derivative is nonzero.

■ **THEOREM 1**

If $f(z)$ is analytic in a region Ω, then $w = f(z)$ is conformal at all points z_0 in Ω for which $f'(z_0) \neq 0$.

PROOF
Consider the curve $C_1 : z = z(t)$, $0 \le t \le b$, with $z_0 = z(0)$. Let $w(t) = f(z(t))$ denote the image of C_1 in the w-plane.

From the chain rule, we have

$$w'(t) = \frac{d}{dt}(f(z(t)) = f'(z(t))z'(t)$$

and, at $t = 0$,

$$w'(0) = f'(z(0))z'(0) = f'(z_0)z'(0). \tag{1}$$

Thus, from formula (15.2.11) on p. 885, if both sides of (1) are nonzero

$$\arg w'(0) = \arg f'(z_0) + \arg z'(0)$$

or

$$\arg w'(0) - \arg z'(0) = \arg f'(z_0). \tag{2}$$

But $f'(z_0)$ does not depend on the particular curve C. Thus if $C_2 : z = z^*(t)$, $0 \le t \le b$, is another curve with $z^*(0) = z_0$, and we denote its image in the w-plane by $w^*(t) = f(z^*(t))$, then

$$\arg w^{*\prime}(0) - \arg z^{*\prime}(0) = \arg f'(z_0). \tag{3}$$

Equating (2) and (3) yields

$$\arg w'(0) - \arg z'(0) = \arg w^{*\prime}(0) - \arg z^{*\prime}(0)$$

or

$$\arg z^{*\prime}(0) - \arg z'(0) = \arg w^{*\prime}(0) - \arg w'(0).$$

This proves the theorem because $\arg z^{*\prime}(0) - \arg z'(0)$ is the angle between the tangents lines to C_1 and C_2 at z_0. ∎

EXAMPLE 1

The mapping $w = e^z$ is conformal at all points in the complex plane $\mathbb{C}$, since its derivative is nonzero. As we saw in Example 19.1.2, this function maps the real axis in the z-plane onto the positive reals in the w-plane. The imaginary axis in the z-plane is mapped repeatedly on the unit circle in the w-plane because $|e^{iy}| = 1$. Thus, the right angle between the coordinate axes in the first quadrant of the z-plane is transformed into the right angle between the positive real axis and the unit circle in the first quadrant of the w-plane (see Fig. 4).

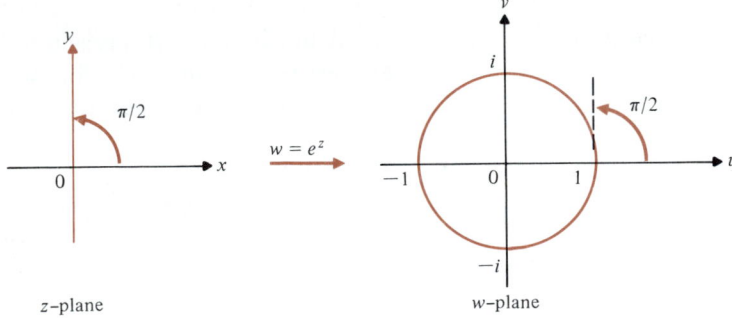

z–plane   w–plane

FIGURE 4 The conformal mapping $w = e^z$

Let $w = f(z)$ be conformal in a region Ω containing the point z_0. Consider the effect of this mapping on a disk centered at z_0 lying in G (see Fig. 5). The

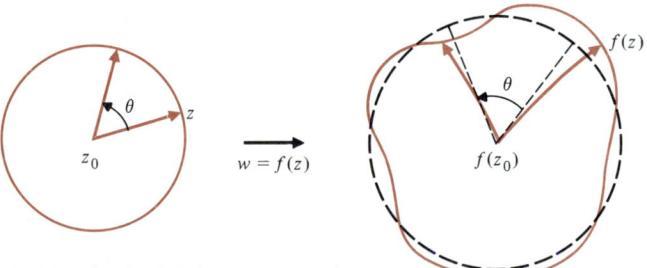

FIGURE 5 Mapping a disk centered at z_0

angles between radial lines are preserved, although their lengths are not. However, since

$$|f'(z_0)| = \lim_{z \to z_0} \frac{|f(z) - f(z_0)|}{|z - z_0|},$$

the radial lines are subject to approximately the same change of scale $|f'(z_0)|$ when the radius is small. Roughly, *"small" circles about z_0 are mapped onto* *"small" circles about $f(z_0)$ with change of scale $|f'(z_0)|$.* Moreover, this indicates that the mapping is *locally* one-to-one, although it is clear nothing can be said about its *global* behavior. For example, $f(z) = e^z$ is locally one-to-one, since $f'(z) = e^z \neq 0$, but $f(0) = f(2\pi i)$, so it is not one-to-one in the complex plane. Indeed, the following theorem can be proved.†

■ **THEOREM 2**

Let $f(z)$ be analytic in an open set Ω containing the point z_0. If $f'(z_0) \neq 0$, then there exists a neighborhood of z_0 on which $w = f(z)$ is one-to-one. ■

Critical Point

DEFINITION

If $w = f(z)$ is analytic at z_0 but $f'(z_0) = 0$, then z_0 is a **critical point** of the mapping.

Angles are magnified at all critical points. For example, $f(z) = z^2$ has a derivative that is zero at the origin. Since $f(1) = f(-1) = 1$ and $f(i) = f(-i) = -1$, the right angles between the axes are mapped into 180° angles. This doubling of angles causes circles around the origin to be mapped to circular curves that wind around the origin twice. This motivates the following theorem.

■ **THEOREM 3**

Let $f(z)$ be analytic in a region Ω containing the point z_0 at which $f'(z)$ has a zero of order k. Then all angles at z_0 are magnified by the factor $k + 1$.

PROOF

We can write $f'(z) = (z - z_0)^k g(z)$, with g analytic and nonzero in an ϵ-neighborhood of z_0. Thus, the terms $f'(z_0), f''(z_0), \ldots, f^{(k)}(z_0)$ all vanish in the Taylor series for $f'(z)$. Hence, the Taylor series of $f(z)$ is

$$f(z) = f(z_0) + \frac{f^{(k+1)}(z_0)}{(k+1)!}(z - z_0)^{k+1} + \cdots.$$

Therefore

$$f(z) - f(z_0) = (z - z_0)^{k+1}\left[\frac{f^{(k+1)}(z_0)}{(k+1)!} + \frac{f^{(k+2)}(z_0)}{(k+2)!}(z - z_0) + \cdots\right]$$

and, from formula (15.2.11),

$$\arg[f(z) - f(z_0)] = (k+1)\arg(z - z_0) + \arg\left[\frac{f^{(k+1)}(z_0)}{(k+1)!} + \cdots\right].$$

---

† The Jacobian $J_f(z_0) = \begin{vmatrix} u_x(z_0) & u_y(z_0) \\ v_x(z_0) & v_y(z_0) \end{vmatrix} = u_x v_y - v_x u_y = u_x^2(z_0) + v_x^2(z_0) = |f'(z_0)|^2 \neq 0$, so the result follows from the inverse function theorem. For a proof of the inverse function theorem, see S. I. Grossman, *Multivariable Calculus, Linear Algebra, and Differential Equations,* 2nd ed., Harcourt Brace Jovanovich, San Diego, CA, 1986, p. 610.

The first two arguments compare the angles between the horizontal direction and the vector pointing from $f(z_0)$ to $f(z)$ and from z_0 to z. If z tends to z_0 along a fixed vector making an angle θ with the horizontal direction, the angle of the vector from $f(z_0)$ to $f(z)$ with the horizontal tends to

$$(k+1)\theta + \arg\left[\frac{f^{(k+1)}(z_0)}{(k+1)!}\right],$$

the last argument being independent of θ. Thus, the angle between the tangents of two smooth arcs intersecting at z_0 is magnified by the factor $k + 1$. ■

EXAMPLE 2

The mapping $w = 1 - \cos z$ is entire and conformal except at the zeros $0, \pm\pi, \pm 2\pi,$ ... of the derivative $(1 - \cos z)' = \sin z$. To examine the behavior of this mapping at $z = 0$, note that $\sin z$ has a zero of order 1 at $z = 0$ since

$$\sin z = z - \frac{z^3}{3!} + \frac{z^5}{5!} - \cdots = z\left(1 - \frac{z^2}{3!} + \cdots\right).$$

Theorem 3 implies that $w = 1 - \cos z$ will magnify all angles at $z = 0$ by 2. Observe in Fig. 6 that the right angle between the coordinate axes in the first quadrant is transformed into an $180°$ angle, since $1 - \cos x > 0$ for $0 < x < \pi/2$ and $1 - \cos iy = 1 - \cosh y < 0$ for $0 < y < \pi/2$.

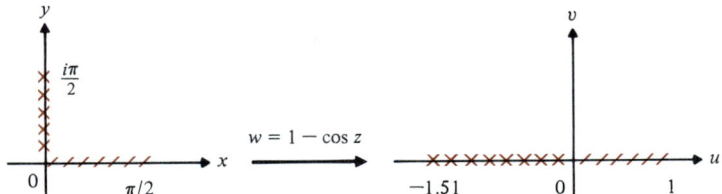

FIGURE 6 Local behavior of $w = 1 - \cos z$ at $z = 0$

Turning to global properties, it is reasonable to ask when a given region Ω_1 can be mapped conformally onto a given region Ω_2. The next result, whose proof is beyond the scope of this book, is the fundamental result in this direction.

■ **THEOREM 4: Riemann Mapping Theorem**
Let z_0 be a point in a simply connected region Ω ($\neq \mathbb{C}$). Then there is a unique analytic function $w = f(z)$ mapping Ω one-to-one onto the disk $|w| < 1$ such that $f(z_0) = 0$ and $f'(z_0) > 0$. ■

Now suppose Ω_1 and Ω_2 are two simply connected regions different from $\mathbb{C}$. The theorem yields the existence of analytic functions f, g mapping Ω_1, Ω_2 onto the unit disk. Thus $g^{-1}f$ is a one-to-one mapping of Ω_1 onto Ω_2. If we can show that g^{-1}, and thus the composition, is analytic, we then have a conformal mapping of Ω_1 onto Ω_2, proving that *any two simply connected regions different from the complex plane can be mapped conformally onto each other.* Since g is conformal (it is one-to-one and analytic), so is g^{-1}. The inverse function theorem of calculus (see p. 1059) shows that g^{-1} has continuous first partial derivatives, and these satisfy the Cauchy-Riemann equations, since

$$\begin{pmatrix} x_u & y_u \\ x_v & y_v \end{pmatrix} = \begin{pmatrix} u_x & v_x \\ u_y & v_y \end{pmatrix}^{-1} = \begin{pmatrix} u_x & v_x \\ -v_x & u_x \end{pmatrix}^{-1} = \frac{1}{u_x^2 + v_x^2} \begin{pmatrix} u_x & -v_x \\ v_x & u_x \end{pmatrix}$$

so that

$$x_u = \frac{u_x}{u_x^2 + v_x^2} = y_v \quad \text{and} \quad y_u = \frac{-v_x}{u_x^2 + v_x^2} = -x_v.$$

Hence, g^{-1} is analytic.

The conditions $f(z_0) = 0$ and $f'(z_0) > 0$ imply that the image of any smooth arc C through z_0 will have the same slope at 0 as the arc C does at z_0, since $\arg f'(z_0) = 0$. This is not a limitation; instead, it is a normalization indicating that there are three "degrees of freedom" in choosing the mapping: the x- and y-coordinates of the point z_0 and the change of direction of angles. Should we wish to change the direction by the angle θ, we need only multiply the mapping by the constant of unit length $e^{i\theta}$.

Although the Riemann mapping theorem asserts the existence of a function mapping a given region conformally onto a disk, it does not show how to find it. Construction of the function can be a matter of great difficulty. The rest of this chapter is devoted to the construction of *specific* conformal mappings and their application in fluid flow.

Problems 19.2

In Problems 1–10 find all points at which the given mapping is not conformal.

1. $w = e^z$

2. $w = \sin z$

3. $w = \dfrac{1}{z}$

4. $w = \sin z^2$

5. $w = z^2 e^z$

6. $w = z^2 - z$

7. $w = z^5$

8. $w = z^7 - z^3$

9. $w = z + \dfrac{1}{z}$

10. $w = \cosh z$

In Problems 11–18 describe what each mapping does to the angle at the origin given by the rays $\arg z = 0$ and $\arg z = \pi/6$.

11. $w = z^3$

12. $w = z^3 \sin z$

13. $w = z - \sin z$

14. $w = e^z - z$

15. $w = z^4$

16. $w = \dfrac{z^3}{\sin z}$

17. $w = e^{z^2} - \cos z$

18. $w = \tan z$

19. Show that the image under the mapping $w = z^2$ of the circle $|z - r| = r$, $r > 0$, is the cardioid with polar equation

$$\rho = 2r^2(1 + \cos\theta).$$

20. Show that the mapping $w = z + 1/z$, maps circles $|z| = r$ onto ellipses

$$\frac{u^2}{(r + 1/r)^2} + \frac{v^2}{(r - 1/r)^2} = 1.$$

21. Why is $\Omega \neq \mathbb{C}$ part of the hypothesis of the Riemann mapping theorem? Explain.

19.3 Linear Fractional Transformations†

A common and important type of conformal mapping is given by the expression

$$w = w(z) = \frac{az+b}{cz+d}, \qquad ad - bc \neq 0, \tag{1}$$

Linear Fractional Transformation

where a, b, c, d are complex constants. Such a mapping is called a **linear fractional transformation.** The condition $ad - bc \neq 0$ prevents its derivative

$$w' = \frac{ad - bc}{(cz + d)^2}$$

from vanishing, as otherwise the function is constant. We can solve for z, obtaining

$$z = \frac{-dw + b}{cw - a},$$

and using the convention that $w(-d/c) = \infty$ and $w(\infty) = a/c$, it follows that w maps $\mathcal{M}$ (the Riemann sphere) one-to-one onto itself. Moreover, the mapping is conformal except at $z = \infty$, $-d/c$, because at these points $w' = 0$ or ∞.

We begin our discussion of linear fractional transformations by examining some special cases of the mapping (1).

EXAMPLE 1: Translation

The mapping

$$w = z + z_0 \tag{2}$$

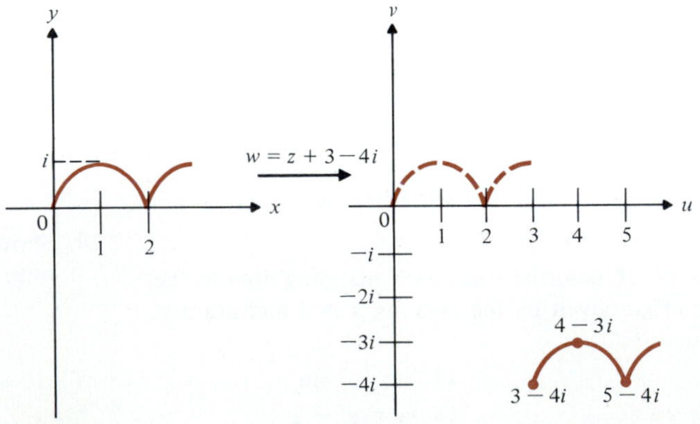

FIGURE 1

† Linear fractional transformations are also called **linear transformations, bilinear transformations,** and **Möbius transformations.**

is called a **translation mapping.** If $z_0 = x_0 + iy_0$, then Re z is shifted or **translated** x_0 units to the right, if $x_0 > 0$, and $|x_0|$ units to the left if $x_0 < 0$. Similarly, Im z is shifted y_0 units up or $|y_0|$ units down. This is illustrated in Fig. 1.

EXAMPLE 2: Rotation

The mapping

$$w = e^{i\alpha}z \tag{3}$$

is called a **rotation mapping** since

$$w = |z|e^{i[\arg z + \alpha]}$$

and

$$\arg w = \arg z + \alpha.$$

Thus z is rotated through an angle α. One such rotation is illustrated in Fig. 2.

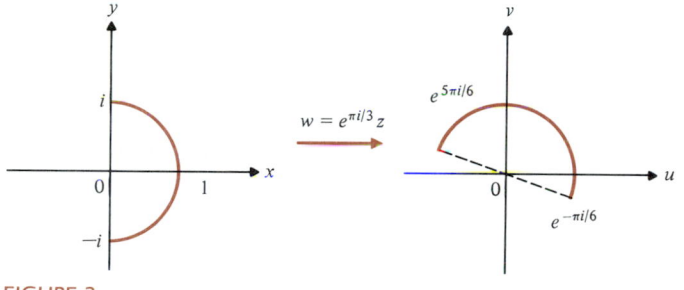

FIGURE 2

EXAMPLE 3: Magnification

The mapping

$$w = Mz, \qquad M > 0, \tag{4}$$

is called a **magnification** because

$$|w| = M|z|.$$

That is, z is magnified by a factor of M. (See Fig. 3.)

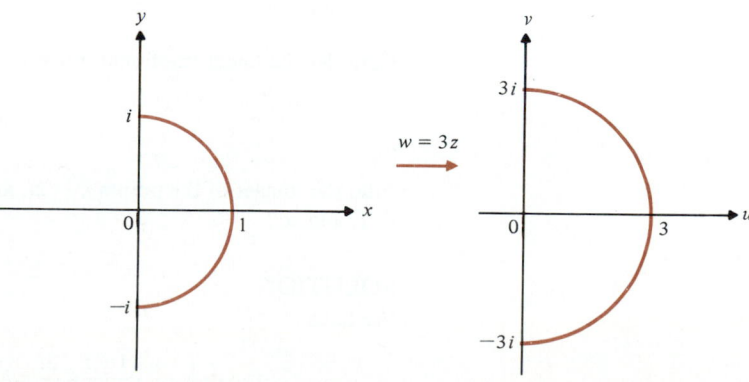

FIGURE 3

EXAMPLE 4: Inverse

The mapping $w = 1/z$ is called the **inverse mapping.** We discussed this mapping in Example 19.1.3.

Any linear fractional transformation is a composition of the four types of mappings given in Examples 1–4.

If $c \neq 0$, we can write

$$\frac{az+b}{cz+d} = \frac{bc-ad}{c^2(z+d/c)} + \frac{a}{c},$$

showing that the transformation may be decomposed into a translation by d/c, followed by a rotation by $e^{2i \arg c}$, a magnification $|c|^2$, an inversion, a rotation, a magnification, and a translation.

If $c = 0$,

$$\frac{az+b}{d} = \frac{a}{d}\left(z + \frac{b}{a}\right),$$

proving that the decomposition consists of a translation, a rotation, and a magnification.

A composition of two linear fractional transformations is again a linear fractional transformation, since

$$\frac{a\left(\dfrac{\alpha z + \beta}{\gamma z + \delta}\right) + b}{c\left(\dfrac{\alpha z + \beta}{\gamma z + \delta}\right) + d} = \frac{(a\alpha + b\gamma)z + (a\beta + b\delta)}{(c\alpha + d\gamma)z + (c\beta + d\delta)}$$

with

$$(a\alpha + b\gamma)(c\beta + d\delta) - (a\beta + b\delta)(c\alpha + d\gamma) = (ad - bc)(\alpha\delta - \beta\gamma) \neq 0.$$

EXAMPLE 5

Consider the linear fractional transformation

$$w = \frac{z-1}{z+1}.$$

Find the images of the points $i, -2i$, and ∞. What points are mapped onto the points 0, 1, and ∞?

SOLUTION
We have

$$i \to \frac{i-1}{i+1} \cdot \frac{1-i}{1-i} = \frac{2i}{2} = i \quad \text{and} \quad -2i \to \frac{-2i-1}{-2i+1} = \frac{3-4i}{5}.$$

Writing

$$w = \frac{1 - (1/z)}{1 + (1/z)},$$

we see that ∞ is mapped onto 1. To find what point is mapped onto 0, we observe that $z = 1$ causes the numerator of the right side of the linear fractional transformation to vanish. Hence, 1 is mapped onto 0. Similarly, -1 causes the denominator to vanish, so -1 is mapped onto ∞.

EXAMPLE 6

Find a linear fractional transformation that maps the circle $|z - i| = 1$ onto the circle $|w - 1| = 2$.

SOLUTION

Consider the sequence of linear fractional transformations shown in Fig. 4: a translation

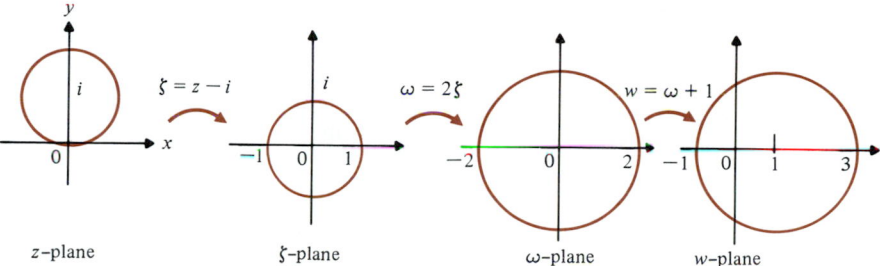

FIGURE 4

$\zeta = z - i$, followed by a magnification $\omega = 2\zeta$, and followed by another translation $w = \omega + 1$. The composition of these three mappings is

$$w = \omega + 1 = 2\zeta + 1 = 2(z - i) + 1$$

or

$$w = 2z + (1 - 2i),$$

and this linear fractional transformation maps $|z - i| = 1$ onto $|w - 1| = 2$.

The fundamental property of linear fractional transformations is that *they map circles onto circles in $\mathcal{M}$*. A circle in $\mathcal{M}$ corresponds to a circle or a straight line in $\mathbb{C}$, as lines in the plane correspond to circles through ∞ on the Riemann sphere (see p. 893). Geometrically, it is clear that translations and rotations carry circles onto circles. Before considering the other two transformations, we make an observation.

Any line not parallel to the y-axis can be written as

$$y = (\tan \theta)x + b, \qquad |\theta| < \pi/2,$$

or

$$y \cos \theta - x \sin \theta = b \cos \theta.$$

Let $z = x + iy$. Then

$$\text{Re}(-ie^{-i\theta}z) = \text{Re}[-i(\cos\theta - i\sin\theta)(x+iy)]$$

$$= y\cos\theta - x\sin\theta = b\cos\theta. \tag{5}$$

Thus, under the magnification $w = Mz$ the line $\text{Re}(-ie^{-i\theta}z) = b\cos\theta$ is mapped into the line

$$\text{Re}(-ie^{-i\theta}w/M) = \frac{\text{Re}(-ie^{-i\theta}z)}{M} = \frac{b}{M}\cos\theta. \tag{6}$$

The magnification $w = Mz$, $M > 0$, maps (by substitution) circles $|z - z_0| = r$ onto circles $|w - Mz_0| = Mr$. The circle $|z - z_0| = r$ (>0) satisfies

$$0 = |z - z_0|^2 - r^2 = |z|^2 + |z_0|^2 - 2\,\text{Re}\,\bar{z}z_0 - r^2. \tag{7}$$

If we substitute $z = 1/w$, then (7) becomes (since $1/\bar{w} = w/|w|^2$)

$$0 = \frac{1}{|w|^2} + |z_0|^2 - r^2 - \frac{2}{|w|^2}\,\text{Re}\,z_0 w. \tag{8}$$

If $|z_0| = r$, indicating that the circle passes through the origin, we obtain the equation

$$0 = \frac{1 - 2\,\text{Re}\,z_0 w}{|w|^2}$$

or

$$\text{Re}\,z_0 w = \tfrac{1}{2}, \tag{9}$$

which by (5) and (6) is the equation of a line in the w-plane with slope $\tan(\pi/2 - \arg z_0) = \cot(\arg z_0)$. If $|z_0| \neq r$, the origin does not lie on the circle, so multiplying equation (8) by the nonzero quantity $|w|^2/(|z_0|^2 - r^2)$, we have

$$0 = \frac{1}{|z_0|^2 - r^2} + |w|^2 - \frac{2}{|z_0|^2 - r^2}\,\text{Re}\,z_0 w$$

$$= \left| w - \frac{\bar{z_0}}{|z_0|^2 - r^2} \right|^2 - \frac{r^2}{(|z_0|^2 - r^2)^2}, \tag{10}$$

a circle. That lines map onto circles through the origin follows by reversing the steps leading to equation (8).

Since any linear fractional transformation is a composition of these special transformations, we have proved the following theorem.

THEOREM 1
Linear fractional transformations map circles onto circles in $\mathcal{M}$.

EXAMPLE 7

Map the intersection of the disks $|z - 1| < 1$ and $|z - i| < 1$ conformally onto the first quadrant.

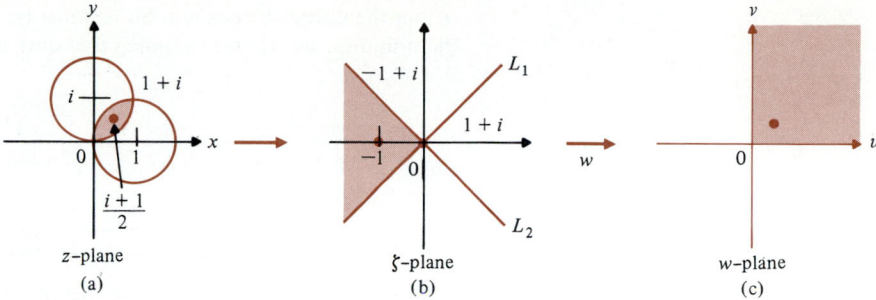

z-plane
(a)

ζ-plane
(b)

w-plane
(c)

FIGURE 5

SOLUTION

The region of interest is sketched in Fig. 5a. How do we proceed? We first note that, at the origin, the circles intersect at right angles (the tangent line to $|z - 1| = 1$ is the y-axis and the tangent line to $|z - i| = 1$ is the x-axis). Thus, we seek a conformal mapping $f(z)$ such that $f(0) = 0$ and the image of each given circle is a straight line, that is, a circle through ∞ in $\mathcal{M}$. In that case the two lines obtained will also intersect at right angles and these lines will divide the plane into four "quadrants." It will then be a simple matter to rotate through an appropriate angle to get the first quadrant.

Since we want both circles to map into straight lines intersecting at 0, we map their other point of intersection, $1 + i$, to ∞. Hence we want a mapping that has the property that $f(0) = 0$ and $f(1 + i) = \infty$. To get $f(1 + i) = \infty$, we put $z - (1 + i)$ in the denominator. To get $f(0) = 0$, we put z in the numerator. Thus

$$\zeta = f(z) = \frac{z}{z - (1 + i)}$$

sends 0 to 0 and $1 + i$ to ∞. Moreover, the images of the circles are straight lines that intersect at right angles. Finding the lines is easy as we need only find two points on each.

L_1 = image of $|z - 1| = 1 : \zeta(0) = 0$ and $\zeta(2) = \dfrac{2}{1 - i} = 1 + i$, so the slope of the line is 1.

L_2 = image of $|z - i| = 1 : \zeta(0) = 0$ and $\zeta(2i) = 1 - i$, so the slope of the line is -1.

Since $\zeta((1 + i)/2) = -1$, the point -1 in the ζ-plane must be in the image of each open disk. Thus the shaded region in Fig. 5b is the image of the shaded region in Fig. 5a. Finally, the rotation

$$w = e^{-3\pi i/4} \zeta = \frac{e^{-3\pi i/4} z}{z - (1 + i)}$$

yields the desired mapping.

EXAMPLE 8

Map the right half plane onto the unit disk $|z| < 1$ so that the point 1 is mapped onto the origin.

SOLUTION

Since we want $f(1) = 0$, we put $z - 1$ in the numerator. Also, the half plane contains

∞ but the unit disk does not. So ∞ must be mapped onto a point on the boundary of the unit disk, say 1. One mapping that does this is

$$w = \frac{z-1}{z+1} \tag{11}$$

since

$$\lim_{z \to \infty} \frac{z-1}{z+1} = \lim_{z \to \infty} \frac{1 - 1/z}{1 + 1/z} = 1.$$

Note that ∞ is not in the range of this mapping because $z + 1$ does not vanish in the right half plane. Furthermore, $\pm i$ are mapped onto themselves [such points are called **fixed points** of the mapping (11)]. Since three points determine a circle, it follows that the imaginary axis is mapped onto the unit circle (see Fig. 6).

Fixed Points

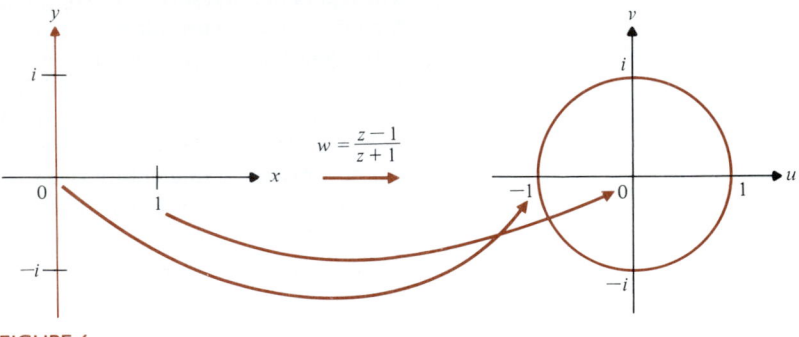

FIGURE 6

EXAMPLE 9

Find the number of roots of the equation

$$p(z) = 11z^4 - 10z^3 - 4z^2 + 10z + 9 = 0 \tag{12}$$

lying in the right half plane.

SOLUTION
In Example 8 we found that the mapping $w = (z - 1)/(z + 1)$ maps the right half plane onto the unit disk $|w| < 1$. If $w = (z - 1)/(z + 1)$, then $z = (1 + w)/(1 - w)$. Substituting this into (12) yields

$$11\left(\frac{1+w}{1-w}\right)^4 - 10\left(\frac{1+w}{1-w}\right)^3 - 4\left(\frac{1+w}{1-w}\right)^2 + 10\left(\frac{1+w}{1-w}\right) + 9 = 0.$$

Multiplying each term by $(1 - w)^4$, we get (after some algebra)

$$16w^4 + 48w^3 + 128w^2 - 32w + 16 = 0.$$

Dividing by 16, we obtain the equivalent problem of finding the number of roots of the equation

$$g(w) = w^4 + 3w^3 + 8w^2 - 2w + 1 = 0$$

lying in $|w| < 1$. Letting $f(w) = 8w^2$, we find that, on $|w| = 1$,

$$|g(w) - f(w)| = |w^4 + 3w^3 - 2w + 1| \le 7 < 8|w|^2 = |f(w)|.$$

But the equation $8w^2 = 0$ has two roots inside $|w| < 1$ (a zero of order 2 at 0). By Rouché's theorem (see p. 1045), $g(w) = 0$ also has two roots inside $|w| < 1$, and, therefore, the answer to our problem is two.

Cross Ratio

Given three distinct points z_1, z_2, z_3 in $\mathcal{M}$, there is a linear fractional transformation carrying them onto 0, 1, ∞, respectively. If none of the points is ∞, the transformation is given by the **cross ratio**

$$w = \frac{(z - z_1)(z_2 - z_3)}{(z - z_3)(z_2 - z_1)} \tag{13}$$

and becomes

$$\frac{z_2 - z_3}{z - z_3}, \quad \frac{z - z_1}{z - z_3}, \quad \text{or} \quad \frac{z - z_1}{z_2 - z_1}, \tag{14}$$

if z_1, z_2, or $z_3 = \infty$. If w^* is another linear fractional transformation with the same property, then the composition $w^* w^{-1}$ keeps the points 0, 1, ∞ fixed. (For example, $w^* w^{-1}(0) = w^*(z_1) = 0$.)

Thus, we have a linear fractional transformation

$$\zeta = \frac{az + b}{cz + d}, \qquad ad - bc \neq 0,$$

satisfying the equations

$$0 = \frac{b}{d}, \qquad 1 = \frac{a + b}{c + d}, \qquad \infty = \frac{a}{c},$$

But then $b = c = 0$ and $a = d$, implying $w^* w^{-1} = I$, the identity mapping, and hence $w^* = w$. Therefore, w is the only linear fractional transformation mapping the points z_1, z_2, z_3 onto 0, 1, ∞, respectively.

Since a circle is determined by three of its points, we can now easily determine a linear fractional transformation carrying a given circle in the z-plane onto a given circle in the w-plane. We select distinct points z_1, z_2, z_3 on the first circle and w_1, w_2, w_3 on the second circle. Then

$$\frac{(w - w_1)(w_2 - w_3)}{(w - w_3)(w_2 - w_1)} = \frac{(z - z_1)(z_2 - z_3)}{(z - z_3)(z_2 - z_1)} \tag{15}$$

maps z_1, z_2, z_3 onto w_1, w_2, w_3, as the right-hand side of the equation maps z_1, z_2, z_3 onto 0, 1, ∞, and the inverse of the left-hand side maps 0, 1, ∞ onto w_1, w_2, w_3.

EXAMPLE 10

Find the linear fractional transformation mapping the points 1, i, -1, onto the points 2, 3, 4, respectively.

SOLUTION
Solve the equation

$$\frac{(w-2)(3-4)}{(w-4)(3-2)} = \frac{(z-1)(i+1)}{(z+1)(i-1)}$$

for w, obtaining

$$w = \frac{(2-4i)z+(2+4i)}{(1-i)z+(1+i)}.$$

Problems 19.3

In Problems 1–3 write each linear fractional transformation as the composition of the four elementary types of mappings: translation, rotation, magnification, and inverse.

1. $\dfrac{z-1}{z+1}$

2. $\dfrac{iz+2}{2iz-4i}$

3. $\dfrac{z-3i+2}{(1+i)z-2+3i}$

In Problems 4–14 find and sketch the image of each point or region under the mapping $f(z) = (z - i)/(z + 1)$.

4. $z_0 = i$ **5.** $z_0 = -i$

6. $z_0 = 1$ **7.** $z_0 = -1$

8. The right half plane   **9.** The left half plane

10. $|z| < 1$ **11.** $|z - 1| < 1$

12. $|z| > 2$ **13.** Re $z \geq 3$

14. Im $z \leq -1$

15. Show that the mapping $w = e^{i\theta}(z - z_0)/(1 - \bar{z}_0 z)$, $|z_0| < 1$, θ real, maps the unit circle onto the unit circle.

In Problems 16–19 find a mapping of the unit circle in the z-plane onto the unit circle in the w-plane such that $f(z_0) = w_0$.

16. $z_0 = i$, $w_0 = 1$ **17.** $z_0 = i$, $w_0 = -i$

18. $z_0 = \dfrac{1+i}{2}$, $w_0 = 0$ **19.** $z_0 = \dfrac{1-i}{2}$, $w_0 = \dfrac{1+i}{2}$

In Problems 20–23, describe the image of the region indicated under the given mapping.

20. The disk $|z| < 1$; $w = i\dfrac{z-1}{z+1}$

21. The quadrant $x > 0$, $y > 0$; $w = \dfrac{z-i}{z+i}$

22. The angular sector $|\arg z| < \dfrac{\pi}{4}$; $w = \dfrac{z}{z-1}$

23. The strip $0 < x < 1$; $w = \dfrac{z}{z-1}$

24. Find the number of roots of the equation

$$11z^4 - 20z^3 + 6z^2 + 20z - 1 = 0$$

lying in the right half plane.

25. How many roots of the equation

$$17z^4 + 26z^3 + 56z^2 + 38z + 7 = 0$$

lie in the first quadrant?

26. Under what conditions will $w = (az + b)/(cz + d)$ map the half plane Im $z \geq 0$ onto the half plane Im $w \geq 0$?

In Problems 27–33 determine the image of the circle $|z + 1| = 1$ and the line Re $z = 2$ under the given mapping.

27. $w = iz$

28. $w = -4z$

29. $w = z - 1 - i$

30. $w = \dfrac{z-1}{z+1}$

31. $w = \dfrac{z+i}{z-i}$

32. $w = \dfrac{1}{z}$

33. $w = \dfrac{1-z}{z}$

34. Map the half plane Im $z \le 0$ onto the unit disk $|w| < 1$ so that the point $-i$ is mapped onto the origin.

35. Map the intersection of the disks $|z + 1| < 1$ and $|z - i| < 1$ conformally onto the third quadrant.

36. Show that the fixed points of the mapping

$$w = \frac{az+b}{cz+d}, \qquad ad - bc \neq 0,$$

are the roots of $cz^2 + (d - a)z - b = 0$.

In Problems 37–39 find the fixed points of the given mapping.

37. $w = \dfrac{z-1}{z+1}$

38. $w = \dfrac{iz+2}{2iz-8i}$

39. $w = \dfrac{z-3i+2}{(1+i)z-2+3i}$

40. Map the intersection of the half planes Re $z < 1$, Im $z > -1$ onto the intersection of two circles in the w-plane.

* 41. Find the image of the region $|z| > 1$ under the mapping $w = z + 1/z$.

Find the linear fractional transformation mapping the points $-1, i, 1 + i$, respectively, onto the points given in Problems 42–45.

42. $0, 1, \infty$

43. $1, \infty, 0$

44. $2, 3, 4$

45. $0, 1, i$

46. Is $w = \bar{z}$ a linear fractional transformation?

* 47. Show that any four distinct points can be mapped by a linear fractional transformation to the points $1, -1, k, -k$, where k depends on the original points.

* 48. Using the exponential function, map the region lying inside $|z| = 2$ and outside $|z - 1| = 1$ onto the upper half plane.

* 49. Map the region $|z - 1| < 1$, Im $z < 0$, onto the upper half plane.

* 50. Map the sector $|\arg z| < \pi/4$ onto the set $|\text{Re } w| < 1$, Im $w > 0$. [*Hint:* Use the sine function.]

19.4 Compositions of Elementary Conformal Mappings

In Section 19.2 we proved that the elementary functions e^z, cos z, sin z, log z, and z^α are conformal in the regions of their domains of definition where their derivative is nonzero. In this section we illustrate how compositions of these functions with linear fractional transformations can be used to map certain regions conformally onto each other. First, let us examine some properties of these mappings.

The Mapping $w = e^z$

This mapping is conformal everywhere because $e^z \neq 0$. We saw in Example 19.1.2 that the images of vertical lines are circles centered at the origin and the images of horizontal lines are rays (see Figs. 19.1.8 and 19.1.9). Thus the images of rectangles under this mapping consist of rectangular wedges as in Fig. 1.

The Mapping $w = \log z$

If $w = \log z$, then $z = e^w$. Thus images under the mapping $w = \log z$ are inverses of images under the exponential mapping. For example, since $\{z : \text{Re } z = c\}$ is mapped into $\{w : |w| = e^c\}$ by the mapping $w = e^z$, we have the following:

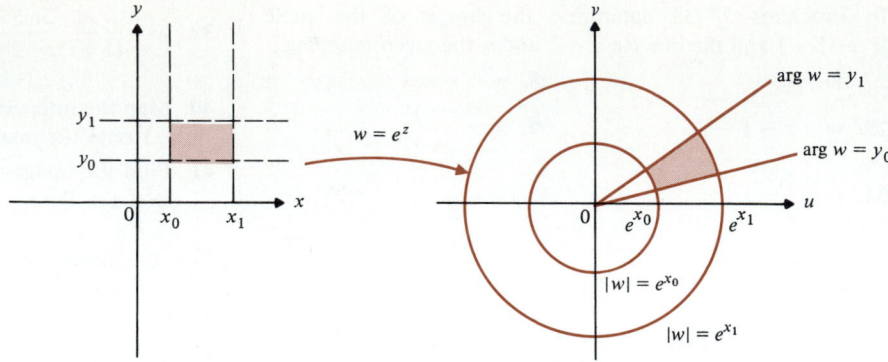

FIGURE 1

The image of the circle $|z| = k > 0$ under the mapping $w = \log z$ is the vertical line $\{w : \operatorname{Re} w = \ln k\}$.

EXAMPLE 1

Find the image of the first quadrant under the mapping $w = \operatorname{Log} z$.

SOLUTION
If z is in the first quadrant, then $z = re^{i\theta}$ where $0 \le \theta \le \pi/2$. Then $w = \operatorname{Log} z = \ln r + i\theta$. $\ln r$ can take on all real values and $0 \le \theta \le \pi/2$, so the image is the infinite strip $\{w : 0 \le \operatorname{Im} w \le \pi/2\}$. This is sketched in Fig. 2.

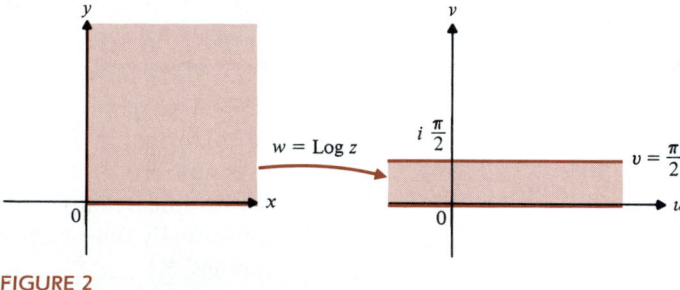

FIGURE 2

The Mapping $w = \cos z$

From formula (15.8.8) we have

$$\cos z = \cos x \cosh y - i \sin x \sinh y, \tag{1}$$

which is periodic of period 2π in x. Moreover, $(\cos z)' = -\sin z$, which is 0 at $n\pi$, $n = 0, \pm 1, \pm 2, \ldots$. Thus the mapping is conformal except at $z = n\pi$, $n = 0, \pm 1, \pm 2, \ldots$. In particular, $\cos z$ is conformal in the infinite strip $0 < \operatorname{Re} z < \pi$.

EXAMPLE 2

What is the image of the rectangle $0 < \text{Re } z < \pi$, $0 \le \text{Im } z \le 1$ under the mapping $w = \cos z$?

SOLUTION

From (1), the line $x = 0$ maps into $w = \cosh y$. Since $0 \le y \le 1$, $1 \le \cosh y \le \cosh 1$. So the set $w = \cosh y$ is the segment $1 \le u \le \cosh 1$. The line $x = \pi$ maps into the set $-\cosh 1 \le u \le -1$. The image of $y = 0$ is the segment $[-1, 1]$. The image of $y = 1$ is the ellipse

$$\frac{u^2}{\cosh^2 1} + \frac{v^2}{\sinh^2 1} = 1.$$

Thus we obtain the sketch in Fig. 3.

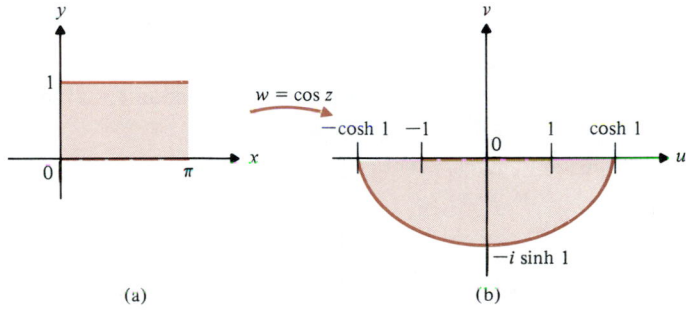

(a)   (b)

FIGURE 3

EXAMPLE 3

Find a conformal mapping of the infinite strip $|\text{Im } z| < \pi/2$ onto the unit disk.

SOLUTION

If we apply the conformal mapping $\zeta = e^z$ to the infinite strip $|\text{Im } z| < \pi/2$, we obtain the right half plane $\text{Re } \zeta > 0$, because for any x we have

$$e^{x \pm i\pi/2} = \pm e^x i \quad \text{and} \quad e^0 = 1.$$

Recall Example 19.3.8, where we constructed a mapping of the right half plane onto the unit disk. This mapping is

$$w = \frac{\zeta - 1}{\zeta + 1}.$$

Hence, the composition of these two mappings $w = w(\zeta(z))$ given by

$$w = \frac{\zeta - 1}{\zeta + 1} = \frac{e^z - 1}{e^z + 1} = \frac{e^{z/2} - e^{-z/2}}{e^{z/2} + e^{-z/2}} = \tanh\left(\frac{z}{2}\right)$$

maps the strip $|\text{Im } z| < \pi/2$ onto the unit disk (see Fig. 4). Note that $z = 0$ is mapped into $\zeta = 1$ and then into $w = 0$.

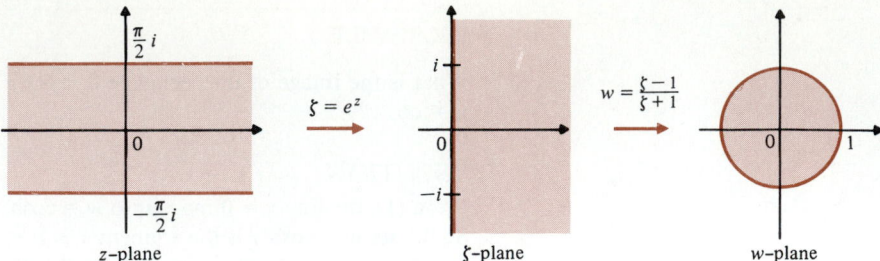

FIGURE 4

EXAMPLE 4

Map the right half plane, with the line $\{z : x \geq 1,\ y = 0\}$ missing, onto the upper half plane.

SOLUTION

First apply the function $\zeta = z^2$ to obtain the plane minus two slits shown in the ζ-plane in Fig. 5.

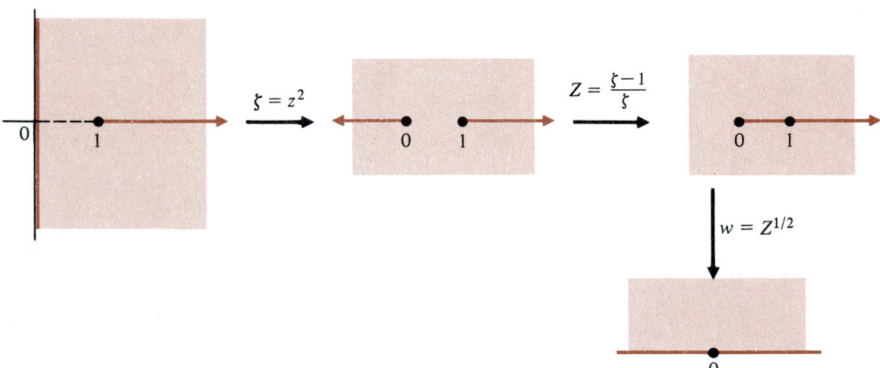

FIGURE 5

Now use the linear fractional transformation that maps $1, \infty, 0$ onto $0, 1, \infty$ [see equation (19.3.14)]

$$Z = \frac{\zeta - 1}{\zeta}.$$

This gives us a plane with the nonnegative real axis removed. Finally, use the principal branch of the square root function $w = Z^{1/2}$ to halve the angle at the origin, obtaining the upper half plane. Thus the desired mapping is

$$w = \sqrt{\frac{\zeta - 1}{\zeta}} = \sqrt{\frac{z^2 - 1}{z^2}}.$$

Problems 19.4

In Problems 1–12 find the image of the line Re $z = 1$ under each mapping.

1. $w = z^2$
2. $w = z^3$
3. $w = e^z$
4. $w = \cos z$
5. $w = \sin z$
6. $w = \cos^2 z$
7. $w = 1/z^2$
8. $w = z + \sin z$
9. $w = \sqrt{z}$
10. $w = z^{1/3}$
11. $w = (z - 2)^2$
12. $w = (z + 1)^2$

13. Find the image of $|z| = 4$ under the mapping $w = \text{Log } z$.

14. Show that the image of the line Re $z = c$ under the mapping $w = e^{(2+3i)z}$ is the curve (given in polar form) $r = e^{-(3/2)\theta + k}$. Find the constant k. This curve is called a **logarithmic spiral**.

15. Show that the image of the line Im $z = c$ under the mapping of Problem 14 is the logarithmic spiral $r = e^{(2/3)\theta - k}$

16. What is the image of the lines Re $z = 2$ and Im $z = 5$ under the mapping $w = e^{(x_0 + iy_0)z}$?

In Problems 17–19 find and sketch the image of the given region under the mapping $w = e^z$.

17. $0 < \text{Re } z < 1, 0 < \text{Im } z < \dfrac{\pi}{2}$

18. $1 < \text{Re } z < 2, -\dfrac{\pi}{2} < \text{Im } z < 0$

19. $-2 < \text{Re } z < 1, \dfrac{\pi}{6} < \text{Im } z < \dfrac{\pi}{3}$

In Problems 20–23 find and sketch the image of the given region under the mapping $w = \cos z$.

20. $0 < \text{Re } z < \dfrac{\pi}{2}, 0 < \text{Im } z < 1$

21. $0 < \text{Re } z < \pi, -1 < \text{Im } z < 1$

22. $0 < \text{Re } z < \dfrac{\pi}{4}, -1 < \text{Im } z < 1$

23. $-\dfrac{\pi}{2} < \text{Re } z < 0$, Im z unrestricted

24. Find a conformal mapping of the unit disk onto the infinite strip $|\text{Re } z| < 1$. [*Hint:* Consider the inverse mapping in Example 3.]

25. Show that

$$w = \frac{z}{\sqrt{z^2 + 1}}$$

conformally maps the upper half plane with the line $\{z : x = 0, y \geq 1\}$ deleted onto the upper half plane.

26. Find a mapping that carries the upper half plane onto the complement of the line segment from -1 to 1.

27. Find a conformal mapping of the square $\{z : |x| \leq 1, |y| \leq 1\}$ onto the annulus $1 < |w| < e^{2\pi}$ with the negative real axis deleted.

28. What is the image of the disk $|z - a| < a$ under the mapping $w = z^2$?

29. Show that the transformation

$$\left(\frac{w - 1}{w + 1}\right)^2 = i\left(\frac{z - 1}{z + 1}\right)$$

conformally maps the upper half of the unit disk onto the unit disk.

* 30. Describe the image of the hyperbola $x^2 - y^2 = \frac{1}{2}$ under the mapping $w = \sqrt{1 - z^2}$.

* 31. Map the complement of the line segment $\{z : y = 0, |x| \leq 1\}$ onto the unit disk.

* 32. Map the outside of the parabola $y^2 = 4x$ onto the unit disk so that $0, -1$ are sent onto $1, 0$.

* 33. Map the region to the left of the right-hand branch of the hyperbola Re$(z^2) = 1$ onto the unit disk. [*Hint:* Consider the mapping $w = z + 1/z$.]

* 34. Prove that the mapping

$$w = \frac{Az^2 + Bz + C}{az^2 + bz + c}$$

can be decomposed into the three successive transformations,

$$\zeta = \frac{\alpha z + \beta}{\gamma z + \delta}, \qquad Z = \frac{1}{2}\left(\zeta + \frac{1}{\zeta}\right), \qquad w = \mu Z + \nu,$$

or into the two successive transformations

$$\zeta = \frac{\alpha z + \beta}{\gamma z + \delta} \quad \text{and} \quad w = \zeta^2 + \nu.$$

19.5 Fluid Flow

In this section we shall discuss a physical problem that can be analyzed with the help of analytic functions. Since a complex function can be decomposed into two real functions, the theory of analytic functions is very useful in solving problems involving two variables in two-dimensional space. However, since this book is not a treatise on mathematical physics, much of what follows is a heuristic outline of the physical theory.

A complete description of the motion of a fluid requires knowledge of the velocity vector at all points of the fluid at any given time. Suppose the fluid is **incompressible** (that is, of constant density) and the flow is **steady** (independent of time) and **two-dimensional** (the same in all planes parallel to the xy-plane in three-dimensional space). Conditions of this type occur, for instance, when the fluid flows past a long cylindrical object whose axis is perpendicular to the direction of the flow. The **velocity vector** can then be given as a continuous complex-valued function of a complex variable $V = V(z)$ for all z in a region Ω. We also assume in this section that no **sources** or **sinks** (points at which fluid is being created or destroyed) lie in the region Ω.

Sources and Sinks

The assumptions that the fluid is incompressible and there are no sources or sinks in Ω imply that a simply connected region in Ω always contains the same amount of fluid. Thus, the quantity of fluid per unit time passing by a length element ds on a pws Jordan curve C lying together with its inside in Ω, is $V_n \, ds$, where V_n is the component (a real number) of V in the outward normal direction to the curve (see Fig. 1). Hence, the total outward flow is

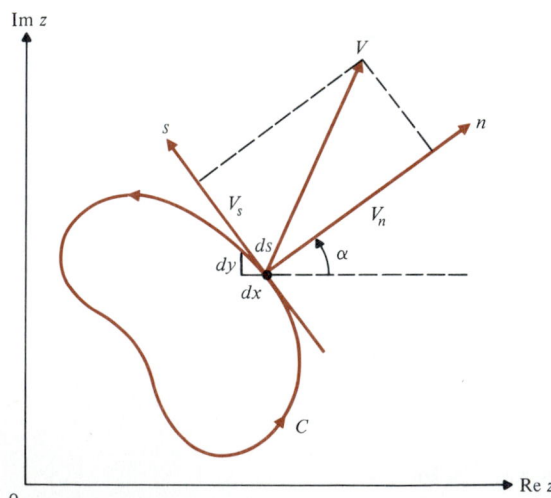

FIGURE 1 Components of the velocity vector

$$Q = \int_C V_n \, ds = 0. \tag{1}$$

The line integral of the tangential component V_s of the velocity V around the curve C,

$$\Gamma = \int_\gamma V_s\,ds, \tag{2}$$

Circulation

is called the **circulation** of V along C. If the circulation is not zero on some curve C, then the tangential components having one sign dominate those having the other sign in the integral (2). Roughly, this means that the fluid rotates around C. The flow is said to be **irrotational** if the circulation is zero along all closed curves in C. We assume the flow is irrotational so that $\Gamma = 0$.

Consider Fig. 1, where the outward normal and tangential directions to the curve C are indicated at a point z. Let $\alpha = \alpha(z)$ be the angle between the (positive) horizontal direction and the outward normal to C at z, and suppose the velocity vector V at z is as indicated.

Rotating the ns-coordinate system about the point z through an angle of $-\alpha$ yields the normal and tangential components of the velocity vector V

$$V_n = \mathrm{Re}(e^{-i\alpha}V), \qquad V_s = \mathrm{Im}(e^{-i\alpha}V).$$

In particular, we have

$$e^{-i\alpha}V = V_n + iV_s. \tag{3}$$

The length element ds is related (see Fig. 1) to the element dx and dy by the identities

$$dx = \cos\!\left(\frac{\pi}{2} + \alpha\right)ds, \qquad dy = \sin\!\left(\frac{\pi}{2} + \alpha\right)ds,$$

implying that

$$dz = dx + i\,dy = e^{i(\pi/2+\alpha)}\,ds = ie^{i\alpha}\,ds. \tag{4}$$

Now, if C is any pws Jordan curve contained together with its inside in Ω, we have, by equations (1)–(4),

$$\int_C \overline{V(z)}\,dz = i\int_C \overline{(e^{-i\alpha}V)}\,ds = i\int_C \overline{(V_n + iV_s)}\,ds = \int_C (V_s + iV_n)\,ds = 0,$$

Complex Potential

Potential and Stream Functions

Streamlines

implying that $\overline{V(z)}$ is analytic by Morera's theorem. If Ω is simply connected, the antiderivative of $\overline{V(z)}$ is an analytic function $w(z) = u(z) + iv(z)$, called the **complex potential** of the flow; u is known as the **potential function** and v as the **stream function**. Individual particles of the fluid move along curves whose direction at each point coincides with that of the velocity vector. Such curves are called **streamlines** and are characterized by the equation $v(z) = $ constant, since the tangent to such a curve has slope

$$\frac{dy}{dx} = -\frac{v_x}{v_y} = -\frac{v_x}{u_x} = -\tan \arg w' = \tan \arg V$$

by the Cauchy-Riemann equations, since $V = \bar{w}'$.

The curves $u(z)$ = constant are called **equipotential lines** and are normal to the streamlines, since

$$\frac{dy}{dx} = -\frac{u_x}{u_y} = \frac{u_x}{v_x} = \frac{-1}{\tan \arg V}.$$

Stagnation Points

Points at which $V(z) = 0$, and consequently $w'(z) = 0$, are known as **stagnation points** of the flow.

EXAMPLE 1

Suppose we have a uniform flow of velocity A (>0) in the positive x-direction in the upper half plane. This approximates fluid flow in extremely wide channels (see Fig. 2).

Since $V(z) = A$, it follows that $w'(z) = A$, so the complex potential is $w(z) = Az + c$, where $c = c_1 + ic_2$ is a complex constant. Thus, $u(z) = Ax + c_1$ and $v(z) = Ay + c_2$, so the equipotential lines are vertical and the streamlines are horizontal (neglecting the effect of viscosity on the real axis). Setting $c = 0$, the streamline $v = 0$ coincides with the real axis.

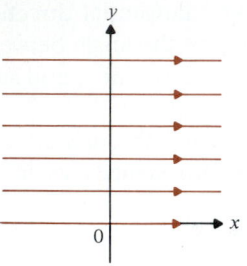

FIGURE 2

Suppose the function $\zeta = f(z)$ maps the region Ω conformally onto the upper half plane $f(\Omega)$. If the complex potential $w(\zeta)$ of the fluid flow in $f(\Omega)$ is known, then the complex potential of the flow in Ω is given by the analytic function $w(f(z))$. For example, if the complex potential in $f(\Omega)$ is that given in Example 1, then the streamlines in Ω are those curves that are mapped by the composite function $w \circ f$ onto the straight lines v = constant in the upper half plane. The determination of such composite functions is the fundamental procedure in the solution of problems in fluid dynamics.

EXAMPLE 2

If we are interested in finding the streamlines along a right angle in a wide channel, we can approximate this situation by studying the flow in the first quadrant. The mapping $\zeta = z^2$ maps the quadrant onto the upper half plane. Thus, if we know the complex potential $w = w(\zeta)$ of the flow in the upper half plane, then $w = w(z^2)$ is the complex potential of the flow in the first quadrant. For example, if we assume that the flow is uniform and of velocity A (>0) in the upper half of the ζ-plane (and $c = 0$), then the complex potential in the ζ-plane is $w = A\zeta$. Thus, the complex potential in the first quadrant satisfies $w = Az^2$, the streamlines are given by the hyperbolas $2Axy$ = constant, and the velocity vector is $V(z) = 2A\bar{z}$. The origin is a stagnation point. (See Fig. 3.)

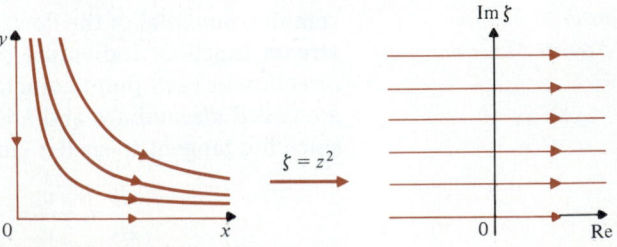

FIGURE 3 Streamlines along a corner

EXAMPLE 3

The mapping $\zeta = z + a^2/z$ has important applications in two-dimensional fluid flow. Rewriting the transformation in the form

$$\frac{(z \pm a)^2}{z} = \zeta \pm 2a,$$

we see that the image ζ of each point on the circle $|z| = b$, $b > a$, satisfies

$$|\zeta - 2a| + |\zeta + 2a| = \frac{|z - a|^2 + |z + a|^2}{b}.$$

By the law of cosines (see Fig. 4), it follows that

$$|z - a|^2 = a^2 + b^2 - 2ab \cos \theta,$$

$$|z + a|^2 = a^2 + b^2 - 2ab \cos(\pi - \theta),$$

where $\theta = \arg z$. Hence,

$$|\zeta - 2a| + |\zeta + 2a| = \frac{2(a^2 + b^2)}{b},$$

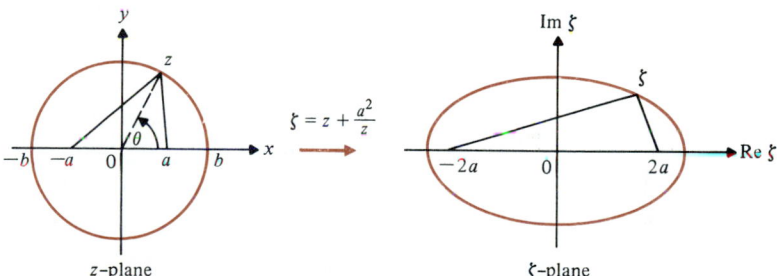

FIGURE 4 The mapping $\zeta = z + \dfrac{a^2}{z}$

and since the right side is constant for fixed b, it follows that the image of the circle $|z| = b$ is an ellipse with foci at $\pm 2a$. Thus, concentric circles of radius $b > a$ centered at the origin of the z-plane map onto confocal ellipses in the ζ-plane. Moreover, the circle $|z| = a$ maps onto the straight line segment joining $-2a$ to $2a$ in the ζ-plane, since $z = ae^{i\theta}$ implies that

$$\zeta = z + \frac{a^2}{z} = ae^{i\theta} + ae^{-i\theta} = 2a \cos \theta, \qquad 0 \leqslant \theta < 2\pi.$$

Noting that $(z + a^2/z)' = 1 - (a/z)^2$, it follows that $\zeta = z + a^2/z$ conformally maps the exterior of the circle $|z| = a$ onto the exterior of the line segment joining $-2a$ to $2a$. Hence, assuming that the motion of the fluid flow in the ζ-plane is uniform and of velocity A (>0) parallel to the real axis, we obtain the complex potential

$$w = A\left(z + \frac{a^2}{z}\right)$$

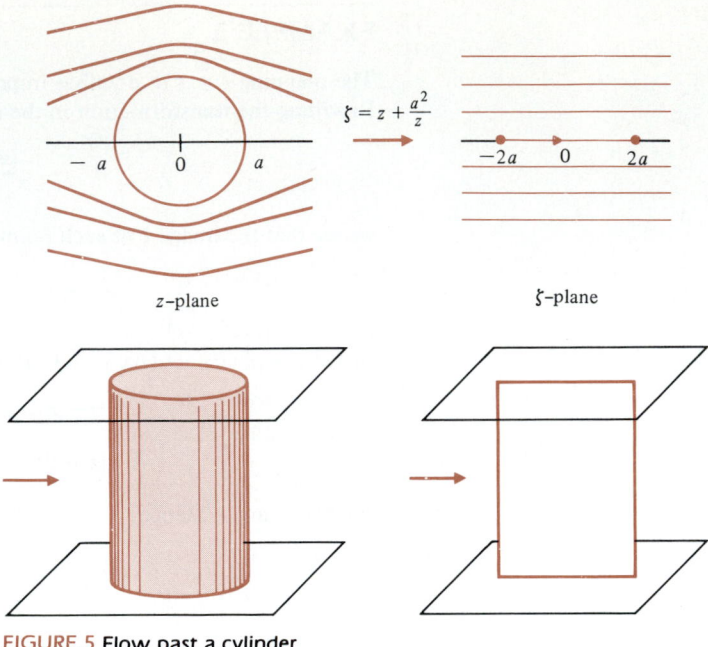

FIGURE 5 Flow past a cylinder

for the flow past a circular cylinder of radius a (see Fig. 5). The stream function is obtained by setting $z = re^{i\theta}$, obtaining

$$v = A\left(r - \frac{a^2}{r}\right)\sin\theta;$$

the streamline $v = 0$ consists of the circle $|z| = a$ and the real axis with $|x| \geq a$. The velocity of the fluid flow is

$$V = \overline{w'} = A\left[1 - \left(\frac{a}{\bar{z}}\right)^2\right],$$

with stagnation points at $z = \pm a$. Note that $V \to A$ as $|z| \to \infty$, implying that although the stream is disturbed by the presence of the cylinder, this disturbance is negligible at great distances from the cylinder and the flow for large $|z|$ is essentially uniform and of velocity A parallel to the x-axis.

Problems 19.5

Find the streamlines for an incompressible fluid flow without sources or sinks in each of the regions given in Problems 1–4. Assume, as in Example 2, that the flow has velocity $A > 0$ as $z \to \infty$. Are there any stagnation points?

1. $0 < \arg z < \dfrac{3\pi}{4}$

2. $0 < \arg z < \dfrac{5\pi}{4}$

3. $0 < \arg z < \dfrac{\pi}{4}$

4. $\dfrac{-\pi}{4} < \arg z < \dfrac{5\pi}{4}$

For Problems 5–8, calculate the speed $|V|$ at $z = 0$, 1, and i for the flow in the upper half plane given by each of the complex potentials. Are there any stagnation points?

5. $w = z + z^3$

6. $w = z + 2iz^2$

7. $w = 3z - iz^2$

8. $w = \sin z$

9. Find the equations of the streamlines for the complex potentials given in Problems 5–8.

10. Find the equations of the equipotential lines for the complex potentials given in Problems 5–8.

* 11. Suppose the complex potential for a flow in the z-plane is given by $w = \cosh^{-1}(z/a)$. Describe the streamlines for this flow. [*Hint:* Consider $z = a \cosh w$.]

* 12. Use Problem 11 to describe a flow through an aperture bounded by the hyperbola $x^2 - y^2 = 1$.

* 13. Use Problem 11 to describe a flow through an aperture of breadth $2a$ in a flat plate. Is this flow physically realizable? [*Hint:* Find the speed at the edges.]

* 14. Use the mapping $\zeta = \sin^{-1} z$ to determine the streamlines of an incompressible fluid flow in the region shown in Fig. 6. Assume that the flow in the ζ-plane is uniform and parallel to the imaginary axis. Is this flow physically realizable?

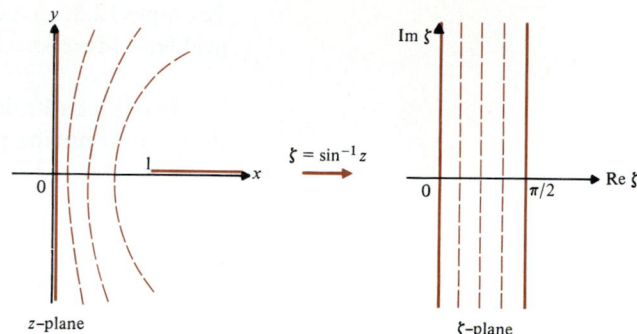

FIGURE 6 Vertical uniform flow in the ζ-plane

* 15. **Bernoulli's theorem** states that in a steady motion of an incompressible fluid, the quantity

$$\frac{p}{\rho} + \frac{1}{2}|V|^2$$

has a constant value at each point of any streamline of the flow, where p, ρ, and $|V|$ are the pressure, density, and speed, respectively. Show in Example 3 that if $A^2 > \frac{2}{3}p(\infty)/\rho$, then there will be points at which the pressure is negative. At these points a vacuum will form, causing the phenomenon of **cavitation.** Cavitation occurs, for example, near the tips of a rapidly moving propeller.

19.6 Dirichlet's Problem

One of the most commonly occurring equations of applied mathematics is **Laplace's equation.** If u is a function of x and y, then Laplace's equation is given by

$$\frac{\partial^2 u}{\partial x^2} + \frac{\partial^2 u}{\partial y^2} = 0.$$

We discussed this equation in Sections 12.8 and 12.9.

In Section 16.7 we stated that any real-valued function $u(x, y)$ with continuous second partial derivatives that satisfies Laplace's equation on a region Ω is called **harmonic** in Ω. We also showed, in Theorem 16.7.1, that the real and imaginary parts of analytic functions are harmonic. Thus, it is easy to find many solutions to Laplace's equation. We get two for every analytic function we choose.

However, solving Laplace's equation is generally only part of the physical problem. Often we are required to solve Laplace's equation in a region Ω with given values specified on the boundary of Ω. (One such example was given in

Example 12.8.1.) As we have seen, any such problem is called a **boundary value problem.** More specifically, we have the following:

Dirichlet's Problem†: Given an arbitrary region Ω, is there a function harmonic in Ω having preassigned values on the boundary of Ω?

EXAMPLE 1

Find a function that is harmonic in the first quadrant and that has the boundary values 0 on the real axis and 100 on the imaginary axis.

SOLUTION

In Example 19.4.1 we saw that the mapping $w = \text{Log } z$ maps the first quadrant onto the strip $0 \le \text{Im } z \le \pi/2$. Thus the mapping

$$w = \frac{200}{\pi} \text{Log } z$$

maps the first quadrant conformally onto the strip $0 \le v \le 100$, where $w = u + iv$ (see Fig. 1). Note that the positive x-axis is transformed into the line $v = 0$, while the positive y-axis becomes the line $v = 100$. Since

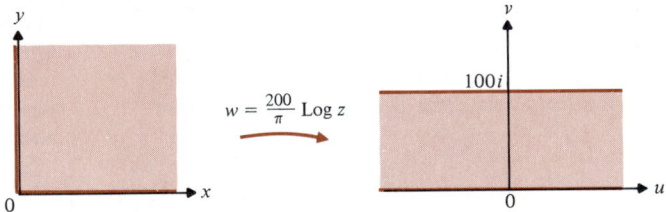

FIGURE 1 Graphical solution of a Dirichlet problem

$$u + iv = \frac{200}{\pi}(\log|z| + i \text{ Arg } z),$$

the function

$$v = \frac{200}{\pi} \text{Arg } z,$$

being the imaginary part of an analytic function, is harmonic in the first quadrant and satisfies the required boundary conditions.

Before continuing, we prove a useful result.

THEOREM 1

Let $w = f(z)$ be a conformal mapping of the simply connected region Ω_1 onto the simply connected region Ω_2, and let $h(w)$ be harmonic on Ω_2. Then the composition mapping $h \circ f$ is harmonic on Ω_1.

† See the accompanying biographical sketch.

LEJEUNE DIRICHLET
(1805–1859)

Courtesy of the Granger Collection

Lejeune Dirichlet was born at Düren, Germany, in 1805 and successively held professorships at Breslau and Berlin. At Gauss's death in 1855 he was appointed Gauss's successor at Göttingen, a fitting honor for so talented a mathematician who was a former student of Gauss and a lifelong admirer of his mentor. While at Göttingen he had hoped to finish Gauss's incomplete works, but his early death in 1859 prevented this.

Fluent in both German and French, Dirichlet served admirably as a liaison between the mathematics and the mathematicians of the two nationalities. Perhaps his most celebrated mathematical accomplishment was his penetrating analysis of the convergence of Fourier series, an undertaking that led him to generalize the function concept. He did much to facilitate the comprehension of some of Gauss's more abstruse methods, and he himself contributed notably to number theory.

Dirichlet was a close friend, expositor, and son-in-law of Jacobi (see p. 656). His name is met by college mathematics students in connection with *Dirichlet's series,* the *Dirichlet function,* the *Dirichlet principle,* and, of course, the *Dirichlet problem.*

Dirichlet has been described as possessing a noble, sincere, human, and modest disposition, but, unlike Jacobi, he seemed unable to communicate with young minds. When a schoolmate expressed envy because Dirichlet's son could always receive help from his gifted father, the son gave this lamentable but memorable reply: "Oh! My father doesn't know the little things anymore." Dirichlet's waggish nephew, Sebastian Hensel, wrote in his memoirs that the mathematics instruction he received in his sixth and seventh years at the gymnasium from his uncle was the most dreadful experience of his life.

Dirichlet was very lax in maintaining family correspondence. When his first child arrived he failed to write of the event to his father-in-law, who was living in London at the time. The father-in-law, when he finally found out, commented that he thought Dirichlet "should have at least been able to write 2 + 1 = 3." This witty father-in-law was none other than Abraham Mendelssohn, a son of the philosopher Moses Mendelssohn, and father of the composer Felix Mendelssohn.

Dirichlet's brain, and also that of Gauss, are preserved in the department of physiology at Göttingen University.

PROOF

Since Ω_2 is simply connected we can use Theorem 16.7.2, p. 967 to construct an analytic function $g(w)$ on Ω_2 such that Re $g = h$. Now the composition of analytic functions is analytic (see p. 904) so $g \circ f$ is analytic on Ω_1. But Re$(g \circ f) = h \circ f$, so $h \circ f$ is harmonic on Ω_1. ∎

Not all Dirichlet problems have a solution. The existence of a solution depends on the *geometry* of the region: *a solution exists whenever no component of the complement of the region can be continuously shrunk to a point.* The proof of this assertion is beyond the scope of this book. Problem 17 provides an example of a region for which Dirichlet's problem need not have a solution.

Dirichlet's problem always has a solution for a simply connected region Ω ($\neq \mathbb{C}$). To see how to obtain an explicit expression for the solution u at any point z_0 in Ω, let g be the function that maps Ω conformally onto the unit disk $|\zeta| < 1$ with $g(z_0) = 0$. (The existence of such a function is guaranteed by the Riemann mapping theorem.) For simplicity, assume g is analytic in an open set containing Ω and its boundary. The composite function $u \circ g^{-1}$ (see Fig. 2) is, by Theorem 1, harmonic on $|\zeta| \le 1$, since g^{-1} is analytic. By the mean value theorem for harmonic functions (Theorem 16.7.4 on p. 970), we can represent $u \circ g^{-1}(0)$ as the integral average of the values of $u \circ g^{-1}$ on $|\zeta| = 1$ by writing (with $z_0 = 0$)

$$u \circ g^{-1}(0) = \frac{1}{2\pi} \int_0^{2\pi} u \circ g^{-1}(e^{i\theta}) \, d\theta.$$

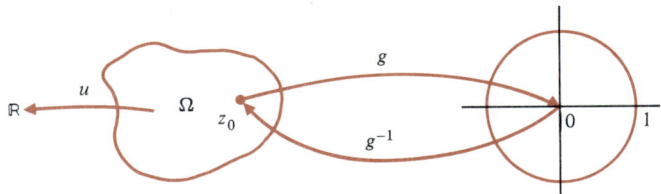

FIGURE 2

Setting $\zeta = e^{i\theta}$, we have $d\zeta/\zeta = i \, d\theta$ so that

$$u \circ g^{-1}(0) = \frac{1}{2\pi i} \int_{|\zeta|=1} u \circ g^{-1}(\zeta) \frac{d\zeta}{\zeta}.$$

Since $\zeta = g(z)$, the integral becomes

$$u(z_0) = \frac{1}{2\pi i} \int_{\partial\Omega} u(z) \frac{g'(z)}{g(z)} \, dz. \tag{1}$$

This equation indicates that the value of a harmonic function u at points interior to the region Ω can be determined as an integral of the boundary values of u.

Observe the similarity of this situation with the Cauchy integral formula. The following examples illustrate the use of this integral.

EXAMPLE 2: Poisson's Integral Formula

Dirichlet's problem can be solved on a disk centered at the origin. Let $\Omega = \{z : |z| < R\}$. The linear fractional transformation

$$g(z) = \frac{R(z - z_0)}{R^2 - \bar{z}_0 z}$$

maps Ω conformally onto the open unit disk with $g(z_0) = 0$. To check this, pick a point on the boundary of Ω, $z = Re^{i\phi}$, and let $z_0 = re^{i\theta}$. Then

$$\left| \frac{R(z - z_0)}{R^2 - \bar{z}_0 z} \right| = \left| \frac{R(Re^{i\phi} - re^{i\theta})}{R^2 - re^{-i\theta}Re^{i\phi}} \right| = \left| \frac{Re^{i\phi} - re^{i\theta}}{R - re^{i(\phi - \theta)}} \right| = |e^{i\phi}| \left| \frac{R - re^{i(\theta - \phi)}}{R - re^{-i(\theta - \phi)}} \right| = 1,$$

since the denominator of the last term is the conjugate of the numerator. Now

$$g'(z) = \frac{(R^2 - \bar{z}_0 z)R + R(z - z_0)\bar{z}_0}{(R^2 - \bar{z}_0 z)^2} = \frac{R^3 - R|z_0|^2}{(R^2 - \bar{z}_0 z)^2}$$

and

$$\frac{g'(z)}{g(z)} = \frac{R^3 - R|z_0|^2}{R(z - z_0)(R^2 - \bar{z}_0 z)} = \frac{R^2 - |z_0|^2}{(z - z_0)(R^2 - \bar{z}_0 z)}.$$

On the boundary of the disk, $|z| = R$, so that

$$\frac{g'(z)}{g(z)} = \frac{|z|^2 - |z_0|^2}{(z - z_0)(z\bar{z} - \bar{z}_0 z)} = \frac{|z|^2 - |z_0|^2}{z(z - z_0)(\bar{z} - \bar{z}_0)} = \frac{|z|^2 - |z_0|^2}{z|z - z_0|^2}.$$

Thus equation (1) becomes

$$u(z_0) = \frac{1}{2\pi i} \int_{|z|=R} u(z) \frac{|z|^2 - |z_0|^2}{|z - z_0|^2} \frac{dz}{z}. \tag{2}$$

Setting $z = Re^{i\phi}$ and $z_0 = re^{i\theta}$ yields **Poisson's integral formula** for the disk $|z| < R$:

$$u(re^{i\theta}) = \frac{1}{2\pi} \int_0^{2\pi} u(Re^{i\phi}) \frac{R^2 - r^2}{R^2 + r^2 - 2Rr \cos(\phi - \theta)} d\phi. \tag{3}$$

Equation (3) is valid for $r < R$.

EXAMPLE 3

Let Ω be the right half plane, and let z_0 be any point interior to Ω. Then the function

$$g(z) = (z - z_0)/(z + \bar{z}_0)$$

maps Ω conformally onto the unit disk with $g(z_0) = 0$. If $z_0 = x_0 + iy_0$ then

$$\frac{g'(z)}{g(z)} = \frac{2x_0}{z^2 - 2iy_0 z - |z_0|^2}.$$

Poisson's integral formula (1) for the right half plane is

$$u(z_0) = \frac{x_0}{\pi i} \int_{\partial\Omega} \frac{u(z)\,dz}{z^2 - 2iy_0 z - |z_0|^2},$$

and setting $z = it$ we have

$$u(z_0) = \frac{x_0}{\pi} \int_{-\infty}^{\infty} \frac{u(it)\,dt}{(t - y_0)^2 + x_0^2}.$$

The following example contrasts the use of conformal mappings with Poisson's integral formula in finding solutions to Dirichlet's problem.

EXAMPLE 4

Find a function u harmonic in the unit disk having the boundary value 1 in the right half of the unit circle and 0 in the left half of the unit circle.

SOLUTION

The linear fractional transformation

$$\zeta = i\left(\frac{i - z}{i + z}\right)$$

maps the points $0, i, -i, 1$ onto $i, 0, \infty, -1$, so it maps the unit disk conformally onto the upper half plane. Following this map with $w = (1/\pi i) \,\mathrm{Log}\, \zeta$ transforms the upper half plane onto the strip $0 < u < 1$, where $w = u + iv$ (see Fig. 3). Hence, the required harmonic function is

$$u(z) = \mathrm{Re}\, w = \frac{1}{\pi} \,\mathrm{Arg}\, \zeta = \frac{1}{\pi} \,\mathrm{Arg}\left[i\left(\frac{i - z}{i + z}\right) \right].$$

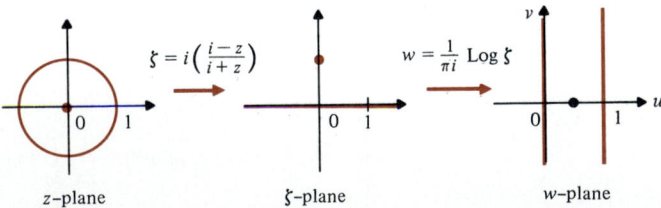

z–plane ζ–plane   w–plane

FIGURE 3

We now solve this Dirichlet problem using Poisson's integral formula. Since $u = 0$ on the boundary of the unit disk in the left half plane, the integral in (2) extends from $-\pi/2$ to $\pi/2$. In that interval $u = 1$ so (2) becomes

$$u(z) = \frac{1}{2\pi} \int_{-\pi/2}^{\pi/2} \frac{1 - |z|^2}{|e^{i\phi} - z|^2} \, d\phi.$$

Setting $z = re^{i\theta}$, we obtain

$$u(z) = \frac{1-r^2}{2\pi} \int_{-\pi/2}^{\pi/2} \frac{d(\phi - \theta)}{1 + r^2 - 2r\cos(\phi - \theta)}$$

$$= \frac{1}{\pi} \tan^{-1}\left(\frac{1+r}{1-r} \tan \frac{\phi - \theta}{2} \right) \Big|_{-\pi/2}^{\pi/2}.$$

Hence,

$$\tan \pi u(z) = \frac{[(1+r)/(1-r)][\tan(\pi/4 - \theta/2) + \tan(\pi/4 + \theta/2)]}{1 - [(1+r)/(1-r)]^2 \tan(\pi/4 - \theta/2)\tan(\pi/4 + \theta/2)}$$

$$= \frac{1-r^2}{2r}\left(\frac{\tan^2 \theta/2 + 1}{\tan^2 \theta/2 - 1} \right) = \frac{1-r^2}{-2r\cos\theta}.$$

To check that these answers are equivalent, observe that

$$\mathrm{Arg}\left[i\left(\frac{i-z}{i+z} \right) \right] = \mathrm{Arg}\left[\frac{-(z+\bar z) + i(1 - |z|^2)}{|i+z|^2} \right]$$

and, using the identity $\mathrm{Arg}(x + iy) = \tan^{-1}(y/x)$ in the right half plane,

$$\mathrm{Arg}\left[i\left(\frac{i-z}{i+z} \right) \right] = \tan^{-1}\left[\frac{1 - |z|^2}{-(z+\bar z)} \right] = \tan^{-1}\left(\frac{1-r^2}{-2r\cos\theta} \right).$$

Problems 19.6

In Problems 1–11 find a harmonic function in the unit disk that takes the given values on the unit circle.

1. $u(\phi) = \sin\phi$

2. $u(\phi) = \cos 2\phi$

3. $u(\phi) = \sin^2\phi$

4. $u(\phi) = 1 - \cos\phi$

5. $u(\phi) = \cos^3\phi$

6. $u(\phi) = \begin{cases} 0, & 0 \le \phi \le \pi \\ 1, & \pi < \phi < 2\pi \end{cases}$

7. $u(\phi) = \begin{cases} -1, & 0 \le \phi \le \pi \\ 1, & \pi < \phi < 2\pi \end{cases}$

8. $u(\phi) = \begin{cases} 0, & 0 \le \phi \le \pi \\ \phi, & \pi < \phi < 2\pi \end{cases}$

9. $u(\phi) = |\phi|$

10. $u(\phi) = \begin{cases} \phi, & 0 \le \phi \le \pi \\ \phi - \pi, & \pi < \phi < 2\pi \end{cases}$

11. $u(\phi) = \begin{cases} -1, & 0 \le \phi \le \pi - 1 \\ \phi, & \pi - 1 < \phi \le \pi + 1 \\ 1, & \pi + 1 < \phi < 2\pi \end{cases}$

* 12. Show that the solution to Dirichlet's problem on the upper half plane with boundary values $u(t)$ is

$$u(z) = \frac{y}{\pi} \int_{-\infty}^{\infty} \frac{u(t)\,dt}{(t-x)^2 + y^2}, \quad z = x + iy.$$

[*Hint:* Look at Example 3.]

* 13. Show that Poisson's integral formula for the first quadrant is

$$u(z) = \frac{4xy}{\pi}\Bigg[\int_0^{\infty} \frac{tu(t)\,dt}{t^4 - 2t^2(x^2 - y^2) + (x^2 + y^2)^2}$$

$$+ \int_0^{\infty} \frac{tu(it)\,dt}{t^4 + 2t^2(x^2 - y^2) + (x^2 + y^2)^2} \Bigg].$$

14. Find the temperature at any point z in the upper half plane if the temperature (in degrees) on the real axis is given by

$$u(t) = \begin{cases} 1 - |t|, & |t| \le 1, \\ 0, & |t| \ge 1. \end{cases}$$

15. Prove that any solution of Dirichlet's problem continuous on the closure of a simply connected region Ω must be unique. [*Hint:* Use the maximum principle.]

16. Suppose $u(z)$ and $v(z)$ are harmonic on a region Ω, continuous on its closure, and satisfy $u(z) \le v(z)$ on the boundary of Ω. Show that $u(z) \le v(z)$ for every z in Ω.

* 17. Let Ω be the region $0 < |z| < 1$. Show that there is no function $u(z)$ harmonic in Ω with boundary values $u(e^{i\theta}) = 0$, $u(0) = a > 0$. (*Hint:* Apply the result of Problem 16 to the functions

$$u_r(z) = a\frac{\ln|z|}{\ln r},$$

harmonic in $r < |z| < 1$.)

18. Use Problem 12 to find a function $u(z)$ that is harmonic in the upper half plane and that satisfies the boundary values

$$u(t) = \begin{cases} 1 & \text{if } |t| < 1, \\ 0 & \text{if } |t| \geq 1. \end{cases}$$

19. Repeat Problem 18 for the boundary values

$$u(t) = \begin{cases} 1 & \text{if } -1 \leq t \leq 0, \\ 1-t & \text{if } 0 \leq t \leq 1, \\ 0 & \text{elsewhere.} \end{cases}$$

Review Exercises for Chapter 19

In Exercises 1–14 find the image in the w-plane of the given set in the z-plane under the given transformation.

1. $y = x + 2$, $w = z^2$ **2.** $x = 3\pi/2$, $w = \cos z$

3. $x = 1$, $w = \cos z$ **4.** $y = 1$, $w = \cos z$

5. $y = 3$, $w = z^3$ **6.** $\{z : \operatorname{Re} z > 2\}$, $w = z^2$

7. $\{z : \operatorname{Re} z < -1 \text{ and } \operatorname{Im} z > 0\}$, $w = z^2$

8. $\{z : \operatorname{Im} z < 1\}$, $w = e^z$

9. $\{z : |z| < 4\}$, $w = e^z$

10. $\{z : \operatorname{Im} z > 1\}$, $w = \sin z$

11. $\{z : \operatorname{Re} z < 2\}$, $w = \sin z$

12. $\{z : |z| = 2\}$, $w = 1/z$

13. $\{z : \operatorname{Re} z = 2 \operatorname{Im} z\}$, $w = 1/z$

14. $|z| = 1$, $w = 4iz - 5$

In Exercises 15–18 find all points at which the given mapping is not conformal.

15. $w = \sin 2z$

16. $w = z - 1/z$

17. $w = z^{100}$

18. $w = z^3 e^z$

19. What happens to the angle at the origin between the rays $\arg z = 0$ and $\arg z = \pi/4$ under the mapping $w = z^6$?

20. Write $w = (2z - i)/(4iz + 5)$ as the composition of translations, rotations, magnifications, and inverses.

In Exercises 21–26 find and sketch the image of each point or region under the mapping $f(z) = (z + 1)/(z - i)$.

21. $z = -3i$ **22.** $z = 1 + 2i$

23. $|z + i| < 1$ **24.** $|z| > 2$

25. $\operatorname{Re} z \geq 2$ **26.** $\operatorname{Im} z \leq 1$

27. Find a mapping of the unit circles in the z- and w-planes such that $w(1) = i$.

28. What is the image of the unit disk under the mapping $w = i(z + 1)/(z - 1)$?

29. Determine the image of the circle $|z + i| = 1$ under the mapping $w = (2 + z)/3z$.

30. Map the half plane $\operatorname{Re} z \leq 0$ onto the unit disk so that the point -1 is mapped onto the origin.

31. Find the image of $|z| < 2$ under the mapping $w = z + 1/z$.

32. Find a linear fractional transformation that maps -1, 1, i to i, 1, ∞, respectively.

In Exercises 33–38 find the image of the line $\operatorname{Im} z = 1$ under the given mapping.

33. $w = z^2$ **34.** $w = e^z$

35. $w = \sin z$ **36.** $w = z^3$

37. $w = z^{1/2}$ **38.** $w = (z - i)^2$

39. Find the image of $\begin{cases} 2 < \operatorname{Re} z < 3 \\ 0 < \operatorname{Im} z < \dfrac{\pi}{2} \end{cases}$ under the mapping $w = e^z$.

40. Find the image of $\begin{cases} 0 < \operatorname{Re} z < \dfrac{\pi}{2} \\ -1 < \operatorname{Im} z < 1 \end{cases}$ under the mapping $w = \cos z$.

41. Find a conformal mapping of the infinite strip $|\operatorname{Im} z| < 1$ onto the unit disk.

42. Find a conformal mapping of the unit disk onto the infinite strip $|\operatorname{Im} z| < 1$.

43. Describe the image of the hyperbola $y^2 - x^2 = \frac{1}{2}$ under the mapping $w = \sqrt{1 - z^2}$.

In Problems 44–48 find a harmonic function that takes the given values on the unit circle.

44. $u(\phi) = \cos \phi$

45. $u(\phi) = \sin 2\phi$

46. $u(\phi) = \begin{cases} -1, & 0 \le \phi \le \pi \\ 2, & \pi \le \phi \le 2\pi \end{cases}$

47. $u(\phi) = |\phi/2|$

48. $u(\phi) = \begin{cases} 0, & 0 < \phi < \pi - 1 \\ \phi, & \pi - 1 < \phi < \pi + 1 \\ 1, & \pi + 1 < \phi < 2\pi \end{cases}$

49. Find a function that is harmonic on the upper half plane and satisfies

$$u(t) = \begin{cases} 1, & \text{if } |t| < 2, \\ 0, & \text{if } |t| > 2. \end{cases}$$

50. Find a conformal mapping from the upper half plane onto the semi-infinite strip $0 < \text{Re } w < 1$, $\text{Im } w < 0$.

appendix 1

Table of Integrals†

Standard Forms

1. $\displaystyle\int a\,dx = ax + C$

2. $\displaystyle\int af(x)\,dx = a\int f(x)\,dx$

3. $\displaystyle\int u\,dv = uv - \int v\,du$ (integration by parts)

4. $\displaystyle\int u^n\,du = \frac{u^{n+1}}{n+1} + C,\ n \neq -1$

5. $\displaystyle\int \frac{du}{u} = \ln|u| + C$

6. $\displaystyle\int e^u\,du = e^u + C$

7. $\displaystyle\int a^u\,du = \int e^{u\ln a}\,du = \frac{e^{u\ln a}}{\ln a} + C$

$\displaystyle = \frac{a^u}{\ln a} + C, \quad a > 0,\ a \neq 1$

8. $\displaystyle\int \sin u\,du = -\cos u + C$

9. $\displaystyle\int \cos u\,du = \sin u + C$

10. $\displaystyle\int \tan u\,du = \ln|\sec u| + C$

$\displaystyle = -\ln|\cos u| + C$

11. $\displaystyle\int \cot u\,du = \ln|\sin u| + C$

12. $\displaystyle\int \sec u\,du = \ln|\sec u + \tan u| + C$

$\displaystyle = \ln\left|\tan\left|\frac{u}{2} + \frac{\pi}{4}\right|\right| + C$

13. $\displaystyle\int \csc u\,du = \ln|\csc u - \cot u| + C$

$\displaystyle = \ln\left|\tan\frac{u}{2}\right| + C$

14. $\displaystyle\int \sec^2 u\,du = \tan u + C$

15. $\displaystyle\int \csc^2 u\,du = -\cot u + C$

16. $\displaystyle\int \sec u \tan u\,du = \sec u + C$

† All angles are measured in radians.

17. $\int \csc u \cot u \, du = -\csc u + C$

18. $\int \dfrac{du}{u^2 + a^2} = \dfrac{1}{a} \tan^{-1} \dfrac{u}{a} + C$

19. $\int \dfrac{du}{u^2 - a^2} = \dfrac{1}{2a} \ln \left| \dfrac{u - a}{u + a} \right| + C$

20. $\int \dfrac{du}{a^2 - u^2} = \dfrac{1}{2a} \ln \left| \dfrac{a + u}{a - u} \right| + C$

$\qquad\qquad = -\dfrac{1}{a} \coth^{-1} \dfrac{u}{a} + C, \quad u^2 > a^2$

$\qquad\qquad = \dfrac{1}{a} \tanh^{-1} \dfrac{u}{a} + C, \quad u^2 < a^2$

21. $\int \dfrac{du}{\sqrt{a^2 - u^2}} = \sin^{-1} \dfrac{u}{|a|} + C$

22. $\int \dfrac{du}{\sqrt{u^2 + a^2}} = \ln(u + \sqrt{u^2 + a^2}) + C$

23. $\int \dfrac{du}{\sqrt{u^2 - a^2}} = \ln|u + \sqrt{u^2 - a^2}| + C$

24. $\int \dfrac{du}{u\sqrt{u^2 - a^2}} = \dfrac{1}{|a|} \sec^{-1} \left| \dfrac{u}{a} \right| + C$

25. $\int \dfrac{du}{u\sqrt{u^2 + a^2}} = -\dfrac{1}{a} \ln \left| \dfrac{a + \sqrt{u^2 + a^2}}{u} \right| + C$

26. $\int \dfrac{du}{u\sqrt{a^2 - u^2}} = -\dfrac{1}{a} \ln \left| \dfrac{a + \sqrt{a^2 - u^2}}{u} \right| + C$

Integrals Involving $au + b$

27. $\int \dfrac{du}{au + b} = \dfrac{1}{a} \ln|au + b| + C$

28. $\int \dfrac{u \, du}{au + b} = \dfrac{u}{a} - \dfrac{b}{a^2} \ln|au + b| + C$

29. $\int \dfrac{u^2 \, du}{au + b} = \dfrac{(au + b)^2}{2a^3} - \dfrac{2b(au + b)}{a^3} + \dfrac{b^2}{a^3} \ln|au + b| + C$

30. $\int \dfrac{du}{u(au + b)} = \dfrac{1}{b} \ln \left| \dfrac{u}{au + b} \right| + C$

31. $\int \dfrac{du}{u^2(au + b)} = -\dfrac{1}{bu} + \dfrac{a}{b^2} \ln \left| \dfrac{au + b}{u} \right| + C$

32. $\int \dfrac{du}{(au + b)^2} = \dfrac{-1}{a(au + b)} + C$

33. $\int \dfrac{u \, du}{(au + b)^2} = \dfrac{b}{a^2(au + b)} + \dfrac{1}{a^2} \ln|au + b| + C$

34. $\int \dfrac{du}{u(au + b)^2} = \dfrac{1}{b(au + b)} + \dfrac{1}{b^2} \ln \left| \dfrac{u}{au + b} \right| + C$

35. $\int (au + b)^n \, du = \dfrac{(au + b)^{n+1}}{(n + 1)a} + C, \quad n \ne -1$

36. $\int u(au + b)^n \, du = \dfrac{(au + b)^{n+2}}{(n + 2)a^2} - \dfrac{b(au + b)^{n+1}}{(n + 1)a^2} + C, \quad n \ne -1, -2$

37. $\int u^m(au + b)^n \, du = \begin{cases} \dfrac{u^{m+1}(au + b)^n}{m + n + 1} + \dfrac{nb}{m + n + 1} \displaystyle\int u^m(au + b)^{n-1} \, du \\[2ex] \dfrac{u^m(au + b)^{n+1}}{(m + n + 1)a} - \dfrac{mb}{(m + n + 1)a} \displaystyle\int u^{m-1}(au + b)^n \, du \\[2ex] \dfrac{-u^{m+1}(au + b)^{n+1}}{(n + 1)b} + \dfrac{m + n + 2}{(n + 1)b} \displaystyle\int u^m(au + b)^{n+1} \, du \end{cases}$

Integrals Involving $\sqrt{au + b}$

38. $\int \dfrac{du}{\sqrt{au + b}} = \dfrac{2\sqrt{au + b}}{a} + C$

39. $\int \dfrac{u \, du}{\sqrt{au + b}} = \dfrac{2(au - 2b)}{3a^2} \sqrt{au + b} + C$

40. $\displaystyle\int \frac{du}{u\sqrt{au+b}} = \begin{cases} \dfrac{1}{\sqrt{b}} \ln\left|\dfrac{\sqrt{au+b}-\sqrt{b}}{\sqrt{au+b}+\sqrt{b}}\right| + C, & b>0 \\[3mm] \dfrac{2}{\sqrt{-b}} \tan^{-1}\sqrt{\dfrac{au+b}{-b}} + C, & b<0 \end{cases}$

41. $\displaystyle\int \sqrt{au+b}\,du = \frac{2\sqrt{(au+b)^3}}{3a} + C$

42. $\displaystyle\int u\sqrt{au+b}\,du = \frac{2(3au-2b)}{15a^2}\sqrt{(au+b)^3} + C$

43. $\displaystyle\int \frac{\sqrt{au+b}}{u}\,du = 2\sqrt{au+b} + b\int \frac{du}{u\sqrt{au+b}}$ (See 40.)

Integrals Involving $u^2 + a^2$

44. $\displaystyle\int \frac{du}{u^2+a^2} = \frac{1}{a}\tan^{-1}\frac{u}{a} + C$

45. $\displaystyle\int \frac{u\,du}{u^2+a^2} = \frac{1}{2}\ln(u^2+a^2) + C$

46. $\displaystyle\int \frac{u^2\,du}{u^2+a^2} = u - a\tan^{-1}\frac{u}{a} + C$

47. $\displaystyle\int \frac{du}{u(u^2+a^2)} = \frac{1}{2a^2}\ln\left(\frac{u^2}{u^2+a^2}\right) + C$

48. $\displaystyle\int \frac{du}{u^2(u^2+a^2)} = -\frac{1}{a^2u} - \frac{1}{a^3}\tan^{-1}\frac{u}{a} + C$

49. $\displaystyle\int \frac{du}{(u^2+a^2)^n} = \frac{u}{2(n-1)a^2(u^2+a^2)^{n-1}} + \frac{2n-3}{(2n-2)a^2}\int \frac{du}{(u^2+a^2)^{n-1}}$

50. $\displaystyle\int \frac{u\,du}{(u^2+a^2)^n} = \frac{-1}{2(n-1)(u^2+a^2)^{n-1}} + C, \quad n\neq 1$

51. $\displaystyle\int \frac{du}{u(u^2+a^2)^n} = \frac{1}{2(n-1)a^2(u^2+a^2)^{n-1}} + \frac{1}{a^2}\int \frac{du}{u(u^2+a^2)^{n-1}}, \quad n\neq 1$

Integrals Involving $u^2 - a^2$, $u^2 > a^2$

52. $\displaystyle\int \frac{du}{u^2-a^2} = \frac{1}{2a}\ln\left|\frac{u-a}{u+a}\right| + C$

53. $\displaystyle\int \frac{u\,du}{u^2-a^2} = \frac{1}{2}\ln(u^2-a^2) + C$

54. $\displaystyle\int \frac{u^2\,du}{u^2-a^2} = u + \frac{a}{2}\ln\left|\frac{u-a}{u+a}\right| + C$

55. $\displaystyle\int \frac{du}{u(u^2-a^2)} = \frac{1}{2a^2}\ln\left|\frac{u^2-a^2}{u^2}\right| + C$

56. $\displaystyle\int \frac{du}{u^2(u^2-a^2)} = \frac{1}{a^2u} + \frac{1}{2a^3}\ln\left|\frac{u-a}{u+a}\right| + C$

57. $\displaystyle\int \frac{du}{(u^2-a^2)^2} = \frac{-u}{2a^2(u^2-a^2)} - \frac{1}{4a^3}\ln\left|\frac{u-a}{u+a}\right| + C$

58. $\displaystyle\int \frac{du}{(u^2-a^2)^n} = \frac{-u}{2(n-1)a^2(u^2-a^2)^{n-1}} - \frac{2n-3}{(2n-2)a^2}\int \frac{du}{(u^2-a^2)^{n-1}}$

59. $\displaystyle\int \frac{u\,du}{(u^2-a^2)^n} = \frac{-1}{2(n-1)(u^2-a^2)^{n-1}} + C$

60. $\displaystyle\int \frac{du}{u(u^2-a^2)^n} = \frac{-1}{2(n-1)a^2(u^2-a^2)^{n-1}} - \frac{1}{a^2}\int \frac{du}{u(u^2-a^2)^{n-1}}$

Integrals Involving $a^2 - u^2$, $u^2 < a^2$

61. $\displaystyle \int \frac{du}{a^2 - u^2} = \frac{1}{2a} \ln \left| \frac{a+u}{a-u} \right| + C = \frac{1}{a} \tanh^{-1} \frac{u}{a} + C$

62. $\displaystyle \int \frac{u\,du}{a^2 - u^2} = -\frac{1}{2} \ln |a^2 - u^2| + C$

63. $\displaystyle \int \frac{u^2\,du}{a^2 - u^2} = -u + \frac{a}{2} \ln \left| \frac{a+u}{a-u} \right| + C$

64. $\displaystyle \int \frac{du}{u(a^2 - u^2)} = \frac{1}{2a^2} \ln \left| \frac{u^2}{a^2 - u^2} \right| + C$

65. $\displaystyle \int \frac{du}{(a^2 - u^2)^2} = \frac{u}{2a^2(a^2 - u^2)} + \frac{1}{4a^3} \ln \left| \frac{a+u}{a-u} \right| + C$

66. $\displaystyle \int \frac{u\,du}{(a^2 - u^2)^2} = \frac{1}{2(a^2 - u^2)} + C$

Integrals Involving $\sqrt{u^2 + a^2}$

67. $\displaystyle \int \frac{du}{\sqrt{u^2 + a^2}} = \ln(u + \sqrt{u^2 + a^2}) + C = \sinh^{-1} \frac{u}{|a|} + C$

68. $\displaystyle \int \frac{u\,du}{\sqrt{u^2 + a^2}} = \sqrt{u^2 + a^2} + C$

69. $\displaystyle \int \frac{u^2\,du}{\sqrt{u^2 + a^2}} = \frac{u\sqrt{u^2 + a^2}}{2} - \frac{a^2}{2} \ln(u + \sqrt{u^2 + a^2}) + C$

70. $\displaystyle \int \frac{du}{u\sqrt{u^2 + a^2}} = -\frac{1}{a} \ln \left| \frac{a + \sqrt{u^2 + a^2}}{u} \right| + C$

71. $\displaystyle \int \sqrt{u^2 + a^2}\,du = \frac{u\sqrt{u^2 + a^2}}{2} + \frac{a^2}{2} \ln(u + \sqrt{u^2 + a^2}) + C$

72. $\displaystyle \int u\sqrt{u^2 + a^2}\,du = \frac{(u^2 + a^2)^{3/2}}{3} + C$

73. $\displaystyle \int u^2\sqrt{u^2 + a^2}\,du = \frac{u(u^2 + a^2)^{3/2}}{4} - \frac{a^2 u\sqrt{u^2 + a^2}}{8} - \frac{a^4}{8} \ln(u + \sqrt{u^2 + a^2}) + C$

74. $\displaystyle \int \frac{\sqrt{u^2 + a^2}}{u}\,du = \sqrt{u^2 + a^2} - a \ln \left| \frac{a + \sqrt{u^2 + a^2}}{u} \right| + C$

75. $\displaystyle \int \frac{\sqrt{u^2 + a^2}}{u^2}\,du = -\frac{\sqrt{u^2 + a^2}}{u} + \ln(u + \sqrt{u^2 + a^2}) + C$

Integrals Involving $\sqrt{u^2 - a^2}$

76. $\displaystyle \int \frac{du}{\sqrt{u^2 - a^2}} = \ln|u + \sqrt{u^2 - a^2}| + C$

77. $\displaystyle \int \frac{u\,du}{\sqrt{u^2 - a^2}} = \sqrt{u^2 - a^2} + C$

78. $\displaystyle \int \frac{u^2\,du}{\sqrt{u^2 - a^2}} = \frac{u\sqrt{u^2 - a^2}}{2} + \frac{a^2}{2} \ln|u + \sqrt{u^2 - a^2}| + C$

79. $\displaystyle \int \frac{du}{u\sqrt{u^2 - a^2}} = \frac{1}{|a|} \sec^{-1} \left| \frac{u}{a} \right| + C$

80. $\displaystyle \int \sqrt{u^2 - a^2}\,du = \frac{u\sqrt{u^2 - a^2}}{2} - \frac{a^2}{2} \ln|u + \sqrt{u^2 - a^2}| + C$

81. $\displaystyle \int u\sqrt{u^2 - a^2}\,du = \frac{(u^2 - a^2)^{3/2}}{3} + C$

82. $\displaystyle\int u^2\sqrt{u^2-a^2}\,du = \frac{u(u^2-a^2)^{3/2}}{4} + \frac{a^2u\sqrt{u^2-a^2}}{8} - \frac{a^4}{8}\ln|u+\sqrt{u^2-a^2}| + C$

83. $\displaystyle\int \frac{\sqrt{u^2-a^2}}{u}\,du = \sqrt{u^2-a^2} - |a|\sec^{-1}\left|\frac{u}{a}\right| + C$

84. $\displaystyle\int \frac{\sqrt{u^2-a^2}}{u^2}\,du = -\frac{\sqrt{u^2-a^2}}{u} + \ln|u+\sqrt{u^2-a^2}| + C$

85. $\displaystyle\int \frac{du}{(u^2-a^2)^{3/2}} = -\frac{u}{a^2\sqrt{u^2-a^2}} + C$

Integrals Involving $\sqrt{a^2-u^2}$

86. $\displaystyle\int \frac{du}{\sqrt{a^2-u^2}} = \sin^{-1}\frac{u}{|a|} + C$

87. $\displaystyle\int \frac{u\,du}{\sqrt{a^2-u^2}} = -\sqrt{a^2-u^2} + C$

88. $\displaystyle\int \frac{u^2\,du}{\sqrt{a^2-u^2}} = -\frac{u\sqrt{a^2-u^2}}{2} + \frac{a^2}{2}\sin^{-1}\frac{u}{|a|} + C$

89. $\displaystyle\int \frac{du}{u\sqrt{a^2-u^2}} = -\frac{1}{a}\ln\left|\frac{a+\sqrt{a^2-u^2}}{u}\right| + C$

90. $\displaystyle\int \frac{du}{u^2\sqrt{a^2-u^2}} = -\frac{\sqrt{a^2-u^2}}{a^2u} + C$

91. $\displaystyle\int \sqrt{a^2-u^2}\,du = \frac{u\sqrt{a^2-u^2}}{2} + \frac{a^2}{2}\sin^{-1}\frac{u}{|a|} + C$

92. $\displaystyle\int u\sqrt{a^2-u^2}\,du = -\frac{(a^2-u^2)^{3/2}}{3} + C$

93. $\displaystyle\int u^2\sqrt{a^2-u^2}\,du = -\frac{u(a^2-u^2)^{3/2}}{4} + \frac{a^2u\sqrt{a^2-u^2}}{8} + \frac{a^4}{8}\sin^{-1}\frac{u}{|a|} + C$

94. $\displaystyle\int \frac{\sqrt{a^2-u^2}}{u}\,du = \sqrt{a^2-u^2} - a\ln\left|\frac{a+\sqrt{a^2-u^2}}{u}\right| + C$

95. $\displaystyle\int \frac{\sqrt{a^2-u^2}}{u^2}\,du = -\frac{\sqrt{a^2-u^2}}{u} - \sin^{-1}\frac{u}{|a|} + C$

Integrals Involving the Trigonometric Functions

96. $\displaystyle\int \sin au\,du = -\frac{\cos au}{a} + C$

97. $\displaystyle\int u\sin au\,du = \frac{\sin au}{a^2} - \frac{u\cos au}{a} + C$

98. $\displaystyle\int u^2\sin au\,du = \frac{2u}{a^2}\sin au + \left(\frac{2}{a^3} - \frac{u^2}{a}\right)\cos au + C$

99. $\displaystyle\int \frac{du}{\sin au} = \frac{1}{a}\ln|\csc au - \cot au| + C$

$\displaystyle = \frac{1}{a}\ln\left|\tan\frac{au}{2}\right| + C$

100. $\displaystyle\int \sin^2 au\,du = \frac{u}{2} - \frac{\sin 2au}{4a} + C$

101. $\displaystyle\int u\sin^2 au\,du = \frac{u^2}{4} - \frac{u\sin 2au}{4a} - \frac{\cos 2au}{8a^2} + C$

102. $\displaystyle\int \frac{du}{\sin^2 au} = -\frac{1}{a}\cot au + C$

103. $\displaystyle\int \sin pu \sin qu\, du = \frac{\sin(p-q)u}{2(p-q)} - \frac{\sin(p+q)u}{2(p+q)} + C, \quad p \neq \pm q$

104. $\displaystyle\int \frac{du}{1-\sin au} = \frac{1}{a}\tan\left(\frac{\pi}{4} + \frac{au}{2}\right) + C$

105. $\displaystyle\int \frac{u\,du}{1-\sin au} = \frac{u}{a}\tan\left(\frac{\pi}{4} + \frac{au}{2}\right) + \frac{2}{a^2}\ln\left|\sin\left(\frac{\pi}{4} - \frac{au}{2}\right)\right| + C$

106. $\displaystyle\int \frac{du}{1+\sin au} = -\frac{1}{a}\tan\left(\frac{\pi}{4} - \frac{au}{2}\right) + C$

107. $\displaystyle\int \frac{du}{p+q\sin au} = \begin{cases} \dfrac{2}{a\sqrt{p^2-q^2}}\tan^{-1}\dfrac{p\tan\frac{1}{2}au+q}{\sqrt{p^2-q^2}} + C, & |p| > |q| \\[3mm] \dfrac{1}{a\sqrt{q^2-p^2}}\ln\left|\dfrac{p\tan\frac{1}{2}au+q-\sqrt{q^2-p^2}}{p\tan\frac{1}{2}au+q+\sqrt{q^2-p^2}}\right| + C, & |p| < |q| \end{cases}$

108. $\displaystyle\int u^m \sin au\, du = -\frac{u^m \cos au}{a} + \frac{mu^{m-1}\sin au}{a^2} - \frac{m(m-1)}{a^2}\int u^{m-2}\sin au\, du$

109. $\displaystyle\int \sin^n au\, du = -\frac{\sin^{n-1} au \cos au}{an} + \frac{n-1}{n}\int \sin^{n-2} au\, du$

110. $\displaystyle\int \frac{du}{\sin^n au} = \frac{-\cos au}{a(n-1)\sin^{n-1} au} + \frac{n-2}{n-1}\int \frac{du}{\sin^{n-2} au}, \quad n \neq 1$

111. $\displaystyle\int \cos au\, du = \frac{\sin au}{a} + C$

112. $\displaystyle\int u \cos au\, du = \frac{\cos au}{a^2} + \frac{u\sin au}{a} + C$

113. $\displaystyle\int u^2 \cos au\, du = \frac{2u}{a^2}\cos au + \left(\frac{u^2}{a} - \frac{2}{a^3}\right)\sin au + C$

114. $\displaystyle\int \frac{du}{\cos au} = \frac{1}{a}\ln|\sec au + \tan au| + C = \frac{1}{a}\ln\left|\tan\left(\frac{\pi}{4} + \frac{au}{2}\right)\right| + C$

115. $\displaystyle\int \cos^2 au\, du = \frac{u}{2} + \frac{\sin 2au}{4a} + C$

116. $\displaystyle\int u\cos^2 au\, du = \frac{u^2}{4} + \frac{u\sin 2au}{4a} + \frac{\cos 2au}{8a^2} + C$

117. $\displaystyle\int \frac{du}{\cos^2 au} = \frac{\tan au}{a} + C$

118. $\displaystyle\int \cos qu \cos pu\, du = \frac{\sin(q-p)u}{2(q-p)} + \frac{\sin(q+p)u}{2(q+p)} + C, \quad q \neq \pm p$

119. $\displaystyle\int \frac{du}{p+q\cos au} = \begin{cases} \dfrac{2}{a\sqrt{p^2-q^2}}\tan^{-1}\left[\sqrt{(p-q)/(p+q)}\,\tan\frac{1}{2}au\right] + C, & |p| > |q| \\[3mm] \dfrac{1}{a\sqrt{q^2-p^2}}\ln\left[\dfrac{\tan\frac{1}{2}au + \sqrt{(q+p)/(q-p)}}{\tan\frac{1}{2}au - \sqrt{(q+p)/(q-p)}}\right] + C, & |p| < |q| \end{cases}$

120. $\displaystyle\int u^m \cos au\, du = \frac{u^m \sin au}{a} + \frac{mu^{m-1}}{a^2}\cos au - \frac{m(m-1)}{a^2}\int u^{m-2}\cos au\, du$

121. $\displaystyle\int \cos^n au\, du = \frac{\sin au \cos^{n-1} au}{an} + \frac{n-1}{n}\int \cos^{n-2} au\, du$

122. $\displaystyle\int \frac{du}{\cos^n au} = \frac{\sin au}{a(n-1)\cos^{n-1} au} + \frac{n-2}{n-1}\int \frac{du}{\cos^{n-2} au}$

123. $\displaystyle\int \sin au \cos au\, du = \frac{\sin^2 au}{2a} + C$

124. $\displaystyle\int \sin pu \cos qu\, du = -\frac{\cos(p-q)u}{2(p-q)} - \frac{\cos(p+q)u}{2(p+q)} + C, \quad p \neq \pm q$

125. $\displaystyle\int \sin^n au \cos au\, du = \frac{\sin^{n+1} au}{(n+1)a} + C, \quad n \neq -1$ **126.** $\displaystyle\int \cos^n au \sin au\, du = -\frac{\cos^{n+1} au}{(n+1)a} + C, \quad n \neq -1$

127. $\displaystyle\int \sin^2 au \cos^2 au\, du = \frac{u}{8} - \frac{\sin 4au}{32a} + C$ **128.** $\displaystyle\int \frac{du}{\sin au \cos au} = \frac{1}{a}\ln|\tan au| + C$

129. $\displaystyle\int \frac{du}{\cos au(1 \pm \sin au)} = \mp\frac{1}{2a(1 \pm \sin au)} + \frac{1}{2a}\ln\left|\tan\left(\frac{au}{2} + \frac{\pi}{4}\right)\right| + C$

130. $\displaystyle\int \frac{du}{\sin au(1 \pm \cos au)} = \pm\frac{1}{2a(1 \pm \cos au)} + \frac{1}{2a}\ln\left|\tan\frac{au}{2}\right| + C$

131. $\displaystyle\int \frac{du}{\sin au \pm \cos au} = \frac{1}{a\sqrt{2}}\ln\left|\tan\left(\frac{au}{2} \pm \frac{\pi}{8}\right)\right| + C$

132. $\displaystyle\int \frac{\sin au\, du}{\sin au \pm \cos au} = \frac{u}{2} \mp \frac{1}{2a}\ln|\sin au \pm \cos au| + C$

133. $\displaystyle\int \frac{\cos au\, du}{\sin au \pm \cos au} = \pm\left[\frac{u}{2} \pm \frac{1}{2a}\ln|\sin au \pm \cos au|\right] + C$

134. $\displaystyle\int \frac{\sin au\, du}{p + q \cos au} = -\frac{1}{aq}\ln|p + q \cos au| + C$ **135.** $\displaystyle\int \frac{\cos au\, du}{p + q \sin au} = \frac{1}{aq}\ln|p + q \sin au| + C$

136. $\displaystyle\int \sin^m au \cos^n au\, du = \begin{cases} -\dfrac{\sin^{m-1} au \cos^{n+1} au}{a(m+n)} + \dfrac{m-1}{m+n}\displaystyle\int \sin^{m-2} au \cos^n au\, du, & m \neq -n \\[3mm] \dfrac{\sin^{m+1} au \cos^{n-1} au}{a(m+n)} + \dfrac{n-1}{m+n}\displaystyle\int \sin^m au \cos^{n-2} au\, du, & m \neq -n \end{cases}$

137. $\displaystyle\int \tan au\, du = \frac{1}{a}\ln|\sec au| + C$ **138.** $\displaystyle\int \tan^2 au\, du = \frac{\tan au}{a} - u + C$

139. $\displaystyle\int \tan^n au \sec^2 au\, du = \frac{\tan^{n+1} au}{(n+1)a} + C, \quad n \neq -1$

140. $\displaystyle\int \tan^n au\, du = \frac{\tan^{n-1} au}{(n-1)a} - \int \tan^{n-2} au\, du, \quad n \neq 1$

141. $\displaystyle\int \cot au\, du = \frac{1}{a}\ln|\sin au| + C$ **142.** $\displaystyle\int \cot^2 au\, du = -\frac{\cot au}{a} - u + C$

143. $\displaystyle\int \cot^n au \csc^2 au \, du = -\frac{\cot^{n+1} au}{(n+1)a} + C, \quad n \neq -1$

144. $\displaystyle\int \cot^n au \, du = -\frac{\cot^{n-1} au}{(n-1)a} - \int \cot^{n-2} au \, du, \quad n \neq 1$

145. $\displaystyle\int \sec au \, du = \frac{1}{a} \ln|\sec au + \tan au| + C = \frac{1}{a} \ln\left|\tan\left(\frac{au}{2} + \frac{\pi}{4}\right)\right| + C$

146. $\displaystyle\int \sec^2 au \, du = \frac{\tan au}{a} + C$

147. $\displaystyle\int \sec^3 au \, du = \frac{\sec au \tan au}{2a} + \frac{1}{2a} \ln|\sec au + \tan au| + C$

148. $\displaystyle\int \sec^n au \tan au \, du = \frac{\sec^n au}{na} + C$

149. $\displaystyle\int \sec^n au \, du = \frac{\sec^{n-2} au \tan au}{a(n-1)} + \frac{n-2}{n-1} \int \sec^{n-2} au \, du, \quad n \neq 1$

150. $\displaystyle\int \csc au \, du = \frac{1}{a} \ln|\csc au - \cot au| + C = \frac{1}{a} \ln\left|\tan\frac{au}{2}\right| + C$

151. $\displaystyle\int \csc^2 au \, du = -\frac{\cot au}{a} + C$
152. $\displaystyle\int \csc^n au \cot au \, du = -\frac{\csc^n au}{na} + C$

153. $\displaystyle\int \csc^n au \, du = -\frac{\csc^{n-2} au \cot au}{a(n-1)} + \frac{n-2}{n-1} \int \csc^{n-2} au \, du, \quad n \neq 1$

Integrals Involving Inverse Trigonometric Functions

154. $\displaystyle\int \sin^{-1}\frac{u}{a} \, du = u \sin^{-1}\frac{u}{a} + \sqrt{a^2 - u^2} + C$

155. $\displaystyle\int u \sin^{-1}\frac{u}{a} \, du = \left(\frac{u^2}{2} - \frac{a^2}{4}\right) \sin^{-1}\frac{u}{a} + \frac{u\sqrt{a^2 - u^2}}{4} + C$

156. $\displaystyle\int \cos^{-1}\frac{u}{a} \, du = u \cos^{-1}\frac{u}{a} - \sqrt{a^2 - u^2} + C$

157. $\displaystyle\int u \cos^{-1}\frac{u}{a} \, du = \left(\frac{u^2}{2} - \frac{a^2}{4}\right) \cos^{-1}\frac{u}{a} - \frac{u\sqrt{a^2 - u^2}}{4} + C$

158. $\displaystyle\int \tan^{-1}\frac{u}{a} \, du = u \tan^{-1}\frac{u}{a} - \frac{a}{2} \ln(u^2 + a^2) + C$
159. $\displaystyle\int u \tan^{-1}\frac{u}{a} \, du = \frac{1}{2}(u^2 + a^2) \tan^{-1}\frac{u}{a} - \frac{au}{2} + C$

160. $\displaystyle\int u^m \sin^{-1}\frac{u}{a} \, du = \frac{u^{m+1}}{m+1} \sin^{-1}\frac{u}{a} - \frac{1}{m+1} \int \frac{u^{m+1}}{\sqrt{a^2 - u^2}} \, du$

161. $\displaystyle\int u^m \cos^{-1}\frac{u}{a} \, du = \frac{u^{m+1}}{m+1} \cos^{-1}\frac{u}{a} + \frac{1}{m+1} \int \frac{u^{m+1}}{\sqrt{a^2 - u^2}} \, du$

162. $\displaystyle\int u^m \tan^{-1}\frac{u}{a}\,du = \frac{u^{m+1}}{m+1}\tan^{-1}\frac{u}{a} - \frac{a}{m+1}\int\frac{u^{m+1}}{u^2+a^2}\,du$

Integrals Involving e^{au}

163. $\displaystyle\int e^{au}\,du = \frac{e^{au}}{a} + C$

164. $\displaystyle\int ue^{au}\,du = \frac{e^{au}}{a}\left(u-\frac{1}{a}\right) + C$

165. $\displaystyle\int u^2 e^{au}\,du = \frac{e^{au}}{a}\left(u^2 - \frac{2u}{a} + \frac{2}{a^2}\right) + C$

166. $\displaystyle\int u^n e^{au}\,du = \frac{u^n e^{au}}{a} - \frac{n}{a}\int u^{n-1}e^{au}\,du$

$\displaystyle\qquad\qquad = \frac{e^{au}}{a}\left[u^n - \frac{nu^{n-1}}{a} + \frac{n(n-1)u^{n-2}}{a^2} - \cdots + \frac{(-1)^n n!}{a^n}\right] + C,$ if n is a positive integer

167. $\displaystyle\int\frac{du}{p+qe^{au}} = \frac{u}{p} - \frac{1}{ap}\ln|p + qe^{au}| + C$

168. $\displaystyle\int e^{au}\sin bu\,du = \frac{e^{au}(a\sin bu - b\cos bu)}{a^2+b^2} + C$

169. $\displaystyle\int e^{au}\cos bu\,du = \frac{e^{au}(a\cos bu + b\sin bu)}{a^2+b^2} + C$

170. $\displaystyle\int ue^{au}\sin bu\,du = \frac{ue^{au}(a\sin bu - b\cos bu)}{a^2+b^2} - \frac{e^{au}[(a^2-b^2)\sin bu - 2ab\cos bu]}{(a^2+b^2)^2} + C$

171. $\displaystyle\int ue^{au}\cos bu\,du = \frac{ue^{au}(a\cos bu + b\sin bu)}{a^2+b^2} - \frac{e^{au}[(a^2-b^2)\cos bu + 2ab\sin bu]}{(a^2+b^2)^2} + C$

172. $\displaystyle\int e^{au}\sin^n bu\,du = \frac{e^{au}\sin^{n-1}bu}{a^2+n^2b^2}(a\sin bu - nb\cos bu) + \frac{n(n-1)b^2}{a^2+n^2b^2}\int e^{au}\sin^{n-2}bu\,du$

173. $\displaystyle\int e^{au}\cos^n bu\,du = \frac{e^{au}\cos^{n-1}bu}{a^2+n^2b^2}(a\cos bu + nb\sin bu) + \frac{n(n-1)b^2}{a^2+n^2b^2}\int e^{au}\cos^{n-2}bu\,du$

Integrals Involving $\ln u$

174. $\displaystyle\int \ln u\,du = u\ln u - u + C$

175. $\displaystyle\int u\ln u\,du = \frac{u^2}{2}(\ln u - \tfrac{1}{2}) + C$

176. $\displaystyle\int u^m \ln u\,du = \frac{u^{m+1}}{m+1}\left(\ln u - \frac{1}{m+1}\right) + C \quad$ if $m \neq -1$

177. $\displaystyle\int\frac{\ln u}{u}\,du = \frac{1}{2}\ln^2 u + C$

178. $\displaystyle\int\frac{\ln^n u\,du}{u} = \frac{\ln^{n+1}u}{n+1} + C \quad$ if $n \neq -1$

179. $\displaystyle\int\frac{du}{u\ln u} = \ln|\ln u| + C$

180. $\displaystyle\int \ln^n u\,du = u\ln^n u - n\int \ln^{n-1}u\,du, \quad n \neq -1$

181. $\displaystyle\int u^m \ln^n u\,du = \frac{u^{m+1}\ln^n u}{m+1} - \frac{n}{m+1}\int u^m \ln^{n-1}u\,du \quad$ if $m \neq -1, n \neq -1$

182. $\displaystyle\int \ln(u^2 + a^2)\, du = u \ln(u^2 + a^2) - 2u + 2a \tan^{-1}\frac{u}{a} + C$

183. $\displaystyle\int \ln|u^2 - a^2|\, du = u \ln|u^2 - a^2| - 2u + a \ln\left|\frac{u+a}{u-a}\right| + C$

Integrals Involving Hyperbolic Functions

184. $\displaystyle\int \sinh au\, du = \frac{\cosh au}{a} + C$

185. $\displaystyle\int u \sinh au\, du = \frac{u \cosh au}{a} - \frac{\sinh au}{a^2} + C$

186. $\displaystyle\int \cosh au\, du = \frac{\sinh au}{a} + C$

187. $\displaystyle\int u \cosh au\, du = \frac{u \sinh au}{a} - \frac{\cosh au}{a^2} + C$

188. $\displaystyle\int \cosh^2 au\, du = \frac{u}{2} + \frac{\sinh au \cosh au}{2a} + C$

189. $\displaystyle\int \sinh^2 au\, du = \frac{\sinh au \cosh au}{2a} - \frac{u}{2} + C$

190. $\displaystyle\int \sinh^n au\, du = \frac{\sinh^{n-1} au \cosh au}{an} - \frac{n-1}{n}\int \sinh^{n-2} au\, du$

191. $\displaystyle\int \cosh^n au\, du = \frac{\cosh^{n-1} au \sinh au}{an} + \frac{n-1}{n}\int \cosh^{n-2} au\, du$

192. $\displaystyle\int \sinh au \cosh au\, du = \frac{\sinh^2 au}{2a} + C$

193. $\displaystyle\int \sinh pu \cosh qu\, du = \frac{\cosh(p+q)u}{2(p+q)} + \frac{\cosh(p-q)u}{2(p-q)} + C$

194. $\displaystyle\int \tanh au\, du = \frac{1}{a} \ln \cosh au + C$

195. $\displaystyle\int \tanh^2 au\, du = u - \frac{\tanh au}{a} + C$

196. $\displaystyle\int \tanh^n au\, du = \frac{-\tanh^{n-1} au}{a(n-1)} + \int \tanh^{n-2} au\, du$

197. $\displaystyle\int \coth au\, du = \frac{1}{a} \ln|\sinh au| + C$

198. $\displaystyle\int \coth^2 au\, du = u - \frac{\coth au}{a} + C$

199. $\displaystyle\int \operatorname{sech} au\, du = \frac{2}{a} \tan^{-1} e^{au} + C$

200. $\displaystyle\int \operatorname{sech}^2 au\, du = \frac{\tanh au}{a} + C$

201. $\displaystyle\int \operatorname{sech}^n au\, du = \frac{\operatorname{sech}^{n-2} au \tanh au}{a(n-1)} + \frac{n-2}{n-1}\int \operatorname{sech}^{n-2} au\, du$

202. $\displaystyle\int \operatorname{csch} au\, du = \frac{1}{a} \ln\left|\tanh \frac{au}{2}\right| + C$

203. $\displaystyle\int \operatorname{csch}^2 au\, du = -\frac{\coth au}{a} + C$

204. $\displaystyle\int \operatorname{sech} u \tanh u\, du = -\operatorname{sech} u + C$

205. $\displaystyle\int \operatorname{csch} u \coth u\, du = -\operatorname{csch} u + C$

Some Definite Integrals

Unless otherwise stated, the letters a, b, m, n, or p stand for positive constants.

206. $\displaystyle\int_0^\infty \frac{dx}{x^2 + a^2} = \frac{\pi}{2a}$

207. $\displaystyle\int_0^\infty \frac{x^{p-1}}{1 + x} \, dx = \frac{\pi}{\sin p\pi}, \quad 0 < p < 1$

208. $\displaystyle\int_0^a \frac{dx}{\sqrt{a^2 - x^2}} = \frac{\pi}{2}$

209. $\displaystyle\int_0^a \sqrt{a^2 - x^2} \, dx = \frac{\pi a^2}{4}$

210. $\displaystyle\int_0^\pi \sin mx \sin nx \, dx = \begin{cases} 0, & \text{if } m, n \text{ integers and } m \neq n \\[2mm] \dfrac{\pi}{2}, & \text{if } m, n \text{ integers and } m = n \end{cases}$

211. $\displaystyle\int_0^\pi \cos mx \cos nx \, dx = \begin{cases} 0, & \text{if } m, n \text{ integers and } m \neq n \\[2mm] \dfrac{\pi}{2}, & \text{if } m, n \text{ integers and } m = n \end{cases}$

212. $\displaystyle\int_0^\pi \sin mx \cos nx \, dx = \begin{cases} 0, & \text{if } m, n \text{ integers and } m + n \text{ is even} \\[2mm] \dfrac{2m}{(m^2 - n^2)}, & \text{if } m, n \text{ integers and } m + n \text{ is odd} \end{cases}$

213. $\displaystyle\int_0^{\pi/2} \sin^2 x \, dx = \int_0^{\pi/2} \cos^2 x \, dx = \frac{\pi}{4}$

214. $\displaystyle\int_0^\infty e^{-ax} \cos bx \, dx = \frac{a}{a^2 + b^2}$

215. $\displaystyle\int_0^\infty e^{-ax} \sin bx \, dx = \frac{b}{a^2 + b^2}$

216. $\displaystyle\int_0^\infty e^{-a^2 x^2} \, dx = \frac{\sqrt{\pi}}{2a}$

217. $\displaystyle\int_0^{\pi/2} \sin^{2m} x \, dx = \int_0^{\pi/2} \cos^{2m} x \, dx = \frac{1 \cdot 3 \cdot 5 \cdot \ \cdots \ \cdot (2m-1)}{2 \cdot 4 \cdot 6 \cdot \ \cdots \ \cdot 2m} \frac{\pi}{2}, \ m = 1, 2, 3, \ldots$

218. $\displaystyle\int_0^{\pi/2} \sin^{2m+1} x \, dx = \int_0^{\pi/2} \cos^{2m+1} x \, dx = \frac{2 \cdot 4 \cdot 6 \cdot \ \cdots \ \cdot 2m}{1 \cdot 3 \cdot 5 \cdot \ \cdots \ \cdot (2m+1)}, \ m = 1, 2, 3, \ldots$

219. $\displaystyle\int_0^\infty \frac{e^{-x}}{\sqrt{x}} \, dx = \sqrt{\pi}$

220. $\displaystyle\int_0^1 x^m (\ln x)^n \, dx = \frac{(-1)^n n!}{(m+1)^{n+1}}, \quad m, n = 0, 1, 2, \ldots$

221. $\displaystyle\int_0^\infty x^{n-1} e^{-x} \, dx = (n-1)!, \quad n = 1, 2, 3, \ldots$

222. $\displaystyle\int_0^1 x^m (1 - x)^n \, dx = \frac{m! \, n!}{(m+n+1)!} \, m, n = 0, 1, 2, \ldots$

223. $\displaystyle\int_0^\infty \frac{x^{m-1} \, dx}{1 + x^n} = \frac{\pi}{n \sin \dfrac{m\pi}{n}}, \quad 0 < m < n$

224. $\displaystyle\int_0^\infty \frac{\tan x \, dx}{x} = \frac{\pi}{2}$

225. $\displaystyle\int_0^\infty \frac{\sin^2 x \, dx}{x^2} = \frac{\pi}{2}$

226. $\displaystyle\int_0^\infty \sin(x^2) \, dx = \frac{1}{2} \sqrt{\frac{\pi}{2}}$

227. $\displaystyle\int_0^{\pi/2} \ln(\sin x) \, dx = -\frac{\pi}{2} \ln 2$

228. $\displaystyle\int_0^{\pi/2} \ln(\tan x) \, dx = 0$

appendix 2

Table of Laplace Transforms[†]

$F(s) = \mathscr{L}\{f(t)\}$	$f(t) = \mathscr{L}^{-1}\{F(s)\}$
1. $\dfrac{1}{s^r}, \quad r > 0$	$\dfrac{t^{r-1}}{\Gamma(r)}, \quad \Gamma(n+1) = n!$
2. $\dfrac{1}{(s-a)^r}, \quad r > 0$	$\dfrac{t^{r-1}e^{at}}{\Gamma(r)}$
3. $\dfrac{1}{(s-a)(s-b)}, \quad a \neq b$	$\dfrac{e^{at} - e^{bt}}{a-b}$
4. $\dfrac{s}{(s-a)(s-b)}, \quad a \neq b$	$\dfrac{ae^{at} - be^{bt}}{a-b}$
5. $\dfrac{1}{s^2 + k^2}$	$\dfrac{1}{k}\sin kt$
6. $\dfrac{s}{s^2 + k^2}$	$\cos kt$
7. $\dfrac{1}{s^2 - k^2}$	$\dfrac{1}{k}\sinh kt$
8. $\dfrac{s}{s^2 - k^2}$	$\cosh kt$
9. $\dfrac{1}{(s-a)^2 + k^2}$	$\dfrac{1}{k}e^{at}\sin kt$
10. $\dfrac{s-a}{(s-a)^2 + k^2}$	$e^{at}\cos kt$

[†] For a more extensive list of Laplace transforms and their inverses, see A. Erdelyi et al., *Tables of Integral Transforms*, 2 vols. (New York: McGraw-Hill, 1954).

$F(s) = \mathcal{L}\{f(t)\}$	$f(t) = \mathcal{L}^{-1}\{F(s)\}$
11. $\dfrac{1}{(s-a)^2 - k^2}$	$\dfrac{1}{k}e^{at}\sinh kt$
12. $\dfrac{s-a}{(s-a)^2 - k^2}$	$e^{at}\cosh kt$
13. $\dfrac{1}{s(s^2 + k^2)}$	$\dfrac{1}{k^2}(1 - \cos kt)$
14. $\dfrac{1}{s^2(s^2 + k^2)}$	$\dfrac{1}{k^3}(kt - \sin kt)$
15. $\dfrac{1}{(s^2 + k^2)^2}$	$\dfrac{1}{2k^3}(\sin kt - kt\cos kt)$
16. $\dfrac{s}{(s^2 + k^2)^2}$	$\dfrac{t}{2k}\sin kt$
17. $\dfrac{s^2}{(s^2 + k^2)^2}$	$\dfrac{1}{2k}(\sin kt + kt\cos kt)$
18. $\dfrac{1}{(s^2 + a^2)(s^2 + b^2)}, \quad a^2 \neq b^2$	$\dfrac{a\sin bt - b\sin at}{ab(a^2 - b^2)}$
19. $\dfrac{s}{(s^2 + a^2)(s^2 + b^2)}, \quad a^2 \neq b^2$	$\dfrac{\cos bt - \cos at}{a^2 - b^2}$
20. $\dfrac{1}{s^4 - k^4}$	$\dfrac{1}{2k^3}(\sinh kt - \sin kt)$
21. $\dfrac{s}{s^4 - k^4}$	$\dfrac{1}{2k^2}(\cosh kt - \cos kt)$
22. $\dfrac{1}{s^4 + 4k^4}$	$\dfrac{1}{4k^3}(\sin kt\cosh kt - \cos kt\sinh kt)$
23. $\dfrac{s}{s^4 + 4k^4}$	$\dfrac{1}{2k^2}\sin kt\sinh kt$
24. $\sqrt{s-a} - \sqrt{s-b}$	$\dfrac{e^{bt} - e^{at}}{2\sqrt{\pi t^3}}$
25. $\dfrac{s}{(s-a)^{3/2}}$	$\dfrac{e^{at}}{\sqrt{\pi t}}(1 + 2at)$
26. $\dfrac{1}{\sqrt{s+a}\,\sqrt{s+b}}$	$e^{-(a+b)t/2}I_0\!\left(\dfrac{a-b}{2}t\right)$ [see p. 281]
27. $\dfrac{(\sqrt{s^2 + k^2} - s)^r}{\sqrt{s^2 + k^2}}, \quad r > -1$	$k^r J_r(kt)$ [see p. 273]

$F(s) = \mathcal{L}\{f(t)\}$	$f(t) = \mathcal{L}^{-1}\{F(s)\}$
28. $\dfrac{1}{(s^2 + k^2)^r}, \quad r > 0$	$\dfrac{\sqrt{\pi}}{\Gamma(r)}\left(\dfrac{t}{2k}\right)^{r-1/2} J_{r-1/2}(kt)$
29. $(\sqrt{s^2 + k^2} - s)^r, \quad r > 0$	$\dfrac{rk^r}{t} J_r(kt)$ [see p. 273]
30. $\dfrac{(s - \sqrt{s^2 - k^2})^r}{\sqrt{s^2 - k^2}}, \quad r > -1$	$k^r I_r(kt)$ [see p. 281]
31. $\dfrac{1}{(s^2 - k^2)^r}, \quad r > 0$	$\dfrac{\sqrt{\pi}}{\Gamma(r)}\left(\dfrac{t}{2k}\right)^{r-1/2} I_{r-1/2}(kt)$
32. $\dfrac{e^{-k/s}}{s^r}, \quad r > 0$	$\left(\dfrac{t}{k}\right)^{(r-1)/2} J_{r-1}(2\sqrt{kt})$
33. $\dfrac{e^{-k/s}}{\sqrt{s}}$	$\dfrac{1}{\sqrt{\pi t}} \cos 2\sqrt{kt}$
34. $\dfrac{e^{k/s}}{s^r}, \quad r > 0$	$\left(\dfrac{t}{k}\right)^{(r-1)/2} I_{r-1}(2\sqrt{kt})$
35. $\dfrac{e^{k/s}}{\sqrt{s}}$	$\dfrac{1}{\sqrt{\pi t}} \cosh 2\sqrt{kt}$
36. $\dfrac{1}{s} \ln s$	$-\ln t - \gamma, \quad \gamma \approx 0.5772$
37. $\ln \dfrac{s-a}{s-b}$	$\dfrac{e^{bt} - e^{at}}{t}$
38. $\ln\left(1 + \dfrac{k^2}{s^2}\right)$	$\dfrac{2}{t}(1 - \cos kt)$
39. $\ln\left(1 - \dfrac{k^2}{s^2}\right)$	$\dfrac{2}{t}(1 - \cosh kt)$
40. $\tan^{-1}\left(\dfrac{k}{s}\right)$	$\dfrac{\sin kt}{t}$
41. $\dfrac{1}{s} \tan^{-1}\left(\dfrac{k}{s}\right)$	$\mathrm{Si}(kt) = \displaystyle\int_0^{kt} \dfrac{\sin u}{u}\, du$
42. $e^{-r\sqrt{s}}, \quad r > 0$	$\dfrac{r}{2\sqrt{\pi t^3}} \exp\left(-\dfrac{r^2}{4t}\right)$
43. $\dfrac{e^{-r\sqrt{s}}}{s}, \quad r \geq 0$	$1 - \mathrm{erf}\left(\dfrac{r}{2\sqrt{t}}\right) = 1 - \dfrac{2}{\sqrt{\pi}} \displaystyle\int_0^{r/2\sqrt{t}} e^{-u^2}\, du$
44. $e^{r^2 s^2}(1 - \mathrm{erf}(rs)), \quad r > 0$	$\dfrac{1}{r\sqrt{\pi}} \exp\left(-\dfrac{t^2}{4r^2}\right)$

$F(s) = \mathscr{L}\{f(t)\}$	$f(t) = \mathscr{L}^{-1}\{F(s)\}$
45. $\dfrac{1}{s}e^{r^2s^2}(1-\text{erf}(rs)), \quad r>0$	$\text{erf}\left(\dfrac{t}{2r}\right)$
46. $\text{erf}\left(\dfrac{r}{\sqrt{s}}\right)$	$\dfrac{1}{\pi t}\sin(2k\sqrt{t})$
47. $e^{rs}(1-\text{erf}\sqrt{rs}), \quad r>0$	$\dfrac{\sqrt{r}}{\pi\sqrt{t}(t+r)}$
48. $\dfrac{1}{\sqrt{s}}e^{rs}(1-\text{erf}\sqrt{rs}), \quad r>0$	$\dfrac{1}{\sqrt{\pi(t+r)}}$

appendix 3

Mathematical Induction

Mathematical induction† is the name given to an elementary logical principle that can be used to prove a certain type of mathematical statement. Typically, we use mathematical induction to prove that a certain statement or equation holds for every positive integer. For example, we may need to prove that $2^n > n$ for all integers $n \geq 1$.

To do so, we proceed in two steps:

> **(i)** We prove that the statement is true for some integer N (usually $N = 1$).
>
> **(ii)** We *assume* that the statement is true for an integer k ($> N$) and then *prove* based on this assumption that it is true for the integer $k + 1$.

If we can complete these two steps, then we will have demonstrated the validity of the statement for *all* positive integers greater than or equal to N. To convince you of this fact, we reason as follows: Since the statement is true for N [by step (i)] it is true for the integer $N + 1$ [by step (ii)]. Then it is also true for the integer $(N + 1) + 1 = N + 2$ [again by step (ii)], and so on. We now demonstrate the procedure with some examples.

EXAMPLE 1

Show that $2^n > n$ for all integers $n \geq 1$.

---

† This technique was first used in a mathematical proof by the great French mathematician Pierre de Fermat (1601–1665).

SOLUTION

(i) If $n = 1$, then $2^n = 2^1 = 2 > 1 = n$, so $2^n > n$ for $n = 1$.

(ii) Assume that $2^k > k$, where $k \geq 1$ is an integer. Then

$$\text{Since } 2^k > k$$
$$\downarrow$$
$$2^{k+1} = 2 \cdot 2^k = 2^k + 2^k > k + k \geq k + 1.$$

This completes the proof since we have shown that $2^1 > 1$, which implies, by step (ii), that $2^2 > 2$, so that, again by step (ii), $2^3 > 3$, so that $2^4 > 4$, and so on.

EXAMPLE 2

Use mathematical induction to prove the formula for the sum of the first n positive integers:

$$1 + 2 + 3 + \cdots + n = \frac{n(n+1)}{2}. \tag{1}$$

SOLUTION

(i) If $n = 1$, then the sum of the first one integer is 1. But $(1)(1 + 1)/2 = 1$, so that equation (1) holds in the case in which $n = 1$.

(ii) Assume that (1) holds for $n = k$; that is,

$$1 + 2 + 3 + \cdots + k = \frac{k(k+1)}{2}.$$

We must now show that it holds for $n = k + 1$. That is, we must show that

$$1 + 2 + 3 + \cdots + k + (k+1) = \frac{(k+1)(k+2)}{2}.$$

But

$$1 + 2 + 3 + \cdots + k + (k+1) = (1 + 2 + 3 + \cdots + k) + (k+1)$$
$$= \frac{k(k+1)}{2} + (k+1)$$
$$= \frac{k(k+1) + 2(k+1)}{2} = \frac{(k+1)(k+2)}{2},$$

and the proof is complete.

You may wish to try a few examples to illustrate that formula (1) really works. For example,

$$1 + 2 + 3 + 4 + 5 + 6 + 7 + 8 + 9 + 10 = \frac{10(11)}{2} = 55.$$

EXAMPLE 3

Use mathematical induction to prove the formula for the sum of the squares of the first n positive integers:

$$1^2 + 2^2 + 3^2 + \cdots + n^2 = \frac{n(n+1)(2n+1)}{6}. \tag{2}$$

SOLUTION

(i) Since $1(1 + 1)(2 \cdot 1 + 1)/6 = 1 = 1^2$, equation (2) is valid for $n = 1$.

(ii) Suppose that equation (2) holds for $n = k$; that is,

$$1^2 + 2^2 + 3^2 + \cdots + k^2 = \frac{k(k+1)(2k+1)}{6}.$$

Then to prove that (2) is true for $n = k + 1$, we have

$$1^2 + 2^2 + 3^2 + \cdots + k^2 + (k+1)^2 = \frac{k(k+1)(2k+1)}{6} + (k+1)^2$$

$$= \frac{k(k+1)(2k+1) + 6(k+1)^2}{6}$$

$$= \frac{k+1}{6}[k(2k+1) + 6(k+1)]$$

$$= \frac{k+1}{6}(2k^2 + 7k + 6) = \frac{k+1}{6}[(k+2)(2k+3)]$$

$$= \frac{(k+1)(k+2)[2(k+1)+1]}{6},$$

which is equation (2) for $n = k + 1$, and the proof is complete.

Again you may wish to experiment with this formula. For example,

$$1^2 + 2^2 + 3^2 + 4^2 + 5^2 + 6^2 + 7^2 = \frac{7(7+1)(2\cdot7+1)}{6} = \frac{7\cdot8\cdot15}{6} = 140.$$

EXAMPLE 4

For $a \neq 1$, use mathematical induction to prove the formula for the sum of a geometric progression:

$$1 + a + a^2 + \cdots + a^n = \frac{1 - a^{n+1}}{1 - a}. \tag{3}$$

SOLUTION

(i) If $n = 0$, then

$$\frac{1 - a^{0+1}}{1 - a} = \frac{1 - a}{1 - a} = 1.$$

Thus equation (3) holds for $n = 0$. (We use $n = 0$ instead of $n = 1$ since $a^0 = 1$ is the first term.)

(ii) Assume that (3) holds for $n = k$; that is

$$1 + a + a^2 + \cdots + a^k = \frac{1 - a^{k+1}}{1 - a}.$$

Then

$$1 + a + a^2 + \cdots + a^k + a^{k+1} = \frac{1 - a^{k+1}}{1 - a} + a^{k+1}$$

$$= \frac{1 - a^{k+1} + (1 - a)a^{k+1}}{1 - a} = \frac{1 - a^{k+2}}{1 - a},$$

so that equation (3) also holds for $n = k + 1$, and the proof is complete.

Problems

1. Use mathematical induction to prove that the sum of the cubes of the first n positive integers is given by

$$1^3 + 2^3 + 3^3 + \cdots + n^3 = \frac{n^2(n+1)^2}{4}. \tag{4}$$

2. Let the functions $f_1, f_2, \ldots, f_n$ be integrable on $[0, 1]$. Show that $f_1 + f_2 + \cdots + f_n$ is integrable on $[0, 1]$ and that

$$\int_0^1 [f_1(x) + f_2(x) + \cdots + f_n(x)] \, dx$$

$$= \int_0^1 f_1(x) \, dx + \int_0^1 f_2(x) \, dx + \cdots + \int_0^1 f_n(x) \, dx.$$

3. Use mathematical induction to prove that the nth derivative of the nth-order polynomial

$$P_n(x) = x^n + a_{n-1}x^{n-1} + a_{n-2}x^{n-2} + \cdots + a_1 x^1 + a_0$$

is equal to $n![n! = n(n - 1)(n - 2) \cdots 3 \cdot 2 \cdot 1]$.

4. Show that if $a \neq 1$,

$$1 + 2a + 3a^2 + \cdots + na^{n-1} = \frac{1 - (n+1)a^n + na^{n+1}}{(1 - a)^2}.$$

* 5. Prove, using mathematical induction, that there are exactly 2^n subsets of a set containing n elements.

6. Use mathematical induction to prove that

$$\ln(a_1 a_2 a_3 \ldots a_n) = \ln a_1 + \ln a_2 + \cdots + \ln a_n,$$

if $a_k > 0$ for $k = 1, 2, \ldots, n$.

7. Let $\mathbf{u}, \mathbf{v}_1, \mathbf{v}_2, \ldots, \mathbf{v}_n$ be $n + 1$ vectors in $\mathbb{R}^2$. Prove that

$$\mathbf{u} \cdot (\mathbf{v}_1 + \mathbf{v}_2 + \cdots + \mathbf{v}_n) = \mathbf{u} \cdot \mathbf{v}_1 + \mathbf{u} \cdot \mathbf{v}_2 + \cdots + \mathbf{u} \cdot \mathbf{v}_n.$$

Answers to Odd-Numbered Problems

Chapter 1

Problems 1.1, page 7

1. 62,500 after 20 days; 156,250 after 30 days
3. $250,000(1.06)^{10} \approx 447,712$ in 1980;
   $250,000(1.06)^{30} \approx 1,435,873$ in 2000
5. **a.** $150(0.16)^2 \approx 3.84°C$ **b.** $-\ln 15/\ln 0.16 \approx 1.48$
   minutes
7. $5730 \ln 0.7/\ln 0.5 \approx 2949$ years
9. **a.** $25(0.6)^{2.4} \approx 7.34$ kg **b.** $10 \ln 0.02/\ln 0.6 \approx 76.6$ hours
11. $\ln 0.5/(-\alpha) \approx 4.62 \times 10^6$ years
13. $\beta = (1/1500)\ln 845.6/1013.25 \approx -1.205812 \times 10^{-4}$
    a. 625.526 mbar **b.** 303.416 mbar **c.** 429.033 mbar
    d. 348.632 mbar **e.** $-(1/\beta)\ln 1013.25 \approx 57.4$ km
15. **a.** $\approx 10,000$ **b.** 7 **c.** $214.20 \approx 8.9$ days
17. 55.1 minutes
19. 38.7 minutes

Problems 1.2, page 16

1. $y = \pm\sqrt{e^x + c}$
3. $y = \frac{1}{2}\ln(x^2 + 1) + c$
5. $z = \tan\left(\frac{r^3}{3} + c\right)$
7. $P = ce^{\sin Q - \cos Q}$
9. $s = e^{(t^3/3) - 2t}$
11. $y = \dfrac{2401}{(1 + x)^3}$
13. $y = c \cos x - 3$
15. If $|x| < 1$ and $|y| < 1$, $y = \sin(c - \sin^{-1} x)$.
    If $|x| > 1$ and $|y| > 1$,
    $y = \pm\cosh(c - \cosh^{-1} x)$.

17. $y = \dfrac{3x}{4x - 3}$
19. $x = \ln\left(1 - \dfrac{1 - e}{e^t}\right)$
21. $y = \pm\sqrt{2e^x + c}$
23. $y = cx^3/e^x$
25. $2y^3 + 3y^2 = 3x^2 + 5$
27. $x = x_0 + v_0(\cos \theta)t$, $y = y_0 + v_0(\sin \theta)t - \frac{1}{2}gt^2$
29. $h' = v = [(cv_0 + g)e^{-ct} - g]/c$, terminal velocity $= -g/c$.
31. Revolve $y = kx^4$ about the y-axis to get the cistern's shape.
33. At $t = 9$, predicted mass is 447.2; error is 1.4%.
    At $t = 18$, predicted mass is 660.4; error is 0.2%.
35. $t = \dfrac{\pi d}{r}\left[\dfrac{\sqrt{i}}{2(k\pi)^{3/2}}\ln\left|\dfrac{\sqrt{i} + \sqrt{k\pi u}}{\sqrt{i} - \sqrt{k\pi u}}\right| - \dfrac{u}{k\pi}\right]$ where $u = \left(\dfrac{3rv}{\pi d}\right)^{1/3}$;
    to prevent overflow, $i = k\pi u^2$

37. 9:30 a.m.

Problems 1.3, page 27

1. first
3. third
5. second
7. third
9. initial-value
11. initial-value
13. boundary value
19. $y_1 = e^{2x} \cos x$, $y_2 = e^{2x} \sin x$
25. $\phi''(-1) = -3$, $\phi'''(-1) = 2$

27.

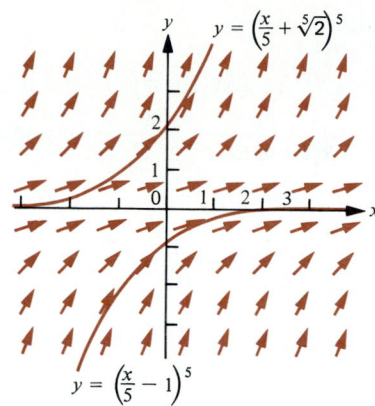

$$y = \left(\frac{x}{5} + \sqrt[5]{2}\right)^5$$

$$y = \left(\frac{x}{5} - 1\right)^5$$

29.

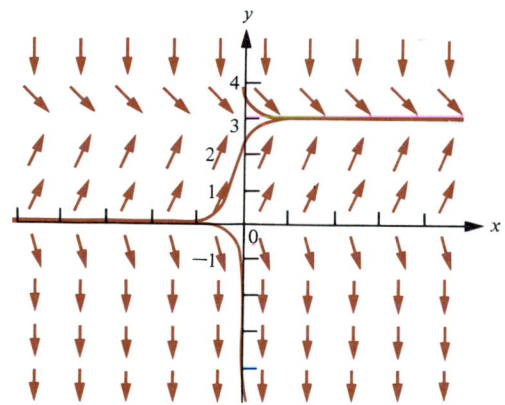

31.

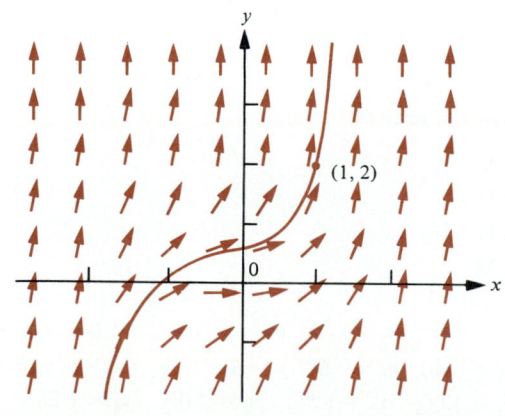

(1, 2)

33.

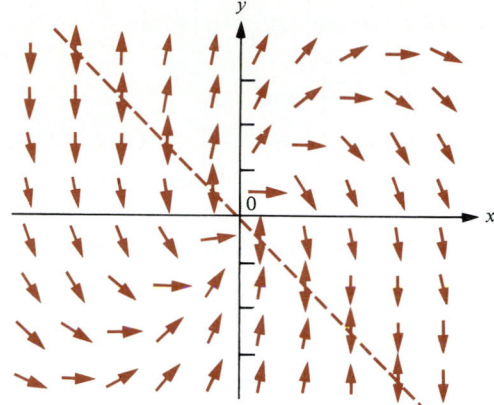

35. a.

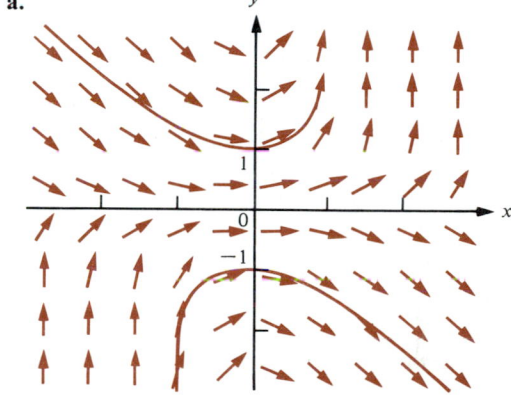

b. It is the constant solution $y = 0$.

c. Their graphs are symmetric with respect to the origin.

Problems 1.4, page 35

1. $x = ce^{3t}$

3. $x = 2e^t - 1$

5. $x = y^y(1 + ce^{-y})$

7. $y = x^4 + 3x^3$

9. $x = ce^{2t} + \dfrac{t^3 e^{2t}}{3}$

11. $s = \left(c + \dfrac{u^2}{2}\right)e^{-u} + 1$

13. $x = e^{-y}(c + y)$

15. $x = \dfrac{1}{k} + ce^{-kt}$

17. $y = B + ce^{-kt}$

19. $P = \dfrac{\beta}{\delta + (\beta/P_0 - \delta)e^{-\beta t}}$

21. a. $\dfrac{dN}{dt} = rN\left(1 - \dfrac{N}{K}\right) - h$

$\qquad = -\dfrac{r}{K}\left(N^2 - KN + \dfrac{hK}{r}\right)$

b. $N = \frac{1}{2}K + \frac{1}{2}D \tanh\left[\frac{1}{2}D\frac{r}{K}t + c\right]$ where

$$D = \sqrt{K^2 - \frac{4Kh}{r}} \text{ and}$$

$$c = \tanh^{-1}[2N_0 - K/D]$$

23. Graph $y = rN(1 - N/K)$ and observe maximum $rK/4$ occurs at $N = K/2$.

25. The object satisfies $m\dot{v} = -mg + kv^2$. Thus $\lim_{t\to\infty} v = -\sqrt{mg/k}$.

27. $y = \pm \sqrt{\dfrac{x}{6 + ce^{-x}}}$

29. $y = \dfrac{1}{x\sqrt{2 - x^2}}, \quad 0 < x < \sqrt{2}$

31. $y = \dfrac{1}{x^3(\frac{1}{2} - \ln x)}, \quad 0 < x < \sqrt[3]{e}$

Problems 1.5, page 41

1. $x^2y + y = c$

3. $x^4y^3 + \ln\left|\dfrac{x}{y}\right| = e^3 - 1$

5. $x^2 - 2y \sin x = \dfrac{\pi^2}{4} - 2$

7. $e^{xy} + 4xy^3 - y^2 + 3 = 0$

9. $x^2y + xe^y = c$

11. $\tan^{-1}\dfrac{y}{x} + \ln\left|\dfrac{x}{y}\right| = c$

13. $e^{2x}(x^2 + y^2) = c$

15. $x^2 \ln|x| - y = cx^2$ and $x = 0$

17. $y^2(x^2 + y^2 + 2) = c$

19. $x^2y = c$

Problems 1.6, page 45

1. $I = \frac{6}{5}(1 - e^{-10t})$

3. $I = 2(1 - e^{-25t})$

5. $I = \frac{1}{20}(e^t - e^{-t})$

7. $Q = \frac{1}{680}[3 \sin 60t + 5 \cos 60t - 5e^{-100t}]$

9. $Q = \frac{1}{100}(1 + 99e^{-100t})$

11. $1000\dfrac{dQ}{dt} + 10^6 Q = 0; \quad Q(0) = 10, \quad Q = 10e^{-1000t}$

13. $Q(60) = \frac{1}{2000}[1 - e^{-10^5(600+3600)}]$

15. $I(t) = \dfrac{E_0 C}{1 + (\omega RC)^2}$
$$\times \left[RC\omega^2 \cos \omega t - \omega \sin \omega t + \dfrac{1}{RC}e^{-t/RC}\right];$$

$$Q(t) = \dfrac{E_0 C}{1 + (\omega RC)^2}$$
$$\times [(\cos \omega t - e^{-t/RC}) + \omega RC \sin \omega t]$$

17. $I_{\text{transient}}(0) = \dfrac{E_0}{R[1 + (\omega RC)^2]} \approx \dfrac{E_0}{R}$ for very small R

19. $I(t) = \begin{cases} \dfrac{3}{50}(1 - e^{-50t}), & 0 \le t \le 10 \\[2mm] \left[\dfrac{e^{500}}{(70)^2} - \dfrac{3}{50}\right]e^{-50t} + \dfrac{7}{100} - \dfrac{e^{10-t}}{98}, & t \ge 10 \end{cases}$

Problems 1.7, page 53

1. 2000-year-old wood has $(\frac{1}{2})^{200/573} \approx 0.785$ the C^{14}-concentration of freshly cut wood.

3. $I(t) = \dfrac{1 + N}{1 + Ne^{-k(1+N)t}}$

5. $t = 25 \ln 2 \approx 17.3$ min

7. $47\frac{43}{91}$ g/liter

9. $p(t) = 400 + \dfrac{2}{\pi} - \dfrac{1}{3}t + \dfrac{2}{\pi}\sin\dfrac{\pi}{12}(t - 6)$

11. $x(t) = \dfrac{A}{k} + \dfrac{B}{k^2 + \omega^2}(k \sin \omega t - \omega \cos \omega t)$
$$+ \left(x_0 + \dfrac{B\omega}{\omega^2 + k^2} - \dfrac{A}{k}\right)e^{-kt}$$

13. distance is $\sqrt{x^2 + (vt - y)^2} = x\sqrt{1 + (y')^2} =$
$$\dfrac{x}{2}\left[\left(\dfrac{x}{a}\right)^{v/w} + \left(\dfrac{a}{x}\right)^{v/w}\right]$$

15. At the point $(0, -b/2)$

17. $r = 2e^{(\theta + \pi/2)/\sqrt{8}}$

19. $r = ae^{-\sqrt{3}\theta}/\sqrt{3}$

21. The eagle's speed is $\dfrac{3 + \sqrt{73}}{8}$ times the pigeon's. The pigeon flies $\frac{200}{3}$ ft, the hawk flies $\frac{400}{3}$ ft, and the eagle flies $\frac{25}{3}(3 + \sqrt{73})$ ft

Problem 1.8, page 60

1. 2.98

3. 0.71

5. 8.31

7. 0.34

9. 156.45

11. $y_1 = 1.02, \ y_2 = 1.04, \ y_3 = 1.07, \ y_4 = 1.09, \ y_5 = 1.12$

13. $y_1 = 0.40, \ y_2 = 0.68, \ y_3 = 0.79, \ y_4 = 0.95, \ y_5 = 1.20$

15. $y_1 = 1.00, \quad y_2 = 1.56, \quad y_3 = 2.06, \quad y_4 = 1.28,$
$y_5 = 0.67, \quad y_6 = 1.05, \quad y_7 = 0.28, \quad y_8 = 1.07$

17. $y_1 = 1.10,$ $y_2 = 1.21,$ $y_3 = 1.33,$ $y_4 = 1.46,$
  $y_5 = 1.60,$ $y_6 = 1.75,$ $y_7 = 1.91,$ $y_8 = 2.10,$
  $y_9 = 2.29,$ $y_{10} = 2.50$

19. $y_1 = 1.55,$ $y_2 = 1.19,$ $y_3 = 0.91,$ $y_4 = 0.69,$
  $y_5 = 0.52$

Problems 1.9, page 64

1. $y = e^{-x}$

3. $y = e^x - 1 - x$

Review Exercises for Chapter 1, page 64

1. $y = \frac{3}{2}x^2 + C$

3. $x = -3e^{t/2}$

5. $P(5) = 10,000(1.15)^5 \approx 20,114;$
  $P(10) = 10,000(1.15)^{10} \approx 40,456$

7. a. $T(20) = 23 + 102(\frac{57}{102})^2 \approx 54.9°C$

  b. $t = \dfrac{10 \ln(2/102)}{\ln(57/102)} \approx 67.6$ minutes

9. $h = \ln 0.5 / \ln 0.8 \approx 3.11$ weeks

11.

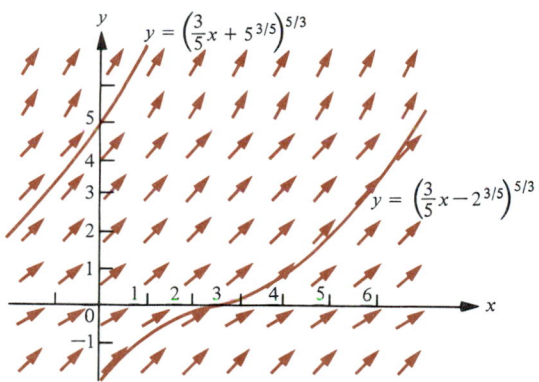

$y = \left(\frac{3}{5}x + 5^{3/5}\right)^{5/3}$

$y = \left(\frac{3}{5}x - 2^{3/5}\right)^{5/3}$

13. Let I = light incident on layer of thickness Δx; then $\Delta I = -kI \, \Delta x$ or $I' = -kI$. Thus, $I(x) = I(0)e^{-kx}$.

15. $x = \dfrac{n^2 kt}{nkt + 1}$

17. $y = \dfrac{1}{1 - \ln|x|}$

19. $y = \sin(\ln|x| + c)$

21. $y = \dfrac{2}{c - x^2}$

23. $y = \sqrt{2} \sin x$

25. $y = [1 + ce^{-x^3/3}]^{-1}$

27. $y = e^{(x^2-1)/2}$

29. $y = ce^{-\cos x} - 1$

31. $y = \dfrac{x + c}{\sqrt{1 + x^2}}$

33. $y = x^2(1 + \ln|x|)$

35. $y = \begin{cases} e^{x^2/2}(c - \int_x^0 e^{-t^2/2} \, dt), & x \le 0 \\ ce^{x^2/2}, & x > 0 \end{cases}$

37. $y = \frac{1}{2}(\sqrt{3x^2 + 4} - x)$

39. $y = \dfrac{(x + c)^2 - 4}{4x}$

41. $y = \dfrac{1}{1 - x + ce^{-x}}$

43. $xy + c = e^y \tan x$

45. $\dfrac{y}{x^3} + \dfrac{3}{x} + \dfrac{x}{y^2} = c$

47. $x = \dfrac{1 - e^{-bt}}{1001}$ where $b = 1.001 \times 10^{-6}$; $x(36,000) \approx 3.536$
  $\times 10^{-5}$; $x(360,000) \approx 3.023 \times 10^{-4}$; $x(31,536,000) \approx$
  9.99×10^{-4}

49. Approximate by assuming that pollutants are dumped continuously at the rate $\frac{1}{24}$ m^3/sec; $x \approx (1 - e^{-bt})/24,024$; maximum concentration is $1/24,024$.

51. $6 - \cosh(\sqrt{g/6}t)$ as long as this is nonnegative

53. $(e^x)'' - 3(e^x)' + 2e^x = e^x - 3e^x + 2e^x = 0,$
  $(e^{2x})'' - 3(e^{2x})' + 2e^{2x} = 4e^{2x} - 6e^{2x} + 2e^{2x} = 0.$

55. $y'' + (a + b)y' + aby = z' + bz = 0$ so $y' + ay = z = ce^{-bx}$.
  Thus $y = \dfrac{ce^{-bx}}{a - b} + ke^{-ax}$.

57. $Q/c = E_0 \sin \omega t$, so $I/c = E_0 \omega \cos \omega t$

59. 1.18 hr

61. $P + \exp \dfrac{1}{b} [a - e^{-b(t+c)}]$; $\lim_{t \to \infty} P(t) = e^{a/b}$

63. 3.125 ft.

65. $y(3) \approx 6.076$

67. $y(3) \approx 5.445$

69. $y(3) \approx 0.625$

71. $y = e^x - 1$

73. $y = x$

Chapter 2

Problems 2.1, page 74

1. linear, homogeneous, variable coefficients

3. nonlinear

5. linear, nonhomogeneous, constant coefficients

7. linear, nonhomogeneous, variable coefficients

9. linear, homogeneous, constant coefficients

11. 3

13. 3

15. $y = c_1 + c_2 x$

17. $y = c_1 + c_2 x^2$

19. At that point $y_1(x_1) = y_1'(x_1) = 0$. But $y \equiv 0$ also satisfies that initial condition.

21. a. $xy_1'' - y_1' + 4x^3 y_1 = x(2 \cos x^2 - 4x^2 \sin x^2)$
  $- 2x \cos x^2 + 4x^3 \sin x^2 = 0$

$$xy_2'' - y_2' + 4x^3 y_2 = x(-2 \sin x^2 - 4x^2 \cos x^2)$$
$$+ 2x \sin x^2 + 4x^3 \cos x^2 = 0$$

  b. $\dfrac{y_1}{y_2} = \tan x^2$ is not constant; also $W(y_1, y_2) = -2x$; no

    since $a(x) = 1/x$ is not continuous.

23. $y_3 = 2y_1 - \frac{1}{2}y_2$

27. $W = -2/x$

Problems 2.2, page 79

1. $y_2(x) = xe^x$

3. $y_2(x) = \dfrac{3x^2 - 1}{4} \ln\left|\dfrac{1+x}{1-x}\right| - \dfrac{3x}{2}$

5. $y_2(x) = x^{-2}$

7. $y_2(x) = 2x - 1$

9. $y_2(x) = \cos x^2$

11. $y_2(x) = \cos x/\sqrt{x}$

13. a. $e^{-ax^2/2}[(a^2 x^2 - a) + ax(-ax) + a] = 0$

   b. $y_2(x) = e^{-ax^2/2} \int e^{ax^2/2}\, dx$

Problems 2.3, page 84

1. $y = c_1 e^{2x} + c_2 e^{-2x}$

3. $y = c_1 e^x + c_2 e^{2x}$

5. $x = (1 + \frac{9}{2}t)e^{-5t/2}$

7. $x = -\frac{1}{5}(e^{3t} + 4e^{-2t})$

9. $y = c_1 + c_2 e^{5x}$

11. $y = (c_1 + c_2 x)e^{-\pi x}$

13. $z = c_1 e^{-5x} + c_2 e^{3x}$

15. $y = (1 + 2x)e^{4x}$

17. $y = c_1 e^{\sqrt{2}x} + c_2 e^{-\sqrt{2}x}$

19. $y = e^{\sqrt{5}x} + 2e^{-\sqrt{5}x}$

21. $y = e^{-x}(c_1 \cos x + c_2 \sin x)$

23. $x = e^{-t/2}\left(c_1 \cos \dfrac{3\sqrt{3}}{2} t + c_2 \sin \dfrac{3\sqrt{3}}{2} t\right)$

25. $x = -\frac{3}{2} \cos 2\theta + \sin 2\theta$

27. $y = \sin \dfrac{x}{2} + 2 \cos \dfrac{x}{2}$

29. $y = e^{-x}(c_1 \cos 2x + c_2 \sin 2x)$

31. $y = e^{-x}(\sin x - \cos x)$

33. If $c_1 \neq 0$, then

$$\frac{z'}{z} = \frac{c_1 z_1' + c_2 z_2'}{c_1 z_1 + c_2 z_2} = \frac{z_1' + (c_2/c_1)z_2'}{z_1 + (c_2/c_1)z_2}$$

  and c_2/c_1 is a single constant.

35. $y = 1 - \dfrac{4}{e^{2x} + 2}$

37. $y = 1 + \dfrac{1}{c + x}$

39. $y = 2 - \dfrac{3c}{e^{3x} + c}$

Problems 2.4, page 92

1. $y = c_1 \sin 2x + c_2 \cos 2x + \sin x$

3. $y = c_1 e^x + c_2 e^{2x} + 3e^{3x}$

5. $y = e^x(c_1 + c_2 x - 2x^2)$

7. $y = 3e^{5x} - 10e^{2x} + 10x + 7$

9. $y = c_1 + c_2 e^{-x} + \dfrac{x^4}{4} - \dfrac{4}{3}x^3 + 4x^2 - 8x$

11. $y = e^{2x}(c_1 \cos x + c_2 \sin x) + xe^{2x} \sin x$

13. $y = e^{-3x}(c_1 + c_2 x + 5x^2)$

15. $y = c_1 e^{3x} + c_2 e^{-x} + \frac{20}{27} - \frac{7}{9}x + \frac{1}{3}x^2 - \frac{1}{4}e^x$

17. $y = (c_1 + c_2 x)e^{-2x} + (\frac{1}{9}x - \frac{2}{27})e^x + \frac{3}{25} \sin x - \frac{4}{25} \cos x$

21. $y = (-\frac{1}{6}x^3 - \frac{1}{4}x^2 + \frac{1}{4}x + c_1)\cos x + (\frac{1}{4}x^2 + \frac{1}{4}x + c_2)\sin x$

23. $y = (\frac{1}{12}x^4 + c_1 x + c_2)e^x$

25. $y = \begin{cases} x, & 0 \le x \le 1 \\ 1, & x \ge 1 \end{cases}$

Problems 2.5, page 96

1. $y = c_1 \cos 2x + c_2 \sin 2x + \sin x$

3. $y = c_1 e^x + c_2 e^{2x} + 3e^{3x}$

5. $y = (c_1 + c_2 x - 2x^2)e^x$

7. $y = c_1 e^{2x} + c_2 e^{5x} + 10x + 7$

9. $y = c_1 e^{-x} + c_2 + \dfrac{x^4}{4} - \dfrac{4}{3}x^3 + 4x^2 - 8x$

11. $y = c_1 + c_2 e^x - \ln|\cos x|$

13. $y = c_1 \cos 2x + c_2 \sin 2x + \frac{1}{2}x \sin 2x$
    $+ \frac{1}{4} \cos 2x \ln|\cos 2x|$

15. $y = e^x[c_1 + c_2 x - \ln|1 - x|]$

17. $y = c_1 e^x + c_2 e^{-x} + e^x \ln|x|$

19. $y = e^{2x}[c_1 + c_2 x + (x + 1)\ln|x + 1|]$

21. $y_p = \frac{1}{2}\left[-1 - x \ln x + \left(x - \dfrac{1}{x}\right)\ln(1 + x)\right]$

25. $y_1 = e^{\omega x}$ and $y_2 = e^{-\omega x}$ are solutions of the homogeneous equation. By Problem 23.e

$$y_p(x) = \frac{1}{\omega} \int_0^x f(t) \sinh \omega(x - t)\,dt,$$

$$y_p'(x) = \int_0^x f(t) \cosh \omega(x - t)\,dt.$$

Problems 2.6, page 101

1. $y = c_1 x + c_2 x^{-1}$

3. $y = x[c_1 \cos(\ln x) + c_2 \sin(\ln x)]$

5. $y = x^{3/2} - x^{1/2}$

7. $y = c_1 x + c_2 x^3$

9. $y = x^{-2}[c_1 \cos(\ln x) + c_2 \sin(\ln x)]$

11. $y = c_1 x^3 + c_2 x^{-4}$

13. $y = c_1 x^3 + c_2 x^{-5} - \dfrac{1}{16x}$

15. $y = x^3(c_1 + c_2 \ln x + \frac{1}{2} \ln^2 x)$

17. $y = c_1 \cos(\ln x) + c_2 \sin(\ln x) + 10$

19. $y = c_1 x^{-5} + c_2 x^{-1} + x/12$

21. $y = c_1 x^2 + c_2 x^{-1} + \frac{1}{4} - \frac{1}{2} \ln x$

23. $y = c_1 + c_2 x^{-3}$

Problems 2.7, page 104

1. $I(t) = 6(e^{-5t} - e^{-20t}); \quad Q(t) = \frac{1}{10}(9 - 12e^{-5t} + 3e^{-20t})$

3. $I_{\text{steady state}} = \cos t + 2 \sin t$

5. $I_{\text{steady state}} = \frac{1}{13}(70 \sin 10t - 90 \cos 10t)$

7. $I_{\text{transient}} = e^{-t}(-\cos t - 3 \sin t)$

9. $I_{\text{transient}} = \frac{40}{3} e^{-5t} - \frac{250}{39} e^{-2t}$

11. $I_{\text{transient}} = \dfrac{e^{-t}}{481}(-245 \cos 7t - \frac{255}{7} \sin 7t)$

13. $I_{\text{transient}} = \dfrac{e^{-600t}}{2320}[-153 \sin 800t - 96 \cos 800t];$

$I_{\text{steady state}} = \dfrac{1}{580}[24 \cos 600t + 27 \sin 600t]$

15. $Q = \begin{cases} c_1 \cos \dfrac{t}{\sqrt{LC}} + c_2 \sin \dfrac{t}{\sqrt{LC}} \\ \qquad + \left(\dfrac{CE_0}{1 - CL\omega^2}\right) \sin \omega t, \quad \omega \neq \dfrac{1}{\sqrt{LC}} \\ \left(c_1 - \dfrac{E_0 t}{2\omega L}\right) \cos \omega t + c_2 \sin \omega t, \quad \omega = \dfrac{1}{\sqrt{LC}} \end{cases}$

17. b. $\omega = \dfrac{\sqrt{(4L/C) - R^2}}{2L}$

19. b. No ω produces resonance.

Problems 2.8, page 108

1. $x = \cos 10t$

3. $x = 3 \cos t + 4 \sin t$

5. $x = \frac{3}{5} \sin 5t$

7. $v = -0.2\sqrt{g}$ m/sec

9. $\sqrt{\dfrac{8\pi}{g}}$ seconds; $\quad x(t) = \left(1 - \dfrac{2}{\pi}\right) \cos\left(\sqrt{\dfrac{\pi g}{2}} \, t\right)$ m

11. $\theta = \frac{1}{2} \cos \sqrt{2g} t, \quad f = \dfrac{\sqrt{2g}}{2\pi} \approx 1.28$ Hz

13. $f = \dfrac{\sqrt{96g}}{2\pi} \approx 8.85$ Hz

15. 844 sec $= 14.1$ min

17. $x'' + \dfrac{T_0 g}{L\omega} x = 0, \quad 2\pi \sqrt{\dfrac{L\omega}{T_0 g}}$

Problems 2.9, page 113

1. $x(t) = (1 + 10t)e^{-10t}$

3. $x(t) = e^{-\sqrt{5}t/2}[3 \cosh \frac{1}{2}t + (8 + 3\sqrt{5})\sinh \frac{1}{2}t] =$
$\left(\dfrac{3\sqrt{5}+11}{2}\right)e^{[(1-\sqrt{5})t/2]} - \left(\dfrac{3\sqrt{5}+5}{2}\right)e^{[(-1-\sqrt{5})t/2]}$

5. $x(t) = e^{-4t} \sin 3t$

7. $k = \dfrac{25}{4}\left(1 + \dfrac{\pi^2}{10\,000}\right)$

In Problems 9–13, $p = \sqrt{g/96} \approx 0.579$

9. $x(t) = e^{-3pt}\left(\frac{1}{6} \cos \sqrt{91}pt - \dfrac{1 - \frac{1}{2}p}{\sqrt{91}p} \sin \sqrt{91}pt\right)$

11. $x(t) = e^{-10pt}[\frac{1}{6} - (1 - \frac{5}{3}p)t]$

13. $x(t) = e^{-3pt/\sqrt{2}}\left(-\dfrac{24}{25} \cos \frac{1}{2}\sqrt{182}pt - \dfrac{72}{25\sqrt{91}} \sin \frac{1}{2}\sqrt{182}pt\right)$

15. No

Problems 2.10, page 117

1. $x(t) = \frac{2001}{2000}(1 + 10t)e^{-10t} - \frac{1}{2000} \cos 10t$

3. $x(t) = \frac{1}{50}\{e^{-\sqrt{5}t/2}[(150 + \sqrt{5}) \cosh \frac{1}{2}t$
$\qquad + (405 + 150\sqrt{5}) \sinh \frac{1}{2}t] - \sqrt{5} \cos t\}$
$\quad = \frac{1}{100}[(555 + 151\sqrt{5})e^{(1-\sqrt{5})t/2}$
$\qquad - (255 + 149\sqrt{5})e^{(-1-\sqrt{5})t/2} - 2\sqrt{5} \cos t]$

5. $x(t) = \frac{1}{104}[e^{-4t}(3 \cos 3t + 106 \sin 3t) - 3 \cos 3t + 2 \sin 3t]$

7. $x(t) = -0.1419e^{-2t} + 0.0029e^{-19.62t} - 0.1390 \cos 2t$
$\quad + 0.1133 \sin 2t$

9. Use $c^2 = \dfrac{144(240)}{g(4\pi^2 + 0.36)}$ and solve the system

$\left(\dfrac{240}{100}g - 144\right)A + \dfrac{12}{100}cgB = \dfrac{g}{5}$

$\dfrac{-12}{100}cgA + \left(\dfrac{240}{100}g - 144\right)B = 0.$

11. b. Let $A = \dfrac{F_0}{k - m\omega^2}$. With $x(0) = x'(0) = 0$,

$x = -\dfrac{A\omega}{\omega_0} \sin \omega_0 t + A \sin \omega t.$

If $\omega \approx \omega_0$, $x \approx A(\sin \omega t - \sin \omega_0 t) = 2A \sin \frac{1}{2}(\omega - \omega_0)t$
$\times \cos \frac{1}{2}(\omega + \omega_0)t$, the product of a slowly varying component and a rapidly varying one.

Problems 2.11, page 122

1. $y = (c_1 + c_2 x) \cos x + (c_3 + c_4 x) \sin x$

3. $y = (1 + x)e^x$

5. $y = 1 + e^{3x} + e^{-3x}$

7. $y = c_1 + c_2 x + c_3 x^2 + c_4 x^3$

9. $y = c_1 e^x + c_2 e^{-x} + c_3 e^{2x} + c_4 e^{-2x}$

11. $y = 1 - x + e^{2x} - e^{-2x}$

13. $y = c_1 e^x + c_2 \cos x + c_3 \sin x$

15. $y = c_1 e^{3x} + e^{-3x/2}\left(c_2 \cos \dfrac{3\sqrt{3}}{2}x + c_3 \sin \dfrac{3\sqrt{3}}{2}x\right)$

17. If $c_1 y_1 + c_2 y_2 + c_3 y_3 \equiv 0$, then we also must have
$$c_1 y_1' + c_2 y_2' + c_3 y_3' \equiv 0,$$
$$c_1 y_1'' + c_2 y_2'' + c_3 y_3'' \equiv 0.$$

In order for all three of these identities to hold at x_0, it must be that $c_1 = c_2 = c_3 = 0$.

19. **a.** $y_1(v')'' + (3y_1' + ay_1)(v')' + (3y_1'' + 2ay_1' + by_1)v' = 0$

   c. $v' = \dfrac{W(y_1, y_2)}{y_1^2} \displaystyle\int \dfrac{y_1 e^{-\int a(x)\,dx}}{W^2(y_1, y_2)}\,dx$

25. $y = c_1 e^x + c_2 x e^x + c_3 e^{-x} + \frac{1}{4}x^2 e^x$

27. $y = c_1 e^{2x} + c_2 e^{-3x} + c_3 e^{4x} + \frac{1}{24}x + \frac{41}{288}$

29. **b.** $y_p = \frac{1}{10}e^{2x}\int e^{-3x}\tan x\,dx + \frac{1}{10}e^{-x}[(3\cos x - \sin x)\ln|\sec x + \tan x| + 1]$

31. $y = c_1 x^{-1} + (c_2 + c_3 \ln|x|)x$

33. $y = c_1 x^{-1} + c_2 \cos(\ln|x|) + c_3 \sin(\ln|x|)$

Review Exercises for Chapter 2, page 123

1. $y_2 = \cos 2x$

3. $y_2 = x^{-2}$

5. $y_2 = -2 + x \ln\left|\dfrac{1+x}{1-x}\right|$

7. $y = 13e^{4x} - 10e^{5x}$

9. $y = \dfrac{2}{\sqrt{7}}e^{(3/2)x}\sin\dfrac{\sqrt{7}}{2}x$

11. $y = (c_1 + c_2 x)e^{-x/2}$

13. $y = e^x(c_1 \cos \sqrt{6}x + c_2 \sin \sqrt{6}x)$

15. $y = c_1 e^x + c_2 e^{2x} + c_3 e^{3x}$

17. $y = -3e^x + x^2 + 4x + 5$

19. $y = e^x(c_1 + c_2 x + x^{-1})$

21. $y = (c_1 + c_2 \ln x)x^{-2}$

23. $y = (c_1 + c_2 x)e^{-2x} + c_3 e^{3x}$

25. **a.** $x = \cos \sqrt{50}t$

   b. $x = e^{-5t}(\cos 5t + \sin 5t)$

   c. $x = \frac{1}{1000}[1000 \cos \sqrt{50}t + \sqrt{2}\sin \sqrt{50}t - \sin 10t]$

   d. $x = \frac{1}{2500}[e^{-5t}(2501 \cos 5t + 2502 \sin 5t) - \cos 10t - \frac{1}{2}\sin 10t]$

27. **a.** $x = 3 \cos \dfrac{2\sqrt{10}}{5}t + \sqrt{10}\sin \dfrac{2\sqrt{10}}{5}t$

   b. $x = e^{-t/\sqrt{5}}\left(3 \cos \sqrt{\dfrac{7}{5}}t + \dfrac{3 + 4\sqrt{5}}{\sqrt{7}}\sin \sqrt{\dfrac{7}{5}}t\right)$

   c. $x = 3 \cos \dfrac{2\sqrt{10}}{5}t + \dfrac{59\sqrt{10}}{60}\sin \dfrac{2\sqrt{10}}{5}t + \dfrac{1}{15}\sin t$

   d. $x = \dfrac{1}{145}\left[e^{-t/\sqrt{5}}\left((435 + 2\sqrt{5})\cos \sqrt{\dfrac{7}{5}}t \right.\right.$
   $\left.\left. + \left(579\sqrt{\dfrac{5}{7}} + \dfrac{435}{\sqrt{7}}\right)\sin \sqrt{\dfrac{7}{5}}t\right) - 2\sqrt{5}\cos t + 3 \sin t\right]$

29. **a.** $x = \frac{6}{5}\sin \frac{5}{2}t$

   b. $x = \frac{2}{7}\sqrt{21}e^{-t}\sin \dfrac{\sqrt{21}}{2}t$

   c. $x = \frac{72}{55}\sin \frac{5}{2}t - \frac{1}{11}\sin 3t$

   d. $x = \dfrac{1}{697}\left[e^{-t}\left(24 \cos \dfrac{\sqrt{21}}{2}t + \dfrac{4296}{\sqrt{21}}\sin \dfrac{\sqrt{21}}{2}t\right)\right.$
   $\left. - 24 \cos 3t - 11 \sin 3t\right]$

31. $Q_{ss} = \frac{1}{10}$,
$$Q_{tran} = -\frac{1}{10}e^{-25t/2}\left[\cos \dfrac{25\sqrt{31}}{2}t + \dfrac{1}{\sqrt{31}}\sin \dfrac{25\sqrt{31}}{2}t\right]$$

33. $I_{ss} = \frac{50}{14841}[120 \cos 5t + 21 \sin 5t]$

35. $I_{ss} = 4 \cos 5t + 2 \sin 5t$

37. $x = c_1 \cos \sqrt{2g}t + c_2 \sin \sqrt{2g}t + \dfrac{g}{10(g - 32)}\cos 8t$

39. $x = e^{-gt/2000}(c_1 \cos rt + c_2 \sin rt) - \dfrac{13g^2}{8000D}\cos \dfrac{65}{8}t$
$$+ \left(\dfrac{2g^2 - 4225g/64}{5D}\right)\sin \dfrac{65t}{8} \text{ where } r^2 = 2g - g^2/4 \times 10^6$$

and $D = (2g - 4225/64)^2 + (65g/8000)^2$

Chapter 3

Problems 3.1, page 134

1. $x = c_1 e^t + c_2 e^{-t}$
   $y = c_1 e^t + \frac{5}{3}c_2 e^{-t}$

3. $x = e^t(c_1 \cos t + c_2 \sin t)$
   $y = e^t((2c_1 + c_2)\cos t + (2c_2 - c_1)\sin t)$

5. $x = 2(1 + t)e^{-3t}$
   $y = -2(2 + t)e^{-3t}$

7. $x = 3 - 2e^{-2t}$
   $y = 4 - 4e^{-2t}$

9. $x = (c_1 + c_2 t)e^{10t}$
$y = -(2c_1 + c_2 + 2c_2 t)e^{10t}$

11. $x = \frac{4}{5}e^{-t} + \frac{6}{5}e^{4t} + 3t - 2$
$y = \dfrac{-4}{5}e^{-t} + \dfrac{9}{5}e^{4t} - 2t + 3$

13. $x = e^{2t}\left(c_1 \cos 2t + c_2 \sin 2t + \dfrac{t}{4}\right)$
$y = e^{2t}(2c_2 \cos 2t - 2c_1 \sin 2t - \frac{11}{4})$

15. $x = e^{6(t-\pi/4)}(-\frac{1}{2}\cos 2t)$
$y = e^{6(t-\pi/4)}(\sin 2t - \cos 2t)$

17. $x_1' = x_2$, $\quad x_2' = 6tx_2 - 3t^3 x_1 + \cos(t)$

19. $x_1' = x_2$, $\quad x_2' = x_3$, $\quad x_3' = x_4$, $\quad x_4' = \cos x_1 + t$

21. $x_1' = x_2$, $\quad x_2' = x_3$, $\quad x_3' = t^5/x_1 x_2 x_3$

23. $x' = x_2$, $\quad y' = y_2$, $\quad z' = z_2$
$x_2' = f(t, x, y, z)/m$
$y_2' = g(t, x, y, z)/m$
$z_2' = h(t, x, y, z)/m$

25. $x_1 = ae^t$
$x_2 = e^t(b \sin 2t + c \cos 2t - \frac{3}{2}a)$
$x_3 = e^t(c \sin 2t - b \cos 2t + a)$

27. $t = \dfrac{100}{2\sqrt{3}} \ln\!\left(\dfrac{3+\sqrt{3}}{3-\sqrt{3}}\right) = \dfrac{50}{\sqrt{3}} \ln(2 + \sqrt{3})$
$y_{\max} = 50(3 + \sqrt{3})(2 + \sqrt{3})^{(-1-\sqrt{3})/2}$

29. $y = 500(e^{-0.04t} - e^{-0.12t})$ so $t = (\ln 3)/0.08 \approx 13.7$ min and
$y_{\max} = 500\left[\dfrac{1}{\sqrt{3}} - \dfrac{1}{3\sqrt{3}}\right] = \dfrac{1000}{3\sqrt{3}} \approx 192.45$ lbs.

31. $m_1 x_1'' = -k_1 x_1 + k_2(x_2 - x_1)$
$m_2 x_2'' = -k_2(x_2 - x_1) + k(x_3 - x_2)$
$m_3 x_3'' = -k_3(x_3 - x_2)$

33. Let $x_1 = x$, $x_2 = x'$, ..., $x_n = x^{(n-1)}$. Then $x_1' = x_2$, $x_2' = x_3$, ..., $x_{n-1}' = x_n$ and $x_n' = g(t, x_1, x_2, \dots, x_n)$

35. the concentration of H_2S approaches $\dfrac{\gamma}{\alpha}$; the concentration of SO_2 approaches $\dfrac{\gamma + \delta}{\beta}$

37. $I_L = e^{-100t}((b - B) \sin 100t - B \cos 100t) + A \sin t$
$+ B \cos t$
$I_R = e^{-100t}((b - 2B) \sin 100t - b \cos 100t) + a \sin t$
$+ b \cos t$, where
$a = \dfrac{4000(19999)}{20000(19999) + 20001}$, $B = \dfrac{-20001a}{200(19999)}$
$b = \dfrac{-200a}{19999}$, $A = -100b$. Set $t = 1$.

Problems 3.2, page 142

1. (e^t, e^t) and $(e^{-t}, \frac{5}{3}e^{-t})$

3. $(e^t \cos t, e^t(2 \cos t - \sin t))$ and $(e^t \sin t, e^t(2 \sin t + \cos t))$

5. $(e^{-3t}, -e^{-3t})$ and $(te^{-3t}, e^{-3t}(-1 - t))$

7. $(1, \frac{4}{3})$ and $(e^{-2t}, 2e^{-2t})$

9. $(\frac{1}{4}te^{2t}, -\frac{11}{4}e^{2t})$

11. $(-\sin t, -\cos t - 2 \sin t)$

13. $(-t + 3e^t, \frac{1}{2} - t + 3e^t)$

Problems 3.3, page 147

1. $(I_L, I_R) = (1 - 1.025e^{-0.05}, 1 - 1.05e^{-0.05})$

3. $(100(k_1 e^{\lambda_1 t} + k_2 e^{\lambda_2 t}) + 1, -8(k_1\lambda_1 e^{\lambda_1 t} + k_2\lambda_2 e^{\lambda_2 t}) + 1)$
where $\begin{aligned}k_1 \\ k_2\end{aligned}\Big\} = \dfrac{\mp 3\sqrt{2} - 4}{800}$, $\begin{aligned}\lambda_1 \\ \lambda_2\end{aligned}\Big\} = -50 \pm 25\sqrt{2}$

5. The general solution $(I_L, I_R) = (I_L, I_R)_h + (I_L, I_R)_p$ is given by
$$(I_L, I_R)_p = (A \sin \omega t + B \cos \omega t, C \sin \omega t + D \cos \omega t),$$
where $\omega = 60\pi$, $\Delta = \omega^2 + 2500$, $A = (2500/\Delta)^2$, $B = -25\omega(\omega^2 + 7500)/\Delta^2$, $C = -2500(\omega^2 - 2500)/\Delta^2$, $D = -250{,}000\omega/\Delta^2$, and

a. $(I_L, I_R)_h = \left(\dfrac{e^{-50t}}{2}\left[k_1 + k_2\left(t + \dfrac{1}{25}\right)\right]\right.$,
$\left. e^{-50t}\left[k_1 + k_2\left(t + \dfrac{1}{25}\right)\right]\right)$

b. $(I_L, I_R)_h = (e^{-50t}[k_1 \cos 50\sqrt{3}t + k_2 \sin 50\sqrt{3}t],$
$e^{-50t}[k_2 \cos 50\sqrt{3}t + k_4 \sin 50\sqrt{3}t])$

c. $(I_L, I_R)_h = (e^{-50t}(k_1 e^{25\sqrt{2}t} + k_2 e^{-25\sqrt{2}t}),$
$e^{-50t}[(2 - \sqrt{2})k_1 e^{25\sqrt{2}t} + (2 + \sqrt{2})k_2 e^{-25\sqrt{2}t}])$

7. $I_1 = \frac{1}{20}(8 \sin t - 6 \cos t + 5e^{-t} + e^{-3t})$;
$I_2 = \frac{1}{20}(2 \sin t - 4 \cos t + 5e^{-t} - e^{-3t})$

9. $I = 50t$

11. $E_k = A\,\dfrac{\sinh a(n - k)}{\sinh an}\cos \omega t$

13. Natural frequencies are
$$\omega_N = \left[2\sqrt{CL}\, \sin \dfrac{N\pi}{2(n + 1)}\right]^{-1}.$$
Cut-off frequency is $\omega_n \approx \dfrac{1}{2\sqrt{CL}}$.

15. See the reference.

Problems 3.4, page 152

1. When $t = \sqrt{5} \ln\left(\dfrac{3+\sqrt{5}}{3-\sqrt{5}}\right)$,

$$y_{max} = \frac{40}{3-\sqrt{5}}\left(\frac{3+\sqrt{5}}{3-\sqrt{5}}\right)^{(-5-3\sqrt{5})/10}$$

3. When $t = \dfrac{150}{\sqrt{3}} \ln\dfrac{7+\sqrt{13}}{7-\sqrt{13}}$,

$$y_{max} = \frac{3000}{7-\sqrt{13}}\left(\frac{7-\sqrt{13}}{7+\sqrt{13}}\right)^{(7+\sqrt{13})/(2\sqrt{13})}$$

5. Let $k = AB - (a+b)^2/4$. If $k > 0$

$x = e^{(a-b)t/2}[x_0 \cos\sqrt{k}t + (\frac{1}{2}x_0(a+b) - Ay_0)\sin\sqrt{k}t/\sqrt{k}]$
$y = e^{(a-b)t/2}[y_0 \cos\sqrt{k}t - (\frac{1}{2}y_0(a+b) - Bx_0)\sin\sqrt{k}t/\sqrt{k}]$.

If $k < 0$, replace cos, sin, $\sqrt{k}$ with cosh, sinh, $\sqrt{-k}$.

7. Let $D = a_{11}a_{22} - a_{12}a_{21} \neq 0$, λ_1 and λ_2 be the roots of $\lambda^2 - (a_{11} + a_{22})\lambda + D = 0$. Then

$x = c_1 e^{\lambda_1 t} + c_2 e^{\lambda_2 t} + (b_2 a_{12} - b_1 a_{22})/D$
$y = c_1(\lambda_1 - a_{11})e^{\lambda_1 t}/a_{12} + c_2(\lambda_2 - a_{11})e^{\lambda_2 t}/a_{12}$
  $+ (b_1 a_{21} - b_2 a_{11})/D$.

9. Let $D = a_1 b_2 - a_2 b_1 \neq 0$, $\omega = \pi/4$, λ_1 and λ_2 be the roots of $\lambda^2 + (a_1 + b_2)\lambda + D = 0$, and A_1, A_2, A_3, A_4 be the solution of the system

$$\begin{array}{rl}
a_1 A_1 + \omega A_2 - a_2 A_3 & = -a \\
-\omega A_1 + a_1 A_2 \quad\quad - a_2 A_4 & = 0 \\
-b_1 A_1 \quad\quad + b_2 A_3 + \omega A_4 & = 0 \\
b_1 A_2 + \omega A_3 - b_2 A_4 & = 0.
\end{array}$$

$x = c_1 e^{\lambda_1 t} + c_2 e^{\lambda_2 t} + ab_2/D + A_1 \cos\omega t + A_2 \sin\omega t$
$y = (c_1\lambda_1 + a_1)e^{\lambda_1 t}/a_2 + (c_2\lambda_2 + a_2)e^{\lambda_2 t}/a_2$
  $+ ab_1/D + A_3 \cos\omega t + A_4 \sin\omega t$.

Problems 3.5, page 155

1. If $\tan\theta = B/A$, then $A \cos\omega t + B \sin\omega t = \sqrt{A^2 + B^2}\cos(\omega t - \theta)$.

3. $x_1 = (\frac{1}{2}x_{10} + \frac{1}{4}x_{20})\cos t + (\frac{1}{2}x_{30} + \frac{1}{4}x_{40})\sin t$
  $+ (\frac{1}{2}x_{10} - \frac{1}{4}x_{20})\cos\sqrt{3}t$
  $+ (\frac{1}{2}x_{30} - \frac{1}{4}x_{40})\sin\sqrt{3}t/\sqrt{3}$
  $x_2 = (x_{10} + \frac{1}{2}x_{20})\cos t + (x_{30} + \frac{1}{2}x_{40})\sin t$
  $- (x_{10} - \frac{1}{2}x_{20})\cos\sqrt{3}t$
  $- (x_{30} - \frac{1}{2}x_{40})\sin\sqrt{3}t/\sqrt{3}$

5. $\omega_n = \dfrac{n^2\pi^2}{64}\sqrt{\dfrac{4.32\times10^9}{24(8)^4/\pi g}} \approx 325n^2$ rev/sec. The lowest frequency is 325 rev/sec.

7. $l_1(m_1 + m_2)\theta'' + m_2 l_2\phi'' \cos(\theta - \phi) + m_2 l_2\phi'^2 \sin(\theta - \phi) + g(m_1 + m_2)\sin\theta = 0$
  $l_1\theta'' \cos(\theta - \phi) + l_2\phi'' - l_1\theta'^2 \sin(\theta - \phi) + g \sin\phi = 0$

9. The characteristic equation is $m_1 l_1 l_2 \lambda^4 + (m_1 + m_2)(l_1 + l_2)g\lambda^2 + (m_1 + m_2)g^2 = 0$. The discriminant as a quadratic in λ^2 is positive, so the roots are pure imaginary, and the solution is a superposition of two simple harmonics.

Problems 3.6, page 163

1. a. $x' = x$; b. $x = 0, x = 1$;

  c. $x(t) = \dfrac{x(0)e^t}{x(0)e^t + 1 - x(0)}$.

3. a. $x' = 2x$; b. $x = 0, x = -\frac{2}{3}$;

  c. $x(t) = \dfrac{2x(0)e^{2t}}{2 + 3x(0)(1 - e^{2t})}$.

5. a. $x' = 2x$; b. $x = 0, x = 1, x = 2$;

  c. $\dfrac{x^2 - 2x}{(x-1)^2} = ce^{2t}$ where $c = \dfrac{x(0)^2 - 2x(0)}{(x(0) - 1)^2}$.

7. a. The orbits are the circles $x^2 + y^2 = a^2 + b^2$.

9. The orbits are $x = \exp[\frac{1}{2}(\ln^2 y - 2t_0 \ln y)]$.

11. At (x_0, y_0) we have $x' = y' = 0$ so the orbit is stationary. If the other orbit passes through this point, the two orbits would meet, contradicting Problem 10.

Problems 3.7, page 175

1. saddle point

3. unstable focus

5. stable focus

7. unstable focus

9. Consider $\lambda = (T \pm \sqrt{c})/2$.

11. $\lambda = 0$ if and only if $D = 0$.

15. a. $x = c_1, y = c_1 + c_2 e^t$;

  c. (c, c) is a critical point for any real number c.

Problems 3.8, page 184

1. $(0, 0)$ is a saddle point (unstable).

3. $(0, 0)$ is unstable.

5. $(0, 0)$ is an unstable node.

7. $(0, 0)$ is the only critical point and is an unstable node or focus.

9. $(0, 0)$ is an unstable node or focus, $(0, \frac{1}{3})$ and $(1, 0)$ are saddle points and $(\frac{8}{7}, \frac{1}{7})$ is a stable focus.

11. $(0, 0)$, $(3, -9)$ and $(-2, -4)$ are saddle points; $(1, -1)$ is a stable focus and $(-1, -1)$ is an unstable focus.

13. $(0, 0)$ is the only critical point and is a stable focus.

15. Let $y = x' + \epsilon(x^3/3 - x)$. Then $y' = x'' + \epsilon(x^2 - 1)x' = -x$, so we have

$$\begin{cases} x' = y - \epsilon(x^3/3 - x) \\ y' = -x \end{cases}$$

The associated linear system has the characteristic equation $\lambda^2 - \epsilon\lambda + 1 = 0$. If $\epsilon < 0$, the origin is a stable focus. Apply Theorem 1(i)d and Theorem 2(i).

Review Exercises for Chapter 3, page 185

1. $x_1' = x_2,\ x_2' = -4x_1$

3. $x_1' = x_2,\quad x_2' = x_3,\quad x_3' = 6x_3 - 2x_2 + 5x_1$

5. $x_1' = x_2,\quad x_2' = x_3,\quad x_3' = (\ln t - x_1 x_3)/x_2$

7. $x = c_1 e^{5t} + c_2 e^{-t},\quad y = 2c_1 e^{5t} - c_2 e^{-t}$

9. $x = e^{2t}(c_1 \cos 3t + c_2 \sin 3t)$
$\quad y = \frac{1}{2}e^{2t}((3c_2 - c_1)\cos 3t - (c_2 + 3c_1)\sin 3t)$

11. $x = c_1 e^{-t} + 3e^{-2t},\quad y = (c_2 - 2c_1 t)e^{-t} + 12e^{-2t}$

13. a. $\dfrac{dE}{dx} = -IR,\quad \dfrac{dI}{dx} = -EG$

 b. $E = E_0 e^{-\sqrt{RG}x},\quad I = \sqrt{\dfrac{G}{R}}\, E_0 e^{-\sqrt{RG}x}$

15. a. $x = 30(1 - e^{-9t/75}),\quad y = 30(1 + e^{-9t/75})$

 b. y decreases steadily to 30 lb, x increases steadily to 30 lb.

 c. There is no maximum.

17. $y = \dfrac{bd}{(a+c)(a+d)}\,[a + ce^{-(a+c)t}]$

19. $\begin{cases} x' = \dfrac{2}{50}y - \dfrac{6}{100}x, & x(0) = 100 \\[2mm] y' = \dfrac{6x}{100} + \dfrac{z}{100} - \dfrac{7}{50}y, & y(0) = 20 \\[2mm] z' = \dfrac{5}{50}y - \dfrac{5}{100}z, & z(0) = 0 \end{cases}$

21. $(0, 0)$ is a saddle point (unstable).

23. $(0, 0)$ is a center (stable).

25. $(0, 0)$ is a saddle point (unstable).

27. $(0, 0)$ is a saddle point; $\left(\dfrac{1}{\sqrt{2}}, 0\right)$ and $\left(-\dfrac{1}{\sqrt{2}}, 0\right)$ are unstable nodes.

29. $(0, 0)$ is a saddle point; $(3, 4)$ is a saddle point; $(0, 5/2)$ is a stable node; $(1, 0)$ is an unstable node.

Chapter 4

Problems 4.1, page 198

1. $\dfrac{5}{s^2} + \dfrac{2}{s},\ s > 0$

3. $\dfrac{18}{s^3} - \dfrac{7}{s},\ s > 0$

5. $\dfrac{2}{s^3} + \dfrac{8}{s^2} - \dfrac{16}{s},\ s > 0$

7. $\dfrac{3}{4s^4} + \dfrac{1}{2s^3} + \dfrac{1}{2s^2} + \dfrac{1}{s},\ s > 0$

9. $\dfrac{a}{s^2} + \dfrac{b}{s},\ s > 0$

11. $\dfrac{e^2}{s - 5},\ s > 5$

13. $\dfrac{1}{s - 1/2},\ s > 1/2$

15. $\dfrac{e^{-1/2}}{s + 1},\ s > -1$

17. $\dfrac{3}{s^2 + 9},\ s > 0$

19. $\dfrac{s}{s^2 + 49},\ s > 0$

21. $\dfrac{5\cos 2}{s^2 + 25} + \dfrac{s\sin 2}{s^2 + 25},\ s > 0$

23. $\dfrac{s\cos b}{s^2 + a^2} - \dfrac{a\sin b}{s^2 + a^2},\ s > 0$

25. $\dfrac{s}{s^2 - 1/4},\ s > 1/2$

27. $\dfrac{s\cosh 2}{s^2 - 25} - \dfrac{5\sinh 2}{s^2 - 25},\ s > 5$

29. $\dfrac{a\cosh b}{s^2 - a^2} + \dfrac{s\sinh b}{s^2 - a^2},\ s > |a|$

31. $\dfrac{1}{(s-1)^2},\ s > 1$

33. $\dfrac{6}{(s+1)^4} - \dfrac{1}{s+1},\ s > -1$

35. $\dfrac{1}{(s-1)^2 + 1},\ s > 1$

37. $\dfrac{s-4}{(s-4)^2 + 4},\ s > 4$

39. $\dfrac{s+2}{(s+1)^2 + 1},\ s > -1$

43. $9t^2 + 7$

45. $\cos t + \sin t$

47. $\cosh \sqrt{2}t - \sqrt{2}\sinh \sqrt{2}t$

49. te^t

51. $\dfrac{3}{\sqrt{5}}\,e^{-2t}\sin \sqrt{5}t$

53. $2e^{-t}\cos \sqrt{7}t - \dfrac{3}{\sqrt{7}}\,e^{-t}\sin \sqrt{7}t$

55. $ce^{-at}\cos \sqrt{b - a^2}t + \dfrac{d - ac}{\sqrt{b - a^2}}\,e^{-at}\sin \sqrt{b - a^2}t$

63. $\dfrac{2abs}{(s^2 - a^2 - b^2)^2 - 4a^2 b^2}$

65. $\dfrac{s(s^2 - a^2 + b^2)}{(s^2 + a^2 + b^2)^2 - 4a^2 s^2}$

67. $\dfrac{a(s^2 - a^2 - b^2)}{(s^2 + a^2 + b^2)^2 - 4a^2 s^2}$

71. $\dfrac{6}{(s+1)^4}$

73. $\dfrac{2s^3 - 54s}{(s^2 + 9)^3}$

75. $\dfrac{(s-a)^2 - b^2}{[(s-a)^2 + b^2]^2}$

77. $\dfrac{3(s+1)^2+3}{[(s+1)^2-1]^2}$

81. a. $\sqrt{\pi/s}$ **b.** $\dfrac{\sqrt{\pi}}{2s^{3/2}}$ **c.** $\dfrac{15\sqrt{\pi}}{8s^{7/2}}$

Problems 4.2, page 212

1. $\cos t$

3. $A\cosh at + \dfrac{B}{a}\sinh at$

5. $e^{-t}(\cos 2t + \sin 2t)$

7. $\frac{1}{3} - e^t + \frac{5}{3}e^{3t}$

9. $-\dfrac{t}{9} + \dfrac{23}{27}e^{3t} + \dfrac{4}{27}e^{-3t}$

11. e^{-t}

13. $\sinh t$

15. $\left(1 + \dfrac{t}{2k}\right)\sin kt$

17. $\left(a + \dfrac{1}{2a^2}\right)\sin at + \left(a - \dfrac{t}{2a}\right)\cos at$

19. $\frac{5}{4}\cosh t - \frac{1}{4}(\cos t + t\sin t)$

21. $\dfrac{s^2 + 2a^2}{s(s^2 + 4a^2)}$

23. $\dfrac{2a(3s^2 - a^2)}{(s^2 + a^2)^3}$

25. $\dfrac{8 + 6s^2}{s^2(s^2 + 4)^2}$

27. $\dfrac{2a^2(4a^2 + 3s^2)}{s^2(s^2 + 4a^2)^2}$

29. $\dfrac{16}{s^2 + 16}\left[\dfrac{1}{s^3} + \dfrac{1}{s(s^2 + 16)} + \dfrac{4s}{(s^2 + 16)^2}\right]$

33. $\dfrac{\pi}{2} - \tan^{-1}s = \tan^{-1}(1/s)$

35. $\dfrac{\pi}{2} - \tan^{-1}(s/3) = \tan^{-1}(3/s)$

37. $\tan^{-1}(k/s)$

39. $\dfrac{1}{2}\ln\dfrac{s^2 - a^2}{s^2}$

41. $\dfrac{1}{2s}\ln\dfrac{s^2 - a^2}{s^2}$

43. $\dfrac{e^{s^2/4}}{s}\left(1 - \operatorname{erf}\left(\dfrac{s}{2}\right)\right)$

45. $\dfrac{2}{t}(1 - \cos at)$

47. $\dfrac{\sin t}{t}$

49. a. $Y' - s^2 Y = -1$
  b. $Y(s) = [Y(0) - \int_0^s e^{-u^3/3}\,du]e^{s^3/3}$

51. $I(t) = \frac{1}{676}[12\cos t + 5\sin t - (12 + 65t)e^{-5t}]$

53. $Q(t) = \frac{1}{125}[-4\cos 10t - 3\sin 10t + (4 + 50t)e^{-5t}]$

Problems 4.3, page 225

9. $\dfrac{b}{s(e^{as} + 1)}$

11. $\dfrac{b(1 - e^{-as})}{s(1 + e^{-2as})}$

13. $\dfrac{1}{s^2(1 + e^{-as})}$

15. $\dfrac{1}{(s^2 + 1)(1 - e^{-\pi s})}$

17. $\dfrac{1 + se^{-4\pi s}}{s^2 + 1}$

19. $\sin(t - \pi)H(t - \pi)$

21. $\cos t(1 + H(t - \pi))$

23. $\dfrac{1}{4!}[t^4 + (t - 4)^4 H(t - 4)]$

25. $f(t) = 1 + 2H(t - 1) + 2H(t - 7);$
  $\mathcal{L}\{f\} = \dfrac{1}{s}[1 + 2e^{-s} + 2e^{-7s}]$

27. $y = t - \sin t - H(t - \pi)(t + \sin t + \pi\cos t)$

29. $y = \frac{1}{10}[1 - e^{-t}(\cos 3t + \frac{1}{3}\sin 3t)]$
  $- \frac{1}{10}[1 - e^{-(t-1)}(\cos 3(t - 1)$
  $+ \frac{1}{3}\sin 3(t - 1))]H(t - 1)$

31. $y = \frac{1}{10} + \frac{9}{10}e^{-t}\cos 3t + \frac{3}{10}e^{-t}\sin 3t$
  $- \frac{1}{10}H(t - 1)[1 - e^{-(t-1)}\cos 3(t - 1)$
  $- \frac{1}{3}e^{-(t-1)}\sin 3(t - 1)]$

33. $y = e^{2t}\sin 3t + e^{2(t-1)}\sin 3(t - 1)H(t - 1)$

Problems 4.4, page 231

1. $I(t) = \dfrac{10}{\sqrt{6}}e^{-20(t-30)}\sin\sqrt{600}(t - 30)H(t - 30)$

3. $I(t) = 1 - \cos 4t + H(t - 5)[\cos 4(t - 5) - 1]$

5. $I(t) = 1 - \cos\sqrt{10}t - H(t - 2)[1 - \cos\sqrt{10}(t - 2)]$
  $+ 2H(t - 4)[1 - \cos\sqrt{10}(t - 4)]$
  $+ \dfrac{60}{\sqrt{10}}H(t - 4)\sin\sqrt{10}(t - 4)$

7. a. $x(t) = \dfrac{t}{2} - \dfrac{\sin 2t}{4} - \dfrac{1}{4}H\left(t - \dfrac{\pi}{2}\right)$
  $\times [2t - \pi\cos(2t - \pi) - \sin(2t - \pi)]$

  b. $x(t) = \dfrac{t}{2} - \dfrac{\sin 2t}{4} + \cos 2t$
  $- \dfrac{1}{4}H\left(t - \dfrac{\pi}{2}\right)[2t - \pi\cos(2t - \pi) - \sin(2t - \pi)]$

Problems 4.5, page 237

1. $\dfrac{3!}{s^4(s^2 + 1)}$

3. $\dfrac{3!5!}{s^{10}}$

5. $\dfrac{19!}{s^{20}(s - 17)}$

7. $\frac{1}{2}\displaystyle\int_0^t (t - u)^3\sin u\,du = \dfrac{t^3}{2} - 3t + 3\sin t$

9. $\dfrac{1}{a^2}(1 - \cos at)$

11. $\frac{1}{2}(t-3)^2 H(t-3)$

13. $e^{-t}(1-t)^2$

15. $\frac{1}{13}[1 - e^{-2t}(\cos 3t + \frac{2}{3}\sin 3t)]$
$- \frac{1}{13}[1 - e^{-2(t-\pi)}(\cos 3(t-\pi) + \frac{2}{3}\sin 3(t-\pi)]H(t-\pi)$

17. $\mathcal{L}\{1*f\} = \mathcal{L}\{1\}\mathcal{L}\{f\}$

Problems 4.6, page 242

1. $x = \cosh t, \quad y = \sinh t$

3. $x = e^{3t}(2\cos 3t + 2\sin 3t),$
$y = e^{3t}(-2\cos 3t + 4\sin 3t)$

5. $x = \frac{3}{2}\cos t + \frac{1}{2}\sin t + \frac{7}{2}e^t - 5e^{-t} - 4t,$
$y = \cos t + \frac{7}{2}e^t - \frac{5}{2}e^{-t} - 1 - 3t$

7. $x = \dfrac{e^{3t}}{4(13)^3}(6887\cos 2t + 2637\sin 2t)$
$+ \dfrac{e^t}{4} - \dfrac{5t^2}{13} - \dfrac{34t}{(13)^2} - \dfrac{74}{(13)^3},$

$y = \dfrac{e^{3t}}{4(13)^3}(-9524\cos 2t + 4250\sin 2t)$
$+ \dfrac{4}{13}t^2 + \dfrac{48t}{(13)^2} + \dfrac{184}{(13)^3}$

9. $x = \cos t, \quad y = -\cos t - \sin t$

11. $x = \cos t - \sin t, \quad y \equiv 1, \quad z = \sin t + \cos t$

Review Exercises for Chapter 4, page 242

1. $\dfrac{3}{s^2} - \dfrac{2}{s}$

3. $\dfrac{1}{e(s-2)}$

5. $\dfrac{1}{(s+1)^2}$

7. $\dfrac{3\cosh 4 - s(\sinh 4)}{s^2 - 9}$

9. $\dfrac{(s-1)^2 - 1}{[(s-1)^2 + 1]^2}$

11. $\dfrac{s^2 + 2}{s(s^2 + 4)}$

13. $\dfrac{2s(s^2 - 12)}{(s^2 + 4)^3}$

15. e^{-3s}

17. $\dfrac{1 + se^{-2\pi s}}{s^2 + 1}$

19. -14

21. te^{2t}

23. $e^{-2t}\sin t$

25. $2e^{2t} - e^t$

27. $-2e^{-t} - 5te^{-t} + 2\cos t + 4\sin t$

29. $\dfrac{2}{t}(1 - \cos 2t)$

31. $\cos t - \cos\left(t - \dfrac{\pi}{2}\right)H\left(t - \dfrac{\pi}{2}\right)$
$= \begin{cases} \cos t, & t < \dfrac{\pi}{2} \\ \\ \cos t - \sin t, & t > \dfrac{\pi}{2} \end{cases}$

33. $y = \cos t + 3\sin t$

35. $y = 5e^{2t} - 3e^{3t}$

37. $y = \dfrac{e^{2t}}{9}(3t - 4) + e^t - \dfrac{5}{9}e^{-t}$

39. $y = \frac{1}{2}[1 - e^{-(t-3)}(\cos(t-3) - \sin(t-3))]H(t-3)$

41. $y = te^{2t} - (t-1)e^{2(t-1)}H(t-1)$

43. $\dfrac{4!7!}{s^{13}}$

45. a. $\int_0^t \sin(t-u)u^2\,du = t^2 + 2\cos t - 2;$

b. $\int_0^t \frac{1}{2}\sin 2u\,du = \frac{1}{4}(1 - \cos 2t)$

47. a. $\int_0^t g(t-u)f'(u)\,du = g(t-u)f(u)|_0^t + \int_0^t g'(t-u)f(u)\,du$

b. $\dfrac{d}{dt}\int_0^t f(u)g(t-u)\,du = f(t)g(0) + \int_0^t f(u)g'(t-u)\,du$

Chapter 5

Problems 5.1, page 251

1. $e^x = e\left[1 + (x-1) + \dfrac{(x-1)^2}{2!} + \dfrac{(x-1)^3}{3!} + \cdots\right]$

3. $\cos x =$
$\dfrac{\sqrt{2}}{2}\left[1 + \left(x - \dfrac{\pi}{4}\right) - \dfrac{\left(x - \dfrac{\pi}{4}\right)^2}{2!} - \dfrac{\left(x - \dfrac{\pi}{4}\right)^3}{3!} + \cdots\right]$

5. $e^{bx} =$
$e^{-b}\left[1 + b(x+1) + \dfrac{b^2}{2!}(x+1)^2 + \dfrac{b^3}{3!}(x+1)^3 + \cdots\right]$

7. $x^2 e^{-x^2} = \displaystyle\sum_{n=0}^{\infty} \dfrac{(-1)^n x^{2n+2}}{n!}$

9. $(x-1)\ln x = (x-1)^2 - \frac{1}{2}(x-1)^3 + \frac{1}{3}(x-1)^4 - \cdots$

19. $y = e^x + x + 1$

21. $y = c_0\cos x + (c_1 - 1)\sin x + x$

23. $y = c_0(1 + x\tan^{-1}x) + c_1 x$

25. $y = xe^x$

27. $y = 1 - 2x^2$

29. $y = x + \sin x$

31. $y = \dfrac{1}{1-x}$

33. $y = e^{x^2} - \dfrac{x}{4}$

35. $y = c_0\left[1 + \displaystyle\sum_{n=1}^{\infty} \dfrac{1\cdot 4\cdot\;\cdots\;\cdot(3n-2)}{(3n)!}x^{3n}\right]$
$+ c_1\left[x + \displaystyle\sum_{n=1}^{\infty} \dfrac{2\cdot 5\cdot\;\cdots\;\cdot(3n-1)}{(3n+1)!}x^{3n+1}\right]$

Problems 5.2, page 258

1. $y_1 = x$;

$$y_2 = 1 - \frac{x^2}{2!} - \frac{x^4}{4!} - \frac{3x^6}{6!} - \frac{3 \cdot 5}{8!} x^8 - \cdots$$

3. $y_1 = 1 + \frac{x^3}{2} + \frac{1}{2!(5 \cdot 2)} x^6 + \frac{1}{3!(8 \cdot 5 \cdot 2)} x^9 + \cdots$

$$= 1 + \sum_{n=1}^{\infty} \frac{x^{3n}}{n![2 \cdot 5 \cdot \cdots \cdot (3n-1)]}$$

$$y_2 = x + \frac{1}{4} x^4 + \frac{1}{2!(7 \cdot 4)} x^7 + \frac{1}{3!(10 \cdot 7 \cdot 4)} x^{10} + \cdots$$

$$= \sum_{n=0}^{\infty} \frac{x^{3n+1}}{n![1 \cdot 4 \cdot \cdots \cdot (3n+1)]}$$

5. $y_1 = (x-2)e^x$; $y_2 = (x-2)e^{x-2} \int \frac{e^{(x-2)^2/2}}{(x-2)^2} \, dx$

7. $y_1 = e^{-x^3/3}$; $y_2 = e^{-x^3/3} \int e^{x^3/3} \, dx$

9. $y_1 = x$, $y_2 = 1 + x \tan^{-1} x$

11. $y_1 = 1 - x$, $y_2 = \dfrac{1}{1-x}$

13. $y_1 = x + \frac{8}{3} x^3$,
$y_2 = 1 + 9x^2 + \frac{15}{2} x^4 - \frac{7}{2} x^6 + \frac{27}{8} x^8 - \cdots$

15. $y = c_1 x + c_0 \left(1 - \frac{1}{2} x^2 - \frac{1}{4!} x^4 - \frac{3}{6!} x^6 - \cdots \right) + \sin x$

$$= c_1 x - c_0 x \int \frac{e^{x^2/2}}{x^2} \, dx + \sin x$$

17. $y = 1 + \frac{x^4}{4!} + \frac{1 \cdot 5 x^8}{8!} + \frac{1 \cdot 5 \cdot 9}{12!} x^{12} + \cdots$

19. $y = e^{x^2/2}$

21. $y = 1 + x + \frac{x^2}{2!} + \frac{2}{3!} x^3 + \frac{4}{4!} x^4 + \cdots$

23. $y = x - \frac{x^3}{3!} + \frac{4x^5}{5!} - \frac{19x^7}{7!} + \cdots$

25. $y = e^x(c_1 + c_2 \ln|x|)$

27. $y = c_1 x + c_2 x \int \frac{e^{-x}}{x} \, dx$

$$= c_1 x + c_2 \left(x \ln|x| + \sum_{n=1}^{\infty} \frac{(-1)^n x^{n+1}}{(n!)n} \right)$$

29. $y = c_1 x^2 + c_2 x^2 \int \frac{e^{-x}}{x^3} \, dx$

$$= c_1 x^2 + c_2 \left(-\frac{1}{2} + x + \frac{1}{2} x^2 \ln|x| + \sum_{n=3}^{\infty} \frac{(-1)^n x^n}{n!(n-2)} \right)$$

31. $y = c_1(x^2 + 2x + 3) + \dfrac{c_2 x^4}{(1-x)^2}$

33. No; the solutions $y = ce^{-1/x}$ and $y = ce^{-1/2x^2}$ do not have power series about $x = 0$

35. Substitute $y = \sum_{n=0}^{\infty} c_n x^n$ to get:

$$c_0 + \sum_{n=1}^{\infty} [c_{n-1}(n-1)(n-2) + c_n(n+1)] x^n = 0.$$

Then $0 = c_0 = c_1 = \cdots$, so method fails.

Problems 5.3, page 270

1. $y_0 = c_0 \cos \sqrt{x} + c_1 \sin \sqrt{x}$ if $x > 0$;
$c_0 \cosh \sqrt{-x} + c_1 \sinh \sqrt{-x}$ if $x < 0$

3. $y = c_0 \dfrac{\sinh x}{x^3} + c_1 \dfrac{\cosh x}{x^3}$

5. $y = c_0 \dfrac{\sin x^2}{x^2} + c_1 \dfrac{\cos x^2}{x^2}$

7. $y = c_0 x + \dfrac{c_1}{x^2}$

9. $y = c_0 \dfrac{\sin x}{x} + c_1 \dfrac{\cos x}{x}$

11. $y = c_0 e^x + c_1 e^x \ln|x|$

13. $y = \sqrt{|x|} e^x (c_0 + c_1 \ln|x|)$

15. $y = \dfrac{c_0}{1-x} + c_1 \dfrac{\ln|x|}{1-x}$

17. $y = (c_0 + c_1 \ln|x|) \left(1 + \frac{x^2}{2^2} + \frac{x^4}{(2 \cdot 4)^2} + \cdots \right)$

$$- c_1 \left(\frac{x^2}{4} + \frac{3x^4}{8 \cdot 16} + \cdots \right)$$

19. $y = c_0 \sqrt{x} (1 - \frac{7}{6} x + \frac{21}{40} x^2 + \cdots)$
$+ c_1(1 - 3x + 2x^2 + \cdots)$

21. $y = c_0 x + c_1 \left(x \ln|x| + \sum_{n=1}^{\infty} \frac{(-1)^n}{n!n} x^{n+1} \right)$

23. $y = \sqrt{|x|} (c_0 + c_1 \ln|x|)$

25. b. $y_2 = xe^{-1/x}$

Problems 5.4, page 280

1. $\left(\dfrac{384}{x^4} - \dfrac{72}{x^2} + 1 \right) J_1(x) - \left(\dfrac{192}{x^3} - \dfrac{12}{x} \right) J_0(x)$

3. a. $4J_p'' = 2(2J_p')' = 2J_{p-1}' - 2J_{p+1}'$

5. Expand in series using (6) & (9) $x^p Y_p = (\cot p\pi) x^p J_p$
$- (\csc p\pi) x^p J_{-p}$ and differentiate term by term. Use $\csc p\pi = -\csc(p-1)\pi$.

7. By Problems 5 and 6

$$xY_p' = xY_{p-1} - pY_p$$
$$xY_p' = -xY_{p+1} + pY_p$$

9. a. $\int x^2 J_0 \, dx = \int x(xJ_1)' \, dx = x^2 J_1 - \int xJ_1 \, dx$
$= x^2 J_1 + \int xJ_0 \, dx$

11. Differentiate the right side and use equations (14) and (15).

13. Use equation (17).

15. Set $z = \sqrt{x}$; then $y = AJ_p(\sqrt{x}) + BY_p(\sqrt{x})$.

17. Set $y = xu$; then $y = x[AJ_1(x) + BY_1(x)]$.

19. Set $y = ux^{-k}$; then $y = x^{-k}(AJ_k(x) + BY_k(x))$.

21. Set $y = \sqrt{x}u$, $z = kx^3/3$;
then $y = \sqrt{x}(AJ_{1/6}(kx^3/3) + BY_{1/6}(kx^3/3))$.

27. Multiply the first equation in Problem 25.d by $\cos n\theta$ and the second by $\sin n\theta$. Show that

$$\frac{1}{\pi}\int_0^\pi \cos(x\sin\theta)\cos n\theta \, d\theta = \begin{cases} J_n(x), & n \text{ even} \\ 0, & n \text{ odd} \end{cases}$$

and

$$\frac{1}{\pi}\int_0^\pi \sin(x\sin\theta)\sin n\theta \, d\theta = \begin{cases} 0, & n \text{ even} \\ J_n(x), & n \text{ odd.} \end{cases}$$

Then add the two integrals.

33. b. $u^2\dfrac{d^2x}{du^2} + \left(u^2 + \dfrac{4k_1}{ma^2}\right)x = 0$

c. Substitute $x = \sqrt{u}z$ getting

$$u^2 z'' + uz' + \left(u^2 - \left(\frac{1}{4} - \frac{4k_1}{ma^2}\right)\right)z = 0.$$

Problems 5.5, page 287

1. $P_5 = \dfrac{63x^5 - 70x^3 + 15x}{8}$,

$P_6 = \dfrac{231x^6 - 315x^4 + 105x^2 - 5}{16}$,

$P_7 = \dfrac{429x^7 - 693x^5 + 315x^3 - 35x}{16}$,

$P_8 = \dfrac{429(15x^8 - 28x^6) + 630(11x^4 - 2x^2) + 35}{128}$

3. $\left|\dfrac{c_{2n+1}x^{2n+1}}{c_{2n-1}x^{2n-1}}\right| = \left|\dfrac{(p-2n+1)(p+2n)x^2}{2n(2n+1)}\right| =$

$|x|^2 \left|\dfrac{\left(1+\dfrac{p}{2n}\right)\left(\dfrac{p+1}{2n}-1\right)}{\left(1+\dfrac{1}{2n}\right)}\right| \to |x|^2$ as $n \to \infty$

5. Using Rodrigues' formula, observe that $(x^2-1)^{2p+1}$ consists of only even powers of x which must be differentiated an odd number of times, and then evaluated at $x = 0$.

7. By (12) $P'_{2p} = xP'_{2p-1} + 2pP_{2p-1}$. Set $x = 0$ and use Problem 5.

9. $H_0(x) = 1$, $\quad H_1(x) = 2x$, $\quad H_2(x) = 4x^2 - 2$,
$H_3(x) = 8x^3 - 12x$, $\quad H_4(x) = 16x^4 - 48x^2 + 12$

11. $[1 - (2xz - z^2)]^{-1/2} = 1 + \frac{1}{2}(2xz - z^2)$
$+ \dfrac{(\frac{1}{2})(\frac{3}{2})}{2!}(2xz - z^2)^2 + \cdots$

Then the coefficient of z^p occurs only in the terms from $(2xz - z^2)^p = z^p(2x - z)^p$ down. Verify this is $P_p(x)$.

Review Exercises for Chapter 5, page 288

1. Use Taylor's theorem with $a = 1$.

3. Use Maclaurin's series. **7.** $y = x^2 + e^x$

9. $y = x(\sin x + 1)$ **11.** $y = e^{-x^2}$

13. $y = c_0 e^x + c_1\sqrt{x}$ **15.** $y = \dfrac{c_0}{x} + \dfrac{c_1}{1-x}$

17. $y = c_1\dfrac{e^{x/2}}{\sqrt{x}} + c_2 x\left(1 + \sum_{n=1}^\infty \dfrac{x^n}{5\cdot7\cdot\cdots\cdot(2n+3)}\right)$

19. a. $F(-a, b, b; -x) = 1 + ax + \dfrac{a(a-1)}{2!}x^2 + \cdots = (1+x)^a$ (use a Maclaurin series)

b. $xF(1, 1, 2; -x) = x\left[1 - \dfrac{x}{2} + \dfrac{x^2}{3} - \dfrac{x^3}{4} + \cdots\right]$

21. a. Note that $J_0(ax)$ is a solution of $x^2 u'' + xu' + a^2 x^2 u = 0$. Consider

$$\begin{cases} xu'' + u' + a^2 xu = 0 \\ xv'' + v' + b^2 xv = 0 \end{cases}$$

Then

$$[x(uv' - vu')]' = (a^2 - b^2)xuv$$

and

$$\int_0^1 xuv \, dx = x(uv' - vu')|_0^1 = 0.$$

23. $y = c_0 xJ_0(x) + c_1 xY_0(x)$

25. $y = P_n(\cos x)$

Chapter 6

Problems 6.1, page 301

1. $|\mathbf{v}| = 4\sqrt{2}$, $\quad \theta = \pi/4$

3. $|\mathbf{v}| = 4\sqrt{2}$, $\quad \theta = 7\pi/4$

5. $|\mathbf{v}| = 2$, $\quad \theta = \pi/6$

7. $|\mathbf{v}| = 2$, $\quad \theta = 2\pi/3$

9. $|\mathbf{v}| = 2$, $\quad \theta = 4\pi/3$

11. $|\mathbf{v}| = \sqrt{89}$, $\quad \theta = \pi + \tan^{-1}(-\frac{8}{5}) \approx 2.13$ (in the second quadrant)

13. $|\mathbf{v}| = 3$, $\cos \alpha = 0$, $\cos \beta = 1$, $\cos \gamma = 0$

15. $|\mathbf{v}| = 14$, $\cos \alpha = 0$, $\cos \beta = 0$, $\cos \gamma = 1$

17. $|\mathbf{v}| = \sqrt{5}$, $\cos \alpha = 1/\sqrt{5}$, $\cos \beta = 0$, $\cos \gamma = 2/\sqrt{5}$

19. $|\mathbf{v}| = \sqrt{3}$, $\cos \alpha = 1/\sqrt{3}$, $\cos \beta = -1/\sqrt{3}$, $\cos \gamma = 1/\sqrt{3}$

21. $|\mathbf{v}| = \sqrt{3}$, $\cos \alpha = \cos \beta = \cos \gamma = -1/\sqrt{3}$

23. $|\mathbf{v}| = \sqrt{82}$, $\cos \alpha = -3/\sqrt{82}$, $\cos \beta = 3/\sqrt{82}$, $\cos \gamma = 8/\sqrt{82}$.

25. $\mathbf{v} = \mathbf{j}$

27. $\mathbf{v} = -6\mathbf{i} + \mathbf{j}$

29. $\mathbf{v} = \mathbf{i} + \mathbf{j} + 4\mathbf{k}$

31. a. $3\mathbf{u} = (6, 9)$.

　　b. $\mathbf{u} + \mathbf{v} = (-3, 7)$

　　c. $\mathbf{v} - \mathbf{u} = (-7, 1)$

　　d. $2\mathbf{u} - 7\mathbf{v} = (39, -22)$

33. $(1/\sqrt{2})\mathbf{i} - (1/\sqrt{2})\mathbf{j}$

35. $(1/\sqrt{3})\mathbf{i} + (1/\sqrt{3})\mathbf{j} + (1/\sqrt{3})\mathbf{k}$

37. $(4/\sqrt{97})\mathbf{i} - (9/\sqrt{97})\mathbf{k}$

39. $\mathbf{F}_1 + \mathbf{F}_2 = -5\sqrt{3}\mathbf{j}$, $\mathbf{F} = 5\sqrt{3}\mathbf{j}$

41. $-(1/\sqrt{2})\mathbf{i} - (1/\sqrt{2})\mathbf{j}$

43. $\frac{3}{5}\mathbf{i} - \frac{4}{5}\mathbf{j}$

45. $(1/\sqrt{26})\mathbf{i} - (3/\sqrt{26})\mathbf{j} + (4/\sqrt{26})\mathbf{k}$

47. $4\mathbf{i} + 4\sqrt{3}\mathbf{j}$

49. If $\mathbf{u} = (a, b)$ and $\mathbf{v} = (c, d)$, then $\mathbf{u} + \mathbf{v} = (a + c, b + d)$,
$$|\mathbf{u} + \mathbf{v}| = \sqrt{(a+c)^2 + (b+c)^2},$$
$$|\mathbf{u}| = \sqrt{a^2 + b^2} \quad \text{and} \quad |\mathbf{v}| = \sqrt{c^2 + d^2}.$$

But
$$0 \le (ad - bc)^2 = a^2d^2 - 2abcd + b^2c^2,$$
and
$$(ac + bd)^2 \le a^2c^2 + a^2d^2 + b^2c^2 + b^2d^2$$
$$= (a^2 + b^2)(c^2 + d^2) = |\mathbf{u}|^2|\mathbf{v}|^2.$$

Hence
$$|\mathbf{u} + \mathbf{v}|^2 = |\mathbf{u}|^2 + 2(ac + bd) + |\mathbf{v}|^2$$
$$\le |\mathbf{u}|^2 + 2|\mathbf{u}||\mathbf{v}| + |\mathbf{v}|^2 = (|\mathbf{u}| + |\mathbf{v}|)^2.$$

Problems 6.2, page 306

1. 0;　0　　　　　　　　**3.** 0;　0

5. 20;　20/29　　　　　**7.** -2;　$-\sqrt{2}/3$

9. 11;　$11/(2\sqrt{57})$　　**11.** $\mathbf{u} \cdot \mathbf{v} = \alpha\beta - \beta\alpha = 0$

13. parallel　　　　　　**15.** neither

17. orthogonal　　　　　**19.** parallel

21. a. $-\frac{3}{4}$
　　b. $\frac{4}{3}$
　　c. $\frac{1}{7}$
　　d. $(-96 \pm \sqrt{7500})/78 \approx -0.12, -2.34$

23. $(-1, 0, 0) = -\mathbf{i}$

25. $\overrightarrow{PQ} = -11\mathbf{i}$. From Problem 24, $R = (-3, y, z)$ and
$$\overrightarrow{PR} = (y - 1)\mathbf{j} + (z - 7)\mathbf{k}.$$

$|\overrightarrow{PR}| = 1$ means that $(y - 1)^2 + (z - 7)^2 = 1$. This is the equation of the circle lying in the plane $x = -3$ with center at $(-3, 1, 7)$ and radius 1.

27. $10\mathbf{i} + 3\mathbf{j} - 7\mathbf{k}$

29. $\mathbf{i} - \mathbf{j} - 6\mathbf{k}$

31. $-13\mathbf{i} + 28\mathbf{j} + 12\mathbf{k}$

33. 25

35. 45

37. $\cos \theta = \dfrac{-10}{\sqrt{59}\sqrt{50}} \approx -0.1841$, $\theta \approx 1.756$ rad $\approx 100.6°$

39. $\overrightarrow{PQ} = (-2, -3, 1)$, $\overrightarrow{PR} = (3, -4, -6)$, $\overrightarrow{PQ} \cdot \overrightarrow{PR} = 0$,

41. If either $\mathbf{u} = \mathbf{0}$ or $\mathbf{v} = \mathbf{0}$ both sides are zero. Otherwise $\left| \dfrac{\mathbf{u} \cdot \mathbf{v}}{|\mathbf{u}||\mathbf{v}|} \right| = |\cos \phi| \le 1$, so $|\mathbf{u} \cdot \mathbf{v}| \le |\mathbf{u}||\mathbf{v}|$.

43. $\frac{3}{2}\mathbf{i} + \frac{3}{2}\mathbf{j}$　　　　　　**45.** 0

47. $-\frac{2}{13}\mathbf{i} + \frac{3}{13}\mathbf{j}$　　　**49.** $(\frac{14}{13})\mathbf{i} - (\frac{4}{13})\mathbf{j} - (\frac{10}{13})\mathbf{k}$

51. $W = -3$ N·m.　　　**53.** $W = -6\sqrt{2}$ N·m.

55. $W = \frac{21}{2}(1 + \sqrt{3}) \approx 28.687$ N·m.

57. $W = 14\sqrt{3}$ N·m.　　**59.** neither

61. parallel　　　　　　**63.** orthogonal

65. parallel　　　　　　**67.** neither

69. parallel

71. The line $ax + by + c = 0$ has slope $-a/b$. A vector parallel to the line is $\mathbf{u} = \mathbf{i} - (a/b)\mathbf{j}$, and $\mathbf{u} \cdot \mathbf{v} = 1 \cdot a - (a/b) \cdot b = 0$.

73. $52/5\sqrt{113} \approx 0.9783$;　$61/\sqrt{34}\sqrt{113} \approx 0.9841$;　$-27/5\sqrt{34} \approx -0.9261$

81. $\overrightarrow{FE} = \mathbf{u} + \mathbf{v}$, $\overrightarrow{AE} = \mathbf{u} + 2\mathbf{v}$, $\overrightarrow{AG} = \mathbf{u} + \mathbf{v} + \mathbf{w}$

Problems 6.3, page 313

1. $-6\mathbf{i} - 3\mathbf{j}$　　　　　　**3.** $-\mathbf{i} - \mathbf{j} + \mathbf{k}$

5. $12\mathbf{i} + 8\mathbf{j} - 21\mathbf{k}$　　**7.** $(bc - ad)\mathbf{j}$

9. $-5\mathbf{i} - \mathbf{j} + 7\mathbf{k}$　　　**11.** 0

13. $42\mathbf{i} + 6\mathbf{j}$

15. $\pm[-(9/\sqrt{181})\mathbf{i} - (6/\sqrt{181})\mathbf{j} + (8/\sqrt{181})\mathbf{k}]$

17. $\sqrt{5/29} \approx 0.415$

19. $5\sqrt{5}$

21. $\sqrt{523}$

23. $\sqrt{a^2b^2 + a^2c^2 + b^2c^2}$

25. See Figure 2. The triangle PQR has half the area of the parallelogram.

27. $\frac{1}{2}\sqrt{595}$

29. coplanar

31. not coplanar

33. not coplanar

35. -1

37. 32

39. -5

41. 60

43. $-\mathbf{i} + \mathbf{j}$

45. $-3\mathbf{i} - 6\mathbf{j} + 14\mathbf{k}$

47. $-5\mathbf{i} - 10\mathbf{j} - 5\mathbf{k}$

49. 14

51. 5

53. $(\mathbf{u} \times \mathbf{v}) \times \mathbf{w} = -\mathbf{i} - 6\mathbf{j} - 6\mathbf{k}$
$(\mathbf{u} \cdot \mathbf{w})\mathbf{v} = -2\mathbf{v} = -4\mathbf{i} - 6\mathbf{k}$
$(\mathbf{v} \cdot \mathbf{w})\mathbf{u} = -3\mathbf{u} = -3\mathbf{i} + 6\mathbf{j}$

55. $(\mathbf{u} \times \mathbf{v}) \times (\mathbf{w} \times \mathbf{y}) = -7\mathbf{i} + 14\mathbf{j}$
$\mathbf{u} \cdot (\mathbf{v} \times \mathbf{y})\mathbf{w} = 7\mathbf{w} = 7\mathbf{j} - 7\mathbf{k}$
$\mathbf{u} \cdot (\mathbf{v} \times \mathbf{w})\mathbf{y} = -7\mathbf{y} = 7\mathbf{i} - 7\mathbf{j} - 7\mathbf{k}$

57. Interchanging rows changes sign of determinant.

59.
$$\begin{vmatrix} \mathbf{i} & \mathbf{j} & \mathbf{k} \\ u_1 & u_2 & u_3 \\ v_1 + w_1 & v_2 + w_2 & v_3 + w_3 \end{vmatrix}$$
$$= \begin{vmatrix} \mathbf{i} & \mathbf{j} & \mathbf{k} \\ u_1 & u_2 & u_3 \\ v_1 & v_2 & v_3 \end{vmatrix} + \begin{vmatrix} \mathbf{i} & \mathbf{j} & \mathbf{k} \\ u_1 & u_2 & u_3 \\ w_1 & w_2 & w_3 \end{vmatrix}$$

61. By (7), we have
$$(\mathbf{u} \times \mathbf{v}) \times (\mathbf{w} \times \mathbf{y}) = [(\mathbf{u} \times \mathbf{v}) \cdot \mathbf{y}]\mathbf{w} - [(\mathbf{u} \times \mathbf{v}) \cdot \mathbf{w}]\mathbf{y}.$$
Applying Theorem 2(v)
$$(\mathbf{u} \times \mathbf{v}) \cdot \mathbf{y} = \mathbf{u} \cdot (\mathbf{v} \times \mathbf{y}) \quad \text{and}$$
$$(\mathbf{u} \times \mathbf{v}) \cdot \mathbf{w} = \mathbf{u} \cdot (\mathbf{v} \times \mathbf{w}).$$

Problems 6.4, page 319

In the answers to Problems 1–4 we assume that the first point is P and the second point is Q. The vector equations are of the form $\overrightarrow{QR} = \overrightarrow{QP} + t\mathbf{v}$. Only $\mathbf{v}$ is given in the answers.

1. $\mathbf{v} = -\mathbf{i} + \mathbf{j} - 4\mathbf{k}$;
$x = 2 - t, y = 1 + t, z = 3 - 4t$;
$(x - 2)/(-1) = y - 1 = (z - 3)/(-4)$

3. $\mathbf{v} = -\mathbf{j} - 2\mathbf{k}$; $x = -4$, $y = 1 - t$, $z = 3 - 2t$;
$x = -4$ and $z = 1 + 2y$

In Problems 5 and 6, $\mathbf{v}$ is already given.

5. $x = 2 + 2t, \quad y = 2 - t, \quad z = 1 - t$;
$(x - 2)/2 = (y - 2)/(-1) = (z - 1)/(-1)$

7. $x = a + dt, \quad y = b + et, \quad z = c$; $(x - a)/d = (y - b)/e$
and $z = c$

9. $\mathbf{v} = 3\mathbf{i} + 6\mathbf{j} + 2\mathbf{k}$; $x = 4 + 3t$, $y = 1 + 6t$,
$z = -6 + 2t$; $(x - 4)/3 = (y - 1)/6 = (z + 6)/2$

11. Vectors on L_1 are multiples of $(2, 4, -1)$, on L_2 of $(5, -2, 2)$. Then $(2, 4, -1) \cdot (5, -2, 2) = 0$ so orthogonal.

13. $3\mathbf{i} + 6\mathbf{j} + 9\mathbf{k} = 3(\mathbf{i} + 2\mathbf{j} + 3\mathbf{k})$ so the direction vectors of the lines are parallel. Note that they are not coincident since, for example, the point $(1, -3, -3)$ is on L_1 but not on L_2.

15. If they had a point in common, we would have
$$2 - t = 1 + s$$
$$1 + t = -2s$$
$$-2t = 3 + 2s.$$
The unique solution of the first two of these equations is $s = -2, t = 3$; but this pair does not satisfy the third equation.

17. a. $(\sqrt{186}/3), (t = \frac{1}{3})$
b. $\sqrt{1518}/11 = \sqrt{138/11}, (t = -\frac{4}{11})$
c. $\sqrt{750}/6 = 5\sqrt{30}/6, (t = -\frac{1}{6})$

19. $(x + 4)/26 = (y - 7)/1 = (z - 3)/37$

21. $(x - 4)/(-4) = (y - 6)/16 = z/24$

23. 3

25. $y = 0$ (xz-plane)

27. $x + y = 3$

29. $y + z = 5$

31. $-3x - 4y + z = 45$

33. $2x - 7y - 8z = -20$

35. $-12x - 21y + 22z = 63$

37. $2x + y = 7$

39. $33/\sqrt{59}$

41. $3/\sqrt{68}$

43. One point P on the plane (obtained by setting $y = z = 0$) is $(d/a, 0, 0)$. (If $a = 0$, set $x = z = 0$ or $x = y = 0$.) Then
$$\overrightarrow{PQ} = (x_0 - d/a, y_0, z_0), \mathbf{n} = (a, b, c), |\mathbf{n}| = \sqrt{a^2 + b^2 + c^2} \text{ and}$$
$$|\overrightarrow{PQ} \cdot \mathbf{n}| = \left| a\left(x_0 - \frac{d}{a}\right) + by_0 + cz_0 \right|$$
$$= |ax_0 + by_0 + cz_0 - d|.$$
The result then follows from Problem 38.

45. $13/\sqrt{830}$

47. $\dfrac{x}{1} = \dfrac{y + 11}{-37} = \dfrac{z + 2}{-10}$

49. $\cos^{-1}(9/\sqrt{3}\sqrt{29}) = \cos^{-1}(0.9649) \approx 0.2657 \approx 15.23°$

51. $\cos^{-1}|20/\sqrt{294}\sqrt{6}| = \cos^{-1}(\frac{20}{42}) \approx 1.074 \approx 61.56°$

53. $x - 22y - 17z = 0$

55. not coplanar

57. not coplanar

Problems 6.5, page 326

1. No, (iv) or (vi) when $\alpha < 0$

3. yes   **5.** yes

7. No; (iii), (iv), and (vi)   **9.** yes

11. No; (vii), (viii)   **13.** yes

15. yes

17. Let **y** and **z** be additive inverses for **x**. Then $\mathbf{y} = \mathbf{y} + \mathbf{0} = \mathbf{y} + (\mathbf{x} + \mathbf{z}) = (\mathbf{y} + \mathbf{x}) + \mathbf{z} = \mathbf{0} + \mathbf{z} = \mathbf{z}.$

19. (i), (vi): If x and y are positive and real, then xy and x^α are positive and real so that (i) and (vi) are satisfied.

(ii): $(x + y) + z = xy + z = xyz = x(yz) = x(y + z)$
$= x + (y + z).$

(iii): 1 is the additive identity: $x + 1 = x1 = 1x = x.$

(iv): $x + \dfrac{1}{x} = x\dfrac{1}{x} = 1$ so that $\dfrac{1}{x}$ is the additive inverse.

(v): $x + y = xy = yx = y + x.$

(vii): $\alpha(x + y) = \alpha(xy) = (xy)^\alpha = x^\alpha y^\alpha = (\alpha x)(\alpha y) = \alpha x + \alpha y.$

(viii): $(\alpha + \beta)x = x^{\alpha+\beta} = x^\alpha x^\beta = x^\alpha + x^\beta = \alpha x + \beta x.$

(ix): $\alpha(\beta x) = (x^\beta)^\alpha = x^{\beta\alpha} = x^{\alpha\beta} = (\alpha\beta)x.$

(x): 1 is also the multiplicative identity: $1x = x^1 = x.$

Problems 6.6, page 329

1. no, because $\alpha(x, y) \notin H$ if $\alpha < 0$

3. yes

5. no; for example $x^4 \in H$ and $x^3 - x^4 \in H$, but $x^4 + (x^3 - x^4) = x^3 \notin H.$

7. yes

9. yes

11. yes

13. No. If $f \in H$ and $\alpha \neq 1$, then $\int_a^b \alpha f(x)\, dx = \alpha \int_a^b f(x)\, dx = \alpha \cdot 1 = \alpha$ so $\alpha f \notin H.$

15. If **x** and $\mathbf{y} \in H_i$, $i = 1, 2$ so does $\alpha\mathbf{x} + \beta\mathbf{y}.$

Problems 6.7, page 338

1. independent

3. dependent

5. dependent

7. dependent

9.
$$\begin{pmatrix} 0 \\ 1 \\ 0 \end{pmatrix} = \begin{pmatrix} 1 \\ 1 \\ 0 \end{pmatrix} - \begin{pmatrix} 1 \\ 0 \\ 0 \end{pmatrix} \quad \text{and} \quad \begin{pmatrix} 0 \\ 0 \\ 1 \end{pmatrix} = \begin{pmatrix} 1 \\ 1 \\ 1 \end{pmatrix} - \begin{pmatrix} 1 \\ 1 \\ 0 \end{pmatrix}$$

and we know

$$\begin{pmatrix} 1 \\ 0 \\ 0 \end{pmatrix}, \begin{pmatrix} 0 \\ 1 \\ 0 \end{pmatrix}, \begin{pmatrix} 0 \\ 0 \\ 1 \end{pmatrix}$$

are linearly independent by Example 9.

11. independent

13. independent

15. dependent

17. independent

19. Let $p(x) = a_0 + a_1 x + a_2 x^2$ and $q(x) = b_0 + b_1 x + b_2 x^2$ be two polynomials in P_2. Let $r(x) = c_0 + c_1 x + c_2 x^2$ be a third polynomial in P_2. If p and q span P_2, then there are scalars α and β such that $r = \alpha p + \beta q$; that is,

$$c_0 = \alpha a_0 + \beta b_0,$$
$$c_1 = \alpha a_1 + \beta b_1,$$
$$c_2 = \alpha a_2 + \beta b_2.$$

This "overdetermined" system of 3 equations in 2 unknowns will have a solution if and only if the third equation is a linear combination of the first two. Since c_0, c_1, and c_2 are arbitrary, this will rarely be the case. Thus p and q cannot span P_2.

29. Let $\alpha_1\mathbf{v}_1 + \alpha_2(\mathbf{v}_1 + \mathbf{v}_2) + \cdots + \alpha_n(\mathbf{v}_1 + \cdots + \mathbf{v}_n) = \mathbf{0}.$ Then $(\alpha_1 + \alpha_2 + \cdots + \alpha_n)\mathbf{v}_1 + (\alpha_2 + \alpha_3 + \cdots + \alpha_n)\mathbf{v}_2 + \cdots + (\alpha_{n-1} + \alpha_n)\mathbf{v}_{n-1} + \alpha_n\mathbf{v}_n = \mathbf{0}.$ Since $\mathbf{v}_1, \mathbf{v}_2, \ldots, \mathbf{v}_n$ are independent, we have

$$\alpha_1 + \alpha_2 + \alpha_3 + \cdots + \alpha_{n-1} + \alpha_n = 0,$$
$$\alpha_2 + \alpha_3 + \cdots + \alpha_{n-1} + \alpha_n = 0,$$
$$\vdots$$
$$\alpha_{n-1} + \alpha_n = 0,$$
$$\alpha_n = 0.$$

Starting with the last equation and working backward, we find that

$$\alpha_n = \alpha_{n-1} = \cdots = \alpha_3 = \alpha_2 = \alpha_1 = 0.$$

31. There exist scalars c_1, c_2 not both zero such that $c_1\mathbf{v}_1 + c_2\mathbf{v}_2 = \mathbf{0}.$ If $\mathbf{v}_2 = \mathbf{0}$, then either $c_1 = 0$ or $\mathbf{v}_1 = \mathbf{0}.$ In either case, $\mathbf{v}_2$ is a (zero) multiple of $\mathbf{v}_1.$ If $c_2 \neq 0$, then $\mathbf{v}_2 = (c_1/c_2)\mathbf{v}_1.$ The same argument shows that $\mathbf{v}_i$ is a multiple of $\mathbf{v}_1$ for $i = 2, 3, \ldots, n.$

33. no   **35.** no   **37.** yes

39. no   **41.** yes   **43.** yes   **45.** yes

47. $\left\{ \begin{pmatrix} 0 \\ 1 \\ -1 \end{pmatrix}, \begin{pmatrix} 1 \\ 0 \\ 2 \end{pmatrix} \right\}$ **49.** $\left\{ \begin{pmatrix} 2 \\ 3 \\ 4 \end{pmatrix} \right\}$

51. Since dim $\mathbb{R}^2 = 2$, a proper subspace H must have dimension 1. Let $\{(x_0, y_0)\}$ be a basis for H. If $(x, y) \in H$, then $(x, y) = c(x_0, y_0)$ for some number c. This means that $x = cx_0$, $y = cy_0$ or $c = \dfrac{x}{x_0} = \dfrac{y}{y_0}$ and $y = \left(\dfrac{y_0}{x_0}\right)x$, which is the equation of a straight line through the origin with slope $\dfrac{y_0}{x_0}$ if $x_0 \neq 0$. If $x_0 = 0$, then the line is the y-axis.

53. Since dim $H = 1$, H is spanned by a single vector $\mathbf{v}$ with coordinates (a, b, c). Then $H = \{(x, y, z) : (x, y, z) = t\mathbf{v}\}$.

55. a. Show that if $\mathbf{x}, \mathbf{y} \in H$ so does $\mathbf{x} + \mathbf{y}$ and $\alpha\mathbf{x}$.

b. $\left\{ \begin{pmatrix} 1 \\ 0 \\ 0 \\ -a/d \end{pmatrix}, \begin{pmatrix} 0 \\ 1 \\ 0 \\ -b/d \end{pmatrix}, \begin{pmatrix} 0 \\ 0 \\ 1 \\ -c/d \end{pmatrix} \right\}$.

63. Any basis of V must span H.

65. a. If $y_1, y_2 \in H$ then $c_1 y_1 + c_2 y_2 \in V$ and $c_1 y_1(0) + c_2 y_2(0) = c_1 \cdot 0 + c_2 \cdot 0 = 0$ so it also belongs to H.

b. dim $H = 1$.

67. $\{e^t, e^{3t}\}$

69. $\{e^{2t}, te^{2t}\}$

71. $\{\cos 2t, \sin 2t\}$

73. $\{e^t, e^{2t}, e^{3t}\}$

Problems 6.8, page 348

1. 13

3. 31

5. n

7. $\sqrt{205}/12$

9. $\sqrt{a^2 + b^2 + c^2 + d^2 + e^2}$

11. $\left\{ \begin{pmatrix} 1/\sqrt{2} \\ 1/\sqrt{2} \end{pmatrix}, \begin{pmatrix} -1/\sqrt{2} \\ 1/\sqrt{2} \end{pmatrix} \right\}$

13. $\left\{ \begin{pmatrix} 1/\sqrt{2} \\ -1/\sqrt{2} \end{pmatrix} \right\}$

15. $\left\{ \begin{pmatrix} a/\sqrt{a^2 + b^2} \\ b/\sqrt{a^2 + b^2} \end{pmatrix}, \begin{pmatrix} -b/\sqrt{a^2 + b^2} \\ a/\sqrt{a^2 + b^2} \end{pmatrix} \right\}$

17. $\left\{ \begin{pmatrix} 1/\sqrt{5} \\ 2/\sqrt{5} \\ 0 \end{pmatrix}, \begin{pmatrix} 2/\sqrt{5} \\ -1/\sqrt{5} \\ 0 \end{pmatrix}, \begin{pmatrix} 0 \\ 0 \\ 1 \end{pmatrix} \right\}$

19. $\{(1/\sqrt{5}, 0, 2/\sqrt{5}), (2/\sqrt{30}, 5/\sqrt{30}, -1/\sqrt{30})\}$

21. $2i$

23. $6 + 6i$

25. a. $\sqrt{7}$ **b.** $\sqrt{17}$

27. a. 1 **b.** $\sqrt{\tfrac{1}{3}}, \sqrt{\dfrac{e^2 - 1}{2}}$

29. a. $\dfrac{e(\sin 1 - \cos 1) + 1}{2}$

b. $\sqrt{\dfrac{e^2 - 1}{2}}, \dfrac{\sqrt{2}\sin 2}{2}$

31. $1, \sqrt{12}(x - \tfrac{1}{2}), \sqrt{180}(x^2 - x + \tfrac{1}{6})$

33. a. $(\mathbf{x}, \mathbf{y})_* = (\mathbf{y}, \mathbf{x})_*$

$(\mathbf{x} + \mathbf{y}, \mathbf{z})_* = (x_1 + y_1)z_1 + 3(x_2 + y_2)z_2$

$\qquad = (x_1 z_1 + 3x_2 z_2) + (y_1 z_1 + 3y_2 z_2)$

$\qquad = (\mathbf{x}, \mathbf{z})_* + (\mathbf{y}, \mathbf{z})_*$, etc.

b. $\left| \begin{pmatrix} 2 \\ 3 \end{pmatrix} \right|_*^2 = \left(\begin{pmatrix} 2 \\ 3 \end{pmatrix}, \begin{pmatrix} 2 \\ 3 \end{pmatrix} \right)_* = 4 + 3(9) = 31$

so $\left| \begin{pmatrix} 2 \\ 3 \end{pmatrix} \right|_* = \sqrt{31}$

35. a. $a^2 + b^2 = 1$

b. Begin by assuming $\left\{ \begin{pmatrix} a \\ b \end{pmatrix}, \begin{pmatrix} c \\ d \end{pmatrix} \right\}$ is an orthonormal basis and determine c and d in terms of a and b.

Review Exercises for Chapter 6, page 350

1. $|\mathbf{v}| = 3\sqrt{2}$, $\theta = \pi/4$

3. $|\mathbf{v}| = 4$, $\theta = 5\pi/3$

5. $|\mathbf{v}| = 12\sqrt{2}$, $\theta = 5\pi/4$

7. $\sqrt{5}$, $2/\sqrt{5}$, $-1/\sqrt{5}$

9. $\sqrt{14}$, $1/\sqrt{14}$, $-2/\sqrt{14}$, $-3/\sqrt{14}$

11. $(1/\sqrt{2})\mathbf{i} + (1/\sqrt{2})\mathbf{j}$

13. $(2/\sqrt{29})\mathbf{i} + (5/\sqrt{29})\mathbf{j}$

15. $\tfrac{3}{5}\mathbf{i} + \tfrac{4}{5}\mathbf{j}$

17. $\dfrac{-7}{\sqrt{78}}\mathbf{i} + \dfrac{2}{\sqrt{78}}\mathbf{j} + \dfrac{5}{\sqrt{78}}\mathbf{k}$

19. a. $-\tfrac{8}{3}$

b. 6

c. $\tfrac{4}{5}$ and -20

d. $32 \pm 52/\sqrt{3}$

21. $7\mathbf{i} - 7\mathbf{j}$

23. $-\tfrac{3}{10}\mathbf{i} + \tfrac{9}{10}\mathbf{j}$

25. -3

27. $-8(3 + 5\sqrt{3})$

29. $\mathbf{i} - 14\mathbf{j} + 20\mathbf{k}$

31. $\tfrac{26}{21}\mathbf{i} - \tfrac{52}{21}\mathbf{j} + \tfrac{13}{21}\mathbf{k}$

33. 22

35. $\cos^{-1}(-9/\sqrt{798}) \approx 1.895 \approx 108.6°$

37. $68/\sqrt{3}$

39. $-7\mathbf{i} - 7\mathbf{k}$

41. $-26\mathbf{i} - 8\mathbf{j} + 7\mathbf{k}$

43. $\sqrt{2065}$

45. 4

47. not coplanar

49. -5

51. $3\mathbf{i} - 8\mathbf{j} + 3\mathbf{k}$

53. $-4\mathbf{i} + \mathbf{j} + (7\mathbf{i} - \mathbf{j} + 7\mathbf{k})t$
$x = -4 + 7t, \quad y = 1 - t, \quad z = 7t$
$$\frac{x+4}{7} = \frac{y-1}{-1} = \frac{z}{7}$$

55. $\overrightarrow{OR} = \mathbf{i} - 2\mathbf{j} - 3\mathbf{k} + t(5\mathbf{i} - 3\mathbf{j} + 2\mathbf{k})$;
$x = 1 + 5t, y = -2 - 3t, z = -3 + 2t$;
$(x - 1)/5 = (y + 2)/(-3) = (z + 3)/2$

57. $\dfrac{x+1}{14} = \dfrac{y-2}{-26} = \dfrac{z-4}{-11}$

59. $2y - 3z = -26$

61. $46x + 14y - 19z = -55$

63. yes; dimension 2; basis $\{(1, 0, 1), (0, 1, 2)\}$

65. yes; dimension 3; basis $\{(1, 0, 0, -1), (0, 1, 0, -1), (0, 0, 1, -1)\}$

67. yes; dimension 5, basis $\{1, x, x^2, x^3, x^4\}$

69. no

71. yes; dimension 3, basis $\{x, x^2, x^3\}$.

73. independent   **75.** dependent

77. independent   **79.** independent

81. dependent   **83.** 28

85. $-\pi$

87. $\left\{ \left(\dfrac{1}{\sqrt{2}}, 0, \dfrac{1}{\sqrt{2}}\right), \left(\dfrac{1}{\sqrt{6}}, \dfrac{2}{\sqrt{6}}, \dfrac{-1}{\sqrt{6}}\right) \right\}$

89. $\left\{ \left(\dfrac{1}{\sqrt{2}}, 0, \dfrac{1}{\sqrt{2}}, 0\right), \left(0, \dfrac{1}{\sqrt{2}}, 0, \dfrac{1}{\sqrt{2}}\right) \right\}$

Chapter 7

Problems 7.1, page 355

1. $\begin{pmatrix} 3 & 9 \\ 6 & 15 \\ -3 & 6 \end{pmatrix}$

3. $\begin{pmatrix} 2 & 2 \\ -2 & -1 \\ 6 & -1 \end{pmatrix}$

5. $\begin{pmatrix} 0 & 0 \\ 0 & 0 \\ 0 & 0 \end{pmatrix}$

7. $\begin{pmatrix} -2 & 4 \\ 7 & 15 \\ -15 & 10 \end{pmatrix}$

9. $\begin{pmatrix} 4 & 10 \\ 17 & 22 \\ -9 & 1 \end{pmatrix}$

11. $\begin{pmatrix} 0 & 6 \\ 5 & 14 \\ -9 & 9 \end{pmatrix}$

13. $\begin{pmatrix} 1 & -5 & 0 \\ -3 & 4 & -5 \\ -14 & 13 & -1 \end{pmatrix}$

15. $\begin{pmatrix} 1 & 1 & 5 \\ 9 & 5 & 10 \\ 7 & -7 & 3 \end{pmatrix}$

17. $\begin{pmatrix} -1 & -1 & -1 \\ -3 & -3 & -10 \\ -7 & 3 & 5 \end{pmatrix}$

19. $\begin{pmatrix} -1 & -1 & -5 \\ -9 & -5 & -10 \\ -7 & 7 & -3 \end{pmatrix}$

25. $\begin{pmatrix} 0 & 1 & 1 & 0 \\ 1 & 0 & 1 & 0 \\ 1 & 1 & 0 & 1 \\ 0 & 0 & 1 & 0 \end{pmatrix}$

27. $\begin{pmatrix} -1 & 0 & 0 & 0 & 1 & 0 & 0 \\ 1 & 0 & 0 & 0 & -1 & 0 & 0 \\ 0 & 0 & 0 & 0 & 0 & 1 & -1 \\ 0 & 0 & 0 & 1 & 0 & -1 & 1 \\ 0 & 0 & 1 & -1 & 0 & 0 & 0 \\ 0 & -1 & -1 & 0 & 0 & 0 & 0 \end{pmatrix}$

Problems 7.2, page 360

1. $\begin{pmatrix} 8 & 20 \\ -4 & 11 \end{pmatrix}$

3. $\begin{pmatrix} -3 & -3 \\ 1 & 3 \end{pmatrix}$

5. $\begin{pmatrix} 13 & 35 & 18 \\ 20 & 26 & 20 \end{pmatrix}$

7. $\begin{pmatrix} 19 & -17 & 34 \\ 8 & -12 & 20 \\ -8 & -11 & 7 \end{pmatrix}$

9. $\begin{pmatrix} 18 & 15 & 35 \\ 9 & 21 & 13 \\ 10 & 9 & 9 \end{pmatrix}$

11. $(7 \quad 16)$

13. $\begin{pmatrix} 3 & -2 & 1 \\ 4 & 0 & 6 \\ 5 & 1 & 9 \end{pmatrix}$

15. $\begin{pmatrix} a & b & c \\ d & e & f \\ g & h & j \end{pmatrix}$

17. If $D = a_{11}a_{22} - a_{12}a_{21}$, then
$$\begin{pmatrix} b_{11} & b_{21} \\ b_{21} & b_{22} \end{pmatrix} = \begin{pmatrix} a_{22}/D & -a_{12}/D \\ -a_{21}/D & a_{11}/D \end{pmatrix}$$

21. $\begin{pmatrix} 0 & -8 \\ 32 & 32 \end{pmatrix}$

23. $\begin{pmatrix} 11 & 38 \\ 57 & 106 \end{pmatrix}$

31. The simplest example is $A = B = \begin{pmatrix} 0 & 1 \\ 0 & 0 \end{pmatrix}$

33. **a.** $\begin{pmatrix} U_2 \\ I_2 \end{pmatrix} = \begin{pmatrix} U_1 - ZI_2 \\ I_1 \end{pmatrix}$

b. $\begin{pmatrix} U_2 \\ I_2 \end{pmatrix} = \begin{pmatrix} U_1 \\ -U_1/Z + I_1 \end{pmatrix}$

35. a. $X_1^2 = X_2^2 = X_3^2 = \begin{pmatrix} 1/4 & 0 \\ 0 & 1/4 \end{pmatrix}$

b. $X_i X_j + X_j X_i = \begin{pmatrix} 0 & 0 \\ 0 & 0 \end{pmatrix}$ if $i \neq j$

$X_i^2 + X_i^2 = \begin{pmatrix} 1/2 & 0 \\ 0 & 1/2 \end{pmatrix}$ if $i = j$

Problems 7.3, page 375

1. $x_1 = \frac{-13}{5}, x_2 = \frac{-11}{5}$

3. no solutions

5. $x_1 = \frac{11}{2}, \quad x_2 = -30; \quad \det = -2$

7. infinite number of solutions; $x_2 = \frac{2}{3}x_1$, where x_1 is arbitrary

9. $x_1 = -1, \quad x_2 = 2$

Note: Where there were an infinite number of solutions, we wrote the solutions with the last variable chosen arbitrarily. The solutions can be written in other ways as well.

11. $(2, -3, 1)$

13. $(3 + \frac{2}{9}x_3, \frac{8}{9}x_3, x_3), \quad x_3$ arbitrary.

15. $(-9, 30, 14)$

17. no solution

19. $(-\frac{4}{5}x_3, \frac{9}{5}x_3, x_3), \quad x_3$ arbitrary

21. $(-1, \frac{5}{2} + \frac{1}{2}x_3, x_3), \quad x_3$ arbitrary

23. no solution

25. $(\frac{20}{13} - \frac{4}{13}x_4, \frac{-28}{13} + \frac{3}{13}x_4, \frac{-45}{13} + \frac{9}{13}x_4, x_4), \quad x_4$ arbitrary

27. $(18 - 4x_4, \frac{-15}{2} + 2x_4, -31 + 7x_4, x_4), \quad x_4$ arbitrary

29. no solution

31. row echelon form

33. reduced row echelon form

35. neither

37. reduced row echelon form

39. neither

41. row echelon form: $\begin{pmatrix} 1 & -6 \\ 0 & 1 \end{pmatrix}$;

reduced row echelon form: $\begin{pmatrix} 1 & 0 \\ 0 & 1 \end{pmatrix}$

43. row echelon form: $\begin{pmatrix} 1 & -2 & 4 \\ 0 & 1 & -4/11 \\ 0 & 0 & 1 \end{pmatrix}$;

reduced row echelon form: $\begin{pmatrix} 1 & 0 & 0 \\ 0 & 1 & 0 \\ 0 & 0 & 1 \end{pmatrix}$

45. row echelon form: $\begin{pmatrix} 1 & -7/2 \\ 0 & 1 \\ 0 & 0 \end{pmatrix}$;

reduced row echelon form: $\begin{pmatrix} 1 & 0 \\ 0 & 1 \\ 0 & 0 \end{pmatrix}$

47. 131,161 units of FN, 120,324 units of FM, 79,194 units of BM, 178,936 units of BN, 66,703 units of E, 426,542 units of S

49. $\begin{pmatrix} 1 & -1 & 3 \\ 4 & 1 & -1 \\ 2 & -1 & 3 \end{pmatrix} \begin{pmatrix} x_1 \\ x_2 \\ x_3 \end{pmatrix} = \begin{pmatrix} 11 \\ -4 \\ 10 \end{pmatrix}$

51. $\begin{pmatrix} 4 & -1 & 1 & -1 \\ 3 & 1 & -5 & 6 \\ 2 & -1 & 1 & 0 \end{pmatrix} \begin{pmatrix} x_1 \\ x_2 \\ x_3 \\ x_4 \end{pmatrix} = \begin{pmatrix} -7 \\ 8 \\ 9 \end{pmatrix}$

53. $\begin{pmatrix} 2 & 3 & -1 \\ -4 & 2 & 1 \\ 7 & 3 & -9 \end{pmatrix} \begin{pmatrix} x_1 \\ x_2 \\ x_3 \end{pmatrix} = \begin{pmatrix} 0 \\ 0 \\ 0 \end{pmatrix}$

55. Let $D = R_1 R_2 + R_2 R_3 + R_2 R_4 + R_1 R_3$. Then
$I_1 = \frac{(R_2 + R_3)E}{D}, I_2 = \frac{R_3 E}{D}, I_3 = \frac{R_2 E}{D}.$

Problems 7.4, page 385

1. $(0, 0)$

3. $(0, 0, 0)$

5. $(\frac{1}{6}x_3, \frac{5}{6}x_3, x_3), \quad x_3$ arbitrary

7. $(0, 0)$

9. $(-4x_4, 2x_4, 7x_4, x_4), \quad x_4$ arbitrary.

11. $(0, 0)$ **13.** $(0, 0, 0)$

15. $k = \frac{95}{11}$ **17.** independent

19. independent   **21.** dependent

23. dependent for every α (the second vector is -2 times the first)

25. $(6, 0, 0) + x_2(1, 1, 0) + x_3(-1, 0, 1)$

27. No solution; system is inconsistent.

29. $(5 + \frac{1}{2}x_4, 4 - \frac{1}{4}x_4, -3 + \frac{1}{4}x_4, x_4)$

Problems 7.5, page 399

1. $\begin{pmatrix} 2 & -1 \\ -3 & 2 \end{pmatrix}$ **3.** $\begin{pmatrix} 0 & 1 \\ 1 & 0 \end{pmatrix}$

5. not invertible   **7.** $\begin{pmatrix} 1/3 & -1/3 & -1/3 \\ 0 & 1/2 & 1 \\ 0 & 0 & -1 \end{pmatrix}$

9. not invertible   **11.** not invertible

13.
$$\begin{pmatrix} 7/3 & -1/3 & -1/3 & -2/3 \\ 4/9 & -1/9 & -4/9 & 1/9 \\ -1/9 & -2/9 & 1/9 & 2/9 \\ -5/3 & 2/3 & 2/3 & 1/3 \end{pmatrix}$$

15.
$$\begin{pmatrix} 0 & 1 & 0 & 2 \\ 1 & -1 & -2 & 2 \\ 0 & 1 & 3 & -3 \\ -2 & 2 & 3 & -2 \end{pmatrix}$$

27.
$$\begin{pmatrix} 1/2 & -1/6 & 7/30 \\ 0 & 1/3 & -4/15 \\ 0 & 0 & 1/5 \end{pmatrix}$$

31. any nonzero multiple of $(1, 2)$

33. 3 chairs and 2 tables

35. 4 units of A and 5 units of B

37. $(I - A)^{-1}$
$$\approx \begin{pmatrix} 1.234 & 0.014 & 0.006 & 0.064 & 0.007 & 0.018 \\ 0.017 & 1.436 & 0.057 & 0.012 & 0.020 & 0.032 \\ 0.071 & 0.465 & 1.877 & 0.019 & 0.045 & 0.031 \\ 0.751 & 0.134 & 0.100 & 1.740 & 0.066 & 0.124 \\ 0.061 & 0.045 & 0.130 & 0.082 & 1.578 & 0.059 \\ 0.339 & 0.236 & 0.307 & 0.312 & 0.376 & 1.349 \end{pmatrix}$$

(all computations rounded to 3 decimal places)
The solution is given in the answer to Problem 7.3.47.

39. a. $A = \begin{pmatrix} 0.293 & 0 & 0 \\ 0.014 & 0.207 & 0.017 \\ 0.044 & 0.010 & 0.216 \end{pmatrix}$;

$I - A = \begin{pmatrix} 0.707 & 0 & 0 \\ -0.014 & 0.793 & -0.017 \\ -0.044 & -0.010 & 0.784 \end{pmatrix}$

b. $\begin{pmatrix} 18,639 \\ 22,597 \\ 3,613 \end{pmatrix}$

41. $\begin{pmatrix} 1 & 1/2 \\ 0 & 0 \end{pmatrix}$; yes

43. $\begin{pmatrix} 1 & 2/3 & 1/3 \\ 0 & 1 & 1 \\ 0 & 0 & 1 \end{pmatrix}$; yes

45. $\begin{pmatrix} 1 & -1/2 & 2 \\ 0 & 1 & -14 \\ 0 & 0 & 0 \end{pmatrix}$; no

47. $\begin{pmatrix} 1 & 0 & 2 & 3 \\ 0 & 1 & 2 & 7 \\ 0 & 0 & 1 & 10/7 \\ 0 & 0 & 0 & 0 \end{pmatrix}$; no

Problems 7.6, page 403

1. $\begin{pmatrix} -1 & 6 \\ 4 & 5 \end{pmatrix}$

3. $\begin{pmatrix} 2 & -1 & 1 \\ 3 & 2 & 4 \end{pmatrix}$

5. $\begin{pmatrix} 1 & -1 & 1 \\ 2 & 0 & 5 \\ 3 & 4 & 5 \end{pmatrix}$

7. $\begin{pmatrix} 1 & 0 \\ 0 & 1 \\ 1 & 0 \\ 0 & 1 \end{pmatrix}$

9. $\begin{pmatrix} a & d & g \\ b & e & h \\ c & f & j \end{pmatrix}$

Problems 7.7, page 410

1. -10 **3.** 47

5. 4   **7.** -36

9. -260 **11.** -183

13. 24   **15.** -296

17. 138   **19.** $abcde$

23. Almost any example will work. For instance,

$$\det\begin{pmatrix} 1 & 0 \\ 0 & 1 \end{pmatrix} = 1, \text{ but } \det\begin{pmatrix} 1 & 0 \\ 0 & 0 \end{pmatrix} + \det\begin{pmatrix} 0 & 0 \\ 0 & 1 \end{pmatrix} = 0 + 0 \neq 1.$$

Problems 7.8, page 422

1. 28   **3.** 2

5. 32   **7.** -36

9. -50 **11.** -18

13. -260 **15.** -183

17. -24 **19.** 138

21. -8 **23.** 16

25. -16 **27.** -16

Problems 7.9, page 431

1. $\begin{pmatrix} 1/2 & -1/2 \\ -1/4 & 3/4 \end{pmatrix}$

3. $\begin{pmatrix} 0 & 1 \\ 1 & 0 \end{pmatrix}$

5. $\begin{pmatrix} 1/3 & -1/4 & -1/6 \\ 0 & 1/4 & 1/2 \\ 0 & 1/4 & -1/2 \end{pmatrix}$

7. $\begin{pmatrix} 0 & 1 & -1 \\ 2 & -2 & -1 \\ -1 & 1 & 1 \end{pmatrix}$

9. not invertible

11. $\begin{pmatrix} 7/3 & -1/3 & -1/3 & -2/3 \\ 4/9 & -1/9 & -4/9 & 1/9 \\ -1/9 & -2/9 & 1/9 & 2/9 \\ -5/3 & 2/3 & 2/3 & 1/3 \end{pmatrix}$

13. $x_1 = 1, \quad x_2 = 3$

15. $x_1 = 1, \quad x_2 = -1, \quad x_3 = 1$

17. $x_1 = \frac{45}{13}, \quad x_2 = -\frac{11}{13}, \quad x_3 = \frac{23}{13}$

19. $x_1 = \frac{3}{2}, \quad x_2 = \frac{3}{2}, \quad x_3 = \frac{1}{2}$

21. $x_1 = \frac{21}{29}, \quad x_2 = \frac{171}{29}, \quad x_3 = -\frac{284}{29}, \quad x_4 = -\frac{182}{29}$

23. $\det A = 3; \quad \det A^{-1} = \begin{vmatrix} 5/3 & -1/3 \\ -2/3 & 1/3 \end{vmatrix} = \frac{1}{3}$

25. $\alpha = 4$ or $\alpha = -3$

29. independent ($\det = 13$)

31. independent ($\det = -3$)

Problems 7.10, page 441

1. $\rho = 2, \quad \nu = 0$ 3. $\rho = 1, \quad \nu = 2$

5. $\rho = 2, \quad \nu = 1$ 7. $\rho = 2, \quad \nu = 2$

9. $\rho = 2, \quad \nu = 0$ 11. $\rho = 2, \quad \nu = 2$

13. $\rho = 3, \quad \nu = 1$ 15. $\rho = 2, \quad \nu = 1$

17. row space basis $= \{(1, -1, 2), (0, 2, -1)\}$

     kernel $= \text{span}\left\{ \begin{pmatrix} -3 \\ 1 \\ 2 \end{pmatrix} \right\}$

19. row space basis $= \{(1, -1, 2, 3), (0, 1, 4, 3), (0, 0, 0, 1)\}$

     kernel $= \text{span}\left\{ \begin{pmatrix} -6 \\ -4 \\ 1 \\ 0 \end{pmatrix} \right\}$

21. row space basis $= \{(1, -1, 2, 3)\};$

     kernel $= \text{span}\left\{ \begin{pmatrix} 1 \\ 1 \\ 0 \\ 0 \end{pmatrix}, \begin{pmatrix} 0 \\ 2 \\ 1 \\ 0 \end{pmatrix}, \begin{pmatrix} 0 \\ 3 \\ 0 \\ 1 \end{pmatrix} \right\}$

23. no

25. yes

Problems 7.11, page 449

1. linear; $A_T = \begin{pmatrix} 1 & 0 \\ 0 & 0 \end{pmatrix}$

3. linear; $A_T = \begin{pmatrix} 1 & 0 & 0 \\ 0 & 1 & 0 \end{pmatrix}$ 5. not linear

7. linear; $A_T = \begin{pmatrix} 0 & 1 \\ 1 & 0 \end{pmatrix}$ 9. not linear

11. linear; $A_T = \begin{pmatrix} 1 \\ 1 \\ \vdots \\ 1 \end{pmatrix}$ 13. not linear

15. not linear   17. not linear

19. linear   21. linear; $A_T = \begin{pmatrix} 1 & 0 & 0 \\ 0 & 1 & 0 \end{pmatrix}$

23. not linear   25. linear

27. linear   29. not linear

31. a. $\begin{pmatrix} -14 \\ 4 \\ 26 \end{pmatrix}$

     b. $\begin{pmatrix} -31 \\ -6 \\ 26 \end{pmatrix}$

33. It rotates a vector counterclockwise around the z-axis through an angle of θ in a plane parallel to the xy-plane.

45. $T\begin{pmatrix} x \\ y \end{pmatrix} = \begin{pmatrix} c - ap & p \\ bc - aq & q \end{pmatrix}\begin{pmatrix} x \\ y \end{pmatrix},$

where c, p, and q are real numbers and $c \neq 0$.

Review Exercises for Chapter 7, page 450

1. $\left(\frac{1}{7}, \frac{10}{7}\right)$

3. no solution

5. $(0, 0, 0)$

7. $\left(-\frac{1}{2}, 0, \frac{5}{2}\right)$

9. $\left(\frac{1}{3}x_3, \frac{7}{3}x_3, x_3\right), \quad x_3$ arbitrary

11. no solution

13. $(0, 0, 0, 0)$

15. $1/\sqrt{5}$

17. $(4 - 2x_2 + 7x_3, x_2, x_3), \quad x_2, x_3$ arbitrary

19. $(0, 0, 0)$ 21. $\left(1, -\frac{1}{3}, \frac{1}{2}, 4\right)$

23. 20 units of each   25. $\begin{pmatrix} 3 & 0 & 7 \\ 0 & 4 & 14 \end{pmatrix}$

27. $\begin{pmatrix} 16 & 19 \\ 3 & 29 \end{pmatrix}$ 29. $\begin{pmatrix} -26 & 16 & 35 \\ -10 & 19 & 30 \\ -42 & 17 & 32 \end{pmatrix}$

31. $\begin{pmatrix} 7 \\ 29 \\ 5 \end{pmatrix}$

33. $\begin{pmatrix} 1 & 3/2 \\ 0 & 1 \end{pmatrix}$; inverse is $\begin{pmatrix} 4/11 & -3/11 \\ 1/11 & 2/11 \end{pmatrix}$

35. $\begin{pmatrix} 1 & 2 & 0 \\ 0 & 1 & 1/3 \\ 0 & 0 & 1 \end{pmatrix}$; inverse is $\begin{pmatrix} -1/4 & 1/4 & 1/4 \\ 5/8 & -1/8 & -1/8 \\ 1/8 & -5/8 & 3/8 \end{pmatrix}$

37. $\begin{pmatrix} 1 & 0 & 2 \\ 0 & 1 & 1 \\ 0 & 0 & 1 \end{pmatrix}$; inverse is $\begin{pmatrix} 5/6 & 2/3 & -2 \\ 1/3 & 2/3 & -1 \\ -1/6 & -1/3 & 1 \end{pmatrix}$

39. $\begin{pmatrix} 1 & 2 & 0 \\ 2 & 1 & -1 \\ 3 & 1 & 1 \end{pmatrix}\begin{pmatrix} x_1 \\ x_2 \\ x_3 \end{pmatrix} = \begin{pmatrix} 3 \\ -1 \\ 7 \end{pmatrix}$; A^{-1} is given in Exercise 35;

$x_1 = \frac{3}{4}, \quad x_2 = \frac{9}{8}, \quad x_3 = \frac{29}{8}$

41. $\begin{pmatrix} 2 & -1 \\ 3 & 0 \\ 1 & 2 \end{pmatrix}$; neither

43. $\begin{pmatrix} 2 & 3 & 1 \\ 3 & -6 & -5 \\ 1 & -5 & 9 \end{pmatrix}$; symmetric

45. $\begin{pmatrix} 1 & -1 & 4 & 6 \\ -1 & 2 & 5 & 7 \\ 4 & 5 & 3 & -8 \\ 6 & 7 & -8 & 9 \end{pmatrix}$; symmetric

47. $x_1 = \frac{11}{7}, \quad x_2 = \frac{1}{7}$

49. $x_1 = \frac{1}{4}, \quad x_2 = \frac{5}{4}, \quad x_3 = -\frac{3}{4}$

51. -4 **53.** 24

55. 60 **57.** 34

59. $\begin{pmatrix} -1/11 & 4/11 \\ 2/11 & 3/11 \end{pmatrix}$ **61.** not invertible

63. $\begin{pmatrix} 1/11 & 1/11 & 0 & 3/11 \\ 9/11 & -2/11 & 0 & -6/11 \\ 3/11 & 3/11 & 0 & -2/11 \\ 1/22 & 1/22 & -1/2 & 3/22 \end{pmatrix}$

65. **a.** independent
   b. dependent

67. ker A = multiples of $\begin{pmatrix} -2 \\ 1 \\ 1 \end{pmatrix}$, $\nu(A) = 1$, $\rho(A) = 2$

69. ker A = span $\left\{ \begin{pmatrix} -2 \\ 1 \\ 0 \end{pmatrix}, \begin{pmatrix} 1 \\ 0 \\ 1 \end{pmatrix} \right\}$, $\nu(A) = 2$, $\rho(A) = 1$

71. ker A = span $\left\{ \begin{pmatrix} -1 \\ 1 \\ 1 \\ 0 \end{pmatrix}, \begin{pmatrix} -3 \\ 0 \\ 0 \\ 1 \end{pmatrix} \right\}$,

$\nu(A) = \rho(A) = 2$

73. not linear

75. linear

77. linear

Chapter 8

Problems 8.1, page 463

In the following answers E_i denotes the eigenspace corresponding to the eigenvalue $\lambda = i$ (see Problem 32). The geometric multiplicity (see Problem 33) is also given.

1. $-4, 3$; $E_{-4} = \text{span}\left\{ \begin{pmatrix} 1 \\ 1 \end{pmatrix} \right\}$; $E_3 = \text{span}\left\{ \begin{pmatrix} 2 \\ -5 \end{pmatrix} \right\}$

3. $i, -i$; $E_i = \text{span}\left\{ \begin{pmatrix} 2 + i \\ 5 \end{pmatrix} \right\}$; $E_{-i} = \text{span}\left\{ \begin{pmatrix} 2 - i \\ 5 \end{pmatrix} \right\}$

5. $-3, -3$; $E_{-3} = \text{span}\left\{ \begin{pmatrix} 1 \\ 0 \end{pmatrix} \right\}$; geom. mult. is 1

7. $0, 1, 3$; $E_0 = \text{span}\left\{ \begin{pmatrix} 1 \\ 1 \\ 1 \end{pmatrix} \right\}$; $E_1 = \text{span}\left\{ \begin{pmatrix} -1 \\ 0 \\ 1 \end{pmatrix} \right\}$;

$E_3 = \text{span}\left\{ \begin{pmatrix} 1 \\ -2 \\ 1 \end{pmatrix} \right\}$

9. $1, 1, 10$; $E_1 = \text{span}\left\{ \begin{pmatrix} 1 \\ 0 \\ -2 \end{pmatrix}, \begin{pmatrix} 0 \\ 1 \\ -2 \end{pmatrix} \right\}$;

$E_{10} = \text{span}\left\{ \begin{pmatrix} 2 \\ 2 \\ 1 \end{pmatrix} \right\}$; geom. mult. of 1 is 2

11. $1, 1, 1$; $E_1 = \text{span}\left\{ \begin{pmatrix} 1 \\ 1 \\ 1 \end{pmatrix} \right\}$;

geom. mult. is 1 (alg. mult. is 3)

13. $-1, i, -i$; $E_{-1} = \text{span}\left\{ \begin{pmatrix} 0 \\ -1 \\ 1 \end{pmatrix} \right\}$;

$E_i = \text{span}\left\{ \begin{pmatrix} 1 + i \\ 1 \\ 1 \end{pmatrix} \right\}$; $E_{-i} = \text{span}\left\{ \begin{pmatrix} 1 - i \\ 1 \\ 1 \end{pmatrix} \right\}$

15. 1, 2, 2; $E_1 = \text{span}\left\{\begin{pmatrix} 4 \\ 1 \\ -3 \end{pmatrix}\right\}$;

$E_2 = \text{span}\left\{\begin{pmatrix} 3 \\ 1 \\ -2 \end{pmatrix}\right\}$; geom. mult. of 2 is 1

17. a, a, a, a; $E_a = \mathbb{R}^4$; geom. mult. of $a = $ alg. mult. of $a = 4$

19. a, a, a, a;

$$E_a = \text{span}\left\{\begin{pmatrix} 1 \\ 0 \\ 0 \\ 0 \end{pmatrix}, \begin{pmatrix} 0 \\ 0 \\ 0 \\ 1 \end{pmatrix}\right\};$$

alg. mult. of $a = 4$; geom. mult. of $a = 2$.

35. $E_{-1} = \text{span}\left\{\begin{pmatrix} 3 \\ -6 \\ 2 \end{pmatrix}\right\}$

41. Each matrix has the eigenvalues $\pm\frac{1}{2}$.

Problems 8.2, page 469

1.

n	$p_{j,n}$	$p_{a,n}$	T_n	$p_{j,n}/p_{a,n}$	T_n/T_{n-1}
0	0	12	12	0	—
1	36	7	43	5.14	3.58
2	21	19	40	1.11	0.930
5	104	45	149	2.31	—
10	600	291	891	2.06	—
19	16,090	7737	23827	2.08	—
20	23,170	11140	34310	2.08	1.44

Note that the eigenvalues are 1.44 and -0.836. The corresponding eigenvectors are $\begin{pmatrix} 2.08 \\ 1 \end{pmatrix}$ and $\begin{pmatrix} -3.57 \\ 1 \end{pmatrix}$.

3.

n	$p_{j,n}$	$p_{a,n}$	T_n	$p_{j,n}/p_{a,n}$	T_n/T_{n-1}
0	0	20	20	0	—
1	80	16	96	5	4.8
2	64	69	133	0.928	1.39
5	1092	498	1590	2.19	—
10	3114	1970	5084	1.58	—
19	3.69×10^7	1.95×10^7	5.64×10^7	1.89	—
20	7.82×10^7	4.14×10^7	11.96×10^7	1.89	2.12

The eigenvalues are 2.12 and -1.32 with corresponding eigenvectors $\begin{pmatrix} 1.89 \\ 1 \end{pmatrix}$ and $\begin{pmatrix} -3.03 \\ 1 \end{pmatrix}$.

Problems 8.3, page 474

1. yes; $C = \begin{pmatrix} 1 & 2 \\ 1 & -5 \end{pmatrix}$, $C^{-1}AC = \begin{pmatrix} -4 & 0 \\ 0 & 3 \end{pmatrix}$

3. yes; $C = \begin{pmatrix} 1 & 1 \\ 2-i & 2+i \end{pmatrix}$; $C^{-1}AC = \begin{pmatrix} i & 0 \\ 0 & -i \end{pmatrix}$

5. yes; $C = \begin{pmatrix} 2 & 2 \\ -1+3i & -1-3i \end{pmatrix}$;

$C^{-1}AC = \begin{pmatrix} 2+3i & 0 \\ 0 & 2-3i \end{pmatrix}$

7. yes; $C = \begin{pmatrix} 3 & 1 & 1 \\ 2 & 3 & 0 \\ 1 & 1 & 1 \end{pmatrix}$; $C^{-1}AC = \begin{pmatrix} 1 & 0 & 0 \\ 0 & 2 & 0 \\ 0 & 0 & -1 \end{pmatrix}$

9. yes; $C = \begin{pmatrix} 0 & 0 & 1 \\ 1 & 1 & 0 \\ 0 & 2 & 0 \end{pmatrix}$; $C^{-1}AC = \begin{pmatrix} 0 & 0 & 0 \\ 0 & 2 & 0 \\ 0 & 0 & 3 \end{pmatrix}$

11. $C = \begin{pmatrix} 1 & 0 & 2 \\ 3 & -2 & 1 \\ 0 & 1 & 2 \end{pmatrix}$; $C^{-1}AC = \begin{pmatrix} 1 & 0 & 0 \\ 0 & 1 & 0 \\ 0 & 0 & 2 \end{pmatrix}$

13. No, since 1 is an eigenvalue of algebraic multiplicity 3 and geometric multiplicity 1.

15. yes; $C = \begin{pmatrix} 0 & -1 & 1 & 1 \\ 0 & 1 & 1 & 1 \\ 1 & 0 & 1 & -1 \\ 0 & 1 & -1 & 1 \end{pmatrix}$;

$C^{-1}AC = \begin{pmatrix} 2 & 0 & 0 & 0 \\ 0 & 2 & 0 & 0 \\ 0 & 0 & 4 & 0 \\ 0 & 0 & 0 & 6 \end{pmatrix}$

21. $\begin{pmatrix} 1 & 0 \\ 0 & 1 \end{pmatrix}$

Problems 8.4, page 484

1. $Q = \begin{pmatrix} 2/\sqrt{5} & 1/\sqrt{5} \\ 1/\sqrt{5} & -2/\sqrt{5} \end{pmatrix}$, $D = \begin{pmatrix} 5 & 0 \\ 0 & -5 \end{pmatrix}$

3. $Q = \begin{pmatrix} 1/\sqrt{2} & 1/\sqrt{2} \\ 1/\sqrt{2} & -1/\sqrt{2} \end{pmatrix}$, $D = \begin{pmatrix} 0 & 0 \\ 0 & 2 \end{pmatrix}$

5. $Q = \begin{pmatrix} 1/\sqrt{2} & 1/2 & 1/2 \\ -1/\sqrt{2} & 1/2 & 1/2 \\ 0 & 1/\sqrt{2} & -1/\sqrt{2} \end{pmatrix}$,

$D = \begin{pmatrix} -3 & 0 & 0 \\ 0 & 1+2\sqrt{2} & 0 \\ 0 & 0 & 1-2\sqrt{2} \end{pmatrix}$

7. $Q = \begin{pmatrix} -2/3 & 1/3 & 2/3 \\ 2/3 & 2/3 & 1/3 \\ 1/3 & -2/3 & 2/3 \end{pmatrix}$, $D = \begin{pmatrix} 0 & 0 & 0 \\ 0 & 3 & 0 \\ 0 & 0 & 6 \end{pmatrix}$

15. $A^* = \begin{pmatrix} 1-i & 5-2i \\ 2 & 3 \end{pmatrix}$

17. $A^* = \begin{pmatrix} 2 & i & 1-i \\ 3 & -2i & -1-i \\ 5 & -5i & 2+3i \end{pmatrix}$

25. $U = \dfrac{1}{\sqrt{3}} \begin{pmatrix} -1+i & 1 \\ 1 & 1+i \end{pmatrix}$; $U^*AU = \begin{pmatrix} -1 & 0 \\ 0 & 8 \end{pmatrix}$

Problems 8.5, page 491

1. $\begin{pmatrix} 3 & -1 \\ -1 & 0 \end{pmatrix}\begin{pmatrix} x \\ y \end{pmatrix} \cdot \begin{pmatrix} x \\ y \end{pmatrix} = 5$;

$Q = \begin{pmatrix} \dfrac{2}{\sqrt{26-6\sqrt{13}}} & \dfrac{2}{\sqrt{26+6\sqrt{13}}} \\ \dfrac{3-\sqrt{13}}{\sqrt{26-6\sqrt{13}}} & \dfrac{3+\sqrt{13}}{\sqrt{26+6\sqrt{13}}} \end{pmatrix}$

$\approx \begin{pmatrix} 0.9571 & 0.2898 \\ -0.2898 & 0.9571 \end{pmatrix}$; $\dfrac{x'^2}{\left(\dfrac{10}{\sqrt{13}+3}\right)} - \dfrac{y'^2}{\left(\dfrac{10}{\sqrt{13}-3}\right)} = 1$;

hyperbola; $\theta \approx 5.989 \approx 343°$

3. $\begin{pmatrix} 4 & 2 \\ 2 & -1 \end{pmatrix}\begin{pmatrix} x \\ y \end{pmatrix} \cdot \begin{pmatrix} x \\ y \end{pmatrix} = 9$;

$Q = \begin{pmatrix} \dfrac{5+\sqrt{41}}{\sqrt{82+10\sqrt{41}}} & \dfrac{5-\sqrt{41}}{\sqrt{82-10\sqrt{41}}} \\ \dfrac{4}{\sqrt{82+10\sqrt{41}}} & \dfrac{4}{\sqrt{82-10\sqrt{41}}} \end{pmatrix}$

$\approx \begin{pmatrix} 0.9436 & -0.3310 \\ 0.3310 & 0.9436 \end{pmatrix}$; $\dfrac{x'^2}{\left(\dfrac{18}{\sqrt{41}+3}\right)} - \dfrac{y'^2}{\left(\dfrac{18}{\sqrt{41}-3}\right)} = 1$;

hyperbola; $\theta \approx 0.3374 \approx 19.33°$

5. $\begin{pmatrix} 0 & 1/2 \\ 1/2 & 0 \end{pmatrix}\begin{pmatrix} x \\ y \end{pmatrix} \cdot \begin{pmatrix} x \\ y \end{pmatrix} = a > 0$;

$Q = \begin{pmatrix} 1/\sqrt{2} & 1/\sqrt{2} \\ -1/\sqrt{2} & 1/\sqrt{2} \end{pmatrix}$; $\dfrac{x'^2}{2a} - \dfrac{y'^2}{2a} = 1$;

hyperbola; $\theta = 7\pi/4 = 315°$.

7. Same as Problem 5 except that now we have a hyperbola with the roles of x' and y' reversed; since $a < 0$, we have

$$\dfrac{y'^2}{(-2a)} - \dfrac{x'^2}{(-2a)} = 1.$$

9. $\begin{pmatrix} -1 & 1 \\ 1 & -1 \end{pmatrix}\begin{pmatrix} x \\ y \end{pmatrix} \cdot \begin{pmatrix} x \\ y \end{pmatrix} = 0$;

$Q = \begin{pmatrix} 1/\sqrt{2} & -1/\sqrt{2} \\ 1/\sqrt{2} & 1/\sqrt{2} \end{pmatrix}$;

$y'^2 = 0$, which is the equation of a straight line through the origin; $\theta = \pi/4 = 45°$.

11. $\begin{pmatrix} 3 & -3 \\ -3 & 5 \end{pmatrix}\begin{pmatrix} x \\ y \end{pmatrix} \cdot \begin{pmatrix} x \\ y \end{pmatrix} = 36$;

$Q = \begin{pmatrix} \dfrac{1+\sqrt{10}}{\sqrt{20+2\sqrt{10}}} & \dfrac{1-\sqrt{10}}{\sqrt{20-2\sqrt{10}}} \\ \dfrac{3}{\sqrt{20+2\sqrt{10}}} & \dfrac{3}{\sqrt{20-2\sqrt{10}}} \end{pmatrix}$

$\approx \begin{pmatrix} 0.8112 & -0.5847 \\ 0.5847 & 0.8112 \end{pmatrix}$

$\dfrac{x'^2}{\left(\dfrac{36}{4-\sqrt{10}}\right)} + \dfrac{y'^2}{\left(\dfrac{36}{4+\sqrt{10}}\right)} = 1$;

ellipse; $\theta \approx 0.6245 \approx 35.78°$

13. $\begin{pmatrix} 6 & 5/2 \\ 5/2 & -6 \end{pmatrix}\begin{pmatrix} x \\ y \end{pmatrix} \cdot \begin{pmatrix} x \\ y \end{pmatrix} = -7$;

$Q = \begin{pmatrix} 5/\sqrt{26} & -1/\sqrt{26} \\ 1/\sqrt{26} & 5/\sqrt{26} \end{pmatrix}$;

$\dfrac{y'^2}{(14/13)} - \dfrac{x'^2}{(14/13)} = 1$;

hyperbola; $\theta \approx 1.377 \approx 78.91°$

15. $\begin{pmatrix} 1 & -1 & -1 \\ -1 & 1 & -1 \\ -1 & -1 & 1 \end{pmatrix}\begin{pmatrix} x \\ y \\ z \end{pmatrix} \cdot \begin{pmatrix} x \\ y \\ z \end{pmatrix}$;

$Q = \begin{pmatrix} 1/\sqrt{3} & 1/\sqrt{2} & 1/\sqrt{6} \\ 1/\sqrt{3} & -1/\sqrt{2} & 1/\sqrt{6} \\ 1/\sqrt{3} & 0 & -2/\sqrt{6} \end{pmatrix}$;

$-x'^2 + 2y'^2 + 2z'^2$

17. $\begin{pmatrix} 3 & 2 & 2 \\ 2 & 2 & 0 \\ 2 & 0 & 4 \end{pmatrix}\begin{pmatrix} x \\ y \\ z \end{pmatrix} \cdot \begin{pmatrix} x \\ y \\ z \end{pmatrix}$;

$Q = \begin{pmatrix} -2/3 & 1/3 & 2/3 \\ 2/3 & 2/3 & 1/3 \\ 1/3 & -2/3 & 2/3 \end{pmatrix}$;

$3y'^2 + 6z'^2$

19. $\begin{pmatrix} 1 & 1 & 2 & 7/2 \\ 1 & 1 & 3 & -1 \\ 2 & 3 & 3 & 0 \\ 7/2 & -1 & 0 & 1 \end{pmatrix}$

21. $\begin{pmatrix} 3 & -7/2 & 1/2 & -1 & 3/2 \\ -7/2 & -2 & -1/2 & 1/2 & 0 \\ 1/2 & -1/2 & 3 & -2 & -5/2 \\ -1 & 1/2 & -2 & -6 & 1/2 \\ 3/2 & 0 & -5/2 & 1/2 & -1 \end{pmatrix}$

29. negative definite

31. positive definite

33. indefinite

35. negative definite

Problems 8.6, page 500

1. $\begin{pmatrix} c_1 e^{-4t} + 2c_2 e^{3t} \\ c_1 e^{-4t} - 5c_2 e^{3t} \end{pmatrix}$

3. $\begin{pmatrix} 2c_1 \sin t + c_1 \cos t - c_2 \sin t \\ 5c_1 \sin t - 2c_2 \sin t + c_2 \cos t \end{pmatrix}$

5. $\begin{pmatrix} 2c_1 e^{t/2} + 3c_2 e^{t/3} \\ c_1 e^{t/2} + 2c_2 e^{t/3} \end{pmatrix}$

7. $\begin{pmatrix} c_1 e^{-4t} + 4c_2 e^{-6t} \\ c_1 e^{-4t} - 3c_2 e^{-6t} \end{pmatrix}$

9. $c_1 \begin{pmatrix} 4e^t - 3e^{2t} + 6te^{2t} \\ e^t - e^{2t} + 2te^{2t} \\ -3e^t + 3e^{2t} - 4te^{2t} \end{pmatrix} + c_2 \begin{pmatrix} -12e^t + 12e^{2t} - 6te^{2t} \\ -3e^t + 4e^{2t} - 2te^{2t} \\ 9e^t - 9e^{2t} + 4te^{2t} \end{pmatrix}$
$+ c_3 \begin{pmatrix} 6te^{2t} \\ 2te^{2t} \\ -4te^{2t} + e^{2t} \end{pmatrix}$

11. $\begin{pmatrix} I_L \\ I_R \end{pmatrix} \approx e^{-6.25t} \begin{pmatrix} -0.688 \sin(14.52t) \\ 0.5 \cos(14.52t) - 0.215 \sin(14.52t) \end{pmatrix}$

13. $\mathbf{x}(t) \approx \begin{pmatrix} 0.28 \cos(2.91t) + 0.72 \cos(1.24t) \\ -0.10 \cos(2.91t) + 3.10 \cos(1.24t) \\ -0.64 \sin(2.91t) - 0.89 \sin(1.24t) \\ 0.25 \sin(2.91t) - 3.84 \sin(1.24t) \end{pmatrix}$

15. a. $x_1' = x_2, \quad x_2' = x_3, \quad x_3' = x_4, \quad x_4' \frac{MQ^2}{EI} x_1$

c. $\omega_n = \frac{n^2 \pi^2}{64} \sqrt{\frac{4.32 \times 10^9}{24(8)^4/\pi g}} \approx 325n^2$ rev/sec. The lowest frequency is 325 rev/sec.

17. $x_1(t) = (600 - 400e^{-t}); \quad x_2(t) = (300 + 200e^{-t})$

23. $x(t) = \frac{1}{4}(7e^{-t} - 3e^{-5t})$

25. b. $\mathbf{v}_1 = \begin{pmatrix} 1 \\ 3 \end{pmatrix}$

c. $\mathbf{u} = \begin{pmatrix} 1 \\ 1 \end{pmatrix}, \quad \mathbf{w} = \begin{pmatrix} 2 \\ 6 \end{pmatrix}$

d. $\mathbf{x}(t) = e^{4t} \left[c_1 \begin{pmatrix} 1 \\ 3 \end{pmatrix} + c_2 \begin{pmatrix} 1 + 2t \\ 1 + 6t \end{pmatrix} \right]$

27. $\mathbf{x}(t) = e^{-5t} \left[c_1 \begin{pmatrix} 1 \\ 1 \end{pmatrix} + c_2 \begin{pmatrix} -1 + 7t \\ 7t \end{pmatrix} \right]$

Problems 8.7, page 505

1. a. $p(\lambda) = \lambda^2 + \lambda - 12 = 0;$

c. $A^{-1} = \frac{1}{12} \begin{pmatrix} -1 & -2 \\ -5 & 2 \end{pmatrix}$

3. a. $p(\lambda) = -\lambda^3 + 4\lambda^2 - 3\lambda;$

c. A^{-1} does not exist.

5. a. $p(\lambda) = -\lambda^3 + 3\lambda^2 - 3\lambda + 1 = 0$

c. $A^{-1} = \begin{pmatrix} 3 & -3 & 1 \\ 1 & 0 & 0 \\ 0 & 1 & 0 \end{pmatrix}$

7. a. $p(\lambda) = -\lambda^3 + 6\lambda^2 + 18\lambda + 9 = 0$

c. $A^{-1} = \frac{1}{9} \begin{pmatrix} -27 & 18 & -9 \\ -6 & 3 & 0 \\ 19 & -11 & 6 \end{pmatrix}$

9. a. $p(\lambda) = (a - \lambda)^4$

c. $A^{-1} = \begin{pmatrix} 1/a & -b/a^2 & cb/a^3 & -bcd/a^4 \\ 0 & 1/a & -c/a^2 & cd/a^3 \\ 0 & 0 & 1/a & -d/a^2 \\ 0 & 0 & 0 & 1/a \end{pmatrix}$

Review Exercises for Chapter 8, page 505

1. $4, -2;$

$E_4 = \text{span} \left\{ \begin{pmatrix} 1 \\ 1 \end{pmatrix} \right\};$

$E_{-2} = \text{span} \left\{ \begin{pmatrix} 2 \\ 1 \end{pmatrix} \right\}$

3. $1, 7, -5;$

$E_1 = \text{span} \left\{ \begin{pmatrix} -6 \\ 3 \\ 4 \end{pmatrix} \right\};$

$E_7 = \text{span} \left\{ \begin{pmatrix} 0 \\ 3 \\ 1 \end{pmatrix} \right\};$

$E_{-5} = \text{span} \left\{ \begin{pmatrix} 0 \\ 0 \\ 1 \end{pmatrix} \right\}$

5. $1, 3, 3 + \sqrt{2}i, 3 - \sqrt{2}i;$

$E_1 = \text{span} \left\{ \begin{pmatrix} 1 \\ 2 \\ 0 \\ 0 \end{pmatrix} \right\};$

$$E_3 = \text{span} \left\{ \begin{pmatrix} 1 \\ 1 \\ 0 \\ 0 \end{pmatrix} \right\} ;$$

$$E_{3+\sqrt{2}i} = \text{span} \left\{ \begin{pmatrix} 0 \\ 0 \\ -1 \\ \sqrt{2}i \end{pmatrix} \right\} ;$$

$$E_{3-\sqrt{2}i} = \text{span} \left\{ \begin{pmatrix} 0 \\ 0 \\ 1 \\ \sqrt{2}i \end{pmatrix} \right\}$$

7. $C = \begin{pmatrix} -3 & 1 \\ 4 & -1 \end{pmatrix};$ $C^{-1}AC = \begin{pmatrix} 2 & 0 \\ 0 & -3 \end{pmatrix}$

9. $C = \begin{pmatrix} 0 & -1-i & -1+i \\ 1 & 1 & 1 \\ -1 & 1 & 1 \end{pmatrix};$

$$C^{-1}AC = \begin{pmatrix} -1 & 0 & 0 \\ 0 & i & 0 \\ 0 & 0 & -i \end{pmatrix}$$

11. not diagonalizable

13. $Q = \begin{pmatrix} 1/\sqrt{2} & 0 & 1/\sqrt{2} \\ 1/\sqrt{2} & 0 & -1/\sqrt{2} \\ 0 & 1 & 0 \end{pmatrix};$ $Q^t A Q = \begin{pmatrix} 4 & 0 & 0 \\ 0 & -3 & 0 \\ 0 & 0 & 0 \end{pmatrix}$

15. $C = \begin{pmatrix} 1 & 1 & 1 & -1 \\ -1 & 0 & 0 & 0 \\ 0 & 1 & 0 & 0 \\ -1 & -1 & -1 & 2 \end{pmatrix};$ $C^{-1}AC = \begin{pmatrix} -1 & 0 & 0 & 0 \\ 0 & -1 & 0 & 0 \\ 0 & 0 & 3 & 0 \\ 0 & 0 & 0 & 3 \end{pmatrix}$

17. $\dfrac{x'^2}{8/(3+\sqrt{2})} + \dfrac{y'^2}{8/(3-\sqrt{2})} = 1$: ellipse

19. $\dfrac{x'^2}{10/(\sqrt{13}+3)} - \dfrac{y'^2}{10/(\sqrt{13}-3)} = 1$; hyperbola

21. $4x'^2 - 3y'^2$

23. $x_1 = c_1 e^{4t} + c_2 e^{-2t};$ $x_2 = c_1 e^{4t} + \frac{1}{2} c_2 e^{-2t};$

25. $\begin{pmatrix} x_1 \\ x_2 \\ x_3 \end{pmatrix} = \begin{pmatrix} -c_1 e^t + c_2 e^{-2t} + c_3 e^{3t} \\ 4c_1 e^t - c_2 e^{-2t} + 2c_3 e^{3t} \\ c_1 e^t - c_2 e^{-2t} + c_3 e^{3t} \end{pmatrix}$

Chapter 9

Problems 9.1, page 512

1. $\mathbb{R}^2$; $[0, \infty)$

3. $\{(x, y): y \neq 0\}$; $\mathbb{R}$

5. $\{(x, y): x^2 - 4y^2 \leq 1\}$; $[0, \infty)$

7. $\mathbb{R}^2$; $(0, \infty)$

9. $\{(x, y): x - y \neq (n + \frac{1}{2})\pi, n = 0, \pm 1, \pm 2, \ldots\}$; $\mathbb{R}$

11. $\{(x, y): |x| \geq |y|$ and $x \neq -y\}$; $[0, \infty)$

13. $\{(x, y): |x - y| \leq 1\}$: $[0, \pi]$

15. $\{(x, y): x \neq -y\}$; $\mathbb{R}$

17. $\{(x, y): x \neq 0$ and $y \neq 0\}$; $(-\infty, -2] \cup [2, \infty)$ (The range is obtained by using the fact that if $u = x/(2y)$ and $u > 0$, then $f(x, y) = u + (1/u)$ which has minimum value 2.)

19. $\{(x, y, z): x + y + z \geq 0\}$—this is the half-space "in front of" the plane $x + y + z = 0$; $[0, \infty)$

21. $\{(x, y, z): y^2 < x^2 + z^2\}$—this is the region "outside" the cone $y^2 = x^2 + z^2$; $(0, \infty)$

23. $\{(0, 0, 0)\}$; $\{0\}$

25. $\{(x, y, z): z \neq 0\}$; $\mathbb{R}$

27. $\{(x, y, z): |x + y - z| \leq 1\}$—this is the part of $\mathbb{R}^3$ between the planes $x + y - z = -1$ and $x + y - z = 1$; $[-\pi/2, \pi/2]$

29. $\{(x, y, z): y \neq 0$ and $(x + z)/y \neq (n + \frac{1}{2})\pi,$ $n = 0, \pm 1, \pm 2, \ldots\}$; $\mathbb{R}$

31. $\mathbb{R}^3$; $(0, \infty)$

33. $\{(x, y, z): x \neq 0, y \neq 0$ and $z \neq 0\}$; $\mathbb{R} - \{0\}$

35. $\mathbb{R}^3$; $[-3, 3]$

37. elliptic paraboloid opening around the y-axis; cross sections parallel to the xz-plane are ellipses

39. hyperbolic paraboloid; cross sections parallel to the xy-plane are hyperbolas

41. two half planes parallel to the xy-plane; $\{(x, y, z): z = 1$ and $x \geq 0\}$ and $\{(x, y, z); z = -1$ and $x \geq 0\}$

43. these are the parallel straight lines (with slopes of -1) $y = -x + (z^2 - 1)$

45. these are concentric ellipses (centered at the origin) with equations $x^2 + 4y^2 = 1 - z^2$, $|z| \leq 1$; for $z = 1$ we obtain the single point $(0, 0, 0)$

47. parallel straight lines (with slopes of 1) $y = x - \cos z$

49. for each value of z we get a family of straight lines all of which have a slope of -1: for $z = 0$ we obtain $x + y = n\pi$, n an integer; for $z = 1$ we obtain $x + y = (\pi/4) + n\pi$; for $z = -1$ we obtain $x + y = -(\pi/4) + n\pi$; for $z = \sqrt{3}$ we obtain $x + y = (\pi/3) + n\pi$

51. concentric ellipses (centered at the origin); $a = \sqrt{T - 20}$, $b = \frac{1}{2}\sqrt{T - 20}$, for $T > 20$

53. concentric ellipses (centered at the origin); $a = \sqrt{(P - 100)/2}$, $b = \sqrt{(P - 100)/3}$, for $P > 100$; for $P = 100$ we obtain the single point $(0, 0)$

55. (h)

57. (a)

59. (b)

61. (i)

63. (g)

Problems 9.2, page 516

1. $2(x + y)(\mathbf{i} + \mathbf{j})$

3. $\sin(x - y)(-\mathbf{i} + \mathbf{j})$

5. $(x/\sqrt{x^2 + y^3})\mathbf{i} + (3y^2/2\sqrt{x^2 + y^3})\mathbf{j}$

7. $-y \sec^2(y - x)\mathbf{i} + [\tan(y - x) + y \sec^2(y - x)]\mathbf{j}$

9. $\sec(x + 3y)\tan(x + 3y)(\mathbf{i} + 3\mathbf{j})$

11. $[4xy/(x^2 + y^2)^2](y\mathbf{i} - x\mathbf{j})$

13. $yz\mathbf{i} + xz\mathbf{j} + xy\mathbf{k}$

15. $\left(\dfrac{1}{3xy}\right)\left[\left(\dfrac{x^2 + y^2 - z^2}{x}\right)\mathbf{i} - \left(\dfrac{x^2 + y^2 + z^2}{y}\right)\mathbf{j} + 2z\mathbf{k}\right]$

17. $y^2\mathbf{i} + 2y(x + z^3)\mathbf{j} + 3y^2z^2\mathbf{k}$

19. $\sin y \ln z\,\mathbf{i} + x \cos y \ln z\,\mathbf{j} + (x/z)\sin y\,\mathbf{k}$

21. $(\cosh z - y \cos x)\mathbf{i} - \sin x\,\mathbf{j} + x \sinh z\,\mathbf{k}$

23. $-\omega y\mathbf{i} + \omega x\mathbf{j}$

Problems 9.3, page 526

1. $\mathbb{R} - \{0, 1\}$

3. $\mathbb{R} - \{-1, 1\}$

5. $(0, 1)$

7. $\mathbb{R} - \{n\pi/2: n = 0, \pm 1, \pm 2, \ldots\}$

9. $y^2 = 4x$ **11.** $y = \pm x^{3/2}$

13. $y = 2x + 5$ **15.** $x = y^2 + y + 1, \quad y \geq 0$

17. $y = x^3 - 1$ **19.** $y = (\ln x)^2, \; x > 0$

21. $\mathbf{f}' = \mathbf{i} - 5t^4\mathbf{j}; \quad \mathbf{f}'' = -20t^3\mathbf{j}$

23. $\mathbf{f}' = (2 \cos 2t)\mathbf{i} - 3(\sin 3t)\mathbf{j};$
$\mathbf{f}'' = (-4 \sin 2t)\mathbf{i} - (9 \cos 3t)\mathbf{j}$

25. $\mathbf{f}' = (1/t)\mathbf{i} + 3e^{3t}\mathbf{j}; \quad \mathbf{f}'' = -(1/t^2)\mathbf{i} + 9e^{3t}\mathbf{j}$

27. $\mathbf{f}' = (\sec^2 t)\mathbf{i} + (\sec t)(\tan t)\mathbf{j};$
$\mathbf{f}'' = 2(\sec^2 t)(\tan t)\mathbf{i} + [\sec^3 t + (\sec t)(\tan^2 t)]\mathbf{j}$

29. $\mathbf{f}' = -(\tan t)\mathbf{i} + (\cot t)\mathbf{j}; \quad \mathbf{f}'' = -(\sec^2 t)\mathbf{i} - (\csc^2 t)\mathbf{j}$

31. $(2/\sqrt{13})\mathbf{i} + (3/\sqrt{13})\mathbf{j}$

33. $\mathbf{j}$

35. $-(1/\sqrt{2})\mathbf{i} + (1/\sqrt{2})\mathbf{j}$

37. $\mathbf{i}$

39. $(4/\sqrt{97})\mathbf{i} - (9/\sqrt{97})\mathbf{j}$

43. $(650\sqrt{2}/4)^2 = 52{,}812.5$ ft (using the value $g = 32$ ft/sec²)

45. $x(t) = 50\sqrt{3}t; \quad y(t) = -(9.81t^2/2) - 50t + 150$

47. $x = \frac{1}{2}\alpha - \frac{1}{4} \sin \alpha$ m, $\quad y = \frac{1}{2}\alpha - \frac{1}{4} \cos \alpha$ m

49. $-(a/\sqrt{a^2 + b^2})\mathbf{i} + (b/\sqrt{a^2 + b^2})\mathbf{j}$

51. $(1/\sqrt{2})\mathbf{i} + (1/\sqrt{2})\mathbf{j}$

53. $(0, 1/\sqrt{14}, 2/\sqrt{14}, 3/\sqrt{14})$

55. $(1/2, -1/2\sqrt{3}, 1/2, -1/2\sqrt{3}, 1/2, -1/2\sqrt{3})$

Problems 9.4, page 535

1. $\frac{1}{27}(148^{3/2} - 13^{3/2})$

3. $8(2^{3/2} - 1)$

5. $\displaystyle\int_{1/\sqrt{5}}^{1} \sqrt{1 + x^2}\, dx = \int_{\tan^{-1}(1/\sqrt{5})}^{\pi/4} \sec^3 \theta\, d\theta =$
$\dfrac{1}{2}\left[\sqrt{2} - \left(\dfrac{\sqrt{6}}{5}\right) + \ln\left(\dfrac{\sqrt{5}(1 + \sqrt{2})}{1 + \sqrt{6}}\right)\right] \approx 0.6861$

7. $\sqrt{2}$ **9.** $8a$ (see Example 3)

11. $\frac{2}{27}[13^{3/2} - 8]$ **13.** $|a|\pi$

15. $\sqrt{2}(e^3 - 1)$ **17.** $6\sqrt{2}$

19. $\sqrt{2}(e^{\pi/2} - 1)$

21. $t = 0.$ $\mathbf{f} = (\cos 3t)\mathbf{i} + (\sin 3t)\mathbf{j}.$ Circle $x^2 + y^2 = 1.$
$\mathbf{f}' = -3(\sin 3t)\mathbf{i} + 3(\cos 3t)\mathbf{j}.$
$\mathbf{T} = -(\sin 3t)\mathbf{i} + (\cos 3t)\mathbf{j} = \mathbf{j}.$
$\mathbf{T}' = -3(\cos 3t)\mathbf{i} - 3(\sin 3t)\mathbf{j}.$
$\mathbf{n} = -(\cos 3t)\mathbf{i} - (\sin 3t)\mathbf{j} = -\mathbf{i}$

23. $t = \pi/4.$ $\mathbf{f} = 2 \cos 4t\mathbf{i} + 2 \sin 4t\mathbf{j}.$ Circle $x^2 + y^2 = 4.$ $\mathbf{f}' = -8(\sin 4t)\mathbf{i} + 8(\cos 4t)\mathbf{j}.$ $\mathbf{T} = -(\sin 4t)\mathbf{i} + (\cos 4t)\mathbf{j} = -\mathbf{j}$ $\mathbf{T}' = -4(\cos 4t)\mathbf{i} - 4(\sin 4t)\mathbf{j}.$ $\mathbf{n} = -(\cos 4t)\mathbf{i} - (\sin 4t)\mathbf{j} = \mathbf{i}.$

25. $t = \pi/4.$ $\mathbf{f} = 8(\cos t)\mathbf{i} + 8(\sin t)\mathbf{j}.$ Circle $x^2 + y^2 = 64.$ $\mathbf{f}' = -8 \sin t\mathbf{i} + 8 \cos t\mathbf{j}.$ $\mathbf{T} = -\sin t\mathbf{i} + \cos t\mathbf{j} = (-\mathbf{i} + \mathbf{j})/\sqrt{2}.$ $\mathbf{T}' = -(\cos t)\mathbf{i} - (\sin t)\mathbf{j} = \mathbf{n} = -(\mathbf{i} + \mathbf{j})/\sqrt{2}.$

27. $t = 3.$ $\mathbf{f} = (2 + 3t)\mathbf{i} + (8 - 5t)\mathbf{j}.$ Line $(x - 2)/3 = (y - 8)/-5.$ $\mathbf{f}' = 3\mathbf{i} - 5\mathbf{j}.$ $\mathbf{T} = (3\mathbf{i} - 5\mathbf{j})/\sqrt{34}.$ $\mathbf{T}' = \mathbf{0}.$ $\mathbf{n} = \pm(5\mathbf{i} + 3\mathbf{j})/\sqrt{34}.$

29. $t = t_0.$ $\mathbf{f} = (a + bt)\mathbf{i} + (c + dt)\mathbf{j}.$ Line $(x - a)/b = (y - c)/d.$ $\mathbf{f}' = b\mathbf{i} + d\mathbf{j}.$ $\mathbf{T} = (b\mathbf{i} + d\mathbf{j})/(b^2 + d^2)^{1/2}.$ $\mathbf{T}' = \mathbf{0}.$ $\mathbf{n} = \pm(d\mathbf{i} - b\mathbf{j})/(b^2 - d^2)^{1/2}.$

31. and **33.** $\mathbf{f} = (t - \cos t)\mathbf{i} + (1 - \sin t)\mathbf{j}.$ Inverted cycloid
$\mathbf{f}' = (1 + \sin t)\mathbf{i} - (\cos t)\mathbf{j} = (1 + \sin t)\mathbf{i} - s(1 - \sin^2 t)^{1/2}\mathbf{j}$
$|\mathbf{f}'| = \sqrt{2}(1 + \sin t)^{1/2}.$ $\mathbf{T} = [(1 + \sin t)^{1/2}\mathbf{i} - s(1 - \sin t)^{1/2}\mathbf{j}]/\sqrt{2}.$ $\mathbf{T}' = [\frac{1}{2}(1 + \sin t)^{-1/2} \cos t\mathbf{i} + \frac{1}{2}s(1 - \sin t)^{-1/2} \cos t\mathbf{j}]/\sqrt{2} = \frac{1}{2}[s(1 - \sin t)^{1/2}\mathbf{i} + (1 + \sin t)^{1/2}\mathbf{j}]/\sqrt{2} = \frac{1}{2}\mathbf{n}\ (s = \text{sgn}(\cos t)).$

31. $\mathbf{f}\left(\dfrac{\pi}{2}\right) = \dfrac{\pi}{2}\mathbf{j}.$ $\mathbf{T}\left(\dfrac{\pi}{2}\right) = \mathbf{i}.$ $\mathbf{n}\left(\dfrac{\pi}{2}\right) = \mathbf{j}.$

33. $\mathbf{f}\left(\dfrac{\pi}{4}\right) = \left(\dfrac{\pi}{4} - \dfrac{1}{2}\sqrt{2}\right)\mathbf{i} + \left(1 - \dfrac{1}{2}\sqrt{2}\right)\mathbf{j} \approx 0.078\mathbf{i} + 0.293\mathbf{j}.$

$\mathbf{T}\left(\dfrac{\pi}{4}\right) = [(1 + \tfrac{1}{2}\sqrt{2})^{1/2}\mathbf{i} - (1 - \tfrac{1}{2}\sqrt{2})^{1/2}\mathbf{j}]/\sqrt{2}$

$$\approx 0.924\mathbf{i} - 0.383\mathbf{j}.$$

$$\mathbf{n}\left(\frac{\pi}{4}\right) = [(1 - \tfrac{1}{2}\sqrt{2})^{1/2}\mathbf{i} + (1 + \tfrac{1}{2}\sqrt{2})^{1/2}\mathbf{j}]/\sqrt{2}$$

$$\approx 0.383\mathbf{i} + 0.924\mathbf{j}.$$

35. $(1/\sqrt{14})\mathbf{i} + (2/\sqrt{14})\mathbf{j} + (3/\sqrt{14})\mathbf{k}$

37. $(1/\sqrt{3})\mathbf{i} + (1/\sqrt{3})\mathbf{j} - (1/\sqrt{3})\mathbf{k}$

39. $(-8/\sqrt{65})\mathbf{i} + (1/\sqrt{65})\mathbf{k}$

41. $\int_0^{10} \sqrt{36 + 4t^2}\, dt = 18 \int_0^{\tan^{-1}(10/3)} \sec^3 \theta\, d\theta = 10\sqrt{109} + 9 \ln[(\sqrt{109} + 10)/3]$

43. $\frac{4}{729}(328^{3/2} - 8) \approx 32.55$

45. $(4/\sqrt{6})\sqrt{n(n+1)(2n+1)}$

Problems 9.5, page 543

1. $\kappa = \frac{1}{2}, \quad \rho = 2$

3. $\kappa = 2/5^{3/2}, \quad \rho = 5^{3/2}/2$

5. $\kappa = \frac{3}{16}, \quad \rho = \frac{16}{3}$

7. $\kappa = 6/\pi, \quad \rho = \pi/6$

9. $\kappa = 2/5^{3/2}, \quad \rho = 5^{3/2}/2$

11. $\kappa = 1/2\sqrt{2}, \quad \rho = 2\sqrt{2}$

13. $\kappa = 1/2\sqrt{2}, \quad \rho = 2\sqrt{2}$

15. $\kappa = 2a/(1 + b^2)^{3/2}, \quad \rho = (1 + b^2)^{3/2}/2a$

17. $\kappa = 1, \quad \rho = 1$

19. $\kappa = 0, \quad \rho$ is undefined

21. at the origin

23. minimum for $t = \pi/2$, maximum for $t = \pi/4$

Note: In 25, 27, 29, and 31, we used the formula

$$\tau = \frac{\mathbf{f}''' \cdot (\mathbf{f}' \times \mathbf{f}'')}{|\mathbf{f}' \times \mathbf{f}''|^2}.$$

25. $\mathbf{T} = (\tfrac{1}{2}a\sqrt{2}\mathbf{i} - \tfrac{1}{2}a\sqrt{2}\mathbf{j} + \mathbf{k})/\sqrt{a^2 + 1}; \quad \mathbf{n} = -\frac{\sqrt{2}}{2}\mathbf{i} - \frac{\sqrt{2}}{2}\mathbf{j};$

$$\mathbf{B} = \frac{1}{\sqrt{a^2 + 1}}\left(\frac{\sqrt{2}}{2}\mathbf{i} - \frac{\sqrt{2}}{2}\mathbf{j} - a\mathbf{k}\right); \quad \tau = -1/(a^2 + 1)$$

27. $\mathbf{T} = \frac{1}{\sqrt{b^2 + 1}}(b\mathbf{j} + \mathbf{k}); \quad \mathbf{n} = -\mathbf{i}; \quad \mathbf{B} = \frac{1}{\sqrt{b^2 + 1}}(-\mathbf{j} + b\mathbf{k});$

$$\tau = b/[a(1 + b^2)]$$

29. $\mathbf{T} = \mathbf{i}; \quad \mathbf{n} = \mathbf{j}; \quad \mathbf{B} = \mathbf{k}; \quad \tau = 3$

31. $\mathbf{T} = \frac{1}{\sqrt{6}}(-2\mathbf{i} + \mathbf{j} + \mathbf{k}); \quad \mathbf{n} = \frac{1}{\sqrt{5}}(-\mathbf{i} - 2\mathbf{j});$

$$\mathbf{B} = \frac{1}{\sqrt{30}}(2\mathbf{i} - \mathbf{j} + 5\mathbf{k}); \quad \tau = 1/3e^{\pi/4}$$

Problems 9.6, page 552

1. $\mathbf{v} = 3\mathbf{j}, \quad |\mathbf{v}| = 3, \quad \mathbf{a} = -9\mathbf{i}, \quad |\mathbf{a}| = 9$

3. $\mathbf{v} = -4\sqrt{3}\mathbf{i} - 4\mathbf{j}; \quad |\mathbf{v}| = 8; \quad \mathbf{a} = 16\mathbf{i} - 16\sqrt{3}\mathbf{j}; \quad |\mathbf{a}| = 32$

5. $\mathbf{v} = 4\mathbf{i} + 4\mathbf{j}; \quad |\mathbf{v}| = 4\sqrt{2}; \quad \mathbf{a} = 4\mathbf{j}; \quad |\mathbf{a}| = 4$

7. $\mathbf{v} = -7\mathbf{i} + 5\mathbf{j}; \quad |\mathbf{v}| = \sqrt{74}; \quad \mathbf{a} = 0; \quad |\mathbf{a}| = 0$

9. $\mathbf{v} = 2\mathbf{i}; \quad |\mathbf{v}| = 2; \quad \mathbf{a} = 2\mathbf{j}; \quad |\mathbf{a}| = 2$

11. $\mathbf{v} = -(\sin 1)\mathbf{i} + (\cos 1)\mathbf{j} + 4\mathbf{k}; \quad |\mathbf{v}| = \sqrt{17};$
$\mathbf{a} = -(\cos 1)\mathbf{i} - (\sin 1)\mathbf{j} + 12\mathbf{k}; \quad |\mathbf{a}| = \sqrt{145}$

13. $\mathbf{v} = \mathbf{j} + \mathbf{k}; \quad |\mathbf{v}| = \sqrt{2}, \quad \mathbf{a} = \mathbf{i}; \quad |\mathbf{a}| = 1$

15. $\mathbf{v} = \tfrac{1}{2}\mathbf{i} + \tfrac{1}{3}\mathbf{j} + \tfrac{1}{4}\mathbf{k}; \quad |\mathbf{v}| = \sqrt{61}/12;$
$\mathbf{a} = -\tfrac{1}{4}\mathbf{i} - \tfrac{2}{9}\mathbf{j} - \tfrac{3}{16}\mathbf{k}; \quad |\mathbf{a}| = \sqrt{3049}/144$

17. a. $3000^2\left[\sqrt{3} + \tfrac{1}{4}\ln\left(\frac{1 + \sqrt{3}/2}{1 - \sqrt{3}/2}\right)\right]\Big/ 64 \approx 336{,}168.25$ ft $\approx$

   66.96 miles;

   b. $140{,}625\sqrt{3} \approx 243{,}569$ ft ≈ 48.5 miles

   c. $105{,}468.75$ ft ≈ 21.0 miles;

   d. $1500\sqrt{3}/2 \approx 1837.1$ ft/sec.

19. a. $\sqrt{3000/g} \approx 17.5$ sec

   b. $125 t_f \approx 2186$ m

   c. $-g t_f \approx -171.6$ m/sec

21. $a_T = 0, \quad a_n = 4$

23. $a_T = (4t/\sqrt{1 + 4t^2}), \quad a_n = (2/\sqrt{1 + 4t^2})$

25. $a_T = (18t^3 - 14t)/\sqrt{9t^4 - 14t^2 + 9},$
$a_n = 6(1 + t^2)/\sqrt{9t^4 - 14t^2 + 9}$

27. $a_T = (4 + 18t^2)/\sqrt{4 + 9t^2}, \quad a_n = 6t/\sqrt{4 + 9t^2}$

31. $[(10{,}000)(80{,}000)^2/(3600)^2] \cdot (1/\sqrt{2}) \approx 3{,}491{,}885.3$ N

33. $v = 150/\sqrt{2} \approx 106$ km/hr (reduce speed by a factor of $\sqrt{2}$)

37. $4v_0 t \mathbf{r}' = 4v_0\omega t(-\sin \omega t \mathbf{i} + \cos \omega t \mathbf{j})$

39. $[-2v_0/(1 + t^2)]\mathbf{r}' = [-2v_0\omega/(1 + t^2)](-\sin \omega t \mathbf{i} + \cos \omega t \mathbf{j})$

Problems 9.7, page 559

5. The function is not defined along the line $y = x$.

7. Along the line $y = x$ the limit is 0; along the line $y = x^2$ the limit is 1.

9. $g(x) = 2x$

11. $\partial z/\partial x = 2xy; \quad \partial z/\partial y = x^2$

13. $\partial z/\partial x = y/2\sqrt{xy + 2y^3}; \quad \partial z/\partial y = (x + 6y^2)/2\sqrt{xy + 2y^3}$

15 b. no (neither $f(x, 0)$ nor $f(0, y)$ is continuous at $(0, 0)$)

17. $3e^{3t}$

19. 1

21. $\partial z/\partial r = 2r; \quad \partial z/\partial s = -2s$

23. $\partial z/\partial r = -e^{s-r}; \quad \partial z/\partial s = e^{s-r}$

25. $w_r = f_x \cos \theta + f_y \sin \theta; \quad w_\theta = -f_x r \sin \theta + f_y r \cos \theta;$
$w_t = f_z$

27. 122.248 cm³/sec, increasing

Problems 9.8, page 570

1. $9/\sqrt{10}$

3. $\sqrt{2}/7$

5. $-5/(4\sqrt{13})$

7. $(2e^2 + 3e)/\sqrt{2}$

9. $2\sqrt{3}$

11. $1/(5\sqrt{6})$

13. $61/\sqrt{26}$

15. 0

17. a. at the origin

   b. $\nabla T = 200e^{-(x^2+y^2+z^2)}(-x\mathbf{i} - y\mathbf{j} - z\mathbf{k})$

19. $(x(t), y(t)) = (ae^{-2t/a^2}, be^{-2t/b^2}); \quad (x/a)^{a^2} = (y/b)^{b^2}$

21. a. descend, because $\nabla h \cdot \mathbf{u} < 0$ when $\mathbf{u} = (1/\sqrt{2})\mathbf{i} - (1/\sqrt{2})\mathbf{j}$

   b. $(-120\mathbf{i} + 40\mathbf{j})/\sqrt{120^2 + 40^2} = (-3\mathbf{i} + \mathbf{j})/\sqrt{10}$; i.e., $\approx 18.4°$ (east of north)

25. $f(x, y) = x^2/2 + y^2/2 + C$

29. $F(x, y, z) = (\alpha/2) \ln(x^2 + y^2 + z^2) = \alpha \ln|\mathbf{x}|$

Problems 9.9, page 571

1. $x = 1; \quad y = 0, \quad z = 0$

3. $z = 1; \quad x = 0, \quad y = 0$

5. $(1/a)(x - a) + (1/b)(y - b) + (1/c)(z - c) = 0;$ $a(x - a) = b(y - b) = c(z - c)$

7. $\frac{1}{4}(x - 4) + \frac{1}{2}(y - 1) + \frac{1}{6}(z - 9) = 0;$ $4(x - 4) = 2(y - 1) = 6(z - 9)$

9. $2(x - 1) + (y - 2) + (z - 2) = 0;$ $(x - 1)/2 = y - 2 = z - 2$

11. $24(x - 3) - 2(y - 1) + 20(z + 2) = 0;$ $(x - 3)/24 = (y - 1)/(-2) = (z + 2)/20$

13. $(\pi/\sqrt{3})(y - 1) + \sqrt{3}[z - (\pi/3)] = 0; \quad x = \pi/2,$ $(\sqrt{3}/\pi)(y - 1) = (1/\sqrt{3})[z - (\pi/3)]$

15. $(\ln 5)(x - 1) + (\ln 5)(y - 1) + (z - \ln 5) = 0;$ $(x - 1)/(\ln 5) = (y - 1)/(\ln 5) = z - \ln 5$

17. $z = 1 + (x - 1) + 2(y - 1);$ $x - 1 = (y - 1)/2 = (z - 1)/(-1)$

19. $z = 1; \quad x = \pi/8, \quad y = \pi/20$

21. $z = -(\pi/4) - \frac{1}{4}(x + 2) - \frac{1}{4}(y - 2);$ $4(x + 2) = 4(y - 2) = z + \pi/4$

23. $z = 2 + 2\sqrt{3}(x - \pi/2) - 2\sqrt{3}(y - \pi/6);$ $(x - \pi/2)/2\sqrt{3} = (y - \pi/6)/(-2\sqrt{3}) = (z - 2)/(-1)$

25. $(0, 2, 4), (1, 1, 2)$

29. All tangent planes pass through the origin. (Note that the surface cannot pass through the origin since $f(y/x)$ is not defined when $x = 0$.)

33. $[(x - x_0)\mathbf{i} + (y - y_0)\mathbf{j}] \cdot \nabla F(x_0, y_0) = 0$

35. $3\mathbf{i} - 8\mathbf{j}; \quad 3(x - 1) - 8(y + 2) = 0$

37. $\mathbf{i} + \mathbf{j}; \quad (x - 1) + y = 0$

39. $2\mathbf{i} + 2\mathbf{j}; \quad (x - \pi/4) + y = 0$

41. $\sqrt{5}$

43. $1/\sqrt{2}$

Problems 9.10, page 583

1. $x + y$

3. $1 + x_1 - 4x_2 + x_3$

5. $2 + \frac{1}{4}(x - 2) + \frac{1}{4}(y - 2)$

7. $\frac{5}{2} + \frac{3}{4}(x - 2) - \frac{3}{2}(y - 1)$

9. $x^2 + y^2$

11. $xyz - (x^3y^3z^3/3!)$

13. $y + xy$

15. $\dfrac{e^2\sqrt{2}}{2} + \dfrac{e^2\sqrt{2}}{2}(x - 2) + \dfrac{e^2\sqrt{2}}{2}\left(y - \dfrac{\pi}{4}\right) + \dfrac{e^2\sqrt{2}}{4}(x - 2)^2$

$+ \dfrac{e^2\sqrt{2}}{2}(x - 2)\left(y - \dfrac{\pi}{4}\right) - \dfrac{e^2\sqrt{2}}{4}\left(y - \dfrac{\pi}{4}\right)^2$

19. b. It is still equal to f. This may be hard to see, however. For example, the second-degree Taylor polynomial of $f(x) = x^2$ around $x = 3$ is $9 + 6(x - 3) + (x - 3)^2 = x^2$.

21. $\dfrac{1}{5!}[f_{xxxxx}(x_0, y_0)(x - x_0)^5 + 5f_{xxxxy}(x_0, y_0)(x - x_0)^4(y - y_0)$

$+ 10f_{xxxyy}(x_0, y_0)(x - x_0)^3(y - y_0)^2$

$+ 10f_{xxyyy}(x_0, y_0)(x - x_0)^2(y - y_0)^3$

$+ 5f_{xyyyy}(x_0, y_0)(x - x_0)(y - y_0)^4$

$+ f_{yyyyy}(x_0, y_0)(y - y_0)^5]$

Review Exercises for Chapter 9, page 584

1. $y = 2x$

3. $x = [(y/2) + 3]^2$

5. $x^2 + y^2 = 1$

7. $x = y^3, \quad x \geq 0$

9. $2; \text{V}: (0, 0); \text{no H}$

11. $\sqrt{3}; \text{V}: (1, 0), (-1, 0); \text{H}: (0, 1), (0, -1)$

13. undefined; V: $(1, 0)$; no H

15. $-4/3\sqrt{3}; \text{V}: (3, 0), (-3, 0); \quad \text{H}: (0, 4), (0, -4)$

17. $1/\sqrt{3}$

19. π

21. $\mathbf{f}' = 2\mathbf{i} - 3t^2\mathbf{j}; \quad \mathbf{f}'' = -6t\mathbf{j}$

23. $\mathbf{f}' = (-5 \sin 5t)\mathbf{i} + (8 \cos 4t)\mathbf{j};$ $\mathbf{f}'' = (-25 \cos 5t)\mathbf{i} - (32 \sin 4t)\mathbf{j}$

25. $\mathbf{T} = (4/\sqrt{41})\mathbf{i} + (5/\sqrt{41})\mathbf{j}; \quad \mathbf{n} = (-5/\sqrt{41})\mathbf{i} + (4/\sqrt{41})\mathbf{j}$

27. $\mathbf{T} = (\sqrt{3}/2)\mathbf{i} - \frac{1}{2}\mathbf{j}; \quad \mathbf{n} = -\frac{1}{2}\mathbf{i} - (\sqrt{3}/2)\mathbf{j}$

29. $\mathbf{T} = (2/\sqrt{5})\mathbf{i} + (1/\sqrt{5})\mathbf{j}; \quad \mathbf{n} = -(1/\sqrt{5})\mathbf{i} + (2/\sqrt{5})\mathbf{j}$

31. $\pi/3$

33. 16

35. $\frac{1}{3}(29^{3/2} - 5^{3/2})$

37. $\mathbf{f} = \frac{3}{4}[(2s + 1)^{2/3} - 1]\mathbf{i} + [(2s + 1)^{2/3} - 1]^{3/2}\mathbf{j}$

39. $\mathbf{f} = 2 \cos(s/2)\mathbf{i} + 2 \sin(s/2)\mathbf{j}$

41. $f_{xx} = 0, \quad f_{xy} = f_{yx} = 3y^2, \quad f_{yy} = 6xy$

43. $f_{xx} = \dfrac{4y}{(x-y)^3}$; $f_{xy} = \dfrac{-2(x+y)}{(x-y)^3} = f_{yx}$; $f_{yy} = \dfrac{4x}{(x-y)^3}$

45. $\frac{1}{2}\mathbf{i} - \frac{1}{2}\mathbf{j}$

47. $-\frac{1}{2}\mathbf{i} + \mathbf{j}$

49. $\frac{1}{12}\mathbf{i} - \frac{1}{12}\mathbf{j} - \frac{1}{48}\mathbf{k}$

51. $-3\mathbf{j} - 4\mathbf{k}$

53. $(t^2 + 2t)/(1+t)^2\sqrt{1 - t^4/(1+t)^2}$

55. $2r/(r-s)^2$

57. $[\cos(e^{r+s} - e^{r-s})][e^{r+s} + e^{r-s}]$

59. $\partial w/\partial s = 4r^4 s^3 t^2 + 2r^4 st^{-2}$; $\partial w/\partial t = 2r^4 s^4 t - 2r^4 s^2 t^{-3}$

61. along $y = kx$ limit is $k/(k^2 - 1)$

63. Since $0 \le (x - y^2)^2 = x^2 - 2xy^2 + y^4$, $|2xy^2| \le x^2 + y^4$
so that $|4xy^3/(x^2 + y^4)| \le |4xy^3/2xy^2| = 2|y|$

65. $\kappa = \rho = 1$

67. $\kappa = 36/(97/2)^{3/2}$, $\rho = (97/2)^{3/2}/36$

69. $\kappa = 16/17^{3/2}$, $\rho = 17^{3/2}/16$

71. $\kappa = 2/17^{3/2}$, $\rho = 17^{3/2}/2$

73. $\kappa = \frac{3}{4}$, $\rho = \frac{4}{3}$

75. $a_T = 0$, $a_n = 2$

77. $a_T = (6t + 12t^3)/\sqrt{t^2 + t^4}$; $a_n = 6t^2/\sqrt{t^2 + t^4}$

79. $[1300(175,000)^2/(3600)^2] \cdot (1/65) \approx 47,261$ N

81. a. $(1500)(40) - (9.81)(2500) = 35,475$ N
    b. $(1500)(40) - (9.81)(1500) = 45,285$ N

83. $-3/\sqrt{2}$

85. $-1/(2\sqrt{13})$

87. $9/(\sqrt{14})(6^{3/2})$

89. $-1/|\mathbf{x}|^3$ is a potential function

91. $x + y + z = 3$; $x = y = z$

93. $3x - y + 5z = 15$; $(x+1)/3 = (y-2)/(-1) = (z-4)/5$

95. $-3x + 6y - 2z = 18$;

97. $1 - \dfrac{(x+2y)^2}{2!}$

Chapter 10

Problems 10.1, page 593

1. $\mathbf{x} = (t, 2t)$; $I = \int_0^2 9t^2\, dt = 24$

3. $\mathbf{x} = (t, 2t - 4)$; $I = \int_1^2 (2t^2 - 2t - 8)\, dt = -\frac{19}{3}$

5. $\mathbf{x} = (\cos\theta, \sin\theta)$; $I = \int_0^{2\pi} [(\cos\theta)(\sin\theta)(-\sin\theta) + (\sin\theta - \cos\theta)\cos\theta]\, d\theta = -\pi$

7. $\int_0^1 0\, dt + \int_0^1 (t-1)\, dt + \int_0^1 t^2\, dt = -\frac{5}{6}$

9. $\mathbf{x} = (3\cos\theta, -\sin\theta)$; $I = \int_{\pi/2}^{3\pi/2} [(9\cos^2\theta - 2\sin\theta) \times (-3\sin\theta) + (-\sin^2\theta)(-\cos\theta)]\, d\theta = 3\pi - \frac{2}{3}$

11. $\int_0^1 e^t\, dt + \int_0^1 (-e + e^{1-2t})\, dt + \int_1^0 e^{-1}\, dt = \frac{1}{2}(e + e^{-1}) - 2 = \cosh 1 - 2$

13. $\mathbf{x} = (2 + 2t, 1 + 5t)$;
$$I = \int_0^1 \left\{ \frac{1+5t}{[2(1+t)]^2} \cdot 2 + \frac{2(1+t)}{[1+5t]^2} \cdot 5 \right\} dt$$
$$= \frac{5}{2}\ln 2 + \frac{2}{5}\ln 6 + \frac{1}{3}$$

15. $W = \int_0^{\pi/2} [\sin^3 t \cos t + e^{2t} \sin t]\, dt = \frac{9}{20} + (2e^\pi/5) \approx 9.71$ J

17. $W = -\int_0^{2\pi} [(\cos\theta \sin\theta)(-\sin\theta) + (2\cos^3\theta - \sin\theta)(\cos\theta)]\, d\theta = -3\pi/2$ J

19. $W = \int_0^1 0\, dt + \int_0^1 t^2\, dt + \int_1^0 (2t^2 + t^2)\, dt = -\frac{2}{3}$ J

Problems 10.2, page 601

1. $x^2 y + y + C$

3. not conservative

5. not conservative

7. $(x^2/2) - y\sin x + C$

9. $(3x^2/2)\ln x - (3x^2/4) + (x^6/6) - xy + C$

11. $\tan x - x + \tan^{-1}(y/x) - e^y + C$

13. not conservative

15. $x + y + z + C$

17. $(xy/z) + (x^3/3) + \cos y + \sin z + C$

19. $\mathbf{F} = \nabla(x^2 y + y)$; $I = 14$

21. $\mathbf{F} = \nabla x\sin(x+y)$; $I = \pi/6$

23. $\mathbf{F} = \nabla x^2 \cos y$; $I = \pi^2/4$

25. $\mathbf{F} = \nabla xe^y$; $I = 5e^7$

Problems 10.3, page 608

1. $\frac{2}{3}$

3. $e^{-5} - e^{-1} - e^{-2} + e^2$

5. -31

7. $\frac{162}{5}$

9. $\frac{16}{3}$

11. $\frac{20}{3}$

13. $\frac{1}{2}(e^{19} - e^{17} - e^3 + e)$

15. $\frac{1}{3} + \pi/16$

17. $\int_0^{1/2} \int_x^{1-x} (x + 2y)\, dy\, dx = \frac{7}{24}$

19. $\int_0^{1/\sqrt{2}} \int_{x^2}^{1-x^2} (x^2 + y)\, dy\, dx = \sqrt{2}/5$

21. $\int_1^2 \int_1^y (y/\sqrt{x^2 + y^2})\, dx\, dy = \int_1^2 \int_x^2 (y/\sqrt{x^2 + y^2})\, dy\, dx = \int_1^2 (\sqrt{x^2 + 4} - x\sqrt{2})\, dx = 1/\sqrt{2} - \frac{1}{2}\sqrt{5} + 2\ln[(2 + 2\sqrt{2})/(1 + \sqrt{5})]$

23. $\int_0^\infty \int_0^\infty (x + y)e^{-(x+y)}\, dy\, dx = \int_0^\infty (1 + y)e^{-y}\, dy = 2$

25. a. $\int_{-5}^8 \int_0^4 (x + y)\, dx\, dy$
    b. 182
    c. region is a rectangle

27. a. $\int_0^1 \int_y^1 dx\, dy$

   b. $\frac{1}{2}$

   c. triangle with vertices at $(0, 0)$, $(1, 0)$ and $(1,1)$

29. a. $\int_0^1 \int_0^{\cos^{-1} x} y\, dy\, dx$

   b. $(\pi/2) - 1$

   c. region in first quadrant bounded by x-axis, y-axis, and curve $y = \cos^{-1} x$ (vertices are $(0, 0)$, $(\pi/2, 0)$, $(1, 0)$)

31. a. $\int_0^1 \int_{y^3}^{y^2} (1 + y^6)\, dx\, dy$

   b. $\frac{17}{180}$

   c. region bounded by $y = x^{1/2}$ and $y = x^{1/3}$ (the curves meet at $(0, 0)$, and $(1, 1)$)

33. a. $\int_0^\infty \int_0^y (1 + y^2)^{-7/5}\, dx\, dy$

   b. $\frac{5}{4}$

   c. region is the "triangular" part of the first quadrant above the line $y = x$

35. 36

37. 2π

39. $\pi(18 - 22\sqrt{5}/3)$

Problems 10.4, page 614

1. 2

3. $\int_0^1 \int_0^{1-y} 2e^x \sin y\, dx\, dy = \cos 1 - \sin 1 + e - 2$

5. $2(e^2 - 1)(1 - \cos 1)$

7. $\iint_{\text{disk}} (-4y)\, dx\, dy = 0$

9. $(\pi/8) - (\pi/3) + (\pi/3\sqrt{2}) = \pi[(4\sqrt{2} - 5)/24]$

11. $\frac{242}{5} - \frac{26}{3} = \frac{596}{15}$

13. $\iint_\Omega (b - a)\, dA = (b - a)(\text{area of } \Omega)$

15. 0 (conservative with $f = -2x^2/\sqrt{1 + y^2}$)

17. Use (14). Note that the line from (a_1, b_1) to (a_2, b_2) is parametrized as $\mathbf{x} = (a_1 + t(a_2 - a_1), b_1 + t(b_2 - b_1))$ for $0 \le t \le 1$. This yields area as $\frac{1}{2}|(a_1 + a_2)(b_2 - b_1) + (a_2 + a_3)(b_3 - b_2) + (a_3 + a_1)(b_1 - b_3)|$.

19. $\frac{1}{2}|(a_1 + a_2)(b_2 - b_1) + (a_2 + a_3)(b_3 - b_2) + (a_3 + a_4)(b_4 - b_3) + (a_4 + a_1)(b_1 - b_4)|$
(See answer to Problem 17.)

21. a. $2x - 2y$

   b. 0

   c. 0

   d. 0

23. a. $3(x^2 - y^2)$

   b. 0

   c. 0

   d. 0

25. a. -2

   b. -2π

   c. 0

   d. 0

27. curl $\mathbf{F} = 0$

29. curl $\mathbf{F}$ is an odd function

31. curl $\mathbf{F} = 0$

33. div $\mathbf{F} = 0$

35. c. $\mathbf{F}(0, 0)$ is undefined

Problems 10.5, page 631

1. $2\sqrt{6}$

3. $(\pi/2)\sqrt{1 + a^2 + b^2}$

5. $2 \int_0^1 \int_1^2 \sqrt{1 + \frac{4}{9}x^{-2/3}}\, dy\, dx = \frac{2}{27}(13^{3/2} - 8)$ (Note that $\int_{-1}^1 \int_1^2 \sqrt{1 + \frac{4}{9}x^{-2/3}}\, dy\, dx$ is *wrong* since the integrand is not defined at $x = 0$; this improper integral is treated by breaking it into two parts.)

7. 132

9. $\int_0^1 \int_0^2 \sqrt{1 + 9x}\, dy\, dx = \frac{4}{27}(10^{3/2} - 1)$ (This is the surface area for $z \ge -1$.)

11. $(5^{3/2} - 1)/6$

13. $(9\sqrt{5}/16) - (\frac{1}{32})\ln(2 + \sqrt{5})$

15. $\frac{14}{3}$

17. 0

19. $-7\sqrt{14}/108$

21. $76\sqrt{2}\pi$

23. $64 + 32\sqrt{21}/3$; note that a tetrahedron has four faces. The integral over the sloping face is $32\sqrt{21}/3$. The other three integrals are 0, $\frac{64}{3}$; and $\frac{128}{3}$.

25. 0

27. 54π

29. $\sqrt{3}/12$

33. $-\frac{32}{3}$

35. 2π

37. 0

39. $-\pi/3$

43. $\frac{419}{3360}$

Problems 10.6, page 638

1. div $\mathbf{F} = 2(x + y + z)$;   curl $\mathbf{F} = 0$

3. div $\mathbf{F} = 0$;   curl $\mathbf{F} = 0$

5. div $\mathbf{F} = x + y + z$;   curl $\mathbf{F} = -y\mathbf{i} - z\mathbf{j} - x\mathbf{k}$

7. div $\mathbf{F} = 0$;   curl $\mathbf{F} = x(e^{xy} - e^{xz})\mathbf{i} + y(e^{yz} - e^{xy})\mathbf{j} + z(e^{xz} - e^{yz})\mathbf{k}$

9. div $\mathbf{F} = (1/y) + (1/z) + (1/x)$;   curl $\mathbf{F} = (y/z^2)\mathbf{i} + (z/x^2)\mathbf{j} + (x/y^2)\mathbf{k}$

19. 0

21. 0

23. only the functions in Problems 19 and 21

29. b. $\mathbf{F} = \nabla(2x^2yz)$

Problems 10.7, page 644

1. -3π

3. -6

5. 108π

7. $21 - \sin 3$

11. the integral of $\mathbf{F}$ along the straight line segment from $(6, 0, 0)$ to $(0, 3, 0)$ is -18, from $(0, 3, 0)$ to $(0, 0, 2)$ is -4, from $(0, 0, 2)$ to $(6, 0, 0)$ is -24; the integral of $\mathbf{F}$ along the triangle is -46

13. the integral equals -18π

Problems 10.8, page 652

1. 4π

3. 36π

5. 0

7. 2

9. 108π

11. $\frac{1}{6}$

13. $\frac{1}{2}$

15. $\frac{184}{35}$

Problems 10.9, page 662

1. $[\partial(x, y)/\partial(u, v)] = -2$

3. $[\partial(x, y)/\partial(u, v)] = 4(u^2 + v^2)$

5. $[\partial(x, y)/\partial(u, v)] = -2$

7. $[\partial(x, y)/\partial(u, v)] = -(a^2 + b^2)$

9. $[\partial(x, y)/\partial(u, v)] = (1 - uv)e^{u+v}$

11. $[\partial(x, y)/\partial(u, v)] = [1/(u + v)][(1/v) - (1/u)]$

13. $[\partial(x, y)/\partial(u, v)] = \sec u \csc v(1 + uv \cot u \tan v)$

15. $[\partial(x, y, z)/\partial(u, v, w)] = 0$

17. $[\partial(x, y, z)/\partial(u, v, w)] = 2(u^2v - uv^2 + v^2w - vw^2 + w^2u - wu^2)$

19. $[\partial(x, y, z)/\partial(u, v, w)] = e^{u+v+w}$

21. $\int_0^1 \int_y^1 xy \, dx \, dy = \int_{-1/2}^0 \int_{-v}^{1+v} (u^2 - v^2)(2) \, du \, dv = \frac{1}{8}$

23. 2π

25. $\frac{128}{15}$

27. $4\pi abc/3$

Review Exercises for Chapter 10, page 662

1. $3(x + y)^2(\mathbf{i} + \mathbf{j})$

3. $\frac{1}{2}\sqrt{y/x}\mathbf{i} + \frac{1}{2}\sqrt{x/y}\mathbf{j}$

5. $2x\mathbf{i} + 2y\mathbf{j} + 2z\mathbf{k}$

9. $\mathbf{F} = \nabla((x^3/3) + (y^3/3));\quad I = \frac{2}{3}$

11. $-\frac{1}{2}$

13. $\frac{5}{3}$

15. a. $\mathbf{F} = \nabla(xe^{xy})$

b. 1

17. $\int_0^1 \int_x^1 (y^2 - x^2) \, dy \, dx = \frac{1}{6}$ (using Green's theorem)

19. $\int_{-1}^1 \int_{x^2}^1 (x/\sqrt{1+x^2}) \, dy \, dx = 0.$

21. a. $\operatorname{curl} \mathbf{F} = 0$

b. 0

c. $\operatorname{div} \mathbf{F} = y^2 + x^2$

d. 8π

23. $\operatorname{curl}[(\cos x^2)\mathbf{i} + e^y\mathbf{j}] = 0$

25. $\frac{3}{2}$

27. $-\frac{5}{3}$

29. $(\frac{91}{48})\sqrt{5} + (\frac{61}{96})\ln(2 + \sqrt{5})$

31. 0

33. $567\sqrt{2}\pi/4$

35. $\alpha\pi^2/2$ (α is proportionality constant)

37. 4π

39. π

41. $\operatorname{div} \mathbf{F} = 3;\quad \operatorname{curl} \mathbf{F} = \mathbf{i} + \mathbf{j} + \mathbf{k}$

43. $\operatorname{div} \mathbf{F} = (1/x) + (1/y) + (1/z);\quad \operatorname{curl} \mathbf{F} = \mathbf{0}$

45. $\operatorname{div} \mathbf{F} = -\sin z;\ \operatorname{curl} \mathbf{F} = (-\sin x + \sin y)\mathbf{k}$

47. 0

49. 0

51. $4(a + b + c)\pi/3$

53. 0

55. $\frac{5}{4}$ if the unit cube is $(0, 1)^3$; 0 if it is $(-\frac{1}{2}, \frac{1}{2})^3$

57. $[\partial(x, y)/\partial(u, v)] = 18u^2v^2$

59. $[\partial(x, y)/\partial(u, v)] = (uv - 1)e^{u+v}$

61. $[\partial(x, y)/\partial(u, v)] = v[\sec^2(uv)][u \sec^2 u - \tan u]$

63. $[\partial(x, y, z)/\partial(u, v, w)] = 2uvw$

65. $\int_2^3 \int_{-v}^v e^{(u/v)} \frac{1}{2} \, du \, dv = \frac{5}{4}(e - e^{-1}) = \frac{5}{2}\sinh 1$

Chapter 11

Problems 11.1, page 670

1. eigenvalues: $\left(\dfrac{n\pi}{T}\right)^2$;

eigenfunctions: $A \sin \dfrac{n\pi t}{T}, \quad n = 1, 2, 3, \ldots$

3. eigenvalues: $-\left(\dfrac{n\pi}{T}\right)^2$;

eigenfunctions: $A \sin \dfrac{n\pi t}{T}, \quad n = 1, 2, 3, \ldots$

5. eigenvalues: $\left(\dfrac{n\pi}{2T}\right)^2$;

eigenfunctions: $A \sin \dfrac{n\pi(t + T)}{2T}, \quad n = 1, 2, 3, \ldots$

7. eigenvalues: $\left[\left(n+\dfrac{1}{2}\right)\dfrac{\pi}{T}\right]^2$; eigenfunctions:

$A \sin\left(n+\dfrac{1}{2}\right)\dfrac{\pi t}{T}$, $n = 0, 1, 2, 3, \ldots$

9. eigenvalues: $(n\pi)^2$; eigenfunctions: $A \cos n\pi t$,
$n = 0, 1, 2, 3, \ldots$

11. eigenvalues: all values of λ such that if $\lambda > \frac{1}{4}$ and $\beta = \dfrac{\sqrt{4\lambda-1}}{2}$, then β satisfies $2\beta + \tan \beta = 0$; eigenfunctions: $Ax^{-1/2}[2\beta \cos(\beta \ln|x|) + \sin(\beta \ln|x|)]$.

Problems 11.2, page 674

1. $\dfrac{1}{\sqrt{2\pi}}, \dfrac{\cos x}{\sqrt{\pi}}, \dfrac{\cos 2x}{\sqrt{\pi}}, \ldots$

3. $\dfrac{1}{\sqrt{T}}, \sqrt{\dfrac{2}{T}} \cos \dfrac{2\pi x}{T}, \sqrt{\dfrac{2}{T}} \cos \dfrac{4\pi x}{T}, \ldots$

5. $\sqrt{\dfrac{1}{\pi}}, \sqrt{\dfrac{2}{\pi}} \cos 2x, \sqrt{\dfrac{2}{\pi}} \cos 4x, \ldots$

7. $\dfrac{1}{\sqrt{2\pi}}, \dfrac{1}{\sqrt{\pi}} \cos 3x, \dfrac{1}{\sqrt{\pi}} \cos 6x, \dfrac{1}{\sqrt{\pi}} \cos 9x, \ldots$

9. b. $H'_n(x) = \dfrac{d}{dx}\left[(-1)^n e^{x^2} \dfrac{d^n}{dx^n} e^{-x^2}\right]$

$= (-1)^n 2x e^{x^2} \dfrac{d^n}{dx^n} e^{x^2} + (-1)^n e^{x^2} \dfrac{d^{n+1}}{dx^{n+1}} e^{-x^2}$

$= 2x H_n(x) - H_{n+1}(x)$.

c. Show $H_{n+1}(x) = 2x H_n(x) - 2n H_{n-1}(x)$ by using equation (5.5.11). Then $H'_n = 2n H_{n-1}$ and $H''_n = 2n H'_{n-1} = 2n[2x H_{n-1} - H_n] = 2x H'_n - 2n H_n$.

d. $\displaystyle\int_{-\infty}^{\infty} e^{-x^2} H_n(x)\, dx = (-1)^n \dfrac{d^{n-1}}{dx^n} (e^{-x^2})\Big|_{-\infty}^{\infty} = -e^{x^2} H_{n-1}(x)|_{-\infty}^{\infty} = 0$, since H_{n-1} is a polynomial.

Problems 11.3, page 688

1. $\pi, 2, T/n, 1/k$

3. bT/a

5. $2(\sin x - \frac{1}{2} \sin 2x + \frac{1}{3} \sin 3x - \frac{1}{4} \sin 4x + \cdots)$

7. $\dfrac{\pi^2}{3} - 4(\cos x - \frac{1}{4} \cos 2x + \frac{1}{9} \cos 3x - \frac{1}{16} \cos 4x + \cdots)$

9. $\dfrac{\pi}{2} - \dfrac{4}{\pi}\left[\cos x + \dfrac{1}{3^2} \cos 3x + \dfrac{1}{5^2} \cos 5x + \cdots\right]$

11. $\dfrac{2}{\pi}[(1 + \sin(\pi - 1))\sin x - \frac{1}{2}(1 + \frac{1}{2} \sin 2(\pi - 1))\sin 2x + \frac{1}{3}(1 + \frac{1}{3} \sin 3(\pi - 1))\sin 3x - \cdots]$

13. $\dfrac{\sinh \pi}{\pi}\left[1 + 2 \sum_{n=1}^{\infty} \dfrac{(-1)^n}{1+n^2} \cos nx - 2 \sum_{n=1}^{\infty} \dfrac{n(-1)^n}{1+n^2} \sin nx\right]$

15. $1 - \dfrac{2}{\pi} \sum_{n=1}^{\infty} \dfrac{\sin n\pi x}{n}$

17. $\dfrac{1}{3} + \dfrac{4}{\pi^2} \sum_{n=1}^{\infty} \dfrac{(-1)^n}{n^2} \cos n\pi x$

19. $\dfrac{1}{2} - \dfrac{2}{\pi}\left(\sin \pi x + \dfrac{1}{3} \sin 3\pi x + \dfrac{1}{5} \sin 5\pi x + \cdots\right)$

21. $\dfrac{3}{4} - \dfrac{1}{\pi} \sum_{n=1}^{\infty} \dfrac{1}{n} \sin n\pi x$

$- \dfrac{2}{\pi^2}\left(\cos \pi x + \dfrac{1}{3^2} \cos 3\pi x + \dfrac{1}{5^2} \cos 5\pi x + \cdots\right)$

Problems 11.4, page 696

1. even

3. even

5. odd

7. neither

9. neither

11. $\dfrac{\pi^2}{3} - 4(\cos x - \frac{1}{4} \cos 2x + \frac{1}{9} \cos 3x - \frac{1}{16} \cos 4x + \cdots)$

13. $\dfrac{\pi^3}{4} - 6\pi\left(\cos x - \dfrac{\cos 2x}{4} + \dfrac{\cos 3x}{9} - \dfrac{\cos 4x}{16} + \cdots\right)$

$+ \dfrac{24}{\pi}\left(\cos x + \dfrac{\cos 3x}{3^4} + \dfrac{\cos 5x}{5^4} + \cdots\right)$

15. $\dfrac{T}{2} + \dfrac{4T}{\pi^2}\left(\cos \dfrac{\pi x}{T} + \dfrac{1}{9} \cos \dfrac{3\pi x}{T} + \dfrac{1}{25} \cos \dfrac{5\pi x}{T} + \cdots\right)$

17. $\dfrac{2T}{\pi}\left(\sin \dfrac{\pi x}{T} + \dfrac{1}{2} \sin \dfrac{2\pi x}{T} + \dfrac{1}{3} \sin \dfrac{3\pi x}{T} + \dfrac{1}{4} \sin \dfrac{4\pi x}{T} + \cdots\right)$

19. $\dfrac{1}{3} + \dfrac{4}{\pi^2}(\cos \pi x + \frac{1}{4} \cos 2\pi x + \frac{1}{9} \cos 3\pi x + \frac{1}{16} \cos 4\pi x + \cdots)$

21. even: k

odd: $\dfrac{4k}{\pi}(\sin \pi x + \frac{1}{3} \sin 3\pi x + \frac{1}{5} \sin 5\pi x + \cdots)$

23. even: $\dfrac{1}{4} - \dfrac{6}{\pi^2}(\cos \pi x - \frac{1}{4} \cos 2\pi x + \frac{1}{9} \cos 3\pi x - \frac{1}{16} \cos 4\pi x + \cdots) + \dfrac{24}{\pi^4}\left(\cos \pi x + \dfrac{\cos 3\pi x}{3^4} + \dfrac{\cos 5\pi x}{5^4} + \cdots\right)$

odd: $\dfrac{2}{\pi}[\sin \pi x - \frac{1}{2} \sin 2\pi x + \frac{1}{3} \sin 3\pi x$

$$-\tfrac{1}{4}\sin 4\pi x + \cdots) - \frac{12}{\pi^3}\left[\sin \pi x - \frac{\sin 2\pi x}{2^3} + \frac{\sin 3\pi x}{3^3}\right.$$

$$\left. - \frac{\sin 4\pi x}{4^3} + \cdots\right]$$

25. even: $\dfrac{1}{2} + \dfrac{2}{\pi}\left(\cos\dfrac{\pi x}{2} - \dfrac{1}{3}\cos\dfrac{3\pi x}{2}\right.$

$$+ \frac{1}{5}\cos\frac{5\pi x}{2} - \frac{1}{7}\cos\frac{7\pi x}{2} + \cdots\bigg)$$

odd: $\dfrac{2}{\pi}\left(\sin\dfrac{\pi x}{2} + \sin \pi x + \dfrac{1}{3}\sin\dfrac{3\pi x}{2} + \dfrac{1}{5}\sin\dfrac{5\pi x}{2}\right.$

$$+ \frac{1}{3}\sin 3\pi x + \frac{1}{7}\sin\frac{7\pi x}{2} + \frac{1}{9}\sin\frac{9\pi x}{2}$$

$$+ \frac{1}{5}\sin 5\pi x + \cdots\bigg)$$

27. even: $\dfrac{3}{4} + \dfrac{4}{\pi^2}\left(\cos\dfrac{\pi x}{2} - \dfrac{2}{2^2}\cos \pi x + \dfrac{1}{3^2}\cos\dfrac{3\pi x}{2}\right.$

$$+ \frac{1}{5^2}\cos\frac{5\pi x}{2} - \frac{2}{6^2}\cos 3\pi x + \frac{1}{7^2}\cos\frac{7\pi x}{2}$$

$$+ \frac{1}{9^2}\cos\frac{9\pi x}{2} - \frac{2}{10^2}\cos 5\pi x + \cdots\bigg)$$

odd: $\dfrac{2}{\pi}\left(\sin\dfrac{\pi x}{2} + \dfrac{1}{2}\sin \pi x + \dfrac{1}{3}\sin\dfrac{3\pi x}{2}\right.$

$$+ \frac{1}{4}\sin 2\pi x + \frac{1}{5}\sin\frac{5\pi x}{2} + \cdots\bigg)$$

$$+ \frac{4}{\pi^2}\left(\sin\frac{\pi x}{2} - \frac{1}{3^2}\sin\frac{3\pi x}{2} + \frac{1}{5^2}\sin\frac{5\pi x}{2} - \cdots\right)$$

29. even:

$$\frac{2}{3} - \frac{16}{\pi^3}\left(\cos\frac{\pi x}{2} - \frac{1}{3^3}\cos\frac{3\pi x}{2} + \frac{1}{5^3}\cos\frac{5\pi x}{2} - \cdots\right)$$

$$- \frac{8}{\pi^2}\left(\frac{1}{2^2}\cos \pi x - \frac{1}{4^2}\cos 2\pi x + \frac{1}{6^2}\cos 3\pi x - \cdots\right)$$

odd: $\dfrac{2}{\pi}\left(\sin\dfrac{\pi x}{2} - \dfrac{1}{2}\sin \pi x + \dfrac{1}{3}\sin\dfrac{3\pi x}{2} - \dfrac{1}{4}\sin 2\pi x\right.$

$$+ \frac{1}{5}\sin\frac{5\pi x}{2} - \cdots\bigg) + \frac{8}{\pi^2}\left(\sin\frac{\pi x}{2} - \frac{1}{3^2}\sin\frac{3\pi x}{2}\right.$

$$+ \frac{1}{5^2}\sin\frac{5\pi x}{2} - \cdots\bigg) - \frac{16}{\pi^3}\left(\sin\frac{\pi x}{2} + \frac{2}{2^3}\sin \pi x\right.$

$$+ \frac{1}{3^3}\sin\frac{3\pi x}{2} + \frac{1}{5^3}\sin\frac{5\pi x}{2} + \frac{2}{6^3}\sin 3\pi x + \frac{1}{7^3}\sin\frac{7\pi x}{2}$$

$$+ \frac{1}{9^3}\sin\frac{9\pi x}{2} + \cdots\bigg)$$

Problems 11.5, page 702

1. $\dfrac{2}{\pi}\displaystyle\int_0^\infty (\sin t - t\cos t)\sin tx\,dt$

3. $\dfrac{2}{\pi}\displaystyle\int_0^\infty \dfrac{(1-\cos t)}{t}\sin tx\,dt$

5. $\displaystyle\int_0^\infty e^{-t}\cos tx\,dt$

7. $\dfrac{2}{\pi}\displaystyle\int_0^\infty \dfrac{\sin \pi t}{1-t^2}\sin tx\,dt$

9. $\dfrac{2}{\pi}\displaystyle\int_0^\infty \left[\dfrac{\sin tT\cosh T - 2t\cos tT\sinh T}{1+t^2}\right]\sin tx\,dt$

11. $\dfrac{2}{\pi}\displaystyle\int_0^\infty \left[\dfrac{\cos t - \cos 2t}{t^2}\right]\cos tx\,dt$

17. $F(u)\sin tu$ is odd; apply Theorem 11.4.1 (c).

19. The integral $\int_0^\infty \cos tu\,du$ diverges

Problems 11.6, page 707

1. $\lambda_k = k^2$, $y_k = A\sin kx$, $k = 1, 2, 3, \ldots$

3. $\lambda_k = k^2$, $y_k = A\sin k\left(x + \dfrac{\pi}{2}\right)$, $k = 1, 2, 3, \ldots$

5. $\lambda_k = \dfrac{k^2\pi^2}{4}$, $y_k = A\sin\left(\dfrac{k\pi \ln x}{2}\right)$, $k = 1, 2, 3, \ldots$

7. $\lambda_k = k^2\pi^2$, $y_k = A\sin\dfrac{k\pi}{x}$, $k = 1, 2, 3, \ldots$

Problems 11.7, page 711

1. $\lambda_k = k^2$ are real and simple, $y_k = A\sin kx$, $k = 1, 2, 3, \ldots$, are orthogonal.

3. $\lambda_k = k^2$ are real and simple, $y_k = A\sin k\left(x + \dfrac{\pi}{2}\right)$, $k = 1, 2, 3, \ldots$, are orthogonal.

5. $\lambda_k = k^2\pi^2$ are real and simple, $y_k = A\sin\dfrac{k\pi}{x}$, $k = 1, 2, 3, \ldots$, are orthogonal with respect to $1/x^2$ on $1/2 \le x \le 1$.

7. b. Let $r(x) = q(x) = e^{-x^2}$ and note that $\lim_{|x|\to\infty} r(x) = 0$. Then apply the remark on p. 709.

9. The eigenvalues of this Euler equation are $1 \pm \dfrac{2k\pi}{\ln 2}\,i$, $k = 1, 2, 3, \ldots$

Problems 11.8, page 717

1. No solution since zero is an eigenvalue and $\int_0^\pi X \times \sin X\,dx = \pi \ne 0$.

3. $4 \sum_{k=1}^{\infty} \dfrac{\left[1+(-1)^k\left(\dfrac{2k-1}{2}\pi\right)\right]}{\left(\dfrac{2k-1}{2}\pi\right)^3\left[\left(\dfrac{2k-1}{4}\pi\right)^2-1\right]} \sin\left(\dfrac{2k-1}{2}\right)\pi x$

5. $2(\ln 2)^2 \sin(\ln 2)\pi$

$\times \sum_{k=1}^{\infty} \dfrac{(-1)^k k}{[3(\ln 2)^2-(k\pi)^2][(\ln 2)^2-(k\pi)^2]} \sin\left(\dfrac{k\pi}{\ln 2}\ln x\right)$

13. $K(x, t) = \begin{cases} -\dfrac{\sinh(1-t)\sinh x}{\sinh(1)}, & x<t, \\[2mm] -\dfrac{\sinh(1-x)\sinh t}{\sinh(1)}, & t<x \end{cases}$

17. $K(x, t) = \begin{cases} -\dfrac{\sin 2x \cos 2(1-t)}{2\cos 2}, & x<t, \\[2mm] -\dfrac{\cos 2(1-x)\sin 2t}{2\cos 2}, & t<x \end{cases}$

Problems 11.9, page 722

1. $a_0 = 2, \quad a_1 = a_2 = 0$

3. $a_1 = 2/\pi, \quad a_2 = -1/\pi, \quad a_3 = 2/(3\pi)$.

5. $a_0 = 2/3, \quad a_n = \dfrac{4(-1)^n}{n^2\pi^2}, \quad n = 1, 2, 3$.

7. $\frac{1}{5} P_0 + \frac{4}{7} P_2 + \frac{8}{35} P_4$

9. $\frac{18}{5} P_3 - \frac{16}{3} P_2 + \frac{62}{5} P_1 - \frac{26}{3} P_0$

11. $\sin x \approx 3(\sin 1 - \cos 1)x$

13. $\sinh x \approx 3e^{-1}x$

15. $\int_{-1}^{1} x^k p(x)\, dx = 0$, for $k = 0, 1, \ldots, n-1$ implies $p(x)$ is orthogonal to $P_0, P_1, \ldots, P_{n-1}$. Since deg $p = n$, p must be a multiple of P_n.

17. a. Integrate by parts with $u = P_m$ and

$v = \dfrac{d^{n-1}}{dx^{n-1}}[(x^2-1)^n]$.

b. $x^2 = 1$ is a factor of v.

19. The nth degree term of P_n has the coefficient $(2n)!/2^n(n!)^2$. Thus $P_n^{(n)} = (2n)!/2^n n!$ so that $\|P_n\|^2 =$
$\dfrac{(2n)!}{2^{2n}(n!)^2} \int_{-1}^{1} (1-x^2)^n\, dx$. Substituting $x = \sin\theta$, we get
by entry 218 $\|P_n\|^2 = \dfrac{(2n)!}{2^{2n}(n!)^2} \int_{-\pi/2}^{\pi/2} \cos^{2n+1}\theta\, d\theta = \dfrac{2}{2n+1}$.

Review Exercises for Chapter 11, p. 723

1. $\left\{ \sqrt{\dfrac{2}{\pi}} \sin nx \right\}$

3. $\{\sqrt{2} \sin 2n\pi x, \ \sqrt{2} \cos 2\, n\pi x\}$

5. $-1 + \pi \sin x - \dfrac{1}{2} \cos x + 2 \sum_{n=2}^{\infty} \dfrac{\cos nx}{n^2-1}$

7. $\dfrac{2}{\pi}\left\{1 - 2 \sum_{n=1}^{\infty} \dfrac{\cos 2nx}{4n^2-1}\right\}$

9. $2\pi \sum_{n=1}^{\infty} \left\{\left(\dfrac{1}{2n-1} - \dfrac{4}{(2n-1)^{2n-1}\pi^2}\right)\sin(2n-1)x - \dfrac{1}{2^n} \sin 2nx\right\}; \dfrac{\pi^2}{3} + 4 \sum_{n=1}^{\infty} \dfrac{(-1)}{n^2} \cos nx$

11. $\dfrac{2}{\pi} \sum_{n=1}^{\infty} \dfrac{(-1)^{n-1}}{n} \sin n\pi x; \dfrac{-4}{\pi^2} \sum_{n=1}^{\infty} \dfrac{1}{(2n-1)^2} \cos(2n-1)\pi x$

13. $\dfrac{-32}{\pi^3}\left(\sin\dfrac{\pi x}{2} + \dfrac{\pi^2}{8} \sin\dfrac{2\pi x}{2} + \dfrac{1}{3^3} \sin\dfrac{3\pi x}{2} + \dfrac{\pi^2}{16} \sin\dfrac{4\pi x}{2} + \cdots\right);$

$\pi^3 = 32\left(1 - \dfrac{1}{3^3} + \dfrac{1}{5^3} - \dfrac{1}{7^3} + \cdots\right)$

15. $\lambda_n = \dfrac{(2n-1)^2}{4}, \quad y_n(x) = \cos\dfrac{(2n-1)x}{2}, n = 1, 2, 3, \ldots$

17. $y_n(x) = \sin\sqrt{\lambda_n} x$, where λ_n are the roots of $\tan k = -k$. For large n, $\sqrt{\lambda_n} \approx \dfrac{(2n-1)\pi}{2}$.

19. Same as Exercise 17.

21. $(\sqrt{1-x^2}\, y')' + \dfrac{n^2}{\sqrt{1-x^2}} y = 0$ with $q(x) = 1/\sqrt{1-x^2}$ on $[-1, 1]$.

23. $y = \sum_{n=0}^{\infty} c_n \cos\dfrac{(2n+1)\pi x}{2}$ where

$c_n = \dfrac{32}{(2n+1)^2\pi^2+4}\left[\dfrac{2}{(2n+1)^2\pi^2} - \dfrac{(-1)^n}{(2n+1)\pi}\right]$

Chapter 12

Problems 12.1, page 732

1. $z(x, y) = (x + y)e^{\sqrt{2}y}$

3. $z(x, y) = (x - y)e^{-\sqrt{2}y} + e^{x-(1+\sqrt{2})y}$

5. $z(x, y) = \frac{7}{2} + e^{x+(5/4)y} - \frac{7}{2} e^{y/2}$

7. $z(x, y) = ye^x$

9. $z(x, y) = (x + y)e^{\sqrt{2}y}$

11. $z(x, y) = (x - y)e^{-\sqrt{2}y} + e^{x-(1+\sqrt{2})y}$

13. $z(x, y) = \frac{7}{2} + e^{x+(5/4)y} - \frac{7}{2} e^{y/2}$

15. $z(x, y) = ye^x$

17. $z(x, y) = \ln x + xe^{-y}$

19. $z(x, y) = y^2(x - \ln y)e^{-x}$

21. $z(x, y) = \sin(x - y^2)e^{y^2} - 2$

23. $dx = dy = \dfrac{dz}{z}$ so $x + c_1 = y$ and $z = c_2 e^x$. If $c_2 = f(c_1)$, then $ze^{-x} = f(y - x)$ or $z(x, y) = e^x f(y - x)$. Setting $x = y$, we get $e^y = z(y, y) = e^y f(0)$, implying that $f(0) = 1$. Hence, any such function f will do. For example, $z(x, y) = e^x(y - x + 1)$ is a solution.

25. $z - x = f(x - y)$ so $f(0) = 7$. Hence $z = x + f(x - y)$ for any differentiable function satisfying $f(0) = 7$ will do. For example $z(x, y) = (n + 1)x - ny + 7$, for all n.

Problems 12.2, page 737

1. $z(x, y) = f(x - y)$ (for any differentiable function f)

3. $z(x, y) = f(xe^{-y})$

5. $z^2 - 2y = f(xe^{-z})$

7. $z = \cos(x - y/z^2)$

9. $z = \left(1 - \dfrac{1}{x} + \dfrac{2}{y}\right)^{-1}$

11. $z = y\sqrt{3(z^2 - x^2)} - x$

13. $z = \dfrac{x}{1 + y}$

15. $\dfrac{1}{z} = f\left(\dfrac{x}{y}\right) - \ln y$

17. $z = \dfrac{y^2(1 + e^{x/y}) - (1 - e^{x/y})}{y^2(1 + e^{x/y}) + (1 - e^{x/y})}$

19. $y^2 - 2xz + \dfrac{2}{3}x^3 = f\left(\dfrac{x^2}{2} - z\right)$

21. $z = f(x - (1 + \frac{3}{2}z)y)$

Problems 12.3, page 742

1. no; $u = \dfrac{x}{1 + t}$

3. $u = \dfrac{1 + 2tx - \sqrt{1 + 4tx}}{2t^2}$

$= \dfrac{1 + 2tx - (1 + 2tx - 2t^2x^2 + \cdots)}{2t^2} \to x^2$

as $t \to 0$ so no shocks develop.

5. The shock (x_s, t_s) lies on the line $x - tf(x_1) = x_1$ where $f'(x - tu) = f'(x_1)$. Hence $t_s = 1/f'(x_1)$ and $x_s = x_1 - f(x_1)/f'(x_1)$.

7. $u = f(x - F'(u)t)$ so $u_x = f'/(1 + tf'F'')$ and shocks develop whenever $1 + tf'(x - F'(u)t)F''(u) = 0$.

Problems 12.4, page 747

1. hyperbolic

3. parabolic

5. elliptic

7. elliptic

9. elliptic

11. $\dfrac{\partial^2 z}{\partial x^2} + \dfrac{\partial^2 z}{\partial y^2} - z = 0$

13. $\dfrac{\partial^2 z}{\partial x^2} - \dfrac{\partial^2 z}{\partial y^2} + \dfrac{57}{4}z = 0$

15. $\dfrac{\partial^2 z}{\partial x^2} + \dfrac{z}{2} = 0$

17. $\dfrac{\partial^2 z}{\partial x^2} + \dfrac{\partial^2 z}{\partial y^2} + 5z = 0$

19. $\dfrac{\partial^2 z}{\partial x^2} - \dfrac{\partial^2 z}{\partial y^2} - \dfrac{1}{12}z = 0$

21. Multiply out $-AC$ to get $\left[b^2 - \dfrac{(a - c)^2}{4}\right]\sin^2 2\alpha + b(a - c)\sin 2\alpha \cos 2\alpha - ac$. Then substitute $a - c = 2b/\tan 2\alpha$ to obtain $b^2 - ac$.

23. Elliptic in second and fourth quadrants, hyperbolic in first and third quadrants, and parabolic on axes.

25. Elliptic for $|y| < |x|$, hyperbolic for $|y| > |x|$, and parabolic for $|y| = |x|$.

27. Elliptic for $y > e^{-x}$, hyperbolic for $y < e^{-x}$, and parabolic for $y = e^{-x}$.

Problems 12.5, page 754

1. -0.0075 m

3. 0.0 m

5. 0.1 m

7. 0.2 m

9. By (14) $y(x, t) = h(x + ct) + g(x - ct)$. Since $y(x, 0) = 0$ we get $h(x) = -g(x)$. Thus $f'(x) = y_t(x, 0) = c[h'(x + ct) + h'(x - ct)]|_{t=0} = 2ch'(x)$. Thus $h(x) = \dfrac{1}{2c}[f(x) + k]$ from which the result follows. Furthermore, since $y(0, t) = 0$, it follows that f is even with period $2L$.

11. $-\dfrac{0.045}{\pi}$ m

13. $\dfrac{0.03}{\pi}\left[\dfrac{\pi}{6} - \dfrac{1}{4}\right]$ m

15. 0.002 m

17. $y(x, t) = \sin\dfrac{\pi x}{2}\left[\cos\dfrac{\pi t}{2} + \dfrac{2}{\pi}\sin\dfrac{\pi t}{2}\right]$

19. If $RC = LG$, let $v(x, t) = u(x, t)e^{-Rt/L}$. Then u satisfies the wave equation $u_{tt} = (1/LC)u_{xx}$, so that v has the form

$$v(x, t) = e^{-Rt/L}\left[g\left(x - \dfrac{1}{\sqrt{LC}}t\right) + h\left(x + \dfrac{1}{\sqrt{LC}}t\right)\right].$$

A much more complicated solution involving integral equations exists when $RC \neq LG$ (see Zachmanoglou and Thoe, *Introduction to Partial Differential Equations*, Williams & Wilkins, Baltimore, MD (1976), p. 375).

Problems 12.6, page 759

1. $y(x, t) = 0.1\cos\dfrac{\pi ct}{L}\sin\dfrac{\pi x}{L}$

3. $y(x, t) = \dfrac{0.8L^2}{\pi^3} \displaystyle\sum_{n=0}^{\infty} \dfrac{1}{(2n+1)^3}$

$\times \cos \dfrac{(2n+1)\pi ct}{L} \sin \dfrac{(2n+1)\pi x}{L}$

5. $y(x, t) = \dfrac{2.4}{\pi^2} \displaystyle\sum_{n=1}^{\infty} \dfrac{1}{n^2} \sin\!\left(\dfrac{n\pi}{2}\right)\cos\!\left(\dfrac{n\pi ct}{6}\right)\sin\!\left(\dfrac{n\pi x}{6}\right)$

7. $T' = kT$ and $X'' + kX = 0$

9. $X' + X = 0$ or $T' = 0$; hence $y(x, t) = c_1 te^{-x} + c_2$

11. $Y'' - kyY = 0$ and $xX' + kX = 0$

13. $\Theta'' + k\Theta = 0$ and $r^2 R'' + rR' - kR = 0$

15. a. $y(x, t) = \sqrt{x}(b_1 e^{\sqrt{k}t} + b_2 e^{-\sqrt{k}t})$

$\times (c_1 x^{\sqrt{1-4k}/2} + c_2 x^{-\sqrt{1-4k}/2})$, all k

b. $y(x, t) = c_1 \sqrt{x}\!\left(x^{\sqrt{1-4k}/2} - \dfrac{\pi^{\sqrt{1-4k}}}{x^{\sqrt{1-4k}/2}}\right)$

$\times (b_1 e^{\sqrt{k}t} + b_2 e^{-\sqrt{k}t}), \quad k > 0$

17. a. $y(x, t) = (b_1 e^{\sqrt{k}t} + b_2 e^{-\sqrt{k}t})$

$\times (c_1 \cos \sqrt{k}x + c_2 \sin \sqrt{k}x)$, all k

b. $c_1 = 0$ and $k = 1^2, 2^2, 3^2, \ldots$, so

$$y(x, t) = \sin nx(b_1 e^{nt} + b_2 e^{-nt})$$

c. Use a Fourier sine series.

19. a. $y(x, t) = (b_1 e^{\sqrt{k}x} + b_2 e^{-\sqrt{k}x})(c_1 e^{\sqrt{k-1}t} + c_2 e^{-\sqrt{k-1}t})$

b. Let $k < 0$, then $b_1 = 0$ and $k = -1^2, -2^2, \ldots$, so that

$y(x, t) = \sin nx(c_1 \cos \sqrt{n^2 + 1}t + c_2 \sin \sqrt{n^2 + 1}t)$.

c. Use a Fourier sine series.

Problems 12.7, page 764

1. $T(x, t) = \dfrac{4}{\pi} \displaystyle\sum_{n=1}^{\infty} \dfrac{[1 - (-1)^n]}{n^3} (\sin nx)e^{-n^2\delta t}$

3. $T(x, t) = 4 \displaystyle\sum_{n=1}^{\infty} \dfrac{[2(-1)^{n+1} - 1]}{n^3} (\sin nx)e^{-n^2\delta t}$

5. $T(x, t) = \dfrac{2}{\pi} \displaystyle\sum_{n=1}^{\infty} \dfrac{1}{n^2}\!\left[\sin \dfrac{n\pi}{2} - \dfrac{n\pi}{2} \cos \dfrac{n\pi}{2}\right]\!(\sin nx)e^{-n^2\delta t}$

7. $[(L - x)T_1 + xT_2]/L$

9. $f(x) = \dfrac{c_0^2}{T_1} \cos cx + c_0 c_2 \sin cx$, where

$c_2 = \dfrac{c_0(1 - \cos cL)}{T \sin cL}$. If cL is a multiple of π, then c_2 is arbitrary.

Problems 12.8, page 768

1. $T(x, y) = \dfrac{100}{\pi} \displaystyle\sum_{n=1}^{\infty} \dfrac{[1 - (-1)^n]}{n \sinh n\pi} \sin \dfrac{n\pi x}{L} \sinh \dfrac{n\pi y}{L}$

Evaluate at $\left(\dfrac{L}{2}, \dfrac{L}{2}\right)$ and $\left(\dfrac{L}{4}, \dfrac{L}{4}\right)$.

3. $T(x, y) = \dfrac{100}{\sinh \pi} \sin \dfrac{\pi x}{L} \sinh \dfrac{\pi y}{L}$

5. $T(x, y) = \dfrac{-4L^3}{\pi^3} \displaystyle\sum_{n=1}^{\infty} \dfrac{[1 + 2(-1)^n]}{n^3 \sinh n\pi} \sin \dfrac{n\pi x}{L} \sinh \dfrac{n\pi y}{L}$.

Evaluate at $\left(\dfrac{L}{4}, \dfrac{L}{4}\right)$ and $\left(\dfrac{L}{4}, \dfrac{3L}{4}\right)$.

7. $T(x, y) = \dfrac{-160L}{\pi^2} \displaystyle\sum_{n=1}^{\infty} \dfrac{n \sin \dfrac{2n\pi x}{L} \sinh \dfrac{2n\pi y}{L}}{(4n^2 - 1)^2 \sinh(2n\pi)}$

9. The solution is the superposition of the solutions T_1 and T_2 of two problems: first where $T_1 = 50°$ on the top edge and $0°$ on the other edges; second where $T_2 = 50 \sin \pi y/2L°$ on the side edges and $0°$ on the horizontal edges. Thus

$$T(x, y) = T_1(x, y) + T_2(x, y)$$

$$= \dfrac{200}{\pi} \sum_{n=1}^{\infty} \dfrac{(-1)^n}{(2n+1)\sinh[(2n+1)\pi/2]}$$

$$\times \cos\!\left(\dfrac{2n+1}{2}\right)\dfrac{\pi x}{L} \sinh\!\left(\dfrac{2n+1}{2}\right)\dfrac{\pi y}{L}$$

$$- \dfrac{400}{\pi} \sum_{n=1}^{\infty} \dfrac{(-1)^n n}{(4n^2 - 1)\cosh n\pi} \cosh \dfrac{n\pi x}{L} \sin \dfrac{n\pi y}{L}.$$

11. $T(x, y) = 200 \displaystyle\sum_{n=0}^{\infty} \dfrac{(-1)^n}{\sinh[(2n+1)\pi/2]}$

$$\times \left[\dfrac{1}{(2n+1)\pi} - \left(\dfrac{2}{(2n+1)\pi}\right)^3\right]$$

$$\times \cos\!\left(\dfrac{2n+1}{2}\right)\dfrac{\pi x}{L} \sinh\!\left(\dfrac{2n+1}{2}\right)\dfrac{\pi y}{L}$$

$$- \dfrac{100}{\pi} \sum_{n=1}^{\infty} \dfrac{(-1)^n}{n \cosh n\pi} \cosh \dfrac{n\pi x}{L} \sin \dfrac{n\pi y}{L}$$

Problems 12.9, page 778

1. $T(r, \theta) = 2 \displaystyle\sum_{n=1}^{\infty} \dfrac{r^n(-1)^n}{n} \sin n\theta$

3. $T(r, \theta) = \dfrac{1 - r^2}{2\pi} \displaystyle\int_{-\pi}^{\pi} \dfrac{(\pi - \phi)\, d\phi}{r^2 + 1 - 2r \cos(\theta - \phi)}\bigg|_{(1/2, 0)} =$

$\dfrac{3}{8\pi} \displaystyle\int_{-\pi}^{\pi} \dfrac{(\pi - \phi)\, d\theta}{\frac{5}{4} - \cos \theta}$

5. a. $A_0 = 1$, $A_2 = -\frac{1}{2}$; all other coefficients are zero, so that $T(r, \theta) = \frac{1}{2}(1 - r^2 \cos 2\theta)$. Hence $T(\frac{1}{2}, 0) = \frac{3}{8}°$.

b. $T(\frac{1}{2}, 0) = \dfrac{3}{8\pi} \displaystyle\int_{-\pi}^{\pi} \dfrac{\sin^2 \phi}{\frac{5}{4} - \cos \phi}\, d\phi = \frac{3}{8}°$. (The last equality can be proved using the calculus of residues).

7. a. $T(r, \theta) = -\dfrac{\pi^2}{3} - 4 \displaystyle\sum_{n=1}^{\infty} (-r)^n \left(\dfrac{\cos n\theta}{n^2} + \dfrac{\pi \sin n\theta}{n}\right)$

b. $T(r, \theta) = \dfrac{1 - r^2}{2\pi} \displaystyle\int_{-\pi}^{\pi} \dfrac{\phi(2\pi - \phi)\, d\phi}{r^2 + 1 - 2r\cos(\theta - \phi)}$

9. Integrating by parts, we get

$$|a_n| \leq \dfrac{T}{n^2\pi^2} \left| \int_{-T}^{T} F''(x)\cos\dfrac{n\pi x}{T}\, dx \right| < \dfrac{2MT^2}{n^2\pi^2}.$$

And compare with $\sum 1/n^2$.

11. a. If u attains a maximal value at (x_0, y_0), the cross sections of the surface with the planes $x = x_0$ and $y = y_0$ have slope zero at (x_0, y_0) and are concave down. Thus $u_{xx} \leq 0$ and $u_{yy} \leq 0$ at (x_0, y_0).

c. $w_{xx} + w_{yy} = u_{xx} + u_{yy} + 4\epsilon > 0$

13. Apply equation (2).

15. $\dfrac{V''}{c^2 V} = \dfrac{U'' + \dfrac{1}{r} U'}{U} = -\lambda^2$

17. a. Because drum is fixed at the boundary and $u(R, t) = 0$.

b. Let $\lambda_n = s_n/R$, $z = r\lambda_n$, then $U''_n + \dfrac{1}{r} U'_n + \lambda_n^2 U_n =$

$\lambda_n^2 \left[J''_0 + \dfrac{1}{z} J'_0 + J_0 \right] = \dfrac{\lambda_n^2}{z^2} [z^2 J''_0 + z J'_0 + z^2 J_0] = 0.$

19. b. Observe that

$$f_1(r) = \sum_{n=1}^{\infty} a_n J_0\left(\dfrac{s_n r}{R}\right),$$

$$f_2(r) = \sum_{n=1}^{\infty} b_n k_n J_0\left(\dfrac{s_n r}{R}\right).$$

Now rewrite (18) as

$$(rU')' + \lambda_n^2 rU = 0$$

and apply the orthogonality theorem (11.7.1) to construct a generalized Fourier series in $\{J_0(s_n r/R)\}$, all orthogonal with respect to the weight function r. Finally apply equation (11.3.15) to determine the coefficients a_n and b_n.

Problems 12.10, page 784

1. $y(x, t) = t$

3. $T(x, t) = 1 - \text{erf}\left(\dfrac{x}{2\sqrt{\delta t}}\right)$

5. $T(x, t) = \displaystyle\int_0^t \sin\omega(t - u) \dfrac{x}{2\sqrt{\pi \delta u^3}} e^{-x^2/4\delta u}\, du$

Review Exercises for Chapter 12, page 785

1. $z(x, y) = xe^y$

3. $z(x, y) = x + e^y$

5. $T(x, t) = \dfrac{4a^2}{\pi^3} \displaystyle\sum_{n=1}^{\infty} \dfrac{[1 - (-1)^n]}{n^3} \sin\dfrac{n\pi x}{a} e^{-(n\pi/a)^2 \delta t}$

7. $T(x, t) = \dfrac{4a}{\pi^2} \displaystyle\sum_{n=0}^{\infty} \dfrac{(-1)^n}{(2n+1)^2} \sin\dfrac{(2n+1)\pi x}{a} e^{-[(2n+1)\pi/a]^2 \delta t}$

9. $y = X(x)T(t)$ where the general solutions to X and T are

$$X(x) = \begin{cases} c_1 x^{\lambda_1} + c_2 x^{\lambda_2}, & k > -\frac{1}{4} \\ c_1\sqrt{x} + c_2\sqrt{x}\ln x, & k = -\frac{1}{4} \\ \sqrt{x}[c_1\cos(\ln|x|^\beta) + c_2\sin(\ln|x|^\beta)], & k < -\frac{1}{4}, \end{cases}$$

$$T(t) = \begin{cases} b_1 t^{\lambda_1} + b_2 t^{\lambda_2}, & k > -\frac{1}{4} \\ b_1\sqrt{t} + b_2\sqrt{t}\ln t, & k = -\frac{1}{4} \\ \sqrt{t}[b_1\cos(\ln|t|^\beta) + b_2\sin(\ln|t|^\beta)], & k < -\frac{1}{4}, \end{cases}$$

where $\begin{aligned}\lambda_1 \\ \lambda_2\end{aligned}\bigg\} = \dfrac{1 \pm \sqrt{1 + 4k}}{2}$, respectively, and

$$\beta = \dfrac{\sqrt{-(1 + 4k)}}{2}.$$

11. $y = X(x)T(t)$ where the general solutions to X and T are

$$X(x) = \begin{cases} c_1 e^{\sqrt{k+1}x} + c_2 e^{-\sqrt{k+1}x}, & k > -1 \\ c_1 + c_2 x, & k = -1 \\ c_1\cos\sqrt{-(k+1)}x + c_2\sin\sqrt{-(k+1)}x, & k < -1, \end{cases}$$

$$T(t) = \begin{cases} b_1 e^{\sqrt{k}t} + b_2 e^{-\sqrt{k}t}, & k > 0 \\ b_1 + b_1 t, & k = 0 \\ b_1\cos\sqrt{-k}t + b_2\sin\sqrt{-k}t, & k < 0. \end{cases}$$

13. $y(x, t) = \dfrac{1}{2}\left(3\sin\dfrac{\pi x}{a}\cos\dfrac{c\pi t}{a} - \sin\dfrac{3\pi x}{a}\cos\dfrac{3c\pi t}{a}\right)$

15. a. $\dfrac{\partial^2 y}{\partial t^2} = c^2\dfrac{\partial^2 y}{\partial x^2} - 2b\dfrac{\partial y}{\partial t}$, where b is a constant of proportionality;

b. $y(x, t) = \dfrac{e^{-bt}}{4}\left[3\left(\cos k_1 t + \dfrac{b}{k_1}\sin k_1 t\right)\sin\dfrac{\pi x}{a}\right.$
$\left. - \left(\cos k_3 t + \dfrac{b}{k_3}\sin k_3 t\right)\sin\dfrac{3\pi x}{a}\right];$

c. $y(x, t) = e^{-bt}\displaystyle\sum_{n=1}^{\infty} A_n\left(\cos k_n t + \dfrac{b}{k_n}\sin k_n t\right)\sin\dfrac{n\pi x}{a}$, where

$k_n = \sqrt{n^2\pi^2 c^2 - b^2 a^2}/a$ and $A_n = \dfrac{2}{a}\displaystyle\int_0^a f(x)\sin\dfrac{n\pi x}{a}\, dx.$

Chapter 13

Problems 13.2, page 796

1. [9, 10]

3. $f(x) = (x + \frac{5}{2})^2 + \frac{3}{4} \geq \frac{3}{4}$

5. $x_0 = 9, \quad \sqrt{90} \approx 9.486832981$

7. $4\sqrt{25} \approx 2.236067978$

9. 6.1925824, 0.8074176

11. 7.984792473 (other roots are complex)

13. b. 0.7390851332

15. 0.904788218

17. $x_{n+1} = x_n - \dfrac{F(x_n)(x_n - x_{n-1})}{F(x_n) - F(x_{n-1})} = \dfrac{x_{n-1}F(x_n) - x_n F(x_{n-1})}{F(x_n) - F(x_{n-1})}$

Problems 13.3, page 806

In Problems 1–15 the answers are given in the order requested in the text. The last two numbers give the actual error in the trapezoidal and Simpson's estimates, respectively. The calculations were made with a hand calculator with ten decimal place precision.

1. 0.5; 0.5; 0; 0; 0.5; 0; 0

3. $\frac{11}{32} = 0.34375$; $\frac{1}{3}$, $\frac{1}{96} = 0.0104167$; 0, $\frac{1}{3}$; $-\frac{1}{96}$; 0

5. 6.448104763; 6.389488576; 0.1368343722;
0.0010135879; 6.389056099; 0.0590486641;
0.0004324773

7. 1.218760835; 1.218951005; 0.0003255208;
0.0000012716; 1.218951416; −0.0001905814;
−0.0000004111

9. 0.9871158010; 1.000134585; 0.0201863780;
0.0002075329; 1.0; −0.012884199; 0.000134585

11. 0.6973999653; 0.6932588905; 0.0207193081;
0.0010413912; ln 2 ≈ 0.6931471806; 0.0042527848;
0.0001117099

13. 0.9956971321; 0.9999360657; 0.0117435512;
0.0003852527; 1.0; −0.0043028736; −0.000063343

15. 0.912905114; 0.9093257681; 0.0040387583;
0.0000124097; 0.9093306736; 0.0035898377;
−0.0000049056

In Problems 17–29 the trapezoidal approximation is given first.

17. 0.6590191268; 0.6593301635

19. 1.987795499; 1.994503740

21. 0.6903257237; 0.8367702680

23. 1.488736680; 1.493674110

25. 0.9091616587; 0.9096068101

27. 0.9841199229; 0.9838189106

29. 0.3604819483; 0.3579282259

31. $|y''| = |3(2x + 3x^4)e^{x^3}| \leq 15e$ on [0, 1] and $|y^{(4)}| = |9(9x^8 + 36x^5 + 20x^2)e^{x^3}| \leq 585e$. Thus $|\epsilon_8^T| \leq 15e/(12 \cdot 8^2) \approx 0.05309$ and $|\epsilon_8^S| \leq 585e/(180 \cdot 8^4) \approx 0.00216$

33. $|y''| = |-2 + 4x^2|e^{-x^2} \leq 2$ on [−1, 1] and $|y^{(4)}| = |12 - 48x^2 + 16x^4|e^{-x^2} \leq 12$. Thus $|\epsilon_{10}^T| \leq 2^4/(12)10^2 \approx 0.01333$ and $|\epsilon_{10}^S| \leq 12 \cdot 2^5/(180 \cdot 10^4) \approx 0.00021$

35. $|y''| = |3(4x^2 + 3x^5)e^{x^3}| \leq 21e$ on [0, 1]; $|y^{(4)}| = |3(8 + 132x^3 + 144x^6 + 27x^9)e^{x^3}| \leq 933e$. Thus $|\epsilon_{10}^T| \leq 21e/(12 \cdot 10^2) \approx 0.04757$ and $|\epsilon_{10}^S| \leq 933e/(180 \cdot 10^4) \approx 0.00141$

37. $|y^{(4)}| \leq 3$ so we need $(3/\sqrt{2\pi})1^5/(180 \cdot n^4) \leq 0.005$ (for half the integral—using the hint); $n = 2$ will do; we obtain 0.6830581043 (the "true" value is 0.6826894921 giving an error of 0.0003686122)

39. a. On [0, 50] need $(3/\sqrt{2\pi})50^5/(180 \cdot n^4) \leq 0.05$ or $n \geq 82$; this leads to the estimate $(1/\sqrt{2\pi}) \int_{-50}^{50} = (2\sqrt{2\pi}) \int_{-50}^{50} \approx 2(0.4999994266) = 0.9999988532$

b. $\lim_{N\to\infty} (1/\sqrt{2\pi}) \int_{-N}^{N} e^{-x^2/2}\, dx = 1$

41. $|y''| = 2/x^3 \leq 2$ on [1, 2]; need $2/12n^2 \leq 10^{-10}$ or $12n^2 > 2 \cdot 10^{10}$ which implies that $n > 40{,}824.8$

43. It is not difficult to verify that on $[\frac{1}{2}, 1]$, $|J''_{1/2}(x)| \leq 8$. Then, with the trapezoidal rule, we need $8(\frac{1}{2})^3/12n^2 \leq 0.01$ or $n > 2.88$. Using $n = 3$ gives $\int_{1/2}^{1} J_{1/2}(x)\, dx \approx 0.3095670957$.

45. $\displaystyle\int_{x_i}^{x_i+2\Delta x} (ax^2 + bx + c)\, dx = \frac{\Delta x}{3}[a(6x_i^2 + 12x_i\,\Delta x + 8(\Delta x)^2)$
$+ 6b(x_i + \Delta x) + 6c]; \dfrac{\Delta x}{3}[f(x_i) + 4f(x_{i+1}) + f(x_{i+2})] =$

$\dfrac{\Delta x}{3}[f(x_i) + 4f(x_i + \Delta x) + f(x_i + 2\Delta x)] =$

$\dfrac{\Delta x}{3}[a(6x_i^2 + 12x_i\,\Delta x + 8(\Delta x)^2) + 6b(x_i + \Delta x) + 6c]$

Problems 13.4, page 814

1. $q_1(x) = (2/\pi)x$

3. $q_2(x) = -\frac{1}{60}x^2 + \frac{23}{60}x + 1$

5. $q_2(x) = (\cosh 1 - 1)x^2 + (\sinh 1)x + 1 = \{[(e + e^{-1})/2] - 1\}x^2 + \{[(e - e^{-1})/2]\}x + 1$

7. $q_2(x) = (1/\pi^2)[48 - (72/\sqrt{3})]x^2 + (1/\pi)[(18/\sqrt{3}) - 8]x$

9. $q_2(x) = \frac{7}{4}x^2 - \frac{3}{4}x$

11. $q_1(x) = -0.117779 + 0.405463x$; $q_1(2.3) = 0.8147859$;
max error $\leq (0.3)(0.7)/(2)(2^2) = 0.02625$ (actual error $\approx$ 0.018123)

13. $q_3(x) = -0.937499 + 1.220053x - 0.246195x^2$
$+ 0.021915x^3$; $q_3(2.3) \approx 0.832891155$; max error $\leq [(\frac{3}{10})(\frac{1}{30})(\frac{11}{30})(\frac{7}{10})/4!](6/2^4) \approx 0.0000401$; actual error $\approx$ 0.00001797

15. a. max value occurs at $x - x_3 = -\sqrt{1.5 + \sqrt{145}/10}\,\Delta x$
and is approximately equal to $(3.6314)(\Delta x)^5$.
b. $|f(x) - q(x)| < M\,(3.6314)(\Delta x)^5/5!$

17. $q_4(x) = \frac{1}{3}(0.97509 + 2.75045x - 0.9561x^2 + 0.26224x^3 - 0.03168x^4)$; $q_4(1.2) = 1.095435$; max error $\approx$
$(0.2)(0.05)(0.3)(0.55)(0.8)(\frac{1}{2})(\frac{1}{2})(\frac{3}{2})(\frac{5}{2})(\frac{7}{2})/5! \approx 0.000036$
(actual error ≈ 0.000001); $q_4(1.7) = 1.303839$;
max error ≈ 0.0000258 (actual error ≈ 0.0000016)

21. $q_1(x) = 0.63662x$ (error on $[0, \pi/2] < 0.308$); $q_2(x) = -0.335749x^2 + 1.16401x$ (error on $[0, \pi/2] < 0.2486$)
$q_3(x) = -0.113872x^3 - 0.065471x^2 + 1.02043x$ (error on $[0, \pi/2] < 0.1427$)

	Bound for this	
Estimate	estimate	Actual error
$q_1(0.5) = 0.31831$	0.2677	0.16112
$q_2(0.5) = 0.49807$	0.02547	0.01860
$q_3(0.5) = 0.47961$	0.000576	0.000184
$q_1(1) = 0.63662$	0.285	0.20485
$q_2(1) = 0.82826$	0.02042	0.01321
$q_3(1) = 0.84109$	0.000535	0.00038

Problems 13.5, page 820

1. $y = 2e^x - x - 1$, $y(1) = 2(e - 1) \approx 3.4366$;
a. $y_E = 2.98$;
b. $y_{IE} = 3.405$
3. $y = \sqrt{2x^2 + 1} - x$; $y(1) = \sqrt{3} - 1 \approx 0.73205$
a. $y_E = 0.71$;
b. $y_{IE} = 0.73207$
5. $y = \sinh(\frac{1}{2}x^2 - \frac{1}{2})$; $y(3) \approx 27.2899$
a. $y_E = 8.31$;
b. $y_{IE} = 21.671$
7. $y = (x^3 + x^{-2})^{-1/2}$, $y(2) = 2/\sqrt{33} \approx 0.3481553$;
a. $y_E = 0.343$
b. $y_{IE} = 0.34939$
9. $y = 2e^{e^x - 1}$, $y(2) = 2e^{e^2 - 1} \approx 1190.59$;
a. $y_E = 156$;
b. $y_{IE} = 781.56$

In 11–19 answers are given for the improved Euler method. Part (a) is given in the answers for Section 1.3.

11. $y_1 = 1.02$, $y_2 = 1.04$, $y_3 = 1.07$, $y_4 = 1.09$,
$y_5 = 1.12$
13. $y_1 = 0.34$, $y_2 = 0.56$, $y_3 = 0.76$, $y_4 = 0.98$,
$y_5 = 1.34$
15. $y_1 = 1.28$, $y_2 = 1.65$, $y_3 = 1.46$, $y_4 = 1.09$,
$y_5 = 0.87$, $y_6 = 0.79$, $y_7 = 0.93$, $y_8 = 0.95$
17. $y_1 = 1.10$, $y_2 = 1.22$, $y_3 = 1.35$, $y_4 = 1.48$,
$y_5 = 1.63$, $y_6 = 1.79$, $y_7 = 1.97$, $y_8 = 2.16$,
$y_9 = 2.37$, $y_{10} = 2.60$
19. $y_1 = 1.60$, $y_2 = 1.27$, $y_3 = 1.00$, $y_4 = 0.80$,
$y_5 = 0.64$
21. No method will provide a correct answer since the solution to the differential equation is the hyperbola
$x^2 - 2xy - y^2 = 4$ or $y = \sqrt{2x^2 - 4} - x$, which is not defined if $x = 1$.
23. The solution to the differential equation is $y = [x^2(2 - x^2)]^{-1/2}$, which is not defined at $x = 3$.

Problems 13.6, page 825

1. a. $|e_n| \leq 0.859h$
b. $h = 0.1$, $|e_n| \leq 0.086$; $h = 0.2$, $|e_n| \leq 0.172$.
3. $|e_n| \leq 6.06h$
a. 1,212,000
b. 12,120,000.

Problems 13.7, page 831

1. $y = 2e^x - x - 1$; $y(1) = 2(e - 1) \approx 3.4365637$;
$y_{RK} = 3.436502$
3. $y = \sqrt{2x^2 + 1} - x$; $y(1) = \sqrt{3} - 1 \approx 0.732050807$;
$y_{RK} = 0.73205044$
5. $y = \sinh(\frac{1}{2}x^2 - \frac{1}{2})$; $y(3) \approx 27.2899$; $y_{RK} = 27.0275$
7. $y = (x^3 + x^{-2})^{-1/2}$; $y(2) = 2/\sqrt{33} \approx 0.3481553$;
$y_{RK} = 0.348161$
9. $y = 2e^{e^x - 1}$, $y(2) = 2e^{e^2 - 1} \approx 1190.59$; $y_{RK} = 1164.76$
11. $y_1 = 1.02$, $y_2 = 1.04$, $y_3 = 1.07$, $y_4 = 1.10$,
$y_5 = 1.12$.
13. $y_1 = 0.38$, $y_2 = 0.61$, $y_3 = 0.78$, $y_4 = 0.98$,
$y_5 = 1.35$.
15. $y_1 = 1.33$, $y_2 = 1.91$, $y_3 = 1.61$, $y_4 = 1.16$,
$y_5 = 0.92$, $y_6 = 0.79$, $y_7 = 0.54$, $y_8 = 0.77$.
17. $y_1 = 1.11$, $y_2 = 1.22$, $y_3 = 1.35$, $y_4 = 1.48$,
$y_5 = 1.63$, $y_6 = 1.79$, $y_7 = 1.97$, $y_8 = 2.16$,
$y_9 = 2.37$, $y_{10} = 2.60$.
19. $y_1 = 1.59$, $y_2 = 1.26$, $y_3 = 1.00$, $y_4 = 0.79$,
$y_5 = 0.64$.

21. See answer to Problem 13.5.21.

23. See answer to Problem 13.5.23.

25. $a = \frac{1}{4}$, $b = 0$, $c = \frac{3}{4}$, $d = 0$

27. $n = p = \frac{2}{3}$

29. Because the determinant of the resulting 4×4 system is zero. There are an infinite number of solutions; no.

Problems 13.9, page 843

1. $u_{i,j+1} = \frac{1}{4}[u_{i+1,j} + 2u_{i,j} + u_{i-1,j}]$ so we obtain the values in the table below:

0	$\frac{2875}{2048}$	$\frac{4625}{2048}$	$\frac{4625}{2048}$	$\frac{2875}{2048}$	0
0	$\frac{25}{16}$	$\frac{1275}{512}$	$\frac{1275}{512}$	$\frac{25}{16}$	0
0	$\frac{225}{128}$	$\frac{350}{128}$	$\frac{350}{128}$	$\frac{225}{128}$	0
0	$\frac{65}{32}$	$\frac{95}{32}$	$\frac{95}{32}$	$\frac{65}{32}$	0
0	5/2	25/8	25/8	5/2	0
0	7/2	3	3	7/2	0
0	6	2	2	6	0

3. Use

$$u_{i,j+1} - u_{i,j} = \frac{1}{h}(u_{i+1,j} - 2u_{i,j} + u_{i-1,j}) + (u_{i+1,j} - u_{i-1,j})$$

or

$$u_{i,j+1} = \left(1 + \frac{1}{h}\right)u_{i+1,j} + \left(1 - \frac{2}{h}\right)u_{i,j} + \left(-1 + \frac{1}{h}\right)u_{i-1,j}$$

Since $h = 4$ we get

$$u_{i,j+1} = \frac{5}{4}u_{i+1,j} + \frac{1}{2}u_{i,j} - \frac{3}{4}u_{i-1,j}$$

yielding the table below:

0	-2.16	-36.2	-2.49	13.8	0
0	13.7	-7.2	-17.8	0.82	0
0	5.906	8.594	-5.66	-6.84	0
0	1.5	4.125	6.125	-4.5	0
0	5.5	-1	7	1.5	0
0	6	2	2	6	0

5.

0	20.77	-12.1	7.69	12.52	10
0	27.94	5.44	4.94	7.44	10
0	21.5	13.75	11.75	7.5	10
0	13	12	14	11	10
0	6	8	10	12	10
0	2	4	6	8	10

7.

0	0	0	0	0
0	1/16	1/8	3/16	1/4
0	1/8	1/4	3/8	1/2
0	3/16	3/8	9/16	3/4
0	1/4	1/2	3/4	1

9. (a) $z_{i,j+1} = z_{i,j} + \frac{2}{h}[z_{i+1,j} - 2z_{i,j} + z_{i-1,j}]$

Review Exercises for Chapter 13, page 843

1. $y^2 = 2(e^x + 1)$, $y(3) = \sqrt{2(e^3 + 1)} \approx 6.4939$

  a. $y_{IE} = 6.56$

  b. $y_{RK} = 6.4942$

3. $y = x + \sqrt{1 + x^2}$, $y(3) = 3 + \sqrt{10} \approx 6.1623$

  a. $y_{IE} = 5.96$

  b. $y_{RK} = 6.1605$

5. $y^{-2} = (2x - 1 + 3e^{-2x})/2$, $y(3) = 0.6320$

  a. $y_{IE} = 0.6356$

  b. $y_{RK} = 0.6329$

7. The exact solution is $y^2(y + 1) = x^2$. If we graph this curve and begin moving to the right from the initial point, we cannot proceed *on that branch* beyond $x = 2/3\sqrt{3}$.

9. Solution $y = \tan x$ becomes infinite at $x = \pi/2$, so no numerical method will yield $y(2)$.

11. 0.632121

13. 1.34557

15. 0.927056

17. -0.218412

31. $z_{ij} = \frac{1}{4 + h^2}[z_{i+1,j} + z_{i-1,j} + z_{i,j+1} + z_{i,j-1}]$

0	0	0	0
0	0.07051	0.15181	0.26174	0.4375
0	0.13464	0.28448	0.47399	0.75
0	0.19198	0.39528	0.62936	0.9375
0	0.25	0.5	0.75	1

Chapter 14

Problems 14.1, page 851

1. Exact solution is $x_1 = 1.6$, $x_2 = -0.8$, $x_3 = -3.7$

3. Exact solution is $(0, -2.61, 4.3)$.

5. a. with pivoting; $x_1 = 5.99$, $x_2 = -2$, $x_3 = 3.99$

  b. without pivoting: $x_1 = 6$, $x_2 = -2$, and $x_3 = 4$ (Yes, sometimes it's better to follow the simplest path. In Problem 6, pivoting gives much more accurate answers.) The relative errors with pivoting are $\frac{1}{600} = 0.0017$, 0, and $\frac{1}{400} = 0.0025$.

7. A solution with rounding to 3 significant figures is $x_1 = 1050$ and $x_2 = -1000$. The exact solution is $x_1 = \frac{15650}{13} \approx 1204$ and $x_2 = -\frac{15000}{13} \approx -1154$. The relative errors are $0.1465 \approx 15\%$ and $0.1333 \approx 13\%$.

Problems 14.2, page 859

1. yes

3. no

5. yes

7. Jacobi: $x_1 = 2.9757$, $x_2 = -0.9919$ (8 iterations)
   Gauss–Seidel: $x_1 = 3.0041$, $x_2 = -1.0025$ (5 iterations)
   Exact solution is $(3, -1)$.

9. Jacobi: $x_1 = 1.9999$, $x_2 = -2.988$, $x_3 = 7.0146$ (7 iterations)
   Gauss–Seidel: $x_1 = -1.9974$, $x_2 = -2.9993$, $x_3 = 6.999$ (6 iterations) Exact solution is $(-2, -3, 7)$.

11. Jacobi: $x_1 = -8.2864$, $x_2 = 14.386$, $x_3 = -0.10296$ (13 iterations); Gauss–Seidel: $x_1 = -8.2989$, $x_2 = 14.3995$, $x_3 = -0.10039$ (7 iterations) Exact solution is $(-8.3, 14.4, -0.1)$.

13. **a.** $|a_{ii}| = 1$ and
$$|a_{12}| + |a_{13}| = |a_{21}| + |a_{23}|$$
$$= |a_{31}| + |a_{32}|$$
$$= 1;$$

thus $|a_{ii}| = \sum_{\substack{j=1 \\ j\neq i}}^{3} |a_{ij}|$ so that $a_{ii} \not> \sum_{\substack{j=1 \\ j\neq i}}^{3} |a_{ij}|$

b. *Jacobi*

n	$x_1^{(n)}$	$x_2^{(n)}$	$x_3^{(n)}$
0	0.8	0.8	0.8
1	1.2	1.2	1.2
2	0.8	0.8	0.8
3	1.2	1.2	1.2
4	0.8	0.8	0.8
5	1.2	1.2	1.2

c. *Gauss–Seidel*

n	$x_1^{(n)}$	$x_2^{(n)}$	$x_3^{(n)}$
0	0.8	0.8	0.8
1	1.2	1	0.9
2	1.05	1.025	0.9625
3	1.0063	1.0156	0.98906
4	0.99766	1.0066	0.99785
5	0.99775	1.0022	1
6	0.99889	1.0005	1.0003
7	0.99958	1.0001	1.0002
8	0.99988	0.99997	1.0001

d. (i) $D^{-1}(L + U)$

$$= \begin{pmatrix} 0 & 1/2 & 1/2 \\ 1/2 & 0 & 1/2 \\ 1/2 & 1/2 & 0 \end{pmatrix}, \text{ which has the eigenvalue 1}$$

so, by Theorem 2(i), the Jacobi iterates fail to converge.

(ii) $(D + L)^{-1}U$

$$= \begin{pmatrix} -3/8 & -1/8 & 0 \\ 1/4 & -1/4 & 0 \\ 1/2 & 1/2 & 0 \end{pmatrix},$$

whose characteristic polynomial is
$\lambda(\lambda^2 + \frac{5}{8}\lambda + \frac{1}{8}) = 0$ with roots 0, $(-5 \pm \sqrt{7}i)/16$.
$|(-5 + \sqrt{7}i)/16|$
$\quad = |(-5 - \sqrt{7}i)/16|$
$\quad = \sqrt{(\frac{5}{16})^2 + (\sqrt{7}/16)^2}$
$\quad \approx 0.3536.$

Thus $r[(D + L)^{-1}U] \approx 0.3536 < 1$ and, by Theorem 2(ii), the Gauss–Seidel iterates converge.

15. If A is diagonal, then $L = U = 0$. Then $r[D^{-1}(L + U)] = r(0) = 0$, and $r[(D + L)^{-1}U] = r(D0) = r(0) = 0$.

17. $D = \begin{pmatrix} a & 0 \\ 0 & d \end{pmatrix}$, $L = \begin{pmatrix} 0 & b \\ 0 & 0 \end{pmatrix}$, and

$U = \begin{pmatrix} 0 & 0 \\ c & 0 \end{pmatrix}$ so that

$$D^{-1}(L + U) = \begin{pmatrix} 0 & \dfrac{b}{a} \\ \dfrac{c}{d} & 0 \end{pmatrix} \text{ and}$$

$$(D + L)^{-1}U = \begin{pmatrix} -\dfrac{bc}{ad} & 0 \\ \dfrac{c}{d} & 0 \end{pmatrix}.$$

The eigenvalues of $D^{-1}(L + U)$ are $\pm\sqrt{\dfrac{bc}{ad}}$, and the eigenvalues of $(D + L)^{-1}U$ are 0 and $\dfrac{bc}{ad}$. These are all less than one in absolute value if and only if $\left|\dfrac{bc}{ad}\right| < 1$.

Problems 14.4, page 869

1. $-4,\ \begin{pmatrix} 1 \\ 1 \end{pmatrix}$

3. $-6.3,\ \begin{pmatrix} 1 \\ -0.5 \end{pmatrix}$

5. $8, \begin{pmatrix} 1 \\ 0.5 \\ 1 \end{pmatrix}$

7. a. $\lambda \approx 8.5536$ (without scaling);
$\epsilon_r \approx 0.00076374$

b. $\lambda = 2 + \sqrt{43} = 8.5574$. The actual relative error is 0.000444.

9. Starting with $\mathbf{x}_0 = \begin{pmatrix} 1 \\ 1 \end{pmatrix}$ and without scaling, we obtain

$\begin{pmatrix} 1 \\ 1 \end{pmatrix}, \begin{pmatrix} 2 \\ 1 \end{pmatrix}, \begin{pmatrix} -1 \\ -1 \end{pmatrix}, \begin{pmatrix} -2 \\ -1 \end{pmatrix}, \begin{pmatrix} 1 \\ 1 \end{pmatrix}, \begin{pmatrix} 2 \\ 1 \end{pmatrix}, \ldots$. That is, the

method doesn't converge. Note that the eigenvalues of A are $\pm i$, both of which have absolute value 1. That is, A does not have a dominant eigenvalue.

11. The second eigenvalue is $\lambda_2 = 3$.

13. The other eigenvalues are -2 and 1.

Review Exercises for Chapter 14, page 870

1. $\begin{pmatrix} 1 & -2 & 3 \\ 0 & 1 & -2 \\ 0 & 0 & 1 \end{pmatrix}$

3. $\begin{pmatrix} 1 & 3 & 5 \\ 0 & 1 & 2 \\ 0 & 0 & 0 \end{pmatrix}$

5. $x_1 = -4.29246$, $x_2 = 2.89408$, $x_3 = 1.09666$. The exact solutions are -4.3, 2.9, and 1.1.

7. yes

9. yes

11. *Jacobi*: $x_1 = 1.29961$, $x_2 = -2.70014$, $x_3 = 0.799741$ (after 10 iterations)
Gauss–Seidel: $x_1 = 1.30028$, $x_2 = -2.69984$, $x_3 = 0.799936$ (after 8 iterations)
Exact solutions: 1.3, -2.7, 0.8

13. $6, \begin{pmatrix} 1 \\ 1 \end{pmatrix}$

15. $3, \begin{pmatrix} -1/2 \\ 1 \\ -1/2 \end{pmatrix}$

Chapter 15

Problems 15.1, page 879

1. $9 - 7i$

3. 2

5. $-27 + 5i$

7. i

9. $\frac{2}{25} + \frac{11}{25}i$

11. $\frac{10}{17} - \frac{11}{17}i$

13. 5

15. $-5 + 10i$

17. $1 + 2i$

19. $-2 + 2i$

21. $-2i$

23. $-3 - 4i$

25. $-\frac{7}{10} + \frac{19}{10}i$

27. -10

29. $-i$

31. $3 + 8i$

33. $5 - 6i$

35. $x + iy = x - iy$ if and only if $y = 0$.

37. $iz = ix - y$

39. If $z_2 \neq 0$, then $z_2\bar{z}_2 = (\text{length } z_2)^2 \neq 0$, But $z_1z_2\bar{z}_2 = 0$, so divide both sides by $(\text{length } z_2)^2$ to get $z_1 = 0$.

41. $\text{Im } z_1 = -\text{Im } z_2$ so $\text{Im } z_1z_2 = \text{Im } z_2$. $(\text{Re } z_1 - \text{Re } z_2) = 0$. If $\text{Re } z_1 = \text{Re } z_2$, then $\text{Re } z_1z_2 > 0$. Hence $\text{Im } z_2 = 0$.

43. $(x_1 - iy_1)(x_2 - iy_2) = (x_1x_2 - y_1y_2) - i(x_1y_2 + x_2y_1) = \overline{(x_1x_2 - y_1y_2) + i(x_1y_2 + x_2y_1)}$

45. $0 = \overline{p(z)} = \bar{z}^n + \bar{a}_{n-1}\bar{z}^{n-1} + \cdots + \bar{a}_1\bar{z} + \bar{a}_0$
$= \bar{z}^n + a_{n-1}\bar{z}^{n-1} + a_1\bar{z} + a_0$
$= p(\bar{z})$, since $a_0, a_1, \ldots, a_{n-1}$ real

47. Let $z_j = x_j + iy_j$, $j = 1, 2, 3$. Then show that both $(z_1 + z_2) + z_3$ and $z_1 + (z_2 + z_3)$ equal $(x_1 + x_2 + x_3) + i(y_1 + y_2 + y_3)$. Proceed similarly for the products.

49. $w^3 = z^3 + pz^2 + \frac{1}{3}p^2z + \frac{p^3}{27}$ so
$w^3 + (q - \frac{1}{3}p^2)w + \left(r - \frac{pq}{3} + \frac{2p^3}{27}\right) = 0.$

51. $A = \sqrt[3]{-15 + D}$, $B = \sqrt[3]{-15 - D}$, where
$D = \sqrt{\frac{900}{4} - \frac{6859}{27}} \approx \sqrt{-29.037} \approx 5.3886i$

Problems 15.2, page 886

1. $|i| = 1$, $\arg i = \frac{\pi}{2} + 2k\pi$,
$i = \cos\left(\frac{\pi}{2} + 2k\pi\right) + i\sin\left(\frac{\pi}{2} + 2k\pi\right)$

3. $|1 - i| = \sqrt{2}$, $\arg(1 - i) = \frac{-\pi}{4} + 2k\pi$,
$1 - i = \sqrt{2}\left[\cos\left(\frac{-\pi}{4} + 2k\pi\right) + i\sin\left(\frac{-\pi}{4} + 2k\pi\right)\right]$

5. $|-1 - i| = \sqrt{2}$, $\arg(-1 - i) = \frac{-3\pi}{4} + 2k\pi$,
$-1 - i = \sqrt{2}\left[\cos\left(\frac{-3\pi}{4} + 2k\pi\right) + i\sin\left(\frac{-3\pi}{4} + 2k\pi\right)\right]$

7. $|-1 + \sqrt{3}i| = 2$, $\arg(-1 + \sqrt{3}i) = \frac{2\pi}{3} + 2k\pi$,
$-1 + \sqrt{3}i = 2\left[\cos\left(\frac{2\pi}{3} + 2k\pi\right) + i\sin\left(\frac{2\pi}{3} + 2k\pi\right)\right]$

9. $|-\sqrt{3} - i| = 2$, $\arg(-\sqrt{3} - i) = \dfrac{-5\pi}{6} + 2k\pi$,

$$-\sqrt{3} - i = 2\left[\cos\left(\frac{-5\pi}{6} + 2k\pi\right) + i\sin\left(\frac{-5\pi}{6} + 2k\pi\right)\right]$$

11. $|4 + 3i| = 5$, $\arg(4 + 3i) \approx 0.6435 + 2k\pi$,
$4 + 3i \approx 5[\cos(0.6435 + 2k\pi) + i\sin(0.6435 + 2k\pi)]$

13. $|2 + 7i| = \sqrt{53}$, $\arg(2 + 7i) \approx 1.2925 + 2k\pi$,
$2 + 7i \approx \sqrt{53}[\cos(1.2925 + 2k\pi) + i\sin(1.2925 + 2k\pi)]$

15. $|5 + 2i| = \sqrt{29}$, $\arg(5 + 2i) \approx 0.3805 + 2k\pi$,
$5 + 2i \approx \sqrt{29}[\cos(0.3805 + 2k\pi) + i\sin(0.3805 + 2k\pi)]$

17. $|7 - 2i| = \sqrt{53}$, $\arg(7 - 2i) \approx -0.2783 + 2k\pi$,
$7 - 2i \approx \sqrt{53}[\cos(-0.2783 + 2k\pi)$
$+ i\sin(-0.2783 + 2k\pi)]$

19. $|-100 + 100i| = 100\sqrt{2}$, $\arg(-100 + 100i) =$
$\dfrac{3\pi}{4} + 2k\pi$, $-100 + 100i = 100\sqrt{2}\left[\cos\left(\dfrac{3\pi}{4} + 2k\pi\right)\right.$

$\left. + i\sin\left(\dfrac{3\pi}{4} + 2k\pi\right)\right]$

21. Circle centered at z_0 of radius a

23. $|z_1 + z_2|^2 = |z_1|^2 + |z_2|^2 + 2\,\mathrm{Re}\,z_1\bar{z}_2 \leqslant |z_1|^2 + |z_2|^2$
$+ 2|z_1\bar{z}_2| = (|z_1| + |z_2|)^2$ since $|z_1\bar{z}_2| = |z_1|\,|\bar{z}_2| = |z_1|\,|z_2|$
and $\mathrm{Re}\,z \leqslant |z|$.

25. $|z - a|^2 = |z|^2 + |a|^2 - 2\,\mathrm{Re}\,z\bar{a}$; $|1 - az|^2 = 1 + |a|^2\,|z|^2$
$- 2\,\mathrm{Re}\,z\bar{a}$, and $0 < (1 - |z|^2)(1 - |a|^2) =$
$1 - |z|^2 - |a|^2 + |a|^2|z|^2$

27. See the proof of Problem 15.1.45.

Problems 15.3, page 890

1. $(1 + i)^{29} = (\sqrt{2})^{29}\left[\cos\left(\dfrac{\pi}{4} + 2k\pi\right) + i\sin\left(\dfrac{\pi}{4} + 2k\pi\right)\right]^{29} =$
$-2^{14}(1 + i)$

3. $(-1 - i)^{36} = 2^{18}\left[\cos\left(\dfrac{\pi}{4} + 2k\pi\right) + i\sin\left(\dfrac{\pi}{4} + 2k\pi\right)\right]^{36} = -2^{18}$

5. $(\sqrt{3} + i)^{15} = 2^{15}i$

7. $(-1 - \sqrt{3}i)^{10} = -2^9(1 + \sqrt{3}i)$

9. $(-3 + 4i)^8 = 164833 - 354144i$

11. $\dfrac{1}{\sqrt{2}} + \dfrac{1}{\sqrt{2}}i, \dfrac{-1}{\sqrt{2}} - \dfrac{1}{\sqrt{2}}i$

13. $1.45535 - 0.34356i$, $-1.45535 + 0.34356i$

15. $1.29207 + 0.20129i$, $-0.82036 + 1.01832i$,
$-0.47171 - 1.21962i$

17. $0.92388 + 0.38268i$, $-0.38268 + 0.92388i$,
$-0.92388 - 0.38268i$, $0.38268 - 0.92388i$

19. $z_k = \cos\dfrac{k\pi}{4} + i\sin\dfrac{k\pi}{4}$, $k = 0, 1, \ldots, 7$

or $1, \dfrac{1+i}{\sqrt{2}}, i, \dfrac{-1+i}{\sqrt{2}}, -1, \dfrac{-1-i}{\sqrt{2}}, -i, \dfrac{1-i}{\sqrt{2}}$

21. $\cos 4\theta + i\sin 4\theta = (\cos\theta + i\sin\theta)^4 = \cos^4\theta$
$+ 4i\cos^3\theta\sin\theta - 6\cos^2\theta\sin^2\theta - 4i\cos\theta\sin^3\theta$
$+ \sin^4\theta$ hence

$$\cos 4\theta = \cos^4\theta - 6\cos^2\theta\sin^2\theta + \sin^4\theta$$
$$\sin 4\theta = 4\cos^3\theta\sin\theta - 4\cos\theta\sin^3\theta$$

23. $|z|^n(\cos n\theta + i\sin n\theta) = z^n = 1 = \cos 2\pi k + i\sin 2\pi k$ so $|z|$
$= 1$ and $\theta = 2\pi k/n$. But

$$\frac{2\pi(k + n)}{n} = \frac{2\pi k}{n} + 2\pi \text{ so } z_{k+n} = z_k.$$

25.
$$1 + z + \cdots + z^{n-1} = \frac{1 - z^n}{1 - z} = 0$$

for any nth root of unity z_k since $z_k^n = 1$. Since the first n
of the nth roots of unity are all different (see Problem 23)
and the polynomial $1 - z^n$ has exactly n roots, one of
which is 1, $1 + z + \cdots + z^{n-1}$ must have each of the
other nth roots of unity as a root.

27. $\sqrt{3i} = \pm\dfrac{\sqrt{3}}{2}(1 + i)$

29. $\sqrt{-1 + i} = \pm\left[\sqrt{\dfrac{\sqrt{2} - 1}{2}} + i\sqrt{\dfrac{\sqrt{2} + 1}{2}}\right]$

31. $\sqrt{2 - i} = \pm\left[\sqrt{\dfrac{\sqrt{5} + 2}{2}} - i\sqrt{\dfrac{\sqrt{5} - 2}{2}}\right]$

33. $\left|\displaystyle\sum_{k=1}^{n} a_k z_k\right|^2 = \left(\displaystyle\sum_{k=1}^{n} a_k z_k\right)\left(\displaystyle\sum_{j=1}^{n} \bar{a}_j \bar{z}_j\right)$

$$= \sum_{k=1}^{n} |a_k|^2|z_k|^2 + \sum_{j\neq k} a_k\bar{a}_j\bar{z}_k z_j$$

$$= \left(\sum_{k=1}^{n} |a_k|^2\right)\left(\sum_{k=1}^{n} |z_k|^2\right) + \sum_{j\neq k} a_k\bar{a}_j z_k\bar{z}_j$$

$$- \sum_{j\neq k} |a_k|^2|z_j|^2$$

$$= \left(\sum_{k=1}^{n} |a_k|^2\right)\left(\sum_{k=1}^{n} |z_k|^2\right) - \sum_{1\leqslant j < k \leqslant n} [|a_j|^2|z_k|^2$$

$$- a_j\bar{a}_k z_j\bar{z}_k - a_k\bar{a}_j z_k\bar{z}_j + |a_k|^2|z_j|^2]$$

$$= \sum_{k=1}^{n} |a_k|^2 \sum_{k=1}^{n} |z_k|^2 - \sum_{1\leqslant j < k \leqslant n} |a_j\bar{z}_k - a_k\bar{z}_j|^2.$$

Problems 15.4, page 894

1. open; bounded; simply connected, circle $|z + 3| = 2$;

3. open; unbounded; not connected, lines $\mathrm{Im}\,z = \pm 1$;

5. closed; unbounded; simply connected, parabola $|z| = \mathrm{Re}\,z + 2$;

7. closed; unbounded; connected but not simply connected
ellipse $|z + 1| + |z + i| = 2$;

9. open; bounded; connected but not simply connected, ellipses $|z - 1| + |z + 1| = 3$ and $|z - 1| + |z + 1| = 2\sqrt{2}$

11. Assume the closure $\bar{S}$ is disconnected. Then there are open sets G and H such that $\bar{S} \subset G \cup H$, $G \cap H$ is empty, and $G \cap \bar{S}$ and $H \cap \bar{S}$ are nonempty. Since S is connected it lies in only one of these sets, say G. Then S is contained in the closed set $\mathbb{C} - H$, so $\bar{S} \cap H$ is empty, a contradiction.

Problems 15.5, page 901

1. $|2z - 2| = 2|z - 1| < 2\delta$ so set $\epsilon = 2\delta$

3. $|z + i| < \delta$ so set $\epsilon = \delta$

5. $|(2z - 3) - (-1 + 2i)| = |2z - 2 - 2i| = 2|z - (1 + i)| < 2\delta = \epsilon$

7. $\left| \dfrac{z^2 - 4}{z - 2} - 4 \right| = \dfrac{|z^2 - 4z + 4|}{|z - 2|} = |z - 2| < \delta = \epsilon$

9. $\left| \dfrac{z^3 - 1}{z - 1} - 3 \right| = \dfrac{|z^3 - 3z + 2|}{|z - 1|} = |z^2 + z - 2| = |(z - 1)^2$
$+ 3(z - 1)| < \delta(\delta + 3)$. Let $\delta < 1$ and select $\epsilon = 4\delta$.

11. If approach zero along positive real axis, $z = x$ and
$$\lim_{z \to 0} \frac{z}{|z|} = \lim_{x \to 0} \frac{x}{x} = 1.$$ If approach with $z = iy$, $y > 0$, then
$$\lim_{z \to 0} \frac{z}{|z|} = \lim_{y \to 0} \frac{iy}{y} = i.$$

13. $\lim_{z \to 0} \dfrac{|z|^2}{z} = \lim_{z \to 0} \dfrac{z\bar{z}}{z} = \lim_{z \to 0} \bar{z} = 0$

15. $\lim_{z \to i} (z^3 - 2z + 3)$
$= \lim_{z \to i} z^3 - \lim_{z \to i} 2z + \lim_{z \to i} 3$ (by rule (i))
$= (\lim_{z \to i} z)^3 - 2 \lim_{z \to i} z + 3$ (by rule (ii))
$= i^3 - 2 \cdot i + 3 = 3 - 3i$

17. $\dfrac{-2i}{2 - i} = \frac{2}{5} - \frac{4}{5}i$

19. $\dfrac{i}{7 + 6i} = \frac{6}{85} + \frac{7}{85}i$

21. $|\text{Re } z - \text{Re } a| = |\text{Re } (z - a)| \leqslant |z - a| < \delta = \epsilon$ so $\lim_{z \to a} \text{Re } z = \text{Re } a$

23. $|\bar{z} - \bar{a}| = |z - a| < \delta = \epsilon$ so $\lim_{z \to a} \bar{z} = \bar{a}$.

25. $\lim_{z \to a} \text{Re } f(z) = \text{Re } (\lim_{z \to a} f(z)) = \text{Re } f(a)$

27. $||f(z)| - |f(a)|| < |f(z) - f(a)| < \epsilon$

29. $\lim_{z \to \pm 1} \dfrac{z^3 - 1}{z^2 - 1} = \lim_{z \to \pm 1} \dfrac{z^2 + z + 1}{z + 1} = \frac{3}{2}$, if $z = 1$; undefined, if $z = -1$

31. $f(z) = \bar{z}$ for $z \neq 0$ so is continuous in $\mathbb{C} - \{0\}$ by Problem 23. Since $\lim_{z \to 0} \bar{z} = 0$ define $f(0) = 0$ to make it continuous in $\mathbb{C}$.

33. $f(z) = \dfrac{x^2 - y^2}{x^2 + y^2}$ is continuous in $\mathbb{C} - \{0\}$ since quotient of continuous functions with nonzero denominator, $\lim_{z \to 0} f(z)$ doesn't exist because we get 1 if we approach along real axis and -1 along imaginary axis. Hence function cannot be made continuous at $z = 0$.

35. $czw + dw = az + b$ so $z = \dfrac{b - dw}{cw - a}$ for $z \neq -d/c$ and $w \neq a/c$. The point $-d/c$ maps to ∞ and ∞ maps to a/c.

37. $|P(z)| > |a_0| - [|a_1| + \cdots + |a_n|] \geqslant 0$ in $|z| < 1$

Problems 15.6, page 910

1. $f'(z) = 3z^2 - 2$; entire

3. $f'(z) = -1/z^2$; $z = 0$

5. $f'(z) = \dfrac{-2z^3 + 3z^2 + 1}{(z^3 + 1)^2}$; $-1, \frac{1}{2} \pm \dfrac{\sqrt{3}i}{2}$

7. $f'(z) = \dfrac{-4z}{(z^2 - 1)^2}$; $z = \pm 1$

9. $f'(z) = 8(1 - z)^{-3}$; $z = 1$

11. $f'(z) = 1 - i$, since $f(z) = (1 - i)z$

13. $f(z) = z^2 + 2z$, so $f'(z) = 2(z + 1)$

15. $u_x = \cos x \cosh y = v_y$
$-u_y = -\sin x \sinh y = v_x$ and partials are continuous.
$f'(z) = f_x(z) = u_x + iv_x = \cos x \cosh y - i \sin x \sinh y$

17. $f(z) = z^3$, so $f'(z) = 3z^2$

19. $u + iv = x - iy$, so $u = x$, $v = -y$ and $u_x = 1 \neq -1 = v_y$, so Cauchy-Riemann equations do not hold.

21. $u + iv = y$, so $u = y$, $v = 0$ and $-u_y = -1 \neq 0 = v_x$, so Cauchy-Riemann equations do not hold.

23. $u + iv = (x - iy)^2 = (x^2 - y^2) - 2xyi$, so $u_x = 2x \neq -2x = v_y$ for $x \neq 0$, and $-u_y = 2y \neq -2y = v_x$ for $y \neq 0$. Hence the only point Cauchy-Riemann equations hold is the origin, where $f'(0) = \lim_{z \to 0} \dfrac{\bar{z}^2}{z} = 0$, since
$$\left| \frac{\bar{z}^2}{z} \right| = |z|.$$

25. $u + iv = x(x - iy) = x^2 - iyx$, so
$u_x = 2x \neq -x = v_y$ for $x \neq 0$,
$-u_y = 0 \neq -y = v_x$ for $y \neq 0$.
Thus we need only check the origin:
$f'(0) = \lim_{z \to 0} \dfrac{\bar{z} \, \text{Re } z}{z} = 0$, since $\left| \dfrac{\bar{z} \, \text{Re } z}{z} \right| = |\text{Re } z|$

27. $u_x = \dfrac{y^2 - x^2}{(x^2 + y^2)^2} = v_y$ for $z \neq 0$,
and $-u_y = \dfrac{2xy}{(x^2 + y^2)^2} = v_x$ for $z \neq 0$,

with all first partials continuous in $\mathbb{C} - \{0\}$. Thus f is analytic in $\mathbb{C} - \{0\}$. Note that $f(z) = \dfrac{\bar{z}}{|z|^2} = \dfrac{\bar{z}}{z\bar{z}} = \dfrac{1}{z}$.

29. $u_x = \left(\cos\dfrac{x}{x^2+y^2}\cosh\dfrac{y}{x^2+y^2}\right)\left(\dfrac{y^2-x^2}{(x^2+y^2)^2}\right)$

$\qquad + (\sin\dfrac{x}{x^2+y^2}\sinh\dfrac{y}{x^2+y^2})\left(\dfrac{-2xy}{(x^2+y^2)^2}\right) = v_y$

$\quad -u_y = \left(\cos\dfrac{x}{x^2+y^2}\cosh\dfrac{y}{x^2+y^2}\right)\left(\dfrac{2xy}{(x^2+y^2)^2}\right)$

$\qquad + \left(\sin\dfrac{x}{x^2+y^2}\sinh\dfrac{y}{x^2+y^2}\right)\left(\dfrac{y^2-x^2}{(x^2+y^2)^2}\right) = v_x$

for $z \neq 0$, implying f is analytic in $\mathbb{C} - \{0\}$.

31. $\dfrac{f(z)}{z} = \left(\dfrac{\bar{z}}{z}\right)^2$ has value 1 on the real axis and -1 on the imaginary axis; thus no derivative at $z = 0$. But

$$u = \dfrac{x^3 - 3xy^2}{x^2+y^2}, \quad v = \dfrac{y^3 - 3x^2y}{x^2+y^2}$$

satisfy $u_x = 1 = v_y$ and $-u_y = 0 = v_x$ at $z = 0$.

33. $rv_r = r(v_x x_r + v_y y_r) = xv_x + yv_y = -xu_y + yu_x = -u_\theta$, since $x_\theta = -y$ and $y_\theta = x$. Similarly for the other identity.

35. Let $P(z) = A \prod_{k=1}^{n} (z - a_k)$, where $\operatorname{Re} a_k < 0$. Then $\dfrac{P'(z)}{P(z)} = \sum_{k=1}^{n} \dfrac{1}{z - a_k}$ for $z \neq a_k$. Let $P'(z_0) = 0$ and suppose $\operatorname{Re} z_0 \geqslant 0$. Then $\operatorname{Re}(z_0 - a_k) > 0$ and therefore, $\operatorname{Re}(z_0 - a_k)^{-1} > 0$, since $1/z = \bar{z}/|z|^2$, implying that $\operatorname{Re}(P'(z_0)/P(z_0)) > 0$, a contradiction. Hence, $\operatorname{Re} z_0 < 0$.

37. $\operatorname{Im} f^2 = 2uv$ is constant, so $|f|^2$ is constant. Apply the zero derivative theorem 15.6.3.

39. $\operatorname{Re} f^2 = u^2 - v^2 = 0$, so $|f|^2$ is constant.

41. $\dfrac{f(z_2) - f(z_1)}{z_2 - z_1} = \dfrac{-i - 1}{i - 1} = i$ and

$$f'(z) = 3z^2 = 3[t + i(1 - t)]^2$$
$$= 3[-1 + 2t + 2it(1 - t)].$$

Hence, f' is pure imaginary only when $t = \frac{1}{2}$, with $f'((1 + i)/2) = 3i/2$.

Problems 15.7, page 915

1. -1

3. $\frac{1}{2} + \dfrac{\sqrt{3}}{2}i$

5. $-\dfrac{1}{2} - \dfrac{\sqrt{3}}{2}i$

7. $e^3 i$

9. $\dfrac{1}{e}\left(\dfrac{\sqrt{3}}{2} + \dfrac{1}{2}i\right)$

11. $\dfrac{1}{e}\left(\dfrac{1}{\sqrt{2}} + \dfrac{1}{\sqrt{2}}i\right)$

13. $-e^3 i$

15. 1

17. $\dfrac{1}{e^2}\left(\dfrac{\sqrt{3}}{2} - \dfrac{1}{2}i\right)$

19. $z = \dfrac{\pi}{2}i + 2k\pi i, \quad k = 0, \pm 1, \pm 2, \ldots$

21. $z = 2k\pi - i \ln 2, \quad k = 0, \pm 1, \pm 2, \ldots$

23. $z = \ln 2 + \pi(\frac{5}{6} + 2k)i, \quad k = 0, \pm 1, \pm 2, \ldots$

25. $z = \frac{1}{3} \ln(6\sqrt{2}) + \pi i\left(\dfrac{2k}{3} - \dfrac{1}{12}\right), \quad k = 0, \pm 1, \pm 2, \ldots$

27. $\pm 1, \pm i$

29. $-2^{14}(1 + i)$

31. -2^{18}

33. $2^{15} i$

35. $-2^{13}(1 + \sqrt{3}i)$

37. Use the identity in the solution to Problem 25 of section 15.3 to get

$$\sin\dfrac{(n+1)x}{2} \cos\dfrac{nx}{2} \bigg/ \sin\dfrac{x}{2}.$$

39. $\sin\dfrac{(n+1)x}{2} \sin\dfrac{nx}{2} \bigg/ \sin\dfrac{x}{2}$

41. Use the chain rule to get $(e^{f(z)})' = e^{f(z)}f'(z)$.

43. $e^{2k\pi i/5}, \quad k = 0, 1, 2, 3, 4$

45. Let $f(z)$ be any entire solution to $f'(z) = f(z), f(0) = 1$. Then $f(z)/e^z$ is entire and

$$\left(\dfrac{f(z)}{e^z}\right)' = \dfrac{e^z f(z) - e^z f(z)}{e^{2z}} = 0.$$

Thus, by the zero derivative theorem, $f(z)/e^z$ is constant and

$$\dfrac{f(z)}{e^z} = \dfrac{f(0)}{e^0} = \dfrac{1}{1} = 1.$$

Problem 15.8, page 920

1. $i \sinh 1 \approx 1.175i$

3. $\cosh 1 \cos 1 - i \sinh 1 \sin 1 \approx 0.834 - 0.989i$

5. $i \tanh 2 \approx 0.964i$

7. $\sin 2 \cosh 5 + i \cos 2 \sinh 5 \approx 67.479 - 30.879i$

9. $\cosh 1 \cos 1 + i \sinh 1 \sin 1 \approx 0.834 + 0.989i$

11. $\cos 2\pi = 1$

13. $\cos\left(\dfrac{\pi}{4}\right) = \dfrac{1}{\sqrt{2}} \approx 0.707$

15. $i \tanh\dfrac{\pi}{4} \approx 0.656i$

17. $\sqrt{\sin^2 x \cosh^2 y + \cos^2 x \sinh^2 y}$

$\quad = \sqrt{\cosh^2 y (\sin^2 x + \cos^2 x) - \cos^2 x (\cosh^2 y - \sinh^2 y)}$

$\quad = \sqrt{\cosh^2 y - \cos^2 x}$

19. $\sqrt{\dfrac{\cosh^2 y - \cos^2 x}{\cosh^2 y - \sin^2 x}}$

21. $\left(\dfrac{\sin x \cos x}{\cosh^2 y - \sin^2 x}\right) + \left(\dfrac{\sinh y \cosh y}{\cosh^2 y - \sin^2 x}\right)i$

23. $e^{2iz}(1 + i) = (-1 + i)$ so $e^{2iz} = i$ and $z = \dfrac{\pi}{4} + k\pi$,

$\quad k = 0, \pm 1, \pm 2, \dots$

25. $\{z : z = x + yi, x = \dfrac{\pi}{2} + 2k\pi, k = 0, \pm 1, \pm 2, \dots,$

$\quad y = \ln(2 \pm \sqrt{3});$

27. $z = \pi k i, \quad k = 0, \pm 1, \pm 2, \dots$

29. $z = \ln(2 \pm \sqrt{3}) + 2\pi k i, \quad k = 0, \pm 1, \pm 2, \dots$

31. No, because $e^{-z} \neq 0$

33. $\frac{1}{2}[e^{iz} + e^{-iz}] = \frac{1}{2}[\overline{e^{iz} + e^{-iz}}] = \frac{1}{2}[e^{-iz} + e^{iz}]$

35. $\dfrac{e^{iz_1} + e^{-iz_1}}{2} \dfrac{e^{iz_2} + e^{-iz_2}}{2} + \dfrac{e^{iz_1} - e^{-iz_1}}{2i} \dfrac{e^{iz_2} - e^{-iz_2}}{2i}$

$\quad\quad = \dfrac{e^{iz_1}e^{-iz_2} + e^{-iz_1}e^{iz_2}}{2}$

37. $\dfrac{e^{2iz} - e^{-2iz}}{2i} = 2\dfrac{(e^{iz} - e^{-iz})}{2i}\dfrac{(e^{iz} + e^{-iz})}{2}$

39. $\dfrac{\sin 2z}{\cos 2z} = \dfrac{2 \sin z \cos z}{\cos^2 z - \sin^2 z}$

and divide numerator and denominator by $\cos^2 z$.

41. $|\cos z|^2 = \cos^2 x \cosh^2 y + \sin^2 x \sinh^2 y$

$\quad\quad = \cosh^2 y - \sin^2 x$

43. $\sinh z_1 \cosh z_2 + \cosh z_1 \sinh z_2$

$\quad = \frac{1}{4}[(e^{z_1} - e^{-z_1})(e^{z_2} + e^{-z_2}) + (e^{z_1} + e^{-z_1})(e^{z_2} - e^{-z_2})]$

$\quad = \frac{1}{2}[e^{z_1+z_2} - e^{-(z_1+z_2)}] = \sinh(z_1 + z_2)$

45. $i \sinh z = i \dfrac{e^z - e^{-z}}{2} = \dfrac{e^{iz} - e^{-iz}}{2i} = \sin(iz),$

$\quad \cosh z = \dfrac{e^z + e^{-z}}{2} = \dfrac{e^{iz} + e^{-iz}}{2} = \cos(iz),$ so

$\quad i \tanh z = \tan(iz)$

47. $(\sinh z)' = \left(\dfrac{e^z - e^{-z}}{2}\right)' = \dfrac{e^z + e^{-z}}{2} = \cosh z,$

$\quad (\cosh z)' = \left(\dfrac{e^z + e^{-z}}{2}\right)' = \dfrac{e^z - e^{-z}}{2} = \sinh z$

49. $(\operatorname{sech} z)' = \left(\dfrac{2}{e^z + e^{-z}}\right)' = \dfrac{-2(e^z - e^{-z})}{(e^z + e^{-z})^2}$

$\quad = \dfrac{-\sinh z}{\cosh^2 z} = -\tanh z \operatorname{sech} z,$

$(\operatorname{csch} z)' = \left(\dfrac{2}{e^z - e^{-z}}\right)' = \dfrac{-2(e^z + e^{-z})}{(e^z - e^{-z})^2}$

$\quad = \dfrac{-\cosh z}{\sinh^2 z} = -\coth z \operatorname{csch} z$

51. $\cosh z + \sinh z = \dfrac{e^z + e^{-z}}{2} + \dfrac{e^z - e^{-z}}{2} = e^z.$

Problem 15.9, page 929

1. a. $(\frac{1}{2} + 2k)\pi i, \quad k = 0, \pm 1, \pm 2, \dots$

b. $\pi i / 2$

3. a. $2k\pi i, \quad k = 0, \pm 1, \pm 2, \dots$

b. 0

5. a. $\frac{1}{2} \ln 2 + (\frac{3}{4} + 2k)\pi i, \quad k = 0, \pm 1, \pm 2, \dots$

b. $\frac{1}{2} \ln 2 + \frac{3}{4}\pi i$

7. a. $\ln 2 + (\frac{1}{6} + 2k)\pi i, \quad k = 0, \pm 1, \pm 2, \dots$

b. $\ln 2 + \pi i / 6$

9. a. $\ln 2 + (2k - \frac{1}{3})\pi i, \quad k = 0, \pm 1, \pm 2, \dots$

b. $\ln 2 - \pi i / 3$

11. a. $\frac{1}{2} \ln 13 + (\tan^{-1}(\frac{3}{2}) + 2k\pi)i, \quad k = 0, \pm 1, \pm 2, \dots$

b. $\frac{1}{2} \ln 13 + i \tan^{-1}(\frac{3}{2}) \approx 1.282 + 0.983i$

13. a. $-\ln 4 + (\frac{1}{2} + 2k)\pi i, \quad k = 0, \pm 1, \pm 2, \dots$

b. $-\ln 4 + \pi i / 2$

15. a. $e^{-2k\pi}, \quad k = 0, \pm 1, \pm 2, \dots$

b. 1

17. a. $e^{-2k\pi + i \ln 2}, \quad k = 0, \pm 1, \pm 2, \dots$

b. $e^{i \ln 2} = \cos(\ln 2) + i \sin(\ln 2) \approx 0.769 + 0.639i$

19. a. $e^{-(1/4 + 2k)\pi + (i/2)\ln 2}, \quad k = 0, \pm 1, \pm 2, \dots$

b. $e^{-(\pi/4) + (i/2)\ln 2}$

21. a. $e^{((1/2)\ln 2 - \pi/4 - 2k\pi) + i((1/2)\ln 2 + \pi/4 + 2k\pi)}$

b. $e^{((1/2)\ln 2 - \pi/4) + i((1/2)\ln 2 + \pi/4)}$

23. a. $e^{(\ln 2 + (1/3 - 2k)\pi) + i(\ln 2 + (2k - 1/3)\pi)}$

b. $e^{(\ln 2 + \pi/3) + i(\ln 2 - \pi/3)}$

25. $\operatorname{Log}(-1 - i) = \ln \sqrt{2} - \dfrac{3\pi i}{4}$, $\operatorname{Log} i = \dfrac{\pi i}{2}$ but

$\quad \operatorname{Log} \dfrac{-1 - i}{i} = \operatorname{Log}(-1 + i) = \ln \sqrt{2} + \dfrac{3\pi i}{4}$

27. $\log z^a = \log(e^{a \log z}) = a \log z$ since the exponential and logarithm are inverse functions.

29. Let $z = \cos w = (e^{iw} + e^{-iw})/2$, then $e^{2iw} - 2ze^{iw} + 1 = 0$ has roots $e^{iw} = z + (z^2 - 1)^{1/2}$ where the square root maps $[\mathbb{C} - \{0\}]^2$ onto $\mathbb{C} - \{0\}$.

31. Let $z = \sinh w = (e^w - e^{-w})/2 = (e^{2w} - 1)/2e^w$, so $e^{2w} - 2ze^w - 1 = 0$. Hence,

$$e^w = \frac{2z + \sqrt{4z^2 + 4}}{2} = z + \sqrt{z^2 + 1},$$

where the square root is two-valued, and we get the desired result by taking the logarithm of both sides of this equation.

33. Let $z = \tanh w = \dfrac{e^w - e^{-w}}{e^w + e^{-w}} = \dfrac{e^{2w} - 1}{e^{2w} + 1}$, so that

$$e^{2w} = \frac{1 + z}{1 - z},$$

from which the result follows by taking the logarithm.

35. Let $z = \cos w$, then $1 = -\sin w \cdot \dfrac{dw}{dz}$, so that

$$\frac{dw}{dz} = \frac{-1}{\sin w} = \frac{-1}{\sqrt{1 - \cos^2 w}} = \frac{-1}{\sqrt{1 - z^2}}, z \neq \pm 1.$$

37. Let $z = \sinh w$, then $1 = \cosh w \cdot \dfrac{dw}{dz}$, so that

$$\frac{dw}{dz} = \frac{1}{\cosh w} = \frac{1}{\sqrt{1 + \sinh^2 w}} = \frac{1}{\sqrt{1 + z^2}}, z \neq \pm i.$$

39. $z^{k/2}$ maps $[\mathbb{C} - \{0\}]^2$ onto $[\mathbb{C} - \{0\}]^k$, for $k = 1, 3$ so the range spaces are different for $z^{1/2}$ and $z^{3/2}$. Hence $(-1)^{1/2} \neq (-1)^{3/2}$.

Review Exercises for Chapter 15, page 929

1. $11 - 3i$

3. $\frac{7}{6} + \frac{17}{6}i$

5. $\frac{1}{10} + \frac{4}{5}i$

7. -10

9. $4, \pi, 4e^{\pi i}$

11. $4, \dfrac{\pi}{6}, 4e^{\pi i/6}$

13. $-2^{25}i$

15. $\frac{1}{2} + \dfrac{\sqrt{3}}{2}i$

17. $1, i, -1, -i$

19. Approximately $\pm(1.674 - 0.896i)$

21. Open disk of radius 3 centered at 4; boundary $= \{z : |z - 4| = 3\}$

23. Horizontal strip between $-4i$ and $4i$; boundary $= \{z : z = x + iy, \ y = \pm 4\}$

25. $\frac{3}{10} - \frac{2}{5}i$

27. $u = (x + 2)^2 - y^2, \quad v = 2y(x + 2),$

$u_x = 2(x + 2) = v_y, \quad u_y = -2y = -v_x,$

$f'(z) = 2(z + 2)$

29. $u_x = 2e^{2x} \cos 2y = v_y$

$u_y = -2e^{2x} \sin 2y = -v_x$

$f'(z) = 2f(z) = 2e^{2z}$

31. $u_x = 4x \neq -4x = v_y \quad$ for $x \neq 0$,

$u_y = -4y \neq 4y = -v_x \quad$ for $y \neq 0$,

so not differentiable for $z \neq 0$. Note that $f(z) = 2\bar{z}^2$ and

$$\lim_{z \to 0} \frac{f(z)}{z} = 2 \lim_{z \to 0} \frac{\bar{z}^2}{z} = 0.$$

33. 1

35. 1

37. $\dfrac{1}{e^3}\left(-\dfrac{\sqrt{3}}{2} + \dfrac{1}{2}i\right)$

39. $\sin 1 \cosh 1 - i \cos 1 \sinh 1 \approx 1.298 - 0.635i$

41. $\cos\left(\dfrac{\pi}{6}\right) = \dfrac{\sqrt{3}}{2}$

43. $k\pi i, \quad k = 0, \pm 1, \pm 2, \ldots$

45. $\frac{1}{2} \ln 2 + (\frac{3}{4} + 2k)\pi i$

47. $(2k + \frac{1}{2})\pi$

49. $2k\pi + i \ln(3 \pm 2\sqrt{2})$

51. $\ln(1 + \sqrt{2}) + 2k\pi i \quad$ and $\quad \ln(\sqrt{2} - 1) + (2k + 1)\pi i$

53. a. $\ln 2 + (2k + \frac{1}{2})\pi i$

    b. $\ln 2 + \frac{1}{2}\pi i$

55. a. $\ln 4 + (2k - \frac{5}{6})\pi i$

    b. $\ln 4 - \dfrac{5\pi}{6}i$

57. a. $e^{-6k\pi}$

    b. 1

59. a. $e^{i \ln 8 - (4k + 1/2)\pi}$

    b. $e^{i \ln 8 - \pi/2}$

Chapter 16

Problem 16.1, page 937

1. $z(t) = t, \quad 0 \leq t \leq 1$

3. $z(t) = it, \quad -1 \leq t \leq 1$

5. $z(t) = t + i(1 - t), \quad 0 \leq t \leq 1$

7. $z(t) = 1 + it, \quad -1 \leq t \leq 1$

9. $z(t) = (-2 + 5t) + (5 - 12t)i, \quad 0 \leq t \leq 1$

11. $x = a \cos t, \quad y = b \sin t$, so that

$$\frac{x^2}{a^2} + \frac{y^2}{b^2} = \cos^2 t + \sin^2 t = 1$$

13. $z(t) = \begin{cases} t, & 0 \leq t \leq 1 \\ 2 - t + i(t - 1), & 1 \leq t \leq 2 \\ i(3 - t), & 2 \leq t \leq 3 \end{cases}$

15. $z(t) = \begin{cases} -1 + t, & 0 \leq t \leq 2 \\ 1 + i(t - 2), & 2 \leq t \leq 3 \\ (4 - t) + i, & 3 \leq t \leq 5 \\ -1 + i(6 - t), & 5 \leq t \leq 6 \end{cases}$

17. $z(t) = \begin{cases} (1-t)+2it, & 0 \le t \le 1 \\ (1-t)+2i(2-t), & 1 \le t \le 2 \\ (t-3)+2i(2-t), & 2 \le t \le 3 \\ (t-3)+2i(t-4), & 3 \le t \le 4 \end{cases}$

19. $z(t) = (3-2i) + 5e^{it}, \ -\dfrac{\pi}{2} \le t \le \dfrac{\pi}{2}$

21. C is parametrized by $z(t) = z_0 + re^{it}, 0 \le t \le 2\pi$. Then

$dz = rie^{it} dt$ so $\displaystyle\int_C \frac{dz}{z - z_0} = \int_0^{2\pi} \frac{rie^{it} dt}{re^{it}} = it \Big|_0^{2\pi} = 2\pi i.$

23. The tangent to the arc $\gamma: z = z(t)$ at any point has slope

$\dfrac{dy}{dx} = \dfrac{y'(t)}{x'(t)} = \tan \arg z'(t).$

25. $\displaystyle\int_{|z|=1} \bar{z}\, dz = \int_0^{2\pi} e^{-it}(ie^{it} dt) = 2\pi i,$

$\displaystyle\int_{|z|=1} x\, dz = \int_0^{2\pi} \cos t(ie^{it} dt)$

$= \displaystyle\int_0^{2\pi} (-\cos t \sin t + i \cos^2 t)\, dt$

$= \tfrac{1}{2}[\cos^2 t + i(t + \sin t \cos t)]\big|_0^{2\pi} = \pi i$

$\displaystyle\int_{|z|=1} y\, dz = \int_0^{2\pi} \sin t(ie^{it} dt)$

$= i\displaystyle\int_0^{2\pi} (\sin t \cos t + i \sin^2 t)\, dt$

$= \dfrac{i}{2}[\sin^2 t + i(t - \sin t \cos t)]\big|_0^{2\pi} = -\pi.$

27. Let $\gamma: z(t) = 1 - t + ti, 0 \le t \le 1$. Then

$\displaystyle\int_\gamma y\, dz = \int_0^1 t(-1+i)\, dt = (-1+i)\frac{t^2}{2}\Big|_0^1 = \frac{-1+i}{2}.$

29. $\displaystyle\int_\gamma y\, dz = \int_1^0 0 \cdot dx + i\int_0^1 y\, dy = \frac{iy^2}{2}\Big|_0^1 = \frac{i}{2}.$

Problems 16.2, page 942

1. $-\tfrac{1}{3}(1+i)$

3. $\tfrac{1}{2}(e^{2i} - e^2)$

5. 0

7. $-1 + 3i$

9. $\cosh(3 + 3i) - \cosh(3 - 7i)$

11. Integral is zero for $n \ne -1$. For $n = -1$ see the solution to Problem 16.1.21.

13. $\displaystyle\int_\gamma e^z\, dz = \int_0^{\pi/2} e^{e^{it}}ie^{it}\, dt = e^{e^{it}}\Big|_0^{\pi/2} = e^i - e.$

15. $\displaystyle\int_{-i}^i e^{\pi z}\, dz = \frac{e^{\pi z}}{\pi}\Big|_{-i}^i = 0$

17. $\displaystyle\int_1^i (z-1)^3\, dz = \frac{(z-1)^4}{4}\Big|_1^i = \frac{(i-1)^4}{4} = -1$

19. $\displaystyle\int_{\gamma_1} \frac{dz}{z^2} = \int_0^\pi e^{-2it}(ie^{it}\, dt) = -e^{-it}\Big|_0^\pi = 2,$

$\displaystyle\int_{\gamma_2} \frac{dz}{z^2} = \int_0^\pi e^{2it}(-ie^{-it}\, dt) = -e^{it}\Big|_0^\pi = 2$

21. $\displaystyle\int_{\gamma_1} \sqrt{z}\, dz = \int_0^\pi e^{it/2}(ie^{it}\, dt) = \frac{-2}{3}(1+i),$

$\displaystyle\int_{\gamma_2} \sqrt{z}\, dz = \int_0^\pi e^{-it/2}(-ie^{-it}\, dt) = \frac{-2}{3}(1-i)$

Problems 16.3, page 951

1. $-\tfrac{3}{4} + i$ **3.** $i \sinh 1$

5. $\dfrac{1}{2}\left(\dfrac{1}{\sqrt{2}} - 1\right)$ **7.** $\tfrac{1}{2}[e^{-4\pi^2/9} - 1]$

9. 0   **11.** See Problem 16.1.21.

13. 0   **15.** 0

17. 0   **19.** 0

21. Set $z = e^{it}$ to get $-\displaystyle\int_{-\pi}^\pi t\, dt = 0$. For $0 \le \arg z \le 2\pi$ we

get $-2\pi^2$.

23. $|\text{Log } z| \le \ln R + \dfrac{\pi}{2}$ on γ, so that

$\left|\displaystyle\int_\gamma \frac{\text{Log } z}{z^2}\, dz\right| \le \left(\ln R + \frac{\pi}{2}\right)\int_{-\pi/2}^{\pi/2} \frac{dt}{R} = \frac{\pi}{R}\left(\ln R + \frac{\pi}{2}\right).$

25. See page 1022.

Problems 16.4, page 955

1. $0 = \displaystyle\int_\gamma e^z\, dz = \int_0^a e^x\, dx + ai\int_0^{\pi/2} e^{ae^{it}}e^{it}\, dt - i\int_0^a e^{iy}\, dy,$

where $e^{it} = \cos t + i \sin t$, and consider the imaginary parts of the last two integrals

3. $\text{Sin } z = \sin x \cosh y + i \cos x \sinh y$, where $x = at$ and $y = bt, 0 \le t \le T$

5. $0 = \displaystyle\int_\gamma \frac{dz}{1+z^2} = \int_{-a}^a \frac{dx}{1+x^2} + i\int_0^b \frac{dy}{(1+a^2-y^2)+2iay}$

$-i\displaystyle\int_0^b \frac{dy}{(1+a^2-y^2)-2iay} - \int_{-a}^a \frac{dx}{(1+x^2-b^2)+2ibx}$

Multiply numerator and denominator of last integral by complex conjugate and take the limit as $a \to \infty$.

7. Use the rectangle in Example 2 and the function $f(z) = e^{ikz}/(1+z^2)$.

9. Integrate $f(z) = e^{-z^2}$ on the boundary of the rectangle $0 \le x \le a, 0 \le y \le b$, and let $a \to \infty$.

11. Show that

$$\left| Ri \int_0^{\pi/8} e^{-R^2 e^{2it}} e^{it} \, dt \right| \leqslant R \int_0^{\pi/8} e^{-R^2 \cos 2t} \, dt \to 0,$$

as $R \to \infty$.

Problems 16.5, page 962

1. $-2\pi i$

3. $2\pi i e$

5. $\sqrt{2}\pi i$

7. $2\pi i$

9. 0

11. $2\pi i$

13. $\dfrac{2\pi i}{5}(1 - e^{-5})$

15. $6\pi i$

17. $-\pi i$

19. $4\pi/e$

21. $\pi i(1 + \sqrt{3})/4!$

23. $-\pi i$

25. $\pi i/e$

27. $-\pi i/e$

29. 0

31. $-2\pi i/(b - a)^2$

33. Apply Cauchy's estimate to f with $M = (1 - r)^{-1}$ and minimize the result for all $0 \leqslant r \leqslant 1$.

35. Let $z = e^{i\theta}$, $0 \leqslant \theta \leqslant 2\pi$, and equate imaginary terms.

37. Let $P(z) = (z - r_1)^{k_1} \cdot \cdots \cdot (z - r_n)^{k_n} Q(z)$, where $r_1, \ldots,$ r_n lie inside γ and the roots of $Q(z)$ lie outside of γ. Then

$$\frac{1}{2\pi i}\int_\gamma \frac{P'(z)\,dz}{P(z)} = \frac{1}{2\pi i}\int_\gamma \left[\frac{k_1}{z - r_1} + \cdots + \frac{k_n}{z - r} + \frac{Q'(z)}{Q(z)}\right] dz$$
$$= k_1 + \cdots + k_n,$$

since Q'/Q is analytic inside and on γ.

Problems 16.6, page 966

1. If not, f is bounded, contradicting Liouville's theorem.

3. Let $|z_0| < r$, then by Cauchy's formula for derivatives

$$|f''(z_0)| \leqslant \frac{1}{\pi}\int_{|z - z_0| = r} \frac{|f(z)|}{|z - z_0|^3}|dz|$$

$$\leqslant \frac{1}{\pi r^3}\int_{|z - z_0| = r} A|z|^2|dz| \leqslant \frac{A(2r)^2 2\pi r}{\pi r^3} = 8A.$$

Since f'' is entire and bounded (by the inequality above), Liouville's theorem implies f'' is constant. Integrate.

5. Let

$$F(0) = \lim_{z \to 0} \frac{f(z)}{z} = f'(0).$$

Then F is continuous on $|z| < 1$. Analyticity follows since, for any $r < 1$,

$$\frac{1}{2\pi i}\int_{|z|=r} \frac{F(z)}{z - \zeta}\,dz = \begin{cases} \dfrac{1}{2\pi i \zeta}\int_{|z|=r} f(z)\left[\dfrac{1}{z - \zeta} - \dfrac{1}{z}\right]dz = \dfrac{f(\zeta)}{\zeta} \\ \qquad = F(\zeta) \\ \dfrac{1}{2\pi i}\int_{|z|=r} \dfrac{f(z)}{z^2}\,dz = f'(0), \ \zeta \neq 0. \end{cases}$$

7. By Cauchy's estimate, $|f'(0)| \leqslant r^{-1}$ for all $0 < r < 1$.

9. Let $f(z) = z$, and let Ω be the open unit disk.

11. $|f| \neq 0$ on the boundary of Ω and if f has no zeros in Ω, the maximum and minimum principles imply f is constant in Ω.

Problems 16.7, page 970

1. $u_{xx} = u_{yy} = 0$

3. $u_{xx} = u_{yy} = 0$

5. $u_{xx} = u_{yy} = 0$

7. $u_{xx} = 6x = -u_{yy}$

9. $u = \text{Re} \sin z$

11. $u_{xx} = 0$, $u_{yy} = 2x/y^3$ so not harmonic. Thus not real or imaginary part of an analytic function.

13. $u_{xx} + u_{yy} \neq 0$, so no.

15. $u_{xx} + u_{yy} = 5u \neq 0$, so no.

17. $\frac{1}{2}(x^2 - y^2) + k$

19. $5x + 2y + k$

21. $\arg z + k$

23. $2x(1 - y) + k$

25. $-\ln(x^2 + y^2) + k$

27. $\Delta(au + bv) = a\,\Delta u + b\,\Delta v$
$\Delta(uv) = u\,\Delta v + v\,\Delta u + 2(u_x v_x + u_y v_y)$ and Cauchy-Riemann equations hold for conjugate harmonic functions.

29. Cut the region into simply connected regions and apply maximum principle on each of these. Hence maximum is on their boundaries. But cut lines are arbitrary.

Review Exercises for Chapter 16, page 971

1. $z(t) = 2(1 - t) - 2it$, $0 \leqslant t \leqslant 1$

3. $z(t) = e^{it}$, $-\pi/2 \leqslant t \leqslant \pi/2$

5. $z(t) = \begin{cases} t, & -1 \leqslant t \leqslant 2 \\ 2 + i(t - 2), & 2 \leqslant t \leqslant 5 \\ (7 - t) + 3i, & 5 \leqslant t \leqslant 8 \\ -1 + (11 - t)i, & 8 \leqslant t \leqslant 11 \end{cases}$

7. $1 - i$

9. 0

11. $(1 + i)/2$

13. 0

15. 0

17. 0

19. πi

21. 0

23. $-8i$

25. $\dfrac{1}{2}\left[\dfrac{1}{e} - 1\right]$

27. 0

29. 0

31. $2\pi i$

33. $22\pi i$

35. $-26\pi i$

37. $\frac{125}{3}\pi i e^{5i}$

39. $-2\pi i[\sinh 1 + 2\cosh 1]$

41. $\int_0^{(a+ib)T} \cos z \, dz = \sin(a+ib)T$

$$= \sin aT \cosh bT + i \cos aT \sinh bT,$$

and

$$\int_0^{(a+ib)T} \cos z \, dz = (a+ib) \int_0^T (\cos at \cosh bt$$
$$- i \sin at \sinh bt) dt.$$

Solve for the real part of the last integral.

43. b. $5iz^2$

45. b. $-iz^4/4$

47. b. $6 \log z$

Chapter 17

Problems 17.1, page 977

1. $8, 11\frac{1}{2}, 16, 20\frac{3}{4}$; unbounded; divergent

3. $e^{-\pi i/6}, e^{-\pi i/3}, e^{-\pi i/2}, e^{-2\pi i/3}$; bounded; divergent

5. $0, -1, 0, 1$; bounded; divergent

7. $-i, -\dfrac{1}{\sqrt{2}}, \dfrac{i}{\sqrt{3}}, \dfrac{1}{2}$; bounded; convergent, $L = 0$

9. $-i, -\dfrac{1}{4}, \dfrac{i}{27}, \dfrac{1}{256}$; bounded; convergent, $L = 0$

11. $\dfrac{1}{2} - \dfrac{i}{2}, \dfrac{8}{5} - \dfrac{4}{5}i, \dfrac{27}{10} - \dfrac{9}{10}i, \dfrac{64}{17} - \dfrac{16}{17}i$;
unbounded, divergent.

13. $\dfrac{i}{2}, \dfrac{2i}{3}, \dfrac{3i}{4}, \dfrac{4i}{5}$; bounded; convergent, $L = i$

15. $i \sinh 1, \dfrac{i}{2} \sinh 2, \dfrac{i}{3} \sinh 3, \dfrac{i}{4} \sinh 4$;
unbounded divergent

17. If $z_n \to a$ and $z_n \to b$, then there exists an N such that $n \geq N$ implies $|z_n - a| < \epsilon$ and an M such that $n \geq M$ implies $|z_n - b| < \epsilon$. If $|a - b| > 2\epsilon$, then when $n \geq \max(N, M)$ we have by the triangle inequality

$$2\epsilon < |a - b| = |a - z_n + z_n - b|$$
$$\leq |a - z_n| + |z_n - b| < 2\epsilon,$$

a contradiction.

19. $|z_n - z_m| \leq \dfrac{1}{n} + \dfrac{1}{m} < \dfrac{2}{n} \leq \dfrac{2}{N} < \epsilon$ for $n \geq N > 2/\epsilon$.
Hence Cauchy; $L = 0$.

21. $|z_n - z_m| \leq \dfrac{2}{n} + \dfrac{2}{m} < \dfrac{4}{n} \leq \dfrac{4}{N} < \epsilon$ for $n \geq N > \dfrac{4}{\epsilon}$.
Cauchy; $L = 0$

Problems 17.2, page 982

1. $1 + i$

3. $1 - i$

5. $\dfrac{(18 - 3\sqrt{3}) + 3i}{20 - 6\sqrt{3}}$

7. converges

9. converges

11. diverges

13. converges

15. converges

17. diverges, since $\left(\dfrac{k}{k+20i}\right)^k > \left(1 - \dfrac{20}{k+20}\right)^k >$
$\left(1 - \dfrac{20}{k+20}\right)^{k+20} \to e^{-20}$

19. $\dfrac{1}{(k+1)^2} < \dfrac{1}{k(k+1)}$ and $\displaystyle\sum_{k=1}^{\infty} \dfrac{1}{k(k+1)}$ converges.

21. Take the real and imaginary parts of the result in problem 20:

$$1 + \cos\theta + \cos 2\theta + \cdots = \dfrac{1}{2}$$

$$\sin\theta + \sin 2\theta + \cdots = \dfrac{1}{2} \cot\dfrac{\theta}{2}$$

Problems 17.3, page 993

1. $1 - z^2 + z^4 - z^6 + \cdots$

3. $\dfrac{\pi}{4} = \tan^{-1} 1 = 1 - \dfrac{1}{3} + \dfrac{1}{5} - \dfrac{1}{7} + \cdots,$

so $\pi = 4 \displaystyle\sum_{k=0}^{\infty} \dfrac{(-1)^k}{2k+1} \approx 4 \sum_{k=0}^{118} \dfrac{(-1)^k}{2k+1} \approx 3.149996$

5. $R = 6$

7. $R = \dfrac{1}{4}$

9. $R = 1$

11. $R = 1$

13. $R = \infty$

15. $R = \infty$

17. $R = 16$

19. $R = 0$

21. $R = 1/e$

23. $R = 1$

25. $2 \displaystyle\sum_{k=0}^{\infty} z^{2k}; R = 1$

27. $\displaystyle\sum_{k=0}^{\infty} (k+1)(-z)^k; R = 1$

29. $-\dfrac{1}{3} \displaystyle\sum_{k=0}^{\infty} \left[1 - \left(\dfrac{-1}{2}\right)^{k+1}\right] z^k; R = 1$

31. $R = 1$ and apply Theorem 4

33. $R = 6$, apply Theorem 4

35. $R = 1$, apply Theorem 4

37. $R = \infty$, apply Theorem 4

39. $\displaystyle\sum_{k=0}^{\infty} (e^{-z})^k = \dfrac{1}{1 - e^{-z}}$ if $|e^{-z}| < 1$
But $|e^{-z}| = e^{-\text{Re } z} \leq e^{-\delta} < 1$, so convergence is uniform.

41. $R = 1$, apply Theorem 4

43. a. $R = \infty$

b. $J_0' = \sum\limits_{k=1}^{\infty} \dfrac{k(-1)^k}{(k!)^2}\left(\dfrac{z}{2}\right)^{2k-1}$

$J_0'' = \sum\limits_{k=1}^{\infty} \dfrac{k(k-\frac{1}{2})(-1)^k}{(k!)^2}\left(\dfrac{z}{2}\right)^{2k-2}$

so

$z^2 J_0'' + z J_0' = \sum\limits_{k=1}^{\infty} \dfrac{4k^2(-1)^k}{(k!)^2}\left(\dfrac{z}{2}\right)^{2k}$

$= z^2 \sum\limits_{k=1}^{\infty} \dfrac{(-1)^k}{[(k-1)!]^2}\left(\dfrac{z}{2}\right)^{2k-2} = -z^2 J_0.$

45. $f(z) = \sum\limits_{n=0}^{\infty}\left(\int_0^1 t^n g(t)\,dt\right)\dfrac{z^n}{n!}$ and $\left|\int_0^1 t^n g(t)\,dt\right| \le \dfrac{M}{n+1}$,

where $|g(t)| \le M$ on $0 \le t \le 1$. Hence,

$R^{-1} = \lim\limits_{n} \sqrt[n]{\dfrac{M}{(n+1)!}} = 0$, so $R = \infty$.

47. $\sin zt = \sum\limits_{n=0}^{\infty} \dfrac{(-1)^n(zt)^{2n+1}}{(2n+1)!}$

converges uniformly on any closed bounded subset of $\mathscr{C}$. Hence,

$f(z) = \sum\limits_{n=0}^{\infty}(-1)^n\left(\int_0^1 g(t)t^{2n+1}\,dt\right)\dfrac{z^{2n+1}}{(2n+1)!}$

is entire, since

$\left|\int_0^1 g(t)t^{2n+1}\,dt\right| \le \dfrac{M}{2n+2}$,

where

$|g(t)| \le M$ on $0 \le t \le 1$, and

$R^{-1} = \lim\limits_{n} \sqrt[n]{\dfrac{M}{(2n+2)!}} = 0.$

Finally,

$f'(z) = \sum\limits_{n=0}^{\infty}(-1)^n\left(\int_0^1 g(t)t^{2n+1}\,dt\right)\dfrac{z^{2n}}{(2n)!}.$

Problems 17.4, page 1002

1. $\sum\limits_{k=0}^{\infty}\dfrac{(-1)^k 2^{2k+1}}{(2k+1)!}z^{2k+1};\quad R=\infty$

3. $\sum\limits_{k=0}^{\infty}\dfrac{z^{2k+1}}{(2k+1)!};\quad R=\infty$

5. $\sum\limits_{k=0}^{\infty} z^{4k};\quad R=1$

7. $\sum\limits_{k=0}^{\infty}\dfrac{(-1)^k}{k!}z^{3k};\quad R=\infty$

9. $\sum\limits_{k=1}^{\infty}\dfrac{(-1)^{k+1}}{k}(iz)^k;\quad R=1$

11. $\sum\limits_{k=0}^{\infty}\dfrac{(-1)^k z^{2k}}{(2k+1)!};\quad R=\infty$

13. $\sum\limits_{k=0}^{\infty}\dfrac{z^{2(k+1)}}{k!};\quad R=\infty$

15. $\cos^2 z = \dfrac{1+\cos 2z}{2}$

$= 1 + \sum\limits_{k=1}^{\infty}\dfrac{(-1)^k 2^{2k-1}}{(2k)!}z^{2k};\quad R=\infty$

17. $\dfrac{1}{1-i-(z-i)} = \dfrac{1}{1-i}\left[1 + \dfrac{z-i}{1-i} + \left(\dfrac{z-i}{1-i}\right)^2 + \cdots\right]$,

$|z-i| < \sqrt{2}$

19. $\sum\limits_{n=0}^{\infty}(-1)^n\dfrac{(z-\pi/2)^{2n}}{(2n)!},\quad |z| < \infty$

21. $\dfrac{i\pi}{2} - \sum\limits_{n=1}^{\infty}\dfrac{(i)^n}{n}(z-i)^n,\quad |z-i| < 1$

23. $\ln|2| + 3\pi i - \sum\limits_{n=1}^{\infty}\dfrac{(z-2e^{3\pi i})^n}{2^n n},\ |z-2e^{3\pi i}| < 2$

25. $\sum\limits_{k=2}^{\infty}\dfrac{(-1)^k(z-1)^k}{(k-1)},\quad |z-1| < 1$

27. $-\dfrac{z^2}{2!} - \dfrac{2}{4!}z^4 - \dfrac{16}{6!}z^6$

29. Expand $e^{\alpha\,\mathrm{Log}(1+z)}$ in a Maclaurin series.

31. $1 + \sum\limits_{k=1}^{\infty}\dfrac{\frac{1}{4}(-\frac{3}{4})\cdot\cdots\cdot(\frac{5}{4}-k)}{k!}z^k$

33. 0.5038

35. a. $\mathrm{erf}(z) = \dfrac{2}{\sqrt{\pi}}\sum\limits_{k=0}^{\infty}\dfrac{(-1)^k z^{2k+1}}{k!(2k+1)}$

b. $\mathrm{erf}(1) \approx 0.8427$
$\mathrm{erf}(\frac{1}{2}) \approx 0.5205$

37. b. $\mathrm{Si}(z) = \sum\limits_{k=0}^{\infty}\dfrac{(-1)^k z^{2k+1}}{(2k+1)!(2k+1)}$

c. $\mathrm{Si}(1) \approx 0.9461$
$\mathrm{Si}(\frac{1}{2}) \approx 0.4931$

39. c. $B_0 = 1,\quad B_1 = -\frac{1}{2},\quad B_2 = \frac{1}{6},\quad B_3 = 0,\quad B_4 = \dfrac{-1}{6!},$

$B_5 = 0,\quad B_6 = \frac{1}{30240},\quad B_7 = 0.$

41. a. $\tan z$ can be factored from each odd derivative

b. $E_0 = 1,\quad E_2 = -1,\quad E_4 = 5,\quad E_6 = -61\quad E_8 = 1385$

43. $f(0) = 0,\quad f'(z) = 1 + zf(z)|_{z=0} = 1,$

$f''(z) = f(z) + zf'(z)|_{z=0} = 0,$

$f'''(z) = 2f'(z) + zf''(z)|_{z=0} = 2,$

$f^{(4)}(z) = 3f''(z) + zf'''(z)|_{z=0} = 0$, and in general,

$f^{(n)}(z) = (n-1)f^{(n-2)}(z) + zf^{(n-1)}(z)|_{z=0}$

$$= \begin{cases} 0, & n \text{ even,} \\ \left[\dfrac{n}{2}\right]! 2^{[n/2]}, & n \text{ odd,} \end{cases}$$

where $[x]$ is largest integer $\leqslant x$, so

$$f(z) = \sum_{n=0}^{\infty} \frac{2^n n! z^{2n+1}}{(2n+1)!}, \quad R = \infty.$$

45. $(1-z^2) \sum_{k=2}^{\infty} k(k-1)a_k z^{k-2} - 2z \sum_{k=1}^{\infty} k a_k z^{k-1}$

$+ n(n+1) \sum_{k=0}^{\infty} a_k z^k = 0,$

so that

$$\sum_{k=0}^{\infty} [(k+2)(k+1)a_{k+2} - (k(k-1)+2k-n(n+1))a_k]z^k = 0$$

and

$$(k+2)(k+1)a_{k+2} = [k(k+1) - n(n+1)]a_k,$$

or

$$a_{k+2} = \frac{-(n+k+1)(n-k)a_k}{(k+2)(k+1)}.$$

Thus

$$f(z) = a_0\left[1 - \frac{n(n+1)z^2}{2!} + \frac{(n-2)n(n+1)(n+3)z^4}{4!} \right.$$

$$\left. - \cdots \right] + a_1\left[z - \frac{(n-1)(n+2)z^3}{3!} \right.$$

$$\left. + \frac{(n-3)(n-1)(n+2)(n+4)z^5}{5!} - \cdots \right].$$

47. Use the Taylor series of $f(z)$ and $g(z)$ centered at z_0 to get

$$f(z)/g(z) = \frac{f'(z_0)(z-z_0) + f''(z_0)(z-z_0)^2/2 + \cdots}{g'(z_0)(z-z_0) + g''(z_0)(z-z_0)^2/2 + \cdots},$$

Then divide numerator and denominator by $z - z_0$ and take the limit as $z \to z_0$.

Problems 17.5, page 1008

1. $\displaystyle\sum_{k=-1}^{\infty} \frac{z^k}{(k+1)!}$

3. $\displaystyle\sum_{k=-\infty}^{0} \frac{(-1)^k z^{1+4k}}{(-2k)!}$

5. $\displaystyle\sum_{k=0}^{\infty} \frac{z^{2k-10}}{k!} = \sum_{k=-5}^{\infty} \frac{z^{2k}}{(k+5)!}$

7. $\displaystyle\sum_{k=0}^{\infty} \frac{z^{1-k}}{k!} = \sum_{k=-\infty}^{1} \frac{z^k}{(1-k)!}$

9. $\dfrac{1}{z} - 1 + z - z^2 + z^3 - \cdots = \displaystyle\sum_{n=0}^{\infty} (-1)^n z^{n-1}$

11. $-\displaystyle\sum_{n=0}^{\infty} (1-z)^{-(n+1)} - \frac{1}{2} \sum_{n=0}^{\infty} \left(\frac{1-z}{2}\right)^n$

13. $\displaystyle\sum_{n=1}^{\infty} z^{-2n-1}$

15. $\dfrac{1}{4} \displaystyle\sum_{n=-1}^{\infty} \left(\frac{1-z}{2}\right)^n + \sum_{n=-\infty}^{-2} (1-z)^n$

17. $\dfrac{1/3}{z-1} + \dfrac{2}{9} \displaystyle\sum_{n=0}^{\infty} \left(\frac{1-z}{3}\right)^n$

19. $\dfrac{1}{3} \displaystyle\sum_{n=1}^{\infty} \left[\left(\frac{-z}{2}\right)^n + z^{-n}\right] + \frac{1}{3}$

21. $\dfrac{1}{z+2} + \displaystyle\sum_{n=1}^{\infty} \frac{3^{n-1}}{(z+2)^{n+1}}$

23. $\left(\displaystyle\sum_{n=0}^{\infty} (1-z)^n\right) \cdot \left(\sum_{n=0}^{\infty} \frac{(-1)^n (z-1)^{-(2n+1)}}{(2n+1)!}\right) =$

$\displaystyle\sum_{n=-\infty}^{\infty} (-1)^{n+1} c_n (z-1)^n \quad$ where $\quad c_n = \displaystyle\sum_{k=q}^{\infty} \frac{(-1)^k}{(2k+1)!}$,

with $q = 0$, for $n \geqslant -1$, $q = [|n/2|]$, for $n \leqslant -2$, where $[m]$ is the largest integer $\leqslant m$.

25. $\displaystyle\sum_{n=-\infty}^{\infty} c_n z^n$, where $c_n = c_{-n}$ and $c_n = \displaystyle\sum_{k=0}^{\infty} [k!(n+k)!]^{-1}$

27. $|a_n| = \left|\dfrac{1}{2\pi i} \displaystyle\int_{|\zeta-z_0|=\rho} \frac{f(\zeta)\,d\zeta}{(\zeta-z_0)^{n+1}}\right| \leqslant \dfrac{M 2\pi\rho}{2\pi\rho^{n+1}} = \dfrac{M}{\rho^n}$

for $r < \rho < R$. Then let ρ tend to r or R. Setting $r = 0$ implies $a_{-n} = 0$ for $n = 1, 2, \ldots$, so z_0 is a removable singularity and $f(z_0) = a_0$.

29. $\displaystyle\int_{|z|=1} z^{n-1} e^{1/z}\,dz = \int_{|z|=1} z^{n-1}\left(1 + \frac{1}{z} + \frac{1}{2!z^2} + \cdots\right)dz = \frac{2\pi i}{n!}.$

Let $z = e^{i\theta}$, getting

$$\frac{2\pi i}{n!} = \int_0^{2\pi} e^{\cos\theta - i\sin\theta} e^{i(n-1)\theta} i e^{i\theta}\,d\theta$$

$$= i\int_0^{2\pi} e^{\cos\theta - i(\sin\theta - n\theta)}\,d\theta.$$

Divide by i and take the real part of both sides.

31. Let $\pi < \rho < 2\pi$ and consider

$$a_n = \frac{1}{2\pi i} \int_{|z|=\rho} \frac{dz}{z^{n+1}\sin z}$$

where $z^{n+1}\sin z$ has simple zeros at $z = \pm\pi$ and a zero of order $n + 2$ at $z = 0$. By Cauchy's theorem,

$$2\pi i a_n = \int_{|z-\pi|=\epsilon} \frac{-dz}{z^{n+1} \sin(z-\pi)} + \int_{|z|=\epsilon} \frac{dz}{z^{n+1} \sin z}$$

$$+ \int_{|z+\pi|=\epsilon} \frac{-dz}{z^{n+1} \sin(z+\pi)}.$$

But $\dfrac{1}{\sin \zeta} = \dfrac{1}{\zeta} + \dfrac{\zeta}{3!} + \dfrac{7}{360} \zeta^3 + \cdots$, so that

$$\int_{|z-\pi|=\epsilon} \frac{-dz}{z^{n+1} \sin(z-\pi)} = \int_{|z-\pi|=\epsilon} \frac{-z^{-n-1} dz}{z-\pi} = -2\pi i \pi^{-n-1}.$$

Similarly for $|z + \pi| = \epsilon$. For $|z| = \epsilon$, note that the integral

is zero if n is even. Thus, $a_0 = 0$, $a_1 = \dfrac{1}{3!} - \dfrac{2}{\pi^2}$, $a_{-1} = 1$,

and so on.

Problems 17.6, page 1015

1. $z = 0$ of order 3
3. $z = \pm i$ of order 4
5. $z = 0$ of order 4 and $z = k\pi$, $k \neq 0$, of order 1
7. $z = 1$, $e^{2\pi i/5}$, $e^{4\pi i/5}$, $e^{6\pi i/5}$, $e^{8\pi i/5}$ of order 1
9. $z = 0$ of order 10
11. The difference of the two functions has a nonisolated zero in G. Since the difference is analytic, it must be zero.
13. $f(z) = 0$ for $z = 1/2n$, but $f(z) = z$ for $z = 1/(2n + 1)$. Since f coincides with **both** 0 and z on a subset of $|z| < 2$ having a limit point at 0, f must equal **both** 0 and z. This is impossible.
15. $f(z) = \dfrac{2z}{3z + 1}$ for $z = 1, \frac{1}{3}, \frac{1}{5}, \ldots$, while

$f(z) = \dfrac{-2z}{2z + 1}$ for $z = \frac{1}{2}, \frac{1}{4}, \ldots$,

so that f coincides with two different analytic functions on subsets of $|z| < \frac{1}{3}$ that contain a limit point at 0. This is impossible.
17. $g(z) = f(z) - \alpha$ is nonconstant and analytic in Ω and its zeros are isolated.
19. simple pole at $z = 0$
21. simple poles at $z = \pm i$
23. simple poles at $z = k\pi$, $k = 0, \pm 1, \pm 2, \ldots$
25. essential singularity at $z = 0$
27. essential singularity at $z = 0$
29. pole of order 4 at $z = 0$
31. simple poles at $z = \dfrac{1}{k\pi}$, $k = \pm 1, \pm 2, \ldots$; nonisolated

singularity at $z = 0$
33. simple poles at $z = k\pi$, $k = 0, \pm 1, \pm 2, \ldots$

35. pole of order 9 at $z = 0$
37. simple poles at $z = (k + \frac{1}{2})\pi i$, $k = 0, \pm 1, \ldots$
39. poles of order 2 at $z = 0, \pm \pi, \pm 2\pi, \ldots$
41. removable singularity at $z = 0$, simple poles at $z = k\pi$, $k = \pm 1, \pm 2, \ldots$
43. $z = 2/(2k + 1)\pi$, $k = 0, \pm 1, \ldots$, are essential singularities, $z = 0$ is a limit point of essential singularities, $z = \infty$ is removable.
45. If z_0 is removable, then $\lim_{z \to z_0} f(z)$ exists so $\lim_{z \to z_0}(z - z_0) f(z) = 0$. If z_0 is a pole or an essential singularity, then $\lim_{z \to z_0}(z - z_0) f(z)$ is nonzero or doesn't exist.
47. $\text{Si}(z) = \displaystyle\int_0^z \left(\sum_{n=0}^{\infty} (-1)^n \frac{\zeta^{2n}}{(2n + 1)!} \right) d\zeta$

$= \displaystyle\sum_{n=0}^{\infty} \frac{(-1)^n z^{2n+1}}{(2n + 1)(2n + 1)!}$,

$\text{Si}(0) = 0$.
49. $E(z) = \displaystyle\int_0^z \left(\sum_{n=0}^{\infty} \frac{\zeta^n}{(n + 1)!} \right) d\zeta = \sum_{n=0}^{\infty} \frac{z^{n+1}}{(n + 1)(n + 1)!}$,

$E(0) = 0$.
51. Let $w = z^{-1}$ and observe that e^w maps each strip $(2k - 1)\pi < |\text{Im } w| < (2k + 1)\pi$ onto $\mathbb{C} - \{0\}$.
53. If $f(z)$ is meromorphic in $\mathscr{M}$, it has finitely many isolated poles $z_1, \ldots, z_n$ of orders $r_1, \ldots, r_n$. Then,

$$f(z) = \frac{g(z)}{(z - z_1)^{r_1} \cdot \cdots \cdot (z - z_n)^{r_n}},$$

where g is entire and has a nonessential singularity at ∞. Hence, g is a polynomial by Problem 52.

Review Exercises for Chapter 17, page 1016

1. converges
3. converges
5. converges
7. $R = 4$
9. $R = 1$
11. $R = \infty$
13. $\frac{9}{13} + \frac{6}{13}i$
15. $\frac{6}{5} + \frac{3}{5}i$
17. $1 + z^4 + \dfrac{z^8}{2!} + \cdots = \displaystyle\sum_{k=0}^{\infty} \frac{z^{4k}}{k!}$
19. $i \displaystyle\sum_{k=0}^{\infty} \frac{z^{2k+1}}{(2k + 1)!} = i \sinh z$
21. $\displaystyle\sum_{k=1}^{\infty} \frac{(-iz)^k}{k} = -iz + \frac{z^2}{2} + \frac{iz^3}{3} - \frac{z^4}{4} + \cdots$
23. $-\displaystyle\sum_{k=0}^{\infty} \frac{(-1)^k (z - \pi)^{2k+1}}{(2k + 1)!}$

25. $c_1 = -1$, $c_3 = \dfrac{-1}{3}$, $c_5 = \dfrac{-2}{15}$, $c_7 = \dfrac{-17}{315}$

27. $R = \infty$ and apply Theorem 17.3.4

29. $\left| \displaystyle\sum_{k=n+1}^{\infty} \dfrac{e^{-2i|z|}}{k^3} \right| \le \displaystyle\sum_{k=n+1}^{\infty} \dfrac{1}{k^3} \le \displaystyle\int_n^{\infty} x^{-3}\, dx = \dfrac{1}{2n^2} < \epsilon$

for n sufficiently large

31. $\displaystyle\sum_{k=0}^{\infty} \dfrac{z^{2k-5}}{k!}$

33. $\displaystyle\sum_{k=0}^{\infty} \dfrac{(-1)^k z^{2-k}}{k!}$

35. $\displaystyle\sum_{k=0}^{\infty} \dfrac{(-1)^k z^{k-1}}{4^{k+1}}$

37. $\dfrac{1}{16} \displaystyle\sum_{k=0}^{\infty} (-1)^k \left(\dfrac{4}{z-4} \right)^{k+1} - \dfrac{1}{32} \displaystyle\sum_{k=0}^{\infty} (-1)^k \left(\dfrac{z-4}{8} \right)^k$

39. $\dfrac{-2}{z-2} - 3 \displaystyle\sum_{k=0}^{\infty} (z-2)^k$

41. $2 \displaystyle\sum_{k=0}^{\infty} \dfrac{(-1)^k}{(2k)!(z-2)^{2k}} + \displaystyle\sum_{k=0}^{\infty} \dfrac{(-1)^k}{(2k)!(z-2)^{2k-1}}$

43. $z = 2$ of order 3

45. $z = 0$ of order 7; $z = 2k\pi$ of order 2

47. simple pole at $z = 0$

49. essential singularity at $z = 0$

51. simple pole at $z = 1$, pole of order 2 at $z = 2$, pole of order 10 at $z = \pm i$

53. pole of order 5 at $z = 0$

Chapter 18

Problems 18.1, page 1031

1. $\mathrm{Res}_{-1} f = 1$

3. $\mathrm{Res}_{-1} f = -\frac{1}{2}$, $\mathrm{Res}_1 f = \frac{1}{2}$

5. $\mathrm{Res}_0 f = 1$

7. $\mathrm{Res}_0 f = 0$

9. $\mathrm{Res}_0 f = -1$, $\mathrm{Res}_1 f = \dfrac{e}{2}$, $\mathrm{Res}_{-1} f = \dfrac{1}{2e}$

11. $\mathrm{Res}_0 f = 1$

13. $\mathrm{Res}_4 f = \frac{4}{3}$, $\mathrm{Res}_1 f = -\frac{4}{3}$

15. $\mathrm{Res}_0 f = \frac{5}{2}$, $\mathrm{Res}_1 f = -8$, $\mathrm{Res}_2 f = \frac{17}{2}$

17. $\mathrm{Res}_{\pm i} f = 0$

19. $\mathrm{Res}_0 f = -\frac{1}{2}$

21. $2\pi i$

23. $2\pi i$

25. $2\pi i$

27. $2\pi i \cos(1)$

29. $\pi i/16$

31. $2\pi i$

33. $2\pi i(1 - \cos 2)$

35. 0

37. $-4\pi i$

39. 0

41. $(50 \cos 1 - 4 \sin 1 + 3 \sinh 2)\pi i/25$

43. 0

45. $-3\pi i/64$

47. $-4i$

49. 0

51. Let $\gamma : |z| = R$ be sufficiently large so that all poles of F and C lie inside this circle. Then

$$2\pi i \sum_{S^*} \mathrm{Res}\, F = \int_{\gamma - C} F(z)\, dz$$

$$= \int_{|z|=R} F(z)\, dz - \int_C F(z)\, dz.$$

But $\lim_{R \to \infty} \int_{|z|=R} F(z)\, dz = 0$ because

$$\left| \int_{|z|=R} F(z)\, dz \right| \le \int_{|z|=R} \dfrac{|z^2 F(z)|}{|z|^2} |dz| \le \dfrac{2\pi R M}{R^2}.$$

53. 0

55. πi^n

57. Remove all common factors of P and Q. Then

$$\left(\dfrac{P}{(z-r)^a Q_1} \right)' = \dfrac{1}{(z-r)^a} \left(\dfrac{P}{Q_1} \right)' - \dfrac{1}{(z-r)^{a+1}} \cdot \left(\dfrac{aP}{Q_1} \right)$$

so $\mathrm{Res}_r \left(\dfrac{P}{Q} \right)' = \lim_{z \to r} \dfrac{d^a}{dz^a} \left\{ (z-r) \left(\dfrac{P}{Q_1} \right)' - \dfrac{aP}{Q_1} \right\} = 0.$

Problems 18.2, page 1035

1. $2\pi/\sqrt{3}$

3. $-35\pi/24$

5. $\pi(3 - 2\sqrt{2})$

7. $67\pi/24$

9. $\dfrac{1}{4} \displaystyle\int_0^{2\pi} \dfrac{d\theta}{a + \sin^2 \theta} = i \displaystyle\int_{|z|=1} \dfrac{z\, dz}{(z^2 - 1)^2 - 4az^2}$

and $|\sqrt{a} - \sqrt{a+1}| < 1$

11. $-4i \displaystyle\int_{|z|=1} \dfrac{z\, dz}{(a^2 - b^2)z^4 + 2(a^2 + b^2)z^2 + (a^2 - b^2)}$

$$= \dfrac{-4i}{a^2 - b^2} \cdot \displaystyle\int_{|z|=1} \dfrac{z\, dz}{\left(z^2 + \dfrac{a+b}{a-b} \right)\left(z^2 + \dfrac{a-b}{a+b} \right)}$$

and $|a - b| < a + b$

13. $i \displaystyle\int_{|z|=1} \dfrac{dz}{(az - 1)(z - a)}$

15. $\dfrac{-i}{2^n} \cdot \displaystyle\int_{|z|=1} \dfrac{[(a-ib)z^2 + (a+ib)]^n \, dz}{z^{n+1}}$

and calculate residue at $z = 0$.

17. Use $\cos ib = \cosh b$ and $\sin ib = i \sinh b$.

Problems 18.3, page 1040

1. $\dfrac{\pi}{2}$ **3.** 0

5. 0   **7.** $\dfrac{-\pi}{4e^2}$

9. $\dfrac{\pi}{2} e^{-1/\sqrt{2}} \cos\left(\dfrac{1}{\sqrt{2}} + \dfrac{\pi}{4}\right)$ **11.** $\dfrac{\pi}{2}$

13. $\dfrac{\pi}{\sqrt{3}} e^{\sqrt{3}/\sqrt{2}} \cos\left(\dfrac{\sqrt{3}}{\sqrt{2}} + \dfrac{\pi}{4}\right)$ **15.** $\dfrac{\pi}{3}$

17. Set $a = 0$ in Theorem 1 and evaluate the residue of $z/(z^2 + 2z + 2)^2$ at $-1 + i$.

19. $\dfrac{1}{2} \displaystyle\int_{-\infty}^{\infty} \dfrac{x^2 \, dx}{(x^2 + a^2)^2} = \dfrac{1}{2} \operatorname{Re}\left\{2\pi i \operatorname{Res}_{ai} \dfrac{z^2}{(z^2 + a^2)^2}\right\}$

21. $\operatorname{Re}\{2\pi i \operatorname{Res}_i(z^2 + 1)^{-n-1}\}$

23. $\operatorname{Im}\left\{2\pi i \operatorname{Res}_{ib} \dfrac{z^3 e^{iaz}}{(z^2 + b^2)^2}\right\}$

25. $\dfrac{1}{2} \operatorname{Im}\left\{2\pi i\left[\operatorname{Res}_{(1+i)b/\sqrt{2}} \dfrac{z\, e^{iaz}}{z^4 + b^4} + \operatorname{Res}_{(-1+i)b/\sqrt{2}} \dfrac{z\, e^{iaz}}{z^4 + b^4}\right]\right\}$

Problems 18.4, page 1043

1. $\dfrac{1}{8} + \dfrac{\pi}{4} \coth 2\pi$

3. $\dfrac{1}{2} + \dfrac{\pi}{2\sqrt{2}}\left[\dfrac{\sin \pi/\sqrt{2} \cos \pi/\sqrt{2} + \sinh \pi/\sqrt{2} \cosh \pi/\sqrt{2}}{\sin^2 \pi/\sqrt{2} + \sinh^2 \pi/\sqrt{2}}\right]$

5. $\dfrac{1}{2} + \dfrac{\pi}{2} \operatorname{csch} \pi$

7. $4\pi^2/27$

9. $-\dfrac{\pi}{4} \sqrt[4]{2}[\cot(\pi \sqrt[4]{2}) + \coth(\pi \sqrt[4]{2})]$

11. $\pi^2 \csc^2 \pi a$

Problems 18.5, page 1047

1. $10\pi i$ **3.** $2\pi i$

5. $4\pi i$ **7.** 0

9. 0   **11.** (i) 5, (ii) 8, (iii) 5, (iv) 6

13. 3. Let $f(z) = (z^2 - 1) \cdot (z^4 - 5z^2 + 5)$ and note that $|z^2 - 1| > |z|$ on the semicircle $|z| = R > 2$, $x > 0$, and on the line segment $z = iy$, $|y| \leqslant R$.

15. 0. Multiply the equation by $z^2 + 2$.

17. $f(C)$ does not wind around any number off the real line. Hence $\operatorname{Im} f = 0$, and apply the zero-derivative theorem on page 909.

Review Exercises for Chapter 18, page 1047

1. 1   **3.** 0

5. $\cos 2$ **7.** $\pm i/3$

9. $-\dfrac{3}{2} \pm \dfrac{i}{\sqrt{2}}, \dfrac{3}{2} \pm i$ **11.** 57, -7

13. $2\pi i$ **15.** 0

17. 0   **19.** 0

21. 0

23. $\dfrac{2\pi i}{169}\left[13 \cos 2 - 4 \sin 2 - \dfrac{5}{3} \sinh 3\right]$

25. $2\pi/\sqrt{3}$

27. $\pi\left(\dfrac{180 - 104\sqrt{3}}{24 - 14\sqrt{3}}\right)$

29. $\pi/16\sqrt{2}$

31. $\pi/2e^2$

33. $\dfrac{\pi}{4}\left[\dfrac{1}{\sqrt{2}} - e^{-2\sqrt{2}} \cos\left(2\sqrt{2} + \dfrac{\pi}{4}\right)\right]$

35. $\dfrac{\pi}{4} \operatorname{csch}(2\pi) + \dfrac{1}{8}$

37. $\pi^2/9$

39. $2\pi i$

Chapter 19

Problems 19.1, page 1055

1. $16u^2 - 24uv + 9v^2 + 12u + 16v = 4$

3. a. $\cos 0 = \cos 2\pi$

   b. The strip $0 \leqslant x \leqslant \pi$ maps onto $\mathbb{C}$

5. $u \geqslant 1$

7. $\{(u, v) : v = 0, \ |u| \leqslant 1\}$

9. Parametrically: $u = 8 - 6t^2$, $v = 12t - t^3$, $\ |t| < \infty$

11. $\mathbb{C}$

13. $v \geqslant 0$

15. $v \leqslant 0$

17. $\left\{u > \dfrac{v^2}{16} - 4\right\} \cap \left\{u > 9 - \dfrac{v^2}{36}\right\}$

19. $|w| \leqslant 4$ **21.** $|w| < e$

23. $|w| > 0$ **25.** $|w| > 1$

27. $0 < |w| < 1$ **29.** $|w| \geqslant e^2$

31. Parametrically: the interior of

$$u = e^{\cos \theta} \cos(\sin \theta)$$
$$v = e^{\cos \theta} \sin(\sin \theta)$$
$$0 \leqslant \theta < 2\pi$$

33. $\begin{cases} v \geqslant 1, & k \text{ even} \\ v \leqslant -1, & k \text{ odd} \end{cases}$

35. $\dfrac{u^2}{\cosh^2\left[(2k+1)\dfrac{\pi}{2}\right]} + \dfrac{v^2}{\sinh^2\left[(2k+1)\dfrac{\pi}{2}\right]} = 1$

37. $\dfrac{u^2}{\sin^2 c} - \dfrac{v^2}{\cos^2 c} = 1$

39. $\dfrac{u^2}{\cosh^2 2} + \dfrac{v^2}{\sinh^2 2} \geqslant 1$

41. Use $\sin z = \sin x \cosh y + i \cos x \sinh y$ and set $y = c$.

43. See page 1066.

45. $\left| z + \dfrac{i}{2} \right| > \dfrac{1}{2}$

47. $(u + 1)^2 + (v + \frac{1}{2})^2 = \frac{5}{4}$

49. $0 < |w| \leqslant \frac{1}{2}$

51. $u = -\frac{1}{2}$

53. no

55. yes

57. no

59. no

61. no

63. $f(z) = 1 + i + z$

65. $u + av = 2 - 3b$

Problems 19.2, page 1061

1. none

3. $z = 0$

5. $z = 0, -2$

7. $z = 0$

9. $z = 0, \pm 1$

11. triples angle, $0 < \arg w < \pi/2$

13. triples angle, $0 < \arg w < \pi/2$

15. quadruples angle, $0 < \arg w < 2\pi/3$

17. doubles angle, $0 < \arg w < \pi/3$

19. Set $z = r(1 + e^{i\theta})$ and square.

21. Look at Liouville's theorem.

Problems 19.3, page 1070

1. $1 - \dfrac{2}{z+1}$ $\therefore$ translate by 1, invert, magnify by 2, rotate by π, and translate by 1.

3. $\dfrac{(5-i)/2}{\sqrt{2}e^{\pi i/4}\left(z - \dfrac{1+5i}{2}\right)} + \dfrac{1-i}{2}$

5. $1 - i$

7. ∞

9. $\left| w - \dfrac{1-i}{2} \right| > \dfrac{1}{\sqrt{2}}$

11. $\left| w - \dfrac{1-2i}{3} \right| < \dfrac{\sqrt{2}}{3}$

13. $\left| w - \dfrac{7-i}{8} \right| \leqslant \dfrac{\sqrt{2}}{8}$

15. $\dfrac{z - z_0}{1 - \bar{z}_0 z}$ maps $0, z_0, z_0/|z_0|$ onto $-z_0, 0, z_0/|z_0|, e^{i\theta}$ rotates

17. $w = -z$

19. $w = \dfrac{(i + \frac{1}{2})z - 1}{z + (i - \frac{1}{2})}$

21. lower half of unit disk

23. $\operatorname{Re} w < 1$ and $|w - \frac{1}{2}| > \frac{1}{2}$

25. two

27. $|w + i| = 1$ and $\operatorname{Im} w = 2$

29. $|w + 2 + i| = 1$ and $\operatorname{Re} w = 1$

31. $|w + 1 + 2i| = 2$ and $\left| w - \left(1 + \dfrac{i}{2}\right) \right| = \dfrac{1}{2}$

33. $\operatorname{Re} w = -\frac{3}{2}$ and $|w + \frac{3}{4}| = \frac{1}{4}$

35. $w = \dfrac{e^{\pi i/4}z}{z + 1 - i}$

37. $z = \pm i$

39. $z = \dfrac{-3i}{4} \pm \dfrac{\sqrt{-11 - 10i}}{2}$

41. $\mathbb{C} - \{w : |u| \leqslant 2, v = 0\}$

43. $w = \dfrac{(1+i)z - 2i}{(2+i)z + (1-2i)}$

45. $w = \dfrac{i(z+1)}{3z - 1 - 2i}$

47. Map a, b, c to $1, -1, k$ by

$$\left(\dfrac{z-a}{z-c}\right)\left(\dfrac{b-c}{b-a}\right) = \left(\dfrac{w-1}{w-k}\right)\left(\dfrac{-1-k}{-1-1}\right)$$
$$= \dfrac{1+k}{2}\left(\dfrac{w-1}{w-k}\right)$$

and solve

$$\left(\dfrac{d-a}{d-c}\right)\left(\dfrac{b-c}{b-a}\right) = \dfrac{(1+k)^2}{4k}, \text{ for } k.$$

49. $w = \left(\dfrac{2-z}{z}\right)^2$

Problems 19.4, page 1075

1. $u = 1 - \dfrac{v^2}{4}$

3. $|w| = e$

5. $\dfrac{u^2}{\sin^2 1} - \dfrac{v^2}{\cos^2 1} = 1$

7. parametrically for $|t| < \infty$:

$$u = \frac{1 - t^2}{(1 + t^2)^2}, \quad v = \frac{-2t}{(1 + t^2)^2}$$

9. right branch of $u^2 - v^2 = 1$

11. $u = 1 - \dfrac{v^2}{4}$

13. $u = \ln 4, \; -\pi \leqslant v < \pi$

15. $k = \frac{13}{3}c$

17. portion of $1 < |w| < e$ in the first quadrant

19. sector satisfying

$$\frac{\pi}{6} < \arg w < \frac{\pi}{3} \quad \text{and} \quad e^{-2} < |w| < e$$

21. interior of $\dfrac{u^2}{\cosh^2 1} + \dfrac{v^2}{\sinh^2 1} = 1$ with lines $\{w : v = 0,$

$|u| \geqslant 1\}$ excluded

23. Re $w > 0$ with the line $\{w : v = 0, \quad u \geqslant 1\}$ excluded

25. Consider the mappings $\zeta = z^2$, $Z = \dfrac{\zeta}{\zeta + 1}$, and $w = \sqrt{Z}$.

27. $w = e^{\pi(z+1)}$

29. Observe that $\zeta = i\left(\dfrac{1 - z}{1 + z}\right)$ maps the unit disk onto the upper half-plane. Find where the upper half of the unit disk is mapped.

31. Look at Problem 19.3.41.

33. Consider the mappings:

$$z = \frac{1}{\sqrt{2}}\left(\zeta + \frac{1}{\zeta}\right),$$

$$Z = -\zeta,$$

$$W = Z^{2/3},$$

$$\mu = \frac{1}{2}\left(W + \frac{1}{W}\right),$$

$$w = \frac{\mu - 1}{\mu + 1}.$$

Then check that the mapping is one-to-one on the line segment $-\sqrt{2} < x \leqslant 1, \; y = 0$.

Problem 19.5, page 1080

1. $\text{Im}(Az^{4/3}) = A \, \text{Im}(z^{4/3}) = $ constant. The origin is a stagnation point.

3. $\text{Im}(Az^4) = $ constant or $x^3y - xy^3 = $ constant. The origin is a stagnation point.

5. $V = \bar{w}' = 1 - 3\bar{z}^2$, so $|V| = 1, 2, 4$ at $0, 1, i,$ respectively. $w' = 0$ when $z = \pm(3)^{-1/2}$.

7. $V = 3 + 2i\bar{z}$, so $|V| = 3, \sqrt{13}, 1$ at $0, 1, i,$ respectively. $w' = 0$ when $z = -3i/2$.

9. $y + 3x^2y - y^3 = $ constant;
$y + 2(x^2 - y^2) = $ constant;
$3y - (x^2 - y^2) = $ constant;
$\cos x \sinh y = $ constant

11. $e^w = (z \pm \sqrt{z^2 - a^2})/a$ so $\arg[(z \pm \sqrt{z^2 - a^2})/a] = $ constant.

13. Streamlines are the confocal hyperbolas

$$\frac{x^2}{a^2 \cos^2 v} - \frac{y^2}{a^2 \sin^2 v} = 1,$$

$v = $ constant. For $v = 0, \pi$ get the flow on the edges of the aperture, but this flow is not physically realizable because

$$\frac{1}{\bar{V}}\frac{dz}{dw} = a \sinh w = a\sqrt{\cosh^2 w - 1} = \sqrt{z^2 - a^2}, \text{ so that}$$

$|V|$ is infinite at $z = \pm a$.

15. $\bar{V} = w' = A(1 - a^2/z^2)$ so for $z = ae^{i\theta}$ we get $|\bar{V}(ae^{i\theta})| = |A(1 - e^{-2i\theta})| = 2A|\sin \theta|$. But

$$\frac{p(\infty)}{\rho} + \frac{1}{2}A^2 = \frac{p(z)}{\rho} + 2A^2 \sin^2 \theta$$

implying that

$$\frac{p(z)}{\rho} = \frac{p(\infty)}{\rho} + \frac{1}{2}A^2[1 - 4\sin^2 \theta].$$

If $A^2 > \dfrac{2p(\infty)}{3\rho}$ and $\theta = \pm\pi/2$, cavitation occurs.

Problem 19.6, page 1087

1. $u = r \sin \phi$ for $z = re^{i\theta}$

3. $u = \frac{1}{2}(1 - r^2 \cos 2\phi)$

5. $u = \frac{1}{4}(r^3 \cos 3\phi - 3r \cos \phi)$

7. $u = \dfrac{-2}{\pi} \text{Arg}\left(\dfrac{1 + z}{1 - z}\right)$

9. $u = \pi + 2 \, \text{Arg}(1 - z)$

11. $u = 1 + \sum_{n=1}^{\infty}\left[\dfrac{2(-r)^n}{n} \sin n \cos n\phi \right.$

$\left. + \dfrac{2}{\pi}r^n\left(\dfrac{(-1)^n \sin n}{n^2} - \dfrac{1}{n}\right)\sin n\phi\right]$

13. Use $g(z) = (z^2 - z_0^2)/(z^2 - \bar{z}_0^2)$.

Then $u(z) = \dfrac{1}{2\pi i} \int_{\partial G} u(\zeta) \dfrac{2\zeta(z^2 - \bar{z}^2)}{(\zeta^2 - \bar{z}^2)(\zeta^2 - z^2)} d\zeta$

$= \dfrac{4xy}{\pi}\left[\int_0^{\infty} t\left(\dfrac{u(it)}{|t^2 + z^2|^2} + \dfrac{u(t)}{|t^2 - z^2|^2}\right)dt\right]$

15. If $u(z), U(z)$ are both solutions of Dirichlet's problem that are continuous on $\bar{G}$, then $U(z) - u(z)$ is harmonic on the

simply connected region G and continuous on $\bar{G}$. The maximum principle implies that $U - u$ attains both its maximum and minimum on ∂G. But $U - u = 0$ on ∂G, so u is unique.

17. $u_r(e^{i\theta}) = 0 = u(e^{i\theta})$ and $u_r(re^{i\theta}) = a \geqslant u(re^{i\theta})$ by the maximum principle. Hence, by Problem 16 for $r < k < 1$

$$u(ke^{i\theta}) \leqslant u_r(ke^{i\theta}) = a \frac{\ln k}{\ln r} \to 0$$

as $r \to 0+$. Since k can be made arbitrarily small, u is not continuous at 0.

19. $u = \dfrac{1}{\pi}\left[\text{Arg}\left(\dfrac{z}{z+1}\right) + (1-x)\text{Arg}\left(\dfrac{z-1}{z}\right)\right] - \dfrac{y}{\pi}\ln\left|\dfrac{1-z}{z}\right|$

Review Exercises for Chapter 19, page 1088

1. $8v = u^2 - 16$

3. $\dfrac{u^2}{\cos^2 1} - \dfrac{v^2}{\sin^2 1} = 1$; right branch

5. $u^2 = \left(\dfrac{v}{9} + 3\right)\left(\dfrac{v}{9} - 24\right)^2$

7. $u > 1 - \dfrac{v^2}{4}, \quad v < 0$

9. inside $\begin{cases} u = e^{4\cos\theta}\cos(4\sin\theta) \\ v = e^{4\cos\theta}\sin(4\sin\theta) \end{cases}$
$0 \leqslant \theta < 2\pi.$

11. $\mathbb{C}$

13. $u = -2v$

15. $z = \dfrac{(2k+1)}{4}\pi$, k an integer

17. $z = 0$

19. magnified by six: $0 \leqslant \arg w \leqslant \dfrac{3\pi}{2}$

21. $(3 + i)/4$

23. $\left|w - \dfrac{1 + 2i}{3}\right| < \dfrac{\sqrt{2}}{3}$

25. $\left|w - \dfrac{5 + i}{4}\right| \leqslant \dfrac{1}{2\sqrt{2}}$

27. $w = iz$

29. $v = \tfrac{1}{3}$

31. $|w - 2| + |w + 2| < 5$

33. $u = \dfrac{v^2}{4} - 1$

35. $\dfrac{u^2}{\cosh^2 1} + \dfrac{v^2}{\sinh^2 1} = 1$

37. $uv = \tfrac{1}{2}$

39. $\left\{w : 0 < \arg w < \dfrac{\pi}{2}, \quad e^2 < |w| < e^3\right\}$

41. $w = -\tanh\left(\dfrac{\pi z}{4}\right)$

43. $u^2 - v^2 = \tfrac{3}{2}$

45. $u = r^2\sin 2\phi = \text{Im}(z^2)$

47. $u = \dfrac{\pi}{2} + \text{Arg}(1 - z)$

49. $u = \dfrac{1}{\pi}\left[\tan^{-1}\left(\dfrac{2 - x}{y}\right) + \tan^{-1}\left(\dfrac{2 + x}{y}\right)\right]$

Problems Appendix 3, page A19

1. $1^3 = \dfrac{1(1+1)^2}{4} = 1$, so true for $n = 1$. If true for $n = k$, then

$$1^3 + 2^3 + \cdots + k^3 + (k+1)^3 = \dfrac{k^2(k+1)^2}{4} + (k+1)^3 =$$

$$\dfrac{(k+1)^2}{4}[k^2 + 4(k+1)] = \dfrac{(k+1)^2(k+2)^2}{4}.$$

3. $\dfrac{d}{dx}(x + a_0) = 1 = 1!$, so true for $n = 1$. Assume true for $n = k$. Let $P_{n+1}(x) = x^{n+1} + a_n x^n + \cdots + a_1 x + a_0 = xP_n(x) + a_0$. Then $P'_{n+1} = xP'_n(x) + P_n(x)$. But $xP'_n(x)$ is a polynomial of degree n with leading coefficient n; $\left[\dfrac{d}{dx}(x^n + \cdots) = nx^{n-1} + \cdots\right]$. Thus $\dfrac{d^n}{dx^n}[xP'_n(x)] = n \times n!$, so $\dfrac{d^{n+1}}{dx^{n+1}}P_{n+1}(x) = \dfrac{d^n}{dx^n}P'_{n+1}(x) = \dfrac{d^n}{dx^n}[xP'_n(x) + P_n(x)] = n \times n! + n! = (n+1)n! = (n+1)!$

5. $\varnothing$ (the empty set) is the only subset of $\varnothing$, so a set with 0 elements has $2^0 = 1$ subset. Assume true for $n = k$. Suppose S has $k + 1$ elements. Take out one element and call it x. The set $S - \{x\}$ has k elements and, by assumption, 2^k subsets. For each such subset, there are two subsets of S that can be obtained by either adding or not adding x. Thus S has $2 \times 2^k = 2^{k+1}$ subsets.

7. $\mathbf{u} \cdot (\mathbf{v}_1 + \mathbf{v}_2) = \mathbf{u} \cdot \mathbf{v}_1 + \mathbf{u} \cdot \mathbf{v}_2$ by Theorem 6.2.1 on page 302. Assume true for $n = k$. Then $\mathbf{u} \cdot (\mathbf{v}_1 + \mathbf{v}_2 + \cdots + \mathbf{v}_k + \mathbf{v}_{k+1}) = \mathbf{u} \cdot (\mathbf{v}_1 + \mathbf{v}_2 + \cdots + \mathbf{v}_k) + \mathbf{u} \cdot \mathbf{v}_{k+1}$ (by the case $n = 2$) $= \mathbf{u} \cdot \mathbf{v}_1 + \mathbf{u} \cdot \mathbf{v}_2 + \cdots + \mathbf{u} \cdot \mathbf{v}_k + \mathbf{u} \cdot \mathbf{v}_{k+1}$ (by the induction assumption).

Index